299.95 per copy (in United States).
rice subject to change without prior notice.

0206

RSMeans

Facilities Construction Cost Data

21st Annual Edition

2006

RSMeans
Construction Publishers & Consultants
63 Smiths Lane
Kingston, MA 02364-0800
(781) 422-5000

Printed in the United States of America.
ISSN 0888-6709
ISBN 0-87629-799-8

Senior Editors
Melville J. Mossman, PE
Stephen C. Plotner

Contributing Editors
Christopher Babbitt
Ted Baker
Barbara Balboni
Robert A. Bastoni
John H. Chiang, PE
John Kane
Robert J. Kuchta
Robert C. McNichols
Robert W. Mewis, CCC
John J. Moylan
Jeannene D. Murphy
Eugene R. Spencer
Marshall J. Stetson
Phillip R. Waier, PE

Senior Engineering Operations Manager
John H. Ferguson, PE

Senior Vice President & General Manager
John Ware

Vice President of Sales
John M. Shea

Production Manager
Michael Kokernak

Technical Support
Wayne D. Anderson
Thomas J. Dion
Jonathan Forgit
Mary Lou Geary
Gary L. Hoitt
Genevieve Medeiros
Paula Reale-Camelio
Kathryn S. Rodriguez
Sheryl A. Rose
Laurie Thom

Book & Cover Design
Norman R. Forgit

Reed Construction Data®

This book is recyclable.

This book is printed on recycled stock.

First Printing

Foreword

RSMeans is a product line of Reed Construction Data, Inc., a leading provider of construction information, products, and services in North America and globally. Reed Construction Data's project information products include more than 100 regional editions, national construction data, sales leads, and local plan rooms in major business centers. Reed Construction Data's PlansDirect provides surveys, plans, and specifications. The First Source suite of products consists of *First Source for Products,* SPEC-DATA™, MANU-SPEC™, CADBlocks, Manufacturer Catalogs, and First Source Exchange (www.firstsourceexchange.com) for the selection of nationally available building products. Reed Construction Data also publishes ProFile, a database of more than 20,000 U.S. architectural firms. RSMeans provides construction cost data, training, and consulting services in print, CD-ROM, and online. Reed Construction Data, headquartered in Atlanta, is owned by Reed Business Information (www.reedconstructiondata.com), a leading provider of critical information and marketing solutions to business professionals in the media, manufacturing, electronics, construction, and retail industries. Its market-leading properties include more than 135 business-to-business publications and over 125 Webzines and Web portals as well as online services, custom publishing, directories, research, and direct-marketing lists. Reed Business Information is a member of the Reed Elsevier plc group (NYSE: RUK and ENL)—a world-leading publisher and information provider operating in the science and medical, legal, education, and business-to-business industry sectors.

Our Mission

Since 1942, RSMeans has been actively engaged in construction cost publishing and consulting throughout North America.

Today, over 60 years after RSMeans began, our primary objective remains the same: to provide you, the construction and facilities professional, with the most current and comprehensive construction cost data possible.

Whether you are a contractor, an owner, an architect, an engineer, a facilities manager, or anyone else who needs a fast and reliable construction cost estimate, you'll find this publication to be a highly useful and necessary tool.

Today, with the constant flow of new construction methods and materials, it's difficult to find the time to look at and evaluate all the different construction cost possibilities. In addition, because labor and material costs keep changing, last year's cost information is not a reliable basis for today's estimate or budget.

That's why so many construction professionals turn to RSMeans. We keep track of the costs for you, along with a wide range of other key information, from city cost indexes . . . to productivity rates . . . to crew composition . . . to contractor's overhead and profit rates.

RSMeans performs these functions by collecting data from all facets of the industry and organizing it in a format that is instantly accessible to you. From the preliminary budget to the detailed unit price estimate, you'll find the data in this book useful for all phases of construction cost determination.

The Staff, the Organization, and Our Services

When you purchase one of RSMeans' publications, you are, in effect, hiring the services of a full-time staff of construction and engineering professionals.

Our thoroughly experienced and highly qualified staff works daily at collecting, analyzing, and disseminating comprehensive cost information for your needs. These staff members have years of practical construction experience and engineering training prior to joining the firm. As a result, you can count on them not only for the cost figures, but also for additional background reference information that will help you create a realistic estimate.

The RSMeans organization is always prepared to help you solve construction problems through its five major divisions: Construction and Cost Data Publishing, Electronic Products and Services, Consulting Services, Insurance Services, and Educational Services.

Besides a full array of construction cost estimating books, RSMeans also publishes a number of other reference works for the construction industry. Subjects include construction estimating and project and business management; special topics such as HVAC, roofing, plumbing, and hazardous waste remediation; and a library of facility management references.

In addition, you can access all of our construction cost data through your computer with *Means CostWorks 2006* CD-ROM, an electronic tool that offers over 50,000 lines of RSMeans detailed construction cost data, along with assembly and whole building cost data. You can also access RSMeans cost information from our Web site at www.rsmeans.com.

What's more, you can increase your knowledge and improve your construction estimating and management performance with an RSMeans Construction Seminar or In-House Training Program. These two-day seminar programs offer unparalleled opportunities for everyone in your organization to get updated on a wide variety of construction-related issues.

RSMeans also is a worldwide provider of construction cost management and analysis services for commercial and government owners and of claims and valuation services for insurers.

In short, RSMeans can provide you with the tools and expertise for constructing accurate and dependable construction estimates and budgets in a variety of ways.

Robert Snow Means Established a Tradition of Quality That Continues Today

Robert Snow Means spent years building RSMeans, making certain he always delivered a quality product.

Today, at RSMeans, we do more than talk about the quality of our data and the usefulness of our books. We stand behind all of our data, from historical cost indexes to construction materials and techniques to current costs.

If you have any questions about our products or services, please call us toll-free at 1-800-334-3509. Our customer service representatives will be happy to assist you. You can also visit our Web site at www.rsmeans.com.

Table of Contents

Related RSMeans Products and Services

Melville J. Mossman, P.E. and Stephen C. Plotner, Senior Editors of this cost data book, suggest the following RSMeans products and services as **companion** information resources to *Facilities Construction Cost Data*:

Construction Cost Data Books

Interior Cost Data 2006
Building Construction Cost Data 2006
Facilities Maintenance & Repair Cost Data 2006

Reference Books

Facilities Planning and Relocation
Life Cycle Costing for Facilities
Cost Planning & Estimating for Facilities Maintenance
Job Order Contracting
Estimating Building Costs
Estimating Handbook, 2nd Edition
ADA Compliance Pricing Guide, 2nd Edition
Building Security: Strategies & Costs

Seminars and In-House Training

Facilities Maintenance & Repair Estimating
Repair & Remodeling Estimating
Unit Price Estimating
Advanced Project Management
Means CostWorks Training
Means Data for Job Order Contracting (JOC)

RSMeans on the Internet

Visit RSMeans at **http://www.rsmeans.com.** The site contains useful ***interactive*** cost and reference material. Request or download ***FREE*** estimating software demos. Visit our bookstore for convenient ordering and to learn more about new publications and companion products.

RSMeans Data on CD-ROM

Get the information found in RSMeans traditional cost books on RSMeans' new, easy-to-use CD-ROM, *CostWorks 2006. Means CostWorks* users can now enhance their Costlist with the *Means CostWorks Estimator*. For more information see the special full-color brochure inserted in this book.

RSMeans Business Solutions

RSMeans Business Solutions provides construction cost-related services including: Customized Databases, Benchmark Studies, Estimating Audits, Feasibility Studies, Litigation Support, and Predictive Cost Models.

New! RSMeans for Job Order Contracting (JOC)

New methods and "best practices" for estimating and project management to reduce administrative costs for a $147 billion+ renovation market in the U.S. Finish renovation projects faster in schools and universities, on government building projects, and in medical facilities.

* RSMeans Job Order Contracting (JOC) Cost Data
* JOCWorks Software (Basic, Advanced, PRO)
* Consultation in Contracting Methods
* RSMeans Data™ Licensing Program

Construction Costs for Software Applications

Over 25 unit price and assemblies cost databases are available through a number of leading estimating and facilities management software providers (listed below). For more information see the "yellow" pages at the back of this publication.

RSMeansData™ is also available to federal, state, and local government agencies as multi-year, multi-seat licenses.

- 3D International
- 4Clicks-Solutions, LLC
- Aepco, Inc.
- Applied Flow Technology
- ArenaSoft Estimating
- Best Software–Timberline
- BSD – Building Systems Design, Inc.
- CMS – Construction Management Software
- Corecon Technologies, Inc.
- CorVet Systems
- Estimating Systems, Inc.
- Maximus Asset Solutions
- MC^2 – Management Computer Controls
- Shaw Beneco Enterprises, Inc.
- US Cost, Inc.
- VFA–Vanderweil Facility Advisers
- WinEstimator, Inc.

How the Book Is Built: An Overview

A Powerful Construction Tool

You have in your hands one of the most powerful construction tools available today. A successful project is built on the foundation of an accurate and dependable estimate. This book will enable you to construct just such an estimate.

For the casual user the book is designed to be:

- quickly and easily understood so you can get right to your estimate.
- filled with valuable information so you can understand the necessary factors that go into the cost estimate.

For the regular user, the book is designed to be:

- a handy desk reference that can be quickly referred to for key costs.
- a comprehensive, fully reliable source of current construction costs and productivity rates so you'll be prepared to estimate any project.
- a source book for preliminary project cost, product selections, and alternate materials and methods.

To meet all of these requirements we have organized the book into the following clearly defined sections.

How To Use the Book: The Details

This section contains an in-depth explanation of how the book is arranged . . . and how you can use it to determine a reliable construction cost estimate. It includes information about how we develop our cost figures and how to completely prepare your estimate.

Unit Price Section

All cost data has been divided into the 16 divisions according to the MasterFormat system of classification and numbering as developed by the Construction Specifications Institute (CSI) and Construction Specifications Canada (CSC). For a listing of these divisions and an outline of their subdivisions, see the Unit Price Section Table of Contents.

Estimating tips are included at the beginning of each division.

Assemblies Section

The cost data in this section has been organized in an "Assemblies" format. These assemblies are the functional elements of a building and are arranged according to the 7 divisions of the UNIFORMAT II classification system. For a complete explanation of a typical "Assemblies" page, see "How To Use the Assemblies Cost Tables."

Reference Section

This section includes information on Equipment Rental Costs, Crew Listings, Historical Cost Indexes, City Cost Indexes, Location Factors, Reference Tables, Change Orders, Square Foot Costs, and a listing of Abbreviations.

Equipment Rental Costs: This section contains the average costs to rent and operate hundreds of pieces of construction equipment.

Crew Listings: This section lists all the crews referenced in the book. For the purposes of this book, a crew is composed of more than one trade classification and/or the addition of power equipment to any trade classification. Power equipment is included in the cost of the crew. Costs are shown both with the bare labor rates and with the installing contractor's overhead and profit added. For each, the total crew cost per eight-hour day and the composite cost per labor-hour are listed.

Historical Cost Indexes: These indexes provide you with data to adjust construction costs over time.

City Cost Indexes: All costs in this book are U.S. national averages. Costs vary because of the regional economy. You can adjust costs by CSI Division to over 700 locations throughout the U.S. and Canada by using the data in this section.

Location Factors: You can adjust total project costs to over 900 locations throughout the U.S. and Canada by using the data in this section.

Reference Tables: At the beginning of selected major classifications in the Unit Price and Assemblies sections are "reference numbers" shown in a bold box. These numbers refer you to related information in the Reference Section. In this section, you'll find reference tables, explanations, and estimating information that support how we develop the unit price data, technical data, and estimating procedures.

Change Orders: This section includes information on the factors that influence the pricing of change orders.

Square Foot Costs: Formerly Division 17, this section contains costs for 59 different building types that allow you to make a rough estimate for the overall cost of a project or its major components.

Abbreviations: A listing of abbreviations used throughout this book, along with the terms they represent, is included in this section.

Index

A comprehensive listing of all terms and subjects in this book will help you quickly find what you need when you are not sure where it falls in MasterFormat.

The Scope of This Book

This book is designed to be as comprehensive and as easy to use as possible. To that end we have made certain assumptions and limited its scope in three key ways:

1. We have established material prices based on a national average.
2. We have computed labor costs based on a 30-city national average of union wage rates.
3. We have targeted the data for projects of a certain size range.

For a more detailed explanation of how the cost data is developed, see "How To Use the Book: The Details."

Project Size

This book is aimed primarily at commercial and industrial projects costing $10,000 to $1,000,000.

With reasonable exercise of judgment the figures can be used for any building work. *However, for civil engineering structures such as bridges, dams, highways, or the like, please refer to **RSMeans Heavy Construction Cost Data.***

How to Use the Book: The Details

What's Behind the Numbers? The Development of Cost Data

The staff at RSMeans continuously monitors developments in the construction industry in order to ensure reliable, thorough, and up-to-date cost information.

While *overall* construction costs may vary relative to general economic conditions, price fluctuations within the industry are dependent upon many factors. Individual price variations may, in fact, be opposite to overall economic trends. Therefore, costs are continually monitored and complete updates are published yearly. Also, new items are frequently added in response to changes in materials and methods.

Costs—$ (U.S.)

All costs represent U.S. national averages and are given in U.S. dollars. The RSMeans City Cost Indexes can be used to adjust costs to a particular location. The City Cost Indexes for Canada can be used to adjust U.S. national averages to local costs in Canadian dollars. No exchange rate conversion is necessary.

Material Costs

The RSMeans staff contacts manufacturers, dealers, distributors, and contractors all across the U.S. and Canada to determine national average material costs. If you have access to current material costs for your specific location, you may wish to make adjustments to reflect differences from the national average. Included within material costs are fasteners for a normal installation. RSMeans engineers use manufacturers' recommendations, written specifications, and/or standard construction practice for size and spacing of fasteners. Adjustments to material costs may be required for your specific application or location. Material costs do not include sales tax.

Labor Costs

Labor costs are based on the average of wage rates from 30 major U.S. cities. Rates are determined from labor union agreements or prevailing wages for construction trades for the current year. Rates, along with overhead and profit markups, are listed on the inside back cover of this book.

- If wage rates in your area vary from those used in this book, or if rate increases are expected within a given year, labor costs should be adjusted accordingly.

Labor costs reflect productivity based on actual working conditions. These figures include time spent during a normal workday on tasks other than actual installation, such as material receiving and handling, mobilization at site, site movement, breaks, and cleanup.

Productivity data is developed over an extended period so as not to be influenced by abnormal variations and reflects a typical average.

Equipment Costs

Equipment costs include not only rental, but also operating costs for equipment under normal use. The operating costs include parts and labor for routine servicing such as repair and replacement of pumps, filters, and worn lines. Normal operating expendables, such as fuel, lubricants, tires, and electricity (where applicable), are also included. Extraordinary operating expendables with highly variable wear patterns, such as diamond bits and blades, are excluded. These costs are included under materials. Equipment rental rates are obtained from industry sources throughout North America—contractors, suppliers, dealers, manufacturers, and distributors.

Crew Equipment Cost/Day—The power equipment required for each crew is included in the crew cost. The daily cost for crew equipment is based on dividing the weekly bare rental rate by 5 (number of working days per week), and then adding the hourly operating cost times 8 (hours per day). This "Crew Equipment Cost/Day" is listed in the Reference Section.

Mobilization/Demobilization—The cost to move construction equipment from an equipment yard or rental company to the job site and back again is not included in equipment costs. Mobilization (to the site) and demobilization (from the site) costs can be found in Section 02305-250. If a piece of equipment is already at the job site, it is not appropriate to utilize mob/demob costs again in an estimate.

General Conditions

Cost data in this book is presented in two ways: Bare Costs and Total Cost including O&P (Overhead and Profit). General Conditions, when applicable, should also be added to the Total Cost including O&P. The costs for General Conditions are listed in Division 1 of the Unit Price Section and the Reference Section of this book. General Conditions for the *Installing Contractor* may range from 0% to 10% of the Total Cost including O&P. For the *General* or *Prime Contractor*, costs for General Conditions may range from 5% to 15% of the Total Cost including O&P, with a figure of 10% as the most typical allowance.

Overhead and Profit

Total Cost including O&P for the *Installing Contractor* is shown in the last column on both the Unit Price and Assemblies pages of this book. This figure is the sum of the bare material cost plus 10% for profit, the base labor cost plus total overhead and profit, and the bare equipment cost plus 10% for profit. Details for the calculation of Overhead and Profit on labor are shown on the inside back cover and in the Reference Section of this book. (See the "How to Use the Unit Price Pages" for an example of this calculation.)

Factors Affecting Costs

Costs can vary depending upon a number of variables. Here's how we have handled the main factors affecting costs.

Quality—The prices for materials and the workmanship upon which productivity is based represent sound construction work. They are also in line with U.S. government specifications.

Overtime—We have made no allowance for overtime. If you anticipate premium time or work beyond normal working hours, be sure to make an appropriate adjustment to your labor costs.

Productivity—The productivity, daily output, and labor-hour figures for each line item are based on working an eight-hour day in daylight hours in moderate temperatures. For work that extends beyond normal work hours or is performed under adverse conditions, productivity may decrease. (See the section in "How To Use the Unit Price Pages" for more on productivity.)

Size of Project — The size, scope of work, and type of construction project will have a significant impact on cost. Economies of scale can reduce costs for large projects. Unit costs can often run higher for small projects. Costs in this book are intended for the size and type of project as previously described in "How the Book Is Built: An Overview." Costs for projects of a significantly different size or type should be adjusted accordingly.

Location — Material prices in this book are for metropolitan areas. However, in dense urban areas, traffic and site storage limitations may increase costs. Beyond a 20-mile radius of large cities, extra trucking or transportation charges may also increase the material costs slightly. On the other hand, lower wage rates may be in effect. Be sure to consider both of these factors when preparing an estimate, particularly if the job site is located in a central city or remote rural location.

In addition, highly specialized subcontract items may require travel and per-diem expenses for mechanics.

Other Factors—
- season of year
- contractor management
- weather conditions
- local union restrictions
- building code requirements
- availability of:
    - adequate energy
    - skilled labor
    - building materials
- owner's special requirements/restrictions
- safety requirements
- environmental considerations

Unpredictable Factors — General business conditions influence "in-place" costs of all items. Substitute materials and construction methods may have to be employed. These may affect the installed cost and/or life cycle costs. Such factors may be difficult to evaluate and cannot necessarily be predicted on the basis of the job's location in a particular section of the country. Thus, where these factors apply, you may find significant, but unavoidable cost variations for which you will have to apply a measure of judgment to your estimate.

Rounding of Costs

In general, all unit prices in excess of $5.00 have been rounded to make them easier to use and still maintain adequate precision of the results. The rounding rules we have chosen are in the following table.

Prices from . . .	Rounded to the nearest . . .
$.01 to $5.00	$.01
$5.01 to $20.00	$.05
$20.01 to $100.00	$.50
$100.01 to $300.00	$1.00
$300.01 to $1,000.00	$5.00
$1,000.01 to $10,000.00	$25.00
$10,000.01 to $50,000.00	$100.00
$50,000.01 and above	$500.00

Important Estimating Considerations

One reason for listing a job size "minimum" is to ensure that the construction craftsmen are productive for 8 hours a day on a continuous basis. Otherwise, tasks that require only 6 to 7 hours could be billed as an 8-hour day, thus providing an erroneous productivity estimate. The "productivity," or daily output, of each craftsman includes mobilization and cleanup time, break time and plan layout time, as well as an allowance to carry stock from the storage trailer or location on the job site up to 200′ into the building and to the first or second floor. If material has to be transported over greater distances or to higher floors, an additional allowance should be considered by the estimator. An allowance has also been included in the piping and fittings installation time for a leak check and minor tightening.

Equipment installation time includes the following applicable items: positioning, leveling and securing the unit in place, connecting all associated piping, ducts, vents, etc., which shall have been estimated separately, connecting to an adjacent power source, filling/bleeding, startup, adjusting the controls up and down to ensure proper response, setting the integral controls/valves/regulators/thermostats for proper operation (does not include external building type control systems, DDC systems, etc.), explaining/training owner's operator, and warranty. A reasonable breakdown of the labor costs is as follows:

1	Movement into building, installation/setting of equipment	35%
2.	Connection to piping/duct/power, etc.	25%
3.	Filling/flushing/cleaning/touchup, etc.	15%
4.	Startup/running adjustments	5%
5.	Training owner's rep.	5%
6.	Warranty/call back/service	15%

Note that cranes or other lifting equipment are not included on any lines in Division 15. For example, if a crane is required to lift a heavy piece of pipe into place high above a gym floor, or to put a rooftop unit on the roof of a four-story building, etc., it must be added. Due to the potential for extreme variation—from nothing additional required to a major crane or helicopter—we feel that including a nominal amount for "lifting contingency" would be useless and detract from the accuracy of the estimate. When using equipment rental from Means, remember to include the cost of the operator(s).

Final Checklist

Estimating can be a straightforward process provided you remember the basics. Here's a checklist of some of the steps you should remember to complete before finalizing your estimate.

Did you remember to . . .

- factor in the City Cost Index for your locale?
- take into consideration which items have been marked up and by how much?
- mark up the entire estimate sufficiently for your purposes?
- read the background information on techniques and technical matters that could impact your project time span and cost?
- include all components of your project in the final estimate?
- double check your figures for accuracy?
- call RSMeans if you have any questions about your estimate or the data you've found in our publications?

For more information, please see "Tips for Accurate Estimating," R01100-005 in the Reference Section.

Remember, RSMeans stands behind its publications. If you have any questions about your estimate . . . about the costs you've used from our books . . . or even about the technical aspects of the job that may affect your estimate, feel free to call the RSMeans editors at 1-800-334-3509.

Using Minimum Labor/Equipment Charges for Small Quantities

Estimating small construction or repair tasks often creates situations in which the quantity of work to be performed is very small. When this occurs, the labor and/or equipment costs to perform the work may be too low to allow for the crew to get to the job, receive instructions, find materials, get set up, perform the work, clean up, and get to the next job. In these situations, the estimator should compare the developed labor and/or equipment costs for performing the work (e.g., quantity x labor and/or equipment costs) with the *"minimum labor/equipment charge"* within that Unit Price section of the book.

If the labor and/or equipment costs developed by the estimator are LOWER THAN the *"minimum labor/equipment charge"* listed at the bottom of specific sections of Unit Price costs, the estimator should adjust the developed costs upward to the *"minimum labor/equipment charge."* The proper use of a *"minimum labor/equipment charge"* results in having enough money in the estimate to cover the contractor's higher cost of performing a very small amount of work during a partial workday.

A *"minimum labor/equipment charge"* should be used only when the task being estimated is the only task the crew will perform at the job site that day. If, however, the crew will be able to perform other tasks at the job site that day, the use of a *"minimum labor/equipment charge"* is not appropriate.

08500 | Windows

	08550 | Wood Windows	CREW	DAILY OUTPUT	LABOR-HOURS	UNIT	2006 BARE COSTS MAT.	LABOR	EQUIP.	TOTAL	TOTAL INCL O&P	
200	0010 **CASEMENT WINDOW** Including frame, screen, and grills										200
	0100 Avg. quality, bldrs. model, 2'-0" x 3'-0" H, dbl. insulated glass	1 Carp	10	.800	Ea.	186	28.50		214.50	252	
	0150 Low E glass		10	.800		225	28.50		253.50	295	
	0200 2'-0" x 4'-6" high, double insulated glass		9	.889		242	31.50		273.50	320	
	0250 Low E glass		9	.889		320	31.50		351.50	410	
	0300 2'-4" x 6'-0" high, double insulated glass		8	1		355	35.50		390.50	450	
	0350 Low E glass		8	1		385	35.50		420.50	485	
	0522 Vinyl clad, premium, double insulated glass, 2'-0" x 3'-0"		10	.800		264	28.50		292.50	335	
	8100 Metal clad, deluxe, dbl. insul. glass, 2'-0" x 3'-0" high		10	.800		206	28.50		234.50	274	
	8120 2'-0" x 4'-0" high		9	.889		248	31.50		279.50	325	
	8140 2'-0" x 5'-0" high		8	1		282	35.50		317.50	370	
	8300 For installation, add per leaf						15%				
	9000 Minimum labor/equipment charge	1 Carp	3	2.667	Job		95		95	157	

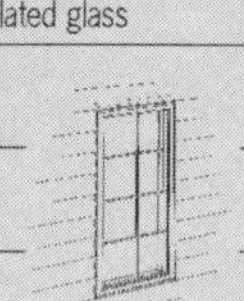

Example:

Establish the bid price to install two casement windows. Assume installation of 2′ x 4′ metal clad windows with insulating glass [Unit Price line number 08550-200-8120], and that this is the only task this crew will perform at the job site that day.

Solution:

Step One —Develop the Bare Labor Cost for this task:

Bare Labor Cost = 2 windows @ \$31.50/each = \$63.00

Step Two —Evaluate the *"minimum labor/equipment charge"* for this Unit Price section against the developed Bare Labor Cost for this task:

"minimum labor/equipment charge" = \$95.00 (compare with \$63.00)

Step Three —Choose to adjust the developed labor cost upward to the *"minimum labor/equipment charge."*

Step Four —Develop the bid price for this task (including O&P):

Add together the marked-up Bare Material Cost for this task and the marked-up *"minimum labor/equipment charge"* for this Unit Price section.

2 x (\$248.00 + 10%) + (\$95.00 + 65.7%)
= 2 x (\$248.00 + \$24.80) + (\$95.00 + \$62.42)
= 2 x (\$272.80) + \$157.42
= \$545.60 + \$157.42
= \$703.02

ANSWER: \$703.02 is the correct bid price to use. This sum takes into consideration the Material Cost (with 10% for profit) for these two windows, plus the *"minimum labor/equipment charge"* (with O&P included) for this section of the Unit Price book.

Unit Price Section

Table of Contents

How to Use the Unit Price Pages

The following is a detailed explanation of a sample entry in the Unit Price Section. Next to each bold number below is the described item with the appropriate component of the sample entry following in parentheses. Some prices are listed as bare costs; others as costs that include overhead and profit of the installing contractor. In most cases, if the work is to be subcontracted, the general contractor will need to add an additional markup (RSMeans suggests using 10%) to the figures in the column "Total Incl. O&P."

1 *Division Number/Title (03300/Cast-In-Place Concrete)*

Use the Unit Price Section Table of Contents to locate specific items. The sections are classified according to the CSI MasterFormat (1995 Edition).

2 *Line Numbers (03310 240 3920)*

Each unit price line item has been assigned a unique 12-digit code based on the CSI MasterFormat classification.

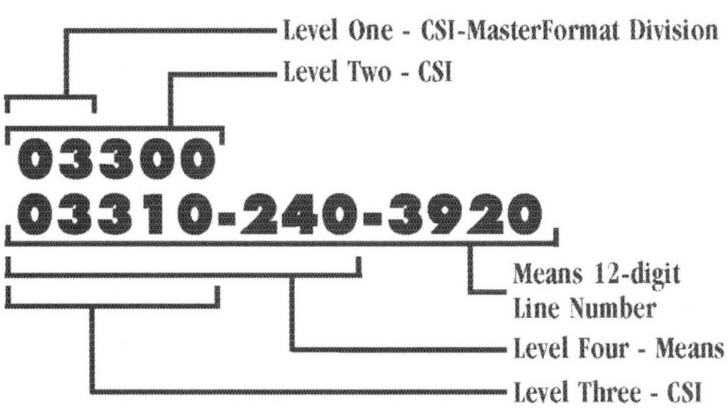

3 *Description (Concrete-In-Place, etc.)*

Each line item is described in detail. Sub-items and additional sizes are indented beneath the appropriate line items. The first line or two after the main item (in boldface) may contain descriptive information that pertains to all line items beneath this boldface listing.

4 *Reference Number Information*

R033053 -10 You'll see reference numbers shown in bold rectangles at the beginning of some sections. These refer to related items in the Reference Section, visually identified by a vertical gray bar on the page edges.

The relation may be: (1) an estimating procedure that should be read before estimating, (2) an alternate pricing method, or (3) technical information.

The "R" designates the Reference Section. The numbers refer to the MasterFormat classification system.

It is strongly recommended that you review all reference numbers that appear within the section in which you are working.

Note: Not all reference numbers appear in all Means publications.

03300 | Cast-In-Place Concrete

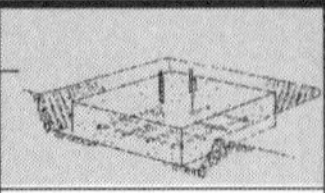

		03310 | Structural Concrete		CREW	DAILY OUTPUT	LABOR-HOURS	UNIT	2006 BARE COSTS MAT.	LABOR	EQUIP.	TOTAL	TOTAL INCL O&P	
240	0010	**CONCRETE IN PLACE**	R033053 -10										240
	0020	Including forms (4 uses), co... placeme...t, rei...g											
	0050	steel and finishing unless othe... indicated	R033053 -50										240
	0300	Beams, 5 kip per L.F., 10′ span		C-14A	15.62	12.804	C.Y.	287	455		788	1,125	
	3800	Footings, spread under 1 C.Y.		C-14C	38.07	2.942	C	175	99.50	.56	275.06	[illegible]	
	3850	Over 5 C.Y.			81.04	1.382		242	47	.26	289.26	[illegible]	
	3900	Footings, strip, 18″ x 9″, unreinfo...			40	2.800		109	95	.53	204.53	278	
	3920	18″ x 9″, reinforced			35	3.200		130	108	.61	238.61	325	
	3925	20″ x 10″, unreinforced			45	2.489		106	84.50	.47	190.97	257	
	3930	20″ x 10″, reinforced	R033053 -10	C-14C	40	2.800	C.Y.	122	95	.53	217.53	293	
	3935	24″ x 12″, unreinforced			55	2.036		103	69	.39	172.39	228	
	3940	24″ x 12″, reinforced	R033053 -50		48	2.333		121	79	.44	200.44	264	
	3945	36″ x 12″, unreinforced			70	1.600		99.50	54	.30	153.80	199	

5 *Crew (C-14C)*

The "Crew" column designates the typical trade or crew used to install the item. If an installation can be accomplished by one trade and requires no power equipment, that trade and the number of workers are listed (for example, "2 Carpenters"). If an installation requires a composite crew, a crew code designation is listed (for example, "C-14C"). You'll find full details on all composite crews in the Crew Listings.

- For a complete list of all trades utilized in this book and their abbreviations, see the inside back cover.

Crews

Crew No.	Bare Costs		Incl. Subs O & P		Cost Per Labor-Hour	
Crew C-14C	Hr.	Daily	Hr.	Daily	Bare Costs	Incl. O&P
1 Carpenter Foreman (out)	$37.55	$300.40	$62.20	$497.60	$33.85	$56.21
6 Carpenters	35.55	1706.40	58.90	2827.20		
2 Rodmen (reinf.)	39.50	632.00	67.80	1084.80		
4 Laborers	27.40	876.80	45.40	1452.80		
1 Cement Finisher	34.40	275.20	54.20	433.60		
1 Gas Engine Vibrator		21.40		23.55	.19	.21
112 L.H., Daily Totals		$3812.20		$6319.55	$34.04	$56.42

6 *Productivity: Daily Output (35)/Labor-Hours (3.200)*

The "Daily Output" represents the typical number of units the designated crew will install in a normal 8-hour day. To find out the number of days the given crew would require to complete the installation, divide your quantity by the daily output. For example:

Quantity	÷	Daily Output	=	Duration
100 C.Y.	÷	35/ Crew Day	=	2.86 Crew Days

The "Labor-Hours" figure represents the number of labor-hours required to install one unit of work. To find out the number of labor-hours required for your particular task, multiply the quantity of the item times the number of labor-hours shown. For example:

Quantity	x	Productivity Rate	=	Duration
100 C.Y.	x	3.200 Labor-Hours/ C.Y.	=	320 Labor-Hours

7 *Unit (C.Y.)*

The abbreviated designation indicates the unit of measure upon which the price, production, and crew are based (C.Y. = Cubic Yard). For a complete listing of abbreviations refer to the Abbreviations Listing in the Reference Section of this book.

8 *Bare Costs:*

Mat. (Bare Material Cost) (130.00)

The unit material cost is the "bare" material cost with no overhead and profit included. *Costs shown reflect national average material prices for January of the current year and include delivery to the job site. No sales taxes are included.*

Labor (108.00)

The unit labor cost is derived by multiplying bare labor-hour costs for Crew C-14C by labor-hour units. The bare labor-hour cost is found in the Crew Section under C-14C. (If a trade is listed, the hourly labor cost—the wage rate—is found on the inside back cover.)

Labor-Hour Cost Crew C-14C	x	Labor-Hour Units	=	Labor
$33.85	x	3.200	=	$108.45

Equip. (Equipment) (.61)

Equipment costs for each crew are listed in the description of each crew. Tools or equipment whose value justifies purchase or ownership by a contractor are considered overhead as shown on the inside back cover. The unit equipment cost is derived by multiplying the bare equipment hourly cost by the labor-hour units.

Equipment Cost Crew C-14C	x	Labor-Hour Units	=	Equip.
.19	x	3.20	=	.61

Total (238.61)

The total of the bare costs is the arithmetic total of the three previous columns: mat., labor, and equip.

Material	+	Labor	+	Equip.	=	Total
$130	+	$108.00	+	$.61	=	$238.61

9 *Total Costs Including O&P*

This figure is the sum of the bare material cost plus 10% for profit; the bare labor cost plus total overhead and profit (per the inside back cover or, if a crew is listed, from the crew listings); and the bare equipment cost plus 10% for profit.

Material is Bare Material Cost + 10% = 130 + 13.00	=	$143.00
Labor for Crew C-14C = Labor-Hour Cost (56.21) x Labor-Hour Units (3.20)	=	$179.87
Equip. is Bare Equip. Cost + 10% = .61 + .06	=	$ .67
Total (Rounded)	=	$325

Division 1 General Requirements

Estimating Tips

The General Requirements of any contract are very important to both the bidder and the owner. These lay the ground rules under which the contract will be executed and have a significant influence on the cost of operations. Therefore, it is extremely important to thoroughly read and understand the General Requirements, both before preparing an estimate and when the estimate is complete, to ascertain that nothing in the contract is overlooked. Caution should be exercised when applying items listed in Division 1 to an estimate. Many of the items are included in the unit prices listed in the other divisions such as markups on labor and company overhead.

01200 Price & Payment Procedures

- When estimating historic preservation projects (depending on the condition of the existing structure and the owner's requirements), a 15%-20% contingency or allowance is recommended, regardless of the stage of the drawings.

01300 Administrative Requirements

- Before determining a final cost estimate, it is a good practice to review all the items listed in Subdivision 01300 to make final adjustments for items that may need customizing to specific job conditions.
- Historic preservation projects may require specialty labor and methods, as well as extra time, to protect existing materials that must be preserved and/or restored. Some additional expenses may be incurred in architectural fees for facility surveys and other special inspections and analyses.
- Requirements for initial and periodic submittals can represent a significant cost to the General Requirements of a job. Thoroughly check the submittal specifications when estimating a project to determine any costs that should be included.

01400 Quality Requirements

- All projects will require some degree of Quality Control. This cost is not included in the unit cost of construction listed in each division. Depending upon the terms of the contract, the various costs of inspection and testing can be the responsibility of either the owner or the contractor. Be sure to include the required costs in your estimate.

01500 Temporary Facilities & Controls

- Barricades, access roads, safety nets, scaffolding, security, and many more requirements for the execution of a safe project are elements of direct cost. These costs can easily be overlooked when preparing an estimate. When looking through the major classifications of this subdivision, determine which items apply to each division in your estimate.

01740 Execution Requirements

- When preparing an estimate, thoroughly read the specifications to determine the requirements for Contract Closeout. Final cleaning, record documentation, operation and maintenance data, warranties and bonds, and spare parts and maintenance materials can all be elements of cost for the completion of a contract. Do not overlook these in your estimate.

01830 Operations & Maintenance

- If maintenance and repair are included in your contract, they require special attention. Estimating the cost to remove and replace any unit usually requires a site visit to determine the accessibility and the specific difficulty at that location. Obstructions, dust control, safety, and often, overtime hours must be considered when preparing your estimate.

Reference Numbers

Reference numbers are shown in bold squares at the beginning of some major classifications. These numbers refer to related items in the Reference Section. The reference information may be an estimating procedure, an alternate pricing method, or technical information.

Note: Not all subdivisions listed here necessarily appear in this publication.

01100 | Summary

01103 | Models & Renderings

		01103 Models & Renderings	CREW	DAILY OUTPUT	LABOR-HOURS	UNIT	2006 BARE COSTS MAT.	LABOR	EQUIP.	TOTAL	TOTAL INCL O&P	
200	0010	**MODELS**										200
	0020	Cardboard & paper, 1 building, minimum				Ea.	640			640	705	
	0050	Maximum					1,475			1,475	1,600	
	0100	2 buildings, minimum					860			860	945	
	0150	Maximum					1,950			1,950	2,125	
	0200	Plexiglass and metal, basic layout				SF Flr.	.06			.06	.07	
	0210	Including equipment and personnel				"	.29			.29	.32	
	0300	Site plan layout, minimum				Ea.	1,225			1,225	1,350	
	0350	Maximum				"	2,050			2,050	2,250	
500	0010	**RENDERINGS** Color, matted, 20" x 30", eye level,										500
	0020	1 building, minimum				Ea.	1,850			1,850	2,050	
	0050	Average					2,650			2,650	2,925	
	0100	Maximum					4,250			4,250	4,675	
	1000	5 buildings, minimum					3,725			3,725	4,075	
	1100	Maximum					7,425			7,425	8,175	
	2000	Aerial perspective, color, 1 building, minimum					2,650			2,650	2,925	
	2100	Maximum					7,425			7,425	8,175	
	3000	5 buildings, minimum					5,300			5,300	5,825	
	3100	Maximum					10,600			10,600	11,700	

01107 | Professional Consultant

		01107 Professional Consultant	CREW	DAILY OUTPUT	LABOR-HOURS	UNIT	MAT.	LABOR	EQUIP.	TOTAL	TOTAL INCL O&P	
100	0011	**ARCHITECTURAL FEES** R011110-10										100
	0020	For new construction										
	0060	Minimum				Project					4.90%	
	0090	Maximum									16%	
	0100	For alteration work, to $500,000, add to fee									50%	
	0150	Over $500,000, add to fee									25%	
200	0010	**CONSTRUCTION MANAGEMENT FEES**										200
	0060	For work to $10,000				Project					10%	
	0070	To $25,000									9%	
	0090	To $100,000									6%	
	0100	To $500,000									5%	
	0110	To $1,000,000									4%	
	0300	$5,000,000 job, minimum									2.50%	
	0350	Maximum									4%	
300	0010	**ENGINEERING FEES** R011110-30										300
	0020	Educational planning consultant, minimum				Project					.50%	
	0100	Maximum				"					2.50%	
	0200	Electrical, minimum				Contrct					4.10%	
	0300	Maximum									10.10%	
	0400	Elevator & conveying systems, minimum									2.50%	
	0500	Maximum									5%	
	0600	Food service & kitchen equipment, minimum									8%	
	0700	Maximum									12%	
	0800	Landscaping & site development, minimum									2.50%	
	0900	Maximum									6%	
	1000	Mechanical (plumbing & HVAC), minimum									4.10%	
	1100	Maximum									10.10%	
	1200	Structural, minimum				Project					1%	
	1300	Maximum				"					2.50%	
700	0010	**TOPOGRAPHICAL SURVEYS**										700
	0020	Minimum	A-7	3.30	7.273	Acre	16.80	283	17.65	317.45	495	
	0100	Maximum	A-8	.60	53.333		50.50	2,000	97	2,147.50	3,425	
	0300	Lot location and lines, minimum, for large quantities	A-7	2	12		26.50	465	29	520.50	815	

01100 | Summary

	01107 Professional Consultant		CREW	DAILY OUTPUT	LABOR-HOURS	UNIT	2006 BARE COSTS MAT.	LABOR	EQUIP.	TOTAL	TOTAL INCL O&P	
700	0320	Average	A-7	1.25	19.200	Acre	47.50	745	46.50	839	1,300	700
	0400	Maximum, for small quantities	A-8	1	32	↓	75.50	1,200	58	1,333.50	2,100	
	0600	Monuments, 3′ long	A-7	10	2.400	Ea.	20.50	93	5.85	119.35	180	
	0800	Property lines, perimeter, cleared land	"	1000	.024	L.F.	.03	.93	.06	1.02	1.60	
	1100	Crew for layout of building, trenching or pipe laying, 2 person crew	A-6	1	16	Day		565	58	623	985	
	1200	3 person crew	A-7	1	24			930	58.50	988.50	1,575	
	1400	Crew for roadway layout, 4 person crew	A-8	1	32	↓		1,200	58	1,258	2,025	
	1500	Aerial surveying, including ground control, minimum fee, 10 acres				Total					5,700	
	1510	100 acres									9,500	
	1550	From existing photography, deduct				↓					1,370	
	1600	2′ contours, 10 acres				Acre					460	
	1650	20 acres									315	
	1800	50 acres									95	
	1850	100 acres				↓					85	
	2150	For 1′ contours and										
	2160	dense urban areas, add to above				Acre					40%	
	3000	Inertial guidance system for										
	3010	locating coordinates, rent per day				Ea.					4,000	

01200 | Price & Payment Procedures

	01250 Contract Modification Procedures		CREW	DAILY OUTPUT	LABOR-HOURS	UNIT	2006 BARE COSTS MAT.	LABOR	EQUIP.	TOTAL	TOTAL INCL O&P	
200	0010	**CONTINGENCIES**										200
	0020	For estimate at conceptual stage				Project					25%	
	0050	Schematic stage									20%	
	0100	Preliminary working drawing stage (Design Dev.)									15%	
	0150	Final working drawing stage				↓					8%	
400	0010	**FACTORS** Cost adjustments R012153-10										400
	0100	Add to construction costs for particular job requirements										
	0500	Cut & patch to match existing construction, add, minimum				Costs	2%	3%				
	0550	Maximum					5%	9%				
	0800	Dust protection, add, minimum					1%	2%				
	0850	Maximum					4%	11%				
	1100	Equipment usage curtailment, add, minimum					1%	1%				
	1150	Maximum					3%	10%				
	1400	Material handling & storage limitation, add, minimum					1%	1%				
	1450	Maximum					6%	7%				
	1700	Protection of existing work, add, minimum					2%	2%				
	1750	Maximum					5%	7%				
	2000	Shift work requirements, add, minimum						5%				
	2050	Maximum						30%				
	2300	Temporary shoring and bracing, add, minimum					2%	5%				
	2350	Maximum					5%	12%				
	2400	Work inside prisons and high security areas, add, minimum						30%				
	2450	Maximum				↓		50%				
500	0010	**JOB CONDITIONS** Modifications to total										500
	0020	project cost summaries										
	0100	Economic conditions, favorable, deduct				Project					2%	
	0200	Unfavorable, add				↓					5%	

01200 | Price & Payment Procedures

		01250 | Contract Modification Procedures	CREW	DAILY OUTPUT	LABOR-HOURS	UNIT	2006 BARE COSTS MAT.	LABOR	EQUIP.	TOTAL	TOTAL INCL O&P	
500	0300	Hoisting conditions, favorable, deduct				Project					2%	500
	0400	Unfavorable, add									5%	
	0500	General Contractor management, experienced, deduct									2%	
	0600	Inexperienced, add									10%	
	0700	Labor availability, surplus, deduct									1%	
	0800	Shortage, add									10%	
	0900	Material storage area, available, deduct									1%	
	1000	Not available, add									2%	
	1100	Subcontractor availability, surplus, deduct									5%	
	1200	Shortage, add									12%	
	1300	Work space, available, deduct									2%	
	1400	Not available, add									5%	
600	0010	**OVERTIME** For early completion of projects or where (R012909-90)										600
	0020	labor shortages exist, add to usual labor, up to				Costs		100%				

		01255 | Cost Indexes	CREW	DAILY OUTPUT	LABOR-HOURS	UNIT	MAT.	LABOR	EQUIP.	TOTAL	TOTAL INCL O&P	
400	0010	**HISTORICAL COST INDEXES** (Reference) Back to 1956										400

		01290 | Payment Procedures	CREW	DAILY OUTPUT	LABOR-HOURS	UNIT	MAT.	LABOR	EQUIP.	TOTAL	TOTAL INCL O&P	
800	0010	**TAXES** (R012909-80)										800
	0020	Sales tax, state, average				%	4.83%					
	0050	Maximum (R012909-85)					7%					
	0200	Social Security, on first $90,000 of wages						7.65%				
	0300	Unemployment, combined Federal and State, minimum						.80%				
	0350	Average						6.20%				
	0400	Maximum						11.76%				

01300 | Administrative Requirements

		01310 | Project Management/Coordination	CREW	DAILY OUTPUT	LABOR-HOURS	UNIT	2006 BARE COSTS MAT.	LABOR	EQUIP.	TOTAL	TOTAL INCL O&P	
150	0010	**PERMITS**										150
	0020	Rule of thumb, most cities, minimum				Job					.50%	
	0100	Maximum				"					2%	
200	0010	**PERFORMANCE BOND** (R013113-80)										200
	0020	For buildings, minimum				Job					.60%	
	0100	Maximum				"					2.50%	
300	0010	**CONSTRUCTION TIME** Requirements (R012157-20)										300
350	0010	**INSURANCE** (R013113-40)										350
	0020	Builders risk, standard, minimum				Job					.24%	
	0050	Maximum (R013113-60)									.64%	
	0200	All-risk type, minimum									.25%	
	0250	Maximum									.62%	
	0400	Contractor's equipment floater, minimum				Value					.50%	
	0450	Maximum				"					1.50%	
	0800	Workers' compensation & employer's liability, average										

01300 | Administrative Requirements

01310 | Project Management/Coordination

		01310 Project Management/Coordination	CREW	DAILY OUTPUT	LABOR-HOURS	UNIT	2006 BARE COSTS MAT.	LABOR	EQUIP.	TOTAL	TOTAL INCL O&P	
350	0850	by trade, carpentry, general R013113-40				Payroll		18.39%				350
	0900	Clerical						.65%				
	0950	Concrete R013113-60						15.82%				
	1000	Electrical						6.84%				
	1050	Excavation						10.63%				
	1100	Glazing						13.87%				
	1150	Insulation						15.76%				
	1200	Lathing						11.44%				
	1250	Masonry						14.92%				
	1300	Painting & decorating						13.29%				
	1350	Pile driving						22.07%				
	1400	Plastering						14%				
	1450	Plumbing						8.22%				
	1500	Roofing						32.40%				
	1550	Sheet metal work (HVAC)						11.76%				
	1600	Steel erection, structural						39.65%				
	1650	Tile work, interior ceramic						9.52%				
	1700	Waterproofing, brush or hand caulking						7.04%				
	1800	Wrecking						40.06%				
	2000	Range of 35 trades in 50 states, excl. wrecking, min.						2.40%				
	2100	Average						16.30%				
	2200	Maximum						137.60%				
400	0010	**MAIN OFFICE EXPENSE** Average for General Contractors R013113-50										400
	0020	As a percentage of their annual volume										
	0030	Annual volume to $50,000, minimum				% Vol.				20%		
	0040	Maximum								30%		
	0060	To $100,000, minimum								17%		
	0070	Maximum								22%		
	0080	To $250,000, minimum								16%		
	0090	Maximum								19%		
	0110	To $500,000, minimum								14%		
	0120	Maximum								16%		
	0130	To $1,000,000, minimum								8%		
	0140	Maximum								10%		
500	0010	**MARK-UP** For General Contractors for change										500
	0100	of scope of job as bid										
	0200	Extra work, by subcontractors, add				%					10%	
	0250	By General Contractor, add									15%	
	0400	Omitted work, by subcontractors, deduct all but									5%	
	0450	By General Contractor, deduct all but									7.50%	
	0600	Overtime work, by subcontractors, add									15%	
	0650	By General Contractor, add									10%	
610	0010	**OVERHEAD** As a percent of installing contractors direct costs R013113-50										610
	0040	Includes an allowance for home office expenses, FICA,										
	0060	Risk & public liability insur. and unemploy.; minimum				%				5%		
	0080	Average								15%		
	0100	Maximum								30%		
	0120	With profit allowance, by size of project; under $50,000								40%		
	0140	$50,000 to $100,000								35%		
	0160	$100,000 to $500,000								25%		
	0180	$500,000 to $1,000,000								20%		
	0200	Mark-up on general contractors total incl. O&P for										
	0220	Handling subcontract; minimum				%					5%	
	0240	Average									10%	
	0260	Maximum									15%	

01300 | Administrative Requirements

	01310 | Project Management/Coordination		CREW	DAILY OUTPUT	LABOR-HOURS	UNIT	2006 BARE COSTS MAT.	LABOR	EQUIP.	TOTAL	TOTAL INCL O&P	
700	0010	FIELD PERSONNEL										700
	0020	Clerk, average				Week		335		335	515	
	0100	Field engineer, minimum						800		800	1,225	
	0120	Average						1,040		1,040	1,600	
	0140	Maximum						1,200		1,200	1,850	
	0160	General purpose laborer, average						1,100		1,100	1,675	
	0180	Project manager, minimum						1,500		1,500	2,300	
	0200	Average						1,700		1,700	2,600	
	0220	Maximum						1,950		1,950	3,000	
	0240	Superintendent, minimum						1,450		1,450	2,225	
	0260	Average						1,575		1,575	2,425	
	0280	Maximum						1,800		1,800	2,750	
	0290	Timekeeper, average						930		930	1,425	

	01320 | Const. Progress Documentation		CREW	DAILY OUTPUT	LABOR-HOURS	UNIT	MAT.	LABOR	EQUIP.	TOTAL	TOTAL INCL O&P	
200	0010	SCHEDULING										200
	0020	Critical path, as % of architectural fee, minimum				%					.50%	
	0100	Maximum				"					1%	
	0300	Computer update, micro, no plots, minimum				Ea.					450	
	0400	Including plots, maximum				"					1,550	
	0600	Rule of thumb, CPM scheduling, small job ($10 Million)				Job					.05%	
	0650	Large job ($50 Million +)									.03%	
	0700	Including cost control, small job									.08%	
	0750	Large job									.04%	

	01321 | Construction Photos		CREW	DAILY OUTPUT	LABOR-HOURS	UNIT	MAT.	LABOR	EQUIP.	TOTAL	TOTAL INCL O&P	
500	0010	PHOTOGRAPHS										500
	0020	8″ x 10″, 4 shots, 2 prints ea., std. mounting				Set	300			300	330	
	0100	Hinged linen mounts					325			325	360	
	0200	8″ x 10″, 4 shots, 2 prints each, in color					350			350	385	
	0300	For I.D. slugs, add to all above					4.21			4.21	4.63	
	0500	Aerial photos, initial fly-over, 6 shots, 1 print ea., 8″ x 10″					710			710	780	
	0550	11″ x 14″ prints					790			790	870	
	0600	16″ x 20″ prints					1,025			1,025	1,125	
	0700	For full color prints, add					40%				40%	
	0750	Add for traffic control area					294			294	325	
	0900	For over 30 miles from airport, add per				Mile	5.30			5.30	5.85	
	1000	Vertical photography, 4 to 6 shots with										
	1010	different scales, 1 print each				Set	1,100			1,100	1,200	
	1500	Time lapse equipment, camera and projector, buy					3,775			3,775	4,175	
	1550	Rent per month					565			565	620	
	1700	Cameraman and film, including processing, B.&W.				Day	1,375			1,375	1,525	
	1720	Color				"	1,375			1,375	1,525	

01450 | Quality Control

		01450 Quality Control	CREW	DAILY OUTPUT	LABOR-HOURS	UNIT	2006 BARE COSTS MAT.	LABOR	EQUIP.	TOTAL	TOTAL INCL O&P	
500	0010	**TESTING** and Inspectional Services										500
	0015	For concrete building costing $1,000,000, minimum				Project					5,200	
	0020	Maximum									41,800	
	0050	Steel building, minimum									5,200	
	0070	Maximum									16,300	
	0100	For building costing, $10,000,000, minimum									33,100	
	0150	Maximum									53,000	
	0200	Asphalt testing, compressive strength Marshall stability, set of 3				Ea.					165	
	0220	Density, set of 3									95	
	0250	Extraction, individual tests on sample									150	
	0300	Penetration									45	
	0350	Mix design, 5 specimens									200	
	0360	Additional specimen									40	
	0400	Specific gravity									45	
	0420	Swell test									70	
	0450	Water effect and cohesion, set of 6									200	
	0470	Water effect and plastic flow									70	
	0600	Concrete testing, aggregates, abrasion, ASTM C 131									150	
	0650	Absorption, ASTM C 127									46	
	0800	Petrographic analysis, ASTM C 295									850	
	0900	Specific gravity, ASTM C 127									55	
	1000	Sieve analysis, washed, ASTM C 136									65	
	1050	Unwashed									65	
	1200	Sulfate soundness									125	
	1300	Weight per cubic foot									40	
	1500	Cement, physical tests, ASTM C 150									350	
	1600	Chemical tests, ASTM C 150									270	
	1800	Compressive test, cylinder, delivered to lab, ASTM C 39									13	
	1900	Picked up by lab, minimum									15	
	1950	Average									20	
	2000	Maximum									30	
	2200	Compressive strength, cores (not incl. drilling), ASTM C 42									40	
	2250	Core drilling, 4" diameter (plus technician)				Inch					25	
	2260	Technician for core drilling				Hr.					50	
	2300	Patching core holes				Ea.					24	
	2400	Drying shrinkage at 28 days									260	
	2500	Flexural test beams, ASTM C 78									65	
	2600	Mix design, one batch mix									285	
	2650	Added trial batches									132	
	2800	Modulus of elasticity, ASTM C 469									180	
	2900	Tensile test, cylinders, ASTM C 496									50	
	3000	Water-Cement ratio curve, 3 batches									155	
	3100	4 batches									205	
	3300	Masonry testing, absorption, per 5 brick, ASTM C 67									50	
	3350	Chemical resistance, per 2 brick									55	
	3400	Compressive strength, per 5 brick, ASTM C 67									75	
	3420	Efflorescence, per 5 brick, ASTM C 67									75	
	3440	Imperviousness, per 5 brick									96	
	3470	Modulus of rupture, per 5 brick									95	
	3500	Moisture, block only									35	
	3550	Mortar, compressive strength, set of 3									25	
	4100	Reinforcing steel, bend test									61	
	4200	Tensile test, up to #8 bar									40	
	4220	#9 to #11 bar									45	
	4240	#14 bar and larger									70	
	4400	Soil testing, Atterberg limits, liquid and plastic limits									65	

01400 | Quality Requirements

	01450 | Quality Control		CREW	DAILY OUTPUT	LABOR-HOURS	UNIT	2006 BARE COSTS MAT.	LABOR	EQUIP.	TOTAL	TOTAL INCL O&P	
500	4510	Hydrometer analysis				Ea.					120	500
	4530	Specific gravity, ASTM D 354									48	
	4600	Sieve analysis, washed, ASTM D 422									60	
	4700	Unwashed, ASTM D 422									65	
	4710	Consolidation test (ASTM D2435), minimum									275	
	4715	Maximum									475	
	4720	Density and classification of undisturbed sample									80	
	4735	Soil density, nuclear method, ASTM D2922									38.67	
	4740	Sand cone method ASTM D1556									30.17	
	4750	Moisture content, ASTM D 2216									10	
	4780	Permeability test, double ring infiltrometer									550	
	4800	Permeability, var. or constant head, undist., ASTM D 2434									250	
	4850	Recompacted									275	
	4900	Proctor compaction, 4" standard mold, ASTM D 698									135	
	4950	6" modified mold									75	
	5100	Shear tests, triaxial, minimum									450	
	5150	Maximum									600	
	5300	Direct shear, minimum, ASTM D 3080									350	
	5350	Maximum									450	
	5550	Technician for inspection, per day, earthwork									220	
	5570	Concrete									270	
	5650	Bolting									280	
	5750	Roofing									255	
	5790	Welding									270	
	5820	Non-destructive testing, dye penetrant				Day					330	
	5840	Magnetic particle									330	
	5860	Radiography									495	
	5880	Ultrasonic									340	
	6000	Welding certification, minimum				Ea.					100	
	6100	Maximum				"					275	
	7000	Underground storage tank										
	7500	Volumetric tightness test, <=12,000 gal				Ea.					400	
	7510	<=30,000 gal				"					660	
	7600	Vadose zone (soil gas) sampling, 10-40 samples, min.				Day					1,500	
	7610	Maximum				"					2,500	
	7700	Ground water monitoring incl. drilling 3 wells, min.				Total					5,000	
	7710	Maximum				"					7,000	
	8000	X-ray concrete slabs				Ea.					200	

01500 | Temporary Facilities & Controls

	01510 | Temporary Utilities		CREW	DAILY OUTPUT	LABOR-HOURS	UNIT	2006 BARE COSTS MAT.	LABOR	EQUIP.	TOTAL	TOTAL INCL O&P	
800	0010	TEMPORARY UTILITIES										800
	0100	Heat, incl. fuel and operation, per week, 12 hrs. per day	1 Skwk	100	.080	CSF Flr	7.20	2.92		10.12	12.75	
	0200	24 hrs. per day	"	60	.133		10.80	4.87		15.67	19.85	
	0350	Lighting, incl. service lamps, wiring & outlets, minimum	1 Elec	34	.235		2.39	9.90		12.29	17.85	
	0360	Maximum	"	17	.471		5.20	19.75		24.95	36.50	
	0400	Power for temp lighting only, per month, min/month 6.6 KWH								.75	1.18	
	0450	Maximum/month 23.6 KWH								2.85	3.14	
	0600	Power for job duration incl. elevator, etc., minimum								47	51.70	

01500 | Temporary Facilities & Controls

		01510 | Temporary Utilities	CREW	DAILY OUTPUT	LABOR-HOURS	UNIT	2006 BARE COSTS MAT.	LABOR	EQUIP.	TOTAL	TOTAL INCL O&P	
800	0650	Maximum				CSF Flr				110	121	800
	0700	Temporary construction water bill per mo. average				Month	60			60	66	
	1000	Toilet, portable, see also Reference Section										
		01520 | Construction Facilities										
500	0010	OFFICE										500
	0020	Trailer, furnished, no hookups, 20′ x 8′, buy	2 Skwk	1	16	Ea.	7,550	585		8,135	9,250	
	0250	Rent per month					167			167	183	
	0300	32′ x 8′, buy	2 Skwk	.70	22.857		10,400	835		11,235	12,800	
	0350	Rent per month					183			183	201	
	0400	50′ x 10′, buy	2 Skwk	.60	26.667		18,100	975		19,075	21,600	
	0450	Rent per month					293			293	320	
	0500	50′ x 12′, buy	2 Skwk	.50	32		23,300	1,175		24,475	27,500	
	0550	Rent per month					340			340	375	
	0700	For air conditioning, rent per month, add					41			41	45	
	0800	For delivery, add per mile				Mile	3.50			3.50	3.85	
	1000	Portable buildings, prefab, on skids, economy, 8′ x 8′	2 Carp	265	.060	S.F.	82	2.15		84.15	93.50	
	1100	Deluxe, 8′ x 12′	″	150	.107	″	90	3.79		93.79	105	
	1200	Storage boxes, 20′ x 8′, buy	2 Skwk	1.80	8.889	Ea.	3,250	325		3,575	4,100	
	1250	Rent per month					75			75	82.50	
	1300	40′ x 8′, buy	2 Skwk	1.40	11.429		4,275	415		4,690	5,375	
	1350	Rent per month					109			109	120	
	5000	Air supported structures, see Division 13011-200										
550	0010	FIELD OFFICE EXPENSE										550
	0100	Field office expense, office equipment rental average				Month	145			145	160	
	0120	Office supplies, average				″	90			90	99	
	0125	Office trailer rental, see Division 01520-500										
	0140	Telephone bill; avg. bill/month incl. long dist.				Month	210			210	231	
	0160	Field office lights & HVAC				″	110			110	121	
900	0010	WEATHER STATION										900
	0020	Remote recording, minimum				Ea.	5,500			5,500	6,050	
	0100	Maximum				″	27,000			27,000	29,700	
		01530 | Temporary Construction										
700	0010	PROTECTION										700
	0020	Stair tread, 2″ x 12″ planks, 1 use	1 Carp	75	.107	Tread	4.80	3.79		8.59	11.60	
	0100	Exterior plywood, 1/2″ thick, 1 use		65	.123		1.84	4.38		6.22	9.30	
	0200	3/4″ thick, 1 use		60	.133		2.46	4.74		7.20	10.55	
900	0010	WINTER PROTECTION Reinforced plastic on wood										900
	0100	framing to close openings	2 Clab	750	.021	S.F.	.39	.58		.97	1.40	
	0200	Tarpaulins hung over scaffolding, 8 uses, not incl. scaffolding		1500	.011		.19	.29		.48	.69	
	0250	Tarpaulin polyester reinf. w/integral fastening system 11 mils thick		1600	.010		.77	.27		1.04	1.30	
	0300	Prefab. fiberglass panels, steel frame, 8 uses		1200	.013		.75	.37		1.12	1.44	
		01540 | Construction Aids										
500	0010	PERSONNEL PROTECTIVE EQUIPMENT										500
	0015	Hazardous waste protection										
	0020	Respirator mask only, full face, silicone				Ea.	197			197	217	
	0030	Half face, silicone					26.50			26.50	29.50	
	0040	Respirator cartridges, 2 req'd/mask, dust or asbestos					4.80			4.80	5.30	
	0050	Chemical vapor					4.61			4.61	5.05	
	0060	Combination vapor and dust					10			10	11	
	0100	Emergency escape breathing apparatus, 5 min.					430			430	475	

01540 | Construction Aids

		01540 Construction Aids	CREW	DAILY OUTPUT	LABOR-HOURS	UNIT	2006 BARE COSTS MAT.	LABOR	EQUIP.	TOTAL	TOTAL INCL O&P	
500	0110	10 min.				Ea.	490			490	535	500
	0150	Self contained breathing apparatus with full face piece, 30 min.					1,725			1,725	1,900	
	0160	60 min.					3,125			3,125	3,450	
	0200	Encapsulating suits, limited use, level A					895			895	980	
	0210	Level B					230			230	253	
	0300	Over boots, latex				Pr.	4.20			4.20	4.62	
	0310	PVC					12.50			12.50	13.75	
	0320	Neoprene					51.50			51.50	57	
	0400	Gloves, nitrile/PVC					5.40			5.40	5.95	
	0410	Neoprene coated					24			24	26.50	
550	0010	**PUMP STAGING**, Aluminum R015423-20										550
	0200	24′ long pole section, buy				Ea.	360			360	400	
	0300	18′ long pole section, buy					278			278	305	
	0400	12′ long pole section, buy					191			191	210	
	0500	6′ long pole section, buy					100			100	110	
	0600	6′ long splice joint section, buy					73.50			73.50	81	
	0700	Pump jack					119			119	131	
	0900	Foldable brace					51.50			51.50	56.50	
	1000	Workbench/back safety rail support					63			63	69	
	1100	Scaffolding planks/workbench, 14″ wide x 24′ long					580			580	640	
	1200	Plank end safety rail					196			196	215	
	1250	Safety net, 22′ long					285			285	315	
	1300	System in place, 50′ working height, per use based on 50 uses	2 Carp	84.80	.189	C.S.F.	5.30	6.70		12	16.95	
	1400	100 uses		84.80	.189		2.66	6.70		9.36	14	
	1500	150 uses		84.80	.189		1.78	6.70		8.48	13.05	
700	0010	**SAFETY NETS**										700
	0020	No supports, stock sizes, nylon, 4″ mesh				S.F.	1.03			1.03	1.13	
	0100	Polypropylene, 6″ mesh					1.59			1.59	1.75	
	0200	Small mesh debris nets, 1/4″ & 3/4″ mesh, stock sizes					.74			.74	.81	
	0220	Combined 4″ mesh and 1/4″ mesh, stock sizes					2.05			2.05	2.26	
	0300	Monthly rental, 4″ mesh, stock sizes, 1st month					.40			.40	.44	
	0320	2nd month rental					.20			.20	.22	
	0340	Maximum rental/year					.90			.90	.99	
750	0010	**SCAFFOLDING** R015423-10										750
	0015	Steel tube, reg., rent/mo., no plank, incl. erect or dismantle										
	0090	Building exterior, wall face, 1 to 5 stories, 6′-4″ x 5′ frames	3 Carp	24	1	C.S.F.	25.50	35.50		61	87	
	0200	6 to 12 stories	4 Carp	21.20	1.509		25.50	53.50		79	117	
	0310	13 to 20 stories	5 Carp	20	2		25.50	71		96.50	146	
	0460	Building interior, wall face area, up to 16′ high	3 Carp	25	.960		25.50	34		59.50	84.50	
	0560	16′ to 40′ high		23	1.043		25.50	37		62.50	89.50	
	0800	Building interior floor area, up to 30′ high		312	.077	C.C.F.	2.70	2.73		5.43	7.50	
	0900	Over 30′ high	4 Carp	275	.116	″	2.70	4.14		6.84	9.80	
	0910	Steel tubular, heavy duty shoring, buy										
	0920	Frames 5′ high 2′ wide				Ea.	86.50			86.50	95.50	
	0925	5′ high 4′ wide					98			98	108	
	0930	6′ high 2′ wide					99.50			99.50	109	
	0935	6′ high 4′ wide					117			117	128	
	0940	Accessories										
	0945	Cross braces				Ea.	18.50			18.50	20.50	
	0950	U-head, 8″ x 8″					20			20	22	
	0955	J-head, 4″ x 8″					14.80			14.80	16.25	
	0960	Base plate, 8″ x 8″					16.40			16.40	18.05	
	0965	Leveling jack					35			35	39	
	1000	Steel tubular, regular, buy										
	1100	Frames 3′ high 5′ wide				Ea.	67			67	73.50	

01540 | Construction Aids

R015423-10

		01540 Construction Aids	CREW	DAILY OUTPUT	LABOR-HOURS	UNIT	2006 BARE COSTS MAT.	LABOR	EQUIP.	TOTAL	TOTAL INCL O&P	
750	1150	5′ high 5′ wide				Ea.	77.50			77.50	85	750
	1200	6′-4″ high 5′ wide					97			97	107	
	1350	7′-6″ high 6′ wide					167			167	184	
	1500	Accessories cross braces					17.35			17.35	19.05	
	1550	Guardrail post					17.35			17.35	19.05	
	1600	Guardrail 7′ section					8.40			8.40	9.20	
	1650	Screw jacks & plates					27.50			27.50	30.50	
	1700	Sidearm brackets					32.50			32.50	35.50	
	1750	8″ casters					38			38	42	
	1800	Plank 2″ x 10″ x 16′-0″					49			49	54	
	1900	Stairway section					283			283	310	
	1910	Stairway starter bar					33.50			33.50	37	
	1920	Stairway inside handrail					61			61	67.50	
	1930	Stairway outside handrail					84.50			84.50	93	
	1940	Walk-thru frame guardrail					42.50			42.50	46.50	
	2000	Steel tubular, regular, rent/mo.										
	2100	Frames 3′ high 5′ wide				Ea.	3.94			3.94	4.33	
	2150	5′ high 5′ wide					3.94			3.94	4.33	
	2200	6′-4″ high 5′ wide					3.94			3.94	4.33	
	2250	7′-6″ high 6′ wide					7.35			7.35	8.10	
	2500	Accessories, cross braces					.63			.63	.69	
	2550	Guardrail post					1.05			1.05	1.16	
	2600	Guardrail 7′ section					.79			.79	.87	
	2650	Screw jacks & plates					1.58			1.58	1.74	
	2700	Sidearm brackets					1.58			1.58	1.74	
	2750	8″ casters					6.30			6.30	6.95	
	2800	Outrigger for rolling tower					3.15			3.15	3.47	
	2850	Plank 2″ x 10″ x 16′-0″					5.25			5.25	5.80	
	2900	Stairway section					10.50			10.50	11.55	
	2910	Stairway starter bar					.11			.11	.12	
	2920	Stairway inside handrail					5.25			5.25	5.80	
	2930	Stairway outside handrail					5.25			5.25	5.80	
	2940	Walk-thru frame guardrail					2.10			2.10	2.31	
	3000	Steel tubular, heavy duty shoring, rent/mo.										
	3250	5′ high 2′ & 4′ wide				Ea.	5.25			5.25	5.80	
	3300	6′ high 2′ & 4′ wide					5.25			5.25	5.80	
	3500	Accessories, cross braces					1.05			1.05	1.16	
	3600	U - head, 8″ x 8″					1.05			1.05	1.16	
	3650	J - head, 4″ x 8″					1.05			1.05	1.16	
	3700	Base plate, 8″ x 8″					1.05			1.05	1.16	
	3750	Leveling jack					2.10			2.10	2.31	
	4000	Scaffolding, regular, rent/mo, no plank, w/ erect or dismantle										
	4100	Building exterior 2 stories	3 Carp	25	.960	C.S.F.	25.50	34		59.50	84.50	
	4150	4 stories	"	23	1.043		25.50	37		62.50	89.50	
	4200	6 stories	4 Carp	24	1.333		25.50	47.50		73	107	
	4250	8 stories		22	1.455		25.50	51.50		77	114	
	4300	10 stories		20	1.600		25.50	57		82.50	122	
	4350	12 stories		18	1.778		25.50	63		88.50	133	
	4500	1st tier, 3′ high x 5′ wide frames, 1 use/mo, erect or dismantle	1 Carp	2.38	3.361		39.50	120		159.50	241	
	4550	2 uses/month		2.38	3.361		19.65	120		139.65	220	
	4600	4 uses/month		2.38	3.361		9.80	120		129.80	209	
	4650	8 uses/month		2.38	3.361		4.91	120		124.91	203	
	5000	5′ high x 5′ wide frames, 1 use/month		3.97	2.015		28.50	71.50		100	151	
	5050	2 uses/month		3.97	2.015		14.35	71.50		85.85	135	
	5100	4 uses/month		3.97	2.015		7.15	71.50		78.65	127	
	5150	8 uses/month		3.97	2.015		3.58	71.50		75.08	123	

	01540 | Construction Aids		CREW	DAILY OUTPUT	LABOR-HOURS	UNIT	2006 BARE COSTS MAT.	LABOR	EQUIP.	TOTAL	TOTAL INCL O&P	
750	5500	6′-4″ high x 5′ wide frames, 1 use/month (R015423-10)	1 Carp	5.02	1.594	C.S.F.	25.50	56.50		82	122	750
	5550	2 uses/month		5.02	1.594		12.75	56.50		69.25	108	
	5600	4 uses/month		5.02	1.594		6.40	56.50		62.90	101	
	5650	8 uses/month		5.02	1.594		3.19	56.50		59.69	97.50	
	5700	Planks, 2x10x16′-0″, labor only, erect/remove to 50′ H	3 Carp	144	.167	Ea.		5.95		5.95	9.80	
	5800	Over 50′ high	4 Carp	160	.200	"		7.10		7.10	11.80	
	6000	Heavy duty shoring for elevated slab forms to 8′-2″ high, floor area										
	6010	incl. erection or dismantle labor										
	6100	1 use/month	4 Carp	36	.889	C.S.F.	31	31.50		62.50	86.50	
	6150	2 uses/month		36	.889		15.55	31.50		47.05	69.50	
	6200	3 uses/month		36	.889		10.35	31.50		41.85	64	
	6500	To 14′-8″ high										
	6600	1 use/month	4 Carp	18	1.778	C.S.F.	45.50	63		108.50	155	
	6650	2 uses/month		18	1.778		22.50	63		85.50	130	
	6700	3 uses/month		18	1.778		15.10	63		78.10	122	
755	0010	**SCAFFOLDING SPECIALTIES**										755
	1200	Sidewalk bridge, heavy duty steel posts & beams, including										
	1210	parapet protection & waterproofing										
	1220	8′ to 10′ wide, 2 posts	3 Carp	15	1.600	L.F.	33.50	57		90.50	131	
	1230	3 posts	"	10	2.400	"	51.50	85.50		137	198	
	1500	Sidewalk bridge using tubular steel										
	1512	scaffold frames, including planking	3 Carp	55	.436	L.F.	4.96	15.50		20.46	31	
	1600	For 2 uses per month, deduct from all above					50%					
	1700	For 1 use every 2 months, add to all above					100%					
	1900	Catwalks, 20″ wide, no guardrails, 7′ span, buy				Ea.	144			144	159	
	2000	10′ span, buy				"	185			185	203	
	2800	Hand winch-operated masons scaffolding, no plank										
	2810	plank moving not required										
	2900	98′ long, 10′-6″ high, buy				Ea.	30,300			30,300	33,300	
	3000	Rent per month					1,200			1,200	1,325	
	3100	28′-6″ high, buy					37,200			37,200	40,900	
	3200	Rent per month					1,475			1,475	1,625	
	3400	196′ long, 28′-6″ high, buy					72,000			72,000	79,000	
	3500	Rent per month					2,875			2,875	3,150	
	3600	64′-6″ high, buy					98,500			98,500	108,500	
	3700	Rent per month					3,950			3,950	4,325	
	3720	Putlog, standard, 8′ span, with hangers, buy					70.50			70.50	77.50	
	3730	Rent per month					10.50			10.50	11.55	
	3750	12′ span, buy					106			106	117	
	3755	Rent per month					15.75			15.75	17.35	
	3760	Trussed type, 16′ span, buy					243			243	267	
	3770	Rent per month					21			21	23	
	3790	22′ span, buy					291			291	320	
	3795	Rent per month					31.50			31.50	34.50	
	3800	Rolling ladders with handrails, 30″ wide, buy, 2 step					197			197	216	
	4000	7 step					660			660	725	
	4050	10 step					870			870	955	
	4100	Rolling towers, buy, 5′ wide, 7′ long, 10′ high					1,325			1,325	1,450	
	4200	For 5′ high added sections, to buy, add					219			219	241	
	4300	Complete incl. wheels, railings, outriggers,										
	4350	21′ high, to buy				Ea.	2,225			2,225	2,450	
	4400	Rent/month				"	167			167	184	
760	0010	**STAGING AIDS** and fall protection equipment										760
	0100	Sidewall staging bracket, tubular, buy				Ea.	34			34	37	

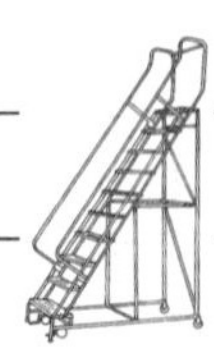

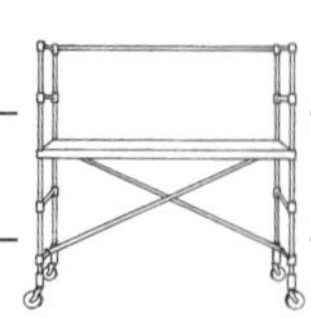

01500 | Temporary Facilities & Controls

		01540 | Construction Aids	CREW	DAILY OUTPUT	LABOR-HOURS	UNIT	2006 BARE COSTS MAT.	LABOR	EQUIP.	TOTAL	TOTAL INCL O&P	
760	0110	Cost each per day, based on 250 days' use				Day	.14			.14	.15	760
	0200	Guard post, buy				Ea.	16.50			16.50	18.15	
	0210	Cost each per day, based on 250 days' use				Day	.07			.07	.07	
	0300	End guard chains, buy per pair				Pair	27.50			27.50	30.50	
	0310	Cost per set per day, based on 250 days' use				Day	.14			.14	.15	
	1000	Roof shingling bracket, steel, buy				Ea.	7.15			7.15	7.85	
	1010	Cost each per day, based on 250 days' use				Day	.03			.03	.03	
	1100	Wood bracket, buy				Ea.	14.20			14.20	15.60	
	1110	Cost each per day, based on 250 days' use				Day	.06			.06	.06	
	2000	Ladder jack, aluminum, buy per pair				Pair	89.50			89.50	98.50	
	2010	Cost per pair per day, based on 250 days' use				Day	.36			.36	.39	
	2100	Steel siderail jack, buy per pair				Pair	69.50			69.50	76	
	2110	Cost per pair per day, based on 250 days' use				Day	.28			.28	.30	
	3000	Laminated wood plank, 2x10x16', buy				Ea.	47			47	51.50	
	3010	Cost each per day, based on 250 days' use				Day	.19			.19	.21	
	3100	Aluminum scaffolding plank, 20" wide x 24' long, buy				Ea.	725			725	795	
	3110	Cost each per day, based on 250 days' use				Day	2.89			2.89	3.18	
	4000	Nylon full-body harness, lanyard and rope grab				Ea.	218			218	239	
	4010	Cost each per day, based on 250 days' use				Day	.87			.87	.96	
	4100	Rope for safety line, 5/8" x 100' nylon, buy				Ea.	45			45	49.50	
	4110	Cost each per day, based on 250 days' use				Day	.18			.18	.20	
	4200	Permanent U-Bolt roof anchor, buy				Ea.	32.50			32.50	36	
	4300	Temporary (one use) roof ridge anchor, buy				"	26.50			26.50	29	
	5000	Installation (setup and removal) of staging aids										
	5010	Sidewall staging bracket	2 Carp	64	.250	Ea.		8.90		8.90	14.75	
	5020	Guard post with 2 wood rails	"	64	.250			8.90		8.90	14.75	
	5030	End guard chains, set	1 Carp	64	.125			4.44		4.44	7.35	
	5100	Roof shingling bracket		96	.083			2.96		2.96	4.91	
	5200	Ladder jack		64	.125			4.44		4.44	7.35	
	5300	Wood plank, 2x10x16'	2 Carp	80	.200			7.10		7.10	11.80	
	5310	Aluminum scaffold plank, 20" x 24'	"	40	.400			14.20		14.20	23.50	
	5410	Safety rope	1 Carp	40	.200			7.10		7.10	11.80	
	5420	Permanent U-Bolt roof anchor (install only)	2 Carp	40	.400			14.20		14.20	23.50	
	5430	Temporary roof ridge anchor (install only)	1 Carp	64	.125			4.44		4.44	7.35	
780	0010	**SWING STAGING**, 500 lb cap., 2' wide to 24' long, hand-operated hoist										780
	0020	steel cable type, with 60' cables, buy				Ea.	4,675			4,675	5,150	
	0030	Rent per month				"	470			470	515	
	0600	Lightweight (not for masons) 24' long for 150' height										
	0610	manual type, buy				Ea.	4,950			4,950	5,450	
	0620	Rent per month					495			495	545	
	0700	Powered, electric or air, to 150' high, buy					17,500			17,500	19,300	
	0710	Rent per month					1,225			1,225	1,350	
	0780	To 300' high, buy					20,600			20,600	22,600	
	0800	Rent per month					1,450			1,450	1,575	
	1000	Bosun's chair or work basket 3' x 3.5', to 300' high, electric, buy					7,725			7,725	8,500	
	1010	Rent per month					540			540	595	
	2200	Move swing staging (setup and remove)	E-4	2	16	Move		645	44.50	689.50	1,250	
790	0010	**SURVEYOR STAKES**										790
	0020	Hardwood, 1" x 1" x 48" long				C	48			48	53	
	0100	2" x 2" x 18" long					63			63	69.50	
	0150	2" x 2" x 24" long					75			75	82.50	
	0200	2" x 2" x 30" long					74			74	81.50	
800	0010	**TARPAULINS**										800
	0020	Cotton duck, 10 oz. to 13.13 oz. per S.Y., minimum				S.F.	.50			.50	.55	

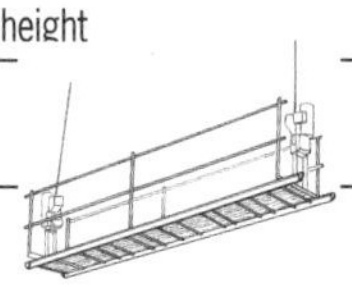

01540 | Construction Aids

		01540 Construction Aids	CREW	DAILY OUTPUT	LABOR-HOURS	UNIT	2006 BARE COSTS MAT.	LABOR	EQUIP.	TOTAL	TOTAL INCL O&P	
800	0050	Maximum				S.F.	.58			.58	.64	800
	0100	Polyvinyl-coated nylon, 14 oz. to 18 oz., minimum					.48			.48	.53	
	0150	Maximum					.68			.68	.75	
	0200	Reinforced polyethylene 3 mils thick, white					.11			.11	.12	
	0300	4 mils thick, white, clear or black					.14			.14	.15	
	0400	5.5 mils thick, clear					.20			.20	.22	
	0500	White, fire retardant					.18			.18	.20	
	0600	7.5 mils, oil resistant, fire retardant					.19			.19	.21	
	0700	8.5 mils, black					.24			.24	.26	
	0710	Woven polyethylene, 6 mils thick					.48			.48	.53	
	0720	Steel reinforced polyethylene, 4 mils thick					.53			.53	.58	
	0730	Polyester reinforced w/integral fastening system 11 mils thick					1.07			1.07	1.18	
	0740	Mylar polyester, non-reinforced, 7 mils thick				↓	1.17			1.17	1.29	
820	0010	**SMALL TOOLS** R013113-50										820
	0020	As % of contractor's work, minimum				Total					.50%	
	0100	Maximum				"					2%	

01550 | Vehicular Access & Parking

		01550 Vehicular Access & Parking	CREW	DAILY OUTPUT	LABOR-HOURS	UNIT	MAT.	LABOR	EQUIP.	TOTAL	TOTAL INCL O&P	
700	0010	**ROADS AND SIDEWALKS** Temporary										700
	0050	Roads, gravel fill, no surfacing, 4" gravel depth	B-14	715	.067	S.Y.	3.30	1.95	.32	5.57	7.20	
	0100	8" gravel depth	"	615	.078	"	6.60	2.27	.37	9.24	11.35	
	1000	Ramp, 3/4" plywood on 2" x 6" joists, 16" O.C.	2 Carp	300	.053	S.F.	1.41	1.90		3.31	4.69	
	1100	On 2" x 10" joists, 16" O.C.	"	275	.058		2.10	2.07		4.17	5.75	
	2200	Sidewalks, 2" x 12" planks, 2 uses	1 Carp	350	.023		.80	.81		1.61	2.23	
	2300	Exterior plywood, 2 uses, 1/2" thick		750	.011		.31	.38		.69	.97	
	2500	3/4" thick	↓	600	.013	↓	.41	.47		.88	1.24	

01560 | Barriers & Enclosures

		01560 Barriers & Enclosures	CREW	DAILY OUTPUT	LABOR-HOURS	UNIT	MAT.	LABOR	EQUIP.	TOTAL	TOTAL INCL O&P	
100	0010	**BARRICADES**										100
	0020	5' high, 3 rail @ 2" x 8", fixed	2 Carp	20	.800	L.F.	5.10	28.50		33.60	52.50	
	0150	Movable	"	30	.533	"	4.38	18.95		23.33	36.50	
	0300	Stock units, 6' high, 8' wide, plain, buy				Ea.	435			435	480	
	0350	With reflective tape, buy				"	525			525	580	
	0400	Break-a-way 3" PVC pipe barricade										
	0410	with 3 ea. 1' x 4' reflectorized panels, buy				Ea.	305			305	335	
	0500	Plywood with steel legs, 32" wide					72			72	79	
	0600	Telescoping Christmas tree, 9' high, 5 flags, buy					122			122	134	
	0800	Traffic cones, PVC, 18" high					6.25			6.25	6.85	
	0850	28" high				↓	19			19	21	
	1000	Guardrail, wooden, 3' high, 1" x 6", on 2" x 4" posts	2 Carp	200	.080	L.F.	1.04	2.84		3.88	5.85	
	1100	2" x 6", on 4" x 4" posts	"	165	.097		1.91	3.45		5.36	7.80	
	1200	Portable metal with base pads, buy					15.95			15.95	17.55	
	1250	Typical installation, assume 10 reuses	2 Carp	600	.027	↓	1.65	.95		2.60	3.39	
	1300	Barricade tape, polyethelyne, 7 mil, 3" wide x 500' long roll				Ea.	25			25	27.50	
	5000	Barricades, see also Reference Section										
250	0010	**TEMPORARY FENCING**										250
	0020	Chain link, 11 ga., 5' high	2 Clab	400	.040	L.F.	4.79	1.10		5.89	7.05	
	0100	6' high		300	.053		4.62	1.46		6.08	7.50	
	0200	Rented chain link, 6' high, to 1,000' (up to 12 mo.)		400	.040		3.03	1.10		4.13	5.15	
	0250	Over 1,000' (up to 12 mo.)	↓	300	.053		2.19	1.46		3.65	4.83	
	0350	Plywood, painted, 2" x 4" frame, 4' high	A-4	135	.178		5.15	6.10		11.25	15.70	
	0400	4" x 4" frame, 8' high	"	110	.218		9.85	7.50		17.35	23	
	0500	Wire mesh on 4" x 4" posts, 4' high	2 Carp	100	.160	↓	8.15	5.70		13.85	18.35	

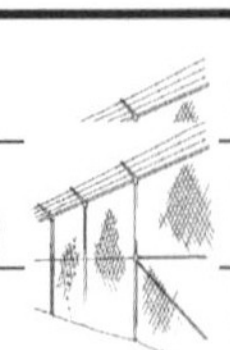

01500 | Temporary Facilities & Controls

	01560 | Barriers & Enclosures		CREW	DAILY OUTPUT	LABOR-HOURS	UNIT	2006 BARE COSTS MAT.	LABOR	EQUIP.	TOTAL	TOTAL INCL O&P	
250	0550	8′ high	2 Carp	80	.200	L.F.	12.60	7.10		19.70	25.50	250
400	0010	**TEMPORARY CONSTRUCTION** See also Division 01530										400
800	0010	**WATCHMAN**										800
	0020	Service, monthly basis, uniformed person, minimum				Hr.					20	
	0100	Maximum									50	
	0200	Person and command dog, minimum									28	
	0300	Maximum				↓					60	
	0500	Sentry dog, leased, with job patrol (yard dog), 1 dog				Week					250	
	0600	2 dogs				"					320	
	0800	Purchase, trained sentry dog, minimum				Ea.					1,500	
	0900	Maximum				"					3,000	
	01580 | Project Signs											
700	0010	**SIGNS**										700
	0020	Hi Intensity reflectorized, no posts, buy				S.F.	16.55			16.55	18.20	

01700 | Execution Requirements

	01740 | Cleaning		CREW	DAILY OUTPUT	LABOR-HOURS	UNIT	2006 BARE COSTS MAT.	LABOR	EQUIP.	TOTAL	TOTAL INCL O&P	
500	0010	**CLEANING UP**										500
	0020	After job completion, allow, minimum				Job					.30%	
	0040	Maximum				"					1%	
	0042	Rubbish removal, see division 02220-350										
	0052	Cleanup of floor area, continuous, per day, during const.	A-5	16	1.125	M.S.F.	1.70	31	2.01	34.71	55	
	0100	Final by GC at end of job	"	11.50	1.565	"	2.71	43	2.80	48.51	77	

01800 | Facility Operation

	01810 | Commissioning		CREW	DAILY OUTPUT	LABOR-HOURS	UNIT	2006 BARE COSTS MAT.	LABOR	EQUIP.	TOTAL	TOTAL INCL O&P	
100	0010	**COMMISSIONING** Including documentation of design intent										100
	0100	performance verification, O&M, training, min				Project					.50%	
	0150	Maximum				"					.75%	
	01830 | Operation & Maintenance											
500	0010	**FACILITIES MAINTENANCE EQUIPMENT** Purchase										500
	0090	Carpet care equipment										
	0110	Dual motor vac, 1 H.P., 16″ brush				Ea.	214			214	235	
	0120	Upright vacuum, 12″ brush					257			257	282	
	0130	14″ brush					420			420	460	
	0140	Soil extractor, hot water 5′ wand, 12″ head				↓	1,450			1,450	1,600	

	01830 | Operation & Maintenance	CREW	DAILY OUTPUT	LABOR-HOURS	UNIT	2006 BARE COSTS MAT.	LABOR	EQUIP.	TOTAL	TOTAL INCL O&P	
500	0150 Dry foam, 13″ brush				Ea.	975			975	1,075	500
	0160 24″ brush				↓	2,125			2,125	2,325	
	0240 Floor care equipment										
	0260 Polishing, buffing, waxing machine, 175 RPM										
	0270 .33 H.P., 11″ dia. brush				Ea.	500			500	550	
	0280 1 H.P., 16″ dia. brush					830			830	910	
	0290 18″ dia. brush					885			885	975	
	0300 1.5 H.P., 20″ dia. brush					955			955	1,050	
	0310 24″ dia. brush				↓	1,125			1,125	1,250	
	0330 Scrubber, automatic, 2 stage 1 H.P. vacuum motor										
	0340 20″ dia. brush				Ea.	4,775			4,775	5,275	
	0350 28″ dia. brush				″	7,550			7,550	8,300	
	1200 Plumbing maintenance equipment										
	1220 Kinetic water ram				Ea.	271			271	298	
	1240 Cable pipe snake, 104 ft., self feed, electric, .5 H.P.				″	1,650			1,650	1,825	
	1500 Specialty equipment										
	1550 Litterbuggy, 8 H.P. gasoline				Ea.	2,075			2,075	2,300	
	1560 Propane					2,675			2,675	2,925	
	1570 Pressure cleaner, hot water, minimum					2,750			2,750	3,025	
	1580 Maximum				↓	5,650			5,650	6,200	
	1590 Vacuum cleaners, steel canister										
	1600 Dry only, two stage, 1 H.P.				Ea.	305			305	340	
	1610 Wet/dry, two stage, 2 H.P.					365			365	400	
	1620 4 H.P.					540			540	590	
	1630 Wet/dry, three stage, 1.5 H.P.					340			340	375	
	1670 Squeegee wet only, 2 stage, 1 H.P.					765			765	840	
	1680 2 H.P.					1,425			1,425	1,550	
	1690 Asbestos and hazardous waste dry vacs, 1 H.P.					1,800			1,800	1,975	
	1700 2 H.P.					1,850			1,850	2,025	
	1710 Three stage, 1 H.P.					2,175			2,175	2,400	
	1800 Upholstery soil extractor, dry-foam				↓	2,350			2,350	2,575	
	1900 Snow removal equipment										
	1920 Thrower, 3 H.P., 20″, single stage, gas				Ea.	575			575	635	
	1930 Electric start					680			680	750	
	1940 5 H.P., 24″, 2 stage, gas					1,000			1,000	1,100	
	1950 11 H.P., 36″				↓	2,225			2,225	2,450	
	1970 Trash receptacles and closures; see div. 02870-520										
	2000 Tube cleaning equip., for heat exchangers, condensers, evap.										
	2010 Tubes/pipes 1/4″-1″, 115v, .5 H.P. drive				Ea.	1,475			1,475	1,625	
	2020 Air powered					1,550			1,550	1,700	
	2040 1″ and up, 115 volt, 1 H.P. drive				↓	1,650			1,650	1,800	
	01832 | Facilities Maintenance										
220	0010 **SITE WORK FACILITIES MAINTENANCE** See Div. 02985-700										220
230	0010 **CONCRETE FACILITIES MAINTENANCE**										230
	0200 Epoxy floor - resurface concrete floor with aggregated epoxy										
	0210 Clean, patch, prime, sand/agg. mix, power trowel, sealer										
	0220 Minimum cost				S.F.					4.50	
	0230 Maximum cost				″					7.50	
	1000 Patching concrete										
	1010 Floors, 1/4″ thick, small areas, regular grout	1 Cefi	170	.047	S.F.	1.17	1.62		2.79	3.84	
	1015 Epoxy grout	″	100	.080	″	8.40	2.75		11.15	13.60	
	1020 Walls, including chipping, cleaning and epoxy grout										
	1025 Minimum	1 Cefi	65	.123	S.F.	3.50	4.23		7.73	10.50	

	01832 | Facilities Maintenance		CREW	DAILY OUTPUT	LABOR-HOURS	UNIT	2006 BARE COSTS MAT.	LABOR	EQUIP.	TOTAL	TOTAL INCL O&P	
230	1030	Average	1 Cefi	50	.160	S.F.	8.15	5.50		13.65	17.60	230
	1035	Maximum		40	.200		11.40	6.90		18.30	23.50	
	1037	Stair cleaning, sweep	1 Clab	50	.160	Flight		4.38		4.38	7.25	
	1039	Mop or scrub	"	5	1.600	"	.07	44		44.07	72.50	
	1040	Sealing concrete floor, 2 coats, oil or urethane base, min. cost				S.F.				.50	.60	
	1050	Maximum cost				"				.75	.90	
	1080	Sweep and damp mop concrete floor	1 Clab	20	.400	M.S.F.	1.57	10.95		12.52	19.85	
240	0010	**MASONRY FACILITIES MAINTENANCE**										240
	0040	Caulking masonry, no staging included										
	0050	Re-caulk only, oil base, 1/2" x 1/2"	1 Bric	225	.036	L.F.	.47	1.30		1.77	2.63	
	0060	Butyl		205	.039		.30	1.43		1.73	2.64	
	0070	Poly sulfide		200	.040		.60	1.46		2.06	3.03	
	0080	Silicone		195	.041		.46	1.50		1.96	2.94	
	0100	Cut out and re-caulk, oil base		145	.055		.47	2.02		2.49	3.79	
	0110	Butyl		130	.062		.30	2.25		2.55	3.98	
	0120	Polysulfide		125	.064		.30	2.34		2.64	4.13	
	0130	Silicone		125	.064		.46	2.34		2.80	4.31	
	0800	Cleaning masonry, no staging included										
	0850	Chemical cleaning, brush and wash, minimum cost	D-1	800	.020	S.F.	.03	.64		.67	1.07	
	0860	Average cost		400	.040		.04	1.29		1.33	2.13	
	0870	Maximum cost		330	.048		.07	1.56		1.63	2.61	
	0880	High pressure water only, minimum cost	B-9	2000	.020			.56	.07	.63	1	
	0900	Average cost		1500	.027			.74	.10	.84	1.34	
	0910	Maximum cost		1000	.040			1.11	.15	1.26	2	
	0920	High pressure water and chemical, minimum cost		1800	.022		.03	.62	.08	.73	1.14	
	0930	Average cost		1200	.033		.04	.93	.12	1.09	1.72	
	0940	Maximum cost		800	.050		.07	1.39	.18	1.64	2.58	
	0950	Sandblast, wet system, minimum cost		1700	.024		.22	.65	.09	.96	1.42	
	0960	Average cost		1100	.036		.27	1.01	.13	1.41	2.12	
	0970	Maximum cost		700	.057		.33	1.59	.21	2.13	3.22	
	0980	Dry system, minimum cost		3000	.013		.22	.37	.05	.64	.90	
	0990	Average cost		1750	.023		.27	.64	.08	.99	1.44	
	1000	Maximum cost		1500	.027		.33	.74	.10	1.17	1.70	
	1010	For walnut shells, add					.37			.37	.40	
	1020	For corn chips, add					.27			.27	.30	
	1030	Steam cleaning, minimum cost	B-9	3000	.013			.37	.05	.42	.66	
	1040	Average cost		2500	.016			.44	.06	.50	.80	
	1050	Maximum cost		1500	.027			.74	.10	.84	1.34	
	1060	Add for masking doors & windows									.80	
	1070	Add for pedestrian protection				Job					10%	
	1080	Add for wire cut face brick				S.F.					.32	
	3000	Pointing masonry										
	3030	Cut and repoint brick, hard mortar, running bond	1 Bric	80	.100	S.F.	.23	3.66		3.89	6.20	
	3040	Common bond		77	.104		.23	3.80		4.03	6.40	
	3050	Flemish bond		70	.114		.24	4.18		4.42	7.05	
	3060	English bond		65	.123		.24	4.50		4.74	7.55	
	3070	Soft old mortar, running bond		100	.080		.25	2.92		3.17	5	
	3080	Common bond		96	.083		.25	3.05		3.30	5.20	
	3090	Flemish bond		90	.089		.26	3.25		3.51	5.55	
	3100	English bond		82	.098		.26	3.57		3.83	6.10	
	3110	Stonework, hard mortar		140	.057	L.F.	.35	2.09		2.44	3.78	
	3120	Soft old mortar		160	.050	"	.39	1.83		2.22	3.40	
	3130	Repoint, mask and grout method, running bond		95	.084	S.F.	.35	3.08		3.43	5.40	
	3140	Common bond		90	.089		.35	3.25		3.60	5.65	
	3150	Flemish bond		86	.093		.35	3.40		3.75	5.90	

		01832 | Facilities Maintenance	Crew	Daily Output	Labor-Hours	Unit	2006 Bare Costs Mat.	Labor	Equip.	Total	Total Incl O&P	
240	3160	English bond	1 Bric	77	.104	S.F.	.35	3.80		4.15	6.55	240
	3170	Scrub coat, sand grout on walls, minimum		120	.067		.35	2.44		2.79	4.34	
	3180	Maximum	↓	98	.082	↓	.38	2.98		3.36	5.25	
250	0010	**METALS FACILITIES MAINTENANCE**										250
	2000	Steel surface treat., sand blast & paint, see Div. 05910-500										
	2010	Steel siding painting, see Div. 09910-700 & 09910-710										
270	0010	**MOISTURE-THERMAL CONTROL FACILITIES MAINTENANCE**										270
	0100	Caulking and sealants see Division 07920-800										
	0110	Caulking around exterior doors and windows, silicone	1 Carp	400	.020	L.F.	.12	.71		.83	1.31	
	2000	Roofing & siding demolition see Division 07060-110										
	2500	Single-ply or built-up roofing maintenance										
	2510	Walk roof, clean drain trap, pick-up trash										
	2520	Allow per year, minimum				S.F.					.02	
	2530	Maximum				"					.06	
	3000	Roof non-destructive testing, infrared analysis system										
	3040	Minimum cost				S.F.					.04	
	3050	Maximum cost				"					.06	
280	0010	**DOOR & WINDOW FACILITIES MAINTENANCE**										280
	0200	Caulking & sealants see Division 07920-800										
	0250	Cylinder dead bolt lock set, repair	1 Skwk	11	.727	Ea.		26.50		26.50	43.50	
	0300	Demolition, windows see Division 08060-120										
	0500	Door closer, overhaul	1 Skwk	5	1.600	Ea.		58.50		58.50	95.50	
	0510	Repair concealed door check or closer		4	2			73		73	119	
	0520	Adjust	↓	26	.308	↓		11.25		11.25	18.35	
	0530	Replace, see Division 08710-300										
	1500	Remove and reset window, minimum cost (labor)	1 Carp	4	2	Ea.		71		71	118	
	1510	Average cost		2	4			142		142	236	
	1520	Maximum cost	↓	1.33	6.015			214		214	355	
	1900	Washing door, 3′ x 7′, damp wipe	1 Clab	145	.055	↓	.16	1.51		1.67	2.67	
	2000	Washing windows, at ground level w/ladder, sponge & squeegee										
	2010	Both sides, 4′ x 7′ 2 pane	1 Clab	60	.133	Ea.	.17	3.65		3.82	6.25	
	2020	4′ x 6′, 4 pane		55	.145		.15	3.99		4.14	6.75	
	2030	3.5′ x 5.5′, 8 pane		45	.178		.12	4.87		4.99	8.20	
	2040	2.5′ x 5.7′, 12 pane		40	.200		.09	5.50		5.59	9.20	
	2050	4′ x 6′, 16 pane		35	.229		.15	6.25		6.40	10.55	
	2060	4′ x 7′, industrial, 20 pane		30	.267		.17	7.30		7.47	12.30	
	2070	6′ x 7′, austral casement, 6 pane		35	.229	↓	.26	6.25		6.51	10.70	
	2080	Clear glass partition, 8 sq. ft. per unit		3	2.667	M.S.F.	3.10	73		76.10	124	
	2090	Opaque, 20 sq. ft. per unit	↓	3	2.667	"	3.10	73		76.10	124	
	2100	Alternate pricing method by window area to be washed										
	2110	Minimum productivity	1 Clab	1	8	M.S.F.	.29	219		219.29	365	
	2120	Average productivity		2.50	3.200		.29	87.50		87.79	145	
	2130	Maximum productivity	↓	4	2	↓	.29	55		55.29	91.50	
	2200	Weatherstripping see Division 08720-300										
290	0010	**FINISHES, FACILITIES MAINTENANCE** R019313-10										290
	0100	Ceiling maintenance										
	0120	Acoustical tile cleaning, chemical R019313-20				S.F.					.25	
	0130	Clean and apply accoustical coating				"					.60	
	0400	Floor maintenance R019313-30										
	0500	Carpet fiber sealant, anti-soilant, commercial				S.F.					.28	
	0510	Residential				"					.37	
	0530	Carpet cleaning, vacuum, dry pick-up										

01832 | Facilities Maintenance

			CREW	DAILY OUTPUT	LABOR-HOURS	UNIT	2006 BARE COSTS MAT.	LABOR	EQUIP.	TOTAL	TOTAL INCL O&P	
290	0540	Unobstructed R019313-10	1 Clab	30	.267	M.S.F.		7.30		7.30	12.10	290
	0550	Obstructed		22	.364			9.95		9.95	16.50	
	0560	Wet pick-up, unobstructed R019313-20		21	.381		7.15	10.45		17.60	25	
	0570	Obstructed		14	.571		7.15	15.65		22.80	34	
	0580	Steam clean, minimum R019313-30		9760	.001	S.F.		.02		.02	.04	
	0590	Maximum cost		3250	.002	"		.07		.07	.11	
	0700	Composition, resilient or wood flooring										
	0720	Dust mop, unobstructed	1 Clab	60	.133	M.S.F.		3.65		3.65	6.05	
	0730	Obstructed		35	.229			6.25		6.25	10.40	
	0740	Damp mop, unobstructed		26	.308		2.91	8.45		11.36	17.15	
	0750	Obstructed		16	.500		2.91	13.70		16.61	25.50	
	0780	Hand scrub, unobstructed		1.80	4.444		2.91	122		124.91	205	
	0790	Obstructed		1.30	6.154		2.91	169		171.91	282	
	0800	Machine scrub, unobstructed		17	.471		2.91	12.90		15.81	24.50	
	0810	Obstructed		12	.667		2.91	18.25		21.16	33.50	
	0820	Machine polish, unobstructed		28	.286			7.85		7.85	12.95	
	0830	Obstructed		14	.571			15.65		15.65	26	
	0850	Refinish old wood floors, minimum	1 Carp	400	.020	S.F.	.02	.71		.73	1.20	
	0860	Maximum cost	"	130	.062	"	.03	2.19		2.22	3.65	
	0870	Strip & rewax/polish, unobstructed	1 Clab	4	2	M.S.F.	18.95	55		73.95	112	
	0880	Obstructed		3	2.667		18.95	73		91.95	142	
	0890	Sweeping, unobstructed		47	.170			4.66		4.66	7.75	
	0900	Obstructed		35	.229			6.25		6.25	10.40	
	0950	Painting concrete or wood floors, see division 09910-630										
	1200	Paint removal, see division 13281-440										
	1300	Scrape after fire damage, see division 09990-500										
	2000	Wall maintenance										
	2400	Washing enamel finish walls with mild cleanser	1 Clab	16	.500	M.S.F.	3.73	13.70		17.43	26.50	
300	0010	**SPECIALTIES, FACILITIES MAINTENANCE**										300
	0100	Bathroom accessories										
	0120	Accessory, replacement see Division 10810-100										
	0130	All purpose cleaner, concentrate				Gal.	14.90			14.90	16.40	
	0150	Clean mirror, 36″ x 24″	1 Clab	840	.010	Ea.	.02	.26		.28	.45	
	0160	48″ x 24″		630	.013		.03	.35		.38	.61	
	0170	72″ x 24″		420	.019		.04	.52		.56	.91	
	0200	General cleaning of fixtures (basins, water closets, urinals)										
	0210	Including shelves, partitions and dispenser servicing	1 Clab	80	.100	Fixture	.02	2.74		2.76	4.56	
	1500	Partition replacement, hospital, office, toilet see Division 10160-100										
310	0010	**ARCHITECTURAL EQUIPMENT, FACILITIES MAINTENANCE**										310
	0100	Bird control needle strips										
	0110	Anchor clip mounted	2 Clab	500	.032	L.F.	5.40	.88		6.28	7.40	
	0120	Adhesive mounted	"	550	.029	"	5.40	.80		6.20	7.25	
	0500	Laundry equipment repair and maintenance										
	0520	Dryer, domestic, remove or reinstall basket	1 Skwk	6	1.333	Ea.		48.50		48.50	79.50	
	0530	Commercial, disassemble, clean & reassemble		2	4			146		146	239	
	0550	Extractor, replace belts (2)		6	1.333			48.50		48.50	79.50	
	0560	Replace pressure pad		2	4			146		146	239	
	0570	Flatwork ironer, replace feed ribbons		2	4			146		146	239	
	0580	Press, overhaul air system		4	2			73		73	119	
	0590	Tumbler, extractor, replace bearings and seal		2	4			146		146	239	
	0600	Washer, domestic, replace bearings and seal		3	2.667			97.50		97.50	159	
	0610	Remove or reinstall tub		4	2			73		73	119	
	0700	Medical and ward equipment repair and maintenance										
	0710	Aspirator-overhaul	1 Skwk	6	1.333	Ea.		48.50		48.50	79.50	
	0720	Autoclave-overhaul		4	2	"		73		73	119	
	0730	Bed rails, assemble or repair		10	.800	Pr.		29		29	48	

01832 | Facilities Maintenance

		01832 Facilities Maintenance	CREW	DAILY OUTPUT	LABOR-HOURS	UNIT	2006 BARE COSTS MAT.	LABOR	EQUIP.	TOTAL	TOTAL INCL O&P	
310	0770	Cart, medication, ice, etc. - replace wheels	1 Skwk	6	1.333	Carton		48.50		48.50	79.50	310
	0780	Replace rubber bumper		6	1.333	"		48.50		48.50	79.50	
	0800	Centrifuge, repair or replace timer or switch		11	.727	Ea.		26.50		26.50	43.50	
	0850	Litter, attach safety belt		16	.500			18.25		18.25	30	
	0880	Operating table, overhaul and lubricate		6	1.333			48.50		48.50	79.50	
	0900	Oxygen tent apparatus, overhaul		2	4			146		146	239	
	0950	Still, clean		3	2.667			97.50		97.50	159	
	0970	Striker saw, overhaul		5	1.600			58.50		58.50	95.50	
	1100	Wheelchair, complete overhaul		3	2.667			97.50		97.50	159	
	1110	Replace wheel		11	.727			26.50		26.50	43.50	
	2000	Other equipment repair and maintenance										
	2040	Cart or dolly, assemble or disassemble axles and wheels	1 Skwk	6	1.333	Carton		48.50		48.50	79.50	
	2090	Forge and harden tool		13	.615	Ea.		22.50		22.50	36.50	
	3100	Office furniture, repair		7	1.143			41.50		41.50	68	
	3120	Portable power tool, assemble or disassemble		6	1.333			48.50		48.50	79.50	
	3130	Repair or replace part		6	1.333			48.50		48.50	79.50	
	3180	Replace tool handle		16	.500			18.25		18.25	30	
	3200	Sharpen hand tool		16	.500			18.25		18.25	30	
320	0010	FURNISHINGS FACILITIES MAINTENANCE										320
	1000	Reupholster, side chair	1 Skwk	2.50	3.200	Ea.		117		117	191	
	1040	Tubular frame chair seat and back		3	2.667			97.50		97.50	159	
	1080	Sofa		2	4			146		146	239	
	1500	Upholstery fabric protection/cleaner, minimum cost				S.Y.					10	
	1510	Maximum cost				"					26	
	1540	Vacuum, large divan	1 Clab	80	.100	Ea.		2.74		2.74	4.54	
340	0010	CONVEYING SYSTEMS FACILITIES MAINTENANCE										340
	0900	Elevator, general cleaning, vacuum, dust, mop, polish	1 Clab	16	.500	Ea.	.68	13.70		14.38	23.50	
	0920	Motor generator set, service	1 Elec	8	1			42		42	64.50	
	0930	Repair, minor		2	4			168		168	259	
	0940	Limit switch, repair		11	.727			30.50		30.50	47	
	0950	Troubleshoot, main control panel	2 Elec	2.50	6.400			269		269	415	
	1000	Escalator cleaning, dusting, polishing	1 Clab	20	.400	Flight	.75	10.95		11.70	18.95	
	1500	Vehicle maintenance, grease, light vehicle	1 Skwk	20	.400	Ea.		14.60		14.60	24	
	1510	Heavy vehicle		7	1.143			41.50		41.50	68	
	1520	Wash, light vehicle		8	1		.58	36.50		37.08	60	
	1530	Heavy vehicle		6	1.333		1.04	48.50		49.54	80.50	
	1540	Change oil and filter, light vehicle		10	.800		12.30	29		41.30	61.50	
	1570	Rotate tires, light vehicle		18	.444			16.20		16.20	26.50	
350	0010	MECHANICAL FACILITIES MAINTENANCE										350
	0100	Air conditioning system maintenance										
	0130	Belt, replace	1 Stpi	15	.533	Ea.		23		23	35.50	
	0170	Fan, clean		16	.500			21.50		21.50	33.50	
	0180	Filter, remove, clean, replace		12	.667			28.50		28.50	44.50	
	0190	Flexible coupling alignment, inspect		40	.200			8.60		8.60	13.40	
	0200	Gas leak locate and repair		4	2			86		86	134	
	0250	Pump packing gland, remove and replace		11	.727			31.50		31.50	48.50	
	0270	Tighten		32	.250			10.75		10.75	16.75	
	0290	Pump disassemble and assemble		4	2			86		86	134	
	0300	Air pressure regulator, disassemble, clean, assemble	1 Skwk	4	2			73		73	119	
	0310	Repair or replace part	"	6	1.333			48.50		48.50	79.50	
	0320	Purging system	1 Stpi	16	.500			21.50		21.50	33.50	
	0400	Compressor, air, remove or install fan wheel	1 Skwk	20	.400			14.60		14.60	24	
	0410	Disassemble or assemble 2 cylinder, 2 stage		4	2			73		73	119	
	0420	4 cylinder, 4 stage		1	8			292		292	480	

Important: See the Reference Section for supporting data - Crews, Rental Equipment, City Cost Indexes and Reference Data

01832 | Facilities Maintenance

		Description	Crew	Daily Output	Labor-Hours	Unit	2006 Bare Costs Mat.	2006 Bare Costs Labor	2006 Bare Costs Equip.	2006 Bare Costs Total	Total Incl O&P
350	0430	Repair or replace part	1 Skwk	2	4	Ea.		146		146	239
	0700	Demolition, for mech. demolition see Div. 15055-300 or 15055-600									
	0800	Ductwork, clean									
	0810	Rectangular									
	0820	6"	1 Shee	187.50	.043	L.F.		1.80		1.80	2.86
	0830	8"		140.63	.057			2.40		2.40	3.81
	0840	10"		112.50	.071			3		3	4.77
	0850	12"		93.75	.085			3.60		3.60	5.70
	0860	14"		80.36	.100			4.20		4.20	6.65
	0870	16"		70.31	.114			4.80		4.80	7.65
	0900	Round									
	0910	4"	1 Shee	358.10	.022	L.F.		.94		.94	1.50
	0920	6"		238.73	.034			1.41		1.41	2.25
	0930	8"		179.05	.045			1.88		1.88	3
	0940	10"		143.24	.056			2.35		2.35	3.74
	0950	12"		119.37	.067			2.82		2.82	4.49
	0960	16"		89.52	.089			3.77		3.77	6
	1000	Expansion joint, not screwed, install or remove	1 Stpi	3	2.667	Ea.		115		115	179
	1010	Repack	"	6	1.333	"		57.50		57.50	89.50
	1200	Fire protection equipment									
	1220	Fire hydrant, replace	Q-1	3	5.333	Ea.		205		205	320
	1230	Service, lubricate, inspect, flush, clean	1 Plum	7	1.143			49		49	76
	1240	Test	"	11	.727			31		31	48.50
	1310	Inspect valves, pressure, nozzle,	1 Spri	4	2	System		83.50		83.50	130
	1500	Instrumentation preventive maintenance, rule of thumb									
	1510	Annual cost per instrument for calibration & trouble shooting									
	1520	Includes flow meters, analytical transmitters, flow control									
	1530	chemical feed systems, recorders and distrib. control sys.				Ea.					325
	1800	Plumbing fixtures, for installation see Division 15400									
	1840	Open drain with snake	1 Plum	13	.615	Ea.		26.50		26.50	41
	1850	Open drain with toilet auger		16	.500			21.50		21.50	33
	1860	Grease trap, clean		2	4			171		171	266
	1870	Plaster trap, clean		6	1.333			57		57	88.50
	1900	Relief valve, test and adjust	1 Stpi	20	.400			17.20		17.20	27
	2000	Repair or replace-steam trap		8	1			43		43	67
	2020	Y-type or bell strainer		6	1.333			57.50		57.50	89.50
	2040	Water trap or vacuum breaker, screwed joints		13	.615			26.50		26.50	41
	2100	Steam specialties, clean									
	2120	Air separator with automatic trap, 1" fittings	1 Stpi	12	.667	Ea.		28.50		28.50	44.50
	2130	Bucket trap, 2" pipe		7	1.143			49		49	76.50
	2150	Drip leg, 2" fitting		45	.178			7.65		7.65	11.90
	2200	Thermodynamic trap, 1" fittings		50	.160			6.90		6.90	10.70
	2210	Thermostatic		65	.123			5.30		5.30	8.25
	2240	Screen and seat in y-type strainer, plug type.		25	.320			13.80		13.80	21.50
	2242	Screen and seat in y-type strainer, flange type.		12	.667			28.50		28.50	44.50
	2500	Valve, replace broken handwheel		24	.333			14.35		14.35	22.50
	3000	Valve, overhaul, regulator, relief, flushometer, mixing									
	3040	Cold water, gas	1 Stpi	5	1.600	Ea.		69		69	107
	3050	Hot water, steam		3	2.667			115		115	179
	3080	Globe, gate, check up to 4" cold water, gas		10	.800			34.50		34.50	53.50
	3090	Hot water, steam		5	1.600			69		69	107
	3100	Over 4" ID hot or cold line		1.40	5.714			246		246	385
	3120	Remove and replace, gate, globe or check up to 4"	Q-5	6	2.667			103		103	161
	3130	Over 4"	"	2	8			310		310	480
	3150	Repack up to 4"	1 Stpi	13	.615			26.50		26.50	41
	3160	Over 4"	"	4	2			86		86	134

01800 | Facility Operation

01832 | Facilities Maintenance

			CREW	DAILY OUTPUT	LABOR-HOURS	UNIT	2006 BARE COSTS MAT.	LABOR	EQUIP.	TOTAL	TOTAL INCL O&P	
360	0010	**ELECTRICAL FACILITIES MAINTENANCE**										360
	0700	Cathodic protection systems										
	0720	Check and adjust reading on rectifier	1 Elec	20	.400	Ea.		16.80		16.80	26	
	0730	Check pipe to soil potential		20	.400			16.80		16.80	26	
	0740	Replace lead connection		4	2			84		84	129	
	0800	Control device, install		5.70	1.404			59		59	91	
	0810	Disassemble, clean and reinstall		7	1.143			48		48	74	
	0820	Replace		10.70	.748			31.50		31.50	48.50	
	0830	Trouble shoot		10	.800			33.50		33.50	52	
	0900	Demolition, for electrical demolition see Division 16055-300										
	1000	Distribution systems and equipment install or repair a breaker										
	1010	In power panels up to 200 amps	1 Elec	7	1.143	Ea.		48		48	74	
	1020	Over 200 amps		2	4			168		168	259	
	1030	Reset breaker or replace fuse		20	.400			16.80		16.80	26	
	1100	Megger test MCC (each stack)		4	2			84		84	129	
	1110	MCC vacuum and clean (each stack)		5.30	1.509			63.50		63.50	97.50	
	2500	Remove/replace or maint, road fixture & lamp		3	2.667		775	112		887	1,025	
	2510	Fluorescent fixture		7	1.143		62.50	48		110.50	143	
	2515	Relamp (fluor.) facility area each tube		60	.133		4.88	5.60		10.48	14	
	2518	Fluorescent fixture, clean (area)		44	.182		.09	7.65		7.74	11.85	
	2520	Incandescent fixture		11	.727		49	30.50		79.50	101	
	2530	Lamp (incadescent or fluorescent)		60	.133		2.81	5.60		8.41	11.75	
	2535	Replace cord in socket lamp		13	.615		2.10	26		28.10	42.50	
	2540	Ballast electronic type for two tubes		8	1		34.50	42		76.50	103	
	2541	Starter		30	.267		1	11.20		12.20	18.35	
	2545	Replace other lighting parts		11	.727		12.75	30.50		43.25	61	
	2550	Switch		11	.727		5.20	30.50		35.70	53	
	2555	Receptacle		11	.727		5.15	30.50		35.65	52.50	
	2560	Floodlight		4	2		250	84		334	405	
	2570	Christmas lighting, indoor, per string		16	.500			21		21	32.50	
	2580	Outdoor		13	.615			26		26	40	
	2590	Test battery operated emergency lights		40	.200			8.40		8.40	12.95	
	2600	Repair/replace component in communication system		6	1.333		53.50	56		109.50	145	
	2700	Repair misc. appliances (incl. clocks, vent fan, blower, etc.)		6	1.333			56		56	86.50	
	2710	Reset clocks & timers		50	.160			6.70		6.70	10.35	
	2720	Adjust time delay relays		16	.500			21		21	32.50	
	2730	Test specific gravity of lead-acid batteries		80	.100			4.20		4.20	6.45	
	3000	Motors and generators										
	3020	Disassemble, clean and reinstall motor, up to 1/4 HP	1 Elec	4	2	Ea.		84		84	129	
	3030	Up to 3/4 HP		3	2.667			112		112	173	
	3040	Up to 10 HP		2	4			168		168	259	
	3050	Replace part, up to 1/4 HP		6	1.333			56		56	86.50	
	3060	Up to 3/4 HP		4	2			84		84	129	
	3070	Up to 10 HP		3	2.667			112		112	173	
	3080	Megger test motor windings		5.33	1.501			63		63	97	
	3082	Motor vibration check		16	.500			21		21	32.50	
	3084	Oil motor bearings		25	.320			13.45		13.45	20.50	
	3086	Run test emergency generator for 30 minutes		11	.727			30.50		30.50	47	
	3090	Rewind motor, up to 1/4 HP		3	2.667			112		112	173	
	3100	Up to 3/4 HP		2	4			168		168	259	
	3110	Up to 10 HP		1.50	5.333			224		224	345	
	3150	Generator, repair or replace part		4	2			84		84	129	
	3160	Repair DC generator		2	4			168		168	259	
	4000	Stub pole, install or remove		3	2.667			112		112	173	
	4500	Transformer maintenance up to 15 kVA		2.70	2.963			124		124	192	

01840 | Moving Equipment

		Description	CREW	DAILY OUTPUT	LABOR-HOURS	UNIT	2006 BARE COSTS MAT.	LABOR	EQUIP.	TOTAL	TOTAL INCL O&P	
100	0010	**MOVING EQUIPMENT**, Remove and reset, 100′ distance,										100
	0020	No obstructions, no assembly or leveling unless noted										
	0100	Annealing furnace, 24′ overall	B-67	4	4	Ea.		145	60.50	205.50	295	
	0200	Annealing oven, small		14	1.143			41.50	17.30	58.80	84	
	0240	Very large		1	16			580	242	822	1,175	
	0400	Band saw, small		12	1.333			48	20	68	98	
	0440	Large		8	2			72.50	30.50	103	148	
	0500	Blue print copy machine		7	2.286			82.50	34.50	117	168	
	0600	Bonding mill, 6″		7	2.286			82.50	34.50	117	168	
	0620	12″		6	2.667			96.50	40.50	137	197	
	0640	18″		4	4			145	60.50	205.50	295	
	0660	24″		2	8			289	121	410	590	
	0700	Boring machine (jig)	B-68	7	3.429			125	34.50	159.50	235	
	0800	Bridgeport mill, standard	B-67	14	1.143			41.50	17.30	58.80	84	
	1000	Calibrator, 6 unit	″	14	1.143			41.50	17.30	58.80	84	
	1100	Comparitor, bench top	2 Clab	14	1.143			31.50		31.50	52	
	1140	Floor mounted	B-67	7	2.286			82.50	34.50	117	168	
	1200	Computer, desk top	2 Clab	25	.640			17.55		17.55	29	
	1300	Copy machine	″	25	.640			17.55		17.55	29	
	1500	Deflasher	B-67	14	1.143			41.50	17.30	58.80	84	
	1600	Degreaser, small		14	1.143			41.50	17.30	58.80	84	
	1640	Large 24′ overall		1	16			580	242	822	1,175	
	1700	Desk with chair	2 Clab	25	.640			17.55		17.55	29	
	1800	Dial press	B-67	7	2.286			82.50	34.50	117	168	
	1900	Drafting table	2 Clab	14	1.143			31.50		31.50	52	
	2000	Drill press, bench top	″	14	1.143			31.50		31.50	52	
	2040	Floor mounted	B-67	14	1.143			41.50	17.30	58.80	84	
	2080	Industrial radial	″	7	2.286			82.50	34.50	117	168	
	2100	Dust collector, portable	2 Clab	25	.640			17.55		17.55	29	
	2140	Stationary, small	B-67	7	2.286			82.50	34.50	117	168	
	2180	Stationary, large	″	2	8			289	121	410	590	
	2300	Electric discharge machine	B-68	7	3.429			125	34.50	159.50	235	
	2400	Environmental chamber walls, including assembly	4 Clab	18	1.778	L.F.		48.50		48.50	80.50	
	2600	File cabinet	2 Clab	25	.640	Ea.		17.55		17.55	29	
	2800	Grinder/sander, pedestal mount	B-67	14	1.143			41.50	17.30	58.80	84	
	3000	Hack saw, power	2 Clab	24	.667			18.25		18.25	30.50	
	3100	Hydraulic press	B-67	14	1.143			41.50	17.30	58.80	84	
	3500	Laminar flow tables	″	14	1.143			41.50	17.30	58.80	84	
	3600	Lathe, bench	2 Clab	14	1.143			31.50		31.50	52	
	3640	6″	B-67	14	1.143			41.50	17.30	58.80	84	
	3680	10″		13	1.231			44.50	18.65	63.15	90.50	
	3720	12″		12	1.333			48	20	68	98	
	4000	Milling machine		8	2			72.50	30.50	103	148	
	4100	Molding press, 25 ton		5	3.200			116	48.50	164.50	237	
	4140	60 ton		4	4			145	60.50	205.50	295	
	4180	100 ton		2	8			289	121	410	590	
	4220	150 ton		1.50	10.667			385	162	547	790	
	4260	200 ton		1	16			580	242	822	1,175	
	4300	300 ton		.75	21.333			770	325	1,095	1,575	
	4700	Oil pot stand		14	1.143			41.50	17.30	58.80	84	
	5000	Press, 10 ton		14	1.143			41.50	17.30	58.80	84	
	5040	15 ton		12	1.333			48	20	68	98	
	5080	20 ton		10	1.600			58	24	82	118	
	5120	30 ton		8	2			72.50	30.50	103	148	
	5160	45 ton		6	2.667			96.50	40.50	137	197	
	5200	60 ton		4	4			145	60.50	205.50	295	

01800 | Facility Operation

	01840 Moving Equipment		CREW	DAILY OUTPUT	LABOR-HOURS	UNIT	2006 BARE COSTS MAT.	LABOR	EQUIP.	TOTAL	TOTAL INCL O&P	
100	5240	75 ton	B-67	2.50	6.400	Ea.		231	97	328	470	100
	5280	100 ton	↓	2	8	↓		289	121	410	590	
	5500	Raised floor, including assembly	2 Carp	250	.064	S.F.		2.28		2.28	3.77	
	5600	Rolling mill, 6″	B-67	7	2.286	Ea.		82.50	34.50	117	168	
	5640	9″		6	2.667			96.50	40.50	137	197	
	5680	12″		4	4			145	60.50	205.50	295	
	5720	13″		3.50	4.571			165	69.50	234.50	335	
	5760	18″		2	8			289	121	410	590	
	5800	25″		1	16			580	242	822	1,175	
	6000	Sander, floor stand		14	1.143			41.50	17.30	58.80	84	
	6100	Screw machine		7	2.286			82.50	34.50	117	168	
	6200	Shaper, 16″		14	1.143			41.50	17.30	58.80	84	
	6300	Shear, power assist		4	4			145	60.50	205.50	295	
	6400	Slitter, 6″		14	1.143			41.50	17.30	58.80	84	
	6440	8″		13	1.231			44.50	18.65	63.15	90.50	
	6480	10″		12	1.333			48	20	68	98	
	6520	12″		11	1.455			52.50	22	74.50	108	
	6560	16″		10	1.600			58	24	82	118	
	6600	20″		8	2			72.50	30.50	103	148	
	6640	24″		6	2.667			96.50	40.50	137	197	
	6800	Snag and tap machine		7	2.286			82.50	34.50	117	168	
	6900	Solder machine (auto)	↓	7	2.286			82.50	34.50	117	168	
	7000	Storage cabinet metal, small	2 Clab	36	.444			12.20		12.20	20	
	7040	Large		25	.640			17.55		17.55	29	
	7100	Storage rack open, small		14	1.143			31.50		31.50	52	
	7140	Large	↓	7	2.286			62.50		62.50	104	
	7200	Surface bench, small	B-67	14	1.143			41.50	17.30	58.80	84	
	7240	Large	″	5	3.200			116	48.50	164.50	237	
	7300	Surface grinder, large wet	B-68	5	4.800			175	48.50	223.50	330	
	7500	Time check machine	2 Clab	14	1.143			31.50		31.50	52	
	8000	Welder, 30 KVA (bench)		14	1.143			31.50		31.50	52	
	8100	Work bench with chair	↓	25	.640	↓		17.55		17.55	29	

For information about Means Estimating Seminars, see yellow pages 12 and 13 in back of book

Division 2 Site Construction

Estimating Tips

02200 Site Preparation

- If possible visit the site and take an inventory of the type, quantity, and size of the trees. Certain trees may have a landscape resale value or firewood value. Stump disposal can be very expensive, particularly if they cannot be buried at the site. Consider using a bulldozer in lieu of hand cutting trees.
- Estimators should visit the site to determine the need for haul road access, storage of materials, and security considerations. When estimating for access roads on unstable soil, consider using a geotextile stabilization fabric. It can greatly reduce the quantity of crushed stone or gravel. Sites of limited size and access can cause cost overruns due to lost productivity. Theft and damage is another consideration if the location is isolated. A temporary fence or security guards may be required. Investigate the site thoroughly.

02210 Subsurface Investigation

In preparing estimates on structures involving earthwork or foundations, all information concerning soil characteristics should be obtained. Look particularly for hazardous waste, evidence of prior dumping of debris, and previous stream beds.

02220 Selective Demolition

The costs shown for selective demolition do not include rubbish handling or disposal. These items should be estimated separately using Means data or other sources.

- Historic preservation often requires that the contractor remove materials from the existing structure, rehab them, and replace them. The estimator must be aware of any related measures and precautions that must be taken when doing selective demolition and cutting and patching. Requirements may include special handling and storage as well as security.
- In addition to Section 02220, you can find selective demolition items in each division. Example: Roofing demolition is in Division 7.

02300 Earthwork

- Estimating the actual cost of performing earthwork requires careful consideration of the variables involved. This includes items such as type of soil, whether water will be encountered, dewatering, whether banks need bracing, disposal of excavated earth, and length of haul to fill or spoil sites, etc. If the project has large quantities of cut or fill, consider raising or lowering the site to reduce costs, while paying close attention to the effect on site drainage and utilities.
- If the project has large quantities of fill, creating a borrow pit on the site can significantly lower the costs.
- It is very important to consider what time of year the project is scheduled for completion. Bad weather can create large cost overruns from dewatering, site repair, and lost productivity from cold weather.

02500 Utility Services
02600 Drainage & Containment

- Never assume that the water, sewer, and drainage lines will go in at the early stages of the project. Consider the site access needs before dividing the site in half with open trenches, loose pipe, and machinery obstructions. Always inspect the site to establish that the site drawings are complete. Check off all existing utilities on your drawing as you locate them. If you find any discrepancies, mark up the site plan for further research. Differing site conditions can be very costly if discovered later in the project.
- See also Section 02955 for restoration of pipe where removal/replacement may be undesirable. Use of new types of piping materials can reduce the overall project cost. Owners/design engineers should consider the installing construction as a valuable source of current information on piping products that could lead to significant utility cost savings.

02700 Bases, Ballasts, Pavements/Appurtenances

- When estimating paving, keep in mind the project schedule. If an asphaltic paving project is in a colder climate and runs through to the spring, consider placing the base course in the autumn and then topping it in the spring just prior to completion. This could save considerable costs in spring repair. Keep in mind that prices for asphalt and concrete are generally higher in the cold seasons.
- See also Sections 02960/02965.

02900 Planting

- The timing of planting and guarantee specifications often dictate the costs for establishing tree and shrub growth and a stand of grass or ground cover. Establish the work performance schedule to coincide with the local planting season. Maintenance and growth guarantees can add from 20% to 100% to the total landscaping cost. The cost to replace trees and shrubs can be as high as 5% of the total cost depending on the planting zone, soil conditions, and time of year.

02960 & 02965 Flexible Pavement Surfacing Recovery

- Recycling of asphalt pavement is becoming very popular and is an alternative to removal and replacement of asphalt pavement. It can be a good value engineering proposal if removed pavement can be recycled either at the site or another site that is reasonably close to the project site.

Reference Numbers

Reference numbers are shown in bold squares at the beginning of some major classifications. These numbers refer to related items in the Reference Section. The reference information may be an estimating procedure, an alternate pricing method, or technical information.

Note: Not all subdivisions listed here necessarily appear in this publication.

	02065 | Cement & Concrete		CREW	DAILY OUTPUT	LABOR-HOURS	UNIT	2006 BARE COSTS MAT.	LABOR	EQUIP.	TOTAL	TOTAL INCL O&P	
300	0010	**PLANT MIXED BITUMINOUS CONCRETE**										300
	0020	Asphaltic concrete plant mix (145 LB per c.f.)				Ton	38			38	41.50	
	0040	Asphaltic concrete less than 300 tons add trucking costs										
	0050	See Div. 02315-490 for hauling costs										
	0200	All weather patching mix, hot					38			38	41.50	
	0250	Cold patch					41			41	45	
	0300	Berm mix					38			38	41.50	
	0400	Base mix					38			38	41.50	
	0500	Binder mix					38			38	41.50	
	0600	Sand or sheet mix					41			41	45	
400	0010	**RECYCLED PLANT MIXED BITUMINOUS CONCRETE**										400
	0200	Reclaimed pavement in stockpile				Ton	14.35			14.35	15.75	
	0400	Recycled pavement, at plant, ratio old: new, 70:30					21.50			21.50	23.50	
	0600	Ratio old: new, 30:70					31			31	34	

	02080 | Utility Materials		CREW	DAILY OUTPUT	LABOR-HOURS	UNIT	MAT.	LABOR	EQUIP.	TOTAL	TOTAL INCL O&P	
100	0010	**FIRE HYDRANTS**										100
	0020	Mechanical joints unless otherwise noted										
	1000	Fire hydrants, two way; excavation and backfill not incl.										
	1100	4-1/2" valve size, depth 2'-0"	B-21	10	2.800	Ea.	1,100	90	15.25	1,205.25	1,400	
	1200	4'-6"		9	3.111		1,250	100	16.95	1,366.95	1,575	
	1260	6'-0" 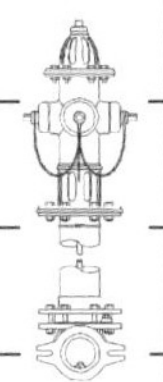		7	4		1,225	128	22	1,375	1,575	
	1340	8'-0"		6	4.667		1,400	150	25.50	1,575.50	1,800	
	1420	10'-0"		5	5.600		1,475	180	30.50	1,685.50	1,950	
	2000	5-1/4" valve size, depth 2'-0"		10	2.800		880	90	15.25	985.25	1,125	
	2080	4'-0"		9	3.111		995	100	16.95	1,111.95	1,275	
	2160	6'-0"		7	4		1,100	128	22	1,250	1,450	
	2240	8'-0"		6	4.667		1,225	150	25.50	1,400.50	1,625	
	2320	10'-0"		5	5.600		1,350	180	30.50	1,560.50	1,800	
	2350	For threeway valves, add					7%					
	2400	Lower barrel extensions with stems, 1'-0"	B-20	14	1.714		350	53.50		403.50	475	
	2440	2'-0"		13	1.846		430	57.50		487.50	565	
	2480	3'-0"		12	2		505	62		567	660	
	2520	4'-0"		10	2.400		580	74.50		654.50	765	
	5000	Indicator post										
	5020	Adjustable, valve size 4" to 14", 4' bury	B-21	10	2.800	Ea.	385	90	15.25	490.25	590	
	5060	8' bury 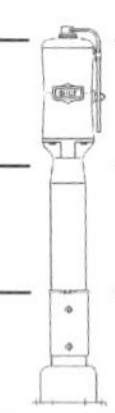		7	4		495	128	22	645	775	
	5080	10' bury		6	4.667		505	150	25.50	680.50	830	
	5100	12' bury		5	5.600		760	180	30.50	970.50	1,175	
	5120	14' bury		4	7		830	225	38	1,093	1,325	
	5500	Non-adjustable, valve size 4" to 14", 3' bury		10	2.800		385	90	15.25	490.25	590	
	5520	3'-6" bury		10	2.800		385	90	15.25	490.25	590	
	5540	4' bury		9	3.111		385	100	16.95	501.95	605	
400	0010	**UTILITY BOXES** Precast concrete, 6" thick										400
	0040	4' x 6' x 6' high, I.D.	B-13	2	28	Ea.	1,250	840	325	2,415	3,100	
	0050	5' x 10' x 6' high, I.D.		2	28		1,550	840	325	2,715	3,450	
	0100	6' x 10' x 6' high, I.D.		2	28		1,600	840	325	2,765	3,500	
	0150	5' x 12' x 6' high, I.D.		2	28		1,700	840	325	2,865	3,600	
	0200	6' x 12' x 6' high, I.D.		1.80	31.111		1,925	930	360	3,215	4,025	
	0250	6' x 13' x 6' high, I.D.		1.50	37.333		2,525	1,125	435	4,085	5,075	
	0300	8' x 14' x 7' high, I.D.		1	56		2,725	1,675	650	5,050	6,450	
	0350	Hand hole, precast concrete, 1-1/2" thick										
	0400	1'-0" x 2'-0" x 1'-9", I.D., light duty	B-1	4	6	Ea.	265	168		433	570	
	0450	4'-6" x 3'-2" x 2'-0", O.D., heavy duty	B-6	3	8		815	240	75.50	1,130.50	1,375	
	0460	Meter pit, 4' x 4', 4' deep		2	12		695	360	113	1,168	1,475	

02050 | Basic Site Materials & Methods

02080 | Utility Materials

		02080 Utility Materials	CREW	DAILY OUTPUT	LABOR-HOURS	UNIT	2006 BARE COSTS MAT.	LABOR	EQUIP.	TOTAL	TOTAL INCL O&P	
400	0470	6′ deep	B-6	1.60	15	Ea.	990	450	141	1,581	1,950	400
	0480	8′ deep		1.40	17.143		1,300	515	162	1,977	2,475	
	0490	10′ deep		1.20	20		1,650	600	189	2,439	3,000	
	0500	15′ deep		1	24		2,425	720	226	3,371	4,100	
	0510	6′ x 6′, 4′ deep		1.40	17.143		1,025	515	162	1,702	2,150	
	0520	6′ deep		1.20	20		1,550	600	189	2,339	2,875	
	0530	8′ deep		1	24		2,050	720	226	2,996	3,675	
	0540	10′ deep		.80	30		2,575	900	283	3,758	4,600	
	0550	15′ deep		.60	40		3,900	1,200	375	5,475	6,675	
500	0010	**VALVES** Water distribution, see also div. 15110										500
	3000	Butterfly valves with boxes, cast iron, mech. jt.										
	3100	4″ diameter	B-20	6	4	Ea.	218	124		342	445	
	3180	8″ diameter	B-21	4	7		305	225	38	568	745	
	3340	12″ diameter		3	9.333		585	300	51	936	1,175	
	3400	14″ diameter		2	14		1,000	450	76.50	1,526.50	1,925	
	3460	18″ diameter		1.50	18.667		1,675	600	102	2,377	2,950	
	3480	20″ diameter		1	28		2,100	900	153	3,153	3,975	
	3500	24″ diameter		.50	56		2,950	1,800	305	5,055	6,500	
	3600	With lever operator										
	3610	4″ diameter	B-20	6	4	Ea.	118	124		242	335	
	3616	8″ diameter	B-21	4	7		240	225	38	503	675	
	3620	12″ diameter		3	9.333		715	300	51	1,066	1,325	
	3624	16″ diameter		2	14		1,550	450	76.50	2,076.50	2,525	
	3630	24″ diameter		.50	56		2,550	1,800	305	4,655	6,075	
	3700	Check valves, flanged										
	3710	4″ diameter	B-20	6	4	Ea.	435	124		559	680	
	3714	6″ diameter	″	5	4.800		735	149		884	1,050	
	3716	8″ diameter	B-21	4	7		1,500	225	38	1,763	2,050	
	3720	12″ diameter		3	9.333		3,650	300	51	4,001	4,575	
	3724	16″ diameter		2	14		8,050	450	76.50	8,576.50	9,675	
	3726	18″ diameter		1.50	18.667		11,900	600	102	12,602	14,200	
	3730	24″ diameter		.50	56		19,900	1,800	305	22,005	25,200	
	3800	Gate valves, C.I., 250 PSI, mechanical joint, w/boxes										
	3810	4″ diameter	B-21A	8	5	Ea.	169	172	67	408	535	
	3814	6″ diameter		6.80	5.882		219	202	79	500	650	
	3816	8″ diameter		5.60	7.143		345	245	96	686	875	
	3820	12″ diameter		4.08	9.804		740	335	131	1,206	1,500	
	3824	16″ diameter	B-21	1	28		915	900	153	1,968	2,650	
	3828	20″ diameter		.80	35		1,425	1,125	191	2,741	3,625	
	3830	24″ diameter		.50	56		1,750	1,800	305	3,855	5,200	
	3900	Globe valves, flanged, iron body, class 125										
	3910	4″ diameter	B-20	10	2.400	Ea.	1,125	74.50		1,199.50	1,375	
	3914	6″ diameter	″	9	2.667		1,850	83		1,933	2,175	
	3916	8″ diameter	B-21	6	4.667		3,450	150	25.50	3,625.50	4,075	
	3920	12″ diameter		4	7		6,425	225	38	6,688	7,475	
	3924	16″ diameter		2	14		8,025	450	76.50	8,551.50	9,650	
	3928	20″ diameter		.60	46.667		8,700	1,500	254	10,454	12,300	
	3930	24″ diameter		.50	56		10,300	1,800	305	12,405	14,700	
600	0010	**UTILITY ACCESSORIES**										600
	0400	Underground tape, detectable, reinforced, alum. foil core, 2″	1 Clab	150	.053	C.L.F.	1.46	1.46		2.92	4.03	
	0500	6″		140	.057	″	3.65	1.57		5.22	6.60	
	9000	Minimum labor/equipment charge		4	2	Job		55		55	91	

2 SITE CONSTRUCTION

02100 | Site Remediation

	02115 | Underground Storage Tank Removal		CREW	DAILY OUTPUT	LABOR-HOURS	UNIT	2006 BARE COSTS MAT.	LABOR	EQUIP.	TOTAL	TOTAL INCL O&P	
200	0010	REMOVAL OF UNDERGROUND STORAGE TANKS R026510-20										200
	0011	Petroleum storage tanks, non-leaking										
	0100	Excavate & load onto trailer										
	0110	3000 gal. to 5000 gal. tank	B-14	4	12	Ea.		350	56.50	406.50	630	
	0120	6000 gal to 8000 gal tank	B-3A	3	13.333			390	241	631	905	
	0130	9000 gal to 12000 gal tank	"	2	20	↓		585	360	945	1,350	
	0190	Known leaking tank add				%				100%	100%	
	0200	Remove sludge, water and remaining product from tank bottom										
	0201	of tank with vacuum truck										
	0300	3000 gal to 5000 gal tank	A-13	5	1.600	Ea.		56.50	100	156.50	199	
	0310	6000 gal to 8000 gal tank		4	2			70.50	125	195.50	249	
	0320	9000 gal to 12000 gal tank	↓	3	2.667	↓		94	167	261	330	
	0390	Dispose of sludge off-site, average				Gal.					4.40	
	0400	Insert inert solid CO_2 "dry ice" into tank										
	0401	For cleaning/transporting tanks (1.5 lbs./100 gal. cap)	1 Clab	500	.016	Lb.	1.37	.44		1.81	2.24	
	1020	Haul tank to certified salvage dump, 100 miles round trip										
	1023	3000 gal. to 5000 gal. tank				Ea.				550	690	
	1026	6000 gal. to 8000 gal. tank								650	825	
	1029	9,000 gal. to 12,000 gal. tank				↓				875	1,100	
	1100	Disposal of contaminated soil to landfill										
	1110	Minimum				C.Y.					110	
	1111	Maximum				"					330	
	1120	Disposal of contaminated soil to										
	1121	bituminous concrete batch plant										
	1130	Minimum				C.Y.					55	
	1131	Maximum				"					110	
	2010	Decontamination of soil on site incl poly tarp on top/bottom										
	2011	Soil containment berm, and chemical treatment										
	2020	Minimum	B-11C	100	.160	C.Y.	5.70	5.15	2.26	13.11	17.05	
	2021	Maximum	"	100	.160		7.40	5.15	2.26	14.81	18.90	
	2050	Disposal of decontaminated soil, minimum									66	
	2055	Maximum				↓					135	

02200 | Site Preparation

	02210 | Subsurface Investigation		CREW	DAILY OUTPUT	LABOR-HOURS	UNIT	2006 BARE COSTS MAT.	LABOR	EQUIP.	TOTAL	TOTAL INCL O&P	
120	0010	BORING AND EXPLORATORY DRILLING										120
	0020	Borings, initial field stake out & determination of elevations	A-6	1	16	Day		565	58	623	985	
	0100	Drawings showing boring details				Total		185		185	270	
	0200	Report and recommendations from P.E.						415		415	595	
	0300	Mobilization and demobilization, minimum	B-55	4	6	↓		165	202	367	495	
	0350	For over 100 miles, per added mile		450	.053	Mile		1.47	1.80	3.27	4.38	
	0600	Auger holes in earth, no samples, 2-1/2" diameter		78.60	.305	L.F.		8.40	10.30	18.70	25	
	0650	4" diameter		67.50	.356			9.75	11.95	21.70	29.50	
	0800	Cased borings in earth, with samples, 2-1/2" diameter		55.50	.432		15.30	11.90	14.55	41.75	52.50	
	0850	4" diameter	↓	32.60	.736		24	20	25	69	87.50	
	1000	Drilling in rock, "BX" core, no sampling	B-56	34.90	.458			14.35	30.50	44.85	56.50	
	1050	With casing & sampling		31.70	.505		15.30	15.80	33.50	64.60	79.50	
	1200	"NX" core, no sampling		25.92	.617			19.30	41	60.30	76.50	
	1250	With casing and sampling	↓	25	.640	↓	18.75	20	42.50	81.25	100	

	02210 | Subsurface Investigation		CREW	DAILY OUTPUT	LABOR-HOURS	UNIT	2006 BARE COSTS MAT.	LABOR	EQUIP.	TOTAL	TOTAL INCL O&P	
120	1400	Drill rig and crew with truck mounted auger	B-55	1	24	Day		660	810	1,470	1,975	120
	1450	With crawler type drill	B-56	1	16	″		500	1,075	1,575	1,975	
	1500	For inner city borings add, minimum									10%	
	1510	Maximum									20%	
200	0010	CORE DRILLING										200
	0020	Reinf. conc slab, up to 6″ thick, incl. bit, layout & set up										
	0100	1″ diameter core	B-89A	28	.571	Ea.	2.59	18.25	4.13	24.97	37.50	
	0150	Each added inch thick, add		300	.053		.46	1.70	.39	2.55	3.73	
	0300	3″ diameter core		23	.696		5.75	22	5	32.75	48.50	
	0350	Each added inch thick, add		186	.086		1.04	2.75	.62	4.41	6.35	
	0500	4″ diameter core		19	.842		5.75	27	6.10	38.85	57.50	
	0550	Each added inch thick, add		170	.094		1.31	3.01	.68	5	7.15	
	0700	6″ diameter core		14	1.143		9.50	36.50	8.25	54.25	79.50	
	0750	Each added inch thick, add		140	.114		1.61	3.65	.83	6.09	8.70	
	0900	8″ diameter core		11	1.455		12.95	46.50	10.50	69.95	102	
	0950	Each added inch thick, add		95	.168		2.18	5.40	1.22	8.80	12.60	
	1100	10″ diameter core		10	1.600		17.30	51	11.55	79.85	116	
	1150	Each added inch thick, add		80	.200		2.86	6.40	1.44	10.70	15.25	
	1300	12″ diameter core		9	1.778		21	57	12.85	90.85	131	
	1350	Each added inch thick, add		68	.235		3.43	7.50	1.70	12.63	18	
	1500	14″ diameter core		7	2.286		25	73	16.50	114.50	166	
	1550	Each added inch thick, add		55	.291		4.36	9.30	2.10	15.76	22.50	
	1700	18″ diameter core		4	4		32.50	128	29	189.50	278	
	1750	Each added inch thick, add		28	.571		5.70	18.25	4.13	28.08	41	
	1760	For horizontal holes, add to above								30%	30%	
	1770	Prestressed hollow core plank, 6″ thick										
	1780	1″ diameter core	B-89A	52	.308	Ea.	1.72	9.85	2.22	13.79	20.50	
	1790	Each added inch thick, add		350	.046		.30	1.46	.33	2.09	3.09	
	1800	3″ diameter core		50	.320		3.79	10.20	2.31	16.30	23.50	
	1810	Each added inch thick, add		240	.067		.63	2.13	.48	3.24	4.72	
	1820	4″ diameter core		48	.333		5.05	10.65	2.41	18.11	25.50	
	1830	Each added inch thick, add		216	.074		.87	2.37	.53	3.77	5.45	
	1840	6″ diameter core		44	.364		6.25	11.60	2.63	20.48	29	
	1850	Each added inch thick, add		175	.091		1.04	2.92	.66	4.62	6.65	
	1860	8″ diameter core		32	.500		8.35	16	3.61	27.96	39.50	
	1870	Each added inch thick, add		118	.136		1.45	4.33	.98	6.76	9.85	
	1880	10″ diameter core		28	.571		11.30	18.25	4.13	33.68	47	
	1890	Each added inch thick, add		99	.162		1.56	5.15	1.17	7.88	11.50	
	1900	12″ diameter core		22	.727		13.75	23	5.25	42	59	
	1910	Each added inch thick, add		85	.188		2.29	6	1.36	9.65	13.90	
	1950	Minimum charge for above, 3″ diameter core		7	2.286	Total		73	16.50	89.50	138	
	2000	4″ diameter core		6.80	2.353			75	17	92	143	
	2050	6″ diameter core		6	2.667			85	19.25	104.25	161	
	2100	8″ diameter core		5.50	2.909			93	21	114	176	
	2150	10″ diameter core		4.75	3.368			108	24.50	132.50	204	
	2200	12″ diameter core		3.90	4.103			131	29.50	160.50	249	
	2250	14″ diameter core		3.38	4.734			151	34	185	287	
	2300	18″ diameter core		3.15	5.079			162	36.50	198.50	310	
	3010	Bits for core drill, diamond, premium, 1″ diameter				Ea.	116			116	127	
	3020	3″ diameter					286			286	315	
	3040	4″ diameter					320			320	350	
	3050	6″ diameter					510			510	560	
	3080	8″ diameter					695			695	765	
	3120	12″ diameter					1,100			1,100	1,200	
	3180	18″ diameter					2,275			2,275	2,500	
	3240	24″ diameter					3,025			3,025	3,325	

SITE CONSTRUCTION 2

02210 | Subsurface Investigation

			CREW	DAILY OUTPUT	LABOR-HOURS	UNIT	2006 BARE COSTS MAT.	LABOR	EQUIP.	TOTAL	TOTAL INCL O&P
700	0010	TEST PITS									
	0020	Hand digging, light soil	1 Clab	4.50	1.778	C.Y.		48.50		48.50	80.50
	0100	Heavy soil	″	2.50	3.200			87.50		87.50	145
	0120	Loader-backhoe, light soil	B-11M	28	.571			18.30	9.90	28.20	40.50
	0130	Heavy soil	″	20	.800			25.50	13.90	39.40	57
	1000	Subsurface exploration, mobilization				Mile				5.50	6.30
	1010	Difficult access for rig, add				Hr.				110	126
	1020	Auger borings, drill rig, incl. samples				L.F.				12.65	14.60
	1030	Hand auger								18.70	21.50
	1050	Drill and sample every 5′, split spoon								16.50	19
	1060	Extra samples				Ea.				22	25.50

02220 | Site Demolition

			CREW	DAILY OUTPUT	LABOR-HOURS	UNIT	MAT.	LABOR	EQUIP.	TOTAL	TOTAL INCL O&P
110	0010	**BUILDING DEMOLITION** Large urban projects, incl. 20 mi. haul (R024119-10)									
	0011	No foundation or dump fees, C.F. is vol. of building standing									
	0012	Steel	B-8	21500	.003	C.F.		.09	.11	.20	.27
	0050	Concrete		15300	.004			.13	.16	.29	.38
	0080	Masonry		20100	.003			.10	.12	.22	.29
	0100	Mixture of types, average		20100	.003			.10	.12	.22	.29
	0500	Small bldgs, or single bldgs, no salvage included, steel	B-3	14800	.003			.10	.12	.22	.29
	0600	Concrete		11300	.004			.13	.15	.28	.38
	0650	Masonry		14800	.003			.10	.12	.22	.29
	0700	Wood		14800	.003			.10	.12	.22	.29
	1000	Single family, one story house, wood, minimum				Ea.				2,525	2,975
	1020	Maximum								4,400	5,275
	1200	Two family, two story house, wood, minimum								3,300	3,950
	1220	Maximum								6,375	7,700
	1300	Three family, three story house, wood, minimum								4,400	5,275
	1320	Maximum								7,700	9,250
	5000	For buildings with no interior walls, deduct								50%	

			CREW	DAILY OUTPUT	LABOR-HOURS	UNIT	MAT.	LABOR	EQUIP.	TOTAL	TOTAL INCL O&P
130	0010	BLDG. FOOTINGS AND FOUNDATIONS DEMOLITION (R024119-10)									
	0200	Floors, concrete slab on grade,									
	0240	4″ thick, plain concrete	B-9C	500	.080	S.F.		2.22	.29	2.51	4
	0280	Reinforced, wire mesh		470	.085			2.37	.31	2.68	4.26
	0300	Rods		400	.100			2.78	.37	3.15	5
	0400	6″ thick, plain concrete		375	.107			2.97	.39	3.36	5.35
	0420	Reinforced, wire mesh		340	.118			3.27	.43	3.70	5.90
	0440	Rods		300	.133			3.71	.49	4.20	6.70
	1000	Footings, concrete, 1′ thick, 2′ wide	B-5	300	.187	L.F.		5.65	3.14	8.79	12.70
	1080	1′-6″ thick, 2′ wide		250	.224			6.80	3.77	10.57	15.25
	1120	3′ wide		200	.280			8.50	4.71	13.21	19.05
	1140	2′ thick, 3′ wide		175	.320			9.70	5.40	15.10	22
	1200	Average reinforcing, add								10%	10%
	1220	Heavy reinforcing, add								20%	20%
	2000	Walls, block, 4″ thick	1 Clab	180	.044	S.F.		1.22		1.22	2.02
	2040	6″ thick		170	.047			1.29		1.29	2.14
	2080	8″ thick		150	.053			1.46		1.46	2.42
	2100	12″ thick		150	.053			1.46		1.46	2.42
	2200	For horizontal reinforcing, add								10%	10%
	2220	For vertical reinforcing, add								20%	20%
	2400	Concrete, plain concrete, 6″ thick	B-9	160	.250			6.95	.92	7.87	12.50
	2420	8″ thick		140	.286			7.95	1.05	9	14.30
	2440	10″ thick		120	.333			9.25	1.23	10.48	16.70
	2500	12″ thick		100	.400			11.10	1.47	12.57	20

02220 | Site Demolition

		02220 Site Demolition	Crew	Daily Output	Labor-Hours	Unit	Mat.	Labor	Equip.	Total	Total Incl O&P	
							2006 Bare Costs					
130	2600	For average reinforcing, add (R024119-10)				S.F.				10%	10%	130
	2620	For heavy reinforcing, add								20%	20%	
	9000	Minimum labor/equipment charge	A-1	2	4	Job		110	24.50	134.50	209	
220	0010	FENCING DEMOLITION (R024119-10)										220
	1600	Fencing, barbed wire, 3 strand	2 Clab	430	.037	L.F.		1.02		1.02	1.69	
	1650	5 strand	″	280	.057			1.57		1.57	2.59	
	1700	Chain link, posts & fabric, remove only, 8′ to 10′ high	B-6	445	.054			1.62	.51	2.13	3.19	
	1750	Remove and reset	″	70	.343			10.30	3.23	13.53	20.50	
230	0010	HYDRODEMOLITION (R024119-10)										230
	0015	Hydrodemolition, concrete pavement, 4000 PSI, 2″ depth	B-5	500	.112	S.F.		3.40	1.88	5.28	7.60	
	0120	4″ depth		450	.124			3.78	2.09	5.87	8.45	
	0130	6″ depth		400	.140			4.25	2.35	6.60	9.50	
	0410	6000 PSI, 2″ depth		410	.137			4.14	2.30	6.44	9.30	
	0420	4″ depth		350	.160			4.85	2.69	7.54	10.85	
	0430	6″ depth		300	.187			5.65	3.14	8.79	12.70	
	0510	8000 PSI, 2″ depth		330	.170			5.15	2.85	8	11.55	
	0520	4″ depth		280	.200			6.05	3.36	9.41	13.60	
	0530	6″ depth		240	.233			7.10	3.92	11.02	15.85	
240	0010	MINOR SITE DEMOLITION (R024119-10)										240
	0015	No hauling, abandon catch basin or manhole	B-6	7	3.429	Ea.		103	32.50	135.50	203	
	0020	Remove existing catch basin or manhole, masonry		4	6			180	56.50	236.50	355	
	0030	Catch basin or manhole frames and covers, stored		13	1.846			55.50	17.40	72.90	109	
	0040	Remove and reset		7	3.429			103	32.50	135.50	203	
	0100	Roadside delineators, remove only	B-80	175	.183			5.45	3.04	8.49	12.25	
	0110	Remove and reset	″	100	.320			9.55	5.30	14.85	21.50	
	0400	Minimum labor/equipment charge	B-6	4	6	Job		180	56.50	236.50	355	
	0800	Guiderail, corrugated steel, remove only	B-80A	100	.240	L.F.		6.60	1.76	8.36	12.85	
	0850	Remove and reset	″	40	.600	″		16.45	4.40	20.85	32	
	0860	Guide posts, remove only	B-80B	120	.267	Ea.		7.85	1.96	9.81	14.95	
	0870	Remove and reset	B-55	50	.480	″		13.20	16.15	29.35	39.50	
	0890	Minimum labor/equipment charge	2 Clab	4	4	Job		110		110	182	
	0900	Hydrants, fire, remove only	B-21A	5	8	Ea.		275	107	382	560	
	0950	Remove and reset	″	2	20	″		685	268	953	1,400	
	0990	Minimum labor/equipment charge	2 Plum	2	8	Job		340		340	530	
	1000	Masonry walls, block or tile, solid, remove	B-5	1800	.031	C.F.		.94	.52	1.46	2.12	
	1100	Cavity wall		2200	.025			.77	.43	1.20	1.73	
	1200	Brick, solid		900	.062			1.89	1.05	2.94	4.23	
	1300	With block back-up		1130	.050			1.50	.83	2.33	3.37	
	1400	Stone, with mortar		900	.062			1.89	1.05	2.94	4.23	
	1500	Dry set		1500	.037			1.13	.63	1.76	2.54	
	1600	Median barrier, precast concrete, remove and store	B-3	430	.112	L.F.		3.30	4.06	7.36	9.85	
	1610	Remove and reset	″	390	.123	″		3.64	4.48	8.12	10.90	
	1650	Minimum labor/equipment charge	A-1	4	2	Job		55	12.25	67.25	105	
	2900	Pipe removal, sewer/water, no excavation, 12″ diameter	B-6	175	.137	L.F.		4.11	1.29	5.40	8.10	
	2930	15″-18″ diameter		150	.160			4.80	1.51	6.31	9.45	
	2960	21″-24″ diameter		120	.200			6	1.89	7.89	11.80	
	3000	27″-36″ diameter		90	.267			8	2.51	10.51	15.75	
	3200	Steel, welded connections, 4″ diameter		160	.150			4.50	1.41	5.91	8.85	
	3300	10″ diameter		80	.300			9	2.83	11.83	17.75	
	3390	Minimum labor/equipment charge		3	8	Job		240	75.50	315.50	475	
	3500	Railroad track removal, ties and track	B-13	330	.170	L.F.		5.10	1.97	7.07	10.45	
	3600	Ballast	B-14	500	.096	C.Y.		2.79	.45	3.24	5.05	
	3700	Remove and re-install, ties & track using new bolts & spikes		50	.960	L.F.		28	4.52	32.52	50.50	
	3800	Turnouts using new bolts and spikes		1	48	Ea.		1,400	226	1,626	2,525	

	02220 | Site Demolition		CREW	DAILY OUTPUT	LABOR-HOURS	UNIT	2006 BARE COSTS MAT.	LABOR	EQUIP.	TOTAL	TOTAL INCL O&P
240	3850	Runways, remove rubber skid marks, 4-6 passes (R024119-10)	B-59A	35	.686	M.S.F.	27.50	19	9.30	55.80	72
	3860	6-10 passes	"	35	.686	"	41.50	19	9.30	69.80	87
	3890	Minimum labor/equipment charge	B-14	5	9.600	Job		279	45	324	505
	4000	Sidewalk removal, bituminous, 2-1/2" thick	B-6	325	.074	S.Y.		2.22	.70	2.92	4.37
	4050	Brick, set in mortar		185	.130			3.89	1.22	5.11	7.70
	4100	Concrete, plain, 4"		160	.150			4.50	1.41	5.91	8.85
	4200	Mesh reinforced	↓	150	.160	↓		4.80	1.51	6.31	9.45
	4290	Minimum labor/equipment charge	B-39	12	4	Job		116	12.30	128.30	205
250	0010	**DEMOLISH, REMOVE PAVEMENT AND CURB** (R024119-10)									
	5010	Pavement removal, bituminous roads, 3" thick	B-38	690	.058	S.Y.		1.81	1.21	3.02	4.26
	5050	4" to 6" thick		420	.095			2.97	1.99	4.96	7
	5100	Bituminous driveways		640	.063			1.95	1.31	3.26	4.60
	5200	Concrete to 6" thick, hydraulic hammer, mesh reinforced		255	.157			4.90	3.28	8.18	11.55
	5300	Rod reinforced		200	.200	↓		6.25	4.18	10.43	14.70
	5400	Concrete, 7" to 24" thick, plain		33	1.212	C.Y.		38	25.50	63.50	89.50
	5500	Reinforced		24	1.667	"		52	35	87	123
	5590	Minimum labor/equipment charge	↓	6	6.667	Job		208	139	347	490
	5600	With hand held air equipment, bituminous, to 6" thick	B-39	1900	.025	S.F.		.73	.08	.81	1.29
	5700	Concrete to 6" thick, no reinforcing		1600	.030			.87	.09	.96	1.53
	5800	Mesh reinforced		1400	.034			1	.11	1.11	1.75
	5900	Rod reinforced	↓	765	.063	↓		1.82	.19	2.01	3.20
	5990	Minimum labor/equipment charge	B-38	6	6.667	Job		208	139	347	490
	6000	Curbs, concrete, plain	B-6	360	.067	L.F.		2	.63	2.63	3.94
	6100	Reinforced		275	.087			2.62	.82	3.44	5.15
	6200	Granite		360	.067			2	.63	2.63	3.94
	6300	Bituminous		528	.045	↓		1.36	.43	1.79	2.69
	6390	Minimum labor/equipment charge		6	4	Job		120	37.50	157.50	237
	6500	Site demo, berms under 4" in height, bituminous		528	.045	L.F.		1.36	.43	1.79	2.69
	6600	4" or over in height	↓	300	.080	"		2.40	.75	3.15	4.73
310	0010	**SELECTIVE DEMOLITION, CUTOUT** (R024119-10)									
	0020	Concrete, elev. slab, light reinforcement, under 6 CF	B-9C	65	.615	C.F.		17.10	2.26	19.36	31
	0050	Light reinforcing, over 6 C.F.	"	75	.533	"		14.85	1.96	16.81	26.50
	0200	Slab on grade to 6" thick, not reinforced, under 8 S.F.	B-9	85	.471	S.F.		13.10	1.73	14.83	23.50
	0250	8 - 16 S.F.	"	175	.229	"		6.35	.84	7.19	11.50
	0255	For over 16 SF see 02220-130-0400									
	0600	Walls, not reinforced, under 6 C.F.	B-9	60	.667	C.F.		18.55	2.45	21	33
	0650	6 - 12 C.F.	"	80	.500	"		13.90	1.84	15.74	25
	0655	For over 12 CF see 02220-130-2500									
	1000	Concrete, elevated slab, bar reinforced, under 6 C.F.	B-9C	45	.889	C.F.		24.50	3.27	27.77	44.50
	1050	Bar reinforced, over 6 C.F.	"	50	.800	"		22	2.94	24.94	40
	1200	Slab on grade to 6" thick, bar reinforced, under 8 S.F.	B-9	75	.533	S.F.		14.85	1.96	16.81	26.50
	1250	8 - 16 S.F.	"	150	.267	"		7.40	.98	8.38	13.40
	1255	For over 16 SF see 02220-130-0440									
	1400	Walls, bar reinforced, under 6 C.F.	B-9C	50	.800	C.F.		22	2.94	24.94	40
	1450	6 - 12 CF	"	70	.571	"		15.90	2.10	18	29
	1455	For over 12 CF see 02220-130-2500 & 2600									
	2000	Brick, to 4 S.F. opening, not including toothing									
	2040	4" thick	B-9C	30	1.333	Ea.		37	4.91	41.91	67
	2060	8" thick		18	2.222			62	8.20	70.20	111
	2080	12" thick		10	4			111	14.70	125.70	200
	2400	Concrete block, to 4 S.F. opening, 2" thick		35	1.143			32	4.21	36.21	57
	2420	4" thick		30	1.333			37	4.91	41.91	67
	2440	8" thick		27	1.481			41	5.45	46.45	74
	2460	12" thick	↓	24	1.667			46.50	6.15	52.65	84
	2600	Gypsum block, to 4 S.F. opening, 2" thick	B-9	80	.500	↓		13.90	1.84	15.74	25

02220 | Site Demolition

	Line	Description	Crew	Daily Output	Labor-Hours	Unit	2006 Bare Costs Mat.	Labor	Equip.	Total	Total Incl O&P	
310	2620	4″ thick	B-9	70	.571	Ea.		15.90	2.10	18	29	310
	2640	8″ thick (R024119-10)		55	.727			20	2.68	22.68	36.50	
	2800	Terra cotta, to 4 S.F. opening, 4″ thick		70	.571			15.90	2.10	18	29	
	2840	8″ thick		65	.615			17.10	2.26	19.36	31	
	2880	12″ thick		50	.800			22	2.94	24.94	40	
	4000	For toothing masonry, see Division 04910-800										
	6000	Walls, interior, not including re-framing,										
	6010	openings to 5 S.F.										
	6100	Drywall to 5/8″ thick	1 Clab	24	.333	Ea.		9.15		9.15	15.15	
	6200	Paneling to 3/4″ thick		20	.400			10.95		10.95	18.15	
	6300	Plaster, on gypsum lath		20	.400			10.95		10.95	18.15	
	6340	On wire lath		14	.571			15.65		15.65	26	
	7000	Wood frame, not including re-framing, openings to 5 S.F.										
	7200	Floors, sheathing and flooring to 2″ thick	1 Clab	5	1.600	Ea.		44		44	72.50	
	7310	Roofs, sheathing to 1″ thick, not including roofing		6	1.333			36.50		36.50	60.50	
	7410	Walls, sheathing to 1″ thick, not including siding		7	1.143			31.50		31.50	52	
	8500	Minimum labor/equipment charge		4	2	Job		55		55	91	
320	0010	**SELECTIVE DEMOLITION, DISPOSAL ONLY**										320
	0015	Urban bldg w/salvage value allowed (R024119-10)										
	0020	Including loading and 5 mile haul to dump										
	0200	Steel frame	B-3	430	.112	C.Y.		3.30	4.06	7.36	9.85	
	0300	Concrete frame		365	.132			3.89	4.79	8.68	11.60	
	0400	Masonry construction		445	.108			3.19	3.93	7.12	9.50	
	0500	Wood frame		247	.194			5.75	7.05	12.80	17.20	
330	0010	**SELECTIVE DEMOLITION, DUMP CHARGES**										330
	0020	Dump charges, typical urban city, tipping fees only (R024119-10)										
	0100	Building construction materials				Ton					70	
	0200	Trees, brush, lumber									50	
	0300	Rubbish only									60	
	0500	Reclamation station, usual charge									85	
340	0010	**SELECTIVE DEMOLITION, GUTTING**										340
	0020	Building interior, including disposal, dumpter fees not incl. (R024119-10)										
	0500	Residential building										
	0560	Minimum	B-16	400	.080	SF Flr.		2.25	1.19	3.44	5	
	0580	Maximum	″	360	.089	″		2.50	1.32	3.82	5.60	
	0900	Commercial building										
	1000	Minimum	B-16	350	.091	SF Flr.		2.57	1.36	3.93	5.75	
	1020	Maximum		250	.128	″		3.60	1.91	5.51	8.05	
	3000	Minimum labor/equipment charge		4	8	Job		225	119	344	500	
350	0010	**SELECTIVE DEMOLITION, RUBBISH HANDLING**										350
	0020	The following are to be added to the demolition prices (R024119-10)										
	0400	Chute, circular, prefabricated steel, 18″ diameter	B-1	40	.600	L.F.	28.50	16.85		45.35	59.50	
	0440	30″ diameter	″	30	.800	″	38.50	22.50		61	79	
	0725	Dumpster, weekly rental, 1 dump/week, 20 C.Y. capacity (8 Tons)				Week	420			420	462	
	0800	30 C.Y. capacity (10 Tons)					605			605	665	
	0840	40 C.Y. capacity (13 Tons)					775			775	825	
	0900	Alternate pricing for dumpsters										
	0910	Delivery, average for all sizes				Ea.	50			50	55	
	0920	Haul, average for all sizes					150			150	165	
	0930	Rent per day, average for all sizes					5			5	5.50	
	0940	Rent per month, average for all sizes					45			45	49.50	
	0950	Disposal fee per ton, average for all sizes				Ton	40			40	44	
	1000	Dust partition, 6 mil polyethylene, 1″ x 3″ frame	2 Carp	2000	.008	S.F.	.18	.28		.46	.67	

	02220 | Site Demolition		Crew	Daily Output	Labor-Hours	Unit	2006 Bare Costs				Total Incl O&P	
							Mat.	Labor	Equip.	Total		
350	1080	2" x 4" frame (R024119-10)	2 Carp	2000	.008	S.F.	.32	.28		.60	.82	350
	2000	Load, haul, and dump, 50' haul	2 Clab	24	.667	C.Y.		18.25		18.25	30.50	
	2040	100' haul		16.50	.970			26.50		26.50	44	
	2080	Over 100' haul, add per 100 L.F.		35.50	.451			12.35		12.35	20.50	
	2120	In elevators, per 10 floors, add		140	.114			3.13		3.13	5.20	
	3000	Loading & trucking, including 2 mile haul, chute loaded	B-16	45	.711			20	10.60	30.60	44.50	
	3040	Hand loading truck, 50' haul	"	48	.667			18.75	9.95	28.70	42	
	3080	Machine loading truck	B-17	120	.267			7.90	4.61	12.51	17.90	
	5000	Haul, per mile, up to 8 C.Y. truck	B-34B	1165	.007			.19	.41	.60	.77	
	5100	Over 8 C.Y. truck	"	1550	.005			.15	.31	.46	.58	
360	0010	**SELECTIVE DEMOLITION, SAW CUTTING** (R024119-10)										360
	0015	Asphalt, up to 3" deep	B-89	1050	.015	L.F.	.28	.48	.28	1.04	1.39	
	0020	Each additional inch of depth		1800	.009		.06	.28	.17	.51	.70	
	0400	Concrete slabs, mesh reinforcing, up to 3" deep		980	.016		.38	.51	.30	1.19	1.57	
	0420	Each additional inch of depth		1600	.010		.13	.31	.19	.63	.84	
	0800	Concrete walls, hydraulic saw, plain, per inch of depth	B-89B	250	.064		.34	2.01	1.86	4.21	5.65	
	0820	Rod reinforcing, per inch of depth		150	.107		.47	3.35	3.09	6.91	9.25	
	1200	Masonry walls, hydraulic saw, brick, per inch of depth		300	.053		.34	1.67	1.55	3.56	4.75	
	1220	Block walls, solid, per inch of depth		250	.064		.36	2.01	1.86	4.23	5.65	
	2000	Brick or masonry w/hand held saw, per inch of depth	A-1	125	.064		.29	1.75	.39	2.43	3.66	
	5000	Wood sheathing to 1" thick, on walls	1 Carp	200	.040			1.42		1.42	2.36	
	5020	On roof	"	250	.032			1.14		1.14	1.88	
	9000	Minimum labor/equipment charge	A-1	2	4	Job		110	24.50	134.50	209	
	9950	See also Div. 02210-200 core drilling										
370	0010	**SELECTIVE DEMOLITION, TORCH CUTTING** (R024119-10)										370
	0020	Steel, 1" thick plate	1 Clab	360	.022	L.F.	.18	.61		.79	1.21	
	0040	1" diameter bar	"	210	.038	Ea.		1.04		1.04	1.73	
	1000	Oxygen lance cutting, reinforced concrete walls										
	1040	12" to 16" thick walls	1 Clab	10	.800	L.F.		22		22	36.50	
	1080	24" thick walls	"	6	1.333	"		36.50		36.50	60.50	
	1090	Minimum labor/equipment charge	E-25	2	4	Job		168	40.50	208.50	360	
	1100	See also division 05090-920										
372	0010	**SELECTIVE DEMOLITION, UTILITY MATERIALS** (R024119-10)										372
	0020	See other utility items in 02220-240-0010										
	0100	Fire Hydrant extensions	B-20	14	1.714	Ea.		53.50		53.50	88	
	0200	Precast Utility boxes up to 8'x14'x7'	B-13	2	28			840	325	1,165	1,725	
	0300	Handholes and meter pits	B-6	2	12			360	113	473	710	
	0400	Utility valves 4"-12"	B-20	4	6			187		187	310	
	0500	14"-24"	B-21	2	14			450	76.50	526.50	820	
374	0010	**SELECTIVE DEMOLITION, RIP-RAP & ROCK LINING** (R024119-10)										374
	0100	Slope protection, broken stone	B-13	62	.903	C.Y.		27	10.50	37.50	55.50	
	0200	3/8 to 1/4 C.Y. pieces		60	.933	S.Y.		28	10.85	38.85	57.50	
	0300	18 inch depth		60	.933	"		28	10.85	38.85	57.50	
	0400	Dumped stone		93	.602	Ton		18.05	7	25.05	37	
	0500	Gabions, 6 - 12 inches deep		60	.933	S.Y.		28	10.85	38.85	57.50	
	0600	18 - 36 inches deep		30	1.867	"		56	21.50	77.50	115	
376	0010	**SELECTIVE DEMOLITION, SHORE PROTECT/MOORING STRUCTURES**										376
	0100	Breakwaters, bulkheads, concrete, maximun	B-9	12	3.333	L.F.		92.50	12.25	104.75	168	
	0200	Breakwaters, bulkheads, concrete, 12', minimun		10	4			111	14.70	125.70	200	
	0300	Maximum		9	4.444			124	16.35	140.35	223	
	0400	Steel, shore driven	B-40B	54	.889			27	21	48	67	
	0500	Barge driven	B-76A	30	2.133			63	51.50	114.50	160	
	0600	Jetties, docks, floating	B-21B	600	.067	S.F.		2	1.06	3.06	4.44	
	0700	Pier supported, 3" - 4" decking		300	.133			3.99	2.12	6.11	8.90	

	02220 | Site Demolition		Crew	Daily Output	Labor-Hours	Unit	2006 Bare Costs Mat.	Labor	Equip.	Total	Total Incl O&P	
376	0800	Floating, prefab, small boat, minimum	B-21B	600	.067	S.F.		2	1.06	3.06	4.44	376
	0900	Maximum		300	.133			3.99	2.12	6.11	8.90	
	1000	Floating, prefab, per slip, minimum		3	13.333	Ea.		400	212	612	890	
	1010	Maximum		3	13.333	"		400	212	612	890	
378	0010	**SELECTIVE DEMOLITION, PILES**										378
	0100	Cast in place piles, 8"-10", corrugated	B-19	600	.107	V.L.F.		3.79	2.62	6.41	9.15	
	0200	12"-14" corrugated		500	.128			4.55	3.14	7.69	10.95	
	0300	16" corrugated		400	.160			5.70	3.93	9.63	13.70	
	0400	12" fluted		600	.107			3.79	2.62	6.41	9.15	
	0500	14"-18" fluted		500	.128			4.55	3.14	7.69	10.95	
	0600	12" end bearing		600	.107			3.79	2.62	6.41	9.15	
	0700	14"-18" end bearing		500	.128			4.55	3.14	7.69	10.95	
	0800	Precast prestressed piles, 12"-14" dia		700	.091			3.25	2.24	5.49	7.80	
	0900	16"-24" dia		600	.107			3.79	2.62	6.41	9.15	
	1000	36"-66" dia		300	.213			7.60	5.25	12.85	18.25	
	1100	10"-14" thick		600	.107			3.79	2.62	6.41	9.15	
	1200	16"-24" thick		500	.128			4.55	3.14	7.69	10.95	
	1300	Pressure grouted pile, 5"		150	.427			15.15	10.45	25.60	36.50	
	1400	Steel piles, 8"-12" tip		600	.107			3.79	2.62	6.41	9.15	
	1500	H sections HP8 to HP12		600	.107			3.79	2.62	6.41	9.15	
	1600	H section HP14	B-19A	600	.107			3.79	3.35	7.14	9.95	
	1700	Steel pipe piles, 8"-12"	B-19	600	.107			3.79	2.62	6.41	9.15	
	1800	14"-18" plain		500	.128			4.55	3.14	7.69	10.95	
	1900	14"-18" concrete filled		400	.160			5.70	3.93	9.63	13.70	
	2000	Timber piles to 14" dia		600	.107			3.79	2.62	6.41	9.15	
380	0010	**SELECTIVE DEMOLITION, UTILITY VALVES & ACCESSORIES**										380
	0015	Excludes excavation										
	0100	Utility valves 4"-12" dia	B-20	4	6	Ea.		187		187	310	
	0200	14"-24" dia	B-21	2	14			450	76.50	526.50	820	
	0300	Crosses 4"-12"	B-20	8	3			93.50		93.50	154	
	0400	14"-24"	B-21	4	7			225	38	263	410	
	0500	Utility cut-in valves 4"-12" dia	B-20	20	1.200			37.50		37.50	61.50	
	0600	Curb boxes	"	20	1.200			37.50		37.50	61.50	
381	0010	**SELECTIVE DEMOLITION, WATER & SEWER PIPING AND FITTINGS**										381
	0015	Excludes excavation										
	0020	See other utility items in 02220-240										
	0090	Concrete pipe 4"-10" diameter	B-6	250	.096	L.F.		2.88	.91	3.79	5.70	
	0100	42"-48" diameter	B-13B	96	.583			17.45	9.90	27.35	39.50	
	0200	60"-84" diameter	"	80	.700			21	11.90	32.90	47	
	0300	96" diameter	B-13C	80	.700			21	19.85	40.85	56	
	0400	108"-144" diameter	"	64	.875			26	25	51	70	
	0450	Concrete fittings 12" diameter	B-6	24	1	Ea.		30	9.45	39.45	59.50	
	0480	Concrete end pieces 12" diameter		200	.120	L.F.		3.60	1.13	4.73	7.10	
	0485	15" diameter		150	.160			4.80	1.51	6.31	9.45	
	0490	18" diameter		150	.160			4.80	1.51	6.31	9.45	
	0500	24" -36" diameter		100	.240			7.20	2.26	9.46	14.20	
	0600	Concrete fittings 24"-36" diameter		12	2	Ea.		60	18.85	78.85	118	
	0700	48"-84" diameter	B-13B	12	4.667			140	79.50	219.50	315	
	0800	96" diameter	"	8	7			210	119	329	470	
	0900	108"-144" diameter	B-13C	4	14			420	395	815	1,125	
	1000	Ductile iron pipe 4" diameter	B-21B	200	.200	L.F.		6	3.18	9.18	13.30	
	1100	6"-12" diameter		175	.229			6.85	3.64	10.49	15.20	
	1200	14"-24" diameter		120	.333			10	5.30	15.30	22	
	1300	Ductile iron fittings 4"-12" diametrer		24	1.667	Ea.		50	26.50	76.50	111	
	1400	14"-16" diameter		18	2.222			66.50	35.50	102	148	

02220 | Site Demolition

		02220 Site Demolition	CREW	DAILY OUTPUT	LABOR-HOURS	UNIT	2006 BARE COSTS MAT.	LABOR	EQUIP.	TOTAL	TOTAL INCL O&P	
381	1500	18"-24" diameter	B-21B	12	3.333	Ea.		100	53	153	222	381
	1600	Plastic pipe 3/4"-4" diameter	B-20	700	.034	L.F.		1.07		1.07	1.76	
	1700	6"-8" diameter		500	.048			1.49		1.49	2.46	
	1800	10"-18" diameter		300	.080			2.49		2.49	4.10	
	1900	20"-36" diameter		200	.120			3.73		3.73	6.15	
	1910	42"-48" diameter		180	.133			4.15		4.15	6.85	
	1920	54"-60" diameter		160	.150			4.67		4.67	7.70	
	2000	Plastic fittings 4"- 8" diameter	B-6	75	.320	Ea.		9.60	3.02	12.62	18.90	
	2100	10"-14" diameter		50	.480			14.40	4.53	18.93	28.50	
	2200	16"-24" diameter		20	1.200			36	11.30	47.30	71	
	2210	30"-36" diameter		15	1.600			48	15.10	63.10	94.50	
	2220	42 "- 48" diameter		12	2			60	18.85	78.85	118	
	2300	Copper pipe 3/4"-2" diameter	Q-1	500	.032	L.F.		1.23		1.23	1.91	
	2400	2 1/2" - 3" diameter		300	.053			2.05		2.05	3.19	
	2500	4"- 6" diameter		200	.080			3.07		3.07	4.78	
	2600	Copper fittings 3/4"- 2" diameter		15	1.067	Ea.		41		41	63.50	
	2700	Cast iron pipe 4" diameter		200	.080	L.F.		3.07		3.07	4.78	
	2800	5"- 6" diameter	Q-2	200	.120			4.78		4.78	7.45	
	2900	8"- 12" diameter	Q-3	200	.160			6.50		6.50	10.10	
	3000	Cast iron fittings 4" diameter	Q-1	30	.533	Ea.		20.50		20.50	32	
	3100	5"- 6" diameter	Q-2	30	.800			32		32	49.50	
	3200	8"- 15" diameter	Q-3	30	1.067			43.50		43.50	67.50	
	3300	Vent cast iron pipe 4"- 8" diameter	Q-1	200	.080	L.F.		3.07		3.07	4.78	
	3400	10"- 15" diameter	Q-3	200	.160	"		6.50		6.50	10.10	
	3500	Vent cast iron fittings 4"- 8" diameter	Q-2	30	.800	Ea.		32		32	49.50	
	3600	10"- 15" diameter	"	20	1.200	"		48		48	74.50	
382	0010	**SELECTIVE DEMOLITION, METAL DRAINAGE PIPING**										382
	0015	Excludes excavation										
	0020	See other utility items in 02220-240, 02220-380										
	0025	02220-381										
	0100	CMP pipe, aluminium, 6"-10" dia	B-21	800	.035	L.F.		1.12	.19	1.31	2.05	
	0110	12" dia		600	.047			1.50	.25	1.75	2.73	
	0120	18" dia		600	.047			1.50	.25	1.75	2.73	
	0140	Steel, 6"-10" dia		800	.035			1.12	.19	1.31	2.05	
	0150	12" dia		600	.047			1.50	.25	1.75	2.73	
	0160	18" dia		400	.070			2.25	.38	2.63	4.10	
	0170	24" dia	B-13	300	.187			5.60	2.17	7.77	11.50	
	0180	30" - 36" dia		250	.224			6.70	2.60	9.30	13.80	
	0190	48" - 60" dia		200	.280			8.40	3.25	11.65	17.25	
	0200	steel, 72" dia	B-13B	100	.560			16.75	9.50	26.25	38	
	0210	CMP end sections, steel, 10"-18" dia	B-21	40	.700	Ea.		22.50	3.82	26.32	41	
	0220	24"-36" dia	B-13	30	1.867			56	21.50	77.50	115	
	0230	48" dia		20	2.800			84	32.50	116.50	173	
	0240	60" dia		10	5.600			168	65	233	345	
	0250	72" dia	B-13B	10	5.600			168	95	263	380	
	0260	CMP fittings, 8"-12" dia	B-21	60	.467			15	2.54	17.54	27.50	
	0270	18" dia	"	40	.700			22.50	3.82	26.32	41	
	0280	24"-48" dia	B-13	30	1.867			56	21.50	77.50	115	
	0290	60" dia	"	20	2.800			84	32.50	116.50	173	
	0300	72" dia	B-13B	10	5.600			168	95	263	380	
	0310	Oval arch 17" x 13", 21" x 15", 15-18" equivalent	B-21	400	.070	L.F.		2.25	.38	2.63	4.10	
	0320	28" x 20", 24" equivalent	B-13	300	.187			5.60	2.17	7.77	11.50	
	0330	35" x 24", 42" x 29", 30-36" equivalent		250	.224			6.70	2.60	9.30	13.80	
	0340	49" x 33", 57" x 38", 42-48" equivalent		200	.280			8.40	3.25	11.65	17.25	
	0350	Oval arch 17" x 13" end piece, 15" equivalent	B-21	40	.700	Ea.		22.50	3.82	26.32	41	
	0360	42" x 29" end piece, 36" equivalent	B-13	30	1.867	"		56	21.50	77.50	115	

	02220 | Site Demolition		CREW	DAILY OUTPUT	LABOR-HOURS	UNIT	2006 BARE COSTS MAT.	LABOR	EQUIP.	TOTAL	TOTAL INCL O&P	
383	0010	**SELECTIVE DEMOLITION, MANHOLES & CATCH BASINS**										383
	0015	Excludes excavation										
	0020	See other utility items in 02220-240										
	0100	Manholes, precast or brick over 8' deep	B-6	8	3	V.L.F.		90	28.50	118.50	177	
	0200	Cast in place 4'-8' deep	B-9	127	.315	SF Face		8.75	1.16	9.91	15.80	
	0300	Over 8' deep	"	100	.400	"		11.10	1.47	12.57	20	
	0400	Top, precast, 8" thick, 4'-6" dia	B-6	8	3	Ea.		90	28.50	118.50	177	
	0500	Steps	1 Clab	60	.133	"		3.65		3.65	6.05	
384	0010	**SELECTIVE DEMOLITION, BOX CULVERT**										384
	0015	Excludes excavation										
	0100	Box culvert 8' x 6' x 3' to 8' x 8' x 8'	B-69	300	.160	L.F.		4.86	3.53	8.39	11.80	
	0200	8' x 10' x 3' to 8' x 12' x 8'	"	200	.240	"		7.30	5.30	12.60	17.65	
386	0010	**SELECTIVE DEMOLITION, WATER WELLS**										386
	0100	Well, 40' deep with casing & gravel pack, 24"-36" dia	B-23	.25	160	Ea.		4,450	14,200	18,650	23,000	
	0200	Riser pipe, 1 1/4", for observation well	"	300	.133	L.F.		3.71	11.85	15.56	19.20	
	0300	Pump, 1/2 to 5 HP up to 100' depth	Q-1	3	5.333	Ea.		205		205	320	
	0400	Up to 150' well 25 HP pump	Q-22	1.50	10.667			410	425	835	1,100	
	0500	Up to 500' well 30 HP pump	"	1	16	↓		615	635	1,250	1,650	
	0600	Well screen 2" to 8"	B-23	300	.133	V.L.F.		3.71	11.85	15.56	19.20	
	0700	10" to 16"		200	.200			5.55	17.80	23.35	29	
	0800	18" to 26"		150	.267			7.40	23.50	30.90	38.50	
	0900	Slotted PVC for wells 1 1/4"-8"		600	.067			1.85	5.95	7.80	9.55	
	1000	Well screen and casing 6" to 16"		300	.133			3.71	11.85	15.56	19.20	
	1100	18" to 26"		150	.267			7.40	23.50	30.90	38.50	
	1200	30" to 36"	↓	100	.400	↓		11.10	35.50	46.60	57.50	
388	0010	**SELECTIVE DEMOLITION, SEPTIC TANKS & RELATED COMPONENTS**										388
	0020	Excludes excavation										
	0100	Septic tanks, precast,1000-1250 gal	B-21	8	3.500	Ea.		112	19.10	131.10	205	
	0200	1500 gal		7	4			128	22	150	234	
	0300	2000-2500 gal		5	5.600			180	30.50	210.50	330	
	0400	4000 gal	↓	4	7			225	38	263	410	
	0500	Precast, 5000 gal, 4 piece	B-13	3	18.667			560	217	777	1,150	
	0600	15,000 gal	B-13B	1.70	32.941			985	560	1,545	2,225	
	0700	25,000 gal		1.10	50.909			1,525	865	2,390	3,425	
	0800	40,000 gal	↓	.80	70			2,100	1,200	3,300	4,725	
	0900	Precast, 50,000 gal, 5 piece	B-13C	.60	93.333			2,800	2,650	5,450	7,475	
	1000	Cast-in-place, 75,000 gal	B-9	.06	666			18,500	2,450	20,950	33,400	
	1100	100,000 gal	"	.05	800			22,200	2,950	25,150	40,100	
	1200	HDPE, 1000 gal	B-21	9	3.111			100	16.95	116.95	182	
	1300	1500 gal		8	3.500			112	19.10	131.10	205	
	1400	Galley, 4' x 4' x 4'	↓	16	1.750			56	9.55	65.55	103	
	1500	Distribution boxes, concrete, 7 outlets	2 Clab	16	1			27.50		27.50	45.50	
	1600	9 outlets	"	8	2			55		55	91	
	1700	Leaching chambers 13' x 3'-7" x 1'-4", standard	B-13	16	3.500			105	40.50	145.50	216	
	1800	8' x 4' x 1'-6", heavy duty		14	4			120	46.50	166.50	246	
	1900	13' x 3'-9" x 1'-6"		12	4.667			140	54	194	288	
	2100	20' x 4' x 1'-6"	↓	5	11.200			335	130	465	690	
	2200	Leaching pit 6'-6" x 6' deep	B-21	5	5.600			180	30.50	210.50	330	
	2300	6'-6" x 8' deep		4	7			225	38	263	410	
	2400	8' x 6' deep H20		4	7			225	38	263	410	
	2500	8' x 8' deep H20		3	9.333			300	51	351	545	
	2600	Velocity reducing pit, precast 6' x 3' deep	↓	4.70	5.957	↓		191	32.50	223.50	350	
390	0010	**SELECTIVE DEMOLITION, STEEL PIPE WITH INSULATION**										390
	0020	Excludes excavation										

	02220 | Site Demolition		CREW	DAILY OUTPUT	LABOR-HOURS	UNIT	2006 BARE COSTS MAT.	LABOR	EQUIP.	TOTAL	TOTAL INCL O&P	
390	0100	Steel pipe, with insulation, 3/4"-4"	B-1A	400	.060	L.F.		1.68	.41	2.09	3.24	390
	0200	5"-10"	B-1B	360	.089			2.72	2.22	4.94	6.90	
	0300	12"-16"		240	.133			4.08	3.33	7.41	10.30	
	0400	18"-24"		160	.200			6.10	4.99	11.09	15.50	
	0450	26"-36"		100	.320			9.80	8	17.80	25	
	0500	Steel gland seal, with insulation, 3/4"-4"	B-1A	100	.240	Ea.		6.75	1.62	8.37	12.95	
	0600	5"-10"	B-1B	75	.427			13.05	10.65	23.70	33	
	0700	12"-16"		60	.533			16.30	13.30	29.60	41	
	0800	18"-24"		50	.640			19.55	15.95	35.50	49.50	
	0850	26"-36"		40	.800			24.50	19.95	44.45	62	
	0900	Demo steel fittings with insulation 3/4"-4"	B-1A	60	.400			11.25	2.70	13.95	21.50	
	1000	5"-10"	B-1B	40	.800			24.50	19.95	44.45	62	
	1100	12"-16"		30	1.067			32.50	26.50	59	82.50	
	1200	18"-24"		20	1.600			49	40	89	124	
	1300	26"-36"		15	2.133			65	53.50	118.50	165	
	1400	Steel pipe anchors, 5"-10 "		40	.800			24.50	19.95	44.45	62	
	1500	12"-16"		30	1.067			32.50	26.50	59	82.50	
	1600	18"-24"		20	1.600			49	40	89	124	
	1700	26"-36"		15	2.133			65	53.50	118.50	165	
392	0010	**SELECTIVE DEMOLITION, GASOLINE CONTAINMENT PIPING**										392
	0020	Excludes excavation										
	0030	Excludes environmental site remediation										
	0100	Gasoline plastic primary containment piping 2" to 4"	Q-6	800	.030	L.F.		1.21		1.21	1.87	
	0200	Fittings 2" to 4"		40	.600	Ea.		24		24	37.50	
	0300	Gasoline plastic secondary containment piping 3" to 6"		800	.030	L.F.		1.21		1.21	1.87	
	0400	Fittings 3" to 6"		40	.600	Ea.		24		24	37.50	
394	0010	**SELECTIVE DEMOLITION, NATURAL GAS, PE PIPE**										394
	0020	Excludes excavation										
	0100	Natural gas coils, PE, 1 1/4 to 3 inch	Q-6	800	.030	L.F.		1.21		1.21	1.87	
	0200	Joints, 40', PE, 3-4 "		800	.030			1.21		1.21	1.87	
	0300	6 - 8 "		600	.040			1.61		1.61	2.50	
396	0010	**SELECTIVE DEMOLITION, NATURAL GAS, STEEL PIPE**										396
	0020	Excludes excavation										
	0100	Natural gas steel pipe 1- 4 "	B-1A	800	.030	L.F.		.84	.20	1.04	1.62	
	0200	5 - 10"	B-1B	360	.089			2.72	2.22	4.94	6.90	
	0300	12 - 16"		240	.133			4.08	3.33	7.41	10.30	
	0400	18 - 24"		160	.200			6.10	4.99	11.09	15.50	
	0500	Natural gas steel fittings 1 - 4 "	B-1A	160	.150	Ea.		4.21	1.01	5.22	8.10	
	0600	5 - 10"	B-1B	160	.200			6.10	4.99	11.09	15.50	
	0700	12 - 16"		108	.296			9.05	7.40	16.45	23	
	0800	18 - 24"		70	.457			14	11.40	25.40	35.50	
398	0010	**SELECTIVE DEMO, NATURAL GAS, VALVES, FITTINGS, REGULATORS**										398
	0100	Gas stops 1 1/4"-2"	1 Plum	22	.364	Ea.		15.55		15.55	24	
	0200	Gas regulator 1 1/2"-2"	"	22	.364			15.55		15.55	24	
	0300	3 - 4"	Q-1	22	.727			28		28	43.50	
	0400	Gas plug valve, 3/4 - 2"	1 Plum	22	.364			15.55		15.55	24	
	0500	2 1/2 - 3"	Q-1	10	1.600			61.50		61.50	95.50	
401	0010	**SELECTIVE DEMOLITION, RADIO TOWERS**										401
	0100	Radio tower,guyed, 50'	2 Skwk	16	1	Ea.		36.50		36.50	59.50	
	0200	190', 40 Lb section	K-2	.70	34.286			1,250	252	1,502	2,550	
	0300	200', 70 Lb section		.70	34.286			1,250	252	1,502	2,550	
	0400	300', 70 Lb section		.40	60			2,200	440	2,640	4,425	
	0500	270', 90 Lb section		.40	60			2,200	440	2,640	4,425	

		02220 | Site Demolition	CREW	DAILY OUTPUT	LABOR-HOURS	UNIT	2006 BARE COSTS MAT.	LABOR	EQUIP.	TOTAL	TOTAL INCL O&P	
401	0600	400′	K-2	.30	80	Ea.		2,925	585	3,510	5,925	401
	0700	Self supported, 60′		.90	26.667			975	196	1,171	1,975	
	0800	120′		.80	30			1,100	220	1,320	2,225	
	0900	190′		.40	60			2,200	440	2,640	4,425	
402	0010	**SELECTIVE DEMOLITION, ELECTRIC DUCTS & FITTINGS**										402
	0020	Excludes excavation										
	0100	Plastic conduit, 1/2″ - 2 ″	1 Elec	600	.013	L.F.		.56		.56	.86	
	0200	3″ - 6 ″	2 Elec	400	.040	″		1.68		1.68	2.59	
	0300	Fittings, 1/2″ - 2 ″	1 Elec	50	.160	Ea.		6.70		6.70	10.35	
	0400	3″ - 6 ″	″	40	.200	″		8.40		8.40	12.95	
404	0010	**SELECTIVE DEMOLITION, ELECTRIC DUCT BANKS**										404
	0020	Excludes excavation										
	0100	Hand holes sized to 4′ x 4′ x 4′	R-3	7	2.857	Ea.		118	22	140	207	
	0200	Manholes sized to 6′ x 10′ x 7′	B-13	6	9.333	″		279	108	387	575	
	0300	Conduit 1 @ 2″ diameter, EB plastic, no concrete	2 Elec	1000	.016	L.F.		.67		.67	1.04	
	0400	2 @ 2″ diameter		500	.032			1.34		1.34	2.07	
	0500	4 @ 2″ diameter		250	.064			2.69		2.69	4.14	
	0600	1 @ 3″ diameter		800	.020			.84		.84	1.29	
	0700	2 @ 3″ diameter		400	.040			1.68		1.68	2.59	
	0800	4 @ 3″ diameter		200	.080			3.36		3.36	5.20	
	0900	1 @ 4″ diameter		800	.020			.84		.84	1.29	
	1000	2 @ 4″ diameter		400	.040			1.68		1.68	2.59	
	1100	4 @ 4″ diameter		200	.080			3.36		3.36	5.20	
	1200	6 @ 4″ diameter		100	.160			6.70		6.70	10.35	
	1300	1 @ 5″ diameter		500	.032			1.34		1.34	2.07	
	1400	2 @ 5″ diameter		250	.064			2.69		2.69	4.14	
	1500	4 @ 5″ diameter		160	.100			4.20		4.20	6.45	
	1600	6 @ 5″ diameter		100	.160			6.70		6.70	10.35	
	1700	1 @ 6″ diameter		500	.032			1.34		1.34	2.07	
	1800	2 @ 6″ diameter		250	.064			2.69		2.69	4.14	
	1900	4 @ 6″ diameter		160	.100			4.20		4.20	6.45	
	2000	6 @ 6″ diameter		100	.160			6.70		6.70	10.35	
	2100	Conduit 1 EB plastic, with concrete, 0.92 CF/LF	B-9	150	.267			7.40	.98	8.38	13.40	
	2200	2 EB plastic 1.52 CF/LF		100	.400			11.10	1.47	12.57	20	
	2300	2 x 2 EB plastic 2.51 CF/LF		80	.500			13.90	1.84	15.74	25	
	2400	2 x 3 EB plastic 3.56 CF/LF		60	.667			18.55	2.45	21	33	
	2500	Conduit 2 @ 2″ diameter, steel, no concrete	2 Elec	400	.040			1.68		1.68	2.59	
	2600	4 @ 2″ diameter		200	.080			3.36		3.36	5.20	
	2700	2 @ 3″ diameter		200	.080			3.36		3.36	5.20	
	2800	4 @ 3″ diameter		100	.160			6.70		6.70	10.35	
	2900	2 @ 4″ diameter		200	.080			3.36		3.36	5.20	
	3000	4 @ 4″ diameter		100	.160			6.70		6.70	10.35	
	3100	6 @ 4″ diameter		50	.320			13.45		13.45	20.50	
	3200	2 @ 5″ diameter		120	.133			5.60		5.60	8.65	
	3300	4 @ 5″ diameter		60	.267			11.20		11.20	17.25	
	3400	6 @ 5″ diameter		40	.400			16.80		16.80	26	
	3500	2 @ 6″ diameter		120	.133			5.60		5.60	8.65	
	3600	4 @ 6″ diameter		60	.267			11.20		11.20	17.25	
	3700	6 @ 6″ diameter		40	.400			16.80		16.80	26	
	3800	Conduit 2 steel, with concrete, 1.52 CF/LF	B-9	80	.500			13.90	1.84	15.74	25	
	3900	2 x 2 EB steel 2.51 CF/LF		60	.667			18.55	2.45	21	33	
	4000	2 x 3 steel 3.56 CF/LF		50	.800			22	2.94	24.94	40	
	4100	Conduit fittings, PVC type EB, 2″ - 3″	1 Elec	30	.267	Ea.		11.20		11.20	17.25	
	4200	4″ - 6″	″	20	.400	″		16.80		16.80	26	
406	0010	**SELECTIVE DEMOLITION, UTILITY POLES & CROSS ARMS**										406
	0100	Utility poles, wood, 20′ - 30′ high	R-3	6	3.333	Ea.		138	25.50	163.50	242	

	02220 | Site Demolition		CREW	DAILY OUTPUT	LABOR-HOURS	UNIT	2006 BARE COSTS MAT.	LABOR	EQUIP.	TOTAL	TOTAL INCL O&P	
406	0200	35′ - 45′ high	R-3	5	4	Ea.		166	30.50	196.50	290	406
	0300	Cross arms, wood,4′ - 6′ long	1 Elec	5	1.600	↓		67		67	104	
408	0010	**SELECTIVE REMOVAL, PAVEMENT LINES & MARKINGS**										408
	0015	Does not include traffic control costs										
	0020	See other items in 02760-300										
	0100	Remove painted traffic lines and markings permanent	B-78A	500	.016	C.L.F.		.56	1.59	2.15	2.64	
	0200	Temporary traffic line tape	2 Clab	1500	.011	L.F.		.29		.29	.48	
	0300	Thermoplastic traffic lines and markings	B-79A	500	.024	C.L.F.		.84	2.26	3.10	3.81	
	0400	Painted pavement markings	B-78B	500	.036	S.F.		1.02	.49	1.51	2.22	
410	0010	**SELECTIVE DEMOLITION, WALKS, STEPS AND PAVERS**										410
	0020	See other utility items in 02220-240										
	0100	Splash blocks	1 Clab	300	.027	S.F.		.73		.73	1.21	
	0200	Tree grates	"	50	.160	Ea.		4.38		4.38	7.25	
	0300	Walks, limestone pavers	2 Clab	150	.107	S.F.		2.92		2.92	4.84	
	0400	Redwood sections		600	.027			.73		.73	1.21	
	0500	Redwood planks		480	.033			.91		.91	1.51	
	0600	Shale paver		300	.053			1.46		1.46	2.42	
	0700	Tile thinset paver	↓	675	.024	↓		.65		.65	1.08	
	0800	Wood round	B-1	350	.069	Ea.		1.92		1.92	3.19	
	0900	Asphalt block	2 Clab	450	.036	S.F.		.97		.97	1.61	
	1000	Bluestone		450	.036			.97		.97	1.61	
	1100	Slate, 1″ or thinner		675	.024			.65		.65	1.08	
	1200	Granite blocks		300	.053			1.46		1.46	2.42	
	1300	Precast patio blocks		450	.036			.97		.97	1.61	
	1400	Planter blocks		600	.027			.73		.73	1.21	
	1500	Brick paving, dry set		300	.053			1.46		1.46	2.42	
	1600	Mortar set		180	.089			2.44		2.44	4.04	
	1700	Dry set on edge		240	.067	↓		1.83		1.83	3.03	
	1800	Steps, brick		200	.080	L.F.		2.19		2.19	3.63	
	1900	Railroad tie		150	.107			2.92		2.92	4.84	
	2000	Bluestone		180	.089			2.44		2.44	4.04	
	2100	Wood/steel edging for steps		1000	.016			.44		.44	.73	
	2200	Timber or railroad tie edging for steps	↓	400	.040	↓		1.10		1.10	1.82	
414	0010	**SELECTIVE DEMOLITION, ATHLETIC SURFACES**										414
	0100	Synthetic grass	2 Clab	2000	.008	S.F.		.22		.22	.36	
	0200	Surface coat latex rubber	"	2000	.008	"		.22		.22	.36	
	0300	Tennis court posts	B-11C	16	1	Ea.		32	14.15	46.15	67	
418	0010	**SELECTIVE DEMOLITION, LAWN SPRINKLER SYSTEMS**										418
	0100	Golf course sprinkler system, 9 hole	4 Skwk	.10	320	Ea.		11,700		11,700	19,100	
	0200	Sprinkler system, 24′ diam. @ 15′ O.C., per head	B-20	110	.218	Head		6.80		6.80	11.20	
	0300	60′ diam. @ 24′ O.C., per head	"	52	.462	"		14.35		14.35	23.50	
	0400	Sprinkler heads, plastic	2 Skwk	150	.107	Ea.		3.89		3.89	6.35	
	0500	Impact circle pattern, 28-76′ diam.		75	.213			7.80		7.80	12.75	
	0600	Pop-up, 42′-76′ diam.		50	.320			11.70		11.70	19.10	
	0700	39′-99′ diameter		50	.320			11.70		11.70	19.10	
	0800	Sprinkler valves		40	.400			14.60		14.60	24	
	0900	Valve boxes		40	.400			14.60		14.60	24	
	1000	Controls		2	8			292		292	480	
	1100	Backflow preventer		4	4			146		146	239	
	1200	Vacuum breaker	↓	4	4	↓		146		146	239	
420	0010	**SELECTIVE DEMOLITION, CHAIN LINK FENCES & GATES**										420
	0020	See other fence items in 02220-220										
	0100	Chain link, gates, 3′ - 4′ width	B-6	30	.800	Ea.		24	7.55	31.55	47.50	
	0200	10-12′ width	↓	16	1.500	↓		45	14.15	59.15	88.50	

	02220 | Site Demolition		CREW	DAILY OUTPUT	LABOR-HOURS	UNIT	2006 BARE COSTS MAT.	LABOR	EQUIP.	TOTAL	TOTAL INCL O&P	
420	0300	14′ width	B-6	15	1.600	Ea.		48	15.10	63.10	94.50	420
	0400	20′ width		10	2.400			72	22.50	94.50	142	
	0500	18′ width overhead & cantilever		80	.300	L.F.		9	2.83	11.83	17.75	
	0510	Sliding		80	.300			9	2.83	11.83	17.75	
	0520	Cantilever to 40′ wide		80	.300			9	2.83	11.83	17.75	
	0530	Motor operators	2 Skwk	1	16	Ea.		585		585	955	
	0540	Transmitter systems	"	15	1.067	"		39		39	63.50	
	0600	Chain link, fence, 5′ high	B-6	890	.027	L.F.		.81	.25	1.06	1.60	
	0650	3-4′ high		1000	.024			.72	.23	.95	1.42	
	0675	12′ high		400	.060			1.80	.57	2.37	3.55	
	0700	Snow fence, 4′ high		1000	.024			.72	.23	.95	1.42	
	0800	Chain link, fence, braces	B-1	2000	.012	Ea.		.34		.34	.56	
	0900	Privacy slats	"	2000	.012			.34		.34	.56	
	1000	Fence posts, steel, in concrete	B-6	80	.300			9	2.83	11.83	17.75	
	1100	Fence fabric & accessories, fabric to 8′ high		800	.030	L.F.		.90	.28	1.18	1.77	
	1200	Barbed wire		5000	.005	"		.14	.05	.19	.28	
	1300	Extension arms & eye tops		300	.080	Ea.		2.40	.75	3.15	4.73	
	1400	Fence rails		2000	.012	L.F.		.36	.11	.47	.71	
	1500	Reinforcing wire		5000	.005	"		.14	.05	.19	.28	
421	0010	**SELECTIVE DEMOLITION, VINYL FENCES & GATES**										421
	0100	Vinyl fence up to 6 ′ high	B-6	1000	.024	L.F.		.72	.23	.95	1.42	
	0200	Gates, up to 6′ high	"	40	.600	Ea.		18	5.65	23.65	35.50	
422	0010	**SELECTIVE DEMOLITION, MISC METAL FENCES & GATES**										422
	0020	See other fence items in 02220-220										
	0100	Misc steel mesh fences, 4′ - 6′ high	B-6	600	.040	L.F.		1.20	.38	1.58	2.36	
	0200	Kennels, 6′ - 12′ long	2 Clab	8	2	Ea.		55		55	91	
	0300	Tops, 6′ - 12′ long	"	30	.533	"		14.60		14.60	24	
	0400	Security fences, 12′ - 16′ high	B-6	100	.240	L.F.		7.20	2.26	9.46	14.20	
	0500	Metal tubular picket fences 4′ - 6′ high		500	.048	"		1.44	.45	1.89	2.84	
	0600	Gates 3′ - 4′ wide		20	1.200	Ea.		36	11.30	47.30	71	
424	0010	**SELECTIVE DEMOLITION, WOOD FENCES & GATES**										424
	0020	See other fence items in 02220-220										
	0100	Wood fence gates 3′ - 4′ wide	2 Clab	20	.800	Ea.		22		22	36.50	
	0200	Wood fence, open rail, to 4′ high		2560	.006	L.F.		.17		.17	.28	
	0300	Wood fences 8′ high		368	.043	"		1.19		1.19	1.97	
	0400	Post, in concrete		50	.320	Ea.		8.75		8.75	14.55	
426	0010	**SELECTIVE DEMOLITION, RETAINING WALLS**										426
	0020	See other retaining wall items in 02220-240										
	0100	Concrete retaining wall, 6′ high, no reinforcing	B-9	12.70	3.150	L.F.		87.50	11.60	99.10	158	
	0200	8′ high		10	4			111	14.70	125.70	200	
	0300	10′ high		7.80	5.128			143	18.85	161.85	257	
	0400	With reinforcing, 6′ high		11.50	3.478			96.50	12.80	109.30	174	
	0500	8′ high		9	4.444			124	16.35	140.35	223	
	0600	10′ high		7	5.714			159	21	180	286	
	0700	20′ high		4	10			278	37	315	500	
	0800	Concrete cribbing, 12′ high, open/closed face		126	.317	S.F.		8.85	1.17	10.02	15.90	
	0900	Interlocking segmental retaining wall	B-62	800	.030			.90	.19	1.09	1.67	
	1000	Wall caps	"	600	.040			1.20	.26	1.46	2.23	
	1100	Metal bin retaining wall, 10′ wide, 4-12′ high	B-13	1200	.047			1.40	.54	1.94	2.88	
	1200	10′ wide, 16-28′ high		1000	.056			1.68	.65	2.33	3.45	
	1300	Stone gabions, 6′ x 3′ x 1′, stone filled		170	.329	Ea.		9.85	3.83	13.68	20.50	
	1400	6′ x 3′ x 1′-6″		75	.747			22.50	8.70	31.20	46	
	1500	6′ x 3′ x 3′		25	2.240			67	26	93	138	
	1600	9′ x 3′ x 1′		75	.747			22.50	8.70	31.20	46	

	02220 | Site Demolition	CREW	DAILY OUTPUT	LABOR-HOURS	UNIT	2006 BARE COSTS MAT.	LABOR	EQUIP.	TOTAL	TOTAL INCL O&P	
426 1700	9′ x 3′ x 1′-6″	B-13	33	1.697	Ea.		51	19.70	70.70	105	426
1800	9′ x 3′ x 3′		12	4.667			140	54	194	288	
1900	12′ x 3′ x 1′		42	1.333			40	15.50	55.50	82	
2000	12′ x 3′ x 1′-6″		20	2.800			84	32.50	116.50	173	
2100	12′ x 3′ x 3′		6	9.333			279	108	387	575	
428 0010	**SELECTIVE DEMOLITION, HIGHWAY GUARD RAILS & BARRIERS**										428
0100	Guard rail, corrugated steel	B-6	600	.040	L.F.		1.20	.38	1.58	2.36	
0200	End sections		40	.600	Ea.		18	5.65	23.65	35.50	
0300	Wrap around		40	.600	″		18	5.65	23.65	35.50	
0400	Timber 4″ x 8″		600	.040	L.F.		1.20	.38	1.58	2.36	
0500	Three 3/4″ cables		600	.040	″		1.20	.38	1.58	2.36	
0600	Wood posts		240	.100	Ea.		3	.94	3.94	5.90	
0700	Guide rail, 6″ x 6″ box beam	B-80B	120	.267	L.F.		7.85	1.96	9.81	14.95	
0800	Median barrier, 6″ x 8″ box beam		240	.133			3.91	.98	4.89	7.50	
0850	Precast concrete 3′-6″ high x 2′ wide		300	.107			3.13	.78	3.91	5.95	
0900	Impact barrier, UTMCD, barrel type	B-16	60	.533	Ea.		15	7.95	22.95	34	
1000	Resilient guide fence and light shield 6′ high	″	120	.267	L.F.		7.50	3.97	11.47	16.75	
1100	Concrete posts, 6′-5″ triangular	B-6	200	.120	Ea.		3.60	1.13	4.73	7.10	
1200	Speed bumps 10 1/2 x 2 1/4 x 48 ″		300	.080	L.F.		2.40	.75	3.15	4.73	
1300	Pavement marking channelizing		200	.120	Ea.		3.60	1.13	4.73	7.10	
1400	Barrier and curb delineators		300	.080			2.40	.75	3.15	4.73	
1500	Rumble strips 24″ x 3 1/2″ x 1/2 ″		150	.160			4.80	1.51	6.31	9.45	
430 0010	**SELECTIVE DEMOLITION, PARKING APPURTENANCES**										430
0100	Bumper rails, garage, 6″ wide	B-6	300	.080	L.F.		2.40	.75	3.15	4.73	
0200	12″ channel rail		300	.080			2.40	.75	3.15	4.73	
0300	Parking bumper, timber		1000	.024			.72	.23	.95	1.42	
0400	Folding, with locks	B-1	100	.240	Ea.		6.75		6.75	11.15	
0500	Flexible fixed garage stanchion	B-6	150	.160			4.80	1.51	6.31	9.45	
0600	Wheel stops, precast concrete		120	.200			6	1.89	7.89	11.80	
0700	Thermoplastic		120	.200			6	1.89	7.89	11.80	
0800	Pipe bollards, 6″ - 12″ dia		80	.300			9	2.83	11.83	17.75	
432 0010	**SELECTIVE DEMOLITION, PREFAB PEDESTRIAN BRIDGES**										432
0100	Bridges, pedestrian, precast, 60′-150′ long	B-21C	250	.224	S.F.		6.70	6.35	13.05	17.95	
0200	Steel, 50′-160′ long, 8-10′ wide	″	500	.112			3.35	3.18	6.53	8.95	
0300	Laminated wood, 80′ - 130′ long	C-12	300	.160			5.60	2.12	7.72	11.55	
434 0010	**SELECTIVE DEMOLITION, SITE FURNISHINGS**										434
0100	Benches, all types	2 Clab	10	1.600	Ea.		44		44	72.50	
0200	Trash receptacles, all types		80	.200			5.50		5.50	9.10	
0300	Trash enclosures, steel or wood		10	1.600			44		44	72.50	
435 0010	**SELECTIVE DEMOLITION, ATHLETIC SCREENING**										435
0020	See other items in 02220-220 & 02220-420										
0100	Baseball backstops	B-6	2	12	Ea.		360	113	473	710	
0200	Basketball goal, single		6	4			120	37.50	157.50	237	
0300	Double		4	6			180	56.50	236.50	355	
0400	Tennis wire mesh, pair ends		5	4.800	Set		144	45.50	189.50	284	
0500	Enclosed court		3	8	Ea.		240	75.50	315.50	475	
0600	Wood or masonry handball court		1	24	″		720	226	946	1,425	
438 0010	**SELECTIVE DEMOLITION, ATHLETIC/PLAYGROUND EQUIPMENT**										438
0020	See other items in 02220-442										
0100	Bike rack	B-6	32	.750	Ea.		22.50	7.05	29.55	44.50	
0200	Climber arch		16	1.500			45	14.15	59.15	88.50	
0300	Fitness trail, 9 to 10 stations wood or metal		.50	48			1,450	455	1,905	2,850	
0400	16 to 20 stations wood or metal		.25	96			2,875	905	3,780	5,675	

02220 | Site Demolition

		02220 Site Demolition	CREW	DAILY OUTPUT	LABOR-HOURS	UNIT	2006 BARE COSTS MAT.	LABOR	EQUIP.	TOTAL	TOTAL INCL O&P
438	0500	Monkey bars 14' long	B-6	8	3	Ea.		90	28.50	118.50	177
	0600	Parallel bars 10' long		8	3			90	28.50	118.50	177
	0700	Post tether ball		32	.750			22.50	7.05	29.55	44.50
	0800	Poles 10'-6" long		32	.750			22.50	7.05	29.55	44.50
	0900	Pole ground sockets		80	.300			9	2.83	11.83	17.75
	1000	See-saw, 2 units		12	2			60	18.85	78.85	118
	1100	4 units		10	2.400			72	22.50	94.50	142
	1200	6 units		8	3			90	28.50	118.50	177
	1300	Shelter, fiberglass, 3 person		10	2.400			72	22.50	94.50	142
	1400	Slides, 12' long		8	3			90	28.50	118.50	177
	1500	20' long		6	4			120	37.50	157.50	237
	1600	Swings, 4 seats		6	4			120	37.50	157.50	237
	1700	8 seats		4	6			180	56.50	236.50	355
	1800	Whirlers, 8 - 10' diameter		6	4			120	37.50	157.50	237
	1900	Football or football/soccer goal posts		3	8	Pair		240	75.50	315.50	475
	2000	Soccer goal posts		4	6	"		180	56.50	236.50	355
	2100	Playground surfacing, 4 " depth	2 Clab	900	.018	S.F.		.49		.49	.81
	2200	2" topping	"	1800	.009	"		.24		.24	.40
	2300	Platform/paddle tennis court, alum. deck & frame	B-1	.20	120	Court		3,375		3,375	5,575
	2400	Aluminum deck and wood frame		.25	96	"		2,700		2,700	4,475
	2500	Heater		2	12	Ea.		335		335	560
	2600	Wood deck & frame		.25	96	Court		2,700		2,700	4,475
	2700	Wood deck & steel frame		.25	96			2,700		2,700	4,475
	2800	Steel deck & wood frame		.25	96			2,700		2,700	4,475
442	0010	**SELECTIVE DEMOLITION, MODULAR PLAYGROUND EQUIPMENT**									
	0100	Modular playground, 48" deck, square or triangle shape	B-1	2	12	Ea.		335		335	560
	0200	Various posts		18	1.333			37.50		37.50	62
	0300	Roof 54" x 54"		18	1.333			37.50		37.50	62
	0400	Wheel chair transfer module		6	4			112		112	186
	0500	Guardrail pipe		90	.267	L.F.		7.50		7.50	12.40
	0600	Steps, 3 risers		16	1.500	Ea.		42		42	70
	0700	Activity panel crawl through		8	3			84		84	140
	0800	Alphabet/spelling panel		8	3			84		84	140
	0900	Crawl tunnel, each 4' section		8	3			84		84	140
	1000	Slide tunnel		16	1.500			42		42	70
	1100	Spiral slide tunnel		12	2			56		56	93
	1200	Ladder, horizontal or vertical		10	2.400			67.50		67.50	112
	1300	Corkscrew climber		6	4			112		112	186
	1400	Fire pole		12	2			56		56	93
	1500	Bridge, ring chamber		8	3			84		84	140
	1600	Bridge, suspension		8	3			84		84	140
446	0010	**SELECTIVE DEMOLITION, TRAFFIC LIGHT SYSTEMS**									
	0100	Traffic signal, pedestrian cross walk	R-2	.50	112	Total		4,150	610	4,760	7,175
	0200	8 signal intersection, 2 each direction	"	.25	224			8,300	1,225	9,525	14,400
	0300	Each traffic phase controller	L-9	3	12			385		385	650
	0400	Each semi-actuated detector		1.50	24			765		765	1,300
	0500	Each fully-actuated detector		1.50	24			765		765	1,300
	0600	Pedestrian push button system		2	18			575		575	975
	0650	Each optically programmed head		3	12			385		385	650
	0700	School signal flashing system		1	36	Signal		1,150		1,150	1,950
	0800	Complete traffic light intersection system, without lane control	R-2	.25	224	Ea.		8,300	1,225	9,525	14,400
	0900	With lane control		.20	280			10,400	1,525	11,925	18,000
	1000	Traffic light, each left turn protection system		.66	84.848			3,150	460	3,610	5,425
448	0010	**SELECTIVE DEMOLITION, SOD, EDGING, PLANTERS & TREE GUYING**									
	0100	Remove sod 1" deep, 3 M.S.F. or greater, level ground	B-63	40	1	M.S.F.		29	3.85	32.85	51.50

For expanded coverage of these items see *Means Site Work & Landscape Cost Data 2006*

02200 | Site Preparation

02220 | Site Demolition

	Line	Description	Crew	Daily Output	Labor-Hours	Unit	Mat. (2006 Bare Costs)	Labor (2006 Bare Costs)	Equip. (2006 Bare Costs)	Total (2006 Bare Costs)	Total Incl O&P
448	0200	Less than 3 M.S.F., level ground	B-63	27	1.481	M.S.F.		43	5.70	48.70	77
	0300	3 M.S.F. or greater, sloped ground		12	3.333			96.50	12.85	109.35	172
	0400	Less than 3 M.S.F., sloped ground		8	5			145	19.25	164.25	258
	0500	Edging, aluminum	B-1	800	.030	L.F.		.84		.84	1.40
	0600	Brick horizontal		400	.060			1.68		1.68	2.79
	0700	Brick vertical		200	.120			3.37		3.37	5.60
	0800	Corrugated aluminum		1200	.020			.56		.56	.93
	0900	Granite 5″ x 16″		600	.040			1.12		1.12	1.86
	1000	Polyethylene grass barrier		1200	.020			.56		.56	.93
	1100	Precast 2″ x 8″ x 16″		800	.030			.84		.84	1.40
	1200	Railroad ties		200	.120			3.37		3.37	5.60
	1300	Wood		400	.060			1.68		1.68	2.79
	1400	Steel strips including stakes		600	.040			1.12		1.12	1.86
	1500	Planters, concrete, 48″ dia	2 Clab	30	.533	Ea.		14.60		14.60	24
	1600	Concrete 7′ dia		20	.800			22		22	36.50
	1700	Fiberglass 36″ dia		30	.533			14.60		14.60	24
	1800	Fiberglass 60″ dia		20	.800			22		22	36.50
	1900	Fiberglass 24″ square		30	.533			14.60		14.60	24
	2000	Fiberglass 48″ square		30	.533			14.60		14.60	24
	2100	Wood 48″ square		30	.533			14.60		14.60	24
	2200	Wood 48″ dia		20	.800			22		22	36.50
	2300	Wood 72″ dia		20	.800			22		22	36.50
	2400	Planter/bench, fiberglass 72″-96″ square		10	1.600			44		44	72.50
	2500	Wood, 72″ square		10	1.600			44		44	72.50
	2600	Tree guying, 3″-4″ caliper using stakes		40	.400			10.95		10.95	18.15
	2700	3″ caliper using anchors		40	.400			10.95		10.95	18.15
	2800	3″- 6″ caliper using anchors		30	.533			14.60		14.60	24
	2900	6″ caliper using anchors		20	.800			22		22	36.50
	3000	8″ caliper using anchors		15	1.067			29		29	48.50

02230 | Site Clearing

	Line	Description	Crew	Daily Output	Labor-Hours	Unit	Mat. (2006 Bare Costs)	Labor (2006 Bare Costs)	Equip. (2006 Bare Costs)	Total (2006 Bare Costs)	Total Incl O&P
100	0010	**CLEAR AND GRUB**									
	0020	Cut & chip light trees to 6″ diam.	B-7	1	48	Acre		1,400	1,025	2,425	3,425
	0150	Grub stumps and remove	B-30	2	12			375	840	1,215	1,525
	0200	Cut & chip medium, trees to 12″ diam.	B-7	.70	68.571			2,000	1,475	3,475	4,925
	0250	Grub stumps and remove	B-30	1	24			745	1,675	2,420	3,050
	0300	Cut & chip heavy, trees to 24″ diam.	B-7	.30	160			4,675	3,450	8,125	11,500
	0350	Grub stumps and remove	B-30	.50	48			1,500	3,350	4,850	6,075
	0400	If burning is allowed, reduce cut & chip									40%
	3000	Chipping stumps, to 18″ deep, 12″ diam.	B-86	20	.400	Ea.		14.70	3.18	17.88	26.50
	3040	18″ diameter		16	.500			18.35	3.97	22.32	33.50
	3080	24″ diameter		14	.571			21	4.54	25.54	38
	3100	30″ diameter		12	.667			24.50	5.30	29.80	44.50
	3120	36″ diameter		10	.800			29.50	6.35	35.85	53.50
	3160	48″ diameter		8	1			36.50	7.95	44.45	67
	5000	Tree thinning, feller buncher, conifer									
	5080	Up to 8″ diameter	B-93	240	.033	Ea.		1.22	2.05	3.27	4.18
	5120	12″ diameter		160	.050			1.84	3.07	4.91	6.30
	5240	Hardwood, up to 4″ diameter		240	.033			1.22	2.05	3.27	4.18
	5280	8″ diameter		180	.044			1.63	2.73	4.36	5.60
	5320	12″ diameter		120	.067			2.45	4.10	6.55	8.35
	7000	Tree removal, congested area, aerial lift truck									
	7040	8″ diameter	B-85	7	5.714	Ea.		168	99.50	267.50	385
	7080	12″ diameter		6	6.667			196	116	312	450
	7120	18″ diameter		5	8			236	139	375	540

Important: See the Reference Section for supporting data - Crews, Rental Equipment, City Cost Indexes and Reference Data

		02230 | Site Clearing	CREW	DAILY OUTPUT	LABOR-HOURS	UNIT	MAT. (2006 BARE COSTS)	LABOR	EQUIP.	TOTAL	TOTAL INCL O&P	
100	7160	24″ diameter	B-85	4	10	Ea.		295	174	469	670	100
	7240	36″ diameter		3	13.333			395	232	627	895	
	7280	48″ diameter		2	20			590	350	940	1,350	
300	0010	**SELECTIVE TREE REMOVAL** With tractor, large tract, firm										300
	0020	level terrain, no boulders, less than 12″ diam. trees										
	0300	300 HP dozer, up to 400 trees/acre, 0 to 25% hardwoods	B-10M	.75	16	Acre		540	1,600	2,140	2,600	
	0340	25% to 50% hardwoods		.60	20			670	2,000	2,670	3,275	
	0370	75% to 100% hardwoods		.45	26.667			895	2,650	3,545	4,350	
	0400	500 trees/acre, 0% to 25% hardwoods		.60	20			670	2,000	2,670	3,275	
	0440	25% to 50% hardwoods		.48	25			840	2,500	3,340	4,100	
	0470	75% to 100% hardwoods		.36	33.333			1,125	3,325	4,450	5,450	
	0500	More than 600 trees/acre, 0 to 25% hardwoods		.52	23.077			775	2,300	3,075	3,775	
	0540	25% to 50% hardwoods		.42	28.571			960	2,850	3,810	4,650	
	0570	75% to 100% hardwoods		.31	38.710			1,300	3,850	5,150	6,325	
	0900	Large tract clearing per tree										
	1500	300 HP dozer, to 12″ diameter, softwood	B-10M	320	.038	Ea.		1.26	3.74	5	6.15	
	1550	Hardwood		100	.120			4.03	11.95	15.98	19.60	
	1600	12″ to 24″ diameter, softwood		200	.060			2.02	6	8.02	9.85	
	1650	Hardwood		80	.150			5.05	14.95	20	24.50	
	1700	24″ to 36″ diameter, softwood		100	.120			4.03	11.95	15.98	19.60	
	1750	Hardwood		50	.240			8.05	24	32.05	39.50	
	1800	36″ to 48″ diameter, softwood		70	.171			5.75	17.10	22.85	28	
	1850	Hardwood		35	.343			11.50	34	45.50	56	
	2000	Stump removal on site by hydraulic backhoe, 1-1/2 C.Y.										
	2040	4″ to 6″ diameter	B-17	60	.533	Ea.		15.80	9.20	25	35.50	
	2050	8″ to 12″ diameter	B-30	33	.727			22.50	51	73.50	92.50	
	2100	14″ to 24″ diameter		25	.960			30	67	97	122	
	2150	26″ to 36″ diameter		16	1.500			46.50	105	151.50	190	
	3000	Remove selective trees, on site using chain saws and chipper,										
	3050	not incl. stumps, up to 6″ diameter	B-7	18	2.667	Ea.		78	57.50	135.50	191	
	3100	8″ to 12″ diameter		12	4			117	86	203	287	
	3150	14″ to 24″ diameter		10	4.800			141	103	244	345	
	3200	26″ to 36″ diameter		8	6			176	129	305	430	
	3300	Machine load, 2 mile haul to dump, 12″ diam. tree, add								160	240	
500	0010	**STRIPPING & STOCKPILING OF SOIL**										500
	0020	200 H.P. dozer, ideal conditions	B-10B	2300	.005	C.Y.		.18	.40	.58	.72	
	0100	Adverse conditions	″	1150	.010			.35	.80	1.15	1.44	
	0200	300 HP dozer, ideal conditions	B-10M	3000	.004			.13	.40	.53	.66	
	0300	Adverse conditions	″	1650	.007			.24	.73	.97	1.19	
	0400	400 HP dozer, ideal conditions	B-10X	3900	.003			.10	.39	.49	.60	
	0500	Adverse conditions	″	2000	.006			.20	.77	.97	1.16	
	0600	Clay, dry and soft, 200 HP dozer, ideal conditions	B-10B	1600	.008			.25	.58	.83	1.03	
	0700	Adverse conditions	″	800	.015			.50	1.15	1.65	2.08	
	1000	Medium hard, 300 HP dozer, ideal conditions	B-10M	2000	.006			.20	.60	.80	.98	
	1100	Adverse conditions	″	1100	.011			.37	1.09	1.46	1.79	
	1200	Very hard, 400 HP dozer, ideal conditions	B-10X	2600	.005			.16	.59	.75	.90	
	1300	Adverse conditions	″	1340	.009			.30	1.15	1.45	1.74	

		02240 | Dewatering	CREW	DAILY OUTPUT	LABOR-HOURS	UNIT	MAT.	LABOR	EQUIP.	TOTAL	TOTAL INCL O&P	
500	0010	**DEWATERING**										500
	0020	Excavate drainage trench, 2′ wide, 2′ deep	B-11C	90	.178	C.Y.		5.70	2.51	8.21	11.95	
	0100	2′ wide, 3′ deep, with backhoe loader	″	135	.119			3.80	1.68	5.48	8	
	0200	Excavate sump pits by hand, light soil	1 Clab	7.10	1.127			31		31	51	

For expanded coverage of these items see *Means Site Work & Landscape Cost Data 2006*

	02240 | Dewatering		Crew	Daily Output	Labor-Hours	Unit	2006 Bare Costs Mat.	Labor	Equip.	Total	Total Incl O&P	
500	0300	Heavy soil	1 Clab	3.50	2.286	C.Y.		62.50		62.50	104	500
	0500	Pumping 8 hr., attended 2 hrs. per day, including 20 L.F.										
	0550	of suction hose & 100 L.F. discharge hose										
	0600	2" diaphragm pump used for 8 hours	B-10H	4	3	Day		101	14.50	115.50	177	
	0650	4" diaphragm pump used for 8 hours	B-10I	4	3			101	21	122	184	
	0800	8 hrs. attended, 2" diaphragm pump	B-10H	1	12			405	58	463	710	
	0900	3" centrifugal pump	B-10J	1	12			405	64.50	469.50	715	
	1000	4" diaphragm pump	B-10I	1	12			405	83	488	735	
	1100	6" centrifugal pump	B-10K	1	12			405	267	672	940	
	1300	CMP, incl. excavation 3' deep, 12" diameter	B-6	115	.209	L.F.	10.05	6.25	1.97	18.27	23.50	
	1400	18" diameter		100	.240	"	12.50	7.20	2.26	21.96	28	
	1600	Sump hole construction, incl. excavation and gravel, pit		1250	.019	C.F.	.75	.58	.18	1.51	1.97	
	1700	With 12" gravel collar, 12" pipe, corrugated, 16 ga.		70	.343	L.F.	16.30	10.30	3.23	29.83	38.50	
	1800	15" pipe, corrugated, 16 ga.		55	.436		21	13.10	4.11	38.21	49	
	1900	18" pipe, corrugated, 16 ga.		50	.480		24.50	14.40	4.53	43.43	55.50	
	2000	24" pipe, corrugated, 14 ga.		40	.600		29.50	18	5.65	53.15	68	
	2200	Wood lining, up to 4' x 4', add		300	.080	SFCA	13.80	2.40	.75	16.95	19.95	
	9950	See Div. 02240-900 for wellpoints										
	9960	See Div. 02240-700 for deep well systems										
700	0010	**WELLS** For dewatering 10' to 20' deep, 2' diameter										700
	0020	with steel casing, minimum	B-6	165	.145	V.L.F.	2.54	4.36	1.37	8.27	11.40	
	0050	Average		98	.245		5.05	7.35	2.31	14.71	20	
	0100	Maximum		49	.490		13.50	14.70	4.62	32.82	44	
900	0010	**WELLPOINTS** For wellpoint equip. rental, see Reference Section										900
	0100	Installation and removal of single stage system										
	0110	Labor only, .75 labor-hours per L.F., minimum	1 Clab	10.70	.748	LF Hdr		20.50		20.50	34	
	0200	2.0 labor-hours per L.F., maximum	"	4	2	"		55		55	91	
	0400	Pump operation, 4 @ 6 hr. shifts										
	0410	Per 24 hour day	4 Eqlt	1.27	25.197	Day		885		885	1,400	
	0500	Per 168 hour week, 160 hr. straight, 8 hr. double time		.18	177	Week		6,250		6,250	9,875	
	0550	Per 4.3 week month		.04	800	Month		28,200		28,200	44,500	
	0600	Complete installation, operation, equipment rental, fuel &										
	0610	removal of system with 2" wellpoints 5' O.C.										
	0700	100' long header, 6" diameter, first month	4 Eqlt	3.23	9.907	LF Hdr	135	350		485	700	
	0800	Thereafter, per month		4.13	7.748		108	273		381	550	
	1000	200' long header, 8" diameter, first month		6	5.333		119	188		307	430	
	1100	Thereafter, per month		8.39	3.814		60.50	134		194.50	279	
	1300	500' long header, 8" diameter, first month		10.63	3.010		47	106		153	219	
	1400	Thereafter, per month		20.91	1.530		33.50	54		87.50	122	
	1600	1,000' long header, 10" diameter, first month		11.62	2.754		40.50	97		137.50	198	
	1700	Thereafter, per month		41.81	.765		20	27		47	64.50	
	1900	Note: above figures include pumping 168 hrs. per week										
	1910	and include the pump operator and one stand-by pump.										

	02250 | Shoring & Underpinning		Crew	Daily Output	Labor-Hours	Unit	Mat.	Labor	Equip.	Total	Total Incl O&P	
100	0010	**GROUTING, PRESSURE**										100
	0020	Grouting, pressure, cement & sand, 1:1 mix, minimum	B-61	124	.323	Bag	9.50	9.45	2.10	21.05	28.50	
	0100	Maximum		51	.784	"	9.50	23	5.10	37.60	53.50	
	0200	Cement and sand, 1:1 mix, minimum		250	.160	C.F.	19.05	4.70	1.04	24.79	30	
	0300	Maximum		100	.400		28.50	11.75	2.60	42.85	53.50	
	0400	Epoxy cement grout, minimum		137	.292		141	8.55	1.90	151.45	172	
	0500	Maximum		57	.702		141	20.50	4.57	166.07	195	
	0600	Structural epoxy grout				Gal.	49.50			49.50	54.50	
	0700	Alternate pricing method: (Add for materials)										
	0710	5 person crew and equipment	B-61	1	40	Day		1,175	260	1,435	2,200	

	02250 | Shoring & Underpinning		Crew	Daily Output	Labor-Hours	Unit	2006 Bare Costs Mat.	Labor	Equip.	Total	Total Incl O&P	
400	0010	**SHEET PILING**										400
	0020	Sheet piling steel, not incl. wales, 22 psf, 15′ excav., left in place	B-40	10.81	5.920	Ton	935	210	255	1,400	1,650	
	0100	Drive, extract & salvage		6	10.667		415	380	460	1,255	1,575	
	0300	20′ deep excavation, 27 psf, left in place		12.95	4.942		935	176	213	1,324	1,550	
	0400	Drive, extract & salvage		6.55	9.771		415	345	420	1,180	1,500	
	0600	25′ deep excavation, 38 psf, left in place		19	3.368		935	120	145	1,200	1,375	
	0700	Drive, extract & salvage		10.50	6.095		415	217	263	895	1,100	
	0900	40′ deep excavation, 38 psf, left in place		21.20	3.019		935	107	130	1,172	1,350	
	1000	Drive, extract & salvage		12.25	5.224		415	186	225	826	1,000	
	1200	15′ deep excavation, 22 psf, left in place		983	.065	S.F.	10.85	2.31	2.80	15.96	18.85	
	1300	Drive, extract & salvage		545	.117		4.66	4.17	5.05	13.88	17.60	
	1500	20′ deep excavation, 27 psf, left in place		960	.067		13.65	2.37	2.87	18.89	22	
	1600	Drive, extract & salvage		485	.132		6.05	4.69	5.70	16.44	20.50	
	1800	25′ deep excavation, 38 psf, left in place		1000	.064		20	2.27	2.76	25.03	29	
	1900	Drive, extract & salvage		553	.116		8.30	4.11	4.99	17.40	21.50	
	2100	Rent steel sheet piling and wales, first month				Ton	222			222	244	
	2200	Per added month					22			22	24.50	
	2300	Rental piling left in place, add to rental					740			740	815	
	2500	Wales, connections & struts, 2/3 salvage					227			227	249	
	2700	High strength piling, 50,000 psi, add					50.50			50.50	55.50	
	2800	55,000 psi, add					53.50			53.50	59	
	3000	Tie rod, not upset, 1-1/2″ to 4″ diameter with turnbuckle					1,650			1,650	1,825	
	3100	No turnbuckle					1,275			1,275	1,400	
	3300	Upset, 1-3/4″ to 4″ diameter with turnbuckle					1,850			1,850	2,050	
	3400	No turnbuckle					1,600			1,600	1,775	
	3600	Lightweight, 18″ to 28″ wide, 7 ga., 9.22 psf, and										
	3610	9 ga., 8.6 psf, minimum				Lb.	.68			.68	.75	
	3700	Average					.74			.74	.81	
	3750	Maximum					.85			.85	.94	
	3900	Wood, solid sheeting, incl. wales, braces and spacers,										
	3910	drive, extract & salvage, 8′ deep excavation	B-31	330	.121	S.F.	1.78	3.57	.41	5.76	8.30	
	4000	10′ deep, 50 S.F./hr. in & 150 S.F./hr. out		300	.133		1.83	3.92	.45	6.20	9	
	4100	12′ deep, 45 S.F./hr. in & 135 S.F./hr. out		270	.148		1.88	4.36	.51	6.75	9.85	
	4200	14′ deep, 42 S.F./hr. in & 126 S.F./hr. out		250	.160		1.94	4.71	.55	7.20	10.55	
	4300	16′ deep, 40 S.F./hr. in & 120 S.F./hr. out		240	.167		2	4.91	.57	7.48	11	
	4400	18′ deep, 38 S.F./hr. in & 114 S.F./hr. out		230	.174		2.06	5.10	.59	7.75	11.40	
	4500	20′ deep, 35 S.F./hr. in & 105 S.F./hr. out		210	.190		2.13	5.60	.65	8.38	12.35	
	4520	Left in place, 8′ deep, 55 S.F./hr.		440	.091		3.20	2.68	.31	6.19	8.30	
	4540	10′ deep, 50 S.F./hr.		400	.100		3.37	2.94	.34	6.65	8.95	
	4560	12′ deep, 45 S.F./hr.		360	.111		3.56	3.27	.38	7.21	9.75	
	4565	14′ deep, 42 S.F./hr.		335	.119		3.76	3.51	.41	7.68	10.40	
	4570	16′ deep, 40 S.F./hr.		320	.125		4	3.68	.43	8.11	10.95	
	4580	18′ deep, 38 S.F./hr.		305	.131		4.27	3.86	.45	8.58	11.60	
	4590	20′ deep, 35 S.F./hr.		280	.143		4.57	4.20	.49	9.26	12.55	
	4700	Alternate pricing, left in place, 8′ deep		1.76	22.727	M.B.F.	720	670	77.50	1,467.50	1,975	
	4800	Drive, extract and salvage, 8′ deep		1.32	30.303	"	640	890	103	1,633	2,300	
	4990	Minimum labor/equipment charge		2	20	Job		590	68	658	1,050	
	5000	For treated lumber add cost of treatment to lumber										
	5010	See Division 06070-400										
500	0010	**SHORING**										500
	0020	Shoring, existing building, with timber, no salvage allowance	B-51	2.20	21.818	M.B.F.	800	605	58.50	1,463.50	1,950	
	1000	On cribbing with 35 ton screw jacks, per box and jack		3.60	13.333	Jack	48	370	35.50	453.50	705	
	1090	Minimum labor/equipment charge		2	24	Ea.		665	64.50	729.50	1,175	
800	0010	**UNDERPINNING FOUNDATIONS** Including excavation,										800
	0020	forming, reinforcing, concrete and equipment										

02200 | Site Preparation

02250 | Shoring & Underpinning

			CREW	DAILY OUTPUT	LABOR-HOURS	UNIT	2006 BARE COSTS MAT.	LABOR	EQUIP.	TOTAL	TOTAL INCL O&P	
800	0100	5′ to 16′ below grade, 100 to 500 C.Y.	B-52	2.30	24.348	C.Y.	240	790	173	1,203	1,750	800
	0200	Over 500 C.Y.		2.50	22.400		216	730	159	1,105	1,625	
	0400	16′ to 25′ below grade, 100 to 500 C.Y.		2	28		264	910	199	1,373	2,000	
	0500	Over 500 C.Y.		2.10	26.667		250	870	189	1,309	1,900	
	0700	26′ to 40′ below grade, 100 to 500 C.Y.		1.60	35		288	1,150	248	1,686	2,475	
	0800	Over 500 C.Y.		1.80	31.111		264	1,000	221	1,485	2,200	
	0900	For under 50 C.Y., add					10%	40%				
	1000	For 50 C.Y. to 100 C.Y., add					5%	20%				
900	0010	**VIBROFLOTATION**										900
	0900	Vibroflotation compacted sand cylinder, minimum	B-60	750	.075	V.L.F.		2.40	1.62	4.02	5.65	
	0950	Maximum		325	.172			5.55	3.73	9.28	13	
	1100	Vibro replacement compacted stone cylinder, minimum		500	.112			3.60	2.42	6.02	8.45	
	1150	Maximum		250	.224			7.20	4.85	12.05	16.95	
	1300	Mobilization and demobilization, minimum		.47	119	Total		3,825	2,575	6,400	9,000	
	1400	Maximum		.14	400	″		12,900	8,650	21,550	30,200	

02260 | Excavation Support/Protection

			CREW	DAILY OUTPUT	LABOR-HOURS	UNIT	MAT.	LABOR	EQUIP.	TOTAL	TOTAL INCL O&P	
200	0010	**COFFERDAMS**, incl. mobilization and temporary sheeting										200
	0080	Soldier beams & lagging H piles with 3″ wood sheeting										
	0090	horizontal between piles, including removal of wales & braces										
	0100	No hydrostatic head, 15′ deep, 1 line of braces, minimum	B-50	545	.206	S.F.	8	6.95	3.21	18.16	24	
	0200	Maximum		495	.226		8.90	7.65	3.53	20.08	26.50	
	0400	15′ to 22′ deep with 2 lines of braces, 10″ H, minimum		360	.311		9.40	10.50	4.86	24.76	33	
	0500	Maximum		330	.339		10.65	11.45	5.30	27.40	36.50	
	0700	23′ to 35′ deep with 3 lines of braces, 12″ H, minimum		325	.345		12.30	11.65	5.40	29.35	39	
	0800	Maximum		295	.380		13.35	12.80	5.95	32.10	42.50	
	1000	36′ to 45′ deep with 4 lines of braces, 14″ H, minimum		290	.386		13.80	13.05	6.05	32.90	43.50	
	1100	Maximum		265	.423		14.55	14.25	6.60	35.40	47	
	1300	No hydrostatic head, left in place, 15′ dp., 1 line of braces, min.		635	.176		10.65	5.95	2.76	19.36	24.50	
	1400	Maximum		575	.195		11.45	6.55	3.04	21.04	27	
	1600	15′ to 22′ deep with 2 lines of braces, minimum		455	.246		16	8.30	3.84	28.14	35.50	
	1700	Maximum		415	.270		17.80	9.10	4.22	31.12	39.50	
	1900	23′ to 35′ deep with 3 lines of braces, minimum		420	.267		19.05	9	4.17	32.22	40.50	
	2000	Maximum		380	.295		21	9.95	4.60	35.55	44.50	
	2200	36′ to 45′ deep with 4 lines of braces, minimum		385	.291		23	9.80	4.54	37.34	46.50	
	2300	Maximum		350	.320		26.50	10.80	5	42.30	53	
	2350	Lagging only, 3″ thick wood between piles 8′ O.C., minimum	B-46	400	.120		1.78	3.77	.08	5.63	8.35	
	2370	Maximum		250	.192		2.67	6.05	.13	8.85	13.20	
	2400	Open sheeting no bracing, for trenches to 10′ deep, min.		1736	.028		.80	.87	.02	1.69	2.36	
	2450	Maximum		1510	.032		.89	1	.02	1.91	2.67	
	2500	Tie-back method, add to open sheeting, add, minimum								20%	20%	
	2550	Maximum								60%	60%	
	2700	Tie-backs only, based on tie-backs total length, minimum	B-46	86.80	.553	L.F.	10.70	17.35	.39	28.44	41	
	2750	Maximum		38.50	1.247	″	18.80	39	.87	58.67	87	
	3500	Tie-backs only, typical average, 25′ long		2	24	Ea.	470	755	16.80	1,241.80	1,800	
	3600	35′ long		1.58	30.380	″	625	955	21.50	1,601.50	2,325	
	4500	Trench box, 7′ deep, 16′ x 8′, see reference section				Day				175	193	
	4600	20′ x 10′, see reference section				″				295	320	
	5200	Wood sheeting, in trench, jacks at 4′ O.C., 8′ deep	B-1	800	.030	S.F.	.62	.84		1.46	2.08	
	5250	12′ deep		700	.034		.73	.96		1.69	2.39	
	5300	15′ deep		600	.040		1	1.12		2.12	2.96	
720	0010	**ROCK BOLTING**										720
	2020	Hollow core, prestressable anchor, 1″ diameter, 5′ long	2 Skwk	32	.500	Ea.	81	18.25		99.25	119	
	2025	10′ long		24	.667		162	24.50		186.50	218	
	2060	2″ diameter, 5′ long		32	.500		300	18.25		318.25	365	

02200 | Site Preparation

	02260 | Excavation Support/Protection	CREW	DAILY OUTPUT	LABOR-HOURS	UNIT	MAT.	LABOR	EQUIP.	TOTAL	TOTAL INCL O&P	
720	2065 10' long	2 Skwk	24	.667	Ea.	605	24.50		629.50	705	720
	2100 Super high-tensile, 3/4" diameter, 5' long		32	.500		18.35	18.25		36.60	50	
	2105 10' long		24	.667		36.50	24.50		61	80.50	
	2160 2" diameter, 5' long		32	.500		164	18.25		182.25	210	
	2165 10' long	↓	24	.667		325	24.50		349.50	400	
	4400 Drill hole for rock bolt, 1-3/4" diam., 5' long (for 3/4" bolt)	B-56	17	.941			29.50	63	92.50	117	
	4405 10' long		9	1.778			55.50	119	174.50	220	
	4420 2" diameter, 5' long (for 1" bolt)		13	1.231			38.50	82	120.50	153	
	4425 10' long		7	2.286			71.50	153	224.50	283	
	4460 3-1/2" diameter, 5' long (for 2" bolt)		10	1.600			50	107	157	198	
	4465 10' long	↓	5	3.200	↓		100	214	314	395	

SITE CONSTRUCTION 2

02300 | Earthwork

	02305 | Equipment	CREW	DAILY OUTPUT	LABOR-HOURS	UNIT	MAT.	LABOR	EQUIP.	TOTAL	TOTAL INCL O&P	
250	0010 **MOBILIZATION OR DEMOB.** (One or the other, unless noted) R015433-10										250
	0015 Up to 25 mi haul dist (50 mi RT for mob/demob crew)										
	0020 Dozer, loader, backhoe, excav., grader, paver, roller, 70 to 150 H.P.	B-34N	4	2	Ea.		56.50	112	168.50	216	
	0100 Above 150 HP	B-34K	3	2.667			75.50	175	250.50	315	
	0300 Scraper, towed type (incl. tractor), 6 C.Y. capacity		3	2.667			75.50	175	250.50	315	
	0400 10 C.Y.		2.50	3.200			90.50	210	300.50	380	
	0600 Self-propelled scraper, 15 C.Y.		2.50	3.200			90.50	210	300.50	380	
	0700 24 C.Y.		2	4			113	262	375	475	
	0900 Shovel or dragline, 3/4 C.Y.		3.60	2.222			63	146	209	263	
	1000 1-1/2 C.Y.	↓	3	2.667			75.50	175	250.50	315	
	1100 Small equipment, placed in rear of, or towed by pickup truck	A-3A	8	1			27.50	10.65	38.15	56.50	
	1150 Equip up to 70 HP, on flatbed trailer behind pickup truck	A-3D	4	2			55	43.50	98.50	138	
	2000 Mob & demob truck-mounted crane up to 75 ton, driver only	1 Eqhv	3.60	2.222			84.50		84.50	134	
	2100 Crane, truck-mounted, over 75 ton	A-3E	2.50	6.400			213	34	247	380	
	2200 Crawler-mounted, up to 75 ton	A-3F	2	8			266	278	544	730	
	2300 Over 75 ton	A-3G	1.50	10.667	↓		355	405	760	1,000	
	2500 For each additional 5 miles haul distance, add						10%	10%			
	3000 For large pieces of equipment, allow for assembly/knockdown										
	3001 For mob/demob of vibrofloatation equip, see section 02250-900										
	3100 For mob/demob of micro-tunneling equip, see section 02441-400										
	3200 For mob/demob of pile driving equip, see section 02455-650										
	3300 For mob/demob of caisson drilling equip, see 02465-950										
	02310 | Grading										
100	0010 **FINISH GRADING**										100
	0012 Finish grading area to be paved with grader, small area	B-11L	400	.040	S.Y.		1.28	1.14	2.42	3.33	
	0100 Large area		2000	.008			.26	.23	.49	.66	
	0200 Grade subgrade for base course, roadways	↓	3500	.005			.15	.13	.28	.38	
	1020 For large parking lots	B-32C	5000	.010			.31	.32	.63	.85	
	1050 For small irregular areas	"	2000	.024			.78	.79	1.57	2.12	
	1100 Fine grade for slab on grade, machine	B-11L	1040	.015			.49	.44	.93	1.27	
	1150 Hand grading	B-18	700	.034			.96	.05	1.01	1.64	
	1200 Fine grade granular base for sidewalks and bikeways	B-62	1200	.020	↓		.60	.13	.73	1.12	
	2550 Hand grade select gravel	2 Clab	60	.267	C.S.F.		7.30		7.30	12.10	

(Cost columns MAT., LABOR, EQUIP., TOTAL are 2006 BARE COSTS.)

02310 | Grading

	Line	Description	Crew	Daily Output	Labor-Hours	Unit	Mat. (2006 Bare Costs)	Labor (2006 Bare Costs)	Equip. (2006 Bare Costs)	Total (2006 Bare Costs)	Total Incl O&P	
100	3000	Hand grade select gravel, including compaction, 4″ deep	B-18	555	.043	S.Y.		1.21	.06	1.27	2.08	100
	3100	6″ deep		400	.060			1.68	.09	1.77	2.89	
	3120	8″ deep		300	.080			2.25	.12	2.37	3.85	
	3300	Finishing grading slopes, gentle	B-11L	8900	.002			.06	.05	.11	.15	
	3310	Steep slopes	"	7100	.002			.07	.06	.13	.19	
	9000	Minimum labor/equipment charge, hand grading	1 Clab	2	4	Job		110		110	182	
	9100	Minimum labor/equipment charge, machine grading	B-11L	2	8	"		256	229	485	665	

02315 | Excavation and Fill

	Line	Description	Crew	Daily Output	Labor-Hours	Unit	Mat. (2006 Bare Costs)	Labor (2006 Bare Costs)	Equip. (2006 Bare Costs)	Total (2006 Bare Costs)	Total Incl O&P	
110	0010	**BACKFILL, GENERAL**										110
	0015	By hand, no compaction, light soil	1 Clab	14	.571	L.C.Y.		15.65		15.65	26	
	0100	Heavy soil		11	.727	"		19.95		19.95	33	
	0300	Compaction in 6″ layers, hand tamp, add to above		20.60	.388	E.C.Y.		10.65		10.65	17.65	
	0400	Roller compaction operator walking, add	B-10A	100	.120			4.03	1.26	5.29	7.85	
	0500	Air tamp, add	B-9D	190	.211			5.85	.97	6.82	10.75	
	0600	Vibrating plate, add	A-1D	60	.133			3.65	.44	4.09	6.55	
	0800	Compaction in 12″ layers, hand tamp, add to above	1 Clab	34	.235			6.45		6.45	10.70	
	0900	Roller compaction operator walking, add	B-10A	150	.080			2.69	.84	3.53	5.20	
	1000	Air tamp, add	B-9	285	.140			3.90	.52	4.42	7	
	1100	Vibrating plate, add	A-1E	90	.089			2.44	.38	2.82	4.46	
120	0010	**BACKFILL, STRUCTURAL** Dozer or F.E. loader										120
	0020	From existing stockpile, no compaction										
	2000	80 H.P., 50′ haul, sand & gravel	B-10L	1100	.011	L.C.Y.		.37	.29	.66	.91	
	2020	Common earth		975	.012			.41	.33	.74	1.02	
	2040	Clay		850	.014			.47	.38	.85	1.17	
	2200	150′ haul, sand & gravel		550	.022			.73	.58	1.31	1.81	
	2220	Common earth		490	.024			.82	.65	1.47	2.04	
	2240	Clay		425	.028			.95	.75	1.70	2.35	
	2400	300′ haul, sand & gravel		370	.032			1.09	.86	1.95	2.69	
	2420	Common earth		330	.036			1.22	.97	2.19	3.02	
	2440	Clay		290	.041			1.39	1.10	2.49	3.44	
	3000	105 H.P., 50′ haul, sand & gravel	B-10W	1350	.009			.30	.34	.64	.86	
	3020	Common earth		1225	.010			.33	.38	.71	.95	
	3040	Clay		1100	.011			.37	.42	.79	1.05	
	3200	150′ haul, sand & gravel		670	.018			.60	.69	1.29	1.72	
	3220	Common earth		610	.020			.66	.76	1.42	1.89	
	3240	Clay		550	.022			.73	.84	1.57	2.10	
	3300	300′ haul, sand & gravel		465	.026			.87	1	1.87	2.48	
	3320	Common earth		415	.029			.97	1.12	2.09	2.79	
	3340	Clay		370	.032			1.09	1.25	2.34	3.12	
	4000	200 H.P., 50′ haul, sand & gravel	B-10B	2500	.005			.16	.37	.53	.67	
	4020	Common earth		2200	.005			.18	.42	.60	.75	
	4040	Clay		1950	.006			.21	.47	.68	.85	
	4200	150′ haul, sand & gravel		1225	.010			.33	.75	1.08	1.36	
	4220	Common earth		1100	.011			.37	.84	1.21	1.51	
	4240	Clay		975	.012			.41	.94	1.35	1.70	
	4400	300′ haul, sand & gravel		805	.015			.50	1.14	1.64	2.06	
	4420	Common earth		735	.016			.55	1.25	1.80	2.26	
	4440	Clay		660	.018			.61	1.40	2.01	2.51	
	5000	300 H.P., 50′ haul, sand & gravel	B-10M	3170	.004			.13	.38	.51	.62	
	5020	Common earth		2900	.004			.14	.41	.55	.67	
	5040	Clay		2700	.004			.15	.44	.59	.73	
	5200	150′ haul, sand & gravel		2200	.005			.18	.54	.72	.89	
	5220	Common earth		1950	.006			.21	.61	.82	1	

02315 | Excavation and Fill

		02315 Excavation and Fill	CREW	DAILY OUTPUT	LABOR-HOURS	UNIT	2006 BARE COSTS MAT.	LABOR	EQUIP.	TOTAL	TOTAL INCL O&P	
120	5240	Clay	B-10M	1700	.007	L.C.Y.		.24	.70	.94	1.15	120
	5400	300' haul, sand & gravel		1500	.008			.27	.80	1.07	1.31	
	5420	Common earth		1350	.009			.30	.89	1.19	1.46	
	5440	Clay		1225	.010			.33	.98	1.31	1.61	
210	0010	**BORROW, LOADING AND/OR SPREADING**										210
	0020	Material only, bank run gravel				Ton	13			13	14.30	
	0100	Crushed stone, 1-1/2" to 3/4" size					14.20			14.20	15.60	
	0150	3/8" size					18.70			18.70	20.50	
	0200	Dead or bank run sand					6.10			6.10	6.70	
	0500	Haul 2 mi. spread, 200HP dozer, bank run gravel	B-15	1100	.025			.78	1.70	2.48	3.13	
	0540	Crushed stone, 1-1/2" to 3/4" size		1150	.024			.75	1.63	2.38	2.99	
	0560	3/8" size		1150	.024			.75	1.63	2.38	2.99	
	0600	Bank run or dead sand		1000	.028			.86	1.87	2.73	3.44	
	1000	Hand spread, bank run gravel	A-5	33	.545			14.95	.98	15.93	25.50	
	1040	Crushed stone, 1-1/2" to 3/4" size		38	.474			13	.85	13.85	22.50	
	1060	3/8" size		40	.450			12.35	.81	13.16	21.50	
	1100	Bank run or dead sand		33	.545			14.95	.98	15.93	25.50	
	1800	Delivery charge, minimum 20 tons, 1 hr round trip, add	B-34B	130	.062			1.74	3.67	5.41	6.85	
	1820	1-1/2 hr. round trip, add		93	.086			2.44	5.15	7.59	9.60	
	1840	2 hr. round trip, add		65	.123			3.49	7.35	10.84	13.75	
	7000	Topsoil or loam from stockpile, shovel, 1 C.Y. bucket	B-12N	840	.019	B.C.Y.	21	.62	1.12	22.74	25	
	7010	1-1/2 C.Y. bucket	B-12O	1135	.014		21	.46	.88	22.34	24.50	
	7020	3 C.Y. bucket	B-12T	1800	.009		21	.29	.73	22.02	24.50	
	7030	Front end loader, wheel mounted										
	7050	3/4 C.Y. bucket	B-10R	550	.022	B.C.Y.	21	.73	.35	22.08	24.50	
	7060	1-1/2 C.Y. bucket	B-10S	970	.012		21	.42	.25	21.67	24	
	7070	3 C.Y. bucket	B-10T	1575	.008		21	.26	.19	21.45	23.50	
	7080	5 C.Y. bucket	B-10U	2600	.005		21	.16	.27	21.43	23.50	
310	0010	**COMPACTION, GENERAL**										310
	5000	Riding, vibrating roller, 6" lifts, 2 passes	B-10Y	3000	.004	E.C.Y.		.13	.14	.27	.37	
	5020	3 passes		2300	.005			.18	.18	.36	.48	
	5040	4 passes		1900	.006			.21	.22	.43	.58	
	5060	12" lifts, 2 passes		5200	.002			.08	.08	.16	.21	
	5080	3 passes		3500	.003			.12	.12	.24	.31	
	5100	4 passes		2600	.005			.16	.16	.32	.43	
	5600	Sheepsfoot or wobbly wheel roller, 6" lifts, 2 passes	B-10G	2400	.005			.17	.35	.52	.66	
	5620	3 passes		1735	.007			.23	.49	.72	.91	
	5640	4 passes		1300	.009			.31	.65	.96	1.22	
	5680	12" lifts, 2 passes		5200	.002			.08	.16	.24	.30	
	5700	3 passes		3500	.003			.12	.24	.36	.45	
	5720	4 passes		2600	.005			.16	.33	.49	.61	
	7000	Walk behind, vibrating plate 18" wide, 6" lifts, 2 passes	A-1D	200	.040			1.10	.13	1.23	1.97	
	7020	3 passes		185	.043			1.18	.14	1.32	2.12	
	7040	4 passes		140	.057			1.57	.19	1.76	2.80	
	7200	12" lifts, 2 passes	A-1E	560	.014			.39	.06	.45	.72	
	7220	3 passes		375	.021			.58	.09	.67	1.07	
	7240	4 passes		280	.029			.78	.12	.90	1.44	
	7500	Vibrating roller 24" wide, 6" lifts, 2 passes	B-10A	420	.029			.96	.30	1.26	1.87	
	7520	3 passes		280	.043			1.44	.45	1.89	2.79	
	7540	4 passes		210	.057			1.92	.60	2.52	3.73	
	7600	12" lifts, 2 passes		840	.014			.48	.15	.63	.93	
	7620	3 passes		560	.021			.72	.22	.94	1.40	
	7640	4 passes		420	.029			.96	.30	1.26	1.87	
	8000	Rammer tamper, 6" to 11", 4" lifts, 2 passes	A-1F	130	.062			1.69	.28	1.97	3.10	
	8050	3 passes		97	.082			2.26	.38	2.64	4.16	
	8100	4 passes		65	.123			3.37	.57	3.94	6.25	

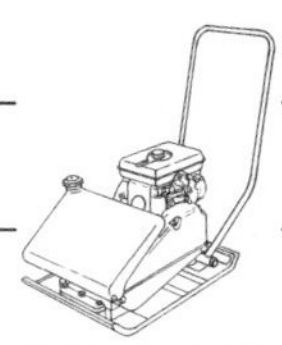

02315 | Excavation and Fill

		02315 | Excavation and Fill	CREW	DAILY OUTPUT	LABOR-HOURS	UNIT	2006 BARE COSTS MAT.	LABOR	EQUIP.	TOTAL	TOTAL INCL O&P	
310	8200	8" lifts, 2 passes	A-1F	260	.031	E.C.Y.		.84	.14	.98	1.56	310
	8250	3 passes		195	.041			1.12	.19	1.31	2.07	
	8300	4 passes		130	.062			1.69	.28	1.97	3.10	
	8400	13" to 18", 4" lifts, 2 passes	A-1G	390	.021			.56	.10	.66	1.03	
	8450	3 passes		290	.028			.76	.13	.89	1.39	
	8500	4 passes		195	.041			1.12	.19	1.31	2.07	
	8600	8" lifts, 2 passes		780	.010			.28	.05	.33	.52	
	8650	3 passes		585	.014			.37	.06	.43	.69	
	8700	4 passes		390	.021			.56	.10	.66	1.03	
	9000	Water, 3000 gal. truck, 3 mile haul	B-45	1888	.008		.20	.28	.28	.76	.97	
	9010	6 mile haul		1444	.011		.20	.36	.37	.93	1.21	
	9020	12 mile haul		1000	.016		.20	.52	.53	1.25	1.64	
	9030	6000 gal. wagon, 3 mile haul	B-59	2000	.004		.20	.11	.16	.47	.58	
	9040	6 mile haul	"	1600	.005		.20	.14	.20	.54	.67	
	9900	Minimum labor/equipment charge	1 Clab	4	2	Job		55		55	91	
320	0010	**COMPACTION, STRUCTURAL**										320
	0020	Steel wheel tandem roller, 5 tons	B-10E	8	1.500	Hr.		50.50	14.30	64.80	96	
412	0010	**DRILLING ONLY ROCK**										412
	0015	Drilling only rock, 2" hole for rock bolts	B-47	316	.076	L.F.		2.33	3.44	5.77	7.55	
	0800	2-1/2" hole for pre-splitting		600	.040			1.23	1.81	3.04	3.99	
	4600	Quarry operations, 2-1/2" to 3-1/2" diameter		715	.034			1.03	1.52	2.55	3.35	
	4610	6" diameter drill holes	B-47A	1350	.018			.59	.61	1.20	1.62	
416	0010	**DRILLING AND BLASTING ROCK**										416
	0020	Rock, open face, under 1500 CY	B-47	225	.107	B.C.Y.	2.04	3.27	4.83	10.14	12.85	
	0100	Over 1500 C.Y.		300	.080		2.04	2.45	3.62	8.11	10.20	
	0200	Areas where blasting mats are required, under 1500 C.Y.		175	.137		2.04	4.21	6.20	12.45	15.95	
	0250	Over 1500 C.Y.		250	.096		2.04	2.94	4.35	9.33	11.80	
	0300	Bulk drilling and blasting, can vary greatly, average									5.50	
	0500	Pits, average									22	
	1300	Deep hole method, up to 1500 C.Y.	B-47	50	.480		2.04	14.70	21.50	38.24	50	
	1400	Over 1500 C.Y.		66	.364		2.04	11.15	16.45	29.64	38.50	
	1900	Restricted areas, up to 1500 C.Y.		13	1.846		2.04	56.50	83.50	142.04	186	
	2000	Over 1500 C.Y.		20	1.200		2.04	37	54.50	93.54	122	
	2200	Trenches, up to 1500 C.Y.		22	1.091		5.90	33.50	49.50	88.90	116	
	2300	Over 1500 C.Y.		26	.923		5.90	28.50	42	76.40	98.50	
	2500	Pier holes, up to 1500 C.Y.		22	1.091		2.04	33.50	49.50	85.04	111	
	2600	Over 1500 C.Y.		31	.774		2.04	23.50	35	60.54	79	
	2800	Boulders under 1/2 C.Y., loaded on truck, no hauling	B-100	80	.150			5.05	6.95	12	15.70	
	2900	Boulders, drilled, blasted	B-47	100	.240		2.04	7.35	10.85	20.24	26	
	3100	Jackhammer operators with foreman compressor, air tools	B-9	1	40	Day		1,100	147	1,247	2,000	
	3300	Track drill, compressor, operator and foreman	B-47	1	24	"		735	1,075	1,810	2,400	
	3500	Blasting caps				Ea.	3.70			3.70	4.07	
	3700	Explosives					.31			.31	.34	
	3900	Blasting mats, rent, for first day					102			102	112	
	4000	Per added day					34			34	37.50	
	4200	Preblast survey for 6 room house, individual lot, minimum	A-6	2.40	6.667			236	24.50	260.50	410	
	4300	Maximum	"	1.35	11.852			420	43	463	730	
	4500	City block within zone of influence, minimum	A-8	25200	.001	S.F.		.05		.05	.08	
	4600	Maximum	"	15100	.002	"		.08		.08	.13	
	5000	Excavate and load boulders, less than 0.5 C.Y.	B-10T	80	.150	B.C.Y.		5.05	3.82	8.87	12.25	
	5020	0.5 C.Y. to 1 C.Y.	B-10U	100	.120			4.03	7	11.03	14.15	
	5200	Excavate and load blasted rock, 3 C.Y. power shovel	B-12T	1530	.010			.34	.86	1.20	1.50	
	5400	Haul boulders, 25 Ton off-highway dump, 1 mile round trip	B-34E	330	.024			.69	2.76	3.45	4.16	
	5420	2 mile round trip		275	.029			.82	3.32	4.14	4.99	

02315 | Excavation and Fill

			CREW	DAILY OUTPUT	LABOR-HOURS	UNIT	2006 BARE COSTS MAT.	LABOR	EQUIP.	TOTAL	TOTAL INCL O&P	
416	5440	3 mile round trip	B-34E	225	.036	B.C.Y.		1.01	4.05	5.06	6.10	416
	5460	4 mile round trip	↓	200	.040	↓		1.13	4.56	5.69	6.85	
	5600	Bury boulders on site, less than 0.5 C.Y., 300 H.P. dozer										
	5620	150′ haul	B-10M	310	.039	B.C.Y.		1.30	3.86	5.16	6.35	
	5640	300′ haul		210	.057			1.92	5.70	7.62	9.30	
	5800	0.5 to 1 C.Y., 300 H.P. dozer, 150′ haul		300	.040			1.34	3.99	5.33	6.55	
	5820	300′ haul	↓	200	.060	↓		2.02	6	8.02	9.85	
418	0010	**RIPPING**										418
	0020	Ripping, trap rock, soft, 300 HP dozer, ideal conditions	B-11S	700	.017	B.C.Y.		.58	1.81	2.39	2.92	
	1400	Ideal conditions		710	.017			.57	1.79	2.36	2.88	
	1500	Adverse conditions		660	.018			.61	1.92	2.53	3.10	
	1600	Medium hard, 300 HP dozer, ideal conditons		600	.020			.67	2.12	2.79	3.41	
	1700	Adverse conditions	↓	540	.022			.75	2.35	3.10	3.78	
	2000	Very hard, 410 HP dozer, ideal conditions	B-11T	350	.034			1.15	4.62	5.77	6.95	
	2100	Adverse conditions	″	310	.039			1.30	5.20	6.50	7.85	
	2200	Shale, soft, 300 HP dozer, ideal conditons	B-11S	1500	.008			.27	.85	1.12	1.36	
	2300	Adverse conditions		1350	.009			.30	.94	1.24	1.52	
	2400	Medium hard, 300 HP dozer, ideal conditons		1200	.010			.34	1.06	1.40	1.70	
	2500	Adverse conditions	↓	1080	.011			.37	1.18	1.55	1.89	
	2600	Very hard, 410 HP dozer, ideal conditons	B-11T	800	.015			.50	2.02	2.52	3.03	
	2700	Adverse conditions	″	720	.017			.56	2.25	2.81	3.37	
	3000	Dozing ripped material, 200 HP, 100′ haul	B-10B	700	.017			.58	1.32	1.90	2.37	
	3050	300′ haul	″	250	.048			1.61	3.68	5.29	6.65	
	3200	300 HP, 100′ haul	B-10M	1150	.010			.35	1.04	1.39	1.70	
	3250	300′ haul	″	400	.030			1.01	2.99	4	4.90	
	3400	410 HP, 100′ haul	B-10X	1680	.007			.24	.91	1.15	1.39	
	3450	300′ haul	″	600	.020	↓		.67	2.56	3.23	3.90	
424	0010	**EXCAVATING, BULK BANK MEASURE** Common earth piled										424
	0020	For loading onto trucks, add								15%	15%	
	0050	For mobilization and demobilization, see Division 02305-250										
	0100	For hauling, see Division 02315-490										
	0200	Backhoe, hydraulic, crawler mtd., 1 C.Y. cap. = 75 C.Y./hr.	B-12A	600	.027	B.C.Y.		.87	.93	1.80	2.44	
	0250	1-1/2 C.Y. cap. = 100 C.Y./hr.	B-12B	800	.020			.66	.90	1.56	2.05	
	0260	2 C.Y. cap. = 130 C.Y./hr.	B-12C	1040	.015			.50	.88	1.38	1.78	
	0300	3 C.Y. cap. = 160 C.Y./hr.	B-12D	1280	.013			.41	1.58	1.99	2.40	
	0310	Wheel mounted, 1/2 C.Y. cap. = 30 C.Y./hr.	B-12E	240	.067			2.18	1.40	3.58	5.05	
	0360	3/4 C.Y. cap. = 45 C.Y./hr.	B-12F	360	.044			1.46	1.33	2.79	3.81	
	0500	Clamshell, 1/2 C.Y. cap. = 20 C.Y./hr.	B-12G	160	.100			3.28	3.55	6.83	9.20	
	0550	1 C.Y. cap. = 35 C.Y./hr.	B-12H	280	.057			1.87	3.30	5.17	6.65	
	0950	Dragline, 1/2 C.Y. cap. = 30 C.Y./hr.	B-12I	240	.067			2.18	2.84	5.02	6.65	
	1000	3/4 C.Y. cap. = 35 C.Y./hr.	″	280	.057			1.87	2.44	4.31	5.70	
	1050	1-1/2 C.Y. cap. = 65 C.Y./hr.	B-12P	520	.031			1.01	1.86	2.87	3.66	
	1100	3 C.Y. cap. = 112 C.Y./hr.	B-12V	900	.018			.58	1.42	2	2.50	
	1200	Front end loader, track mtd., 1-1/2 C.Y. cap. = 70 C.Y./hr.	B-10N	560	.021			.72	.56	1.28	1.77	
	1250	2-1/2 C.Y. cap. = 95 C.Y./hr.	B-10O	760	.016			.53	.73	1.26	1.66	
	1300	3 C.Y. cap. = 130 C.Y./hr.	B-10P	1040	.012			.39	.76	1.15	1.46	
	1350	5 C.Y. cap. = 160 C.Y./hr.	B-10Q	1280	.009			.32	.87	1.19	1.45	
	1500	Wheel mounted, 3/4 C.Y. cap. = 45 C.Y./hr.	B-10R	360	.033			1.12	.54	1.66	2.38	
	1550	1-1/2 C.Y. cap. = 80 C.Y./hr.	B-10S	640	.019			.63	.38	1.01	1.42	
	1601	3 C.Y. cap. = 140 C.Y./hr.	B-10T	1120	.011			.36	.27	.63	.88	
	1650	5 C.Y. cap. = 185 C.Y./hr.	B-10U	1480	.008			.27	.47	.74	.96	
	1800	Hydraulic excavator, truck mtd, 1/2 C.Y. = 30 C.Y./hr.	B-12J	240	.067			2.18	3.50	5.68	7.35	
	1850	48 inch bucket, 1 C.Y. = 45 C.Y./hr.	B-12K	360	.044			1.46	2.71	4.17	5.35	
	3700	Shovel, 1/2 C.Y. capacity = 55 C.Y./hr.	B-12L	440	.036			1.19	1.33	2.52	3.38	
	3750	3/4 C.Y. capacity = 85 C.Y./hr.	B-12M	680	.024	↓		.77	1.05	1.82	2.40	

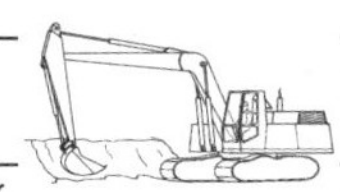

02315 | Excavation and Fill

		02315 Excavation and Fill	Crew	Daily Output	Labor-Hours	Unit	Mat.	Labor	Equip.	Total	Total Incl O&P	
							2006 Bare Costs					
424	3800	1 C.Y. capacity = 120 C.Y./hr.	B-12N	960	.017	B.C.Y.		.55	.98	1.53	1.96	424
	3850	1-1/2 C.Y. capacity = 160 C.Y./hr.	B-12O	1280	.013			.41	.78	1.19	1.52	
	3900	3 C.Y. cap. = 250 C.Y./hr.	B-12T	2000	.008			.26	.66	.92	1.14	
	4000	For soft soil or sand, deduct								15%	15%	
	4100	For heavy soil or stiff clay, add								60%	60%	
	4200	For wet excavation with clamshell or dragline, add								100%	100%	
	4250	All other equipment, add								50%	50%	
	4400	Clamshell in sheeting or cofferdam, minimum	B-12H	160	.100			3.28	5.75	9.03	11.65	
	4450	Maximum	″	60	.267			8.75	15.40	24.15	31	
	8000	For hauling excavated material, see Div. 02315-490										
	9000	Minimum labor/equipment charge	B-10L	2	6	Job		202	159	361	500	
432	0010	**EXCAVATING, BULK, DOZER** Open site										432
	2000	80 H.P., 50′ haul, sand & gravel	B-10L	460	.026	B.C.Y.		.88	.69	1.57	2.16	
	2020	Common earth		400	.030			1.01	.80	1.81	2.49	
	2040	Clay		250	.048			1.61	1.28	2.89	3.98	
	2200	150′ haul, sand & gravel		230	.052			1.75	1.39	3.14	4.33	
	2220	Common earth		200	.060			2.02	1.59	3.61	4.98	
	2240	Clay		125	.096			3.23	2.55	5.78	7.95	
	2400	300′ haul, sand & gravel		120	.100			3.36	2.66	6.02	8.30	
	2420	Common earth		100	.120			4.03	3.19	7.22	9.95	
	2440	Clay		65	.185			6.20	4.91	11.11	15.35	
	3000	105 H.P., 50′ haul, sand & gravel	B-10W	700	.017			.58	.66	1.24	1.65	
	3020	Common earth		610	.020			.66	.76	1.42	1.89	
	3040	Clay		385	.031			1.05	1.20	2.25	3	
	3200	150′ haul, sand & gravel		310	.039			1.30	1.49	2.79	3.72	
	3220	Common earth		270	.044			1.49	1.71	3.20	4.28	
	3240	Clay		170	.071			2.37	2.72	5.09	6.80	
	3300	300′ haul, sand & gravel		140	.086			2.88	3.31	6.19	8.25	
	3320	Common earth		120	.100			3.36	3.86	7.22	9.65	
	3340	Clay		100	.120			4.03	4.63	8.66	11.55	
	4000	200 H.P., 50′ haul, sand & gravel	B-10B	1400	.009			.29	.66	.95	1.18	
	4020	Common earth		1230	.010			.33	.75	1.08	1.34	
	4040	Clay		770	.016			.52	1.20	1.72	2.16	
	4200	150′ haul, sand & gravel		595	.020			.68	1.55	2.23	2.78	
	4220	Common earth		516	.023			.78	1.78	2.56	3.21	
	4240	Clay		325	.037			1.24	2.83	4.07	5.10	
	4400	300′ haul, sand & gravel		310	.039			1.30	2.97	4.27	5.35	
	4420	Common earth		270	.044			1.49	3.41	4.90	6.15	
	4440	Clay		170	.071			2.37	5.40	7.77	9.75	
	5000	300 H.P., 50′ haul, sand & gravel	B-10M	1900	.006			.21	.63	.84	1.03	
	5020	Common earth		1650	.007			.24	.73	.97	1.19	
	5040	Clay		1025	.012			.39	1.17	1.56	1.91	
	5200	150′ haul, sand & gravel		920	.013			.44	1.30	1.74	2.13	
	5220	Common earth		800	.015			.50	1.50	2	2.46	
	5240	Clay		500	.024			.81	2.39	3.20	3.92	
	5400	300′ haul, sand & gravel		470	.026			.86	2.55	3.41	4.17	
	5420	Common earth		410	.029			.98	2.92	3.90	4.78	
	5440	Clay		250	.048			1.61	4.79	6.40	7.85	
452	0010	**EXCAVATION, BULK, SCRAPERS**										452
	0100	Elev. scraper 11 C.Y., sand & gravel 1500′ haul, 1/4 dozer	B-33F	690	.020	B.C.Y.		.69	1.63	2.32	2.89	
	0150	3000′ haul		610	.023			.78	1.84	2.62	3.27	
	0200	5000′ haul		505	.028			.94	2.22	3.16	3.95	
	0300	Common earth, 1500′ haul		600	.023			.79	1.87	2.66	3.33	
	0350	3000′ haul		530	.026			.90	2.12	3.02	3.77	

02315 | Excavation and Fill

		02315 Excavation and Fill	CREW	DAILY OUTPUT	LABOR-HOURS	UNIT	2006 BARE COSTS MAT.	2006 BARE COSTS LABOR	2006 BARE COSTS EQUIP.	2006 BARE COSTS TOTAL	TOTAL INCL O&P
452	0400	5000′ haul	B-33F	440	.032	B.C.Y.		1.08	2.55	3.63	4.53
	0500	Clay, 1500′ haul		375	.037			1.27	2.99	4.26	5.30
	0550	3000′ haul		330	.042			1.44	3.40	4.84	6.05
	0600	5000′ haul		275	.051			1.73	4.08	5.81	7.25
	1000	Self propelled scraper, 14 C.Y. 1/4 push dozer, sand									
	1050	Sand and gravel, 1500′ haul	B-33D	920	.015	B.C.Y.		.52	2.04	2.56	3.07
	1100	3000′ haul		805	.017			.59	2.33	2.92	3.51
	1200	5000′ haul		645	.022			.74	2.91	3.65	4.38
	1300	Common earth, 1500′ haul		800	.018			.60	2.35	2.95	3.53
	1350	3000′ haul		700	.020			.68	2.68	3.36	4.04
	1400	5000′ haul		560	.025			.85	3.35	4.20	5.05
	1500	Clay, 1500′ haul		500	.028			.95	3.75	4.70	5.65
	1550	3000′ haul		440	.032			1.08	4.26	5.34	6.40
	1600	5000′ haul		350	.040			1.36	5.35	6.71	8.05
	2000	21 C.Y., 1/4 push dozer, sand & gravel, 1500′ haul	B-33E	1180	.012			.40	2.24	2.64	3.11
	2100	3000′ haul		910	.015			.52	2.91	3.43	4.04
	2200	5000′ haul		750	.019			.64	3.53	4.17	4.90
	2300	Common earth, 1500′ haul		1030	.014			.46	2.57	3.03	3.57
	2350	3000′ haul		790	.018			.60	3.35	3.95	4.65
	2400	5000′ haul		650	.022			.73	4.08	4.81	5.65
	2500	Clay, 1500′ haul		645	.022			.74	4.11	4.85	5.70
	2550	3000′ haul		495	.028			.96	5.35	6.31	7.45
	2600	5000′ haul		405	.035			1.18	6.55	7.73	9.10
	2700	Towed, 10 C.Y., 1/4 push dozer, sand & gravel, 1500′ haul	B-33B	560	.025			.85	2.99	3.84	4.65
	2720	3000′ haul		450	.031			1.06	3.73	4.79	5.80
	2730	5000′ haul		365	.038			1.31	4.59	5.90	7.15
	2750	Common earth, 1500′ haul		420	.033			1.13	3.99	5.12	6.20
	2770	3000′ haul		400	.035			1.19	4.19	5.38	6.50
	2780	5000′ haul		310	.045			1.54	5.40	6.94	8.40
	2800	Clay, 1500′ haul		315	.044			1.51	5.30	6.81	8.25
	2820	3000′ haul		300	.047			1.59	5.60	7.19	8.70
	2840	5000′ haul		225	.062			2.12	7.45	9.57	11.60
	2900	15 C.Y., 1/4 push dozer, sand & gravel, 1500′ haul	B-33C	800	.018			.60	2.10	2.70	3.26
	2920	3000′ haul		640	.022			.74	2.62	3.36	4.07
	2940	5000′ haul		520	.027			.92	3.22	4.14	5
	2960	Common earth, 1500′ haul		600	.023			.79	2.79	3.58	4.34
	2980	3000′ haul		560	.025			.85	2.99	3.84	4.65
	3000	5000′ haul		440	.032			1.08	3.81	4.89	5.90
	3020	Clay, 1500′ haul		450	.031			1.06	3.73	4.79	5.80
	3040	3000′ haul		420	.033			1.13	3.99	5.12	6.20
	3060	5000′ haul		320	.044			1.49	5.25	6.74	8.15
462	0010	**EXCAVATION, STRUCTURAL**									
	0015	Hand, pits to 6′ deep, sandy soil	1 Clab	8	1	B.C.Y.		27.50		27.50	45.50
	0100	Heavy soil or clay		4	2			55		55	91
	0300	Pits 6′ to 12′ deep, sandy soil		5	1.600			44		44	72.50
	0500	Heavy soil or clay		3	2.667			73		73	121
	0700	Pits 12′ to 18′ deep, sandy soil		4	2			55		55	91
	0900	Heavy soil or clay		2	4			110		110	182
	1100	Hand loading trucks from stock pile, sandy soil		12	.667			18.25		18.25	30.50
	1300	Heavy soil or clay		8	1			27.50		27.50	45.50
	1500	For wet or muck hand excavation, add to above				%				50%	50%
	1550	Excavation rock by hand/air tool	B-9	3.40	11.765	B.C.Y.		325	43.50	368.50	590
	9000	Minimum labor/equipment charge	1 Clab	4	2	Job		55		55	91
490	0010	**HAULING**, excavated or borrow, loose cubic yards									
	0012	no loading included, highway haulers									

02315 | Excavation and Fill

		02315 Excavation and Fill	CREW	DAILY OUTPUT	LABOR-HOURS	UNIT	2006 BARE COSTS MAT.	LABOR	EQUIP.	TOTAL	TOTAL INCL O&P	
490	0020	6 C.Y. dump truck, 1/4 mile round trip, 5.0 loads/hr.	B-34A	195	.041	L.C.Y.		1.16	1.68	2.84	3.73	490
	0030	1/2 mile round trip, 4.1 loads/hr.		160	.050			1.42	2.04	3.46	4.56	
	0040	1 mile round trip, 3.3 loads/hr.		130	.062			1.74	2.51	4.25	5.60	
	0100	2 mile round trip, 2.6 loads/hr.		100	.080			2.27	3.27	5.54	7.30	
	0150	3 mile round trip, 2.1 loads/hr.		80	.100			2.84	4.08	6.92	9.10	
	0200	4 mile round trip, 1.8 loads/hr.		70	.114			3.24	4.67	7.91	10.40	
	0310	12 C.Y. dump truck, 1/4 mile round trip 3.7 loads/hr.	B-34B	288	.028			.79	1.66	2.45	3.10	
	0320	1/2 mile round trip, 3.2 loads/hr.		250	.032			.91	1.91	2.82	3.58	
	0330	1 mile round trip 2.7, loads/hr.		210	.038			1.08	2.27	3.35	4.26	
	0400	2 mile round trip, 2.2 loads/hr.		180	.044			1.26	2.65	3.91	4.96	
	0450	3 mile round trip, 1.9 loads/hr.		170	.047			1.33	2.80	4.13	5.25	
	0500	4 mile round trip, 1.6 loads/hr.		125	.064			1.81	3.81	5.62	7.15	
	0540	5 mile round trip, 1 load/hr.		78	.103			2.91	6.10	9.01	11.45	
	0550	10 mile round trip, 0.60 load/hr.		58	.138			3.91	8.20	12.11	15.40	
	0560	20 mile round trip, 0.4 load/hr.		39	.205			5.80	12.20	18	23	
	0600	16.5 C.Y. dump trailer, 1 mile round trip, 2.6 loads/hr.	B-34C	280	.029			.81	1.45	2.26	2.91	
	0700	2 mile round trip, 2.1 loads/hr.		225	.036			1.01	1.80	2.81	3.62	
	1000	3 mile round trip, 1.8 loads/hr.		193	.041			1.18	2.10	3.28	4.22	
	1100	4 mile round trip, 1.6 loads/hr.		172	.047			1.32	2.35	3.67	4.74	
	1110	5 mile round trip, 1 load/hr.		108	.074			2.10	3.75	5.85	7.55	
	1120	10 mile round trip, .60 load/hr.		80	.100			2.84	5.05	7.89	10.15	
	1130	20 mile round trip, .4 load/hr.		54	.148			4.20	7.50	11.70	15.10	
	1150	20 C.Y. dump trailer, 1 mile round trip, 2.5 loads/hr.	B-34D	325	.025			.70	1.28	1.98	2.55	
	1200	2 mile round trip, 2 loads/hr.		260	.031			.87	1.61	2.48	3.19	
	1220	3 mile round trip, 1.7 loads/hr.		221	.036			1.03	1.89	2.92	3.75	
	1240	4 mile round trip, 1.5 loads/hr.		195	.041			1.16	2.14	3.30	4.24	
	1245	5 mile round trip, 1.1 load/hr.		143	.056			1.59	2.92	4.51	5.80	
	1250	10 mile round trip, .75 load/hr.		110	.073			2.06	3.80	5.86	7.55	
	1255	20 mile round trip, .5 load/hr.		78	.103			2.91	5.35	8.26	10.65	
	1300	Hauling in medium traffic, add								20%	20%	
	1400	Heavy traffic, add								30%	30%	
	1600	Grading at dump, or embankment if required, by dozer	B-10B	1000	.012			.40	.92	1.32	1.66	
	1800	Spotter at fill or cut, if required	1 Clab	8	1	Hr.		27.50		27.50	45.50	
	4700	Highway hauling beyond 20 miles, per loaded mile, minimum				Mile				1.20	1.32	
	4750	Maximum				"				2.40	2.64	
510	0010	**FILL BY BORROW**, load, 1 mile haul, spread with dozer										510
	0020	for embankments	B-15	1200	.023	L.C.Y.	6.10	.71	1.56	8.37	9.60	
	0100	Select fill for shoulders & embankments	"	1200	.023	"	8.05	.71	1.56	10.32	11.70	
	0201	For hauling over 1 mile, add to above per C.Y., see Div 02315-490				Mile				.69	.92	
520	0010	**FILL**, spread dumped material, no compaction										520
	0020	By dozer, no compaction	B-10B	1000	.012	L.C.Y.		.40	.92	1.32	1.66	
	0100	By hand	1 Clab	12	.667	"		18.25		18.25	30.50	
	0150	Spread fill, from stockpile with 2-1/2 C.Y. F.E. loader										
	0170	130 H.P., 300′ haul	B-10P	600	.020	L.C.Y.		.67	1.32	1.99	2.54	
	0190	With dozer 300 H.P., 300′ haul	B-10M	600	.020	"		.67	2	2.67	3.27	
	0500	Gravel fill, compacted, under floor slabs, 4″ deep	B-37	10000	.005	S.F.	.16	.14	.01	.31	.42	
	0600	6″ deep		8600	.006		.24	.16	.01	.41	.54	
	0700	9″ deep		7200	.007		.40	.19	.02	.61	.78	
	0800	12″ deep		6000	.008		.56	.23	.02	.81	1.02	
	1000	Alternate pricing method, 4″ deep		120	.400	E.C.Y.	12.05	11.60	.95	24.60	33.50	
	1100	6″ deep		160	.300		12.05	8.70	.71	21.46	28.50	
	1200	9″ deep		200	.240		12.05	6.95	.57	19.57	25.50	
	1300	12″ deep		220	.218		12.05	6.35	.52	18.92	24	
	1500	For fill under exterior paving, see Div. 02720-200										
	1600	For flowable fill, see Division 03310-220										

	02315 | Excavation and Fill		CREW	DAILY OUTPUT	LABOR-HOURS	UNIT	2006 BARE COSTS MAT.	LABOR	EQUIP.	TOTAL	TOTAL INCL O&P	
520	9000	Minimum labor/equipment charge	1 Clab	4	2	Job		55		55	91	520
610	0010	**EXCAVATING, TRENCH** or continuous footing, common earth										610
	0020	No sheeting or dewatering included										
	0050	1′ to 4′ deep, 3/8 C.Y. tractor loader/backhoe	B-11C	150	.107	B.C.Y.		3.42	1.51	4.93	7.15	
	0060	1/2 C.Y. tractor loader/backhoe	B-11M	200	.080			2.56	1.39	3.95	5.65	
	0090	4′ to 6′ deep, 1/2 C.Y. tractor loader/backhoe	"	200	.080			2.56	1.39	3.95	5.65	
	0100	5/8 C.Y. hydraulic backhoe	B-12Q	250	.064			2.10	1.74	3.84	5.30	
	0300	1/2 C.Y. hydraulic excavator, truck mounted	B-12J	200	.080			2.62	4.20	6.82	8.85	
	0500	6′ to 10′ deep, 3/4 C.Y. hydraulic backhoe, 6′ to 10′ deep	B-12F	225	.071			2.33	2.12	4.45	6.10	
	0600	1 C.Y. hydraulic excavator, truck mounted	B-12K	400	.040			1.31	2.44	3.75	4.79	
	0900	10′ to 14′ deep, 3/4 C.Y. hydraulic backhoe	B-12F	200	.080			2.62	2.39	5.01	6.85	
	1000	1-1/2 C.Y. hydraulic backhoe	B-12B	540	.030			.97	1.34	2.31	3.03	
	1300	14′ to 20′ deep, 1 C.Y. hydraulic backhoe	B-12A	320	.050			1.64	1.75	3.39	4.57	
	1400	By hand with pick and shovel 2′ to 6′ deep, light soil	1 Clab	8	1			27.50		27.50	45.50	
	1500	Heavy soil	"	4	2	↓		55		55	91	
	1700	For tamping backfilled trenches, air tamp, add	A-1G	100	.080	E.C.Y.		2.19	.37	2.56	4.04	
	1900	Vibrating plate, add	B-18	180	.133	"		3.74	.19	3.93	6.40	
	2100	Trim sides and bottom for concrete pours, common earth		1500	.016	S.F.		.45	.02	.47	.77	
	2300	Hardpan	↓	600	.040	"		1.12	.06	1.18	1.92	
	2400	Pier and spread footing excavation, add to above				B.C.Y.				30%	30%	
	3000	Backfill trench, F.E. loader, wheel mtd., 1 C.Y. bucket										
	3020	Minimal haul	B-10R	400	.030	L.C.Y.		1.01	.49	1.50	2.15	
	3040	100′ haul		200	.060			2.02	.97	2.99	4.30	
	3060	200′ haul	↓	100	.120	↓		4.03	1.95	5.98	8.60	
	9000	Minimum labor/equipment charge	1 Clab	4	2	Job		55		55	91	
620	0010	**EXCAVATING, UTILITY TRENCH** Common earth										620
	0050	Trenching with chain trencher, 12 H.P., operator walking										
	0100	4″ wide trench, 12″ deep	B-53	800	.010	L.F.		.35	.06	.41	.63	
	0150	18″ deep		750	.011			.38	.06	.44	.66	
	0200	24″ deep		700	.011			.40	.07	.47	.71	
	0300	6″ wide trench, 12″ deep		650	.012			.43	.07	.50	.76	
	0350	18″ deep		600	.013			.47	.08	.55	.83	
	0400	24″ deep		550	.015			.51	.09	.60	.91	
	0450	36″ deep		450	.018			.63	.11	.74	1.11	
	0600	8″ wide trench, 12″ deep		475	.017			.59	.10	.69	1.05	
	0650	18″ deep		400	.020			.70	.12	.82	1.24	
	0700	24″ deep		350	.023			.80	.14	.94	1.42	
	0750	36″ deep		300	.027	↓		.94	.16	1.10	1.65	
	0900	Minimum labor/equipment charge	↓	2	4	Job		141	24	165	248	
	1000	Backfill by hand including compaction, add										
	1050	4″ wide trench, 12″ deep	A-1G	800	.010	L.F.		.27	.05	.32	.50	
	1100	18″ deep		530	.015			.41	.07	.48	.77	
	1150	24″ deep		400	.020			.55	.09	.64	1.01	
	1300	6″ wide trench, 12″ deep		540	.015			.41	.07	.48	.75	
	1350	18″ deep		405	.020			.54	.09	.63	1	
	1400	24″ deep		270	.030			.81	.14	.95	1.50	
	1450	36″ deep		180	.044			1.22	.21	1.43	2.25	
	1600	8″ wide trench, 12″ deep		400	.020			.55	.09	.64	1.01	
	1650	18″ deep		265	.030			.83	.14	.97	1.52	
	1700	24″ deep		200	.040			1.10	.19	1.29	2.02	
	1750	36″ deep	↓	135	.059	↓		1.62	.27	1.89	2.99	
	2000	Chain trencher, 40 H.P. operator riding										
	2050	6″ wide trench and backfill, 12″ deep	B-54	1200	.007	L.F.		.23	.18	.41	.57	
	2100	18″ deep		1000	.008			.28	.22	.50	.68	
	2150	24″ deep	↓	975	.008	↓		.29	.22	.51	.70	

For expanded coverage of these items see *Means Site Work & Landscape Cost Data 2006*

02300 | Earthwork

2 SITE CONSTRUCTION

02315 | Excavation and Fill

		02315 Excavation and Fill	CREW	DAILY OUTPUT	LABOR-HOURS	UNIT	2006 BARE COSTS MAT.	LABOR	EQUIP.	TOTAL	TOTAL INCL O&P	
620	2200	36" deep	B-54	900	.009	L.F.		.31	.24	.55	.76	620
	2250	48" deep		750	.011			.38	.29	.67	.91	
	2300	60" deep		650	.012			.43	.33	.76	1.05	
	2400	8" wide trench and backfill, 12" deep		1000	.008			.28	.22	.50	.68	
	2450	18" deep		950	.008			.30	.23	.53	.72	
	2500	24" deep		900	.009			.31	.24	.55	.76	
	2550	36" deep		800	.010			.35	.27	.62	.86	
	2600	48" deep		650	.012			.43	.33	.76	1.05	
	2700	12" wide trench and backfill, 12" deep		975	.008			.29	.22	.51	.70	
	2750	18" deep		860	.009			.33	.25	.58	.80	
	2800	24" deep		800	.010			.35	.27	.62	.86	
	2850	36" deep		725	.011			.39	.30	.69	.94	
	3000	16" wide trench and backfill, 12" deep		835	.010			.34	.26	.60	.82	
	3050	18" deep		750	.011			.38	.29	.67	.91	
	3100	24" deep		700	.011			.40	.31	.71	.98	
	3200	Compaction with vibratory plate, add								50%	50%	
	5100	Hand excavate and trim for pipe bells after trench excavation										
	5200	8" pipe	1 Clab	155	.052	L.F.		1.41		1.41	2.34	
	5300	18" pipe	"	130	.062	"		1.69		1.69	2.79	
	9000	Minimum labor/equipment charge	A-1G	4	2	Job		55	9.25	64.25	101	
640	0010	**UTILITY BEDDING** For pipe and conduit, not incl. compaction										640
	0050	Crushed or screened bank run gravel	B-6	150	.160	L.C.Y.	20.50	4.80	1.51	26.81	32	
	0100	Crushed stone 3/4" to 1/2"		150	.160		27.50	4.80	1.51	33.81	40	
	0200	Sand, dead or bank		150	.160		4.21	4.80	1.51	10.52	14.10	
	0500	Compacting bedding in trench	A-1D	90	.089	E.C.Y.		2.44	.29	2.73	4.36	
	0600	If material source exceeds 2 miles, add for extra mileage.										
	0610	See 02315-490 for hauling mileage add.										

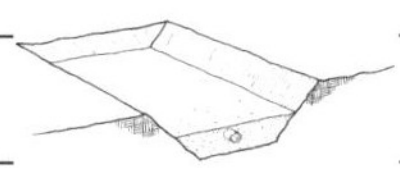

02325 | Dredging

		02325 Dredging	CREW	DAILY OUTPUT	LABOR-HOURS	UNIT	MAT.	LABOR	EQUIP.	TOTAL	TOTAL INCL O&P	
250	0010	**DREDGING**										250
	0020	Dredging mobilization and demobilization., add to below, minimum	B-8	.53	120	Total		3,725	4,525	8,250	11,000	
	0100	Maximum	"	.10	640	"		19,700	24,000	43,700	58,500	
	0300	Barge mounted clamshell excavation into scows,										
	0310	Dumped 20 miles at sea, minimum	B-57	310	.155	B.C.Y.		4.90	4.28	9.18	12.60	
	0400	Maximum	"	213	.225	"		7.15	6.20	13.35	18.35	
	0500	Barge mounted dragline or clamshell, hopper dumped,										
	0510	pumped 1000' to shore dump, minimum	B-57	340	.141	B.C.Y.		4.47	3.90	8.37	11.50	
	0525	All pumping uses 2000 gallons of water per cubic yard										
	0600	Maximum	B-57	243	.198	B.C.Y.		6.25	5.45	11.70	16.10	
	1000	Hydraulic method, pumped 1000' to shore dump, minimum		460	.104			3.30	2.88	6.18	8.50	
	1100	Maximum		310	.155			4.90	4.28	9.18	12.60	
	1400	Into scows dumped 20 miles, minimum		425	.113			3.58	3.12	6.70	9.20	
	1500	Maximum		243	.198			6.25	5.45	11.70	16.10	
	1600	For inland rivers and canals in South, deduct								30%	30%	

02340 | Soil Stabilization

		02340 Soil Stabilization	CREW	DAILY OUTPUT	LABOR-HOURS	UNIT	MAT.	LABOR	EQUIP.	TOTAL	TOTAL INCL O&P	
100	0010	**ASPHALT SOIL STABILIZATION** Including scarifying and compaction										100
	0020	Asphalt, 1-1/2" deep, 1/2 gal/S.Y.	B-75	4000	.014	S.Y.	1.04	.46	.88	2.38	2.84	
	0040	3/4 gal/S.Y.		4000	.014		1.55	.46	.88	2.89	3.41	
	0100	3" deep, 1 gal/S.Y.		3500	.016		2.07	.53	1	3.60	4.23	
	0140	1-1/2 gal/S.Y.		3500	.016		3.11	.53	1	4.64	5.35	
	0200	6" deep, 2 gal/S.Y.		3000	.019		4.14	.62	1.17	5.93	6.80	
	0240	3 gal/S.Y.		3000	.019		6.20	.62	1.17	7.99	9.10	
	0300	8" deep, 2-2/3 gal/S.Y.		2800	.020		5.55	.66	1.25	7.46	8.55	

02340 | Soil Stabilization

			CREW	DAILY OUTPUT	LABOR-HOURS	UNIT	2006 BARE COSTS MAT.	LABOR	EQUIP.	TOTAL	TOTAL INCL O&P	
100	0340	4 gal/S.Y.	B-75	2800	.020	S.Y.	8.30	.66	1.25	10.21	11.55	100
	0500	12″ deep, 4 gal/S.Y.		5000	.011		8.30	.37	.70	9.37	10.45	
	0540	6 gal/S.Y.		2600	.022		12.40	.71	1.35	14.46	16.25	
200	0010	**CEMENT SOIL STABILIZATION** Including scarifying and compaction										200
	1020	Cement, 4% mix, by volume, 6″ deep	B-74	1100	.058	S.Y.	1.02	1.89	3.46	6.37	7.95	
	1030	8″ deep		1050	.061		1.33	1.98	3.63	6.94	8.65	
	1060	12″ deep		960	.067		1.99	2.17	3.97	8.13	10.05	
	1100	6% mix, 6″ deep		1100	.058		1.46	1.89	3.46	6.81	8.45	
	1120	8″ deep		1050	.061		1.90	1.98	3.63	7.51	9.25	
	1160	12″ deep		960	.067		2.87	2.17	3.97	9.01	11	
	1200	9% mix, 6″ deep		1100	.058		2.21	1.89	3.46	7.56	9.30	
	1220	8″ deep		1050	.061		2.87	1.98	3.63	8.48	10.35	
	1260	12″ deep		960	.067		4.33	2.17	3.97	10.47	12.60	
	1300	12% mix, 6″ deep		1100	.058		2.87	1.89	3.46	8.22	10	
	1320	8″ deep		1050	.061		3.85	1.98	3.63	9.46	11.40	
	1360	12″ deep		960	.067		5.75	2.17	3.97	11.89	14.15	
300	0010	**GEOTEXTILE SOIL STABILIZATION**										300
	1500	Geotextile fabric, woven, 200 lb. tensile strength	2 Clab	2500	.006	S.Y.	2.07	.18		2.25	2.57	
	1510	Heavy Duty, 600 lb. tensile strength		2400	.007		1.85	.18		2.03	2.34	
	1550	Non-woven, 120 lb. tensile strength		2500	.006		1.05	.18		1.23	1.45	
500	0010	**LIME SOIL STABILIZATION** Including scarifying and compaction										500
	2020	Hydrated lime, for base, 2% mix by weight, 6″ deep	B-74	1800	.036	S.Y.	2.72	1.16	2.12	6	7.20	
	2030	8″ deep		1700	.038		3.62	1.23	2.24	7.09	8.40	
	2060	12″ deep		1550	.041		2.73	1.34	2.46	6.53	7.85	
	2100	4% mix, 6″ deep		1800	.036		2.77	1.16	2.12	6.05	7.25	
	2120	8″ deep		1700	.038		3.75	1.23	2.24	7.22	8.55	
	2160	12″ deep		1550	.041		5.60	1.34	2.46	9.40	11	
	2200	6% mix, 6″ deep		1800	.036		4.20	1.16	2.12	7.48	8.80	
	2220	8″ deep		1700	.038		5.60	1.23	2.24	9.07	10.60	
	2260	12″ deep		1550	.041		8.35	1.34	2.46	12.15	14	
700	0010	**CALCIUM CHLORIDE**										700
	0020	Calcium chloride delivered, 100 LB bags, truckload lots				Ton	495			495	545	
	0030	Solution, 4 lb. flake per gallon, tank truck delivery				Gal.	1.06			1.06	1.17	

02360 | Soil Treatment

			CREW	DAILY OUTPUT	LABOR-HOURS	UNIT	MAT.	LABOR	EQUIP.	TOTAL	TOTAL INCL O&P	
200	0010	**TERMITE PRETREATMENT**										200
	0020	Slab and walls, residential	1 Skwk	1200	.007	SF Flr.	.29	.24		.53	.72	
	0100	Commercial, minimum		2496	.003		.31	.12		.43	.53	
	0200	Maximum		1645	.005		.47	.18		.65	.80	
	0390	Minimum labor/equipment charge		4	2	Job		73		73	119	
	0400	Insecticides for termite control, minimum		14.20	.563	Gal.	11.85	20.50		32.35	46.50	
	0500	Maximum		11	.727	″	20.50	26.50		47	66	
	3000	Soil poisoning (sterilization)	1 Clab	4496	.002	S.F.	.28	.05		.33	.39	
	3100	Herbicide application from truck	B-59	19000	.001	S.Y.		.01	.02	.03	.04	

02370 | Erosion & Sedimentation Control

			CREW	DAILY OUTPUT	LABOR-HOURS	UNIT	MAT.	LABOR	EQUIP.	TOTAL	TOTAL INCL O&P	
450	0010	**RIP-RAP & ROCK LINING**, Random, broken stone										450
	0100	Machine placed for slope protection	B-12G	62	.258	L.C.Y.	25.50	8.45	9.15	43.10	51.50	
	0110	3/8 to 1/4 C.Y. pieces, grouted	B-13	80	.700	S.Y.	49.50	21	8.15	78.65	97.50	
	0200	18″ minimum thickness, not grouted	″	53	1.057	″	15.85	31.50	12.30	59.65	82.50	
	0300	Dumped, 50 lb. average	B-11A	800	.020	Ton	18.25	.64	1.15	20.04	22.50	
	0350	100 lb. average		700	.023		26	.73	1.32	28.05	31	

02300 | Earthwork

02370 | Erosion & Sedimentation Control

		02370 Erosion & Sedimentation Control	CREW	DAILY OUTPUT	LABOR-HOURS	UNIT	2006 BARE COSTS MAT.	LABOR	EQUIP.	TOTAL	TOTAL INCL O&P	
450	0370	300 lb. average	B-11A	600	.027	Ton	30.50	.85	1.53	32.88	36.50	450
	0400	Gabions, galvanized steel mesh mats or boxes, stone filled, 6″ deep	B-13	200	.280	S.Y.	17.20	8.40	3.25	28.85	36	
	0500	9″ deep		163	.344		26.50	10.30	3.99	40.79	50	
	0600	12″ deep		153	.366		28	10.95	4.25	43.20	53	
	0700	18″ deep		102	.549		35.50	16.45	6.40	58.35	73	
	0800	36″ deep		60	.933		60.50	28	10.85	99.35	124	
700	0010	**SYNTHETIC EROSION CONTROL**										700
	0020	Jute mesh, 100 SY per roll, 4′ wide, stapled	B-80A	2400	.010	S.Y.	.75	.27	.07	1.09	1.36	
	0060	Nylon, 3 dimensional geomatrix, 9 mil thick		700	.034		4.37	.94	.25	5.56	6.65	
	0062	12 mil thick		515	.047		5.95	1.28	.34	7.57	9	
	0064	18 mil thick		460	.052		6.25	1.43	.38	8.06	9.65	
	0100	Plastic netting, stapled, 2″ x 1″ mesh, 20 mil	B-1	2500	.010		.68	.27		.95	1.20	
	0120	Revegetation mat, webbed	2 Clab	1000	.016		5.45	.44		5.89	6.75	
	0200	Polypropylene mesh, stapled, 6.5 oz./S.Y.	B-1	2500	.010		1.40	.27		1.67	1.99	
	0300	Tobacco netting, or jute mesh #2, stapled	″	2500	.010		.08	.27		.35	.54	
	1000	Silt fence, polypropylene, 3′ high, ideal conditions	2 Clab	1600	.010	L.F.	.34	.27		.61	.82	
	1100	Adverse conditions	″	950	.017	″	.34	.46		.80	1.13	
	1130	Cellular confinement, poly, 3-dimen, 8′ x 20′ panels, 4″ deep cell	B-6	1600	.015	S.F.	1.82	.45	.14	2.41	2.89	
	1140	8″ deep cells	″	1200	.020	″	2.51	.60	.19	3.30	3.95	
	1200	Place and remove hay bales	A-2	3	8	Ton	56	220	43	319	470	
	1250	Hay bales, staked	″	2500	.010	L.F.	2.25	.26	.05	2.56	2.96	

02390 | Shore Protect/Mooring Structures

		02390 Shore Protect/Mooring Structures	CREW	DAILY OUTPUT	LABOR-HOURS	UNIT	MAT.	LABOR	EQUIP.	TOTAL	TOTAL INCL O&P	
110	0010	**BREAKWATERS, BULKHEADS** Reinforced concrete,										110
	0015	include footing and tie-backs										
	0060	Maximum	C-17C	24.25	3.423	L.F.	69.50	126	16.35	211.85	300	
	0100	12′ high, minimum		20	4.150		113	153	19.85	285.85	395	
	0160	Maximum		18.50	4.486		131	166	21.50	318.50	440	
	0200	Steel sheeting, with 4′ x 4′ x 8″ concrete deadmen, @ 10′ O.C.										
	0210	12′ high, shore driven	B-40	27	2.370	L.F.	62.50	84	102	248.50	320	
	0260	Barge driven	B-76	15	4.800	″	93.50	170	155	418.50	555	
340	0010	**JETTIES, DOCKS** Floating, recreational, prefabricated galvanized steel with										340
	0020	polyethylene encased polystyrene, no pilings included	F-3	330	.121	S.F.	29.50	4.37	1.93	35.80	42	
	0200	Pile supported, shore constructed, bare, 3″ decking		130	.308		20.50	11.10	4.90	36.50	46	
	0250	4″ decking		120	.333		19.60	12	5.30	36.90	47	
	0400	Floating, small boat, prefab, no shore facilities, minimum		250	.160		10	5.75	2.55	18.30	23.50	
	0500	Maximum		150	.267		34	9.60	4.25	47.85	58	
	0700	Per slip, minimum (180 S.F. each)		1.59	25.157	Ea.	2,150	905	400	3,455	4,300	
	0800	Maximum		1.40	28.571	″	6,375	1,025	455	7,855	9,200	
500	0010	**REVETMENTS, CONCRETE**										500
	0100	Concrete revetment matt 8′ x 20′ x 4 1/2 ″ excluding site preparation										
	0110	Includes all labor, material and equip. for complete installation				Ea.	2,150			2,150	2,375	

02400 | Tunneling, Boring & Jacking

02441 | Microtunneling

		02441 Microtunneling	CREW	DAILY OUTPUT	LABOR-HOURS	UNIT	2006 BARE COSTS MAT.	LABOR	EQUIP.	TOTAL	TOTAL INCL O&P	
400	0010	**MICROTUNNELING** Not including excavation, backfill, shoring,										400
	0020	or dewatering, average 50′/day, slurry method										

02400 | Tunneling, Boring & Jacking

	02441 | Microtunneling		Crew	Daily Output	Labor-Hours	Unit	2006 Bare Costs Mat.	Labor	Equip.	Total	Total Incl O&P	
400	0100	24" to 48" outside diameter, minimum				L.F.					700	400
	0110	Adverse conditions, add				%					50%	
	1000	Rent microtunneling machine, average monthly lease				Month					94,000	
	1010	Operating technician				Day					700	
	1100	Mobilization and demobilization, minimum				Job					47,000	
	1110	Maximum				"					450,000	

	02445 | Boring or Jacking Conduits		Crew	Daily Output	Labor-Hours	Unit	Mat.	Labor	Equip.	Total	Total Incl O&P	
300	0010	**HORIZONTAL BORING** Casing only, 100′ minimum,										300
	0020	not incl. jacking pits or dewatering										
	0100	Roadwork, 1/2" thick wall, 24" diameter casing	B-42	20	3.200	L.F.	72.50	100	56.50	229	310	
	0200	36" diameter		16	4		115	125	71	311	415	
	0300	48" diameter		15	4.267		169	133	75.50	377.50	490	
	0500	Railroad work, 24" diameter		15	4.267		72.50	133	75.50	281	385	
	0600	36" diameter		14	4.571		115	143	81	339	455	
	0700	48" diameter		12	5.333		169	166	94.50	429.50	570	
	0900	For ledge, add								155	190	
	1000	Small diameter boring, 3", sandy soil	B-82	900	.018		19.95	.56	.07	20.58	23	
	1040	Rocky soil	"	500	.032		19.95	1	.13	21.08	24	

02450 | Foundation & Load Bearing Elements

	02455 | Driven Piles		Crew	Daily Output	Labor-Hours	Unit	2006 Bare Costs Mat.	Labor	Equip.	Total	Total Incl O&P	
100	0010	**CAST IN PLACE CONCRETE PILES**, 200 piles, 60′ long										100
	0020	unless specified otherwise, not incl. pile caps or mobilization										
	0100	Cast in place, thin wall shell pile, straight sided,										
	0110	not incl. reinforcing, 8" diam., 16 ga., 5.8 lb./L.F.	B-19	700	.091	V.L.F.	6.10	3.25	2.24	11.59	14.50	
	0200	10" diameter, 16 ga. corrugated, 7.3 lb./L.F.		650	.098		8	3.50	2.42	13.92	17.20	
	0300	12" diameter, 16 ga. corrugated, 8.7 lb./L.F.		600	.107		10.40	3.79	2.62	16.81	20.50	
	0400	14" diameter, 16 ga. corrugated, 10.0 lb./L.F.		550	.116		12.20	4.13	2.86	19.19	23.50	
	0500	16" diameter, 16 ga. corrugated, 11.6 lb./L.F.		500	.128		14.95	4.55	3.14	22.64	27.50	
	0800	Cast in place friction pile, 50′ long, fluted,										
	0810	tapered steel, 4000 psi concrete, no reinforcing										
	0900	12" diameter, 7 ga.	B-19	600	.107	V.L.F.	18.35	3.79	2.62	24.76	29	
	1000	14" diameter, 7 ga.		560	.114		20	4.06	2.81	26.87	32	
	1100	16" diameter, 7 ga.		520	.123		23.50	4.37	3.02	30.89	36.50	
	1200	18" diameter, 7 ga.		480	.133		27.50	4.74	3.27	35.51	42	
	1300	End bearing, fluted, constant diameter,										
	1320	4000 psi concrete, no reinforcing										
	1340	12" diameter, 7 ga.	B-19	600	.107	V.L.F.	19.15	3.79	2.62	25.56	30	
	1360	14" diameter, 7 ga.		560	.114		24	4.06	2.81	30.87	36.50	
	1380	16" diameter, 7 ga.		520	.123		28	4.37	3.02	35.39	41	
	1400	18" diameter, 7 ga.		480	.133		30.50	4.74	3.27	38.51	45	
	1500	For reinforcing steel, add				Lb.	.86			.86	.94	
	1700	For ball or pedestal end, add	B-19	11	5.818	C.Y.	98.50	207	143	448.50	605	
	1900	For lengths above 60′, concrete, add	"	11	5.818	"	102	207	143	452	610	
	2000	For steel thin shell, pipe only				Lb.	.69			.69	.76	
450	0010	**PRESTRESSED CONCRETE PILES**, 200 piles										450
	0020	unless specified otherwise, not incl. pile caps or mobilization										

	02455 | Driven Piles		Crew	Daily Output	Labor-Hours	Unit	2006 Bare Costs Mat.	Labor	Equip.	Total	Total Incl O&P	
450	2200	Precast, prestressed, 50′ long, 12″ diam., 2-3/8″ wall	B-19	720	.089	V.L.F.	12.55	3.16	2.18	17.89	21.50	450
	2300	14″ diameter, 2-1/2″ wall		680	.094		16.55	3.34	2.31	22.20	26	
	2500	16″ diameter, 3″ wall		640	.100		23	3.55	2.46	29.01	33.50	
	2600	18″ diameter, 3″ wall	B-19A	600	.107		29	3.79	3.35	36.14	41.50	
	2800	20″ diameter, 3-1/2″ wall		560	.114		33	4.06	3.59	40.65	47	
	2900	24″ diameter, 3-1/2″ wall		520	.123		40.50	4.37	3.86	48.73	56	
	2920	36″ diameter, 4-1/2″ wall	B-19	400	.160		57.50	5.70	3.93	67.13	76.50	
	2940	54″ diameter, 5″ wall		340	.188		76.50	6.70	4.62	87.82	101	
	2960	66″ diameter, 6″ wall		220	.291		115	10.35	7.15	132.50	152	
	3100	Precast, prestressed, 40′ long, 10″ thick, square		700	.091		9.10	3.25	2.24	14.59	17.80	
	3200	12″ thick, square		680	.094		11.45	3.34	2.31	17.10	20.50	
	3400	14″ thick, square		600	.107		13.55	3.79	2.62	19.96	24	
	3500	Octagonal		640	.100		18.70	3.55	2.46	24.71	29	
	3700	16″ thick, square		560	.114		21.50	4.06	2.81	28.37	33.50	
	3800	Octagonal		600	.107		22.50	3.79	2.62	28.91	33.50	
	4000	18″ thick, square	B-19A	520	.123		25.50	4.37	3.86	33.73	40	
	4100	Octagonal	B-19	560	.114		26.50	4.06	2.81	33.37	39	
	4300	20″ thick, square	B-19A	480	.133		31.50	4.74	4.19	40.43	47.50	
	4400	Octagonal	B-19	520	.123		29.50	4.37	3.02	36.89	43	
	4600	24″ thick, square	B-19A	440	.145		45.50	5.15	4.57	55.22	63.50	
	4700	Octagonal	B-19	480	.133		42.50	4.74	3.27	50.51	58	
	4730	Precast, prestressed, 60′ long, 10″ thick, square		700	.091		9.55	3.25	2.24	15.04	18.30	
	4740	12″ thick, square (60′ long)		680	.094		11.90	3.34	2.31	17.55	21	
	4750	Mobilization for 10,000 L.F. pile job, add		3300	.019			.69	.48	1.17	1.66	
	4800	25,000 L.F. pile job, add		8500	.008			.27	.18	.45	.64	
	5000	Pressure grouted pin pile, 5″ diam., cased, up to 50 ton,										
	5040	End bearing, less than 20′	B-48	90	.622	V.L.F.	28.50	19.30	40.50	88.30	107	
	5080	More than 40′		135	.415		16.55	12.90	27	56.45	68.50	
	5120	Friction, loose sand and gravel		107	.523		28.50	16.25	34	78.75	95	
	5160	Dense sand and gravel		135	.415		16.55	12.90	27	56.45	68.50	
	5200	Uncased, up to 10 ton capacity, 20′		135	.415		16.75	12.90	27	56.65	69	
600	0010	**STEEL PILES**, Not including mobilization or demobilization										600
	0100	Step tapered, round, concrete filled										
	0110	8″ tip, 60 ton capacity, 30′ depth	B-19	760	.084	V.L.F.	7.60	2.99	2.07	12.66	15.55	
	0120	60′ depth		740	.086		8.55	3.07	2.12	13.74	16.80	
	0130	80′ depth		700	.091		8.85	3.25	2.24	14.34	17.55	
	0150	10″ tip, 90 ton capacity, 30′ depth		700	.091		9.35	3.25	2.24	14.84	18.05	
	0160	60′ depth		690	.093		9.60	3.29	2.28	15.17	18.50	
	0170	80′ depth		670	.096		10.35	3.39	2.35	16.09	19.60	
	0190	12″ tip, 120 ton capacity, 30′ depth		660	.097		12.80	3.44	2.38	18.62	22.50	
	0200	60′ depth, 12″ diameter		630	.102		12.80	3.61	2.49	18.90	23	
	0210	80′ depth		590	.108		11.40	3.85	2.66	17.91	22	
	0250	"H" Sections, 50′ long, HP8 x 36		640	.100		12.80	3.55	2.46	18.81	22.50	
	0400	HP10 X 42		610	.105		14.95	3.73	2.58	21.26	25.50	
	0500	HP10 X 57		610	.105		20.50	3.73	2.58	26.81	31.50	
	0700	HP12 X 53		590	.108		19.10	3.85	2.66	25.61	30.50	
	0800	HP12 X 74	B-19A	590	.108		26.50	3.85	3.40	33.75	39.50	
	1000	HP14 X 73		540	.119		26.50	4.21	3.72	34.43	40	
	1100	HP14 X 89		540	.119		32	4.21	3.72	39.93	46.50	
	1300	HP14 X 102		510	.125		37	4.46	3.94	45.40	52	
	1400	HP14 X 117		510	.125		42.50	4.46	3.94	50.90	58	
	1600	Splice on standard points, not in leads, 8″ or 10″	1 Sswl	5	1.600	Ea.	84	64		148	212	
	1700	12″ or 14″		4	2		122	80		202	283	
	1900	Heavy duty points, not in leads, 10″ wide		4	2		130	80		210	291	
	2100	14″ wide		3.50	2.286		168	91.50		259.50	355	

02455 | Driven Piles

		02455 Driven Piles	CREW	DAILY OUTPUT	LABOR-HOURS	UNIT	2006 BARE COSTS MAT.	LABOR	EQUIP.	TOTAL	TOTAL INCL O&P	
600	2600	Pipe piles, 50′ lg. 8″ diam., 29 lb. per L.F., no concrete	B-19	500	.128	V.L.F.	13.85	4.55	3.14	21.54	26	600
	2700	Concrete filled		460	.139		14.70	4.94	3.42	23.06	28	
	2900	10″ diameter, 34 lb. per L.F., no concrete		500	.128		17.25	4.55	3.14	24.94	30	
	3000	Concrete filled		450	.142		19	5.05	3.49	27.54	33	
	3200	12″ diameter, 44 lb. per L.F., no concrete		475	.135		21.50	4.79	3.31	29.60	35	
	3300	Concrete filled		415	.154		22	5.50	3.79	31.29	37.50	
	3500	14″ diameter, 46 lb. per L.F., no concrete		430	.149		23	5.30	3.65	31.95	37.50	
	3600	Concrete filled		355	.180		24.50	6.40	4.43	35.33	42.50	
	3800	16″ diameter, 52 lb. per L.F., no concrete		385	.166		25.50	5.90	4.08	35.48	42	
	3900	Concrete filled		335	.191		28	6.80	4.69	39.49	47.50	
	4100	18″ diameter, 59 lb. per L.F., no concrete		355	.180		33.50	6.40	4.43	44.33	52	
	4200	Concrete filled		310	.206		33	7.35	5.05	45.40	54	
	4400	Splices for pipe piles, not in leads, 8″ diameter	1 Sswl	4.67	1.713	Ea.	62	68.50		130.50	197	
	4500	14″ diameter		3.79	2.111		81.50	84.50		166	248	
	4600	16″ diameter		3.03	2.640		101	105		206	310	
	4650	18″ diameter		4.50	1.778		133	71		204	279	
	4800	Points, standard, 8″ diameter		4.61	1.735		69.50	69.50		139	207	
	4900	14″ diameter		4.05	1.975		96.50	79		175.50	254	
	5000	16″ diameter		3.37	2.374		118	95		213	305	
	5050	18″ diameter		5	1.600		184	64		248	320	
	5200	Points, heavy duty, 10″ diameter		2.89	2.768		48	111		159	260	
	5300	14″ or 16″ diameter		2.02	3.960		77	158		235	380	
	5500	For reinforcing steel, add		1150	.007	Lb.	.57	.28		.85	1.15	
	5700	For thick wall sections, add				″	.65			.65	.72	
	6020	Steel pipe pile end plates, 8″ diameter	1 Sswl	14	.571	Ea.	53	23		76	101	
	6050	10″ diameter		14	.571		55.50	23		78.50	104	
	6100	12″ diameter		12	.667		59	26.50		85.50	115	
	6150	14″ diameter		10	.800		61	32		93	127	
	6200	16″ diameter		9	.889		65	35.50		100.50	138	
	6250	18″ diameter		8	1		73	40		113	155	
	6300	Steel pipe pile shoes, 8″ diameter		12	.667		113	26.50		139.50	174	
	6350	10″ diameter		12	.667		117	26.50		143.50	178	
	6400	12″ diameter		10	.800		125	32		157	197	
	6450	14″ diameter		9	.889		144	35.50		179.50	225	
	6500	16″ diameter		8	1		160	40		200	251	
	6550	18″ diameter		6	1.333		179	53.50		232.50	297	
650	0010	**TIMBER PILES**, Friction or end bearing, not including										650
	0100	Untreated piles, up to 30′ long, 12″ butts, 8″ points	B-19	625	.102	V.L.F.	6.55	3.64	2.51	12.70	15.95	
	0200	30′ to 39′ long, 12″ butts, 8″ points		700	.091		6.55	3.25	2.24	12.04	15	
	0300	40′ to 49′ long, 12″ butts, 7″ points		720	.089		6.55	3.16	2.18	11.89	14.80	
	0400	50′ to 59′ long, 13″butts, 7″ points		800	.080		6.65	2.84	1.96	11.45	14.15	
	0500	60′ to 69′ long, 13″ butts, 7″ points		840	.076		7.45	2.71	1.87	12.03	14.75	
	0600	70′ to 80′ long, 13″ butts, 6″ points		840	.076		8.30	2.71	1.87	12.88	15.70	
	0800	Treated piles, 12 lb. per C.F.,										
	0810	friction or end bearing, ASTM class B										
	1000	Up to 30′ long, 12″ butts, 8″ points	B-19	625	.102	V.L.F.	9.70	3.64	2.51	15.85	19.40	
	1100	30′ to 39′ long, 12″ butts, 8″ points		700	.091		9.45	3.25	2.24	14.94	18.20	
	1200	40′ to 49′ long, 12″ butts, 7″ points		720	.089		10.15	3.16	2.18	15.49	18.75	
	1300	50′ to 59′ long, 13″ butts, 7″ points		800	.080		11.20	2.84	1.96	16	19.15	
	1400	60′ to 69′ long, 13″ butts, 6″ points	B-19A	840	.076		15.30	2.71	2.39	20.40	24	
	1500	70′ to 80′ long, 13″ butts, 6″ points	″	840	.076		19.20	2.71	2.39	24.30	28	
	1600	Treated piles, C.C.A., 2.5# per C.F.										
	1610	8″ butts, 10′ long	B-19	400	.160	V.L.F.	7.05	5.70	3.93	16.68	21.50	
	1620	11′ to 16′ long		500	.128		7.05	4.55	3.14	14.74	18.70	
	1630	17′ to 20′ long		575	.111		7.05	3.95	2.73	13.73	17.25	
	1640	10″ butts, 10′ to 16′ long		500	.128		8.05	4.55	3.14	15.74	19.80	

02450 | Foundation & Load Bearing Elements

02455 | Driven Piles

		02455 Driven Piles	CREW	DAILY OUTPUT	LABOR-HOURS	UNIT	2006 BARE COSTS MAT.	LABOR	EQUIP.	TOTAL	TOTAL INCL O&P	
650	1650	17′ to 20′ long	B-19	575	.111	V.L.F.	8.05	3.95	2.73	14.73	18.35	650
	1660	21′ to 40′ long		700	.091		8.05	3.25	2.24	13.54	16.65	
	1670	12″ butts, 10′ to 20′ long		575	.111		8.70	3.95	2.73	15.38	19.05	
	1680	21′ to 35′ long		650	.098		8.70	3.50	2.42	14.62	17.95	
	1690	36′ to 40′ long		700	.091		8.70	3.25	2.24	14.19	17.35	
	1695	14″ butts. to 40′ long	↓	700	.091	↓	12.45	3.25	2.24	17.94	21.50	
	1700	Boot for pile tip, minimum	1 Pile	27	.296	Ea.	19.85	10.30		30.15	39.50	
	1800	Maximum		21	.381		59.50	13.20		72.70	88	
	2000	Point for pile tip, minimum		20	.400		19.80	13.90		33.70	45.50	
	2100	Maximum	↓	15	.533		71.50	18.50		90	110	
	2300	Splice for piles over 50′ long, minimum	B-46	35	1.371		49.50	43	.96	93.46	128	
	2400	Maximum		20	2.400	↓	59.50	75.50	1.68	136.68	193	
	2600	Concrete encasement with wire mesh and tube	↓	331	.145	V.L.F.	9.30	4.55	.10	13.95	18	
	2700	Mobilization for 10,000 L.F. pile job, add	B-19	3300	.019			.69	.48	1.17	1.66	
	2800	25,000 L.F. pile job, add	″	8500	.008	↓		.27	.18	.45	.64	
800	0010	**PILING SPECIAL COSTS**										800
	0011	Pile caps, see Division 03310-240										
	0500	Cutoffs, concrete piles, plain	1 Pile	5.50	1.455	Ea.		50.50		50.50	85.50	
	0600	With steel thin shell, add		38	.211			7.30		7.30	12.40	
	0700	Steel pile or "H" piles		19	.421			14.60		14.60	25	
	0800	Wood piles	↓	38	.211	↓		7.30		7.30	12.40	
	0900	Pre-augering up to 30′ deep, average soil, 24″ diameter	B-43	180	.267	L.F.		8.10	18.75	26.85	33.50	
	0920	36″ diameter		115	.417			12.65	29.50	42.15	53	
	0960	48″ diameter		70	.686			21	48.50	69.50	87	
	0980	60″ diameter	↓	50	.960	↓		29	67.50	96.50	122	
	1000	Testing, any type piles, test load is twice the design load										
	1050	50 ton design load, 100 ton test				Ea.				15,000	16,050	
	1100	100 ton design load, 200 ton test								19,250	20,300	
	1150	150 ton design load, 300 ton test								24,100	25,700	
	1200	200 ton design load, 400 ton test								26,200	28,800	
	1250	400 ton design load, 800 ton test				↓				30,300	33,700	
	1500	Wet conditions, soft damp ground										
	1600	Requiring mats for crane, add								40%	40%	
	1700	Barge mounted driving rig, add								30%	30%	
900	0010	**MOBILIZATION**										900
	0020	Set up & remove, air compressor, 600 C.F.M.	A-5	3.30	5.455	Ea.		150	9.75	159.75	258	
	0100	1200 C.F.M.	″	2.20	8.182			224	14.65	238.65	385	
	0200	Crane, with pile leads and pile hammer, 75 ton	B-19	.60	106			3,800	2,625	6,425	9,125	
	0300	150 ton	″	.36	177			6,325	4,375	10,700	15,200	
	0500	Drill rig, for caissons, to 36″, minimum	B-43	2	24			730	1,700	2,430	3,025	
	0520	Maximum		.50	96			2,925	6,750	9,675	12,200	
	0600	Up to 84″	↓	1	48			1,450	3,375	4,825	6,100	
	0800	Auxiliary boiler, for steam small	A-5	1.66	10.843			297	19.40	316.40	510	
	0900	Large	″	.83	21.687			595	39	634	1,025	
	1100	Rule of thumb: complete pile driving set up, small	B-19	.45	142			5,050	3,500	8,550	12,200	
	1200	Large	″	.27	237	↓		8,425	5,825	14,250	20,300	

02465 | Bored Piles

		02465 Bored Piles	CREW	DAILY OUTPUT	LABOR-HOURS	UNIT	MAT.	LABOR	EQUIP.	TOTAL	TOTAL INCL O&P	
800	0010	**DRILLED CAISSONS** Incl. excav., concrete, 50 lbs. reinf.										800
	0020	per C.Y., not incl. mobiliz., boulder removal, disposal										
	0100	Open style, machine drilled, to 50′ deep, in stable ground, no										
	0110	casings or ground water, 18″ diam., 0.065 C.Y./L.F.	B-43	200	.240	V.L.F.	7	7.30	16.90	31.20	38	
	0200	24″ diameter, 0.116 C.Y./L.F.		190	.253		12.55	7.65	17.80	38	46	
	0300	30″ diameter, 0.182 C.Y./L.F.	↓	150	.320	↓	19.70	9.70	22.50	51.90	62.50	

02450 | Foundation & Load Bearing Elements

	02465 | Bored Piles	CREW	DAILY OUTPUT	LABOR-HOURS	UNIT	2006 BARE COSTS MAT.	LABOR	EQUIP.	TOTAL	TOTAL INCL O&P	
800	0400 36" diameter, 0.262 C.Y./L.F.	B-43	125	.384	V.L.F.	28.50	11.65	27	67.15	79.50	800
	0500 48" diameter, 0.465 C.Y./L.F.		100	.480		50.50	14.55	34	99.05	116	
	0600 60" diameter, 0.727 C.Y./L.F.		90	.533		79	16.20	37.50	132.70	155	
	0700 72" diameter, 1.05 C.Y./L.F.		80	.600		114	18.20	42	174.20	201	
	0800 84" diameter, 1.43 C.Y./L.F.		75	.640		155	19.45	45	219.45	252	
	1000 For bell excavation and concrete, add										
	1020 4' bell diameter, 24" shaft, 0.444 C.Y.	B-43	20	2.400	Ea.	38.50	73	169	280.50	350	
	1040 6' bell diameter, 30" shaft, 1.57 C.Y.		5.70	8.421		137	256	595	988	1,225	
	1060 8' bell diameter, 36" shaft, 3.72 C.Y.		2.40	20		325	605	1,400	2,330	2,900	
	1080 9' bell diameter, 48" shaft, 4.48 C.Y.		2	24		390	730	1,700	2,820	3,450	
	1100 10' bell diameter, 60" shaft, 5.24 C.Y.		1.70	28.235		455	855	2,000	3,310	4,075	
	1120 12' bell diameter, 72" shaft, 8.74 C.Y.		1	48		760	1,450	3,375	5,585	6,925	
	1140 14' bell diameter, 84" shaft, 13.6 C.Y.		.70	68.571		1,175	2,075	4,825	8,075	9,975	
	1200 Open style, machine drilled, to 50' deep, in wet ground, pulled										
	1300 casing and pumping, 18" diameter, 0.065 C.Y./L.F.	B-48	160	.350	V.L.F.	7	10.85	22.50	40.35	50.50	
	1400 24" diameter, 0.116 C.Y./L.F.		125	.448		12.55	13.90	29	55.45	68.50	
	1500 30" diameter, 0.182 C.Y./L.F.		85	.659		19.70	20.50	43	83.20	102	
	1600 36" diameter, 0.262 C.Y./L.F.		60	.933		28.50	29	60.50	118	145	
	1700 48" diameter, 0.465 C.Y./L.F.	B-49	55	1.600		50.50	52	78	180.50	226	
	1800 60" diameter, 0.727 C.Y./L.F.		35	2.514		79	81.50	123	283.50	355	
	1900 72" diameter, 1.05 C.Y./L.F.		30	2.933		114	95.50	143	352.50	435	
	2000 84" diameter, 1.43 C.Y./L.F.		25	3.520		155	114	172	441	545	
	2100 For bell excavation and concrete, add										
	2120 4' bell diameter, 24" shaft, 0.444 C.Y.	B-48	19.80	2.828	Ea.	38.50	88	184	310.50	385	
	2140 6' bell diameter, 30" shaft, 1.57 C.Y.		5.70	9.825		137	305	640	1,082	1,350	
	2160 8' bell diameter, 36" shaft, 3.72 C.Y.		2.40	23.333		325	725	1,525	2,575	3,200	
	2180 9' bell diameter, 48" shaft, 4.48 C.Y.	B-49	3.30	26.667		390	865	1,300	2,555	3,250	
	2200 10' bell diameter, 60" shaft, 5.24 C.Y.		2.80	31.429		455	1,025	1,525	3,005	3,825	
	2220 12' bell diameter, 72" shaft, 8.74 C.Y.		1.60	55		760	1,775	2,675	5,210	6,675	
	2240 14' bell diameter, 84" shaft, 13.6 C.Y.		1	88		1,175	2,850	4,300	8,325	10,700	
	2300 Open style, machine drilled, to 50' deep, in soft rocks and										
	2400 medium hard shales, 18" diameter, 0.065 C.Y./L.F.	B-49	50	1.760	V.L.F.	7	57	86	150	195	
	2500 24" diameter, 0.116 C.Y./L.F.		30	2.933		12.55	95.50	143	251.05	325	
	2600 30" diameter, 0.182 C.Y./L.F.		20	4.400		19.70	143	214	376.70	490	
	2700 36" diameter, 0.262 C.Y./L.F.		15	5.867		28.50	191	286	505.50	655	
	2800 48" diameter, 0.465 C.Y./L.F.		10	8.800		50.50	286	430	766.50	990	
	2900 60" diameter, 0.727 C.Y./L.F.		7	12.571		79	410	615	1,104	1,425	
	3000 72" diameter, 1.05 C.Y./L.F.		6	14.667		114	475	715	1,304	1,675	
	3100 84" diameter, 1.43 C.Y./L.F.		5	17.600		155	570	860	1,585	2,050	
	3200 For bell excavation and concrete, add										
	3220 4' bell diameter, 24" shaft, 0.444 C.Y.	B-49	10.90	8.073	Ea.	38.50	262	395	695.50	905	
	3240 6' bell diameter, 30" shaft, 1.57 C.Y.		3.10	28.387		137	920	1,375	2,432	3,175	
	3260 8' bell diameter, 36" shaft, 3.72 C.Y.		1.30	67.692		325	2,200	3,300	5,825	7,550	
	3280 9' bell diameter, 48" shaft, 4.48 C.Y.		1.10	80		390	2,600	3,900	6,890	8,950	
	3300 10' bell diameter, 60" shaft, 5.24 C.Y.		.90	97.778		455	3,175	4,775	8,405	10,900	
	3320 12' bell diameter, 72" shaft, 8.74 C.Y.		.60	146		760	4,775	7,150	12,685	16,500	
	3340 14' bell diameter, 84" shaft, 13.6 C.Y.		.40	220		1,175	7,150	10,700	19,025	24,700	
	3600 For rock excavation, sockets, add, minimum		120	.733	C.F.		24	35.50	59.50	78	
	3650 Average		95	.926			30	45	75	98.50	
	3700 Maximum		48	1.833			59.50	89.50	149	196	
	3900 For 50' to 100' deep, add				V.L.F.				7%	7%	
	4000 For 100' to 150' deep, add								25%	25%	
	4100 For 150' to 200' deep, add								30%	30%	
	4200 For casings left in place, add				Lb.	.72			.72	.79	
	4300 For other than 50 lb. reinf. per C.Y., add or deduct				"	.77			.77	.85	
	4400 For steel "I" beam cores, add	B-49	8.30	10.602	Ton	1,525	345	515	2,385	2,800	

02450 | Foundation & Load Bearing Elements

	02465 | Bored Piles		CREW	DAILY OUTPUT	LABOR-HOURS	UNIT	2006 BARE COSTS MAT.	LABOR	EQUIP.	TOTAL	TOTAL INCL O&P	
800	4500	Load and haul excess excavation, 2 miles	B-34B	178	.045	L.C.Y.		1.27	2.68	3.95	5	800
950	0010	**DRILLED CONCRETE PIERS AND SHAFTS** or Displacement										950
	0100	Caissons, incl. mobilization and demobilization, up to 50 miles										
	0200	Uncased shafts, 30 to 80 tons cap., 17″ diam., 10′ depth	B-44	88	.727	V.L.F.	17.40	25.50	11.20	54.10	73.50	
	0300	25′ depth		165	.388		12.45	13.55	6	32	42.50	
	0400	80-150 ton capacity, 22″ diameter, 10′ depth		80	.800		22	28	12.35	62.35	84	
	0500	20′ depth		130	.492		17.40	17.20	7.60	42.20	56	
	0700	Cased shafts, 10 to 30 ton capacity, 10-5/8″ diam., 20′ depth		175	.366		12.45	12.75	5.65	30.85	41	
	0800	30′ depth		240	.267		11.60	9.30	4.11	25.01	32.50	
	0850	30 to 60 ton capacity, 12″ diameter, 20′ depth		160	.400		17.40	13.95	6.15	37.50	49	
	0900	40′ depth		230	.278		13.40	9.70	4.29	27.39	35.50	
	1000	80 to 100 ton capacity, 16″ diameter, 20′ depth		160	.400		25	13.95	6.15	45.10	57.50	
	1100	40′ depth		230	.278		23	9.70	4.29	36.99	46.50	
	1200	110 to 140 ton capacity, 17-5/8″ diameter, 20′ depth		160	.400		27	13.95	6.15	47.10	59.50	
	1300	40′ depth		230	.278		25	9.70	4.29	38.99	48.50	
	1400	140 to 175 ton capacity, 19″ diameter, 20′ depth		130	.492		29	17.20	7.60	53.80	69	
	1500	40′ depth	↓	210	.305	↓	27	10.65	4.70	42.35	52.50	
	1700	Over 30′ long, L.F. cost tends to be lower										
	1900	Maximum depth is about 90′										

02500 | Utility Services

	02510 | Water Distribution		CREW	DAILY OUTPUT	LABOR-HOURS	UNIT	2006 BARE COSTS MAT.	LABOR	EQUIP.	TOTAL	TOTAL INCL O&P	
600	0010	**VALVES**										600
	9000	Valves, gate valve, N.R.S. post type, 4″ diameter	B-21	32	.875	Ea.	365	28	4.77	397.77	455	
	9040	8″ diameter		16	1.750		705	56	9.55	770.55	880	
	9080	12″ diameter		13	2.154		1,425	69	11.75	1,505.75	1,675	
	9120	O.S.&Y., 4″ diameter		32	.875		291	28	4.77	323.77	370	
	9160	8″ diameter		16	1.750		695	56	9.55	760.55	870	
	9200	12″ diameter		13	2.154		1,475	69	11.75	1,555.75	1,750	
	9220	14″ diameter	↓	11	2.545		4,150	81.50	13.85	4,245.35	4,725	
	9400	Check valves, rubber disc, 2-1/2″ diameter	B-20	44	.545		350	16.95		366.95	415	
	9440	4″ diameter	B-21	32	.875		430	28	4.77	462.77	525	
	9500	8″ diameter		16	1.750		1,050	56	9.55	1,115.55	1,250	
	9540	12″ diameter		13	2.154		2,675	69	11.75	2,755.75	3,075	
	9700	Detector check valves, reducing, 4″ diameter		32	.875		705	28	4.77	737.77	825	
	9740	8″ diameter		16	1.750		1,775	56	9.55	1,840.55	2,050	
	9800	Galvanized, 4″ diameter		32	.875		920	28	4.77	952.77	1,075	
	9840	8″ diameter	↓	16	1.750	↓	2,150	56	9.55	2,215.55	2,450	
710	0010	**TAPPING, CROSSES AND SLEEVES**										710
	4000	Drill and tap pressurized main (labor only)										
	4100	6″ main, 1″ to 2″ service	Q-1	3	5.333	Ea.		205		205	320	
	4150	8″ main, 1″ to 2″ service	″	2.75	5.818	″		224		224	350	
	4500	Tap and insert gate valve										
	4600	8″ main, 4″ branch	B-21	3.20	8.750	Ea.		281	47.50	328.50	515	
	4650	6″ branch		2.70	10.370			335	56.50	391.50	605	
	4651	Piping, drill, tap & insert gate valve, 8″ main, 6″ branch		2.70	10.370			335	56.50	391.50	605	
	4700	10″ Main, 4″ branch		2.70	10.370			335	56.50	391.50	605	
	4750	6″ branch	↓	2.35	11.915	↓		380	65	445	695	

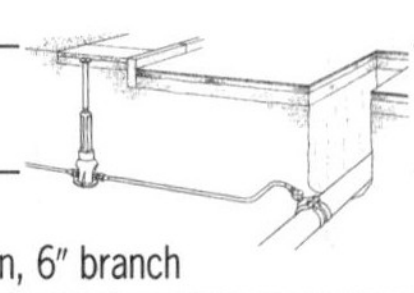

	02510 | Water Distribution		CREW	DAILY OUTPUT	LABOR-HOURS	UNIT	2006 BARE COSTS MAT.	LABOR	EQUIP.	TOTAL	TOTAL INCL O&P	
710	4800	12″ main, 6″ branch	B-21	2.35	11.915	Ea.		380	65	445	695	710
	4850	8″ branch		2.35	11.915			380	65	445	695	
	7000	Tapping crosses, sleeves, valves; with rubber gaskets										
	7020	Crosses, 4″ x 4″	B-21	37	.757	Ea.	460	24.50	4.12	488.62	555	
	7060	8″ x 6″		21	1.333		675	43	7.25	725.25	820	
	7080	8″ x 8″		21	1.333		730	43	7.25	780.25	885	
	7100	10″ x 6″		21	1.333		1,325	43	7.25	1,375.25	1,550	
	7160	12″ x 12″		18	1.556		1,675	50	8.50	1,733.50	1,950	
	7180	14″ x 6″		16	1.750		3,325	56	9.55	3,390.55	3,775	
	7240	16″ x 10″		14	2		3,625	64	10.90	3,699.90	4,100	
	7280	18″ x 6″		10	2.800		5,350	90	15.25	5,455.25	6,050	
	7320	18″ x 18″		10	2.800		4,925	90	15.25	5,030.25	5,575	
	7340	20″ x 6″		8	3.500		4,325	112	19.10	4,456.10	4,950	
	7360	20″ x 12″		8	3.500		4,625	112	19.10	4,756.10	5,300	
	7420	24″ x 12″		6	4.667		5,650	150	25.50	5,825.50	6,500	
	7440	24″ x 18″		6	4.667		8,450	150	25.50	8,625.50	9,550	
	7600	Cut-in sleeves with rubber gaskets, 4″		18	1.556		153	50	8.50	211.50	260	
	7640	8″		10	2.800		265	90	15.25	370.25	455	
	7680	12″		9	3.111		435	100	16.95	551.95	660	
	7800	Cut-in valves with rubber gaskets, 4″		18	1.556		390	50	8.50	448.50	515	
	7840	8″		10	2.800		810	90	15.25	915.25	1,050	
	7880	12″		9	3.111		1,225	100	16.95	1,341.95	1,525	
	7900	Tapping Valve 4 inch, MJ, ductile iron		18	1.556		410	50	8.50	468.50	545	
	7920	Tapping Valve 6 inch, MJ, ductile iron		12	2.333		490	75	12.70	577.70	675	
	8000	Sleeves with rubber gaskets, 4″ x 4″		37	.757		815	24.50	4.12	843.62	940	
	8030	8″ x 4″		21	1.333		505	43	7.25	555.25	635	
	8040	8″ x 6″		21	1.333		540	43	7.25	590.25	675	
	8060	8″ x 8″		21	1.333		620	43	7.25	670.25	765	
	8070	10″ x 4″		21	1.333		505	43	7.25	555.25	635	
	8080	10″ x 6″		21	1.333		565	43	7.25	615.25	705	
	8090	10″ x 8″		21	1.333		905	43	7.25	955.25	1,075	
	8140	12″ x 12″		18	1.556		1,075	50	8.50	1,133.50	1,275	
	8160	14″ x 6″		16	1.750		2,600	56	9.55	2,665.55	2,975	
	8220	16″ x 10″		14	2		2,850	64	10.90	2,924.90	3,250	
	8260	18″ x 6″		10	2.800		4,175	90	15.25	4,280.25	4,775	
	8300	18″ x 18″		10	2.800		4,325	90	15.25	4,430.25	4,950	
	8320	20″ x 6″		8	3.500		3,375	112	19.10	3,506.10	3,925	
	8340	20″ x 12″		8	3.500		3,625	112	19.10	3,756.10	4,200	
	8400	24″ x 12″		6	4.667		4,450	150	25.50	4,625.50	5,150	
	8420	24″ x 18″		6	4.667		6,625	150	25.50	6,800.50	7,550	
	8800	Curb box, 6′ long	B-20	20	1.200		140	37.50		177.50	216	
	8820	8′ long	″	18	1.333		144	41.50		185.50	228	
720	0010	**WATER SUPPLY, CONCRETE PIPE**										720
	0020	Not including excavation or backfill, without gaskets										
	3000	Conc. cylinder pipe (CCP), 150 PSI, 40′ L, 12 diam.	B-13	192	.292	L.F.	55	8.75	3.39	67.14	79	
	3010	24″ diameter	″	128	.438		68.50	13.10	5.10	86.70	103	
	3040	36″ diameter	B-13B	96	.583		85	17.45	9.90	112.35	133	
	3050	48″ diameter		64	.875		116	26	14.85	156.85	187	
	3060	Prestressed (PCCP), 150 PSI, 24′ L, 60″ diameter		60	.933		191	28	15.85	234.85	273	
	3070	72″ diameter		60	.933		239	28	15.85	282.85	325	
	3080	84″ diameter		40	1.400		310	42	24	376	435	
	3090	96″ diameter	B-13C	40	1.400		460	42	40	542	615	
	3100	108″ diameter		32	1.750		630	52.50	49.50	732	835	
	3102	120″ diameter		16	3.500		935	105	99.50	1,139.50	1,300	
	3104	144″ diameter		16	3.500		1,150	105	99.50	1,354.50	1,550	
	3110	Conc. cylinder pipe (CCP), 150 PSI, elbow, 90°, 12″ diameter	B-13	32	1.750	Ea.	192	52.50	20.50	265	320	

	02510 | Water Distribution		CREW	DAILY OUTPUT	LABOR-HOURS	UNIT	2006 BARE COSTS MAT.	LABOR	EQUIP.	TOTAL	TOTAL INCL O&P	
720	3140	24″ diameter	B-13	6	9.333	Ea.	345	279	108	732	955	720
	3150	36″ diameter	B-13B	15	3.733		610	112	63.50	785.50	925	
	3160	48″ diameter		12	4.667		675	140	79.50	894.50	1,050	
	3170	Prestressed (PCCP), 150 PSI, elbow, 90°, 60″ diameter		10	5.600		1,050	168	95	1,313	1,525	
	3180	72″ diameter		4	14		10,700	420	238	11,358	12,700	
	3190	84″ diameter		3.20	17.500		15,200	525	297	16,022	17,900	
	3200	96″ diameter		2	28		29,600	840	475	30,915	34,500	
	3210	108″ diameter	B-13C	1.20	46.667		35,000	1,400	1,325	37,725	42,200	
	3220	120″ diameter		.40	140		36,300	4,200	3,975	44,475	51,000	
	3225	144″ diameter		.30	184		49,400	5,525	5,225	60,150	69,500	
	3230	Concrete cylinder pipe (CCP), 150 PSI, elbow, 45°, 12″ diameter	B-13	24	2.333		161	70	27	258	320	
	3250	24″ diameter	"	6	9.333		345	279	108	732	955	
	3260	36″ diameter	B-13B	4	14		1,850	420	238	2,508	3,000	
	3270	48″ diameter		3	18.667		3,350	560	315	4,225	4,950	
	3280	Prestressed, (PCCP), 150 PSI, elbow, 45°, 60″ diameter		2	28		4,675	840	475	5,990	7,050	
	3290	72″ diameter		1.60	35		8,300	1,050	595	9,945	11,500	
	3300	84″ diameter		1.30	42.945		10,400	1,275	730	12,405	14,300	
	3310	96″ diameter		1	56		12,400	1,675	950	15,025	17,400	
	3320	108″ diameter	B-13C	.66	84.337		15,600	2,525	2,400	20,525	24,000	
	3330	120″ diameter		.40	140		17,600	4,200	3,975	25,775	30,500	
	3340	144″ diameter		.30	184		17,700	5,525	5,225	28,450	34,300	
730	0010	**WATER SUPPLY, DUCTILE IRON PIPE** cement lined R331113-80										730
	0020	Not including excavation or backfill										
	2000	Pipe, class 50 water piping, 18′ lengths										
	2020	Mechanical joint, 4″ diameter	B-21A	200	.200	L.F.	13.50	6.85	2.68	23.03	29	
	2040	6″ diameter		160	.250		15.65	8.60	3.35	27.60	34.50	
	2060	8″ diameter		133.33	.300		17.35	10.30	4.02	31.67	40	
	2080	10″ diameter		114.29	.350		23.50	12	4.69	40.19	50.50	
	2100	12″ diameter		105.26	.380		29	13.05	5.10	47.15	58	
	2120	14″ diameter		100	.400		37	13.75	5.35	56.10	68.50	
	2140	16″ diameter		72.73	.550		40	18.90	7.40	66.30	82	
	2160	18″ diameter		68.97	.580		50.50	19.90	7.80	78.20	96	
	2170	20″ diameter		57.14	.700		59	24	9.40	92.40	114	
	2180	24″ diameter		47.06	.850		75.50	29	11.40	115.90	142	
	3000	Tyton, push-on joint, 4″ diameter		400	.100		7.60	3.44	1.34	12.38	15.35	
	3020	6″ diameter		333.33	.120		8.90	4.12	1.61	14.63	18.10	
	3040	8″ diameter		200	.200		12.05	6.85	2.68	21.58	27	
	3060	10″ diameter		181.82	.220		18.50	7.55	2.95	29	36	
	3080	12″ diameter		160	.250		19.45	8.60	3.35	31.40	39	
	3100	14″ diameter		133.33	.300		21.50	10.30	4.02	35.82	44.50	
	3120	16″ diameter		114.29	.350		30.50	12	4.69	47.19	58	
	3140	18″ diameter		100	.400		34	13.75	5.35	53.10	65	
	3160	20″ diameter		88.89	.450		37	15.45	6.05	58.50	72	
	3180	24″ diameter		76.92	.520		49.50	17.85	6.95	74.30	90.50	
	8000	Fittings, mechanical joint										
	8006	90° bend, 4″ diameter	B-20A	16	2	Ea.	174	67		241	298	
	8020	6″ diameter		12.80	2.500		210	83.50		293.50	365	
	8040	8″ diameter		10.67	2.999		340	100		440	530	
	8060	10″ diameter	B-21A	11.43	3.500		355	120	47	522	640	
	8080	12″ diameter		10.53	3.799		610	130	51	791	940	
	8100	14″ diameter		10	4		965	137	53.50	1,155.50	1,325	
	8120	16″ diameter		7.27	5.502		1,150	189	74	1,413	1,650	
	8140	18″ diameter		6.90	5.797		1,150	199	77.50	1,426.50	1,675	
	8160	20″ diameter		5.71	7.005		2,475	241	94	2,810	3,225	
	8180	24″ diameter		4.70	8.511		4,625	292	114	5,031	5,675	

	02510 Water Distribution		Crew	Daily Output	Labor-Hours	Unit	2006 Bare Costs Mat.	Labor	Equip.	Total	Total Incl O&P	
730	8200	Wye or tee, 4" diameter (R331113-80)	B-20A	10.67	2.999	Ea.	268	100		368	455	730
	8220	6" diameter		8.53	3.751		360	125		485	595	
	8240	8" diameter		7.11	4.501		510	150		660	805	
	8260	10" diameter	B-21A	7.62	5.249		880	180	70.50	1,130.50	1,325	
	8280	12" diameter		7.02	5.698		1,125	196	76.50	1,397.50	1,625	
	8300	14" diameter		6.67	5.997		1,250	206	80.50	1,536.50	1,800	
	8320	16" diameter		4.85	8.247		1,300	283	111	1,694	2,000	
	8340	18" diameter		4.60	8.696		3,375	299	117	3,791	4,325	
	8360	20" diameter		3.81	10.499		5,300	360	141	5,801	6,550	
	8380	24" diameter		3.14	12.739		6,275	440	171	6,886	7,775	
	8398	45° bends, 4" diameter	B-20A	16	2		157	67		224	280	
	8400	6" diameter	"	12.80	2.500		213	83.50		296.50	370	
	8410	12" diameter	B-21A	10.53	3.799		585	130	51	766	910	
	8420	16" diameter		7.27	5.502		585	189	74	848	1,025	
	8430	20" diameter		5.71	7.005		2,375	241	94	2,710	3,125	
	8440	24" diameter		4.70	8.511		3,350	292	114	3,756	4,275	
	8450	Decreaser, 6" x 4" diameter	B-20A	14.22	2.250		191	75		266	330	
	8460	8" x 6" diameter	"	11.64	2.749		279	92		371	450	
	8470	10" x 6 " diameter	B-21A	13.33	3.001		340	103	40	483	585	
	8480	12" x 6" diameter		12.70	3.150		450	108	42	600	715	
	8490	16" x 6" diameter		10	4		850	137	53.50	1,040.50	1,225	
	8500	20" x 6" diameter		8.42	4.751		1,675	163	63.50	1,901.50	2,175	
	8550	Butterfly valves with boxes, cast iron										
	8560	4" diameter	B-20	6	4	Ea.	760	124		884	1,050	
	8580	8" diameter	B-21	4	7		760	225	38	1,023	1,250	
	8590	10" diameter		3.50	8		1,050	257	43.50	1,350.50	1,625	
	8600	12" diameter		3	9.333		1,350	300	51	1,701	2,025	
	8610	14" diameter		2	14		1,875	450	76.50	2,401.50	2,875	
	8620	16" diameter		2	14		2,400	450	76.50	2,926.50	3,475	
	9600	Steel sleeve and tap, 4" diameter	B-20	3	8		430	249		679	880	
	9620	6" diameter		2	12		505	375		880	1,175	
	9630	8" diameter		2	12		680	375		1,055	1,375	
740	0010	**WATER SUPPLY, POLYETHYLENE PIPE, C901** (R331113-80)										740
	0020	Not including excavation or backfill										
	1000	Piping, 160 P.S.I., 3/4" diameter	B-20	525	.046	L.F.	.31	1.42		1.73	2.68	
	1120	1" diameter		485	.049		.71	1.54		2.25	3.32	
	1140	1-1/2" diameter		450	.053		.99	1.66		2.65	3.82	
	1160	2" diameter		365	.066		1.55	2.04		3.59	5.10	
750	0010	**WATER SUPPLY, POLYVINYL CHLORIDE PIPE** (R331113-80)										750
	0020	Not including excavation or backfill, unless specified										
	2100	AWWA Class 160, S.D.R. 26, 1-1/2" diameter	B-20	750	.032	L.F.	.46	1		1.46	2.15	
	2120	2" diameter		686	.035		1.11	1.09		2.20	3.01	
	2140	2-1/2" diameter		500	.048		1.65	1.49		3.14	4.28	
	2160	3" diameter		430	.056		2.36	1.74		4.10	5.45	
	2180	4" diameter		375	.064		3.85	1.99		5.84	7.50	
	2200	6" diameter		316	.076		8.30	2.36		10.66	13.05	
	2210	8" diameter		260	.092		14.10	2.87		16.97	20	
	3010	AWWA C905, PR 100, DR 41										
	3030	14" diameter	B-20A	213	.150	L.F.	12.85	5		17.85	22	
	3040	16" diameter		200	.160		17	5.35		22.35	27.50	
	3050	18" diameter		160	.200		21.50	6.70		28.20	34	
	3060	20" diameter		133	.241		26.50	8.05		34.55	42.50	
	3070	24" diameter		107	.299		37.50	10		47.50	57	
	3080	30" diameter		80	.400		58	13.35		71.35	85	
	3090	36" diameter		80	.400		82	13.35		95.35	112	
	3100	42" diameter		60	.533		112	17.80		129.80	152	

02510 | Water Distribution

	Line	Description	Crew	Daily Output	Labor-Hours	Unit	2006 Bare Costs Mat.	Labor	Equip.	Total	Total Incl O&P
750	3200	48″ diameter (R331113-80)	B-20A	60	.533	L.F.	145	17.80		162.80	189
	4520	Class 150, SDR 18, AWWA C900, 4″		380	.084		2.57	2.81		5.38	7.35
	4530	6″		316	.101		5.10	3.38		8.48	11
	4540	8″		264	.121		8.80	4.05		12.85	16.10
	4550	10″		220	.145		13.30	4.86		18.16	22.50
	4560	12″		186	.172		18.80	5.75		24.55	29.50
	8000	Fittings with rubber gasket									
	8003	Class 150, D.R. 18									
	8006	90° Bend , 4″ diameter	B-20	100	.240	Ea.	51	7.45		58.45	69
	8020	6″ diameter		90	.267		91	8.30		99.30	114
	8040	8″ diameter		80	.300		176	9.35		185.35	209
	8060	10″ diameter		50	.480		269	14.95		283.95	320
	8080	12″ diameter		30	.800		410	25		435	490
	8100	Tee, 4″ diameter		90	.267		70.50	8.30		78.80	91
	8120	6″ diameter		80	.300		158	9.35		167.35	189
	8140	8″ diameter		70	.343		257	10.65		267.65	300
	8160	10″ diameter		40	.600		440	18.65		458.65	515
	8180	12″ diameter		20	1.200		615	37.50		652.50	740
	8200	45° Bend, 4″ diameter		100	.240		50.50	7.45		57.95	68
	8220	6″ diameter		90	.267		89	8.30		97.30	111
	8240	8″ diameter		50	.480		270	14.95		284.95	320
	8260	10″ diameter		50	.480		410	14.95		424.95	480
	8280	12″ diameter		30	.800		545	25		570	640
	8300	Reducing tee 6″x 4″		100	.240		153	7.45		160.45	181
	8320	8″ x 6″		90	.267		249	8.30		257.30	288
	8330	10″ x 6″		90	.267		288	8.30		296.30	330
	8340	10″x 8″		90	.267		310	8.30		318.30	355
	8350	12″ x 6″		90	.267		385	8.30		393.30	440
	8360	12″ x 8″		90	.267		430	8.30		438.30	485
	8400	Tapped service tee (threaded type) 6″ x 6″ x 3/4		100	.240		54.50	7.45		61.95	72.50
	8420	6″ x 6″ x 3/4″		90	.267		54.50	8.30		62.80	73.50
	8430	6″ x 6″ x 1″		90	.267		54.50	8.30		62.80	73.50
	8440	6″ x 6″ x 1 1/2″		90	.267		54.50	8.30		62.80	73.50
	8450	6″ x 6″ x 2″		90	.267		54.50	8.30		62.80	73.50
	8460	8″ x 8″ x 3/4″		90	.267		230	8.30		238.30	267
	8470	8″ x 8″ x 1″		90	.267		230	8.30		238.30	267
	8480	8″ x 8″ x 1 1/2″		90	.267		230	8.30		238.30	267
	8490	8″ x 8″ x 2″		90	.267		230	8.30		238.30	267
	8500	Repair coupling 4″		100	.240		32.50	7.45		39.95	48
	8520	6″ diameter		90	.267		50	8.30		58.30	68.50
	8540	8″ diameter		50	.480		93.50	14.95		108.45	128
	8560	10″ diameter		50	.480		143	14.95		157.95	182
	8580	12″ diameter		50	.480		208	14.95		222.95	254
	8600	Plug end 4″		100	.240		27.50	7.45		34.95	43
	8620	6″ diameter		90	.267		49.50	8.30		57.80	68
	8640	8″ diameter		50	.480		70.50	14.95		85.45	102
	8660	10″ diameter		50	.480		91.50	14.95		106.45	126
	8680	12″ diameter		50	.480		118	14.95		132.95	154
760	0010	**WATER SUPPLY, HDPE**, butt fusion joints, SDR 21, 40′ lengths (R331113-80)									
	0100	4″ diameter	B-22A	400	.095	L.F.	1.99	2.99	1.52	6.50	8.75
	0200	6″ diameter		380	.100		2.86	3.14	1.60	7.60	10.05
	0300	8″ diameter		320	.119		4.84	3.73	1.90	10.47	13.50
	0400	10″ diameter		300	.127		7.50	3.98	2.02	13.50	17
	0500	12″ diameter		260	.146		10.60	4.59	2.34	17.53	21.50
	0600	14″ diameter		220	.173		12.75	5.45	2.76	20.96	26
	0700	16″ diameter		180	.211		16.65	6.65	3.37	26.67	33

02510 | Water Distribution

			Crew	Daily Output	Labor-Hours	Unit	2006 Bare Costs Mat.	Labor	Equip.	Total	Total Incl O&P	
760	0800	18" diameter (R331113-80)	B-22A	140	.271	L.F.	21	8.55	4.34	33.89	41.50	760
	0900	24" diameter		100	.380		37.50	11.95	6.05	55.50	68	
	1000	Fittings										
	1100	Elbows, 90 degrees										
	1200	4" diameter	B-22B	32	.500	Ea.	27.50	16	4.64	48.14	61.50	
	1300	6" diameter		28	.571		72.50	18.25	5.30	96.05	116	
	1400	8" diameter		24	.667		182	21.50	6.20	209.70	242	
	1500	10" diameter		18	.889		315	28.50	8.25	351.75	400	
	1600	12" diameter		12	1.333		530	42.50	12.35	584.85	670	
	1700	14" diameter		9	1.778		530	57	16.50	603.50	690	
	1800	16" diameter		6	2.667		635	85	25	745	865	
	1900	18" diameter		4	4		860	128	37	1,025	1,200	
	2000	24" diameter		3	5.333		1,750	170	49.50	1,969.50	2,250	
	2100	Tees										
	2200	4" diameter	B-22B	30	.533	Ea.	38	17.05	4.95	60	75.50	
	2300	6" diameter		26	.615		101	19.65	5.70	126.35	150	
	2400	8" diameter		22	.727		255	23	6.75	284.75	325	
	2500	10" diameter		15	1.067		335	34	9.90	378.90	435	
	2600	12" diameter		10	1.600		460	51	14.85	525.85	605	
	2700	14" diameter		8	2		545	64	18.55	627.55	720	
	2800	16" diameter		6	2.667		640	85	25	750	870	
	2900	18" diameter		4	4		860	128	37	1,025	1,200	
	3000	24" diameter		2	8		1,775	256	74	2,105	2,475	
780	0010	**WATER SUPPLY, COPPER PIPE**										780
	0020	Not including excavation or backfill										
	2000	Tubing, type K, 20' joints, 3/4" diameter	Q-1	400	.040	L.F.	3.22	1.54		4.76	5.95	
	2200	1" diameter		320	.050		4.57	1.92		6.49	8.05	
	3000	1-1/2" diameter		265	.060		7.05	2.32		9.37	11.35	
	3020	2" diameter		230	.070		10.90	2.67		13.57	16.15	
	3040	2-1/2" diameter		146	.110		16.25	4.21		20.46	24.50	
	3060	3" diameter		134	.119		22.50	4.59		27.09	32	
	4012	4" diameter		95	.168		37.50	6.45		43.95	51.50	
	4016	6" diameter	Q-2	95	.253		114	10.05		124.05	141	
	5000	Tubing, type L										
	5108	2" diameter	Q-1	105	.152	L.F.	9.05	5.85		14.90	19.05	
	6010	3" diameter		140	.114		18.40	4.39		22.79	27.50	
	6012	4" diameter		95	.168		30.50	6.45		36.95	44	
	6016	6" diameter	Q-2	100	.240		106	9.55		115.55	131	
	7020	Fittings, brass, corporation stops, 3/4" diameter	1 Plum	19	.421	Ea.	16.65	18		34.65	46.50	
	7040	1" diameter		16	.500		29	21.50		50.50	65	
	7060	1-1/2" diameter		13	.615		86.50	26.50		113	137	
	7080	2" diameter		11	.727		144	31		175	207	
	7100	Curb stops, 3/4" diameter		19	.421		41	18		59	73	
	7120	1" diameter		16	.500		56.50	21.50		78	95.50	
	7140	1-1/2" diameter		13	.615		126	26.50		152.50	179	
	7160	2" diameter		11	.727		206	31		237	275	
	7180	Curb box, cast iron, 1/2" to 1" curb stops		12	.667		59	28.50		87.50	110	
	7200	1-1/4" to 2" curb stops		8	1		108	42.50		150.50	186	
	7220	Saddles, 3/4" diameter, add					27.50			27.50	30.50	
	7240	2" diameter, add					32.50			32.50	36	

02520 | Wells

			Crew	Daily Output	Labor-Hours	Unit	Mat.	Labor	Equip.	Total	Total Incl O&P	
510	0010	**WELLS & ACCESSORIES**, domestic										510
	0100	Drilled, 4" to 6" diameter	B-23	120	.333	L.F.		9.25	29.50	38.75	48	

02520 | Wells

510		CREW	DAILY OUTPUT	LABOR-HOURS	UNIT	2006 BARE COSTS MAT.	LABOR	EQUIP.	TOTAL	TOTAL INCL O&P
0200	8″ diameter	B-23	95.20	.420	L.F.		11.70	37.50	49.20	60.50
0400	Gravel pack well, 40′ deep, incl. gravel & casing, complete									
0500	24″ diameter casing x 18″ diameter screen	B-23	.13	307	Total	24,800	8,550	27,300	60,650	71,500
0600	36″ diameter casing x 18″ diameter screen		.12	333	″	26,600	9,275	29,600	65,475	77,500
0800	Observation wells, 1-1/4″ riser pipe	↓	163	.245	V.L.F.	13.65	6.80	22	42.45	50.50
0900	For flush Buffalo roadway box, add	1 Skwk	16.60	.482	Ea.	37	17.60		54.60	70
1200	Test well, 2-1/2″ diameter, up to 50′ deep (15 to 50 GPM)	B-23	1.51	26.490	″	555	735	2,350	3,640	4,450
1300	Over 50′ deep, add	″	121.80	.328	L.F.	14.85	9.15	29	53	63.50
1500	Pumps, installed in wells to 100′ deep, 4″ submersible									
1510	1/2 H.P.	Q-1	3.22	4.969	Ea.	345	191		536	675
1520	3/4 H.P.		2.66	6.015		400	231		631	800
1600	1 H.P.	↓	2.29	6.987		425	269		694	880
1700	1-1/2 H.P.	Q-22	1.60	10		1,100	385	400	1,885	2,250
1800	2 H.P.		1.33	12.030		1,275	460	480	2,215	2,650
1900	3 H.P.		1.14	14.035		1,475	540	560	2,575	3,075
2000	5 H.P.		1.14	14.035		2,000	540	560	3,100	3,650
2050	Remove and install motor only, 4 H.P.		1.14	14.035		755	540	560	1,855	2,275
3000	Pump, 6″ submersible, 25′ to 150′ deep, 25 H.P., 249 to 297 GPM		.89	17.978		4,600	690	715	6,005	6,900
3100	25′ to 500′ deep, 30 H.P., 100 to 300 GPM	↓	.73	21.918	↓	5,225	840	870	6,935	8,000
8000	Steel well casing	B-23A	3020	.008	Lb.	.63	.25	1.15	2.03	2.35
8110	Well screen assembly, stainless steel, 2″ diameter		273	.088	L.F.	52	2.74	12.65	67.39	75.50
8120	3″ diameter		253	.095		72	2.96	13.65	88.61	99
8130	4″ diameter		200	.120		82.50	3.74	17.30	103.54	116
8140	5″ diameter		168	.143		96.50	4.45	20.50	121.45	136
8150	6″ diameter		126	.190		115	5.95	27.50	148.45	167
8160	8″ diameter		98.50	.244		150	7.60	35	192.60	216
8170	10″ diameter		73	.329		189	10.25	47.50	246.75	277
8180	12″ diameter		62.50	.384		222	11.95	55.50	289.45	325
8190	14″ diameter		54.30	.442		252	13.80	63.50	329.30	370
8200	16″ diameter		48.30	.497		278	15.50	71.50	365	410
8210	18″ diameter		39.20	.612		345	19.10	88	452.10	510
8220	20″ diameter		31.20	.769		395	24	111	530	595
8230	24″ diameter		23.80	1.008		490	31.50	145	666.50	745
8240	26″ diameter		21	1.143		550	35.50	165	750.50	845
8300	Slotted PVC, 1-1/4″ diameter		521	.046		1.83	1.44	6.65	9.92	11.65
8310	1-1/2″ diameter		488	.049		2.66	1.53	7.10	11.29	13.20
8320	2″ diameter		273	.088		3.68	2.74	12.65	19.07	22.50
8330	3″ diameter		253	.095		3.91	2.96	13.65	20.52	24
8340	4″ diameter		200	.120		4.52	3.74	17.30	25.56	30
8350	5″ diameter		168	.143		4.74	4.45	20.50	29.69	35
8360	6″ diameter		126	.190		6.55	5.95	27.50	40	47
8370	8″ diameter	↓	98.50	.244		9.95	7.60	35	52.55	62
8400	Artificial gravel pack, 2″ screen, 6″ casing	B-23B	174	.138		2.99	4.30	21.50	28.79	34
8405	8″ casing		111	.216		4.09	6.75	33.50	44.34	52
8410	10″ casing		74.50	.322		5.15	10.05	49.50	64.70	76.50
8415	12″ casing		60	.400		6.95	12.45	61.50	80.90	96
8420	14″ casing		50.20	.478		8.95	14.90	73.50	97.35	115
8425	16″ casing		40.70	.590		12.35	18.40	91	121.75	144
8430	18″ casing		36	.667		14.35	21	103	138.35	163
8435	20″ casing		29.50	.814		16.55	25.50	125	167.05	197
8440	24″ casing		25.70	.934		18.35	29	144	191.35	226
8445	26″ casing		24.60	.976		20.50	30.50	150	201	237
8450	30″ casing		20	1.200		23.50	37.50	185	246	290
8455	36″ casing		16.40	1.463	↓	25.50	45.50	225	296	350
8500	Develop well		8	3	Hr.	224	93.50	460	777.50	910
8550	Pump test well	↓	8	3	↓	59.50	93.50	460	613	730

02520 | Wells

		02520 Wells	CREW	DAILY OUTPUT	LABOR-HOURS	UNIT	2006 BARE COSTS MAT.	LABOR	EQUIP.	TOTAL	TOTAL INCL O&P	
510	8560	Standby well	B-23A	8	3	Hr.	58	93.50	430	581.50	690	510
	8570	Standby, drill rig	↓	8	3	↓		93.50	430	523.50	625	
	8580	Surface seal well, concrete filled	↓	1	24	Ea.	565	750	3,450	4,765	5,650	
	8590	Well test pump, install & remove	B-23	1	40			1,100	3,550	4,650	5,750	
	8600	Well sterilization, chlorine	2 Clab	1	16	↓	415	440		855	1,175	
	9000	Minimum labor/equipment charge	B-21	1.80	15.556	Job		500	85	585	910	
	9950	See Div. 02240-900 for wellpoints										
	9960	See Div. 02240-700 for drainage wells										
520	0010	**WATER SUPPLY WELLS, PUMPS** with pressure control										520
	1000	Deep well, jet, 42 gal. galvanized tank										
	1040	3/4 HP	1 Plum	.80	10	Ea.	635	425		1,060	1,350	
	3000	Shallow well, jet, 30 gal. galvanized tank										
	3040	1/2 HP	1 Plum	2	4	Ea.	400	171		571	705	

02530 | Sanitary Sewerage

		02530 Sanitary Sewerage	CREW	DAILY OUTPUT	LABOR-HOURS	UNIT	MAT.	LABOR	EQUIP.	TOTAL	TOTAL INCL O&P	
720	0010	**SEWAGE COLLECTION, VENT CAST IRON PIPE**										720
	0020	Not including excavation or backfill										
	2022	Sewage vent cast iron, B & S, 4″ diameter	Q-1	66	.242	L.F.	8.35	9.30		17.65	23.50	
	2024	5″ diameter	Q-2	88	.273		11.85	10.85		22.70	30	
	2026	6″ diameter	″	84	.286		14.25	11.40		25.65	33.50	
	2028	8″ diameter	Q-3	70	.457		23	18.60		41.60	54	
	2030	10″ diameter		66	.485		37.50	19.75		57.25	72	
	2032	12″ diameter		57	.561		54	23		77	95	
	2034	15″ diameter	↓	49	.653	↓	78	26.50		104.50	128	
	8001	Fittings, bends and elbows										
	8110	4″ diameter	Q-1	13	1.231	Ea.	27	47.50		74.50	104	
	8112	5″ diameter	Q-2	18	1.333		39	53		92	125	
	8114	6″ diameter	″	17	1.412		46.50	56.50		103	139	
	8116	8″ diameter	Q-3	11	2.909		133	118		251	330	
	8118	10″ diameter		10	3.200		196	130		326	420	
	8120	12″ diameter		9	3.556		263	145		408	515	
	8122	15″ diameter	↓	7	4.571	↓	765	186		951	1,125	
	8500	Wyes and tees										
	8510	4″ diameter	Q-1	8	2	Ea.	44	77		121	169	
	8512	5″ diameter	Q-2	12	2		72.50	79.50		152	204	
	8514	6″ diameter	″	11	2.182		93	87		180	237	
	8516	8″ diameter	Q-3	7	4.571		223	186		409	535	
	8518	10″ diameter		6	5.333		355	217		572	730	
	8520	12″ diameter		4	8		585	325		910	1,150	
	8522	15″ diameter	↓	3	10.667	↓	1,175	435		1,610	1,975	
730	0010	**SEWAGE COLLECTION, CONCRETE PIPE**										730
	0020	See 02630-530 for sewage/drainage collection, concrete pipe										
770	0010	**SEWAGE COLLECTION, PLASTIC PIPE**										770
	0020	Not including excavation & backfill										
	1100	Piping, DWV Sch 40 ABS, 4″ diameter	B-20	375	.064	L.F.	1.77	1.99		3.76	5.20	
	1110	6″ diameter		350	.069	″	15.10	2.13		17.23	20	
	1120	Fitting, 1/4 bend, 4″		19	1.263	Ea.	22.50	39.50		62	89.50	
	1130	6″		15	1.600		15.10	50		65.10	98.50	
	1140	Tee, 4″		12	2	↓	22.50	62		84.50	128	
	3000	Piping, HDPE Corrugated Type S with watertight gaskets, 4″ diameter		425	.056	L.F.	.90	1.76		2.66	3.89	
	3020	6″ diameter		400	.060		2.06	1.87		3.93	5.35	
	3040	8″ diameter		380	.063		3.94	1.96		5.90	7.60	
	3060	10″ diameter		370	.065		5.45	2.02		7.47	9.35	
	3080	12″ diameter	↓	340	.071	↓	6.10	2.20		8.30	10.30	

02530 | Sanitary Sewerage

		02530 Sanitary Sewerage	CREW	DAILY OUTPUT	LABOR-HOURS	UNIT	2006 BARE COSTS MAT.	LABOR	EQUIP.	TOTAL	TOTAL INCL O&P	
770	3100	15″ diameter	B-20	300	.080	L.F.	8.25	2.49		10.74	13.20	770
	3120	18″ diameter	B-21	275	.102		11.80	3.27	.55	15.62	18.90	
	3140	24″ diameter		250	.112		18.25	3.60	.61	22.46	26.50	
	3160	30″ diameter		200	.140		29	4.49	.76	34.25	40	
	3180	36″ diameter		180	.156		36.50	4.99	.85	42.34	49.50	
	3200	42″ diameter		175	.160		51.50	5.15	.87	57.52	66	
	3220	48″ diameter		170	.165		66.50	5.30	.90	72.70	82.50	
	3240	54″ diameter		160	.175		103	5.60	.95	109.55	123	
	3260	60″ diameter	↓	150	.187	↓	120	6	1.02	127.02	143	
	3300	Watertight elbows 12″ diam	B-20	11	2.182	Ea.	67.50	68		135.50	186	
	3320	15″ diam	″	9	2.667		104	83		187	251	
	3340	18″ diam	B-21	9	3.111		171	100	16.95	287.95	370	
	3360	24″ diam		9	3.111		365	100	16.95	481.95	580	
	3380	30″ diam		8	3.500		580	112	19.10	711.10	845	
	3400	36″ diam		8	3.500		745	112	19.10	876.10	1,025	
	3420	42″ diam		6	4.667		945	150	25.50	1,120.50	1,325	
	3440	48″ diam	↓	6	4.667		1,450	150	25.50	1,625.50	1,850	
	3460	Watertight tee 12″ diam	B-20	7	3.429		151	107		258	340	
	3480	15″ diam	″	6	4		180	124		304	405	
	3500	18″ diam	B-21	6	4.667		264	150	25.50	439.50	565	
	3520	24″ diam		5	5.600		345	180	30.50	555.50	710	
	3540	30″ diam		5	5.600		655	180	30.50	865.50	1,050	
	3560	36″ diam		4	7		850	225	38	1,113	1,350	
	3580	42″ diam		4	7		1,475	225	38	1,738	2,000	
	3600	48″ diam	↓	4	7	↓	2,425	225	38	2,688	3,050	
780	0010	**SEWAGE COLLECTION, POLYVINYL CHLORIDE PIPE**										780
	0020	Not including excavation or backfill										
	2000	10′ lengths, S.D.R. 35, B&S, 4″ diameter	B-20	375	.064	L.F.	2.53	1.99		4.52	6.05	
	2040	6″ diameter		350	.069		5.30	2.13		7.43	9.35	
	2080	8″ diameter	↓	335	.072		8.05	2.23		10.28	12.50	
	2120	10″ diameter	B-21	330	.085		11.65	2.72	.46	14.83	17.75	
	2160	12″ diameter		320	.088		11.55	2.81	.48	14.84	17.85	
	2200	15″ diameter	↓	190	.147	↓	16.95	4.73	.80	22.48	27.50	
	3040	Fittings, bends or elbows, 4″ diameter	B-20	19	1.263	Ea.	5.35	39.50		44.85	71	
	3080	6″ diameter		15	1.600		16.95	50		66.95	101	
	3120	Tees, 4″ diameter	↓	12	2		5.55	62		67.55	109	
	3160	6″ diameter	2 Skwk	10	1.600		23	58.50		81.50	121	
	3200	Wyes, 4″ diameter	″	12	1.333		7.35	48.50		55.85	87.50	
	3240	6″ diameter	B-20	10	2.400	↓	31	74.50		105.50	158	
	4000	Piping, DWV PVC,no exc/bkfill, 10′ L, Sch 40, 4″ dia		375	.064	L.F.	2.05	1.99		4.04	5.55	
	4010	6″ dia		350	.069		4.43	2.13		6.56	8.40	
	4020	8″ dia	↓	335	.072	↓	13.15	2.23		15.38	18.10	

02540 | Septic Tank Systems

		02540 Septic Tank Systems	CREW	DAILY OUTPUT	LABOR-HOURS	UNIT	MAT.	LABOR	EQUIP.	TOTAL	TOTAL INCL O&P	
400	0010	**SEPTIC TANKS**										400
	0015	Septic tanks, not incl exc or piping, precast, 1,000 gal	B-21	8	3.500	Ea.	585	112	19.10	716.10	850	
	0020	1,250 gallon		8	3.500		820	112	19.10	951.10	1,100	
	0060	1,500 gallon		7	4		965	128	22	1,115	1,275	
	0100	2,000 gallon		5	5.600		1,175	180	30.50	1,385.50	1,600	
	0140	2,500 gallon		5	5.600		1,550	180	30.50	1,760.50	2,025	
	0180	4,000 gallon	↓	4	7		4,525	225	38	4,788	5,375	
	0220	5,000 gal., 4 piece	B-13	3	18.667		7,225	560	217	8,002	9,100	
	0300	15,000 gallon, 4 piece	B-13B	1.70	32.941		16,400	985	560	17,945	20,200	
	0400	25,000 gallon, 4 piece	↓	1.10	50.909	↓	34,600	1,525	865	36,990	41,500	

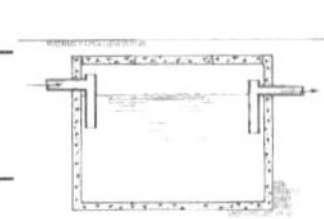

02500 | Utility Services

02540 | Septic Tank Systems

			CREW	DAILY OUTPUT	LABOR-HOURS	UNIT	2006 BARE COSTS MAT.	LABOR	EQUIP.	TOTAL	TOTAL INCL O&P	
400	0500	40,000 gallon, 4 piece	B-13B	.80	70	Ea.	40,800	2,100	1,200	44,100	49,500	400
	0520	50,000 gallon, 5 piece	B-13C	.60	93.333		46,900	2,800	2,650	52,350	59,000	
	0540	75,000 gallon, cast in place	C-14C	.25	448		57,000	15,200	85	72,285	88,500	
	0560	100,000 gallon	"	.15	746		70,500	25,300	142	95,942	119,500	
	0600	High density polyethylene, 1,000 gallon	B-21	6	4.667		910	150	25.50	1,085.50	1,275	
	0700	1,500 gallon		4	7		1,175	225	38	1,438	1,675	
	0900	Galley, 4′ x 4′ x 4′	↓	16	1.750		216	56	9.55	281.55	340	
	1000	Distribution boxes, concrete, 7 outlets	2 Clab	16	1		109	27.50		136.50	166	
	1100	9 outlets	"	8	2		295	55		350	415	
	1150	Leaching field chambers, 13′ x 3′-7″ x 1′-4″, standard	B-13	16	3.500		590	105	40.50	735.50	865	
	1200	Heavy duty, 8′ x 4′ x 1′-6″		14	4		410	120	46.50	576.50	695	
	1300	13′ x 3′-9″ x 1′-6″		12	4.667		1,450	140	54	1,644	1,875	
	1350	20′ x 4′ x 1′-6″	↓	5	11.200		955	335	130	1,420	1,750	
	1600	Leaching pit, 6′-6″ diameter, 6′ deep	B-21	5	5.600		655	180	30.50	865.50	1,050	
	1620	8′ deep		4	7		770	225	38	1,033	1,250	
	1700	8′ diameter, H-20 load, 6′ deep		4	7		1,050	225	38	1,313	1,575	
	1720	8′ deep		3	9.333		1,300	300	51	1,651	2,000	
	2000	Velocity reducing pit, precast conc., 6′ diameter, 3′ deep	↓	4.70	5.957	↓	325	191	32.50	548.50	710	
	2200	Excavation for septic tank, 3/4 C.Y. backhoe	B-12F	145	.110	C.Y.		3.61	3.29	6.90	9.40	
	2400	4′ trench for disposal field, 3/4 C.Y. backhoe	"	335	.048	L.F.		1.56	1.43	2.99	4.09	
	2600	Gravel fill, run of bank	B-6	150	.160	C.Y.	16.50	4.80	1.51	22.81	27.50	
	2800	Crushed stone, 3/4″	"	150	.160	"	23	4.80	1.51	29.31	34.50	

02550 | Piped Energy Distribution

			CREW	DAILY OUTPUT	LABOR-HOURS	UNIT	MAT.	LABOR	EQUIP.	TOTAL	TOTAL INCL O&P	
200	0010	PIPE CONDUIT, PREFABRICATED / PREINSULATED										200
	0020	Does not include trenching, fittings or crane.										
	0300	For cathodic protection, add 12 to 14%										
	0310	of total built-up price (casing plus service pipe)										
	0580	Polyurethane insulated system, 250° F. max. temp.										
	0620	Black steel service pipe, standard wt., 1/2″ insulation										
	0660	3/4″ diam. pipe size	Q-17	54	.296	L.F.	26.50	11.50	1.28	39.28	48.50	
	0670	1″ diam. pipe size		50	.320		29	12.40	1.38	42.78	53	
	0680	1-1/4″ diam. pipe size		47	.340		32.50	13.20	1.47	47.17	57.50	
	0690	1-1/2″ diam. pipe size 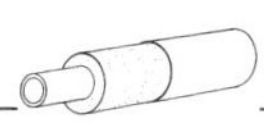		45	.356		35	13.80	1.54	50.34	61.50	
	0700	2″ diam. pipe size		42	.381		36.50	14.75	1.65	52.90	65	
	0710	2-1/2″ diam. pipe size		34	.471		37	18.25	2.03	57.28	71	
	0720	3″ diam. pipe size		28	.571		43	22	2.47	67.47	84	
	0730	4″ diam. pipe size		22	.727		54	28	3.14	85.14	107	
	0740	5″ diam. pipe size	↓	18	.889		69	34.50	3.84	107.34	134	
	0750	6″ diam. pipe size	Q-18	23	1.043		80	42	3.01	125.01	156	
	0760	8″ diam. pipe size		19	1.263		117	51	3.64	171.64	212	
	0770	10″ diam. pipe size		16	1.500		155	60.50	4.32	219.82	268	
	0780	12″ diam. pipe size		13	1.846		192	74	5.30	271.30	330	
	0790	14″ diam. pipe size		11	2.182		214	87.50	6.30	307.80	380	
	0800	16″ diam. pipe size		10	2.400		246	96.50	6.90	349.40	430	
	0810	18″ diam. pipe size		8	3		283	121	8.65	412.65	505	
	0820	20″ diam. pipe size		7	3.429		315	138	9.85	462.85	570	
	0830	24″ diam. pipe size	↓	6	4		385	161	11.50	557.50	690	
	0900	For 1″ thick insulation, add					10%					
	0940	For 1-1/2″ thick insulation, add					13%					
	0980	For 2″ thick insulation, add				↓	20%					
	1500	Gland seal for system, 3/4″ diam. pipe size	Q-17	32	.500	Ea.	430	19.40	2.16	451.56	500	
	1510	1″ diam. pipe size		32	.500		430	19.40	2.16	451.56	500	
	1540	1-1/4″ diam. pipe size		30	.533		460	20.50	2.30	482.80	540	
	1550	1-1/2″ diam. pipe size		30	.533		460	20.50	2.30	482.80	540	
	1560	2″ diam. pipe size	↓	28	.571	↓	545	22	2.47	569.47	635	

02550 | Piped Energy Distribution		CREW	DAILY OUTPUT	LABOR-HOURS	UNIT	2006 BARE COSTS MAT.	LABOR	EQUIP.	TOTAL	TOTAL INCL O&P
200 1570	2-1/2" diam. pipe size	Q-17	26	.615	Ea.	585	24	2.66	611.66	685 200
1580	3" diam. pipe size		24	.667		630	26	2.88	658.88	740
1590	4" diam. pipe size		22	.727		745	28	3.14	776.14	865
1600	5" diam. pipe size	↓	19	.842		915	32.50	3.64	951.14	1,050
1610	6" diam. pipe size	Q-18	26	.923		975	37	2.66	1,014.66	1,125
1620	8" diam. pipe size		25	.960		1,125	38.50	2.76	1,166.26	1,325
1630	10" diam. pipe size		23	1.043		1,375	42	3.01	1,420.01	1,600
1640	12" diam. pipe size		21	1.143		1,525	46	3.29	1,574.29	1,750
1650	14" diam. pipe size		19	1.263		1,700	51	3.64	1,754.64	1,950
1660	16" diam. pipe size		18	1.333		2,000	53.50	3.84	2,057.34	2,300
1670	18" diam. pipe size		16	1.500		2,125	60.50	4.32	2,189.82	2,450
1680	20" diam. pipe size		14	1.714		2,425	69	4.94	2,498.94	2,775
1690	24" diam. pipe size	↓	12	2	↓	2,700	80.50	5.75	2,786.25	3,100
2000	Elbow, 45° for system									
2020	3/4" diam. pipe size	Q-17	14	1.143	Ea.	270	44.50	4.94	319.44	370
2040	1" diam. pipe size		13	1.231		278	47.50	5.30	330.80	385
2050	1-1/4" diam. pipe size		11	1.455		315	56.50	6.30	377.80	440
2060	1-1/2" diam. pipe size		9	1.778		330	69	7.70	406.70	480
2070	2" diam. pipe size		6	2.667		345	103	11.50	459.50	555
2080	2-1/2" diam. pipe size		4	4		375	155	17.30	547.30	670
2090	3" diam. pipe size		3.50	4.571		430	177	19.75	626.75	770
2100	4" diam. pipe size		3	5.333		500	207	23	730	895
2110	5" diam. pipe size	↓	2.80	5.714		645	221	24.50	890.50	1,075
2120	6" diam. pipe size	Q-18	4	6		730	241	17.30	988.30	1,200
2130	8" diam. pipe size		3	8		1,050	320	23	1,393	1,700
2140	10" diam. pipe size		2.40	10		1,350	400	29	1,779	2,125
2150	12" diam. pipe size		2	12		1,775	480	34.50	2,289.50	2,750
2160	14" diam. pipe size		1.80	13.333		2,200	535	38.50	2,773.50	3,300
2170	16" diam. pipe size		1.60	15		2,625	605	43	3,273	3,875
2180	18" diam. pipe size		1.30	18.462		3,300	740	53	4,093	4,825
2190	20" diam. pipe size		1	24		4,150	965	69	5,184	6,150
2200	24" diam. pipe size	↓	.70	34.286		5,225	1,375	98.50	6,698.50	8,000
2260	For elbow, 90°, add					25%				
2300	For tee, straight, add					85%	30%			
2340	For tee, reducing, add					170%	30%			
2380	For weldolet, straight, add				↓	50%				
2800	Calcium silicate insulated system, high temp. (1200° F)									
2840	Steel casing with protective exterior coating									
2850	6-5/8" diameter	Q-18	52	.462	L.F.	49.50	18.55	1.33	69.38	85
2860	8-5/8" diameter 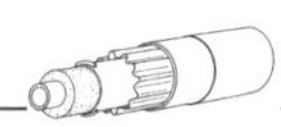		50	.480		54	19.30	1.38	74.68	91
2870	10-3/4" diameter		47	.511		63	20.50	1.47	84.97	103
2880	12-3/4" diameter		44	.545		69	22	1.57	92.57	111
2890	14" diameter		41	.585		77.50	23.50	1.69	102.69	123
2900	16" diameter		39	.615		83	24.50	1.77	109.27	132
2910	18" diameter		36	.667		91.50	27	1.92	120.42	145
2920	20" diameter		34	.706		103	28.50	2.03	133.53	160
2930	22" diameter 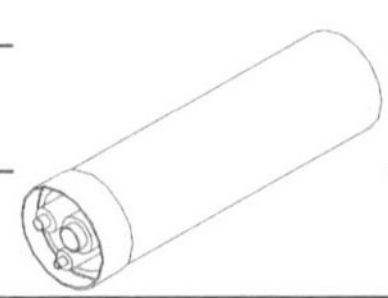		32	.750		143	30	2.16	175.16	207
2940	24" diameter		29	.828		163	33.50	2.38	198.88	234
2950	26" diameter		26	.923		186	37	2.66	225.66	265
2960	28" diameter		23	1.043		232	42	3.01	277.01	325
2970	30" diameter		21	1.143		247	46	3.29	296.29	345
2980	32" diameter		19	1.263		278	51	3.64	332.64	390
2990	34" diameter		18	1.333		281	53.50	3.84	338.34	400
3000	36" diameter	↓	16	1.500		300	60.50	4.32	364.82	430
3040	For multi-pipe casings, add					10%				
3060	For oversize casings, add				↓	2%				

200	02550 | Piped Energy Distribution	CREW	DAILY OUTPUT	LABOR-HOURS	UNIT	2006 BARE COSTS MAT.	LABOR	EQUIP.	TOTAL	TOTAL INCL O&P
3400	Steel casing gland seal, single pipe									
3420	6-5/8" diameter	Q-18	25	.960	Ea.	690	38.50	2.76	731.26	820
3440	8-5/8" diameter		23	1.043		805	42	3.01	850.01	955
3450	10-3/4" diameter		21	1.143		905	46	3.29	954.29	1,075
3460	12-3/4" diameter		19	1.263		1,075	51	3.64	1,129.64	1,250
3470	14" diameter		17	1.412		1,175	56.50	4.07	1,235.57	1,400
3480	16" diameter		16	1.500		1,375	60.50	4.32	1,439.82	1,625
3490	18" diameter		15	1.600		1,525	64.50	4.61	1,594.11	1,775
3500	20" diameter		13	1.846		1,700	74	5.30	1,779.30	1,975
3510	22" diameter		12	2		1,900	80.50	5.75	1,986.25	2,225
3520	24" diameter		11	2.182		2,125	87.50	6.30	2,218.80	2,500
3530	26" diameter		10	2.400		2,425	96.50	6.90	2,528.40	2,825
3540	28" diameter		9.50	2.526		2,750	102	7.30	2,859.30	3,200
3550	30" diameter		9	2.667		2,800	107	7.70	2,914.70	3,275
3560	32" diameter		8.50	2.824		3,150	113	8.15	3,271.15	3,650
3570	34" diameter		8	3		3,450	121	8.65	3,579.65	3,975
3580	36" diameter	↓	7	3.429		3,650	138	9.85	3,797.85	4,250
3620	For multi-pipe casings, add				↓	5%				
4000	Steel casing anchors, single pipe									
4020	6-5/8" diameter	Q-18	8	3	Ea.	615	121	8.65	744.65	875
4040	8-5/8" diameter		7.50	3.200		645	129	9.20	783.20	920
4050	10-3/4" diameter		7	3.429		845	138	9.85	992.85	1,150
4060	12-3/4" diameter		6.50	3.692		905	148	10.65	1,063.65	1,250
4070	14" diameter		6	4		1,050	161	11.50	1,222.50	1,450
4080	16" diameter		5.50	4.364		1,225	175	12.55	1,412.55	1,625
4090	18" diameter		5	4.800		1,400	193	13.80	1,606.80	1,850
4100	20" diameter		4.50	5.333		1,525	214	15.35	1,754.35	2,025
4110	22" diameter		4	6		1,700	241	17.30	1,958.30	2,275
4120	24" diameter		3.50	6.857		1,875	276	19.75	2,170.75	2,500
4130	26" diameter		3	8		2,100	320	23	2,443	2,850
4140	28" diameter		2.50	9.600		2,300	385	27.50	2,712.50	3,175
4150	30" diameter		2	12		2,450	480	34.50	2,964.50	3,500
4160	32" diameter		1.50	16		2,950	645	46	3,641	4,275
4170	34" diameter		1	24		3,300	965	69	4,334	5,200
4180	36" diameter	↓	1	24		3,575	965	69	4,609	5,525
4220	For multi-pipe, add				↓	5%	20%			
4800	Steel casing elbow									
4820	6-5/8" diameter	Q-18	15	1.600	Ea.	845	64.50	4.61	914.11	1,025
4830	8-5/8" diameter		15	1.600		905	64.50	4.61	974.11	1,100
4850	10-3/4" diameter		14	1.714		1,100	69	4.94	1,173.94	1,300
4860	12-3/4" diameter		13	1.846		1,300	74	5.30	1,379.30	1,550
4870	14" diameter		12	2		1,375	80.50	5.75	1,461.25	1,650
4880	16" diameter		11	2.182		1,500	87.50	6.30	1,593.80	1,800
4890	18" diameter		10	2.400		1,725	96.50	6.90	1,828.40	2,050
4900	20" diameter		9	2.667		1,825	107	7.70	1,939.70	2,200
4910	22" diameter		8	3		1,975	121	8.65	2,104.65	2,375
4920	24" diameter		7	3.429		2,175	138	9.85	2,322.85	2,625
4930	26" diameter		6	4		2,375	161	11.50	2,547.50	2,900
4940	28" diameter		5	4.800		2,575	193	13.80	2,781.80	3,150
4950	30" diameter		4	6		2,575	241	17.30	2,833.30	3,250
4960	32" diameter		3	8		2,950	320	23	3,293	3,750
4970	34" diameter		2	12		3,225	480	34.50	3,739.50	4,350
4980	36" diameter	↓	2	12	↓	3,450	480	34.50	3,964.50	4,575
5500	Black steel service pipe, std. wt., 1" thick insulation									
5510	3/4" diameter pipe size	Q-17	54	.296	L.F.	21.50	11.50	1.28	34.28	43
5540	1" diameter pipe size	↓	50	.320	↓	22.50	12.40	1.38	36.28	45.50

	02550 | Piped Energy Distribution		CREW	DAILY OUTPUT	LABOR-HOURS	UNIT	2006 BARE COSTS MAT.	LABOR	EQUIP.	TOTAL	TOTAL INCL O&P	
200	5550	1-1/4" diameter pipe size	Q-17	47	.340	L.F.	25	13.20	1.47	39.67	49.50	200
	5560	1-1/2" diameter pipe size		45	.356		27.50	13.80	1.54	42.84	53	
	5570	2" diameter pipe size		42	.381		30.50	14.75	1.65	46.90	58.50	
	5580	2-1/2" diameter pipe size		34	.471		32	18.25	2.03	52.28	65.50	
	5590	3" diameter pipe size		28	.571		36.50	22	2.47	60.97	77	
	5600	4" diameter pipe size		22	.727		46.50	28	3.14	77.64	99	
	5610	5" diameter pipe size		18	.889		63	34.50	3.84	101.34	127	
	5620	6" diameter pipe size	Q-18	23	1.043		69	42	3.01	114.01	144	
	6000	Black steel service pipe, std. wt., 1-1/2" thick insul.										
	6010	3/4" diameter pipe size	Q-17	54	.296	L.F.	22	11.50	1.28	34.78	43.50	
	6040	1" diameter pipe size		50	.320		24.50	12.40	1.38	38.28	48	
	6050	1-1/4" diameter pipe size		47	.340		27.50	13.20	1.47	42.17	52	
	6060	1-1/2" diameter pipe size		45	.356		30	13.80	1.54	45.34	56	
	6070	2" diameter pipe size		42	.381		32.50	14.75	1.65	48.90	60.50	
	6080	2-1/2" diameter pipe size		34	.471		34.50	18.25	2.03	54.78	68.50	
	6090	3" diameter pipe size		28	.571		38.50	22	2.47	62.97	79.50	
	6100	4" diameter pipe size		22	.727		49.50	28	3.14	80.64	102	
	6110	5" diameter pipe size		18	.889		63	34.50	3.84	101.34	127	
	6120	6" diameter pipe size	Q-18	23	1.043		71.50	42	3.01	116.51	147	
	6130	8" diameter pipe size		19	1.263		103	51	3.64	157.64	197	
	6140	10" diameter pipe size		16	1.500		135	60.50	4.32	199.82	246	
	6150	12" diameter pipe size		13	1.846		161	74	5.30	240.30	298	
	6190	For 2" thick insulation, add					15%					
	6220	For 2-1/2" thick insulation, add					25%					
	6260	For 3" thick insulation, add					30%					
	6800	Black steel service pipe, ex. hvy. wt., 1" thick insul.										
	6820	3/4" diameter pipe size	Q-17	50	.320	L.F.	22.50	12.40	1.38	36.28	46	
	6840	1" diameter pipe size		47	.340		24	13.20	1.47	38.67	48.50	
	6850	1-1/4" diameter pipe size		44	.364		28	14.10	1.57	43.67	54	
	6860	1-1/2" diameter pipe size		42	.381		29.50	14.75	1.65	45.90	57.50	
	6870	2" diameter pipe size		40	.400		31.50	15.50	1.73	48.73	60.50	
	6880	2-1/2" diameter pipe size		31	.516		38.50	20	2.23	60.73	76	
	6890	3" diameter pipe size		27	.593		43.50	23	2.56	69.06	86.50	
	6900	4" diameter pipe size		21	.762		57.50	29.50	3.29	90.29	113	
	6910	5" diameter pipe size		17	.941		80.50	36.50	4.07	121.07	149	
	6920	6" diameter pipe size	Q-18	22	1.091		89	44	3.14	136.14	169	
	7400	Black steel service pipe, ex. hvy. wt., 1-1/2" thick insul.										
	7420	3/4" diameter pipe size	Q-17	50	.320	L.F.	22.50	12.40	1.38	36.28	46	
	7440	1" diameter pipe size		47	.340		26	13.20	1.47	40.67	50.50	
	7450	1-1/4" diameter pipe size		44	.364		30	14.10	1.57	45.67	56.50	
	7460	1-1/2" diameter pipe size		42	.381		33	14.75	1.65	49.40	61	
	7470	2" diameter pipe size		40	.400		33	15.50	1.73	50.23	62.50	
	7480	2-1/2" diameter pipe size		31	.516		39.50	20	2.23	61.73	76.50	
	7490	3" diameter pipe size		27	.593		46	23	2.56	71.56	89.50	
	7500	4" diameter pipe size		21	.762		60	29.50	3.29	92.79	116	
	7510	5" diameter pipe size		17	.941		83	36.50	4.07	123.57	152	
	7520	6" diameter pipe size	Q-18	22	1.091		91.50	44	3.14	138.64	172	
	7530	8" diameter pipe size		18	1.333		138	53.50	3.84	195.34	239	
	7540	10" diameter pipe size		15	1.600		163	64.50	4.61	232.11	285	
	7550	12" diameter pipe size		13	1.846		201	74	5.30	280.30	340	
	7590	For 2" thick insulation, add					13%					
	7640	For 2-1/2" thick insulation, add					18%					
	7680	For 3" thick insulation, add					24%					
464	0010	**PIPING, GAS SERVICE & DISTRIBUTION, POLYETHYLENE**										464
	0020	not including excavation or backfill										

02550 | Piped Energy Distribution

		02550 Piped Energy Distribution	CREW	DAILY OUTPUT	LABOR-HOURS	UNIT	MAT.	LABOR	EQUIP.	TOTAL	TOTAL INCL O&P	
							2006 BARE COSTS					
464	1000	60 psi coils, comp cplg @ 100′, 1/2″ diameter, SDR 9.3	B-20A	608	.053	L.F.	.51	1.76		2.27	3.37	464
	1040	1-1/4″ diameter, SDR 11		544	.059		.95	1.97		2.92	4.18	
	1100	2″ diameter, SDR 11		488	.066		1.17	2.19		3.36	4.79	
	1160	3″ diameter, SDR 11		408	.078		2.43	2.62		5.05	6.85	
	1500	60 PSI 40′ joints with coupling, 3″ diameter, SDR 11	B-21A	408	.098		2.43	3.37	1.31	7.11	9.45	
	1540	4″ diameter, SDR 11		352	.114		5.60	3.90	1.52	11.02	14.05	
	1600	6″ diameter, SDR 11		328	.122		17.35	4.19	1.64	23.18	27.50	
	1640	8″ diameter, SDR 11		272	.147		23.50	5.05	1.97	30.52	36	
	9000	Minimum labor/equipment charge	B-20	2	12	Job		375		375	615	
466	0010	**PIPING, GAS SERVICE & DISTRIBUTION, STEEL**										466
	0020	not including excavation or backfill, tar coated and wrapped										
	4000	Schedule 40, plain end										
	4040	1″ diameter	Q-4	300	.107	L.F.	2.85	4.34	.23	7.42	10.15	
	4080	2″ diameter		280	.114		4.47	4.65	.25	9.37	12.45	
	4120	3″ diameter		260	.123		7.40	5	.27	12.67	16.25	
	4160	4″ diameter	B-35	255	.188		9.65	6.50	2.14	18.29	23.50	
	4200	5″ diameter		220	.218		13.95	7.50	2.49	23.94	30	
	4240	6″ diameter		180	.267		17.10	9.20	3.04	29.34	37	
	4280	8″ diameter		140	.343		27	11.80	3.91	42.71	53.50	
	4320	10″ diameter		100	.480		93.50	16.55	5.45	115.50	136	
	4360	12″ diameter		80	.600		117	20.50	6.85	144.35	170	
	4400	14″ diameter		75	.640		125	22	7.30	154.30	181	
	4440	16″ diameter		70	.686		137	23.50	7.80	168.30	197	
	4480	18″ diameter		65	.738		176	25.50	8.40	209.90	243	
	4520	20″ diameter		60	.800		273	27.50	9.10	309.60	355	
	4560	24″ diameter		50	.960		310	33	10.95	353.95	410	
	5000	Threaded and coupled										
	5002	4″ diameter	B-20	144	.167	L.F.	27	5.20		32.20	38.50	
	5008	8″ diameter	B-21	108	.259		87.50	8.30	1.41	97.21	112	
	5012	12″ diameter	″	72	.389		126	12.50	2.12	140.62	162	
	6000	Schedule 80, plain end										
	6002	4″ diameter	B-35	144	.333	L.F.	42	11.50	3.80	57.30	69	
	6006	6″ diameter		126	.381		70	13.10	4.34	87.44	103	
	6008	8″ diameter		108	.444		93.50	15.30	5.05	113.85	133	
	6012	12″ diameter		72	.667		187	23	7.60	217.60	251	
	8008	Elbow, weld joint, standard weight										
	8020	4″ diameter	Q-16	6.80	3.529	Ea.	62.50	141	10.15	213.65	299	
	8026	8″ diameter		3.40	7.059		259	281	20.50	560.50	745	
	8030	12″ diameter		2.30	10.435		580	415	30	1,025	1,325	
	8034	16″ diameter		1.50	16		1,100	640	46	1,786	2,250	
	8038	20″ diameter		1.20	20		2,075	795	57.50	2,927.50	3,625	
	8040	24″ diameter		1.02	23.529		2,750	940	68	3,758	4,575	
	8100	Extra heavy										
	8102	4″ diameter	Q-16	5.30	4.528	Ea.	125	180	13.05	318.05	435	
	8108	8″ diameter		2.60	9.231		390	370	26.50	786.50	1,025	
	8112	12″ diameter		1.80	13.333		775	530	38.50	1,343.50	1,725	
	8116	16″ diameter		1.20	20		1,475	795	57.50	2,327.50	2,950	
	8120	20″ diameter		.94	25.532		2,775	1,025	73.50	3,873.50	4,700	
	8122	24″ diameter		.80	30		3,675	1,200	86.50	4,961.50	6,000	
	8200	Malleable, standard weight										
	8202	4″ diameter	B-20	12	2	Ea.	228	62		290	355	
	8208	8″ diameter		6	4		625	124		749	895	
	8212	12″ diameter		4	6		690	187		877	1,075	
	8300	Extra heavy										
	8302	4″ diameter	B-20	12	2	Ea.	455	62		517	610	

	02550 | Piped Energy Distribution		CREW	DAILY OUTPUT	LABOR-HOURS	UNIT	2006 BARE COSTS MAT.	LABOR	EQUIP.	TOTAL	TOTAL INCL O&P	
466	8308	8" diameter	B-21	6	4.667	Ea.	785	150	25.50	960.50	1,150	466
	8312	12" diameter	"	4	7	↓	920	225	38	1,183	1,400	
	8500	Tee weld, standard weight										
	8510	4" diameter	Q-16	4.50	5.333	Ea.	101	213	15.35	329.35	460	
	8514	6" diameter		3	8		173	320	23	516	710	
	8516	8" diameter		2.30	10.435		305	415	30	750	1,025	
	8520	12" diameter		1.50	16		840	640	46	1,526	1,975	
	8524	16" diameter		1	24		1,650	955	69	2,674	3,375	
	8528	20" diameter		.80	30		4,050	1,200	86.50	5,336.50	6,400	
	8530	24" diameter	↓	.70	34.286	↓	5,225	1,375	98.50	6,698.50	7,975	
	8810	Malleable, standard weight										
	8812	4" diameter	B-20	8	3	Ea.	390	93.50		483.50	585	
	8818	8" diameter	B-21	4	7		670	225	38	933	1,150	
	8822	12" diameter	"	2.70	10.370	↓	1,275	335	56.50	1,666.50	2,000	
	8900	Extra heavy										
	8902	4" diameter	B-20	8	3	Ea.	585	93.50		678.50	800	
	8908	8" diameter	B-21	4	7		835	225	38	1,098	1,325	
	8912	12" diameter	"	2.70	10.370	↓	1,275	335	56.50	1,666.50	2,000	
468	0010	**PIPING, VALVES & METERS, GAS DISTRIBUTION**										468
	0020	not including excavation or backfill										
	0100	Gas stops, with or without checks										
	0140	1-1/4" size	1 Plum	12	.667	Ea.	49	28.50		77.50	98.50	
	0180	1-1/2" size		10	.800		66	34		100	126	
	0200	2" size	↓	8	1	↓	99.50	42.50		142	176	
	0600	Pressure regulator valves, iron and bronze										
	0640	1-1/2" diameter	1 Plum	13	.615	Ea.	116	26.50		142.50	169	
	0680	2" diameter	"	11	.727		224	31		255	295	
	0700	3" diameter	Q-1	13	1.231		455	47.50		502.50	575	
	0740	4" diameter	"	8	2	↓	1,300	77		1,377	1,550	
	2000	Lubricated semi-steel plug valve										
	2040	3/4" diameter	1 Plum	16	.500	Ea.	84.50	21.50		106	126	
	2080	1" diameter		14	.571		82.50	24.50		107	129	
	2100	1-1/4" diameter		12	.667		85	28.50		113.50	138	
	2140	1-1/2" diameter		11	.727		85	31		116	142	
	2180	2" diameter	↓	8	1		79.50	42.50		122	154	
	2300	2-1/2" diameter	Q-1	5	3.200		127	123		250	330	
	2340	3" diameter	"	4.50	3.556	↓	196	137		333	430	
550	0010	**GAS STATION PRODUCT LINE**										550
	0020	Primary containment pipe, fiberglass-reinforced										
	0030	Plastic pipe 15' & 30' lengths										
	0040	2" diameter	Q-6	425	.056	L.F.	3.49	2.27		5.76	7.35	
	0050	3" diameter		400	.060		4.57	2.41		6.98	8.80	
	0060	4" diameter	↓	375	.064	↓	5.90	2.57		8.47	10.50	
	0100	Fittings										
	0110	Elbows, 90° & 45°, bell-ends, 2"	Q-6	24	1	Ea.	35.50	40		75.50	102	
	0120	3" diameter		22	1.091		37	44		81	109	
	0130	4" diameter		20	1.200		49	48		97	129	
	0200	Tees, bell ends, 2"		21	1.143		43	46		89	119	
	0210	3" diameter		18	1.333		43	53.50		96.50	131	
	0220	4" diameter		15	1.600		59	64.50		123.50	165	
	0230	Flanges bell ends, 2"		24	1		14.15	40		54.15	78	
	0240	3" diameter		22	1.091		17.85	44		61.85	87.50	
	0250	4" diameter		20	1.200		24.50	48		72.50	102	
	0260	Sleeve couplings, 2"		21	1.143		9	46		55	81.50	
	0270	3" diameter	↓	18	1.333	↓	12.80	53.50		66.30	97.50	

02550 | Piped Energy Distribution

			CREW	DAILY OUTPUT	LABOR-HOURS	UNIT	2006 BARE COSTS MAT.	LABOR	EQUIP.	TOTAL	TOTAL INCL O&P	
550	0280	4″ diameter	Q-6	15	1.600	Ea.	17.65	64.50		82.15	119	550
	0290	Threaded adapters 2″		21	1.143		11.85	46		57.85	84.50	
	0300	3″ diameter		18	1.333		21	53.50		74.50	107	
	0310	4″ diameter		15	1.600		28	64.50		92.50	131	
	0320	Reducers, 2″		27	.889		15.75	35.50		51.25	73	
	0330	3″ diameter		22	1.091		18.25	44		62.25	88	
	0340	4″ diameter		20	1.200		23.50	48		71.50	101	
	1010	Gas station product line for secondary containment (double wall)										
	1100	Fiberglass reinforced plastic pipe 25′ lengths										
	1120	Pipe, plain end, 3″	Q-6	375	.064	L.F.	6	2.57		8.57	10.60	
	1130	4″ diameter		350	.069		9.85	2.76		12.61	15.10	
	1140	5″ diameter		325	.074		12.80	2.97		15.77	18.65	
	1150	6″ diameter		300	.080		13.15	3.21		16.36	19.50	
	1200	Fittings										
	1230	Elbows, 90° & 45°, 3″	Q-6	18	1.333	Ea.	43	53.50		96.50	131	
	1240	4″ diameter		16	1.500		75.50	60.50		136	177	
	1250	5″ diameter		14	1.714		175	69		244	299	
	1260	6″ diameter		12	2		177	80.50		257.50	320	
	1270	Tees, 3″		15	1.600		64	64.50		128.50	171	
	1280	4″ diameter		12	2		94	80.50		174.50	228	
	1290	5″ diameter		9	2.667		190	107		297	375	
	1300	6″ diameter		6	4		199	161		360	470	
	1310	Couplings, 3″		18	1.333		30.50	53.50		84	117	
	1320	4″ diameter		16	1.500		78.50	60.50		139	180	
	1330	5″ diameter		14	1.714		163	69		232	286	
	1340	6″ diameter		12	2		169	80.50		249.50	310	
	1350	Cross-over nipples, 3″		18	1.333		6.90	53.50		60.40	91	
	1360	4″ diameter		16	1.500		8.10	60.50		68.60	102	
	1370	5″ diameter		14	1.714		12.05	69		81.05	120	
	1380	6″ diameter		12	2		12.65	80.50		93.15	139	
	1400	Telescoping, reducers, concentric 4″ x 3″		18	1.333		23	53.50		76.50	109	
	1410	5″ x 4″		17	1.412		60	56.50		116.50	154	
	1420	6″ x 5″		16	1.500		144	60.50		204.50	253	

02580 | Elec/Communication Structures

			CREW	DAILY OUTPUT	LABOR-HOURS	UNIT	2006 BARE COSTS MAT.	LABOR	EQUIP.	TOTAL	TOTAL INCL O&P	
100	0010	**RADIO TOWERS**										100
	0020	Guyed, 50′h, 40 lb. sec., 70 MPH basic wind spd.	2 Sswk	1	16	Ea.	2,400	640		3,040	3,850	
	0100	Wind load 90 MPH basic wind speed	″	1	16		2,400	640		3,040	3,850	
	0300	190′ high, 40 lb. section, wind load 70 MPH basic wind speed	K-2	.33	72.727		6,500	2,650	535	9,685	12,500	
	0400	200′ high, 70 lb. section, wind load 90 MPH basic wind speed		.33	72.727		13,100	2,650	535	16,285	19,800	
	0600	300′ high, 70 lb. section, wind load 70 MPH basic wind speed		.20	120		18,500	4,375	880	23,755	29,300	
	0700	270′ high, 90 lb. section, wind load 90 MPH basic wind speed		.20	120		21,400	4,375	880	26,655	32,400	
	0800	400′ high, 100 lb. section, wind load 70 MPH basic wind speed		.14	171		31,200	6,250	1,250	38,700	47,000	
	0900	Self-supporting, 60′ high, wind load 70 MPH basic wind speed		.80	30		5,000	1,100	220	6,320	7,725	
	0910	60′ high, wind load 90 MPH basic wind speed		.45	53.333		9,025	1,950	390	11,365	13,900	
	1000	120′ high, wind load 70 MPH basic wind speed		.40	60		12,300	2,200	440	14,940	17,900	
	1200	190′ high, wind load 90 MPH basic wind speed		.20	120		29,800	4,375	880	35,055	41,700	
	2000	For states west of Rocky Mountains, add for shipping					10%					
410	0010	**UNDERGROUND DUCTS AND MANHOLES**, In slab or duct bank										410
	0011	Not including excavation, backfill and cast in place concrete										
	1000	Direct burial										
	1010	PVC, schedule 40, w/coupling, 1/2″ diameter	1 Elec	340	.024	L.F.	.38	.99		1.37	1.94	
	1020	3/4″ diameter		290	.028		.50	1.16		1.66	2.34	
	1030	1″ diameter		260	.031		.74	1.29		2.03	2.80	

02580 | Elec/Communication Structures

		02580 Elec/Communication Structures	CREW	DAILY OUTPUT	LABOR-HOURS	UNIT	2006 BARE COSTS MAT.	LABOR	EQUIP.	TOTAL	TOTAL INCL O&P	
410	1040	1-1/2" diameter	1 Elec	210	.038	L.F.	1.20	1.60		2.80	3.79	410
	1050	2" diameter		180	.044		1.58	1.87		3.45	4.62	
	1060	3" diameter	2 Elec	240	.067		3.01	2.80		5.81	7.60	
	1070	4" diameter		160	.100		4.22	4.20		8.42	11.10	
	1080	5" diameter		120	.133		6.15	5.60		11.75	15.40	
	1090	6" diameter		90	.178		8.10	7.45		15.55	20.50	
	1110	Elbows, 1/2" diameter	1 Elec	48	.167	Ea.	1.58	7		8.58	12.55	
	1120	3/4" diameter		38	.211		1.59	8.85		10.44	15.35	
	1130	1" diameter		32	.250		2.69	10.50		13.19	19.15	
	1140	1-1/2" diameter		21	.381		5.20	16		21.20	30	
	1150	2" diameter		16	.500		7.55	21		28.55	41	
	1160	3" diameter		12	.667		23	28		51	68.50	
	1170	4" diameter		9	.889		40	37.50		77.50	102	
	1180	5" diameter		8	1		70	42		112	142	
	1190	6" diameter		5	1.600		119	67		186	235	
	1210	Adapters, 1/2" diameter		52	.154		.51	6.45		6.96	10.50	
	1220	3/4" diameter		43	.186		.94	7.80		8.74	13.10	
	1230	1" diameter		39	.205		1.19	8.60		9.79	14.55	
	1240	1-1/2" diameter		35	.229		1.85	9.60		11.45	16.85	
	1250	2" diameter		26	.308		2.66	12.90		15.56	23	
	1260	3" diameter		20	.400		6.75	16.80		23.55	33.50	
	1270	4" diameter		14	.571		11.80	24		35.80	50	
	1280	5" diameter		12	.667		23	28		51	68.50	
	1290	6" diameter		9	.889		28	37.50		65.50	88.50	
	1340	Bell end & cap, 1-1/2" diameter		35	.229		9	9.60		18.60	24.50	
	1350	Bell end & plug, 2" diameter		26	.308		9.50	12.90		22.40	30.50	
	1360	3" diameter		20	.400		11.90	16.80		28.70	39	
	1370	4" diameter		14	.571		13.90	24		37.90	52.50	
	1380	5" diameter		12	.667		21	28		49	66	
	1390	6" diameter		9	.889		24	37.50		61.50	83.50	
	1450	Base spacer, 2" diameter		56	.143		1.61	6		7.61	11	
	1460	3" diameter		46	.174		1.78	7.30		9.08	13.20	
	1470	4" diameter		41	.195		1.88	8.20		10.08	14.65	
	1480	5" diameter		37	.216		2.31	9.10		11.41	16.55	
	1490	6" diameter		34	.235		3.24	9.90		13.14	18.75	
	1550	Intermediate spacer, 2" diameter		60	.133		1.40	5.60		7	10.20	
	1560	3" diameter		46	.174		1.78	7.30		9.08	13.20	
	1570	4" diameter		41	.195		1.88	8.20		10.08	14.65	
	1580	5" diameter		37	.216		2.31	9.10		11.41	16.55	
	1590	6" diameter		34	.235		3.24	9.90		13.14	18.75	
	4010	PVC, schedule 80, w/coupling, 1/2" diameter		215	.037	L.F.	.45	1.56		2.01	2.91	
	4020	3/4" diameter		180	.044		.61	1.87		2.48	3.55	
	4030	1" diameter		145	.055		.86	2.32		3.18	4.52	
	4040	1-1/2" diameter		120	.067		1.44	2.80		4.24	5.90	
	4050	2" diameter		100	.080		1.98	3.36		5.34	7.40	
	4060	3" diameter	2 Elec	130	.123		4.06	5.15		9.21	12.40	
	4070	4" diameter		90	.178		5.90	7.45		13.35	18	
	4080	5" diameter		70	.229		8.50	9.60		18.10	24	
	4090	6" diameter		50	.320		11.70	13.45		25.15	33.50	
	4110	Elbows, 1/2" diameter	1 Elec	29	.276	Ea.	3.05	11.60		14.65	21	
	4120	3/4" diameter		23	.348		4.79	14.60		19.39	28	
	4130	1" diameter		20	.400		6.25	16.80		23.05	33	
	4140	1-1/2" diameter		16	.500		12.30	21		33.30	46	
	4150	2" diameter		12	.667		16.85	28		44.85	61.50	
	4160	3" diameter		9	.889		57	37.50		94.50	120	
410	4170	4" diameter		7	1.143		93	48		141	176	410

02580 | Elec/Communication Structures

			CREW	DAILY OUTPUT	LABOR-HOURS	UNIT	2006 BARE COSTS MAT.	LABOR	EQUIP.	TOTAL	TOTAL INCL O&P	
410	4180	5″ diameter	1 Elec	6	1.333	Ea.	135	56		191	236	410
	4190	6″ diameter		4	2		265	84		349	420	
	4210	Adapter, 1/2″ diameter		39	.205		.51	8.60		9.11	13.80	
	4220	3/4″ diameter		33	.242		.94	10.20		11.14	16.75	
	4230	1″ diameter		29	.276		1.19	11.60		12.79	19.15	
	4240	1-1/2″ diameter		26	.308		1.85	12.90		14.75	22	
	4250	2″ diameter		23	.348		2.66	14.60		17.26	25.50	
	4260	3″ diameter		18	.444		6.75	18.65		25.40	36.50	
	4270	4″ diameter		13	.615		11.80	26		37.80	53	
	4280	5″ diameter		11	.727		23	30.50		53.50	72.50	
	4290	6″ diameter		8	1		28	42		70	95.50	
	4310	Bell end & cap, 1-1/2″ diameter		26	.308		9	12.90		21.90	30	
	4320	Bell end & plug, 2″ diameter		23	.348		9.50	14.60		24.10	33	
	4330	3″ diameter		18	.444		11.90	18.65		30.55	42	
	4340	4″ diameter		13	.615		13.90	26		39.90	55.50	
	4350	5″ diameter		11	.727		21	30.50		51.50	70	
	4360	6″ diameter		8	1		24	42		66	90.50	
	4370	Base spacer, 2″ diameter		42	.190		1.61	8		9.61	14.05	
	4380	3″ diameter		33	.242		1.78	10.20		11.98	17.65	
	4390	4″ diameter		29	.276		1.88	11.60		13.48	19.90	
	4400	5″ diameter		26	.308		2.31	12.90		15.21	22.50	
	4410	6″ diameter		25	.320		3.24	13.45		16.69	24	
	4420	Intermediate spacer, 2″ diameter		45	.178		1.40	7.45		8.85	13.05	
	4430	3″ diameter		34	.235		1.78	9.90		11.68	17.15	
	4440	4″ diameter		31	.258		1.88	10.85		12.73	18.75	
	4450	5″ diameter		28	.286		2.31	12		14.31	21	
	4460	6″ diameter		25	.320		3.24	13.45		16.69	24	
420	0010	**ELECTRIC & TELEPHONE UNDERGROUND**, Not including excavation,										420
	0200	backfill and cast in place concrete										
	0380	Underground marking tape, 6″ wide	R-19	2530	.008	L.F.	.31	.33		.64	.85	
	0400	Hand holes, precast concrete, with concrete cover										
	0600	2′ x 2′ x 3′ deep	R-3	2.40	8.333	Ea.	268	345	63.50	676.50	900	
	0800	3′ x 3′ x 3′ deep		1.90	10.526		350	435	80.50	865.50	1,150	
	1000	4′ x 4′ x 4′ deep		1.40	14.286		695	590	109	1,394	1,800	
	1200	Manholes, precast with iron racks & pulling irons, C.I. frame										
	1400	and cover, 4′ x 6′ x 7′ deep	B-13	2	28	Ea.	1,325	840	325	2,490	3,200	
	1600	6′ x 8′ x 7′ deep		1.90	29.474		1,625	880	340	2,845	3,625	
	1800	6′ x 10′ x 7′ deep		1.80	31.111		1,825	930	360	3,115	3,950	
	4200	Underground duct, banks ready for concrete fill, min. of 7.5″										
	4400	between conduits, ctr. to ctr.(for wire & cable see Div. 16120)										
	4580	PVC, type EB, 1 @ 2″ diameter	2 Elec	480	.033	L.F.	.75	1.40		2.15	2.99	
	4600	2 @ 2″ diameter		240	.067		1.50	2.80		4.30	5.95	
	4800	4 @ 2″ diameter		120	.133		3	5.60		8.60	11.95	
	4900	1 @ 3″ diameter		400	.040		1.03	1.68		2.71	3.72	
	5000	2 @ 3″ diameter		200	.080		2.06	3.36		5.42	7.45	
	5200	4 @ 3″ diameter		100	.160		4.11	6.70		10.81	14.85	
	5300	1 @ 4″ diameter		320	.050		1.58	2.10		3.68	4.98	
	5400	2 @ 4″ diameter		160	.100		3.16	4.20		7.36	9.90	
	5600	4 @ 4″ diameter		80	.200		6.30	8.40		14.70	19.90	
	5800	6 @ 4″ diameter		54	.296		9.45	12.45		21.90	29.50	
	5810	1 @ 5″ diameter		260	.062		2.35	2.58		4.93	6.55	
	5820	2 @ 5″ diameter		130	.123		4.70	5.15		9.85	13.10	
	5840	4 @ 5″ diameter		70	.229		9.40	9.60		19	25	
	5860	6 @ 5″ diameter		50	.320		14.10	13.45		27.55	36	
	5870	1 @ 6″ diameter		200	.080		3.36	3.36		6.72	8.90	

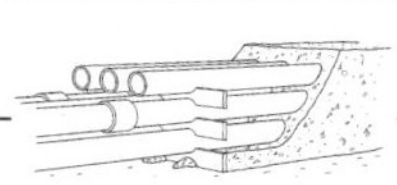

02580 | Elec/Communication Structures		CREW	DAILY OUTPUT	LABOR-HOURS	UNIT	2006 BARE COSTS MAT.	LABOR	EQUIP.	TOTAL	TOTAL INCL O&P
420 5880	2 @ 6" diameter	2 Elec	100	.160	L.F.	6.75	6.70		13.45	17.75 420
5900	4 @ 6" diameter		50	.320		13.45	13.45		26.90	35.50
5920	6 @ 6" diameter		30	.533		20	22.50		42.50	56.50
6200	Rigid galvanized steel, 2 @ 2" diameter		180	.089		14.80	3.73		18.53	22
6400	4 @ 2" diameter		90	.178		29.50	7.45		36.95	44
6800	2 @ 3" diameter		100	.160		35	6.70		41.70	49
7000	4 @ 3" diameter		50	.320		69.50	13.45		82.95	97
7200	2 @ 4" diameter		70	.229		48.50	9.60		58.10	68
7400	4 @ 4" diameter		34	.471		96.50	19.75		116.25	137
7600	6 @ 4" diameter		22	.727		145	30.50		175.50	206
7620	2 @ 5" diameter		60	.267		105	11.20		116.20	132
7640	4 @ 5" diameter		30	.533		209	22.50		231.50	265
7660	6 @ 5" diameter		18	.889		315	37.50		352.50	405
7680	2 @ 6" diameter		40	.400		152	16.80		168.80	193
7700	4 @ 6" diameter		20	.800		305	33.50		338.50	385
7720	6 @ 6" diameter		14	1.143		455	48		503	575
7800	For Cast-in-place Concrete - Add									
7810	Under 1 C.Y.	C-6	16	3	C.Y.	139	86.50	2.67	228.17	298
7820	1 C.Y. - 5 C.Y.		19.20	2.500		126	72.50	2.23	200.73	260
7830	Over 5 C.Y.		24	2		104	58	1.78	163.78	212
7850	For Reinforcing Rods - Add									
7860	#4 to #7	2 Rodm	1.10	14.545	Ton	855	575		1,430	1,925
7870	#8 to #14	"	1.50	10.667	"	855	420		1,275	1,675
8000	Fittings, PVC type EB, elbow, 2" diameter	1 Elec	16	.500	Ea.	11	21		32	44.50
8200	3" diameter		14	.571		11.60	24		35.60	50
8400	4" diameter		12	.667		16.45	28		44.45	61
8420	5" diameter		10	.800		39	33.50		72.50	95
8440	6" diameter		9	.889		74	37.50		111.50	139
8500	Coupling, 2" diameter					.79			.79	.87
8600	3" diameter					2.74			2.74	3.01
8700	4" diameter					4.30			4.30	4.73
8720	5" diameter					7.75			7.75	8.55
8740	6" diameter					22.50			22.50	25
8800	Adapter, 2" diameter	1 Elec	26	.308		1.40	12.90		14.30	21.50
9000	3" diameter		20	.400		3.90	16.80		20.70	30.50
9200	4" diameter		16	.500		5.20	21		26.20	38.50
9220	5" diameter		13	.615		13	26		39	54.50
9240	6" diameter		10	.800		17.15	33.50		50.65	71
9400	End bell, 2" diameter		16	.500		6.80	21		27.80	40
9600	3" diameter		14	.571		8.15	24		32.15	46
9800	4" diameter		12	.667		9.60	28		37.60	53.50
9810	5" diameter		10	.800		14.50	33.50		48	68
9820	6" diameter		8	1		27.50	42		69.50	95
9830	5° angle coupling, 2" diameter		26	.308		11.45	12.90		24.35	32.50
9840	3" diameter		20	.400		14.50	16.80		31.30	42
9850	4" diameter		16	.500		17.15	21		38.15	51.50
9860	5" diameter		13	.615		18.80	26		44.80	60.50
9870	6" diameter		10	.800		19.25	33.50		52.75	73
9880	Expansion joint, 2" diameter		16	.500		33.50	21		54.50	69.50
9890	3" diameter		18	.444		59	18.65		77.65	94
9900	4" diameter		12	.667		85.50	28		113.50	137
9910	5" diameter		10	.800		133	33.50		166.50	198
9920	6" diameter		8	1		179	42		221	262
9930	Heat bender, 2" diameter					440			440	485
9940	6" diameter					1,225			1,225	1,350
9950	Cement, quart					14.70			14.70	16.15

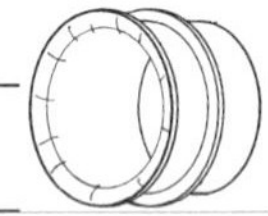

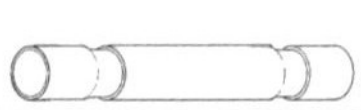

02500 | Utility Services

	02580 | Elec/Communication Structures		CREW	DAILY OUTPUT	LABOR-HOURS	UNIT	2006 BARE COSTS MAT.	LABOR	EQUIP.	TOTAL	TOTAL INCL O&P	
420	9960	Nylon polyethylene pull rope, 1/4″	2 Elec	2000	.008	L.F.	.10	.34		.44	.63	420
	9990	Minimum labor/equipment charge	1 Elec	3.50	2.286	Job		96		96	148	
500	0010	UTILITY POLES										500
	6200	Poles, wood, preservative treatment, 20′ high, see also Div. 16520	R-3	3.10	6.452	Ea.	251	267	49	567	745	
	6400	25′ high		2.90	6.897		265	286	52.50	603.50	790	
	6600	30′ high		2.60	7.692		292	320	58.50	670.50	880	
	6800	35′ high		2.40	8.333		380	345	63.50	788.50	1,025	
	7000	40′ high		2.30	8.696		465	360	66.50	891.50	1,150	
	7200	45′ high		1.70	11.765		570	485	90	1,145	1,475	
	7400	Cross arms with hardware & insulators										
	7600	4′ long	1 Elec	2.50	3.200	Ea.	119	134		253	340	
	7800	5′ long		2.40	3.333		138	140		278	370	
	8000	6′ long		2.20	3.636		159	153		312	410	

02600 | Drainage & Containment

	02620 | Subdrainage		CREW	DAILY OUTPUT	LABOR-HOURS	UNIT	2006 BARE COSTS MAT.	LABOR	EQUIP.	TOTAL	TOTAL INCL O&P	
300	0010	GEOTEXTILES FOR SUBSURFACE DRAINAGE										300
	0100	Fabric, laid in trench, polypropylene, ideal conditions	2 Clab	2400	.007	S.Y.	1.38	.18		1.56	1.82	
	0110	Adverse conditions		1600	.010	″	1.38	.27		1.65	1.97	
	0170	Fabric ply bonded to 3 dimen. nylon mat, 0.4″ thk, ideal conditions		2000	.008	S.F.	.96	.22		1.18	1.42	
	0180	Adverse conditions		1200	.013	″	1.20	.37		1.57	1.93	
	0185	Soil drainage mat on vertical wall, 0.44″ thick		265	.060	S.Y.	16.85	1.65		18.50	21	
	0188	0.25″ thick		300	.053	″	10.25	1.46		11.71	13.65	
	0190	0.8″ thick, ideal conditions		2400	.007	S.F.	1.51	.18		1.69	1.96	
	0200	Adverse conditions		1600	.010	″	1.99	.27		2.26	2.64	
	0300	Drainage material, 3/4″ gravel fill in trench	B-6	260	.092	C.Y.	54.50	2.77	.87	58.14	65.50	
	0400	Pea stone	″	260	.092	″	19.15	2.77	.87	22.79	26.50	
610	0010	PIPING, SUBDRAINAGE, CONCRETE										610
	0021	Not including excavation and backfill										
	3000	Porous wall concrete underdrain, std. strength, 4″ diameter	B-20	335	.072	L.F.	2.05	2.23		4.28	5.95	
	3020	6″ diameter	″	315	.076		2.67	2.37		5.04	6.85	
	3040	8″ diameter	B-21	310	.090		3.29	2.90	.49	6.68	8.90	
	3060	12″ diameter		285	.098		6.95	3.15	.54	10.64	13.40	
	3080	15″ diameter		230	.122		8	3.91	.66	12.57	15.95	
	3100	18″ diameter		165	.170		10.55	5.45	.92	16.92	21.50	
	4000	Extra strength, 6″ diameter	B-20	315	.076		2.70	2.37		5.07	6.90	
	4020	8″ diameter	B-21	310	.090		4.05	2.90	.49	7.44	9.75	
	4040	10″ diameter		285	.098		8.10	3.15	.54	11.79	14.65	
	4060	12″ diameter		230	.122		8.80	3.91	.66	13.37	16.80	
	4080	15″ diameter		200	.140		9.70	4.49	.76	14.95	18.90	
	4100	18″ diameter		165	.170		14.20	5.45	.92	20.57	25.50	
620	0010	PIPING, SUBDRAINAGE, CORRUGATED METAL										620
	0021	Not including excavation and backfill										
	2010	Aluminum, perforated										
	2020	6″ diameter, 18 ga.	B-14	380	.126	L.F.	3.04	3.67	.60	7.31	10	
	2200	8″ diameter, 16 ga.		370	.130		4.34	3.77	.61	8.72	11.65	
	2220	10″ diameter, 16 ga.		360	.133		5.45	3.87	.63	9.95	13	

02600 | Drainage & Containment

02620 | Subdrainage

		02620 Subdrainage	CREW	DAILY OUTPUT	LABOR-HOURS	UNIT	2006 BARE COSTS MAT.	LABOR	EQUIP.	TOTAL	TOTAL INCL O&P	
620	2240	12″ diameter, 16 ga.	B-14	285	.168	L.F.	6.10	4.89	.79	11.78	15.60	620
	2260	18″ diameter, 16 ga.	↓	205	.234	↓	9.10	6.80	1.10	17	22.50	
	3000	Uncoated galvanized, perforated										
	3020	6″ diameter, 18 ga.	B-20	380	.063	L.F.	4.81	1.96		6.77	8.55	
	3200	8″ diameter, 16 ga.	″	370	.065		6.60	2.02		8.62	10.60	
	3220	10″ diameter, 16 ga.	B-21	360	.078		9.90	2.50	.42	12.82	15.45	
	3240	12″ diameter, 16 ga.		285	.098		10.40	3.15	.54	14.09	17.15	
	3260	18″ diameter, 16 ga.	↓	205	.137	↓	15.85	4.38	.74	20.97	25.50	
	4000	Steel, perforated, asphalt coated										
	4020	6″ diameter 18 ga.	B-20	380	.063	L.F.	3.84	1.96		5.80	7.45	
	4030	8″ diameter 18 ga	″	370	.065		6	2.02		8.02	9.95	
	4040	10″ diameter 16 ga	B-21	360	.078		6.90	2.50	.42	9.82	12.15	
	4050	12″ diameter 16 ga		285	.098		7.95	3.15	.54	11.64	14.45	
	4060	18″ diameter 16 ga	↓	205	.137	↓	10.80	4.38	.74	15.92	19.90	
630	0010	**PIPING, SUBDRAINAGE, PLASTIC**										630
	0020	Not including excavation and backfill										
	1110	Piping, subdrainage, corr polthn, coupler, 10″				Ea.	3.51			3.51	3.86	
	2100	Perforated PVC, 4″ diameter	B-14	314	.153	L.F.	.83	4.44	.72	5.99	9	
	2110	6″ diameter		300	.160		1.54	4.64	.75	6.93	10.10	
	2120	8″ diameter		290	.166		1.74	4.81	.78	7.33	10.70	
	2130	10″ diameter		280	.171		2.70	4.98	.81	8.49	12	
	2140	12″ diameter	↓	270	.178	↓	3.74	5.15	.84	9.73	13.50	

02630 | Storm Drainage

		02630 Storm Drainage	CREW	DAILY OUTPUT	LABOR-HOURS	UNIT	MAT.	LABOR	EQUIP.	TOTAL	TOTAL INCL O&P	
110	0010	**CATCH BASIN GRATES AND FRAMES** not including footing, excavation										110
	1600	Frames & covers, C.I., 24″ square, 500 lb.	B-6	7.80	3.077	Ea.	268	92.50	29	389.50	475	
	1700	26″ D shape, 600 lb.		7	3.429		460	103	32.50	595.50	710	
	1800	Light traffic, 18″ diameter, 100 lb.		10	2.400		150	72	22.50	244.50	305	
	1900	24″ diameter, 300 lb.		8.70	2.759		232	83	26	341	420	
	2000	36″ diameter, 900 lb.		5.80	4.138		465	124	39	628	755	
	2100	Heavy traffic, 24″ diameter, 400 lb.		7.80	3.077		224	92.50	29	345.50	430	
	2200	36″ diameter, 1150 lb.		3	8		740	240	75.50	1,055.50	1,300	
	2300	Mass. State standard, 26″ diameter, 475 lb.		7	3.429		560	103	32.50	695.50	820	
	2400	30″ diameter, 620 lb.		7	3.429		355	103	32.50	490.50	595	
	2500	Watertight, 24″ diameter, 350 lb.		7.80	3.077		380	92.50	29	501.50	600	
	2600	26″ diameter, 500 lb.		7	3.429		360	103	32.50	495.50	600	
	2700	32″ diameter, 575 lb.	↓	6	4	↓	805	120	37.50	962.50	1,125	
	2800	3 piece cover & frame, 10″ deep,										
	2900	1200 lbs., for heavy equipment	B-6	3	8	Ea.	1,150	240	75.50	1,465.50	1,725	
	3000	Raised for paving 1-1/4″ to 2″ high,										
	3100	4 piece expansion ring										
	3200	20″ to 26″ diameter	1 Clab	3	2.667	Ea.	126	73		199	259	
	3300	30″ to 36″ diameter	″	3	2.667	″	175	73		248	315	
	3320	Frames and covers, existing, raised for paving, 2″, including										
	3340	row of brick, concrete collar, up to 12″ wide frame	B-6	18	1.333	Ea.	39.50	40	12.55	92.05	122	
	3360	20″ to 26″ wide frame		11	2.182		62	65.50	20.50	148	197	
	3380	30″ to 36″ wide frame	↓	9	2.667		76.50	80	25	181.50	242	
	3400	Inverts, single channel brick	D-1	3	5.333		86	171		257	375	
	3500	Concrete		5	3.200		67.50	103		170.50	241	
	3600	Triple channel, brick		2	8		131	257		388	560	
	3700	Concrete	↓	3	5.333	↓	115	171		286	405	
400	0010	**STORM DRAINAGE MANHOLES, FRAMES & COVERS** not including										400
	0020	footing, excavation, backfill (See line items for frame & cover)										
	0050	Brick, 4′ inside diameter, 4′ deep	D-1	1	16	Ea.	360	515		875	1,225	
	0100	6′ deep	↓	.70	22.857	↓	505	735		1,240	1,750	

	02630 | Storm Drainage		CREW	DAILY OUTPUT	LABOR-HOURS	UNIT	2006 BARE COSTS MAT.	LABOR	EQUIP.	TOTAL	TOTAL INCL O&P	
400	0150	8′ deep	D-1	.50	32	Ea.	645	1,025		1,670	2,375	400
	0200	For depths over 8′, add		4	4	V.L.F.	182	129		311	410	
	0400	Concrete blocks (radial), 4′ I.D., 4′ deep		1.50	10.667	Ea.	300	345		645	885	
	0500	6′ deep		1	16		400	515		915	1,275	
	0600	8′ deep		.70	22.857		500	735		1,235	1,750	
	0700	For depths over 8′, add		5.50	2.909	V.L.F.	51.50	93.50		145	209	
	0800	Concrete, cast in place, 4′ x 4′, 8″ thick, 4′ deep	C-14H	2	24	Ea.	470	840	10.80	1,320.80	1,925	
	0900	6′ deep		1.50	32		685	1,125	14.40	1,824.40	2,625	
	1000	8′ deep		1	48		975	1,675	21.50	2,671.50	3,875	
	1100	For depths over 8′, add		8	6	V.L.F.	113	210	2.70	325.70	470	
	1110	Precast, 4′ I.D., 4′ deep	B-22	4.10	7.317	Ea.	710	238	56	1,004	1,225	
	1120	6′ deep		3	10		915	325	76.50	1,316.50	1,625	
	1130	8′ deep		2	15		1,075	490	114	1,679	2,100	
	1140	For depths over 8′, add		16	1.875	V.L.F.	150	61	14.30	225.30	280	
	1150	5′ I.D., 4′ deep	B-6	3	8	Ea.	755	240	75.50	1,070.50	1,300	
	1160	6′ deep		2	12		1,025	360	113	1,498	1,825	
	1170	8′ deep		1.50	16		1,275	480	151	1,906	2,350	
	1180	For depths over 8′, add		12	2	V.L.F.	167	60	18.85	245.85	300	
	1190	6′ I.D., 4′ deep		2	12	Ea.	1,225	360	113	1,698	2,050	
	1200	6′ deep		1.50	16		1,600	480	151	2,231	2,725	
	1210	8′ deep		1	24		1,975	720	226	2,921	3,600	
	1220	For depths over 8′, add		8	3	V.L.F.	258	90	28.50	376.50	460	
	1250	Slab tops, precast, 8″ thick										
	1300	4′ diameter manhole	B-6	8	3	Ea.	173	90	28.50	291.50	370	
	1400	5′ diameter manhole		7.50	3.200		340	96	30	466	565	
	1500	6′ diameter manhole		7	3.429		425	103	32.50	560.50	675	
	3800	Steps, heavyweight cast iron, 7″ x 9″	1 Bric	40	.200		13.95	7.30		21.25	27	
	3900	8″ x 9″		40	.200		21	7.30		28.30	35	
	3928	12″ x 10-1/2″		40	.200		19.80	7.30		27.10	34	
	4000	Standard sizes, galvanized steel		40	.200		16.85	7.30		24.15	30.50	
	4100	Aluminum		40	.200		23	7.30		30.30	37	
510	0010	**PIPING, STORM DRAINAGE, CORRUGATED METAL**										510
	0020	Not including excavation or backfill										
	2000	Corrugated metal pipe, galvanized and coated										
	2020	Bituminous coated with paved invert, 20′ lengths										
	2040	8″ diameter, 16 ga.	B-14	330	.145	L.F.	9.30	4.22	.69	14.21	17.95	
	2060	10″ diameter, 16 ga.		260	.185		11.15	5.35	.87	17.37	22	
	2080	12″ diameter, 16 ga.		210	.229		13.40	6.65	1.08	21.13	27	
	2100	15″ diameter, 16 ga.		200	.240		16.30	6.95	1.13	24.38	30.50	
	2120	18″ diameter, 16 ga.		190	.253		21	7.35	1.19	29.54	36.50	
	2140	24″ diameter, 14 ga.		160	.300		25.50	8.70	1.41	35.61	44	
	2160	30″ diameter, 14 ga.	B-13	120	.467		33.50	13.95	5.40	52.85	66	
	2180	36″ diameter, 12 ga.		120	.467		49.50	13.95	5.40	68.85	83.50	
	2200	48″ diameter, 12 ga.		100	.560		75	16.75	6.50	98.25	117	
	2220	60″ diameter, 10 ga.	B-13B	75	.747		97.50	22.50	12.70	132.70	157	
	2240	72″ diameter, 8 ga.	″	45	1.244		145	37.50	21	203.50	244	
	2250	End sections, 8″ diameter, 16 ga.	B-14	20	2.400	Ea.	54.50	69.50	11.30	135.30	186	
	2255	10″ diameter, 16 ga.		20	2.400		62	69.50	11.30	142.80	195	
	2260	12″ diameter, 16 ga.		18	2.667		76.50	77.50	12.55	166.55	225	
	2265	15″ diameter, 16 ga.		18	2.667		94.50	77.50	12.55	184.55	245	
	2270	18″ diameter, 16 ga.		16	3		120	87	14.15	221.15	290	
	2275	24″ diameter, 16 ga.	B-13	16	3.500		171	105	40.50	316.50	405	
	2280	30″ diameter, 16 ga.		14	4		325	120	46.50	491.50	600	
	2285	36″ diameter, 14 ga.		14	4		425	120	46.50	591.50	715	
	2290	48″ diameter, 14 ga.		10	5.600		885	168	65	1,118	1,325	

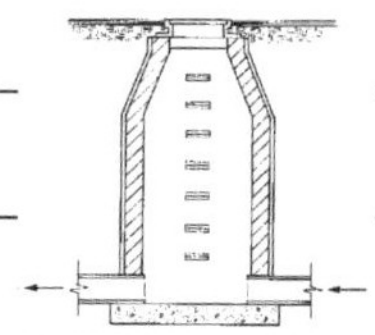

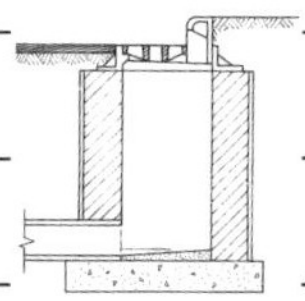

SITE CONSTRUCTION 2

02630 | Storm Drainage

		02630 Storm Drainage	CREW	DAILY OUTPUT	LABOR-HOURS	UNIT	2006 BARE COSTS MAT.	LABOR	EQUIP.	TOTAL	TOTAL INCL O&P	
510	2292	60″ diameter, 14 ga.	B-13	6	9.333	Ea.	1,200	279	108	1,587	1,900	510
	2294	72″ diameter, 14 ga.	B-13B	5	11.200		2,125	335	190	2,650	3,075	
	2300	Bends or elbows, 8″ diameter	B-14	28	1.714		87.50	50	8.05	145.55	186	
	2320	10″ diameter		25	1.920		108	55.50	9.05	172.55	220	
	2340	12″ diameter, 16 ga.		30	1.600		125	46.50	7.55	179.05	222	
	2342	18″ diameter, 16 ga.		20	2.400		181	69.50	11.30	261.80	325	
	2344	24″ diameter, 14 ga.		16	3		252	87	14.15	353.15	435	
	2346	30″ diameter, 14 ga.		15	3.200		315	93	15.05	423.05	515	
	2348	36″ diameter, 14 ga.	B-13	15	3.733		425	112	43.50	580.50	695	
	2350	48″ diameter, 12 ga.	″	12	4.667		605	140	54	799	955	
	2352	60″ diameter, 10 ga.	B-13B	10	5.600		810	168	95	1,073	1,275	
	2354	72″ diameter, 10 ga.	″	6	9.333		1,000	279	159	1,438	1,725	
	2360	Wyes or tees, 8″ diameter	B-14	25	1.920		122	55.50	9.05	186.55	235	
	2380	10″ diameter		21	2.286		152	66.50	10.75	229.25	288	
	2400	12″ diameter, 16 ga.		22.48	2.135		127	62	10.05	199.05	252	
	2410	18″ diameter, 16 ga.		16	3		241	87	14.15	342.15	425	
	2412	24″ diameter, 14 ga.		16	3		365	87	14.15	466.15	560	
	2414	30″ diameter, 14 ga.	B-13	12	4.667		470	140	54	664	810	
	2416	36″ diameter, 14 ga.		11	5.091		580	152	59	791	955	
	2418	48″ diameter, 12 ga.		10	5.600		880	168	65	1,113	1,325	
	2420	60″ diameter, 10 ga.	B-13B	8	7		1,275	210	119	1,604	1,900	
	2422	72″ diameter, 10 ga.	″	5	11.200		1,550	335	190	2,075	2,450	
	2500	Galvanized, uncoated, 20′ lengths										
	2520	8″ diameter, 16 ga.	B-14	355	.135	L.F.	7.60	3.93	.64	12.17	15.50	
	2540	10″ diameter, 16 ga.		280	.171		8.35	4.98	.81	14.14	18.25	
	2560	12″ diameter, 16 ga.		220	.218		9.55	6.35	1.03	16.93	22	
	2580	15″ diameter, 16 ga.		220	.218		12.10	6.35	1.03	19.48	25	
	2600	18″ diameter, 16 ga.		205	.234		16.05	6.80	1.10	23.95	30	
	2620	24″ diameter, 14 ga.		175	.274		23	7.95	1.29	32.24	40	
	2640	30″ diameter, 14 ga.	B-13	130	.431		29.50	12.90	5	47.40	59	
	2660	36″ diameter, 12 ga.		130	.431		48.50	12.90	5	66.40	80	
	2680	48″ diameter, 12 ga.		110	.509		65	15.25	5.90	86.15	103	
	2690	60″ diameter, 10 ga.	B-13B	78	.718		102	21.50	12.20	135.70	160	
	2695	72″ diameter, 10 ga.	″	60	.933		129	28	15.85	172.85	205	
	2711	Bends or elbows, 12″ diameter, 16 ga.	B-14	30	1.600	Ea.	108	46.50	7.55	162.05	203	
	2712	15″ diameter, 16 ga.		25.04	1.917		134	55.50	9.05	198.55	248	
	2714	18″ diameter, 16 ga.		20	2.400		150	69.50	11.30	230.80	291	
	2716	24″ diameter, 14 ga.		16	3		219	87	14.15	320.15	400	
	2718	30″ diameter, 14 ga.		15	3.200		274	93	15.05	382.05	470	
	2720	36″ diameter, 14 ga.	B-13	15	3.733		375	112	43.50	530.50	645	
	2722	48″ diameter, 12 ga.		12	4.667		515	140	54	709	855	
	2724	60″ diameter, 10 ga.		10	5.600		790	168	65	1,023	1,225	
	2726	72″ diameter, 10 ga.		6	9.333		980	279	108	1,367	1,650	
	2728	Wyes or tees, 12″ diameter, 16 ga.	B-14	22.48	2.135		144	62	10.05	216.05	271	
	2730	18″ diameter, 16 ga.		15	3.200		209	93	15.05	317.05	400	
	2732	24″ diameter, 14 ga.		15	3.200		325	93	15.05	433.05	525	
	2734	30″ diameter, 14 ga.		14	3.429		425	99.50	16.15	540.65	645	
	2736	36″ diameter, 14 ga.	B-13	14	4		550	120	46.50	716.50	850	
	2738	48″ diameter, 12 ga.		12	4.667		810	140	54	1,004	1,175	
	2740	60″ diameter, 10 ga.		10	5.600		1,175	168	65	1,408	1,650	
	2742	72″ diameter, 10 ga.		6	9.333		1,400	279	108	1,787	2,125	
	2780	End sections, 8″ diameter	B-14	24	2		74	58	9.40	141.40	187	
	2785	10″ diameter		22	2.182		76	63.50	10.30	149.80	199	
	2790	12″ diameter		35	1.371		86	40	6.45	132.45	167	
	2800	18″ diameter		30	1.600		102	46.50	7.55	156.05	196	
	2810	24″ diameter	B-13	25	2.240		149	67	26	242	300	

	02630 | Storm Drainage		CREW	DAILY OUTPUT	LABOR-HOURS	UNIT	2006 BARE COSTS MAT.	LABOR	EQUIP.	TOTAL	TOTAL INCL O&P	
510	2820	30″ diameter	B-13	25	2.240	Ea.	282	67	26	375	450	510
	2825	36″ diameter		20	2.800		375	84	32.50	491.50	585	
	2830	48″ diameter		10	5.600		775	168	65	1,008	1,200	
	2835	60″ diameter	B-13B	5	11.200		1,000	335	190	1,525	1,850	
	2840	72″ diameter	″	4	14		1,825	420	238	2,483	2,950	
	2850	Couplings, 12″ diameter					16.60			16.60	18.30	
	2855	18″ diameter					24.50			24.50	26.50	
	2860	24″ diameter					33			33	36	
	2865	30″ diameter					40			40	44.50	
	2870	36″ diameter					56.50			56.50	62	
	2875	48″ diameter					76.50			76.50	84	
	2880	60″ diameter					120			120	132	
	2885	72″ diameter					166			166	183	
520	0010	**PIPING, DRAINAGE & SEWAGE, CORRUGATED HDPE TYPE S**										520
	0020	Not including excavation & backfill, bell & spigot										
	1000	With gaskets, 4″ diameter	B-20	425	.056	L.F.	.78	1.76		2.54	3.76	
	1010	6″ diameter		400	.060		1.79	1.87		3.66	5.05	
	1020	8″ diameter		380	.063		3.43	1.96		5.39	7	
	1030	10″ diameter		370	.065		4.74	2.02		6.76	8.55	
	1040	12″ diameter		340	.071		5.30	2.20		7.50	9.40	
	1050	15″ diameter		300	.080		7.20	2.49		9.69	12	
	1060	18″ diameter	B-21	275	.102		10.25	3.27	.55	14.07	17.20	
	1070	24″ diameter		250	.112		15.90	3.60	.61	20.11	24	
	1080	30″ diameter		200	.140		25	4.49	.76	30.25	35.50	
	1090	36″ diameter		180	.156		32	4.99	.85	37.84	44	
	1100	42″ diameter		175	.160		44.50	5.15	.87	50.52	58.50	
	1110	48″ diameter		170	.165		58	5.30	.90	64.20	73	
	1120	54″ diameter		160	.175		89.50	5.60	.95	96.05	108	
	1130	60″ diameter		150	.187		104	6	1.02	111.02	126	
	1135	Add 15% to material pipe cost for water tight connection bell & spigot										
	1140	HDPE type s, elbows 12″ diam	B-20	11	2.182	Ea.	58.50	68		126.50	177	
	1150	HDPE type s, elbows 15″ diam	″	9	2.667		90.50	83		173.50	237	
	1160	HDPE type s, elbows 18″ diam	B-21	9	3.111		149	100	16.95	265.95	345	
	1170	HDPE type s, elbows 24″ diam		9	3.111		315	100	16.95	431.95	530	
	1180	HDPE type s, elbows 30″ diam		8	3.500		505	112	19.10	636.10	760	
	1190	HDPE type s, elbows 36″ diam		8	3.500		650	112	19.10	781.10	920	
	1240	HDPE type s, Tee 12″ diam	B-20	7	3.429		132	107		239	320	
	1260	HDPE type s, Tee, 15″ diam	″	6	4		156	124		280	375	
	1280	HDPE type s, Tee, 18″ diam	B-21	6	4.667		229	150	25.50	404.50	525	
	1300	HDPE type s, Tee, 24″ diam		5	5.600		300	180	30.50	510.50	660	
	1320	HDPE type s, Tee, 30″ diam		5	5.600		565	180	30.50	775.50	955	
	1400	Add to basic installation cost for each split coupling joint										
	1402	HDPE type s, split coupling, 12″ diam	B-20	17	1.412	Ea.	4.27	44		48.27	77	
	1420	HDPE type s, split coupling, 15″ diam	″	15	1.600	″	5.65	50		55.65	88.50	
530	0010	**SEWAGE/DRAINAGE COLLECTION, CONCRETE PIPE**										530
	0020	Not including excavation or backfill										
	0100	Box culvert, precast, base price, 8′ long, 6′ x 3′	B-69	140	.343	L.F.	168	10.40	7.55	185.95	210	
	0150	6′ x 7′		125	.384		255	11.65	8.45	275.10	310	
	0200	8′ x 3′		133	.361		232	10.95	7.95	250.90	282	
	0250	8′ x 8′		100	.480		315	14.55	10.60	340.15	380	
	0300	10′ x 3′		110	.436		345	13.25	9.65	367.90	410	
	0350	10′ x 8′		80	.600		390	18.20	13.25	421.45	475	
	0400	12′ x 3′		100	.480		335	14.55	10.60	360.15	400	
	0450	12′ x 8′		67	.716		575	22	15.80	612.80	685	

For expanded coverage of these items see *Means Site Work & Landscape Cost Data 2006*

	02630 | Storm Drainage		CREW	DAILY OUTPUT	LABOR-HOURS	UNIT	2006 BARE COSTS MAT.	LABOR	EQUIP.	TOTAL	TOTAL INCL O&P	
530	0500	Set up charge at plant, add to base price				Job	3,200			3,200	3,525	530
	0510	Inserts and keyway, add				Ea.	281			281	310	
	0520	Sloped or skewed end, add				"	445			445	490	
	1000	Non-reinforced pipe, extra strength, B&S or T&G joints										
	1010	6" diameter	B-14	265.04	.181	L.F.	4.46	5.25	.85	10.56	14.50	
	1020	8" diameter		224	.214		4.91	6.20	1.01	12.12	16.70	
	1030	10" diameter		216	.222		5.45	6.45	1.05	12.95	17.75	
	1040	12" diameter		200	.240		6.70	6.95	1.13	14.78	20	
	1050	15" diameter		180	.267		7.80	7.75	1.26	16.81	22.50	
	1060	18" diameter		144	.333		9.60	9.70	1.57	20.87	28	
	1070	21" diameter		112	.429		11.80	12.45	2.02	26.27	35.50	
	1080	24" diameter	↓	100	.480	↓	14.50	13.95	2.26	30.71	41.50	
	1560	Reinforced culvert, class 2, no gaskets										
	1590	27" diameter	B-21	88	.318	L.F.	29	10.20	1.73	40.93	50.50	
	1592	30" diameter	B-13	80	.700		31.50	21	8.15	60.65	77.50	
	1594	36" diameter	"	72	.778	↓	36.50	23.50	9.05	69.05	88.50	
	2000	Reinforced culvert, class 3, no gaskets										
	2010	12" diameter	B-14	150	.320	L.F.	12.10	9.30	1.51	22.91	30.50	
	2020	15" diameter		150	.320		15.50	9.30	1.51	26.31	34	
	2030	18" diameter		132	.364		16.30	10.55	1.71	28.56	37	
	2035	21" diameter		120	.400		21	11.60	1.88	34.48	44.50	
	2040	24" diameter	↓	100	.480		24	13.95	2.26	40.21	52	
	2045	27" diameter	B-13	92	.609		30.50	18.20	7.05	55.75	71	
	2050	30" diameter		88	.636		33	19.05	7.40	59.45	75.50	
	2060	36" diameter	↓	72	.778		47	23.50	9.05	79.55	99.50	
	2070	42" diameter	B-13B	72	.778		64	23.50	13.20	100.70	123	
	2080	48" diameter		64	.875		79.50	26	14.85	120.35	146	
	2090	60" diameter		48	1.167		127	35	19.80	181.80	219	
	2100	72" diameter		40	1.400		180	42	24	246	293	
	2120	84" diameter		32	1.750		305	52.50	29.50	387	455	
	2140	96" diameter	↓	24	2.333		365	70	39.50	474.50	560	
	2200	With gaskets, class 3, 12" diameter	B-21	168	.167		13.25	5.35	.91	19.51	24.50	
	2220	15" diameter		160	.175		15.90	5.60	.95	22.45	28	
	2230	18" diameter		152	.184		19.90	5.90	1	26.80	33	
	2240	24" diameter	↓	136	.206		30	6.60	1.12	37.72	45	
	2260	30" diameter	B-13	88	.636		40	19.05	7.40	66.45	82.50	
	2270	36" diameter	"	72	.778		59.50	23.50	9.05	92.05	113	
	2290	48" diameter	B-13B	64	.875		97.50	26	14.85	138.35	166	
	2310	72" diameter	"	40	1.400		252	42	24	318	370	
	2330	Flared ends, 6'-1" long, 12" diameter	B-21	190	.147		36.50	4.73	.80	42.03	48.50	
	2340	15" diameter		155	.181		41.50	5.80	.98	48.28	56.50	
	2400	6'-2" long, 18" diameter		122	.230		43.50	7.35	1.25	52.10	61.50	
	2420	24" diameter	↓	88	.318		50	10.20	1.73	61.93	73.50	
	2440	36" diameter	B-13	60	.933	↓	91	28	10.85	129.85	157	
	2500	Class 4										
	2510	12" diameter	B-21	168	.167	L.F.	14.25	5.35	.91	20.51	25.50	
	2512	15" diameter		160	.175		16.85	5.60	.95	23.40	29	
	2514	18" diameter		152	.184		19.95	5.90	1	26.85	33	
	2516	21" diameter		144	.194		24.50	6.25	1.06	31.81	38.50	
	2518	24" diameter		136	.206		29	6.60	1.12	36.72	44	
	2520	27" diameter	↓	120	.233		36.50	7.50	1.27	45.27	53.50	
	2522	30" diameter	B-13	88	.636		40	19.05	7.40	66.45	83	
	2524	36" diameter	"	72	.778	↓	59	23.50	9.05	91.55	113	
	2600	Class 5										
	2610	12" diameter	B-21	168	.167	L.F.	14.55	5.35	.91	20.81	26	
	2612	15" diameter	↓	160	.175	↓	17.10	5.60	.95	23.65	29	

02600 | Drainage & Containment

		02630 | Storm Drainage	CREW	DAILY OUTPUT	LABOR-HOURS	UNIT	2006 BARE COSTS MAT.	LABOR	EQUIP.	TOTAL	TOTAL INCL O&P	
530	2614	18″ diameter	B-21	152	.184	L.F.	20.50	5.90	1	27.40	33.50	530
	2616	21″ diameter		144	.194		25.50	6.25	1.06	32.81	39.50	
	2618	24″ diameter		136	.206		30	6.60	1.12	37.72	44.50	
	2620	27″ diameter		120	.233		38	7.50	1.27	46.77	55.50	
	2622	30″ diameter	B-13	88	.636		42	19.05	7.40	68.45	85.50	
	2624	36″ diameter	″	72	.778		59	23.50	9.05	91.55	113	
	2800	Add for rubber joints,					12%					
	3080	Radius pipe, add to pipe prices, 12″ to 60″ diameter					50%					
	3090	Over 60″ diameter, add					20%					
	3500	Reinforced elliptical, 8′ lengths, C507 class 3										
	3520	14″ x 23″ inside, round equivalent 18″ diameter	B-21	82	.341	L.F.	28.50	10.95	1.86	41.31	51.50	
	3530	24″ x 38″ inside, round equivalent 30″ diameter	B-13	58	.966		50.50	29	11.20	90.70	115	
	3540	29″ x 45″ inside, round equivalent 36″ diameter		52	1.077		65	32	12.50	109.50	138	
	3550	38″ x 60″ inside, round equivalent 48″ diameter		38	1.474		85	44	17.10	146.10	184	
	3560	48″ x 76″ inside, round equivalent 60″ diameter		26	2.154		130	64.50	25	219.50	276	
	3570	58″ x 91″ inside, round equivalent 72″ diameter		22	2.545		185	76	29.50	290.50	360	
	3780	Concrete slotted pipe, class 4 mortar joint										
	3800	12″ diameter	B-21	168	.167	L.F.	14.20	5.35	.91	20.46	25.50	
	3840	18″ diameter	″	152	.184	″	22	5.90	1	28.90	35	
	3900	Class 4 O-ring										
	3940	12″ diameter	B-21	168	.167	L.F.	14.85	5.35	.91	21.11	26	
	3960	18″ diameter	″	152	.184	″	19.95	5.90	1	26.85	33	
	6200	Gasket, conc. pipe joint, 12″				Ea.	3.41			3.41	3.75	
	6220	24″					6.25			6.25	6.85	
	6240	36″					9.05			9.05	10	
	6260	48″					14.35			14.35	15.75	
	6265	54″					15.90			15.90	17.50	
	6270	60″					20.50			20.50	22.50	
	6280	72″					22.50			22.50	25	
		02640 | Culverts/Manuf Const										
810	0010	OVAL ARCH CULVERTS										810
	3000	Corrugated galvanized or aluminum, coated & paved										
	3020	17″ x 13″, 16 ga., 15″ equivalent	B-14	200	.240	L.F.	29	6.95	1.13	37.08	44.50	
	3040	21″ x 15″, 16 ga., 18″ equivalent		150	.320		37.50	9.30	1.51	48.31	58	
	3060	28″ x 20″, 14 ga., 24″ equivalent		125	.384		53.50	11.15	1.81	66.46	79.50	
	3080	35″ x 24″, 14 ga., 30″ equivalent		100	.480		65.50	13.95	2.26	81.71	97.50	
	3100	42″ x 29″, 12 ga., 36″ equivalent	B-13	100	.560		97.50	16.75	6.50	120.75	142	
	3120	49″ x 33″, 12 ga., 42″ equivalent		90	.622		117	18.65	7.25	142.90	167	
	3140	57″ x 38″, 12 ga., 48″ equivalent		75	.747		131	22.50	8.70	162.20	190	
	3160	Steel, plain oval arch culverts, plain										
	3180	17″ x 13″, 16 ga., 15″ equivalent	B-14	225	.213	L.F.	15.75	6.20	1	22.95	28.50	
	3200	21″ x 15″, 16 ga., 18″ equivalent		175	.274		18.65	7.95	1.29	27.89	35	
	3220	28″ x 20″, 14 ga., 24″ equivalent		150	.320		30	9.30	1.51	40.81	50	
	3240	35″ x 24″, 14 ga., 30″ equivalent	B-13	108	.519		37.50	15.50	6.05	59.05	73.50	
	3260	42″ x 29″, 12 ga., 36″ equivalent		108	.519		62	15.50	6.05	83.55	100	
	3280	49″ x 33″, 12 ga., 42″ equivalent		92	.609		73.50	18.20	7.05	98.75	118	
	3300	57″ x 38″, 12 ga., 48″ equivalent		75	.747		64.50	22.50	8.70	95.70	117	
	3320	End sections, 17″ x 13″		22	2.545	Ea.	86	76	29.50	191.50	252	
	3340	42″ x 29″		17	3.294	″	355	98.50	38.50	492	595	
	3360	Multi-plate arch, steel	B-20	1690	.014	Lb.	.99	.44		1.43	1.82	
		02660 | Ponds & Reservoirs										
600	0010	**SLURRY TRENCH** Excavated slurry trench in wet soils										600
	0020	backfilled with 3000 PSI concrete, no reinforcing steel										

02600 | Drainage & Containment

02660 | Ponds & Reservoirs

			CREW	DAILY OUTPUT	LABOR-HOURS	UNIT	2006 BARE COSTS MAT.	LABOR	EQUIP.	TOTAL	TOTAL INCL O&P	
600	0050	Minimum	C-7	333	.216	C.F.	6.45	6.50	3.04	15.99	21	600
	0100	Maximum		200	.360	"	10.80	10.80	5.05	26.65	35	
	0200	Alternate pricing method, minimum		150	.480	S.F.	12.90	14.40	6.75	34.05	45	
	0300	Maximum		120	.600		19.35	18	8.40	45.75	60.50	
	0500	Reinforced slurry trench, minimum	B-48	177	.316		9.65	9.80	20.50	39.95	49	
	0600	Maximum	"	69	.812		32	25	52.50	109.50	135	
	0800	Haul for disposal, 2 mile haul, excavated material, add	B-34B	99	.081	C.Y.		2.29	4.81	7.10	9.05	
	0900	Haul bentonite castings for disposal, add	"	40	.200	"		5.65	11.90	17.55	22.50	

02700 | Bases, Ballasts, Pavements & Appurtenances

02710 | Bound Base Courses

			CREW	DAILY OUTPUT	LABOR-HOURS	UNIT	2006 BARE COSTS MAT.	LABOR	EQUIP.	TOTAL	TOTAL INCL O&P	
100	0010	**AGGREGATE-BITUMINOUS BASE COURSE** for roadways										100
	0020	and large paved areas										
	0500	Bituminous concrete, 4″ thick	B-25	4545	.019	S.Y.	8.50	.58	.47	9.55	10.80	
	0550	6″ thick		3700	.024		12.50	.72	.57	13.79	15.55	
	0560	8″ thick		3000	.029		16.65	.88	.71	18.24	20.50	
	0570	10″ thick		2545	.035		20.50	1.04	.83	22.37	25	
	8900	For small and irregular areas, add						50%	50%			
200	0010	**ASPHALT-TREATED PERMEABLE BASE COURSE** for roadways										200
	0020	and large paved areas										
	0700	Liquid application to gravel base, asphalt emulsion	B-45	6000	.003	Gal.	3	.09	.09	3.18	3.54	
	0800	Prime and seal, cut back asphalt		6000	.003	"	3.54	.09	.09	3.72	4.13	
	1000	Macadam penetration crushed stone, 2 gal. per S.Y., 4″ thick		6000	.003	S.Y.	6	.09	.09	6.18	6.85	
	1100	6″ thick, 3 gal. per S.Y.		4000	.004		9	.13	.13	9.26	10.25	
	1200	8″ thick, 4 gal. per S.Y.		3000	.005		12	.17	.18	12.35	13.70	
	8900	For small and irregular areas, add						50%	50%			

02720 | Unbound Base Courses & Ballasts

			CREW	DAILY OUTPUT	LABOR-HOURS	UNIT	2006 BARE COSTS MAT.	LABOR	EQUIP.	TOTAL	TOTAL INCL O&P	
200	0010	**AGGREGRATE BASE COURSE** For roadways and large areas										200
	0050	Crushed 3/4″ stone base, compacted, 3″ deep	B-36C	5200	.008	S.Y.	2.69	.26	.51	3.46	3.93	
	0100	6″ deep		5000	.008		5.40	.27	.53	6.20	6.90	
	0200	9″ deep		4600	.009		8.05	.29	.58	8.92	10	
	0300	12″ deep		4200	.010		10.75	.32	.63	11.70	13.05	
	0301	Crushed 1-1/2″ stone base, compacted to 4″ deep	B-36B	6000	.011		4.25	.35	.50	5.10	5.80	
	0302	6″ deep		5400	.012		6.40	.38	.56	7.34	8.25	
	0303	8″ deep		4500	.014		8.50	.46	.67	9.63	10.85	
	0304	12″ deep		3800	.017		12.75	.55	.79	14.09	15.80	
	0310	Minimum labor/equipment charge	B-36	1	40	Job		1,250	1,150	2,400	3,325	
	0350	Bank run gravel, spread and compacted										
	0370	6″ deep	B-32	6000	.005	S.Y.	3.54	.18	.26	3.98	4.47	
	0390	9″ deep		4900	.007		5.30	.22	.32	5.84	6.55	
	0400	12″ deep		4200	.008		7.05	.26	.38	7.69	8.65	
	1500	Alternate method to figure base course										
	1510	Crushed stone, 3/4″ , compacted, 3″ deep	B-36C	435	.092	E.C.Y.	27.50	3.09	6.10	36.69	42	
	1511	6″ deep	B-36B	835	.077		27.50	2.49	3.62	33.61	38.50	
	1512	9″ deep		1150	.056		27.50	1.80	2.63	31.93	36.50	
	1513	12″ deep		1400	.046		27.50	1.48	2.16	31.14	35.50	
	1520	Crushed stone, 1-1/2″ , compacted 4″ deep		665	.096		27.50	3.12	4.54	35.16	40.50	

02700 | Bases, Ballasts, Pavements & Appurtenances

02720 | Unbound Base Courses & Ballasts

			CREW	DAILY OUTPUT	LABOR-HOURS	UNIT	2006 BARE COSTS MAT.	LABOR	EQUIP.	TOTAL	TOTAL INCL O&P
200	1521	6" deep	B-36B	900	.071	E.C.Y.	27.50	2.31	3.36	33.17	38
	1522	8" deep		1000	.064		27.50	2.07	3.02	32.59	37
	1523	12" deep		1265	.051		27.50	1.64	2.39	31.53	36
	1530	Gravel, bank run, compacted, 6" deep	B-36C	835	.048		18.15	1.61	3.17	22.93	26
	1531	9" deep		1150	.035		18.15	1.17	2.30	21.62	24.50
	1532	12" deep		1400	.029		18.15	.96	1.89	21	23.50
	2000	Alternate method to figure base course									
	2005	Bituminous concrete, 4" thick	B-25	1000	.088	Ton	38	2.65	2.12	42.77	48
	2006	6" thick		1220	.072		38	2.17	1.74	41.91	47
	2007	8" thick		1320	.067		38	2.01	1.60	41.61	46.50
	2008	10" thick		1400	.063		38	1.89	1.51	41.40	46.50
	2010	Crushed stone, 3/4" maximum size, 3" deep	B-36	540	.074		16.70	2.33	2.13	21.16	24.50
	2011	6" deep		1625	.025		16.70	.78	.71	18.19	20.50
	2012	9" deep		1785	.022		16.70	.71	.64	18.05	20
	2013	12" deep		1950	.021		16.70	.65	.59	17.94	20
	2020	Crushed stone, 1-1/2" maximum size, 4" deep		720	.056		16.70	1.75	1.60	20.05	23
	2021	6" deep		815	.049		16.70	1.55	1.41	19.66	22.50
	2022	8" deep		835	.048		16.70	1.51	1.38	19.59	22.50
	2023	12" deep		975	.041		16.70	1.29	1.18	19.17	22
	2030	Bank run gravel, 6" deep	B-32A	875	.027		12.20	.92	1.12	14.24	16.15
	2031	9" deep		970	.025		12.20	.83	1.01	14.04	15.90
	2032	12" deep		1060	.023		12.20	.76	.92	13.88	15.70
	6000	Stabilization fabric, polypropylene, 6 oz./S.Y.	B-6	10000	.002	S.Y.	.94	.07	.02	1.03	1.17
	6900	For small and irregular areas, add						50%	50%		
	6990	Minimum labor/equipment charge	1 Clab	3	2.667	Job		73		73	121
	7000	Prepare and roll sub-base, small areas to 2500 S.Y.	B-32A	1500	.016	S.Y.		.54	.65	1.19	1.58
	8000	Large areas over 2500 S.Y.	B-32	3700	.009	"		.30	.43	.73	.94
	9000	Minimum labor/equipment charge	1 Clab	4	2	Job		55		55	91

02740 | Flexible Pavement

			CREW	DAILY OUTPUT	LABOR-HOURS	UNIT	MAT.	LABOR	EQUIP.	TOTAL	TOTAL INCL O&P
310	0010	**ASPHALTIC CONCRETE PAVEMENT, HIGHWAYS**									
	0020	and large paved areas									
	0080	Binder course, 1-1/2" thick	B-25	7725	.011	S.Y.	2.89	.34	.27	3.50	4.04
	0120	2" thick		6345	.014		3.85	.42	.33	4.60	5.30
	0160	3" thick		4905	.018		5.70	.54	.43	6.67	7.65
	0200	4" thick		4140	.021		7.65	.64	.51	8.80	10
	0300	Wearing course, 1" thick	B-25B	10575	.009		2.04	.28	.22	2.54	2.94
	0340	1-1/2" thick		7725	.012		3.11	.38	.30	3.79	4.37
	0380	2" thick		6345	.015		4.18	.46	.36	5	5.75
	0420	2-1/2" thick		5480	.018		5.15	.54	.42	6.11	7
	0460	3" thick		4900	.020		6.15	.60	.47	7.22	8.25
	0500	Open graded friction course	B-25C	5000	.010		1.85	.30	.37	2.52	2.93
	0800	Alternate method of figuring paving costs									
	0810	Binder course, 1-1/2" thick	B-25	630	.140	Ton	38	4.21	3.36	45.57	52
	0811	2" thick		690	.128		38	3.84	3.07	44.91	51
	0812	3" thick		800	.110		38	3.31	2.65	43.96	50
	0813	4" thick		850	.104		38	3.12	2.49	43.61	49.50
	0850	Wearing course, 1" thick	B-25B	575	.167		41	5.10	4.03	50.13	57.50
	0851	1-1/2" thick		630	.152		41	4.67	3.67	49.34	56.50
	0852	2" thick		690	.139		41	4.27	3.35	48.62	55.50
	0853	2-1/2" thick		745	.129		41	3.95	3.11	48.06	55
	0854	3" thick		800	.120		41	3.68	2.89	47.57	54
	1000	Pavement replacement over trench, 2" thick	B-37	90	.533	S.Y.	4.25	15.50	1.27	21.02	31.50
	1050	4" thick		70	.686		8.40	19.90	1.63	29.93	43.50
	1080	6" thick		55	.873		13.40	25.50	2.08	40.98	58.50
	1200	For paving projects 300 tons or less add for trucking									

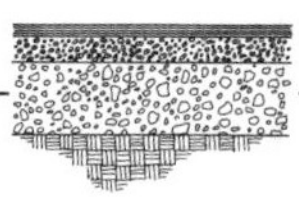

02740 | Flexible Pavement

			CREW	DAILY OUTPUT	LABOR-HOURS	UNIT	2006 BARE COSTS MAT.	LABOR	EQUIP.	TOTAL	TOTAL INCL O&P	
310	1300	See 02315-490-0010 for hauling costs										310
315	0011	**ASPHALTIC CONCRETE PAVEMENT, LOTS & DRIVEWAYS**										315
	0020	6" stone base, 2" binder course, 1" topping	B-25C	9000	.005	S.F.	1.42	.16	.21	1.79	2.06	
	0300	Binder course, 1-1/2" thick		35000	.001		.34	.04	.05	.43	.51	
	0400	2" thick		25000	.002		.45	.06	.07	.58	.67	
	0500	3" thick		15000	.003		.69	.10	.12	.91	1.06	
	0600	4" thick		10800	.004		.90	.14	.17	1.21	1.40	
	0800	Sand finish course, 3/4" thick		41000	.001		.18	.04	.05	.27	.31	
	0900	1" thick		34000	.001		.22	.04	.05	.31	.37	
	1000	Fill pot holes, hot mix, 2" thick	B-16	4200	.008		.47	.21	.11	.79	.98	
	1100	4" thick		3500	.009		.68	.26	.14	1.08	1.32	
	1120	6" thick		3100	.010		.92	.29	.15	1.36	1.66	
	1140	Cold patch, 2" thick	B-51	3000	.016		.47	.44	.04	.95	1.30	
	1160	4" thick		2700	.018		.90	.49	.05	1.44	1.86	
	1180	6" thick		1900	.025		1.40	.70	.07	2.17	2.77	

02750 | Rigid Pavement

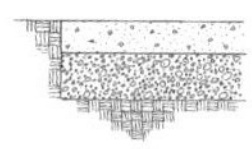

			CREW	DAILY OUTPUT	LABOR-HOURS	UNIT	MAT.	LABOR	EQUIP.	TOTAL	TOTAL INCL O&P	
300	0010	**PLAIN CEMENT CONCRETE PAVEMENT**										300
	0015	Including joints, finishing and curing										
	0020	Fixed form, 12' pass, unreinforced, 6" thick	B-26	3000	.029	S.Y.	21.50	.91	.90	23.31	26	
	0030	7" thick		2850	.031		27	.96	.95	28.91	32	
	0100	8" thick		2750	.032		29.50	.99	.98	31.47	35	
	0110	8" thick, small area		1375	.064		29.50	1.98	1.96	33.44	38	
	0200	9" thick		2500	.035		34.50	1.09	1.08	36.67	41	
	0300	10" thick		2100	.042		37	1.30	1.28	39.58	44.50	
	0310	10" thick, small area		1050	.084		37	2.60	2.57	42.17	48	
	0400	12" thick		1800	.049		40	1.52	1.50	43.02	48	
	0500	15" thick		1500	.059		44.50	1.82	1.80	48.12	54	
	0505	Minimum labor/equipment charge	1 Cefi	1	8	Job		275		275	435	
	0510	For small irregular areas, add				%		100%				
	0610	For under 10' pass, add								7.30	8.55	
	0650	For 24' pass, deduct						20%	20%	20%		
	0655	For slipforming, deduct						10%	10%	10%		
	0700	Finishing, broom finish small areas	2 Cefi	120	.133	S.Y.		4.59		4.59	7.25	
	0710	Transverse joint support dowels	C-1	350	.091	Ea.	3.24	3.06		6.30	8.65	
	0720	Transverse contraction joints, saw cut & grind	A-1B	120	.067	L.F.		1.83	.89	2.72	4.01	
	0730	Transverse expansion joints, incl. premolded bit. jt. filler	C-1	150	.213		1.57	7.15		8.72	13.60	
	0740	Transverse construction joint using bulkhead	"	73	.438		2.28	14.70		16.98	27	
	0750	Longitudinal joint tie bars, grouted	B-23	70	.571	Ea.	3.53	15.90	51	70.43	86.50	
	1000	Curing, with sprayed membrane by hand	2 Clab	1500	.011	S.Y.	.50	.29		.79	1.03	
	1650	For integral coloring, see div. 03310-220										

02760 | Paving Specialties

			CREW	DAILY OUTPUT	LABOR-HOURS	UNIT	MAT.	LABOR	EQUIP.	TOTAL	TOTAL INCL O&P	
300	0010	**PAINTED TRAFFIC LINES AND MARKINGS**										300
	0020	Acrylic waterborne, white or yellow, 4" wide	B-78	20000	.002	L.F.	.17	.07	.02	.26	.32	
	0200	6" wide		11000	.004		.16	.12	.04	.32	.42	
	0500	8" wide		10000	.005		.22	.13	.04	.39	.50	
	0600	12" wide		4000	.012		.41	.33	.10	.84	1.11	
	0620	Arrows or gore lines		2300	.021	S.F.	.64	.58	.17	1.39	1.85	
	0640	Temporary paint, white or yellow		15000	.003	L.F.	.19	.09	.03	.31	.39	
	0660	Removal	1 Clab	300	.027			.73		.73	1.21	
	0680	Temporary tape	2 Clab	1500	.011		1.85	.29		2.14	2.52	
	0710	Thermoplastic, white or yellow, 4" wide	B-79	15000	.003		.74	.07	.06	.87	1	
	0730	6" wide		14000	.003		1.07	.08	.06	1.21	1.38	
	0740	8" wide		12000	.003		1.44	.09	.07	1.60	1.82	

02760 | Paving Specialties

			CREW	DAILY OUTPUT	LABOR-HOURS	UNIT	2006 BARE COSTS MAT.	LABOR	EQUIP.	TOTAL	TOTAL INCL O&P	
300	0750	12" wide	B-79	6000	.007	L.F.	2.15	.19	.15	2.49	2.83	300
	0760	Arrows		660	.061	S.F.	1.98	1.69	1.34	5.01	6.45	
	0770	Gore lines		2500	.016		1.31	.45	.36	2.12	2.57	
	0780	Letters		660	.061		1.64	1.69	1.34	4.67	6.05	
	0782	Thermoplastic material				Ton	1,050			1,050	1,150	
	0784	Glass beads, add				M.L.F.	9.30			9.30	10.25	
500	0010	PAVEMENT MARKINGS										500
	0790	Layout of pavement marking	A-2	25000	.001	L.F.		.03	.01	.04	.05	
	0800	Parking stall, paint, white	B-78	440	.109	Stall	3.43	3.03	.88	7.34	9.75	
	1000	Street letters and numbers	"	1600	.030	S.F.	.66	.83	.24	1.73	2.38	
	1100	Pavement marking letter, 6"	2 Pord	400	.040	Ea.	1.32	1.27		2.59	3.49	
	1110	12" letter		272	.059		1.93	1.86		3.79	5.10	
	1120	24" letter		160	.100		3.22	3.17		6.39	8.65	
	1130	36" letter		84	.190		6.50	6.05		12.55	16.85	
	1140	42" letter		84	.190		15.75	6.05		21.80	27	
	1150	72" letter		40	.400		44.50	12.70		57.20	69	
	1200	Handicap symbol		40	.400		27.50	12.70		40.20	50.50	

02770 | Curbs and Gutters

			CREW	DAILY OUTPUT	LABOR-HOURS	UNIT	MAT.	LABOR	EQUIP.	TOTAL	TOTAL INCL O&P	
100	0010	BITUMINOUS CONCRETE CURBS										100
	0012	Curbs, asphaltic, machine formed, 8" wide, 6" high, 40 L.F./ton	B-27	1000	.032	L.F.	.82	.89	.23	1.94	2.64	
	0100	8" wide, 8" high, 30 L.F. per ton		900	.036		.94	.99	.26	2.19	2.96	
	0150	Asphaltic berm, 12" W, 3"-6" H, 35 L.F./ton, before pavement		700	.046		1.07	1.28	.33	2.68	3.65	
	0200	12" W, 1-1/2" to 4" H, 60 L.F. per ton, laid with pavement	B-2	1050	.038		.65	1.06		1.71	2.47	
300	0010	CEMENT CONCRETE CURBS										300
	0300	Concrete, wood forms, 6" x 18", straight	C-2A	500	.096	L.F.	2.89	3.30		6.19	8.55	
	0400	6" x 18", radius	"	200	.240		3	8.25		11.25	16.85	
	0415	Machine formed, 6" x 18", straight	B-69A	2000	.024		3.80	.73	.30	4.83	5.70	
	0416	6" x 18", radius	"	900	.053		3.96	1.62	.66	6.24	7.75	
	0421	Curb and gutter, straight										
	0422	with 6" high curb and 6" thick gutter, wood forms										
	0430	24" wide, .055 C.Y. per L.F.	C-2A	375	.128	L.F.	14.30	4.39		18.69	23	
	0435	30" wide, .066 C.Y. per L.F.	"	340	.141		15.60	4.85		20.45	25	
	0550	Precast, 6" x 18", straight	B-29	700	.080		7.95	2.40	1.20	11.55	13.95	
	0600	6" x 18", radius	"	325	.172		9.05	5.15	2.59	16.79	21.50	
500	0010	STONE CURBS										500
	1000	Granite, split face, straight, 5" x 16"	D-13	500	.096	L.F.	10.20	3.27	1.07	14.54	17.70	
	1300	Radius curbing, 6" x 18", over 10' radius	B-29	260	.215	"	16.40	6.45	3.24	26.09	32	
	1400	Corners, 2' radius		80	.700	Ea.	55	21	10.50	86.50	106	
	1600	Edging, 4-1/2" x 12", straight		300	.187	L.F.	5.10	5.60	2.80	13.50	17.80	
	1800	Curb inlets, (guttermouth) straight		41	1.366	Ea.	122	41	20.50	183.50	224	
	2000	Indian granite (belgian block)										
	2100	Jumbo, 10-1/2" x 7-1/2" x 4", grey	D-1	150	.107	L.F.	1.86	3.43		5.29	7.60	
	2150	Pink		150	.107		2.41	3.43		5.84	8.20	
	2200	Regular, 9" x 4-1/2" x 4-1/2", grey		160	.100		1.68	3.22		4.90	7.05	
	2250	Pink		160	.100		2.31	3.22		5.53	7.75	
	2300	Cubes, 4" x 4" x 4", grey		175	.091		1.60	2.94		4.54	6.55	
	2350	Pink		175	.091		1.68	2.94		4.62	6.60	
	2400	6" x 6" x 6", pink		155	.103		4.17	3.32		7.49	10	
	2500	Alternate pricing method for indian granite										
	2550	Jumbo, 10-1/2" x 7-1/2" x 4" (30 lb), grey				Ton	106			106	116	
	2600	Pink					140			140	154	
	2650	Regular, 9" x 4-1/2" x 4-1/2" (20 lb), grey					119			119	131	

		02770 | Curbs and Gutters	CREW	DAILY OUTPUT	LABOR-HOURS	UNIT	2006 BARE COSTS MAT.	LABOR	EQUIP.	TOTAL	TOTAL INCL O&P	
500	2700	Pink				Ton	162			162	178	500
	2750	Cubes, 4″ x 4″ x 4″ (15 lb), grey					194			194	214	
	2800	Pink					216			216	238	
	2850	6″ x 6″ x 6″ (25 lb), pink					162			162	178	
	2900	For pallets, add					16.50			16.50	18.15	

		02775 | Sidewalks	CREW	DAILY OUTPUT	LABOR-HOURS	UNIT	MAT.	LABOR	EQUIP.	TOTAL	TOTAL INCL O&P	
275	0010	**SIDEWALKS, DRIVEWAYS, & PATIOS** No base										275
	0020	Asphaltic concrete, 2″ thick	B-37	720	.067	S.Y.	4.01	1.94	.16	6.11	7.75	
	0100	2-1/2″ thick	"	660	.073	"	5.10	2.11	.17	7.38	9.25	
	0110	Bedding for brick or stone, mortar, 1″ thick	D-1	300	.053	S.F.	.65	1.71		2.36	3.50	
	0120	2″ thick	"	200	.080		1.63	2.57		4.20	5.95	
	0130	Sand, 2″ thick	B-18	8000	.003		.16	.08		.24	.32	
	0140	4″ thick	"	4000	.006		.32	.17	.01	.50	.65	
	0300	Concrete, 3000 psi, CIP, 6 x 6 - W1.4 x W1.4 mesh,										
	0310	broomed finish, no base, 4″ thick	B-24	600	.040	S.F.	1.45	1.30		2.75	3.71	
	0350	5″ thick		545	.044		1.93	1.43		3.36	4.46	
	0400	6″ thick 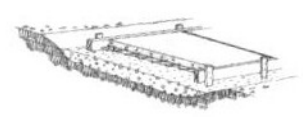		510	.047		2.26	1.53		3.79	4.97	
	0450	For bank run gravel base, 4″ thick, add	B-18	2500	.010		.42	.27	.01	.70	.94	
	0520	8″ thick, add	"	1600	.015		.85	.42	.02	1.29	1.66	
	0550	Exposed aggregate finish, add to above, minimum	B-24	1875	.013		.11	.42		.53	.80	
	0600	Maximum	"	455	.053		.35	1.71		2.06	3.18	
	0850	Splash block, precast concrete	1 Clab	150	.053	Ea.	6	1.46		7.46	9	
	0950	Concrete tree grate, 5′ square	B-6	25	.960		345	29	9.05	383.05	435	
	0960	Cast iron tree grate with frame, 2 piece, round, 5′ diameter		25	.960		950	29	9.05	988.05	1,100	
	0980	Square, 5′ side		25	.960		980	29	9.05	1,018.05	1,125	
	1000	Crushed stone, 1″ thick, white marble	2 Clab	1700	.009	S.F.	.20	.26		.46	.65	
	1050	Bluestone		1700	.009		.21	.26		.47	.67	
	1070	Granite chips		1700	.009		.21	.26		.47	.66	
	1660	Limestone pavers, 3″ thick	D-1	72	.222		8.05	7.15		15.20	20.50	
	1670	4″ thick		70	.229		10.65	7.35		18	23.50	
	1680	5″ thick		68	.235		13.35	7.55		20.90	27	
	1700	Redwood, prefabricated, 4′ x 4′ sections	2 Carp	316	.051		8.10	1.80		9.90	11.90	
	1750	Redwood planks, 1″ thick, on sleepers	"	240	.067		5.65	2.37		8.02	10.20	
	1830	1-1/2″ thick	B-28	167	.144		3.93	4.72		8.65	12.10	
	1840	2″ thick		167	.144		5.05	4.72		9.77	13.35	
	1850	3″ thick		150	.160		7.45	5.25		12.70	16.90	
	1860	4″ thick		150	.160		9.80	5.25		15.05	19.45	
	1870	5″ thick		150	.160		12.45	5.25		17.70	22.50	
	2100	River or beach stone, stock	B-1	18	1.333	Ton	27	37.50		64.50	91.50	
	2150	Quarried	"	18	1.333	"	47	37.50		84.50	114	
	2160	Load, dump, and spread stone with skid steer, 100′ haul	B-62	24	1	C.Y.		30	6.40	36.40	56	
	2165	200′ haul		18	1.333			40	8.55	48.55	74.50	
	2168	300′ haul		12	2			60	12.85	72.85	112	
	2170	Shale paver, 2-1/4″ thick	D-1	200	.080	S.F.	2.76	2.57		5.33	7.20	
	2200	Coarse washed sand bed, 1″	B-62	1350	.018	S.Y.	1.21	.53	.11	1.85	2.33	
	2250	Stone dust, 4″ thick	"	900	.027	"	2.94	.80	.17	3.91	4.72	
	2300	Tile thinset pavers, 3/8″ thick	D-1	300	.053	S.F.	3.03	1.71		4.74	6.10	
	2350	3/4″ thick	"	280	.057	"	4.83	1.84		6.67	8.30	
	2400	Wood rounds, cypress	B-1	175	.137	Ea.	9.10	3.85		12.95	16.40	
	9000	Minimum labor/equipment charge	D-1	2	8	Job		257		257	415	

		02780 | Unit Pavers	CREW	DAILY OUTPUT	LABOR-HOURS	UNIT	MAT.	LABOR	EQUIP.	TOTAL	TOTAL INCL O&P	
100	0010	**ASPHALT BLOCKS**										100
	0020	Rectangular, 6″ x 12″ x 1-1/4″, w/bed & neopr. adhesive	D-1	135	.119	S.F.	4.64	3.81		8.45	11.30	

		02780 | Unit Pavers	CREW	DAILY OUTPUT	LABOR-HOURS	UNIT	2006 BARE COSTS MAT.	LABOR	EQUIP.	TOTAL	TOTAL INCL O&P	
100	0100	3" thick	D-1	130	.123	S.F.	6.50	3.96		10.46	13.55	100
	0300	Hexagonal tile, 8" wide, 1-1/4" thick		135	.119		4.64	3.81		8.45	11.30	
	0400	2" thick		130	.123		6.50	3.96		10.46	13.55	
	0500	Square, 8" x 8", 1-1/4" thick		135	.119		4.64	3.81		8.45	11.30	
	0600	2" thick		130	.123		6.50	3.96		10.46	13.55	
	0900	For exposed aggregate (ground finish) add					.32			.32	.35	
	0910	For colors, add					.24			.24	.26	
	9000	Minimum labor/equipment charge	1 Bric	2	4	Job		146		146	237	
200	0010	**BRICK PAVING**										200
	0012	4" x 8" x 1-1/2", without joints (4.5 brick/S.F.)	D-1	110	.145	S.F.	2.57	4.68		7.25	10.45	
	0100	Grouted, 3/8" joint (3.9 brick/S.F.)		90	.178		3.01	5.70		8.71	12.55	
	0200	4" x 8" x 2-1/4", without joints (4.5 bricks/S.F.)		110	.145		3.33	4.68		8.01	11.25	
	0300	Grouted, 3/8" joint (3.9 brick/S.F.)		90	.178		3.07	5.70		8.77	12.65	
	0500	Bedding, asphalt, 3/4" thick	B-25	5130	.017		.38	.52	.41	1.31	1.71	
	0540	Course washed sand bed, 1" thick 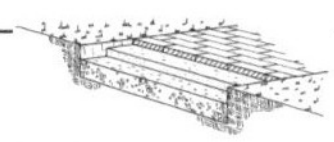	B-18	5000	.005		.18	.13	.01	.32	.43	
	0580	Mortar, 1" thick	D-1	300	.053		.54	1.71		2.25	3.38	
	0620	2" thick		200	.080		1.09	2.57		3.66	5.35	
	2000	Brick pavers, laid on edge, 7.2 per S.F.		70	.229		2.48	7.35		9.83	14.60	
	2500	For 4" thick concrete bed and joints, add		595	.027		1.03	.86		1.89	2.53	
	2800	For steam cleaning, add	A-1H	950	.008		.05	.23	.06	.34	.51	
	9000	Minimum labor/equipment charge	1 Bric	2	4	Job		146		146	237	
400	0010	**PRECAST CONCRETE PAVING SLABS**										400
	0710	Precast concrete patio blocks, 2-3/8" thick, colors, 8" x 16"	D-1	265	.060	S.F.	1.24	1.94		3.18	4.51	
	0715	12" x 12"		300	.053		1.61	1.71		3.32	4.55	
	0720	16" x 16"		335	.048		1.77	1.54		3.31	4.44	
	0730	24" x 24"		510	.031		2.21	1.01		3.22	4.07	
	0740	Green, 8" x 16"		265	.060		1.60	1.94		3.54	4.91	
	0750	Exposed local aggregate, natural	2 Bric	250	.064		5.30	2.34		7.64	9.65	
	0800	Colors		250	.064		5.80	2.34		8.14	10.15	
	0850	Exposed granite or limestone aggregate		250	.064		6.40	2.34		8.74	10.85	
	0900	Exposed white tumblestone aggregate		250	.064		3.81	2.34		6.15	8	
410	0010	**PLANTER BLOCKS** Precast concrete, interlocking										410
	0020	"V" blocks for retaining soil	D-1	205	.078	S.F.	4.61	2.51		7.12	9.10	
600	0010	**STONE PAVERS**										600
	1100	Flagging, bluestone, irregular, 1" thick,	D-1	81	.198	S.F.	4.72	6.35		11.07	15.50	
	1110	1-1/2" thick		90	.178		5.55	5.70		11.25	15.40	
	1120	Pavers, 1/2" thick		110	.145		7.85	4.68		12.53	16.25	
	1130	3/4" thick		95	.168		10	5.40		15.40	19.80	
	1140	1" thick		81	.198		10.70	6.35		17.05	22	
	1150	Snapped random rectangular, 1" thick		92	.174		7.15	5.60		12.75	16.90	
	1200	1-1/2" thick		85	.188		8.60	6.05		14.65	19.25	
	1250	2" thick		83	.193		10	6.20		16.20	21	
	1300	Slate, natural cleft, irregular, 3/4" thick		92	.174		5.25	5.60		10.85	14.85	
	1310	1" thick		85	.188		6.15	6.05		12.20	16.55	
	1350	Random rectangular, gauged, 1/2" thick		105	.152		11.40	4.90		16.30	20.50	
	1400	Random rectangular, butt joint, gauged, 1/4" thick		150	.107		12.25	3.43		15.68	19.05	
	1450	For sand rubbed finish, add					5.70			5.70	6.30	
	1500	For interior setting, add								25%	25%	
	1550	Granite blocks, 3-1/2" x 3-1/2" x 3-1/2"	D-1	92	.174	S.F.	6.40	5.60		12	16.05	
	1560	4" x 4" x 4"		95	.168		6.75	5.40		12.15	16.20	
	1600	4" to 12" long, 3" to 5" wide, 3" to 5" thick		98	.163		5.30	5.25		10.55	14.35	
	1650	6" to 15" long, 3" to 6" wide, 3" to 5" thick		105	.152		2.84	4.90		7.74	11.05	
700	0010	**STEPS** Incl. excav., borrow & concrete base, where applicable										700
	0100	Brick steps	B-24	35	.686	LF Riser	10.35	22.50		32.85	47.50	

02780 | Unit Pavers

			CREW	DAILY OUTPUT	LABOR-HOURS	UNIT	2006 BARE COSTS MAT.	LABOR	EQUIP.	TOTAL	TOTAL INCL O&P	
700	0200	Railroad ties	2 Clab	25	.640	LF Riser	3.04	17.55		20.59	32.50	700
	0300	Bluestone treads, 12" x 2" or 12" x 1-1/2"	B-24	30	.800	↓	23	26		49	68	
	0490	Minimum labor/equipment charge	D-1	2	8	Job		257		257	415	
	0600	Precast concrete, see Division 03480-800										
	4000	Edging, redwood, 2" x 4"	2 Carp	330	.048	L.F.	1.89	1.72		3.61	4.94	
	4025	Steel edge strips, incl. stakes, 1/4" x 5"	B-1	390	.062		3.62	1.73		5.35	6.85	
	4050	Edging, landscape timber or railroad ties, 6" x 8"	2 Carp	170	.094	↓	2.69	3.35		6.04	8.50	

02785 | Flexible Pavement Coating

			CREW	DAILY OUTPUT	LABOR-HOURS	UNIT	MAT.	LABOR	EQUIP.	TOTAL	TOTAL INCL O&P	
250	0010	**FOG SEAL**										250
	0012	Sealcoating, 2 coat coal tar pitch emulsion over 10,000 SY	B-45	5000	.003	S.Y.	.53	.10	.11	.74	.87	
	0030	1000 to 10,000 S.Y.	"	3000	.005		.53	.17	.18	.88	1.06	
	0100	Under 1000 S.Y.	B-1	1050	.023		.53	.64		1.17	1.64	
	0300	Petroleum resistant, over 10,000 S.Y.	B-45	5000	.003		.66	.10	.11	.87	1.02	
	0320	1000 to 10,000 S.Y.	"	3000	.005		.66	.17	.18	1.01	1.21	
	0400	Under 1000 S.Y.	B-1	1050	.023		.66	.64		1.30	1.79	
	0600	Non-skid pavement renewal, over 10,000 S.Y.	B-45	5000	.003		.77	.10	.11	.98	1.14	
	0620	1000 to 10,000 S.Y.	"	3000	.005		.77	.17	.18	1.12	1.33	
	0700	Under 1000 S.Y.	B-1	1050	.023		.77	.64		1.41	1.91	
	0800	Prepare and clean surface for above	A-2	8545	.003	↓		.08	.02	.10	.15	
	1000	Hand seal asphalt curbing	B-1	4420	.005	L.F.	.39	.15		.54	.68	
	1900	Asphalt surface treatment, single course, small area										
	1901	0.30 gal/S.Y. asphalt material, 20#/S.Y. aggregate	B-91	5000	.013	S.Y.	.91	.42	.30	1.63	1.99	
	1910	Roadway or large area		10000	.006		.84	.21	.15	1.20	1.41	
	1950	Asphalt surface treatment, dbl. course for small area		3000	.021		1.68	.69	.49	2.86	3.50	
	1960	Roadway or large area		6000	.011		1.51	.35	.25	2.11	2.49	
	1980	Asphalt surface treatment,single course, for shoulders	↓	7500	.009	↓	.97	.28	.20	1.45	1.74	
300	0010	**LATEX MODIFIED EMULSION**										300
	3600	Waterproofing, membrane, tar and fabric, small area	B-63	233	.172	S.Y.	6.20	4.97	.66	11.83	15.75	
	3640	Large area		1435	.028		5.70	.81	.11	6.62	7.75	
	3680	Preformed rubberized asphalt, small area		100	.400		8.40	11.60	1.54	21.54	30	
	3720	Large area	↓	367	.109		7.65	3.16	.42	11.23	14	
	3780	Rubberized asphalt (latex) seal	B-45	5000	.003	↓	1.21	.10	.11	1.42	1.62	
	9000	Minimum labor/equipment charge	1 Clab	2	4	Job		110		110	182	
400	0010	**SAND SEAL**										400
	2080	Sand sealing, sharp sand, asphalt emulsion, small area	B-91	10000	.006	S.Y.	.66	.21	.15	1.02	1.22	
	2120	Roadway or large area	"	18000	.004	"	.57	.12	.08	.77	.90	
	3000	Sealing random cracks, min 1/2" wide, to 1-1/2", 1,000 L.F.	B-77	2800	.014	L.F.	.52	.40	.14	1.06	1.38	
	3040	10,000 L.F.		4000	.010	"	.40	.28	.09	.77	1	
	3080	Alternate method, 1,000 L.F.		200	.200	Gal.	7.85	5.55	1.90	15.30	19.95	
	3120	10,000 L.F.	↓	325	.123	"	6.30	3.43	1.17	10.90	13.90	
	3200	Multi-cracks (flooding), 1 coat, small area	B-92	460	.070	S.Y.	2.36	1.94	.71	5.01	6.60	
	3240	Large area		2850	.011		2.10	.31	.11	2.52	2.96	
	3280	2 coat, small area		230	.139		7.75	3.88	1.42	13.05	16.55	
	3320	Large area		1425	.022	↓	7.05	.63	.23	7.91	9.05	
	3360	Alternate method, small area		115	.278	Gal.	8.10	7.75	2.84	18.69	25	
	3400	Large area	↓	715	.045	"	7.60	1.25	.46	9.31	10.95	
500	0010	**SLURRY SEAL**										500
	0100	Slurry seal, type I, 8 lbs agg./S.Y., 1 coat, small or irregular area	B-90	2800	.023	S.Y.	.90	.68	.68	2.26	2.85	
	0150	Roadway or large area		10000	.006		.90	.19	.19	1.28	1.51	
	0200	Type II, 12 lbs aggregate/S.Y., 2 coats,small or irregular area		2000	.032		1.80	.95	.95	3.70	4.58	
	0250	Roadway or large area		8000	.008		1.80	.24	.24	2.28	2.63	
	0300	Type III, 20 lbs aggregate/S.Y., 2 coats, small or irregular area	↓	1800	.036	↓	2.12	1.06	1.06	4.24	5.25	

		02785 | Flexible Pavement Coating	CREW	DAILY OUTPUT	LABOR-HOURS	UNIT	2006 BARE COSTS MAT.	LABOR	EQUIP.	TOTAL	TOTAL INCL O&P	
500	0350	Roadway or large area	B-90	6000	.011	S.Y.	2.12	.32	.32	2.76	3.21	500
	0400	Slurry seal, thermoplastic coal-tar, type I, small or irregular area		2400	.027		1.87	.80	.79	3.46	4.22	
	0450	Roadway or large area		8000	.008		1.87	.24	.24	2.35	2.71	
	0500	Type II, small or irregular area		2400	.027		2.39	.80	.79	3.98	4.79	
	0550	Roadway or large area		7800	.008		2.39	.25	.24	2.88	3.30	
	0600	Average mobilization cost				Ea.				3,750	3,750	
600	0010	SURFACE TREATMENT										600
	3000	Pavement overlay, polypropylene										
	3040	6 oz. per S.Y., ideal conditions	B-63	10000	.004	S.Y.	1.25	.12	.02	1.39	1.59	
	3080	Adverse conditions		1000	.040		1.68	1.16	.15	2.99	3.91	
	3120	4 oz. per S.Y., ideal conditions		10000	.004		.85	.12	.02	.99	1.15	
	3160	Adverse conditions		1000	.040		1.10	1.16	.15	2.41	3.28	
	3200	Tack coat, emulsion, .05 gal. per S.Y., 1000 S.Y	B-45	2500	.006		.24	.21	.21	.66	.82	
	3240	10,000 S.Y.		10000	.002		.19	.05	.05	.29	.35	
	3270	.10 gal per S.Y., 1000 S.Y.		2500	.006		.45	.21	.21	.87	1.06	
	3275	.10 gal per S.Y., 10,000 S.Y.		10000	.002		.45	.05	.05	.55	.64	
	3280	.15 gal per S.Y., 1000 S.Y.		2500	.006		.66	.21	.21	1.08	1.29	
	3320	10,000 S.Y.		10000	.002		.53	.05	.05	.63	.72	
		02790 | Athletic/Recreational Surfaces										
400	0010	SYNTHETIC GRASS SURFACING										400
	0015	Not including asphalt base or drainage,										
	0020	but including cushion pad, over 50,000 S.F.										
	0200	1/2" pile and 5/16" cushion pad, standard	C-17	3200	.025	S.F.	7.25	.92		8.17	9.50	
	0300	Deluxe		2560	.031		8.60	1.15		9.75	11.35	
	0500	1/2" pile and 5/8" cushion pad, standard		2844	.028		10.55	1.04		11.59	13.35	
	0600	Deluxe		2327	.034		11.60	1.27		12.87	14.85	
	0800	For asphaltic concrete base, 2-1/2" thick,										
	0900	with 6" crushed stone sub-base, add	B-25	12000	.007	S.F.	1.26	.22	.18	1.66	1.93	
500	0010	SYNTHETIC RUNNING TRACK SURFACING										500
	0020	Running track, asphalt, incl base, 3" thick	B-37	300	.160	S.Y.	13.80	4.64	.38	18.82	23	
	0100	Surface, latex rubber system, 3/8" thick, black	B-20	125	.192		6.20	5.95		12.15	16.70	
	0150	Colors		125	.192		11.05	5.95		17	22	
	0300	Urethane rubber system, 3/8" thick, black		120	.200		16.60	6.20		22.80	28.50	
	0400	Color coating		115	.209		19.65	6.50		26.15	32	
600	0010	TENNIS COURT SURFACING										600
	0020	Tennis court, asphalt, incl. base, 2-1/2" thick, one court	B-37	450	.107	S.Y.	16.15	3.10	.25	19.50	23	
	0200	Two courts		675	.071		8.65	2.06	.17	10.88	13.10	
	0300	Clay courts		360	.133		31.50	3.87	.32	35.69	41	
	0400	Pulverized natural greenstone with 4" base, fast dry		250	.192		28.50	5.55	.46	34.51	41	
	0800	Rubber-acrylic base resilient pavement		600	.080		32	2.32	.19	34.51	39	
	1000	Colored sealer, acrylic emulsion, 3 coats	2 Clab	800	.020		4.17	.55		4.72	5.50	
	1100	3 coat, 2 colors	"	900	.018		5.80	.49		6.29	7.15	
	1200	For preparing old courts, add	1 Clab	825	.010			.27		.27	.44	
	1400	Posts for nets, 3-1/2" diameter with eye bolts	B-1	3.40	7.059	Pr.	175	198		373	525	
	1500	With pulley & reel		3.40	7.059	"	249	198		447	605	
	1700	Net, 42' long, nylon thread with binder		50	.480	Ea.	231	13.45		244.45	277	
	1800	All metal		6.50	3.692	"	415	104		519	625	
	2000	Paint markings on asphalt, 2 coats	1 Pord	1.78	4.494	Court	69.50	142		211.50	305	
	2200	Complete court with fence, etc., asphaltic conc., minimum	B-37	.20	240		14,500	6,975	570	22,045	27,900	
	2300	Maximum		.16	300		18,000	8,700	715	27,415	34,900	
	2800	Clay courts, minimum		.20	240		17,000	6,975	570	24,545	30,700	
	2900	Maximum		.16	300		20,400	8,700	715	29,815	37,600	

02800 | Site Improvements and Amenities

02810 | Irrigation System

		Description	CREW	DAILY OUTPUT	LABOR-HOURS	UNIT	2006 BARE COSTS MAT.	LABOR	EQUIP.	TOTAL	TOTAL INCL O&P	
300	0010	**SPRINKLER IRRIGATION SYSTEM** For lawns										300
	0100	Golf course with fully automatic system	C-17	.05	1600	9 holes	86,000	59,000		145,000	191,000	
	0200	24' diam. head at 15' O.C incl. piping, auto oper., minimum	B-20	70	.343	Head	18.95	10.65		29.60	38.50	
	0300	Maximum		40	.600		43.50	18.65		62.15	79	
	0500	60' diam. head at 40' O.C. incl. piping, auto oper., minimum		28	.857		57.50	26.50		84	107	
	0600	Maximum		23	1.043		161	32.50		193.50	231	
	0800	Residential system, custom, 1" supply		2000	.012	S.F.	.28	.37		.65	.93	
	0900	1-1/2" supply		1800	.013	"	.33	.41		.74	1.04	
	1020	Pop up spray head w/risers, hi-pop, full circle pattern, 4"	2 Skwk	76	.211	Ea.	3.42	7.70		11.12	16.30	
	1030	1/2 circle pattern		76	.211		3.50	7.70		11.20	16.40	
	1040	6", full circle pattern		76	.211		8.75	7.70		16.45	22	
	1050	1/2 circle pattern		76	.211		8.75	7.70		16.45	22	
	1060	12", full circle pattern		76	.211		12.50	7.70		20.20	26.50	
	1070	1/2 circle pattern		76	.211		12.55	7.70		20.25	26.50	
	1080	Pop up bubbler head w/risers, hi-pop bubbler head, 4"		76	.211		3.57	7.70		11.27	16.50	
	1090	6"		76	.211		8.75	7.70		16.45	22	
	1100	12"		76	.211		12.50	7.70		20.20	26.50	
	1110	Impact full/part circle sprinklers, 28'-54' 25-60 PSI		37	.432		11.70	15.80		27.50	39	
	1120	Spaced 37'-49' @ 25-50 PSI		37	.432		25.50	15.80		41.30	54	
	1130	Spaced 43'-61' @ 30-60 PSI		37	.432		46.50	15.80		62.30	77	
	1140	Spaced 54'-78' @ 40-80 PSI		37	.432		107	15.80		122.80	143	
	1145	Impact rotor pop-up full/part commercial circle sprinklers										
	1150	Spaced 42'-65' 35-80 PSI	2 Skwk	25	.640	Ea.	23.50	23.50		47	64	
	1160	Spaced 48'-76' 45-85 PSI	"	25	.640	"	9.75	23.50		33.25	48.50	
	1165	Impact rotor pop-up part. circle comm., 53'-75', 55-100 PSI, w/ acc										
	1170	Plastic case, metal cover	2 Skwk	25	.640	Ea.	163	23.50		186.50	218	
	1180	Rubber cover		25	.640		123	23.50		146.50	173	
	1190	Iron case, metal cover		22	.727		149	26.50		175.50	208	
	1200	Rubber cover		22	.727		162	26.50		188.50	222	
	1250	Plastic case, 2 nozzle, metal cover		25	.640		119	23.50		142.50	169	
	1260	Rubber cover		25	.640		126	23.50		149.50	177	
	1270	Iron case, 2 nozzle, metal cover		22	.727		173	26.50		199.50	235	
	1280	Rubber cover		22	.727		177	26.50		203.50	239	
	1282	Impact rotor pop-up full circle comm., 39'-99', 30-100 PSI										
	1284	Plastic case, metal cover	2 Skwk	25	.640	Ea.	122	23.50		145.50	172	
	1286	Rubber cover		25	.640		135	23.50		158.50	187	
	1288	Iron case, metal cover		22	.727		175	26.50		201.50	237	
	1290	Rubber cover		22	.727		184	26.50		210.50	247	
	1292	Plastic case, 2 nozzle, metal cover		22	.727		131	26.50		157.50	188	
	1294	Rubber cover		22	.727		136	26.50		162.50	194	
	1296	Iron case, 2 nozzle, metal cover		20	.800		169	29		198	234	
	1298	Rubber cover		20	.800		181	29		210	247	
	1305	Electric remote control valve, plastic, 3/4"		18	.889		18.40	32.50		50.90	73.50	
	1310	1"		18	.889		32.50	32.50		65	88.50	
	1320	1-1/2"		18	.889		62.50	32.50		95	122	
	1330	2"		18	.889		92.50	32.50		125	155	
	1335	Quick coupling valves, brass, locking cover										
	1340	Inlet coupling valve, 3/4"	2 Skwk	18.75	.853	Ea.	45	31		76	101	
	1350	1"		18.75	.853		55	31		86	112	
	1360	Controller valve boxes, 6" round boxes		18.75	.853		4.02	31		35.02	55.50	
	1370	10" round boxes		14.25	1.123		8.80	41		49.80	76.50	
	1380	12" square box		9.75	1.641		17.70	60		77.70	118	
	1388	Electromech. control, 14 day 3-60 min, auto start to 23/day										
	1390	4 station	2 Skwk	1.04	15.385	Ea.	191	560		751	1,125	
	1400	7 station		.64	25		247	915		1,162	1,775	
300	1410	12 station		.40	40		305	1,450		1,755	2,725	

02810 | Irrigation System

		02810 Irrigation System	Crew	Daily Output	Labor-Hours	Unit	2006 Bare Costs Mat.	Labor	Equip.	Total	Total Incl O&P	
300	1420	Dual programs, 18 station	2 Skwk	.24	66.667	Ea.	1,525	2,425		3,950	5,650	300
	1430	23 station		.16	100		1,975	3,650		5,625	8,150	
	1435	Backflow preventer, bronze, 0-175 PSI, w/valves, test cocks										
	1440	3/4"	2 Skwk	2	8	Ea.	125	292		417	615	
	1450	1"		2	8		130	292		422	625	
	1460	1-1/2"		2	8		258	292		550	765	
	1470	2"		2	8		269	292		561	775	
	1475	Pressure vacuum breaker, brass, 15-150 PSI										
	1480	3/4"	2 Skwk	2	8	Ea.	75	292		367	565	
	1490	1"		2	8		98.50	292		390.50	590	
	1500	1-1/2"		2	8		194	292		486	695	
	1510	2"		2	8		180	292		472	680	

02820 | Fences & Gates

		02820 Fences & Gates	Crew	Daily Output	Labor-Hours	Unit	Mat.	Labor	Equip.	Total	Total Incl O&P	
110	0010	**CHAIN LINK FENCES AND GATES**										110
	4750	Gate, transom for 10' fence, galv. steel, single, 3' x 7' x 3'	B-80A	52	.462	Ea.	335	12.65	3.39	351.04	390	
	4752	4' x 7' x 3'		10	2.400		350	66	17.60	433.60	515	
	4754	3' x 10'		8	3		335	82	22	439	530	
	4756	4' x 10'		10	2.400		375	66	17.60	458.60	545	
	4758	Double transom, 10' x 7' x 3'	B-80B	10	3.200		610	94	23.50	727.50	855	
	4760	12' x 7' x 3'		6	5.333		665	157	39	861	1,025	
	4762	14' x 7' x 3'		5	6.400		735	188	47	970	1,175	
	4764	10' x 10'		4	8		785	235	59	1,079	1,325	
	4766	12' x 10'		7	4.571		840	134	33.50	1,007.50	1,175	
	4768	14' x 10'		7	4.571		905	134	33.50	1,072.50	1,250	
	4780	Vinyl clad, single transom, 3' x 7' x 3'		6	5.333		505	157	39	701	855	
	4782	4' x 7' x 3'		10	3.200		535	94	23.50	652.50	765	
	4784	3' x 10'		8	4		515	117	29.50	661.50	790	
	4786	4' x 10'		10	3.200		580	94	23.50	697.50	815	
	4788	Double transom, 10' x 7' x 3'		10	3.200		1,125	94	23.50	1,242.50	1,425	
	4790	12' x 7' x 3'		6	5.333		1,225	157	39	1,421	1,650	
	4792	14' x 7' x 3'		5	6.400		1,375	188	47	1,610	1,875	
	4794	10' x 10'		4	8		1,225	235	59	1,519	1,800	
	4798	12' x 12'		7	4.571		1,325	134	33.50	1,492.50	1,725	
	4799	14' x 14'		7	4.571		1,425	134	33.50	1,592.50	1,825	
130	0010	**FENCE, CHAIN LINK INDUSTRIAL**, schedule 40										130
	0020	3 strands barb wire, 2" post @ 10' O.C., set in concrete, 6' H										
	0200	9 ga. wire, galv. steel	B-80C	240	.100	L.F.	12.75	2.75	.56	16.06	19.15	
	0300	Aluminized steel		240	.100		16.35	2.75	.56	19.66	23	
	0500	6 ga. wire, galv. steel		240	.100		20	2.75	.56	23.31	27	
	0600	Aluminized steel		240	.100		22.50	2.75	.56	25.81	30	
	0800	6 ga. wire, 6' high but omit barbed wire, galv. steel		250	.096		19.40	2.64	.54	22.58	26.50	
	0900	Aluminized steel		250	.096		27	2.64	.54	30.18	35	
	0920	8' H, 6 ga. wire, 2-1/2" line post, galv. steel		180	.133		31	3.66	.75	35.41	41	
	0940	Aluminized steel		180	.133		38	3.66	.75	42.41	48.50	
	1400	Gate for 6' high fence, 1-5/8" frame, 3' wide, galv. steel		10	2.400	Ea.	141	66	13.40	220.40	280	
	1500	Aluminized steel		10	2.400	"	174	66	13.40	253.40	315	
	2000	5'-0" high fence, 9 ga., no barbed wire, 2" line post,										
	2010	10' O.C., 1-5/8" top rail										
	2100	Galvanized steel	B-80C	300	.080	L.F.	11.05	2.20	.45	13.70	16.25	
	2200	Aluminized steel		300	.080	"	13.05	2.20	.45	15.70	18.45	
	2400	Gate, 4' wide, 5' high, 2" frame, galv. steel		10	2.400	Ea.	161	66	13.40	240.40	300	
	2500	Aluminized steel		10	2.400	"	181	66	13.40	260.40	325	
	3100	Overhead slide gate, chain link, 6' high, to 18' wide		38	.632	L.F.	142	17.35	3.53	162.88	188	
	3110	Cantilever type	B-80	48	.667		61	19.95	11.10	92.05	112	

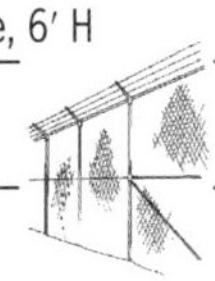

02820 | Fences & Gates

		02820 Fences & Gates	CREW	DAILY OUTPUT	LABOR-HOURS	UNIT	2006 BARE COSTS MAT.	LABOR	EQUIP.	TOTAL	TOTAL INCL O&P	
130	3120	8′ high	B-80	24	1.333	L.F.	88.50	40	22	150.50	187	130
	3130	10′ high		18	1.778		105	53	29.50	187.50	234	
	5000	Double swing gates, incl. posts & hardware										
	5010	5′ high, 12′ opening	B-80C	3.40	7.059	Opng.	430	194	39.50	663.50	840	
	5020	20′ opening		2.80	8.571		590	235	48	873	1,100	
	5060	6′ high, 12′ opening		3.20	7.500		725	206	42	973	1,175	
	5070	20′ opening		2.60	9.231		1,000	254	51.50	1,305.50	1,575	
	5080	8′ high, 12′ opening	B-80	2.13	15.002		1,125	450	250	1,825	2,250	
	5090	20′ opening		1.45	22.069		1,475	660	365	2,500	3,100	
	5100	10′ high, 12′ opening		1.31	24.427		1,300	730	405	2,435	3,075	
	5110	20′ opening		1.03	31.068		1,925	930	515	3,370	4,200	
	5120	12′ high, 12′ opening		1.05	30.476		1,875	910	505	3,290	4,100	
	5130	20′ opening		.85	37.647		2,425	1,125	625	4,175	5,175	
	5190	For aluminized steel add					20%					
	7001	Snow fence on steel posts 10′ O.C., 4′ high	B-1	500	.048	L.F.	2.38	1.35		3.73	4.85	
	7055	Braces, galv. steel	B-80A	960	.025		1.79	.69	.18	2.66	3.31	
	7056	Aluminized steel	"	960	.025		2.15	.69	.18	3.02	3.70	
	7071	Privacy slats, vertical, vinyl	1 Clab	500	.016	S.F.	1.50	.44		1.94	2.38	
	7072	Redwood		450	.018		2.62	.49		3.11	3.69	
	7073	Diagonal, aluminum		300	.027		3.41	.73		4.14	4.96	
	9000	Minimum labor/equipment charge	B-80	2	16	Job		480	266	746	1,075	
140	0010	**FENCE, CHAIN LINK RESIDENTIAL**, sch. 20, 11 ga. wire, 1-5/8″ post										140
	0020	10′ O.C., 1-3/8″ top rail, 2″ corner post, galv. stl. 3′ high	B-80C	500	.048	L.F.	3.86	1.32	.27	5.45	6.70	
	0050	4′ high		400	.060		4.38	1.65	.34	6.37	7.90	
	0100	6′ high		200	.120		5.30	3.30	.67	9.27	12.05	
	0150	Add for gate 3′ wide, 1-3/8″ frame, 3′ high		12	2	Ea.	57.50	55	11.20	123.70	166	
	0170	4′ high		10	2.400		66.50	66	13.40	145.90	197	
	0190	6′ high		10	2.400		91	66	13.40	170.40	224	
	0200	Add for gate 4′ wide, 1-3/8″ frame, 3′ high		9	2.667		65	73.50	14.90	153.40	209	
	0220	4′ high		9	2.667		73.50	73.50	14.90	161.90	218	
	0240	6′ high		8	3		86	82.50	16.75	185.25	249	
	0350	Aluminized steel, 11 ga. wire, 3′ high		500	.048	L.F.	5.10	1.32	.27	6.69	8.05	
	0380	4′ high		400	.060		6.55	1.65	.34	8.54	10.30	
	0400	6′ high		200	.120		9.20	3.30	.67	13.17	16.35	
	0450	Add for gate 3′ wide, 1-3/8″ frame, 3′ high		12	2	Ea.	70.50	55	11.20	136.70	180	
	0470	4′ high		10	2.400		116	66	13.40	195.40	251	
	0490	6′ high		10	2.400		146	66	13.40	225.40	284	
	0500	Add for gate 4′ wide, 1-3/8″ frame, 3′ high		10	2.400		80	66	13.40	159.40	212	
	0520	4′ high		9	2.667		88.50	73.50	14.90	176.90	235	
	0540	6′ high		8	3		102	82.50	16.75	201.25	266	
	0620	Vinyl covered, 9 ga. wire, 3′ high		500	.048	L.F.	4.17	1.32	.27	5.76	7.05	
	0640	4′ high		400	.060		4.87	1.65	.34	6.86	8.45	
	0660	6′ high		200	.120		6.15	3.30	.67	10.12	13	
	0720	Add for gate 3′ wide, 1-3/8″ frame, 3′ high		12	2	Ea.	80.50	55	11.20	146.70	191	
	0740	4′ high		10	2.400		91	66	13.40	170.40	224	
	0760	6′ high		10	2.400		141	66	13.40	220.40	279	
	0780	Add for gate 4′ wide, 1-3/8″ frame, 3′ high		10	2.400		92.50	66	13.40	171.90	226	
	0800	4′ high		9	2.667		101	73.50	14.90	189.40	248	
	0820	6′ high		8	3		114	82.50	16.75	213.25	279	
150	0010	**FENCE, CHAIN LINK, GATES & POSTS** (1/3 post length in ground)										150
	6580	Line posts, galvanized, 2-1/2″ OD, set in conc., 4′	B-80	80	.400	Ea.	22	11.95	6.65	40.60	51	
	6585	5′		76	.421		26.50	12.60	7	46.10	57	
	6590	6′		74	.432		27	12.95	7.20	47.15	59	
	6595	7′		72	.444		29.50	13.30	7.40	50.20	62	
	6600	8′		69	.464		32.50	13.85	7.70	54.05	66.50	

02820 | Fences & Gates

			CREW	DAILY OUTPUT	LABOR-HOURS	UNIT	2006 BARE COSTS MAT.	LABOR	EQUIP.	TOTAL	TOTAL INCL O&P	
150	6610	H-beam, 1-7/8″, 4′	B-80	83	.386	Ea.	25	11.55	6.40	42.95	53.50	150
	6615	5′		81	.395		28.50	11.80	6.55	46.85	58	
	6620	6′		78	.410		33.50	12.25	6.85	52.60	64.50	
	6625	7′		75	.427		36.50	12.75	7.10	56.35	69.50	
	6630	8′		73	.438		40	13.10	7.30	60.40	73.50	
	6635	Vinyl coated, 2-1/2″ OD, set in conc., 4′		79	.405		43	12.10	6.75	61.85	74.50	
	6640	5′		77	.416		49.50	12.45	6.90	68.85	82	
	6645	6′		74	.432		62	12.95	7.20	82.15	97.50	
	6650	7′		72	.444		66.50	13.30	7.40	87.20	103	
	6655	8′		69	.464		72.50	13.85	7.70	94.05	111	
	6660	End gate post, steel, 3″ OD, set in conc.,4′		68	.471		50.50	14.05	7.85	72.40	87	
	6665	5′		65	.492		64	14.70	8.20	86.90	104	
	6670	6′		63	.508		65	15.20	8.45	88.65	105	
	6675	7′		61	.525		87	15.70	8.75	111.45	131	
	6685	Vinyl, 4′		68	.471		84.50	14.05	7.85	106.40	125	
	6690	5′		65	.492		93	14.70	8.20	115.90	135	
	6695	6′		63	.508		115	15.20	8.45	138.65	161	
	6705	8′		59	.542		137	16.20	9.05	162.25	187	
	6710	Corner post, galv. steel, 4″ OD, set in conc., 4′		65	.492		85	14.70	8.20	107.90	127	
	6715	6′		63	.508		120	15.20	8.45	143.65	166	
	6720	7′		61	.525		135	15.70	8.75	159.45	183	
	6725	8′		65	.492		155	14.70	8.20	177.90	204	
	6730	Vinyl, 5′		65	.492		148	14.70	8.20	170.90	196	
	6735	6′		63	.508		197	15.20	8.45	220.65	251	
	6740	7′		61	.525		180	15.70	8.75	204.45	232	
	6745	8′		59	.542		235	16.20	9.05	260.25	294	
	7031	For corner, end, & pull post bracing, add					20%	15%				
	7770	Gates, sliding w/overhead support, 4′ high	B-80B	35	.914	L.F.	45.50	27	6.70	79.20	101	
	7775	5′ high		32	1		49.50	29.50	7.35	86.35	111	
	7780	6′ high		28	1.143		53	33.50	8.40	94.90	123	
	7785	7′ high		25	1.280		70	37.50	9.40	116.90	149	
	7790	8′ high		23	1.391		87	41	10.25	138.25	173	
	7795	Cantilever, manual, exp. roller, (pr), 40′ wide x 8′ high	B-22	1	30	Ea.	3,575	975	229	4,779	5,775	
	7800	30′ wide x 8′ high		1	30		3,000	975	229	4,204	5,150	
	7805	24′ wide x 8′ high		1	30		2,225	975	229	3,429	4,300	
	7810	Motor operators for gates, (no elec wiring), 3′ wide swing	2 Skwk	.50	32		1,225	1,175		2,400	3,225	
	7815	Up to 20′ wide swing		.50	32		4,050	1,175		5,225	6,350	
	7820	Up to 45′ sliding		.50	32		4,125	1,175		5,300	6,425	
	7825	Overhead gate, 6′ to 18′ wide, sliding/cantilever		45	.356	L.F.	144	13		157	179	
	7830	Gate operators, digital receiver		7	2.286	Ea.	435	83.50		518.50	610	
	7835	Two button transmitter		24	.667		55.50	24.50		80	101	
	7840	3 button station		14	1.143		68.50	41.50		110	144	
	7845	Master slave system		4	4		244	146		390	510	
	7900	Auger fence post hole, 3′ deep, medium soil, by hand	1 Clab	30	.267			7.30		7.30	12.10	
	7925	By machine	B-80	175	.183			5.45	3.04	8.49	12.25	
	7950	Rock, with jackhammer	B-9C	32	1.250			35	4.60	39.60	62.50	
	7975	With rock drill	B-47C	65	.246			7.70	17.05	24.75	31	
170	0010	**FENCE, FABRIC & ACCESSORIES**										170
	1000	Fabric, 9 ga., galv., 1.2 oz. coat, 2″ chain link, 4′	B-80A	304	.079	L.F.	3.45	2.16	.58	6.19	8	
	1150	5′		285	.084		4.17	2.31	.62	7.10	9.10	
	1200	6′		266	.090		6.60	2.47	.66	9.73	12.10	
	1250	7′		247	.097		9.70	2.66	.71	13.07	15.90	
	1300	8′		228	.105		12.10	2.88	.77	15.75	18.95	
	1400	9 ga., fused, 4′		304	.079		5.65	2.16	.58	8.39	10.45	
	1450	5′		285	.084		7.05	2.31	.62	9.98	12.25	

02820 | Fences & Gates

			CREW	DAILY OUTPUT	LABOR-HOURS	UNIT	2006 BARE COSTS MAT.	LABOR	EQUIP.	TOTAL	TOTAL INCL O&P	
170	1500	6′	B-80A	266	.090	L.F.	8.50	2.47	.66	11.63	14.20	170
	1550	7′		247	.097		9.90	2.66	.71	13.27	16.10	
	1600	8′		228	.105		14.15	2.88	.77	17.80	21	
	1650	Barbed wire, galv., cost per strand		2280	.011		.11	.29	.08	.48	.69	
	1700	Vinyl coated		2280	.011		.26	.29	.08	.63	.86	
	1750	Extension arms, 3 strands		143	.168	Ea.	7.85	4.60	1.23	13.68	17.60	
	1800	6 strands, 2-3/8″		119	.202		9.95	5.55	1.48	16.98	21.50	
	1850	Eye tops, 2-3/8″		143	.168		3.26	4.60	1.23	9.09	12.55	
	1900	Top rail, incl. tie wires, 1-5/8″, galv.		912	.026	L.F.	3.10	.72	.19	4.01	4.81	
	1950	Vinyl coated		912	.026		4.88	.72	.19	5.79	6.75	
	2100	Rail, middle/bottom, w/tie wire, 1-5/8″, galv.		912	.026		2.62	.72	.19	3.53	4.28	
	2150	Vinyl coated		912	.026		3.53	.72	.19	4.44	5.30	
	2200	Reinforcing wire, coiled spring, 7 ga. galv.		2279	.011		.16	.29	.08	.53	.75	
	2250	9 ga., vinyl coated		2282	.011		.24	.29	.08	.61	.83	
210	0010	**FENCE, SECURITY**										210
	0100	Fence, chain link, security, 7′ high, no gates/signs	B-80C	480	.050	L.F.	43	1.37	.28	44.65	50	
	0200	Fence, barbed wire, security, 7′ high, with 3 wire barbed wire arm	″	400	.060	″	5.85	1.65	.34	7.84	9.55	
	0300	Taunt wire fence detection system										
	0310	Includes all labor, material and equip. for complete installation				M.L.F.	15,000			15,000	16,500	
	0400	Microwave fence detection system										
	0410	Includes all labor, material and equip. for complete installation				M.L.F.	25,000			25,000	27,500	
	0500	Passive magnetic fence detection system										
	0510	Includes all labor, material and equip. for complete installation				M.L.F.	12,000			12,000	13,200	
	0600	Infrared fence detection system										
	0610	Includes all labor, material and equip. for complete installation				M.L.F.	8,000			8,000	8,800	
	0700	Strain relief fence detection system										
	0710	Includes all labor, material and equip. for complete installation				M.L.F.	15,000			15,000	16,500	
	0800	Electro shock fence detection system										
	0810	Includes all labor, material and equip. for complete installation				M.L.F.	22,000			22,000	24,200	
	0900	Photo electric fence detection system										
	0910	Includes all labor, material and equip. for complete installation				M.L.F.	10,000			10,000	11,000	
300	0010	**FENCE, VINYL**, white, steel reinforced, stainless steel fasteners										300
	0020	Picket, 4″ x 4″ posts @ 6′ - 0″ OC, 3′ high	B-1	140	.171	L.F.	17.15	4.81		21.96	27	
	0030	4′ high		130	.185		19.75	5.20		24.95	30.50	
	0040	5′ high		120	.200		22.50	5.60		28.10	34.50	
	0100	Board (semi-privacy), 5″ x 5″ posts @ 7′ - 6″ OC, 5′ high		130	.185		23.50	5.20		28.70	34.50	
	0120	6′ high		125	.192		27	5.40		32.40	38.50	
	0200	Basketweave, 5″ x 5″ posts @ 7′ - 6″ OC, 5′ high		160	.150		21	4.21		25.21	30	
	0220	6′ high		150	.160		24.50	4.49		28.99	34.50	
	0300	Privacy, 5″ x 5″ posts @ 7′ - 6″ OC, 5′ high		130	.185		25.50	5.20		30.70	36.50	
	0320	6′ high		150	.160		29	4.49		33.49	39.50	
	0350	Gate, 5′ high		9	2.667	Ea.	360	75		435	520	
	0360	6′ high		9	2.667		365	75		440	530	
	0400	For posts set in concrete, add		25	.960		7.30	27		34.30	52.50	
410	0010	**FENCES, MISC. METAL**										410
	0012	Chicken wire, posts @ 4′, 1″ mesh, 4′ high	B-80C	410	.059	L.F.	1.76	1.61	.33	3.70	4.95	
	0100	2″ mesh, 6′ high		350	.069		1.59	1.88	.38	3.85	5.25	
	0200	Galv. steel, 12 ga., 2″ x 4″ mesh, posts 5′ O.C., 3′ high		300	.080		2.39	2.20	.45	5.04	6.75	
	0300	5′ high		300	.080		3.19	2.20	.45	5.84	7.60	
	0400	14 ga., 1″ x 2″ mesh, 3′ high		300	.080		2.54	2.20	.45	5.19	6.90	
	0500	5′ high		300	.080		3.51	2.20	.45	6.16	7.95	
	1000	Kennel fencing, 1-1/2″ mesh, 6′ long, 3′-6″ wide, 6′-2″ high	2 Clab	4	4	Ea.	400	110		510	620	
	1050	12′ long		4	4		480	110		590	705	
	1200	Top covers, 1-1/2″ mesh, 6′ long		15	1.067		81	29		110	138	

	02820 | Fences & Gates		CREW	DAILY OUTPUT	LABOR-HOURS	UNIT	2006 BARE COSTS MAT.	LABOR	EQUIP.	TOTAL	TOTAL INCL O&P	
410	1250	12′ long	2 Clab	12	1.333	Ea.	130	36.50		166.50	204	410
	1300	For kennel doors, see Division 08310-400										
	4500	Security fence, prison grade, set in concrete, 12′ high	B-80	25	1.280	L.F.	35	38.50	21.50	95	125	
	4600	16′ high	"	20	1.600		42	48	26.50	116.50	154	
	5300	Tubular picket, steel, 6′ sections, 1-9/16″ posts, 4′ high	B-80C	300	.080		25	2.20	.45	27.65	31	
	5400	2″ posts, 5′ high		240	.100		34.50	2.75	.56	37.81	42.50	
	5600	2″ posts, 6′ high		200	.120		39	3.30	.67	42.97	48.50	
	5700	Staggered picket 1-9/16″ posts, 4′ high		300	.080		22.50	2.20	.45	25.15	28.50	
	5800	2″ posts, 5′ high		240	.100		36.50	2.75	.56	39.81	45.50	
	5900	2″ posts, 6′ high		200	.120		38	3.30	.67	41.97	48	
	6200	Gates, 4′ high, 3′ wide	B-1	10	2.400	Ea.	215	67.50		282.50	350	
	6300	5′ high, 3′ wide		10	2.400		279	67.50		346.50	415	
	6400	6′ high, 3′ wide		10	2.400		287	67.50		354.50	425	
	6500	4′ wide		10	2.400		335	67.50		402.50	480	
420	0010	**WIRE FENCING, GENERAL**										420
	0015	Barbed wire, galvanized, domestic steel, hi-tensile 15-1/2 ga.				M.L.F.	29			29	32	
	0020	Standard, 12-3/4 ga.					39			39	43	
	0210	Barbless wire, 2-strand galvanized, 12-1/2 ga.					39			39	42.50	
	0500	Helical razor ribbon, stainless steel, 18″ dia x 18″ spacing				C.L.F.	113			113	124	
	0600	Hardware cloth galv., 1/4″ mesh, 23 ga., 2′ wide				C.S.F.	50.50			50.50	55.50	
	0700	3′ wide					49.50			49.50	54	
	0900	1/2″ mesh, 19 ga., 2′ wide					44.50			44.50	49	
	1000	4′ wide					44			44	48	
	1200	Chain link fabric, steel, 2″ mesh, 6 ga, galvanized					167			167	184	
	1300	9 ga, galvanized					82.50			82.50	91	
	1350	Vinyl coated					69			69	76	
	1360	Aluminized					107			107	118	
	1400	2-1/4″ mesh, 11.5 ga, galvanized					56			56	61.50	
	1600	1-3/4″ mesh (tennis courts), 11.5 ga (core), vinyl coated					79			79	87	
	1700	9 ga, galvanized					71.50			71.50	78.50	
	2100	Welded wire fabric, galvanized, 1″ x 2″, 14 ga.					47.50			47.50	52	
	2200	2″ x 4″, 12-1/2 ga.					16.15			16.15	17.80	
510	0010	**FENCE, WOOD** Basket weave, 3/8″ x 4″ boards, 2″ x 4″										510
	0020	stringers on spreaders, 4″ x 4″ posts										
	0050	No. 1 cedar, 6′ high	B-80C	160	.150	L.F.	8.45	4.12	.84	13.41	16.95	
	0070	Treated pine, 6′ high	"	150	.160	"	10.25	4.40	.89	15.54	19.55	
	0200	Board fence, 1″ x 4″ boards, 2″ x 4″ rails, 4″ x 4″ post										
	0220	Preservative treated, 2 rail, 3′ high	B-80C	145	.166	L.F.	6.25	4.55	.93	11.73	15.40	
	0240	4′ high		135	.178		6.90	4.88	.99	12.77	16.70	
	0260	3 rail, 5′ high		130	.185		7.75	5.05	1.03	13.83	18.05	
	0300	6′ high		125	.192		8.90	5.25	1.07	15.22	19.70	
	0320	No. 2 grade western cedar, 2 rail, 3′ high		145	.166		6.85	4.55	.93	12.33	16	
	0340	4′ high		135	.178		8.10	4.88	.99	13.97	18.05	
	0360	3 rail, 5′ high		130	.185		9.35	5.05	1.03	15.43	19.75	
	0400	6′ high		125	.192		10.25	5.25	1.07	16.57	21	
	0420	No. 1 grade cedar, 2 rail, 3′ high		145	.166		10.30	4.55	.93	15.78	19.80	
	0440	4′ high		135	.178		11.70	4.88	.99	17.57	22	
	0460	3 rail, 5′ high		130	.185		13.50	5.05	1.03	19.58	24.50	
	0500	6′ high		125	.192		15.05	5.25	1.07	21.37	26.50	
	0540	Shadow box, 1″ x 6″ board, 2″ x 4″ rail, 4″ x 4″post										
	0560	Pine, pressure treated, 3 rail, 6′ high	B-80C	150	.160	L.F.	11.50	4.40	.89	16.79	21	
	0600	Gate, 3′-6″ wide		8	3	Ea.	66	82.50	16.75	165.25	227	
	0620	No. 1 cedar, 3 rail, 4′ high		130	.185	L.F.	14.15	5.05	1.03	20.23	25	
	0640	6′ high		125	.192		17.45	5.25	1.07	23.77	29	
	0860	Open rail fence, split rails, 2 rail 3′ high, no. 1 cedar		160	.150		5.70	4.12	.84	10.66	13.95	
	0870	No. 2 cedar		160	.150		4.42	4.12	.84	9.38	12.60	

2 SITE CONSTRUCTION

	02820 | Fences & Gates		CREW	DAILY OUTPUT	LABOR-HOURS	UNIT	2006 BARE COSTS MAT.	LABOR	EQUIP.	TOTAL	TOTAL INCL O&P	
510	0880	3 rail, 4′ high, no. 1 cedar	B-80C	150	.160	L.F.	7.65	4.40	.89	12.94	16.65	510
	0890	No. 2 cedar 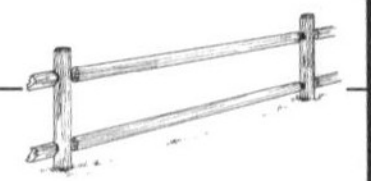		150	.160		5.05	4.40	.89	10.34	13.80	
	0920	Rustic rails, 2 rail 3′ high, no. 1 cedar		160	.150		3.54	4.12	.84	8.50	11.60	
	0930	No. 2 cedar		160	.150		3.40	4.12	.84	8.36	11.45	
	0940	3 rail, 4′ high		150	.160		4.75	4.40	.89	10.04	13.50	
	0950	No. 2 cedar		150	.160		3.59	4.40	.89	8.88	12.20	
	0960	Picket fence, gothic, pressure treated pine										
	1000	2 rail, 3′ high	B-80C	140	.171	L.F.	4.37	4.71	.96	10.04	13.60	
	1020	3 rail, 4′ high		130	.185	″	5.15	5.05	1.03	11.23	15.15	
	1040	Gate, 3′-6″ wide		9	2.667	Ea.	46	73.50	14.90	134.40	188	
	1060	No. 2 cedar, 2 rail, 3′ high		140	.171	L.F.	5.45	4.71	.96	11.12	14.80	
	1100	3 rail, 4′ high		130	.185	″	5.60	5.05	1.03	11.68	15.65	
	1120	Gate, 3′-6″ wide		9	2.667	Ea.	54.50	73.50	14.90	142.90	197	
	1140	No. 1 cedar, 2 rail 3′ high		140	.171	L.F.	10.90	4.71	.96	16.57	21	
	1160	3 rail, 4′ high		130	.185	″	12.70	5.05	1.03	18.78	23.50	
	1170	Gate, 3′-6″ wide		9	2.667	Ea.	121	73.50	14.90	209.40	270	
	1200	Rustic picket, molded pine, 2 rail, 3′ high		140	.171	L.F.	5	4.71	.96	10.67	14.30	
	1220	No. 1 cedar, 2 rail, 3′ high		140	.171		6.80	4.71	.96	12.47	16.30	
	1240	Stockade fence, no. 1 cedar, 3-1/4″ rails, 6′ high		160	.150		10.30	4.12	.84	15.26	19.05	
	1260	8′ high		155	.155		13.30	4.25	.87	18.42	22.50	
	1300	No. 2 cedar, treated wood rails, 6′ high		160	.150		10.30	4.12	.84	15.26	19.05	
	1320	Gate, 3′-6″ wide		8	3	Ea.	60.50	82.50	16.75	159.75	221	
	1360	Treated pine, treated rails, 6′ high		160	.150	L.F.	10.05	4.12	.84	15.01	18.80	
	1400	8′ high		150	.160	″	15.15	4.40	.89	20.44	25	
520	0010	**FENCE, WOOD RAIL**										520
	0012	Picket, No. 2 cedar, Gothic, 2 rail, 3′ high	B-1	160	.150	L.F.	5.40	4.21		9.61	12.95	
	0050	Gate, 3′-6″ wide	B-80C	9	2.667	Ea.	46.50	73.50	14.90	134.90	188	
	0400	3 rail, 4′ high		150	.160	L.F.	6.20	4.40	.89	11.49	15.10	
	0500	Gate, 3′-6″ wide		9	2.667	Ea.	56	73.50	14.90	144.40	199	
	5000	Fence rail, redwood, 2″ x 4″, merch grade 8′	B-1	2400	.010	L.F.	1.06	.28		1.34	1.64	
	5050	Select grade, 8′		2400	.010	″	3.28	.28		3.56	4.08	
	6000	Fence post, select redwood, earthpacked & treated, 4″ x 4″ x 6′		96	.250	Ea.	10.50	7		17.50	23	
	6010	4″ x 4″ x 8′		96	.250		12.45	7		19.45	25.50	
	6020	Set in concrete, 4″ x 4″ x 6′		50	.480		13.50	13.45		26.95	37.50	
	6030	4″ x 4″ x 8′		50	.480		16.15	13.45		29.60	40.50	
	6040	Wood post, 4′ high, set in concrete, incl. concrete		50	.480		7.90	13.45		21.35	31	
	6050	Earth packed		96	.250		5.40	7		12.40	17.55	
	6060	6′ high, set in concrete, incl. concrete		50	.480		10.10	13.45		23.55	33.50	
	6070	Earth packed		96	.250		7.35	7		14.35	19.70	
	9000	Minimum labor/equipment charge	1 Clab	2	4	Job		110		110	182	

	02830 | Retaining Walls		CREW	DAILY OUTPUT	LABOR-HOURS	UNIT	MAT.	LABOR	EQUIP.	TOTAL	TOTAL INCL O&P	
100	0010	**CAST IN PLACE RETAINING WALLS**										100
	1800	Concrete gravity wall with vertical face including excavation & backfill										
	1850	No reinforcing										
	1900	6′ high, level embankment 	C-17C	36	2.306	L.F.	64	85	11	160	221	
	2000	33° slope embankment		32	2.594		58	96	12.40	166.40	235	
	2200	8′ high, no surcharge		27	3.074		79	114	14.70	207.70	289	
	2300	33° slope embankment		24	3.458		95.50	128	16.55	240.05	330	
	2500	10′ high, level embankment		19	4.368		113	161	21	295	410	
	2600	33° slope embankment		18	4.611		157	170	22	349	475	
	2800	Reinforced concrete cantilever, incl. excavation, backfill & reinf.										
	2900	6′ high, 33° slope embankment	C-17C	35	2.371	L.F.	58	87.50	11.35	156.85	219	
	3000	8′ high, 33° slope embankment		29	2.862		67	106	13.70	186.70	262	
	3100	10′ high, 33° slope embankment		20	4.150		87	153	19.85	259.85	370	
	3200	20′ high, 500 lb. per L.F. surcharge		7.50	11.067		261	410	53	724	1,025	

02830 | Retaining Walls

		02830 Retaining Walls	CREW	DAILY OUTPUT	LABOR-HOURS	UNIT	2006 BARE COSTS MAT.	LABOR	EQUIP.	TOTAL	TOTAL INCL O&P	
100	3500	Concrete cribbing, incl. excavation and backfill										100
	3700	12′ high, open face	B-13	210	.267	S.F.	29	8	3.10	40.10	48.50	
	3900	Closed face	"	210	.267	"	27	8	3.10	38.10	46.50	
	4100	Concrete filled slurry trench, see Div. 02660-600										
200	0010	**INTERLOCKING SEGMENTAL RETAINING WALLS**										200
	7100	Segmental Retaining Wall system, incl pins, and void fill										
	7120	base not included										
	7140	Large unit, 8″ high x 18″ wide x 20″ deep, 3 plane split	B-62	300	.080	S.F.	9.25	2.40	.51	12.16	14.65	
	7150	Straight split		300	.080		9.25	2.40	.51	12.16	14.65	
	7160	Medium, ltwt, 8″ high x 18″ wide x 12″ deep, 3 plane split		400	.060		8.15	1.80	.39	10.34	12.30	
	7170	Straight split		400	.060		8.15	1.80	.39	10.34	12.30	
	7180	Small unit, 4″ x 18″ x 10″ deep, 3 plane split		400	.060		10.20	1.80	.39	12.39	14.60	
	7190	Straight split		400	.060		10.20	1.80	.39	12.39	14.60	
	7200	Cap unit, 3 plane split		300	.080		11.45	2.40	.51	14.36	17.05	
	7210	Cap unit, st split	↓	300	.080		11.45	2.40	.51	14.36	17.05	
	7260	For reinforcing, add				↓				2.20	2.75	
	8000	For higher walls, add components as necessary										
600	0010	**METAL BIN RETAINING WALLS**, Aluminized steel bin, excavation										600
	0020	and backfill not included, 10′ wide										
	0100	4′ high, 5.5′ deep	B-13	650	.086	S.F.	17.95	2.58	1	21.53	25	
	0200	8′ high, 5.5′ deep		615	.091		20.50	2.73	1.06	24.29	28	
	0300	10′ high, 7.7′ deep		580	.097		21.50	2.89	1.12	25.51	30	
	0400	12′ high, 7.7′ deep		530	.106		23.50	3.16	1.23	27.89	32.50	
	0500	16′ high, 7.7′ deep		515	.109		25	3.26	1.26	29.52	33.50	
	0600	16′ high, 9.9′ deep		500	.112		26	3.35	1.30	30.65	35.50	
	0700	20′ high, 9.9′ deep		470	.119		29.50	3.57	1.38	34.45	40	
	0800	20′ high, 12.1′ deep		460	.122		31.50	3.64	1.41	36.55	42	
	0900	24′ high, 12.1′ deep		455	.123		33.50	3.69	1.43	38.62	44	
	1000	24′ high, 14.3′ deep		450	.124		35	3.73	1.45	40.18	45.50	
	1100	28′ high, 14.3′ deep	↓	440	.127	↓	36	3.81	1.48	41.29	47.50	
	1300	For plain galvanized bin type walls, deduct					10%					
700	0010	**STONE GABION RETAINING WALLS**										700
	4300	Stone filled gabions, not incl. excavation,										
	4310	Stone, delivered, 3′ wide										
	4340	Galvanized, 6′ long, 1′ high	B-13	113	.496	Ea.	63.50	14.85	5.75	84.10	100	
	4400	1′-6″ high		50	1.120		80.50	33.50	13	127	157	
	4490	3′-0″ high		13	4.308		133	129	50	312	410	
	4590	9′ long, 1′ high		50	1.120		90.50	33.50	13	137	168	
	4650	1′-6″ high		22	2.545		119	76	29.50	224.50	288	
	4690	3′-0″ high		6	9.333		203	279	108	590	795	
	4890	12′ long, 1′ high		28	2		121	60	23	204	256	
	4950	1′-6″ high		13	4.308		155	129	50	334	435	
	4990	3′-0″ high		3	18.667		267	560	217	1,044	1,450	
	5200	PVC coated, 6′ long, 1′ high		113	.496		72.50	14.85	5.75	93.10	110	
	5250	1′-6″ high		50	1.120		88.50	33.50	13	135	166	
	5300	3′ high		13	4.308		148	129	50	327	430	
	5500	9′ long, 1′ high		50	1.120		95.50	33.50	13	142	174	
	5550	1′-6″ high		22	2.545		129	76	29.50	234.50	299	
	5600	3′ high		6	9.333		219	279	108	606	815	
	5800	12′ long, 1′ high		28	2		133	60	23	216	269	
	5850	1′-6″ high		13	4.308		170	129	50	349	450	
	5900	3′ high		3	18.667	↓	287	560	217	1,064	1,475	
	6000	Galvanized, 6′ long, 1′ high		75	.747	C.Y.	63	22.50	8.70	94.20	115	
	6010	1′-6″ high		50	1.120		80.50	33.50	13	127	157	
	6020	3′-0″ high	↓	25	2.240	↓	66.50	67	26	159.50	211	

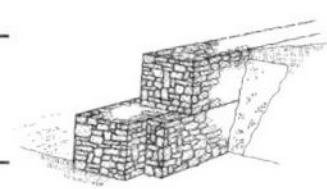

SITE CONSTRUCTION 2

02830 | Retaining Walls

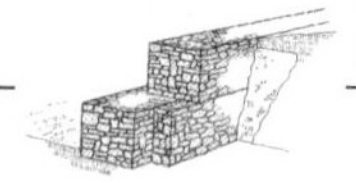

		02830 Retaining Walls	Crew	Daily Output	Labor-Hours	Unit	2006 Bare Costs Mat.	Labor	Equip.	Total	Total Incl O&P	
700	6030	9′ long, 1′ high	B-13	50	1.120	C.Y.	90.50	33.50	13	137	168	700
	6040	1′-6″ high		33.30	1.682		79	50.50	19.55	149.05	191	
	6050	3′-0″ high		16.70	3.353		67.50	100	39	206.50	282	
	6060	12′ long, 1′ high		37.50	1.493		91	44.50	17.35	152.85	192	
	6070	1′-6″ high		25	2.240		77.50	67	26	170.50	223	
	6080	3′-0″ high		12.50	4.480		67	134	52	253	350	
	6100	PVC coated, 6′ long, 1′ high		75	.747		108	22.50	8.70	139.20	165	
	6110	1′-6″ high		50	1.120		88.50	33.50	13	135	166	
	6120	3′ high		25	2.240		74	67	26	167	219	
	6130	9′ long, 1′ high		50	1.120		95.50	33.50	13	142	174	
	6140	1′-6″ high		33.30	1.682		86	50.50	19.55	156.05	198	
	6150	3′ high		16.67	3.359		73	101	39	213	287	
	6160	12′ long, 1′ high		37.50	1.493		99.50	44.50	17.35	161.35	201	
	6170	1′-6″ high		25	2.240		77.50	67	26	170.50	223	
	6180	3′ high	↓	12.50	4.480	↓	71.50	134	52	257.50	355	
800	0010	**STONE RETAINING WALLS**										800
	0015	Including excavation, concrete footing and										
	0020	stone 3′ below grade. Price is exposed face area.										
	0200	Decorative random stone, to 6′ high, 1′-6″ thick, dry set	D-1	35	.457	S.F.	35.50	14.70		50.20	63.50	
	0300	Mortar set		40	.400		35.50	12.85		48.35	60.50	
	0500	Cut stone, to 6′ high, 1′-6″ thick, dry set		35	.457		35.50	14.70		50.20	63.50	
	0600	Mortar set		40	.400		35.50	12.85		48.35	60.50	
	0800	Random stone, 6′ to 10′ high, 2′ thick, dry set		45	.356		35.50	11.45		46.95	58	
	0900	Mortar set		50	.320		35.50	10.30		45.80	56	
	1100	Cut stone, 6′ to 10′ high, 2′ thick, dry set		45	.356		35.50	11.45		46.95	58	
	1200	Mortar set		50	.320	↓	35.50	10.30		45.80	56	
	5100	Setting stone, dry		100	.160	C.F.		5.15		5.15	8.35	
	5600	With mortar		120	.133	″		4.29		4.29	6.95	
	9000	Minimum labor/equipment charge	↓	2	8	Job		257		257	415	

02840 | Walk/Road/Parking Appurtenances

		02840 Walk/Road/Parking Appurtenances	Crew	Daily Output	Labor-Hours	Unit	2006 Bare Costs Mat.	Labor	Equip.	Total	Total Incl O&P	
200	0010	**GUIDE/GUARD RAIL**										200
	0012	Corrugated stl, galv. stl posts, 6′-3″ O.C.	B-80	850	.038	L.F.	16.45	1.13	.63	18.21	20.50	
	0200	End sections, galvanized, flared		50	.640	Ea.	68	19.15	10.65	97.80	118	
	0300	Wrap around end		50	.640	″	102	19.15	10.65	131.80	156	
	0400	Timber guide rail, 4″ x 8″ with 6″ x 8″ wood posts, treated		960	.033	L.F.	21.50	1	.55	23.05	26	
	0600	Cable guide rail, 3 at 3/4″ cables, steel posts, single face		900	.036		7.25	1.06	.59	8.90	10.40	
	0700	Wood posts		950	.034		10.10	1.01	.56	11.67	13.35	
	0900	Guide rail, steel box beam, 6″ x 6″		120	.267		25	7.95	4.44	37.39	45.50	
	1100	Median barrier, steel box beam, 6″ x 8″	↓	215	.149		30.50	4.45	2.48	37.43	44	
	1400	Resilient guide fence and light shield, 6′ high	B-2	130	.308	↓	24	8.55		32.55	40.50	
	1500	Concrete posts, individual, 6′-5″, triangular	B-80	110	.291	Ea.	51	8.70	4.84	64.54	75.50	
	1550	Square	″	110	.291	″	54.50	8.70	4.84	68.04	79.50	
	2000	Median, precast concrete, 3′-6″ high, 2′ wide, single face	B-29	380	.147	L.F.	39.50	4.41	2.21	46.12	53	
	2200	Double face	″	340	.165	″	45.50	4.93	2.47	52.90	61	
	2400	Speed bumps, thermoplastic, 10-1/2″ x 2-1/4″ x 48″ long	B-2	120	.333	Ea.	97	9.25		106.25	121	
	3030	Impact barrier, UTMCD, barrel type	B-16	30	1.067	″	330	30	15.90	375.90	430	
210	0010	**SECURITY VEHICLE BARRIERS**										210
	0020	Security planters excludes filling material										
	0100	Concrete security planter, exposed aggregate 36″ dia x 30″ high	B-11M	8	2	Ea.	420	64	34.50	518.50	600	
	0200	48″ dia x 36″ high		8	2		685	64	34.50	783.50	895	
	0300	53″ dia x 18″ high		8	2		510	64	34.50	608.50	700	
	0400	72″ dia x 18″ high with seats		8	2		1,050	64	34.50	1,148.50	1,300	
	0450	84″ dia x 36″ high		8	2		1,025	64	34.50	1,123.50	1,275	
	0500	36″ x 36″ x 24″ high square	↓	8	2	↓	370	64	34.50	468.50	545	

02840	Walk/Road/Parking Appurtenances	CREW	DAILY OUTPUT	LABOR-HOURS	UNIT	2006 BARE COSTS MAT.	LABOR	EQUIP.	TOTAL	TOTAL INCL O&P
210 0600	36″ x 36″ x 30″ high square	B-11M	8	2	Ea.	420	64	34.50	518.50	600
0700	48″ L x 24″ W x 30″ H rectangle		8	2		345	64	34.50	443.50	520
0800	72″ L x 24″ W x 30″ H rectangle		8	2		410	64	34.50	508.50	590
0900	96″ L x 24″ W x 30″ H rectangle		8	2		570	64	34.50	668.50	765
0950	Decorative geometric concrete barrier, 96″ L x 24″ W x 36″ H		8	2		850	64	34.50	948.50	1,075
1000	Concrete security planter, filling with 1 C.Y. washed sand/gravel		8	2		25.50	64	34.50	124	169
1050	Concrete security planter, filling with 2 C.Y. washed sand/gravel		6	2.667		51	85.50	46.50	183	246
1200	Jersey concete barrier, 10′ L x 2′ by 0.5′ W x 30″ H	B-21B	16	2.500		248	75	40	363	440
1300	10′ L x 2′ by 0.5′ W x 30″ H, 10 or more same site		24	1.667		248	50	26.50	324.50	385
1400	Jersey concete barrier, 10′ L x 2′ by 0.5′ W x 32″ H		16	2.500		258	75	40	373	450
1500	10′ L x 2′ by 0.5′ W x 32″ H, 10 or more same site		24	1.667		258	50	26.50	334.50	395
1600	Jersey concete barrier, 20′ L x 2′ by 0.5′ W x 30″ H		12	3.333		625	100	53	778	905
1700	20′ L x 2′ by 0.5′ W x 30″ H, 10 or more same site		18	2.222		625	66.50	35.50	727	835
1800	Jersey concete barrier, 20′ L x 2′ by 0.5′ W x 32″ H		12	3.333		635	100	53	788	920
1900	20′ L x 2′ by 0.5′ W x 32″ H, 10 or more same site		18	2.222		635	66.50	35.50	737	850
2000	GFRC decorative security barrier per 10 feet section including concrete		4	10		1,650	299	159	2,108	2,500
2100	GFRC decorative security barrier per 12 feet section including concrete		4	10		2,000	299	159	2,458	2,875
2210	GFRC decorative security barrier will stop 30 MPH, 4000 lb vehicle									
2300	High security barrier base prep per 12 feet section on bare ground	B-11C	4	4	Ea.	25	128	56.50	209.50	297
2310	GFRC barrier base prep does not include haul away of excavated matl									
2400	GFRC decorative high security barrier per 12 feet section w/concrete	B-21B	3	13.333	Ea.	4,375	400	212	4,987	5,700
2410	GFRC decorative high security barrier will stop 50 MPH, 15000 lb vehl									
2500	GFRC decorative impaler security barrier per 10 feet section w/prep	B-6	4	6	Ea.	1,675	180	56.50	1,911.50	2,175
2600	GFRC decorative impaler security barrier per 12 feet section w/prep	″	4	6	″	2,000	180	56.50	2,236.50	2,550
2610	Impaler barrier should stop 50 MPH, 15000 lb vehl w/some penetr									
2700	Pipe bollards, steel, concrete filled/painted, 8′ L x 4′ D hole, 8″ diam.	B-6	10	2.400	Ea.	695	72	22.50	789.50	905
2710	Schedule 80 concrete bollards will stop 4000 lb vehicle @ 30 MPH									
2800	GFRC decorative jersey barrier cover per 10 feet section excludes soil	B-6	8	3	Ea.	750	90	28.50	868.50	1,000
2900	GFRC decorative jersey barrier cover per 12 feet section excludes soil		8	3		900	90	28.50	1,018.50	1,175
3000	GFRC decorative 8″ diameter bollard cover		12	2		375	60	18.85	453.85	535
3100	GFRC decorative barrier face 10 foot section excludes earth backing		10	2.400		700	72	22.50	794.50	910
3200	Drop arm crash barrier, 20′ width or less, 15000 lb vehl @ 50 MPH									
3210	Includes all labor, material and equip. for complete installation				Ea.	66,000			66,000	72,500
3300	Drop arm crash barrier, 24′ width, 15000 lb vehicle @ 50 MPH									
3310	Includes all labor, material and equip. for complete installation				Ea.	71,500			71,500	78,500
3400	Portable wedge crash barrier, 10′ width, 15000 lb vehl @ 50 MPH									
3410	Includes all labor, material and equip. for complete installation				Ea.	65,000			65,000	71,500
3500	Portable wedge crash barrier, 12′ width, 15000 lb vehl @ 50 MPH									
3510	Includes all labor, material and equip. for complete installation				Ea.	71,500			71,500	78,500
3600	Sliding crash barrier, 12′ width, 15000 lb vehl @ 50 MPH									
3610	Includes all labor, material and equip. for complete installation				Ea.	92,500			92,500	101,500
3700	Sliding roller crash barrier, 20′ width, 15000 lb vehl @ 50 MPH									
3710	Includes all labor, material and equip. for complete installation				Ea.	66,000			66,000	72,500
3800	Sliding cantilever crash barrier, 20′ width, 15000 lb vehl @ 50 MPH									
3810	Includes all labor, material and equip. for complete installation				Ea.	71,500			71,500	78,500
3890	Note: Raised bollard crash barriers should be used w/tire shredders									
3900	Raised bollard crash barrier, 10′ width, 15000 lb vehicle @ 50 MPH									
3910	Includes all labor, material and equip. for complete installation				Ea.	22,000			22,000	24,200
4000	Raised bollard crash barrier, 12′ width, 15000 lb vehl @ 50 MPH									
4010	Includes all labor, material and equip. for complete installation				Ea.	25,300			25,300	27,800
4100	Raised bollard crash barrier, 10′ width with solar power									
4110	Includes all labor, material and equip. for complete installation				Ea.	28,000			28,000	30,800
4200	Raised bollard crash barrier, 12′ width with solar power									
4210	Includes all labor, material and equip. for complete installation				Ea.	31,300			31,300	34,400
4300	In ground tire shredder, 16′ width									
4310	Includes all labor, material and equip. for complete installation				Ea.	29,000			29,000	31,900

	02840	Walk/Road/Parking Appurtenances	CREW	DAILY OUTPUT	LABOR-HOURS	UNIT	2006 BARE COSTS MAT.	LABOR	EQUIP.	TOTAL	TOTAL INCL O&P	
400	0010	**FENDERS**										400
	0015	Bumper rails for garages, 12 Ga. rail, 6″ wide, with steel										
	0020	posts 12′-6″ O.C., minimum	E-4	190	.168	L.F.	13.15	6.80	.47	20.42	27.50	
	0030	Average		165	.194		16.45	7.85	.54	24.84	33.50	
	0100	Maximum		140	.229		19.70	9.25	.64	29.59	39.50	
	0300	12″ channel rail, minimum		160	.200		16.45	8.10	.56	25.11	34	
	0400	Maximum		120	.267		24.50	10.80	.74	36.04	48	
	9000	Minimum labor/equipment charge		2	16	Job		645	44.50	689.50	1,250	
800	0010	**PARKING BUMPERS**										800
	0020	Parking barriers, timber w/saddles, treated type										
	0100	4″ x 4″ for cars	B-2	520	.077	L.F.	2.54	2.14		4.68	6.35	
	0200	6″ x 6″ for trucks		520	.077	″	5.30	2.14		7.44	9.40	
	0600	Flexible fixed stanchion, 2′ high, 3″ diameter		100	.400	Ea.	25.50	11.10		36.60	46.50	
	1000	Wheel stops, precast concrete incl. dowels, 6″ x 10″ x 6′-0″		120	.333		40	9.25		49.25	59.50	
	1100	8″ x 13″ x 6′-0″		120	.333		46.50	9.25		55.75	66.50	
	1200	Thermoplastic, 6″ x 10″ x 6′-0″		120	.333		74.50	9.25		83.75	97.50	
	1300	Pipe bollards, conc filled/paint, 8′ L x 4′ D hole, 6″ diam.	B-6	20	1.200		300	36	11.30	347.30	400	
	1400	8″ diam.		15	1.600		455	48	15.10	518.10	595	
	1500	12″ diam.		12	2		595	60	18.85	673.85	770	
	1592	Bollards, steel, 3′ H, retractable, incl hydraulic controls, min		4	6		27,800	180	56.50	28,036.50	31,000	
	1594	Max		2	12		31,000	360	113	31,473	34,800	
	2030	Folding with individual padlocks	B-2	50	.800		730	22		752	835	
	9000	Parking lot control, see Div. 11156-600										
910	0010	**TRAFFIC BARRIERS, TRAFFIC CONTROL DEVICES**										910
	0020	Crash barriers										
	0100	Traffic channelizing pavement markers, layout only	A-7	2000	.012	Ea.		.47	.03	.50	.78	
	0110	13″ x 7-1/2″ x 2-1/2″ high, non-plowable install	2 Clab	96	.167		19.85	4.57		24.42	29.50	
	0200	8″ x 8″x 3-1/4″ high, non-plowable, install		96	.167		18.05	4.57		22.62	27.50	
	0230	4″ x 4″ x 3/4″ high, non-plowable, install		120	.133		2.41	3.65		6.06	8.70	
	0240	9-1/4″ x 5-7/8″ x 1/4″ high, plowable, concrete pav't	A-2A	70	.343		14.45	9.40	3.37	27.22	35	
	0250	9-1/4″ x 5-7/8″ x 1/4″ high, plowable, asphalt pav't	″	120	.200		2.76	5.50	1.97	10.23	14.25	
	0300	Barrier and curb delineators,reflectorized, 2″ x 4″	2 Clab	150	.107		1.57	2.92		4.49	6.55	
	0310	3″ x 5″	″	150	.107		3.19	2.92		6.11	8.35	
	0500	Rumble strip, polycarbonate										
	0510	24″ x 3-1/2″ x 1/2″ high	2 Clab	50	.320	Ea.	6	8.75		14.75	21	

	02850	Prefabricated Bridges	CREW	DAILY OUTPUT	LABOR-HOURS	UNIT	MAT.	LABOR	EQUIP.	TOTAL	TOTAL INCL O&P	
210	0010	**BRIDGES, PEDESTRIAN** spans over streams, roadways, etc.										210
	0020	including erection, not including foundations										
	0050	Precast concrete, complete in place, 8′ wide, 60′ span	E-2	215	.260	S.F.	52	10.15	6.65	68.80	82.50	
	0100	100′ span		185	.303		56.50	11.80	7.70	76	92	
	0150	120′ span		160	.350		61.50	13.60	8.90	84	102	
	0200	150′ span		145	.386		64	15	9.85	88.85	108	
	0300	Steel, trussed or arch spans, compl. in place, 8′ wide, 40′ span		320	.175		67.50	6.80	4.46	78.76	91	
	0400	50′ span		395	.142		60.50	5.50	3.61	69.61	80.50	
	0500	60′ span		465	.120		60.50	4.68	3.07	68.25	78.50	
	0600	80′ span		570	.098		72	3.82	2.50	78.32	88.50	
	0700	100′ span		465	.120		101	4.68	3.07	108.75	123	
	0800	120′ span		365	.153		128	5.95	3.91	137.86	156	
	0900	150′ span		310	.181		136	7.05	4.60	147.65	168	
	1000	160′ span		255	.220		136	8.55	5.60	150.15	172	
	1100	10′ wide, 80′ span		640	.088		75	3.40	2.23	80.63	91	
	1200	120′ span		415	.135		97	5.25	3.44	105.69	120	
	1300	150′ span		445	.126		109	4.90	3.21	117.11	132	
	1400	200′ span		205	.273		116	10.65	6.95	133.60	155	

		02850 | Prefabricated Bridges	CREW	DAILY OUTPUT	LABOR-HOURS	UNIT	2006 BARE COSTS MAT.	LABOR	EQUIP.	TOTAL	TOTAL INCL O&P	
210	1600	Wood, laminated type, complete in place, 80′ span	C-12	203	.236	S.F.	44.50	8.25	3.14	55.89	66	210
	1700	130′ span	″	153	.314		46.50	10.95	4.16	61.61	73.50	
		02870 | Site Furnishings										
310	0010	**BENCHES**										310
	0012	Seating, benches, park, precast conc, w/backs, wood rails, 4′ long	2 Clab	5	3.200	Ea.	335	87.50		422.50	515	
	0100	8′ long		4	4		700	110		810	950	
	0300	Fiberglass, without back, one piece, 4′ long		10	1.600		440	44		484	560	
	0400	8′ long		7	2.286		905	62.50		967.50	1,100	
	0500	Steel barstock pedestals w/backs, 2″ x 3″ wood rails, 4′ long		10	1.600		835	44		879	990	
	0510	8′ long		7	2.286		985	62.50		1,047.50	1,175	
	0520	3″ x 8″ wood plank, 4′ long		10	1.600		840	44		884	1,000	
	0530	8′ long		7	2.286		875	62.50		937.50	1,075	
	0540	Backless, 4″ x 4″ wood plank, 4′ square		10	1.600		815	44		859	970	
	0550	8′ long		7	2.286		775	62.50		837.50	955	
	0600	Aluminum pedestals, with backs, aluminum slats, 8′ long		8	2		405	55		460	535	
	0610	15′ long		5	3.200		400	87.50		487.50	585	
	0620	Portable, aluminum slats, 8′ long		8	2		355	55		410	480	
	0630	15′ long		5	3.200		520	87.50		607.50	715	
	0800	Cast iron pedestals, back & arms, wood slats, 4′ long		8	2		310	55		365	430	
	0820	8′ long		5	3.200		895	87.50		982.50	1,125	
	0840	Backless, wood slats, 4′ long		8	2		535	55		590	680	
	0860	8′ long		5	3.200		530	87.50		617.50	730	
	1700	Steel frame, fir seat, 10′ long		10	1.600		190	44		234	282	
	9000	Minimum labor/equipment charge		2	8	Job		219		219	365	
510	0010	**TRASH RECEPTACLE**										510
	0020	Trash receptacle, fiberglass, 2′ square, 18″ high	2 Clab	30	.533	Ea.	266	14.60		280.60	315	
	0100	2′ square, 2′-6″ high		30	.533		370	14.60		384.60	435	
	0300	Circular , 2′ diameter, 18″ high		30	.533		239	14.60		253.60	287	
	0400	2′ diameter, 2′-6″ high		30	.533		320	14.60		334.60	375	
	1000	Alum. frame, hardboard panels, steel drum base,										
	1020	30 gal. capacity, silk screen on plastic finish	2 Clab	25	.640	Ea.	570	17.55		587.55	660	
	1040	Aggregate finish		25	.640		445	17.55		462.55	520	
	1100	50 gal. capacity, silk screen on plastic finish		20	.800		585	22		607	680	
	1140	Aggregate finish		20	.800		690	22		712	795	
	1200	Formed plastic liner, 14 gal., silk screen on plastic finish		40	.400		253	10.95		263.95	296	
	1240	Aggregate finish		40	.400		289	10.95		299.95	340	
	1300	30 gal. capacity, silk screen on plastic finish		35	.457		345	12.55		357.55	395	
	1340	Aggregate finish		35	.457		395	12.55		407.55	455	
	1400	Redwood slats, plastic liner, leg base, 14 gal. capacity,										
	1420	Varnish w/routed message	2 Clab	40	.400	Ea.	300	10.95		310.95	350	
	2000	Concrete, precast, 2′ to 2-1/2′ wide, 3′ high, sandblasted	″	15	1.067	″	630	29		659	740	
	3000	Galv. steel frame and panels, leg base, poly bag retainer,										
	3020	40 gal. capacity, silk screen on enamel finish	2 Clab	25	.640	Ea.	355	17.55		372.55	420	
	3040	Aggregate finish		25	.640		465	17.55		482.55	540	
	3200	Formed plastic liner, 50 gal., silk screen on enamel finish		20	.800		530	22		552	620	
	3240	Aggregate finish		20	.800		620	22		642	715	
	4000	Perforated steel, pole mounted, 12″ diam., 10 gal., painted		25	.640		110	17.55		127.55	150	
	4040	Redwood slats		25	.640		285	17.55		302.55	345	
	4100	22 gal. capacity, painted		25	.640		162	17.55		179.55	207	
	4140	Redwood slats		25	.640		345	17.55		362.55	405	
	4500	Galvanized steel street basket, 52 gal. capacity, unpainted		40	.400		233	10.95		243.95	274	
520	0010	**TRASH CLOSURE** Steel with pullover cover										520
	0020	2′-3″ wide, 4′-7″ high, 6′-2″ long	2 Clab	5	3.200	Ea.	950	87.50		1,037.50	1,200	

02800 | Site Improvements and Amenities

02870 | Site Furnishings

			Crew	Daily Output	Labor-Hours	Unit	2006 Bare Costs Mat.	Labor	Equip.	Total	Total Incl O&P	
520	0100	10′-1″ long	2 Clab	4	4	Ea.	1,250	110		1,360	1,550	520
	0300	Wood, 10′ wide, 6′ high, 10′ long		1.20	13.333		990	365		1,355	1,700	

02880 | Playfield Equipment

			Crew	Daily Output	Labor-Hours	Unit	2006 Bare Costs Mat.	Labor	Equip.	Total	Total Incl O&P	
100	0010	**ATHLETIC OR RECREATIONAL SCREENING**										100
	0015	Backstops, baseball, prefabricated, 30′ wide, 12′ high & 1 overhang	B-1	1	24	Ea.	2,675	675		3,350	4,075	
	0100	40′ wide, 12′ high & 2 overhangs	″	.75	32		3,350	900		4,250	5,200	
	0110	Regulation, galvanized	B-13	2.40	23.333		6,500	700	271	7,471	8,600	
	0120	Vinyl coated	″	2.40	23.333		13,900	700	271	14,871	16,700	
	0180	Softball, prefabricated, no overhang	B-1	1.60	15		2,350	420		2,770	3,300	
	0200	Softball, regulation, galv., 3″ posts @8′, 6 & 9 ga. mesh, 14′ H	B-13	8.80	6.364		8,050	191	74	8,315	9,250	
	0205	18′ high		.75	74.667		9,625	2,225	870	12,720	15,200	
	0210	20′ high		.72	77.778		10,100	2,325	905	13,330	15,900	
	0215	22′ high		.70	80		11,900	2,400	930	15,230	18,000	
	0220	24′ high		.60	92.838		10,700	2,775	1,075	14,550	17,500	
	0250	Vinyl coated, 14′ high		1.20	46.667		12,200	1,400	540	14,140	16,300	
	0255	18′ high		.75	74.667		14,100	2,225	870	17,195	20,100	
	0260	20′ high		.72	77.778		14,900	2,325	905	18,130	21,100	
	0265	22′ high		.70	80		17,200	2,400	930	20,530	23,800	
	0270	24′ high		.60	93.333		19,900	2,800	1,075	23,775	27,700	
	0300	Basketball, steel, single goal		3.04	18.421		1,050	550	214	1,814	2,275	
	0400	Double goal		1.92	29.167		640	875	340	1,855	2,500	
	0600	Tennis, wire mesh with pair of ends	B-1	2.48	9.677	Set	1,650	272		1,922	2,250	
	0700	Enclosed court	″	1.30	18.462	Ea.	4,675	520		5,195	6,000	
	0900	Handball or squash court, outdoor, wood	2 Carp	.50	32		3,850	1,150		5,000	6,100	
	1000	Masonry handball/squash court	D-1	.30	53.333		22,400	1,725		24,125	27,500	
110	0010	**TENNIS COURT FENCES**										110
	0860	Tennis courts, 11 ga. wire, 2-1/2″ post set										
	0870	in concrete, 10′ O.C., 1-5/8″ top rail										
	0900	10′ high	B-80	190	.168	L.F.	12.55	5.05	2.80	20.40	25	
	0920	12′ high		170	.188	″	14.75	5.65	3.13	23.53	29	
	1000	Add for gate 4′ wide, 1-5/8″ frame 7′ high		10	3.200	Ea.	118	95.50	53.50	267	345	
	1040	Aluminized steel, 11 ga. wire 10′ high		190	.168	L.F.	15.20	5.05	2.80	23.05	28	
	1100	12′ high		170	.188	″	18.75	5.65	3.13	27.53	33	
	1140	Add for gate 4′ wide, 1-5/8″ frame, 7′ high		10	3.200	Ea.	138	95.50	53.50	287	365	
	1250	Vinyl covered, 9 ga. wire, 10′ high		190	.168	L.F.	14.70	5.05	2.80	22.55	27.50	
	1300	12′ high		170	.188	″	24.50	5.65	3.13	33.28	39.50	
	1310	Fence, CL, tennis ct, transom gate, single, galv, 4′ x 7′ x 3′	B-80A	8.72	2.752	Ea.	256	75.50	20	351.50	430	
	1400	Add for gate 4′ wide, 1-5/8″ frame, 7′ high	B-80	10	3.200	″	138	95.50	53.50	287	365	
210	0010	**PLAYGROUND EQUIPMENT** See also individual items										210
	0200	Bike rack, 10′ long, permanent	B-1	12	2	Ea.	505	56		561	655	
	0240	Climber, arch, 6′ high, 12′ long, 5′ wide		4	6		780	168		948	1,125	
	0260	Fitness trail, with signs, 9 to 10 stations, treated pine, minimum		.25	96		4,150	2,700		6,850	9,025	
	0270	Maximum		.17	141		9,000	3,975		12,975	16,500	
	0280	Metal, minimum		.25	96		4,175	2,700		6,875	9,075	
	0285	Maximum		.17	141		6,350	3,975		10,325	13,600	
	0300	Redwood, minimum		.25	96		12,400	2,700		15,100	18,100	
	0310	Maximum		.17	141		9,000	3,975		12,975	16,500	
	0320	16 to 20 station, treated pine, minimum		.17	141		6,125	3,975		10,100	13,300	
	0330	Maximum		.13	184		6,050	5,175		11,225	15,200	
	0340	Metal, minimum		.17	141		12,800	3,975		16,775	20,700	
	0350	Maximum		.13	184		16,700	5,175		21,875	27,000	
	0360	Redwood, minimum		.17	141		7,450	3,975		11,425	14,800	
	0370	Maximum		.13	184		18,000	5,175		23,175	28,400	
	0400	Horizontal monkey ladder, 14′ long, 6′ high		4	6		660	168		828	1,000	

	02880 | Playfield Equipment		CREW	DAILY OUTPUT	LABOR-HOURS	UNIT	2006 BARE COSTS MAT.	LABOR	EQUIP.	TOTAL	TOTAL INCL O&P	
210	0590	Parallel bars, 10′ long	B-1	4	6	Ea.	284	168		452	595	210
	0600	Posts, tether ball set, 2-3/8″ O.D.		12	2		188	56		244	300	
	0800	Poles, multiple purpose, 10′-6″ long		12	2	Pr.	141	56		197	248	
	1000	Ground socket for movable posts, 2-3/8″ post		10	2.400		99.50	67.50		167	221	
	1100	3-1/2″ post		10	2.400		139	67.50		206.50	265	
	1300	See-saw, spring, steel, 2 units		6	4	Ea.	980	112		1,092	1,250	
	1400	4 units		4	6		1,250	168		1,418	1,650	
	1500	6 units		3	8		2,500	225		2,725	3,125	
	1700	Shelter, fiberglass golf tee, 3 person		4.60	5.217		2,750	146		2,896	3,275	
	1900	Slides, stainless steel bed, 12′ long, 6′ high		3	8		2,450	225		2,675	3,075	
	2000	20′ long, 10′ high		2	12		3,575	335		3,910	4,500	
	2200	Swings, plain seats, 8′ high, 4 seats		2	12		1,100	335		1,435	1,775	
	2300	8 seats		1.30	18.462		2,025	520		2,545	3,075	
	2500	12′ high, 4 seats		2	12		1,500	335		1,835	2,175	
	2600	8 seats		1.30	18.462		1,950	520		2,470	3,000	
	2800	Whirlers, 8′ diameter		3	8		2,700	225		2,925	3,325	
	2900	10′ diameter		3	8		3,275	225		3,500	3,975	
220	0010	**MODULAR PLAYGROUND** Basic components										220
	0100	Deck, square, steel, 48″ x 48″	B-1	1	24	Ea.	770	675		1,445	1,975	
	0110	Recycled polyurethane		1	24		695	675		1,370	1,900	
	0120	Triangular, steel, 48″ side		1	24		565	675		1,240	1,750	
	0130	Post, steel, 5″ square		18	1.333	L.F.	25.50	37.50		63	90	
	0140	Aluminum, 2-3/8″ square		20	1.200		23	33.50		56.50	81.50	
	0150	5″ square		18	1.333		31	37.50		68.50	96	
	0160	Roof, square poly, 54″ side		18	1.333	Ea.	965	37.50		1,002.50	1,100	
	0170	Wheelchair transfer module, for 3′ high deck		3	8	″	2,525	225		2,750	3,150	
	0180	Guardrail, pipe, 36″ high		60	.400	L.F.	289	11.25		300.25	340	
	0190	Steps, deck-to-deck, 3 - 8″ steps		8	3	Ea.	1,375	84		1,459	1,650	
	0200	Activity panel, crawl through panel		2	12		395	335		730	990	
	0210	Alphabet/spelling panel		2	12		745	335		1,080	1,375	
	0360	With guardrails		3	8		2,125	225		2,350	2,700	
	0370	Crawl tunnel, straight, 56″ long		4	6		1,250	168		1,418	1,650	
	0380	90°, 4′ long		4	6		1,525	168		1,693	1,950	
	1200	Slide, tunnel, for 56″ high deck		8	3		1,725	84		1,809	2,050	
	1210	Straight, poly		8	3		245	84		329	410	
	1220	Stainless steel, 54″ high deck		6	4		253	112		365	465	
	1230	Curved, poly, 40″ high deck		6	4		1,275	112		1,387	1,575	
	1240	Spiral slide, 56″ - 72″ high		5	4.800		4,450	135		4,585	5,125	
	1300	Ladder, vertical, for 24″ - 72″ high deck		5	4.800		565	135		700	845	
	1310	Horizontal, 8′ long		5	4.800		785	135		920	1,075	
	1320	Corkscrew climber, 6′ high		3	8		605	225		830	1,025	
	1330	Fire pole for 72″ high deck		6	4		288	112		400	500	
	1340	Bridge, ring climber, 8′ long		4	6		1,150	168		1,318	1,525	
	1350	Suspension		4	6	L.F.	325	168		493	640	
330	0010	**GOAL POSTS**										330
	0020	Goal posts, steel, football, double post	B-1	1.50	16	Pr.	1,650	450		2,100	2,575	
	0100	Deluxe, single post		1.50	16		2,675	450		3,125	3,675	
	0300	Football, convertible to soccer		1.50	16		2,650	450		3,100	3,675	
	0500	Soccer, regulation		2	12		2,100	335		2,435	2,850	
350	0010	**PLATFORM/PADDLE TENNIS COURT** Complete with lighting, etc.										350
	0100	Aluminum slat deck with aluminum frame	B-1	.08	300	Court	43,600	8,425		52,025	62,000	
	0500	Aluminum slat deck and wood frame	C-1	.12	266		43,600	8,925		52,525	62,500	
	0800	Aluminum deck heater, add	B-1	1.18	20.339		4,250	570		4,820	5,625	

02800 | Site Improvements and Amenities

02880 | Playfield Equipment

		02880 Playfield Equipment	CREW	DAILY OUTPUT	LABOR-HOURS	UNIT	2006 BARE COSTS MAT.	LABOR	EQUIP.	TOTAL	TOTAL INCL O&P
350	0900	Douglas fir planking and wood frame 2″ x 6″ x 30′	C-1	.12	266	Court	42,800	8,925		51,725	62,000
	1000	Plywood deck with steel frame		.12	266		42,800	8,925		51,725	62,000
	1100	Steel slat deck with wood frame		.12	266		30,100	8,925		39,025	47,900

02890 | Traffic Signs & Signals

		02890 Traffic Signs & Signals	CREW	DAILY OUTPUT	LABOR-HOURS	UNIT	2006 BARE COSTS MAT.	LABOR	EQUIP.	TOTAL	TOTAL INCL O&P
100	0010	**SIGNS**									
	0012	Stock, 24″ x 24″, no posts, .080″ alum. reflectorized	B-80	70	.457	Ea.	55	13.65	7.60	76.25	91
	0100	High intensity		70	.457		55	13.65	7.60	76.25	91
	0300	30″ x 30″, reflectorized		70	.457		112	13.65	7.60	133.25	153
	0400	High intensity		70	.457		112	13.65	7.60	133.25	153
	0600	Guide and directional signs, 12″ x 18″, reflectorized		70	.457		35.50	13.65	7.60	56.75	70
	0700	High intensity		70	.457		34	13.65	7.60	55.25	68
	0900	18″ x 24″, stock signs, reflectorized		70	.457		40	13.65	7.60	61.25	74.50
	1000	High intensity		70	.457		40	13.65	7.60	61.25	74.50
	1200	24″ x 24″, stock signs, reflectorized		70	.457		49.50	13.65	7.60	70.75	85
	1300	High intensity		70	.457		49.50	13.65	7.60	70.75	85
	1500	Add to above for steel posts, galvanized, 10′-0″ upright, bolted		200	.160		17.90	4.78	2.66	25.34	30.50
	1600	12′-0″ upright, bolted		140	.229		24	6.85	3.80	34.65	41.50
	1800	Highway road signs, aluminum, over 20 S. F., reflectorized		350	.091	S.F.	23	2.73	1.52	27.25	31
	2000	High intensity		350	.091		23	2.73	1.52	27.25	31
	2200	Highway, suspended over road, 80 S.F. min., reflectorized		165	.194		23	5.80	3.23	32.03	38
	2300	High intensity		165	.194		23	5.80	3.23	32.03	38
	5300	Remove traffic posts to 12′-0″ high	B-6	100	.240	Ea.		7.20	2.26	9.46	14.20
300	0010	**TRAFFIC SIGNALS** Mid block pedestrian crosswalk,									
	0020	with pushbutton and mast arms	R-11	.30	186	Total	57,000	7,350	2,575	66,925	77,000
	0600	Traffic signals, School flashing system, solar powered, remote control'd	″	1	56	Signal	14,000	2,200	770	16,970	19,700
	1000	Intersection traffic signals, LED, mast, programmable, no lane control									
	1010	Includes all labor, material and equip. for complete installation				Ea.	157,500			157,500	173,500
	1100	Intersection traffic signals, LED, mast, programmable, R/L lane control									
	1110	Includes all labor, material and equip. for complete installation				Ea.	210,000			210,000	231,000
	1200	Add protective/permissive left turns to existing traffic light									
	1210	Includes all labor, material and equip. for complete installation				Ea.	63,000			63,000	69,500
	1300	Replace existing light heads with LED Heads wire hung									
	1310	Includes all labor, material and equip. for complete installation				Ea.	42,000			42,000	46,200
	1400	Replace existing light heads with LED Heads mast arm hung									
	1410	Includes all labor, material and equip. for complete installation				Ea.	84,000			84,000	92,500

02900 | Planting

02905 | Plants, Planting, Transplanting

		02905 Plants, Planting, Transplanting	CREW	DAILY OUTPUT	LABOR-HOURS	UNIT	2006 BARE COSTS MAT.	LABOR	EQUIP.	TOTAL	TOTAL INCL O&P
725	0010	**PLANTING**									
	0012	Moving shrubs on site, 12″ ball	B-62	28	.857	Ea.		25.50	5.50	31	48
	0100	24″ ball	″	22	1.091			32.50	7	39.50	60.50
	0300	Moving trees on site, 36″ ball	B-6	3.75	6.400			192	60.50	252.50	375
	0400	60″ ball	″	1	24			720	226	946	1,425
925	0010	**TREE REMOVAL**									
	0100	Dig & lace, shrubs, broadleaf evergreen, 18″-24″	B-1	55	.436	Ea.		12.25		12.25	20.50

02900 | Planting

SITE CONSTRUCTION 2

	02905 | Plants, Planting, Transplanting		CREW	DAILY OUTPUT	LABOR-HOURS	UNIT	2006 BARE COSTS MAT.	LABOR	EQUIP.	TOTAL	TOTAL INCL O&P	
925	0200	2′-3′	B-1	35	.686	Ea.		19.25		19.25	32	925
	0300	3′-4′	B-6	30	.800			24	7.55	31.55	47.50	
	0400	4′-5′	"	20	1.200			36	11.30	47.30	71	
	1000	Deciduous, 12″-15″	B-1	110	.218			6.10		6.10	10.15	
	1100	18″-24″		65	.369			10.35		10.35	17.15	
	1200	2′-3′		55	.436			12.25		12.25	20.50	
	1300	3′-4′	B-6	50	.480			14.40	4.53	18.93	28.50	
	2000	Evergreeen, 18″-24″	B-1	55	.436			12.25		12.25	20.50	
	2100	2′-0″ to 2′-6″		50	.480			13.45		13.45	22.50	
	2200	2′-6″ to 3′-0″		35	.686			19.25		19.25	32	
	2300	3′-0″ to 3′-6″		20	1.200			33.50		33.50	56	
	3000	Trees, deciduous, small, 2′-3′		55	.436			12.25		12.25	20.50	
	3100	3′-4′	B-6	50	.480			14.40	4.53	18.93	28.50	
	3200	4′-5′		35	.686			20.50	6.45	26.95	40.50	
	3300	5′-6′		30	.800			24	7.55	31.55	47.50	
	4000	Shade, 5′-6′		50	.480			14.40	4.53	18.93	28.50	
	4100	6′-8′		35	.686			20.50	6.45	26.95	40.50	
	4200	8′-10′		25	.960			29	9.05	38.05	57	
	4300	2″ caliper		12	2			60	18.85	78.85	118	
	5000	Evergreen, 4′-5′		35	.686			20.50	6.45	26.95	40.50	
	5100	5′-6′		25	.960			29	9.05	38.05	57	
	5200	6′-7′		19	1.263			38	11.90	49.90	74.50	
	5300	7′-8′		15	1.600			48	15.10	63.10	94.50	
	5400	8′-10′		11	2.182			65.50	20.50	86	129	

	02910 | Plant Preparation		CREW	DAILY OUTPUT	LABOR-HOURS	UNIT	MAT.	LABOR	EQUIP.	TOTAL	TOTAL INCL O&P	
500	0010	**MULCHING**										500
	0100	Aged barks, 3″ deep, hand spread	1 Clab	100	.080	S.Y.	2.07	2.19		4.26	5.90	
	0150	Skid steer loader	B-63	13.50	2.963	M.S.F.	230	86	11.40	327.40	405	
	0200	Hay, 1″ deep, hand spread	1 Clab	475	.017	S.Y.	.57	.46		1.03	1.39	
	0250	Power mulcher, small	B-64	180	.089	M.S.F.	63.50	2.44	1.35	67.29	75	
	0350	Large	B-65	530	.030	"	63.50	.83	.59	64.92	71.50	
	0400	Humus peat, 1″ deep, hand spread	1 Clab	700	.011	S.Y.	2.17	.31		2.48	2.91	
	0450	Push spreader	"	2500	.003	"	2.17	.09		2.26	2.54	
	0550	Tractor spreader	B-66	700	.011	M.S.F.	241	.40	.26	241.66	266	
	0600	Oat straw, 1″ deep, hand spread	1 Clab	475	.017	S.Y.	.33	.46		.79	1.12	
	0650	Power mulcher, small	B-64	180	.089	M.S.F.	36.50	2.44	1.35	40.29	46	
	0700	Large	B-65	530	.030	"	36.50	.83	.59	37.92	42.50	
	0750	Add for asphaltic emulsion	B-45	1770	.009	Gal.	1.77	.29	.30	2.36	2.75	
	0800	Peat moss, 1″ deep, hand spread	1 Clab	900	.009	S.Y.	1.72	.24		1.96	2.29	
	0850	Push spreader	"	2500	.003	"	1.72	.09		1.81	2.04	
	0950	Tractor spreader	B-66	700	.011	M.S.F.	191	.40	.26	191.66	211	
	1000	Polyethylene film, 6 mil.	2 Clab	2000	.008	S.Y.	.17	.22		.39	.55	
	1100	Redwood nuggets, 3″ deep, hand spread	1 Clab	150	.053	"	4.61	1.46		6.07	7.45	
	1150	Skid steer loader	B-63	13.50	2.963	M.S.F.	510	86	11.40	607.40	720	
	1200	Stone mulch, hand spread, ceramic chips, economy	1 Clab	125	.064	S.Y.	6.25	1.75		8	9.75	
	1250	Deluxe	"	95	.084	"	9.60	2.31		11.91	14.35	
	1300	Granite chips	B-1	10	2.400	C.Y.	31	67.50		98.50	147	
	1400	Marble chips		10	2.400		117	67.50		184.50	241	
	1600	Pea gravel		28	.857		58	24		82	104	
	1700	Quartz		10	2.400		150	67.50		217.50	278	
	1800	Tar paper, 15 Lb. felt	1 Clab	800	.010	S.Y.	.31	.27		.58	.79	
	1900	Wood chips, 2″ deep, hand spread	"	220	.036	"	1.77	1		2.77	3.60	
	1950	Skid steer loader	B-63	20.30	1.970	M.S.F.	197	57	7.60	261.60	320	
710	0010	**LAWN BED PREPARATION**										710
	0100	Rake topsoil, site material, harley rock rake, ideal	B-6	33	.727	M.S.F.		22	6.85	28.85	43	

02900 | Planting

02910 | Plant Preparation		CREW	DAILY OUTPUT	LABOR-HOURS	UNIT	2006 BARE COSTS MAT.	LABOR	EQUIP.	TOTAL	TOTAL INCL O&P
710	0200 Adverse	B-6	7	3.429	M.S.F.		103	32.50	135.50	203
	0300 Screened loam, york rake and finish, ideal	B-62	24	1			30	6.40	36.40	56
	0400 Adverse	"	20	1.200			36	7.70	43.70	67
	1000 Remove topsoil & stock pile on site, 75 HP dozer, 6" deep, 50' haul	B-10L	30	.400			13.45	10.65	24.10	33
	1050 300' haul		6.10	1.967			66	52.50	118.50	164
	1100 12" deep, 50' haul		15.50	.774			26	20.50	46.50	64
	1150 300' haul		3.10	3.871			130	103	233	320
	1200 200 HP dozer, 6" deep, 50' haul	B-10B	125	.096			3.23	7.35	10.58	13.25
	1250 300' haul		30.70	.391			13.15	30	43.15	54
	1300 12" deep, 50' haul		62	.194			6.50	14.85	21.35	27
	1350 300' haul		15.40	.779			26	60	86	108
	1400 Alternate method, 75 HP dozer, 50' haul	B-10L	860	.014	C.Y.		.47	.37	.84	1.16
	1450 300' haul	"	114	.105			3.54	2.80	6.34	8.75
	1500 200 HP dozer, 50' haul	B-10B	2660	.005			.15	.35	.50	.62
	1600 300' haul	"	570	.021			.71	1.62	2.33	2.91
	1800 Rolling topsoil, hand push roller	1 Clab	3200	.003	S.F.		.07		.07	.11
	1850 Tractor drawn roller	B-66	10666	.001	"		.03	.02	.05	.06
	2000 Root raking and loading, residential, no boulders	B-6	53.30	.450	M.S.F.		13.50	4.25	17.75	26.50
	2100 With boulders		32	.750			22.50	7.05	29.55	44.50
	2200 Municipal, no boulders		200	.120			3.60	1.13	4.73	7.10
	2300 With boulders		120	.200			6	1.89	7.89	11.80
	2400 Large commercial, no boulders	B-10B	400	.030			1.01	2.30	3.31	4.14
	2500 With boulders	"	240	.050			1.68	3.84	5.52	6.90
	3000 Scarify subsoil, residential, skid steer loader w/scarifiers, 50 HP	B-66	32	.250			8.80	5.60	14.40	20
	3050 Municipal, skid steer loader w/scarifiers, 50 HP	"	120	.067			2.35	1.49	3.84	5.35
	3100 Large commercial, 75 HP, dozer w/scarifier	B-10L	240	.050			1.68	1.33	3.01	4.15
	3500 Screen topsoil from stockpile, vibrating screen, wet material (organic)	B-10P	200	.060	C.Y.		2.02	3.97	5.99	7.60
	3550 Dry material	"	300	.040			1.34	2.65	3.99	5.05
	3600 Mixing with conditioners, manure and peat	B-10R	550	.022			.73	.35	1.08	1.56
	3650 Mobilization add for 2 days or less operation	B-34K	3	2.667	Job		75.50	175	250.50	315
	3800 Spread conditioned topsoil, 6" deep, by hand	B-1	360	.067	S.Y.	4.65	1.87		6.52	8.20
	3850 300 HP dozer	B-10M	27	.444	M.S.F.	505	14.95	44.50	564.45	630
	4000 Spread soil conditioners, alum. sulfate, 1#/S.Y., hand push spreader	1 Clab	17500	.001	S.Y.	15.30	.01		15.31	16.85
	4050 Tractor spreader	B-66	700	.011	M.S.F.	1,700	.40	.26	1,700.66	1,875
	4100 Fertilizer, 0.2#/S.Y., push spreader	1 Clab	17500	.001	S.Y.	.08	.01		.09	.11
	4150 Tractor spreader	B-66	700	.011	M.S.F.	8.90	.40	.26	9.56	10.70
	4200 Ground limestone, 1#/S.Y., push spreader	1 Clab	17500	.001	S.Y.	.09	.01		.10	.12
	4250 Tractor spreader	B-66	700	.011	M.S.F.	10	.40	.26	10.66	11.90
	4300 Lusoil, 3#/S.Y., push spreader	1 Clab	17500	.001	S.Y.	.47	.01		.48	.54
	4350 Tractor spreader	B-66	700	.011	M.S.F.	52	.40	.26	52.66	58.50
	4400 Manure, 18#/S.Y., push spreader	1 Clab	2500	.003	S.Y.	2.79	.09		2.88	3.22
	4450 Tractor spreader	B-66	280	.029	M.S.F.	310	1.01	.64	311.65	340
	4500 Perlite, 1" deep, push spreader	1 Clab	17500	.001	S.Y.	8.60	.01		8.61	9.50
	4550 Tractor spreader	B-66	700	.011	M.S.F.	960	.40	.26	960.66	1,050
	4600 Vermiculite, push spreader	1 Clab	17500	.001	S.Y.	2.63	.01		2.64	2.91
	4650 Tractor spreader	B-66	700	.011	M.S.F.	292	.40	.26	292.66	320
	5000 Spread topsoil, skid steer loader and hand dress	B-62	270	.089	C.Y.	21	2.67	.57	24.24	28
	5100 Articulated loader and hand dress	B-100	320	.038		21	1.26	1.74	24	27
	5200 Articulated loader and 75HP dozer	B-10M	500	.024		21	.81	2.39	24.20	27
	5300 Road grader and hand dress	B-11L	1000	.016		21	.51	.46	21.97	24.50
	6000 Tilling topsoil, 20 HP tractor, disk harrow, 2" deep	B-66	450	.018	M.S.F.		.63	.40	1.03	1.43
	6050 4" deep		360	.022			.78	.50	1.28	1.79
	6100 6" deep		270	.030			1.04	.66	1.70	2.38
	6150 26" rototiller, 2" deep	A-1J	1250	.006	S.Y.		.18	.08	.26	.38
	6200 4" deep		1000	.008			.22	.10	.32	.47
	6250 6" deep		750	.011			.29	.14	.43	.63

02900 | Planting

	02910 | Plant Preparation		Crew	Daily Output	Labor-Hours	Unit	2006 Bare Costs Mat.	Labor	Equip.	Total	Total Incl O&P	
710	7000	Lawn maintenance see Division 02985-700										710
720	0010	**PLANT BED PREPARATION, SHRUB & TREE**										720
	0100	Backfill planting pit, by hand, on site topsoil	2 Clab	18	.889	C.Y.		24.50		24.50	40.50	
	0200	Prepared planting mix	"	24	.667			18.25		18.25	30.50	
	0300	Skid steer loader, on site topsoil	B-62	340	.071			2.12	.45	2.57	3.94	
	0400	Prepared planting mix	"	410	.059			1.76	.38	2.14	3.27	
	1000	Excavate planting pit, by hand, sandy soil	2 Clab	16	1			27.50		27.50	45.50	
	1100	Heavy soil or clay	"	8	2			55		55	91	
	1200	1/2 C.Y. backhoe, sandy soil	B-11C	150	.107			3.42	1.51	4.93	7.15	
	1300	Heavy soil or clay	"	115	.139			4.46	1.97	6.43	9.35	
	2000	Mix planting soil, incl. loam, manure, peat, by hand	2 Clab	60	.267		36.50	7.30		43.80	52.50	
	2100	Skid steer loader	B-62	150	.160		36.50	4.80	1.03	42.33	49.50	
	3000	Pile sod, skid steer loader	"	2800	.009	S.Y.		.26	.06	.32	.48	
	3100	By hand	2 Clab	400	.040			1.10		1.10	1.82	
	4000	Remove sod, F.E. loader	B-10S	2000	.006			.20	.12	.32	.45	
	4100	Sod cutter	B-12K	3200	.005			.16	.30	.46	.59	
	4200	By hand	2 Clab	240	.067			1.83		1.83	3.03	
	02915 | Shrub and Tree Transplanting											
300	0010	**GROUND COVER AND VINES** Planting only, no preparation										300
	0100	Ajuga, 1 year, bare root	B-1	9	2.667	C	65	75		140	196	
	0150	Potted, 2 year		6	4		239	112		351	450	
	0200	Berberis, potted, 2 year		6	4		266	112		378	480	
	0250	Cotoneaster, 15″ to 18″, shady areas, B & B		.60	40		760	1,125		1,885	2,675	
	0300	Boston ivy, on bank, 1 year, bare root		6	4		120	112		232	320	
	0350	Potted, 2 year		6	4		299	112		411	515	
	0400	English ivy, 1 year, bare root		9	2.667		68.50	75		143.50	200	
	0450	Potted, 2 year		6	4		111	112		223	310	
	0500	Halls honeysuckle, 1 year, bare root		5	4.800		121	135		256	355	
	0550	Potted, 2 year		4	6		385	168		553	705	
	0600	Memorial rose, 9″ to 12″, 1 gallon container		3	8		182	225		407	570	
	0650	Potted, 2 gallon container		2	12		330	335		665	920	
	0700	Pachysandra, 1 year, bare root		10	2.400		13.85	67.50		81.35	127	
	0750	Potted, 2 year		6	4		98	112		210	294	
	0800	Vinca minor, 1 year, bare root		10	2.400		69.50	67.50		137	189	
	0850	Potted, 2 year		6	4		92.50	112		204.50	288	
	0900	Woodbine, on bank, 1/2 year, bare root		6	4		109	112		221	305	
	0950	Potted, 2 year		4	6		285	168		453	595	
	2000	Alternate method of figuring										
	2100	Ajuga, field division, 4000/M.S.F.	B-1	.23	104	M.S.F.	2,225	2,925		5,150	7,300	
	2300	Boston ivy, 1 year, 60/M.S.F.		10	2.400		71	67.50		138.50	190	
	2400	English ivy, 1 yr., 500/M.S.F.		1.80	13.333		305	375		680	955	
	2500	Halls honeysuckle, 1 yr., 333/M.S.F.		1.50	16		171	450		621	935	
	2600	Memorial rose, 9″-12″, 1 gal., 333/M.S.F.		.90	26.667		565	750		1,315	1,875	
	2700	Pachysandra, 1 yr., 4000/M.S.F.		.25	96		595	2,700		3,295	5,125	
	2800	Vinca minor, rooted cutting, 2000/M.S.F.		1	24		1,225	675		1,900	2,475	
	2900	Woodbine, 1 yr., 60/M.S.F.		10	2.400		71	67.50		138.50	190	
400	0010	**PLANTING** Trees, shrubs and ground cover										400
	0100	Light soil										
	0110	Bare root seedlings, 3″ to 5″	1 Clab	960	.008	Ea.		.23		.23	.38	
	0120	6″ to 10″		520	.015			.42		.42	.70	
	0130	11″ to 16″		370	.022			.59		.59	.98	
	0140	17″ to 24″		210	.038			1.04		1.04	1.73	
	0200	Potted, 2-1/4″ diameter		840	.010			.26		.26	.43	
	0210	3″ diameter		700	.011			.31		.31	.52	

For expanded coverage of these items see *Means Site Work & Landscape Cost Data 2006*

2 SITE CONSTRUCTION

	02915 | Shrub and Tree Transplanting	CREW	DAILY OUTPUT	LABOR-HOURS	UNIT	2006 BARE COSTS MAT.	LABOR	EQUIP.	TOTAL	TOTAL INCL O&P	
400	0220 4" diameter	1 Clab	620	.013	Ea.		.35		.35	.59	400
	0300 Container, 1 gallon	2 Clab	84	.190			5.20		5.20	8.65	
	0310 2 gallon		52	.308			8.45		8.45	13.95	
	0320 3 gallon		40	.400			10.95		10.95	18.15	
	0330 5 gallon		29	.552			15.10		15.10	25	

	02920 | Lawns & Grasses	CREW	DAILY OUTPUT	LABOR-HOURS	UNIT	MAT.	LABOR	EQUIP.	TOTAL	TOTAL INCL O&P	
320	0010 SEEDING, ATHLETIC FIELDS R329219-50										320
	0020 Seeding, athletic fields, athletic field mix, 8#/MSFpush spreader	1 Clab	8	1	M.S.F.	22	27.50		49.50	70	
	0100 Tractor spreader	B-66	52	.154		22	5.40	3.44	30.84	37	
	0200 Hydro or air seeding, with mulch & fertil.	B-81	80	.300		24.50	9.25	5.05	38.80	47.50	
	0400 Birdsfoot trefoil, .45#/M.S.F., push spreader	1 Clab	8	1		7.15	27.50		34.65	53.50	
	0500 Tractor spreader	B-66	52	.154		7.15	5.40	3.44	15.99	20	
	0600 Hydro or air seeding,with mulch & fertil.	B-81	80	.300		13.80	9.25	5.05	28.10	35.50	
	0800 Bluegrass, 4#/M.S.F., common, push spreader	1 Clab	8	1		15.10	27.50		42.60	62	
	0900 Tractor spreader	B-66	52	.154		15.10	5.40	3.44	23.94	29	
	1000 Hydro or air seeding, with mulch & fertil.	B-81	80	.300		25	9.25	5.05	39.30	48	
	1100 Baron, push spreader	1 Clab	8	1		19.95	27.50		47.45	67.50	
	1200 Tractor spreader	B-66	52	.154		19.95	5.40	3.44	28.79	34.50	
	1300 Hydro or air seeding, with mulch & fertil.	B-81	80	.300		27.50	9.25	5.05	41.80	50.50	
	1500 Clover, 0.67#/M.S.F., white, push spreader	1 Clab	8	1		1.31	27.50		28.81	47	
	1600 Tractor spreader	B-66	52	.154		1.31	5.40	3.44	10.15	13.75	
	1700 Hydro or air seeding, with mulch and fertil.	B-81	80	.300		7.20	9.25	5.05	21.50	28.50	
	1800 Ladino, push spreader	1 Clab	8	1		5.15	27.50		32.65	51	
	1900 Tractor spreader	B-66	52	.154		5.15	5.40	3.44	13.99	18.05	
	2000 Hydro or air seeding, with mulch and fertil.	B-81	80	.300		23	9.25	5.05	37.30	45.50	
	2200 Fescue 5.5#/M.S.F., tall, push spreader	1 Clab	8	1		10.25	27.50		37.75	57	
	2300 Tractor spreader	B-66	52	.154		10.25	5.40	3.44	19.09	23.50	
	2400 Hydro or air seeding, with mulch and fertilizer	B-81	80	.300		34	9.25	5.05	48.30	57.50	
	2500 Chewing, push spreader	1 Clab	8	1		10.25	27.50		37.75	57	
	2600 Tractor spreader	B-66	52	.154		10.25	5.40	3.44	19.09	23.50	
	2700 Hydro or air seeding, with mulch and fertil.	B-81	80	.300		34	9.25	5.05	48.30	57.50	
	2900 Crown vetch, 4#/M.S.F., push spreader	1 Clab	8	1		37.50	27.50		65	87	
	3000 Tractor spreader	B-66	52	.154		37.50	5.40	3.44	46.34	54	
	3100 Hydro or air seeding, with mulch and fertilizer	B-81	80	.300		52	9.25	5.05	66.30	77.50	
	3300 Rye, 10#/M.S.F., annual, push spreader	1 Clab	8	1		6.90	27.50		34.40	53	
	3400 Tractor spreader	B-66	52	.154		6.90	5.40	3.44	15.74	19.95	
	3500 Hydro or air seeding, with mulch and fertilizer	B-81	80	.300		15.20	9.25	5.05	29.50	37	
	3600 Fine textured, push spreader	1 Clab	8	1		6.90	27.50		34.40	53	
	3700 Tractor spreader	B-66	52	.154		6.90	5.40	3.44	15.74	19.95	
	3800 Hydro or air seeding, with mulch and fertilizer	B-81	80	.300		15.20	9.25	5.05	29.50	37	
	4000 Shade mix, 6#/M.S.F., push spreader	1 Clab	8	1		9.85	27.50		37.35	56.50	
	4100 Tractor spreader	B-66	52	.154		9.85	5.40	3.44	18.69	23	
	4200 Hydro or air seeding, with mulch and fertilizer	B-81	80	.300		21.50	9.25	5.05	35.80	44.50	
	4400 Slope mix, 6#/M.S.F., push spreader	1 Clab	8	1		9.85	27.50		37.35	56.50	
	4500 Tractor spreader	B-66	52	.154		9.85	5.40	3.44	18.69	23	
	4600 Hydro or air seeding, with mulch and fertilizer	B-81	80	.300		24.50	9.25	5.05	38.80	47.50	
	4800 Turf mix, 4#/M.S.F., push spreader	1 Clab	8	1		6.55	27.50		34.05	53	
	4900 Tractor spreader	B-66	52	.154		6.55	5.40	3.44	15.39	19.60	
	5000 Hydro or air seeding, with mulch and fertilizer	B-81	80	.300		16.45	9.25	5.05	30.75	38.50	
	5200 Utility mix, 7#/M.S.F., push spreader	1 Clab	8	1		11.50	27.50		39	58	
	5300 Tractor spreader	B-66	52	.154		11.50	5.40	3.44	20.34	25	
	5400 Hydro or air seeidng, with mulch and fertilizer	B-81	80	.300		43	9.25	5.05	57.30	68	
	5600 Wildflower, .10#/M.S.F., push spreader	1 Clab	8	1		4.03	27.50		31.53	50	
	5700 Tractor spreader	B-66	52	.154		4.03	5.40	3.44	12.87	16.75	

02900 | Planting

02920 | Lawns & Grasses

		02920 | Lawns & Grasses	CREW	DAILY OUTPUT	LABOR-HOURS	UNIT	2006 BARE COSTS MAT.	LABOR	EQUIP.	TOTAL	TOTAL INCL O&P	
320	5800	Hydro or air seeding, with mulch and fertilizer (R329219-50)	B-81	80	.300	M.S.F.	22	9.25	5.05	36.30	45	320
	7025	Limestone, mechanical spread	1 Clab	1.75	4.571	Acre	3.46	125		128.46	212	
	7060	Limestone, mechanical spread		1.74	4.598	"	3.46	126		129.46	213	
	9000	Minimum labor/equipment charge		4	2	Job		55		55	91	
400	0010	**SODDING**										400
	0020	Sodding, 1" deep, bluegrass sod, on level ground, over 8 MSF	B-63	22	1.818	M.S.F.	223	52.50	7	282.50	340	
	0200	4 M.S.F.		17	2.353		249	68	9.05	326.05	395	
	0300	1000 S.F.		13.50	2.963		271	86	11.40	368.40	455	
	0500	Sloped ground, over 8 M.S.F.		6	6.667		223	193	25.50	441.50	590	
	0600	4 M.S.F.		5	8		249	232	31	512	690	
	0700	1000 S.F.		4	10		271	290	38.50	599.50	815	
	1000	Bent grass sod, on level ground, over 6 M.S.F.		20	2		500	58	7.70	565.70	655	
	1100	3 M.S.F.		18	2.222		555	64.50	8.55	628.05	725	
	1200	Sodding 1000 S.F. or less		14	2.857		630	82.50	11	723.50	845	
	1500	Sloped ground, over 6 M.S.F.		15	2.667		500	77	10.25	587.25	690	
	1600	3 M.S.F.		13.50	2.963		555	86	11.40	652.40	765	
	1700	1000 S.F.		12	3.333		630	96.50	12.85	739.35	865	

02930 | Exterior Plants

		02930 | Exterior Plants	CREW	DAILY OUTPUT	LABOR-HOURS	UNIT	MAT.	LABOR	EQUIP.	TOTAL	TOTAL INCL O&P	
050	0010	**TRAVEL** To all nursery items, for 10 to 20 miles, add				All					5%	050
	0100	30 to 50 miles, add				"					10%	
310	0010	**SHRUBS AND TREES** Evergreen, in prepared beds, B & B										310
	0100	Arborvitae pyramidal, 4'-5'	B-17	30	1.067	Ea.	44	31.50	18.45	93.95	120	
	0150	Globe, 12"-15"	B-1	96	.250		11.05	7		18.05	24	
	0200	Balsam, fraser, 6' - 7'	B-17	30	1.067		105	31.50	18.45	154.95	188	
	0300	Cedar, blue, 8'-10'		18	1.778		183	52.50	30.50	266	320	
	0350	Japanese, 4' - 5'		55	.582		91	17.20	10.05	118.25	139	
	0400	Cypress, hinoki, 15" - 18"	B-1	80	.300		58	8.40		66.40	78	
	0500	Hemlock, canadian, 2-1/2'-3'		36	.667		23	18.70		41.70	56.50	
	0550	Holly, Savannah, 8' - 10' H		9.68	2.479		530	69.50		599.50	695	
	0600	Juniper, andorra, 18"-24"		80	.300		15.90	8.40		24.30	31.50	
	0620	Wiltoni, 15"-18"		80	.300		15.25	8.40		23.65	30.50	
	0640	Skyrocket, 4-1/2'-5'	B-17	55	.582		48.50	17.20	10.05	75.75	92.50	
	0660	Blue pfitzer, 2'-2-1/2'	B-1	44	.545		25	15.30		40.30	53	
	0680	Ketleerie, 2-1/2'-3'		50	.480		32	13.45		45.45	57.50	
	0700	Pine, black, 2-1/2'-3'		50	.480		39	13.45		52.45	65.50	
	0720	Mugo, 18"-24"		60	.400		34	11.25		45.25	56	
	0740	White, 4'-5'	B-17	75	.427		50.50	12.65	7.35	70.50	84	
	0800	Spruce, blue, 18"-24"	B-1	60	.400		38	11.25		49.25	60.50	
	0820	Dwarf alberta, 18" - 24"	"	60	.400		45.50	11.25		56.75	68.50	
	0840	Norway, 4'-5'	B-17	75	.427		85.50	12.65	7.35	105.50	123	
	0900	Yew, denisforma, 12"-15"	B-1	60	.400		23	11.25		34.25	43.50	
	1000	Capitata, 18"-24"		30	.800		19.65	22.50		42.15	58.50	
	1100	Hicksi, 2'-2-1/2'		30	.800		29	22.50		51.50	69	
320	0010	**SHRUBS** Broadleaf evergreen, planted in prepared beds										320
	0100	Andromeda, 15"-18", container	B-1	96	.250	Ea.	22.50	7		29.50	36.50	
	0200	Azalea, 15" - 18", container		96	.250		25	7		32	39	
	0300	Barberry, 9"-12", container		130	.185		10.05	5.20		15.25	19.65	
	0400	Boxwood, 15"-18", B & B		96	.250		27	7		34	41	
	0500	Euonymus, emerald gaiety, 12" to 15", container		115	.209		16.20	5.85		22.05	27.50	
	0600	Holly, 15"-18", B & B		96	.250		15.80	7		22.80	29	
	0700	Leucothoe, 15" - 18", container		96	.250		14.10	7		21.10	27	
	0800	Mahonia, 18" - 24", container		80	.300		24.50	8.40		32.90	41	
	0900	Mount laurel, 18" - 24", B & B		80	.300		50	8.40		58.40	69	

SITE CONSTRUCTION 2

	02930 | Exterior Plants		CREW	DAILY OUTPUT	LABOR-HOURS	UNIT	2006 BARE COSTS MAT.	LABOR	EQUIP.	TOTAL	TOTAL INCL O&P	
320	1000	Paxistema, 9 - 12" high	B-1	130	.185	Ea.	16.20	5.20		21.40	26.50	320
	1100	Rhododendron, 18"-24", container		48	.500		28	14.05		42.05	54	
	1200	Rosemary, 1 gal container		600	.040		59	1.12		60.12	67	
	2000	Deciduous, amelanchier, 2'-3', B & B		57	.421		80.50	11.80		92.30	108	
	2100	Azalea, 15"-18", B & B		96	.250		21.50	7		28.50	35	
	2200	Barberry, 2' - 3', B & B		57	.421		21	11.80		32.80	42.50	
	2300	Bayberry, 2'-3', B & B		57	.421		24.50	11.80		36.30	46.50	
	2400	Boston ivy, 2 year, container		600	.040		15.65	1.12		16.77	19.10	
	2500	Corylus, 3' - 4', B & B	B-17	75	.427		34	12.65	7.35	54	66	
	2600	Cotoneaster, 15"-18", B & B	B-1	80	.300		14.70	8.40		23.10	30	
	2700	Deutzia, 12" - 15", B & B	"	96	.250		9.15	7		16.15	21.50	
	2800	Dogwood, 3'-4', B & B	B-17	40	.800		23	23.50	13.80	60.30	79	
	2900	Euonymus, alatus compacta, 15" to 18", container	B-1	80	.300		19.30	8.40		27.70	35	
	3000	Flowering almond, 2' - 3', container	"	36	.667		17	18.70		35.70	49.50	
	3100	Flowering currant, 3' - 4', container	B-17	75	.427		19.50	12.65	7.35	39.50	50	
	3200	Forsythia, 2'-3', container	B-1	60	.400		17	11.25		28.25	37.50	
	3300	Hibiscus, 3'-4', B & B	B-17	75	.427		13.25	12.65	7.35	33.25	43	
	3400	Honeysuckle, 3'-4', B & B	B-1	60	.400		18.95	11.25		30.20	39.50	
	3500	Hydrangea, 2'-3', B & B	"	57	.421		22	11.80		33.80	43.50	
	3600	Lilac, 3'-4', B & B	B-17	40	.800		23	23.50	13.80	60.30	78.50	
	3700	Mockorange, 3' - 4', B & B	B-1	36	.667		14.70	18.70		33.40	47	
	3800	Osier willow, 2' - 3', B & B		57	.421		18.30	11.80		30.10	39.50	
	3900	Privet, bare root, 18"-24"		80	.300		11.75	8.40		20.15	27	
	4000	Pyracantha, 2' - 3', container		80	.300		36	8.40		44.40	53.50	
	4100	Quince, 2'-3', B & B		57	.421		19.50	11.80		31.30	41	
	4200	Russian olive, 3'-4', B & B	B-17	75	.427		21	12.65	7.35	41	52	
	4300	Snowberry, 2' - 3', B & B	B-1	57	.421		11.60	11.80		23.40	32.50	
	4400	Spirea, 3'-4', B & B	"	70	.343		24	9.60		33.60	42.50	
	4500	Viburnum, 3'-4', B & B	B-17	40	.800		25	23.50	13.80	62.30	81	
	4600	Weigela, 3' - 4', B & B	B-1	70	.343		18.20	9.60		27.80	36	
410	0010	**TREES** Deciduous, in prep. beds, balled & burlapped (B&B)										410
	0100	Ash, 2" caliper	B-17	8	4	Ea.	113	118	69	300	395	
	0200	Beech, 5'-6'		50	.640		221	18.95	11.05	251	287	
	0300	Birch, 6'-8', 3 stems		20	1.600		122	47.50	27.50	197	242	
	0400	Cherry, 6' - 8', 1" caliper		24	1.333		81	39.50	23	143.50	179	
	0500	Crabapple, 6'-8'		20	1.600		159	47.50	27.50	234	283	
	0600	Dogwood, 4'-5'		40	.800		68	23.50	13.80	105.30	128	
	0700	Eastern redbud 4'-5'		40	.800		133	23.50	13.80	170.30	200	
	0800	Elm, 8'-10'		20	1.600		112	47.50	27.50	187	231	
	0900	Ginkgo, 6'-7'		24	1.333		164	39.50	23	226.50	271	
	1000	Hawthorn, 8'-10', 1" caliper		20	1.600		125	47.50	27.50	200	246	
	1100	Honeylocust, 10'-12', 1-1/2" caliper		10	3.200		149	94.50	55.50	299	380	
	1200	Laburnum, 6' - 8', 1" caliper		24	1.333		56.50	39.50	23	119	152	
	1300	Larch, 8'		32	1		96.50	29.50	17.30	143.30	173	
	1400	Linden, 8'-10', 1" caliper		20	1.600		104	47.50	27.50	179	222	
	1500	Magnolia, 4'-5'		20	1.600		71.50	47.50	27.50	146.50	186	
	1600	Maple, red, 8'-10', 1-1/2" caliper		10	3.200		159	94.50	55.50	309	390	
	1700	Mountain ash, 8'-10', 1" caliper		16	2		164	59	34.50	257.50	315	
	1800	Oak, 2-1/2"-3" caliper		6	5.333		248	158	92	498	630	
	1900	Pagoda, 6' - 8'		20	1.600		82.50	47.50	27.50	157.50	198	
	2000	Pear, 6' - 8', 1" caliper		20	1.600		133	47.50	27.50	208	254	
	2100	Planetree, 9'-11', 1-1/4" caliper		10	3.200		102	94.50	55.50	252	330	
	2200	Plum, 6'-8', 1" caliper		20	1.600		90.50	47.50	27.50	165.50	207	
	2300	Poplar, 9'-11', 1-1/4" caliper		10	3.200		47	94.50	55.50	197	267	
	2400	Shadbush, 4' - 5'		60	.533		66.50	15.80	9.20	91.50	109	
	2500	Sumac, 2'-3'		75	.427		22.50	12.65	7.35	42.50	53.50	

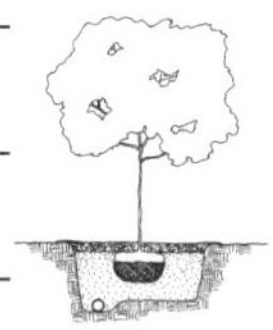

02900 | Planting

	02930	Exterior Plants	CREW	DAILY OUTPUT	LABOR-HOURS	UNIT	2006 BARE COSTS MAT.	LABOR	EQUIP.	TOTAL	TOTAL INCL O&P	
410	2600	Tupelo, 5′ - 6′	B-17	40	.800	Ea.	48	23.50	13.80	85.30	107	410
	2700	Tulip, 5′-6′		40	.800		49.50	23.50	13.80	86.80	108	
	2800	Willow, 6′-8′, 1″ caliper		20	1.600		60.50	47.50	27.50	135.50	175	
	9000	Minimum labor/equipment charge	1 Clab	4	2	Job		55		55	91	
	02935	**Plant Maintenance**										
100	0010	**FERTILIZE**										100
	0100	Dry granular, 4#/M.S.F., hand spread	1 Clab	24	.333	M.S.F.	2.06	9.15		11.21	17.40	
	0110	Push rotary		140	.057	"	2.06	1.57		3.63	4.86	
	0112	Push rotary, per 1076 feet squared		130	.062	Ea.	2.06	1.69		3.75	5.05	
	0120	Tractor towed spreader, 8′	B-66	500	.016	M.S.F.	2.06	.56	.36	2.98	3.55	
	0130	12′ spread		800	.010		2.06	.35	.22	2.63	3.08	
	0140	Truck whirlwind spreader		1200	.007		2.06	.23	.15	2.44	2.80	
	0180	Water soluable, hydro spread, 1.5 # /MSF	B-64	600	.027		2.16	.73	.40	3.29	4.02	
	0190	Add for weed control					.32			.32	.35	
300	0010	**MOWING**										300
	1650	Mowing brush, tractor with rotary mower										
	1660	Light density	B-84	22	.364	M.S.F.		13.35	10.50	23.85	32.50	
	1670	Medium density		13	.615			22.50	17.75	40.25	55	
	1680	Heavy density		9	.889			32.50	25.50	58	79.50	
	4000	Lawn mowing, improved areas, 16″ hand push	1 Clab	48	.167			4.57		4.57	7.55	
	4050	Power mower, 18″ - 22″		65	.123			3.37		3.37	5.60	
	4100	22″ - 30″		110	.073			1.99		1.99	3.30	
	4150	30″ - 32″		140	.057			1.57		1.57	2.59	
	4160	Riding mower, 36″ - 44″	B-66	300	.027			.94	.60	1.54	2.14	
	4170	48″ - 58″	"	480	.017			.59	.37	.96	1.34	
	4175	Mowing with tractor & attachments										
	4180	3 gang reel, 7′	B-66	930	.009	M.S.F.		.30	.19	.49	.69	
	4190	5 gang reel, 12′		1200	.007			.23	.15	.38	.53	
	4200	Cutter or sickle-bar, 5′, rough terrain		210	.038			1.34	.85	2.19	3.06	
	4210	Cutter or sickle-bar, 5′, smooth terrain		340	.024			.83	.53	1.36	1.89	
	4220	Drainage channel, 5′ sickle bar		5	1.600	Mile		56.50	36	92.50	129	
	4250	Lawnmower, rotary type, sharpen (all sizes)	1 Clab	10	.800	Ea.		22		22	36.50	
	4260	Repair or replace part		7	1.143	"		31.50		31.50	52	
	5000	Edge trimming with weed whacker		5760	.001	L.F.		.04		.04	.06	
410	0010	**TREE PRUNING**										410
	0020	1-1/2″ caliper	1 Clab	84	.095	Ea.		2.61		2.61	4.32	
	0030	2″ caliper		70	.114			3.13		3.13	5.20	
	0040	2-1/2″ caliper		50	.160			4.38		4.38	7.25	
	0050	3″ caliper		30	.267			7.30		7.30	12.10	
	0060	4″ caliper, by hand	2 Clab	21	.762			21		21	34.50	
	0070	Aerial lift equipment	B-85	38	1.053			31	18.30	49.30	70.50	
	0100	6″ caliper, by hand	2 Clab	12	1.333			36.50		36.50	60.50	
	0110	Aerial lift equipment	B-85	20	2			59	35	94	135	
	0200	9″ caliper, by hand	2 Clab	7.50	2.133			58.50		58.50	97	
	0210	Aerial lift equipment	B-85	12.50	3.200			94	55.50	149.50	216	
	0300	12″ caliper, by hand	2 Clab	6.50	2.462			67.50		67.50	112	
	0310	Aerial lift equipment	B-85	10.80	3.704			109	64.50	173.50	249	
	0400	18″ caliper by hand	2 Clab	5.60	2.857			78.50		78.50	130	
	0410	Aerial lift equipment	B-85	9.30	4.301			127	75	202	290	
	0500	24″ caliper, by hand	2 Clab	4.60	3.478			95.50		95.50	158	
	0510	Aerial lift equipment	B-85	7.70	5.195			153	90.50	243.50	350	
	0600	30″ caliper, by hand	2 Clab	3.70	4.324			118		118	196	
	0610	Aerial lift equipment	B-85	6.20	6.452			190	112	302	435	
	0700	36″ caliper, by hand	2 Clab	2.70	5.926			162		162	269	

02900 | Planting

	02935 | Plant Maintenance		CREW	DAILY OUTPUT	LABOR-HOURS	UNIT	2006 BARE COSTS MAT.	LABOR	EQUIP.	TOTAL	TOTAL INCL O&P	
410	0710	Aerial lift equipment	B-85	4.50	8.889	Ea.		262	155	417	595	410
	0800	48″ caliper, by hand	2 Clab	1.70	9.412			258		258	425	
	0810	Aerial lift equipment	B-85	2.80	14.286			420	249	669	960	
420	0010	SHRUB PRUNING										420
	6700	Prune, shrub bed	1 Clab	7	1.143	M.S.F.		31.50		31.50	52	
	6710	Shrub under 3′ height		190	.042	Ea.		1.15		1.15	1.91	
	6720	4′ height		90	.089			2.44		2.44	4.04	
	6730	Over 6′		50	.160			4.38		4.38	7.25	
	7350	Prune trees from ground		20	.400			10.95		10.95	18.15	
	7360	High work		8	1			27.50		27.50	45.50	
500	0010	WATERING										500
	4900	Water lawn or planting bed with hose, 1″ of water	1 Clab	16	.500	M.S.F.		13.70		13.70	22.50	
	4910	50′ soaker hoses, in place		82	.098			2.67		2.67	4.43	
	4920	60′ soaker hoses, in place		89	.090			2.46		2.46	4.08	
	7500	Water trees or shrubs, under 1″ caliper		32	.250	Ea.		6.85		6.85	11.35	
	7550	1″ - 3″ caliper		17	.471			12.90		12.90	21.50	
	7600	3″ - 4″ caliper		12	.667			18.25		18.25	30.50	
	7650	Over 4″ caliper		10	.800			22		22	36.50	
	9000	For sprinkler irrigation systems, see Div. 02810										
610	0010	WEEDING										610
	0100	Weed planting bed	1 Clab	800	.010	S.Y.		.27		.27	.45	

	02945 | Planting Accessories		CREW	DAILY OUTPUT	LABOR-HOURS	UNIT	MAT.	LABOR	EQUIP.	TOTAL	TOTAL INCL O&P	
120	0010	EDGING										120
	0050	Aluminum alloy, including stakes, 1/8″ x 4″, mill finish	B-1	390	.062	L.F.	2.43	1.73		4.16	5.55	
	0051	Black paint		390	.062		2.82	1.73		4.55	5.95	
	0052	Black anodized		390	.062		3.26	1.73		4.99	6.45	
	0060	3/16″ x 4″, mill finish		380	.063		3.50	1.77		5.27	6.80	
	0061	Black paint		380	.063		4.03	1.77		5.80	7.40	
	0062	Black anodized		380	.063		4.71	1.77		6.48	8.15	
	0070	1/8″ x 5-1/2″ mill finish		370	.065		3.52	1.82		5.34	6.90	
	0071	Black paint		370	.065		4.18	1.82		6	7.60	
	0072	Black anodized		370	.065		4.79	1.82		6.61	8.25	
	0080	3/16″ x 5-1/2″ mill finish		360	.067		4.71	1.87		6.58	8.30	
	0081	Black paint		360	.067		5.30	1.87		7.17	8.95	
	0082	Black anodized		360	.067		6.15	1.87		8.02	9.85	
	0100	Brick, set horizontally, 1-1/2 bricks per L.F.	D-1	370	.043		1.03	1.39		2.42	3.39	
	0150	Set vertically, 3 bricks per L.F.	″	135	.119		2.76	3.81		6.57	9.25	
	0200	Corrugated aluminum, roll, 4″ wide	1 Carp	650	.012		.37	.44		.81	1.14	
	0250	6″ wide	″	550	.015		.46	.52		.98	1.37	
	0300	Concrete, cast in place, see 03310-240										
	0350	Granite, 5″ x 16″, straight	B-29	300	.187	L.F.	10.20	5.60	2.80	18.60	23.50	
	0400	Polyethylene grass barrier, 5″ x 1/8″	D-1	400	.040		.61	1.29		1.90	2.76	
	0500	Precast scallops, green, 2″ x 8″ x 16″		400	.040		.88	1.29		2.17	3.06	
	0550	2″ x 8″ x 16″ other than green		400	.040		.70	1.29		1.99	2.86	
	0600	Railroad ties, 6″ x 8″	2 Carp	170	.094		2.69	3.35		6.04	8.50	
	0650	7″ x 9″	″	136	.118		2.99	4.18		7.17	10.25	
	0700	Redwood										
	0750	2″ x 4″	2 Carp	330	.048	L.F.	2.20	1.72		3.92	5.30	
	0800	Steel edge strips, incl. stakes, 1/4″ x 5″	B-1	390	.062		3.62	1.73		5.35	6.85	
	0850	3/16″ x 4″	″	390	.062		2.86	1.73		4.59	6	
	0900	Hardwood, pressure treated, 4″ x 6″	2 Carp	250	.064		2.17	2.28		4.45	6.15	
	0940	6″ x 6″		200	.080		3.10	2.84		5.94	8.10	

02900 | Planting

02945 | Planting Accessories

		02945 Planting Accessories	CREW	DAILY OUTPUT	LABOR-HOURS	UNIT	2006 BARE COSTS MAT.	LABOR	EQUIP.	TOTAL	TOTAL INCL O&P	
120	0980	6″ x 8″	2 Carp	170	.094	L.F.	3.54	3.35		6.89	9.45	120
	1000	Pine, pressure treated, 1″ x 4″		500	.032		.42	1.14		1.56	2.34	
	1040	2″ x 4″		330	.048		.71	1.72		2.43	3.64	
	1080	4″ x 6″		250	.064		2.35	2.28		4.63	6.35	
	1100	6″ x 6″		200	.080		3.53	2.84		6.37	8.60	
	1140	6″ x 8″		170	.094		4.62	3.35		7.97	10.65	
	9000	Minimum labor/equipment charge	1 Carp	4	2	Job		71		71	118	
300	0010	PLANTERS										300
	0012	Concrete, sandblasted, precast, 48″ diameter, 24″ high	2 Clab	15	1.067	Ea.	585	29		614	690	
	0100	Fluted, precast, 7′ diameter, 36″ high	″	10	1.600	″	985	44		1,029	1,150	
	0280	Jersey barriers, concrete, 42″H, for bldg perimeter security, excl freight	B-29	350	.160	L.F.	39.50	4.79	2.40	46.69	54	
	0282	Jersey barriers, plastic, water-filled, 42″H x 8′L x 24″ W base, incl conn	2 Clab	40	.400	Ea.	480	10.95		490.95	550	
	0300	Fiberglass, circular, 36″ diameter, 24″ high		15	1.067		415	29		444	510	
	0320	36″ diameter, 27″ high		12	1.333		330	36.50		366.50	425	
	0330	33″ high		15	1.067		435	29		464	525	
	0335	24″ diameter, 36″ high		15	1.067		385	29		414	470	
	0340	60″ diameter, 39″ high		8	2		990	55		1,045	1,200	
	0400	60″ diameter, 24″ high		10	1.600		705	44		749	850	
	0600	Square, 24″ side, 36″ high		15	1.067		405	29		434	495	
	0610	24″ side, 27″ high		12	1.333		265	36.50		301.50	350	
	0620	24″ side, 16″ high		20	.800		293	22		315	355	
	0700	48″ side, 36″ high		15	1.067		905	29		934	1,050	
	0900	Planter/bench, 72″ square, 36″ high		5	3.200		1,125	87.50		1,212.50	1,375	
	1000	96″ square, 27″ high		5	3.200		1,775	87.50		1,862.50	2,100	
	1200	Wood, square, 48″ side, 24″ high		15	1.067		905	29		934	1,050	
	1300	Circular, 48″ diameter, 30″ high		10	1.600		745	44		789	895	
	1500	72″ diameter, 30″ high		10	1.600		1,325	44		1,369	1,525	
	1600	Planter/bench, 72″		5	3.200		2,925	87.50		3,012.50	3,350	
	9000	Minimum labor/equipment charge	1 Clab	2	4	Job		110		110	182	
510	0010	TREE GUYING										510
	0015	Tree guying Including stakes, guy wire and wrap										
	0100	Less than 3″ caliper, 2 stakes	2 Clab	35	.457	Ea.	15.50	12.55		28.05	38	
	0200	3″ to 4″ caliper, 3 stakes	″	21	.762	″	18.30	21		39.30	54.50	
	1000	Including arrowhead anchor, cable, turnbuckles and wrap										
	1100	Less than 3″ caliper, 3″ anchors	2 Clab	20	.800	Ea.	48.50	22		70.50	90	
	1200	3″ to 6″ caliper, 4″ anchors		15	1.067		69.50	29		98.50	125	
	1300	6″ caliper, 6″ anchors		12	1.333		86	36.50		122.50	156	
	1400	8″ caliper, 8″ anchors		9	1.778		98.50	48.50		147	190	

02950 | Site Restoration & Rehabilitation

02955 | Restoration of Underground Piping

		02955 Restoration of Underground Piping	CREW	DAILY OUTPUT	LABOR-HOURS	UNIT	2006 BARE COSTS MAT.	LABOR	EQUIP.	TOTAL	TOTAL INCL O&P	
210	0010	PIPE INTERNAL CLEANING & INSPECTION										210
	0100	Cleaning, pressure pipe systems										
	0120	Pig method, lengths 1000′ to 10,000′										
	0140	4″ diameter thru 24″ diameter, minimum				L.F.				2.14	2.46	
	0160	Maximum				″				5.35	6.40	
	6000	Sewage/sanitary systems										

	02955	Restoration of Underground Piping	CREW	DAILY OUTPUT	LABOR-HOURS	UNIT	2006 BARE COSTS MAT.	LABOR	EQUIP.	TOTAL	TOTAL INCL O&P
210	6100	Power rodder with header & cutters									
	6110	Mobilization charge, minimum				Total				320	375
	6120	Mobilization charge, maximum				"				750	850
	6140	4″ diameter				L.F.				1.07	1.23
	6150	6″ diameter								1.34	1.55
	6160	8″ diameter								1.60	1.87
	6170	10″ diameter								1.87	2.14
	6180	12″ diameter								2.03	2.35
	6190	14″ diameter								2.14	2.46
	6200	16″ diameter								2.41	2.78
	6210	18″ diameter								2.68	3.05
	6220	20″ diameter								2.78	3.21
	6230	24″ diameter								3.10	3.58
	6240	30″ diameter								3.75	4.28
	6250	36″ diameter								4.28	4.92
	6260	48″ diameter								4.60	5.30
	6270	60″ diameter								6	6.90
	6280	72″ diameter								6.65	7.65
	9000	Inspection, television camera with film									
	9060	500 linear feet				Total				1,075	1,230
220	0010	**PIPEBURSTING**, 300′ runs, replace with HDPE pipe									
	0020	Not including excavation, backfill, shoring, or dewatering									
	0100	6″ to 15″ diameter, minimum				L.F.					80
	0200	Maximum									160
	0300	18″ to 36″ diameter, minimum									187
	0400	Maximum									320
	0500	Mobilize and demobilize, minimum				Job					2,675
	0600	Maximum				"					26,800
230	0010	**LINING PIPE** with cement, incl. bypass and cleaning									
	0020	Less than 10,000 L.F., urban, 6″ to 10″	C-17E	130	.615	L.F.	7.20	22.50	.62	30.32	45.50
	0050	10″ to 12″		125	.640		8.85	23.50	.65	33	49
	0070	12″ to 16″		115	.696		9.10	25.50	.70	35.30	53
	0100	16″ to 20″		95	.842		10.70	31	.85	42.55	63.50
	0200	24″ to 36″		90	.889		11.50	33	.90	45.40	67
	0300	48″ to 72″		80	1		18.35	37	1.01	56.36	81.50
	0500	Rural, 6″ to 10″		180	.444		7.20	16.40	.45	24.05	35.50
	0550	10″ to 12″		175	.457		8.85	16.85	.46	26.16	38
	0570	12″ to 16″		160	.500		9.25	18.45	.51	28.21	41
	0600	16″ to 20″		135	.593		9.80	22	.60	32.40	47.50
	0700	24″ to 36″		125	.640		11.70	23.50	.65	35.85	52
	0800	48″ to 72″		100	.800		18.35	29.50	.81	48.66	69.50
	1000	Greater than 10,000 L.F., urban, 6″ to 10″		160	.500		7.20	18.45	.51	26.16	38.50
	1050	10″ to 12″		155	.516		8.75	19.05	.52	28.32	41
	1070	12″ to 16″		140	.571		9.10	21	.58	30.68	45
	1100	16″ to 20″		120	.667		9.80	24.50	.67	34.97	51.50
	1200	24″ to 36″		115	.696		11.70	25.50	.70	37.90	55.50
	1300	48″ to 72″		95	.842		18.35	31	.85	50.20	72
	1500	Rural, 6″ to 10″		215	.372		7.20	13.75	.38	21.33	31
	1550	10″ to 12″		210	.381		8.85	14.05	.38	23.28	33
	1570	12″ to 16″		185	.432		9.10	15.95	.44	25.49	36.50
	1600	16″ to 20″		150	.533		9.80	19.70	.54	30.04	43.50
	1700	24″ to 36″		140	.571		11.70	21	.58	33.28	48
	1800	48″ to 72″		120	.667		18.50	24.50	.67	43.67	61
	2000	Cured in place pipe, non-pressure, flexible felt resin, 400′ runs									
	2100	6″ diameter				L.F.					32
	2200	8″ diameter									43

		02955 | Restoration of Underground Piping	CREW	DAILY OUTPUT	LABOR-HOURS	UNIT	2006 BARE COSTS MAT.	LABOR	EQUIP.	TOTAL	TOTAL INCL O&P	
230	2300	10″ diameter				L.F.					59	230
	2400	12″ diameter									75	
	2500	15″ diameter									102	
	2600	18″ diameter									123	
	2700	21″ diameter									144	
	2800	24″ diameter									161	
	2900	30″ diameter									203	
	3000	36″ diameter									268	
	3100	48″ diameter									360	
	3200	60″ diameter									450	
	3300	72″ diameter									535	
240	0010	**HDPE PIPE LINING**, excludes cleaning and video inspection										240
	0020	Pipe relined with one pipe size smaller than original (4″ for 6 ″)										
	0100	6″ diameter, original size	B-6B	600	.080	L.F.	1.99	2.25	1.30	5.54	7.35	
	0150	8″ diameter, original size		600	.080		2.86	2.25	1.30	6.41	8.30	
	0200	10″ diameter, original size		600	.080		4.84	2.25	1.30	8.39	10.45	
	0250	12″ diameter, original size		400	.120		7.50	3.37	1.95	12.82	16	
	0300	14″ diameter, original size		400	.120		10.60	3.37	1.95	15.92	19.40	
310	0010	**CORROSION RESISTANCE**										310
	0012	Wrap & coat, add to pipe, 4″ dia.				L.F.	1.58			1.58	1.74	
	0020	5″ diameter					1.65			1.65	1.82	
	0040	6″ diameter					1.65			1.65	1.82	
	0060	8″ diameter					2.56			2.56	2.82	
	0080	10″ diameter					3.82			3.82	4.20	
	0100	12″ diameter					3.82			3.82	4.20	
	0120	14″ diameter					5.60			5.60	6.15	
	0140	16″ diameter					5.60			5.60	6.15	
	0160	18″ diameter					5.65			5.65	6.25	
	0180	20″ diameter					5.90			5.90	6.50	
	0200	24″ diameter					7.25			7.25	7.95	
	0220	Small diameter pipe, 1″ diameter, add					1.19			1.19	1.31	
	0240	2″ diameter					1.31			1.31	1.44	
	0260	2-1/2″ diameter					1.44			1.44	1.58	
	0280	3″ diameter					1.44			1.44	1.58	
	0300	Fittings, field covered, add				S.F.	7.55			7.55	8.30	
	0500	Coating, bituminous, per diameter inch, 1 coat, add				L.F.	.29			.29	.32	
	0540	3 coat					.45			.45	.50	
	0560	Coal tar epoxy, per diameter inch, 1 coat, add					.17			.17	.19	
	0600	3 coat					.37			.37	.41	
	1000	Polyethylene H.D. extruded, .025″ thk., 1/2″ diameter add					.26			.26	.29	
	1020	3/4″ diameter					.27			.27	.30	
	1040	1″ diameter					.30			.30	.33	
	1060	1-1/4″ diameter					.37			.37	.41	
	1080	1-1/2″ diameter					.41			.41	.45	
	1100	.030″ thk., 2″ diameter					.44			.44	.48	
	1120	2-1/2″ diameter					.52			.52	.57	
	1140	.035″ thk., 3″ diameter					.65			.65	.72	
	1160	3-1/2″ diameter					.73			.73	.80	
	1180	4″ diameter					.82			.82	.90	
	1200	.040″ thk, 5″ diameter					1.13			1.13	1.24	
	1220	6″ diameter					1.18			1.18	1.30	
	1240	8″ diameter					1.54			1.54	1.69	
	1260	10″ diameter					1.88			1.88	2.07	
	1280	12″ diameter					2.25			2.25	2.48	

For expanded coverage of these items see *Means Site Work & Landscape Cost Data 2006*

	02955 | Restoration of Underground Piping		CREW	DAILY OUTPUT	LABOR-HOURS	UNIT	2006 BARE COSTS MAT.	LABOR	EQUIP.	TOTAL	TOTAL INCL O&P	
310	1300	.060" thk., 14" diameter				L.F.	3.10			3.10	3.41	310
	1320	16" diameter					3.63			3.63	3.99	
	1340	18" diameter					4.29			4.29	4.72	
	1360	20" diameter					4.82			4.82	5.30	
	1380	Fittings, field wrapped, add				S.F.	4.98			4.98	5.50	
400	0010	**PIPE REPAIR**										400
	0020	Not including excavation or backfill										
	0100	Clamp, stainless steel, lightweight, for steel pipe										
	0110	3" long, 1/2" diameter pipe	1 Plum	34	.235	Ea.	11.30	10.05		21.35	28	
	0120	3/4" diameter pipe		32	.250		11.65	10.70		22.35	29.50	
	0130	1" diameter pipe		30	.267		12.50	11.40		23.90	31.50	
	0140	1-1/4" diameter pipe		28	.286		13.30	12.20		25.50	33.50	
	0150	1-1/2" diameter pipe		26	.308		13.70	13.15		26.85	35.50	
	0160	2" diameter pipe		24	.333		15	14.25		29.25	38.50	
	0170	2-1/2" diameter pipe		23	.348		16.30	14.85		31.15	41	
	0180	3" diameter pipe		22	.364		18	15.55		33.55	44	
	0190	3-1/2" diameter pipe		21	.381		19.30	16.25		35.55	46.50	
	0200	4" diameter pipe	B-20	56	.429		20	13.35		33.35	44	
	0210	5" diameter pipe		53	.453		23	14.10		37.10	48.50	
	0220	6" diameter pipe		48	.500		25.50	15.55		41.05	54	
	0230	8" diameter pipe		30	.800		30	25		55	74	
	0240	10" diameter pipe		28	.857		77	26.50		103.50	129	
	0250	12" diameter pipe		24	1		82	31		113	142	
	0260	14" diameter pipe		22	1.091		85.50	34		119.50	150	
	0270	16" diameter pipe		20	1.200		92.50	37.50		130	164	
	0280	18" diameter pipe		18	1.333		97.50	41.50		139	176	
	0290	20" diameter pipe		16	1.500		108	46.50		154.50	196	
	0300	24" diameter pipe		14	1.714		120	53.50		173.50	220	
	0360	For 6" long, add					100%	40%				
	0370	For 9" long, add					200%	100%				
	0380	For 12" long, add					300%	150%				
	0390	For 18" long, add					500%	200%				
	0400	Pipe Freezing for live repairs of systems 3/8 inch to 6 inch										
	0410	Note: Pipe Freezing can also used to install a valve into a live system										
	0420	Pipe Freezing each side 3/8 inch	2 Skwk	8	2	Ea.	520	73		593	690	
	0425	Pipe Freezing each side 3/8 inch , second location same kit		8	2		19.75	73		92.75	141	
	0430	Pipe Freezing each side 3/4 inch		8	2		445	73		518	610	
	0435	Pipe Freezing each side 3/4 inch , second location same kit		8	2		19.75	73		92.75	141	
	0440	Pipe Freezing each side 1 1/2 inch		6	2.667		445	97.50		542.50	650	
	0445	Pipe Freezing each side 1 1/2 inch , second location same kit		6	2.667		19.75	97.50		117.25	181	
	0450	Pipe Freezing each side 2 inch		6	2.667		660	97.50		757.50	885	
	0455	Pipe Freezing each side 2 inch , second location same kit		6	2.667		19.75	97.50		117.25	181	
	0460	Pipe Freezing each side 2 1/2 inch-3 inch		6	2.667		925	97.50		1,022.50	1,175	
	0465	Pipe Frz each side 2 1/2-3 inch , second loc same kit		6	2.667		39.50	97.50		137	203	
	0470	Pipe Freezing each side 4 inch		4	4		1,500	146		1,646	1,900	
	0475	Pipe Freezing each side 4 inch , second location same kit		4	4		62.50	146		208.50	310	
	0480	Pipe Freezing each side 5-6 inch		4	4		3,625	146		3,771	4,225	
	0485	Pipe Freezing each side 5-6 inch , second location same kit		4	4		187	146		333	445	
	0490	Pipe Frz extra 20 lb CO_2 cylinders (3/8 to 2 " - 1 ea, 3 " -2 ea)					197			197	216	
	0500	Pipe Freezing extra 50 lb CO_2 cylinders (4 " - 2 ea, 5-6 " -6 ea)					440			440	485	
	1000	Clamp, stainless steel, with threaded service tap										
	1040	Full seal for iron, steel, PVC pipe										
	1100	6" long, 2" diameter pipe	1 Plum	17	.471	Ea.	146	20		166	192	
	1110	2-1/2" diameter pipe		16	.500		154	21.50		175.50	203	
	1120	3" diameter pipe		15.60	.513		164	22		186	215	

	02955	Restoration of Underground Piping	CREW	DAILY OUTPUT	LABOR-HOURS	UNIT	2006 BARE COSTS MAT.	LABOR	EQUIP.	TOTAL	TOTAL INCL O&P	
400	1130	3-1/2″ diameter pipe	1 Plum	15	.533	Ea.	171	23		194	224	400
	1140	4″ diameter pipe	B-20	40	.600		188	18.65		206.65	238	
	1150	6″ diameter pipe		34	.706		218	22		240	275	
	1160	8″ diameter pipe		21	1.143		271	35.50		306.50	355	
	1170	10″ diameter pipe		20	1.200		300	37.50		337.50	390	
	1180	12″ diameter pipe		17	1.412		345	44		389	450	
	1205	For 9″ long, add					20%	45%				
	1210	For 12″ long, add					40%	80%				
	1220	For 18″ long, add					70%	110%				
	1600	Clamp, stainless steel, single section										
	1640	Full seal for iron, steel, PVC pipe										
	1700	6″ long, 2″ diameter pipe	1 Plum	17	.471	Ea.	75.50	20		95.50	115	
	1710	2-1/2″ diameter pipe		16	.500		82	21.50		103.50	124	
	1720	3″ diameter pipe		15.60	.513		85.50	22		107.50	128	
	1730	3-1/2″ diameter pipe		15	.533		94	23		117	140	
	1740	4″ diameter pipe	B-20	40	.600		103	18.65		121.65	144	
	1750	6″ diameter pipe		34	.706		128	22		150	177	
	1760	8″ diameter pipe		21	1.143		171	35.50		206.50	247	
	1770	10″ diameter pipe		20	1.200		197	37.50		234.50	279	
	1780	12″ diameter pipe		17	1.412		214	44		258	310	
	1800	For 9″ long, add					40%	45%				
	1810	For 12″ long, add					60%	80%				
	1820	For 18″ long, add					120%	110%				
	2000	Clamp, stainless steel, two section										
	2040	Full seal, for iron, steel, PVC pipe										
	2100	6″ long, 4″ diameter pipe	B-20	24	1	Ea.	146	31		177	212	
	2110	6″ diameter pipe		20	1.200		171	37.50		208.50	250	
	2120	8″ diameter pipe		13	1.846		188	57.50		245.50	300	
	2130	10″ diameter pipe		12	2		274	62		336	405	
	2140	12″ diameter pipe		10	2.400		300	74.50		374.50	455	
	2200	9″ long, 4″ diameter pipe		16	1.500		188	46.50		234.50	284	
	2210	6″ diameter pipe		13	1.846		214	57.50		271.50	330	
	2220	8″ diameter pipe		9	2.667		223	83		306	380	
	2230	10″ diameter pipe		8	3		310	93.50		403.50	495	
	2240	12″ diameter pipe		7	3.429		345	107		452	550	
	2250	14″ diameter pipe		6.40	3.750		515	117		632	755	
	2260	16″ diameter pipe		6	4		540	124		664	800	
	2270	18″ diameter pipe		5	4.800		555	149		704	855	
	2280	20″ diameter pipe		4.60	5.217		600	162		762	930	
	2290	24″ diameter pipe		4	6		1,025	187		1,212	1,425	
	2320	For 12″ long, add to 9″					15%	25%				
	2330	For 18″ long, add to 9″					70%	55%				
	8000	For internal cleaning and inspection, see Div. 02955-210										
	8100	For pipe testing, see Div. 15955-700										

	02960	Flex. Pavement Surfacing Recovery	CREW	DAILY OUTPUT	LABOR-HOURS	UNIT	MAT.	LABOR	EQUIP.	TOTAL	TOTAL INCL O&P	
100	0010	PAVEMENT MILLING AND PAVEMENT COLD PLANING										100
	5200	Cold planing & cleaning, 1″ to 3″ asphalt pavmt., over 25,000 S.Y.	B-71	6000	.009	S.Y.		.30	.90	1.20	1.47	
	5280	5,000 S.Y. to 10,000 S.Y.	″	4000	.014	″		.44	1.35	1.79	2.20	
	5300	Asphalt pavement removal from conc. base, no haul										
	5320	Rip, load & sweep 1″ to 3″	B-70	8000	.007	S.Y.		.22	.16	.38	.53	
	5330	3″ to 6″ deep	″	5000	.011			.35	.25	.60	.85	
	5340	Profile grooving, asphalt pavement load & sweep, 1″ deep	B-71	12500	.004			.14	.43	.57	.70	
	5350	3″ deep		9000	.006			.20	.60	.80	.98	
	5360	6″ deep		5000	.011			.35	1.08	1.43	1.76	

02950 | Site Restoration & Rehabilitation

		02965 | Flex./Bit. Pavement Recycling	CREW	DAILY OUTPUT	LABOR-HOURS	UNIT	2006 BARE COSTS MAT.	LABOR	EQUIP.	TOTAL	TOTAL INCL O&P	
200	0010	**COLD IN-PLACE RECYCLED BITUMINOUS PAVEMENT COURSES**										200
	5000	Reclamation, pulverizing and blending with existing base										
	5040	Aggregate base, 4" thick pavement, over 15,000 S.Y.	B-73	2400	.027	S.Y.		.89	1.69	2.58	3.29	
	5080	5,000 S.Y. to 15,000 S.Y.		2200	.029			.97	1.84	2.81	3.59	
	5120	8" thick pavement, over 15,000 S.Y.		2200	.029			.97	1.84	2.81	3.59	
	5160	5,000 S.Y. to 15,000 S.Y.		2000	.032			1.07	2.03	3.10	3.95	
	5180	Add for mobilization and demobilization for crew B-73				Ea.				1,700	2,150	
400	0010	**HOT IN-PLACE RECYCLED BITUMINOUS PAVEMENT COURSES**										400
	5500	Recycle asphalt pavement at site										
	5520	Remove, rejuvenate and spread 4" deep	B-72	2500	.026	S.Y.	2.55	.83	3.56	6.94	8.05	
	5521	6" deep	"	2000	.032	"	3.75	1.03	4.45	9.23	10.70	
		02985 | Site Maintenance										
700	0010	**SITE MAINTENANCE** R019313-10										700
	0800	Flower bed maintenance										
	0810	Cultivate bed, no mulch	1 Clab	14	.571	M.S.F.		15.65		15.65	26	
	0830	Fall clean-up of flower bed, including pick-up mulch for re-use		1	8			219		219	365	
	0840	Fertilize flower bed, dry granular 3 lb./c.s.f.		85	.094		9.95	2.58		12.53	15.15	
	1000	Mulch - see Division 02910-500										
	1100	Plant bed preparation see Division 02910-720										
	1110	Plant from flats	1 Clab	800	.010	S.F.		.27		.27	.45	
	1120	Planting, ground cover see Division 02915-300										
	1130	Police, hand pickup	1 Clab	30	.267	M.S.F.		7.30		7.30	12.10	
	1140	Vacuum (outside)		48	.167			4.57		4.57	7.55	
	1200	Spring prepare		2	4			110		110	182	
	1300	Weed mulched bed		20	.400			10.95		10.95	18.15	
	1310	Unmulched bed		8	1			27.50		27.50	45.50	
	1550	General site work maintenance										
	1560	Clearing brush with brush saw & rake	1 Clab	565	.014	S.Y.		.39		.39	.64	
	1570	By hand	"	280	.029			.78		.78	1.30	
	1580	With dozer, ball and chain, light clearing	B-11A	3675	.004			.14	.25	.39	.50	
	1590	Medium clearing	"	3110	.005			.16	.30	.46	.60	
	1600	Ground cover planting see Division 02915-300										
	1610	Grounds, policing, incl. picking up trash & other debris	1 Clab	300	.027	M.S.F.		.73		.73	1.21	
	1750	Site improvements, (bleachers, playground equipment,										
	1760	fencing, planters, etc.) see Div. 02800										
	1800	Soil stabilization, see Division 02340										
	3000	Lawn maintenance										
	3010	Aerate lawn, 18" cultivating width, walk behind	A-1K	95	.084	M.S.F.		2.31	.24	2.55	4.08	
	3040	48" cultivating width	B-66	750	.011			.38	.24	.62	.85	
	3060	72" cultivating width	"	1100	.007			.26	.16	.42	.58	
	3100	Edge lawn, by hand at walks	1 Clab	16	.500	C.L.F.		13.70		13.70	22.50	
	3150	At planting beds		7	1.143			31.50		31.50	52	
	3200	Using gas powered edger at walks		88	.091			2.49		2.49	4.13	
	3250	At planting beds		24	.333			9.15		9.15	15.15	
	3260	Vacuum, 30" gas, outdoors with hose	A-1L	96	.083	M.L.F.		2.28	.24	2.52	4.04	
	3350	Edging material for lawns & planting beds see Div. 02945-120										
	3400	Weed lawn, by hand	1 Clab	3	2.667	M.S.F.		73		73	121	
	3800	Lawn bed preparation see Division 02910-710										
	4500	Rake leaves or lawn, by hand, (cavex)	1 Clab	7.50	1.067	M.S.F.		29		29	48.50	
	4510	Power rake	A-1L	45	.178	"		4.87	.50	5.37	8.60	
	4700	Seeding lawn, see Division 02920-320										
	4750	Sodding, see Division 02920-400										
	5900	Road & walk maintenance										
	5910	Asphaltic concrete paving, cold patch, 2" thick	B-37	350	.137	S.Y.	4.74	3.98	.33	9.05	12.10	

	02985 | Site Maintenance		CREW	DAILY OUTPUT	LABOR-HOURS	UNIT	2006 BARE COSTS MAT.	LABOR	EQUIP.	TOTAL	TOTAL INCL O&P
700	5913	3" thick R019313-10	B-37	260	.185	S.Y.	6.80	5.35	.44	12.59	16.75
	5915	De-icing roads and walks									
	5920	Calcium Chloride in truckload lots see Division 02340-700									
	6000	Ice melting comp., 90% Calc. Chlor., effec. to -30° F									
	6010	50-80 lb. poly bags, med. applic. 19 lbs./M.S.F., by hand	1 Clab	60	.133	M.S.F.	15.90	3.65		19.55	23.50
	6050	With hand operated rotary spreader		110	.073		15.90	1.99		17.89	21
	6100	Rock salt, med. applic. on road & walkway, by hand		60	.133		3.99	3.65		7.64	10.45
	6110	With hand operated rotary spreader		110	.073		3.99	1.99		5.98	7.70
	6130	Hosing, sidewalks & other paved areas		30	.267			7.30		7.30	12.10
	6150	Painting lines on pavement see Division 02760-300									
	6250	Seal coating bituminous surfaces see Division 02785-300									
	6260	Sidewalk, brick pavers, steam cleaning	A-1H	950	.008	S.F.	.05	.23	.06	.34	.51
	6400	Sweep walk by hand	1 Clab	15	.533	M.S.F.		14.60		14.60	24
	6410	Power vacuum	A-1L	100	.080			2.19	.23	2.42	3.88
	6420	Drives & parking areas with power vacuum	1 Clab	120	.067			1.83		1.83	3.03
	6600	Shrub maintenance									
	6630	Mulch-see Div. 02910-500									
	6640	Shrub bed fertilize dry granular 3 lbs./M.S.F.	1 Clab	85	.094	M.S.F.	.95	2.58		3.53	5.30
	6800	Weed, by handhoe		8	1			27.50		27.50	45.50
	6810	Spray out		32	.250			6.85		6.85	11.35
	6820	Spray after mulch		48	.167			4.57		4.57	7.55
	6840	Shrub purchase and installation see Div. 02915, 02930									
	7100	Tree maintenance									
	7140	Clear and grub trees, see Div. 02230-100									
	7160	Cutting and piling trees, see Div. 02230-300									
	7200	Fertilize, tablets, slow release, 30 gram/tree	1 Clab	100	.080	Ea.	.34	2.19		2.53	4
	7280	Guying, including stakes, guy wire & wrap, see Div. 02945-510									
	7300	Planting, trees, Deciduous, in prep. beds, see Div. 02915-400									
	7400	Removal, trees see Div. 02905-925									
	7420	Pest control, spray	1 Clab	24	.333	Ea.	15.45	9.15		24.60	32
	7430	Systemic	"	48	.167	"	14.35	4.57		18.92	23.50
750	0010	**SNOW REMOVAL**									
	0020	Plowing, 12 ton truck, 2"-4" deep	A-3	250	.032	M.S.F.		.91	1.31	2.22	2.92
	0040	4"-10" deep		200	.040			1.13	1.63	2.76	3.65
	0060	10"-15" deep		150	.053			1.51	2.18	3.69	4.86
	0080	Pickup truck, 2"-4" deep	A-3A	175	.046			1.26	.49	1.75	2.59
	0100	4"-10" deep		130	.062			1.70	.66	2.36	3.48
	0120	10"-15" deep		75	.107			2.94	1.14	4.08	6.05
	0140	Load and haul snow, 1 mile round trip	A-3B	230	.070	C.Y.		2.26	3.40	5.66	7.35
	0160	2 mile round trip		175	.091			2.97	4.47	7.44	9.70
	0180	3 mile round trip		150	.107			3.47	5.20	8.67	11.30
	0200	4 mile round trip		130	.123			4	6	10	13
	0220	5 mile round trip		120	.133			4.34	6.50	10.84	14.10
	0240	Clearing with wheeled skid steer loader, 1 C.Y.	A-3C	80	.100			3.52	2.53	6.05	8.35
	0260	Spread sand and salt mix	A-3	375	.021	M.S.F.	3.66	.60	.87	5.13	5.95
	0280	Sidewalks and drives, by hand	1 Clab	780	.010	C.F.		.28		.28	.47
	0300	Power, 24" blower	A-1L	1200	.007	"		.18	.02	.20	.32
	0320	2"-4" deep, single driveway (10' x 50')	A-1M	8	1	Ea.		27.50	12.95	40.45	60
	0340	Double driveway (20' x 50')		4.50	1.778			48.50	23	71.50	106
	0360	4"-10" deep, single driveway		6	1.333			36.50	17.25	53.75	79.50
	0380	Double driveway		3.25	2.462			67.50	32	99.50	147
	0400	10"-15" deep, single driveway		4	2			55	26	81	120
	0420	Double driveway		2.25	3.556			97.50	46	143.50	212
	0440	For heavy wet snow, add								20%	
	9000	Minimum labor and equipment charge	A-3A	2	4	Job		110	42.50	152.50	227

	02990 | Structure Moving		CREW	DAILY OUTPUT	LABOR-HOURS	UNIT	2006 BARE COSTS MAT.	LABOR	EQUIP.	TOTAL	TOTAL INCL O&P	
300	0010	**MOVING BUILDINGS** One day move, up to 24′ wide										300
	0020	Reset on new foundation, patch & hook-up, average move				Total					9,300	
	0040	Wood or steel frame bldg., based on ground floor area	B-4	185	.259	S.F.		7.25	1.77	9.02	13.90	
	0060	Masonry bldg., based on ground floor area	"	137	.350	↓		9.75	2.39	12.14	18.80	
	0200	For 24′ to 42′ wide, add									15%	
	0220	For each additional day on road, add	B-4	1	48	Day		1,350	330	1,680	2,550	
	0240	Construct new basement, move building, 1 day										
	0300	move, patch & hook-up, based on ground floor area	B-3	155	.310	S.F.	6.45	9.15	11.25	26.85	34.50	

For information about Means Estimating Seminars, see yellow pages 12 and 13 in back of book

Division 3 Concrete

Estimating Tips

General

- Carefully check all the plans and specifications. Concrete often appears on drawings other than structural drawings, including mechanical and electrical drawings for equipment pads. The cost of cutting and patching is often difficult to estimate. See Subdivisions 02220 and 03055 for demolition costs.
- Always obtain concrete prices from suppliers near the job site. A volume discount can often be negotiated depending upon competition in the area. Remember to add for waste, particularly for slabs and footings on grade.

03100 Concrete Forms & Accessories

- A primary cost for concrete construction is forming. Most jobs today are constructed with prefabricated forms. The selection of the forms best suited for the job and the total square feet of forms required for efficient concrete forming and placing are key elements in estimating concrete construction. Enough forms must be available for erection to make efficient use of the concrete placing equipment and crew.
- Concrete accessories for forming and placing depend upon the systems used. Study the plans and specifications to assure that all special accessory requirements have been included in the cost estimate such as anchor bolts, inserts, and hangers.

03200 Concrete Reinforcement

- Ascertain that the reinforcing steel supplier has included all accessories, cutting, bending, and an allowance for lapping, splicing, and waste. A good rule of thumb is 10% for lapping, splicing, and waste. Also, 10% waste should be allowed for welded wire fabric.

03300 Cast-in-Place Concrete

- When estimating structural concrete, pay particular attention to requirements for concrete additives, curing methods, and surface treatments. Special consideration for climate, hot or cold, must be included in your estimate. Be sure to include requirements for concrete placing equipment and concrete finishing.

03400 Precast Concrete
03500 Cementitious Decks & Toppings

- The cost of hauling precast concrete structural members is often an important factor. For this reason, it is important to get a quote from the nearest supplier. It may become economically feasible to set up precasting beds on the site if the hauling costs are prohibitive.

Reference Numbers

Reference numbers are shown in bold squares at the beginning of some major classifications. These numbers refer to related items in the Reference Section. The reference information may be an estimating procedure, an alternate pricing method, or technical information.

Note: Not all subdivisions listed here necessarily appear in this publication.

03055 | Selective Demolition

		03055 Selective Demolition	CREW	DAILY OUTPUT	LABOR-HOURS	UNIT	2006 BARE COSTS MAT.	LABOR	EQUIP.	TOTAL	TOTAL INCL O&P	
110	0010	**SELECTIVE CONCRETE DEMOLITION** R024119-10										110
	0012	Excludes saw cutting, torch cutting, loading or hauling										
	0050	Break up into small pieces, minimum reinforcing	B-9	24	1.667	C.Y.		46.50	6.15	52.65	84	
	0060	Average reinforcing		16	2.500			69.50	9.20	78.70	125	
	0070	Maximum reinforcing		8	5			139	18.40	157.40	251	
	0150	Remove whole pieces, up to 2 tons per piece	E-18	36	1.111	Ea.		44	26.50	70.50	109	
	0160	2 - 5 tons per piece		30	1.333			53	31.50	84.50	131	
	0170	5 - 10 tons per piece		24	1.667			66	39.50	105.50	164	
	0180	10 - 15 tons per piece		18	2.222			88	52.50	140.50	218	
	0250	Precast unit embedded in masonry, up to 1 CF	D-1	16	1			32		32	52	
	0260	1 - 2 CF		12	1.333			43		43	69.50	
	0270	2 - 5 CF		10	1.600			51.50		51.50	83.50	
	0280	5 - 10 CF		8	2			64.50		64.50	104	
	1910	Minimum labor/equipment charge	B-9	2	20	Job		555	73.50	628.50	1,000	

03060 | Basic Concrete Materials

		03060 Basic Concrete Materials	CREW	DAILY OUTPUT	LABOR-HOURS	UNIT	MAT.	LABOR	EQUIP.	TOTAL	TOTAL INCL O&P	
100	0010	**CONCRETE ADMIXTURES & SURFACE TREATMENTS**										100
	0040	Abrasives, aluminum oxide, over 20 tons				Lb.	1.18			1.18	1.30	
	0070	Under 1 ton					1.28			1.28	1.41	
	0100	Silicon carbide, black, over 20 tons					1.64			1.64	1.80	
	0120	Under 1 ton					1.77			1.77	1.95	
	0200	Air entraining agent, .7 to 1.5 oz. per bag, 55 gallon lots				Gal.	8.95			8.95	9.85	
	0220	5 gallon lots					8.25			8.25	9.05	
	0300	Bonding agent, acrylic latex, 250 S.F. per gallon					17.80			17.80	19.55	
	0320	Epoxy resin, 80 S.F. per gallon					39			39	43	
	0400	Calcium chloride, 50 lb. bags				Ton	415			415	455	
	0420	Less than truckload lots				Bag	29.50			29.50	32.50	
	0500	Carbon black, liquid, 2 to 8 lbs. per bag of cement				Lb.	4.27			4.27	4.70	
	0600	Colors, integral, 2 to 10 lb. per bag of cement, minimum					2.32			2.32	2.55	
	0610	Average					3.12			3.12	3.43	
	0620	Maximum					3.95			3.95	4.35	
	0700	Curing compound, solvent based, 400 S.F./gal, 55 gal. lots				Gal.	12.35			12.35	13.60	
	0720	5 gallon lots					14.20			14.20	15.60	
	0800	Water based, 250 S.F./gal, 55 gallon lots					11.35			11.35	12.50	
	0820	5 gallon lots					13			13	14.30	
	0900	Dustproofing compound, (200-600 S.F./gal.), 55 gallon lots					3.84			3.84	4.22	
	0920	5 gallon lots					5.70			5.70	6.30	
	1000	Epoxy dustproof coating, colors, (300-400 S.F. per coat),										
	1010	or transparent, (400-600 S.F. per coat)				Gal.	29.50			29.50	32	
	1100	Hardeners, metallic, 55 lb. bags, natural (grey)				Lb.	.70			.70	.77	
	1200	Colors					.97			.97	1.07	
	1300	Non-metallic, 55 lb. bags, natural grey					.30			.30	.33	
	1320	Colors					.55			.55	.61	
	1550	Release agent, for tilt slabs				Gal.	9.25			9.25	10.15	
	1570	For forms, average					5.65			5.65	6.25	
	1600	Sealer, hardener and dustproofer, epoxy-based, 200 SF/gal, min					29.50			29.50	32	
	1620	Maximum					32			32	35.50	
	1630	Solvent-based, 250 S.F./gal., minimum					12.35			12.35	13.60	
	1640	Maximum					14.20			14.20	15.60	
	1650	Water based, 350 S.F., minimum					11.35			11.35	12.50	
	1660	Maximum					13			13	14.30	
	1700	Colors (300-400 S.F. per gallon)					49			49	53.50	
	1800	Set accelerator for below freezing, 1 to 1-1/2 gal. per C.Y.					10			10	11	
	1900	Set retarder, 2 to 4 fl. oz. per bag of cement					13.30			13.30	14.65	
	2000	Waterproofing, integral 1 lb. per bag of cement				Lb.	.98			.98	1.08	
	2100	Powdered metallic, 40 lbs. per 100 S.F., minimum					1.22			1.22	1.34	

	03060 | Basic Concrete Materials		CREW	DAILY OUTPUT	LABOR-HOURS	UNIT	2006 BARE COSTS MAT.	LABOR	EQUIP.	TOTAL	TOTAL INCL O&P	
100	2120	Maximum				Lb.	2.44			2.44	2.68	100
110	0010	**AGGREGATE** R033105-20										110
	0020	Expanded shale, C.L. lots, 52 lb. per C.F., minimum				Ton	36			36	39.50	
	0050	Maximum R033105-30				"	48			48	53	
	0100	Lightweight vermiculite or perlite, 4 C.F. bag, C.L. lots				Bag	6.80			6.80	7.50	
	0150	L.C.L. lots R033105-40				"	7.50			7.50	8.25	
	0250	Sand & stone, loaded at pit, crushed bank gravel				Ton	18.85			18.85	20.50	
	0350	Sand, washed, for concrete R033105-50					25.50			25.50	28	
	0400	For plaster or brick					12.75			12.75	14.05	
	0450	Stone, 3/4" to 1-1/2"					18.40			18.40	20	
	0500	3/8" roofing stone & 1/2" pea stone					15.95			15.95	17.50	
	0550	For trucking 10 miles (5 mile R/T), add to the above	B-34B	117	.068			1.94	4.07	6.01	7.65	
	0600	30 miles (15 mi R/T), add to the above	"	72	.111			3.15	6.60	9.75	12.45	
	0850	Sand & stone, loaded at pit, crushed bank gravel				C.Y.	21			21	23.50	
	0950	Sand, washed, for concrete					12.95			12.95	14.25	
	1000	For plaster or brick					37			37	41	
	1050	Stone, 3/4" to 1-1/2"					23			23	25	
	1100	3/8" roofing stone & 1/2" pea stone					33.50			33.50	36.50	
	1150	For trucking 10 miles (5 mi R/T), add to the above	B-34B	78	.103			2.91	6.10	9.01	11.45	
	1200	30 miles (15 mi R/T), add to the above	"	48	.167			4.73	9.95	14.68	18.60	
	1310	Quartz chips, 50 lb. bags				Cwt.	17			17	18.70	
	1330	Silica chips, 50 lb. bags					9.35			9.35	10.25	
	1410	White marble, 3/8" to 1/2", 50 lb. bags					14.05			14.05	15.45	
	1430	3/4"				Ton	71			71	78	
200	0010	**CEMENT** Material only R033105-20										200
	0240	Portland, Type I/II, TL lots, 94 lb bags				Bag	8.25			8.25	9.10	
	0250	LTL/LCL lots R033105-30				"	10.45			10.45	11.50	
	0300	Trucked in bulk, per cwt				Cwt.	4.42			4.42	4.86	
	0400	Type III, high early strength, TL lots, 94 lb bags R033105-40				Bag	9.50			9.50	10.45	
	0420	L.T.L. or L.C.L. lots					10.30			10.30	11.35	
	0500	White, type III, high early strength, T.L. or C.L. lots, bags R033105-50					21.50			21.50	24	
	0520	L.T.L. or L.C.L. lots					23			23	25	
	0600	White, type I, T.L. or C.L. lots, bags					20.50			20.50	22.50	
	0620	L.T.L. or L.C.L. lots					21.50			21.50	23.50	
210	0010	**CRIBBING** See under Retaining Walls, division 02830-100										210
250	0010	**DAMPPROOFING** See division 07100										250
300	0010	**EQUIPMENT** For placing concrete see Reference Section R015433-10										300
400	0010	**LIFT SLAB** See division 03310-240 R033105-85										400
700	0010	**SAWING CONCRETE** See division 02220-360										700
850	0010	**WATERPROOFING AND DAMPPROOFING** See division 07100										850
	0050	Integral waterproofing, add to cost of regular concrete				C.Y.	5.90			5.90	6.45	
870	0010	**WINTER PROTECTION**										870
	0012	For heated ready mix, add, minimum				C.Y.	4.73			4.73	5.20	
	0050	Maximum				"	5.90			5.90	6.50	
	0100	Temporary heat to protect concrete, 24 hours, minimum	2 Clab	50	.320	M.S.F.	108	8.75		116.75	134	
	0150	Maximum	"	25	.640	"	141	17.55		158.55	184	
	0200	Temporary shelter for slab on grade, wood frame/polyethylene sheeting										
	0201	Build or remove, minimum	2 Carp	10	1.600	M.S.F.	284	57		341	405	
	0210	Maximum	"	3	5.333	"	340	190		530	690	

03050 | Basic Concrete Materials & Methods

		03060 | Basic Concrete Materials	CREW	DAILY OUTPUT	LABOR-HOURS	UNIT	2006 BARE COSTS MAT.	LABOR	EQUIP.	TOTAL	TOTAL INCL O&P	
870	0300	See also Division 03390-200										870

03100 | Concrete Forms & Accessories

		03110 | Structural C.I.P. Forms	CREW	DAILY OUTPUT	LABOR-HOURS	UNIT	2006 BARE COSTS MAT.	LABOR	EQUIP.	TOTAL	TOTAL INCL O&P	
300	0010	**EXPANSION JOINT** See division 03150-250										300
405	0010	**FORMS IN PLACE, BEAMS AND GIRDERS** R031113-40										405
	0500	Exterior spandrel, job-built plywood, 12" wide, 1 use	C-2	225	.213	SFCA	3.76	7.35		11.11	16.35	
	0550	2 use R031113-60		275	.175		1.99	6.05		8.04	12.20	
	0650	4 use		310	.155		1.22	5.35		6.57	10.20	
	1000	18" wide, 1 use		250	.192		3.22	6.65		9.87	14.55	
	1050	2 use		275	.175		1.77	6.05		7.82	11.95	
	1100	3 use		305	.157		1.29	5.45		6.74	10.40	
	1150	4 use		315	.152		1.05	5.25		6.30	9.85	
	1500	24" wide, 1 use		265	.181		2.95	6.25		9.20	13.60	
	1550	2 use		290	.166		1.66	5.70		7.36	11.30	
	1600	3 use		315	.152		1.18	5.25		6.43	10	
	1650	4 use		325	.148		.96	5.10		6.06	9.50	
	2000	Interior beam, job-built plywood, 12" wide, 1 use		300	.160		4.10	5.50		9.60	13.65	
	2050	2 use		340	.141		2.03	4.87		6.90	10.35	
	2100	3 use		364	.132		1.64	4.55		6.19	9.35	
	2150	4 use		377	.127		1.33	4.40		5.73	8.75	
	2500	24" wide, 1 use		320	.150		3.01	5.20		8.21	11.90	
	2550	2 use		365	.132		1.69	4.54		6.23	9.35	
	2600	3 use		385	.125		1.20	4.31		5.51	8.45	
	2650	4 use		395	.122		.97	4.20		5.17	8	
	3000	Encasing steel beam, hung, job-built plywood, 1 use		325	.148		2.94	5.10		8.04	11.70	
	3050	2 use		390	.123		1.62	4.25		5.87	8.85	
	3100	3 use		415	.116		1.18	3.99		5.17	7.90	
	3150	4 use		430	.112		.96	3.85		4.81	7.45	
	3500	Bottoms only, to 30" wide, job-built plywood, 1 use		230	.209		4.48	7.20		11.68	16.90	
	3550	2 use		265	.181		2.51	6.25		8.76	13.10	
	3600	3 use		280	.171		1.79	5.90		7.69	11.75	
	3650	4 use		290	.166		1.46	5.70		7.16	11.05	
	4000	Sides only, vertical, 36" high, job-built plywood, 1 use		335	.143		4.69	4.95		9.64	13.35	
	4050	2 use		405	.119		2.58	4.09		6.67	9.65	
	4100	3 use		430	.112		1.88	3.85		5.73	8.45	
	4150	4 use		445	.108		1.52	3.72		5.24	7.85	
	4500	Sloped sides, 36" high, 1 use		305	.157		4.59	5.45		10.04	14.05	
	4550	2 use		370	.130		2.56	4.48		7.04	10.20	
	4600	3 use		405	.119		1.84	4.09		5.93	8.80	
	4650	4 use		425	.113		1.49	3.90		5.39	8.10	
	5000	Upstanding beams, 36" high, 1 use		225	.213		5.60	7.35		12.95	18.40	
	5050	2 use		255	.188		3.14	6.50		9.64	14.20	
	5100	3 use		275	.175		2.27	6.05		8.32	12.50	
	5150	4 use	↓	280	.171	↓	1.84	5.90		7.74	11.80	
	9000	Minimum labor/equipment charge	2 Carp	2	8	Job		284		284	470	

03110 | Structural C.I.P. Forms

	03110 | Structural C.I.P. Forms	CREW	DAILY OUTPUT	LABOR-HOURS	UNIT	2006 BARE COSTS MAT.	LABOR	EQUIP.	TOTAL	TOTAL INCL O&P
410	0010 FORMS IN PLACE, COLUMNS R031113-40									
	0500 Round fiberglass, 4 use per mo., rent, 12″ diameter	C-1	160	.200	L.F.	6.40	6.70		13.10	18.10
	0550 16″ diameter R031113-60		150	.213		7.65	7.15		14.80	20.50
	0600 18″ diameter		140	.229		8.50	7.65		16.15	22
	0650 24″ diameter		135	.237		10.60	7.95		18.55	25
	0700 28″ diameter		130	.246		11.80	8.25		20.05	26.50
	0800 30″ diameter		125	.256		12.35	8.60		20.95	28
	0850 36″ diameter		120	.267		16.45	8.95		25.40	33
	1500 Round fiber tube, 1 use, 8″ diameter		155	.206		1.51	6.90		8.41	13.10
	1550 10″ diameter		155	.206		1.94	6.90		8.84	13.60
	1600 12″ diameter		150	.213		2.32	7.15		9.47	14.40
	1650 14″ diameter		145	.221		3.03	7.40		10.43	15.60
	1700 16″ diameter		140	.229		3.54	7.65		11.19	16.60
	1750 20″ diameter		135	.237		5.65	7.95		13.60	19.35
	1800 24″ diameter		130	.246		7.30	8.25		15.55	21.50
	1850 30″ diameter		125	.256		10.40	8.60		19	25.50
	1900 36″ diameter		115	.278		13.15	9.30		22.45	30
	1950 42″ diameter		100	.320		31.50	10.70		42.20	53
	2000 48″ diameter		85	.376		40	12.60		52.60	65
	2200 For seamless type, add					15%				
	3000 Round, steel, 4 use per mo., rent, regular duty, 12″ diameter	C-1	145	.221	L.F.	10.45	7.40		17.85	24
	3050 16″ diameter		125	.256		11.75	8.60		20.35	27
	3100 Heavy duty, 20″ diameter		105	.305		12.90	10.20		23.10	31
	3150 24″ diameter		85	.376		14.15	12.60		26.75	36.50
	3200 30″ diameter		70	.457		16.25	15.30		31.55	43.50
	3250 36″ diameter		60	.533		17.40	17.85		35.25	48.50
	3300 48″ diameter		50	.640		24	21.50		45.50	62
	3350 60″ diameter		45	.711		32	24		56	74.50
	4000 Column capitals, steel, 4 uses/mo., 24″ col, 4′ cap diameter		12	2.667	Ea.	198	89.50		287.50	365
	4050 5′ cap diameter		11	2.909		226	97.50		323.50	410
	4100 6′ cap diameter		10	3.200		310	107		417	520
	4150 7′ cap diameter		9	3.556		380	119		499	610
	4500 For second and succeeding months, deduct					50%				
	5000 Job-built plywood, 8″ x 8″ columns, 1 use	C-1	165	.194	SFCA	2.18	6.50		8.68	13.15
	5050 2 use		195	.164		1.25	5.50		6.75	10.50
	5100 3 use		210	.152		.87	5.10		5.97	9.40
	5150 4 use		215	.149		.72	4.99		5.71	9.05
	5500 12″ x 12″ columns, 1 use		180	.178		2.19	5.95		8.14	12.25
	5550 2 use		210	.152		1.20	5.10		6.30	9.75
	5600 3 use		220	.145		.87	4.87		5.74	9.05
	5650 4 use		225	.142		.71	4.77		5.48	8.70
	6000 16″ x 16″ columns, 1 use		185	.173		2.23	5.80		8.03	12.05
	6050 2 use		215	.149		1.17	4.99		6.16	9.55
	6100 3 use		230	.139		.89	4.66		5.55	8.75
	6150 4 use		235	.136		.73	4.56		5.29	8.35
	6500 24″ x 24″ columns, 1 use		190	.168		2.54	5.65		8.19	12.15
	6550 2 use		216	.148		1.40	4.96		6.36	9.80
	6600 3 use		230	.139		1.02	4.66		5.68	8.85
	6650 4 use		238	.134		.83	4.51		5.34	8.35
	7000 36″ x 36″ columns, 1 use		200	.160		2.26	5.35		7.61	11.40
	7050 2 use		230	.139		1.27	4.66		5.93	9.15
	7100 3 use		245	.131		.90	4.38		5.28	8.25
	7150 4 use		250	.128		.74	4.29		5.03	7.90
	7500 Steel framed plywood, 4 use per mo., rent, 8″ x 8″		340	.094		3.26	3.15		6.41	8.85
	7550 10″ x 10″		350	.091		2.45	3.06		5.51	7.80
	7600 12″ x 12″		370	.086		3.11	2.90		6.01	8.20

3 CONCRETE

03110	Structural C.I.P. Forms		CREW	DAILY OUTPUT	LABOR-HOURS	UNIT	2006 BARE COSTS MAT.	LABOR	EQUIP.	TOTAL	TOTAL INCL O&P	
410	7650	16" x 16" R031113-40	C-1	400	.080	SFCA	3.41	2.68		6.09	8.20	410
	7700	20" x 20"		420	.076		1.64	2.55		4.19	6.05	
	7750	24" x 24" R031113-60		440	.073		1.54	2.44		3.98	5.75	
	7755	30" x 30"		440	.073		1.38	2.44		3.82	5.55	
	9000	Minimum labor/equipment charge	2 Carp	2	8	Job		284		284	470	
415	0010	FORMS IN PLACE, CULVERT R031113-40										415
	0015	5' to 8' square or rectangular, 1 use	C-1	170	.188	SFCA	5.25	6.30		11.55	16.20	
	0050	2 use R031113-60		180	.178		1.81	5.95		7.76	11.85	
	0100	3 use		190	.168		1.43	5.65		7.08	10.90	
	0150	4 use		200	.160		1.24	5.35		6.59	10.25	
420	0010	FORMS IN PLACE, ELEVATED SLABS R031113-40										420
	0050	See also corrugated form deck, Division 05310-300										
	1000	Flat plate, job-built plywood, to 15' high, 1 use R031113-60	C-2	470	.102	S.F.	4.31	3.53		7.84	10.60	
	1050	2 use		520	.092		2.37	3.19		5.56	7.90	
	1100	3 use		545	.088		1.72	3.04		4.76	6.95	
	1150	4 use		560	.086		1.40	2.96		4.36	6.45	
	1500	15' to 20' high ceilings, 4 use		495	.097		1.66	3.35		5.01	7.40	
	1600	21' to 35' high ceilings, 4 use		450	.107		2.33	3.68		6.01	8.65	
	2000	Flat slab, drop panels, job-built plywood, to 15' high, 1 use		449	.107		4.79	3.69		8.48	11.35	
	2050	2 use		509	.094		2.63	3.26		5.89	8.30	
	2100	3 use		532	.090		1.91	3.12		5.03	7.25	
	2150	4 use		544	.088		1.56	3.05		4.61	6.75	
	2250	15' to 20' high ceilings, 4 use		480	.100		4.20	3.45		7.65	10.30	
	2350	20' to 35' high ceilings, 4 use		435	.110		4.87	3.81		8.68	11.65	
	3000	Floor slab hung from steel beams, 1 use		485	.099		2.30	3.42		5.72	8.20	
	3050	2 use		535	.090		1.65	3.10		4.75	6.95	
	3100	3 use		550	.087		1.43	3.01		4.44	6.55	
	3150	4 use		565	.085		1.32	2.93		4.25	6.30	
	3500	Floor slab, with 20" metal pans, 1 use		415	.116		6.10	3.99		10.09	13.30	
	3550	2 use		445	.108		3.94	3.72		7.66	10.50	
	3600	3 use		475	.101		3.22	3.49		6.71	9.35	
	3650	4 use		500	.096		2.86	3.31		6.17	8.65	
	3700	Floor slab with 30" pans, 1 use		418	.115		6	3.97		9.97	13.15	
	3720	2 use		455	.105		3.86	3.64		7.50	10.30	
	3740	3 use		470	.102		3.14	3.53		6.67	9.30	
	3760	4 use		480	.100		2.78	3.45		6.23	8.75	
	3801	Rent 30" metal pans, 1 use					4.88			4.88	5.35	
	3821	2 use					4.88			4.88	5.35	
	3841	3 use					4.88			4.88	5.35	
	3861	4 use					4.88			4.88	5.35	
	4000	Floor slab with 19" metal domes, 1 use	C-2	405	.119		6.45	4.09		10.54	13.90	
	4050	2 use		435	.110		4.29	3.81		8.10	11	
	4100	3 use		465	.103		3.57	3.56		7.13	9.85	
	4150	4 use		495	.097		3.21	3.35		6.56	9.10	
	4500	With 30" fiberglass domes, 1 use		405	.119		6.30	4.09		10.39	13.70	
	4520	2 use		450	.107		4.12	3.68		7.80	10.65	
	4530	3 use		460	.104		3.40	3.60		7	9.70	
	4550	4 use		470	.102		3.04	3.53		6.57	9.20	
	4701	Rent 30" fiberglass domes, 1 use					3.29			3.29	3.62	
	4721	2 use					3.29			3.29	3.62	
	4741	3 use					3.29			3.29	3.62	
	4761	4 use					3.29			3.29	3.62	
	5000	Box out for slab openings, over 16" deep, 1 use	C-2	190	.253	SFCA	3.22	8.70		11.92	18	
	5050	2 use		240	.200	"	1.77	6.90		8.67	13.40	

	03110 | Structural C.I.P. Forms		Crew	Daily Output	Labor-Hours	Unit	2006 Bare Costs Mat.	Labor	Equip.	Total	Total Incl O&P	
420	5500	Shallow slab box outs, to 10 S.F. R031113-40	C-2	42	1.143	Ea.	10.70	39.50		50.20	77.50	420
	5550	Over 10 S.F. (use perimeter)		600	.080	L.F.	1.43	2.76		4.19	6.15	
	6000	Bulkhead forms for slab, with keyway, 1 use, 2 piece R031113-60		500	.096		1.59	3.31		4.90	7.25	
	6100	3 piece (see also edge forms)		460	.104		1.74	3.60		5.34	7.85	
	6200	Bulkhead form, 4-1/2" high, exp. metal, keyway & stakes	C-1	1200	.027		.84	.89		1.73	2.40	
	6210	5-1/2" high		1100	.029		.90	.97		1.87	2.61	
	6215	7-1/2" high		960	.033		1.08	1.12		2.20	3.04	
	6220	9-1/2" high		840	.038		1.88	1.28		3.16	4.19	
	6500	Curb forms, wood, 6" to 12" high, on elevated slabs, 1 use		180	.178	SFCA	1.54	5.95		7.49	11.55	
	6550	2 use		205	.156		.85	5.25		6.10	9.60	
	6600	3 use		220	.145		.62	4.87		5.49	8.80	
	6650	4 use		225	.142		.50	4.77		5.27	8.45	
	7000	Edge forms to 6" high, on elevated slab, 4 use		500	.064	L.F.	.19	2.14		2.33	3.76	
	7070	7" to 12" high, 1 use		162	.198	SFCA	1.17	6.60		7.77	12.25	
	7080	2 use		198	.162		.64	5.40		6.04	9.65	
	7090	3 use		222	.144		.47	4.83		5.30	8.50	
	7101	4 use		350	.091		.38	3.06		3.44	5.50	
	7500	Depressed area forms to 12" high, 4 use		300	.107	L.F.	.72	3.57		4.29	6.70	
	7550	12" to 24" high, 4 use		175	.183		.97	6.15		7.12	11.20	
	8000	Perimeter deck and rail for elevated slabs, straight		90	.356		14.60	11.90		26.50	36	
	8050	Curved		65	.492		20	16.50		36.50	49.50	
	8500	Void forms, round fiber, 3" diameter		450	.071		.64	2.38		3.02	4.65	
	8550	4" diameter		425	.075		.75	2.52		3.27	5	
	8600	6" diameter		400	.080		1.13	2.68		3.81	5.70	
	8650	8" diameter		375	.085		1.82	2.86		4.68	6.75	
	8700	10" diameter		350	.091		2.35	3.06		5.41	7.70	
	8750	12" diameter		300	.107		3.04	3.57		6.61	9.25	
	9000	Minimum labor/equipment charge	2 Carp	2	8	Job		284		284	470	
425	0010	**FORMS IN PLACE, EQUIPMENT FOUNDATIONS** job built R031113-40										425
	0020	1 use	C-2	160	.300	SFCA	2.55	10.35		12.90	19.95	
	0050	2 use R031113-60		190	.253		1.40	8.70		10.10	16	
	0100	3 use		200	.240		1.02	8.30		9.32	14.85	
	0150	4 use		205	.234		.83	8.10		8.93	14.30	
	9000	Minimum labor/equipment charge	1 Carp	3	2.667	Job		95		95	157	
430	0010	**FORMS IN PLACE, FOOTINGS** R031113-40										430
	0020	Continuous wall, plywood, 1 use	C-1	375	.085	SFCA	2.62	2.86		5.48	7.60	
	0050	2 use R031113-60		440	.073		1.44	2.44		3.88	5.60	
	0100	3 use		470	.068		1.05	2.28		3.33	4.93	
	0150	4 use		485	.066		.85	2.21		3.06	4.60	
	0500	Dowel supports for footings or beams, 1 use		500	.064	L.F.	.84	2.14		2.98	4.47	
	1000	Integral starter wall, to 4" high, 1 use		400	.080		.89	2.68		3.57	5.40	
	1500	Keyway, 4 use, tapered wood, 2" x 4"	1 Carp	530	.015		.20	.54		.74	1.11	
	1550	2" x 6"		500	.016		.30	.57		.87	1.27	
	2000	Tapered plastic, 2" x 3"		530	.015		.64	.54		1.18	1.59	
	2050	2" x 4"		500	.016		.82	.57		1.39	1.84	
	2250	For keyway hung from supports, add		150	.053		.84	1.90		2.74	4.06	
	3000	Pile cap, square or rectangular, job-built plywood, 1 use	C-1	290	.110	SFCA	2.45	3.70		6.15	8.85	
	3050	2 use		346	.092		1.35	3.10		4.45	6.65	
	3100	3 use		371	.086		.98	2.89		3.87	5.85	
	3150	4 use		383	.084		.80	2.80		3.60	5.50	
	4000	Triangular or hexagonal, 1 use		225	.142		2.89	4.77		7.66	11.10	
	4050	2 use		280	.114		1.59	3.83		5.42	8.10	
	4100	3 use		305	.105		1.16	3.52		4.68	7.10	
	4150	4 use		315	.102		.94	3.40		4.34	6.70	
	5000	Spread footings, job-built lumber, 1 use		305	.105		1.83	3.52		5.35	7.85	
	5050	2 use		371	.086		1.01	2.89		3.90	5.90	

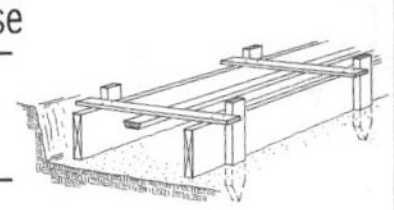

		03110 | Structural C.I.P. Forms	CREW	DAILY OUTPUT	LABOR-HOURS	UNIT	2006 BARE COSTS MAT.	LABOR	EQUIP.	TOTAL	TOTAL INCL O&P
430	5100	3 use (R031113-40)	C-1	401	.080	SFCA	.73	2.67		3.40	5.25
	5150	4 use		414	.077		.59	2.59		3.18	4.94
	6000	Supports for dowels, plinths or templates, 2′ x 2′ ftgs (R031113-60)		25	1.280	Ea.	5.45	43		48.45	77
	6050	4′ x 4′ footing		22	1.455		10.95	48.50		59.45	93
	6100	8′ x 8′ footing		20	1.600		22	53.50		75.50	113
	6150	12′ x 12′ footing		17	1.882		35	63		98	144
	7000	Plinths, job-built plywood, 1 use		250	.128	SFCA	3.63	4.29		7.92	11.10
	7100	4 use		270	.119	"	1.19	3.97		5.16	7.90
	9000	Minimum labor/equipment charge	1 Carp	3	2.667	Job		95		95	157
435	0010	**FORMS IN PLACE, GRADE BEAM** (R031113-40)									
	0020	Job-built plywood, 1 use	C-2	530	.091	SFCA	1.74	3.13		4.87	7.10
	0050	2 use (R031113-60)		580	.083		.96	2.86		3.82	5.80
	0100	3 use		600	.080		.70	2.76		3.46	5.35
	0150	4 use		605	.079		.57	2.74		3.31	5.15
	9000	Minimum labor/equipment charge	2 Carp	2	8	Job		284		284	470
440	0010	**FORMS IN PLACE, MAT FOUNDATION** (R031113-40)									
	0020	Job-built plywood, 1 use	C-2	290	.166	SFCA	2.54	5.70		8.24	12.25
	0050	2 use (R031113-60)		310	.155		1.01	5.35		6.36	9.95
	0100	3 use		330	.145		.71	5		5.71	9.10
	0120	4 use		350	.137		.59	4.74		5.33	8.50
445	0010	**FORMS IN PLACE, SLAB ON GRADE** (R031113-40)									
	1000	Bulkhead forms w/keyway, wood, 6″ high, 1 use	C-1	510	.063	L.F.	.97	2.10		3.07	4.55
	1050	2 uses (R031113-60)		400	.080		.53	2.68		3.21	5.05
	1100	4 uses		350	.091		.32	3.06		3.38	5.45
	1400	Bulkhead form, 4-1/2″ high, exp metal, incl keyway & stakes (R031113-80)		1200	.027		.84	.89		1.73	2.40
	1410	5-1/2″ high		1100	.029		.90	.97		1.87	2.61
	1420	7-1/2″ high		960	.033		1.08	1.12		2.20	3.04
	1430	9-1/2″ high		840	.038		1.88	1.28		3.16	4.19
	2000	Curb forms, wood, 6″ to 12″ high, on grade, 1 use		215	.149	SFCA	2.23	4.99		7.22	10.70
	2050	2 use		250	.128		1.23	4.29		5.52	8.45
	2100	3 use		265	.121		.89	4.05		4.94	7.70
	2150	4 use		275	.116		.72	3.90		4.62	7.25
	3000	Edge forms, wood, 4 use, on grade, to 6″ high		600	.053	L.F.	.29	1.79		2.08	3.28
	3050	7″ to 12″ high		435	.074	SFCA	.76	2.47		3.23	4.92
	3500	For depressed slabs, 4 use, to 12″ high		300	.107	L.F.	.69	3.57		4.26	6.65
	3550	To 24″ high		175	.183		.91	6.15		7.06	11.15
	4000	For slab blockouts, to 12″ high, 1 use		200	.160		.74	5.35		6.09	9.70
	4050	To 24″ high, 1 use		120	.267		.94	8.95		9.89	15.85
	4100	Plastic (extruded), to 6″ high, multiple use, on grade		800	.040		.37	1.34		1.71	2.63
	5000	Screed, 24 ga. metal key joint, see Div 03150-250									
	5020	Wood, incl. wood stakes, 1″ x 3″	C-1	900	.036	L.F.	.57	1.19		1.76	2.59
	5050	2″ x 4″		900	.036	"	.66	1.19		1.85	2.69
	6000	Trench forms in floor, wood, 1 use		160	.200	SFCA	2.50	6.70		9.20	13.85
	6050	2 use		175	.183		1.27	6.15		7.42	11.55
	6100	3 use		180	.178		.93	5.95		6.88	10.85
	6150	4 use		185	.173		.75	5.80		6.55	10.45
	8760	Void form, corrugated fiberboard, 6″ x 12″, 10′ long		240	.133	S.F.	.77	4.47		5.24	8.25
	9000	Minimum labor/equipment charge	1 Carp	2	4	Job		142		142	236
450	0010	**FORMS IN PLACE, STAIRS** (R031113-40)									
	0015	(Slant length x width), 1 use	C-2	165	.291	S.F.	5.90	10.05		15.95	23
	0050	2 use (R031113-60)		170	.282		3.48	9.75		13.23	20
	0100	3 use		180	.267		2.67	9.20		11.87	18.20
	0150	4 use		190	.253		2.27	8.70		10.97	16.95
	1000	Alternate pricing method (1.0 L.F./S.F.), 1 use		100	.480	LF Rsr	5.90	16.55		22.45	34

3 CONCRETE

	03110 | Structural C.I.P. Forms		CREW	DAILY OUTPUT	LABOR-HOURS	UNIT	2006 BARE COSTS MAT.	LABOR	EQUIP.	TOTAL	TOTAL INCL O&P	
450	1050	2 use (R031113-40)	C-2	105	.457	LF Rsr	3.48	15.80		19.28	30	450
	1100	3 use		110	.436		2.67	15.05		17.72	28	
	1150	4 use (R031113-60)		115	.417		2.27	14.40		16.67	26.50	
	2000	Stairs, cast on sloping ground (length x width), 1 use		220	.218	S.F.	2.16	7.55		9.71	14.85	
	2100	4 use		240	.200	"	.70	6.90		7.60	12.20	
455	0010	**FORMS IN PLACE, WALLS** (R031113-10)										455
	0100	Box out for wall openings, to 16" thick, to 10 S.F.	C-2	24	2	Ea.	23.50	69		92.50	140	
	0150	Over 10 S.F. (use perimeter) (R031113-40)	"	280	.171	L.F.	2.02	5.90		7.92	12	
	0250	Brick shelf, 4" w, add to wall forms, use wall area abv shelf										
	0260	1 use (R031113-60)	C-2	240	.200	SFCA	2.20	6.90		9.10	13.85	
	0300	2 use		275	.175		1.21	6.05		7.26	11.35	
	0350	4 use		300	.160		.88	5.50		6.38	10.10	
	0500	Bulkhead, with keyway, 1 use, 2 piece		265	.181	L.F.	3.17	6.25		9.42	13.85	
	0550	3 piece		175	.274		4	9.45		13.45	20	
	0600	Bulkhead forms with keyway, 1 Pc. expanded metal, 8" wall	C-1	1000	.032		1.08	1.07		2.15	2.97	
	0610	10" wall		800	.040		1.88	1.34		3.22	4.29	
	0620	12" wall		525	.061		2.28	2.04		4.32	5.90	
	0700	Buttress, to 8' high, 1 use	C-2	350	.137	SFCA	4.01	4.74		8.75	12.25	
	0750	2 use		430	.112		2.20	3.85		6.05	8.80	
	0800	3 use		460	.104		1.61	3.60		5.21	7.70	
	0850	4 use		480	.100		1.32	3.45		4.77	7.15	
	1000	Corbel or haunch, to 12" wide, add to wall forms, 1 use		150	.320	L.F.	2.14	11.05		13.19	20.50	
	1050	2 use		170	.282		1.17	9.75		10.92	17.45	
	1100	3 use		175	.274		.85	9.45		10.30	16.65	
	1150	4 use		180	.267		.69	9.20		9.89	16	
	2000	Wall, below grade, job-built plywood, to 8' high, 1 use		300	.160	SFCA	2.47	5.50		7.97	11.85	
	2050	2 use		365	.132		1.57	4.54		6.11	9.25	
	2150	4 use		435	.110		.93	3.81		4.74	7.30	
	2400	Over 8' to 16' high, 1 use		280	.171		4.97	5.90		10.87	15.25	
	2420	2 use		345	.139		2	4.80		6.80	10.15	
	2430	3 use		375	.128		1.68	4.42		6.10	9.15	
	2440	4 use		395	.122		1.51	4.20		5.71	8.60	
	2445	Exterior wall, 8' to 16' high, 1 use		280	.171		2.17	5.90		8.07	12.20	
	2450	2 use		345	.139		1.19	4.80		5.99	9.25	
	2500	3 use		375	.128		.85	4.42		5.27	8.25	
	2550	4 use		395	.122		.70	4.20		4.90	7.70	
	2700	Over 16' high, 1 use		235	.204		2.42	7.05		9.47	14.35	
	2750	2 use		290	.166		1.33	5.70		7.03	10.90	
	2800	3 use		315	.152		.97	5.25		6.22	9.75	
	2850	4 use		330	.145		.79	5		5.79	9.15	
	3000	For architectural finish, add		1820	.026		5.45	.91		6.36	7.50	
	4000	Radial, smooth curved, job-built plywood, 1 use		245	.196		2.27	6.75		9.02	13.70	
	4050	2 use		300	.160		1.25	5.50		6.75	10.50	
	4100	3 use		325	.148		.91	5.10		6.01	9.45	
	4150	4 use		335	.143		.74	4.95		5.69	9	
	4200	Below grade, job-built plywood, 1 use		225	.213		3.13	7.35		10.48	15.65	
	4210	2 use		225	.213		1.73	7.35		9.08	14.10	
	4220	3 use		225	.213		1.40	7.35		8.75	13.75	
	4230	4 use		225	.213		1.02	7.35		8.37	13.30	
	4300	Curved, 2' chords, job-built plywood, 1 use		290	.166		1.92	5.70		7.62	11.55	
	4350	2 use		355	.135		1.06	4.67		5.73	8.90	
	4400	3 use		385	.125		.77	4.31		5.08	8	
	4450	4 use		400	.120		.62	4.14		4.76	7.55	
	4500	Over 8' high, 1 use		290	.166		.82	5.70		6.52	10.35	
	4525	2 use		355	.135		.45	4.67		5.12	8.25	

455	03110 | Structural C.I.P. Forms		CREW	DAILY OUTPUT	LABOR-HOURS	UNIT	2006 BARE COSTS MAT.	LABOR	EQUIP.	TOTAL	TOTAL INCL O&P	455
	4550	3 use	R031113-10	C-2	385	.125	SFCA	.33	4.31		4.64	7.50
	4575	4 use			400	.120		.27	4.14		4.41	7.15
	4600	Retaining wall, battered, job-built plyw'd, to 8' h, 1 use	R031113-40		300	.160		1.81	5.50		7.31	11.15
	4650	2 use			355	.135		1	4.67		5.67	8.85
	4700	3 use	R031113-60		375	.128		.72	4.42		5.14	8.10
	4750	4 use			390	.123		.54	4.25		4.79	7.65
	4900	Over 8' to 16' high, 1 use			240	.200		1.98	6.90		8.88	13.65
	4950	2 use			295	.163		1.09	5.60		6.69	10.50
	5000	3 use			305	.157		.79	5.45		6.24	9.85
	5050	4 use			320	.150		.64	5.20		5.84	9.30
	5500	For gang wall forming, 192 S.F. sections, deduct						10%	10%			
	5550	384 S.F. sections, deduct						20%	20%			
	5750	Liners for forms (add to wall forms), A.B.S. plastic										
	5800	Aged wood, 4" wide, 1 use		1 Carp	250	.032	SFCA	6.45	1.14		7.59	8.95
	5820	2 use			400	.020		3.54	.71		4.25	5.05
	5840	4 use			750	.011		2.09	.38		2.47	2.93
	5900	Fractured rope rib, 1 use			250	.032		3.84	1.14		4.98	6.10
	6000	4 use			750	.011		1.25	.38		1.63	2
	6100	Ribbed look, 1/2" & 3/4" deep, 1 use			300	.027		5.45	.95		6.40	7.55
	6200	4 use			800	.010		1.77	.36		2.13	2.54
	6300	Rustic brick pattern, 1 use			250	.032		3.84	1.14		4.98	6.10
	6400	4 use			750	.011		1.25	.38		1.63	2
	6500	Striated, random, 3/8" x 3/8" deep, 1 use			300	.027		3.84	.95		4.79	5.80
	6600	4 use			800	.010		1.25	.36		1.61	1.96
	6800	Rustication strips, A.B.S. plastic, 2 piece snap-on										
	6850	1" deep x 1-3/8" wide, 1 use		C-2	400	.120	L.F.	4.26	4.14		8.40	11.55
	6900	2 use			600	.080		2.34	2.76		5.10	7.15
	6950	4 use			800	.060		1.38	2.07		3.45	4.95
	7050	Wood, beveled edge, 3/4" deep, 1 use			600	.080		.10	2.76		2.86	4.69
	7100	1" deep, 1 use			450	.107		.15	3.68		3.83	6.25
	7200	For solid board finish, uniform, 1 use, add to wall forms			300	.160	SFCA	.90	5.50		6.40	10.15
	7300	Non-uniform finish			250	.192		.84	6.65		7.49	11.90
	7500	Lintel or sill forms, 1 use		1 Carp	30	.267		2.93	9.50		12.43	18.95
	7520	2 use			34	.235		1.61	8.35		9.96	15.60
	7540	3 use			36	.222		1.17	7.90		9.07	14.40
	7560	4 use			37	.216		.95	7.70		8.65	13.80
	7800	Modular prefabricated plywood, to 8' high, 1 use		C-2	1180	.041		1.91	1.40		3.31	4.43
	7820	2 use			1200	.040		1.05	1.38		2.43	3.45
	7840	3 use			1240	.039		.76	1.34		2.10	3.05
	7860	4 use			1260	.038		.63	1.32		1.95	2.87
	8000	To 16' high, 1 use			715	.067		2.50	2.32		4.82	6.60
	8020	2 use			740	.065		1.38	2.24		3.62	5.20
	8040	3 use			770	.062		1	2.15		3.15	4.67
	8060	4 use			790	.061		.83	2.10		2.93	4.39
	8100	Over 16' high, 1 use			715	.067		3	2.32		5.32	7.15
	8120	2 use			740	.065		1.65	2.24		3.89	5.55
	8140	3 use			770	.062		1.20	2.15		3.35	4.89
	8160	4 use			790	.061		1	2.10		3.10	4.58
	8600	Pilasters, 1 use			270	.178		2.87	6.15		9.02	13.30
	8620	2 use			330	.145		1.58	5		6.58	10.05
	8640	3 use			370	.130		1.15	4.48		5.63	8.65
	8660	4 use			385	.125		.93	4.31		5.24	8.20
	9010	Steel framed plywood, based on 100 uses of purchased										
	9020	forms, and 4 uses of bracing lumber										
	9060	To 8' high		C-2	600	.080	SFCA	.40	2.76		3.16	5.05
	9260	Over 8' to 16' high			450	.107		.40	3.68		4.08	6.55

		03110 | Structural C.I.P. Forms	CREW	DAILY OUTPUT	LABOR-HOURS	UNIT	2006 BARE COSTS MAT.	LABOR	EQUIP.	TOTAL	TOTAL INCL O&P	
455	9460	Over 16′ to 20′ high (R031113-10)	C-2	400	.120	SFCA	.40	4.14		4.54	7.30	455
	9475	For elevated walls, add						10%				
	9900	Minimum labor/equipment charge (R031113-40)	2 Carp	2	8	Job		284		284	470	
460	0010	**FORMS IN PLACE, INSULATING CONCRETE**										460
	0020	Forms left in place, S.F. is for both sides										
	1000	Panel system, flat cavity, minimum	2 Carp	960	.017	S.F.	1.84	.59		2.43	3	
	1010	Maximum		960	.017		2.76	.59		3.35	4.02	
	1020	Grid cavity, minimum		960	.017		2	.59		2.59	3.18	
	1030	Maximum		960	.017		3	.59		3.59	4.28	
	1040	Post and beam cavity, minimum		960	.017		2.20	.59		2.79	3.40	
	1050	Maximum		960	.017		3.30	.59		3.89	4.61	
	1060	Plank system, flat cavity, minimum		1920	.008		2.40	.30		2.70	3.13	
	1070	Maximum		1920	.008		3.60	.30		3.90	4.45	
	1120	Block system, flat cavity, minimum		480	.033		2.42	1.18		3.60	4.62	
	1130	Maximum		480	.033		3.62	1.18		4.80	5.95	
	1140	Grid cavity, minimum		480	.033		2.42	1.18		3.60	4.62	
	1150	Maximum		480	.033		3.62	1.18		4.80	5.95	
	1160	Post and beam cavity, minimum		480	.033		2.42	1.18		3.60	4.62	
	1170	Maximum		480	.033		3.62	1.18		4.80	5.95	
500	0010	**GAS STATION FORMS** Curb fascia, with template,										500
	0050	12 ga. steel, left in place, 9″ high	1 Carp	50	.160	L.F.	9.95	5.70		15.65	20.50	
	1000	Sign or light bases, 18″ diameter, 9″ high		9	.889	Ea.	62.50	31.50		94	122	
	1050	30″ diameter, 13″ high		8	1	″	99.50	35.50		135	169	
	1990	Minimum labor/equipment charge		2	4	Job		142		142	236	
	2000	Island forms, 10′ long, 9″ high, 3′- 6″ wide	C-1	10	3.200	Ea.	278	107		385	485	
	2050	4′ wide		9	3.556		287	119		406	510	
	2500	20′ long, 9″ high, 4′ wide		6	5.333		460	179		639	805	
	2550	5′ wide		5	6.400		480	214		694	885	
	9000	Minimum labor/equipment charge	1 Carp	3	2.667	Job		95		95	157	
800	0010	**SCAFFOLDING** See Division 01540-750										800
820	0010	**CONCRETE SLIP FORMING** (R031113-30)										820
	0020	Silos, minimum	C-17E	3885	.021	SFCA	3.15	.76	.02	3.93	4.73	
	0050	Maximum		1095	.073		4.52	2.70	.07	7.29	9.45	
	1000	Buildings, minimum		3660	.022		2.10	.81	.02	2.93	3.65	
	1050	Maximum		875	.091		3.89	3.37	.09	7.35	9.90	
		03150 | Concrete Accessories										
085	0013	**ANCHOR BOLTS** See Divisions 04080-070 and 05090-080										085
160	0010	**ACCESSORIES, CHAMFER STRIPS**										160
	2000	Polyvinyl chloride, 1/2″ wide with leg	1 Carp	535	.015	L.F.	.23	.53		.76	1.13	
	2200	3/4″ wide with leg		525	.015		.32	.54		.86	1.25	
	2400	1″ radius with leg		515	.016		.67	.55		1.22	1.65	
	2800	1-1/2″ radius with leg		500	.016		2.20	.57		2.77	3.36	
	5000	Wood, 1/2″ wide		535	.015		.10	.53		.63	.99	
	5200	3/4″ wide		525	.015		.10	.54		.64	1.01	
	5400	1″ wide		515	.016		.15	.55		.70	1.08	
250	0010	**EXPANSION JOINTS**										250
	0020	Keyed, cold, 24 ga, incl. stakes, 3-1/2″ high	1 Carp	200	.040	L.F.	.69	1.42		2.11	3.12	
	0050	4-1/2″ high		200	.040		.84	1.42		2.26	3.28	
	0100	5-1/2″ high		195	.041		.90	1.46		2.36	3.41	
	0150	7-1/2″ high		190	.042		1.08	1.50		2.58	3.67	
	0300	Poured asphalt, plain, 1/2″ x 1″	1 Clab	450	.018		.39	.49		.88	1.24	

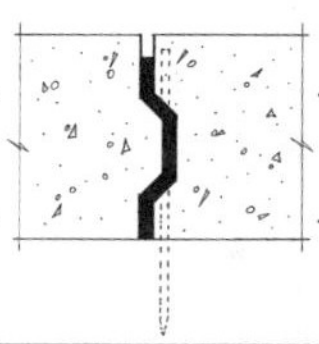

03150 | Concrete Accessories

			CREW	DAILY OUTPUT	LABOR-HOURS	UNIT	2006 BARE COSTS MAT.	LABOR	EQUIP.	TOTAL	TOTAL INCL O&P	
250	0350	1″ x 2″	1 Clab	400	.020	L.F.	1.55	.55		2.10	2.62	250
	0500	Neoprene, liquid, cold applied, 1/2″ x 1″		450	.018		1.85	.49		2.34	2.85	
	0550	1″ x 2″		400	.020		6.25	.55		6.80	7.80	
	0700	Polyurethane, poured, 2 part, 1/2″ x 1″		400	.020		1.27	.55		1.82	2.31	
	0750	1″ x 2″		350	.023		5.10	.63		5.73	6.65	
	0900	Rubberized asphalt, hot or cold applied, 1/2″ x 1″		450	.018		.37	.49		.86	1.22	
	0950	1″ x 2″		400	.020		1.36	.55		1.91	2.41	
	1100	Hot applied, fuel resistant, 1/2″ x 1″		450	.018		1.04	.49		1.53	1.95	
	1150	1″ x 2″		400	.020		2	.55		2.55	3.11	
	2000	Premolded, bituminous fiber, 1/2″ x 6″	1 Carp	375	.021		.37	.76		1.13	1.67	
	2050	1″ x 12″		300	.027		1.51	.95		2.46	3.23	
	2250	Cork with resin binder, 1/2″ x 6″		375	.021		1.68	.76		2.44	3.11	
	2300	1″ x 12″		300	.027		6.10	.95		7.05	8.30	
	2500	Neoprene sponge, closed cell, 1/2″ x 6″		375	.021		1.42	.76		2.18	2.82	
	2550	1″ x 12″		300	.027		6.50	.95		7.45	8.70	
	2750	Polyethylene foam, 1/2″ x 6″		375	.021		.47	.76		1.23	1.78	
	2800	1″ x 12″		300	.027		1.69	.95		2.64	3.43	
	3000	Polyethylene backer rod, 3/8″ diameter		460	.017		.03	.62		.65	1.05	
	3050	3/4″ diameter		460	.017		.07	.62		.69	1.09	
	3100	1″ diameter		460	.017		.11	.62		.73	1.14	
	3500	Polyurethane foam, with polybutylene, 1/2″ x 1/2″		475	.017		.92	.60		1.52	2	
	3550	1″ x 1″		450	.018		1.82	.63		2.45	3.05	
	3750	Polyurethane foam, regular, closed cell, 1/2″ x 6″		375	.021		.62	.76		1.38	1.94	
	3800	1″ x 12″		300	.027		1.24	.95		2.19	2.93	
	4000	Polyvinyl chloride foam, closed cell, 1/2″ x 6″		375	.021		2.42	.76		3.18	3.92	
	4050	1″ x 12″		300	.027		6.15	.95		7.10	8.35	
	4250	Rubber, gray sponge, 1/2″ x 6″		375	.021		2.65	.76		3.41	4.18	
	4300	1″ x 12″		300	.027		10.25	.95		11.20	12.80	
	4500	Lead wool for joints, 1 ton lots				Lb.	1.90			1.90	2.09	
	5000	For installation in walls, add						75%				
	5250	For installation in boxouts, add						25%				
600	0010	**SHORES** Erect and strip, by hand, horizontal members										600
	0500	Aluminum joists and stringers	2 Carp	60	.267	Ea.		9.50		9.50	15.70	
	0600	Steel, adjustable beams		45	.356			12.65		12.65	21	
	0700	Wood joists		50	.320			11.40		11.40	18.85	
	0800	Wood stringers		30	.533			18.95		18.95	31.50	
	1000	Vertical members to 10′ high		55	.291			10.35		10.35	17.15	
	1050	To 13′ high		50	.320			11.40		11.40	18.85	
	1100	To 16′ high		45	.356			12.65		12.65	21	
	1500	Reshoring		1400	.011	S.F.	.42	.41		.83	1.13	
	1600	Flying truss system	C-17D	9600	.009	SFCA		.32	.06	.38	.59	
	1760	Horizontal, aluminum joists, 6-1/4″ high x 5′ to 21′ span, buy				L.F.	33.50			33.50	37	
	1770	Beams, 7-1/4″ high x 4′ to 30′ span				″	47			47	51.50	
	1810	Horizontal, steel beam, adjustable, 4′ to 7′ span				Ea.	535			535	590	
	1830	6′ to 10′ span					690			690	760	
	1920	9′ to 15′ span					875			875	965	
	1940	12′ to 20′ span					895			895	985	
	1970	Steel stringer, W8x10, 4′ to 16′ span, buy				L.F.	19.25			19.25	21	
	3000	Rent for job duration, aluminum joist @ 2′ O.C., per mo				SF Flr.	.84			.84	.93	
	3050	Steel W8x10					.48			.48	.53	
	3060	Steel adjustable					1.73			1.73	1.90	
	3500	#1 post shore, steel, 5′-7″ to 9′-6″ high, 10000# cap., buy				Ea.	355			355	390	
	3550	#2 post shore, 7′-3″ to 12′-10″ high, 7800# capacity					380			380	420	
	3600	#3 post shore, 8′-10″ to 16′-1″ high, 3800# capacity					415			415	455	
	5010	Frame shoring systems, steel, 12000#/leg, buy										

	03150 | Concrete Accessories		CREW	DAILY OUTPUT	LABOR-HOURS	UNIT	2006 BARE COSTS MAT.	LABOR	EQUIP.	TOTAL	TOTAL INCL O&P	
600	5040	Frame, 2′ wide x 6′ high				Ea.	330			330	365	600
	5250	X-brace					52.50			52.50	57.50	
	5550	Base plate					41.50			41.50	45.50	
	5600	Screw jack					132			132	145	
	5650	U-head, 8″ x 8″					21			21	23	
660	0010	**STAIR TREAD INSERTS**										660
	0015	Cast iron, abrasive, 3″ wide	1 Carp	90	.089	L.F.	8.05	3.16		11.21	14.10	
	0020	4″ wide		80	.100		10.70	3.56		14.26	17.70	
	0040	6″ wide		75	.107		16.10	3.79		19.89	24	
	0050	9″ wide		70	.114		24	4.06		28.06	33.50	
	0100	12″ wide		65	.123		32	4.38		36.38	43	
	0300	Cast aluminum, compared to cast iron, deduct					10%					
	0500	Extruded aluminum safety tread, 3″ wide	1 Carp	75	.107		9.15	3.79		12.94	16.35	
	0550	4″ wide		75	.107		12.20	3.79		15.99	19.75	
	0600	6″ wide		75	.107		18.30	3.79		22.09	26.50	
	0650	9″ wide to resurface stairs		70	.114		27.50	4.06		31.56	37	
	1700	Cement fill for pan-type metal treads, plain	1 Cefi	115	.070	S.F.	1.61	2.39		4	5.55	
	1750	Non-slip	″	100	.080	″	1.77	2.75		4.52	6.30	
860	0010	**WATERSTOP**										860
	0020	PVC, ribbed 3/16″ thick, 4″ wide	1 Carp	155	.052	L.F.	.79	1.83		2.62	3.91	
	0050	6″ wide		145	.055		1.30	1.96		3.26	4.68	
	0500	Ribbed, PVC, with center bulb, 6″ wide, 3/16″ thick		135	.059		1.19	2.11		3.30	4.80	
	0550	3/8″ thick		130	.062		1.86	2.19		4.05	5.65	
	0800	Dumbbell type, PVC, 6″ wide, 3/16″ thick		150	.053		1.15	1.90		3.05	4.41	
	0850	3/8″ thick		145	.055		2.27	1.96		4.23	5.75	
	1000	9″ wide, 3/8″ thick, PVC, plain		130	.062		3.26	2.19		5.45	7.20	
	1050	Center bulb		130	.062		6.65	2.19		8.84	10.90	
	1250	Split PVC, 3/8″ thick, 6″ wide		145	.055		1.31	1.96		3.27	4.69	
	1300	9″ wide		130	.062		2.53	2.19		4.72	6.40	
	2000	Rubber, flat dumbbell, 3/8″ thick, 6″ wide		145	.055		6.25	1.96		8.21	10.15	
	2050	9″ wide		135	.059		9.35	2.11		11.46	13.80	
	2500	Flat dumbbell split, 3/8″ thick, 6″ wide		145	.055		1.31	1.96		3.27	4.69	
	2550	9″ wide		135	.059		2.53	2.11		4.64	6.25	
	3000	Center bulb, 1/4″ thick, 6″ wide		145	.055		5.45	1.96		7.41	9.25	
	3050	9″ wide		135	.059		10.70	2.11		12.81	15.30	
	3500	Center bulb split, 3/8″ thick, 6″ wide		145	.055		6.60	1.96		8.56	10.55	
	3550	9″ wide		135	.059		11.45	2.11		13.56	16.10	
	5000	Waterstop fittings, rubber, flat										
	5010	Dumbbell or center bulb, 3/8″ thick,										
	5200	Field union, 6″ wide	1 Carp	50	.160	Ea.	21	5.70		26.70	32.50	
	5250	9″ wide		50	.160		25	5.70		30.70	36.50	
	5500	Flat cross, 6″ wide		30	.267		43	9.50		52.50	63	
	5550	9″ wide		30	.267		61.50	9.50		71	83	
	6000	Flat tee, 6″ wide		30	.267		40	9.50		49.50	59.50	
	6050	9″ wide		30	.267		52.50	9.50		62	73	
	6500	Flat ell, 6″ wide		40	.200		36	7.10		43.10	51.50	
	6550	9″ wide		40	.200		48.50	7.10		55.60	65	
	7000	Vertical tee, 6″ wide		25	.320		35	11.40		46.40	57.50	
	7050	9″ wide		25	.320		44.50	11.40		55.90	68	
	7500	Vertical ell, 6″ wide		35	.229		31	8.15		39.15	47.50	
	7550	9″ wide		35	.229		42.50	8.15		50.65	60.50	

		03210 | Reinforcing Steel	CREW	DAILY OUTPUT	LABOR-HOURS	UNIT	2006 BARE COSTS MAT.	LABOR	EQUIP.	TOTAL	TOTAL INCL O&P	
100	0010	ACCESSORIES										100
	0020	See also Form Accessories, Division 03150										
	0100	Beam bolsters, (BB) standard, lower, up to 1-1/2" high, plain				C.L.F.	62.50			62.50	69	
	0102	Galvanized					96.50			96.50	106	
	0104	Stainless					182			182	200	
	0106	Plastic					109			109	120	
	0108	Epoxy					149			149	164	
	0110	2-1/2" to 3" high, plain					82			82	90	
	0120	Galvanized					109			109	120	
	0140	Stainless					249			249	274	
	0160	Plastic					114			114	125	
	0162	Epoxy					159			159	175	
	0200	Upper, standard (BBU) to 1-1/2" high, plain					180			180	198	
	0210	2-1/2" to 3" high					325			325	360	
	0300	Beam bolster with plate (BBP) to 1-1/2" high, plain					207			207	228	
	0310	2-1/2" to 3" high					400			400	440	
	0500	Slab bolsters, continuous, plain (SB) 3/4" to 1" high, plain					51.50			51.50	56.50	
	0502	Galvanized					59.50			59.50	65.50	
	0504	Stainless					83.50			83.50	92	
	0506	Plastic					70.50			70.50	78	
	0510	1" to 2" high, plain					64.50			64.50	70.50	
	0515	Galvanized					75.50			75.50	83	
	0520	Stainless					143			143	157	
	0525	Plastic					85			85	93.50	
	0530	For bolsters with wire runners (SBR), add					72.50			72.50	79.50	
	0540	For bolsters with plates (SBP), add				↓	175			175	193	
	0700	Clip or bar ties, 16 ga., plain, 3" long				C	14.30			14.30	15.75	
	0710	4" long					15.10			15.10	16.60	
	0720	6" long					17.70			17.70	19.45	
	0730	8" long				↓	18.50			18.50	20.50	
	0900	Flange clips, expandable flanges, 10 ga., 12" O.C., continuous,										
	0910	galvanized, over 500 L.F., 4" to 8"				C.L.F.	53			53	58.50	
	0920	9" to 12"					72.50			72.50	79.50	
	0930	17" to 24"				↓	74			74	81.50	
	1200	High chairs, individual, no plates (1 HC), to 3" high, plain				C	77.50			77.50	85	
	1202	Galvanized					108			108	119	
	1204	Stainless					221			221	243	
	1206	Plastic					103			103	113	
	1210	5" high, plain					119			119	131	
	1212	Galvanized					155			155	171	
	1214	Stainless					335			335	370	
	1216	Plastic					144			144	159	
	1220	8" high, plain					257			257	283	
	1222	Galvanized					345			345	380	
	1224	Stainless					565			565	625	
	1226	Plastic					325			325	355	
	1230	12" high, plain					535			535	585	
	1232	Galvanized					680			680	750	
	1234	Stainless					1,025			1,025	1,125	
	1236	Plastic					600			600	660	
	1240	15" high, plain					1,000			1,000	1,100	
	1242	Galvanized					1,175			1,175	1,275	
	1244	Stainless					1,725			1,725	1,900	
	1246	Plastic					1,075			1,075	1,200	
	1250	For each added inch up to 24" high, plain, add					74			74	81.50	
100	1252	Galvanized, add				↓	86			86	94.50	100

	03210 | Reinforcing Steel	CREW	DAILY OUTPUT	LABOR-HOURS	UNIT	2006 BARE COSTS MAT.	LABOR	EQUIP.	TOTAL	TOTAL INCL O&P	
100	1254 Stainless, add				C	96			96	106	100
	1256 Plastic, add					86			86	94.50	
	1400 Individual high chairs, with plates, (HCP), to 5" high, add					340			340	375	
	1410 Over 5" high, add					390			390	430	
	1500 Bar chair (BC) for up to 1-3/4" high, plain					44			44	48	
	1520 Galvanized					50.50			50.50	55.50	
	1530 Stainless					135			135	148	
	1540 Plastic					89.50			89.50	98	
	1550 Joist chair (JC), joists up to 6", plain					53			53	58.50	
	1580 Galvanized					67.50			67.50	74	
	1600 Stainless					96			96	106	
	1620 Plastic					91.50			91.50	101	
	1630 Epoxy				▼	222			222	245	
	1700 Continuous high chairs, legs 8" O.C. (CHC) to 4" high, plain				C.L.F.	103			103	113	
	1705 Galvanized					126			126	139	
	1710 Stainless					270			270	297	
	1715 Plastic					127			127	140	
	1718 Epoxy					192			192	211	
	1720 6" high, plain					148			148	163	
	1725 Galvanized					212			212	234	
	1730 Stainless					315			315	345	
	1735 Plastic					202			202	223	
	1738 Epoxy					258			258	284	
	1740 8" high, plain					215			215	237	
	1745 Galvanized					298			298	330	
	1750 Stainless					370			370	410	
	1755 Plastic					281			281	310	
	1758 Epoxy					390			390	430	
	1760 12" high, plain					530			530	580	
	1765 Galvanized					635			635	700	
	1770 Stainless					890			890	975	
	1775 Plastic					605			605	665	
	1778 Epoxy					925			925	1,025	
	1780 15" high, plain					595			595	650	
	1785 Galvanized					715			715	785	
	1790 Stainless					710			710	780	
	1795 Plastic					675			675	745	
	1798 Epoxy					1,225			1,225	1,325	
	1800 For each added 1" up to 24" high, plain, add					50.50			50.50	55.50	
	1820 Galvanized, add					62.50			62.50	68.50	
	1840 Stainless, add					55.50			55.50	61	
	1860 Plastic, add					55.50			55.50	61	
	1900 For continuous bottom plate, (CHCP), add					243			243	267	
	1940 For upper continuous high chairs, (CHCU), add					243			243	267	
	1960 For galvanized wire runners, add					227			227	250	
	2100 Paper tubing, 4' lengths, for #2 & #3 bar					91			91	100	
	2120 For #6 bar				▼	113			113	125	
	2200 Screed base, 1/2" diameter, 2-1/2" high, plain				C	235			235	259	
	2210 Galvanized					244			244	268	
	2220 5-1/2" high, plain					284			284	310	
	2250 Galvanized					300			300	330	
	2300 3/4" diameter, 2-1/2" high, plain					292			292	320	
	2310 Galvanized					310			310	340	
	2320 5-1/2" high, plain					360			360	395	
	2350 Galvanized					385			385	425	
100	2400 Screed holder, 1/2" diam. for 1" I.D. pipe, plain, 6" long				▼	244			244	268	100

03200 | Concrete Reinforcement

		03210 | Reinforcing Steel	CREW	DAILY OUTPUT	LABOR-HOURS	UNIT	2006 BARE COSTS MAT.	LABOR	EQUIP.	TOTAL	TOTAL INCL O&P	
100	2420	12" long				C	405			405	445	100
	2500	3/4" diameter, for 1-1/2" I.D. pipe, 6" long					440			440	480	
	2520	12" long					720			720	795	
	2700	Screw anchor for bolts, plain, 1/2" diameter					160			160	176	
	2720	1" diameter					480			480	525	
	2740	1-1/2" diameter					790			790	870	
	2800	Screw eye bolts, 1/2" x 5" long					182			182	200	
	2820	1" x 9" long					720			720	795	
	2840	1-1/2" x 14" long					1,750			1,750	1,925	
	2900	Screw anchor bolts, 1/2" x up to 7" long					730			730	805	
	2920	1" x up to 12" long					2,375			2,375	2,600	
	3000	Slab lifting inserts, single, 3/4" dia., galv., 4" high					490			490	540	
	3010	6" high					600			600	660	
	3030	7" high					685			685	755	
	3100	1" diameter, 5" high					775			775	850	
	3120	7" high					815			815	900	
	3200	Double lifting inserts, 1" diameter, 5" high					1,525			1,525	1,675	
	3220	7" high					1,625			1,625	1,775	
	3330	1-1/4" diameter, 5" high					1,675			1,675	1,850	
	3500	Sleeper clips for wood sleepers, 20 ga., galv., 2" wide				M	585			585	645	
	3520	4" wide					720			720	795	
	3600	Spacers, plastic for 1" bar clearance, average					89.50			89.50	98.50	
	3620	For 2" bar clearance, average					108			108	119	
	3800	Subgrade chairs, 1/2" diameter, 3-1/2" high				C	480			480	530	
	3850	12" high					1,350			1,350	1,475	
	3900	3/4" diameter, 3-1/2" high					620			620	680	
	3950	12" high					1,450			1,450	1,600	
	4200	Subgrade stakes, 3/4" diameter, 12" long					500			500	550	
	4250	24" long					670			670	735	
	4300	1" diameter, 12" long					750			750	820	
	4350	24" long					1,100			1,100	1,200	
	4500	Tie wire, 16 ga. annealed steel, under 500 lbs.				Cwt.	144			144	159	
	4520	2,000 to 4,000 lbs.				"	136			136	149	
	4550	Tie wire holder, plastic case				Ea.	56.50			56.50	62.50	
	4600	Aluminum case				"	68.50			68.50	75.50	
200	0010	**COATED REINFORCING** Add to fabricated & delivered price										200
	0100	Epoxy coated, A775				Ton	365			365	400	
	0150	Galvanized, #3					730			730	800	
	0200	#4					730			730	800	
	0250	#5					715			715	785	
	0300	#6 or over					715			715	785	
	1000	For over 20 tons, #6 or larger, minimum					665			665	730	
	1500	Maximum					795			795	875	
600	0010	**REINFORCING IN PLACE** A615 Grade 60, incl. access. labor R032110-10										600
	0102	Beams & Girders, #3 to #7	4 Rodm	3200	.010	Lb.	.44	.40		.84	1.17	
	0152	#8 to #18 R032110-20		5400	.006		.44	.23		.67	.89	
	0202	Columns, #3 to #7		3000	.011		.44	.42		.86	1.21	
	0252	#8 to #18 R032110-25		4600	.007		.44	.27		.71	.96	
	0300	Spirals, hot rolled, 8" to 15" diameter		2.20	14.545	Ton	1,125	575		1,700	2,225	
	0320	15" to 24" diameter R032110-40		2.20	14.545		1,075	575		1,650	2,175	
	0330	24" to 36" diameter		2.30	13.913		1,025	550		1,575	2,075	
	0340	36" to 48" diameter R032110-50		2.40	13.333		975	525		1,500	1,975	
	0360	48" to 64" diameter		2.50	12.800		1,075	505		1,580	2,075	
	0380	64" to 84" diameter R032110-70		2.60	12.308		1,125	485		1,610	2,075	
	0390	84" to 96" diameter		2.70	11.852		1,175	470		1,645	2,100	

03210 | Reinforcing Steel

		03210 Reinforcing Steel		CREW	DAILY OUTPUT	LABOR-HOURS	UNIT	2006 BARE COSTS MAT.	LABOR	EQUIP.	TOTAL	TOTAL INCL O&P	
600	0402	Elevated slabs, #4 to #7	R032110 -10	4 Rodm	5800	.006	Lb.	.47	.22		.69	.88	600
	0502	Footings, #4 to #7			4200	.008		.44	.30		.74	1.01	
	0552	#8 to #18	R032110 -20		7200	.004		.44	.18		.62	.79	
	0602	Slab on grade, #3 to #7			4200	.008		.43	.30		.73	.99	
	0702	Walls, #3 to #7	R032110 -25		6000	.005		.44	.21		.65	.85	
	0752	#8 to #18			8000	.004		.44	.16		.60	.76	
	0900	Use the following for a rough estimate guide	R032110 -40										
	1000	Typical in place, average, under 10 ton job, #3 to #7		4 Rodm	1.80	17.778	Ton	885	700		1,585	2,175	
	1010	#8 to #18	R032110 -50		2.70	11.852		900	470		1,370	1,800	
	1050	10 - 50 ton job, #3 to #7			2.10	15.238		870	600		1,470	1,975	
	1060	#8 to #18	R032110 -70		3	10.667		885	420		1,305	1,700	
	1100	50 - 100 ton job, #3 - #7			2.20	14.545		845	575		1,420	1,925	
	1110	#8 to #18	R032110 -80		3.10	10.323		865	410		1,275	1,650	
	1150	Over 100 ton job, #3 - #7			2.30	13.913		835	550		1,385	1,875	
	1160	#8 - #18			3.20	10		855	395		1,250	1,625	
	1200	High strength steel, Grade 75, #14 bars only, add						62			62	68.50	
	2000	Unloading & sorting, add to above		C-5	100	.560			21.50	6.50	28	43.50	
	2200	Crane cost for handling, add to above, minimum			135	.415			16	4.82	20.82	32.50	
	2210	Average			92	.609			23.50	7.05	30.55	47.50	
	2220	Maximum			35	1.600			61.50	18.60	80.10	125	
	2300	For epoxy coated rebar, add						45%					
	2350	For stainless steel rebar, add						300%					
	2400	Dowels, 2 feet long, deformed, #3		2 Rodm	520	.031	Ea.	.35	1.22		1.57	2.48	
	2410	#4			480	.033		.63	1.32		1.95	2.95	
	2420	#5			435	.037		.98	1.45		2.43	3.57	
	2430	#6			360	.044		1.41	1.76		3.17	4.56	
	2450	Longer and heavier dowels, add			725	.022	Lb.	.47	.87		1.34	2.02	
	2500	Smooth dowels, 12" long, 1/4" or 3/8" diameter			140	.114	Ea.	.72	4.51		5.23	8.55	
	2520	5/8" diameter			125	.128		1.25	5.05		6.30	10.10	
	2530	3/4" diameter			110	.145		1.55	5.75		7.30	11.55	
	2600	Dowel sleeves for CIP concrete, 2-part system											
	2610	Sleeve base, plastic, for #5 bar, fasten to edge form		1 Rodm	200	.040	Ea.	.36	1.58		1.94	3.11	
	2615	Sleeve, plastic, for #5 bar x 9" long, snap onto base			400	.020		1.03	.79		1.82	2.49	
	2620	Sleeve base, for #6 bar			175	.046		.36	1.81		2.17	3.50	
	2625	Sleeve, for #6 bar			350	.023		1.06	.90		1.96	2.72	
	2630	Sleeve base, for #8 bar			150	.053		.44	2.11		2.55	4.10	
	2635	Sleeve, for #8 bar			300	.027		1.18	1.05		2.23	3.11	
	2700	Dowel caps, visual warning only, plastic, #3 to #8		2 Rodm	800	.020		.27	.79		1.06	1.66	
	2720	#7 to #14			750	.021		.35	.84		1.19	1.83	
	2750	Impalement protective, plastic, #3 to #7			800	.020		1.59	.79		2.38	3.11	
	2760	#7 to #11			775	.021		2.21	.82		3.03	3.83	
	2770	#11 to #16			750	.021		2.25	.84		3.09	3.93	
	3000	For epoxy dowel anchoring, see Div 05090-300											
	9000	Minimum labor/equipment charge		1 Rodm	4	2	Job		79		79	136	
700	0010	**SPLICING REINFORCING BARS**	R032110 -70										700
	0020	Including holding bars in place while splicing											
	0100	Butt weld columns #4 bars		C-5	190	.295	Ea.	4.09	11.35	3.42	18.86	27.50	
	0110	#6 bars			150	.373		5.75	14.40	4.34	24.49	35	
	0130	#10 bars			95	.589		14.10	22.50	6.85	43.45	61	
	0150	#14 bars			65	.862		22.50	33	10	65.50	91.50	
	0280	Column splices, bar to bar end bearing											
	0300	#7 or #8 bars		C-5	190	.295	Ea.	37	11.35	3.42	51.77	63.50	
	0310	#9 or #10 bars			170	.329		40.50	12.70	3.83	57.03	70	
	0320	#11 bars			160	.350		49	13.50	4.07	66.57	81	
	0330	#14 bars			150	.373		61.50	14.40	4.34	80.24	96.50	
	0340	#18 bars			140	.400		88.50	15.45	4.65	108.60	129	

	03210 | Reinforcing Steel		CREW	DAILY OUTPUT	LABOR-HOURS	UNIT	2006 BARE COSTS MAT.	LABOR	EQUIP.	TOTAL	TOTAL INCL O&P	
700	0500	Transition bar to bar end bearing, #14 to #18 bar (R032110-70)				Ea.	87			87	96	700
	0520	#14 to #11 bar					67			67	74	
	0540	#11 to #10 bar					50.50			50.50	55.50	
	0550	#10 to #9 bar					42			42	46.50	
	0560	#9 to #8 bar					39			39	43	
	0580	#8 to #7 bar					38			38	42	
	0600	For bolted speed sleeve type, deduct						15%				
	0800	Mechanical butt splice, sleeve type w/ filler metal, compression										
	0810	only, all grades, columns only #11 bars	C-5	68	.824	Ea.	43	32	9.55	84.55	112	
	0900	#14 bars		62	.903		70	35	10.50	115.50	147	
	0920	#18 bars		62	.903		128	35	10.50	173.50	211	
	1000	125% yield point, grade 60, columns only, #6 bars		68	.824		34.50	32	9.55	76.05	102	
	1020	#7 or #8 bars		68	.824		37	32	9.55	78.55	105	
	1030	#9 bars		68	.824		38	32	9.55	79.55	106	
	1040	#10 bars		68	.824		40.50	32	9.55	82.05	109	
	1050	#11 bars		68	.824		49	32	9.55	90.55	118	
	1060	#14 bars		62	.903		61.50	35	10.50	107	138	
	1070	#18 bars		62	.903		88.50	35	10.50	134	168	
	1200	Full tension, grade 60 steel, columns,										
	1220	slabs or beams, #6, #7, #8 bars	C-5	68	.824	Ea.	11.65	32	9.55	53.20	77	
	1230	#9 bars		68	.824		12.70	32	9.55	54.25	78	
	1240	#10 bars		68	.824		14.10	32	9.55	55.65	79.50	
	1250	#11 bars		68	.824		15	32	9.55	56.55	80.50	
	1260	#14 bars		62	.903		22.50	35	10.50	68	94.50	
	1270	#18 bars		62	.903		34	35	10.50	79.50	108	
	1400	If equipment handling not required, deduct						50%				
	1600	Mechanical threaded type, bar threading not included,										
	1700	Straight bars, #10 & #11	C-5	140	.400	Ea.	15	15.45	4.65	35.10	47.50	
	1750	#14 bars		130	.431		22.50	16.60	5	44.10	58	
	1800	#18 bars		75	.747		34	29	8.70	71.70	95.50	
	2100	#11 to #18 & #14 to #18 transition		75	.747		36.50	29	8.70	74.20	98	
	2400	Bent bars, #10 & #11		105	.533		15	20.50	6.20	41.70	58	
	2500	#14		90	.622		25.50	24	7.25	56.75	76.50	
	2600	#18		70	.800		42	31	9.30	82.30	109	
	2800	#11 to #14 transition		75	.747		24.50	29	8.70	62.20	84.50	
	2900	#11 to #18 & #14 to #18 transition		70	.800		40	31	9.30	80.30	106	

	03220 | Welded Wire Fabric		CREW	DAILY OUTPUT	LABOR-HOURS	UNIT	MAT.	LABOR	EQUIP.	TOTAL	TOTAL INCL O&P	
200	0010	**WELDED WIRE FABRIC** ASTM A185 (R032205-30)										200
	0050	Sheets										
	0100	6 x 6 - W1.4 x W1.4 (10 x 10) 21 lb. per C.S.F.	2 Rodm	35	.457	C.S.F.	12	18.05		30.05	44	
	0200	6 x 6 - W2.1 x W2.1 (8 x 8) 30 lb. per C.S.F.		31	.516		14.30	20.50		34.80	51	
	0300	6 x 6 - W2.9 x W2.9 (6 x 6) 42 lb. per C.S.F.		29	.552		17.15	22		39.15	56.50	
	0400	6 x 6 - W4 x W4 (4 x 4) 58 lb. per C.S.F.		27	.593		23.50	23.50		47	65.50	
	0500	4 x 4 - W1.4 x W1.4 (10 x 10) 31 lb. per C.S.F.		31	.516		17.70	20.50		38.20	54.50	
	0600	4 x 4 - W2.1 x W2.1 (8 x 8) 44 lb. per C.S.F.		29	.552		21	22		43	60.50	
	0650	4 x 4 - W2.9 x W2.9 (6 x 6) 61 lb. per C.S.F.		27	.593		27.50	23.50		51	70.50	
	0700	4 x 4 - W4 x W4 (4 x 4) 85 lb. per C.S.F.		25	.640		39	25.50		64.50	86	
	0750	Rolls										
	0800	2 x 2 - #14 galv., 21 lb/C.S.F., beam & column wrap	2 Rodm	6.50	2.462	C.S.F.	30.50	97		127.50	201	
	0900	2 x 2 - #12 galv. for gunite reinforcing	"	6.50	2.462	"	35	97		132	206	
	9000	Minimum labor/equipment charge	1 Rodm	4	2	Job		79		79	136	

	03230 | Stressing Tendons		CREW	DAILY OUTPUT	LABOR-HOURS	UNIT	MAT.	LABOR	EQUIP.	TOTAL	TOTAL INCL O&P	
600	0010	**PRESTRESSING STEEL** (R034136-90)										600
	0100	Grouted strand, post-tensioned in field, 50' span, 100 kip	C-3	1200	.053	Lb.	1.44	1.93	.09	3.46	4.94	

03200 | Concrete Reinforcement

	03230 | Stressing Tendons		CREW	DAILY OUTPUT	LABOR-HOURS	UNIT	2006 BARE COSTS MAT.	LABOR	EQUIP.	TOTAL	TOTAL INCL O&P	
600	0150	300 kip R034136-90	C-3	2700	.024	Lb.	.71	.86	.04	1.61	2.28	600
	0300	100′ span, 100 kip		1700	.038		1.44	1.36	.06	2.86	3.95	
	0350	300 kip		3200	.020		1.26	.72	.03	2.01	2.65	
	0500	200′ span, 100 kip		2700	.024		1.43	.86	.04	2.33	3.07	
	0550	300 kip		3500	.018		1.26	.66	.03	1.95	2.53	
	0800	Grouted bars, 50′ span, 42 kip		2600	.025		.70	.89	.04	1.63	2.32	
	0850	143 kip		3200	.020		.68	.72	.03	1.43	2	
	1000	75′ span, 42 kip		3200	.020		.70	.72	.03	1.45	2.03	
	1050	143 kip		4200	.015		.62	.55	.03	1.20	1.64	
	1200	Ungrouted strand, 50′ span, 100 kip	C-4	1275	.025		.47	1	.02	1.49	2.27	
	1250	300 kip		1475	.022		.47	.87	.02	1.36	2.03	
	1400	100′ span, 100 kip		1500	.021		.47	.85	.02	1.34	2	
	1450	300 kip		1650	.019		.47	.78	.02	1.27	1.87	
	1600	200′ span, 100 kip		1500	.021		.47	.85	.02	1.34	2	
	1650	300 kip		1700	.019		.47	.75	.02	1.24	1.83	
	1800	Ungrouted bars, 50′ span, 42 kip		1400	.023		.43	.91	.02	1.36	2.06	
	1850	143 kip		1700	.019		.43	.75	.02	1.20	1.78	
	2000	75′ span, 42 kip		1800	.018		.43	.71	.02	1.16	1.71	
	2050	143 kip		2200	.015		.43	.58	.01	1.02	1.49	
	2220	Ungrouted single strand, 100′ slab, 25 kip		1200	.027		.47	1.07	.03	1.57	2.38	
	2250	35 kip		1475	.022		.47	.87	.02	1.36	2.03	

	03240 | Fibrous Reinforcing		CREW	DAILY OUTPUT	LABOR-HOURS	UNIT	MAT.	LABOR	EQUIP.	TOTAL	TOTAL INCL O&P	
300	0010	**FIBROUS REINFORCING**										300
	0100	Synthetic fibers, add to concrete				Lb.	3.97			3.97	4.37	
	0110	1-1/2 lb. per C.Y.				C.Y.	6.15			6.15	6.75	
	0150	Steel fibers, add to concrete				Lb.	.46			.46	.51	
	0155	25 lb. per C.Y.				C.Y.	11.50			11.50	12.65	
	0160	50 lb. per C.Y.					23			23	25.50	
	0170	75 lb. per C.Y.					35.50			35.50	39	
	0180	100 lb. per C.Y.					46			46	50.50	

03300 | Cast-In-Place Concrete

	03310 | Structural Concrete		CREW	DAILY OUTPUT	LABOR-HOURS	UNIT	2006 BARE COSTS MAT.	LABOR	EQUIP.	TOTAL	TOTAL INCL O&P	
200	0010	**CONCRETE, FIELD MIX** R033105-65										200
	0015	FOB forms 2250 psi				C.Y.	78			78	85.50	
	0020	3000 psi				"	81			81	89	
220	0010	**CONCRETE, READY MIX** Normal weight R033105-10										220
	0012	Includes local aggregate, sand, portland cement, and water										
	0015	Excludes all additives and treatments R033105-20										
	0020	2000 psi				C.Y.	83.50			83.50	92	
	0100	2500 psi R033105-30					85			85	93.50	
	0150	3000 psi					87			87	95.50	
	0200	3500 psi R033105-40					89			89	98	
	0300	4000 psi					91			91	100	
	0350	4500 psi R033105-50					93			93	102	
	0400	5000 psi					96			96	106	

	03310 | Structural Concrete			CREW	DAILY OUTPUT	LABOR-HOURS	UNIT	2006 BARE COSTS MAT.	LABOR	EQUIP.	TOTAL	TOTAL INCL O&P	
220	0411	6000 psi	R033105-10				C.Y.	109			109	120	220
	0412	8000 psi						179			179	196	
	0413	10,000 psi	R033105-20					253			253	279	
	0414	12,000 psi						305			305	335	
	0700	Lightweight, 110#/CF	R033105-30										
	1000	For high early strength cement, add					C.Y.	10%					
	1010	For structural lightweight with regular sand, add	R033105-40					25%					
	2000	For all lightweight aggregate, add						45%					
	3000	For integral colors, 2500 psi (5 bag mix)	R033105-50										
	3100	Red, yellow or brown, 1.8 lb. per bag, add					C.Y.	21			21	23	
	3200	9.4 lb. per bag, add						109			109	120	
	3400	Black, 1.8 lb. per bag, add						28			28	31	
	3500	7.5 lb. per bag, add						117			117	129	
	3700	Green, 1.8 lb. per bag, add						35.50			35.50	39	
	3800	7.5 lb. per bag, add						148			148	163	
	4000	Flowable fill: ash, cement, aggregate, water											
	4100	40 - 80 psi					C.Y.	69			69	76	
	4150	Structural: ash, cement, aggregate, water & sand											
	4200	50 psi					C.Y.	73.50			73.50	80.50	
	4250	140 psi						76.50			76.50	84	
	4300	500 psi						79.50			79.50	87.50	
	4350	1000 psi						84			84	92.50	
240	0010	**CONCRETE IN PLACE**	R033053-10										240
	0020	Including forms (4 uses), concrete, placement, reinforcing											
	0050	steel and finishing unless otherwise indicated	R033053-50										
	0300	Beams, 5 kip per L.F., 10′ span		C-14A	15.62	12.804	C.Y.	287	455	46	788	1,125	
	0350	25′ span	R033053-60	"	18.55	10.782		298	385	39	722	1,000	
	0500	Chimney foundations, industrial, minimum		C-14C	32.22	3.476		129	118	.66	247.66	340	
	0510	Maximum	R033105-80	"	23.71	4.724		152	160	.90	312.90	435	
	0700	Columns, square, 12″ x 12″, minimum reinforcing		C-14A	11.96	16.722		305	595	60.50	960.50	1,400	
	0720	Average reinforcing	R033105-85		10.13	19.743		485	705	71.50	1,261.50	1,775	
	0740	Maximum reinforcing			9.03	22.148		725	790	80	1,595	2,200	
	0800	16″ x 16″, minimum reinforcing			16.22	12.330		243	440	44.50	727.50	1,050	
	0820	Average reinforcing			12.57	15.911		410	565	57.50	1,032.50	1,450	
	0840	Maximum reinforcing			10.25	19.512		640	695	70.50	1,405.50	1,925	
	0900	24″ x 24″, minimum reinforcing			23.66	8.453		207	300	30.50	537.50	760	
	0920	Average reinforcing			17.71	11.293		370	400	41	811	1,125	
	0940	Maximum reinforcing			14.15	14.134		585	505	51	1,141	1,525	
	1000	36″ x 36″, minimum reinforcing			33.69	5.936		182	211	21.50	414.50	575	
	1020	Average reinforcing			23.32	8.576		325	305	31	661	895	
	1040	Maximum reinforcing			17.82	11.223		545	400	40.50	985.50	1,300	
	1200	16″ diameter, minimum reinforcing			31.49	6.351		236	226	23	485	660	
	1220	Average reinforcing			19.12	10.460		415	370	38	823	1,125	
	1240	Maximum reinforcing			13.77	14.524		630	515	52.50	1,197.50	1,600	
	1300	20″ diameter, minimum reinforcing			41.04	4.873		238	174	17.60	429.60	570	
	1320	Average reinforcing			24.05	8.316		400	296	30	726	965	
	1340	Maximum reinforcing			17.01	11.758		630	420	42.50	1,092.50	1,425	
	1400	24″ diameter, minimum reinforcing			51.85	3.857		223	137	13.90	373.90	490	
	1420	Average reinforcing			27.06	7.391		400	263	26.50	689.50	905	
	1440	Maximum reinforcing			18.29	10.935		620	390	39.50	1,049.50	1,375	
	1500	36″ diameter, minimum reinforcing			75.04	2.665		224	95	9.60	328.60	415	
	1520	Average reinforcing			37.49	5.335		380	190	19.25	589.25	755	
	1540	Maximum reinforcing			22.84	8.757		600	310	31.50	941.50	1,225	
	1900	Elevated slabs, flat slab with drops, 125 psf Sup. Load, 20′ span		C-14B	38.45	5.410		242	192	18.75	452.75	610	
	1950	30′ span			50.99	4.079		250	145	14.15	409.15	530	
	2100	Flat plate, 125 psf Sup. Load, 15′ span			30.24	6.878		220	245	24	489	675	

03310 | Structural Concrete

240

	03310 Structural Concrete	CREW	DAILY OUTPUT	LABOR-HOURS	UNIT	2006 BARE COSTS MAT.	LABOR	EQUIP.	TOTAL	TOTAL INCL O&P
2150	25′ span (R033053-10)	C-14B	49.60	4.194	C.Y.	226	149	14.55	389.55	510
2300	Waffle const., 30″ domes, 125 psf Sup. Load, 20′ span		37.07	5.611		330	200	19.45	549.45	715
2350	30′ span (R033053-50)		44.07	4.720		294	168	16.40	478.40	620
2500	One way joists, 30″ pans, 125 psf Sup. Load, 15′ span		27.38	7.597		410	270	26.50	706.50	930
2550	25′ span (R033053-60)		31.15	6.677		375	237	23	635	830
2700	One way beam & slab, 125 psf Sup. Load, 15′ span		20.59	10.102		245	360	35	640	905
2750	25′ span (R033105-80)		28.36	7.334		225	261	25.50	511.50	705
2900	Two way beam & slab, 125 psf Sup. Load, 15′ span		24.04	8.652		232	310	30	572	800
2950	25′ span (R033105-85)	↓	35.87	5.799	↓	196	206	20	422	575
3100	Elevated slabs including finish, not									
3110	including forms or reinforcing									
3150	Regular concrete, 4″ slab	C-8	2613	.021	S.F.	1.18	.66	.27	2.11	2.67
3200	6″ slab		2585	.022		1.74	.67	.27	2.68	3.31
3250	2-1/2″ thick floor fill		2685	.021		.76	.65	.26	1.67	2.17
3300	Lightweight, 110# per C.F., 2-1/2″ thick floor fill		2585	.022		1.04	.67	.27	1.98	2.54
3400	Cellular concrete, 1-5/8″ fill, under 5000 S.F.		2000	.028		.70	.87	.35	1.92	2.57
3450	Over 10,000 S.F.		2200	.025		.66	.79	.32	1.77	2.36
3500	Add per floor for 3 to 6 stories high		31800	.002			.05	.02	.07	.11
3520	For 7 to 20 stories high	↓	21200	.003	↓		.08	.03	.11	.17
3800	Footings, spread under 1 C.Y.	C-14C	38.07	2.942	C.Y.	175	99.50	.56	275.06	360
3850	Over 5 C.Y. 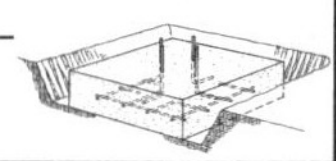		81.04	1.382		242	47	.26	289.26	345
3900	Footings, strip, 18″ x 9″, unreinforced		40	2.800		109	95	.53	204.53	278
3920	18″ x 9″, reinforced		35	3.200		130	108	.61	238.61	325
3925	20″ x 10″, unreinforced		45	2.489		106	84.50	.47	190.97	257
3930	20″ x 10″, reinforced		40	2.800		122	95	.53	217.53	293
3935	24″ x 12″, unreinforced		55	2.036		103	69	.39	172.39	228
3940	24″ x 12″, reinforced		48	2.333		121	79	.44	200.44	264
3945	36″ x 12″, unreinforced		70	1.600		99.50	54	.30	153.80	199
3950	36″ x 12″, reinforced		60	1.867		115	63	.35	178.35	231
4000	Foundation mat, under 10 C.Y.		38.67	2.896		179	98	.55	277.55	360
4050	Over 20 C.Y.	↓	56.40	1.986		156	67	.38	223.38	283
4200	Grade walls, 8″ thick, 8′ high	C-14D	45.83	4.364		157	154	15.75	326.75	445
4250	14′ high		27.26	7.337		203	259	26.50	488.50	680
4260	12″ thick, 8′ high		64.32	3.109		140	110	11.25	261.25	350
4270	14′ high		40.01	4.999		159	176	18.05	353.05	485
4300	15″ thick, 8′ high		80.02	2.499		132	88	9	229	300
4350	12′ high		51.26	3.902		140	138	14.10	292.10	400
4500	18′ high	↓	48.85	4.094	↓	155	144	14.80	313.80	425
4520	Handicap access ramp, railing both sides, 3′ wide	C-14H	14.58	3.292	L.F.	204	115	1.48	320.48	415
4525	5′ wide		12.22	3.928		212	137	1.77	350.77	460
4530	With 6″ curb and rails both sides, 3′ wide		8.55	5.614		211	196	2.53	409.53	560
4535	5′ wide	↓	7.31	6.566	↓	215	230	2.95	447.95	620
4650	Slab on grade, not including finish, 4″ thick	C-14E	60.75	1.449	C.Y.	106	50.50	.35	156.85	202
4700	6″ thick	″	92	.957	″	102	33.50	.23	135.73	168
4751	Slab on grade, incl. troweled finish, not incl. forms									
4760	or reinforcing, over 10,000 S.F., 4″ thick	C-14F	3425	.021	S.F.	1.16	.68	.01	1.85	2.37
4820	6″ thick		3350	.021		1.69	.69	.01	2.39	2.98
4840	8″ thick		3184	.023		2.31	.73	.01	3.05	3.73
4900	12″ thick		2734	.026		3.47	.85	.01	4.33	5.20
4950	15″ thick	↓	2505	.029	↓	4.36	.93	.01	5.30	6.30
5000	Slab on grade, incl. textured finish, not incl. forms									
5001	or reinforcing, 4″ thick	C-14G	2873	.019	S.F.	1.13	.62	.01	1.76	2.24
5010	6″ thick		2590	.022		1.77	.69	.01	2.47	3.06
5020	8″ thick	↓	2320	.024	↓	2.31	.77	.01	3.09	3.78
5200	Lift slab in place above the foundation, incl. forms,									
5210	reinforcing, concrete and columns, minimum	C-14B	2113	.098	S.F.	5.35	3.50	.34	9.19	12.05

03310 | Structural Concrete

			Crew	Daily Output	Labor-Hours	Unit	2006 Bare Costs Mat.	Labor	Equip.	Total	Total Incl O&P	
240	5250	Average (R033053-10)	C-14B	1650	.126	S.F.	5.80	4.48	.44	10.72	14.30	240
	5300	Maximum		1500	.139		6.25	4.93	.48	11.66	15.65	
	5500	Lightweight, ready mix, including screed finish only, (R033053-50)										
	5510	not including forms or reinforcing										
	5550	1:4 for structural roof decks (R033053-60)	C-14B	260	.800	C.Y.	122	28.50	2.78	153.28	184	
	5600	1:6 for ground slab with radiant heat	C-14F	92	.783		123	25.50	.23	148.73	176	
	5650	1:3:2 with sand aggregate, roof deck (R033105-80)	C-14B	260	.800		120	28.50	2.78	151.28	182	
	5700	Ground slab	C-14F	107	.673		120	21.50	.20	141.70	167	
	5900	Pile caps, incl. forms and reinf., sq. or rect., under 5 C.Y. (R033105-85)	C-14C	54.14	2.069		151	70	.39	221.39	282	
	5950	Over 10 C.Y.		75	1.493		139	50.50	.28	189.78	237	
	6000	Triangular or hexagonal, under 5 C.Y.		53	2.113		107	71.50	.40	178.90	236	
	6050	Over 10 C.Y.		85	1.318		120	44.50	.25	164.75	206	
	6200	Retaining walls, gravity, 4′ high see division 02830-100	C-14D	66.20	3.021		125	107	10.90	242.90	325	
	6250	10′ high		125	1.600		117	56.50	5.80	179.30	229	
	6300	Cantilever, level backfill loading, 8′ high		70	2.857		136	101	10.30	247.30	330	
	6350	16′ high		91	2.198		130	77.50	7.95	215.45	280	
	6800	Stairs, not including safety treads, free standing, 3′-6″ wide	C-14H	83	.578	LF Nose	5.30	20	.26	25.56	39.50	
	6850	Cast on ground		125	.384	"	4.03	13.45	.17	17.65	26.50	
	7000	Stair landings, free standing		200	.240	S.F.	4.36	8.40	.11	12.87	18.80	
	7050	Cast on ground		475	.101	"	3.05	3.54	.05	6.64	9.25	
	9000	Minimum labor/equipment charge	2 Carp	1	16	Job		570		570	940	
450	0010	**INSULATING CONCRETE** See division 03520-250										450
700	0010	**PLACING CONCRETE** (R033105-70)										700
	0020	Includes labor and equipment to place and vibrate										
	0050	Beams, elevated, small beams, pumped	C-20	60	1.067	C.Y.		31.50	12.40	43.90	65	
	0100	With crane and bucket	C-7	45	1.600			48	22.50	70.50	103	
	0200	Large beams, pumped	C-20	90	.711			21	8.25	29.25	43.50	
	0250	With crane and bucket	C-7	65	1.108			33	15.55	48.55	71	
	0400	Columns, square or round, 12″ thick, pumped	C-20	60	1.067			31.50	12.40	43.90	65	
	0450	With crane and bucket	C-7	40	1.800			54	25.50	79.50	116	
	0600	18″ thick, pumped	C-20	90	.711			21	8.25	29.25	43.50	
	0650	With crane and bucket	C-7	55	1.309			39.50	18.40	57.90	84	
	0800	24″ thick, pumped	C-20	92	.696			20.50	8.10	28.60	42.50	
	0850	With crane and bucket	C-7	70	1.029			31	14.45	45.45	66	
	1000	36″ thick, pumped	C-20	140	.457			13.55	5.30	18.85	28	
	1050	With crane and bucket	C-7	100	.720			21.50	10.10	31.60	46	
	1400	Elevated slabs, less than 6″ thick, pumped	C-20	140	.457			13.55	5.30	18.85	28	
	1450	With crane and bucket	C-7	95	.758			22.50	10.65	33.15	48.50	
	1500	6″ to 10″ thick, pumped	C-20	160	.400			11.90	4.65	16.55	24.50	
	1550	With crane and bucket	C-7	110	.655			19.65	9.20	28.85	42	
	1600	Slabs over 10″ thick, pumped	C-20	180	.356			10.55	4.13	14.68	22	
	1650	With crane and bucket	C-7	130	.554			16.60	7.80	24.40	35.50	
	1900	Footings, continuous, shallow, direct chute	C-6	120	.400			11.55	.36	11.91	19.35	
	1950	Pumped	C-20	150	.427			12.65	4.96	17.61	26	
	2000	With crane and bucket	C-7	90	.800			24	11.25	35.25	51.50	
	2100	Footings, continuous, deep, direct chute	C-6	140	.343			9.90	.31	10.21	16.60	
	2150	Pumped	C-20	160	.400			11.90	4.65	16.55	24.50	
	2200	With crane and bucket	C-7	110	.655			19.65	9.20	28.85	42	
	2400	Footings, spread, under 1 C.Y., direct chute	C-6	55	.873			25	.78	25.78	42.50	
	2450	Pumped	C-20	65	.985			29	11.45	40.45	60	
	2500	With crane and bucket	C-7	45	1.600			48	22.50	70.50	103	
	2600	Over 5 C.Y., direct chute	C-6	120	.400			11.55	.36	11.91	19.35	
	2650	Pumped	C-20	150	.427			12.65	4.96	17.61	26	
	2700	With crane and bucket	C-7	100	.720			21.50	10.10	31.60	46	

03300 | Cast-In-Place Concrete

03310 | Structural Concrete

			CREW	DAILY OUTPUT	LABOR-HOURS	UNIT	2006 BARE COSTS MAT.	LABOR	EQUIP.	TOTAL	TOTAL INCL O&P	
700	2900	Foundation mats, over 20 C.Y., direct chute (R033105-70)	C-6	350	.137	C.Y.		3.96	.12	4.08	6.65	700
	2950	Pumped	C-20	400	.160			4.75	1.86	6.61	9.80	
	3000	With crane and bucket	C-7	300	.240			7.20	3.37	10.57	15.40	
	3200	Grade beams, direct chute	C-6	150	.320			9.25	.28	9.53	15.45	
	3250	Pumped	C-20	180	.356			10.55	4.13	14.68	22	
	3300	With crane and bucket	C-7	120	.600			18	8.40	26.40	39	
	3500	High rise, for more than 5 stories, pumped, add per story	C-20	2100	.030			.91	.35	1.26	1.87	
	3510	With crane and bucket, add per story	C-7	2100	.034			1.03	.48	1.51	2.20	
	3700	Pile caps, under 5 C.Y., direct chute	C-6	90	.533			15.40	.47	15.87	26	
	3750	Pumped	C-20	110	.582			17.25	6.75	24	35.50	
	3800	With crane and bucket	C-7	80	.900			27	12.65	39.65	58	
	3850	Pile cap, 5 C.Y. to 10 C.Y., direct chute	C-6	175	.274			7.95	.24	8.19	13.25	
	3900	Pumped	C-20	200	.320			9.50	3.72	13.22	19.60	
	3950	With crane and bucket	C-7	150	.480			14.40	6.75	21.15	31	
	4000	Over 10 C.Y., direct chute	C-6	215	.223			6.45	.20	6.65	10.80	
	4050	Pumped	C-20	240	.267			7.90	3.10	11	16.35	
	4100	With crane and bucket	C-7	185	.389			11.65	5.45	17.10	25	
	4300	Slab on grade, 4" thick, direct chute	C-6	110	.436			12.60	.39	12.99	21	
	4350	Pumped	C-20	130	.492			14.60	5.70	20.30	30.50	
	4400	With crane and bucket	C-7	110	.655			19.65	9.20	28.85	42	
	4600	Over 6" thick, direct chute	C-6	165	.291			8.40	.26	8.66	14.10	
	4650	Pumped	C-20	185	.346			10.25	4.02	14.27	21	
	4700	With crane and bucket	C-7	145	.497			14.90	6.95	21.85	31.50	
	4900	Walls, 8" thick, direct chute	C-6	90	.533			15.40	.47	15.87	26	
	4950	Pumped	C-20	100	.640			19	7.45	26.45	39	
	5000	With crane and bucket	C-7	80	.900			27	12.65	39.65	58	
	5050	12" thick, direct chute	C-6	100	.480			13.85	.43	14.28	23.50	
	5100	Pumped	C-20	110	.582			17.25	6.75	24	35.50	
	5200	With crane and bucket	C-7	90	.800			24	11.25	35.25	51.50	
	5300	15" thick, direct chute	C-6	105	.457			13.20	.41	13.61	22	
	5350	Pumped	C-20	120	.533			15.85	6.20	22.05	33	
	5400	With crane and bucket	C-7	95	.758			22.50	10.65	33.15	48.50	
	5600	Wheeled concrete dumping, add to placing costs above										
	5610	Walking cart, 50' haul, add	C-18	32	.281	C.Y.		7.75	1.63	9.38	14.65	
	5620	150' haul, add		24	.375			10.35	2.18	12.53	19.55	
	5700	250' haul, add		18	.500			13.80	2.90	16.70	26	
	5800	Riding cart, 50' haul, add	C-19	80	.113			3.11	.93	4.04	6.15	
	5810	150' haul, add		60	.150			4.14	1.24	5.38	8.20	
	5900	250' haul, add		45	.200			5.50	1.65	7.15	10.95	
	9000	Minimum labor/equipment charge	C-6	2	24	Job		695	21.50	716.50	1,175	

03350 | Concrete Finishing

			CREW	DAILY OUTPUT	LABOR-HOURS	UNIT	MAT.	LABOR	EQUIP.	TOTAL	TOTAL INCL O&P	
300	0010	FINISHING FLOORS										300
	0020	Monolithic, screed finish	1 Cefi	900	.009	S.F.		.31		.31	.48	
	0100	Screed and bull float (darby) finish		725	.011			.38		.38	.60	
	0150	Screed, float, and broom finish		630	.013			.44		.44	.69	
	0200	Screed, float, and hand trowel		600	.013			.46		.46	.72	
	0250	Machine trowel		550	.015			.50		.50	.79	
	0370	Minimum labor/equipment charge		4	2	Job		69		69	108	
	0400	Integral topping and finish, using 1:1:2 mix, 3/16" thick	C-10B	1000	.040	S.F.	.06	1.21	.18	1.45	2.22	
	0450	1/2" thick		950	.042		.17	1.27	.19	1.63	2.45	
	0500	3/4" thick		850	.047		.26	1.42	.21	1.89	2.82	
	0600	1" thick		750	.053		.35	1.61	.23	2.19	3.25	
	0800	Granolithic topping, laid after, 1:1:1-1/2 mix, 1/2" thick		590	.068		.19	2.05	.30	2.54	3.86	
	0820	3/4" thick		580	.069		.29	2.08	.30	2.67	4.02	
	0850	1" thick		575	.070		.39	2.10	.31	2.80	4.17	

03350 | Concrete Finishing

		03350 Concrete Finishing	CREW	DAILY OUTPUT	LABOR-HOURS	UNIT	2006 BARE COSTS MAT.	LABOR	EQUIP.	TOTAL	TOTAL INCL O&P
300	0950	2″ thick	C-10B	500	.080	S.F.	.77	2.42	.35	3.54	5.15
	1200	Heavy duty, 1:1:2, 3/4″ thick, preshrunk, gray, 20 MSF		320	.125		.26	3.78	.55	4.59	7
	1300	100 MSF	↓	380	.105		.26	3.18	.46	3.90	5.95
	1600	Exposed local aggregate finish, minimum	1 Cefi	625	.013		.16	.44		.60	.87
	1650	Maximum		465	.017		.24	.59		.83	1.19
	1800	Floor abrasives, .25 psf, aluminum oxide		850	.009		.32	.32		.64	.86
	1850	Silicon carbide		850	.009		.44	.32		.76	1
	2000	Floor hardeners, metallic, light service, .50 psf, add		850	.009		.38	.32		.70	.93
	2050	Medium service, .75 psf		750	.011		.57	.37		.94	1.20
	2100	Heavy service, 1.0 psf		650	.012		.75	.42		1.17	1.50
	2150	Extra heavy, 1.5 psf		575	.014		1.13	.48		1.61	2
	2300	Non-metallic, light service, .50 psf		850	.009		.13	.32		.45	.65
	2350	Medium service, .75 psf		750	.011		.19	.37		.56	.79
	2400	Heavy service, 1.00 psf		650	.012		.25	.42		.67	.95
	2450	Extra heavy, 1.50 psf	↓	575	.014	↓	.38	.48		.86	1.17
	2800	Trap rock wearing surface for monolithic floors									
	2810	2.0 psf	C-10B	1250	.032	S.F.	.02	.97	.14	1.13	1.74
	3000	Floor coloring, dusted on, minimum (0.6 psf), add to above	1 Cefi	1300	.006		.60	.21		.81	.99
	3050	Maximum (1.0 psf), add to above	″	625	.013	↓	1	.44		1.44	1.79
	3100	Colored powder only				Lb.	1			1	1.10
	3600	1/2″ topping using 0.6 psf powdered color	C-10B	590	.068	S.F.	4.76	2.05	.30	7.11	8.90
	3650	1/2″ topping using 1.0 psf powdered color	″	590	.068		5.15	2.05	.30	7.50	9.35
	3800	Dustproofing, solvent-based, 1 coat	1 Cefi	1900	.004		.24	.14		.38	.49
	3850	2 coats		1300	.006		.86	.21		1.07	1.28
	4000	Epoxy-based, 1 coat		1500	.005		.15	.18		.33	.45
	4050	2 coats		1500	.005		.29	.18		.47	.61
	4400	Stair finish, float		275	.029			1		1	1.58
	4500	Steel trowel finish		200	.040			1.38		1.38	2.17
	4600	Silicon carbide finish, .25 psf	↓	150	.053	↓	.32	1.83		2.15	3.24
	9100	Minimum labor/equipment charge	C-10	2	12	Job		385		385	615
325	0010	**CONTROL JOINTS, SAW CUT**									
	0100	Sawcut in green concrete									
	0120	1″ depth	C-27	2000	.008	L.F.	.08	.28	.05	.41	.58
	0140	1-1/2″ depth		1800	.009		.13	.31	.06	.50	.69
	0160	2″ depth	↓	1600	.010		.17	.34	.07	.58	.79
	0200	Clean out control joint of debris	C-28	6000	.001	↓		.05		.05	.07
	0300	Joint sealant									
	0320	Backer rod, polyethylene, 1/4″ diameter	1 Cefi	460	.017	L.F.	.02	.60		.62	.96
	0340	Sealant, polyurethane									
	0360	1/4″ x 1/4″ (308 LF/Gal)	1 Cefi	270	.030	L.F.	.16	1.02		1.18	1.79
	0380	1/4″ x 1/2″ (154 LF/Gal)	″	255	.031	″	.32	1.08		1.40	2.05
350	0010	**FINISHING WALLS**									
	0020	Break ties and patch voids	1 Cefi	540	.015	S.F.	.03	.51		.54	.83
	0050	Burlap rub with grout		450	.018		.03	.61		.64	.99
	0100	Carborundum rub, dry		270	.030			1.02		1.02	1.61
	0150	Wet rub	↓	175	.046			1.57		1.57	2.48
	0300	Bush hammer, green concrete	B-39	1000	.048			1.39	.15	1.54	2.45
	0350	Cured concrete	″	650	.074			2.14	.23	2.37	3.77
	0600	Float finish, 1/16″ thick	1 Cefi	300	.027		.24	.92		1.16	1.71
	0700	Sandblast, light penetration	C-10	1100	.022		.24	.70		.94	1.39
	0750	Heavy penetration		375	.064	↓	.48	2.05		2.53	3.81
	9000	Minimum labor/equipment charge	↓	2	12	Job		385		385	615
600	0010	**SLAB TEXTURE STAMPING**									
	0020	Approx. 3 S.F.- 5 S.F. each, buy, minimum				Ea.	99			99	109

03300 | Cast-In-Place Concrete

03350 | Concrete Finishing

			CREW	DAILY OUTPUT	LABOR-HOURS	UNIT	2006 BARE COSTS MAT.	LABOR	EQUIP.	TOTAL	TOTAL INCL O&P	
600	0030	Average				Ea.	139			139	153	600
	0120	Maximum					169			169	186	
	0200	Commonly used chemicals for texture systems										
	0210	Hardener, colored powder				S.F.	.51			.51	.56	
	0220	Release agent, colored powder					.08			.08	.09	
	0230	Curing & sealing compound, solvent based					.05			.05	.05	
	0300	Broadcasting hardener & release agent, stamping	2 Cefi	1000	.016			.55		.55	.87	

03370 | Specially Placed Concrete

			CREW	DAILY OUTPUT	LABOR-HOURS	UNIT	MAT.	LABOR	EQUIP.	TOTAL	TOTAL INCL O&P	
300	0010	**GUNITE (DRY-MIX)**										300
	0020	Applied in 1″ layers, no mesh included	C-8	2000	.028	S.F.	.27	.87	.35	1.49	2.09	
	0100	Mesh for gunite 2 x 2, #12, to 3″ thick	2 Rodm	800	.020		.35	.79		1.14	1.75	
	0150	Over 3″ thick	″	500	.032		.35	1.26		1.61	2.56	
	0300	Typical in place, including mesh, 2″ thick, minimum	C-16	1000	.072		.88	2.37	.70	3.95	5.65	
	0350	Maximum		500	.144		.88	4.74	1.40	7.02	10.30	
	0500	4″ thick, minimum		750	.096		1.41	3.16	.93	5.50	7.80	
	0550	Maximum		350	.206		1.41	6.75	2	10.16	14.90	
	0900	Prepare old walls, no scaffolding, minimum	C-10	1000	.024			.77		.77	1.23	
	0950	Maximum	″	275	.087			2.80		2.80	4.47	
	1100	For high finish requirement or close tolerance, add, minimum						50%				
	1150	Maximum						110%				
	9000	Minimum labor/equipment charge	C-10	1	24	Job		770		770	1,225	

03390 | Concrete Curing

			CREW	DAILY OUTPUT	LABOR-HOURS	UNIT	MAT.	LABOR	EQUIP.	TOTAL	TOTAL INCL O&P	
200	0010	**WATER CURING**										200
	0015	With burlap, 4 uses assumed, 7.5 oz.	2 Clab	55	.291	C.S.F.	7.05	7.95		15	21	
	0100	10 oz.		55	.291		12.65	7.95		20.60	27	
	0200	Curing blanket, burlap/poly, 2-ply		70	.229		15.05	6.25		21.30	27	
	0300	Sprayed membrane curing compound		95	.168		5.55	4.61		10.16	13.75	
	0400	Curing blankets, 1″ to 2″ thick, buy, minimum				S.F.	.27			.27	.30	
	0450	Maximum					.41			.41	.45	
	0500	Electrically heated pads, 110 volts, 15 watts per S.F., buy					5.40			5.40	5.90	
	0600	20 watts per S.F., buy					7.15			7.15	7.85	
	0710	Electrically, heated pads, 15 watts/S.F., 20 uses, minimum					.19			.19	.21	
	0800	Maximum					.32			.32	.35	
	9000	Minimum labor/equipment charge	1 Clab	5	1.600	Job		44		44	72.50	

03400 | Precast Concrete

03410 | Plant-Precast Structural Concrete

			CREW	DAILY OUTPUT	LABOR-HOURS	UNIT	2006 BARE COSTS MAT.	LABOR	EQUIP.	TOTAL	TOTAL INCL O&P	
100	0010	**PRECAST BEAMS** R034105-30										100
	0011	L-shaped, 20′ span, 12″ x 20″	C-11	32	2.250	Ea.	1,050	88	48	1,186	1,375	
	1000	Inverted tee beams, add to above, small beams					15%					
	1050	Large beams					20%					
	1200	Rectangular, 20′ span, 12″ x 20″	C-11	32	2.250		780	88	48	916	1,075	
	1250	18″ x 36″		24	3		885	117	64	1,066	1,250	
	1300	24″ x 44″		22	3.273		1,100	128	70	1,298	1,525	
	1400	30′ span, 12″ x 36″		24	3		1,200	117	64	1,381	1,575	

	03410 | Plant-Precast Structural Concrete		Ref.	Crew	Daily Output	Labor-Hours	Unit	2006 Bare Costs Mat.	Labor	Equip.	Total	Total Incl O&P
100	1450	18″ x 44″	R034105-30	C-11	20	3.600	Ea.	1,500	141	77	1,718	2,000
	1500	24″ x 52″			16	4.500		1,925	176	96	2,197	2,550
	1600	40′ span, 12″ x 52″			20	3.600		2,200	141	77	2,418	2,775
	1650	18″ x 52″			16	4.500		2,625	176	96	2,897	3,325
	1700	24″ x 52″			12	6		2,875	235	128	3,238	3,750
	2000	"T" shaped, 20′ span, 12″ x 20″			32	2.250		1,325	88	48	1,461	1,675
	2050	18″ x 36″			24	3		1,500	117	64	1,681	1,925
	2100	24″ x 44″			22	3.273		1,875	128	70	2,073	2,375
	2200	30′ span, 12″ x 36″			24	3		2,025	117	64	2,206	2,500
	2250	18″ x 44″			20	3.600		2,550	141	77	2,768	3,150
	2300	24″ x 52″			16	4.500		3,275	176	96	3,547	4,025
	2500	40′ span, 12″ x 52″			20	3.600		3,750	141	77	3,968	4,450
	2550	18″ x 52″			16	4.500		4,475	176	96	4,747	5,350
	2600	24″ x 52″			12	6		4,900	235	128	5,263	5,975
210	0010	**PRECAST COLUMNS**	R034105-30									
	0020	Rectangular to 12′ high, small columns		C-11	120	.600	L.F.	42.50	23.50	12.85	78.85	104
	0050	Large columns			96	.750		74.50	29.50	16.05	120.05	153
	0300	24′ high, small columns			192	.375		42.50	14.65	8	65.15	82.50
	0350	Large columns			144	.500		74.50	19.55	10.70	104.75	129
	0700	24′ high, 1 haunch, 12″ x 12″			32	2.250	Ea.	1,025	88	48	1,161	1,350
	0800	20″ x 20″			28	2.571	″	1,800	101	55	1,956	2,225
400	0010	**PRECAST JOISTS**	R034105-30									
	0015	40 psf L.L., 6″ deep for 12′ spans		C-12	600	.080	L.F.	7.10	2.80	1.06	10.96	13.55
	0050	8″ deep for 16′ spans			575	.083		11.85	2.92	1.11	15.88	19
	0100	10″ deep for 20′ spans			550	.087		20.50	3.05	1.16	24.71	29.50
	0150	12″ deep for 24′ spans			525	.091		28.50	3.20	1.21	32.91	37.50
620	0010	**PRECAST SLAB PLANKS**	R034105-30									
	0020	Prestressed roof/floor members, grouted, solid, 4″ thick		C-11	2400	.030	S.F.	4.44	1.17	.64	6.25	7.70
	0050	6″ thick			2800	.026		5.50	1.01	.55	7.06	8.45
	0100	Hollow, 8″ thick			3200	.023		5.75	.88	.48	7.11	8.50
	0150	10″ thick			3600	.020		6.10	.78	.43	7.31	8.60
	0200	12″ thick			4000	.018		6.90	.70	.38	7.98	9.30
650	0010	**PRESTRESSED CONCRETE** post-tensioned in place	R034105-30									
	0020	See also Division 03230-600										
	0100	Post-tensioned in place, small job	R034136-90	C-17B	8.50	9.647	C.Y.	585	355	32	972	1,250
	0200	Large job		″	10	8.200	″	440	305	27	772	1,000
660	0010	**PRESTRESSED** roof and floor members	R034105-30									
750	0010	**TEES** Prestressed	R034105-30									
	0020	Quad tee, short spans, roof		C-11	7200	.010	S.F.	5.65	.39	.21	6.25	7.15
	0050	Floor			7200	.010		5.65	.39	.21	6.25	7.15
	0200	Double tee, floor members, 60′ span			8400	.009		7.35	.34	.18	7.87	8.90
	0250	80′ span			8000	.009		8.60	.35	.19	9.14	10.35
	0300	Roof members, 30′ span			4800	.015		6.25	.59	.32	7.16	8.25
	0350	50′ span			6400	.011		7.15	.44	.24	7.83	8.90
	0400	Wall members, up to 55′ high			3600	.020		9.45	.78	.43	10.66	12.30
	0500	Single tee roof members, 40′ span			3200	.023		7.50	.88	.48	8.86	10.40
	0550	80′ span			5120	.014		9.20	.55	.30	10.05	11.50
	0600	100′ span			6000	.012		13.85	.47	.26	14.58	16.40
	0650	120′ span			6000	.012		15.10	.47	.26	15.83	17.80
	1000	Double tees, floor members										
	1100	Lightweight, 20″ x 8′ wide, 45′ span		C-11	20	3.600	Ea.	2,425	141	77	2,643	3,000
	1150	24″ x 8′ wide, 50′ span			18	4		2,625	157	85.50	2,867.50	3,250
	1200	32″ x 10′ wide, 60′ span			16	4.500		4,425	176	96	4,697	5,300

03400 | Precast Concrete

		03410 | Plant-Precast Structural Concrete	CREW	DAILY OUTPUT	LABOR-HOURS	UNIT	2006 BARE COSTS MAT.	LABOR	EQUIP.	TOTAL	TOTAL INCL O&P
750	1250	Standard weight, 12″ x 8′ wide, 20′ span R034105-30	C-11	22	3.273	Ea.	905	128	70	1,103	1,300
	1300	16″ x 8′ wide, 25′ span		20	3.600		1,200	141	77	1,418	1,675
	1350	18″ x 8′ wide, 30′ span		20	3.600		1,525	141	77	1,743	2,025
	1400	20″ x 8′ wide, 45′ span		18	4		1,725	157	85.50	1,967.50	2,275
	1450	24″ x 8′ wide, 50′ span		16	4.500		2,225	176	96	2,497	2,875
	1500	32″ x 10′ wide, 60′ span		14	5.143		4,075	201	110	4,386	4,950
	2000	Roof members									
	2050	Lightweight, 20″ x 8′ wide, 40′ span	C-11	20	3.600	Ea.	2,025	141	77	2,243	2,575
	2100	24″ x 8′ wide, 50′ span		18	4		2,700	157	85.50	2,942.50	3,350
	2150	32″ x 10′ wide, 60′ span		16	4.500		4,475	176	96	4,747	5,350
	2200	Standard weight, 12″ x 8′ wide, 30′ span		22	3.273		1,350	128	70	1,548	1,800
	2250	16″ x 8′ wide, 30′ span		20	3.600		1,425	141	77	1,643	1,925
	2300	18″ x 8′ wide, 30′ span		20	3.600		1,575	141	77	1,793	2,100
	2350	20″ x 8′ wide, 40′ span		18	4		1,625	157	85.50	1,867.50	2,175
	2400	24″ x 8′ wide, 50′ span		16	4.500		2,175	176	96	2,447	2,800
	2450	32″ x 10′ wide, 60′ span		14	5.143		3,800	201	110	4,111	4,650
		03450 | Plant-Precast Architectural Concrete									
850	0011	**WALL PANELS** R034513-10									
	0050	Uninsulated 4″ thick, smooth gray									
	0150	Low rise, 4′ x 8′x 4″ thick	C-11	320	.225	S.F.	12.55	8.80	4.81	26.16	35
	0200	8′x 8′ x 4″ thick		576	.125		12.40	4.89	2.67	19.96	25.50
	0250	8′x 16′x 4″ thick		1024	.070		12.30	2.75	1.50	16.55	20
	0400	8′x 8′, 4″ thick, smooth gray		576	.125		12.40	4.89	2.67	19.96	25.50
	0500	Exposed aggregate		576	.125		15.85	4.89	2.67	23.41	29
	0600	High rise, 4′ x 8′ x 4″ thick		288	.250		12.55	9.80	5.35	27.70	37.50
	0650	8′ x 8′x 4″ thick		512	.141		12.40	5.50	3.01	20.91	27
	0700	8′ x 16′ x 4″ thick		768	.094		12.30	3.67	2	17.97	22.50
	0800	Insulated panel, 2″ polystyrene, add					.93			.93	1.02
	0850	2″ urethane, add					.74			.74	.81
	1200	Finishes, white, add					2.29			2.29	2.52
	1250	Exposed aggregate, add					1.13			1.13	1.24
	1300	Granite faced, domestic, add					24.50			24.50	27
	2200	Fiberglass reinforced cement with urethane core									
	2210	R20, 8′ x 8′, minimum	E-2	750	.075	S.F.	16.55	2.90	1.90	21.35	25.50
	2220	Maximum	″	600	.093	″	23	3.63	2.38	29.01	34.50
855	0010	**PRECAST WINDOW SILLS**									
	0600	Precast concrete, 4″ tapers to 3″, 9″ wide	D-1	70	.229	L.F.	9.35	7.35		16.70	22
	0650	11″ wide		60	.267		12.50	8.55		21.05	27.50
	0700	13″ wide, 3 1/2″ tapers to 2 1/2″, 12″ wall		50	.320		13.50	10.30		23.80	31.50
		03470 | Tilt-Up Precast Concrete									
600	0010	**TILT-UP WALL PANELS** R034713-20									
	0015	Wall panel construction, walls only, 5-1/2″ thick	C-14	1600	.090	S.F.	4.52	3.11	.66	8.29	10.85
	0100	7-1/2″ thick		1550	.093		5.70	3.21	.68	9.59	12.30
	0500	Walls and columns, 5-1/2″ thick walls, 12″ x 12″ columns		1565	.092		7	3.18	.68	10.86	13.70
	0550	7-1/2″ thick wall, 12″ x 12″ columns		1370	.105		8.65	3.63	.77	13.05	16.35
	0800	Columns only, site precast, minimum		200	.720	L.F.	11.90	25	5.30	42.20	60
	0850	Maximum		105	1.371	″	14.90	47.50	10.10	72.50	106
	3010	Burlap rub with grout	1 Cefi	450	.018	S.F.	.03	.61		.64	.99
		03480 | Precast Concrete Specialties									
200	0010	**CURBS** Roadway type, see Division 02770-300									
400	0010	**LINTELS**									
	0800	Precast concrete, 4″ wide, 8″ high, to 5′ long	D-10	28	1.429	Ea.	26	48	19.15	93.15	128

03400 | Precast Concrete

03480 | Precast Concrete Specialties

			CREW	DAILY OUTPUT	LABOR-HOURS	UNIT	2006 BARE COSTS MAT.	LABOR	EQUIP.	TOTAL	TOTAL INCL O&P	
400	0850	5'-12' long	D-10	24	1.667	Ea.	66.50	56	22.50	145	188	400
	1000	6" wide, 8" high, to 5' long		26	1.538		32.50	52	20.50	105	142	
	1050	5'-12' long		22	1.818		90	61.50	24.50	176	225	
	1200	8" wide, 8" high, to 5' long		24	1.667		31.50	56	22.50	110	150	
	1250	5'-12' long		20	2		80.50	67.50	27	175	227	
	1400	10" wide, 8" high U-Shape, to 14' long		18	2.222		168	75	30	273	340	
	1450	12" wide, 8" high U-Shape, to 19' long	↓	16	2.500	↓	219	84.50	33.50	337	415	
800	0010	**PRECAST STAIRS**										800
	0020	Precast concrete treads on steel stringers, 3' wide	C-12	75	.640	Riser	101	22.50	8.50	132	157	
	0300	Front entrance, 5' wide with 48" platform, 2 risers		16	3	Flight	315	105	40	460	565	
	0350	5 risers		12	4		515	140	53	708	855	
	0500	6' wide, 2 risers		15	3.200		335	112	42.50	489.50	600	
	0550	5 risers		11	4.364		530	153	58	741	900	
	0700	7' wide, 2 risers		14	3.429		420	120	45.50	585.50	705	
	0750	5 risers	↓	10	4.800		680	168	63.50	911.50	1,100	
	1200	Basement entrance stairs, steel bulkhead doors, minimum	B-51	22	2.182		700	60.50	5.85	766.35	875	
	1250	Maximum	"	11	4.364	↓	1,200	121	11.70	1,332.70	1,550	

03500 | Cementitious Decks & Underlayments

03510 | Cementitious Roof Deck

			CREW	DAILY OUTPUT	LABOR-HOURS	UNIT	2006 BARE COSTS MAT.	LABOR	EQUIP.	TOTAL	TOTAL INCL O&P	
200	0010	**WOOD FIBER** Lightweight cement system (R051223-50)										200
	0050	Plank, beveled, 1" thick	2 Carp	1000	.016	S.F.	1.85	.57		2.42	2.98	
	0100	Plank, T & G, 1-1/2" thick		975	.016		1.97	.58		2.55	3.14	
	0150	2" thick		950	.017		2.61	.60		3.21	3.86	
	0200	2-1/2" thick		925	.017		2.86	.62		3.48	4.17	
	0250	3" thick		900	.018		3.27	.63		3.90	4.65	
	0300	3-1/2" thick		875	.018		4.79	.65		5.44	6.35	
	0350	4" thick	↓	850	.019		5.35	.67		6.02	7	
	1000	Bulb tee, sub-purlin and grout, 6' span, add	E-1	5000	.005		1.65	.19	.02	1.86	2.16	
	1100	8' span	"	4200	.006	↓	1.87	.22	.02	2.11	2.48	
250	0010	**PRECAST CONCRETE CHANNEL SLABS**										250
	0020	Lightweight channel slab, 2.75" or 3.5" thk, straight	C-12	1575	.030	S.F.	4.65	1.07	.40	6.12	7.30	
	0050	Chopped up		785	.061		4.90	2.14	.81	7.85	9.80	
	0200	6" thick, span to 20'		1300	.037		5.15	1.29	.49	6.93	8.35	
	0300	8" thick, span to 24'	↓	1100	.044	↓	5.95	1.53	.58	8.06	9.70	
270	0010	**PRECAST LIGHTWEIGHT CONCRETE PLANK**										270
	0015	Lightweight plank, nailable, T&G, 2" thick	C-12	1800	.027	S.F.	3.20	.93	.35	4.48	5.45	
	0050	2-3/4" thick		1575	.030		3.29	1.07	.40	4.76	5.80	
	0100	3-3/4" thick	↓	1375	.035		4.65	1.22	.46	6.33	7.60	
	0150	For premium ceiling finish, add				↓	10%					
	0200	For sloping roofs, slope over 4 in 12, add						25%				
	0250	Slope over 6 in 12, add						150%				
350	0010	**FORMBOARD** Including sub-purlins (R051223-50)										350
	0050	Non-asbestos fiber cement, 1/8" thick	C-13	2950	.008	S.F.	2.15	.31	.03	2.49	2.97	
	0070	1/4" thick		2950	.008		2.48	.31	.03	2.82	3.33	
	0100	Fiberglass, 1" thick, economy	↓	2700	.009	↓	2.68	.34	.03	3.05	3.61	

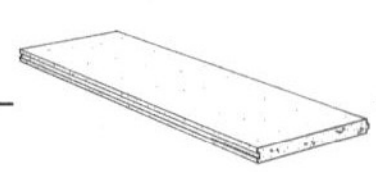

03500 | Cementitious Decks & Underlayments

	03510 | Cementitious Roof Deck		CREW	DAILY OUTPUT	LABOR-HOURS	UNIT	2006 BARE COSTS MAT.	LABOR	EQUIP.	TOTAL	TOTAL INCL O&P	
350	1000	Poured gypsum, 2″ thick, add to formboard above (R051223-50)	C-8	6000	.009	S.F.	1.60	.29	.12	2.01	2.36	350
	1100	3″ thick	″	4800	.012		2.41	.36	.15	2.92	3.40	
900	0010	**WOOD PLANK** Roof decks, see Divisions 06150-600 & 06170-550										900

	03520 | Lightweight Concrete Roof Insulation		CREW	DAILY OUTPUT	LABOR-HOURS	UNIT	MAT.	LABOR	EQUIP.	TOTAL	TOTAL INCL O&P	
250	0010	**INSULATING ROOF FILL**, lightweight cellular concrete (R035216-10)										250
	0020	Portland cement and foaming agent	C-8	50	1.120	C.Y.	84.50	34.50	14	133	164	
	0100	Poured vermiculite or perlite, field mix,										
	0110	1:6 field mix	C-8	50	1.120	C.Y.	92.50	34.50	14	141	173	
	0200	Ready mix, 1:6 mix, roof fill, 2″ thick		10000	.006	S.F.	.51	.17	.07	.75	.92	
	0250	3″ thick		7700	.007		.77	.23	.09	1.09	1.31	
	0400	Expanded volcanic glass rock, with binder, minimum	2 Carp	1500	.011		.29	.38		.67	.95	
	0450	Maximum	″	1200	.013		.88	.47		1.35	1.76	

03600 | Grouts

	03610 | Construction Grout		CREW	DAILY OUTPUT	LABOR-HOURS	UNIT	2006 BARE COSTS MAT.	LABOR	EQUIP.	TOTAL	TOTAL INCL O&P	
400	0010	**GROUT**										400
	0020	Column & machine bases, non-shrink, metallic, 1″ deep	1 Cefi	35	.229	S.F.	5.75	7.85		13.60	18.75	
	0050	2″ deep		25	.320		11.50	11		22.50	30	
	0300	Non-shrink, non-metallic, 1″ deep		35	.229		4.32	7.85		12.17	17.15	
	0350	2″ deep		25	.320		8.65	11		19.65	27	

03900 | Concrete Restoration & Cleaning

	03920 | Concrete Resurfacing		CREW	DAILY OUTPUT	LABOR-HOURS	UNIT	2006 BARE COSTS MAT.	LABOR	EQUIP.	TOTAL	TOTAL INCL O&P	
600	0010	**PATCHING CONCRETE**										600
	2510	Underlayment, P.C based self-leveling, 4100 psi, pumped, 1/4″	C-8	20000	.003	S.F.	1.35	.09	.04	1.48	1.66	
	2520	1/2″		19000	.003		2.69	.09	.04	2.82	3.15	
	2530	3/4″		18000	.003		4.04	.10	.04	4.18	4.64	
	2540	1″		17000	.003		5.40	.10	.04	5.54	6.10	
	2550	1-1/2″		15000	.004		8.05	.12	.05	8.22	9.15	
	2560	Hand mix, 1/2″	C-18	4000	.002		2.69	.06	.01	2.76	3.07	
	2610	Topping, P.C. based self-level/dry 6100 psi, pumped, 1/4″	C-8	20000	.003		2.08	.09	.04	2.21	2.47	
	2620	1/2″		19000	.003		4.16	.09	.04	4.29	4.77	
	2630	3/4″		18000	.003		6.25	.10	.04	6.39	7.05	
	2660	1″		17000	.003		8.35	.10	.04	8.49	9.35	
	2670	1-1/2″		15000	.004		12.50	.12	.05	12.67	14	
	2680	Hand mix, 1/2″	C-18	4000	.002		4.16	.06	.01	4.23	4.69	

	03930 | Concrete Rehabilitation	CREW	DAILY OUTPUT	LABOR-HOURS	UNIT	2006 BARE COSTS MAT.	LABOR	EQUIP.	TOTAL	TOTAL INCL O&P	
300	0010 **CRACK REPAIR**, including chipping, sand blasting and cleaning										300
	0100 Epoxy injection, 1/8" wide, 12" deep	B-9	80	.500	L.F.	4.36	13.90	1.84	20.10	30	
	0110 1/4" wide, 12" deep		60	.667		8.75	18.55	2.45	29.75	43	
	0200 Latex injection, 1/8" wide, 12" deep		100	.400		1.75	11.10	1.47	14.32	22	
	0210 1/4" wide, 12" deep	▼	75	.533	▼	3.50	14.85	1.96	20.31	30.50	

For information about Means Estimating Seminars, see yellow pages 12 and 13 in back of book

Division 4 Masonry

Estimating Tips

04050 Basic Masonry Materials & Methods

- The terms *mortar* and *grout* are often used interchangeably, and incorrectly. Mortar is used to bed masonry units, seal the entry of air and moisture, provide architectural appearance, and allow for size variations in the units. Grout is used primarily in reinforced masonry construction and is used to bond the masonry to the reinforcing steel. Common mortar types are M(2500 psi), S(1800 psi), N(750 psi), and O(350 psi), and conform to ASTM C270. Grout is either fine or coarse and conforms to ASTM C476, and in-place strengths generally exceed 2500 psi. Mortar and grout are different components of masonry construction and are placed by entirely different methods. An estimator should be aware of their unique uses and costs.
- Waste, specifically the loss/droppings of mortar and the breakage of brick and block, is included in all masonry assemblies in this division. A factor of 25% is added for mortar and 3% for brick and concrete masonry units.
- Scaffolding or staging is not included in any of the Division 4 costs. Refer to section 01540 for scaffolding and staging costs.

04800 Masonry Assemblies

- The most common types of unit masonry are brick and concrete masonry. The major classifications of brick are building brick (ASTM C62), facing brick (ASTM C216), glazed brick, fire brick, and pavers. Many varieties of texture and appearance can exist within these classifications, and the estimator would be wise to check local custom and availability within the project area. For repair and remodeling jobs, matching the existing brick may be the most important criteria.
- Brick and concrete block are priced by the piece and then converted into a price per square foot of wall. Openings less than two square feet are generally ignored by the estimator because any savings in units used is offset by the cutting and trimming required.
- It is often difficult and expensive to find and purchase small lots of historic brick. Costs can vary widely. Many design issues affect costs, selection of mortar mix, and repairs or replacement of masonry materials. Cleaning techniques must be reflected in the estimate.
- All masonry walls, whether interior or exterior, require bracing. The cost of bracing walls during construction should be included by the estimator, and this bracing must remain in place until permanent bracing is complete. Permanent bracing of masonry walls is accomplished by masonry itself, in the form of pilasters or abutting wall corners, or by anchoring the walls to the structural frame. Accessories in the form of anchors, anchor slots, and ties are used, but their supply and installation can be by different trades. For instance, anchor slots on spandrel beams and columns are supplied and welded in place by the steel fabricator, but the ties from the slots into the masonry are installed by the bricklayer. Regardless of the installation method the estimator must be certain that these accessories are accounted for in pricing.

Reference Numbers

Reference numbers are shown in bold squares at the beginning of some major classifications. These numbers refer to related items in the Reference Section. The reference information may be an estimating procedure, an alternate pricing method, or technical information.

Note: Not all subdivisions listed here necessarily appear in this publication.

		04055 | Selective Demolition	CREW	DAILY OUTPUT	LABOR-HOURS	UNIT	2006 BARE COSTS MAT.	LABOR	EQUIP.	TOTAL	TOTAL INCL O&P	
110	0010	**SELECTIVE DEMOLITION, MASONRY** R024119-10										110
	0300	Concrete block walls, unreinforced, 2" thick	2 Clab	1200	.013	S.F.		.37		.37	.61	
	0310	4" thick		1150	.014			.38		.38	.63	
	0320	6" thick		1100	.015			.40		.40	.66	
	0330	8" thick		1050	.015			.42		.42	.69	
	0340	10" thick		1000	.016			.44		.44	.73	
	0360	12" thick		950	.017			.46		.46	.76	
	0380	Reinforced alternate courses, 2" thick		1130	.014			.39		.39	.64	
	0390	4" thick		1080	.015			.41		.41	.67	
	0400	6" thick		1035	.015			.42		.42	.70	
	0410	8" thick		990	.016			.44		.44	.73	
	0420	10" thick		940	.017			.47		.47	.77	
	0430	12" thick		890	.018			.49		.49	.82	
	0440	Reinforced alternate courses & vertically 48" OC, 4" thick		900	.018			.49		.49	.81	
	0450	6" thick		850	.019			.52		.52	.85	
	0460	8" thick		800	.020			.55		.55	.91	
	0480	10" thick		750	.021			.58		.58	.97	
	0490	12" thick		700	.023			.63		.63	1.04	
	1000	Chimney, 16" x 16", soft old mortar	1 Clab	55	.145	C.F.		3.99		3.99	6.60	
	1020	Hard mortar		40	.200			5.50		5.50	9.10	
	1030	16" x 20", soft old mortar		55	.145			3.99		3.99	6.60	
	1040	Hard mortar		40	.200			5.50		5.50	9.10	
	1050	16" x 24", soft old mortar		55	.145			3.99		3.99	6.60	
	1060	Hard mortar		40	.200			5.50		5.50	9.10	
	1080	20" x 20", soft old mortar		55	.145			3.99		3.99	6.60	
	1100	Hard mortar		40	.200			5.50		5.50	9.10	
	1110	20" x 24", soft old mortar		55	.145			3.99		3.99	6.60	
	1120	Hard mortar		40	.200			5.50		5.50	9.10	
	1140	20" x 32", soft old mortar		55	.145			3.99		3.99	6.60	
	1160	Hard mortar		40	.200			5.50		5.50	9.10	
	1200	48" x 48", soft old mortar		55	.145			3.99		3.99	6.60	
	1220	Hard mortar		40	.200			5.50		5.50	9.10	
	1250	Metal, high temp steel jacket, 24" diameter	E-2	130	.431	V.L.F.		16.75	11	27.75	42	
	1260	60" diameter	"	60	.933			36.50	24	60.50	91	
	1280	Flue lining, up to 12" x 12"	1 Clab	200	.040			1.10		1.10	1.82	
	1282	Up to 24" x 24"		150	.053			1.46		1.46	2.42	
	2000	Columns, 8" x 8", soft old mortar		48	.167			4.57		4.57	7.55	
	2020	Hard mortar		40	.200			5.50		5.50	9.10	
	2060	16" x 16", soft old mortar		16	.500			13.70		13.70	22.50	
	2100	Hard mortar		14	.571			15.65		15.65	26	
	2140	24" x 24", soft old mortar		8	1			27.50		27.50	45.50	
	2160	Hard mortar		6	1.333			36.50		36.50	60.50	
	2200	36" x 36", soft old mortar		4	2			55		55	91	
	2220	Hard mortar		3	2.667			73		73	121	
	2230	Alternate pricing method, soft old mortar		30	.267	C.F.		7.30		7.30	12.10	
	2240	Hard mortar		23	.348	"		9.55		9.55	15.80	
	3000	Copings, precast or masonry, to 8" wide										
	3020	Soft old mortar	1 Clab	180	.044	L.F.		1.22		1.22	2.02	
	3040	Hard mortar	"	160	.050	"		1.37		1.37	2.27	
	3100	To 12" wide										
	3120	Soft old mortar	1 Clab	160	.050	L.F.		1.37		1.37	2.27	
	3140	Hard mortar	"	140	.057	"		1.57		1.57	2.59	
	4000	Fireplace, brick, 30" x 24" opening										
	4020	Soft old mortar	1 Clab	2	4	Ea.		110		110	182	
	4040	Hard mortar		1.25	6.400			175		175	291	
	4100	Stone, soft old mortar		1.50	5.333			146		146	242	

		04055 | Selective Demolition	CREW	DAILY OUTPUT	LABOR-HOURS	UNIT	2006 BARE COSTS MAT.	LABOR	EQUIP.	TOTAL	TOTAL INCL O&P	
110	4120	Hard mortar (R024119-10)	1 Clab	1	8	Ea.		219		219	365	110
	5000	Veneers, brick, soft old mortar		140	.057	S.F.		1.57		1.57	2.59	
	5020	Hard mortar		125	.064			1.75		1.75	2.91	
	5100	Granite and marble, 2" thick		180	.044			1.22		1.22	2.02	
	5120	4" thick		170	.047			1.29		1.29	2.14	
	5140	Stone, 4" thick		180	.044			1.22		1.22	2.02	
	5160	8" thick		175	.046			1.25		1.25	2.08	
	5400	Alternate pricing method, stone, 4" thick		60	.133	C.F.		3.65		3.65	6.05	
	5420	8" thick		85	.094	"		2.58		2.58	4.27	
	9000	Minimum labor/equipment charge		2	4	Job		110		110	182	

		04060 | Masonry Mortar	CREW	DAILY OUTPUT	LABOR-HOURS	UNIT	MAT.	LABOR	EQUIP.	TOTAL	TOTAL INCL O&P	
200	0010	**CEMENT**										200
	0020	Gypsum, 80 lb. bag, T.L. lots				Bag	16.70			16.70	18.35	
	0050	L.T.L. lots					22.50			22.50	25	
	0100	Masonry, 70 lb. bag, T.L. lots					6.45			6.45	7.10	
	0150	L.T.L. lots					6.85			6.85	7.50	
	0200	White, 70 lb. bag, T.L. lots					13.40			13.40	14.75	
	0250	L.T.L. lots					14.30			14.30	15.75	
400	0010	**LIME**										400
	0020	Masons, hydrated, 50 lb. bag, T.L. lots				Bag	6.55			6.55	7.20	
	0050	L.T.L. lots					7.40			7.40	8.15	
	0200	Finish, double hydrated, 50 lb. bag, T.L. lots					8.40			8.40	9.25	
	0250	L.T.L. lots					9.30			9.30	10.25	
500	0010	**MORTAR** (R042110-50)										500
	0100	Type M, 1:1:6 mix	1 Brhe	143	.056	C.F.	3.44	1.55		4.99	6.30	
	0200	Type N, 1:3 mix		143	.056		3.12	1.55		4.67	5.95	
	0300	Type O, 1:3 mix		143	.056		5.25	1.55		6.80	8.30	
	0400	Type PM, 1:1:6 mix, 2500 psi		143	.056		5.55	1.55		7.10	8.65	
	0500	Type S, 1/2:1:4 mix		143	.056		5.90	1.55		7.45	9	
	2000	With portland cement and lime										
	2100	Type M, 1:1/4:3 mix	1 Brhe	143	.056	C.F.	6.60	1.55		8.15	9.75	
	2200	Type N, 1:1:6 mix, 750 psi		143	.056		5.55	1.55		7.10	8.60	
	2300	Type O, 1:2:9 mix (Pointing Mortar)		143	.056		5.90	1.55		7.45	9	
	2400	Type PL, 1:1/2:4 mix, 2500 psi		143	.056		6.05	1.55		7.60	9.15	
	2500	Type K, 1:3:12 mix, 75 psi		143	.056		5.25	1.55		6.80	8.25	
	2600	Type S, 1:1/2:4 mix, 1800 psi		143	.056		6.95	1.55		8.50	10.15	
	2650	Pre-mixed, type S or N					4.44			4.44	4.89	
	2700	Mortar for glass block	1 Brhe	143	.056		8.15	1.55		9.70	11.45	
	2800	Gypsum cement mortar					8.15			8.15	9	
	2900	Mortar for Fire Brick, 80 lb. bag, T.L. Lots				Bag	21			21	23	
520	0010	**MORTAR** (See also Division 04060-500)										520
	0020	Masonry restoration mix				Lb.	1.60			1.60	1.76	
	0050	White				"	1.84			1.84	2.02	
540	0010	**MORTAR PIGMENTS**, 50 lb. bags (2 bags per M bricks),										540
	0020	range 2 to 10 lb. per bag of cement, minimum				Lb.	6.20			6.20	6.80	
	0050	Average					8.15			8.15	8.95	
	0100	Maximum					16.15			16.15	17.80	
750	0010	**SAND**, screened and washed at the pit										750
	0020	For mortar, per ton				Ton	12.40			12.40	13.65	
	0050	With 10 mile haul					17.30			17.30	19	
	0100	With 30 mile haul					19.30			19.30	21.50	

04050 | Basic Masonry Materials and Methods

04060 | Masonry Mortar

		04060 Masonry Mortar	CREW	DAILY OUTPUT	LABOR-HOURS	UNIT	2006 BARE COSTS MAT.	LABOR	EQUIP.	TOTAL	TOTAL INCL O&P	
750	0200	Screened and washed, at the pit				C.Y.	17.25			17.25	19	750
	0250	With 10 mile haul					24			24	26.50	
	0300	With 30 mile haul					27			27	29.50	
770	0010	**SURFACE BONDING** CMU walls with fiberglass mortar,										770
	0020	gray or white colors, not incl. block work	1 Bric	540	.015	S.F.	.11	.54		.65	1	
900	0010	**WATERPROOFING ADMIXTURE**										900
	0020	Per quart (1 qt. to 2 bags of masonry cement)				Qt.	7.20			7.20	7.90	

04070 | Masonry Grout

		04070 Masonry Grout	CREW	DAILY OUTPUT	LABOR-HOURS	UNIT	MAT.	LABOR	EQUIP.	TOTAL	TOTAL INCL O&P	
420	0010	**GROUTING** Bond bms. & lintels, 8" dp., pumped, not incl. block										420
	0020	8" thick, 0.2 C.F. per L.F.	D-4	1400	.023	L.F.	.82	.73	.10	1.65	2.18	
	0050	10" thick, 0.25 C.F. per L.F.		1200	.027		.91	.85	.12	1.88	2.50	
	0060	12" thick, 0.3 C.F. per L.F.		1040	.031		1.09	.98	.14	2.21	2.93	
	0200	Concrete block cores, solid, 4" thk., by hand, 0.067 C.F./S.F. of wall	D-8	1100	.036	S.F.	.24	1.20		1.44	2.22	
	0210	6" thick, pumped, 0.175 C.F. per S.F.	D-4	720	.044		.64	1.41	.20	2.25	3.20	
	0250	8" thick, pumped, 0.258 C.F. per S.F.		680	.047		.94	1.50	.21	2.65	3.67	
	0300	10" thick, pumped, 0.340 C.F. per S.F.		660	.048		1.24	1.54	.22	3	4.08	
	0350	12" thick, pumped, 0.422 C.F. per S.F.		640	.050		1.54	1.59	.23	3.36	4.50	
	0500	Cavity walls, 2" space, pumped, 0.167 C.F./S.F. of wall		1700	.019		.61	.60	.09	1.30	1.72	
	0550	3" space, 0.250 C.F./S.F.		1200	.027		.91	.85	.12	1.88	2.50	
	0600	4" space, 0.333 C.F. per S.F.		1150	.028		1.21	.89	.13	2.23	2.91	
	0700	6" space, 0.500 C.F. per S.F.		800	.040		1.82	1.27	.18	3.27	4.26	
	0800	Door frames, 3' x 7' opening, 2.5 C.F. per opening		60	.533	Opng.	9.10	16.95	2.42	28.47	40	
	0850	6' x 7' opening, 3.5 C.F. per opening		45	.711	"	12.75	22.50	3.22	38.47	54	
	2000	Grout, C476, for bond beams, lintels and CMU cores		350	.091	C.F.	3.65	2.91	.41	6.97	9.15	
	9000	Minimum labor/equipment charge	1 Bric	2	4	Job		146		146	237	

04080 | Anchorage & Reinforcement

		04080 Anchorage & Reinforcement	CREW	DAILY OUTPUT	LABOR-HOURS	UNIT	MAT.	LABOR	EQUIP.	TOTAL	TOTAL INCL O&P	
070	0010	**ANCHOR BOLTS**										070
	0020	Hooked, with nut and washer, 1/2" diam., 8" long	1 Bric	200	.040	Ea.	.68	1.46		2.14	3.12	
	0030	12" long		190	.042		1.29	1.54		2.83	3.92	
	0040	5/8" diameter, 8" long		180	.044		1.01	1.62		2.63	3.75	
	0050	12" long		170	.047		1.11	1.72		2.83	4.01	
	0060	3/4" diameter, 8" long		160	.050		1.54	1.83		3.37	4.66	
	0070	12" long		150	.053		1.92	1.95		3.87	5.25	

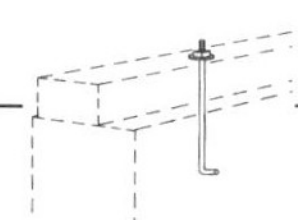

		04080 Anchorage & Reinforcement	CREW	DAILY OUTPUT	LABOR-HOURS	UNIT	MAT.	LABOR	EQUIP.	TOTAL	TOTAL INCL O&P	
200	0010	**REINFORCING** R040519-50										200
	0015	Steel bars A615, placed horiz., #3 & #4 bars	1 Bric	450	.018	Lb.	.43	.65		1.08	1.52	
	0020	#5 & #6 bars		800	.010		.43	.37		.80	1.06	
	0050	Placed vertical, #3 & #4 bars		350	.023		.43	.84		1.27	1.83	
	0060	#5 & #6 bars		650	.012		.43	.45		.88	1.20	
	0500	Joint reinforcing, ladder type, mill std galvanized										
	0600	9 ga. sides, 9 ga. ties, 4" wall	1 Bric	30	.267	C.L.F.	9.40	9.75		19.15	26	
	0650	6" wall		30	.267		10.65	9.75		20.40	27.50	
	0700	8" wall		25	.320		11	11.70		22.70	31	
	0750	10" wall		20	.400		11.70	14.60		26.30	36.50	
	0800	12" wall		20	.400		12.35	14.60		26.95	37	
	1000	Truss type										
	1100	9 ga. sides, 9 ga. ties, 4" wall	1 Bric	30	.267	C.L.F.	11.50	9.75		21.25	28.50	
	1150	6" wall		30	.267		11.95	9.75		21.70	29	
	1200	8" wall		25	.320		12.60	11.70		24.30	33	
	1250	10" wall		20	.400		13.25	14.60		27.85	38	
	1300	12" wall		20	.400		14.05	14.60		28.65	39	
	1500	3/16" sides, 9 ga. ties, 4" wall		30	.267		16.85	9.75		26.60	34.50	

04080 | Anchorage & Reinforcement

		04080 Anchorage & Reinforcement	CREW	DAILY OUTPUT	LABOR-HOURS	UNIT	2006 BARE COSTS MAT.	LABOR	EQUIP.	TOTAL	TOTAL INCL O&P	
200	1550	6″ wall (R040519-50)	1 Bric	30	.267	C.L.F.	17	9.75		26.75	34.50	200
	1600	8″ wall		25	.320		17.60	11.70		29.30	38.50	
	1650	10″ wall		20	.400		18.30	14.60		32.90	43.50	
	1700	12″ wall		20	.400		19.15	14.60		33.75	44.50	
	2000	3/16″ sides, 3/16″ ties, 4″ wall		30	.267		19.70	9.75		29.45	37.50	
	2050	6″ wall		30	.267		19.80	9.75		29.55	38	
	2100	8″ wall		25	.320		20.50	11.70		32.20	41.50	
	2150	10″ wall		20	.400		21.50	14.60		36.10	47	
	2200	12″ wall		20	.400		22.50	14.60		37.10	48	
	2500	Cavity truss type, galvanized										
	2600	9 ga. sides, 9 ga. ties, 4″ wall	1 Bric	25	.320	C.L.F.	28.50	11.70		40.20	50.50	
	2650	6″ wall		25	.320		29	11.70		40.70	50.50	
	2700	8″ wall		20	.400		29.50	14.60		44.10	56	
	2750	10″ wall		15	.533		30.50	19.50		50	65	
	2800	12″ wall		15	.533		31.50	19.50		51	66	
	3000	3/16″ sides, 9 ga. ties, 4″ wall		25	.320		35	11.70		46.70	57.50	
	3050	6″ wall		25	.320		35.50	11.70		47.20	58	
	3100	8″ wall		20	.400		36	14.60		50.60	63	
	3150	10″ wall		15	.533		36.50	19.50		56	72	
	3200	12″ wall		15	.533		38	19.50		57.50	73	
	3500	For hot dip galvanizing, add					80%					
650	0010	**WALL TIES**										650
	0020	For brick veneer, galv., corrugated, 7/8″ x 7″, 22 Ga.	1 Bric	10.50	.762	C	8.35	28		36.35	54	
	0100	24 Ga.		10.50	.762		7.20	28		35.20	53	
	0150	16 Ga.		10.50	.762		18.25	28		46.25	65	
	0200	Buck anchors, galv., corrugated, 16 gauge, 2″ bend, 8″ x 2″		10.50	.762		121	28		149	178	
	0250	8″ x 3″ 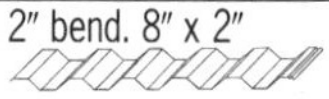		10.50	.762		107	28		135	163	
	0300	Adjustable, rectangular, 4-1/8″ wide										
	0350	Anchor and tie, 3/16″ wire, mill galv.										
	0400	2-3/4″ eye, 3-1/4″ tie	1 Bric	1.05	7.619	M	430	278		708	920	
	0500	4-3/4″ tie		1.05	7.619		475	278		753	970	
	0520	5-1/2″ tie		1.05	7.619		500	278		778	1,000	
	0550	4-3/4″ eye, 3-1/4″ tie		1.05	7.619		480	278		758	975	
	0570	4-3/4″ tie		1.05	7.619		540	278		818	1,050	
	0580	5-1/2″ tie		1.05	7.619		565	278		843	1,075	
	0680	1/4″ diameter		10.50	.762	C	33	28		61	81.50	
	0850	8″ long, 3/16″ diameter		10.50	.762		28.50	28		56.50	76.50	
	0855	1/4″ diameter		10.50	.762		78.50	28		106.50	132	
	1000	Rectangular type, galvanized, 1/4″ diameter, 2″ x 6″		10.50	.762		35.50	28		63.50	84	
	1050	4″ x 6″		10.50	.762		39.50	28		67.50	88.50	
	1100	3/16″ diameter, 2″ x 6″		10.50	.762		28.50	28		56.50	76	
	1150	4″ x 6″		10.50	.762		30.50	28		58.50	78.50	
	1200	Mesh wall tie, 1/2″ mesh, hot dip galvanized										
	1400	16 gauge, 12″ long, 3″ wide	1 Bric	9	.889	C	73.50	32.50		106	134	
	1420	6″ wide		9	.889		112	32.50		144.50	176	
	1440	12″ wide		8.50	.941		180	34.50		214.50	254	
	1500	Rigid partition anchors, plain, 8″ long, 1″ x 1/8″		10.50	.762		69	28		97	121	
	1550	1″ x 1/4″		10.50	.762		114	28		142	170	
	1580	1-1/2″ x 1/8″		10.50	.762		92	28		120	146	
	1600	1-1/2″ x 1/4″		10.50	.762		187	28		215	251	
	1650	2″ x 1/8″		10.50	.762		119	28		147	176	
	1700	2″ x 1/4″		10.50	.762		219	28		247	286	
	2000	Column flange ties, wire, galvanized										
	2300	3/16″ diameter, up to 3″ wide	1 Bric	10.50	.762	C	73.50	28		101.50	126	
	2350	To 5″ wide		10.50	.762		80	28		108	133	

MASONRY 4

	04080 | Anchorage & Reinforcement		CREW	DAILY OUTPUT	LABOR-HOURS	UNIT	2006 BARE COSTS MAT.	LABOR	EQUIP.	TOTAL	TOTAL INCL O&P	
650	2400	To 7" wide	1 Bric	10.50	.762	C	85.50	28		113.50	139	650
	2600	To 9" wide		10.50	.762		92	28		120	146	
	2650	1/4" diameter, up to 3" wide		10.50	.762		99	28		127	154	
	2700	To 5" wide		10.50	.762		109	28		137	165	
	2800	To 7" wide		10.50	.762		118	28		146	174	
	2850	To 9" wide		10.50	.762		128	28		156	186	
	2900	For hot dip galvanized, add					35%					
	4000	Channel slots, 1-3/8" x 1/2" x 8"										
	4100	12 gauge, plain	1 Bric	10.50	.762	C	188	28		216	251	
	4150	16 gauge, galvanized	"	10.50	.762	"	104	28		132	159	
	4200	Channel slot anchors										
	4300	16 gauge, galvanized, 1-1/4" x 3-1/2"				C	41.50			41.50	45.50	
	4350	1-1/4" x 5-1/2"					48			48	53	
	4400	1-1/4" x 7-1/2"					54.50			54.50	60	
	4500	1/8" plain, 1-1/4" x 3-1/2"					67			67	73.50	
	4550	1-1/4" x 5-1/2"					72			72	79	
	4600	1-1/4" x 7-1/2"					78			78	86	
	4700	For corrugation, add					39			39	43	
	4750	For hot dip galvanized, add					35%					
	5000	Dowels										
	5100	Plain, 1/4" diameter, 3" long				C	28			28	31	
	5150	4" long					31			31	34	
	5200	6" long					38			38	42	
	5300	3/8" diameter, 3" long					37			37	40.50	
	5350	4" long					46			46	50.50	
	5400	6" long					53			53	58.50	
	5500	1/2" diameter, 3" long					54			54	59.50	
	5550	4" long					64			64	70.50	
	5600	6" long					85			85	93.50	
	5700	5/8" diameter, 3" long					74			74	81.50	
	5750	4" long					91			91	100	
	5800	6" long					125			125	138	
	6000	3/4" diameter, 3" long					92			92	101	
	6100	4" long					118			118	130	
	6150	6" long					167			167	184	
	6300	For hot dip galvanized, add					35%					

	04090 | Masonry Accessories		CREW	DAILY OUTPUT	LABOR-HOURS	UNIT	MAT.	LABOR	EQUIP.	TOTAL	TOTAL INCL O&P	
170	0010	CONTROL JOINT										170
	0020	Rubber, 4" and wider wall	1 Bric	400	.020	L.F.	1.80	.73		2.53	3.17	
	0050	PVC, 4" wall		400	.020		1.10	.73		1.83	2.40	
	0100	Rubber, 6" wall		320	.025		2.31	.91		3.22	4.02	
	0120	PVC, 6" wall		320	.025		1.48	.91		2.39	3.11	
	0140	Rubber, 8" and wider wall		280	.029		2.99	1.04		4.03	4.98	
	0160	PVC, 8" wall		280	.029		1.68	1.04		2.72	3.54	
	0180	12" wall		240	.033		3.46	1.22		4.68	5.80	
860	0010	VENT BOX										860
	0020	Extruded aluminum, 4" deep, 2-3/8" x 8-1/8"	1 Bric	30	.267	Ea.	22.50	9.75		32.25	40.50	
	0050	5" x 8-1/8"		25	.320		28.50	11.70		40.20	50	
	0100	2-1/4" x 25"		25	.320		58	11.70		69.70	82.50	
	0150	5" x 16-1/2"		22	.364		52.50	13.30		65.80	79.50	
	0200	6" x 16-1/2"		22	.364		68	13.30		81.30	96.50	
	0250	7-3/4" x 16-1/2"		20	.400		66.50	14.60		81.10	97	
	0400	For baked enamel finish, add					35%					
	0500	For cast aluminum, painted, add					60%					
	1000	Stainless steel ventilators, 6" x 6"	1 Bric	25	.320		98.50	11.70		110.20	127	

04050 | Basic Masonry Materials and Methods

04090 | Masonry Accessories

		Description	Crew	Daily Output	Labor-Hours	Unit	2006 Bare Costs Mat.	Labor	Equip.	Total	Total Incl O&P	
860	1050	8″ x 8″	1 Bric	24	.333	Ea.	104	12.20		116.20	134	860
	1100	12″ x 12″		23	.348		120	12.70		132.70	153	
	1150	12″ x 6″		24	.333		105	12.20		117.20	135	
	1200	Foundation block vent, galv., 1-1/4″ thk, 8″ high, 16″ long, no damper		30	.267		19.25	9.75		29	37	
	1250	For damper, add					6.40			6.40	7.05	
900	0010	**WALL PLUGS** (for nailing to brickwork)										900
	0020	26 ga., galvanized, plain	1 Bric	10.50	.762	C	27	28		55	74.50	
	0050	Wood filled	″	10.50	.762	″	90	28		118	144	

04200 | Masonry Units

04210 | Clay Masonry Units

		Description	Crew	Daily Output	Labor-Hours	Unit	2006 Bare Costs Mat.	Labor	Equip.	Total	Total Incl O&P	
100	0010	**COMMON BUILDING BRICK** C62, TL lots, material only R042110-20										100
	0020	Standard, minimum				M	315			315	345	
	0050	Average (select)				″	385			385	425	
300	0010	**FACE BRICK** C216, TL lots, material only R042110-20										300
	0300	Standard modular, 4″ x 2-2/3″ x 8″, minimum				M	400			400	440	
	0350	Maximum R042110-50					515			515	565	
	0450	Economy, 4″ x 4″ x 8″, minimum					775			775	855	
	0500	Maximum					1,000			1,000	1,100	
	0510	Economy, 4″ x 4″ x 12″, minimum					1,075			1,075	1,200	
	0520	Maximum					1,400			1,400	1,550	
	0550	Jumbo, 6″ x 4″ x 12″, minimum					1,225			1,225	1,350	
	0600	Maximum					1,550			1,550	1,700	
	0610	Jumbo, 8″ x 4″ x 12″, minimum					1,225			1,225	1,350	
	0620	Maximum					1,550			1,550	1,700	
	0650	Norwegian, 4″ x 3-1/5″ x 12″, minimum					785			785	860	
	0700	Maximum					1,025			1,025	1,125	
	0710	Norwegian, 6″ x 3-1/5″ x 12″, minimum					1,175			1,175	1,300	
	0720	Maximum					1,525			1,525	1,675	
	0850	Standard glazed, plain colors, 4″ x 2-2/3″ x 8″, minimum					1,225			1,225	1,350	
	0900	Maximum					1,600			1,600	1,750	
	1000	Deep trim shades, 4″ x 2-2/3″ x 8″, minimum					1,250			1,250	1,375	
	1050	Maximum					1,400			1,400	1,525	
	1080	Jumbo utility, 4″ x 4″ x 12″					1,125			1,125	1,250	
	1120	4″ x 8″ x 8″					1,175			1,175	1,275	
	1140	4″ x 8″ x 16″					2,450			2,450	2,700	
	1260	Engineer, 4″ x 3-1/5″ x 8″, minimum					500			500	550	
	1270	Maximum					530			530	580	
	1350	King, 4″ x 2-3/4″ x 10″, minimum					465			465	515	
	1360	Maximum					500			500	550	
	1400	Norman, 4″ x 2-3/4″ x 12″					465			465	515	
	1450	Roman, 4″ x 2″ x 12″					465			465	515	
	1500	SCR, 6″ x 2-2/3″ x 12″					465			465	515	
	1550	Double, 4″ x 5-1/3″ x 8″					465			465	515	
	1600	Triple, 4″ x 5-1/3″ x 12″					465			465	515	
	1770	Standard modular, double glazed, 4″ x 2-2/3″ x 8″					1,050			1,050	1,150	

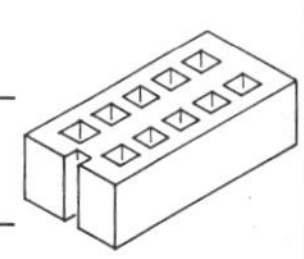

		04210 | Clay Masonry Units		Crew	Daily Output	Labor-Hours	Unit	2006 Bare Costs Mat.	Labor	Equip.	Total	Total Incl O&P	
300	1850	Jumbo, colored glazed ceramic, 6" x 4" x 12"	R042110 -20				M	1,550			1,550	1,700	300
	2050	Jumbo utility, glazed, 4" x 4" x 12"						1,175			1,175	1,275	
	2100	4" x 8" x 8"	R042110 -50					1,600			1,600	1,775	
	2150	4" x 16" x 8"						2,775			2,775	3,050	
	2170	For less than truck load lots, add						10			10	11	
	2180	For buff or gray brick, add						15			15	16.50	
	3050	Used brick, minimum						375			375	410	
	3100	Maximum						490			490	540	
	3150	Add for brick to match existing work, minimum						5%					
	3200	Maximum						50%					

		04220 | Concrete Masonry Units		Crew	Daily Output	Labor-Hours	Unit	Mat.	Labor	Equip.	Total	Total Incl O&P	
210	0010	**CONCRETE BLOCK** material only	R042210 -20										210
	0020	2" x 8" x 16" solid, normal-weight, 2,000 psi					Ea.	.67			.67	.74	
	0050	3,500 psi	R042110 -50					.71			.71	.78	
	0100	5,000 psi						.84			.84	.92	
	0150	Lightweight, std.						.80			.80	.88	
	0300	3" x 8" x 16" solid, normal-weight, 2000 psi						.70			.70	.77	
	0350	3,500 psi						1.27			1.27	1.40	
	0400	5,000 psi						1.29			1.29	1.42	
	0450	Lightweight, std.						.83			.83	.91	
	0600	4" x 8" x 16" hollow, normal-weight, 2000 psi						.76			.76	.84	
	0650	3,500 psi						.82			.82	.90	
	0700	5000 psi						.93			.93	1.02	
	0750	Lightweight, std.						.94			.94	1.03	
	1300	Solid, normal-weight, 2,000 psi						1.01			1.01	1.11	
	1350	3,500 psi						1.15			1.15	1.27	
	1400	5,000 psi						1.05			1.05	1.16	
	1450	Lightweight, std.						1.43			1.43	1.57	
	1600	6" x 8" x 16" hollow, normal-weight, 2,000 psi						1.16			1.16	1.28	
	1650	3,500 psi						1.24			1.24	1.36	
	1700	5,000 psi						1.22			1.22	1.34	
	1750	Lightweight, std.						1.29			1.29	1.42	
	2300	Solid, normal-weight, 2,000 psi						1.22			1.22	1.34	
	2350	3,500 psi						1.34			1.34	1.47	
	2400	5,000 psi						1.70			1.70	1.87	
	2450	Lightweight, std.						1.60			1.60	1.76	
	2600	8" x 8" x 16" hollow, normal-weight, 2000 psi						1.21			1.21	1.33	
	2650	3500 psi						1.41			1.41	1.55	
	2700	5,000 psi						1.45			1.45	1.60	
	2750	Lightweight, std.						1.57			1.57	1.73	
	3200	Solid, normal-weight, 2,000 psi						1.76			1.76	1.94	
	3250	3,500 psi						2.06			2.06	2.27	
	3300	5,000 psi						2.34			2.34	2.57	
	3350	Lightweight, std.						2.14			2.14	2.35	
	3400	10" x 8" x 16" hollow, normal-weight, 2000 psi						1.21			1.21	1.33	
	3410	3500 psi						1.41			1.41	1.55	
	3420	5,000 psi						1.45			1.45	1.60	
	3430	Lightweight, std.						1.57			1.57	1.73	
	3480	Solid, normal-weight, 2,000 psi						1.76			1.76	1.94	
	3490	3,500 psi						2.06			2.06	2.27	
	3500	5,000 psi						2.34			2.34	2.57	
	3510	Lightweight, std.						2.14			2.14	2.35	
	3600	12" x 8" x 16" hollow, normal-weight, 2000 psi						1.78			1.78	1.96	
	3650	3,500 psi						1.83			1.83	2.01	
	3700	5,000 psi						2.02			2.02	2.22	

04200 | Masonry Units

04220 | Concrete Masonry Units

			CREW	DAILY OUTPUT	LABOR-HOURS	UNIT	2006 BARE COSTS MAT.	LABOR	EQUIP.	TOTAL	TOTAL INCL O&P	
210	3750	Lightweight, std. (R042210-20)				Ea.	2.06			2.06	2.27	210
	4300	Solid, normal-weight, 2,000 psi					2.74			2.74	3.01	
	4350	3,500 psi (R042110-50)					2.48			2.48	2.73	
	4400	5,000 psi					3.52			3.52	3.87	
	4500	Lightweight, std.					3.74			3.74	4.11	

04500 | Refractories

04550 | Flue Liners

			CREW	DAILY OUTPUT	LABOR-HOURS	UNIT	2006 BARE COSTS MAT.	LABOR	EQUIP.	TOTAL	TOTAL INCL O&P	
250	0010	**FLUE LINING**, including mortar joints										250
	0020	8″ x 8″	D-1	125	.128	V.L.F.	3.41	4.12		7.53	10.45	
	0100	8″ x 12″		103	.155		5.30	4.99		10.29	13.95	
	0200	12″ x 12″		93	.172		6.70	5.55		12.25	16.30	
	0300	12″ x 18″		84	.190		11.90	6.10		18	23	
	0400	18″ x 18″		75	.213		17.10	6.85		23.95	30	
	0500	20″ x 20″		66	.242		31	7.80		38.80	47	
	0600	24″ x 24″		56	.286		38	9.20		47.20	57	
	1000	Round, 18″ diameter		66	.242		25	7.80		32.80	40	
	1100	24″ diameter		47	.340		49	10.95		59.95	72	

04580 | Refractory Brick

			CREW	DAILY OUTPUT	LABOR-HOURS	UNIT	2006 BARE COSTS MAT.	LABOR	EQUIP.	TOTAL	TOTAL INCL O&P	
250	0010	**FIRE BRICK**										250
	0012	Low duty, 2000° F, 9″ x 2-1/2″ x 4-1/2″	D-1	.60	26.667	M	860	855		1,715	2,350	
	0050	High duty, 3000° F	″	.60	26.667	″	1,625	855		2,480	3,200	
260	0010	**FIRE CLAY**										260
	0020	Gray, high duty, 100 lb. bag				Bag	46			46	50.50	
	0050	100 lb. drum, premixed (400 brick per drum)				Drum	56.50			56.50	62.50	

04700 | Simulated Masonry

04710 | Simulated Brick

			CREW	DAILY OUTPUT	LABOR-HOURS	UNIT	2006 BARE COSTS MAT.	LABOR	EQUIP.	TOTAL	TOTAL INCL O&P	
600	0010	**SIMULATED BRICK**										600
	0020	Aluminum, baked on colors	1 Carp	200	.040	S.F.	2.73	1.42		4.15	5.35	
	0050	Fiberglass panels		200	.040		3	1.42		4.42	5.65	
	0100	Urethane pieces cemented in mastic		150	.053		5.20	1.90		7.10	8.85	
	0150	Vinyl siding panels		200	.040		2.07	1.42		3.49	4.64	
	0160	Cement base, brick, incl. mastic	D-1	100	.160		3.27	5.15		8.42	11.95	
	0170	Corner		50	.320	V.L.F.	8.20	10.30		18.50	25.50	
	0180	Stone face, incl. mastic		100	.160	S.F.	7.90	5.15		13.05	17.05	
	0190	Corner		50	.320	V.L.F.	8.70	10.30		19	26.50	

04700 | Simulated Masonry

	04730 | Simulated Stone		Crew	Daily Output	Labor-Hours	Unit	2006 Bare Costs Mat.	Labor	Equip.	Total	Total Incl O&P	
600	0010	**SIMULATED STONE**										600
	0100	Insulated fiberglass panels, 5/8" ply backer	L-4	200	.120	S.F.	9.80	3.98		13.78	17.30	

04800 | Masonry Assemblies

	04810 | Unit Masonry Assemblies		Crew	Daily Output	Labor-Hours	Unit	2006 Bare Costs Mat.	Labor	Equip.	Total	Total Incl O&P	
040	0010	**ADOBE BRICK** Semi-stabilized, with cement mortar										040
	0060	Brick, 10" x 4" x 14" , 2.6/S.F.	D-8	560	.071	S.F.	2.22	2.36		4.58	6.25	
	0080	12" x 4" x 16", 2.3/S.F.		580	.069		5.70	2.28		7.98	10	
	0100	10" x 4" x 16", 2.3/S.F.		590	.068		3.72	2.24		5.96	7.70	
	0120	8" x 4" x 16", 2.3/S.F.		560	.071		3.96	2.36		6.32	8.20	
	0140	4" x 4" x 16", 2.3/S.F.		540	.074		2.97	2.45		5.42	7.25	
	0160	6" x 4" x 16", 2.3/S.F.		540	.074		2.44	2.45		4.89	6.65	
	0180	4" x 4" x 12", 3.0/S.F.		520	.077		2.42	2.54		4.96	6.80	
	0200	8" x 4" x 12", 3.0/S.F.		520	.077		2.82	2.54		5.36	7.20	
050	0010	**AUTOCLAVED AERATED CONCRETE BLOCK** Scaffolding not incl										050
	0050	Solid, 4" x 12" x 24", incl mortar	D-8	600	.067	S.F.	1.87	2.20		4.07	5.60	
	0060	6" x 12" x 24" R042110-50		600	.067		2.51	2.20		4.71	6.35	
	0070	8" x 8" x 24"		575	.070		3.15	2.30		5.45	7.20	
	0080	10" x 12" x 24"		575	.070		4.11	2.30		6.41	8.25	
	0090	12" x 12" x 24"		550	.073		4.78	2.40		7.18	9.15	
100	0010	**BRICK VENEER** Scaffolding not included, truck load lots										100
	0015	Material costs incl. 3% brick and 25% mortar waste										
	0020	Standard, select common, 4" x 2-2/3" x 8" (6.75/S.F.)	D-8	1.50	26.667	M	470	880		1,350	1,950	
	0050	Red, 4" x 2-2/3" x 8", running bond		1.50	26.667		485	880		1,365	1,950	
	0100	Full header every 6th course (7.88/S.F.)		1.45	27.586		485	910		1,395	2,000	
	0150	English, full header every 2nd course (10.13/S.F.)		1.40	28.571		485	945		1,430	2,050	
	0200	Flemish, alternate header every course (9.00/S.F.)		1.40	28.571		485	945		1,430	2,050	
	0250	Flemish, alt. header every 6th course (7.13/S.F.)		1.45	27.586		485	910		1,395	2,000	
	0300	Full headers throughout (13.50/S.F.)		1.40	28.571		485	945		1,430	2,050	
	0350	Rowlock course (13.50/S.F.)		1.35	29.630		485	980		1,465	2,125	
	0400	Rowlock stretcher (4.50/S.F.)		1.40	28.571		490	945		1,435	2,075	
	0450	Soldier course (6.75/S.F.)		1.40	28.571		485	945		1,430	2,050	
	0500	Sailor course (4.50/S.F.)		1.30	30.769		490	1,025		1,515	2,200	
	0601	Buff or gray face, running bond, (6.75/S.F.)		1.50	26.667		485	880		1,365	1,950	
	0700	Glazed face, 4" x 2-2/3" x 8", running bond		1.40	28.571		1,400	945		2,345	3,075	
	0750	Full header every 6th course (7.88/S.F.)		1.35	29.630		1,350	980		2,330	3,075	
	1000	Jumbo, 6" x 4" x 12",(3.00/S.F.)		1.30	30.769		1,475	1,025		2,500	3,275	
	1051	Norman, 4" x 2-2/3" x 12" (4.50/S.F.)		1.45	27.586		945	910		1,855	2,500	
	1100	Norwegian, 4" x 3-1/5" x 12" (3.75/S.F.)		1.40	28.571		915	945		1,860	2,525	
	1150	Economy, 4" x 4" x 8" (4.50 per S.F.)		1.40	28.571		885	945		1,830	2,500	
	1201	Engineer, 4" x 3-1/5" x 8", (5.63/S.F.)		1.45	27.586		595	910		1,505	2,125	
	1251	Roman, 4" x 2" x 12", (6.00/S.F.)		1.50	26.667		870	880		1,750	2,375	
	1300	S.C.R. 6" x 2-2/3" x 12" (4.50/S.F.)		1.40	28.571		1,050	945		1,995	2,675	
	1350	Utility, 4" x 4" x 12" (3.00/S.F.)		1.35	29.630		1,275	980		2,255	3,000	
	1360	For less than truck load lots, add					10			10	11	
	1400	For battered walls, add						30%				
	1450	For corbels, add						75%				
	1500	For curved walls, add						30%				

04810 | Unit Masonry Assemblies

		04810 Unit Masonry Assemblies	CREW	DAILY OUTPUT	LABOR-HOURS	UNIT	2006 BARE COSTS MAT.	LABOR	EQUIP.	TOTAL	TOTAL INCL O&P	
100	1550	For pits and trenches, deduct						20%				100
	1999	Alternate method of figuring by square foot										
	2000	Standard, sel. common, 4" x 2-2/3" x 8", (6.75/S.F.)	D-8	230	.174	S.F.	3.28	5.75		9.03	12.90	
	2020	Standard, red, 4" x 2-2/3" x 8", running bond (6.75/SF)		220	.182		3.28	6		9.28	13.35	
	2050	Full header every 6th course (7.88/S.F.)		185	.216		3.82	7.15		10.97	15.80	
	2100	English, full header every 2nd course (10.13/S.F.)		140	.286		4.90	9.45		14.35	20.50	
	2150	Flemish, alternate header every course (9.00/S.F.)		150	.267		4.36	8.80		13.16	19.10	
	2200	Flemish, alt. header every 6th course (7.13/S.F.)		205	.195		3.46	6.45		9.91	14.25	
	2250	Full headers throughout (13.50/S.F.)		105	.381		6.50	12.60		19.10	27.50	
	2300	Rowlock course (13.50/S.F.)		100	.400		6.50	13.20		19.70	28.50	
	2350	Rowlock stretcher (4.50/S.F.)		310	.129		2.20	4.26		6.46	9.30	
	2400	Soldier course (6.75/S.F.)		200	.200		3.28	6.60		9.88	14.30	
	2450	Sailor course (4.50/S.F.)		290	.138		2.20	4.56		6.76	9.80	
	2600	Buff or gray face, running bond, (6.75/S.F.)		220	.182		3.47	6		9.47	13.55	
	2700	Glazed face brick, running bond		210	.190		9.05	6.30		15.35	20	
	2750	Full header every 6th course (7.88/S.F.)		170	.235		10.55	7.75		18.30	24	
	3000	Jumbo, 6" x 4" x 12" running bond (3.00/S.F.)		435	.092		4.03	3.04		7.07	9.35	
	3050	Norman, 4" x 2-2/3" x 12" running bond, (4.5/S.F.)		320	.125		4.43	4.13		8.56	11.55	
	3100	Norwegian, 4" x 3-1/5" x 12" (3.75/S.F.)		375	.107		3.32	3.52		6.84	9.35	
	3150	Economy, 4" x 4" x 8" (4.50/S.F.)		310	.129		3.94	4.26		8.20	11.25	
	3200	Engineer, 4" x 3-1/5" x 8" (5.63/S.F.)		260	.154		3.33	5.10		8.43	11.90	
	3250	Roman, 4" x 2" x 12" (6.00/S.F.)		250	.160		5.10	5.30		10.40	14.15	
	3300	SCR, 6" x 2-2/3" x 12" (4.50/S.F.)		310	.129		4.63	4.26		8.89	12	
	3350	Utility, 4" x 4" x 12" (3.00/S.F.)	↓	450	.089	↓	3.74	2.94		6.68	8.90	
	3400	For cavity wall construction, add						15%				
	3450	For stacked bond, add						10%				
	3500	For interior veneer construction, add						15%				
	3550	For curved walls, add						30%				
	9000	Minimum labor/equipment charge	D-1	2	8	Job		257		257	415	
110	0010	**OVERSIZED BRICK**, scaffolding not included										110
	0100	Veneer, 4" x 2.25" x 16"	D-8	387	.103	S.F.	4.48	3.41		7.89	10.50	
	0105	4" x 2.75" x 16"		412	.097		4.05	3.21		7.26	9.65	
	0110	4" x 4" x 16"		460	.087		3.96	2.87		6.83	9	
	0120	4" x 8" x 16"		533	.075		9.35	2.48		11.83	14.25	
	0125	Loadbearing, 6" x 4" x 16", grouted and reinforced		387	.103		6.85	3.41		10.26	13.10	
	0130	8" x 4" x 16", grouted and reinforced		327	.122		7.20	4.04		11.24	14.45	
	0135	6" x 8" x 16", grouted and reinforced		440	.091		6.55	3		9.55	12.05	
	0140	8" x 8" x 16", grouted and reinforced		400	.100		7.30	3.30		10.60	13.35	
	0145	Curtainwall / reinforced veneer, 6" x 4" x 16"		387	.103		10.45	3.41		13.86	17.05	
	0150	8" x 4" x 16"		327	.122		12.60	4.04		16.64	20.50	
	0155	6" x 8" x 16"		440	.091		10.55	3		13.55	16.45	
	0160	8" x 8" x 16"	↓	400	.100		12.70	3.30		16	19.30	
	0200	For 1 to 3 slots in face, add					25%					
	0210	For 4 to 7 slots in face, add					15%					
	0220	For bond beams, add					20%					
	0230	For bullnose shapes, add					20%					
	0240	For open end knockout, add					10%					
	0250	For white or gray color group, add					10%					
	0260	For 135 degree corner, add				↓	250%					
160	0010	**CHIMNEY** See Div. 03310-240 for foundation, add to prices below										160
	0100	Brick, 16" x 16", 8" flue, scaff. not incl.	D-1	18.20	.879	V.L.F.	17.45	28.50		45.95	65	
	0150	16" x 20" with one 8" x 12" flue		16	1		27	32		59	81.50	
	0200	16" x 24" with two 8" x 8" flues		14	1.143		38.50	36.50		75	102	
	0250	20" x 20" with one 12" x 12" flue		13.70	1.168		31.50	37.50		69	95.50	
	0300	20" x 24" with two 8" x 12" flues	↓	12	1.333	↓	44	43		87	118	

	04810 | Unit Masonry Assemblies		CREW	DAILY OUTPUT	LABOR-HOURS	UNIT	2006 BARE COSTS MAT.	LABOR	EQUIP.	TOTAL	TOTAL INCL O&P	
160	0350	20″ x 32″ with two 12″ x 12″ flues	D-1	10	1.600	V.L.F.	55	51.50		106.50	144	160
	1800	Metal, high temp. steel jacket, factory lining, 24″ diam.	E-2	65	.862		191	33.50	22	246.50	294	
	1900	60″ diameter	″	30	1.867		695	72.50	47.50	815	950	
	2100	Poured concrete, brick lining, 200′ high x 10′ diam.					6,100			6,100	6,725	
	2800	500′ x 20′ diameter					10,700			10,700	11,800	
165	0010	**CHIMNEY BLOCK** Scaffolding not included										165
	0220	1 piece, with 8″ x 8″ flue, 16″ x 16″	D-1	28	.571	V.L.F.	10.75	18.35		29.10	42	
	0230	2 piece, 16″ x 16″		26	.615		12.45	19.80		32.25	45.50	
	0240	2 piece, with 8″ x 12″ flue, 16″ x 20″		24	.667		22.50	21.50		44	59.50	
170	0010	**COLUMNS** Face brick, includes mortar, scaffolding not included										170
	0050	8″ x 8″, 9 brick per course	D-1	56	.286	V.L.F.	4.21	9.20		13.41	19.55	
	0100	12″ x 8″, 13.5 brick		37	.432		6.30	13.90		20.20	29.50	
	0200	12″ x 12″, 20 brick		25	.640		9.35	20.50		29.85	44	
	0300	16″ x 12″, 27 brick		19	.842		12.65	27		39.65	58	
	0400	16″ x 16″, 36 brick		14	1.143		16.85	36.50		53.35	78	
	0500	20″ x 16″, 45 brick		11	1.455		21	47		68	99	
	0600	20″ x 20″, 56 brick		9	1.778		26	57		83	122	
	0700	24″ x 20″, 68 brick		7	2.286		32	73.50		105.50	154	
	0800	24″ x 24″, 81 brick		6	2.667		38	85.50		123.50	181	
	1000	36″ x 36″, 182 brick		3	5.333		85	171		256	370	
	9000	Minimum labor/equipment charge		2	8	Job		257		257	415	
172	0010	**CONCRETE BLOCK, BACK-UP,** C90, 2000 psi R042210-20										172
	0020	Normal weight, 8″ x 16″ units, tooled joint 1 side										
	0050	Not-reinforced, 2000 psi, 2″ thick R042110-50	D-8	475	.084	S.F.	1.01	2.78		3.79	5.60	
	0200	4″ thick		460	.087		1.21	2.87		4.08	6	
	0300	6″ thick		440	.091		1.76	3		4.76	6.80	
	0350	8″ thick		400	.100		1.93	3.30		5.23	7.45	
	0400	10″ thick		330	.121		2.41	4		6.41	9.15	
	0450	12″ thick	D-9	310	.155		2.78	4.98		7.76	11.10	
	1000	Reinforced, alternate courses, 4″ thick	D-8	450	.089		1.29	2.94		4.23	6.20	
	1100	6″ thick		430	.093		1.85	3.07		4.92	7	
	1150	8″ thick		395	.101		2.02	3.34		5.36	7.65	
	1200	10″ thick		320	.125		2.51	4.13		6.64	9.45	
	1250	12″ thick	D-9	300	.160		2.88	5.15		8.03	11.50	
	2000	Lightweight, not reinforced, 4″ thick	D-8	460	.087		1.41	2.87		4.28	6.20	
	2100	6″ thick		445	.090		1.91	2.97		4.88	6.90	
	2150	8″ thick		435	.092		2.33	3.04		5.37	7.50	
	2200	10″ thick		410	.098		3.02	3.22		6.24	8.60	
	2250	12″ thick	D-9	390	.123		3.09	3.96		7.05	9.80	
	3000	Lightweight, reinforced, alternate courses, 4″ thick	D-8	450	.089		1.50	2.94		4.44	6.40	
	3100	6″ thick		430	.093		2	3.07		5.07	7.20	
	3150	8″ thick		420	.095		2.43	3.15		5.58	7.75	
	3200	10″ thick		400	.100		3.12	3.30		6.42	8.80	
	3250	12″ thick	D-9	380	.126		3.20	4.06		7.26	10.10	
	9000	Minimum labor/equipment charge	D-1	2	8	Job		257		257	415	
175	0010	**CONCRETE BLOCK BOND BEAM** C90, 2000 psi										175
	0020	Not including grout or reinforcing										
	0130	8″ high, 8″ thick	D-8	565	.071	L.F.	2	2.34		4.34	6	
	0150	12″ thick	D-9	510	.094		2.78	3.03		5.81	7.95	
	0525	Lightweight, 6″ thick	D-8	592	.068		1.84	2.23		4.07	5.65	
	0530	8″ high, 8″ thick	″	575	.070		1.87	2.30		4.17	5.80	
	0550	12″ thick	D-9	520	.092		3.10	2.97		6.07	8.20	
	2000	Including grout and 2 #5 bars										
	2100	Regular block, 8″ high, 8″ thick	D-8	300	.133	L.F.	3.77	4.40		8.17	11.30	
	2150	12″ thick	D-9	250	.192		5	6.15		11.15	15.50	

04810 | Unit Masonry Assemblies

		04810 Unit Masonry Assemblies	CREW	DAILY OUTPUT	LABOR-HOURS	UNIT	2006 BARE COSTS MAT.	LABOR	EQUIP.	TOTAL	TOTAL INCL O&P	
175	2500	Lightweight, 8″ high, 8″ thick	D-8	305	.131	L.F.	4.03	4.33		8.36	11.50	175
	2550	12″ thick	D-9	255	.188		5.35	6.05		11.40	15.65	
	9000	Minimum labor/equipment charge	D-1	2	8	Job		257		257	415	
180	0010	**CONCRETE BLOCK COLUMN** or pilaster, scaffolding not included										180
	0050	Including vertical reinforcing (4-#4 bars) and grout										
	0160	1 piece unit, 16″ x 16″	D-1	26	.615	V.L.F.	15.95	19.80		35.75	49.50	
	0170	2 piece units, 16″ x 20″		24	.667		21.50	21.50		43	58.50	
	0180	20″ x 20″		22	.727		31	23.50		54.50	72.50	
	0190	22″ x 24″		18	.889		43.50	28.50		72	94.50	
	0200	20″ x 32″		14	1.143		47	36.50		83.50	112	
182	0010	**CONCRETE BLOCK, DECORATIVE** C90, 2000 psi										182
	0020	Embossed, simulated brick face										
	0100	8″ x 16″ units, 4″ thick	D-8	400	.100	S.F.	2.71	3.30		6.01	8.35	
	0200	8″ thick		340	.118		3.74	3.89		7.63	10.40	
	0250	12″ thick		300	.133		4.92	4.40		9.32	12.55	
	0400	Embossed both sides										
	0500	8″ thick	D-8	300	.133	S.F.	4.19	4.40		8.59	11.75	
	0550	12″ thick	″	275	.145	″	5.30	4.80		10.10	13.60	
	1000	Fluted high strength										
	1100	8″ x 16″ x 4″ thick, flutes 1 side,	D-8	345	.116	S.F.	3.21	3.83		7.04	9.75	
	1150	Flutes 2 sides		335	.119		3.90	3.94		7.84	10.70	
	1200	8″ thick		300	.133		5.05	4.40		9.45	12.75	
	1250	For special colors, add					.32			.32	.35	
	1400	Deep grooved, smooth face										
	1450	8″ x 16″ x 4″ thick	D-8	345	.116	S.F.	2.10	3.83		5.93	8.50	
	1500	8″ thick	″	300	.133	″	3.63	4.40		8.03	11.15	
	2000	Formblock, incl. inserts & reinforcing										
	2100	8″ x 16″ x 8″ thick	D-8	345	.116	S.F.	3.68	3.83		7.51	10.25	
	2150	12″ thick	″	310	.129	″	4.59	4.26		8.85	11.95	
	2500	Ground face										
	2600	8″ x 16″ x 4″ thick	D-8	345	.116	S.F.	3.58	3.83		7.41	10.15	
	2650	6″ thick		325	.123		4.45	4.07		8.52	11.50	
	2700	8″ thick		300	.133		5.15	4.40		9.55	12.85	
	2750	12″ thick	D-9	265	.181		6.55	5.80		12.35	16.65	
	2900	For special colors, add, minimum					15%					
	2950	For special colors, add, maximum					45%					
	4000	Slump block										
	4100	4″ face height x 16″ x 4″ thick	D-1	165	.097	S.F.	3.42	3.12		6.54	8.80	
	4150	6″ thick		160	.100		4.82	3.22		8.04	10.50	
	4200	8″ thick		155	.103		5.20	3.32		8.52	11.10	
	4250	10″ thick		140	.114		9.75	3.67		13.42	16.65	
	4300	12″ thick		130	.123		10.15	3.96		14.11	17.60	
	4400	6″ face height x 16″ x 6″ thick		155	.103		4.30	3.32		7.62	10.15	
	4450	8″ thick		150	.107		5.80	3.43		9.23	11.95	
	4500	10″ thick		130	.123		8.65	3.96		12.61	15.90	
	4550	12″ thick		120	.133		9.25	4.29		13.54	17.10	
	5000	Split rib profile units, 1″ deep ribs, 8 ribs										
	5100	8″ x 16″ x 4″ thick	D-8	345	.116	S.F.	2.47	3.83		6.30	8.90	
	5150	6″ thick		325	.123		2.87	4.07		6.94	9.75	
	5200	8″ thick		300	.133		3.31	4.40		7.71	10.80	
	5250	12″ thick	D-9	275	.175		3.94	5.60		9.54	13.45	
	5400	For special deeper colors, 4″ thick, add					.88			.88	.97	
	5450	12″ thick, add					.76			.76	.83	
	5600	For white, 4″ thick, add					.88			.88	.97	

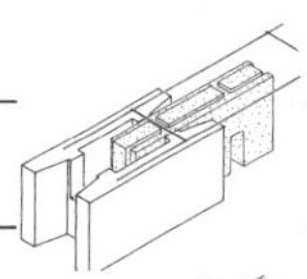

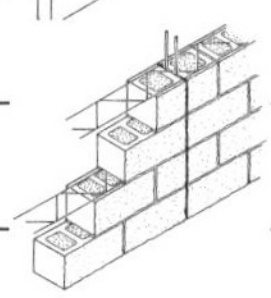

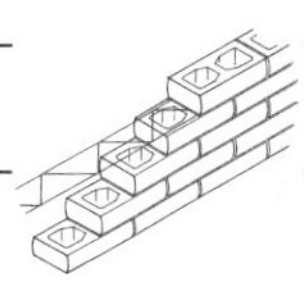

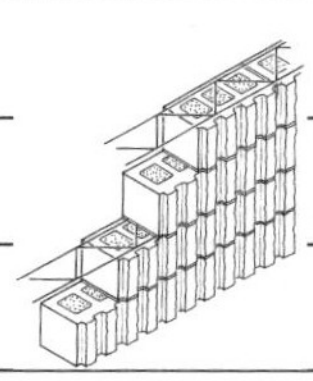

	04810 | Unit Masonry Assemblies		CREW	DAILY OUTPUT	LABOR-HOURS	UNIT	2006 BARE COSTS MAT.	LABOR	EQUIP.	TOTAL	TOTAL INCL O&P
182	5650	6″ thick, add				S.F.	.88			.88	.97
	5700	8″ thick, add					.82			.82	.91
	5750	12″ thick, add					.76			.76	.83
	6000	Split face									
	6100	8″ x 16″ x 4″ thick	D-8	350	.114	S.F.	2.36	3.78		6.14	8.70
	6150	6″ thick		325	.123		2.73	4.07		6.80	9.60
	6200	8″ thick		300	.133		3.20	4.40		7.60	10.65
	6250	12″ thick	D-9	270	.178		3.82	5.70		9.52	13.45
	6300	For scored, add					.25			.25	.28
	6400	For special deeper colors, 4″ thick, add					.51			.51	.56
	6450	6″ thick, add					.51			.51	.56
	6500	8″ thick, add					.45			.45	.50
	6550	12″ thick, add					.38			.38	.42
	6650	For white, 4″ thick, add					.88			.88	.97
	6700	6″ thick, add					.88			.88	.97
	6750	8″ thick, add					.82			.82	.91
	6800	12″ thick, add					.76			.76	.83
	7000	Scored ground face, 2 to 5 scores									
	7100	8″ x 16″ x 4″ thick	D-8	340	.118	S.F.	4.11	3.89		8	10.80
	7150	6″ thick		310	.129		4.76	4.26		9.02	12.15
	7200	8″ thick		290	.138		5.50	4.56		10.06	13.45
	7250	12″ thick	D-9	265	.181		6.55	5.80		12.35	16.65
	8000	Hexagonal face profile units, 8″ x 16″ units									
	8100	4″ thick, hollow	D-8	340	.118	S.F.	2.57	3.89		6.46	9.15
	8200	Solid		340	.118		3.46	3.89		7.35	10.10
	8300	6″ thick, hollow		310	.129		2.89	4.26		7.15	10.10
	8350	8″ thick, hollow		290	.138		3.93	4.56		8.49	11.70
	8500	For stacked bond, add						26%			
	8550	For high rise construction, add per story	D-8	67.80	.590	M.S.F.		19.50		19.50	31.50
	8600	For scored block, add					10%				
	8650	For honed or ground face, per face, add				Ea.	.32			.32	.35
	8700	For honed or ground end, per end, add				″	2.49			2.49	2.73
	8750	For bullnose block, add					10%				
	8800	For special color, add					13%				
	9000	Minimum labor/equipment charge	D-1	2	8	Job		257		257	415
184	0010	**CONCRETE BLOCK, EXTERIOR** C90, 2000 psi									
	0020	Reinforced alt courses, tooled joints 2 sides									
	0100	Normal weight, 8″ x 16″ x 6″ thick	D-8	395	.101	S.F.	1.97	3.34		5.31	7.60
	0200	8″ thick		360	.111		2.90	3.67		6.57	9.15
	0250	10″ thick		290	.138		3.51	4.56		8.07	11.25
	0300	12″ thick	D-9	250	.192		3.59	6.15		9.74	13.95
	0500	Lightweight, 8″ x 16″ x 6″ thick	D-8	450	.089		2.27	2.94		5.21	7.25
	0600	8″ thick		430	.093		2.99	3.07		6.06	8.25
	0650	10″ thick		395	.101		3.50	3.34		6.84	9.30
	0700	12″ thick	D-9	350	.137		5	4.41		9.41	12.65
	9000	Minimum labor/equipment charge	D-1	2	8	Job		257		257	415
186	0010	**CONCRETE BLOCK FOUNDATION WALL** C90/C145									
	0050	Normal-weight, cut joints, horiz joint reinf, no vert reinf									
	0200	Hollow, 8″ x 16″ x 6″ thick	D-8	455	.088	S.F.	2.13	2.90		5.03	7.05
	0250	8″ thick		425	.094		2.31	3.11		5.42	7.60
	0300	10″ thick		350	.114		2.81	3.78		6.59	9.20
	0350	12″ thick	D-9	300	.160		3.18	5.15		8.33	11.85
	0500	Solid, 8″ x 16″ block, 6″ thick	D-8	440	.091		2.20	3		5.20	7.30
	0550	8″ thick	″	415	.096		2.93	3.18		6.11	8.40

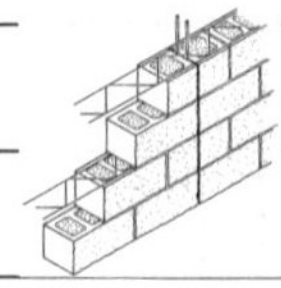

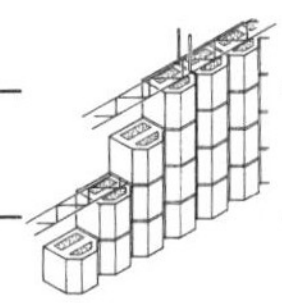

04810 | Unit Masonry Assemblies

			CREW	DAILY OUTPUT	LABOR-HOURS	UNIT	2006 BARE COSTS MAT.	LABOR	EQUIP.	TOTAL	TOTAL INCL O&P	
186	0600	12″ thick	D-9	350	.137	S.F.	4.27	4.41		8.68	11.85	186
	1000	Reinforced, #4 vert @ 48″										
	1100	Hollow, 8″ x 16″ block, 4″ thick	D-8	455	.088	S.F.	2.02	2.90		4.92	6.95	
	1125	6″ thick		445	.090		2.86	2.97		5.83	7.95	
	1150	8″ thick		415	.096		3.34	3.18		6.52	8.80	
	1200	10″ thick		340	.118		4.13	3.89		8.02	10.85	
	1250	12″ thick	D-9	290	.166		4.81	5.30		10.11	13.95	
	1500	Solid, 8″ x 16″ block, 6″ thick	D-8	430	.093		2.20	3.07		5.27	7.40	
	1600	8″ thick	″	405	.099		2.93	3.26		6.19	8.55	
	1650	12″ thick	D-9	340	.141		4.27	4.54		8.81	12.05	
	9000	Minimum labor/equipment charge	D-1	2	8	Job		257		257	415	
187	0010	**CONCRETE BLOCK, HIGH STRENGTH** Normal weight (R042210-20)										187
	0050	Hollow, reinforced alternate courses, 8″ x 16″ units										
	0200	3500 psi, 4″ thick	D-8	440	.091	S.F.	1.61	3		4.61	6.65	
	0250	6″ thick		395	.101		1.91	3.34		5.25	7.55	
	0300	8″ thick		360	.111		2.83	3.67		6.50	9.05	
	0350	12″ thick	D-9	250	.192		3.50	6.15		9.65	13.85	
	0500	5000 psi, 4″ thick	D-8	440	.091		1.53	3		4.53	6.55	
	0550	6″ thick		395	.101		2.38	3.34		5.72	8.05	
	0600	8″ thick		360	.111		3.21	3.67		6.88	9.50	
	0650	12″ thick	D-9	300	.160		4.76	5.15		9.91	13.60	
	1000	For 75% solid block, add					30%					
	1050	For 100% solid block, add					50%					
188	0010	**CONCRETE BLOCK INSULATION INSERTS**										188
	0100	Inserts, styrofoam, plant installed, add to block prices										
	0200	8″ x 16″ units, 6″ thick				S.F.	1.02			1.02	1.12	
	0250	8″ thick					1.02			1.02	1.12	
	0300	10″ thick					1.20			1.20	1.32	
	0350	12″ thick					1.26			1.26	1.39	
	0500	8″ x 8″ units, 8″ thick					.84			.84	.92	
	0550	12″ thick					1.02			1.02	1.12	
189	0010	**CONCRETE BLOCK, INTERLOCKING**										189
	0100	Not including grout or reinforcing										
	0200	8″ x 16″ units, 2,000 psi, 8″ thick	D-1	245	.065	S.F.	2.14	2.10		4.24	5.75	
	0300	12″ thick		220	.073		3.17	2.34		5.51	7.25	
	0350	16″ thick		185	.086		4.78	2.78		7.56	9.75	
	0400	Including grout & reinforcing, 8″ thick	D-4	245	.131		6.15	4.15	.59	10.89	14.10	
	0450	12″ thick		220	.145		7.35	4.63	.66	12.64	16.25	
	0500	16″ thick		185	.173		9.15	5.50	.78	15.43	19.75	
	9000	Minimum labor/equipment charge	D-1	2	8	Job		257		257	415	
190	0010	**CONCRETE BLOCK, LINTELS** C90, normal weight										190
	0100	Including grout and horizontal reinforcing										
	0200	8″ x 8″ x 8″, 1 #4 bar	D-4	300	.107	L.F.	3.99	3.39	.48	7.86	10.35	
	0250	2 #4 bars		295	.108		4.18	3.45	.49	8.12	10.70	
	0400	8″ x 16″ x 8″, 1 #4 bar		275	.116		6.85	3.70	.53	11.08	14.10	
	0450	2 #4 bars		270	.119		7.05	3.77	.54	11.36	14.40	
	1000	12″ x 8″ x 8″, 1 #4 bar		275	.116		5.60	3.70	.53	9.83	12.70	
	1100	2 #4 bars		270	.119		5.80	3.77	.54	10.11	13	
	1150	2 #5 bars		270	.119		6	3.77	.54	10.31	13.25	
	1200	2 #6 bars		265	.121		6.25	3.84	.55	10.64	13.70	
	1500	12″ x 16″ x 8″, 1 #4 bar		250	.128		8.85	4.07	.58	13.50	16.95	
	1600	2 #3 bars		245	.131		8.90	4.15	.59	13.64	17.15	
	1650	2 #4 bars		245	.131		9.05	4.15	.59	13.79	17.30	
	1700	2 #5 bars		240	.133		9.30	4.24	.60	14.14	17.70	

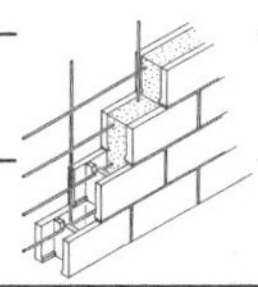

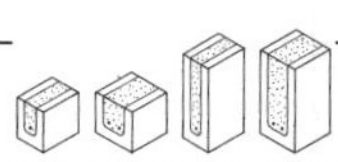

	04810 | Unit Masonry Assemblies		CREW	DAILY OUTPUT	LABOR-HOURS	UNIT	2006 BARE COSTS MAT.	LABOR	EQUIP.	TOTAL	TOTAL INCL O&P	
190	9000	Minimum labor/equipment charge	D-1	2	8	Job		257		257	415	190
210	0010	**CONCRETE BLOCK, PARTITIONS**, scaffolding not included R042210-20										210
	0100	Acoustical slotted block										
	0200	4" thick, type A-1	D-8	315	.127	S.F.	2.68	4.19		6.87	9.75	
	0210	8" thick		275	.145		3.96	4.80		8.76	12.15	
	0250	8" thick, type Q		275	.145		5.40	4.80		10.20	13.75	
	0260	4" thick, type RSC		315	.127		3.96	4.19		8.15	11.15	
	0270	6" thick		295	.136		3.96	4.48		8.44	11.60	
	0280	8" thick		275	.145		3.96	4.80		8.76	12.15	
	0290	12" thick		250	.160		3.96	5.30		9.26	12.90	
	0300	8" thick, type RSR		275	.145		3.96	4.80		8.76	12.15	
	0400	8" thick, type RSC/RF		275	.145		4.91	4.80		9.71	13.20	
	0410	10" thick		260	.154		5.30	5.10		10.40	14.10	
	0500	NRC .60 type R, 8" thick		265	.151		4.50	4.99		9.49	13.05	
	0600	NRC .65 type RR, 8" thick		265	.151		5.05	4.99		10.04	13.65	
	0700	NRC .65 type 4R-RF, 8" thick		265	.151		5.75	4.99		10.74	14.45	
	0710	NRC .70 type R, 12" thick		245	.163		6.85	5.40		12.25	16.30	
	1000	Lightweight block, tooled joints, 2 sides, hollow										
	1100	Not reinforced, 8" x 16" x 4" thick	D-8	440	.091	S.F.	1.34	3		4.34	6.35	
	1150	6" thick		410	.098		1.84	3.22		5.06	7.25	
	1200	8" thick		385	.104		2.26	3.43		5.69	8.05	
	1250	10" thick		370	.108		2.95	3.57		6.52	9.05	
	1300	12" thick	D-9	350	.137		3.02	4.41		7.43	10.45	
	1500	Reinforced alternate courses, 4" thick	D-8	435	.092		1.96	3.04		5	7.10	
	1600	6" thick		405	.099		1.92	3.26		5.18	7.40	
	1650	8" thick		380	.105		2.34	3.48		5.82	8.25	
	1700	10" thick		365	.110		3.04	3.62		6.66	9.20	
	1750	12" thick	D-9	345	.139		3.11	4.47		7.58	10.65	
	2000	Not reinforced, 8" x 24" x 4" thick, hollow		460	.104		.98	3.35		4.33	6.55	
	2100	6" thick		440	.109		1.33	3.51		4.84	7.15	
	2150	8" thick		415	.116		1.67	3.72		5.39	7.90	
	2200	10" thick		385	.125		2.16	4.01		6.17	8.85	
	2250	12" thick		365	.132		2.24	4.23		6.47	9.30	
	2400	Reinforced alternate courses, 4" thick		455	.105		1.05	3.39		4.44	6.65	
	2500	6" thick		435	.110		1.43	3.55		4.98	7.30	
	2550	8" thick		410	.117		1.75	3.76		5.51	8	
	2600	10" thick		380	.126		2.25	4.06		6.31	9.05	
	2650	12" thick		360	.133		2.33	4.29		6.62	9.50	
	2800	Solid, not reinforced, 8" x 16" x 2" thick	D-8	440	.091		1.08	3		4.08	6.05	
	2900	4" thick		420	.095		1.89	3.15		5.04	7.20	
	2950	6" thick		390	.103		2.19	3.39		5.58	7.90	
	3000	8" thick		365	.110		2.91	3.62		6.53	9.05	
	3050	10" thick		350	.114		3.41	3.78		7.19	9.85	
	3100	12" thick	D-9	330	.145		4.92	4.68		9.60	13	
	3300	Solid, reinforced alternate courses, 4" thick	D-8	415	.096		1.96	3.18		5.14	7.30	
	3400	6" thick		385	.104		2.37	3.43		5.80	8.15	
	3450	8" thick		360	.111		2.99	3.67		6.66	9.25	
	3500	10" thick		345	.116		3.50	3.83		7.33	10.05	
	3550	12" thick	D-9	325	.148		5	4.75		9.75	13.20	
	4000	Regular block, tooled joints, 2 sides, hollow										
	4100	Not reinforced, 8" x 16" x 4" thick	D-8	430	.093	S.F.	1.14	3.07		4.21	6.25	
	4150	6" thick		400	.100		1.69	3.30		4.99	7.20	
	4200	8" thick		375	.107		1.85	3.52		5.37	7.75	
	4250	10" thick		360	.111		2.34	3.67		6.01	8.55	
	4300	12" thick	D-9	340	.141		2.70	4.54		7.24	10.30	

04810 | Unit Masonry Assemblies

		04810 Unit Masonry Assemblies	CREW	DAILY OUTPUT	LABOR-HOURS	UNIT	2006 BARE COSTS MAT.	LABOR	EQUIP.	TOTAL	TOTAL INCL O&P
210	4500	Reinforced alternate courses, 8" x 16" x 4" thick R042210-20	D-8	425	.094	S.F.	1.22	3.11		4.33	6.40
	4550	6" thick		395	.101		1.78	3.34		5.12	7.40
	4600	8" thick		370	.108		1.95	3.57		5.52	7.95
	4650	10" thick		355	.113		2.70	3.72		6.42	9
	4700	12" thick	D-9	335	.143		2.81	4.61		7.42	10.55
	4900	Solid, not reinforced, 2" thick	D-8	435	.092		.97	3.04		4.01	6
	5000	3" thick		430	.093		1.03	3.07		4.10	6.10
	5050	4" thick		415	.096		1.42	3.18		4.60	6.70
	5100	6" thick		385	.104		1.76	3.43		5.19	7.50
	5150	8" thick		360	.111		2.48	3.67		6.15	8.65
	5200	12" thick	D-9	325	.148		3.79	4.75		8.54	11.85
	5500	Solid, reinforced alternate courses, 4" thick	D-8	420	.095		1.49	3.15		4.64	6.75
	5550	6" thick		380	.105		1.84	3.48		5.32	7.65
	5600	8" thick		355	.113		2.56	3.72		6.28	8.85
	5650	12" thick	D-9	320	.150		3.62	4.82		8.44	11.80
	9000	Minimum labor/equipment charge	D-1	2	8	Job		257		257	415
225	0010	**CONCRETE BRICK** C55, grade N, type I									
	0100	Regular, 4 x 2-1/4 x 8	D-8	220	.182	Ea.	.35	6		6.35	10.15
	0125	Rusticated, 4 x 2-1/4 x 8		220	.182		.39	6		6.39	10.20
	0150	Frog, 4 x 2-1/4 x 8		220	.182		.38	6		6.38	10.15
	0200	Double, 4 x 4-7/8 x 8		180	.222		.61	7.35		7.96	12.55
240	0010	**CONCRETE SCREEN BLOCK**									
	0200	8" x 16", 4" thick	D-8	330	.121	S.F.	1.80	4		5.80	8.50
	0300	8" thick		270	.148		2.73	4.89		7.62	10.95
	0350	12" x 12", 4" thick		290	.138		2.15	4.56		6.71	9.75
	0500	8" thick		250	.160		2.77	5.30		8.07	11.60
250	0010	**COPING** Stock units									
	0050	Precast concrete, 10" wide, 4" tapers to 3-1/2", 8" wall	D-1	75	.213	L.F.	15	6.85		21.85	27.50
	0100	12" wide, 3-1/2" tapers to 3", 10" wall		70	.229		10.50	7.35		17.85	23.50
	0110	14" wide, 4" tapers to 3-1/2", 12" wall		65	.246		19.85	7.90		27.75	35
	0150	16" wide, 4" tapers to 3-1/2", 14" wall		60	.267		14.50	8.55		23.05	30
	0250	Precast concrete corners		40	.400	Ea.	24	12.85		36.85	47.50
	0300	Limestone for 12" wall, 4" thick		90	.178	L.F.	13.50	5.70		19.20	24
	0350	6" thick		80	.200		15.75	6.45		22.20	28
	0500	Marble, to 4" thick, no wash, 9" wide		90	.178		19.25	5.70		24.95	30.50
	0550	12" wide		80	.200		29	6.45		35.45	42.50
	0700	Terra cotta, 9" wide		90	.178		4.75	5.70		10.45	14.50
	0750	12" wide		80	.200		7.80	6.45		14.25	19.05
	0800	Aluminum, for 12" wall		80	.200		11.25	6.45		17.70	23
	9000	Minimum labor/equipment charge		2	8	Job		257		257	415
260	0010	**CORNICES** Brick, on existing building									
	0110	Face bricks, 12 brick/S.F., minimum	D-1	30	.533	SF Face	5	17.15		22.15	33.50
	0150	15 brick/S.F., maximum		23	.696	"	6	22.50		28.50	43
	9000	Minimum labor/equipment charge		1.50	10.667	Job		345		345	555
325	0010	**GLASS BLOCK**									
	0100	Plain, 4" thick, under 1,000 S.F., 6" x 6"	D-8	115	.348	S.F.	17.95	11.50		29.45	38.50
	0150	8" x 8"		160	.250		11.95	8.25		20.20	26.50
	0160	end block		160	.250		34	8.25		42.25	50.50
	0170	90 deg corner		160	.250		32.50	8.25		40.75	49.50
	0180	45 deg corner		160	.250		14.05	8.25		22.30	29
	0210	4" x 8"		160	.250		8	8.25		16.25	22
	0220	6" x 8"		160	.250		9.40	8.25		17.65	24
	0300	1,000 to 5,000 S.F., 6" x 6"		135	.296		17.60	9.80		27.40	35.50
	0350	8" x 8"		190	.211		11.75	6.95		18.70	24

04810 | Unit Masonry Assemblies

		04810 Unit Masonry Assemblies	CREW	DAILY OUTPUT	LABOR-HOURS	UNIT	2006 BARE COSTS MAT.	LABOR	EQUIP.	TOTAL	TOTAL INCL O&P	
325	0400	12" x 12"	D-8	215	.186	S.F.	14	6.15		20.15	25.50	325
	0410	4" x 8"		215	.186		3.73	6.15		9.88	14.05	
	0420	6" x 8"		215	.186		4.13	6.15		10.28	14.50	
	0500	Over 5,000 S.F., 6" x 6"		145	.276		17.05	9.10		26.15	33.50	
	0550	8" x 8"		215	.186		11.35	6.15		17.50	22.50	
	0600	12" x 12"		240	.167		13.55	5.50		19.05	24	
	0610	4" x 8"		240	.167		3.75	5.50		9.25	13.10	
	0620	6" x 8"		240	.167		3.85	5.50		9.35	13.20	
	0700	For solar reflective blocks, add					100%					
	1000	Thinline, plain, 3-1/8" thick, under 1,000 S.F., 6" x 6"	D-8	115	.348	S.F.	19.70	11.50		31.20	40	
	1050	8" x 8"		160	.250		9.90	8.25		18.15	24.50	
	1200	Over 5,000 S.F., 6" x 6"		145	.276		12.50	9.10		21.60	28.50	
	4000	Accessories										
	4100	Anchors, 20 ga. galv., 1-3/4" wide x 24" long				Ea.	2.57			2.57	2.83	
	4200	Emulsion asphalt				Gal.	9.60			9.60	10.55	
	4300	Expansion joint, fiberglass				L.F.	.61			.61	.67	
	4400	Steel mesh, double galvanized				"	.25			.25	.28	
	9000	Minimum labor/equipment charge	D-1	2	8	Job		257		257	415	
350	0010	**GLAZED CONCRETE BLOCK** C744										350
	0100	Single face, 8" x 16" units, 2" thick	D-8	360	.111	S.F.	6.45	3.67		10.12	13.05	
	0200	4" thick		345	.116		6.55	3.83		10.38	13.45	
	0250	6" thick		330	.121		7.05	4		11.05	14.25	
	0300	8" thick		310	.129		7.55	4.26		11.81	15.25	
	0350	10" thick		295	.136		8.50	4.48		12.98	16.60	
	0400	12" thick	D-9	280	.171		8.90	5.50		14.40	18.70	
	0700	Double face, 8" x 16" units, 4" thick	D-8	340	.118		10.25	3.89		14.14	17.55	
	0750	6" thick		320	.125		11.25	4.13		15.38	19.10	
	0800	8" thick		300	.133		11.75	4.40		16.15	20	
	1000	Jambs, bullnose or square, single face, 8" x 16", 2" thick		315	.127	Ea.	11.55	4.19		15.74	19.50	
	1050	4" thick		285	.140	"	12.75	4.64		17.39	21.50	
	1200	Caps, bullnose or square, 8" x 16", 2" thick		420	.095	L.F.	10.30	3.15		13.45	16.40	
	1250	4" thick		380	.105		14	3.48		17.48	21	
	1500	Cove base, 8" x 16", 2" thick		315	.127		6.45	4.19		10.64	13.90	
	1550	4" thick		285	.140		6.20	4.64		10.84	14.35	
	1600	6" thick		265	.151		6.80	4.99		11.79	15.55	
	1650	8" thick		245	.163		7.45	5.40		12.85	16.90	
	9000	Minimum labor/equipment charge	D-1	2	8	Job		257		257	415	
500	0010	**TERRA COTTA COPING**										500
	0020	Split type, not glazed, 9" wide	D-1	90	.178	L.F.	7.50	5.70		13.20	17.50	
	0100	13" wide		80	.200		11.30	6.45		17.75	23	
	0200	Split type, glazed, 9" wide		90	.178		12.70	5.70		18.40	23.50	
	0250	13" wide		80	.200		16.60	6.45		23.05	28.50	
	0500	Partition or back-up blocks, scored, in C.L. lots										
	0700	Non-load bearing 12" x 12", 3" thick, special order	D-8	550	.073	S.F.	12.35	2.40		14.75	17.45	
	0750	4" thick, standard		500	.080		4.87	2.64		7.51	9.65	
	0800	6" thick		450	.089		5.75	2.94		8.69	11.10	
	0850	8" thick		400	.100		7.45	3.30		10.75	13.55	
	1000	Load bearing, 12" x 12", 4" thick, in walls		500	.080		5.10	2.64		7.74	9.90	
	1050	In floors		750	.053		5.10	1.76		6.86	8.45	
	1200	6" thick, in walls		450	.089		5.75	2.94		8.69	11.10	
	1250	In floors		675	.059		5.75	1.96		7.71	9.55	
	1400	8" thick, in walls		400	.100		7.45	3.30		10.75	13.55	
	1450	In floors		575	.070		7.45	2.30		9.75	11.95	
	1600	10" thick, in walls, special order		350	.114		9.25	3.78		13.03	16.30	
	1650	In floors, special order		500	.080		9.25	2.64		11.89	14.50	

	04810 | Unit Masonry Assemblies		CREW	DAILY OUTPUT	LABOR-HOURS	UNIT	2006 BARE COSTS MAT.	LABOR	EQUIP.	TOTAL	TOTAL INCL O&P	
500	1800	12" thick, in walls, special order	D-8	300	.133	S.F.	18.50	4.40		22.90	27.50	500
	1850	In floors, special order		450	.089		18.50	2.94		21.44	25.50	
	2000	For reinforcing with steel rods, add to above					15%	5%				
	2100	For smooth tile instead of scored, add					2.40			2.40	2.64	
	2200	For L.C.L. quantities, add					10%	10%				
	9000	Minimum labor/equipment charge	D-1	2	8	Job		257		257	415	
510	0010	**TERRA COTTA TILE** On walls, dry set, 1/2" thick										510
	0100	Square, hexagonal or lattice shapes, unglazed	1 Tilf	135	.059	S.F.	4.79	2.03		6.82	8.45	
	0300	Glazed, plain colors		130	.062		6.60	2.11		8.71	10.55	
	0400	Intense colors		125	.064		7.70	2.19		9.89	11.90	
650	0010	**WALLS** Building brick, including mortar R042110-20										650
	0140	4" thick, facing, 4" x 2-2/3" x 8"	D-8	1.45	27.586	M	390	910		1,300	1,900	
	0150	4" thick, as back-up, 6.75 bricks per S.F. R042110-50		1.60	25		390	825		1,215	1,775	
	0204	8" thick, 13.50 bricks per S.F.		1.80	22.222		405	735		1,140	1,650	
	0250	12" thick, 20.25 bricks per S.F.		1.90	21.053		410	695		1,105	1,575	
	0304	16" thick, 27.00 bricks per S.F.		2	20		415	660		1,075	1,525	
	0500	Reinforced, 4" wall, 4" x 2-2/3" x 8"		1.40	28.571		390	945		1,335	1,950	
	0550	8" thick, 13.50 bricks per S.F.		1.75	22.857		480	755		1,235	1,750	
	0600	12" thick, 20.25 bricks per S.F.		1.85	21.622		480	715		1,195	1,675	
	0650	16" thick, 27.00 bricks per S.F.		1.95	20.513		485	680		1,165	1,625	
	0660	4" thick, select common, face, 4" x 2-2/3" x 8"		1.45	27.586		465	910		1,375	1,975	
	0790	Alternate method of figuring by square foot										
	0800	4" wall, face, 4" x 2-2/3" x 8"	D-8	215	.186	S.F.	3.23	6.15		9.38	13.50	
	0850	4" thick, as back up, 6.75 bricks per S.F.		240	.167		2.64	5.50		8.14	11.85	
	0900	8" thick wall, 13.50 brick per S.F.		135	.296		5.50	9.80		15.30	22	
	1000	12" thick wall, 20.25 bricks per S.F.		95	.421		8.25	13.90		22.15	31.50	
	1050	16" thick wall, 27.00 bricks per S.F.		75	.533		11.20	17.60		28.80	41	
	1200	Reinforced, 4" x 2-2/3" x 8" , 4" wall		205	.195		2.64	6.45		9.09	13.35	
	1250	8" thick wall, 13.50 brick per S.F.		130	.308		5.50	10.15		15.65	22.50	
	1300	12" thick wall, 20.25 bricks per S.F.		90	.444		8.25	14.70		22.95	33	
	1350	16" thick wall, 27.00 bricks per S.F.		70	.571		11.20	18.85		30.05	43	
	9000	Minimum labor/equipment charge	D-1	2	8	Job		257		257	415	
670	0010	**STEPS**, with select common bricks										670
	0012	Entry steps, brick	D-1	.30	53.333	M	385	1,725		2,110	3,200	
700	0010	**STRUCTURAL BRICK** C652, Grade SW, incl mortar, scaffolding not incl										700
	0100	Standard unit, 4-5/8" x 2-3/4" x 9-5/8"	D-8	245	.163	S.F.	4.67	5.40		10.07	13.90	
	0120	Bond beam		225	.178		8.85	5.85		14.70	19.30	
	0140	V cut bond beam		225	.178		9.15	5.85		15	19.60	
	0160	Stretcher quoin, 5-5/8" x 2-3/4" x 9-5/8"		245	.163		8.80	5.40		14.20	18.45	
	0180	Corner quoin		245	.163		9.90	5.40		15.30	19.65	
	0200	Corner, 45 deg, 4-5/8" x 2-3/4" x 10-7/16"		235	.170		10.25	5.60		15.85	20.50	
750	0010	**STRUCTURAL FACING TILE** Scaffolding not incl, standard colors										750
	0020	6T series, 5-1/3" x 12", 2.3 pieces per S.F., glazed 1 side, 2" thick	D-8	225	.178	S.F.	6	5.85		11.85	16.15	
	0100	4" thick		220	.182		7.05	6		13.05	17.50	
	0150	Glazed 2 sides		195	.205		11.20	6.80		18	23.50	
	0250	6" thick		210	.190		10.95	6.30		17.25	22	
	0300	Glazed 2 sides		185	.216		12.80	7.15		19.95	25.50	
	0400	8" thick		180	.222		13.65	7.35		21	27	
	0500	Special shapes, group 1		400	.100	Ea.	5.40	3.30		8.70	11.25	
	0550	Group 2		375	.107		8.20	3.52		11.72	14.70	
	0600	Group 3		350	.114		10.15	3.78		13.93	17.25	
	0650	Group 4		325	.123		17.50	4.07		21.57	26	
	0700	Group 5		300	.133		23	4.40		27.40	32	
	0750	Group 6		275	.145		27.50	4.80		32.30	38.50	
	1000	Fire rated, 4" thick, 1 hr. rating		210	.190	S.F.	12.05	6.30		18.35	23.50	

04810 | Unit Masonry Assemblies

			CREW	DAILY OUTPUT	LABOR-HOURS	UNIT	2006 BARE COSTS MAT.	LABOR	EQUIP.	TOTAL	TOTAL INCL O&P	
750	1300	Acoustic, 4″ thick	D-8	210	.190	S.F.	27	6.30		33.30	39.50	750
	1400	For designer colors, add					25%					
	2000	8W series, 8″ x 16″, 1.125 pieces per S.F.										
	2050	2″ thick, glazed 1 side	D-8	360	.111	S.F.	6.50	3.67		10.17	13.10	
	2100	4″ thick, glazed 1 side		345	.116		7.10	3.83		10.93	14	
	2150	Glazed 2 sides		325	.123		10.60	4.07		14.67	18.25	
	2200	6″ thick, glazed 1 side		330	.121		10	4		14	17.55	
	2250	8″ thick, glazed 1 side		310	.129		11.65	4.26		15.91	19.70	
	2500	Special shapes, group 1		300	.133	Ea.	8.70	4.40		13.10	16.70	
	2550	Group 2		280	.143		11.45	4.72		16.17	20.50	
	2600	Group 3		260	.154		11.95	5.10		17.05	21.50	
	2650	Group 4		250	.160		26.50	5.30		31.80	38	
	2700	Group 5		240	.167		43.50	5.50		49	57	
	2750	Group 6		230	.174		51	5.75		56.75	65.50	
	3000	4″ thick, glazed 1 side		345	.116	S.F.	8.30	3.83		12.13	15.30	
	3100	Acoustic, 4″ thick		345	.116	″	11	3.83		14.83	18.30	
	3120	4W series, 8″ x 8″, 2.25 pieces per S.F.										
	3125	2″ thick, glazed 1 side	D-8	360	.111	S.F.	7.65	3.67		11.32	14.35	
	3130	4″ thick, glazed 1 side		345	.116		7.65	3.83		11.48	14.60	
	3135	Glazed 2 sides		325	.123		10.95	4.07		15.02	18.65	
	3140	6″ thick, glazed 1 side		330	.121		10.20	4		14.20	17.75	
	3150	8″ thick, glazed 1 side		310	.129		14.75	4.26		19.01	23	
	3155	Special shapes, group I		300	.133	Ea.	5.40	4.40		9.80	13.10	
	3160	Group II		280	.143	″	6.55	4.72		11.27	14.85	
	3200	For designer colors, add					25%					
	3300	For epoxy mortar joints, add				S.F.	1.25			1.25	1.38	
	9000	Minimum labor/equipment charge	D-1	2	8	Job		257		257	415	

04840 | Prefabricated Masonry Panels

			CREW	DAILY OUTPUT	LABOR-HOURS	UNIT	MAT.	LABOR	EQUIP.	TOTAL	TOTAL INCL O&P	
900	0010	**WALL PANELS**										900
	0020	Prefabricated, 4″ thick, minimum	C-11	775	.093	S.F.	9.10	3.64	1.99	14.73	18.80	
	0100	Maximum	″	500	.144		11.70	5.65	3.08	20.43	26.50	
	0200	4″ brick & 2″ concrete back-up, add					50%					
	0300	4″ brick & 1″ urethane & 3″ concrete back-up, add					70%					

04850 | Stone Assemblies

			CREW	DAILY OUTPUT	LABOR-HOURS	UNIT	MAT.	LABOR	EQUIP.	TOTAL	TOTAL INCL O&P	
050	0011	**ASHLAR VENEER** 4″ + or - thk, random or random rectangular										050
	0150	Sawn face, split joints, low priced stone	D-8	140	.286	S.F.	10.85	9.45		20.30	27	
	0200	Medium priced stone		130	.308		13.35	10.15		23.50	31	
	0300	High priced stone		120	.333		18.85	11		29.85	38.50	
	0600	Seam face, split joints, medium price stone		125	.320		14	10.55		24.55	32.50	
	0700	High price stone		120	.333		20	11		31	40	
	1000	Split or rock face, split joints, medium price stone		125	.320		14	10.55		24.55	32.50	
	1100	High price stone		120	.333		20	11		31	40	
100	0010	**BLUESTONE** Cut to size										100
	0100	Paving, natural cleft, to 4′, 1″ thick	D-8	150	.267	S.F.	7.05	8.80		15.85	22	
	0150	1-1/2″ thick		145	.276		6.40	9.10		15.50	22	
	0200	Smooth finish, 1″ thick		150	.267		7.55	8.80		16.35	22.50	
	0250	1-1/2″ thick		145	.276		8.10	9.10		17.20	23.50	
	0300	Thermal finish, 1″ thick		150	.267		9.25	8.80		18.05	24.50	
	0350	1-1/2″ thick		145	.276		10.40	9.10		19.50	26.50	
	0500	Sills, natural cleft, 10″ wide to 6′ long, 1-1/2″ thick	D-11	70	.343	L.F.	12.35	11.75		24.10	32.50	
	0550	2″ thick		63	.381		14.05	13.05		27.10	36.50	
	0600	Smooth finish, 1-1/2″ thick		70	.343		18.50	11.75		30.25	39.50	

		04850 | Stone Assemblies	CREW	DAILY OUTPUT	LABOR-HOURS	UNIT	2006 BARE COSTS MAT.	LABOR	EQUIP.	TOTAL	TOTAL INCL O&P
100	0650	2″ thick	D-11	63	.381	L.F.	22	13.05		35.05	45
	0800	Thermal finish, 1-1/2″ thick		70	.343		20	11.75		31.75	41
	0850	2″ thick	↓	63	.381		22	13.05		35.05	45
	1000	Stair treads, natural cleft, 12″ wide, 6′ long, 1-1/2″ thick	D-10	115	.348		18.20	11.75	4.66	34.61	44
	1050	2″ thick		105	.381		19.10	12.85	5.10	37.05	47
	1100	Smooth finish, 1-1/2″ thick		115	.348		22	11.75	4.66	38.41	48
	1150	2″ thick		105	.381		22	12.85	5.10	39.95	50
	1300	Thermal finish		115	.348		27.50	11.75	4.66	43.91	54
	1350	2″ thick	↓	105	.381	↓	31	12.85	5.10	48.95	60
	2000	Coping, finished top & 2 sides, 12″ to 6′									
	2100	Natural cleft, 1-1/2″ thick	D-10	115	.348	L.F.	19.10	11.75	4.66	35.51	45
	2150	2″ thick		105	.381		22	12.85	5.10	39.95	50
	2200	Smooth finish, 1-1/2″ thick		115	.348		27.50	11.75	4.66	43.91	54
	2250	2″ thick		105	.381		28.50	12.85	5.10	46.45	57.50
	2300	Thermal finish, 1-1/2″ thick		115	.348		31	11.75	4.66	47.41	58
	2350	2″ thick	↓	105	.381	↓	31	12.85	5.10	48.95	60
	9000	Minimum labor/equipment charge	D-1	2.50	6.400	Job		206		206	335
300	0010	**GRANITE** Cut to size									
	0050	Veneer, polished face, 3/4″ to 1-1/2″ thick									
	0150	Low price, gray, light gray, etc.	D-10	130	.308	S.F.	22.50	10.40	4.13	37.03	46
	0180	Medium price, pink, brown, etc.		130	.308		24.50	10.40	4.13	39.03	48.50
	0220	High price, red, black, etc.	↓	130	.308	↓	37.50	10.40	4.13	52.03	63
	0300	1-1/2″ to 2-1/2″ thick, veneer									
	0350	Low price, gray, light gray, etc.	D-10	130	.308	S.F.	24.50	10.40	4.13	39.03	48.50
	0500	Medium price, pink, brown, etc.		130	.308		31	10.40	4.13	45.53	55.50
	0550	High price, red, black, etc.	↓	130	.308	↓	42.50	10.40	4.13	57.03	68.50
	0700	2-1/2″ to 4″ thick, veneer									
	0750	Low price, gray, light gray, etc.	D-10	110	.364	S.F.	31	12.25	4.88	48.13	59
	0850	Medium price, pink, brown, etc.		110	.364		37	12.25	4.88	54.13	65.50
	0950	High price, red, black, etc.	↓	110	.364		49.50	12.25	4.88	66.63	79.50
	1000	For bush hammered finish, deduct					5%				
	1050	Coarse rubbed finish, deduct					10%				
	1100	Honed finish, deduct					5%				
	1150	Thermal finish, deduct				↓	18%				
	1800	Carving or bas-relief, from templates or plaster molds									
	1850	Minimum	D-10	80	.500	C.F.	140	16.85	6.70	163.55	188
	1900	Maximum	″	80	.500	″	420	16.85	6.70	443.55	500
	2000	Intricate or hand finished pieces									
	2010	Mouldings, radius cuts, bullnose edges, etc.									
	2050	Add, minimum					30%				
	2100	Add, maximum					300%				
	2500	Steps, copings, etc., finished on more than one surface									
	2550	Minimum	D-10	50	.800	C.F.	84	27	10.75	121.75	148
	2600	Maximum	″	50	.800	″	135	27	10.75	172.75	203
	2700	Pavers, Belgian block, 8″-13″ long, 4″-6″ wide, 4″-6″ deep	D-11	120	.200	S.F.	21.50	6.85		28.35	34.50
	2800	Pavers, 4″ x 4″ x 4″ blocks, split face and joints									
	2850	Minimum	D-11	80	.300	S.F.	12.35	10.30		22.65	30.50
	2900	Maximum	″	80	.300	″	24.50	10.30		34.80	43.50
	3000	Pavers, 4″ x 4″ x 4″, thermal face, sawn joints									
	3050	Minimum	D-11	65	.369	S.F.	22.50	12.65		35.15	45
	3100	Maximum	″	65	.369	″	29.50	12.65		42.15	53
	3500	Curbing, city street type, See Division 02770-300									
	3700	Slope, 4-1/2″ x 12″, split face,									
	3710	sawn top, 2′ to 6′ lengths	D-10	300	.133	L.F.	9.55	4.50	1.79	15.84	19.70
	3800	Radius curbs, over 5′ radius, add				↓	50%				

	04850 | Stone Assemblies		CREW	DAILY OUTPUT	LABOR-HOURS	UNIT	2006 BARE COSTS MAT.	LABOR	EQUIP.	TOTAL	TOTAL INCL O&P	
300	3850	Under 5′ radius, add				L.F.	100%					300
	4000	Soffits, 2″ thick, minimum	D-13	35	1.371	S.F.	35	46.50	15.35	96.85	131	
	4100	Maximum		35	1.371		73	46.50	15.35	134.85	172	
	4200	4″ thick, minimum		35	1.371		50.50	46.50	15.35	112.35	148	
	4300	Maximum		35	1.371		94.50	46.50	15.35	156.35	196	
	9000	Minimum labor/equipment charge	D-1	2	8	Job		257		257	415	
350	0011	**LIGHTWEIGHT NATURAL STONE** Lava type										350
	0100	Veneer, rubble face, sawed back, irregular shapes	D-10	130	.308	S.F.	6.05	10.40	4.13	20.58	28	
	0200	Sawed face and back, irregular shapes	″	130	.308	″	6.05	10.40	4.13	20.58	28	
400	0010	**LIMESTONE,** Cut to size										400
	0020	Veneer facing panels										
	0750	5″ thick, 5′ x 14′ panels	D-10	275	.145	S.F.	50.50	4.91	1.95	57.36	65.50	
	1000	Sugarcube finish, 2″ Thick, 3′ x 5′ panels		275	.145		25	4.91	1.95	31.86	37.50	
	1050	3″ Thick, 4′ x 9′ panels		275	.145		37.50	4.91	1.95	44.36	51.50	
	1200	4″ Thick, 5′ x 11′ panels		275	.145		39.50	4.91	1.95	46.36	53.50	
	1400	Sugarcube, textured finish, 4-1/2″ thick, 5′ x 12′		275	.145		43.50	4.91	1.95	50.36	58	
	1450	5″ thick, 5′ x 14′ panels		275	.145		51.50	4.91	1.95	58.36	66.50	
	2000	Coping, sugarcube finish, top & 2 sides		30	1.333	C.F.	125	45	17.90	187.90	230	
	2100	Sills, lintels, jambs, trim, stops, sugarcube finish, average		20	2		125	67.50	27	219.50	277	
	2150	Detailed		20	2		125	67.50	27	219.50	277	
	2300	Steps, extra hard, 14″ wide, 6″ rise		50	.800	L.F.	22.50	27	10.75	60.25	80.50	
	3000	Quoins, plain finish, 6″x12″x12″	D-12	25	1.280	Ea.	125	42		167	206	
	3050	6″x16″x24″	″	25	1.280	″	167	42		209	251	
	9000	Minimum labor/equipment charge	D-1	2	8	Job		257		257	415	
500	0011	**MARBLE**, ashlar, split face, 4″ + or - thick, random										500
	0040	lengths 1′ to 4′ & heights 2″ to 7-1/2″, average	D-8	175	.229	S.F.	14.55	7.55		22.10	28.50	
	0100	Base, polished, 3/4″ or 7/8″ thick, polished, 6″ high	D-10	65	.615	L.F.	13.25	21	8.25	42.50	57	
	0300	Carvings or bas relief, from templates, average		80	.500	S.F.	118	16.85	6.70	141.55	164	
	0350	Maximum		80	.500	″	276	16.85	6.70	299.55	340	
	0600	Columns, cornices, mouldings, etc.										
	0650	Hand or special machine cut, average	D-10	35	1.143	C.F.	118	38.50	15.35	171.85	209	
	0700	Maximum	″	35	1.143	″	253	38.50	15.35	306.85	360	
	1000	Facing, polished finish, cut to size, 3/4″ to 7/8″ thick										
	1050	Average	D-10	130	.308	S.F.	19.45	10.40	4.13	33.98	43	
	1100	Maximum		130	.308		45	10.40	4.13	59.53	71	
	1300	1-1/4″ thick, average		125	.320		28	10.80	4.29	43.09	53	
	1350	Maximum		125	.320		56.50	10.80	4.29	71.59	84	
	1500	2″ thick, average		120	.333		32.50	11.25	4.47	48.22	59	
	1550	Maximum		120	.333		56.50	11.25	4.47	72.22	85	
	1700	Rubbed finish, cut to size, 4″ thick										
	1740	Average	D-10	100	.400	S.F.	32.50	13.50	5.35	51.35	64	
	1780	Maximum	″	100	.400	″	63	13.50	5.35	81.85	97.50	
	2500	Flooring, polished tiles, 12″ x 12″ x 3/8″ thick										
	2510	Thin set, average	D-11	90	.267	S.F.	9.75	9.15		18.90	25.50	
	2600	Maximum		90	.267		150	9.15		159.15	180	
	2700	Mortar bed, average		65	.369		9.90	12.65		22.55	31.50	
	2740	Maximum		65	.369		150	12.65		162.65	186	
	2780	Travertine, 3/8″ thick, average	D-10	130	.308		12.40	10.40	4.13	26.93	35	
	2790	Maximum	″	130	.308		30.50	10.40	4.13	45.03	55	
	2800	Patio tile, non-slip, 1/2″ thick, flame finish	D-11	75	.320		12.65	10.95		23.60	32	
	2900	Shower or toilet partitions, 7/8″ thick partitions										
	3050	3/4″ or 1-1/4″ thick stiles, polished 2 sides, average	D-11	75	.320	S.F.	37.50	10.95		48.45	59.50	
	3201	Soffits, add to above prices				″	20%	100%				
	3210	Stairs, risers, 7/8″ thick x 6″ high	D-10	115	.348	L.F.	12.75	11.75	4.66	29.16	38	

	04850 | Stone Assemblies		CREW	DAILY OUTPUT	LABOR-HOURS	UNIT	2006 BARE COSTS MAT.	LABOR	EQUIP.	TOTAL	TOTAL INCL O&P	
500	3360	Treads, 12" wide x 1-1/4" thick	D-10	115	.348	L.F.	18.55	11.75	4.66	34.96	44.50	500
	3500	Thresholds, 3' long, 7/8" thick, 4" to 5" wide, plain	D-12	24	1.333	Ea.	14.05	43.50		57.55	86	
	3550	Beveled		24	1.333	"	16.30	43.50		59.80	88.50	
	3700	Window stools, polished, 7/8" thick, 5" wide		85	.376	L.F.	13.10	12.30		25.40	34.50	
	9000	Minimum labor/equipment charge	D-1	2	8	Job		257		257	415	
600	0011	**ROUGH STONE WALL**, Dry										600
	0150	Over 18" thick	D-12	63	.508	C.F.	10.70	16.60		27.30	39	
	9000	Minimum labor/equipment charge	D-1	2	8	Job		257		257	415	
700	0011	**SANDSTONE OR BROWNSTONE**										700
	0100	Sawed face veneer, 2-1/2" thick, to 2' x 4' panels	D-10	130	.308	S.F.	16.25	10.40	4.13	30.78	39	
	0150	4' thick, to 3'-6" x 8'panels		100	.400		16.25	13.50	5.35	35.10	46	
	0300	Split face, random sizes		100	.400		11.70	13.50	5.35	30.55	41	
	0350	Cut stone trim (limestone)										
	0360	Ribbon stone, 4" thick, 5' pieces	D-8	120	.333	Ea.	144	11		155	176	
	0370	Cove stone, 4" thick, 5' pieces		105	.381		144	12.60		156.60	179	
	0380	Cornice stone, 10" to 12" wide		90	.444		178	14.70		192.70	220	
	0390	Band stone, 4" thick, 5' pieces		145	.276		92	9.10		101.10	116	
	0410	Window and door trim, 3" to 4" wide		160	.250		78	8.25		86.25	99.50	
	0420	Key stone, 18" long		60	.667		82	22		104	126	
	9000	Minimum labor/equipment charge	D-1	2.50	6.400	Job		206		206	335	
800	0010	**SLATE** Pennsylvania, blue gray to gray black; Vermont,										800
	0050	Unfading green, mottled green & purple, gray & purple										
	0100	Virginia, blue black										
	0200	Exterior paving, natural cleft, 1" thick										
	0250	6" x 6" Pennsylvania	D-12	100	.320	S.F.	6.75	10.45		17.20	24.50	
	0300	Vermont		100	.320		8.80	10.45		19.25	26.50	
	0350	Virginia		100	.320		11.70	10.45		22.15	30	
	0500	24" x 24", Pennsylvania		120	.267		10.85	8.70		19.55	26	
	0550	Vermont		120	.267		13.65	8.70		22.35	29	
	0600	Virginia		120	.267		15.60	8.70		24.30	31.50	
	0700	18" x 30" Pennsylvania		120	.267		11.75	8.70		20.45	27	
	0750	Vermont		120	.267		13.65	8.70		22.35	29	
	0800	Virginia		120	.267		13.90	8.70		22.60	29.50	
	1000	Interior flooring, natural cleft, 1/2" thick										
	1100	6" x 6" Pennsylvania	D-12	100	.320	S.F.	3.75	10.45		14.20	21	
	1150	Vermont		100	.320		8.80	10.45		19.25	26.50	
	1200	Virginia		100	.320		9.25	10.45		19.70	27	
	1300	24" x 24" Pennsylvania		120	.267		7	8.70		15.70	22	
	1350	Vermont		120	.267		12	8.70		20.70	27.50	
	1400	Virginia		120	.267		11.25	8.70		19.95	26.50	
	1500	18" x 24" Pennsylvania		120	.267		7	8.70		15.70	22	
	1550	Vermont		120	.267		12	8.70		20.70	27.50	
	1600	Virginia		120	.267		11.40	8.70		20.10	26.50	
	2000	Facing panels, 1-1/4" thick, to 4' x 4' panels										
	2100	Natural cleft finish, Pennsylvania	D-10	180	.222	S.F.	26	7.50	2.98	36.48	44	
	2110	Vermont		180	.222		23	7.50	2.98	33.48	41	
	2120	Virginia		180	.222		25.50	7.50	2.98	35.98	43.50	
	2150	Sand rubbed finish, surface, add					5			5	5.50	
	2200	Honed finish, add					2.50			2.50	2.75	
	2500	Ribbon, natural cleft finish, 1" thick, to 9 S.F.	D-10	80	.500		10	16.85	6.70	33.55	45.50	
	2550	Sand rubbed finish		80	.500		13.75	16.85	6.70	37.30	49.50	
	2600	Honed finish		80	.500		12.75	16.85	6.70	36.30	48.50	
	2700	1-1/2" thick		78	.513		13	17.30	6.90	37.20	50	
	2750	Sand rubbed finish		78	.513		18	17.30	6.90	42.20	55.50	

04800 | Masonry Assemblies

	04850 | Stone Assemblies		CREW	DAILY OUTPUT	LABOR-HOURS	UNIT	2006 BARE COSTS MAT.	LABOR	EQUIP.	TOTAL	TOTAL INCL O&P	
800	2800	Honed finish	D-10	78	.513	S.F.	16.50	17.30	6.90	40.70	53.50	800
	2850	2″ thick		76	.526		15.60	17.75	7.05	40.40	53.50	
	2900	Sand rubbed finish		76	.526		22	17.75	7.05	46.80	60.50	
	2950	Honed finish		76	.526		20.50	17.75	7.05	45.30	59	
	3000	Roofing, see division 07310-800										
	3500	Stair treads, sand finish, 1″ thick x 12″ wide										
	3550	Under 3 L.F.	D-10	85	.471	L.F.	17.75	15.90	6.30	39.95	52	
	3600	3 L.F. to 6 L.F.	″	120	.333	″	17.90	11.25	4.47	33.62	43	
	3700	Ribbon, sand finish, 1″ thick x 12″ wide										
	3750	To 6 L.F.	D-10	120	.333	L.F.	14.75	11.25	4.47	30.47	39.50	
	4000	Stools or sills, sand finish, 1″ thick, 6″ wide	D-12	160	.200		9.25	6.55		15.80	21	
	4100	Honed finish		160	.200		8.75	6.55		15.30	20.50	
	4200	10″ wide		90	.356		14	11.60		25.60	34.50	
	4250	Honed finish		90	.356		13.25	11.60		24.85	33.50	
	4400	2″ thick, 6″ wide		140	.229		15.25	7.45		22.70	29	
	4450	Honed finish		140	.229		15	7.45		22.45	28.50	
	4600	10″ wide		90	.356		23.50	11.60		35.10	44.50	
	4650	Honed finish		90	.356		22.50	11.60		34.10	43.50	
	4800	For lengths over 3′, add					25%					
	9000	Minimum labor/equipment charge	D-1	2.50	6.400	Job		206		206	335	
900	0010	**WINDOW SILL**										900
	0020	Bluestone, thermal top, 10″ wide, 1-1/2″ thick	D-1	85	.188	S.F.	14.85	6.05		20.90	26	
	0050	2″ thick		75	.213	″	17.35	6.85		24.20	30	
	0100	Cut stone, 5″ x 8″ plain		48	.333	L.F.	10.20	10.70		20.90	28.50	
	0200	Face brick on edge, brick, 8″ wide		80	.200		2.15	6.45		8.60	12.80	
	0400	Marble, 9″ wide, 1″ thick		85	.188		7.50	6.05		13.55	18.05	
	0900	Slate, colored, unfading, honed, 12″ wide, 1″ thick		85	.188		15.25	6.05		21.30	26.50	
	0950	2″ thick		70	.229		21.50	7.35		28.85	35.50	
	9000	Minimum labor/equipment charge	1 Bric	2	4	Job		146		146	237	

	04880 | Masonry Fireplaces		CREW	DAILY OUTPUT	LABOR-HOURS	UNIT	MAT.	LABOR	EQUIP.	TOTAL	TOTAL INCL O&P	
600	0010	**FIREPLACE** For prefabricated fireplace, see Div. 10305-100										600
	0100	Brick fireplace, not incl. foundations or chimneys										
	0110	30″ x 29″ opening, incl. chamber, plain brickwork	D-1	.40	40	Ea.	415	1,275		1,690	2,525	
	0200	Fireplace box only (110 brick)	″	2	8	″	136	257		393	565	
	0300	For elaborate brickwork and details, add					35%	35%				
	0400	For hearth, brick & stone, add	D-1	2	8	Ea.	153	257		410	585	
	0410	For steel angle, damper, cleanouts, add		4	4		107	129		236	325	
	0600	Plain brickwork, incl. metal circulator		.50	32		790	1,025		1,815	2,550	
	0800	Face brick only, standard size, 8″ x 2-2/3″ x 4″		.30	53.333	M	415	1,725		2,140	3,225	
	0900	Stone fireplace, fieldstone, add				SF Face	10.90			10.90	12	
	1000	Cut stone, add				″	12			12	13.20	
	9000	Minimum labor/equipment charge	D-1	2	8	Job		257		257	415	

04900 | Masonry Restoration and Cleaning

	04910 | Unit Masonry Restoration		CREW	DAILY OUTPUT	LABOR-HOURS	UNIT	2006 BARE COSTS MAT.	LABOR	EQUIP.	TOTAL	TOTAL INCL O&P	
600	0010	**NEEDLE BEAM MASONRY** Incl. wood shoring 10′ x 10′ opening										600
	0400	Block, concrete, 8″ thick	B-9	7.10	5.634	Ea.	36.50	157	20.50	214	325	

04900 | Masonry Restoration and Cleaning

	04910 | Unit Masonry Restoration		Crew	Daily Output	Labor-Hours	Unit	2006 Bare Costs Mat.	Labor	Equip.	Total	Total Incl O&P
600	0420	12" thick	B-9	6.70	5.970	Ea.	42.50	166	22	230.50	345
	0800	Brick, 4" thick with 8" backup block		5.70	7.018		42.50	195	26	263.50	400
	1000	Brick, solid, 8" thick		6.20	6.452		36.50	179	23.50	239	365
	1040	12" thick		4.90	8.163		42.50	227	30	299.50	455
	1080	16" thick		4.50	8.889		53.50	247	32.50	333	505
	2000	Add for additional floors of shoring	B-1	6	4		36.50	112		148.50	227
	9000	Minimum labor/equipment charge	"	2	12	Job		335		335	560
720	0010	**POINTING MASONRY**									
	0300	Cut and repoint brick, hard mortar, running bond	1 Bric	80	.100	S.F.	.43	3.66		4.09	6.40
	0320	Common bond		77	.104		.43	3.80		4.23	6.60
	0360	Flemish bond		70	.114		.45	4.18		4.63	7.30
	0400	English bond		65	.123		.45	4.50		4.95	7.80
	0600	Soft old mortar, running bond		100	.080		.44	2.92		3.36	5.25
	0620	Common bond		96	.083		.43	3.05		3.48	5.40
	0640	Flemish bond		90	.089		.45	3.25		3.70	5.75
	0680	English bond		82	.098		.45	3.57		4.02	6.30
	0700	Stonework, hard mortar		140	.057	L.F.	.57	2.09		2.66	4.02
	0720	Soft old mortar		160	.050	"	.57	1.83		2.40	3.60
	1000	Repoint, mask and grout method, running bond		95	.084	S.F.	.57	3.08		3.65	5.60
	1020	Common bond		90	.089		.57	3.25		3.82	5.90
	1040	Flemish bond		86	.093		.57	3.40		3.97	6.15
	1060	English bond		77	.104		.57	3.80		4.37	6.80
	2000	Scrub coat, sand grout on walls, minimum		120	.067		2.77	2.44		5.21	7
	2020	Maximum		98	.082		1.99	2.98		4.97	7.05
	9000	Minimum labor/equipment charge		3	2.667	Job		97.50		97.50	158
750	0010	**SAWING**									
	0050	Brick or block by hand, per inch depth	A-1	125	.064	L.F.		1.75	.39	2.14	3.34
800	0010	**TOOTHING MASONRY**									
	0500	Brickwork, soft old mortar	1 Clab	40	.200	V.L.F.		5.50		5.50	9.10
	0520	Hard mortar		30	.267			7.30		7.30	12.10
	0700	Blockwork, soft old mortar		70	.114			3.13		3.13	5.20
	0720	Hard mortar		50	.160			4.38		4.38	7.25
	9000	Minimum labor/equipment charge		4	2	Job		55		55	91
	04930 | Unit Masonry Cleaning										
220	0010	**MASONRY CLEANING**									
	0200	Chemical cleaning, new construction, brush and wash, minimum	D-1	1000	.016	S.F.	.04	.51		.55	.87
	0220	Average		800	.020		.06	.64		.70	1.10
	0240	Maximum		600	.027		.08	.86		.94	1.47
	0260	Light restoration, minimum		800	.020		.07	.64		.71	1.12
	0270	Average		400	.040		.11	1.29		1.40	2.21
	0280	Maximum		330	.048		.15	1.56		1.71	2.69
	0300	Heavy restoration, minimum		600	.027		.08	.86		.94	1.48
	0310	Average		400	.040		.12	1.29		1.41	2.22
	0320	Maximum		250	.064		.16	2.06		2.22	3.51
	0400	High pressure water only, minimum	B-9	2000	.020			.56	.07	.63	1
	0420	Average		1500	.027			.74	.10	.84	1.34
	0440	Maximum		1000	.040			1.11	.15	1.26	2
	0800	High pressure water and chemical, minimum		1800	.022		.09	.62	.08	.79	1.21
	0820	Average		1200	.033		.13	.93	.12	1.18	1.82
	0840	Maximum		800	.050		.18	1.39	.18	1.75	2.69
	1200	Sandblast, wet system, minimum		1750	.023		.16	.64	.08	.88	1.32
	1220	Average		1100	.036		.24	1.01	.13	1.38	2.09
	1240	Maximum		700	.057		.32	1.59	.21	2.12	3.22
	1400	Dry system, minimum		2500	.016		.16	.44	.06	.66	.98

04900 | Masonry Restoration and Cleaning

	04930 Unit Masonry Cleaning		Crew	Daily Output	Labor-Hours	Unit	2006 Bare Costs Mat.	Labor	Equip.	Total	Total Incl O&P	
220	1420	Average	B-9	1750	.023	S.F.	.24	.64	.08	.96	1.41	220
	1440	Maximum		1000	.040		.32	1.11	.15	1.58	2.36	
	1800	For walnut shells, add					.41			.41	.45	
	1820	For corn chips, add					.41			.41	.45	
	2000	Steam cleaning, minimum	B-9	3000	.013			.37	.05	.42	.66	
	2020	Average		2500	.016			.44	.06	.50	.80	
	2040	Maximum		1500	.027			.74	.10	.84	1.34	
	4000	Add for masking doors and windows									.80	
	4200	Add for pedestrian protection				Job					10%	
	4400	Add for wire cut face brick				S.F.	.15			.15	.16	
	9000	Minimum labor/equipment charge	D-4	2	16	Job		510	72.50	582.50	900	
900	0010	**WASHING MASONRY**										900
	0012	Acid cleanser, smooth brick surface	1 Bric	560	.014	S.F.	.03	.52		.55	.88	
	0050	Rough brick		400	.020		.03	.73		.76	1.23	
	0060	Stone, acid wash		600	.013		.04	.49		.53	.84	
	1000	Muriatic acid, price per gallon in 5 gallon lots				Gal.	5.25			5.25	5.75	

For information about Means Estimating Seminars, see yellow pages 12 and 13 in back of book

Division 5 Metals

Estimating Tips

05050 Basic Metal Materials & Methods

- Nuts, bolts, washers, connection angles, and plates can add a significant amount to both the tonnage of a structural steel job as well as the estimated cost. As a rule of thumb, add 10% to the total weight to account for these accessories.
- Type 2 steel construction, commonly referred to as "simple construction," consists generally of field-bolted connections with lateral bracing supplied by other elements of the building, such as masonry walls or x-bracing. The estimator should be aware, however, that shop connections may be accomplished by welding or bolting. The method may be particular to the fabrication shop and may have an impact on the estimated cost.

05200 Metal Joists

- In any given project the total weight of open web steel joists is determined by the loads to be supported and the design. However, economies can be realized in minimizing the amount of labor used to place the joists. This is done by maximizing the joist spacing and therefore minimizing the number of joists required to be installed on the job. Certain spacings and locations may be required by the design, but in other cases maximizing the spacing and keeping it as uniform as possible will keep the costs down.

05300 Metal Deck

- The takeoff and estimating of metal deck involves more than simply the area of the floor or roof and the type of deck specified or shown on the drawings. Many different sizes and types of openings may exist. Small openings for individual pipes or conduits may be drilled after the floor/roof is installed, but larger openings may require special deck lengths as well as reinforcing or structural support. The estimator should determine who will be supplying this reinforcing. Additionally, some deck terminations are part of the deck package, such as screed angles and pour stops, and others will be part of the steel contract, such as angles attached to structural members and cast-in-place angles and plates. The estimator must ensure that all pieces are accounted for in the complete estimate.

05500 Metal Fabrications

- The most economical steel stairs are those that use common materials, standard details, and most importantly, a uniform and relatively simple method of field assembly. Commonly available A36 channels and plates are very good choices for the main stringers of the stairs, as are angles and tees for the carrier members. Risers and treads are usually made by specialty shops, and it is most economical to use a typical detail in as many places as possible. The stairs should be pre-assembled and shipped directly to the site. The field connections should be simple and straightforward to be accomplished efficiently and with a minimum of equipment and labor.

Reference Numbers

Reference numbers are shown in bold squares at the beginning of some major classifications. These numbers refer to related items in the Reference Section. The reference information may be an estimating procedure, an alternate pricing method, or technical information.

Note: Not all subdivisions listed here necessarily appear in this publication.

05060 | Selective Demolition

			CREW	DAILY OUTPUT	LABOR-HOURS	UNIT	2006 BARE COSTS MAT.	LABOR	EQUIP.	TOTAL	TOTAL INCL O&P	
110	0010	**SELECTIVE METALS DEMOLITION** R024119-10										110
	0015	Excludes shores, bracing, cutting, loading, hauling, dumping										
	0020	Remove nuts only up to 3/4" diameter	1 Sswk	480	.017	Ea.		.67		.67	1.25	
	0030	7/8" to 1-1/4" diameter		240	.033			1.33		1.33	2.49	
	0040	1-3/8" to 2" diameter		160	.050			2		2	3.74	
	0060	Unbolt and remove structural bolts up to 3/4" diameter		240	.033			1.33		1.33	2.49	
	0070	7/8" to 2" diameter		160	.050			2		2	3.74	
	0140	Light weight framing members, remove whole or cut up, up to 20 lb.		240	.033			1.33		1.33	2.49	
	0150	21 - 40 lb.	2 Sswk	210	.076			3.04		3.04	5.70	
	0160	41 - 80 lb.	3 Sswk	180	.133			5.35		5.35	9.95	
	0170	81 - 120 lb.	4 Sswk	150	.213			8.50		8.50	15.95	
	0230	Structural members, remove whole or cut up, up to 500 lb.	E-19	48	.500			19.50	19.85	39.35	57	
	0240	1/4 - 2 tons	E-18	36	1.111			44	26.50	70.50	109	
	0250	2 - 5 tons	E-24	30	1.067			42	21.50	63.50	99	
	0260	5 - 10 tons	E-20	24	2.667			104	47.50	151.50	240	
	0270	10 - 15 tons	E-2	18	3.111			121	79.50	200.50	305	
	0340	Fabricated item, remove whole or cut up, up to 20 lb.	1 Sswk	96	.083			3.33		3.33	6.20	
	0350	21 - 40 lb.	2 Sswk	84	.190			7.60		7.60	14.25	
	0360	41 - 80 lb.	3 Sswk	72	.333			13.30		13.30	25	
	0370	81 - 120 lb.	4 Sswk	60	.533			21.50		21.50	40	
	0380	121 - 500 lb.	E-19	48	.500			19.50	19.85	39.35	57	
	0390	501 - 1000 lb.	"	36	.667			26	26.50	52.50	75.50	
	2950	Minimum labor/equipment charge	B-13	2	28	Job		840	325	1,165	1,725	

05090 | Metal Fastenings

			CREW	DAILY OUTPUT	LABOR-HOURS	UNIT	MAT.	LABOR	EQUIP.	TOTAL	TOTAL INCL O&P	
080	0010	**ANCHOR BOLTS**										080
	0020	See also Division 04080-070										
	0100	J-type, incl. hex nut & washer, 1/2" diameter x 6" long	2 Carp	70	.229	Ea.	1.03	8.15		9.18	14.60	
	0110	12" long		65	.246		1.29	8.75		10.04	15.90	
	0120	18" long		60	.267		1.68	9.50		11.18	17.55	
	0130	3/4" diameter x 8" long		50	.320		1.54	11.40		12.94	20.50	
	0140	12" long		45	.356		1.92	12.65		14.57	23	
	0150	18" long		40	.400		2.50	14.20		16.70	26.50	
	0160	1" diameter x 12" long		35	.457		3.65	16.25		19.90	31	
	0170	18" long		30	.533		4.40	18.95		23.35	36.50	
	0180	24" long		25	.640		5.35	23		28.35	43.50	
	0190	36" long		20	.800		7.35	28.50		35.85	55	
	0200	1-1/2" diameter x 18" long		22	.727		13	26		39	57.50	
	0210	24" long		16	1		15.50	35.50		51	76	
	0300	L-type, incl. hex nut & washer, 3/4" diameter x 12" long		45	.356		1.44	12.65		14.09	22.50	
	0310	18" long		40	.400		1.85	14.20		16.05	25.50	
	0320	24" long		35	.457		2.25	16.25		18.50	29.50	
	0330	30" long		30	.533		2.86	18.95		21.81	34.50	
	0340	36" long		25	.640		3.26	23		26.26	41	
	0350	1" diameter x 12" long		35	.457		2.50	16.25		18.75	30	
	0360	18" long		30	.533		3.12	18.95		22.07	35	
	0370	24" long		25	.640		3.85	23		26.85	41.50	
	0380	30" long		23	.696		4.54	24.50		29.04	46	
	0390	36" long		20	.800		5.20	28.50		33.70	52.50	
	0400	42" long		18	.889		6.30	31.50		37.80	59.50	
	0410	48" long		15	1.067		7.10	38		45.10	71	
	0420	1-1/4" diameter x 18" long		25	.640		5.60	23		28.60	43.50	
	0430	24" long		22	.727		6.65	26		32.65	50.50	
	0440	30" long		20	.800		7.75	28.50		36.25	55.50	
	0450	36" long		18	.889		8.80	31.50		40.30	62	

05090 | Metal Fastenings

			CREW	DAILY OUTPUT	LABOR-HOURS	UNIT	2006 BARE COSTS MAT.	LABOR	EQUIP.	TOTAL	TOTAL INCL O&P
080	0460	42" long	2 Carp	16	1	Ea.	9.90	35.50		45.40	70
	0470	48" long		14	1.143		11.35	40.50		51.85	80
	0480	54" long		12	1.333		13.35	47.50		60.85	93
	0490	60" long		10	1.600		14.65	57		71.65	110
	0500	1-1/2" diameter x 18" long		22	.727		8.65	26		34.65	52.50
	0510	24" long		19	.842		10.10	30		40.10	60.50
	0520	30" long		17	.941		11.40	33.50		44.90	68
	0530	36" long		16	1		13.10	35.50		48.60	73.50
	0540	42" long		15	1.067		14.90	38		52.90	79.50
	0550	48" long		13	1.231		16.75	44		60.75	91
	0560	54" long		11	1.455		20.50	51.50		72	108
	0570	60" long		9	1.778		22.50	63		85.50	130
	0580	1-3/4" diameter x 18" long		20	.800		13.05	28.50		41.55	61.50
	0590	24" long		18	.889		15.30	31.50		46.80	69.50
	0600	30" long		17	.941		17.80	33.50		51.30	75
	0610	36" long		16	1		20.50	35.50		56	81.50
	0620	42" long		14	1.143		23	40.50		63.50	92.50
	0630	48" long		12	1.333		25	47.50		72.50	106
	0640	54" long		10	1.600		31	57		88	128
	0650	60" long		8	2		33.50	71		104.50	155
	0660	2" diameter x 24" long		17	.941		19.50	33.50		53	77
	0670	30" long		15	1.067		22	38		60	87
	0680	36" long		13	1.231		24	44		68	99
	0690	42" long		11	1.455		27	51.50		78.50	115
	0700	48" long		10	1.600		31	57		88	128
	0710	54" long		9	1.778		37	63		100	146
	0720	60" long		8	2		39.50	71		110.50	162
	0730	66" long		7	2.286		42.50	81.50		124	182
	0740	72" long		6	2.667		46.50	95		141.50	208
	0990	For galvanized, add					75%				
150	0010	**BOLTS & HEX NUTS** Steel, A307									
	0100	1/4" diameter, 1/2" long	1 Sswk	140	.057	Ea.	.07	2.28		2.35	4.35
	0200	1" long		140	.057		.08	2.28		2.36	4.36
	0300	2" long		130	.062		.11	2.46		2.57	4.72
	0400	3" long		130	.062		.16	2.46		2.62	4.78
	0500	4" long		120	.067		.18	2.66		2.84	5.15
	0600	3/8" diameter, 1" long		130	.062		.12	2.46		2.58	4.73
	0700	2" long		130	.062		.15	2.46		2.61	4.77
	0800	3" long		120	.067		.20	2.66		2.86	5.20
	0900	4" long		120	.067		.25	2.66		2.91	5.25
	1000	5" long		115	.070		.31	2.78		3.09	5.55
	1100	1/2" diameter, 1-1/2" long		120	.067		.22	2.66		2.88	5.20
	1200	2" long		120	.067		.25	2.66		2.91	5.25
	1300	4" long		115	.070		.38	2.78		3.16	5.60
	1400	6" long		110	.073		.52	2.91		3.43	6
	1500	8" long		105	.076		.68	3.04		3.72	6.45
	1600	5/8" diameter, 1-1/2" long		120	.067		.45	2.66		3.11	5.45
	1700	2" long		120	.067		.49	2.66		3.15	5.50
	1800	4" long		115	.070		.68	2.78		3.46	5.95
	1900	6" long		110	.073		.85	2.91		3.76	6.40
	2000	8" long		105	.076		1.23	3.04		4.27	7.05
	2100	10" long		100	.080		1.53	3.20		4.73	7.70
	2200	3/4" diameter, 2" long		120	.067		.70	2.66		3.36	5.75
	2300	4" long		110	.073		.97	2.91		3.88	6.50
	2400	6" long		105	.076		1.23	3.04		4.27	7.05
	2500	8" long		95	.084		1.83	3.36		5.19	8.30

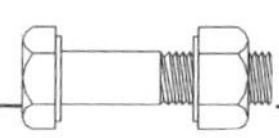

05090 | Metal Fastenings

		05090 Metal Fastenings	CREW	DAILY OUTPUT	LABOR-HOURS	UNIT	2006 BARE COSTS MAT.	LABOR	EQUIP.	TOTAL	TOTAL INCL O&P	
150	2600	10" long	1 Sswk	85	.094	Ea.	2.38	3.76		6.14	9.65	150
	2700	12" long		80	.100		2.78	4		6.78	10.50	
	2800	1" diameter, 3" long		105	.076		1.82	3.04		4.86	7.70	
	2900	6" long		90	.089		2.80	3.55		6.35	9.75	
	3000	12" long		75	.107		5.25	4.26		9.51	13.75	
	3100	For galvanized, add					75%					
	3200	For stainless, add					350%					
300	0010	**CHEMICAL ANCHORS**, Includes layout & drilling										300
	1430	Chemical anchor, w/rod & epoxy cartridge, 3/4" diam. x 9-1/2" long	B-89A	27	.593	Ea.	12.80	18.95	4.28	36.03	50	
	1435	1" diameter x 11-3/4" long		24	.667		25	21.50	4.81	51.31	68	
	1440	1-1/4" diameter x 14" long		21	.762		47	24.50	5.50	77	98	
	1445	1-3/4" diameter x 15" long		20	.800		89.50	25.50	5.80	120.80	147	
	1450	18" long		17	.941		107	30	6.80	143.80	175	
	1455	2" diameter x 18" long		16	1		137	32	7.20	176.20	210	
	1460	24" long		15	1.067		179	34	7.70	220.70	261	
	1500	Chemical anchoring, epoxy cartridge, excludes layout, drilling, fastener										
	1530	For fastener 3/4" dia. x 6" embedment	B-89A	27	.593	Ea.	5.05	18.95	4.28	28.28	41.50	
	1535	1" dia. x 8" embedment		24	.667		7.55	21.50	4.81	33.86	48.50	
	1540	1-1/4" dia. x 10" embedment		21	.762		15.10	24.50	5.50	45.10	62.50	
	1545	1-3/4" dia. x 12" embedment		20	.800		25	25.50	5.80	56.30	76	
	1550	14" embedment		17	.941		30	30	6.80	66.80	90.50	
	1555	2" dia. x 12" embedment		16	1		40.50	32	7.20	79.70	105	
	1560	18" embedment		15	1.067		50.50	34	7.70	92.20	120	
340	0010	**DRILLING**										340
	0050	Up to 4" deep in conc/brick floor/wall, incl. bit & layout, no anchor										
	0100	Holes, 1/4" diameter	1 Carp	75	.107	Ea.	.09	3.79		3.88	6.40	
	0150	For each additional inch of depth, add		430	.019		.02	.66		.68	1.13	
	0200	3/8" diameter		63	.127		.09	4.51		4.60	7.60	
	0250	For each additional inch of depth, add		340	.024		.02	.84		.86	1.41	
	0300	1/2" diameter		50	.160		.09	5.70		5.79	9.50	
	0350	For each additional inch of depth, add		250	.032		.02	1.14		1.16	1.90	
	0400	5/8" diameter		48	.167		.16	5.95		6.11	10	
	0450	For each additional inch of depth, add		240	.033		.04	1.18		1.22	2	
	0500	3/4" diameter		45	.178		.19	6.30		6.49	10.65	
	0550	For each additional inch of depth, add		220	.036		.05	1.29		1.34	2.19	
	0600	7/8" diameter		43	.186		.23	6.60		6.83	11.20	
	0650	For each additional inch of depth, add		210	.038		.06	1.35		1.41	2.30	
	0700	1" diameter		40	.200		.27	7.10		7.37	12.10	
	0750	For each additional inch of depth, add		190	.042		.07	1.50		1.57	2.55	
	0800	1-1/4" diameter		38	.211		.38	7.50		7.88	12.80	
	0850	For each additional inch of depth, add		180	.044		.10	1.58		1.68	2.72	
	0900	1-1/2" diameter		35	.229		.58	8.15		8.73	14.10	
	0950	For each additional inch of depth, add		165	.048		.14	1.72		1.86	3.02	
	1000	For ceiling installations, add						40%				
	1100	Drilling & layout for drywall/plaster walls, up to 1" deep, no anchor										
	1200	Holes, 1/4" diameter	1 Carp	150	.053	Ea.	.01	1.90		1.91	3.15	
	1300	3/8" diameter		140	.057		.01	2.03		2.04	3.38	
	1400	1/2" diameter		130	.062		.01	2.19		2.20	3.63	
	1500	3/4" diameter		120	.067		.02	2.37		2.39	3.96	
	1600	1" diameter		110	.073		.03	2.59		2.62	4.32	
	1700	1-1/4" diameter		100	.080		.05	2.84		2.89	4.76	
	1800	1-1/2" diameter		90	.089		.07	3.16		3.23	5.35	
	1900	For ceiling installations, add						40%				
	1910	Drilling & layout for steel, up to 1/4" deep, no anchor										
	1920	Holes, 1/4" diameter	1 Sswk	112	.071	Ea.	.12	2.85		2.97	5.50	

05050 | Basic Metal Materials & Methods

05090 | Metal Fastenings

		05090 Metal Fastenings	CREW	DAILY OUTPUT	LABOR-HOURS	UNIT	2006 BARE COSTS MAT.	LABOR	EQUIP.	TOTAL	TOTAL INCL O&P	
340	1925	For each additional 1/4" depth, add	1 Sswk	336	.024	Ea.	.12	.95		1.07	1.92	340
	1930	3/8" diameter		104	.077		.14	3.07		3.21	5.90	
	1935	For each additional 1/4" depth, add		312	.026		.14	1.02		1.16	2.07	
	1940	1/2" diameter		96	.083		.16	3.33		3.49	6.35	
	1945	For each additional 1/4" depth, add		288	.028		.16	1.11		1.27	2.25	
	1950	5/8" diameter		88	.091		.26	3.63		3.89	7.10	
	1955	For each additional 1/4" depth, add		264	.030		.26	1.21		1.47	2.54	
	1960	3/4" diameter		80	.100		.29	4		4.29	7.75	
	1965	For each additional 1/4" depth, add		240	.033		.29	1.33		1.62	2.81	
	1970	7/8" diameter		72	.111		.34	4.44		4.78	8.70	
	1975	For each additional 1/4" depth, add		216	.037		.34	1.48		1.82	3.15	
	1980	1" diameter		64	.125		.39	4.99		5.38	9.80	
	1985	For each additional 1/4" depth, add	↓	192	.042		.39	1.66		2.05	3.54	
	1990	For drilling up, add				↓		40%				
	2000	Minimum labor/equipment charge	1 Carp	4	2	Job		71		71	118	
380	0010	**EXPANSION ANCHORS**										380
	0100	Anchors for concrete, brick or stone, no layout and drilling										
	0200	Expansion shields, zinc, 1/4" diameter, 1-5/16" long, single	1 Carp	90	.089	Ea.	1.06	3.16		4.22	6.40	
	0300	1-3/8" long, double		85	.094		1.16	3.35		4.51	6.85	
	0400	3/8" diameter, 1-1/2" long, single		85	.094		1.74	3.35		5.09	7.45	
	0500	2" long, double 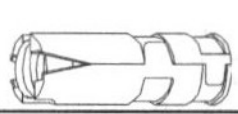		80	.100		2.15	3.56		5.71	8.25	
	0600	1/2" diameter, 2-1/16" long, single		80	.100		2.88	3.56		6.44	9.05	
	0700	2-1/2" long, double 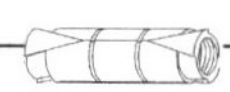		75	.107		2.78	3.79		6.57	9.35	
	0800	5/8" diameter, 2-5/8" long, single		75	.107		4.12	3.79		7.91	10.85	
	0900	2-3/4" long, double		70	.114		4.12	4.06		8.18	11.30	
	1000	3/4" diameter, 2-3/4" long, single		70	.114		6.10	4.06		10.16	13.50	
	1100	3-15/16" long, double		65	.123		8.15	4.38		12.53	16.25	
	1500	Self drilling anchor, snap-off, for 1/4" diameter bolt		26	.308		.91	10.95		11.86	19.10	
	1600	3/8" diameter bolt		23	.348		1.32	12.35		13.67	22	
	1700	1/2" diameter bolt 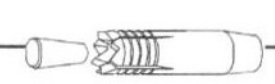		20	.400		2.03	14.20		16.23	25.50	
	1800	5/8" diameter bolt		18	.444		3.39	15.80		19.19	29.50	
	1900	3/4" diameter bolt	↓	16	.500	↓	5.70	17.80		23.50	36	
	2100	Hollow wall anchors for gypsum wall board, plaster or tile										
	2300	1/8" diameter, short	1 Carp	160	.050	Ea.	.29	1.78		2.07	3.27	
	2400	Long 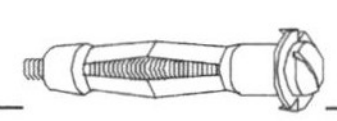		150	.053		.34	1.90		2.24	3.51	
	2500	3/16" diameter, short		150	.053		.61	1.90		2.51	3.81	
	2600	Long		140	.057		.65	2.03		2.68	4.09	
	2700	1/4" diameter, short		140	.057		.74	2.03		2.77	4.18	
	2800	Long		130	.062		.84	2.19		3.03	4.54	
	3000	Toggle bolts, bright steel, 1/8" diameter, 2" long		85	.094		.27	3.35		3.62	5.85	
	3100	4" long		80	.100		.41	3.56		3.97	6.35	
	3200	3/16" diameter, 3" long		80	.100		.46	3.56		4.02	6.40	
	3300	6" long 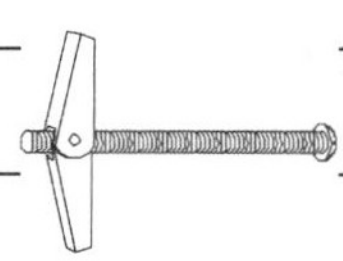		75	.107		.64	3.79		4.43	7	
	3400	1/4" diameter, 3" long		75	.107		.51	3.79		4.30	6.85	
	3500	6" long		70	.114		.73	4.06		4.79	7.55	
	3600	3/8" diameter, 3" long		70	.114		.99	4.06		5.05	7.85	
	3700	6" long		60	.133		1.73	4.74		6.47	9.75	
	3800	1/2" diameter, 4" long		60	.133		2.57	4.74		7.31	10.70	
	3900	6" long	↓	50	.160	↓	4.23	5.70		9.93	14.05	
	4000	Nailing anchors										
	4100	Nylon nailing anchor, 1/4" diameter, 1" long	1 Carp	3.20	2.500	C	19.80	89		108.80	169	
	4200	1-1/2" long		2.80	2.857		25.50	102		127.50	196	
	4300	2" long 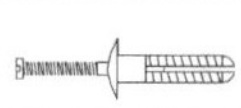		2.40	3.333		42.50	119		161.50	243	
	4400	Metal nailing anchor, 1/4" diameter, 1" long		3.20	2.500		30.50	89		119.50	181	
	4500	1-1/2" long	↓	2.80	2.857	↓	41.50	102		143.50	214	

METALS 5

5 METALS

05090 | Metal Fastenings

			CREW	DAILY OUTPUT	LABOR-HOURS	UNIT	2006 BARE COSTS MAT.	LABOR	EQUIP.	TOTAL	TOTAL INCL O&P	
380	4600	2″ long	1 Carp	2.40	3.333	C	52.50	119		171.50	254	380
	5000	Screw anchors for concrete, masonry,										
	5100	stone & tile, no layout or drilling included										
	5200	Jute fiber, #6, #8, & #10, 1″ long	1 Carp	240	.033	Ea.	.25	1.18		1.43	2.24	
	5300	#12, 1-1/2″ long		200	.040		.37	1.42		1.79	2.77	
	5400	#14, 2″ long 		160	.050		.57	1.78		2.35	3.58	
	5500	#16, 2″ long		150	.053		.60	1.90		2.50	3.80	
	5600	#20, 2″ long		140	.057		.96	2.03		2.99	4.43	
	5700	Lag screw shields, 1/4″ diameter, short		90	.089		.46	3.16		3.62	5.75	
	5800	Long		85	.094		.54	3.35		3.89	6.15	
	5900	3/8″ diameter, short 		85	.094		.84	3.35		4.19	6.45	
	6000	Long		80	.100		.99	3.56		4.55	7	
	6100	1/2″ diameter, short		80	.100		1.17	3.56		4.73	7.20	
	6200	Long		75	.107		1.46	3.79		5.25	7.90	
	6300	3/4″ diameter, short		70	.114		3.28	4.06		7.34	10.35	
	6400	Long		65	.123		3.97	4.38		8.35	11.60	
	6600	Lead, #6 & #8, 3/4″ long		260	.031		.16	1.09		1.25	1.99	
	6700	#10 - #14, 1-1/2″ long 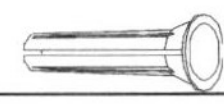		200	.040		.24	1.42		1.66	2.62	
	6800	#16 & #18, 1-1/2″ long		160	.050		.33	1.78		2.11	3.31	
	6900	Plastic, #6 & #8, 3/4″ long		260	.031		.11	1.09		1.20	1.93	
	7000	#8 & #10, 7/8″ long 		240	.033		.05	1.18		1.23	2.02	
	7100	#10 & #12, 1″ long		220	.036		.14	1.29		1.43	2.29	
	7200	#14 & #16, 1-1/2″ long	↓	160	.050	↓	.08	1.78		1.86	3.04	
	8000	Wedge anchors, not including layout or drilling										
	8050	Carbon steel, 1/4″ diameter, 1-3/4″ long	1 Carp	150	.053	Ea.	.46	1.90		2.36	3.65	
	8100	3 1/4″ long		140	.057		.61	2.03		2.64	4.04	
	8150	3/8″ diameter, 2-1/4″ long 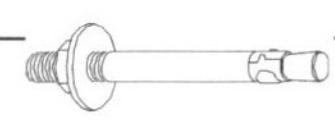		145	.055		.69	1.96		2.65	4.01	
	8200	5″ long		140	.057		1.21	2.03		3.24	4.70	
	8250	1/2″ diameter, 2-3/4″ long		140	.057		1.06	2.03		3.09	4.53	
	8300	7″ long		125	.064		1.81	2.28		4.09	5.75	
	8350	5/8″ diameter, 3-1/2″ long		130	.062		2.09	2.19		4.28	5.90	
	8400	8-1/2″ long		115	.070		4.45	2.47		6.92	9	
	8450	3/4″ diameter, 4-1/4″ long		115	.070		2.52	2.47		4.99	6.90	
	8500	10″ long		95	.084		5.75	2.99		8.74	11.25	
	8550	1″ diameter, 6″ long		100	.080		8.45	2.84		11.29	14	
	8575	9″ long		85	.094		11	3.35		14.35	17.65	
	8600	12″ long		75	.107		11.85	3.79		15.64	19.35	
	8650	1-1/4″ diameter, 9″ long		70	.114		15.40	4.06		19.46	23.50	
	8700	12″ long	↓	60	.133	↓	19.70	4.74		24.44	29.50	
	8750	For type 303 stainless steel, add					350%					
	8800	For type 316 stainless steel, add					450%					
	8950	Self-drilling concrete screw, hex washer head, 3/16″ dia. x 1-3/4″ long	1 Carp	300	.027	Ea.	.35	.95		1.30	1.96	
	8960	2-1/4″ long		250	.032		.54	1.14		1.68	2.47	
	8970	Phillips flat head, 3/16″ dia. x 1-3/4″ long		300	.027		.36	.95		1.31	1.97	
	8980	2-1/4″ long		250	.032	↓	.53	1.14		1.67	2.46	
	9000	Minimum labor/equipment charge	↓	4	2	Job		71		71	118	
420	0010	**HIGH STRENGTH BOLTS** R050523-10										420
	0020	A325 Type 1, structural steel, bolt-nut-washer set										
	0100	1/2″ diameter x 1-1/2″ long	1 Sswk	130	.062	Ea.	.36	2.46		2.82	4.99	
	0120	2″ long		125	.064		.38	2.56		2.94	5.20	
	0150	3″ long		120	.067		.49	2.66		3.15	5.50	
	0170	5/8″ diameter x 1-1/2″ long		125	.064		.55	2.56		3.11	5.40	
	0180	2″ long		120	.067		.58	2.66		3.24	5.60	
	0190	3″ long		115	.070		.68	2.78		3.46	5.95	
	0200	3/4″ diameter x 2″ long		120	.067		.81	2.66		3.47	5.85	
	0220	3″ long	↓	115	.070	↓	.93	2.78		3.71	6.20	

	05090 | Metal Fastenings		CREW	DAILY OUTPUT	LABOR-HOURS	UNIT	2006 BARE COSTS MAT.	LABOR	EQUIP.	TOTAL	TOTAL INCL O&P	
420	0250	4″ long R050523-10	1 Sswk	110	.073	Ea.	1.09	2.91		4	6.65	420
	0300	6″ long		105	.076		1.36	3.04		4.40	7.20	
	0350	8″ long		95	.084		2.48	3.36		5.84	9.05	
	0360	7/8″ diameter x 2″ long		115	.070		1.21	2.78		3.99	6.55	
	0365	3″ long		110	.073		1.38	2.91		4.29	6.95	
	0370	4″ long		105	.076		1.61	3.04		4.65	7.45	
	0380	6″ long		100	.080		1.97	3.20		5.17	8.15	
	0390	8″ long		90	.089		2.98	3.55		6.53	9.95	
	0400	1″ diameter x 2″ long		105	.076		1.69	3.04		4.73	7.55	
	0420	3″ long		100	.080		1.85	3.20		5.05	8.05	
	0450	4″ long		95	.084		2.04	3.36		5.40	8.55	
	0500	6″ long		90	.089		2.57	3.55		6.12	9.45	
	0550	8″ long		85	.094		4.13	3.76		7.89	11.60	
	0600	1-1/4″ diameter x 3″ long		85	.094		3.81	3.76		7.57	11.25	
	0650	4″ long		80	.100		1.73	4		5.73	9.35	
	0700	6″ long		75	.107		4.96	4.26		9.22	13.40	
	0750	8″ long		70	.114		6.05	4.57		10.62	15.20	
	1020	A490, bolt-nut-washer set										
	1170	5/8″ diameter x 1-1/2″ long	1 Sswk	125	.064	Ea.	.79	2.56		3.35	5.65	
	1180	2″ long		120	.067		.90	2.66		3.56	5.95	
	1190	3″ long		115	.070		1.05	2.78		3.83	6.35	
	1200	3/4″ diameter x 2″ long		120	.067		1.05	2.66		3.71	6.15	
	1220	3″ long		115	.070		1.19	2.78		3.97	6.50	
	1250	4″ long		110	.073		1.34	2.91		4.25	6.95	
	1300	6″ long		105	.076		1.85	3.04		4.89	7.75	
	1350	8″ long		95	.084		2.97	3.36		6.33	9.55	
	1360	7/8″ diameter x 2″ long		115	.070		1.42	2.78		4.20	6.75	
	1365	3″ long		110	.073		1.61	2.91		4.52	7.20	
	1370	4″ long		105	.076		1.91	3.04		4.95	7.80	
	1380	6″ long		100	.080		2.54	3.20		5.74	8.80	
	1390	8″ long		90	.089		3.53	3.55		7.08	10.55	
	1400	1″ diameter x 2″ long		105	.076		1.93	3.04		4.97	7.85	
	1420	3″ long		100	.080		2.21	3.20		5.41	8.45	
	1450	4″ long		95	.084		2.46	3.36		5.82	9	
	1500	6″ long		90	.089		3.09	3.55		6.64	10.05	
	1550	8″ long		85	.094		4.55	3.76		8.31	12.05	
	1600	1-1/4″ diameter x 3″ long		85	.094		4.44	3.76		8.20	11.95	
	1650	4″ long		80	.100		4.92	4		8.92	12.85	
	1700	6″ long		75	.107		6.30	4.26		10.56	14.90	
	1750	8″ long		70	.114		7.85	4.57		12.42	17.20	
460	0010	**LAG SCREWS**										460
	0020	Steel, 1/4″ diameter, 2″ long	1 Carp	200	.040	Ea.	.09	1.42		1.51	2.46	
	0100	3/8″ diameter, 3″ long		150	.053		.26	1.90		2.16	3.43	
	0200	1/2″ diameter, 3″ long 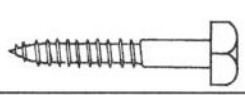		130	.062		.42	2.19		2.61	4.08	
	0300	5/8″ diameter, 3″ long		120	.067		.82	2.37		3.19	4.83	
500	0010	**MACHINE SCREWS**										500
	0020	Steel, round head, #8 x 1″ long	1 Carp	4.80	1.667	C	2.51	59.50		62.01	101	
	0110	#8 x 2″ long		2.40	3.333		5.45	119		124.45	202	
	0200	#10 x 1″ long		4	2		3.57	71		74.57	122	
	0300	#10 x 2″ long		2	4		6.70	142		148.70	243	
540	0010	**MACHINERY ANCHORS**, heavy duty, incl. sleeve, floating base nut,										540
	0020	lower stud & coupling nut, fiber plug, connecting stud, washer & nut.										
	0030	For flush mounted embedment in poured concrete heavy equip. pads.										
	0200	1/2″ diameter stud & bolt	E-16	40	.400	Ea.	58	16.40	2.23	76.63	97	

	05090 | Metal Fastenings		Crew	Daily Output	Labor-Hours	Unit	2006 Bare Costs Mat.	Labor	Equip.	Total	Total Incl O&P	
540	0300	5/8″ diameter	E-16	35	.457	Ea.	64	18.70	2.55	85.25	108	540
	0500	3/4″ diameter		30	.533		74	22	2.98	98.98	126	
	0600	7/8″ diameter		25	.640		81	26	3.57	110.57	142	
	0800	1″ diameter		20	.800		85	33	4.46	122.46	160	
	0900	1-1/4″ diameter		15	1.067		113	43.50	5.95	162.45	212	
580	0010	**POWDER ACTUATED TOOLS & FASTENERS**										580
	0020	Stud driver, .22 caliber, buy, minimum				Ea.	340			340	370	
	0100	Maximum				″	545			545	600	
	0300	Powder charges for above, low velocity				C	17.40			17.40	19.15	
	0400	Standard velocity					25			25	27.50	
	0600	Drive pins & studs, 1/4″ & 3/8″ diam., to 3″ long, minimum	1 Carp	4.80	1.667		13.05	59.50		72.55	112	
	0700	Maximum	″	4	2		51	71		122	174	
	0800	Pneumatic stud driver for 1/8″ diameter studs				Ea.	2,350			2,350	2,600	
	0900	Drive pins for above, 1/2″ to 3/4″ long	1 Carp	1	8	M	540	284		824	1,075	
600	0010	**RIVETS**										600
	0100	Aluminum rivet & mandrel, 1/2″ grip length x 1/8″ diameter	1 Carp	4.80	1.667	C	5.25	59.50		64.75	104	
	0200	3/16″ diameter		4	2		8.10	71		79.10	127	
	0300	Aluminum rivet, steel mandrel, 1/8″ diameter		4.80	1.667		8.05	59.50		67.55	107	
	0400	3/16″ diameter		4	2		7.35	71		78.35	126	
	0500	Copper rivet, steel mandrel, 1/8″ diameter		4.80	1.667		7.30	59.50		66.80	106	
	0600	Monel rivet, steel mandrel, 1/8″ diameter		4.80	1.667		26	59.50		85.50	127	
	0700	3/16″ diameter		4	2		75	71		146	201	
	0800	Stainless rivet & mandrel, 1/8″ diameter		4.80	1.667		13.05	59.50		72.55	112	
	0900	3/16″ diameter		4	2		24.50	71		95.50	145	
	1000	Stainless rivet, steel mandrel, 1/8″ diameter		4.80	1.667		10.20	59.50		69.70	109	
	1100	3/16″ diameter		4	2		18.45	71		89.45	139	
	1200	Steel rivet and mandrel, 1/8″ diameter		4.80	1.667		6.30	59.50		65.80	105	
	1300	3/16″ diameter		4	2		9.50	71		80.50	128	
	1400	Hand riveting tool, minimum				Ea.	122			122	134	
	1500	Maximum					233			233	257	
	1600	Power riveting tool, minimum					880			880	970	
	1700	Maximum					2,225			2,225	2,450	
820	0010	**VIBRATION PADS**										820
	0300	Laminated synthetic rubber impregnated cotton duck, 1/2″ thick	2 Sswk	24	.667	S.F.	59	26.50		85.50	115	
	0400	1″ thick		20	.800		119	32		151	191	
	0600	Neoprene bearing pads, 1/2″ thick		24	.667		22.50	26.50		49	74.50	
	0700	1″ thick		20	.800		45.50	32		77.50	110	
	0900	Fabric reinforced neoprene, 5000 psi, 1/2″ thick		24	.667		10	26.50		36.50	61	
	1000	1″ thick		20	.800		20	32		52	82	
	1200	Felt surfaced vinyl pads, cork and sisal, 5/8″ thick		24	.667		26	26.50		52.50	78.50	
	1300	1″ thick		20	.800		47	32		79	112	
	1500	Teflon bonded to 10 ga. carbon steel, 1/32″ layer		24	.667		46.50	26.50		73	101	
	1600	3/32″ layer		24	.667		69.50	26.50		96	127	
	1800	Bonded to 10 ga. stainless steel, 1/32″ layer		24	.667		82.50	26.50		109	141	
	1900	3/32″ layer		24	.667		107	26.50		133.50	168	
	2100	Circular machine leveling pad & stud				Kip	6.75			6.75	7.40	
840	0010	**WELD SHEAR CONNECTORS**										840
	0020	3/4″ diameter, 3-3/16″ long	E-10	960	.017	Ea.	.41	.68	.28	1.37	2.03	
	0030	3-3/8″ long		950	.017		.43	.69	.28	1.40	2.07	
	0200	3-7/8″ long		945	.017		.46	.69	.28	1.43	2.12	
	0300	4-3/16″ long		935	.017		.48	.70	.28	1.46	2.15	
	0500	4-7/8″ long		930	.017		.54	.70	.29	1.53	2.22	
	0600	5-3/16″ long		920	.017		.56	.71	.29	1.56	2.27	
	0800	5-3/8″ long		910	.018		.57	.72	.29	1.58	2.29	

	05090 | Metal Fastenings		CREW	DAILY OUTPUT	LABOR-HOURS	UNIT	2006 BARE COSTS MAT.	LABOR	EQUIP.	TOTAL	TOTAL INCL O&P
840	0900	6-3/16″ long	E-10	905	.018	Ea.	.62	.72	.29	1.63	2.35
	1000	7-3/16″ long		895	.018		.77	.73	.30	1.80	2.55
	1100	8-3/16″ long		890	.018		.85	.74	.30	1.89	2.64
	1500	7/8″ diameter, 3-11/16″ long		920	.017		.67	.71	.29	1.67	2.38
	1600	4-3/16″ long		910	.018		.71	.72	.29	1.72	2.46
	1700	5-3/16″ long		905	.018		.81	.72	.29	1.82	2.56
	1800	6-3/16″ long		895	.018		.91	.73	.30	1.94	2.70
	1900	7-3/16″ long		890	.018		1.01	.74	.30	2.05	2.82
	2000	8-3/16″ long		880	.018		1.10	.74	.30	2.14	2.93
	9000	Minimum labor/equipment charge	1 Sswk	2	4	Job		160		160	299
860	0010	WELD STUDS									
	0020	1/4″ diameter, 2-11/16″ long	E-10	1120	.014	Ea.	.27	.59	.24	1.10	1.64
	0100	4-1/8″ long		1080	.015		.25	.61	.25	1.11	1.68
	0200	3/8″ diameter, 4-1/8″ long		1080	.015		.29	.61	.25	1.15	1.72
	0300	6-1/8″ long		1040	.015		.38	.63	.26	1.27	1.87
	0400	1/2″ diameter, 2-1/8″ long		1040	.015		.27	.63	.26	1.16	1.76
	0500	3-1/8″ long		1025	.016		.33	.64	.26	1.23	1.85
	0600	4-1/8″ long		1010	.016		.39	.65	.26	1.30	1.93
	0700	5-5/16″ long		990	.016		.48	.66	.27	1.41	2.06
	0800	6-1/8″ long		975	.016		.52	.67	.27	1.46	2.13
	0900	8-1/8″ long		960	.017		.73	.68	.28	1.69	2.39
	1000	5/8″ diameter, 2-11/16″ long		1000	.016		.47	.66	.27	1.40	2.04
	1010	4-3/16″ long		990	.016		.58	.66	.27	1.51	2.17
	1100	6-9/16″ long		975	.016		.76	.67	.27	1.70	2.40
	1200	8-3/16″ long		960	.017		1.02	.68	.28	1.98	2.70
	9000	Minimum labor/equipment charge	1 Sswk	2	4	Job		160		160	299
880	0010	WELD ROD									
	0020	Steel, type 6011, 1/8″ dia., less than 500#				Lb.	1.90			1.90	2.09
	0100	500# to 2,000#					1.71			1.71	1.88
	0200	2,000# to 5,000#					1.61			1.61	1.77
	0300	5/32″ diameter, less than 500#					1.82			1.82	2
	0310	500# to 2,000#					1.64			1.64	1.80
	0320	2,000# to 5,000#					1.54			1.54	1.70
	0400	3/16″ dia., less than 500#					1.83			1.83	2.01
	0500	500# to 2,000#					1.65			1.65	1.82
	0600	2,000# to 5,000#					1.55			1.55	1.71
	0620	Steel, type 6010, 1/8″ dia., less than 500#					2.09			2.09	2.30
	0630	500# to 2,000#					1.88			1.88	2.07
	0640	2,000# to 5,000#					1.77			1.77	1.94
	0650	Steel, type 7018 Low Hydrogen, 1/8″ dia., less than 500#					1.79			1.79	1.97
	0660	500# to 2,000#					1.61			1.61	1.77
	0670	2,000# to 5,000#					1.51			1.51	1.66
	0700	Steel, type 7024 Jet Weld, 1/8″ dia., less than 500#					1.89			1.89	2.08
	0710	500# to 2,000#					1.70			1.70	1.87
	0720	2,000# to 5,000#					1.60			1.60	1.76
	1550	Aluminum, type 4043 TIG, 1/8″ dia., less than 10#					6.55			6.55	7.20
	1560	10# to 60#					5.90			5.90	6.50
	1570	Over 60#					5.55			5.55	6.10
	1600	Aluminum, type 5356 TIG, 1/8″ dia., less than 10#					7.25			7.25	8
	1610	10# to 60#					6.55			6.55	7.20
	1620	Over 60#					6.15			6.15	6.75
	1900	Cast iron, type 8 Nickel, 1/8″ dia., less than 500#					22.50			22.50	24.50
	1910	500# to 1,000#					20			20	22
	1920	Over 1,000#					19			19	21
	2000	Stainless steel, type 316/316L, 1/8″ dia., less than 500#					10.30			10.30	11.35
	2100	500# to 1000#					9.30			9.30	10.25

METALS 5

05050 | Basic Metal Materials & Methods

5 METALS

05090 | Metal Fastenings

			CREW	DAILY OUTPUT	LABOR-HOURS	UNIT	2006 BARE COSTS MAT.	LABOR	EQUIP.	TOTAL	TOTAL INCL O&P	
880	2220	Over 1000#				Lb.	8.75			8.75	9.60	880
900	0010	WELDING STRUCTURAL R050521-20										900
	0020	Field welding, 1/8″ E6011, cost per welder, no oper. engr	E-14	8	1	Hr.	3.80	42	11.15	56.95	95	
	0200	With 1/2 operating engineer	E-13	8	1.500		3.80	59.50	11.15	74.45	122	
	0300	With 1 operating engineer	E-12	8	2		3.80	77	11.15	91.95	150	
	0500	With no operating engineer, 2# weld rod per ton	E-14	8	1	Ton	3.80	42	11.15	56.95	95	
	0600	8# E6011 per ton	″	2	4		15.20	168	44.50	227.70	380	
	0800	With one operating engineer per welder, 2# E6011 per ton	E-12	8	2		3.80	77	11.15	91.95	150	
	0900	8# E6011 per ton	″	2	8		15.20	310	44.50	369.70	600	
	1200	Continuous fillet, stick welding, incl. equipment										
	1300	Single pass, 1/8″ thick, 0.1#/L.F.	E-14	150	.053	L.F.	.19	2.24	.59	3.02	5.05	
	1400	3/16″ thick, 0.2#/L.F.		75	.107		.38	4.47	1.19	6.04	10.10	
	1500	1/4″ thick, 0.3#/L.F.		50	.160		.57	6.70	1.78	9.05	15.15	
	1610	5/16″ thick, 0.4#/L.F.		38	.211		.76	8.85	2.35	11.96	19.90	
	1800	3 passes, 3/8″ thick, 0.5#/L.F.		30	.267		.95	11.20	2.97	15.12	25.50	
	2010	4 passes, 1/2″ thick, 0.7#/L.F.		22	.364		1.33	15.25	4.05	20.63	34.50	
	2200	5 to 6 passes, 3/4″ thick, 1.3#/L.F.		12	.667		2.47	28	7.45	37.92	63.50	
	2400	8 to 11 passes, 1″ thick, 2.4#/L.F.		6	1.333		4.56	56	14.85	75.41	126	
	2600	For all position welding, add, minimum						20%				
	2700	Maximum						300%				
	2900	For semi-automatic welding, deduct, minimum						5%				
	3000	Maximum						15%				
	4000	Cleaning and welding plates, bars, or rods										
	4010	to existing beams, columns, or trusses	E-14	12	.667	L.F.	.95	28	7.45	36.40	61.50	
	9000	Minimum labor/equipment charge	″	4	2	Job		84	22.50	106.50	182	
920	0010	STEEL CUTTING										920
	0020	Hand burning, incl. preparation, torch cutting & grinding, no staging										
	0100	Steel to 1/2″ thick	E-25	320	.025	L.F.		1.05	.25	1.30	2.24	
	0150	3/4″ thick		260	.031			1.29	.31	1.60	2.75	
	0200	1″ thick		200	.040			1.68	.41	2.09	3.59	
	9000	Minimum labor/equipment charge		2	4	Job		168	40.50	208.50	360	

05100 | Structural Metal Framing

05120 | Structural Steel

			CREW	DAILY OUTPUT	LABOR-HOURS	UNIT	2006 BARE COSTS MAT.	LABOR	EQUIP.	TOTAL	TOTAL INCL O&P	
140	0010	SUBPURLINS R051223-50										140
	0020	Bulb tees, painted, 32-5/8″ O.C., 40 psf L.L.										
	0100	Type 178, max 8′-9″ span, 2.15 plf, 2″ high x 1-5/8″ wide	E-1	4200	.006	S.F.	1.38	.22	.02	1.62	1.94	
	0200	Type 218, max 10′-2″ span, 3.19 plf, 2-1/8″ high x 2-1/8″ wide	″	3100	.008		1.60	.30	.03	1.93	2.33	
	1420	For 24-5/8″ spacing, add					33%	33%				
	1430	For 48-5/8″ spacing, deduct					50%	50%				
180	0010	CANOPY FRAMING										180
	0020	6″ and 8″ members	E-4	3000	.011	Lb.	1.14	.43	.03	1.60	2.09	
	9000	Minimum labor/equipment charge	1 Sswk	1	8	Job		320		320	600	
220	0010	CEILING SUPPORTS										220
	1000	Entrance door/folding partition supports	E-4	60	.533	L.F.	19	21.50	1.49	41.99	63	
	1100	Linear accelerator door supports		14	2.286		86.50	92.50	6.40	185.40	275	
	1200	Lintels or shelf angles, hung, exterior hot dipped galv.		267	.120		12.95	4.85	.33	18.13	23.50	

05120 | Structural Steel

		05120 Structural Steel	CREW	DAILY OUTPUT	LABOR-HOURS	UNIT	2006 BARE COSTS MAT.	LABOR	EQUIP.	TOTAL	TOTAL INCL O&P	
220	1250	Two coats primer paint instead of galv.	E-4	267	.120	L.F.	11.25	4.85	.33	16.43	22	220
	1400	Monitor support, ceiling hung, expansion bolted		4	8	Ea.	300	325	22.50	647.50	960	
	1450	Hung from pre-set inserts		6	5.333		325	216	14.90	555.90	775	
	1600	Motor supports for overhead doors		4	8		153	325	22.50	500.50	800	
	1700	Partition support for heavy folding partitions, without pocket		24	1.333	L.F.	43	54	3.72	100.72	153	
	1750	Supports at pocket only		12	2.667		86.50	108	7.45	201.95	305	
	2000	Rolling grilles & fire door supports		34	.941		37	38	2.63	77.63	115	
	2100	Spider-leg light supports, expansion bolted to ceiling slab		8	4	Ea.	124	162	11.15	297.15	455	
	2150	Hung from pre-set inserts		12	2.667	"	133	108	7.45	248.45	355	
	2400	Toilet partition support		36	.889	L.F.	43	36	2.48	81.48	117	
	2500	X-ray travel gantry support		12	2.667	"	148	108	7.45	263.45	375	
250	0010	**COLUMNS, LIGHTWEIGHT**										250
	1000	Lightweight units (lally), 3-1/2" diameter	E-2	780	.072	L.F.	2.99	2.79	1.83	7.61	10.30	
	1050	4" diameter	"	900	.062	"	4.40	2.42	1.59	8.41	10.90	
	5800	Adjustable jack post, 8' maximum height, 2-3/4" diameter				Ea.	32			32	35	
	5850	4" diameter				"	51			51	56	
260	0010	**COLUMNS, STRUCTURAL** (R051223-20)										260
	0020	Shop fab'd for 100-ton, 1-2 story project, bolted conn's.										
	0800	Steel, concrete filled, extra strong pipe, 3-1/2" diameter	E-2	660	.085	L.F.	31.50	3.30	2.16	36.96	43	
	0830	4" diameter		780	.072		35	2.79	1.83	39.62	45.50	
	0890	5" diameter		1020	.055		41.50	2.14	1.40	45.04	51.50	
	0930	6" diameter		1200	.047		55	1.82	1.19	58.01	65	
	0940	8" diameter		1100	.051		55	1.98	1.30	58.28	65.50	
	1100	For galvanizing, add				Lb.	.22			.22	.25	
	1300	For web ties, angles, etc., add per added lb.	1 Sswk	945	.008		.95	.34		1.29	1.68	
	1500	Steel pipe, extra strong, no concrete, 3" to 5" diameter	E-2	16000	.004		.95	.14	.09	1.18	1.39	
	1600	6" to 12" diameter		14000	.004		.95	.16	.10	1.21	1.44	
	1700	Steel pipe, extra strong, no concrete, 3" diameter x 12'-0"		60	.933	Ea.	117	36.50	24	177.50	220	
	1750	4" diameter x 12'-0"		58	.966		171	37.50	24.50	233	283	
	1800	6" diameter x 12'-0"		54	1.037		325	40.50	26.50	392	460	
	1850	8" diameter x 14'-0"		50	1.120		575	43.50	28.50	647	745	
	1900	10" diameter x 16'-0"		48	1.167		830	45.50	29.50	905	1,025	
	1950	12" diameter x 18'-0"		45	1.244		1,125	48.50	31.50	1,205	1,350	
	3300	Structural tubing, square, A500GrB, 4" to 6" square, light section		11270	.005	Lb.	.95	.19	.13	1.27	1.54	
	3600	Heavy section		32000	.002	"	.95	.07	.04	1.06	1.22	
	4000	Concrete filled, add				L.F.	3.47			3.47	3.81	
	4500	Structural tubing, sq, 4" x 4" x 1/4" x 12'-0"	E-2	58	.966	Ea.	157	37.50	24.50	219	267	
	4550	6" x 6" x 1/4" x 12'-0"		54	1.037		257	40.50	26.50	324	385	
	4600	8" x 8" x 3/8" x 14'-0"		50	1.120		555	43.50	28.50	627	720	
	4650	10" x 10" x 1/2" x 16'-0"		48	1.167		1,025	45.50	29.50	1,100	1,250	
	5100	Structural tubing, rect, 5" to 6" wide, light section		8000	.007	Lb.	.95	.27	.18	1.40	1.74	
	5200	Heavy section		12000	.005		.95	.18	.12	1.25	1.51	
	5300	7" to 10" wide, light section		15000	.004		.95	.15	.10	1.20	1.41	
	5400	Heavy section		18000	.003		.95	.12	.08	1.15	1.36	
	5500	Structural tubing, rect, 5" x 3" x 1/4" x 12'-0"		58	.966	Ea.	152	37.50	24.50	214	262	
	5550	6" x 4" x 5/16" x 12'-0"		54	1.037		238	40.50	26.50	305	365	
	5600	8" x 4" x 3/8" x 12'-0"		54	1.037		345	40.50	26.50	412	480	
	5650	10" x 6" x 3/8" x 14'-0"		50	1.120		555	43.50	28.50	627	720	
	5700	12" x 8" x 1/2" x 16'-0"		48	1.167		1,025	45.50	29.50	1,100	1,250	
	6800	W Shape, A992 steel, 2 tier, W8 x 24		1080	.052	L.F.	25	2.02	1.32	28.34	32.50	
	6850	W8 x 31		1080	.052		32.50	2.02	1.32	35.84	40.50	
	6900	W8 x 48		1032	.054		50	2.11	1.38	53.49	60.50	
	6950	W8 x 67		984	.057		70	2.21	1.45	73.66	82.50	
	7000	W10 x 45		1032	.054		47	2.11	1.38	50.49	57	

5 METALS

	05120 | Structural Steel		Crew	Daily Output	Labor-Hours	Unit	2006 Bare Costs Mat.	Labor	Equip.	Total	Total Incl O&P	
260	7050	W10 x 68 (R051223-20)	E-2	984	.057	L.F.	71	2.21	1.45	74.66	83.50	260
	7100	W10 x 112		960	.058		117	2.27	1.49	120.76	135	
	7150	W12 x 50		1032	.054		52.50	2.11	1.38	55.99	63	
	7200	W12 x 87		984	.057		91	2.21	1.45	94.66	106	
	7250	W12 x 120		960	.058		125	2.27	1.49	128.76	144	
	7300	W12 x 190		912	.061		199	2.39	1.56	202.95	224	
	7350	W14 x 74		984	.057		77.50	2.21	1.45	81.16	90.50	
	7400	W14 x 120		960	.058		125	2.27	1.49	128.76	144	
	7450	W14 x 176		912	.061		184	2.39	1.56	187.95	208	
	8090	For projects 75 to 99 tons, add				All	10%					
	8092	50 to 74 tons, add					20%					
	8094	25 to 49 tons, add					30%	10%				
	8096	10 to 24 tons, add					50%	25%				
	8098	2 to 9 tons, add					75%	50%				
	8099	Less than 2 tons, add					100%	100%				
	9000	Minimum labor/equipment charge	1 Sswk	1	8	Job		320		320	600	
300	0010	CURB EDGING										300
	0020	Steel angle w/anchors, on forms, 1″ x 1″, 0.8#/L.F.	E-4	350	.091	L.F.	1.61	3.70	.26	5.57	8.95	
	0100	2″ x 2″ angles, 3.92#/L.F.		330	.097		5.15	3.92	.27	9.34	13.35	
	0200	3″ x 3″ angles, 6.1#/L.F.		300	.107		8.15	4.31	.30	12.76	17.40	
	0300	4″ x 4″ angles, 8.2#/L.F.		275	.116		10.55	4.71	.32	15.58	21	
	1000	6″ x 4″ angles, 12.3#/L.F.		250	.128		15.25	5.20	.36	20.81	27	
	1050	Steel channels with anchors, on forms, 3″ channel, 5#/L.F.		290	.110		6.40	4.46	.31	11.17	15.75	
	1100	4″ channel, 5.4#/L.F.		270	.119		6.85	4.79	.33	11.97	16.85	
	1200	6″ channel, 8.2#/L.F.		255	.125		10.55	5.10	.35	16	21.50	
	1300	8″ channel, 11.5#/L.F.		225	.142		14.35	5.75	.40	20.50	27	
	1400	10″ channel, 15.3#/L.F.		180	.178		18.65	7.20	.50	26.35	34.50	
	1500	12″ channel, 20.7#/L.F.		140	.229		25	9.25	.64	34.89	45.50	
	2000	For curved edging, add					35%	10%				
	9000	Minimum labor/equipment charge	E-4	4	8	Job		325	22.50	347.50	630	
440	0010	LIGHTWEIGHT FRAMING (R051223-35)										440
	0200	For load-bearing steel studs see Division 05410-400										
	0400	Angle framing, field fabricated, 4″ and larger (R051223-45)	E-3	440	.055	Lb.	.55	2.22	.20	2.97	4.97	
	0450	Less than 4″ angles		265	.091	″	.57	3.68	.34	4.59	7.90	
	0460	1/2″ x 1/2″ x 1/8″		200	.120	L.F.	.11	4.87	.45	5.43	9.70	
	0462	3/4″ x 3/4″ x 1/8″		160	.150		.32	6.10	.56	6.98	12.35	
	0464	1″ x 1″ x 1/8″		135	.178		.46	7.20	.66	8.32	14.75	
	0466	1-1/4″ x 1-1/4″ x 3/16″		115	.209		.84	8.50	.78	10.12	17.65	
	0468	1-1/2″ x 1-1/2″ x 3/16″		100	.240		1.03	9.75	.89	11.67	20.50	
	0470	2″ x 2″ x 1/4″		90	.267		1.82	10.85	.99	13.66	23.50	
	0472	2-1/2″ x 2-1/2″ x 1/4″		72	.333		2.34	13.55	1.24	17.13	29.50	
	0474	3″ x 2″ x 3/8″		65	.369		3.36	15	1.37	19.73	33	
	0476	3″ x 3″ x 3/8″		57	.421		4.10	17.10	1.57	22.77	38	
	0600	Channel framing, field fabricated, 8″ and larger		500	.048	Lb.	.57	1.95	.18	2.70	4.48	
	0650	Less than 8″ channels		335	.072	″	.57	2.91	.27	3.75	6.35	
	0660	C2 x 1.78		115	.209	L.F.	1.01	8.50	.78	10.29	17.80	
	0662	C3 x 4.1		80	.300		2.34	12.20	1.12	15.66	27	
	0664	C4 x 5.4		66	.364		3.08	14.75	1.35	19.18	32.50	
	0666	C5 x 6.7		57	.421		3.82	17.10	1.57	22.49	38	
	0668	C6 x 8.2		55	.436		4.52	17.70	1.62	23.84	40	
	0670	C7 x 9.8		40	.600		5.60	24.50	2.23	32.33	54	
	0672	C8 x 11.5		36	.667		6.55	27	2.48	36.03	60.50	
	0710	Structural bar tee, field fabricated, 3/4″ x 3/4″ x 1/8″		160	.150		.32	6.10	.56	6.98	12.35	
	0712	1″ x 1″ x 1/8″		135	.178		.46	7.20	.66	8.32	14.75	
	0714	1-1/2″ x 1-1/2″ x 1/4″		114	.211		1.33	8.55	.78	10.66	18.35	
	0716	2″ x 2″ x 1/4″		89	.270		1.82	10.95	1	13.77	23.50	

		05120 | Structural Steel	CREW	DAILY OUTPUT	LABOR-HOURS	UNIT	2006 BARE COSTS MAT.	LABOR	EQUIP.	TOTAL	TOTAL INCL O&P	
440	0718	2-1/2" x 2-1/2" x 3/8" R051223-35	E-3	72	.333	L.F.	3.36	13.55	1.24	18.15	30.50	440
	0720	3" x 3" x 3/8"		57	.421		4.10	17.10	1.57	22.77	38	
	0730	Structural zee, field fabricated, 1-1/4" x 1-3/4" x 1-3/4" R051223-45		114	.211		.43	8.55	.78	9.76	17.35	
	0732	2-11/16" x 3" x 2-11/16"		114	.211		1.01	8.55	.78	10.34	18	
	0734	3-1/16" x 4" x 3-1/16"		133	.180		1.53	7.35	.67	9.55	16.15	
	0736	3-1/4" x 5" x 3-1/4"		133	.180		2.09	7.35	.67	10.11	16.75	
	0738	3-1/2" x 6" x 3-1/2"		160	.150		3.15	6.10	.56	9.81	15.50	
	0740	Junior beam, field fabricated, 3"		80	.300		3.25	12.20	1.12	16.57	28	
	0742	4"		72	.333		4.39	13.55	1.24	19.18	31.50	
	0744	5"		67	.358		5.70	14.55	1.33	21.58	34.50	
	0746	6"		62	.387		7.15	15.70	1.44	24.29	39	
	0748	7"		57	.421		8.70	17.10	1.57	27.37	43.50	
	0750	8"		53	.453		10.50	18.40	1.68	30.58	48	
	1000	Continuous slotted channel framing system, shop fab, min.	2 Sswk	2400	.007	Lb.	2.95	.27		3.22	3.74	
	1200	Maximum	"	1600	.010		3.33	.40		3.73	4.41	
	1250	Plate & bar stock for reinforcing beams and trusses					1.05			1.05	1.15	
	1300	Cross bracing, rods, shop fabricated, 3/4" diameter	E-3	700	.034		1.14	1.39	.13	2.66	3.99	
	1310	7/8" diameter		850	.028		1.14	1.15	.11	2.40	3.51	
	1320	1" diameter		1000	.024		1.14	.97	.09	2.20	3.17	
	1330	Angle, 5" x 5" x 3/8"		2800	.009		1.14	.35	.03	1.52	1.94	
	1350	Hanging lintels, shop fabricated, average		850	.028		1.14	1.15	.11	2.40	3.51	
	1380	Roof frames, shop fabricated, 3'-0" square, 5' span	E-2	4200	.013		1.14	.52	.34	2	2.55	
	1400	Tie rod, not upset, 1-1/2" to 4" diameter, with turnbuckle	2 Sswk	800	.020		1.24	.80		2.04	2.85	
	1420	No turnbuckle		700	.023		1.19	.91		2.10	3.02	
	1500	Upset, 1-3/4" to 4" diameter, with turnbuckle		800	.020		1.24	.80		2.04	2.85	
	1520	No turnbuckle		700	.023		1.19	.91		2.10	3.02	
	9000	Minimum labor/equipment charge		2	8	Job		320		320	600	
480	0010	**LINTELS**										480
	0020	Plain steel angles, under 500 lb.	1 Bric	550	.015	Lb.	.73	.53		1.26	1.66	
	0100	500 to 1000 lb.		640	.013		.71	.46		1.17	1.52	
	0200	1,000 to 2,000 lb.		640	.013		.69	.46		1.15	1.50	
	0300	2,000 to 4,000 lb.		640	.013		.67	.46		1.13	1.48	
	0500	For built-up angles and plates, add to above					.24			.24	.26	
	0700	For engineering, add to above					.10			.10	.10	
	0900	For galvanizing, add to above, under 500 lb.					.30			.30	.33	
	0950	500 to 2,000 lb.					.27			.27	.30	
	1000	Over 2,000 lb.					.22			.22	.25	
	2000	Steel angles, 3-1/2" x 3", 1/4" thick, 2'-6" long	1 Bric	47	.170	Ea.	10.25	6.20		16.45	21.50	
	2100	4'-6" long		26	.308		18.45	11.25		29.70	39	
	2500	3-1/2" x 3-1/2" x 5/16", 5'-0" long		18	.444		27.50	16.25		43.75	56.50	
	2600	4" x 3-1/2", 1/4" thick, 5'-0" long		21	.381		23.50	13.90		37.40	48.50	
	2700	9'-0" long		12	.667		42.50	24.50		67	86	
	2800	4" x 3-1/2" x 5/16", 7'-0" long		12	.667		41	24.50		65.50	84.50	
	2900	5" x 3-1/2" x 5/16", 10'-0" long		8	1		66	36.50		102.50	132	
	3500	For precast concrete lintels, see Div. 03480-400										
	9000	Minimum labor/equipment charge	1 Bric	4	2	Job		73		73	119	
520	0010	**PIPE SUPPORT FRAMING**										520
	0020	Under 10#/L.F.	E-4	3900	.008	Lb.	1.27	.33	.02	1.62	2.05	
	0200	10.1 to 15#/L.F.		4300	.007		1.25	.30	.02	1.57	1.96	
	0400	15.1 to 20#/L.F.		4800	.007		1.24	.27	.02	1.53	1.88	
	0600	Over 20#/L.F.		5400	.006		1.22	.24	.02	1.48	1.81	
560	0010	**PLATES** R051223-80										560
	0020	For connections & stiffener plates, shop fabricated										

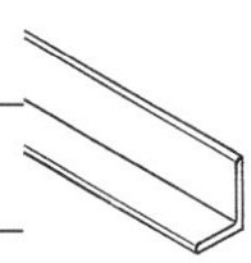

05100 | Structural Metal Framing

	05120	Structural Steel	CREW	DAILY OUTPUT	LABOR-HOURS	UNIT	2006 BARE COSTS MAT.	LABOR	EQUIP.	TOTAL	TOTAL INCL O&P	
560	0050	1/8″ thick (5.1 Lb./S.F.) R051223-80				S.F.	4.85			4.85	5.35	560
	0100	1/4″ thick (10.2 Lb./S.F.)					9.70			9.70	10.65	
	0300	3/8″ thick (15.3 Lb./S.F.)					14.55			14.55	16	
	0400	1/2″ thick (20.4 Lb./S.F.)					19.40			19.40	21.50	
	0450	3/4″ thick (30.6 Lb./S.F.)					29			29	32	
	0500	1″ thick (40.8 Lb.S.F.)					39			39	42.50	
600	0010	**STRESSED SKIN ROOF & CEILING SYSTEM**										600
	0020	Double panel flat roof, spans to 100′	E-2	1150	.049	S.F.	7.60	1.89	1.24	10.73	13.10	
	0100	Double panel convex roof, spans to 200′		960	.058		12.35	2.27	1.49	16.11	19.30	
	0200	Double panel arched roof, spans to 300′		760	.074		19	2.87	1.88	23.75	28	
640	0010	**STRUCTURAL STEEL MEMBERS** R051223-10										640
	0020	Shop fab'd for 100-ton, 1-2 story project, bolted conn's.										
	0102	W 6 x 9 R051223-15	E-2	600	.093	L.F.	9.40	3.63	2.38	15.41	19.45	
	0302	W 8 x 10		600	.093		10.45	3.63	2.38	16.46	20.50	
	0502	x 31		550	.102		32.50	3.96	2.59	39.05	45.50	
	0702	W 10 x 22		600	.093		23	3.63	2.38	29.01	34.50	
	0902	x 49		550	.102		51	3.96	2.59	57.55	66.50	
	1102	W 12 x 14		880	.064		14.65	2.48	1.62	18.75	22.50	
	1302	x 22		880	.064		23	2.48	1.62	27.10	31.50	
	1502	x 26		880	.064		27	2.48	1.62	31.10	36	
	1702	x 72		640	.088		75	3.40	2.23	80.63	91.50	
	1902	W 14 x 26		990	.057		27	2.20	1.44	30.64	35.50	
	2102	x 30		900	.062		31.50	2.42	1.59	35.51	40.50	
	2302	x 34		810	.069		35.50	2.69	1.76	39.95	46	
	2502	x 120		720	.078		125	3.03	1.98	130.01	146	
	2702	W 16 x 26		1000	.056		27	2.18	1.43	30.61	35.50	
	2902	x 31		900	.062		32.50	2.42	1.59	36.51	41.50	
	3102	x 40		800	.070		42	2.72	1.78	46.50	53	
	3302	W 18 x 35	E-5	960	.083		36.50	3.28	1.58	41.36	47.50	
	3502	x 40		960	.083		42	3.28	1.58	46.86	53.50	
	3702	x 50		912	.088		52.50	3.46	1.66	57.62	65.50	
	3902	x 55		912	.088		57.50	3.46	1.66	62.62	71	
	4102	W 21 x 44		1064	.075		46	2.96	1.42	50.38	57.50	
	4302	x 50		1064	.075		52.50	2.96	1.42	56.88	64.50	
	4502	x 62		1036	.077		65	3.04	1.46	69.50	78.50	
	4702	x 68		1036	.077		71	3.04	1.46	75.50	85	
	4902	W 24 x 55		1110	.072		57.50	2.84	1.37	61.71	69.50	
	5102	x 62		1110	.072		65	2.84	1.37	69.21	78	
	5302	x 68		1110	.072		71	2.84	1.37	75.21	84.50	
	5502	x 76		1110	.072		79.50	2.84	1.37	83.71	94	
	5702	x 84		1080	.074		88	2.92	1.40	92.32	103	
	5902	W 27 x 94		1190	.067		98	2.65	1.27	101.92	114	
	6102	W 30 x 99		1200	.067		103	2.63	1.26	106.89	120	
	6302	x 108		1200	.067		113	2.63	1.26	116.89	130	
	6502	x 116		1160	.069		121	2.72	1.31	125.03	139	
	6702	W 33 x 118		1176	.068		123	2.68	1.29	126.97	142	
	6902	x 130		1134	.071		136	2.78	1.34	140.12	156	
	7102	x 141		1134	.071		147	2.78	1.34	151.12	169	
	7302	W 36 x 135		1170	.068		141	2.70	1.30	145	161	
	7502	x 150		1170	.068		157	2.70	1.30	161	178	
	7702	x 194		1125	.071		203	2.80	1.35	207.15	230	
	7902	x 230		1125	.071		240	2.80	1.35	244.15	271	
	8102	x 300		1035	.077		315	3.05	1.46	319.51	350	
	8490	For projects 75 to 99 tons, add					10%					
	8492	50 to 74 tons, add					20%					
	8494	25 to 49 tons, add					30%	10%				

5 METALS

	05120 | Structural Steel			Crew	Daily Output	Labor-Hours	Unit	2006 Bare Costs Mat.	Labor	Equip.	Total	Total Incl O&P	
640	8496	10 to 24 tons, add	R051223 -10				L.F.	50%	25%				640
	8498	2 to 9 tons, add						75%	50%				
	8499	Less than 2 tons, add	R051223 -15					100%	100%				
	9000	Minimum labor/equipment charge		E-2	2	28	Job		1,100	715	1,815	2,725	
680	0010	**STRUCTURAL STEEL PROJECTS**	R050516 -30										680
	0020	Shop fab'd for 100-ton, 1-2 story project, bolted conn's.											
	0200	Apartments, nursing homes, etc., 1 to 2 stories	R050523 -10	E-5	10.30	7.767	Ton	1,900	305	147	2,352	2,825	
	0300	3 to 6 stories		"	10.10	7.921		1,950	310	150	2,410	2,850	
	0400	7 to 15 stories	R051223 -10	E-6	14.20	9.014		1,975	355	115	2,445	2,950	
	0500	Over 15 stories		"	13.90	9.209		2,050	365	118	2,533	3,050	
	0700	Offices, hospitals, etc., steel bearing, 1 to 2 stories	R051223 -15	E-5	10.30	7.767		1,900	305	147	2,352	2,825	
	0800	3 to 6 stories		E-6	14.40	8.889		1,950	350	114	2,414	2,900	
	0900	7 to 15 stories	R051223 -20		14.20	9.014		1,975	355	115	2,445	2,950	
	1000	Over 15 stories			13.90	9.209		2,050	365	118	2,533	3,050	
	1100	For multi-story masonry wall bearing construction, add	R051223 -25						30%				
	1300	Industrial bldgs., 1 story, beams & girders, steel bearing		E-5	12.90	6.202		1,900	244	118	2,262	2,675	
	1400	Masonry bearing		"	10	8		1,900	315	152	2,367	2,850	
	1500	Industrial bldgs., 1 story, under 10 tons,											
	1510	steel from warehouse, trucked		E-2	7.50	7.467	Ton	2,275	290	190	2,755	3,225	
	1600	1 story with roof trusses, steel bearing		E-5	10.60	7.547		2,250	298	143	2,691	3,175	
	1700	Masonry bearing		"	8.30	9.639		2,250	380	183	2,813	3,375	
	1900	Monumental structures, banks, stores, etc., minimum		E-6	13	9.846		1,900	390	126	2,416	2,950	
	2000	Maximum		"	9	14.222		3,150	560	182	3,892	4,700	
	2200	Churches, minimum		E-5	11.60	6.897		1,775	272	131	2,178	2,600	
	2300	Maximum		"	5.20	15.385		2,350	605	292	3,247	4,025	
	2800	Power stations, fossil fuels, minimum		E-6	11	11.636		1,900	460	149	2,509	3,100	
	2900	Maximum			5.70	22.456		2,850	885	287	4,022	5,075	
	2950	Nuclear fuels, non-safety steel, minimum			7	18.286		1,900	720	234	2,854	3,675	
	3000	Maximum			5.50	23.273		2,850	920	298	4,068	5,125	
	3040	Safety steel, minimum			2.50	51.200		2,775	2,025	655	5,455	7,450	
	3070	Maximum			1.50	85.333		3,650	3,375	1,100	8,125	11,400	
	3100	Roof trusses, minimum		E-5	13	6.154		2,650	243	117	3,010	3,500	
	3200	Maximum			8.30	9.639		3,225	380	183	3,788	4,450	
	3210	Schools, minimum			14.50	5.517		1,900	217	105	2,222	2,600	
	3220	Maximum			8.30	9.639		2,775	380	183	3,338	3,950	
	3400	Welded construction, simple commercial bldgs., 1 to 2 stories		E-7	7.60	10.526		1,950	415	211	2,576	3,100	
	3500	7 to 15 stories		E-9	8.30	15.422		2,250	610	226	3,086	3,825	
	3700	Welded rigid frame, 1 story, minimum		E-7	15.80	5.063		1,975	200	102	2,277	2,650	
	3800	Maximum		"	5.50	14.545		2,575	575	292	3,442	4,200	
	3810	Fabrication shop costs (included in project material cost, above)											
	3820	Mini mill base price, A992					Ton	525			525	580	
	3830	Mill extras and delivery to shop						185			185	204	
	3840	Shop extra, add for shop drawings and detailing						225			225	248	
	3850	Add for shop fabricating and handling						760			760	835	
	3860	Add for sandblasting and primer coat of paint						120			120	132	
	3870	Add for shop delivery to the job site						85			85	93.50	
	3880	Total material cost, shop fabricated, primed, delivered						1,900			1,900	2,100	
	3900	High strength steel mill spec extras: A242, A441,											
	3950	A529, A572 (42 ksi) and A992: same as A36 steel											
	4000	Add to A992 price for A572 (50, 60, 65 ksi)					Ton	30			30	33	
	4100	A588 Weathering					"	60			60	66	
	4200	Mill size extras for W-Shapes: 0 to 30 plf: no extra charge											
	4210	Member sizes 31 to 65 plf, add					Ton	1			1	1.10	
	4220	Member sizes 66 to 100 plf, add						1			1	1.10	
	4230	Member sizes 101 to 387 plf, add						30			30	33	
	4300	Column base plates, light, up to 150 lb.		2 Sswk	2000	.008	Lb.	1.05	.32		1.37	1.75	

METALS 5

05100 | Structural Metal Framing

05120 | Structural Steel

			Ref.	Crew	Daily Output	Labor-Hours	Unit	2006 Bare Costs Mat.	Labor	Equip.	Total	Total Incl O&P	
680	4400	Heavy, over 150 lb.	R050516-30	E-2	7500	.007	Lb.	1.09	.29	.19	1.57	1.93	680
	4600	Castellated beams, light sections, to 50#/L.F., minimum			10.70	5.234	Ton	2,000	204	133	2,337	2,700	
	4700	Maximum	R050523-10		7	8		2,175	310	204	2,689	3,175	
	4900	Heavy sections, over 50# per L.F., minimum			11.70	4.786		2,100	186	122	2,408	2,775	
	5000	Maximum	R051223-10		7.80	7.179		2,275	279	183	2,737	3,200	
	5390	For projects 75 to 99 tons, add						10%					
	5392	50 to 74 tons, add	R051223-15					20%					
	5394	25 to 49 tons, add						30%	10%				
	5396	10 to 24 tons, add	R051223-20					50%	25%				
	5398	2 to 9 tons, add						75%	50%				
	5399	Less than 2 tons, add	R051223-25					100%	100%				

05140 | Structural Aluminum

			Crew	Daily Output	Labor-Hours	Unit	Mat.	Labor	Equip.	Total	Total Incl O&P	
080	0010	**ALUMINUM**										080
	0020	Structural shapes, 1" to 10" members, under 1 ton	E-2	1050	.053	Lb.	2.30	2.07	1.36	5.73	7.75	
	0050	1 to 5 tons		1330	.042		2.19	1.64	1.07	4.90	6.55	
	0100	Over 5 tons		1330	.042		2.13	1.64	1.07	4.84	6.45	
	0300	Extrusions, over 5 tons, stock shapes		1330	.042		2.30	1.64	1.07	5.01	6.65	
	0400	Custom shapes		1330	.042		2.35	1.64	1.07	5.06	6.70	
	0600	For formed aluminum columns, see Div. 05580-200										

05150 | Wire Rope Assemblies

			Crew	Daily Output	Labor-Hours	Unit	Mat.	Labor	Equip.	Total	Total Incl O&P	
800	0010	**STEEL WIRE ROPE**										800
	0020	6 x 19, bright, fiber core, 5000' rolls, 1/2" diameter				L.F.	.84			.84	.93	
	0050	Steel core					1.11			1.11	1.22	
	0100	Fiber core, 1" diameter					2.84			2.84	3.13	
	0150	Steel core					3.24			3.24	3.57	
	0300	6 x 19, galvanized, fiber core, 1/2" diameter					1.24			1.24	1.37	
	0350	Steel core					1.42			1.42	1.56	
	0400	Fiber core, 1" diameter					3.64			3.64	4	
	0450	Steel core					3.82			3.82	4.20	
	0500	6 x 7, bright, IPS, fiber core, <500 L.F. w/acc., 1/4" diameter	E-17	6400	.003		.63	.10		.73	.88	
	0510	1/2" diameter		2100	.008		1.53	.31		1.84	2.26	
	0520	3/4" diameter		960	.017		2.77	.68	.01	3.46	4.34	
	0550	6 x 19, bright, IPS, IWRC, <500 L.F. w/acc., 1/4" diameter		5760	.003		.93	.11		1.04	1.24	
	0560	1/2" diameter		1730	.009		1.51	.38		1.89	2.37	
	0570	3/4" diameter		770	.021		2.62	.85	.01	3.48	4.48	
	0580	1" diameter		420	.038		4.44	1.56	.01	6.01	7.80	
	0590	1-1/4" diameter		290	.055		7.35	2.26	.02	9.63	12.35	
	0600	1-1/2" diameter		192	.083		9.05	3.41	.03	12.49	16.40	
	0610	1-3/4" diameter	E-18	240	.167		14.45	6.60	3.95	25	32	
	0620	2" diameter		160	.250		18.50	9.95	5.90	34.35	45	
	0630	2-1/4" diameter		160	.250		25	9.95	5.90	40.85	52	
	0650	6 x 37, bright, IPS, IWRC, <500 L.F. w/acc., 1/4" diameter	E-17	6400	.003		1.18	.10		1.28	1.49	
	0660	1/2" diameter		1730	.009		2	.38		2.38	2.92	
	0670	3/4" diameter		770	.021		3.24	.85	.01	4.10	5.15	
	0680	1" diameter		430	.037		5.15	1.52	.01	6.68	8.50	
	0690	1-1/4" diameter		290	.055		7.75	2.26	.02	10.03	12.80	
	0700	1-1/2" diameter		190	.084		11.10	3.45	.03	14.58	18.70	
	0710	1-3/4" diameter	E-18	260	.154		17.65	6.10	3.64	27.39	34.50	
	0720	2" diameter		200	.200		23	7.95	4.74	35.69	44.50	
	0730	2-1/4" diameter		160	.250		30	9.95	5.90	45.85	58	
	0800	6 x 19 & 6 x 37, swaged, 1/2" diameter	E-17	1220	.013		3.60	.54		4.14	4.96	
	0810	9/16" diameter		1120	.014		4.18	.59		4.77	5.70	

05150 | Wire Rope Assemblies

			CREW	DAILY OUTPUT	LABOR-HOURS	UNIT	2006 BARE COSTS MAT.	LABOR	EQUIP.	TOTAL	TOTAL INCL O&P	
800	0820	5/8″ diameter	E-17	930	.017	L.F.	4.96	.70	.01	5.67	6.80	800
	0830	3/4″ diameter		640	.025		6.30	1.02	.01	7.33	8.85	
	0840	7/8″ diameter		480	.033		7.95	1.36	.01	9.32	11.30	
	0850	1″ diameter		350	.046		9.70	1.87	.02	11.59	14.20	
	0860	1-1/8″ diameter		288	.056		11.95	2.28	.02	14.25	17.40	
	0870	1-1/4″ diameter		230	.070		14.50	2.85	.02	17.37	21.50	
	0880	1-3/8″ diameter	↓	192	.083		16.70	3.41	.03	20.14	25	
	0890	1-1/2″ diameter	E-18	300	.133	↓	20.50	5.30	3.16	28.96	35.50	
	1500	Thimbles, heavy duty, 1/4″	E-17	160	.100	Ea.	.54	4.10	.03	4.67	8.30	
	1510	1/2″		160	.100		2.39	4.10	.03	6.52	10.30	
	1520	3/4″		105	.152		5.45	6.25	.05	11.75	17.70	
	1530	1″		52	.308		10.85	12.60	.10	23.55	35.50	
	1540	1-1/4″		38	.421		16.75	17.25	.14	34.14	50.50	
	1550	1-1/2″		13	1.231		47	50.50	.42	97.92	146	
	1560	1-3/4″		8	2		97	82	.68	179.68	261	
	1570	2″		6	2.667		141	109	.91	250.91	360	
	1580	2-1/4″		4	4		191	164	1.36	356.36	515	
	1600	Clips, 1/4″ diameter		160	.100		2.45	4.10	.03	6.58	10.40	
	1610	3/8″ diameter		160	.100		2.68	4.10	.03	6.81	10.65	
	1620	1/2″ diameter		160	.100		4.31	4.10	.03	8.44	12.45	
	1630	3/4″ diameter		102	.157		7	6.40	.05	13.45	19.75	
	1640	1″ diameter		64	.250		11.65	10.25	.09	21.99	32	
	1650	1-1/4″ diameter		35	.457		19.10	18.70	.16	37.96	56	
	1670	1-1/2″ diameter		26	.615		26	25	.21	51.21	75.50	
	1680	1-3/4″ diameter		16	1		60	41	.34	101.34	143	
	1690	2″ diameter		12	1.333		67	54.50	.45	121.95	176	
	1700	2-1/4″ diameter		10	1.600		98.50	65.50	.54	164.54	232	
	2200	Jaw & jaw turnbuckles, 1/4″ x 4″		160	.100		19.50	4.10	.03	23.63	29	
	2250	1/2″ x 6″		96	.167		24.50	6.85	.06	31.41	40	
	2260	1/2″ x 9″		77	.208		33	8.50	.07	41.57	52	
	2270	1/2″ x 12″		66	.242		37	9.95	.08	47.03	59	
	2300	3/4″ x 6″		38	.421		48	17.25	.14	65.39	85	
	2310	3/4″ x 9″		30	.533		53.50	22	.18	75.68	99.50	
	2320	3/4″ x 12″		28	.571		68.50	23.50	.19	92.19	120	
	2330	3/4″ x 18″		23	.696		82	28.50	.24	110.74	144	
	2350	1″ x 6″		17	.941		93.50	38.50	.32	132.32	175	
	2360	1″ x 12″		13	1.231		103	50.50	.42	153.92	208	
	2370	1″ x 18″		10	1.600		154	65.50	.54	220.04	293	
	2380	1″ x 24″		9	1.778		169	73	.60	242.60	325	
	2400	1-1/4″ x 12″		7	2.286		172	93.50	.78	266.28	365	
	2410	1-1/4″ x 18″		6.50	2.462		213	101	.84	314.84	425	
	2420	1-1/4″ x 24″		5.60	2.857		287	117	.97	404.97	535	
	2450	1-1/2″ x 12″		5.20	3.077		289	126	1.05	416.05	555	
	2460	1-1/2″ x 18″		4	4		310	164	1.36	475.36	645	
	2470	1-1/2″ x 24″		3.20	5		415	205	1.70	621.70	840	
	2500	1-3/4″ x 18″		3.20	5		625	205	1.70	831.70	1,075	
	2510	1-3/4″ x 24″		2.80	5.714		715	234	1.94	950.94	1,225	
	2550	2″ x 24″	↓	1.60	10	↓	965	410	3.40	1,378.40	1,825	

05210 | Steel Joists

			CREW	DAILY OUTPUT	LABOR-HOURS	UNIT	2006 BARE COSTS MAT.	LABOR	EQUIP.	TOTAL	TOTAL INCL O&P
600	0010	**OPEN WEB JOISTS**									
	0020	K series, 40-ton job lots, horiz. bridging, spans to 30′, minimum	E-7	15	5.333	Ton	1,175	210	107	1,492	1,800
	0050	Average		12	6.667		1,325	263	134	1,722	2,075
	0080	Maximum		9	8.889		1,575	350	178	2,103	2,575
	0130	8K1, 5.1 Lb/LF		1200	.067	L.F.	3.36	2.63	1.34	7.33	9.95
	0140	10K1, 5.0 Lb/LF		1200	.067		3.30	2.63	1.34	7.27	9.90
	0160	12K3, 5.7 Lb/LF		1500	.053		3.76	2.10	1.07	6.93	9.15
	0180	14K3, 6.0 Lb/LF		1500	.053		3.96	2.10	1.07	7.13	9.35
	0200	16K3, 6.3 Lb/LF		1800	.044		4.16	1.75	.89	6.80	8.75
	0220	16K6, 8.1 Lb/LF		1800	.044		5.35	1.75	.89	7.99	10.05
	0240	18K5, 7.7 Lb/LF		2000	.040		5.10	1.58	.80	7.48	9.35
	0260	18K9, 10.2 Lb/LF		2000	.040		6.75	1.58	.80	9.13	11.15
	0410	Span 30′ to 50′, minimum		17	4.706	Ton	1,150	186	94.50	1,430.50	1,725
	0440	Average		17	4.706		1,300	186	94.50	1,580.50	1,875
	0460	Maximum		10	8		1,375	315	161	1,851	2,275
	0500	20K5, 8.2 Lb/LF		2000	.040	L.F.	5.35	1.58	.80	7.73	9.60
	0520	20K9, 10.8 Lb/LF		2000	.040		7	1.58	.80	9.38	11.45
	0540	22K5, 8.8 Lb/LF		2000	.040		5.70	1.58	.80	8.08	10.05
	0560	22K9, 11.3 Lb/LF		2000	.040		7.35	1.58	.80	9.73	11.85
	0580	24K6, 9.7 Lb/LF		2200	.036		6.30	1.43	.73	8.46	10.35
	0600	24K10, 13.1 Lb/LF		2200	.036		8.50	1.43	.73	10.66	12.75
	0620	26K6, 10.6 Lb/LF		2200	.036		6.90	1.43	.73	9.06	11
	0640	26K10, 13.8 Lb/LF		2200	.036		8.95	1.43	.73	11.11	13.25
	0660	28K8, 12.7 Lb/LF		2400	.033		8.25	1.31	.67	10.23	12.25
	0680	28K12, 17.1 Lb/LF		2400	.033		11.10	1.31	.67	13.08	15.40
	0700	30K8, 13.2 Lb/LF		2400	.033		8.60	1.31	.67	10.58	12.60
	0720	30K12, 17.6 Lb/LF		2400	.033		11.45	1.31	.67	13.43	15.75
	1010	CS series, horizontal bridging									
	1020	Spans to 30′, minimum	E-7	15	5.333	Ton	1,225	210	107	1,542	1,850
	1040	Average		12	6.667		1,350	263	134	1,747	2,125
	1060	Maximum		9	8.889		1,600	350	178	2,128	2,575
	1100	10CS2, 7.5 Lb/LF		1200	.067	L.F.	5.10	2.63	1.34	9.07	11.85
	1120	12CS2, 8.0 Lb/LF		1500	.053		5.40	2.10	1.07	8.57	10.95
	1140	14CS2, 8.0Lb/LF		1500	.053		5.40	2.10	1.07	8.57	10.95
	1160	16CS2, 8.5 Lb/LF		1800	.044		5.75	1.75	.89	8.39	10.50
	1180	16CS4, 14.5 Lb/LF		1800	.044		9.80	1.75	.89	12.44	14.95
	1200	18CS2, 9.0 Lb/LF		2000	.040		6.10	1.58	.80	8.48	10.45
	1220	18CS4, 15.0 Lb/LF		2000	.040		10.15	1.58	.80	12.53	14.90
	1240	20CS2, 9.5 Lb/LF		2000	.040		6.45	1.58	.80	8.83	10.85
	1260	20CS4, 16.5 Lb/LF		2000	.040		11.15	1.58	.80	13.53	16.05
	1280	22CS2, 10.0 Lb/LF		2000	.040		6.75	1.58	.80	9.13	11.20
	1300	22CS4, 16.5 Lb/LF		2000	.040		11.15	1.58	.80	13.53	16.05
	1320	24CS2, 10.0 Lb/LF		2200	.036		6.75	1.43	.73	8.91	10.85
	1340	24CS4, 16.5 Lb/LF		2200	.036		11.15	1.43	.73	13.31	15.70
	1360	26CS2, 10.0 Lb/LF		2200	.036		6.75	1.43	.73	8.91	10.85
	1380	26CS4, 16.5 Lb/LF		2200	.036		11.15	1.43	.73	13.31	15.70
	1400	28CS2, 10.5 Lb/LF		2400	.033		7.10	1.31	.67	9.08	10.95
	1420	28CS4, 16.5 Lb/LF		2400	.033		11.15	1.31	.67	13.13	15.45
	1440	30CS2, 11.0 Lb/LF		2400	.033		7.45	1.31	.67	9.43	11.35
	1460	30CS4, 16.5 Lb/LF		2400	.033		11.15	1.31	.67	13.13	15.45
	2000	LH series, bolted cross bridging									
	2020	Spans to 96′, minimum	E-7	16	5	Ton	1,325	197	100	1,622	1,950
	2040	Average		13	6.154		1,450	243	124	1,817	2,175
	2080	Maximum		11	7.273		1,725	287	146	2,158	2,575
	2200	18LH04, 12 Lb/LF		1400	.057	L.F.	8.75	2.25	1.15	12.15	15
600	2220	18LH08, 19 Lb/LF		1400	.057		13.90	2.25	1.15	17.30	20.50

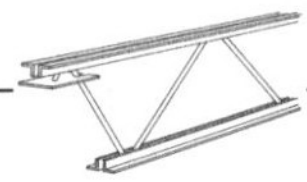

05210 | Steel Joists

		Description	CREW	DAILY OUTPUT	LABOR-HOURS	UNIT	2006 BARE COSTS MAT.	LABOR	EQUIP.	TOTAL	TOTAL INCL O&P	
600	2240	20LH04, 12 Lb/LF	E-7	1400	.057	L.F.	8.75	2.25	1.15	12.15	15	600
	2260	20LH08, 19 Lb/LF		1400	.057		13.90	2.25	1.15	17.30	20.50	
	2280	24LH05, 13 Lb/LF		1400	.057		9.50	2.25	1.15	12.90	15.80	
	2300	24LH10, 23 Lb/LF		1400	.057		16.80	2.25	1.15	20.20	24	
	2320	28LH06, 16 Lb/LF		1800	.044		11.70	1.75	.89	14.34	17	
	2340	28LH11, 25 Lb/LF		1800	.044		18.25	1.75	.89	20.89	24	
	2360	32LH08, 17 Lb/LF		1800	.044		12.40	1.75	.89	15.04	17.80	
	2380	32LH13, 30 Lb/LF		1800	.044		22	1.75	.89	24.64	28	
	2400	36LH09, 21 Lb/LF		1800	.044		15.35	1.75	.89	17.99	21	
	2420	36LH14, 36 Lb/LF		1800	.044		26.50	1.75	.89	29.14	33	
	2440	40LH10, 21 Lb/LF		2200	.036		15.35	1.43	.73	17.51	20.50	
	2460	40LH15, 36 Lb/LF		2200	.036		26.50	1.43	.73	28.66	32.50	
	2480	44LH11, 22 Lb/LF		2200	.036		16.10	1.43	.73	18.26	21	
	2500	44LH16, 42 Lb/LF		2200	.036		30.50	1.43	.73	32.66	37.50	
	2520	48LH11, 22 Lb/LF		2200	.036		16.10	1.43	.73	18.26	21	
	2540	48LH16, 42 Lb/LF		2200	.036		30.50	1.43	.73	32.66	37.50	
	3010	DLH series, bolted cross bridging										
	3020	Spans to 144′ (shipped in 2 pieces), minimum	E-7	16	5	Ton	1,450	197	100	1,747	2,075	
	3040	Average		13	6.154		1,550	243	124	1,917	2,275	
	3100	Maximum		11	7.273		1,875	287	146	2,308	2,750	
	3200	52DLH11, 26 Lb/LF		2000	.040	L.F.	19.45	1.58	.80	21.83	25.50	
	3220	52DLH16, 45 Lb/LF		2000	.040		35	1.58	.80	37.38	42.50	
	3240	56DLH11, 26 Lb/LF		2000	.040		20	1.58	.80	22.38	26	
	3260	56DLH16, 46 Lb/LF		2000	.040		36	1.58	.80	38.38	43.50	
	3280	60DLH12, 29 Lb/LF		2000	.040		22.50	1.58	.80	24.88	29	
	3300	60DLH17, 52 Lb/LF		2000	.040		40.50	1.58	.80	42.88	48.50	
	3320	64DLH12, 31 Lb/LF		2200	.036		24	1.43	.73	26.16	30	
	3340	64DLH17, 52 Lb/LF		2200	.036		40.50	1.43	.73	42.66	48	
	3360	68DLH13, 37 Lb/LF		2200	.036		29	1.43	.73	31.16	35	
	3380	68DLH18, 61 Lb/LF		2200	.036		47.50	1.43	.73	49.66	55.50	
	3400	72DLH14, 41 Lb/LF		2200	.036		32	1.43	.73	34.16	38.50	
	3420	72DLH19, 70 Lb/LF		2200	.036		54.50	1.43	.73	56.66	63.50	
	4010	SLH series, bolted cross bridging										
	4020	Spans to 200′ (shipped in 3 pieces), minimum	E-7	16	5	Ton	1,425	197	100	1,722	2,050	
	4040	Average		13	6.154		1,600	243	124	1,967	2,325	
	4060	Maximum		11	7.273		1,900	287	146	2,333	2,775	
	4200	80SLH15, 40 Lb/LF		1500	.053	L.F.	32	2.10	1.07	35.17	40	
	4220	80SLH20, 75 Lb/LF		1500	.053		60	2.10	1.07	63.17	71	
	4240	88SLH16, 46 Lb/LF		1500	.053		36.50	2.10	1.07	39.67	45.50	
	4260	88SLH21, 89 Lb/LF		1500	.053		71	2.10	1.07	74.17	83	
	4280	96SLH17, 52 Lb/LF		1500	.053		41.50	2.10	1.07	44.67	50.50	
	4300	96SLH22, 102 Lb/LF		1500	.053		81.50	2.10	1.07	84.67	94.50	
	4320	104SLH18, 59 Lb/LF		1800	.044		47	1.75	.89	49.64	56	
	4340	104SLH23, 109 Lb/LF		1800	.044		87	1.75	.89	89.64	100	
	4360	112SLH19, 67 Lb/LF		1800	.044		53.50	1.75	.89	56.14	63	
	4380	112SLH24, 131 Lb/LF		1800	.044		105	1.75	.89	107.64	119	
	4400	120SLH20, 77 Lb/LF		1800	.044		61.50	1.75	.89	64.14	71.50	
	4420	120SLH25, 152 Lb/LF		1800	.044		121	1.75	.89	123.64	138	
	6000	For welded cross bridging, add						30%				
	6100	For less than 40-ton job lots										
	6102	For 30 to 39 tons, add					10%					
	6104	20 to 29 tons, add					20%					
	6106	10 to 19 tons, add					30%					
	6107	5 to 9 tons, add					50%	25%				
	6108	1 to 4 tons, add					75%	50%				
	6109	Less than 1 ton, add					100%	100%				

05200 | Metal Joists

	05210 | Steel Joists		CREW	DAILY OUTPUT	LABOR-HOURS	UNIT	2006 BARE COSTS MAT.	LABOR	EQUIP.	TOTAL	TOTAL INCL O&P	
600	6200	For shop prime paint other than mfrs. standard, add					20%					600
	6300	For bottom chord extensions, add per chord				Ea.	28.50			28.50	31.50	
	7000	Joist girders, minimum	E-5	15	5.333	Ton	1,200	210	101	1,511	1,825	
	7020	Average		13	6.154		1,325	243	117	1,685	2,025	
	7040	Maximum		11	7.273		1,375	287	138	1,800	2,200	
	8000	Trusses, factory fabricated WT chords, average	↓	11	7.273	↓	4,325	287	138	4,750	5,425	

5 METALS

05300 | Metal Deck

	05310 | Steel Deck		CREW	DAILY OUTPUT	LABOR-HOURS	UNIT	2006 BARE COSTS MAT.	LABOR	EQUIP.	TOTAL	TOTAL INCL O&P	
300	0010	**METAL DECKING** Steel decking R053100-10										300
	0200	Cellular units, galv, 2″ deep, 20-20 gauge, over 15 squares	E-4	1460	.022	S.F.	5.85	.89	.06	6.80	8.20	
	0250	18-20 gauge		1420	.023		6.65	.91	.06	7.62	9.05	
	0300	18-18 gauge 		1390	.023		6.85	.93	.06	7.84	9.30	
	0320	16-18 gauge		1360	.024		8.15	.95	.07	9.17	10.80	
	0340	16-16 gauge		1330	.024		9.05	.97	.07	10.09	11.85	
	0400	3″ deep, galvanized, 20-20 gauge		1375	.023		6.45	.94	.06	7.45	8.95	
	0500	18-20 gauge		1350	.024		7.80	.96	.07	8.83	10.40	
	0600	18-18 gauge		1290	.025		7.75	1	.07	8.82	10.50	
	0700	16-18 gauge		1230	.026		8.75	1.05	.07	9.87	11.65	
	0800	16-16 gauge		1150	.028		9.55	1.13	.08	10.76	12.70	
	1000	4-1/2″ deep, galvanized, 20-18 gauge		1100	.029		9	1.18	.08	10.26	12.20	
	1100	18-18 gauge		1040	.031		8.95	1.24	.09	10.28	12.25	
	1200	16-18 gauge		980	.033		10.05	1.32	.09	11.46	13.60	
	1300	16-16 gauge	↓	935	.034	↓	10.95	1.38	.10	12.43	14.75	
	1500	For acoustical deck, add					15%					
	1700	For cells used for ventilation, add					15%					
	1900	For multi-story or congested site, add						50%				
	2100	Open type, galv., 1-1/2″ deep wide rib, 22 gauge, under 50 squares	E-4	4500	.007	S.F.	1.47	.29	.02	1.78	2.17	
	2400	Over 500 squares		5100	.006		1.06	.25	.02	1.33	1.65	
	2600	20 gauge, under 50 squares 		3865	.008		1.73	.33	.02	2.08	2.56	
	2700	Over 500 squares		4300	.007		1.24	.30	.02	1.56	1.95	
	2900	18 gauge, under 50 squares		3800	.008		2.24	.34	.02	2.60	3.13	
	3000	Over 500 squares		4300	.007		1.61	.30	.02	1.93	2.35	
	3050	16 gauge, under 50 squares		3700	.009		3.01	.35	.02	3.38	3.99	
	3100	Over 500 squares		4200	.008		2.17	.31	.02	2.50	2.99	
	3200	3″ deep, 22 gauge, under 50 squares		3600	.009		2.02	.36	.02	2.40	2.92	
	3300	20 gauge, under 50 squares		3400	.009		2.36	.38	.03	2.77	3.33	
	3400	18 gauge, under 50 squares		3200	.010		3.05	.40	.03	3.48	4.15	
	3500	16 gauge, under 50 squares		3000	.011		4.02	.43	.03	4.48	5.25	
	3700	4-1/2″ deep, long span roof, over 50 squares, 20 gauge		2700	.012		3.78	.48	.03	4.29	5.10	
	3800	18 gauge		2460	.013		4.88	.53	.04	5.45	6.35	
	3900	16 gauge		2350	.014		3.64	.55	.04	4.23	5.05	
	4100	6″ deep, long span, 18 gauge		2000	.016		7	.65	.04	7.69	8.95	
	4200	16 gauge		1930	.017		5.20	.67	.05	5.92	7	
	4300	14 gauge		1860	.017		6.70	.70	.05	7.45	8.70	
	4500	7-1/2″ deep, long span, 18 gauge		1690	.019		7.65	.77	.05	8.47	9.95	
	4600	16 gauge		1590	.020		5.70	.81	.06	6.57	7.90	
	4700	14 gauge	↓	1490	.021	↓	7.35	.87	.06	8.28	9.80	
	4800	For painted instead of galvanized, deduct					2%					

05300 | Metal Deck

	05310 Steel Deck		CREW	DAILY OUTPUT	LABOR-HOURS	UNIT	2006 BARE COSTS MAT.	LABOR	EQUIP.	TOTAL	TOTAL INCL O&P	
300	5000	For acoustical perforated, with fiberglass, add R053100-10				S.F.	1			1	1.10	300
	5200	Non-cellular composite deck, galv., 2" deep, 22 gauge	E-4	3860	.008		1.40	.34	.02	1.76	2.20	
	5300	20 gauge		3600	.009		1.55	.36	.02	1.93	2.41	
	5400	18 gauge 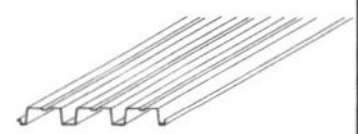		3380	.009		1.97	.38	.03	2.38	2.92	
	5500	16 gauge		3200	.010		2.46	.40	.03	2.89	3.50	
	5700	3" deep, galv., 22 gauge		3200	.010		1.53	.40	.03	1.96	2.47	
	5800	20 gauge		3000	.011		1.71	.43	.03	2.17	2.72	
	5900	18 gauge		2850	.011		2.10	.45	.03	2.58	3.19	
	6000	16 gauge		2700	.012		2.80	.48	.03	3.31	4.02	
	6100	Slab form, steel, 28 gauge, 9/16" deep, uncoated		4000	.008		.98	.32	.02	1.32	1.71	
	6200	Galvanized		4000	.008		.87	.32	.02	1.21	1.59	
	6220	24 gauge, 1" deep, uncoated 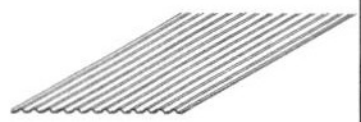		3900	.008		1.07	.33	.02	1.42	1.83	
	6240	Galvanized		3900	.008		1.26	.33	.02	1.61	2.04	
	6300	24 gauge, 1-5/16" deep, uncoated		3800	.008		1.14	.34	.02	1.50	1.92	
	6400	Galvanized		3800	.008		1.34	.34	.02	1.70	2.14	
	6500	22 gauge, 1-5/16" deep, uncoated		3700	.009		1.43	.35	.02	1.80	2.25	
	6600	Galvanized		3700	.009		1.46	.35	.02	1.83	2.29	
	6700	22 gauge, 2" deep uncoated		3600	.009		1.89	.36	.02	2.27	2.78	
	6800	Galvanized		3600	.009		1.85	.36	.02	2.23	2.74	
	7000	Sheet metal edge closure form, 12" wide with 2 bends, galv										
	7100	18 gauge	E-14	360	.022	L.F.	3.04	.93	.25	4.22	5.35	
	7200	16 gauge	"	360	.022	"	4.12	.93	.25	5.30	6.55	
	8000	Metal deck and trench, 2" thick, 20 gauge, combination										
	8010	60% cellular, 40% non-cellular, inserts and trench	R-4	1100	.036	S.F.	11.70	1.48	.08	13.26	15.60	
	9000	Minimum labor/equipment charge	1 Sswk	1	8	Job		320		320	600	

05400 | Cold-Formed Metal Framing

	05410 Load-Bearing Metal Studs		CREW	DAILY OUTPUT	LABOR-HOURS	UNIT	2006 BARE COSTS MAT.	LABOR	EQUIP.	TOTAL	TOTAL INCL O&P	
100	0010	**BRACING**, shear wall X-bracing, per 10' x 10' bay, one face										100
	0120	Metal strap, 20 ga. x 4" wide	2 Carp	18	.889	Ea.	18	31.50		49.50	72.50	
	0130	6" wide		18	.889		28	31.50		59.50	83	
	0160	18 ga. x 4" wide		16	1		26	35.50		61.50	87.50	
	0170	6" wide		16	1		38.50	35.50		74	101	
	0410	Continuous strap bracing, per horizontal row on both faces										
	0420	Metal strap, 20 ga. x 2" wide, studs 12" O.C.	1 Carp	7	1.143	C.L.F.	46	40.50		86.50	119	
	0430	16" O.C.		8	1		46	35.50		81.50	110	
	0440	24" O.C.		10	.800		46	28.50		74.50	98	
	0450	18 ga. x 2" wide, studs 12" O.C.		6	1.333		64	47.50		111.50	149	
	0460	16" O.C.		7	1.143		64	40.50		104.50	138	
	0470	24" O.C.		8	1		64	35.50		99.50	129	
120	0010	**BRIDGING**, solid between studs w/ 1-1/4" leg track, per stud bay										120
	0200	Studs 12" O.C., 18 ga. x 2-1/2" wide	1 Carp	125	.064	Ea.	.77	2.28		3.05	4.62	
	0210	3-5/8" wide		120	.067		.93	2.37		3.30	4.95	
	0220	4" wide		120	.067		.99	2.37		3.36	5	
	0230	6" wide		115	.070		1.29	2.47		3.76	5.50	
	0240	8" wide		110	.073		1.63	2.59		4.22	6.10	
	0300	16 ga. x 2-1/2" wide		115	.070		.97	2.47		3.44	5.15	
	0310	3-5/8" wide		110	.073		1.18	2.59		3.77	5.60	

	05410 | Load-Bearing Metal Studs		CREW	DAILY OUTPUT	LABOR-HOURS	UNIT	2006 BARE COSTS MAT.	LABOR	EQUIP.	TOTAL	TOTAL INCL O&P	
120	0320	4″ wide	1 Carp	110	.073	Ea.	1.26	2.59		3.85	5.65	120
	0330	6″ wide		105	.076		1.61	2.71		4.32	6.25	
	0340	8″ wide		100	.080		2.06	2.84		4.90	7	
	1200	Studs 16″ O.C., 18 ga. x 2-1/2″ wide		125	.064		.99	2.28		3.27	4.86	
	1210	3-5/8″ wide		120	.067		1.19	2.37		3.56	5.25	
	1220	4″ wide		120	.067		1.27	2.37		3.64	5.35	
	1230	6″ wide		115	.070		1.65	2.47		4.12	5.90	
	1240	8″ wide		110	.073		2.10	2.59		4.69	6.60	
	1300	16 ga. x 2-1/2″ wide		115	.070		1.24	2.47		3.71	5.45	
	1310	3-5/8″ wide		110	.073		1.52	2.59		4.11	5.95	
	1320	4″ wide		110	.073		1.62	2.59		4.21	6.05	
	1330	6″ wide		105	.076		2.06	2.71		4.77	6.75	
	1340	8″ wide		100	.080		2.64	2.84		5.48	7.60	
	2200	Studs 24″ O.C., 18 ga. x 2-1/2″ wide		125	.064		1.43	2.28		3.71	5.35	
	2210	3-5/8″ wide		120	.067		1.72	2.37		4.09	5.80	
	2220	4″ wide		120	.067		1.84	2.37		4.21	5.95	
	2230	6″ wide		115	.070		2.39	2.47		4.86	6.75	
	2240	8″ wide		110	.073		3.03	2.59		5.62	7.60	
	2300	16 ga. x 2-1/2″ wide		115	.070		1.79	2.47		4.26	6.05	
	2310	3-5/8″ wide		110	.073		2.20	2.59		4.79	6.70	
	2320	4″ wide		110	.073		2.34	2.59		4.93	6.85	
	2330	6″ wide		105	.076		2.98	2.71		5.69	7.75	
	2340	8″ wide	↓	100	.080	↓	3.82	2.84		6.66	8.90	
	3000	Continuous bridging, per row										
	3100	16 ga. x 1-1/2″ channel thru studs 12″ O.C.	1 Carp	6	1.333	C.L.F.	43	47.50		90.50	126	
	3110	16″ O.C.		7	1.143		43	40.50		83.50	115	
	3120	24″ O.C.		8.80	.909		43	32.50		75.50	101	
	4100	2″ x 2″ angle x 18 ga., studs 12″ O.C.		7	1.143		62.50	40.50		103	137	
	4110	16″ O.C.		9	.889		62.50	31.50		94	122	
	4120	24″ O.C.		12	.667		62.50	23.50		86	109	
	4200	16 ga., studs 12″ O.C.		5	1.600		80.50	57		137.50	183	
	4210	16″ O.C.		7	1.143		80.50	40.50		121	156	
	4220	24″ O.C.	↓	10	.800	↓	80.50	28.50		109	136	
300	0010	**FRAMING, BOXED HEADERS/BEAMS**										300
	0200	Double, 18 ga. x 6″ deep	2 Carp	220	.073	L.F.	4.49	2.59		7.08	9.20	
	0210	8″ deep		210	.076		4.99	2.71		7.70	10	
	0220	10″ deep		200	.080		6.05	2.84		8.89	11.35	
	0230	12 ″ deep		190	.084		6.65	2.99		9.64	12.25	
	0300	16 ga. x 8″ deep		180	.089		5.75	3.16		8.91	11.55	
	0310	10″ deep		170	.094		6.85	3.35		10.20	13.10	
	0320	12 ″ deep		160	.100		7.50	3.56		11.06	14.15	
	0400	14 ga. x 10″ deep		140	.114		8	4.06		12.06	15.55	
	0410	12 ″ deep		130	.123		8.75	4.38		13.13	16.90	
	1210	Triple, 18 ga. x 8″ deep		170	.094		7.20	3.35		10.55	13.50	
	1220	10″ deep		165	.097		8.65	3.45		12.10	15.20	
	1230	12 ″ deep		160	.100		9.55	3.56		13.11	16.40	
	1300	16 ga. x 8″ deep		145	.110		8.30	3.92		12.22	15.65	
	1310	10″ deep		140	.114		9.90	4.06		13.96	17.65	
	1320	12 ″ deep		135	.119		10.85	4.21		15.06	18.90	
	1400	14 ga. x 10″ deep		115	.139		10.90	4.95		15.85	20	
	1410	12 ″ deep	↓	110	.145	↓	12.10	5.15		17.25	22	
400	0010	**FRAMING, STUD WALLS** w/ top & bottom track, no openings,										400
	0020	headers, beams, bridging or bracing										
	4100	8′ high walls, 18 ga. x 2-1/2″ wide, studs 12″ O.C.	2 Carp	54	.296	L.F.	7.55	10.55		18.10	26	
	4110	16″ O.C.	↓	77	.208	↓	6.05	7.40		13.45	18.90	

	05410 | Load-Bearing Metal Studs	CREW	DAILY OUTPUT	LABOR-HOURS	UNIT	2006 BARE COSTS MAT.	LABOR	EQUIP.	TOTAL	TOTAL INCL O&P	
400	4120 24" O.C.	2 Carp	107	.150	L.F.	4.51	5.30		9.81	13.75	400
	4130 3-5/8" wide, studs 12" O.C.		53	.302		8.95	10.75		19.70	27.50	
	4140 16" O.C.		76	.211		7.15	7.50		14.65	20.50	
	4150 24" O.C.		105	.152		5.35	5.40		10.75	14.90	
	4160 4" wide, studs 12" O.C.		52	.308		9.40	10.95		20.35	28.50	
	4170 16" O.C.		74	.216		7.55	7.70		15.25	21	
	4180 24" O.C.		103	.155		5.65	5.50		11.15	15.35	
	4190 6" wide, studs 12" O.C.		51	.314		11.90	11.15		23.05	31.50	
	4200 16" O.C.		73	.219		9.55	7.80		17.35	23.50	
	4210 24" O.C.		101	.158		7.15	5.65		12.80	17.25	
	4220 8" wide, studs 12" O.C.		50	.320		14.55	11.40		25.95	35	
	4230 16" O.C.		72	.222		11.70	7.90		19.60	26	
	4240 24" O.C.		100	.160		8.85	5.70		14.55	19.10	
	4300 16 ga. x 2-1/2" wide, studs 12" O.C.		47	.340		8.85	12.10		20.95	29.50	
	4310 16" O.C.		68	.235		7	8.35		15.35	21.50	
	4320 24" O.C.		94	.170		5.15	6.05		11.20	15.70	
	4330 3-5/8" wide, studs 12" O.C.		46	.348		10.55	12.35		22.90	32	
	4340 16" O.C.		66	.242		8.35	8.60		16.95	23.50	
	4350 24" O.C.		92	.174		6.15	6.20		12.35	17.05	
	4360 4" wide, studs 12" O.C.		45	.356		11.15	12.65		23.80	33.50	
	4370 16" O.C.		65	.246		8.85	8.75		17.60	24.50	
	4380 24" O.C.		90	.178		6.55	6.30		12.85	17.65	
	4390 6" wide, studs 12" O.C.		44	.364		14.05	12.95		27	37	
	4400 16" O.C.		64	.250		11.15	8.90		20.05	27	
	4410 24" O.C.		88	.182		8.25	6.45		14.70	19.80	
	4420 8" wide, studs 12" O.C.		43	.372		17.25	13.25		30.50	41	
	4430 16" O.C.		63	.254		13.75	9.05		22.80	30	
	4440 24" O.C.		86	.186		10.20	6.60		16.80	22	
	5100 10' high walls, 18 ga. x 2-1/2" wide, studs 12" O.C.		54	.296		9.05	10.55		19.60	27.50	
	5110 16" O.C.		77	.208		7.15	7.40		14.55	20	
	5120 24" O.C.		107	.150		5.25	5.30		10.55	14.60	
	5130 3-5/8" wide, studs 12" O.C.		53	.302		10.75	10.75		21.50	29.50	
	5140 16" O.C.		76	.211		8.50	7.50		16	22	
	5150 24" O.C.		105	.152		6.25	5.40		11.65	15.90	
	5160 4" wide, studs 12" O.C.		52	.308		11.30	10.95		22.25	30.50	
	5170 16" O.C.		74	.216		8.95	7.70		16.65	22.50	
	5180 24" O.C.		103	.155		6.60	5.50		12.10	16.40	
	5190 6" wide, studs 12" O.C.		51	.314		14.25	11.15		25.40	34	
	5200 16" O.C.		73	.219		11.30	7.80		19.10	25.50	
	5210 24" O.C.		101	.158		8.35	5.65		14	18.55	
	5220 8" wide, studs 12" O.C.		50	.320		17.40	11.40		28.80	38	
	5230 16" O.C.		72	.222		13.85	7.90		21.75	28.50	
	5240 24" O.C.		100	.160		10.25	5.70		15.95	20.50	
	5300 16 ga. x 2-1/2" wide, studs 12" O.C.		47	.340		10.65	12.10		22.75	32	
	5310 16" O.C.		68	.235		8.35	8.35		16.70	23	
	5320 24" O.C.		94	.170		6.05	6.05		12.10	16.75	
	5330 3-5/8" wide, studs 12" O.C.		46	.348		12.75	12.35		25.10	34.50	
	5340 16" O.C.		66	.242		10	8.60		18.60	25.50	
	5350 24" O.C.		92	.174		7.25	6.20		13.45	18.25	
	5360 4" wide, studs 12" O.C.		45	.356		13.50	12.65		26.15	36	
	5370 16" O.C.		65	.246		10.60	8.75		19.35	26	
	5380 24" O.C.		90	.178		7.70	6.30		14	18.90	
	5390 6" wide, studs 12" O.C.		44	.364		16.95	12.95		29.90	40	
	5400 16" O.C.		64	.250		13.35	8.90		22.25	29.50	
	5410 24" O.C.		88	.182		9.70	6.45		16.15	21.50	
400	5420 8" wide, studs 12" O.C.	↓	43	.372	↓	21	13.25		34.25	45	400

	05410 | Load-Bearing Metal Studs		CREW	DAILY OUTPUT	LABOR-HOURS	UNIT	2006 BARE COSTS MAT.	LABOR	EQUIP.	TOTAL	TOTAL INCL O&P	
400	5430	16″ O.C.	2 Carp	63	.254	L.F.	16.40	9.05		25.45	33	400
	5440	24″ O.C.		86	.186		11.95	6.60		18.55	24	
	6190	12′ high walls, 18 ga. x 6″ wide, studs 12″ O.C.		41	.390		16.60	13.85		30.45	41.50	
	6200	16″ O.C.		58	.276		13.05	9.80		22.85	30.50	
	6210	24″ O.C.		81	.198		9.55	7		16.55	22	
	6220	8″ wide, studs 12″ O.C.		40	.400		20.50	14.20		34.70	46	
	6230	16″ O.C.		57	.281		16	10		26	34	
	6240	24″ O.C.		80	.200		11.70	7.10		18.80	24.50	
	6390	16 ga. x 6″ wide, studs 12″ O.C.		35	.457		19.85	16.25		36.10	49	
	6400	16″ O.C.		51	.314		15.50	11.15		26.65	35.50	
	6410	24″ O.C.		70	.229		11.15	8.15		19.30	25.50	
	6420	8″ wide, studs 12″ O.C.		34	.471		24.50	16.75		41.25	54.50	
	6430	16″ O.C.		50	.320		19.05	11.40		30.45	40	
	6440	24″ O.C.		69	.232		13.75	8.25		22	29	
	6530	14 ga. x 3-5/8″ wide, studs 12″ O.C.		34	.471		18.80	16.75		35.55	48	
	6540	16″ O.C.		48	.333		14.65	11.85		26.50	36	
	6550	24″ O.C.		65	.246		10.55	8.75		19.30	26	
	6560	4″ wide, studs 12″ O.C.		33	.485		19.90	17.25		37.15	50.50	
	6570	16″ O.C.		47	.340		15.55	12.10		27.65	37	
	6580	24″ O.C.		64	.250		11.15	8.90		20.05	27	
	6730	12 ga. x 3-5/8″ wide, studs 12″ O.C.		31	.516		26	18.35		44.35	59	
	6740	16″ O.C.		43	.372		20	13.25		33.25	44	
	6750	24″ O.C.		59	.271		14.15	9.65		23.80	31.50	
	6760	4″ wide, studs 12″ O.C.		30	.533		28	18.95		46.95	62.50	
	6770	16″ O.C.		42	.381		21.50	13.55		35.05	46	
	6780	24″ O.C.		58	.276		15.20	9.80		25	33	
	7390	16′ high walls, 16 ga. x 6″ wide, studs 12″ O.C.		33	.485		25.50	17.25		42.75	56.50	
	7400	16″ O.C.		48	.333		19.85	11.85		31.70	41.50	
	7410	24″ O.C.		67	.239		14.05	8.50		22.55	29.50	
	7420	8″ wide, studs 12″ O.C.		32	.500		31.50	17.80		49.30	64	
	7430	16″ O.C.		47	.340		24.50	12.10		36.60	47	
	7440	24″ O.C.		66	.242		17.25	8.60		25.85	33.50	
	7560	14 ga. x 4″ wide, studs 12″ O.C.		31	.516		26	18.35		44.35	59	
	7570	16″ O.C.		45	.356		19.90	12.65		32.55	43	
	7580	24″ O.C.		61	.262		14.10	9.30		23.40	31	
	7590	6″ wide, studs 12″ O.C.		30	.533		32.50	18.95		51.45	67	
	7600	16″ O.C.		44	.364		25	12.95		37.95	49	
	7610	24″ O.C.		60	.267		17.70	9.50		27.20	35	
	7760	12 ga. x 4″ wide, studs 12″ O.C.		29	.552		36.50	19.60		56.10	72.50	
	7770	16″ O.C.		40	.400		28	14.20		42.20	54.50	
	7780	24″ O.C.		55	.291		19.45	10.35		29.80	38.50	
	7790	6″ wide, studs 12″ O.C.		28	.571		46	20.50		66.50	84	
	7800	16″ O.C.		39	.410		35	14.60		49.60	62.50	
	7810	24″ O.C.		54	.296		24.50	10.55		35.05	44.50	
	8590	20′ high walls, 14 ga. x 6″ wide, studs 12″ O.C.		29	.552		39.50	19.60		59.10	76	
	8600	16″ O.C.		42	.381		30.50	13.55		44.05	56	
	8610	24″ O.C.		57	.281		21.50	10		31.50	40	
	8620	8″ wide, studs 12″ O.C.		28	.571		48.50	20.50		69	87	
	8630	16″ O.C.		41	.390		37.50	13.85		51.35	64	
	8640	24″ O.C.		56	.286		26	10.15		36.15	46	
	8790	12 ga. x 6″ wide, studs 12″ O.C.		27	.593		56.50	21		77.50	97.50	
	8800	16″ O.C.		37	.432		43.50	15.35		58.85	73	
	8810	24″ O.C.		51	.314		30	11.15		41.15	51.50	
	8820	8″ wide, studs 12″ O.C.		26	.615		69	22		91	113	
	8830	16″ O.C.		36	.444		53	15.80		68.80	84	
400	8840	24″ O.C.	↓	50	.320	↓	36.50	11.40		47.90	59	400

05400 | Cold-Formed Metal Framing

		05410 | Load-Bearing Metal Studs	Crew	Daily Output	Labor-Hours	Unit	2006 Bare Costs Mat.	Labor	Equip.	Total	Total Incl O&P	
400	9000	Minimum labor/equipment charge	2 Carp	4	4	Job		142		142	236	400

		05420 | Cold-Formed Metal Joists	Crew	Daily Output	Labor-Hours	Unit	2006 Bare Costs Mat.	Labor	Equip.	Total	Total Incl O&P	
100	0010	**BRACING**, continuous, per row, top & bottom										100
	0120	Flat strap, 20 ga.x 2" wide, joists at 12" O.C.	1 Carp	4.67	1.713	C.L.F.	48.50	61		109.50	154	
	0130	16" O.C.		5.33	1.501		46.50	53.50		100	140	
	0140	24" O.C.		6.66	1.201		45	42.50		87.50	121	
	0150	18 ga. x 2" wide, joists at 12" O.C.		4	2		63	71		134	188	
	0160	16" O.C.		4.67	1.713		62	61		123	170	
	0170	24" O.C.		5.33	1.501		61	53.50		114.50	156	
120	0010	**BRIDGING**, solid between joists w/ 1-1/4" leg track, per joist bay										120
	0230	Joists 12" O.C., 18 ga .track x 6" wide	1 Carp	80	.100	Ea.	1.29	3.56		4.85	7.30	
	0240	8" wide		75	.107		1.63	3.79		5.42	8.10	
	0250	10" wide		70	.114		2.02	4.06		6.08	8.95	
	0260	12" wide		65	.123		2.33	4.38		6.71	9.80	
	0330	16 ga. track x 6" wide		70	.114		1.61	4.06		5.67	8.50	
	0340	8" wide		65	.123		2.06	4.38		6.44	9.50	
	0350	10" wide		60	.133		2.54	4.74		7.28	10.65	
	0360	12" wide		55	.145		2.91	5.15		8.06	11.75	
	0440	14 ga. track x 8" wide		60	.133		2.59	4.74		7.33	10.70	
	0450	10" wide		55	.145		3.18	5.15		8.33	12.05	
	0460	12" wide		50	.160		3.68	5.70		9.38	13.45	
	0550	12 ga. track x 10" wide		45	.178		4.67	6.30		10.97	15.60	
	0560	12" wide		40	.200		5.30	7.10		12.40	17.60	
	1230	16" O.C., 18 ga. track x 6" wide		80	.100		1.65	3.56		5.21	7.70	
	1240	8" wide		75	.107		2.10	3.79		5.89	8.60	
	1250	10" wide		70	.114		2.59	4.06		6.65	9.60	
	1260	12" wide		65	.123		2.99	4.38		7.37	10.55	
	1330	16 ga. track x 6" wide		70	.114		2.06	4.06		6.12	9	
	1340	8" wide		65	.123		2.64	4.38		7.02	10.15	
	1350	10" wide		60	.133		3.25	4.74		7.99	11.45	
	1360	12" wide		55	.145		3.73	5.15		8.88	12.65	
	1440	14 ga. track x 8" wide		60	.133		3.32	4.74		8.06	11.50	
	1450	10" wide		55	.145		4.08	5.15		9.23	13.05	
	1460	12" wide		50	.160		4.72	5.70		10.42	14.60	
	1550	12 ga. track x 10" wide		45	.178		6	6.30		12.30	17.05	
	1560	12" wide		40	.200		6.80	7.10		13.90	19.25	
	2230	24" O.C., 18 ga. track x 6" wide		80	.100		2.39	3.56		5.95	8.55	
	2240	8" wide		75	.107		3.03	3.79		6.82	9.65	
	2250	10" wide		70	.114		3.75	4.06		7.81	10.85	
	2260	12" wide		65	.123		4.32	4.38		8.70	12	
	2330	16 ga. track x 6" wide		70	.114		2.98	4.06		7.04	10.05	
	2340	8" wide		65	.123		3.82	4.38		8.20	11.45	
	2350	10" wide		60	.133		4.70	4.74		9.44	13	
	2360	12" wide		55	.145		5.40	5.15		10.55	14.50	
	2440	14 ga. track x 8" wide		60	.133		4.80	4.74		9.54	13.15	
	2450	10" wide		55	.145		5.90	5.15		11.05	15.05	
	2460	12" wide		50	.160		6.85	5.70		12.55	16.90	
	2550	12 ga. track x 10" wide		45	.178		8.65	6.30		14.95	20	
	2560	12" wide		40	.200		9.80	7.10		16.90	22.50	
200	0010	**FRAMING, BAND JOIST** (track) fastened to bearing wall										200
	0220	18 ga. track x 6" deep	2 Carp	1000	.016	L.F.	1.05	.57		1.62	2.10	
	0230	8" deep		920	.017		1.33	.62		1.95	2.49	
	0240	10" deep		860	.019		1.65	.66		2.31	2.91	

METALS 5

05420	Cold-Formed Metal Joists		Crew	Daily Output	Labor-Hours	Unit	2006 Bare Costs Mat.	Labor	Equip.	Total	Total Incl O&P	
200	0320	16 ga. track x 6" deep	2 Carp	900	.018	L.F.	1.31	.63		1.94	2.49	200
	0330	8" deep		840	.019		1.68	.68		2.36	2.97	
	0340	10" deep		780	.021		2.07	.73		2.80	3.49	
	0350	12" deep		740	.022		2.37	.77		3.14	3.88	
	0430	14 ga. track x 8" deep		750	.021		2.11	.76		2.87	3.58	
	0440	10" deep		720	.022		2.59	.79		3.38	4.16	
	0450	12" deep		700	.023		3	.81		3.81	4.65	
	0540	12 ga. track x 10" deep		670	.024		3.81	.85		4.66	5.60	
	0550	12" deep		650	.025		4.32	.88		5.20	6.20	
300	0010	**FRAMING, BOXED HEADERS/BEAMS**										300
	0200	Double, 18 ga. x 6" deep	2 Carp	220	.073	L.F.	4.49	2.59		7.08	9.20	
	0210	8" deep		210	.076		4.99	2.71		7.70	10	
	0220	10" deep		200	.080		6.05	2.84		8.89	11.35	
	0230	12" deep		190	.084		6.65	2.99		9.64	12.25	
	0300	16 ga. x 8" deep		180	.089		5.75	3.16		8.91	11.55	
	0310	10" deep		170	.094		6.85	3.35		10.20	13.10	
	0320	12" deep		160	.100		7.50	3.56		11.06	14.15	
	0400	14 ga. x 10" deep		140	.114		8	4.06		12.06	15.55	
	0410	12" deep		130	.123		8.75	4.38		13.13	16.90	
	0500	12 ga. x 10" deep		110	.145		10.55	5.15		15.70	20	
	0510	12" deep		100	.160		11.70	5.70		17.40	22.50	
	1210	Triple, 18 ga .x 8" deep		170	.094		7.20	3.35		10.55	13.50	
	1220	10" deep		165	.097		8.65	3.45		12.10	15.20	
	1230	12" deep		160	.100		9.55	3.56		13.11	16.40	
	1300	16 ga. x 8" deep		145	.110		8.30	3.92		12.22	15.65	
	1310	10" deep		140	.114		9.90	4.06		13.96	17.65	
	1320	12" deep		135	.119		10.85	4.21		15.06	18.90	
	1400	14 ga. x 10" deep		115	.139		11.60	4.95		16.55	21	
	1410	12" deep		110	.145		12.75	5.15		17.90	22.50	
	1500	12 ga. x 10" deep		90	.178		15.45	6.30		21.75	27.50	
	1510	12" deep		85	.188		17.15	6.70		23.85	30	
410	0010	**FRAMING, JOISTS**, no band joists (track), web stiffeners, headers,										410
	0020	beams, bridging or bracing										
	0030	Joists (2" flange) and fasteners, materials only										
	0220	18 ga. x 6" deep				L.F.	1.39			1.39	1.52	
	0230	8" deep					1.65			1.65	1.81	
	0240	10" deep					1.93			1.93	2.13	
	0320	16 ga. x 6" deep					1.70			1.70	1.87	
	0330	8" deep					2.04			2.04	2.24	
	0340	10" deep					2.37			2.37	2.61	
	0350	12" deep					2.70			2.70	2.97	
	0430	14 ga. x 8" deep					2.57			2.57	2.83	
	0440	10" deep					2.96			2.96	3.26	
	0450	12" deep					3.37			3.37	3.71	
	0540	12 ga. x 10" deep					4.32			4.32	4.75	
	0550	12" deep					4.91			4.91	5.40	
	1010	Installation of joists to band joists, beams & headers, labor only										
	1220	18 ga. x 6" deep	2 Carp	110	.145	Ea.		5.15		5.15	8.55	
	1230	8" deep		90	.178			6.30		6.30	10.45	
	1240	10" deep		80	.200			7.10		7.10	11.80	
	1320	16 ga. x 6" deep		95	.168			6		6	9.90	
	1330	8" deep		70	.229			8.15		8.15	13.45	
	1340	10" deep		60	.267			9.50		9.50	15.70	
	1350	12" deep		55	.291			10.35		10.35	17.15	
	1430	14 ga. x 8" deep		65	.246			8.75		8.75	14.50	

05400 | Cold-Formed Metal Framing

	05420 | Cold-Formed Metal Joists		CREW	DAILY OUTPUT	LABOR-HOURS	UNIT	2006 BARE COSTS MAT.	LABOR	EQUIP.	TOTAL	TOTAL INCL O&P	
410	1440	10″ deep	2 Carp	45	.356	Ea.		12.65		12.65	21	410
	1450	12″ deep		35	.457			16.25		16.25	27	
	1540	12 ga. x 10″ deep		40	.400			14.20		14.20	23.50	
	1550	12″ deep		30	.533			18.95		18.95	31.50	
	9000	Minimum labor/equipment charge		4	4	Job		142		142	236	
500	0010	**FRAMING, WEB STIFFENERS** at joist bearing, fabricated from										500
	0020	stud piece (1-5/8″ flange) to stiffen joist (2″ flange)										
	2120	For 6″ deep joist, with 18 ga. x 2-1/2″ stud	1 Carp	120	.067	Ea.	1.67	2.37		4.04	5.75	
	2130	3-5/8″ stud		110	.073		1.85	2.59		4.44	6.30	
	2140	4″ stud		105	.076		1.79	2.71		4.50	6.45	
	2150	6″ stud		100	.080		1.95	2.84		4.79	6.85	
	2160	8″ stud		95	.084		2	2.99		4.99	7.15	
	2220	8″ deep joist, with 2-1/2″ stud		120	.067		1.83	2.37		4.20	5.95	
	2230	3-5/8″ stud		110	.073		1.99	2.59		4.58	6.45	
	2240	4″ stud		105	.076		1.95	2.71		4.66	6.65	
	2250	6″ stud		100	.080		2.13	2.84		4.97	7.05	
	2260	8″ stud		95	.084		2.30	2.99		5.29	7.50	
	2320	10″ deep joist, with 2-1/2″ stud		110	.073		2.59	2.59		5.18	7.10	
	2330	3-5/8″ stud		100	.080		2.84	2.84		5.68	7.85	
	2340	4″ stud		95	.084		2.81	2.99		5.80	8.05	
	2350	6″ stud		90	.089		3.04	3.16		6.20	8.60	
	2360	8″ stud		85	.094		3.09	3.35		6.44	8.95	
	2420	12″ deep joist, with 2-1/2″ stud		110	.073		2.74	2.59		5.33	7.30	
	2430	3-5/8″ stud		100	.080		2.97	2.84		5.81	8	
	2440	4″ stud		95	.084		2.91	2.99		5.90	8.15	
	2450	6″ stud		90	.089		3.19	3.16		6.35	8.75	
	2460	8″ stud		85	.094		3.43	3.35		6.78	9.35	
	3130	For 6″ deep joist, with 16 ga. x 3-5/8″ stud		100	.080		1.93	2.84		4.77	6.85	
	3140	4″ stud		95	.084		1.91	2.99		4.90	7.05	
	3150	6″ stud		90	.089		2.10	3.16		5.26	7.55	
	3160	8″ stud		85	.094		2.21	3.35		5.56	8	
	3230	8″ deep joist, with 3-5/8″ stud		100	.080		2.14	2.84		4.98	7.05	
	3240	4″ stud		95	.084		2.10	2.99		5.09	7.25	
	3250	6″ stud		90	.089		2.33	3.16		5.49	7.80	
	3260	8″ stud		85	.094		2.49	3.35		5.84	8.30	
	3330	10″ deep joist, with 3-5/8″ stud		85	.094		2.92	3.35		6.27	8.75	
	3340	4″ stud		80	.100		2.98	3.56		6.54	9.20	
	3350	6″ stud		75	.107		3.25	3.79		7.04	9.85	
	3360	8″ stud		70	.114		3.38	4.06		7.44	10.45	
	3430	12″ deep joist, with 3-5/8″ stud		85	.094		3.19	3.35		6.54	9.05	
	3440	4″ stud		80	.100		3.13	3.56		6.69	9.35	
	3450	6″ stud		75	.107		3.48	3.79		7.27	10.15	
	3460	8″ stud		70	.114		3.72	4.06		7.78	10.85	
	4230	For 8″ deep joist, with 14 ga. x 3-5/8″ stud		90	.089		2.77	3.16		5.93	8.30	
	4240	4″ stud		85	.094		2.84	3.35		6.19	8.65	
	4250	6″ stud		80	.100		3.07	3.56		6.63	9.25	
	4260	8″ stud		75	.107		3.29	3.79		7.08	9.90	
	4330	10″ deep joist, with 3-5/8″ stud		75	.107		3.89	3.79		7.68	10.60	
	4340	4″ stud		70	.114		3.88	4.06		7.94	11	
	4350	6″ stud		65	.123		4.25	4.38		8.63	11.95	
	4360	8″ stud		60	.133		4.44	4.74		9.18	12.75	
	4430	12″ deep joist, with 3-5/8″ stud		75	.107		4.14	3.79		7.93	10.85	
	4440	4″ stud		70	.114		4.23	4.06		8.29	11.40	
	4450	6″ stud		65	.123		4.58	4.38		8.96	12.30	
	4460	8″ stud		60	.133		4.91	4.74		9.65	13.25	

05400 | Cold-Formed Metal Framing

							2006 BARE COSTS					
	05420 | Cold-Formed Metal Joists		CREW	DAILY OUTPUT	LABOR-HOURS	UNIT	MAT.	LABOR	EQUIP.	TOTAL	TOTAL INCL O&P	
500	5330	For 10″ deep joist, with 12 ga. x 3-5/8″ stud	1 Carp	65	.123	Ea.	4.11	4.38		8.49	11.75	500
	5340	4″ stud		60	.133		4.24	4.74		8.98	12.50	
	5350	6″ stud		55	.145		4.67	5.15		9.82	13.70	
	5360	8″ stud		50	.160		5.15	5.70		10.85	15.05	
	5430	12″ deep joist, with 3-5/8″ stud		65	.123		4.55	4.38		8.93	12.25	
	5440	4″ stud		60	.133		4.47	4.74		9.21	12.75	
	5450	6″ stud		55	.145		5.10	5.15		10.25	14.15	
	5460	8″ stud		50	.160		5.85	5.70		11.55	15.85	
	05425 | Cold-Formed Roof Framing											
100	0010	**FRAMING, BRACING**										100
	0020	Continuous bracing, per row										
	0100	16 ga. x 1-1/2″ channel thru rafters/trusses @ 16″ O.C.	1 Carp	4.50	1.778	C.L.F.	43	63		106	153	
	0120	24″ O.C.		6	1.333		43	47.50		90.50	126	
	0300	2″ x 2″ angle x 18 ga., rafters/trusses @ 16″ O.C.		6	1.333		62.50	47.50		110	148	
	0320	24″ O.C.		8	1		62.50	35.50		98	128	
	0400	16 ga., rafters/trusses @ 16″ O.C.		4.50	1.778		80.50	63		143.50	194	
	0420	24″ O.C.		6.50	1.231		80.50	44		124.50	161	
200	0010	**FRAMING, BRIDGING**										200
	0020	Solid, between rafters w/ 1-1/4″ leg track, per rafter bay										
	1200	Rafters 16″ O.C., 18 ga. x 4″ deep	1 Carp	60	.133	Ea.	1.27	4.74		6.01	9.25	
	1210	6″ deep		57	.140		1.65	4.99		6.64	10.05	
	1220	8″ deep		55	.145		2.10	5.15		7.25	10.85	
	1230	10″ deep		52	.154		2.59	5.45		8.04	11.90	
	1240	12″ deep		50	.160		2.99	5.70		8.69	12.70	
	2200	24″ O.C., 18 ga. x 4″ deep		60	.133		1.84	4.74		6.58	9.85	
	2210	6″ deep		57	.140		2.39	4.99		7.38	10.90	
	2220	8″ deep		55	.145		3.03	5.15		8.18	11.90	
	2230	10″ deep		52	.154		3.75	5.45		9.20	13.15	
	2240	12″ deep		50	.160		4.32	5.70		10.02	14.15	
500	0010	**FRAMING, PARAPETS**										500
	0100	3′ high installed on 1st story, 18 ga. x 4″ wide studs, 12″ O.C.	2 Carp	100	.160	L.F.	4.71	5.70		10.41	14.60	
	0110	16″ O.C.		150	.107		4	3.79		7.79	10.70	
	0120	24″ O.C.		200	.080		3.30	2.84		6.14	8.35	
	0200	6″ wide studs, 12″ O.C.		100	.160		6	5.70		11.70	16	
	0210	16″ O.C.		150	.107		5.10	3.79		8.89	11.90	
	0220	24″ O.C.		200	.080		4.22	2.84		7.06	9.35	
	1100	Installed on 2nd story, 18 ga. x 4″ wide studs, 12″ O.C.		95	.168		4.71	6		10.71	15.10	
	1110	16″ O.C.		145	.110		4	3.92		7.92	10.90	
	1120	24″ O.C.		190	.084		3.30	2.99		6.29	8.60	
	1200	6″ wide studs, 12″ O.C.		95	.168		6	6		12	16.50	
	1210	16″ O.C.		145	.110		5.10	3.92		9.02	12.10	
	1220	24″ O.C.		190	.084		4.22	2.99		7.21	9.60	
	2100	Installed on gable, 18 ga. x 4″ wide studs, 12″ O.C.		85	.188		4.71	6.70		11.41	16.30	
	2110	16″ O.C.		130	.123		4	4.38		8.38	11.65	
	2120	24″ O.C.		170	.094		3.30	3.35		6.65	9.20	
	2200	6″ wide studs, 12″ O.C.		85	.188		6	6.70		12.70	17.70	
	2210	16″ O.C.		130	.123		5.10	4.38		9.48	12.85	
	2220	24″ O.C.		170	.094		4.22	3.35		7.57	10.20	
550	0010	**FRAMING, ROOF RAFTERS**										550
	0100	Boxed ridge beam, double, 18 ga. x 6″ deep	2 Carp	160	.100	L.F.	4.49	3.56		8.05	10.85	
	0110	8″ deep		150	.107		4.99	3.79		8.78	11.80	
	0120	10″ deep		140	.114		6.05	4.06		10.11	13.40	
	0130	12″ deep		130	.123		6.65	4.38		11.03	14.55	
	0200	16 ga. x 6″ deep		150	.107		5.10	3.79		8.89	11.90	

	05425 | Cold-Formed Roof Framing		CREW	DAILY OUTPUT	LABOR-HOURS	UNIT	2006 BARE COSTS MAT.	LABOR	EQUIP.	TOTAL	TOTAL INCL O&P	
550	0210	8″ deep	2 Carp	140	.114	L.F.	5.75	4.06		9.81	13.05	550
	0220	10″ deep		130	.123		6.85	4.38		11.23	14.80	
	0230	12″ deep		120	.133		7.50	4.74		12.24	16.10	
	1100	Rafters, 2″ flange, material only, 18 ga. x 6″ deep					1.39			1.39	1.52	
	1110	8″ deep					1.65			1.65	1.81	
	1120	10″ deep					1.93			1.93	2.13	
	1130	12″ deep					2.25			2.25	2.47	
	1200	16 ga. x 6″ deep					1.70			1.70	1.87	
	1210	8″ deep					2.04			2.04	2.24	
	1220	10″ deep					2.37			2.37	2.61	
	1230	12″ deep					2.70			2.70	2.97	
	2100	Installation only, ordinary rafter to 4:12 pitch, 18 ga .x 6″ deep	2 Carp	35	.457	Ea.		16.25		16.25	27	
	2110	8″ deep		30	.533			18.95		18.95	31.50	
	2120	10″ deep		25	.640			23		23	37.50	
	2130	12″ deep		20	.800			28.50		28.50	47	
	2200	16 ga. x 6″ deep		30	.533			18.95		18.95	31.50	
	2210	8″ deep		25	.640			23		23	37.50	
	2220	10″ deep		20	.800			28.50		28.50	47	
	2230	12″ deep		15	1.067			38		38	63	
	8100	Add to labor, ordinary rafters on steep roofs						25%				
	8110	Dormers & complex roofs						50%				
	8200	Hip & valley rafters to 4:12 pitch						25%				
	8210	Steep roofs						50%				
	8220	Dormers & complex roofs						75%				
	8300	Hip & valley jack rafters to 4:12 pitch						50%				
	8310	Steep roofs						75%				
	8320	Dormers & complex roofs						100%				
	9000	Minimum labor/equipment charge	2 Carp	4	4	Job		142		142	236	
600	0010	**FRAMING, ROOF TRUSSES**										600
	0020	Fabrication of trusses on ground, Fink (W) or King Post, to 4:12 pitch										
	0120	18 ga. x 4″ chords, 16′ span	2 Carp	12	1.333	Ea.	52.50	47.50		100	137	
	0130	20′ span		11	1.455		66	51.50		117.50	158	
	0140	24′ span		11	1.455		79	51.50		130.50	173	
	0150	28′ span		10	1.600		92	57		149	195	
	0160	32′ span		10	1.600		105	57		162	210	
	0250	6″ chords, 28′ span		9	1.778		116	63		179	232	
	0260	32′ span		9	1.778		132	63		195	250	
	0270	36′ span		8	2		149	71		220	282	
	0280	40′ span		8	2		165	71		236	300	
	1120	5:12 to 8:12 pitch, 18 ga. x 4″ chords, 16′ span		10	1.600		60	57		117	160	
	1130	20′ span		9	1.778		75	63		138	188	
	1140	24′ span		9	1.778		90	63		153	205	
	1150	28′ span		8	2		105	71		176	234	
	1160	32′ span		8	2		120	71		191	250	
	1250	6″ chords, 28′ span		7	2.286		132	81.50		213.50	280	
	1260	32′ span		7	2.286		151	81.50		232.50	300	
	1270	36′ span		6	2.667		170	95		265	345	
	1280	40′ span		6	2.667		189	95		284	365	
	2120	9:12 to 12:12 pitch, 18 ga. x 4″ chords, 16′ span		8	2		75	71		146	201	
	2130	20′ span		7	2.286		94	81.50		175.50	238	
	2140	24′ span		7	2.286		113	81.50		194.50	259	
	2150	28′ span		6	2.667		132	95		227	300	
	2160	32′ span		6	2.667		150	95		245	320	
	2250	6″ chords, 28′ span		5	3.200		165	114		279	370	
	2260	32′ span		5	3.200		189	114		303	395	
	2270	36′ span		4	4		212	142		354	470	

	05425 | Cold-Formed Roof Framing		CREW	DAILY OUTPUT	LABOR-HOURS	UNIT	2006 BARE COSTS MAT.	LABOR	EQUIP.	TOTAL	TOTAL INCL O&P	
600	2280	40′ span	2 Carp	4	4	Ea.	236	142		378	495	600
	4900	Minimum labor/equipment charge		4	4	Job		142		142	236	
	5120	Erection only of roof trusses, to 4:12 pitch, 16′ span	F-6	48	.833	Ea.		27.50	13.25	40.75	59.50	
	5130	20′ span		46	.870			28.50	13.85	42.35	62	
	5140	24′ span		44	.909			30	14.45	44.45	65	
	5150	28′ span		42	.952			31	15.15	46.15	67.50	
	5160	32′ span		40	1			33	15.90	48.90	71.50	
	5170	36′ span		38	1.053			34.50	16.75	51.25	75	
	5180	40′ span		36	1.111			36.50	17.70	54.20	79	
	5220	5:12 to 8:12 pitch, 16′ span		42	.952			31	15.15	46.15	67.50	
	5230	20′ span		40	1			33	15.90	48.90	71.50	
	5240	24′ span		38	1.053			34.50	16.75	51.25	75	
	5250	28′ span		36	1.111			36.50	17.70	54.20	79	
	5260	32′ span		34	1.176			38.50	18.75	57.25	83.50	
	5270	36′ span		32	1.250			41	19.90	60.90	89	
	5280	40′ span		30	1.333			43.50	21	64.50	95	
	5320	9:12 to 12:12 pitch, 16′ span		36	1.111			36.50	17.70	54.20	79	
	5330	20′ span		34	1.176			38.50	18.75	57.25	83.50	
	5340	24′ span		32	1.250			41	19.90	60.90	89	
	5350	28′ span		30	1.333			43.50	21	64.50	95	
	5360	32′ span		28	1.429			47	22.50	69.50	102	
	5370	36′ span		26	1.538			50.50	24.50	75	110	
	5380	40′ span		24	1.667			54.50	26.50	81	119	
	9000	Minimum labor/equipment charge		2	20	Job		655	320	975	1,425	
650	0010	**FRAMING, SOFFITS & CANOPIES**										650
	0130	Continuous ledger track @ wall, studs @ 16″ O.C., 18 ga .x 4″ wide	2 Carp	535	.030	L.F.	.85	1.06		1.91	2.69	
	0140	6″ wide		500	.032		1.10	1.14		2.24	3.09	
	0150	8″ wide		465	.034		1.40	1.22		2.62	3.57	
	0160	10″ wide		430	.037		1.73	1.32		3.05	4.09	
	0230	Studs @ 24″ O.C., 18 ga. x 4″ wide		800	.020		.81	.71		1.52	2.07	
	0240	6″ wide		750	.021		1.05	.76		1.81	2.42	
	0250	8″ wide		700	.023		1.33	.81		2.14	2.82	
	0260	10″ wide		650	.025		1.65	.88		2.53	3.26	
	1000	Horizontal soffit and canopy members, material only										
	1030	1-5/8″ flange studs, 18 ga. x 4″ deep				L.F.	1.13			1.13	1.24	
	1040	6″ deep					1.42			1.42	1.56	
	1050	8″ deep					1.72			1.72	1.89	
	1140	2″ flange joists, 18 ga. x 6″ deep					1.58			1.58	1.74	
	1150	8″ deep					1.88			1.88	2.07	
	1160	10″ deep					2.21			2.21	2.43	
	4030	Installation only, 18 ga., 1-5/8″ flange x 4″ deep	2 Carp	130	.123	Ea.		4.38		4.38	7.25	
	4040	6″ deep		110	.145			5.15		5.15	8.55	
	4050	8″ deep		90	.178			6.30		6.30	10.45	
	4140	2″ flange, 18 ga. x 6″ deep		110	.145			5.15		5.15	8.55	
	4150	8″ deep		90	.178			6.30		6.30	10.45	
	4160	10″ deep		80	.200			7.10		7.10	11.80	
	6010	Clips to attach facia to rafter tails, 2″ x 2″ x 18 ga. angle	1 Carp	120	.067		.74	2.37		3.11	4.75	
	6020	16 ga. angle	″	100	.080		.95	2.84		3.79	5.75	
	9000	Minimum labor/equipment charge	2 Carp	4	4	Job		142		142	236	

05500 | Metal Fabrications

05514 | Ladders

		05514 Ladders	CREW	DAILY OUTPUT	LABOR-HOURS	UNIT	2006 BARE COSTS MAT.	LABOR	EQUIP.	TOTAL	TOTAL INCL O&P	
500	0010	**LADDER**, shop fabricated										500
	0020	Steel, 20" wide, bolted to concrete, with cage	E-4	50	.640	V.L.F.	68	26	1.79	95.79	125	
	0100	Without cage		85	.376		31.50	15.25	1.05	47.80	64.50	
	0300	Aluminum, bolted to concrete, with cage		50	.640		91.50	26	1.79	119.29	151	
	0400	Without cage		85	.376		53	15.25	1.05	69.30	88	
	1350	Alternating tread stair, 56/68°, steel, standard paint color	2 Sswk	50	.320		151	12.80		163.80	190	
	1360	Non-standard paint color		50	.320		171	12.80		183.80	212	
	1370	Galvanized steel		50	.320		170	12.80		182.80	211	
	1380	Stainless steel		50	.320		253	12.80		265.80	305	
	1390	68°, aluminum		50	.320		185	12.80		197.80	228	
	9000	Minimum labor/equipment charge	E-4	2	16	Job		645	44.50	689.50	1,250	

05517 | Metal Stairs

		05517 Metal Stairs	CREW	DAILY OUTPUT	LABOR-HOURS	UNIT	MAT.	LABOR	EQUIP.	TOTAL	TOTAL INCL O&P	
300	0010	**FIRE ESCAPE**, shop fabricated										300
	0200	2' wide balcony, 1" x 1/4" bars 1-1/2" O.C.	1 Sswk	5	1.600	L.F.	45.50	64		109.50	170	
	0400	1st story cantilevered stair, standard	"	.09	88.889	Ea.	1,900	3,550		5,450	8,750	
	0500	Cable counterweight				"	1,775			1,775	1,950	
	0700	Platform & fixed stair, 36" x 40"	1 Sswk	.17	47.059	Flight	845	1,875		2,720	4,450	
	0900	For 3'-6" wide escapes, add to above					100%	150%				
350	0010	**FIRE ESCAPE STAIRS**										350
	0020	One story, disappearing, stainless steel	2 Sswk	20	.800	V.L.F.	184	32		216	262	
	0100	Portable ladder				Ea.	55.50			55.50	61.50	
700	0010	**STAIR**, shop fabricated, steel stringers, safety nosing on treads										700
	0020	Grating tread and pipe railing, 3'-6" wide	E-4	35	.914	Riser	218	37	2.55	257.55	310	
	0100	4'-0" wide		30	1.067		283	43	2.98	328.98	395	
	0200	Cement fill metal pan, picket rail, 3'-6" wide		35	.914		325	37	2.55	364.55	430	
	0300	4'-0" wide		30	1.067		370	43	2.98	415.98	490	
	0350	Wall rail, both sides, 3'-6" wide		53	.604		250	24.50	1.68	276.18	320	
	0400	Cast iron tread and pipe rail, 3'-6" wide		35	.914		350	37	2.55	389.55	455	
	0500	Checkered plate tread, industrial, 3'-6" wide		28	1.143		218	46	3.19	267.19	330	
	0550	Circular, for tanks, 3'-0" wide		33	.970		240	39	2.71	281.71	340	
	0600	For isolated stairs, add						100%				
	0800	Custom steel stairs, 3'-6" wide, minimum	E-4	35	.914		325	37	2.55	364.55	430	
	0810	Average		30	1.067		435	43	2.98	480.98	565	
	0900	Maximum		20	1.600		545	64.50	4.46	613.96	725	
	1100	For 4' wide stairs, add					5%	5%				
	1300	For 5' wide stairs, add					10%	10%				
	1500	Landing, steel pan, conventional	E-4	160	.200	S.F.	43.50	8.10	.56	52.16	64	
	1600	Pre-erected	"	255	.125	"	76.50	5.10	.35	81.95	94	
	1700	Pre-erected, steel pan tread, 3'-6" wide, 2 line pipe rail	E-2	87	.644	Riser	360	25	16.40	401.40	460	
	1810	Spiral aluminum, 5'-0" diameter, stock units	E-4	45	.711		395	29	1.98	425.98	490	
	1820	Custom units		45	.711		745	29	1.98	775.98	875	
	1900	Spiral, cast iron, 4'-0" diameter, ornamental, minimum		45	.711		350	29	1.98	380.98	445	
	1920	Maximum		25	1.280		480	52	3.57	535.57	630	
	2000	Spiral, steel, industrial checkered plate, 4' diameter		45	.711		350	29	1.98	380.98	445	
	2200	Stock units, 6'-0" diameter		40	.800		425	32.50	2.23	459.73	535	
	3110	Spiral steel, stock units, primed, flat metal tread, 3'-6" dia.	2 Carp	1.60	10	Flight	1,050	355		1,405	1,775	
	3120	4'-0" dia.		1.45	11.034		1,225	390		1,615	2,000	
	3130	4'-6" dia.		1.35	11.852		1,350	420		1,770	2,175	
	3140	5'-0" dia.		1.25	12.800		1,450	455		1,905	2,350	
	3210	Galvanized, 3'-6" dia.		1.60	10		1,850	355		2,205	2,650	
	3220	4'-0" dia.		1.45	11.034		2,075	390		2,465	2,950	
	3230	4'-6" dia.		1.35	11.852		2,275	420		2,695	3,200	
	3240	5'-0" dia.		1.25	12.800		2,450	455		2,905	3,450	

05517 | Metal Stairs

			Crew	Daily Output	Labor-Hours	Unit	2006 Bare Costs Mat.	Labor	Equip.	Total	Total Incl O&P	
700	3310	Checkered plate tread, 3'-6" dia.	2 Carp	1.45	11.034	Flight	1,300	390		1,690	2,075	700
	3320	4'-0" dia.		1.35	11.852		1,475	420		1,895	2,325	
	3330	4'-6" dia.		1.25	12.800		1,625	455		2,080	2,525	
	3340	5'-0" dia.		1.15	13.913		1,750	495		2,245	2,750	
	3410	Galvanized, 3'-6" dia.		1.45	11.034		2,125	390		2,515	2,975	
	3420	4'-0" dia.		1.35	11.852		2,375	420		2,795	3,325	
	3430	4'-6" dia.		1.25	12.800		2,575	455		3,030	3,600	
	3440	5'-0" dia.		1.15	13.913		2,775	495		3,270	3,900	
	3510	Red oak tread on flat metal, 3'-6" dia.		1.35	11.852		1,775	420		2,195	2,650	
	3520	4'-0" dia.		1.25	12.800		1,975	455		2,430	2,900	
	3530	4'-6" dia.		1.15	13.913		2,125	495		2,620	3,175	
	3540	5'-0" dia.		1.05	15.238		2,300	540		2,840	3,425	
	3900	Industrial ships ladder, 3' W, grating treads, 2 line pipe rail	E-4	30	1.067	Riser	142	43	2.98	187.98	240	
	4000	Aluminum		30	1.067	"	218	43	2.98	263.98	325	
	9000	Minimum labor/equipment charge		2	16	Job		645	44.50	689.50	1,250	

05520 | Handrails & Railings

			Crew	Daily Output	Labor-Hours	Unit	Mat.	Labor	Equip.	Total	Total Incl O&P	
700	0010	**RAILING, PIPE**, shop fabricated										700
	0020	Aluminum, 2 rail, satin finish, 1-1/4" diameter	E-4	160	.200	L.F.	19.75	8.10	.56	28.41	37.50	
	0030	Clear anodized		160	.200		24.50	8.10	.56	33.16	43	
	0040	Dark anodized		160	.200		27.50	8.10	.56	36.16	46.50	
	0080	1-1/2" diameter, satin finish		160	.200		23.50	8.10	.56	32.16	42	
	0090	Clear anodized		160	.200		26.50	8.10	.56	35.16	45	
	0100	Dark anodized		160	.200		29	8.10	.56	37.66	48	
	0140	Aluminum, 3 rail, 1-1/4" diam., satin finish		137	.234		30	9.45	.65	40.10	52	
	0150	Clear anodized		137	.234		38	9.45	.65	48.10	60	
	0160	Dark anodized		137	.234		42	9.45	.65	52.10	64.50	
	0200	1-1/2" diameter, satin finish		137	.234		36.50	9.45	.65	46.60	58.50	
	0210	Clear anodized		137	.234		41	9.45	.65	51.10	63.50	
	0220	Dark anodized		137	.234		45	9.45	.65	55.10	68	
	0500	Steel, 2 rail, on stairs, primed, 1-1/4" diameter		160	.200		16.20	8.10	.56	24.86	33.50	
	0520	1-1/2" diameter		160	.200		17.80	8.10	.56	26.46	35.50	
	0540	Galvanized, 1-1/4" diameter		160	.200		22.50	8.10	.56	31.16	40.50	
	0560	1-1/2" diameter		160	.200		25	8.10	.56	33.66	43.50	
	0580	Steel, 3 rail, primed, 1-1/4" diameter		137	.234		24	9.45	.65	34.10	45	
	0600	1-1/2" diameter		137	.234		25.50	9.45	.65	35.60	46.50	
	0620	Galvanized, 1-1/4" diameter		137	.234		34	9.45	.65	44.10	56	
	0640	1-1/2" diameter		137	.234		40	9.45	.65	50.10	62.50	
	0700	Stainless steel, 2 rail, 1-1/4" diam. #4 finish		137	.234		58.50	9.45	.65	68.60	82.50	
	0720	High polish		137	.234		94	9.45	.65	104.10	121	
	0740	Mirror polish		137	.234		118	9.45	.65	128.10	148	
	0760	Stainless steel, 3 rail, 1-1/2" diam., #4 finish		120	.267		88	10.80	.74	99.54	118	
	0770	High polish		120	.267		145	10.80	.74	156.54	181	
	0780	Mirror finish		120	.267		177	10.80	.74	188.54	216	
	0900	Wall rail, alum. pipe, 1-1/4" diam., satin finish		213	.150		11.30	6.10	.42	17.82	24	
	0905	Clear anodized		213	.150		13.75	6.10	.42	20.27	27	
	0910	Dark anodized		213	.150		16.65	6.10	.42	23.17	30	
	0915	1-1/2" diameter, satin finish		213	.150		12.50	6.10	.42	19.02	25.50	
	0920	Clear anodized		213	.150		15.70	6.10	.42	22.22	29	
	0925	Dark anodized		213	.150		19.40	6.10	.42	25.92	33.50	
	0930	Steel pipe, 1-1/4" diameter, primed		213	.150		9.85	6.10	.42	16.37	22.50	
	0935	Galvanized		213	.150		14.25	6.10	.42	20.77	27.50	
	0940	1-1/2" diameter		176	.182		10.10	7.35	.51	17.96	25.50	
	0945	Galvanized		213	.150		14.30	6.10	.42	20.82	27.50	
	0955	Stainless steel pipe, 1-1/2" diam., #4 finish		107	.299		46.50	12.10	.83	59.43	74.50	

05500 | Metal Fabrications

	05520 | Handrails & Railings		CREW	DAILY OUTPUT	LABOR-HOURS	UNIT	2006 BARE COSTS MAT.	LABOR	EQUIP.	TOTAL	TOTAL INCL O&P	
700	0960	High polish	E-4	107	.299	L.F.	95	12.10	.83	107.93	128	700
	0965	Mirror polish	↓	107	.299	↓	112	12.10	.83	124.93	146	
	9000	Minimum labor/equipment charge	1 Sswk	2	4	Job		160		160	299	
780	0010	**RAILINGS, INDUSTRIAL** Welded, shop fabricated										780
	0020	2 rail, 3'-6" high, 1-1/2" pipe	E-4	255	.125	L.F.	21.50	5.10	.35	26.95	33.50	
	0100	2" angle rail	"	255	.125		19.60	5.10	.35	25.05	31.50	
	0200	For 4" high kick plate, 10 gauge, add					4.48			4.48	4.93	
	0300	1/4" thick, add					5.75			5.75	6.35	
	0500	For curved rails, add				↓	30%	30%				
	9000	Minimum labor/equipment charge	1 Sswk	2	4	Job		160		160	299	
	05530 | Gratings											
300	0010	**FLOOR GRATING, ALUMINUM**, field fabricated from panels										300
	0110	Bearing bars @ 1-3/16" O.C., cross bars @ 4" O.C.,										
	0111	Up to 300 S.F., 1" x 1/8" bar	E-4	900	.036	S.F.	9.70	1.44	.10	11.24	13.50	
	0112	Over 300 S.F.		850	.038		8.85	1.52	.11	10.48	12.65	
	0113	1-1/4" x 1/8" bar, up to 300 S.F.		800	.040		11.05	1.62	.11	12.78	15.30	
	0114	Over 300 S.F.		1000	.032		10.05	1.29	.09	11.43	13.55	
	0122	1-1/4" x 3/16" bar, up to 300 S.F.		750	.043		14.15	1.73	.12	16	18.95	
	0124	Over 300 S.F.		1000	.032		12.90	1.29	.09	14.28	16.65	
	0132	1-1/2" x 1/8" bar, up to 300 S.F.		700	.046		8.85	1.85	.13	10.83	13.35	
	0134	Over 300 S.F.		1000	.032		8.05	1.29	.09	9.43	11.35	
	0136	1-3/4" x 3/16" bar, up to 300 S.F.		500	.064		19	2.59	.18	21.77	26	
	0138	Over 300 S.F.		1000	.032		17.30	1.29	.09	18.68	21.50	
	0146	2-1/4" x 3/16" bar, up to 300 S.F.		600	.053		22.50	2.16	.15	24.81	28.50	
	0148	Over 300 S.F.		1000	.032		20.50	1.29	.09	21.88	25	
	0162	Cross bars @ 2" O.C., 1" x 1/8", up to 300 S.F.		600	.053		12.25	2.16	.15	14.56	17.65	
	0164	Over 300 S.F.		1000	.032		11.15	1.29	.09	12.53	14.75	
	0172	1-1/4" x 3/16" bar, up to 300 S.F.		600	.053		16.95	2.16	.15	19.26	23	
	0174	Over 300 S.F.		1000	.032		15.45	1.29	.09	16.83	19.45	
	0182	1-1/2" x 1/8" bar, up to 300 S.F.		600	.053		12.15	2.16	.15	14.46	17.55	
	0184	Over 300 S.F.		1000	.032		11.05	1.29	.09	12.43	14.65	
	0186	1-3/4" x 3/16" bar, up to 300 S.F.		600	.053		17.65	2.16	.15	19.96	23.50	
	0188	Over 300 S.F.	↓	1000	.032	↓	16.05	1.29	.09	17.43	20	
	0200	For straight cuts, add				L.F.	2.45			2.45	2.70	
	0212	Close mesh, 1" x 1/8", up to 300 S.F.	E-4	520	.062	S.F.	31	2.49	.17	33.66	39	
	0214	Over 300 S.F.		920	.035		28	1.41	.10	29.51	33.50	
	0222	1-1/4" x 3/16", up to 300 S.F.		520	.062		38.50	2.49	.17	41.16	47.50	
	0224	Over 300 S.F.		920	.035		35	1.41	.10	36.51	41	
	0232	1-1/2" x 1/8", up to 300 S.F.		520	.062		15.60	2.49	.17	18.26	22	
	0234	Over 300 S.F.	↓	920	.035	↓	14.20	1.41	.10	15.71	18.35	
	0600	For aluminum checkered plate nosings, add				L.F.	4.03			4.03	4.43	
	0700	For straight toe plate, add					7.30			7.30	8.05	
	0800	For curved toe plate, add					9.15			9.15	10.05	
	1000	For cast aluminum abrasive nosings, add				↓	4.90			4.90	5.40	
	1200	Expanded aluminum, .65# per S.F.	E-4	1050	.030	S.F.	4.27	1.23	.09	5.59	7.10	
	1600	Heavy duty, all extruded plank, 3/4" deep, 1.8 # per S.F.		1100	.029		11.95	1.18	.08	13.21	15.45	
	1700	1-1/4" deep, 2.9# per S.F.		1000	.032		12.75	1.29	.09	14.13	16.50	
	1800	1-3/4" deep, 4.2# per S.F.		925	.035		15.20	1.40	.10	16.70	19.45	
	1900	2-1/4" deep, 5.0# per S.F.	↓	875	.037	↓	17.40	1.48	.10	18.98	22	
	2100	For safety serrated surface, add					15%					
320	0010	**FLOOR GRATING PLANKS**, field fabricated from planks										320
	0020	Aluminum, 9-1/2" wide, 14 ga., 2" rib	E-4	950	.034	L.F.	14.10	1.36	.09	15.55	18.15	

	05530 | Gratings		CREW	DAILY OUTPUT	LABOR-HOURS	UNIT	2006 BARE COSTS MAT.	LABOR	EQUIP.	TOTAL	TOTAL INCL O&P	
320	0200	Galvanized steel, 9-1/2" wide, 14 ga., 2-1/2" rib	E-4	950	.034	L.F.	9.10	1.36	.09	10.55	12.65	320
	0300	4" rib		950	.034		10.90	1.36	.09	12.35	14.65	
	0500	12 gauge, 2-1/2" rib		950	.034		12.10	1.36	.09	13.55	15.95	
	0600	3" rib		950	.034		14.15	1.36	.09	15.60	18.25	
	0800	Stainless steel, type 304, 16 ga., 2" rib		950	.034		24.50	1.36	.09	25.95	29.50	
	0900	Type 316		950	.034		29.50	1.36	.09	30.95	35	
340	0010	**FLOOR GRATING, STEEL**, field fabricated from panels										340
	0050	Labor for installing, from ground/floor	E-4	845	.038	S.F.		1.53	.11	1.64	2.98	
	0100	Elevated		460	.070	"		2.81	.19	3	5.45	
	0300	Platforms, to 12' high, rectangular		3150	.010	Lb.	1.39	.41	.03	1.83	2.33	
	0400	Circular		2300	.014	"	1.54	.56	.04	2.14	2.78	
	0410	Painted bearing bars @ 1-3/16"										
	0412	Cross bars @ 4" O.C., 3/4" x 1/8" bar, up to 300 S.F.	E-2	500	.112	S.F.	5.85	4.36	2.85	13.06	17.40	
	0414	Over 300 S.F.		750	.075		5.30	2.90	1.90	10.10	13.15	
	0422	1-1/4" x 3/16", up to 300 S.F.		400	.140		7.90	5.45	3.57	16.92	22.50	
	0424	Over 300 S.F.		600	.093		7.20	3.63	2.38	13.21	17	
	0432	1-1/2" x 1/8", up to 300 S.F.		400	.140		7.05	5.45	3.57	16.07	21.50	
	0434	Over 300 S.F.		600	.093		6.45	3.63	2.38	12.46	16.15	
	0436	1-3/4" x 3/16", up to 300 S.F.		400	.140		10.25	5.45	3.57	19.27	25	
	0438	Over 300 S.F.		600	.093		9.35	3.63	2.38	15.36	19.35	
	0452	2-1/4" x 3/16", up to 300 S.F.		300	.187		12.20	7.25	4.76	24.21	31.50	
	0454	Over 300 S.F.		450	.124		11.10	4.84	3.17	19.11	24.50	
	0462	Cross bars @ 2" O.C., 3/4" x 1/8", up to 300 S.F.		500	.112		11.30	4.36	2.85	18.51	23.50	
	0464	Over 300 S.F.		750	.075		9.40	2.90	1.90	14.20	17.65	
	0472	1-1/4" x 3/16", up to 300 S.F.		400	.140		16.90	5.45	3.57	25.92	32.50	
	0474	Over 300 S.F.		600	.093		14.10	3.63	2.38	20.11	24.50	
	0482	1-1/2" x 1/8", up to 300 S.F.		400	.140		13.05	5.45	3.57	22.07	28	
	0484	Over 300 S.F.		600	.093		10.90	3.63	2.38	16.91	21	
	0486	1-3/4" x 3/16", up to 300 S.F.		400	.140		17.60	5.45	3.57	26.62	33	
	0488	Over 300 S.F.		600	.093		14.70	3.63	2.38	20.71	25.50	
	0502	2-1/4" x 3/16", up to 300 S.F.		300	.187		21	7.25	4.76	33.01	41.50	
	0504	Over 300 S.F.		450	.124		17.50	4.84	3.17	25.51	31.50	
	0601	Painted bearing bars @ 15/16" O.C., cross bars @ 4" O.C.,										
	0612	Up to 300 S.F., 3/4" x 1/8" bars	E-4	850	.038	S.F.	7.30	1.52	.11	8.93	11	
	0622	1-1/4" x 3/16" bars		600	.053		10.30	2.16	.15	12.61	15.55	
	0632	1-1/2" x 1/8" bars		550	.058		9.40	2.35	.16	11.91	14.95	
	0636	1-3/4" x 3/16" bars		450	.071		13.55	2.88	.20	16.63	20.50	
	0652	2-1/4" x 3/16" bars	E-2	300	.187		17.75	7.25	4.76	29.76	38	
	0662	Cross bars @ 2" O.C., up to 300 S.F., 3/4" x 1/8"		500	.112		12.40	4.36	2.85	19.61	24.50	
	0672	1-1/4" x 3/16" bars		400	.140		13.40	5.45	3.57	22.42	28.50	
	0682	1-1/2" x 1/8" bars		400	.140		12.75	5.45	3.57	21.77	27.50	
	0686	1-3/4" x 3/16" bars		300	.187		17.35	7.25	4.76	29.36	37.50	
	0690	For galvanized grating, add					25%					
	0800	For straight cuts, add				L.F.	2.61			2.61	2.87	
	0900	For curved cuts, add					3.58			3.58	3.94	
	1000	For straight banding, add					2.98			2.98	3.28	
	1100	For curved banding, add					4.61			4.61	5.05	
	1200	For checkered plate nosings, add					5.25			5.25	5.80	
	1300	For straight toe or kick plate, add					7.40			7.40	8.15	
	1400	For curved toe or kick plate, add					9.05			9.05	9.95	
	1500	For abrasive nosings, add					5.45			5.45	6	
	1600	For safety serrated surface, minimum, add					15%					
	1700	Maximum, add					25%					
	2000	Stainless steel gratings, close spaced, 1" x 1/8" bars, up to 300 S.F.	E-4	450	.071	S.F.	33	2.88	.20	36.08	42	
	2100	Standard spacing, 3/4" x 1/8" bars		500	.064		26	2.59	.18	28.77	33.50	
	2200	1-1/4" x 3/16" bars		400	.080		41.50	3.24	.22	44.96	52	

05500 | Metal Fabrications

05530 | Gratings

		Description	CREW	DAILY OUTPUT	LABOR-HOURS	UNIT	2006 BARE COSTS MAT.	LABOR	EQUIP.	TOTAL	TOTAL INCL O&P	
340	2400	Expanded steel grating, at ground, 3.0# per S.F.	E-4	900	.036	S.F.	3.26	1.44	.10	4.80	6.40	340
	2500	3.14# per S.F.		900	.036		3.45	1.44	.10	4.99	6.60	
	2600	4.0# per S.F.		850	.038		4.26	1.52	.11	5.89	7.65	
	2650	4.27# per S.F.		850	.038		4.61	1.52	.11	6.24	8	
	2700	5.0# per S.F.		800	.040		5.50	1.62	.11	7.23	9.20	
	2800	6.25# per S.F.		750	.043		6.95	1.73	.12	8.80	10.95	
	2900	7.0# per S.F.		700	.046		7.80	1.85	.13	9.78	12.20	
	3100	For flattened expanded steel grating, add					8%					
	3300	For elevated installation above 15′, add						15%				
360	0010	**GRATING FRAME**, field fabricated										360
	0020	Aluminum, for gratings 1″ to 1-1/2″ deep	1 Sswk	70	.114	L.F.	7.40	4.57		11.97	16.70	
	0100	For each corner, add				Ea.	5.85			5.85	6.45	

05540 | Floor Plates

		Description	CREW	DAILY OUTPUT	LABOR-HOURS	UNIT	MAT.	LABOR	EQUIP.	TOTAL	TOTAL INCL O&P	
200	0010	**CHECKERED PLATE**, field fabricated										200
	0020	1/4″ & 3/8″, 2000 to 5000 S.F., bolted	E-4	2900	.011	Lb.	.80	.45	.03	1.28	1.74	
	0100	Welded		4400	.007	″	.75	.29	.02	1.06	1.40	
	0300	Pit or trench cover and frame, 1/4″ plate, 2′ to 3′ wide		100	.320	S.F.	21	12.95	.89	34.84	48	
	0400	For galvanizing, add				Lb.	.43			.43	.47	
	0500	Platforms, 1/4″ plate, no handrails included, rectangular	E-4	4200	.008		1.27	.31	.02	1.60	2	
	0600	Circular	″	2500	.013		1.77	.52	.04	2.33	2.96	
700	0010	**TRENCH COVER**, field fabricated										700
	0020	Cast iron grating with bar stops and angle frame, to 18″ wide	1 Sswk	20	.400	L.F.	56	16		72	91.50	
	0100	Frame only (both sides of trench), 1″ grating		45	.178		12.20	7.10		19.30	27	
	0150	2″ grating		35	.229		19	9.15		28.15	38	
	0200	Aluminum, stock units, including frames and										
	0210	3/8″ plain cover plate, 4″ opening	E-4	205	.156	L.F.	35	6.30	.44	41.74	51	
	0300	6″ opening		185	.173		43.50	7	.48	50.98	61.50	
	0400	10″ opening		170	.188		60	7.60	.53	68.13	80.50	
	0500	16″ opening		155	.206		82	8.35	.58	90.93	106	
	0700	Add per inch for additional widths to 24″					3.27			3.27	3.60	
	0900	For custom fabrication, add					50%					
	1100	For 1/4″ plain cover plate, deduct					12%					
	1500	For cover recessed for tile, 1/4″ thick, deduct					12%					
	1600	3/8″ thick, add					5%					
	1800	For checkered plate cover, 1/4″ thick, deduct					12%					
	1900	3/8″ thick, add					2%					
	2100	For slotted or round holes in cover, 1/4″ thick, add					3%					
	2200	3/8″ thick, add					4%					
	2300	For abrasive cover, add					12%					

05550 | Stair Treads & Nosings

		Description	CREW	DAILY OUTPUT	LABOR-HOURS	UNIT	MAT.	LABOR	EQUIP.	TOTAL	TOTAL INCL O&P	
700	0010	**STAIR TREADS**										700
	0020	Aluminum grating, 3′ long, 1-1/2″ x 3/16″ rect. bars, 6″ wide	1 Sswk	24	.333	Ea.	49.50	13.30		62.80	79.50	
	0100	12″ wide		22	.364		73	14.55		87.55	107	
	0200	1-1/2″ x 3/16″ I-bars, 6″ wide		24	.333		41.50	13.30		54.80	70.50	
	0300	12″ wide		22	.364		67	14.55		81.55	101	
	0400	For abrasive nosings, add					12.95			12.95	14.25	
	0500	For narrow mesh, add					60%					
	0600	Stair treads, not incl. stringers. See also Div. 03150-660										
	0700	Cast aluminum, abrasive, 3′ long x 12″ wide, 5/16″ thick	1 Sswk	15	.533	Ea.	110	21.50		131.50	161	
	0800	3/8″ thick		15	.533		117	21.50		138.50	168	

05550 | Stair Treads & Nosings

			CREW	DAILY OUTPUT	LABOR-HOURS	UNIT	2006 BARE COSTS MAT.	LABOR	EQUIP.	TOTAL	TOTAL INCL O&P	
700	0900	1/2″ thick	1 Sswk	15	.533	Ea.	128	21.50		149.50	181	700
	1000	Cast bronze, abrasive, 3/8″ thick		8	1		475	40		515	600	
	1100	1/2″ thick		8	1		625	40		665	765	
	1200	Cast iron, abrasive, 3′ long x 12″ wide, 3/8″ thick		15	.533		96.50	21.50		118	146	
	1300	1/2″ thick		15	.533		111	21.50		132.50	163	
	1400	Fiberglass reinforced plastic with safety nosing,										
	1500	1-1/2″ thick, 12″ wide, 24″ long	1 Sswk	22	.364	Ea.	104	14.55		118.55	142	
	1600	30″ long		22	.364		109	14.55		123.55	147	
	1700	36″ long		22	.364		119	14.55		133.55	158	
	2000	Steel grating, painted, 3′ long, 1-1/4″ x 3/16″ bars, 6″ wide		20	.400		42	16		58	76	
	2010	12″ wide		18	.444		48.50	17.75		66.25	86.50	
	2022	Plain, 1″ x 3/16″, 6″ wide		17	.471		38.50	18.80		57.30	77	
	2024	12″ wide		15	.533		46.50	21.50		68	91	
	2032	1-1/4″ x 3/16″, 6″ wide		17	.471		41	18.80		59.80	80.50	
	2034	12″ wide		15	.533		51.50	21.50		73	96.50	
	2062	Serrated, 1-1/4″ x 3/16″, 6″ wide		17	.471		29	18.80		47.80	67	
	2064	12″ wide		15	.533		48.50	21.50		70	93.50	
	2072	1-1/2″ x 3/16″, 6″ wide		17	.471		32	18.80		50.80	70	
	2074	12″ wide		15	.533		57	21.50		78.50	103	
	2092	For galvanizing, 1″ x 3/16″, add				S.F.	1.50			1.50	1.65	
	2094	1-1/4″ x 3/16″, add				″	3.33			3.33	3.66	
	2100	Add for abrasive nosing, 3′ long, painted				Ea.	13.10			13.10	14.40	
	2200	Galvanized					17.60			17.60	19.35	
	2500	Expanded steel, 2-1/2″ deep, 9″ x 3′ long, 18 gauge	1 Sswk	20	.400		23	16		39	55.50	
	2600	14 gauge	″	20	.400		28.50	16		44.50	61.50	

05560 | Metal Castings

			CREW	DAILY OUTPUT	LABOR-HOURS	UNIT	MAT.	LABOR	EQUIP.	TOTAL	TOTAL INCL O&P	
200	0010	**CONSTRUCTION CASTINGS**										200
	0020	Manhole covers and frames see Division 02630-110										
	0100	Column bases, cast iron, 16″ x 16″, approx. 65 lbs.	E-4	46	.696	Ea.	107	28	1.94	136.94	173	
	0200	32″ x 32″, approx. 256 lbs.	″	23	1.391		400	56.50	3.88	460.38	550	
	0400	Cast aluminum for wood columns, 8″ x 8″	1 Carp	32	.250		34.50	8.90		43.40	53	
	0500	12″ x 12″	″	32	.250		74.50	8.90		83.40	97	
	0600	Miscellaneous C.I. castings, light sections, less than 150 lbs.	E-4	3200	.010	Lb.	1.66	.40	.03	2.09	2.62	
	1100	Heavy sections, more than 150 lb.		4200	.008		.81	.31	.02	1.14	1.49	
	1300	Special low volume items		3200	.010		2.89	.40	.03	3.32	3.97	
	1500	For ductile iron, add					100%					

05580 | Formed Metal Fabrications

			CREW	DAILY OUTPUT	LABOR-HOURS	UNIT	MAT.	LABOR	EQUIP.	TOTAL	TOTAL INCL O&P	
150	0010	**ALLOY STEEL CHAIN**, Grade 80										150
	0015	Self-colored, cut lengths, w/accessories, 1/4″	E-17	4	4	C.L.F.	470	164	1.36	635.36	820	
	0020	3/8″		2	8		605	330	2.72	937.72	1,275	
	0030	1/2″		1.20	13.333		950	545	4.53	1,499.53	2,075	
	0040	5/8″		.72	22.222		1,500	910	7.55	2,417.55	3,350	
	0050	3/4″		.48	33.333		2,200	1,375	11.35	3,586.35	4,950	
	0060	7/8″		.40	40		3,225	1,650	13.60	4,888.60	6,650	
	0070	1″		.35	45.714		5,300	1,875	15.55	7,190.55	9,375	
	0080	1-1/4″		.24	66.667		9,900	2,725	22.50	12,647.50	16,000	
	0110	Clevis slip hook, 1/4″				Ea.	13.60			13.60	14.95	
	0120	3/8″					20.50			20.50	22.50	
	0130	1/2″					34.50			34.50	38	
	0140	5/8″					51.50			51.50	57	
	0150	3/4″					108			108	119	
	0160	Eye/sling hook w/ hammerlock coupling, 7/8″					305			305	335	
	0170	1″					430			430	470	

05500 | Metal Fabrications

	05580 | Formed Metal Fabrications		CREW	DAILY OUTPUT	LABOR-HOURS	UNIT	2006 BARE COSTS MAT.	LABOR	EQUIP.	TOTAL	TOTAL INCL O&P	
150	0180	1-1/4"				Ea.	645			645	710	150
200	0010	**ALUMINUM COLUMNS**										200
	0020	Aluminum, extruded, stock units, no cap or base, 6" diameter	E-4	240	.133	L.F.	8.70	5.40	.37	14.47	20	
	0100	8" diameter		170	.188		11.70	7.60	.53	19.83	27.50	
	0200	10" diameter		150	.213		15.45	8.65	.60	24.70	34	
	0300	12" diameter		140	.229		27	9.25	.64	36.89	47.50	
	0400	15" diameter	↓	120	.267	↓	32.50	10.80	.74	44.04	56.50	
	0410	Caps and bases, plain, 6" diameter				Set	18.95			18.95	21	
	0420	8" diameter					23.50			23.50	25.50	
	0430	10" diameter					31			31	34	
	0440	12" diameter					57.50			57.50	63	
	0450	15" diameter					86			86	95	
	0460	Caps, ornamental, minimum					218			218	239	
	0470	Maximum				↓	950			950	1,050	
	0500	For square columns, add to column prices above				L.F.	50%					
	0700	Residential, flat, 8' high, plain	E-4	20	1.600	Ea.	65	64.50	4.46	133.96	197	
	0720	Fancy		20	1.600		132	64.50	4.46	200.96	271	
	0740	Corner type, plain		20	1.600		115	64.50	4.46	183.96	253	
	0760	Fancy	↓	20	1.600	↓	232	64.50	4.46	300.96	380	
600	0010	**LAMP POSTS**										600
	0020	Aluminum, 7' high, stock units, post only	1 Carp	16	.500	Ea.	32	17.80		49.80	65	
	0100	Mild steel, plain		16	.500	"	28	17.80		45.80	60.50	
	9000	Minimum labor/equipment charge	↓	4	2	Job		71		71	118	
900	0010	**WINDOW GUARDS**, shop fabricated										900
	0015	Expanded metal, steel angle frame, permanent	E-4	350	.091	S.F.	19.65	3.70	.26	23.61	28.50	
	0025	Steel bars, 1/2" x 1/2", spaced 5" O.C.	"	290	.110	"	13.60	4.46	.31	18.37	23.50	
	0030	Hinge mounted, add				Opng.	39.50			39.50	43.50	
	0040	Removable type, add				"	25			25	27.50	
	0050	For galvanized guards, add				S.F.	35%					
	0070	For pivoted or projected type, add					105%	40%				
	0100	Mild steel, stock units, economy	E-4	405	.079		5.35	3.20	.22	8.77	12.10	
	0200	Deluxe		405	.079	↓	10.95	3.20	.22	14.37	18.30	
	0400	Woven wire, stock units, 3/8" channel frame, 3' x 5' opening		40	.800	Opng.	144	32.50	2.23	178.73	222	
	0500	4' x 6' opening	↓	38	.842		231	34	2.35	267.35	320	
	0800	Basket guards for above, add					197			197	217	
	1000	Swinging guards for above, add				↓	68			68	74.50	
	9000	Minimum labor/equipment charge	1 Sswk	2	4	Job		160		160	299	

05650 | Railroad Track & Accessories

	05655 | Railroad Trackwork		CREW	DAILY OUTPUT	LABOR-HOURS	UNIT	2006 BARE COSTS MAT.	LABOR	EQUIP.	TOTAL	TOTAL INCL O&P	
700	0010	**RAILROAD SIDINGS** R347216-10										700
	0020	Car bumpers, standard	B-14	2	24	Ea.	2,650	695	113	3,458	4,175	
	0100	Heavy duty R347216-20		2	24		5,000	695	113	5,808	6,775	
	0200	Derails hand throw (sliding)		10	4.800		835	139	22.50	996.50	1,175	
	0300	Hand throw with standard timbers, open stand & target		8	6	↓	910	174	28.50	1,112.50	1,325	
	0400	Resurface and realign existing track		200	.240	L.F.		6.95	1.13	8.08	12.70	
	0600	For crushed stone ballast, add	↓	500	.096	"	12.75	2.79	.45	15.99	19.10	
	0800	Siding, yard spur, level grade										

05650 | Railroad Track & Accessories

05655 | Railroad Trackwork

			CREW	DAILY OUTPUT	LABOR-HOURS	UNIT	2006 BARE COSTS MAT.	LABOR	EQUIP.	TOTAL	TOTAL INCL O&P	
700	0810	100 lb. rail, new material on wood ties R347216-10	B-14	57	.842	L.F.	92.50	24.50	3.97	120.97	146	700
	1000	Steel ties in concrete w/100 lb. new rail, fasteners & plates		22	2.182	"	128	63.50	10.30	201.80	256	
	1200	Switch timber, for a #8 switch, pressure treated R347216-20		3.70	12.973	M.B.F.	2,250	375	61	2,686	3,150	
	1300	Complete set of timbers, 3.7 M.B.F. for #8 switch		1	48	Total	8,650	1,400	226	10,276	12,000	
	1400	Ties, concrete, 8'-6" long, 30" O.C.		80	.600	Ea.	89.50	17.40	2.83	109.73	130	
	1600	Wood, pressure treated, 6" x 8" x 8'-6", C.L. lots		90	.533		37.50	15.50	2.51	55.51	69.50	
	1700	L.C.L. lots		90	.533		39.50	15.50	2.51	57.51	71.50	
	1900	Heavy duty, 7" x 9" x 8'-6", C.L. lots		70	.686		41	19.90	3.23	64.13	81.50	
	2000	L.C.L. lots	↓	70	.686	↓	41	19.90	3.23	64.13	81.50	
	2200	Turnouts, #8, incl. 100 lb. rails, plates, bars, frog, switch pt.										
	2300	Timbers and ballast 6" below bottom of tie	B-14	.50	96	Ea.	10,000	2,775	450	13,225	16,200	
	2400	Wheel stops, fixed		18	2.667	Pr.	595	77.50	12.55	685.05	795	
	2450	Hinged	↓	14	3.429	"	800	99.50	16.15	915.65	1,050	

			CREW	DAILY OUTPUT	LABOR-HOURS	UNIT	MAT.	LABOR	EQUIP.	TOTAL	TOTAL INCL O&P	
750	0010	**RAILROAD TRACK** R347216-10										750
	0020	Track bolts				Ea.	2.08			2.08	2.29	
	0100	Joint bars				Pr.	63			63	69.50	
	0200	Spikes				Ea.	.48			.48	.53	
	0300	Tie plates				"	10.50			10.50	11.55	
	1000	Rail, 100 lb. prime grade				L.F.	23.50			23.50	26	
	1500	Relay rail				"	11.80			11.80	13	

05700 | Ornamental Metal

05720 | Ornamental Handrails & Railings

			CREW	DAILY OUTPUT	LABOR-HOURS	UNIT	2006 BARE COSTS MAT.	LABOR	EQUIP.	TOTAL	TOTAL INCL O&P	
700	0010	**RAILINGS, ORNAMENTAL**, shop fabricated										700
	0020	Aluminum, bronze or stainless, minimum	1 Sswk	24	.333	L.F.	27.50	13.30		40.80	55.50	
	0100	Maximum		9	.889		261	35.50		296.50	355	
	0200	Aluminum ornamental rail, minimum		15	.533		27.50	21.50		49	70.50	
	0300	Maximum		8	1		82.50	40		122.50	166	
	0400	Hand-forged wrought iron, minimum		12	.667		86	26.50		112.50	145	
	0500	Maximum		8	1		258	40		298	360	
	0600	Composite metal/wood/glass, minimum		6	1.333		165	53.50		218.50	281	
	0700	Maximum		5	1.600	↓	330	64		394	485	
	9000	Minimum labor/equipment charge	↓	2	4	Job		160		160	299	

05730 | Ornamental Formed Metal

			CREW	DAILY OUTPUT	LABOR-HOURS	UNIT	MAT.	LABOR	EQUIP.	TOTAL	TOTAL INCL O&P	
200	0010	**COLUMNS, ORNAMENTAL**, shop fabricated										200
	6400	Mild steel, flat, 9" wide, stock units, painted, plain	E-4	160	.200	L.F.	6.10	8.10	.56	14.76	22.50	
	6450	Fancy		160	.200		12.35	8.10	.56	21.01	29.50	
	6500	Corner columns, painted, plain		160	.200		10.85	8.10	.56	19.51	27.50	
	6550	Fancy	↓	160	.200	↓	22	8.10	.56	30.66	40	

05800 | Expansion Control

	05810 | Exp. Joint Cover Assemblies		CREW	DAILY OUTPUT	LABOR-HOURS	UNIT	2006 BARE COSTS MAT.	LABOR	EQUIP.	TOTAL	TOTAL INCL O&P	
350	0010	**EXPANSION JOINT ASSEMBLIES** Custom units										350
	0200	Floor cover assemblies, 1″ space, aluminum	1 Sswk	38	.211	L.F.	16.20	8.40		24.60	33.50	
	0300	Bronze		38	.211		33	8.40		41.40	52	
	0500	2″ space, aluminum		38	.211		19.60	8.40		28	37.50	
	0600	Bronze		38	.211		35.50	8.40		43.90	55	
	0800	Wall and ceiling assemblies, 1″ space, aluminum		38	.211		9.70	8.40		18.10	26.50	
	0900	Bronze		38	.211		31	8.40		39.40	50	
	1100	2″ space, aluminum		38	.211		16.40	8.40		24.80	34	
	1200	Bronze		38	.211		38	8.40		46.40	57.50	
	1400	Floor to wall assemblies, 1″ space, aluminum		38	.211		14.75	8.40		23.15	32	
	1500	Bronze or stainless		38	.211		37.50	8.40		45.90	57	
	1700	Gym floor angle covers, aluminum, 3″ x 3″ angle		46	.174		12.80	6.95		19.75	27	
	1800	3″ x 4″ angle		46	.174		15.10	6.95		22.05	29.50	
	2000	Roof closures, aluminum, flat roof, low profile, 1″ space		57	.140		29.50	5.60		35.10	43	
	2100	High profile		57	.140		36	5.60		41.60	50.50	
	2300	Roof to wall, low profile, 1″ space		57	.140		16.30	5.60		21.90	28.50	
	2400	High profile		57	.140		21	5.60		26.60	33.50	
	9000	Minimum labor/equipment charge		2	4	Job		160		160	299	

05900 | Metal Restoration & Cleaning

	05910 | Metal Cleaning		CREW	DAILY OUTPUT	LABOR-HOURS	UNIT	2006 BARE COSTS MAT.	LABOR	EQUIP.	TOTAL	TOTAL INCL O&P	
500	0010	**METAL CLEANING**										500
	6125	Steel surface treatments										
	6170	Wire brush, hand (SSPC-SP2)	1 Psst	240	.033	S.F.	.02	1.08		1.10	2.15	
	6180	Power tool (SSPC-SP3)	″	600	.013		.05	.43		.48	.90	
	6215	Pressure washing, 2800-6000 S.F./day	1 Pord	3500	.002			.07		.07	.12	
	6220	Steam cleaning, 2800-4000 S.F./day		2400	.003			.11		.11	.17	
	6225	Water blasting		3200	.003			.08		.08	.13	
	6230	Brush-off blast (SSPC-SP7)	E-11	6000	.005		.08	.17	.02	.27	.42	
	6235	Com'l blast (SSPC-SP6), loose scale, fine pwder rust, 2.0 #/S.F. sand		4000	.008		.16	.26	.04	.46	.68	
	6240	Tight mill scale, little/no rust, 3.0 #/S.F. sand		3200	.010		.24	.32	.05	.61	.89	
	6245	Exist coat blistered/pitted, 4.0 #/S.F. sand		2100	.015		.32	.49	.07	.88	1.31	
	6250	Exist coat badly pitted/nodules, 6.7 #/S.F. sand		1300	.025		.54	.78	.11	1.43	2.13	
	6255	Near white blast (SSPC-SP10), loose scale, fine rust, 5.6 #/S.F. sand		1700	.019		.45	.60	.09	1.14	1.67	
	6260	Tight mill scale, little/no rust, 6.9 #/S.F. sand		1400	.023		.56	.73	.11	1.40	2.04	
	6265	Exist coat blistered/pitted, 9.0 #/S.F. sand		1100	.029		.73	.93	.13	1.79	2.61	
	6270	Exist coat badly pitted/nodules, 11.3 #/S.F. sand		800	.040		.91	1.28	.18	2.37	3.48	

	05950 | Paints & Protective Coatings		CREW	DAILY OUTPUT	LABOR-HOURS	UNIT	MAT.	LABOR	EQUIP.	TOTAL	TOTAL INCL O&P	
650	0010	**PAINTS & PROTECTIVE COATINGS** R050516-30										650
	5900	Galvanizing structural steel in shop, under 1 ton				Ton	540			540	595	
	5950	1 ton to 20 tons					450			450	490	
	6000	Over 20 tons					395			395	435	
	6100	Cold galvanizing, brush in field	1 Psst	1100	.007	S.F.	.07	.24		.31	.53	
	6510	Paints & protective coatings, sprayed in field										
	6520	Alkyds, primer	2 Psst	3600	.004	S.F.	.06	.14		.20	.35	
	6540	Gloss topcoats		3200	.005		.05	.16		.21	.37	
	6560	Silicone alkyd		3200	.005		.10	.16		.26	.43	
	6610	Epoxy, primer		3000	.005		.17	.17		.34	.53	

		05950 | Paints & Protective Coatings	CREW	DAILY OUTPUT	LABOR-HOURS	UNIT	2006 BARE COSTS MAT.	LABOR	EQUIP.	TOTAL	TOTAL INCL O&P	
650	6630	Intermediate or topcoat R050516-30	2 Psst	2800	.006	S.F.	.14	.19		.33	.51	650
	6650	Enamel coat		2800	.006		.17	.19		.36	.55	
	6700	Epoxy ester, primer		2800	.006		.34	.19		.53	.74	
	6720	Topcoats		2800	.006		.11	.19		.30	.49	
	6810	Latex primer		3600	.004		.05	.14		.19	.34	
	6830	Topcoats		3200	.005		.06	.16		.22	.39	
	6910	Universal primers, one part, phenolic, modified alkyd		2000	.008		.08	.26		.34	.60	
	6940	Two part, epoxy spray		2000	.008		.20	.26		.46	.73	
	7000	Zinc rich primers, self cure, spray, inorganic		1800	.009		.51	.29		.80	1.13	
	7010	Epoxy, spray, organic	↓	1800	.009	↓	.14	.29		.43	.72	
	7020	Above one story, spray painting simple structures, add						25%				
	7030	Intricate structures, add						50%				

For information about Means Estimating Seminars, see yellow pages 12 and 13 in back of book

Division 6
Wood & Plastics

Estimating Tips

06050 Basic Wood & Plastic Materials & Methods

- Common to any wood-framed structure are the accessory connector items such as screws, nails, adhesives, hangers, connector plates, straps, angles, and holdowns. For typical wood-framed buildings, such as residential projects, the aggregate total for these items can be significant, especially in areas where seismic loading is a concern. For floor and wall framing, the material cost is based on 10 to 25 lbs. per MBF. Holdowns, hangers, and other connectors should be taken off by the piece.

06100 Rough Carpentry

- Lumber is a traded commodity and therefore sensitive to supply and demand in the marketplace. Even in "budgetary" estimating of wood-framed projects, it is advisable to call local suppliers for the latest market pricing.
- Common quantity units for wood framed projects are "thousand board feet" (MBF). A board foot is a volume of wood, 1″ x 1′ x 1′, or 144 cubic inches. Board-foot quantities are generally calculated using nominal material dimensions—dressed sizes are ignored. Board foot per lineal foot of any stick of lumber can be calculated by dividing the nominal cross sectional area by 12. As an example, 2,000 lineal feet of 2 x 12 equates to 4 MBF by dividing the nominal area, 2 x 12, by 12, which equals 2, and multiplying by 2,000 to give 4,000 board feet. This simple rule applies to all nominal dimensioned lumber.
- Waste is an issue of concern at the quantity takeoff for any area of construction. Framing lumber is sold in even foot lengths, i.e., 10′, 12′, 14′, 16′, and depending on spans, wall heights and the grade of lumber, waste is inevitable. A rule of thumb for lumber waste is 5% to 10% depending on material quality and the complexity of the framing.
- Wood in various forms and shapes is used in many projects, even where the main structural framing is steel, concrete, or masonry. Plywood as a back-up partition material and 2x boards used as blocking and cant strips around roof edges are two common examples. The estimator should ensure that the costs of all wood materials are included in the final estimate.

06200 Finish Carpentry

- It is necessary to consider the grade of workmanship when estimating labor costs for erecting millwork and interior finish. In practice, there are three grades: premium, custom, and economy. The Means daily output for base and case moldings is in the range of 200 to 250 L.F. per carpenter per day. This is appropriate for most average custom-grade projects. For premium projects an adjustment to productivity of 25% to 50% should be made depending on the complexity of the job.

Reference Numbers

Reference numbers are shown in bold squares at the beginning of some major classifications. These numbers refer to related items in the Reference Section. The reference information may be an estimating procedure, an alternate pricing method, or technical information.

Note: Not all subdivisions listed here necessarily appear in this publication.

	06052 | Selective Demolition	CREW	DAILY OUTPUT	LABOR-HOURS	UNIT	2006 BARE COSTS MAT.	LABOR	EQUIP.	TOTAL	TOTAL INCL O&P	
110	0010 SELECTIVE DEMOLITION, WOOD FRAMING R024119-10										110
	0100 Timber connector, nailed, small	1 Clab	96	.083	Ea.		2.28		2.28	3.78	
	0110 Medium		60	.133			3.65		3.65	6.05	
	0120 Large		48	.167			4.57		4.57	7.55	
	0130 Bolted, small		48	.167			4.57		4.57	7.55	
	0140 Medium		32	.250			6.85		6.85	11.35	
	0150 Large		24	.333			9.15		9.15	15.15	
	2958 Beams, 2″ x 6″	2 Clab	1100	.015	L.F.		.40		.40	.66	
	2960 2″ x 8″		825	.019			.53		.53	.88	
	2965 2″ x 10″		665	.024			.66		.66	1.09	
	2970 2″ x 12″		550	.029			.80		.80	1.32	
	2972 2″ x 14″		470	.034			.93		.93	1.55	
	2975 4″ x 8″	B-1	413	.058			1.63		1.63	2.70	
	2980 4″ x 10″		330	.073			2.04		2.04	3.38	
	2985 4″ x 12″		275	.087			2.45		2.45	4.06	
	3000 6″ x 8″		275	.087			2.45		2.45	4.06	
	3040 6″ x 10″		220	.109			3.06		3.06	5.05	
	3080 6″ x 12″		185	.130			3.64		3.64	6.05	
	3120 8″ x 12″		140	.171			4.81		4.81	7.95	
	3160 10″ x 12″		110	.218			6.10		6.10	10.15	
	3162 Alternate pricing method		1.10	21.818	M.B.F.		610		610	1,025	
	3170 Blocking, in 16″ OC wall framing, 2″ x 4″	1 Clab	600	.013	L.F.		.37		.37	.61	
	3172 2″ x 6″		400	.020			.55		.55	.91	
	3174 In 24″ OC wall framing, 2″ x 4″		600	.013			.37		.37	.61	
	3176 2″ x 6″		400	.020			.55		.55	.91	
	3178 Alt method, wood blocking removal from wood framimg		.40	20	M.B.F.		550		550	910	
	3179 Wood blocking removal from steel framimg		.36	22.222	″		610		610	1,000	
	3180 Bracing, let in, 1″ x 3″, studs 16″ OC		1050	.008	L.F.		.21		.21	.35	
	3181 Studs 24″ OC		1080	.007			.20		.20	.34	
	3182 1″ x 4″, studs 16″ OC		1050	.008			.21		.21	.35	
	3183 Studs 24″ OC		1080	.007			.20		.20	.34	
	3184 1″ x 6″, studs 16″ OC		1050	.008			.21		.21	.35	
	3185 Studs 24″ OC		1080	.007			.20		.20	.34	
	3186 2″ x 3″, studs 16″ OC		800	.010			.27		.27	.45	
	3187 Studs 24″ OC		830	.010			.26		.26	.44	
	3188 2″ x 4″, studs 16″ OC		800	.010			.27		.27	.45	
	3189 Studs 24″ OC		830	.010			.26		.26	.44	
	3190 2″ x 6″, studs 16″ OC		800	.010			.27		.27	.45	
	3191 Studs 24″ OC		830	.010			.26		.26	.44	
	3192 2″ x 8″, studs 16″ OC		800	.010			.27		.27	.45	
	3193 Studs 24″ OC		830	.010			.26		.26	.44	
	3194 "T" shaped metal bracing, studs at 16″ OC		1060	.008			.21		.21	.34	
	3195 Studs at 24″ OC		1200	.007			.18		.18	.30	
	3196 Metal straps, studs at 16″ OC		1200	.007			.18		.18	.30	
	3197 Studs at 24″ OC		1240	.006			.18		.18	.29	
	3200 Columns, round, 8′ to 14′ tall		40	.200	Ea.		5.50		5.50	9.10	
	3202 Dimensional lumber sizes	2 Clab	1.10	14.545	M.B.F.		400		400	660	
	3250 Blocking, between joists	1 Clab	320	.025	Ea.		.69		.69	1.14	
	3252 Bridging, metal strap, between joists		320	.025	Pr.		.69		.69	1.14	
	3260 Door buck, studs, header & access, 8′ high 2″ x 4″ wall, 3′ wide		32	.250	Ea.		6.85		6.85	11.35	
	3261 4′ wide		32	.250			6.85		6.85	11.35	
	3262 5′ wide		32	.250			6.85		6.85	11.35	
	3263 6′ wide		32	.250			6.85		6.85	11.35	
	3264 8′ wide		30	.267			7.30		7.30	12.10	
	3265 10′ wide		30	.267			7.30		7.30	12.10	
	3266 12′ wide		30	.267			7.30		7.30	12.10	

06052 | Selective Demolition

			CREW	DAILY OUTPUT	LABOR-HOURS	UNIT	2006 BARE COSTS MAT.	LABOR	EQUIP.	TOTAL	TOTAL INCL O&P	
110	3267	2″ x 6″ wall, 3′ wide R024119-10	1 Clab	32	.250	Ea.		6.85		6.85	11.35	110
	3268	4′ wide		32	.250			6.85		6.85	11.35	
	3269	5′ wide		32	.250			6.85		6.85	11.35	
	3270	6′ wide		32	.250			6.85		6.85	11.35	
	3271	8′ wide		30	.267			7.30		7.30	12.10	
	3272	10′ wide		30	.267			7.30		7.30	12.10	
	3273	12′ wide		30	.267			7.30		7.30	12.10	
	3274	Window buck, studs, header & access, 8′ high 2″ x 4″ wall, 2′ wide		24	.333			9.15		9.15	15.15	
	3275	3′ wide		24	.333			9.15		9.15	15.15	
	3276	4′ wide		24	.333			9.15		9.15	15.15	
	3277	5′ wide		24	.333			9.15		9.15	15.15	
	3278	6′ wide		24	.333			9.15		9.15	15.15	
	3279	7′ wide		24	.333			9.15		9.15	15.15	
	3280	8′ wide		22	.364			9.95		9.95	16.50	
	3281	10′ wide		22	.364			9.95		9.95	16.50	
	3282	12′ wide		22	.364			9.95		9.95	16.50	
	3283	2″ x 6″ wall, 2′ wide		24	.333			9.15		9.15	15.15	
	3284	3′ wide		24	.333			9.15		9.15	15.15	
	3285	4′ wide		24	.333			9.15		9.15	15.15	
	3286	5′ wide		24	.333			9.15		9.15	15.15	
	3287	6′ wide		24	.333			9.15		9.15	15.15	
	3288	7′ wide		24	.333			9.15		9.15	15.15	
	3289	8′ wide		22	.364			9.95		9.95	16.50	
	3290	10′ wide		22	.364			9.95		9.95	16.50	
	3291	12′ wide		22	.364			9.95		9.95	16.50	
	3400	Fascia boards, 1″ x 6″		500	.016	L.F.		.44		.44	.73	
	3440	1″ x 8″		450	.018			.49		.49	.81	
	3480	1″ x 10″		400	.020			.55		.55	.91	
	3490	2″ x 6″		450	.018			.49		.49	.81	
	3500	2″ x 8″		400	.020			.55		.55	.91	
	3510	2″ x 10″		350	.023			.63		.63	1.04	
	3610	Furring, on wood walls or ceiling		4000	.002	S.F.		.05		.05	.09	
	3620	On masonry or concrete walls or ceiling		1200	.007	″		.18		.18	.30	
	3800	Headers over openings, 2 @ 2″ x 6″		110	.073	L.F.		1.99		1.99	3.30	
	3840	2 @ 2″ x 8″		100	.080			2.19		2.19	3.63	
	3880	2 @ 2″ x 10″		90	.089			2.44		2.44	4.04	
	3885	Alternate pricing method		.26	30.651	M.B.F.		840		840	1,400	
	3920	Joists, 1″ x 4″		1250	.006	L.F.		.18		.18	.29	
	3930	1″ x 6″		1135	.007			.19		.19	.32	
	3940	1″ x 8″		1000	.008			.22		.22	.36	
	3950	1″ x 10″		895	.009			.25		.25	.41	
	3960	1″ x 12″		765	.010			.29		.29	.47	
	4200	2″ x 4″	2 Clab	1000	.016			.44		.44	.73	
	4230	2″ x 6″		970	.016			.45		.45	.75	
	4240	2″ x 8″		940	.017			.47		.47	.77	
	4250	2″ x 10″		910	.018			.48		.48	.80	
	4280	2″ x 12″		880	.018			.50		.50	.83	
	4281	2″ x 14″		850	.019			.52		.52	.85	
	4282	Composite joists, 9-1/2″		960	.017			.46		.46	.76	
	4283	11-7/8″		930	.017			.47		.47	.78	
	4284	14″		897	.018			.49		.49	.81	
	4285	16″		865	.019			.51		.51	.84	
	4290	Wood joists, alternate pricing method		1.50	10.667	M.B.F.		292		292	485	
	4500	Open web joist, 12″ deep		500	.032	L.F.		.88		.88	1.45	
	4505	14″ deep		475	.034			.92		.92	1.53	
	4510	16″ deep		450	.036			.97		.97	1.61	

6 WOOD & PLASTICS

	06052 | Selective Demolition		CREW	DAILY OUTPUT	LABOR-HOURS	UNIT	2006 BARE COSTS MAT.	LABOR	EQUIP.	TOTAL	TOTAL INCL O&P	
110	4520	18" deep R024119-10	2 Clab	425	.038	L.F.		1.03		1.03	1.71	110
	4560	1" x 3"	1 Clab	1200	.007			.18		.18	.30	
	4580	2" x 2"		1100	.007			.20		.20	.33	
	4590	2" x 4"		1000	.008			.22		.22	.36	
	4600	2" x 6"		1000	.008			.22		.22	.36	
	4601	2" x 8 or 2" x 10"		800	.010			.27		.27	.45	
	4602	4" x 6"		600	.013			.37		.37	.61	
	4604	4" x 8"		450	.018			.49		.49	.81	
	5400	Posts, 4" x 4"	2 Clab	800	.020			.55		.55	.91	
	5405	4" x 6"		550	.029			.80		.80	1.32	
	5410	4" x 8"		440	.036			1		1	1.65	
	5425	4" x 10"		390	.041			1.12		1.12	1.86	
	5430	4" x 12"		350	.046			1.25		1.25	2.08	
	5440	6" x 6"		400	.040			1.10		1.10	1.82	
	5445	6" x 8"		350	.046			1.25		1.25	2.08	
	5450	6" x 10"		320	.050			1.37		1.37	2.27	
	5455	6" x 12"		290	.055			1.51		1.51	2.50	
	5480	8" x 8"		300	.053			1.46		1.46	2.42	
	5500	10" x 10"		240	.067			1.83		1.83	3.03	
	5660	Tongue and groove floor planks		2	8	M.B.F.		219		219	365	
	5750	Rafters, ordinary, 16" OC, 2" x 4"		880	.018	S.F.		.50		.50	.83	
	5755	2" x 6"		840	.019			.52		.52	.86	
	5760	2" x 8"		820	.020			.53		.53	.89	
	5770	2" x 10"		820	.020			.53		.53	.89	
	5780	2" x 12"		810	.020			.54		.54	.90	
	5785	24" OC, 2" x 4"		1170	.014			.37		.37	.62	
	5786	2" x 6"		1117	.014			.39		.39	.65	
	5787	2" x 8"		1091	.015			.40		.40	.67	
	5788	2" x 10"		1091	.015			.40		.40	.67	
	5789	2" x 12"		1077	.015			.41		.41	.67	
	5795	Rafters, ordinary, 2" x 4" (alternate method)		862	.019	L.F.		.51		.51	.84	
	5800	2" x 6" (alternate method)		850	.019			.52		.52	.85	
	5840	2" x 8" (alternate method)		837	.019			.52		.52	.87	
	5855	2" x 10" (alternate method)		825	.019			.53		.53	.88	
	5865	2" x 12" (alternate method)		812	.020			.54		.54	.89	
	5870	Sill plate, 2" x 4"	1 Clab	1170	.007			.19		.19	.31	
	5871	2" x 6"		780	.010			.28		.28	.47	
	5872	2" x 8"		586	.014			.37		.37	.62	
	5873	Alternate pricing method		.78	10.256	M.B.F.		281		281	465	
	5885	Ridge board, 1" x 4"	2 Clab	900	.018	L.F.		.49		.49	.81	
	5886	1" x 6"		875	.018			.50		.50	.83	
	5887	1" x 8"		850	.019			.52		.52	.85	
	5888	1" x 10"		825	.019			.53		.53	.88	
	5889	1" x 12"		800	.020			.55		.55	.91	
	5890	2" x 4"		900	.018			.49		.49	.81	
	5892	2" x 6"		875	.018			.50		.50	.83	
	5894	2" x 8"		850	.019			.52		.52	.85	
	5896	2" x 10"		825	.019			.53		.53	.88	
	5898	2" x 12"		800	.020			.55		.55	.91	
	5900	Hip & valley rafters, 2" x 6"		500	.032			.88		.88	1.45	
	5940	2" x 8"		420	.038			1.04		1.04	1.73	
	6050	Rafter tie, 1" x 4"		1250	.013			.35		.35	.58	
	6052	1" x 6"		1135	.014			.39		.39	.64	
	6054	2" x 4"		1000	.016			.44		.44	.73	
	6056	2" x 6"		970	.016			.45		.45	.75	
	6070	Sleepers, on concrete, 1" x 2"	1 Clab	4700	.002			.05		.05	.08	

	06052 | Selective Demolition		Crew	Daily Output	Labor-Hours	Unit	2006 Bare Costs Mat.	Labor	Equip.	Total	Total Incl O&P	
110	6075	1″ x 3″ R024119-10	1 Clab	4000	.002	L.F.		.05		.05	.09	110
	6080	2″ x 4″		3000	.003			.07		.07	.12	
	6085	2″ x 6″		2600	.003			.08		.08	.14	
	6086	Sheathing from roof, 5/16″	2 Clab	1600	.010	S.F.		.27		.27	.45	
	6088	3/8″		1525	.010			.29		.29	.48	
	6090	1/2″		1400	.011			.31		.31	.52	
	6092	5/8″		1300	.012			.34		.34	.56	
	6094	3/4″		1200	.013			.37		.37	.61	
	6096	Board sheathing from roof		1400	.011			.31		.31	.52	
	6100	Sheathing, from walls, 1/4″		1200	.013			.37		.37	.61	
	6110	5/16″		1175	.014			.37		.37	.62	
	6120	3/8″		1150	.014			.38		.38	.63	
	6130	1/2″		1125	.014			.39		.39	.65	
	6140	5/8″		1100	.015			.40		.40	.66	
	6150	3/4″		1075	.015			.41		.41	.68	
	6152	Board sheathing from walls		1500	.011			.29		.29	.48	
	6158	Subfloor, with boards		1050	.015			.42		.42	.69	
	6160	Plywood, 1/2″ thick		768	.021			.57		.57	.95	
	6162	5/8″ thick		760	.021			.58		.58	.96	
	6164	3/4″thick		750	.021			.58		.58	.97	
	6165	1-1/8″ thick		720	.022			.61		.61	1.01	
	6166	Underlayment, particle board, 3/8″ thick	1 Clab	780	.010			.28		.28	.47	
	6168	1/2″ thick		768	.010			.29		.29	.47	
	6170	5/8″ thick		760	.011			.29		.29	.48	
	6172	3/4″ thick		750	.011			.29		.29	.48	
	6200	Stairs and stringers, minimum	2 Clab	40	.400	Riser		10.95		10.95	18.15	
	6240	Maximum	″	26	.615	″		16.85		16.85	28	
	6300	Components, tread	1 Clab	110	.073	Ea.		1.99		1.99	3.30	
	6320	Riser		80	.100	″		2.74		2.74	4.54	
	6390	Stringer, 2″ x 10″		260	.031	L.F.		.84		.84	1.40	
	6400	2″ x 12″		260	.031			.84		.84	1.40	
	6410	3″ x 10″		250	.032			.88		.88	1.45	
	6420	3″ x 12″		250	.032			.88		.88	1.45	
	6590	Wood studs, 2″ x 3″	2 Clab	3076	.005			.14		.14	.24	
	6600	2″ x 4″		2000	.008			.22		.22	.36	
	6640	2″ x 6″		1600	.010			.27		.27	.45	
	6720	Wall framing, including studs plates and blocking, 2″ x 4″	1 Clab	600	.013	S.F.		.37		.37	.61	
	6740	2″ x 6″		480	.017	″		.46		.46	.76	
	6750	Headers, 2″ x 4″		1125	.007	L.F.		.19		.19	.32	
	6755	2″ x 6″		1125	.007			.19		.19	.32	
	6760	2″ x 8″		1050	.008			.21		.21	.35	
	6765	2″ x 10″		1050	.008			.21		.21	.35	
	6770	2″ x 12″		1000	.008			.22		.22	.36	
	6780	4″ x 10″		525	.015			.42		.42	.69	
	6785	4″ x 12″		500	.016			.44		.44	.73	
	6790	6″ x 8″		560	.014			.39		.39	.65	
	6795	6″ x 10″		525	.015			.42		.42	.69	
	6797	6″ x 12″		500	.016			.44		.44	.73	
	8000	Soffit, T & G wood		520	.015	S.F.		.42		.42	.70	
	8010	Hardboard, vinyl or aluminum		640	.013			.34		.34	.57	
	8030	Plywood	2 Carp	315	.051			1.81		1.81	2.99	
	9000	Minimum labor/equipment charge	1 Clab	4	2	Job		55		55	91	
	9500	See Div. 02220-350 for rubbish handling										
120	0010	**SELECTIVE DEMOLITION, MILLWORK AND TRIM** R024119-10										120
	1000	Cabinets, wood, base cabinets, per L.F.	2 Clab	80	.200	L.F.		5.50		5.50	9.10	

06050 | Basic Wood / Plastic Materials / Methods

06052 | Selective Demolition

		06052 Selective Demolition	CREW	DAILY OUTPUT	LABOR-HOURS	UNIT	2006 BARE COSTS MAT.	LABOR	EQUIP.	TOTAL	TOTAL INCL O&P	
120	1020	Wall cabinets, per L.F. R024119-10	2 Clab	80	.200	L.F.		5.50		5.50	9.10	120
	1060	Remove and reset, base cabinets	2 Carp	18	.889	Ea.		31.50		31.50	52.50	
	1070	Wall cabinets	″	20	.800	″		28.50		28.50	47	
	1100	Steel, painted, base cabinets	2 Clab	60	.267	L.F.		7.30		7.30	12.10	
	1120	Wall cabinets		60	.267	″		7.30		7.30	12.10	
	1200	Casework, large area		320	.050	S.F.		1.37		1.37	2.27	
	1220	Selective		200	.080	″		2.19		2.19	3.63	
	1500	Counter top, minimum		200	.080	L.F.		2.19		2.19	3.63	
	1510	Maximum		120	.133			3.65		3.65	6.05	
	1550	Remove and reset, minimum	2 Carp	50	.320			11.40		11.40	18.85	
	1560	Maximum	″	40	.400			14.20		14.20	23.50	
	2000	Paneling, 4′ x 8′ sheets	2 Clab	2000	.008	S.F.		.22		.22	.36	
	2100	Boards, 1″ x 4″		700	.023			.63		.63	1.04	
	2120	1″ x 6″		750	.021			.58		.58	.97	
	2140	1″ x 8″		800	.020			.55		.55	.91	
	3000	Trim, baseboard, to 6″ wide		1200	.013	L.F.		.37		.37	.61	
	3040	Greater than 6″ and up to 12″ wide		1000	.016			.44		.44	.73	
	3080	Remove and reset, minimum	2 Carp	400	.040			1.42		1.42	2.36	
	3090	Maximum	″	300	.053			1.90		1.90	3.14	
	3100	Ceiling trim	2 Clab	1000	.016			.44		.44	.73	
	3120	Chair rail		1200	.013			.37		.37	.61	
	3140	Railings with balusters		240	.067			1.83		1.83	3.03	
	3160	Wainscoting		700	.023	S.F.		.63		.63	1.04	
	9000	Minimum labor/equipment charge	1 Clab	4	2	Job		55		55	91	

06070 | Lumber Treatment

		06070 Lumber Treatment	CREW	DAILY OUTPUT	LABOR-HOURS	UNIT	MAT.	LABOR	EQUIP.	TOTAL	TOTAL INCL O&P	
400	0011	**LUMBER TREATMENT**										400
	0400	Fire retardant, wet				M.B.F.	310			310	340	
	0500	KDAT					297			297	325	
	0700	Salt treated, water borne, .40 lb. retention					142			142	156	
	0800	Oil borne, 8 lb. retention					166			166	183	
	1000	Kiln dried lumber, 1″ & 2″ thick, softwoods					95			95	104	
	1100	Hardwoods					101			101	111	
	1500	For small size 1″ stock, add					12.80			12.80	14.10	
	1700	For full size rough lumber, add					20%					
600	0010	**PLYWOOD TREATMENT**										600
	0020	Fire retardant, 1/4″ thick				M.S.F.	237			237	261	
	0030	3/8″ thick					261			261	287	
	0050	1/2″ thick					279			279	305	
	0070	5/8″ thick					297			297	325	
	0100	3/4″ thick					325			325	360	
	0200	For KDAT, add					71			71	78	
	0500	Salt treated water borne, .25 lb., wet, 1/4″ thick					130			130	143	
	0530	3/8″ thick					136			136	150	
	0550	1/2″ thick					142			142	156	
	0570	5/8″ thick					155			155	170	
	0600	3/4″ thick					161			161	177	
	0800	For KDAT add					71			71	78	
	0900	For .40 lb., per C.F. retention, add					59.50			59.50	65.50	
	1000	For certification stamp, add					35			35	38.50	

06090 | Wood & Plastic Fastenings

		06090 Wood & Plastic Fastenings	CREW	DAILY OUTPUT	LABOR-HOURS	UNIT	MAT.	LABOR	EQUIP.	TOTAL	TOTAL INCL O&P	
600	0010	**NAILS**, material only, based upon 50# box purchase										600
	0020	Copper nails, plain				Lb.	6.60			6.60	7.25	

	06090 | Wood & Plastic Fastenings		CREW	DAILY OUTPUT	LABOR-HOURS	UNIT	2006 BARE COSTS MAT.	LABOR	EQUIP.	TOTAL	TOTAL INCL O&P	
600	0400	Stainless steel, plain				Lb.	5.95			5.95	6.55	600
	0500	Box, 3d to 20d, bright					.94			.94	1.03	
	0520	Galvanized					1.38			1.38	1.52	
	0600	Common, 3d to 60d, plain 					.78			.78	.86	
	0700	Galvanized					1.02			1.02	1.12	
	0800	Aluminum					4.45			4.45	4.90	
	1000	Annular or spiral thread, 4d to 60d, plain					1.88			1.88	2.07	
	1200	Galvanized					1.95			1.95	2.15	
	1400	Drywall nails, plain					.79			.79	.87	
	1600	Galvanized					1.54			1.54	1.69	
	1800	Finish nails, 4d to 10d, plain 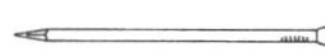					1.05			1.05	1.16	
	2000	Galvanized					1.32			1.32	1.45	
	2100	Aluminum					4.10			4.10	4.51	
	2300	Flooring nails, hardened steel, 2d to 10d, plain 					1.61			1.61	1.77	
	2400	Galvanized					2.25			2.25	2.48	
	2500	Gypsum lath nails, 1-1/8", 13 ga. flathead, blued					1.54			1.54	1.69	
	2600	Masonry nails, hardened steel, 3/4" to 3" long, plain					1.48			1.48	1.63	
	2700	Galvanized 					1.55			1.55	1.71	
	2900	Roofing nails, threaded, galvanized					1.32			1.32	1.45	
	3100	Aluminum					4.80			4.80	5.30	
	3300	Compressed lead head, threaded, galvanized					1.50			1.50	1.65	
	3600	Siding nails, plain shank, galvanized					1.45			1.45	1.60	
	3800	Aluminum					4.11			4.11	4.52	
	5000	Add to prices above for cement coating					.10			.10	.11	
	5200	Zinc or tin plating					.13			.13	.14	
	5500	Vinyl coated sinkers, 8d to 16d				↓	.57			.57	.63	
650	0010	**NAILS**, pneumatic tools										650
	0020	Framing, per carton of 5000, 2"				Ea.	37			37	41	
	0100	2-3/8"					42.50			42.50	46.50	
	0200	Per carton of 4000, 3"					37.50			37.50	41.50	
	0300	3-1/4"					40			40	44	
	0400	Per carton of 5000, 2-3/8", galv.					57.50			57.50	63.50	
	0500	Per carton of 4000, 3", galv.					65			65	71.50	
	0600	3-1/4", galv.					80.50			80.50	88.50	
	0700	Roofing, per carton of 7200, 1"					35			35	38.50	
	0800	1-1/4"					32.50			32.50	36	
	0900	1-1/2"					37.50			37.50	41.50	
	1000	1-3/4"				↓	45.50			45.50	50	
700	0010	**SHEET METAL SCREWS**										700
	0020	Steel, standard, #8 x 3/4", plain				C	3.29			3.29	3.62	
	0100	Galvanized					3.44			3.44	3.78	
	0300	#10 x 1", plain					4.49			4.49	4.94	
	0400	Galvanized					4.35			4.35	4.79	
	0600	With washers, #14 x 1", plain					10.20			10.20	11.20	
	0700	Galvanized					10.50			10.50	11.55	
	0900	#14 x 2", plain					18			18	19.80	
	1000	Galvanized					18.15			18.15	19.95	
	1500	Self-drilling, with washers, (pinch point) #8 x 3/4", plain					7.10			7.10	7.80	
	1600	Galvanized					7.20			7.20	7.90	
	1800	#10 x 3/4", plain					9.30			9.30	10.20	
	1900	Galvanized					9.40			9.40	10.35	
	3000	Stainless steel w/aluminum or neoprene washers, #14 x 1", plain					18.35			18.35	20	
	3100	#14 x 2", plain				↓	25			25	27.50	
750	0010	**WOOD SCREWS**										750
	0020	Steel, #8 x 1" long				C	3.29			3.29	3.62	

For expanded coverage of these items see *Means Interior Cost Data 2006*

	06090 | Wood & Plastic Fastenings		CREW	DAILY OUTPUT	LABOR-HOURS	UNIT	2006 BARE COSTS MAT.	LABOR	EQUIP.	TOTAL	TOTAL INCL O&P	
750	0100	Brass				C	11.50			11.50	12.65	750
	0200	#8, 2" long, steel					5.60			5.60	6.15	
	0300	Brass					19.50			19.50	21.50	
	0400	#10, 1" long, steel					5.90			5.90	6.50	
	0500	Brass					14			14	15.40	
	0600	#10, 2" long, steel					6.30			6.30	6.90	
	0700	Brass					23			23	25.50	
	0800	#10, 3" long, steel					10.50			10.50	11.55	
	1000	#12, 2" long, steel					7.50			7.50	8.25	
	1100	Brass					30			30	33	
	1500	#12, 3" long, steel					11.50			11.50	12.65	
	2000	#12, 4" long, steel					43			43	47.50	
800	0010	**TIMBER CONNECTORS** Add up cost of each part for total										800
	0020	cost of connection										
	0100	Connector plates, steel, with bolts, straight	2 Carp	75	.213	Ea.	22.50	7.60		30.10	37.50	
	0110	Tee		50	.320		33	11.40		44.40	55.50	
	0120	T- Strap, 14 gauge, 12" x 8" x 2"		50	.320		33	11.40		44.40	55.50	
	0150	Anchor plate, 7 gauge, 9" x 7"		75	.213		22.50	7.60		30.10	37.50	
	0200	Bolts, machine, sq. hd. with nut & washer, 1/2" diameter, 4" long	1 Carp	140	.057		.38	2.03		2.41	3.79	
	0300	7-1/2" long		130	.062		.67	2.19		2.86	4.36	
	0500	3/4" diameter, 7-1/2" long		130	.062		1.83	2.19		4.02	5.65	
	0610	Machine bolts, w/ nut, washer, 3/4" dia, 15" L, HD's & beam hangers		95	.084		3.36	2.99		6.35	8.65	
	0800	Drilling bolt holes in timber, 1/2" diameter		450	.018	Inch		.63		.63	1.05	
	0900	1" diameter		350	.023	"		.81		.81	1.35	
	1100	Framing anchors, 2 or 3 dimensional, 10 gauge, no nails incl.		175	.046	Ea.	.56	1.63		2.19	3.31	
	1150	Framing anchors, 18 gauge, 4 1/2" x 2 3/4"		175	.046		.56	1.63		2.19	3.31	
	1160	Framing anchors, 18 gauge, 4 1/2" x 3"		175	.046		.56	1.63		2.19	3.31	
	1170	Clip anchors plates, 18 gauge, 12" x 1 1/8"		175	.046		.56	1.63		2.19	3.31	
	1250	Holdowns, 3 gauge base, 10 gauge body		8	1		18.75	35.50		54.25	79.50	
	1260	Holdowns, 7 gauge 11 1/16" x 3 1/4"		8	1		18.75	35.50		54.25	79.50	
	1270	Holdowns, 7 gauge 14 3/8" x 3 1/8"		8	1		18.75	35.50		54.25	79.50	
	1275	Holdowns, 12 gauge 8" x 2 1/2"		8	1		18.75	35.50		54.25	79.50	
	1300	Joist and beam hangers, 18 ga. galv., for 2" x 4" joist		175	.046		.70	1.63		2.33	3.46	
	1400	2" x 6" to 2" x 10" joist		165	.048		.80	1.72		2.52	3.74	
	1600	16 ga. galv., 3" x 6" to 3" x 10" joist 		160	.050		3.16	1.78		4.94	6.45	
	1700	3" x 10" to 3" x 14" joist		160	.050		3.67	1.78		5.45	7	
	1800	4" x 6" to 4" x 10" joist		155	.052		2.78	1.83		4.61	6.10	
	1900	4" x 10" to 4" x 14" joist		155	.052		3.77	1.83		5.60	7.20	
	2000	Two-2" x 6" to two-2" x 10" joists		150	.053		3.04	1.90		4.94	6.50	
	2100	Two-2" x 10" to two-2" x 14" joists		150	.053		3.04	1.90		4.94	6.50	
	2300	3/16" thick, 6" x 8" joist		145	.055		6.60	1.96		8.56	10.50	
	2400	6" x 10" joist		140	.057		7.80	2.03		9.83	11.90	
	2500	6" x 12" joist		135	.059		9.40	2.11		11.51	13.80	
	2700	1/4" thick, 6" x 14" joist		130	.062		11.65	2.19		13.84	16.40	
	2900	Plywood clips, extruded aluminum H clip, for 3/4" panels					.18			.18	.20	
	3000	Galvanized 18 ga. back-up clip					.17			.17	.19	
	3200	Post framing, 16 ga. galv. for 4" x 4" base, 2 piece	1 Carp	130	.062		6.55	2.19		8.74	10.80	
	3300	Cap		130	.062		3.21	2.19		5.40	7.15	
	3500	Rafter anchors, 18 ga. galv., 1-1/2" wide, 5-1/4" long		145	.055		.55	1.96		2.51	3.86	
	3600	10-3/4" long 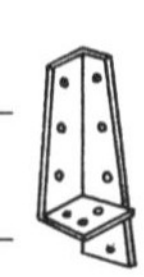		145	.055		1.09	1.96		3.05	4.45	
	3800	Shear plates, 2-5/8" diameter		120	.067		1.96	2.37		4.33	6.10	
	3900	4" diameter		115	.070		4.50	2.47		6.97	9.05	
	4000	Sill anchors, embedded in concrete or block, 18-5/8" long		115	.070		1.32	2.47		3.79	5.55	
	4100	Spike grids, 4" x 4", flat or curved		120	.067		.49	2.37		2.86	4.47	
	4400	Split rings, 2-1/2" diameter		120	.067		1.58	2.37		3.95	5.65	
	4500	4" diameter		110	.073		2.45	2.59		5.04	7	

06050 | Basic Wood / Plastic Materials / Methods

06090 | Wood & Plastic Fastenings

			Crew	Daily Output	Labor-Hours	Unit	2006 Bare Costs Mat.	Labor	Equip.	Total	Total Incl O&P	
800	4550	Tie plate, 20 gauge, 7″ x 3 1/8″	1 Carp	110	.073	Ea.	2.45	2.59		5.04	7	800
	4560	Tie plate, 20 gauge, 5″ x 4 1/8″		110	.073		2.45	2.59		5.04	7	
	4575	Twist straps, 18 gauge, 12″ x 1 1/4″		110	.073		2.45	2.59		5.04	7	
	4580	Twist straps, 18 gauge, 16″ x 1 1/4″		110	.073		2.45	2.59		5.04	7	
	4600	Strap ties, 20 ga., 2 -1/16″ wide, 12 13/16″ long		180	.044		1.43	1.58		3.01	4.19	
	4700	Strap ties, 16 ga., 1-3/8″ wide, 12″ long		180	.044		1.43	1.58		3.01	4.19	
	4800	24″ long		160	.050		1.97	1.78		3.75	5.10	
	5000	Toothed rings, 2-5/8″ or 4″ diameter		90	.089		1.35	3.16		4.51	6.75	
	5200	Truss plates, nailed, 20 gauge, up to 32′ span		17	.471	Truss	9.75	16.75		26.50	38.50	
	5400	Washers, 2″ x 2″ x 1/8″				Ea.	.31			.31	.34	
	5500	3″ x 3″ x 3/16″				″	.80			.80	.88	
	9000	Minimum labor/equipment charge	1 Carp	4	2	Job		71		71	118	
825	0010	**ROUGH HARDWARE**, average % of carpentry material										825
	0020	Minimum					.50%					
	0200	Maximum					1.50%					
	0210	In seismic or hurricane areas, up to					10%					
850	0010	**BRACING**										850
	0300	Let-in, "T" shaped, 22 ga. galv. steel, studs at 16″ O.C.	1 Carp	5.80	1.379	C.L.F.	53	49		102	140	
	0400	Studs at 24″ O.C.		6	1.333		53	47.50		100.50	137	
	0500	16 ga. galv. steel straps, studs at 16″ O.C.		6	1.333		80	47.50		127.50	167	
	0600	Studs at 24″ O.C.		6.20	1.290		80	46		126	164	

6 WOOD & PLASTICS

06100 | Rough Carpentry

06110 | Wood Framing

			Crew	Daily Output	Labor-Hours	Unit	2006 Bare Costs Mat.	Labor	Equip.	Total	Total Incl O&P	
100	0010	**BLOCKING**										100
	2600	Miscellaneous, to wood construction										
	2620	2″ x 4″	1 Carp	.17	47.059	M.B.F.	560	1,675		2,235	3,400	
	2625	Pneumatic nailed		.21	38.095		560	1,350		1,910	2,875	
	2660	2″ x 8″		.27	29.630		640	1,050		1,690	2,450	
	2665	Pneumatic nailed		.33	24.242		640	860		1,500	2,125	
	2720	To steel construction										
	2740	2″ x 4″	1 Carp	.14	57.143	M.B.F.	560	2,025		2,585	4,000	
	2780	2″ x 8″		.21	38.095	″	640	1,350		1,990	2,950	
	9000	Minimum labor/equipment charge		4	2	Job		71		71	118	
150	0010	**BRACING**										150
	0011	Let-in, with 1″ x 6″ boards, studs 16″ OC	1 Carp	1.50	5.333	C.L.F.	55	190		245	375	
	0202	Studs @ 24″ O.C.	″	230	.035	L.F.	.55	1.24		1.79	2.66	
200	0010	**BRIDGING**										200
	0011	Wood, for joists 16″ OC, 1″ x 3″	1 Carp	1.30	6.154	C.Pr.	42.50	219		261.50	405	
	0017	Pneumatic nailed		170	.047	Pr.	.42	1.67		2.09	3.24	
	0102	2″ x 3″ bridging		130	.062		.57	2.19		2.76	4.25	
	0107	Pneumatic nailed		170	.047		.57	1.67		2.24	3.40	
	0302	Steel, galvanized, 18 ga., for 2″ x 10″ joists at 12″ O.C.		130	.062		1.18	2.19		3.37	4.91	
	0402	24″ O.C.		140	.057		1.42	2.03		3.45	4.93	
	0902	Compression type, 16″ O.C., 2″ x 8″ joists		200	.040		1.38	1.42		2.80	3.88	
	1002	2″ x 12″ joists		200	.040		1.38	1.42		2.80	3.88	

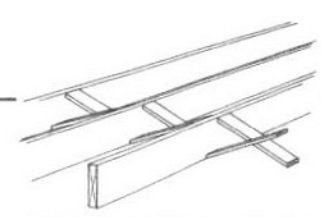

		06110 | Wood Framing	CREW	DAILY OUTPUT	LABOR-HOURS	UNIT	2006 BARE COSTS MAT.	LABOR	EQUIP.	TOTAL	TOTAL INCL O&P	
505	0010	**FRAMING, BEAMS & GIRDERS** R061110-30										505
	1002	Single, 2" x 6"	2 Carp	700	.023	L.F.	.59	.81		1.40	2	
	1007	Pneumatic nailed		812	.020		.59	.70		1.29	1.81	
	1022	2" x 8"		650	.025		.85	.88		1.73	2.39	
	1027	Pneumatic nailed		754	.021		.85	.75		1.60	2.19	
	1042	2" x 10"		600	.027		1.29	.95		2.24	2.99	
	1047	Pneumatic nailed		696	.023		1.29	.82		2.11	2.77	
	1062	2" x 12"		550	.029		1.60	1.03		2.63	3.47	
	1067	Pneumatic nailed		638	.025		1.60	.89		2.49	3.24	
	1082	2" x 14"		500	.032		2.18	1.14		3.32	4.28	
	1087	Pneumatic nailed		580	.028		2.18	.98		3.16	4.03	
	1102	3" x 8"		550	.029		2.59	1.03		3.62	4.56	
	1122	3" x 10"		500	.032		3.26	1.14		4.40	5.45	
	1142	3" x 12"		450	.036		3.92	1.26		5.18	6.40	
	1162	3" x 14"		400	.040		4.73	1.42		6.15	7.55	
	1170	4" x 6"	F-3	1100	.036		1.87	1.31	.58	3.76	4.85	
	1182	4" x 8"		1000	.040		2.59	1.44	.64	4.67	5.90	
	1202	4" x 10"		950	.042		3.26	1.52	.67	5.45	6.80	
	1222	4" x 12"		900	.044		4.04	1.60	.71	6.35	7.85	
	1242	4" x 14"		850	.047		4.95	1.70	.75	7.40	9.05	
	2002	Double, 2" x 6"	2 Carp	625	.026		1.18	.91		2.09	2.81	
	2007	Pneumatic nailed		725	.022		1.18	.78		1.96	2.60	
	2022	2" x 8"		575	.028		1.71	.99		2.70	3.52	
	2027	Pneumatic nailed		667	.024		1.71	.85		2.56	3.29	
	2042	2" x 10"		550	.029		2.58	1.03		3.61	4.55	
	2047	Pneumatic nailed		638	.025		2.58	.89		3.47	4.32	
	2062	2" x 12"		525	.030		3.20	1.08		4.28	5.30	
	2067	Pneumatic nailed		610	.026		3.20	.93		4.13	5.05	
	2082	2" x 14"		475	.034		4.36	1.20		5.56	6.75	
	2087	Pneumatic nailed		551	.029		4.36	1.03		5.39	6.50	
	3002	Triple, 2" x 6"		550	.029		1.78	1.03		2.81	3.66	
	3007	Pneumatic nailed		638	.025		1.78	.89		2.67	3.43	
	3022	2" x 8"		525	.030		2.56	1.08		3.64	4.62	
	3027	Pneumatic nailed		609	.026		2.56	.93		3.49	4.37	
	3042	2" x 10"		500	.032		3.87	1.14		5.01	6.15	
	3047	Pneumatic nailed		580	.028		3.87	.98		4.85	5.90	
	3062	2" x 12"		475	.034		4.80	1.20		6	7.30	
	3067	Pneumatic nailed		551	.029		4.80	1.03		5.83	7	
	3082	2" x 14"		450	.036		6.55	1.26		7.81	9.30	
	3087	Pneumatic nailed		522	.031		6.55	1.09		7.64	9	
	9000	Minimum labor/equipment charge	1 Carp	2	4	Job		142		142	236	
510	0010	**FRAMING, CEILINGS**										510
	6002	Suspended, 2" x 3"	2 Carp	1000	.016	L.F.	.38	.57		.95	1.36	
	6052	2" x 4"		900	.018		.37	.63		1	1.46	
	6102	2" x 6"		800	.020		.59	.71		1.30	1.83	
	6152	2" x 8"		650	.025		.85	.88		1.73	2.39	
	9000	Minimum labor/equipment charge	1 Carp	4	2	Job		71		71	118	
515	0010	**FRAMING, COLUMNS**										515
	0100	4" x 4"	2 Carp	390	.041	L.F.	1.25	1.46		2.71	3.80	
	0150	4" x 6"		275	.058		1.87	2.07		3.94	5.50	
	0200	4" x 8"		220	.073		2.59	2.59		5.18	7.15	
	0250	6" x 6"		215	.074		4.54	2.65		7.19	9.35	
	0300	6" x 8"		175	.091		6.05	3.25		9.30	12.05	
	0350	6" x 10"		150	.107		10.90	3.79		14.69	18.30	
	9000	Minimum labor/equipment charge	1 Carp	2	4	Job		142		142	236	

06110 | Wood Framing

		06110 Wood Framing	CREW	DAILY OUTPUT	LABOR-HOURS	UNIT	2006 BARE COSTS MAT.	LABOR	EQUIP.	TOTAL	TOTAL INCL O&P	
520	0010	**HEAVY MILL TIMBER FRAMING**										520
	0020	Beams, single 6″ x 10″	2 Carp	1.10	14.545	M.B.F.	2,150	515		2,665	3,225	
	0100	Single 8″ x 16″		1.20	13.333	″	2,700	475		3,175	3,750	
	0202	Built from 2″ lumber, multiple 2″ x 14″		900	.018	B.F.	.93	.63		1.56	2.08	
	0212	Built from 3″ lumber, multiple 3″ x 6″		700	.023		1.28	.81		2.09	2.76	
	0222	Multiple 3″ x 8″		800	.020		1.30	.71		2.01	2.60	
	0232	Multiple 3″ x 10″		900	.018		1.30	.63		1.93	2.48	
	0242	Multiple 3″ x 12″		1000	.016		1.31	.57		1.88	2.38	
	0252	Built from 4″ lumber, multiple 4″ x 6″		800	.020		.94	.71		1.65	2.21	
	0262	Multiple 4″ x 8″		900	.018		.97	.63		1.60	2.12	
	0272	Multiple 4″ x 10″		1000	.016		.98	.57		1.55	2.02	
	0282	Multiple 4″ x 12″		1100	.015		1.01	.52		1.53	1.97	
	0292	Columns, structural grade, 1500f, 4″ x 4″		450	.036	L.F.	2.34	1.26		3.60	4.66	
	0302	6″ x 6″		225	.071		6.40	2.53		8.93	11.25	
	0402	8″ x 8″		240	.067		12.45	2.37		14.82	17.65	
	0502	10″ x 10″		90	.178		19.85	6.30		26.15	32.50	
	0602	12″ x 12″		70	.229		28.50	8.15		36.65	45	
	0802	Floor planks, 2″ thick, T & G, 2″ x 6″		1050	.015	B.F.	1.88	.54		2.42	2.97	
	0902	2″ x 10″		1100	.015		1.88	.52		2.40	2.93	
	1102	3″ thick, 3″ x 6″		1050	.015		1.35	.54		1.89	2.39	
	1202	3″ x 10″		1100	.015		1.35	.52		1.87	2.35	
	1402	Girders, structural grade, 12″ x 12″		800	.020		1.94	.71		2.65	3.31	
	1502	10″ x 16″		1000	.016		1.89	.57		2.46	3.02	
	2050	Roof planks, see division 06150-600										
	2302	Roof purlins, 4″ thick, structural grade	2 Carp	1050	.015	B.F.	1.32	.54		1.86	2.35	
	2502	Roof trusses, add timber connectors, division 06090-800	″	450	.036	″	1.29	1.26		2.55	3.51	
	9000	Minimum labor/equipment charge	1 Carp	2	4	Job		142		142	236	
530	0010	**FRAMING, JOISTS** R061110-30										530
	2002	Joists, 2″ x 4″	2 Carp	1250	.013	L.F.	.37	.46		.83	1.16	
	2007	Pneumatic nailed		1438	.011		.37	.40		.77	1.07	
	2100	2″ x 6″		1250	.013		.59	.46		1.05	1.40	
	2105	Pneumatic nailed		1438	.011		.59	.40		.99	1.31	
	2152	2″ x 8″		1100	.015		.85	.52		1.37	1.80	
	2157	Pneumatic nailed		1265	.013		.85	.45		1.30	1.69	
	2202	2″ x 10″		900	.018		1.29	.63		1.92	2.47	
	2207	Pneumatic nailed		1035	.015		1.29	.55		1.84	2.33	
	2252	2″ x 12″		875	.018		1.60	.65		2.25	2.84	
	2257	Pneumatic nailed		1006	.016		1.60	.57		2.17	2.70	
	2302	2″ x 14″		770	.021		2.18	.74		2.92	3.62	
	2307	Pneumatic nailed		886	.018		2.18	.64		2.82	3.46	
	2352	3″ x 6″		925	.017		1.92	.62		2.54	3.13	
	2402	3″ x 10″		780	.021		3.26	.73		3.99	4.80	
	2452	3″ x 12″		600	.027		3.92	.95		4.87	5.90	
	2502	4″ x 6″		800	.020		1.87	.71		2.58	3.24	
	2552	4″ x 10″		600	.027		3.26	.95		4.21	5.15	
	2602	4″ x 12″		450	.036		4.04	1.26		5.30	6.55	
	2607	Sister joist, 2″ x 6″		800	.020		.59	.71		1.30	1.83	
	2608	Pneumatic nailed		960	.017		.59	.59		1.18	1.63	
	2612	2″ x 8″		640	.025		.85	.89		1.74	2.41	
	2613	Pneumatic nailed		768	.021		.85	.74		1.59	2.17	
	2617	2″ x 10″		535	.030		1.29	1.06		2.35	3.18	
	2618	Pneumatic nailed		642	.025		1.29	.89		2.18	2.89	
	2622	2″ x 12″		455	.035		1.60	1.25		2.85	3.83	
	2627	Pneumatic nailed		546	.029		1.60	1.04		2.64	3.49	
	3000	Composite wood joist 9-1/2″ deep		.90	17.778	M.L.F.	1,825	630		2,455	3,050	

06110 | Wood Framing

		06110 Wood Framing	CREW	DAILY OUTPUT	LABOR-HOURS	UNIT	2006 BARE COSTS MAT.	LABOR	EQUIP.	TOTAL	TOTAL INCL O&P	
530	3010	11-1/2" deep (R061110-30)	2 Carp	.88	18.182	M.L.F.	1,950	645		2,595	3,225	530
	3020	14" deep		.82	19.512		2,250	695		2,945	3,625	
	3030	16" deep		.78	20.513		2,550	730		3,280	4,000	
	4000	Open web joist 12" deep		.88	18.182		1,825	645		2,470	3,075	
	4010	14" deep		.82	19.512		2,125	695		2,820	3,475	
	4020	16" deep		.78	20.513		2,200	730		2,930	3,625	
	4030	18" deep	↓	.74	21.622	↓	2,250	770		3,020	3,750	
	9000	Minimum labor/equipment charge	1 Carp	4	2	Job		71		71	118	
535	0010	**FRAMING, LIGHT**										535
	0020	Average cost for all light framing	2 Carp	1.05	15.238	M.B.F.	675	540		1,215	1,650	
	6500	Sills, 4" x 4", treated		.78	20.513	"	1,050	730		1,780	2,350	
	6802	Sleepers on concrete, treated, 1" x 2"	↓	2350	.007	L.F.	.10	.24		.34	.51	
545	0010	**FRAMING, MISCELLANEOUS**										545
	2002	Firestops, 2" x 4"	2 Carp	780	.021	L.F.	.37	.73		1.10	1.62	
	2007	Pneumatic nailed		952	.017		.37	.60		.97	1.40	
	2102	2" x 6"		600	.027		.59	.95		1.54	2.22	
	2107	Pneumatic nailed		732	.022		.59	.78		1.37	1.94	
	5002	Nailers, treated, wood construction, 2" x 4"		800	.020		.49	.71		1.20	1.71	
	5007	Pneumatic nailed		960	.017		.49	.59		1.08	1.51	
	5102	2" x 6"		750	.021		.75	.76		1.51	2.08	
	5107	Pneumatic nailed		900	.018		.75	.63		1.38	1.87	
	5122	2" x 8"		700	.023		.98	.81		1.79	2.42	
	5127	Pneumatic nailed		840	.019		.98	.68		1.66	2.19	
	5202	Steel construction, 2" x 4"		750	.021		.49	.76		1.25	1.79	
	5222	2" x 6"		700	.023		.75	.81		1.56	2.17	
	5242	2" x 8"		650	.025		.98	.88		1.86	2.52	
	7002	Rough bucks, treated, for doors or windows, 2" x 6"		400	.040		.75	1.42		2.17	3.18	
	7007	Pneumatic nailed		480	.033		.75	1.18		1.93	2.78	
	7102	2" x 8"		380	.042		.98	1.50		2.48	3.55	
	7107	Pneumatic nailed		456	.035		.98	1.25		2.23	3.14	
	8000	Stair stringers, 2" x 10"		130	.123		1.29	4.38		5.67	8.65	
	8100	2" x 12"		130	.123		1.60	4.38		5.98	9	
	8150	3" x 10"		125	.128		3.26	4.55		7.81	11.15	
	8200	3" x 12"	↓	125	.128	↓	3.92	4.55		8.47	11.85	
	9000	Minimum labor/equipment charge	1 Carp	4	2	Job		71		71	118	
550	0010	**PARTITIONS** Wood stud with single bottom plate and										550
	0020	double top plate, no waste, std. & better lumber										
	0182	2" x 4" studs, 8' high, studs 12" O.C.	2 Carp	80	.200	L.F.	4.50	7.10		11.60	16.75	
	0187	12" O.C., pneumatic nailed		96	.167		4.50	5.95		10.45	14.75	
	0202	16" O.C.		100	.160		3.68	5.70		9.38	13.45	
	0207	16" O.C., pneumatic nailed		120	.133		3.68	4.74		8.42	11.90	
	0302	24" O.C.		125	.128		2.86	4.55		7.41	10.70	
	0307	24" O.C., pneumatic nailed		150	.107		2.86	3.79		6.65	9.45	
	0382	10' high, studs 12" O.C.		80	.200		5.30	7.10		12.40	17.65	
	0387	12" O.C., pneumatic nailed		96	.167		5.30	5.95		11.25	15.65	
	0402	16" O.C.		100	.160		4.30	5.70		10	14.15	
	0407	16" O.C., pneumatic nailed		120	.133		4.30	4.74		9.04	12.60	
	0502	24" O.C.		125	.128		3.27	4.55		7.82	11.15	
	0507	24" O.C., pneumatic nailed		150	.107		3.27	3.79		7.06	9.90	
	0582	12' high, studs 12" O.C.		65	.246		6.15	8.75		14.90	21.50	
	0587	12" O.C., pneumatic nailed		78	.205		6.15	7.30		13.45	18.85	
	0602	16" O.C.		80	.200		4.91	7.10		12.01	17.20	
	0607	16" O.C., pneumatic nailed		96	.167		4.91	5.95		10.86	15.20	
	0701	24" O.C.		100	.160		3.68	5.70		9.38	13.45	
	0706	24" O.C., pneumatic nailed	↓	120	.133	↓	3.68	4.74		8.42	11.90	

	06110 | Wood Framing		CREW	DAILY OUTPUT	LABOR-HOURS	UNIT	2006 BARE COSTS MAT.	LABOR	EQUIP.	TOTAL	TOTAL INCL O&P	
550	0782	2" x 6" studs, 8' high, studs 12" O.C.	2 Carp	70	.229	L.F.	7.15	8.15		15.30	21.50	550
	0787	12" O.C., pneumatic nailed		84	.190		7.15	6.75		13.90	19.10	
	0802	16" O.C.		90	.178		5.85	6.30		12.15	16.90	
	0807	16" O.C., pneumatic nailed		108	.148		5.85	5.25		11.10	15.20	
	0902	24" O.C.		115	.139		4.56	4.95		9.51	13.20	
	0907	24" O.C., pneumatic nailed		138	.116		4.56	4.12		8.68	11.85	
	0982	10' high, studs 12" O.C.		70	.229		8.45	8.15		16.60	23	
	0987	12" O.C., pneumatic nailed		84	.190		8.45	6.75		15.20	20.50	
	1002	16" O.C.		90	.178		6.85	6.30		13.15	17.95	
	1007	16" O.C., pneumatic nailed		108	.148		6.85	5.25		12.10	16.25	
	1102	24" O.C.		115	.139		5.20	4.95		10.15	13.95	
	1107	24" O.C., pneumatic nailed		138	.116		5.20	4.12		9.32	12.60	
	1182	12' high, studs 12" O.C.		55	.291		9.75	10.35		20.10	28	
	1187	12" O.C., pneumatic nailed		66	.242		9.75	8.60		18.35	25	
	1202	16" O.C.		70	.229		7.80	8.15		15.95	22	
	1207	16" O.C., pneumatic nailed		84	.190		7.80	6.75		14.55	19.80	
	1302	24" O.C.		90	.178		5.85	6.30		12.15	16.90	
	1307	24" O.C., pneumatic nailed		108	.148		5.85	5.25		11.10	15.20	
	1402	For horizontal blocking, 2" x 4", add		600	.027		.41	.95		1.36	2.02	
	1502	2" x 6", add		600	.027		.65	.95		1.60	2.29	
	1600	For openings, add		250	.064			2.28		2.28	3.77	
	1702	Headers for above openings, material only, add				B.F.	.70			.70	.77	
	9000	Minimum labor/equipment charge	1 Carp	4	2	Job		71		71	118	
552	0010	**FRAMING, PORCH OR DECK**										552
	0100	Treated lumber, posts or columns, 4" x 4"	2 Carp	390	.041	L.F.	1.41	1.46		2.87	3.97	
	0110	4" x 6"		275	.058		2.49	2.07		4.56	6.15	
	0120	4" x 8"		220	.073		3.32	2.59		5.91	7.95	
	0130	Girder, single, 4" x 4"		675	.024		1.41	.84		2.25	2.95	
	0140	4" x 6"		600	.027		2.49	.95		3.44	4.31	
	0150	4" x 8"		525	.030		3.32	1.08		4.40	5.45	
	0160	Double, 2" x 4"		625	.026		.99	.91		1.90	2.60	
	0170	2" x 6"		600	.027		1.53	.95		2.48	3.25	
	0180	2" x 8"		575	.028		1.99	.99		2.98	3.83	
	0190	2" x 10"		550	.029		3.03	1.03		4.06	5.05	
	0200	2" x 12"		525	.030		3.96	1.08		5.04	6.15	
	0210	Triple, 2" x 4"		575	.028		1.49	.99		2.48	3.28	
	0220	2" x 6"		550	.029		2.29	1.03		3.32	4.23	
	0230	2" x 8"		525	.030		2.99	1.08		4.07	5.10	
	0240	2" x 10"		500	.032		4.55	1.14		5.69	6.90	
	0250	2" x 12"		475	.034		5.95	1.20		7.15	8.55	
	0260	Ledger, bolted 4' O.C., 2" x 4"		400	.040		.59	1.42		2.01	3.01	
	0270	2" x 6"		395	.041		.85	1.44		2.29	3.33	
	0280	2" x 8"		390	.041		1.08	1.46		2.54	3.61	
	0300	2" x 12"		380	.042		2.06	1.50		3.56	4.74	
	0310	Joists, 2" x 4"		1250	.013		.50	.46		.96	1.30	
	0320	2" x 6"		1250	.013		.77	.46		1.23	1.60	
	0330	2" x 8"		1100	.015		1	.52		1.52	1.96	
	0340	2" x 10"		900	.018		1.53	.63		2.16	2.73	
	0350	2" x 12"		875	.018		1.83	.65		2.48	3.09	
	0440	Balusters, square, 2" x 2"		220	.073		.31	2.59		2.90	4.62	
	0450	Turned, 2" x 2"		140	.114		.41	4.06		4.47	7.20	
	0460	Stair stringer, 2" x 10"		130	.123		1.53	4.38		5.91	8.95	
	0470	2" x 12"		130	.123		1.83	4.38		6.21	9.25	
	0480	Stair treads, 1" x 4"		140	.114		.64	4.06		4.70	7.45	
	0490	2" x 4"		140	.114		.50	4.06		4.56	7.30	

		06110 | Wood Framing	CREW	DAILY OUTPUT	LABOR-HOURS	UNIT	2006 BARE COSTS MAT.	LABOR	EQUIP.	TOTAL	TOTAL INCL O&P	
552	0500	2″ x 6″	2 Carp	160	.100	L.F.	.75	3.56		4.31	6.75	552
	0510	5/4″ x 6″		160	.100		.60	3.56		4.16	6.55	
	0520	Turned handrail post, 4″ x 4″		64	.250	Ea.	13.05	8.90		21.95	29	
	0530	Lattice panel, 4′ x 8′		1600	.010	S.F.	.70	.36		1.06	1.36	
	0540	Cedar, posts or columns, 4″ x 4″		390	.041	L.F.	3.18	1.46		4.64	5.90	
	0550	4″ x 6″		275	.058		4.83	2.07		6.90	8.75	
	0560	4″ x 8″		220	.073		7.25	2.59		9.84	12.25	
	0800	Decking, 1″ x 4″		550	.029		1.36	1.03		2.39	3.21	
	0810	2″ x 4″		600	.027		2.63	.95		3.58	4.46	
	0820	2″ x 6″		640	.025		4.24	.89		5.13	6.15	
	0830	5/4″ x 6″		640	.025		3.79	.89		4.68	5.65	
	0840	Railings and trim, 1″ x 4″		600	.027		1.36	.95		2.31	3.07	
	0860	2″ x 4″		600	.027		2.63	.95		3.58	4.46	
	0870	2″ x 6″		600	.027		4.24	.95		5.19	6.25	
	0920	Stair treads, 1″ x 4″		140	.114		1.36	4.06		5.42	8.25	
	0930	2″ x 4″		140	.114		2.63	4.06		6.69	9.65	
	0940	2″ x 6″		160	.100		4.24	3.56		7.80	10.55	
	0950	5/4″ x 6″		160	.100		3.79	3.56		7.35	10.05	
	0980	Redwood, posts or columns, 4″ x 4″		390	.041		6.35	1.46		7.81	9.40	
	0990	4″ x 6″		275	.058		10.65	2.07		12.72	15.20	
	1000	4″ x 8″		220	.073		18.55	2.59		21.14	25	
	1280	Railings and trim, 1″ x 4″		600	.027		1.48	.95		2.43	3.20	
	1310	2″ x 6″		600	.027		5.75	.95		6.70	7.90	
	1420	Alternative decking, wood / plastic composite, 5/4″ x 6″		640	.025		1.93	.89		2.82	3.59	
	1430	Vinyl, 1-1/2″ x 5-1/2″		640	.025		3.48	.89		4.37	5.30	
	1440	1″ x 4″ square edge fir		550	.029		1.30	1.03		2.33	3.14	
	1450	1″ x 4″ tongue and groove fir		450	.036		1.30	1.26		2.56	3.52	
	1460	1″ x 4″ mahogany		550	.029		1.30	1.03		2.33	3.14	
	1470	Accessories, joist hangers, 2″ x 4″	1 Carp	160	.050	Ea.	.70	1.78		2.48	3.72	
	1480	2″ x 6″ through 2″ x 12″	″	150	.053		.80	1.90		2.70	4.02	
	1530	Post footing, incl excav, backfill, tube form & concrete, 4′ deep, 8″ dia	F-7	12	2.667		10.40	84		94.40	150	
	1540	10″ diameter		11	2.909		14.70	91.50		106.20	168	
	1550	12″ diameter		10	3.200		19.70	101		120.70	189	
555	0010	**FRAMING, ROOFS**										555
	2000	Fascia boards, 2″ x 8″	2 Carp	225	.071	L.F.	.85	2.53		3.38	5.15	
	2100	2″ x 10″		180	.089		1.29	3.16		4.45	6.65	
	5000	Rafters, to 4 in 12 pitch, 2″ x 6″, ordinary		1000	.016		.59	.57		1.16	1.59	
	5060	2″ x 8″, ordinary		950	.017		.85	.60		1.45	1.93	
	5300	Hip and valley rafters, 2″ x 6″, ordinary		760	.021		.59	.75		1.34	1.89	
	5360	2″ x 8″, ordinary		720	.022		.85	.79		1.64	2.25	
	5540	Hip and valley jacks, 2″ x 6″, ordinary		600	.027		.59	.95		1.54	2.22	
	5600	2″ x 8″, ordinary		490	.033		.85	1.16		2.01	2.86	
	5761	For slopes steeper than 4 in 12, add						30%				
	5770	For dormers or complex roofs, add						50%				
	5780	Rafter tie, 1″ x 4″, #3	2 Carp	800	.020	L.F.	.35	.71		1.06	1.57	
	5800	Ridge board, #2 or better, 1″ x 6″		600	.027		.70	.95		1.65	2.34	
	5820	1″ x 8″		550	.029		.89	1.03		1.92	2.68	
	5840	1″ x 10″		500	.032		1.30	1.14		2.44	3.31	
	5860	2″ x 6″		500	.032		.59	1.14		1.73	2.53	
	5880	2″ x 8″		450	.036		.85	1.26		2.11	3.03	
	5900	2″ x 10″		400	.040		1.29	1.42		2.71	3.78	
	5920	Roof cants, split, 4″ x 4″		650	.025		1.25	.88		2.13	2.83	
	5940	6″ x 6″		600	.027		4.54	.95		5.49	6.55	
	5960	Roof curbs, untreated, 2″ x 6″		520	.031		.59	1.09		1.68	2.46	
	5980	2″ x 12″		400	.040		1.60	1.42		3.02	4.12	

		06110 | Wood Framing	Crew	Daily Output	Labor-Hours	Unit	2006 Bare Costs Mat.	Labor	Equip.	Total	Total Incl O&P	
555	6000	Sister rafters, 2″ x 6″	2 Carp	800	.020	L.F.	.59	.71		1.30	1.83	555
	6020	2″ x 8″		640	.025		.85	.89		1.74	2.41	
	6040	2″ x 10″		535	.030		1.29	1.06		2.35	3.18	
	6060	2″ x 12″		455	.035		1.60	1.25		2.85	3.83	
	9000	Minimum labor/equipment charge	1 Carp	4	2	Job		71		71	118	
560	0010	**FRAMING, SILLS**										560
	2002	Ledgers, nailed, 2″ x 4″	2 Carp	755	.021	L.F.	.37	.75		1.12	1.66	
	2052	2″ x 6″		600	.027		.59	.95		1.54	2.22	
	2102	Bolted, not including bolts, 3″ x 6″		325	.049		1.92	1.75		3.67	5	
	2152	3″ x 12″		233	.069		3.92	2.44		6.36	8.35	
	2602	Mud sills, redwood, construction grade, 2″ x 4″		895	.018		2.20	.64		2.84	3.47	
	2622	2″ x 6″		780	.021		3.30	.73		4.03	4.84	
	4002	Sills, 2″ x 4″		600	.027		.37	.95		1.32	1.98	
	4052	2″ x 6″		550	.029		.59	1.03		1.62	2.36	
	4082	2″ x 8″		500	.032		.85	1.14		1.99	2.82	
	4202	Treated, 2″ x 4″		550	.029		.49	1.03		1.52	2.24	
	4222	2″ x 6″		500	.032		.75	1.14		1.89	2.70	
	4242	2″ x 8″		450	.036		.98	1.26		2.24	3.16	
	4402	4″ x 4″		450	.036		1.39	1.26		2.65	3.62	
	4422	4″ x 6″		350	.046		2.46	1.63		4.09	5.40	
	4462	4″ x 8″		300	.053		3.27	1.90		5.17	6.75	
	4481	4″ x 10″		260	.062		4.10	2.19		6.29	8.15	
	9000	Minimum labor/equipment charge	1 Carp	4	2	Job		71		71	118	
565	0010	**FRAMING, SLEEPERS**										565
	0100	On concrete, treated, 1″ x 2″	2 Carp	2350	.007	L.F.	.10	.24		.34	.51	
	0150	1″ x 3″		2000	.008		.19	.28		.47	.68	
	0200	2″ x 4″		1500	.011		.49	.38		.87	1.16	
	0250	2″ x 6″		1300	.012		.75	.44		1.19	1.55	
	9000	Minimum labor/equipment charge	1 Carp	4	2	Job		71		71	118	
570	0010	**FRAMING, SOFFITS & CANOPIES**										570
	1002	Canopy or soffit framing , 1″ x 4″	2 Carp	900	.018	L.F.	.46	.63		1.09	1.56	
	1042	1″ x 8″		750	.021		.89	.76		1.65	2.23	
	1102	2″ x 4″		620	.026		.37	.92		1.29	1.93	
	1142	2″ x 8″		500	.032		.85	1.14		1.99	2.82	
	1202	3″ x 4″		500	.032		1.09	1.14		2.23	3.08	
	1242	3″ x 10″		300	.053		3.26	1.90		5.16	6.75	
	9000	Minimum labor/equipment charge	1 Carp	4	2	Job		71		71	118	
575	0010	**FRAMING, TREATED LUMBER**										575
	0100	2″ x 4″				M.B.F.	730			730	800	
	0110	2″ x 6″					750			750	825	
	0120	2″ x 8″					730			730	805	
	0130	2″ x 10″					895			895	985	
	0140	2″ x 12″					975			975	1,075	
	0200	4″ x 4″					1,050			1,050	1,150	
	0210	4″ x 6″					1,225			1,225	1,350	
	0220	4″ x 8″					1,225			1,225	1,350	
590	0010	**FRAMING, WALLS** R061110-30										590
	0100	Door buck, studs, header, access, 8′ H, 2″x4″ wall, 3′ W	1 Carp	32	.250	Ea.	15.75	8.90		24.65	32	
	0110	4′ wide		32	.250		16.95	8.90		25.85	33.50	
	0120	5′ wide		32	.250		21	8.90		29.90	38	
	0130	6′ wide		32	.250		22.50	8.90		31.40	40	
	0140	8′ wide		30	.267		33	9.50		42.50	52	
	0150	10′ wide		30	.267		44.50	9.50		54	64.50	
	0160	12′ wide		30	.267		65.50	9.50		75	87.50	

06110 | Wood Framing

			CREW	DAILY OUTPUT	LABOR-HOURS	UNIT	2006 BARE COSTS MAT.	LABOR	EQUIP.	TOTAL	TOTAL INCL O&P	
590	0170	2″ x 6″ wall, 3′ wide R061110-30	1 Carp	32	.250	Ea.	23	8.90		31.90	40	590
	0180	4′ wide		32	.250		24	8.90		32.90	41.50	
	0190	5′ wide		32	.250		28	8.90		36.90	45.50	
	0200	6′ wide		32	.250		29.50	8.90		38.40	47.50	
	0210	8′ wide		30	.267		40	9.50		49.50	60	
	0220	10′ wide		30	.267		51.50	9.50		61	72.50	
	0230	12′ wide		30	.267		72.50	9.50		82	95	
	0240	Window buck, studs, header & access, 8′ high 2″ x 4″ wall, 2′ wide		24	.333		16.55	11.85		28.40	38	
	0250	3′ wide		24	.333		19.45	11.85		31.30	41	
	0260	4′ wide		24	.333		21.50	11.85		33.35	43	
	0270	5′ wide		24	.333		25.50	11.85		37.35	47.50	
	0280	6′ wide		24	.333		28	11.85		39.85	50.50	
	0300	8′ wide		22	.364		41	12.95		53.95	66.50	
	0310	10′ wide		22	.364		53.50	12.95		66.45	80.50	
	0320	12′ wide		22	.364		76.50	12.95		89.45	106	
	0330	2″ x 6″ wall, 2′ wide		24	.333		25.50	11.85		37.35	47.50	
	0340	3′ wide		24	.333		28.50	11.85		40.35	51	
	0350	4′ wide		24	.333		31	11.85		42.85	53.50	
	0360	5′ wide		24	.333		35	11.85		46.85	58	
	0370	6′ wide		24	.333		38.50	11.85		50.35	62	
	0380	7′ wide		24	.333		48.50	11.85		60.35	72.50	
	0390	8′ wide		22	.364		52.50	12.95		65.45	79.50	
	0400	10′ wide		22	.364		66	12.95		78.95	94	
	0410	12′ wide		22	.364		90	12.95		102.95	121	
	2002	Headers over openings, 2″ x 6″	2 Carp	360	.044	L.F.	.59	1.58		2.17	3.27	
	2007	2″ x 6″, pneumatic nailed		432	.037		.59	1.32		1.91	2.83	
	2052	2″ x 8″		340	.047		.85	1.67		2.52	3.71	
	2057	2″ x 8″, pneumatic nailed		408	.039		.85	1.39		2.24	3.25	
	2100	2″ x 10″		320	.050		1.29	1.78		3.07	4.37	
	2105	2″ x 10″, pneumatic nailed		384	.042		1.29	1.48		2.77	3.87	
	2152	2″ x 12″		300	.053		1.60	1.90		3.50	4.90	
	2157	2″ x 12″, pneumatic nailed		360	.044		1.60	1.58		3.18	4.38	
	2202	4″ x 12″		190	.084		4.04	2.99		7.03	9.40	
	2207	4″ x 12″, pneumatic nailed		228	.070		4.04	2.49		6.53	8.55	
	2252	6″ x 12″		140	.114		9.60	4.06		13.66	17.30	
	5002	Plates, untreated, 2″ x 3″		850	.019		.38	.67		1.05	1.53	
	5007	2″ x 3″, pneumatic nailed		1020	.016		.38	.56		.94	1.34	
	5022	2″ x 4″		800	.020		.37	.71		1.08	1.59	
	5027	2″ x 4″, pneumatic nailed		960	.017		.37	.59		.96	1.39	
	5041	2″ x 6″		750	.021		.59	.76		1.35	1.91	
	5046	2″ x 6″, pneumatic nailed		900	.018		.59	.63		1.22	1.70	
	5122	Studs, 8′ high wall, 2″ x 3″		1200	.013		.38	.47		.85	1.21	
	5127	2″ x 3″, pneumatic nailed		1440	.011		.38	.40		.78	1.07	
	5142	2″ x 4″		1100	.015		.37	.52		.89	1.27	
	5147	2″ x 4″, pneumatic nailed		1320	.012		.37	.43		.80	1.12	
	5162	2″ x 6″		1000	.016		.59	.57		1.16	1.59	
	5167	2″ x 6″, pneumatic nailed		1200	.013		.59	.47		1.06	1.44	
	5182	3″ x 4″		800	.020		1.09	.71		1.80	2.38	
	5187	3″ x 4″, pneumatic nailed		960	.017		1.09	.59		1.68	2.18	
	8200	For 12′ high walls, deduct						5%				
	8220	For stub wall, 6′ high, add						20%				
	8240	3′ high, add						40%				
	8250	For second story & above, add						5%				
	8300	For dormer & gable, add						15%				
	9000	Minimum labor/equipment charge	1 Carp	4	2	Job		71		71	118	

06100 | Rough Carpentry

	06110 | Wood Framing		CREW	DAILY OUTPUT	LABOR-HOURS	UNIT	2006 BARE COSTS MAT.	LABOR	EQUIP.	TOTAL	TOTAL INCL O&P	
600	0010	**FURRING**										600
	0012	Wood strips, 1" x 2", on walls, on wood	1 Carp	550	.015	L.F.	.19	.52		.71	1.07	
	0017	Pneumatic nailed		710	.011		.19	.40		.59	.87	
	0302	On masonry		495	.016		.19	.57		.76	1.16	
	0402	On concrete		260	.031		.19	1.09		1.28	2.02	
	0602	1" x 3", on walls, on wood		550	.015		.28	.52		.80	1.17	
	0607	Pneumatic nailed		710	.011		.28	.40		.68	.97	
	0702	On masonry		495	.016		.28	.57		.85	1.26	
	0802	On concrete		260	.031		.28	1.09		1.37	2.12	
	0852	On ceilings, on wood		350	.023		.28	.81		1.09	1.66	
	0857	Pneumatic nailed		450	.018		.28	.63		.91	1.36	
	0902	On masonry		320	.025		.28	.89		1.17	1.78	
	0952	On concrete		210	.038		.28	1.35		1.63	2.55	
	9000	Minimum labor/equipment charge		4	2	Job		71		71	118	
700	0010	**GROUNDS**										700
	0020	For casework, 1" x 2" wood strips, on wood	1 Carp	330	.024	L.F.	.19	.86		1.05	1.64	
	0102	On masonry		285	.028		.19	1		1.19	1.86	
	0202	On concrete		250	.032		.19	1.14		1.33	2.09	
	0402	For plaster, 3/4" deep, on wood		450	.018		.19	.63		.82	1.26	
	0502	On masonry		225	.036		.19	1.26		1.45	2.30	
	0602	On concrete		175	.046		.19	1.63		1.82	2.90	
	0702	On metal lath		200	.040		.19	1.42		1.61	2.57	
	9000	Minimum labor/equipment charge		4	2	Job		71		71	118	

	06150 | Wood Decking		CREW	DAILY OUTPUT	LABOR-HOURS	UNIT	MAT.	LABOR	EQUIP.	TOTAL	TOTAL INCL O&P	
600	0010	**ROOF DECKS**										600
	0020	For laminated decks, see division 06170-550										
	0200	For cementitious decks, see division 03510-200										
	0400	Cedar planks, 3" thick	2 Carp	320	.050	S.F.	6.40	1.78		8.18	10	
	0500	4" thick		250	.064		8.65	2.28		10.93	13.25	
	0702	Douglas fir, 3" thick		320	.050		2.40	1.78		4.18	5.60	
	0802	4" thick		250	.064		3.21	2.28		5.49	7.30	
	1002	Hemlock, 3" thick		320	.050		2.40	1.78		4.18	5.60	
	1102	4" thick		250	.064		3.20	2.28		5.48	7.30	
	1302	Western white spruce, 3" thick		320	.050		2.31	1.78		4.09	5.50	
	1402	4" thick		250	.064		3.08	2.28		5.36	7.15	
	9000	Minimum labor/equipment charge	1 Carp	2	4	Job		142		142	236	

	06160 | Sheathing		CREW	DAILY OUTPUT	LABOR-HOURS	UNIT	MAT.	LABOR	EQUIP.	TOTAL	TOTAL INCL O&P	
800	0010	**SHEATHING** Plywood on roof, CDX (R061636-20)										800
	0032	5/16" thick	2 Carp	1600	.010	S.F.	.62	.36		.98	1.27	
	0037	Pneumatic nailed (R061110-30)		1952	.008		.62	.29		.91	1.16	
	0052	3/8" thick		1525	.010		.46	.37		.83	1.12	
	0057	Pneumatic nailed		1860	.009		.46	.31		.77	1.01	
	0102	1/2" thick		1400	.011		.61	.41		1.02	1.35	
	0103	Pneumatic nailed		1708	.009		.61	.33		.94	1.23	
	0202	5/8" thick		1300	.012		.66	.44		1.10	1.46	
	0207	Pneumatic nailed		1586	.010		.66	.36		1.02	1.32	
	0302	3/4" thick		1200	.013		.82	.47		1.29	1.69	
	0307	Pneumatic nailed		1464	.011		.82	.39		1.21	1.54	
	0502	Plywood on walls with exterior CDX, 3/8" thick		1200	.013		.46	.47		.93	1.29	
	0507	Pneumatic nailed		1488	.011		.46	.38		.84	1.13	
	0602	1/2" thick		1125	.014		.61	.51		1.12	1.52	
	0607	Pneumatic nailed		1395	.011		.61	.41		1.02	1.36	
	0702	5/8" thick		1050	.015		.66	.54		1.20	1.63	

	06160 | Sheathing		CREW	DAILY OUTPUT	LABOR-HOURS	UNIT	2006 BARE COSTS MAT.	LABOR	EQUIP.	TOTAL	TOTAL INCL O&P
800	0707	Pneumatic nailed (R061636-20)	2 Carp	1302	.012	S.F.	.66	.44		1.10	1.45
	0802	3/4″ thick		975	.016		.82	.58		1.40	1.87
	0807	Pneumatic nailed (R061110-30)	↓	1209	.013	↓	.82	.47		1.29	1.68
	1000	For shear wall construction, add						20%			
	1200	For structural 1 exterior plywood, add				S.F.	10%				
	1402	With boards, on roof 1″ x 6″ boards, laid horizontal	2 Carp	725	.022		1.20	.78		1.98	2.62
	1502	Laid diagonal		650	.025		1.20	.88		2.08	2.77
	1702	1″ x 8″ boards, laid horizontal		875	.018		1.11	.65		1.76	2.31
	1802	Laid diagonal	↓	725	.022		1.11	.78		1.89	2.53
	2000	For steep roofs, add						40%			
	2200	For dormers, hips and valleys, add					5%	50%			
	2402	Boards on walls, 1″ x 6″ boards, laid regular	2 Carp	650	.025		1.20	.88		2.08	2.77
	2502	Laid diagonal		585	.027		1.20	.97		2.17	2.93
	2702	1″ x 8″ boards, laid regular		765	.021		1.11	.74		1.85	2.46
	2802	Laid diagonal		650	.025		1.11	.88		1.99	2.68
	2852	Gypsum, weatherproof, 1/2″ thick		1050	.015		.62	.54		1.16	1.58
	2902	Sealed, 4/10″ thick		1100	.015		.45	.52		.97	1.36
	3000	Wood fiber, regular, no vapor barrier, 1/2″ thick		1200	.013		.55	.47		1.02	1.40
	3100	5/8″ thick		1200	.013		.72	.47		1.19	1.58
	3300	No vapor barrier, in colors, 1/2″ thick		1200	.013		.78	.47		1.25	1.65
	3400	5/8″ thick		1200	.013		.96	.47		1.43	1.85
	3600	With vapor barrier one side, white, 1/2″ thick		1200	.013		.55	.47		1.02	1.40
	3700	Vapor barrier 2 sides, 1/2″ thick		1200	.013		.77	.47		1.24	1.64
	3800	Asphalt impregnated, 25/32″ thick		1200	.013		.28	.47		.75	1.10
	3850	Intermediate, 1/2″ thick	↓	1200	.013	↓	.18	.47		.65	.99
	9000	Minimum labor/equipment charge	1 Carp	2	4	Job		142		142	236
850	0010	**SUBFLOOR**									
	0011	Plywood, CDX, 1/2″ thick	2 Carp	1500	.011	SF Flr.	.61	.38		.99	1.31
	0017	Pneumatic nailed		1860	.009		.61	.31		.92	1.19
	0102	5/8″ thick		1350	.012		.66	.42		1.08	1.43
	0107	Pneumatic nailed		1674	.010		.66	.34		1	1.29
	0202	3/4″ thick		1250	.013		.82	.46		1.28	1.65
	0207	Pneumatic nailed		1550	.010		.82	.37		1.19	1.51
	0302	1-1/8″ thick, 2-4-1 including underlayment		1050	.015		1.19	.54		1.73	2.21
	0452	1″ x 8″ S4S, laid regular		1000	.016		1.11	.57		1.68	2.17
	0462	Laid diagonal		850	.019		1.11	.67		1.78	2.34
	0502	1″ x 10″ S4S, laid regular		1100	.015		1.36	.52		1.88	2.36
	0602	Laid diagonal	↓	900	.018	↓	1.36	.63		1.99	2.55
	9000	Minimum labor/equipment charge	1 Carp	4	2	Job		71		71	118
900	0010	**UNDERLAYMENT** (R061636-20)									
	0030	Plywood, underlayment grade, 3/8″ thick	2 Carp	1500	.011	SF Flr.	.78	.38		1.16	1.49
	0080	Pneumatic nailed		1860	.009		.78	.31		1.09	1.37
	0102	1/2″ thick		1450	.011		.87	.39		1.26	1.61
	0107	Pneumatic nailed		1798	.009		.87	.32		1.19	1.48
	0202	5/8″ thick		1400	.011		1.22	.41		1.63	2.01
	0207	Pneumatic nailed		1736	.009		1.22	.33		1.55	1.88
	0302	3/4″ thick		1300	.012		1.32	.44		1.76	2.18
	0306	Pneumatic nailed		1612	.010		1.32	.35		1.67	2.03
	0502	Particle board, 3/8″ thick		1500	.011		.33	.38		.71	.99
	0507	Pneumatic nailed		1860	.009		.33	.31		.64	.87
	0602	1/2″ thick		1450	.011		.37	.39		.76	1.06
	0607	Pneumatic nailed		1798	.009		.37	.32		.69	.93
	0802	5/8″ thick		1400	.011		.41	.41		.82	1.12
	0807	Pneumatic nailed		1736	.009		.41	.33		.74	.99
	0902	3/4″ thick	↓	1300	.012	↓	.56	.44		1	1.35

06160 | Sheathing

		06160 Sheathing	CREW	DAILY OUTPUT	LABOR-HOURS	UNIT	2006 BARE COSTS MAT.	LABOR	EQUIP.	TOTAL	TOTAL INCL O&P	
900	0907	Pneumatic nailed R061636-20	2 Carp	1612	.010	SF Flr.	.56	.35		.91	1.20	900
	1102	Hardboard, underlayment grade, 4' x 4', .215" thick		1500	.011		.39	.38		.77	1.06	
	9000	Minimum labor/equipment charge	1 Carp	4	2	Job		71		71	118	

06170 | Prefabricated Structural Wood

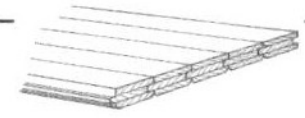

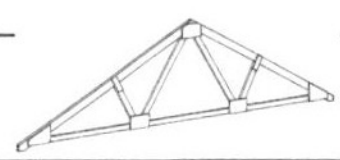

		06170 Prefabricated Structural Wood	CREW	DAILY OUTPUT	LABOR-HOURS	UNIT	MAT.	LABOR	EQUIP.	TOTAL	TOTAL INCL O&P	
550	0010	**LAMINATED ROOF DECK**										550
	0020	Pine or hemlock, 3" thick	2 Carp	425	.038	S.F.	3.02	1.34		4.36	5.55	
	0100	4" thick		325	.049		4.02	1.75		5.77	7.30	
	0300	Cedar, 3" thick		425	.038		3.69	1.34		5.03	6.30	
	0400	4" thick		325	.049		4.70	1.75		6.45	8.05	
	0600	Fir, 3" thick		425	.038		3.08	1.34		4.42	5.60	
	0700	4" thick		325	.049		3.85	1.75		5.60	7.15	
	9000	Minimum labor/equipment charge	1 Carp	3	2.667	Job		95		95	157	
600	0010	**STRUCTURAL JOISTS** Fabricated "I" joists with wood flanges,										600
	0100	Plywood webs, incl. bridging & blocking, panels 24" O.C.										
	1200	15' to 24' span, 50 psf live load	F-5	2400	.013	SF Flr.	2.07	.48		2.55	3.08	
	1300	55 psf live load		2250	.014		2.23	.51		2.74	3.30	
	1400	24' to 30' span, 45 psf live load		2600	.012		2.57	.44		3.01	3.56	
	1500	55 psf live load		2400	.013		2.91	.48		3.39	4	
	1600	Tubular steel open webs, 45 psf, 24" O.C., 40' span	F-3	6250	.006		2.07	.23	.10	2.40	2.77	
	1700	55' span		7750	.005		2.01	.19	.08	2.28	2.61	
	1800	70' span		9250	.004		2.61	.16	.07	2.84	3.21	
	1900	85 psf live load, 26' span		2300	.017		2.43	.63	.28	3.34	4	
980	0010	**ROOF TRUSSES**										980
	0020	For timber connectors, see div. 06090-800										
	0100	Fink (W) or King post type, 2'-0" O.C.										
	0200	Metal plate connected, 4 in 12 slope										
	0210	24' to 29' span	F-3	3000	.013	SF Flr.	1.84	.48	.21	2.53	3.04	
	0300	30' to 43' span		3000	.013		2.04	.48	.21	2.73	3.26	
	0400	44' to 60' span		3000	.013		2.25	.48	.21	2.94	3.50	
	0700	Glued and nailed, add					50%					
	0800	Flat wood truss 16' to 29' span	F-3	3000	.013		2.20	.48	.21	2.89	3.44	

06180 | Glued-Laminated Construction

		06180 Glued-Laminated Construction	CREW	DAILY OUTPUT	LABOR-HOURS	UNIT	MAT.	LABOR	EQUIP.	TOTAL	TOTAL INCL O&P	
400	0010	**LAMINATED FRAMING** Not including decking										400
	0200	Straight roof beams, 20' clear span, beams 8' O.C.	F-3	2560	.016	SF Flr.	1.78	.56	.25	2.59	3.15	
	0300	Beams 16' O.C.		3200	.013		1.28	.45	.20	1.93	2.37	
	0500	40' clear span, beams 8' O.C.		3200	.013		3.40	.45	.20	4.05	4.70	
	0600	Beams 16' O.C.		3840	.010		2.77	.38	.17	3.32	3.85	
	0800	60' clear span, beams 8' O.C.	F-4	2880	.017		5.85	.59	.33	6.77	7.70	
	0900	Beams 16' O.C.	"	3840	.013		4.35	.44	.25	5.04	5.80	
	1100	Tudor arches, 30' to 40' clear span, frames 8' O.C.	F-3	1680	.024		7.60	.86	.38	8.84	10.20	
	1200	Frames 16' O.C.	"	2240	.018		5.95	.64	.28	6.87	7.90	
	1400	50' to 60' clear span, frames 8' O.C.	F-4	2200	.022		8.20	.77	.43	9.40	10.80	
	1500	Frames 16' O.C.		2640	.018		7	.64	.36	8	9.15	
	1700	Radial arches, 60' clear span, frames 8' O.C.		1920	.025		7.70	.89	.50	9.09	10.45	
	1800	Frames 16' O.C.		2880	.017		5.90	.59	.33	6.82	7.80	
	2000	100' clear span, frames 8' O.C.		1600	.030		7.95	1.06	.59	9.60	11.15	
	2100	Frames 16' O.C.		2400	.020		7	.71	.40	8.11	9.30	
	2300	120' clear span, frames 8' O.C.		1440	.033		10.55	1.18	.66	12.39	14.30	
	2400	Frames 16' O.C.		1920	.025		9.65	.89	.50	11.04	12.60	
	2600	Bowstring trusses, 20' O.C., 40' clear span	F-3	2400	.017		4.76	.60	.27	5.63	6.55	
	2700	60' clear span	F-4	3600	.013		4.27	.47	.26	5	5.75	
	2800	100' clear span		4000	.012		6.05	.43	.24	6.72	7.60	

06180 | Glued-Laminated Construction

		06180 Glued-Laminated Construction	CREW	DAILY OUTPUT	LABOR-HOURS	UNIT	2006 BARE COSTS MAT.	LABOR	EQUIP.	TOTAL	TOTAL INCL O&P	
400	2900	120′ clear span	F-4	3600	.013	SF Flr.	6.50	.47	.26	7.23	8.20	400
	3100	For premium appearance, add to S.F. prices					5%					
	3300	For industrial type, deduct					15%					
	3500	For stain and varnish, add					5%					
	3900	For 3/4″ laminations, add to straight					25%					
	4100	Add to curved					15%					
	4300	Alternate pricing method: (use nominal footage of										
	4310	components). Straight beams, camber less than 6″	F-3	3.50	11.429	M.B.F.	2,625	410	182	3,217	3,775	
	4400	Columns, including hardware		2	20		2,825	720	320	3,865	4,650	
	4600	Curved members, radius over 32′		2.50	16		2,900	575	255	3,730	4,425	
	4700	Radius 10′ to 32′		3	13.333		2,875	480	212	3,567	4,175	
	4900	For complicated shapes, add maximum					100%					
	5100	For pressure treating, add to straight					35%					
	5200	Add to curved					45%					
	6000	Laminated veneer members, southern pine or western species										
	6050	1-3/4″ wide x 5-1/2″ deep	2 Carp	480	.033	L.F.	3.02	1.18		4.20	5.30	
	6100	9-1/2″ deep		480	.033		3.76	1.18		4.94	6.10	
	6150	14″ deep		450	.036		5.60	1.26		6.86	8.25	
	6200	18″ deep		450	.036		7.60	1.26		8.86	10.45	
	6300	Parallel strand members, southern pine or western species										
	6350	1-3/4″ wide x 9-1/4″ deep	2 Carp	480	.033	L.F.	3.56	1.18		4.74	5.90	
	6400	11-1/4″ deep		450	.036		4.38	1.26		5.64	6.90	
	6450	14″ deep		400	.040		5.20	1.42		6.62	8.10	
	6500	3-1/2″ wide x 9-1/4″ deep		480	.033		8.65	1.18		9.83	11.45	
	6550	11-1/4″ deep		450	.036		10.70	1.26		11.96	13.90	
	6600	14″ deep		400	.040		12.70	1.42		14.12	16.35	
	6650	7″ wide x 9-1/4″ deep		450	.036		18.05	1.26		19.31	22	
	6700	11-1/4″ deep		420	.038		22.50	1.35		23.85	27	
	6750	14″ deep		400	.040		27	1.42		28.42	32	
	9000	Minimum labor/equipment charge	F-3	2.50	16	Job		575	255	830	1,225	

06200 | Finish Carpentry

06220 | Millwork

		06220 Millwork	CREW	DAILY OUTPUT	LABOR-HOURS	UNIT	2006 BARE COSTS MAT.	LABOR	EQUIP.	TOTAL	TOTAL INCL O&P	
200	0010	**MOLDINGS, BASE**										200
	0500	Base, stock pine, 9/16″ x 3-1/2″	1 Carp	240	.033	L.F.	2.21	1.18		3.39	4.39	
	0550	9/16″ x 4-1/2″		200	.040	″	2.40	1.42		3.82	5	
	9000	Minimum labor/equipment charge		4	2	Job		71		71	118	
400	0010	**MOLDINGS, CASINGS**										400
	0090	Apron, stock pine, 5/8″ x 2″	1 Carp	250	.032	L.F.	1.04	1.14		2.18	3.02	
	0110	5/8″ x 3-1/2″		220	.036		1.25	1.29		2.54	3.52	
	0300	Band, stock pine, 11/16″ x 1-1/8″		270	.030		.57	1.05		1.62	2.38	
	0350	11/16″ x 1-3/4″		250	.032		1.12	1.14		2.26	3.11	
	0700	Casing, stock pine, 11/16″ x 2-1/2″		240	.033		1.28	1.18		2.46	3.37	
	0750	11/16″ x 3-1/2″		215	.037		1.83	1.32		3.15	4.20	
	0760	Door & window casing, exterior, 1-1/4″ x 2″		200	.040		2.13	1.42		3.55	4.70	
	0770	Finger jointed, 1-1/4″ x 2″		200	.040		1.04	1.42		2.46	3.50	
	9000	Minimum labor/equipment charge		4	2	Job		71		71	118	
450	0010	**MOLDINGS, CEILINGS**										450
	0600	Bed, stock pine, 9/16″ x 1-3/4″	1 Carp	270	.030	L.F.	.88	1.05		1.93	2.72	

06200 | Finish Carpentry

	06220 | Millwork		CREW	DAILY OUTPUT	LABOR-HOURS	UNIT	2006 BARE COSTS MAT.	LABOR	EQUIP.	TOTAL	TOTAL INCL O&P	
450	0650	9/16" x 2"	1 Carp	240	.033	L.F.	1.19	1.18		2.37	3.27	450
	1200	Cornice molding, stock pine, 9/16" x 1-3/4"		330	.024		.82	.86		1.68	2.33	
	1300	9/16" x 2-1/4"		300	.027		1.02	.95		1.97	2.69	
	2400	Cove scotia, stock pine, 9/16" x 1-3/4"		270	.030		.65	1.05		1.70	2.47	
	2500	11/16" x 2-3/4"		255	.031		1.25	1.12		2.37	3.23	
	2600	Crown, stock pine, 9/16" x 3-5/8"		250	.032		1.86	1.14		3	3.93	
	2700	11/16" x 4-5/8"		220	.036	↓	3.11	1.29		4.40	5.55	
	9000	Minimum labor/equipment charge	↓	4	2	Job		71		71	118	
500	0010	**MOLDINGS, EXTERIOR**										500
	1500	Cornice, boards, pine, 1" x 2"	1 Carp	330	.024	L.F.	.36	.86		1.22	1.83	
	1600	1" x 4"		250	.032		.71	1.14		1.85	2.66	
	1700	1" x 6"		250	.032		1.24	1.14		2.38	3.24	
	1800	1" x 8"		200	.040		1.44	1.42		2.86	3.94	
	1900	1" x 10"		180	.044		1.68	1.58		3.26	4.47	
	2000	1" x 12"		180	.044		1.89	1.58		3.47	4.70	
	2200	Three piece, built-up, pine, minimum		80	.100		2.31	3.56		5.87	8.45	
	2300	Maximum		65	.123		5	4.38		9.38	12.75	
	3000	Corner board, sterling pine, 1" x 4"		200	.040		.69	1.42		2.11	3.12	
	3100	1" x 6"		200	.040		1.50	1.42		2.92	4.01	
	3200	2" x 6"		165	.048		2.12	1.72		3.84	5.20	
	3300	2" x 8"		165	.048		2.87	1.72		4.59	6	
	3350	Fascia, sterling pine, 1" x 6"		250	.032		1.50	1.14		2.64	3.53	
	3370	1" x 8"		225	.036		1.90	1.26		3.16	4.18	
	3395	Grounds, 1" x 1" redwood		300	.027		.19	.95		1.14	1.78	
	3400	Trim, exterior, sterling pine, back band		250	.032		.74	1.14		1.88	2.69	
	3500	Casing		250	.032		1.96	1.14		3.10	4.04	
	3600	Crown		250	.032		1.89	1.14		3.03	3.96	
	3700	Porch rail with balusters		22	.364		15.40	12.95		28.35	38.50	
	3800	Screen		395	.020		1.19	.72		1.91	2.50	
	4100	Verge board, sterling pine, 1" x 4"		200	.040		.78	1.42		2.20	3.22	
	4200	1" x 6"		200	.040		1.08	1.42		2.50	3.55	
	4300	2" x 6"		165	.048		1.75	1.72		3.47	4.79	
	4400	2" x 8"	↓	165	.048	↓	2.32	1.72		4.04	5.40	
	4700	For redwood trim, add					200%					
	9000	Minimum labor/equipment charge	1 Carp	4	2	Job		71		71	118	
700	0010	**MOLDINGS, TRIM**										700
	0200	Astragal, stock pine, 11/16" x 1-3/4"	1 Carp	255	.031	L.F.	1.16	1.12		2.28	3.13	
	0250	1-5/16" x 2-3/16"		240	.033		2.71	1.18		3.89	4.94	
	0800	Chair rail, stock pine, 5/8" x 2-1/2"		270	.030		1.44	1.05		2.49	3.33	
	0900	5/8" x 3-1/2"		240	.033		2.20	1.18		3.38	4.38	
	1000	Closet pole, stock pine, 1-1/8" diameter		200	.040		.87	1.42		2.29	3.32	
	1100	Fir, 1-5/8" diameter		200	.040		1.60	1.42		3.02	4.12	
	3300	Half round, stock pine, 1/4" x 1/2"		270	.030		.23	1.05		1.28	2	
	3350	1/2" x 1"	↓	255	.031	↓	.60	1.12		1.72	2.51	
	3400	Handrail, fir, single piece, stock, hardware not included										
	3450	1-1/2" x 1-3/4"	1 Carp	80	.100	L.F.	1.77	3.56		5.33	7.85	
	3470	Pine, 1-1/2" x 1-3/4"		80	.100		1.77	3.56		5.33	7.85	
	3500	1-1/2" x 2-1/2"		76	.105		1.63	3.74		5.37	8	
	3600	Lattice, stock pine, 1/4" x 1-1/8"		270	.030		.41	1.05		1.46	2.20	
	3700	1/4" x 1-3/4"		250	.032		.45	1.14		1.59	2.38	
	3800	Miscellaneous, custom, pine, 1" x 1"		270	.030		.34	1.05		1.39	2.12	
	3900	1" x 3"		240	.033		.69	1.18		1.87	2.72	
	4100	Birch or oak, nominal 1" x 1"		240	.033		.53	1.18		1.71	2.54	
	4200	Nominal 1" x 3"		215	.037		1.81	1.32		3.13	4.18	
	4400	Walnut, nominal 1" x 1"	↓	215	.037	↓	.87	1.32		2.19	3.15	

For expanded coverage of these items see *Means Interior Cost Data 2006*

06200 | Finish Carpentry

	06220 | Millwork		CREW	DAILY OUTPUT	LABOR-HOURS	UNIT	2006 BARE COSTS MAT.	LABOR	EQUIP.	TOTAL	TOTAL INCL O&P	
700	4500	Nominal 1″ x 3″	1 Carp	200	.040	L.F.	2.61	1.42		4.03	5.25	700
	4700	Teak, nominal 1″ x 1″		215	.037		1.23	1.32		2.55	3.54	
	4800	Nominal 1″ x 3″		200	.040		3.52	1.42		4.94	6.25	
	4900	Quarter round, stock pine, 1/4″ x 1/4″		275	.029		.22	1.03		1.25	1.95	
	4950	3/4″ x 3/4″		255	.031		.64	1.12		1.76	2.55	
	5600	Wainscot moldings, 1-1/8″ x 9/16″, 2′ high, minimum		76	.105	S.F.	10.50	3.74		14.24	17.75	
	5700	Maximum		65	.123	″	19.65	4.38		24.03	29	
	9000	Minimum labor/equipment charge		4	2	Job		71		71	118	
800	0010	**MOLDINGS, WINDOW AND DOOR**										800
	2800	Door moldings, stock, decorative, 1-1/8″ wide, plain	1 Carp	17	.471	Set	43.50	16.75		60.25	75.50	
	2900	Detailed		17	.471	″	87.50	16.75		104.25	124	
	2960	Clear pine door jamb, no stops, 11/16″ x 4-9/16″		240	.033	L.F.	2.30	1.18		3.48	4.49	
	3150	Door trim set, 1 head and 2 sides, pine, 2-1/2 wide		5.90	1.356	Opng.	22	48		70	104	
	3170	3-1/2″ wide		5.30	1.509	″	31	53.50		84.50	123	
	3250	Glass beads, stock pine, 3/8″ x 1/2″		275	.029	L.F.	.31	1.03		1.34	2.05	
	3270	3/8″ x 7/8″		270	.030		.41	1.05		1.46	2.20	
	4850	Parting bead, stock pine, 3/8″ x 3/4″		275	.029		.35	1.03		1.38	2.10	
	4870	1/2″ x 3/4″		255	.031		.44	1.12		1.56	2.33	
	5000	Stool caps, stock pine, 11/16″ x 3-1/2″		200	.040		2.20	1.42		3.62	4.78	
	5100	1-1/16″ x 3-1/4″		150	.053		3.48	1.90		5.38	6.95	
	5300	Threshold, oak, 3′ long, inside, 5/8″ x 3-5/8″		32	.250	Ea.	7.75	8.90		16.65	23.50	
	5400	Outside, 1-1/2″ x 7-5/8″		16	.500	″	35	17.80		52.80	68	
	5900	Window trim sets, including casings, header, stops,										
	5910	stool and apron, 2-1/2″ wide, minimum	1 Carp	13	.615	Opng.	30	22		52	69	
	5950	Average		10	.800		35	28.50		63.50	85.50	
	6000	Maximum		6	1.333		62.50	47.50		110	147	
	9000	Minimum labor/equipment charge		4	2	Job		71		71	118	
900	0010	**SOFFITS**										900
	0200	Soffits, pine, 1″ x 4″	2 Carp	420	.038	L.F.	.35	1.35		1.70	2.63	
	0210	1″ x 6″		420	.038		.55	1.35		1.90	2.85	
	0220	1″ x 8″		420	.038		.70	1.35		2.05	3.01	
	0230	1″ x 10″		400	.040		1.08	1.42		2.50	3.55	
	0240	1″ x 12″		400	.040		1.10	1.42		2.52	3.56	
	0250	STK cedar, 1″ x 4″		420	.038		.56	1.35		1.91	2.86	
	0260	1″ x 6″		420	.038		.83	1.35		2.18	3.15	
	0270	1″ x 8″		420	.038		1.09	1.35		2.44	3.44	
	0280	1″ x 10″		400	.040		2.13	1.42		3.55	4.70	
	0290	1″ x 12″		400	.040		2.89	1.42		4.31	5.55	
	1000	Exterior AC plywood, 1/4″ thick		420	.038	S.F.	.75	1.35		2.10	3.07	
	1050	3/8″ thick		420	.038		.78	1.35		2.13	3.10	
	1100	1/2″ thick		420	.038		.87	1.35		2.22	3.20	
	1150	Polyvinyl chloride, white, solid	1 Carp	230	.035		.82	1.24		2.06	2.95	
	1160	Perforated	″	230	.035		.82	1.24		2.06	2.95	
	9000	Minimum labor/equipment charge	2 Carp	5	3.200	Job		114		114	188	

	06250 | Prefinished Paneling		CREW	DAILY OUTPUT	LABOR-HOURS	UNIT	MAT.	LABOR	EQUIP.	TOTAL	TOTAL INCL O&P	
200	0010	**PANELING, HARDBOARD**										200
	0050	Not incl. furring or trim, hardboard, tempered, 1/8″ thick	2 Carp	500	.032	S.F.	.32	1.14		1.46	2.23	
	0100	1/4″ thick		500	.032		.48	1.14		1.62	2.41	
	0300	Tempered pegboard, 1/8″ thick		500	.032		.39	1.14		1.53	2.31	
	0400	1/4″ thick		500	.032		.51	1.14		1.65	2.44	
	0600	Untempered hardboard, natural finish, 1/8″ thick		500	.032		.35	1.14		1.49	2.27	
	0700	1/4″ thick		500	.032		.34	1.14		1.48	2.25	
	0900	Untempered pegboard, 1/8″ thick		500	.032		.35	1.14		1.49	2.27	

06250 | Prefinished Paneling

		06250 Prefinished Paneling	CREW	DAILY OUTPUT	LABOR-HOURS	UNIT	2006 BARE COSTS MAT.	LABOR	EQUIP.	TOTAL	TOTAL INCL O&P
200	1000	1/4″ thick	2 Carp	500	.032	S.F.	.39	1.14		1.53	2.31
	1200	Plastic faced hardboard, 1/8″ thick		500	.032		.58	1.14		1.72	2.52
	1300	1/4″ thick		500	.032		.77	1.14		1.91	2.73
	1500	Plastic faced pegboard, 1/8″ thick		500	.032		.55	1.14		1.69	2.49
	1600	1/4″ thick		500	.032		.68	1.14		1.82	2.63
	1800	Wood grained, plain or grooved, 1/4″ thick, minimum		500	.032		.51	1.14		1.65	2.44
	1900	Maximum		425	.038		1.08	1.34		2.42	3.41
	2100	Moldings for hardboard, wood or aluminum, minimum		500	.032	L.F.	.35	1.14		1.49	2.27
	2200	Maximum		425	.038	″	.98	1.34		2.32	3.30
	9000	Minimum labor/equipment charge	1 Carp	2	4	Job		142		142	236
500	0010	**PANELING, PLYWOOD**									
	2400	Plywood, prefinished, 1/4″ thick, 4′ x 8′ sheets									
	2410	with vertical grooves. Birch faced, minimum	2 Carp	500	.032	S.F.	.84	1.14		1.98	2.80
	2420	Average		420	.038		1.28	1.35		2.63	3.65
	2430	Maximum		350	.046		1.87	1.63		3.50	4.75
	2600	Mahogany, African		400	.040		2.39	1.42		3.81	4.99
	2700	Philippine (Lauan)		500	.032		1.03	1.14		2.17	3.01
	2900	Oak or Cherry, minimum		500	.032		2	1.14		3.14	4.08
	3000	Maximum		400	.040		3.07	1.42		4.49	5.75
	3200	Rosewood		320	.050		4.36	1.78		6.14	7.75
	3400	Teak		400	.040		3.07	1.42		4.49	5.75
	3600	Chestnut		375	.043		4.54	1.52		6.06	7.50
	3800	Pecan		400	.040		1.96	1.42		3.38	4.52
	3900	Walnut, minimum		500	.032		2.62	1.14		3.76	4.76
	3950	Maximum		400	.040		4.96	1.42		6.38	7.80
	4000	Plywood, prefinished, 3/4″ thick, stock grades, minimum		320	.050		1.19	1.78		2.97	4.26
	4100	Maximum		224	.071		5.10	2.54		7.64	9.85
	4300	Architectural grade, minimum		224	.071		3.78	2.54		6.32	8.35
	4400	Maximum		160	.100		5.80	3.56		9.36	12.25
	4600	Plywood, "A" face, birch, V.C., 1/2″ thick, natural		450	.036		1.79	1.26		3.05	4.06
	4700	Select		450	.036		1.96	1.26		3.22	4.25
	4900	Veneer core, 3/4″ thick, natural		320	.050		1.89	1.78		3.67	5.05
	5000	Select		320	.050		2.13	1.78		3.91	5.30
	5200	Lumber core, 3/4″ thick, natural		320	.050		2.84	1.78		4.62	6.05
	5500	Plywood, knotty pine, 1/4″ thick, A2 grade		450	.036		1.55	1.26		2.81	3.80
	5600	A3 grade		450	.036		1.96	1.26		3.22	4.25
	5800	3/4″ thick, veneer core, A2 grade		320	.050		2.01	1.78		3.79	5.15
	5900	A3 grade		320	.050		2.26	1.78		4.04	5.45
	6100	Aromatic cedar, 1/4″ thick, plywood		400	.040		1.98	1.42		3.40	4.54
	6200	1/4″ thick, particle board		400	.040		.96	1.42		2.38	3.42
	9000	Minimum labor/equipment charge	1 Carp	2	4	Job		142		142	236

06260 | Board Paneling

		06260 Board Paneling	CREW	DAILY OUTPUT	LABOR-HOURS	UNIT	MAT.	LABOR	EQUIP.	TOTAL	TOTAL INCL O&P
400	0010	**PANELING, BOARDS**									
	6400	Wood board paneling, 3/4″ thick, knotty pine	2 Carp	300	.053	S.F.	1.40	1.90		3.30	4.68
	6500	Rough sawn cedar		300	.053		1.79	1.90		3.69	5.10
	6700	Redwood, clear, 1″ x 4″ boards		300	.053		4.18	1.90		6.08	7.75
	6900	Aromatic cedar, closet lining, boards		275	.058		3.22	2.07		5.29	6.95
	9000	Minimum labor/equipment charge	1 Carp	2	4	Job		142		142	236

06270 | Closet/Utility Wood Shelving

		06270 Closet/Utility Wood Shelving	CREW	DAILY OUTPUT	LABOR-HOURS	UNIT	MAT.	LABOR	EQUIP.	TOTAL	TOTAL INCL O&P
200	0010	**SHELVING**									
	0020	Pine, clear grade, no edge band, 1″ x 8″	1 Carp	115	.070	L.F.	1.81	2.47		4.28	6.10
	0100	1″ x 10″		110	.073		2.31	2.59		4.90	6.80
	0200	1″ x 12″		105	.076		3.10	2.71		5.81	7.90

06200 | Finish Carpentry

	06270 Closet/Utility Wood Shelving		CREW	DAILY OUTPUT	LABOR-HOURS	UNIT	2006 BARE COSTS MAT.	LABOR	EQUIP.	TOTAL	TOTAL INCL O&P	
200	0600	Plywood, 3/4" thick with lumber edge, 12" wide	1 Carp	75	.107	L.F.	1.66	3.79		5.45	8.15	200
	0700	24" wide		70	.114		2.98	4.06		7.04	10.05	
	0900	Bookcase, clear grade pine, shelves 12" O.C., 8" deep, /SF shelf		70	.114	S.F.	5.90	4.06		9.96	13.20	
	1000	12" deep shelves		65	.123	"	10.10	4.38		14.48	18.35	
	1200	Adjustable closet rod and shelf, 12" wide, 3' long		20	.400	Ea.	8.70	14.20		22.90	33	
	1300	8' long		15	.533	"	22	18.95		40.95	55.50	
	1500	Prefinished shelves with supports, stock, 8" wide		75	.107	L.F.	3.83	3.79		7.62	10.50	
	1600	10" wide		70	.114	"	4.26	4.06		8.32	11.45	
	9000	Minimum labor/equipment charge		4	2	Job		71		71	118	

06400 | Architectural Woodwork

	06410 Custom Cabinets		CREW	DAILY OUTPUT	LABOR-HOURS	UNIT	2006 BARE COSTS MAT.	LABOR	EQUIP.	TOTAL	TOTAL INCL O&P	
100	0010	**CABINETS** Corner china cabinets, stock pine,										100
	0020	80" high, unfinished, minimum	2 Carp	6.60	2.424	Ea.	455	86		541	645	
	0100	Maximum	"	4.40	3.636	"	1,000	129		1,129	1,325	
	0700	Kitchen base cabinets, hardwood, not incl. counter tops,										
	0710	24" deep, 35" high, prefinished										
	0800	One top drawer, one door below, 12" wide	2 Carp	24.80	.645	Ea.	138	23		161	190	
	0820	15" wide		24	.667		188	23.50		211.50	247	
	0840	18" wide		23.30	.687		205	24.50		229.50	266	
	0860	21" wide		22.70	.705		213	25		238	276	
	0880	24" wide		22.30	.717		245	25.50		270.50	315	
	1000	Four drawers, 12" wide		24.80	.645		325	23		348	400	
	1020	15" wide		24	.667		251	23.50		274.50	315	
	1040	18" wide		23.30	.687		280	24.50		304.50	350	
	1060	24" wide		22.30	.717		305	25.50		330.50	380	
	1200	Two top drawers, two doors below, 27" wide		22	.727		272	26		298	340	
	1220	30" wide		21.40	.748		291	26.50		317.50	365	
	1240	33" wide		20.90	.766		300	27		327	375	
	1260	36" wide		20.30	.788		315	28		343	390	
	1280	42" wide		19.80	.808		335	28.50		363.50	420	
	1300	48" wide		18.90	.847		360	30		390	445	
	1500	Range or sink base, two doors below, 30" wide		21.40	.748		240	26.50		266.50	310	
	1520	33" wide		20.90	.766		258	27		285	330	
	1540	36" wide		20.30	.788		270	28		298	345	
	1560	42" wide		19.80	.808		288	28.50		316.50	365	
	1580	48" wide		18.90	.847		300	30		330	380	
	1800	For sink front units, deduct					52			52	57.50	
	2000	Corner base cabinets, 36" wide, standard	2 Carp	18	.889		395	31.50		426.50	490	
	2100	Lazy Susan with revolving door	"	16.50	.970		385	34.50		419.50	475	
	4000	Kitchen wall cabinets, hardwood, 12" deep with two doors										
	4050	12" high, 30" wide	2 Carp	24.80	.645	Ea.	151	23		174	204	
	4100	36" wide		24	.667		176	23.50		199.50	233	
	4400	15" high, 30" wide		24	.667		158	23.50		181.50	214	
	4420	33" wide		23.30	.687		177	24.50		201.50	236	
	4440	36" wide		22.70	.705		179	25		204	239	
	4450	42" wide		22.70	.705		208	25		233	270	
	4700	24" high, 30" wide		23.30	.687		197	24.50		221.50	257	

06400 | Architectural Woodwork

06410 | Custom Cabinets

		06410 Custom Cabinets	CREW	DAILY OUTPUT	LABOR-HOURS	UNIT	2006 BARE COSTS MAT.	LABOR	EQUIP.	TOTAL	TOTAL INCL O&P
100	4720	36″ wide	2 Carp	22.70	.705	Ea.	218	25		243	281
	4740	42″ wide		22.30	.717		240	25.50		265.50	305
	5000	30″ high, one door, 12″ wide		22	.727		133	26		159	189
	5020	15″ wide		21.40	.748		150	26.50		176.50	209
	5040	18″ wide		20.90	.766		164	27		191	226
	5060	24″ wide		20.30	.788		185	28		213	251
	5300	Two doors, 27″ wide		19.80	.808		229	28.50		257.50	300
	5320	30″ wide		19.30	.829		222	29.50		251.50	293
	5340	36″ wide		18.80	.851		253	30.50		283.50	330
	5360	42″ wide		18.50	.865		276	31		307	355
	5380	48″ wide		18.40	.870		310	31		341	390
	6000	Corner wall, 30″ high, 24″ wide		18	.889		152	31.50		183.50	220
	6050	30″ wide		17.20	.930		180	33		213	253
	6100	36″ wide		16.50	.970		195	34.50		229.50	272
	6500	Revolving Lazy Susan		15.20	1.053		298	37.50		335.50	385
	7000	Broom cabinet, 84″ high, 24″ deep, 18″ wide		10	1.600		405	57		462	540
	7500	Oven cabinets, 84″ high, 24″ deep, 27″ wide		8	2		590	71		661	770
	7750	Valance board trim		396	.040	L.F.	8.40	1.44		9.84	11.65
	9000	For deluxe models of all cabinets, add					40%				
	9500	For custom built in place, add					25%	10%			
	9550	Rule of thumb, kitchen cabinets not including									
	9560	appliances & counter top, minimum	2 Carp	30	.533	L.F.	95.50	18.95		114.45	137
	9600	Maximum	″	25	.640	″	250	23		273	315
	9610	For metal cabinets, see division 12310-750									
	9700	Minimum labor/equipment charge	1 Carp	3	2.667	Job		95		95	157
210	0010	**CASEWORK, FRAMES**									
	0050	Base cabinets, counter storage, 36″ high, one bay									
	0100	18″ wide	1 Carp	2.70	2.963	Ea.	107	105		212	293
	0200	24″ wide		2.50	3.200		126	114		240	325
	0300	36″ wide		2.30	3.478		149	124		273	370
	0400	Two bay, 36″ wide		2.20	3.636		163	129		292	395
	1000	48″ wide		2	4		176	142		318	430
	1050	72″ wide		1.80	4.444		208	158		366	490
	1100	Three bay, 54″ wide		1.50	5.333		194	190		384	530
	1200	72″ wide		1.30	6.154		239	219		458	625
	1800	108″ wide		1.10	7.273		277	259		536	735
	2000	Four bay, 72″ wide		1.10	7.273		271	259		530	730
	2100	96″ wide		1	8		300	284		584	800
	2500	144″ wide		.90	8.889		365	315		680	925
	2800	Book cases, one bay, 7′ high, 18″ wide		2.40	3.333		126	119		245	335
	3000	24″ wide		2.30	3.478		145	124		269	365
	3100	36″ wide		2.20	3.636		176	129		305	410
	3500	Two bay, 36″ wide		1.60	5		183	178		361	495
	3800	48″ wide		1.50	5.333		227	190		417	565
	4000	72″ wide		1.40	5.714		278	203		481	640
	4100	Three bay, 54″ wide		1.20	6.667		300	237		537	725
	4200	72″ wide		1.10	7.273		335	259		594	795
	4500	108″ wide		1	8		430	284		714	940
	4600	Four bay, 72″ wide		.95	8.421		340	299		639	870
	4700	96″ wide		.90	8.889		415	315		730	980
	5000	144″ wide		.85	9.412		565	335		900	1,175
	5100	Coat racks, one bay, 7′ high, 24″ wide		4.50	1.778		126	63		189	243
	5200	36″ wide		4.30	1.860		138	66		204	261
	5300	Two bay, 48″ wide		2.75	2.909		175	103		278	365
	5600	72″ wide		2.50	3.200		201	114		315	410

	06410 | Custom Cabinets		CREW	DAILY OUTPUT	LABOR-HOURS	UNIT	2006 BARE COSTS MAT.	LABOR	EQUIP.	TOTAL	TOTAL INCL O&P	
210	5800	Three bay, 72" wide	1 Carp	2.10	3.810	Ea.	258	135		393	505	210
	6000	108" wide		1.90	4.211		278	150		428	555	
	6100	Wall mounted cabinet, one bay, 24" high, 18" wide		3.60	2.222		69.50	79		148.50	207	
	6400	24" wide		3.50	2.286		75.50	81.50		157	218	
	6600	36" wide		3.40	2.353		95	83.50		178.50	243	
	6800	Two bay, 36" wide		2.20	3.636		101	129		230	325	
	7000	48" wide		2.10	3.810		113	135		248	350	
	7200	72" wide		2	4		145	142		287	395	
	7400	Three bay, 54" wide		1.70	4.706		126	167		293	415	
	7600	72" wide		1.60	5		151	178		329	460	
	7800	108" wide		1.50	5.333		189	190		379	525	
	8000	Four bay, 72" wide		1.40	5.714		151	203		354	500	
	8100	96" wide		1.30	6.154		176	219		395	555	
	8200	144" wide		1.20	6.667		237	237		474	655	
	8400	30" high, one bay, 18" wide		3.60	2.222		75.50	79		154.50	214	
	8600	24" wide		3.40	2.353		87	83.50		170.50	235	
	8800	36" wide		3.20	2.500		100	89		189	257	
	9000	Two bay, 36" wide		2.15	3.721		100	132		232	330	
	9100	48" wide		2	4		118	142		260	365	
	9200	72" wide		1.85	4.324		143	154		297	415	
	9400	Three bay, 54" wide		1.60	5		125	178		303	430	
	9600	72" wide		1.50	5.333		150	190		340	480	
	9650	108" wide		1.40	5.714		330	203		533	700	
	9700	Four bay, 72" wide		1.30	6.154		156	219		375	530	
	9750	96" wide		1.22	6.557		187	233		420	590	
	9780	144" wide		1.15	6.957		252	247		499	685	
	9800	Wardrobe, 7' high, single, 24" wide		2.70	2.963		139	105		244	325	
	9850	36" wide		2.50	3.200		139	114		253	340	
	9880	Partition & adjustable shelves, 48" wide		1.70	4.706		176	167		343	470	
	9890	72" wide		1.55	5.161		283	183		466	615	
	9950	Partition, adjustable shelves & drawers, 48" wide		1.40	5.714		265	203		468	625	
	9960	72" wide		1.25	6.400	↓	345	228		573	755	
	9970	Minimum labor/equipment charge	↓	4	2	Job		71		71	118	
220	0010	CABINET DOORS										220
	2000	Glass panel, hardwood frame										
	2200	12" wide, 18" high	1 Carp	34	.235	Ea.	22	8.35		30.35	38.50	
	2400	24" high		33	.242		24	8.60		32.60	40.50	
	2600	30" high		32	.250		25.50	8.90		34.40	43	
	2800	36" high		30	.267		29.50	9.50		39	48	
	3000	48" high		23	.348		37.50	12.35		49.85	62	
	3200	60" high		17	.471		46	16.75		62.75	78	
	3400	72" high		15	.533		55	18.95		73.95	92	
	3600	15" wide x 18" high		33	.242		23	8.60		31.60	39.50	
	3800	24" high		32	.250		24.50	8.90		33.40	42	
	4000	30" high		30	.267		26	9.50		35.50	44.50	
	4250	36" high		28	.286		30.50	10.15		40.65	50.50	
	4300	48" high		22	.364		38.50	12.95		51.45	64	
	4350	60" high		16	.500		54.50	17.80		72.30	89.50	
	4400	72" high		14	.571		63.50	20.50		84	104	
	4450	18" wide, 18" high		32	.250		24	8.90		32.90	41	
	4500	24" high		30	.267		24.50	9.50		34	42.50	
	4550	30" high		29	.276		26.50	9.80		36.30	45.50	
	4600	36" high		27	.296		31.50	10.55		42.05	52	
	4650	48" high		21	.381		39.50	13.55		53.05	65.50	
	4700	60" high	↓	15	.533	↓	55.50	18.95		74.45	92.50	

06410 | Custom Cabinets

		06410 Custom Cabinets	CREW	DAILY OUTPUT	LABOR-HOURS	UNIT	2006 BARE COSTS MAT.	LABOR	EQUIP.	TOTAL	TOTAL INCL O&P
220	4750	72" high	1 Carp	13	.615	Ea.	64.50	22		86.50	108
	5000	Hardwood, raised panel									
	5100	12" wide, 18" high	1 Carp	16	.500	Ea.	33.50	17.80		51.30	66.50
	5150	24" high		15.50	.516		36.50	18.35		54.85	70.50
	5200	30" high		15	.533		50	18.95		68.95	86.50
	5250	36" high		14	.571		43	20.50		63.50	81
	5300	48" high		11	.727		52.50	26		78.50	101
	5320	60" high		8	1		70.50	35.50		106	137
	5340	72" high		7	1.143		81	40.50		121.50	157
	5360	15" wide x 18" high		15.50	.516		35	18.35		53.35	68.50
	5380	24" high		15	.533		38.50	18.95		57.45	74
	5400	30" high		14.50	.552		42.50	19.60		62.10	79
	5420	36" high		13.50	.593		48.50	21		69.50	88.50
	5440	48" high		10.50	.762		59	27		86	110
	5460	60" high		7.50	1.067		86	38		124	158
	5480	72" high		6.50	1.231		97	44		141	180
	5500	18" wide, 18" high		15	.533		37	18.95		55.95	72
	5550	24" high		14.50	.552		41	19.60		60.60	77.50
	5600	30" high		14	.571		45	20.50		65.50	83
	5650	36" high		13	.615		51.50	22		73.50	93
	5700	48" high		10	.800		62	28.50		90.50	115
	5750	60" high		7	1.143		89.50	40.50		130	166
	5800	72" high	↓	6	1.333	↓	101	47.50		148.50	190
	6000	Plastic laminate on particle board									
	6100	12" wide, 18" high	1 Carp	25	.320	Ea.	10.35	11.40		21.75	30.50
	6120	24" high		24	.333		13.80	11.85		25.65	35
	6140	30" high		23	.348		17.25	12.35		29.60	39.50
	6160	36" high		21	.381		20.50	13.55		34.05	45.50
	6200	48" high		16	.500		27.50	17.80		45.30	60
	6250	60" high		13	.615		34.50	22		56.50	74.50
	6300	72" high		12	.667		41.50	23.50		65	85
	6320	15" wide x 18" high		24.50	.327		13.75	11.60		25.35	34.50
	6340	24" high		23.50	.340		18.35	12.10		30.45	40
	6360	30" high		22.50	.356		23	12.65		35.65	46.50
	6380	36" high		20.50	.390		27.50	13.85		41.35	53.50
	6400	48" high		15.50	.516		36.50	18.35		54.85	71
	6450	60" high		12.50	.640		46	23		69	88
	6480	72" high		11.50	.696		55	24.50		79.50	102
	6500	18" wide, 18" high		24	.333		15.50	11.85		27.35	36.50
	6550	24" high		23	.348		20.50	12.35		32.85	43.50
	6600	30" high		22	.364		26	12.95		38.95	50
	6650	36" high		20	.400		31	14.20		45.20	57.50
	6700	48" high		15	.533		41.50	18.95		60.45	77
	6750	60" high		12	.667		52	23.50		75.50	96.50
	6800	72" high		11	.727	↓	62	26		88	112
	9000	Minimum labor/equipment charge	↓	4	2	Job		71		71	118
230	0010	**CABINET HARDWARE**									
	1000	Catches, minimum	1 Carp	235	.034	Ea.	.84	1.21		2.05	2.93
	1020	Average		119.40	.067		2.75	2.38		5.13	7
	1040	Maximum	↓	80	.100	↓	5.20	3.56		8.76	11.65
	2000	Door/drawer pulls, handles									
	2200	Handles and pulls, projecting, metal, minimum	1 Carp	160	.050	Ea.	3.90	1.78		5.68	7.25
	2220	Average		95.24	.084		5.20	2.99		8.19	10.70
	2240	Maximum		68	.118		7.15	4.18		11.33	14.85
	2300	Wood, minimum		160	.050		3.90	1.78		5.68	7.25
	2320	Average	↓	95.24	.084	↓	5.20	2.99		8.19	10.70

	06410 | Custom Cabinets		CREW	DAILY OUTPUT	LABOR-HOURS	UNIT	MAT.	LABOR	EQUIP.	TOTAL	TOTAL INCL O&P	
230	2340	Maximum	1 Carp	68	.118	Ea.	7.15	4.18		11.33	14.85	230
	2600	Flush, metal, minimum		160	.050		3.90	1.78		5.68	7.25	
	2620	Average		95.24	.084		5.20	2.99		8.19	10.70	
	2640	Maximum		68	.118		7.15	4.18		11.33	14.85	
	3000	Drawer tracks/glides, minimum		48	.167	Pr.	6.60	5.95		12.55	17.10	
	3020	Average		32	.250		11.20	8.90		20.10	27	
	3040	Maximum		24	.333		19.25	11.85		31.10	40.50	
	4000	Cabinet hinges, minimum		160	.050		2.25	1.78		4.03	5.45	
	4020	Average		95.24	.084		3.25	2.99		6.24	8.55	
	4040	Maximum		68	.118		7.85	4.18		12.03	15.60	
	5000	Cabinet locks, minimum		47.90	.167	Ea.	4.21	5.95		10.16	14.50	
	5020	Average		23.95	.334		6.60	11.85		18.45	27	
	5040	Maximum		16	.500		16	17.80		33.80	47	
240	0010	**DRAWERS**										240
	0100	Solid hardwood front										
	1000	4″ high, 12″ wide	1 Carp	17	.471	Ea.	3.50	16.75		20.25	31.50	
	1200	18″ wide		16	.500		4.95	17.80		22.75	35	
	1400	24″ wide		15	.533		6.60	18.95		25.55	39	
	1600	6″ high, 12″ wide		16	.500		5	17.80		22.80	35	
	1800	18″ wide		15	.533		7.50	18.95		26.45	40	
	2000	24″ wide		14	.571		10	20.50		30.50	44.50	
	2200	9″ high, 12″ wide		15	.533		7.50	18.95		26.45	40	
	2400	18″ wide		14	.571		11.25	20.50		31.75	46	
	2600	24″ wide		13	.615		15	22		37	53	
	2800	Plastic laminate on particle board front										
	3000	4″ high, 12″ wide	1 Carp	17	.471	Ea.	4.39	16.75		21.14	32.50	
	3200	18″ wide		16	.500		6.60	17.80		24.40	37	
	3600	24″ wide		15	.533		8.75	18.95		27.70	41	
	3800	6″ high, 12″ wide		16	.500		6.65	17.80		24.45	37	
	4000	18″ wide		15	.533		10	18.95		28.95	42.50	
	4500	24″ wide		14	.571		13.30	20.50		33.80	48	
	4800	9″ high, 12″ wide		15	.533		10	18.95		28.95	42.50	
	5000	18″ wide		14	.571		15	20.50		35.50	50	
	5200	24″ wide		13	.615		19.95	22		41.95	58.50	
	5400	Plywood, flush panel front										
	6000	4″ high, 12″ wide	1 Carp	17	.471	Ea.	5.95	16.75		22.70	34	
	6200	18″ wide		16	.500		8.90	17.80		26.70	39.50	
	6400	24″ wide		15	.533		11.90	18.95		30.85	44.50	
	6600	6″ high, 12″ wide		16	.500		9	17.80		26.80	39.50	
	7000	18″ wide		15	.533		13.50	18.95		32.45	46.50	
	7500	24″ wide		14	.571		18	20.50		38.50	53.50	
	8000	9″ high, 12″ wide		15	.533		13.50	18.95		32.45	46.50	
	8500	18″ wide		14	.571		20.50	20.50		41	56	
	8600	24″ wide		13	.615		27	22		49	66	
	9000	Minimum labor/equipment charge		4	2	Job		71		71	118	
400	0010	**VANITIES**										400
	8000	Vanity bases, 2 doors, 30″ high, 21″ deep, 24″ wide	2 Carp	20	.800	Ea.	184	28.50		212.50	249	
	8050	30″ wide		16	1		211	35.50		246.50	291	
	8100	36″ wide		13.33	1.200		282	42.50		324.50	380	
	8150	48″ wide		11.43	1.400		335	50		385	455	
	9000	For deluxe models of all vanities, add to above					40%					
	9500	For custom built in place, add to above					25%	10%				

06400 | Architectural Woodwork

		06415 | Countertops	CREW	DAILY OUTPUT	LABOR-HOURS	UNIT	2006 BARE COSTS MAT.	LABOR	EQUIP.	TOTAL	TOTAL INCL O&P	
100	0010	**COUNTER TOP**										100
	0020	Stock plastic laminate, 24" wide w/ backsplash, minimum	1 Carp	30	.267	L.F.	8.85	9.50		18.35	25.50	
	0100	Maximum	"	25	.320	"	16	11.40		27.40	36.50	
	1700	For end splash, add				Ea.	15.15			15.15	16.65	
	9000	Minimum labor/equipment charge	1 Carp	3.75	2.133	Job		76		76	126	
300	0010	**TOPS** Counter, plastic laminate										300
	1000	Counter top, 24" wide, 1-1/2" thick edging										
	1500	Plastic laminate edging, minimum	1 Carp	25	.320	L.F.	3.03	11.40		14.43	22	
	1520	Average		24	.333		5.55	11.85		17.40	26	
	1540	Maximum		22	.364		8.35	12.95		21.30	30.50	
	2500	Hardwood edging, minimum		20	.400		3.74	14.20		17.94	27.50	
	2520	Average		18	.444		6	15.80		21.80	32.50	
	2540	Maximum		16	.500		8.60	17.80		26.40	39	
	2600	Backsplash, add to above, minimum		36	.222		1.21	7.90		9.11	14.45	
	2620	Average		35	.229		2.02	8.15		10.17	15.65	
	2640	Maximum		34	.235		2.83	8.35		11.18	16.95	
	2700	For metal cove, add					.71			.71	.78	
	2900	Postformed backsplash, add to above										
	2920	Minimum	1 Carp	21	.381	L.F.	1.09	13.55		14.64	23.50	
	2940	Average		20	.400		2.18	14.20		16.38	26	
	2960	Maximum		19	.421		3.27	14.95		18.22	28.50	
	3500	Well openings (for typewriters etc.)		2.50	3.200	Ea.		114		114	188	
	3900	Cutouts for sinks, lavatories		12	.667	"		23.50		23.50	39.50	

		06430 | Stairs & Railings	CREW	DAILY OUTPUT	LABOR-HOURS	UNIT	MAT.	LABOR	EQUIP.	TOTAL	TOTAL INCL O&P	
500	0010	**RAILINGS**										500
	0020	Custom design, architectural grade, hardwood, minimum	1 Carp	38	.211	L.F.	5.60	7.50		13.10	18.55	
	0100	Maximum		30	.267		46.50	9.50		56	67	
	0300	Stock interior railing with spindles 6" O.C., 4' long		40	.200		42	7.10		49.10	58	
	0400	8' long		48	.167		21	5.95		26.95	33	
	9000	Minimum labor/equipment charge		3	2.667	Job		95		95	157	
620	0010	**STAIRS, PREFABRICATED**										620
	0100	Box stairs, prefabricated, 3'-0" wide										
	0110	Oak treads, up to 14 risers	2 Carp	39	.410	Riser	71.50	14.60		86.10	103	
	0600	With pine treads for carpet, up to 14 risers	"	39	.410	"	46	14.60		60.60	74.50	
	1100	For 4' wide stairs, add				Flight	25%					
	1550	Stairs, prefabricated stair handrail with balusters	1 Carp	30	.267	L.F.	64	9.50		73.50	85.50	
	1700	Basement stairs, prefabricated, pine treads										
	1710	Pine risers, 3' wide, up to 14 risers	2 Carp	52	.308	Riser	46	10.95		56.95	68.50	
	4000	Residential, wood, oak treads, prefabricated		1.50	10.667	Flight	930	380		1,310	1,650	
	4200	Built in place		.44	36.364	"	1,925	1,300		3,225	4,275	
	4400	Spiral, oak, 4'-6" diameter, unfinished, prefabricated,										
	4500	incl. railing, 9' high	2 Carp	1.50	10.667	Flight	4,400	380		4,780	5,475	
	9000	Minimum labor/equipment charge	"	3	5.333	Job		190		190	315	
630	0010	**STAIR PARTS**										630
	0020	Balusters, turned, 30" high, pine, minimum	1 Carp	28	.286	Ea.	3.70	10.15		13.85	21	
	0100	Maximum		26	.308		19	10.95		29.95	39	
	0300	30" high birch balusters, minimum		28	.286		6.30	10.15		16.45	24	
	0400	Maximum		26	.308		27.50	10.95		38.45	48	
	0600	42" high, pine balusters, minimum		27	.296		4.90	10.55		15.45	23	
	0700	Maximum		25	.320		27.50	11.40		38.90	49	
	0900	42" high birch balusters, minimum		27	.296		10.70	10.55		21.25	29	
	1000	Maximum		25	.320		38.50	11.40		49.90	61.50	
	1050	Baluster, stock pine, 1-1/4" x 1-1/4"		240	.033	L.F.	3.12	1.18		4.30	5.40	

06400 | Architectural Woodwork

06430 | Stairs & Railings

		06430 Stairs & Railings	Crew	Daily Output	Labor-Hours	Unit	2006 Bare Costs Mat.	Labor	Equip.	Total	Total Incl O&P	
630	1100	1-3/4″ x 1-3/4″	1 Carp	220	.036	L.F.	8.90	1.29		10.19	11.95	630
	1200	Newels, 3-1/4″ wide, starting, minimum		7	1.143	Ea.	38.50	40.50		79	110	
	1300	Maximum		6	1.333		320	47.50		367.50	430	
	1500	Landing, minimum		5	1.600		104	57		161	208	
	1600	Maximum		4	2		350	71		421	505	
	1800	Railings, oak, built-up, minimum		60	.133	L.F.	31.50	4.74		36.24	43	
	1900	Maximum		55	.145		43.50	5.15		48.65	56.50	
	2100	Add for sub rail		110	.073		5.30	2.59		7.89	10.15	
	2300	Risers, beech, 3/4″ x 7-1/2″ high		64	.125		5.80	4.44		10.24	13.70	
	2400	Fir, 3/4″ x 7-1/2″ high		64	.125		1.60	4.44		6.04	9.10	
	2600	Oak, 3/4″ x 7-1/2″ high		64	.125		6.10	4.44		10.54	14.05	
	2800	Pine, 3/4″ x 7-1/2″ high		66	.121		3.05	4.31		7.36	10.50	
	2850	Skirt board, pine, 1″ x 10″		55	.145		2.90	5.15		8.05	11.75	
	2900	1″ x 12″		52	.154		3.45	5.45		8.90	12.85	
	3000	Treads, oak, 1-1/4″ x 10″ wide, 3′ long		18	.444	Ea.	29	15.80		44.80	58	
	3100	4′ long, oak		17	.471		116	16.75		132.75	156	
	3300	1-1/4″ x 11-1/2″ wide, 3′ long, oak		18	.444		31	15.80		46.80	60	
	3400	6′ long, oak		14	.571		186	20.50		206.50	239	
	3600	Beech treads, add					40%					
	3800	For mitered return nosings, add				L.F.	3.90			3.90	4.29	
	9000	Minimum labor/equipment charge	1 Carp	3	2.667	Job		95		95	157	

06440 | Wood Ornaments

		06440 Wood Ornaments	Crew	Daily Output	Labor-Hours	Unit	Mat.	Labor	Equip.	Total	Total Incl O&P	
150	0010	**BEAMS, DECORATIVE**										150
	0020	Rough sawn cedar, non-load bearing, 4″ x 4″	2 Carp	180	.089	L.F.	1.35	3.16		4.51	6.75	
	0100	4″ x 6″		170	.094		2.60	3.35		5.95	8.40	
	0200	4″ x 8″		160	.100		3.34	3.56		6.90	9.55	
	0300	4″ x 10″		150	.107		4.64	3.79		8.43	11.40	
	0400	4″ x 12″		140	.114		5.60	4.06		9.66	12.95	
	0500	8″ x 8″		130	.123		7.85	4.38		12.23	15.90	
	1100	Beam connector plates see div. 06090-800										
	9000	Minimum labor/equipment charge	1 Carp	3	2.667	Job		95		95	157	
350	0010	**GRILLES** and panels, hardwood, sanded										350
	0020	2′ x 4′ to 4′ x 8′, custom designs, unfinished, minimum	1 Carp	38	.211	S.F.	12.10	7.50		19.60	26	
	0050	Average		30	.267		26.50	9.50		36	44.50	
	0100	Maximum		19	.421		40.50	14.95		55.45	69.50	
	0300	As above, but prefinished, minimum		38	.211		12.10	7.50		19.60	26	
	0400	Maximum		19	.421		45.50	14.95		60.45	75	
	9000	Minimum labor/equipment charge		2	4	Job		142		142	236	
400	0010	**LOUVERS**										400
	0020	Redwood, 2′-0″ diameter, full circle	1 Carp	16	.500	Ea.	136	17.80		153.80	180	
	0100	Half circle		16	.500		130	17.80		147.80	173	
	0200	Octagonal 		16	.500		104	17.80		121.80	144	
	0300	Triangular, 5/12 pitch, 5′-0″ at base		16	.500		220	17.80		237.80	272	
	9000	Minimum labor/equipment charge		3.50	2.286	Job		81.50		81.50	135	
500	0010	**FIREPLACE MANTELS**										500
	0015	6″ molding, 6′ x 3′-6″ opening, minimum	1 Carp	5	1.600	Opng.	139	57		196	247	
	0100	Maximum		5	1.600		173	57		230	284	
	0300	Prefabricated pine, colonial type, stock, deluxe		2	4		2,175	142		2,317	2,625	
	0400	Economy		3	2.667		375	95		470	565	
	9000	Minimum labor/equipment charge		3	2.667	Job		95		95	157	
550	0010	**FIREPLACE MANTEL BEAMS**										550
	0020	Rough texture wood, 4″ x 8″	1 Carp	36	.222	L.F.	5.15	7.90		13.05	18.75	

06440 | Wood Ornaments

		06440 Wood Ornaments	CREW	DAILY OUTPUT	LABOR-HOURS	UNIT	2006 BARE COSTS MAT.	LABOR	EQUIP.	TOTAL	TOTAL INCL O&P	
550	0100	4″ x 10″	1 Carp	35	.229	L.F.	6.20	8.15		14.35	20.50	550
	0300	Laminated hardwood, 2-1/4″ x 10-1/2″ wide, 6′ long		5	1.600	Ea.	98.50	57		155.50	202	
	0400	8′ long		5	1.600	″	137	57		194	244	
	0600	Brackets for above, rough sawn		12	.667	Pr.	9.05	23.50		32.55	49.50	
	0700	Laminated		12	.667	″	13.70	23.50		37.20	54.50	
	9000	Minimum labor/equipment charge		4	2	Job		71		71	118	
700	0010	**COLUMNS**										700
	0050	Aluminum, round colonial, 6″ diameter	2 Carp	80	.200	V.L.F.	18.10	7.10		25.20	31.50	
	0100	8″ diameter		62.25	.257		25	9.15		34.15	42.50	
	0200	10″ diameter		55	.291		30	10.35		40.35	50	
	0250	Fir, stock units, hollow round, 6″ diameter		80	.200		20.50	7.10		27.60	34.50	
	0300	8″ diameter		80	.200		24.50	7.10		31.60	38.50	
	0350	10″ diameter		70	.229		31	8.15		39.15	47.50	
	0400	Solid turned, to 8′ high, 3-1/2″ diameter		80	.200		8.90	7.10		16	21.50	
	0500	4-1/2″ diameter		75	.213		12.75	7.60		20.35	26.50	
	0600	5-1/2″ diameter		70	.229		17.80	8.15		25.95	33	
	0800	Square columns, built-up, 5″ x 5″		65	.246		16.55	8.75		25.30	32.50	
	0900	Solid, 3-1/2″ x 3-1/2″		130	.123		7.65	4.38		12.03	15.65	
	1600	Hemlock, tapered, T & G, 12″ diam, 10′ high		100	.160		38	5.70		43.70	51.50	
	1700	16′ high		65	.246		67.50	8.75		76.25	88.50	
	1900	10′ high, 14″ diameter		100	.160		98	5.70		103.70	117	
	2000	18′ high		65	.246		93	8.75		101.75	117	
	2200	18″ diameter, 12′ high		65	.246		131	8.75		139.75	159	
	2300	20′ high		50	.320		126	11.40		137.40	158	
	2500	20″ diameter, 14′ high		40	.400		155	14.20		169.20	195	
	2600	20′ high		35	.457		160	16.25		176.25	203	
	2800	For flat pilasters, deduct					33%					
	3000	For splitting into halves, add				Ea.	114			114	125	
	4000	Rough sawn cedar posts, 4″ x 4″	2 Carp	250	.064	V.L.F.	3.70	2.28		5.98	7.85	
	4100	4″ x 6″		235	.068		6.70	2.42		9.12	11.35	
	4200	6″ x 6″		220	.073		12.50	2.59		15.09	18.05	
	4300	8″ x 8″		200	.080		28.50	2.84		31.34	36	
	9000	Minimum labor/equipment charge	1 Carp	3	2.667	Job		95		95	157	

06445 | Simulated Wood Ornaments

		06445 Simulated Wood Ornaments	CREW	DAILY OUTPUT	LABOR-HOURS	UNIT	MAT.	LABOR	EQUIP.	TOTAL	TOTAL INCL O&P	
100	0010	**MILLWORK, HIGH DENSITY POLYMER**										100
	0100	Base, 9/16″ x 3-3/16″	1 Carp	230	.035	L.F.	1.34	1.24		2.58	3.52	
	0200	Casing, fluted, 5/8″ x 3-1/4″		215	.037		3.12	1.32		4.44	5.60	
	0300	Chair rail, 9/16″ x 2-1/4″		260	.031		2.04	1.09		3.13	4.05	
	0600	Cove, 13/16″ x 3-3/4″		260	.031		3.90	1.09		4.99	6.10	
	0700	Crown, 3/4″ x 3-13/16″		260	.031		5.85	1.09		6.94	8.25	
	0800	Half round, 15/16″ x 2″		240	.033		17.50	1.18		18.68	21	

06470 | Screen, Blinds & Shutters

		06470 Screen, Blinds & Shutters	CREW	DAILY OUTPUT	LABOR-HOURS	UNIT	MAT.	LABOR	EQUIP.	TOTAL	TOTAL INCL O&P	
100	0010	**SHUTTERS, EXTERIOR**										100
	0012	Aluminum, louvered, 1′-4″ wide, 3′-0″ long	1 Carp	10	.800	Pr.	45.50	28.50		74	97	
	0200	4′-0″ long		10	.800		54.50	28.50		83	107	
	0300	5′-4″ long		10	.800		72	28.50		100.50	126	
	0400	6′-8″ long		9	.889		91.50	31.50		123	154	
	1000	Pine, louvered, primed, each 1′-2″ wide, 3′-3″ long		10	.800		86.50	28.50		115	143	
	1100	4′-7″ long		10	.800		117	28.50		145.50	176	
	1250	Each 1′-4″ wide, 3′-0″ long		10	.800		88.50	28.50		117	145	
	1350	5′-3″ long		10	.800		133	28.50		161.50	193	
	1500	Each 1′-6″ wide, 3′-3″ long		10	.800		92	28.50		120.50	148	

06400 | Architectural Woodwork

06470 | Screen, Blinds & Shutters

			CREW	DAILY OUTPUT	LABOR-HOURS	UNIT	2006 BARE COSTS MAT.	LABOR	EQUIP.	TOTAL	TOTAL INCL O&P	
100	1600	4'-7" long	1 Carp	10	.800	Pr.	129	28.50		157.50	189	100
	1620	Hemlock, louvered, 1'-2" wide, 5'-7" long		10	.800		142	28.50		170.50	203	
	1630	Each 1'-4" wide, 2'-2" long		10	.800		88.50	28.50		117	145	
	1640	3'-0" long		10	.800		88.50	28.50		117	145	
	1650	3'-3" long		10	.800		95.50	28.50		124	152	
	1660	3'-11" long		10	.800		108	28.50		136.50	166	
	1670	4'-3" long		10	.800		106	28.50		134.50	163	
	1680	5'-3" long		10	.800		132	28.50		160.50	192	
	1690	5'-11" long		10	.800		149	28.50		177.50	211	
	1700	Door blinds, 6'-9" long, each 1'-3" wide		9	.889		150	31.50		181.50	218	
	1710	1'-6" wide		9	.889		161	31.50		192.50	230	
	1720	Hemlock, solid raised panel, each 1'-4" wide, 3'-3" long		10	.800		143	28.50		171.50	204	
	1730	3'-11" long		10	.800		167	28.50		195.50	231	
	1740	4'-3" long		10	.800		181	28.50		209.50	246	
	1750	4'-7" long		10	.800		195	28.50		223.50	262	
	1760	4'-11" long		10	.800		212	28.50		240.50	281	
	1770	5'-11" long		10	.800		242	28.50		270.50	315	
	1800	Door blinds, 6'-9" long, each 1'-3" wide		9	.889		272	31.50		303.50	355	
	1900	1'-6" wide		9	.889		297	31.50		328.50	380	
	2500	Polystyrene, solid raised panel, each 1'-4" wide, 3'-3" long		10	.800		38.50	28.50		67	89	
	2600	3'-11" long		10	.800		44	28.50		72.50	95.50	
	2700	4'-7" long		10	.800		48	28.50		76.50	100	
	2800	5'-3" long		10	.800		52	28.50		80.50	104	
	2900	6'-8" long		9	.889		62	31.50		93.50	121	
	4500	Polystyrene, louvered, each 1'-2" wide, 3'-3" long		10	.800		26	28.50		54.50	75.50	
	4600	4'-7" long		10	.800		32	28.50		60.50	82.50	
	4750	5'-3" long		10	.800		37	28.50		65.50	87.50	
	4850	6'-8" long		9	.889		44	31.50		75.50	101	
	6000	Vinyl, louvered, each 1'-2" x 4'-7" long		10	.800		32	28.50		60.50	82	
	6200	Each 1'-4" x 6'-8" long		9	.889	↓	45	31.50		76.50	102	
	9000	Minimum labor/equipment charge	↓	4	2	Job		71		71	118	
200	0010	**SHUTTERS, INTERIOR** Wood, louvered,										200
	0200	Two panel, 27" wide, 36" high	1 Carp	5	1.600	Set	108	57		165	213	
	0300	33" wide, 36" high		5	1.600		142	57		199	250	
	0500	47" wide, 36" high		5	1.600		189	57		246	300	
	1000	Four panel, 27" wide, 36" high		5	1.600		128	57		185	235	
	1100	33" wide, 36" high		5	1.600		157	57		214	267	
	1300	47" wide, 36" high	↓	5	1.600	↓	223	57		280	340	

06500 | Structural Plastics

06520 | Plastic Struct Assemblies

			CREW	DAILY OUTPUT	LABOR-HOURS	UNIT	2006 BARE COSTS MAT.	LABOR	EQUIP.	TOTAL	TOTAL INCL O&P	
175	0010	**FLOOR GRATING, FIBERGLASS**										175
	0100	Reinforced polyester, fire retardant, 1" x 4" grid, 1" thick	E-4	510	.063	S.F.	13.35	2.54	.18	16.07	19.65	
	0200	1-1/2" x 6" mesh, 1-1/2" thick		500	.064		15.65	2.59	.18	18.42	22	
	0300	With grit surface, 1-1/2" x 6" grid, 1-1/2" thick	↓	500	.064	↓	15.95	2.59	.18	18.72	22.50	

	06620	Non-Structural Plastics	CREW	DAILY OUTPUT	LABOR-HOURS	UNIT	2006 BARE COSTS MAT.	LABOR	EQUIP.	TOTAL	TOTAL INCL O&P	
810	0010	**SOLID SURFACE COUNTERTOPS**, Acrylic polymer										810
	0020	Pricing for orders of 100 L.F. or greater										
	0100	25" wide, solid colors	2 Carp	28	.571	L.F.	44	20.50		64.50	82	
	0200	Patterned colors		28	.571		56	20.50		76.50	95	
	0300	Premium patterned colors		28	.571		70	20.50		90.50	111	
	0400	With silicone attached 4" backsplash, solid colors		27	.593		48.50	21		69.50	88.50	
	0500	Patterned colors		27	.593		61.50	21		82.50	103	
	0600	Premium patterned colors		27	.593		76.50	21		97.50	119	
	0700	With hard seam attached 4" backsplash, solid colors		23	.696		48.50	24.50		73	94.50	
	0800	Patterned colors		23	.696		61.50	24.50		86	109	
	0900	Premium patterned colors		23	.696		76.50	24.50		101	125	
	1000	Pricing for order of 51 - 99 L.F.										
	1100	25" wide, solid colors	2 Carp	24	.667	L.F.	51	23.50		74.50	95.50	
	1200	Patterned colors		24	.667		64.50	23.50		88	111	
	1300	Premium patterned colors		24	.667		80.50	23.50		104	128	
	1400	With silicone attached 4" backsplash, solid colors		23	.696		55.50	24.50		80	103	
	1500	Patterned colors		23	.696		70.50	24.50		95	119	
	1600	Premium patterned colors		23	.696		88	24.50		112.50	138	
	1700	With hard seam attached 4" backsplash, solid colors		20	.800		55.50	28.50		84	109	
	1800	Patterned colors		20	.800		70.50	28.50		99	125	
	1900	Premium patterned colors		20	.800		88	28.50		116.50	144	
	2000	Pricing for order of 1 - 50 L.F.										
	2100	25" wide, solid colors	2 Carp	20	.800	L.F.	59.50	28.50		88	113	
	2200	Patterned colors		20	.800		75.50	28.50		104	130	
	2300	Premium patterned colors		20	.800		94.50	28.50		123	151	
	2400	With silicone attached 4" backsplash, solid colors		19	.842		65.50	30		95.50	122	
	2500	Patterned colors		19	.842		83	30		113	141	
	2600	Premium patterned colors		19	.842		103	30		133	164	
	2700	With hard seam attached 4" backsplash, solid colors		15	1.067		65.50	38		103.50	135	
	2800	Patterned colors		15	1.067		83	38		121	154	
	2900	Premium patterned colors		15	1.067		103	38		141	177	
	3000	Sinks, pricing for order of 100 or greater units										
	3100	Single bowl, hard seamed, solid colors, 13" x 17"	1 Carp	3	2.667	Ea.	298	95		393	485	
	3200	10" x 15"		7	1.143		138	40.50		178.50	220	
	3300	Cutouts for sinks		8	1			35.50		35.50	59	
	3400	Sinks, pricing for order of 51 - 99 units										
	3500	Single bowl, hard seamed, solid colors, 13" x 17"	1 Carp	2.55	3.137	Ea.	345	112		457	560	
	3600	10" x 15"		6	1.333		159	47.50		206.50	253	
	3700	Cutouts for sinks		7	1.143			40.50		40.50	67.50	
	3800	Sinks, pricing for order of 1 - 50 units										
	3900	Single bowl, hard seamed, solid colors, 13" x 17"	1 Carp	2	4	Ea.	405	142		547	680	
	4000	10" x 15"		4.55	1.758		186	62.50		248.50	310	
	4100	Cutouts for sinks		5.25	1.524			54		54	90	
	4200	Cooktop cutouts, pricing for 100 or greater units		4	2		22	71		93	142	
	4300	51 - 99 units		3.40	2.353		25.50	83.50		109	167	
	4400	1 - 50 units		3	2.667		29.50	95		124.50	190	
850	0010	**VANITY TOPS**										850
	0015	Solid surface, center bowl, 17" x 19"	1 Carp	12	.667	Ea.	182	23.50		205.50	240	
	0020	19" x 25"		12	.667		220	23.50		243.50	282	
	0030	19" x 31"		12	.667		267	23.50		290.50	335	
	0040	19" x 37"		12	.667		310	23.50		333.50	380	
	0050	22" x 25"		10	.800		194	28.50		222.50	260	
	0060	22" x 31"		10	.800		227	28.50		255.50	296	
	0070	22" x 37"		10	.800		264	28.50		292.50	335	
	0080	22" x 43"		10	.800		300	28.50		328.50	375	
	0090	22" x 49"		10	.800		335	28.50		363.50	410	

	06620 | Non-Structural Plastics	CREW	DAILY OUTPUT	LABOR-HOURS	UNIT	2006 BARE COSTS MAT.	LABOR	EQUIP.	TOTAL	TOTAL INCL O&P	
850	0110 22" x 55"	1 Carp	8	1	Ea.	380	35.50		415.50	475	850
	0120 22" x 61"		8	1		430	35.50		465.50	535	
	0220 Double bowl, 22" x 61"		8	1		490	35.50		525.50	595	
	0230 Double bowl, 22" x 73"	↓	8	1	↓	675	35.50		710.50	805	
	0240 For aggregate colors, add					35%					
	0250 For faucets and fittings see 15410-300										

For information about Means Estimating Seminars, see yellow pages 12 and 13 in back of book

Division 7
Thermal & Moisture Protection

Estimating Tips

07100 Dampproofing & Waterproofing

- Be sure of the job specifications before pricing this subdivision. The difference in cost between waterproofing and dampproofing can be great. Waterproofing will hold back standing water. Dampproofing prevents the transmission of water vapor. Also included in this section are vapor retarding membranes.

07200 Thermal Protection

- Insulation and fireproofing products are measured by area, thickness, volume or R value. Specifications may give only what the specific R value should be in a certain situation. The estimator may need to choose the type of insulation to meet that R value.

07300 Shingles, Roof Tiles & Roof Coverings
07400 Roofing & Siding Panels

- Many roofing and siding products are bought and sold by the square. One square is equal to an area that measures 100 square feet.

  This simple change in unit of measure could create a large error if the estimator is not observant. Accessories necessary for a complete installation must be figured into any calculations for both material and labor.

07500 Membrane Roofing
07600 Flashing & Sheet Metal
07700 Roof Specialties & Accessories

- The items in these subdivisions compose a roofing system. No one component completes the installation, and all must be estimated. Built-up or single-ply membrane roofing systems are made up of many products and installation trades. Wood blocking at roof perimeters or penetrations, parapet coverings, reglets, roof drains, gutters, downspouts, sheet metal flashing, skylights, smoke vents, and roof hatches all need to be considered along with the roofing material. Several different installation trades will need to work together on the roofing system. Inherent difficulties in the scheduling and coordination of various trades must be accounted for when estimating labor costs.

07900 Joint Sealers

- To complete the weather-tight shell the sealants and caulkings must be estimated. Where different materials meet—at expansion joints, at flashing penetrations, and at hundreds of other locations throughout a construction project—they provide another line of defense against water penetration. Often, an entire system is based on the proper location and placement of caulking or sealants. The detailed drawings that are included as part of a set of architectural plans show typical locations for these materials. When caulking or sealants are shown at typical locations, this means the estimator must include them for all the locations where this detail is applicable. Be careful to keep different types of sealants separate, and remember to consider backer rods and primers if necessary.

Reference Numbers

Reference numbers are shown in bold squares at the beginning of some major classifications. These numbers refer to related items in the Reference Section. The reference information may be an estimating procedure, an alternate pricing method, or technical information.

Note: Not all subdivisions listed here necessarily appear in this publication.

07050 | Basic Thermal & Moisture Protection Materials & Methods

07060 | Selective Demolition

		07060 Selective Demolition	CREW	DAILY OUTPUT	LABOR-HOURS	UNIT	2006 BARE COSTS MAT.	LABOR	EQUIP.	TOTAL	TOTAL INCL O&P	
110	0010	**SELECTIVE DEMO, THERMAL & MOISTURE PROTECTION** R024119-10										110
	0200	Waterproofing demo, scrape off, to 1/2" thick	2 Clab	2000	.008	S.F.		.22		.22	.36	
	0210	Over 1/2" thick		1750	.009	″		.25		.25	.42	
	0250	Protection / drain board		3900	.004	B.F.		.11		.11	.19	
	1000	Deck, roof, concrete plank	B-13	1680	.033	S.F.		1	.39	1.39	2.06	
	1100	Gypsum plank		3900	.014			.43	.17	.60	.88	
	1150	Metal decking		3500	.016			.48	.19	.67	.98	
	1200	Wood, boards, tongue and groove, 2" x 6"	2 Clab	960	.017			.46		.46	.76	
	1220	2" x 10"		1040	.015			.42		.42	.70	
	1280	Standard planks, 1" x 6"		1080	.015			.41		.41	.67	
	1320	1" x 8"		1160	.014			.38		.38	.63	
	1340	1" x 12"		1200	.013			.37		.37	.61	
	1350	Plywood, to 1" thick		2000	.008			.22		.22	.36	
	2000	Gutters, aluminum or wood, edge hung	1 Clab	240	.033	L.F.		.91		.91	1.51	
	2100	Built-in		100	.080	″		2.19		2.19	3.63	
	2200	Insulation removal, loose fitting		3000	.003	C.F.		.07		.07	.12	
	2250	Air barrier		3500	.002	S.F.		.06		.06	.10	
	2300	Batts or blankets		1400	.006	C.F.		.16		.16	.26	
	2350	Rigid board		3450	.002	B.F.		.06		.06	.11	
	2500	Roof accessories, plumbing vent flashing		14	.571	Ea.		15.65		15.65	26	
	2600	Adjustable metal chimney flashing		9	.889	″		24.50		24.50	40.50	
	2650	Coping, sheet metal, up to 12" wide		240	.033	L.F.		.91		.91	1.51	
	2660	Concrete, up to 12" wide	2 Clab	160	.100	″		2.74		2.74	4.54	
	3000	Roofing, built-up, 5 ply roof, no gravel	B-2	1600	.025	S.F.		.70		.70	1.15	
	3001	Including gravel		890	.045			1.25		1.25	2.07	
	3100	Gravel removal, minimum		5000	.008			.22		.22	.37	
	3120	Maximum		2000	.020			.56		.56	.92	
	3400	Roof insulation board, up to 2" thick		3900	.010			.29		.29	.47	
	3405	Over 2" thick		7800	.005	B.F.		.14		.14	.24	
	3450	Roll roofing, cold adhesive	1 Clab	12	.667	Sq.		18.25		18.25	30.50	
	4000	Shingles, asphalt strip, 1 layer	B-2	3500	.011	S.F.		.32		.32	.53	
	4100	Slate		2500	.016			.44		.44	.74	
	4300	Wood		2200	.018			.51		.51	.84	
	4500	Skylight to 10 S.F.	1 Clab	8	1	Ea.		27.50		27.50	45.50	
	5000	Siding, metal, horizontal		444	.018	S.F.		.49		.49	.82	
	5020	Vertical		400	.020			.55		.55	.91	
	5200	Wood, boards, vertical		400	.020			.55		.55	.91	
	5220	Clapboards, horizontal		380	.021			.58		.58	.96	
	5240	Shingles		350	.023			.63		.63	1.04	
	5260	Textured plywood		725	.011			.30		.30	.50	
	9000	Minimum labor/equipment charge		2	4	Job		110		110	182	

07100 | Dampproofing and Waterproofing

07110 | Dampproofing

		07110 Dampproofing	CREW	DAILY OUTPUT	LABOR-HOURS	UNIT	2006 BARE COSTS MAT.	LABOR	EQUIP.	TOTAL	TOTAL INCL O&P	
100	0010	**BITUMINOUS ASPHALT COATING** For foundation										100
	0030	Brushed on, below grade, 1 coat	1 Rofc	665	.012	S.F.	.09	.37		.46	.76	
	0100	2 coat		500	.016		.18	.49		.67	1.07	
	0300	Sprayed on, below grade, 1 coat, 25.6 S.F./gal.		830	.010		.09	.30		.39	.63	

07100 | Dampproofing and Waterproofing

07110 | Dampproofing

			CREW	DAILY OUTPUT	LABOR-HOURS	UNIT	2006 BARE COSTS MAT.	LABOR	EQUIP.	TOTAL	TOTAL INCL O&P	
100	0400	2 coat, 20.5 S.F./gal.	1 Rofc	500	.016	S.F.	.17	.49		.66	1.07	100
	0500	Asphalt coating, with fibers				Gal.	4.18			4.18	4.60	
	0600	Troweled on, asphalt with fibers, 1/16″ thick	1 Rofc	500	.016	S.F.	.18	.49		.67	1.08	
	0700	1/8″ thick		400	.020		.32	.61		.93	1.45	
	1000	1/2″ thick		350	.023		1.05	.70		1.75	2.41	
	9000	Minimum labor/equipment charge		3	2.667	Job		81.50		81.50	147	
200	0010	CEMENT PARGING										200
	0020	Portland cement, 2 coats, 1/2″ thick	D-1	250	.064	S.F.	.18	2.06		2.24	3.54	
	0100	Waterproofed Portland cement	″	250	.064	″	1.85	2.06		3.91	5.35	

07130 | Sheet Waterproofing

			CREW	DAILY OUTPUT	LABOR-HOURS	UNIT	MAT.	LABOR	EQUIP.	TOTAL	TOTAL INCL O&P	
200	0010	ELASTOMERIC WATERPROOFING										200
	0090	EPDM, plain, 45 mils thick	2 Rofc	580	.028	S.F.	.99	.84		1.83	2.61	
	0100	60 mils thick		570	.028		1.07	.86		1.93	2.72	
	0300	Nylon reinforced sheets, 45 mils thick		580	.028		1.50	.84		2.34	3.17	
	0400	60 mils thick		570	.028		1.89	.86		2.75	3.62	
	0600	Vulcanizing splicing tape for above, 2″ wide				C.L.F.	41			41	45	
	0700	4″ wide				″	55			55	60.50	
	0900	Adhesive, bonding, 60 SF per gal				Gal.	14.60			14.60	16.05	
	1000	Splicing, 75 SF per gal				″	31			31	34.50	
	1200	Neoprene sheets, plain, 45 mils thick	2 Rofc	580	.028	S.F.	1.41	.84		2.25	3.07	
	1300	60 mils thick		570	.028		2.21	.86		3.07	3.97	
	1500	Nylon reinforced, 45 mils thick		580	.028		1.80	.84		2.64	3.50	
	1600	60 mils thick		570	.028		2.05	.86		2.91	3.80	
	1800	120 mils thick		500	.032		3.37	.98		4.35	5.45	
	1900	Adhesive, splicing, 150 S.F. per gal. per coat				Gal.	18.15			18.15	19.95	
	2100	Fiberglass reinforced, fluid applied, 1/8″ thick	2 Rofc	500	.032	S.F.	1.62	.98		2.60	3.54	
	2200	Polyethylene and rubberized asphalt sheets, 1/8″ thick		550	.029		.54	.89		1.43	2.19	
	2210	Asphaltic hardboard protection board, 1/8″ thick		500	.032		.31	.98		1.29	2.10	
	2220	1/4″ thick		450	.036		.55	1.09		1.64	2.57	
	2400	Polyvinyl chloride sheets, plain, 10 mils thick		580	.028		.16	.84		1	1.70	
	2500	20 mils thick		570	.028		.25	.86		1.11	1.82	
	2700	30 mils thick		560	.029		.35	.87		1.22	1.96	
	3000	Adhesives, trowel grade, 40-100 SF per gal				Gal.	23			23	25.50	
	3100	Brush grade, 100-250 SF per gal.				″	23			23	25.50	
	3300	Bitumen modified polyurethane, fluid applied, 55 mils thick	2 Rofc	665	.024	S.F.	.67	.74		1.41	2.06	
	3600	Vinyl plastic, sprayed on, 25 to 40 mils thick		475	.034	″	1	1.03		2.03	2.95	
	9000	Minimum labor/equipment charge		2	8	Job		245		245	440	

			CREW	DAILY OUTPUT	LABOR-HOURS	UNIT	MAT.	LABOR	EQUIP.	TOTAL	TOTAL INCL O&P	
500	0010	MEMBRANE WATERPROOFING										500
	0012	On slabs, 1 ply, felt, mopped	G-1	3000	.019	S.F.	.19	.53	.11	.83	1.29	
	0100	On slabs, 1 ply, glass fiber fabric, mopped		2100	.027		.21	.76	.16	1.13	1.78	
	0300	On slabs, 2 ply, felt, mopped		2500	.022		.39	.64	.13	1.16	1.73	
	0400	On slabs, 2 ply, glass fiber fabric, mopped		1650	.034		.46	.97	.20	1.63	2.47	
	0600	On slabs, 3 ply, felt, mopped		2100	.027		.58	.76	.16	1.50	2.19	
	0700	On slabs, 3 ply, glass fiber fabric, mopped		1550	.036		.62	1.03	.22	1.87	2.78	
	0710	Asphaltic hardboard protection board, 1/8″ thick	2 Rofc	500	.032		.31	.98		1.29	2.10	
	1000	For adhered 1/4″ EPS protection board, add		3500	.005		.16	.14		.30	.43	
	1050	3/8″ thick, add		3500	.005		.18	.14		.32	.45	
	1060	1/2″ thick, add		3500	.005		.20	.14		.34	.47	
	1070	Fiberglass fabric, black, 20/10 mesh		116	.138	Sq.	9.95	4.22		14.17	18.55	
	1080	White, 20/10 mesh		116	.138	″	10.25	4.22		14.47	18.90	
	1100	1/16″ urethane, troweled		200	.080	S.F.	.66	2.45		3.11	5.15	
	1200	Roller applied		120	.133	″	.60	4.08		4.68	8	
	9000	Minimum labor/equipment charge		2	8	Job		245		245	440	

07100 | Dampproofing and Waterproofing

	07160 | Cement. & Reactive Waterproofing		CREW	DAILY OUTPUT	LABOR-HOURS	UNIT	2006 BARE COSTS MAT.	LABOR	EQUIP.	TOTAL	TOTAL INCL O&P	
150	0010	**CEMENTITIOUS WATERPROOFING** One coat cement base										150
	0020	1/8" application, sprayed on	G-2	1000	.024	S.F.	1.55	.70	.12	2.37	2.98	
	0030	2 coat, cementitious/metallic slurry, troweled, 1/4" thick	1 Cefi	2.48	3.226	C.S.F.	36	111		147	215	
	0040	3 coat, 3/8" thick		1.84	4.348		54	150		204	296	
	0050	4 coat, 1/2" thick	↓	1.20	6.667	↓	72	229		301	440	

	07170 | Bentonite Waterproofing		CREW	DAILY OUTPUT	LABOR-HOURS	UNIT	MAT.	LABOR	EQUIP.	TOTAL	TOTAL INCL O&P	
700	0010	**BENTONITE**										700
	0020	Panels, 4' x 4', 3/16" thick	1 Rofc	625	.013	S.F.	.76	.39		1.15	1.54	
	0100	Rolls, 3/8" thick, with geotextile fabric both sides	"	550	.015	"	.97	.45		1.42	1.87	
	0300	Granular bentonite, 50 lb. bags (.625 C.F.)				Bag	19.20			19.20	21	
	0400	3/8" thick, troweled on	1 Rofc	475	.017	S.F.	.96	.52		1.48	1.99	
	0500	Drain board, expanded polystyrene, binder encapsulated, 1-1/2" thick	1 Rohe	1600	.005		1.07	.11		1.18	1.38	
	0510	2" thick		1600	.005		1.42	.11		1.53	1.76	
	0520	3" thick		1600	.005		2.13	.11		2.24	2.54	
	0530	4" thick		1600	.005		2.83	.11		2.94	3.31	
	0600	With filter fabric, 1-1/2" thick		1600	.005		1.29	.11		1.40	1.62	
	0625	2" thick		1600	.005		1.70	.11		1.81	2.07	
	0650	3" thick		1600	.005		2.41	.11		2.52	2.85	
	0675	4" thick	↓	1600	.005	↓	3.12	.11		3.23	3.63	
	0700	Vapor retarder, see polyethelene, Div. 07260-100										

	07190 | Water Repellents		CREW	DAILY OUTPUT	LABOR-HOURS	UNIT	MAT.	LABOR	EQUIP.	TOTAL	TOTAL INCL O&P	
700	0010	**RUBBER COATING**										700
	0020	Water base liquid, roller applied	2 Rofc	7000	.002	S.F.	.57	.07		.64	.76	
	0200	Silicone or stearate, sprayed on CMU, 1 coat	1 Rofc	4000	.002		.32	.06		.38	.47	
	0300	2 coats		3000	.003	↓	.65	.08		.73	.86	
	9000	Minimum labor/equipment charge	↓	3	2.667	Job		81.50		81.50	147	

07200 | Thermal Protection

	07210 | Building Insulation		CREW	DAILY OUTPUT	LABOR-HOURS	UNIT	2006 BARE COSTS MAT.	LABOR	EQUIP.	TOTAL	TOTAL INCL O&P	
150	0010	**BLOWN-IN INSULATION** Ceilings, with open access										150
	0020	Cellulose, 3-1/2" thick, R13	G-4	5000	.005	S.F.	.19	.13	.05	.37	.48	
	0030	5-3/16" thick, R19		3800	.006		.28	.18	.06	.52	.66	
	0050	6-1/2" thick, R22		3000	.008		.35	.22	.08	.65	.84	
	0100	8-11/16" thick, R30		2600	.009		.48	.26	.09	.83	1.05	
	0120	10-7/8" thick, R38		1800	.013		.61	.37	.13	1.11	1.43	
	1000	Fiberglass, 5" thick, R11		3800	.006		.17	.18	.06	.41	.54	
	1050	6" thick, R13		3000	.008		.21	.22	.08	.51	.68	
	1100	8-1/2" thick, R19		2200	.011		.29	.31	.10	.70	.94	
	1200	10" thick, R22		1800	.013		.34	.37	.13	.84	1.13	
	1300	12" thick, R26		1500	.016		.41	.45	.15	1.01	1.36	
	2000	Mineral wool, 4" thick, R12		3500	.007		.19	.19	.06	.44	.60	
	2050	6" thick, R17		2500	.010		.21	.27	.09	.57	.78	
	2100	9" thick, R23	↓	1750	.014	↓	.30	.38	.13	.81	1.11	
	2500	Wall installation, incl. drilling & patching from outside, two 1"										
	2510	diam. holes @ 16" O.C., top & mid-point of wall, add to above										

07200 | Thermal Protection

07210 | Building Insulation

				DAILY	LABOR-		2006 BARE COSTS				TOTAL	
			CREW	OUTPUT	HOURS	UNIT	MAT.	LABOR	EQUIP.	TOTAL	INCL O&P	
150	2700	For masonry	G-4	415	.058	S.F.	.06	1.62	.55	2.23	3.36	150
	2800	For wood siding		840	.029		.06	.80	.27	1.13	1.70	
	2900	For stucco/plaster		665	.036		.06	1.01	.34	1.41	2.12	
	9000	Minimum labor/equipment charge		4	6	Job		168	56.50	224.50	340	
500	0010	**POURED INSULATION**, Cellulose fiber										500
	0020	R3.8 per inch	1 Carp	200	.040	C.F.	.58	1.42		2	3	
	0040	Ceramic type (perlite), R3.2 per inch		200	.040		1.70	1.42		3.12	4.23	
	0080	Fiberglass wool, R4 per inch		200	.040		.41	1.42		1.83	2.81	
	0100	Mineral wool, R3 per inch		200	.040		.35	1.42		1.77	2.75	
	0300	Polystyrene, R4 per inch		200	.040		2.89	1.42		4.31	5.55	
	0400	Vermiculite or perlite, R2.7 per inch		200	.040		1.70	1.42		3.12	4.23	
	0700	Wood fiber, R3.85 per inch		200	.040		.65	1.42		2.07	3.08	
	9000	Minimum labor/equipment charge		4	2	Job		71		71	118	
550	0010	**MASONRY INSULATION** Vermiculite or perlite, poured										550
	0100	In cores of concrete block, 4″ thick wall, .115 CF/SF	D-1	4800	.003	S.F.	.20	.11		.31	.39	
	0200	6″ thick wall, .175 CF/SF		3000	.005		.30	.17		.47	.61	
	0300	8″ thick wall, .258 CF/SF		2400	.007		.44	.21		.65	.83	
	0400	10″ thick wall, .340 CF/SF		1850	.009		.58	.28		.86	1.09	
	0500	12″ thick wall, .422 CF/SF		1200	.013		.72	.43		1.15	1.49	
	0550	For sand fill, deduct from above					70%					
	0600	Poured cavity wall, vermiculite or perlite, water repellant	D-1	250	.064	C.F.	1.70	2.06		3.76	5.20	
	0700	Foamed in place, urethane in 2-5/8″ cavity	G-2	1035	.023	S.F.	.41	.68	.11	1.20	1.67	
	0800	For each 1″ added thickness, add	″	2372	.010	″	.12	.30	.05	.47	.66	
600	0010	**PERIMETER INSULATION**										600
	0600	Polystyrene, expanded, 1″ thick, R4	1 Carp	680	.012	S.F.	.25	.42		.67	.97	
	0700	2″ thick, R8		675	.012	″	.53	.42		.95	1.28	
	9000	Minimum labor/equipment charge		4	2	Job		71		71	118	
700	0010	**REFLECTIVE INSULATION**										700
	0020	Aluminum foil on reinforced scrim	1 Carp	19	.421	C.S.F.	14.30	14.95		29.25	41	
	0102	Reinforced with woven polyolefin		1900	.004	S.F.	.17	.15		.32	.44	
	0502	With single bubble air space, R8.8		1500	.005		.28	.19		.47	.62	
	0602	With double bubble air space, R9.8		1500	.005		.30	.19		.49	.64	
	9000	Minimum labor/equipment charge		4	2	Job		71		71	118	
800	0010	**SPRAYED INSULATION**										800
	0020	Fibrous/cementitious, finished wall, 1″ thick, R3.7	G-2	2050	.012	S.F.	.23	.34	.06	.63	.87	
	0100	Attic, 5.2″ thick, R19	″	1550	.015	″	.37	.45	.08	.90	1.23	
	0300	Foam type, incl. preparation										
	0600	3 #/CF, 1″ thick, R3.8	G-2	770	.031	S.F.	.50	.91	.15	1.56	2.20	
	0700	2″ thick, R7.5		475	.051	″	1.02	1.48	.25	2.75	3.79	
	9000	Minimum labor/equipment charge		2	12	Job		350	58.50	408.50	635	
900	0010	**WALL INSULATION, RIGID**										900
	0040	Fiberglass, 1.5#/CF, unfaced, 1″ thick, R4.1	1 Carp	1000	.008	S.F.	.38	.28		.66	.89	
	0060	1-1/2″ thick, R6.2		1000	.008		.49	.28		.77	1.01	
	0080	2″ thick, R8.3		1000	.008		.59	.28		.87	1.12	
	0120	3″ thick, R12.4		800	.010		.68	.36		1.04	1.34	
	0370	3#/CF, unfaced, 1″ thick, R4.3		1000	.008		.43	.28		.71	.94	
	0390	1-1/2″ thick, R6.5		1000	.008		.82	.28		1.10	1.37	
	0400	2″ thick, R8.7		890	.009		.99	.32		1.31	1.62	
	0420	2-1/2″ thick, R10.9		800	.010		1.21	.36		1.57	1.92	
	0440	3″ thick, R13		800	.010		1.44	.36		1.80	2.17	
	0520	Foil faced, 1″ thick, R4.3		1000	.008		.95	.28		1.23	1.52	
	0540	1-1/2″ thick, R6.5		1000	.008		1.29	.28		1.57	1.89	
	0560	2″ thick, R8.7		890	.009		1.61	.32		1.93	2.30	
	0580	2-1/2″ thick, R10.9		800	.010		1.90	.36		2.26	2.68	

07210		Building Insulation	CREW	DAILY OUTPUT	LABOR-HOURS	UNIT	2006 BARE COSTS MAT.	LABOR	EQUIP.	TOTAL	TOTAL INCL O&P	
900	0600	3" thick, R13	1 Carp	800	.010	S.F.	2.07	.36		2.43	2.87	900
	0670	6#/CF, unfaced, 1" thick, R4.3		1000	.008		.92	.28		1.20	1.48	
	0690	1-1/2" thick, R6.5		890	.009		1.42	.32		1.74	2.09	
	0700	2" thick, R8.7		800	.010		2	.36		2.36	2.79	
	0721	2-1/2" thick, R10.9		800	.010		2.19	.36		2.55	3	
	0741	3" thick, R13		730	.011		2.62	.39		3.01	3.53	
	0821	Foil faced, 1" thick, R4.3		1000	.008		1.30	.28		1.58	1.90	
	0840	1-1/2" thick, R6.5		890	.009		1.87	.32		2.19	2.59	
	0850	2" thick, R8.7		800	.010		2.44	.36		2.80	3.27	
	0880	2-1/2" thick, R10.9		800	.010		2.93	.36		3.29	3.81	
	0900	3" thick, R13		730	.011		3.50	.39		3.89	4.50	
	1500	Foamglass, 1-1/2" thick, R4.5		800	.010		1.25	.36		1.61	1.97	
	1550	3" thick, R9		730	.011		3.01	.39		3.40	3.96	
	1700	Perlite, 1" thick, R2.77		800	.010		.28	.36		.64	.90	
	1750	2" thick, R5.55		730	.011		.54	.39		.93	1.24	
	1900	Extruded polystyrene, 25 PSI compressive strength, 1" thick, R5		800	.010		.47	.36		.83	1.11	
	1940	2" thick R10		730	.011		.93	.39		1.32	1.67	
	1960	3" thick, R15		730	.011		1.30	.39		1.69	2.08	
	2100	Expanded polystyrene, 1" thick, R3.85		800	.010		.22	.36		.58	.83	
	2120	2" thick, R7.69		730	.011		.55	.39		.94	1.26	
	2140	3" thick, R11.49		730	.011		.68	.39		1.07	1.40	
	9000	Minimum labor/equipment charge		4	2	Job		71		71	118	
950	0010	**WALL OR CEILING INSUL., NON-RIGID**										950
	0040	Fiberglass, kraft faced, batts or blankets										
	0060	3-1/2" thick, R11, 11" wide	1 Carp	1150	.007	S.F.	.29	.25		.54	.73	
	0080	15" wide		1600	.005		.29	.18		.47	.61	
	0100	23" wide		1600	.005		.29	.18		.47	.61	
	0140	6" thick, R19, 11" wide		1000	.008		.42	.28		.70	.93	
	0160	15" wide		1350	.006		.42	.21		.63	.81	
	0180	23" wide		1600	.005		.42	.18		.60	.75	
	0200	9" thick, R30, 15" wide		1150	.007		.71	.25		.96	1.19	
	0220	23" wide		1350	.006		.71	.21		.92	1.13	
	0240	12" thick, R38, 15" wide		1000	.008		.85	.28		1.13	1.41	
	0260	23" wide		1350	.006		.85	.21		1.06	1.29	
	0400	Fiberglass, foil faced, batts or blankets										
	0420	3-1/2" thick, R11, 15" wide	1 Carp	1600	.005	S.F.	.47	.18		.65	.81	
	0440	23" wide		1600	.005		.47	.18		.65	.81	
	0460	6" thick, R19, 15" wide		1350	.006		.51	.21		.72	.91	
	0480	23" wide		1600	.005		.51	.18		.69	.85	
	0500	9" thick, R30, 15" wide		1150	.007		.78	.25		1.03	1.27	
	0550	23" wide		1350	.006		.78	.21		.99	1.21	
	0800	Fiberglass, unfaced, batts or blankets										
	0820	3-1/2" thick, R11, 15" wide	1 Carp	1350	.006	S.F.	.33	.21		.54	.71	
	0830	23" wide		1600	.005		.33	.18		.51	.65	
	0860	6" thick, R19, 15" wide		1150	.007		.56	.25		.81	1.03	
	0880	23" wide		1350	.006		.56	.21		.77	.97	
	0900	9" thick, R30, 15" wide		1000	.008		.66	.28		.94	1.20	
	0920	23" wide		1150	.007		.66	.25		.91	1.14	
	0940	12" thick, R38, 15" wide		1000	.008		.83	.28		1.11	1.38	
	0960	23" wide		1150	.007		.83	.25		1.08	1.32	
	1300	Mineral fiber batts, kraft faced										
	1320	3-1/2" thick, R12	1 Carp	1600	.005	S.F.	.30	.18		.48	.62	
	1340	6" thick, R19		1600	.005		.40	.18		.58	.73	
	1380	10" thick, R30		1350	.006		.59	.21		.80	1	
	1850	Friction fit wire insulation supports, 16" O.C.		960	.008	Ea.	.07	.30		.37	.57	
	1900	For foil backing, add				S.F.	.04			.04	.04	

		07210 | Building Insulation	CREW	DAILY OUTPUT	LABOR-HOURS	UNIT	2006 BARE COSTS MAT.	LABOR	EQUIP.	TOTAL	TOTAL INCL O&P	
950	9000	Minimum labor/equipment charge	1 Carp	4	2	Job		71		71	118	950

		07220 | Roof and Deck Insulation	CREW	DAILY OUTPUT	LABOR-HOURS	UNIT	MAT.	LABOR	EQUIP.	TOTAL	TOTAL INCL O&P	
700	0010	ROOF DECK INSULATION										700
	0020	Fiberboard low density, 1/2″ thick R1.39	1 Rofc	1000	.008	S.F.	.19	.24		.43	.65	
	0030	1″ thick R2.78		800	.010		.38	.31		.69	.97	
	0080	1 1/2″ thick R4.17		800	.010		.57	.31		.88	1.18	
	0100	2″ thick R5.56		800	.010		.76	.31		1.07	1.39	
	0110	Fiberboard high density, 1/2″ thick R1.3		1000	.008		.20	.24		.44	.66	
	0120	1″ thick R2.5		800	.010		.40	.31		.71	.99	
	0130	1-1/2″ thick R3.8		800	.010		.60	.31		.91	1.21	
	0200	Fiberglass, 3/4″ thick R2.78		1000	.008		.50	.24		.74	.99	
	0400	15/16″ thick R3.70		1000	.008		.67	.24		.91	1.18	
	0460	1-1/16″ thick R4.17		1000	.008		.84	.24		1.08	1.36	
	0600	1-5/16″ thick R5.26		1000	.008		1.15	.24		1.39	1.71	
	0650	2-1/16″ thick R8.33		800	.010		1.23	.31		1.54	1.90	
	0700	2-7/16″ thick R10		800	.010		1.40	.31		1.71	2.09	
	1500	Foamglass, 1-1/2″ thick R4.5		800	.010		1.30	.31		1.61	1.98	
	1530	3″ thick R9		700	.011		2.65	.35		3	3.55	
	1600	Tapered for drainage		600	.013	B.F.	1.13	.41		1.54	1.97	
	1650	Perlite, 1/2″ thick R1.32		1050	.008	S.F.	.28	.23		.51	.73	
	1655	3/4″ thick R2.08		800	.010		.30	.31		.61	.88	
	1660	1″ thick R2.78		800	.010		.32	.31		.63	.90	
	1670	1-1/2″ thick R4.17		800	.010		.40	.31		.71	.99	
	1680	2″ thick R5.56		700	.011		.64	.35		.99	1.33	
	1685	2-1/2″ thick R6.67		700	.011		.79	.35		1.14	1.50	
	1690	Tapered for drainage		800	.010	B.F.	.61	.31		.92	1.22	
	1700	Polyisocyanurate, 2#/CF density, 3/4″ thick, R5.1		1500	.005	S.F.	.60	.16		.76	.95	
	1705	1″ thick R7.14		1400	.006		.65	.17		.82	1.03	
	1715	1-1/2″ thick R10.87		1250	.006		.69	.20		.89	1.11	
	1725	2″ thick R14.29		1100	.007		.91	.22		1.13	1.40	
	1735	2-1/2″ thick R16.67		1050	.008		1	.23		1.23	1.52	
	1745	3″ thick R21.74		1000	.008		1.46	.24		1.70	2.05	
	1755	3-1/2″ thick R25		1000	.008		1.48	.24		1.72	2.07	
	1765	Tapered for drainage		1400	.006	B.F.	.76	.17		.93	1.15	
	1900	Extruded Polystyrene										
	1910	15 PSI compressive strength, 1″ thick, R5	1 Rofc	1500	.005	S.F.	.43	.16		.59	.76	
	1920	2″ thick, R10		1250	.006		.66	.20		.86	1.08	
	1930	3″ thick R15		1000	.008		.86	.24		1.10	1.39	
	1932	4″ thick R20		1000	.008		1.34	.24		1.58	1.91	
	1934	Tapered for drainage		1500	.005	B.F.	.44	.16		.60	.77	
	1940	25 PSI compressive strength, 1″ thick R5		1500	.005	S.F.	.63	.16		.79	.98	
	1942	2″ thick R10		1250	.006		1.21	.20		1.41	1.68	
	1944	3″ thick R15		1000	.008		1.84	.24		2.08	2.46	
	1946	4″ thick R20		1000	.008		2.45	.24		2.69	3.14	
	1948	Tapered for drainage		1500	.005	B.F.	.50	.16		.66	.84	
	1950	40 psi compressive strength, 1″ thick R5		1500	.005	S.F.	.44	.16		.60	.77	
	1952	2″ thick R10		1250	.006		.86	.20		1.06	1.30	
	1954	3″ thick R15		1000	.008		1.26	.24		1.50	1.83	
	1956	4″ thick R20		1000	.008		1.69	.24		1.93	2.30	
	1958	Tapered for drainage		1400	.006	B.F.	.63	.17		.80	1	
	1960	60 PSI compressive strength, 1″ thick R5		1450	.006	S.F.	.53	.17		.70	.88	
	1962	2″ thick R10		1200	.007		.94	.20		1.14	1.40	
	1964	3″ thick R15		975	.008		1.40	.25		1.65	1.99	
	1966	4″ thick R20		950	.008		1.95	.26		2.21	2.61	

07220 | Roof and Deck Insulation

		07220 Roof and Deck Insulation	CREW	DAILY OUTPUT	LABOR-HOURS	UNIT	2006 BARE COSTS MAT.	LABOR	EQUIP.	TOTAL	TOTAL INCL O&P	
700	1968	Tapered for drainage	1 Rofc	1400	.006	B.F.	.76	.17		.93	1.15	700
	2010	Expanded polystyrene, 1#/CF density, 3/4" thick R2.89		1500	.005	S.F.	.25	.16		.41	.57	
	2020	1" thick R3.85		1500	.005		.25	.16		.41	.57	
	2100	2" thick R7.69		1250	.006		.53	.20		.73	.93	
	2110	3" thick R11.49		1250	.006		.61	.20		.81	1.02	
	2120	4" thick R15.38		1200	.007		.74	.20		.94	1.18	
	2130	5" thick R19.23		1150	.007		.92	.21		1.13	1.39	
	2140	6" thick R23.26		1150	.007		1.08	.21		1.29	1.57	
	2150	Tapered for drainage		1500	.005	B.F.	.44	.16		.60	.77	
	2400	Composites with 2" EPS										
	2410	1" fiberboard	1 Rofc	950	.008	S.F.	.98	.26		1.24	1.54	
	2420	7/16" oriented strand board		800	.010		1.16	.31		1.47	1.83	
	2430	1/2" plywood		800	.010		1.25	.31		1.56	1.93	
	2440	1" perlite		800	.010		1.03	.31		1.34	1.68	
	2450	Composites with 1-1/2" polyisocyanurate										
	2460	1" fiberboard	1 Rofc	800	.010	S.F.	1.34	.31		1.65	2.02	
	2470	1" perlite		850	.009		1.41	.29		1.70	2.07	
	2480	7/16" oriented strand board		800	.010		1.62	.31		1.93	2.33	
	9000	Minimum labor/equipment charge		3.25	2.462	Job		75.50		75.50	135	

07240 | Ext. Insulation Finish Systems (EIFS)

		07240 Ext. Insulation Finish Systems (EIFS)	CREW	DAILY OUTPUT	LABOR-HOURS	UNIT	MAT.	LABOR	EQUIP.	TOTAL	TOTAL INCL O&P	
100	0010	EXTERIOR INSULATION FINISH SYSTEM										100
	0095	Field applied, 1" EPS insulation	J-1	295	.136	S.F.	2.07	4.15	.36	6.58	9.35	
	0100	With 1/2" cement board sheathing		220	.182		2.83	5.55	.48	8.86	12.65	
	0105	2" EPS insulation		295	.136		2.40	4.15	.36	6.91	9.75	
	0110	With 1/2" cement board sheathing		220	.182		3.16	5.55	.48	9.19	13	
	0115	3" EPS insulation		295	.136		2.53	4.15	.36	7.04	9.90	
	0120	With 1/2" cement board sheathing		220	.182		3.29	5.55	.48	9.32	13.15	
	0125	4" EPS insulation		295	.136		2.95	4.15	.36	7.46	10.35	
	0130	With 1/2" cement board sheathing		220	.182		4.47	5.55	.48	10.50	14.45	
	0140	Premium finish add		1265	.032		.29	.97	.08	1.34	1.97	
	0150	Heavy duty reinforcement add		914	.044		1.73	1.34	.12	3.19	4.19	
	0160	2.5#/S.Y. metal lath substrate add	1 Lath	75	.107	S.Y.	2.22	3.50		5.72	8	
	0170	3.4#/S.Y. metal lath substrate add	"	75	.107	"	2.32	3.50		5.82	8.10	
	0180	Color or texture change,	J-1	1265	.032	S.F.	.75	.97	.08	1.80	2.48	
	0190	With substrate leveling base coat	1 Plas	530	.015		.75	.49		1.24	1.62	
	0210	With substrate sealing base coat	1 Pord	1224	.007		.07	.21		.28	.41	
	0370	V groove shape in panel face				L.F.	.54			.54	.59	
	0380	U groove shape in panel face				"	.71			.71	.78	
	0433	Crack repair, acrylic rubber, fluid applied, 20 mils thick	1 Plas	350	.023	S.F.	1.36	.74		2.10	2.70	
	0437	50 mils thick, reinforced	"	200	.040	"	2.53	1.30		3.83	4.87	
	0440	For higher than one story, add						25%				

07260 | Vapor Retarders

		07260 Vapor Retarders	CREW	DAILY OUTPUT	LABOR-HOURS	UNIT	MAT.	LABOR	EQUIP.	TOTAL	TOTAL INCL O&P	
100	0010	BUILDING PAPER										100
	0020	Aluminum and kraft laminated, foil 1 side	1 Carp	37	.216	Sq.	4.78	7.70		12.48	18	
	0100	Foil 2 sides		37	.216		7.95	7.70		15.65	21.50	
	0400	Asphalt felt sheathing paper, 15#		37	.216		3.41	7.70		11.11	16.50	
	0450	Housewrap, exterior, spun bonded polypropylene										
	0470	Small roll	1 Carp	3800	.002	S.F.	.18	.08		.26	.32	
	0480	Large roll	"	4000	.002	"	.11	.07		.18	.24	
	0500	Material only, 3' x 111.1' roll				Ea.	61			61	67	
	0520	9' x 111.1' roll				"	110			110	121	
	0600	Polyethylene vapor barrier, standard, .002" thick	1 Carp	37	.216	Sq.	.96	7.70		8.66	13.80	

07200 | Thermal Protection

	07260 | Vapor Retarders		Crew	Daily Output	Labor-Hours	Unit	2006 Bare Costs Mat.	Labor	Equip.	Total	Total Incl O&P	
100	0700	.004″ thick	1 Carp	37	.216	Sq.	2.74	7.70		10.44	15.75	100
	0900	.006″ thick		37	.216		4.09	7.70		11.79	17.25	
	1200	.010″ thick		37	.216		5.10	7.70		12.80	18.35	
	1300	Clear reinforced, fire retardant, .008″ thick		37	.216		8.55	7.70		16.25	22	
	1350	Cross laminated type, .003″ thick		37	.216		6.65	7.70		14.35	20	
	1400	.004″ thick		37	.216		7.40	7.70		15.10	21	
	1500	Red rosin paper, 5 sq rolls, 4 lb per square		37	.216		1.73	7.70		9.43	14.65	
	1600	5 lbs. per square		37	.216		2.19	7.70		9.89	15.15	
	1800	Reinf. waterproof, .002″ polyethylene backing, 1 side		37	.216		5	7.70		12.70	18.25	
	1900	2 sides		37	.216		6.65	7.70		14.35	20	
	2100	Roof deck vapor barrier, class 1 metal decks	1 Rofc	37	.216		11.75	6.60		18.35	25	
	2200	For all other decks	″	37	.216		8.35	6.60		14.95	21	
	2400	Waterproofed kraft with sisal or fiberglass fibers, minimum	1 Carp	37	.216		5.40	7.70		13.10	18.70	
	2500	Maximum		37	.216		13.50	7.70		21.20	27.50	
	9950	Minimum labor/equipment charge		4	2	Job		71		71	118	

07300 | Shingles, Roof Tiles and Roof Coverings

	07310 | Shingles		Crew	Daily Output	Labor-Hours	Unit	2006 Bare Costs Mat.	Labor	Equip.	Total	Total Incl O&P	
050	0010	**ALUMINUM SHINGLES**										050
	0020	Mill finish, .019 thick	1 Carp	5	1.600	Sq.	158	57		215	268	
	0100	.020″ thick	″	5	1.600		177	57		234	288	
	0300	For colors, add					15.15			15.15	16.65	
	0600	Ridge cap, .024″ thick	1 Carp	170	.047	L.F.	1.82	1.67		3.49	4.77	
	0700	End wall flashing, .024″ thick		170	.047		1.61	1.67		3.28	4.54	
	0900	Valley section, .024″ thick		170	.047		2.48	1.67		4.15	5.50	
	1000	Starter strip, .024″ thick		400	.020		1.35	.71		2.06	2.67	
	1200	Side wall flashing, .024″ thick		170	.047		1.62	1.67		3.29	4.55	
	9000	Minimum labor/equipment charge		3	2.667	Job		95		95	157	
100	0010	**ASPHALT SHINGLES**										100
	0100	Standard strip shingles										
	0150	Inorganic, class A, 210-235 lb/sq	1 Rofc	5.50	1.455	Sq.	38.50	44.50		83	122	
	0155	Pneumatic nailed		7	1.143		38.50	35		73.50	105	
	0200	Organic, class C, 235-240 lb/sq		5	1.600		43	49		92	136	
	0205	Pneumatic nailed		6.25	1.280		43	39		82	118	
	0250	Standard, laminated multi-layered shingles										
	0300	Class A, 240-260 lb/sq	1 Rofc	4.50	1.778	Sq.	47	54.50		101.50	150	
	0305	Pneumatic nailed		5.63	1.422		47	43.50		90.50	130	
	0350	Class C, 260-300 lb/square, 4 bundles/square		4	2		50	61		111	165	
	0355	Pneumatic nailed		5	1.600		50	49		99	143	
	0400	Premium, laminated multi-layered shingles										
	0450	Class A, 260-300 lb, 4 bundles/sq	1 Rofc	3.50	2.286	Sq.	63	70		133	195	
	0455	Pneumatic nailed		4.37	1.831		63	56		119	170	
	0500	Class C, 300-385 lb/square, 5 bundles/square		3	2.667		75.50	81.50		157	230	
	0505	Pneumatic nailed		3.75	2.133		75.50	65.50		141	200	
	0800	#15 felt underlayment		64	.125		3.41	3.83		7.24	10.65	
	0825	#30 felt underlayment		58	.138		7.20	4.22		11.42	15.50	
	0850	Self adhering polyethylene and rubberized asphalt underlayment		22	.364		46	11.15		57.15	70.50	
	0900	Ridge shingles		330	.024	L.F.	1.33	.74		2.07	2.79	

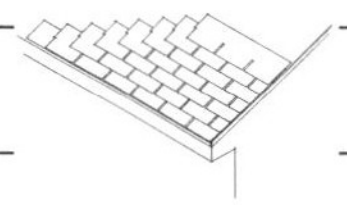

	07310 | Shingles		CREW	DAILY OUTPUT	LABOR-HOURS	UNIT	2006 BARE COSTS MAT.	LABOR	EQUIP.	TOTAL	TOTAL INCL O&P	
100	0905	Pneumatic nailed	1 Rofc	412.50	.019	L.F.	1.33	.59		1.92	2.53	100
	1000	For steep roofs (7 to 12 pitch or greater), add						50%				
	9000	Minimum labor/equipment charge	1 Rofc	3	2.667	Job		81.50		81.50	147	
500	0010	**FIBER CEMENT SHINGLES**										500
	0012	Field shingles, 16" x 9.35", 500 lb per square	1 Rofc	2.20	3.636	Sq.	292	111		403	520	
	0110	Starters, 16" x 9.35"		3	2.667	C.L.F.	99.50	81.50		181	256	
	0120	Hip & ridge, 4.75" x 14"		1	8	"	720	245		965	1,225	
	0200	Shakes, 16" x 9.35", 550 lb per square		2.20	3.636	Sq.	265	111		376	490	
	0300	Hip & ridge, 4.75 x 14"		1	8	C.L.F.	720	245		965	1,225	
	0400	Hexagonal, 16" x 16"		3	2.667	Sq.	198	81.50		279.50	365	
	0500	Square, 16" x 16"		3	2.667		178	81.50		259.50	340	
	2000	For steep roofs (7/12 pitch or greater), add						50%				
	9000	Minimum labor/equipment charge	1 Rofc	3	2.667	Job		81.50		81.50	147	
800	0010	**SLATE**, Buckingham, Virginia, black										800
	0100	3/16" - 1/4" thick	1 Rots	1.75	4.571	Sq.	585	140		725	890	
	0200	1/4" thick		1.75	4.571		615	140		755	925	
	0900	Pennsylvania black, Bangor, #1 clear		1.75	4.571		475	140		615	775	
	1200	Vermont, unfading, green, mottled green		1.75	4.571		365	140		505	655	
	1300	Semi-weathering green & gray		1.75	4.571		350	140		490	635	
	1400	Purple		1.75	4.571		390	140		530	680	
	1500	Black or gray		1.75	4.571		360	140		500	645	
	2500	Slate roof repair, extensive replacement		1	8		575	245		820	1,075	
	2600	Repair individual pieces, scattered		19	.421	Ea.	5.70	12.90		18.60	29.50	
	9000	Minimum labor/equipment charge		3	2.667	Job		81.50		81.50	147	
900	0010	**STEEL SHINGLES**										900
	0012	Galvanized, 26 gauge	1 Rots	2.20	3.636	Sq.	183	111		294	400	
	0200	24 gauge	"	2.20	3.636		192	111		303	410	
	0300	For colored galvanized shingles, add					50			50	55	
	0500	For 1" factory applied polystyrene insulation, add					36.50			36.50	40.50	
	9000	Minimum labor/equipment charge	1 Rots	3	2.667	Job		81.50		81.50	147	
980	0010	**WOOD SHINGLES**										980
	0012	16" No. 1 red cedar shingles, 5" exposure, on roof	1 Carp	2.50	3.200	Sq.	187	114		301	395	
	0015	Pneumatic nailed		3.25	2.462		187	87.50		274.50	350	
	0200	7-1/2" exposure, on walls		2.05	3.902		125	139		264	365	
	0205	Pneumatic nailed		2.67	2.996		125	107		232	315	
	0300	18" No. 1 red cedar perfections, 5-1/2" exposure, on roof		2.75	2.909		168	103		271	355	
	0305	Pneumatic nailed		3.57	2.241		168	79.50		247.50	315	
	0600	Resquared, and rebutted, 5-1/2" exposure, on roof		3	2.667		210	95		305	390	
	0605	Pneumatic nailed		3.90	2.051		210	73		283	350	
	0900	7-1/2" exposure, on walls		2.45	3.265		154	116		270	360	
	0905	Pneumatic nailed		3.18	2.516		154	89.50		243.50	315	
	1000	Add to above for fire retardant shingles, 16" long					35			35	38.50	
	1050	18" long					35			35	38.50	
	1060	Preformed ridge shingles	1 Carp	400	.020	L.F.	1.65	.71		2.36	3	
	1100	Hand-split red cedar shakes, 1/2" thick x 24" long, 10" exp. on roof		2.50	3.200	Sq.	116	114		230	315	
	1105	Pneumatic nailed		3.25	2.462		116	87.50		203.50	272	
	1110	3/4" thick x 24" long, 10" exp. on roof		2.25	3.556		116	126		242	335	
	1115	Pneumatic nailed		2.92	2.740		116	97.50		213.50	288	
	1200	1/2" thick, 18" long, 8-1/2" exp. on roof		2	4		89.50	142		231.50	335	
	1205	Pneumatic nailed		2.60	3.077		89.50	109		198.50	280	
	1210	3/4" thick x 18" long, 8 1/2" exp. on roof		1.80	4.444		89.50	158		247.50	360	
	1215	Pneumatic nailed		2.34	3.419		89.50	122		211.50	300	
	1255	10" exp. on walls		2	4		101	142		243	345	
	1260	10" exposure on walls, pneumatic nailed		2.60	3.077		101	109		210	292	

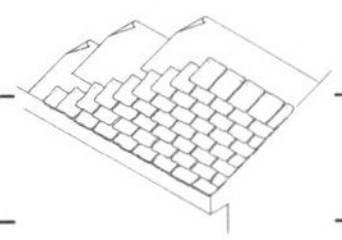

07300 | Shingles, Roof Tiles and Roof Coverings

07310 | Shingles

		07310 Shingles	Crew	Daily Output	Labor-Hours	Unit	2006 Bare Costs Mat.	Labor	Equip.	Total	Total Incl O&P
980	1700	Add to above for fire retardant shakes, 24″ long				Sq.	50			50	55
	1800	18″ long					45			45	49.50
	1810	Ridge shakes	1 Carp	350	.023	L.F.	2.35	.81		3.16	3.94
	2000	White cedar shingles, 16″ long, extras, 5″ exposure, on roof		2.40	3.333	Sq.	163	119		282	375
	2005	Pneumatic nailed		3.12	2.564		163	91		254	330
	2050	5″ exposure on walls		2	4		163	142		305	415
	2055	Pneumatic nailed		2.60	3.077		163	109		272	360
	2100	7-1/2″ exposure, on walls		2	4		116	142		258	365
	2105	Pneumatic nailed		2.60	3.077		116	109		225	310
	2150	"B" grade, 5″ exposure on walls		2	4		146	142		288	395
	2155	Pneumatic nailed		2.60	3.077		146	109		255	340
	2300	For 15# organic felt underlayment on roof, 1 layer, add		64	.125		3.41	4.44		7.85	11.10
	2400	2 layers, add		32	.250		6.80	8.90		15.70	22.50
	2600	For steep roofs (7/12 pitch or greater), add to above						50%			
	2700	Panelized systems, No.1 cedar shingles on 5/16″ CDX plywood									
	2800	On walls, 8′ strips, 7″ or 14″ exposure	2 Carp	700	.023	S.F.	3.20	.81		4.01	4.87
	3500	On roofs, 8′ strips, 7″ or 14″ exposure	1 Carp	3	2.667	Sq.	320	95		415	505
	3505	Pneumatic nailed		4	2	"	320	71		391	470
	9000	Minimum labor/equipment charge		3	2.667	Job		95		95	157

07320 | Roof Tiles

		07320 Roof Tiles	Crew	Daily Output	Labor-Hours	Unit	Mat.	Labor	Equip.	Total	Total Incl O&P
100	0010	**ALUMINUM ROOF TILES**									
	0020	Accessories included, .032″ thick, mission tile	1 Carp	2.50	3.200	Sq.	565	114		679	815
	0200	Spanish tiles		3	2.667	"	390	95		485	585
	9000	Minimum labor/equipment charge		3	2.667	Job		95		95	157
200	0010	**CLAY TILE** ASTM C1167, GR 1, severe weathering, acces. incl.									
	0200	Lanai tile or Classic tile, 158 pc per sq	1 Rots	1.65	4.848	Sq.	405	148		553	715
	0300	Americana, 158 pc per sq, most colors		1.65	4.848		525	148		673	840
	0350	Green, gray or brown		1.65	4.848		505	148		653	825
	0400	Blue		1.65	4.848		505	148		653	825
	0600	Spanish tile, 171 pc per sq, red		1.80	4.444		268	136		404	540
	0800	Blend		1.80	4.444		430	136		566	720
	0900	Glazed white		1.80	4.444		515	136		651	810
	1100	Mission tile, 192 pc per sq, machine scored finish, red		1.15	6.957		650	213		863	1,100
	1700	French tile, 133 pc per sq, smooth finish, red		1.35	5.926		590	181		771	975
	1750	Blue or green		1.35	5.926		705	181		886	1,100
	1800	Norman black 317 pc per sq		1	8		830	245		1,075	1,350
	2200	Williamsburg tile, 158 pc per sq, aged cedar		1.35	5.926		505	181		686	880
	2250	Gray or green		1.35	5.926		505	181		686	880
	2510	One piece mission tile, natural red, 75 pc per square		1.65	4.848		137	148		285	420
	2530	Mission Tile, 134 pc per square		1.15	6.957		183	213		396	585
	3000	For steep roofs (7/12 pitch or greater), add to above						50%			
	9000	Minimum labor/equipment charge	1 Rots	3	2.667	Job		81.50		81.50	147
300	0010	**CONCRETE TILE** Including installation of accessories									
	0020	Corrugated, 13″ x 16-1/2″, 90 per sq, 950 lb per sq									
	0050	Earthtone colors, nailed to wood deck	1 Rots	1.35	5.926	Sq.	93.50	181		274.50	430
	0150	Blues		1.35	5.926		94.50	181		275.50	430
	0200	Greens		1.35	5.926		94.50	181		275.50	430
	0250	Premium colors		1.35	5.926		158	181		339	500
	0500	Shakes, 13″ x 16-1/2″, 90 per sq, 950 lb per sq									
	0600	All colors, nailed to wood deck	1 Rots	1.50	5.333	Sq.	191	163		354	505
	1500	Accessory pieces, ridge & hip, 10″ x 16-1/2″, 8 lbs. each				Ea.	2.32			2.32	2.55
	1700	Rake, 6-1/2″ x 16-3/4″, 9 lbs. each					2.32			2.32	2.55
	1800	Mansard hip, 10″ x 16-1/2″, 9.2 lbs. each					2.32			2.32	2.55
	1900	Hip starter, 10″ x 16-1/2″, 10.5 lbs. each					9.80			9.80	10.75

07300 | Shingles, Roof Tiles and Roof Coverings

	07320 | Roof Tiles		Crew	Daily Output	Labor-Hours	Unit	2006 Bare Costs Mat.	Labor	Equip.	Total	Total Incl O&P	
300	2000	3 or 4 way apex, 10″ each side, 11.5 lbs. each				Ea.	10.55			10.55	11.60	300
	9000	Minimum labor/equipment charge	1 Rots	3	2.667	Job		81.50		81.50	147	

07400 | Roofing and Siding Panels

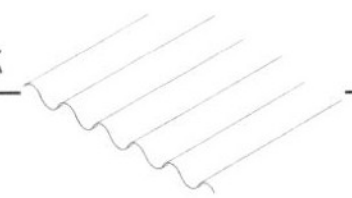

	07410 | Metal Roof and Wall Panels		Crew	Daily Output	Labor-Hours	Unit	2006 Bare Costs Mat.	Labor	Equip.	Total	Total Incl O&P	
100	0010	**ALUMINUM ROOFING PANELS**										100
	0020	Corrugated or ribbed, .0155″ thick, natural	G-3	1200	.027	S.F.	.69	.93		1.62	2.26	
	0300	Painted		1200	.027		1	.93		1.93	2.60	
	0400	Corrugated, .018″ thick, on steel frame, natural finish		1200	.027		.90	.93		1.83	2.49	
	0600	Painted		1200	.027		1.11	.93		2.04	2.72	
	0700	Corrugated, on steel frame, natural, .024″ thick		1200	.027		1.30	.93		2.23	2.93	
	0800	Painted		1200	.027		1.57	.93		2.50	3.23	
	0900	.032″ thick, natural		1200	.027		1.54	.93		2.47	3.19	
	1200	Painted		1200	.027		2.13	.93		3.06	3.84	
	1300	V-Beam, on steel frame construction, .032″ thick, natural		1200	.027		1.73	.93		2.66	3.40	
	1500	Painted		1200	.027		2.16	.93		3.09	3.88	
	1600	.040″ thick, natural		1200	.027		2.12	.93		3.05	3.83	
	1800	Painted		1200	.027		2.58	.93		3.51	4.34	
	1900	.050″ thick, natural		1200	.027		2.57	.93		3.50	4.33	
	2100	Painted		1200	.027		3.10	.93		4.03	4.91	
	2200	For roofing on wood frame, deduct		4600	.007	↓	.06	.24		.30	.46	
	2400	Ridge cap, .032″ thick, natural	↓	800	.040	L.F.	2.03	1.39		3.42	4.48	
	9000	Minimum labor/equipment charge	1 Rofc	3	2.667	Job		81.50		81.50	147	
500	0010	**MANSARD** Colored aluminum, with battens, .032″ thick										500
	0600	Stock units, straight surfaces	1 Shee	115	.070	S.F.	2.39	2.93		5.32	7.30	
	0700	Concave or convex surfaces		75	.107	″	2.63	4.50		7.13	10.05	
	0800	For framing, to 5′ high, add		115	.070	L.F.	2.63	2.93		5.56	7.55	
	0900	Soffits, to 1′ wide		125	.064	S.F.	1.30	2.70		4	5.70	
	9000	Minimum labor/equipment charge	↓	2.50	3.200	Job		135		135	215	
690	0010	**METAL FACING PANELS**										690
	0400	Textured aluminum, 4′ x 8′ x 5/16″ plywood backing, single face	2 Shee	375	.043	S.F.	2.63	1.80		4.43	5.75	
	0600	Double face		375	.043		3.86	1.80		5.66	7.10	
	0700	4′ x 10′ x 5/16″ plywood backing, single face		375	.043		2.73	1.80		4.53	5.85	
	0900	Double face		375	.043		4.01	1.80		5.81	7.25	
	1000	4′ x 12′ x 5/16″ plywood backing, single face		375	.043		3.52	1.80		5.32	6.75	
	1300	Smooth al, 1/4″ panel, fluoropolymer finish, double face		375	.043		6.30	1.80		8.10	9.75	
	1350	Clear anodized finish, double face		375	.043		7	1.80		8.80	10.55	
	1400	Double face textured aluminum, structural panel, 1″ EPS insulation	↓	375	.043	↓	4.23	1.80		6.03	7.50	
	1500	Accessories, outside corner	1 Shee	175	.046	L.F.	1.94	1.93		3.87	5.20	
	1600	Inside corner		175	.046		2.11	1.93		4.04	5.40	
	1800	Batten mounting clip		200	.040		.39	1.69		2.08	3.11	
	1900	Low profile batten		480	.017		.80	.70		1.50	2	
	2100	High profile batten		480	.017		1.36	.70		2.06	2.62	
	2200	Water table		200	.040		1.96	1.69		3.65	4.84	
	2400	Horizontal joint connector		200	.040		1.83	1.69		3.52	4.69	
	2500	Corner cap		200	.040		1.69	1.69		3.38	4.54	
	2700	H - moulding	↓	480	.017	↓	1.17	.70		1.87	2.41	
700	0010	**STEEL ROOFING PANELS**										700
	0012	Corrugated or ribbed, on steel framing, 30 ga galv	G-3	1100	.029	S.F.	1.05	1.01		2.06	2.80	

07400 | Roofing and Siding Panels

07410 | Metal Roof and Wall Panels

		07410 Metal Roof and Wall Panels	CREW	DAILY OUTPUT	LABOR-HOURS	UNIT	2006 BARE COSTS MAT.	LABOR	EQUIP.	TOTAL	TOTAL INCL O&P
700	0100	28 ga	G-3	1050	.030	S.F.	1.10	1.06		2.16	2.92
	0300	26 ga		1000	.032		1.53	1.11		2.64	3.48
	0400	24 ga		950	.034		1.89	1.17		3.06	3.97
	0600	Colored, 28 ga		1050	.030		1.08	1.06		2.14	2.90
	0700	26 ga		1000	.032		1.65	1.11		2.76	3.62
	0710	Flat profile, 1-3/4" standing seams, 10" wide, standard finish, 26 ga		1000	.032		3.22	1.11		4.33	5.35
	0715	24 ga		950	.034		3.74	1.17		4.91	6
	0720	22 ga		900	.036		4.61	1.24		5.85	7.05
	0725	Zinc aluminum alloy finish, 26 ga		1000	.032		2.52	1.11		3.63	4.57
	0730	24 ga		950	.034		3.01	1.17		4.18	5.20
	0735	22 ga		900	.036		3.45	1.24		4.69	5.80
	0740	12" wide, standard finish, 26 ga		1000	.032		3.21	1.11		4.32	5.35
	0745	24 ga		950	.034		4.22	1.17		5.39	6.55
	0750	Zinc aluminum alloy finish, 26 ga		1000	.032		3.65	1.11		4.76	5.80
	0755	24 ga		950	.034		3	1.17		4.17	5.20
	0840	Flat profile, 1" x 3/8" batten, 12" wide, standard finish, 26 ga		1000	.032		2.83	1.11		3.94	4.91
	0845	24 ga		950	.034		3.32	1.17		4.49	5.55
	0850	22 ga		900	.036		3.98	1.24		5.22	6.40
	0855	Zinc aluminum alloy finish, 26 ga		1000	.032		2.72	1.11		3.83	4.79
	0860	24 ga		950	.034		3.03	1.17		4.20	5.20
	0865	22 ga		900	.036		3.50	1.24		4.74	5.85
	0870	16-1/2" wide, standard finish, 24 ga		950	.034		3.27	1.17		4.44	5.50
	0875	22 ga		900	.036		3.66	1.24		4.90	6.05
	0880	Zinc aluminum alloy finish, 24 ga		950	.034		2.85	1.17		4.02	5.05
	0885	22 ga		900	.036		3.19	1.24		4.43	5.50
	0890	Flat profile, 2" x 2" batten, 12" wide, standard finish, 26 ga		1000	.032		3.25	1.11		4.36	5.40
	0895	24 ga		950	.034		3.88	1.17		5.05	6.15
	0900	22 ga		900	.036		4.75	1.24		5.99	7.25
	0905	Zinc aluminum alloy finish, 26 ga		1000	.032		3.03	1.11		4.14	5.15
	0910	24 ga		950	.034		3.46	1.17		4.63	5.70
	0915	22 ga		900	.036		4.03	1.24		5.27	6.45
	0920	16-1/2" wide, standard finish, 24 ga		950	.034		3.58	1.17		4.75	5.85
	0925	22 ga		900	.036		4.18	1.24		5.42	6.60
	0930	Zinc aluminum alloy finish, 24 ga		950	.034		3.24	1.17		4.41	5.45
	0935	22 ga		900	.036	↓	3.68	1.24		4.92	6.05
	1200	Ridge, galvanized, 10" wide		800	.040	L.F.	2.35	1.39		3.74	4.84
	1210	20" wide	↓	750	.043	"	4.13	1.48		5.61	6.95
	9000	Minimum labor/equipment charge	1 Rofc	2	4	Job		122		122	220

07420 | Plastic Roof and Wall Panels

		07420 Plastic Roof and Wall Panels	CREW	DAILY OUTPUT	LABOR-HOURS	UNIT	MAT.	LABOR	EQUIP.	TOTAL	TOTAL INCL O&P
770	0010	**FIBERGLASS CORRUGATED ROOFING PANELS**									
	0012	Corrugated, 8 oz per SF	G-3	1000	.032	S.F.	2.24	1.11		3.35	4.26
	0100	12 oz per SF		1000	.032		3.23	1.11		4.34	5.35
	0300	Corrugated siding, 6 oz per SF		880	.036		1.94	1.26		3.20	4.17
	0400	8 oz per SF		880	.036		2.24	1.26		3.50	4.50
	0500	Fire retardant		880	.036		3.09	1.26		4.35	5.45
	0600	12 oz. siding, textured		880	.036		3.14	1.26		4.40	5.50
	0700	Fire retardant		880	.036		4.14	1.26		5.40	6.60
	0900	Flat panels, 6 oz per SF, clear or colors		880	.036		1.73	1.26		2.99	3.94
	1100	Fire retardant, class A		880	.036		3	1.26		4.26	5.35
	1300	8 oz per SF, clear or colors		880	.036		2.24	1.26		3.50	4.50
	1700	Sandwich panels, fiberglass, 1-9/16" thick, panels to 20 SF		180	.178		21.50	6.20		27.70	34
	1900	As above, but 2-3/4" thick, panels to 100 SF	↓	265	.121	↓	16.10	4.20		20.30	24.50
	9000	Minimum labor/equipment charge	1 Rofc	2	4	Job		122		122	220

07400 | Roofing and Siding Panels

07440 | Faced Panels

		07440 Faced Panels	CREW	DAILY OUTPUT	LABOR-HOURS	UNIT	MAT.	LABOR	EQUIP.	TOTAL	TOTAL INCL O&P	
200	0010	EXPOSED AGGREGATE PANELS										200
	1400	Fiberglass polymer back-up,										
	1500	Small size aggregate	F-3	445	.090	S.F.	4.90	3.24	1.43	9.57	12.25	
	1600	Medium size aggregate		445	.090		5.55	3.24	1.43	10.22	12.95	
	1700	Large size aggregate		445	.090		5.70	3.24	1.43	10.37	13.10	

07460 | Siding

		07460 Siding	CREW	DAILY OUTPUT	LABOR-HOURS	UNIT	MAT.	LABOR	EQUIP.	TOTAL	TOTAL INCL O&P	
100	0010	ALUMINUM SIDING PANELS										100
	0012	Corrugated, on steel framing, .019 thick, natural finish	G-3	775	.041	S.F.	.74	1.44		2.18	3.13	
	0100	Painted		775	.041		.96	1.44		2.40	3.38	
	0400	Farm type, .021″ thick on steel frame, natural		775	.041		.86	1.44		2.30	3.27	
	0600	Painted		775	.041		1.05	1.44		2.49	3.48	
	0700	Industrial type, corrugated, on steel, .024″ thick, mill		775	.041		1.09	1.44		2.53	3.52	
	0900	Painted		775	.041		1.31	1.44		2.75	3.76	
	1000	.032″ thick, mill		775	.041		1.59	1.44		3.03	4.07	
	1200	Painted		775	.041		2.02	1.44		3.46	4.54	
	1300	V-Beam, on steel frame, .032″ thick, mill		775	.041		1.86	1.44		3.30	4.37	
	1500	Painted		775	.041		2.21	1.44		3.65	4.75	
	1600	.040″ thick, mill		775	.041		2.26	1.44		3.70	4.81	
	1800	Painted		775	.041		2.65	1.44		4.09	5.25	
	1900	.050″ thick, mill		775	.041		2.67	1.44		4.11	5.25	
	2100	Painted		775	.041		3.19	1.44		4.63	5.85	
	2200	Ribbed, 3″ profile, on steel frame, .032″ thick, natural		775	.041		1.55	1.44		2.99	4.03	
	2400	Painted		775	.041		2.04	1.44		3.48	4.56	
	2500	.040″ thick, natural		775	.041		1.90	1.44		3.34	4.41	
	2700	Painted		775	.041		2.23	1.44		3.67	4.77	
	2750	.050″ thick, natural		775	.041		2.20	1.44		3.64	4.74	
	2760	Painted		775	.041		2.50	1.44		3.94	5.05	
	3300	For siding on wood frame, deduct from above		2800	.011		.07	.40		.47	.72	
	3400	Screw fasteners, aluminum, self tapping, neoprene washer, 1″				M	141			141	155	
	3600	Stitch screws, self tapping, with neoprene washer, 5/8″				″	105			105	116	
	3630	Flashing, sidewall, .032″ thick	G-3	800	.040	L.F.	2	1.39		3.39	4.45	
	3650	End wall, .040″ thick		800	.040		2.34	1.39		3.73	4.82	
	3670	Closure strips, corrugated, .032″ thick		800	.040		.59	1.39		1.98	2.90	
	3680	Ribbed, 4″ or 8″, .032″ thick		800	.040		.59	1.39		1.98	2.90	
	3690	V-beam, .040″ thick		800	.040		.81	1.39		2.20	3.14	
	3800	Horizontal, colored clapboard, 8″ wide, plain	2 Carp	515	.031	S.F.	1.21	1.10		2.31	3.16	
	3900	Insulated		515	.031		1.56	1.10		2.66	3.55	
	4000	Vertical board & batten, colored, non-insulated		515	.031		1.21	1.10		2.31	3.16	
	4200	For simulated wood design, add					.09			.09	.10	
	4300	Corners for above, outside	2 Carp	515	.031	V.L.F.	1.86	1.10		2.96	3.88	
	4500	Inside corners	″	515	.031	″	1.10	1.10		2.20	3.04	
	4600	Sandwich panels, 1″ insulation, single story	G-3	395	.081	S.F.	5.15	2.82		7.97	10.25	
	4900	Multi-story	″	345	.093		7.25	3.23		10.48	13.20	
	5100	For baked enamel finish 1 side, add					.27			.27	.30	
	5200	See also Metal Facing Panels, division 07410-690										
	9000	Minimum labor/equipment charge	1 Carp	3	2.667	Job		95		95	157	
300	0010	FASCIA Aluminum, reverse board and batten,										300
	0100	.032″ thick, colored, no furring included	1 Shee	145	.055	S.F.	2.95	2.33		5.28	6.95	
	0300	Steel, galv and enameled, stock, no furring, long panels		145	.055		3	2.33		5.33	7	
	0600	Short panels		115	.070		4	2.93		6.93	9.05	
	9000	Minimum labor/equipment charge		4	2	Job		84.50		84.50	134	
500	0010	FIBER CEMENT SIDING										500
	0020	Lap siding, 5/16″ thick, 6″ wide, smooth texture	2 Carp	415	.039	S.F.	1.23	1.37		2.60	3.63	

	07460 | Siding		CREW	DAILY OUTPUT	LABOR-HOURS	UNIT	2006 BARE COSTS MAT.	LABOR	EQUIP.	TOTAL	TOTAL INCL O&P	
500	0025	Woodgrain texture	2 Carp	415	.039	S.F.	1.23	1.37		2.60	3.63	500
	0030	7-1/2" wide, smooth texture		425	.038		1.08	1.34		2.42	3.40	
	0035	Woodgrain texture		425	.038		1.08	1.34		2.42	3.40	
	0040	8" wide, smooth texture		425	.038		1.07	1.34		2.41	3.39	
	0045	Roughsawn texture		425	.038		1.07	1.34		2.41	3.39	
	0050	9-1/2" wide, smooth texture		440	.036		1.03	1.29		2.32	3.28	
	0055	Woodgrain texture		440	.036		1.03	1.29		2.32	3.28	
	0060	12" wide, smooth texture		455	.035		.99	1.25		2.24	3.16	
	0065	Woodgrain texture		455	.035		.99	1.25		2.24	3.16	
	0070	Panel siding, 5/16" thick, smooth texture		750	.021		.89	.76		1.65	2.24	
	0075	Stucco texture		750	.021		.89	.76		1.65	2.24	
	0080	Grooved woodgrain texture		750	.021		.89	.76		1.65	2.24	
	0085	V - grooved woodgrain texture		750	.021		.89	.76		1.65	2.24	
	0090	Wood starter strip		400	.040	L.F.	.19	1.42		1.61	2.57	
600	0010	**VINYL SIDING**										600
	0020	Clapboard profile, woodgrain texture, .048 thick, double 4	1 Carp	255	.031	S.F.	.74	1.12		1.86	2.66	
	0100	with 3/8" insulation		255	.031		.84	1.12		1.96	2.77	
	0200	Soffit and fascia		205	.039		1.60	1.39		2.99	4.06	
	0300	Window and door trim moldings		185	.043	L.F.	.40	1.54		1.94	2.99	
	0500	Corner posts, outside corner		205	.039		1.44	1.39		2.83	3.88	
	0600	Inside corner		205	.039		.67	1.39		2.06	3.04	
	9000	Minimum labor/equipment charge		3	2.667	Job		95		95	157	
750	0010	**SOFFIT**										750
	0012	Aluminum, residential, .020" thick	1 Carp	210	.038	S.F.	1.24	1.35		2.59	3.60	
	0100	Baked enamel on steel, 16 or 18 gauge		105	.076		4.60	2.71		7.31	9.55	
	0300	Polyvinyl chloride, white, solid		230	.035		.82	1.24		2.06	2.95	
	0400	Perforated		230	.035		.82	1.24		2.06	2.95	
	0500	For colors, add					.09			.09	.10	
	9000	Minimum labor/equipment charge	1 Carp	3	2.667	Job		95		95	157	
800	0010	**STEEL SIDING**										800
	0020	Beveled, vinyl coated, 8" wide, including fasteners	1 Carp	265	.030	S.F.	1.50	1.07		2.57	3.43	
	0050	10" wide		275	.029	"	1.60	1.03		2.63	3.47	
	0060	Minimum labor/equipment charge		3	2.667	Job		95		95	157	
	0080	Galv, corrugated or ribbed, on steel frame, 30 gauge	G-3	800	.040	S.F.	1	1.39		2.39	3.35	
	0100	28 gauge		795	.040		1.05	1.40		2.45	3.42	
	0300	26 gauge		790	.041		1.47	1.41		2.88	3.90	
	0400	24 gauge		785	.041		1.48	1.42		2.90	3.92	
	0600	22 gauge		770	.042		1.70	1.45		3.15	4.21	
	0700	Colored, corrugated/ribbed, on steel frame, 10 yr fnsh, 28 ga.		800	.040		1.56	1.39		2.95	3.97	
	0900	26 gauge		795	.040		1.63	1.40		3.03	4.05	
	1000	24 gauge		790	.041		1.89	1.41		3.30	4.36	
	1020	20 gauge		785	.041		2.40	1.42		3.82	4.93	
	1200	Factory sandwich panel, 26 ga., 1" insulation, galvanized		380	.084		3.99	2.93		6.92	9.15	
	1300	Colored 1 side		380	.084		4.88	2.93		7.81	10.10	
	1500	Galvanized 2 sides		380	.084		6.05	2.93		8.98	11.40	
	1600	Colored 2 sides		380	.084		6.30	2.93		9.23	11.65	
	1800	Acrylic paint face, regular paint liner		380	.084		4.73	2.93		7.66	9.95	
	1900	For 2" thick polystyrene, add					.73			.73	.80	
	2000	22 ga, galv, 2" insulation, baked enamel exterior	G-3	360	.089		9.25	3.09		12.34	15.20	
	2100	P.V.F. exterior finish	"	360	.089		9.75	3.09		12.84	15.75	
	9000	Minimum labor/equipment charge	1 Carp	3	2.667	Job		95		95	157	
900	0010	**WOOD SIDING, BOARDS**										900
	3200	Wood, cedar bevel, A grade, 1/2" x 6"	1 Carp	250	.032	S.F.	3.20	1.14		4.34	5.40	
	3300	1/2" x 8"		275	.029		3.11	1.03		4.14	5.15	
	3500	3/4" x 10", clear grade		300	.027		3.63	.95		4.58	5.55	

07400 | Roofing and Siding Panels

07460 | Siding

	07460 Siding		CREW	DAILY OUTPUT	LABOR-HOURS	UNIT	2006 BARE COSTS MAT.	LABOR	EQUIP.	TOTAL	TOTAL INCL O&P	
900	3600	"B" grade	1 Carp	300	.027	S.F.	2.80	.95		3.75	4.65	900
	3800	Cedar, rough sawn, 1" x 4", A grade, natural		240	.033		2.91	1.18		4.09	5.15	
	3900	Stained		240	.033		3.28	1.18		4.46	5.55	
	4100	1" x 12", board & batten, #3 & Btr., natural		260	.031		2.20	1.09		3.29	4.23	
	4200	Stained		260	.031		2.56	1.09		3.65	4.63	
	4400	1" x 8" channel siding, #3 & Btr., natural		250	.032		2.14	1.14		3.28	4.23	
	4500	Stained		250	.032		2.43	1.14		3.57	4.55	
	4700	Redwood, clear, beveled, vertical grain, 1/2" x 4"		200	.040		3.40	1.42		4.82	6.10	
	4750	1/2" x 6"		225	.036		2.86	1.26		4.12	5.25	
	4800	1/2" x 8"		250	.032		2.31	1.14		3.45	4.42	
	5000	3/4" x 10"		300	.027		3.77	.95		4.72	5.70	
	5200	Channel siding, 1" x 10", B grade		285	.028		2.43	1		3.43	4.32	
	5250	Redwood, T&G boards, B grade, 1" x 4"	2 Carp	300	.053		2.90	1.90		4.80	6.35	
	5270	1" x 8"	"	375	.043		2.50	1.52		4.02	5.25	
	5400	White pine, rough sawn, 1" x 8", natural	1 Carp	275	.029		.73	1.03		1.76	2.51	
	5500	Stained		275	.029		1.08	1.03		2.11	2.90	
	9000	Minimum labor/equipment charge		2	4	Job		142		142	236	
950	0010	**WOOD PRODUCT SIDING**										950
	0030	Lap siding, hardboard, 7/16" x 8", primed										
	0051	Wood grain texture finish	L-2	552	.029	S.F.	1.13	.90		2.03	2.73	
	0100	Panels, 7/16" thick, smooth, textured or grooved, primed	2 Carp	700	.023		.84	.81		1.65	2.27	
	0200	Stained		700	.023		.96	.81		1.77	2.41	
	0700	Particle board, overlaid, 3/8" thick		750	.021		.69	.76		1.45	2.02	
	0900	Plywood, medium density overlaid, 3/8" thick		750	.021		.84	.76		1.60	2.18	
	1000	1/2" thick		700	.023		.95	.81		1.76	2.40	
	1100	3/4" thick		650	.025		1.28	.88		2.16	2.86	
	1600	Texture 1-11, cedar, 5/8" thick, natural		675	.024		2.38	.84		3.22	4.02	
	1700	Factory stained		675	.024		2.21	.84		3.05	3.83	
	1900	Texture 1-11, fir, 5/8" thick, natural		675	.024		1.01	.84		1.85	2.51	
	2000	Factory stained		675	.024		1.08	.84		1.92	2.59	
	2050	Texture 1-11, S.Y.P., 5/8" thick, natural		675	.024		.81	.84		1.65	2.29	
	2100	Factory stained		675	.024		.88	.84		1.72	2.37	
	2200	Rough sawn cedar, 3/8" thick, natural		675	.024		1.08	.84		1.92	2.59	
	2300	Factory stained		675	.024		1.20	.84		2.04	2.72	
	2500	Rough sawn fir, 3/8" thick, natural		675	.024		.58	.84		1.42	2.04	
	2600	Factory stained		675	.024		.65	.84		1.49	2.12	
	2800	Redwood, textured siding, 5/8" thick		675	.024		1.80	.84		2.64	3.38	
	3000	Polyvinyl chloride coated, 3/8" thick		750	.021		.88	.76		1.64	2.23	
	9000	Minimum labor/equipment charge	1 Carp	2	4	Job		142		142	236	

07500 | Membrane Roofing

07510 | Built-Up Bituminous Roofing

	07510 Built-Up Bituminous Roofing		CREW	DAILY OUTPUT	LABOR-HOURS	UNIT	2006 BARE COSTS MAT.	LABOR	EQUIP.	TOTAL	TOTAL INCL O&P	
050	0010	**BUILT- UP ROOFING COMPONENTS**										050
	0012	Asphalt saturated felt, #30, 2 square per roll	1 Rofc	58	.138	Sq.	7.20	4.22		11.42	15.50	
	0200	#15, 4 sq per roll, plain or perforated, not mopped		58	.138		3.41	4.22		7.63	11.35	
	0300	Roll roofing, smooth, #65		15	.533		6.85	16.30		23.15	37	
	0500	#90		15	.533		19.05	16.30		35.35	50.50	
	0520	Mineralized		15	.533		16.55	16.30		32.85	47.50	

		07510 | Built-Up Bituminous Roofing	CREW	DAILY OUTPUT	LABOR-HOURS	UNIT	2006 BARE COSTS MAT.	LABOR	EQUIP.	TOTAL	TOTAL INCL O&P
050	0540	D.C. (Double coverage), 19" selvage edge	1 Rofc	10	.800	Sq.	32	24.50		56.50	79
	0580	Adhesive (lap cement)				Gal.	3.97			3.97	4.37
	0600	Steep, flat or dead level asphalt, 10 ton lots, bulk				Ton	280			280	310
	0800	Packaged				"	410			410	450
	9000	Minimum labor/equipment charge	1 Rofc	4	2	Job		61		61	110
300	0010	**BUILT-UP ROOFING**									
	0120	Asphalt flood coat with gravel/slag surfacing, not including									
	0140	Insulation, flashing or wood nailers									
	0200	Asphalt base sheet, 3 plies #15 asphalt felt, mopped	G-1	22	2.545	Sq.	49.50	73	15.30	137.80	202
	0350	On nailable decks		21	2.667		55.50	76	16.05	147.55	216
	0500	4 plies #15 asphalt felt, mopped		20	2.800		71.50	80	16.85	168.35	242
	0550	On nailable decks		19	2.947		65	84.50	17.75	167.25	242
	0700	Coated glass base sheet, 2 plies glass (type IV), mopped		22	2.545		53	73	15.30	141.30	206
	0850	3 plies glass, mopped		20	2.800		63.50	80	16.85	160.35	232
	0950	On nailable decks		19	2.947		60	84.50	17.75	162.25	237
	1100	4 plies glass fiber felt (type IV), mopped		20	2.800		77	80	16.85	173.85	247
	1150	On nailable decks		19	2.947		70	84.50	17.75	172.25	248
	1200	Coated & saturated base sheet, 3 plies #15 asph. felt, mopped		20	2.800		57	80	16.85	153.85	226
	1250	On nailable decks		19	2.947		53.50	84.50	17.75	155.75	230
	1300	4 plies #15 asphalt felt, mopped		22	2.545		66.50	73	15.30	154.80	221
	2000	Asphalt flood coat, smooth surface									
	2200	Asphalt base sheet & 3 plies #15 asphalt felt, mopped	G-1	24	2.333	Sq.	54.50	66.50	14.05	135.05	195
	2400	On nailable decks		23	2.435		51	69.50	14.65	135.15	198
	2600	4 plies #15 asphalt felt, mopped		24	2.333		63.50	66.50	14.05	144.05	205
	2700	On nailable decks		23	2.435		60.50	69.50	14.65	144.65	208
	2900	Coated glass fiber base sheet, mopped, and 2 plies of									
	2910	glass fiber felt (type IV)	G-1	25	2.240	Sq.	48.50	64	13.50	126	183
	3100	On nailable decks		24	2.333		46	66.50	14.05	126.55	186
	3200	3 plies, mopped		23	2.435		59	69.50	14.65	143.15	206
	3300	On nailable decks		22	2.545		55.50	73	15.30	143.80	209
	3800	4 plies glass fiber felt (type IV), mopped		23	2.435		69	69.50	14.65	153.15	217
	3900	On nailable decks		22	2.545		65.50	73	15.30	153.80	220
	4000	Coated & saturated base sheet, 3 plies #15 asph. felt, mopped		24	2.333		52.50	66.50	14.05	133.05	193
	4200	On nailable decks		23	2.435		49.50	69.50	14.65	133.65	195
	4300	4 plies #15 organic felt, mopped		22	2.545		62	73	15.30	150.30	216
	4500	Coal tar pitch with gravel/slag surfacing									
	4600	4 plies #15 tarred felt, mopped	G-1	21	2.667	Sq.	118	76	16.05	210.05	284
	4800	3 plies glass fiber felt (type IV), mopped	"	19	2.947	"	96	84.50	17.75	198.25	277
	5000	Coated glass fiber base sheet, and 2 plies of									
	5010	glass fiber felt, (type IV), mopped	G-1	19	2.947	Sq.	95.50	84.50	17.75	197.75	276
	5300	On nailable decks		18	3.111		85	89	18.75	192.75	274
	5600	4 plies glass fiber felt (type IV), mopped		21	2.667		133	76	16.05	225.05	300
	5800	On nailable decks		20	2.800		123	80	16.85	219.85	298
400	0010	**CANTS**									
	0012	Lumber, treated, 4" x 4" cut diagonally	1 Rofc	325	.025	L.F.	1.38	.75		2.13	2.87
	0100	Foamglass		325	.025		2.15	.75		2.90	3.72
	0300	Mineral or fiber, trapezoidal, 1"x 4" x 48"		325	.025		.17	.75		.92	1.54
	0400	1-1/2" x 5-5/8" x 48"		325	.025		.29	.75		1.04	1.67
	9000	Minimum labor/equipment charge		4	2	Job		61		61	110
700	0010	**ROOFING FELTS**									
	0012	Glass fibered roofing felt, #15, not mopped	1 Rofc	58	.138	Sq.	4.49	4.22		8.71	12.55
	0300	Base sheet, #45, channel vented		58	.138		19.15	4.22		23.37	28.50
	0400	#50, coated		58	.138		10.35	4.22		14.57	19
	0500	Cap, mineral surfaced		58	.138		19.15	4.22		23.37	28.50
	0600	Flashing membrane, #65		16	.500		29.50	15.30		44.80	59.50

07500 | Membrane Roofing

	07510 | Built-Up Bituminous Roofing		CREW	DAILY OUTPUT	LABOR-HOURS	UNIT	2006 BARE COSTS MAT.	LABOR	EQUIP.	TOTAL	TOTAL INCL O&P	
700	0800	Coal tar fibered, #15, no mopping	1 Rofc	58	.138	Sq.	8.70	4.22		12.92	17.15	700
	0900	Asphalt felt, #15, 4 sq per roll, no mopping		58	.138		3.41	4.22		7.63	11.35	
	1100	#30, 2 sq per roll		58	.138		7.20	4.22		11.42	15.50	
	1200	Double coated, #33		58	.138		6.85	4.22		11.07	15.10	
	1400	#40, base sheet		58	.138		9.40	4.22		13.62	17.95	
	1450	Coated and saturated		58	.138		7.50	4.22		11.72	15.85	
	1500	Tarred felt, organic, #15, 4 sq rolls		58	.138		10.85	4.22		15.07	19.50	
	1550	#30, 2 sq roll		58	.138		18.55	4.22		22.77	28	
	1700	Add for mopping above felts, per ply, asphalt, 24 lb per sq	G-1	192	.292		4.92	8.35	1.76	15.03	22.50	
	1800	Coal tar mopping, 30 lb per sq		186	.301		9.25	8.60	1.81	19.66	27.50	
	1900	Flood coat, with asphalt, 60 lb per sq		60	.933		12.30	26.50	5.60	44.40	68	
	2000	With coal tar, 75 lb per sq		56	1		23	28.50	6	57.50	83.50	
	9000	Minimum labor/equipment charge	1 Rofc	4	2	Job		61		61	110	
	07520 | Cold Applied Bituminous Roofing											
200	0010	**COLD APPLIED BUILT UP ROOF**										200
	0020	3 ply system, installation only (components listed below)	G-5	50	.800	Sq.		22	2.93	24.93	43	
	0100	Spunbond poly. fabric, 1.35 oz/SY, 36"W, 10.8 Sq/roll				Ea.	132			132	146	
	0200	49" wide, 14.6 Sq./roll					183			183	201	
	0300	2.10 oz./S.Y., 36" wide, 10.8 Sq./roll					199			199	219	
	0400	49" wide, 14.6 Sq./roll					270			270	297	
	0500	Base & finish coat, 3 gal./Sq., 5 gal./can				Gal.	3.51			3.51	3.86	
	0600	Coating, ceramic granules, 1/2 Sq./bag				Ea.	12.40			12.40	13.65	
	0700	Aluminum, 2 gal./Sq.				Gal.	10			10	11	
	0800	Emulsion, fibered or non-fibered, 4 gal./Sq.				"	4.59			4.59	5.05	
	07530 | Elastomeric Membrane Roofing											
350	0010	**ELASTOMERIC ROOFING**										350
	0100	For Elastomeric waterproofing, see Division 07130-200										
	0110	Acrylic rubber, fluid applied, 20 mils thick	G-5	2000	.020	S.F.	1.80	.56	.07	2.43	3.06	
	0120	50 mils, reinforced		1200	.033		2.80	.93	.12	3.85	4.87	
	0130	For walking surface, add		900	.044		.85	1.23	.16	2.24	3.34	
	0300	Hypalon neoprene, fluid applied, 20 mil thick, not-reinforced	G-1	1135	.049		2.05	1.41	.30	3.76	5.10	
	0600	Non-woven polyester, reinforced		960	.058		2.07	1.67	.35	4.09	5.65	
	0700	5 coat neoprene deck, 60 mil thick, under 10,000 SF		325	.172		4.56	4.93	1.04	10.53	15	
	0900	Over 10,000 SF		625	.090		4.25	2.56	.54	7.35	9.85	
	1300	Vinyl plastic traffic deck, sprayed, 2 to 4 mils thick		625	.090		1.33	2.56	.54	4.43	6.65	
	1500	Vinyl and neoprene membrane traffic deck		1550	.036		1.41	1.03	.22	2.66	3.65	
	9000	Minimum labor/equipment charge	1 Rofc	2	4	Job		122		122	220	
800	0010	**SINGLE-PLY MEMBRANE**										800
	0800	Chlorosulfonated polyethylene-hypalon (CSPE), 45 mils,										
	0900	0.29 P.S.F., fully adhered	G-5	26	1.538	Sq.	123	42.50	5.65	171.15	219	
	1100	Loose-laid & ballasted with stone (10 P.S.F.)		51	.784		130	22	2.87	154.87	185	
	1200	Mechanically attached		35	1.143		125	32	4.18	161.18	200	
	1300	Plates with adhesive attachment		35	1.143		123	32	4.18	159.18	198	
	3500	Ethylene propylene diene monomer (EPDM), 45 mils, 0.28 P.S.F.										
	3600	Loose-laid & ballasted with stone (10 P.S.F.)	G-5	51	.784	Sq.	61	22	2.87	85.87	109	
	3700	Mechanically attached		35	1.143		53.50	32	4.18	89.68	120	
	3800	Fully adhered with adhesive		26	1.538		77	42.50	5.65	125.15	168	
	4500	60 mils, 0.40 P.S.F.										
	4600	Loose-laid & ballasted with stone (10 P.S.F.)	G-5	51	.784	Sq.	75.50	22	2.87	100.37	125	
	4700	Mechanically attached		35	1.143		66.50	32	4.18	102.68	135	
	4800	Fully adhered with adhesive		26	1.538		90	42.50	5.65	138.15	183	
	4810	45 mil, .28 PSF, membrane only					31.50			31.50	34.50	
	4820	60 mil, .40 PSF, membrane only					43.50			43.50	48	

07500 | Membrane Roofing

		07530 | Elastomeric Membrane Roofing	CREW	DAILY OUTPUT	LABOR-HOURS	UNIT	2006 BARE COSTS MAT.	LABOR	EQUIP.	TOTAL	TOTAL INCL O&P	
800	4850	Seam tape for membrane, 3″ x 100′ roll				Ea.	41			41	45	800
	4900	Batten strips, 10′ sections					2.95			2.95	3.25	
	4910	Cover tape for batten strips, 6″ x 100′ roll					144			144	158	
	4930	Plate anchors				M	73.50			73.50	80.50	
	4970	Adhesive for fully adhered systems, 60 S.F./gal.				Gal.	14.60			14.60	16.05	
	7500	Polyisobutylene (PIB), 100 mils, 0.57 P.S.F.										
	7600	Loose-laid & ballasted with stone/gravel (10 P.S.F.)	G-5	51	.784	Sq.	128	22	2.87	152.87	183	
	7700	Partially adhered with adhesive		35	1.143		160	32	4.18	196.18	238	
	7800	Hot asphalt attachment		35	1.143		153	32	4.18	189.18	230	
	7900	Fully adhered with contact cement		26	1.538		165	42.50	5.65	213.15	265	
		07540 | PVC Single Ply Membrane										
800	0010	**POLYVINYL CHLORIDE ROOFING**										800
	8200	Heat welded seams										
	8700	Reinforced, 48 mils, 0.33 P.S.F.										
	8750	Loose-laid & ballasted with stone/gravel (12 P.S.F.)	G-5	51	.784	Sq.	97.50	22	2.87	122.37	149	
	8800	Mechanically attached		35	1.143		88.50	32	4.18	124.68	159	
	8850	Fully adhered with adhesive		26	1.538		119	42.50	5.65	167.15	213	
	8860	Reinforced, 60 mils, .40 P.S.F.										
	8870	Loose-laid & ballasted with stone/gravel (12 P.S.F.)	G-5	51	.784	Sq.	90	22	2.87	114.87	141	
	8880	Mechanically attached		35	1.143		81	32	4.18	117.18	151	
	8890	Fully adhered with adhesive		26	1.538		111	42.50	5.65	159.15	205	
		07550 | Modified Bit. Membrane Roofing										
500	0010	**MODIFIED BITUMEN ROOFING** R075213-30										500
	0020	Base sheet, #15 glass fiber felt, nailed to deck	1 Rofc	58	.138	Sq.	5.25	4.22		9.47	13.35	
	0030	Spot mopped to deck	G-1	295	.190		6.95	5.45	1.14	13.54	18.65	
	0040	Fully mopped to deck	″	192	.292		9.40	8.35	1.76	19.51	27.50	
	0050	#15 organic felt, nailed to deck	1 Rofc	58	.138		4.15	4.22		8.37	12.15	
	0060	Spot mopped to deck	G-1	295	.190		5.85	5.45	1.14	12.44	17.45	
	0070	Fully mopped to deck	″	192	.292		8.35	8.35	1.76	18.46	26	
	0080	SBS modified, granule surf cap sheet, poly rein., mopped										
	0600	150 mils	G-1	2000	.028	S.F.	.49	.80	.17	1.46	2.17	
	1100	160 mils		2000	.028		.71	.80	.17	1.68	2.41	
	1500	Glass fiber reinforced, mopped, 160 mils		2000	.028		.45	.80	.17	1.42	2.13	
	1600	Smooth surface cap sheet, mopped, 145 mils		2100	.027		.45	.76	.16	1.37	2.05	
	1700	Smooth surface flashing, 145 mils		1260	.044		.45	1.27	.27	1.99	3.07	
	1800	150 mils		1260	.044		.44	1.27	.27	1.98	3.05	
	1900	Granular surface flashing, 150 mils		1260	.044		.49	1.27	.27	2.03	3.11	
	2000	160 mils		1260	.044		.71	1.27	.27	2.25	3.35	
	2100	APP mod., smooth surf. cap sheet, poly. reinf., torched, 160 mils	G-5	2100	.019		.44	.53	.07	1.04	1.51	
	2150	170 mils		2100	.019		.50	.53	.07	1.10	1.58	
	2200	Granule surface cap sheet, poly. reinf., torched, 180 mils		2000	.020		.55	.56	.07	1.18	1.69	
	2250	Smooth surface flashing, torched, 160 mils		1260	.032		.44	.88	.12	1.44	2.20	
	2300	170 mils		1260	.032		.50	.88	.12	1.50	2.27	
	2350	Granule surface flashing, torched, 180 mils		1260	.032		.55	.88	.12	1.55	2.33	
	2400	Fibrated aluminum coating	1 Rofc	3800	.002		.10	.06		.16	.23	
350	0010	**FOAMED COATED ROOFING**										350
	1600	Polyurethane spray-on with 20 mil silicone rubber coating applied										
	1700	1″ thick, R7, minimum	G-2A	875	.027	S.F.	1.34	.74	.13	2.21	2.91	
	1800	Maximim		805	.030		1.80	.80	.15	2.75	3.54	
	1900	2″ thick, R14, minimum		685	.035		1.90	.94	.17	3.01	3.93	
	2000	Maximum		575	.042		2.41	1.12	.20	3.73	4.83	
	2100	3″ thick, R21, minimum		500	.048		3.25	1.29	.23	4.77	6.10	
	2200	Maximum		440	.055		4.16	1.46	.27	5.89	7.45	

07500 | Membrane Roofing

	07580 | Roll Roofing		CREW	DAILY OUTPUT	LABOR-HOURS	UNIT	2006 BARE COSTS MAT.	LABOR	EQUIP.	TOTAL	TOTAL INCL O&P	
200	0010	**ROLL ROOFING**										200
	0100	Asphalt, mineral surface										
	0200	1 ply #15 organic felt, 1 ply mineral surfaced										
	0300	Selvage roofing, lap 19", nailed & mopped	G-1	27	2.074	Sq.	42	59.50	12.50	114	167	
	0400	3 plies glass fiber felt (type IV), 1 ply mineral surfaced										
	0500	Selvage roofing, lapped 19", mopped	G-1	25	2.240	Sq.	65	64	13.50	142.50	201	
	0600	Coated glass fiber base sheet, 2 plies of glass fiber										
	0700	Felt (type IV), 1 ply mineral surfaced selvage										
	0800	Roofing, lapped 19", mopped	G-1	25	2.240	Sq.	71	64	13.50	148.50	208	
	0900	On nailable decks	"	24	2.333	"	66	66.50	14.05	146.55	208	
	1000	3 plies glass fiber felt (type III), 1 ply mineral surfaced										
	1100	Selvage roofing, lapped 19", mopped	G-1	25	2.240	Sq.	65	64	13.50	142.50	201	

	07590 | Roof Maintenance and Repairs		CREW	DAILY OUTPUT	LABOR-HOURS	UNIT	MAT.	LABOR	EQUIP.	TOTAL	TOTAL INCL O&P	
300	0010	**ROOF COATINGS**										300
	0012	Asphalt, brush grade, material only				Gal.	2.95			2.95	3.25	
	0200	Asphalt base, fibered aluminum coating					9.80			9.80	10.80	
	0300	Asphalt primer, 5 gallon				↓	4.21			4.21	4.63	
	0600	Coal tar pitch, 200 lb. barrels				Ton	615			615	680	
	0700	Tar roof cement, 5 gal. lots				Gal.	6.50			6.50	7.15	
	0800	Glass fibered roof & patching cement, 5 gallon				"	4.64			4.64	5.10	
	0900	Reinforcing glass membrane, 450 S.F./roll				Ea.	45.50			45.50	50	
	1000	Neoprene roof coating, 5 gal, 2 gal/sq				Gal.	22			22	24	
	1100	Roof patch & flashing cement, 5 gallon					19.15			19.15	21	
	1200	Roof resaturant, glass fibered, 3 gal/sq					7.30			7.30	8.05	
	1300	Mineral rubber, 3 gal/sq				↓	4.71			4.71	5.20	

07600 | Flashing and Sheet Metal

	07610 | Sheet Metal Roofing		CREW	DAILY OUTPUT	LABOR-HOURS	UNIT	2006 BARE COSTS MAT.	LABOR	EQUIP.	TOTAL	TOTAL INCL O&P	
300	0010	**COPPER ROOFING**, field fabricated										300
	0012	Batten seam, over 10 sq, 16 oz, 130 lb/sq	1 Shee	1.10	7.273	Sq.	440	305		745	975	
	0200	18 oz, 145 lb per sq		1	8		490	335		825	1,075	
	0300	20 oz, 160 lb per sq		1	8		545	335		880	1,125	
	0400	Standing seam, over 10 squares, 16 oz, 125 lb per sq		1.30	6.154		425	259		684	880	
	0600	18 oz, 140 lb per sq		1.20	6.667		475	281		756	965	
	0700	20 oz, 150 lb per sq		1.10	7.273		510	305		815	1,050	
	0900	Flat seam, over 10 squares, 16 oz, 115 lb per sq		1.20	6.667		390	281		671	875	
	1000	20 oz, 145 lb per sq	↓	1.10	7.273		490	305		795	1,025	
	1200	For abnormal conditions or small areas, add					25%	100%				
	1300	For lead-coated copper, add				↓	25%					
	9000	Minimum labor/equipment charge	1 Shee	2	4	Job		169		169	268	
500	0010	**LEAD ROOFING**, field fabricated										500
	0020	5 lb. per SF, batten seam	1 Shee	1.20	6.667	Sq.	395	281		676	880	
	0100	Flat seam		1.30	6.154	"	395	259		654	850	
	9000	Minimum labor/equipment charge	↓	2	4	Job		169		169	268	
700	0010	**STAINLESS STEEL ROOFING**										700
	0020	Type 304, batten seam, 28 gauge	1 Shee	1.20	6.667	Sq.	400	281		681	885	
	0100	26 gauge	"	1.15	6.957		495	293		788	1,000	
	0200	For standing seam construction, deduct				↓	2%					

07600 | Flashing and Sheet Metal

	07610 | Sheet Metal Roofing		CREW	DAILY OUTPUT	LABOR-HOURS	UNIT	2006 BARE COSTS MAT.	LABOR	EQUIP.	TOTAL	TOTAL INCL O&P	
700	0500	For flat seam construction, deduct				Sq.	3%					700
	0800	For lead or terne coated stainless, 28 gauge, add					91			91	100	
	0900	For 26 gauge, add					121			121	133	
	9000	Minimum labor/equipment charge	1 Shee	2.75	2.909	Job		123		123	195	
900	0010	**ZINC / COPPER ALLOY ROOFING,** field fabricated										900
	0012	Batten seam, .020 thick	1 Shee	1.20	6.667	Sq.	560	281		841	1,075	
	0100	.027" thick		1.15	6.957		680	293		973	1,200	
	0300	.032" thick		1.10	7.273		765	305		1,070	1,325	
	0400	.040" thick		1.05	7.619		900	320		1,220	1,500	
	0600	For standing seam construction, deduct					2%					
	0700	For flat seam construction, deduct					3%					
	9000	Minimum labor/equipment charge	1 Shee	2.75	2.909	Job		123		123	195	
	07650 | Flexible Flashing											
600	0010	**FLASHING,** including up to 4 bends										600
	0020	Aluminum, mill finish, .013" thick	1 Rofc	145	.055	S.F.	.38	1.69		2.07	3.45	
	0030	.016" thick		145	.055		.56	1.69		2.25	3.65	
	0060	.019" thick		145	.055		.71	1.69		2.40	3.81	
	0100	.032" thick		145	.055		1.17	1.69		2.86	4.32	
	0200	.040" thick		145	.055		1.58	1.69		3.27	4.77	
	0300	.050" thick		145	.055		2.01	1.69		3.70	5.25	
	0325	Mill finish 5" x 7" step flashing, .016" thick		1920	.004	Ea.	.12	.13		.25	.36	
	0350	Mill finish 12" x 12" step flashing, .016" thick		1600	.005	"	.48	.15		.63	.81	
	0400	Painted finish, add				S.F.	.26			.26	.29	
	0500	Fabric-backed 2 sides, .004" thick	1 Rofc	330	.024		1.01	.74		1.75	2.44	
	0700	.005" thick		330	.024		1.19	.74		1.93	2.64	
	0750	Mastic-backed, self adhesive		460	.017		2.46	.53		2.99	3.67	
	0800	Mastic-coated 2 sides, .004" thick		330	.024		1.01	.74		1.75	2.44	
	1000	.005" thick		330	.024		1.19	.74		1.93	2.64	
	1100	.016" thick		330	.024		1.31	.74		2.05	2.77	
	1300	Asphalt flashing cement, 5 gallon				Gal.	4.77			4.77	5.25	
	1600	Copper, 16 oz, sheets, under 1000 lbs.	1 Rofc	115	.070	S.F.	2.76	2.13		4.89	6.85	
	1700	Over 4000 lbs.		155	.052		3.06	1.58		4.64	6.20	
	1900	20 oz sheets, under 1000 lbs.		110	.073		4.10	2.23		6.33	8.50	
	2000	Over 4000 lbs.		145	.055		3.80	1.69		5.49	7.20	
	2200	24 oz sheets, under 1000 lbs.		105	.076		4.92	2.33		7.25	9.60	
	2300	Over 4000 lbs.		135	.059		4.56	1.81		6.37	8.25	
	2500	32 oz sheets, under 1000 lbs.		100	.080		6.55	2.45		9	11.60	
	2600	Over 4000 lbs.		130	.062		6.10	1.88		7.98	10.10	
	2700	W shape for valleys, 16 oz, 24" wide		100	.080	L.F.	6.20	2.45		8.65	11.20	
	2800	Copper, paperbacked 1 side, 2 oz		330	.024	S.F.	.90	.74		1.64	2.32	
	2900	3 oz		330	.024		1.18	.74		1.92	2.63	
	3100	Paperbacked 2 sides, 2 oz		330	.024		.91	.74		1.65	2.33	
	3150	3 oz		330	.024		1.17	.74		1.91	2.62	
	3200	5 oz		330	.024		1.75	.74		2.49	3.26	
	3250	7 oz		330	.024		2.86	.74		3.60	4.48	
	3400	Mastic-backed 2 sides, copper, 2 oz		330	.024		1.07	.74		1.81	2.51	
	3500	3 oz		330	.024		1.32	.74		2.06	2.78	
	3700	5 oz		330	.024		1.94	.74		2.68	3.46	
	3800	Fabric-backed 2 sides, copper, 2 oz		330	.024		1.14	.74		1.88	2.58	
	4000	3 oz		330	.024		1.47	.74		2.21	2.95	
	4100	5 oz		330	.024		2	.74		2.74	3.53	
	4300	Copper-clad stainless steel, .015" thick, under 500 lbs.		115	.070		3.32	2.13		5.45	7.50	
	4400	Over 2000 lbs.		155	.052		3.20	1.58		4.78	6.35	
	4600	.018" thick, under 500 lbs.		100	.080		4.39	2.45		6.84	9.25	
	4700	Over 2000 lbs.		145	.055		3.21	1.69		4.90	6.55	

07650 | Flexible Flashing

			CREW	DAILY OUTPUT	LABOR-HOURS	UNIT	2006 BARE COSTS MAT.	LABOR	EQUIP.	TOTAL	TOTAL INCL O&P	
600	4900	Fabric, asphalt-saturated cotton, specification grade	1 Rofc	35	.229	S.Y.	1.93	7		8.93	14.65	600
	5000	Utility grade		35	.229		1.22	7		8.22	13.90	
	5200	Open-mesh fabric, saturated, 40 oz per S.Y.		35	.229		1.35	7		8.35	14.05	
	5300	Close-mesh fabric, saturated, 17 oz per S.Y.		35	.229		1.42	7		8.42	14.10	
	5500	Fiberglass, resin-coated		35	.229		1.16	7		8.16	13.85	
	5600	Asphalt-coated, 40 oz per S.Y.		35	.229		7.90	7		14.90	21.50	
	5800	Lead, 2.5 lb. per SF, up to 12" wide		135	.059	S.F.	3.03	1.81		4.84	6.60	
	5900	Over 12" wide		135	.059		3.69	1.81		5.50	7.30	
	6100	Lead-coated copper, fabric-backed, 2 oz		330	.024		1.54	.74		2.28	3.02	
	6200	5 oz		330	.024		1.77	.74		2.51	3.28	
	6400	Mastic-backed 2 sides, 2 oz		330	.024		1.20	.74		1.94	2.65	
	6500	5 oz		330	.024		1.49	.74		2.23	2.97	
	6700	Paperbacked 1 side, 2 oz		330	.024		1.04	.74		1.78	2.47	
	6800	3 oz		330	.024		1.22	.74		1.96	2.67	
	7000	Paperbacked 2 sides, 2 oz		330	.024		1.07	.74		1.81	2.51	
	7100	5 oz		330	.024		1.75	.74		2.49	3.26	
	7300	Polyvinyl chloride, black, .010" thick		285	.028		.17	.86		1.03	1.73	
	7400	.020" thick		285	.028		.24	.86		1.10	1.80	
	7600	.030" thick		285	.028		.32	.86		1.18	1.89	
	7700	.056" thick		285	.028		.76	.86		1.62	2.38	
	7900	Black or white for exposed roofs, .060" thick		285	.028		1.66	.86		2.52	3.37	
	8060	PVC tape, 5" x 45 mils, for joint covers, 100 L.F./roll				Ea.	85			85	93.50	
	8100	Rubber, butyl, 1/32" thick	1 Rofc	285	.028	S.F.	.75	.86		1.61	2.37	
	8200	1/16" thick		285	.028		1.12	.86		1.98	2.77	
	8300	Neoprene, cured, 1/16" thick		285	.028		1.58	.86		2.44	3.28	
	8400	1/8" thick		285	.028		3.20	.86		4.06	5.05	
	8500	Shower pan, bituminous membrane, 7 oz		155	.052		1.16	1.58		2.74	4.12	
	8550	3 ply copper and fabric, 3 oz		155	.052		1.74	1.58		3.32	4.75	
	8600	7 oz		155	.052		3.61	1.58		5.19	6.80	
	8650	Copper, 16 oz		100	.080		3.33	2.45		5.78	8.05	
	8700	Lead on copper and fabric, 5 oz		155	.052		1.77	1.58		3.35	4.79	
	8800	7 oz		155	.052		3.14	1.58		4.72	6.30	
	8850	Polyvinyl chloride, .030" thick		160	.050		.32	1.53		1.85	3.10	
	8900	Stainless steel sheets, 32 ga, .010" thick		155	.052		2.70	1.58		4.28	5.80	
	9000	28 ga, .015" thick		155	.052		3.35	1.58		4.93	6.55	
	9100	26 ga, .018" thick		155	.052		4.06	1.58		5.64	7.30	
	9200	24 ga, .025" thick		155	.052		5.30	1.58		6.88	8.65	
	9290	For mechanically keyed flashing, add					40%					
	9300	Stainless steel, paperbacked 2 sides, .005" thick	1 Rofc	330	.024	S.F.	2.38	.74		3.12	3.95	
	9400	Terne coated stainless steel, .015" thick, 28 ga		155	.052		5.05	1.58		6.63	8.40	
	9500	.018" thick, 26 ga		155	.052		5.70	1.58		7.28	9.15	
	9600	Zinc and copper alloy (brass), .020" thick		155	.052		3.55	1.58		5.13	6.75	
	9700	.027" thick		155	.052		4.76	1.58		6.34	8.10	
	9800	.032" thick		155	.052		5.55	1.58		7.13	8.95	
	9900	.040" thick		155	.052		6.80	1.58		8.38	10.30	
	9950	Minimum labor/equipment charge		3	2.667	Job		81.50		81.50	147	

07710 | Manufactured Roof Specialties

		07710 Manufactured Roof Specialties	CREW	DAILY OUTPUT	LABOR-HOURS	UNIT	2006 BARE COSTS MAT.	LABOR	EQUIP.	TOTAL	TOTAL INCL O&P	
400	0010	DOWNSPOUTS										400
	0020	Aluminum 2" x 3", .020" thick, embossed	1 Shee	190	.042	L.F.	.85	1.77		2.62	3.76	
	0100	Enameled		190	.042		1.26	1.77		3.03	4.21	
	0300	Enameled, .024" thick, 2" x 3"		180	.044		1.77	1.87		3.64	4.93	
	0400	3" x 4"		140	.057		2.50	2.41		4.91	6.60	
	0600	Round, corrugated aluminum, 3" diameter, .020" thick		190	.042		1.36	1.77		3.13	4.32	
	0700	4" diameter, .025" thick		140	.057		2.03	2.41		4.44	6.05	
	0900	Wire strainer, round, 2" diameter		155	.052	Ea.	1.85	2.18		4.03	5.50	
	1000	4" diameter		155	.052		2	2.18		4.18	5.65	
	1200	Rectangular, perforated, 2" x 3"		145	.055		2.25	2.33		4.58	6.20	
	1300	3" x 4"		145	.055		3.25	2.33		5.58	7.30	
	1500	Copper, round, 16 oz., stock, 2" diameter		190	.042	L.F.	4.55	1.77		6.32	7.80	
	1600	3" diameter		190	.042		4.40	1.77		6.17	7.65	
	1800	4" diameter		145	.055		5.20	2.33		7.53	9.40	
	1900	5" diameter		130	.062		7.70	2.59		10.29	12.60	
	2100	Rectangular, corrugated copper, stock, 2" x 3"		190	.042		3.88	1.77		5.65	7.10	
	2200	3" x 4"		145	.055		4.60	2.33		6.93	8.75	
	2400	Rectangular, plain copper, stock, 2" x 3"		190	.042		5.25	1.77		7.02	8.60	
	2500	3" x 4"		145	.055		5.95	2.33		8.28	10.25	
	2700	Wire strainers, rectangular, 2" x 3"		145	.055	Ea.	2.85	2.33		5.18	6.85	
	2800	3" x 4"		145	.055		4.45	2.33		6.78	8.60	
	3000	Round, 2" diameter		145	.055		2.65	2.33		4.98	6.60	
	3100	3" diameter		145	.055		3.70	2.33		6.03	7.75	
	3300	4" diameter		145	.055		5.75	2.33		8.08	10.05	
	3400	5" diameter		115	.070		8.30	2.93		11.23	13.80	
	3600	Lead-coated copper, round, stock, 2" diameter		190	.042	L.F.	5.50	1.77		7.27	8.85	
	3700	3" diameter		190	.042		7.35	1.77		9.12	10.90	
	3900	4" diameter		145	.055		9.70	2.33		12.03	14.40	
	4000	5" diameter, corrugated		130	.062		13.80	2.59		16.39	19.30	
	4200	6" diameter, corrugated		105	.076		14.85	3.21		18.06	21.50	
	4300	Rectangular, corrugated, stock, 2" x 3"		190	.042		7.80	1.77		9.57	11.40	
	4500	Plain, stock, 2" x 3"		190	.042		7.35	1.77		9.12	10.90	
	4600	3" x 4"		145	.055		7.95	2.33		10.28	12.45	
	4800	Steel, galvanized, round, corrugated, 2" or 3" diam, 28 ga		190	.042		1.34	1.77		3.11	4.29	
	4900	4" diameter, 28 gauge		145	.055		1.40	2.33		3.73	5.25	
	5100	5" diameter, 28 gauge		130	.062		2.34	2.59		4.93	6.70	
	5200	26 gauge		130	.062		2.35	2.59		4.94	6.70	
	5400	6" diameter, 28 gauge		105	.076		3.32	3.21		6.53	8.75	
	5500	26 gauge		105	.076		3.25	3.21		6.46	8.70	
	5700	Rectangular, corrugated, 28 gauge, 2" x 3"		190	.042		1.26	1.77		3.03	4.21	
	5800	3" x 4"		145	.055		1.26	2.33		3.59	5.10	
	6000	Rectangular, plain, 28 gauge, galvanized, 2" x 3"		190	.042		1.78	1.77		3.55	4.78	
	6100	3" x 4"		145	.055		1.58	2.33		3.91	5.45	
	6300	Epoxy painted, 24 gauge, corrugated, 2" x 3"		190	.042		1.55	1.77		3.32	4.53	
	6400	3" x 4"		145	.055		1.78	2.33		4.11	5.65	
	6600	Wire strainers, rectangular, 2" x 3"		145	.055	Ea.	1.57	2.33		3.90	5.45	
	6700	3" x 4"		145	.055		3.13	2.33		5.46	7.15	
	6900	Round strainers, 2" or 3" diameter		145	.055		1.57	2.33		3.90	5.45	
	7000	4" diameter		145	.055		3.13	2.33		5.46	7.15	
	7200	5" diameter		145	.055		2.20	2.33		4.53	6.10	
	7300	6" diameter		115	.070		2.63	2.93		5.56	7.55	
	7500	Steel pipe, black, extra heavy, 4" diameter		20	.400	L.F.	18.15	16.85		35	47	
	7600	6" diameter		18	.444		38.50	18.75		57.25	72.50	
	7800	Stainless steel tubing, schedule 5, 2" x 3" or 3" diameter		190	.042		20.50	1.77		22.27	25.50	
	7900	3" x 4" or 4" diameter		145	.055		26	2.33		28.33	32	
	8100	4" x 5" or 5" diameter		135	.059		53.50	2.50		56	63	

07700 | Roof Specialties and Accessories

		07710 | Manufactured Roof Specialties	CREW	DAILY OUTPUT	LABOR-HOURS	UNIT	2006 BARE COSTS MAT.	LABOR	EQUIP.	TOTAL	TOTAL INCL O&P	
400	8200	Vinyl, rectangular, 2" x 3"	1 Shee	210	.038	L.F.	.72	1.61		2.33	3.34	400
	8300	Round, 2-1/2"		220	.036		.72	1.53		2.25	3.23	
	9000	Minimum labor/equipment charge		4	2	Job		84.50		84.50	134	
450	0010	**DRIP EDGE**										450
	0020	Aluminum, .016" thick, 5" wide, mill finish	1 Carp	400	.020	L.F.	.30	.71		1.01	1.51	
	0100	White finish		400	.020		.35	.71		1.06	1.57	
	0200	8" wide, mill finish		400	.020		.42	.71		1.13	1.64	
	0300	Ice belt, 28" wide, mill finish		100	.080		3.76	2.84		6.60	8.85	
	0310	Vented, mill finish		400	.020		1.32	.71		2.03	2.63	
	0320	Painted finish		400	.020		1.43	.71		2.14	2.75	
	0400	Galvanized, 5" wide		400	.020		.40	.71		1.11	1.62	
	0500	8" wide, mill finish		400	.020		.49	.71		1.20	1.72	
	0510	Rake edge, aluminum, 1-1/2" x 1-1/2"		400	.020		.15	.71		.86	1.35	
	0520	3-1/2" x 1-1/2"		400	.020		.19	.71		.90	1.39	
	9000	Minimum labor/equipment charge		4	2	Job		71		71	118	
500	0010	**DOWNSPOUT ELBOWS**										500
	0020	Aluminum, 2" x 3", embossed	1 Shee	100	.080	Ea.	3.30	3.37		6.67	9	
	0100	Enameled		100	.080		3.30	3.37		6.67	9	
	0200	3" x 4", .025" thick, embossed		100	.080		3.30	3.37		6.67	9	
	0300	Enameled		100	.080		3.20	3.37		6.57	8.85	
	0400	Round corrugated, 3", embossed, .020" thick		100	.080		1.98	3.37		5.35	7.55	
	0500	4", .025" thick		100	.080		4.29	3.37		7.66	10.05	
	0600	Copper, 16 oz. round, 2" diameter		100	.080		11.15	3.37		14.52	17.65	
	0700	3" diameter		100	.080		8.15	3.37		11.52	14.35	
	0800	4" diameter		100	.080		9.50	3.37		12.87	15.80	
	1000	2" x 3" corrugated		100	.080		6.15	3.37		9.52	12.10	
	1100	3" x 4" corrugated		100	.080		7.20	3.37		10.57	13.30	
	1300	Vinyl, 2-1/2" diameter, 45° or 75°		100	.080		2	3.37		5.37	7.55	
	1400	Tee Y junction		75	.107		8.50	4.50		13	16.50	
	9000	Minimum labor/equipment charge		4	2	Job		84.50		84.50	134	
550	0010	**GRAVEL STOP**										550
	0020	Aluminum, .050" thick, 4" face height, mill finish	1 Shee	145	.055	L.F.	4.28	2.33		6.61	8.40	
	0080	Duranodic finish		145	.055		4.13	2.33		6.46	8.25	
	0100	Painted		145	.055		4.77	2.33		7.10	8.95	
	0300	6" face height		135	.059		4.59	2.50		7.09	9	
	0350	Duranodic finish		135	.059		4.87	2.50		7.37	9.30	
	0400	Painted		135	.059		5.65	2.50		8.15	10.20	
	0600	8" face height		125	.064		5.75	2.70		8.45	10.60	
	0650	Duranodic finish		125	.064		5.60	2.70		8.30	10.45	
	0700	Painted		125	.064		5.70	2.70		8.40	10.55	
	0900	12" face height, .080 thick, 2 piece		100	.080		7.65	3.37		11.02	13.75	
	0950	Duranodic finish		100	.080		7.05	3.37		10.42	13.10	
	1000	Painted		100	.080		8.30	3.37		11.67	14.50	
	1350	Galv steel, 24 ga., 4" leg, plain, with continuous cleat, 4" face		145	.055		1.68	2.33		4.01	5.55	
	1360	6" face height		145	.055		2.54	2.33		4.87	6.50	
	1500	Polyvinyl chloride, 6" face height		135	.059		3.67	2.50		6.17	8	
	1600	9" face height		125	.064		4.33	2.70		7.03	9.05	
	1800	Stainless steel, 24 ga., 6" face height		135	.059		8	2.50		10.50	12.75	
	1900	12" face height		100	.080		16.80	3.37		20.17	24	
	2100	20 ga., 6" face height		135	.059		9.05	2.50		11.55	13.95	
	2200	12" face height		100	.080		19.15	3.37		22.52	26.50	
	9000	Minimum labor/equipment charge		3.50	2.286	Job		96.50		96.50	153	
650	0010	**GUTTERS**										650
	0012	Aluminum, stock units, 5" K type, .027" thick, plain	1 Shee	120	.067	L.F.	1.33	2.81		4.14	5.95	

	07710 | Manufactured Roof Specialties		CREW	DAILY OUTPUT	LABOR-HOURS	UNIT	2006 BARE COSTS MAT.	LABOR	EQUIP.	TOTAL	TOTAL INCL O&P	
650	0100	Enameled	1 Shee	120	.067	L.F.	1.25	2.81		4.06	5.85	650
	0300	5″ K type type, .032″ thick, plain 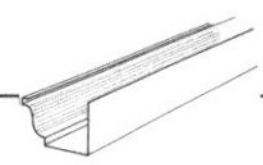		120	.067		2.35	2.81		5.16	7.05	
	0400	Enameled		120	.067		2.36	2.81		5.17	7.05	
	0700	Copper, half round, 16 oz, stock units, 4″ wide		120	.067		4.10	2.81		6.91	9	
	0900	5″ wide		120	.067		5.25	2.81		8.06	10.20	
	1000	6″ wide		115	.070		5.70	2.93		8.63	10.90	
	1200	K type, 16 oz, stock, 4″ wide		120	.067		5.05	2.81		7.86	10	
	1300	5″ wide		120	.067		5.50	2.81		8.31	10.50	
	1500	Lead coated copper, half round, stock, 4″ wide		120	.067		7.50	2.81		10.31	12.70	
	1600	6″ wide		115	.070		11.40	2.93		14.33	17.15	
	1800	K type, stock, 4″ wide		120	.067		8.10	2.81		10.91	13.35	
	1900	5″ wide		120	.067		8.90	2.81		11.71	14.20	
	2100	Stainless steel, half round or box, stock, 4″ wide		120	.067		4.65	2.81		7.46	9.55	
	2200	5″ wide		120	.067		5	2.81		7.81	9.95	
	2400	Steel, galv, half round or box, 28 ga, 5″ wide, plain		120	.067		1.18	2.81		3.99	5.75	
	2500	Enameled		120	.067		1.12	2.81		3.93	5.70	
	2700	26 ga, stock, 5″ wide		120	.067		.95	2.81		3.76	5.50	
	2800	6″ wide	↓	120	.067		1.60	2.81		4.41	6.25	
	3000	Vinyl, O.G., 4″ wide	1 Carp	110	.073		.85	2.59		3.44	5.20	
	3100	5″ wide		110	.073		1	2.59		3.59	5.40	
	3200	4″ half round, stock units	↓	110	.073	↓	.68	2.59		3.27	5.05	
	3250	Joint connectors				Ea.	1.36			1.36	1.50	
	3300	Wood, clear treated cedar, fir or hemlock, 3″ x 4″ 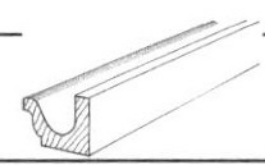	1 Carp	100	.080	L.F.	6.30	2.84		9.14	11.60	
	3400	4″ x 5″	″	100	.080	″	7.30	2.84		10.14	12.70	
	9000	Minimum labor/equipment charge	1 Shee	3.75	2.133	Job		90		90	143	
700	0010	**GUTTER GUARD**										700
	0020	6″ wide strip, aluminum mesh	1 Carp	500	.016	L.F.	.37	.57		.94	1.35	
	0100	Vinyl mesh		500	.016	″	.40	.57		.97	1.38	
	9000	Minimum labor/equipment charge	↓	4	2	Job		71		71	118	
750	0010	**REGLET**										750
	0020	Aluminum, .025″ thick, in concrete parapet	1 Carp	225	.036	L.F.	1.02	1.26		2.28	3.21	
	0100	Copper, 10 oz.		225	.036		1.77	1.26		3.03	4.04	
	0300	16 oz.		225	.036		2.35	1.26		3.61	4.68	
	0400	Galvanized steel, 24 gauge		225	.036		.75	1.26		2.01	2.92	
	0600	Stainless steel, .020″ thick		225	.036		1.74	1.26		3	4	
	0700	Zinc and copper alloy, 20 oz.	↓	225	.036		1.95	1.26		3.21	4.24	
	0900	Counter flashing for above, 12″ wide, .032″ aluminum	1 Shee	150	.053		1.32	2.25		3.57	5.05	
	1000	Copper, 10 oz.		150	.053		3.70	2.25		5.95	7.65	
	1200	16 oz.		150	.053		4.11	2.25		6.36	8.10	
	1300	Galvanized steel, .020″ thick		150	.053		.69	2.25		2.94	4.34	
	1500	Stainless steel, .020″ thick		150	.053		3.07	2.25		5.32	6.95	
	1600	Zinc and copper alloy, 20 oz.	↓	150	.053	↓	3.46	2.25		5.71	7.40	
	9000	Minimum labor/equipment charge	1 Carp	3	2.667	Job		95		95	157	
800	0010	**EXPANSION JOINT**										800
	0300	Butyl or neoprene center with foam insulation, metal flanges										
	0400	Aluminum, .032″ thick for openings to 2-1/2″	1 Rofc	165	.048	L.F.	8.15	1.48		9.63	11.60	
	0600	For joint openings to 3-1/2″		165	.048		9.55	1.48		11.03	13.15	
	0610	For joint openings to 5″		165	.048		11.45	1.48		12.93	15.25	
	0620	For joint openings to 8″		165	.048		18.05	1.48		19.53	22.50	
	0700	Copper, 16 oz. for openings to 2-1/2″		165	.048		11.45	1.48		12.93	15.25	
	0900	For joint openings to 3-1/2″		165	.048		13.15	1.48		14.63	17.15	
	0910	For joint openings to 5″		165	.048		15.40	1.48		16.88	19.60	
	0920	For joint openings to 8″		165	.048		23	1.48		24.48	28	
	1000	Galvanized steel, 26 ga. for openings to 2-1/2″		165	.048		6.90	1.48		8.38	10.25	
	1200	For joint openings to 3-1/2″	↓	165	.048	↓	8.15	1.48		9.63	11.60	

07700 | Roof Specialties and Accessories

07710 | Manufactured Roof Specialties

		07710 Manufactured Roof Specialties	CREW	DAILY OUTPUT	LABOR-HOURS	UNIT	2006 BARE COSTS MAT.	LABOR	EQUIP.	TOTAL	TOTAL INCL O&P	
800	1210	For joint openings to 5"	1 Rofc	165	.048	L.F.	10.20	1.48		11.68	13.85	800
	1220	For joint openings to 8"		165	.048		17.35	1.48		18.83	22	
	1300	Lead-coated copper, 16 oz. for openings to 2-1/2"		165	.048		20.50	1.48		21.98	25	
	1500	For joint openings to 3-1/2"		165	.048		24	1.48		25.48	29	
	1600	Stainless steel, .018", for openings to 2-1/2"		165	.048		10.20	1.48		11.68	13.85	
	1800	For joint openings to 3-1/2"		165	.048		11.60	1.48		13.08	15.45	
	1810	For joint openings to 5"		165	.048		14.35	1.48		15.83	18.45	
	1820	For joint openings to 8"		165	.048		22	1.48		23.48	26.50	
	1900	Neoprene, double-seal type with thick center, 4-1/2" wide		125	.064		9.80	1.96		11.76	14.30	
	1950	Polyethylene bellows, with galv steel flat flanges		100	.080		3.80	2.45		6.25	8.60	
	1960	With galvanized angle flanges		100	.080		4.19	2.45		6.64	9	
	2000	Roof joint with extruded aluminum cover, 2"	1 Shee	115	.070		26	2.93		28.93	33.50	
	2100	Roof joint, plastic curbs, foam center, standard	1 Rofc	100	.080		9.60	2.45		12.05	14.95	
	2200	Large		100	.080		12.80	2.45		15.25	18.45	
	2300	Transitions, regular, minimum		10	.800	Ea.	82.50	24.50		107	135	
	2350	Maximum		4	2		105	61		166	226	
	2400	Large, minimum		9	.889		121	27		148	182	
	2450	Maximum		3	2.667		127	81.50		208.50	286	
	2500	Roof to wall joint with extruded aluminum cover	1 Shee	115	.070	L.F.	22.50	2.93		25.43	29	
	2600	See also Divisions 03150-250 & 05810-350										
	2700	Wall joint, closed cell foam on PVC cover, 9" wide	1 Rofc	125	.064	L.F.	3.41	1.96		5.37	7.25	
	2800	12" wide	"	115	.070	"	3.86	2.13		5.99	8.10	
	9000	Minimum labor/equipment charge	1 Shee	3	2.667	Job		112		112	179	

07720 | Roof Accessories

		07720 Roof Accessories	CREW	DAILY OUTPUT	LABOR-HOURS	UNIT	MAT.	LABOR	EQUIP.	TOTAL	TOTAL INCL O&P	
480	0010	**PITCH POCKETS**										480
	0100	Adjustable, 4" to 7", welded corners, 4" deep	1 Rofc	48	.167	Ea.	10.75	5.10		15.85	21	
	0200	Side extenders, 6"	"	240	.033	"	1.80	1.02		2.82	3.81	
500	0010	**ROOF VENTS**										500
	0020	Mushroom shape, for built-up roofs, aluminum	1 Rofc	30	.267	Ea.	24	8.15		32.15	41	
	0100	PVC, 6" high		30	.267	"	27.50	8.15		35.65	45	
	9000	Minimum labor/equipment charge		2.75	2.909	Job		89		89	160	
550	0010	**RIDGE VENT**										550
	0100	Aluminum strips, mill finish	1 Rofc	160	.050	L.F.	1.21	1.53		2.74	4.08	
	0150	Painted finish		160	.050	"	2.08	1.53		3.61	5.05	
	0200	Connectors		48	.167	Ea.	1.99	5.10		7.09	11.35	
	0300	End caps		48	.167	"	.96	5.10		6.06	10.20	
	0400	Galvanized strips		160	.050	L.F.	2.07	1.53		3.60	5.05	
	0430	Molded polyethylene, shingles not included		160	.050	"	2.70	1.53		4.23	5.70	
	0440	End plugs		48	.167	Ea.	.96	5.10		6.06	10.20	
	0450	Flexible roll, shingles not included		160	.050	L.F.	1.99	1.53		3.52	4.93	
560	0010	**SNOW GUARDS**										560
	0100	Slate & asphalt shingle roofs	1 Rofc	160	.050	Ea.	7.75	1.53		9.28	11.25	
	0200	Standing seam metal roofs		48	.167		12.25	5.10		17.35	22.50	
	0300	Surface mount for metal roofs		48	.167		6.75	5.10		11.85	16.60	
	0400	Double rail pipe type, including pipe		130	.062	L.F.	17.75	1.88		19.63	23	
700	0010	**ROOF HATCHES** With curb, 1" fiberglass insulation, 2'-6" x 3'-0"										700
	0500	Aluminum curb and cover	G-3	10	3.200	Ea.	550	111		661	785	
	0520	Galvanized steel curb and aluminum cover		10	3.200		380	111		491	600	
	0540	Galvanized steel curb and cover		10	3.200		445	111		556	670	
	0600	2'-6" x 4'-6", aluminum curb and cover		9	3.556		650	124		774	915	
	0800	Galvanized steel curb and aluminum cover		9	3.556		540	124		664	795	

07700 | Roof Specialties and Accessories

	07720 | Roof Accessories		CREW	DAILY OUTPUT	LABOR-HOURS	UNIT	2006 BARE COSTS MAT.	LABOR	EQUIP.	TOTAL	TOTAL INCL O&P	
700	0900	Galvanized steel curb and cover	G-3	9	3.556	Ea.	685	124		809	950	700
	1100	4′ x 4′ aluminum curb and cover		8	4		670	139		809	960	
	1120	Galvanized steel curb and aluminum cover		8	4		540	139		679	820	
	1140	Galvanized steel curb and cover		8	4		625	139		764	915	
	1200	2′-6″ x 8′-0″, aluminum curb and cover		6.60	4.848		1,375	169		1,544	1,775	
	1400	Galvanized steel curb and aluminum cover		6.60	4.848		1,050	169		1,219	1,425	
	1500	Galvanized steel curb and cover		6.60	4.848		940	169		1,109	1,300	
	1800	For plexiglass panels, 2′-6″ x 3′-0″, add to above					375			375	410	
	9000	Minimum labor/equipment charge	2 Carp	2	8	Job		284		284	470	
800	0010	**ROOF WALKWAY**										800
	0020	Asphalt impregnated, 3′ x 6′ x 1/2″ thick	1 Rofc	400	.020	S.F.	1.20	.61		1.81	2.42	
	0100	3′ x 3′ x 3/4″ thick	″	400	.020		2.30	.61		2.91	3.63	
	0300	Concrete patio blocks, 2″ thick, natural	1 Clab	115	.070		1.47	1.91		3.38	4.78	
	0400	Colors	″	115	.070		1.86	1.91		3.77	5.20	
	9000	Minimum labor/equipment charge	1 Rofc	2.75	2.909	Job		89		89	160	
850	0010	**SMOKE HATCHES** Unlabeled, not including hand winch operator										850
	0200	For 3′-0″ long, add to roof hatches from Division 07720-700				Ea.	25%	5%				
	0250	For 4′-0″ long, add to roof hatches from Division 07720-700					20%	5%				
	0300	For 8′-0″ long, add to roof hatches from Division 07720-700					10%	5%				
860	0010	**SMOKE VENT**, insulated, 4′ x 4′										860
	0100	Aluminum cover and frame	G-3	13	2.462	Ea.	1,150	85.50		1,235.50	1,400	
	0200	Galvanized steel cover and frame		13	2.462		1,050	85.50		1,135.50	1,300	
	0300	4′ x 8′ aluminum cover and frame		8	4		1,550	139		1,689	1,925	
	0400	Galvanized steel cover and frame		8	4		1,350	139		1,489	1,725	
	9000	Minimum labor/equipment charge	2 Carp	2	8	Job		284		284	470	
870	0010	**VENTS, ONE-WAY**										870
	0020	Plastic, for insulated decks, 1 per M.S.F., minimum	1 Rofc	40	.200	Ea.	12.90	6.10		19	25	
	0100	Maximum		20	.400		29.50	12.25		41.75	54.50	
	0300	Aluminum		30	.267		12.90	8.15		21.05	29	
	0800	Polystyrene baffles, 12″ wide for 16″ O.C. rafter spacing	1 Carp	90	.089		.31	3.16		3.47	5.60	
	0900	For 24″ O.C. rafter spacing		110	.073		.52	2.59		3.11	4.85	
	9000	Minimum labor/equipment charge		3	2.667	Job		95		95	157	

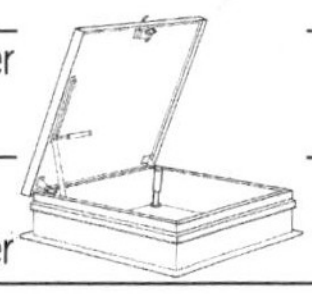

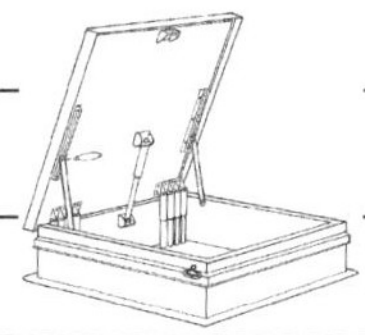

07800 | Fire and Smoke Protection

	07812 | Cementitious Fireproofing		CREW	DAILY OUTPUT	LABOR-HOURS	UNIT	2006 BARE COSTS MAT.	LABOR	EQUIP.	TOTAL	TOTAL INCL O&P	
600	0010	**SPRAYED** Mineral fiber or cementitious for fireproofing,										600
	0050	not incl tamping or canvas protection										
	0100	1″ thick, on flat plate steel	G-2	3000	.008	S.F.	.43	.23	.04	.70	.89	
	0200	Flat decking		2400	.010		.43	.29	.05	.77	1	
	0400	Beams		1500	.016		.43	.47	.08	.98	1.32	
	0500	Corrugated or fluted decks		1250	.019		.64	.56	.09	1.29	1.71	
	0700	Columns, 1-1/8″ thick		1100	.022		.48	.64	.11	1.23	1.69	
	0800	2-3/16″ thick		700	.034		.91	1	.17	2.08	2.81	
	0850	For tamping, add						10%				
	0900	For canvas protection, add	G-2	5000	.005	S.F.	.06	.14	.02	.22	.33	
	1000	Acoustical sprayed, 1″ thick, finished, straight work, minimum		520	.046		.45	1.35	.23	2.03	2.95	
	1100	Maximum		200	.120		.48	3.51	.59	4.58	6.90	

07812 | Cementitious Fireproofing		CREW	DAILY OUTPUT	LABOR-HOURS	UNIT	2006 BARE COSTS MAT.	LABOR	EQUIP.	TOTAL	TOTAL INCL O&P
600 1300	Difficult access, minimum	G-2	225	.107	S.F.	.48	3.12	.52	4.12	6.20 600
1400	Maximum	↓	130	.185	↓	.53	5.40	.90	6.83	10.35
1500	Intumescent epoxy fireproofing on wire mesh, 3/16″ thick									
1550	1 hour rating, exterior use	G-2	136	.176	S.F.	5.55	5.15	.86	11.56	15.50
1600	Magnesium oxychloride, 35# to 40# density, 1/4″ thick		3000	.008		1.17	.23	.04	1.44	1.71
1650	1/2″ thick		2000	.012		2.35	.35	.06	2.76	3.22
1700	60# to 70# density, 1/4″ thick		3000	.008		1.55	.23	.04	1.82	2.13
1750	1/2″ thick		2000	.012		3.12	.35	.06	3.53	4.06
2000	Vermiculite cement, troweled or sprayed, 1/4″ thick		3000	.008		1.06	.23	.04	1.33	1.59
2050	1/2″ thick		2000	.012	↓	2.10	.35	.06	2.51	2.94
9000	Minimum labor/equipment charge	↓	3	8	Job		234	39	273	425

07840 | Firestopping		CREW	DAILY OUTPUT	LABOR-HOURS	UNIT	MAT.	LABOR	EQUIP.	TOTAL	TOTAL INCL O&P
100 0010	FIRESTOPPING R078413-30									100
0100	Metallic piping, non insulated									
0110	Through walls, 2″ diameter	1 Carp	16	.500	Ea.	9.95	17.80		27.75	40.50
0120	4″ diameter		14	.571		15.20	20.50		35.70	50.50
0130	6″ diameter		12	.667		20.50	23.50		44	62
0140	12″ diameter		10	.800		36.50	28.50		65	87
0150	Through floors, 2″ diameter		32	.250		6.05	8.90		14.95	21.50
0160	4″ diameter		28	.286		8.70	10.15		18.85	26.50
0170	6″ diameter		24	.333		11.40	11.85		23.25	32
0180	12″ diameter	↓	20	.400	↓	19.25	14.20		33.45	44.50
0190	Metallic piping, insulated									
0200	Through walls, 2″ diameter	1 Carp	16	.500	Ea.	14.15	17.80		31.95	45
0210	4″ diameter		14	.571		19.40	20.50		39.90	55
0220	6″ diameter		12	.667		24.50	23.50		48	66.50
0230	12″ diameter		10	.800		40.50	28.50		69	91.50
0240	Through floors, 2″ diameter		32	.250		10.20	8.90		19.10	26
0250	4″ diameter		28	.286		12.85	10.15		23	31
0260	6″ diameter		24	.333		15.60	11.85		27.45	37
0270	12″ diameter	↓	20	.400	↓	19.25	14.20		33.45	44.50
0280	Non metallic piping, non insulated									
0290	Through walls, 2″ diameter	1 Carp	12	.667	Ea.	41	23.50		64.50	85
0300	4″ diameter		10	.800		51.50	28.50		80	104
0310	6″ diameter		8	1		72	35.50		107.50	138
0330	Through floors, 2″ diameter		16	.500		32	17.80		49.80	65
0340	4″ diameter		6	1.333		40	47.50		87.50	123
0350	6″ diameter	↓	6	1.333	↓	48	47.50		95.50	131
0370	Ductwork, insulated & non insulated, round									
0380	Through walls, 6″ diameter	1 Carp	12	.667	Ea.	21	23.50		44.50	62.50
0390	12″ diameter		10	.800		41.50	28.50		70	93
0400	18″ diameter		8	1		67.50	35.50		103	134
0410	Through floors, 6″ diameter		16	.500		11.45	17.80		29.25	42
0420	12″ diameter		14	.571		21	20.50		41.50	56.50
0430	18″ diameter	↓	12	.667	↓	36.50	23.50		60	79.50
0440	Ductwork, insulated & non insulated, rectangular									
0450	With stiffener/closure angle, through walls, 6″ x 12″	1 Carp	8	1	Ea.	17.35	35.50		52.85	78
0460	12″ x 24″		6	1.333		23	47.50		70.50	104
0470	24″ x 48″		4	2		65.50	71		136.50	190
0480	With stiffener/closure angle, through floors, 6″ x 12″		10	.800		9.35	28.50		37.85	57.50
0490	12″ x 24″		8	1		16.85	35.50		52.35	77.50
0500	24″ x 48″	↓	6	1.333	↓	33	47.50		80.50	115
0510	Multi trade openings									
0520	Through walls, 6″ x 12″	1 Carp	2	4	Ea.	36.50	142		178.50	276

07800 | Fire and Smoke Protection

	07840 | Firestopping		CREW	DAILY OUTPUT	LABOR-HOURS	UNIT	2006 BARE COSTS MAT.	LABOR	EQUIP.	TOTAL	TOTAL INCL O&P
100	0530	12″ x 24″ R078413-30	1 Carp	1	8	Ea.	147	284		431	630
	0540	24″ x 48″	2 Carp	1	16		585	570		1,155	1,575
	0550	48″ x 96″	″	.75	21.333		2,350	760		3,110	3,850
	0560	Through floors, 6″ x 12″	1 Carp	2	4		36.50	142		178.50	276
	0570	12″ x 24″	″	1	8		147	284		431	630
	0580	24″ x 48″	2 Carp	.75	21.333		585	760		1,345	1,900
	0590	48″ x 96″	″	.50	32	↓	2,350	1,150		3,500	4,475
	0600	Structural penetrations, through walls									
	0610	Steel beams, W8 x 10	1 Carp	8	1	Ea.	23	35.50		58.50	84
	0620	W12 x 14		6	1.333		36.50	47.50		84	119
	0630	W21 x 44		5	1.600		73	57		130	174
	0640	W36 x 135		3	2.667		177	95		272	350
	0650	Bar joists, 18″ deep		6	1.333		33.50	47.50		81	115
	0660	24″ deep		6	1.333		41.50	47.50		89	125
	0670	36″ deep		5	1.600		62.50	57		119.50	163
	0680	48″ deep	↓	4	2	↓	73	71		144	198
	0690	Construction joints, floor slab at exterior wall									
	0700	Precast, brick, block or drywall exterior									
	0710	2″ wide joint	1 Carp	125	.064	L.F.	5.20	2.28		7.48	9.45
	0720	4″ wide joint	″	75	.107	″	10.40	3.79		14.19	17.75
	0730	Metal panel, glass or curtain wall exterior									
	0740	2″ wide joint	1 Carp	40	.200	L.F.	12.30	7.10		19.40	25.50
	0750	4″ wide joint	″	25	.320	″	16.80	11.40		28.20	37.50
	0760	Floor slab to drywall partition									
	0770	Flat joint	1 Carp	100	.080	L.F.	5.10	2.84		7.94	10.30
	0780	Fluted joint		50	.160		10.40	5.70		16.10	21
	0790	Etched fluted joint	↓	75	.107	↓	6.75	3.79		10.54	13.75
	0800	Floor slab to concrete/masonry partition									
	0810	Flat joint	1 Carp	75	.107	L.F.	11.45	3.79		15.24	18.90
	0820	Fluted joint	″	50	.160	″	13.55	5.70		19.25	24.50
	0830	Concrete/CMU wall joints									
	0840	1″ wide	1 Carp	100	.080	L.F.	6.25	2.84		9.09	11.55
	0850	2″ wide		75	.107		11.45	3.79		15.24	18.90
	0860	4″ wide	↓	50	.160	↓	22	5.70		27.70	33.50
	0870	Concrete/CMU floor joints									
	0880	1″ wide	1 Carp	200	.040	L.F.	3.12	1.42		4.54	5.80
	0890	2″ wide		150	.053		5.70	1.90		7.60	9.45
	0900	4″ wide ↓	↓	100	.080	↓	10.90	2.84		13.74	16.70

07900 | Joint Sealers

	07920 | Joint Sealants		CREW	DAILY OUTPUT	LABOR-HOURS	UNIT	2006 BARE COSTS MAT.	LABOR	EQUIP.	TOTAL	TOTAL INCL O&P
800	0010	CAULKING AND SEALANTS									
	0020	Acoustical sealant, elastomeric, cartridges				Ea.	2.28			2.28	2.51
	0032	Backer rod, polyethylene, 1/4″ diameter	1 Bric	460	.017	L.F.	.02	.64		.66	1.05
	0052	1/2″ diameter		460	.017		.04	.64		.68	1.07
	0072	3/4″ diameter		460	.017		.07	.64		.71	1.10
	0092	1″ diameter	↓	460	.017	↓	.11	.64		.75	1.15
	0100	Acrylic latex caulk, white									
	0200	11 fl. oz cartridge				Ea.	1.88			1.88	2.07

	07920 | Joint Sealants	CREW	DAILY OUTPUT	LABOR-HOURS	UNIT	2006 BARE COSTS MAT.	LABOR	EQUIP.	TOTAL	TOTAL INCL O&P	
800 0500	1/4″ x 1/2″	1 Bric	248	.032	L.F.	.15	1.18		1.33	2.08	800
0600	1/2″ x 1/2″		250	.032		.31	1.17		1.48	2.24	
0800	3/4″ x 3/4″		230	.035		.69	1.27		1.96	2.82	
0900	3/4″ x 1″		200	.040		.92	1.46		2.38	3.38	
1000	1″ x 1″		180	.044		1.15	1.62		2.77	3.91	
1400	Butyl based, bulk				Gal.	23			23	25	
1500	Cartridges				″	28			28	30.50	
1700	Bulk, in place 1/4″ x 1/2″, 154 L.F./gal.	1 Bric	230	.035	L.F.	.15	1.27		1.42	2.22	
1800	1/2″ x 1/2″, 77 L.F./gal.	″	180	.044	″	.30	1.62		1.92	2.97	
2000	Latex acrylic based, bulk				Gal.	24			24	26.50	
2100	Cartridges					29.50			29.50	32.50	
2300	Polysulfide compounds, 1 component, bulk					45			45	49.50	
2600	1 or 2 component, in place, 1/4″ x 1/4″, 308 L.F./gal.	1 Bric	145	.055	L.F.	.15	2.02		2.17	3.43	
2700	1/2″ x 1/4″, 154 L.F./gal.		135	.059		.29	2.17		2.46	3.83	
2900	3/4″ x 3/8″, 68 L.F./gal.		130	.062		.66	2.25		2.91	4.38	
3000	1″ x 1/2″, 38 L.F./gal.		130	.062		1.19	2.25		3.44	4.95	
3200	Polyurethane, 1 or 2 component				Gal.	49			49	54	
3500	Bulk, in place, 1/4″ x 1/4″	1 Bric	150	.053	L.F.	.16	1.95		2.11	3.34	
3600	1/2″ x 1/4″		145	.055		.32	2.02		2.34	3.62	
3800	3/4″ x 3/8″, 68 L.F./gal.		130	.062		.72	2.25		2.97	4.44	
3900	1″ x 1/2″		110	.073		1.27	2.66		3.93	5.70	
4100	Silicone rubber, bulk				Gal.	35.50			35.50	39	
4200	Cartridges				″	38			38	42	
4400	Neoprene gaskets, closed cell, adhesive, 1/8″ x 3/8″	1 Bric	240	.033	L.F.	.21	1.22		1.43	2.21	
4500	1/4″ x 3/4″		215	.037		.49	1.36		1.85	2.75	
4700	1/2″ x 1″		200	.040		1.44	1.46		2.90	3.95	
4800	3/4″ x 1-1/2″		165	.048		3	1.77		4.77	6.15	
5500	Resin epoxy coating, 2 component, heavy duty				Gal.	26.50			26.50	29	
5802	Tapes, sealant, P.V.C. foam adhesive, 1/16″ x 1/4″				L.F.	.05			.05	.05	
5902	1/16″ x 1/2″					.07			.07	.08	
5952	1/16″ x 1″					.12			.12	.13	
6002	1/8″ x 1/2″					.08			.08	.09	
6200	Urethane foam, 2 component, handy pack, 1 C.F.				Ea.	28.50			28.50	31	
6300	50.0 C.F. pack				C.F.	14.45			14.45	15.90	
9000	Minimum labor/equipment charge	1 Bric	4	2	Job		73		73	119	

For information about Means Estimating Seminars, see yellow pages 12 and 13 in back of book

Division 8
Doors & Windows

Estimating Tips

08100 Metal Doors & Frames

- Most metal doors and frames look alike, but there may be significant differences among them. When estimating these items be sure to choose the line item that most closely compares to the specification or door schedule requirements regarding:
  - type of metal
  - metal gauge
  - door core material
  - fire rating
  - finish

08200 Wood & Plastic Doors

- Wood and plastic doors vary considerably in price. The primary determinant is the veneer material. Lauan, birch, and oak are the most common veneers. Other variables include the following:
  - hollow or solid core
  - fire rating
  - flush or raised panel
  - finish
- If the specifications require compliance with AWI (Architectural Woodwork Institute) standards or acoustical standards, the cost of the door may increase substantially. All wood doors are priced pre-mortised for hinges and pre-drilled for cylindrical locksets.
- Frequently, doors, frames, and windows are unique in old buildings. Specified replacement units could be stock, custom (similar to the original), or exact reproduction. The estimator should work closely with a window consultant to determine any extra costs that may be associated with the unusual installation requirements.

08300 Specialty Doors

- There are many varieties of special doors, and they are usually priced per each. Add frames, hardware, or operators required for a complete installation.

08510 Steel Windows

- Most metal windows are delivered preglazed. However, some metal windows are priced without glass. Refer to 08800 Glazing for glass pricing. The grade C indicates commercial grade windows, usually ASTM C-35.

08550 Wood Windows

- All wood windows are priced preglazed. The two glazing options priced are single pane float glass and insulating glass 1/2″ thick. Add the cost of screens and grills if required.

08700 Hardware

- Hardware costs add considerably to the cost of a door. The most efficient method to determine the hardware requirements for a project is to review the door schedule. This schedule, in conjunction with the specifications, is all you should need to take off the door hardware.
- Door hinges are priced by the pair, with most doors requiring 1-1/2 pairs per door. The hinge prices do not include installation labor because it is included in door installation. Hinges are classified according to the frequency of use.

08800 Glazing

- Different openings require different types of glass. The three most common types are:
  - float
  - tempered
  - insulating
- Most exterior windows are glazed with insulating glass. Entrance doors and window walls, where the glass is less than 18″ from the floor, are generally glazed with tempered glass. Interior windows and some residential windows are glazed with float glass.
- Energy efficient coatings are also available

08900 Glazed Curtain Wall

- Glazed curtain walls consist of the metal tube framing and the glazing material. The cost data in this subdivision is presented for the metal tube framing alone or the composite wall. If your estimate requires a detailed takeoff of the framing, be sure to add the glazing cost.

Reference Numbers

Reference numbers are shown in bold squares at the beginning of some major classifications. These numbers refer to related items in the Reference Section. The reference information may be an estimating procedure, an alternate pricing method, or technical information.

Note: Not all subdivisions listed here necessarily appear in this publication.

	08060 | Selective Demolition	CREW	DAILY OUTPUT	LABOR-HOURS	UNIT	2006 BARE COSTS MAT.	LABOR	EQUIP.	TOTAL	TOTAL INCL O&P	
110	0010 **SELECTIVE DEMOLITION, DOORS** R024119-10										110
	0200 Doors, exterior, 1-3/4" thick, single, 3' x 7' high	1 Clab	16	.500	Ea.		13.70		13.70	22.50	
	0220 Double, 6' x 7' high		12	.667			18.25		18.25	30.50	
	0500 Interior, 1-3/8" thick, single, 3' x 7' high		20	.400			10.95		10.95	18.15	
	0520 Double, 6' x 7' high		16	.500			13.70		13.70	22.50	
	0700 Bi-folding, 3' x 6'-8" high		20	.400			10.95		10.95	18.15	
	0720 6' x 6'-8" high		18	.444			12.20		12.20	20	
	0900 Bi-passing, 3' x 6'-8" high		16	.500			13.70		13.70	22.50	
	0940 6' x 6'-8" high		14	.571			15.65		15.65	26	
	1500 Remove and reset, minimum	1 Carp	8	1			35.50		35.50	59	
	1520 Maximum		6	1.333			47.50		47.50	78.50	
	2000 Frames, including trim, metal		8	1			35.50		35.50	59	
	2200 Wood	2 Carp	32	.500			17.80		17.80	29.50	
	2201 Alternate pricing method	1 Carp	200	.040	L.F.		1.42		1.42	2.36	
	2950 Minimum labor/equipment charge	1 Clab	4	2	Job		55		55	91	
	3000 Special doors, counter doors	2 Carp	6	2.667	Ea.		95		95	157	
	3100 Double acting		10	1.600			57		57	94	
	3200 Floor door (trap type), or access type		8	2			71		71	118	
	3300 Glass, sliding, including frames		12	1.333			47.50		47.50	78.50	
	3400 Overhead, commercial, 12' x 12' high		4	4			142		142	236	
	3440 up to 20' x 16' high		3	5.333			190		190	315	
	3445 up to 35' x 30' high		1	16			570		570	940	
	3500 Residential, 9' x 7' high		8	2			71		71	118	
	3540 16' x 7' high		7	2.286			81.50		81.50	135	
	3600 Remove and reset, minimum		4	4			142		142	236	
	3620 Maximum		2.50	6.400			228		228	375	
	3700 Roll-up grille		5	3.200			114		114	188	
	3800 Revolving door		2	8			284		284	470	
	3900 Storefront swing door		3	5.333			190		190	315	
	5020 Remove bay/bow window		6	2.667			95		95	157	
	5032 Remove skylight, plstc. domes,flush/curb mtd.	G-3	395	.081	S.F.		2.82		2.82	4.56	
	6600 Demo flexible transparent strip entrance	3 Shee	115	.209	SF Surf		8.80		8.80	14	
	7100 Remove double swing pneumatic doors, openers and sensors	2 Skwk	.50	32	Opng.		1,175		1,175	1,900	
	7110 Remove automatic operators, industrial, sliding doors, to 12' wide	"	.40	40	"		1,450		1,450	2,400	
	7550 Hangar door demo	2 Sswk	220	.073	S.F.		2.91		2.91	5.45	
	7570 Remove shock absorbing door	"	1.90	8.421	Opng.		335		335	630	
	9000 Minimum labor/equipment charge	1 Carp	4	2	Job		71		71	118	
120	0010 **SELECTIVE DEMOLITION, WINDOWS** R024119-10										120
	0200 Aluminum, including trim, to 12 S.F.	1 Clab	16	.500	Ea.		13.70		13.70	22.50	
	0240 To 25 S.F.		11	.727			19.95		19.95	33	
	0280 To 50 S.F.		5	1.600			44		44	72.50	
	0320 Storm windows/scree, to 12 S.F.		27	.296			8.10		8.10	13.45	
	0360 To 25 S.F.		21	.381			10.45		10.45	17.30	
	0400 To 50 S.F.		16	.500			13.70		13.70	22.50	
	0600 Glass, minimum		200	.040	S.F.		1.10		1.10	1.82	
	0620 Maximum		150	.053	"		1.46		1.46	2.42	
	1000 Steel, including trim, to 12 S.F.		13	.615	Ea.		16.85		16.85	28	
	1020 To 25 S.F.		9	.889			24.50		24.50	40.50	
	1040 To 50 S.F.		4	2			55		55	91	
	2000 Wood, including trim, to 12 S.F.		22	.364			9.95		9.95	16.50	
	2020 To 25 S.F.		18	.444			12.20		12.20	20	
	2060 To 50 S.F.		13	.615			16.85		16.85	28	
	2065 To 180 S.F.		8	1			27.50		27.50	45.50	
	5020 Remove and reset window, minimum	1 Carp	6	1.333			47.50		47.50	78.50	
	5040 Average		4	2			71		71	118	

08050 | Basic Door and Window Materials and Methods

		08060 | Selective Demolition	Crew	Daily Output	Labor-Hours	Unit	2006 Bare Costs Mat.	Labor	Equip.	Total	Total Incl O&P	
120	5080	Maximum (R024119-10)	1 Carp	2	4	Ea.		142		142	236	120
	9000	Minimum labor/equipment charge	1 Clab	4	2	Job		55		55	91	

08100 | Metal Doors and Frames

		08110 | Steel Doors and Frames	Crew	Daily Output	Labor-Hours	Unit	2006 Bare Costs Mat.	Labor	Equip.	Total	Total Incl O&P	
200	0010	COMMERCIAL STEEL DOORS (R081313-20)										200
	0015	Flush, full panel, hollow core										
	0020	1-3/8″ thick, 20 ga., 2′-0″x 6′-8″	2 Carp	20	.800	Ea.	246	28.50		274.50	320	
	0040	2′-8″ x 6′-8″		18	.889		254	31.50		285.50	330	
	0060	3′-0″ x 6′-8″		17	.941		259	33.50		292.50	340	
	0100	3′-0″ x 7′-0″		17	.941		241	33.50		274.50	320	
	0120	For vision lite, add					84			84	92.50	
	0140	For narrow lite, add					85.50			85.50	94	
	0320	Half glass, 20 ga., 2′-0″ x 6′-8″	2 Carp	20	.800		365	28.50		393.50	450	
	0340	2′-8″ x 6′-8″		18	.889		375	31.50		406.50	470	
	0360	3′-0″ x 6′-8″		17	.941		380	33.50		413.50	475	
	0400	3′-0″ x 7′-0″		17	.941		380	33.50		413.50	475	
	0410	1-3/8″ thick, 18 ga., 2′-0″x 6′-8″		20	.800		295	28.50		323.50	370	
	0420	3′-0″ x 6′-8″		17	.941		299	33.50		332.50	385	
	0425	3′-0″ x 7′-0″		17	.941		305	33.50		338.50	390	
	0450	For vision lite, add					84			84	92.50	
	0452	For narrow lite, add					85.50			85.50	94	
	0460	Half glass, 18 ga., 2′-0″ x 6′-8″	2 Carp	20	.800		415	28.50		443.50	505	
	0465	2′-8″ x 6′-8″		18	.889		430	31.50		461.50	530	
	0470	3′-0″ x 6′-8″		17	.941		420	33.50		453.50	520	
	0475	3′-0″ x 7′-0″		17	.941		425	33.50		458.50	525	
	0500	Hollow core, 1-3/4″ thick, full panel, 20 ga., 2′-8″ x 6′-8″		18	.889		277	31.50		308.50	360	
	0520	3′-0″ x 6′-8″		17	.941		259	33.50		292.50	340	
	0640	3′-0″ x 7′-0″		17	.941		305	33.50		338.50	395	
	0680	4′-0″ x 7′-0″		15	1.067		430	38		468	535	
	0700	4′-0″ x 8′-0″		13	1.231		490	44		534	615	
	1000	18 ga., 2′-8″ x 6′-8″		17	.941		204	33.50		237.50	281	
	1020	3′-0″ x 6′-8″		16	1		199	35.50		234.50	278	
	1120	3′-0″ x 7′-0″		17	.941		345	33.50		378.50	430	
	1180	4′-0″ x 7′-0″		14	1.143		415	40.50		455.50	525	
	1200	4′-0″ x 8′-0″		17	.941		490	33.50		523.50	595	
	1212	For vision lite, add					84			84	92.50	
	1214	For narrow lite, add					85.50			85.50	94	
	1230	Half glass, 20 ga., 2′-8″ x 6′-8″	2 Carp	20	.800		380	28.50		408.50	465	
	1240	3′-0″ x 6′-8″		18	.889		380	31.50		411.50	475	
	1260	3′-0″ x 7′-0″		18	.889		390	31.50		421.50	485	
	1320	18 ga., 2′-8″ x 6′-8″		18	.889		425	31.50		456.50	525	
	1340	3′-0″ x 6′-8″		17	.941		420	33.50		453.50	520	
	1360	3′-0″ x 7′-0″		17	.941		430	33.50		463.50	530	
	1380	4′-0″ x 7′-0″		15	1.067		535	38		573	655	
	1400	4′-0″ x 8′-0″		14	1.143		615	40.50		655.50	745	
	1720	Insulated, 1-3/4″ thick, full panel, 18 ga., 3′-0″ x 6′-8″		15	1.067		370	38		408	475	
	1740	2′-8″ x 7′-0″		16	1		390	35.50		425.50	485	
	1760	3′-0″ x 7′-0″		15	1.067		380	38		418	485	

	08110 | Steel Doors and Frames		Crew	Daily Output	Labor-Hours	Unit	2006 Bare Costs Mat.	Labor	Equip.	Total	Total Incl O&P	
200	1800	4′-0″ x 8′-0″ (R081313-20)	2 Carp	13	1.231	Ea.	565	44		609	695	200
	1805	For vision lite, add					84			84	92.50	
	1810	For narrow lite, add					85.50			85.50	94	
	1820	Half glass, 18 ga., 3′-0″ x 6′-8″	2 Carp	16	1		495	35.50		530.50	605	
	1840	2′-8″ x 7′-0″		17	.941		510	33.50		543.50	615	
	1860	3′-0″ x 7′-0″		16	1		565	35.50		600.50	680	
	1900	4′-0″ x 8′-0″		14	1.143		685	40.50		725.50	825	
	2000	For bottom louver, add					138			138	152	
	2020	For baked enamel finish, add					30%	15%				
	2040	For galvanizing, add					15%					
	9000	Minimum labor/equipment charge	1 Carp	4	2	Job		71		71	118	
250	0010	DOOR FRAMES										250
	0020	Steel channels with anchors and bar stops										
	0100	6″ channel @ 8.2#/L.F., 3′ x 7′ door, weighs 150#	E-4	13	2.462	Ea.	171	99.50	6.85	277.35	380	
	0200	8″ channel @ 11.5#/L.F., 6′ x 8′ door, weighs 275#		9	3.556		315	144	9.90	468.90	625	
	0300	8′ x 12′ door, weighs 400#		6.50	4.923		455	199	13.75	667.75	885	
	0400	10″ channel @ 15.3#/L.F., 10′ x 10′ door, weighs 500#		6	5.333		570	216	14.90	800.90	1,050	
	0500	12′ x 12′ door, weighs 600#		5.50	5.818		685	235	16.25	936.25	1,200	
	0600	12″ channel @ 20.7#/L.F., 12′ x 12′ door, weighs 825#		4.50	7.111		940	288	19.85	1,247.85	1,575	
	0700	12′ x 16′ door, weighs 1000#		4	8		1,150	325	22.50	1,497.50	1,875	
	0800	For frames without bar stops, light sections, deduct					15%					
	0900	Heavy sections, deduct					10%					
	9000	Minimum labor/equipment charge	E-4	4	8	Job		325	22.50	347.50	630	
300	0010	FIRE DOOR (R081313-20)										300
	0015	Steel, flush, "B" label, 90 minute										
	0020	Full panel, 20 ga., 2′-0″ x 6′-8″	2 Carp	20	.800	Ea.	310	28.50		338.50	385	
	0040	2′-8″ x 6′-8″		18	.889		320	31.50		351.50	405	
	0060	3′-0″ x 6′-8″		17	.941		325	33.50		358.50	410	
	0080	3′-0″ x 7′-0″		17	.941		335	33.50		368.50	425	
	0140	18 ga., 3′-0″ x 6′-8″		16	1		365	35.50		400.50	460	
	0160	2′-8″ x 7′-0″		17	.941		385	33.50		418.50	475	
	0180	3′-0″ x 7′-0″		16	1		365	35.50		400.50	465	
	0200	4′-0″ x 7′-0″		15	1.067		480	38		518	590	
	0220	For "A" label, 3 hour, 18 ga., use same price as "B" label										
	0240	For vision lite, add				Ea.	104			104	115	
	0520	Flush, "B" label 90 min., composite, 20 ga., 2′-0″ x 6′-8″	2 Carp	18	.889		400	31.50		431.50	495	
	0540	2′-8″ x 6′-8″		17	.941		410	33.50		443.50	505	
	0560	3′-0″ x 6′-8″		16	1		410	35.50		445.50	515	
	0580	3′-0″ x 7′-0″		16	1		425	35.50		460.50	525	
	0640	Flush, "A" label 3 hour, composite, 18 ga., 3′-0″ x 6′-8″		15	1.067		345	38		383	445	
	0660	2′-8″ x 7′-0″		16	1		365	35.50		400.50	460	
	0680	3′-0″ x 7′-0″		15	1.067		360	38		398	460	
	0700	4′-0″ x 7′-0″		14	1.143		465	40.50		505.50	580	
	9000	Minimum labor/equipment charge	1 Carp	4	2	Job		71		71	118	
600	0010	RESIDENTIAL STEEL DOOR										600
	0020	Prehung, insulated, exterior										
	0030	Embossed, full panel, 2′-8″ x 6′-8″	2 Carp	17	.941	Ea.	216	33.50		249.50	293	
	0040	3′-0″ x 6′-8″		15	1.067		217	38		255	300	
	0060	3′-0″ x 7′-0″		15	1.067		280	38		318	375	
	0070	5′-4″ x 6′-8″, double		8	2		440	71		511	605	
	0220	Half glass, 2′-8″ x 6′-8″		17	.941		261	33.50		294.50	345	
	0240	3′-0″ x 6′-8″		16	1		261	35.50		296.50	345	
	0260	3′-0″ x 7′-0″		16	1		315	35.50		350.50	410	
	0270	5′-4″ x 6′-8″, double		8	2		540	71		611	715	

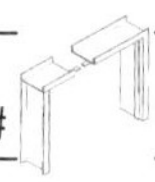

	08110 Steel Doors and Frames	CREW	DAILY OUTPUT	LABOR-HOURS	UNIT	2006 BARE COSTS MAT.	LABOR	EQUIP.	TOTAL	TOTAL INCL O&P	
600	0720 Raised plastic face, full panel, 2'-8" x 6'-8"	2 Carp	16	1	Ea.	256	35.50		291.50	340	600
	0740 3'-0" x 6'-8"		15	1.067		258	38		296	345	
	0760 3'-0" x 7'-0"		15	1.067		261	38		299	350	
	0780 5'-4" x 6'-8", double		8	2		485	71		556	650	
	0820 Half glass, 2'-8" x 6'-8"		17	.941		287	33.50		320.50	370	
	0840 3'-0" x 6'-8"		16	1		290	35.50		325.50	380	
	0860 3'-0" x 7'-0"		16	1		320	35.50		355.50	410	
	0880 5'-4" x 6'-8", double		8	2		630	71		701	815	
	1320 Flush face, full panel, 2'-6" x 6'-8"		16	1		216	35.50		251.50	297	
	1340 3'-0" x 6'-8"		15	1.067		216	38		254	300	
	1360 3'-0" x 7'-0"		15	1.067		278	38		316	370	
	1380 5'-4" x 6'-8", double		8	2		405	71		476	565	
	1420 Half glass, 2'-8" x 6'-8"		17	.941		270	33.50		303.50	355	
	1440 3'-0" x 6'-8"		16	1		274	35.50		309.50	360	
	1460 3'-0" x 7'-0"		16	1		315	35.50		350.50	405	
	1480 5'-4" x 6'-8", double	↓	8	2		530	71		601	700	
	1500 Sidelight, full lite, 1'-0" x 6'-8" with grille					211			211	232	
	1510 1'-0" x 6'-8", low e					242			242	267	
	1520 1'-0" x 6'-8", half lite					189			189	208	
	1530 1'-0" x 6'-8", half lite, low e					230			230	253	
	2300 Interior, residential, closet, bi-fold, 6'-8" x 2'-0" wide	2 Carp	16	1		140	35.50		175.50	213	
	2330 3'-0" wide		16	1		157	35.50		192.50	232	
	2360 4'-0" wide		15	1.067		238	38		276	325	
	2400 5'-0" wide		14	1.143		276	40.50		316.50	375	
	2420 6'-0" wide	↓	13	1.231	↓	310	44		354	415	
	9000 Minimum labor/equipment charge	1 Carp	4	2	Job		71		71	118	
820	0010 **STEEL FRAMES, KNOCK DOWN**										820
	0020 16 ga., up to 5-3/4" jamb depth										
	0025 6'-8" high, 3'-0" wide, single	2 Carp	16	1	Ea.	121	35.50		156.50	192	
	0028 3'-6" wide, single		16	1		107	35.50		142.50	177	
	0030 4'-0" wide, single		16	1		107	35.50		142.50	177	
	0040 6'-0" wide, double		14	1.143		143	40.50		183.50	226	
	0045 8'-0" wide, double		14	1.143		149	40.50		189.50	232	
	0100 7'-0" high, 3'-0" wide, single		16	1		124	35.50		159.50	195	
	0110 3'-6" wide, single		16	1		110	35.50		145.50	180	
	0112 4'-0" wide, single		16	1		188	35.50		223.50	266	
	0140 6'-0" wide, double		14	1.143		150	40.50		190.50	233	
	0145 8'-0" wide, double		14	1.143		140	40.50		180.50	222	
	1000 16 ga., up to 4-7/8" deep, 7'-0" H, 3'-0" W, single		16	1		121	35.50		156.50	192	
	1140 6'-0" wide, double		14	1.143		148	40.50		188.50	231	
	2800 14 ga., up to 3-7/8" deep, 7'-0" high, 3'-0" wide, single		16	1		124	35.50		159.50	195	
	2840 6'-0" wide, double		14	1.143		152	40.50		192.50	235	
	3000 14 ga., up to 5-3/4" deep, 6'-8" high, 3'-0" wide, single		16	1		134	35.50		169.50	207	
	3002 3'-6" wide, single		16	1		131	35.50		166.50	203	
	3005 4'-0" wide, single		16	1		131	35.50		166.50	203	
	3600 up to 5-3/4" jamb depth, 7'-0" high, 4'-0" wide, single		15	1.067		103	38		141	177	
	3620 6'-0" wide, double		12	1.333		162	47.50		209.50	257	
	3640 8'-0" wide, double		12	1.333		136	47.50		183.50	229	
	3700 8'-0" high, 4'-0" wide, single		15	1.067		136	38		174	213	
	3740 8'-0" wide, double		12	1.333		169	47.50		216.50	265	
	4000 6-3/4" deep, 7'-0" high, 4'-0" wide, single		15	1.067		144	38		182	222	
	4020 6'-0" wide, double		12	1.333		165	47.50		212.50	261	
	4040 8'-0", wide double		12	1.333		191	47.50		238.50	289	
	4100 8'-0" high, 4'-0" wide, single		15	1.067		162	38		200	241	
	4140 8'-0" wide, double		12	1.333		189	47.50		236.50	287	
	4400 8-3/4" deep, 7'-0" high, 4'-0" wide, single	↓	15	1.067	↓	152	38		190	230	

	08110 | Steel Doors and Frames		CREW	DAILY OUTPUT	LABOR-HOURS	UNIT	2006 BARE COSTS MAT.	LABOR	EQUIP.	TOTAL	TOTAL INCL O&P	
820	4440	8'-0" wide, double	2 Carp	12	1.333	Ea.	200	47.50		247.50	299	820
	4500	8'-0" high, 4'-0" wide, single		15	1.067		178	38		216	258	
	4540	8'-0" wide, double		12	1.333		208	47.50		255.50	310	
	4900	For welded frames, add					44.50			44.50	49	
	5400	14 ga., "B" label, up to 5-3/4" deep, 7'-0" high, 4'-0" wide, single	2 Carp	15	1.067		160	38		198	239	
	5440	8'-0" wide, double		12	1.333		187	47.50		234.50	285	
	5800	6-3/4" deep, 7'-0" high, 4'-0" wide, single		15	1.067		140	38		178	217	
	5840	8'-0" wide, double		12	1.333		207	47.50		254.50	305	
	6200	8-3/4" deep, 7'-0" high, 4'-0" wide, single		15	1.067		170	38		208	250	
	6240	8'-0" wide, double		12	1.333		220	47.50		267.50	320	
	6300	For "A" label use same price as "B" label										
	6400	For baked enamel finish, add					30%	15%				
	6500	For galvanizing, add					15%					
	6600	For hospital stop, add				Ea.	263			263	289	
	7900	Transom lite frames, fixed, add	2 Carp	155	.103	S.F.	42.50	3.67		46.17	53	
	8000	Movable, add	"	130	.123	"	51.50	4.38		55.88	64	
	9000	Minimum labor/equipment charge	1 Carp	4	2	Job		71		71	118	
840	0010	**WRAP AROUND DRYWALL FRAMES**										840
	0400	Wrap-around drywall frame, 16 ga., 6-1/4" x 7'-0" x 2'-6" wide	2 Carp	16	1	Ea.	125	35.50		160.50	197	
	0430	2'-8" wide		16	1		125	35.50		160.50	197	
	0490	3'-6" wide		15	1.067		132	38		170	208	
	0520	5'-0" wide		14	1.143		139	40.50		179.50	221	
	0580	6'-0" wide		14	1.143		146	40.50		186.50	229	
	0610	7'-0" wide		13	1.231		159	44		203	248	
	2000	18 ga., 4-3/4" x 7'-0 x 2'-6" wide		16	1		112	35.50		147.50	182	
	2030	2'-8" wide		16	1		112	35.50		147.50	182	
	2060	3'-0" wide		15	1.067		112	38		150	186	
	2090	3'-6" wide		15	1.067		117	38		155	191	
	2120	5'-0" wide		14	1.143		127	40.50		167.50	207	
	2150	5'-4" wide		14	1.143		138	40.50		178.50	220	
	2200	6'-0" wide		14	1.143		138	40.50		178.50	220	
	2250	7'-0" wide		13	1.231		141	44		185	228	
	3700	Fire door frames, 6-1/4" x 7'-0" x 2'-6" wide		16	1		143	35.50		178.50	216	
	3730	2'-8" wide		16	1		143	35.50		178.50	216	
	3760	3'-0" wide		15	1.067		143	38		181	220	
	3820	5'-0" wide		14	1.143		162	40.50		202.50	246	
	3850	5'-4" wide		14	1.143		171	40.50		211.50	256	
	3880	6'-0" wide		14	1.143		171	40.50		211.50	256	
	9000	Minimum labor/equipment charge	1 Carp	2	4	Job		142		142	236	
	08160 | Sliding Metal Doors and Grilles											
300	0010	**STEEL, SLIDING**										300
	0020	Up to 50' x 18', electric, standard duty, minimum	L-5	360	.156	S.F.	22	6.20	1.81	30.01	38	
	0100	Maximum		340	.165		37	6.60	1.91	45.51	54.50	
	0500	Heavy duty, minimum		297	.189		30	7.55	2.19	39.74	49	
	0600	Maximum		277	.202		79	8.10	2.35	89.45	104	
	08180 | Metal Screen and Storm Doors											
100	0010	**STORM DOORS & FRAMES** Aluminum, residential,										100
	0020	combination storm and screen										
	0400	Clear anodic coating, 6'-8" x 2'-6" wide	2 Carp	15	1.067	Ea.	162	38		200	242	
	0420	2'-8" wide		14	1.143		186	40.50		226.50	273	
	0440	3'-0" wide		14	1.143		186	40.50		226.50	273	
	0500	For 7' door height, add					5%					

08100 | Metal Doors and Frames

08180 | Metal Screen and Storm Doors

			Crew	Daily Output	Labor-Hours	Unit	2006 Bare Costs Mat.	2006 Bare Costs Labor	2006 Bare Costs Equip.	2006 Bare Costs Total	Total Incl O&P	
100	1000	Mill finish, 6′-8″ x 2′-6″ wide	2 Carp	15	1.067	Ea.	216	38		254	300	100
	1020	2′-8″ wide		14	1.143		216	40.50		256.50	305	
	1040	3′-0″ wide		14	1.143		234	40.50		274.50	325	
	1100	For 7′-0″ door, add					5%					
	1500	White painted, 6′-8″ x 2′-6″ wide	2 Carp	15	1.067		216	38		254	300	
	1520	2′-8″ wide		14	1.143		220	40.50		260.50	310	
	1540	3′-0″ wide		14	1.143		229	40.50		269.50	320	
	1600	For 7′-0″ door, add					5%					
	2000	Wood door & screen, see Division 08210-930										
	9000	Minimum labor/equipment charge	1 Carp	4	2	Job		71		71	118	

08200 | Wood and Plastic Doors

08210 | Wood Doors

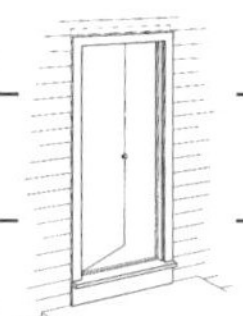

			Crew	Daily Output	Labor-Hours	Unit	2006 Bare Costs Mat.	2006 Bare Costs Labor	2006 Bare Costs Equip.	2006 Bare Costs Total	Total Incl O&P	
450	0010	**KALAMEIN**										450
	0020	Interior, flush type, 3′ x 7′	2 Carp	4.30	3.721	Opng.	176	132		308	410	
	9000	Minimum labor/equipment charge	1 Carp	2	4	Job		142		142	236	
720	0010	**PRE-HUNG DOORS**										720
	0300	Exterior, wood, comb. storm & screen, 6′-9″ x 2′-6″ wide	2 Carp	15	1.067	Ea.	283	38		321	375	
	0320	2′-8″ wide		15	1.067		283	38		321	375	
	0340	3′-0″ wide		15	1.067		291	38		329	385	
	0360	For 7′-0″ high door, add					24.50			24.50	27	
	0370	For aluminum storm doors, see Division 08180-100										
	1600	Entrance door, flush, birch, solid core										
	1620	4-5/8″ solid jamb, 1-3/4″ x 6′-8″ x 2′-8″ wide	2 Carp	16	1	Ea.	288	35.50		323.50	375	
	1640	3′-0″ wide	″	16	1		296	35.50		331.50	385	
	1680	For 7′-0″ high door, add					17.50			17.50	19.25	
	2000	Entrance door, colonial, 6 panel pine										
	2020	4-5/8″ solid jamb, 1-3/4″ x 6′-8″ x 2′-8″ wide	2 Carp	16	1	Ea.	520	35.50		555.50	630	
	2040	3′-0″ wide	″	16	1		520	35.50		555.50	630	
	2060	For 7′-0″ high door, add					48			48	52.50	
	2200	For 5-5/8″ solid jamb, add					38.50			38.50	42	
	4000	Interior, passage door, 4-5/8″ solid jamb										
	4400	Lauan, flush, solid core, 1-3/8″ x 6′-8″ x 2′-6″ wide	2 Carp	20	.800	Ea.	178	28.50		206.50	242	
	4420	2′-8″ wide		20	.800		178	28.50		206.50	242	
	4440	3′-0″ wide		19	.842		191	30		221	260	
	4600	Hollow core, 1-3/8″ x 6′-8″ x 2′-6″ wide		20	.800		120	28.50		148.50	179	
	4620	2′-8″ wide		20	.800		118	28.50		146.50	177	
	4640	3′-0″ wide		19	.842		121	30		151	183	
	4700	For 7′-0″ high door, add					23			23	25.50	
	5000	Birch, flush, solid core, 1-3/8″ x 6′-8″ x 2′-6″ wide	2 Carp	20	.800		166	28.50		194.50	229	
	5020	2′-8″ wide		20	.800		187	28.50		215.50	253	
	5040	3′-0″ wide		19	.842		200	30		230	270	
	5200	Hollow core, 1-3/8″ x 6′-8″ x 2′-6″ wide		20	.800		137	28.50		165.50	198	
	5220	2′-8″ wide		20	.800		144	28.50		172.50	205	
	5240	3′-0″ wide		19	.842		144	30		174	208	
	5280	For 7′-0″ high door, add					19.95			19.95	22	
	5500	Hardboard paneled, 1-3/8″ x 6′-8″ x 2′-6″ wide	2 Carp	20	.800		138	28.50		166.50	199	
	5520	2′-8″ wide		20	.800		145	28.50		173.50	206	

		08210 | Wood Doors	CREW	DAILY OUTPUT	LABOR-HOURS	UNIT	2006 BARE COSTS MAT.	LABOR	EQUIP.	TOTAL	TOTAL INCL O&P	
720	5540	3'-0" wide	2 Carp	19	.842	Ea.	143	30		173	207	720
	6000	Pine paneled, 1-3/8" x 6'-8" x 2'-6" wide		20	.800		240	28.50		268.50	310	
	6020	2'-8" wide		20	.800		261	28.50		289.50	335	
	6040	3'-0" wide		19	.842		266	30		296	340	
	6500	For 5-5/8" solid jamb, add					11.80			11.80	12.95	
	6520	For split jamb, deduct					13.95			13.95	15.35	
	9000	Minimum labor/equipment charge	1 Carp	4	2	Job		71		71	118	
850	0010	TIN CLAD										850
	0020	3 ply, 6' x 7', double sliding, doors only	2 Carp	1	16	Opng.	1,375	570		1,945	2,475	
	1000	For electric operator, add	1 Elec	2	4	"	2,450	168		2,618	2,950	
	9000	Minimum labor/equipment charge	2 Carp	1	16	Job		570		570	940	
900	0010	WOOD DOOR, ARCHITECTURAL										900
	0015	Flush, int., 1-3/8", 7 ply, hollow core,										
	0020	Lauan face, 2'-0" x 6'-8"	2 Carp	17	.941	Ea.	29	33.50		62.50	87.50	
	0040	2'-6" x 6'-8"		17	.941		33	33.50		66.50	92	
	0080	3'-0" x 6'-8"		17	.941		39	33.50		72.50	98.50	
	0100	4'-0" x 6'-8"		16	1		69.50	35.50		105	136	
	0120	Birch face, 2'-0" x 6'-8"		17	.941		46	33.50		79.50	106	
	0140	2'-6" x 6'-8"		17	.941		44.50	33.50		78	105	
	0180	3'-0" x 6'-8"		17	.941		50	33.50		83.50	111	
	0200	4'-0" x 6'-8"		16	1		95	35.50		130.50	163	
	0220	Oak face, 2'-0" x 6'-8"		17	.941		72.50	33.50		106	136	
	0240	2'-6" x 6'-8"		17	.941		77.50	33.50		111	141	
	0280	3'-0" x 6'-8"		17	.941		83	33.50		116.50	147	
	0300	4'-0" x 6'-8"		16	1		105	35.50		140.50	175	
	0320	Walnut face, 2'-0" x 6'-8"		17	.941		147	33.50		180.50	217	
	0340	2'-6" x 6'-8"		17	.941		150	33.50		183.50	220	
	0380	3'-0" x 6'-8"		17	.941		155	33.50		188.50	227	
	0400	4'-0" x 6'-8"		16	1		176	35.50		211.50	253	
	0430	For 7'-0" high, add					14.45			14.45	15.90	
	0440	For 8'-0" high, add					20.50			20.50	22.50	
	0480	For prefinishing, clear, add					32			32	35	
	0500	For prefinishing, stain, add					43.50			43.50	47.50	
	1320	M.D. overlay on hardboard, 2'-0" x 6'-8"	2 Carp	17	.941		89.50	33.50		123	154	
	1340	2'-6" x 6'-8"		17	.941		89.50	33.50		123	154	
	1380	3'-0" x 6'-8"		17	.941		106	33.50		139.50	173	
	1400	4'-0" x 6'-8"		16	1		146	35.50		181.50	220	
	1420	For 7'-0" high, add					7.95			7.95	8.75	
	1440	For 8'-0" high, add					21			21	23.50	
	1720	H.P. plastic laminate, 2'-0" x 6'-8"	2 Carp	16	1		213	35.50		248.50	294	
	1740	2'-6" x 6'-8"		16	1		213	35.50		248.50	294	
	1780	3'-0" x 6'-8"		15	1.067		246	38		284	335	
	1800	4'-0" x 6'-8"		14	1.143		340	40.50		380.50	445	
	1820	For 7'-0" high, add					8.35			8.35	9.20	
	1840	For 8'-0" high, add					21.50			21.50	23.50	
	2020	5 ply particle core, lauan face, 2'-6" x 6'-8"	2 Carp	15	1.067		73	38		111	144	
	2040	3'-0" x 6'-8"		14	1.143		76	40.50		116.50	151	
	2080	3'-0" x 7'-0"		13	1.231		85	44		129	166	
	2100	4'-0" x 7'-0"		12	1.333		98	47.50		145.50	187	
	2120	Birch face, 2'-6" x 6'-8"		15	1.067		83	38		121	155	
	2140	3'-0" x 6'-8"		14	1.143		91	40.50		131.50	168	
	2180	3'-0" x 7'-0"		13	1.231		93	44		137	175	
	2200	4'-0" x 7'-0"		12	1.333		113	47.50		160.50	204	
	2220	Oak face, 2'-6" x 6'-8"		15	1.067		92	38		130	164	
	2240	3'-0" x 6'-8"		14	1.143		101	40.50		141.50	179	

08210 | Wood Doors

		08210 Wood Doors	CREW	DAILY OUTPUT	LABOR-HOURS	UNIT	2006 BARE COSTS MAT.	LABOR	EQUIP.	TOTAL	TOTAL INCL O&P	
900	2280	3′-0″ x 7′-0″	2 Carp	13	1.231	Ea.	104	44		148	187	900
	2300	4′-0″ x 7′-0″		12	1.333		127	47.50		174.50	219	
	2320	Walnut face, 2′-0″ x 6′-8″		15	1.067		102	38		140	175	
	2340	2′-6″ x 6′-8″		14	1.143		116	40.50		156.50	196	
	2380	3′-0″ x 6′-8″		13	1.231		131	44		175	217	
	2400	4′-0″ x 6′-8″		12	1.333		171	47.50		218.50	267	
	2440	For 8′-0″ high, add					25			25	27.50	
	2460	For 8′-0″ high walnut, add					13.65			13.65	15.05	
	2480	For solid wood core, add					30.50			30.50	33.50	
	2720	For prefinishing, clear, add					19.95			19.95	22	
	2740	For prefinishing, stain, add					44.50			44.50	49	
	3320	M.D. overlay on hardboard, 2′-6″ x 6′-8″	2 Carp	14	1.143		89.50	40.50		130	166	
	3340	3′-0″ x 6′-8″		13	1.231		93.50	44		137.50	176	
	3380	3′-0″ x 7′-0″		12	1.333		95.50	47.50		143	184	
	3400	4′-0″ x 7′-0″		10	1.600		117	57		174	222	
	3440	For 8′-0″ height, add					26			26	28.50	
	3460	For solid wood core, add					34			34	37.50	
	3720	H.P. plastic laminate, 2′-6″ x 6′-8″	2 Carp	13	1.231		131	44		175	217	
	3740	3′-0″ x 6′-8″		12	1.333		148	47.50		195.50	242	
	3780	3′-0″ x 7′-0″		11	1.455		154	51.50		205.50	255	
	3800	4′-0″ x 7′-0″		8	2		187	71		258	325	
	3840	For 8′-0″ height, add					26			26	28.50	
	3860	For solid wood core, add					32			32	35	
	4000	Exterior, flush, solid wood stave core, birch, 1-3/4″ x 7′-0″ x 2′-6″	2 Carp	15	1.067		158	38		196	236	
	4020	2′-8″ wide		15	1.067		165	38		203	244	
	4040	3′-0″ wide		14	1.143		176	40.50		216.50	262	
	4100	Oak faced 1-3/4″ x 7′-0″ x 2′-6″ wide		15	1.067		174	38		212	254	
	4120	2′-8″ wide		15	1.067		186	38		224	267	
	4140	3′-0″ wide		14	1.143		198	40.50		238.50	286	
	4200	Walnut faced, 1-3/4″ x 7′-0″ x 2′-6″ wide		15	1.067		255	38		293	345	
	4220	2′-8″ wide		15	1.067		267	38		305	355	
	4240	3′-0″ wide		14	1.143		278	40.50		318.50	375	
	4300	For 6′-8″ high door, deduct from 7′-0″ door					14.20			14.20	15.60	
	9000	Minimum labor/equipment charge	1 Carp	4	2	Job		71		71	118	
910	0010	**WOOD DOORS, DECORATOR**										910
	3000	Solid wood, 1-3/4″ thick stile and rail										
	3020	Mahogany, 3′-0″ x 7′-0″, minimum	2 Carp	14	1.143	Ea.	945	40.50		985.50	1,125	
	3030	Maximum		10	1.600		1,300	57		1,357	1,550	
	3040	3′-6″ x 8′-0″, minimum		10	1.600		895	57		952	1,075	
	3050	Maximum		8	2		1,625	71		1,696	1,925	
	3100	Pine, 3′-0″ x 7′-0″, minimum		14	1.143		425	40.50		465.50	540	
	3110	Maximum		10	1.600		705	57		762	870	
	3120	3′-6″ x 8′-0″, minimum		10	1.600		655	57		712	815	
	3130	Maximum		8	2		1,075	71		1,146	1,325	
	3200	Red oak, 3′-0″ x 7′-0″, minimum		14	1.143		1,475	40.50		1,515.50	1,700	
	3210	Maximum		10	1.600		1,750	57		1,807	2,025	
	3220	3′-6″ x 8′-0″, minimum		10	1.600		1,600	57		1,657	1,875	
	3230	Maximum		8	2		2,850	71		2,921	3,275	
	4000	Hand carved door, mahogany										
	4020	3′-0″ x 7′-0″, minimum	2 Carp	14	1.143	Ea.	1,450	40.50		1,490.50	1,650	
	4030	Maximum		11	1.455		2,825	51.50		2,876.50	3,200	
	4040	3′-6″ x 8′-0″, minimum		10	1.600		1,800	57		1,857	2,100	
	4050	Maximum		8	2		2,775	71		2,846	3,200	
	4200	Red oak, 3′-0″ x 7′-0″, minimum		14	1.143		4,400	40.50		4,440.50	4,925	
	4210	Maximum		11	1.455		12,100	51.50		12,151.50	13,400	
	4220	3′-6″ x 8′-0″, minimum		10	1.600		4,950	57		5,007	5,550	

	08210 | Wood Doors		CREW	DAILY OUTPUT	LABOR-HOURS	UNIT	2006 BARE COSTS MAT.	LABOR	EQUIP.	TOTAL	TOTAL INCL O&P	
910	4280	For 6'-8" high door, deduct from 7'-0" door				Ea.	31.50			31.50	34.50	910
	4400	For custom finish, add					335			335	370	
	4600	Side light, mahogany, 7'-0" x 1'-6" wide, minimum	2 Carp	18	.889		810	31.50		841.50	945	
	4610	Maximum		14	1.143		2,375	40.50		2,415.50	2,700	
	4620	8'-0" x 1'-6" wide, minimum		14	1.143		1,525	40.50		1,565.50	1,750	
	4630	Maximum		10	1.600		1,725	57		1,782	2,000	
	4640	Side light, oak, 7'-0" x 1'-6" wide, minimum		18	.889		935	31.50		966.50	1,075	
	4650	Maximum		14	1.143		1,675	40.50		1,715.50	1,900	
	4660	8'-0" x 1-6" wide, minimum		14	1.143		880	40.50		920.50	1,050	
	4670	Maximum		10	1.600		1,675	57		1,732	1,925	
	6520	Interior cafe doors, 2'-6" opening, stock, panel pine		16	1		188	35.50		223.50	265	
	6540	3'-0" opening	↓	16	1	↓	196	35.50		231.50	274	
	6550	Louvered pine										
	6560	2'-6" opening	2 Carp	16	1	Ea.	164	35.50		199.50	239	
	8000	3'-0" opening		16	1		175	35.50		210.50	251	
	8010	2'-6" opening, hardwood		16	1		282	35.50		317.50	370	
	8020	3'-0" opening	↓	16	1	↓	310	35.50		345.50	405	
	8800	Pre-hung doors see Division 08210-720										
	9000	Minimum labor/equipment charge	1 Carp	4	2	Job		71		71	118	
920	0010	**WOOD DOORS, PANELED**										920
	0020	Interior, six panel, hollow core, 1-3/8" thick										
	0040	Molded hardboard, 2'-0" x 6'-8"	2 Carp	17	.941	Ea.	47.50	33.50		81	108	
	0060	2'-6" x 6'-8"		17	.941		51	33.50		84.50	112	
	0070	2'-8" x 6'-8"		17	.941		53.50	33.50		87	115	
	0080	3'-0" x 6'-8"		17	.941		56.50	33.50		90	118	
	0140	Embossed print, molded hardboard, 2'-0" x 6'-8"		17	.941		51	33.50		84.50	112	
	0160	2'-6" x 6'-8"		17	.941		51	33.50		84.50	112	
	0180	3'-0" x 6'-8"		17	.941		56.50	33.50		90	118	
	0540	Six panel, solid, 1-3/8" thick, pine, 2'-0" x 6'-8"		15	1.067		124	38		162	199	
	0560	2'-6" x 6'-8"		14	1.143		139	40.50		179.50	221	
	0580	3'-0" x 6'-8"		13	1.231		160	44		204	249	
	1020	Two panel, bored rail, solid, 1-3/8" thick, pine, 1'-6" x 6'-8"		16	1		226	35.50		261.50	310	
	1040	2'-0" x 6'-8"		15	1.067		297	38		335	390	
	1060	2'-6" x 6'-8"		14	1.143		340	40.50		380.50	445	
	1340	Two panel, solid, 1-3/8" thick, fir, 2'-0" x 6'-8"		15	1.067		124	38		162	199	
	1360	2'-6" x 6'-8"		14	1.143		139	40.50		179.50	221	
	1380	3'-0" x 6'-8"		13	1.231		340	44		384	450	
	1740	Five panel, solid, 1-3/8" thick, fir, 2'-0" x 6'-8"		15	1.067		222	38		260	305	
	1760	2'-6" x 6'-8"		14	1.143		355	40.50		395.50	460	
	1780	3'-0" x 6'-8"	↓	13	1.231	↓	355	44		399	465	
	9000	Minimum labor/equipment charge	1 Carp	4	2	Job		71		71	118	
930	0010	**WOOD DOORS, RESIDENTIAL**										930
	0200	Exterior, combination storm & screen, pine										
	0220	Cross buck, 6'-9" x 2'-6" wide	2 Carp	11	1.455	Ea.	310	51.50		361.50	425	
	0260	2'-8" wide		10	1.600		272	57		329	395	
	0280	3'-0" wide		9	1.778		278	63		341	410	
	0300	7'-1" x 3'-0" wide		9	1.778		298	63		361	435	
	0400	Full lite, 6'-9" x 2'-6" wide		11	1.455		290	51.50		341.50	405	
	0420	2'-8" wide		10	1.600		290	57		347	415	
	0440	3'-0" wide		9	1.778		299	63		362	435	
	0500	7'-1" x 3'-0" wide		9	1.778		320	63		383	460	
	0700	Dutch door, pine, 1-3/4" x 6'-8" x 2'-8" wide, minimum		12	1.333		675	47.50		722.50	820	
	0720	Maximum		10	1.600		710	57		767	880	
	0800	3'-0" wide, minimum		12	1.333		685	47.50		732.50	835	
	0820	Maximum	↓	10	1.600	↓	750	57		807	920	

08210 | Wood Doors		CREW	DAILY OUTPUT	LABOR-HOURS	UNIT	2006 BARE COSTS MAT.	LABOR	EQUIP.	TOTAL	TOTAL INCL O&P
930 1000	Entrance door, colonial, 1-3/4" x 6'-8" x 2'-8" wide	2 Carp	16	1	Ea.	355	35.50		390.50	455
1020	6 panel pine, 3'-0" wide		15	1.067		400	38		438	505
1100	8 panel pine, 2'-8" wide		16	1		595	35.50		630.50	715
1120	3'-0" wide		15	1.067		535	38		573	655
1200	For tempered safety glass lites, (min of 2) add					62.50			62.50	68.50
1300	Flush, birch, solid core, 1-3/4" x 6'-8" x 2'-8" wide	2 Carp	16	1		94.50	35.50		130	163
1320	3'-0" wide		15	1.067		98	38		136	171
1350	7'-0" x 2'-8" wide		16	1		104	35.50		139.50	173
1360	3'-0" wide		15	1.067		111	38		149	185
1380	For tempered safety glass lites, add					93			93	102
2700	Interior, closet, bi-fold, w/hardware, no frame or trim incl.									
2720	Flush, birch, 6'-6" or 6'-8" x 2'-6" wide	2 Carp	13	1.231	Ea.	48.50	44		92.50	126
2740	3'-0" wide		13	1.231		52.50	44		96.50	131
2760	4'-0" wide		12	1.333		96	47.50		143.50	184
2780	5'-0" wide		11	1.455		97	51.50		148.50	193
2800	6'-0" wide		10	1.600		103	57		160	208
3000	Raised panel pine, 6'-6" or 6'-8" x 2'-6" wide		13	1.231		154	44		198	242
3020	3'-0" wide		13	1.231		196	44		240	288
3040	4'-0" wide		12	1.333		300	47.50		347.50	410
3060	5'-0" wide		11	1.455		360	51.50		411.50	480
3080	6'-0" wide		10	1.600		395	57		452	530
3200	Louvered, pine 6'-6" or 6'-8" x 2'-6" wide		13	1.231		104	44		148	187
3220	3'-0" wide		13	1.231		157	44		201	246
3240	4'-0" wide		12	1.333		194	47.50		241.50	292
3260	5'-0" wide		11	1.455		220	51.50		271.50	330
3280	6'-0" wide		10	1.600		243	57		300	360
4400	Bi-passing closet, incl. hardware and frame, no trim incl.									
4420	Flush, lauan, 6'-8" x 4'-0" wide	2 Carp	12	1.333	Opng.	164	47.50		211.50	260
4440	5'-0" wide		11	1.455		180	51.50		231.50	284
4460	6'-0" wide		10	1.600		193	57		250	305
4600	Flush, birch, 6'-8" x 4'-0" wide		12	1.333		201	47.50		248.50	300
4620	5'-0" wide		11	1.455		203	51.50		254.50	310
4640	6'-0" wide		10	1.600		241	57		298	360
4800	Louvered, pine, 6'-8" x 4'-0" wide		12	1.333		400	47.50		447.50	520
4820	5'-0" wide		11	1.455		385	51.50		436.50	505
4840	6'-0" wide		10	1.600		490	57		547	635
5000	Paneled, pine, 6'-8" x 4'-0" wide		12	1.333		380	47.50		427.50	500
5020	5'-0" wide		11	1.455		400	51.50		451.50	525
5040	6'-0" wide		10	1.600		470	57		527	615
6100	Folding accordion, closet, including track and frame									
6120	Vinyl, 2 layer, stock (see also Division 10651-100)	2 Carp	400	.040	S.F.	3.09	1.42		4.51	5.75
6180	Economy vinyl, stock		400	.040		1.74	1.42		3.16	4.27
6200	Rigid PVC		400	.040		4.63	1.42		6.05	7.45
6220	For custom partition, add					25%	10%			
7310	Passage doors, flush, no frame included									
7320	Hardboard, hollow core, 1-3/8" x 6'-8" x 1'-6" wide	2 Carp	18	.889	Ea.	40	31.50		71.50	96.50
7330	2'-0" wide		18	.889		40.50	31.50		72	97
7340	2'-6" wide		18	.889		44.50	31.50		76	102
7350	2'-8" wide		18	.889		47	31.50		78.50	105
7360	3'-0" wide		17	.941		49.50	33.50		83	110
7420	Lauan, hollow core, 1-3/8" x 6'-8" x 1'-6" wide		18	.889		28	31.50		59.50	83.50
7440	2'-0" wide		18	.889		27	31.50		58.50	82.50
7450	2'-4" wide		18	.889		30.50	31.50		62	86
7460	2'-6" wide		18	.889		30.50	31.50		62	86
7480	2'-8" wide		18	.889		32	31.50		63.50	88
7500	3'-0" wide		17	.941		33.50	33.50		67	92.50

	08210 | Wood Doors		CREW	DAILY OUTPUT	LABOR-HOURS	UNIT	2006 BARE COSTS MAT.	LABOR	EQUIP.	TOTAL	TOTAL INCL O&P	
930	7700	Birch, hollow core, 1-3/8" x 6'-8" x 1'-6" wide	2 Carp	18	.889	Ea.	35.50	31.50		67	91.50	930
	7720	2'-0" wide		18	.889		40	31.50		71.50	96.50	
	7740	2'-6" wide		18	.889		44.50	31.50		76	102	
	7760	2'-8" wide		18	.889		46	31.50		77.50	103	
	7780	3'-0" wide		17	.941		50	33.50		83.50	111	
	8000	Pine louvered, 1-3/8" x 6'-8" x 1'-6" wide		19	.842		100	30		130	160	
	8020	2'-0" wide		18	.889		125	31.50		156.50	191	
	8040	2'-6" wide		18	.889		137	31.50		168.50	203	
	8060	2'-8" wide		18	.889		144	31.50		175.50	211	
	8080	3'-0" wide		17	.941		154	33.50		187.50	226	
	8300	Pine paneled, 1-3/8" x 6'-8" x 1'-6" wide		19	.842		108	30		138	169	
	8320	2'-0" wide		18	.889		125	31.50		156.50	191	
	8330	2'-4" wide		18	.889		137	31.50		168.50	203	
	8340	2'-6" wide		18	.889		140	31.50		171.50	207	
	8360	2'-8" wide		18	.889		151	31.50		182.50	220	
	8380	3'-0" wide	↓	17	.941	↓	158	33.50		191.50	230	
	8550	For over 20 doors, deduct					15%					
	9900	Minimum labor/equipment charge	1 Carp	4	2	Job		71		71	118	
950	0010	**WOOD FIRE DOORS**										950
	0020	Particle core, 7 face plys, "B" label,										
	0040	1 hour, birch face, 1-3/4" x 2'-6" x 6'-8"	2 Carp	14	1.143	Ea.	298	40.50		338.50	400	
	0080	3'-0" x 6'-8"		13	1.231		310	44		354	415	
	0090	3'-0" x 7'-0"		12	1.333		320	47.50		367.50	435	
	0100	4'-0" x 7'-0"		12	1.333		430	47.50		477.50	550	
	0140	Oak face, 2'-6" x 6'-8"		14	1.143		298	40.50		338.50	400	
	0180	3'-0" x 6'-8"		13	1.231		310	44		354	415	
	0190	3'-0" x 7'-0"		12	1.333		325	47.50		372.50	435	
	0200	4'-0" x 7'-0"		12	1.333		420	47.50		467.50	545	
	0240	Walnut face, 2'-6" x 6'-8"		14	1.143		390	40.50		430.50	500	
	0280	3'-0" x 6'-8"		13	1.231		400	44		444	515	
	0290	3'-0" x 7'-0"		12	1.333		420	47.50		467.50	540	
	0300	4'-0" x 7'-0"		12	1.333		565	47.50		612.50	700	
	0440	M.D. overlay on hardboard, 2'-6" x 6'-8"		15	1.067		262	38		300	350	
	0480	3'-0" x 6'-8"		14	1.143		272	40.50		312.50	370	
	0490	3'-0" x 7'-0"		13	1.231		287	44		331	390	
	0500	4'-0" x 7'-0"		12	1.333		350	47.50		397.50	465	
	0540	H.P. plastic laminate, 2'-6" x 6'-8"		13	1.231		350	44		394	460	
	0580	3'-0" x 6'-8"		12	1.333		370	47.50		417.50	485	
	0590	3'-0" x 7'-0"		11	1.455		370	51.50		421.50	495	
	0600	4'-0" x 7'-0"		10	1.600		475	57		532	615	
	0740	90 minutes, birch face, 1-3/4" x 2'-6" x 6'-8"		14	1.143		256	40.50		296.50	350	
	0780	3'-0" x 6'-8"		13	1.231		262	44		306	360	
	0790	3'-0" x 7'-0"		12	1.333		315	47.50		362.50	430	
	0800	4'-0" x 7'-0"		12	1.333		380	47.50		427.50	495	
	0840	Oak face, 2'-6" x 6'-8"		14	1.143		269	40.50		309.50	365	
	0880	3'-0" x 6'-8"		13	1.231		279	44		323	380	
	0890	3'-0" x 7'-0"		12	1.333		293	47.50		340.50	400	
	0900	4'-0" x 7'-0"		12	1.333		405	47.50		452.50	525	
	0940	Walnut face, 2'-6" x 6'-8"		14	1.143		370	40.50		410.50	475	
	0980	3'-0" x 6'-8"		13	1.231		380	44		424	490	
	0990	3'-0" x 7'-0"		12	1.333		395	47.50		442.50	515	
	1000	4'-0" x 7'-0"		12	1.333		570	47.50		617.50	705	
	1140	M.D. overlay on hardboard, 2'-6" x 6'-8"		15	1.067		292	38		330	385	
	1180	3'-0" x 6'-8"		14	1.143		300	40.50		340.50	400	
	1190	3'-0" x 7'-0"		13	1.231		310	44		354	420	
	1200	4'-0" x 7'-0"	↓	12	1.333	↓	425	47.50		472.50	545	

08200 | Wood and Plastic Doors

	08210 | Wood Doors		CREW	DAILY OUTPUT	LABOR-HOURS	UNIT	2006 BARE COSTS MAT.	LABOR	EQUIP.	TOTAL	TOTAL INCL O&P	
950	1240	For 8'-0" height, add				Ea.	53.50			53.50	59	950
	1260	For 8'-0" height walnut, add					71			71	78	
	1340	H.P. plastic laminate, 2'-6" x 6'-8"	2 Carp	13	1.231		365	44		409	475	
	1380	3'-0" x 6'-8"		12	1.333		380	47.50		427.50	500	
	1390	3'-0" x 7'-0"		11	1.455		385	51.50		436.50	510	
	1400	4'-0" x 7'-0"		10	1.600		500	57		557	645	
	2200	Custom architectural "B" label, flush, 1-3/4" thick, birch,										
	2210	Solid core										
	2220	2'-6" x 7'-0"	2 Carp	15	1.067	Ea.	248	38		286	335	
	2260	3'-0" x 7'-0"		14	1.143		257	40.50		297.50	350	
	2300	4'-0" x 7'-0"		13	1.231		360	44		404	470	
	2420	4'-0" x 8'-0"		11	1.455		450	51.50		501.50	580	
	2480	For oak veneer, add					50%					
	2500	For walnut veneer, add					75%					
	9000	Minimum labor/equipment charge	1 Carp	4	2	Job		71		71	118	
960	0010	**WOOD FRAMES**										960
	0400	Exterior frame, incl. ext. trim, pine, 5/4 x 4-9/16" deep	2 Carp	375	.043	L.F.	5.10	1.52		6.62	8.10	
	0420	5-3/16" deep		375	.043		7.95	1.52		9.47	11.25	
	0440	6-9/16" deep		375	.043		7.90	1.52		9.42	11.15	
	0600	Oak, 5/4 x 4-9/16" deep		350	.046		9.45	1.63		11.08	13.05	
	0620	5-3/16" deep		350	.046		10.60	1.63		12.23	14.35	
	0640	6-9/16" deep		350	.046		11.80	1.63		13.43	15.70	
	0800	Walnut, 5/4 x 4-9/16" deep		350	.046		11.10	1.63		12.73	14.90	
	0820	5-3/16" deep		350	.046		16	1.63		17.63	20.50	
	0840	6-9/16" deep		350	.046		18.95	1.63		20.58	23.50	
	1000	Sills, 8/4 x 8" deep, oak, no horns		100	.160		17.55	5.70		23.25	29	
	1020	2" horns		100	.160		15.45	5.70		21.15	26.50	
	1040	3" horns		100	.160		16	5.70		21.70	27	
	1100	8/4 x 10" deep, oak, no horns		90	.178		22.50	6.30		28.80	35	
	1120	2" horns		90	.178		21	6.30		27.30	33.50	
	1140	3" horns		90	.178		21	6.30		27.30	33.50	
	2000	Exterior, colonial, frame & trim, 3' opng., in-swing, minimum		22	.727	Ea.	310	26		336	385	
	2010	Average		21	.762		460	27		487	555	
	2020	Maximum		20	.800		1,050	28.50		1,078.50	1,200	
	2100	5'-4" opening, in-swing, minimum		17	.941		355	33.50		388.50	445	
	2120	Maximum		15	1.067		1,050	38		1,088	1,225	
	2140	Out-swing, minimum		17	.941		360	33.50		393.50	450	
	2160	Maximum		15	1.067		1,100	38		1,138	1,275	
	2400	6'-0" opening, in-swing, minimum		16	1		340	35.50		375.50	430	
	2420	Maximum		10	1.600		1,100	57		1,157	1,300	
	2460	Out-swing, minimum		16	1		365	35.50		400.50	460	
	2480	Maximum		10	1.600		1,275	57		1,332	1,525	
	2600	For two sidelights, add, minimum		30	.533	Opng.	345	18.95		363.95	410	
	2620	Maximum		20	.800	"	1,125	28.50		1,153.50	1,275	
	2700	Custom birch frame, 3'-0" opening		16	1	Ea.	199	35.50		234.50	278	
	2750	6'-0" opening		16	1		300	35.50		335.50	390	
	2900	Exterior, modern, plain trim, 3' opng., in-swing, minimum		26	.615		33.50	22		55.50	73.50	
	2920	Average		24	.667		40	23.50		63.50	83.50	
	2940	Maximum		22	.727		48.50	26		74.50	96.50	
	3000	Interior frame, pine, 11/16" x 3-5/8" deep		375	.043	L.F.	4.31	1.52		5.83	7.25	
	3020	4-9/16" deep		375	.043		5.80	1.52		7.32	8.90	
	3200	Oak, 11/16" x 3-5/8" deep		350	.046		4	1.63		5.63	7.10	
	3220	4-9/16" deep		350	.046		4.31	1.63		5.94	7.45	
	3240	5-3/16" deep		350	.046		4.47	1.63		6.10	7.60	
	3400	Walnut, 11/16" x 3-5/8" deep		350	.046		6.75	1.63		8.38	10.15	

08200 | Wood and Plastic Doors

	08210 | Wood Doors		CREW	DAILY OUTPUT	LABOR-HOURS	UNIT	2006 BARE COSTS MAT.	LABOR	EQUIP.	TOTAL	TOTAL INCL O&P	
960	3420	4-9/16" deep	2 Carp	350	.046	L.F.	7.10	1.63		8.73	10.50	960
	3440	5-3/16" deep		350	.046		7.40	1.63		9.03	10.85	
	3600	Pocket door frame		16	1	Ea.	62.50	35.50		98	128	
	3800	Threshold, oak, 5/8" x 3-5/8" deep		200	.080	L.F.	2.60	2.84		5.44	7.55	
	3820	4-5/8" deep		190	.084		3.20	2.99		6.19	8.50	
	3840	5-5/8" deep		180	.089		5.65	3.16		8.81	11.45	
	4000	For casing see division 06220-400 & 06220-800										
	9000	Minimum labor/equipment charge	1 Carp	4	2	Job		71		71	118	

	08220 | Plastic Doors		CREW	DAILY OUTPUT	LABOR-HOURS	UNIT	MAT.	LABOR	EQUIP.	TOTAL	TOTAL INCL O&P	
100	0010	FIBERGLASS DOORS										100
	0020	Exterior, fiberglass, door, 2'-8" wide x 6'-8" high	2 Carp	15	1.067	Ea.	305	38		343	400	
	0040	3'-0" wide x 6'-8" high		15	1.067		305	38		343	400	
	0060	3'-0" wide x 7'-0" high		15	1.067		375	38		413	480	
	0080	3'-0" wide x 6'-8" high, with two lites		15	1.067		370	38		408	470	
	0100	3'-0" wide x 7'-0" high, with two lites		15	1.067		440	38		478	550	
	0110	Half glass, 3'-0" wide x 6'-8" high		15	1.067		415	38		453	520	
	0120	3'-0" wide x 6'-8" high, low e		15	1.067		440	38		478	545	
	0130	3'-0" wide x 7'-0" high		15	1.067		485	38		523	595	
	0140	3'-0" wide x 7'-0" high, low e		15	1.067		510	38		548	625	
	0150	Side lights, 1'-0" wide x 6'-8" high,					281			281	310	
	0160	1'-0" wide x 6'-8" high, low e					293			293	325	
	0180	1'-0" wide x 6'-8" high, full glass					320			320	355	
	0190	1'-0" wide x 6'-8" high, low e					345			345	380	

	08260 | Sliding Wood and Plastic Doors		CREW	DAILY OUTPUT	LABOR-HOURS	UNIT	MAT.	LABOR	EQUIP.	TOTAL	TOTAL INCL O&P	
700	0010	GLASS, SLIDING, VINYL										700
	0012	Vinyl clad, 1" insul. glass, 6'-0" x 6'-10" high	2 Carp	4	4	Opng.	1,250	142		1,392	1,625	
	0030	6'-0" x 8'-0" high		4	4	Ea.	1,900	142		2,042	2,325	
	0100	8'-0" x 6'-10" high		4	4	Opng.	1,950	142		2,092	2,375	
	0500	3 leaf, 9'-0" x 6'-10" high		3	5.333		1,775	190		1,965	2,275	
	0600	12'-0" x 6'-10" high		3	5.333		2,200	190		2,390	2,750	
	9000	Minimum labor/equipment charge	1 Carp	4	2	Job		71		71	118	

08300 | Specialty Doors

	08310 | Access Doors and Panels		CREW	DAILY OUTPUT	LABOR-HOURS	UNIT	2006 BARE COSTS MAT.	LABOR	EQUIP.	TOTAL	TOTAL INCL O&P	
100	0010	ACCESS DOORS										100
	1000	Fire rated door with lock										
	1100	Metal, 12" x 12"	1 Carp	10	.800	Ea.	143	28.50		171.50	204	
	1150	18" x 18"		9	.889		186	31.50		217.50	258	
	1200	24" x 24"		9	.889		225	31.50		256.50	300	
	1250	24" x 36"		8	1		300	35.50		335.50	390	
	1300	24" x 48"		8	1		375	35.50		410.50	470	
	1350	36" x 36"		7.50	1.067		450	38		488	560	
	1400	48" x 48"		7.50	1.067		575	38		613	700	
	1600	Stainless steel, 12" x 12"		10	.800		255	28.50		283.50	325	
	1650	18" x 18"		9	.889		370	31.50		401.50	460	
	1700	24" x 24"		9	.889		455	31.50		486.50	555	

 Important: See the Reference Section for supporting data - Crews, Rental Equipment, City Cost Indexes and Reference Data

	08310 | Access Doors and Panels		CREW	DAILY OUTPUT	LABOR-HOURS	UNIT	2006 BARE COSTS MAT.	LABOR	EQUIP.	TOTAL	TOTAL INCL O&P	
100	1750	24″ x 36″	1 Carp	8	1	Ea.	580	35.50		615.50	700	100
	2000	Flush door for finishing										
	2100	Metal 8″ x 8″	1 Carp	10	.800	Ea.	41	28.50		69.50	92	
	2150	12″ x 12″	″	10	.800	″	46	28.50		74.50	97.50	
	3000	Recessed door for acoustic tile										
	3100	Metal, 12″ x 12″	1 Carp	4.50	1.778	Ea.	63	63		126	174	
	3150	12″ x 24″		4.50	1.778		82	63		145	196	
	3200	24″ x 24″		4	2		110	71		181	239	
	3250	24″ x 36″		4	2		141	71		212	274	
	4000	Recessed door for drywall										
	4100	Metal 12″ x 12″	1 Carp	6	1.333	Ea.	70	47.50		117.50	156	
	4150	12″ x 24″		5.50	1.455		104	51.50		155.50	200	
	4200	24″ x 36″		5	1.600		166	57		223	276	
	6000	Standard door										
	6100	Metal, 8″ x 8″	1 Carp	10	.800	Ea.	36.50	28.50		65	87	
	6150	12″ x 12″		10	.800		41	28.50		69.50	92	
	6200	18″ x 18″		9	.889		57	31.50		88.50	115	
	6250	24″ x 24″		9	.889		73.50	31.50		105	134	
	6300	24″ x 36″		8	1		109	35.50		144.50	179	
	6350	36″ x 36″		8	1		133	35.50		168.50	205	
	6500	Stainless steel, 8″ x 8″		10	.800		72.50	28.50		101	127	
	6550	12″ x 12″		10	.800		96.50	28.50		125	153	
	6600	18″ x 18″		9	.889		179	31.50		210.50	250	
	6650	24″ x 24″		9	.889		234	31.50		265.50	310	
	7010	Aluminum cover	G-3	11	2.909		535	101		636	750	
	9000	Minimum labor/equipment charge	1 Carp	4	2	Job		71		71	118	
150	0010	**BULKHEAD CELLAR DOORS**										150
	0020	Steel, not incl. sides, 44″ x 62″	1 Carp	5.50	1.455	Ea.	214	51.50		265.50	320	
	0100	52″ x 73″		5.10	1.569		238	56		294	355	
	0500	With sides and foundation plates, 57″ x 45″ x 24″		4.70	1.702		280	60.50		340.50	410	
	0600	42″ x 49″ x 51″		4.30	1.860		335	66		401	480	
	9000	Minimum labor/equipment charge		2	4	Job		142		142	236	
300	0010	**FLOOR, COMMERCIAL**										300
	0020	Aluminum tile, steel frame, one leaf, 2′ x 2′ opng.	2 Sswk	3.50	4.571	Opng.	385	183		568	760	
	0050	3′-6″ x 3′-6″ opening		3.50	4.571		695	183		878	1,100	
	0500	Double leaf, 4′ x 4′ opening		3	5.333		1,025	213		1,238	1,525	
	0550	5′ x 5′ opening		3	5.333		1,500	213		1,713	2,050	
	9000	Minimum labor/equipment charge		2	8	Job		320		320	600	
350	0010	**FLOOR, INDUSTRIAL**										350
	0020	Steel 300 psf L.L., single leaf, 2′ x 2′, 175#	2 Sswk	6	2.667	Opng.	565	107		672	820	
	0050	3′ x 3′ opening, 300#		5.50	2.909		780	116		896	1,075	
	0300	Double leaf, 4′ x 4′ opening, 455#		5	3.200		1,175	128		1,303	1,550	
	0350	5′ x 5′ opening, 645#		4.50	3.556		1,525	142		1,667	1,975	
	1000	Aluminum, 300 psf L.L., single leaf, 2′ x 2′, 60#		6	2.667		555	107		662	810	
	1050	3′ x 3′ opening, 100#		5.50	2.909		865	116		981	1,175	
	1500	Double leaf, 4′ x 4′ opening, 160#		5	3.200		1,375	128		1,503	1,750	
	1550	5′ x 5′ opening, 235#		4.50	3.556		1,800	142		1,942	2,275	
	2000	Aluminum, 150 psf L.L., single leaf, 2′ x 2′, 60#		6	2.667		510	107		617	760	
	2050	3′ x 3′ opening, 95#		5.50	2.909		755	116		871	1,050	
	2500	Double leaf, 4′ x 4′ opening, 150#		5	3.200		1,200	128		1,328	1,575	
	2550	5′ x 5′ opening, 230#		4.50	3.556		1,625	142		1,767	2,075	
	9000	Minimum labor/equipment charge		2	8	Job		320		320	600	
400	0010	**KENNEL DOORS**										400
	0020	2 way, swinging type, 13″ x 19″ opening	2 Carp	11	1.455	Opng.	207	51.50		258.50	315	
	0100	17″ x 29″ opening	″	11	1.455	″	223	51.50		274.50	330	

For expanded coverage of these items see *Means Interior Cost Data 2006*

08300 | Specialty Doors

	08320 | Detention Doors and Frames		CREW	DAILY OUTPUT	LABOR-HOURS	UNIT	2006 BARE COSTS MAT.	LABOR	EQUIP.	TOTAL	TOTAL INCL O&P	
900	0010	**SECURITY GATES** See Division 10610-100										900
950	0010	**VAULT FRONT** See Division 11021-800										950

08330 | Coiling Doors and Grilles

			CREW	DAILY OUTPUT	LABOR-HOURS	UNIT	MAT.	LABOR	EQUIP.	TOTAL	TOTAL INCL O&P	
130	0010	**COUNTER DOORS** (security item)										130
	0020	Manual, incl. frm. and hdwe., galv. stl., 4′ roll-up, 6′ long	2 Carp	2	8	Opng.	950	284		1,234	1,525	
	0300	Galvanized steel, UL label		1.80	8.889		1,050	315		1,365	1,675	
	0600	Stainless steel, 4′ high roll-up, 6′ long		2	8		1,450	284		1,734	2,075	
	0700	10′ long		1.80	8.889		2,100	315		2,415	2,825	
	2000	Aluminum, 4′ high, 4′ long		2.20	7.273		920	259		1,179	1,425	
	2020	6′ long		2	8		1,025	284		1,309	1,600	
	2040	8′ long		1.90	8.421		1,175	299		1,474	1,800	
	2060	10′ long		1.80	8.889		1,400	315		1,715	2,075	
	2080	14′ long		1.40	11.429		2,025	405		2,430	2,925	
	2100	6′ high, 4′ long		2	8		1,025	284		1,309	1,600	
	2120	6′ long		1.60	10		1,200	355		1,555	1,925	
	2140	10′ long		1.40	11.429		1,625	405		2,030	2,475	
	9000	Minimum labor/equipment charge	1 Carp	2	4	Job		142		142	236	
640	0010	**COILING GRILLE**										640
	0015	Aluminum, manual, incl. frame, mill finish										
	0020	Top coiling, 4′ high, 4′ long	2 Sswk	3.20	5	Opng.	1,175	200		1,375	1,675	
	0030	6′ long		3.20	5		1,225	200		1,425	1,725	
	0040	8′ long		2.40	6.667		1,400	266		1,666	2,050	
	0050	12′ long		2.40	6.667		1,650	266		1,916	2,300	
	0060	16′ long		1.60	10		2,050	400		2,450	3,000	
	0070	6′ high, 4′ long		3.20	5		1,250	200		1,450	1,750	
	0080	6′ long		3.20	5		1,300	200		1,500	1,825	
	0090	8′ long		2.40	6.667		1,475	266		1,741	2,125	
	0100	12′ long		1.60	10		1,725	400		2,125	2,650	
	0110	16′ long		1.20	13.333		2,200	535		2,735	3,425	
	0200	Side coiling, 8′ high, 12′ long		.60	26.667		1,850	1,075		2,925	4,025	
	0220	18′ long		.50	32		2,850	1,275		4,125	5,525	
	0240	24′ long		.40	40		2,825	1,600		4,425	6,125	
	0260	12′ high, 12′ long		.50	32		2,675	1,275		3,950	5,350	
	0280	18′ long		.40	40		3,950	1,600		5,550	7,350	
	0300	24′ long		.28	57.143		5,275	2,275		7,550	10,100	
	9000	Minimum labor/equipment charge		1	16	Job		640		640	1,200	
700	0010	**ROLLING GRILLE SUPPORTS**										700
	0020	Rolling grille supports, overhead framed	E-4	36	.889	L.F.	18.40	36	2.48	56.88	89.50	
720	0010	**ROLLING SERVICE DOORS** Steel, manual, 20 ga., incl. hardware										720
	0050	8′ x 8′ high	2 Sswk	1.60	10	Ea.	810	400		1,210	1,625	
	0100	10′ x 10′ high		1.40	11.429		1,075	455		1,530	2,025	
	0200	20′ x 10′ high		1	16		2,250	640		2,890	3,675	
	0300	12′ x 12′ high		1.20	13.333		1,375	535		1,910	2,525	
	0400	20′ x 12′ high		.90	17.778		2,300	710		3,010	3,875	
	0500	14′ x 14′ high		.80	20		1,800	800		2,600	3,500	
	0600	20′ x 16′ high		.60	26.667		2,700	1,075		3,775	4,975	
	0700	10′ x 20′ high		.50	32		1,625	1,275		2,900	4,200	
	1000	12′ x 12′, crank operated, crank on door side		.80	20		1,275	800		2,075	2,900	
	1100	Crank thru wall		.70	22.857		1,425	915		2,340	3,250	
	1300	For vision panel, add					232			232	255	
	1400	For 22 ga., deduct				S.F.	.71			.71	.78	
	1600	3′ x 7′ pass door within rolling steel door, new construction				Ea.	1,250			1,250	1,375	

08300 | Specialty Doors

08330 | Coiling Doors and Grilles

			CREW	DAILY OUTPUT	LABOR-HOURS	UNIT	2006 BARE COSTS MAT.	LABOR	EQUIP.	TOTAL	TOTAL INCL O&P	
720	1700	Existing construction	2 Sswk	2	8	Ea.	1,250	320		1,570	1,975	720
	2000	Class A fire doors, manual, 20 ga., 8′ x 8′ high		1.40	11.429		1,100	455		1,555	2,050	
	2100	10′ x 10′ high		1.10	14.545		1,450	580		2,030	2,675	
	2200	20′ x 10′ high		.80	20		2,900	800		3,700	4,700	
	2300	12′ x 12′ high		1	16		1,925	640		2,565	3,325	
	2400	20′ x 12′ high		.80	20		3,350	800		4,150	5,175	
	2500	14′ x 14′ high		.60	26.667		2,500	1,075		3,575	4,750	
	2600	20′ x 16′ high		.50	32		4,175	1,275		5,450	7,000	
	2700	10′ x 20′ high		.40	40		2,650	1,600		4,250	5,900	
	3000	For 18 ga. doors, add				S.F.	.82			.82	.90	
	3300	For enamel finish, add				"	.98			.98	1.08	
	3600	For safety edge bottom bar, pneumatic, add				L.F.	12.85			12.85	14.15	
	3700	Electric, add					24			24	26.50	
	4000	For weatherstripping, extruded rubber, jambs, add					8.35			8.35	9.15	
	4100	Hood, add					5.30			5.30	5.85	
	4200	Sill, add					3.03			3.03	3.33	
	4500	Motor operators, to 14′ x 14′ opening	2 Sswk	5	3.200	Ea.	780	128		908	1,100	
	4600	Over 14′ x 14′, jack shaft type	"	5	3.200		770	128		898	1,075	
	4700	For fire door, additional fusible link, add					15.15			15.15	16.65	
	9000	Minimum labor/equipment charge	2 Sswk	1	16	Job		640		640	1,200	

08340 | Special Function Doors

			CREW	DAILY OUTPUT	LABOR-HOURS	UNIT	MAT.	LABOR	EQUIP.	TOTAL	TOTAL INCL O&P	
100	0010	**COLD STORAGE**										100
	0020	Single, 20 ga. galvanized steel										
	0300	Horizontal sliding, 5′ x 7′, manual operation, 3.5″ thick	2 Carp	2	8	Ea.	2,600	284		2,884	3,350	
	0400	4″ thick		2	8		3,175	284		3,459	3,950	
	0500	6″ thick		2	8		2,850	284		3,134	3,625	
	0800	5′ x 7′, power operation, 2″ thick		1.90	8.421		4,750	299		5,049	5,725	
	0900	4″ thick		1.90	8.421		4,825	299		5,124	5,800	
	1000	6″ thick		1.90	8.421		5,500	299		5,799	6,550	
	1300	9′ x 10′, manual operation, 2″ insulation		1.70	9.412		3,825	335		4,160	4,775	
	1400	4″ insulation		1.70	9.412		3,950	335		4,285	4,900	
	1500	6″ insulation		1.70	9.412		4,775	335		5,110	5,800	
	1800	Power operation, 2″ insulation		1.60	10		6,600	355		6,955	7,850	
	1900	4″ insulation		1.60	10		6,750	355		7,105	8,025	
	2000	6″ insulation		1.70	9.412		7,650	335		7,985	8,975	
	2300	For stainless steel face, add					20%					
	3000	Hinged, lightweight, 3′ x 7′-0″, galvanized 1 face, 2″ thick	2 Carp	2	8	Ea.	1,200	284		1,484	1,800	
	3050	4″ thick		1.90	8.421		1,400	299		1,699	2,025	
	3300	Aluminum doors, 3′ x 7′-0″, 4″ thick		1.90	8.421		1,125	299		1,424	1,750	
	3350	6″ thick		1.40	11.429		2,025	405		2,430	2,900	
	3600	Stainless steel, 3′ x 7′-0″, 4″ thick		1.90	8.421		1,450	299		1,749	2,100	
	3650	6″ thick		1.40	11.429		2,425	405		2,830	3,350	
	3900	Painted, 3′ x 7′-0″, 4″ thick		1.90	8.421		1,050	299		1,349	1,650	
	3950	6″ thick		1.40	11.429		1,975	405		2,380	2,850	
	5000	Bi-parting, electric operated										
	5010	6′ x 8′ opening, galv. faces, 4″ thick for cooler	2 Carp	.80	20	Opng.	6,200	710		6,910	8,000	
	5050	For freezer, 4″ thick		.80	20		6,825	710		7,535	8,700	
	5300	For door buck framing and door protection, add		2.50	6.400		505	228		733	935	
	6000	Galvanized batten door, galvanized hinges, 4′ x 7′		2	8		1,525	284		1,809	2,150	
	6050	6′ x 8′		1.80	8.889		2,075	315		2,390	2,825	
	6500	Fire door, 3 hr., 6′ x 8′, single slide		.80	20		7,175	710		7,885	9,050	
	6550	Double, bi-parting		.70	22.857		10,700	815		11,515	13,200	
	9000	Minimum labor/equipment charge	1 Carp	2	4	Job		142		142	236	
200	0010	**HANGAR DOOR**										200
	0020	Bi-fold, ovhd., 20 PSF wind load, incl. elec. oper.										

DOORS & WINDOWS 8

		08340 | Special Function Doors	CREW	DAILY OUTPUT	LABOR-HOURS	UNIT	2006 BARE COSTS MAT.	LABOR	EQUIP.	TOTAL	TOTAL INCL O&P	
200	0100	Without sheeting, 12′ high x 40′	2 Sswk	240	.067	S.F.	13.55	2.66		16.21	19.90	200
	0200	16′ high x 60′		230	.070		17.25	2.78		20.03	24	
	0300	20′ high x 80′		220	.073		17.55	2.91		20.46	25	
610	0010	**ACOUSTICAL DOORS**										610
	0020	Including framed seals, 3′ x 7′, wood, 27 STC rating	2 Carp	1.50	10.667	Ea.	385	380		765	1,050	
	0100	Steel, 40 STC rating		1.50	10.667		1,450	380		1,830	2,225	
	0200	45 STC rating		1.50	10.667		1,925	380		2,305	2,750	
	0300	48 STC rating		1.50	10.667		2,425	380		2,805	3,300	
	0400	52 STC rating		1.50	10.667		2,925	380		3,305	3,825	
	9000	Minimum labor/equipment charge	1 Carp	4	2	Job		71		71	118	
700	0010	**DOUBLE ACTING, SWING**										700
	1000	.063″ aluminum, 7′-0″ high, 4′-0″ wide	2 Carp	4.20	3.810	Pr.	1,825	135		1,960	2,225	
	1050	6′-0″ wide	″	4	4	″	2,400	142		2,542	2,875	
	2000	Solid core wood, 3/4″ thick, metal frame, stainless steel										
	2010	base plate, 7′ high opening, 4′ wide	2 Carp	4	4	Pr.	2,175	142		2,317	2,625	
	2050	7′ wide		3.80	4.211	″	2,400	150		2,550	2,900	
	9000	Minimum labor/equipment charge		2	8	Job		284		284	470	
710	0010	**GLASS DOOR, SWING**										710
	0020	Including hardware, 1/2″ thick, tempered, 3′ x 7′ opening	2 Glaz	2	8	Opng.	1,800	274		2,074	2,425	
	0100	6′ x 7′ opening		1.40	11.429	″	3,500	390		3,890	4,475	
	9000	Minimum labor/equipment charge		2	8	Job		274		274	440	
720	0010	**SHOCK ABSORBING DOORS**										720
	0020	Rigid, no frame, 1-1/2″ thick, 5′ x 7′	2 Sswk	1.90	8.421	Opng.	1,275	335		1,610	2,025	
	0100	8′ x 8′		1.80	8.889		1,800	355		2,155	2,650	
	0500	Flexible, no frame, insulated, .16″ thick, economy, 5′ x 7′		2	8		1,575	320		1,895	2,325	
	0600	Deluxe		1.90	8.421		2,375	335		2,710	3,225	
	1000	8′ x 8′ opening, economy		2	8		2,450	320		2,770	3,275	
	1100	Deluxe		1.90	8.421		3,150	335		3,485	4,075	
	9000	Minimum labor/equipment charge		2	8	Job		320		320	600	
730	0010	**SWING DOORS**, aluminum entrance door including hardware & operator										730
	0015	Aluminum entrance 6′ x 7′	2 Sswk	.70	22.857	Opng.	5,200	915		6,115	7,425	
	0020	For anodized finish, add				″	229			229	252	
	9000	Minimum labor/equipment charge	2 Sswk	1	16	Job		640		640	1,200	
740	0010	**TUBULAR STEEL SWING DOORS**										740
	0020	Tubular steel, 7′ high, single, 3′-4″ opening	2 Sswk	2.50	6.400	Ea.	580	256		836	1,125	
	0100	Double, 6′-0″ opening	″	2	8	Pr.	735	320		1,055	1,400	
		08360 | Overhead Doors										
550	0010	**OVERHEAD, COMMERCIAL** Frames not included										550
	1000	Stock, sectional, heavy duty, wood, 1-3/4″ thick, 8′ x 8′ high	2 Carp	2	8	Ea.	605	284		889	1,125	
	1100	10′ x 10′ high		1.80	8.889		910	315		1,225	1,525	
	1200	12′ x 12′ high		1.50	10.667		1,325	380		1,705	2,075	
	1300	Chain hoist, 14′ x 14′ high		1.30	12.308		1,925	440		2,365	2,850	
	1400	12′ x 16′ high		1	16		1,700	570		2,270	2,825	
	1500	20′ x 8′ high		1.30	12.270		1,575	435		2,010	2,450	
	1600	20′ x 16′ high		.65	24.615		3,975	875		4,850	5,800	
	1800	Center mullion openings, 8′ high		4	4		690	142		832	995	
	1900	20′ high		2	8		1,300	284		1,584	1,900	
	2100	For medium duty custom door, deduct					5%	5%				
	2150	For medium duty stock doors, deduct					10%	5%				
	2300	Fiberglass and aluminum, heavy duty, sectional, 12′ x 12′ high	2 Carp	1.50	10.667	Ea.	1,925	380		2,305	2,725	
	2450	Chain hoist, 20′ x 20′ high		.50	32		4,775	1,150		5,925	7,125	

		08360 | Overhead Doors	Crew	Daily Output	Labor-Hours	Unit	2006 Bare Costs Mat.	Labor	Equip.	Total	Total Incl O&P	
550	2600	Steel, 24 ga. sectional, manual, 8′ x 8′ high	2 Carp	2	8	Ea.	530	284		814	1,050	550
	2650	10′ x 10′ high		1.80	8.889		735	315		1,050	1,325	
	2700	12′ x 12′ high		1.50	10.667		1,000	380		1,380	1,725	
	2800	Chain hoist, 20′ x 14′ high		.70	22.857		2,975	815		3,790	4,600	
	2850	For 1-1/4″ rigid insulation and 26 ga. galv.										
	2860	back panel, add				S.F.	2.51			2.51	2.76	
	2900	For electric trolley operator, 1/3 H.P., to 12′ x 12′, add	1 Carp	2	4	Ea.	700	142		842	1,000	
	2950	Over 12′ x 12′, 1/2 H.P., add	"	1	8	"	785	284		1,069	1,325	
	9000	Minimum labor/equipment charge	2 Carp	1.50	10.667	Job		380		380	630	
600	0010	**RESIDENTIAL GARAGE DOORS** Including hardware, no frame										600
	0050	Hinged, wood, custom, double door, 9′ x 7′	2 Carp	4	4	Ea.	390	142		532	660	
	0070	16′ x 7′		3	5.333		660	190		850	1,050	
	0200	Overhead, sectional, incl. hardware, fiberglass, 9′ x 7′, standard		5.28	3.030		610	108		718	850	
	0220	Deluxe		5.28	3.030		775	108		883	1,025	
	0300	16′ x 7′, standard		6	2.667		1,100	95		1,195	1,350	
	0320	Deluxe		6	2.667		1,350	95		1,445	1,650	
	0500	Hardboard, 9′ x 7′, standard		8	2		405	71		476	565	
	0520	Deluxe		8	2		540	71		611	715	
	0600	16′ x 7′, standard		6	2.667		790	95		885	1,025	
	0620	Deluxe		6	2.667		920	95		1,015	1,175	
	0700	Metal, 9′ x 7′, standard		5.28	3.030		475	108		583	700	
	0720	Deluxe		8	2		635	71		706	820	
	0800	16′ x 7′, standard		3	5.333		605	190		795	980	
	0820	Deluxe		6	2.667		965	95		1,060	1,225	
	0900	Wood, 9′ x 7′, standard		8	2		500	71		571	670	
	0920	Deluxe		8	2		1,425	71		1,496	1,700	
	1000	16′ x 7′, standard		6	2.667		1,000	95		1,095	1,250	
	1020	Deluxe		6	2.667		2,100	95		2,195	2,450	
	1800	Door hardware, sectional	1 Carp	4	2		220	71		291	360	
	1810	Door tracks only		4	2		102	71		173	230	
	1820	One side only		7	1.143		71	40.50		111.50	146	
	3000	Swing-up, including hardware, fiberglass, 9′ x 7′, standard	2 Carp	8	2		690	71		761	875	
	3020	Deluxe		8	2		720	71		791	915	
	3100	16′ x 7′, standard		6	2.667		870	95		965	1,100	
	3120	Deluxe		6	2.667		895	95		990	1,150	
	3200	Hardboard, 9′ x 7′, standard		8	2		315	71		386	465	
	3220	Deluxe		8	2		420	71		491	580	
	3300	16′ x 7′, standard		6	2.667		440	95		535	640	
	3320	Deluxe		6	2.667		655	95		750	875	
	3400	Metal, 9′ x 7′, standard		8	2		345	71		416	500	
	3420	Deluxe		8	2		610	71		681	790	
	3500	16′ x 7′, standard		6	2.667		540	95		635	750	
	3520	Deluxe		6	2.667		870	95		965	1,125	
	3600	Wood, 9′ x 7′, standard		8	2		380	71		451	535	
	3620	Deluxe		8	2		665	71		736	850	
	3700	16′ x 7′, standard		6	2.667		655	95		750	875	
	3720	Deluxe		6	2.667		925	95		1,020	1,175	
	3900	Door hardware only, swing up	1 Carp	4	2		109	71		180	238	
	3920	One side only		7	1.143		61.50	40.50		102	135	
	4000	For electric operator, economy, add		8	1		287	35.50		322.50	375	
	4100	Deluxe, including remote control		8	1		420	35.50		455.50	520	
	4500	For transmitter/receiver control , add to operator				Total	86.50			86.50	95.50	
	4600	Transmitters, additional				"	31.50			31.50	35	
	9000	Minimum labor/equipment charge	1 Carp	2.50	3.200	Job		114		114	188	

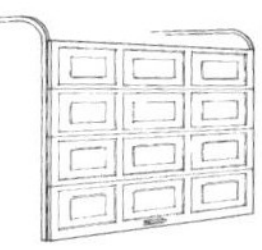

08300 | Specialty Doors

08360 | Overhead Doors

		08360 Overhead Doors	CREW	DAILY OUTPUT	LABOR-HOURS	UNIT	2006 BARE COSTS MAT.	LABOR	EQUIP.	TOTAL	TOTAL INCL O&P
800	0010	**TELESCOPING STEEL DOORS** Baked enamel finish									
	1000	Overhead, .03" thick, electric operated, 10' x 10'	E-3	.80	30	Ea.	10,100	1,225	112	11,437	13,500
	2000	20' x 10'		.60	40		13,400	1,625	149	15,174	17,900
	3000	20' x 16'	↓	.40	60	↓	16,300	2,425	223	18,948	22,800

08370 | Vertical Lift Doors

		08370 Vertical Lift Doors	CREW	DAILY OUTPUT	LABOR-HOURS	UNIT	MAT.	LABOR	EQUIP.	TOTAL	TOTAL INCL O&P
950	0010	**VERTICAL LIFT DOORS**									
	0020	Motorized, 14 ga.. stl., incl., frm. and ctrl. pnl.									
	0050	16' x 16' high	L-10	.50	48	Ea.	17,600	1,925	1,275	20,800	24,200
	0100	10' x 20' high		1.30	18.462		23,900	740	490	25,130	28,200
	0120	15' x 20' high		1.30	18.462		29,300	740	490	30,530	34,100
	0140	20' x 20' high		1	24		34,400	960	635	35,995	40,200
	0160	25' x 20' high		1	24		38,100	960	635	39,695	44,300
	0170	32' x 24' high		.75	32		35,100	1,275	850	37,225	41,800
	0180	20' x 25' high		1	24		39,400	960	635	40,995	45,700
	0200	25' x 25' high		.70	34.286		44,400	1,375	910	46,685	52,500
	0220	25' x 30' high		.70	34.286		52,000	1,375	910	54,285	60,500
	0240	30' x 30' high		.70	34.286		59,500	1,375	910	61,785	69,000
	0260	35' x 30' high	↓	.70	34.286	↓	65,500	1,375	910	67,785	75,500

08380 | Traffic Doors

		08380 Traffic Doors	CREW	DAILY OUTPUT	LABOR-HOURS	UNIT	MAT.	LABOR	EQUIP.	TOTAL	TOTAL INCL O&P
480	0010	**FLEXIBLE TRANSPARENT STRIP ENTRANCE**									
	0100	12" strip width, 2/3 overlap	3 Shee	135	.178	SF Surf	5.20	7.50		12.70	17.60
	0200	Full overlap		115	.209		5.45	8.80		14.25	20
	0220	8" strip width, 1/2 overlap		140	.171		5.05	7.25		12.30	17.05
	0240	Full overlap	↓	120	.200	↓	6.65	8.45		15.10	21
	0300	Add for suspension system, header mount				L.F.	7.70			7.70	8.45
	0400	Wall mount				"	7.75			7.75	8.55

08400 | Entrances and Storefronts

08410 | Metal-Framed Storefronts

		08410 Metal-Framed Storefronts	CREW	DAILY OUTPUT	LABOR-HOURS	UNIT	2006 BARE COSTS MAT.	LABOR	EQUIP.	TOTAL	TOTAL INCL O&P
110	0010	**ALUMINUM-FRAMED STOREFRONTS,** No glazing									
	0020	Entrance, 3' x 7' opening, clear anodized finish	2 Sswk	7	2.286	Opng.	254	91.50		345.50	450
	0100	Bronze finish		7	2.286		340	91.50		431.50	540
	0500	6' x 7' opening, clear finish		6	2.667		310	107		417	540
	0520	Bronze finish		6	2.667		350	107		457	585
	1000	With 3' high transoms, 3' x 10' opening, clear finish		6.50	2.462		370	98.50		468.50	595
	1050	Bronze finish		6.50	2.462		410	98.50		508.50	635
	1100	Black finish		6.50	2.462		485	98.50		583.50	715
	1500	With 3' high transoms, 6' x 10' opening, clear finish		5.50	2.909		450	116		566	710
	1550	Bronze finish		5.50	2.909		485	116		601	745
	1600	Black finish		5.50	2.909	↓	575	116		691	845
	9000	Minimum labor/equipment charge	↓	4	4	Job		160		160	299
120	0010	**ALUMINUM DOORS** Commercial entrance, no glazing									
	0020	Incl. hinges, push/pull, deadlock, cyl., threshold									
	0800	Narrow stile, no glazing, standard hardware, pair of 2'-6" x 7'-0"	2 Carp	1.70	9.412	Pr.	770	335		1,105	1,400
	1000	3'-0" x 7'-0", single	↓	3	5.333	Ea.	510	190		700	875

08410 | Metal-Framed Storefronts

		08410 Metal-Framed Storefronts	CREW	DAILY OUTPUT	LABOR-HOURS	UNIT	2006 BARE COSTS MAT.	LABOR	EQUIP.	TOTAL	TOTAL INCL O&P	
120	1200	Pair of 3'-0" x 7'-0"	2 Carp	1.70	9.412	Pr.	1,025	335		1,360	1,675	120
	1500	3'-6" x 7'-0", single		3	5.333	Ea.	545	190		735	915	
	2000	Medium stile, pair of 2'-6" x 7'-0"		1.70	9.412	Pr.	955	335		1,290	1,600	
	2100	3'-0" x 7'-0", single		3	5.333	Ea.	670	190		860	1,050	
	2200	Pair of 3'-0" x 7'-0"		1.70	9.412	Pr.	1,300	335		1,635	2,000	
	2300	3'-6" x 7'-0", single		5.33	3.002	Ea.	790	107		897	1,050	
	5000	Flush panel doors, pair of 2'-6" x 7'-0"	2 Sswk	2	8	Pr.	945	320		1,265	1,650	
	5050	3'-0" x 7'-0", single		2.50	6.400	Ea.	470	256		726	1,000	
	5100	Pair of 3'-0" x 7'-0"		2	8	Pr.	945	320		1,265	1,650	
	5150	3'-6" x 7'-0", single		2.50	6.400	Ea.	560	256		816	1,100	
130	0010	**ALUMINUM DOORS & FRAMES** Entrance, narrow stile, including										130
	0015	Standard hardware, clear finish, not incl. glass, 2'-6" x 7'-0" opng.	2 Sswk	2	8	Ea.	695	320		1,015	1,350	
	0020	3'-0" x 7'-0" opening		2	8		560	320		880	1,225	
	0030	3'-6" x 7'-0" opening		2	8		500	320		820	1,150	
	0100	3'-0" x 10'-0" opening, 3' high transom		1.80	8.889		785	355		1,140	1,525	
	0200	3'-6" x 10'-0" opening, 3' high transom		1.80	8.889		770	355		1,125	1,525	
	0280	5'-0" x 7'-0" opening		2	8		850	320		1,170	1,550	
	0300	6'-0" x 7'-0" opening		1.30	12.308	Pr.	865	490		1,355	1,875	
	0400	6'-0" x 10'-0" opening, 3' high transom		1.10	14.545		730	580		1,310	1,875	
	0420	7'-0" x 7'-0" opening		1	16		830	640		1,470	2,125	
	0500	Wide stile, 2'-6" x 7'-0" opening		2	8	Ea.	680	320		1,000	1,350	
	0520	3'-0" x 7'-0" opening		2	8		695	320		1,015	1,375	
	0540	3'-6" x 7'-0" opening		2	8		740	320		1,060	1,425	
	0560	5'-0" x 7'-0" opening		2	8		1,200	320		1,520	1,925	
	0580	6'-0" x 7'-0" opening		1.30	12.308	Pr.	1,150	490		1,640	2,200	
	0600	7'-0" x 7'-0" opening		1	16	"	1,250	640		1,890	2,575	
	1100	For full vision doors, with 1/2" glass, add				Leaf	55%					
	1200	For non-standard size, add					67%					
	1300	Light bronze finish, add					36%					
	1400	Dark bronze finish, add					18%					
	1500	For black finish, add					36%					
	1600	Concealed panic device, add					1,025			1,025	1,125	
	1700	Electric striker release, add				Opng.	262			262	288	
	1800	Floor check, add				Leaf	775			775	855	
	1900	Concealed closer, add				"	515			515	570	
	9000	Minimum labor/equipment charge	2 Carp	4	4	Job		142		142	236	
140	0010	**STOREFRONT SYSTEMS** Aluminum frame, clear 3/8" plate glass,										140
	0020	incl. 3' x 7' door with hardware (400 sq. ft. max. wall)										
	0500	Wall height to 12' high, commercial grade	2 Glaz	150	.107	S.F.	13.85	3.65		17.50	21	
	0600	Institutional grade		130	.123		17.40	4.22		21.62	26	
	0700	Monumental grade		115	.139		26.50	4.77		31.27	36.50	
	1000	6' x 7' door with hardware, commercial grade		135	.119		14.15	4.06		18.21	22	
	1100	Institutional grade		115	.139		19.35	4.77		24.12	29	
	1200	Monumental grade		100	.160		36	5.50		41.50	48.50	
	1500	For bronze anodized finish, add					15%					
	1600	For black anodized finish, add					30%					
	1700	For stainless steel framing, add to monumental					75%					
	9000	Minimum labor/equipment charge	2 Glaz	1	16	Job		550		550	885	
300	0010	**STAINLESS STEEL AND GLASS** Entrance unit, narrow stiles										300
	0020	3' x 7' opening, including hardware, minimum	2 Sswk	1.60	10	Opng.	4,825	400		5,225	6,050	
	0050	Average		1.40	11.429		5,225	455		5,680	6,600	
	0100	Maximum		1.20	13.333		5,600	535		6,135	7,150	
	1000	For solid bronze entrance units, statuary finish, add					60%					
	1100	Without statuary finish, add					45%					
	2000	Balanced doors, 3' x 7', economy	2 Sswk	.90	17.778	Ea.	6,550	710		7,260	8,525	
	2100	Premium		.70	22.857	"	11,300	915		12,215	14,100	

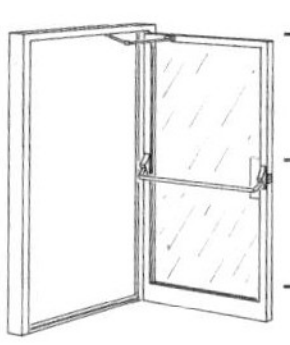

08400 | Entrances and Storefronts

		08410 | Metal-Framed Storefronts	Crew	Daily Output	Labor-Hours	Unit	2006 Bare Costs Mat.	Labor	Equip.	Total	Total Incl O&P	
300	9000	Minimum labor/equipment charge	2 Sswk	2	8	Job		320		320	600	300

		08460 | Automatic Entrance Doors	Crew	Daily Output	Labor-Hours	Unit	Mat.	Labor	Equip.	Total	Total Incl O&P	
600	0010	**SLIDING ENTRANCE** 12′ x 7′-6″ opng., 5′ x 7′ door, 2 way traf.,										600
	0020	mat activated, panic pushout, incl. operator & hardware,										
	0030	not including glass or glazing	2 Glaz	.70	22.857	Opng.	6,200	785		6,985	8,075	
	9000	Minimum labor/equipment charge	″	.70	22.857	Job		785		785	1,250	

		08470 | Revolving Entrance Doors	Crew	Daily Output	Labor-Hours	Unit	Mat.	Labor	Equip.	Total	Total Incl O&P	
600	0010	**REVOLVING DOORS** Aluminum, 6′-6″ to 7′-0″ diameter,										600
	0020	6′-10″ to 7′ high, stock units, minimum	4 Sswk	.75	42.667	Opng.	16,400	1,700		18,100	21,200	
	0050	Average		.60	53.333		20,200	2,125		22,325	26,200	
	0100	Maximum		.45	71.111		26,600	2,850		29,450	34,600	
	1000	Stainless steel		.30	105		30,200	4,225		34,425	41,100	
	1100	Solid bronze		.15	213		35,300	8,525		43,825	54,500	
	1500	For automatic controls, add	2 Elec	2	8		12,300	335		12,635	14,100	

		08480 | Balanced Entrance Doors	Crew	Daily Output	Labor-Hours	Unit	Mat.	Labor	Equip.	Total	Total Incl O&P	
150	0010	**BALANCED DOORS**										150
	0020	Hardware & frame, alum. & glass, 3′ x 7′, econ.	2 Sswk	.90	17.778	Ea.	5,300	710		6,010	7,150	
	0150	Premium		.70	22.857	″	6,600	915		7,515	8,950	
	9000	Minimum labor/equipment charge		1	16	Job		640		640	1,200	

		08490 | Sliding Storefronts	Crew	Daily Output	Labor-Hours	Unit	Mat.	Labor	Equip.	Total	Total Incl O&P	
100	0010	**SLIDING PANELS**										100
	0020	Mall fronts, aluminum & glass, 15′ x 9′ high	2 Glaz	1.30	12.308	Opng.	2,425	420		2,845	3,350	
	0100	24′ x 9′ high		.70	22.857		3,525	785		4,310	5,125	
	0200	48′ x 9′ high, with fixed panels		.90	17.778		6,575	610		7,185	8,200	
	0500	For bronze finish, add					17%					
	9000	Minimum labor/equipment charge	2 Glaz	1	16	Job		550		550	885	

08500 | Windows

		08510 | Steel Windows	Crew	Daily Output	Labor-Hours	Unit	2006 Bare Costs Mat.	Labor	Equip.	Total	Total Incl O&P	
700	0010	**SCREENS**										700
	0020	For metal sash, aluminum or bronze mesh, flat screen	2 Sswk	1200	.013	S.F.	3.61	.53		4.14	4.97	
	0500	Wicket screen, inside window		1000	.016		5.50	.64		6.14	7.25	
	0800	Security screen, aluminum frame with stainless steel cloth		1200	.013		19.45	.53		19.98	22.50	
	0900	Steel grate, painted, on steel frame		1600	.010		10.45	.40		10.85	12.25	
	1000	For solar louvers, add		160	.100		20	4		24	29.50	
	4000	See also Division 05580-900										
750	0010	**STEEL SASH** Custom units, glazing and trim not included										750
	0100	Casement, 100% vented	2 Sswk	200	.080	S.F.	42	3.20		45.20	52	
	0200	50% vented		200	.080		38	3.20		41.20	48	
	0300	Fixed		200	.080		26	3.20		29.20	35	
	1000	Projected, commercial, 40% vented		200	.080		44	3.20		47.20	54	
	1100	Intermediate, 50% vented		200	.080		48	3.20		51.20	58.50	

08500 | Windows

08510 | Steel Windows

		08510 Steel Windows	CREW	DAILY OUTPUT	LABOR-HOURS	UNIT	2006 BARE COSTS MAT.	LABOR	EQUIP.	TOTAL	TOTAL INCL O&P	
750	1500	Industrial, horizontally pivoted	2 Sswk	200	.080	S.F.	45.50	3.20		48.70	56	750
	1600	Fixed		200	.080		26	3.20		29.20	34.50	
	2000	Industrial security sash, 50% vented		200	.080		49	3.20		52.20	59.50	
	2100	Fixed		200	.080		39.50	3.20		42.70	49.50	
	2500	Picture window		200	.080		24.50	3.20		27.70	33	
	3000	Double hung		200	.080		47	3.20		50.20	58	
	5000	Mullions for above, open interior face		240	.067	L.F.	8.45	2.66		11.11	14.25	
	5100	With interior cover		240	.067	"	13.95	2.66		16.61	20.50	
	9000	Minimum labor/equipment charge	1 Sswk	2	4	Job		160		160	299	
770	0010	**STEEL WINDOWS** Stock, including frame, trim and insul. glass										770
	0020	See also Division 13128-700										
	1000	Custom units, double hung, 2'-8" x 4'-6" opening	2 Sswk	12	1.333	Ea.	595	53.50		648.50	755	
	1100	2'-4" x 3'-9" opening		12	1.333		490	53.50		543.50	640	
	1500	Commercial projected, 3'-9" x 5'-5" opening		10	1.600		1,050	64		1,114	1,275	
	1600	6'-9" x 4'-1" opening		7	2.286		1,375	91.50		1,466.50	1,700	
	2000	Intermediate projected, 2'-9" x 4'-1" opening		12	1.333		585	53.50		638.50	740	
	2100	4'-1" x 5'-5" opening		10	1.600		1,175	64		1,239	1,425	
	9000	Minimum labor/equipment charge	1 Sswk	3	2.667	Job		107		107	199	

08520 | Aluminum Windows

		08520 Aluminum Windows	CREW	DAILY OUTPUT	LABOR-HOURS	UNIT	MAT.	LABOR	EQUIP.	TOTAL	TOTAL INCL O&P	
100	0010	**ALUMINUM SASH**										100
	0020	Stock, grade C, glaze & trim not incl., casement	2 Sswk	200	.080	S.F.	29.50	3.20		32.70	38	
	0050	Double hung		200	.080		29.50	3.20		32.70	38.50	
	0100	Fixed casement		200	.080		11.70	3.20		14.90	18.85	
	0150	Picture window		200	.080		12.55	3.20		15.75	19.85	
	0200	Projected window		200	.080		27	3.20		30.20	35.50	
	0250	Single hung		200	.080		13.80	3.20		17	21	
	0300	Sliding		200	.080		17.90	3.20		21.10	25.50	
	1000	Mullions for above, tubular		240	.067	L.F.	4.60	2.66		7.26	10.05	
	2000	Custom aluminum sash, grade HC, glazing not included, minimum		200	.080	S.F.	33	3.20		36.20	42.50	
	2100	Maximum		85	.188	"	43.50	7.50		51	61.50	
	9000	Minimum labor/equipment charge	1 Sswk	2	4	Job		160		160	299	
120	0010	**ALUMINUM WINDOWS** Incl. frame and glazing, Commercial grade										120
	0020	See also Division 05120-440										
	1000	Stock units, casement, 3'-1" x 3'-2" opening	2 Sswk	10	1.600	Ea.	310	64		374	460	
	1050	Add for storms					62.50			62.50	68.50	
	1600	Projected, with screen, 3'-1" x 3'-2" opening	2 Sswk	10	1.600		222	64		286	365	
	1700	Add for storms					59			59	64.50	
	2000	4'-5" x 5'-3" opening	2 Sswk	8	2		310	80		390	495	
	2100	Add for storms					81.50			81.50	90	
	2500	Enamel finish windows, 3'-1" x 3'-2"	2 Sswk	10	1.600		199	64		263	340	
	2600	4'-5" x 5'-3"		8	2		298	80		378	480	
	3000	Single hung, 2' x 3' opening, enameled, standard glazed		10	1.600		149	64		213	284	
	3100	Insulating glass		10	1.600		181	64		245	320	
	3300	2'-8" x 6'-8" opening, standard glazed		8	2		315	80		395	500	
	3400	Insulating glass		8	2		405	80		485	600	
	3700	3'-4" x 5'-0" opening, standard glazed		9	1.778		204	71		275	355	
	3800	Insulating glass		9	1.778		287	71		358	450	
	3890	Awning type, 3' x 3' opening standard glass		14	1.143		365	45.50		410.50	490	
	3900	Insulating glass		14	1.143		390	45.50		435.50	515	
	3910	3' x 4' opening, standard glass		10	1.600		430	64		494	590	
	3920	Insulating glass		10	1.600		495	64		559	660	
	3930	3' x 5'-4" opening, standard glass		10	1.600		515	64		579	690	
	3940	Insulating glass		10	1.600		610	64		674	790	

8 DOORS & WINDOWS

For expanded coverage of these items see *Means Interior Cost Data 2006*

	08520 | Aluminum Windows		CREW	DAILY OUTPUT	LABOR-HOURS	UNIT	2006 BARE COSTS MAT.	LABOR	EQUIP.	TOTAL	TOTAL INCL O&P
120	3950	4′ x 5′-4″ opening, standard glass	2 Sswk	9	1.778	Ea.	565	71		636	760
	3960	Insulating glass		9	1.778		680	71		751	880
	4000	Sliding aluminum, 3′ x 2′ opening, standard glazed		10	1.600		168	64		232	305
	4100	Insulating glass		10	1.600		186	64		250	325
	4300	5′ x 3′ opening, standard glazed		9	1.778		213	71		284	365
	4400	Insulating glass		9	1.778		298	71		369	465
	4600	8′ x 4′ opening, standard glazed		6	2.667		305	107		412	535
	4700	Insulating glass		6	2.667		495	107		602	740
	5000	9′ x 5′ opening, standard glazed		4	4		465	160		625	810
	5100	Insulating glass		4	4		740	160		900	1,125
	5500	Sliding, with thermal barrier and screen, 6′ x 4′, 2 track		8	2		630	80		710	845
	5700	4 track		8	2		770	80		850	1,000
	6000	For above units with bronze finish, add					12%				
	6200	For installation in concrete openings, add					5%				
	9000	Minimum labor/equipment charge	1 Sswk	3	2.667	Job		107		107	199
500	0010	**JALOUSIES**									
	0020	Aluminum incl. glazing & screens, stock, 1′-7″ x 3′-2″	2 Sswk	10	1.600	Ea.	136	64		200	270
	0100	2′-3″ x 4′-0″		10	1.600		194	64		258	335
	0200	3′-1″ x 2′-0″		10	1.600		153	64		217	289
	0300	3′-1″ x 5′-3″		10	1.600		277	64		341	425
	1000	Mullions for above, 2′-0″ long		80	.200		10.45	8		18.45	26.50
	1100	5′-3″ long		80	.200		17.90	8		25.90	34.50
	9000	Minimum labor/equipment charge	1 Sswk	3.50	2.286	Job		91.50		91.50	171

	08550 | Wood Windows		CREW	DAILY OUTPUT	LABOR-HOURS	UNIT	MAT.	LABOR	EQUIP.	TOTAL	TOTAL INCL O&P
100	0010	**AWNING WINDOW** Including frame, screens and grills									
	0100	Average quality, builders model, 34″ x 22″, double insulated glass	1 Carp	10	.800	Ea.	233	28.50		261.50	305
	0200	Low E glass		10	.800		246	28.50		274.50	315
	0300	40″ x 28″, double insulated glass		9	.889		294	31.50		325.50	380
	0400	Low E Glass		9	.889		310	31.50		341.50	400
	0500	48″ x 36″, double insulated glass		8	1		430	35.50		465.50	530
	0600	Low E glass		8	1		450	35.50		485.50	555
	1000	34″ x 22″		10	.800		247	28.50		275.50	320
	1100	40″ x 22″		10	.800		270	28.50		298.50	345
	1200	36″ x 28″		9	.889		287	31.50		318.50	370
	1300	36″ x 36″		9	.889		320	31.50		351.50	410
	1400	48″ x 28″		8	1		345	35.50		380.50	440
	1500	60″ x 36″		8	1		500	35.50		535.50	610
	2000	Metal clad, deluxe, double insulated glass, 34″ x 22″		10	.800		231	28.50		259.50	300
	2100	40″ x 22″		10	.800		271	28.50		299.50	345
	2200	36″ x 25″		9	.889		251	31.50		282.50	330
	2300	40″ x 30″		9	.889		315	31.50		346.50	400
	2400	48″ x 28″		8	1		320	35.50		355.50	410
	2500	60″ x 36″		8	1		340	35.50		375.50	435
	9000	Minimum labor/equipment charge		4	2	Job		71		71	118
150	0010	**BOW-BAY WINDOW** Including frame, screens and grills,									
	0020	end panels operable									
	1000	Bow type, casement, wood, bldrs mdl, 8′ x 5′ dbl insltd glass, 4 panel	2 Carp	10	1.600	Ea.	1,500	57		1,557	1,750
	1050	Low E glass		10	1.600		1,200	57		1,257	1,425
	1100	10′-0″ x 5′-0″ , double insulated glass, 6 panels		6	2.667		1,250	95		1,345	1,525
	1200	Low E glass, 6 panels		6	2.667		1,325	95		1,420	1,600
	1300	Vinyl clad, bldrs model, double insulated glass, 6′-0″ x 4′-0″, 3 panel		10	1.600		1,375	57		1,432	1,600
	1340	9′-0″ x 4′-0″, 4 panel		8	2		1,375	71		1,446	1,625
	1380	10′-0″ x 6′-0″, 5 panels		7	2.286		2,200	81.50		2,281.50	2,550
	1420	12′-0″ x 6′-0″, 6 panels		6	2.667		2,775	95		2,870	3,200

	08550 | Wood Windows		CREW	DAILY OUTPUT	LABOR-HOURS	UNIT	2006 BARE COSTS MAT.	LABOR	EQUIP.	TOTAL	TOTAL INCL O&P	
150	1600	Metal clad, casement, bldrs mdl, 6'-0" x 4'-0", dbl insltd gls, 3 panels	2 Carp	10	1.600	Ea.	855	57		912	1,025	150
	1640	9'-0" x 4'-0" , 4 panels		8	2		1,200	71		1,271	1,450	
	1680	10'-0" x 5'-0", 5 panels		7	2.286		1,650	81.50		1,731.50	1,950	
	1720	12'-0" x 6'-0", 6 panels		6	2.667		2,300	95		2,395	2,675	
	2000	Bay window, casement, builders model, 8' x 5' dbl insul glass, 4 panels		10	1.600		1,675	57		1,732	1,925	
	2050	Low E glass,		10	1.600		2,025	57		2,082	2,325	
	2100	12'-0" x 6'-0" , double insulated glass, 6 panels		6	2.667		2,075	95		2,170	2,450	
	2200	Low E glass		6	2.667		2,050	95		2,145	2,400	
	2280	6'-0" x 4'-0"		11	1.455		1,075	51.50		1,126.50	1,275	
	2300	Vinyl clad, premium, double insulated glass, 8'-0" x 5'-0"		10	1.600		1,275	57		1,332	1,500	
	2340	10'-0" x 5'-0" 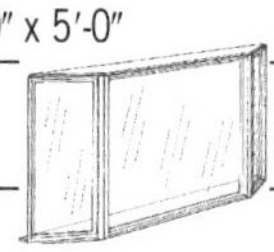		8	2		1,800	71		1,871	2,125	
	2380	10'-0" x 6'-0"		7	2.286		1,900	81.50		1,981.50	2,200	
	2420	12'-0" x 6'-0"		6	2.667		2,250	95		2,345	2,625	
	2600	Metal clad, deluxe, dbl insul. glass, 8'-0" x 5'-0" high, 4 panels		10	1.600		1,475	57		1,532	1,725	
	2640	10'-0" x 5'-0" high, 5 panels		8	2		1,575	71		1,646	1,875	
	2680	10'-0" x 6'-0" high, 5 panels		7	2.286		1,875	81.50		1,956.50	2,175	
	2720	12'-0" x 6'-0" high, 6 panels		6	2.667		2,600	95		2,695	3,000	
	3000	Double hung, bldrs. model, bay, 8' x 4' high,dbl insulated glass		10	1.600		1,175	57		1,232	1,375	
	3050	Low E glass		10	1.600		1,250	57		1,307	1,475	
	3100	9'-0" x 5'-0" high, doublel insulated glass		6	2.667		1,250	95		1,345	1,550	
	3200	Low E glass		6	2.667		1,325	95		1,420	1,625	
	3300	Vinyl clad, premium, double insulated glass, 7'-0" x 4'-6"		10	1.600		1,225	57		1,282	1,425	
	3340	8'-0" x 4'-6"		8	2		1,250	71		1,321	1,500	
	3380	8'-0" x 5'-0"		7	2.286		1,300	81.50		1,381.50	1,550	
	3420	9'-0" x 5'-0"		6	2.667		1,325	95		1,420	1,625	
	3600	Metal clad, deluxe, dbl insul. glass, 7'-0" x 4'-0" high		10	1.600		1,125	57		1,182	1,325	
	3640	8'-0" x 4'-0" high		8	2		1,175	71		1,246	1,400	
	3680	8'-0" x 5'-0" high		7	2.286		1,200	81.50		1,281.50	1,450	
	3720	9'-0" x 5'-0" high		6	2.667		1,275	95		1,370	1,550	
	9000	Minimum labor/equipment charge		2.50	6.400	Job		228		228	375	
200	0010	**CASEMENT WINDOW** Including frame, screen, and grills										200
	0100	Avg. quality, bldrs. model, 2'-0" x 3'-0" H, dbl. insulated glass	1 Carp	10	.800	Ea.	186	28.50		214.50	252	
	0150	Low E glass		10	.800		225	28.50		253.50	295	
	0200	2'-0" x 4'-6" high, double insulated glass		9	.889		242	31.50		273.50	320	
	0250	Low E glass 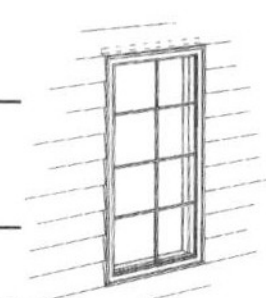		9	.889		320	31.50		351.50	410	
	0300	2'-4" x 6'-0" high, double insulated glass		8	1		355	35.50		390.50	450	
	0350	Low E glass		8	1		385	35.50		420.50	485	
	0522	Vinyl clad, premium, double insulated glass, 2'-0" x 3'-0"		10	.800		264	28.50		292.50	335	
	0524	2'-0" x 4'-0"		9	.889		310	31.50		341.50	395	
	0525	2'-0" x 5'-0"		8	1		355	35.50		390.50	450	
	0528	2'-0" x 6'-0"		8	1		400	35.50		435.50	500	
	8100	Metal clad, deluxe, dbl. insul. glass, 2'-0" x 3'-0" high		10	.800		206	28.50		234.50	274	
	8120	2'-0" x 4'-0" high		9	.889		248	31.50		279.50	325	
	8140	2'-0" x 5'-0" high		8	1		282	35.50		317.50	370	
	8160	2'-0" x 6'-0" high		8	1		325	35.50		360.50	415	
	8200	For multiple leaf units, deduct for stationary sash										
	8220	2' high				Ea.	21.50			21.50	23.50	
	8240	4'-6" high					24.50			24.50	27	
	8260	6' high					33			33	36.50	
	8300	For installation, add per leaf						15%				
	9000	Minimum labor/equipment charge	1 Carp	3	2.667	Job		95		95	157	
250	0010	**DOUBLE HUNG** Including frame, screens, and grills										250
	0100	Avg. quality, bldrs. model, 2'-0" x 3'-0" high, dbl insul. glass	1 Carp	10	.800	Ea.	204	28.50		232.50	272	
	0150	Low E glass		10	.800		186	28.50		214.50	252	
	0200	3'-0" x 4'-0" high, double insulated glass		9	.889		266	31.50		297.50	345	

	08550 | Wood Windows		CREW	DAILY OUTPUT	LABOR-HOURS	UNIT	2006 BARE COSTS MAT.	LABOR	EQUIP.	TOTAL	TOTAL INCL O&P	
250	0250	Low E glass	1 Carp	9	.889	Ea.	249	31.50		280.50	325	250
	0300	4'-0" x 4'-6" high, double insulated glass		8	1		310	35.50		345.50	400	
	0350	Low E glass		8	1		335	35.50		370.50	425	
	1000	Vinyl clad, premium, double insulated glass, 2'-6" x 3'-0"		10	.800		218	28.50		246.50	287	
	1100	3'-0" x 3'-6"		10	.800		256	28.50		284.50	330	
	1200	3'-0" x 4'-0"		9	.889		350	31.50		381.50	440	
	1300	3'-0" x 4'-6"		9	.889		296	31.50		327.50	380	
	1400	3'-0" x 5'-0"		8	1		310	35.50		345.50	400	
	1500	3'-6" x 6'-0"		8	1		360	35.50		395.50	460	
	2000	Metal clad, deluxe, dbl. insul. glass, 2'-6" x 3'-0" high		10	.800		219	28.50		247.50	287	
	2100	3'-0" x 3'-6" high		10	.800		259	28.50		287.50	330	
	2200	3'-0" x 4'-0" high		9	.889		275	31.50		306.50	360	
	2300	3'-0" x 4'-6" high		9	.889		298	31.50		329.50	380	
	2400	3'-0" x 5'-0" high		8	1		320	35.50		355.50	410	
	2500	3'-6" x 6'-0" high		8	1		385	35.50		420.50	485	
	9000	Minimum labor/equipment charge		3	2.667	Job		95		95	157	
650	0010	**PALLADIAN WINDOWS**										650
	0020	Vinyl clad, double insulated glass, including frame and grills										
	0040	3'-2" x 2'-6" high	2 Carp	11	1.455	Ea.	1,200	51.50		1,251.50	1,400	
	0060	3'-2" x 4'-10"		11	1.455		1,450	51.50		1,501.50	1,675	
	0080	3'-2" x 6'-4"		10	1.600		1,625	57		1,682	1,900	
	0100	4'-0" x 4'-0"		10	1.600		1,450	57		1,507	1,700	
	0120	4'-0" x 5'-4"	3 Carp	10	2.400		1,750	85.50		1,835.50	2,075	
	0140	4'-0" x 6'-0"		9	2.667		1,950	95		2,045	2,300	
	0160	4'-0" x 7'-4"		9	2.667		1,950	95		2,045	2,300	
	0180	5'-5" x 4'-10"		9	2.667		2,050	95		2,145	2,425	
	0200	5'-5" x 6'-10"		9	2.667		2,375	95		2,470	2,750	
	0220	5'-5" x 7'-9"		9	2.667		2,575	95		2,670	2,975	
	0240	6'-0" x 7'-11"		8	3		3,225	107		3,332	3,725	
	0260	8'-0" x 6'-0"		8	3		2,750	107		2,857	3,200	
670	0010	**PICTURE WINDOW** Including frame and grills										670
	0100	Average quality, bldrs. model, 3'-6" x 4'-0" high, dbl insulated glass	2 Carp	12	1.333	Ea.	284	47.50		331.50	390	
	0150	Low E glass		12	1.333		310	47.50		357.50	420	
	0200	4'-0" x 4'-6" high, double insulated glass		11	1.455		315	51.50		366.50	430	
	0250	Low E glass		11	1.455		345	51.50		396.50	465	
	0300	5'-0" x 4'-0" high, double insulated glass		11	1.455		390	51.50		441.50	510	
	0350	Low E glass		11	1.455		435	51.50		486.50	565	
	0400	6'-0" x 4'-6" high, double insulated glass		10	1.600		495	57		552	640	
	0450	Low E glass		10	1.600		560	57		617	710	
	1000	Vinyl clad, premium, dbl. insul. glass, 4'-0" x 4'-0"		12	1.333		455	47.50		502.50	580	
	1100	4'-0" x 6'-0"		11	1.455		675	51.50		726.50	825	
	1200	5'-0" x 6'-0"		10	1.600		970	57		1,027	1,175	
	1300	6'-0" x 6'-0"		10	1.600		990	57		1,047	1,175	
	2000	Metal clad, deluxe, dbl. insul. glass, 4'-0" x 4'-0" high		12	1.333		325	47.50		372.50	435	
	2100	4'-0" x 6'-0" high		11	1.455		480	51.50		531.50	610	
	2200	5'-0" x 6'-0" high		10	1.600		525	57		582	675	
	2300	6'-0" x 6'-0" high		10	1.600		605	57		662	760	
	9000	Minimum labor/equipment charge		2.75	5.818	Job		207		207	345	
750	0010	**SLIDING WINDOW** Including frame, screen, and grills										750
	0100	Average quality, bldrs. model, 3'-0" x 3'-0" high, double insulated	1 Carp	10	.800	Ea.	153	28.50		181.50	215	
	0120	Low E glass		10	.800		193	28.50		221.50	259	
	0200	4'-0" x 3'-6" high, double insulated		9	.889		182	31.50		213.50	253	
	0220	Low E glass		9	.889		227	31.50		258.50	305	
	0300	6'-0" x 5'-0" high, double insulated		8	1		335	35.50		370.50	425	
	0320	Low E glass		8	1		400	35.50		435.50	500	
	1000	Vinyl clad, premium, dbl. insulated glass, 3'-0" x 3'-0"		10	.800		510	28.50		538.50	605	

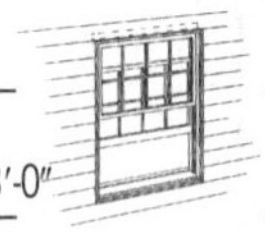

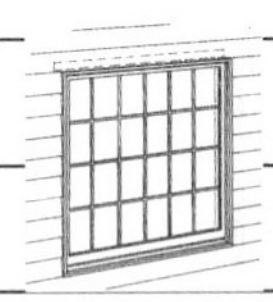

08500 | Windows

08550 | Wood Windows

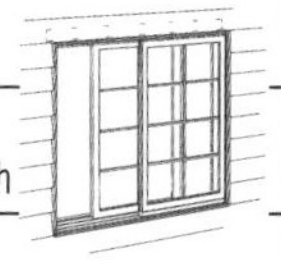

		08550 Wood Windows	Crew	Daily Output	Labor-Hours	Unit	2006 Bare Costs Mat.	Labor	Equip.	Total	Total Incl O&P	
750	1050	4'-0" x 3'-6"	1 Carp	9	.889	Ea.	630	31.50		661.50	745	750
	1100	5'-0" x 4'-0"		9	.889		760	31.50		791.50	890	
	1150	6'-0" x 5'-0"		8	1		965	35.50		1,000.50	1,100	
	2000	Metal clad, deluxe, double insul. glass, 3'-0" x 3'-0" h		10	.800		290	28.50		318.50	365	
	2050	4'-0" x 3'-6" high		9	.889		355	31.50		386.50	445	
	2100	5'-0" x 4'-0" high		9	.889		430	31.50		461.50	525	
	2150	6'-0" x 5'-0" high		8	1		655	35.50		690.50	780	
	9000	Minimum labor/equipment charge		3	2.667	Job		95		95	157	
800	0010	**WINDOW GRILLE OR MUNTIN** Snap-in type										800
	0020	Standard pattern interior grills										
	2000	Wood, awning window, glass size 28" x 16" high	1 Carp	30	.267	Ea.	20	9.50		29.50	38	
	2060	44" x 24" high		32	.250		29.50	8.90		38.40	47.50	
	2100	Casement, glass size, 20" x 36" high		30	.267		25	9.50		34.50	43	
	2180	20" x 56" high		32	.250		36	8.90		44.90	54.50	
	2200	Double hung, glass size, 16" x 24" high		24	.333	Set	44	11.85		55.85	68	
	2280	32" x 32" high		34	.235	″	123	8.35		131.35	150	
	2500	Picture, glass size, 48" x 48" high		30	.267	Ea.	141	9.50		150.50	171	
	2580	60" x 68" high		28	.286	″	108	10.15		118.15	136	
	2600	Sliding, glass size, 14" x 36" high		24	.333	Set	25	11.85		36.85	47	
	2680	36" x 36" high		22	.364	″	38.50	12.95		51.45	63.50	
	9000	Minimum labor/equipment charge		5	1.600	Job		57		57	94	
820	0010	**WOOD SASH** Including glazing but not including trim										820
	0050	Custom, 5'-0" x 4'-0", 1" dbl. glazed, 3/16" thick lites	2 Carp	3.20	5	Ea.	164	178		342	475	
	0100	1/4" thick lites		5	3.200		169	114		283	375	
	0200	1" thick, triple glazed		5	3.200		385	114		499	615	
	0300	7'-0" x 4'-6" high, 1" double glazed, 3/16" thick lites		4.30	3.721		395	132		527	650	
	0400	1/4" thick lites		4.30	3.721		445	132		577	705	
	0500	1" thick, triple glazed		4.30	3.721		505	132		637	775	
	0600	8'-6" x 5'-0" high, 1" double glazed, 3/16" thick lites		3.50	4.571		530	163		693	855	
	0700	1/4" thick lites		3.50	4.571		580	163		743	910	
	0800	1" thick, triple glazed		3.50	4.571		585	163		748	915	
	0900	Window frames only, based on perimeter length				L.F.	3.20			3.20	3.52	
	1200	Window sill, stock, per lineal foot					7.20			7.20	7.90	
	1250	Casing, stock					2.58			2.58	2.84	
	3000	Replacement sash, double hung, double glazing, to 12 S.F.	1 Carp	64	.125	S.F.	17.15	4.44		21.59	26	
	3100	12 S.F. to 20 S.F.		94	.085		17.35	3.03		20.38	24	
	3200	20 S.F. and over		106	.075		15.10	2.68		17.78	21	
	3800	Triple glazing for above, add					2.40			2.40	2.64	
	7000	Sash, single lite, 2'-0" x 2'-0" high	1 Carp	20	.400	Ea.	34	14.20		48.20	60.50	
	7050	2'-6" x 2'-0" high		19	.421		39	14.95		53.95	68	
	7100	2'-6" x 2'-6" high		18	.444		45	15.80		60.80	75.50	
	7150	3'-0" x 2'-0" high		17	.471		48	16.75		64.75	80.50	
	9000	Minimum labor/equipment charge		4	2	Job		71		71	118	
840	0010	**WOOD SCREENS**										840
	0020	Over 3 S.F., 3/4" frames	2 Carp	375	.043	S.F.	3.69	1.52		5.21	6.55	
	0100	1-1/8" frames	″	375	.043	″	6.45	1.52		7.97	9.55	
	9000	Minimum labor/equipment charge	1 Carp	4	2	Job		71		71	118	

08560 | Plastic Windows

		08560 Plastic Windows	Crew	Daily Output	Labor-Hours	Unit	Mat.	Labor	Equip.	Total	Total Incl O&P	
100	0010	**VINYL SINGLE HUNG WINDOWS**										100
	0100	Grids, low E, J fin, ext. jambs, 21" x 53"	2 Carp	18	.889	Ea.	146	31.50		177.50	214	
	0110	21" x 57"		17	.941		150	33.50		183.50	221	
	0120	21" x 65"		16	1		156	35.50		191.50	230	
	0130	25" x 41"		20	.800		138	28.50		166.50	199	
	0140	25" x 49"		18	.889		152	31.50		183.50	221	

08560 | Plastic Windows

		08560 Plastic Windows	CREW	DAILY OUTPUT	LABOR-HOURS	UNIT	2006 BARE COSTS MAT.	LABOR	EQUIP.	TOTAL	TOTAL INCL O&P	
100	0150	25″ x 57″	2 Carp	17	.941	Ea.	156	33.50		189.50	227	100
	0160	25″ x 65″		16	1		162	35.50		197.50	237	
	0170	29″ x 41″		18	.889		147	31.50		178.50	214	
	0180	29″ x 53″		18	.889		157	31.50		188.50	226	
	0190	29″ x 57″		17	.941		161	33.50		194.50	233	
	0200	29″ x 65″		16	1		167	35.50		202.50	243	
	0210	33″ x 41″		20	.800		152	28.50		180.50	214	
	0220	33″ x 53″		18	.889		163	31.50		194.50	233	
	0230	33″ x 57″		17	.941		167	33.50		200.50	240	
	0240	33″ x 65″		16	1		174	35.50		209.50	250	
	0250	37″ x 41″		20	.800		160	28.50		188.50	223	
	0260	37″ x 53″		18	.889		172	31.50		203.50	242	
	0270	37″ x 57″		17	.941		175	33.50		208.50	249	
	0280	37″ x 65″		16	1		182	35.50		217.50	260	
200	0010	**VINYL DOUBLE HUNG WINDOWS**										200
	0100	Grids, low E, J fin, ext. jambs, 21″ x 53″	2 Carp	18	.889	Ea.	167	31.50		198.50	237	
	0102	21″ x 37″		18	.889		149	31.50		180.50	217	
	0104	21″ x 41″		18	.889		153	31.50		184.50	221	
	0106	21″ x 49″		18	.889		160	31.50		191.50	229	
	0110	21″ x 57″		17	.941		171	33.50		204.50	244	
	0120	21″ x 65″		16	1		177	35.50		212.50	254	
	0128	25″ x 37″		20	.800		157	28.50		185.50	220	
	0130	25″ x 41″		20	.800		161	28.50		189.50	224	
	0140	25″ x 49″		18	.889		166	31.50		197.50	235	
	0145	25″ x 53″		18	.889		172	31.50		203.50	243	
	0150	25″ x 57″		17	.941		173	33.50		206.50	246	
	0160	25″ x 65″		16	1		184	35.50		219.50	261	
	0162	25″ x 69″		16	1		191	35.50		226.50	269	
	0164	25″ x 77″		16	1		202	35.50		237.50	281	
	0168	29″ x 37″		18	.889		162	31.50		193.50	231	
	0170	29″ x 41″		18	.889		166	31.50		197.50	235	
	0172	29″ x 49″		18	.889		174	31.50		205.50	244	
	0180	29″ x 53″		18	.889		178	31.50		209.50	248	
	0190	29″ x 57″		17	.941		181	33.50		214.50	255	
	0200	29″ x 65″		16	1		188	35.50		223.50	266	
	0202	29″ x 69″		16	1		195	35.50		230.50	274	
	0205	29″ x 77″		16	1		207	35.50		242.50	286	
	0208	33″ x 37″		20	.800		167	28.50		195.50	230	
	0210	33″ x 41″		20	.800		170	28.50		198.50	234	
	0215	33″ x 49″		20	.800		179	28.50		207.50	244	
	0220	33″ x 53″		18	.889		183	31.50		214.50	254	
	0230	33″ x 57″		17	.941		187	33.50		220.50	262	
	0240	33″ x 65″		16	1		192	35.50		227.50	270	
	0242	33″ x 69″		16	1		204	35.50		239.50	283	
	0246	33″ x 77″		16	1		214	35.50		249.50	294	
	0250	37″ x 41″		20	.800		174	28.50		202.50	239	
	0255	37″ x 49″		20	.800		183	28.50		211.50	249	
	0260	37″ x 53″		18	.889		191	31.50		222.50	263	
	0270	37″ x 57″		17	.941		195	33.50		228.50	270	
	0280	37″ x 65″		16	1		200	35.50		235.50	279	
	0282	37″ x 69″		16	1		262	35.50		297.50	345	
	0286	37″ x 77″		16	1		274	35.50		309.50	360	
	0300	Solid vinyl, average quality, double insulated glass, 2′-0″ x 3′-0″	1 Carp	10	.800		137	28.50		165.50	198	
	0310	3′-0″ x 4′-0″		9	.889		165	31.50		196.50	234	
	0320	4′-0″ x 4′-6″		8	1		197	35.50		232.50	276	
	0330	Premium, double insulated glass, 2′-6″ x 3′-0″		10	.800		153	28.50		181.50	215	

	08560 | Plastic Windows		CREW	DAILY OUTPUT	LABOR-HOURS	UNIT	2006 BARE COSTS MAT.	LABOR	EQUIP.	TOTAL	TOTAL INCL O&P	
200	0340	3'-0" x 3'-6"	1 Carp	9	.889	Ea.	177	31.50		208.50	248	200
	0350	3'-0" x 4'-0"		9	.889		188	31.50		219.50	260	
	0360	3'-0" x 4'-6"		9	.889		193	31.50		224.50	265	
	0370	3'-0" x 5'-0"		8	1		198	35.50		233.50	277	
	0380	3'-6" x 6'-0"	↓	8	1	↓	218	35.50		253.50	298	
300	0010	VINYL CASEMENT WINDOWS										300
	0100	Grids, low E, J fin, ext. jambs, 1 lt, 21" x 41"	2 Carp	20	.800	Ea.	216	28.50		244.50	285	
	0110	21" x 47"		20	.800		236	28.50		264.50	305	
	0120	21" x 53"		20	.800		255	28.50		283.50	325	
	0128	24" x 35"		19	.842		208	30		238	278	
	0130	24" x 41"		19	.842		226	30		256	298	
	0140	24" x 47"		19	.842		244	30		274	320	
	0150	24" x 53"		19	.842		263	30		293	340	
	0158	28" x 35"		19	.842		221	30		251	294	
	0160	28" x 41"		19	.842		239	30		269	315	
	0170	28" x 47"		19	.842		258	30		288	335	
	0180	28" x 53"		19	.842		284	30		314	365	
	0184	28" x 59"		19	.842		290	30		320	370	
	0188	Two lites, 33" x 35"		18	.889		355	31.50		386.50	445	
	0190	33" x 41"		18	.889		380	31.50		411.50	475	
	0200	33" x 47"		18	.889		410	31.50		441.50	505	
	0210	33" x 53"		18	.889		440	31.50		471.50	535	
	0212	33" x 59"		18	.889		465	31.50		496.50	565	
	0215	33" x 72"		18	.889		480	31.50		511.50	585	
	0220	41" x 41"		18	.889		415	31.50		446.50	515	
	0230	41" x 47"		18	.889		445	31.50		476.50	545	
	0240	41" x 53"		17	.941		475	33.50		508.50	575	
	0242	41" x 59"		17	.941		500	33.50		533.50	605	
	0246	41" x 72"		17	.941		520	33.50		553.50	630	
	0250	47" x 41"		17	.941		420	33.50		453.50	520	
	0260	47" x 47"		17	.941		450	33.50		483.50	550	
	0270	47" x 53"		17	.941		475	33.50		508.50	575	
	0272	47" x 59"		17	.941		520	33.50		553.50	625	
	0280	56" x 41"		15	1.067		450	38		488	560	
	0290	56" x 47"		15	1.067		475	38		513	585	
	0300	56" x 53"		15	1.067		520	38		558	635	
	0302	56" x 59"		15	1.067		540	38		578	660	
	0310	56" x 72"	↓	15	1.067		590	38		628	710	
	0340	Solid vinyl, premium, double insulated glass, 2'-0" x 3'-0" high	1 Carp	10	.800		185	28.50		213.50	250	
	0360	2'-0" x 4'-0" high		9	.889		227	31.50		258.50	300	
	0380	2'-0" x 5'-0" high	↓	8	1	↓	258	35.50		293.50	345	
400	0010	VINYL PICTURE WINDOWS										400
	0100	Grids, low E, J fin, ext. jambs, 33" x 47"	2 Carp	12	1.333	Ea.	217	47.50		264.50	320	
	0110	35" x 71"		12	1.333		230	47.50		277.50	330	
	0120	41" x 47"		12	1.333		252	47.50		299.50	355	
	0130	41" x 71"		12	1.333		273	47.50		320.50	380	
	0140	47" x 47"		12	1.333		285	47.50		332.50	395	
	0150	47" x 71"		11	1.455		299	51.50		350.50	415	
	0160	53" x 47"		11	1.455		280	51.50		331.50	395	
	0170	53" x 71"		11	1.455		293	51.50		344.50	410	
	0180	59" x 47"		11	1.455		320	51.50		371.50	435	
	0190	59" x 71"		11	1.455		340	51.50		391.50	460	
	0200	71" x 47"		10	1.600		350	57		407	485	
	0210	71" x 71"	↓	10	1.600	↓	370	57		427	505	

08580 | Special Function Windows

	Line	Description	Crew	Daily Output	Labor-Hours	Unit	2006 Bare Costs Mat.	Labor	Equip.	Total	Total Incl O&P	
900	0010	**STORM WINDOWS** Aluminum, residential										900
	0300	Basement, mill finish, incl. fiberglass screen										
	0320	1′-10″ x 1′-0″ high	2 Carp	30	.533	Ea.	29.50	18.95		48.45	64	
	0340	2′-9″ x 1′-6″ high	″	30	.533	″	32	18.95		50.95	67	
	1600	Double-hung, combination, storm & screen										
	2000	Average quality, clear anodic coating, 2′-0″ x 3′-5″ high	2 Carp	30	.533	Ea.	77.50	18.95		96.45	117	
	2020	2′-6″ x 5′-0″ high		28	.571		95	20.50		115.50	138	
	2040	4′-0″ x 6′-0″ high		25	.640		115	23		138	165	
	2400	White painted, 2′-0″ x 3′-5″ high		30	.533		76.50	18.95		95.45	116	
	2420	2′-6″ x 5′-0″ high		28	.571		84.50	20.50		105	127	
	2440	4′-0″ x 6′-0″ high		25	.640		92.50	23		115.50	140	
	2600	Mill finish, 2′-0″ x 3′-5″ high		30	.533		69.50	18.95		88.45	108	
	2620	2′-6″ x 5′-0″ high		28	.571		77.50	20.50		98	119	
	2640	4′-0″ x 6-8″ high		25	.640		87	23		110	134	
	9410	Minimum labor/equipment charge	1 Carp	4	2	Job		71		71	118	

08590 | Window Restoration & Replace

	Line	Description	Crew	Daily Output	Labor-Hours	Unit	2006 Bare Costs Mat.	Labor	Equip.	Total	Total Incl O&P	
600	0010	**SOLID VINYL REPLACEMENT WINDOWS** (R085313-20)										600
	0020	Double hung, insulated glass, up to 83 united inches	2 Carp	8	2	Ea.	218	71		289	360	
	0040	84 to 93		8	2		243	71		314	385	
	0060	94 to 101		6	2.667		277	95		372	460	
	0080	102 to 111		6	2.667		305	95		400	495	
	0100	112 to 120		6	2.667		345	95		440	535	
	0120	For each united inch over 120 , add		800	.020	Inch	3.65	.71		4.36	5.20	
	0140	Casement windows, one operating sash , 42 to 60 united inches		8	2	Ea.	191	71		262	330	
	0160	61 to 70		8	2		217	71		288	355	
	0180	71 to 80		8	2		236	71		307	375	
	0200	81 to 96		8	2		250	71		321	395	
	0220	Two operating sash, 58 to 78 united inches		8	2		380	71		451	540	
	0240	79 to 88		8	2		405	71		476	570	
	0260	89 to 98		8	2		445	71		516	610	
	0280	99 to 108		6	2.667		465	95		560	665	
	0300	109 to 121		6	2.667		500	95		595	705	
	0320	Three operating sash, 73 to 108 united inches		8	2		600	71		671	780	
	0340	109 to 118		8	2		635	71		706	815	
	0360	119 to 128		6	2.667		650	95		745	870	
	0380	129 to 138		6	2.667		695	95		790	920	
	0400	139 to 156		6	2.667		735	95		830	965	
	0420	Four operating sash, 98 to 118 united inches		8	2		865	71		936	1,075	
	0440	119 to 128		8	2		925	71		996	1,150	
	0460	129 to 138		6	2.667		980	95		1,075	1,225	
	0480	139 to 148		6	2.667		1,025	95		1,120	1,275	
	0500	149 to 168		6	2.667		1,100	95		1,195	1,350	
	0520	169 to 178		6	2.667		1,200	95		1,295	1,450	
	0540	For venting unit to fixed unit, deduct					17.40			17.40	19.15	
	0560	Fixed picture window, up to 63 united inches	2 Carp	8	2		138	71		209	270	
	0580	64 to 83		8	2		163	71		234	298	
	0600	84 to 101		8	2		210	71		281	350	
	0620	For each united inch over 101, add		900	.018	Inch	2.53	.63		3.16	3.83	
	0640	Picture window opt., low E glazing, up to 101 united inches				Ea.	19.65			19.65	21.50	
	0660	102 to 124					25.50			25.50	28	
	0680	124 and over					38			38	42	
	0700	Options, low E glazing, up to 101 united inches					9.80			9.80	10.80	
	0720	102 to 124					12.65			12.65	13.90	
	0740	124 and over					19.05			19.05	21	

08500 | Windows

	08590 Window Restoration & Replace	CREW	DAILY OUTPUT	LABOR-HOURS	UNIT	2006 BARE COSTS MAT.	LABOR	EQUIP.	TOTAL	TOTAL INCL O&P	
600	0760 Muntins, between glazing, square, per lite R085313-20				Ea.	1.85			1.85	2.04	600
	0780 Diamond shape, per full or partial diamond				↓	2.92			2.92	3.21	
	0800 Celluose fiber insulation, poured into sash balance cavity	1 Carp	36	.222	C.F.	.58	7.90		8.48	13.75	
	0820 Silicone caulking at perimeter	"	800	.010	L.F.	.12	.36		.48	.73	

08600 | Skylights

	08620 Unit Skylights	CREW	DAILY OUTPUT	LABOR-HOURS	UNIT	2006 BARE COSTS MAT.	LABOR	EQUIP.	TOTAL	TOTAL INCL O&P	
400	0010 **PREFABRICATED**, Glass block with metal frame										400
	0020 Minimum	G-3	265	.121	S.F.	43.50	4.20		47.70	54.50	
	0100 Maximum	"	160	.200	"	86.50	6.95		93.45	107	
800	0010 **SKYLIGHT** Plastic domes, flush or curb mounted, ten or										800
	0100 more units, curb not included										
	0300 Nominal size under 10 S.F., double	G-3	130	.246	S.F.	19.10	8.55		27.65	35	
	0400 Single		160	.200		13.95	6.95		20.90	26.50	
	0600 10 S.F. to 20 S.F., double		315	.102		16.85	3.53		20.38	24.50	
	0700 Single		395	.081		8.70	2.82		11.52	14.10	
	0900 20 S.F. to 30 S.F., double		395	.081		15.40	2.82		18.22	21.50	
	1000 Single		465	.069		10.95	2.39		13.34	15.90	
	1200 30 S.F. to 65 S.F., double		465	.069		11.75	2.39		14.14	16.75	
	1300 Single	↓	610	.052		15	1.82		16.82	19.45	
	1500 For insulated 4" curbs, double, add					25%					
	1600 Single, add				↓	30%					
	1800 For integral insulated 9" curbs, double, add					30%					
	1900 Single, add					40%					
	2120 Ventilating insulated plexiglass dome with										
	2130 curb mounting, 36" x 36"	G-3	12	2.667	Ea.	380	93		473	570	
	2150 52" x 52"		12	2.667		570	93		663	775	
	2160 28" x 52"		10	3.200		445	111		556	670	
	2170 36" x 52"	↓	10	3.200		480	111		591	710	
	2180 For electric opening system, add				↓	285			285	315	
	2200 Field fabricated, factory type, aluminum and wire glass	G-3	120	.267	S.F.	14.70	9.25		23.95	31	
	2300 Insulated safety glass with aluminum frame		160	.200		85.50	6.95		92.45	106	
	2400 Sandwich panels, fiberglass, for walls, 1-9/16" thick, to 250 S.F.		200	.160		15.55	5.55		21.10	26	
	2500 250 S.F. and up		265	.121		13.95	4.20		18.15	22	
	2700 As above, but for roofs, 2-3/4" thick, to 250 S.F.		295	.108		22.50	3.77		26.27	30.50	
	2800 250 S.F. and up	↓	330	.097	↓	18.35	3.37		21.72	25.50	

08700 | Hardware

	08710 Door Hardware	CREW	DAILY OUTPUT	LABOR-HOURS	UNIT	2006 BARE COSTS MAT.	LABOR	EQUIP.	TOTAL	TOTAL INCL O&P	
100	0010 **AUTOMATIC OPENERS COMMERCIAL**										100
	0020 Pneumatic, incl opener, motion sens, control box, tubing, compressor										

	08710 | Door Hardware		CREW	DAILY OUTPUT	LABOR-HOURS	UNIT	2006 BARE COSTS MAT.	LABOR	EQUIP.	TOTAL	TOTAL INCL O&P	
100	0050	For single swing door, per opening	2 Skwk	.80	20	Ea.	3,725	730		4,455	5,300	100
	0100	Pair, per opening		.50	32	Opng.	6,200	1,175		7,375	8,725	
	1000	For single sliding door, per opening		.60	26.667		4,125	975		5,100	6,125	
	1300	Bi-parting pair		.50	32		6,225	1,175		7,400	8,750	
	1420	Electronic door opener incl motion sens, 12V control box, motor										
	1450	For single swing door, per opening	2 Skwk	.80	20	Opng.	3,250	730		3,980	4,775	
	1500	Pair, per opening		.50	32		5,325	1,175		6,500	7,750	
	1600	For single sliding door, per opening		.60	26.667		3,525	975		4,500	5,475	
	1700	Bi-parting pair		.50	32		5,400	1,175		6,575	7,850	
	1750	Handicap actuator buttons, 2, including 12V DC wiring, add	1 Carp	1.50	5.333	Pr.	375	190		565	725	
120	0010	**AUTOMATIC OPENERS INDUSTRIAL**										120
	0015	Sliding doors up to 6′ wide	2 Skwk	.60	26.667	Opng.	4,875	975		5,850	6,950	
	0200	To 12′ wide	"	.40	40	"	5,925	1,450		7,375	8,925	
	0400	Over 12′ wide, add per L.F. of excess				L.F.	650			650	715	
	1000	Swing doors, to 5′ wide	2 Skwk	.80	20	Ea.	2,850	730		3,580	4,325	
	1860	Add for controls, wall pushbutton, 3 button		4	4		174	146		320	430	
	1870	Ceiling pull cord		4.30	3.721		149	136		285	385	
150	0010	**AVERAGE** Percentage for hardware, total job cost										150
	0025	Minimum									.75%	
	0050	Maximum									3.50%	
	0500	Total hardware for building, average distribution					85%	15%				
	1000	Door hardware, apartment, interior				Door	127			127	140	
	1500	Hospital bedroom, minimum					284			284	315	
	2000	Maximum					625			625	685	
	2100	Pocket door				Ea.	127			127	140	
	2250	School, single exterior, incl. lever, not incl. panic device				Door	420			420	460	
	2500	Single interior, regular use, no lever included					280			280	310	
	2550	Including handicap lever					380			380	420	
	2600	Heavy use, incl. lever and closer					490			490	540	
	2850	Stairway, single interior					700			700	770	
	3100	Double exterior, with panic device				Pr.	985			985	1,075	
	3600	Toilet, public, single interior				Door	154			154	170	
200	0010	**BOLTS, FLUSH** (security item)										200
	0020	Standard, concealed	1 Carp	7	1.143	Ea.	20	40.50		60.50	89.50	
	0800	Automatic fire exit	"	5	1.600		271	57		328	390	
	1600	For electric release, add	1 Elec	3	2.667		110	112		222	294	
	3000	Barrel, brass, 2″ long	1 Carp	40	.200		4.30	7.10		11.40	16.55	
	3020	4″ long		40	.200		2.92	7.10		10.02	15	
	3060	6″ long		40	.200		8.70	7.10		15.80	21.50	
220	0010	**DOOR PROTECTORS**										220
	0020	1-3/4″ x 3/4″ U channel	2 Carp	80	.200	L.F.	18	7.10		25.10	31.50	
	0021	1-3/4″ x 1-1/4″ U channel		80	.200	"	8.45	7.10		15.55	21	
	1000	Tear drop, spring-stl., 8″ high x 19″ long		15	1.067	Ea.	78	38		116	149	
	1010	Tear drop, spring-stl., 8″ high x 32″ long		15	1.067		96.50	38		134.50	169	
	1100	8″ high x 19″ long		15	1.067		199	38		237	282	
	1200	8″ high x 32″ long		15	1.067		248	38		286	335	
300	0010	**DOOR CLOSERS**										300
	0015	Door closer Rack and pinion	1 Carp	6.50	1.231	Ea.	134	44		178	220	
	0020	Adjustable backcheck, 3 way mount, all sizes, regular arm		6	1.333		137	47.50		184.50	230	
	0040	Hold open arm		6	1.333		157	47.50		204.50	252	
	0100	Fusible link		6.50	1.231		122	44		166	207	
	0200	Non sized, regular arm		6	1.333		136	47.50		183.50	229	

From the planning stages through the punch list . . .

maximize your estimating & budgeting efforts quickly & efficiently with a *Means CostWorks 2006* subscription.

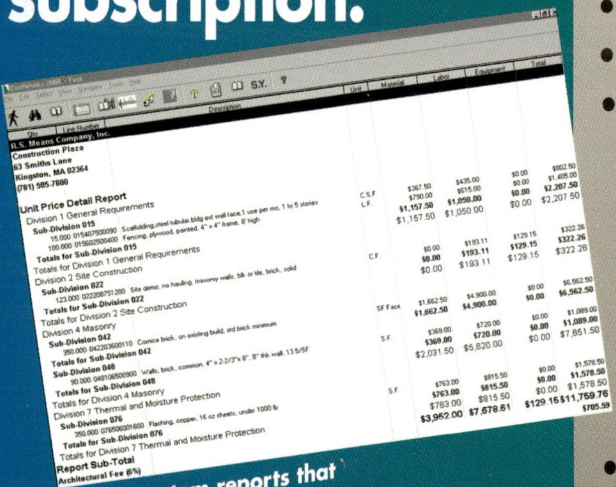

Design custom reports that meet your specifications

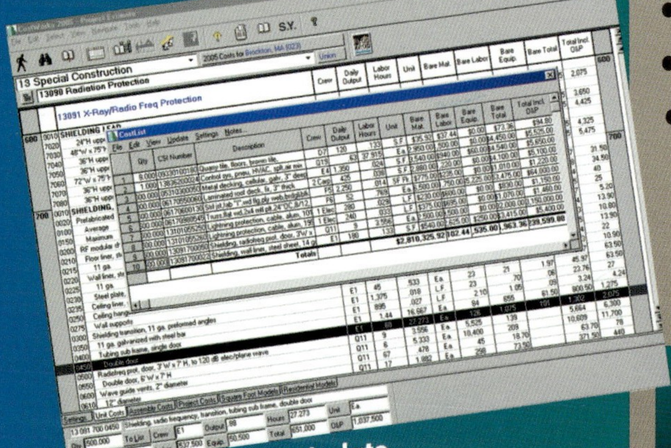

Select items and calculate project costs with the CostList

As building requirements grow more complex, the estimating and budgeting process has never been more demanding. Tighter deadlines and more stringent cost controls mean you must find ways to work more efficiently. Let ***Means CostWorks 2006***, available in CD format, help you quickly and easily perform in-depth cost research . . . and take your estimating and budgeting to the next level.

Purchase your titles separately or in value-priced packages. With your purchase you receive access to regular cost updates, access to the Means HOTLINE, and the Wage Rate Utility (with selected titles).

Your *Means CostWorks 2006* subscription now includes, Free:

- Access to Means Hotline
- The *RSMeans Quarterly Update Service*
- Wage Rate Utility – allows you to reference Union and Open Shop rates (included with selected titles)
- Means Estimator with your purchase of a *Means CostWorks 2006* package

The same Cost Data you know and trust . . . delivered electronically in a CD format!

- Localize costs to your geographic area to ensure accuracy
- Easily search and organize detailed cost data
- Determine best-cost solutions for each project
- Change and calculate results automatically
- Receive free online cost-factor updates with the *RSMeans Quarterly Update Service*
- Available as a 12-month subscription

Quick and easy . . . Get this powerful tool before you estimate your next project!

	08710 Door Hardware		Crew	Daily Output	Labor-Hours	Unit	2006 Bare Costs Mat.	Labor	Equip.	Total	Total Incl O&P
300	0240	Hold open arm	1 Carp	6	1.333	Ea.	170	47.50		217.50	266
	0400	4 way mount, non sized, regular arm		6	1.333		187	47.50		234.50	285
	0440	Hold open arm		6	1.333		201	47.50		248.50	300
	2000	Backcheck and adjustable power, hinge face mount									
	2010	All sizes, regular arm	1 Carp	6.50	1.231	Ea.	172	44		216	262
	2040	Hold open arm		6.50	1.231		185	44		229	277
	2400	Top jamb mount, all sizes, regular arm		6	1.333		172	47.50		219.50	268
	2440	Hold open arm		6	1.333		185	47.50		232.50	283
	2800	Top face mount, all sizes, regular arm		6.50	1.231		172	44		216	262
	2840	Hold open arm		6.50	1.231		185	44		229	276
	4000	Backcheck, overhead concealed, all sizes, regular arm		5.50	1.455		182	51.50		233.50	286
	4040	Concealed arm		5	1.600		194	57		251	310
	4400	Compact overhead, concealed, all sizes, regular arm		5.50	1.455		330	51.50		381.50	450
	4440	Concealed arm		5	1.600		345	57		402	475
	4800	Concealed in door, all sizes, regular arm		5.50	1.455		123	51.50		174.50	221
	4840	Concealed arm		5	1.600		132	57		189	239
	4900	Floor concealed, all sizes, single acting		2.20	3.636		156	129		285	385
	4940	Double acting		2.20	3.636		201	129		330	435
	5000	For cast aluminum cylinder, deduct					16.50			16.50	18.15
	5040	For delayed action, add					29			29	32
	5080	For fusible link arm, add					11.95			11.95	13.15
	5120	For shock absorbing arm, add					36			36	39.50
	5160	For spring power adjustment, add					27.50			27.50	30.50
	6000	Closer-holder, hinge face mount, all sizes, exposed arm	1 Carp	6.50	1.231		127	44		171	212
	7000	Electronic closer-holder, hinge facemount, concealed arm		5	1.600		193	57		250	305
	7400	With built-in detector		5	1.600		580	57		637	735
	8000	Surface mounted, stand. duty, parallel arm, primed, traditional		6	1.333		134	47.50		181.50	226
	8030	Light duty		6	1.333		79.50	47.50		127	166
	8050	Heavy duty		6	1.333		160	47.50		207.50	255
	8100	Standard duty, parallel arm, modern		6	1.333		150	47.50		197.50	244
	8150	Heavy duty		6	1.333		169	47.50		216.50	265
	9000	Minimum labor/equipment charge		4	2	Job		71		71	118
320	0010	**DEADLOCKS**									
	0011	Mortise heavy duty outside key(security item)	1 Carp	9	.889	Ea.	131	31.50		162.50	197
	0020	Double cylinder		9	.889		145	31.50		176.50	212
	0100	Medium duty, outside key		10	.800		102	28.50		130.50	159
	0110	Double cylinder		10	.800		127	28.50		155.50	187
	1000	Tubular, standard duty, outside key		10	.800		55	28.50		83.50	107
	1010	Double cylinder		10	.800		70.50	28.50		99	125
	1200	Night latch, outside key		10	.800		69	28.50		97.50	123
340	0010	**DOORSTOPS**									
	0020	Holder & bumper, floor or wall	1 Carp	32	.250	Ea.	31.50	8.90		40.40	49.50
	1300	Wall bumper, 4" diameter, with rubber pad, aluminum		32	.250		9.35	8.90		18.25	25
	1600	Door bumper, floor type, aluminum		32	.250		4.82	8.90		13.72	20
	1900	Plunger type, door mounted		32	.250		25.50	8.90		34.40	43
	9000	Minimum labor/equipment charge		6	1.333	Job		47.50		47.50	78.50
400	0010	**ENTRANCE LOCKS**									
	0015	Cylinder, grip handle deadlocking latch	1 Carp	9	.889	Ea.	120	31.50		151.50	185
	0020	Deadbolt		8	1		146	35.50		181.50	220
	0100	Push and pull plate, dead bolt		8	1		139	35.50		174.50	212
	0200	Push bar and pull, dead bolt, bronze		7	1.143		253	40.50		293.50	345
	0240	Push bar and pull bar, dead bolt, bronze		7	1.143		395	40.50		435.50	505
	0900	For handicapped lever, add ♿					152			152	167
450	0010	**FLOOR CHECKS**									
	0020	For over 3′ wide doors single acting	1 Carp	2.50	3.200	Ea.	460	114		574	695

For expanded coverage of these items see *Means Interior Cost Data 2006*

	08710	Door Hardware	CREW	DAILY OUTPUT	LABOR-HOURS	UNIT	2006 BARE COSTS MAT.	LABOR	EQUIP.	TOTAL	TOTAL INCL O&P
450	0500	Double acting	1 Carp	2.50	3.200	Ea.	585	114		699	835
500	0010	**HASP**									
	0015	3″	1 Carp	26	.308	Ea.	2.68	10.95		13.63	21
	0020	4-1/2″		13	.615		3.43	22		25.43	40.50
	0040	6″		12.50	.640		5.20	23		28.20	43
510	0010	**MORTISED HINGES**									
	0200	Average frequency, steel plated, ball bearing				Pr.	23.50			23.50	26
	0300	Bronze, ball bearing					26.50			26.50	29.50
	1100	Bronze, ball bearing					74			74	81.50
	1300	Average frequency, steel plated, ball bearing					29			29	32
	1500	Bronze, ball bearing, to 36″ wide					31			31	34
	1700	Low frequency, steel, plated, plain bearing					13.05			13.05	14.35
	1900	Bronze, plain bearing					14.85			14.85	16.35
520	0010	**HINGES**									
	0012	Full mortise, avg. freq., steel base, USP, 4-1/2″ x 4-1/2″				Pr.	21			21	23.50
	0040	US26D					26.50			26.50	29
	0080	US10A					31			31	34
	0100	5″ x 5″, USP					35.50			35.50	39
	0200	6″ x 6″, USP					75.50			75.50	83
	0400	Brass base, 4-1/2″ x 4-1/2″, US10					43.50			43.50	48
	0440	US26D					51.50			51.50	57
	0480	US10B					57.50			57.50	63.50
	0500	5″ x 5″, US10					63.50			63.50	70
	0600	6″ x 6″, US10					108			108	119
	0800	Stainless steel base, 4-1/2″ x 4-1/2″, US32					65.50			65.50	72
	0900	For non removable pin, add (security item)				Ea.	2.38			2.38	2.62
	0910	For floating pin, driven tips, add					2.73			2.73	3
	0930	For hospital type tip on pin, add					11.80			11.80	12.95
	0940	For steeple type tip on pin, add					10.30			10.30	11.35
	0950	Full mortise, high frequency, steel base, 3-1/2″ x 3-1/2″, US26D				Pr.	19.30			19.30	21
	1000	4-1/2″ x 4-1/2″, USP					51.50			51.50	56.50
	1040	US26D					52			52	57.50
	1080	US26					67			67	73.50
	1100	5″ x 5″, USP					48			48	53
	1200	6″ x 6″, USP					117			117	129
	1300	8″ x 8″, USP					197			197	216
	1400	Brass base, 3-1/2″ x 3-1/2″, US4					40.50			40.50	44.50
	1430	4-1/2″ x 4-1/2″, US10					69			69	76
	1440	US26D					84			84	92.50
	1480	US10B					96			96	105
	1500	5″ x 5″, US10					103			103	113
	1600	6″ x 6″, US10					149			149	164
	1700	8″ x 8″, US10					275			275	305
	1800	Stainless steel base, 4-1/2″ x 4-1/2″, US32					111			111	122
	1810	5″ x 4-1/2″, US32					154			154	170
	1930	For hospital type tip on pin, add				Ea.	6.95			6.95	7.65
	1950	Full mortise, low frequency, steel base, 3-1/2″ x 3-1/2″, US26D				Pr.	10.10			10.10	11.10
	2000	4-1/2″ x 4-1/2″, USP					9.40			9.40	10.35
	2040	US26D					13.65			13.65	15.05
	2080	US10A					17.10			17.10	18.80
	2100	5″ x 5″, USP					26			26	28.50
	2200	6″ x 6″, USP					52			52	57
	2300	4-1/2″ x 4-1/2″, US3					15.20			15.20	16.70
	2310	5″ x 5″, US3					37.50			37.50	41.50
	2400	Brass bass, 4-1/2″ x 4-1/2″, US10					36.50			36.50	40

	08710 | Door Hardware		Crew	Daily Output	Labor-Hours	Unit	2006 Bare Costs Mat.	Labor	Equip.	Total	Total Incl O&P	
520	2440	US26D				Pr.	42.50			42.50	47	520
	2480	US10A					39			39	43	
	2500	5″ x 5″, US10					55.50			55.50	61	
	2800	Stainless steel base, 4-1/2″ x 4-1/2″, US32					63			63	69	
	3000	Half surface, half mortise, or full surface, average frequency										
	3010	Steel base, 4-1/2″ x 4-1/2″, USP				Pr.	33.50			33.50	36.50	
	3040	US26D					53			53	58	
	3080	US10A					70.50			70.50	77.50	
	3100	5″ x 5″, USP					65.50			65.50	72	
	3400	Brass base, 4-1/2″ x 4-1/2″, US10					125			125	137	
	3440	US26D					134			134	148	
	3480	US10B					147			147	161	
	3500	5″ x 5″, US10					158			158	174	
	3800	Stainless steel base, 4-1/2″ x 4-1/2″, US32					171			171	188	
	4000	Half surface, half mortise or full surface, high frequency										
	4010	Steel base, 4-1/2″ x 4-1/2″, USP				Pr.	65.50			65.50	72	
	4040	US26D					121			121	133	
	4080	US10A					147			147	161	
	4100	5″ x 5″, USP					129			129	142	
	4400	Brass base, 4-1/2″ x 4-1/2″, US10					150			150	165	
	4440	US26D					157			157	173	
	4480	US10B					179			179	197	
	4500	5″ x 5″, US10					178			178	195	
	4800	Stainless steel base, 4-1/2″ x 4-1/2″, US32					197			197	217	
	5000	Half surface, half mortise, or full surface, low frequency										
	5010	Steel base, 4-1/2″ x 4-1/2″, USP				Pr.	24.50			24.50	27	
	5040	US26D					30			30	33	
	5080	US10A					49.50			49.50	54.50	
550	0010	**KICK PLATE**										550
	0020	Stainless steel	1 Carp	15	.533	Ea.	27.50	18.95		46.45	62	
	0500	Bronze		15	.533		35	18.95		53.95	69.50	
	2000	Aluminum, .050, with 3 beveled edges, 10″ x 28″		15	.533		19.80	18.95		38.75	53.50	
	2010	10″ x 30″		15	.533		21.50	18.95		40.45	55	
	2020	10″ x 34″		15	.533		22.50	18.95		41.45	56	
	2040	10″ x 38″		15	.533		25	18.95		43.95	59	
	9000	Minimum labor/equipment charge		6	1.333	Job		47.50		47.50	78.50	
650	0010	**LOCKSET** Standard duty, cylindrical, with sectional trim (security item)										650
	0020	Non-keyed, passage	1 Carp	12	.667	Ea.	43	23.50		66.50	86.50	
	0100	Privacy		12	.667		53.50	23.50		77	98	
	0400	Keyed, single cylinder function		10	.800		74.50	28.50		103	129	
	0420	Hotel		8	1		106	35.50		141.50	176	
	0500	Lever handled, keyed, single cylinder function		10	.800		132	28.50		160.50	192	
	1000	Heavy duty with sectional trim, non-keyed, passages		12	.667		123	23.50		146.50	176	
	1100	Privacy		12	.667		156	23.50		179.50	212	
	1400	Keyed, single cylinder function		10	.800		184	28.50		212.50	250	
	1420	Hotel		8	1		275	35.50		310.50	360	
	1600	Communicating		10	.800		218	28.50		246.50	287	
	1690	For re-core cylinder, add					31			31	34	
	1700	Residential, interior door, minimum	1 Carp	16	.500		14.90	17.80		32.70	46	
	1720	Maximum		8	1		39.50	35.50		75	103	
	1800	Exterior, minimum		14	.571		33	20.50		53.50	70	
	1810	Average		8	1		65	35.50		100.50	131	
	1820	Maximum		8	1		140	35.50		175.50	212	
	9000	Minimum labor/equipment charge		6	1.333	Job		47.50		47.50	78.50	
700	0010	**MORTISE LOCKSET** Comm., wrought knobs & full escutcheon trim										700
	0020	Non-keyed, passage, minimum	1 Carp	9	.889	Ea.	160	31.50		191.50	228	

08710 | Door Hardware

		08710 Door Hardware	CREW	DAILY OUTPUT	LABOR-HOURS	UNIT	2006 BARE COSTS MAT.	LABOR	EQUIP.	TOTAL	TOTAL INCL O&P	
700	0030	Maximum	1 Carp	8	1	Ea.	258	35.50		293.50	340	700
	0040	Privacy, minimum		9	.889		170	31.50		201.50	240	
	0050	Maximum		8	1		278	35.50		313.50	365	
	0100	Keyed, office/entrance/apartment, minimum		8	1		195	35.50		230.50	273	
	0110	Maximum		7	1.143		335	40.50		375.50	440	
	0120	Single cylinder, typical, minimum		8	1		167	35.50		202.50	243	
	0130	Maximum		7	1.143		310	40.50		350.50	410	
	0200	Hotel, minimum		7	1.143		201	40.50		241.50	289	
	0210	Maximum		6	1.333		325	47.50		372.50	440	
	0300	Communication, double cylinder, minimum		8	1		201	35.50		236.50	280	
	0310	Maximum		7	1.143		260	40.50		300.50	355	
	1000	Wrought knobs and sectional trim, non-keyed, passage, minimum		10	.800		106	28.50		134.50	163	
	1010	Maximum		9	.889		209	31.50		240.50	283	
	1040	Privacy, minimum		10	.800		124	28.50		152.50	183	
	1050	Maximum		9	.889		223	31.50		254.50	298	
	1100	Keyed, entrance, office/apartment, minimum		9	.889		184	31.50		215.50	255	
	1110	Maximum		8	1		265	35.50		300.50	350	
	1120	Single cylinder, typical, minimum		9	.889		177	31.50		208.50	247	
	1130	Maximum		8	1		257	35.50		292.50	340	
	2000	Cast knobs and full escutcheon trim										
	2010	Non-keyed, passage, minimum	1 Carp	9	.889	Ea.	225	31.50		256.50	300	
	2020	Maximum		8	1		365	35.50		400.50	460	
	2040	Privacy, minimum		9	.889		271	31.50		302.50	350	
	2050	Maximum		8	1		390	35.50		425.50	490	
	2120	Keyed, single cylinder, typical, minimum		8	1		271	35.50		306.50	355	
	2130	Maximum		7	1.143		425	40.50		465.50	540	
	2200	Hotel, minimum		7	1.143		300	40.50		340.50	400	
	2210	Maximum		6	1.333		540	47.50		587.50	670	
	3000	Cast knob and sectional trim, non-keyed, passage, minimum		10	.800		175	28.50		203.50	240	
	3010	Maximum		10	.800		350	28.50		378.50	430	
	3040	Privacy, minimum		10	.800		199	28.50		227.50	266	
	3050	Maximum		10	.800		350	28.50		378.50	430	
	3100	Keyed, office/entrance/apartment, minimum		9	.889		226	31.50		257.50	300	
	3110	Maximum		9	.889		355	31.50		386.50	445	
	3120	Single cylinder, typical, minimum		9	.889		226	31.50		257.50	300	
	3130	Maximum		9	.889		440	31.50		471.50	535	
	3190	For re-core cylinder, add					30.50			30.50	33.50	
	3800	Cipher lockset (security item)	1 Carp	13	.615		715	22		737	820	
	3900	Keyless, pushbutton type										
	4000	Residential/light commercial, deadbolt, standard	1 Carp	9	.889	Ea.	102	31.50		133.50	165	
	4010	Heavy duty		9	.889		121	31.50		152.50	186	
	4020	Industrial, heavy duty, with deadbolt		9	.889		238	31.50		269.50	315	
	4030	Key override		9	.889		264	31.50		295.50	345	
	4040	Lever activated handle		9	.889		290	31.50		321.50	375	
	4050	Key override		9	.889		320	31.50		351.50	410	
	4060	Double sided pushbutton type		8	1		530	35.50		565.50	645	
	4070	Key override		8	1		570	35.50		605.50	690	
750	0010	**PANIC DEVICE**										750
	0015	For rim locks, single door exit only	1 Carp	6	1.333	Ea.	375	47.50		422.50	490	
	0020	Outside key and pull		5	1.600		425	57		482	560	
	0200	Bar and vertical rod, exit only		5	1.600		540	57		597	690	
	0210	Outside key and pull		4	2		645	71		716	830	
	0400	Bar and concealed rod		4	2		545	71		616	715	
	0600	Touch bar, exit only		6	1.333		430	47.50		477.50	550	
	0610	Outside key and pull		5	1.600		515	57		572	665	

	08710 | Door Hardware		CREW	DAILY OUTPUT	LABOR-HOURS	UNIT	2006 BARE COSTS MAT.	LABOR	EQUIP.	TOTAL	TOTAL INCL O&P	
750	0700	Touch bar and vertical rod, exit only	1 Carp	5	1.600	Ea.	585	57		642	740	750
	0710	Outside key and pull		4	2		685	71		756	870	
	0800	Touch bar, low profile, exit only		6	1.333		330	47.50		377.50	440	
	0810	Outside key and pull		5	1.600		385	57		442	520	
	0900	Touch bar and vertical rod, low profile, exit only		5	1.600		475	57		532	620	
	0910	Outside key and pull		4	2		520	71		591	690	
	1000	Mortise, bar, exit only		4	2		480	71		551	650	
	1600	Touch bar, exit only		4	2		550	71		621	725	
	2000	Narrow stile, rim mounted, bar, exit only		6	1.333		580	47.50		627.50	715	
	2010	Outside key and pull		5	1.600		630	57		687	785	
	2200	Bar and vertical rod, exit only		5	1.600		595	57		652	750	
	2210	Outside key and pull		4	2		595	71		666	775	
	2400	Bar and concealed rod, exit only		3	2.667		695	95		790	920	
	3000	Mortise, bar, exit only		4	2		505	71		576	675	
	3600	Touch bar, exit only		4	2		735	71		806	930	
	6000	Trim, rim mounted, cylinder and pull		25	.320		105	11.40		116.40	134	
	6100	Cylinder, pull and thumb piece		20	.400		103	14.20		117.20	138	
	6200	Pull only		30	.267		60	9.50		69.50	81.50	
	6400	Mortise, cylinder and pull		25	.320		122	11.40		133.40	153	
	6500	Cylinder, pull and thumb piece		20	.400		118	14.20		132.20	154	
	6600	Pull only		30	.267		105	9.50		114.50	131	
	9000	Minimum labor/equipment charge		2.50	3.200	Job		114		114	188	
770	0010	**DOOR AND WINDOW ACCESSORIES**										770
	0050	Door closing coordinator, 36" (for paired openings up to 56")	1 Carp	8	1	Ea.	87.50	35.50		123	155	
	0060	48" (for paired openings up to 84")		8	1		93.50	35.50		129	162	
	0070	56" (for paired openings up to 96")		8	1		103	35.50		138.50	172	
780	0010	**PUSH-PULL PLATE**										780
	0100	Push plate, .050 thick, 4" x 16", aluminum	1 Carp	12	.667	Ea.	6.85	23.50		30.35	47	
	0500	Bronze		12	.667		17.35	23.50		40.85	58.50	
	1500	Pull handle and push bar, aluminum		11	.727		121	26		147	176	
	2000	Bronze		10	.800		157	28.50		185.50	219	
	3000	Push plate both sides, aluminum		14	.571		14.80	20.50		35.30	50	
	3500	Bronze		13	.615		37	22		59	77	
	4000	Door pull, designer style, cast aluminum, minimum		12	.667		64.50	23.50		88	111	
	5000	Maximum		8	1		325	35.50		360.50	420	
	6000	Cast bronze, minimum		12	.667		76	23.50		99.50	123	
	7000	Maximum		8	1		355	35.50		390.50	450	
	8000	Walnut, minimum		12	.667		58	23.50		81.50	104	
	9000	Maximum		8	1		325	35.50		360.50	415	
	9800	Minimum labor/equipment charge		5	1.600	Job		57		57	94	
	9897	Bronze		12	.667	Ea.	17.35	23.50		40.85	58.50	
790	0010	**PEEPHOLE**										790
	2010	Peephole	1 Carp	32	.250	Ea.	14.10	8.90		23	30.50	
800	0010	**SPECIAL HINGES**										800
	0015	Paumelle, high frequency										
	0020	Steel base, 6" x 4-1/2", US10				Pr.	134			134	147	
	0040	US26D					166			166	183	
	0080	US10A					174			174	192	
	0100	Bronze base, 5" x 4-1/2", US10					170			170	187	
	0140	US26D					187			187	206	
	0180	US3					197			197	217	
	0200	Paumelle, average frequency, steel base, 4-1/2" x 3-1/2", US10					90.50			90.50	99.50	
	0240	US26D					94.50			94.50	104	
	0280	US10A					97.50			97.50	107	
	0400	Olive knuckle, low frequency, brass base, 6" x 4-1/2", US10					155			155	171	

08710 | Door Hardware		CREW	DAILY OUTPUT	LABOR-HOURS	UNIT	2006 BARE COSTS MAT.	LABOR	EQUIP.	TOTAL	TOTAL INCL O&P
800 0440	US26D				Pr.	174			174	192 800
0480	US3					176			176	193
0800	Emergency door pivot, average frequency									
0810	Brass base, 4-7/8″ jamb plate, USP				Pr.	59.50			59.50	65.50
0840	US26D					66			66	72.50
0880	For emergency door stop and hold back, add					56.50			56.50	62
1000	Electric hinge with concealed conductor, average frequency									
1010	Steel base, 4-1/2″ x 4-1/2″, US26D				Pr.	286			286	315
1100	Bronze base, 4-1/2″ x 4-1/2″, US26D				″	300			300	330
1200	Electric hinge with concealed conductor, high frequency									
1210	Steel base, 4-1/2″ x 4-1/2″, US26D				Pr.	213			213	235
1400	Non template, full mortise, low frequency									
1410	Steel base, 4″ x 4″, USP				Pr.	15.25			15.25	16.80
1440	US26D					17.15			17.15	18.85
1480	US10A					21			21	23.50
1500	Non template, full mortise, average frequency									
1510	Steel base, 4″ x 4″, USP				Pr.	33			33	36.50
1540	US26D					32			32	35
1580	US10A					41			41	45
1600	Double weight, 800 lb., steel base, removable pin, 5″ x 6″, USP					125			125	137
1700	Steel base-welded pin, 5″ x 6″, USP					141			141	155
1800	Triple weight, 2000 lb., steel base, welded pin, 5″ x 6″, USP					143			143	158
2000	Pivot reinf., high frequency, steel base, 7-3/4″ door plate, USP					166			166	183
2040	US26D					143			143	158
2080	US10A					174			174	191
2200	Bronze base, 7-3/4″ door plate, US10					198			198	218
2240	US26D					207			207	227
2280	US10A					237			237	261
3000	Swing clear, full mortise, full or half surface, high frequency,									
3010	Steel base, 5″ high, USP				Pr.	143			143	157
3040	US26D					138			138	151
3080	US10A					173			173	191
3200	Swing clear, full mortise, average frequency									
3210	Steel base, 4-1/2″ high, USP				Pr.	113			113	125
3280	US10A				″	139			139	153
3400	Swing clear, half mortise, high frequency									
3410	Steel base, 4-1/2″ high, USP				Pr.	143			143	158
3440	US26D					148			148	163
3480	US10A					175			175	192
4000	Wide throw, average frequency, steel base, 4-1/2″ x 6″, USP					86.50			86.50	95.50
4040	US26D					92.50			92.50	102
4080	US10A					99			99	109
4100	5″ x 7″, USP					106			106	116
4200	High frequency, steel base, 4-1/2″ x 6″, USP					133			133	146
4240	US26D					153			153	168
4280	US10A					171			171	188
4300	5″ x 7″, USP					135			135	149
4400	Wide throw, low frequency, steel base, 4-1/2″ x 6″, USP					65			65	71.50
4480	US10A					85.50			85.50	94
4500	5″ x 7″, USP					92.50			92.50	102
4600	Spring hinge, single acting, 6″ flange, steel				Ea.	49			49	54
4700	Brass					86			86	94.50
4900	Double acting, 6″ flange, steel					86			86	95
4950	Brass					141			141	155
9000	Continuous hinge, steel, full mortise, heavy duty	2 Carp	64	.250	L.F.	11.10	8.90		20	27

	08720 | Weatherstripping & Seals	CREW	DAILY OUTPUT	LABOR-HOURS	UNIT	2006 BARE COSTS MAT.	LABOR	EQUIP.	TOTAL	TOTAL INCL O&P	
100	0010 **ASTRAGALS** One piece overlapping										100
	0400 Cadmium plated steel, flat, 3/16" x 2"	1 Carp	90	.089	L.F.	3.06	3.16		6.22	8.60	
	0600 Prime coated steel, flat, 1/8" x 3"		90	.089		4.25	3.16		7.41	9.95	
	0800 Stainless steel, flat, 3/32" x 1-5/8"		90	.089		15.95	3.16		19.11	23	
	1000 Aluminum, flat, 1/8" x 2"		90	.089		3.07	3.16		6.23	8.65	
	1200 Nail on, "T" extrusion		120	.067		.68	2.37		3.05	4.68	
	1300 Vinyl bulb insert		105	.076		1.10	2.71		3.81	5.70	
	1600 Screw on, "T" extrusion		90	.089		4.47	3.16		7.63	10.15	
	1700 Vinyl insert		75	.107		3.02	3.79		6.81	9.60	
	2000 "L" extrusion, neoprene bulbs		75	.107		1.74	3.79		5.53	8.20	
	2100 Neoprene sponge insert		75	.107		5.30	3.79		9.09	12.15	
	2200 Magnetic		75	.107		8.70	3.79		12.49	15.90	
	2400 Spring hinged security seal, with cam		75	.107		5.60	3.79		9.39	12.45	
	2600 Spring loaded locking bolt, vinyl insert		45	.178		7.65	6.30		13.95	18.85	
	2800 Neoprene sponge strip, "Z" shaped, aluminum		60	.133		3.63	4.74		8.37	11.85	
	2900 Solid neoprene strip, nail on aluminum strip	↓	90	.089	↓	3.05	3.16		6.21	8.60	
	3000 One piece stile protection										
	3020 Neoprene fabric loop, nail on aluminum strips	1 Carp	60	.133	L.F.	.54	4.74		5.28	8.45	
	3110 Flush mounted aluminum extrusion, 1/2" x 1-1/4"		60	.133		2.89	4.74		7.63	11.05	
	3140 3/4" x 1-3/8"		60	.133		3.67	4.74		8.41	11.90	
	3160 1-1/8" x 1-3/4"		60	.133		5.85	4.74		10.59	14.30	
	3300 Mortise, 9/16" x 3/4"		60	.133		3.11	4.74		7.85	11.25	
	3320 13/16" x 1-3/8"		60	.133		3.37	4.74		8.11	11.55	
	3600 Spring bronze strip, nail on type		105	.076		2.65	2.71		5.36	7.40	
	3620 Screw on, with retainer		75	.107		2.17	3.79		5.96	8.70	
	3800 Flexible stainless steel housing, pile insert, 1/2" door		105	.076		6	2.71		8.71	11.10	
	3820 3/4" door		105	.076		6.75	2.71		9.46	11.95	
	4000 Extruded aluminum retainer, flush mount, pile insert		105	.076		2.10	2.71		4.81	6.80	
	4080 Mortise, felt insert		90	.089		3.74	3.16		6.90	9.35	
	4160 Mortise with spring, pile insert		90	.089		2.92	3.16		6.08	8.45	
	4400 Rigid vinyl retainer, mortise, pile insert		105	.076		2.09	2.71		4.80	6.80	
	4600 Wool pile filler strip, aluminum backing	↓	105	.076	↓	2.10	2.71		4.81	6.80	
	5000 Two piece overlapping astragal, extruded aluminum retainer										
	5010 Pile insert	1 Carp	60	.133	L.F.	2.72	4.74		7.46	10.85	
	5020 Vinyl bulb insert		60	.133		1.76	4.74		6.50	9.80	
	5040 Vinyl flap insert		60	.133		5.45	4.74		10.19	13.85	
	5060 Solid neoprene flap insert		60	.133		5.40	4.74		10.14	13.80	
	5080 Hypalon rubber flap insert		60	.133		5.55	4.74		10.29	13.95	
	5090 Snap on cover, pile insert		60	.133		6.30	4.74		11.04	14.80	
	5400 Magnetic aluminum, surface mounted		60	.133		21.50	4.74		26.24	31.50	
	5500 Interlocking aluminum, 5/8" x 1" neoprene bulb insert		45	.178		3.39	6.30		9.69	14.20	
	5600 Adjustable aluminum, 9/16" x 21/32", pile insert	↓	45	.178		16.10	6.30		22.40	28	
	5790 For vinyl bulb, deduct					.41			.41	.45	
	5800 Magnetic, adjustable, 9/16" x 21/32"	1 Carp	45	.178	↓	20.50	6.30		26.80	33	
	6000 Two piece stile protection										
	6010 Cloth backed rubber loop, 1" gap, nail on aluminum strips	1 Carp	45	.178	L.F.	3.47	6.30		9.77	14.25	
	6040 Screw on aluminum strips		45	.178		5.40	6.30		11.70	16.40	
	6100 1-1/2" gap, screw on aluminum extrusion		45	.178		4.86	6.30		11.16	15.80	
	6240 Vinyl fabric loop, slotted aluminum extrusion, 1" gap		45	.178		1.72	6.30		8.02	12.35	
	6300 1-1/4" gap	↓	45	.178	↓	5.10	6.30		11.40	16.10	
300	0010 **WEATHERSTRIPPING**, Window, double hung, 3' X 5'										300
	0020 Zinc	1 Carp	7.20	1.111	Opng.	12.05	39.50		51.55	79	
	0100 Bronze		7.20	1.111		23.50	39.50		63	91.50	
	0500 As above but heavy duty, zinc		4.60	1.739		15.35	62		77.35	119	
	0600 Bronze		4.60	1.739		26.50	62		88.50	132	
	1000 Doors, wood frame, interlocking, for 3' x 7' door, zinc	↓	3	2.667	↓	13.75	95		108.75	172	

DOORS & WINDOWS 8

08700 | Hardware

08720 | Weatherstripping & Seals

		08720 Weatherstripping & Seals	CREW	DAILY OUTPUT	LABOR-HOURS	UNIT	2006 BARE COSTS MAT.	LABOR	EQUIP.	TOTAL	TOTAL INCL O&P	
300	1100	Bronze	1 Carp	3	2.667	Opng.	21.50	95		116.50	181	300
	1300	6′ x 7′ opening, zinc		2	4		15.05	142		157.05	253	
	1400	Bronze		2	4		28.50	142		170.50	267	
	1700	Wood frame, spring type, bronze										
	1800	3′ x 7′ door	1 Carp	7.60	1.053	Opng.	17.70	37.50		55.20	81.50	
	1900	6′ x 7′ door	″	7	1.143	″	21	40.50		61.50	90.50	
	2200	Metal frame, spring type, bronze										
	2300	3′ x 7′ door	1 Carp	3	2.667	Opng.	30	95		125	190	
	2400	6′ x 7′ door	″	2.50	3.200	″	41	114		155	233	
	2500	For stainless steel, spring type, add					133%					
	2700	Metal frame, extruded sections, 3′ x 7′ door, aluminum	1 Carp	2	4	Opng.	40	142		182	280	
	2800	Bronze		2	4		101	142		243	345	
	3100	6′ x 7′ door, aluminum		1.20	6.667		51	237		288	450	
	3200	Bronze		1.20	6.667		119	237		356	525	
	3500	Threshold weatherstripping										
	3650	Door sweep, flush mounted, aluminum	1 Carp	25	.320	Ea.	11.80	11.40		23.20	32	
	3700	Vinyl		25	.320		14	11.40		25.40	34.50	
	5000	Garage door bottom weatherstrip, 12′ aluminum, clear		14	.571		18.90	20.50		39.40	54.50	
	5010	Bronze		14	.571		72	20.50		92.50	113	
	5050	Bottom protection, 12′ aluminum, clear		14	.571		21	20.50		41.50	56.50	
	5100	Bronze		14	.571		89.50	20.50		110	132	
	9000	Minimum labor/equipment charge		3	2.667	Job		95		95	157	
800	0010	**THRESHOLDS**										800
	0011	Threshold 3′ long saddles aluminum	1 Carp	48	.167	L.F.	3.62	5.95		9.57	13.80	
	0100	Aluminum, 8″ wide, 1/2″ thick		12	.667	Ea.	30.50	23.50		54	73	
	0500	Bronze		60	.133	L.F.	32	4.74		36.74	43	
	0600	Bronze, panic threshold, 5″ wide, 1/2″ thick		12	.667	Ea.	60	23.50		83.50	106	
	0700	Rubber, 1/2″ thick, 5-1/2″ wide		20	.400		33.50	14.20		47.70	60.50	
	0800	2-3/4″ wide		20	.400		14.05	14.20		28.25	39	
	9000	Minimum labor/equipment charge		4	2	Job		71		71	118	

08770 | Door/Window Accessories

		08770 Door/Window Accessories	CREW	DAILY OUTPUT	LABOR-HOURS	UNIT	MAT.	LABOR	EQUIP.	TOTAL	TOTAL INCL O&P	
100	0010	**AREA WINDOW WELL**, Galvanized steel 20 ga.										100
	0020	3′-2″ wide, 1′ deep	1 Sswk	29	.276	Ea.	12.95	11		23.95	35	
	0100	2′ deep		23	.348		23	13.90		36.90	51.50	
	0300	16 ga., galv., 3′-2″ wide, 1′ deep		29	.276		18.25	11		29.25	40.50	
	0400	3′ deep		23	.348		37	13.90		50.90	66.50	
	0600	Welded grating for above, 15 lbs., painted		45	.178		33.50	7.10		40.60	50.50	
	0700	Galvanized		45	.178		81	7.10		88.10	102	
	0900	Translucent plastic cap for above		60	.133		13.40	5.35		18.75	24.50	
200	0010	**DOOR PROTECTION** Acrylic and neoprene cover for door										200
	0020	frame, high impact type	1 Carp	100	.080	L.F.	7.50	2.84		10.34	12.95	
550	0010	**DETECTION SYSTEMS** See Division 13720										550

08800 | Glazing

08810 | Glass

		08810 Glass	CREW	DAILY OUTPUT	LABOR-HOURS	UNIT	2006 BARE COSTS MAT.	LABOR	EQUIP.	TOTAL	TOTAL INCL O&P	
100	0010	**ACOUSTICAL GLASS UNITS**, one lite										100
	0020	For 1″ thick	2 Glaz	100	.160	S.F.	23.50	5.50		29	35	

	08810 | Glass		CREW	DAILY OUTPUT	LABOR-HOURS	UNIT	2006 BARE COSTS MAT.	LABOR	EQUIP.	TOTAL	TOTAL INCL O&P	
100	0100	For 4″ thick	2 Glaz	80	.200	S.F.	40.50	6.85		47.35	55.50	100
160	0010	**BEVELED GLASS**, with design patterns										160
	0020	Minimum	2 Glaz	150	.107	S.F.	40.50	3.65		44.15	50.50	
	0050	Average		125	.128		95	4.38		99.38	112	
	0100	Maximum		100	.160		168	5.50		173.50	194	
250	0010	**FACETED**, Color tinted glass										250
	0020	Minimum	2 Glaz	95	.168	S.F.	34	5.75		39.75	47	
	0100	Maximum	″	75	.213	″	60	7.30		67.30	78	
260	0010	**FLOAT GLASS**										260
	0020	3/16″ thick, plain	2 Glaz	130	.123	S.F.	4.67	4.22		8.89	11.95	
	0200	Tempered, clear		130	.123		5.45	4.22		9.67	12.80	
	0300	Tinted		130	.123		6.95	4.22		11.17	14.45	
	0600	1/4″ thick, clear, plain		120	.133		5.55	4.57		10.12	13.45	
	0700	Tinted		120	.133		5.45	4.57		10.02	13.35	
	0800	Tempered, clear		120	.133		6.80	4.57		11.37	14.80	
	0900	Tinted		120	.133		9.35	4.57		13.92	17.65	
	1600	3/8″ thick, clear, plain		75	.213		8.80	7.30		16.10	21.50	
	1700	Tinted		75	.213		10.90	7.30		18.20	24	
	1800	Tempered, clear		75	.213		13.65	7.30		20.95	27	
	1900	Tinted		75	.213		16.95	7.30		24.25	30.50	
	2200	1/2″ thick, clear, plain		55	.291		17.60	9.95		27.55	35.50	
	2300	Tinted		55	.291		19.05	9.95		29	37	
	2400	Tempered, clear		55	.291		20.50	9.95		30.45	38.50	
	2500	Tinted		55	.291		25.50	9.95		35.45	44	
	2800	5/8″ thick, clear, plain		45	.356		19.05	12.20		31.25	40.50	
	2900	Tempered, clear		45	.356		22	12.20		34.20	43.50	
	3200	3/4″ thick, clear, plain		35	.457		24.50	15.65		40.15	52	
	3300	Tempered, clear		35	.457		28.50	15.65		44.15	56.50	
	3600	1″ thick, clear, plain		30	.533		41	18.25		59.25	74.50	
	8900	For low emissivity coating for 3/16″ & 1/4″ only, add to above					15%					
	9000	Minimum labor/equipment charge	1 Glaz	2	4	Job		137		137	221	
270	0010	**FULL VISION**										270
	0020	Up to 10′ high	H-2	130	.185	S.F.	55	5.90		60.90	70	
	0100	10′ to 20′ high, minimum		110	.218		58.50	7		65.50	75.50	
	0150	Average		100	.240		63	7.65		70.65	81.50	
	0200	Maximum		80	.300		70.50	9.60		80.10	93.50	
	9000	Minimum labor/equipment charge	1 Glaz	2	4	Job		137		137	221	
300	0010	**GLAZING VARIABLES**										300
	0500	For high rise glazing, exterior, add per S.F. per story				S.F.					.09	
	0600	For glass replacement, add				″		100%				
	0700	For gasket settings, add				L.F.	4.08			4.08	4.49	
	0900	For sloped glazing, add				S.F.		25%				
	2000	Fabrication, polished edges, 1/4″ thick				Inch	.35			.35	.39	
	2100	1/2″ thick					.89			.89	.98	
	2500	Mitered edges, 1/4″ thick					.89			.89	.98	
	2600	1/2″ thick					1.44			1.44	1.58	
460	0010	**INSULATING GLASS** 2 lites 1/8″ float, 1/2″ thk, under 15 S.F.										460
	0020	Clear	2 Glaz	95	.168	S.F.	7.85	5.75		13.60	17.95	
	0100	Tinted		95	.168		11.55	5.75		17.30	22	
	0200	2 lites 3/16″ float, for 5/8″ thk unit, 15 to 30 S.F., clear		90	.178		9.55	6.10		15.65	20.50	
	0300	Tinted		90	.178		9.80	6.10		15.90	20.50	
	0400	1″ thk, dbl. glazed, 1/4″ float, 30-70 S.F., clear		75	.213		13.45	7.30		20.75	26.50	

	08810 | Glass		CREW	DAILY OUTPUT	LABOR-HOURS	UNIT	2006 BARE COSTS MAT.	LABOR	EQUIP.	TOTAL	TOTAL INCL O&P	
460	0500	Tinted	2 Glaz	75	.213	S.F.	16.35	7.30		23.65	30	460
	0600	1" thick double glazed, 1/4" float, 1/4" wire		75	.213		19.95	7.30		27.25	34	
	0700	1/4" float, 1/4" tempered		75	.213		19.95	7.30		27.25	34	
	0800	1/4" wire, 1/4" tempered		75	.213		27.50	7.30		34.80	42	
	0900	Both lites, 1/4" wire		75	.213		26	7.30		33.30	40.50	
	2000	Both lites, light & heat reflective		85	.188		22	6.45		28.45	34.50	
	2500	Heat reflective, film inside, 1" thick unit, clear		85	.188		19.05	6.45		25.50	31.50	
	2600	Tinted		85	.188		20.50	6.45		26.95	33	
	3000	Film on weatherside, clear, 1/2" thick unit		95	.168		13.60	5.75		19.35	24.50	
	3100	5/8" thick unit		90	.178		16.60	6.10		22.70	28	
	3200	1" thick unit		85	.188		18.75	6.45		25.20	31	
	9000	Minimum labor/equipment charge	1 Glaz	2	4	Job		137		137	221	
500	0010	**LAMINATED GLASS** (security item)										500
	0020	Clear float .03" vinyl 1/4"	2 Glaz	90	.178	S.F.	9.45	6.10		15.55	20	
	0100	3/8" thick		78	.205		14.90	7.05		21.95	27.50	
	0200	.06" vinyl, 1/2" thick		65	.246		17.40	8.45		25.85	33	
	1000	5/8" thick		90	.178		20.50	6.10		26.60	32.50	
	2000	Bullet-resisting, 1-3/16" thick, to 15 S.F.		16	1		56	34.50		90.50	117	
	2100	Over 15 S.F.		16	1		50.50	34.50		85	111	
	2500	2-1/4" thick, to 15 S.F.		12	1.333		67	45.50		112.50	147	
	2600	Over 15 S.F.		12	1.333		59.50	45.50		105	139	
550	0010	**LEAD GLASS** For X-ray, see Division 13091-600										550
600	0010	**OBSCURE GLASS**										600
	0020	Minimum	2 Glaz	140	.114	S.F.	7.70	3.91		11.61	14.75	
	0100	Maximum		125	.128		9.20	4.38		13.58	17.20	
	0300	7/32" thick, minimum		120	.133		8.55	4.57		13.12	16.75	
	0400	Maximum		105	.152		10.90	5.20		16.10	20.50	
650	0010	**PATTERNED GLASS**										650
	0020	Minimum	2 Glaz	140	.114	S.F.	8.60	3.91		12.51	15.75	
	0100	Maximum		125	.128		10.55	4.38		14.93	18.70	
	0300	7/32" thick, minimum		120	.133		10.15	4.57		14.72	18.50	
	0400	Maximum		105	.152		11.35	5.20		16.55	21	
675	0010	**REFLECTIVE GLASS**										675
	0100	1/4" float with fused metallic oxide fixed	2 Glaz	115	.139	S.F.	11.35	4.77		16.12	20	
	0500	1/4" float glass with reflective applied coating		115	.139		9.45	4.77		14.22	18.10	
	2000	Solar film on glass, not including glass, minimum		180	.089		4.57	3.04		7.61	9.95	
	2010	Window protection film, controls blast damage		80	.200		4.57	6.85		11.42	16.10	
	2050	Solar film on glass, not including glass, maximum		225	.071		10.55	2.44		12.99	15.60	
740	0010	**SANDBLASTED GLASS**, plate glass										740
	0020	1/8" thick	2 Glaz	160	.100	S.F.	7.55	3.43		10.98	13.80	
	0100	3/16" thick		130	.123		8.35	4.22		12.57	16	
	0500	1/4" thick		120	.133		8.70	4.57		13.27	16.90	
	0600	3/8" thick		75	.213		9.45	7.30		16.75	22	
760	0010	**SHEET GLASS**										760
	0020	1/8" thick	2 Glaz	160	.100	S.F.	3.94	3.43		7.37	9.85	
	0200	1/4" thick	"	130	.123	"	5.15	4.22		9.37	12.50	
780	0010	**SPANDREL GLASS**										780
	0020	Up to 100 S.F.	2 Glaz	110	.145	S.F.	13.25	4.98		18.23	22.50	
	0200	1,000 to 2,000 S.F.	"	120	.133	"	12.40	4.57		16.97	21	
	0300	For custom colors, add				Total	10%					

08800 | Glazing

08810 | Glass

		08810 Glass	Crew	Daily Output	Labor-Hours	Unit	2006 Bare Costs Mat.	Labor	Equip.	Total	Total Incl O&P	
780	0500	For 3/8" thick, add				S.F.	7.85			7.85	8.60	780
	1000	For double coated, 1/4" thick, add					2.92			2.92	3.21	
	1200	For insulation on panels, add					4.83			4.83	5.30	
	2000	Panels, insulated, with aluminum backed fiberglass, 1" thick	2 Glaz	120	.133		11.65	4.57		16.22	20	
	2100	2" thick	"	120	.133		13.55	4.57		18.12	22.50	
	2500	With galvanized steel backing, add					4.03			4.03	4.43	
850	0010	**WINDOW GLASS**										850
	0015	1/8" thick, clear float	2 Glaz	480	.033	S.F.	3.78	1.14		4.92	6	
	0500	3/16" thick, clear		480	.033		4.66	1.14		5.80	7	
	0600	Tinted		480	.033		5.25	1.14		6.39	7.65	
	0700	Tempered		480	.033		6.35	1.14		7.49	8.85	
	9000	Minimum labor/equipment charge		5	3.200	Job		110		110	177	
900	0010	**WIRE GLASS** (chicken wire) (security item)										900
	0012	1/4" thick rough obscure	2 Glaz	135	.119	S.F.	12.45	4.06		16.51	20.50	
	1000	Polished wire, 1/4" thick, diamond, clear		135	.119		15.45	4.06		19.51	23.50	
	1500	Pinstripe, obscure		135	.119		15.50	4.06		19.56	23.50	

08830 | Mirrors

		08830 Mirrors	Crew	Daily Output	Labor-Hours	Unit	Mat.	Labor	Equip.	Total	Total Incl O&P	
100	0010	**MIRRORS** No frames, wall type, 1/4" plate glass, polished edge										100
	0100	Up to 5 S.F.	2 Glaz	125	.128	S.F.	6.70	4.38		11.08	14.40	
	0200	Over 5 S.F.		160	.100		6.45	3.43		9.88	12.60	
	0500	Door type, 1/4" plate glass, up to 12 S.F.		160	.100		6.95	3.43		10.38	13.15	
	1000	Float glass, up to 10 S.F., 1/8" thick		160	.100		3.91	3.43		7.34	9.80	
	1100	3/16" thick		150	.107		4.54	3.65		8.19	10.90	
	1500	12" x 12" wall tiles, square edge, clear		195	.082		1.59	2.81		4.40	6.30	
	1600	Veined		195	.082		4.33	2.81		7.14	9.30	
	2000	1/4" thick, stock sizes, one way transparent		125	.128		15.10	4.38		19.48	23.50	
	2010	Bathroom, unframed, laminated		160	.100		11.10	3.43		14.53	17.75	
	2500	Tempered		160	.100		12.25	3.43		15.68	19	

08840 | Plastic Glazing

		08840 Plastic Glazing	Crew	Daily Output	Labor-Hours	Unit	Mat.	Labor	Equip.	Total	Total Incl O&P	
600	0010	**PLEXIGLASS ACRYLIC**, Masked										600
	0020	1/8" thick, cut sheets	2 Glaz	170	.094	S.F.	3.45	3.22		6.67	9	
	0200	Full sheets		195	.082		1.80	2.81		4.61	6.50	
	0500	1/4" thick, cut sheets		165	.097		6.10	3.32		9.42	12.05	
	0600	Full sheets		185	.086		3.31	2.96		6.27	8.40	
	0900	3/8" thick, cut sheets		155	.103		11.15	3.54		14.69	17.95	
	1000	Full sheets		180	.089		6	3.04		9.04	11.50	
	1300	1/2" thick, cut sheets		135	.119		12.85	4.06		16.91	20.50	
	1400	Full sheets		150	.107		12.45	3.65		16.10	19.60	
	1700	3/4" thick, cut sheets		115	.139		45.50	4.77		50.27	57.50	
	1800	Full sheets		130	.123		26.50	4.22		30.72	36	
	2100	1" thick, cut sheets		105	.152		51.50	5.20		56.70	65	
	2200	Full sheets		125	.128		32	4.38		36.38	42	
	3000	Colored, 1/8" thick, cut sheets		170	.094		10.60	3.22		13.82	16.85	
	3200	Full sheets		195	.082		6.85	2.81		9.66	12.10	
	3500	1/4" thick, cut sheets		165	.097		11.85	3.32		15.17	18.40	
	3600	Full sheets		185	.086		8.10	2.96		11.06	13.65	
	4000	Mirrors, untinted, cut sheets, 1/8" thick		185	.086		4.98	2.96		7.94	10.25	
	4200	1/4" thick		180	.089		8	3.04		11.04	13.70	
650	0010	**POLYCARBONATE** (security item)										650
	0020	1/8" thick	2 Glaz	170	.094	S.F.	6	3.22		9.22	11.80	
	0500	3/16" thick		165	.097		7.25	3.32		10.57	13.30	
	1000	1/4" thick		155	.103		8	3.54		11.54	14.50	

For expanded coverage of these items see *Means Interior Cost Data 2006*

08800 | Glazing

		08840 | Plastic Glazing	CREW	DAILY OUTPUT	LABOR-HOURS	UNIT	2006 BARE COSTS MAT.	LABOR	EQUIP.	TOTAL	TOTAL INCL O&P	
650	1500	3/8" thick	2 Glaz	150	.107	S.F.	14.75	3.65		18.40	22	650
	9000	Minimum labor/equipment charge	1 Glaz	2	4	Job		137		137	221	
700	0010	**SECURITY FILM** clear, 32000 psi tensile strength, adh to glass										700
	0100	.002" thick, daylight installation	H-2	950	.025	S.F.	.65	.81		1.46	2.03	
	0150	.004" thick, daylight installation		800	.030		.77	.96		1.73	2.41	
	0200	.006" thick, daylight installation		700	.034		.82	1.10		1.92	2.68	
	0210	Install for anchorage		600	.040		.91	1.28		2.19	3.08	
	0400	.007" thick, daylight installation		600	.040		.87	1.28		2.15	3.04	
	0410	Install for anchorage		500	.048		.97	1.53		2.50	3.55	
	0500	.008" thick, daylight installation		500	.048		1.20	1.53		2.73	3.81	
	0510	Install for anchorage		500	.048		1.33	1.53		2.86	3.96	
	0600	.015" thick, daylight installation		400	.060		2.49	1.92		4.41	5.85	
	0610	Install for anchorage	↓	400	.060	↓	1.33	1.92		3.25	4.59	
	0900	Security Film Anchorage, mechanical attachment and cover plate	H-3	370	.043	L.F.	6.55	1.31		7.86	9.35	
	0950	Security film anchorage, wet glaze structural caulking	1 Glaz	225	.036	"	1.20	1.22		2.42	3.28	
	1000	Adhered security film removal	1 Clab	275	.029	S.F.		.80		.80	1.32	
900	0010	**VINYL GLASS**										900
	0020	.090" thick	2 Glaz	170	.094	S.F.	8.10	3.22		11.32	14.10	
	0500	.120" thick		170	.094		10.25	3.22		13.47	16.50	
	1000	.250" thick	↓	155	.103		14.30	3.54		17.84	21.50	
	1500	For non-standard sizes, add				↓	15%					

08900 | Glazed Curtain Wall

		08910 | Glazed Alumimum Curtain Wall	CREW	DAILY OUTPUT	LABOR-HOURS	UNIT	2006 BARE COSTS MAT.	LABOR	EQUIP.	TOTAL	TOTAL INCL O&P	
200	0010	**CURTAIN WALLS**										200
	0020	Minimum	H-1	205	.156	S.F.	23.50	5.80		29.30	35.50	
	0050	Average, single glazed		195	.164		30	6.10		36.10	43.50	
	0150	Average, double glazed		180	.178		42	6.60		48.60	57.50	
	0200	Maximum	↓	160	.200	↓	112	7.40		119.40	136	
700	0010	**TUBE FRAMING** For window walls and store fronts, aluminum, stock										700
	0050	Plain tube frame, mill finish, 1-3/4" x 1-3/4"	2 Glaz	103	.155	L.F.	6.55	5.30		11.85	15.75	
	0150	1-3/4" x 4"		98	.163		8.75	5.60		14.35	18.60	
	0200	1-3/4" x 4-1/2"		95	.168		10.20	5.75		15.95	20.50	
	0250	2" x 6"		89	.180		14.65	6.15		20.80	26	
	0350	4" x 4"		87	.184		14.35	6.30		20.65	26	
	0400	4-1/2" x 4-1/2"		85	.188		16.65	6.45		23.10	28.50	
	0450	Glass bead		240	.067		1.86	2.28		4.14	5.75	
	1000	Flush tube frame, mill finish, 1/4" glass, 1-3/4" x 4", open header		80	.200		8.60	6.85		15.45	20.50	
	1050	Open sill		82	.195		7.45	6.70		14.15	18.95	
	1100	Closed back header		83	.193		12	6.60		18.60	24	
	1150	Closed back sill		85	.188		11.40	6.45		17.85	23	
	1200	Vertical mullion, one piece		75	.213		12.70	7.30		20	26	
	1250	Two piece		73	.219		13.65	7.50		21.15	27	
	1300	90° or 180° vertical corner post		75	.213		21.50	7.30		28.80	35.50	
	1400	1-3/4" x 4-1/2", open header		80	.200		10.45	6.85		17.30	22.50	
	1450	Open sill		82	.195		8.65	6.70		15.35	20.50	
	1500	Closed back header	↓	83	.193	↓	12.75	6.60		19.35	24.50	

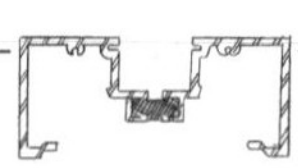

08910 | Glazed Alumimum Curtain Wall

		08910 Glazed Alumimum Curtain Wall	CREW	DAILY OUTPUT	LABOR-HOURS	UNIT	2006 BARE COSTS MAT.	LABOR	EQUIP.	TOTAL	TOTAL INCL O&P	
700	1550	Closed back sill	2 Glaz	85	.188	L.F.	12.30	6.45		18.75	24	700
	1600	Vertical mullion, one piece		75	.213		13.75	7.30		21.05	27	
	1650	Two piece		73	.219		14.55	7.50		22.05	28	
	1700	90° or 180° vertical corner post		75	.213		14.90	7.30		22.20	28	
	2000	Flush tube frame, mill fin. for ins. glass, 2" x 4-1/2", open header		75	.213		11.45	7.30		18.75	24.50	
	2050	Open sill		77	.208		10.05	7.10		17.15	22.50	
	2100	Closed back header		78	.205		12.50	7.05		19.55	25	
	2150	Closed back sill		80	.200		12.35	6.85		19.20	24.50	
	2200	Vertical mullion, one piece		70	.229		13.85	7.85		21.70	28	
	2250	Two piece		68	.235		14.80	8.05		22.85	29.50	
	2300	90° or 180° vertical corner post		70	.229		14	7.85		21.85	28	
	5000	Flush tube frame, mill fin., thermal brk., 2-1/4" x 4-1/2", open header		74	.216		12.65	7.40		20.05	26	
	5050	Open sill		75	.213		11.05	7.30		18.35	24	
	5100	Vertical mullion, one piece		69	.232		15.25	7.95		23.20	29.50	
	5150	Two piece		67	.239		16.30	8.20		24.50	31	
	5200	90° or 180° vertical corner post		69	.232		14.65	7.95		22.60	29	
	6980	Door stop (snap in)		380	.042		2.60	1.44		4.04	5.20	
	7000	For joints, 90°, clip type, add				Ea.	20			20	22	
	7050	Screw spline joint, add					15.35			15.35	16.85	
	7100	For joint other than 90°, add					32			32	35.50	
	8000	For bronze anodized aluminum, add					15%					
	8020	For black finish, add					27%					
	8050	For stainless steel materials, add					350%					
	8100	For monumental grade, add					50%					
	8150	For steel stiffener, add	2 Glaz	200	.080	L.F.	7.85	2.74		10.59	13	
	8200	For 2 to 5 stories, add per story				Story		5%				
	9000	Minimum labor/equipment charge	2 Glaz	2	8	Job		274		274	440	
900	0010	**WINDOW WALLS**, Aluminum stock including glazing										900
	0020	Minimum	H-2	160	.150	S.F.	28	4.80		32.80	39	
	0050	Average		140	.171		34.50	5.50		40	47	
	0100	Maximum		110	.218		108	7		115	130	
	0500	For translucent sandwich wall systems, see Div. 07420-770										
	0850	Cost of the above walls depends on material,										
	0860	finish, repetition, and size of units.										
	0870	The larger the opening, the lower the S.F. cost										
	1200	Double glazed acoustical window wall for airports,										
	1220	including 1" thick glass with 2" x 4-1/2" tube frame	H-2	40	.600	S.F.	62.50	19.20		81.70	99.50	

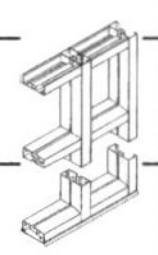

08950 | Translucent Wall & Roof Assemblies

		08950 Translucent Wall & Roof Assemblies	CREW	DAILY OUTPUT	LABOR-HOURS	UNIT	MAT.	LABOR	EQUIP.	TOTAL	TOTAL INCL O&P	
100	0010	**SKYROOFS**, Translucent panels, 2-3/4" thick										100
	0020	Under 500 S.F.	G-3	395	.081	SF Hor.	20.50	2.82		23.32	27	
	0100	Over 5000 S.F.		465	.069		18.10	2.39		20.49	24	
	0300	Continuous vaulted, semi-circular, to 8' wide, double glazed		145	.221		45.50	7.70		53.20	62.50	
	0400	Single glazed		160	.200		30.50	6.95		37.45	45.50	
	0600	To 20' wide, single glazed		175	.183		34.50	6.35		40.85	48.50	
	0700	Over 20' wide, single glazed		200	.160		39.50	5.55		45.05	52	
	0900	Motorized opening type, single glazed, 1/3 opening		145	.221		42	7.70		49.70	59	
	1000	Full opening		130	.246		48	8.55		56.55	67	
	1200	Pyramid type units, self-supporting, to 30' clear opening,										
	1300	square or circular, single glazed, minimum	G-3	200	.160	SF Hor.	22	5.55		27.55	33.50	
	1310	Average		165	.194		31.50	6.75		38.25	45.50	
	1400	Maximum		130	.246		45	8.55		53.55	63.50	
	1500	Grid type, 4' to 10' modules, single glass glazed, minimum		200	.160		29	5.55		34.55	41	
	1550	Maximum		128	.250		47.50	8.70		56.20	66	
	1600	Preformed acrylic, minimum		300	.107		35	3.71		38.71	44.50	

08900 | Glazed Curtain Wall

08950 | Translucent Wall & Roof Assemblies			CREW	DAILY OUTPUT	LABOR-HOURS	UNIT	2006 BARE COSTS MAT.	LABOR	EQUIP.	TOTAL	TOTAL INCL O&P	
100	1650	Maximum	G-3	175	.183	SF Hor.	48.50	6.35		54.85	64	100
	1800	Skyroofs,dome type, self-supporting, to 100′ clear opening, circular										
	1900	Rise to span ratio = 0.20										
	1920	Minimum	G-3	197	.162	SF Hor.	14.45	5.65		20.10	25	
	1950	Maximum		113	.283		48	9.85		57.85	69	
	2100	Rise to span ratio = 0.33, minimum		169	.189		29	6.60		35.60	42	
	2150	Maximum		101	.317		57	11		68	80.50	
	2200	Rise to span ratio = 0.50, minimum		148	.216		43.50	7.50		51	60	
	2250	Maximum		87	.368		63.50	12.80		76.30	90	
	2400	Ridge units, continuous, to 8′ wide, double		130	.246		111	8.55		119.55	136	
	2500	Single		200	.160		76	5.55		81.55	92.50	
	2700	Ridge and furrow units, over 4′ O.C., double, minimum		200	.160		24	5.55		29.55	35.50	
	2750	Maximum		120	.267		44.50	9.25		53.75	63.50	
	2800	Single, minimum		214	.150		20.50	5.20		25.70	31	
	2850	Maximum		153	.209	▼	43	7.25		50.25	59.50	
	3000	Rolling roof, translucent panels, flat roof, mininum		253	.126	S.F.	18.40	4.40		22.80	27	
	3030	Maximum		160	.200		37	6.95		43.95	52	
	3100	Lean-to skyroof, long span, double, minimum		197	.162		24.50	5.65		30.15	36	
	3150	Maximum		101	.317		49	11		60	72	
	3300	Single, minimum		321	.100		18.45	3.47		21.92	26	
	3350	Maximum	▼	160	.200	▼	30.50	6.95		37.45	45.50	
	9000	Minimum labor/equipment charge	2 Carp	8	2	Job		71		71	118	

For information about Means Estimating Seminars, see yellow pages 12 and 13 in back of book

Division 9 Finishes

Estimating Tips

General

- Room Finish Schedule: A complete set of plans should contain a room finish schedule. If one is not available, it would be well worth the time and effort to put one together. A room finish schedule should contain the room number, room name (for clarity), floor materials, base materials, wainscot materials, wainscot height, wall materials (for each wall), ceiling materials, ceiling height, and special instructions.
- Surplus Finishes: Review the specifications to determine if there is any requirement to provide certain amounts of extra materials for the owner's maintenance department. In some cases the owner may require a substantial amount of materials, especially when it is a special order item or long lead time item.

09200 Plaster & Gypsum Board

- Lath is estimated by the square yard for both gypsum and metal lath, plus usually 5% allowance for waste. Furring, channels, and accessories are measured by the linear foot. An extra foot should be allowed for each accessory miter or stop.
- Plaster is also estimated by the square yard. Deductions for openings vary by preference, from zero deduction to 50% of all openings over 2 feet in width. Some estimators deduct a percentage of the total yardage for openings. The estimator should allow one extra square foot for each linear foot of horizontal interior or exterior angle located below the ceiling level. Also, double the areas of small radius work.
- Each room should be measured, perimeter times maximum wall height. Floors and ceiling areas are equal to length times width.
- Drywall accessories, studs, track, and acoustical caulking are all measured by the linear foot. Drywall taping is figured by the square foot. Gypsum wallboard is estimated by the square foot. No material deductions should be made for door or window openings under 32 S.F. Coreboard can be obtained in a 1″ thickness for solid wall and shaft work. Additions should be made to price out the inside or outside corners.
- Different types of partition construction should be listed separately on the quantity sheets. There may be walls with studs of various widths, double studded, and similar or dissimilar surface materials. Shaft work is usually different construction from surrounding partitions requiring separate quantities and pricing of the work.

09300 Tile
09400 Terrazzo

- Tile and terrazzo areas are taken off on a square foot basis. Trim and base materials are measured by the linear foot. Accent tiles are listed per each. Two basic methods of installation are used. Mud set is approximately 30% more expensive than the thin set. In terrazzo work, be sure to include the linear footage of embedded decorative strips, grounds, machine rubbing, and power cleanup.

09600 Flooring

- Wood flooring is available in strip, parquet, or block configuration. The latter two types are set in adhesives with quantities estimated by the square foot. The laying pattern will influence labor costs and material waste. In addition to the material and labor for laying wood floors, the estimator must make allowances for sanding and finishing these areas unless the flooring is prefinished.
- Most of the various types of flooring are all measured on a square foot basis. Base is measured by the linear foot. If adhesive materials are to be quantified, they are estimated at a specified coverage rate by the gallon depending upon the specified type and the manufacturer's recommendations.
- Sheet flooring is measured by the square yard. Roll widths vary, so consideration should be given to use the most economical width, as waste must be figured into the total quantity. Consider also the installation methods available, direct glue down or stretched.

09700 Wall Finishes

- Wall coverings are estimated by the square foot. The area to be covered is measured, length by height of wall above baseboards, to calculate the square footage of each wall. This figure is divided by the number of square feet in the single roll which is being used. Deduct, in full, the areas of openings such as doors and windows. Where a pattern match is required allow 25%-30% waste. One gallon of paste should be sufficient to hang 12 single rolls of light- to medium-weight paper.

09800 Acoustical Treatment

- Acoustical systems fall into several categories. The takeoff of these materials should be by the square foot of area with a 5% allowance for waste. Do not forget about scaffolding, if applicable, when estimating these systems.

09900 Paints & Coatings

- A major portion of the work in painting involves surface preparation. Be sure to include cleaning, sanding, filling, and masking costs in the estimate.
- Painting is one area where bids vary to a greater extent than almost any other section of a project. This arises from the many methods of measuring surfaces to be painted. The estimator should check the plans and specifications carefully to be sure of the required number of coats.
- Protection of adjacent surfaces is not included in painting costs. When considering the method of paint application, an important factor is the amount of protection and masking required. These must be estimated separately and may be the determining factor in choosing the method of application.

Reference Numbers

Reference numbers are shown in bold squares at the beginning of some major classifications. These numbers refer to related items in the Reference Section. The reference information may be an estimating procedure, an alternate pricing method, or technical information.

Note: Not all subdivisions listed here necessarily appear in this publication.

09050 | Basic Material Finishes and Methods

9 FINISHES

09060 | Selective Demolition

		09060 Selective Demolition	CREW	DAILY OUTPUT	LABOR-HOURS	UNIT	2006 BARE COSTS MAT.	LABOR	EQUIP.	TOTAL	TOTAL INCL O&P	
110	0010	**SELECTIVE DEMOLITION, CEILINGS** R024119-10										110
	0200	Ceiling, drywall, furred and nailed or screwed	2 Clab	800	.020	S.F.		.55		.55	.91	
	0220	On metal frame		760	.021			.58		.58	.96	
	0240	On suspension system, including system		720	.022			.61		.61	1.01	
	1000	Plaster, lime and horse hair, on wood lath, incl. lath		700	.023			.63		.63	1.04	
	1020	On metal lath		570	.028			.77		.77	1.27	
	1100	Gypsum, on gypsum lath		720	.022			.61		.61	1.01	
	1120	On metal lath		500	.032			.88		.88	1.45	
	1200	Suspended ceiling, mineral fiber, 2′ x 2′ or 2′ x 4′		1500	.011			.29		.29	.48	
	1250	On suspension system, incl. system		1200	.013			.37		.37	.61	
	1500	Tile, wood fiber, 12″ x 12″, glued		900	.018			.49		.49	.81	
	1540	Stapled		1500	.011			.29		.29	.48	
	1580	On suspension system, incl. system		760	.021			.58		.58	.96	
	2000	Wood, tongue and groove, 1″ x 4″		1000	.016			.44		.44	.73	
	2040	1″ x 8″		1100	.015			.40		.40	.66	
	2400	Plywood or wood fiberboard, 4′ x 8′ sheets		1200	.013			.37		.37	.61	
	9000	Minimum labor/equipment charge	1 Clab	2	4	Job		110		110	182	
120	0010	**SELECTIVE DEMOLITION, FLOORING** R024119-10										120
	0200	Brick with mortar	2 Clab	475	.034	S.F.		.92		.92	1.53	
	0400	Carpet, bonded, including surface scraping		2000	.008			.22		.22	.36	
	0480	Tackless		9000	.002			.05		.05	.08	
	0600	Composition, acrylic or epoxy		400	.040			1.10		1.10	1.82	
	0700	Concrete, scarify skin	A-1A	225	.036			1.30	1.39	2.69	3.64	
	0800	Resilient, sheet goods	2 Clab	1400	.011			.31		.31	.52	
	0820	For gym floors		900	.018			.49		.49	.81	
	0900	Vinyl composition tile, 12″ x 12″		1000	.016			.44		.44	.73	
	2000	Tile, ceramic, thin set		675	.024			.65		.65	1.08	
	2020	Mud set		625	.026			.70		.70	1.16	
	2200	Marble, slate, thin set		675	.024			.65		.65	1.08	
	2220	Mud set		625	.026			.70		.70	1.16	
	2600	Terrazzo, thin set		450	.036			.97		.97	1.61	
	2620	Mud set		425	.038			1.03		1.03	1.71	
	2640	Cast in place		300	.053			1.46		1.46	2.42	
	3000	Wood, block, on end	1 Carp	400	.020			.71		.71	1.18	
	3200	Parquet		450	.018			.63		.63	1.05	
	3400	Strip flooring, interior, 2-1/4″ x 25/32″ thick		325	.025			.88		.88	1.45	
	3500	Exterior, porch flooring, 1″ x 4″		220	.036			1.29		1.29	2.14	
	3800	Subfloor, tongue and groove, 1″ x 6″		325	.025			.88		.88	1.45	
	3820	1″ x 8″		430	.019			.66		.66	1.10	
	3840	1″ x 10″		520	.015			.55		.55	.91	
	4000	Plywood, nailed		600	.013			.47		.47	.79	
	4100	Glued and nailed		400	.020			.71		.71	1.18	
	8000	Remove flooring, bead blast, minimum	A-1A	1000	.008			.29	.31	.60	.82	
	8100	Maximum		400	.020			.73	.78	1.51	2.05	
	8150	Mastic only		1500	.005			.19	.21	.40	.55	
	9000	Minimum labor/equipment charge	1 Clab	4	2	Job		55		55	91	
130	0010	**SELECTIVE DEMOLITION, WALLS AND PARTITIONS** R024119-10										130
	0100	Brick, 4″ to 12″ thick	B-9C	220	.182	C.F.		5.05	.67	5.72	9.10	
	0200	Concrete block, 4″ thick		1000	.040	S.F.		1.11	.15	1.26	2	
	0280	8″ thick		810	.049			1.37	.18	1.55	2.47	
	0300	Exterior stucco 1″ thick over mesh	B-9	3200	.013			.35	.05	.40	.63	
	1000	Drywall, nailed or screwed	1 Clab	1000	.008			.22		.22	.36	
	1020	Glued and nailed		900	.009			.24		.24	.40	
	1500	Fiberboard, nailed		900	.009			.24		.24	.40	

09050 | Basic Material Finishes and Methods

	09060 Selective Demolition		CREW	DAILY OUTPUT	LABOR-HOURS	UNIT	2006 BARE COSTS MAT.	LABOR	EQUIP.	TOTAL	TOTAL INCL O&P	
130	1520	Glued and nailed R024119-10	1 Clab	800	.010	S.F.		.27		.27	.45	130
	1568	Plenum barrier, sheet lead		300	.027			.73		.73	1.21	
	2000	Movable walls, metal, 5′ high		300	.027			.73		.73	1.21	
	2020	8′ high		400	.020			.55		.55	.91	
	2200	Metal or wood studs, finish 2 sides, fiberboard	B-1	520	.046			1.30		1.30	2.15	
	2250	Lath and plaster		260	.092			2.59		2.59	4.29	
	2300	Plasterboard (drywall)		520	.046			1.30		1.30	2.15	
	2350	Plywood		450	.053			1.50		1.50	2.48	
	3000	Plaster, lime and horsehair, on wood lath	1 Clab	400	.020			.55		.55	.91	
	3020	On metal lath		335	.024			.65		.65	1.08	
	3400	Gypsum or perlite, on gypsum lath		410	.020			.53		.53	.89	
	3420	On metal lath		300	.027			.73		.73	1.21	
	3750	Terra cotta block and plaster, to 6″ thick	B-1	175	.137			3.85		3.85	6.40	
	3800	Toilet partitions, slate or marble	1 Clab	5	1.600	Ea.		44		44	72.50	
	3820	Hollow metal	"	8	1	"		27.50		27.50	45.50	
	5000	Wallcovering, vinyl	1 Pape	700	.011	S.F.		.36		.36	.58	
	5040	Designer	"	480	.017	"		.53		.53	.85	
	9000	Minimum labor/equipment charge	1 Clab	4	2	Job		55		55	91	

09100 | Metal Support Assemblies

	09110 Non-Load Bearing Wall Framing		CREW	DAILY OUTPUT	LABOR-HOURS	UNIT	2006 BARE COSTS MAT.	LABOR	EQUIP.	TOTAL	TOTAL INCL O&P	
100	0010	**METAL STUDS AND TRACK**										100
	1600	Non-load bearing, galv, 8′ high, 25 ga. 1-5/8″ wide, 16″ O.C.	1 Carp	619	.013	S.F.	.24	.46		.70	1.02	
	1610	24″ O.C.		950	.008		.18	.30		.48	.70	
	1620	2-1/2″ wide, 16″ O.C.		613	.013		.29	.46		.75	1.08	
	1630	24″ O.C.		938	.009		.22	.30		.52	.74	
	1640	3-5/8″ wide, 16″ O.C.		600	.013		.33	.47		.80	1.15	
	1650	24″ O.C.		925	.009		.25	.31		.56	.78	
	1660	4″ wide, 16″ O.C.		594	.013		.35	.48		.83	1.18	
	1670	24″ O.C.		925	.009		.26	.31		.57	.80	
	1680	6″ wide, 16″ O.C.		588	.014		.51	.48		.99	1.36	
	1690	24″ O.C.		906	.009		.38	.31		.69	.94	
	1700	20 ga. studs, 1-5/8″ wide, 16″ O.C.		494	.016		.39	.58		.97	1.38	
	1710	24″ O.C.		763	.010		.29	.37		.66	.94	
	1720	2-1/2″ wide, 16″ O.C.		488	.016		.46	.58		1.04	1.48	
	1730	24″ O.C.		750	.011		.35	.38		.73	1.01	
	1740	3-5/8″ wide, 16″ O.C.		481	.017		.55	.59		1.14	1.58	
	1750	24″ O.C.		738	.011		.41	.39		.80	1.09	
	1760	4″ wide, 16″ O.C.		475	.017		.60	.60		1.20	1.65	
	1770	24″ O.C.		738	.011		.45	.39		.84	1.14	
	1780	6″ wide, 16″ O.C.		469	.017		.76	.61		1.37	1.84	
	1790	24″ O.C.		725	.011		.57	.39		.96	1.28	
	2000	Non-load bearing, galv, 10′ high, 25 ga. 1-5/8″ wide, 16″ O.C.		495	.016		.22	.57		.79	1.20	
	2100	24″ O.C.		760	.011		.17	.37		.54	.80	
	2200	2-1/2″ wide, 16″ O.C.		490	.016		.27	.58		.85	1.26	
	2250	24″ O.C.		750	.011		.20	.38		.58	.85	
	2300	3-5/8″ wide, 16″ O.C.		480	.017		.31	.59		.90	1.32	
	2350	24″ O.C.		740	.011		.23	.38		.61	.89	
	2400	4″ wide, 16″ O.C.		475	.017		.33	.60		.93	1.35	

09110 | Non-Load Bearing Wall Framing

			CREW	DAILY OUTPUT	LABOR-HOURS	UNIT	2006 BARE COSTS MAT.	LABOR	EQUIP.	TOTAL	TOTAL INCL O&P	
100	2450	24" O.C.	1 Carp	740	.011	S.F.	.25	.38		.63	.91	100
	2500	6" wide, 16" O.C.		470	.017		.48	.61		1.09	1.53	
	2550	24" O.C.		725	.011		.36	.39		.75	1.04	
	2600	20 ga. studs, 1-5/8" wide, 16" O.C.		395	.020		.37	.72		1.09	1.59	
	2650	24" O.C.		610	.013		.27	.47		.74	1.07	
	2700	2-1/2" wide, 16" O.C.		390	.021		.44	.73		1.17	1.69	
	2750	24" O.C.		600	.013		.32	.47		.79	1.14	
	2800	3-5/8" wide, 16" O.C.		385	.021		.52	.74		1.26	1.79	
	2850	24" O.C.		590	.014		.38	.48		.86	1.22	
	2900	4" wide, 16" O.C.		380	.021		.57	.75		1.32	1.87	
	2950	24" O.C.		590	.014		.42	.48		.90	1.26	
	3000	6" wide, 16" O.C.		375	.021		.72	.76		1.48	2.05	
	3050	24" O.C.		580	.014		.53	.49		1.02	1.40	
	3060	Non-load bearing, galv, 12' high, 25 ga. 1-5/8" wide, 16" O.C.		413	.019		.21	.69		.90	1.37	
	3070	24" O.C.		633	.013		.16	.45		.61	.91	
	3080	2-1/2" wide, 16" O.C.		408	.020		.26	.70		.96	1.44	
	3090	24" O.C.		625	.013		.19	.46		.65	.96	
	3100	3-5/8" wide, 16" O.C.		400	.020		.30	.71		1.01	1.50	
	3110	24" O.C.		617	.013		.22	.46		.68	1	
	3120	4" wide, 16" O.C.		396	.020		.32	.72		1.04	1.54	
	3130	24" O.C.		617	.013		.23	.46		.69	1.01	
	3140	6" wide, 16" O.C.		392	.020		.46	.73		1.19	1.71	
	3150	24" O.C.		604	.013		.34	.47		.81	1.15	
	3160	20 ga. studs, 1-5/8" wide, 16" O.C.		329	.024		.35	.86		1.21	1.81	
	3170	24" O.C.		508	.016		.26	.56		.82	1.21	
	3180	2-1/2" wide, 16" O.C.		325	.025		.42	.88		1.30	1.91	
	3190	24" O.C.		500	.016		.30	.57		.87	1.27	
	3200	3-5/8" wide, 16" O.C.		321	.025		.50	.89		1.39	2.02	
	3210	24" O.C.		492	.016		.36	.58		.94	1.36	
	3220	4" wide, 16" O.C.		317	.025		.54	.90		1.44	2.09	
	3230	24" O.C.		492	.016		.40	.58		.98	1.40	
	3240	6" wide, 16" O.C.		313	.026		.69	.91		1.60	2.27	
	3250	24" O.C.	↓	483	.017	↓	.50	.59		1.09	1.53	
	5000	Load bearing studs, see division 05410-400										
	9000	Minimum labor/equipment charge	1 Carp	4	2	Job		71		71	118	

09120 | Ceiling Suspension

			CREW	DAILY OUTPUT	LABOR-HOURS	UNIT	MAT.	LABOR	EQUIP.	TOTAL	TOTAL INCL O&P	
100	0010	**CEILING SUSPENSION SYSTEMS** For gypsum board or plaster										100
	8000	Suspended ceilings, including carriers										
	8200	1-1/2" carriers, 24" O.C. with:										
	8300	7/8" channels, 16" O.C.	1 Lath	165	.048	S.F.	.39	1.59		1.98	2.95	
	8320	24" O.C.		200	.040		.32	1.31		1.63	2.43	
	8400	1-5/8" channels, 16" O.C.		155	.052		.47	1.69		2.16	3.21	
	8420	24" O.C.	↓	190	.042	↓	.37	1.38		1.75	2.60	
	8600	2" carriers, 24" O.C. with:										
	8700	7/8" channels, 16" O.C.	1 Lath	155	.052	S.F.	.44	1.69		2.13	3.17	
	8720	24" O.C.		190	.042		.37	1.38		1.75	2.60	
	8800	1-5/8" channels, 16" O.C.		145	.055		.52	1.81		2.33	3.44	
	8820	24" O.C.	↓	180	.044	↓	.42	1.46		1.88	2.77	

09130 | Acoustical Suspension

			CREW	DAILY OUTPUT	LABOR-HOURS	UNIT	MAT.	LABOR	EQUIP.	TOTAL	TOTAL INCL O&P	
100	0010	**CEILING SUSPENSION SYSTEMS** For boards and tile										100
	0050	Class A suspension system, 15/16" T bar, 2' x 4' grid	1 Carp	800	.010	S.F.	.52	.36		.88	1.17	
	0300	2' x 2' grid	"	650	.012		.66	.44		1.10	1.45	
	0350	For 9/16" grid, add				↓	.15			.15	.17	

09100 | Metal Support Assemblies

	09130 | Acoustical Suspension		CREW	DAILY OUTPUT	LABOR-HOURS	UNIT	2006 BARE COSTS MAT.	LABOR	EQUIP.	TOTAL	TOTAL INCL O&P	
100	0360	For fire rated grid, add				S.F.	.09			.09	.10	100
	0370	For colored grid, add					.19			.19	.21	
	0400	Concealed Z bar suspension system, 12″ module	1 Carp	520	.015		.47	.55		1.02	1.43	
	0600	1-1/2″ carrier channels, 4′ O.C., add	″	470	.017		.10	.61		.71	1.11	
	0700	Carrier channels for ceilings with										
	0900	recessed lighting fixtures, add	1 Carp	460	.017	S.F.	.18	.62		.80	1.22	
	1040	Hanging wire, 12 ga., 4′ long		65	.123	C.S.F.	5.65	4.38		10.03	13.45	
	1080	8′ long		65	.123	″	11.30	4.38		15.68	19.65	
	2402	For strip lighting, see division 16510-440										
	9000	Minimum labor/equipment charge	1 Carp	5	1.600	Job		57		57	94	

09200 | Plaster & Gypsum Board

	09205 | Furring & Lathing		CREW	DAILY OUTPUT	LABOR-HOURS	UNIT	2006 BARE COSTS MAT.	LABOR	EQUIP.	TOTAL	TOTAL INCL O&P	
530	0010	**FURRING** Beams & columns, 7/8″ galvanized channels,										530
	0030	12″ O.C.	1 Lath	155	.052	S.F.	.25	1.69		1.94	2.97	
	0050	16″ O.C.		170	.047		.21	1.54		1.75	2.68	
	0070	24″ O.C.		185	.043		.14	1.42		1.56	2.40	
	0100	Ceilings, on steel, 7/8″ channels, galvanized, 12″ O.C.		210	.038		.23	1.25		1.48	2.23	
	0300	16″ O.C.		290	.028		.21	.91		1.12	1.67	
	0400	24″ O.C.		420	.019		.14	.62		.76	1.14	
	0600	1-5/8″ channels, galvanized, 12″ O.C.		190	.042		.32	1.38		1.70	2.54	
	0700	16″ O.C.		260	.031		.29	1.01		1.30	1.92	
	0900	24″ O.C.		390	.021		.19	.67		.86	1.28	
	1000	Walls, 7/8″ channels, galvanized, 12″ O.C.		235	.034		.23	1.12		1.35	2.02	
	1200	16″ O.C.		265	.030		.21	.99		1.20	1.80	
	1300	24″ O.C.		350	.023		.14	.75		.89	1.34	
	1500	1-5/8″ channels, galvanized, 12″ O.C.		210	.038		.32	1.25		1.57	2.33	
	1600	16″ O.C.		240	.033		.29	1.09		1.38	2.05	
	1800	24″ O.C.		305	.026		.19	.86		1.05	1.58	
	9000	Minimum labor/equipment charge		4	2	Job		65.50		65.50	104	
540	0010	**GYPSUM LATH**										540
	0012	Plain or perforated, nailed, 3/8″ thick	1 Lath	85	.094	S.Y.	5.40	3.09		8.49	10.85	
	0100	1/2″ thick, nailed		80	.100		4.23	3.28		7.51	9.85	
	0300	Clipped to steel studs, 3/8″ thick		75	.107		5.40	3.50		8.90	11.50	
	0400	1/2″ thick		70	.114		4.23	3.75		7.98	10.60	
	0600	Firestop gypsum base, to steel studs, 3/8″ thick		70	.114		3.69	3.75		7.44	10	
	0700	1/2″ thick		65	.123		4.50	4.04		8.54	11.35	
	0900	Foil back, to steel studs, 3/8″ thick		75	.107		3.96	3.50		7.46	9.90	
	1000	1/2″ thick		70	.114		4.14	3.75		7.89	10.50	
	1500	For ceiling installations, add		216	.037			1.21		1.21	1.93	
	1600	For columns and beams, add		170	.047			1.54		1.54	2.45	
	9000	Minimum labor/equipment charge		4.25	1.882	Job		61.50		61.50	98	
560	0010	**METAL LATH**										560
	0020	Diamond, expanded, 2.5 lb. per S.Y., painted				S.Y.	2.85			2.85	3.14	
	0100	Galvanized, 2.5 lb. per S.Y.					2.37			2.37	2.61	
	0300	3.4 lb. per S.Y., painted					3.71			3.71	4.08	
	0400	Galvanized					3.11			3.11	3.42	
	0600	For 15# asphalt sheathing paper, add					.31			.31	.34	

For expanded coverage of these items see *Means Interior Cost Data 2006*

09200 | Plaster & Gypsum Board

	09205 Furring & Lathing		CREW	DAILY OUTPUT	LABOR-HOURS	UNIT	2006 BARE COSTS MAT.	LABOR	EQUIP.	TOTAL	TOTAL INCL O&P	
560	0900	Flat rib, 1/8" high, 2.75 lb., painted				S.Y.	3.04			3.04	3.34	560
	1000	Foil backed					3.65			3.65	4.02	
	1200	3.4 lb. per S.Y., painted					3.28			3.28	3.61	
	1300	Galvanized					4.13			4.13	4.54	
	1500	For 15# asphalt sheating paper, add					.31			.31	.34	
	1800	High rib, 3/8" high, 3.4 lb. per S.Y., painted					4.40			4.40	4.84	
	1900	Galvanized					3.67			3.67	4.04	
	2400	High rib, 3/4" high, painted, .60 lb. per S.F.				S.F.	.42			.42	.46	
	2500	.75 lb. per S.F.				"	.90			.90	.99	
	2800	Stucco mesh, painted, 3.6 lb.				S.Y.	3.32			3.32	3.65	
	3000	K-lath, perforated, absorbent paper, regular					3.21			3.21	3.53	
	3100	Heavy duty					3.79			3.79	4.17	
	3300	Waterproof, heavy duty, grade B backing					3.71			3.71	4.08	
	3400	Fire resistant backing					4.10			4.10	4.51	
	3600	2.5 lb. diamond painted, on wood framing, on walls	1 Lath	85	.094		2.85	3.09		5.94	8.05	
	3700	On ceilings		75	.107		2.85	3.50		6.35	8.70	
	3900	3.4 lb. diamond painted, on wood framing, on walls		80	.100		3.28	3.28		6.56	8.80	
	4000	On ceilings		70	.114		3.28	3.75		7.03	9.55	
	4200	3.4 lb. diamond painted, wired to steel framing		75	.107		3.28	3.50		6.78	9.15	
	4300	On ceilings		60	.133		3.28	4.37		7.65	10.55	
	4500	Columns and beams, wired to steel		40	.200		3.28	6.55		9.83	14	
	4600	Cornices, wired to steel		35	.229		3.28	7.50		10.78	15.50	
	4800	Screwed to steel studs, 2.5 lb.		80	.100		2.85	3.28		6.13	8.35	
	4900	3.4 lb.		75	.107		3.71	3.50		7.21	9.65	
	5100	Rib lath, painted, wired to steel, on walls, 2.5 lb.		75	.107		3.04	3.50		6.54	8.90	
	5200	3.4 lb.		70	.114		4.40	3.75		8.15	10.80	
	5400	4.0 lb.		65	.123		4.52	4.04		8.56	11.35	
	5500	For self-furring lath, add					.08			.08	.09	
	5700	Suspended ceiling system, incl. 3.4 lb. diamond lath, painted	1 Lath	15	.533		11.90	17.50		29.40	41	
	5800	Galvanized	"	15	.533		12.25	17.50		29.75	41.50	
	6000	Hollow metal stud partitions, 3.4 lb. painted lath both sides										
	6010	Non-load bearing, 25 ga., w/rib lath 2-1/2" studs, 12" O.C.	1 Lath	20.30	.394	S.Y.	11.85	12.95		24.80	33.50	
	6300	16" O.C.		21.10	.379		11.15	12.45		23.60	32	
	6350	24" O.C.		22.70	.352		10.60	11.55		22.15	30	
	6400	3-5/8" studs, 16" O.C.		19.50	.410		11.60	13.45		25.05	34.50	
	6600	24" O.C.		20.40	.392		10.85	12.85		23.70	32.50	
	6700	4" studs, 16" O.C.		20.40	.392		11.80	12.85		24.65	33.50	
	6900	24" O.C.		21.60	.370		11	12.15		23.15	31.50	
	7000	6" studs, 16" O.C.		19.50	.410		13.15	13.45		26.60	36	
	7100	24" O.C.		21.10	.379		12	12.45		24.45	33	
	7200	L.B. partitions, 16 ga., w/rib lath, 2-1/2" studs, 16" O.C.		20	.400		11.35	13.10		24.45	33.50	
	7300	3-5/8" studs, 16 ga.		19.70	.406		12.85	13.30		26.15	35	
	7500	4" studs, 16 ga.		19.50	.410		13.35	13.45		26.80	36	
	7600	6" studs, 16 ga.		18.70	.428		15.80	14.05		29.85	40	
	9000	Minimum labor/equipment charge		4.25	1.882	Job		61.50		61.50	98	
600	0010	**SECURITY MESH**, expanded metal, flat, screwed to framing										600
	0100	On walls, 3/4", 1.76 lb/S.F.	2 Carp	1500	.011	S.F.	1.56	.38		1.94	2.35	
	0110	1-1/2", 1.14 lb/S.F.		1600	.010		1.29	.36		1.65	2.01	
	0200	On ceilings, 3/4", 1.76 lb/S.F.		1350	.012		1.56	.42		1.98	2.42	
	0210	1-1/2", 1.14 lb/S.F.		1450	.011		1.29	.39		1.68	2.07	
700	0010	**ACCESSORIES, PLASTER**										700
	0020	Casing bead, expanded flange, galvanized	1 Lath	2.70	2.963	C.L.F.	36.50	97		133.50	195	
	0900	Channels, cold rolled, 16 ga., 3/4" deep, galvanized					23			23	25.50	
	1200	1-1/2" deep, 16 ga., galvanized					32			32	35	

	09205 | Furring & Lathing		CREW	DAILY OUTPUT	LABOR-HOURS	UNIT	2006 BARE COSTS MAT.	LABOR	EQUIP.	TOTAL	TOTAL INCL O&P	
700	1620	Corner bead, expanded bullnose, 3/4" radius, #10, galvanized	1 Lath	2.60	3.077	C.L.F.	30	101		131	193	700
	1650	#1, galvanized		2.55	3.137		47.50	103		150.50	215	
	1670	Expanded wing, 2-3/4" wide, galv. #1		2.65	3.019		30	99		129	190	
	1700	Inside corner, (corner rite) 3" x 3", painted		2.60	3.077		26.50	101		127.50	189	
	1750	Strip-ex, 4" wide, painted		2.55	3.137		23	103		126	188	
	1800	Expansion joint, 3/4" grounds, limited expansion, galv., 1 piece		2.70	2.963		95	97		192	259	
	2100	Extreme expansion, galvanized, 2 piece	↓	2.60	3.077	↓	162	101		263	340	

	09210 | Gypsum Plaster		CREW	DAILY OUTPUT	LABOR-HOURS	UNIT	MAT.	LABOR	EQUIP.	TOTAL	TOTAL INCL O&P	
100	0010	GYPSUM PLASTER										100
	0020	80# bag, less than 1 ton				Bag	14.45			14.45	15.85	
	0100	Over 1 ton				"	13.20			13.20	14.50	
	0300	2 coats, no lath included, on walls	J-1	105	.381	S.Y.	3.29	11.65	1.01	15.95	23.50	
	0400	On ceilings	"	92	.435		3.29	13.30	1.15	17.74	26.50	
	0600	2 coats on and incl. 3/8" gypsum lath on steel, on walls	J-2	97	.495		8.70	15.35	1.09	25.14	35.50	
	0700	On ceilings	"	83	.578		8.70	17.90	1.27	27.87	40	
	0900	3 coats, no lath included, on walls	J-1	87	.460		4.66	14.10	1.21	19.97	29	
	1000	On ceilings	"	78	.513		4.66	15.70	1.35	21.71	32	
	1200	3 coats on and including painted metal lath, on wood studs	J-2	86	.558		8.85	17.30	1.23	27.38	39	
	1300	On ceilings	"	76.50	.627	↓	8.85	19.45	1.38	29.68	43	
	1600	For irregular or curved surfaces, add						30%				
	1800	For columns & beams, add						50%				
	9000	Minimum labor/equipment charge	1 Plas	1	8	Job		260		260	420	
200	0010	GAUGING PLASTER										200
	0020	100 lb. bags, less than 1 ton				Bag	22.50			22.50	25	
	0100	Over 1 ton				"	18.75			18.75	20.50	
300	0010	KEENES CEMENT										300
	0020	In 100 lb. bags, less than 1 ton				Bag	23.50			23.50	26	
	0100	Over 1 ton				"	23			23	25.50	
	0300	Finish only, add to plaster prices, standard	J-1	215	.186	S.Y.	2.23	5.70	.49	8.42	12.20	
	0400	High quality	"	144	.278	"	2.25	8.50	.73	11.48	17.05	
500	0010	PERLITE OR VERMICULITE PLASTER										500
	0020	In 100 lb. bags, under 200 bags				Bag	14.55			14.55	16	
	0100	Over 200 bags				"	13.10			13.10	14.40	
	0300	2 coats, no lath included, on walls	J-1	92	.435	S.Y.	3.24	13.30	1.15	17.69	26.50	
	0400	On ceilings	"	79	.506		3.24	15.50	1.34	20.08	30	
	0600	2 coats, on and incl. 3/8" gypsum lath, on metal studs	J-2	84	.571		6.90	17.70	1.26	25.86	37.50	
	0700	On ceilings	"	70	.686		6.90	21.50	1.51	29.91	43.50	
	0900	3 coats, no lath included, on walls	J-1	74	.541		5.85	16.55	1.43	23.83	34.50	
	1000	On ceilings	"	63	.635		5.85	19.45	1.68	26.98	40	
	1200	3 coats, on and incl. painted metal lath, on metal studs	J-2	72	.667		10.25	20.50	1.47	32.22	46	
	1300	On ceilings		61	.787		10.25	24.50	1.73	36.48	52	
	1500	3 coats, on and incl. suspended metal lath ceiling	↓	37	1.297		17.75	40	2.85	60.60	87	
	1700	For irregular or curved surfaces, add to above						30%				
	1800	For columns and beams, add to above						50%				
	1900	For soffits, add to ceiling prices				↓		40%				
	9000	Minimum labor/equipment charge	1 Plas	1	8	Job		260		260	420	
650	0010	PLASTER PARTITION WALL										650
	0400	Stud walls, 3.4 lb. metal lath, 3 coat gypsum plaster, 2 sides										
	0600	2" x 4" wood studs, 16" O.C.	J-2	315	.152	S.F.	3.21	4.72	.34	8.27	11.50	
	0700	2-1/2" metal studs, 25 ga., 12" O.C.		325	.148		2.97	4.58	.32	7.87	11	
	0800	3-5/8" metal studs, 25 ga., 16" O.C.	↓	320	.150	↓	2.99	4.65	.33	7.97	11.15	
	0900	Gypsum lath, 2 coat vermiculite plaster, 2 sides										

FINISHES 9

		09210 | Gypsum Plaster	CREW	DAILY OUTPUT	LABOR-HOURS	UNIT	2006 BARE COSTS MAT.	LABOR	EQUIP.	TOTAL	TOTAL INCL O&P	
650	1000	2" x 4" wood studs, 16" O.C.	J-2	355	.135	S.F.	3.58	4.19	.30	8.07	11	650
	1200	2-1/2" metal studs, 25 ga., 12" O.C.		365	.132		3.17	4.08	.29	7.54	10.35	
	1300	3-5/8" metal studs, 25 ga., 16" O.C.		360	.133		3.26	4.13	.29	7.68	10.55	
900	0010	THIN COAT										900
	0012	1 coat veneer, not incl. lath	J-1	3600	.011	S.F.	.07	.34	.03	.44	.66	
	1000	In 50 lb. bags				Bag	9.85			9.85	10.80	
		09220 | Portland Cement Plaster										
200	0010	STUCCO										200
	0015	3 coats 1" thick, float finish, with mesh, on wood frame	J-2	63	.762	S.Y.	5.35	23.50	1.68	30.53	45.50	
	0100	On masonry construction, no mesh incl.	J-1	67	.597		2.05	18.30	1.58	21.93	33.50	
	0300	For trowel finish, add	1 Plas	170	.047			1.53		1.53	2.46	
	0400	For 3/4" thick, on masonry, deduct	J-1	880	.045		.57	1.39	.12	2.08	3.01	
	0600	For coloring and special finish, add, minimum		685	.058		.37	1.79	.15	2.31	3.47	
	0700	Maximum		200	.200		1.29	6.15	.53	7.97	11.90	
	0900	For soffits, add	J-2	155	.310		1.99	9.60	.68	12.27	18.40	
	1000	Exterior stucco, with bonding agent, 3 coats, on walls, no mesh incl.	J-1	200	.200		3.33	6.15	.53	10.01	14.15	
	1200	Ceilings		180	.222		3.33	6.80	.59	10.72	15.30	
	1300	Beams		80	.500		3.33	15.30	1.32	19.95	29.50	
	1500	Columns		100	.400		3.33	12.25	1.06	16.64	24.50	
	1550	Minimum labor/equipment charge	1 Plas	1	8	Job		260		260	420	
	1600	Mesh, painted, nailed to wood, 1.8 lb.	1 Lath	60	.133	S.Y.	4.17	4.37		8.54	11.55	
	1800	3.6 lb.		55	.145		3.32	4.77		8.09	11.20	
	1900	Wired to steel, painted, 1.8 lb.		53	.151		4.17	4.95		9.12	12.45	
	2100	3.6 lb.		50	.160		3.32	5.25		8.57	12	
	9000	Minimum labor/equipment charge		4	2	Job		65.50		65.50	104	
		09250 | Gypsum Board										
200	0010	CEMENTITIOUS BACKERBOARD										200
	0070	Cementitious backerboard, on floor, 3' x 4'x 1/2" sheets	2 Carp	525	.030	S.F.	1.04	1.08		2.12	2.95	
	0080	3' x 5' x 1/2" sheets		525	.030		1.04	1.08		2.12	2.95	
	0090	3' x 6' x 1/2" sheets		525	.030		.76	1.08		1.84	2.63	
	0100	3' x 4'x 5/8" sheets		525	.030		1.08	1.08		2.16	2.99	
	0110	3' x 5' x 5/8" sheets		525	.030		1.06	1.08		2.14	2.97	
	0120	3' x 6' x 5/8" sheets		525	.030		1.07	1.08		2.15	2.97	
	0150	On wall, 3' x 4'x 1/2" sheets		350	.046		1.04	1.63		2.67	3.84	
	0160	3' x 5' x 1/2" sheets		350	.046		1.04	1.63		2.67	3.84	
	0170	3' x 6' x 1/2" sheets		350	.046		.76	1.63		2.39	3.52	
	0180	3' x 4'x 5/8" sheets		350	.046		1.08	1.63		2.71	3.88	
	0190	3' x 5' x 5/8" sheets		350	.046		1.06	1.63		2.69	3.86	
	0200	3' x 6' x 5/8" sheets		350	.046		1.07	1.63		2.70	3.86	
	0250	On counter, 3' x 4'x 1/2" sheets		180	.089		1.04	3.16		4.20	6.40	
	0260	3' x 5' x 1/2" sheets		180	.089		1.04	3.16		4.20	6.40	
	0270	3' x 6' x 1/2" sheets		180	.089		.76	3.16		3.92	6.10	
	0300	3' x 4'x 5/8" sheets		180	.089		1.08	3.16		4.24	6.45	
	0310	3' x 5' x 5/8" sheets		180	.089		1.06	3.16		4.22	6.40	
	0320	3' x 6' x 5/8" sheets		180	.089		1.07	3.16		4.23	6.40	
300	0010	BLUEBOARD For use with thin coat										300
	0100	plaster application (see division 09210-900)										
	1000	3/8" thick, on walls or ceilings, standard, no finish included	2 Carp	1900	.008	S.F.	.23	.30		.53	.75	
	1100	With thin coat plaster finish		875	.018		.30	.65		.95	1.41	
	1400	On beams, columns, or soffits, standard, no finish included		675	.024		.26	.84		1.10	1.69	
	1450	With thin coat plaster finish		475	.034		.34	1.20		1.54	2.35	

	09250 | Gypsum Board		CREW	DAILY OUTPUT	LABOR-HOURS	UNIT	2006 BARE COSTS MAT.	LABOR	EQUIP.	TOTAL	TOTAL INCL O&P	
300	3000	1/2" thick, on walls or ceilings, standard, no finish included	2 Carp	1900	.008	S.F.	.27	.30		.57	.80	300
	3100	With thin coat plaster finish		875	.018		.34	.65		.99	1.46	
	3300	Fire resistant, no finish included		1900	.008		.27	.30		.57	.80	
	3400	With thin coat plaster finish		875	.018		.34	.65		.99	1.46	
	3450	On beams, columns, or soffits, standard, no finish included		675	.024		.31	.84		1.15	1.74	
	3500	With thin coat plaster finish		475	.034		.38	1.20		1.58	2.40	
	3700	Fire resistant, no finish included		675	.024		.31	.84		1.15	1.74	
	3800	With thin coat plaster finish		475	.034		.38	1.20		1.58	2.40	
	5000	5/8" thick, on walls or ceilings, fire resistant, no finish included		1900	.008		.29	.30		.59	.82	
	5100	With thin coat plaster finish		875	.018		.36	.65		1.01	1.48	
	5500	On beams, columns, or soffits, no finish included		675	.024		.33	.84		1.17	1.77	
	5600	With thin coat plaster finish		475	.034		.41	1.20		1.61	2.43	
	6000	For high ceilings, over 8' high, add		3060	.005			.19		.19	.31	
	6500	For over 3 stories high, add per story		6100	.003			.09		.09	.15	
	9000	Minimum labor/equipment charge	1 Carp	2	4	Job		142		142	236	
500	0010	**CEILINGS** Gypsum drywall, fire rated, finished										500
	0100	Screwed to grid, channel or joists, 1/2" thick	2 Carp	765	.021	S.F.	.30	.74		1.04	1.56	
	0200	5/8" thick		765	.021		.30	.74		1.04	1.56	
	0300	Over 8' high, 1/2" thick		615	.026		.30	.93		1.23	1.86	
	0400	5/8" thick		615	.026		.30	.93		1.23	1.86	
	0600	Grid suspension system, direct hung										
	0700	1-1/2" C.R.C., with 7/8" hi hat furring channel, 16" O.C.	2 Carp	600	.027	S.F.	1	.95		1.95	2.67	
	0800	24" O.C.		900	.018		.91	.63		1.54	2.05	
	0900	3-5/8" C.R.C., with 7/8" hi hat furring channel, 16" O.C.		600	.027		1.06	.95		2.01	2.74	
	1000	24" O.C.		900	.018		.92	.63		1.55	2.06	
700	0010	**DRYWALL** Gypsum plasterboard, nailed or screwed	R092910-10									700
	0100	to studs unless otherwise noted										
	0150	3/8" thick, on walls, standard, no finish included	2 Carp	2000	.008	S.F.	.24	.28		.52	.73	
	0200	On ceilings, standard, no finish included		1800	.009		.24	.32		.56	.78	
	0250	On beams, columns, or soffits, no finish included		675	.024		.24	.84		1.08	1.66	
	0300	1/2" thick, on walls, standard, no finish included		2000	.008		.27	.28		.55	.77	
	0350	Taped and finished (level 4 finish)		965	.017		.30	.59		.89	1.32	
	0390	With compound skim coat (level 5 finish)		775	.021		.34	.73		1.07	1.60	
	0400	Fire resistant, no finish included		2000	.008		.30	.28		.58	.80	
	0450	Taped and finished (level 4 finish)		965	.017		.33	.59		.92	1.35	
	0490	With compound skim coat (level 5 finish)		775	.021		.37	.73		1.10	1.63	
	0500	Water resistant, no finish included		2000	.008		.25	.28		.53	.75	
	0550	Taped and finished (level 4 finish)		965	.017		.28	.59		.87	1.29	
	0590	With compound skim coat (level 5 finish)		775	.021		.32	.73		1.05	1.58	
	0600	Prefinished, vinyl, clipped to studs		900	.018		.56	.63		1.19	1.67	
	1000	On ceilings, standard, no finish included		1800	.009		.27	.32		.59	.82	
	1050	Taped and finished (level 4 finish)		765	.021		.30	.74		1.04	1.57	
	1090	With compound skim coat (level 5 finish)		610	.026		.34	.93		1.27	1.92	
	1100	Fire resistant, no finish included		1800	.009		.30	.32		.62	.85	
	1150	Taped and finished (level 4 finish)		765	.021		.33	.74		1.07	1.60	
	1195	With compound skim coat (level 5 finish)		610	.026		.37	.93		1.30	1.95	
	1200	Water resistant, no finish included		1800	.009		.25	.32		.57	.80	
	1250	Taped and finished (level 4 finish)		765	.021		.28	.74		1.02	1.54	
	1290	With compound skim coat (level 5 finish)		610	.026		.32	.93		1.25	1.90	
	1500	On beams, columns, or soffits, standard, no finish included		675	.024		.31	.84		1.15	1.74	
	1550	Taped and finished (level 4 finish)		475	.034		.30	1.20		1.50	2.32	
	1590	With compound skim coat (level 5 finish)		540	.030		.34	1.05		1.39	2.13	
	1600	Fire resistant, no finish included		675	.024		.35	.84		1.19	1.78	
	1650	Taped and finished (level 4 finish)		475	.034		.33	1.20		1.53	2.35	
	1690	With compound skim coat (level 5 finish)		540	.030		.37	1.05		1.42	2.16	

09250 | Gypsum Board

		09250 Gypsum Board	CREW	DAILY OUTPUT	LABOR-HOURS	UNIT	2006 BARE COSTS MAT.	LABOR	EQUIP.	TOTAL	TOTAL INCL O&P
700	1700	Water resistant, no finish included (R092910-10)	2 Carp	675	.024	S.F.	.29	.84		1.13	1.72
	1750	Taped and finished (level 4 finish)		475	.034		.28	1.20		1.48	2.29
	1790	With compound skim coat (level 5 finish)		540	.030		.32	1.05		1.37	2.11
	2000	5/8" thick, on walls, standard, no finish included		2000	.008		.30	.28		.58	.80
	2050	Taped and finished (level 4 finish)		965	.017		.33	.59		.92	1.35
	2090	With compound skim coat (level 5 finish)		775	.021		.37	.73		1.10	1.63
	2100	Fire resistant, no finish included		2000	.008		.30	.28		.58	.80
	2150	Taped and finished (level 4 finish)		965	.017		.33	.59		.92	1.35
	2195	With compound skim coat (level 5 finish)		775	.021		.37	.73		1.10	1.63
	2200	Water resistant, no finish included		2000	.008		.27	.28		.55	.77
	2250	Taped and finished (level 4 finish)		965	.017		.30	.59		.89	1.32
	2290	With compound skim coat (level 5 finish)		775	.021		.34	.73		1.07	1.60
	2300	Prefinished, vinyl, clipped to studs		900	.018		.65	.63		1.28	1.77
	3000	On ceilings, standard, no finish included		1800	.009		.30	.32		.62	.85
	3050	Taped and finished (level 4 finish)		765	.021		.33	.74		1.07	1.60
	3090	With compound skim coat (level 5 finish)		615	.026		.37	.93		1.30	1.94
	3100	Fire resistant, no finish included		1800	.009		.30	.32		.62	.85
	3150	Taped and finished (level 4 finish)		765	.021		.33	.74		1.07	1.60
	3190	With compound skim coat (level 5 finish)		615	.026		.37	.93		1.30	1.94
	3200	Water resistant, no finish included		1800	.009		.27	.32		.59	.82
	3250	Taped and finished (level 4 finish)		765	.021		.30	.74		1.04	1.57
	3290	With compound skim coat (level 5 finish)		615	.026		.34	.93		1.27	1.91
	3500	On beams, columns, or soffits, no finish included		675	.024		.35	.84		1.19	1.78
	3550	Taped and finished (level 4 finish)		475	.034		.38	1.20		1.58	2.40
	3590	With compound skim coat (level 5 finish)		380	.042		.43	1.50		1.93	2.95
	3600	Fire resistant, no finish included		675	.024		.35	.84		1.19	1.78
	3650	Taped and finished (level 4 finish)		475	.034		.38	1.20		1.58	2.40
	3690	With compound skim coat (level 5 finish)		380	.042		.37	1.50		1.87	2.89
	3700	Water resistant, no finish included		675	.024		.31	.84		1.15	1.74
	3750	Taped and finished (level 4 finish)		475	.034		.35	1.20		1.55	2.37
	3790	With compound skim coat (level 5 finish)		380	.042		.34	1.50		1.84	2.86
	4000	Fireproofing, beams or columns, 2 layers, 1/2" thick, incl finish		330	.048		.63	1.72		2.35	3.56
	4050	5/8" thick		300	.053		.67	1.90		2.57	3.88
	4100	3 layers, 1/2" thick		225	.071		.93	2.53		3.46	5.20
	4150	5/8" thick		210	.076		1	2.71		3.71	5.60
	5050	For 1" thick coreboard on columns		480	.033		.48	1.18		1.66	2.49
	5100	For foil-backed board, add					.10			.10	.11
	5200	For work over 8' high, add	2 Carp	3060	.005			.19		.19	.31
	5270	For textured spray, add	2 Lath	1600	.010		.05	.33		.38	.58
	5300	For over 3 stories high, add per story	2 Carp	6100	.003			.09		.09	.15
	5350	For finishing inner corners, add		950	.017	L.F.	.08	.60		.68	1.07
	5355	For finishing outer corners, add		1250	.013		.20	.46		.66	.97
	5500	For acoustical sealant, add per bead	1 Carp	500	.016		.03	.57		.60	.97
	5550	Sealant, 1 quart tube				Ea.	4.89			4.89	5.40
	5600	Sound deadening board, 1/4" gypsum	2 Carp	1800	.009	S.F.	.27	.32		.59	.82
	5650	1/2" wood fiber	"	1800	.009	"	.38	.32		.70	.94
	9000	Minimum labor/equipment charge	1 Carp	2	4	Job		142		142	236
800	0010	**HIGH ABUSE GYPSUM BOARD**, fiber reinforced, nailed or									
	0100	screwed to studs unless otherwise noted									
	0110	1/2" thick, on walls, no finish included	2 Carp	1800	.009	S.F.	.53	.32		.85	1.10
	0120	Taped and finished (level 4 finish)		870	.018		.56	.65		1.21	1.70
	0130	With compound skim coat (level 5 finish)		700	.023		.60	.81		1.41	2.01
	0150	On ceilings, no finish included		1620	.010		.53	.35		.88	1.16
	0160	Taped and finished (level 4 finish)		690	.023		.56	.82		1.38	1.99
	0170	With compound skim coat (level 5 finish)		550	.029		.60	1.03		1.63	2.37

09250 | Gypsum Board

		09250 Gypsum Board	CREW	DAILY OUTPUT	LABOR-HOURS	UNIT	2006 BARE COSTS MAT.	LABOR	EQUIP.	TOTAL	TOTAL INCL O&P	
800	0210	5/8" thick, on walls, no finish included	2 Carp	1800	.009	S.F.	.63	.32		.95	1.21	800
	0220	Taped and finished (level 4 finish)		870	.018		.66	.65		1.31	1.81	
	0230	With compound skim coat (level 5 finish)		700	.023		.70	.81		1.51	2.12	
	0250	On ceilings, no finish included		1620	.010		.63	.35		.98	1.27	
	0260	Taped and finished (level 4 finish)		690	.023		.66	.82		1.48	2.10	
	0270	With compound skim coat (level 5 finish)		550	.029		.70	1.03		1.73	2.48	
	0310	5/8" thick, on walls, very high impact, no finish included		1800	.009		.73	.32		1.05	1.32	
	0320	Taped and finished (level 4 finish)		870	.018		.76	.65		1.41	1.92	
	0330	With compound skim coat (level 5 finish)		700	.023		.80	.81		1.61	2.23	
	0350	On ceilings, no finish included		1620	.010		.73	.35		1.08	1.38	
	0360	Taped and finished (level 4 finish)		690	.023		.76	.82		1.58	2.21	
	0370	With compound skim coat (level 5 finish)		550	.029		.80	1.03		1.83	2.59	
	0400	High abuse, gypsum core, paper face										
	0410	1/2" thick, on walls, no finish included	2 Carp	1800	.009	S.F.	.57	.32		.89	1.15	
	0420	Taped and finished (level 4 finish)		870	.018		.60	.65		1.25	1.75	
	0430	With compound skim coat (level 5 finish)		700	.023		.64	.81		1.45	2.06	
	0450	On ceilings, no finish included		1620	.010		.57	.35		.92	1.21	
	0460	Taped and finished (level 4 finish)		690	.023		.60	.82		1.42	2.04	
	0470	With compound skim coat (level 5 finish)		550	.029		.64	1.03		1.67	2.42	
	0510	5/8" thick, on walls, no finish included		1800	.009		.63	.32		.95	1.21	
	0520	Taped and finished (level 4 finish)		870	.018		.66	.65		1.31	1.81	
	0530	With compound skim coat (level 5 finish)		700	.023		.70	.81		1.51	2.12	
	0550	On ceilings, no finish included		1620	.010		.63	.35		.98	1.27	
	0560	Taped and finished (level 4 finish)		690	.023		.66	.82		1.48	2.10	
	0570	With compound skim coat (level 5 finish)		550	.029		.70	1.03		1.73	2.48	
	1000	For high ceilings, over 8' high, add		2750	.006			.21		.21	.34	
	1010	For over 3 stories high, add per story		5500	.003			.10		.10	.17	

09260 | Gypsum Board Systems

		09260 Gypsum Board Systems	CREW	DAILY OUTPUT	LABOR-HOURS	UNIT	MAT.	LABOR	EQUIP.	TOTAL	TOTAL INCL O&P	
100	0010	**PARTITION WALL** Stud wall, 8' to 12' high										100
	0050	1/2", interior, gypsum board, std, tape & finish 2 sides										
	0500	Installed on and incl., 2" x 4" wood studs, 16" O.C.	2 Carp	310	.052	S.F.	1.03	1.83		2.86	4.17	
	1000	Metal studs, NLB, 25 ga., 16" O.C., 3-5/8" wide		350	.046		.92	1.63		2.55	3.70	
	1200	6" wide		330	.048		1.09	1.72		2.81	4.06	
	1400	Water resistant, on 2" x 4" wood studs, 16" O.C.		310	.052		.99	1.83		2.82	4.13	
	1600	Metal studs, NLB, 25 ga., 16" O.C., 3-5/8" wide		350	.046		.88	1.63		2.51	3.66	
	1800	6" wide		330	.048		1.05	1.72		2.77	4.02	
	2000	Fire res., 2 layers, 1-1/2 hr., on 2" x 4" wood studs, 16" O.C.		210	.076		1.69	2.71		4.40	6.35	
	2200	Metal studs, NLB, 25 ga., 16" O.C., 3-5/8" wide		250	.064		1.58	2.28		3.86	5.50	
	2400	6" wide		230	.070		1.75	2.47		4.22	6.05	
	2600	Fire & water res., 2 layers, 1-1/2 hr., 2" x 4" studs, 16" O.C.		210	.076		1.69	2.71		4.40	6.35	
	2800	Metal studs, NLB, 25 ga., 16" O.C., 3-5/8" wide		250	.064		1.58	2.28		3.86	5.50	
	3000	6" wide		230	.070		1.75	2.47		4.22	6.05	
	3200	5/8", interior, gypsum board, std, tape & finish 2 sides										
	3400	Installed on and including 2" x 4" wood studs, 16" O.C.	2 Carp	300	.053	S.F.	1.09	1.90		2.99	4.34	
	3600	24" O.C.		330	.048		.99	1.72		2.71	3.95	
	3800	Metal studs, NLB, 25 ga., 16" O.C., 3-5/8" wide		340	.047		.98	1.67		2.65	3.85	
	4000	6" wide		320	.050		1.15	1.78		2.93	4.22	
	4200	24" O.C., 3-5/8" wide		360	.044		.90	1.58		2.48	3.61	
	4400	6" wide		340	.047		1.03	1.67		2.70	3.90	
	4800	Water resistant, on 2" x 4" wood studs, 16" O.C.		300	.053		1.03	1.90		2.93	4.27	
	5000	24" O.C.		330	.048		.93	1.72		2.65	3.89	
	5200	Metal studs, NLB, 25 ga. 16" O.C., 3-5/8" wide		340	.047		.92	1.67		2.59	3.78	
	5400	6" wide		320	.050		1.09	1.78		2.87	4.15	
	5600	24" O.C., 3-5/8" wide		360	.044		.84	1.58		2.42	3.54	

		09260 | Gypsum Board Systems	CREW	DAILY OUTPUT	LABOR-HOURS	UNIT	2006 BARE COSTS MAT.	LABOR	EQUIP.	TOTAL	TOTAL INCL O&P	
100	5800	6″ wide	2 Carp	340	.047	S.F.	.97	1.67		2.64	3.83	100
	6000	Fire res., 2 layers, 2 hr., on 2″ x 4″ wood studs, 16″ O.C.		205	.078		1.59	2.77		4.36	6.35	
	6200	24″ O.C.		235	.068		1.59	2.42		4.01	5.75	
	6400	Metal studs, NLB, 25 ga., 16″ O.C., 3-5/8″ wide		245	.065		1.60	2.32		3.92	5.60	
	6600	6″ wide		225	.071		1.75	2.53		4.28	6.10	
	6800	24″ O.C., 3-5/8″ wide		265	.060		1.50	2.15		3.65	5.20	
	7000	6″ wide		245	.065		1.63	2.32		3.95	5.65	
	7200	Fire & water res., 2 layers, 2 hr., 2″ x 4″ studs, 16″ O.C.		205	.078		1.69	2.77		4.46	6.45	
	7400	24″ O.C.		235	.068		1.59	2.42		4.01	5.75	
	7600	Metal studs, NLB, 25 ga., 16″ O.C., 3-5/8″ wide		245	.065		1.58	2.32		3.90	5.60	
	7800	6″ wide		225	.071		1.75	2.53		4.28	6.10	
	8000	24″ O.C., 3-5/8″ wide		265	.060		1.50	2.15		3.65	5.20	
	8200	6″ wide		245	.065		1.63	2.32		3.95	5.65	
	8600	1/2″ blueboard, mesh tape both sides										
	8620	Installed on and including 2″ x 4″ wood studs, 16″ O.C.	2 Carp	300	.053	S.F.	1.09	1.90		2.99	4.34	
	8640	Metal studs, NLB, 25 ga., 16″ O.C., 3-5/8″ wide		340	.047		.98	1.67		2.65	3.85	
	8660	6″ wide		320	.050		1.15	1.78		2.93	4.22	
	9000	Exterior, 1/2″ gypsum sheathing, 1/2″ gypsum finished, interior,										
	9100	including foil faced insulation, metal studs, 20 ga.										
	9200	16″ O.C., 3-5/8″ wide	2 Carp	290	.055	S.F.	1.91	1.96		3.87	5.35	
	9400	6″ wide	"	270	.059	"	2.12	2.11		4.23	5.80	
800	0010	**SHAFT WALL** Cavity type on 25 ga. J track & C-H studs, 24″ O.C.										800
	0030	1″ thick coreboard wall liner on shaft side										
	0040	2-hour assembly with double layer										
	0060	5/8″ fire rated gypsum board on room side	2 Carp	220	.073	S.F.	1.08	2.59		3.67	5.45	
	0100	3-hour assembly with triple layer										
	0300	5/8″ fire rated gypsum board on room side	2 Carp	180	.089	S.F.	1.38	3.16		4.54	6.75	
	0400	4-hour assembly, 1″ coreboard, 5/8″ fire rated gypsum board										
	0600	and 3/4″ galv. metal furring channels, 24″ O.C., with										
	0700	Double layer 5/8″ fire rated gypsum board on room side	2 Carp	110	.145	S.F.	1.22	5.15		6.37	9.90	
	0900	For taping & finishing, add per side	1 Carp	1050	.008	"	.03	.27		.30	.49	
	1000	For insulation, see div. 07210										

		09270 | Drywall Accessories	CREW	DAILY OUTPUT	LABOR-HOURS	UNIT	MAT.	LABOR	EQUIP.	TOTAL	TOTAL INCL O&P	
100	0010	**ACCESSORIES, DRYWALL**										100
	0020	Casing bead, galvanized steel	1 Carp	2.90	2.759	C.L.F.	18.10	98		116.10	182	
	0100	Vinyl		3	2.667		18.35	95		113.35	177	
	0300	Corner bead, galvanized steel, 1″ x 1″		4	2		13.95	71		84.95	133	
	0400	Corner bead, galvanized steel, 1-1/4″ x 1-1/4″		3.50	2.286		20.50	81.50		102	158	
	0600	Vinyl corner bead		4	2		18.55	71		89.55	139	
	0900	Furring channel, galv. steel, 7/8″ deep, standard		2.60	3.077		24	109		133	207	
	1000	Resilient		2.55	3.137		23.50	112		135.50	211	
	1100	J trim, galvanized steel, 1/2″ wide		3	2.667		18.45	95		113.45	178	
	1120	5/8″ wide		2.95	2.712		16.65	96.50		113.15	178	
	1140	L trim, galvanized		3	2.667		15.85	95		110.85	174	
	1150	U trim, galvanized		2.95	2.712		16.45	96.50		112.95	178	
	1160	Screws #6 x 1″ A				M	6.80			6.80	7.50	
	1170	#6 x 1-5/8″ A				"	9.50			9.50	10.45	
	1200	For stud partitions, see Divisions 05410-400 and 09110-100										
	1500	Z stud, galvanized steel, 1-1/2″ wide	1 Carp	2.60	3.077	C.L.F.	34.50	109		143.50	219	
	1600	2″ wide		2.55	3.137	"	44	112		156	234	
	9000	Minimum labor/equipment charge		3	2.667	Job		95		95	157	

09200 | Plaster & Gypsum Board

	09280 | Gypsum Wallboard Repairs		CREW	DAILY OUTPUT	LABOR-HOURS	UNIT	2006 BARE COSTS MAT.	LABOR	EQUIP.	TOTAL	TOTAL INCL O&P	
100	0010	GYPSUM WALLBOARD REPAIRS										100
	0100	Fill and sand, pin / nail holes	1 Carp	960	.008	Ea.		.30		.30	.49	
	0110	Screw head pops		480	.017			.59		.59	.98	
	0120	Dents, up to 2″ square		48	.167		.01	5.95		5.96	9.80	
	0130	2″ to 4″ square		24	.333		.03	11.85		11.88	19.70	
	0140	Cut square, patch, sand and finish, holes, up to 2″ square		12	.667		.03	23.50		23.53	39.50	
	0150	2″ to 4″ square		11	.727		.07	26		26.07	43	
	0160	4″ to 8″ square		10	.800		.19	28.50		28.69	47	
	0170	8″ to 12″ square		8	1		.37	35.50		35.87	59.50	
	9000	Minimum labor/equipment charge		2	4	Job		142		142	236	

09300 | Tile

	09310 | Ceramic Tile		CREW	DAILY OUTPUT	LABOR-HOURS	UNIT	2006 BARE COSTS MAT.	LABOR	EQUIP.	TOTAL	TOTAL INCL O&P	
100	0010	CERAMIC TILE										100
	0050	Base, using 1′ x 4″ high pc. with 1″ x 1″ tiles, mud set	D-7	82	.195	L.F.	4.07	5.95		10.02	13.80	
	0100	Thin set	″	128	.125		3.87	3.80		7.67	10.20	
	0300	For 6″ high base, 1″ x 1″ tile face, add					.64			.64	.70	
	0400	For 2″ x 2″ tile face, add to above					.34			.34	.37	
	0600	Cove base, 4-1/4″ x 4-1/4″ high, mud set	D-7	91	.176		3.19	5.35		8.54	11.85	
	0700	Thin set		128	.125		3.21	3.80		7.01	9.50	
	0900	6″ x 4-1/4″ high, mud set		100	.160		2.94	4.86		7.80	10.85	
	1000	Thin set		137	.117		2.94	3.55		6.49	8.80	
	1200	Sanitary cove base, 6″ x 4-1/4″ high, mud set		93	.172		3.26	5.25		8.51	11.80	
	1300	Thin set		124	.129		3.70	3.92		7.62	10.20	
	1500	6″ x 6″ high, mud set		84	.190		4.06	5.80		9.86	13.50	
	1600	Thin set		117	.137		4.06	4.15		8.21	10.95	
	1800	Bathroom accessories, average		82	.195	Ea.	9.70	5.95		15.65	19.95	
	1900	Bathtub, 5′, rec. 4-1/4″ x 4-1/4″ tile wainscot, adhesive set 6′ high		2.90	5.517		140	168		308	415	
	2100	7′ high wainscot		2.50	6.400		160	194		354	480	
	2200	8′ high wainscot		2.20	7.273		170	221		391	530	
	2400	Bullnose trim, 4-1/4″ x 4-1/4″, mud set		82	.195	L.F.	2.82	5.95		8.77	12.40	
	2500	Thin set		128	.125		2.61	3.80		6.41	8.80	
	2700	6″ x 4-1/4″ bullnose trim, mud set		84	.190		2.32	5.80		8.12	11.60	
	2800	Thin set		124	.129		2.32	3.92		6.24	8.70	
	3000	Floors, natural clay, random or uniform, thin set, color group 1		183	.087	S.F.	3.69	2.66		6.35	8.20	
	3100	Color group 2		183	.087		3.98	2.66		6.64	8.55	
	3255	Floors, glazed, thin set, 6″ x 6″, color group 1		200	.080		3.14	2.43		5.57	7.25	
	3260	8″ x 8″ tile		250	.064		3.14	1.94		5.08	6.50	
	3270	12″ x 12″ tile		325	.049		3.94	1.50		5.44	6.65	
	3280	16″ x 16″ tile		550	.029		5.40	.88		6.28	7.35	
	3285	Border, 6″ x 12″ tile		275	.058		10.25	1.77		12.02	14	
	3290	3″ x 12″ tile		200	.080		30	2.43		32.43	37	
	3300	Porcelain type, 1 color, color group 2, 1″ x 1″		183	.087		4.27	2.66		6.93	8.85	
	3310	2″ x 2″ or 2″ x 1″, thin set		190	.084		4.48	2.56		7.04	8.95	
	3350	For random blend, 2 colors, add					.79			.79	.87	
	3360	4 colors, add					1.12			1.12	1.23	
	3370	For color group 3, add					.46			.46	.51	
	3380	For abrasive non-slip tile, add					.45			.45	.50	
	4300	Specialty tile, 4-1/4″ x 4-1/4″ x 1/2″, decorator finish	D-7	183	.087		8.95	2.66		11.61	14	

09300 | Tile

		09310 | Ceramic Tile	CREW	DAILY OUTPUT	LABOR-HOURS	UNIT	2006 BARE COSTS MAT.	LABOR	EQUIP.	TOTAL	TOTAL INCL O&P	
100	4500	Add for epoxy grout, 1/16" joint, 1" x 1" tile	D-7	800	.020	S.F.	.55	.61		1.16	1.56	100
	4600	2" x 2" tile		820	.020		.50	.59		1.09	1.48	
	4800	Pregrouted sheets, walls, 4-1/4" x 4-1/4", 6" x 4-1/4"										
	4810	and 8-1/2" x 4-1/4", 4 S.F. sheets, silicone grout	D-7	240	.067	S.F.	4.23	2.03		6.26	7.85	
	5100	Floors, unglazed, 2 S.F. sheets,										
	5110	urethane adhesive	D-7	180	.089	S.F.	4.21	2.70		6.91	8.85	
	5400	Walls, interior, thin set, 4-1/4" x 4-1/4" tile		190	.084		1.99	2.56		4.55	6.20	
	5500	6" x 4-1/4" tile		190	.084		2.35	2.56		4.91	6.60	
	5700	8-1/2" x 4-1/4" tile		190	.084		3.32	2.56		5.88	7.65	
	5800	6" x 6" tile		200	.080		2.73	2.43		5.16	6.80	
	5810	8" x 8" tile		225	.071		3.64	2.16		5.80	7.40	
	5820	12" x 12" tile		300	.053		3.04	1.62		4.66	5.90	
	5830	16" x 16" tile		500	.032		3.29	.97		4.26	5.15	
	6000	Decorated wall tile, 4-1/4" x 4-1/4", minimum		270	.059		3.65	1.80		5.45	6.85	
	6100	Maximum		180	.089		39	2.70		41.70	47	
	6300	Exterior walls, frostproof, mud set, 4-1/4" x 4-1/4"		102	.157		4.02	4.77		8.79	11.85	
	6400	1-3/8" x 1-3/8"		93	.172		3.86	5.25		9.11	12.45	
	6600	Crystalline glazed, 4-1/4" x 4-1/4", mud set, plain		100	.160		3.33	4.86		8.19	11.25	
	6700	4-1/4" x 4-1/4", scored tile		100	.160		4.13	4.86		8.99	12.15	
	6900	6"x 6" plain		93	.172		4.40	5.25		9.65	13.05	
	7000	For epoxy grout, 1/16" joints, 4-1/4" tile, add		800	.020		.33	.61		.94	1.31	
	7200	For tile set in dry mortar, add		1735	.009			.28		.28	.44	
	7300	For tile set in portland cement mortar, add		290	.055			1.68		1.68	2.63	
	9500	Minimum labor/equipment charge		3.25	4.923	Job		150		150	234	
300	0010	**CERAMIC TILE PANELS**										300
	0020	Insulated, over 1000 S.F., 1-1/2" thick	D-7	220	.073	S.F.	8.85	2.21		11.06	13.15	
	0100	2-1/2" thick	"	220	.073	"	9.45	2.21		11.66	13.85	
		09330 | Quarry Tile										
100	0010	**QUARRY TILE** Base, cove or sanitary, 2" or 5" high, mud set										100
	0100	1/2" thick	D-7	110	.145	L.F.	4.59	4.42		9.01	12	
	0300	Bullnose trim, red, mud set, 6" x 6" x 1/2" thick		120	.133		3.92	4.05		7.97	10.65	
	0400	4" x 4" x 1/2" thick		110	.145		4.43	4.42		8.85	11.80	
	0600	4" x 8" x 1/2" thick, using 8" as edge		130	.123		3.89	3.74		7.63	10.15	
	0700	Floors, mud set, 1,000 S.F. lots, red, 4" x 4" x 1/2" thick		120	.133	S.F.	3.89	4.05		7.94	10.65	
	0900	6" x 6" x 1/2" thick		140	.114		3.14	3.47		6.61	8.90	
	1000	4" x 8" x 1/2" thick		130	.123		3.89	3.74		7.63	10.15	
	1300	For waxed coating, add					.62			.62	.68	
	1500	For colors other than green, add					.37			.37	.41	
	1600	For abrasive surface, add					.44			.44	.48	
	1800	Brown tile, imported, 6" x 6" x 3/4"	D-7	120	.133		4.62	4.05		8.67	11.45	
	1900	8" x 8" x 1"		110	.145		5.25	4.42		9.67	12.70	
	2100	For thin set mortar application, deduct		700	.023			.69		.69	1.09	
	2200	For epoxy grout & mortar, 6" x 6" x 1/2", add		350	.046		1.69	1.39		3.08	4.04	
	2700	Stair tread, 6" x 6" x 3/4", plain		50	.320		4.82	9.70		14.52	20.50	
	2800	Abrasive		47	.340		4.91	10.35		15.26	21.50	
	3000	Wainscot, 6" x 6" x 1/2", thin set, red		105	.152		3.67	4.63		8.30	11.30	
	3100	Colors other than green		105	.152		4.09	4.63		8.72	11.75	
	3300	Window sill, 6" wide, 3/4" thick		90	.178	L.F.	4.60	5.40		10	13.50	
	3400	Corners		80	.200	Ea.	5.15	6.10		11.25	15.25	
	9000	Minimum labor/equipment charge		3.25	4.923	Job		150		150	234	
		09350 | Glass Mosaics										
100	0010	**GLASS MOSAICS** 3/4" tile on 12" sheets, standard grout										100
	0300	Color group 1 & 2	D-7	73	.219	S.F.	15.70	6.65		22.35	28	

09300 | Tile

09350 | Glass Mosaics

		09350 Glass Mosaics	CREW	DAILY OUTPUT	LABOR-HOURS	UNIT	2006 BARE COSTS MAT.	LABOR	EQUIP.	TOTAL	TOTAL INCL O&P	
100	0350	Color group 3	D-7	73	.219	S.F.	16.55	6.65		23.20	28.50	100
	0400	Color group 4		73	.219		22.50	6.65		29.15	35.50	
	0450	Color group 5		73	.219		25.50	6.65		32.15	38.50	
	0500	Color group 6		73	.219		30.50	6.65		37.15	44.50	
	0600	Color group 7		73	.219		35.50	6.65		42.15	49.50	
	0700	Color group 8, golds, silvers & specialties		64	.250		51.50	7.60		59.10	68.50	

09370 | Metal Tile

		09370 Metal Tile	CREW	DAILY OUTPUT	LABOR-HOURS	UNIT	MAT.	LABOR	EQUIP.	TOTAL	TOTAL INCL O&P	
100	0010	**METAL TILE** 4′ x 4′ sheet, 24 ga., tile pattern, nailed										100
	0200	Stainless steel	2 Carp	512	.031	S.F.	23	1.11		24.11	27.50	
	0400	Aluminized steel	″	512	.031	″	12.50	1.11		13.61	15.60	
	9000	Minimum labor/equipment charge	1 Carp	4	2	Job		71		71	118	

09400 | Terrazzo

09410 | Portland Cement Terrazzo

		09410 Portland Cement Terrazzo	CREW	DAILY OUTPUT	LABOR-HOURS	UNIT	2006 BARE COSTS MAT.	LABOR	EQUIP.	TOTAL	TOTAL INCL O&P	
100	0010	**PORTLAND CEMENT TERRAZZO**, cast-in-place										100
	0020	Cove base, 6″ high, 16ga. zinc	1 Mstz	20	.400	L.F.	3	13.65		16.65	25	
	0100	Curb, 6″ high and 6″ wide		6	1.333		4.78	45.50		50.28	77	
	0300	Divider strip for floors, 14 ga., 1-1/4″ deep, zinc		375	.021		1.08	.73		1.81	2.33	
	0400	Brass		375	.021		1.94	.73		2.67	3.27	
	0600	Heavy top strip 1/4″ thick, 1-1/4″ deep, zinc		300	.027		1.53	.91		2.44	3.11	
	0900	Galv. bottoms, brass		300	.027		2.64	.91		3.55	4.33	
	1200	For thin set floors, 16 ga., 1/2″ x 1/2″, zinc		350	.023		.66	.78		1.44	1.95	
	1300	Brass		350	.023		1.32	.78		2.10	2.67	
	1500	Floor, bonded to concrete, 1-3/4″ thick, gray cement	J-3	130	.123	S.F.	2.55	3.79	1.81	8.15	10.75	
	1600	White cement, mud set		130	.123		2.90	3.79	1.81	8.50	11.15	
	1800	Not bonded, 3″ total thickness, gray cement		115	.139		3.19	4.29	2.05	9.53	12.45	
	1900	White cement, mud set		115	.139		3.48	4.29	2.05	9.82	12.80	
	2100	For Venetian terrazzo, 1″ topping, add					50%	50%				
	2200	For heavy duty abrasive terrazzo, add					50%	50%				
	2700	Monolithic terrazzo, 1/2″ thick										
	2710	10′ panels	J-3	125	.128	S.F.	2.32	3.94	1.88	8.14	10.80	
	3000	Stairs, cast in place, pan filled treads		30	.533	L.F.	2.33	16.45	7.85	26.63	37	
	3100	Treads and risers		14	1.143	″	4.83	35	16.80	56.63	79	
	3300	For stair landings, add to floor prices						50%				
	3400	Stair stringers and fascia	J-3	30	.533	S.F.	3.88	16.45	7.85	28.18	39	
	3600	For abrasive metal nosings on stairs, add		150	.107	L.F.	6.70	3.29	1.57	11.56	14.25	
	3700	For abrasive surface finish, add		600	.027	S.F.	.92	.82	.39	2.13	2.73	
	3900	For raised abrasive strips, add		150	.107	L.F.	.88	3.29	1.57	5.74	7.85	
	4000	Wainscot, bonded, 1-1/2″ thick		30	.533	S.F.	2.97	16.45	7.85	27.27	38	
	4200	Epoxy terrazzo, 1/4″ thick		40	.400	″	4.34	12.30	5.90	22.54	30.50	
	4300	Stone chips, onyx gemstone, per 50 lb. bag				Bag	13.50			13.50	14.85	
	9000	Minimum labor/equipment charge	1 Mstz	1	8	Job		273		273	430	

09420 | Precast Terrazzo

		09420 Precast Terrazzo	CREW	DAILY OUTPUT	LABOR-HOURS	UNIT	MAT.	LABOR	EQUIP.	TOTAL	TOTAL INCL O&P	
900	0010	**TERRAZZO, PRECAST**										900
	0020	Base, 6″ high, straight	1 Mstz	35	.229	L.F.	9.65	7.80		17.45	23	
	0100	Cove		30	.267		10.90	9.10		20	26.50	
	0300	8″ high base, straight		30	.267		9.75	9.10		18.85	25	

09400 | Terrazzo

	09420 | Precast Terrazzo		Crew	Daily Output	Labor-Hours	Unit	2006 Bare Costs Mat.	Labor	Equip.	Total	Total Incl O&P	
900	0400	Cove	1 Mstz	25	.320	L.F.	14.35	10.90		25.25	33	900
	0600	For white cement, add					.39			.39	.43	
	0700	For 16 ga. zinc toe strip, add					1.43			1.43	1.57	
	0900	Curbs, 4″ x 4″ high	1 Mstz	19	.421		27.50	14.35		41.85	53	
	1000	8″ x 8″ high	″	15	.533		31.50	18.20		49.70	63	
	1200	Floor tiles, non-slip, 1″ thick, 12″ x 12″	D-1	29	.552	S.F.	16.20	17.75		33.95	47	
	1300	1-1/4″ thick, 12″ x 12″		29	.552		18.10	17.75		35.85	49	
	1500	16″ x 16″		23	.696		19.70	22.50		42.20	58	
	1600	1-1/2″ thick, 16″ x 16″		21	.762		18	24.50		42.50	59.50	
	1800	For Venetian terrazzo, add					5.40			5.40	5.95	
	1900	For white cement, add					.50			.50	.55	
	2400	Stair treads, 1-1/2″ thick, non-slip, three line pattern	2 Mstz	70	.229	L.F.	36.50	7.80		44.30	52	
	2500	Nosing and two lines		70	.229		36.50	7.80		44.30	52	
	2700	2″ thick treads, straight		60	.267		38.50	9.10		47.60	57	
	2800	Curved		50	.320		50.50	10.90		61.40	72.50	
	3000	Stair risers, 1″ thick, to 6″ high, straight sections		60	.267		8.80	9.10		17.90	24	
	3100	Cove		50	.320		12.85	10.90		23.75	31.50	
	3300	Curved, 1″ thick, to 6″ high, vertical		48	.333		17.55	11.35		28.90	37	
	3400	Cove		38	.421		33.50	14.35		47.85	59.50	
	3600	Stair tread and riser, single piece, straight, minimum		60	.267		46.50	9.10		55.60	65.50	
	3700	Maximum		40	.400		60	13.65		73.65	87.50	
	3900	Curved tread and riser, minimum		40	.400		65	13.65		78.65	93	
	4000	Maximum		32	.500		81.50	17.05		98.55	117	
	4200	Stair stringers, notched, 1″ thick		25	.640		26.50	22		48.50	63	
	4300	2″ thick		22	.727		31.50	25		56.50	73.50	
	4500	Stair landings, structural, non-slip, 1-1/2″ thick		85	.188	S.F.	29	6.40		35.40	42	
	4600	3″ thick		75	.213		41	7.25		48.25	56.50	
	4800	Wainscot, 12″ x 12″ x 1″ tiles	1 Mstz	12	.667		5.85	22.50		28.35	42	
	4900	16″ x 16″ x 1-1/2″ tiles	″	8	1		12.65	34		46.65	67.50	
	9500	Minimum labor/equipment charge	1 Tilf	2	4	Job		137		137	215	

	09430 | Conductive Terrazzo		Crew	Daily Output	Labor-Hours	Unit	Mat.	Labor	Equip.	Total	Total Incl O&P	
100	0010	**CONDUCTIVE TERRAZZO**										100
	2400	Bonded conductive floor for hospitals	J-3	90	.178	S.F.	3.48	5.50	2.62	11.60	15.30	
	2500	Epoxy terrazzo, 1/4″ thick, minimum		100	.160		3.94	4.93	2.36	11.23	14.60	
	2550	Average		75	.213		4.34	6.55	3.14	14.03	18.50	
	2600	Maximum		60	.267		4.76	8.20	3.93	16.89	22.50	

	09450 | Cast-in-Place Terrazzo		Crew	Daily Output	Labor-Hours	Unit	Mat.	Labor	Equip.	Total	Total Incl O&P	
200	0010	**TILE OR TERRAZZO BASE**										200
	0020	Scratch coat only	1 Mstz	150	.053	S.F.	.38	1.82		2.20	3.27	
	0500	Scratch and brown coat only		75	.107	″	.72	3.64		4.36	6.50	
	9000	Minimum labor/equipment charge		1	8	Job		273		273	430	

09500 | Ceilings

	09510 | Acoustical Ceilings		Crew	Daily Output	Labor-Hours	Unit	2006 Bare Costs Mat.	Labor	Equip.	Total	Total Incl O&P	
700	0010	**SUSPENDED ACOUSTIC CEILING TILES**, Not including										700
	0100	suspension system										

9 FINISHES

09510 | Acoustical Ceilings

		09510 Acoustical Ceilings	Crew	Daily Output	Labor-Hours	Unit	2006 Bare Costs Mat.	Labor	Equip.	Total	Total Incl O&P	
700	0300	Fiberglass boards, film faced, 2′ x 2′ or 2′ x 4′, 5/8″ thick	1 Carp	625	.013	S.F.	.52	.46		.98	1.32	700
	0400	3/4″ thick		600	.013		1.19	.47		1.66	2.10	
	0500	3″ thick, thermal, R11		450	.018		1.31	.63		1.94	2.49	
	0600	Glass cloth faced fiberglass, 3/4″ thick		500	.016		1.72	.57		2.29	2.83	
	0700	1″ thick		485	.016		1.90	.59		2.49	3.06	
	0820	1-1/2″ thick, nubby face		475	.017		2.36	.60		2.96	3.59	
	1110	Mineral fiber tile, lay-in, 2′ x 2′ or 2′ x 4′, 5/8″ thick, fine texture		625	.013		.45	.46		.91	1.25	
	1115	Rough textured		625	.013		1.10	.46		1.56	1.96	
	1125	3/4″ thick, fine textured		600	.013		1.22	.47		1.69	2.13	
	1130	Rough textured		600	.013		1.53	.47		2	2.47	
	1135	Fissured		600	.013		1.81	.47		2.28	2.78	
	1150	Tegular, 5/8″ thick, fine textured		470	.017		1.08	.61		1.69	2.19	
	1155	Rough textured		470	.017		1.40	.61		2.01	2.54	
	1165	3/4″ thick, fine textured		450	.018		1.53	.63		2.16	2.73	
	1170	Rough textured		450	.018		1.73	.63		2.36	2.95	
	1175	Fissured		450	.018		2.69	.63		3.32	4.01	
	1180	For aluminum face, add					4.91			4.91	5.40	
	1185	For plastic film face, add					.81			.81	.89	
	1190	For fire rating, add					.36			.36	.40	
	1300	Mirror faced panels, 15/16″ thick, 2′ x 2′	1 Carp	500	.016		10.20	.57		10.77	12.20	
	1900	Eggcrate, acrylic, 1/2″ x 1/2″ x 1/2″ cubes		500	.016		1.48	.57		2.05	2.57	
	2100	Polystyrene eggcrate, 3/8″ x 3/8″ x 1/2″ cubes		510	.016		1.24	.56		1.80	2.28	
	2200	1/2″ x 1/2″ x 1/2″ cubes		500	.016		1.66	.57		2.23	2.77	
	2400	Luminous panels, prismatic, acrylic		400	.020		1.80	.71		2.51	3.16	
	2500	Polystyrene		400	.020		.92	.71		1.63	2.19	
	2700	Flat white acrylic		400	.020		3.13	.71		3.84	4.62	
	2800	Polystyrene		400	.020		2.14	.71		2.85	3.53	
	3000	Drop pan, white, acrylic		400	.020		4.59	.71		5.30	6.25	
	3100	Polystyrene		400	.020		3.84	.71		4.55	5.40	
	3600	Perforated aluminum sheets, .024″ thick, corrugated, painted		490	.016		1.83	.58		2.41	2.97	
	3700	Plain		500	.016		3.10	.57		3.67	4.35	
	3720	Mineral fiber, 24″ x 24″ or 48″, reveal edge, painted, 5/8″ thick		600	.013		1.02	.47		1.49	1.91	
	3740	3/4″ thick		575	.014		1.66	.49		2.15	2.65	
	9000	Minimum labor/equipment charge		4	2	Job		71		71	118	
760	0010	**SUSPENDED CEILINGS, COMPLETE** Including standard										760
	0100	suspension system but not incl. 1-1/2″ carrier channels										
	0600	Fiberglass ceiling board, 2′ x 4′ x 5/8″, plain faced,	1 Carp	500	.016	S.F.	1.04	.57		1.61	2.09	
	0700	Offices, 2′ x 4′ x 3/4″		380	.021		1.71	.75		2.46	3.12	
	0800	Mineral fiber, on 15/16″ T bar susp. 2′ x 2′ x 3/4″ lay-in board		345	.023		1.88	.82		2.70	3.43	
	0810	2′ x 4′ x 5/8″ tile		380	.021		.97	.75		1.72	2.31	
	0820	Tegular, 2′ x 2′ x 5/8″ tile on 9/16″ grid		250	.032		1.89	1.14		3.03	3.95	
	0830	2′ x 4′ x 3/4″ tile		275	.029		2.20	1.03		3.23	4.13	
	0900	Luminous panels, prismatic, acrylic		255	.031		2.32	1.12		3.44	4.41	
	1200	Metal pan with acoustic pad, steel		75	.107		3.44	3.79		7.23	10.10	
	1300	Painted aluminum		75	.107		2.35	3.79		6.14	8.90	
	1500	Aluminum, degreased finish		75	.107		3.98	3.79		7.77	10.70	
	1600	Stainless steel		75	.107		7.45	3.79		11.24	14.50	
	1800	Tile, Z bar suspension, 5/8″ mineral fiber tile		150	.053		1.61	1.90		3.51	4.91	
	1900	3/4″ mineral fiber tile		150	.053		1.72	1.90		3.62	5.05	
	2402	For strip lighting, see division 16510-440										
	2500	For rooms under 500 S.F., add				S.F.		25%				
	9000	Minimum labor/equipment charge	1 Carp	2	4	Job		142		142	236	
900	0010	**CEILING TILE**, Stapled or cemented										900
	0100	12″ x 12″ or 12″ x 24″, not including furring										
	0600	Mineral fiber, vinyl coated, 5/8″ thick	1 Carp	1000	.008	S.F.	1.47	.28		1.75	2.09	
	0700	3/4″ thick		1000	.008		1.41	.28		1.69	2.02	

09500 | Ceilings

09510 | Acoustical Ceilings

		09510 Acoustical Ceilings	CREW	DAILY OUTPUT	LABOR-HOURS	UNIT	2006 BARE COSTS MAT.	LABOR	EQUIP.	TOTAL	TOTAL INCL O&P	
900	0900	Fire rated, 3/4" thick, plain faced	1 Carp	1000	.008	S.F.	1.27	.28		1.55	1.87	900
	1000	Plastic coated face		1000	.008		1.36	.28		1.64	1.97	
	1200	Aluminum faced, 5/8" thick, plain		1000	.008		1.12	.28		1.40	1.70	
	3700	Wall application of above, add		3100	.003			.09		.09	.15	
	3900	For ceiling primer, add					.12			.12	.13	
	4000	For ceiling cement, add					.34			.34	.37	
	9000	Minimum labor/equipment charge	1 Carp	4	2	Job		71		71	118	

09600 | Flooring

09620 | Specialty Flooring

		09620 Specialty Flooring	CREW	DAILY OUTPUT	LABOR-HOURS	UNIT	2006 BARE COSTS MAT.	LABOR	EQUIP.	TOTAL	TOTAL INCL O&P	
100	0010	**ATHLETIC FLOORING**										100
	3700	Polyethylene, in rolls, no base incl., landscape surfaces	1 Tilf	275	.029	S.F.	2.57	1		3.57	4.39	
	3800	Nylon action surface, 1/8" thick		275	.029		2.76	1		3.76	4.60	
	3900	1/4" thick		275	.029		3.98	1		4.98	5.95	
	4000	3/8" thick		275	.029		5	1		6	7.05	
	4100	Golf tee surface with foam back		235	.034		4.95	1.17		6.12	7.30	
	4200	Practice putting, knitted nylon surface		235	.034		4.20	1.17		5.37	6.45	
	4400	Polyurethane, thermoset, prefabricated in place, indoor										
	4500	3/8" thick for basketball, gyms, etc.	1 Tilf	100	.080	S.F.	3.95	2.74		6.69	8.65	
	4600	1/2" thick for professional sports		95	.084		5.30	2.88		8.18	10.35	
	4700	Outdoor, 1/4" thick, smooth, for tennis		100	.080		4.10	2.74		6.84	8.80	
	4800	Rough, for track, 3/8" thick		95	.084		4.75	2.88		7.63	9.75	
	5000	Poured in place, indoor, with finish, 1/4" thick		80	.100		2.92	3.43		6.35	8.55	
	5050	3/8" thick		65	.123		3.54	4.22		7.76	10.50	
	5100	1/2" thick		50	.160		4.69	5.50		10.19	13.75	
	5500	Polyvinyl chloride, sheet goods for gyms, 1/4" thick		80	.100		5	3.43		8.43	10.85	
	5600	3/8" thick		60	.133		5.65	4.57		10.22	13.35	

09631 | Brick Flooring

		09631 Brick Flooring	CREW	DAILY OUTPUT	LABOR-HOURS	UNIT	2006 BARE COSTS MAT.	LABOR	EQUIP.	TOTAL	TOTAL INCL O&P	
100	0010	**BRICK FLOORING**										100
	0020	Acid proof shales, red, 8" x 3-3/4" x 1-1/4" thick	D-7	.43	37.209	M	760	1,125		1,885	2,600	
	0050	2-1/4" thick	D-1	.40	40		825	1,275		2,100	2,975	
	0200	Acid proof clay brick, 8" x 3-3/4" x 2-1/4" thick	"	.40	40		785	1,275		2,060	2,950	
	0260	Cast ceramic, pressed, 4" x 8" x 1/2", unglazed	D-7	100	.160	S.F.	5.15	4.86		10.01	13.25	
	0270	Glazed		100	.160		6.85	4.86		11.71	15.15	
	0280	Hand molded flooring, 4" x 8" x 3/4", unglazed		95	.168		6.80	5.10		11.90	15.50	
	0290	Glazed		95	.168		8.50	5.10		13.60	17.35	
	0300	8" hexagonal, 3/4" thick, unglazed		85	.188		7.45	5.70		13.15	17.15	
	0310	Glazed		85	.188		13.45	5.70		19.15	24	
	0400	Heavy duty industrial, cement mortar bed, 2" thick, not incl. brick	D-1	80	.200		.70	6.45		7.15	11.20	
	0450	Acid proof joints, 1/4" wide	"	65	.246		1.18	7.90		9.08	14.15	
	0500	Pavers, 8" x 4", 1" to 1-1/4" thick, red	D-7	95	.168		3	5.10		8.10	11.30	
	0510	Ironspot	"	95	.168		4.23	5.10		9.33	12.65	
	0540	1-3/8" to 1-3/4" thick, red	D-1	95	.168		2.89	5.40		8.29	12	
	0560	Ironspot		95	.168		4.18	5.40		9.58	13.40	
	0580	2-1/4" thick, red		90	.178		2.94	5.70		8.64	12.50	
	0590	Ironspot		90	.178		4.56	5.70		10.26	14.25	

09600 | Flooring

09631 | Brick Flooring

		09631 Brick Flooring	CREW	DAILY OUTPUT	LABOR-HOURS	UNIT	2006 BARE COSTS MAT.	LABOR	EQUIP.	TOTAL	TOTAL INCL O&P	
100	0800	For sidewalks and patios with pavers, see division 02780-200										100
	0870	For epoxy joints, add	D-1	600	.027	S.F.	2.24	.86		3.10	3.85	
	0880	For Furan underlayment, add	″	600	.027		1.85	.86		2.71	3.43	
	0890	For waxed surface, steam cleaned, add	A-1H	1000	.008		.16	.22	.06	.44	.60	
	9000	Minimum labor/equipment charge	1 Bric	2	4	Job		146		146	237	

09635 | Marble Flooring

		09635 Marble Flooring	CREW	DAILY OUTPUT	LABOR-HOURS	UNIT	MAT.	LABOR	EQUIP.	TOTAL	TOTAL INCL O&P	
100	0010	**MARBLE**										100
	0020	Thin gauge tile, 12″ x 6″, 3/8″, White Carara	D-7	60	.267	S.F.	9.10	8.10		17.20	22.50	
	0100	Travertine		60	.267		10	8.10		18.10	23.50	
	0200	12″ x 12″ x 3/8″, thin set, floors		60	.267		6.30	8.10		14.40	19.65	
	0300	On walls		52	.308		8.90	9.35		18.25	24.50	
	9000	Minimum labor/equipment charge		3	5.333	Job		162		162	254	

09637 | Stone Flooring

		09637 Stone Flooring	CREW	DAILY OUTPUT	LABOR-HOURS	UNIT	MAT.	LABOR	EQUIP.	TOTAL	TOTAL INCL O&P	
100	0010	**SLATE TILE**										100
	0020	Vermont, 6″ x 6″ x 1/4″ thick, thin set	D-7	180	.089	S.F.	4.26	2.70		6.96	8.90	
	0200	See also division 02780-600										
	9000	Minimum labor/equipment charge	D-7	3	5.333	Job		162		162	254	
200	0010	**SLATE & STONE FLOORS** See division 02780-600										200
300	0010	**CONCRETE FLOORS** And toppings, see division 03350-300										300

09642 | Wood Athletic Flooring

		09642 Wood Athletic Flooring	CREW	DAILY OUTPUT	LABOR-HOURS	UNIT	MAT.	LABOR	EQUIP.	TOTAL	TOTAL INCL O&P	
100	0010	**WOOD ATHLETIC FLOORING**										100
	0600	Gym floor, in mastic, over 2 ply felt, #2 & better										
	0700	25/32″ thick maple	1 Carp	100	.080	S.F.	3.55	2.84		6.39	8.60	
	0900	33/32″ thick maple		98	.082		4.10	2.90		7	9.30	
	1000	For 1/2″ corkboard underlayment, add		750	.011		.79	.38		1.17	1.50	
	1300	For #1 grade maple, add					.42			.42	.46	
	1600	Maple flooring, over sleepers, #2 & better										
	1700	25/32″ thick	1 Carp	85	.094	S.F.	3.84	3.35		7.19	9.75	
	1900	33/32″ thick	″	83	.096		4.47	3.43		7.90	10.60	
	2000	For #1 grade, add					.45			.45	.50	
	2200	For 3/4″ subfloor, add	1 Carp	350	.023		1	.81		1.81	2.45	
	2300	With two 1/2″ subfloors, 25/32″ thick	″	69	.116		4.81	4.12		8.93	12.15	
	2500	Maple, incl. finish, #2 & btr., 25/32″ thick, on rubber										
	2600	Sleepers, with two 1/2″ subfloors	1 Carp	76	.105	S.F.	5.15	3.74		8.89	11.90	
	2800	With steel spline, double connection to channels	″	73	.110		5.50	3.90		9.40	12.50	
	2900	For 33/32″ maple, add					.60			.60	.66	
	3100	For #1 grade maple, add					.45			.45	.50	
	3500	For termite proofing all of the above, add					.23			.23	.25	
	3700	Portable hardwood, prefinished panels	1 Carp	83	.096		6.90	3.43		10.33	13.25	
	3720	Insulated with polystyrene, 1″ thick, add		165	.048		.60	1.72		2.32	3.52	
	3750	Running tracks, Sitka spruce surface, 25/32″ x 2-1/4″		62	.129		12.60	4.59		17.19	21.50	
	3770	3/4″ plywood surface, finished		100	.080		2.99	2.84		5.83	8	
	3800	See also resilient gym floors, division 09620-100										

09643 | Wood Block Flooring

		09643 Wood Block Flooring	CREW	DAILY OUTPUT	LABOR-HOURS	UNIT	MAT.	LABOR	EQUIP.	TOTAL	TOTAL INCL O&P	
100	0010	**WOOD BLOCK FLOORING**										100
	0020	End grain flooring, coated, 2″ thick	1 Carp	295	.027	S.F.	2.98	.96		3.94	4.88	
	0400	Natural finish, 1″ thick, fir		125	.064		3.08	2.28		5.36	7.15	
	0600	1-1/2″ thick, pine		125	.064		3.03	2.28		5.31	7.10	

09643 | Wood Block Flooring

			CREW	DAILY OUTPUT	LABOR-HOURS	UNIT	2006 BARE COSTS MAT.	LABOR	EQUIP.	TOTAL	TOTAL INCL O&P	
100	0700	2″ thick, pine	1 Carp	125	.064	S.F.	3.55	2.28		5.83	7.70	100
	9000	Minimum labor/equipment charge		2	4	Job		142		142	236	

09644 | Wood Comp. Flooring

			CREW	DAILY OUTPUT	LABOR-HOURS	UNIT	MAT.	LABOR	EQUIP.	TOTAL	TOTAL INCL O&P	
100	0010	**WOOD COMPOSITION** Gym floors										100
	0100	2-1/4″ x 6-7/8″ x 3/8″, on 2″ grout setting bed	D-7	150	.107	S.F.	5.20	3.24		8.44	10.85	
	0200	Thin set, on concrete	″	250	.064		4.77	1.94		6.71	8.30	
	0300	Sanding and finishing, add	1 Carp	200	.040		.70	1.42		2.12	3.13	

09647 | Wood Parquet Flooring

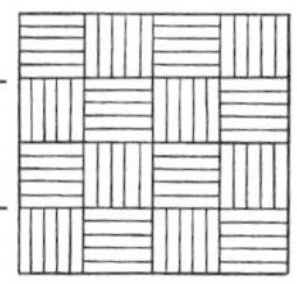

			CREW	DAILY OUTPUT	LABOR-HOURS	UNIT	MAT.	LABOR	EQUIP.	TOTAL	TOTAL INCL O&P	
100	0010	**WOOD PARQUET** flooring										100
	5200	Parquetry, standard, 5/16″ thick, not incl. finish, oak, minimum	1 Carp	160	.050	S.F.	3.43	1.78		5.21	6.70	
	5300	Maximum		100	.080		5.30	2.84		8.14	10.55	
	5500	Teak, minimum		160	.050		4.53	1.78		6.31	7.95	
	5600	Maximum		100	.080		7.90	2.84		10.74	13.40	
	5650	13/16″ thick, select grade oak, minimum		160	.050		8.75	1.78		10.53	12.55	
	5700	Maximum		100	.080		13.25	2.84		16.09	19.30	
	5800	Custom parquetry, including finish, minimum		100	.080		14.65	2.84		17.49	21	
	5900	Maximum		50	.160		19.40	5.70		25.10	31	
	6700	Parquetry, prefinished white oak, 5/16″ thick, minimum		160	.050		3.50	1.78		5.28	6.80	
	6800	Maximum		100	.080		7.15	2.84		9.99	12.60	
	7000	Walnut or teak, parquetry, minimum		160	.050		4.91	1.78		6.69	8.35	
	7100	Maximum		100	.080		8.55	2.84		11.39	14.15	
	7200	Acrylic wood parquet blocks, 12″ x 12″ x 5/16″,										
	7210	irradiated, set in epoxy	1 Carp	160	.050	S.F.	7.15	1.78		8.93	10.80	

09648 | Wood Strip Flooring

			CREW	DAILY OUTPUT	LABOR-HOURS	UNIT	MAT.	LABOR	EQUIP.	TOTAL	TOTAL INCL O&P	
100	0010	**WOOD**										100
	0020	Fir, vertical grain, 1″ x 4″, not incl. finish, B & better	1 Carp	255	.031	S.F.	2.51	1.12		3.63	4.61	
	0100	C grade & better		255	.031		2.36	1.12		3.48	4.45	
	4000	Maple, strip, 25/32″ x 2-1/4″, not incl. finish, select		170	.047		4.78	1.67		6.45	8	
	4100	#2 & better		170	.047		2.95	1.67		4.62	6	
	4300	33/32″ x 3-1/4″, not incl. finish, #1 grade		170	.047		3.73	1.67		5.40	6.85	
	4400	#2 & better		170	.047		3.32	1.67		4.99	6.40	
	4600	Oak, white or red, 25/32″ x 2-1/4″, not incl. finish										
	4700	#1 common	1 Carp	170	.047	S.F.	3	1.67		4.67	6.05	
	4900	Select quartered, 2-1/4″ wide		170	.047		2.78	1.67		4.45	5.85	
	5000	Clear		170	.047		3.72	1.67		5.39	6.85	
	6100	Prefinished, white oak, prime grade, 2-1/4″ wide		170	.047		6.20	1.67		7.87	9.55	
	6200	3-1/4″ wide		185	.043		7.95	1.54		9.49	11.30	
	6400	Ranch plank		145	.055		7.70	1.96		9.66	11.70	
	6500	Hardwood blocks, 9″ x 9″, 25/32″ thick		160	.050		5.30	1.78		7.08	8.75	
	7400	Yellow pine, 3/4″ x 3-1/8″, T & G, C & better, not incl. finish		200	.040		2.28	1.42		3.70	4.87	
	7550	Refinish old floors, see division 01832-290										
	7800	Sanding and finishing, 2 coats polyurethane	1 Clab	295	.027	S.F.	.71	.74		1.45	2.01	
	7900	Subfloor and underlayment, see division 06160										
	8015	Transition molding, 2 1/4″ wide, 5′ long	1 Carp	19.20	.417	Ea.	12.90	14.80		27.70	38.50	
	8300	Floating floor, wood composition strip, complete.	1 Clab	133	.060	S.F.	3.98	1.65		5.63	7.10	
	8310	Floating floor components, T & G wood composite strips					3.45			3.45	3.80	
	8320	Film					.14			.14	.15	
	8330	Foam					.23			.23	.25	
	8340	Adhesive					.21			.21	.23	
	8350	Installation kit					.16			.16	.18	
	8360	Trim, 2″ wide x 3′ long				L.F.	2.31			2.31	2.54	
	8370	Reducer moulding				″	3.99			3.99	4.39	

	09648 | Wood Strip Flooring		Crew	Daily Output	Labor-Hours	Unit	2006 Bare Costs Mat.	Labor	Equip.	Total	Total Incl O&P	
100	9000	Minimum labor/equipment charge	1 Carp	2	4	Job		142		142	236	100

	09651 | Resilient Base & Access.		Crew	Daily Output	Labor-Hours	Unit	Mat.	Labor	Equip.	Total	Total Incl O&P	
100	0010	**STAIR TREADS AND RISERS** See index for materials other										100
	0100	than rubber and vinyl										
	0300	Rubber, molded tread, 12" wide, 5/16" thick, black	1 Tilf	115	.070	L.F.	9.45	2.38		11.83	14.15	
	0400	Colors		115	.070		8.95	2.38		11.33	13.60	
	0600	1/4" thick, black		115	.070		8.75	2.38		11.13	13.40	
	0700	Colors		115	.070		8.75	2.38		11.13	13.40	
	0900	Grip strip safety tread, colors, 5/16" thick		115	.070		13	2.38		15.38	18.05	
	1000	3/16" thick		120	.067		9.15	2.28		11.43	13.70	
	1200	Landings, smooth sheet rubber, 1/8" thick		120	.067	S.F.	4.20	2.28		6.48	8.20	
	1300	3/16" thick		120	.067	"	5.35	2.28		7.63	9.50	
	1500	Nosings, 3" wide, 3/16" thick, black		140	.057	L.F.	2.90	1.96		4.86	6.25	
	1600	Colors		140	.057		2.79	1.96		4.75	6.15	
	1800	Risers, 7" high, 1/8" thick, flat		250	.032		3.39	1.10		4.49	5.45	
	1900	Coved		250	.032		3.03	1.10		4.13	5.05	
	2100	Vinyl, molded tread, 12" wide, colors, 1/8" thick		115	.070		3.85	2.38		6.23	8	
	2200	1/4" thick		115	.070		5.65	2.38		8.03	9.95	
	2300	Landing material, 1/8" thick		200	.040	S.F.	3.95	1.37		5.32	6.50	
	2400	Riser, 7" high, 1/8" thick, coved		175	.046	L.F.	2.25	1.57		3.82	4.93	
	2500	Tread and riser combined, 1/8" thick		80	.100	"	6.55	3.43		9.98	12.55	
	9000	Minimum labor/equipment charge		3	2.667	Job		91.50		91.50	143	
200	0010	**RESILIENT BASE**										200
	0800	Base, cove, rubber or vinyl, .080" thick										
	1100	Standard colors, 2-1/2" high	1 Tilf	315	.025	L.F.	.52	.87		1.39	1.93	
	1150	4" high		315	.025		.53	.87		1.40	1.94	
	1200	6" high		315	.025		.92	.87		1.79	2.37	
	1450	1/8" thick, standard colors, 2-1/2" high		315	.025		.60	.87		1.47	2.02	
	1500	4" high		315	.025		.60	.87		1.47	2.02	
	1550	6" high		315	.025		.95	.87		1.82	2.41	
	1600	Corners, 2-1/2" high		315	.025	Ea.	1.17	.87		2.04	2.65	
	1630	4" high		315	.025		1.75	.87		2.62	3.29	
	1660	6" high		315	.025		2	.87		2.87	3.56	

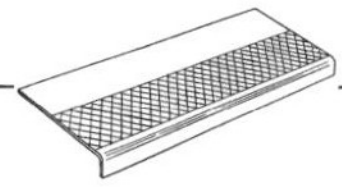

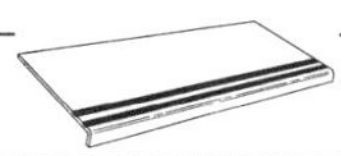

	09653 | Resilient Sheet Flooring		Crew	Daily Output	Labor-Hours	Unit	Mat.	Labor	Equip.	Total	Total Incl O&P	
100	0010	**RESILIENT SHEET FLOORING**										100
	5900	Rubber, sheet goods, 36" wide, 1/8" thick	1 Tilf	120	.067	S.F.	4.25	2.28		6.53	8.25	
	5950	3/16" thick		100	.080		6	2.74		8.74	10.90	
	6000	1/4" thick		90	.089		7	3.04		10.04	12.45	
	8000	Vinyl sheet goods, backed, .065" thick, minimum		250	.032		2.21	1.10		3.31	4.15	
	8050	Maximum		200	.040		2.71	1.37		4.08	5.15	
	8100	.080" thick, minimum		230	.035		2.39	1.19		3.58	4.50	
	8150	Maximum		200	.040		3.40	1.37		4.77	5.90	
	8200	.125" thick, minimum		230	.035		2.69	1.19		3.88	4.83	
	8250	Maximum		200	.040		4.25	1.37		5.62	6.85	
	8700	Adhesive cement, 1 gallon does 200 to 300 S.F.				Gal.	16.75			16.75	18.40	
	8800	Asphalt primer, 1 gallon per 300 S.F.					10.30			10.30	11.35	
	8900	Emulsion, 1 gallon per 140 S.F.					13.10			13.10	14.40	
	8950	Latex underlayment, liquid, fortified					33			33	36.50	

	09658 | Resilient Tile Flooring		Crew	Daily Output	Labor-Hours	Unit	Mat.	Labor	Equip.	Total	Total Incl O&P	
100	0010	**RESILIENT TILE FLOORING**										100
	2200	Cork tile, standard finish, 1/8" thick	1 Tilf	315	.025	S.F.	4.24	.87		5.11	6	

09658 | Resilient Tile Flooring

		09658 Resilient Tile Flooring	Crew	Daily Output	Labor-Hours	Unit	Mat.	Labor	Equip.	Total	Total Incl O&P	
100	2250	3/16″ thick	1 Tilf	315	.025	S.F.	4.91	.87		5.78	6.75	100
	2300	5/16″ thick		315	.025		5.45	.87		6.32	7.35	
	2350	1/2″ thick		315	.025		6.45	.87		7.32	8.45	
	2500	Urethane finish, 1/8″ thick		315	.025		5.20	.87		6.07	7.05	
	2550	3/16″ thick		315	.025		5.60	.87		6.47	7.50	
	2600	5/16″ thick		315	.025		7.05	.87		7.92	9.10	
	2650	1/2″ thick		315	.025		9.85	.87		10.72	12.20	
	6050	Tile, marbleized colors, 12″ x 12″, 1/8″ thick		400	.020		4.86	.69		5.55	6.40	
	6100	3/16″ thick		400	.020		6.50	.69		7.19	8.20	
	6300	Special tile, plain colors, 1/8″ thick		400	.020		5	.69		5.69	6.55	
	6350	3/16″ thick		400	.020		6.80	.69		7.49	8.55	
	6410	Raised, radial or square, minimum		400	.020		7.90	.69		8.59	9.75	
	6430	Maximum		400	.020		7.90	.69		8.59	9.75	
	6450	For golf course, skating rink, etc., 1/4″ thick		275	.029		7.90	1		8.90	10.25	
	6700	Synthetic turf, 3/8″ thick		90	.089		3.62	3.04		6.66	8.75	
	6750	Interlocking 2′ x 2′ squares, 1/2″ thick, not										
	6810	cemented, for playgrounds, minimum	1 Tilf	210	.038	S.F.	2.99	1.30		4.29	5.35	
	6850	Maximum		190	.042		7.70	1.44		9.14	10.70	
	7000	Vinyl composition tile, 12″ x 12″, 1/16″ thick		500	.016		.86	.55		1.41	1.81	
	7050	Embossed		500	.016		1.35	.55		1.90	2.35	
	7100	Marbleized		500	.016		1.35	.55		1.90	2.35	
	7150	Solid		500	.016		1.50	.55		2.05	2.51	
	7200	3/32″ thick, embossed		500	.016		1.03	.55		1.58	1.99	
	7250	Marbleized		500	.016		1.50	.55		2.05	2.51	
	7300	Solid		500	.016		1.96	.55		2.51	3.02	
	7350	1/8″ thick, marbleized		500	.016		1.17	.55		1.72	2.15	
	7400	Solid		500	.016		2.05	.55		2.60	3.12	
	7450	Conductive		500	.016		3.89	.55		4.44	5.15	
	7500	Vinyl tile, 12″ x 12″, .050″ thick, minimum		500	.016		2.10	.55		2.65	3.17	
	7550	Maximum		500	.016		3.95	.55		4.50	5.20	
	7600	1/8″ thick, minimum		500	.016		2.75	.55		3.30	3.89	
	7650	Solid colors		500	.016		5.05	.55		5.60	6.40	
	7700	Marbleized or Travertine pattern		500	.016		3.95	.55		4.50	5.20	
	7750	Florentine pattern		500	.016		4.42	.55		4.97	5.70	
	7800	Maximum		500	.016		9.10	.55		9.65	10.85	
	9500	Minimum labor/equipment charge		4	2	Job		68.50		68.50	107	

09662 | Static Control Flooring

		09662 Static Control Flooring	Crew	Daily Output	Labor-Hours	Unit	Mat.	Labor	Equip.	Total	Total Incl O&P	
100	0010	CONDUCTIVE RESILIENT FLOORING										100
	1700	Conductive flooring, rubber tile, 1/8″ thick	1 Tilf	315	.025	S.F.	3.50	.87		4.37	5.20	
	1800	Homogeneous vinyl tile, 1/8″ thick	″	315	.025	″	4.75	.87		5.62	6.60	

09670 | Fluid Applied Flooring

		09670 Fluid Applied Flooring	Crew	Daily Output	Labor-Hours	Unit	Mat.	Labor	Equip.	Total	Total Incl O&P	
100	0010	RESILIENT TILE UNDERLAYMENT										100
	3600	Latex underlayment, 1/8″ thk., cementitious for resilient flooring	1 Tilf	160	.050	S.F.	1.55	1.71		3.26	4.39	

09673 | Composition Flooring

		09673 Composition Flooring	Crew	Daily Output	Labor-Hours	Unit	Mat.	Labor	Equip.	Total	Total Incl O&P	
100	0010	COMPOSITION FLOORING										100
	0020	Cementitous acrylic, 1/4″ thick	C-6	520	.092	S.F.	1.30	2.67	.08	4.05	5.90	
	0100	3/8″ thick	″	450	.107		1.71	3.08	.09	4.88	7.05	
	0200	Methyl methachrylate, 1/4″ thick	C-8A	3000	.016		5	.48		5.48	6.30	
	0210	1/8″ thick	″	3000	.016		2.75	.48		3.23	3.81	
	0300	Cupric oxychloride, on bond coat, minimum	C-6	480	.100		2.91	2.89	.09	5.89	8.05	
	0400	Maximum		420	.114		4.86	3.30	.10	8.26	10.85	
	0600	Epoxy, with colored quartz chips, broadcast, minimum		675	.071		2.30	2.06	.06	4.42	5.95	

Table column headings: DAILY OUTPUT, LABOR-HOURS, and 2006 BARE COSTS (Mat., Labor, Equip., Total); TOTAL INCL O&P.

	09673 | Composition Flooring		CREW	DAILY OUTPUT	LABOR-HOURS	UNIT	2006 BARE COSTS MAT.	LABOR	EQUIP.	TOTAL	TOTAL INCL O&P	
100	0700	Maximum	C-6	490	.098	S.F.	2.79	2.83	.09	5.71	7.80	100
	0900	Trowelled, minimum		560	.086		2.97	2.48	.08	5.53	7.40	
	1000	Maximum		480	.100		4.33	2.89	.09	7.31	9.60	
	1200	Heavy duty epoxy topping, 1/4" thick,										
	1300	500 to 1,000 S.F.	C-6	420	.114	S.F.	5.15	3.30	.10	8.55	11.15	
	1500	1,000 to 2,000 S.F.		450	.107		4.06	3.08	.09	7.23	9.60	
	1600	Over 10,000 S.F.		480	.100		3.76	2.89	.09	6.74	9	
	1800	Epoxy terrazzo, 1/4" thick, chemical resistant, minimum	J-3	200	.080		5.45	2.46	1.18	9.09	11.15	
	1900	Maximum		150	.107		8.35	3.29	1.57	13.21	16.10	
	2100	Conductive, minimum		210	.076		6.75	2.35	1.12	10.22	12.35	
	2200	Maximum		160	.100		9.15	3.08	1.47	13.70	16.50	
	2400	Mastic, hot laid, 2 coat, 1-1/2" thick, standard, minimum	C-6	690	.070		3.37	2.01	.06	5.44	7.10	
	2500	Maximum		520	.092		4.33	2.67	.08	7.08	9.25	
	2700	Acidproof, minimum		605	.079		4.33	2.29	.07	6.69	8.60	
	2800	Maximum		350	.137		6	3.96	.12	10.08	13.25	
	3000	Neoprene, trowelled on, 1/4" thick, minimum		545	.088		3.32	2.55	.08	5.95	7.90	
	3100	Maximum		430	.112		4.57	3.23	.10	7.90	10.45	
	3150	Polyacrylate terrazzo, 1/4" thick, minimum		735	.065		3.04	1.89	.06	4.99	6.50	
	3170	Maximum		480	.100		3.94	2.89	.09	6.92	9.15	
	3200	3/8" thick, minimum		620	.077		4	2.24	.07	6.31	8.15	
	3220	Maximum		480	.100		5.90	2.89	.09	8.88	11.30	
	3300	Conductive terrazzo, 1/4" thick, minimum		450	.107		7.10	3.08	.09	10.27	13	
	3330	Maximum		305	.157		9.10	4.55	.14	13.79	17.65	
	3350	3/8" thick, minimum		365	.132		9.50	3.80	.12	13.42	16.85	
	3370	Maximum		255	.188		11.75	5.45	.17	17.37	22	
	3450	Granite, conductive, 1/4" thick, minimum		695	.069		7.40	2	.06	9.46	11.50	
	3470	Maximum		420	.114		9.60	3.30	.10	13	16.05	
	3500	3/8" thick, minimum		695	.069		10.85	2	.06	12.91	15.25	
	3520	Maximum		380	.126		13.10	3.65	.11	16.86	20.50	
	3600	Polyester, with colored quartz chips, 1/16" thick, minimum		1065	.045		2.95	1.30	.04	4.29	5.45	
	3700	Maximum		560	.086		4.09	2.48	.08	6.65	8.65	
	3900	1/8" thick, minimum		810	.059		3.24	1.71	.05	5	6.45	
	4000	Maximum		675	.071		4.57	2.06	.06	6.69	8.50	
	4200	Polyester, heavy duty, compared to epoxy, add		2590	.019		1.21	.54	.02	1.77	2.23	
	4300	Polyurethane, with suspended vinyl chips, minimum		1065	.045		6.45	1.30	.04	7.79	9.30	
	4500	Maximum		860	.056		9.30	1.61	.05	10.96	12.95	
	09680 | Carpet											
600	0010	**CARPET PAD**, commercial grade										600
	9000	Sponge rubber pad, minimum	1 Tilf	150	.053	S.Y.	3.58	1.83		5.41	6.80	
	9100	Maximum		150	.053		8.25	1.83		10.08	11.95	
	9200	Felt pad, minimum		150	.053		3.57	1.83		5.40	6.80	
	9300	Maximum		150	.053		6.70	1.83		8.53	10.25	
	9400	Bonded urethane pad, minimum		150	.053		3.95	1.83		5.78	7.20	
	9500	Maximum		150	.053		6.75	1.83		8.58	10.30	
	9600	Prime urethane pad, minimum		150	.053		2.27	1.83		4.10	5.35	
	9700	Maximum		150	.053		4.19	1.83		6.02	7.45	
800	0010	**CARPET** Commercial grades, direct cement										800
	0700	Nylon, level loop, 26 oz., light to medium traffic	1 Tilf	75	.107	S.Y.	17.60	3.65		21.25	25	
	0720	28 oz., light to medium traffic		75	.107		16.55	3.65		20.20	24	
	0900	32 oz., medium traffic		75	.107		25	3.65		28.65	33.50	
	1100	40 oz., medium to heavy traffic		75	.107		37	3.65		40.65	46.50	
	2920	Nylon plush, 30 oz., medium traffic		57	.140		18.40	4.81		23.21	27.50	
	3000	36 oz., medium traffic		75	.107		24	3.65		27.65	32.50	
	3100	42 oz., medium to heavy traffic		70	.114		22	3.91		25.91	30.50	

09600 | Flooring

	09680 | Carpet		Crew	Daily Output	Labor-Hours	Unit	2006 Bare Costs Mat.	Labor	Equip.	Total	Total Incl O&P	
800	3200	46 oz., medium to heavy traffic	1 Tilf	70	.114	S.Y.	33	3.91		36.91	42.50	800
	3300	54 oz., heavy traffic		70	.114		37.50	3.91		41.41	47	
	3340	60 oz., heavy traffic		70	.114		53	3.91		56.91	64	
	3665	Olefin, 24 oz., light to medium traffic		75	.107		11.15	3.65		14.80	18	
	3670	26 oz., medium traffic		75	.107		10.90	3.65		14.55	17.75	
	3680	28 oz., medium to heavy traffic		75	.107		13.50	3.65		17.15	20.50	
	3700	32 oz., medium to heavy traffic		75	.107		16.60	3.65		20.25	24	
	3730	42 oz., heavy traffic		70	.114		24.50	3.91		28.41	33	
	4110	Wool, level loop, 40 oz., medium traffic		75	.107		83	3.65		86.65	97	
	4500	50 oz., medium to heavy traffic		75	.107		84.50	3.65		88.15	99	
	4700	Patterned, 32 oz., medium to heavy traffic		70	.114		83.50	3.91		87.41	98	
	4900	48 oz., heavy traffic		70	.114		85	3.91		88.91	99.50	
	5000	For less than full roll, add					25%					
	5100	For small rooms, less than 12′ wide, add						25%				
	5200	For large open areas (no cuts), deduct						25%				
	5600	For bound carpet baseboard, add	1 Tilf	300	.027	L.F.	1.35	.91		2.26	2.92	
	5610	For stairs, not incl. price of carpet, add	″	30	.267	Riser		9.15		9.15	14.30	
	5620	For borders and patterns, add to labor						18%				
	8950	For tackless, stretched installation, add padding to above										
	9850	For "branded" fiber, add				S.Y.	25%					
	9910	Minimum labor/equipment charge	1 Tilf	3	2.667	Job		91.50		91.50	143	
900	0010	**CARPET TILE**										900
	0100	Tufted nylon, 18″ x 18″, hard back, 20 oz.	1 Tilf	150	.053	S.Y.	20.50	1.83		22.33	25.50	
	0110	26 oz.		150	.053		35.50	1.83		37.33	42	
	0200	Cushion back, 20 oz.		150	.053		26	1.83		27.83	31.50	
	0210	26 oz.		150	.053		40.50	1.83		42.33	47.50	
910	0011	**CARPET MAINTENANCE** See Division 01832-290										910

09700 | Wall Finishes

	09720 | Wall Coverings		Crew	Daily Output	Labor-Hours	Unit	2006 Bare Costs Mat.	Labor	Equip.	Total	Total Incl O&P	
100	0010	**WALL COVERING** Including sizing, add 10-30% waste @ takeoff										100
	0050	Aluminum foil	1 Pape	275	.029	S.F.	.87	.93		1.80	2.45	
	0100	Copper sheets, .025″ thick, vinyl backing		240	.033		4.66	1.06		5.72	6.85	
	0300	Phenolic backing		240	.033		6.05	1.06		7.11	8.35	
	0600	Cork tiles, light or dark, 12″ x 12″ x 3/16″		240	.033		3.74	1.06		4.80	5.80	
	0700	5/16″ thick		235	.034		3.19	1.08		4.27	5.25	
	0900	1/4″ basketweave		240	.033		4.90	1.06		5.96	7.10	
	1000	1/2″ natural, non-directional pattern		240	.033		6.25	1.06		7.31	8.60	
	1100	3/4″ natural, non-directional pattern		240	.033		10.25	1.06		11.31	13	
	1200	Granular surface, 12″ x 36″, 1/2″ thick		385	.021		1.06	.66		1.72	2.23	
	1300	1″ thick		370	.022		1.37	.69		2.06	2.62	
	1500	Polyurethane coated, 12″ x 12″ x 3/16″ thick		240	.033		3.31	1.06		4.37	5.35	
	1600	5/16″ thick		235	.034		4.71	1.08		5.79	6.95	
	1800	Cork wallpaper, paperbacked, natural		480	.017		1.88	.53		2.41	2.92	
	1900	Colors		480	.017		2.33	.53		2.86	3.41	
	2100	Flexible wood veneer, 1/32″ thick, plain woods		100	.080		1.98	2.55		4.53	6.25	

9 FINISHES

		09720 | Wall Coverings	CREW	DAILY OUTPUT	LABOR-HOURS	UNIT	2006 BARE COSTS MAT.	LABOR	EQUIP.	TOTAL	TOTAL INCL O&P	
100	2200	Exotic woods	1 Pape	95	.084	S.F.	3.01	2.68		5.69	7.60	100
	2400	Gypsum-based, fabric-backed, fire										
	2500	resistant for masonry walls, minimum, 21 oz./S.Y.	1 Pape	800	.010	S.F.	.68	.32		1	1.26	
	2600	Average		720	.011		1.03	.35		1.38	1.70	
	2700	Maximum, (small quantities)	↓	640	.013		1.14	.40		1.54	1.89	
	2750	Acrylic, modified, semi-rigid PVC, .028″ thick	2 Carp	330	.048		.95	1.72		2.67	3.91	
	2800	.040″ thick	″	320	.050		1.25	1.78		3.03	4.33	
	3000	Vinyl wall covering, fabric-backed, lightweight, (12-15 oz./S.Y.)	1 Pape	640	.013		.59	.40		.99	1.29	
	3300	Medium weight, type 2, (20-24 oz./S.Y.)		480	.017		.73	.53		1.26	1.65	
	3400	Heavy weight, type 3, (28 oz./S.Y.)	↓	435	.018	↓	1.17	.59		1.76	2.23	
	3600	Adhesive, 5 gal. lots, (18SY/Gal.)				Gal.	8.95			8.95	9.80	
	3700	Wallpaper, average workmanship, solid pattern, low cost paper	1 Pape	640	.013	S.F.	.30	.40		.70	.97	
	3900	basic patterns (matching required), avg. cost paper		535	.015		.65	.48		1.13	1.48	
	4000	Paper at $85 per double roll, quality workmanship	↓	435	.018	↓	1.53	.59		2.12	2.62	
	4100	Linen wall covering, paper backed										
	4150	Flame treatment, minimum				S.F.	.70			.70	.77	
	4200	Grass cloths with lining paper, minimum	1 Pape	400	.020		.65	.64		1.29	1.74	
	4300	Maximum		350	.023		2.09	.73		2.82	3.47	
	5990	Wallpaper removal, 1 layer, minimum		800	.010		.06	.32		.38	.58	
	6000	Wallpaper removal, 3 layer, maximum		400	.020	↓	.11	.64		.75	1.14	
	9000	Minimum labor/equipment charge	↓	2	4	Job		127		127	205	

		09770 | Special Wall Surfaces	CREW	DAILY OUTPUT	LABOR-HOURS	UNIT	MAT.	LABOR	EQUIP.	TOTAL	TOTAL INCL O&P	
400	0010	**FIBERGLASS REINFORCED PLASTIC** panels, .090″ thick, on walls										400
	0020	Adhesive mounted, embossed surface	2 Carp	640	.025	S.F.	1.52	.89		2.41	3.14	
	0030	Smooth surface		640	.025		1.78	.89		2.67	3.43	
	0040	Fire rated, embossed surface		640	.025		1.98	.89		2.87	3.65	
	0050	Nylon rivet mounted, on drywall, embossed surface		480	.033		1.41	1.18		2.59	3.51	
	0060	Smooth surface		480	.033		1.66	1.18		2.84	3.79	
	0070	Fire rated, embossed surface		480	.033		1.85	1.18		3.03	4	
	0080	On masonry, embossed surface		320	.050		1.41	1.78		3.19	4.50	
	0090	Smooth surface		320	.050		1.78	1.78		3.56	4.91	
	0100	Fire rated, embossed surface		320	.050		1.88	1.78		3.66	5	
	0110	Nylon rivet and adhesive mounted, on drywall, embossed surface		240	.067		1.66	2.37		4.03	5.75	
	0120	Smooth surface		240	.067		1.88	2.37		4.25	6	
	0130	Fire rated, embossed surface		240	.067		2.09	2.37		4.46	6.25	
	0140	On masonry, embossed surface		190	.084		1.66	2.99		4.65	6.80	
	0150	Smooth surface		190	.084		1.88	2.99		4.87	7.05	
	0160	Fire rated, embossed surface	↓	190	.084	↓	2.09	2.99		5.08	7.25	
	0170	For moldings add	1 Carp	250	.032	L.F.	.28	1.14		1.42	2.19	
	0180	On ceilings, for lay in grid system, embossed surface		400	.020	S.F.	1.52	.71		2.23	2.85	
	0190	Smooth surface		400	.020		1.78	.71		2.49	3.14	
	0200	Fire rated, embossed surface	↓	400	.020	↓	1.98	.71		2.69	3.36	
700	0010	**PANEL SYSTEM**										700
	0100	Raised panel, eng. wood core w/ wood veneer, std., paint grade	2 Carp	300	.053	S.F.	10.20	1.90		12.10	14.35	
	0110	Oak veneer		300	.053		16.90	1.90		18.80	21.50	
	0120	Maple veneer		300	.053		21.50	1.90		23.40	27	
	0130	Cherry veneer		300	.053		27.50	1.90		29.40	33	
	0300	Class I fire rated, paint grade		300	.053		12.20	1.90		14.10	16.55	
	0310	Oak veneer		300	.053		20.50	1.90		22.40	25.50	
	0320	Maple veneer		300	.053		26	1.90		27.90	31.50	
	0330	Cherry veneer		300	.053		33	1.90		34.90	39	
	0510	Beadboard, 5/8″ MDF, standard, primed		300	.053		7.20	1.90		9.10	11.05	
	0520	Oak veneer, unfinished		300	.053		12.50	1.90		14.40	16.90	
	0530	Maple veneer, unfinished	↓	300	.053	↓	14.80	1.90		16.70	19.45	

09700 | Wall Finishes

	09770 | Special Wall Surfaces		CREW	DAILY OUTPUT	LABOR-HOURS	UNIT	2006 BARE COSTS MAT.	LABOR	EQUIP.	TOTAL	TOTAL INCL O&P	
700	0610	Rustic paneling, 5/8" MDF, standard, maple veneer, unfinished	2 Carp	300	.053	S.F.	19	1.90		20.90	24	700
	5000	For prefinished paneling, see division 06250-500 & 06250-200										
750	0010	**SLATWALL PANELS AND ACCESSORIES**										750
	0100	Slatwall panel, 4' x 8' x 3/4" T, MDF, paint grade	1 Carp	500	.016	S.F.	1.41	.57		1.98	2.49	
	0110	Melamine finish		500	.016		2.14	.57		2.71	3.29	
	0120	High pressure plastic laminate finish		500	.016		3.42	.57		3.99	4.70	
	0130	Aluminum channel inserts, add					2.68			2.68	2.95	
	0200	Accessories, corner forms, 8' L				L.F.	3.82			3.82	4.20	
	0210	T-connector, 8' L					5.30			5.30	5.80	
	0220	J-mold, 8' L					1.35			1.35	1.49	
	0230	Edge cap, 8' L					.88			.88	.97	
	0240	Finish end cap, 8' L					3.23			3.23	3.55	
	0300	Display hook, 4" L				Ea.	1			1	1.10	
	0310	6" L					1.11			1.11	1.22	
	0320	8" L					1.18			1.18	1.30	
	0330	10" L					1.30			1.30	1.43	
	0340	12" L					1.41			1.41	1.55	
	0350	Acrylic, 4" L					.82			.82	.90	
	0360	6" L					.95			.95	1.05	
	0370	8" L					1			1	1.10	
	0380	10" L					1.12			1.12	1.23	
	0400	Waterfall hanger, metal, 12" - 16"					6.65			6.65	7.30	
	0410	Acrylic					9.50			9.50	10.45	
	0500	Shelf bracket, metal, 8"					5.15			5.15	5.70	
	0510	10"					5.55			5.55	6.10	
	0520	12"					5.90			5.90	6.50	
	0530	14"					6.65			6.65	7.30	
	0540	16"					7.40			7.40	8.15	
	0550	Acrylic, 8"					3.08			3.08	3.39	
	0560	10"					3.45			3.45	3.80	
	0570	12"					3.82			3.82	4.20	
	0580	14"					4.31			4.31	4.74	
	0600	Shelf, acrylic, 12" x 16" x 1/4"					26			26	28.50	
	0610	12" x 24" x 1/4"					43			43	47.50	

09800 | Acoustical Treatment

	09820 | Acoustical Insul/Sealants		CREW	DAILY OUTPUT	LABOR-HOURS	UNIT	2006 BARE COSTS MAT.	LABOR	EQUIP.	TOTAL	TOTAL INCL O&P	
500	0010	SOUND ATTENUATION										500
	0020	Blanket, 1" thick	1 Carp	925	.009	S.F.	.25	.31		.56	.79	
	0500	1-1/2" thick		920	.009		.25	.31		.56	.79	
	1000	2" thick		915	.009		.31	.31		.62	.85	
	1500	3" thick		910	.009		.42	.31		.73	.98	
	3000	Thermal or acoustical batt above ceiling, 2" thick		900	.009		.46	.32		.78	1.03	
	3100	3" thick		900	.009		.69	.32		1.01	1.28	
	3200	4" thick		900	.009		.85	.32		1.17	1.46	
	3400	Urethane plastic foam, open cell, on wall, 2" thick	2 Carp	2050	.008		2.82	.28		3.10	3.56	
	3500	3" thick		1550	.010		3.74	.37		4.11	4.72	
	3600	4" thick		1050	.015		5.25	.54		5.79	6.70	
	3700	On ceiling, 2" thick		1700	.009		2.81	.33		3.14	3.64	

09800 | Acoustical Treatment

		09820 | Acoustical Insul/Sealants	CREW	DAILY OUTPUT	LABOR-HOURS	UNIT	2006 BARE COSTS MAT.	LABOR	EQUIP.	TOTAL	TOTAL INCL O&P	
500	3800	3″ thick	2 Carp	1300	.012	S.F.	3.74	.44		4.18	4.84	500
	3900	4″ thick		900	.018		5.25	.63		5.88	6.85	
	4000	Nylon matting 0.4″ thick, with carbon black spinerette										
	4010	plus polyester fabric, on floor	J-4	4000	.004	S.F.	2.09	.12		2.21	2.49	
	4200	Fiberglass reinf. backer board underlayment, 7/16″ thick, on floor	″	800	.020	″	1.75	.61		2.36	2.88	
	9000	Minimum labor/equipment charge	1 Carp	5	1.600	Job		57		57	94	

		09830 | Acoustical Barriers	CREW	DAILY OUTPUT	LABOR-HOURS	UNIT	MAT.	LABOR	EQUIP.	TOTAL	TOTAL INCL O&P	
100	0011	**BARRIERS** Plenum										100
	0600	Aluminum foil, fiberglass reinf., parallel with joists	1 Carp	275	.029	S.F.	.72	1.03		1.75	2.50	
	0700	Perpendicular to joists		180	.044		.89	1.58		2.47	3.60	
	0900	Aluminum mesh, kraft paperbacked		275	.029		.71	1.03		1.74	2.49	
	0970	Fiberglass batts, kraft faced, 3-1/2″ thick		1400	.006		.26	.20		.46	.63	
	0980	6″ thick		1300	.006		.43	.22		.65	.83	
	1000	Sheet lead, 1 lb., 1/64″ thick, perpendicular to joists		150	.053		2.75	1.90		4.65	6.15	
	1100	Vinyl foam reinforced, 1/8″ thick, 1.0 lb. per S.F.		150	.053		3.98	1.90		5.88	7.50	

		09840 | Acoustical Wall Treatment	CREW	DAILY OUTPUT	LABOR-HOURS	UNIT	MAT.	LABOR	EQUIP.	TOTAL	TOTAL INCL O&P	
100	0010	**SOUND ABSORBING PANELS** Perforated steel facing, painted with										100
	0100	fiberglass or mineral filler, no backs, 2-1/4″ thick, modular										
	0200	space units, ceiling or wall hung, white or colored	1 Carp	100	.080	S.F.	9.15	2.84		11.99	14.80	
	0300	Fiberboard sound deadening panels, 1/2″ thick	″	600	.013	″	.30	.47		.77	1.12	
	0500	Fiberglass panels, 4′ x 8′ x 1″ thick, with										
	0600	glass cloth face for walls, cemented	1 Carp	155	.052	S.F.	6.20	1.83		8.03	9.85	
	0700	1-1/2″ thick, dacron covered, inner aluminum frame,										
	0710	wall mounted	1 Carp	300	.027	S.F.	7.60	.95		8.55	9.90	
	0900	Mineral fiberboard panels, fabric covered, 30″x 108″,										
	1000	3/4″ thick, concealed spline, wall mounted	1 Carp	150	.053	S.F.	5.45	1.90		7.35	9.15	
	9000	Minimum labor/equipment charge	″	4	2	Job		71		71	118	

FINISHES 9

09900 | Paints & Coatings

		09910 | Paints	CREW	DAILY OUTPUT	LABOR-HOURS	UNIT	2006 BARE COSTS MAT.	LABOR	EQUIP.	TOTAL	TOTAL INCL O&P	
100	0010	**CABINETS AND CASEWORK**										100
	1000	Primer coat, oil base, brushwork	1 Pord	650	.012	S.F.	.05	.39		.44	.68	
	2000	Paint, oil base, brushwork, 1 coat		650	.012		.06	.39		.45	.70	
	2500	2 coats		400	.020		.12	.63		.75	1.15	
	3000	Stain, brushwork, wipe off		650	.012		.05	.39		.44	.69	
	4000	Shellac, 1 coat, brushwork		650	.012		.06	.39		.45	.69	
	4500	Varnish, 3 coats, brushwork, sand after 1st coat		325	.025		.17	.78		.95	1.44	
	5000	For latex paint, deduct					10%					
300	0010	**DOORS AND WINDOWS, EXTERIOR** R099100-20										300
	0100	Door frames & trim, only										
	0110	Brushwork, primer	1 Pord	512	.016	L.F.	.05	.50		.55	.86	
	0120	Finish coat, exterior latex		512	.016		.06	.50		.56	.86	
	0130	Primer & 1 coat, exterior latex		300	.027		.11	.85		.96	1.48	
	0135	2 coats, exterior latex, both sides		15	.533	Ea.	5.20	16.90		22.10	32.50	
	0140	Primer & 2 coats, exterior latex		265	.030	L.F.	.17	.96		1.13	1.73	
	0150	Doors, flush, both sides, incl. frame & trim										

	09910 | Paints	CREW	DAILY OUTPUT	LABOR-HOURS	UNIT	2006 BARE COSTS MAT.	LABOR	EQUIP.	TOTAL	TOTAL INCL O&P	
300	0160 Roll & brush, primer (R099100-20)	1 Pord	10	.800	Ea.	4.01	25.50		29.51	45	300
	0170 Finish coat, exterior latex		10	.800		4.42	25.50		29.92	45.50	
	0180 Primer & 1 coat, exterior latex		7	1.143		8.45	36		44.45	67.50	
	0190 Primer & 2 coats, exterior latex		5	1.600		12.85	50.50		63.35	95.50	
	0200 Brushwork, stain, sealer & 2 coats polyurethane	↓	4	2	↓	17	63.50		80.50	121	
	0210 Doors, French, both sides, 10-15 lite, incl. frame & trim										
	0220 Brushwork, primer	1 Pord	6	1.333	Ea.	2	42.50		44.50	70	
	0230 Finish coat, exterior latex		6	1.333		2.21	42.50		44.71	70.50	
	0240 Primer & 1 coat, exterior latex		3	2.667		4.21	84.50		88.71	141	
	0250 Primer & 2 coats, exterior latex		2	4		6.30	127		133.30	211	
	0260 Brushwork, stain, sealer & 2 coats polyurethane	↓	2.50	3.200	↓	6.10	101		107.10	170	
	0270 Doors, louvered, both sides, incl. frame & trim										
	0280 Brushwork, primer	1 Pord	7	1.143	Ea.	4.01	36		40.01	62.50	
	0290 Finish coat, exterior latex		7	1.143		4.42	36		40.42	63	
	0300 Primer & 1 coat, exterior latex		4	2		8.45	63.50		71.95	111	
	0310 Primer & 2 coats, exterior latex		3	2.667		12.60	84.50		97.10	150	
	0320 Brushwork, stain, sealer & 2 coats polyurethane	↓	4.50	1.778	↓	17	56.50		73.50	109	
	0330 Doors, panel, both sides, incl. frame & trim										
	0340 Roll & brush, primer	1 Pord	6	1.333	Ea.	4.01	42.50		46.51	72.50	
	0350 Finish coat, exterior latex		6	1.333		4.42	42.50		46.92	73	
	0360 Primer & 1 coat, exterior latex		3	2.667		8.45	84.50		92.95	145	
	0370 Primer & 2 coats, exterior latex		2.50	3.200		12.60	101		113.60	177	
	0380 Brushwork, stain, sealer & 2 coats polyurethane	↓	3	2.667	↓	17	84.50		101.50	155	
	0400 Windows, per ext. side, based on 15 SF										
	0410 1 to 6 lite										
	0420 Brushwork, primer	1 Pord	13	.615	Ea.	.79	19.50		20.29	32.50	
	0430 Finish coat, exterior latex		13	.615		.87	19.50		20.37	32.50	
	0440 Primer & 1 coat, exterior latex		8	1		1.66	31.50		33.16	53	
	0450 Primer & 2 coats, exterior latex		6	1.333		2.49	42.50		44.99	70.50	
	0460 Stain, sealer & 1 coat varnish	↓	7	1.143	↓	2.42	36		38.42	60.50	
	0470 7 to 10 lite										
	0480 Brushwork, primer	1 Pord	11	.727	Ea.	.79	23		23.79	38	
	0490 Finish coat, exterior latex		11	.727		.87	23		23.87	38	
	0500 Primer & 1 coat, exterior latex		7	1.143		1.66	36		37.66	60	
	0510 Primer & 2 coats, exterior latex		5	1.600		2.49	50.50		52.99	84	
	0520 Stain, sealer & 1 coat varnish	↓	6	1.333	↓	2.42	42.50		44.92	70.50	
	0530 12 lite										
	0540 Brushwork, primer	1 Pord	10	.800	Ea.	.79	25.50		26.29	41.50	
	0550 Finish coat, exterior latex		10	.800		.87	25.50		26.37	41.50	
	0560 Primer & 1 coat, exterior latex		6	1.333		1.66	42.50		44.16	70	
	0570 Primer & 2 coats, exterior latex		5	1.600		2.49	50.50		52.99	84	
	0580 Stain, sealer & 1 coat varnish	↓	6	1.333		2.40	42.50		44.90	70.50	
	0590 For oil base paint, add				↓	10%					
310	0010 **DOORS & WINDOWS, INTERIOR LATEX** (R099100-20)										310
	0100 Doors flush, both sides, incl. frame & trim										
	0110 Roll & brush, primer	1 Pord	10	.800	Ea.	3.63	25.50		29.13	44.50	
	0120 Finish coat, latex		10	.800		3.66	25.50		29.16	44.50	
	0130 Primer & 1 coat latex		7	1.143		7.30	36		43.30	66	
	0140 Primer & 2 coats latex		5	1.600		10.75	50.50		61.25	93.50	
	0160 Spray, both sides, primer		20	.400		3.82	12.70		16.52	24.50	
	0170 Finish coat, latex		20	.400		3.84	12.70		16.54	24.50	
	0180 Primer & 1 coat latex		11	.727		7.70	23		30.70	45.50	
	0190 Primer & 2 coats latex	↓	8	1	↓	11.35	31.50		42.85	63.50	
	0200 Doors, French, both sides, 10-15 lite, incl. frame & trim										
	0210 Roll & brush, primer	1 Pord	6	1.333	Ea.	1.81	42.50		44.31	70	

9 FINISHES

09900 | Paints & Coatings

		09910 | Paints	CREW	DAILY OUTPUT	LABOR-HOURS	UNIT	2006 BARE COSTS MAT.	LABOR	EQUIP.	TOTAL	TOTAL INCL O&P	
310	0220	Finish coat, latex (R099100-20)	1 Pord	6	1.333	Ea.	1.83	42.50		44.33	70	310
	0230	Primer & 1 coat latex		3	2.667		3.64	84.50		88.14	140	
	0240	Primer & 2 coats latex		2	4		5.35	127		132.35	210	
	0260	Doors, louvered, both sides, incl. frame & trim										
	0270	Roll & brush, primer	1 Pord	7	1.143	Ea.	3.63	36		39.63	62	
	0280	Finish coat, latex		7	1.143		3.66	36		39.66	62	
	0290	Primer & 1 coat, latex		4	2		7.05	63.50		70.55	110	
	0300	Primer & 2 coats, latex		3	2.667		10.95	84.50		95.45	148	
	0320	Spray, both sides, primer		20	.400		3.82	12.70		16.52	24.50	
	0330	Finish coat, latex		20	.400		3.84	12.70		16.54	24.50	
	0340	Primer & 1 coat, latex		11	.727		7.70	23		30.70	45.50	
	0350	Primer & 2 coats, latex		8	1		11.60	31.50		43.10	64	
	0360	Doors, panel, both sides, incl. frame & trim										
	0370	Roll & brush, primer	1 Pord	6	1.333	Ea.	3.82	42.50		46.32	72	
	0380	Finish coat, latex		6	1.333		3.66	42.50		46.16	72	
	0390	Primer & 1 coat, latex		3	2.667		7.30	84.50		91.80	144	
	0400	Primer & 2 coats, latex		2.50	3.200		10.95	101		111.95	175	
	0420	Spray, both sides, primer		10	.800		3.82	25.50		29.32	44.50	
	0430	Finish coat, latex		10	.800		3.84	25.50		29.34	44.50	
	0440	Primer & 1 coat, latex		5	1.600		7.70	50.50		58.20	90	
	0450	Primer & 2 coats, latex		4	2		11.60	63.50		75.10	115	
	0460	Windows, per interior side, based on 15 SF										
	0470	1 to 6 lite										
	0480	Brushwork, primer	1 Pord	13	.615	Ea.	.72	19.50		20.22	32.50	
	0490	Finish coat, enamel		13	.615		.72	19.50		20.22	32.50	
	0500	Primer & 1 coat enamel		8	1		1.44	31.50		32.94	52.50	
	0510	Primer & 2 coats enamel		6	1.333		2.16	42.50		44.66	70.50	
	0530	7 to 10 lite										
	0540	Brushwork, primer	1 Pord	11	.727	Ea.	.72	23		23.72	38	
	0550	Finish coat, enamel		11	.727		.72	23		23.72	38	
	0560	Primer & 1 coat enamel		7	1.143		1.44	36		37.44	59.50	
	0570	Primer & 2 coats enamel		5	1.600		2.16	50.50		52.66	84	
	0590	12 lite										
	0600	Brushwork, primer	1 Pord	10	.800	Ea.	.72	25.50		26.22	41.50	
	0610	Finish coat, enamel		10	.800		.72	25.50		26.22	41.50	
	0620	Primer & 1 coat enamel		6	1.333		1.44	42.50		43.94	69.50	
	0630	Primer & 2 coats enamel		5	1.600		2.16	50.50		52.66	84	
	0650	For oil base paint, add					10%					
320	0010	**DOORS AND WINDOWS, INTERIOR ALKYD (OIL BASE)**										320
	0500	Flush door & frame, 3′ x 7′, oil, primer, brushwork	1 Pord	10	.800	Ea.	2.20	25.50		27.70	43	
	1000	Paint, 1 coat		10	.800		2.04	25.50		27.54	42.50	
	1200	2 coats		6	1.333		3.43	42.50		45.93	72	
	1220	3 coats		5	1.600		7.95	50.50		58.45	90.50	
	1400	Stain, brushwork, wipe off		18	.444		1.06	14.10		15.16	23.50	
	1600	Shellac, 1 coat, brushwork		25	.320		1.18	10.15		11.33	17.60	
	1800	Varnish, 3 coats, brushwork, sand after 1st coat		9	.889		3.61	28		31.61	49	
	2000	Panel door & frame, 3′ x 7′, oil, primer, brushwork		6	1.333		1.82	42.50		44.32	70	
	2200	Paint, 1 coat		6	1.333		2.04	42.50		44.54	70	
	2400	2 coats		3	2.667		5.90	84.50		90.40	143	
	2420	3 coats		2	4		7.70	127		134.70	212	
	2600	Stain, brushwork, panel door, 3′ x 7′, not incl. frame		16	.500		1.06	15.85		16.91	26.50	
	2800	Shellac, 1 coat, brushwork		22	.364		1.18	11.55		12.73	19.80	
	3000	Varnish, 3 coats, brushwork, sand after 1st coat		7.50	1.067		3.61	34		37.61	58.50	
	4400	Windows, including frame and trim, per side										
	4600	Colonial type, 6/6 lites, 2′ x 3′, oil, primer, brushwork	1 Pord	14	.571	Ea.	.29	18.10		18.39	29.50	
	5800	Paint, 1 coat		14	.571		.32	18.10		18.42	29.50	

	09910 | Paints	CREW	DAILY OUTPUT	LABOR-HOURS	UNIT	2006 BARE COSTS MAT.	LABOR	EQUIP.	TOTAL	TOTAL INCL O&P	
320	6000 2 coats	1 Pord	9	.889	Ea.	.63	28		28.63	45.50	320
	6010 3 coats		7	1.143		.93	36		36.93	59	
	6200 3′ x 5′ opening, 6/6 lites, primer coat, brushwork		12	.667		.72	21		21.72	35	
	6400 Paint, 1 coat		12	.667		.80	21		21.80	35	
	6600 2 coats		7	1.143		1.56	36		37.56	59.50	
	6610 3 coats		6	1.333		2.32	42.50		44.82	70.50	
	6800 4′ x 8′ opening, 6/6 lites, primer coat, brushwork		8	1		1.54	31.50		33.04	52.50	
	7000 Paint, 1 coat		8	1		1.72	31.50		33.22	53	
	7200 2 coats		5	1.600		3.34	50.50		53.84	85	
	7210 3 coats		4	2		4.95	63.50		68.45	107	
	7500 Standard, 6/6 lites, 2′ x 3′, primer		14	.571		.29	18.10		18.39	29.50	
	7520 Paint 1 coat		14	.571		.32	18.10		18.42	29.50	
	7540 2 coats		9	.889		.63	28		28.63	45.50	
	7560 3 coats		7	1.143		.93	36		36.93	59	
	7580 3′ x 5′, 6/6 lites, primer		12	.667		.72	21		21.72	35	
	7600 Paint 1 coat		12	.667		.80	21		21.80	35	
	7620 2 coats		7	1.143		1.56	36		37.56	59.50	
	7640 3 coats		6	1.333		2.32	42.50		44.82	70.50	
	7660 4′ x 8′, primer		8	1		1.54	31.50		33.04	52.50	
	7680 Paint 1 coat		8	1		1.72	31.50		33.22	53	
	7700 2 coats		5	1.600		3.34	50.50		53.84	85	
	7720 3 coats		4	2		4.95	63.50		68.45	107	
	8000 Single lite type, 2′ x 3′, oil base, primer coat, brushwork		33	.242		.29	7.70		7.99	12.65	
	8200 Paint, 1 coat		33	.242		.32	7.70		8.02	12.70	
	8400 2 coats		20	.400		.63	12.70		13.33	21	
	8410 3 coats		16	.500		.93	15.85		16.78	26.50	
	8600 3′ x 5′ opening, primer coat, brushwork		20	.400		.72	12.70		13.42	21.50	
	8800 Paint, 1 coat		20	.400		.80	12.70		13.50	21.50	
	9000 2 coats		13	.615		1.56	19.50		21.06	33	
	9010 3 coats		10	.800		2.32	25.50		27.82	43	
	9200 4′ x 8′ opening, primer coat, brushwork		14	.571		1.54	18.10		19.64	30.50	
	9400 Paint, 1 coat		14	.571		1.72	18.10		19.82	31	
	9600 2 coats		8	1		3.34	31.50		34.84	54.50	
	9610 3 coats		7	1.143		4.95	36		40.95	63.50	
400	0010 FENCES (R099100-20)										400
	0100 Chain link or wire metal, one side, water base										
	0110 Roll & brush, first coat	1 Pord	960	.008	S.F.	.06	.26		.32	.48	
	0120 Second coat		1280	.006		.05	.20		.25	.38	
	0130 Spray, first coat		2275	.004		.06	.11		.17	.24	
	0140 Second coat		2600	.003		.06	.10		.16	.22	
	0150 Picket, water base										
	0160 Roll & brush, first coat	1 Pord	865	.009	S.F.	.06	.29		.35	.54	
	0170 Second coat		1050	.008		.06	.24		.30	.46	
	0180 Spray, first coat		2275	.004		.06	.11		.17	.25	
	0190 Second coat		2600	.003		.06	.10		.16	.23	
	0200 Stockade, water base										
	0210 Roll & brush, first coat	1 Pord	1040	.008	S.F.	.06	.24		.30	.46	
	0220 Second coat		1200	.007		.06	.21		.27	.41	
	0230 Spray, first coat		2275	.004		.06	.11		.17	.25	
	0240 Second coat		2600	.003		.06	.10		.16	.23	
500	0010 FLOORS, INTERIOR										500
	0100 Concrete										
	0110 Brushwork, latex, block filler										
	0120 1st coat	1 Pord	975	.008	S.F.	.12	.26		.38	.56	
	0130 2nd coat		1150	.007		.08	.22		.30	.44	
	0140 3rd coat		1300	.006		.07	.20		.27	.38	

	09910 | Paints		Crew	Daily Output	Labor-Hours	Unit	2006 Bare Costs Mat.	Labor	Equip.	Total	Total Incl O&P
500	0150	Roll, latex, block filler									
	0160	1st coat	1 Pord	2600	.003	S.F.	.17	.10		.27	.34
	0170	2nd coat		3250	.002		.10	.08		.18	.24
	0180	3rd coat		3900	.002		.07	.07		.14	.18
	0190	Spray, latex, block filler									
	0200	1st coat	1 Pord	2600	.003	S.F.	.14	.10		.24	.32
	0210	2nd coat		3250	.002		.08	.08		.16	.22
	0220	3rd coat		3900	.002		.06	.07		.13	.17
620	0010	**MISCELLANEOUS, EXTERIOR** R099100-20									
	0015	For painting metals, see Div. 05950-650									
	0100	Railing, ext., decorative wood, incl. cap & baluster									
	0110	newels & spindles @ 12″ O.C.									
	0120	Brushwork, stain, sand, seal & varnish									
	0130	First coat	1 Pord	90	.089	L.F.	.48	2.82		3.30	5.05
	0140	Second coat	″	120	.067	″	.48	2.11		2.59	3.92
	0150	Rough sawn wood, 42″ high, 2″ x 2″ verticals, 6″ O.C.									
	0160	Brushwork, stain, each coat	1 Pord	90	.089	L.F.	.15	2.82		2.97	4.69
	0170	Wrought iron, 1″ rail, 1/2″ sq. verticals									
	0180	Brushwork, zinc chromate, 60″ high, bars 6″ O.C.									
	0190	Primer	1 Pord	130	.062	L.F.	.51	1.95		2.46	3.69
	0200	Finish coat		130	.062		.16	1.95		2.11	3.30
	0210	Additional coat		190	.042		.19	1.33		1.52	2.34
	0220	Shutters or blinds, single panel, 2′ x 4′, paint all sides									
	0230	Brushwork, primer	1 Pord	20	.400	Ea.	.58	12.70		13.28	21
	0240	Finish coat, exterior latex		20	.400		.47	12.70		13.17	21
	0250	Primer & 1 coat, exterior latex		13	.615		.92	19.50		20.42	32.50
	0260	Spray, primer		35	.229		.84	7.25		8.09	12.60
	0270	Finish coat, exterior latex		35	.229		.99	7.25		8.24	12.75
	0280	Primer & 1 coat, exterior latex		20	.400		.91	12.70		13.61	21.50
	0290	For louvered shutters, add				S.F.	10%				
	0300	Stair stringers, exterior, metal									
	0310	Roll & brush, zinc chromate, to 14″, each coat	1 Pord	320	.025	L.F.	.05	.79		.84	1.33
	0320	Rough sawn wood, 4″ x 12″									
	0330	Roll & brush, exterior latex, each coat	1 Pord	215	.037	L.F.	.07	1.18		1.25	1.97
	0340	Trellis/lattice, 2″ x 2″ @ 3″ O.C. with 2″ x 8″ supports									
	0350	Spray, latex, per side, each coat	1 Pord	475	.017	S.F.	.07	.53		.60	.94
	0450	Decking, Ext., sealer, alkyd, brushwork, sealer coat		1140	.007		.06	.22		.28	.42
	0460	1st coat		1140	.007		.06	.22		.28	.42
	0470	2nd coat		1300	.006		.04	.20		.24	.36
	0500	Paint, alkyd, brushwork, primer coat		1140	.007		.07	.22		.29	.44
	0510	1st coat		1140	.007		.07	.22		.29	.44
	0520	2nd coat		1300	.006		.05	.20		.25	.37
	0600	Sand paint, alkyd, brushwork, 1 coat		150	.053		.09	1.69		1.78	2.81
630	0010	**MISCELLANEOUS, INTERIOR** R099100-20									
	2400	Floors, conc./wood, oil base, primer/sealer coat, brushwork	2 Pord	1950	.008	S.F.	.06	.26		.32	.49
	2450	Roller		5200	.003		.06	.10		.16	.23
	2600	Spray		6000	.003		.06	.08		.14	.21
	2650	Paint 1 coat, brushwork		1950	.008		.06	.26		.32	.48
	2800	Roller		5200	.003		.06	.10		.16	.22
	2850	Spray		6000	.003		.06	.08		.14	.21
	3000	Stain, wood floor, brushwork, 1 coat		4550	.004		.05	.11		.16	.24
	3200	Roller		5200	.003		.05	.10		.15	.22
	3250	Spray		6000	.003		.05	.08		.13	.20
	3400	Varnish, wood floor, brushwork		4550	.004		.06	.11		.17	.24
	3450	Roller		5200	.003		.06	.10		.16	.23

		09910 | Paints	CREW	DAILY OUTPUT	LABOR-HOURS	UNIT	2006 BARE COSTS MAT.	LABOR	EQUIP.	TOTAL	TOTAL INCL O&P
630	3600	Spray R099100-20	2 Pord	6000	.003	S.F.	.06	.08		.14	.21
	3650	For dust proofing or anti skid, see division 03350-300									
	3800	Grilles, per side, oil base, primer coat, brushwork	1 Pord	520	.015	S.F.	.10	.49		.59	.89
	3850	Spray		1140	.007		.10	.22		.32	.47
	3880	Paint 1 coat, brushwork		520	.015		.11	.49		.60	.90
	3900	Spray		1140	.007		.12	.22		.34	.49
	3920	Paint 2 coats, brushwork		325	.025		.21	.78		.99	1.48
	3940	Spray		650	.012		.24	.39		.63	.89
	3950	Prime & paint 1 coat		325	.025		.20	.78		.98	1.47
	3960	Prime & paint 2 coats		270	.030		.20	.94		1.14	1.73
	4250	Paint 1 coat, brushwork	2 Pord	1300	.012	L.F.	.11	.39		.50	.75
	4500	Louvers, one side, primer, brushwork	1 Pord	524	.015	S.F.	.06	.48		.54	.85
	4520	Paint one coat, brushwork		520	.015		.06	.49		.55	.84
	4530	Spray		1140	.007		.06	.22		.28	.43
	4540	Paint two coats, brushwork		325	.025		.11	.78		.89	1.37
	4550	Spray		650	.012		.12	.39		.51	.76
	4560	Paint three coats, brushwork		270	.030		.16	.94		1.10	1.69
	4570	Spray		500	.016		.18	.51		.69	1.01
	5000	Pipe, thru 4" diameter, primer or sealer coat, oil base, brushwork	2 Pord	1250	.013	L.F.	.06	.41		.47	.72
	5100	Spray		2165	.007		.06	.23		.29	.45
	5200	Paint 1 coat, brushwork		1250	.013		.06	.41		.47	.72
	5300	Spray		2165	.007		.06	.23		.29	.44
	5350	Paint 2 coats, brushwork		775	.021		.11	.65		.76	1.17
	5400	Spray		1240	.013		.13	.41		.54	.80
	5420	Paint 3 coats, brushwork		775	.021		.17	.65		.82	1.24
	5450	Thru 8" diameter, primer or sealer coat, brushwork		620	.026		.13	.82		.95	1.45
	5500	Spray		1085	.015		.21	.47		.68	.98
	5550	Paint 1 coat, brushwork		620	.026		.17	.82		.99	1.50
	5600	Spray		1085	.015		.19	.47		.66	.96
	5650	Paint 2 coats, brushwork		385	.042		.23	1.32		1.55	2.37
	5700	Spray		620	.026		.25	.82		1.07	1.59
	5720	Paint 3 coats, brushwork		385	.042		.34	1.32		1.66	2.49
	5750	Thru 12" diameter, primer or sealer coat, brushwork		415	.039		.19	1.22		1.41	2.17
	5800	Spray		725	.022		.21	.70		.91	1.35
	5850	Paint 1 coat, brushwork		415	.039		.17	1.22		1.39	2.15
	6000	Spray		725	.022		.19	.70		.89	1.33
	6200	Paint 2 coats, brushwork		260	.062		.34	1.95		2.29	3.50
	6250	Spray		415	.039		.38	1.22		1.60	2.37
	6270	Paint 3 coats, brushwork		260	.062		.50	1.95		2.45	3.68
	6300	Thru 16" diameter, primer or sealer coat, brushwork		310	.052		.25	1.64		1.89	2.91
	6350	Spray		540	.030		.28	.94		1.22	1.82
	6400	Paint 1 coat, brushwork		310	.052		.23	1.64		1.87	2.89
	6450	Spray		540	.030		.26	.94		1.20	1.80
	6500	Paint 2 coats, brushwork		195	.082		.45	2.60		3.05	4.68
	6550	Spray		310	.052		.50	1.64		2.14	3.18
	6600	Radiators, per side, primer, brushwork	1 Pord	520	.015	S.F.	.06	.49		.55	.85
	6620	Paint one coat, brushwork		520	.015		.06	.49		.55	.84
	6640	Paint two coats, brushwork		340	.024		.11	.75		.86	1.32
	6660	Paint three coats, brushwork		283	.028		.16	.90		1.06	1.62
	7000	Trim, wood, incl. puttying, under 6" wide									
	7200	Primer coat, oil base, brushwork	1 Pord	650	.012	L.F.	.02	.39		.41	.66
	7250	Paint, 1 coat, brushwork		650	.012		.03	.39		.42	.66
	7400	2 coats		400	.020		.05	.63		.68	1.08
	7450	3 coats		325	.025		.08	.78		.86	1.34
	7500	Over 6" wide, primer coat, brushwork		650	.012		.05	.39		.44	.68
	7550	Paint, 1 coat, brushwork		650	.012		.05	.39		.44	.69

		09910 | Paints	CREW	DAILY OUTPUT	LABOR-HOURS	UNIT	2006 BARE COSTS MAT.	LABOR	EQUIP.	TOTAL	TOTAL INCL O&P	
630	7600	2 coats R099100-20	1 Pord	400	.020	L.F.	.10	.63		.73	1.13	630
	7650	3 coats		325	.025		.15	.78		.93	1.42	
	8000	Cornice, simple design, primer coat, oil base, brushwork		650	.012	S.F.	.05	.39		.44	.68	
	8250	Paint, 1 coat		650	.012		.05	.39		.44	.69	
	8300	2 coats		400	.020		.10	.63		.73	1.13	
	8350	Ornate design, primer coat		350	.023		.05	.72		.77	1.21	
	8400	Paint, 1 coat		350	.023		.05	.72		.77	1.22	
	8450	2 coats		400	.020		.10	.63		.73	1.13	
	8600	Balustrades, primer coat, oil base, brushwork		520	.015		.05	.49		.54	.83	
	8650	Paint, 1 coat		520	.015		.05	.49		.54	.84	
	8700	2 coats		325	.025		.10	.78		.88	1.36	
	8900	Trusses and wood frames, primer coat, oil base, brushwork		800	.010		.05	.32		.37	.56	
	8950	Spray		1200	.007		.05	.21		.26	.40	
	9000	Paint 1 coat, brushwork		750	.011		.05	.34		.39	.60	
	9200	Spray		1200	.007		.06	.21		.27	.41	
	9220	Paint 2 coats, brushwork		500	.016		.10	.51		.61	.92	
	9240	Spray		600	.013		.12	.42		.54	.81	
	9260	Stain, brushwork, wipe off		600	.013		.05	.42		.47	.74	
	9280	Varnish, 3 coats, brushwork		275	.029		.17	.92		1.09	1.67	
	9350	For latex paint, deduct					10%					
650	0010	**COATINGS & PAINTS** In 5 gallon lots R099100-20										650
	0050	For 100 gallons or more, deduct					10%				10%	
	0100	Paint, Exterior alkyd (oil base)										
	0200	Flat				Gal.	24			24	26.50	
	0300	Gloss					23.50			23.50	26	
	0400	Primer					24			24	26.50	
	0500	Latex (water base)										
	0600	Acrylic stain				Gal.	20.50			20.50	22.50	
	0700	Gloss enamel					24.50			24.50	27	
	0800	Flat					19.80			19.80	22	
	0900	Primer					21			21	23	
	1000	Semi-gloss					27			27	29.50	
	1100	Interior, alkyd (oil base)										
	1200	Enamel undercoat				Gal.	23			23	25.50	
	1300	Flat					23			23	25	
	1400	Gloss					27			27	29.50	
	1500	Primer sealer					19.20			19.20	21	
	1600	Semi-gloss					23			23	25	
	1700	Latex (water base)										
	1800	Enamel undercoat				Gal.	16.65			16.65	18.30	
	1900	Flat					17.35			17.35	19.10	
	2000	Floor and deck					22			22	24.50	
	2100	Gloss					25.50			25.50	28	
	2200	Primer sealer					19.10			19.10	21	
	2300	Semi-gloss					20.50			20.50	22.50	
	2400	Masonry, Exterior										
	2500	Alkali resistant primer				Gal.	20			20	22	
	2600	Block filler, epoxy					22			22	24	
	2700	Latex					10.60			10.60	11.65	
	2800	Latex, flat					21			21	23	
	2900	Semi-gloss					21.50			21.50	24	
	3000	Masonry, Interior										
	3100	Alkali resistant primer				Gal.	17.30			17.30	19.05	
	3200	Block filler, epoxy					25			25	27.50	
	3300	Latex					10.85			10.85	11.95	
	3400	Floor, alkyd					23.50			23.50	26	

	09910 | Paints	CREW	DAILY OUTPUT	LABOR-HOURS	UNIT	2006 BARE COSTS MAT.	LABOR	EQUIP.	TOTAL	TOTAL INCL O&P	
650	3500 Latex (R099100-20)				Gal.	25			25	27.50	650
	3600 Latex, flat acrylic					14.05			14.05	15.50	
	3700 Flat emulsion					11.95			11.95	13.15	
	3800 Sealer					14			14	15.40	
	3900 Semi-gloss					19.65			19.65	21.50	
	4000 Metal										
	4100 Galvanizing paint				Gal.	26			26	28.50	
	4200 High heat					46.50			46.50	51	
	4300 Heat resistant					28			28	31	
	4400 Machinery enamel, alkyd					27.50			27.50	30	
	4500 Metal pretreatment (polyvinyl butyral)					25			25	27.50	
	4600 Rust inhibitor, ferrous metal					24			24	26.50	
	4700 Zinc chromate					21			21	23.50	
	4800 Zinc rich primer					101			101	112	
	4900 Varnish and Stain										
	5000 Alkyd, clear				Gal.	23			23	25	
	5100 Polyurethane, clear					25			25	27.50	
	5200 Primer sealer					19.15			19.15	21	
	5300 Stain, semi-transparent					20			20	22	
	5400 Solid color					19.95			19.95	22	
	5500 Coatings										
	5600 Heavy duty										
	5700 Acrylic urethane				Gal.	47.50			47.50	52	
	5800 Chlorinated rubber					32			32	35.50	
	5900 Coal tar epoxy					40.50			40.50	44.50	
	6000 Polyamide epoxy, finish					34			34	37.50	
	6100 Primer					33.50			33.50	37	
	6200 Silicone alkyd					34			34	37.50	
	6300 2 component solvent based acrylic epoxy					49			49	54	
	6400 Polyester epoxy					69			69	76	
	6500 Vinyl					29			29	31.50	
	6600 Special/Miscellaneous										
	6700 Aluminum				Gal.	24.50			24.50	27	
	6900 Dry fall out, flat					13.05			13.05	14.35	
	7000 Fire retardant, intumescent					39			39	43	
	7100 Linseed oil					12.15			12.15	13.40	
	7200 Shellac					22.50			22.50	25	
	7300 Swimming pool, epoxy or urethane base					39			39	42.50	
	7400 Rubber base					29			29	32	
	7500 Texture paint					14.30			14.30	15.70	
	7600 Turpentine					11.40			11.40	12.55	
	7700 Water repellent, 5% silicone					19			19	21	
700	0010 **SIDING EXTERIOR**, Alkyd (oil base) (R099100-20)										700
	0450 Steel siding, oil base, paint 1 coat, brushwork	2 Pord	2015	.008	S.F.	.06	.25		.31	.46	
	0500 Spray		4550	.004		.08	.11		.19	.27	
	0800 Paint 2 coats, brushwork		1300	.012		.11	.39		.50	.75	
	1000 Spray		4550	.004		.14	.11		.25	.33	
	1200 Stucco, rough, oil base, paint 2 coats, brushwork		1300	.012		.11	.39		.50	.75	
	1400 Roller		1625	.010		.12	.31		.43	.63	
	1600 Spray		2925	.005		.12	.17		.29	.42	
	1800 Texture 1-11 or clapboard, oil base, primer coat, brushwork		1300	.012		.10	.39		.49	.74	
	2000 Spray		4550	.004		.10	.11		.21	.29	
	2100 Paint 1 coat, brushwork		1300	.012		.08	.39		.47	.72	
	2200 Spray		4550	.004		.08	.11		.19	.27	
	2400 Paint 2 coats, brushwork		810	.020		.16	.63		.79	1.19	
	2600 Spray		2600	.006		.18	.20		.38	.51	

9 FINISHES

	09910 | Paints	CREW	DAILY OUTPUT	LABOR-HOURS	UNIT	2006 BARE COSTS MAT.	LABOR	EQUIP.	TOTAL	TOTAL INCL O&P	
700	3000 Stain 1 coat, brushwork (R099100-20)	2 Pord	1520	.011	S.F.	.05	.33		.38	.60	700
	3200 Spray		5320	.003		.06	.10		.16	.21	
	3400 Stain 2 coats, brushwork		950	.017		.10	.53		.63	.97	
	4000 Spray		3050	.005		.11	.17		.28	.39	
	4200 Wood shingles, oil base primer coat, brushwork		1300	.012		.09	.39		.48	.73	
	4400 Spray		3900	.004		.08	.13		.21	.30	
	4600 Paint 1 coat, brushwork		1300	.012		.07	.39		.46	.70	
	4800 Spray		3900	.004		.08	.13		.21	.30	
	5000 Paint 2 coats, brushwork		810	.020		.13	.63		.76	1.16	
	5200 Spray		2275	.007		.13	.22		.35	.50	
	5800 Stain 1 coat, brushwork		1500	.011		.05	.34		.39	.60	
	6000 Spray		3900	.004		.05	.13		.18	.27	
	6500 Stain 2 coats, brushwork		950	.017		.10	.53		.63	.97	
	7000 Spray		2660	.006		.14	.19		.33	.46	
	8000 For latex paint, deduct					10%					
	8100 For work over 12′ H, from pipe scaffolding, add						15%				
	8200 For work over 12′ H, from extension ladder, add						25%				
	8300 For work over 12′ H, from swing staging, add						35%				
710	0010 **SIDING, MISC.**, latex paint (R099100-20)										710
	0100 Aluminum siding										
	0110 Brushwork, primer	2 Pord	2275	.007	S.F.	.05	.22		.27	.42	
	0120 Finish coat, exterior latex		2275	.007		.04	.22		.26	.41	
	0130 Primer & 1 coat exterior latex		1300	.012		.10	.39		.49	.74	
	0140 Primer & 2 coats exterior latex		975	.016		.14	.52		.66	1	
	0150 Mineral Fiber shingles										
	0160 Brushwork, primer	2 Pord	1495	.011	S.F.	.10	.34		.44	.65	
	0170 Finish coat, industrial enamel		1495	.011		.10	.34		.44	.65	
	0180 Primer & 1 coat enamel		810	.020		.20	.63		.83	1.23	
	0190 Primer & 2 coats enamel		540	.030		.30	.94		1.24	1.84	
	0200 Roll, primer		1625	.010		.11	.31		.42	.62	
	0210 Finish coat, industrial enamel		1625	.010		.11	.31		.42	.62	
	0220 Primer & 1 coat enamel		975	.016		.22	.52		.74	1.08	
	0230 Primer & 2 coats enamel		650	.025		.33	.78		1.11	1.61	
	0240 Spray, primer		3900	.004		.08	.13		.21	.30	
	0250 Finish coat, industrial enamel		3900	.004		.09	.13		.22	.31	
	0260 Primer & 1 coat enamel		2275	.007		.17	.22		.39	.55	
	0270 Primer & 2 coats enamel		1625	.010		.27	.31		.58	.79	
	0280 Waterproof sealer, first coat		4485	.004		.07	.11		.18	.25	
	0290 Second coat		5235	.003		.06	.10		.16	.23	
	0300 Rough wood incl. shingles, shakes or rough sawn siding										
	0310 Brushwork, primer	2 Pord	1280	.013	S.F.	.12	.40		.52	.77	
	0320 Finish coat, exterior latex		1280	.013		.07	.40		.47	.72	
	0330 Primer & 1 coat exterior latex		960	.017		.19	.53		.72	1.06	
	0340 Primer & 2 coats exterior latex		700	.023		.26	.72		.98	1.45	
	0350 Roll, primer		2925	.005		.16	.17		.33	.45	
	0360 Finish coat, exterior latex		2925	.005		.09	.17		.26	.38	
	0370 Primer & 1 coat exterior latex		1790	.009		.24	.28		.52	.73	
	0380 Primer & 2 coats exterior latex		1300	.012		.33	.39		.72	1	
	0390 Spray, primer		3900	.004		.13	.13		.26	.36	
	0400 Finish coat, exterior latex		3900	.004		.07	.13		.20	.29	
	0410 Primer & 1 coat exterior latex		2600	.006		.20	.20		.40	.53	
	0420 Primer & 2 coats exterior latex		2080	.008		.27	.24		.51	.69	
	0430 Waterproof sealer, first coat		4485	.004		.12	.11		.23	.31	
	0440 Second coat		4485	.004		.07	.11		.18	.25	
	0450 Smooth wood incl. butt, T&G, beveled, drop or B&B siding										
	0460 Brushwork, primer	2 Pord	2325	.007	S.F.	.08	.22		.30	.44	

	09910 | Paints		CREW	DAILY OUTPUT	LABOR-HOURS	UNIT	2006 BARE COSTS MAT.	LABOR	EQUIP.	TOTAL	TOTAL INCL O&P	
710	0470	Finish coat, exterior latex (R099100-20)	2 Pord	1280	.013	S.F.	.07	.40		.47	.72	710
	0480	Primer & 1 coat exterior latex		800	.020		.16	.63		.79	1.19	
	0490	Primer & 2 coats exterior latex		630	.025		.23	.81		1.04	1.54	
	0500	Roll, primer		2275	.007		.09	.22		.31	.46	
	0510	Finish coat, exterior latex		2275	.007		.08	.22		.30	.45	
	0520	Primer & 1 coat exterior latex		1300	.012		.17	.39		.56	.82	
	0530	Primer & 2 coats exterior latex		975	.016		.25	.52		.77	1.12	
	0540	Spray, primer		4550	.004		.07	.11		.18	.26	
	0550	Finish coat, exterior latex		4550	.004		.07	.11		.18	.26	
	0560	Primer & 1 coat exterior latex		2600	.006		.14	.20		.34	.47	
	0570	Primer & 2 coats exterior latex		1950	.008		.21	.26		.47	.65	
	0580	Waterproof sealer, first coat		5230	.003		.07	.10		.17	.23	
	0590	Second coat	↓	5980	.003		.07	.09		.16	.21	
	0600	For oil base paint, add				↓	10%					
800	0010	**TRIM, EXTERIOR** (R099100-20)										800
	0100	Door frames & trim (see Doors, interior or exterior)										
	0110	Fascia, latex paint, one coat coverage										
	0120	1" x 4", brushwork	1 Pord	640	.013	L.F.	.02	.40		.42	.66	
	0130	Roll		1280	.006		.02	.20		.22	.34	
	0140	Spray		2080	.004		.01	.12		.13	.22	
	0150	1" x 6" to 1" x 10", brushwork		640	.013		.06	.40		.46	.71	
	0160	Roll		1230	.007		.07	.21		.28	.40	
	0170	Spray		2100	.004		.05	.12		.17	.24	
	0180	1" x 12", brushwork		640	.013		.06	.40		.46	.71	
	0190	Roll		1050	.008		.07	.24		.31	.46	
	0200	Spray	↓	2200	.004	↓	.05	.12		.17	.24	
	0210	Gutters & downspouts, metal, zinc chromate paint										
	0220	Brushwork, gutters, 5", first coat	1 Pord	640	.013	L.F.	.06	.40		.46	.70	
	0230	Second coat		960	.008		.05	.26		.31	.48	
	0240	Third coat		1280	.006		.04	.20		.24	.37	
	0250	Downspouts, 4", first coat		640	.013		.06	.40		.46	.70	
	0260	Second coat		960	.008		.05	.26		.31	.48	
	0270	Third coat	↓	1280	.006	↓	.04	.20		.24	.37	
	0280	Gutters & downspouts, wood										
	0290	Brushwork, gutters, 5", primer	1 Pord	640	.013	L.F.	.05	.40		.45	.70	
	0300	Finish coat, exterior latex		640	.013		.05	.40		.45	.70	
	0310	Primer & 1 coat exterior latex		400	.020		.11	.63		.74	1.14	
	0320	Primer & 2 coats exterior latex		325	.025		.17	.78		.95	1.44	
	0330	Downspouts, 4", primer		640	.013		.05	.40		.45	.70	
	0340	Finish coat, exterior latex		640	.013		.05	.40		.45	.70	
	0350	Primer & 1 coat exterior latex		400	.020		.11	.63		.74	1.14	
	0360	Primer & 2 coats exterior latex	↓	325	.025	↓	.08	.78		.86	1.34	
	0370	Molding, exterior, up to 14" wide										
	0380	Brushwork, primer	1 Pord	640	.013	L.F.	.06	.40		.46	.71	
	0390	Finish coat, exterior latex		640	.013		.06	.40		.46	.71	
	0400	Primer & 1 coat exterior latex		400	.020		.13	.63		.76	1.17	
	0410	Primer & 2 coats exterior latex		315	.025		.13	.81		.94	1.44	
	0420	Stain & fill		1050	.008		.06	.24		.30	.46	
	0430	Shellac		1850	.004		.07	.14		.21	.29	
	0440	Varnish	↓	1275	.006	↓	.07	.20		.27	.40	
910	0350	**WALLS, MASONRY (CMU), EXTERIOR**										910
	0360	Concrete masonry units (CMU), smooth surface										
	0370	Brushwork, latex, first coat	1 Pord	640	.013	S.F.	.04	.40		.44	.68	
	0380	Second coat		960	.008		.03	.26		.29	.45	
	0390	Waterproof sealer, first coat		736	.011		.24	.34		.58	.81	
	0400	Second coat	↓	1104	.007	↓	.24	.23		.47	.63	

09910 | Paints

		09910 | Paints	CREW	DAILY OUTPUT	LABOR-HOURS	UNIT	2006 BARE COSTS MAT.	LABOR	EQUIP.	TOTAL	TOTAL INCL O&P	
910	0410	Roll, latex, paint, first coat	1 Pord	1465	.005	S.F.	.04	.17		.21	.33	910
	0420	Second coat		1790	.004		.03	.14		.17	.26	
	0430	Waterproof sealer, first coat		1680	.005		.24	.15		.39	.50	
	0440	Second coat		2060	.004		.24	.12		.36	.46	
	0450	Spray, latex, paint, first coat		1950	.004		.03	.13		.16	.25	
	0460	Second coat		2600	.003		.03	.10		.13	.19	
	0470	Waterproof sealer, first coat		2245	.004		.24	.11		.35	.44	
	0480	Second coat		2990	.003		.24	.09		.33	.40	
	0490	Concrete masonry unit (CMU), porous										
	0500	Brushwork, latex, first coat	1 Pord	640	.013	S.F.	.07	.40		.47	.72	
	0510	Second coat		960	.008		.04	.26		.30	.46	
	0520	Waterproof sealer, first coat		736	.011		.24	.34		.58	.81	
	0530	Second coat		1104	.007		.24	.23		.47	.63	
	0540	Roll latex, first coat		1465	.005		.05	.17		.22	.34	
	0550	Second coat		1790	.004		.03	.14		.17	.27	
	0560	Waterproof sealer, first coat		1680	.005		.24	.15		.39	.50	
	0570	Second coat		2060	.004		.24	.12		.36	.46	
	0580	Spray latex, first coat		1950	.004		.04	.13		.17	.25	
	0590	Second coat		2600	.003		.03	.10		.13	.19	
	0600	Waterproof sealer, first coat		2245	.004		.24	.11		.35	.44	
	0610	Second coat		2990	.003		.24	.09		.33	.40	
920	0010	**WALLS AND CEILINGS, INTERIOR** R099100-20										920
	0100	Concrete, dry wall or plaster, oil base, primer or sealer coat										
	0200	Smooth finish, brushwork	1 Pord	1150	.007	S.F.	.05	.22		.27	.40	
	0240	Roller		1350	.006		.05	.19		.24	.35	
	0280	Spray		2750	.003		.03	.09		.12	.19	
	0300	Sand finish, brushwork		975	.008		.05	.26		.31	.47	
	0340	Roller		1150	.007		.05	.22		.27	.41	
	0380	Spray		2275	.004		.04	.11		.15	.22	
	0400	Paint 1 coat, smooth finish, brushwork		1200	.007		.05	.21		.26	.40	
	0440	Roller		1300	.006		.05	.20		.25	.37	
	0480	Spray		2275	.004		.05	.11		.16	.23	
	0500	Sand finish, brushwork		1050	.008		.05	.24		.29	.45	
	0540	Roller		1600	.005		.05	.16		.21	.31	
	0580	Spray		2100	.004		.05	.12		.17	.24	
	0800	Paint 2 coats, smooth finish, brushwork		680	.012		.10	.37		.47	.71	
	0840	Roller		800	.010		.11	.32		.43	.63	
	0880	Spray		1625	.005		.09	.16		.25	.35	
	0900	Sand finish, brushwork		605	.013		.10	.42		.52	.78	
	0940	Roller		1020	.008		.11	.25		.36	.52	
	0980	Spray		1700	.005		.09	.15		.24	.34	
	1200	Paint 3 coats, smooth finish, brushwork		510	.016		.15	.50		.65	.97	
	1240	Roller		650	.012		.16	.39		.55	.81	
	1280	Spray		1625	.005		.14	.16		.30	.40	
	1300	Sand finish, brushwork		454	.018		.15	.56		.71	1.07	
	1340	Roller		680	.012		.16	.37		.53	.78	
	1380	Spray		1133	.007		.14	.22		.36	.51	
	1600	Glaze coating, 5 coats, spray, clear		900	.009		.60	.28		.88	1.11	
	1640	Multicolor		900	.009		.85	.28		1.13	1.39	
	1700	For latex paint, deduct					10%					
	1800	For ceiling installations, add						25%				
	2000	Masonry or concrete block, oil base, primer or sealer coat										
	2100	Smooth finish, brushwork	1 Pord	1224	.007	S.F.	.05	.21		.26	.38	
	2180	Spray		2400	.003		.07	.11		.18	.24	
	2200	Sand finish, brushwork		1089	.007		.07	.23		.30	.45	

	09910 | Paints		CREW	DAILY OUTPUT	LABOR-HOURS	UNIT	2006 BARE COSTS MAT.	LABOR	EQUIP.	TOTAL	TOTAL INCL O&P	
920	2280	Spray (R099100-20)	1 Pord	2400	.003	S.F.	.07	.11		.18	.24	920
	2400	Paint 1 coat, smooth finish, brushwork		1100	.007		.07	.23		.30	.45	
	2480	Spray		2400	.003		.07	.11		.18	.24	
	2500	Sand finish, brushwork		979	.008		.07	.26		.33	.50	
	2580	Spray		2400	.003		.07	.11		.18	.24	
	2800	Paint 2 coats, smooth finish, brushwork		756	.011		.15	.34		.49	.70	
	2880	Spray		1360	.006		.14	.19		.33	.45	
	2900	Sand finish, brushwork		672	.012		.15	.38		.53	.77	
	2980	Spray		1360	.006		.14	.19		.33	.45	
	3200	Paint 3 coats, smooth finish, brushwork		560	.014		.22	.45		.67	.97	
	3280	Spray		1088	.007		.20	.23		.43	.59	
	3300	Sand finish, brushwork		498	.016		.22	.51		.73	1.06	
	3380	Spray		1088	.007		.20	.23		.43	.59	
	3600	Glaze coating, 5 coats, spray, clear		900	.009		.60	.28		.88	1.11	
	3620	Multicolor		900	.009		.85	.28		1.13	1.39	
	4000	Block filler, 1 coat, brushwork		425	.019		.12	.60		.72	1.10	
	4100	Silicone, water repellent, 2 coats, spray	↓	2000	.004		.25	.13		.38	.48	
	4120	For latex paint, deduct					10%					
	8200	For work 8 - 15′ H, add						10%				
	8300	For work over 15′ H, add				↓		20%				
940	0010	**DRY FALL PAINTING** (R099100-20)										940
	0100	Walls										
	0200	Wallboard and smooth plaster, one coat, brush	1 Pord	910	.009	S.F.	.04	.28		.32	.50	
	0210	Roll		1560	.005		.04	.16		.20	.31	
	0220	Spray		2600	.003		.04	.10		.14	.21	
	0230	Two coats, brush		520	.015		.09	.49		.58	.88	
	0240	Roll		877	.009		.09	.29		.38	.56	
	0250	Spray		1560	.005		.09	.16		.25	.36	
	0260	Concrete or textured plaster, one coat, brush		747	.011		.04	.34		.38	.60	
	0270	Roll		1300	.006		.04	.20		.24	.36	
	0280	Spray		1560	.005		.04	.16		.20	.31	
	0290	Two coats, brush		422	.019		.09	.60		.69	1.07	
	0300	Roll		747	.011		.09	.34		.43	.65	
	0310	Spray		1300	.006		.09	.20		.29	.41	
	0320	Concrete block, one coat, brush		747	.011		.04	.34		.38	.60	
	0330	Roll		1300	.006		.04	.20		.24	.36	
	0340	Spray		1560	.005		.04	.16		.20	.31	
	0350	Two coats, brush		422	.019		.09	.60		.69	1.07	
	0360	Roll		747	.011		.09	.34		.43	.65	
	0370	Spray		1300	.006		.09	.20		.29	.41	
	0380	Wood, one coat, brush		747	.011		.04	.34		.38	.60	
	0390	Roll		1300	.006		.04	.20		.24	.36	
	0400	Spray		877	.009		.04	.29		.33	.51	
	0410	Two coats, brush		487	.016		.09	.52		.61	.94	
	0420	Roll		747	.011		.09	.34		.43	.65	
	0430	Spray	↓	650	.012	↓	.09	.39		.48	.73	
	0440	Ceilings										
	0450	Wallboard and smooth plaster, one coat, brush	1 Pord	600	.013	S.F.	.04	.42		.46	.73	
	0460	Roll		1040	.008		.04	.24		.28	.44	
	0470	Spray		1560	.005		.04	.16		.20	.31	
	0480	Two coats, brush		341	.023		.09	.74		.83	1.29	
	0490	Roll		650	.012		.09	.39		.48	.73	
	0500	Spray		1300	.006		.09	.20		.29	.41	
	0510	Concrete or textured plaster, one coat, brush		487	.016		.04	.52		.56	.89	
	0520	Roll		877	.009		.04	.29		.33	.51	
	0530	Spray	↓	1560	.005	↓	.04	.16		.20	.31	

		09910 | Paints	CREW	DAILY OUTPUT	LABOR-HOURS	UNIT	2006 BARE COSTS MAT.	LABOR	EQUIP.	TOTAL	TOTAL INCL O&P	
940	0540	Two coats, brush (R099100-20)	1 Pord	276	.029	S.F.	.09	.92		1.01	1.58	940
	0550	Roll		520	.015		.09	.49		.58	.88	
	0560	Spray		1300	.006		.09	.20		.29	.41	
	0570	Structural steel, bar joists or metal deck, one coat, spray		1560	.005		.04	.16		.20	.31	
	0580	Two coats, spray		1040	.008		.09	.24		.33	.49	

		09930 | Stains/Transp. Finishes	CREW	DAILY OUTPUT	LABOR-HOURS	UNIT	MAT.	LABOR	EQUIP.	TOTAL	TOTAL INCL O&P	
100	0010	**VARNISH**										100
	0012	1 coat + sealer, on wood trim, no sanding included	1 Pord	400	.020	S.F.	.07	.63		.70	1.10	
	0100	Hardwood floors, 2 coats, no sanding included, roller		1890	.004	"	.13	.13		.26	.36	
	9000	Minimum labor/equipment charge		4	2	Job		63.50		63.50	102	

		09963 | Glazed Coatings	CREW	DAILY OUTPUT	LABOR-HOURS	UNIT	MAT.	LABOR	EQUIP.	TOTAL	TOTAL INCL O&P	
200	0010	**WALL COATINGS**										200
	0100	Acrylic glazed coatings, minimum	1 Pord	525	.015	S.F.	.26	.48		.74	1.07	
	0200	Maximum		305	.026		.54	.83		1.37	1.93	
	0300	Epoxy coatings, minimum		525	.015		.33	.48		.81	1.14	
	0400	Maximum		170	.047		1.02	1.49		2.51	3.52	
	0600	Exposed aggregate, troweled on, 1/16" to 1/4", minimum		235	.034		.50	1.08		1.58	2.28	
	0700	Maximum (epoxy or polyacrylate)		130	.062		1.09	1.95		3.04	4.33	
	0900	1/2" to 5/8" aggregate, minimum		130	.062		1.01	1.95		2.96	4.24	
	1000	Maximum		80	.100		1.72	3.17		4.89	7	
	1200	1" aggregate size, minimum		90	.089		1.75	2.82		4.57	6.45	
	1300	Maximum		55	.145		2.68	4.61		7.29	10.35	
	1500	Exposed aggregate, sprayed on, 1/8" aggregate, minimum		295	.027		.47	.86		1.33	1.90	
	1600	Maximum		145	.055		.87	1.75		2.62	3.77	
	1800	High build epoxy, 50 mil, minimum		390	.021		.56	.65		1.21	1.66	
	1900	Maximum		95	.084		.96	2.67		3.63	5.35	
	2100	Laminated epoxy with fiberglass, minimum		295	.027		.61	.86		1.47	2.05	
	2200	Maximum		145	.055		1.08	1.75		2.83	4	
	2400	Sprayed perlite or vermiculite, 1/16" thick, minimum		2935	.003		.22	.09		.31	.38	
	2500	Maximum		640	.013		.62	.40		1.02	1.32	
	2700	Vinyl plastic wall coating, minimum		735	.011		.28	.34		.62	.86	
	2800	Maximum		240	.033		.68	1.06		1.74	2.45	
	3000	Urethane on smooth surface, 2 coats, minimum		1135	.007		.21	.22		.43	.59	
	3100	Maximum		665	.012		.46	.38		.84	1.12	
	3300	3 coat, minimum		840	.010		.28	.30		.58	.79	
	3400	Maximum		470	.017		.62	.54		1.16	1.55	
	3600	Ceramic-like glazed coating, cementitious, minimum		440	.018		.40	.58		.98	1.37	
	3700	Maximum		345	.023		.67	.74		1.41	1.92	
	3900	Resin base, minimum		640	.013		.28	.40		.68	.95	
	4000	Maximum		330	.024		.45	.77		1.22	1.73	

		09990 | Paint Restoration	CREW	DAILY OUTPUT	LABOR-HOURS	UNIT	MAT.	LABOR	EQUIP.	TOTAL	TOTAL INCL O&P	
500	0010	**SCRAPE AFTER FIRE DAMAGE**										500
	0050	Boards, 1" x 4"	1 Pord	336	.024	L.F.		.75		.75	1.21	
	0060	1" x 6"		260	.031			.98		.98	1.57	
	0070	1" x 8"		207	.039			1.23		1.23	1.97	
	0080	1" x 10"		174	.046			1.46		1.46	2.34	
	0500	Framing, 2" x 4"		265	.030			.96		.96	1.54	
	0510	2" x 6"		221	.036			1.15		1.15	1.84	
	0520	2" x 8"		190	.042			1.33		1.33	2.14	
	0530	2" x 10"		165	.048			1.54		1.54	2.47	
	0540	2" x 12"		144	.056			1.76		1.76	2.83	

	09990 | Paint Restoration		CREW	DAILY OUTPUT	LABOR-HOURS	UNIT	2006 BARE COSTS MAT.	LABOR	EQUIP.	TOTAL	TOTAL INCL O&P	
500	1000	Heavy framing, 3″ x 4″	1 Pord	226	.035	L.F.		1.12		1.12	1.80	500
	1010	4″ x 4″		210	.038			1.21		1.21	1.94	
	1020	4″ x 6″		191	.042			1.33		1.33	2.13	
	1030	4″ x 8″		165	.048			1.54		1.54	2.47	
	1040	4″ x 10″		144	.056			1.76		1.76	2.83	
	1060	4″ x 12″		131	.061			1.94		1.94	3.11	
	2900	For sealing, minimum		825	.010	S.F.	.13	.31		.44	.63	
	2920	Maximum		460	.017	"	.26	.55		.81	1.18	
	9000	Minimum labor/equipment charge		3	2.667	Job		84.50		84.50	136	
800	0010	**SANDING** and puttying interior trim, compared to										800
	0100	Painting 1 coat, on quality work				L.F.		100%				
	0300	Medium work						50%				
	0400	Industrial grade						25%				
	0500	Surface protection, placement and removal										
	0510	Basic drop cloths	1 Pord	6400	.001	S.F.		.04		.04	.06	
	0520	Masking with paper		800	.010		.03	.32		.35	.54	
	0530	Volume cover up (using plastic sheathing, or building paper)		16000	.001			.02		.02	.03	
900	0010	**SURFACE PREPARATION, EXTERIOR**										900
	0015	Doors, per side, not incl. frames or trim										
	0020	Scrape & sand										
	0030	Wood, flush	1 Pord	616	.013	S.F.		.41		.41	.66	
	0040	Wood, detail		496	.016			.51		.51	.82	
	0050	Wood, louvered		280	.029			.91		.91	1.45	
	0060	Wood, overhead		616	.013			.41		.41	.66	
	0070	Wire brush										
	0080	Metal, flush	1 Pord	640	.013	S.F.		.40		.40	.64	
	0090	Metal, detail		520	.015			.49		.49	.78	
	0100	Metal, louvered		360	.022			.70		.70	1.13	
	0110	Metal or fibr., overhead		640	.013			.40		.40	.64	
	0120	Metal, roll up		560	.014			.45		.45	.73	
	0130	Metal, bulkhead		640	.013			.40		.40	.64	
	0140	Power wash, based on 2500 lb. operating pressure										
	0150	Metal, flush	B-9	2240	.018	S.F.		.50	.07	.57	.89	
	0160	Metal, detail		2120	.019			.52	.07	.59	.95	
	0170	Metal, louvered		2000	.020			.56	.07	.63	1	
	0180	Metal or fibr., overhead		2400	.017			.46	.06	.52	.84	
	0190	Metal, roll up		2400	.017			.46	.06	.52	.84	
	0200	Metal, bulkhead		2200	.018			.51	.07	.58	.91	
	0400	Windows, per side, not incl. trim										
	0410	Scrape & sand										
	0420	Wood, 1-2 lite	1 Pord	320	.025	S.F.		.79		.79	1.27	
	0430	Wood, 3-6 lite		280	.029			.91		.91	1.45	
	0440	Wood, 7-10 lite		240	.033			1.06		1.06	1.70	
	0450	Wood, 12 lite		200	.040			1.27		1.27	2.04	
	0460	Wood, Bay / Bow		320	.025			.79		.79	1.27	
	0470	Wire brush										
	0480	Metal, 1-2 lite	1 Pord	480	.017	S.F.		.53		.53	.85	
	0490	Metal, 3-6 lite		400	.020			.63		.63	1.02	
	0500	Metal, Bay / Bow		480	.017			.53		.53	.85	
	0510	Power wash, based on 2500 lb. operating pressure										
	0520	1-2 lite	B-9	4400	.009	S.F.		.25	.03	.28	.46	
	0530	3-6 lite		4320	.009			.26	.03	.29	.47	
	0540	7-10 lite		4240	.009			.26	.03	.29	.47	
	0550	12 lite		4160	.010			.27	.04	.31	.48	
	0560	Bay / Bow		4400	.009			.25	.03	.28	.46	

		09990 | Paint Restoration	CREW	DAILY OUTPUT	LABOR-HOURS	UNIT	2006 BARE COSTS MAT.	LABOR	EQUIP.	TOTAL	TOTAL INCL O&P	
900	0600	Siding, scrape and sand, light=10-30%, med.=30-70%										900
	0610	Heavy=70-100%, % of surface to sand										
	0650	Texture 1-11, light	1 Pord	480	.017	S.F.		.53		.53	.85	
	0660	Med.		440	.018			.58		.58	.93	
	0670	Heavy		360	.022			.70		.70	1.13	
	0680	Wood shingles, shakes, light		440	.018			.58		.58	.93	
	0690	Med.		360	.022			.70		.70	1.13	
	0700	Heavy		280	.029			.91		.91	1.45	
	0710	Clapboard, light		520	.015			.49		.49	.78	
	0720	Med.		480	.017			.53		.53	.85	
	0730	Heavy		400	.020			.63		.63	1.02	
	0740	Wire brush										
	0750	Aluminum, light	1 Pord	600	.013	S.F.		.42		.42	.68	
	0760	Med.		520	.015			.49		.49	.78	
	0770	Heavy		440	.018			.58		.58	.93	
	0780	Pressure wash, based on 2500 lb.. operating pressure										
	0790	Stucco	B-9	3080	.013	S.F.		.36	.05	.41	.65	
	0800	Aluminum or vinyl		3200	.013			.35	.05	.40	.63	
	0810	Siding, masonry, brick & block		2400	.017			.46	.06	.52	.84	
	1300	Miscellaneous, wire brush										
	1310	Metal, pedestrian gate	1 Pord	100	.080	S.F.		2.54		2.54	4.07	
	8000	For Chemical Washing, see Division 04930										
	8010	For Steam Cleaning, see Division 04930										
	8020	For Sand Blasting, see Division 05910-500 and 03350-350										
910	0010	**SURFACE PREPARATION, INTERIOR**										910
	0020	Doors										
	0030	Scrape & sand										
	0040	Wood, flush	1 Pord	616	.013	S.F.		.41		.41	.66	
	0050	Wood, detail		496	.016			.51		.51	.82	
	0060	Wood, louvered		280	.029			.91		.91	1.45	
	0070	Wire brush										
	0080	Metal, flush	1 Pord	640	.013	S.F.		.40		.40	.64	
	0090	Metal, detail		520	.015			.49		.49	.78	
	0100	Metal, louvered		360	.022			.70		.70	1.13	
	0110	Hand wash										
	0120	Wood, flush	1 Pord	2160	.004	S.F.		.12		.12	.19	
	0130	Wood, detailed		2000	.004			.13		.13	.20	
	0140	Wood, louvered		1360	.006			.19		.19	.30	
	0150	Metal, flush		2160	.004			.12		.12	.19	
	0160	Metal, detail		2000	.004			.13		.13	.20	
	0170	Metal, louvered		1360	.006			.19		.19	.30	
	0400	Windows, per side, not incl. trim										
	0410	Scrape & sand										
	0420	Wood, 1-2 lite	1 Pord	360	.022	S.F.		.70		.70	1.13	
	0430	Wood, 3-6 lite		320	.025			.79		.79	1.27	
	0440	Wood, 7-10 lite		280	.029			.91		.91	1.45	
	0450	Wood, 12 lite		240	.033			1.06		1.06	1.70	
	0460	Wood, Bay / Bow		360	.022			.70		.70	1.13	
	0470	Wire brush										
	0480	Metal, 1-2 lite	1 Pord	520	.015	S.F.		.49		.49	.78	
	0490	Metal, 3-6 lite		440	.018			.58		.58	.93	
	0500	Metal, Bay / Bow		520	.015			.49		.49	.78	
	0600	Walls, sanding, light=10-30%										
	0610	Med.=30-70%, heavy=70-100%, % of surface to sand										
	0650	Walls, sand										
	0660	Drywall, gypsum, plaster, light	1 Pord	3077	.003	S.F.		.08		.08	.13	

	09990 | Paint Restoration		CREW	DAILY OUTPUT	LABOR-HOURS	UNIT	MAT.	LABOR	EQUIP.	TOTAL	TOTAL INCL O&P	
910	0670	Drywall, gypsum, plaster, med.	1 Pord	2160	.004	S.F.		.12		.12	.19	910
	0680	Drywall, gypsum, plaster, heavy		923	.009			.27		.27	.44	
	0690	Wood, T&G, light		2400	.003			.11		.11	.17	
	0700	Wood, T&G, med.		1600	.005			.16		.16	.25	
	0710	Wood, T&G, heavy	↓	800	.010	↓		.32		.32	.51	
	0720	Walls, wash										
	0730	Drywall, gypsum, plaster	1 Pord	3200	.003	S.F.		.08		.08	.13	
	0740	Wood, T&G		3200	.003			.08		.08	.13	
	0750	Masonry, brick & block, smooth		2800	.003			.09		.09	.15	
	0760	Masonry, brick & block, coarse	↓	2000	.004	↓		.13		.13	.20	
	8000	For Chemical Washing, see Division 04930										
	8010	For Steam Cleaning, see Division 04930										
	8020	For Sand Blasting, see Division 03350-350 and 05910-500										

(2006 BARE COSTS: MAT., LABOR, EQUIP., TOTAL)

For information about Means Estimating Seminars, see yellow pages 12 and 13 in back of book

Division 10 Specialties

Estimating Tips

General

- The items in this division are usually priced per square foot or each.
- Many items in Division 10 require some type of support system or special anchors that are not usually furnished with the item. The required anchors must be added to the estimate in the appropriate division.
- Some items in Division 10, such as lockers, may require assembly before installation. Verify the amount of assembly required. Assembly can often exceed installation time.

10150 Compartments & Cubicles

- Support angles and blocking are not included in the installation of toilet compartments, shower/dressing compartments, or cubicles. Appropriate line items from Divisions 5 or 6 may need to be added to support the installations.
- Toilet partitions are priced by the stall. A stall consists of a side wall, pilaster, and door with hardware. Toilet tissue holders and grab bars are extra.

10600 Partitions

- The required acoustical rating of a folding partition can have a significant impact on costs. Verify the sound transmission coefficient rating of the panel priced to the specification requirements.

10800 Toilet/Bath/Laundry Accessories

- Grab bar installation does not include supplemental blocking or backing to support the required load. When grab bars are installed at an existing facility, provisions must be made to attach the grab bars to solid structure.

Reference Numbers

Reference numbers are shown in bold squares at the beginning of some major classifications. These numbers refer to related items in the Reference Section. The reference information may be an estimating procedure, an alternate pricing method, or technical information.

Note: Not all subdivisions listed here necessarily appear in this publication.

10050 | Basic Specialties Materials and Methods

10060 | Selective Demolition

		10060 Selective Demolition	CREW	DAILY OUTPUT	LABOR-HOURS	UNIT	2006 BARE COSTS MAT.	LABOR	EQUIP.	TOTAL	TOTAL INCL O&P	
100	0010	**SELECTIVE DEMOLITION, SPECIALTIES** R024119-10										100
	1100	Boards and panels, wall mounted	2 Clab	15	1.067	Ea.		29		29	48.50	
	1200	Cases, for directory and/or bulletin boards, incl doors	↓	24	.667	↓		18.25		18.25	30.50	
	1850	Shower partitons, cabinet or stall, incl base and door	↓	8	2	↓		55		55	91	
	1855	Shower receptor, terrazo or concrete	1 Clab	14	.571	↓		15.65		15.65	26	
	1900	Curtain track or rod, hospital type, ceiling mounted or suspended	↓	220	.036	L.F.		1		1	1.65	
	2650	Wall guard, misc. wall or corner protection	↓	320	.025	"		.69		.69	1.14	
	2750	Access floor, metal panel system, incl pedestals, covering	2 Clab	850	.019	S.F.		.52		.52	.85	
	3050	Fireplace, prefab, freestanding or wall hung, incl hood and screen	1 Clab	2	4	Ea.		110		110	182	
	3054	Chimney top, simulated brick, 4' high	"	15	.533			14.60		14.60	24	
	3200	Stove, woodburning, cast iron	2 Clab	2	8			219		219	365	
	3440	Weathervane, residential	1 Clab	12	.667			18.25		18.25	30.50	
	3500	Flagpole, groundset, to 70' high, excl base/fndn	K-1	1	16			505	176	681	1,025	
	3555	To 30' high	"	2.50	6.400			202	70.50	272.50	410	
	4300	Letter, signs or plaques, exterior on wall	1 Clab	20	.400			10.95		10.95	18.15	
	4310	Signs, street, reflective aluminum, incl post and bracket	"	60	.133			3.65		3.65	6.05	
	4550	Turnstiles, manual or electric	2 Clab	2	8	↓		219		219	365	
	5050	Lockers	"	15	1.067	Opng.		29		29	48.50	
	5250	Cabinets, recessed	Q-12	12	1.333	Ea.		50		50	78	
	5350	Awning, fabric, including frame	2 Clab	100	.160	S.F.		4.38		4.38	7.25	
	6050	Partition, woven wire		1400	.011			.31		.31	.52	
	6100	Folding gate, security, door or window		500	.032			.88		.88	1.45	
	6580	Acoustic air wall		650	.025	↓		.67		.67	1.12	
	7550	Telephone enclosure, exterior, post mounted		3	5.333	Ea.		146		146	242	
	8850	Scale, platform, excludes foundation or pit	↓	.25	64	"		1,750		1,750	2,900	

10100 | Visual Display Boards

10110 | Chalkboards

		10110 Chalkboards	CREW	DAILY OUTPUT	LABOR-HOURS	UNIT	2006 BARE COSTS MAT.	LABOR	EQUIP.	TOTAL	TOTAL INCL O&P	
240	0010	**CHALKBOARDS** Porcelain enamel steel										240
	0100	Freestanding, reversible										
	0120	Economy, wood frame, 4' x 6'										
	0140	Chalkboard both sides				Ea.	545			545	600	
	0160	Chalkboard one side, cork other side				"	455			455	500	
	0200	Standard, lightweight satin finished aluminum, 4' x 6'										
	0220	Chalkboard both sides				Ea.	565			565	620	
	0240	Chalkboard one side, cork other side				"	490			490	540	
	0300	Deluxe, heavy duty extruded aluminum, 4' x 6'										
	0320	Chalkboard both sides				Ea.	1,175			1,175	1,300	
	0340	Chalkboard one side, cork other side				"	1,075			1,075	1,175	
	0400	Sliding chalkboards										
	0450	Vertical, one sliding board with back panel, wall mounted										
	0500	8' x 4'	2 Carp	8	2	Ea.	1,525	71		1,596	1,800	
	0520	8' x 8'	↓	7.50	2.133	↓	2,125	76		2,201	2,475	
	0540	8' x 12'	↓	7	2.286	↓	2,650	81.50		2,731.50	3,025	
	0600	Two sliding boards, with back panel										
	0620	8' x 4'	2 Carp	8	2	Ea.	2,325	71		2,396	2,675	
	0640	8' x 8'	↓	7.50	2.133	↓	3,200	76		3,276	3,650	
	0660	8' x 12'	↓	7	2.286	↓	3,975	81.50		4,056.50	4,500	

10100 | Visual Display Boards

10110 | Chalkboards

240

	10110 Chalkboards	CREW	DAILY OUTPUT	LABOR-HOURS	UNIT	2006 BARE COSTS MAT.	LABOR	EQUIP.	TOTAL	TOTAL INCL O&P
0700	Horizontal, two track									
0800	4' x 8', 2 sliding panels	2 Carp	8	2	Ea.	1,575	71		1,646	1,850
0820	4' x 12', 2 sliding panels	↓	7.50	2.133	↓	1,975	76		2,051	2,300
0840	4' x 16', 4 sliding panels	↓	7	2.286	↓	2,675	81.50		2,756.50	3,075
0900	Four track, four sliding panels									
0920	4' x 8'	2 Carp	8	2	Ea.	2,525	71		2,596	2,900
0940	4' x 12'	↓	7.50	2.133	↓	3,200	76		3,276	3,625
0960	4' x 16'	↓	7	2.286	↓	4,050	81.50		4,131.50	4,575
1200	Vertical, motor operated									
1400	One sliding panel with back panel									
1450	10' x 4'	2 Carp	4	4	Ea.	5,775	142		5,917	6,575
1500	10' x 10'	↓	3.75	4.267	↓	6,700	152		6,852	7,625
1550	10' x 16'	↓	3.50	4.571	↓	7,650	163		7,813	8,675
1700	Two sliding panels with back panel									
1750	10' x 4'	2 Carp	4	4	Ea.	9,250	142		9,392	10,400
1800	10' x 10'	↓	3.75	4.267	↓	10,500	152		10,652	11,800
1850	10' x 16'	↓	3.50	4.571	↓	11,800	163		11,963	13,300
2000	Three sliding panels with back panel									
2100	10' x 4'	2 Carp	4	4	Ea.	12,100	142		12,242	13,500
2150	10' x 10'	↓	3.75	4.267	↓	14,000	152		14,152	15,700
2200	10' x 16'	↓	3.50	4.571	↓	15,800	163		15,963	17,700
2400	For projection screen, glass beaded, add				S.F.	10.80			10.80	11.90
2500	For remote control, 1 panel control, add				Ea.	455			455	500
2600	2 panel control, add				"	785			785	865
2800	For units without back panels, deduct				S.F.	4.99			4.99	5.50
2850	For liquid chalk porcelain panels, add				"	4.21			4.21	4.63
3000	Swing leaf, any comb. of chalkboard & cork, aluminum frame									
3100	Floor style, 6 panels									
3150	30" x 40" panels				Ea.	880			880	965
3200	48" x 40" panels				"	2,275			2,275	2,525
3300	Wall mounted, 6 panels									
3400	30" x 40" panels	2 Carp	16	1	Ea.	810	35.50		845.50	955
3450	48" x 40" panels	"	16	1	"	1,700	35.50		1,735.50	1,900
3600	Extra panels for swing leaf units									
3700	30" x 40" panels				Ea.	263			263	290
3750	48" x 40" panels				"	315			315	345
3900	Wall hung									
4000	Aluminum frame and chalktrough									
4200	3' x 4'	2 Carp	16	1	Ea.	150	35.50		185.50	224
4300	3' x 5'	↓	15	1.067	↓	218	38		256	305
4500	4' x 8'	↓	14	1.143	↓	335	40.50		375.50	435
4600	4' x 12'	↓	13	1.231	↓	480	44		524	605
4700	Wood frame and chalktrough									
4800	3' x 4'	2 Carp	16	1	Ea.	155	35.50		190.50	230
5000	3' x 5'	↓	15	1.067	↓	217	38		255	300
5100	4' x 5'	↓	14	1.143	↓	219	40.50		259.50	310
5300	4' x 8'	↓	13	1.231	↓	310	44		354	415
5400	Liquid chalk, white porcelain enamel, wall hung									
5420	Deluxe units, aluminum trim and chalktrough									
5450	4' x 4'	2 Carp	16	1	Ea.	191	35.50		226.50	269
5500	4' x 8'	↓	14	1.143	↓	335	40.50		375.50	435
5550	4' x 12'	↓	12	1.333	↓	455	47.50		502.50	580
5700	Wood trim and chalktrough									
5900	4' x 4'	2 Carp	16	1	Ea.	440	35.50		475.50	545
6000	4' x 6'	↓	15	1.067	↓	530	38		568	645
6200	4' x 8'	↓	14	1.143	↓	615	40.50		655.50	745

240

10100 | Visual Display Boards

10110 | Chalkboards

			CREW	DAILY OUTPUT	LABOR-HOURS	UNIT	2006 BARE COSTS MAT.	LABOR	EQUIP.	TOTAL	TOTAL INCL O&P	
240	6250	Dry erase board, 36″ tall, incl. placement	2 Carp	104	.154	S.F.	5.55	5.45		11	15.15	240
	6300	Liquid chalk, felt tip markers				Ea.	1.50			1.50	1.65	
	6500	Erasers					1.53			1.53	1.68	
	6600	Board cleaner, 8 oz. bottle					4.13			4.13	4.54	
	9000	Minimum labor/equipment charge	2 Carp	3	5.333	Job		190		190	315	

10120 | Tack & Visual Aid Boards

			CREW	DAILY OUTPUT	LABOR-HOURS	UNIT	MAT.	LABOR	EQUIP.	TOTAL	TOTAL INCL O&P	
350	0010	CONTROL BOARDS										350
	0020	Magnetic, porcelain finish, 18″ x 24″, framed	2 Carp	8	2	Ea.	169	71		240	305	
	0100	24″ x 36″		7.50	2.133		235	76		311	385	
	0200	36″ x 48″		7	2.286		295	81.50		376.50	460	
	0300	48″ x 72″		6	2.667		625	95		720	845	
	0400	48″ x 96″		5	3.200		745	114		859	1,000	
940	0010	BULLETIN BOARD										940
	0020	Cork sheets, unbacked, no frame, 1/4″ thick	2 Carp	290	.055	S.F.	3.60	1.96		5.56	7.20	
	0100	1/2″ thick		290	.055		6.15	1.96		8.11	10	
	0300	Fabric-face, no frame, on 7/32″ cork underlay		290	.055		5.10	1.96		7.06	8.85	
	0400	On 1/4″ cork on 1/4″ hardboard		290	.055		6.15	1.96		8.11	10.05	
	0600	With edges wrapped		290	.055		6.60	1.96		8.56	10.50	
	0700	On 7/16″ fire retardant core		290	.055		4.45	1.96		6.41	8.15	
	0900	With edges wrapped		290	.055		6.60	1.96		8.56	10.50	
	1000	Designer fabric only, cut to size					1.90			1.90	2.09	
	1200	1/4″ vinyl cork, on 1/4″ hardboard, no frame	2 Carp	290	.055		6.20	1.96		8.16	10.10	
	1300	On 1/4″ coreboard		290	.055		6.20	1.96		8.16	10.05	
	2000	For map and display rail, economy, add		385	.042	L.F.	2.08	1.48		3.56	4.74	
	2100	Deluxe, add		350	.046	″	3.04	1.63		4.67	6.05	
	2120	Prefabricated, 1/4″ cork, 3′ x 5′ with aluminum frame		16	1	Ea.	114	35.50		149.50	184	
	2140	Wood frame		16	1		147	35.50		182.50	221	
	2160	4′ x 4′ with aluminum frame		16	1		94.50	35.50		130	163	
	2180	Wood frame		16	1		130	35.50		165.50	202	
	2200	4′ x 8′ with aluminum frame		14	1.143		182	40.50		222.50	268	
	2210	With wood frame		14	1.143		94.50	40.50		135	172	
	2220	4′ x 12′ with aluminum frame		12	1.333		262	47.50		309.50	370	
	2230	Bulletin board case, single glass door, with lock										
	2240	36″ x 24″, economy	2 Carp	12	1.333	Ea.	198	47.50		245.50	296	
	2250	Deluxe		12	1.333		450	47.50		497.50	575	
	2260	42″ x 30″, economy		12	1.333		365	47.50		412.50	480	
	2270	Deluxe		12	1.333		465	47.50		512.50	590	
	2300	Glass enclosed cabinets, alum., cork panel, hinged doors										
	2400	3′ x 3′, 1 door	2 Carp	12	1.333	Ea.	510	47.50		557.50	640	
	2500	4′ x 4′, 2 door		11	1.455		850	51.50		901.50	1,025	
	2600	4′ x 7′, 3 door		10	1.600		1,400	57		1,457	1,650	
	2800	4′ x 10′, 4 door		8	2		2,050	71		2,121	2,375	
	2900	For lights, add per door opening	1 Elec	13	.615		154	26		180	210	
	3100	Horizontal sliding units, 4 doors, 4′ x 8′, 8′ x 4′	2 Carp	9	1.778		1,575	63		1,638	1,825	
	3200	4′ x 12′		7	2.286		1,975	81.50		2,056.50	2,300	
	3400	8 doors, 4′ x 16′		5	3.200		2,675	114		2,789	3,150	
	3500	4′ x 24′		4	4		3,600	142		3,742	4,200	
	9000	Minimum labor/equipment charge		4	4	Job		142		142	236	

10150 | Compartments & Cubicles

10160 | Metal Toilet Compartments

			CREW	DAILY OUTPUT	LABOR-HOURS	UNIT	2006 BARE COSTS MAT.	LABOR	EQUIP.	TOTAL	TOTAL INCL O&P	
100	0010	TOILET PARTITIONS, METAL										100
	0110	Cubicles, ceiling hung										
	0200	Painted metal	2 Carp	4	4	Ea.	475	142		617	755	
	0500	Stainless steel	″	4	4		1,250	142		1,392	1,600	
	0600	For handicap units, incl. 52″ grab bars, add					415			415	455	
	0900	Floor and ceiling anchored										
	1000	Painted metal	2 Carp	5	3.200	Ea.	470	114		584	705	
	1300	Stainless steel	″	5	3.200		1,450	114		1,564	1,800	
	1400	For handicap units, incl. 52″ grab bars, add					287			287	315	
	1610	Floor mounted										
	1700	Painted metal	2 Carp	7	2.286	Ea.	530	81.50		611.50	715	
	2000	Stainless steel	″	7	2.286		1,550	81.50		1,631.50	1,825	
	2100	For handicap units, incl. 52″ grab bars, add					287			287	315	
	2200	For juvenile units, deduct					39.50			39.50	43.50	
	2450	Floor mounted, headrail braced										
	2500	Painted metal	2 Carp	6	2.667	Ea.	470	95		565	670	
	2800	Stainless steel	″	6	2.667		1,450	95		1,545	1,725	
	2900	For handicap units, incl. 52″ grab bars, add					305			305	335	
	3000	Wall hung partitions, painted metal	2 Carp	7	2.286		595	81.50		676.50	790	
	3300	Stainless steel	″	7	2.286		1,425	81.50		1,506.50	1,675	
	3400	For handicap units, incl. 52″ grab bars, add					305			305	335	
	4000	Screens, entrance, floor mounted, 58″ high, 48″ wide										
	4200	Painted metal	2 Carp	15	1.067	Ea.	212	38		250	297	
	4500	Stainless steel	″	15	1.067	″	765	38		803	910	
	4650	Urinal screen, 18″ wide										
	4700	Painted metal	2 Carp	8	2	Ea.	204	71		275	345	
	5000	Stainless steel	″	8	2	″	550	71		621	725	
	5100	Floor mounted, head rail braced										
	5300	Painted metal	2 Carp	8	2	Ea.	198	71		269	335	
	5600	Stainless steel	″	8	2	″	640	71		711	825	
	5750	Pilaster, flush										
	5800	Painted metal	2 Carp	10	1.600	Ea.	269	57		326	390	
	6100	Stainless steel		10	1.600		485	57		542	625	
	6300	Urinal screen, post braced, painted metal		10	1.600		294	57		351	420	
	6600	Stainless steel		10	1.600		480	57		537	625	
	6700	Wall hung, bracket supported										
	6800	Painted metal	2 Carp	10	1.600	Ea.	281	57		338	405	
	7100	Stainless steel		10	1.600		440	57		497	580	
	7400	Flange supported, painted metal		10	1.600		218	57		275	335	
	7700	Stainless steel		10	1.600		510	57		567	655	
	7800	Wedge type, painted metal		10	1.600		257	57		314	375	
	8100	Stainless steel		10	1.600		525	57		582	670	
	9000	Minimum labor/equipment charge	1 Carp	2.50	3.200	Job		114		114	188	

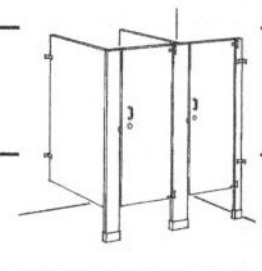

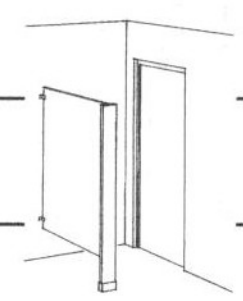

10165 | Plastic Laminate Toilet Compartment

			CREW	DAILY OUTPUT	LABOR-HOURS	UNIT	MAT.	LABOR	EQUIP.	TOTAL	TOTAL INCL O&P	
100	0010	TOILET PARTITIONS, PLASTIC LAMINATE										100
	0110	Cubicles, ceiling hung										
	0300	Plastic laminate on particle board	2 Carp	4	4	Ea.	575	142		717	865	
	0600	For handicap units, incl. 52″ grab bars, add				″	415			415	455	
	0900	Floor and ceiling anchored										
	1100	Plastic laminate on particle board	2 Carp	5	3.200	Ea.	665	114		779	920	
	1400	For handicap units, incl. 52″ grab bars, add				″	287			287	315	
	1610	Floor mounted										
	1800	Plastic laminate on particle board	2 Carp	7	2.286	Ea.	570	81.50		651.50	760	
	2450	Floor mounted, headrail braced										

SPECIALTIES 10

10165 | Plastic Laminate Toilet Compartment

		10165 Plastic Laminate Toilet Compartment	CREW	DAILY OUTPUT	LABOR-HOURS	UNIT	2006 BARE COSTS MAT.	LABOR	EQUIP.	TOTAL	TOTAL INCL O&P	
100	2600	Plastic laminate on particle board	2 Carp	6	2.667	Ea.	770	95		865	1,000	100
	3400	For handicap units, incl. 52" grab bars, add					305			305	335	
	4300	Entrance screen, floor mtd., plas. lam., 58" high, 48" wide	2 Carp	15	1.067		435	38		473	540	
	4800	Urinal screen, 18" wide, ceiling braced, plastic laminate		8	2		293	71		364	440	
	5400	Floor mounted, headrail braced		8	2		355	71		426	510	
	5900	Pilaster, flush, plastic laminate		10	1.600		360	57		417	490	
	6400	Post braced, plastic laminate		10	1.600		360	57		417	490	
	6700	Wall hung, bracket supported										
	6900	Plastic laminate on particle board	2 Carp	10	1.600	Ea.	167	57		224	277	
	7450	Flange supported										
	7500	Plastic laminate on particle board	2 Carp	10	1.600	Ea.	395	57		452	530	

10170 | Plastic Toilet Compartments

		10170 Plastic Toilet Compartments	CREW	DAILY OUTPUT	LABOR-HOURS	UNIT	MAT.	LABOR	EQUIP.	TOTAL	TOTAL INCL O&P	
100	0010	**TOILET PARTITIONS, PLASTIC**										100
	0110	Cubicles, ceiling hung										
	0250	Phenolic	2 Carp	4	4	Ea.	955	142		1,097	1,275	
	0600	For handicap units, incl. 52" grab bars, add				"	415			415	455	
	0900	Floor and ceiling anchored										
	1050	Phenolic	2 Carp	5	3.200	Ea.	985	114		1,099	1,275	
	1400	For handicap units, incl. 52" grab bars, add				"	287			287	315	
	1610	Floor mounted										
	1750	Phenolic	2 Carp	7	2.286	Ea.	1,075	81.50		1,156.50	1,325	
	2100	For handicap units, incl. 52" grab bars, add					287			287	315	
	2200	For juvenile units, deduct					39.50			39.50	43.50	
	2450	Floor mounted, headrail braced										
	2550	Phenolic	2 Carp	6	2.667	Ea.	950	95		1,045	1,200	

10180 | Stone Toilet Compartments

		10180 Stone Toilet Compartments	CREW	DAILY OUTPUT	LABOR-HOURS	UNIT	MAT.	LABOR	EQUIP.	TOTAL	TOTAL INCL O&P	
100	0010	**TOILET PARTITIONS, STONE**										100
	0100	Cubicles, ceiling hung, marble	2 Marb	2	8	Ea.	1,550	279		1,829	2,150	
	0600	For handicap units, incl. 52" grab bars, add					415			415	455	
	0800	Floor & ceiling anchored, marble	2 Marb	2.50	6.400		1,675	223		1,898	2,200	
	1400	For handicap units, incl. 52" grab bars, add					287			287	315	
	1600	Floor mounted, marble	2 Marb	3	5.333		995	186		1,181	1,400	
	2400	Floor mounted, headrail braced, marble	"	3	5.333		950	186		1,136	1,350	
	2900	For handicap units, incl. 52" grab bars, add					305			305	335	
	4100	Entrance screen, floor mounted marble, 58" high, 48" wide	2 Marb	9	1.778		625	62		687	785	
	4600	Urinal screen, 18" wide, ceiling braced, marble	D-1	6	2.667		625	85.50		710.50	825	
	5100	Floor mounted, head rail braced										
	5200	Marble	D-1	6	2.667	Ea.	545	85.50		630.50	740	
	5700	Pilaster, flush, marble		9	1.778		695	57		752	860	
	6200	Post braced, marble		9	1.778		690	57		747	850	
	9000	Minimum labor/equipment charge	1 Carp	2.50	3.200	Job		114		114	188	

10185 | Shower/Dressing Compartments

		10185 Shower/Dressing Compartments	CREW	DAILY OUTPUT	LABOR-HOURS	UNIT	MAT.	LABOR	EQUIP.	TOTAL	TOTAL INCL O&P	
100	0010	**PARTITIONS, SHOWER** Floor mounted, no plumbing										100
	0100	Cabinet, incl. base, no door, painted steel, 1" thick walls	2 Shee	5	3.200	Ea.	765	135		900	1,050	
	0300	With door, fiberglass		4.50	3.556		620	150		770	920	
	0600	Galvanized and painted steel, 1" thick walls		5	3.200		805	135		940	1,100	
	0800	Stall, 1" thick wall, no base, enameled steel		5	3.200		875	135		1,010	1,175	
	1200	Stainless steel		5	3.200		1,400	135		1,535	1,775	
	1400	For double entry type, no doors, deduct					10%					
	1500	Circular fiberglass, cabinet 36" diameter,	2 Shee	4	4		635	169		804	970	

	10185	Shower/Dressing Compartments	CREW	DAILY OUTPUT	LABOR-HOURS	UNIT	2006 BARE COSTS MAT.	LABOR	EQUIP.	TOTAL	TOTAL INCL O&P	
100	1700	One piece, 36″ diameter, less door	2 Shee	4	4	Ea.	535	169		704	860	100
	1800	With door		3.50	4.571		880	193		1,073	1,275	
	2000	Curved shell shower, no door needed		3	5.333		770	225		995	1,200	
	2300	For fiberglass seat, add to both above					109			109	120	
	2400	Glass stalls, with doors, no receptors, chrome on brass	2 Shee	3	5.333		1,250	225		1,475	1,725	
	2700	Anodized aluminum	"	4	4		870	169		1,039	1,225	
	2900	Marble shower stall, stock design, with shower door	2 Marb	1.20	13.333		1,925	465		2,390	2,875	
	3000	With curtain		1.30	12.308		1,700	430		2,130	2,575	
	3200	Receptors, precast terrazzo, 32″ x 32″		14	1.143		236	40		276	325	
	3300	48″ x 34″		9.50	1.684		405	59		464	540	
	3500	Plastic, simulated terrazzo receptor, 32″ x 32″		14	1.143		100	40		140	175	
	3600	32″ x 48″		12	1.333		147	46.50		193.50	238	
	3800	Precast concrete, colors, 32″ x 32″		14	1.143		197	40		237	282	
	3900	48″ x 48″		8	2		210	70		280	345	
	4100	Shower doors, economy plastic, 24″ wide	1 Shee	9	.889		109	37.50		146.50	180	
	4200	Tempered glass door, economy		8	1		184	42		226	269	
	4400	Folding, tempered glass, aluminum frame		6	1.333		330	56		386	450	
	4500	Sliding, tempered glass, 48″ opening		6	1.333		217	56		273	330	
	4700	Deluxe, tempered glass, chrome on brass frame, minimum		8	1		269	42		311	365	
	4800	Maximum		1	8		735	335		1,070	1,350	
	4850	On anodized aluminum frame, minimum		2	4		128	169		297	410	
	4900	Maximum		1	8		430	335		765	1,000	
	5100	Shower enclosure, tempered glass, anodized alum. frame										
	5120	2 panel & door, corner unit, 32″ x 32″	1 Shee	2	4	Ea.	430	169		599	745	
	5140	Neo-angle corner unit, 16″ x 24″ x 16″	"	2	4		795	169		964	1,150	
	5200	Shower surround, 3 wall, polypropylene, 32″ x 32″	1 Carp	4	2		300	71		371	450	
	5220	PVC, 32″ x 32″		4	2		272	71		343	415	
	5240	Fiberglass		4	2		315	71		386	465	
	5250	2 wall, polypropylene, 32″ x 32″		4	2		225	71		296	365	
	5270	PVC		4	2		280	71		351	430	
	5290	Fiberglass		4	2		310	71		381	465	
	5300	Tub doors, tempered glass & frame, minimum	1 Shee	8	1		184	42		226	269	
	5400	Maximum		6	1.333		425	56		481	555	
	5600	Chrome plated, brass frame, minimum		8	1		240	42		282	330	
	5700	Maximum		6	1.333		475	56		531	615	
	5900	Tub/shower enclosure, temp. glass, alum. frame, minimum		2	4		325	169		494	625	
	6200	Maximum		1.50	5.333		635	225		860	1,050	
	6500	On chrome-plated brass frame, minimum		2	4		450	169		619	765	
	6600	Maximum		1.50	5.333		925	225		1,150	1,375	
	6800	Tub surround, 3 wall, polypropylene	1 Carp	4	2		182	71		253	320	
	6900	PVC		4	2		277	71		348	425	
	7000	Fiberglass, minimum		4	2		310	71		381	460	
	7100	Maximum		3	2.667		530	95		625	740	
	9990	Minimum labor/equipment charge	1 Shee	2.50	3.200	Job		135		135	215	

	10190	Cubicles	CREW	DAILY OUTPUT	LABOR-HOURS	UNIT	MAT.	LABOR	EQUIP.	TOTAL	TOTAL INCL O&P	
200	0010	**PARTITIONS, HOSPITAL**										200
	0020	Curtain track, box channel, ceiling mounted	1 Carp	135	.059	L.F.	4.72	2.11		6.83	8.70	
	0100	Suspended	"	100	.080	"	6.45	2.84		9.29	11.80	
	0300	Curtains, nylon mesh tops, fire resistant, 11 oz. per lineal yard										
	0310	Polyester oxford cloth, 9′ ceiling height	1 Carp	425	.019	L.F.	10.45	.67		11.12	12.60	
	0500	8′ ceiling height		425	.019		8.85	.67		9.52	10.80	
	0700	Designer oxford cloth		425	.019		22.50	.67		23.17	25.50	
	0800	I.V. track systems										
	0820	I.V. track, oval	1 Carp	135	.059	L.F.	4.59	2.11		6.70	8.55	
	0830	I.V. trolley		32	.250	Ea.	31.50	8.90		40.40	49.50	

SPECIALTIES 10

10150 | Compartments & Cubicles

	10190 | Cubicles		CREW	DAILY OUTPUT	LABOR-HOURS	UNIT	2006 BARE COSTS MAT.	LABOR	EQUIP.	TOTAL	TOTAL INCL O&P	
200	0840	I.V. pendent, (tree, 5 hook)	1 Carp	32	.250	Ea.	112	8.90		120.90	139	200

10200 | Louvers & Vents

	10210 | Wall Louvers		CREW	DAILY OUTPUT	LABOR-HOURS	UNIT	2006 BARE COSTS MAT.	LABOR	EQUIP.	TOTAL	TOTAL INCL O&P	
800	0010	LOUVERS										800
	0020	Aluminum with screen, residential, 8″ x 8″	1 Carp	38	.211	Ea.	10.05	7.50		17.55	23.50	
	0100	12″ x 12″		38	.211		11.80	7.50		19.30	25.50	
	0200	12″ x 18″		35	.229		15.35	8.15		23.50	30.50	
	0250	14″ x 24″		30	.267		20	9.50		29.50	37.50	
	0300	18″ x 24″		27	.296		24.50	10.55		35.05	44.50	
	0500	24″ x 30″		24	.333		30.50	11.85		42.35	53	
	0700	Triangle, adjustable, small		20	.400		28	14.20		42.20	54.50	
	0800	Large		15	.533		45.50	18.95		64.45	81.50	
	1200	Extruded aluminum, see division 15850-600										
	2100	Midget, aluminum, 3/4″ deep, 1″ diameter	1 Carp	85	.094	Ea.	.91	3.35		4.26	6.55	
	2150	3″ diameter		60	.133		1.98	4.74		6.72	10.05	
	2200	4″ diameter		50	.160		2.98	5.70		8.68	12.70	
	2250	6″ diameter		30	.267		3.52	9.50		13.02	19.55	
	2300	Ridge vent strip, mill finish	1 Shee	155	.052	L.F.	2.66	2.18		4.84	6.40	
	2330	Soffit vent, continuous, 3″ wide, aluminum, mill finish	1 Carp	200	.040		.51	1.42		1.93	2.92	
	2340	Baked enamel finish		200	.040		.46	1.42		1.88	2.87	
	2400	Under eaves vent, aluminum, mill finish, 16″ x 4″		48	.167	Ea.	1.96	5.95		7.91	11.95	
	2500	16″ x 8″		48	.167		2.17	5.95		8.12	12.20	
	7000	Vinyl gable vent, 8″ x 8″		38	.211		10.90	7.50		18.40	24.50	
	7020	12″ x 12″		38	.211		25	7.50		32.50	40	
	7080	12″ x 18″		35	.229		30	8.15		38.15	46.50	
	7200	18″ x 24″		30	.267		34	9.50		43.50	53	
	9000	Minimum labor/equipment charge		2.50	3.200	Job		114		114	188	

10260 | Wall & Corner Guards

	10265 | Wall & Corner Guards		CREW	DAILY OUTPUT	LABOR-HOURS	UNIT	2006 BARE COSTS MAT.	LABOR	EQUIP.	TOTAL	TOTAL INCL O&P	
200	0010	CORNER GUARDS										200
	0020	Steel angle w/anchors, 1″ x 1″ x 1/4″, 1.5#/L.F.	2 Carp	160	.100	L.F.	5.50	3.56		9.06	11.95	
	0100	2″ x 2″ x 1/4″ angles, 3.2#/L.F.		150	.107		9.95	3.79		13.74	17.25	
	0200	3″ x 3″ x 5/16″ angles, 6.1#/L.F.		140	.114		15.10	4.06		19.16	23.50	
	0300	4″ x 4″ x 5/16″ angles, 8.2#/L.F.		120	.133		18.05	4.74		22.79	27.50	
	0350	For angles drilled and anchored to masonry, add					15%	120%				
	0370	Drilled and anchored to concrete, add					20%	170%				
	0400	For galvanized angles, add					35%					
	0450	For stainless steel angles, add					100%					
	0500	Steel door track/wheel guards, 4′ - 0″ high	E-4	22	1.455	Ea.	122	59	4.06	185.06	248	

10260 | Wall & Corner Guards

	10265 | Wall & Corner Guards		CREW	DAILY OUTPUT	LABOR-HOURS	UNIT	2006 BARE COSTS MAT.	LABOR	EQUIP.	TOTAL	TOTAL INCL O&P	
200	0800	Pipe bumper for truck doors, 8′ long, 6″ diameter, filled	E-4	20	1.600	Ea.	320	64.50	4.46	388.96	480	200
	0900	8″ diameter	↓	20	1.600	↓	490	64.50	4.46	558.96	665	
	9000	Minimum labor/equipment charge	1 Carp	2	4	Job		142		142	236	
250	0010	**CORNER PROTECTION**										250
	0100	Stainless steel, 16 ga., adhesive mount, 3-1/2″ leg	1 Sswk	80	.100	L.F.	19	4		23	28.50	
	0200	12 ga. stainless, adhesive mount	″	80	.100		32.50	4		36.50	43.50	
	0300	For screw mount, add						10%				
	0500	Vinyl acrylic, adhesive mount, 3″ leg	1 Carp	128	.063		6.50	2.22		8.72	10.85	
	0550	1-1/2″ leg		160	.050		4.50	1.78		6.28	7.90	
	0600	Screw mounted, 3″ leg		80	.100		6.80	3.56		10.36	13.35	
	0650	1-1/2″ leg		100	.080		4.40	2.84		7.24	9.55	
	0700	Clear plastic, screw mounted, 2-1/2″		60	.133		2.97	4.74		7.71	11.10	
	1000	Vinyl cover, alum. retainer, surface mount, 3″ x 3″		48	.167		6.25	5.95		12.20	16.70	
	1050	2″ x 2″		48	.167		4.96	5.95		10.91	15.25	
	1100	Flush mounted, 3″ x 3″		32	.250		13.75	8.90		22.65	30	
	1150	2″ x 2″	↓	32	.250	↓	10.15	8.90		19.05	26	
500	0010	**WALLGUARD**										500
	0400	Rub rail, vinyl, adhesive mounted	1 Carp	185	.043	L.F.	4	1.54		5.54	6.95	
	0500	Neoprene, aluminum backing, 1-1/2″ x 2″		110	.073		6.85	2.59		9.44	11.80	
	1000	Trolley rail, PVC, clipped to wall, 5″ high		185	.043		7.90	1.54		9.44	11.20	
	1050	8″ high		180	.044	↓	9.75	1.58		11.33	13.30	
	1200	Bed bumper, vinyl acrylic, alum. retainer, 21″ long		10	.800	Ea.	37.50	28.50		66	88.50	
	1300	53″ long with aligner		9	.889	″	111	31.50		142.50	175	
	1400	Bumper, vinyl cover, alum. retain., cush. mnt., 1-1/2″ x 2-3/4″		80	.100	L.F.	11.35	3.56		14.91	18.35	
	1500	2″ x 4-1/4″		80	.100		15.90	3.56		19.46	23.50	
	1600	Surface mounted, 1-3/4″ x 3-5/8″		80	.100		10.75	3.56		14.31	17.70	
	2000	Crash rail, vinyl cover, alum. retainer, 1″ x 4″		110	.073		11.80	2.59		14.39	17.30	
	2100	1″ x 8″		90	.089		14.75	3.16		17.91	21.50	
	2150	Vinyl inserts, aluminum plate, 1″ x 2-1/2″		110	.073		11.90	2.59		14.49	17.40	
	2200	1″ x 5″	↓	90	.089	↓	16.30	3.16		19.46	23	
	3000	Handrail/bumper, vinyl cover, alum. retainer										
	3010	Bracket mounted, flat rail, 5-1/2″	1 Carp	80	.100	L.F.	14.25	3.56		17.81	21.50	
	3100	6-1/2″		80	.100		22.50	3.56		26.06	31	
	3200	Bronze bracket, 1-3/4″ diam. rail	↓	80	.100	↓	18.30	3.56		21.86	26	

10270 | Access Flooring

	10275 | Access Flooring		CREW	DAILY OUTPUT	LABOR-HOURS	UNIT	2006 BARE COSTS MAT.	LABOR	EQUIP.	TOTAL	TOTAL INCL O&P	
150	0010	**PEDESTAL ACCESS FLOORS** Computer room application, metal										150
	0020	Particle board or steel panels, no covering, under 6,000 S.F.	2 Carp	400	.040	S.F.	14.50	1.42		15.92	18.30	
	0300	Metal covered, over 6,000 S.F.		450	.036		8.40	1.26		9.66	11.35	
	0400	Aluminum, 24″ panels	↓	500	.032		31	1.14		32.14	36	
	0600	For carpet covering, add					6			6	6.60	
	0700	For vinyl floor covering, add					6.05			6.05	6.70	
	0900	For high pressure laminate covering, add					4.05			4.05	4.46	
	0910	For snap on stringer system, add	2 Carp	1000	.016	↓	1.38	.57		1.95	2.46	
	0950	Office applications, to 8″ high, steel panels,										
	0960	no covering, over 6,000 S.F.	2 Carp	500	.032	S.F.	9.70	1.14		10.84	12.55	

10270 | Access Flooring

		10275 | Access Flooring	CREW	DAILY OUTPUT	LABOR-HOURS	UNIT	2006 BARE COSTS MAT.	LABOR	EQUIP.	TOTAL	TOTAL INCL O&P	
150	1000	Machine cutouts after initial installation	1 Carp	10	.800	Ea.	4.42	28.50		32.92	52	150
	1050	Pedestals, 6″ to 12″	2 Carp	85	.188		8.50	6.70		15.20	20.50	
	1100	Air conditioning grilles, 4″ x 12″	1 Carp	17	.471		61	16.75		77.75	94.50	
	1150	4″ x 18″	″	14	.571		83.50	20.50		104	126	
	1200	Approach ramps, minimum	2 Carp	85	.188	S.F.	23	6.70		29.70	36.50	
	1300	Maximum	″	60	.267	″	31	9.50		40.50	49.50	
	1500	Handrail, 2 rail, aluminum	1 Carp	15	.533	L.F.	89.50	18.95		108.45	130	

10300 | Fireplaces & Stoves

		10305 | Manufactured Fireplaces	CREW	DAILY OUTPUT	LABOR-HOURS	UNIT	2006 BARE COSTS MAT.	LABOR	EQUIP.	TOTAL	TOTAL INCL O&P	
100	0010	**FIREPLACE, PREFABRICATED** Free standing or wall hung										100
	0100	with hood & screen, minimum	1 Carp	1.30	6.154	Ea.	980	219		1,199	1,425	
	0150	Average		1	8		1,375	284		1,659	1,975	
	0200	Maximum		.90	8.889		3,350	315		3,665	4,225	
	0500	Chimney dbl. wall, all stainless, over 8′-6″, 7″ diam., add		33	.242	V.L.F.	54	8.60		62.60	74	
	0600	10″ diameter, add		32	.250		55	8.90		63.90	75.50	
	0700	12″ diameter, add		31	.258		75	9.15		84.15	97.50	
	0800	14″ diameter, add		30	.267		99	9.50		108.50	125	
	1000	Simulated brick chimney top, 4′ high, 16″ x 16″		10	.800	Ea.	202	28.50		230.50	269	
	1100	24″ x 24″		7	1.143	″	375	40.50		415.50	485	
	1500	Simulated logs, gas fired, 40,000 BTU, 2′ long, minimum		7	1.143	Set	510	40.50		550.50	630	
	1600	Maximum		6	1.333		680	47.50		727.50	830	
	1700	Electric, 1,500 BTU, 1′-6″ long, minimum		7	1.143		140	40.50		180.50	222	
	1800	11,500 BTU, maximum		6	1.333		300	47.50		347.50	410	
		10310 | Fireplace Specialties & Accessories										
100	0010	**FIREPLACE ACCESSORIES**										100
	0020	Chimney screens, galv., 13″ x 13″ flue	1 Bric	8	1	Ea.	43	36.50		79.50	107	
	0050	Galv., 24″ x 24″ flue		5	1.600		100	58.50		158.50	205	
	0200	Stainless steel, 13″ x 13″ flue		8	1		264	36.50		300.50	350	
	0250	20″ x 20″ flue		5	1.600		360	58.50		418.50	490	
	0400	Cleanout doors and frames, cast iron, 8″ x 8″		12	.667		30	24.50		54.50	72.50	
	0450	12″ x 12″		10	.800		36.50	29		65.50	87.50	
	0500	18″ x 24″		8	1		105	36.50		141.50	175	
	0550	Cast iron frame, steel door, 24″ x 30″		5	1.600		227	58.50		285.50	345	
	0800	Damper, rotary control, steel, 30″ opening		6	1.333		64	48.50		112.50	149	
	0850	Cast iron, 30″ opening		6	1.333		71	48.50		119.50	157	
	0880	36″ opening		6	1.333		77	48.50		125.50	164	
	0900	48″ opening		6	1.333		110	48.50		158.50	200	
	0920	60″ opening		6	1.333		247	48.50		295.50	350	
	0950	72″ opening		5	1.600		297	58.50		355.50	420	
	1000	84″ opening, special order		5	1.600		635	58.50		693.50	795	
	1050	96″ opening, special order		4	2		645	73		718	830	
	1200	Steel plate, poker control, 60″ opening		8	1		225	36.50		261.50	310	
	1250	84″ opening, special opening		5	1.600		410	58.50		468.50	545	
	1400	"Universal" type, chain operated, 32″ x 20″ opening		8	1		175	36.50		211.50	253	
	1450	48″ x 24″ opening		5	1.600		261	58.50		319.50	380	
	1600	Dutch Oven door and frame, cast iron, 12″ x 15″ opening		13	.615		92	22.50		114.50	138	

10300 | Fireplaces & Stoves

10310 | Fireplace Specialties & Accessories

			Crew	Daily Output	Labor-Hours	Unit	2006 Bare Costs Mat.	Labor	Equip.	Total	Total Incl O&P	
100	1650	Copper plated, 12″ x 15″ opening	1 Bric	13	.615	Ea.	177	22.50		199.50	231	100
	1800	Fireplace forms, no accessories, 32″ opening		3	2.667		510	97.50		607.50	720	
	1900	36″ opening		2.50	3.200		650	117		767	905	
	2000	40″ opening		2	4		860	146		1,006	1,175	
	2100	78″ opening		1.50	5.333		1,250	195		1,445	1,700	
	2400	Squirrel and bird screens, galvanized, 8″ x 8″ flue		16	.500		37.50	18.30		55.80	70.50	
	2450	13″ x 13″ flue		12	.667	↓	40.50	24.50		65	84	
	9000	Minimum labor/equipment charge	↓	3.50	2.286	Job		83.50		83.50	136	

10320 | Stoves

			Crew	Daily Output	Labor-Hours	Unit	Mat.	Labor	Equip.	Total	Total Incl O&P	
100	0010	WOODBURNING STOVES										100
	0015	Cast iron, minimum	2 Carp	1.30	12.308	Ea.	910	440		1,350	1,725	
	0020	Average		1	16		1,400	570		1,970	2,500	
	0030	Maximum	↓	.80	20		2,350	710		3,060	3,750	
	0050	For gas log lighter, add				↓	40			40	44	

10340 | Manufactured Exterior Specialties

10344 | Weathervanes

			Crew	Daily Output	Labor-Hours	Unit	2006 Bare Costs Mat.	Labor	Equip.	Total	Total Incl O&P	
800	0010	WEATHERVANES										800
	0020	Residential types, minimum	1 Carp	8	1	Ea.	130	35.50		165.50	202	
	0100	Maximum		2	4	″	2,300	142		2,442	2,775	
	9000	Minimum labor/equipment charge	↓	4	2	Job		71		71	118	

10350 | Flagpoles

10355 | Flagpoles

			Crew	Daily Output	Labor-Hours	Unit	2006 Bare Costs Mat.	Labor	Equip.	Total	Total Incl O&P	
400	0010	FLAGPOLE, Ground set										400
	0050	Not including base or foundation										
	0100	Aluminum, tapered, ground set 20′ high	K-1	2	8	Ea.	790	253	88	1,131	1,375	
	0200	25′ high		1.70	9.412		895	297	104	1,296	1,575	
	0300	30′ high		1.50	10.667		890	335	117	1,342	1,675	
	0400	35′ high		1.40	11.429		1,450	360	126	1,936	2,325	
	0500	40′ high		1.20	13.333		2,050	420	147	2,617	3,100	
	0600	50′ high		1	16		2,700	505	176	3,381	4,000	
	0700	60′ high		.90	17.778		4,350	560	196	5,106	5,950	
	0800	70′ high		.80	20		7,800	630	220	8,650	9,875	
	1100	Counterbalanced, internal halyard, 20′ high		1.80	8.889		2,050	281	98	2,429	2,825	
	1200	30′ high		1.50	10.667		2,275	335	117	2,727	3,175	
	1300	40′ high		1.30	12.308		3,800	390	136	4,326	4,975	
	1400	50′ high		1	16		7,000	505	176	7,681	8,725	
	2820	Aluminum, electronically operated, 30′ high		1.40	11.429		3,975	360	126	4,461	5,100	
	2840	35′ high	↓	1.30	12.308	↓	4,275	390	136	4,801	5,525	

10350 | Flagpoles

	10355 | Flagpoles		CREW	DAILY OUTPUT	LABOR-HOURS	UNIT	2006 BARE COSTS MAT.	LABOR	EQUIP.	TOTAL	TOTAL INCL O&P	
400	2860	39′ high	K-1	1.10	14.545	Ea.	4,675	460	160	5,295	6,050	400
	2880	45′ high		1	16		5,975	505	176	6,656	7,600	
	2900	50′ high		.90	17.778		6,350	560	196	7,106	8,125	
	3000	Fiberglass, tapered, ground set, 23′ high		2	8		1,050	253	88	1,391	1,675	
	3100	29′-7″ high		1.50	10.667		1,275	335	117	1,727	2,075	
	3200	36′-1″ high		1.40	11.429		1,650	360	126	2,136	2,550	
	3300	39′-5″ high		1.20	13.333		2,150	420	147	2,717	3,225	
	3400	49′-2″ high		1	16		4,375	505	176	5,056	5,825	
	3500	59′ high		.90	17.778		6,275	560	196	7,031	8,050	
	4300	Steel, direct imbedded installation										
	4400	Internal halyard, 20′ high	K-1	2.50	6.400	Ea.	1,350	202	70.50	1,622.50	1,875	
	4500	25′ high		2.50	6.400		1,500	202	70.50	1,772.50	2,050	
	4600	30′ high		2.30	6.957		1,775	220	76.50	2,071.50	2,400	
	4700	40′ high		2.10	7.619		2,625	241	84	2,950	3,400	
	4800	50′ high		1.90	8.421		3,900	266	92.50	4,258.50	4,800	
	5000	60′ high		1.80	8.889		5,875	281	98	6,254	7,050	
	5100	70′ high		1.60	10		7,175	315	110	7,600	8,550	
	5200	80′ high		1.40	11.429		9,475	360	126	9,961	11,100	
	5300	90′ high		1.20	13.333		13,000	420	147	13,567	15,200	
	5500	100′ high		1	16		13,400	505	176	14,081	15,800	
	6400	Wood poles, tapered, clear vertical grain fir with tilting										
	6410	base, not incl. foundation, 4″ butt, 25′ high	K-1	1.90	8.421	Ea.	625	266	92.50	983.50	1,225	
	6800	6″ butt, 30′ high	″	1.30	12.308	″	775	390	136	1,301	1,650	
	7300	Foundations for flagpoles, including										
	7400	excavation and concrete, to 35′ high poles	C-1	10	3.200	Ea.	500	107		607	730	
	7600	40′ to 50′ high		3.50	9.143		910	305		1,215	1,500	
	7700	Over 60′ high		2	16		1,025	535		1,560	2,025	
900	0010	**FLAGPOLE**, Structure mounted										900
	0100	Fiberglass, vertical wall set, 19′-8″ long	K-1	1.50	10.667	Ea.	1,225	335	117	1,677	2,025	
	0200	23′ long		1.40	11.429		1,300	360	126	1,786	2,150	
	0300	26′-3″ long		1.30	12.308		1,900	390	136	2,426	2,875	
	0800	19′-8″ long outrigger		1.30	12.308		1,250	390	136	1,776	2,175	
	1300	Aluminum, vertical wall set, tapered, with base, 20′ high		1.20	13.333		925	420	147	1,492	1,875	
	1400	29′-6″ high		1	16		2,175	505	176	2,856	3,425	
	2400	Outrigger poles with base, 12′ long		1.30	12.308		860	390	136	1,386	1,750	
	2500	14′ long		1	16		1,375	505	176	2,056	2,550	

10 SPECIALTIES

10400 | Identification Devices

	10410 | Directories		CREW	DAILY OUTPUT	LABOR-HOURS	UNIT	2006 BARE COSTS MAT.	LABOR	EQUIP.	TOTAL	TOTAL INCL O&P	
100	0010	**DIRECTORY BOARDS**										100
	0050	Plastic, glass covered, 30″ x 20″	2 Carp	3	5.333	Ea.	242	190		432	580	
	0100	36″ x 48″		2	8		770	284		1,054	1,325	
	0300	Grooved cork, 30″ x 20″		3	5.333		355	190		545	705	
	0400	36″ x 48″		2	8		300	284		584	800	
	0600	Black felt, 30″ x 20″		3	5.333		192	190		382	525	
	0700	36″ x 48″		2	8		330	284		614	830	
	0900	Outdoor, weatherproof, black plastic, 36″ x 24″		2	8		685	284		969	1,225	

10400 | Identification Devices

10410 | Directories

		10410 Directories	CREW	DAILY OUTPUT	LABOR-HOURS	UNIT	2006 BARE COSTS MAT.	LABOR	EQUIP.	TOTAL	TOTAL INCL O&P	
100	1000	36″ x 36″	2 Carp	1.50	10.667	Ea.	790	380		1,170	1,500	100
	1800	Indoor, economy, open face, 18″ x 24″		7	2.286		113	81.50		194.50	260	
	1900	24″ x 36″		7	2.286		137	81.50		218.50	285	
	2000	36″ x 24″		6	2.667		156	95		251	330	
	2100	36″ x 48″		6	2.667		234	95		329	415	
	2400	Building directory, alum., black felt panels, 1 door, 24″ x 18″		4	4		281	142		423	545	
	2500	36″ x 24″		3.50	4.571		320	163		483	625	
	2600	48″ x 32″		3	5.333		480	190		670	840	
	2700	36″ x 48″, 2 door		2.50	6.400		590	228		818	1,025	
	2800	36″ x 60″		2	8		845	284		1,129	1,400	
	2900	48″ x 60″		1	16		925	570		1,495	1,975	
	3100	For bronze enamel finish, add					15%					
	3200	For bronze anodized finish, add					25%					
	3400	For illuminated directory, single door unit, add					154			154	170	
	3500	For 6″ header panel, 6 letters per foot, add				L.F.	27			27	30	
	5000	Building directory, illuminated, 7″ sections, bronze finish										
	5100	19″ x 36″, 2 sections, 80 name capacity				Ea.	2,225			2,225	2,425	
	5200	26-1/4″ x 36″, 3 sections, 120 name capacity					2,625			2,625	2,900	
	5300	33-5/8″ x 36″, 4 sections, 160 name capacity					3,025			3,025	3,325	
	5400	48-3/4″ x 36″, 6 sections, 240 name capacity					3,900			3,900	4,300	
	5500	63-1/2″ x 36″, 8 sections, 320 name capacity					4,750			4,750	5,225	
	5600	Engraving charge per namestrip					11.80			11.80	13	
	9000	Minimum labor/equipment charge	1 Carp	1	8	Job		284		284	470	

10430 | Exterior Signage

		10430 Exterior Signage	CREW	DAILY OUTPUT	LABOR-HOURS	UNIT	MAT.	LABOR	EQUIP.	TOTAL	TOTAL INCL O&P	
200	0010	**SIGNS**										200
	0020	Letters, 2″ high, 3/8″ deep, cast bronze	1 Carp	24	.333	Ea.	20.50	11.85		32.35	42	
	0140	1/2″ deep, cast aluminum		18	.444		20.50	15.80		36.30	48.50	
	0160	Cast bronze		32	.250		25.50	8.90		34.40	43.50	
	0300	6″ high, 5/8″ deep, cast aluminum		24	.333		28.50	11.85		40.35	51	
	0400	Cast bronze		24	.333		42.50	11.85		54.35	66.50	
	0600	8″ high, 3/4″ deep, cast aluminum		14	.571		32.50	20.50		53	69	
	0700	Cast bronze		20	.400		72.50	14.20		86.70	103	
	0900	10″ high, 1″ deep, cast aluminum		18	.444		34	15.80		49.80	63.50	
	1000	Bronze		18	.444		87.50	15.80		103.30	123	
	1200	12″ high, 1-1/4″ deep, cast aluminum		12	.667		43.50	23.50		67	87.50	
	1500	Cast bronze		18	.444		114	15.80		129.80	151	
	1600	14″ high, 2-5/16″ deep, cast aluminum		12	.667		69.50	23.50		93	116	
	1800	Fabricated stainless steel, 6″ high, 2″ deep		20	.400		113	14.20		127.20	148	
	1900	12″ high, 3″ deep		18	.444		134	15.80		149.80	173	
	2100	18″ high, 3″ deep		12	.667		196	23.50		219.50	255	
	2200	24″ high, 4″ deep		10	.800		273	28.50		301.50	345	
	2700	Acrylic, on high density foam, 12″ high, 2″ deep		20	.400		17.20	14.20		31.40	42.50	
	2800	18″ high, 2″ deep		18	.444		42	15.80		57.80	72	
	3900	Plaques, custom, 20″ x 30″, for up to 450 letters, cast aluminum	2 Carp	4	4		855	142		997	1,175	
	4000	Cast bronze		4	4		1,125	142		1,267	1,475	
	4200	30″ x 36″, up to 900 letters cast aluminum		3	5.333		1,725	190		1,915	2,225	
	4300	Cast bronze		3	5.333		2,225	190		2,415	2,775	
	4500	36″ x 48″, for up to 1300 letters, cast bronze		2	8		3,375	284		3,659	4,175	
	4800	Signs, reflective alum. street signs, dbl. face, 2-way, w/bracket		30	.533		71	18.95		89.95	110	
	4900	4-way		30	.533		111	18.95		129.95	154	
	5100	Exit signs, 24 ga. alum., 14″ x 12″ surface mounted	1 Carp	30	.267		33.50	9.50		43	52.50	
	5200	10″ x 7″		20	.400		19.55	14.20		33.75	45	
	5400	Bracket mounted, double face, 12″ x 10″		30	.267		33	9.50		42.50	52	
	5500	Sticky back, stock decals, 14″ x 10″	1 Clab	50	.160		8.30	4.38		12.68	16.40	

10400 | Identification Devices

	10430 | Exterior Signage	CREW	DAILY OUTPUT	LABOR-HOURS	UNIT	2006 BARE COSTS MAT.	LABOR	EQUIP.	TOTAL	TOTAL INCL O&P	
200	6000 Interior elec., wall mount, fiberglass panels, 2 lamps, 6″	1 Elec	8	1	Ea.	104	42		146	179	200
	6100 8″	″	8	1		46	42		88	115	
	6400 Replacement sign faces, 6″ or 8″	1 Clab	50	.160	↓	25	4.38		29.38	35	
	9000 Minimum labor/equipment charge	1 Carp	4	2	Job		71		71	118	

	10440 | Interior Signage	CREW	DAILY OUTPUT	LABOR-HOURS	UNIT	2006 BARE COSTS MAT.	LABOR	EQUIP.	TOTAL	TOTAL INCL O&P	
200	0010 **INTERIOR SIGNAGE**										200
	1010 Flexible door sign, adhesive back, w/Braille, 5/8″ letters, 4″ x 4″	1 Clab	32	.250	Ea.	24	6.85		30.85	37.50	
	1050 6″ x 6″		32	.250		34.50	6.85		41.35	49	
	1100 8″ x 2″		32	.250		25	6.85		31.85	38.50	
	1150 8″ x 4″		32	.250		31	6.85		37.85	45.50	
	1200 8″ x 8″		32	.250		46	6.85		52.85	62	
	1250 12″ x 2″		32	.250		32	6.85		38.85	46.50	
	1300 12″ x 6″		32	.250		45	6.85		51.85	61	
	1350 12″ x 12″		32	.250		92	6.85		98.85	112	
	1500 Graphic symbols, 2″ x 2″		32	.250		11	6.85		17.85	23.50	
	1550 6″ x 6″		32	.250		28	6.85		34.85	42.50	
	1600 8″ x 8″	↓	32	.250	↓	28	6.85		34.85	42.50	

10450 | Pedestrian Control Devices

	10455 | Turnstiles	CREW	DAILY OUTPUT	LABOR-HOURS	UNIT	2006 BARE COSTS MAT.	LABOR	EQUIP.	TOTAL	TOTAL INCL O&P	
900	0010 **TURNSTILES**										900
	0020 One way, 4 arm, 46″ diameter, economy, manual	2 Carp	5	3.200	Ea.	550	114		664	795	
	0100 Electric		1.20	13.333		900	475		1,375	1,775	
	0300 High security, galv., 5′-5″ diameter, 7′ high, manual		1	16		2,325	570		2,895	3,525	
	0350 Electric		.60	26.667		3,025	950		3,975	4,900	
	0420 Three arm, 24″ opening, light duty, manual		2	8		865	284		1,149	1,425	
	0450 Heavy duty		1.50	10.667		2,150	380		2,530	2,975	
	0460 Manual, with registering & controls, light duty		2	8		1,875	284		2,159	2,525	
	0470 Heavy duty		1.50	10.667		2,250	380		2,630	3,100	
	0480 Electric, heavy duty	↓	1.10	14.545		2,625	515		3,140	3,725	
	0500 For coin or token operating, add				↓	420			420	460	
	1200 One way gate with horizontal bars, 5′-5″ diameter										
	1300 7′ high, recreation or transit type	2 Carp	.80	20	Ea.	3,025	710		3,735	4,500	
	1500 For electronic counter, add				″	185			185	203	

10500 | Lockers

	10505 | Metal Lockers	CREW	DAILY OUTPUT	LABOR-HOURS	UNIT	2006 BARE COSTS MAT.	LABOR	EQUIP.	TOTAL	TOTAL INCL O&P	
500	0011 **LOCKERS** Steel, baked enamel										500
	0110 Single tier box locker, 12″ x 15″ x 72″	1 Shee	8	1	Ea.	165	42		207	249	
	0120 18″ x 15″ x 72″		8	1		178	42		220	263	
	0130 12″ x 18″ x 72″	↓	8	1	↓	170	42		212	254	

10500 | Lockers

	10505 | Metal Lockers		CREW	DAILY OUTPUT	LABOR-HOURS	UNIT	2006 BARE COSTS MAT.	LABOR	EQUIP.	TOTAL	TOTAL INCL O&P
500	0140	18″ x 18″ x 72″	1 Shee	8	1	Ea.	193	42		235	279
	0410	Double tier, 12″ x 15″ x 36″		21	.381		175	16.05		191.05	219
	0420	18″ x 15″ x 36″		21	.381		237	16.05		253.05	286
	0430	12″ x 18″ x 36″		21	.381		184	16.05		200.05	228
	0440	18″ x 18″ x 36″		21	.381		197	16.05		213.05	243
	0500	Two person, 18″ x 15″ x 72″		8	1		263	42		305	355
	0510	18″ x 18″ x 72″		8	1		270	42		312	365
	0520	Duplex, 15″ x 15″ x 72″		8	1		258	42		300	350
	0530	15″ x 21″ x 72″		8	1		300	42		342	400
	0600	5 tier box lockers, minimum		30	.267	Opng.	37	11.25		48.25	59
	0700	Maximum		24	.333		46.50	14.05		60.55	73.50
	0900	6 tier box lockers, minimum		36	.222		36.50	9.35		45.85	55
	1000	Maximum		30	.267		43.50	11.25		54.75	65.50
	1100	Wire meshed wardrobe, floor. mtd., open front varsity type		7.50	1.067	Ea.	166	45		211	254
	2100	Locker bench, laminated maple, top only		100	.080	L.F.	13.20	3.37		16.57	19.85
	2200	Pedestals, steel pipe		25	.320	Ea.	35.50	13.50		49	60.50
	2400	16-person locker unit with clothing rack									
	2500	72 wide x 15″ deep x 72″ high	1 Shee	8	1	Ea.	470	42		512	585
	2550	18″ deep	″	8	1	″	400	42		442	505
	3000	Wall mounted lockers, 4 person, with coat bar									
	3100	48″ wide x 18″ deep x 12″ high	1 Shee	8	1	Ea.	210	42		252	298
	3250	Rack w/ 24 wire mesh baskets		1.50	5.333	Set	325	225		550	715
	3260	30 baskets		1.25	6.400		256	270		526	710
	3270	36 baskets		.95	8.421		325	355		680	920
	3280	42 baskets		.80	10		560	420		980	1,275
	3300	For built-in lock with 2 keys, add				Ea.	7			7	7.70
	3600	For hanger rods, add				″	1.79			1.79	1.97
	9000	Minimum labor/equipment charge	1 Shee	2.50	3.200	Job		135		135	215

10520 | Fire Protection Specialties

	10525 | Fire Prot. Specialties		CREW	DAILY OUTPUT	LABOR-HOURS	UNIT	2006 BARE COSTS MAT.	LABOR	EQUIP.	TOTAL	TOTAL INCL O&P
200	0010	**FIRE EQUIPMENT CABINETS** Not equipped, 20 ga. steel box,									
	0040	recessed, D.S. glass in door, box size given									
	1000	Portable extinguisher, single, 8″ x 12″ x 27″, alum. door & frame	Q-12	8	2	Ea.	96	75.50		171.50	222
	1100	Steel door and frame		8	2		70	75.50		145.50	194
	1200	Stainless steel door and frame		8	2		149	75.50		224.50	281
	2000	Portable extinguisher, large, 8″ x 12″ x 36″, alum. door & frame		8	2		171	75.50		246.50	305
	2100	Steel door and frame		8	2		103	75.50		178.50	230
	2200	Stainless steel door and frame		8	2		171	75.50		246.50	305
	2500	8″ x 16″ x 38″, aluminum door & frame		8	2		162	75.50		237.50	295
	2600	Steel door and frame		8	2		145	75.50		220.50	276
	3000	Hose rack assy., 1-1/2″ valve & 100′ hose, 24″ x 40″ x 5-1/2″									
	3100	Aluminum door and frame	Q-12	6	2.667	Ea.	235	100		335	415
	3200	Steel door and frame		6	2.667		154	100		254	325
	3300	Stainless steel door and frame		6	2.667		310	100		410	495
	4000	Hose rack assy., 2-1/2″ x 1-1/2″ valve, 100′ hose, 24″ x 40″ x 8″									
	4100	Aluminum door and frame	Q-12	6	2.667	Ea.	237	100		337	415
	4200	Steel door and frame		6	2.667		160	100		260	330
	4300	Stainless steel door and frame		6	2.667		315	100		415	500

10520 | Fire Protection Specialties

	10525 | Fire Prot. Specialties		CREW	DAILY OUTPUT	LABOR-HOURS	UNIT	2006 BARE COSTS MAT.	LABOR	EQUIP.	TOTAL	TOTAL INCL O&P	
200	5000	Hose rack assy., 2-1/2" x 1-1/2" valve, 100' hose										200
	5010	and extinguisher, 30" x 40" x 8"										
	5100	Aluminum door and frame	Q-12	5	3.200	Ea.	300	120		420	520	
	5200	Steel door and frame		5	3.200		176	120		296	380	
	5300	Stainless steel door and frame		5	3.200		335	120		455	560	
	6000	Hose rack assy., 1-1/2" valve, 100' hose										
	6010	and 2-1/2" FD valve, 24" x 44" x 8"										
	6100	Aluminum door and frame	Q-12	5	3.200	Ea.	268	120		388	485	
	6200	Steel door and frame		5	3.200		171	120		291	375	
	6300	Stainless steel door and frame		5	3.200		355	120		475	580	
	7000	Hose rack assy., 1-1/2" valve & 100' hose, 2-1/2" FD valve										
	7010	and extinguisher, 30" x 44" x 8"										
	7100	Aluminum door and frame	Q-12	5	3.200	Ea.	315	120		435	540	
	7200	Steel door and frame		5	3.200		325	120		445	545	
	7300	Stainless steel door and frame		5	3.200		430	120		550	660	
	8000	Valve cabinet for 2-1/2" FD angle valve, 18" x 18" x 8"										
	8100	Aluminum door and frame	Q-12	12	1.333	Ea.	104	50		154	193	
	8200	Steel door and frame		12	1.333		86	50		136	173	
	8300	Stainless steel door and frame		12	1.333		140	50		190	232	
300	0010	**FIRE EXTINGUISHERS**										300
	0120	CO_2, portable with swivel horn, 5 lb.				Ea.	130			130	143	
	0140	With hose and "H" horn, 10 lb.					160			160	176	
	0160	15 lb.					190			190	209	
	0180	20 lb.					220			220	242	
	0360	Wheeled type, cart mounted, 50 lb.					1,200			1,200	1,325	
	0400	100 lb.					1,625			1,625	1,775	
	1000	Dry chemical, pressurized										
	1040	Standard type, portable, painted, 2-1/2 lb.				Ea.	27.50			27.50	30.50	
	1060	5 lb.					60			60	66	
	1080	10 lb.					80			80	88	
	1100	20 lb.					110			110	121	
	1120	30 lb.					157			157	173	
	1300	Standard type, wheeled, 150 lb.					1,600			1,600	1,750	
	2000	ABC all purpose type, portable, 2-1/2 lb.					30			30	33	
	2060	5 lb.					45			45	49.50	
	2080	9-1/2 lb.					68			68	75	
	2100	20 lb.					98			98	108	
	2300	Wheeled, 45 lb.					650			650	715	
	2360	150 lb.					1,600			1,600	1,750	
	3000	Dry chemical, outside cartridge to -65°F, painted, 9 lb.					200			200	220	
	3060	26 lb.					250			250	275	
	5000	Pressurized water, 2-1/2 gallon, stainless steel					85			85	93.50	
	5060	With anti-freeze					135			135	149	
	9400	Installation of extinguishers, 12 or more, on wood	1 Carp	30	.267			9.50		9.50	15.70	
	9420	On masonry or concrete	"	15	.533			18.95		18.95	31.50	

	10535 | Awnings & Canopies		CREW	DAILY OUTPUT	LABOR-HOURS	UNIT	2006 BARE COSTS MAT.	LABOR	EQUIP.	TOTAL	TOTAL INCL O&P	
050	0010	**AWNINGS, FABRIC**										050
	0020	Including acrylic canvas and frame, standard design										
	0100	Door and window, slope, 3′ high, 4′ wide	1 Carp	4.50	1.778	Ea.	585	63		648	750	
	0110	6′ wide		3.50	2.286		755	81.50		836.50	965	
	0120	8′ wide		3	2.667		925	95		1,020	1,175	
	0200	Quarter round convex, 4′ wide		3	2.667		910	95		1,005	1,150	
	0210	6′ wide		2.25	3.556		1,175	126		1,301	1,500	
	0220	8′ wide		1.80	4.444		1,450	158		1,608	1,850	
	0300	Dome, 4′ wide		7.50	1.067		350	38		388	455	
	0310	6′ wide		3.50	2.286		790	81.50		871.50	1,000	
	0320	8′ wide		2	4		1,400	142		1,542	1,775	
	0350	Elongated dome, 4′ wide		1.33	6.015		1,325	214		1,539	1,800	
	0360	6′ wide		1.11	7.207		1,575	256		1,831	2,175	
	0370	8′ wide		1	8		1,850	284		2,134	2,500	
	1000	Entry or walkway, peak, 12′ long, 4′ wide	2 Carp	.90	17.778		4,175	630		4,805	5,625	
	1010	6′ wide		.60	26.667		6,425	950		7,375	8,650	
	1020	8′ wide		.40	40		8,875	1,425		10,300	12,100	
	1100	Radius with dome end, 4′ wide		1.10	14.545		3,175	515		3,690	4,325	
	1110	6′ wide		.70	22.857		5,075	815		5,890	6,950	
	1120	8′ wide		.50	32		7,225	1,150		8,375	9,825	
	2000	Retractable lateral arm awning, manual										
	2010	To 12′ wide, 8′ - 6″ projection	2 Carp	1.70	9.412	Ea.	940	335		1,275	1,575	
	2020	To 14′ wide, 8′ - 6″ projection		1.10	14.545		1,100	515		1,615	2,050	
	2030	To 19′ wide, 8′ - 6″ projection		.85	18.824		1,500	670		2,170	2,750	
	2040	To 24′ wide, 8′ - 6″ projection		.67	23.881		1,875	850		2,725	3,475	
	2050	Motor for above, add	1 Carp	2.67	3		810	107		917	1,075	
	3000	Patio/deck canopy with frame										
	3010	12′ wide, 12′ projection	2 Carp	2	8	Ea.	1,325	284		1,609	1,950	
	3020	16′ wide, 14′ projection	″	1.20	13.333		2,075	475		2,550	3,050	
	9000	For fire retardant canvas, add					7%					
	9010	For lettering or graphics, add					35%					
	9020	For painted or coated acrylic canvas, deduct					8%					
	9030	For translucent or opaque vinyl canvas, add					10%					
	9040	For 6 or more units, deduct					20%	15%				
200	0010	**CANOPIES**										200
	0020	Wall hung, .032″, aluminum, prefinished, 8′ x 10′	K-2	1.30	18.462	Ea.	1,900	675	136	2,711	3,450	
	0300	8′ x 20′		1.10	21.818		3,775	795	160	4,730	5,775	
	0500	10′ x 10′		1.30	18.462		2,125	675	136	2,936	3,700	
	0700	10′ x 20′		1.10	21.818		3,925	795	160	4,880	5,950	
	1000	12′ x 20′		1	24		4,825	875	176	5,876	7,100	
	1360	12′ x 30′		.80	30		7,100	1,100	220	8,420	10,000	
	1700	12′ x 40′		.60	40		8,650	1,450	294	10,394	12,500	
	1900	For free standing units, add					20%	10%				
	2300	Aluminum entrance canopies, flat soffit, .032″										
	2500	3′-6″ x 4′-0″, clear anodized	2 Carp	4	4	Ea.	795	142		937	1,100	
	2700	Bronze anodized		4	4		1,400	142		1,542	1,750	
	3000	Polyurethane painted		4	4		1,125	142		1,267	1,475	
	3300	4′-6″ x 10′-0″, clear anodized		2	8		2,200	284		2,484	2,875	
	3500	Bronze anodized		2	8		2,800	284		3,084	3,550	
	3700	Polyurethane painted		2	8		2,350	284		2,634	3,050	
	4000	Wall downspout, 10 L.F., clear anodized	1 Carp	7	1.143		136	40.50		176.50	218	
	4300	Bronze anodized		7	1.143		237	40.50		277.50	330	
	4500	Polyurethane painted		7	1.143		204	40.50		244.50	292	
	7000	Carport, baked vinyl finish, .032″, 20′ x 10′, no foundations, min.	K-2	4	6	Car	3,400	219	44	3,663	4,200	
	7250	Maximum		2	12	″	6,800	440	88	7,328	8,350	
	7500	Walkway cover, to 12′ wide, stl., vinyl finish, .032″,no fndtns., min.		250	.096	S.F.	19.95	3.50	.70	24.15	29	

10530 | Protective Covers

	10535 | Awnings & Canopies		Crew	Daily Output	Labor-Hours	Unit	2006 Bare Costs Mat.	Labor	Equip.	Total	Total Incl O&P	
200	7750	Maximum	K-2	200	.120	S.F.	21.50	4.38	.88	26.76	33	200
	9000	Minimum labor/equipment charge	2 Carp	2	8	Job		284		284	470	

10550 | Postal Specialties

	10555 | Mail Delivery Systems		Crew	Daily Output	Labor-Hours	Unit	2006 Bare Costs Mat.	Labor	Equip.	Total	Total Incl O&P	
600	0010	**MAIL BOXES**										600
	0020	Horiz., key lock, 5"H x 6"W x 15"D, alum., rear load	1 Carp	34	.235	Ea.	33	8.35		41.35	50	
	0100	Front loading		34	.235		37	8.35		45.35	54.50	
	0200	Double, 5"H x 12"W x 15"D, rear loading		26	.308		60	10.95		70.95	84	
	0300	Front loading		26	.308		63	10.95		73.95	87.50	
	0500	Quadruple, 10"H x 12"W x 15"D, rear loading		20	.400		112	14.20		126.20	147	
	0600	Front loading		20	.400		114	14.20		128.20	149	
	0800	Vertical, front load, 15"H x 5"W x 6"D, alum., per compartment		34	.235		26	8.35		34.35	42.50	
	0900	Bronze, duranodic finish		34	.235		44	8.35		52.35	62.50	
	1000	Steel, enameled		34	.235		32	8.35		40.35	49	
	1700	Alphabetical directories, 120 names		10	.800		114	28.50		142.50	173	
	1800	Letter collection box		6	1.333		600	47.50		647.50	740	
	1900	Letter slot, residential		20	.400		60	14.20		74.20	89.50	
	2000	Post office type		8	1		220	35.50		255.50	300	
	2200	Post office counter window, with grille		2	4		515	142		657	800	
	2250	Key keeper, single key, aluminum		26	.308		37	10.95		47.95	59	
	2300	Steel, enameled		26	.308		100	10.95		110.95	128	
	9000	Minimum labor/equipment charge		5	1.600	Job		57		57	94	
700	0010	**MAIL CHUTES**										700
	0020	Aluminum & glass, 14-1/4" wide, 4-5/8" deep	2 Shee	4	4	Floor	650	169		819	985	
	0100	8-5/8" deep		3.80	4.211		790	177		967	1,150	
	0300	8-3/4" x 3-1/2", aluminum		5	3.200		600	135		735	875	
	0400	Bronze or stainless		4.50	3.556		930	150		1,080	1,275	
	0600	Lobby collection boxes, aluminum		5	3.200	Ea.	1,800	135		1,935	2,200	
	0700	Bronze or stainless		4.50	3.556	"	2,300	150		2,450	2,775	

10600 | Partitions

	10605 | Wire Mesh Partitions		Crew	Daily Output	Labor-Hours	Unit	2006 Bare Costs Mat.	Labor	Equip.	Total	Total Incl O&P	
100	0010	**PARTITIONS, WOVEN WIRE** For tool or stockroom enclosures										100
	0100	Channel frame, 1-1/2" diamond mesh, 10 ga. wire, painted										
	0300	Wall panels, 4'-0" wide, 7' high	2 Carp	25	.640	Ea.	115	23		138	164	
	0400	8' high		23	.696		123	24.50		147.50	177	
	0600	10' high		18	.889		145	31.50		176.50	213	
	0700	For 5' wide panels, add					5%					
	0900	Ceiling panels, 10' long, 2' wide	2 Carp	25	.640		92.50	23		115.50	140	
	1000	4' wide		15	1.067		131	38		169	207	

10600 | Partitions

10605 | Wire Mesh Partitions

				DAILY	LABOR-		2006 BARE COSTS				TOTAL	
			CREW	OUTPUT	HOURS	UNIT	MAT.	LABOR	EQUIP.	TOTAL	INCL O&P	
100	1200	Panel with service window & shelf, 5′ wide, 7′ high	2 Carp	20	.800	Ea.	300	28.50		328.50	375	100
	1300	8′ high		15	1.067		310	38		348	410	
	1500	Sliding doors, full height, 3′ wide, 7′ high		6	2.667		305	95		400	490	
	1600	10′ high		5	3.200		330	114		444	555	
	1800	6′ wide sliding door, 7′ full height		5	3.200		400	114		514	635	
	1900	10′ high		4	4		520	142		662	810	
	2100	Swinging doors, 3′ wide, 7′ high, no transom		6	2.667		222	95		317	400	
	2200	7′ high, 3′ transom		5	3.200		305	114		419	525	

10610 | Folding Gates

			CREW	DAILY OUTPUT	LABOR-HOURS	UNIT	MAT.	LABOR	EQUIP.	TOTAL	TOTAL INCL O&P	
100	0010	**SECURITY GATES** For roll up type, see division 08330-130										100
	0300	Scissors type folding gate, ptd. steel, single, 6-1/2′ high, 5-1/2′wide	2 Sswk	4	4	Opng.	142	160		302	455	
	0350	6-1/2′ wide		4	4		138	160		298	450	
	0400	7-1/2′ wide		4	4		166	160		326	480	
	0600	Double gate, 8′ high, 8′ wide		2.50	6.400		217	256		473	720	
	0650	10′ wide		2.50	6.400		246	256		502	750	
	0700	12′ wide		2	8		340	320		660	975	
	0750	14′ wide		2	8		390	320		710	1,025	
	0900	Door gate, folding steel, 4′ wide, 61″ high		4	4		72.50	160		232.50	380	
	1000	71″ high		4	4		75	160		235	380	
	1200	81″ high		4	4		81.50	160		241.50	390	
	1300	Window gates, 2′ to 4′ wide, 31″ high		4	4		43	160		203	345	
	1500	55″ high		3.75	4.267		86	170		256	415	
	1600	79″ high		3.50	4.571		99.50	183		282.50	450	

10615 | Demountable Partitions

			CREW	DAILY OUTPUT	LABOR-HOURS	UNIT	MAT.	LABOR	EQUIP.	TOTAL	TOTAL INCL O&P	
100	0010	**PARTITIONS, MOVABLE OFFICE** Demountable, add for doors										100
	0100	Do not deduct door openings from total L.F.										
	0900	Demountable gypsum system on 2″ to 2-1/2″										
	1000	steel studs, 9′ high, 3″ to 3-3/4″ thick										
	1200	Vinyl clad gypsum	2 Carp	48	.333	L.F.	48	11.85		59.85	72	
	1300	Fabric clad gypsum		44	.364		119	12.95		131.95	153	
	1500	Steel clad gypsum		40	.400		129	14.20		143.20	166	
	1600	1.75 system, aluminum framing, vinyl clad hardboard,										
	1800	paper honeycomb core panel, 1-3/4″ to 2-1/2″ thick										
	1900	9′ high	2 Carp	48	.333	L.F.	80.50	11.85		92.35	108	
	2100	7′ high		60	.267		72	9.50		81.50	94.50	
	2200	5′ high		80	.200		61	7.10		68.10	79	
	2250	Unitized gypsum system										
	2300	Unitized panel, 9′ high, 2″ to 2-1/2″ thick										
	2350	Vinyl clad gypsum	2 Carp	48	.333	L.F.	103	11.85		114.85	134	
	2400	Fabric clad gypsum	″	44	.364	″	170	12.95		182.95	209	
	2500	Unitized mineral fiber system										
	2510	Unitized panel, 9′ high, 2-1/4″ thick, aluminum frame										
	2550	Vinyl clad mineral fiber	2 Carp	48	.333	L.F.	103	11.85		114.85	133	
	2600	Fabric clad mineral fiber	″	44	.364	″	153	12.95		165.95	191	
	2800	Movable steel walls, modular system										
	2900	Unitized panels, 9′ high, 48″ wide										
	3100	Baked enamel, pre-finished	2 Carp	60	.267	L.F.	117	9.50		126.50	144	
	3200	Fabric clad steel		56	.286	″	168	10.15		178.15	202	
	5310	Trackless wall, cork finish, semi-acoustic, 1-5/8″ thick, minimum		325	.049	S.F.	32.50	1.75		34.25	38.50	
	5320	Maximum		190	.084		30.50	2.99		33.49	38.50	
	5330	Acoustic, 2″ thick, minimum		305	.052		26.50	1.87		28.37	32	
	5340	Maximum		225	.071		45.50	2.53		48.03	54	
	5500	For acoustical partitions, add, minimum					1.88			1.88	2.07	
	5550	Maximum					8.75			8.75	9.65	

10615 | Demountable Partitions

		10615 Demountable Partitions	Crew	Daily Output	Labor-Hours	Unit	2006 Bare Costs Mat.	Labor	Equip.	Total	Total Incl O&P
100	5700	For doors, see Div. 08100 & 08200									
	5800	For door hardware, see Div. 08700									
	6100	In-plant modular office system, w/prehung hollow core door									
	6200	3" thick polystyrene core panels									
	6250	12' x 12', 2 wall	2 Clab	3.80	4.211	Ea.	2,675	115		2,790	3,150
	6300	4 wall		1.90	8.421		3,725	231		3,956	4,475
	6350	16' x 16', 2 wall		3.60	4.444		4,175	122		4,297	4,800
	6400	4 wall	↓	1.80	8.889	↓	5,375	244		5,619	6,300
	9000	Minimum labor/equipment charge	2 Carp	3	5.333	Job		190		190	315

10630 | Port. Partitions/Screens/Panels

		10630 Port. Partitions/Screens/Panels	Crew	Daily Output	Labor-Hours	Unit	2006 Bare Costs Mat.	Labor	Equip.	Total	Total Incl O&P
100	0010	**PARTITIONS, PORTABLE** Divider panels, free standing, fiber core									
	0020	Fabric face straight									
	0100	3'-0" long, 4'-0" high	2 Carp	100	.160	L.F.	106	5.70		111.70	126
	0200	5'-0" high		90	.178		107	6.30		113.30	128
	0500	6'-0" high		75	.213		122	7.60		129.60	147
	0900	5'-0" long, 4'-0" high		175	.091		80.50	3.25		83.75	94
	1000	5'-0" high		150	.107		98.50	3.79		102.29	115
	1500	6"-0" high		125	.128		94.50	4.55		99.05	112
	1600	6'-0" long, 5'-0" high		162	.099		79	3.51		82.51	93
	3100	Curved, 3'-0" long, 5'-0" high		90	.178		83	6.30		89.30	102
	3150	6'-0" high		75	.213		95.50	7.60		103.10	118
	3200	Economical panels, fabric face, 4'-0" long, 5'-0" high		132	.121		39.50	4.31		43.81	50.50
	3250	6'-0" high		112	.143		42.50	5.10		47.60	55
	3300	5'-0" long, 5'-0" high		150	.107		34.50	3.79		38.29	44.50
	3350	6'-0" high		125	.128		36	4.55		40.55	47
	3380	3'-0" curved, 5'-0" high		90	.178		83	6.30		89.30	102
	3390	6'-0" high		75	.213		95.50	7.60		103.10	118
	3450	Acoustical panels, 60 to 90 NRC, 3'-0" long, 5'-0" high		90	.178		95.50	6.30		101.80	115
	3550	6'-0" high		75	.213		91.50	7.60		99.10	114
	3600	5'-0" long, 5'-0" high		150	.107		66.50	3.79		70.29	79.50
	3650	6'-0" high		125	.128		73.50	4.55		78.05	88.50
	3700	6'-0" long, 5'-0" high		162	.099		64.50	3.51		68.01	76.50
	3750	6'-0" high		138	.116		64.50	4.12		68.62	78
	3800	Economy acoustical panels, 40 NRC, 4'-0" long, 5'-0" high		132	.121		39.50	4.31		43.81	50.50
	3850	6'-0" high		112	.143		42	5.10		47.10	54.50
	3900	5'-0" long, 6'-0" high		125	.128		36	4.55		40.55	47
	3950	6'-0" long, 5'-0" high		162	.099		29.50	3.51		33.01	38.50
	4000	Metal chalkboard, 6'-6" high, chalkboard, 1 side		125	.128		90.50	4.55		95.05	107
	4100	Metal chalkboard, 2 sides		120	.133		103	4.74		107.74	122
	4300	Tackboard, both sides		123	.130	↓	82	4.62		86.62	97.50
	9000	Minimum labor/equipment charge	↓	3	5.333	Job		190		190	315

10651 | Accordion Folding Partitions

		10651 Accordion Folding Partitions	Crew	Daily Output	Labor-Hours	Unit	2006 Bare Costs Mat.	Labor	Equip.	Total	Total Incl O&P
100	0010	**PARTITIONS, FOLDING ACCORDION**									
	0100	Vinyl covered, over 150 S.F., frame not included									
	0300	Residential, 1.25 lb. per S.F., 8' maximum height	2 Carp	300	.053	S.F.	17.35	1.90		19.25	22
	0400	Commercial, 1.75 lb. per S.F., 8' maximum height		225	.071		19.85	2.53		22.38	26
	0600	2 lb. per S.F., 17' maximum height		150	.107		20.50	3.79		24.29	29
	0700	Industrial, 4 lb. per S.F., 20' maximum height		75	.213		29.50	7.60		37.10	45
	0900	Acoustical, 3 lb. per S.F., 17' maximum height		100	.160		23	5.70		28.70	35
	1200	5 lb. per S.F., 20' maximum height		95	.168		32	6		38	45
	1300	5.5 lb. per S.F., 17' maximum height		90	.178		37.50	6.30		43.80	51.50
	1400	Fire rated, 4.5 psf, 20' maximum height	↓	160	.100	↓	37.50	3.56		41.06	47

10600 | Partitions

	10651 Accordion Folding Partitions		CREW	DAILY OUTPUT	LABOR-HOURS	UNIT	2006 BARE COSTS MAT.	LABOR	EQUIP.	TOTAL	TOTAL INCL O&P	
100	1500	Vinyl clad wood or steel, electric operation, 5.0 psf	2 Carp	160	.100	S.F.	43.50	3.56		47.06	53.50	100
	1900	Wood, non-acoustic, birch or mahogany, to 10′ high		300	.053		23.50	1.90		25.40	28.50	
	9000	Minimum labor/equipment charge		4	4	Job		142		142	236	

	10653 Folding Panel Partitions		CREW	DAILY OUTPUT	LABOR-HOURS	UNIT	MAT.	LABOR	EQUIP.	TOTAL	TOTAL INCL O&P	
200	0010	**PARTITIONS, FOLDING LEAF** Acoustic, wood										200
	0100	Vinyl faced, to 18′ high, 6 psf, minimum	2 Carp	60	.267	S.F.	44	9.50		53.50	64	
	0150	Average		45	.356		52.50	12.65		65.15	79	
	0200	Maximum		30	.533		68	18.95		86.95	107	
	0400	Formica or hardwood finish, minimum		60	.267		45.50	9.50		55	65.50	
	0500	Maximum		30	.533		48.50	18.95		67.45	85	
	0600	Wood, low acoustical type, 4.5 psf, to 14′ high		50	.320		33	11.40		44.40	55.50	
	1100	Steel, acoustical, 9 to 12 lb. per S.F., vinyl faced, minimum		60	.267		47	9.50		56.50	67	
	1200	Maximum		30	.533		57.50	18.95		76.45	94.50	
	1700	Aluminum framed, acoustical, to 12′ high, 5.5 psf, minimum		60	.267		31.50	9.50		41	50.50	
	1800	Maximum		30	.533		38	18.95		56.95	73.50	
	2000	6.5 lb. per S.F., minimum		60	.267		33	9.50		42.50	52	
	2100	Maximum		30	.533		41	18.95		59.95	76.50	
	9000	Minimum labor/equipment charge		4	4	Job		142		142	236	

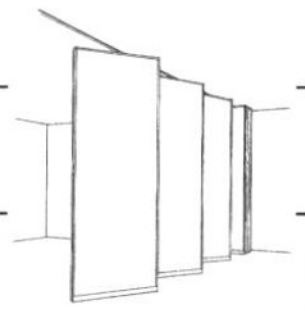

	10658 Acoustic Air Wall		CREW	DAILY OUTPUT	LABOR-HOURS	UNIT	MAT.	LABOR	EQUIP.	TOTAL	TOTAL INCL O&P	
100	0010	**PARTITIONS, OPERABLE**										100
	0020	Acoustic air wall, 1-5/8″ thick, minimum	2 Carp	375	.043	S.F.	27	1.52		28.52	32	
	0100	Maximum		365	.044		46	1.56		47.56	53	
	0300	2-1/4″ thick, minimum		360	.044		30	1.58		31.58	35.50	
	0400	Maximum		330	.048		53	1.72		54.72	61	
	0600	For track type, add to above				L.F.	98.50			98.50	108	
	0700	Overhead track type, acoustical, 3″ thick, 11 psf, minimum	2 Carp	350	.046	S.F.	68	1.63		69.63	77	
	0800	Maximum	″	300	.053	″	81.50	1.90		83.40	93	

10670 | Storage Shelving

	10674 Storage Shelving		CREW	DAILY OUTPUT	LABOR-HOURS	UNIT	2006 BARE COSTS MAT.	LABOR	EQUIP.	TOTAL	TOTAL INCL O&P	
500	0010	**SHELVING**										500
	0020	Metal, industrial, cross-braced, 3′ wide, 12″ deep	1 Sswk	175	.046	SF Shlf	9.65	1.83		11.48	14	
	0100	24″ deep		330	.024		6.90	.97		7.87	9.40	
	0300	4′ wide, 12″ deep		185	.043		9.10	1.73		10.83	13.25	
	0400	24″ deep		380	.021		6.65	.84		7.49	8.85	
	1200	Enclosed sides, cross-braced back, 3′ wide, 12″ deep		175	.046		16.45	1.83		18.28	21.50	
	1300	24″ deep		290	.028		10.65	1.10		11.75	13.75	
	1500	Fully enclosed, sides and back, 3′ wide, 12″ deep		150	.053		14.85	2.13		16.98	20.50	
	1600	24″ deep		255	.031		9.10	1.25		10.35	12.40	
	1800	4′ wide, 12″ deep		150	.053		10.20	2.13		12.33	15.25	
	1900	24″ deep		290	.028		6.55	1.10		7.65	9.25	
	2200	Wide span, 1600 lb. capacity per shelf, 6′ wide, 24″ deep		380	.021		7.25	.84		8.09	9.55	
	2400	36″ deep		440	.018		6.55	.73		7.28	8.55	
	2600	8′ wide, 24″ deep		440	.018		6.75	.73		7.48	8.75	
	2800	36″ deep		520	.015		5.75	.61		6.36	7.45	
	4000	Pallet racks, steel frame 5,000 lb. capacity, 8′ long, 36″ deep	2 Sswk	450	.036		10.80	1.42		12.22	14.55	

10670 | Storage Shelving

	10674	Storage Shelving	CREW	DAILY OUTPUT	LABOR-HOURS	UNIT	2006 BARE COSTS MAT.	LABOR	EQUIP.	TOTAL	TOTAL INCL O&P	
500	4200	42″ deep	2 Sswk	500	.032	SF Shlf	9.45	1.28		10.73	12.80	500
	4400	48″ deep		520	.031		9.25	1.23		10.48	12.45	
	9000	Minimum labor/equipment charge	1 Carp	4	2	Job		71		71	118	
600	0010	**PARTS BINS** Metal, gray baked enamel finish										600
	0100	6′-3″ high, 3′ wide										
	0300	12 bins, 18″ wide x 12″ high, 12″ deep	2 Clab	10	1.600	Ea.	243	44		287	340	
	0400	24″ deep		10	1.600		320	44		364	425	
	0600	72 bins, 6″ wide x 6″ high, 12″ deep		8	2		380	55		435	510	
	0700	18″ deep		8	2		850	55		905	1,025	
	1000	7′-3″ high, 3′ wide										
	1200	14 bins, 18″ wide x 12″ high, 12″ deep	2 Clab	10	1.600	Ea.	255	44		299	355	
	1300	24″ deep		10	1.600		335	44		379	445	
	1500	84 bins, 6″ wide x 6″ high, 12″ deep		8	2		570	55		625	715	
	1600	24″ deep		8	2		745	55		800	910	

10750 | Telephone Specialties

	10755	Telephone Enclosures	CREW	DAILY OUTPUT	LABOR-HOURS	UNIT	2006 BARE COSTS MAT.	LABOR	EQUIP.	TOTAL	TOTAL INCL O&P	
400	0010	TELEPHONE ENCLOSURE										400
	0300	Shelf type, wall hung, minimum	2 Carp	5	3.200	Ea.	1,025	114		1,139	1,325	
	0400	Maximum		5	3.200		2,625	114		2,739	3,100	
	0600	Booth type, painted steel, indoor or outdoor, minimum		1.50	10.667		3,200	380		3,580	4,150	
	0700	Maximum (stainless steel)		1.50	10.667		10,700	380		11,080	12,400	
	1300	Outdoor, acoustical, on post		3	5.333		1,400	190		1,590	1,875	
	1400	Phone carousel, pedestal mounted with dividers		.60	26.667		5,375	950		6,325	7,475	
	1900	Outdoor, drive-up type, wall mounted		4	4		860	142		1,002	1,175	
	2000	Post mounted, stainless steel posts		3	5.333		1,325	190		1,515	1,775	
	2200	Directory shelf, wall mounted, stainless steel										
	2300	3 binders	2 Carp	8	2	Ea.	1,025	71		1,096	1,250	
	2500	4 binders		7	2.286		1,175	81.50		1,256.50	1,400	
	2600	5 binders		6	2.667		1,500	95		1,595	1,800	
	2800	Table type, stainless steel, 4 binders		8	2		1,200	71		1,271	1,450	
	2900	7 binders		7	2.286		1,500	81.50		1,581.50	1,775	
	9000	Minimum labor/equipment charge		4	4	Job		142		142	236	

10800 | Toilet/Bath/Laundry Accessories

	10810	Toilet Accessories	CREW	DAILY OUTPUT	LABOR-HOURS	UNIT	2006 BARE COSTS MAT.	LABOR	EQUIP.	TOTAL	TOTAL INCL O&P	
100	0010	COMMERCIAL TOILET ACCESSORIES										100
	0200	Curtain rod, stainless steel, 5′ long, 1″ diameter	1 Carp	13	.615	Ea.	30.50	22		52.50	70	
	0300	1-1/4″ diameter		13	.615		30	22		52	70	
	0400	Diaper changing station, horizontal, wall mounted, plastic		10	.800		172	28.50		200.50	236	

	10810 | Toilet Accessories		CREW	DAILY OUTPUT	LABOR-HOURS	UNIT	2006 BARE COSTS MAT.	LABOR	EQUIP.	TOTAL	TOTAL INCL O&P	
100	0500	Dispenser units, combined soap & towel dispensers,										100
	0510	mirror and shelf, flush mounted	1 Carp	10	.800	Ea.	350	28.50		378.50	430	
	0600	Towel dispenser and waste receptacle,										
	0610	18 gallon capacity	1 Carp	10	.800	Ea.	269	28.50		297.50	345	
	0800	Grab bar, straight, 1-1/4" diameter, stainless steel, 18" long		24	.333		20	11.85		31.85	41.50	
	0900	24" long		23	.348		19.80	12.35		32.15	42.50	
	1000	30" long		22	.364		22	12.95		34.95	46	
	1100	36" long		20	.400		21	14.20		35.20	46.50	
	1105	42" long		20	.400		21.50	14.20		35.70	47	
	1200	1-1/2" diameter, 24" long		23	.348		52	12.35		64.35	77.50	
	1300	36" long		20	.400		57	14.20		71.20	86	
	1310	42" long		18	.444		48	15.80		63.80	79	
	1500	Tub bar, 1-1/4" diameter, 24" x 36"		14	.571		89.50	20.50		110	132	
	1600	Plus vertical arm		12	.667		90.50	23.50		114	139	
	1900	End tub bar, 1" diameter, 90° angle, 16" x 32"		12	.667		124	23.50		147.50	177	
	2010	Tub/shower/toilet, 2-wall, 36" x 24"		12	.667		80	23.50		103.50	128	
	2300	Hand dryer, surface mounted, electric, 115 volt, 20 amp		4	2		510	71		581	680	
	2400	230 volt, 10 amp		4	2		510	71		581	680	
	2600	Hat and coat strip, stainless steel, 4 hook, 36" long		24	.333		50.50	11.85		62.35	75	
	2700	6 hook, 60" long		20	.400		88	14.20		102.20	121	
	3000	Mirror, with stainless steel 3/4" square frame, 18" x 24"		20	.400		66	14.20		80.20	96	
	3100	36" x 24"		15	.533		147	18.95		165.95	194	
	3200	48" x 24"		10	.800		193	28.50		221.50	259	
	3300	72" x 24"		6	1.333		196	47.50		243.50	295	
	3500	With 5" stainless steel shelf, 18" x 24"		20	.400		157	14.20		171.20	196	
	3600	36" x 24"		15	.533		199	18.95		217.95	251	
	3700	48" x 24"		10	.800		229	28.50		257.50	299	
	3800	72" x 24"		6	1.333		340	47.50		387.50	455	
	4100	Mop holder strip, stainless steel, 5 holders, 48" long		20	.400		86.50	14.20		100.70	119	
	4200	Napkin/tampon dispenser, recessed		15	.533		330	18.95		348.95	390	
	4300	Robe hook, single, regular		36	.222		5.30	7.90		13.20	18.95	
	4400	Heavy duty, concealed mounting		36	.222		12.15	7.90		20.05	26.50	
	4600	Soap dispenser, chrome, surface mounted, liquid		20	.400		47	14.20		61.20	75	
	4700	Powder		20	.400		44	14.20		58.20	72	
	5000	Recessed stainless steel, liquid		10	.800		140	28.50		168.50	201	
	5100	Powder		10	.800		183	28.50		211.50	248	
	5300	Soap tank, stainless steel, 1 gallon		10	.800		167	28.50		195.50	231	
	5400	5 gallon		5	1.600		230	57		287	345	
	5600	Shelf, stainless steel, 5" wide, 18 ga., 24" long		24	.333		45.50	11.85		57.35	69.50	
	5700	48" long		16	.500		89.50	17.80		107.30	128	
	5800	8" wide shelf, 18 ga., 24" long		22	.364		53.50	12.95		66.45	80	
	5900	48" long		14	.571		114	20.50		134.50	159	
	6000	Toilet seat cover dispenser, stainless steel, recessed		20	.400		111	14.20		125.20	146	
	6050	Surface mounted		15	.533		27.50	18.95		46.45	61.50	
	6100	Toilet tissue dispenser, surface mounted, SS, single roll		30	.267		11.70	9.50		21.20	28.50	
	6200	Double roll		24	.333		16.30	11.85		28.15	37.50	
	6400	Towel bar, stainless steel, 18" long		23	.348		35	12.35		47.35	59	
	6500	30" long		21	.381		71.50	13.55		85.05	101	
	6700	Towel dispenser, stainless steel, surface mounted		16	.500		37.50	17.80		55.30	71	
	6800	Flush mounted, recessed		10	.800		209	28.50		237.50	277	
	7000	Towel holder, hotel type, 2 guest size		20	.400		16.15	14.20		30.35	41.50	
	7200	Towel shelf, stainless steel, 24" long, 8" wide		20	.400		62	14.20		76.20	91.50	
	7400	Tumbler holder, tumbler only		30	.267		26	9.50		35.50	44	
	7500	Soap, tumbler & toothbrush		30	.267		23.50	9.50		33	41.50	
	7700	Wall urn ash receiver, surface mount, 11" long		12	.667		117	23.50		140.50	169	
	7800	7-1/2", long	↓	18	.444	↓	76	15.80		91.80	110	100

10800 | Toilet/Bath/Laundry Accessories

		10810 | Toilet Accessories	CREW	DAILY OUTPUT	LABOR-HOURS	UNIT	2006 BARE COSTS MAT.	LABOR	EQUIP.	TOTAL	TOTAL INCL O&P	
100	8000	Waste receptacles, stainless steel, with top, 13 gallon	1 Carp	10	.800	Ea.	231	28.50		259.50	300	100
	8100	36 gallon		8	1	↓	425	35.50		460.50	525	
	9000	Minimum labor/equipment charge	↓	5	1.600	Job		57		57	94	

		10820 | Bath Accessories	CREW	DAILY OUTPUT	LABOR-HOURS	UNIT	MAT.	LABOR	EQUIP.	TOTAL	TOTAL INCL O&P	
400	0010	**MEDICINE CABINETS**										400
	0020	With mirror, st. st. frame, 16" x 22", unlighted	1 Carp	14	.571	Ea.	73	20.50		93.50	114	
	0100	Wood frame		14	.571		101	20.50		121.50	145	
	0300	Sliding mirror doors, 20" x 16" x 4-3/4", unlighted		7	1.143		90.50	40.50		131	167	
	0400	24" x 19" x 8-1/2", lighted		5	1.600		143	57		200	251	
	0600	Triple door, 30" x 32", unlighted, plywood body		7	1.143		214	40.50		254.50	305	
	0700	Steel body		7	1.143		282	40.50		322.50	380	
	0900	Oak door, wood body, beveled mirror, single door		7	1.143		138	40.50		178.50	220	
	1000	Double door		6	1.333		335	47.50		382.50	450	
	1200	Hotel cabinets, stainless, with lower shelf, unlighted		10	.800		181	28.50		209.50	246	
	1300	Lighted		5	1.600	↓	269	57		326	390	
	9000	Minimum labor/equipment charge	↓	4	2	Job		71		71	118	

10880 | Scales

		10885 | Scales	CREW	DAILY OUTPUT	LABOR-HOURS	UNIT	2006 BARE COSTS MAT.	LABOR	EQUIP.	TOTAL	TOTAL INCL O&P	
100	0010	**SCALES** Built-in floor scale, not incl. foundations										100
	0100	Dial type, 5 ton capacity, 8' x 6' platform	3 Carp	.50	48	Ea.	6,525	1,700		8,225	10,000	
	0300	9' x 7' platform		.40	60		8,475	2,125		10,600	12,900	
	0400	10 ton capacity, steel platform, 8' x 6' platform		.40	60		10,300	2,125		12,425	14,800	
	0600	9' x 7' platform	↓	.35	68.571	↓	9,650	2,450		12,100	14,700	
	0700	Truck scales, incl. steel weigh bridge,										
	0800	not including foundation, pits										
	1550	Digital, electronic, 100 ton capacity, steel deck 12' x 10' platform	3 Carp	.20	120	Ea.	10,900	4,275		15,175	19,000	
	1600	40' x 10' platform		.14	171		22,100	6,100		28,200	34,400	
	1640	60' x 10' platform		.13	184		27,300	6,575		33,875	41,000	
	1680	70' x 10' platform	↓	.12	200		30,500	7,100		37,600	45,400	
	2000	For standard automatic printing device, add					2,200			2,200	2,425	
	2100	For remote reading electronic system, add					2,000			2,000	2,200	
	2300	Concrete foundation pits for above, 8' x 6', 5 C.Y. required	C-1	.50	64		830	2,150		2,980	4,475	
	2400	14' x 6' platform, 10 C.Y. required		.35	91.429		1,225	3,075		4,300	6,425	
	2600	50' x 10' platform, 30 C.Y. required		.25	128		1,650	4,300		5,950	8,900	
	2700	70' x 10' platform, 40 C.Y. required	↓	.15	213		3,575	7,150		10,725	15,800	
	2750	Crane scales, dial, 1 ton capacity					885			885	970	
	2780	5 ton capacity					1,025			1,025	1,150	
	2800	Digital, 1 ton capacity					1,850			1,850	2,025	
	2850	10 ton capacity				↓	3,725			3,725	4,100	
	2900	Low profile electronic warehouse scale,										
	3000	not incl. printer, 4' x 4' platform, 10,000 lb. capacity	2 Carp	.30	53.333	Ea.	1,825	1,900		3,725	5,150	
	3300	5' x 7' platform, 10,000 lb. capacity		.25	64		3,575	2,275		5,850	7,725	
	3400	20,000 lb. capacity	↓	.20	80		4,625	2,850		7,475	9,800	
	3500	For printers, incl. time, date & numbering, add					1,150			1,150	1,275	
	3800	Portable, beam type, capacity 1000#, platform 18" x 24"					670			670	735	
	3900	Dial type, capacity 2000#, platform 24" x 24"				↓	1,150			1,150	1,275	

10880 | Scales

	10885 | Scales		CREW	DAILY OUTPUT	LABOR-HOURS	UNIT	2006 BARE COSTS MAT.	LABOR	EQUIP.	TOTAL	TOTAL INCL O&P	
100	4000	Digital type, capacity 1000#, platform 24" x 30"				Ea.	1,850			1,850	2,025	100
	4100	Portable contractor truck scales, 50 ton cap., 40' x 10' platform					28,700			28,700	31,600	
	4200	60' x 10' platform					25,700			25,700	28,300	

10900 | Wardrobe & Closet Specialties

	10905 | Coat Racks/Wardrobes		CREW	DAILY OUTPUT	LABOR-HOURS	UNIT	2006 BARE COSTS MAT.	LABOR	EQUIP.	TOTAL	TOTAL INCL O&P	
500	0010	COAT RACKS & WARDROBES										500
	0020	Floor model hat & coat racks, 6 hangers										
	0050	Standing, beech wood, 21" x 21" x 72", chrome				Ea.	223			223	245	
	0100	18 gauge tubular steel, 21" x 21" x 69", wood walnut				"	237			237	261	
	0500	16 gauge steel frame, 22 gauge steel shelves										
	0650	Single pedestal, 30" x 18" x 63"				Ea.	170			170	187	
	0800	Single face rack, 29" x 18-1/2" x 62"					173			173	190	
	0900	51" x 18-1/2" x 70"					305			305	335	
	0910	Double face rack, 39" x 26" x 70"					201			201	221	
	0920	63" x 26" x 70"					355			355	390	
	0940	For 2" ball casters, add				Set	80.50			80.50	89	
	1400	Utility hook strips, 3/8" x 2-1/2" x 18", 6 hooks	1 Carp	48	.167	Ea.	51.50	5.95		57.45	66.50	
	1500	34" long, 12 hooks	"	48	.167	"	51	5.95		56.95	66	
	1650	Wall mounted racks, 16 gauge steel frame, 22 gauge steel shelves										
	1850	12" x 15" x 26", 6 hangers	1 Carp	32	.250	Ea.	125	8.90		133.90	153	
	2000	12" x 15" x 50", 12 hangers	"	32	.250	"	130	8.90		138.90	158	
	2150	Wardrobe cabinet, steel, baked enamel finish										
	2300	36" x 21" x 78", incl. top shelf & hanger rod				Ea.	261			261	287	

For information about Means Estimating Seminars, see yellow pages 12 and 13 in back of book

Division Notes

		CREW	DAILY OUTPUT	LABOR-HOURS	UNIT	2006 BARE COSTS MAT.	LABOR	EQUIP.	TOTAL	TOTAL INCL O&P

Division 11 Equipment

Estimating Tips

General

- The items in this division are usually priced per square foot or each. Many of these items are purchased by the owner for installation by the contractor. Check the specifications for responsibilities and include time for receiving, storage, installation, and mechanical and electrical hook-ups in the appropriate divisions.
- Many items in Division 11 require some type of support system that is not usually furnished with the item. Examples of these systems include blocking for the attachment of casework and support angles for ceiling-hung projection screens. The required blocking or supports must be added to the estimate in the appropriate division.
- Some items in Division 11 may require assembly or electrical hookups. Verify the amount of assembly required or the need for a hard electrical connection and add the appropriate costs.

Reference Numbers

Reference numbers are shown in bold squares at the beginning of some major classifications. These numbers refer to related items in the Reference Section. The reference information may be an estimating procedure, an alternate pricing method, or technical information.

Note: Not all subdivisions listed here necessarily appear in this publication.

100	11006 | Selective Demolition	Crew	Daily Output	Labor-Hours	Unit	2006 Bare Costs Mat.	Labor	Equip.	Total	Total Incl O&P
0010	SELECTIVE DEMOLITION R024119-10									
0130	Central vacuum, motor unit, residential or commercial	1 Clab	2	4	Ea.	640	110		750	885
0210	Vault door and frame	2 Skwk	2	8		640	292		932	1,175
0215	Day gate, for vault	″	3	5.333		640	195		835	1,025
0380	Bank equipment, teller window, bullet resistant	1 Clab	1.20	6.667			183		183	305
0381	Counter	2 Clab	1.50	10.667	Station		292		292	485
0382	Drive-up window, including drawer and glass		1.50	10.667	″		292		292	485
0383	Thru-wall boxes and chests, selective demolition		2.50	6.400	Ea.		175		175	291
0384	Bullet resistant partitions		20	.800	L.F.		22		22	36.50
0385	Pneumatic tube system, 2 lane drive-up	L-3	.45	35.556	Ea.		1,375		1,375	2,225
0386	Safety deposit box	1 Clab	50	.160	Opng.		4.38		4.38	7.25
0387	Surveillance system, video, complete	2 Elec	2	8	Ea.		335		335	520
0410	Church equipment, misc moveable fixtures	2 Clab	1	16			440		440	725
0412	Church equipment, steeple, to 28′ high	F-3	3	13.333			480	212	692	1,025
0414	40′ to 60′ high	″	.80	50			1,800	795	2,595	3,825
0510	Library equipment, bookshelves, wood, to 90″ high	1 Clab	20	.400	L.F.		10.95		10.95	18.15
0515	Carrels, hardwood, 36″ x 24″	″	9	.889	Ea.		24.50		24.50	40.50
0630	Stage equipment, light control panel	1 Elec	1	8	″		335		335	520
0632	Border lights		40	.200	L.F.		8.40		8.40	12.95
0634	Spot lights		8	1	Ea.		42		42	64.50
0636	Telescoping platforms and risers	2 Clab	175	.091	SF Stg.		2.51		2.51	4.15
1020	Barber equipment, hydraulic chair	1 Clab	40	.200	Ea.		5.50		5.50	9.10
1030	Checkout counter, supermarket or warehouse conveyor	2 Clab	18	.889			24.50		24.50	40.50
1040	Food cases, refrigerated or frozen	Q-5	6	2.667			103		103	161
1190	Laundry equipment, commercial	L-6	3	4			170		170	263
1360	Movie equipment, lamphouse, to 4000 watt, incl. rectifier	1 Elec	4	2			84		84	129
1365	Sound system, incl. amplifier	″	1.25	6.400			269		269	415
1410	Air compressor, to 5 H.P.	2 Clab	2.50	6.400			175		175	291
1412	Lubrication equipment, automotive, 3 reel type, incl. pump, excl piping	L-4	1	24	Set		795		795	1,300
1414	Booth, spray paint, complete, to 26′ long	″	.80	30	″		995		995	1,625
1560	Parking equipment, cashier booth	B-22	2	15	Ea.		490	114	604	920
1600	Loading dock equipment, dock bumpers, rubber	1 Clab	50	.160	″		4.38		4.38	7.25
1610	Door seal for door perimeter	″	50	.160	L.F.		4.38		4.38	7.25
1620	Platform lifter, fixed, 6′ x 8′, 5000 lb. capacity	E-16	1.50	10.667	Ea.		435	59.50	494.50	880
1630	Dock leveller	″	2	8			330	44.50	374.50	665
1640	Lights, single or double arm	1 Elec	8	1			42		42	64.50
1650	Shelter, fabric, truck or train	1 Clab	1.50	5.333			146		146	242
1790	Waste handling equipment, commercial compactor	L-4	2	12			400		400	650
1792	Commercial or municipal incinerator, gas	″	2	12			400		400	650
1795	Crematory, excluding building	Q-3	.25	128			5,200		5,200	8,100
1910	Detection equipment, cell bar front	E-4	4	8			325	22.50	347.50	630
1912	Cell door and frame		8	4			162	11.15	173.15	315
1914	Prefab cell, 1.22m-1.52m w, 2.13m-2.44m high, 2.44m thick		8	4			162	11.15	173.15	315
1916	Cot, bolted, single		40	.800			32.50	2.23	34.73	63
1918	Visitor cubicle		4	8			325	22.50	347.50	630
2850	Hydraulic gates, canal, flap, knife, slide or sluice, to 18″ diameter	L-5A	8	4			160	87	247	385
2852	19″ to 36″ diameter		6	5.333			213	116	329	515
2854	37″ to 48′ diameter		2	16			640	350	990	1,525
2856	49″ to 60″ diameter		1	32			1,275	695	1,970	3,075
2858	Over 60″ diameter		.30	106			4,275	2,325	6,600	10,200
3100	Sewage pumping system, prefabricated, to 1000 GPM	C-17D	.20	420			15,500	2,650	18,150	28,200
3110	Sewage treatment, holding tank for recirc chemical water closet	1 Plum	8	1			42.50		42.50	66.50
3900	Wastewater treatment system, to 1500 gallons	B-21	2	14			450	76.50	526.50	820
4050	Food storage equipment, walk-in refrigerator/freezer	2 Clab	64	.250	S.F.		6.85		6.85	11.35
4052	Shelving, stainless steel, 4 tier or dunnage rack	1 Clab	12	.667	Ea.		18.25		18.25	30.50
4100	Food preparation equipment, small countertop		18	.444			12.20		12.20	20

11006 | Selective Demolition

	11006	Selective Demolition	CREW	DAILY OUTPUT	LABOR-HOURS	UNIT	2006 BARE COSTS MAT.	LABOR	EQUIP.	TOTAL	TOTAL INCL O&P	
100	4150	Food delivery carts, heated cabinets R024119-10	1 Clab	18	.444	Ea.		12.20		12.20	20	100
	4200	Cooking equipment, commercial range	Q-1	12	1.333			51		51	79.50	
	4250	Hood and ventilation equipment, kitchen exhaust hood, excl. fire prot	1 Clab	3	2.667			73		73	121	
	4255	Fire protection system	Q-1	3	5.333			205		205	320	
	4300	Food dispensing equipment, countertop items	1 Clab	15	.533			14.60		14.60	24	
	4310	Serving counter	″	65	.123	L.F.		3.37		3.37	5.60	
	4350	Ice machine, ice cube maker, flakers and storage bins, to 2000 lb/day	Q-1	1.60	10	Ea.		385		385	600	
	4400	Cleaning and disposal, dishwasher, commercial, rack type	L-6	1	12			510		510	790	
	4410	Dishwasher hood	2 Clab	5	3.200			87.50		87.50	145	
	4420	Garbage disposal, commercial, to 5 H.P.	L-1	8	2			84.50		84.50	131	
	4540	Water heater, residential, to 80 gal/day	″	5	3.200			136		136	210	
	4542	Water softener, automatic	2 Plum	10	1.600			68.50		68.50	106	
	4544	Disappearing stairway, to 15′ floor height	2 Clab	6	2.667			73		73	121	
	4710	Darkroom equipment, light	L-7	10	2.800			95.50		95.50	156	
	4712	Heavy	″	1.50	18.667			635		635	1,050	
	4720	Doors	2 Clab	3.50	4.571	Opng.		125		125	208	
	4830	Bowling alley, complete, incl pinsetter, scorer, counters, misc supplies	4 Clab	.40	80	Lane		2,200		2,200	3,625	
	4840	Health club equipment, circuit training apparatus	2 Clab	2	8	Set		219		219	365	
	4842	Squat racks	″	10	1.600	Ea.		44		44	72.50	
	4860	School equipment, basketball backstop	L-2	2	8			248		248	410	
	4862	Table and benches, folding, in wall, 14′ long	L-4	4	6			199		199	325	
	4864	Bleachers, telescoping, to 30 tier	F-5	120	.267	Seat		9.60		9.60	15.95	
	4866	Boxing ring, elevated	L-4	.20	120	Ea.		3,975		3,975	6,525	
	4867	Boxing ring, floor level	″	2	12			400		400	650	
	4868	Exercise equipment	1 Clab	6	1.333			36.50		36.50	60.50	
	4870	Gym divider	L-4	1000	.024	S.F.		.80		.80	1.30	
	4875	Scoreboard	R-3	2	10	Ea.		415	76.50	491.50	725	
	4880	Shooting range, incl. bullet traps, targets, excl. structure	L-9	1	36	Point		1,150		1,150	1,950	
	5200	Vocational shop equipment	2 Clab	8	2	Ea.		55		55	91	
	6200	Fume hood, incl. countertop, excl. HVAC	″	6	2.667	L.F.		73		73	121	
	7100	Medical sterilizing, distiller, water, steam heated, 50 gal. capacity	1 Plum	2.80	2.857	Ea.		122		122	190	
	7200	Medical equipment, surgery table, minor	1 Clab	1	8			219		219	365	
	7210	Surgical lights, doctors office, single or double arm	2 Elec	3	5.333			224		224	345	
	7300	Physical therapy, table	2 Clab	4	4			110		110	182	
	7310	Whirlpool bath, fixed, incl. mixing valves	1 Plum	4	2			85.50		85.50	133	
	7400	Dental equipment, chair, electric or hydraulic	1 Clab	.75	10.667			292		292	485	
	7410	Central suction system	1 Plum	2	4			171		171	266	
	7420	Drill console with accessories	1 Clab	3.20	2.500			68.50		68.50	114	
	7430	X-ray unit	″	4	2			55		55	91	
	7440	X-ray developer	1 Plum	10	.800			34		34	53	

11010 | Maintenance Equipment

	11012	Contractor Equipment	CREW	DAILY OUTPUT	LABOR-HOURS	UNIT	2006 BARE COSTS MAT.	LABOR	EQUIP.	TOTAL	TOTAL INCL O&P	
100	0010	**CONTRACTOR EQUIPMENT** See Reference Section										100
		11013 | Floor/Wall Cleaning Equipment										
800	0010	**VACUUM CLEANING**										800
	0020	Central, 3 inlet, residential	1 Skwk	.90	8.889	Total	640	325		965	1,225	

EQUIPMENT 11

11010 | Maintenance Equipment

	11013 | Floor/Wall Cleaning Equipment	CREW	DAILY OUTPUT	LABOR-HOURS	UNIT	2006 BARE COSTS MAT.	LABOR	EQUIP.	TOTAL	TOTAL INCL O&P	
800	0200 Commercial	1 Skwk	.70	11.429	Total	1,200	415		1,615	2,000	800
	0400 5 inlet system, residential		.50	16		970	585		1,555	2,025	
	0600 7 inlet system, commercial		.40	20		1,075	730		1,805	2,400	
	0800 9 inlet system, residential	↓	.30	26.667		1,375	975		2,350	3,100	
	4010 Rule of thumb: First 1200 S.F., installed				↓					1,180	
	4020 For each additional S.F., add				S.F.					.19	

11020 | Security & Vault Equipment

	11021 | Safes and Vault Doors	CREW	DAILY OUTPUT	LABOR-HOURS	UNIT	2006 BARE COSTS MAT.	LABOR	EQUIP.	TOTAL	TOTAL INCL O&P	
600	0010 **SAFE**										600
	0015 Office, 4 hr. rating, 30″ x 18″ x 18″ inside				Ea.	3,275			3,275	3,600	
	0100 60″ x 36″ x 18″					7,100			7,100	7,825	
	0200 1 hr. rating, 30″ x 18″ x 18″					1,725			1,725	1,900	
	0250 40″ x 18″ x 18″					3,450			3,450	3,775	
	0300 60″ x 36″ x 18″, double door					4,450			4,450	4,900	
	0400 Data, 4 hr. rating, 23-1/2″ x 19-1/2″ x 17″ inside					3,175			3,175	3,500	
	0450 52″ x 19″ x 17″					6,150			6,150	6,775	
	0500 63″ x 34″ x 16″, double door					9,550			9,550	10,500	
	0550 52″ x 34″ x 16″, inside					8,975			8,975	9,875	
	0600 1 hr. rating, 27″ x 19″ x 16″					4,250			4,250	4,675	
	0700 63″ x 34″ x 16″					7,900			7,900	8,700	
	0750 Diskette, 1 hr., 14″ x 12″ x 11″, inside					2,925			2,925	3,225	
	0800 Money, "B" label, 9″ x 14″ x 14″					415			415	455	
	0900 Tool resistive, 24″ x 24″ x 20″					2,800			2,800	3,075	
	1050 Tool and torch resistive, 24″ x 24″ x 20″					6,875			6,875	7,550	
	1150 Jewelers, 23″ x 20″ x 18″					8,775			8,775	9,650	
	1200 63″ x 25″ x 18″					15,300			15,300	16,800	
	1300 For handling into building, add, minimum	A-2	8.50	2.824			77.50	15.15	92.65	145	
	1400 Maximum	"	.78	30.769	↓		845	165	1,010	1,575	
800	0010 **VAULT DOOR**										800
	0020 Door and frame, 32″ x 78″, clear opening										
	0100 1 hour test, 32″ door, weighs 750 lbs.	2 Sswk	1.50	10.667	Opng.	3,550	425		3,975	4,700	
	0200 2 hour test, 32″ door, weighs 950 lbs.		1.30	12.308		3,750	490		4,240	5,050	
	0250 40″ door, weighs 1130 lbs.		1	16		4,275	640		4,915	5,925	
	0300 4 hour test, 32″ door, weighs 1025 lbs.		1.20	13.333		4,100	535		4,635	5,500	
	0350 40″ door, weighs 1140 lbs.	↓	.90	17.778		4,775	710		5,485	6,575	
	0500 For stainless steel front, including frame, add to above					1,750			1,750	1,925	
	0550 Back, add				↓	1,750			1,750	1,925	
	0600 For time lock, two movement, add	1 Elec	2	4	Ea.	1,400	168		1,568	1,800	
	0650 Three movement, add	"	2	4		1,825	168		1,993	2,275	
	0800 Day gate, painted, steel, 32″ wide	2 Sswk	1.50	10.667		1,675	425		2,100	2,650	
	0850 40″ wide		1.40	11.429		1,875	455		2,330	2,900	
	0900 Aluminum, 32″ wide		1.50	10.667		2,550	425		2,975	3,600	
	0950 40″ wide	↓	1.40	11.429	↓	2,825	455		3,280	3,975	
	2050 Security vault door, class I, 3′ wide, 3 1/2″ thick	E-24	.19	166	Opng.	13,400	6,525	3,400	23,325	30,200	
	2100 Class II, 3′ wide, 7″ thick		.19	166		15,500	6,525	3,400	25,425	32,600	
	2150 Class III, 9R, 3′ wide, 10″ thick, minimum		.13	250	↓	19,400	9,775	5,075	34,250	44,600	
	2160 Class V, type 1, 40″ door		2.48	12.903	Ea.	5,575	505	262	6,342	7,325	
	2170 Class V, type 2, 40″ door	↓	2.48	12.903	↓	5,525	505	262	6,292	7,275	

11020 | Security & Vault Equipment

	11021	Safes and Vault Doors	CREW	DAILY OUTPUT	LABOR-HOURS	UNIT	2006 BARE COSTS MAT.	LABOR	EQUIP.	TOTAL	TOTAL INCL O&P	
800	2180	Day gate for class V vault	2 Sswk	2	8	Ea.	1,150	320		1,470	1,850	800

11030 | Teller & Service Equipment

	11038	Bank Equipment	CREW	DAILY OUTPUT	LABOR-HOURS	UNIT	2006 BARE COSTS MAT.	LABOR	EQUIP.	TOTAL	TOTAL INCL O&P	
150	0010	BANK EQUIPMENT										150
	0020	Alarm system, police	2 Elec	1.60	10	Ea.	4,250	420		4,670	5,325	
	0100	With vault alarm	″	.40	40		16,800	1,675		18,475	21,100	
	0400	Bullet resistant teller window, 44″ x 60″	1 Glaz	.60	13.333		2,825	455		3,280	3,825	
	0500	48″ x 60″	″	.60	13.333	↓	3,500	455		3,955	4,600	
	3000	Counters for banks, frontal only	2 Carp	1	16	Station	1,575	570		2,145	2,700	
	3100	Complete with steel undercounter	″	.50	32	″	3,100	1,150		4,250	5,275	
	4600	Door and frame, bullet-resistant, with vision panel, minimum	2 Sswk	1.10	14.545	Ea.	3,500	580		4,080	4,925	
	4700	Maximum		1.10	14.545		4,750	580		5,330	6,300	
	4800	Drive-up window, drawer & mike, not incl. glass, minimum		1	16		4,725	640		5,365	6,375	
	4900	Maximum		.50	32		7,950	1,275		9,225	11,100	
	5000	Night depository, with chest, minimum		1	16		6,500	640		7,140	8,350	
	5100	Maximum		.50	32		9,225	1,275		10,500	12,500	
	5200	Package receiver, painted		3.20	5		1,175	200		1,375	1,675	
	5300	Stainless steel	↓	3.20	5	↓	2,000	200		2,200	2,575	
	5400	Partitions, bullet-resistant, 1-3/16″ glass, 8′ high	2 Carp	10	1.600	L.F.	171	57		228	282	
	5450	Acrylic	″	10	1.600	″	325	57		382	450	
	5500	Pneumatic tube systems, 2 lane drive-up, complete	L-3	.25	64	Total	22,300	2,475		24,775	28,500	
	5550	With T.V. viewer	″	.20	80	″	43,200	3,100		46,300	52,500	
	5570	Safety deposit boxes, minimum	1 Sswk	44	.182	Opng.	49.50	7.25		56.75	68	
	5580	Maximum, 10″ x 15″ opening		19	.421		105	16.80		121.80	148	
	5590	Teller locker, average	↓	15	.533	↓	1,350	21.50		1,371.50	1,550	
	5600	Pass thru, bullet-res. window, painted steel, 24″ x 36″	2 Sswk	1.60	10	Ea.	1,925	400		2,325	2,875	
	5700	48″ x 48″		1.20	13.333		2,300	535		2,835	3,525	
	5800	72″ x 40″	↓	.80	20	↓	2,875	800		3,675	4,650	
	5900	For stainless steel frames, add					20%					
	6100	Surveillance system, video camera, complete	2 Elec	1	16	Ea.	13,100	670		13,770	15,400	
	6110	For each additional camera, add				″	855			855	940	
	6120	CCTV system, see Div. 16850-600										
	6200	Twenty-four hour teller, single unit,										
	6300	automated deposit, cash and memo	L-3	.25	64	Ea.	40,000	2,475		42,475	48,000	
	7000	Vault front, see Div. 11021-800										

11040 | Ecclesiastical Equipment

	11041	Ecclesiastical Equipment	CREW	DAILY OUTPUT	LABOR-HOURS	UNIT	2006 BARE COSTS MAT.	LABOR	EQUIP.	TOTAL	TOTAL INCL O&P	
250	0010	CHURCH EQUIPMENT										250
	0020	Altar, wood, custom design, plain	1 Carp	1.40	5.714	Ea.	1,925	203		2,128	2,425	

11041 | Ecclesiastical Equipment

		Description	CREW	DAILY OUTPUT	LABOR-HOURS	UNIT	2006 BARE COSTS MAT.	LABOR	EQUIP.	TOTAL	TOTAL INCL O&P	
250	0050	Deluxe	1 Carp	.20	40	Ea.	9,225	1,425		10,650	12,600	250
	0070	Granite or marble, average	2 Marb	.50	32		7,400	1,125		8,525	9,950	
	0090	Deluxe	"	.20	80		24,300	2,800		27,100	31,200	
	0100	Arks, prefabricated, plain	2 Carp	.80	20		7,300	710		8,010	9,200	
	0130	Deluxe, maximum	"	.20	80		91,000	2,850		93,850	104,500	
	0150	Baptistry, fiberglass, 3'-6" deep, x 13'-7" long,										
	0160	steps at both ends, incl. plumbing, minimum	L-8	1	20	Ea.	2,900	740		3,640	4,400	
	0200	Maximum	"	.70	28.571		5,675	1,050		6,725	7,975	
	0250	Add for filter, heater and lights					1,250			1,250	1,375	
	0300	Carillon, 4 octave (48 bells), with keyboard				System	661,500			661,500	727,500	
	0320	2 octave (24 bells)					275,500			275,500	303,000	
	0340	3 to 4 bell peal, minimum					60,500			60,500	66,500	
	0360	Maximum					441,000			441,000	485,000	
	0380	Cast bronze bell, average				Ea.	77,000			77,000	85,000	
	0400	Electronic, digital, minimum					13,800			13,800	15,200	
	0410	With keyboard, maximum					66,000			66,000	73,000	
	0500	Reconciliation room, wood, prefabricated, single, plain	1 Carp	.60	13.333		2,425	475		2,900	3,450	
	0550	Deluxe		.40	20		6,675	710		7,385	8,525	
	0650	Double, plain		.40	20		4,850	710		5,560	6,525	
	0700	Deluxe		.20	40		15,200	1,425		16,625	19,100	
	1000	Lecterns, wood, plain		5	1.600		690	57		747	855	
	1100	Deluxe		2	4		6,925	142		7,067	7,850	
	1500	Pews, bench type, hardwood, minimum		20	.400	L.F.	67	14.20		81.20	97.50	
	1550	Maximum		15	.533		133	18.95		151.95	179	
	1570	For kneeler, add					16			16	17.60	
	2000	Pulpits, hardwood, prefabricated, plain	1 Carp	2	4	Ea.	1,225	142		1,367	1,575	
	2100	Deluxe		1.60	5	"	8,325	178		8,503	9,450	
	2500	Railing, hardwood, average		25	.320	L.F.	151	11.40		162.40	185	
	2700	Safes, see division 11021-600										
	3000	Seating, individual, oak, contour, laminated	1 Carp	21	.381	Person	133	13.55		146.55	170	
	3100	Cushion seat		21	.381		121	13.55		134.55	156	
	3200	Fully upholstered		21	.381		108	13.55		121.55	142	
	3300	Combination, self-rising		21	.381		335	13.55		348.55	390	
	3500	For cherry, add					30%					
	4000	Steeples, translucent fiberglass, 30" square, 15' high	F-3	2	20	Ea.	3,775	720	320	4,815	5,675	
	4150	25' high		1.80	22.222		4,425	800	355	5,580	6,575	
	4350	Opaque fiberglass, 24" square, 14' high		2	20		3,150	720	320	4,190	5,000	
	4500	28' high		1.80	22.222		3,550	800	355	4,705	5,625	
	4600	Aluminum, baked finish, 14' high, 16" square					1,875			1,875	2,050	
	4620	20' high, 3'-6" base					4,575			4,575	5,050	
	4640	35' high, 8' base					17,800			17,800	19,600	
	4660	60' high, 14' base					39,400			39,400	43,300	
	4680	152' high, custom					406,500			406,500	447,000	
	4700	Porcelain enamel steeples, custom, 40' high	F-3	.50	80		11,400	2,875	1,275	15,550	18,700	
	4800	60' high	"	.30	133		19,800	4,800	2,125	26,725	32,000	
	5000	Wall cross, aluminum, extruded, 2" x 2" section	1 Carp	34	.235	L.F.	36.50	8.35		44.85	54.50	
	5150	4" x 4" section		29	.276		52.50	9.80		62.30	74.50	
	5300	Bronze, extruded, 1" x 2" section		31	.258		72	9.15		81.15	94.50	
	5350	2-1/2" x 2-1/2" section		34	.235		109	8.35		117.35	134	
	5450	Solid bar stock, 1/2" x 3" section		29	.276		144	9.80		153.80	174	
	5600	Fiberglass, stock		34	.235		29.50	8.35		37.85	46.50	
	5700	Stainless steel, 4" deep, channel section		29	.276		116	9.80		125.80	144	
	5800	4" deep box section		29	.276		158	9.80		167.80	190	

11050 | Library Equipment

	11051 | Library Equipment		CREW	DAILY OUTPUT	LABOR-HOURS	UNIT	2006 BARE COSTS MAT.	LABOR	EQUIP.	TOTAL	TOTAL INCL O&P
400	0010	LIBRARY EQUIPMENT									
	0020	Bookshelf, mtl., 90″ high, 10″ shelf, dbl. face	1 Carp	11.50	.696	L.F.	144	24.50		168.50	199
	0300	Single face	″	12	.667	″	160	23.50		183.50	216
	0600	For 8″ shelving, subtract from above					10%				
	0700	For 12″ shelving, add to above					10%				
	0800	For 42″ high with countertop, subtract from above					20%				
	1100	Magazine shelving, 82″ high, 12″ deep, single face	1 Carp	11.50	.696	L.F.	177	24.50		201.50	236
	1400	Double face		11.50	.696	″	202	24.50		226.50	263
	2500	Carrels, hardwood, 36″ x 24″, minimum		5	1.600	Ea.	665	57		722	825
	2650	Maximum		4	2		855	71		926	1,050
	2700	Card catalog file, 60 trays, complete					5,325			5,325	5,850
	2720	Alternate method: each tray					88.50			88.50	97.50
	3100	Chairs, wood					145			145	159
	3500	Charging desk, built-in, with counter, plastic laminated top	1 Carp	7	1.143	L.F.	420	40.50		460.50	530
	3700	Reading table, laminated top, 60″ x 36″				Ea.	555			555	610
	3800	Mobile compacted shelving, hand crank, 9′-0″ high									
	3820	Double face, including track, 3′ section				Ea.	1,025			1,025	1,125
	3840	For electrical operation, add					25%				

11060 | Theater & Stage Equipment

	11063 | Stage Equipment		CREW	DAILY OUTPUT	LABOR-HOURS	UNIT	2006 BARE COSTS MAT.	LABOR	EQUIP.	TOTAL	TOTAL INCL O&P
600	0010	STAGE EQUIPMENT									
	0050	Control boards with dimmers and breakers, minimum	1 Elec	1	8	Ea.	11,500	335		11,835	13,200
	0100	Average		.50	16		32,000	670		32,670	36,200
	0150	Maximum		.20	40		106,000	1,675		107,675	119,000
	0500	Curtain track, straight, light duty	2 Carp	20	.800	L.F.	21.50	28.50		50	70.50
	0600	Heavy duty		18	.889		45.50	31.50		77	103
	0700	Curved sections		12	1.333		141	47.50		188.50	234
	1000	Curtains, velour, medium weight		600	.027	S.F.	6.85	.95		7.80	9.05
	1150	Silica based yarn, inherently fire retardant		50	.320		13.15	11.40		24.55	33.50
	1500	Flooring, portable oak parquet, 3′ x 3′ sections					11.55			11.55	12.70
	1600	Cart to carry 225 S.F. of flooring				Ea.	345			345	380
	2000	Lights, border, quartz, reflector, vented,									
	2100	colored or white	1 Elec	20	.400	L.F.	149	16.80		165.80	190
	2500	Spotlight, follow spot, with transformer, 2,100 watt	″	4	2	Ea.	1,200	84		1,284	1,425
	2600	For no transformer, deduct					785			785	865
	3000	Stationary spot, fresnel quartz, 6″ lens	1 Elec	4	2		94	84		178	233
	3100	8″ lens		4	2		188	84		272	335
	3500	Ellipsoidal quartz, 1,000W, 6″ lens		4	2		285	84		369	445
	3600	12″ lens		4	2		470	84		554	650
	4000	Strobe light, 1 to 15 flashes per second, quartz		3	2.667		650	112		762	890
	4500	Color wheel, portable, five hole, motorized		4	2		150	84		234	294
	5000	Stages, portable with steps, folding legs, stock, 8″ high				SF Stg.	22.50			22.50	25
	5100	16″ high					22			22	24
	5200	32″ high					31			31	34
	5300	40″ high					51			51	56.50
	6000	Telescoping platforms, extruded alum., straight, minimum	4 Carp	157	.204		24.50	7.25		31.75	39
	6100	Maximum		77	.416		33	14.75		47.75	61
	6500	Pie-shaped, minimum		150	.213		53	7.60		60.60	71

11060 | Theater & Stage Equipment

	11063 | Stage Equipment		CREW	DAILY OUTPUT	LABOR-HOURS	UNIT	2006 BARE COSTS MAT.	LABOR	EQUIP.	TOTAL	TOTAL INCL O&P	
600	6600	Maximum	4 Carp	70	.457	SF Stg.	59.50	16.25		75.75	92.50	600
	6800	For 3/4" plywood covered deck, deduct					3.36			3.36	3.70	
	7000	Band risers, steel frame, plywood deck, minimum	4 Carp	275	.116		26.50	4.14		30.64	36	
	7100	Maximum	"	138	.232	↓	55.50	8.25		63.75	74.50	
	7500	Chairs for above, self-storing, minimum	2 Carp	43	.372	Ea.	86.50	13.25		99.75	118	
	7600	Maximum	"	40	.400	"	145	14.20		159.20	183	
	8000	Rule of thumb: total stage equipment, minimum	4 Carp	100	.320	SF Stg.	84	11.40		95.40	111	
	8100	Maximum	"	25	1.280	"	475	45.50		520.50	595	

11100 | Mercantile Equipment

	11102 | Barber Shop Equipment		CREW	DAILY OUTPUT	LABOR-HOURS	UNIT	2006 BARE COSTS MAT.	LABOR	EQUIP.	TOTAL	TOTAL INCL O&P	
150	0010	BARBER EQUIPMENT										150
	0020	Chair, hydraulic, movable, minimum	1 Carp	24	.333	Ea.	430	11.85		441.85	495	
	0050	Maximum	"	16	.500		2,750	17.80		2,767.80	3,050	
	0200	Wall hung styling station with mirrors, minimum	L-2	8	2		415	62		477	560	
	0300	Maximum	"	4	4		1,825	124		1,949	2,225	
	0500	Sink, hair washing basin, rough plumbing not incl.	1 Plum	8	1		283	42.50		325.50	375	
	1000	Sterilizer, liquid solution for tools					128			128	140	
	1100	Total equipment, rule of thumb, per chair, minimum	L-8	1	20		1,475	740		2,215	2,825	
	1150	Maximum	"	1	20	↓	4,075	740		4,815	5,700	

	11103 | Cash Register/Checking		CREW	DAILY OUTPUT	LABOR-HOURS	UNIT	MAT.	LABOR	EQUIP.	TOTAL	TOTAL INCL O&P	
200	0010	CHECKOUT COUNTER										200
	0020	Supermarket conveyor, single belt	2 Clab	10	1.600	Ea.	2,300	44		2,344	2,600	
	0100	Double belt, power take-away		9	1.778		3,850	48.50		3,898.50	4,325	
	0400	Double belt, power take-away, incl. side scanning		7	2.286		4,800	62.50		4,862.50	5,375	
	0800	Warehouse or bulk type	↓	6	2.667	↓	5,625	73		5,698	6,300	
	1000	Scanning system, 2 lanes, w/registers, scan gun & memory				System	14,000			14,000	15,400	
	1100	10 lanes, single processor, full scan, with scales				"	133,500			133,500	147,000	
	2000	Register, restaurant, minimum				Ea.	570			570	625	
	2100	Maximum					2,500			2,500	2,750	
	2150	Store, minimum					570			570	625	
	2200	Maximum				↓	2,500			2,500	2,750	

	11104 | Display Cases & Systems		CREW	DAILY OUTPUT	LABOR-HOURS	UNIT	MAT.	LABOR	EQUIP.	TOTAL	TOTAL INCL O&P	
700	0010	REFRIGERATED FOOD CASES										700
	0030	Dairy, multi-deck, 12' long	Q-5	3	5.333	Ea.	8,600	207		8,807	9,775	
	0100	For rear sliding doors, add					1,225			1,225	1,350	
	0200	Delicatessen case, service deli, 12' long, single deck	Q-5	3.90	4.103		5,775	159		5,934	6,625	
	0300	Multi-deck, 18 S.F. shelf display		3	5.333		7,075	207		7,282	8,125	
	0400	Freezer, self-contained, chest-type, 30 C.F.		3.90	4.103		4,225	159		4,384	4,900	
	0500	Glass door, upright, 78 C.F.		3.30	4.848		8,025	188		8,213	9,150	
	0600	Frozen food, chest type, 12' long		3.30	4.848		5,800	188		5,988	6,700	
	0700	Glass door, reach-in, 5 door		3	5.333		11,100	207		11,307	12,500	
	0800	Island case, 12' long, single deck		3.30	4.848		6,575	188		6,763	7,525	
	0900	Multi-deck		3	5.333		13,900	207		14,107	15,600	
	1000	Meat case, 12' long, single deck	↓	3.30	4.848	↓	4,775	188		4,963	5,550	

11 EQUIPMENT

11100 | Mercantile Equipment

	11104 | Display Cases & Systems		CREW	DAILY OUTPUT	LABOR-HOURS	UNIT	2006 BARE COSTS MAT.	LABOR	EQUIP.	TOTAL	TOTAL INCL O&P	
700	1050	Multi-deck	Q-5	3.10	5.161	Ea.	8,225	200		8,425	9,350	700
	1100	Produce, 12′ long, single deck		3.30	4.848		6,325	188		6,513	7,250	
	1200	Multi-deck		3.10	5.161		7,050	200		7,250	8,050	

11110 | Commercial Laundry & Dry Cleaning Equipment

	11119 | Laundry Cleaning		CREW	DAILY OUTPUT	LABOR-HOURS	UNIT	2006 BARE COSTS MAT.	LABOR	EQUIP.	TOTAL	TOTAL INCL O&P	
450	0010	**LAUNDRY EQUIPMENT** Not incl. rough-in										450
	0500	Dryers, gas fired residential, 16 lb. capacity, average	1 Plum	3	2.667	Ea.	650	114		764	890	
	1000	Commercial, 30 lb. capacity, coin operated, single		3	2.667		2,750	114		2,864	3,200	
	1100	Double stacked		2	4		5,825	171		5,996	6,675	
	1500	Industrial, 30 lb. capacity		2	4		2,575	171		2,746	3,100	
	1600	50 lb. capacity		1.70	4.706		3,125	201		3,326	3,725	
	2000	Dry cleaners, electric, 20 lb. capacity	L-1	.20	80		32,800	3,400		36,200	41,400	
	2050	25 lb. capacity		.17	94.118		42,700	3,975		46,675	53,000	
	2100	30 lb. capacity		.15	106		45,000	4,525		49,525	56,500	
	2150	60 lb. capacity		.09	177		70,500	7,525		78,025	89,000	
	3500	Folders, blankets & sheets, minimum	1 Elec	.17	47.059		29,900	1,975		31,875	36,000	
	3700	King size with automatic stacker		.10	80		51,000	3,350		54,350	61,000	
	3800	For conveyor delivery, add		.45	17.778		6,100	745		6,845	7,850	
	4500	Ironers, institutional, 110″, single roll		.20	40		27,500	1,675		29,175	32,900	
	4700	Lint collector, ductwork not included, 8,000 to 10,000 C.F.M.	Q-10	.30	80		8,025	3,150		11,175	13,800	
	4800	Pressers, low capacity air operated	L-6	1.75	6.857		8,375	291		8,666	9,675	
	4820	Hand operated		1.75	6.857		7,475	291		7,766	8,675	
	4830	Extractor, low capacity		1.75	6.857		6,375	291		6,666	7,450	
	4840	Ironer 48″, 240V		3.50	3.429		99,500	146		99,646	109,000	
	4860	Coin dry cleaner 20 lb.		1.75	6.857		24,800	291		25,091	27,800	
	4900	Spreader feeders, 240V, 2 station		.70	17.143		48,400	730		49,130	54,500	
	4920	4 station		.35	34.286		59,500	1,450		60,950	68,000	
	5000	Washers, residential, 4 cycle, average	1 Plum	3	2.667		695	114		809	940	
	5300	Commercial, coin operated, average	″	3	2.667		1,125	114		1,239	1,400	
	6000	Combination washer/extractor, 20 lb. capacity	L-6	1.50	8		4,000	340		4,340	4,925	
	6100	30 lb. capacity		.80	15		7,775	635		8,410	9,525	
	6200	50 lb. capacity		.68	17.647		9,650	750		10,400	11,800	
	6300	75 lb. capacity		.30	40		19,800	1,700		21,500	24,400	
	6350	125 lb. capacity		.16	75		24,100	3,175		27,275	31,400	
	6380	Washer extractor/dryer, 110 lb., 240V		1	12		7,725	510		8,235	9,300	
	6400	Washer extractor, 135 lb, 240V		1	12		23,100	510		23,610	26,300	
	6450	Pass through		1	12		62,500	510		63,010	70,000	
	6500	200 lb.		1	12		60,000	510		60,510	67,000	
	6550	Pass through		1	12		65,000	510		65,510	72,000	
	6600	Hand operated presser		.70	17.143		5,600	730		6,330	7,275	
	6620	Mushroom press 115V		.70	17.143		6,625	730		7,355	8,400	

11136 | Projection Screens

	11136 | Projection Screens		CREW	DAILY OUTPUT	LABOR-HOURS	UNIT	2006 BARE COSTS MAT.	LABOR	EQUIP.	TOTAL	TOTAL INCL O&P	
500	0010	**PROJECTION SCREENS** Wall or ceiling hung, matte white										500
	0100	Manually operated, economy	2 Carp	500	.032	S.F.	5.15	1.14		6.29	7.55	
	0300	Intermediate		450	.036		6	1.26		7.26	8.70	
	0400	Deluxe		400	.040		8.30	1.42		9.72	11.50	
	0600	Electric operated, matte white, 25 S.F., economy		5	3.200	Ea.	750	114		864	1,025	
	0700	Deluxe		4	4		1,575	142		1,717	1,950	
	0900	50 S.F., economy		3	5.333		735	190		925	1,125	
	1000	Deluxe		2	8		1,750	284		2,034	2,400	
	1200	Heavy duty, electric operated, 200 S.F.		1.50	10.667		3,425	380		3,805	4,400	
	1300	400 S.F.		1	16		4,225	570		4,795	5,600	
	1500	Rigid acrylic in wall, for rear projection, 1/4" thick	2 Glaz	30	.533	S.F.	41.50	18.25		59.75	75	
	1600	1/2" thick (maximum size 10' x 20')	"	25	.640	"	73.50	22		95.50	116	
	9000	Minimum labor/equipment charge	2 Carp	3	5.333	Job		190		190	315	
600	0010	**MOVIE EQUIPMENT**										600
	0020	Changeover, minimum				Ea.	410			410	450	
	0100	Maximum					795			795	875	
	0400	Film transport, incl. platters and autowind, minimum					4,425			4,425	4,875	
	0500	Maximum					12,600			12,600	13,800	
	0800	Lamphouses, incl. rectifiers, xenon, 1,000 watt	1 Elec	2	4		5,850	168		6,018	6,700	
	0900	1,600 watt		2	4		6,250	168		6,418	7,125	
	1000	2,000 watt		1.50	5.333		6,700	224		6,924	7,725	
	1100	4,000 watt		1.50	5.333		8,300	224		8,524	9,475	
	1400	Lenses, anamorphic, minimum					1,125			1,125	1,250	
	1500	Maximum					2,525			2,525	2,775	
	1800	Flat 35 mm, minimum					975			975	1,075	
	1900	Maximum					1,525			1,525	1,675	
	2200	Pedestals, for projectors					1,300			1,300	1,425	
	2300	Console type					9,425			9,425	10,400	
	2600	Projector mechanisms, incl. soundhead, 35 mm, minimum					9,750			9,750	10,700	
	2700	Maximum					13,400			13,400	14,700	
	3000	Projection screens, rigid, in wall, acrylic, 1/4" thick	2 Glaz	195	.082	S.F.	37	2.81		39.81	45	
	3100	1/2" thick	"	130	.123	"	43	4.22		47.22	54	
	3300	Electric operated, heavy duty, 400 S.F.	2 Carp	1	16	Ea.	2,600	570		3,170	3,800	
	3320	Theatre projection screens, matte white, including frames	"	200	.080	S.F.	5.85	2.84		8.69	11.15	
	3400	Also see division 11136-500										
	3700	Sound systems, incl. amplifier, mono, minimum	1 Elec	.90	8.889	Ea.	2,900	375		3,275	3,775	
	3800	Dolby/Super Sound, maximum		.40	20		15,900	840		16,740	18,800	
	4100	Dual system, 2 channel, front surround, minimum		.70	11.429		4,075	480		4,555	5,225	
	4200	Dolby/Super Sound, 4 channel, maximum		.40	20		14,600	840		15,440	17,300	
	4500	Sound heads, 35 mm					4,650			4,650	5,125	
	4900	Splicer, tape system, minimum					650			650	715	
	5000	Tape type, maximum					1,175			1,175	1,275	
	5300	Speakers, recessed behind screen, minimum	1 Elec	2	4		935	168		1,103	1,275	
	5400	Maximum	"	1	8		2,725	335		3,060	3,525	
	5700	Seating, painted steel, upholstered, minimum	2 Carp	35	.457		116	16.25		132.25	155	
	5800	Maximum	"	28	.571		370	20.50		390.50	445	
	6100	Rewind tables, minimum					2,325			2,325	2,550	
	6200	Maximum					4,150			4,150	4,575	
	7000	For automation, varying sophistication, minimum	1 Elec	1	8	System	2,100	335		2,435	2,825	
	7100	Maximum	2 Elec	.30	53.333	"	4,900	2,250		7,150	8,850	

11 EQUIPMENT

11140 | Vehicle Service Equipment

	11141	Service Station Equipment	CREW	DAILY OUTPUT	LABOR-HOURS	UNIT	2006 BARE COSTS MAT.	LABOR	EQUIP.	TOTAL	TOTAL INCL O&P	
100	0010	**COMPRESSED AIR EQUIPMENT**										100
	0030	Compressors, electric, 1-1/2 H.P., standard controls	L-4	1.50	16	Ea.	281	530		811	1,175	
	0550	Dual controls	↓	1.50	16	↓	475	530		1,005	1,400	
	0600	5 H.P., 115/230 volt, standard controls		1	24		1,650	795		2,445	3,125	
	0650	Dual controls		1	24		1,775	795		2,570	3,250	
200	0010	**FUEL DISPENSING EQUIPMENT**										200
	1100	Product dispenser with vapor recovery for 6 nozzles, installed, not										
	1110	including piping to storage tanks				Ea.	18,900			18,900	20,800	
300	0010	**LUBRICATION EQUIPMENT**										300
	3000	Lube equipment, 3 reel type, with pumps, not including piping	L-4	.50	48	Set	7,175	1,600		8,775	10,500	
400	0010	**SPRAY PAINTING EQUIPMENT**										400
	4000	Spray painting booth, 26′ long, complete	L-4	.40	60	Ea.	13,300	2,000		15,300	17,900	

11150 | Parking Control Equipment

	11156	Parking Equipment	CREW	DAILY OUTPUT	LABOR-HOURS	UNIT	2006 BARE COSTS MAT.	LABOR	EQUIP.	TOTAL	TOTAL INCL O&P	
600	0010	**PARKING EQUIPMENT**										600
	5000	Barrier gate with programmable controller	2 Elec	3	5.333	Ea.	3,200	224		3,424	3,875	
	5020	Industrial	"	3	5.333	↓	4,350	224		4,574	5,150	
	5100	Card reader	1 Elec	2	4		1,750	168		1,918	2,175	
	5120	Proximity with customer display	2 Elec	1	16		5,300	670		5,970	6,850	
	5200	Cashier booth, average	B-22	1	30		9,225	975	229	10,429	12,100	
	5300	Collector station, pay on foot	2 Elec	.20	80		105,500	3,350		108,850	121,000	
	5320	Credit card only	↓	.50	32		19,500	1,350		20,850	23,500	
	5500	Exit verifier		1	16		16,800	670		17,470	19,400	
	5600	Fee computer	1 Elec	1.50	5.333		12,800	224		13,024	14,400	
	5700	Full sign, 4″ letters	"	2	4		1,175	168		1,343	1,525	
	5800	Inductive loop	2 Elec	4	4		162	168		330	435	
	5900	Ticket spitter with time/date stamp, standard	↓	2	8		6,075	335		6,410	7,225	
	5920	Mag stripe encoding		2	8		17,800	335		18,135	20,100	
	5950	Vehicle detector, microprocessor based	1 Elec	3	2.667		375	112		487	585	
	6000	Parking control software, minimum	↓	.50	16		21,100	670		21,770	24,200	
	6020	Maximum		.20	40		87,500	1,675		89,175	99,000	

11160 | Loading Dock Equipment

	11161	Loading Dock Equipment	CREW	DAILY OUTPUT	LABOR-HOURS	UNIT	2006 BARE COSTS MAT.	LABOR	EQUIP.	TOTAL	TOTAL INCL O&P	
200	0010	**DOCK BUMPERS** Bolts not included										200
	0012	2″ x 6″ to 4″ x 8″, average	1 Carp	300	.027	B.F.	.60	.95		1.55	2.23	
400	0010	**LOADING DOCK**										400
	0020	Bumpers, rubber blocks 4-1/2″ thk., 10″ H, 14″ long	1 Carp	26	.308	Ea.	44.50	10.95		55.45	67	

11160 | Loading Dock Equipment

11161 | Loading Dock Equipment

			CREW	DAILY OUTPUT	LABOR-HOURS	UNIT	2006 BARE COSTS MAT.	LABOR	EQUIP.	TOTAL	TOTAL INCL O&P	
400	0200	24″ long	1 Carp	22	.364	Ea.	84	12.95		96.95	114	400
	0300	36″ long		17	.471		85.50	16.75		102.25	122	
	0500	12″ high, 14″ long		25	.320		88.50	11.40		99.90	116	
	0550	24″ long		20	.400		97.50	14.20		111.70	132	
	0600	36″ long		15	.533		109	18.95		127.95	152	
	0800	Rubber blocks 6″ thick, 10″ high, 14″ long		22	.364		77.50	12.95		90.45	107	
	0850	24″ long		18	.444		107	15.80		122.80	144	
	0900	36″ long		13	.615		125	22		147	175	
	0910	20″ high, 11″ long		13	.615		129	22		151	179	
	0920	Extruded rubber bumpers, T section, 22″ x 22″ x 3″ thick		41	.195		47.50	6.95		54.45	64	
	0940	Molded rubber bumpers, 24″ x 12″ x 3″ thick	↓	20	.400		43.50	14.20		57.70	71.50	
	1000	Welded installation of above bumpers	E-14	8	1		3.30	42	11.15	56.45	94.50	
	1100	For drilled anchors, add per anchor	1 Carp	36	.222	↓	5.90	7.90		13.80	19.60	
	1300	Steel bumpers, see Div. 10265-200										
	1350	Wood bumpers, see Div. 11161-200										
	2200	Dock boards, heavy duty, 60″ x 60″, aluminum, 5,000 lb. cap.				Ea.	1,250			1,250	1,375	
	2700	9,000 lb. capacity					1,175			1,175	1,300	
	3200	15,000 lb. capacity				↓	1,300			1,300	1,425	
	3600	Door seal for door perimeter, 12″ x 12″, vinyl covered	1 Carp	26	.308	L.F.	23.50	10.95		34.45	43.50	
	3900	Folding gates, see Div. 10610-100										
	4200	Platform lifter, 6′ x 6′, portable, 3,000 lb. capacity				Ea.	8,075			8,075	8,875	
	4250	4,000 lb. capacity					9,925			9,925	10,900	
	4400	Fixed, 6′ x 8′, 5,000 lb. capacity	E-16	.70	22.857		8,700	935	128	9,763	11,500	
	4500	Levelers, hinged for trucks, 10 ton capacity, 6′ x 8′		1.08	14.815		4,400	605	82.50	5,087.50	6,050	
	4650	7′ x 8′		1.08	14.815		4,175	605	82.50	4,862.50	5,800	
	4670	Air bag power operated, 10 ton cap., 6′ x 8′		1.08	14.815		5,100	605	82.50	5,787.50	6,825	
	4680	7′ x 8′		1.08	14.815		5,125	605	82.50	5,812.50	6,850	
	4700	Hydraulic, 10 ton capacity, 6′ x 8′		1.08	14.815		7,650	605	82.50	8,337.50	9,625	
	4800	7′ x 8′	↓	1.08	14.815		8,225	605	82.50	8,912.50	10,300	
	5000	Lights for loading docks, single arm, 24″ long	1 Elec	3.80	2.105		127	88.50		215.50	276	
	5700	Double arm, 60″ long	″	3.80	2.105		153	88.50		241.50	305	
	6200	Shelters, fabric, for truck or train, scissor arms, minimum	1 Carp	1	8		1,300	284		1,584	1,900	
	6300	Maximum	″	.50	16	↓	1,900	570		2,470	3,025	

11170 | Solid Waste Handling Equipment

11179 | Waste Handling Equipment

			CREW	DAILY OUTPUT	LABOR-HOURS	UNIT	2006 BARE COSTS MAT.	LABOR	EQUIP.	TOTAL	TOTAL INCL O&P	
150	0010	**WASTE HANDLING**										150
	0020	Compactors, 115 volt, 250#/hr., chute fed	L-4	1	24	Ea.	9,950	795		10,745	12,200	
	0100	Hand fed		2.40	10		7,150	330		7,480	8,425	
	0300	Multi-bag, 230 volt, 600#/hr, chute fed		1	24		8,975	795		9,770	11,200	
	0400	Hand fed		1	24		7,700	795		8,495	9,775	
	0500	Containerized, hand fed, 2 to 6 C.Y. containers, 250#/hr.		1	24		9,425	795		10,220	11,700	
	0550	For chute fed, add per floor		1	24		1,075	795		1,870	2,500	
	1000	Heavy duty industrial compactor, 0.5 C.Y. capacity		1	24		6,375	795		7,170	8,300	
	1050	1.0 C.Y. capacity		1	24		9,800	795		10,595	12,100	
	1100	3 C.Y. capacity		.50	48		13,400	1,600		15,000	17,300	
	1150	5.0 C.Y. capacity		.50	48		16,600	1,600		18,200	20,800	
	1200	Combination shredder/compactor (5,000 lbs./hr.)	↓	.50	48	↓	32,400	1,600		34,000	38,300	

11170 | Solid Waste Handling Equipment

	11179 | Waste Handling Equipment		CREW	DAILY OUTPUT	LABOR-HOURS	UNIT	2006 BARE COSTS MAT.	LABOR	EQUIP.	TOTAL	TOTAL INCL O&P	
150	1400	For handling hazardous waste materials, 55 gallon drum packer, std.				Ea.	15,100			15,100	16,600	150
	1410	55 gallon drum packer w/HEPA filter					18,800			18,800	20,700	
	1420	55 gallon drum packer w/charcoal & HEPA filter					25,100			25,100	27,600	
	1430	All of the above made explosion proof, add					11,500			11,500	12,600	
	1500	Crematory, not including building, 1 place	Q-3	.20	160		53,500	6,500		60,000	69,000	
	1750	2 place		.10	320		76,500	13,000		89,500	104,000	
	4400	Incinerator, gas, not incl. chimney, elec. or pipe, 50#/hr., minimum		.80	40		20,800	1,625		22,425	25,400	
	4420	Maximum		.70	45.714		27,200	1,850		29,050	32,800	
	4440	200 lb. per hr., minimum (batch type)		.60	53.333		27,200	2,175		29,375	33,300	
	4460	Maximum (with feeder)		.50	64		53,000	2,600		55,600	62,000	
	4480	400 lb. per hr., minimum (batch type)		.30	106		32,200	4,350		36,550	42,200	
	4500	Maximum (with feeder)		.25	128		60,500	5,200		65,700	75,000	
	4520	800 lb. per hr., with feeder, minimum		.20	160		79,000	6,500		85,500	97,000	
	4540	Maximum		.17	188		108,000	7,650		115,650	131,000	
	4560	1,200 lb. per hr., with feeder, minimum		.15	213		114,500	8,675		123,175	139,500	
	4580	Maximum		.11	290		137,500	11,800		149,300	170,000	
	4600	2,000 lb. per hr., with feeder, minimum		.10	320		200,000	13,000		213,000	240,000	
	4620	Maximum		.05	640		336,500	26,000		362,500	410,500	
	4700	For heat recovery system, add, minimum		.25	128		65,000	5,200		70,200	79,500	
	4710	Add, maximum		.11	290		208,000	11,800		219,800	247,000	
	4720	For automatic ash conveyer, add		.50	64		27,300	2,600		29,900	34,100	
	4750	Large municipal incinerators, incl. stack, minimum		.25	128	Ton/day	16,600	5,200		21,800	26,400	
	4850	Maximum		.10	320	"	44,100	13,000		57,100	69,000	
	5500	Shredder, municipal use, 35 ton per hour				Ea.	247,000			247,000	272,000	
	5600	60 ton per hour					526,500			526,500	579,000	
	5750	Shredder & baler, 50 ton per day					493,500			493,500	543,000	
	5800	Shredder, industrial, minimum					19,400			19,400	21,400	
	5850	Maximum					104,000			104,000	114,500	
	5900	Baler, industrial, minimum					7,775			7,775	8,575	
	5950	Maximum					455,000			455,000	500,500	
	6000	Transfer station compactor, with power unit										
	6050	and pedestal, not including pit, 50 ton per hour				Ea.	155,500			155,500	171,500	

11190 | Detention Equipment

	11191 | Detention Equipment		CREW	DAILY OUTPUT	LABOR-HOURS	UNIT	2006 BARE COSTS MAT.	LABOR	EQUIP.	TOTAL	TOTAL INCL O&P	
150	0010	DETENTION EQUIPMENT										150
	0500	Bar front, rolling, 7/8" bars, 4" O.C., 7' high, 5' wide, with hardware	E-4	2	16	Ea.	4,825	645	44.50	5,514.50	6,575	
	1000	Doors & frames, 3' x 7', complete, with hardware, single plate		4	8		3,625	325	22.50	3,972.50	4,625	
	1650	Double plate		4	8		4,500	325	22.50	4,847.50	5,550	
	2000	Cells, prefab., 5' to 6' wide, 7' to 8' high, 7' to 8' deep,										
	2010	bar front, cot, not incl. plumbing	E-4	1.50	21.333	Ea.	8,000	865	59.50	8,924.50	10,500	
	2500	Cot, bolted, single, painted steel		20	1.600		300	64.50	4.46	368.96	455	
	2700	Stainless steel		20	1.600		835	64.50	4.46	903.96	1,050	
	3000	Toilet apparatus including wash basin, average	L-8	1.50	13.333		3,000	495		3,495	4,100	
	4000	Visitor cubicle, vision panel, no intercom	E-4	2	16		2,625	645	44.50	3,314.50	4,150	

EQUIPMENT 11

		11285 | Hydraulic Gates	CREW	DAILY OUTPUT	LABOR-HOURS	UNIT	2006 BARE COSTS MAT.	LABOR	EQUIP.	TOTAL	TOTAL INCL O&P	
150	0010	**CANAL GATES** Cast iron body, fabricated frame										150
	0100	12″ diameter	L-5A	4.60	6.957	Ea.	560	278	152	990	1,275	
	0110	18″ diameter		4	8		925	320	174	1,419	1,800	
	0120	24″ diameter		3.50	9.143		1,600	365	199	2,164	2,650	
	0130	30″ diameter		2.80	11.429		2,150	455	249	2,854	3,475	
	0140	36″ diameter		2.30	13.913		2,650	555	305	3,510	4,250	
	0150	42″ diameter		1.70	18.824		3,875	755	410	5,040	6,075	
	0160	48″ diameter		1.20	26.667		5,050	1,075	580	6,705	8,125	
	0170	54″ diameter		.90	35.556		9,150	1,425	775	11,350	13,500	
	0180	60″ diameter		.50	64		10,900	2,550	1,400	14,850	18,100	
	0190	66″ diameter		.50	64		9,075	2,550	1,400	13,025	16,100	
	0200	72″ diameter		.40	80		14,700	3,200	1,750	19,650	23,900	
190	0010	**FLAP GATES**										190
	0100	Aluminum, 18″ diameter	L-5A	5	6.400	Ea.	2,200	256	139	2,595	3,050	
	0110	24″ diameter		4	8		2,350	320	174	2,844	3,350	
	0120	30″ diameter		3.50	9.143		2,825	365	199	3,389	4,000	
	0130	36″ diameter		2.80	11.429		3,550	455	249	4,254	5,000	
	0140	42″ diameter		2.30	13.913		4,875	555	305	5,735	6,675	
	0150	48″ diameter		1.70	18.824		5,875	755	410	7,040	8,275	
	0160	54″ diameter		1.20	26.667		7,200	1,075	580	8,855	10,500	
	0170	60″ diameter		.80	40		9,025	1,600	870	11,495	13,800	
	0180	66″ diameter		.50	64		10,800	2,550	1,400	14,750	18,000	
	0190	72″ diameter		.40	80		12,800	3,200	1,750	17,750	21,800	
400	0010	**KNIFE GATES**										400
	0100	Incl. handwheel operator for hub, 6″ diameter	Q-23	7.70	3.117	Ea.	705	129	124	958	1,100	
	0110	8″ diameter		7.20	3.333		945	138	132	1,215	1,400	
	0120	10″ diameter		4.80	5		1,200	207	198	1,605	1,875	
	0130	12″ diameter		3.60	6.667		1,800	276	265	2,341	2,700	
	0140	14″ diameter		3.40	7.059		3,125	292	280	3,697	4,225	
	0150	16″ diameter		3.20	7.500		3,500	310	298	4,108	4,650	
	0160	18″ diameter		3	8		4,625	330	315	5,270	5,975	
	0170	20″ diameter		2.70	8.889		5,650	370	355	6,375	7,200	
	0180	24″ diameter		2.40	10		6,450	415	395	7,260	8,175	
	0190	30″ diameter		1.80	13.333		13,500	550	530	14,580	16,200	
	0200	36″ diameter		1.20	20		18,500	825	795	20,120	22,500	
600	0010	**SLIDE GATES**										600
	0100	Steel, self contained incl. anchor bolts and grout, 12″ x 12″	L-5A	4.60	6.957	Ea.	2,600	278	152	3,030	3,525	
	0110	18″ x 18″		4	8		2,525	320	174	3,019	3,550	
	0120	24″ x 24″		3.50	9.143		2,725	365	199	3,289	3,875	
	0130	30″ x 30″		2.80	11.429		3,300	455	249	4,004	4,750	
	0140	36″ x 36″		2.30	13.913		3,300	555	305	4,160	4,950	
	0150	42″ x 42″		1.70	18.824		3,850	755	410	5,015	6,025	
	0160	48″ x 48″		1.20	26.667		4,025	1,075	580	5,680	7,000	
	0170	54″ x 54″		.90	35.556		5,125	1,425	775	7,325	9,050	
	0180	60″ x 60″		.55	58.182		5,800	2,325	1,275	9,400	12,000	
	0190	72″ x 72″		.36	88.889		8,375	3,550	1,925	13,850	17,800	
700	0010	**SLUICE GATES** Cast iron, AWWA C501										700
	0100	Heavy duty, self contained w/crank oper. gate, 18″ x 18″	L-5A	1.70	18.824	Ea.	5,075	755	410	6,240	7,375	
	0110	24″ x 24″		1.20	26.667		6,500	1,075	580	8,155	9,725	
	0120	30″ x 30″		1	32		7,850	1,275	695	9,820	11,700	
	0130	36″ x 36″		.90	35.556		9,450	1,425	775	11,650	13,800	
	0140	42″ x 42″		.80	40		11,200	1,600	870	13,670	16,100	
	0150	48″ x 48″		.50	64		11,400	2,550	1,400	15,350	18,600	
	0160	54″ x 54″		.40	80		14,500	3,200	1,750	19,450	23,700	

11280 | Hydraulic Gates & Valves

	11285	Hydraulic Gates	CREW	DAILY OUTPUT	LABOR-HOURS	UNIT	2006 BARE COSTS MAT.	LABOR	EQUIP.	TOTAL	TOTAL INCL O&P	
700	0170	60″ x 60″	L-5A	.30	106	Ea.	16,700	4,275	2,325	23,300	28,600	700
	0180	66″ x 66″		.30	106		20,700	4,275	2,325	27,300	32,900	
	0190	72″ x 72″		.20	160		23,400	6,400	3,475	33,275	41,100	
	0200	78″ x 78″		.20	160		31,600	6,400	3,475	41,475	50,000	
	0210	84″ x 84″		.10	320		29,500	12,800	6,975	49,275	63,000	
	0220	90″ x 90″	E-20	.30	213		35,900	8,325	3,775	48,000	58,500	
	0230	96″ x 96″		.30	213		40,100	8,325	3,775	52,200	63,500	
	0240	108″ x 108″		.20	320		46,200	12,500	5,675	64,375	80,000	
	0250	120″ x 120″		.10	640		54,500	25,000	11,400	90,900	117,500	
	0260	132″ x 132″		.10	640		96,000	25,000	11,400	132,400	163,000	

11300 | Fluid Waste Treatment & Disposal Equipment

	11310	Sewage & Sludge Pumps	CREW	DAILY OUTPUT	LABOR-HOURS	UNIT	2006 BARE COSTS MAT.	LABOR	EQUIP.	TOTAL	TOTAL INCL O&P	
700	0010	**SEWAGE PUMPING STATIONS** Prefabricated steel, concrete										700
	0020	or fiberglass, 200 GPM	C-17D	.17	494	Total	37,500	18,300	3,125	58,925	74,500	
	0200	1,000 GPM		.07	1200		53,000	44,400	7,550	104,950	139,000	
	0500	Add for generator unit, 200 GPM, steel		.34	247		27,100	9,125	1,550	37,775	46,400	
	0600	Concrete		.51	164		17,200	6,100	1,050	24,350	30,000	
	1000	Add for generator unit, 1,000 GPM, steel		.30	280		28,800	10,300	1,775	40,875	50,500	
	1200	Concrete		.38	221		24,300	8,175	1,400	33,875	41,600	
	1500	For drilled water well, if required, add	B-23	.50	80		7,000	2,225	7,100	16,325	19,200	

	11390	Pkg Sewage Treat Plants	CREW	DAILY OUTPUT	LABOR-HOURS	UNIT	MAT.	LABOR	EQUIP.	TOTAL	TOTAL INCL O&P	
200	0010	**SEWAGE TREATMENT PLANTS,** not incl. fencing or external piping										200
	0020	Steel packaged, blown air aeration plants										
	0100	1,000 GPD				Gal.				17.70	20.35	
	0200	5,000 GPD								11.80	13.55	
	0300	15,000 GPD								6.50	7.45	
	0400	30,000 GPD								6.15	7.05	
	0500	50,000 GPD								4.75	5.25	
	0600	100,000 GPD								4.15	4.57	
	0700	200,000 GPD								2.95	3.25	
	0800	500,000 GPD								2.89	3.18	
	1000	Concrete, extended aeration, primary and secondary treatment										
	1010	10,000 GPD				Gal.				14.25	15.70	
	1100	30,000 GPD								6.50	7.15	
	1200	50,000 GPD								5.55	6.10	
	1400	100,000 GPD								4.14	4.55	
	1500	500,000 GPD								2.95	3.26	
	1700	Municipal wastewater treatment facility										
	1720	1.0 MGD				Gal.				5.07	5.60	
	1740	1.5 MGD								5.02	5.55	
	1760	2.0 MGD								4.30	4.75	
	1780	3.0 MGD								3.36	3.71	
	1800	5.0 MGD								3.07	3.38	
	2000	Holding tank system, not incl. excavation or backfill										
	2010	Recirculating chemical water closet	2 Plum	4	4	Ea.	810	171		981	1,150	
	2100	For voltage converter, add	"	16	1		215	42.50		257.50	305	
	2200	For high level alarm, add	1 Plum	7.80	1.026		123	44		167	203	

11300 | Fluid Waste Treatment & Disposal Equipment

	11390 | Pkg Sewage Treat Plants		CREW	DAILY OUTPUT	LABOR-HOURS	UNIT	2006 BARE COSTS MAT.	LABOR	EQUIP.	TOTAL	TOTAL INCL O&P	
900	0010	WASTEWATER TREATMENT SYSTEM										900
	0020	Fiberglass, 1,000 gallon	B-21	1.29	21.705	Ea.	3,075	695	118	3,888	4,650	
	0100	1,500 gallon	"	1.03	27.184	"	6,725	875	148	7,748	9,000	

11400 | Food Service Equipment

	11405 | Food Storage Equipment		CREW	DAILY OUTPUT	LABOR-HOURS	UNIT	2006 BARE COSTS MAT.	LABOR	EQUIP.	TOTAL	TOTAL INCL O&P	
110	0010	FOOD STORAGE EQUIPMENT										110
	2350	Cooler, reach-in, beverage, 6′ long	Q-1	6	2.667	Ea.	3,775	102		3,877	4,300	
	4300	Freezers, reach-in, 44 C.F.		4	4		8,625	154		8,779	9,725	
	4500	68 C.F.		3	5.333		9,825	205		10,030	11,100	
	4600	Freezer, pre-fab, 8′ x 8′ w/refrigeration	2 Carp	.45	35.556		7,950	1,275		9,225	10,900	
	4620	8′ x 12′		.35	45.714		9,775	1,625		11,400	13,500	
	4640	8′ x 16′		.25	64		12,300	2,275		14,575	17,300	
	4660	8′ x 20′		.17	94.118		15,000	3,350		18,350	22,200	
	4680	Reach-in, 1 compartment	Q-1	4	4		2,125	154		2,279	2,600	
	4700	2 compartment		3	5.333		4,325	205		4,530	5,075	
	8300	Refrigerators, reach-in type, 44 C.F.		5	3.200		5,200	123		5,323	5,925	
	8310	With glass doors, 68 C.F.		4	4		7,575	154		7,729	8,600	
	8320	Refrigerator, reach-in, 1 compartment	R-18	7.80	3.333		1,675	108		1,783	2,025	
	8330	2 compartment		6.20	4.194		2,200	136		2,336	2,650	
	8340	3 compartment		5.60	4.643		3,375	151		3,526	3,975	
	8350	Pre-fab, with refrigeration, 8′ x 8′	2 Carp	.45	35.556		5,275	1,275		6,550	7,900	
	8360	8′ x 12′		.35	45.714		6,800	1,625		8,425	10,200	
	8370	8′ x 16′		.25	64		8,650	2,275		10,925	13,300	
	8380	8′ x 20′		.17	94.118		10,600	3,350		13,950	17,300	
	8390	Pass-thru/roll-in, 1 compartment	R-18	7.80	3.333		3,400	108		3,508	3,925	
	8400	2 compartment		6.24	4.167		4,875	135		5,010	5,575	
	8410	3 compartment		5.60	4.643		6,175	151		6,326	7,050	
	8420	Walk-in, alum, door & floor only, no refrig., 6′ x 6′ x 7′-6″	2 Carp	1.40	11.429		6,700	405		7,105	8,050	
	8430	10′ x 6′ x 7′-6″		.55	29.091		9,600	1,025		10,625	12,300	
	8440	12′ x 14′ x 7′-6″		.25	64		13,300	2,275		15,575	18,500	
	8450	12′ x 20′ x 7′-6″		.17	94.118		16,400	3,350		19,750	23,600	
	8460	Refrigerated cabinets, mobile					2,825			2,825	3,100	
	8470	Refrigerator/freezer, reach-in, 1 compartment	R-18	5.60	4.643		4,350	151		4,501	5,025	
	8480	2 compartment	"	4.80	5.417		6,275	176		6,451	7,175	
	8600	Stainless steel shelving, louvered 4-tier, 20″ x 3′	1 Clab	6	1.333		975	36.50		1,011.50	1,125	
	8605	20″ x 4′		6	1.333		1,375	36.50		1,411.50	1,575	
	8610	20″ x 6′		6	1.333		2,000	36.50		2,036.50	2,250	
	8615	24″ x 3′		6	1.333		1,275	36.50		1,311.50	1,450	
	8620	24″ x 4′		6	1.333		1,475	36.50		1,511.50	1,675	
	8625	24″ x 6′		6	1.333		2,150	36.50		2,186.50	2,400	
	8630	Flat 4-tier, 20″ x 3′		6	1.333		1,125	36.50		1,161.50	1,300	
	8635	20″ x 4′		6	1.333		1,350	36.50		1,386.50	1,525	
	8640	20″ x 5′		6	1.333		1,550	36.50		1,586.50	1,750	
	8645	24″ x 3′		6	1.333		1,175	36.50		1,211.50	1,350	
	8650	24″ x 4′		6	1.333		1,400	36.50		1,436.50	1,575	
	8655	24″ x 6′		6	1.333		2,900	36.50		2,936.50	3,250	
	8700	Galvanized shelving, louvered 4-tier, 20″ x 3′		6	1.333		495	36.50		531.50	600	

		11405 | Food Storage Equipment	CREW	DAILY OUTPUT	LABOR-HOURS	UNIT	2006 BARE COSTS MAT.	LABOR	EQUIP.	TOTAL	TOTAL INCL O&P	
110	8705	20″ x 4′	1 Clab	6	1.333	Ea.	550	36.50		586.50	665	110
	8710	20″ x 6′		6	1.333		570	36.50		606.50	690	
	8715	24″ x 3′		6	1.333		450	36.50		486.50	555	
	8720	24″ x 4′		6	1.333		590	36.50		626.50	710	
	8725	24″ x 6′		6	1.333		845	36.50		881.50	990	
	8730	Flat 4-tier, 20″ x 3′		6	1.333		350	36.50		386.50	445	
	8735	20″ x 4′		6	1.333		395	36.50		431.50	495	
	8740	20″ x 6′		6	1.333		710	36.50		746.50	840	
	8745	24″ x 3′		6	1.333		340	36.50		376.50	435	
	8750	24″ x 4′		6	1.333		535	36.50		571.50	645	
	8755	24″ x 6′		6	1.333		755	36.50		791.50	890	
	8760	Stainless steel dunnage rack, 24″ x 3′		8	1		680	27.50		707.50	795	
	8765	24″ x 4′		8	1		705	27.50		732.50	820	
	8770	Galvanized dunnage rack, 24″ x 3′		8	1		111	27.50		138.50	169	
	8775	24″ x 4′		8	1		144	27.50		171.50	204	
800	0010	**WINE CELLAR**, refrigerated, Redwood interior, carpeted, walk-in type										800
	0020	6′-8″ high, including racks										
	0200	80 ″ W x 48″ D for 900 bottles	2 Carp	1.50	10.667	Ea.	3,125	380		3,505	4,050	
	0250	80″ W x 72″ D for 1300 bottles		1.33	12.030		4,125	430		4,555	5,250	
	0300	80″ W x 94″ D for 1900 bottles		1.17	13.675		5,350	485		5,835	6,700	
	0400	80″ W x 124″ D for 2500 bottles		1	16		6,475	570		7,045	8,075	
	0600	Portable cabinets, red oak, reach-in temp.& humidity controlled										
	0650	26-5/8″ W x 26-1/2″ D x 68″ H for 235 bottles				Ea.	2,825			2,825	3,100	
	0660	32″ W x 21-1/2″ D x 73-1/2″ H for 144 bottles					3,275			3,275	3,600	
	0670	32″ W x 29-1/2″ D x 73-1/2″ H for 288 bottles					3,300			3,300	3,625	
	0680	39-1/2″ W x 29-1/2″ D x 86-1/2″ H for 440 bottles					3,750			3,750	4,125	
	0690	52-1/2″ W x 29-1/2″ D x 73-1/2″ H for 468 bottles					4,300			4,300	4,725	
	0700	52-1/2″ W x 29-1/2″ D x 86-1/2″ H for 572 bottles					4,525			4,525	4,975	
	0730	Portable, red oak, can be built-in with glass door										
	0750	23-7/8″ W x 24″ D x 34-1/2″ H for 50 bottles				Ea.	1,025			1,025	1,125	

		11410 | Food Preparation Equipment	CREW	DAILY OUTPUT	LABOR-HOURS	UNIT	MAT.	LABOR	EQUIP.	TOTAL	TOTAL INCL O&P	
110	0010	FOOD PREPARATION EQUIPMENT										110
	1700	Choppers, 5 pounds	R-18	7	3.714	Ea.	2,400	120		2,520	2,850	
	1720	16 pounds		5	5.200		2,700	169		2,869	3,250	
	1740	35 to 40 pounds		4	6.500		3,000	211		3,211	3,625	
	1840	Coffee brewer, 5 burners	1 Plum	3	2.667		1,150	114		1,264	1,425	
	1850	Coffee urn, twin 6 gallon urns		2	4		2,650	171		2,821	3,175	
	1860	Single, 3 gallon		3	2.667		1,675	114		1,789	2,025	
	3000	Fast food equipment, total package, minimum	6 Skwk	.08	600		139,500	21,900		161,400	189,500	
	3100	Maximum	″	.07	685		190,000	25,000		215,000	250,000	
	3800	Food mixers, 20 quarts	L-7	7	4		3,150	137		3,287	3,700	
	3850	40 quarts		5.40	5.185		6,775	177		6,952	7,750	
	3900	60 quarts		5	5.600		9,875	191		10,066	11,200	
	4040	80 quarts		3.90	7.179		11,200	245		11,445	12,800	
	4080	130 quarts		2.20	12.727		17,200	435		17,635	19,600	
	4100	Floor type, 20 quarts		15	1.867		2,550	63.50		2,613.50	2,900	
	4120	60 quarts		14	2		7,225	68.50		7,293.50	8,025	
	4140	80 quarts		12	2.333		9,875	79.50		9,954.50	10,900	
	4160	140 quarts		8.60	3.256		22,200	111		22,311	24,600	
	6700	Peelers, small	R-18	8	3.250		1,150	105		1,255	1,450	
	6720	Large	″	6	4.333		3,650	141		3,791	4,250	
	6800	Pulper/extractor, close coupled, 5 HP	1 Plum	1.90	4.211		3,250	180		3,430	3,850	
	8580	Slicer with table	R-18	9	2.889		4,175	93.50		4,268.50	4,750	

11400 | Food Service Equipment

11415 | Food Delivery Carts and Conveyors

			CREW	DAILY OUTPUT	LABOR-HOURS	UNIT	2006 BARE COSTS MAT.	LABOR	EQUIP.	TOTAL	TOTAL INCL O&P	
110	0010	FOOD DELIVERY CARTS										110
	1650	Cabinet, heated, 1 compartment, reach-in	R-18	5.60	4.643	Ea.	2,325	151		2,476	2,825	
	1655	Pass-thru roll-in		5.60	4.643		3,975	151		4,126	4,600	
	1660	2 compartment, reach-in	↓	4.80	5.417		5,925	176		6,101	6,775	
	1670	Mobile					2,625			2,625	2,875	
	6850	Mobile rack w/pan slide					985			985	1,075	
	9180	Tray and silver dispenser, mobile	1 Clab	16	.500	↓	770	13.70		783.70	870	

11420 | Food Cooking Equipment

			CREW	DAILY OUTPUT	LABOR-HOURS	UNIT	MAT.	LABOR	EQUIP.	TOTAL	TOTAL INCL O&P	
110	0010	COOKING EQUIPMENT										110
	0020	Bake oven, gas, one section	Q-1	8	2	Ea.	4,700	77		4,777	5,275	
	0300	Two sections		7	2.286		9,450	88		9,538	10,500	
	0600	Three sections	↓	6	2.667		14,000	102		14,102	15,600	
	0900	Electric convection, single deck	L-7	4	7		4,875	239		5,114	5,750	
	1300	Broiler, without oven, standard	Q-1	8	2		2,950	77		3,027	3,375	
	1550	Infra-red	L-7	4	7		6,475	239		6,714	7,525	
	4750	Fryer, with twin baskets, modular model	Q-1	7	2.286		1,675	88		1,763	1,975	
	5000	Floor model, on 6" legs	"	5	3.200		1,575	123		1,698	1,950	
	5100	Extra single basket, large					94			94	103	
	5300	Griddle, SS, 24" plate, w/4" legs, elec., 208V, 3 phase, 3' long	Q-1	7	2.286		915	88		1,003	1,125	
	5550	4' long	"	6	2.667		1,225	102		1,327	1,500	
	6200	Iced tea brewer	1 Plum	3.44	2.326		625	99.50		724.50	845	
	6350	Kettle, w/steam jacket, tilting, w/positive lock, SS, 20 gallons	L-7	7	4		5,100	137		5,237	5,850	
	6600	60 gallons	"	6	4.667		7,250	159		7,409	8,225	
	6900	Range, restaurant type, 6 burners and 1 standard oven, 36" wide	Q-1	7	2.286		1,875	88		1,963	2,200	
	6950	Convection		7	2.286		2,650	88		2,738	3,050	
	7150	2 standard ovens, 24" griddle, 60" wide		6	2.667		4,650	102		4,752	5,250	
	7200	1 standard, 1 convection oven		6	2.667		5,625	102		5,727	6,350	
	7450	Heavy duty, single 34" standard oven, open top		5	3.200		4,125	123		4,248	4,750	
	7500	Convection oven		5	3.200		5,675	123		5,798	6,450	
	7700	Griddle top		6	2.667		4,925	102		5,027	5,575	
	7750	Convection oven	↓	6	2.667		6,500	102		6,602	7,300	
	8850	Steamer, electric 27 KW	L-7	7	4		8,975	137		9,112	10,100	
	9100	Electric, 10 KW or gas 100,000 BTU	"	5	5.600		4,325	191		4,516	5,075	
	9150	Toaster, conveyor type, 16-22 slices per minute					1,050			1,050	1,175	
	9160	Pop-up, 2 slot				↓	575			575	630	
	9200	For deluxe models of above equipment, add					75%					
	9400	Rule of thumb: Equipment cost based										
	9410	on kitchen work area										
	9420	Office buildings, minimum	L-7	77	.364	S.F.	62	12.40		74.40	88.50	
	9450	Maximum		58	.483		105	16.50		121.50	142	
	9550	Public eating facilities, minimum		77	.364		81.50	12.40		93.90	110	
	9600	Maximum		46	.609		132	21		153	179	
	9750	Hospitals, minimum		58	.483		83.50	16.50		100	119	
	9800	Maximum	↓	39	.718	↓	140	24.50		164.50	194	

11425 | Hood and Ventilation Equipment

			CREW	DAILY OUTPUT	LABOR-HOURS	UNIT	MAT.	LABOR	EQUIP.	TOTAL	TOTAL INCL O&P	
110	0010	HOOD & VENTILATION EQUIPMENT										110
	2970	Exhaust hood, sst, gutter on all sides, 4' x 4' x 2'	1 Carp	1.80	4.444	Ea.	3,250	158		3,408	3,825	
	2980	4'x 4' x 7'	"	1.60	5		5,050	178		5,228	5,850	
	7950	Hood fire protection system, minimum	Q-1	3	5.333		3,450	205		3,655	4,100	
	8050	Maximum	"	1	16	↓	25,300	615		25,915	28,900	

11430 | Food Dispensing Equipment

			CREW	DAILY OUTPUT	LABOR-HOURS	UNIT	2006 BARE COSTS MAT.	LABOR	EQUIP.	TOTAL	TOTAL INCL O&P	
110	0010	FOOD DISPENSING EQUIPMENT										110
	1050	Butter pat dispenser	1 Clab	13	.615	Ea.	780	16.85		796.85	885	
	1100	Bread dispenser, counter top		13	.615		740	16.85		756.85	845	
	1900	Cup and glass dispenser, drop in		4	2		1,050	55		1,105	1,275	
	1920	Disposable cup, drop in		16	.500		320	13.70		333.70	380	
	2650	Dish dispenser, drop in, 12"		11	.727		1,200	19.95		1,219.95	1,350	
	2660	Mobile		10	.800		2,025	22		2,047	2,250	
	3300	Food warmer, counter, 1.2 KW					525			525	580	
	3550	1.6 KW					1,550			1,550	1,700	
	3600	Well, hot food, built-in, rectangular, 12" x 20"	R-30	10	2.600		360	86		446	535	
	3610	Circular, 7 qt.		10	2.600		325	86		411	500	
	3620	Refrigerated, 2 compartments		10	2.600		1,850	86		1,936	2,175	
	3630	3 compartments		9	2.889		1,850	96		1,946	2,200	
	3640	4 compartments		8	3.250		2,575	108		2,683	3,000	
	4720	Frost cold plate		9	2.889		12,700	96		12,796	14,200	
	5700	Hot chocolate dispenser	1 Plum	4	2		855	85.50		940.50	1,075	
	6250	Jet spray dispenser	R-18	4.50	5.778		2,650	187		2,837	3,200	
	6300	Juice dispenser, concentrate	"	4.50	5.778		925	187		1,112	1,325	
	6690	Milk dispenser, bulk, 2 flavor	R-30	8	3.250		1,175	108		1,283	1,475	
	6695	3 flavor	"	8	3.250		1,625	108		1,733	1,950	
	8800	Serving counter, straight	1 Carp	40	.200	L.F.	620	7.10		627.10	695	
	8820	Curved section	"	30	.267	"	775	9.50		784.50	870	
	8825	Solid surface, see section 06620-810										
	8830	Soft serve ice cream machine, medium	R-18	11	2.364	Ea.	8,300	76.50		8,376.50	9,275	
	8840	Large	"	9	2.889	"	12,300	93.50		12,393.50	13,700	

11435 | Ice Machines

			CREW	DAILY OUTPUT	LABOR-HOURS	UNIT	MAT.	LABOR	EQUIP.	TOTAL	TOTAL INCL O&P	
110	0010	ICE MACHINES										110
	5800	Ice cube maker, 50 pounds per day	Q-1	6	2.667	Ea.	1,400	102		1,502	1,675	
	5900	250 pounds per day		1.20	13.333		1,850	510		2,360	2,825	
	6050	500 pounds per day		4	4		2,700	154		2,854	3,225	
	6060	With bin		1.20	13.333		2,200	510		2,710	3,225	
	6090	1000 pounds per day, with bin		1	16		6,625	615		7,240	8,225	
	6100	Ice flakers, 300 pounds per day		1.60	10		3,025	385		3,410	3,925	
	6120	600 pounds per day		.95	16.842		3,325	645		3,970	4,650	
	6130	1000 pounds per day		.75	21.333		4,950	820		5,770	6,700	
	6140	2000 pounds per day		.65	24.615		14,800	945		15,745	17,700	
	6160	Ice storage bin, 500 pound capacity	Q-5	1	16		1,800	620		2,420	2,950	
	6180	1000 pound	"	.56	28.571		3,150	1,100		4,250	5,175	

11440 | Cleaning and Disposal Equipment

			CREW	DAILY OUTPUT	LABOR-HOURS	UNIT	MAT.	LABOR	EQUIP.	TOTAL	TOTAL INCL O&P	
110	0010	CLEANING & DISPOSAL EQUIPMENT										110
	2700	Dishwasher, commercial, rack type										
	2720	10 to 12 racks per hour	Q-1	3.20	5	Ea.	3,875	192		4,067	4,575	
	2750	Semi-automatic 38 to 50 racks per hour	"	1.30	12.308		6,950	475		7,425	8,350	
	2800	Automatic, 190 to 230 racks per hour	L-6	.35	34.286		8,625	1,450		10,075	11,700	
	2820	235 to 275 racks per hour		.25	48		19,900	2,050		21,950	25,100	
	2840	8,750 to 12,500 dishes per hour		.10	120		43,200	5,100		48,300	55,500	
	2950	Dishwasher hood, canopy type	L-3A	10	1.200	L.F.	465	47		512	585	
	2960	Pant leg type	"	2.50	4.800	Ea.	6,525	188		6,713	7,475	
	5200	Garbage disposal 1.5 HP, 100 GPH	L-1	4.80	3.333		1,400	141		1,541	1,775	
	5210	3 HP, 120 GPH		4.60	3.478		2,250	147		2,397	2,700	
	5220	5 HP, 250 GPH		4.50	3.556		3,225	151		3,376	3,775	
	6750	Pot sink, 3 compartment	1 Plum	7.25	1.103	L.F.	610	47		657	750	
	6760	Pot washer, small		1.60	5	Ea.	16,400	214		16,614	18,400	

11400 | Food Service Equipment

	11440 | Cleaning and Disposal Equipment	CREW	DAILY OUTPUT	LABOR-HOURS	UNIT	2006 BARE COSTS MAT.	LABOR	EQUIP.	TOTAL	TOTAL INCL O&P	
110	6770 Large	1 Plum	1.20	6.667	Ea.	36,900	285		37,185	41,000	110
	9170 Trash compactor, small, up to 125 lb. compacted weight	L-4	4	6		17,500	199		17,699	19,500	
	9175 Large, up to 175 lb. compacted weight	″	3	8		21,300	265		21,565	23,900	

11450 | Residential Equipment

	11454 | Residential Appliances	CREW	DAILY OUTPUT	LABOR-HOURS	UNIT	2006 BARE COSTS MAT.	LABOR	EQUIP.	TOTAL	TOTAL INCL O&P	
500	0010 RESIDENTIAL APPLIANCES										500
	0020 Cooking range, 30″ free standing, 1 oven, minimum	2 Clab	10	1.600	Ea.	249	44		293	345	
	0050 Maximum		4	4		1,550	110		1,660	1,875	
	0150 2 oven, minimum		10	1.600		1,525	44		1,569	1,750	
	0200 Maximum		10	1.600		1,600	44		1,644	1,850	
	0350 Built-in, 30″ wide, 1 oven, minimum	1 Elec	6	1.333		460	56		516	590	
	0400 Maximum	2 Carp	2	8		1,350	284		1,634	1,975	
	0500 2 oven, conventional, minimum		4	4		960	142		1,102	1,275	
	0550 1 conventional, 1 microwave, maximum		2	8		1,525	284		1,809	2,150	
	0700 Free-standing, 1 oven, 21″ wide range, minimum	2 Clab	10	1.600		256	44		300	355	
	0750 21″ wide, maximum	″	4	4		283	110		393	490	
	0900 Counter top cook tops, 4 burner, standard, minimum	1 Elec	6	1.333		192	56		248	298	
	0950 Maximum		3	2.667		535	112		647	765	
	1050 As above, but with grille and griddle attachment, minimum		6	1.333		480	56		536	615	
	1100 Maximum		3	2.667		660	112		772	900	
	1200 Induction cooktop, 30″ wide		3	2.667		535	112		647	765	
	1250 Microwave oven, minimum		4	2		85.50	84		169.50	223	
	1300 Maximum		2	4		405	168		573	705	
	1750 Compactor, residential size, 4 to 1 compaction, minimum	1 Carp	5	1.600		430	57		487	565	
	1800 Maximum	″	3	2.667		485	95		580	685	
	2000 Deep freeze, 15 to 23 C.F., minimum	2 Clab	10	1.600		405	44		449	520	
	2050 Maximum		5	3.200		535	87.50		622.50	730	
	2200 30 C.F., minimum		8	2		775	55		830	945	
	2250 Maximum		3	5.333		880	146		1,026	1,200	
	2450 Dehumidifier, portable, automatic, 15 pint					154			154	170	
	2550 40 pint					172			172	189	
	2750 Dishwasher, built-in, 2 cycles, minimum	L-1	4	4		256	169		425	545	
	2800 Maximum		2	8		299	340		639	855	
	2950 4 or more cycles, minimum		4	4		271	169		440	560	
	2960 Average		4	4		360	169		529	655	
	3000 Maximum		2	8		560	340		900	1,150	
	3200 Dryer, automatic, minimum	L-2	3	5.333		287	165		452	590	
	3250 Maximum	″	2	8		810	248		1,058	1,300	
	3300 Garbage disposal, sink type, minimum	L-1	10	1.600		44	68		112	153	
	3350 Maximum	″	10	1.600		150	68		218	270	
	3550 Heater, electric, built-in, 1250 watt, ceiling type, minimum	1 Elec	4	2		72	84		156	209	
	3600 Maximum		3	2.667		122	112		234	305	
	3700 Wall type, minimum		4	2		106	84		190	245	
	3750 Maximum		3	2.667		140	112		252	325	
	3900 1500 watt wall type, with blower		4	2		131	84		215	273	
	3950 3000 watt		3	2.667		267	112		379	465	
	4150 Hood for range, 2 speed, vented, 30″ wide, minimum	L-3	5	3.200		38.50	124		162.50	243	

	11454 | Residential Appliances		Crew	Daily Output	Labor-Hours	Unit	2006 Bare Costs Mat.	Labor	Equip.	Total	Total Incl O&P	
500	4200	Maximum	L-3	3	5.333	Ea.	615	207		822	1,025	500
	4300	42" wide, minimum		5	3.200		232	124		356	455	
	4330	Custom		5	3.200		645	124		769	910	
	4350	Maximum		3	5.333		785	207		992	1,200	
	4500	For ventless hood, 2 speed, add					15.55			15.55	17.10	
	4650	For vented 1 speed, deduct from maximum					40.50			40.50	44.50	
	4850	Humidifier, portable, 8 gallons per day					154			154	170	
	5000	15 gallons per day					185			185	204	
	5200	Icemaker, automatic, 20 lb. per day	1 Plum	7	1.143		375	49		424	485	
	5350	51 lb. per day	"	2	4		1,050	171		1,221	1,425	
	5450	Refrigerator, no frost, 6 C.F.	2 Clab	15	1.067		295	29		324	375	
	5500	Refrigerator, no frost, 10 C.F. to 12 C.F. minimum		10	1.600		455	44		499	575	
	5600	Maximum		6	2.667		705	73		778	895	
	5750	14 C.F. to 16 C.F., minimum		9	1.778		460	48.50		508.50	585	
	5800	Maximum		5	3.200		500	87.50		587.50	695	
	5950	18 C.F. to 20 C.F., minimum		8	2		525	55		580	670	
	6000	Maximum		4	4		870	110		980	1,125	
	6150	21 C.F. to 29 C.F., minimum		7	2.286		670	62.50		732.50	840	
	6200	Maximum		3	5.333		2,150	146		2,296	2,600	
	6400	Sump pump cellar drainer, pedestal, 1/3 H.P., molded PVC base	1 Plum	3	2.667		88.50	114		202.50	275	
	6450	Solid brass	"	2	4		185	171		356	470	
	6460	Sump pump, see also division 15440-940										
	6650	Washing machine, automatic, minimum	1 Plum	3	2.667	Ea.	288	114		402	490	
	6700	Maximum	"	1	8		1,050	340		1,390	1,675	
	6900	Water heater, electric, glass lined, 30 gallon, minimum	L-1	5	3.200		261	136		397	495	
	6950	Maximum		3	5.333		360	226		586	750	
	7100	80 gallon, minimum		2	8		500	340		840	1,075	
	7150	Maximum		1	16		695	680		1,375	1,825	
	7180	Water heater, gas, glass lined, 30 gallon, minimum	2 Plum	5	3.200		565	137		702	830	
	7220	Maximum		3	5.333		785	228		1,013	1,225	
	7260	50 gallon, minimum		2.50	6.400		595	273		868	1,075	
	7300	Maximum		1.50	10.667		830	455		1,285	1,625	
	7310	Water heater, see also division 15480-200										
	7350	Water softener, automatic, to 30 grains per gallon	2 Plum	5	3.200	Ea.	445	137		582	700	
	7400	To 100 grains per gallon	"	4	4		620	171		791	945	
	7450	Vent kits for dryers	1 Carp	10	.800		12.80	28.50		41.30	61	
550	0010	**DISAPPEARING STAIRWAY** No trim included										550
	0100	Custom grade, pine, 8'-6" ceiling, minimum	1 Carp	4	2	Ea.	117	71		188	247	
	0150	Average		3.50	2.286		118	81.50		199.50	265	
	0200	Maximum		3	2.667		195	95		290	370	
	0500	Heavy duty, pivoted, from 7'-7" to 12'-10" floor to floor		3	2.667		385	95		480	575	
	0600	16'-0" ceiling		2	4		1,275	142		1,417	1,625	
	0800	Economy folding, pine, 8'-6" ceiling		4	2		107	71		178	235	
	0900	9'-6" ceiling		4	2		116	71		187	246	
	1000	Fire escape, galvanized steel, 8'-0" to 10'-4" ceiling	2 Carp	1	16		1,350	570		1,920	2,450	
	1010	10'-6" to 13'-6" ceiling		1	16		1,700	570		2,270	2,825	
	1100	Automatic electric, aluminum, floor to floor height, 8' to 9'		1	16		7,000	570		7,570	8,650	
	1400	11' to 12'		.90	17.778		7,550	630		8,180	9,350	
	1700	14' to 15'		.70	22.857		8,025	815		8,840	10,200	
	9000	Minimum labor/equipment charge	1 Carp	2	4	Job		142		142	236	

	11460 | Unit Kitchens		Crew	Daily Output	Labor-Hours	Unit	Mat.	Labor	Equip.	Total	Total Incl O&P	
100	0010	**UNIT KITCHENS**										100
	1500	Combination range, refrigerator and sink, 30" wide, minimum	L-1	2	8	Ea.	755	340		1,095	1,350	
	1550	Maximum		1	16		1,500	680		2,180	2,700	
	1570	60" wide, average		1.40	11.429		2,500	485		2,985	3,500	

11450 | Residential Equipment

	11460 | Unit Kitchens		CREW	DAILY OUTPUT	LABOR-HOURS	UNIT	2006 BARE COSTS MAT.	LABOR	EQUIP.	TOTAL	TOTAL INCL O&P	
100	1590	72″ wide, average	L-1	1.20	13.333	Ea.	2,825	565		3,390	3,975	100
	1600	Office model, 48″ wide		2	8		2,125	340		2,465	2,875	
	1620	Refrigerator and sink only		2.40	6.667		2,125	282		2,407	2,775	
	1640	Combination range, refrigerator, sink, microwave										
	1660	oven and ice maker	L-1	.80	20	Ea.	4,150	845		4,995	5,875	

11470 | Darkroom Equipment

	11471 | Darkroom Processing		CREW	DAILY OUTPUT	LABOR-HOURS	UNIT	2006 BARE COSTS MAT.	LABOR	EQUIP.	TOTAL	TOTAL INCL O&P	
700	0010	DARKROOM EQUIPMENT										700
	0020	Developing sink, 5″ deep, 24″ x 48″	Q-1	2	8	Ea.	4,450	305		4,755	5,350	
	0050	48″ x 52″		1.70	9.412		4,500	360		4,860	5,500	
	0200	10″ deep, 24″ x 48″		1.70	9.412		5,550	360		5,910	6,650	
	0250	24″ x 108″		1.50	10.667		2,000	410		2,410	2,825	
	0500	Dryers, dehumidified filtered air, 36″ x 25″ x 68″ high	L-7	6	4.667		4,325	159		4,484	5,025	
	0550	48″ x 25″ x 68″ high		5	5.600		7,975	191		8,166	9,100	
	2000	Processors, automatic, color print, minimum		4	7		11,200	239		11,439	12,700	
	2050	Maximum		.60	46.667		11,800	1,600		13,400	15,600	
	2300	Black and white print, minimum		2	14		8,500	480		8,980	10,100	
	2350	Maximum		.80	35		55,500	1,200		56,700	63,000	
	2600	Manual processor, 16″ x 20″ maximum print size		2	14		8,575	480		9,055	10,200	
	2650	20″ x 24″ maximum print size		1	28		8,075	955		9,030	10,500	
	3000	Viewing lites, 20″ x 24″		6	4.667		350	159		509	645	
	3100	20″ x 24″ with color correction		6	4.667		495	159		654	805	
	3500	Washers, round, minimum sheet 11″ x 14″	Q-1	2	8		2,725	305		3,030	3,475	
	3550	Maximum sheet 20″ x 24″		1	16		3,225	615		3,840	4,475	
	3800	Square, minimum sheet 20″ x 24″		1	16		2,875	615		3,490	4,125	
	3900	Maximum sheet 50″ x 56″		.80	20		4,475	770		5,245	6,125	
	4500	Combination tank sink, tray sink, washers, with										
	4510	dry side tables, average	Q-1	.45	35.556	Ea.	8,525	1,375		9,900	11,500	

	11472 | Revolving Darkroom Doors		CREW	DAILY OUTPUT	LABOR-HOURS	UNIT	MAT.	LABOR	EQUIP.	TOTAL	TOTAL INCL O&P	
370	0010	DARKROOM DOORS										370
	0015	Revolving, standard, 2 way, 36″ diameter	2 Carp	3.10	5.161	Opng.	1,850	183		2,033	2,350	
	0020	41″ diameter		3.10	5.161		2,000	183		2,183	2,500	
	0050	3 way, 51″ diameter		1.40	11.429		2,475	405		2,880	3,400	
	1000	4 way, 49″ diameter		1.40	11.429		2,950	405		3,355	3,900	
	2000	Hinged safety, 2 way, 41″ diameter		2.30	6.957		2,375	247		2,622	3,000	
	2500	3 way, 51″ diameter		1.40	11.429		2,975	405		3,380	3,950	
	3000	Pop out safety, 2 way, 41″ diameter		3.10	5.161		2,900	183		3,083	3,500	
	4000	3 way, 51″ diameter		1.40	11.429		2,900	405		3,305	3,875	
	5000	Wheelchair-type, pop out, 51″ diameter		1.40	11.429		3,000	405		3,405	3,975	
	5020	72″ diameter		.90	17.778		6,125	630		6,755	7,800	
	9300	For complete dark rooms, see division 13035-300										

11483 | Bowling Alleys

			CREW	DAILY OUTPUT	LABOR-HOURS	UNIT	2006 BARE COSTS MAT.	LABOR	EQUIP.	TOTAL	TOTAL INCL O&P	
150	0010	**BOWLING ALLEYS** Including alley, pinsetter, scorer,										150
	0020	counters and misc. supplies, minimum	4 Carp	.20	160	Lane	35,100	5,700		40,800	48,000	
	0150	Average		.19	168		40,000	5,975		45,975	54,000	
	0300	Maximum		.18	177		46,000	6,325		52,325	61,000	
	0400	Combo table ball rack, add					995			995	1,100	
	0600	For automatic scorer, add, minimum					6,350			6,350	7,000	
	0700	Maximum					7,200			7,200	7,925	

11484 | Exercise Equipment

			CREW	DAILY OUTPUT	LABOR-HOURS	UNIT	MAT.	LABOR	EQUIP.	TOTAL	TOTAL INCL O&P	
400	0010	**HEALTH CLUB EQUIPMENT**										400
	0020	Abdominal rack, 2 board capacity				Ea.	435			435	480	
	0050	Abdominal board, upholstered					490			490	540	
	0200	Bicycle trainer, minimum					760			760	835	
	0300	Deluxe, electric					3,875			3,875	4,250	
	0400	Bar bell set, chrome plated steel, 25 lbs.					228			228	251	
	0420	100 lbs.					345			345	380	
	0450	200 lbs.					665			665	735	
	0500	Weight plates, cast iron, per lb.				Lb.	4.67			4.67	5.15	
	0520	Storage rack, 10 station				Ea.	735			735	810	
	0600	Circuit training apparatus, 12 machines minimum	2 Clab	1.25	12.800	Set	25,700	350		26,050	28,900	
	0700	Average		1	16		31,600	440		32,040	35,400	
	0800	Maximum		.75	21.333		37,400	585		37,985	42,200	
	0820	Dumbbell set, cast iron, with rack and 5 pair					720			720	790	
	0900	Squat racks	2 Clab	5	3.200	Ea.	715	87.50		802.50	930	
	1200	Multi-station gym machine, 5 station					7,425			7,425	8,175	
	1250	9 station					10,400			10,400	11,500	
	1280	Rowing machine, hydraulic					1,375			1,375	1,500	
	1300	Treadmill, manual					1,350			1,350	1,475	
	1320	Motorized					3,800			3,800	4,175	
	1340	Electronic					5,100			5,100	5,600	
	1360	Cardio-testing					7,250			7,250	7,975	
	1400	Treatment/massage tables, minimum					485			485	535	
	1420	Deluxe, with accessories					640			640	705	
	1600	For saunas, see Div. 13035-800										
	1640	For steam baths, see Div. 13035-940										
	1680	For whirlpool baths, see Div. 15418-100										
	1720	For additional exercise equipment, see Div. 11486-700										

11486 | Gymnasium Equipment

			CREW	DAILY OUTPUT	LABOR-HOURS	UNIT	MAT.	LABOR	EQUIP.	TOTAL	TOTAL INCL O&P	
700	0010	**SCHOOL EQUIPMENT**										700
	0200	For exterior equipment, see Div. 02880										
	0201	For chairs & desks, see Div. 12560-200										
	0300	For chalkboards & bulletin boards, see Div. 10120-940 & 10110-240										
	0400	For lockers, see Div. 10505-500										
	1000	Basketball backstops, wall mtd., 6' extended, fixed, minimum	L-2	1	16	Ea.	1,075	495		1,570	2,000	
	1100	Maximum		1	16		1,575	495		2,070	2,550	
	1200	Swing up, minimum		1	16		1,275	495		1,770	2,225	
	1250	Maximum		1	16		5,000	495		5,495	6,325	
	1300	Portable, manual, heavy duty, spring operated		1.90	8.421		10,600	261		10,861	12,100	
	1400	Ceiling suspended, stationary, minimum		.78	20.513		1,900	635		2,535	3,125	
	1450	Fold up, with accessories, maximum		.40	40		5,700	1,250		6,950	8,325	
	1600	For electrically operated, add	1 Elec	1	8		1,875	335		2,210	2,575	
	2000	Benches, folding, in wall, 14' table, 2 benches	L-4	2	12	Set	600	400		1,000	1,300	
	3000	Bleachers, telescoping, manual to 15 tier, minimum	F-5	65	.492	Seat	71	17.75		88.75	108	
	3100	Maximum		60	.533		107	19.25		126.25	149	

11486 | Gymnasium Equipment

			CREW	DAILY OUTPUT	LABOR-HOURS	UNIT	2006 BARE COSTS MAT.	LABOR	EQUIP.	TOTAL	TOTAL INCL O&P	
700	3300	16 to 20 tier, minimum	F-5	60	.533	Seat	170	19.25		189.25	219	700
	3400	Maximum		55	.582		213	21		234	269	
	3600	21 to 30 tier, minimum		50	.640		177	23		200	233	
	3700	Maximum		40	.800		213	29		242	282	
	3900	For integral power operation, add, minimum	2 Elec	300	.053		35.50	2.24		37.74	42.50	
	4000	Maximum	"	250	.064		57	2.69		59.69	66.50	
	4100	Boxing ring, elevated, 22′ x 22′	L-4	.10	240	Ea.	8,075	7,950		16,025	21,900	
	4110	For cellular plastic foam padding, add		.10	240		2,550	7,950		10,500	15,800	
	4120	Floor level, including posts and ropes only, 20′ x 20′		.80	30		1,500	995		2,495	3,275	
	4130	Canvas, 30′ x 30′		5	4.800		965	159		1,124	1,325	
	4150	Exercise equipment, bicycle trainer					405			405	445	
	4180	Chinning bar, adjustable, wall mounted	1 Carp	5	1.600		287	57		344	410	
	4200	Exercise ladder, 16′ x 1′-7″, suspended	L-2	3	5.333		1,250	165		1,415	1,650	
	4210	High bar, floor plate attached	1 Carp	4	2		1,075	71		1,146	1,325	
	4240	Parallel bars, adjustable		4	2		2,375	71		2,446	2,725	
	4270	Uneven parallel bars, adjustable		4	2		2,825	71		2,896	3,225	
	4280	Wall mounted, adjustable	L-2	1.50	10.667	Set	815	330		1,145	1,450	
	4300	Rope, ceiling mounted, 18′ long	1 Carp	3.66	2.186	Ea.	212	77.50		289.50	360	
	4330	Side horse, vaulting		5	1.600		1,700	57		1,757	1,975	
	4360	Treadmill, motorized, deluxe, training type		5	1.600		5,450	57		5,507	6,075	
	4390	Weight lifting multi-station, minimum	2 Clab	1	16		2,575	440		3,015	3,550	
	4450	Maximum	"	.50	32		14,200	875		15,075	17,100	
	4480	For additional exercise equipment, see Div. 11484-400										
	4500	Gym divider curtain, mesh top, vinyl bottom, manual	L-4	500	.048	S.F.	7.75	1.59		9.34	11.10	
	4700	Electric roll up	L-7	400	.070		11.30	2.39		13.69	16.35	
	5500	Gym mats, 2″ thick, naugahyde covered					3.53			3.53	3.88	
	5600	Vinyl/nylon covered					5.30			5.30	5.85	
	5800	Wall pads, 1-1/2″ thick	2 Carp	640	.025		5.80	.89		6.69	7.85	
	6000	Wrestling mats, 1″ thick, heavy duty					4.73			4.73	5.20	
	7000	Scoreboards, baseball, minimum	R-3	1.30	15.385	Ea.	2,975	635	117	3,727	4,375	
	7200	Maximum		.05	400		14,600	16,600	3,050	34,250	45,100	
	7300	Football, minimum		.86	23.256		3,950	965	177	5,092	6,050	
	7400	Maximum		.20	100		12,600	4,150	765	17,515	21,000	
	7500	Basketball (one side), minimum		2.07	9.662		2,050	400	73.50	2,523.50	2,950	
	7600	Maximum		.30	66.667		14,300	2,750	510	17,560	20,500	
	7700	Hockey-basketball (four sides), minimum		.25	80		5,675	3,325	610	9,610	12,000	
	7800	Maximum		.15	133		23,600	5,525	1,025	30,150	35,700	

11488 | Shooting Ranges

			CREW	DAILY OUTPUT	LABOR-HOURS	UNIT	MAT.	LABOR	EQUIP.	TOTAL	TOTAL INCL O&P	
700	0010	**SHOOTING RANGE** Incl. bullet traps, target provisions, controls,										700
	0100	separators, ceiling system, etc. Not incl. structural shell										
	0200	Commercial	L-9	.64	56.250	Point	13,000	1,800		14,800	17,400	
	0300	Law enforcement		.28	128		26,000	4,100		30,100	35,600	
	0400	National Guard armories		.71	50.704		12,500	1,625		14,125	16,600	
	0500	Reserve training centers		.71	50.704		10,500	1,625		12,125	14,400	
	0600	Schools and colleges		.32	112		22,000	3,600		25,600	30,300	
	0700	Major acadamies		.19	189		40,000	6,050		46,050	54,500	
	0800	For acoustical treatment, add					10%	10%				
	0900	For lighting, add					28%	25%				
	1000	For plumbing, add					5%	5%				
	1100	For ventilating system, add, minimum					40%	40%				
	1200	Add, average					25%	25%				
	1300	Add, maximum					35%	35%				

11520	Industrial Equipment	CREW	DAILY OUTPUT	LABOR-HOURS	UNIT	2006 BARE COSTS MAT.	LABOR	EQUIP.	TOTAL	TOTAL INCL O&P
250 0010	**DUST COLLECTION SYSTEMS** Commercial / industrial									250
0120	Central vacuum units									
0130	Includes stand, filters and motorized shaker									
0200	500 CFM, 10″ inlet, 2 HP	Q-20	2.40	8.333	Ea.	3,650	325		3,975	4,500
0220	1000 CFM, 10″ inlet, 3 HP		2.20	9.091		3,800	350		4,150	4,750
0240	1500 CFM, 10″ inlet, 5 HP		2	10		3,900	385		4,285	4,875
0260	3000 CFM, 13″ inlet, 10 HP		1.50	13.333		10,200	515		10,715	12,000
0280	5000 CFM, 16″ inlet, 2 @ 10 HP		1	20		10,800	775		11,575	13,000
1000	Vacuum tubing, galvanized									
1100	2-1/8″ OD, 16 ga.	Q-9	440	.036	L.F.	1.93	1.38		3.31	4.31
1110	2-1/2″ OD, 16 ga.		420	.038		2.23	1.45		3.68	4.75
1120	3″ OD, 16 ga.		400	.040		2.84	1.52		4.36	5.55
1130	3-1/2″ OD, 16 ga.		380	.042		4.03	1.60		5.63	6.95
1140	4″ OD, 16 ga.		360	.044		4.12	1.69		5.81	7.20
1150	5″ OD, 14 ga.		320	.050		7.65	1.90		9.55	11.40
1160	6″ OD, 14 ga.		280	.057		8.55	2.17		10.72	12.85
1170	8″ OD, 14 ga.		200	.080		12.95	3.03		15.98	19.10
1180	10″ OD, 12 ga.		160	.100		12.95	3.79		16.74	20.50
1190	12″ OD, 12 ga.		120	.133		37	5.05		42.05	49
1200	14″ OD, 12 ga.		80	.200		41	7.60		48.60	57
1940	Hose, flexible wire reinforced rubber									
1956	3″ dia.	Q-9	400	.040	L.F.	5.55	1.52		7.07	8.50
1960	4″ dia.		360	.044		6.30	1.69		7.99	9.60
1970	5″ dia.		320	.050		7.15	1.90		9.05	10.85
1980	6″ dia.		280	.057		7.75	2.17		9.92	11.95
2000	90° Elbow, slip fit									
2110	2-1/8″ dia.	Q-9	70	.229	Ea.	8.05	8.65		16.70	22.50
2120	2-1/2″ dia.		65	.246		11.30	9.35		20.65	27.50
2130	3″ dia.		60	.267		15.30	10.10		25.40	33
2140	3-1/2″ dia.		55	.291		18.90	11.05		29.95	38.50
2150	4″ dia.		50	.320		23.50	12.15		35.65	45.50
2160	5″ dia.		45	.356		44.50	13.50		58	70.50
2170	6″ dia.		40	.400		61.50	15.15		76.65	91.50
2180	8″ dia.		30	.533		117	20		137	160
2400	45° Elbow, slip fit									
2410	2-1/8″ dia.	Q-9	70	.229	Ea.	7.05	8.65		15.70	21.50
2420	2-1/2″ dia.		65	.246		10.40	9.35		19.75	26.50
2430	3″ dia.		60	.267		12.55	10.10		22.65	30
2440	3-1/2″ dia.		55	.291		15.80	11.05		26.85	35
2450	4″ dia.		50	.320		20.50	12.15		32.65	42
2460	5″ dia.		45	.356		34.50	13.50		48	59.50
2470	6″ dia.		40	.400		47	15.15		62.15	75.50
2480	8″ dia.		35	.457		91.50	17.35		108.85	129
2800	90° TY, slip fit thru 6″ dia									
2810	2-1/8″ dia.	Q-9	42	.381	Ea.	15.80	14.45		30.25	40.50
2820	2-1/2″ dia.		39	.410		20.50	15.55		36.05	47.50
2830	3″ dia.		36	.444		27	16.85		43.85	57
2840	3-1/2″ dia.		33	.485		34.50	18.40		52.90	67.50
2850	4″ dia.		30	.533		49.50	20		69.50	86.50
2860	5″ dia.		27	.593		94.50	22.50		117	140
2870	6″ dia.		24	.667		129	25.50		154.50	181
2880	8″ dia., butt end					273			273	300
2890	10″ dia., butt end					495			495	540
2900	12″ dia., butt end					495			495	540
2910	14″ dia., butt end					895			895	985
2920	6″ x 4″ dia., butt end					78.50			78.50	86.50

	11520 | Industrial Equipment		CREW	DAILY OUTPUT	LABOR-HOURS	UNIT	2006 BARE COSTS MAT.	LABOR	EQUIP.	TOTAL	TOTAL INCL O&P	
250	2930	8" x 4" dia., butt end				Ea.	149			149	163	250
	2940	10" x 4" dia., butt end					195			195	214	
	2950	12" x 4" dia., butt end					223			223	246	
	3100	90° Elbow, butt end segmented										
	3110	8" dia., butt end, segmented				Ea.	176			176	194	
	3120	10" dia., butt end, segmented					290			290	320	
	3130	12" dia., butt end, segmented					365			365	405	
	3140	14" dia., butt end, segmented					460			460	505	
	3200	45° Elbow, butt end segmented										
	3210	8" dia., butt end, segmented				Ea.	150			150	165	
	3220	10" dia., butt end, segmented					209			209	230	
	3230	12" dia., butt end, segmented					214			214	236	
	3240	14" dia., butt end, segmented					264			264	290	
	3400	All butt end fittings require one coupling per joint.										
	3410	Labor for fitting included with couplings.										
	3460	Compression coupling, galvanized, neoprene gasket										
	3470	2-1/8" dia.	Q-9	44	.364	Ea.	12.30	13.80		26.10	35.50	
	3480	2-1/2" dia.		44	.364		12.30	13.80		26.10	35.50	
	3490	3" dia.		38	.421		16.15	15.95		32.10	43.50	
	3500	3-1/2" dia.		35	.457		18.05	17.35		35.40	47.50	
	3510	4" dia.		33	.485		19.60	18.40		38	51	
	3520	5" dia.		29	.552		22.50	21		43.50	58	
	3530	6" dia.		26	.615		26	23.50		49.50	65.50	
	3540	8" dia.		22	.727		47.50	27.50		75	96.50	
	3550	10" dia.		20	.800		69.50	30.50		100	125	
	3560	12" dia.		18	.889		87	33.50		120.50	149	
	3570	14" dia.		16	1		123	38		161	196	
	3800	Air gate valves, galvanized										
	3810	2-1/8" dia.	Q-9	30	.533	Ea.	93.50	20		113.50	135	
	3820	2-1/2" dia.		28	.571		98	21.50		119.50	143	
	3830	3" dia.		26	.615		114	23.50		137.50	162	
	3840	4" dia.		23	.696		149	26.50		175.50	206	
	3850	6" dia.		18	.889		206	33.50		239.50	281	
300	0010	EQUIPMENT INSTALLATION										300
	0020	Industrial equipment, minimum	E-2	12	4.667	Ton		182	119	301	455	
	0200	Maximum	"	2	28	"		1,100	715	1,815	2,725	
850	0010	VOCATIONAL SHOP EQUIPMENT										850
	0020	Benches, work, wood, average	2 Carp	5	3.200	Ea.	450	114		564	685	
	0100	Metal, average		5	3.200		385	114		499	615	
	0400	Combination belt & disc sander, 6"		4	4		710	142		852	1,025	
	0700	Drill press, floor mounted, 12", 1/2 H.P.		4	4		350	142		492	620	
	0800	Dust collector, not incl. ductwork, 6" diameter	1 Shee	1.10	7.273		2,725	305		3,030	3,500	
	1000	Grinders, double wheel, 1/2 H.P.	2 Carp	5	3.200		192	114		306	400	
	1300	Jointer, 4", 3/4 H.P.		4	4		1,000	142		1,142	1,325	
	1600	Kilns, 16 C.F., to 2000°		4	4		2,150	142		2,292	2,600	
	1900	Lathe, woodworking, 10", 1/2 H.P.		4	4		915	142		1,057	1,225	
	2200	Planer, 13" x 6"		4	4		1,725	142		1,867	2,125	
	2500	Potter's wheel, motorized		4	4		980	142		1,122	1,300	
	2800	Saws, band, 14", 3/4 H.P.		4	4		750	142		892	1,050	
	3100	Metal cutting band saw, 14"		4	4		1,975	142		2,117	2,400	
	3400	Radial arm saw, 10", 2 H.P.		4	4		695	142		837	1,000	
	3700	Scroll saw, 24"		4	4		1,450	142		1,592	1,825	
	4000	Table saw, 10", 3 H.P.		4	4		1,700	142		1,842	2,100	
	4300	Welder AC arc, 30 amp capacity		4	4		2,125	142		2,267	2,575	

11600 | Laboratory Equipment

11620 | Laboratory Equipment

		11620 Laboratory Equipment	CREW	DAILY OUTPUT	LABOR-HOURS	UNIT	2006 BARE COSTS MAT.	LABOR	EQUIP.	TOTAL	TOTAL INCL O&P	
110	0010	LABORATORY EQUIPMENT										110
	0600	Fume hood, with countertop & base, not including HVAC										
	0610	Simple, minimum	2 Carp	5.40	2.963	L.F.	545	105		650	775	
	0620	Complex, including fixtures		2.40	6.667		1,300	237		1,537	1,825	
	0630	Special, maximum	↓	1.70	9.412	↓	1,350	335		1,685	2,025	
	0650	Ductwork, minimum	2 Shee	1	16	Hood	905	675		1,580	2,075	
	0660	Maximum	″	.50	32	″	4,575	1,350		5,925	7,175	
	0670	Service fixtures, average				Ea.	46.50			46.50	51	
	0680	For sink assembly with hot and cold water, add	1 Plum	1.40	5.714		620	244		864	1,075	
	0700	Glassware washer, undercounter, minimum	L-1	1.80	8.889		6,325	375		6,700	7,525	
	0710	Maximum	″	1	16		8,625	680		9,305	10,500	
	0750	Glove box, fiberglass, bacteriological					13,900			13,900	15,300	
	0760	Controlled atmosphere					9,200			9,200	10,100	
	0770	Radioisotope					11,900			11,900	13,100	
	0780	Carcinogenic					11,900			11,900	13,100	
	1000	Incubators, minimum					3,125			3,125	3,425	
	1010	Maximum					17,400			17,400	19,100	
	1200	Refrigerator, blood bank, 28.6 C.F. emergency signal					6,625			6,625	7,275	
	1210	Reach-in, 16.9 C.F.					6,650			6,650	7,325	
	1400	Safety equipment, eye wash, hand held					310			310	340	
	1450	Deluge shower					595			595	655	
	1600	Sink, one piece plastic, flask wash, hose, free standing	1 Plum	1.60	5		1,575	214		1,789	2,050	
	1610	Epoxy resin sink, 25″ x 16″ x 10″	″	2	4		175	171		346	460	
	1700	Thermometer, electric, portable					325			325	360	
	1800	Titration unit, four 2000 ml reservoirs					9,025			9,025	9,950	
	1850	Utensil washer-sanitizer	1 Plum	2	4	↓	9,425	171		9,596	10,700	
	1950	Utility table, acid resistant top with drawers	2 Carp	30	.533	L.F.	129	18.95		147.95	174	
	8000	Alternate pricing method: as percent of lab furniture										
	8050	Installation, not incl. plumbing & duct work				% Furn.					22%	
	8100	Plumbing, final connections, simple system									10%	
	8110	Moderately complex system									15%	
	8120	Complex system									20%	
	8150	Electrical, simple system									10%	
	8160	Moderately complex system									20%	
	8170	Complex system				↓					35%	

11700 | Medical Equipment

11710 | Medical Sterilizing Equipment

		11710 Medical Sterilizing Equipment	CREW	DAILY OUTPUT	LABOR-HOURS	UNIT	2006 BARE COSTS MAT.	LABOR	EQUIP.	TOTAL	TOTAL INCL O&P	
110	0010	MEDICAL STERILIZING EQUIPMENT										110
	0700	Distiller, water, steam heated, 50 gal. capacity	1 Plum	1.40	5.714	Ea.	15,900	244		16,144	17,900	
	5600	Sterilizers, floor loading, 26″ x 62″ x 42″, single door, steam					137,000			137,000	150,500	
	5650	Double door, steam					176,000			176,000	193,500	
	5800	General purpose, 20″ x 20″ x 38″, single door					13,200			13,200	14,500	
	6000	Portable, counter top, steam, minimum					3,275			3,275	3,600	
	6020	Maximum					5,125			5,125	5,650	
	6050	Portable, counter top, gas, 17″ x 15″ x 32-1/2″					33,900			33,900	37,300	
	6150	Manual washer/sterilizer, 16″ x 16″ x 26″	1 Plum	2	4	↓	46,500	171		46,671	51,500	
	6200	Steam generators, electric 10 KW to 180 KW, freestanding										

11700 | Medical Equipment

	11710 | Medical Sterilizing Equipment		CREW	DAILY OUTPUT	LABOR-HOURS	UNIT	2006 BARE COSTS MAT.	LABOR	EQUIP.	TOTAL	TOTAL INCL O&P	
110	6250	Minimum	1 Elec	3	2.667	Ea.	6,900	112		7,012	7,775	110
	6300	Maximum	"	.70	11.429		24,900	480		25,380	28,100	
	8200	Bed pan washer-sanitizer	1 Plum	2	4		6,150	171		6,321	7,050	

	11720 | Examination & Treatment Equip.		CREW	DAILY OUTPUT	LABOR-HOURS	UNIT	MAT.	LABOR	EQUIP.	TOTAL	TOTAL INCL O&P	
110	0010	EXAMINATION & TREATMENT EQUIPMENT										110
	0300	Blood pressure unit, mercurial, wall				Ea.	130			130	143	
	0400	Diagnostic set, wall					560			560	615	
	4400	Scale, physician's, with height rod					254			254	279	
	6500	Surgery table, minor minimum	1 Sswk	.70	11.429		12,000	455		12,455	14,100	
	6520	Maximum	"	.50	16		22,500	640		23,140	26,000	
	6700	Surgical lights, doctor's office, single arm	2 Elec	2	8		1,125	335		1,460	1,750	
	6750	Dual arm	"	1	16		5,025	670		5,695	6,550	

	11730 | Patient Care Equipment		CREW	DAILY OUTPUT	LABOR-HOURS	UNIT	MAT.	LABOR	EQUIP.	TOTAL	TOTAL INCL O&P	
110	0010	PATIENT CARE EQUIPMENT										110
	0750	Exam room furnishings, average per room				Ea.	5,575			5,575	6,125	
	1800	Heat therapy unit, humidified, 26" x 78" x 28"				"	2,925			2,925	3,225	
	2100	Hubbard tank with accessories, stainless steel,										
	2110	125 GPM at 45 psi water pressure				Ea.	22,700			22,700	25,000	
	2150	For electric overhead hoist, add					2,475			2,475	2,725	
	2900	K-Module for heat therapy, 20 oz. capacity, 75° to 110° F					350			350	385	
	3600	Paraffin bath, 126° F, auto controlled					1,300			1,300	1,425	
	3900	Parallel bars for walking training, 12'-0"					1,050			1,050	1,175	
	4600	Station, dietary, medium, with ice					13,800			13,800	15,200	
	4700	Medicine					6,250			6,250	6,875	
	7000	Tables, physical therapy, walk off, electric	2 Carp	3	5.333		2,775	190		2,965	3,375	
	7150	Standard, vinyl top with base cabinets, minimum		3	5.333		1,125	190		1,315	1,550	
	7200	Maximum		2	8		3,325	284		3,609	4,125	
	8400	Whirlpool bath, mobile, sst, 18" x 24" x 60"					3,950			3,950	4,350	
	8450	Fixed, incl. mixing valves	1 Plum	2	4		7,950	171		8,121	9,025	

	11740 | Dental Equipment		CREW	DAILY OUTPUT	LABOR-HOURS	UNIT	MAT.	LABOR	EQUIP.	TOTAL	TOTAL INCL O&P	
200	0010	DENTAL EQUIPMENT										200
	0020	Central suction system, minimum	1 Plum	1.20	6.667	Ea.	3,025	285		3,310	3,775	
	0100	Maximum	"	.90	8.889		23,400	380		23,780	26,300	
	0300	Air compressor, minimum	1 Skwk	.80	10		3,325	365		3,690	4,250	
	0400	Maximum		.50	16		23,100	585		23,685	26,400	
	0600	Chair, electric or hydraulic, minimum		.50	16		3,800	585		4,385	5,125	
	0700	Maximum		.25	32		8,050	1,175		9,225	10,800	
	0800	Doctor's/assistant's stool, minimum					430			430	475	
	0850	Maximum					845			845	930	
	1000	Drill console with accessories, minimum	1 Skwk	1.60	5		3,425	183		3,608	4,075	
	1100	Maximum		1.60	5		9,800	183		9,983	11,100	
	2000	Light, ceiling mounted, minimum		8	1		1,975	36.50		2,011.50	2,225	
	2100	Maximum		8	1		3,975	36.50		4,011.50	4,425	
	2200	Unit light, minimum	2 Skwk	5.33	3.002		1,475	110		1,585	1,800	
	2210	Maximum		5.33	3.002		4,825	110		4,935	5,500	
	2220	Track light, minimum		3.20	5		2,625	183		2,808	3,175	
	2230	Maximum		3.20	5		5,750	183		5,933	6,625	
	2300	Sterilizers, steam portable, minimum					3,350			3,350	3,675	
	2350	Maximum					11,500			11,500	12,700	
	2600	Steam, institutional					14,400			14,400	15,800	
	2650	Dry heat, electric, portable, 3 trays					1,125			1,125	1,225	
	2700	Ultra-sonic cleaner, portable, minimum					905			905	995	

11700 | Medical Equipment

		11740 | Dental Equipment	CREW	DAILY OUTPUT	LABOR-HOURS	UNIT	2006 BARE COSTS MAT.	LABOR	EQUIP.	TOTAL	TOTAL INCL O&P	
200	2750	Maximum (institutional)				Ea.	2,025			2,025	2,225	200
	3000	X-ray unit, wall, minimum	1 Skwk	4	2		4,575	73		4,648	5,150	
	3010	Maximum		4	2		8,625	73		8,698	9,625	
	3100	Panoramic unit	↓	.60	13.333		20,700	485		21,185	23,600	
	3105	Deluxe, minimum	2 Skwk	1.60	10		29,900	365		30,265	33,500	
	3110	Maximum	"	1.60	10		69,000	365		69,365	76,500	
	3500	Developers, X-ray, average	1 Plum	5.33	1.501		4,025	64		4,089	4,525	
	3600	Maximum	"	5.33	1.501	↓	7,475	64		7,539	8,325	

		11760 | Operating Room Equipment	CREW	DAILY OUTPUT	LABOR-HOURS	UNIT	MAT.	LABOR	EQUIP.	TOTAL	TOTAL INCL O&P	
110	0010	OPERATING ROOM EQUIPMENT										110
	5000	Scrub, surgical, stainless steel, single station, minimum	1 Plum	3	2.667	Ea.	2,050	114		2,164	2,425	
	5100	Maximum					7,225			7,225	7,950	
	6550	Major surgery table, minimum	1 Sswk	.50	16		27,100	640		27,740	31,100	
	6570	Maximum	"	.50	16		89,500	640		90,140	99,500	
	6800	Surgical lights, major operating room, dual head, minimum	2 Elec	1	16		15,100	670		15,770	17,600	
	6850	Maximum	"	1	16	↓	25,100	670		25,770	28,600	

		11770 | Radiology Equipment	CREW	DAILY OUTPUT	LABOR-HOURS	UNIT	MAT.	LABOR	EQUIP.	TOTAL	TOTAL INCL O&P	
110	0010	RADIOLOGY EQUIPMENT										110
	8700	X-ray, mobile, minimum				Ea.	11,600			11,600	12,700	
	8750	Maximum					65,000			65,000	71,500	
	8900	Stationary, minimum					36,100			36,100	39,700	
	8950	Maximum					188,000			188,000	206,500	
	9150	Developing processors, minimum					10,100			10,100	11,100	
	9200	Maximum				↓	26,400			26,400	29,000	

		11781 | Mortuary Equipment	CREW	DAILY OUTPUT	LABOR-HOURS	UNIT	MAT.	LABOR	EQUIP.	TOTAL	TOTAL INCL O&P	
110	0010	MORTUARY EQUIPMENT										110
	0015	Autopsy table, standard	1 Plum	1	8	Ea.	7,650	340		7,990	8,950	
	0020	Deluxe	"	.60	13.333		13,800	570		14,370	16,100	
	3200	Mortuary refrigerator, end operated, 2 capacity					11,900			11,900	13,100	
	3300	6 capacity				↓	21,900			21,900	24,100	

Division Notes

	CREW	DAILY OUTPUT	LABOR-HOURS	UNIT	2006 BARE COSTS MAT.	LABOR	EQUIP.	TOTAL	TOTAL INCL O&P

Division 12 Furnishings

Estimating Tips

General

- The items in this division are usually priced per square foot or each. Most of these items are purchased by the owner and placed by the supplier. Do not assume the items in Division 12 will be purchased and installed by the supplier. Check the specifications for responsibilities and include receiving, storage, installation, and mechanical and electrical hookups in the appropriate divisions.
- Some items in this division require some type of support system that is not usually furnished with the item. Examples of these systems include blocking for the attachment of casework and heavy drapery rods. The required blocking must be added to the estimate in the appropriate division.

Reference Numbers

Reference numbers are shown in bold squares at the beginning of some major classifications. These numbers refer to related items in the Reference Section. The reference information may be an estimating procedure, an alternate pricing method, or technical information.

Note: Not all subdivisions listed here necessarily appear in this publication.

12040 | Selective Demolition

	12041 Selective demolition	CREW	DAILY OUTPUT	LABOR-HOURS	UNIT	2006 BARE COSTS MAT.	LABOR	EQUIP.	TOTAL	TOTAL INCL O&P	
100	0010 CASEWORK R024119-10										100
	3100 Metal base cabinets	2 Clab	20	.800	L.F.		22		22	36.50	
	3110 Cabinet base trim		400	.040			1.10		1.10	1.82	
	3120 Countertop, stainless steel		80	.200			5.50		5.50	9.10	
	3130 Wall cabinets, wood, 84" high		30	.533			14.60		14.60	24	
	3500 Laboratory casework, tall storage cabinets, 84" high		40	.400			10.95		10.95	18.15	
	3510 Wall cabinets, metal		40	.400			10.95		10.95	18.15	
	4830 Floor mats, recessed or link mats	1 Clab	300	.027	S.F.		.73		.73	1.21	
	4832 Skate lock tile		200	.040			1.10		1.10	1.82	
	4834 Duckboard		300	.027			.73		.73	1.21	
	4920 Blinds, interior, horizontal or vertical		1000	.008	L.F.		.22		.22	.36	
	4922 Wood folding panels		35	.229	Pr.		6.25		6.25	10.40	
	4924 Shades, interior		700	.011	S.F.		.31		.31	.52	
	4930 Drapery hardware, traverse rods		35	.229	Ea.		6.25		6.25	10.40	
	5200 Fixed seating, per seat	2 Carp	44	.364	"		12.95		12.95	21.50	
	6400 Booth, restaurant	2 Clab	80	.200	L.F.		5.50		5.50	9.10	
	7400 Office systems furniture, cubicle	"	3000	.005	S.F.		.15		.15	.24	

12050 | Fabrics

	12051 Fabrics	CREW	DAILY OUTPUT	LABOR-HOURS	UNIT	2006 BARE COSTS MAT.	LABOR	EQUIP.	TOTAL	TOTAL INCL O&P	
350	0010 DRAPERIES (Material only)										350
	0050 Lined, minimum				S.F.	4.22			4.22	4.64	
	0100 Maximum					12.05			12.05	13.25	
	0200 Unlined, minimum					2.34			2.34	2.57	
	0300 Maximum					9.30			9.30	10.25	
	0400 Lightproof type, add, minimum					1.17			1.17	1.29	
	0500 Maximum					7			7	7.70	
	0600 Fire resistant, add, minimum					.89			.89	.98	
	0700 Maximum					3.03			3.03	3.33	
	0800 Valances, pleated, 10" to 18" depth, minimum				L.F.	1.31			1.31	1.44	
	0900 Maximum				"	7.55			7.55	8.35	
	2000 Alternate method, lined overlaps and returns, average										
	2020 32" thru 48" wide, 26" to 39" long				Ea.	119			119	130	
	2040 40" to 63" long					124			124	136	
	2060 64" to 72" long					127			127	140	
	2080 73" to 81" long					136			136	149	
	2100 82" to 90" long					144			144	158	
	2200 91" to 99" long					154			154	170	
	2400 100" to 108" long					161			161	177	
	2600 109" to 120" long					170			170	187	
	2800 121" to 130" long					173			173	190	
	3000 48" thru 72" wide, 26" to 39" long					173			173	190	
	3050 40" to 63" long					184			184	203	
	3100 64" to 72" long					193			193	212	
	3150 73" to 81" long					202			202	222	
	3200 82" to 90" long					221			221	243	
	3250 91" to 99" long					229			229	252	
	3300 100" to 108" long					238			238	262	

12050 | Fabrics

	12051 | Fabrics		Crew	Daily Output	Labor-Hours	Unit	2006 Bare Costs Mat.	Labor	Equip.	Total	Total Incl O&P	
350	3350	109″ to 120″ long				Ea.	256			256	281	350
	3400	121″ to 130″ long					274			274	300	
	3450	64″ thru 96″ wide, 26″ to 39″ long					230			230	253	
	3500	40″ to 63″ long					247			247	272	
	3550	64″ to 72″ long					255			255	281	
	3600	73″ to 81″ long					274			274	300	
	3650	82″ to 90″ long					291			291	320	
	3700	91″ to 99″ long					310			310	340	
	3750	100″ to 108″ long					320			320	350	
	3800	109″ to 120″ long					335			335	370	
	3820	121″ to 130″ long					360			360	400	
	3840	80″ thru 120″ wide, 26″ to 39″ long					291			291	320	
	3860	40″ to 63″ long					310			310	340	
	3880	64″to 72″ long					320			320	350	
	4000	73″ to 81″ long					345			345	380	
	4020	82″ to 90″ long					360			360	395	
	4080	91″ to 99″ long					380			380	420	
	4100	100″ to 108″ long					405			405	445	
	4150	109″to 120″ long					420			420	460	
	4200	121″ to 130″ long					440			440	485	
	4250	96″ thru 144″ wide, 26″ to 39″ long					345			345	380	
	4300	40″ to 63″ long					370			370	405	
	4400	64″ to 72″ long					380			380	415	
	4500	73″ to 81″ long					400			400	440	
	4600	82″ to 90″ long					430			430	475	
	4700	91″ to 99″ long					460			460	505	
	4800	100″ to 108″ long					485			485	530	
	4900	109″ to 120″ long					510			510	560	
	5000	121″ to 130″ long					545			545	600	
	5100	112″ thru 168″ wide, 26″ to 39″ long					405			405	445	
	5200	40″ to 63″ long					435			435	475	
	5300	64″ to 72″ long					450			450	495	
	5400	73″ to 81″ long					475			475	525	
	5500	82″ to 90″ long					505			505	555	
	5600	91″ to 99″ long					535			535	590	
	5700	100″ to 108″ long					565			565	620	
	5800	109″ to 120″ long					635			635	700	
	6000	121″ to 130″ long					635			635	700	
	6100	128″ thru 192″ wide, 26″ to 39″ long					460			460	505	
	6200	40″ to 63″ long					495			495	545	
	6300	64″ to 72″ long					510			510	565	
	6400	73″ to 81″ long					545			545	600	
	6450	82″ to 90″ long					575			575	630	
	6500	91″ to 99″ long					610			610	670	
	6550	100″ to 108″ long					790			790	865	
	6600	109″ to 120″ long					680			680	750	
	6650	121″ to 130″ long					725			725	795	
	6700	144″ thru 216″ wide, 26″ to 39″ long					520			520	575	
	6750	40″ to 63″ long					625			625	690	
	6800	64″ to 72″ long					575			575	630	
	6850	73″ to 81″ long					610			610	675	
	6880	82″ to 90″ long					655			655	720	
	6900	91″ to 99″ long					690			690	755	
	6920	100″ to 108″ long					725			725	795	
	6980	109″ to 120″ long					760			760	835	
	7000	121″ to 130″ long				↓	820			820	905	

12050 | Fabrics

	12051 | Fabrics		CREW	DAILY OUTPUT	LABOR-HOURS	UNIT	2006 BARE COSTS MAT.	LABOR	EQUIP.	TOTAL	TOTAL INCL O&P	
350	7100	160″ thru 240″ wide, 26″ to 39″ long				Ea.	575			575	630	350
	7150	40″ to 63″ long					620			620	680	
	7200	64″ to 72″ long					635			635	700	
	7250	73″ to 81″ long					680			680	750	
	7300	82″ to 90″ long					725			725	795	
	7350	91″ to 99″ long					770			770	845	
	7400	100″ to 108″ long					805			805	885	
	7500	109″ to 120″ long					845			845	930	
	7600	121″ to 130″ long					910			910	1,000	
	7700	Bedspreads, unquilted, twin, minimum					28.50			28.50	31	
	7800	Maximum					241			241	265	
	7850	Full, minimum					34			34	37	
	7900	Maximum					345			345	380	
	8000	Queen, minimum					46			46	51	
	8100	Maximum					365			365	405	
	8200	Quilted, twin, minimum					61			61	67.50	
	8300	Maximum					305			305	335	
	8400	Full, minimum					79.50			79.50	87.50	
	8500	Maximum					415			415	455	
	8600	Queen, minimum					75			75	82.50	
	8700	Maximum					445			445	485	
	8800	Drapery installation, hardware & drapes,										
	9000	Labor cost only, minimum	1 Clab	75	.107	L.F.		2.92		2.92	4.84	
	9100	Maximum	″	20	.400	″		10.95		10.95	18.15	
750	0010	**UPHOLSTERY MATERIALS**										750
	1000	Fabrics, fabric blends, minimum				S.Y.	14			14	15.40	
	1020	Maximum					75			75	82.50	
	1200	Nylons, minimum					23			23	25.50	
	1220	Maximum					42.50			42.50	46.50	
	1400	Polyester, minimum					31			31	34	
	1420	Maximum					42.50			42.50	46.50	
	1600	Silk, minimum					46.50			46.50	51	
	1620	Maximum					69			69	75.50	
	1800	Wool, minimum					38.50			38.50	42.50	
	1820	Maximum					77.50			77.50	85	
	3000	Treatment, acoustic, 1/8″ flame retardant					5.65			5.65	6.20	
	3100	Acrylic backing					2.02			2.02	2.22	
	3200	Flame proofing					5.05			5.05	5.55	
	3300	Paper backing					6.25			6.25	6.85	
	3400	Scotchguard or teflon					1.29			1.29	1.42	
	5000	Leather, minimum				S.F.	8.25			8.25	9.10	
	5020	Maximum				″	15.35			15.35	16.90	
	6000	Vinyl, minimum				S.Y.	20.50			20.50	22.50	
	6020	Maximum				″	38.50			38.50	42	

12100 | Art

	12120 | Wall Decorations		CREW	DAILY OUTPUT	LABOR-HOURS	UNIT	2006 BARE COSTS MAT.	LABOR	EQUIP.	TOTAL	TOTAL INCL O&P	
100	0010	**ART WORK** framed										100
	1000	Photography, minimum	1 Carp	36	.222	Ea.	64	7.90		71.90	83.50	

12100 | Art

12120 | Wall Decorations

		12120 Wall Decorations	CREW	DAILY OUTPUT	LABOR-HOURS	UNIT	2006 BARE COSTS MAT.	LABOR	EQUIP.	TOTAL	TOTAL INCL O&P	
100	1050	Maximum	1 Carp	30	.267	Ea.	425	9.50		434.50	480	100
	2000	Posters, minimum		36	.222		34.50	7.90		42.40	51	
	2050	Maximum		30	.267		485	9.50		494.50	550	
	3000	Reproductions, minimum		36	.222		64.50	7.90		72.40	84	
	3050	Maximum		30	.267		595	9.50		604.50	670	

12300 | Manufactured Casework

12310 | Metal Casework

		12310 Metal Casework	CREW	DAILY OUTPUT	LABOR-HOURS	UNIT	2006 BARE COSTS MAT.	LABOR	EQUIP.	TOTAL	TOTAL INCL O&P	
100	0010	KEY CABINETS										100
	0020	Wall mounted, 60 key capacity	1 Carp	20	.400	Ea.	74	14.20		88.20	105	
	0200	Drawer type, 600 key capacity	1 Clab	15	.533		725	14.60		739.60	825	
	0300	2,400 key capacity		20	.400		3,600	10.95		3,610.95	3,975	
	0400	Tray type, 20 key capacity		50	.160		38.50	4.38		42.88	50	
	0500	50 key capacity		40	.200		70	5.50		75.50	86	
200	0010	DISPLAY CASES Free standing, all glass										200
	0020	Aluminum frame, 42″ high x 36″ x 12″ deep	2 Carp	8	2	Ea.	1,075	71		1,146	1,300	
	0100	70″ high x 48″ x 18″ deep	″	6	2.667		1,650	95		1,745	1,950	
	0500	For wood bases, add					9%					
	0600	For hardwood frames, deduct					8%					
	0700	For bronze, baked enamel finish, add					10%					
	2000	Wall mounted, glass front, aluminum frame										
	2010	Non-illuminated, one section 3′ x 4′ x 1′-4″	2 Carp	5	3.200	Ea.	1,150	114		1,264	1,450	
	2100	5′ x 4′ x 1′-4″		5	3.200		1,250	114		1,364	1,575	
	2200	6′ x 4′ x 1′-4″		4	4		1,425	142		1,567	1,800	
	2500	Two sections, 8′ x 4′ x 1′-4″		2	8		1,725	284		2,009	2,375	
	2600	10′ x 4′ x 1′-4″		2	8		1,900	284		2,184	2,550	
	3000	Three sections, 16′ x 4′ x 1′-4″		1.50	10.667		3,275	380		3,655	4,225	
	3500	For fluorescent lights, add				Section	260			260	286	
	4000	Table exhibit cases, 2′ wide, 3′ high, 4′ long, flat top	2 Carp	5	3.200	Ea.	740	114		854	1,000	
	4100	3′ wide, 3′ high, 4′ long, sloping top	″	3	5.333	″	1,125	190		1,315	1,575	
560	0010	IRONING CENTER										560
	0020	Including cabinet, board & light, minimum	1 Carp	2	4	Ea.	291	142		433	555	
	0100	Maximum, see also Div. 11119-450	″	1.50	5.333	″	515	190		705	880	
750	0010	CASEWORK										750
	0102	Laboratory cabinets, see Div. 12350-500										
	0500	Hospital, base cabinets, laminated plastic	2 Carp	10	1.600	L.F.	221	57		278	335	
	0700	Enameled steel		10	1.600		194	57		251	310	
	1000	Stainless steel		10	1.600		305	57		362	430	
	1200	For all drawers, add					11.30			11.30	12.40	
	1300	Cabinet base trim, 4″ high, enameled steel	2 Carp	200	.080		35.50	2.84		38.34	43.50	
	1400	Stainless steel		200	.080		55.50	2.84		58.34	65.50	
	1450	Counter top, laminated plastic, no backsplash		40	.400		38	14.20		52.20	65.50	
	1650	With backsplash		40	.400		47.50	14.20		61.70	75.50	
	1800	For sink cutout, add		12.20	1.311	Ea.		46.50		46.50	77.50	
	1900	Stainless steel counter top		40	.400	L.F.	106	14.20		120.20	140	
	2000	For drop-in stainless 43″ x 21″ sink, add				Ea.	590			590	650	
	2050	For solid surface countertops see Div. 06620-810										

12300 | Manufactured Casework

12310 | Metal Casework

		12310 Metal Casework	CREW	DAILY OUTPUT	LABOR-HOURS	UNIT	2006 BARE COSTS MAT.	LABOR	EQUIP.	TOTAL	TOTAL INCL O&P	
750	2100	Nurses station, door type, laminated plastic	2 Carp	10	1.600	L.F.	256	57		313	375	750
	2200	Enameled steel		10	1.600		233	57		290	350	
	2300	Stainless steel		10	1.600		325	57		382	450	
	2400	For drawer type, add					104			104	114	
	2500	Wall cabinets, laminated plastic	2 Carp	15	1.067		166	38		204	245	
	2600	Enameled steel		15	1.067		194	38		232	276	
	2700	Stainless steel		15	1.067		315	38		353	410	
	2800	For glass doors, add					26.50			26.50	29	
	3500	Kitchen, base cabinets, metal, minimum	2 Carp	30	.533		57	18.95		75.95	94	
	3600	Maximum		25	.640		144	23		167	197	
	3700	Wall cabinets, metal, minimum		30	.533		57	18.95		75.95	94	
	3800	Maximum		25	.640		130	23		153	182	
	5000	School, 24″ deep, metal, 84″ high units		15	1.067		284	38		322	375	
	5150	Counter height units		20	.800		221	28.50		249.50	290	
	5450	Wood, custom fabricated, 32″ high counter		20	.800		184	28.50		212.50	250	
	5600	Add for counter top		56	.286		20.50	10.15		30.65	39.50	
	5800	84″ high wall units		15	1.067		192	38		230	274	
	6000	Laminated plastic finish is same price as wood										

12350 | Specialty Casework

		12350 Specialty Casework	CREW	DAILY OUTPUT	LABOR-HOURS	UNIT	MAT.	LABOR	EQUIP.	TOTAL	TOTAL INCL O&P	
500	0010	LABORATORY CASEWORK										500
	0020	Cabinets, base, door units, metal	2 Carp	18	.889	L.F.	167	31.50		198.50	236	
	0300	Drawer units		18	.889		360	31.50		391.50	450	
	0700	Tall storage cabinets, open, 7′ high		20	.800		350	28.50		378.50	430	
	0900	With glazed doors		20	.800		435	28.50		463.50	520	
	1300	Wall cabinets, metal, 12-1/2″ deep, open		20	.800		109	28.50		137.50	167	
	1500	With doors		20	.800		227	28.50		255.50	297	
	1550	Counter tops, not incl. base cabinets, acidproof, minimum		82	.195	S.F.	26.50	6.95		33.45	40.50	
	1600	Maximum		70	.229		36	8.15		44.15	53	
	1650	Stainless steel		82	.195		78	6.95		84.95	97	
	6300	Rule of thumb: lab furniture including installation & connection										
	6320	High school				S.F.					32	
	6340	College									47.25	
	6360	Clinical, health care									40.50	
	6380	Industrial									65	

12400 | Furnishings & Accessories

12411 | Furniture Accessories

		12411 Furniture Accessories	CREW	DAILY OUTPUT	LABOR-HOURS	UNIT	2006 BARE COSTS MAT.	LABOR	EQUIP.	TOTAL	TOTAL INCL O&P	
200	0010	DESK ACCESSORIES metal or plastic										200
	0300	Bookends, minimum				Ea.	7			7	7.70	
	0320	Maximum					62			62	68	
	1000	Calendar with pad, minimum					8.40			8.40	9.25	
	1020	Maximum					80			80	88	
	1200	Carafe, tray, minimum					57.50			57.50	63.50	
	1220	Maximum					67.50			67.50	74.50	
	1300	Desk pad, minimum					17.85			17.85	19.60	
	1320	Maximum					83			83	91.50	
	1400	Double pen set with pens, minimum					24			24	26.50	

12400 | Furnishings & Accessories

	12411 | Furniture Accessories		CREW	DAILY OUTPUT	LABOR-HOURS	UNIT	2006 BARE COSTS MAT.	LABOR	EQUIP.	TOTAL	TOTAL INCL O&P	
200	1420	Maximum				Ea.	160			160	176	200
	1500	Letter tray, minimum					11.50			11.50	12.65	
	1520	Maximum					76			76	84	
	1600	Memo box, minimum					5.25			5.25	5.75	
	1620	Maximum					76			76	84	

	12430 | Portable Lamps		CREW	DAILY OUTPUT	LABOR-HOURS	UNIT	MAT.	LABOR	EQUIP.	TOTAL	TOTAL INCL O&P	
400	0010	**LAMPS**										400
	1000	Ceramic, desk, minimum				Ea.	42.50			42.50	47	
	1020	Maximum					70			70	77	
	1200	End table, minimum					39			39	43	
	1220	Maximum					105			105	115	
	1300	Night stand, minimum					42.50			42.50	47	
	1320	Maximum					95			95	104	
	1400	Wall, minimum					51.50			51.50	56.50	
	1420	Maximum					249			249	274	
	2000	Glass, desk, minimum					178			178	196	
	2020	Maximum					520			520	570	
	2100	End table, minimum					173			173	190	
	2120	Maximum					685			685	750	
	2200	Floor, minimum					435			435	475	
	2250	Maximum					1,025			1,025	1,125	
	2280	Night stand, minimum					273			273	300	
	2300	Maximum					515			515	570	
	2350	Pendant, minimum					44.50			44.50	48.50	
	2400	Maximum					845			845	925	
	3000	Metal, desk, minimum					39			39	43	
	3100	Maximum					67			67	73.50	
	3200	End table, minimum					39			39	43	
	3250	Maximum					67			67	73.50	
	3300	Wall, minimum					59.50			59.50	65.50	
	3350	Maximum					117			117	129	
	3400	Floor, minimum					44.50			44.50	49	
	3450	Maximum					360			360	395	
	3500	Night stand, minimum					39			39	43	
	3550	Maximum					39			39	43	
	3600	Pendant, minimum					47			47	51.50	
	3650	Maximum					150			150	165	
	4000	Stone, desk, minimum					53			53	58.50	
	4100	Maximum					57			57	62.50	
	4200	End table, minimum					53			53	58.50	
	4300	Maximum					57			57	62.50	
	4400	Night stand, minimum					53			53	58.50	
	4500	Maximum					57			57	62.50	
	5000	Wall, minimum					43			43	47.50	
	6000	Maximum					89			89	98	
	8000	Replacement shades, lamp, desk, minimum					13			13	14.30	
	8020	Maximum					27			27	29.50	
	8040	End table, minimum					13			13	14.30	
	8060	Maximum					27			27	29.50	
	8100	Night stand minimum					13			13	14.30	
	8120	Maximum					27			27	29.50	
	8140	Wall, minimum					13			13	14.30	
	8160	Maximum					27			27	29.50	

12460 | Furnishing Accessories

			CREW	DAILY OUTPUT	LABOR-HOURS	UNIT	2006 BARE COSTS MAT.	LABOR	EQUIP.	TOTAL	TOTAL INCL O&P	
900	0010	**ASH/TRASH RECEIVERS**										900
	1000	Ash urn, cylindrical metal										
	1020	8″ diameter, 20″ high	1 Clab	60	.133	Ea.	116	3.65		119.65	134	
	1040	8″ dia., 25″ high		60	.133		168	3.65		171.65	191	
	1060	10″ diameter, 26″ high		60	.133		163	3.65		166.65	185	
	1080	12″ dia., 30″ high		60	.133		237	3.65		240.65	267	
	2000	Combination ash/trash urn, metal										
	2020	8″ diameter, 20″ high	1 Clab	60	.133	Ea.	108	3.65		111.65	125	
	2040	8″ dia., 25″ high		60	.133		114	3.65		117.65	131	
	2050	10″ diameter, 26″ high		60	.133		182	3.65		185.65	206	
	2060	12″ dia., 30″ high		60	.133		265	3.65		268.65	297	
	4000	Trash receptacle, metal										
	4020	8″ diameter, 15″ high	1 Clab	60	.133	Ea.	62	3.65		65.65	74	
	4040	10″ diameter, 18″ high		60	.133		105	3.65		108.65	121	
	4060	16″ dia., 16″ high		60	.133		126	3.65		129.65	145	
	4100	18″ dia., 32″ high		60	.133		340	3.65		343.65	380	
	5000	Trash receptacle, plastic, fire resistant										
	5020	Rectangular 11″ x 8″ x 12″ high	1 Clab	60	.133	Ea.	11	3.65		14.65	18.15	
	5040	16″ x 8″ x 14″ high	″	60	.133	″	16	3.65		19.65	23.50	
	5500	Trash receptacle, plastic, with lid										
	5520	35 gallon	1 Clab	60	.133	Ea.	185	3.65		188.65	209	
	5540	45 gallon		60	.133		220	3.65		223.65	248	
	5550	Plastic recycling barrel, w/lid & wheels, 32 gal		60	.133		105	3.65		108.65	122	
	5560	65 gal		60	.133		125	3.65		128.65	144	
	5570	95 gal		60	.133		310	3.65		313.65	345	

12483 | Floor Mats & Frames

			CREW	DAILY OUTPUT	LABOR-HOURS	UNIT	MAT.	LABOR	EQUIP.	TOTAL	TOTAL INCL O&P	
200	0010	**FLOOR MATS**										200
	0020	Recessed, in-laid black rubber, 3/8″ thick, solid	1 Clab	155	.052	S.F.	20.50	1.41		21.91	25.50	
	0050	Perforated		155	.052		17.50	1.41		18.91	21.50	
	0100	1/2″ thick, solid		155	.052		19.55	1.41		20.96	24	
	0150	Perforated		155	.052		22.50	1.41		23.91	27.50	
	0200	In colors, 3/8″ thick, solid		155	.052		20	1.41		21.41	24.50	
	0250	Perforated		155	.052		20	1.41		21.41	24.50	
	0300	1/2″ thick, solid		155	.052		29.50	1.41		30.91	35	
	0350	Perforated		155	.052		29.50	1.41		30.91	35	
	0500	Link mats, including nosings, aluminum, 3/8″ thick		155	.052		18.50	1.41		19.91	23	
	0550	Black rubber with galvanized tie rods		155	.052		14.40	1.41		15.81	18.20	
	0600	Steel, galvanized, 3/8″ thick		155	.052		8	1.41		9.41	11.15	
	0650	Vinyl, in colors		155	.052		18.55	1.41		19.96	23	
	0750	Add for nosings, rubber				L.F.	3.48			3.48	3.83	
	0850	Recess frames for above mats, aluminum	1 Carp	100	.080		3.48	2.84		6.32	8.55	
	0870	Bronze	″	100	.080		5.05	2.84		7.89	10.25	
	0900	Skate lock tile, 24″ x 24″ x 1/2″ thick, rubber, black	1 Clab	125	.064	S.F.	14.60	1.75		16.35	18.95	
	0950	Color		125	.064	″	18.60	1.75		20.35	23.50	
	1000	12″ x 24″ border, black		75	.107	L.F.	16.30	2.92		19.22	23	
	1100	Color		75	.107	″	22	2.92		24.92	29	
	1150	12″ x 12″ outside corner, black		100	.080	S.F.	13.80	2.19		15.99	18.85	
	1200	Color		100	.080		17.90	2.19		20.09	23.50	
	1500	Duckboard, aluminum slats		155	.052		15.15	1.41		16.56	19.05	
	1700	Hardwood strips on rubber base, to 54″ wide		155	.052		12.70	1.41		14.11	16.35	
	1800	Assembled with brass rods and vinyl spacers, to 48″ wide		155	.052		16.90	1.41		18.31	21	
	1850	Tire fabric, 3/4″ thick		155	.052		11.05	1.41		12.46	14.50	
	1900	Vinyl, 36″ wide, in colors, hollow top & bottoms		155	.052		8.65	1.41		10.06	11.85	
	1950	Solid top & bottom members		155	.052		8.65	1.41		10.06	11.85	

12492 | Blinds and Shades

		Description	CREW	DAILY OUTPUT	LABOR-HOURS	UNIT	2006 BARE COSTS MAT.	LABOR	EQUIP.	TOTAL	TOTAL INCL O&P	
100	0010	BLINDS, INTERIOR										100
	0020	Horizontal, 1" aluminum slats, solid color, stock	1 Carp	590	.014	S.F.	2.89	.48		3.37	3.98	
	0090	Custom, minimum		590	.014		2.58	.48		3.06	3.64	
	0100	Maximum		440	.018		6.50	.65		7.15	8.20	
	0250	2" aluminum slats, custom, minimum		590	.014		3.80	.48		4.28	4.98	
	0350	Maximum		440	.018		13	.65		13.65	15.35	
	0450	Stock, minimum		590	.014		4.21	.48		4.69	5.45	
	0500	Maximum		440	.018		6.85	.65		7.50	8.55	
	0600	2" steel slats, stock, minimum		590	.014		1.60	.48		2.08	2.56	
	0630	Maximum		440	.018		4.46	.65		5.11	6	
	0750	Custom, minimum		590	.014		1.44	.48		1.92	2.38	
	0850	Maximum		400	.020		7.15	.71		7.86	9.05	
	1000	Alternate method of figuring:										
	1300	1" aluminum slats, 48" wide, 48" high	1 Carp	30	.267	Ea.	50	9.50		59.50	70.50	
	1320	72" high		29	.276		76.50	9.80		86.30	100	
	1340	96" high		28	.286		99.50	10.15		109.65	126	
	1400	72" wide, 72" high		25	.320		114	11.40		125.40	145	
	1420	96" high		23	.348		151	12.35		163.35	187	
	1480	96" wide, 96" high		20	.400		177	14.20		191.20	219	
	1490	For special colors, add					12%					
	1500	Vertical, 3" to 5" PVC or cloth strips, minimum	1 Carp	460	.017	S.F.	7.85	.62		8.47	9.65	
	1600	Maximum		400	.020		9.55	.71		10.26	11.70	
	1800	4" aluminum slats, minimum		460	.017		3.12	.62		3.74	4.45	
	1900	Maximum		400	.020		7.50	.71		8.21	9.45	
	1950	Mylar mirror-finish strips, to 8" wide, minimum		460	.017		12.50	.62		13.12	14.75	
	1970	Maximum		400	.020		18.80	.71		19.51	21.50	
	1990	Alternate method of figuring:										
	2000	2" aluminum slats, 48" wide, 48" high	1 Carp	30	.267	Ea.	91.50	9.50		101	117	
	2050	72" high		29	.276		115	9.80		124.80	142	
	2100	96" high		28	.286		138	10.15		148.15	169	
	2200	72" wide, 72" high		25	.320		118	11.40		129.40	149	
	2250	96" high		23	.348		138	12.35		150.35	173	
	2300	96" wide, 96" high		20	.400		179	14.20		193.20	221	
	2500	Mirror finish, 48" wide, 48" high		30	.267		99.50	9.50		109	125	
	2550	72" high		29	.276		117	9.80		126.80	145	
	2600	96" high		28	.286		138	10.15		148.15	169	
	2650	72" wide, 72" high		25	.320		165	11.40		176.40	201	
	2700	96" high		23	.348		193	12.35		205.35	233	
	2750	96" wide, 96" high		20	.400		248	14.20		262.20	297	
	2800	Decorative printed finish, 48" wide, 48" high		30	.267		103	9.50		112.50	130	
	2850	72" high		29	.276		128	9.80		137.80	157	
	2900	96" high		28	.286		148	10.15		158.15	180	
	2950	72" wide, 72" high		25	.320		176	11.40		187.40	212	
	2980	96" high		23	.348		213	12.35		225.35	256	
	2990	96" wide, 96" high		20	.400		277	14.20		291.20	330	
	3000	Wood folding panels with movable louvers, 7" x 20" each		17	.471	Pr.	42	16.75		58.75	74	
	3300	8" x 28" each		17	.471		61	16.75		77.75	94.50	
	3450	9" x 36" each		17	.471		72.50	16.75		89.25	107	
	3600	10" x 40" each		17	.471		82	16.75		98.75	118	
	4000	Fixed louver type, stock units, 8" x 20" each		17	.471		63	16.75		79.75	97	
	4150	10" x 28" each		17	.471		86.50	16.75		103.25	123	
	4300	12" x 36" each		17	.471		110	16.75		126.75	149	
	4450	18" x 40" each		17	.471		131	16.75		147.75	172	
	5000	Insert panel type, stock, 7" x 20" each		17	.471		15.60	16.75		32.35	44.50	
	5150	8" x 28" each		17	.471		28.50	16.75		45.25	59	
100	5300	9" x 36" each		17	.471		36.50	16.75		53.25	67.50	100

12492 | Blinds and Shades

		12492 Blinds and Shades	CREW	DAILY OUTPUT	LABOR-HOURS	UNIT	2006 BARE COSTS MAT.	LABOR	EQUIP.	TOTAL	TOTAL INCL O&P	
100	5450	10″ x 40″ each	1 Carp	17	.471	Pr.	39	16.75		55.75	70.50	100
	5600	Raised panel type, stock, 10″ x 24″ each		17	.471		109	16.75		125.75	148	
	5650	12″ x 26″ each		17	.471		126	16.75		142.75	167	
	5700	14″ x 30″ each		17	.471		143	16.75		159.75	185	
	5750	16″ x 36″ each		17	.471		160	16.75		176.75	204	
	6000	For custom built pine, add					22%					
	6500	For custom built hardwood blinds, add					42%					
600	0010	**SHADES**										600
	0020	Basswood, roll-up, stain finish, 3/8″ slats	1 Carp	300	.027	S.F.	9.85	.95		10.80	12.35	
	0200	7/8″ slats		300	.027		9.30	.95		10.25	11.75	
	0300	Vertical side slide, stain finish, 3/8″ slats		300	.027		14.85	.95		15.80	17.90	
	0400	7/8″ slats		300	.027		14.85	.95		15.80	17.90	
	0500	For fire retardant finishes, add					16%					
	0600	For "B" rated finishes, add					20%					
	0900	Mylar, single layer, non-heat reflective	1 Carp	685	.012		5.75	.42		6.17	7.05	
	1000	Double layered, heat reflective		685	.012		7.95	.42		8.37	9.40	
	1100	Triple layered, heat reflective		685	.012		9.25	.42		9.67	10.85	
	1200	For metal roller instead of wood, add per				Shade	3.42			3.42	3.76	
	1300	Vinyl coated cotton, standard	1 Carp	685	.012	S.F.	1.94	.42		2.36	2.82	
	1400	Lightproof decorator shades		685	.012		1.75	.42		2.17	2.62	
	1500	Vinyl, lightweight, 4 gauge		685	.012		.43	.42		.85	1.16	
	1600	Heavyweight, 6 gauge		685	.012		1.35	.42		1.77	2.18	
	1700	Vinyl laminated fiberglass, 6 ga., translucent		685	.012		1.95	.42		2.37	2.84	
	1800	Lightproof		685	.012		3.10	.42		3.52	4.10	
	3000	Woven aluminum, 3/8″ thick, lightproof and fireproof		350	.023		4.35	.81		5.16	6.15	
	5011	Insulative shades		125	.064		8.70	2.28		10.98	13.30	
	6011	Solar screening, fiberglass		85	.094		3.83	3.35		7.18	9.75	
	8011	Interior insulative shutter										
	8111	Stock unit, 15″ x 60″	1 Carp	17	.471	Pr.	8.70	16.75		25.45	37	
900	0010	**SHADES, LAMP** See Div. 12430-400										900

12493 | Curtains and Drapes

		12493 Curtains and Drapes	CREW	DAILY OUTPUT	LABOR-HOURS	UNIT	MAT.	LABOR	EQUIP.	TOTAL	TOTAL INCL O&P	
200	0010	**DRAPERY HARDWARE**										200
	0030	Standard traverse, per foot, minimum	1 Carp	59	.136	L.F.	2.54	4.82		7.36	10.80	
	0100	Maximum		51	.157	″	10.35	5.60		15.95	20.50	
	0200	Decorative traverse, 28″-48″, minimum		22	.364	Ea.	14.65	12.95		27.60	37.50	
	0220	Maximum		21	.381		34	13.55		47.55	60	
	0300	48″-84″, minimum		20	.400		19.55	14.20		33.75	45	
	0320	Maximum		19	.421		55.50	14.95		70.45	86	
	0400	66″-120″, minimum		18	.444		22	15.80		37.80	50	
	0420	Maximum		17	.471		87	16.75		103.75	124	
	0500	84″-156″, minimum		16	.500		24.50	17.80		42.30	56.50	
	0520	Maximum		15	.533		91.50	18.95		110.45	133	
	0600	130″-240″, minimum		14	.571		29.50	20.50		50	66	
	0620	Maximum		13	.615		116	22		138	165	
	0700	Slide rings, each, minimum					.68			.68	.75	
	0720	Maximum					2.30			2.30	2.53	
	3000	Ripplefold, snap-a-pleat system, 3′ or less, minimum	1 Carp	15	.533		45.50	18.95		64.45	81.50	
	3020	Maximum	″	14	.571		66	20.50		86.50	107	
	3200	Each additional foot, add, minimum				L.F.	2.09			2.09	2.30	
	3220	Maximum				″	6.35			6.35	6.95	
	4000	Traverse rods, adjustable, 28″ to 48″	1 Carp	22	.364	Ea.	21	12.95		33.95	44.50	
	4020	48″ to 84″		20	.400		24.50	14.20		38.70	50.50	
	4040	66″ to 120″		18	.444		28.50	15.80		44.30	57.50	

12 FURNISHINGS

12400 | Furnishings & Accessories

	12493 | Curtains and Drapes	CREW	DAILY OUTPUT	LABOR-HOURS	UNIT	2006 BARE COSTS MAT.	LABOR	EQUIP.	TOTAL	TOTAL INCL O&P	
200	4060 84" to 156"	1 Carp	16	.500	Ea.	32	17.80		49.80	64.50	200
	4080 100" to 180"		14	.571		34.50	20.50		55	71.50	
	4090 156" to 228"		13	.615		46	22		68	87	
	4100 228" to 312"		13	.615		55.50	22		77.50	97.50	
	4200 Double rods, adjustable, 30" to 48"		9	.889		44	31.50		75.50	101	
	4220 48" to 86"		9	.889		58.50	31.50		90	117	
	4240 86" to 150"		8	1		67.50	35.50		103	134	
	4260 100" to 180"		7	1.143		72.50	40.50		113	147	
	4300 Tray and curtain rod, adjustable, 30" to 48"		9	.889		26	31.50		57.50	81	
	4320 48" to 86"		9	.889		36.50	31.50		68	92.50	
	4340 86" to 150"		8	1		48	35.50		83.50	112	
	4360 100" to 180"	↓	7	1.143		51	40.50		91.50	124	
	5000 Stationary rods, first 2 feet				↓	9.25			9.25	10.20	
	5020 Each additional foot, add				L.F.	3.40			3.40	3.74	
400	0010 **BLAST CURTAINS**, per LF horizontal opening width, off-white or gray fabric										400
	0100 Blast curtains, drapery system, complete, including hardware, minimum				L.F.					125	
	0120 Average									135	
	0140 Maximum				↓					145	

12500 | Furniture

	12510 | Office Furniture	CREW	DAILY OUTPUT	LABOR-HOURS	UNIT	2006 BARE COSTS MAT.	LABOR	EQUIP.	TOTAL	TOTAL INCL O&P	
400	0010 **FILES & FILE SYSTEMS** Metal baked enamel finish										400
	1000 2 drawer, vertical, minimum				Ea.	111			111	122	
	1020 Maximum					335			335	370	
	1060 Lateral, 36" wide, minimum					340			340	375	
	1080 Maximum					395			395	435	
	1100 4 drawer, vertical, minimum					139			139	153	
	1120 Maximum					435			435	480	
	1160 Lateral, 36" wide, minimum					515			515	570	
	1180 Maximum					620			620	680	
	1200 Wood, 2 drawer, lateral, minimum					425			425	470	
	1210 Maximum					625			625	685	
	2010 Flat file, steel, 5 drawer, 40" wide x 27" deep					620			620	680	
	2020 46" wide x 33" deep					440			440	480	
	2030 8 drawer, 40" wide x 27" deep					990			990	1,100	
	2040 46" wide x 33" deep					1,125			1,125	1,225	
	2050 Base, 40" wide x 27" deep x 5" high					163			163	179	
	2060 46" wide x 33" deep					131			131	144	
	2100 Hanging file, stand					310			310	340	
	2110 Clamps, package of 6				↓	141			141	155	
580	0010 **MULTI-MEDIA BOARDS** See Div. 10110-240										580
600	0010 **OFFICE CASE GOODS**										600
	0020 Desks, 29" high, double pedestal, 30" x 60", metal, minimum				Ea.	330			330	360	
	0030 Maximum					1,050			1,050	1,150	
	0160 Wood, minimum					350			350	385	
	0180 Maximum					2,050			2,050	2,250	
	0400 36" x 72", metal, minimum				↓	400			400	440	

12500 | Furniture

	12510 | Office Furniture	CREW	DAILY OUTPUT	LABOR-HOURS	UNIT	2006 BARE COSTS MAT.	LABOR	EQUIP.	TOTAL	TOTAL INCL O&P	
600	0420 Maximum				Ea.	1,325			1,325	1,450	600
	0560 Wood, minimum					340			340	375	
	0580 Maximum					2,750			2,750	3,025	
	0600 Desks, single pedestal, 30" x 60", metal, minimum					280			280	305	
	0620 Maximum					855			855	940	
	0640 Wood, minimum					455			455	500	
	0650 Maximum					640			640	705	
	0670 Executive return, 24" x 42", with box, file, wood, minimum					355			355	390	
	0680 Maximum					640			640	705	
	0720 Desks, secretarial, 30" x 60", metal, minimum					219			219	241	
	0730 Maximum					490			490	540	
	0740 Return, 20" x 42", minimum					282			282	310	
	0750 Maximum					390			390	430	
	0800 Wood, minimum					360			360	395	
	0810 Maximum					2,075			2,075	2,275	
	0820 Return, 20" x 42", minimum					350			350	385	
	0830 Maximum					1,125			1,125	1,225	
	0900 Desktop organizer, 72" x 14" x 36" high, wood, minimum					152			152	167	
	0920 Maximum					294			294	325	
	0940 59" x 12" x 23" high, steel, minimum					170			170	187	
	0960 Maximum					228			228	251	
	0970 Keyboard shelf, standard					78			78	85.50	
	0980 Articulating					305			305	335	
	0990 Center drawer, 18" wide					101			101	111	
	1020 Credenza, 18" to 22" x 60" to 72" metal, minimum					430			430	470	
	1040 Maximum					1,575			1,575	1,725	
	1120 Wood, minimum					540			540	595	
	1140 Maximum					2,625			2,625	2,875	
	1150 Hutch/bookcase, minimum					340			340	375	
	1160 Maximum					2,125			2,125	2,325	
	1200 Bookcase, 36" x 12" x 30" high, wood					153			153	168	
	1220 48" high					244			244	269	
	1240 Metal, 29" high					99.50			99.50	109	
	1260 52" high					152			152	167	
	1300 Computer stand, mobile, wood, 25" x 24" x 38" high					445			445	490	
	1320 Metal					253			253	279	
	1340 Resinite					225			225	247	
	1360 Computer desk with hutch, wood, 47" x 24" x 46" high					405			405	445	
	1370 Steel					300			300	330	
	1380 Resinite					385			385	425	
	1400 Printer stand, 22" x 24" x 34" high					280			280	305	
	2000 Chairs, office type, executive, minimum					226			226	248	
	2150 Maximum					1,800			1,800	1,975	
	2200 Management, minimum					171			171	188	
	2250 Maximum					1,700			1,700	1,850	
	2280 Task, minimum					102			102	112	
	2290 Maximum					635			635	700	
	2300 Arm kit, minimum					50.50			50.50	55.50	
	2320 Maximum					78			78	85.50	
	2340 Ergonomic, executive, minimum					370			370	410	
	2380 Maximum					1,675			1,675	1,850	
	2390 Management, minimum					330			330	365	
	2400 Maximum					1,950			1,950	2,150	
	2450 Task, minimum					104			104	114	
	2500 Maximum					510			510	565	
600	2550 Arm kit, minimum				↓	32.50			32.50	35.50	600

12500 | Furniture

	12510 | Office Furniture		CREW	DAILY OUTPUT	LABOR-HOURS	UNIT	2006 BARE COSTS MAT.	LABOR	EQUIP.	TOTAL	TOTAL INCL O&P	
600	2600	Maximum				Ea.	75			75	82.50	600
	3000	Side/guest chairs, upholstered, sled base, metal, min.					135			135	148	
	3100	Maximum					545			545	600	
	3200	Wood, minimum					177			177	195	
	3250	Maximum					1,100			1,100	1,200	
	3400	Traditional, wood leg, minimum					163			163	179	
	3500	Maximum					790			790	870	
	3600	Conference chair, upholstered, metal leg, minimum					118			118	129	
	3700	Maximum					650			650	710	
	3800	Lounge chair, upholstered, minimum					320			320	350	
	3850	Maximum					1,650			1,650	1,800	
	4000	Sofa, two seat, upholstered, minimum					585			585	645	
	4100	Maximum					2,375			2,375	2,625	
	4650	Three seat, minimum					850			850	935	
	4680	Maximum					3,350			3,350	3,675	
	4700	Modular seating, lounge chair unit, upholstered, minimum					1,350			1,350	1,475	
	4750	Maximum					1,950			1,950	2,150	
	4780	Corner unit, minimum					925			925	1,025	
	4800	Maximum					2,150			2,150	2,375	
	5000	Tables, sled base, laminate top, coffee, minimum					161			161	177	
	5100	Maximum					400			400	440	
	5150	End, minimum					159			159	175	
	5200	Maximum					370			370	410	
	5250	Wood cube, coffee					500			500	550	
	5350	End					420			420	465	
	5400	All metal drum, 14″ diameter, 14″ high					360			360	395	
	5420	21″ high					415			415	455	
	5480	18″ diameter, 18″ high					390			390	430	
	5500	22″ diameter, 15″ high					505			505	560	
	5550	20″ high					560			560	615	
	5600	For 3/4″ glass top, add				S.F.	49.50			49.50	54.50	
	5650	For 1″ marble top, add				″	49.50			49.50	54.50	
	5700	Designer table, Mies Barcelona, 40″ sq., glass top, s.s. legs				Ea.	2,050			2,050	2,250	
	5750	Saarinen, 42″ round, plastic top, metal pedestal base					1,125			1,125	1,225	
	5800	Aulenti coffee table, 45″ square, marble top and leg base					7,000			7,000	7,700	
	5840	Bruer, Laccio, 21-1/2″ x 19″, plastic top, tubular legs					248			248	273	
	5860	F. knoll, 24″ square, glass top, solid steel leg base, chrome					1,325			1,325	1,475	
	6000	Table, conference										
	6010	Segmented, oval, 72″ x 144″				Ea.	3,300			3,300	3,625	
	6050	Boat, 96″ x 42″, minimum					610			610	670	
	6150	Maximum					4,125			4,125	4,525	
	6200	120″ x 48″, minimum					970			970	1,075	
	6250	Maximum					5,700			5,700	6,275	
	6300	144″ x 48″, minimum					1,100			1,100	1,225	
	6350	Maximum					6,525			6,525	7,200	
	6400	168″ x 60″, minimum					1,875			1,875	2,050	
	6450	Maximum					9,750			9,750	10,700	
	6500	192″ x 60″, minimum					2,175			2,175	2,400	
	6550	Maximum					11,300			11,300	12,400	
	6680	240″ x 60″, minimum					3,150			3,150	3,475	
	6700	Maximum					11,800			11,800	13,000	
	6720	Rectangle, 96″ x 42″, minimum					720			720	790	
	6740	Maximum					3,400			3,400	3,725	
	6760	120″ x 48″, minimum					1,400			1,400	1,550	
	6780	Maximum					4,800			4,800	5,275	
600	6800	144″ x 60″, minimum					1,950			1,950	2,150	600

12510 | Office Furniture

		12510 Office Furniture	CREW	DAILY OUTPUT	LABOR-HOURS	UNIT	2006 BARE COSTS MAT.	LABOR	EQUIP.	TOTAL	TOTAL INCL O&P	
600	6820	Maximum				Ea.	6,800			6,800	7,475	600
	6840	168″ x 60″, minimum					2,250			2,250	2,475	
	6860	Maximum					7,425			7,425	8,175	
	6880	192″ x 60″, minimum					2,600			2,600	2,875	
	6900	Maximum					9,450			9,450	10,400	
	6960	240″ x 60″, minimum					2,500			2,500	2,750	
	6980	Maximum					10,000			10,000	11,000	
	7010	Powered, 52″ x 108″					2,375			2,375	2,625	
	7020	52″ x 120″					2,425			2,425	2,650	
	7030	52″ x 132″					2,575			2,575	2,825	
	7500	Table, training, with modesty panel, 30″ x 60″				↓	475			475	520	
	8000	Tables, modular										
	8010	Rectangular, 24″ x 48″				Ea.	395			395	430	
	8020	24″ x 60″					375			375	410	
	8030	24″ x 72″					385			385	425	
	8040	30″ x 60″					570			570	625	
	8050	30″ x 72″					600			600	660	
	8100	Corner triangle, 24″ deep					244			244	269	
	8110	30″ deep					275			275	300	
	8200	Half round, 30″ x 60″					455			455	500	
	8300	Trapezoid, 30″ x 60″					294			294	325	
	8400	Crescent, 30″ x 60″				↓	575			575	635	
650	0010	**PANELS & DIVIDERS** Free standing										650
	0030	Fabric panel, class A fire rated										
	0040	Minimum NRC .95, STC 28, fabric edged										
	0200	40″ high, 24″ wide	2 Clab	30	.533	Ea.	295	14.60		309.60	350	
	0220	36″ wide		28	.571		310	15.65		325.65	370	
	0240	48″ wide		26	.615		355	16.85		371.85	425	
	0260	60″ wide		25	.640		380	17.55		397.55	450	
	0400	50″ high, 24″ wide		30	.533		300	14.60		314.60	355	
	0420	36″ wide		28	.571		320	15.65		335.65	375	
	0440	48″ wide		26	.615		365	16.85		381.85	435	
	0460	60″ wide		25	.640		400	17.55		417.55	470	
	0600	60″ high,24″ wide		29	.552		315	15.10		330.10	370	
	0620	36″ wide		27	.593		320	16.25		336.25	380	
	0640	48″ wide		25	.640		450	17.55		467.55	525	
	0660	60″ wide		24	.667		495	18.25		513.25	575	
	0800	72″ high, 24″ wide		29	.552		325	15.10		340.10	380	
	1000	36″ wide		27	.593		365	16.25		381.25	425	
	1200	48″ wide		25	.640		470	17.55		487.55	550	
	1400	60″ wide		24	.667		475	18.25		493.25	550	
	2000	Hardwood edged, 40″ high, 24″ wide		30	.533		320	14.60		334.60	380	
	2100	36″ wide		28	.571		345	15.65		360.65	405	
	2120	48″ wide		26	.615		375	16.85		391.85	440	
	2180	60″ wide		25	.640		415	17.55		432.55	485	
	2200	50″ high, 24″ wide		30	.533		330	14.60		344.60	390	
	2220	36″ wide		28	.571		355	15.65		370.65	415	
	2240	48″ wide		26	.615		410	16.85		426.85	480	
	2300	60″ wide		25	.640		435	17.55		452.55	510	
	2400	60″ high, 24″ wide		29	.552		340	15.10		355.10	400	
	2420	36″ wide		27	.593		390	16.25		406.25	455	
	2440	48″ wide		25	.640		495	17.55		512.55	575	
	2460	60″ wide		24	.667		540	18.25		558.25	625	
	2800	72″ high, 24″ wide		29	.552		360	15.10		375.10	420	
	2820	36″ wide		27	.593		410	16.25		426.25	475	
	2840	48″ wide	↓	25	.640	↓	525	17.55		542.55	605	

		12510 | Office Furniture	CREW	DAILY OUTPUT	LABOR-HOURS	UNIT	2006 BARE COSTS MAT.	LABOR	EQUIP.	TOTAL	TOTAL INCL O&P	
650	2860	60" wide	2 Clab	24	.667	Ea.	525	18.25		543.25	610	650
	3000	Fabric panel, straight, NRC <= .50										
	3110	60" high, 24" wide	2 Clab	29	.552	Ea.	139	15.10		154.10	178	
	3120	30" wide		28	.571		150	15.65		165.65	190	
	3130	36" wide		27	.593		151	16.25		167.25	193	
	3140	48" wide		26	.615		158	16.85		174.85	201	
	3150	60" wide		25	.640		172	17.55		189.55	219	
	3160	72" wide		24	.667		177	18.25		195.25	225	
	3210	72" high, 24" wide		29	.552		169	15.10		184.10	211	
	3220	30" wide		28	.571		174	15.65		189.65	218	
	3230	36" wide		27	.593		190	16.25		206.25	236	
	3240	48" wide		26	.615		200	16.85		216.85	248	
	3250	60" wide		25	.640		215	17.55		232.55	265	
	3260	72" wide		24	.667		227	18.25		245.25	281	
	3310	NRC => .75, 42" high, 24" wide		30	.533		199	14.60		213.60	243	
	3320	30" wide		29	.552		230	15.10		245.10	278	
	3330	36" wide		28	.571		245	15.65		260.65	296	
	3340	48" wide		27	.593		281	16.25		297.25	335	
	3350	60" wide		26	.615		320	16.85		336.85	380	
	3360	72" wide		25	.640		375	17.55		392.55	445	
	3410	60" high, 24" wide		29	.552		245	15.10		260.10	295	
	3420	30" wide		28	.571		288	15.65		303.65	340	
	3430	36" wide		27	.593		286	16.25		302.25	340	
	3440	48" wide		26	.615		300	16.85		316.85	365	
	3450	60" wide		25	.640		330	17.55		347.55	395	
	3460	72" wide		24	.667		385	18.25		403.25	455	
	3510	72" high, 24" wide		29	.552		240	15.10		255.10	288	
	3520	30" wide		28	.571		254	15.65		269.65	305	
	3530	36" wide		27	.593		274	16.25		290.25	325	
	3540	48" wide		26	.615		355	16.85		371.85	425	
	3550	60" wide		25	.640		370	17.55		387.55	435	
	3560	72" wide		24	.667		390	18.25		408.25	455	
	3610	Curved, 60" high		25	.640		249	17.55		266.55	305	
	3620	72" high		24	.667		286	18.25		304.25	345	
	3710	Plexiglass, straight, 60" high, 30" wide		28	.571		315	15.65		330.65	375	
	3720	36" wide		27	.593		325	16.25		341.25	385	
	3730	48" wide		26	.615		355	16.85		371.85	425	
	3740	72" high, 60" wide		25	.640		405	17.55		422.55	475	
	3760	Curved, 60" high		26	.615		465	16.85		481.85	540	
	3770	72" high		24	.667		465	18.25		483.25	540	
	3810	Fabric and plexiglass, straight, 60" high, 30" wide		28	.571		260	15.65		275.65	310	
	3820	36" wide		27	.593		279	16.25		295.25	330	
	3830	48" wide		26	.615		305	16.85		321.85	365	
	3840	72" high, 60" wide		25	.640		390	17.55		407.55	455	
	3860	Curved, 60" high		26	.615		460	16.85		476.85	535	
	3870	72" high		24	.667		590	18.25		608.25	680	
	3910	Connector kit, straight					12.35			12.35	13.60	
	3920	Corner, 2-way					10.65			10.65	11.70	
	3930	3-way					13.55			13.55	14.90	
	3940	4-way					21			21	23.50	
	3950	T-leg					18.30			18.30	20	
675	0010	**POSTS**										675
	0020	Portable for pedestrian traffic control, standard, minimum				Ea.	88			88	97	
	0100	Maximum					153			153	168	
	0300	Deluxe posts, minimum					176			176	193	

FURNISHINGS 12

12500 | Furniture

12510 | Office Furniture

			CREW	DAILY OUTPUT	LABOR-HOURS	UNIT	2006 BARE COSTS MAT.	LABOR	EQUIP.	TOTAL	TOTAL INCL O&P	
675	0400	Maximum				Ea.	285			285	315	675
	0600	Ropes for above posts, plastic covered, 1-1/2" diameter				L.F.	9.35			9.35	10.30	
	0700	Chain core				"	9.60			9.60	10.55	
	1500	Portable security or safety barrier, black with 7' yellow strap				Ea.	200			200	220	
	1510	12' yellow strap				↓	233			233	256	
	1550	Sign holder, standard design				↓	50			50	55	

12520 | Seating

			CREW	DAILY OUTPUT	LABOR-HOURS	UNIT	MAT.	LABOR	EQUIP.	TOTAL	TOTAL INCL O&P	
250	0010	**CHAIRS, FOLDING & STACK**										250
	2000	Folding, all steel, baked enamel finish,										
	2100	Form fitting seat and backrests				Ea.	16.30			16.30	17.90	
	2200	Upholstered seat and back				↓	40			40	44	
	2300	Polypropylene seat and back w/frame				↓	32			32	35	
	2500	Chair caddy for above				↓	177			177	195	
	5000	Stack chair										
	5300	Hardwood frame, seat and back										
	5320	Minimum				Ea.	34			34	37.50	
	5340	Maximum				"	228			228	251	
	5400	Upholstered seat and back										
	5420	Minimum				Ea.	41			41	45	
	5440	Maximum				"	310			310	340	
	5600	Metal frame, upholstered seat and back										
	5620	Minimum				Ea.	38.50			38.50	42.50	
	5640	Maximum				"	221			221	243	
	5700	Plastic shell, metal legs										
	5720	Minimum				Ea.	38			38	42	
	5740	Maximum				↓	113			113	124	
	5800	Chair caddy for above				↓	127			127	140	
	5900	Tablet arms, minimum				↓	70.50			70.50	77.50	
	5920	Maximum				↓	85.50			85.50	94	
550	0010	**SEATING**										550
	0101	Benches, See Div. 02870, 10505 & 11486										
	0110	Bleachers, See Div. 11486-700										
	0500	Classroom, movable chair & desk type, minimum				Set					73	
	0600	Maximum				"					134.50	
	1000	Lecture hall, pedestal type, minimum	2 Carp	22	.727	Ea.	127	26		153	183	
	1200	Maximum	↓	14.50	1.103	↓	440	39		479	550	
	2000	Auditorium chair, all veneer construction	↓	22	.727	↓	140	26		166	197	
	2200	Veneer back, padded seat	↓	22	.727	↓	176	26		202	237	
	2350	Fully upholstered, spring seat	↓	22	.727	↓	178	26		204	238	
	2450	For tablet arms, add				↓	46			46	51	
	2500	For fire retardancy, CATB-133, add				↓	19			19	21	

12540 | Hospitality Furniture

			CREW	DAILY OUTPUT	LABOR-HOURS	UNIT	MAT.	LABOR	EQUIP.	TOTAL	TOTAL INCL O&P	
100	0010	**TABLES, FOLDING** Laminated plastic tops										100
	1000	Tubular steel legs with glides										
	1020	18" x 60", minimum				Ea.	189			189	207	
	1040	Maximum				↓	800			800	880	
	1100	18" x 72", minimum				↓	206			206	227	
	1120	Maximum				↓	855			855	940	
	1200	18" x 96", minimum				↓	263			263	289	
	1220	Maximum				↓	1,000			1,000	1,100	
	1400	30" x 48", minimum				↓	174			174	191	
	1420	Maximum				↓	252			252	278	
	1500	30" x 60", minimum				↓	139			139	153	
	1520	Maximum				↓	875			875	960	

	12540 | Hospitality Furniture		CREW	DAILY OUTPUT	LABOR-HOURS	UNIT	2006 BARE COSTS MAT.	LABOR	EQUIP.	TOTAL	TOTAL INCL O&P	
100	1600	30″ x 72″, minimum				Ea.	253			253	278	100
	1620	Maximum					985			985	1,075	
	1700	30″ x 96″, minimum					205			205	225	
	1720	Maximum					1,200			1,200	1,325	
	1800	36″ x 72″, minimum					182			182	200	
	1820	Maximum					1,200			1,200	1,325	
	1840	36″ x 96″, minimum					213			213	234	
	1860	Maximum					1,575			1,575	1,725	
	2000	Round, wood stained, plywood top, 60″ diameter, minimum					315			315	345	
	2020	Maximum					1,950			1,950	2,150	
	2200	72″ diameter, minimum					272			272	300	
	2220	Maximum					2,200			2,200	2,425	
	4000	Mobile storage carts										
	4020	For 72″ tables, flat, maximum 10 tables				Ea.	410			410	450	
	4040	96″ tables, flat, maximum 10 tables					355			355	395	
	4060	Rounds, on edge, maximum 10 tables					570			570	630	
500	0010	**FURNITURE, HOTEL**										500
	0020	Standard quality set, minimum				Room	1,500			1,500	1,650	
	0200	Maximum				"	7,900			7,900	8,675	
	0300	Bed frame				Ea.	43			43	47.50	
	0400	Bench, upholstered, 42″ x 18″ x 18″					125			125	137	
	0420	18″ x 18″ x 23″					119			119	131	
	0500	Desk section, one drawer, 34″ x 20″ x 30″					175			175	192	
	0600	Free standing, 42″ x 22″ x 30″					200			200	220	
	0700	Desk chair, upholstered, minimum					99			99	109	
	0720	Maximum					270			270	297	
	1000	Dressers, uniplex, 2 drawer					520			520	570	
	1100	3 drawer					226			226	249	
	2000	Guest tables, 30″ diameter					99			99	109	
	2100	34″ diameter					115			115	126	
	3000	Headboards, free standing, twin					127			127	140	
	3050	Full					245			245	269	
	3100	Queen					255			255	280	
	3200	Wall mounted, twin					117			117	129	
	3250	Full					165			165	181	
	3300	Queen					170			170	187	
	4000	Lounge chair, full upholstered, minimum					135			135	148	
	4050	Maximum					430			430	475	
	4200	Open arms, minimum					110			110	121	
	4250	Maximum					475			475	525	
	5000	Mattress/box springs, twin					220			220	242	
	5050	Full					265			265	291	
	5100	Queen					285			285	315	
	5150	Mirror, framed, 29″ x 45″					150			150	165	
	6000	Sleep sofas, twin, minimum					410			410	450	
	6050	Maximum					525			525	580	
	6100	Full, minimum					440			440	485	
	6150	Maximum					815			815	895	
	6200	Queen, minimum					450			450	495	
	6250	Maximum					830			830	915	
	7000	Table, wood top, cocktail, 54″ x 24″ x 16″ high					265			265	291	
	7020	Corner, 30″ x 30″ x 21″ high					200			200	220	
	7040	End, 22″ x 28″ x 21″ high					160			160	176	
550	0010	**MATTRESS & BOX SPRINGS** per set										550
	1000	Hospital, 34″ x 84″				Ea.	405			405	445	

	12540 Hospitality Furniture		CREW	DAILY OUTPUT	LABOR-HOURS	UNIT	2006 BARE COSTS MAT.	LABOR	EQUIP.	TOTAL	TOTAL INCL O&P	
550	2000	Hotel/motel, twin, minimum				Ea.	195			195	214	550
	2020	Maximum					605			605	665	
	2200	Full, minimum					258			258	284	
	2220	Maximum					535			535	590	
	2400	Queen, minimum					283			283	310	
	2420	Maximum					590			590	650	
	3000	Stow away bed with head board, twin size				↓	335			335	370	
700	0010	**FURNITURE, RESTAURANT**										700
	0020	Bars, built-in, front bar	1 Carp	5	1.600	L.F.	220	57		277	335	
	0200	Back bar	"	5	1.600	"	160	57		217	270	
	0300	Booth seating see Div. 12640-150										
	2000	Chair, bentwood side chair, metal, minimum				Ea.	51			51	56	
	2020	Maximum					82			82	90	
	2100	Wood, minimum					71			71	78.50	
	2120	Maximum					112			112	123	
	2400	Bruer Cesca, cane seat & back, arms, minimum					125			125	137	
	2420	Maximum					340			340	375	
	2500	Side, minimum					91.50			91.50	101	
	2520	Maximum					335			335	370	
	2600	Upholstered seat & back, arms, minimum					164			164	180	
	2620	Maximum					350			350	385	
	2640	Side, minimum					111			111	123	
	2660	Maximum					325			325	360	
	2700	Corbusier, arm chair, cane seat and back, minimum					158			158	173	
	2720	Maximum					171			171	189	
	2740	Fledermaus, fabric seat, ash frame, minimum					145			145	160	
	2760	Maximum					178			178	196	
	2780	Hoffman arm, fabric seat, cane back, minimum					166			166	182	
	2800	Maximum					181			181	199	
	2820	Side, minimum					130			130	143	
	2840	Maximum					139			139	153	
	2860	Lombard, minimum					89			89	98	
	2880	Maximum					116			116	128	
	2900	Mies, side chair, leather seat and back, minimum					385			385	420	
	2920	Maximum					565			565	620	
	2940	Napoleon, upholstered seat, wood back, minimum					64			64	70	
	2960	Maximum					81.50			81.50	89.50	
	3000	Prague, arm, upholstered, minimum					119			119	131	
	3020	Maximum					255			255	281	
	3040	Side, upholstered, minimum					118			118	130	
	3060	Maximum					145			145	160	
	3080	Contemporary leather seat & back, chrome frame, min.					237			237	261	
	3100	Maximum					345			345	380	
	3120	Foam padded seat & back, s. steel barstock, min.					635			635	700	
	3140	Maximum					675			675	740	
	3160	Chrome tube cantilever frame, padded arms, min.					97.50			97.50	107	
	3180	Maximum					251			251	276	
	4200	"Tub" style, minimum					167			167	184	
	4220	Maximum					380			380	420	
	4240	Sled base, minimum					225			225	248	
	4260	Maximum					325			325	355	
	4280	Bent wood arms and legs with back bow, minimum					46.50			46.50	51.50	
	4300	Maximum					310			310	340	
	4320	Sled base, minimum					97.50			97.50	107	
	4340	Maximum				↓	310			310	340	

12500 | Furniture

		12540 | Hospitality Furniture	CREW	DAILY OUTPUT	LABOR-HOURS	UNIT	2006 BARE COSTS MAT.	LABOR	EQUIP.	TOTAL	TOTAL INCL O&P	
700	4360	Armchair foam padded seat & back,carved wood frame min.				Ea.	283			283	310	700
	4380	Maximum					540			540	595	
	4400	Ornate wood frame and legs, minimum					410			410	455	
	4420	Maximum					1,050			1,050	1,150	
	4440	Heavy wood frame and legs, tailored back, minimum					154			154	169	
	4460	Maximum					350			350	385	
	4480	Button tufted back, minimum					315			315	345	
	4500	Maximum					720			720	795	
	4520	"Dining" wood frame and legs, minimum					169			169	186	
	4550	Maximum					315			315	345	
	4600	"Queen Ann" wood frame and legs, minimum					185			185	203	
	4620	Maximum					325			325	360	
	4660	Upholstered arms and wood legs, minimum					186			186	205	
	4680	Maximum					261			261	287	
	4700	Wicker, foam padded seat and back, barrel design, minimum					390			390	430	
	4720	Maximum					445			445	485	
	4740	Couch design, minimum					291			291	320	
	4760	Maximum					655			655	720	
	4780	Chrome tube round cantilever base, minimum					168			168	184	
	4800	Maximum					370			370	405	
	4820	Chrome tube square cantilever base, minimum					161			161	177	
	4840	Maximum					360			360	395	
	4860	Misc. chair, upholst. seat, wood arms, bow back & legs, min.					131			131	144	
	4880	Maximum					182			182	200	
	4900	Open curved wood back, minimum					92			92	101	
	4950	Maximum					120			120	132	
	5000	Upholstered seat and open back, wood frame arms, min.					164			164	180	
	5100	Maximum					205			205	226	
	5150	Wood frame sled base, minimum					152			152	167	
	5200	Maximum					251			251	277	
	5250	Wood saddle seat, "Windsor", wood frame & legs, min.					136			136	149	
	5300	Maximum					162			162	178	
	5350	Brass nail trim leather, minimum					182			182	200	
	5400	Maximum					208			208	228	
	5450	Wood frame, minimum					156			156	171	
	5500	Maximum					184			184	203	
	6000	Stools, upholst. seat & back, chrome tube cantilever frame, min.					245			245	269	
	6150	Swivel on chrome post mount with spread base, minimum					143			143	157	
	6250	"Vienna" wood back and legs, minimum					59.50			59.50	65.50	
	6350	"Napoleon" wood back and legs, minimum					65.50			65.50	72	
	6450	Wood swivel seat and curved spindle back, leg base					114			114	125	
	6600	Upholstered seat no back, bent wood legs					83.50			83.50	92	
	7000	With back & straight wood frame and legs					143			143	158	
	7200	Wood swivel seat, "Captain" wood back and legs					145			145	159	
	7400	Veneer seat and curved wood back with arms, leg base					138			138	151	
	7600	Upholstered swivel seat no back, wood legs					54.50			54.50	60	
	8000	With back & wood legs					134			134	147	
	8200	Upholstered seat & wood back, wood legs					142			142	157	
	8400	Upholstered swivel seat and back, wood legs					154			154	170	
	8600	Square tube frame and legs					39.50			39.50	43.50	
	8800	Metal frame and legs				↓	39.50			39.50	43.50	
800	0010	TABLE BASES										800
	0040	Bases, dining height										
	1000	Metal disk design, minimum				Ea.	44.50			44.50	49	
	1200	Maximum				↓	218			218	240	

12540 | Hospitality Furniture

		12540 Hospitality Furniture	CREW	DAILY OUTPUT	LABOR-HOURS	UNIT	2006 BARE COSTS MAT.	LABOR	EQUIP.	TOTAL	TOTAL INCL O&P	
800	1400	Wood, minimum				Ea.	87.50			87.50	96	800
	1600	Maximum					120			120	132	
	2000	Fluted wheel, minimum					47.50			47.50	52	
	2200	Maximum					250			250	275	
	2400	Heavy cast iron, minimum					32			32	35.50	
	2600	Maximum					218			218	240	
	2800	Hobnail design, minimum					84			84	92.50	
	3000	Maximum					257			257	283	
	3200	Manhole design, minimum					35.50			35.50	39	
	3400	Maximum					196			196	215	
	3600	Ring design, minimum					43			43	47.50	
	3800	Maximum					180			180	198	
	4000	Trumpet design, minimum					63.50			63.50	70	
	4200	Maximum					182			182	200	
	4400	Tubular, round or rectangular shape, minimum					30			30	33	
	4600	Maximum					191			191	210	
	5000	Cocktail height bases, minimum					66			66	72.50	
	5200	Maximum					200			200	220	
	5400	For foot ring, add, minimum					10.15			10.15	11.15	
	5600	Maximum				↓	23			23	25.50	
850	0010	**TABLE TOPS** Laminated plastic top and edge										850
	0040	24" wide, 24" long, minimum				Ea.	48.50			48.50	53.50	
	1000	Maximum					148			148	163	
	1200	30" long, minimum					55.50			55.50	61	
	1400	Maximum					212			212	234	
	1600	36" long, minimum					62.50			62.50	69	
	1800	Maximum					218			218	240	
	2000	48" long, minimum					96			96	106	
	2200	Maximum					222			222	244	
	2400	60" long, minimum					155			155	171	
	2600	Maximum					310			310	340	
	2800	72" long, minimum					190			190	209	
	3000	Maximum					320			320	355	
	3050	30" wide, 30" long, minimum					61.50			61.50	67.50	
	3150	Maximum					189			189	208	
	3200	36" long, minimum					83.50			83.50	91.50	
	3250	Maximum					203			203	223	
	3400	42" long, minimum					80			80	88	
	3600	Maximum					135			135	149	
	3800	48" long, minimum					107			107	118	
	4000	Maximum					192			192	211	
	4200	60" long, minimum					153			153	168	
	4400	Maximum					278			278	305	
	4600	72" long, minimum					188			188	207	
	4800	Maximum					325			325	355	
	5000	36" wide, 36" long, minimum					77			77	85	
	5200	Maximum					211			211	232	
	5400	48" long, minimum					128			128	141	
	5600	Maximum					253			253	279	
	5800	60" long, minimum					153			153	168	
	6000	Maximum					325			325	360	
	6200	72" long, minimum					208			208	228	
	6400	Maximum					385			385	420	
	6500	42" wide, 42" long, minimum					158			158	174	
	6600	Maximum					315			315	350	
	6700	Round, 24" diameter, minimum				↓	52.50			52.50	58	

	12540 | Hospitality Furniture		CREW	DAILY OUTPUT	LABOR-HOURS	UNIT	MAT.	LABOR	EQUIP.	TOTAL	TOTAL INCL O&P	
							2006 BARE COSTS					
850	6800	Maximum				Ea.	185			185	203	850
	6900	30″ diameter, minimum					63.50			63.50	69.50	
	7000	Maximum					211			211	232	
	7200	36″ diameter, minimum					78.50			78.50	86.50	
	7400	Maximum					223			223	246	
	7600	42″ diameter, minimum					162			162	178	
	8000	Maximum					320			320	355	
	8200	48″ diameter, minimum					182			182	201	
	8400	Maximum					365			365	400	
	8800	54″ diameter, minimum					291			291	320	
	9000	Maximum					455			455	500	
	9200	60″ diameter, minimum					320			320	350	
	9400	Maximum				▼	590			590	650	
	9990	For table bases, See Div. 12540-800										
860	0010	**TABLE TOPS** laminate top and hardwood edge										860
	0040	24″ wide, 24″ long, minimum				Ea.	188			188	207	
	1000	Maximum					565			565	625	
	1200	30″ long, minimum					208			208	228	
	1400	Maximum					595			595	655	
	1600	36″ long, minimum					197			197	217	
	1800	Maximum					645			645	710	
	2000	48″ long, minimum					272			272	299	
	2100	Maximum					720			720	795	
	2200	60″ long, minimum					390			390	430	
	2400	Maximum					785			785	860	
	2600	72″ long, minimum					440			440	485	
	2800	Maximum					815			815	900	
	3000	30″ wide, 30″ long, minimum					226			226	249	
	3100	Maximum					635			635	695	
	3200	36″ long, minimum					228			228	250	
	3400	Maximum					695			695	765	
	3600	42″ long, minimum					196			196	216	
	3800	Maximum					730			730	800	
	4000	48″ long, minimum					296			296	325	
	4100	Maximum					765			765	840	
	4200	60″ long, minimum					400			400	440	
	4300	Maximum					885			885	970	
	4400	72″ long, minimum					252			252	277	
	4600	Maximum					910			910	1,000	
	4800	36″ wide, 36″ long, minimum					187			187	206	
	5000	Maximum					710			710	780	
	5200	48″ long, minimum					325			325	355	
	5400	Maximum					845			845	930	
	5600	60″ long, minimum					455			455	500	
	5800	Maximum					900			900	990	
	6000	72″ long, minimum					500			500	550	
	6200	Maximum					925			925	1,025	
	6400	42″ wide, 42″ long, minimum					330			330	365	
	6600	Maximum					870			870	960	
	6800	Round, 24″ diameter, minimum					340			340	375	
	6900	Maximum					440			440	485	
	7000	30″ diameter, minimum					375			375	415	
	7200	Maximum					635			635	700	
	7400	36″ diameter, minimum					415			415	455	
	7600	Maximum					655			655	720	
	7800	42″ diameter, minimum				▼	480			480	530	

12540 | Hospitality Furniture

			CREW	DAILY OUTPUT	LABOR-HOURS	UNIT	2006 BARE COSTS MAT.	LABOR	EQUIP.	TOTAL	TOTAL INCL O&P	
860	7900	Maximum				Ea.	855			855	940	860
	8000	48″ diameter, minimum					515			515	570	
	8150	Maximum					875			875	960	
	8200	54″ diameter, minimum					770			770	850	
	8250	Maximum					950			950	1,050	
	8300	60″ diameter, minimum					780			780	860	
	8350	Maximum					1,250			1,250	1,375	
	9990	For table bases, See Div. 12540-800										
870	0010	**TABLE TOPS** Polyester resin top & edge										870
	0050	Add inlay top material to cost										
	0100	24″ wide, 24″ long				Ea.	253			253	279	
	1200	30″ long					300			300	330	
	1600	36″ long					350			350	385	
	2000	48″ long					445			445	490	
	2400	60″ long					540			540	595	
	2800	72″ long					630			630	695	
	3200	30″ wide, 30″ long					365			365	400	
	3600	36″ long					420			420	460	
	4000	42″ long					480			480	530	
	4400	48′ long					535			535	590	
	4800	60″ long					650			650	715	
	5200	72″ long					775			775	850	
	5600	36″ wide, 36″ long					485			485	535	
	6000	48″ long					630			630	695	
	6400	60″ long					650			650	715	
	6800	72″ long					910			910	1,000	
	7200	42″ wide, 42″ long					645			645	710	
	7600	Round, 24″ diameter					199			199	219	
	8000	30″ diameter					325			325	360	
	8400	36″ diameter					465			465	510	
	8800	42″ diameter					630			630	695	
	9150	48″ diameter					820			820	905	
	9250	54″ diameter					905			905	995	
	9350	60″ diameter					1,200			1,200	1,325	
	9990	For table bases, See Div. 12540-800										
		12560 | Institutional Furniture										
100	0010	**BANK FURNITURE** See Div. 12510-600										100
150	0010	**CHURCH FURNITURE** See Div. 11041-250										150
200	0010	**FURNITURE, SCHOOL**										200
	1000	Chair, molded plastic,										
	1100	Integral tablet arm, minimum				Ea.	71			71	78.50	
	1150	Maximum					235			235	258	
	2000	Desk, single pedestal, top book compartment, minimum					53.50			53.50	59	
	2020	Maximum					84			84	92.50	
	2100	Side book compartment, minimum					82.50			82.50	91	
	2120	Maximum					104			104	114	
	2200	Flip top, minimum					86			86	94.50	
	2220	Maximum					102			102	112	
	3000	Preschool, moulded plastic chairs, minimum					24			24	26.50	
	3020	Maximum					48			48	52.50	
	3800	Tables, plastic laminate top, 24″ wide, 36″ long					68.50			68.50	75.50	
	3820	48″ long					82			82	90	

12560 | Institutional Furniture

		12560 Institutional Furniture	CREW	DAILY OUTPUT	LABOR-HOURS	UNIT	2006 BARE COSTS MAT.	LABOR	EQUIP.	TOTAL	TOTAL INCL O&P
200	3840	60″ long				Ea.	105			105	116
	3900	30″ wide, 48″ long					117			117	129
	3920	60″ long					134			134	147
	3940	72″ long					155			155	170
	4000	36″ wide, 36″ long					93			93	102
	4020	60″ long					130			130	143
	4040	72″ long					142			142	156
	4200	Round, 36″ dia.					90.50			90.50	99.50
	4220	42″ dia.					102			102	113
	4230	48″ dia.					128			128	141
	4240	60″ dia.					173			173	191
300	0010	**FURNITURE, DORMITORY**									
	0200	Bookcase, two shelf				Ea.	171			171	188
	0300	Bunkable bed, twin, minimum					238			238	262
	0320	Maximum					620			620	680
	1000	Chest, four drawer, minimum					400			400	440
	1020	Maximum					525			525	575
	1050	Built-in, minimum	2 Carp	13	1.231	L.F.	120	44		164	205
	1150	Maximum		10	1.600		222	57		279	340
	1200	Desk top, built-in, laminated plastic, 24″ deep, minimum		50	.320		25.50	11.40		36.90	47
	1300	Maximum		40	.400		76	14.20		90.20	107
	1450	30″ deep, minimum		50	.320		32.50	11.40		43.90	54.50
	1550	Maximum		40	.400		141	14.20		155.20	180
	1750	Dressing unit, built-in, minimum		12	1.333		180	47.50		227.50	277
	1850	Maximum		8	2		540	71		611	715
	2000	Desk, single pedestal				Ea.	435			435	480
	2100	Hutch/bookcase, with light					211			211	232
	3000	Ladder					171			171	188
	4000	Mirror, with frame					99			99	109
	5000	Nightstand					210			210	231
	6000	Wall unit, open shelving					505			505	555
	7000	Wardrobe					525			525	575
	8000	Rule of thumb: total cost for furniture, minimum				Student					2,125
	8050	Maximum				″					4,050
400	0010	**FURNITURE, HOSPITAL**									
	0020	Beds, manual, minimum				Ea.	775			775	855
	0100	Maximum					2,375			2,375	2,600
	0300	Manual and electric beds, minimum					990			990	1,100
	0400	Maximum					2,650			2,650	2,900
	0600	All electric hospital beds, minimum					1,250			1,250	1,375
	0700	Maximum					3,825			3,825	4,225
	0900	Manual, nursing home beds, minimum					755			755	830
	1000	Maximum					1,575			1,575	1,725
	1020	Overbed table, laminated top, minimum					320			320	355
	1040	Maximum					800			800	880
	1100	Patient wall systems, not incl. plumbing, minimum				Room	960			960	1,050
	1200	Maximum				″	1,775			1,775	1,950
	2000	Geriatric chairs, minimum				Ea.	410			410	450
	2020	Maximum				″	915			915	1,000
	3022	Hospital mattress-see also Div. 12540-550									
700	0010	**FURNITURE, LIBRARY**									
	0100	Attendant desk, 36″ x 62″ x 29″ high	1 Carp	16	.500	Ea.	2,400	17.80		2,417.80	2,650
	0200	Book display, "A" frame display, both sides, 42″ x 42″ x 60″ high		16	.500		2,075	17.80		2,092.80	2,325
	0220	Table with bulletin board, 42″ x 24″ x 49″ high		16	.500		1,475	17.80		1,492.80	1,650

	12560 | Institutional Furniture	CREW	DAILY OUTPUT	LABOR-HOURS	UNIT	2006 BARE COSTS MAT.	LABOR	EQUIP.	TOTAL	TOTAL INCL O&P	
700	0300 Book trucks, descending platform,										700
	0320 Small, 14″ x 30″ x 35″ high	1 Carp	16	.500	Ea.	810	17.80		827.80	920	
	0340 Large, 14″ x 40″ x 42″ high		16	.500		1,000	17.80		1,017.80	1,150	
	0800 Card catalogue, 30 tray unit		16	.500		2,575	17.80		2,592.80	2,875	
	0840 60 tray unit		16	.500		4,525	17.80		4,542.80	5,025	
	0880 72 tray unit	2 Carp	16	1		5,075	35.50		5,110.50	5,625	
	0960 120 tray unit	″	16	1		8,350	35.50		8,385.50	9,250	
	1000 Carrels, single face, initial unit	1 Carp	16	.500		690	17.80		707.80	790	
	1050 Additional unit	″	16	.500		550	17.80		567.80	635	
	1500 Double face, initial unit	2 Carp	16	1		1,050	35.50		1,085.50	1,200	
	1550 Additional unit		16	1		810	35.50		845.50	950	
	1600 Cloverleaf		11	1.455		1,825	51.50		1,876.50	2,100	
	2000 Chairs, sled base, arms, minimum	1 Carp	24	.333		275	11.85		286.85	320	
	2050 Maximum		16	.500		252	17.80		269.80	305	
	2100 No arms, minimum		24	.333		174	11.85		185.85	212	
	2150 Maximum		16	.500		285	17.80		302.80	345	
	2500 Standard leg base, arms, minimum		24	.333		238	11.85		249.85	282	
	2520 Maximum		16	.500		340	17.80		357.80	405	
	2600 No arms, minimum		24	.333		177	11.85		188.85	214	
	2620 Maximum		16	.500		258	17.80		275.80	315	
	3000 Charge desk, modular unit, 35″ x 27″ x 39″ high										
	3020 Wood front and edges, plastic laminate tops										
	3100 Book return	1 Carp	16	.500	Ea.	520	17.80		537.80	600	
	3150 Book truck port		16	.500		825	17.80		842.80	935	
	3200 Card file drawer, 5 drawers		16	.500		1,400	17.80		1,417.80	1,575	
	3250 10 drawers		16	.500		1,750	17.80		1,767.80	1,950	
	3280 15 drawers		16	.500		1,975	17.80		1,992.80	2,200	
	3300 Card & legal file		16	.500		2,850	17.80		2,867.80	3,150	
	3350 Charging machine		16	.500		2,100	17.80		2,117.80	2,325	
	3400 Corner		16	.500		530	17.80		547.80	610	
	3450 Cupboard		16	.500		1,300	17.80		1,317.80	1,475	
	3500 Detachable end panel		16	.500		405	17.80		422.80	475	
	3550 Gate		16	.500		445	17.80		462.80	520	
	3600 Knee space		16	.500		1,675	17.80		1,692.80	1,875	
	3650 Open storage		16	.500		1,350	17.80		1,367.80	1,500	
	3700 Station charge		16	.500		2,400	17.80		2,417.80	2,675	
	3750 Work station		16	.500		2,275	17.80		2,292.80	2,525	
	4000 Dictionary stand, stationary		16	.500		715	17.80		732.80	815	
	4020 Revolving		16	.500		286	17.80		303.80	345	
	4200 Exhibit case, table style, 60″ x 28″ x 36″		11	.727		4,075	26		4,101	4,525	
	4500 Globe stand		16	.500		3,175	17.80		3,192.80	3,525	
	4800 Magazine rack		16	.500		3,725	17.80		3,742.80	4,125	
	5000 Newspaper rack		16	.500		800	17.80		817.80	910	
	7000 Tables, card catalog reference, 24″ x 60″ x 42″		16	.500		920	17.80		937.80	1,050	
	7050 24″ x 60″ x 72″		16	.500		1,225	17.80		1,242.80	1,375	
	7100 Index, single tier, 48″ x 72″		16	.500		1,925	17.80		1,942.80	2,125	
	7150 Double tier, 48″ x 72″		16	.500		2,075	17.80		2,092.80	2,300	
	8000 Study, panel ends, plastic laminate surfaces 29″ high, 36″ x 60″	2 Carp	16	1		660	35.50		695.50	785	
	8100 36″ x 72″		16	1		720	35.50		755.50	850	
	8150 36″ x 90″		16	1		795	35.50		830.50	930	
	8200 48″ x 72″		16	1		895	35.50		930.50	1,050	
	8500 Parsons table, 29″ high, plastic lam. top, wood legs & edges										
	8550 36″ x 36″	2 Carp	16	1	Ea.	500	35.50		535.50	610	
	8600 36″ x 60″		16	1		550	35.50		585.50	670	
	8650 36″ x 72″		16	1		625	35.50		660.50	750	
	8700 36″ x 84″		16	1		850	35.50		885.50	995	

12500 | Furniture

	12560 | Institutional Furniture		CREW	DAILY OUTPUT	LABOR-HOURS	UNIT	2006 BARE COSTS MAT.	LABOR	EQUIP.	TOTAL	TOTAL INCL O&P	
700	8750	42″ x 90″	2 Carp	16	1	Ea.	900	35.50		935.50	1,050	700
	8800	48″ x 72″		16	1		800	35.50		835.50	940	
	8850	48″ x 120″		16	1		1,050	35.50		1,085.50	1,200	
	8900	Round, leg or pedestal base, 36″ diameter		16	1		565	35.50		600.50	680	
	8950	42″ diameter		16	1		585	35.50		620.50	700	
	9000	48″ diameter		16	1		640	35.50		675.50	765	
	9050	60″ diameter	↓	16	1	↓	955	35.50		990.50	1,100	

12600 | Multiple Seating

	12640 | Booths & Tables		CREW	DAILY OUTPUT	LABOR-HOURS	UNIT	2006 BARE COSTS MAT.	LABOR	EQUIP.	TOTAL	TOTAL INCL O&P	
150	0010	BOOTHS										150
	1000	Banquet, upholstered seat and back, custom										
	1500	Straight, minimum	2 Carp	40	.400	L.F.	138	14.20		152.20	176	
	1520	Maximum		36	.444		268	15.80		283.80	320	
	1600	"L" or "U" shape, minimum		35	.457		140	16.25		156.25	181	
	1620	Maximum	↓	30	.533	↓	249	18.95		267.95	305	
	1800	Upholstered outside finished backs for										
	1810	single booths and custom banquets										
	1820	Minimum	2 Carp	44	.364	L.F.	15.90	12.95		28.85	39	
	1840	Maximum	"	40	.400	"	47.50	14.20		61.70	76	
	3000	Fixed seating, one piece plastic chair and										
	3010	plastic laminate table top										
	3100	Two seat, 24″ x 24″ table, minimum	F-7	30	1.067	Ea.	555	33.50		588.50	665	
	3120	Maximum		26	1.231		790	38.50		828.50	935	
	3200	Four seat, 24″ x 48″ table, minimum		28	1.143		550	36		586	665	
	3220	Maximum		24	1.333		940	42		982	1,100	
	3300	Six seat, 24″ x 76″ table, minimum		26	1.231		1,000	38.50		1,038.50	1,175	
	3320	Maximum		22	1.455		1,375	46		1,421	1,600	
	3400	Eight seat, 24″ x 102″ table, minimum		20	1.600		1,325	50.50		1,375.50	1,525	
	3420	Maximum	↓	18	1.778	↓	1,725	56		1,781	2,000	
	4000	Free standing, wood fiber core with										
	4010	plastic laminate face, single booth										
	4100	24″ wide	2 Carp	38	.421	Ea.	263	14.95		277.95	315	
	4150	48″ wide		34	.471		315	16.75		331.75	380	
	4200	60″ wide		30	.533		390	18.95		408.95	460	
	4300	Double booth, 24″ wide		32	.500		390	17.80		407.80	460	
	4350	48″ wide		28	.571		520	20.50		540.50	605	
	4400	60″ wide	↓	26	.615	↓	655	22		677	755	
	4600	Upholstered seat and back										
	4650	Foursome, single booth, minimum	2 Carp	38	.421	Ea.	297	14.95		311.95	350	
	4700	Maximum		30	.533		1,125	18.95		1,143.95	1,250	
	4800	Double booth, minimum		32	.500		655	17.80		672.80	750	
	4850	Maximum	↓	26	.615	↓	1,750	22		1,772	1,950	
	5000	Mount in floor, wood fiber core with										
	5010	plastic laminate face, single booth										
	5050	24″ wide	F-7	30	1.067	Ea.	213	33.50		246.50	290	
	5100	48″ wide		28	1.143		270	36		306	360	
	5150	60″ wide	↓	26	1.231	↓	390	38.50		428.50	495	

12600 | Multiple Seating

12640 | Booths & Tables

				DAILY	LABOR-		2006 BARE COSTS				TOTAL	
			CREW	OUTPUT	HOURS	UNIT	MAT.	LABOR	EQUIP.	TOTAL	INCL O&P	
150	5200	Double booth, 24" wide	F-7	26	1.231	Ea.	345	38.50		383.50	445	150
	5250	48" wide	↓	24	1.333	↓	440	42		482	550	
	5300	60" wide	↓	22	1.455	↓	620	46		666	760	

12650 | Area Seating

			CREW	DAILY OUTPUT	LABOR-HOURS	UNIT	MAT.	LABOR	EQUIP.	TOTAL	TOTAL INCL O&P	
500	0010	**MULTIPLE SEATING**										500
	1000	Area seating, full upholstered, 3 seat straight unit, minimum				Ea.	830			830	915	
	1020	Maximum					1,250			1,250	1,375	
	1100	Four seat with corner table, minimum					1,225			1,225	1,350	
	1120	Maximum					1,775			1,775	1,950	
	1200	2 seat with in-line table, minimum					680			680	745	
	1220	Maximum					965			965	1,050	
	2000	Individual seat with table, minimum					400			400	440	
	2020	Maximum				↓	550			550	605	

12700 | Systems Furniture

12710 | Panel Hung Systems

				DAILY	LABOR-		2006 BARE COSTS				TOTAL	
			CREW	OUTPUT	HOURS	UNIT	MAT.	LABOR	EQUIP.	TOTAL	INCL O&P	
650	0010	**FURNITURE, OFFICE SYSTEMS**, Panel hung										650
	0100	Acoustic panel, 43" high, 24" wide, NRC .85				Ea.	217			217	239	
	0110	30" wide					242			242	266	
	0120	36" wide					268			268	294	
	0130	48" wide					320			320	350	
	0200	64" high, 30" wide					288			288	315	
	0210	36" wide					310			310	340	
	0220	42" wide					355			355	395	
	0230	48" wide					380			380	415	
	0240	60" wide					415			415	460	
	0400	Bookshelf, 36" wide					100			100	110	
	0410	42" wide					105			105	116	
	0420	48" wide				↓	109			109	119	
	1000	Connectors, brackets and supports										
	1200	Bracket, cantilever, 20" deep, 24" deep				Ea.	34			34	37.50	
	1220	Countertop, per pair					14.10			14.10	15.50	
	1240	Flat, 20" deep, 24" wide					17.40			17.40	19.15	
	1260	Worksurface kit, per pair					14.10			14.10	15.50	
	1300	Connector kit, ell, 43" high, 90" deep					60.50			60.50	66.50	
	1320	64" high					35.50			35.50	39	
	1340	Straight, 43" high					25			25	27.50	
	1360	64" high					25			25	27.50	
	1400	Support column for peninsula					87			87	95.50	
	2000	Countertop, 15" deep, 36" wide					114			114	126	
	2020	48" wide					128			128	140	
	3000	Duplex receptacle circuit 1					12.45			12.45	13.65	
	3010	Circuit 2					12.45			12.45	13.65	
	3100	Electric power harness, 36" wide					81			81	89.50	
	3120	42" wide					171			171	188	
	3140	60" wide				↓	85.50			85.50	94	

12700 | Systems Furniture

	12710 | Panel Hung Systems		CREW	DAILY OUTPUT	LABOR-HOURS	UNIT	2006 BARE COSTS MAT.	LABOR	EQUIP.	TOTAL	TOTAL INCL O&P	
650	4000	End cover, panel, 43″ high				Ea.	25			25	27.50	650
	4020	64″ high					25			25	27.50	
	4100	Finish, variable height, 2-way					37.50			37.50	41	
	4600	Overhead cabinet with door, 42″ wide					235			235	259	
	4620	60″ wide					355			355	390	
	5000	Pedestal spacer, 22″ deep, 15″ wide					40.50			40.50	44.50	
	5100	28″ deep, 15″ wide					49			49	54	
	5200	Box, box, file, 22″ deep, 26″ high					284			284	315	
	5300	28″ deep					297			297	325	
	5400	File, file, 22″ deep, 26″ high					264			264	291	
	5600	Lateral file, 2 drawer, 30″ wide					375			375	410	
	6000	Power, base in-feed cable					90.50			90.50	99.50	
	7400	Task light, recessed, 30″ - 36″ wide					110			110	121	
	7420	42″ - 48″ wide					119			119	131	
	7600	60″ wide					129			129	142	
	8000	Worksurface, radius edge, 24″ deep, 36″ wide					119			119	130	
	8020	42″ wide					146			146	160	
	8040	48″ wide					156			156	171	
	8060	60″ wide					195			195	215	
	8080	72″ wide					220			220	243	
	8200	30″ deep, 42″ wide					187			187	206	
	8220	60″ wide					218			218	240	
	8240	72″ wide					249			249	273	
	8400	Corner, 24″ deep, 36″ wide					286			286	315	
	8420	42″ wide					325			325	355	
	8500	Peninsula, 36″ wide, 66″ long				▼	375			375	410	
	9000	For installation of systems furniture, add 5%										

	12720 | Free Standing Systems		CREW	DAILY OUTPUT	LABOR-HOURS	UNIT	MAT.	LABOR	EQUIP.	TOTAL	TOTAL INCL O&P	
500	0010	**FURNITURE, OFFICE SYSTEMS**, Freestanding										500
	0100	Desk table, 24″ deep x 48″ wide				Ea.	279			279	305	
	0110	30″ deep x 48″ wide					300			300	330	
	0120	30″ deep x 60″ wide					460			460	505	
	0130	30″ deep x 72″ wide					445			445	490	
	0200	File, mobile, 2 drawer, 22″ deep					345			345	380	
	0210	Suspended, 2 file					345			345	380	
	0220	Undercounter					405			405	445	
	0230	2 box, 1 file					395			395	435	
	0400	Keyboard platform with articulating arm					165			165	182	
	0500	End support leg					121			121	133	
	0800	Modesty panel, 48″ wide					68.50			68.50	75.50	
	0810	Return					75			75	82.50	
	0820	Pencil/utility drawer					96			96	106	
	0830	Privacy screen, 48″ wide x 17-20″ high					199			199	219	
	0840	72″ wide					320			320	350	
	0910	Printer stand, 30″ deep x 36″ wide					425			425	465	
	0920	Universal					440			440	485	
	0930	Paper basket					50			50	55	
	1010	Storage unit with doors, 60″ wide					640			640	705	
	1020	72″ wide					810			810	890	
	1210	Table, peninsula, 30″ x 72″					420			420	465	
	1220	Return, 24″ x 48″					380			380	415	
	1230	Round conference return, 42″ diameter					375			375	410	
	1510	Work surface, 24″ deep x 48″ wide					230			230	253	
	1520	30″ deep x 30″ wide				▼	218			218	240	

12700 | Systems Furniture

	12720 | Free Standing Systems		CREW	DAILY OUTPUT	LABOR-HOURS	UNIT	2006 BARE COSTS MAT.	LABOR	EQUIP.	TOTAL	TOTAL INCL O&P	
500	1530	60″ wide				Ea.	350			350	385	500
	1540	72″ wide					365			365	405	
	1550	Corner bridge, 30″ x 42″					320			320	350	
	1560	Peninsula, 30″ x 60″					505			505	555	
	1570	Connector plate				↓	12.45			12.45	13.70	

	12740 | Desk System Furniture		CREW	DAILY OUTPUT	LABOR-HOURS	UNIT	MAT.	LABOR	EQUIP.	TOTAL	TOTAL INCL O&P	
800	0010	**SYSTEMS FURNITURE**										800
	1000	Secretarial work station, minimum				Ea.	2,800			2,800	3,075	
	1020	Maximum					9,550			9,550	10,500	
	1400	Management work station, minimum					3,175			3,175	3,500	
	1420	Maximum					10,300			10,300	11,300	
	1800	Executive work station, minimum					7,075			7,075	7,800	
	1820	Maximum				↓	18,200			18,200	20,000	
	1840	For installation of systems furniture, add 15%										

12800 | Interior Plants & Planters

	12820 | Int. Artificial Plants		CREW	DAILY OUTPUT	LABOR-HOURS	UNIT	2006 BARE COSTS MAT.	LABOR	EQUIP.	TOTAL	TOTAL INCL O&P	
600	0010	**PLANTS** Permanent only, weighted										600
	0020	Preserved or polyester leaf, natural wood										
	0030	Or molded trunk. For pots see Div. 12830-600										
	0040	For fill see Div. 02315										
	0100	Plants, acuba, 5′ high				Ea.	162			162	178	
	0200	Apidistra, 4′ high					152			152	167	
	0300	Beech, variegated, 5′ high					148			148	163	
	0400	Birds nest fern, 6′ high					267			267	294	
	0500	Croton, 4′ high					59.50			59.50	65.50	
	0600	Diffenbachia, 3′ high					59.50			59.50	65.50	
	0700	Ficus Benjamina, 3′ high					71.50			71.50	78.50	
	0720	6′ high					143			143	157	
	0760	Nitida, 3′ high					160			160	176	
	0780	6′ high					320			320	355	
	0800	Helicona					148			148	163	
	1200	Palm green date fan, 5′ high					154			154	170	
	1220	7′ high					166			166	183	
	1280	Green chamdora date, 5′ high					137			137	150	
	1300	7′ high					214			214	235	
	1400	Green giant for 7′ high					335			335	365	
	1600	Rubber plant, 5′ high					113			113	124	
	1800	Schefflera, 3′ high					77			77	85	
	1820	4′ high					95			95	105	
	1840	5′ high					148			148	163	
	1860	6′ high					202			202	222	
	1880	7′ high					310			310	340	
	1900	Spathiphyllum, 3′ high				↓	107			107	118	
	4000	Trees, polyester or preserved, with										
	4020	Natural trunks										
	4100	Acuba, 10′ high				Ea.	845			845	930	

	12820 Int. Artificial Plants		CREW	DAILY OUTPUT	LABOR-HOURS	UNIT	2006 BARE COSTS MAT.	LABOR	EQUIP.	TOTAL	TOTAL INCL O&P	
600	4120	12′ high				Ea.	1,175			1,175	1,300	600
	4140	14′ high					1,675			1,675	1,825	
	4400	Bamboo, 10′ high					370			370	405	
	4420	12′ high					390			390	430	
	4800	Beech, 10′ high					1,125			1,125	1,250	
	4820	12′ high					1,300			1,300	1,425	
	4840	14′ high					1,775			1,775	1,950	
	5000	Birch, 10′ high					1,125			1,125	1,250	
	5020	12′ high					1,300			1,300	1,425	
	5040	14′ high					1,775			1,775	1,950	
	5060	16′ high					1,400			1,400	1,550	
	5080	18′ high					1,575			1,575	1,725	
	5500	Ficus Benjamina, 10′ high					830			830	915	
	5520	12′ high					1,075			1,075	1,175	
	5540	14′ high					1,200			1,200	1,300	
	5560	16′ high					1,675			1,675	1,825	
	5580	18′ high					2,500			2,500	2,750	
	6000	Magnolia, 10′ high					1,225			1,225	1,350	
	6020	12′ high					1,725			1,725	1,900	
	6040	14′ high					2,350			2,350	2,575	
	6500	Maple, 10′ high					119			119	131	
	6520	12′ high					1,550			1,550	1,700	
	6540	14′ high					1,900			1,900	2,100	
	7000	Palms, chamadora, 10′ high					238			238	261	
	7020	12′ high					305			305	335	
	7040	Date, 10′ high					325			325	360	
	7060	12′ high				↓	375			375	410	

	12830 Interior Planters		CREW	DAILY OUTPUT	LABOR-HOURS	UNIT	MAT.	LABOR	EQUIP.	TOTAL	TOTAL INCL O&P	
600	0010	**PLANTERS**										600
	1000	Fiberglass, hanging, 12″ diameter, 7″ high				Ea.	43.50			43.50	48	
	1100	15″ diameter, 7″ high					65.50			65.50	72	
	1200	36″ diameter, 8″ high					288			288	315	
	1500	Rectangular, 48″ long, 16″ high x 15″ wide					256			256	281	
	1550	16″ high x 24″ wide					365			365	400	
	1600	24″ high x 24″ wide					535			535	585	
	1650	60″ long, 30″ high, 28″ wide					650			650	715	
	1700	72″ long, 16″ high, 15″ wide					530			530	580	
	1750	21″ high, 24″ wide					790			790	870	
	1800	30″ high, 24″ wide					715			715	790	
	2000	Round, 12″ diameter, 13″ high					85.50			85.50	94	
	2050	25″ high					94.50			94.50	104	
	2150	14″ diameter, 15″ high					86			86	94.50	
	2200	16″ diameter, 16″ high					97.50			97.50	107	
	2250	18″ diameter, 19″ high					116			116	127	
	2300	23″ high					168			168	185	
	2350	20″ diameter, 16″ high					114			114	125	
	2400	18″ high					116			116	128	
	2450	21″ high					165			165	181	
	2500	22″ diameter, 10″ high					83			83	91	
	2550	24″ diameter, 16″ high					162			162	178	
	2600	19″ high					222			222	244	
	2650	25″ high					310			310	340	
	2700	36″ high					273			273	300	
	2750	48″ high				↓	400			400	445	

For expanded coverage of these items see *Means Interior Cost Data 2006*

		12830 | Interior Planters	CREW	DAILY OUTPUT	LABOR-HOURS	UNIT	2006 BARE COSTS MAT.	LABOR	EQUIP.	TOTAL	TOTAL INCL O&P	
600	2800	30" diameter, 16" high				Ea.	248			248	272	600
	2850	18" high					255			255	281	
	2900	21" high					259			259	285	
	3000	24" high					285			285	315	
	3350	27" high					310			310	340	
	3400	36" diameter, 16" high					320			320	355	
	3450	18" high					350			350	385	
	3500	21" high					360			360	395	
	3550	24" high					385			385	425	
	3600	27" high					420			420	460	
	3650	30" high					460			460	510	
	3700	48" diameter, 16" high					515			515	565	
	3750	21" high					550			550	605	
	3800	24" high					630			630	690	
	3850	27" high					665			665	735	
	3900	30" high					705			705	775	
	3950	36" high					605			605	665	
	4000	60" diameter, 16" high					645			645	710	
	4100	21" high					660			660	725	
	4150	27" high					800			800	880	
	4200	30" high					985			985	1,075	
	4250	33" high					820			820	900	
	4300	36" high					1,000			1,000	1,100	
	4400	39" high					1,025			1,025	1,125	
	5000	Square, 10" side, 20" high					115			115	127	
	5100	14" side, 15" high					120			120	132	
	5200	18" side, 19" high					157			157	173	
	5300	20" side, 16" high					169			169	186	
	5320	18" high					199			199	219	
	5340	21" high					196			196	215	
	5400	24" side, 16" high					248			248	272	
	5420	21" high					283			283	310	
	5440	25" high					300			300	330	
	5460	30" side, 16" high					330			330	365	
	5480	24" high					425			425	470	
	5490	27" high					545			545	600	
	5500	Round, 36" diameter, 16" high					380			380	420	
	5510	18" high					465			465	510	
	5520	21" high					420			420	465	
	5530	24" high					465			465	510	
	5540	27" high					470			470	515	
	5550	30" high					670			670	735	
	5800	48" diameter, 16" high					550			550	605	
	5820	21" high					570			570	630	
	5840	24" high					675			675	740	
	5860	27" high					715			715	790	
	5880	30" high					785			785	865	
	5900	60" diameter, 16" high					740			740	815	
	5920	21" high					810			810	895	
	5940	27" high					895			895	985	
	5960	30" high					1,025			1,025	1,125	
	5980	36" high					1,200			1,200	1,300	
	6000	Metal bowl, 32" diameter, 8" high, minimum					345			345	380	
	6050	Maximum					440			440	485	
	6100	Rectangle, 30" long x 12" wide, 6" high, minimum					330			330	365	
	6200	Maximum				↓	385			385	425	

	12830 | Interior Planters	CREW	DAILY OUTPUT	LABOR-HOURS	UNIT	2006 BARE COSTS MAT.	LABOR	EQUIP.	TOTAL	TOTAL INCL O&P	
600	6300 36" long 12" wide, 6" high, minimum				Ea.	415			415	455	600
	6400 Maximum					415			415	455	
	6500 Square, 15" side, minimum					330			330	365	
	6600 Maximum					410			410	450	
	6700 20" side, minimum					655			655	720	
	6800 Maximum					710			710	785	
	6900 Round, 6" diameter x 6" high, minimum					30.50			30.50	33.50	
	7000 Maximum					36.50			36.50	40	
	7100 8" diameter x 8" high, minimum					40.50			40.50	44.50	
	7200 Maximum					43.50			43.50	48	
	7300 10" diameter x 11" high, minimum					72.50			72.50	79.50	
	7400 Maximum					83.50			83.50	92	
	7420 12" diameter x 13" high, minimum					73			73	80.50	
	7440 Maximum					86			86	94.50	
	7500 14" diameter x 15" high, minimum					95.50			95.50	105	
	7550 Maximum					121			121	133	
	7580 16" diameter x 17" high, minimum					98			98	108	
	7600 Maximum					124			124	136	
	7620 18" diameter x 19" high, minimum					109			109	119	
	7640 Maximum					151			151	166	
	7680 22" diameter x 20" high, minimum					149			149	164	
	7700 Maximum					214			214	235	
	7750 24" diameter x 21" high, minimum					276			276	305	
	7800 Maximum					284			284	310	
	7850 31" diameter x 18" high, minimum					570			570	625	
	7900 Maximum					850			850	935	
	7950 38" diameter x 24" high, minimum					1,000			1,000	1,125	
	8000 Maximum					1,475			1,475	1,625	
	8050 48" diameter x 24" high, minimum					1,000			1,000	1,100	
	8150 Maximum				↓	1,475			1,475	1,600	
	8500 Plastic laminate faced, fiberglass liner, square										
	8520 14" sq., 15" high				Ea.	305			305	335	
	8540 24" sq., 16" high					390			390	425	
	8580 36" sq., 21" high					545			545	600	
	8600 Rectangle 36" long, 12" wide, 10" high					330			330	365	
	8650 48" long, 12" wide, 10" high					370			370	405	
	8700 48" long, 12" wide, 24" high				↓	545			545	600	
	8750 Wood, fiberglass liner, square										
	8780 14" square, 15" high, minimum				Ea.	221			221	243	
	8800 Maximum					272			272	299	
	8820 24" sq., 16" high, minimum					272			272	299	
	8840 Maximum					360			360	395	
	8860 36" sq., 21" high, minimum					350			350	385	
	8880 Maximum					525			525	580	
	9000 Rectangle, 36" long x 12" wide, 10" high, minimum					244			244	268	
	9050 Maximum					315			315	350	
	9100 48" long x 12" wide, 10" high, minimum					263			263	289	
	9120 Maximum					330			330	365	
	9200 48" long x 12" wide, 24" high, minimum					315			315	350	
	9300 Maximum					465			465	510	
	9400 Plastic cylinder, molded, 10" diameter, 10" high					8.95			8.95	9.85	
	9500 11" diameter, 11" high					28.50			28.50	31	
	9600 13" diameter, 12" high					37.50			37.50	41	
	9700 16" diameter, 14" high				↓	39			39	43	

FURNISHINGS 12

For expanded coverage of these items see *Means Interior Cost Data 2006*

Division Notes

			CREW	DAILY OUTPUT	LABOR-HOURS	UNIT	2006 BARE COSTS MAT.	LABOR	EQUIP.	TOTAL	TOTAL INCL O&P

Division 13 Special Construction

Estimating Tips

General

- The items and systems in this division are usually estimated, purchased, supplied, and installed as a unit by one or more subcontractors. The estimator must ensure that all parties are operating from the same set of specifications and assumptions and that all necessary items are estimated and will be provided. Many times the complex items and systems are covered but the more common ones such as excavation or a crane are overlooked for the very reason that everyone assumes nobody could miss them. The estimator should be the central focus and be able to ensure that all systems are complete.
- Another area where problems can develop in this division is at the interface between systems. The estimator must ensure, for instance, that anchor bolts, nuts, and washers are estimated and included for the air-supported structures and pre-engineered buildings to be bolted to their foundations. Utility supply is a common area where essential items or pieces of equipment can be missed or overlooked due to the fact that each subcontractor may feel it is another's responsibility. The estimator should also be aware of certain items which may be supplied as part of a package but installed by others and ensure that the installing contractor's estimate includes the cost of installation. Conversely, the estimator must also ensure that items are not costed by two different subcontractors, resulting in an inflated overall estimate.

13120 Pre-Engineered Structures

- The foundations and floor slab, as well as rough mechanical and electrical, should be estimated, as this work is required for the assembly and erection of the structure. Generally, as noted in the book, the pre-engineered building comes as a shell and additional features, such as windows and doors, must be included by the estimator. Here again, the estimator must have a clear understanding of the scope of each portion of the work and all the necessary interfaces.

13200 Storage Tanks

- The prices in this subdivision for above- and below-ground storage tanks do not include foundations or hold-down slabs. The estimator should refer to Divisions 2 and 3 for foundation system pricing. In addition to the foundations, required tank accessories, such as tank gauges, leak detection devices, and additional manholes and piping, must be added to the tank prices.

Reference Numbers

Reference numbers are shown in bold squares at the beginning of some major classifications. These numbers refer to related items in the Reference Section. The reference information may be an estimating procedure, an alternate pricing method, or technical information.

Note: Not all subdivisions listed here necessarily appear in this publication.

	13005 | Selective Demolition	CREW	DAILY OUTPUT	LABOR-HOURS	UNIT	2006 BARE COSTS MAT.	LABOR	EQUIP.	TOTAL	TOTAL INCL O&P	
011	0010 **SELECTIVE DEMOLITION, AIR SUPPORTED STRUCTURES**										011
	0020 Tank covers, scrim, dbl. layer, vinyl poly w/ hdw., blower & controls										
	0050 Round and rectangular R024119-10	B-2	9000	.004	S.F.		.12		.12	.20	
	0100 Warehouse structures										
	0120 Poly/vinyl fabric, 28 oz, incl. tension cables & inflation system	4 Clab	9000	.004	SF Flr.		.10		.10	.16	
	0150 Reinforced vinyl, 12 oz., 3000 S.F.	″	5000	.006			.18		.18	.29	
	0200 12,000 to 24,000 S.F.	8 Clab	20000	.003			.09		.09	.15	
	0250 Tedlar vinyl fabric, 28 oz. w/liner, to 3000 S.F.	4 Clab	5000	.006			.18		.18	.29	
	0300 12,000 to 24,000 S.F.	8 Clab	20000	.003			.09		.09	.15	
	0350 Greenhouse/shelter, woven polyethylene with liner										
	0400 3000 S.F.	4 Clab	5000	.006	SF Flr.		.18		.18	.29	
	0450 12,000 to 24,000 S.F.	8 Clab	20000	.003			.09		.09	.15	
	0500 Tennis/gymnasium, poly/vinyl fabric, 28 oz., incl. thermal liner	4 Clab	9000	.004			.10		.10	.16	
	0600 Stadium/Convention Ctr., teflon coated fiberglass, incl. thermal liner	9 Clab	40000	.002			.05		.05	.08	
	0700 Doors, air lock, 15′ long, 10′ x 10′	2 Carp	1.50	10.667	Ea.		380		380	630	
	0720 15′ x 15′		.80	20			710		710	1,175	
	0750 Revolving personnel door, 6′ dia. x 6′-6″ high		1.50	10.667			380		380	630	
012	0010 **SELECTIVE DEMOLITION, GARDEN HOUSES** R024119-10										012
	0020 Garden house, prefab., wood, excl. foundation, average	2 Clab	400	.040	SF Flr.		1.10		1.10	1.82	
013	0010 **SELECTIVE DEMOLITION, GREENHOUSES**, excl. foundation R024119-10										013
	0020 Greenhouse, resi-type, free standing, 9′ long x 8′ wide	2 Clab	160	.100	SF Flr.		2.74		2.74	4.54	
	0030 11′ wide		170	.094			2.58		2.58	4.27	
	0040 14′ wide		220	.073			1.99		1.99	3.30	
	0050 17′ wide		320	.050			1.37		1.37	2.27	
	0060 Lean-to type, 4′ wide		64	.250			6.85		6.85	11.35	
	0070 7′ wide		120	.133			3.65		3.65	6.05	
	0080 Geodesic hemisphere, 1/8″ plexiglass glazing, 8′ dia.		4	4	Ea.		110		110	182	
	0090 24′ dia.		.80	20			550		550	910	
	0100 48′ dia.		.40	40			1,100		1,100	1,825	
034	0010 **SELECTIVE DEMOLITION, RESIDENTIAL GARAGES** R024119-10										034
035	0010 **SELECTIVE DEMOLITION, SPECIAL PURPOSE ROOMS** R024119-10										035
	0100 Audiometric rooms, under 500 S.F. surface	4 Carp	200	.160	SF Surf		5.70		5.70	9.40	
	0110 Over 500 S.F. surface	″	240	.133	″		4.74		4.74	7.85	
	0200 Clean rooms, 12′ x 12′ soft wall, class 100	1 Carp	.30	26.667	Ea.		950		950	1,575	
	0210 Class 1000		.30	26.667			950		950	1,575	
	0220 Class 10,000		.35	22.857			815		815	1,350	
	0230 Class 100,000		.35	22.857			815		815	1,350	
	0300 Darkrooms, shell complete, 8′ high	2 Carp	220	.073	SF Flr.		2.59		2.59	4.28	
	0310 12′ high		110	.145	″		5.15		5.15	8.55	
	0350 Darkrooms doors, mini-cylindrical, revolving		4	4	Ea.		142		142	236	
	0400 Music room, practice modular		140	.114	SF Surf		4.06		4.06	6.75	
	0500 Refrigeration structures and finishes										
	0510 Wall finish, 2 coat portland cement plaster, 1/2″ thick	1 Clab	200	.040	S.F.		1.10		1.10	1.82	
	0520 Fiberglass panels, 1/8″ thick		400	.020			.55		.55	.91	
	0530 Ceiling finish, polystyrene plastic, 1″ to 2″ thick		500	.016			.44		.44	.73	
	0540 4″ thick		450	.018			.49		.49	.81	
	0550 Refrigerator, prefab aluminum walk-in, 7′-6″ high, 6′ x 6′ OD	2 Carp	100	.160	SF Flr.		5.70		5.70	9.40	
	0560 10′ x 10′ OD		160	.100			3.56		3.56	5.90	
	0570 Over 150 S.F.		200	.080			2.84		2.84	4.71	
	0600 Sauna, prefabricated, including heater & controls, 7′ high, to 30 S.F.		120	.133			4.74		4.74	7.85	
	0610 To 40 S.F.		140	.114			4.06		4.06	6.75	
	0620 To 60 S.F.		175	.091			3.25		3.25	5.40	
	0630 To 100 S.F.		220	.073			2.59		2.59	4.28	
	0640 To 130 S.F.		250	.064			2.28		2.28	3.77	

	13005 | Selective Demolition		CREW	DAILY OUTPUT	LABOR-HOURS	UNIT	2006 BARE COSTS MAT.	LABOR	EQUIP.	TOTAL	TOTAL INCL O&P	
035	0650	Steam bath, heater, timer, head, single, to 140 C.F. R024119-10	1 Plum	2.20	3.636	Ea.		155		155	241	035
	0660	To 300 C.F.		2.20	3.636			155		155	241	
	0670	Comm. size, w/blow-down assembly, to 800 C.F.		1.80	4.444			190		190	295	
	0680	To 2500 C.F.		1.60	5			214		214	330	
	0690	Multiple, for motels, apts, 500 C.F., 2 baths		2	4			171		171	266	
	0700	1,000 C.F., 4 baths		1.40	5.714			244		244	380	
036	0010	SELECTIVE DEMOLITION, SOUND CONTROL R024119-10										036
	0120	Acoustical enclosure, 4" T walls & ceiling panels, 8 lbs/S.F.	3 Carp	144	.167	SF Surf		5.95		5.95	9.80	
	0130	10.5 lbs/S.F.		128	.188			6.65		6.65	11.05	
	0140	Reverb chamber, parallel walls, 4" thick		120	.200			7.10		7.10	11.80	
	0150	Skewed walls, parallel roof, 4" thick		110	.218			7.75		7.75	12.85	
	0160	Skewed walls/roof, 4" layer/air space		96	.250			8.90		8.90	14.75	
	0170	Sound-absorbing panels, Pntd. metal, 2'-6" x 8', under 1,000 S.F.		430	.056			1.98		1.98	3.29	
	0180	Over 1,000 S.F.		480	.050			1.78		1.78	2.95	
	0190	Flexible transparent curtain, clear	3 Shee	430	.056			2.35		2.35	3.74	
	0192	50% clear, 50% foam		430	.056			2.35		2.35	3.74	
	0194	25% clear, 75% foam		430	.056			2.35		2.35	3.74	
	0196	100% foam		430	.056			2.35		2.35	3.74	
	0200	Audio-masking system, incl. speakers, amplfr., signal gnrtr.										
	0205	Ceiling mounted, 5,000 S.F.	2 Elec	4800	.003	S.F.		.14		.14	.22	
	0210	10,000 S.F.		5600	.003			.12		.12	.19	
	0220	Plenum mounted, 5,000 S.F.		7600	.002			.09		.09	.14	
	0230	10,000 S.F.		8800	.002			.08		.08	.12	
038	0010	SELECTIVE DEMOLITION, X-RAY/RADIO FREQ PROTECTION										038
	0020	Shielding lead, lined door frame, excl hdwe, 1/16" thick	1 Clab	4.80	1.667	Ea.		45.50		45.50	75.50	
	0030	Lead sheets, 1/16" thick	2 Clab	270	.059	S.F.		1.62		1.62	2.69	
	0040	1/8" thick		240	.067			1.83		1.83	3.03	
	0050	Lead shielding, 1/4" thick		270	.059			1.62		1.62	2.69	
	0060	1/2" thick		240	.067			1.83		1.83	3.03	
	0070	Lead glass, 1/4" thick, 2.0 mm LE, 12" x 16"	2 Glaz	26	.615	Ea.		21		21	34	
	0080	24" x 36"		16	1			34.50		34.50	55	
	0090	36" x 60"		4	4			137		137	221	
	0100	Lead glass window frame, w/ 1/16" lead & voice passage, 36" x 60"		4	4			137		137	221	
	0110	Lead glass window frame, 24" x 36"		16	1			34.50		34.50	55	
	0120	Lead gypsum board, 5/8" thick with 1/16" lead	2 Clab	320	.050	S.F.		1.37		1.37	2.27	
	0130	1/8" lead		280	.057			1.57		1.57	2.59	
	0140	1/32" lead		400	.040			1.10		1.10	1.82	
	0150	Butt joints, 1/8" lead or thicker, 2" x 7' long batten strip		480	.033	Ea.		.91		.91	1.51	
	0160	X-ray protection, average radiography room, to 300 S.F., 1/16" lead, min.		.50	32	Total		875		875	1,450	
	0170	Maximum		.30	53.333			1,450		1,450	2,425	
	0180	Deep therapy X-ray room, 250 KV cap, to 300 S.F., 1/4" lead, min.		.20	80			2,200		2,200	3,625	
	0190	Maximum		.12	133			3,650		3,650	6,050	
	0880	Radio frequency shielding, prefab or screen-type copper or steel, min.		360	.044	SF Surf		1.22		1.22	2.02	
	0890	Average		310	.052			1.41		1.41	2.34	
	0895	Maximum		290	.055			1.51		1.51	2.50	
039	0010	SELECTIVE DEMOLITION, GEODESIC DOMES										039
	0050	Shell only, interlocking plywood panels, 30' diameter	F-5	3.20	10	Ea.		360		360	595	
	0060	34' diameter		2.30	13.913			500		500	830	
	0070	39' diameter		2	16			575		575	955	
	0080	45' diameter		2.20	14.545			525		525	870	
	0090	55' diameter		2	16			575		575	955	
	0100	60' diameter		2	16			575		575	955	
	0110	65' diameter		1.60	20			720		720	1,200	
040	0010	SELECTIVE DEMOLITION, TENSION STRUCTURES										040
	0020	Steel/alum frame, fabric shell, 60' clear span, 6,000 S.F.	B-41	2000	.022	SF Flr.		.63	.12	.75	1.16	

	13005 Selective Demolition		CREW	DAILY OUTPUT	LABOR-HOURS	UNIT	2006 BARE COSTS MAT.	LABOR	EQUIP.	TOTAL	TOTAL INCL O&P	
040	0030	12,000 S.F.	B-41	2200	.020	SF Flr.		.57	.11	.68	1.06	040
	0040	80′ clear span, 20,800 S.F.		2440	.018			.51	.10	.61	.96	
	0050	100′ clear span, 10,000 S.F.	L-5	4350	.013			.51	.15	.66	1.10	
	0060	26,000 S.F.		4600	.012			.49	.14	.63	1.05	
	0080	120′ clear span, 24,000 S.F.		6000	.009			.37	.11	.48	.80	
	0090	150′ clear span, 30,000 S.F.		12000	.005			.19	.05	.24	.40	
	0100	200′ clear span, 40,000 S.F.	E-6	16000	.008			.32	.10	.42	.68	
	0110	For roll-up door, 12′ x 14′	L-2	2	8	Ea.		248		248	410	
041	0010	**SELECTIVE DEMOLITION, HANGARS**										041
	0020	Prefab, steel T type, galv roof & walls, incl doors, excl fndtn	E-2	2550	.022	SF Flr.		.85	.56	1.41	2.15	
	0030	Circular type, prefab, steel frame, plastic skin, incl foundation, 80′ dia	"	.50	112	Total		4,350	2,850	7,200	11,000	
042	0010	**SELECTIVE DEMOLITION, SILOS**										042
	0020	Concrete stave, indstrl, conical/sloping bott, excl fndtn, 12′ dia, 35′ h	D-8	.22	181	Ea.		6,000		6,000	9,750	
	0030	16′ dia, 45′ h		.16	250			8,250		8,250	13,400	
	0040	25′ dia, 75′ h		.10	400			13,200		13,200	21,400	
	0050	Steel, factory fabricated, 30,000 gal cap, painted or epoxy lined	L-5	2	28			1,125	325	1,450	2,400	
046	0010	**SELECTIVE DEMOLITION, SWIMMING POOL EQUIPMENT**										046
	0020	Diving stand, stainless steel, 3 meter	2 Clab	3	5.333	Ea.		146		146	242	
	0030	1 meter		5	3.200			87.50		87.50	145	
	0040	Diving board, 16′ long, aluminum		5.40	2.963			81		81	135	
	0050	Fiberglass		5.40	2.963			81		81	135	
	0070	Ladders, heavy duty, stainless steel, 2 tread		14	1.143			31.50		31.50	52	
	0080	4 tread		12	1.333			36.50		36.50	60.50	
	0090	Lifeguard chair, stainless steel, fixed		5	3.200			87.50		87.50	145	
	0100	Slide, tubular, fiberglass, aluminum handrails & ladder, 5′, straight		4	4			110		110	182	
	0110	8′, curved		6	2.667			73		73	121	
	0120	10′, curved		3	5.333			146		146	242	
	0130	12′ straight, with platform		2.50	6.400			175		175	291	
	0140	Removable access ramp, stainless steel		4	4			110		110	182	
	0150	Removable stairs, stainless steel, collapsible		4	4			110		110	182	
047	0010	**SELECTIVE DEMOLITION, LIGHTNING PROTECTION**										047
	0020	Air terminal & base, copper, 3/8″ dia x 10″, to 75′ h	1 Clab	16	.500	Ea.		13.70		13.70	22.50	
	0030	1/2″ dia x 12″, over 75′ h		16	.500			13.70		13.70	22.50	
	0050	Aluminum, 1/2″ dia x 12″, to 75′ h		16	.500			13.70		13.70	22.50	
	0060	5/8″ dia x 12″, over 75′ h		16	.500			13.70		13.70	22.50	
	0070	Cable, copper, 220 lb per thousand feet, to 75′ high		640	.013	L.F.		.34		.34	.57	
	0080	375 lb per thousand feet, over 75′ high		460	.017			.48		.48	.79	
	0090	Aluminum, 101 lb per thousand feet, to 75′ high		560	.014			.39		.39	.65	
	0100	199 lb per thousand feet, over 75′ high		480	.017			.46		.46	.76	
	0110	Arrester, 175 V AC, to ground		16	.500	Ea.		13.70		13.70	22.50	
	0120	650 V AC, to ground		13	.615	"		16.85		16.85	28	
128	0010	**SELECTIVE DEMOLITION, PRE-ENGINEERED STEEL BUILDINGS**										128
	0500	Pre-engd steel bldgs, rigid frame, clear span & multi post, excl. salvage										
	0550	3,500 to 7,500 S.F	L-11	1350	.024	SF Flr.		.76	1.05	1.81	2.45	
	0600	7,501 to 12,500 S.F		2000	.016			.51	.71	1.22	1.65	
	0650	12,500 S.F or greater		2200	.015			.47	.65	1.12	1.50	
	0700	Pre-engd steel building components										
	0710	Entrance canopy, including frame 4′ x 4′	E-24	8	4	Ea.		157	81.50	238.50	370	
	0720	4′ x 8′	"	7	4.571			179	93	272	420	
	0730	H.M doors, self framing, single leaf	2 Skwk	8	2			73		73	119	
	0740	Double leaf		5	3.200			117		117	191	
	0760	Gutter, eave type		600	.027	L.F.		.97		.97	1.59	
	0770	Sash, single slide, double slide or fixed		24	.667	Ea.		24.50		24.50	40	

13005 | Selective Demolition

	13005 Selective Demolition		CREW	DAILY OUTPUT	LABOR-HOURS	UNIT	2006 BARE COSTS MAT.	LABOR	EQUIP.	TOTAL	TOTAL INCL O&P	
128	0780	Skylight, fiberglass, to 30 S.F.	2 Skwk	16	1	Ea.		36.50		36.50	59.50	128
	0785	Roof vents, circular, 12″ to 24″ diameter		12	1.333			48.50		48.50	79.50	
	0790	Continuous, 10′ long		8	2			73		73	119	
	0900	Shelters, aluminum frame										
	0910	Aluminum frame, acrylic glazing, 3′ x 9′ x 8′ high	2 Skwk	2	8	Ea.		292		292	480	
	0920	9′ x 12′ x 8′ high	″	1.50	10.667	″		390		390	635	
	0930	Silos, concrete stave industrial, not incl. foundations										
201	0010	**SELECTIVE DEMOLITION, STORAGE TANKS**										201
	0500	Steel tank, single wall, above ground, not incl. fdn., pumps or piping										
	0510	Single wall, 275 gallon	Q-1	3	5.333	Ea.		205		205	320	
	0520	550 thru 2,000 gallon (R024119-10)	B-34P	2	12			430	201	631	900	
	0530	5,000 thru 10,000 gallon	B-34Q	2	12			435	535	970	1,275	
	0540	15,000 thru 30,000 gallon	B-34N	2	4			113	223	336	430	
	0600	Steel tank, double wall, above ground not incl. fdn., pumps & piping										
	0620	500 thru 2,000 gallon	B-34P	2	12	Ea.		430	201	631	900	

13010 | Air Supported Structures

	13011 Air Supported Structures		CREW	DAILY OUTPUT	LABOR-HOURS	UNIT	2006 BARE COSTS MAT.	LABOR	EQUIP.	TOTAL	TOTAL INCL O&P	
100	0010	**AIR SUPPORTED TANK COVERS**, vinyl polyester										100
	0100	scrim, double layer, with hardware, blower, standby & controls										
	0200	Round, 75′ diameter	B-2	4500	.009	S.F.	6.70	.25		6.95	7.75	
	0300	100′ diameter		5000	.008		6.15	.22		6.37	7.15	
	0400	150′ diameter		5000	.008		4.64	.22		4.86	5.45	
	0500	Rectangular, 20′ x 20′		4500	.009		15.10	.25		15.35	17	
	0600	30′ x 40′		4500	.009		15.10	.25		15.35	17	
	0700	50′ x 60′		4500	.009		15.10	.25		15.35	17	
	0800	For single wall construction, deduct, minimum					.44			.44	.48	
	0900	Maximum					1.43			1.43	1.57	
	1000	For maximum resistance to atmosphere or cold, add					.60			.60	.66	
	1100	For average shipping charges, add				Total	1,200			1,200	1,325	
200	0010	**AIR SUPPORTED STRUCTURES**										200
	0020	Site preparation, incl. anchor placement and utilities	B-11B	1000	.016	SF Flr.	.79	.50	.20	1.49	1.90	
	0030	For concrete curb, see division 03310-240										
	0050	Warehouse, polyester/vinyl fabric, 28 oz., over 10 yr. life, welded										
	0060	Seams, tension cables, primary & auxiliary inflation system,										
	0070	airlock, personnel doors and liner										
	0100	5,000 S.F.	4 Clab	5000	.006	SF Flr.	16.75	.18		16.93	18.75	
	0250	12,000 S.F.	″	6000	.005		12.10	.15		12.25	13.60	
	0400	24,000 S.F.	8 Clab	12000	.005		8.90	.15		9.05	10	
	0500	50,000 S.F.	″	12500	.005		7.65	.14		7.79	8.70	
	0700	12 oz. reinforced vinyl fabric, 5 yr. life, sewn seams,										
	0710	accordian door, including liner										
	0750	3000 S.F.	4 Clab	3000	.011	SF Flr.	8.75	.29		9.04	10.15	
	0800	12,000 S.F.	″	6000	.005		6.50	.15		6.65	7.40	
	0850	24,000 S.F.	8 Clab	12000	.005		6.45	.15		6.60	7.30	
	0950	Deduct for single layer					.70			.70	.77	
	1000	Add for welded seams					1.01			1.01	1.11	
	1050	Add for double layer, welded seams included					2.08			2.08	2.29	

13010 | Air Supported Structures

	13011 | Air Supported Structures		CREW	DAILY OUTPUT	LABOR-HOURS	UNIT	2006 BARE COSTS MAT.	LABOR	EQUIP.	TOTAL	TOTAL INCL O&P	
200	1250	Tedlar/vinyl fabric, 28 oz., with liner, over 10 yr. life,										200
	1260	incl. overhead and personnel doors										
	1300	3000 S.F.	4 Clab	3000	.011	SF Flr.	16.25	.29		16.54	18.40	
	1450	12,000 S.F.	"	6000	.005		11.40	.15		11.55	12.80	
	1550	24,000 S.F.	8 Clab	12000	.005		8.90	.15		9.05	10	
	1700	Deduct for single layer				↓	1.44			1.44	1.58	
	2250	Greenhouse/shelter, woven polyethylene with liner, 2 yr. life,										
	2260	sewn seams, including doors										
	2300	3000 S.F.	4 Clab	3000	.011	SF Flr.	6.60	.29		6.89	7.75	
	2350	12,000 S.F.	"	6000	.005		6.70	.15		6.85	7.65	
	2450	24,000 S.F.	8 Clab	12000	.005		5.60	.15		5.75	6.40	
	2550	Deduct for single layer				↓	.63			.63	.69	
	2600	Tennis/gymnasium, polyester/vinyl fabric, 28 oz., over 10 yr. life,										
	2610	including thermal liner, heat and lights										
	2650	7,200 S.F.	4 Clab	6000	.005	SF Flr.	15.65	.15		15.80	17.45	
	2750	13,000 S.F.	"	6500	.005		11.75	.13		11.88	13.10	
	2850	Over 24,000 S.F.	8 Clab	12000	.005		10.60	.15		10.75	11.95	
	2860	For low temperature conditions, add				↓	.71			.71	.78	
	2870	For average shipping charges, add				Total	3,100			3,100	3,425	
	2900	Thermal liner, translucent reinforced vinyl				SF Flr.	.97			.97	1.07	
	2950	Metalized mylar fabric and mesh, double liner				"	1.56			1.56	1.72	
	3050	Stadium/convention center, teflon coated fiberglass, heavy weight,										
	3060	over 20 yr. life, incl. thermal liner and heating system										
	3100	Minimum	9 Clab	26000	.003	SF Flr.	40	.08		40.08	44	
	3110	Maximum	"	19000	.004	"	46.50	.10		46.60	51.50	
	3400	Doors, air lock, 15′ long, 10′ x 10′	2 Carp	.80	20	Ea.	14,700	710		15,410	17,300	
	3600	15′ x 15′	"	.50	32		21,200	1,150		22,350	25,200	
	3700	For each added 5′ length, add					3,050			3,050	3,350	
	3900	Revolving personnel door, 6′ diameter, 6′-6″ high	2 Carp	.80	20	↓	10,700	710		11,410	13,000	
	4200	Double wall, self supporting, shell only, minimum				SF Flr.					21	
	4300	Maximum				"					39	

13030 | Special Purpose Rooms

	13035 | Special Purpose Rooms		CREW	DAILY OUTPUT	LABOR-HOURS	UNIT	2006 BARE COSTS MAT.	LABOR	EQUIP.	TOTAL	TOTAL INCL O&P	
150	0010	**ANECHOIC CHAMBERS** Standard units, 7′ ceiling heights										150
	0100	Area for pricing is net inside dimensions										
	0300	200 cycles per second cutoff, 25 S.F. floor area				SF Flr.	1,425			1,425	1,575	
	0400	50 S.F.									1,150	
	0600	75 S.F.									1,100	
	0700	100 S.F.					1,075			1,075	1,200	
	0900	For 150 cycles per second cutoff, add to 100 S.F. room									30%	
	1000	For 100 cycles per second cutoff, add to 100 S.F. room				↓					45%	
180	0010	**AUDIOMETRIC ROOMS**										180
	0020	Under 500 S.F. surface	4 Carp	98	.327	SF Surf	46	11.60		57.60	70.50	
	0100	Over 500 S.F. surface	"	120	.267	"	44	9.50		53.50	64	
200	0010	**CLEAN ROOMS**										200
	1100	Clean room, soft wall, 12′ x 12′, Class 100	1 Carp	.18	44.444	Ea.	13,000	1,575		14,575	16,900	

13030 | Special Purpose Rooms

	13035 | Special Purpose Rooms		CREW	DAILY OUTPUT	LABOR-HOURS	UNIT	2006 BARE COSTS MAT.	LABOR	EQUIP.	TOTAL	TOTAL INCL O&P
200	1110	Class 1,000	1 Carp	.18	44.444	Ea.	10,100	1,575		11,675	13,700
	1120	Class 10,000		.21	38.095		8,575	1,350		9,925	11,700
	1130	Class 100,000		.21	38.095		7,925	1,350		9,275	11,000
	2800	Ceiling grid support, slotted channel struts 4′-0″ O.C., ea. way				S.F.					6.50
	3000	Ceiling panel, vinyl coated foil on mineral substrate									
	3020	Sealed, non-perforated				S.F.					1.40
	4000	Ceiling panel seal, silicone sealant, 150 L.F./gal.	1 Carp	150	.053	L.F.	.23	1.90		2.13	3.39
	4100	Two sided adhesive tape	″	240	.033	″	.10	1.18		1.28	2.07
	4200	Clips, one per panel				Ea.	.90			.90	.99
	6000	HEPA filter,2′x 4′, 99.97% eff.,3″ dp. beveled frame (silicone seal)					277			277	305
	6040	6″ deep skirted frame (channel seal)					320			320	350
	6100	99.99% efficient, 3″ deep beveled frame (silicone seal)					298			298	330
	6140	6″ deep skirted frame (channel seal)					340			340	375
	6200	99.999% efficient, 3″ deep beveled frame (silicone seal)					340			340	375
	6240	6″ deep skirted frame (channel seal)					385			385	420
	7000	Wall panel systems, including channel strut framing									
	7020	Polyester coated aluminum, particle board				S.F.					20
	7100	Porcelain coated aluminum, particle board									35
	7400	Wall panel support, slotted channel struts, to 12′ high									18
300	0010	**DARKROOMS**									
	0020	Shell, complete except for door, 64 S.F., 8′ high	2 Carp	128	.125	SF Flr.	72	4.44		76.44	87
	0100	12′ high		64	.250		85	8.90		93.90	108
	0500	120 S.F. floor, 8′ high		120	.133		53	4.74		57.74	66.50
	0600	12′ high		60	.267		64	9.50		73.50	86
	0800	240 S.F. floor, 8′ high		120	.133		37	4.74		41.74	48.50
	0900	12′ high		60	.267		51	9.50		60.50	71.50
	1200	Mini-cylindrical, revolving, unlined, 4′ diameter		3.50	4.571	Ea.	3,525	163		3,688	4,150
	1400	5′-6″ diameter		2.50	6.400		5,575	228		5,803	6,500
	1600	Add for lead lining, inner cylinder, 1/32″ thick					1,425			1,425	1,575
	1700	1/16″ thick					1,675			1,675	1,825
	1800	Add for lead lining, inner and outer cylinder, 1/32″ thick					2,625			2,625	2,900
	1900	1/16″ thick					2,750			2,750	3,025
	2000	For darkroom door, see division 11472-370									
500	0010	**MUSIC ROOM**									
	0020	Practice room, modular, perforated steel, under 500 S.F.	2 Carp	70	.229	SF Surf	28.50	8.15		36.65	44.50
	0100	Over 500 S.F.	″	80	.200	″	24	7.10		31.10	38.50
700	0010	**REFRIGERATION**									
	0020	Curbs, 12″ high, 4″ thick, concrete	2 Carp	58	.276	L.F.	3.52	9.80		13.32	20
	1000	Doors, see division 08340-100									
	2400	Finishes, 2 coat portland cement plaster, 1/2″ thick	1 Plas	48	.167	S.F.	.97	5.40		6.37	9.80
	2500	For galvanized reinforcing mesh, add	1 Lath	335	.024		.65	.78		1.43	1.96
	2700	3/16″ thick latex cement	1 Plas	88	.091		1.66	2.95		4.61	6.60
	2900	For glass cloth reinforced ceilings, add	″	450	.018		.40	.58		.98	1.37
	3100	Fiberglass panels, 1/8″ thick	1 Carp	149.45	.054		2.19	1.90		4.09	5.55
	3200	Polystyrene, plastic finish ceiling, 1″ thick		274	.029		2.02	1.04		3.06	3.94
	3400	2″ thick		274	.029		2.30	1.04		3.34	4.25
	3500	4″ thick		219	.037		2.56	1.30		3.86	4.97
	3800	Floors, concrete, 4″ thick	1 Cefi	93	.086		.96	2.96		3.92	5.70
	3900	6″ thick	″	85	.094		1.44	3.24		4.68	6.70
	4000	Insulation, 1″ to 6″ thick, cork				B.F.	.94			.94	1.03
	4100	Urethane					.92			.92	1.01
	4300	Polystyrene, regular					.63			.63	.69
	4400	Bead board					.47			.47	.52
	4600	Installation of above, add per layer	2 Carp	657.60	.024	S.F.	.30	.86		1.16	1.76

	13035 | Special Purpose Rooms		CREW	DAILY OUTPUT	LABOR-HOURS	UNIT	2006 BARE COSTS MAT.	LABOR	EQUIP.	TOTAL	TOTAL INCL O&P	
700	4700	Wall and ceiling juncture	2 Carp	298.90	.054	L.F.	1.50	1.90		3.40	4.80	700
	4900	Partitions, galvanized sandwich panels, 4" thick, stock		219.20	.073	S.F.	6.35	2.59		8.94	11.25	
	5000	Aluminum or fiberglass		219.20	.073	"	6.90	2.59		9.49	11.90	
	5200	Prefab walk-in, 7'-6" high, aluminum, incl. door & floors,										
	5210	not incl. partitions or refrigeration, 6' x 6' O.D. nominal	2 Carp	54.80	.292	SF Flr.	113	10.40		123.40	141	
	5500	10' x 10' O.D. nominal		82.20	.195		91	6.90		97.90	111	
	5700	12' x 14' O.D. nominal		109.60	.146		81.50	5.20		86.70	98.50	
	5800	12' x 20' O.D. nominal		109.60	.146		71	5.20		76.20	86.50	
	6100	For 8'-6" high, add					5%					
	6300	Rule of thumb for complete units, w/o doors & refrigeration, cooler	2 Carp	146	.110		103	3.90		106.90	119	
	6400	Freezer		109.60	.146		121	5.20		126.20	142	
	6600	Shelving, plated or galvanized, steel wire type		360	.044	SF Hor.	9	1.58		10.58	12.50	
	6700	Slat shelf type		375	.043		11.10	1.52		12.62	14.70	
	6900	For stainless steel shelving, add					300%					
	7000	Vapor barrier, on wood walls	2 Carp	1644	.010	S.F.	.13	.35		.48	.71	
	7200	On masonry walls	"	1315	.012	"	.34	.43		.77	1.09	
	7500	For air curtain doors, see division 15840-200										
800	0010	**SAUNA**										800
	0020	Prefabricated, incl. heater & controls, 7' high, 6' x 4', C/C	L-7	2.20	12.727	Ea.	3,675	435		4,110	4,750	
	0050	6' x 4', C/P		2	14		3,400	480		3,880	4,525	
	0400	6' x 5', C/C		2	14		4,125	480		4,605	5,300	
	0450	6' x 5', C/P		2	14		3,850	480		4,330	5,000	
	0600	6' x 6', C/C		1.80	15.556		4,400	530		4,930	5,700	
	0650	6' x 6', C/P		1.80	15.556		4,100	530		4,630	5,375	
	0800	6' x 9', C/C		1.60	17.500		5,500	595		6,095	7,025	
	0850	6' x 9', C/P		1.60	17.500		5,225	595		5,820	6,700	
	1000	8' x 12', C/C		1.10	25.455		8,525	870		9,395	10,800	
	1050	8' x 12', C/P		1.10	25.455		7,800	870		8,670	10,000	
	1200	8' x 8', C/C		1.40	20		6,475	685		7,160	8,250	
	1250	8' x 8', C/P		1.40	20		6,100	685		6,785	7,825	
	1400	8' x 10', C/C		1.20	23.333		7,200	795		7,995	9,225	
	1450	8' x 10', C/P		1.20	23.333		6,700	795		7,495	8,650	
	1600	10' x 12', C/C		1	28		9,025	955		9,980	11,500	
	1650	10' x 12', C/P		1	28		8,175	955		9,130	10,600	
	1700	Door only, cedar, 2' x 6', with tempered insulated glass window	2 Carp	3.40	4.706		480	167		647	805	
	1800	Prehung, incl. jambs, pulls & hardware	"	12	1.333		490	47.50		537.50	620	
	2500	Heaters only (incl. above), wall mounted, to 200 C.F.					475			475	520	
	2750	To 300 C.F.					585			585	640	
	3000	Floor standing, to 720 C.F., 10,000 watts, w/controls	1 Elec	3	2.667		1,500	112		1,612	1,825	
	3250	To 1,000 C.F., 16,000 watts	"	3	2.667		1,650	112		1,762	2,000	
900	0010	**SPORT COURT**										900
	0020	Floors, No. 2 & better maple, 25/32" thick				SF Flr.					6.65	
	0100	Walls, laminated plastic bonded to galv. steel studs				SF Wall					8.45	
	0150	Laminated fiberglass surfacing, minimum									2.10	
	0180	Maximum									2.37	
	0300	Squash, regulation court in existing building, minimum				Court	15,300			15,300	16,800	
	0400	Maximum				"	28,500			28,500	31,400	
	0450	Rule of thumb for components:										
	0470	Walls	3 Carp	.15	160	Court	11,700	5,700		17,400	22,300	
	0500	Floor	"	.25	96		5,225	3,425		8,650	11,400	
	0550	Lighting	2 Elec	.60	26.667		1,650	1,125		2,775	3,550	
	0600	Handball, racquetball court in existing building, minimum	C-1	.20	160		2,025	5,350		7,375	11,100	
	0800	Maximum	"	.10	320		27,100	10,700		37,800	47,600	
	0900	Rule of thumb for components: walls	3 Carp	.12	200		10,700	7,100		17,800	23,500	

13030 | Special Purpose Rooms

		13035 | Special Purpose Rooms	CREW	DAILY OUTPUT	LABOR-HOURS	UNIT	2006 BARE COSTS MAT.	LABOR	EQUIP.	TOTAL	TOTAL INCL O&P	
900	1000	Floor	3 Carp	.25	96	Court	5,550	3,425		8,975	11,800	900
	1100	Ceiling	↓	.33	72.727		2,325	2,575		4,900	6,825	
	1200	Lighting	2 Elec	.60	26.667	↓	2,875	1,125		4,000	4,875	
940	0010	**STEAM BATH**										940
	0020	Heater, timer & head, single, to 140 C.F.	1 Plum	1.20	6.667	Ea.	1,075	285		1,360	1,625	
	0500	To 300 C.F.		1.10	7.273		1,200	310		1,510	1,800	
	1000	Commercial size, with blow-down assembly, to 800 C.F.		.90	8.889		4,175	380		4,555	5,175	
	1500	To 2500 C.F.	↓	.80	10		6,575	425		7,000	7,900	
	2000	Multiple, motels, apts., 2 baths, w/ blow-down assm., 500 C.F.	Q-1	1.30	12.308		4,125	475		4,600	5,250	
	2500	4 baths	"	.70	22.857		4,500	880		5,380	6,325	
	2700	Conversion unit for residential tub, including door				↓	3,375			3,375	3,725	

		13039 | Vaults	CREW	DAILY OUTPUT	LABOR-HOURS	UNIT	MAT.	LABOR	EQUIP.	TOTAL	TOTAL INCL O&P	
900	0010	**VAULT FRONT** See division 11021-800										900

13080 | Sound, Vibration & Seismic Control

		13081 | Sound Control	CREW	DAILY OUTPUT	LABOR-HOURS	UNIT	2006 BARE COSTS MAT.	LABOR	EQUIP.	TOTAL	TOTAL INCL O&P	
100	0010	**AUDIO MASKING** Acoustical enclosure, 4" thick wall and ceiling panels										100
	0020	8# per S.F., up to 12' span	3 Carp	72	.333	SF Surf	29	11.85		40.85	51.50	
	0300	Better quality panels, 10.5# per S.F.		64	.375		33	13.35		46.35	58	
	0400	Reverb-chamber, 4" thick, parallel walls		60	.400		41	14.20		55.20	68.50	
	0600	Skewed wall, parallel roof, 4" thick panels		55	.436		46.50	15.50		62	77	
	0700	Skewed walls, skewed roof, 4" layers, 4" air space		48	.500		52.50	17.80		70.30	87.50	
	0900	Sound-absorbing panels, painted metal, 2'-6" x 8', under 1,000 S.F.		215	.112		10.85	3.97		14.82	18.55	
	1100	Over 1000 S.F.		240	.100		10.45	3.56		14.01	17.40	
	1200	Fabric faced	↓	240	.100		8.45	3.56		12.01	15.20	
	1500	Flexible transparent curtain, clear	3 Shee	215	.112		6.55	4.71		11.26	14.70	
	1600	50% foam		215	.112		9.15	4.71		13.86	17.55	
	1700	75% foam		215	.112		9.15	4.71		13.86	17.55	
	1800	100% foam	↓	215	.112	↓	9.15	4.71		13.86	17.55	
	3100	Audio masking system, including speakers, amplification										
	3110	and signal generator										
	3200	Ceiling mounted, 5,000 S.F.	2 Elec	2400	.007	S.F.	1.19	.28		1.47	1.74	
	3300	10,000 S.F.		2800	.006		.96	.24		1.20	1.43	
	3400	Plenum mounted, 5,000 S.F.		3800	.004		.95	.18		1.13	1.32	
	3500	10,000 S.F.	↓	4400	.004	↓	.64	.15		.79	.94	

	13091 | X-Ray/Radio Freq Protection	CREW	DAILY OUTPUT	LABOR-HOURS	UNIT	2006 BARE COSTS MAT.	LABOR	EQUIP.	TOTAL	TOTAL INCL O&P
600 0010	SHIELDING LEAD									
0100	Laminated lead in wood doors, 1/16″ thick, no hardware				S.F.	30.50			30.50	33.50
0200	Lead lined door frame, not incl. hardware,									
0210	1/16″ thick lead, butt prepared for hardware	1 Lath	2.40	3.333	Ea.	390	109		499	600
0300	Lead sheets, 1/16″ thick	2 Lath	135	.119	S.F.	5.40	3.89		9.29	12.10
0400	1/8″ thick		120	.133		11.05	4.37		15.42	19.10
0500	Lead shielding, 1/4″ thick		135	.119		19.65	3.89		23.54	27.50
0550	1/2″ thick		120	.133		40	4.37		44.37	51
0600	Lead glass, 1/4″ thick, 2.0 mm LE, 12″ x 16″	2 Glaz	13	1.231	Ea.	269	42		311	365
0700	24″ x 36″		8	2		1,050	68.50		1,118.50	1,275
0800	36″ x 60″		2	8		3,075	274		3,349	3,850
0850	Window frame with 1/16″ lead and voice passage, 36″ x 60″		2	8		1,125	274		1,399	1,700
0870	24″ x 36″ frame		8	2		650	68.50		718.50	825
0900	Lead gypsum board, 5/8″ thick with 1/16″ lead		160	.100	S.F.	5.60	3.43		9.03	11.70
0910	1/8″ lead		140	.114		11.05	3.91		14.96	18.50
0930	1/32″ lead	2 Lath	200	.080		3.50	2.62		6.12	8
0950	Lead headed nails (average 1 lb. per sheet)				Lb.	6.20			6.20	6.80
1000	Butt joints in 1/8″ lead or thicker, 2″ batten strip x 7′ long	2 Lath	240	.067	Ea.	14.90	2.19		17.09	19.85
1200	X-ray protection, average radiography or fluoroscopy									
1210	room, up to 300 S.F. floor, 1/16″ lead, minimum	2 Lath	.25	64	Total	6,000	2,100		8,100	9,925
1500	Maximum, 7′-0″ walls	″	.15	106	″	7,500	3,500		11,000	13,800
1600	Deep therapy X-ray room, 250 KV capacity,									
1800	up to 300 S.F. floor, 1/4″ lead, minimum	2 Lath	.08	200	Total	18,500	6,550		25,050	30,700
1900	Maximum, 7′-0″ walls		.06	266	″	23,900	8,750		32,650	40,200
1999	Minimum labor/equipment charge		4.50	3.556	Job		117		117	185
2000	X-ray viewing panels, clear lead plastic									
2010	7 mm thick, 0.3 mm LE, 2.3 lbs/S.F.	H-3	139	.115	S.F.	147	3.49		150.49	168
2020	12 mm thick, 0.5 mm LE, 3.9 lbs/S.F.		82	.195		199	5.90		204.90	229
2030	18 mm thick, 0.8mm LE, 5.9 lbs/S.F.		54	.296		216	9		225	253
2040	22 mm thick, 1.0 mm LE, 7.2 lbs/S.F.		44	.364		221	11.05		232.05	261
2050	35 mm thick, 1.5 mm LE, 11.5 lbs/S.F.		28	.571		248	17.35		265.35	300
2060	46 mm thick, 2.0 mm LE, 15.0 lbs/S.F.		21	.762		325	23		348	400
2090	For panels 12 S.F. to 48 S.F., add crating charge				Ea.					50
4000	X-ray barriers, modular, panels mounted within framework for									
4002	attaching to floor, wall or ceiling, upper portion is clear lead									
4005	plastic window panels 48″H, lower portion is opaque leaded									
4008	steel panels 36″H, structural supports not incl.									
4010	1-section barrier, 36″W x 84″H overall									
4020	0.5 mm LE panels	H-3	6.40	2.500	Ea.	3,325	76		3,401	3,775
4030	0.8 mm LE panels		6.40	2.500		3,525	76		3,601	4,000
4040	1.0 mm LE panels		5.33	3.002		3,650	91		3,741	4,175
4050	1.5 mm LE panels		5.33	3.002		3,900	91		3,991	4,425
4060	2-section barrier, 72″W x 84″H overall									
4070	0.5 mm LE panels	H-3	4	4	Ea.	6,875	121		6,996	7,775
4080	0.8 mm LE panels		4	4		7,275	121		7,396	8,200
4090	1.0 mm LE panels		3.56	4.494		7,425	136		7,561	8,400
5000	1.5 mm LE panels		3.20	5		8,025	152		8,177	9,075
5010	3-section barrier, 108″W x 84″H overall									
5020	0.5 mm LE panels	H-3	3.20	5	Ea.	9,850	152		10,002	11,000
5030	0.8 mm LE panels		3.20	5		10,300	152		10,452	11,600
5040	1.0 mm LE panels		2.67	5.993		10,600	182		10,782	11,900
5050	1.5 mm LE panels		2.46	6.504		11,500	197		11,697	12,900
7000	X-ray barriers, mobile, mounted within framework w/casters on									
7005	bottom, clear lead plastic window panels on upper portion,									
7010	opaque on lower, 30″W x 75″H overall, incl. framework									
7020	24″H upper w/0.5 mm LE, 48″H lower w/0.8 mm LE	1 Carp	16	.500	Ea.	2,250	17.80		2,267.80	2,500

13090 | Radiation Protection

	13091 | X-Ray/Radio Freq Protection		CREW	DAILY OUTPUT	LABOR-HOURS	UNIT	2006 BARE COSTS MAT.	LABOR	EQUIP.	TOTAL	TOTAL INCL O&P	
600	7030	48″W x 75″H overall, incl. framework										600
	7040	36″H upper w/0.5 mm LE, 36″H lower w/0.8 mm LE	1 Carp	16	.500	Ea.	4,000	17.80		4,017.80	4,425	
	7050	36″H upper w/1.0 mm LE, 36″H lower w/1.5 mm LE	″	16	.500	″	4,825	17.80		4,842.80	5,325	
	7060	72″W x 75″H overall, incl. framework										
	7070	36″H upper w/0.5 mm LE, 36″H lower w/0.8 mm LE	1 Carp	16	.500	Ea.	4,725	17.80		4,742.80	5,225	
	7080	36″H upper w/1.0 mm LE, 36″H lower w/1.5 mm LE	″	16	.500	″	5,975	17.80		5,992.80	6,600	
700	0010	**SHIELDING, RADIO FREQUENCY**										700
	0020	Prefabricated or screen-type copper or steel, minimum	2 Carp	180	.089	SF Surf	24	3.16		27.16	32	
	0100	Average		155	.103		26	3.67		29.67	34.50	
	0150	Maximum	↓	145	.110	↓	31	3.92		34.92	40.50	

13100 | Lightning Protection

	13101 | Lightning Protection		CREW	DAILY OUTPUT	LABOR-HOURS	UNIT	2006 BARE COSTS MAT.	LABOR	EQUIP.	TOTAL	TOTAL INCL O&P	
055	0010	**LIGHTNING PROTECTION FOR BUILDINGS**										055
	0200	Air terminals & base, copper										
	0400	3/8″ diameter x 10″ (to 75′ high)	1 Elec	8	1	Ea.	26	42		68	93.50	
	0500	1/2″ diameter x 12″ (over 75′ high)		8	1		40	42		82	109	
	1000	Aluminum, 1/2″ diameter x 12″ (to 75′ high)		8	1		20	42		62	86.50	
	1100	5/8″ diameter x 12″ (over 75′ high)		8	1	↓	27	42		69	94.50	
	2000	Cable, copper, 220 lb. per thousand ft. (to 75′ high)		320	.025	L.F.	1.20	1.05		2.25	2.94	
	2100	375 lb. per thousand ft. (over 75′ high)		230	.035		1.80	1.46		3.26	4.23	
	2500	Aluminum, 101 lb. per thousand ft. (to 75′ high)		280	.029		.65	1.20		1.85	2.57	
	2600	199 lb. per thousand ft. (over 75′ high)		240	.033	↓	1.05	1.40		2.45	3.32	
	3000	Arrester, 175 volt AC to ground		8	1	Ea.	40	42		82	109	
	3100	650 volt AC to ground	↓	6.70	1.194	″	78	50		128	164	

13120 | Pre-Engineered Structures

	13128 | Pre-Engineered Structures		CREW	DAILY OUTPUT	LABOR-HOURS	UNIT	2006 BARE COSTS MAT.	LABOR	EQUIP.	TOTAL	TOTAL INCL O&P	
045	0010	**GUARD HOUSE**										045
	0100	Prefab conc. w/bullet resistant doors & windows, roof and wiring										
	0110	8′ x 8′, Level III	L-10	1	24	Ea.	20,000	960	635	21,595	24,400	
	0120	8′ x 8′, Level IV	″	1	24	″	25,000	960	635	26,595	29,900	
060	0010	**PORTABLE BOOTHS** Prefab aluminum with doors, windows, ext. roof										060
	0100	lights wiring & insulation, 15 S.F. building, O.D., painted, minimum				S.F.	268			268	295	
	0300	30 S.F. building, minimum					184			184	202	
	0400	50 S.F. building, minimum					137			137	150	
	0600	80 S.F. building, minimum					110			110	121	
	0700	100 S.F. building, minimum				↓	102			102	112	
	0900	Acoustical booth, 27 Db @ 1,000 Hz, 15 S.F. floor				Ea.	3,025			3,025	3,325	
	1000	7′ x 7′-6″, including light & ventilation					6,200			6,200	6,825	
	1200	Ticket booth, galv. steel, not incl. foundations., 4′ x 4′					5,125			5,125	5,650	
	1300	4′ x 6′				↓	6,000			6,000	6,600	

13120 | Pre-Engineered Structures

	13128 | Pre-Engineered Structures		CREW	DAILY OUTPUT	LABOR-HOURS	UNIT	2006 BARE COSTS MAT.	LABOR	EQUIP.	TOTAL	TOTAL INCL O&P	
070	0010	**CONTROL TOWERS**										070
	0020	12′ x 10′, incl. instruments, min.				Ea.	491,500			491,500	540,500	
	0100	Maximum									1,020,000	
	0500	With standard 40′ tower, average					685,000			685,000	753,500	
	1000	Temporary portable control towers, 8′ x 12′,										
	1010	complete with one position communications, minimum				Ea.					280,000	
	2000	For fixed facilities, depending on height, minimum									60,000	
	2010	Maximum									120,000	
160	0010	**TENSION STRUCTURES** Rigid steel/alum. frame, vyl. coated polyester										160
	0100	fabric shell, 60′ clear span, not incl. foundations or floors										
	0200	6,000 S.F.	B-41	1000	.044	SF Flr.	10.90	1.25	.23	12.38	14.35	
	0300	12,000 S.F.		1100	.040		9.55	1.14	.21	10.90	12.60	
	0400	80′ clear span, 20,800 S.F.		1220	.036		9.75	1.03	.19	10.97	12.65	
	0410	100′ clear span, 10,000 S.F.	L-5	2175	.026		10.85	1.03	.30	12.18	14.10	
	0430	26,000 S.F.		2300	.024		9.60	.97	.28	10.85	12.65	
	0450	36,000 S.F.		2500	.022		9.40	.90	.26	10.56	12.30	
	0460	120′ clear span, 24,000 S.F.		3000	.019		11.35	.75	.22	12.32	14.10	
	0470	150′ clear span, 30,000 S.F.		6000	.009		11.65	.37	.11	12.13	13.65	
	0480	200′ clear span, 40,000 S.F.	E-6	8000	.016		14.15	.63	.20	14.98	17	
	0500	For roll-up door, 12′ x 14′, add	L-2	1	16	Ea.	4,275	495		4,770	5,525	
	0600	For personnel doors, add, minimum				SF Flr.	5%					
	0700	Add, maximum					15%					
	0800	For site work, simple foundation, etc., add, minimum								1.25	1.95	
	0900	Add, maximum								2.75	3.05	
200	0010	**COMFORT STATIONS** Prefab., stock, w/doors, windows & fixt.										200
	0100	Not incl. interior finish or electrical										
	0300	Mobile, on steel frame, minimum				S.F.	36.50			36.50	40	
	0350	Maximum					57.50			57.50	63.50	
	0400	Permanent, including concrete slab, minimum	B-12J	50	.320		142	10.50	16.80	169.30	191	
	0500	Maximum	″	43	.372		206	12.20	19.55	237.75	268	
	0600	Alternate pricing method, mobile, minimum				Fixture	1,775			1,775	1,950	
	0650	Maximum					2,650			2,650	2,925	
	0700	Permanent, minimum	B-12J	.70	22.857		10,200	750	1,200	12,150	13,700	
	0750	Maximum	″	.50	32		17,000	1,050	1,675	19,725	22,300	
300	0010	**DOMES**										300
	0020	Revolv alum, elec drv for astronomy observation, shell only, stk units										
	0600	10′-0″ diameter, 800#, dome	2 Carp	.25	64	Ea.	10,200	2,275		12,475	15,000	
	0700	Base		.67	23.881		3,525	850		4,375	5,275	
	0900	18′-0″ diameter, 2,500#, dome		.17	94.118		29,700	3,350		33,050	38,300	
	1000	Base		.33	48.485		9,850	1,725		11,575	13,700	
	1200	24′-0″ diameter, 4,500#, dome		.08	200		55,000	7,100		62,100	72,500	
	1300	Base		.25	64		17,700	2,275		19,975	23,300	
	1500	Bulk storage, shell only, dual radius hemispher. arch, steel										
	1600	framing, corrugated steel covering, 150′ diameter	E-2	550	.102	SF Flr.	28	3.96	2.59	34.55	40.50	
	1700	400′ diameter	″	720	.078		22.50	3.03	1.98	27.51	32.50	
	1800	Wood framing, wood decking, to 400′ diameter	F-4	400	.120		20.50	4.26	2.38	27.14	32	
	1900	Radial framed wood (2″ x 6″), 1/2″ thick										
	2000	plywood, asphalt shingles, 50′ diameter	F-3	2000	.020	SF Flr.	38.50	.72	.32	39.54	44	
	2100	60′ diameter		1900	.021		29.50	.76	.34	30.60	34	
	2200	72′ diameter		1800	.022		26.50	.80	.35	27.65	30.50	
	2300	116′ diameter		1730	.023		23	.83	.37	24.20	27.50	
	2400	150′ diameter		1500	.027		20	.96	.42	21.38	24	
340	0010	**GEODESIC DOME** Shell only, interlocking plywood panels										340
	0400	30′ diameter	F-5	1.60	20	Ea.	19,300	720		20,020	22,400	
	0500	34′ diameter		1.14	28.070		22,800	1,000		23,800	26,700	
	0600	39′ diameter		1	32		25,600	1,150		26,750	30,100	

		13128 | Pre-Engineered Structures	CREW	DAILY OUTPUT	LABOR-HOURS	UNIT	2006 BARE COSTS MAT.	LABOR	EQUIP.	TOTAL	TOTAL INCL O&P	
340	0700	45′ diameter	F-3	1.13	35.556	Ea.	31,500	1,275	565	33,340	37,400	340
	0750	55′ diameter		1	40		45,200	1,450	635	47,285	53,000	
	0800	60′ diameter		1	40		58,500	1,450	635	60,585	67,500	
	0850	65′ diameter		.80	50		71,500	1,800	795	74,095	83,000	
	1100	Aluminum panel, with 6″ insulation										
	1200	100′ diameter				SF Flr.	35			35	38.50	
	1300	500′ diameter				″	30			30	33	
	1600	Aluminum framed, plexiglass closure panels										
	1700	40′ diameter				SF Flr.	94			94	103	
	1800	200′ diameter				″	75			75	82.50	
	2100	Aluminum framed, aluminum closure panels										
	2200	40′ diameter				SF Flr.	28			28	31	
	2300	100′ diameter					21			21	23	
	2400	200′ diameter					20			20	22	
	2500	For VRP faced bonded fiberglass insulation, add									10	
	2700	Aluminum framed, fiberglass sandwich panel closure										
	2800	6′ diameter	2 Carp	150	.107	SF Flr.	33	3.79		36.79	43	
	2900	28′ diameter	″	350	.046	″	30	1.63		31.63	35.50	
360	0010	**GARAGE COSTS**										360
	0020	Public parking, average				Car					15,400	
	0100	See Square Foot Costs in Reference Section										
	0300	Residential, prefab shell, stock, wood, single car, minimum	2 Carp	1	16	Total	3,175	570		3,745	4,450	
	0350	Maximum		.67	23.881		7,250	850		8,100	9,375	
	0400	Two car, minimum		.67	23.881		4,800	850		5,650	6,675	
	0450	Maximum		.50	32		9,575	1,150		10,725	12,400	
380	0010	**GARDEN HOUSE** Prefab wood, no floors or foundations										380
	0100	32 to 200 S.F., minimum	2 Carp	200	.080	SF Flr.	19.05	2.84		21.89	25.50	
	0300	Maximum	″	48	.333	″	34.50	11.85		46.35	57.50	
500	0010	**GRANDSTANDS** Permanent, municipal, including foundation										500
	0050	Steel understructure w/aluminum closed deck, minimum				Seat					130	
	0100	Maximum									225	
	0300	Steel, minimum					30			30	33	
	0400	Maximum					80			80	88	
	0600	Aluminum, extruded, stock design, minimum									70	
	0700	Maximum									115	
	0900	Composite, steel, wood and plastic, stock design, minimum					75			75	82.50	
	1000	Maximum					140			140	154	
510	0010	**BLEACHERS**										510
	0020	Bleachers, outdoor, portable, 3 to 5 tiers, to 300′ long, min.	2 Sswk	120	.133	Seat	36.50	5.35		41.85	50	
	0100	Maximum, less than 15′ long, prefabricated		80	.200		47.50	8		55.50	67.50	
	0200	6 to 20 tiers, minimum, up to 300′ long		120	.133		43.50	5.35		48.85	58	
	0300	Max., under 15′, (highly prefabricated, on wheels)		80	.200		63.50	8		71.50	84.50	
	0500	Permanent grandstands, wood seat, steel frame, 24″ row										
	0600	3 to 15 tiers, minimum	2 Sswk	60	.267	Seat	108	10.65		118.65	139	
	0700	Maximum		48	.333		119	13.30		132.30	156	
	0900	16 to 30 tiers, minimum		60	.267		125	10.65		135.65	157	
	0950	Average		55	.291		156	11.60		167.60	194	
	1000	Maximum		48	.333		187	13.30		200.30	231	
	1200	Seat backs only, 30″ row, fiberglass		160	.100		22	4		26	32	
	1300	Steel and wood		160	.100		27.50	4		31.50	37.50	
	1400	NOTE: average seating is 1.5′ in width										
540	0010	**GREENHOUSE** Shell only, stock units, not incl. 2′ stub walls,										540
	0020	foundation, floors, heat or compartments										

		13128 | Pre-Engineered Structures	Crew	Daily Output	Labor-Hours	Unit	2006 Bare Costs Mat.	Labor	Equip.	Total	Total Incl O&P	
540	0300	Residential type, free standing, 8'-6" long x 7'-6" wide	2 Carp	59	.271	SF Flr.	43.50	9.65		53.15	64	540
	0400	10'-6" wide		85	.188		33.50	6.70		40.20	48	
	0600	13'-6" wide		108	.148		30	5.25		35.25	42	
	0700	17'-0" wide		160	.100		33.50	3.56		37.06	43	
	0900	Lean-to type, 3'-10" wide		34	.471		38.50	16.75		55.25	70	
	1000	6'-10" wide		58	.276		30	9.80		39.80	49.50	
	1500	Commercial, custom, truss frame, incl. equip., plumbing, elec.,										
	1510	benches and controls, under 2,000 S.F., minimum				SF Flr.					30	
	1550	Maximum					43			43	47.50	
	1700	Over 5,000 S.F., minimum					26			26	28.50	
	1750	Maximum									30	
	2000	Institutional, custom, rigid frame, including compartments and										
	2010	multi-controls, under 500 S.F., minimum				SF Flr.					77	
	2050	Maximum					112			112	124	
	2150	Over 2,000 S.F., minimum					44			44	48.50	
	2200	Maximum									65	
	2400	Concealed rigid frame, under 500 S.F., minimum									90	
	2450	Maximum									110	
	2550	Over 2,000 S.F., minimum									68	
	2600	Maximum									79	
	2800	Lean-to type, under 500 S.F., minimum									80	
	2850	Maximum									120	
	3000	Over 2,000 S.F., minimum									44	
	3050	Maximum									73	
	3600	For 1/4" clear plate glass, add				SF Surf	1.74			1.74	1.91	
	3700	For 1/4" tempered glass, add				"	3.93			3.93	4.32	
	3900	For cooling, add, minimum				SF Flr.	2.63			2.63	2.89	
	4000	Maximum					6.50			6.50	7.15	
	4200	For heaters, 13.6 MBH, add					4.99			4.99	5.50	
	4300	60 MBH, add					1.87			1.87	2.06	
	4500	For benches, 2' x 3'-6", add				SF Hor.	22.50			22.50	24.50	
	4600	3' x 10', add				S.F.	12.10			12.10	13.30	
	4800	For controls, add, minimum				Total	2,250			2,250	2,475	
	4900	Maximum				"	13,300			13,300	14,700	
	5100	For humidification equipment, add				M.C.F.	5.75			5.75	6.30	
	5200	For vinyl shading, add				S.F.	1.21			1.21	1.33	
	6000	Geodesic hemisphere, 1/8" plexiglass glazing										
	6050	8' diameter	2 Carp	2	8	Ea.	2,525	284		2,809	3,250	
	6150	24' diameter		.35	45.714		13,000	1,625		14,625	17,000	
	6250	48' diameter		.20	80		35,200	2,850		38,050	43,400	
580	0010	**HANGARS** Prefabricated steel T hangars, Galv. steel roof &										580
	0100	walls, incl. electric bi-folding doors, 4 or more units,										
	0110	not including floors or foundations, minimum	E-2	1275	.044	SF Flr.	10.25	1.71	1.12	13.08	15.60	
	0130	Maximum		1063	.053		10.65	2.05	1.34	14.04	16.85	
	0900	With bottom rolling doors, minimum		1386	.040		9.80	1.57	1.03	12.40	14.70	
	1000	Maximum		966	.058		10.65	2.26	1.48	14.39	17.35	
	1200	Alternate pricing method:										
	1300	Galv. roof and walls, electric bi-folding doors, minimum	E-2	1.06	52.830	Plane	12,700	2,050	1,350	16,100	19,100	
	1500	Maximum		.91	61.538		14,100	2,400	1,575	18,075	21,500	
	1600	With bottom rolling doors, minimum		1.25	44.800		10,100	1,750	1,150	13,000	15,500	
	1800	Maximum		.97	57.732		12,000	2,250	1,475	15,725	19,000	
	2000	Circular type, prefab., steel frame, plastic skin, electric										
	2010	door, including foundations, 80' diameter,										
	2020	for up to 5 light planes, minimum	E-2	.50	112	Total	63,000	4,350	2,850	70,200	80,500	
	2200	Maximum	"	.25	224	"	70,500	8,725	5,700	84,925	99,500	

13128 | Pre-Engineered Structures

		13128 Pre-Engineered Structures	CREW	DAILY OUTPUT	LABOR-HOURS	UNIT	2006 BARE COSTS MAT.	LABOR	EQUIP.	TOTAL	TOTAL INCL O&P	
600	0010	**KIOSKS**										600
	0020	Round, 5′ diameter, 8′ high, 1/4″ fiberglass wall				Ea.	5,600			5,600	6,175	
	0100	1″ insulated double wall, fiberglass					6,375			6,375	7,025	
	0500	Rectangular, 5′ x 9′, 7′-6″ high, 1/4″ fiberglass wall					8,150			8,150	8,975	
	0600	1″ insulated double wall, fiberglass					9,700			9,700	10,700	
700	0010	**PRE-ENGINEERED STEEL BUILDINGS** R133419-10										700
	0100	Clear span rigid frame, 26 ga. colored roofing and siding										
	0150	20 wide, 10′ eave height	E-2	425	.132	SF Flr.	7.25	5.15	3.36	15.76	21	
	0160	14′ eave height 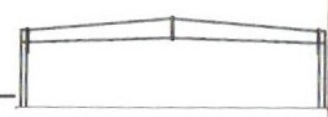		350	.160		7.70	6.20	4.08	17.98	24	
	0170	16′ eave height		320	.175		8.10	6.80	4.46	19.36	26	
	0180	20′ eave height		275	.204		8.90	7.90	5.20	22	29.50	
	0190	24′ eave height		240	.233		10.15	9.10	5.95	25.20	34	
	0200	30′ to 40′ wide, 10′ eave height		535	.105		6	4.07	2.67	12.74	16.85	
	0300	14′ eave height		450	.124		6.35	4.84	3.17	14.36	19.15	
	0400	16′ eave height		415	.135		6.65	5.25	3.44	15.34	20.50	
	0500	20′ eave height		360	.156		7.30	6.05	3.96	17.31	23	
	0600	24′ eave height		320	.175		8.20	6.80	4.46	19.46	26	
	0700	50′ to 100′ wide, 10′ eave height		865	.065		5.15	2.52	1.65	9.32	12.05	
	0800	14′ eave height		770	.073		5.50	2.83	1.85	10.18	13.20	
	0900	16′ eave height		730	.077		5.80	2.98	1.95	10.73	13.90	
	1000	20′ eave height		660	.085		6.25	3.30	2.16	11.71	15.20	
	1100	24′ eave height		605	.093		6.85	3.60	2.36	12.81	16.60	
	1200	Clear span tapered beam frame, 26 ga. colored roofing/siding										
	1300	30′ wide, 10′ eave height	E-2	535	.105	SF Flr.	6.60	4.07	2.67	13.34	17.55	
	1400	14′ eave height 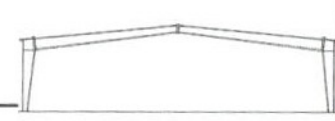		450	.124		7.30	4.84	3.17	15.31	20	
	1500	16′ eave height		415	.135		7.80	5.25	3.44	16.49	22	
	1600	20′ eave height		360	.156		8.60	6.05	3.96	18.61	24.50	
	1700	40′ wide, 10′ eave height		600	.093		6.05	3.63	2.38	12.06	15.75	
	1800	14′ eave height		510	.110		6.60	4.27	2.80	13.67	18	
	1900	16′ eave height		475	.118		6.90	4.59	3	14.49	19.15	
	2000	20′ eave height		415	.135		7.55	5.25	3.44	16.24	21.50	
	2100	50′ to 80′ wide, 10′ eave height		770	.073		5.75	2.83	1.85	10.43	13.50	
	2200	14′ eave height		675	.083		6.15	3.23	2.11	11.49	14.90	
	2300	16′ eave height		635	.088		6.40	3.43	2.25	12.08	15.65	
	2400	20′ eave height		565	.099		7.15	3.86	2.53	13.54	17.55	
	2500	Single post 2-span frame, 26 ga. colored roofing and siding										
	2600	80′ wide, 14′ eave height	E-2	740	.076	SF Flr.	4.95	2.94	1.93	9.82	12.85	
	2700	16′ eave height 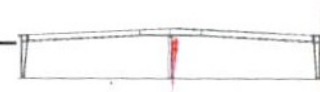		695	.081		5.25	3.13	2.05	10.43	13.70	
	2800	20′ eave height		625	.090		5.75	3.49	2.28	11.52	15.05	
	2900	24′ eave height		570	.098		6.20	3.82	2.50	12.52	16.40	
	3000	100′ wide, 14′ eave height		835	.067		4.80	2.61	1.71	9.12	11.85	
	3100	16′ eave height		795	.070		4.72	2.74	1.79	9.25	12.10	
	3200	20′ eave height		730	.077		5.50	2.98	1.95	10.43	13.55	
	3300	24′ eave height		670	.084		5.95	3.25	2.13	11.33	14.75	
	3400	120′ wide, 14′ eave height		870	.064		4.76	2.50	1.64	8.90	11.55	
	3500	16′ eave height		830	.067		4.96	2.62	1.72	9.30	12.05	
	3600	20′ eave height		765	.073		5.35	2.85	1.87	10.07	13.05	
	3700	24′ eave height		705	.079		5.85	3.09	2.02	10.96	14.25	
	3800	Double post 3-span frame, 26 ga. colored roofing and siding										
	3900	150′ wide, 14′ eave height	E-2	925	.061	SF Flr.	4.42	2.36	1.54	8.32	10.80	
	4000	16′ eave height 		890	.063		4.56	2.45	1.60	8.61	11.15	
	4100	20′ eave height		820	.068		4.84	2.66	1.74	9.24	12	
	4200	24′ eave height		765	.073		5.70	2.85	1.87	10.42	13.40	
	4300	Triple post 4-span frame, 26 ga. colored roofing and siding										
	4400	160′ wide, 14′ eave height	E-2	970	.058	SF Flr.	4.29	2.25	1.47	8.01	10.35	

13128 | Pre-Engineered Structures

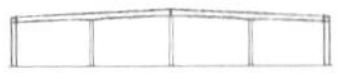

R133419-10

			CREW	DAILY OUTPUT	LABOR-HOURS	UNIT	2006 BARE COSTS MAT.	LABOR	EQUIP.	TOTAL	TOTAL INCL O&P	
700	4500	16′ eave height	E-2	930	.060	SF Flr.	4.49	2.34	1.53	8.36	10.85	700
	4600	20′ eave height		870	.064		4.74	2.50	1.64	8.88	11.50	
	4700	24′ eave height		815	.069		5.60	2.67	1.75	10.02	12.90	
	4800	200′ wide, 14′ eave height		1030	.054		4.32	2.12	1.39	7.83	10.05	
	4900	16′ eave height		995	.056		4.45	2.19	1.43	8.07	10.40	
	5000	20′ eave height		935	.060		4.83	2.33	1.53	8.69	11.15	
	5100	24′ eave height		885	.063		5.60	2.46	1.61	9.67	12.40	
	5200	Accessory items: add to the basic building cost above										
	5250	Eave overhang, 2′ wide, 26 ga., with soffit	E-2	360	.156	L.F.	21.50	6.05	3.96	31.51	38.50	
	5300	4′ wide, without soffit		300	.187		20.50	7.25	4.76	32.51	41	
	5350	With soffit		250	.224		29	8.70	5.70	43.40	53.50	
	5400	6′ wide, without soffit		250	.224		28	8.70	5.70	42.40	52.50	
	5450	With soffit		200	.280		36.50	10.90	7.15	54.55	67.50	
	5500	Entrance canopy, incl. frame, 4′ x 4′		25	2.240	Ea.	277	87	57	421	525	
	5550	4′ x 8′		19	2.947	"	375	115	75	565	705	
	5600	End wall roof overhang, 4′ wide, without soffit		850	.066	L.F.	13.80	2.56	1.68	18.04	21.50	
	5650	With soffit		500	.112	"	20.50	4.36	2.85	27.71	33.50	
	5700	Doors, H.M. self-framing, incl. butts, lockset and trim										
	5750	Single leaf, 3070 (3′ x 7′), economy	2 Sswk	5	3.200	Opng.	430	128		558	715	
	5800	Deluxe		4	4		505	160		665	855	
	5825	Glazed		4	4		515	160		675	870	
	5850	3670 (3′-6″ x 7′)		4	4		530	160		690	885	
	5900	4070 (4′ x 7′)		3	5.333		565	213		778	1,025	
	5950	Double leaf, 6070 (6′ x 7′)		2	8		765	320		1,085	1,450	
	6000	Glazed		2	8		935	320		1,255	1,625	
	6050	Framing only, for openings, 3′ x 7′	E-2	25	2.240		132	87	57	276	365	
	6100	10′ x 10′		21	2.667		435	104	68	607	740	
	6150	For windows below, 2020 (2′ x 2′)		25	2.240		140	87	57	284	375	
	6200	4030 (4′ x 3′)		22	2.545		171	99	65	335	440	
	6250	Flashings, 26 ga., corner or eave, painted	2 Sswk	240	.067	L.F.	3.37	2.66		6.03	8.70	
	6300	Galvanized		240	.067		2.76	2.66		5.42	8	
	6350	Rake flashing, painted		240	.067		3.61	2.66		6.27	8.95	
	6400	Galvanized		240	.067		3.03	2.66		5.69	8.30	
	6450	Ridge flashing, 18″ wide, painted		240	.067		4.47	2.66		7.13	9.90	
	6500	Galvanized		240	.067		3.94	2.66		6.60	9.30	
	6550	Gutter, eave type, 26 ga., painted		320	.050		4.67	2		6.67	8.90	
	6600	Galvanized		320	.050		2.67	2		4.67	6.70	
	6650	Valley type, between buildings, painted		120	.133		7.40	5.35		12.75	18.10	
	6700	Galvanized		120	.133		7.95	5.35		13.30	18.70	
	6750	Insulation, rated .6 lb density, vinyl faced										
	6800	1-1/2″ thick, R5	2 Carp	2300	.007	S.F.	.22	.25		.47	.65	
	6850	3″ thick, R10		2300	.007		.29	.25		.54	.73	
	6900	4″ thick, R13		2300	.007		.39	.25		.64	.84	
	6950	Foil/scrim/kraft (FSK) faced, 1-1/2″ thick, R5		2300	.007		.23	.25		.48	.66	
	7000	2″ thick, R6		2300	.007		.29	.25		.54	.73	
	7050	3″ thick, R10		2300	.007		.36	.25		.61	.81	
	7100	4″ thick, R13		2300	.007		.45	.25		.70	.91	
	7150	Met polyester/scrim/kraft (PSK) facing,1-1/2″ thk,R5		2300	.007		.38	.25		.63	.83	
	7200	2″ thick, R6		2300	.007		.43	.25		.68	.88	
	7250	3″ thick, R11		2300	.007		.46	.25		.71	.92	
	7300	4″ thick, R13		2300	.007		.54	.25		.79	1	
	7350	Vinyl/scrim/foil (VSF), 1-1/2″ thick, R5		2300	.007		.33	.25		.58	.77	
	7400	2″ thick, R6		2300	.007		.41	.25		.66	.86	
	7450	3″ thick, R10		2300	.007		.52	.25		.77	.98	
	7500	4″ thick, R13		2300	.007		.58	.25		.83	1.05	
	7650	Sash, single slide, glazed, with screens, 2020 (2′x 2′)	E-1	22	1.091	Opng.	81	42.50	4.06	127.56	169	

13 SPECIAL CONSTRUCTION

13120 | Pre-Engineered Structures

	13128 | Pre-Engineered Structures		CREW	DAILY OUTPUT	LABOR-HOURS	UNIT	2006 BARE COSTS MAT.	LABOR	EQUIP.	TOTAL	TOTAL INCL O&P	
700	7700	3030 (3′ x 3′) R133419-10	E-1	14	1.714	Opng.	182	67	6.40	255.40	325	700
	7750	4030 (4′ x 3′)		13	1.846		243	72	6.85	321.85	405	
	7800	6040 (6′ x 4′)		12	2		485	78	7.45	570.45	680	
	7850	Double slide sash, 3030 (3′ x 3′)		14	1.714		154	67	6.40	227.40	295	
	7900	6040 (6′ x 4′)		12	2		410	78	7.45	495.45	595	
	7950	Fixed glass, no screens, 3030 (3′ x 3′)		14	1.714		198	67	6.40	271.40	345	
	8000	6040 (6′ x 4′)		12	2		525	78	7.45	610.45	725	
	8050	Prefinished storm sash, 3030 (3′ x 3′)		70	.343		65	13.40	1.28	79.68	97	
	8100	Siding and roofing, see division 07300 & 07400										
	8200	Skylight, fiberglass panels, to 30 S.F.	E-1	10	2.400	Ea.	97	93.50	8.95	199.45	284	
	8250	Larger sizes, add for excess over 30 S.F.	"	300	.080	S.F.	3.23	3.12	.30	6.65	9.45	
	8300	Roof vents, circular with damper, birdscreen										
	8350	and operator hardware, painted										
	8400	26 ga., 12″ diameter	1 Sswk	4	2	Ea.	72	80		152	228	
	8450	20″ diameter		3	2.667		148	107		255	360	
	8500	24 ga., 24″ diameter		2	4		293	160		453	620	
	8550	Galvanized		2	4		245	160		405	570	
	8600	Continuous, 26 ga., 10′ long, 9″ wide	2 Sswk	4	4		20.50	160		180.50	320	
	8650	12″ wide	"	4	4		25	160		185	325	
800	0010	**SHELTERS**										800
	0020	Aluminum frame, acrylic glazing, 3′ x 9′ x 8′ high	2 Sswk	1.14	14.035	Ea.	2,550	560		3,110	3,850	
	0100	9′ x 12′ x 8′ high	"	.73	21.918	"	4,500	875		5,375	6,575	
840	0010	**SILOS** Concrete stave industrial, not incl. foundations, conical or										840
	0100	sloping bottoms, 12′ diameter, 35′ high	D-8	.11	363	Ea.	19,500	12,000		31,500	40,900	
	0200	16′ diameter, 45′ high		.08	500		26,000	16,500		42,500	55,500	
	0400	25′ diameter, 75′ high		.05	800		64,000	26,400		90,400	113,000	
	0500	Steel, factory fab., 30,000 gallon cap., painted, minimum	L-5	1	56		17,200	2,250	650	20,100	23,800	
	0700	Maximum		.50	112		27,400	4,475	1,300	33,175	39,800	
	0800	Epoxy lined, minimum		1	56		28,200	2,250	650	31,100	35,800	
	1000	Maximum		.50	112		35,700	4,475	1,300	41,475	48,900	
880	0010	**SWIMMING POOL ENCLOSURE** Translucent, free standing,										880
	0020	not including foundations, heat or light										
	0200	Economy, minimum	2 Carp	200	.080	SF Hor.	11.90	2.84		14.74	17.80	
	0300	Maximum		100	.160		35	5.70		40.70	48	
	0400	Deluxe, minimum		100	.160		41	5.70		46.70	55	
	0600	Maximum		70	.229		184	8.15		192.15	215	
	0700	For motorized roof, 40% opening, solid roof, add					6.80			6.80	7.45	
	0800	Skylight type roof, add					7.20			7.20	7.90	
	0900	Air-inflated, including blowers and heaters, minimum									3.50	
	1000	Maximum									6.70	

13150 | Swimming Pools

	13151 | Swimming Pools		CREW	DAILY OUTPUT	LABOR-HOURS	UNIT	2006 BARE COSTS MAT.	LABOR	EQUIP.	TOTAL	TOTAL INCL O&P	
200	0010	**SWIMMING POOLS** Residential in-ground, vinyl lined, concrete sides										200
	0020	Sides including equipment, sand bottom	B-52	300	.187	SF Surf	11.65	6.05	1.32	19.02	24.50	
	0100	Metal or polystyrene sides	B-14	410	.117		9.75	3.40	.55	13.70	16.95	
	0200	Add for vermiculite bottom					.75			.75	.83	

	13151 | Swimming Pools		CREW	DAILY OUTPUT	LABOR-HOURS	UNIT	2006 BARE COSTS MAT.	LABOR	EQUIP.	TOTAL	TOTAL INCL O&P	
200	0500	Gunite bottom and sides, white plaster finish										200
	0600	12′ x 30′ pool	B-52	145	.386	SF Surf	19.35	12.55	2.74	34.64	45	
	0720	16′ x 32′ pool		155	.361		17.40	11.75	2.56	31.71	41.50	
	0750	20′ x 40′ pool		250	.224		15.55	7.30	1.59	24.44	31	
	0810	Concrete bottom and sides, tile finish										
	0820	12′ x 30′ pool	B-52	80	.700	SF Surf	19.50	23	4.96	47.46	64.50	
	0830	16′ x 32′ pool		95	.589		16.15	19.20	4.18	39.53	54	
	0840	20′ x 40′ pool		130	.431		12.85	14	3.05	29.90	40.50	
	1100	Motel, gunite with plaster finish, incl. medium										
	1150	capacity filtration & chlorination	B-52	115	.487	SF Surf	24	15.85	3.45	43.30	56	
	1200	Municipal, gunite with plaster finish, incl. high										
	1250	capacity filtration & chlorination	B-52	100	.560	SF Surf	31	18.20	3.97	53.17	68.50	
	1350	Add for formed gutters				L.F.	53.50			53.50	59	
	1360	Add for stainless steel gutters				″	158			158	174	
	1600	For water heating system, see division 15510-880										
	1700	Filtration and deck equipment only, as % of total				Total				20%	20%	
	1800	Deck equipment, rule of thumb, 20′ x 40′ pool				SF Pool					1.30	
	1900	5000 S.F. pool				″					1.90	
	3000	Painting pools, preparation + 3 coats, 20′ x 40′ pool, epoxy	2 Pord	.33	48.485	Total	670	1,525		2,195	3,200	
	3100	Rubber base paint, 18 gallons	″	.33	48.485		505	1,525		2,030	3,025	
	3500	42′ x 82′ pool, 75 gallons, epoxy paint	3 Pord	.14	171		2,825	5,425		8,250	11,800	
	3600	Rubber base paint	″	.14	171		2,175	5,425		7,600	11,100	
700	0010	SWIMMING POOL EQUIPMENT										700
	0020	Diving stand, stainless steel, 3 meter	2 Carp	.40	40	Ea.	5,800	1,425		7,225	8,725	
	0300	1 meter		2.70	5.926		5,175	211		5,386	6,025	
	0600	Diving boards, 16′ long, aluminum		2.70	5.926		2,525	211		2,736	3,125	
	0700	Fiberglass		2.70	5.926		2,025	211		2,236	2,575	
	0900	Filter system, sand or diatomite type, incl. pump, 6,000 gal./hr.	2 Plum	1.80	8.889	Total	1,125	380		1,505	1,825	
	1020	Add for chlorination system, 800 S.F. pool		3	5.333	Ea.	203	228		431	580	
	1040	5,000 S.F. pool		3	5.333	″	1,125	228		1,353	1,600	
	1100	Gutter system, stainless steel, with grating, stock,										
	1110	contains supply and drainage system	E-1	20	1.200	L.F.	170	47	4.46	221.46	275	
	1120	Integral gutter and 5′ high wall system, stainless steel	″	10	2.400	″	254	93.50	8.95	356.45	455	
	1200	Ladders, heavy duty, stainless steel, 2 tread	2 Carp	7	2.286	Ea.	465	81.50		546.50	650	
	1500	4 tread		6	2.667		575	95		670	790	
	1800	Lifeguard chair, stainless steel, fixed		2.70	5.926		1,325	211		1,536	1,800	
	1900	Portable					1,000			1,000	1,100	
	2100	Lights, underwater, 12 volt, with transformer, 300 watt	1 Elec	.40	20		145	840		985	1,450	
	2200	110 volt, 500 watt, standard		.40	20		108	840		948	1,425	
	2400	Low water cutoff type		.40	20		135	840		975	1,450	
	2800	Heaters, see division 15510-880										
	3000	Pool covers, reinforced vinyl	3 Clab	1800	.013	S.F.	.30	.37		.67	.94	
	3050	Automatic, electric									9.65	
	3100	Vinyl water tube	3 Clab	3200	.008		.23	.21		.44	.59	
	3200	Maximum	″	3000	.008		.47	.22		.69	.88	
	3250	Sealed air bubble polyethylene solar blanket					.20			.20	.22	
	3300	Slides, tubular, fiberglass, aluminum handrails & ladder, 5′-0″, straight	2 Carp	1.60	10	Ea.	2,325	355		2,680	3,150	
	3320	8′-0″, curved		3	5.333		9,075	190		9,265	10,300	
	3400	10′-0″, curved		1	16		16,500	570		17,070	19,000	
	3420	12′-0″, straight with platform		1.20	13.333		1,725	475		2,200	2,675	
	4500	Hydraulic lift, movable pool bottom, single ram										
	4520	Under 1,000 S.F. area	L-9	.03	1200	Ea.	80,000	38,300		118,300	153,000	
	4600	Four ram lift, over 1,000 S.F.	″	.02	1800		96,500	57,500		154,000	203,500	
	5000	Removable access ramp, stainless steel	2 Clab	2	8		7,075	219		7,294	8,175	
	5500	Removable stairs, stainless steel, collapsible	″	2	8		2,600	219		2,819	3,225	

13170 | Tubs & Pools

	13171 | Therapeutic Pools		CREW	DAILY OUTPUT	LABOR-HOURS	UNIT	2006 BARE COSTS MAT.	LABOR	EQUIP.	TOTAL	TOTAL INCL O&P	
800	0010	**THERAPEUTIC POOLS** See division 15418										800

13175 | Ice Rinks

	13176 | Ice Rinks		CREW	DAILY OUTPUT	LABOR-HOURS	UNIT	2006 BARE COSTS MAT.	LABOR	EQUIP.	TOTAL	TOTAL INCL O&P	
500	0010	**ICE SKATING** Equipment incl. refrigeration, plumbing & cooling										500
	0020	coils & concrete slab, 85′ x 200′ rink										
	0300	55° system, 5 mos., 100 ton				Total	496,000			496,000	545,500	
	0700	90° system, 12 mos., 135 ton				″	550,500			550,500	605,500	
	1000	Dasher boards, 1/2″ H.D. polyethylene faced steel frame, 3′ acrylic										
	1020	screen at sides, 5′ acrylic ends, 85′ x 200′	F-5	.06	533	Ea.	115,000	19,200		134,200	158,500	
	1100	Fiberglass & aluminum construction, same sides and ends	″	.06	533		160,000	19,200		179,200	208,000	
	1200	Subsoil heating system (recycled from compressor), 85′ x 200′	Q-7	.27	118		25,000	4,875		29,875	35,100	
	1300	Subsoil insulation, 2 lb. polystyrene with vapor barrier, 85′ x 200′	2 Carp	.14	114	↓	30,000	4,075		34,075	39,700	

13200 | Storage Tanks

	13201 | Storage Tanks		CREW	DAILY OUTPUT	LABOR-HOURS	UNIT	2006 BARE COSTS MAT.	LABOR	EQUIP.	TOTAL	TOTAL INCL O&P	
200	0010	**ELEVATED STORAGE TANKS**, not incl pipe, pumps or foundation										200
	3000	Elevated water tanks, 100′ to bottom capacity line, incl. painting										
	3010	50,000 gallons				Ea.					193,764	
	3300	100,000 gallons									267,300	
	3400	250,000 gallons									368,851	
	3600	500,000 gallons									595,298	
	3700	750,000 gallons									820,577	
	3900	1,000,000 gallons				↓					824,180	
300	0010	**PRESTRESSED CONCRETE WATER STORAGE TANKS**										300
	0020	Not including pipe or pumps, 250,000 gallons				Ea.					284,200	
	0100	500,000 gallons									385,700	
	0300	1,000,000 gallons									548,100	
	0400	2,000,000 gallons									828,240	
	0600	4,000,000 gallons									1,309,350	
	0700	6,000,000 gallons									1,786,400	
	0750	8,000,000 gallons									2,273,600	
	0800	10,000,000 gallons									2,740,500	
	0910	Steel, ground level, ht/dia. less than 1, not incl. fdn., 100,000 gallons					137,000			137,000	150,500	
	1000	250,000 gallons									159,910	
	1200	500,000 gallons									243,955	
	1250	750,000 gallons									290,645	
	1300	1,000,000 gallons									444,360	
	1500	2,000,000 gallons									717,858	
	1600	4,000,000 gallons				↓					1,114,723	

13201 | Storage Tanks		CREW	DAILY OUTPUT	LABOR-HOURS	UNIT	2006 BARE COSTS MAT.	LABOR	EQUIP.	TOTAL	TOTAL INCL O&P
300	1800 6,000,000 gallons				Ea.					1,546,606
	1850 8,000,000 gallons									2,025,179
	1910 10,000,000 gallons				↓	2,745,000			2,745,000	3,020,000
	2100 Steel standpipes, hgt/dia. more than 1, 100' to overflow, no fdn.									
	2200 500,000 gallons				Ea.					320,993
	2400 750,000 gallons									391,029
	2500 1,000,000 gallons									476,238
	2700 1,500,000 gallons									665,333
	2800 2,000,000 gallons				↓					817,075
	3000 Steel, storage, above ground, including cradles, coating,									
	3020 fittings, not including fdn., pumps or piping									
	3040 Single wall, interior, 275 gallon	Q-5	5	3.200	Ea.	325	124		449	550
	3060 550 gallon	"	2.70	5.926		1,600	230		1,830	2,100
	3080 1,000 gallon	Q-7	5	6.400		2,500	263		2,763	3,150
	3100 1,500 gallon		4.75	6.737		3,750	276		4,026	4,550
	3120 2,000 gallon		4.60	6.957		4,300	285		4,585	5,200
	3140 5,000 gallon		3.20	10		5,825	410		6,235	7,050
	3150 10,000 gallon		2	16		17,200	655		17,855	19,900
	3160 15,000 gallon		1.70	18.824		18,600	770		19,370	21,600
	3170 20,000 gallon		1.45	22.069		23,200	905		24,105	27,000
	3180 25,000 gallon capacity		1.30	24.615		26,100	1,000		27,100	30,300
	3190 30,000 gallon	↓	1.10	29.091		32,100	1,200		33,300	37,200
	3320 Double wall, 500 gallon capacity	Q-5	2.40	6.667		2,550	258		2,808	3,200
	3330 2,000 gallon capacity	Q-7	4.15	7.711		6,250	315		6,565	7,375
	3340 4,000 gallon capacity		3.60	8.889		11,100	365		11,465	12,800
	3350 6,000 gallon capacity		2.40	13.333		13,100	545		13,645	15,400
	3360 8,000 gallon capacity		2	16		16,800	655		17,455	19,500
	3370 10,000 gallon capacity		1.80	17.778		18,600	730		19,330	21,500
	3380 15,000 gallon capacity		1.50	21.333		28,200	875		29,075	32,500
	3390 20,000 gallon capacity		1.30	24.615		32,100	1,000		33,100	37,000
	3400 25,000 gallon capacity		1.15	27.826		39,100	1,150		40,250	44,800
	3410 30,000 gallon capacity	↓	1	32	↓	42,800	1,325		44,125	49,200
	4000 Fixed roof oil storage tanks, steel, (1 BBL=42 GAL. w/ fdn. 3' D x 1' W)									
	4200 5,000 barrels				Ea.					173,000
	4300 24,000 barrels									296,000
	4500 56,000 barrels									588,500
	4600 110,000 barrels									855,000
	4800 143,000 barrels									1,096,500
	4900 224,000 barrels									1,645,500
	5100 Floating roof gasoline tanks, steel, 5,000 barrels									190,000
	5200 25,000 barrels									325,000
	5400 55,000 barrels									650,000
	5500 100,000 barrels									940,000
	5700 150,000 barrels									1,205,000
	5800 225,000 barrels									1,810,000
	6000 Wood tanks, ground level, 2" cypress, 3,000 gallons	C-1	.19	168		7,475	5,650		13,125	17,600
	6100 2-1/2" cypress, 10,000 gallons		.12	266		19,800	8,925		28,725	36,600
	6300 3" redwood or 3" fir, 20,000 gallons		.10	320		30,300	10,700		41,000	51,000
	6400 30,000 gallons		.08	400		37,400	13,400		50,800	63,500
	6600 45,000 gallons	↓	.07	457	↓	56,500	15,300		71,800	88,000
	6700 Larger sizes, minimum				Gal.					.66
	6900 Maximum				"					.81
	7000 Vinyl coated fabric pillow tanks, freestanding, 5,000 gallons	4 Clab	4	8	Ea.	3,975	219		4,194	4,750
	7100 Supporting embankment not included, 25,000 gallons	6 Clab	2	24		11,600	660		12,260	13,800
	7200 50,000 gallons	8 Clab	1.50	42.667		25,200	1,175		26,375	29,600
	7300 100,000 gallons	9 Clab	.90	80	↓	43,700	2,200		45,900	51,500

	13201 | Storage Tanks		CREW	DAILY OUTPUT	LABOR-HOURS	UNIT	2006 BARE COSTS MAT.	LABOR	EQUIP.	TOTAL	TOTAL INCL O&P	
300	7400	150,000 gallons	9 Clab	.50	144	Ea.	54,500	3,950		58,450	66,500	300
	7500	200,000 gallons		.40	180		54,500	4,925		59,425	68,000	
	7600	250,000 gallons		.30	240		72,500	6,575		79,075	91,000	
800	0010	UNDERGROUND STORAGE TANKS										800
	0210	Fiberglass, underground, single wall, U.L. listed, not including										
	0220	manway or hold-down strap										
	0225	550 gallon capacity	Q-5	2.67	5.993	Ea.	2,350	232		2,582	2,925	
	0230	1,000 gallon capacity	″	2.46	6.504		2,850	252		3,102	3,550	
	0240	2,000 gallon capacity	Q-7	4.57	7.002		4,575	287		4,862	5,475	
	0250	4,000 gallon capacity		3.55	9.014		6,500	370		6,870	7,725	
	0260	6,000 gallon capacity		2.67	11.985		7,200	490		7,690	8,700	
	0270	8,000 gallon capacity		2.29	13.974		7,450	575		8,025	9,100	
	0280	10,000 gallon capacity		2	16		9,700	655		10,355	11,700	
	0282	12,000 gallon capacity		1.88	17.021		9,475	700		10,175	11,500	
	0284	15,000 gallon capacity		1.68	19.048		15,200	780		15,980	17,900	
	0290	20,000 gallon capacity		1.45	22.069		19,400	905		20,305	22,700	
	0300	25,000 gallon capacity		1.28	25		25,000	1,025		26,025	29,100	
	0320	30,000 gallon capacity		1.14	28.070		30,100	1,150		31,250	34,900	
	0340	40,000 gallon capacity		.89	35.955		43,400	1,475		44,875	50,000	
	0360	48,000 gallon capacity		.81	39.506		50,500	1,625		52,125	58,000	
	0500	For manway, fittings and hold-downs, add					20%	15%				
	0600	For manways, add					1,250			1,250	1,375	
	1000	For helical heating coil, add	Q-5	2.50	6.400		2,675	248		2,923	3,300	
	1020	Fiberglass, underground, double wall, U.L. listed										
	1030	includes manways, not incl. hold-down straps										
	1040	600 gallon capacity	Q-5	2.42	6.612	Ea.	5,200	256		5,456	6,125	
	1050	1,000 gallon capacity	″	2.25	7.111		6,800	276		7,076	7,925	
	1060	2,500 gallon capacity	Q-7	4.16	7.692		9,950	315		10,265	11,400	
	1070	3,000 gallon capacity		3.90	8.205		10,700	335		11,035	12,300	
	1080	4,000 gallon capacity		3.64	8.791		12,700	360		13,060	14,500	
	1090	6,000 gallon capacity		2.42	13.223		14,100	545		14,645	16,300	
	1100	8,000 gallon capacity		2.08	15.385		16,100	630		16,730	18,700	
	1110	10,000 gallon capacity		1.82	17.582		18,100	720		18,820	21,000	
	1120	12,000 gallon capacity		1.70	18.824		22,300	770		23,070	25,800	
	1122	15,000 gallon capacity		1.52	21.053		25,900	865		26,765	29,900	
	1124	20,000 gallon capacity		1.33	24.060		31,500	985		32,485	36,200	
	1126	25,000 gallon capacity		1.16	27.586		42,300	1,125		43,425	48,400	
	1128	30,000 gallon capacity		1.03	31.068		51,000	1,275		52,275	58,000	
	1140	For hold-down straps, add					2%	10%				
	1150	For hold-downs 500-4000 gal, add	Q-7	16	2	Set	284	82		366	440	
	1160	For hold-downs 5000-15000 gal, add		8	4		565	164		729	880	
	1170	For hold-downs 20,000 gal, add		5.33	6.004		850	246		1,096	1,325	
	1180	For hold-downs 25,000 gal, add		4	8		1,125	330		1,455	1,750	
	1190	For hold-downs 30,000 gal, add		2.60	12.308		1,700	505		2,205	2,650	
	2210	Fiberglass, underground, single wall, U.L. listed, including										
	2220	hold-down straps, no manways										
	2225	550 gallon capacity	Q-5	2	8	Ea.	2,625	310		2,935	3,350	
	2230	1,000 gallon capacity	″	1.88	8.511		3,125	330		3,455	3,975	
	2240	2,000 gallon capacity	Q-7	3.55	9.014		4,850	370		5,220	5,925	
	2250	4,000 gallon capacity		2.90	11.034		6,775	455		7,230	8,150	
	2260	6,000 gallon capacity		2	16		7,775	655		8,430	9,575	
	2270	8,000 gallon capacity		1.78	17.978		8,025	740		8,765	9,975	
	2280	10,000 gallon capacity		1.60	20		10,300	820		11,120	12,600	
	2282	12,000 gallon capacity		1.52	21.053		10,000	865		10,865	12,500	
	2284	15,000 gallon capacity		1.39	23.022		15,800	945		16,745	18,800	

	13201	Storage Tanks	CREW	DAILY OUTPUT	LABOR-HOURS	UNIT	2006 BARE COSTS MAT.	LABOR	EQUIP.	TOTAL	TOTAL INCL O&P	
800	2290	20,000 gallon capacity	Q-7	1.14	28.070	Ea.	20,200	1,150		21,350	24,100	800
	2300	25,000 gallon capacity		.96	33.333		26,100	1,375		27,475	30,800	
	2320	30,000 gallon capacity		.80	40		31,800	1,650		33,450	37,600	
	3020	Fiberglass, underground, double wall, U.L. listed										
	3030	includes manways and hold-down straps										
	3040	600 gallon capacity	Q-5	1.86	8.602	Ea.	5,500	335		5,835	6,575	
	3050	1,000 gallon capacity	″	1.70	9.412		7,100	365		7,465	8,375	
	3060	2,500 gallon capacity	Q-7	3.29	9.726		10,200	400		10,600	11,900	
	3070	3,000 gallon capacity		3.13	10.224		11,000	420		11,420	12,800	
	3080	4,000 gallon capacity		2.93	10.922		13,000	450		13,450	15,000	
	3090	6,000 gallon capacity		1.86	17.204		14,300	705		15,005	16,900	
	3100	8,000 gallon capacity		1.65	19.394		16,200	795		16,995	19,000	
	3110	10,000 gallon capacity		1.48	21.622		18,700	885		19,585	22,000	
	3120	12,000 gallon capacity		1.40	22.857		22,900	940		23,840	26,700	
	3122	15,000 gallon capacity		1.28	25		26,400	1,025		27,425	30,700	
	3124	20,000 gallon capacity		1.06	30.189		32,400	1,250		33,650	37,500	
	3126	25,000 gallon capacity		.90	35.556		43,500	1,450		44,950	50,000	
	3128	30,000 gallon capacity		.74	43.243		52,500	1,775		54,275	60,500	
	5000	Steel underground, sti-P3, set in place, not incl. hold-down bars.										
	5500	Excavation, pad, pumps and piping not included										
	5510	Single wall, 500 gallon capacity, 7 gauge shell	Q-5	2.70	5.926	Ea.	1,000	230		1,230	1,450	
	5520	1,000 gallon capacity, 7 gauge shell	″	2.50	6.400		1,650	248		1,898	2,175	
	5530	2,000 gallon capacity, 1/4″ thick shell	Q-7	4.60	6.957		3,025	285		3,310	3,775	
	5535	2,500 gallon capacity, 7 gauge shell	Q-5	3	5.333		3,675	207		3,882	4,375	
	5540	5,000 gallon capacity, 1/4″ thick shell	Q-7	3.20	10		6,850	410		7,260	8,175	
	5560	10,000 gallon capacity, 1/4″ thick shell		2	16		11,100	655		11,755	13,300	
	5580	15,000 gallon capacity, 5/16″ thick shell		1.70	18.824		13,200	770		13,970	15,800	
	5600	20,000 gallon capacity, 5/16″ thick shell		1.50	21.333		17,800	875		18,675	20,900	
	5610	25,000 gallon capacity, 3/8″ thick shell		1.30	24.615		21,500	1,000		22,500	25,300	
	5620	30,000 gallon capacity, 3/8″ thick shell		1.10	29.091		28,600	1,200		29,800	33,400	
	5630	40,000 gallon capacity, 3/8″ thick shell		.90	35.556		37,400	1,450		38,850	43,400	
	5640	50,000 gallon capacity, 3/8″ thick shell		.80	40		46,800	1,650		48,450	54,000	
	6200	Steel, underground, 360°, double wall, U.L. listed,										
	6210	with sti-P3 corrosion protection,										
	6220	(dielectric coating, cathodic protection, electrical										
	6230	isolation) 30 year warranty,										
	6240	not incl. manholes or hold-downs.										
	6250	500 gallon capacity	Q-5	2.40	6.667	Ea.	3,025	258		3,283	3,725	
	6260	1,000 gallon capactiy	″	2.25	7.111		4,400	276		4,676	5,275	
	6270	2,000 gallon capacity	Q-7	4.16	7.692		5,800	315		6,115	6,875	
	6280	3,000 gallon capacity		3.90	8.205		6,400	335		6,735	7,575	
	6290	4,000 gallon capacity		3.64	8.791		8,150	360		8,510	9,525	
	6300	5,000 gallon capacity		2.91	10.997		10,700	450		11,150	12,400	
	6310	6,000 gallon capacity		2.42	13.223		12,700	545		13,245	14,700	
	6320	8,000 gallon capacity		2.08	15.385		13,200	630		13,830	15,500	
	6330	10,000 gallon capacity		1.82	17.582		14,900	720		15,620	17,500	
	6340	12,000 gallon capacity		1.70	18.824		18,000	770		18,770	20,900	
	6350	15,000 gallon capacity		1.33	24.060		19,900	985		20,885	23,300	
	6360	20,000 gallon capacity		1.33	24.060		23,900	985		24,885	27,800	
	6370	25,000 gallon capacity		1.16	27.586		40,100	1,125		41,225	45,900	
	6380	30,000 gallon capacity		1.03	31.068		48,300	1,275		49,575	55,000	
	6390	40,000 gallon capacity		.80	40		62,000	1,650		63,650	70,500	
	6395	50,000 gallon capacity		.73	43.836		86,000	1,800		87,800	97,500	
	6400	For hold-downs 500-2000 gal., add		16	2	Set	119	82		201	259	
	6410	For hold-downs 3000-6000 gal., add		12	2.667		241	109		350	435	
	6420	For hold-downs 8000-12,000 gal., add		11	2.909		289	119		408	505	

13200 | Storage Tanks

	13201 Storage Tanks		CREW	DAILY OUTPUT	LABOR-HOURS	UNIT	2006 BARE COSTS MAT.	LABOR	EQUIP.	TOTAL	TOTAL INCL O&P	
800	6430	For hold-downs 15,000 gal., add	Q-7	9	3.556	Set	415	146		561	685	800
	6440	For hold-downs 20,000 gal., add		8	4		480	164		644	780	
	6450	For hold-downs 20,000 gal., plus, add		6	5.333		625	219		844	1,025	
	6500	For manways, add				Ea.	825			825	910	
	6600	In place with hold-downs										
	6652	550 gallon capacity	Q-5	1.84	8.696	Ea.	3,150	335		3,485	3,975	

13280 | Hazardous Material Remediation

	13281 Hazardous Material Remediation		CREW	DAILY OUTPUT	LABOR-HOURS	UNIT	2006 BARE COSTS MAT.	LABOR	EQUIP.	TOTAL	TOTAL INCL O&P	
120	0010	BULK ASBESTOS REMOVAL										120
	0020	Includes disposable tools and 2 suits and 1respirator filter/day/worker										
	0100	Beams, W 10 x 19	A-9	235	.272	L.F.	.76	10.65		11.41	18.25	
	0110	W 12 x 22		210	.305		.85	11.90		12.75	20.50	
	0120	W 14 x 26		180	.356		.99	13.90		14.89	23.50	
	0130	W 16 x 31		160	.400		1.12	15.65		16.77	26.50	
	0140	W 18 x 40		140	.457		1.27	17.90		19.17	30.50	
	0150	W 24 x 55		110	.582		1.62	23		24.62	39	
	0160	W 30 x 108		85	.753		2.10	29.50		31.60	50.50	
	0170	W 36 x 150		72	.889		2.48	35		37.48	59	
	0200	Boiler insulation		480	.133	S.F.	.43	5.20		5.63	8.95	
	0210	With metal lath add				%				50%		
	0300	Boiler breeching or flue insulation	A-9	520	.123	S.F.	.34	4.81		5.15	8.25	
	0310	For active boiler, add				%				100%		
	0400	Duct or AHU insulation	A-10B	440	.073	S.F.	.20	2.85		3.05	4.87	
	0500	Duct vibration isolation joints, up to 24 Sq. In. duct	A-9	56	1.143	Ea.	3.19	44.50		47.69	76.50	
	0520	25 Sq. In. to 48 Sq. In. duct		48	1.333		3.72	52		55.72	89	
	0530	49 Sq. In. to 76 Sq. In. duct		40	1.600		4.46	62.50		66.96	107	
	0600	Pipe insulation, air cell type, up to 4″ diameter pipe		900	.071	L.F.	.20	2.78		2.98	4.76	
	0610	4″ to 8″ diameter pipe		800	.080		.22	3.13		3.35	5.35	
	0620	10″ to 12″ diameter pipe		700	.091		.25	3.58		3.83	6.15	
	0630	14″ to 16″ diameter pipe		550	.116		.32	4.55		4.87	7.75	
	0650	Over 16″ diameter pipe		650	.098	S.F.	.27	3.85		4.12	6.60	
	0700	With glove bag up to 3″ diameter pipe		300	.213	L.F.	3.15	8.35		11.50	17.05	
	1000	Pipe fitting insulation up to 4″ diameter pipe		320	.200	Ea.	.56	7.80		8.36	13.35	
	1100	6″ to 8″ diameter pipe		304	.211		.59	8.25		8.84	14.10	
	1110	10″ to 12″ diameter pipe		192	.333		.93	13.05		13.98	22.50	
	1120	14″ to 16″ diameter pipe		128	.500		1.39	19.55		20.94	33.50	
	1130	Over 16″ diameter pipe		176	.364	S.F.	1.01	14.20		15.21	24	
	1200	With glove bag, up to 8″ diameter pipe		100	.640	L.F.	6.55	25		31.55	48	
	2000	Scrape foam fireproofing from flat surface		2400	.027	S.F.	.07	1.04		1.11	1.78	
	2100	Irregular surfaces		1200	.053		.15	2.09		2.24	3.56	
	3000	Remove cementitious material from flat surface		1800	.036		.10	1.39		1.49	2.38	
	3100	Irregular surface		1400	.046		.13	1.79		1.92	3.06	
	4000	Scrape acoustical coating/fireproofing, from ceiling		3200	.020		.06	.78		.84	1.34	
	5000	Remove VAT from floor by hand		2400	.027		.07	1.04		1.11	1.78	
	5100	By machine	A-11	4800	.013		.04	.52	.01	.57	.90	
	5150	For 2 layers, add				%				50%		
	6000	Remove contaminated soil from crawl space by hand	A-9	400	.160	C.F.	.45	6.25		6.70	10.70	
	6100	With large production vacuum loader	A-12	700	.091	″	.25	3.58	.71	4.54	6.90	

13280 | Hazardous Material Remediation

	13281	Hazardous Material Remediation	CREW	DAILY OUTPUT	LABOR-HOURS	UNIT	2006 BARE COSTS MAT.	LABOR	EQUIP.	TOTAL	TOTAL INCL O&P	
120	7000	Radiator backing, not including radiator removal	A-9	1200	.053	S.F.	.15	2.09		2.24	3.56	120
	8000	Cement-asbestos transite board	2 Asbe	1000	.016		.12	.62		.74	1.16	
	8100	Transite shingle siding	A-10B	750	.043		.20	1.67		1.87	2.95	
	8200	Shingle roofing	"	2000	.016		.07	.63		.70	1.10	
	8250	Built-up, no gravel, non-friable	B-2	1400	.029		.07	.79		.86	1.40	
	8300	Asbestos millboard	2 Asbe	1000	.016		.08	.62		.70	1.11	
	9000	For type C (supplied air) respirator equipment, add				%					10%	
125	0010	**ASBESTOS ABATEMENT WORK AREA** Containment and preparation.										125
	0100	Pre-cleaning, HEPA vacuum and wet wipe, flat surfaces	A-10	12000	.005	S.F.	.01	.21		.22	.36	
	0200	Protect carpeted area, 2 layers 6 mil poly on 3/4" plywood	"	1000	.064		1.50	2.50		4	5.75	
	0300	Separation barrier, 2" x 4" @ 16", 1/2" plywood ea. side, 8' high	2 Carp	400	.040		1.25	1.42		2.67	3.74	
	0310	12' high		320	.050		1.40	1.78		3.18	4.49	
	0320	16' high		200	.080		1.50	2.84		4.34	6.35	
	0400	Personnel decontam. chamber, 2" x 4" @ 16", 3/4" ply ea. side		280	.057		2.50	2.03		4.53	6.10	
	0450	Waste decontam. chamber, 2" x 4" studs @ 16", 3/4" ply ea. side		360	.044		3	1.58		4.58	5.90	
	0500	Cover surfaces with polyethelene sheeting										
	0501	Including glue and tape										
	0550	Floors, each layer, 6 mil	A-10	8000	.008	S.F.	.11	.31		.42	.63	
	0551	4 mil		9000	.007		.06	.28		.34	.52	
	0560	Walls, each layer, 6 mil		6000	.011		.11	.42		.53	.80	
	0561	4 mil		7000	.009		.06	.36		.42	.65	
	0570	For heights above 12', add						20%				
	0575	For heights above 20', add						30%				
	0580	For fire retardant poly, add					100%					
	0590	For large open areas, deduct					10%	20%				
	0600	Seal floor penetrations with foam firestop to 36 Sq. In.	2 Carp	200	.080	Ea.	6.25	2.84		9.09	11.60	
	0610	36 Sq. In. to 72 Sq. In.		125	.128		12.50	4.55		17.05	21.50	
	0615	72 Sq. In. to 144 Sq. In.		80	.200		25	7.10		32.10	39.50	
	0620	Wall penetrations, to 36 square inches		180	.089		6.25	3.16		9.41	12.15	
	0630	36 Sq. In. to 72 Sq. In.		100	.160		12.50	5.70		18.20	23	
	0640	72 Sq. In. to 144 Sq. In.		60	.267		25	9.50		34.50	43	
	0800	Caulk seams with latex	1 Carp	230	.035	L.F.	.15	1.24		1.39	2.22	
	0900	Set up neg. air machine, 1-2k C.F.M. /25 M.C.F. volume	1 Asbe	4.30	1.860	Ea.		72.50		72.50	119	
130	0010	**DEMOLITION IN ASBESTOS CONTAMINATED AREA**										130
	0200	Ceiling, including suspension system, plaster and lath	A-9	2100	.030	S.F.	.08	1.19		1.27	2.03	
	0210	Finished plaster, leaving wire lath		585	.109		.30	4.28		4.58	7.35	
	0220	Suspended acoustical tile		3500	.018		.05	.72		.77	1.23	
	0230	Concealed tile grid system		3000	.021		.06	.83		.89	1.43	
	0240	Metal pan grid system		1500	.043		.12	1.67		1.79	2.85	
	0250	Gypsum board		2500	.026		.07	1		1.07	1.71	
	0260	Lighting fixtures up to 2' x 4'		72	.889	Ea.	2.48	35		37.48	59	
	0400	Partitions, non load bearing										
	0410	Plaster, lath, and studs	A-9	690	.093	S.F.	.81	3.63		4.44	6.80	
	0450	Gypsum board and studs	"	1390	.046	"	.13	1.80		1.93	3.08	
	9000	For type C (supplied air) respirator equipment, add				%					10%	
135	0010	**ASBESTOS ABATEMENT EQUIPMENT** and supplies, buy R028213-20										135
	0200	Air filtration device, 2000 C.F.M.				Ea.	2,500			2,500	2,750	
	0250	Large volume air sampling pump, minimum					365			365	400	
	0260	Maximum					680			680	745	
	0300	Airless sprayer unit, 2 gun					2,000			2,000	2,200	
	0350	Light stand, 500 watt					250			250	275	
	0400	Personal respirators										
	0410	Negative pressure, 1/2 face, dual operation, min.				Ea.	22.50			22.50	25	
	0420	Maximum					24			24	26.50	
	0450	P.A.P.R., full face, minimum					400			400	440	

13280 | Hazardous Material Remediation

	13281 | Hazardous Material Remediation		CREW	DAILY OUTPUT	LABOR-HOURS	UNIT	2006 BARE COSTS MAT.	LABOR	EQUIP.	TOTAL	TOTAL INCL O&P	
135	0460	Maximum (R028213-20)				Ea.	700			700	770	135
	0470	Supplied air, full face, incl. air line, minimum					450			450	495	
	0480	Maximum					600			600	660	
	0500	Personnel sampling pump, minimum					450			450	495	
	0510	Maximum					750			750	825	
	1500	Power panel, 20 unit, incl. G.F.I.					1,800			1,800	1,975	
	1600	Shower unit, including pump and filters					1,125			1,125	1,250	
	1700	Supplied air system (type C)					10,000			10,000	11,000	
	1750	Vacuum cleaner, HEPA, 16 gal., stainless steel, wet/dry					1,000			1,000	1,100	
	1760	55 gallon					2,200			2,200	2,425	
	1800	Vacuum loader, 9-18 ton/hr.					90,000			90,000	99,000	
	1900	Water atomizer unit, including 55 gal. drum					230			230	253	
	2000	Worker protection, whole body, foot, head cover & gloves, plastic					5.60			5.60	6.15	
	2500	Respirator, single use					10.30			10.30	11.35	
	2550	Cartridge for respirator					11.10			11.10	12.20	
	2570	Glove bag, 7 mil, 50″ x 64″					8.50			8.50	9.35	
	2580	10 mil, 44″ x 60″					8.40			8.40	9.25	
	3000	HEPA vacuum for work area, minimun					1,050			1,050	1,150	
	3050	Maximum					2,875			2,875	3,175	
	6000	Disposable polyethelene bags, 6 mil, 3 C.F.					1.15			1.15	1.27	
	6300	Disposable fiber drums, 3 C.F.					6.50			6.50	7.15	
	6400	Pressure sensitive caution lables, 3″ x 5″					1.10			1.10	1.21	
	6450	11″ x 17″					6.50			6.50	7.15	
	6500	Negative air machine, 1800 C.F.M.					775			775	855	
140	0010	**DECONTAMINATION CONTAINMENT AREA DEMOLITION** and clean-up										140
	0100	Spray exposed substrate with surfactant (bridging)										
	0200	Flat surfaces	A-9	6000	.011	S.F.	.36	.42		.78	1.08	
	0250	Irregular surfaces		4000	.016	″	.31	.63		.94	1.36	
	0300	Pipes, beams, and columns		2000	.032	L.F.	.56	1.25		1.81	2.66	
	1000	Spray encapsulate polyethelene sheeting		8000	.008	S.F.	.31	.31		.62	.85	
	1100	Roll down polyethelene sheeting		8000	.008	″		.31		.31	.51	
	1500	Bag polyethelene sheeting		400	.160	Ea.	.77	6.25		7.02	11.05	
	2000	Fine clean exposed substrate, with nylon brush		2400	.027	S.F.		1.04		1.04	1.70	
	2500	Wet wipe substrate		4800	.013			.52		.52	.85	
	2600	Vacuum surfaces, fine brush		6400	.010			.39		.39	.64	
	3000	Structural demolition										
	3100	Wood stud walls	A-9	2800	.023	S.F.		.89		.89	1.46	
	3500	Window manifolds, not incl. window replacement		4200	.015			.60		.60	.97	
	3600	Plywood carpet protection		2000	.032			1.25		1.25	2.04	
	4000	Remove custom decontamination facility	A-10A	8	3	Ea.	15	118		133	209	
	4100	Remove portable decontamination facility	3 Asbe	12	2	″	12.50	78		90.50	141	
	5000	HEPA vacuum, shampoo carpeting	A-9	4800	.013	S.F.	.05	.52		.57	.91	
	9000	Final cleaning of protected surfaces	A-10A	8000	.003	″		.12		.12	.19	
145	0010	**OSHA TESTING**										145
	0100	Certified technician, minimum				Day					300	
	0110	Maximum				″					500	
	0200	Personal sampling, PCM analysis, NIOSH 7400, minimum	1 Asbe	8	1	Ea.	2.75	39		41.75	66.50	
	0210	Maximum	″	4	2	″	3	78		81	130	
	0300	Industrial hygenist, minimum				Day					400	
	0310	Maximum				″					550	
	1000	Cleaned area samples	1 Asbe	8	1	Ea.	2.63	39		41.63	66.50	
	1100	PCM air sample analysis, NIOSH 7400, minimum		8	1		30	39		69	96.50	
	1110	Maximum		4	2		3.09	78		81.09	130	
	1200	TEM air sample analysis, NIOSH 7402, minimum								100	125	
	1210	Maximum								400	500	

	13281	Hazardous Material Remediation	CREW	DAILY OUTPUT	LABOR-HOURS	UNIT	2006 BARE COSTS MAT.	LABOR	EQUIP.	TOTAL	TOTAL INCL O&P	
150	0010	**ENCAPSULATION WITH SEALANTS**										150
	0100	Ceilings and walls, minimum	A-9	21000	.003	S.F.	.27	.12		.39	.49	
	0110	Maximum		10600	.006		.42	.24		.66	.85	
	0200	Columns and beams, minimum		13300	.005		.27	.19		.46	.61	
	0210	Maximum		5325	.012		.47	.47		.94	1.29	
	0300	Pipes to 12″ diameter including minor repairs, minimum		800	.080	L.F.	.37	3.13		3.50	5.50	
	0310	Maximum		400	.160	″	1.04	6.25		7.29	11.35	
155	0010	**WASTE PACKAGING, HANDLING, & DISPOSAL**										155
	0100	Collect and bag bulk material, 3 C.F. bags, by hand	A-9	400	.160	Ea.	1.15	6.25		7.40	11.45	
	0200	Large production vacuum loader	A-12	880	.073		.80	2.84	.57	4.21	6.15	
	1000	Double bag and decontaminate	A-9	960	.067		2.30	2.61		4.91	6.80	
	2000	Containerize bagged material in drums, per 3 C.F. drum	″	800	.080		6.50	3.13		9.63	12.25	
	3000	Cart bags 50′ to dumpster	2 Asbe	400	.040			1.56		1.56	2.55	
	5000	Disposal charges, not including haul, minimum				C.Y.					50	
	5020	Maximum				″					175	
	5100	Remove refrigerant from system	1 Plum	40	.200	Lb.		8.55		8.55	13.30	
	9000	For type C (supplied air) respirator equipment, add				%					10%	
440	0010	**REMOVAL** Existing lead paint, by chemicals, per application R028319-60										440
	0050	Baseboard, to 6″ wide	1 Pord	64	.125	L.F.	1.57	3.96		5.53	8.10	
	0070	To 12″ wide		32	.250	″	3.09	7.95		11.04	16.15	
	0200	Balustrades, one side		28	.286	S.F.	3.50	9.05		12.55	18.40	
	1400	Cabinets, simple design		32	.250		3.06	7.95		11.01	16.10	
	1420	Ornate design		25	.320		3.94	10.15		14.09	20.50	
	1600	Cornice, simple design		60	.133		1.65	4.23		5.88	8.60	
	1620	Ornate design		20	.400		4.86	12.70		17.56	26	
	2800	Doors, one side, flush		84	.095		1.19	3.02		4.21	6.15	
	2820	Two panel		80	.100		1.23	3.17		4.40	6.45	
	2840	Four panel		45	.178		2.17	5.65		7.82	11.45	
	2880	For trim, one side, add		64	.125	L.F.	1.57	3.96		5.53	8.10	
	3000	Fence, picket, one side		30	.267	S.F.	3.29	8.45		11.74	17.15	
	3200	Grilles, one side, simple design		30	.267		3.29	8.45		11.74	17.15	
	3220	Ornate design		25	.320		3.94	10.15		14.09	20.50	
	4400	Pipes, to 4″ diameter		90	.089	L.F.	1.13	2.82		3.95	5.75	
	4420	To 8″ diameter		50	.160		1.96	5.05		7.01	10.30	
	4440	To 12″ diameter		36	.222		2.75	7.05		9.80	14.35	
	4460	To 16″ diameter		20	.400		4.89	12.70		17.59	26	
	4500	For hangers, add		40	.200	Ea.	2.45	6.35		8.80	12.90	
	4800	Siding		90	.089	S.F.	1.13	2.82		3.95	5.75	
	5000	Trusses, open		55	.145	SF Face	1.79	4.61		6.40	9.35	
	6200	Windows, one side only, double hung, 1/1 light, 24″ x 48″ high		4	2	Ea.	24.50	63.50		88	129	
	6220	30″ x 60″ high		3	2.667		33	84.50		117.50	172	
	6240	36″ x 72″ high		2.50	3.200		39.50	101		140.50	207	
	6280	40″ x 80″ high		2	4		49.50	127		176.50	258	
	6400	Colonial window, 6/6 light, 24″ x 48″ high		2	4		49.50	127		176.50	259	
	6420	30″ x 60″ high		1.50	5.333		65.50	169		234.50	345	
	6440	36″ x 72″ high		1	8		98.50	254		352.50	515	
	6480	40″ x 80″ high		1	8		98.50	254		352.50	515	
	6600	8/8 light, 24″ x 48″ high		2	4		49.50	127		176.50	259	
	6620	40″ x 80″ high		1	8		98.50	254		352.50	515	
	6800	12/12 light, 24″ x 48″ high		1	8		98.50	254		352.50	515	
	6820	40″ x 80″ high		.75	10.667		131	340		471	690	
	6840	Window frame & trim items, included in pricing above										
	9000	Minimum labor/equipment charge	1 Pord	3	2.667	Job		84.50		84.50	136	
460	0010	**LEAD PAINT ENCAPSULATION**, water based polymer coating,14 mil DFT										460
	0020	Interior, brushwork, trim, under 6″	1 Pord	240	.033	L.F.	2.37	1.06		3.43	4.31	

	13281 | Hazardous Material Remediation		CREW	DAILY OUTPUT	LABOR-HOURS	UNIT	2006 BARE COSTS MAT.	LABOR	EQUIP.	TOTAL	TOTAL INCL O&P	
460	0030	6″ to 12″ wide	1 Pord	180	.044	L.F.	3.15	1.41		4.56	5.75	460
	0040	Balustrades		300	.027		1.90	.85		2.75	3.45	
	0050	Pipe to 4″ diameter		500	.016		1.14	.51		1.65	2.06	
	0060	To 8″ diameter		375	.021		1.51	.68		2.19	2.75	
	0070	To 12″ diameter		250	.032		2.27	1.01		3.28	4.13	
	0080	To 16″ diameter		170	.047		3.34	1.49		4.83	6.05	
	0090	Cabinets, ornate design		200	.040	S.F.	2.86	1.27		4.13	5.20	
	0100	Simple design		250	.032	"	2.27	1.01		3.28	4.13	
	0110	Doors, 3′x 7′, both sides, incl. frame & trim										
	0120	Flush	1 Pord	6	1.333	Ea.	29	42.50		71.50	100	
	0130	French, 10-15 lite		3	2.667		5.80	84.50		90.30	142	
	0140	Panel		4	2		35	63.50		98.50	141	
	0150	Louvered		2.75	2.909		32	92		124	183	
	0160	Windows, per interior side, per 15 S.F.										
	0170	1 to 6 lite	1 Pord	14	.571	Ea.	20	18.10		38.10	51	
	0180	7 to 10 lite		7.50	1.067		22	34		56	79	
	0190	12 lite		5.75	1.391		30	44		74	104	
	0200	Radiators		8	1		71	31.50		102.50	129	
	0210	Grilles, vents		275	.029	S.F.	2.07	.92		2.99	3.76	
	0220	Walls, roller, drywall or plaster		1000	.008		.57	.25		.82	1.04	
	0230	With spunbonded reinforcing fabric		720	.011		.65	.35		1	1.29	
	0240	Wood		800	.010		.71	.32		1.03	1.29	
	0250	Ceilings, roller, drywall or plaster		900	.009		.65	.28		.93	1.17	
	0260	Wood		700	.011		.81	.36		1.17	1.47	
	0270	Exterior, brushwork, gutters and downspouts		300	.027	L.F.	1.90	.85		2.75	3.45	
	0280	Columns		400	.020	S.F.	1.41	.63		2.04	2.57	
	0290	Spray, siding		600	.013	"	.95	.42		1.37	1.73	
	0300	Miscellaneous										
	0310	Electrical conduit, brushwork, to 2″ diameter	1 Pord	500	.016	L.F.	1.14	.51		1.65	2.06	
	0320	Brick, block or concrete, spray		500	.016	S.F.	1.14	.51		1.65	2.06	
	0330	Steel, flat surfaces and tanks to 12″		500	.016		1.14	.51		1.65	2.06	
	0340	Beams, brushwork		400	.020		1.41	.63		2.04	2.57	
	0350	Trusses		400	.020		1.41	.63		2.04	2.57	
600	0010	**MOLD ABATEMENT WORK AREA** Containment and prep.										600
	0100	Pre-cleaning, HEPA vacuum and wet wipe, flat surfaces	A-10	12000	.005	S.F.	.01	.21		.22	.36	
	0300	Separation barrier, 2″ x 4″ @ 16″, 1/2″ plywood ea. side, 8′ high	2 Carp	400	.040		1.25	1.42		2.67	3.74	
	0310	12′ high		320	.050		1.40	1.78		3.18	4.49	
	0320	16′ high		200	.080		1.50	2.84		4.34	6.35	
	0400	Personnel decontam. chamber, 2″ x 4″ @ 16″, 3/4″ ply ea. side		280	.057		2.50	2.03		4.53	6.10	
	0450	Waste decontam. chamber, 2″ x 4″ studs @ 16″, 3/4″ ply each side		360	.044		3	1.58		4.58	5.90	
	0500	Cover surfaces with polyethelene sheeting										
	0501	Including glue and tape										
	0550	Floors, each layer, 6 mil	A-10	8000	.008	S.F.	.11	.31		.42	.63	
	0551	4 mil		9000	.007		.06	.28		.34	.52	
	0560	Walls, each layer, 6 mil		6000	.011		.09	.42		.51	.78	
	0561	4 mil		7000	.009		.07	.36		.43	.66	
	0570	For heights above 12′, add						20%				
	0575	For heights above 20′, add						30%				
	0580	For fire retardant poly, add					100%					
	0590	For large open areas, deduct					10%	20%				
	0600	Seal floor penetrations with foam firestop to 36 sq. in.	2 Carp	200	.080	Ea.	6.25	2.84		9.09	11.60	
	0610	36 sq in to 72 sq. in.		125	.128		12.50	4.55		17.05	21.50	
	0615	72 sq in to 144 sq. in.		80	.200		25	7.10		32.10	39.50	
	0620	Wall penetrations, to 36 square inches		180	.089		6.25	3.16		9.41	12.15	
	0630	36 sq. in. to 72 sq. in.		100	.160		12.50	5.70		18.20	23	

13280 | Hazardous Material Remediation

13281 | Hazardous Material Remediation

			CREW	DAILY OUTPUT	LABOR-HOURS	UNIT	2006 BARE COSTS MAT.	LABOR	EQUIP.	TOTAL	TOTAL INCL O&P	
600	0640	72 sq. in. to 144 sq. in.	2 Carp	60	.267	Ea.	25	9.50		34.50	43	600
	0800	Caulk seams with latex	1 Carp	230	.035	L.F.	.15	1.24		1.39	2.22	
	0900	Set up neg. air machine, 1-2k C.F.M. /25 M.C.F. volume	1 Asbe	4.30	1.860	Ea.		72.50		72.50	119	
610	0010	DEMOLITION IN MOLD CONTAMINATED AREA										610
	0200	Ceiling, including suspension system, plaster and lath	A-9	2100	.030	S.F.	.08	1.19		1.27	2.03	
	0210	Finished plaster, leaving wire lath		585	.109		.30	4.28		4.58	7.35	
	0220	Suspended acoustical tile		3500	.018		.05	.72		.77	1.23	
	0230	Concealed tile grid system		3000	.021		.06	.83		.89	1.43	
	0240	Metal pan grid system		1500	.043		.12	1.67		1.79	2.85	
	0250	Gypsum board		2500	.026		.07	1		1.07	1.71	
	0255	Plywood		2500	.026		.07	1		1.07	1.71	
	0260	Lighting fixtures up to 2′ x 4′		72	.889	Ea.	2.48	35		37.48	59	
	0400	Partitions, non load bearing										
	0410	Plaster, lath, and studs	A-9	690	.093	S.F.	.81	3.63		4.44	6.80	
	0450	Gypsum board and studs		1390	.046		.13	1.80		1.93	3.08	
	0465	Carpet & pad		1390	.046		.13	1.80		1.93	3.08	
	0600	Pipe insulation, air cell type, up to 4″ diameter pipe		900	.071	L.F.	.20	2.78		2.98	4.76	
	0610	4″ to 8″ diameter pipe		800	.080		.22	3.13		3.35	5.35	
	0620	10″ to 12″ diameter pipe		700	.091		.25	3.58		3.83	6.15	
	0630	14″ to 16″ diameter pipe		550	.116		.32	4.55		4.87	7.75	
	0650	Over 16″ diameter pipe		650	.098	S.F.	.27	3.85		4.12	6.60	
	9000	For type C (supplied air) respirator equipment, add				%					10%	
750	0010	MOLD ABATEMENT METHODS										750
	0020	Initial inspection, average 3 bedroom home				Total					260	
	0030	Initial inspection, average 5 bedroom home									360	
	0040	Air sample test, each									260	
	0050	Swab sample test, each									155	
	0060	Tape sample test, each									155	
	0070	After remediation air test, each									260	
	0080	Mold abatement plan, average 3 bedroom home									1,275	
	0090	Mold abatement plan, average 5 bedroom home									1,550	
	0100	Packup & removal of contents, average 3 bedroom home, excl. storage									10,300	
	0110	Packup & removal of contents, average 5 bedroom home, excl. storage									25,800	
	0120	For demolition in mold contaminated areas, see Div. 13281-130										
	0130	For personal protection equipment, see Div. 13281-135										

13600 | Solar and Wind Energy Equipment

13630 | Solar Collector Components

			CREW	DAILY OUTPUT	LABOR-HOURS	UNIT	2006 BARE COSTS MAT.	LABOR	EQUIP.	TOTAL	TOTAL INCL O&P	
200	0010	SOLAR ENERGY										200
	0020	System/Package prices, not including connecting										
	0030	pipe, insulation, or special heating/plumbing fixtures										
	0150	For solar ultraviolet pipe insulation see Div. 15080-600										
	0500	Hot water, standard package, low temperature										
	0540	1 collector, circulator, fittings, 65 gal. tank	Q-1	.50	32	Ea.	1,625	1,225		2,850	3,675	
	0580	2 collectors, circulator, fittings, 120 gal. tank		.40	40		2,275	1,525		3,800	4,900	
	0620	3 collectors, circulator, fittings, 120 gal. tank		.34	47.059		2,950	1,800		4,750	6,050	

13630 | Solar Collector Components		CREW	DAILY OUTPUT	LABOR-HOURS	UNIT	2006 BARE COSTS MAT.	LABOR	EQUIP.	TOTAL	TOTAL INCL O&P
200 0700	Medium temperature package									
0720	1 collector, circulator, fittings, 80 gal. tank	Q-1	.50	32	Ea.	1,725	1,225		2,950	3,800
0740	2 collectors, circulator, fittings, 120 gal. tank		.40	40		2,525	1,525		4,050	5,175
0780	3 collectors, circulator, fittings, 120 gal. tank		.30	53.333		3,550	2,050		5,600	7,075
0980	For each additional 120 gal. tank, add					880			880	970
2250	Controller, liquid temperature	1 Plum	5	1.600		87.50	68.50		156	202
2300	Circulators, air									
2310	Blowers									
2330	100-300 S.F. system, 1/10 HP	Q-9	16	1	Ea.	189	38		227	269
2340	300-500 S.F. system, 1/5 HP		15	1.067		249	40.50		289.50	340
2350	Two speed, 100-300 S.F., 1/10 HP		14	1.143		136	43.50		179.50	219
2400	Reversible fan, 20″ diameter, 2 speed		18	.889		108	33.50		141.50	173
2520	Space & DHW system, less duct work		.50	32		1,475	1,225		2,700	3,550
2550	Booster fan 6″ diameter, 120 CFM		16	1		35	38		73	99
2570	6″ diameter, 225 CFM		16	1		43	38		81	108
2580	8″ diameter, 150 CFM		16	1		39.50	38		77.50	104
2590	8″ diameter, 310 CFM		14	1.143		60.50	43.50		104	136
2600	8″ diameter, 425 CFM		14	1.143		68	43.50		111.50	144
2650	Rheostat		32	.500		14.80	18.95		33.75	46.50
2660	Shutter/damper		12	1.333		39	50.50		89.50	124
2670	Shutter motor		16	1		92.50	38		130.50	163
2800	Circulators, liquid, 1/25 HP, 5.3 GPM	Q-1	14	1.143		147	44		191	231
2820	1/20 HP, 17 GPM		12	1.333		154	51		205	250
2850	1/20 HP, 17 GPM, stainless steel		12	1.333		243	51		294	345
2870	1/12 HP, 30 GPM		10	1.600		330	61.50		391.50	460
3000	Collector panels, air with aluminum absorber plate									
3010	Wall or roof mount									
3040	Flat black, plastic glazing									
3080	4′ x 8′	Q-9	6	2.667	Ea.	635	101		736	855
3100	4′ x 10′		5	3.200	″	780	121		901	1,050
3200	Flush roof mount, 10′ to 16′ x 22″ wide		96	.167	L.F.	55	6.30		61.30	70.50
3210	Manifold, by L.F. width of collectors		160	.100	″	27	3.79		30.79	35.50
3300	Collector panels, liquid with copper absorber plate									
3320	Black chrome, tempered glass glazing									
3330	Alum. frame, 4′ x 8′, 5/32″ single glazing	Q-1	9.50	1.684	Ea.	630	64.50		694.50	790
3390	Alum. frame, 4′ x 10′, 5/32″ single glazing		6	2.667		745	102		847	980
3450	Flat black, alum. frame, 3.5′ x 7.5′		9	1.778		535	68.50		603.50	695
3500	4′ x 8′		5.50	2.909		605	112		717	840
3520	4′ x 10′		10	1.600		720	61.50		781.50	885
3540	4′ x 12.5′		5	3.200		735	123		858	1,000
3550	Liquid with fin tube absorber plate									
3560	Alum. frame 4′ x 8′ tempered glass	Q-1	10	1.600	Ea.	495	61.50		556.50	640
3580	Liquid with vacuum tubes, 4′ x 6′-10″		9	1.778		880	68.50		948.50	1,075
3600	Liquid, full wetted, plastic, alum. frame, 3′ x 10′		5	3.200		195	123		318	405
3650	Collector panel mounting, flat roof or ground rack		7	2.286		48	88		136	190
3670	Roof clamps		70	.229	Set	2.35	8.80		11.15	16.25
3700	Roof strap, teflon	1 Plum	205	.039	L.F.	9.50	1.67		11.17	13.05
3900	Differential controller with two sensors									
3930	Thermostat, hard wired	1 Plum	8	1	Ea.	95	42.50		137.50	171
3950	Line cord and receptacle		12	.667		103	28.50		131.50	158
4050	Pool valve system		2.50	3.200		215	137		352	450
4070	With 12 VAC actuator		2	4		276	171		447	570
4080	Pool pump system, 2″ pipe size		6	1.333		172	57		229	279
4100	Five station with digital read-out		3	2.667		218	114		332	415
4150	Sensors									
4200	Brass plug, 1/2″ MPT	1 Plum	32	.250	Ea.	17.95	10.70		28.65	36.50

	13630 | Solar Collector Components	CREW	DAILY OUTPUT	LABOR-HOURS	UNIT	2006 BARE COSTS MAT.	LABOR	EQUIP.	TOTAL	TOTAL INCL O&P	
200 4210	Brass plug, reversed	1 Plum	32	.250	Ea.	24.50	10.70		35.20	43.50	200
4220	Freeze prevention		32	.250		22	10.70		32.70	40.50	
4240	Screw attached		32	.250		10.95	10.70		21.65	28.50	
4250	Brass, immersion		32	.250		26.50	10.70		37.20	45.50	
4300	Heat exchanger										
4330	Fluid to air coil, up flow, 45 MBH	Q-1	4	4	Ea.	295	154		449	565	
4380	70 MBH		3.50	4.571		335	176		511	640	
4400	80 MBH		3	5.333		445	205		650	810	
4580	Fluid to fluid package includes two circulating pumps										
4590	expansion tank, check valve, relief valve										
4600	controller, high temperature cutoff and sensors	Q-1	2.50	6.400	Ea.	695	246		941	1,150	
4650	Heat transfer fluid										
4700	Propylene glycol, inhibited anti-freeze	1 Plum	28	.286	Gal.	8.80	12.20		21	28.50	
4800	Solar storage tanks, knocked down										
4810	Air, galvanized steel clad, double wall, 4" fiberglass										
4820	insulation, 20 Mil PVC lining										
4870	4' high, 4' x 4', = 64 C.F./450 gallons	Q-9	2	8	Ea.	2,225	305		2,530	2,925	
4880	4' x 8' = 128 C.F./900 gallons		1.50	10.667		2,850	405		3,255	3,800	
4890	4' x 12' = 190 C.F./1300 gallons		1.30	12.308		3,825	465		4,290	4,950	
4900	8' x 8' = 250 C.F./1700 gallons		1	16		4,275	605		4,880	5,675	
5010	6'-3" high, 7' x 7' = 306 C.F./2000 gallons	Q-10	1.20	20		10,400	785		11,185	12,700	
5020	7' x 10'-6" = 459 C.F./3000 gallons		.80	30		13,100	1,175		14,275	16,300	
5030	7' x 14' = 613 C.F./4000 gallons		.60	40		15,300	1,575		16,875	19,400	
5040	10'-6" x 10'-6" = 689 C.F./4500 gallons		.50	48		15,800	1,900		17,700	20,400	
5050	10'-6" x 14' = 919 C.F./6000 gallons		.40	60		18,700	2,350		21,050	24,300	
5060	14' x 14' = 1225 C.F./8000 gallons	Q-11	.40	80		21,800	3,225		25,025	29,000	
5070	14' x 17'-6" = 1531 C.F./10,000 gallons		.30	106		24,500	4,275		28,775	33,800	
5080	17'-6" x 17'-6" = 1914 C.F./12,500 gallons		.25	128		28,900	5,150		34,050	39,900	
5090	17'-6" x 21' = 2297 C.F./15,000 gallons		.20	160		32,200	6,425		38,625	45,600	
5100	21' x 21' = 2756 C.F./18,000 gallons		.18	177		36,900	7,150		44,050	52,000	
5120	30 Mil reinforced Chemflex lining,										
5140	4' high, 4' x 4' = 64 C.F./450 gallons	Q-9	2	8	Ea.	2,250	305		2,555	2,950	
5150	4' x 8' = 128 C.F./900 gallons		1.50	10.667		3,400	405		3,805	4,375	
5160	4' x 12' = 190 C.F./1300 gallons		1.30	12.308		4,500	465		4,965	5,700	
5170	8' x 8' = 250 C.F./1700 gallons		1	16		5,025	605		5,630	6,500	
5190	6'-3" high, 7' x 7' = 306 C.F./2000 gallons	Q-10	1.20	20		11,100	785		11,885	13,500	
5200	7' x 10'-6" = 459 C.F./3000 gallons		.80	30		13,500	1,175		14,675	16,800	
5210	7' x 14' = 613 C.F./4000 gallons		.60	40		15,800	1,575		17,375	19,800	
5220	10'-6" x 10'-6" = 689 C.F./4500 gallons		.50	48		16,200	1,900		18,100	20,900	
5230	10'-6" x 14' = 919 C.F./6000 gallons		.40	60		19,100	2,350		21,450	24,800	
5240	14' x 14' = 1225 C.F./8000 gallons	Q-11	.40	80		22,400	3,225		25,625	29,700	
5250	14' x 17'-6" = 1531 C.F./10,000 gallons		.30	106		25,000	4,275		29,275	34,300	
5260	17'-6" x 17'-6" = 1914 C.F./12,500 gallons		.25	128		29,400	5,150		34,550	40,600	
5270	17'-6" x 21' = 2297 C.F./15,000 gallons		.20	160		33,100	6,425		39,525	46,600	
5280	21' x 21' = 2756 C.F./18,000 gallons		.18	177		38,000	7,150		45,150	53,000	
5290	30 Mil reinforced Hypalon lining, add					.02%					
7000	Solar control valves and vents										
7050	Air purger, 1" pipe size	1 Plum	12	.667	Ea.	43.50	28.50		72	92	
7070	Air eliminator, automatic 3/4" size		32	.250		28.50	10.70		39.20	48	
7090	Air vent, automatic, 1/8" fitting		32	.250		11.25	10.70		21.95	29	
7100	Manual, 1/8" NPT		32	.250		2.48	10.70		13.18	19.35	
7120	Backflow preventer, 1/2" pipe size		16	.500		58.50	21.50		80	97	
7130	3/4" pipe size		16	.500		62.50	21.50		84	102	
7150	Balancing valve, 3/4" pipe size		20	.400		26	17.10		43.10	55	
7180	Draindown valve, 1/2" copper tube		9	.889		202	38		240	281	
7200	Flow control valve, 1/2" pipe size		22	.364		58.50	15.55		74.05	88.50	

13600 | Solar and Wind Energy Equipment

	13630 | Solar Collector Components	Crew	Daily Output	Labor-Hours	Unit	2006 Bare Costs Mat.	Labor	Equip.	Total	Total Incl O&P	
200	7220 Expansion tank, up to 5 gal.	1 Plum	32	.250	Ea.	61.50	10.70		72.20	84	200
	7250 Hydronic controller (aquastat)		8	1		43.50	42.50		86	115	
	7400 Pressure gauge, 2″ dial		32	.250		22	10.70		32.70	41	
	7450 Relief valve, temp. and pressure 3/4″ pipe size		30	.267		9.10	11.40		20.50	27.50	
	7500 Solenoid valve, normally closed										
	7520 Brass, 3/4″ NPT, 24V	1 Plum	9	.889	Ea.	212	38		250	292	
	7530 1″ NPT, 24V		9	.889		212	38		250	292	
	7750 Vacuum relief valve, 3/4″ pipe size		32	.250		26.50	10.70		37.20	45.50	
	7800 Thermometers										
	7820 Digital temperature monitoring, 4 locations	1 Plum	2.50	3.200	Ea.	128	137		265	355	
	7900 Upright, 1/2″ NPT		8	1		24	42.50		66.50	93	
	7970 Remote probe, 2″ dial		8	1		30.50	42.50		73	100	
	7990 Stem, 2″ dial, 9″ stem		16	.500		20	21.50		41.50	55	
	8250 Water storage tank with heat exchanger and electric element										
	8270 66 gal. with 2″ x 2 lb. density insulation	1 Plum	1.60	5	Ea.	730	214		944	1,125	
	8300 80 gal. with 2″ x 2 lb. density insulation		1.60	5		835	214		1,049	1,250	
	8380 120 gal. with 2″ x 2 lb. density insulation		1.40	5.714		940	244		1,184	1,400	
	8400 120 gal. with 2″ x 2 lb. density insul., 40 S.F. heat coil		1.40	5.714		1,075	244		1,319	1,575	
	8500 Water storage module, plastic										
	8600 Tubular, 12″ diameter, 4′ high	1 Carp	48	.167	Ea.	91	5.95		96.95	110	
	8610 12″ diameter, 8′ high		40	.200		140	7.10		147.10	166	
	8620 18″ diameter, 5′ high		38	.211		153	7.50		160.50	180	
	8630 18″ diameter, 10′ high		32	.250		202	8.90		210.90	237	
	8640 58″ diameter, 5′ high	2 Carp	32	.500		490	17.80		507.80	565	
	8650 Cap, 12″ diameter					16			16	17.60	
	8660 18″ diameter					19			19	21	
	9000 Minimum labor/equipment charge	1 Plum	2	4	Job		171		171	266	

13700 | Security Access and Surveillance

	13710 | Security Access	Crew	Daily Output	Labor-Hours	Unit	2006 Bare Costs Mat.	Labor	Equip.	Total	Total Incl O&P	
300	0010 **ACCESS CONTROL**										300
	0020 Card type, 1 time zone, minimum				Ea.	320			320	350	
	0040 Maximum					1,075			1,075	1,175	
	0060 3 time zones, minimum					785			785	865	
	0080 Maximum					1,800			1,800	2,000	
	0100 System with printer, and control console, 3 zones				Total	8,800			8,800	9,700	
	0120 6 zones				″	11,600			11,600	12,800	
	0140 For each door, minimum, add				Ea.	1,300			1,300	1,425	
	0160 Maximum, add					1,925			1,925	2,125	
	0220 Hand geometry scanner, 512 user memory, excl striker/power	1 Elec	3	2.667		1,600	112		1,712	1,925	
	0230 Memory upgrade, add for 9,700 user profiles		8	1		225	42		267	315	
	0240 Add for 32,500 user profiles		8	1		450	42		492	560	
	0250 Prison type, 256 user profiles, excl strike, power		3	2.667		2,150	112		2,262	2,550	
	0260 Memory upgrade, add for 3,300 user profiles		8	1		160	42		202	241	
	0270 Add for 9,700 user profiles		8	1		350	42		392	450	
	0280 Add for 27,900 user profiles		8	1		450	42		492	560	
	0290 All weather, 512 user memory, excl. striker/power		3	2.667		3,225	112		3,337	3,725	
	0300 Facial & fingerprint scanner, combination unit, excl. striker/power		3	2.667		4,200	112		4,312	4,800	

13700 | Security Access and Surveillance

	13710 | Security Access		CREW	DAILY OUTPUT	LABOR-HOURS	UNIT	MAT. (2006 BARE COSTS)	LABOR (2006 BARE COSTS)	EQUIP. (2006 BARE COSTS)	TOTAL (2006 BARE COSTS)	TOTAL INCL O&P	
300	0310	Access for, for initial setup, excl. striker/power	1 Elec	3	2.667	Ea.	100	112		212	283	300
400	0010	**ACCESS CONTROL**										400
	0200	Video cameras, wireless, hidden in exit signs, clocks, etc., incl. receiver	1 Elec	3	2.667	Ea.	300	112		412	505	
	0210	Video camera accessories, single camera VCR		3	2.667		500	112		612	725	
	0220	multiple cameras VCR		3	2.667		1,500	112		1,612	1,825	
	0230	Video cameras, wireless, for under vehicle searching, complete		2	4		3,500	168		3,668	4,100	
	0240	Metal detector, hand-held, wand type unit									90	
	0250	Metal detector, walk through portal type, single zone	1 Elec	2	4		4,000	168		4,168	4,650	
	0260	Metal detector, walk through portal type, multi-zone		2	4		5,000	168		5,168	5,750	
	0270	Explosives detector, walk through portal type		2	4		3,500	168		3,668	4,100	
	0280	Explosives detector, hand-held, battery operated									28,000	
	0290	X-ray machine, desk top, for mail/small packages/letters	1 Elec	4	2		3,000	84		3,084	3,425	
	0300	X-ray machine, conveyor type, incl monitor, min		2	4		14,000	168		14,168	15,700	
	0310	X-ray machine, conveyor type, incl monitor, max		2	4		25,000	168		25,168	27,800	
	0320	X-ray machine, large unit, for airports, incl. monitor, min.	2 Elec	1	16		35,000	670		35,670	39,500	
	0330	X-ray machine, large unit, for airports, incl monitor, max	"	.50	32		60,000	1,350		61,350	68,000	

	13720 | Detection & Alarm		CREW	DAILY OUTPUT	LABOR-HOURS	UNIT	MAT.	LABOR	EQUIP.	TOTAL	TOTAL INCL O&P	
065	0010	**DETECTION SYSTEMS**, not including wires & conduits										065
	0100	Burglar alarm, battery operated, mechanical trigger	1 Elec	4	2	Ea.	254	84		338	410	
	0200	Electrical trigger		4	2		305	84		389	465	
	0400	For outside key control, add		8	1		72	42		114	144	
	0800	Card reader, flush type, standard		2.70	2.963		850	124		974	1,125	
	1000	Multi-code		2.70	2.963		1,100	124		1,224	1,400	
	1010	Card reader, proximity type		2.70	2.963		470	124		594	705	
	1200	Door switches, hinge switch		5.30	1.509		53.50	63.50		117	157	
	1400	Magnetic switch		5.30	1.509		63	63.50		126.50	167	
	1600	Exit control locks, horn alarm		4	2		315	84		399	480	
	1800	Flashing light alarm		4	2		355	84		439	525	
	2000	Indicating panels, 1 channel		2.70	2.963		335	124		459	560	
	2200	10 channel	2 Elec	3.20	5		1,150	210		1,360	1,575	
	2400	20 channel		2	8		2,250	335		2,585	3,000	
	2600	40 channel		1.14	14.035		4,075	590		4,665	5,400	
	2800	Ultrasonic motion detector, 12 volt	1 Elec	2.30	3.478		210	146		356	455	
	3000	Infrared photoelectric detector		2.30	3.478		173	146		319	415	
	3200	Passive infrared detector		2.30	3.478		259	146		405	510	
	3400	Glass break alarm switch		8	1		43.50	42		85.50	112	
	3420	Switchmats, 30″ x 5′		5.30	1.509		77.50	63.50		141	183	
	3440	30″ x 25′		4	2		186	84		270	335	
	3460	Police connect panel		4	2		223	84		307	375	
	3480	Telephone dialer		5.30	1.509		350	63.50		413.50	485	
	3500	Alarm bell		4	2		71	84		155	207	
	3520	Siren		4	2		134	84		218	276	
	3540	Microwave detector, 10′ to 200′		2	4		610	168		778	935	
	3560	10′ to 350′		2	4		1,775	168		1,943	2,225	
	3594	Fire, alarm control panel										
	3600	4 zone	2 Elec	2	8	Ea.	940	335		1,275	1,550	
	3800	8 zone		1	16		1,400	670		2,070	2,575	
	4000	12 zone		.67	23.988		1,825	1,000		2,825	3,550	
	4020	Alarm device	1 Elec	8	1		122	42		164	199	
	4050	Actuating device	"	8	1		292	42		334	385	
	4160	Alarm command center multiplex, addressable w/o voice large	2 Elec	2	8		2,125	335		2,460	2,875	
	4170	addressable w/ voice large	"	2	8		8,125	335		8,460	9,475	
	4175	Addressable interface device	1 Elec	7.25	1.103		205	46.50		251.50	298	

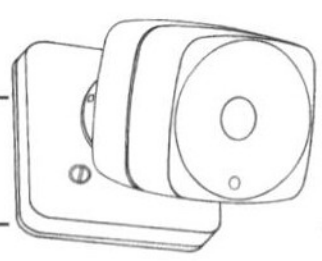

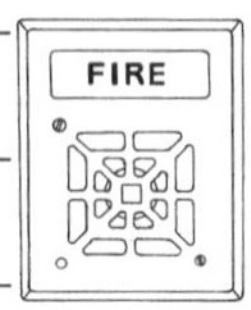

13700 | Security Access and Surveillance

	13720	Detection & Alarm	CREW	DAILY OUTPUT	LABOR-HOURS	UNIT	2006 BARE COSTS MAT.	LABOR	EQUIP.	TOTAL	TOTAL INCL O&P	
065	4200	Battery and rack	1 Elec	4	2	Ea.	690	84		774	890	065
	4400	Automatic charger		8	1		445	42		487	555	
	4600	Signal bell		8	1		49.50	42		91.50	119	
	4800	Trouble buzzer or manual station		8	1		37	42		79	105	
	5000	Detector, rate of rise		8	1		33.50	42		75.50	102	
	5010	Detector, heat (addressable type)		7.25	1.103		150	46.50		196.50	237	
	5100	Fixed temperature		8	1		28	42		70	95.50	
	5200	Smoke detector, ceiling type		6.20	1.290		75	54		129	166	
	5240	Smoke detector (addressable type)		6	1.333		150	56		206	252	
	5400	Duct type		3.20	2.500		250	105		355	435	
	5420	Duct (addressable type)		3.20	2.500		285	105		390	475	
	5600	Strobe and horn		5.30	1.509		95	63.50		158.50	203	
	5610	Strobe and horn (ADA type)		5.30	1.509		95	63.50		158.50	203	
	5620	Visual alarm (ADA type)		6.70	1.194		47.50	50		97.50	130	
	5800	Fire alarm horn		6.70	1.194		36.50	50		86.50	118	
	6000	Door holder, electro-magnetic		4	2		77.50	84		161.50	215	
	6200	Combination holder and closer		3.20	2.500		430	105		535	635	
	6400	Code transmitter		4	2		690	84		774	890	
	6600	Drill switch		8	1		86.50	42		128.50	160	
	6800	Master box		2.70	2.963		3,100	124		3,224	3,600	
	7000	Break glass station		8	1		50	42		92	120	
	7010	Break glass station, addressable		7.25	1.103		68	46.50		114.50	147	
	7800	Remote annunciator, 8 zone lamp	↓	1.80	4.444		175	187		362	480	
	8000	12 zone lamp	2 Elec	2.60	6.154		300	258		558	730	
	8200	16 zone lamp	"	2.20	7.273	↓	300	305		605	800	
	9410	Minimum labor/equipment charge	1 Elec	4	2	Job		84		84	129	

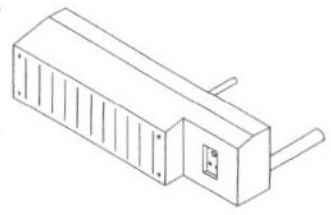

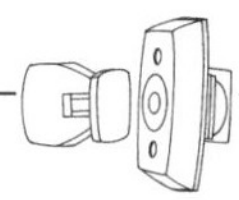

13800 | Building Automation & Control

	13832	Direct Digital Controls	CREW	DAILY OUTPUT	LABOR-HOURS	UNIT	2006 BARE COSTS MAT.	LABOR	EQUIP.	TOTAL	TOTAL INCL O&P	
200	0010	**CONTROL COMPONENTS / DDC SYSTEMS** (Sub's quote incl. M & L)										200
	0100	Analog inputs										
	0110	Sensors (avg. 50′ run in 1/2″ EMT)										
	0120	Duct temperature				Ea.					353.73	
	0130	Space temperature									634.64	
	0140	Duct humidity, +/- 3%									665.85	
	0150	Space humidity, +/- 2%									1,019.59	
	0160	Duct static pressure									541	
	0170	C.F.M./Transducer									728.28	
	0172	Water temp. (see 15120 for well tap add)									624.24	
	0174	Water flow (see 15120 for circuit sensor add)									2,288.88	
	0176	Water pressure differential (see 15120 for tap add)									936.36	
	0177	Steam flow (see 15120 for circuit sensor add)									2,288.88	
	0178	Steam pressure (see 15120 for tap add)									977.97	
	0180	K.W./Transducer									1,300.50	
	0182	K.W.H. totalization (not incl. elec. meter pulse xmtr.)									597.72	
	0190	Space static pressure				↓					1,020	
	1000	Analog outputs (avg. 50′ run in 1/2″ EMT)										
	1010	P/I Transducer				Ea.					603.43	
	1020	Analog output, matl. in MUX				↓					291.31	

13832 | Direct Digital Controls

			CREW	DAILY OUTPUT	LABOR-HOURS	UNIT	2006 BARE COSTS MAT.	LABOR	EQUIP.	TOTAL	TOTAL INCL O&P	
200	1030	Pneumatic (not incl. control device)				Ea.					613.83	200
	1040	Electric (not incl control device)									364.14	
	2000	Status (Alarms)										
	2100	Digital inputs (avg. 50′ run in 1/2″ EMT)										
	2110	Freeze				Ea.					416.16	
	2120	Fire									374.54	
	2130	Differential pressure, (air)									572.22	
	2140	Differential pressure, (water)									811.51	
	2150	Current sensor									416.16	
	2160	Duct high temperature thermostat									546.21	
	2170	Duct smoke detector									676.26	
	2200	Digital output (avg. 50′ run in 1/2″ EMT)										
	2210	Start/stop				Ea.					326.40	
	2220	On/off (maintained contact)				"					561	
	3000	Controller M.U.X. panel, incl. function boards										
	3100	48 point				Ea.					5,045.94	
	3110	128 point				"					6,918.66	
	3200	D.D.C. controller (avg. 50′ run in conduit)										
	3210	Mechanical room										
	3214	16 point controller (incl. 120v/1ph power supply)				Ea.					3,121.20	
	3229	32 point controller (incl. 120v/1ph power supply)				"					5,202	
	3230	Includes software programming and checkout										
	3260	Space										
	3266	V.A.V. terminal box (incl. space temp. sensor)				Ea.					806.31	
	3280	Host computer (avg. 50′ run in conduit)										
	3281	Package complete with PC, keyboard,										
	3282	printer, color CRT, modem, basic software				Ea.					9,363.60	
	4000	Front end costs										
	4100	Computer (P.C.)/software program				Ea.					6,242.40	
	4200	Color graphics software									3,745.44	
	4300	Color graphics slides									468.18	
	4350	Additional dot matrix printer									936.36	
	4400	Communications trunk cable				L.F.					3.64	
	4500	Engineering labor, (not incl. dftg.)				Point					79.07	
	4600	Calibration labor									79.07	
	4700	Start-up, checkout labor									119.64	
	4800	Drafting labor, as req'd.										
	5000	Communications bus (data transmission cable)										
	5010	#18 twisted shielded pair in 1/2″ EMT conduit				C.L.F.					364.14	
	8000	Applications software										
	8050	Basic maintenance manager software (not incl. data base entry)				Ea.					1,872.72	
	8100	Time program				Point					6.55	
	8120	Duty cycle									13.05	
	8140	Optimum start/stop									39.53	
	8160	Demand limiting									19.56	
	8180	Enthalpy program									39.53	
	8200	Boiler optimization				Ea.					1,170.45	
	8220	Chiller optimization				"					1,560.60	
	8240	Custom applications										
	8260	Cost varies with complexity										

13834 | Electric/Electronic Control

			CREW	DAILY OUTPUT	LABOR-HOURS	UNIT	MAT.	LABOR	EQUIP.	TOTAL	TOTAL INCL O&P	
200	0010	CONTROL SYSTEMS, ELECTRONIC										200
	0020	For electronic costs, add to division 13836-200				Ea.					15%	
	9000	Minimum labor/equipment charge	1 Plum	8	1	Job		42.50		42.50	66.50	

	13836 | Pneumatic Controls		CREW	DAILY OUTPUT	LABOR-HOURS	UNIT	2006 BARE COSTS MAT.	LABOR	EQUIP.	TOTAL	TOTAL INCL O&P	
200	0010	**CONTROL SYSTEMS, PNEUMATIC** (Sub's quote incl. mat. & labor)										200
	0011	and nominal 50 Ft. of tubing. Add control panelboard if req'd.										
	0100	Heating and Ventilating, split system										
	0200	Mixed air control, economizer cycle,panel readout,tubing										
	0220	Up to 10 tons	Q-19	.68	35.294	Ea.	2,975	1,400		4,375	5,450	
	0240	For 10 to 20 tons		.63	37.915		3,175	1,500		4,675	5,850	
	0260	For over 20 tons		.58	41.096		3,450	1,625		5,075	6,325	
	0270	Enthalpy cycle, up to 10 tons		.50	48.387		3,275	1,925		5,200	6,600	
	0280	For 10 to 20 tons		.46	52.174		3,550	2,075		5,625	7,125	
	0290	For over 20 tons		.42	56.604		3,850	2,250		6,100	7,725	
	0300	Heating coil, hot water, 3 way valve,										
	0320	Freezestat, limit control on discharge, readout	Q-5	.69	23.088	Ea.	2,200	895		3,095	3,825	
	0500	Cooling coil, chilled water, room										
	0520	Thermostat, 3 way valve	Q-5	2	8	Ea.	985	310		1,295	1,550	
	0600	Cooling tower, fan cycle, damper control,										
	0620	Control system including water readout in/out at panel	Q-19	.67	35.821	Ea.	3,900	1,425		5,325	6,500	
	1000	Unit ventilator, day/night operation,										
	1100	freezestat, ASHRAE, cycle 2	Q-19	.91	26.374	Ea.	2,175	1,050		3,225	4,000	
	2000	Compensated hot water from boiler, valve control,										
	2100	readout and reset at panel, up to 60 GPM	Q-19	.55	43.956	Ea.	4,050	1,750		5,800	7,175	
	2120	For 120 GPM		.51	47.059		4,325	1,875		6,200	7,675	
	2140	For 240 GPM		.49	49.180		4,525	1,950		6,475	8,025	
	3000	Boiler room combustion air, damper to 5 S.F., controls		1.37	17.582		1,950	700		2,650	3,225	
	3500	Fan coil, heating and cooling valves, 4 pipe control system		3	8		880	320		1,200	1,475	
	3600	Heat exchanger system controls		.86	27.907		1,900	1,100		3,000	3,825	
	3900	Multizone control (one per zone), includes thermostat, damper										
	3910	motor and reset of discharge temperature	Q-5	.51	31.373	Ea.	1,950	1,225		3,175	4,050	
	4000	Pneumatic thermostat, including controlling room radiator valve	"	2.43	6.593		590	255		845	1,050	
	4040	Program energy saving optimizer	Q-19	1.21	19.786		4,875	790		5,665	6,600	
	4060	Pump control system	"	3	8		905	320		1,225	1,500	
	4080	Reheat coil control system, not incl. coil	Q-5	2.43	6.593		765	255		1,020	1,225	
	4500	Air supply for pneumatic control system										
	4600	Tank mounted duplex compressor, starter, alternator,										
	4620	piping, dryer, PRV station and filter										
	4630	1/2 HP	Q-19	.68	35.139	Ea.	7,275	1,400		8,675	10,200	
	4640	3/4 HP		.64	37.383		7,625	1,500		9,125	10,700	
	4650	1 HP		.61	39.539		8,325	1,575		9,900	11,600	
	4660	1-1/2 HP		.58	41.739		8,875	1,650		10,525	12,300	
	4680	3 HP		.55	43.956		12,100	1,750		13,850	16,000	
	4690	5 HP		.42	57.143		21,100	2,275		23,375	26,700	
	4800	Main air supply, includes 3/8" copper main and labor	Q-5	1.82	8.791	C.L.F.	244	340		584	800	
	4810	If poly tubing used, deduct									30%	
	7000	Static pressure control for air handling unit, includes pressure										
	7010	sensor, receiver controller, readout and damper motors	Q-19	.64	37.383	Ea.	5,775	1,500		7,275	8,650	
	7020	If return air fan requires control, add									70%	
	8600	VAV boxes, incl. thermostat, damper motor, reheat coil & tubing	Q-5	1.46	10.989		905	425		1,330	1,650	
	8610	If no reheat coil, deduct									204	

	13838 | Pneumatic/Electric Controls		CREW	DAILY OUTPUT	LABOR-HOURS	UNIT	MAT.	LABOR	EQUIP.	TOTAL	TOTAL INCL O&P	
200	0010	**CONTROL COMPONENTS**										200
	0700	Controller, receiver										
	0730	Pneumatic, panel mount, single input	1 Plum	8	1	Ea.	300	42.50		342.50	395	
	0740	With conversion mounting bracket		8	1		310	42.50		352.50	410	
	0750	Dual input, with control point adjustment		7	1.143		415	49		464	530	
	0850	Electric, single snap switch	1 Elec	4	2		360	84		444	525	

SPECIAL CONSTRUCTION 13

	13838 | Pneumatic/Electric Controls	CREW	DAILY OUTPUT	LABOR-HOURS	UNIT	2006 BARE COSTS MAT.	LABOR	EQUIP.	TOTAL	TOTAL INCL O&P
200 0860	Dual snap switches	1 Elec	3	2.667	Ea.	450	112		562	670
1000	Enthalpy control, boiler water temperature control									
1010	governed by outdoor temperature, with timer	1 Elec	3	2.667	Ea.	224	112		336	420
2000	Gauges, pressure or vacuum									
2100	2" diameter dial	1 Stpi	32	.250	Ea.	12.35	10.75		23.10	30.50
2200	2-1/2" diameter dial		32	.250		13.50	10.75		24.25	31.50
2300	3-1/2" diameter dial		32	.250		20	10.75		30.75	39
2400	4-1/2" diameter dial		32	.250		30	10.75		40.75	50
2700	Flanged iron case, black ring									
2800	3-1/2" diameter dial	1 Stpi	32	.250	Ea.	72.50	10.75		83.25	96.50
2900	4-1/2" diameter dial		32	.250		86.50	10.75		97.25	112
3000	6" diameter dial		32	.250		68.50	10.75		79.25	92.50
3300	For compound pressure-vacuum, add					18%				
3350	Humidistat									
3390	Electric operated	1 Shee	8	1	Ea.	59.50	42		101.50	133
3400	Relays									
3430	Pneumatic/electric	1 Plum	16	.500	Ea.	207	21.50		228.50	260
3440	Pneumatic proportioning		8	1		112	42.50		154.50	190
3450	Pneumatic switching		12	.667		99.50	28.50		128	154
3460	Selector, 3 point		6	1.333		71.50	57		128.50	167
3470	Pneumatic time delay		8	1		264	42.50		306.50	355
3500	Sensor, air operated									
3520	Humidity	1 Plum	16	.500	Ea.	233	21.50		254.50	290
3540	Pressure		16	.500		48.50	21.50		70	86
3560	Temperature		12	.667		112	28.50		140.50	168
3600	Electric operated									
3620	Humidity	1 Elec	8	1	Ea.	54.50	42		96.50	124
3650	Pressure		8	1		935	42		977	1,100
3680	Temperature		10	.800		108	33.50		141.50	170
4000	Thermometers									
4100	Dial type, 3-1/2" diameter, vapor type, union connection	1 Stpi	32	.250	Ea.	96.50	10.75		107.25	123
4120	Liquid type, union connection		32	.250		135	10.75		145.75	165
4500	Stem type, 6-1/2" case, 2" stem, 1/2" NPT		32	.250		55	10.75		65.75	77.50
4520	4" stem, 1/2" NPT		32	.250		75.50	10.75		86.25	100
4600	9" case, 3-1/2" stem, 3/4" NPT		28	.286		90	12.30		102.30	118
4620	6" stem, 3/4" NPT		28	.286		96	12.30		108.30	124
4640	8" stem, 3/4" NPT		28	.286		118	12.30		130.30	149
4660	12" stem, 1" NPT		26	.308		130	13.25		143.25	164
5000	Thermostats									
5030	Manual	1 Shee	8	1	Ea.	27	42		69	96.50
5040	1 set back, electric, timed		8	1		81	42		123	156
5050	2 set back, electric, timed		8	1		178	42		220	263
5100	Locking cover					17.95			17.95	19.75
5200	24 hour, automatic, clock	1 Shee	8	1		106	42		148	184
5220	Electric, low voltage, 2 wire	1 Elec	13	.615		18	26		44	60
5230	3 wire	"	10	.800		22	33.50		55.50	76.50
5300	Transmitter, pneumatic									
5320	Temperature averaging element	Q-1	8	2	Ea.	107	77		184	238
5350	Pressure differential	1 Plum	7	1.143		625	49		674	765
5370	Humidity, duct		8	1		210	42.50		252.50	298
5380	Room		12	.667		191	28.50		219.50	255
5390	Temperature, with averaging element		6	1.333		120	57		177	221
5420	Electric operated, humidity	1 Elec	8	1		50.50	42		92.50	120
5430	DPST	"	8	1		71	42		113	143
6000	Valves, motorized zone									
6100	Sweat connections, 1/2" C x C	1 Stpi	20	.400	Ea.	136	17.20		153.20	176 200

13838 | Pneumatic/Electric Controls

			Daily	Labor-		2006 Bare Costs				Total	
	13838 Pneumatic/Electric Controls	Crew	Output	Hours	Unit	Mat.	Labor	Equip.	Total	Incl O&P	
200 6110	3/4" C x C	1 Stpi	20	.400	Ea.	145	17.20		162.20	187	200
6120	1" C x C		19	.421		170	18.15		188.15	215	
6140	1/2" C x C, with end switch, 2 wire		20	.400		143	17.20		160.20	185	
6150	3/4" C x C, with end switch, 2 wire		20	.400		153	17.20		170.20	195	
6160	1" C x C, with end switch, 2 wire		19	.421		173	18.15		191.15	219	
6180	1-1/4" C x C, w/end switch, 2 wire	↓	15	.533	↓	207	23		230	263	
7090	Valves, motor controlled, including actuator										
7100	Electric motor actuated										
7200	Brass, two way, screwed										
7210	1/2" pipe size	L-6	36	.333	Ea.	215	14.15		229.15	258	
7220	3/4" pipe size		30	.400		310	17		327	370	
7230	1" pipe size		28	.429		335	18.20		353.20	400	
7240	1-1/2" pipe size		19	.632		385	27		412	465	
7250	2" pipe size	↓	16	.750	↓	610	32		642	720	
7350	Brass, three way, screwed										
7360	1/2" pipe size	L-6	33	.364	Ea.	238	15.45		253.45	286	
7370	3/4" pipe size		27	.444		267	18.90		285.90	325	
7380	1" pipe size		25.50	.471		235	20		255	290	
7384	1-1/4" pipe size		21	.571		278	24.50		302.50	345	
7390	1-1/2" pipe size		17	.706		340	30		370	420	
7400	2" pipe size	↓	14	.857	↓	525	36.50		561.50	630	
7550	Iron body, two way, flanged										
7560	2-1/2" pipe size	L-6	4	3	Ea.	370	127		497	605	
7570	3" pipe size		3	4		490	170		660	805	
7580	4" pipe size	↓	2	6	↓	740	255		995	1,200	
7850	Iron body, three way, flanged										
7860	2-1/2" pipe size	L-6	3	4	Ea.	865	170		1,035	1,225	
7870	3" pipe size		2.50	4.800		1,050	204		1,254	1,475	
7880	4" pipe size	↓	2	6	↓	1,350	255		1,605	1,875	
8000	Pneumatic, air operated										
8050	Brass, two way, screwed										
8060	1/2" pipe size, class 250	1 Plum	24	.333	Ea.	113	14.25		127.25	146	
8070	3/4" pipe size, class 250		20	.400		149	17.10		166.10	191	
8080	1" pipe size, class 250		19	.421		170	18		188	215	
8090	1-1/4" pipe size, class 125		15	.533		243	23		266	305	
8100	1-1/2" pipe size, class 125		13	.615		282	26.50		308.50	350	
8110	2" pipe size, class 125	↓	11	.727	↓	320	31		351	400	
8180	Brass, three way, screwed										
8190	1/2" pipe size, class 250	1 Plum	22	.364	Ea.	124	15.55		139.55	160	
8200	3/4" pipe size, class 250		18	.444		150	19		169	195	
8210	1" pipe size, class 250		17	.471		172	20		192	222	
8214	1-1/4" pipe size, class 250		14	.571		236	24.50		260.50	298	
8220	1-1/2" pipe size, class 125		11	.727		276	31		307	355	
8230	2" pipe size, class 125	↓	9	.889	↓	345	38		383	440	
8450	Iron body, two way, flanged										
8460	2-1/2" pipe size, 250 lb. flanges	Q-1	5	3.200	Ea.	3,125	123		3,248	3,625	
8470	3" pipe size, 250 lb. flanges		4.50	3.556		3,325	137		3,462	3,850	
8480	4" pipe size, 250 lb. flanges	↓	3	5.333	↓	4,050	205		4,255	4,775	
8560	Iron body, three way, flanged										
8570	2-1/2" pipe size, class 125	Q-1	4.50	3.556	Ea.	765	137		902	1,050	
8580	3" pipe size, class 125		4	4		950	154		1,104	1,300	
8590	4" pipe size, class 125	↓	2.50	6.400		1,900	246		2,146	2,475	
8600	6" pipe size, class 125	Q-2	3	8	↓	2,600	320		2,920	3,350	
9000	Minimum labor/equipment charge	1 Plum	4	2	Job		85.50		85.50	133	

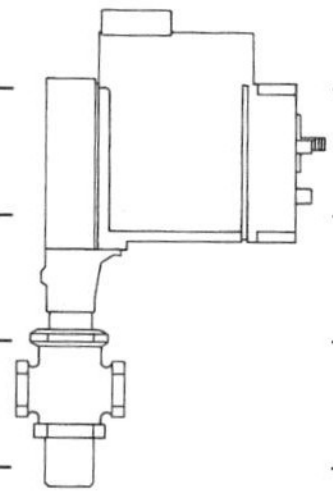

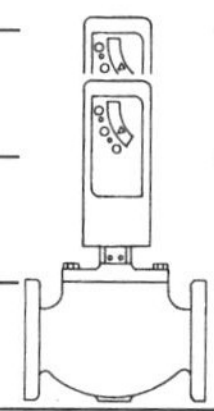

13851 | Detection & Alarm

		13851 Detection & Alarm	CREW	DAILY OUTPUT	LABOR-HOURS	UNIT	2006 BARE COSTS MAT.	LABOR	EQUIP.	TOTAL	TOTAL INCL O&P	
050	0010	**CLOCKS**										050
	0080	12" diameter, single face	1 Elec	8	1	Ea.	75.50	42		117.50	148	
	0100	Double face	"	6.20	1.290	"	144	54		198	242	
055	0010	**CLOCK SYSTEMS**, not including wires & conduits										055
	0100	Time system components, master controller	1 Elec	.33	24.242	Ea.	1,625	1,025		2,650	3,375	
	0200	Program bell		8	1		53.50	42		95.50	124	
	0400	Combination clock & speaker		3.20	2.500		189	105		294	370	
	0600	Frequency generator		2	4		6,350	168		6,518	7,225	
	0800	Job time automatic stamp recorder, minimum		4	2		445	84		529	620	
	1000	Maximum		4	2		680	84		764	880	
	1200	Time stamp for correspondence, hand operated					340			340	375	
	1400	Fully automatic					495			495	545	
	1600	Master time clock system, clocks & bells, 20 room	4 Elec	.20	160		4,075	6,725		10,800	14,900	
	1800	50 room	"	.08	400		9,400	16,800		26,200	36,200	
	2000	Time clock, 100 cards in & out, 1 color	1 Elec	3.20	2.500		950	105		1,055	1,200	
	2200	2 colors		3.20	2.500		1,025	105		1,130	1,275	
	2400	With 3 circuit program device, minimum		2	4		375	168		543	675	
	2600	Maximum		2	4		545	168		713	860	
	2800	Metal rack for 25 cards		7	1.143		60	48		108	140	
	3000	Watchman's tour station		8	1		63	42		105	134	
	3200	Annunciator with zone indication		1	8		230	335		565	775	
	3400	Time clock with tape		1	8		610	335		945	1,200	
350	0010	**TANK LEAK DETECTION SYSTEMS** Liquid and vapor										350
	0100	For hydrocarbons and hazardous liquids/vapors										
	0120	Controller, data acquisition, incl. printer, modem, RS232 port										
	0140	24 channel, for use with all probes				Ea.	2,150			2,150	2,350	
	0160	9 channel, for external monitoring				"	1,650			1,650	1,825	
	0200	Probes										
	0210	Well monitoring										
	0220	Liquid phase detection				Ea.	745			745	820	
	0230	Hydrocarbon vapor, fixed position					685			685	755	
	0240	Hydrocarbon vapor, float mounted					630			630	695	
	0250	Both liquid and vapor hydrocarbon					920			920	1,025	
	0300	Secondary containment, liquid phase										
	0310	Pipe trench/manway sump				Ea.	330			330	365	
	0320	Double wall pipe and manual sump					325			325	355	
	0330	Double wall fiberglass annular space					299			299	330	
	0340	Double wall steel tank annular space					299			299	330	
	0500	Accessories										
	0510	Modem, non-dedicated phone line				Ea.	268			268	295	
	0600	Monitoring, internal										
	0610	Automatic tank gauge, incl. overfill				Ea.	1,075			1,075	1,175	
	0620	Product line				"	1,150			1,150	1,275	
	0700	Monitoring, special										
	0710	Cathodic protection				Ea.	640			640	700	
	0720	Annular space chemical monitor				"	870			870	955	

13900 | Fire Suppression

13910 | Basic Fire Protection Matl/Methd		CREW	DAILY OUTPUT	LABOR-HOURS	UNIT	2006 BARE COSTS MAT.	LABOR	EQUIP.	TOTAL	TOTAL INCL O&P
400 0010	FIRE HOSE AND EQUIPMENT R211226-10									400
0200	Adapters, rough brass, straight hose threads									
0220	One piece, female to male, rocker lugs R211226-20									
0240	1″ x 1″				Ea.	30.50			30.50	33.50
0260	1-1/2″ x 1″					30			30	33
0280	1-1/2″ x 1-1/2″					8.80			8.80	9.70
0300	2″ x 1-1/2″					42			42	46
0320	2″ x 2″					53.50			53.50	59
0340	2-1/2″ x 1-1/2″					9.20			9.20	10.10
0360	2-1/2″ x 2″					24			24	26.50
0380	2-1/2″ x 2-1/2″					16.55			16.55	18.20
0400	3″ x 2-1/2″					79			79	87
0420	3″ x 3″					118			118	130
0500	For polished brass, add					50%				
0520	For polished chrome, add					75%				
0700	One piece, female to male, hexagon									
0740	1-1/2″ x 3/4″				Ea.	28.50			28.50	31
0760	2″ x 1-1/2″					53.50			53.50	59
0780	2-1/2″ x 1″					102			102	112
0800	2-1/2″ x 1-1/2″					35.50			35.50	39
0820	2-1/2″ x 2″					24			24	26.50
0840	3″ x 2-1/2″					44			44	48.50
0900	For polished chrome, add					75%				
1100	Swivel, female to female, pin lugs									
1120	1-1/2″ x 1-1/2″				Ea.	38.50			38.50	42.50
1200	2-1/2″ x 2-1/2″					75.50			75.50	83.50
1260	For polished brass, add					50%				
1280	For polished chrome, add					75%				
1400	Couplings, sngl & dbl jacket, pin lug or rocker lug, cast brass									
1410	1-1/2″				Ea.	33.50			33.50	37
1420	2-1/2″				″	43.50			43.50	48
1500	For polished brass, add					20%				
1520	For polished chrome, add					40%				
1580	Reducing, F x M, interior installation, cast brass									
1590	2″ x 1-1/2″				Ea.	42			42	46
1600	2-1/2″ x 1-1/2″					9.20			9.20	10.10
1680	For polished brass, add					50%				
1720	For polished chrome, add					75%				
1900	Escutcheon plate, for angle valves, polished brass, 1-1/2″					11.05			11.05	12.15
1920	2-1/2″					26			26	29
1940	3″					24			24	26.50
1980	For polished chrome, add					15%				
2200	Hose, less couplings									
2260	Synthetic jacket, lined, 300 lb. test, 1-1/2″ diameter	Q-12	2600	.006	L.F.	1.85	.23		2.08	2.40
2280	2-1/2″ diameter		2200	.007		3.07	.27		3.34	3.81
2360	High strength, 500 lb. test, 1-1/2″ diameter		2600	.006		1.90	.23		2.13	2.45
2380	2-1/2″ diameter		2200	.007		3.30	.27		3.57	4.06
2600	Hose rack, swinging, for 1-1/2″ diameter hose,									
2620	Enameled steel, 50′ & 75′ lengths of hose	Q-12	20	.800	Ea.	33	30		63	83.50
2640	100′ and 125′ lengths of hose		20	.800		33	30		63	83.50
2680	Chrome plated, 50′ and 75′ lengths of hose		20	.800		51	30		81	103
2700	100′ and 125′ lengths of hose		20	.800		51	30		81	103
2780	For hose rack nipple, 1-1/2″ polished brass, add					15.90			15.90	17.50
2820	2-1/2″ polished brass, add					26.50			26.50	29
2840	1-1/2″ polished chrome, add					16.50			16.50	18.15
2860	2-1/2″ polished chrome, add					29			29	31.50

13 SPECIAL CONSTRUCTION

13910 | Basic Fire Protection Matl/Methd

		Description	Crew	Daily Output	Labor-Hours	Unit	2006 Bare Costs Mat.	Labor	Equip.	Total	Total Incl O&P
400	2990	Hose reel, swinging, for 1-1/2" poly. neoprene lined hose (R211226-10)									
	3000	50' long	Q-12	14	1.143	Ea.	83	43		126	159
	3020	100' long (R211226-20)		14	1.143		116	43		159	195
	3060	For 2-1/2" cotton rubber hose, 75' long		14	1.143		135	43		178	215
	3100	150' long		14	1.143		157	43		200	240
	3750	Hydrants, wall, w/caps, single, flush, polished brass									
	3800	2-1/2" x 2-1/2"	Q-12	5	3.200	Ea.	135	120		255	335
	3840	2-1/2" x 3"		5	3.200		180	120		300	385
	3860	3" x 3"		4.80	3.333		221	125		346	440
	3900	For polished chrome, add					20%				
	3950	Double, flush, polished brass									
	4000	2-1/2" x 2-1/2" x 4"	Q-12	5	3.200	Ea.	360	120		480	590
	4040	2-1/2" x 2-1/2" x 6"		4.60	3.478		605	131		736	870
	4080	3" x 3" x 4"		4.90	3.265		775	123		898	1,050
	4120	3" x 3" x 6"		4.50	3.556		800	134		934	1,075
	4200	For polished chrome, add					10%				
	4350	Double, projecting, polished brass									
	4400	2-1/2" x 2-1/2" x 4"	Q-12	5	3.200	Ea.	148	120		268	350
	4450	2-1/2" x 2-1/2" x 6"	"	4.60	3.478	"	286	131		417	520
	4460	Valve control, dbl. flush/projecting hydrant, cap &									
	4470	chain, ext. rod & cplg., escutcheon, polished brass	Q-12	8	2	Ea.	192	75.50		267.50	330
	4480	Four-way square, flush, polished brass									
	4540	2-1/2"(4) x 6"	Q-12	3.60	4.444	Ea.	1,725	167		1,892	2,150
	5000	Nipples, straight hose to tapered iron pipe, brass									
	5060	Female to female, 1-1/2" x 1-1/2"				Ea.	13.45			13.45	14.80
	5080	2" x 2"					18			18	19.80
	5100	2-1/2" x 2-1/2"					21.50			21.50	24
	5200	Double male or male to female, 1" x 1"					27			27	29.50
	5220	1-1/2" x 1"					39			39	43
	5230	1-1/2" x 1-1/2"					8.80			8.80	9.70
	5260	2" x 1-1/2"					42.50			42.50	47
	5270	2" x 2"					53.50			53.50	59
	5280	2-1/2" x 1-1/2"					31			31	34
	5300	2-1/2" x 2"					25			25	27.50
	5310	2-1/2" x 2-1/2"					14.60			14.60	16.10
	5340	For polished chrome, add					75%				
	5600	Nozzles, brass									
	5620	Adjustable fog, 3/4" booster line				Ea.	70.50			70.50	77.50
	5630	1" booster line					87			87	95.50
	5640	1-1/2" leader line					91			91	100
	5660	2-1/2" direct connection					178			178	196
	5680	2-1/2" playpipe nozzle					147			147	162
	5780	For chrome plated, add					8%				
	5850	Electrical fire, adjustable fog, no shock									
	5900	1-1/2"				Ea.	268			268	295
	5920	2-1/2"					365			365	400
	5980	For polished chrome, add					6%				
	6200	Heavy duty, comb. adj. fog and str. stream, with handle									
	6210	1" booster line				Ea.	256			256	282
	6240	1-1/2"					253			253	279
	6260	2-1/2", for playpipe					310			310	340
	6280	2-1/2" direct connection					385			385	420
	6300	2-1/2" playpipe combination					520			520	570
	6480	For polished chrome, add					7%				
	6500	Plain fog, polished brass, 1-1/2"					48.50			48.50	53.50
400	6540	Chrome plated, 1-1/2"					52.50			52.50	58

	13910 | Basic Fire Protection Matl/Methd		CREW	DAILY OUTPUT	LABOR-HOURS	UNIT	2006 BARE COSTS MAT.	LABOR	EQUIP.	TOTAL	TOTAL INCL O&P	
400	6700	Plain stream, polished brass, 1-1/2" x 10" R211226-10				Ea.	30			30	33	400
	6760	2-1/2" x 15" x 7/8" or 1-1/2"				↓	55			55	60.50	
	6860	For polished chrome, add R211226-20					20%					
	7000	Underwriters playpipe, 2-1/2" x 30" with 1-1/8" tip				Ea.	182			182	201	
	7040	Less tip				"	144			144	159	
	7140	Standpipe connections, wall, w/plugs & chains										
	7160	Single, flush, brass, 2-1/2" x 2-1/2"	Q-12	5	3.200	Ea.	101	120		221	299	
	7180	2-1/2" x 3"	"	5	3.200	"	104	120		224	305	
	7240	For polished chrome, add					15%					
	7280	Double, flush, polished brass										
	7300	2-1/2" x 2-1/2" x 4"	Q-12	5	3.200	Ea.	330	120		450	550	
	7330	2-1/2" x 2-1/2" x 6"		4.60	3.478		460	131		591	710	
	7340	3" x 3" x 4"		4.90	3.265		600	123		723	850	
	7370	3" x 3" x 6"	↓	4.50	3.556	↓	720	134		854	1,000	
	7400	For polished chrome, add					15%					
	7440	For sill cock combination, add				Ea.	50			50	55.50	
	7580	Double projecting, polished brass										
	7600	2-1/2" x 2-1/2" x 4"	Q-12	5	3.200	Ea.	273	120		393	490	
	7630	2-1/2" x 2-1/2" x 6"	"	4.60	3.478	"	450	131		581	700	
	7680	For polished chrome, add					15%					
	7900	Three way, flush, polished brass										
	7920	2-1/2" (3) x 4"	Q-12	4.80	3.333	Ea.	930	125		1,055	1,225	
	7930	2-1/2" (3) x 6"	"	4.80	3.333		940	125		1,065	1,225	
	8000	For polished chrome, add				↓	9%					
	8020	Three way, projecting, polished brass										
	8040	2-1/2"(3) x 4"	Q-12	4.80	3.333	Ea.	620	125		745	875	
	8070	2-1/2" (3) x 6"	"	4.60	3.478		620	131		751	885	
	8100	For polished chrome, add				↓	12%					
	8200	Four way, square, flush, polished brass,										
	8240	2-1/2"(4) x 6"	Q-12	3.60	4.444	Ea.	1,625	167		1,792	2,050	
	8300	For polished chrome, add				"	10%					
	8550	Wall, vertical, flush, cast brass										
	8600	Two way, 2-1/2" x 2-1/2" x 4"	Q-12	5	3.200	Ea.	665	120		785	920	
	8660	Four way, 2-1/2" (4) x 6"		3.80	4.211		1,475	158		1,633	1,875	
	8680	Six way, 2-1/2" (6) x 6"	↓	3.40	4.706		1,975	177		2,152	2,450	
	8700	For polished chrome, add				↓	10%					
	8800	Sidewalk siamese unit, polished brass, two way										
	8820	2-1/2" x 2-1/2" x 4"	Q-12	2.50	6.400	Ea.	315	241		556	720	
	8850	2-1/2" x 2-1/2" x 6"		2	8		525	300		825	1,050	
	8860	3" x 3" x 4"		2.50	6.400		575	241		816	1,000	
	8890	3" x 3" x 6"	↓	2	8	↓	775	300		1,075	1,325	
	8940	For polished chrome, add					12%					
	9100	Sidewalk siamese unit, polished brass, three way										
	9120	2-1/2" x 2-1/2" x 2-1/2" x 6"	Q-12	2	8	Ea.	860	300		1,160	1,425	
	9160	For polished chrome, add					15%					
	9200	Storage house, hose only, primed steel					540			540	595	
	9220	Aluminum					1,175			1,175	1,300	
	9280	Hose and hydrant house, primed steel					730			730	800	
	9300	Aluminum					1,050			1,050	1,175	
	9340	Tools, crowbar and brackets	1 Carp	12	.667		74.50	23.50		98	121	
	9360	Combination hydrant wrench and spanner				↓	19.85			19.85	22	
	9380	Fire axe and brackets										
	9400	6 lb.	1 Carp	12	.667	Ea.	76.50	23.50		100	124	
	9900	Minimum labor/equipment charge	1 Plum	2	4	Job		171		171	266	
800	0010	**FIRE VALVES** R211226-10										800
	0020	Angle, combination pressure adjust/restricting, rough brass										

	13910 | Basic Fire Protection Matl/Methd	CREW	DAILY OUTPUT	LABOR-HOURS	UNIT	2006 BARE COSTS MAT.	LABOR	EQUIP.	TOTAL	TOTAL INCL O&P
800 0030	1-1/2" (R211226-20)	1 Spri	12	.667	Ea.	54	28		82	103
0040	2-1/2"	"	7	1.143	"	110	48		158	196
0042	Nonpressure adjustable/restricting, rough brass									
0044	1-1/2"	1 Spri	12	.667	Ea.	33.50	28		61.50	80.50
0046	2-1/2"	"	7	1.143	"	56.50	48		104.50	137
0050	For polished brass, add					30%				
0060	For polished chrome, add					40%				
0080	Wheel handle, 300 lb., 1-1/2"	1 Spri	12	.667	Ea.	30	28		58	76.50
0090	2-1/2"	"	7	1.143	"	51.50	48		99.50	131
0100	For polished brass, add					35%				
0110	For polished chrome, add					50%				
1000	Ball drip, automatic, rough brass, 1/2"	1 Spri	20	.400	Ea.	9.55	16.70		26.25	36.50
1010	3/4"	"	20	.400	"	11.05	16.70		27.75	38
1100	Ball, 175 lb., sprinkler system, FM/UL, threaded, bronze									
1120	Slow close									
1150	1" size	1 Spri	19	.421	Ea.	101	17.60		118.60	139
1160	1-1/4" size		15	.533		109	22.50		131.50	155
1170	1-1/2" size		13	.615		139	25.50		164.50	193
1180	2" size		11	.727		175	30.50		205.50	240
1190	2-1/2" size	Q-12	15	1.067		236	40		276	320
1230	For supervisory switch kit, all sizes									
1240	One circuit, add	1 Spri	48	.167	Ea.	76.50	6.95		83.45	95
1280	Quarter turn for trim									
1300	1/2" size	1 Spri	22	.364	Ea.	13.40	15.20		28.60	38.50
1310	3/4" size		20	.400		14.35	16.70		31.05	42
1320	1" size		19	.421		15.95	17.60		33.55	45
1330	1-1/4" size		15	.533		26	22.50		48.50	64
1340	1-1/2" size		13	.615		33	25.50		58.50	76
1350	2" size		11	.727		39	30.50		69.50	90.50
1400	Caps, polished brass with chain, 3/4"					27			27	29.50
1420	1"					34			34	37
1440	1-1/2"					8.80			8.80	9.70
1460	2-1/2"					13.95			13.95	15.35
1480	3"					21			21	23
3000	Gate, hose, wheel handle, N.R.S., rough brass, 1-1/2"	1 Spri	12	.667		112	28		140	167
3040	2-1/2", 300 lb.	"	7	1.143		149	48		197	238
3080	For polished brass, add					40%				
3090	For polished chrome, add					50%				
3800	Hydrant, screw type, crank handle, brass									
3840	2-1/2" size	Q-12	11	1.455	Ea.	199	54.50		253.50	305
3880	For chrome, same price									
4200	Hydrolator, vent and draining, rough brass, 1-1/2"	1 Spri	12	.667	Ea.	41.50	28		69.50	89.50
4280	For polished brass, add					50%				
4290	For polished chrome, add					90%				
5000	Pressure restricting, adjustable rough brass, 1-1/2"	1 Spri	12	.667		70.50	28		98.50	121
5020	2-1/2"	"	7	1.143		99	48		147	184
5080	For polished brass, add					30%				
5090	For polished chrome, add					45%				
6000	Roof manifold, horiz., brass, without valves & caps									
6040	2-1/2" x 2-1/2" x 4"	Q-12	4.80	3.333	Ea.	103	125		228	310
6060	2-1/2" x 2-1/2" x 6"		4.60	3.478		122	131		253	340
6080	2-1/2" x 2-1/2" x 2-1/2" x 4"		4.60	3.478		166	131		297	385
6090	2-1/2" x 2-1/2" x 2-1/2" x 6"		4.60	3.478		171	131		302	395
7000	Sprinkler line tester, cast brass					17.75			17.75	19.55
8000	Wye, leader line, ball type, swivel female x male x male									
8040	2-1/2" x 1-1/2" x 1-1/2" polished brass				Ea.	252			252	277

13900 | Fire Suppression

	13910 | Basic Fire Protection Matl/Methd		Crew	Daily Output	Labor-Hours	Unit	2006 Bare Costs Mat.	Labor	Equip.	Total	Total Incl O&P	
800	8060	2-1/2" x 1-1/2" x 1-1/2" polished chrome (R211226-10)				Ea.	259			259	285	800
	9990	Minimum labor/equipment charge	1 Spri	4	2	Job		83.50		83.50	130	

	13920 | Fire Pumps		Crew	Daily Output	Labor-Hours	Unit	Mat.	Labor	Equip.	Total	Total Incl O&P	
400	0010	**FIRE PUMPS** Including controller, fittings and relief valve										400
	0030	Diesel										
	0050	500 GPM, 50 psi, 27 HP, 4" pump	Q-13	.64	50	Ea.	53,500	2,000		55,500	62,000	
	0100	500 GPM, 100 psi, 62 HP, 4" pump		.60	53.333		56,500	2,125		58,625	66,000	
	0150	500 GPM, 125 psi, 78 HP, 4" pump		.56	57.143		57,500	2,275		59,775	66,500	
	0200	750 GPM, 50 psi, 44 HP, 5" pump		.60	53.333		57,000	2,125		59,125	66,500	
	0250	750 GPM, 100 psi, 80 HP, 4" pump		.56	57.143		59,500	2,275		61,775	69,000	
	0300	750 GPM, 165 psi, 203 HP, 5" pump		.52	61.538		62,000	2,450		64,450	72,000	
	0350	1000 GPM, 50 psi, 48 HP, 5" pump		.58	55.172		57,500	2,200		59,700	66,500	
	0400	1000 GPM, 100 psi, 89 HP, 4" pump		.56	57.143		61,500	2,275		63,775	71,000	
	0450	1000 GPM, 150 psi, 148 HP, 4" pump		.48	66.667		60,500	2,650		63,150	70,500	
	0470	1000 GPM, 200 psi, 280 HP, 5" pump		.40	80		76,000	3,175		79,175	88,500	
	0480	1250 GPM, 75 psi, 75 HP, 5" pump		.54	59.259		58,000	2,350		60,350	67,500	
	0500	1500 GPM, 50 psi, 66 HP, 6" pump		.50	64		62,000	2,550		64,550	72,000	
	0550	1500 GPM, 100 psi, 140 HP, 6" pump		.46	69.565		63,500	2,775		66,275	74,000	
	0600	1500 GPM, 150 psi, 228 HP, 6" pump		.42	76.190		71,500	3,025		74,525	83,000	
	0650	1500 GPM, 200 psi, 279 HP, 6" pump		.38	84.211		112,000	3,350		115,350	128,500	
	0700	2,000 GPM, 100 psi, 167 HP, 6" pump		.34	94.118		65,000	3,750		68,750	77,500	
	0750	2000 GPM, 150 psi, 284 HP, 6"pump		.30	106		88,000	4,250		92,250	103,000	
	0800	2500 GPM, 100 psi, 213 HP, 8" pump		.32	100		83,000	3,975		86,975	97,000	
	0820	2500 GPM, 150 psi, 365 HP, 8" pump		.26	123		97,000	4,900		101,900	114,500	
	0850	3000 GPM, 100 psi, 250 HP, 8" pump		.28	114		95,500	4,550		100,050	112,500	
	0900	3000 GPM, 150 psi, 384 HP, 10" pump		.20	160		111,500	6,375		117,875	132,500	
	0950	3500 GPM, 100 psi, 300 HP, 10" pump		.24	133		114,500	5,300		119,800	134,000	
	1000	3500 GPM, 150 psi, 518 HP, 10" pump	↓	.20	160	↓	161,500	6,375		167,875	187,500	
	3000	Electric										
	3100	250 GPM, 55 psi, 15 HP, 3,550 RPM, 2" pump	Q-13	.70	45.714	Ea.	12,600	1,825		14,425	16,700	
	3200	500 GPM, 50 psi, 27 HP, 1770 RPM, 4" pump		.68	47.059		15,600	1,875		17,475	20,000	
	3250	500 GPM, 100 psi, 47 HP, 3550 RPM, 3" pump		.66	48.485		18,900	1,925		20,825	23,800	
	3300	500 GPM, 125 psi, 64 HP, 3550 RPM, 3" pump		.62	51.613		20,700	2,050		22,750	26,000	
	3350	750 GPM, 50 psi, 44 HP, 1,770 RPM, 5" pump		.64	50		22,800	2,000		24,800	28,100	
	3400	750 GPM, 100 psi, 66 HP, 3550 RPM, 4" pump		.58	55.172		20,200	2,200		22,400	25,600	
	3450	750 GPM, 165 psi, 120 HP, 3550 RPM, 4" pump		.56	57.143		26,300	2,275		28,575	32,500	
	3500	1000 GPM, 50 psi, 48 HP 1770 RPM, 5" pump		.60	53.333		22,800	2,125		24,925	28,400	
	3550	1000 GPM, 100 psi, 86 HP, 3550 RPM, 5" pump		.54	59.259		25,500	2,350		27,850	31,800	
	3600	1000 GPM, 150 psi, 142 HP, 3550 RPM, 5" pump		.50	64		28,900	2,550		31,450	35,800	
	3650	1000 GPM, 200 psi, 245 HP, 1770 RPM, 6" pump		.36	88.889		62,500	3,550		66,050	74,500	
	3660	1250 GPM, 75 psi, 75 HP, 1770 RPM, 5" pump		.55	58.182		23,400	2,325		25,725	29,300	
	3700	1500 GPM, 50 psi, 66 HP, 1770 RPM, 6" pump		.50	64		24,500	2,550		27,050	31,000	
	3750	1500 GPM, 100 psi, 139 HP, 1770 RPM, 6" pump		.46	69.565		27,700	2,775		30,475	34,800	
	3800	1500 GPM, 150 psi, 200 HP, 1770 RPM, 6" pump		.36	88.889		59,000	3,550		62,550	70,000	
	3850	1500 GPM, 200 psi, 279 HP, 1770 RPM, 6" pump		.32	100		64,500	3,975		68,475	77,000	
	3900	2000 GPM, 100 psi, 167 HP, 1770 RPM, 6" pump		.34	94.118		35,500	3,750		39,250	44,900	
	3950	2000 GPM, 150 psi, 292 HP, 1770 RPM, 6" pump		.28	114		47,500	4,550		52,050	59,000	
	4000	2500 GPM, 100 psi, 213 HP, 1770 RPM, 8" pump		.30	106		38,800	4,250		43,050	49,300	
	4040	2500 GPM, 135 psi, 339 HP, 1770 RPM, 8" pump		.26	123		55,500	4,900		60,400	68,500	
	4100	3000 GPM, 100 psi, 250 HP, 1770 RPM, 8" pump		.28	114		46,500	4,550		51,050	58,000	
	4150	3000 GPM, 140 psi, 428 HP, 1770 RPM, 10" pump		.24	133		74,000	5,300		79,300	89,500	
	4200	3500 GPM, 100 psi, 300 HP, 1770 RPM, 10" pump		.26	123		57,500	4,900		62,400	71,000	
	4250	3500 GPM, 140 psi, 450 HP, 1770 RPM, 10" pump	↓	.24	133		74,000	5,300		79,300	89,500	
	5000	For jockey pump 1", 3 HP, with control, add	Q-12	2	8	↓	2,800	300		3,100	3,550	

13930 | Wet-Pipe Fire Supp. Sprinklers

		Description	Crew	Daily Output	Labor-Hours	Unit	2006 Bare Costs Mat.	2006 Bare Costs Labor	2006 Bare Costs Equip.	2006 Bare Costs Total	Total Incl O&P	
400	0010	SPRINKLER SYSTEM COMPONENTS (R211313-10)										400
	0600	Accelerator	1 Spri	8	1	Ea.	380	42		422	485	
	0800	Air compressor for dry pipe system, automatic, complete (R211313-20)										
	0820	280 gal. system capacity, 3/4 HP	1 Spri	1.30	6.154	Ea.	745	257		1,002	1,225	
	0860	520 gal. system capacity, 1 HP (R211313-30)		1.30	6.154		880	257		1,137	1,375	
	0910	650 gal. system capacity, 1-1/2 HP		1.30	6.154		900	257		1,157	1,400	
	0920	790 gal. system capacity, 2 HP		1.30	6.154		935	257		1,192	1,425	
	0960	Air pressure maintenance control		24	.333		157	13.95		170.95	195	
	1100	Alarm, electric pressure switch (circuit closer)		26	.308		148	12.85		160.85	183	
	1140	For explosion proof, max 20 PSI, contacts close or open		26	.308		296	12.85		308.85	345	
	1220	Water motor, complete with gong		4	2		188	83.50		271.50	335	
	1400	Deluge system, pressured monitoring panel, 120V		18	.444		885	18.60		903.60	1,000	
	1600	Dehydrator package, incl. valves and nipples	↓	12	.667	↓	325	28		353	400	
	1800	Firecycle system, controls, includes panel,										
	1820	batteries, solenoid valves and pressure switches	Q-13	1	32	Ea.	8,800	1,275		10,075	11,700	
	1980	Detector	1 Spri	16	.500		325	21		346	390	
	2000	Release, emergency, manual, for hydraulic or pneumatic system		12	.667		99	28		127	153	
	2060	Release, thermostatic, for hydraulic or pneumatic release line		20	.400		300	16.70		316.70	355	
	2200	Sprinkler cabinets, 6 head capacity		16	.500		35	21		56	71	
	2260	12 head capacity		16	.500		37	21		58	73.50	
	2340	Sprinkler head escutcheons, standard, brass tone, 1" size		40	.200		3.44	8.35		11.79	16.85	
	2360	Chrome, 1" size		40	.200		1.79	8.35		10.14	15	
	2400	Recessed type, brass tone		40	.200		4.44	8.35		12.79	17.95	
	2440	Chrome or white enamel	↓	40	.200	↓	2	8.35		10.35	15.25	
	2600	Sprinkler heads, not including supply piping										
	2640	Dry, pendent, 1/2" orifice, 3/4" or 1" NPT										
	2660	1/2" to 6" length	1 Spri	14	.571	Ea.	44.50	24		68.50	85.50	
	2670	6-1/4" to 9" length		14	.571		47	24		71	88.50	
	2680	9-1/4" to 12" length		14	.571		50	24		74	91.50	
	2690	12-1/4" to 15" length		14	.571		55.50	24		79.50	98	
	2700	15-1/4" to 18" length		14	.571		58.50	24		82.50	102	
	2710	18-1/4" to 21" length		13	.615		61.50	25.50		87	108	
	2720	21-1/4" to 24" length		13	.615		64.50	25.50		90	111	
	2730	24-1/4" to 27" length		13	.615		68	25.50		93.50	115	
	2740	27-1/4" to 30" length		13	.615		70.50	25.50		96	118	
	2750	30-1/4" to 33" length		13	.615		73	25.50		98.50	121	
	2760	33-1/4" to 36" length		13	.615		76.50	25.50		102	124	
	2780	36-1/4" to 39" length		12	.667		79	28		107	130	
	2790	39-1/4" to 42" length	↓	12	.667		81.50	28		109.50	133	
	2800	For each inch or fraction, add					1.03			1.03	1.13	
	3600	Foam-water, pendent or upright, 1/2" NPT	1 Spri	12	.667	↓	61	28		89	111	
	3700	Standard spray, pendent or upright, brass, 135° to 286° F										
	3720	1/2" NPT, 3/8" orifice	1 Spri	16	.500	Ea.	6.75	21		27.75	40	
	3730	1/2" NPT, 7/16" orifice		16	.500		6.75	21		27.75	40	
	3740	1/2" NPT, 1/2" orifice		16	.500		5.10	21		26.10	38	
	3760	1/2" NPT, 17/32" orifice		16	.500		5.85	21		26.85	39	
	3780	3/4" NPT, 17/32" orifice	↓	16	.500	↓	5.55	21		26.55	38.50	
	3800	For open sprinklers, deduct					15%					
	3840	For chrome, add				Ea.	1.39			1.39	1.53	
	3860	For wax and lead coating, add					6.25			6.25	6.85	
	3880	For wax coating, add					3.07			3.07	3.38	
	3900	For lead coating, add				↓	3.69			3.69	4.06	
	3920	For 360° F, same cost										
	3930	For 400° F, add				Ea.	6.25			6.25	6.85	
	3940	For 500° F, add				"	6.25			6.25	6.85	
	4200	Sidewall, vertical brass, 135° to 286° F										

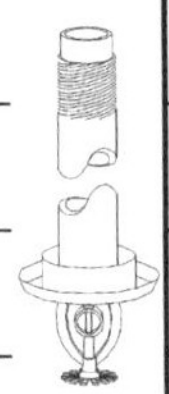

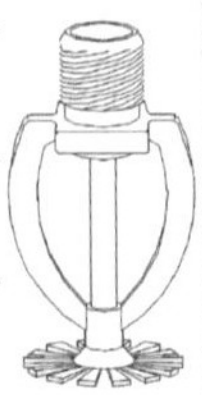

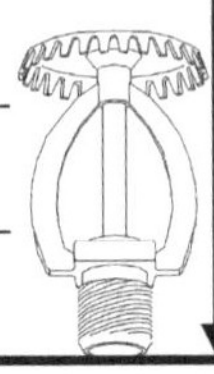

13 SPECIAL CONSTRUCTION

13900 | Fire Suppression

13930 | Wet-Pipe Fire Supp. Sprinklers

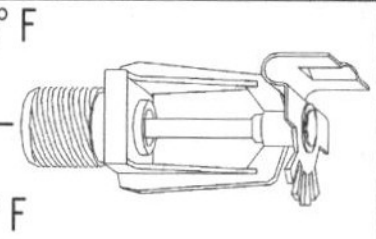

	13930 | Wet-Pipe Fire Supp. Sprinklers		CREW	DAILY OUTPUT	LABOR-HOURS	UNIT	2006 BARE COSTS MAT.	LABOR	EQUIP.	TOTAL	TOTAL INCL O&P	
400	4240	1/2″ NPT, 1/2″ orifice	R211313-10	1 Spri	16	.500	Ea.	8.50	21		29.50	42	400
	4280	3/4″ NPT, 17/32″ orifice		″	16	.500		9.05	21		30.05	42.50	
	4360	For satin chrome, add	R211313-20					1.23			1.23	1.35	
	4400	For 360° F, same cost											
	4410	For 400° F, add	R211313-30				Ea.	6.25			6.25	6.85	
	4420	For 500° F, add					″	8.50			8.50	9.35	
	4500	Sidewall, horizontal, brass, 135° to 286° F											
	4520	1/2″ NPT, 1/2″ orifice		1 Spri	16	.500	Ea.	7.35	21		28.35	40.50	
	4540	For 360° F, same cost											
	4800	Recessed pendent, brass, 135° to 286° F											
	4820	1/2″ NPT, 3/8″ orifice		1 Spri	10	.800	Ea.	17.55	33.50		51.05	71.50	
	4830	1/2″ NPT, 7/16″ orifice			10	.800		12.45	33.50		45.95	65.50	
	4840	1/2″ NPT, 1/2″ orifice			10	.800		14.75	33.50		48.25	68	
	4860	1/2″ NPT, 17/32″ orifice			10	.800		11.90	33.50		45.40	65	
	4900	For satin chrome, add						.92			.92	1.01	
	5000	Recessed-vertical sidewall, brass, 135° to 286° F											
	5020	1/2″ NPT, 3/8″ orifice		1 Spri	10	.800	Ea.	13.60	33.50		47.10	67	
	5030	1/2″ NPT, 7/16″ orifice			10	.800		13.05	33.50		46.55	66.50	
	5040	1/2″ NPT, 1/2″ orifice			10	.800		12.45	33.50		45.95	65.50	
	5100	For bright nickel, same cost											
	5600	Concealed, complete with cover plate											
	5620	1/2″ NPT, 1/2″ orifice, 135° to 212° F		1 Spri	9	.889	Ea.	17	37		54	76.50	
	5800	Window, brass, 1/2″ NPT, 1/4″ orifice			16	.500		15.60	21		36.60	49.50	
	5810	1/2″ NPT, 5/16″ orifice			16	.500		15.60	21		36.60	49.50	
	5820	1/2″ NPT, 3/8″ orifice			16	.500		15.60	21		36.60	49.50	
	5830	1/2″ NPT, 7/16″ orifice			16	.500		15.60	21		36.60	49.50	
	5840	1/2″ NPT, 1/2″ orifice			16	.500		16.60	21		37.60	51	
	5860	For polished chrome, add						2.10			2.10	2.31	
	5880	3/4″ NPT, 5/8″ orifice		1 Spri	16	.500		17.30	21		38.30	51.50	
	5890	3/4 NPT, 3/4″ orifice		″	16	.500		17.30	21		38.30	51.50	
	6000	Sprinkler head guards, bright zinc, 1/2″ NPT						2.80			2.80	3.08	
	6020	Bright zinc, 3/4″ NPT						2.80			2.80	3.08	
	6100	Sprinkler head wrenches, standard head						12.75			12.75	14.05	
	6120	Recessed head						19.85			19.85	22	
	6160	Tamper switch, (valve supervisory switch)		1 Spri	16	.500		80	21		101	121	
	6200	Valves											
	6210	Alarm, includes											
	6220	retard chamber, trim, gauges, alarm line strainer											
	6260	3″ size		Q-12	3	5.333	Ea.	850	201		1,051	1,250	
	6280	4″ size		″	2	8		860	300		1,160	1,425	
	6300	6″ size		Q-13	4	8		955	320		1,275	1,550	
	6320	8″ size		″	3	10.667		1,125	425		1,550	1,900	
	6500	Check, swing, C.I. body, brass fittings, auto. ball drip											
	6520	4″ size		Q-12	3	5.333	Ea.	163	201		364	495	
	6540	6″ size		Q-13	4	8		300	320		620	825	
	6580	8″ size		″	3	10.667		570	425		995	1,275	
	6800	Check, wafer, butterfly type, C.I. body, bronze fittings											
	6820	4″ size		Q-12	4	4	Ea.	207	151		358	460	
	6840	6″ size		Q-13	5.50	5.818		355	232		587	750	
	6860	8″ size			5	6.400		640	255		895	1,100	
	6880	10″ size			4.50	7.111		970	283		1,253	1,525	
	7000	Deluge, assembly, incl. trim, pressure											
	7020	operated relief, emergency release, gauges											
	7040	2″ size		Q-12	2	8	Ea.	1,025	300		1,325	1,600	
	7060	3″ size			1.50	10.667		1,225	400		1,625	1,975	
	7080	4″ size			1	16		1,475	600		2,075	2,575	

13900 | Fire Suppression

	13930 | Wet-Pipe Fire Supp. Sprinklers		CREW	DAILY OUTPUT	LABOR-HOURS	UNIT	2006 BARE COSTS MAT.	LABOR	EQUIP.	TOTAL	TOTAL INCL O&P	
400	7100	6″ size (R211313-10)	Q-13	1.80	17.778	Ea.	1,875	710		2,585	3,150	400
	7800	Pneumatic actuator, bronze, required on all										
	7820	pneumatic release systems, any size deluge (R211313-20)	1 Spri	18	.444	Ea.	196	18.60		214.60	245	
	8000	Dry pipe air check valve, 3″ size	Q-12	2	8		1,175	300		1,475	1,775	
	8200	Dry pipe valve, incl. trim and gauges, 3″ size (R211313-30)		2	8		1,250	300		1,550	1,850	
	8220	4″ size		1	16		1,300	600		1,900	2,375	
	8240	6″ size	Q-13	2	16		1,525	635		2,160	2,675	
	8280	For accelerator trim with gauges, add	1 Spri	8	1		425	42		467	530	
	8400	Firecycle package, includes swing check										
	8420	and flow control valves with required trim										
	8440	2″ size	Q-12	2	8	Ea.	1,325	300		1,625	1,950	
	8460	3″ size		1.50	10.667		2,050	400		2,450	2,875	
	8480	4″ size		1	16		2,300	600		2,900	3,475	
	8500	6″ size	Q-13	1.40	22.857		3,100	910		4,010	4,850	
	8800	Flow control valve, includes trim and gauges, 2″ size	Q-12	2	8		1,175	300		1,475	1,750	
	8820	3″ size	″	1.50	10.667		1,375	400		1,775	2,125	
	8840	4″ size	Q-13	2.80	11.429		1,625	455		2,080	2,500	
	8860	6″ size	″	2	16		1,725	635		2,360	2,900	
	9200	Pressure operated relief valve, brass body	1 Spri	18	.444		128	18.60		146.60	170	
	9600	Waterflow indicator, with recycling retard and										
	9610	two single pole retard switches, 2″ thru 6″ pipe size	1 Spri	8	1	Ea.	118	42		160	195	
	9990	Minimum labor/equipment charge	″	3	2.667	Job		111		111	174	
	13960 | CO2 Fire Extinguishing											
200	0010	AUTOMATIC FIRE SUPPRESSION SYSTEMS										200
	0040	For detectors and control stations, see division 13850										
	0100	Control panel, single zone with batteries (2 zones det., 1 suppr.)	1 Elec	1	8	Ea.	1,275	335		1,610	1,925	
	0150	Multizone (4) with batteries (8 zones det., 4 suppr.)	″	.50	16		2,450	670		3,120	3,725	
	1000	Dispersion nozzle, CO_2, 3″ x 5″	1 Plum	18	.444		50	19		69	84.50	
	1100	FM200, 1-1/2″	″	14	.571		50	24.50		74.50	93	
	2000	Extinguisher, CO_2 system, high pressure, 75 lb. cylinder	Q-1	6	2.667		950	102		1,052	1,200	
	2100	100 lb. cylinder	″	5	3.200		1,025	123		1,148	1,325	
	2400	FM200 system, filled, with mounting bracket										
	2460	26 lb. container	Q-1	8	2	Ea.	1,800	77		1,877	2,100	
	2480	44 lb. container		7	2.286		2,400	88		2,488	2,775	
	2500	63 lb. container		6	2.667		2,800	102		2,902	3,225	
	2520	101 lb. container		5	3.200		3,750	123		3,873	4,325	
	2540	196 lb. container		4	4		5,800	154		5,954	6,625	
	3000	Electro/mechanical release	L-1	4	4		125	169		294	400	
	3400	Manual pull station	1 Plum	6	1.333		45	57		102	138	
	4000	Pneumatic damper release	″	8	1		175	42.50		217.50	260	
	6000	Average FM200 system, minimum				C.F.	1.25			1.25	1.38	
	6020	Maximum				″	2.50			2.50	2.75	

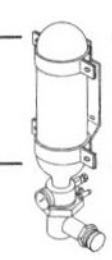

For information about Means Estimating Seminars, see yellow pages 12 and 13 in back of book

13 SPECIAL CONSTRUCTION

Division 14 Conveying Systems

Estimating Tips

General

- Many products in Division 14 will require some type of support or blocking for installation not included with the item itself. Examples are supports for conveyors or tube systems, attachment points for lifts, and footings for hoists or cranes. Add these supports in the appropriate division.

14100 Dumbwaiters
14200 Elevators

- Dumbwaiters and elevators are estimated and purchased in a method similar to buying a car. The manufacturer has a base unit with standard features. Added to this base unit price will be whatever options the owner or specifications require. Increased load capacity, additional vertical travel, additional stops, higher speed, and cab finish options are items to be considered. When developing an estimate for dumbwaiters and elevators, remember that some items needed by the installers may have to be included as part of the general contract.

  Examples are:
  - shaftway
  - rail support brackets
  - machine room
  - electrical supply
  - sill angles
  - electrical connections
  - pits
  - roof penthouses
  - pit ladders

  Check the job specifications and drawings before pricing.
- Installation of elevators and handicapped lifts in historic structures can require significant additional costs. The associated structural requirements may involve cutting into and repairing finishes, mouldings, flooring, etc. The estimator must account for these special conditions.

14300 Escalators & Moving Walks

- Escalators and moving walks are specialty items installed by specialty contractors. There are numerous options associated with these items. For specific options contact a manufacturer or contractor. In a method similar to estimating dumbwaiters and elevators, you should verify the extent of general contract work and add items as necessary.

14400 Lifts
14500 Material Handling
14600 Hoists & Cranes

- Products such as correspondence lifts, conveyors, chutes, pneumatic tube systems, material handling cranes, and hoists, as well as other items specified in this subdivision, may require trained installers. The general contractor might not have any choice as to who will perform the installation or when it will be performed. Long lead times are often required for these products, making early decisions in scheduling necessary.

Reference Numbers

Reference numbers are shown in bold squares at the beginning of some major classifications. These numbers refer to related items in the Reference Section. The reference information may be an estimating procedure, an alternate pricing method, or technical information.

Note: Not all subdivisions listed here necessarily appear in this publication.

14050 | Conveying Systems

	14055 | Conveying Systems		CREW	DAILY OUTPUT	LABOR-HOURS	UNIT	2006 BARE COSTS MAT.	LABOR	EQUIP.	TOTAL	TOTAL INCL O&P	
100	0010	**DUMBWAITER REMOVAL**										100
	0050	Dumbwaiter removal, cab, track and equipment	2 Elev	4	4	Stop		200		200	310	
200	0010	**ELEVATOR REMOVAL**										200
	0050	Elevator removal, cab, track and equipment	2 Elev	2	8	Stop		400		400	615	
300	0010	**ESCALATOR REMOVAL**										300
	0050	Escalator removal, 10′ floor to floor	M-1	.14	228	Ea.		10,800	385	11,185	17,100	
	0100	Escalator removal, 15′ floor to floor		.12	266			12,700	450	13,150	20,000	
	0200	Escalator removal, 20′ floor to floor		.10	320			15,200	540	15,740	24,000	
	0300	Escalator removal, 25′ floor to floor		.08	400			19,000	675	19,675	30,000	
	0400	Moving ramp / walk removal		10	3.200	L.F.		152	5.40	157.40	240	
400	0010	**CHAIR LIFT REMOVAL**										400
	0050	Chair lift removal, minimum	2 Elev	2	8	Ea.		400		400	615	
	0060	Chair lift removal, maximum	"	1	16			800		800	1,225	
	0100	Single post lift removal	L-4	7.50	3.200			106		106	174	
	0200	Double post lift removal		5	4.800			159		159	261	
	0300	Four post lift removal		3	8			265		265	435	
500	0010	**DISTRIBUTION SYSTEM REMOVAL**										500
	0050	Distribution system removal, motorized cart	4 Mill	.30	106	Station		3,950		3,950	6,250	
	0100	Distribution system removal, conveyor belt	2 Mill	80	.200	L.F.		7.40		7.40	11.70	
	0200	Distribution system removal, overhead monorail	1 Elev	50	.160			8		8	12.30	
	0300	Distribution system removal, pneumatic tube	2 Stpi	150	.107			4.59		4.59	7.15	
	0400	Remove linen / trash chute	2 Shee	8	2	Floor		84.50		84.50	134	
600	0010	**CRANE REMOVAL**										600
	0050	Remove crane rail	E-4	300	.107	L.F.		4.31	.30	4.61	8.40	
	0100	Remove overhead bridge crane	M-4	1	36	Ea.		1,425	229	1,654	2,475	

14100 | Dumbwaiters

	14110 | Manual Dumbwaiters		CREW	DAILY OUTPUT	LABOR-HOURS	UNIT	2006 BARE COSTS MAT.	LABOR	EQUIP.	TOTAL	TOTAL INCL O&P	
400	0010	**DUMBWAITERS**, manual										400
	0020	2 stop, minimum	2 Elev	.75	21.333	Ea.	2,450	1,075		3,525	4,350	
	0100	Maximum		.50	32	"	5,600	1,600		7,200	8,625	
	0300	For each additional stop, add		.75	21.333	Stop	890	1,075		1,965	2,625	
	14120 | Electric Dumbwaiters											
400	0010	**DUMBWAITERS**, electric										400
	0020	2 stop, minimum	2 Elev	.13	123	Ea.	6,200	6,150		12,350	16,300	
	0100	Maximum		.11	145	"	18,600	7,275		25,875	31,700	
	0600	For each additional stop, add		.54	29.630	Stop	2,725	1,475		4,200	5,275	
	0750	Correspondence lift, 1 floor, 2 stop, 45 lb capacity	2 Elec	.20	80	Ea.	7,025	3,350		10,375	12,900	

	14210 | Electric Traction Elevators		CREW	DAILY OUTPUT	LABOR-HOURS	UNIT	2006 BARE COSTS MAT.	LABOR	EQUIP.	TOTAL	TOTAL INCL O&P	
100	0010	**ELEVATOR SYSTEMS**										100
	7000	Residential, cab type, 1 floor, 2 stop, minimum	2 Elev	.20	80	Ea.	8,500	4,000		12,500	15,500	
	7100	Maximum		.10	160		14,300	8,000		22,300	28,100	
	7200	2 floor, 3 stop, minimum		.12	133		12,600	6,650		19,250	24,200	
	7300	Maximum		.06	266		20,600	13,300		33,900	43,100	
200	0010	**ELEVATORS** R142000-10										200
	0020	For multi-story buildings, housing project, minimum				% total					2.50%	
	0100	Maximum R142000-20									4.50%	
	0300	Office building, minimum									2.50%	
	0400	Maximum R142000-30									10%	
	0425	Electric freight, base unit, 4000 lb, 200 fpm, 4 stp, std. fin.	2 Elev	.05	320	Ea.	73,500	16,000		89,500	105,000	
	0450	For 5000 lb. capacity, add R142000-40					5,250			5,250	5,775	
	0500	For 6000 lb. capacity, add					9,275			9,275	10,200	
	0525	For 7000 lb. capacity, add					12,400			12,400	13,700	
	0550	For 8000 lb. capacity, add					17,000			17,000	18,700	
	0575	For 10000 lb. capacity, add					20,300			20,300	22,300	
	0600	For 12000 lb. capacity, add					24,800			24,800	27,300	
	0625	For 16000 lb. capacity, add					29,800			29,800	32,800	
	0650	For 20000 lb. capacity, add					32,900			32,900	36,200	
	0675	For increased speed, 250 fpm, add					9,875			9,875	10,900	
	0700	300 fpm, geared electric, add					12,300			12,300	13,500	
	0725	350 fpm, geared electric, add					14,500			14,500	16,000	
	0750	400 fpm, geared electric, add					16,400			16,400	18,100	
	0775	500 fpm, gearless electric, add					21,000			21,000	23,100	
	0800	600 fpm, gearless electric, add					23,200			23,200	25,500	
	0825	700 fpm, gearless electric, add					27,000			27,000	29,700	
	0850	800 fpm, gearless electric, add					30,000			30,000	33,000	
	0875	For class "B" loading, add					1,775			1,775	1,950	
	0900	For class "C-1" loading, add					4,400			4,400	4,850	
	0925	For class "C-2" loading, add					5,250			5,250	5,775	
	0950	For class "C-3" loading, add					7,225			7,225	7,950	
	0975	For travel over 40 V.L.F., add	2 Elev	7.25	2.207	V.L.F.	105	110		215	285	
	1000	For number of stops over 4, add		.27	59.259	Stop	1,875	2,950		4,825	6,625	
	1625	Electric pass., base unit, 2000 lb., 200 fpm, 4 stop, std. fin.		.05	320	Ea.	68,500	16,000		84,500	100,000	
	1650	For 2500 lb. capacity, add					2,800			2,800	3,075	
	1675	For 3000 lb. capacity, add					4,275			4,275	4,700	
	1700	For 3500 lb. capacity, add					5,950			5,950	6,525	
	1725	For 4000 lb. capacity, add					6,350			6,350	7,000	
	1750	For 4500 lb. capacity, add					8,350			8,350	9,200	
	1775	For 5000 lb. capacity, add					10,600			10,600	11,700	
	1800	For increased speed, 250 fpm, geared electric, add					2,350			2,350	2,575	
	1825	300 fpm, geared electric, add					4,850			4,850	5,325	
	1850	350 fpm, geared electric, add					5,750			5,750	6,325	
	1875	400 fpm, geared electric, add					8,150			8,150	8,975	
	1900	500 fpm, gearless electric, add					38,300			38,300	42,100	
	1925	600 fpm, gearless electric, add					40,500			40,500	44,500	
	1950	700 fpm, gearless electric, add					44,300			44,300	48,700	
	1975	800 fpm, gearless electric, add					48,800			48,800	53,500	
	2000	For travel over 40 V.L.F., add	2 Elev	7.25	2.207	V.L.F.	107	110		217	288	
	2025	For number of stops over 4, add		.27	59.259	Stop	2,400	2,950		5,350	7,200	
	2400	Electric hospital, base unit, 4000 lb., 200 fpm, 4 stop, std. fin.		.05	320	Ea.	75,000	16,000		91,000	107,000	
	2425	For 4500 lb. capacity, add					5,175			5,175	5,700	
	2450	For 5000 lb. capacity, add					6,775			6,775	7,450	
	2475	For increased speed, 250 fpm, geared electric, add					2,525			2,525	2,775	
	2500	300 fpm, geared electric, add					4,800			4,800	5,275	

14200 | Elevators

	14210 | Electric Traction Elevators		CREW	DAILY OUTPUT	LABOR-HOURS	UNIT	2006 BARE COSTS MAT.	LABOR	EQUIP.	TOTAL	TOTAL INCL O&P	
200	2525	350 fpm, geared electric, add	R142000-10			Ea.	5,850			5,850	6,450	200
	2550	400 fpm, geared electric, add					8,175			8,175	8,975	
	2575	500 fpm, gearless electric, add	R142000-20				36,600			36,600	40,300	
	2600	600 fpm, gearless electric, add					40,500			40,500	44,500	
	2625	700 fpm, gearless electric, add	R142000-30				44,200			44,200	48,600	
	2650	800 fpm, gearless electric, add					48,800			48,800	53,500	
	2675	For travel over 40 V.L.F., add	R142000-40	2 Elev	7.25	2.207	V.L.F.	107	110		217	288	
	2700	For number of stops over 4, add		"	.27	59.259	Stop	2,975	2,950		5,925	7,850	

	14240 | Hydraulic Elevators	CREW	DAILY OUTPUT	LABOR-HOURS	UNIT	MAT.	LABOR	EQUIP.	TOTAL	TOTAL INCL O&P	
200	0010 HYDRAULIC ELEVATORS										200
	1025 Hydraulic freight, base unit, 2000 lb., 50 fpm, 2 stop, std. fin.	2 Elev	.10	160	Ea.	36,100	8,000		44,100	52,000	
	1050 For 2500 lb. capacity, add					2,625			2,625	2,875	
	1075 For 3000 lb. capacity, add					4,150			4,150	4,550	
	1100 For 3500 lb. capacity, add					6,275			6,275	6,900	
	1125 For 4000 lb. capacity, add					6,700			6,700	7,375	
	1150 For 4500 lb. capacity, add					7,725			7,725	8,500	
	1175 For 5000 lb. capacity, add					10,500			10,500	11,600	
	1200 For 6000 lb. capacity, add					11,100			11,100	12,200	
	1225 For 7000 lb. capacity, add					17,000			17,000	18,700	
	1250 For 8000 lb. capacity, add					19,000			19,000	20,900	
	1275 For 10000 lb. capacity, add					20,200			20,200	22,200	
	1300 For 12000 lb. capacity, add					24,400			24,400	26,800	
	1325 For 16000 lb. capacity, add					31,800			31,800	35,000	
	1350 For 20000 lb. capacity, add					35,300			35,300	38,800	
	1375 For increased speed, 100 fpm, add					755			755	830	
	1400 125 fpm, add					1,450			1,450	1,600	
	1425 150 fpm, add					2,725			2,725	3,000	
	1450 175 fpm, add					4,150			4,150	4,575	
	1475 For class "B" loading, add					1,725			1,725	1,900	
	1500 For class "C-1" loading, add					4,350			4,350	4,800	
	1525 For class "C-2" loading, add					5,225			5,225	5,725	
	1550 For class "C-3" loading, add					7,175			7,175	7,875	
	1575 For travel over 20 V.L.F., add	2 Elev	7.25	2.207	V.L.F.	350	110		460	555	
	1600 For number of stops over 2, add		.27	59.259	Stop	510	2,950		3,460	5,125	
	2050 Hyd. pass., base unit, 1500 lb., 100 fpm, 2 stop, std. fin.		.10	160	Ea.	30,600	8,000		38,600	46,000	
	2075 For 2000 lb. capacity, add					975			975	1,075	
	2100 For 2500 lb. capacity, add					2,000			2,000	2,200	
	2125 For 3000 lb. capacity, add					3,625			3,625	4,000	
	2150 For 3500 lb. capacity, add					5,375			5,375	5,925	
	2175 For 4000 lb. capacity, add					6,475			6,475	7,125	
	2200 For 4500 lb. capacity, add					7,800			7,800	8,600	
	2225 For 5000 lb. capacity, add					11,800			11,800	13,000	
	2250 For increased speed, 125 fpm, add					1,150			1,150	1,275	
	2275 150 fpm, add					2,300			2,300	2,525	
	2300 175 fpm, add					3,725			3,725	4,100	
	2325 200 fpm, add					5,425			5,425	5,975	
	2350 For travel over 12 V.L.F., add	2 Elev	7.25	2.207	V.L.F.	350	110		460	555	
	2375 For number of stops over 2, add		.27	59.259	Stop	3,050	2,950		6,000	7,925	
	2725 Hydraulic hospital, base unit, 4000 lb., 100 fpm, 2 stop, std. fin.		.10	160	Ea.	42,500	8,000		50,500	59,000	
	2750 For 4000 lb. capacity, add					5,125			5,125	5,650	
	2775 For 4500 lb. capacity, add					6,025			6,025	6,625	
	2800 For 5000 lb. capacity, add					8,775			8,775	9,650	
	2825 For increased speed, 125 fpm, add					1,425			1,425	1,575	
	2850 150 fpm, add					2,400			2,400	2,650	
	2875 175 fpm, add					3,875			3,875	4,250	

14 CONVEYING SYSTEMS

14200 | Elevators

	14240 | Hydraulic Elevators		CREW	DAILY OUTPUT	LABOR-HOURS	UNIT	2006 BARE COSTS MAT.	LABOR	EQUIP.	TOTAL	TOTAL INCL O&P	
200	2900	200 fpm, add				Ea.	5,650			5,650	6,200	200
	2925	For travel over 12 V.L.F., add	2 Elev	7.25	2.207	V.L.F.	234	110		344	430	
	2950	For number of stops over 2, add	"	.27	59.259	Stop	2,950	2,950		5,900	7,825	

	14270 | Custom Elevator Cabs		CREW	DAILY OUTPUT	LABOR-HOURS	UNIT	MAT.	LABOR	EQUIP.	TOTAL	TOTAL INCL O&P	
200	0010	CAB FINISHES										200
	3325	Passenger elevator cab finishes (based on 3500 lb. cab size)										
	3350	Acrylic panel ceiling				Ea.	440			440	485	
	3375	Aluminum eggcrate ceiling					510			510	560	
	3400	Stainless steel doors					1,975			1,975	2,175	
	3425	Carpet flooring					370			370	405	
	3450	Epoxy flooring					305			305	335	
	3475	Quarry tile flooring					470			470	515	
	3500	Slate flooring					655			655	720	
	3525	Textured rubber flooring					119			119	131	
	3550	Stainless steel walls					2,650			2,650	2,925	
	3575	Stainless steel returns at door					560			560	615	
	4450	Hospital elevator cab finishes (based on 3500 lb. cab size)										
	4475	Aluminum eggcrate ceiling				Ea.	510			510	560	
	4500	Stainless steel doors					1,625			1,625	1,800	
	4525	Epoxy flooring					305			305	335	
	4550	Quarry tile flooring					465			465	515	
	4575	Textured rubber flooring					119			119	131	
	4600	Stainless steel walls					2,650			2,650	2,925	
	4625	Stainless steel returns at door					560			560	615	

	14280 | Elevator Equipment and Controls		CREW	DAILY OUTPUT	LABOR-HOURS	UNIT	MAT.	LABOR	EQUIP.	TOTAL	TOTAL INCL O&P	
200	0010	ELEVATOR CONTROLS AND DOORS										200
	2975	Passenger elevator options										
	3000	2 car group automatic controls	2 Elev	.66	24.242	Ea.	2,450	1,200		3,650	4,550	
	3025	3 car group automatic controls		.44	36.364		3,675	1,825		5,500	6,825	
	3050	4 car group automatic controls		.33	48.485		6,075	2,425		8,500	10,400	
	3075	5 car group automatic controls		.26	61.538		8,275	3,075		11,350	13,900	
	3100	6 car group automatic controls		.22	72.727		12,500	3,625		16,125	19,400	
	3125	Intercom service		3	5.333		325	266		591	765	
	3150	Duplex car selective collective		.66	24.242		2,700	1,200		3,900	4,825	
	3175	Center opening 1 speed doors		2	8		1,400	400		1,800	2,175	
	3200	Center opening 2 speed doors		2	8		1,675	400		2,075	2,475	
	3225	Rear opening doors (opposite front)		2	8		3,900	400		4,300	4,900	
	3250	Side opening 2 speed doors		2	8		3,700	400		4,100	4,700	
	3275	Automatic emergency power switching		.66	24.242		875	1,200		2,075	2,825	
	3300	Manual emergency power switching		8	2		335	100		435	525	
	3625	Hall finishes, stainless steel doors					925			925	1,025	
	3650	Stainless steel frames					925			925	1,025	
	3675	12 month maintenance contract									2,640	
	3700	Signal devices, hall lanterns	2 Elev	8	2		345	100		445	535	
	3725	Position indicators, up to 3		9.40	1.702		239	85		324	395	
	3750	Position indicators, per each over 3		32	.500		66	25		91	111	
	3775	High speed heavy duty door opener					1,675			1,675	1,850	
	3800	Variable voltage, O.H. gearless machine, min.	2 Elev	.16	100		24,600	5,000		29,600	34,800	
	3815	Maximum		.07	228		54,500	11,400		65,900	77,500	
	3825	Basement installed geared machine		.33	48.485		12,200	2,425		14,625	17,100	
	3850	Freight elevator options										
	3875	Doors, bi-parting	2 Elev	.66	24.242	Ea.	4,175	1,200		5,375	6,475	
	3900	Power operated door and gate	"	.66	24.242		16,900	1,200		18,100	20,500	

14200 | Elevators

14280 | Elevator Equipment and Controls

		14280 Elevator Equipment and Controls	CREW	DAILY OUTPUT	LABOR-HOURS	UNIT	2006 BARE COSTS MAT.	LABOR	EQUIP.	TOTAL	TOTAL INCL O&P	
200	3925	Finishes, steel plate floor				Ea.	735			735	810	200
	3950	14 ga. 1/4" x 4' steel plate walls					1,800			1,800	1,975	
	3975	12 month maintenance contract									1,992	
	4000	Signal devices, hall lanterns	2 Elev	8	2		345	100		445	535	
	4025	Position indicators, up to 3		9.40	1.702		239	85		324	395	
	4050	Position indicators, per each over 3		32	.500		66	25		91	111	
	4075	Variable voltage basement installed geared machine	↓	.66	24.242	↓	13,600	1,200		14,800	16,800	
	4100	Hospital elevator options										
	4125	2 car group automatic controls	2 Elev	.66	24.242	Ea.	2,450	1,200		3,650	4,550	
	4150	3 car group automatic controls		.44	36.364		3,675	1,825		5,500	6,825	
	4175	4 car group automatic controls		.33	48.485		6,075	2,425		8,500	10,400	
	4200	5 car group automatic controls		.26	61.538		8,300	3,075		11,375	13,900	
	4225	6 car group automatic controls		.22	72.727		12,500	3,625		16,125	19,400	
	4250	Intercom service		3	5.333		325	266		591	765	
	4275	Duplex car selective collective		.66	24.242		2,700	1,200		3,900	4,825	
	4300	Center opening 1 speed doors		2	8		1,400	400		1,800	2,175	
	4325	Center opening 2 speed doors		2	8		1,850	400		2,250	2,650	
	4350	Rear opening doors (opposite front)		2	8		3,900	400		4,300	4,900	
	4375	Side opening 2 speed doors		2	8		5,925	400		6,325	7,125	
	4400	Automatic emergency power switching		.66	24.242		875	1,200		2,075	2,825	
	4425	Manual emergency power switching	↓	8	2		335	100		435	525	
	4675	Hall finishes, stainless steel doors					925			925	1,025	
	4700	Stainless steel frames					925			925	1,025	
	4725	12 month maintenance contract									3,985	
	4750	Signal devices, hall lanterns	2 Elev	8	2		345	100		445	535	
	4775	Position indicators, up to 3		9.40	1.702		239	85		324	395	
	4800	Position indicators, per each over 3	↓	32	.500		66	25		91	111	
	4825	High speed heavy duty door opener					1,675			1,675	1,850	
	4850	Variable voltage, O.H. gearless machine, min.	2 Elev	.16	100		24,600	5,000		29,600	34,800	
	4865	Maximum		.07	228		54,500	11,400		65,900	77,500	
	4875	Basement installed geared machine	↓	.33	48.485	↓	12,200	2,425		14,625	17,100	
	5000	Drilling for piston, casing included, 18" diameter	B-48	80	.700	V.L.F.	34	21.50	45.50	101	123	

14300 | Escalators and Moving Walks

14320 | Escalators

		14320 Escalators	CREW	DAILY OUTPUT	LABOR-HOURS	UNIT	2006 BARE COSTS MAT.	LABOR	EQUIP.	TOTAL	TOTAL INCL O&P	
300	0010	ESCALATORS (R143110-10)										300
	1000	Glass, 32" wide x 10' floor to floor height	M-1	.07	457	Ea.	64,000	21,700	775	86,475	105,000	
	1010	48" wide x 10' floor to floor height		.07	457		70,500	21,700	775	92,975	112,000	
	1020	32" wide x 15' floor to floor height		.06	533		67,000	25,300	900	93,200	114,000	
	1030	48" wide x 15' floor to floor height		.06	533		71,000	25,300	900	97,200	118,000	
	1040	32" wide x 20' floor to floor height		.05	653		69,000	31,000	1,100	101,100	125,000	
	1050	48" wide x 20' floor to floor height		.05	653		74,000	31,000	1,100	106,100	130,500	
	1060	32" wide x 25' floor to floor height		.04	800		74,500	38,000	1,350	113,850	142,000	
	1070	48" wide x 25' floor to floor height		.04	800		80,000	38,000	1,350	119,350	148,000	
	1080	Enameled steel, 32" wide x 10' floor to floor height		.07	457		65,500	21,700	775	87,975	106,500	
	1090	48" wide x 10' floor to floor height		.07	457		70,500	21,700	775	92,975	112,000	
	1110	32" wide x 15' floor to floor height		.06	533		67,000	25,300	900	93,200	114,000	
	1120	48" wide x 15' floor to floor height		.06	533		71,000	25,300	900	97,200	118,000	
	1130	32" wide x 20' floor to floor height	↓	.05	653	↓	69,000	31,000	1,100	101,100	125,000	

14300 | Escalators and Moving Walks

14320 | Escalators

			CREW	DAILY OUTPUT	LABOR-HOURS	UNIT	2006 BARE COSTS MAT.	LABOR	EQUIP.	TOTAL	TOTAL INCL O&P	
300	1140	48″ wide x 20′ floor to floor height (R143110-10)	M-1	.05	653	Ea.	74,000	31,000	1,100	106,100	130,500	300
	1150	32″ wide x 25′ floor to floor height		.04	800		74,500	38,000	1,350	113,850	142,000	
	1160	48″ wide x 25′ floor to floor height		.04	800		80,000	38,000	1,350	119,350	148,000	
	1170	Stainless steel, 32″ wide x 10′ floor to floor height		.07	457		72,000	21,700	775	94,475	113,500	
	1180	48″ wide x 10′ floor to floor height		.07	457		75,500	21,700	775	97,975	117,500	
	1500	32″ wide x 15′ floor to floor height		.06	533		73,500	25,300	900	99,700	120,500	
	1700	48″ wide x 15′ floor to floor height		.06	533		76,500	25,300	900	102,700	124,500	
	2300	32″ wide x 25′ floor to floor height		.04	800		83,000	38,000	1,350	122,350	151,500	
	2500	48″ wide x 25′ floor to floor height		.04	800		88,000	38,000	1,350	127,350	157,000	

14360 | Moving Walks

			CREW	DAILY OUTPUT	LABOR-HOURS	UNIT	MAT.	LABOR	EQUIP.	TOTAL	TOTAL INCL O&P	
500	0010	MOVING RAMPS AND WALKS (R143210-20)										500
	0020	Walk, 27″ tread width, minimum	M-1	6.50	4.923	L.F.	715	234	8.30	957.30	1,150	
	0100	300′ to 500′, maximum		4.43	7.223		995	345	12.20	1,352.20	1,650	
	0300	48″ tread width walk, minimum		4.43	7.223		1,625	345	12.20	1,982.20	2,325	
	0400	100′ to 350′, maximum		3.82	8.377		1,900	395	14.15	2,309.15	2,700	
	0600	Ramp, 12° incline, 36″ tread width, minimum		5.27	6.072		1,325	288	10.25	1,623.25	1,900	
	0700	70′ to 90′ maximum		3.82	8.377		1,900	395	14.15	2,309.15	2,700	
	0900	48″ tread width, minimum		3.57	8.964		1,925	425	15.15	2,365.15	2,800	
	1000	40′ to 70′, maximum		2.91	10.997		2,450	520	18.60	2,988.60	3,500	

14400 | Lifts

14420 | Wheelchair Lifts

			CREW	DAILY OUTPUT	LABOR-HOURS	UNIT	2006 BARE COSTS MAT.	LABOR	EQUIP.	TOTAL	TOTAL INCL O&P	
100	0010	CHAIR / WHEELCHAIR LIFT										100
	7700	Stair climber (chair lift), single seat, minimum	2 Elev	1	16	Ea.	4,100	800		4,900	5,750	
	7800	Maximum		.20	80		5,650	4,000		9,650	12,400	
	8000	Wheelchair lift, minimum		1	16		5,625	800		6,425	7,425	
	8500	Maximum		.50	32		13,300	1,600		14,900	17,200	
	8700	Stair lift, minimum		1	16		11,100	800		11,900	13,500	
	8900	Maximum		.20	80		17,600	4,000		21,600	25,600	

14450 | Vehicle Lifts

			CREW	DAILY OUTPUT	LABOR-HOURS	UNIT	MAT.	LABOR	EQUIP.	TOTAL	TOTAL INCL O&P	
700	0010	HYDRAULIC LIFT AUTO/TRUCK										700
	0200	Double post, 9000 lb. frame	L-4	1.15	20.870	Ea.	5,150	690		5,840	6,800	
	2200	Hoists, single post, 8,000# capacity, swivel arms		.40	60		4,525	2,000		6,525	8,225	
	2400	Two posts, adjustable frames, 11,000# capacity		.25	96		5,800	3,175		8,975	11,600	
	2500	24,000# capacity		.15	160		7,750	5,300		13,050	17,200	
	2700	7,500# capacity, frame supports		.50	48		6,450	1,600		8,050	9,700	
	2800	Four post, roll on ramp		.50	48		5,800	1,600		7,400	9,000	
	2810	Hydraulic lifts, above ground, 2 post, clear floor, 6000 lb. cap		2.67	8.989		6,075	298		6,373	7,175	
	2815	9000 lb. capacity		2.29	10.480		14,400	345		14,745	16,400	
	2820	15,000 lb. capacity		2	12		16,300	400		16,700	18,600	
	2825	30,000 lb. capacity		1.60	15		36,000	495		36,495	40,400	
	2830	4 post, ramp style, 25,000 lb. capacity		2	12		14,100	400		14,500	16,200	
	2835	35,000 lb. capacity		1	24		65,500	795		66,295	73,500	
	2840	50,000 lb. capacity		1	24		73,500	795		74,295	82,000	

14400 | Lifts

	14450 | Vehicle Lifts		Crew	Daily Output	Labor-Hours	Unit	2006 Bare Costs Mat.	Labor	Equip.	Total	Total Incl O&P	
700	2845	75,000 lb. capacity	L-4	1	24	Ea.	85,000	795		85,795	95,500	700
	2850	For drive thru tracks, add, minimum					905			905	995	
	2855	Maximum					1,550			1,550	1,700	
	2860	Ramp extensions, 3′ (set of 2)					745			745	815	
	2865	Rolling jack platform					2,575			2,575	2,850	
	2870	Elec/hyd jacking beam					6,900			6,900	7,600	
	2880	Scissor lift, portable, 6000 lb.capacity				↓	6,775			6,775	7,450	

	14460 | Correspondence & Parcel Lifts		Crew	Daily Output	Labor-Hours	Unit	Mat.	Labor	Equip.	Total	Total Incl O&P	
200	0010	**CORRESPONDENCE LIFT**										200
	0020	1 floor, 2 stop, 25 lb. capacity, electric	2 Elev	.20	80	Ea.	5,025	4,000		9,025	11,700	
	0100	Hand, 5 lb. capacity	″	.20	80	″	1,900	4,000		5,900	8,250	
600	0010	**PARCEL LIFT**										600
	0020	20″ x 20″, 100 lb. capacity, electric, per floor	2 Mill	.25	64	Ea.	7,850	2,375		10,225	12,400	

14500 | Material Transport

	14510 | Material Handling		Crew	Daily Output	Labor-Hours	Unit	2006 Bare Costs Mat.	Labor	Equip.	Total	Total Incl O&P	
900	0010	**MOTORIZED CAR** Distribution systems, single track,										900
	0100	20 lb. per car capacity, material handling										
	0200	Minimum, 50′ to 100′	4 Mill	.19	168	Station	25,300	6,250		31,550	37,700	
	0300	Maximum, 100′ to 200′	″	.15	213	″	34,200	7,925		42,125	50,000	
	0400	Larger system, incl. hospital transport, track type,										
	0500	fully automated material handling system										
	0600	Minimum, complete system	4 Mill	.05	640	Ea.	112,000	23,700		135,700	161,000	
	0700	Maximum, with several stations	E-6	.05	2560	″	654,500	101,000	32,800	788,300	940,000	

	14550 | Conveyors		Crew	Daily Output	Labor-Hours	Unit	Mat.	Labor	Equip.	Total	Total Incl O&P	
350	0010	**MATERIAL HANDLING** Conveyors, gravity type 2″ rollers, 3″ O.C.										350
	0050	10′ sections with 2 supports, 600 lb. capacity, 18″ wide				Ea.	375			375	415	
	0100	24″ wide					435			435	480	
	0150	1400 lb. capacity, 18″ wide					345			345	380	
	0200	24″ wide				↓	385			385	425	
	0350	Horizontal belt, center drive and takeup, 60 fpm										
	0400	16″ belt, 26.5′ length	2 Mill	.50	32	Ea.	2,650	1,175		3,825	4,800	
	0450	24″ belt, 41.5′ length		.40	40		4,125	1,475		5,600	6,875	
	0500	61.5′ length	↓	.30	53.333	↓	5,500	1,975		7,475	9,175	
	0600	Inclined belt, 10′ rise with horizontal loader and										
	0620	End idler assembly, 27.5′ length, 18″ belt	2 Mill	.30	53.333	Ea.	7,750	1,975		9,725	11,700	
	0700	24″ belt	″	.15	106	″	8,700	3,950		12,650	15,800	
	3600	Monorail, overhead, manual, channel type										
	3700	125 lb. per L.F.	1 Mill	26	.308	L.F.	8.60	11.40		20	27.50	
	3900	500 lb. per L.F.	″	21	.381	″	7.80	14.15		21.95	31	
	4000	Trolleys for above, 2 wheel, 125 lb. capacity				Ea.	54			54	59.50	
	4200	4 wheel, 250 lb. capacity					164			164	181	
	4300	8 wheel, 1,000 lb. capacity				↓	315			315	350	
900	0010	**VERTICAL CONVEYOR** Automatic selective										900
	0100	to 10 floors, base price	2 Mill	.13	120	Floor	23,100	4,475		27,575	32,600	

	14550 | Conveyors		CREW	DAILY OUTPUT	LABOR-HOURS	UNIT	2006 BARE COSTS MAT.	LABOR	EQUIP.	TOTAL	TOTAL INCL O&P	
900	0200	Add for electrical hook-up and testing	2 Elec	.50	32	Floor		1,350		1,350	2,075	900

	14560 | Chutes		CREW	DAILY OUTPUT	LABOR-HOURS	UNIT	MAT.	LABOR	EQUIP.	TOTAL	TOTAL INCL O&P	
250	0010	**CHUTES** Linen or refuse, incl. sprinklers, 12′ floor height										250
	0050	Aluminized steel, 16 ga., 18″ diameter	2 Shee	3.50	4.571	Floor	870	193		1,063	1,250	
	0100	24″ diameter		3.20	5		945	211		1,156	1,375	
	0200	30″ diameter		3	5.333		1,025	225		1,250	1,500	
	0300	36″ diameter		2.80	5.714		1,175	241		1,416	1,675	
	0400	Galvanized steel, 16 ga., 18″ diameter		3.50	4.571		770	193		963	1,150	
	0500	24″ diameter		3.20	5		870	211		1,081	1,300	
	0600	30″ diameter		3	5.333		975	225		1,200	1,425	
	0700	36″ diameter		2.80	5.714		1,150	241		1,391	1,650	
	0800	Stainless steel, 18″ diameter		3.50	4.571		1,800	193		1,993	2,275	
	0900	24″ diameter		3.20	5		1,900	211		2,111	2,425	
	1000	30″ diameter		3	5.333		2,250	225		2,475	2,825	
	1005	36″ diameter		2.80	5.714		2,500	241		2,741	3,125	
	1200	Linen chute bottom collector, aluminized steel		4	4	Ea.	1,075	169		1,244	1,450	
	1300	Stainless steel		4	4		1,375	169		1,544	1,800	
	1500	Refuse, bottom hopper, aluminized steel, 18″ diameter		3	5.333		705	225		930	1,125	
	1600	24″ diameter		3	5.333		775	225		1,000	1,200	
	1800	36″ diameter		3	5.333		1,325	225		1,550	1,800	
	2900	Package chutes, spiral type, minimum		4.50	3.556	Floor	1,875	150		2,025	2,300	
	3000	Maximum		1.50	10.667	″	4,875	450		5,325	6,100	
	9000	Minimum labor/equipment charge	1 Shee	1	8	Job		335		335	535	

	14580 | Pneumatic Tube Systems		CREW	DAILY OUTPUT	LABOR-HOURS	UNIT	MAT.	LABOR	EQUIP.	TOTAL	TOTAL INCL O&P	
800	0010	**PNEUMATIC TUBE SYSTEM** Single tube, 2 stations, blower										800
	0020	100′ long, stock										
	0100	3″ diameter	2 Stpi	.12	133	Total	5,075	5,750		10,825	14,500	
	0300	4″ diameter	″	.09	177	″	5,725	7,650		13,375	18,200	
	0400	Twin tube, two stations or more, conventional system										
	0600	2-1/2″ round	2 Stpi	62.50	.256	L.F.	10.25	11		21.25	28.50	
	0700	3″ round		46	.348		11.75	14.95		26.70	36.50	
	0900	4″ round		49.60	.323		13	13.90		26.90	36	
	1000	4″ x 7″ oval		37.60	.426		19	18.30		37.30	49.50	
	1050	Add for blower		2	8	System	4,025	345		4,370	4,950	
	1110	Plus for each round station, add		7.50	2.133	Ea.	450	92		542	640	
	1150	Plus for each oval station, add		7.50	2.133	″	450	92		542	640	
	1200	Alternate pricing method: base cost, minimum		.75	21.333	Total	4,925	920		5,845	6,825	
	1300	Maximum		.25	64	″	9,775	2,750		12,525	15,000	
	1500	Plus total system length, add, minimum		93.40	.171	L.F.	6.30	7.35		13.65	18.40	
	1600	Maximum		37.60	.426	″	18.70	18.30		37	49	
	1800	Completely automatic system, 4″ round, 15 to 50 stations		.29	55.172	Station	15,500	2,375		17,875	20,700	
	2200	51 to 144 stations		.32	50		12,000	2,150		14,150	16,600	
	2400	6″ round or 4″ x 7″ oval, 15 to 50 stations		.24	66.667		19,400	2,875		22,275	25,900	
	2800	51 to 144 stations		.23	69.565		16,300	3,000		19,300	22,600	

14610 | Fixed Hoists

		14610 Fixed Hoists	Crew	Daily Output	Labor-Hours	Unit	2006 Bare Costs Mat.	Labor	Equip.	Total	Total Incl O&P	
500	0010	MATERIAL HANDLING										500
	1500	Cranes, portable hydraulic, floor type, 2,000 lb. capacity				Ea.	1,900			1,900	2,100	
	1600	4,000 lb. capacity					3,200			3,200	3,525	
	1800	Movable gantry type, 12′ to 15′ range, 2,000 lb. capacity					2,775			2,775	3,050	
	1900	6,000 lb. capacity					4,025			4,025	4,425	
	2100	Hoists, electric overhead, chain, hook hung, 15′ lift, 1 ton cap.					1,850			1,850	2,025	
	2200	3 ton capacity					2,300			2,300	2,525	
	2500	5 ton capacity					5,100			5,100	5,600	
	2600	For hand-pushed trolley, add					15%					
	2700	For geared trolley, add					30%					
	2800	For motor trolley, add					75%					
	3000	For lifts over 15′, 1 ton, add				L.F.	18			18	19.80	
	3100	5 ton, add				″	40			40	44	
	3300	Lifts, scissor type, portable, electric, 36″ high, 2,000 lb.				Ea.	3,300			3,300	3,625	
	3400	48″ high, 4,000 lb.				″	3,600			3,600	3,950	

14630 | Bridge Cranes

		14630 Bridge Cranes	Crew	Daily Output	Labor-Hours	Unit	Mat.	Labor	Equip.	Total	Total Incl O&P	
300	0010	CRANE RAIL										300
	0020	Box beam bridge, no equipment included	E-4	3400	.009	Lb.	.91	.38	.03	1.32	1.74	
	0210	Running track only, 104 lb. per yard, 20′ piece	″	160	.200	L.F.	16.70	8.10	.56	25.36	34	
700	0010	OVERHEAD BRIDGE CRANES										700
	0100	1 girder, 20′ span, 3 ton	M-3	1	34	Ea.	19,100	1,375	151	20,626	23,300	
	0125	5 ton		1	34		20,900	1,375	151	22,426	25,300	
	0150	7.5 ton		1	34		24,900	1,375	151	26,426	29,700	
	0175	10 ton		.80	42.500		32,900	1,700	189	34,789	39,100	
	0200	15 ton		.80	42.500		42,300	1,700	189	44,189	49,400	
	0225	30′ span, 3 ton		1	34		19,900	1,375	151	21,426	24,200	
	0250	5 ton		1	34		21,900	1,375	151	23,426	26,300	
	0275	7.5 ton		1	34		26,200	1,375	151	27,726	31,100	
	0300	10 ton		.80	42.500		34,100	1,700	189	35,989	40,400	
	0325	15 ton		.80	42.500		44,200	1,700	189	46,089	51,500	
	0350	2 girder, 40′ span, 3 ton	M-4	.50	72		32,900	2,875	460	36,235	41,100	
	0375	5 ton		.50	72		34,400	2,875	460	37,735	42,800	
	0400	7.5 ton		.50	72		37,700	2,875	460	41,035	46,500	
	0425	10 ton		.40	90		44,100	3,575	570	48,245	54,500	
	0450	15 ton		.40	90		60,000	3,575	570	64,145	72,000	
	0475	25 ton		.30	120		71,000	4,775	765	76,540	86,500	
	0500	50′ span, 3 ton		.50	72		37,400	2,875	460	40,735	46,200	
	0525	5 ton		.50	72		38,900	2,875	460	42,235	47,800	
	0550	7.5 ton		.50	72		41,600	2,875	460	44,935	51,000	
	0575	10 ton		.40	90		48,000	3,575	570	52,145	59,000	
	0600	15 ton		.40	90		63,000	3,575	570	67,145	75,000	
	0625	25 ton		.30	120		74,500	4,775	765	80,040	90,000	

For information about Means Estimating Seminars, see yellow pages 12 and 13 in back of book

Division 15 Mechanical

Estimating Tips

15100 Building Services Piping

This subdivision is primarily basic pipe and related materials. The pipe may be used by any of the mechanical disciplines, i.e., plumbing, fire protection, heating, and air conditioning.

- The piping section lists the add to labor for elevated pipe installation. These adds apply to all elevated pipe, fittings, valves, insulation, etc., that are placed above 10′ high. CAUTION: the correct percentage may vary for the same pipe. For example, the percentage add for the basic pipe installation should be based on the maximum height that the craftsman must install for that particular section. If the pipe is to be located 14′ above the floor but it is suspended on threaded rod from beams, the bottom flange of which is 18′ high (4′ rods), then the height is actually 18′ and the add is 20%. The pipe coverer, however, does not have to go above the 14′ and so his or her add should be 10%.
- Most pipe is priced first as straight pipe with a joint (coupling, weld, etc.) every 10′ and a hanger usually every 10′. There are exceptions with hanger spacing such as for cast iron pipe (5′) and plastic pipe (3 per 10′). Following each type of pipe there are several lines listing sizes and the amount to be subtracted to delete couplings and hangers. This is for pipe that is to be buried or supported together on trapeze hangers. The reason that the couplings are deleted is that these runs are usually long and frequently longer lengths of pipe are used. By deleting the couplings the estimator is expected to look up and add back the correct reduced number of couplings.
- When preparing an estimate it may be necessary to approximate the fittings. Fittings usually run between 25% and 50% of the cost of the pipe. The lower percentage is for simpler runs, and the higher number is for complex areas such as mechanical rooms.
- For historic restoration projects, the systems must be as invisible as possible and pathways must be sought for pipes, conduits, and ductwork. While installations in accessible spaces (such as basements and attics) are relatively straightforward to estimate, labor costs may be more difficult to determine when delivery systems must be concealed.

15400 Plumbing Fixtures & Equipment

- Plumbing fixture costs usually require two lines, the fixture itself and its "rough-in, supply and waste."
- In the Assemblies Section (Plumbing D2010) for the desired fixture, the System Components Group at the center of the page shows the fixture on the first line. The rest of the list (fittings, pipe, tubing, etc.) will total up to what we refer to in the Unit Price section as "Rough-in, supply, waste and vent." Note that for most fixtures we allow a nominal 5′ of tubing to reach from the fixture to a main or riser.
- Remember that gas- and oil-fired units need venting.

15500 Heat Generation Equipment

- When estimating the cost of an HVAC system, check to see who is responsible for providing and installing the temperature control system. It is possible to overlook controls, assuming that they would be included in the electrical estimate.
- When looking up a boiler be careful on specified capacity. Some manufacturers rate their products on output while others use input.
- Include HVAC insulation for pipe, boiler, and duct (wrap and liner).
- Be careful when looking up mechanical items to get the correct pressure rating and connection type (thread, weld, flange).

15700 Heating/Ventilation/ Air Conditioning Equipment

- Combination heating and cooling units are sized by the air conditioning requirements. (See Reference No. R236000-20 for preliminary sizing guide.)
- A ton of air conditioning is nominally 400 CFM.
- Rectangular duct is taken off by the linear foot for each size, but its cost is usually estimated by the pound. Remember that SMACNA standards now base duct on internal pressure.
- Prefabricated duct is estimated and purchased like pipe: straight sections and fittings.
- Note that cranes or other lifting equipment are not included on any lines in Division 15. For example, if a crane is required to lift a heavy piece of pipe into place high above a gym floor, or to put a rooftop unit on the roof of a four-story building, etc., it must be added. Due to the potential for extreme variation—from nothing additional required to a major crane or helicopter—we feel that including a nominal amount for "lifting contingency" would be useless and detract from the accuracy of the estimate. When using equipment rental from Means do not forget to include the cost of the operator(s).

Reference Numbers

Reference numbers are shown in bold squares at the beginning of some major classifications. These numbers refer to related items in the Reference Section. The reference information may be an estimating procedure, an alternate pricing method, or technical information.

Note: Not all subdivisions listed here necessarily appear in this publication.

Note: **i2 Trade Service,** *in part, has been used as a reference source for some of the material prices used in Division 15.*

15051 | Mechanical General

		15051 Mechanical General	CREW	DAILY OUTPUT	LABOR-HOURS	UNIT	MAT.	LABOR	EQUIP.	TOTAL	TOTAL INCL O&P	
							2006 BARE COSTS					
100	0010	**BOILERS, GENERAL** Prices do not include flue piping, elec. wiring,										100
	0020	gas or oil piping, boiler base, pad, or tankless unless noted										
	0100	Boiler H.P.: 10 KW = 34 lbs/steam/hr = 33,475 BTU/hr.										
	0120											
	0150	To convert SFR to BTU rating: Hot water, 150 x SFR;										
	0160	Forced hot water, 180 x SFR; steam, 240 x SFR										
700	0010	**PIPING** See also Divisions 02500 & 02600 for site work										700
	1000	Add to labor for elevated installation										
	1080	10′ to 15′ high						10%				
	1100	15′ to 20′ high						20%				
	1120	20′ to 25′ high						25%				
	1140	25′ to 30′ high						35%				
	1160	30′ to 35′ high						40%				
	1180	35′ to 40′ high						50%				
	1200	Over 40′ high						55%				

15055 | Selective Mech Demolition

		15055 Selective Mech Demolition	CREW	DAILY OUTPUT	LABOR-HOURS	UNIT	MAT.	LABOR	EQUIP.	TOTAL	TOTAL INCL O&P	
300	0010	**HVAC DEMOLITION** R220105-10										300
	0100	Air conditioner, split unit, 3 ton	Q-5	2	8	Ea.		310		310	480	
	0150	Package unit, 3 ton R024119-10	Q-6	3	8			320		320	500	
	0190	Rooftop, self contained, up to 5 ton	1 Plum	1.20	6.667			285		285	445	
	0250	Air curtain	Q-9	20	.800	L.F.		30.50		30.50	48.50	
	0254	Air filters, up thru 16,000 CFM		20	.800	Ea.		30.50		30.50	48.50	
	0256	20,000 thru 60,000 CFM		16	1			38		38	60.50	
	0297	Boiler blowdown	Q-5	8	2			77.50		77.50	121	
	0298	Boilers										
	0300	Electric, up thru 148 kW	Q-19	2	12	Ea.		480		480	740	
	0310	150 thru 518 kW	"	1	24			955		955	1,475	
	0320	550 thru 2000 kW	Q-21	.40	80			3,250		3,250	5,050	
	0330	2070 kW and up	"	.30	106			4,325		4,325	6,725	
	0340	Gas and/or oil, up thru 150 MBH	Q-7	2.20	14.545			595		595	930	
	0350	160 thru 2000 MBH		.80	40			1,650		1,650	2,550	
	0360	2100 thru 4500 MBH		.50	64			2,625		2,625	4,075	
	0370	4600 thru 7000 MBH		.30	106			4,375		4,375	6,800	
	0380	7100 thru 12,000 MBH		.16	200			8,200		8,200	12,800	
	0390	12,200 thru 25,000 MBH		.12	266			10,900		10,900	17,000	
	0400	Central station air handler unit, up thru 15 ton	Q-5	1.60	10			390		390	605	
	0410	17.5 thru 30 ton	"	.80	20			775		775	1,200	
	0430	Computer room unit										
	0434	Air cooled split, up thru 10 ton	Q-5	.67	23.881	Ea.		925		925	1,450	
	0436	12 thru 23 ton		.53	30.189			1,175		1,175	1,825	
	0440	Chilled water, up thru 10 ton		1.30	12.308			475		475	740	
	0444	12 thru 23 ton		1	16			620		620	965	
	0450	Glycol system, up thru 10 ton		.53	30.189			1,175		1,175	1,825	
	0454	12 thru 23 ton		.40	40			1,550		1,550	2,400	
	0460	Water cooled, not including condenser, up thru 10 ton		.80	20			775		775	1,200	
	0464	12 thru 23 ton		.60	26.667			1,025		1,025	1,600	
	0600	Condenser, up thru 50 ton		1	16			620		620	965	
	0610	51 thru 100 ton	Q-6	.80	30			1,200		1,200	1,875	
	0620	101 thru 1000 ton	"	.16	150			6,025		6,025	9,375	
	0660	Condensing unit, up thru 10 ton	Q-5	1.25	12.800			495		495	770	
	0670	11 thru 50 ton	"	.40	40			1,550		1,550	2,400	
	0680	60 thru 100 ton	Q-6	.30	80			3,225		3,225	5,000	
	0700	Cooling tower, up thru 400 ton		.80	30			1,200		1,200	1,875	
	0710	450 thru 600 ton		.53	45.283			1,825		1,825	2,825	

15 MECHANICAL

15055 | Selective Mech Demolition

			CREW	DAILY OUTPUT	LABOR-HOURS	UNIT	2006 BARE COSTS MAT.	LABOR	EQUIP.	TOTAL	TOTAL INCL O&P	
300	0720	700 thru 1300 ton (R220105-10)	Q-6	.40	60	Ea.		2,400		2,400	3,750	300
	0780	Dehumidifier, up thru 155 lb./hr.	Q-1	8	2			77		77	120	
	0790	240 lb./hr. and up (R024119-10)	"	2	8	↓		305		305	480	
	1560	Ductwork										
	1570	Metal, steel, sst, fabricated	Q-9	1000	.016	Lb.		.61		.61	.97	
	1580	Aluminum, fabricated		485	.033	"		1.25		1.25	1.99	
	1590	Spiral, prefabricated		400	.040	L.F.		1.52		1.52	2.41	
	1600	Fiberglass, prefabricated		400	.040			1.52		1.52	2.41	
	1610	Flex, prefabricated		500	.032			1.21		1.21	1.93	
	1620	Glass fiber reinforced plastic, prefabricated	↓	280	.057	↓		2.17		2.17	3.45	
	1630	Diffusers, registers or grills, up thru 20" max dimension	1 Shee	50	.160	Ea.		6.75		6.75	10.75	
	1640	21 thru 36" max dimension		36	.222			9.35		9.35	14.90	
	1650	Above 36" max dimension	↓	30	.267			11.25		11.25	17.90	
	1700	Evaporator, up thru 12,000 BTUH	Q-5	5.30	3.019			117		117	182	
	1710	12,500 thru 30,000 BTUH	"	2.70	5.926			230		230	355	
	1720	31,000 BTUH and up	Q-6	1.50	16			645		645	1,000	
	1730	Evaporative cooler, up thru 5 H.P.	Q-9	2.70	5.926			225		225	360	
	1740	10 thru 30 H.P.	"	.67	23.881	↓		905		905	1,450	
	1750	Exhaust systems										
	1760	Exhaust components	1 Shee	8	1	System		42		42	67	
	1770	Weld fume hoods	"	20	.400	Ea.		16.85		16.85	27	
	2120	Fans, up thru 1 H.P. or 2000 CFM	Q-9	8	2			76		76	121	
	2124	1-1/2 thru 10 H.P. or 20,000 CFM		5.30	3.019			115		115	182	
	2128	15 thru 30 H.P. or above 20,000 CFM	↓	4	4			152		152	241	
	2150	Fan coil air conditioner, chilled water, up thru 7.5 ton	Q-5	14	1.143			44.50		44.50	69	
	2154	Direct expansion, up thru 10 ton		8	2			77.50		77.50	121	
	2158	11 thru 30 ton	↓	2	8			310		310	480	
	2170	Flue shutter damper	Q-9	8	2			76		76	121	
	2200	Furnace, electric	Q-20	2	10			385		385	610	
	2300	Gas or oil, under 120 MBH	Q-9	4	4			152		152	241	
	2340	Over 120 MBH	"	3	5.333			202		202	320	
	2730	Heating and ventilating unit	Q-5	2.70	5.926			230		230	355	
	2740	Heater, electric, wall, baseboard and quartz	1 Elec	10	.800			33.50		33.50	52	
	2750	Heater, electric, unit, cabinet, fan and convector	"	8	1			42		42	64.50	
	2760	Heat exchanger, shell and tube type	Q-5	1.60	10			390		390	605	
	2770	Plate type	Q-6	.60	40	↓		1,600		1,600	2,500	
	2810	Heat pump										
	2820	Air source, split, up thru 10 ton	Q-5	.90	17.778	Ea.		690		690	1,075	
	2830	15 thru 25 ton	Q-6	.80	30			1,200		1,200	1,875	
	2850	Single package, up thru 12 ton	Q-5	1	16			620		620	965	
	2860	Water source, up thru 15 ton	"	.90	17.778			690		690	1,075	
	2870	20 thru 50 ton	Q-6	.80	30			1,200		1,200	1,875	
	2910	Heat recovery package, up thru 20,000 CFM	Q-5	2	8			310		310	480	
	2920	25,000 CFM and up		1.20	13.333			515		515	805	
	2930	Heat transfer package, up thru 130 GPM		.80	20			775		775	1,200	
	2934	255 thru 800 GPM		.42	38.095			1,475		1,475	2,300	
	2940	Humidifier		10.60	1.509			58.50		58.50	91	
	2961	Hydronic unit heaters, up thru 200 MBH		14	1.143			44.50		44.50	69	
	2962	Above 200 MBH		8	2			77.50		77.50	121	
	2964	Valance units	↓	32	.500	↓		19.40		19.40	30	
	2966	Radiant floor heating										
	2967	System valves, controls, manifolds	Q-5	16	1	Ea.		39		39	60.50	
	2968	Per room distribution		8	2			77.50		77.50	121	
	2970	Hydronic heating, baseboard radiation		16	1			39		39	60.50	
	2976	Convectors and free standing radiators	↓	18	.889			34.50		34.50	53.50	
	2980	Induced draft fan, up thru 1 H.P.	Q-9	4.60	3.478	↓		132		132	210	

MECHANICAL 15

15055 | Selective Mech Demolition

		15055 Selective Mech Demolition	Ref.	CREW	DAILY OUTPUT	LABOR-HOURS	UNIT	2006 BARE COSTS MAT.	LABOR	EQUIP.	TOTAL	TOTAL INCL O&P	
300	2984	1-1/2 H.P. thru 7-1/2 H.P.	R220105-10	Q-9	2.20	7.273	Ea.		276		276	440	300
	2988	Infra-red unit		Q-5	16	1	↓		39		39	60.50	
	2992	Louvers	R024119-10	1 Shee	46	.174	S.F.		7.35		7.35	11.65	
	3000	Mechanical equip., light items. Unit is weight, not cooling.		Q-5	.90	17.778	Ton		690		690	1,075	
	3600	Heavy items		"	1.10	14.545	"		565		565	875	
	3700	Deduct for salvage (when applicable), minimum					Job					55	
	3710	Maximum					"					420	
	3720	Make up air unit, up thru 6000 CFM		Q-5	3	5.333	Ea.		207		207	320	
	3730	6500 thru 30,000 CFM		"	1.60	10			390		390	605	
	3740	35,000 thru 75,000 CFM		Q-6	1	24			965		965	1,500	
	3800	Mixing boxes, constant and VAV		Q-9	18	.889			33.50		33.50	53.50	
	4000	Packaged terminal air conditioner, up thru 18,000 BTUH		Q-5	8	2			77.50		77.50	121	
	4010	24,000 thru 48,000 BTUH		"	2.80	5.714	↓		221		221	345	
	5000	Refrigerant compressor, reciprocating or scroll											
	5010	Up thru 5 ton		1 Stpi	6	1.333	Ea.		57.50		57.50	89.50	
	5020	5.08 thru 10 ton		Q-5	6	2.667			103		103	161	
	5030	15 thru 50 ton		"	.40	40			1,550		1,550	2,400	
	5040	60 thru 130 ton		Q-6	.48	50			2,000		2,000	3,125	
	5100	Roof top air conditioner, up thru 10 ton		Q-5	1.40	11.429			445		445	690	
	5110	12 thru 40 ton		Q-6	1	24			965		965	1,500	
	5120	50 thru 140 ton			.50	48			1,925		1,925	3,000	
	5130	150 thru 300 ton		↓	.30	80			3,225		3,225	5,000	
	6000	Self contained single package air conditioner, up thru 10 ton		Q-5	1.60	10			390		390	605	
	6010	15 thru 60 ton		Q-6	1.20	20			805		805	1,250	
	6100	Space heaters, up thru 200 MBH		Q-5	10	1.600			62		62	96.50	
	6110	Over 200 MBH			5	3.200			124		124	193	
	6200	Split ductless, both sections		↓	8	2			77.50		77.50	121	
	6300	Steam condensate meter		1 Stpi	11	.727			31.50		31.50	48.50	
	6600	Thru-the-wall air conditioner		L-2	8	2	↓		62		62	103	
	7000	Vent chimney, prefabricated, up thru 12" diameter		Q-9	94	.170	V.L.F.		6.45		6.45	10.25	
	7010	14" thru 36" diameter			40	.400			15.15		15.15	24	
	7020	38" thru 48" diameter		↓	32	.500			18.95		18.95	30	
	7030	54" thru 60" diameter		Q-10	14	1.714	↓		67.50		67.50	107	
	7400	Ventilators, up thru 14" neck diameter		Q-9	58	.276	Ea.		10.45		10.45	16.65	
	7410	16" thru 50" neck diameter			40	.400			15.15		15.15	24	
	7450	Relief vent, up thru 24" x 96"			22	.727			27.50		27.50	44	
	7460	48" x 60" thru 96" x 144"		↓	10	1.600			60.50		60.50	96.50	
	8000	Water chiller up thru 10 ton		Q-5	2.50	6.400			248		248	385	
	8010	15 thru 100 ton		Q-6	.48	50			2,000		2,000	3,125	
	8020	110 thru 500 ton		Q-7	.29	110			4,525		4,525	7,050	
	8030	600 thru 1000 ton			.23	139			5,700		5,700	8,875	
	8040	1100 ton and up		↓	.20	160			6,575		6,575	10,200	
	8400	Window air conditioner		1 Carp	16	.500	↓		17.80		17.80	29.50	
	9000	Minimum labor/equipment charge		Q-6	3	8	Job		320		320	500	
600	0010	**PLUMBING DEMOLITION**	R220105-10										600
	0400	Air compressor, up thru 2 H.P.		Q-1	10	1.600	Ea.		61.50		61.50	95.50	
	0410	3 H.P. thru 7-1/2 H.P.	R024119-10		5.60	2.857			110		110	171	
	0420	10 H.P. thru 15 H.P.		↓	1.40	11.429			440		440	685	
	0430	20 H.P. thru 30 H.P.		Q-2	1.30	18.462			735		735	1,150	
	0500	Backflow preventer, up thru 2" diameter		1 Plum	17	.471			20		20	31.50	
	0510	2-1/2" thru 3" diameter		Q-1	10	1.600			61.50		61.50	95.50	
	0520	4" thru 6" diameter		"	5	3.200			123		123	191	
	0530	8" thru 10" diameter		Q-2	3	8	↓		320		320	495	
	0700	Carriers and supports											
	0710	Fountains, sinks, lavatories and urinals		1 Plum	14	.571	Ea.		24.50		24.50	38	
	0720	Water closets		"	12	.667	"		28.50		28.50	44.50	

15055 | Selective Mech Demolition

600			CREW	DAILY OUTPUT	LABOR-HOURS	UNIT	2006 BARE COSTS MAT.	LABOR	EQUIP.	TOTAL	TOTAL INCL O&P	600
	0730	Grinder pump or sewage ejector system (R220105-10)										
	0732	Simplex	Q-1	7	2.286	Ea.		88		88	137	
	0734	Duplex (R024119-10)	"	2.80	5.714			220		220	340	
	0738	Hot water dispenser	1 Plum	36	.222			9.50		9.50	14.75	
	0740	Hydrant, wall		26	.308			13.15		13.15	20.50	
	0744	Ground		12	.667			28.50		28.50	44.50	
	0760	Cleanouts and drains, up thru 4" pipe diameter		10	.800			34		34	53	
	0764	5" thru 8" pipe diameter	Q-1	10	1.600			61.50		61.50	95.50	
	0780	Industrial safety fixtures	1 Plum	8	1			42.50		42.50	66.50	
	1020	Fixtures, including 10' piping										
	1100	Bath tubs, cast iron	1 Plum	4	2	Ea.		85.50		85.50	133	
	1120	Fiberglass		6	1.333			57		57	88.50	
	1140	Steel		5	1.600			68.50		68.50	106	
	1150	Bidet	Q-1	7	2.286			88		88	137	
	1200	Lavatory, wall hung	1 Plum	10	.800			34		34	53	
	1220	Counter top		8	1			42.50		42.50	66.50	
	1300	Sink, single compartment		8	1			42.50		42.50	66.50	
	1320	Double compartment		7	1.143			49		49	76	
	1340	Shower, stall and receptor	Q-1	6	2.667			102		102	159	
	1350	Group	"	7	2.286			88		88	137	
	1400	Water closet, floor mounted	1 Plum	8	1			42.50		42.50	66.50	
	1420	Wall mounted	"	7	1.143			49		49	76	
	1440	Wash fountain, 36" diameter	Q-2	8	3			120		120	186	
	1442	54" diameter	"	7	3.429			137		137	212	
	1500	Urinal, floor mounted	1 Plum	4	2			85.50		85.50	133	
	1520	Wall mounted	"	7	1.143			49		49	76	
	1590	Whirl pool or hot tub	Q-1	2.60	6.154			236		236	370	
	1600	Water fountains, free standing	1 Plum	8	1			42.50		42.50	66.50	
	1620	Wall or deck mounted		6	1.333			57		57	88.50	
	1800	Medical gas specialties		8	1			42.50		42.50	66.50	
	1900	Piping fittings, single connection, up thru 1-1/2" diameter		30	.267			11.40		11.40	17.70	
	1910	2" thru 4" diameter		14	.571			24.50		24.50	38	
	1980	Pipe hanger/support removal		80	.100			4.27		4.27	6.65	
	1990	Glass pipe with fittings, 1" thru 3" diameter		200	.040	L.F.		1.71		1.71	2.66	
	1992	4" thru 6" diameter		150	.053			2.28		2.28	3.54	
	2000	Piping, metal, up thru 1-1/2" diameter		200	.040			1.71		1.71	2.66	
	2050	2" thru 3-1/2" diameter		150	.053			2.28		2.28	3.54	
	2100	4" thru 6" diameter	2 Plum	100	.160			6.85		6.85	10.60	
	2150	8" thru 14" diameter	"	60	.267			11.40		11.40	17.70	
	2153	16" thru 20 " diameter	Q-18	70	.343			13.80	.99	14.79	22.50	
	2155	24" thru 26 " diameter		55	.436			17.55	1.26	18.81	29	
	2156	30" thru 36 " diameter		40	.600			24	1.73	25.73	39.50	
	2160	Plastic pipe with fittings, up thru 1-1/2" diameter	1 Plum	250	.032			1.37		1.37	2.12	
	2162	2" thru 3" diameter	"	200	.040			1.71		1.71	2.66	
	2164	4" thru 6" diameter	Q-1	200	.080			3.07		3.07	4.78	
	2166	8" thru 14" diameter		150	.107			4.10		4.10	6.35	
	2168	16" diameter		100	.160			6.15		6.15	9.55	
	2170	Prison fixtures, lavatory or sink		18	.889	Ea.		34		34	53	
	2172	Shower		5.60	2.857			110		110	171	
	2174	Urinal or water closet		13	1.231			47.50		47.50	73.50	
	2180	Pumps, all fractional horse-power		12	1.333			51		51	79.50	
	2184	1 H.P. thru 5 H.P.		6	2.667			102		102	159	
	2186	7-1/2 H.P. thru 15 H.P.		2.50	6.400			246		246	380	
	2188	20 H.P. thru 25 H.P.	Q-2	4	6			239		239	370	
	2190	30 H.P. thru 60 H.P.		.80	30			1,200		1,200	1,850	
	2192	75 H.P. thru 100 H.P.		.60	40			1,600		1,600	2,475	

MECHANICAL 15

	15055 | Selective Mech Demolition	CREW	DAILY OUTPUT	LABOR-HOURS	UNIT	2006 BARE COSTS MAT.	LABOR	EQUIP.	TOTAL	TOTAL INCL O&P	
600	2194 150 H.P. (R220105-10)	Q-2	.50	48	Ea.		1,925		1,925	2,975	600
	2198 Pump, sump or submersible	1 Plum	12	.667			28.50		28.50	44.50	
	2200 Receptors and interceptors, up thru 20 GPM (R024119-10)	″	8	1			42.50		42.50	66.50	
	2204 25 thru 100 GPM	Q-1	6	2.667			102		102	159	
	2208 125 thru 300 GPM	″	2.40	6.667			256		256	400	
	2211 325 thru 500 GPM	Q-2	2.60	9.231			370		370	570	
	2230 Temperature maintenance cable	1 Plum	1200	.007	L.F.		.28		.28	.44	
	2250 Water heater, 40 gal.	″	6	1.333	Ea.		57		57	88.50	
	3100 Tanks, water heaters and liquid containers										
	3110 Up thru 45 gallons	Q-1	22	.727	Ea.		28		28	43.50	
	3120 50 thru 120 gallons		14	1.143			44		44	68.50	
	3130 130 thru 240 gallons		7.60	2.105			81		81	126	
	3140 250 thru 500 gallons		5.40	2.963			114		114	177	
	3150 600 thru 1000 gallons	Q-2	1.60	15			600		600	930	
	3160 1100 thru 2000 gallons		.70	34.286			1,375		1,375	2,125	
	3170 2100 thru 4000 gallons		.50	48			1,925		1,925	2,975	
	6000 Remove and reset fixtures, minimum	1 Plum	6	1.333			57		57	88.50	
	6100 Maximum		4	2			85.50		85.50	133	
	9000 Minimum labor/equipment charge		2	4	Job		171		171	266	
	9100 Valve, metal valves, strainers and similiar, up thru 1-1/2″ diameter	1 Stpi	28	.286	Ea.		12.30		12.30	19.15	
	9110 2″ thru 3″ diameter	Q-1	11	1.455			56		56	87	
	9120 4″ thru 6″ diameter	″	8	2			77		77	120	
	9130 8″ thru 14″ diameter	Q-2	8	3			120		120	186	
	9140 16″ thru 20″ diameter		2	12			480		480	745	
	9150 24″ diameter		1.20	20			795		795	1,250	
	9200 Valve, plastic, up thru 1-1/2″ diameter	1 Plum	42	.190			8.15		8.15	12.65	
	9210 2″ thru 3″ diameter		15	.533			23		23	35.50	
	9220 4″ thru 6″ diameter		12	.667			28.50		28.50	44.50	
	9300 Vent flashing and caps		55	.145			6.20		6.20	9.65	
	9350 Water filter, commercial, 1″ thru 1-1/2″	Q-1	2	8			305		305	480	
	9360 2″ thru 2-1/2″	″	1.60	10			385		385	600	
	9400 Water heaters (when tank size given see line 15055 600 3100)										
	9410 Up thru 245 GPH	Q-1	2.40	6.667	Ea.		256		256	400	
	9420 250 thru 756 GPH		1.60	10			385		385	600	
	9430 775 thru 1640 GPH		.80	20			770		770	1,200	
	9440 1650 thru 4000 GPH	Q-2	.50	48			1,925		1,925	2,975	
	9470 Water softener	Q-1	2	8			305		305	480	

	15060 | Hangers & Supports	CREW	DAILY OUTPUT	LABOR-HOURS	UNIT	MAT.	LABOR	EQUIP.	TOTAL	TOTAL INCL O&P	
300	0010 **PIPE HANGERS AND SUPPORTS**										300
	0011 TYPE numbers per MSS-SP58										
	0050 Brackets										
	0060 Beam side or wall, malleable iron, TYPE 34										
	0070 3/8″ threaded rod size	1 Plum	48	.167	Ea.	3.98	7.10		11.08	15.45	
	0080 1/2″ threaded rod size		48	.167		7.65	7.10		14.75	19.50	
	0090 5/8″ threaded rod size		48	.167		10.20	7.10		17.30	22.50	
	0100 3/4″ threaded rod size		48	.167		12.25	7.10		19.35	24.50	
	0110 7/8″ threaded rod size		48	.167		14.30	7.10		21.40	27	
	0120 For concrete installation, add						30%				
	0150 Wall, welded steel, medium, TYPE 32										
	0160 0 size, 12″ wide, 18″ deep	1 Plum	34	.235	Ea.	299	10.05		309.05	345	
	0170 1 size, 18″ wide 24″ deep		34	.235		355	10.05		365.05	405	
	0180 2 size, 24″ wide, 30″ deep		34	.235		470	10.05		480.05	530	
	0300 Clamps										
	0310 C-clamp, for mounting on steel beam flange, w/locknut, TYPE 23										

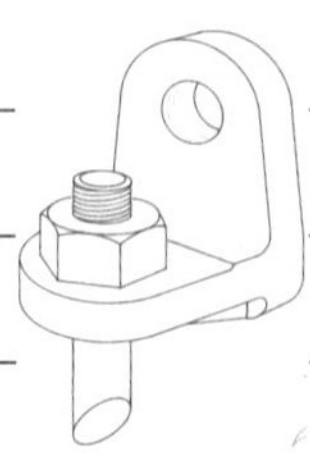

	15060 | Hangers & Supports	CREW	DAILY OUTPUT	LABOR-HOURS	UNIT	2006 BARE COSTS MAT.	LABOR	EQUIP.	TOTAL	TOTAL INCL O&P	
300	0320 3/8" threaded rod size	1 Plum	160	.050	Ea.	2.51	2.14		4.65	6.10	300
	0330 1/2" threaded rod size 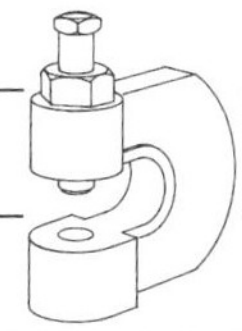		160	.050		3.13	2.14		5.27	6.75	
	0340 5/8" threaded rod size		160	.050		4.79	2.14		6.93	8.55	
	0350 3/4" threaded rod size		160	.050		6.50	2.14		8.64	10.45	
	0400 High temperature to 1050°F, alloy steel										
	0410 4" pipe size	Q-1	106	.151	Ea.	48	5.80		53.80	62	
	0420 6" pipe size 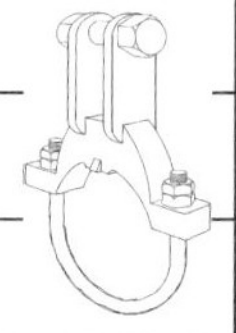		106	.151		81	5.80		86.80	98	
	0430 8" pipe size		97	.165		91.50	6.35		97.85	110	
	0440 10" pipe size		84	.190		124	7.30		131.30	147	
	0450 12" pipe size		72	.222		143	8.55		151.55	171	
	0460 14" pipe size		64	.250		300	9.60		309.60	345	
	0470 16" pipe size		56	.286		340	11		351	390	
	0500 I-beam, for mounting on bottom flange, strap iron, TYPE 21										
	0510 2" flange size	1 Plum	96	.083	Ea.	6.45	3.56		10.01	12.60	
	0520 3" flange size		95	.084		6.10	3.60		9.70	12.35	
	0530 4" flange size		93	.086		6.10	3.67		9.77	12.45	
	0540 5" flange size 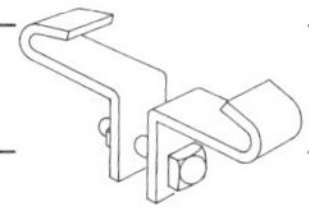		92	.087		6.10	3.71		9.81	12.50	
	0550 6" flange size		90	.089		6.45	3.80		10.25	12.95	
	0560 7" flange size		88	.091		6.45	3.88		10.33	13.10	
	0570 8" flange size		86	.093		12.85	3.97		16.82	20.50	
	0600 One hole, vertical mounting, malleable iron										
	0610 1/2" pipe size	1 Plum	160	.050	Ea.	1.32	2.14		3.46	4.77	
	0620 3/4" pipe size 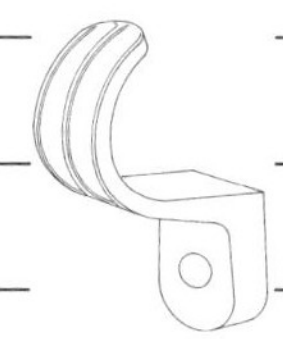		145	.055		1.61	2.36		3.97	5.45	
	0630 1" pipe size		136	.059		2.36	2.51		4.87	6.50	
	0640 1-1/4" pipe size		128	.063		2.65	2.67		5.32	7.05	
	0650 1-1/2" pipe size		120	.067		2.97	2.85		5.82	7.70	
	0660 2" pipe size		112	.071		4.31	3.05		7.36	9.50	
	0670 2-1/2" pipe size		104	.077		8.40	3.28		11.68	14.35	
	0680 3" pipe size		96	.083		8.85	3.56		12.41	15.25	
	0690 3-1/2" pipe size		90	.089		10.60	3.80		14.40	17.55	
	0700 4" pipe size		84	.095		12.50	4.07		16.57	20	
	0750 Riser or extension pipe, carbon steel, TYPE 8										
	0760 3/4" pipe size	1 Plum	48	.167	Ea.	3.83	7.10		10.93	15.25	
	0770 1" pipe size		47	.170		3.87	7.25		11.12	15.55	
	0780 1-1/4" pipe size		46	.174		4.82	7.45		12.27	16.85	
	0790 1-1/2" pipe size		45	.178		5.15	7.60		12.75	17.45	
	0800 2" pipe size		43	.186		5.35	7.95		13.30	18.20	
	0810 2-1/2" pipe size 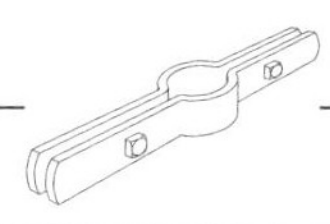		41	.195		5.65	8.35		14	19.15	
	0820 3" pipe size		40	.200		5.95	8.55		14.50	19.85	
	0830 3-1/2" pipe size		39	.205		7.40	8.75		16.15	22	
	0840 4" pipe size		38	.211		7.50	9		16.50	22.50	
	0850 5" pipe size		37	.216		11.20	9.25		20.45	26.50	
	0860 6" pipe size		36	.222		12.85	9.50		22.35	29	
	0870 8" pipe size		34	.235		21	10.05		31.05	38.50	
	0880 10" pipe size		32	.250		31	10.70		41.70	50.50	
	0890 12" pipe size		28	.286		37.50	12.20		49.70	60.50	
	0900 For plastic coating 3/4" to 4", add					190%					
	0910 For copper plating 3/4" to 4", add					58%					
	0950 Two piece, complete, carbon steel, medium weight, TYPE 4										
	0960 1/2" pipe size	Q-1	137	.117	Ea.	4.04	4.49		8.53	11.45	
	0970 3/4" pipe size 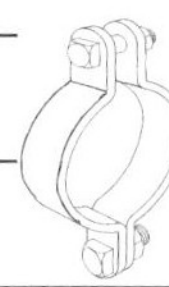		134	.119		4.04	4.59		8.63	11.60	
	0980 1" pipe size		132	.121		4.12	4.66		8.78	11.80	
	0990 1-1/4" pipe size		130	.123		5.35	4.73		10.08	13.25	
	1000 1-1/2" pipe size		126	.127		5.35	4.88		10.23	13.50	
	1010 2" pipe size		124	.129		6.20	4.96		11.16	14.55	
300	1020 2-1/2" pipe size		120	.133		6.55	5.10		11.65	15.20	300

MECHANICAL 15

15060 | Hangers & Supports

	15060 Hangers & Supports	CREW	DAILY OUTPUT	LABOR-HOURS	UNIT	2006 BARE COSTS MAT.	LABOR	EQUIP.	TOTAL	TOTAL INCL O&P	
300	1030 3" pipe size	Q-1	117	.137	Ea.	7.30	5.25		12.55	16.20	300
	1040 3-1/2" pipe size		114	.140		10.20	5.40		15.60	19.60	
	1050 4" pipe size		110	.145		10.20	5.60		15.80	19.90	
	1060 5" pipe size		106	.151		16.75	5.80		22.55	27.50	
	1070 6" pipe size		104	.154		25.50	5.90		31.40	37	
	1080 8" pipe size		100	.160		30.50	6.15		36.65	43	
	1090 10" pipe size		96	.167		55	6.40		61.40	70.50	
	1100 12" pipe size		89	.180		64	6.90		70.90	81.50	
	1110 14" pipe size		82	.195		123	7.50		130.50	147	
	1120 16" pipe size	↓	68	.235		133	9.05		142.05	160	
	1130 For galvanized, add				↓	45%					
	1150 Insert, concrete										
	1160 Wedge type, carbon steel body, malleable iron nut										
	1170 1/4" threaded rod size	1 Plum	96	.083	Ea.	2.15	3.56		5.71	7.90	
	1180 3/8" threaded rod size		96	.083		2.21	3.56		5.77	8	
	1190 1/2" threaded rod size		96	.083		2.33	3.56		5.89	8.10	
	1200 5/8" threaded rod size		96	.083		2.38	3.56		5.94	8.15	
	1210 3/4" threaded rod size		96	.083		2.80	3.56		6.36	8.65	
	1220 7/8" threaded rod size	↓	96	.083		3.52	3.56		7.08	9.40	
	1230 For galvanized, add				↓	.95			.95	1.05	
	1250 Pipe guide sized for insulation										
	1260 No. 1, 1" pipe size, 1" thick insulation	1 Stpi	26	.308	Ea.	200	13.25		213.25	241	
	1270 No. 2, 1-1/4"-2" pipe size, 1" thick insulation		23	.348		236	14.95		250.95	284	
	1280 No. 3, 1-1/4"-2" pipe size, 1-1/2" thick insulation		21	.381		236	16.40		252.40	286	
	1290 No. 4, 2-1/2"-3-1/2" pipe size, 1-1/2" thick insulation		18	.444		236	19.15		255.15	290	
	1300 No. 5, 4"-5" pipe size, 1-1/2" thick insulation	↓	16	.500		262	21.50		283.50	320	
	1310 No. 6, 5"-6" pipe size, 2" thick insulation	Q-5	21	.762		291	29.50		320.50	365	
	1320 No. 7, 8" pipe size, 2" thick insulation		16	1		510	39		549	620	
	1330 No. 8, 10" pipe size, 2" thick insulation	↓	12	1.333		965	51.50		1,016.50	1,125	
	1340 No. 9, 12" pipe size, 2" thick insulation	Q-6	17	1.412		965	56.50		1,021.50	1,150	
	1350 No. 10, 12"-14" pipe size, 2-1/2" thick insulation		16	1.500		1,250	60.50		1,310.50	1,475	
	1360 No. 11, 16" pipe size, 2-1/2" thick insulation		10.50	2.286		1,250	92		1,342	1,525	
	1370 No. 12, 16"-18" pipe size, 3" thick insulation		9	2.667		1,325	107		1,432	1,625	
	1380 No. 13, 20" pipe size, 3" thick insulation		7.50	3.200		1,325	129		1,454	1,650	
	1390 No. 14, 24" pipe size, 3" thick insulation	↓	7	3.429	↓	1,700	138		1,838	2,100	
	1400 Bands										
	1410 Adjustable band, carbon steel, for non-insulated pipe, TYPE 7										
	1420 1/2" pipe size	Q-1	142	.113	Ea.	1.18	4.33		5.51	8.05	
	1430 3/4" pipe size		140	.114		1.18	4.39		5.57	8.15	
	1440 1" pipe size		137	.117		1.36	4.49		5.85	8.50	
	1450 1-1/4" pipe size		134	.119		1.40	4.59		5.99	8.70	
	1460 1-1/2" pipe size		131	.122		1.58	4.69		6.27	9.05	
	1470 2" pipe size		129	.124		2.18	4.77		6.95	9.80	
	1480 2-1/2" pipe size		125	.128		2.28	4.92		7.20	10.15	
	1490 3" pipe size		122	.131		2.51	5.05		7.56	10.60	
	1500 3-1/2" pipe size		119	.134		3.90	5.15		9.05	12.35	
	1510 4" pipe size		114	.140		3.90	5.40		9.30	12.70	
	1520 5" pipe size		110	.145		4.13	5.60		9.73	13.25	
	1530 6" pipe size		108	.148		5.75	5.70		11.45	15.15	
	1540 8" pipe size	↓	104	.154	↓	8.90	5.90		14.80	19	
	1550 For copper plated, add					50%					
	1560 For galvanized, add					30%					
	1570 For plastic coating, add					30%					
	1600 Adjusting nut malleable iron, steel band, TYPE 9										
	1610 1/2" pipe size, galvanized band	Q-1	137	.117	Ea.	4.81	4.49		9.30	12.30	
	1620 3/4" pipe size, galvanized band	↓	135	.119	↓	4.81	4.55		9.36	12.40	300

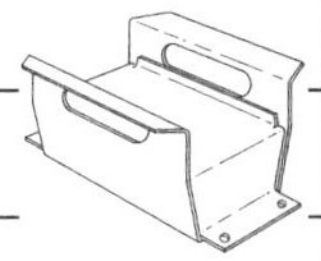

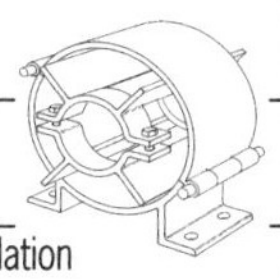

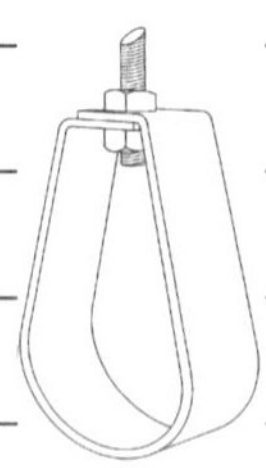

15 MECHANICAL

	15060 | Hangers & Supports	CREW	DAILY OUTPUT	LABOR-HOURS	UNIT	2006 BARE COSTS MAT.	LABOR	EQUIP.	TOTAL	TOTAL INCL O&P
300 1630	1" pipe size, gavanized band	Q-1	132	.121	Ea.	4.81	4.66		9.47	12.55
1640	1-1/4" pipe size, galvanized band		129	.124		5.80	4.77		10.57	13.80
1650	1-1/2" pipe size, galvanized band 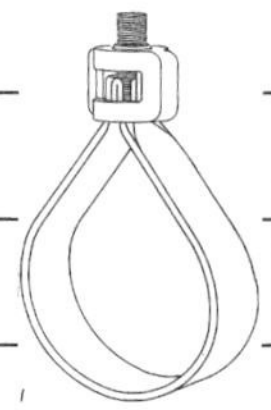		126	.127		5.80	4.88		10.68	14
1660	2" pipe size, galvanized band		124	.129		6.30	4.96		11.26	14.60
1670	2-1/2" pipe size, galvanized band		120	.133		10.55	5.10		15.65	19.55
1680	3" pipe size, galvanized band		117	.137		11.45	5.25		16.70	21
1690	3-1/2" pipe size, galvanized band		114	.140		14.45	5.40		19.85	24.50
1700	4" pipe size, cadmium plated band		110	.145		18.40	5.60		24	29
1740	For plastic coated band, add					35%				
1750	For completely copper coated, add					45%				
1800	Clevis, adjustable, carbon steel, for non-insulated pipe, TYPE 1									
1810	1/2" pipe size	Q-1	137	.117	Ea.	.54	4.49		5.03	7.60
1820	3/4" pipe size		135	.119		.56	4.55		5.11	7.70
1830	1" pipe size		132	.121		.63	4.66		5.29	7.95
1840	1-1/4" pipe size		129	.124		.68	4.77		5.45	8.15
1850	1-1/2" pipe size		126	.127		.77	4.88		5.65	8.45
1860	2" pipe size 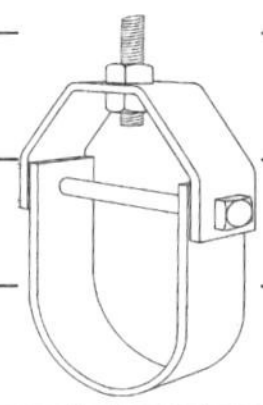		124	.129		.90	4.96		5.86	8.70
1870	2-1/2" pipe size		120	.133		1.62	5.10		6.72	9.75
1880	3" pipe size		117	.137		1.94	5.25		7.19	10.30
1890	3-1/2" pipe size		114	.140		2.07	5.40		7.47	10.70
1900	4" pipe size		110	.145		2.44	5.60		8.04	11.40
1910	5" pipe size		106	.151		3.15	5.80		8.95	12.45
1920	6" pipe size		104	.154		4.04	5.90		9.94	13.65
1930	8" pipe size		100	.160		6.10	6.15		12.25	16.30
1940	10" pipe size 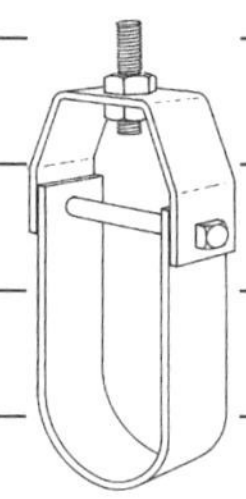		96	.167		31.50	6.40		37.90	45
1950	12" pipe size		89	.180		37.50	6.90		44.40	52.50
1960	14" pipe size		82	.195		60	7.50		67.50	77.50
1970	16" pipe size		68	.235		85.50	9.05		94.55	108
1980	For galvanized, add					66%				
1990	For copper plated 1/2" to 4", add					77%				
2000	For light weight 1/2" to 4", deduct					13%				
2010	Insulated pipe type, 3/4" to 12" pipe, add					180%				
2020	Insulated pipe type, chrome-moly U-strap, add					530%				
2250	Split ring, malleable iron, for non-insulated pipe, TYPE 11									
2260	1/2" pipe size	Q-1	137	.117	Ea.	6.85	4.49		11.34	14.50
2270	3/4" pipe size		135	.119		6.90	4.55		11.45	14.70
2280	1" pipe size		132	.121		7.30	4.66		11.96	15.25
2290	1-1/4" pipe size		129	.124		9.05	4.77		13.82	17.35
2300	1-1/2" pipe size		126	.127		10.95	4.88		15.83	19.65
2310	2" pipe size 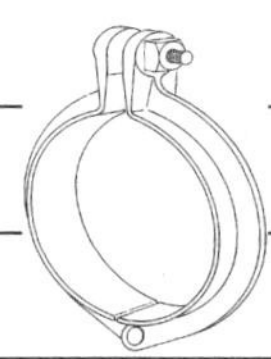		124	.129		12.45	4.96		17.41	21.50
2320	2-1/2" pipe size		120	.133		18	5.10		23.10	28
2330	3" pipe size		117	.137		22.50	5.25		27.75	32.50
2340	3-1/2" pipe size		114	.140		23.50	5.40		28.90	34
2350	4" pipe size		110	.145		23.50	5.60		29.10	34
2360	5" pipe size		106	.151		31	5.80		36.80	43
2370	6" pipe size		104	.154		69.50	5.90		75.40	85.50
2380	8" pipe size		100	.160		123	6.15		129.15	145
2390	For copper plated, add					8%				
2532	Turnbuckle, TYPE 13									
2534	3/8"	1 Plum	80	.100	Ea.	21.50	4.27		25.77	30.50
2535	1/2"		72	.111		27	4.74		31.74	37
2536	5/8"		64	.125		28	5.35		33.35	39.50
2537	3/4"		56	.143		37.50	6.10		43.60	51
2538	7/8"		48	.167		51.50	7.10		58.60	68
2539	1"		40	.200		65	8.55		73.55	85
2540	1-1/8"		32	.250		118	10.70		128.70	146

15060 | Hangers & Supports

	15060 Hangers & Supports	CREW	DAILY OUTPUT	LABOR-HOURS	UNIT	2006 BARE COSTS MAT.	LABOR	EQUIP.	TOTAL	TOTAL INCL O&P
300 2544	Eye rod, welded									
2546	3/8"	1 Plum	56	.143	Ea.	28.50	6.10		34.60	41
2547	1/2"		48	.167		42.50	7.10		49.60	57.50
2548	5/8"		40	.200		64	8.55		72.55	84
2549	3/4"		32	.250		81.50	10.70		92.20	106
2550	7/8"		24	.333		100	14.25		114.25	132
2551	1"		16	.500		126	21.50		147.50	172
2556	Eye rod, welded, linked									
2558	3/8"	1 Plum	56	.143	Ea.	87.50	6.10		93.60	106
2559	1/2"		48	.167		119	7.10		126.10	141
2560	5/8"		40	.200		171	8.55		179.55	201
2561	3/4"		32	.250		221	10.70		231.70	260
2562	7/8"		24	.333		270	14.25		284.25	320
2563	1"		16	.500		345	21.50		366.50	415
2650	Rods, carbon steel									
2660	Continuous thread									
2670	1/4" thread size	1 Plum	144	.056	L.F.	.16	2.37		2.53	3.87
2680	3/8" thread size		144	.056		.23	2.37		2.60	3.94
2690	1/2" thread size		144	.056		.46	2.37		2.83	4.20
2700	5/8" thread size		144	.056		.62	2.37		2.99	4.37
2710	3/4" thread size		144	.056		1.05	2.37		3.42	4.85
2720	7/8" thread size		144	.056		1.35	2.37		3.72	5.20
2725	1/4" thread size, bright finish		144	.056		.70	2.37		3.07	4.46
2726	1/2" thread size, bright finish		144	.056		1.22	2.37		3.59	5.05
2730	For galvanized, add					40%				
2750	Both ends machine threaded 18" length									
2760	3/8" thread size	1 Plum	240	.033	Ea.	2.06	1.42		3.48	4.48
2770	1/2" thread size		240	.033		3.30	1.42		4.72	5.85
2780	5/8" thread size		240	.033		4.75	1.42		6.17	7.45
2790	3/4" thread size		240	.033		7.30	1.42		8.72	10.20
2800	7/8" thread size		240	.033		8.85	1.42		10.27	11.90
2810	1" thread size		240	.033		12.15	1.42		13.57	15.55
2860	Pipe hanger assy., adj. clevis, saddle, rod, clamp, insul. allowance									
2864	1/2" pipe size	Q-5	35	.457	Ea.	25	17.70		42.70	55
2866	3/4" pipe size		34.80	.460		26.50	17.80		44.30	56.50
2868	1" pipe size		34.60	.462		38	17.90		55.90	70
2869	1-1/4" pipe size		34.30	.466		38	18.10		56.10	70
2870	1-1/2" pipe size		33.90	.472		39.50	18.30		57.80	72
2872	2" pipe size		33.30	.480		45	18.60		63.60	78.50
2874	2-1/2" pipe size		32.30	.495		52	19.20		71.20	87
2876	3" pipe size		31.20	.513		58	19.85		77.85	95
2880	4" pipe size		30.70	.521		63.50	20		83.50	102
2884	6" pipe size		29.80	.537		77	21		98	117
2888	8" pipe size		28	.571		124	22		146	171
2892	10" pipe size		25.20	.635		223	24.50		247.50	284
2896	12" pipe size		23.20	.690		315	26.50		341.50	390
2900	Rolls									
2910	Adjustable yoke, carbon steel with CI roll, TYPE 43									
2920	2-1/2" pipe size	Q-1	137	.117	Ea.	19.30	4.49		23.79	28
2930	3" pipe size		131	.122		21.50	4.69		26.19	31
2940	3-1/2" pipe size		124	.129		28.50	4.96		33.46	39
2950	4" pipe size		117	.137		28.50	5.25		33.75	39.50
2960	5" pipe size		110	.145		33.50	5.60		39.10	45
2970	6" pipe size		104	.154		41.50	5.90		47.40	54.50
2980	8" pipe size		96	.167		62	6.40		68.40	78
2990	10" pipe size		80	.200		74	7.70		81.70	93.50

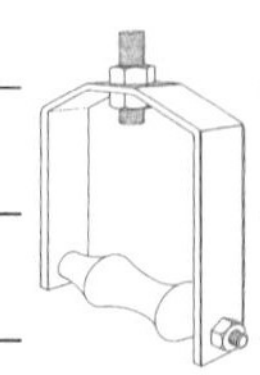

	15060 | Hangers & Supports		CREW	DAILY OUTPUT	LABOR-HOURS	UNIT	2006 BARE COSTS MAT.	LABOR	EQUIP.	TOTAL	TOTAL INCL O&P	
300	3000	12″ pipe size	Q-1	68	.235	Ea.	109	9.05		118.05	134	300
	3010	14″ pipe size		56	.286		300	11		311	345	
	3020	16″ pipe size		48	.333		385	12.80		397.80	445	
	3050	Chair, carbon steel with CI roll										
	3060	2″ pipe size	1 Plum	68	.118	Ea.	25.50	5		30.50	36	
	3070	2-1/2″ pipe size		65	.123		27.50	5.25		32.75	38	
	3080	3″ pipe size		62	.129		28	5.50		33.50	39.50	
	3090	3-1/2″ pipe size		60	.133		36	5.70		41.70	48.50	
	3100	4″ pipe size 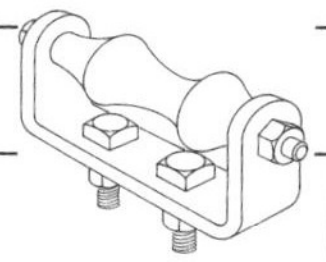		58	.138		40	5.90		45.90	53	
	3110	5″ pipe size		56	.143		43.50	6.10		49.60	57.50	
	3120	6″ pipe size		53	.151		60.50	6.45		66.95	77	
	3130	8″ pipe size		50	.160		82	6.85		88.85	101	
	3140	10″ pipe size		48	.167		103	7.10		110.10	124	
	3150	12″ pipe size		46	.174		150	7.45		157.45	177	
	3170	Single pipe roll, (see line 2650 for rods), TYPE 41, 1″ pipe size	Q-1	137	.117		8.35	4.49		12.84	16.20	
	3180	1-1/4″ pipe size		131	.122		8.45	4.69		13.14	16.60	
	3190	1-1/2″ pipe size		129	.124		8.65	4.77		13.42	16.90	
	3200	2″ pipe size		124	.129		8.80	4.96		13.76	17.40	
	3210	2-1/2″ pipe size 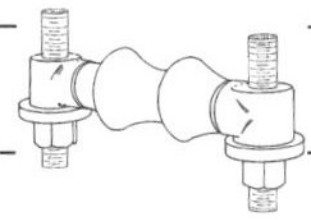		118	.136		10.40	5.20		15.60	19.55	
	3220	3″ pipe size		115	.139		10.80	5.35		16.15	20	
	3230	3-1/2″ pipe size		113	.142		11.30	5.45		16.75	21	
	3240	4″ pipe size		112	.143		14.70	5.50		20.20	24.50	
	3250	5″ pipe size		110	.145		17.10	5.60		22.70	27.50	
	3260	6″ pipe size		101	.158		21	6.10		27.10	32.50	
	3270	8″ pipe size		90	.178		32	6.85		38.85	46	
	3280	10″ pipe size 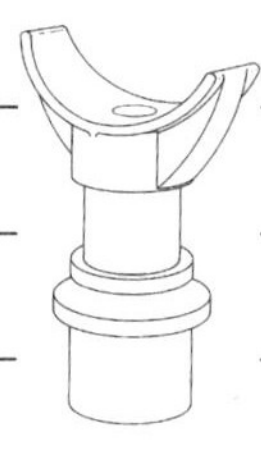		80	.200		38.50	7.70		46.20	54	
	3290	12″ pipe size		68	.235		45	9.05		54.05	63.50	
	3291	14″ pipe size		56	.286		71.50	11		82.50	96	
	3292	16″ pipe size		48	.333		87	12.80		99.80	116	
	3293	18″ pipe size		40	.400		94.50	15.35		109.85	128	
	3294	20″ pipe size		35	.457		137	17.55		154.55	179	
	3296	24″ pipe size		30	.533		217	20.50		237.50	271	
	3297	30″ pipe size		25	.640		520	24.50		544.50	615	
	3298	36″ pipe size		20	.800		545	30.50		575.50	650	
	3300	Saddles (add vertical pipe riser, usually 3″ diameter)										
	3310	Pipe support, complete, adjust., CI saddle, TYPE 36										
	3320	2-1/2″ pipe size	1 Plum	96	.083	Ea.	214	3.56		217.56	241	
	3330	3″ pipe size		88	.091		217	3.88		220.88	245	
	3340	3-1/2″ pipe size		79	.101		218	4.32		222.32	247	
	3350	4″ pipe size		68	.118		283	5		288	320	
	3360	5″ pipe size 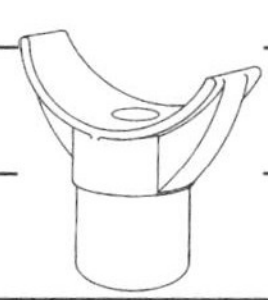		64	.125		291	5.35		296.35	330	
	3370	6″ pipe size		59	.136		291	5.80		296.80	330	
	3380	8″ pipe size		53	.151		310	6.45		316.45	350	
	3390	10″ pipe size		50	.160		340	6.85		346.85	385	
	3400	12″ pipe size		48	.167		355	7.10		362.10	400	
	3450	For standard pipe support, one piece, CI deduct					34%					
	3460	For stanchion support, CI with steel yoke, deduct					60%					
	3500	Multiple pipe support 2′ long strut	1 Plum	30	.267	Ea.	6.15	11.40		17.55	24.50	
	3510	3′ long strut		20	.400		9.20	17.10		26.30	36.50	
	3520	4′ long strut		17	.471		12.25	20		32.25	45	
	3550	Insulation shield 1″ thick, 1/2″ pipe size, TYPE 40	1 Asbe	100	.080		8.70	3.12		11.82	14.70	
	3560	3/4″ pipe size		100	.080		10.85	3.12		13.97	17.05	
	3570	1″ pipe size		98	.082		10.85	3.19		14.04	17.15	
	3580	1-1/4″ pipe size		98	.082		11.60	3.19		14.79	18	
	3590	1-1/2″ pipe size		96	.083		11.60	3.25		14.85	18.10	
	3600	2″ pipe size		96	.083		12	3.25		15.25	18.50	

300	15060 | Hangers & Supports	CREW	DAILY OUTPUT	LABOR-HOURS	UNIT	2006 BARE COSTS MAT.	LABOR	EQUIP.	TOTAL	TOTAL INCL O&P	300
3610	2-1/2" pipe size	1 Asbe	94	.085	Ea.	12.40	3.32		15.72	19.05	
3620	3" pipe size		94	.085		13.25	3.32		16.57	20	
3630	2" thick, 3-1/2" pipe size		92	.087		14.80	3.40		18.20	22	
3640	4" pipe size		92	.087		19.80	3.40		23.20	27.50	
3650	5" pipe size		90	.089		26	3.47		29.47	34	
3660	6" pipe size		90	.089		30.50	3.47		33.97	39	
3670	8" pipe size		88	.091		41.50	3.55		45.05	51.50	
3680	10" pipe size		88	.091		61.50	3.55		65.05	74	
3690	12" pipe size		86	.093		70.50	3.63		74.13	83.50	
3700	14" pipe size		86	.093		141	3.63		144.63	162	
3710	16" pipe size		84	.095		151	3.72		154.72	172	
3720	18" pipe size		84	.095		167	3.72		170.72	190	
3730	20" pipe size		82	.098		184	3.81		187.81	208	
3732	24" pipe size		80	.100		201	3.91		204.91	227	
3733	30" pipe size		78	.103		223	4.01		227.01	252	
3735	36" pipe size	↓	75	.107	↓	234	4.17		238.17	265	
3750	Covering protection saddle, TYPE 39										
3760	1" covering size										
3770	3/4" pipe size	1 Plum	68	.118	Ea.	3.92	5		8.92	12.10	
3780	1" pipe size		68	.118		3.92	5		8.92	12.10	
3790	1-1/4" pipe size		68	.118		3.92	5		8.92	12.10	
3800	1-1/2" pipe size		66	.121		4.25	5.20		9.45	12.75	
3810	2" pipe size		66	.121		4.25	5.20		9.45	12.75	
3820	2-1/2" pipe size		64	.125		6.10	5.35		11.45	15.05	
3830	3" pipe size		64	.125		8.85	5.35		14.20	18.05	
3840	3-1/2" pipe size		62	.129		8.85	5.50		14.35	18.30	
3850	4" pipe size		62	.129		11.20	5.50		16.70	21	
3860	5" pipe size		60	.133		11.20	5.70		16.90	21	
3870	6" pipe size	↓	60	.133	↓	13.30	5.70		19	23.50	
3900	1-1/2" covering size										
3910	3/4" pipe size	1 Plum	68	.118	Ea.	19.40	5		24.40	29.50	
3920	1" pipe size		68	.118		19.40	5		24.40	29.50	
3930	1-1/4" pipe size		68	.118		19.40	5		24.40	29.50	
3940	1-1/2" pipe size		66	.121		19.40	5.20		24.60	29.50	
3950	2" pipe size		66	.121		21	5.20		26.20	31	
3960	2-1/2" pipe size		64	.125		25	5.35		30.35	36	
3970	3" pipe size		64	.125		25	5.35		30.35	36	
3980	3-1/2" pipe size		62	.129		25	5.50		30.50	36	
3990	4" pipe size		62	.129		25	5.50		30.50	36	
4000	5" pipe size		60	.133		25	5.70		30.70	36.50	
4010	6" pipe size		60	.133		36	5.70		41.70	48.50	
4020	8" pipe size	↓	58	.138	↓	45.50	5.90		51.40	59	
4028	2" covering size										
4029	2-1/2" pipe size	1 Plum	62	.129	Ea.	13.40	5.50		18.90	23.50	
4080	10" pipe size		58	.138		50.50	5.90		56.40	64.50	
4090	12" pipe size		56	.143		74.50	6.10		80.60	91.50	
4100	14" pipe size		56	.143		74.50	6.10		80.60	91.50	
4110	16" pipe size		54	.148		105	6.35		111.35	125	
4120	18" pipe size		54	.148		105	6.35		111.35	125	
4130	20" pipe size		52	.154		116	6.55		122.55	138	
4150	24" pipe size		50	.160		136	6.85		142.85	160	
4160	30" pipe size		48	.167		149	7.10		156.10	175	
4180	36" pipe size	↓	45	.178	↓	165	7.60		172.60	193	
4200	Sockets										
4210	Rod end, malleable iron, TYPE 16										
4220	1/4" thread size	1 Plum	240	.033	Ea.	2.14	1.42		3.56	4.56	

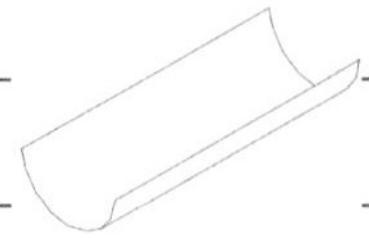

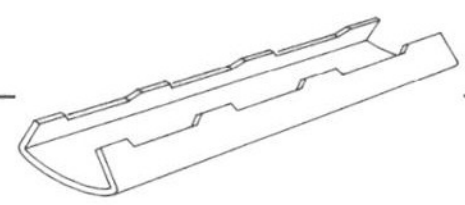

15 MECHANICAL

15060 | Hangers & Supports

300		CREW	DAILY OUTPUT	LABOR-HOURS	UNIT	2006 BARE COSTS MAT.	LABOR	EQUIP.	TOTAL	TOTAL INCL O&P	
4230	3/8" thread size	1 Plum	240	.033	Ea.	2.31	1.42		3.73	4.75	300
4240	1/2" thread size		230	.035		2.75	1.49		4.24	5.35	
4250	5/8" thread size		225	.036		5.45	1.52		6.97	8.35	
4260	3/4" thread size		220	.036		8.90	1.55		10.45	12.15	
4270	7/8" thread size		210	.038		17.40	1.63		19.03	21.50	
4290	Strap, 1/2" pipe size, TYPE 26	Q-1	142	.113		.49	4.33		4.82	7.30	
4300	3/4" pipe size		140	.114		.54	4.39		4.93	7.45	
4310	1" pipe size		137	.117		.55	4.49		5.04	7.60	
4320	1-1/4" pipe size		134	.119		.63	4.59		5.22	7.85	
4330	1-1/2" pipe size		131	.122		.70	4.69		5.39	8.05	
4340	2" pipe size		129	.124		.77	4.77		5.54	8.25	
4350	2-1/2" pipe size		125	.128		1.20	4.92		6.12	8.95	
4360	3" pipe size		122	.131		1.38	5.05		6.43	9.35	
4370	3-1/2" pipe size		119	.134		16.75	5.15		21.90	26.50	
4380	4" pipe size		114	.140		18.75	5.40		24.15	29	
4400	U-bolt, carbon steel										
4410	Standard, with nuts, TYPE 42										
4420	1/2" pipe size	1 Plum	160	.050	Ea.	.58	2.14		2.72	3.96	
4430	3/4" pipe size		158	.051		.58	2.16		2.74	4	
4450	1" pipe size		152	.053		.65	2.25		2.90	4.21	
4460	1-1/4" pipe size		148	.054		.77	2.31		3.08	4.44	
4470	1-1/2" pipe size		143	.056		.85	2.39		3.24	4.65	
4480	2" pipe size		139	.058		.92	2.46		3.38	4.83	
4490	2-1/2" pipe size		134	.060		1.60	2.55		4.15	5.70	
4500	3" pipe size		128	.063		1.70	2.67		4.37	6	
4510	3-1/2" pipe size		122	.066		1.80	2.80		4.60	6.35	
4520	4" pipe size		117	.068		1.83	2.92		4.75	6.55	
4530	5" pipe size		114	.070		2.11	3		5.11	7	
4540	6" pipe size		111	.072		2.11	3.08		5.19	7.10	
4550	8" pipe size		109	.073		3.90	3.13		7.03	9.15	
4560	10" pipe size		107	.075		6.50	3.19		9.69	12.10	
4570	12" pipe size		104	.077		8.50	3.28		11.78	14.45	
4580	For plastic coating on 1/2" thru 6" size, add					150%					
4700	U-hook, carbon steel, requires mounting screws or bolts										
4710	3/4" thru 2" pipe size										
4720	6" long	1 Plum	96	.083	Ea.	.33	3.56		3.89	5.90	
4730	8" long		96	.083		.41	3.56		3.97	6	
4740	10" long		96	.083		.46	3.56		4.02	6.05	
4750	12" long		96	.083		.51	3.56		4.07	6.10	
4760	For copper plated, add					50%					
8000	Pipe clamp, plastic, 1/2" CTS	1 Plum	80	.100		.20	4.27		4.47	6.85	
8010	3/4" CTS		73	.110		.21	4.68		4.89	7.55	
8020	1" CTS		68	.118		.50	5		5.50	8.35	
8080	Economy clamp, 1/4" CTS		175	.046		.05	1.95		2	3.10	
8090	3/8" CTS		168	.048		.07	2.03		2.10	3.24	
8100	1/2" CTS		160	.050		.07	2.14		2.21	3.40	
8110	3/4" CTS		145	.055		.14	2.36		2.50	3.81	
8200	Half clamp, 1/2" CTS		80	.100		.08	4.27		4.35	6.75	
8210	3/4" CTS		73	.110		.12	4.68		4.80	7.45	
8300	Suspension clamp, 1/2" CTS		80	.100		.20	4.27		4.47	6.85	
8310	3/4" CTS		73	.110		.21	4.68		4.89	7.55	
8320	1" CTS		68	.118		.50	5		5.50	8.35	
8400	Insulator, 1/2" CTS		80	.100		.34	4.27		4.61	7	
8410	3/4" CTS		73	.110		.35	4.68		5.03	7.70	
8420	1" CTS		68	.118		.36	5		5.36	8.20	
8800	Wire cable support system										

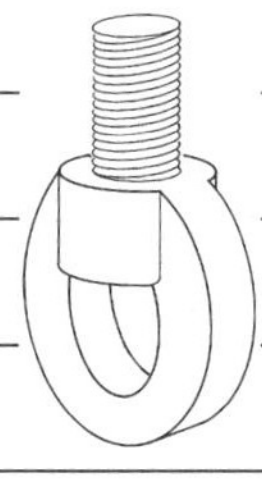

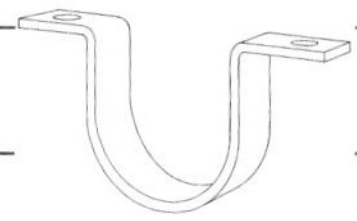

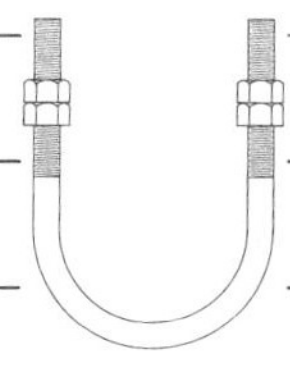

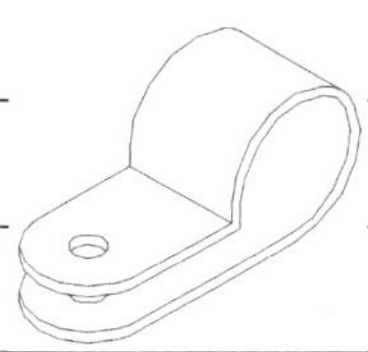

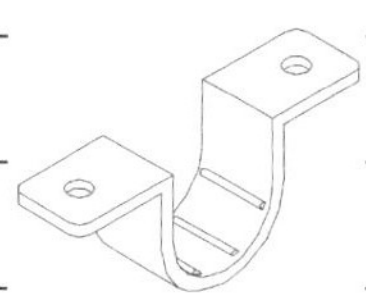

15060 | Hangers & Supports

		15060 Hangers & Supports	CREW	DAILY OUTPUT	LABOR-HOURS	UNIT	2006 BARE COSTS MAT.	LABOR	EQUIP.	TOTAL	TOTAL INCL O&P	
300	8810	Cable with hook terminal and locking device										300
	8830	2 mm, (.079 ″) dia cable, (100 lb. cap.)										
	8840	1 m, (3.3′) length, with hook	1 Shee	96	.083	Ea.	3.63	3.51		7.14	9.60	
	8850	2 m, (6.6′) length, with hook		84	.095		4.19	4.01		8.20	11	
	8860	3 m, (9.9′) length, with hook		72	.111		4.74	4.68		9.42	12.65	
	8870	5 m, (16.4′) length, with hook	Q-9	60	.267		5.95	10.10		16.05	22.50	
	8880	10 m, (32.8′) length, with hook	″	30	.533		8.65	20		28.65	41.50	
	8900	3mm, (.118 ″) dia cable, (200 lb. cap.)										
	8910	1 m, (3.3′) length, with hook	1 Shee	96	.083	Ea.	4.65	3.51		8.16	10.70	
	8920	2 m, (6.6′) length, with hook		84	.095		5.20	4.01		9.21	12.15	
	8930	3 m, (9.9′) length, with hook		72	.111		5.75	4.68		10.43	13.80	
	8940	5 m, (16.4′) length, with hook	Q-9	60	.267		7.10	10.10		17.20	24	
	8950	10 m, (32.8′) length, with hook	″	30	.533		10.25	20		30.25	43.50	
	9000	Cable system accessories										
	9010	Anchor bolt, 3/8″, with nut	1 Shee	140	.057	Ea.	1.32	2.41		3.73	5.30	
	9020	Air duct corner protector		160	.050		.62	2.11		2.73	4.03	
	9030	Air duct support attachment		140	.057		.23	2.41		2.64	4.08	
	9040	Flange clip, hammer-on style										
	9044	For flange thickness 3/32″ - 9/64″, 160 lb. cap.	1 Shee	180	.044	Ea.	.32	1.87		2.19	3.33	
	9048	For flange thickness 1/8″ - 1/4″, 200 lb. cap.		160	.050		.32	2.11		2.43	3.70	
	9052	For flange thickness 5/16″ - 1/2″, 200 lb. cap.		150	.053		.68	2.25		2.93	4.33	
	9056	For flange thickness 9/16″ - 3/4″, 200 lb. cap.		140	.057		.89	2.41		3.30	4.81	
	9060	Wire insulation protection tube		180	.044	L.F.	.33	1.87		2.20	3.34	
	9070	Wire cutter				Ea.	39			39	42.50	

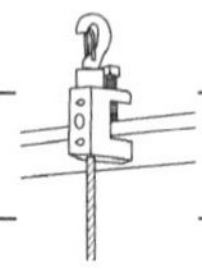

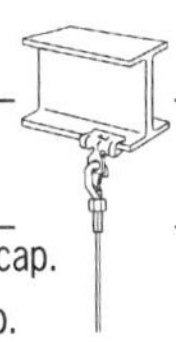

15070 | Mech Sound/Vibration/Seismic Ctrl

		15070 Mech Sound/Vibration/Seismic Ctrl	CREW	DAILY OUTPUT	LABOR-HOURS	UNIT	MAT.	LABOR	EQUIP.	TOTAL	TOTAL INCL O&P	
800	0010	VIBRATION ABSORBERS										800
	0100	Hangers, neoprene flex										
	0200	10 - 120 lb. capacity				Ea.	14.30			14.30	15.75	
	0220	75 - 550 lb. capacity					18.65			18.65	20.50	
	0240	250 - 1100 lb. capacity					39			39	43	
	0260	1000 - 4000 lb. capacity					63			63	69	
	0500	Spring flex, 60 lb. capacity					20			20	22	
	0520	450 lb. capacity					25			25	27.50	
	0540	900 lb. capacity					33			33	36.50	
	0560	1100-1300 lb. capacity					33			33	36.50	
	1000	Mounts, neoprene, 45 - 380 lb. capacity					10.80			10.80	11.90	
	1020	250 - 1100 lb. capacity					28.50			28.50	31	
	1040	1000 - 4000 lb. capacity					63.50			63.50	69.50	
	1100	Spring flex, 60 lb. capacity					39			39	43	
	1120	165 lb. capacity					39			39	43	
	1140	260 lb. capacity					39			39	43	
	1160	450 lb. capacity					51.50			51.50	56.50	
	1180	600 lb. capacity					51.50			51.50	56.50	
	1200	750 lb. capacity					51.50			51.50	56.50	
	1220	900 lb. capacity					53			53	58.50	
	1240	1100 lb. capacity					55			55	60.50	
	1260	1300 lb. capacity					56.50			56.50	62.50	
	1280	1500 lb. capacity					83			83	91.50	
	1300	1800 lb. capacity					89			89	98	
	1320	2200 lb. capacity					87			87	96	
	1340	2600 lb. capacity					89			89	98	
	2000	Pads, cork rib, 18″ x 18″ x 1″, 10-50 psi					102			102	112	
	2020	18″ x 36″ x 1″, 10-50 psi					204			204	224	
	2100	Shear flexible pads, 18″ x 18″ x 3/8″, 20-70 psi					84.50			84.50	93	
	2120	18″ x 36″ x 3/8″, 20-70 psi					157			157	173	

15050 | Basic Materials & Methods

15070 | Mech Sound/Vibration/Seismic Ctrl

		15070 Mech Sound/Vibration/Seismic Ctrl	CREW	DAILY OUTPUT	LABOR-HOURS	UNIT	2006 BARE COSTS MAT.	LABOR	EQUIP.	TOTAL	TOTAL INCL O&P	
800	3000	Note overlap in capacities due to deflections										800

15075 | Mechanical Identification

		15075 Mechanical Identification	CREW	DAILY OUTPUT	LABOR-HOURS	UNIT	MAT.	LABOR	EQUIP.	TOTAL	TOTAL INCL O&P	
400	0010	PIPING SYSTEM IDENTIFICATION LABELS										400
	0100	Indicate contents and flow direction										
	0106	Pipe markers										
	0110	Plastic snap around										
	0114	1/2" pipe	1 Plum	80	.100	Ea.	3.60	4.27		7.87	10.60	
	0116	3/4" pipe		80	.100		3.60	4.27		7.87	10.60	
	0118	1" pipe		80	.100		3.60	4.27		7.87	10.60	
	0120	2" pipe		75	.107		4.60	4.55		9.15	12.15	
	0122	3" pipe		70	.114		8.60	4.88		13.48	17.05	
	0124	4" pipe		60	.133		8.60	5.70		14.30	18.30	
	0126	6" pipe		60	.133		9.50	5.70		15.20	19.30	
	0128	8" pipe		56	.143		12.90	6.10		19	23.50	
	0130	10" pipe		56	.143		12.90	6.10		19	23.50	
	0200	Over 10" pipe size		50	.160		15.80	6.85		22.65	28	
	1110	Self adhesive										
	1114	1" pipe	1 Plum	80	.100	Ea.	2.60	4.27		6.87	9.50	
	1116	2" pipe		75	.107		2.60	4.55		7.15	9.95	
	1118	3" pipe		70	.114		4.20	4.88		9.08	12.20	
	1120	4" pipe		60	.133		4.20	5.70		9.90	13.45	
	1122	6" pipe		60	.133		4.20	5.70		9.90	13.45	
	1124	8" pipe		56	.143		4.20	6.10		10.30	14.10	
	1126	10" pipe		56	.143		5.90	6.10		12	16	
	1200	Over 10" pipe size		50	.160		8.20	6.85		15.05	19.60	
	2000	Valve tags										
	2010	Numbered plus identifying legend										
	2100	Brass, 2" diameter	1 Plum	40	.200	Ea.	2.20	8.55		10.75	15.70	
	2200	Plastic, 1-1/2" Dia.	"	40	.200	"	1.90	8.55		10.45	15.40	

15080 | Mechanical Insulation

		15080 Mechanical Insulation	CREW	DAILY OUTPUT	LABOR-HOURS	UNIT	MAT.	LABOR	EQUIP.	TOTAL	TOTAL INCL O&P	
200	0010	DUCT INSULATION										200
	0100	Rule of thumb, as a percentage of total mechanical costs				Job				10%		
	0110	Insulation req'd is based on the surface size/area to be covered										
	3000	Ductwork										
	3020	Blanket type, fiberglass, flexible										
	3030	Fire resistant liner, black coating one side										
	3050	1/2" thick, 2 lb. density	Q-14	380	.042	S.F.	.43	1.48		1.91	2.88	
	3060	1" thick, 1-1/2 lb. density		350	.046		.56	1.61		2.17	3.24	
	3070	1-1/2" thick, 1-1/2 lb. density		320	.050		.99	1.76		2.75	3.96	
	3080	2" thick, 1-1/2 lb. density		300	.053		1.16	1.87		3.03	4.34	
	3140	FRK vapor barrier wrap, .75 lb. density										
	3160	1" thick	Q-14	350	.046	S.F.	.28	1.61		1.89	2.93	
	3170	1-1/2" thick		320	.050		.30	1.76		2.06	3.20	
	3180	2" thick		300	.053		.43	1.87		2.30	3.53	
	3190	3" thick		260	.062		.53	2.16		2.69	4.11	
	3200	4" thick		242	.066		.83	2.32		3.15	4.70	
	3210	Vinyl jacket, same as FRK										
	3280	Unfaced, 1 lb. density										
	3310	1" thick	Q-14	360	.044	S.F.	.36	1.56		1.92	2.95	
	3320	1-1/2" thick		330	.048		.50	1.70		2.20	3.33	
	3330	2" thick		310	.052		.56	1.81		2.37	3.58	
	3490	Board type, fiberglass liner, 3 lb. density										

15080 | Mechanical Insulation

		15080 Mechanical Insulation	CREW	DAILY OUTPUT	LABOR-HOURS	UNIT	2006 BARE COSTS MAT.	LABOR	EQUIP.	TOTAL	TOTAL INCL O&P	
200	3500	Fire resistant, black pigmented, 1 side										200
	3520	1" thick	Q-14	150	.107	S.F.	1.61	3.75		5.36	7.85	
	3540	1-1/2" thick		130	.123		1.97	4.33		6.30	9.20	
	3560	2" thick		120	.133		2.37	4.69		7.06	10.25	
	3600	FSK vapor barrier										
	3620	1" thick	Q-14	150	.107	S.F.	1.76	3.75		5.51	8.05	
	3630	1-1/2" thick		130	.123		2.27	4.33		6.60	9.55	
	3640	2" thick		120	.133		2.76	4.69		7.45	10.70	
	3680	No finish										
	3700	1" thick	Q-14	170	.094	S.F.	.94	3.31		4.25	6.45	
	3710	1-1/2" thick		140	.114		1.40	4.02		5.42	8.10	
	3720	2" thick		130	.123		1.76	4.33		6.09	9	
	3730	Sheet insulation										
	3760	Polyethylene foam, closed cell, UV resistant										
	3770	Standard temperature (-90° to +212° F)										
	3771	1/4" thick	Q-14	450	.036	S.F.	1.30	1.25		2.55	3.47	
	3772	3/8" thick		440	.036		1.81	1.28		3.09	4.07	
	3773	1/2" thick		420	.038		2.26	1.34		3.60	4.67	
	3774	3/4" thick		400	.040		3.30	1.41		4.71	5.90	
	3775	1" thick		380	.042		4.14	1.48		5.62	6.95	
	3779	Adhesive (see line 7878)										
	3780	Foam, rubber										
	3782	1" thick	1 Stpi	50	.160	S.F.	2.66	6.90		9.56	13.65	
	3795	Finishes										
	3800	1/2" cement over 1" wire mesh, incl. corner bead	Q-14	100	.160	S.F.	1.59	5.60		7.19	10.90	
	3820	Glass cloth, pasted on		170	.094		2.50	3.31		5.81	8.15	
	3900	Weatherproof, non-metallic, 45 mil thick EPDM rubber sheeting		180	.089		2.45	3.12		5.57	7.80	
	7878	Contact cement, quart can				Ea.	22			22	24	
	9600	Minimum labor/equipment charge	1 Stpi	4	2	Job		86		86	134	
400	0010	**EQUIPMENT INSULATION**										400
	0100	Rule of thumb, as a percentage of total mechanical costs				Job				10%		
	0110	Insulation req'd is based on the surface size/area to be covered										
	1000	Boiler, 1-1/2" calcium silicate, 1/2" cement finish	Q-14	50	.320	S.F.	3.55	11.25		14.80	22.50	
	1020	2" fiberglass	"	80	.200	"	2.44	7.05		9.49	14.15	
	2000	Breeching, 2" calcium silicate with 1/2" cement finish, no lath										
	2020	Rectangular	Q-14	42	.381	S.F.	4.65	13.40		18.05	27	
	2040	Round	"	38.70	.413	"	4.85	14.55		19.40	29	
	2300	Calcium silicate block, + 200°F to + 1200°F										
	2310	On irregular surfaces, valves and fittings										
	2340	1" thick	Q-14	30	.533	S.F.	3.15	18.75		21.90	34	
	2360	1-1/2" thick		25	.640		3.25	22.50		25.75	40	
	2380	2" thick		22	.727		3.51	25.50		29.01	45.50	
	2400	3" thick		18	.889		6.15	31		37.15	58	
	2410	On plane surfaces										
	2420	1" thick	Q-14	126	.127	S.F.	3.15	4.46		7.61	10.75	
	2430	1-1/2" thick		120	.133		3.25	4.69		7.94	11.25	
	2440	2" thick		100	.160		3.51	5.60		9.11	13	
	2450	3" thick		70	.229		6.15	8.05		14.20	19.85	
	2900	Domestic water heater wrap kit										
	2920	1-1/2" with vinyl jacket, 20-60 gal.	1 Plum	8	1	Ea.	16.05	42.50		58.55	84	
600	0010	**PIPING INSULATION**										600
	0100	Rule of thumb, as a percentage of total mechanical costs				Job				10%		
	0110	Insulation req'd is based on the surface size/area to be covered										
	2930	Insulated protectors, (ADA)										

15080 | Mechanical Insulation

		15080 Mechanical Insulation	CREW	DAILY OUTPUT	LABOR-HOURS	UNIT	2006 BARE COSTS MAT.	LABOR	EQUIP.	TOTAL	TOTAL INCL O&P	
600	2935	For exposed piping under sinks or lavatories.										600
	2940	Vinyl coated foam, velcro tabs										
	2945	P Trap, 1-1/4″ or 1-1/2″	1 Plum	32	.250	Ea.	16.80	10.70		27.50	35	
	2960	Valve and supply cover										
	2965	1/2″, 3/8″, and 7/16″ pipe size	1 Plum	32	.250	Ea.	16.80	10.70		27.50	35	
	2970	Extension drain cover										
	2975	1-1/4″, or 1-1/2″ pipe size	1 Plum	32	.250	Ea.	17.60	10.70		28.30	36	
	2980	Tailpiece offset (wheelchair)										
	2985	1-1/4″ pipe size	1 Plum	32	.250	Ea.	19.60	10.70		30.30	38	
	4000	Pipe covering (price copper tube one size less than IPS)										
	4280	Cellular glass, closed cell foam, all service jacket, sealant,										
	4281	working temp. (-450°F to +900°F), 0 water vapor xmission										
	4284	1″ wall,										
	4286	1/2″ iron pipe size	Q-14	120	.133	L.F.	1.65	4.69		6.34	9.45	
	4300	1-1/2″ wall,										
	4301	1″ iron pipe size	Q-14	105	.152	L.F.	2.65	5.35		8	11.65	
	4304	2-1/2″ iron pipe size		90	.178		4.46	6.25		10.71	15.10	
	4306	3″ iron pipe size		85	.188		4.51	6.60		11.11	15.75	
	4308	4″ iron pipe size		70	.229		6	8.05		14.05	19.70	
	4310	5″ iron pipe size		65	.246		6.90	8.65		15.55	21.50	
	4320	2″ wall,										
	4322	1″ iron pipe size	Q-14	100	.160	L.F.	3.94	5.60		9.54	13.50	
	4324	2-1/2″ iron pipe size		85	.188		5.85	6.60		12.45	17.20	
	4326	3″ iron pipe size		80	.200		6	7.05		13.05	18.05	
	4328	4″ iron pipe size		65	.246		6.70	8.65		15.35	21.50	
	4330	5″ iron pipe size		60	.267		8.05	9.35		17.40	24	
	4332	6″ iron pipe size		50	.320		9.55	11.25		20.80	29	
	4336	8″ iron pipe size		40	.400		12.30	14.05		26.35	36.50	
	4338	10″ iron pipe size		35	.457		12.75	16.05		28.80	40	
	4350	2-1/2″ wall,										
	4360	12″ iron pipe size	Q-14	32	.500	L.F.	20	17.60		37.60	50.50	
	4362	14″ iron pipe size	″	28	.571	″	21.50	20		41.50	57	
	4370	3″ wall,										
	4378	6″ iron pipe size	Q-14	48	.333	L.F.	12.75	11.70		24.45	33	
	4380	8″ iron pipe size		38	.421		14.85	14.80		29.65	40.50	
	4382	10″ iron pipe size		33	.485		19.65	17.05		36.70	49.50	
	4384	16″ iron pipe size		25	.640		25.50	22.50		48	64.50	
	4386	18″ iron pipe size		22	.727		28	25.50		53.50	72	
	4388	20″ iron pipe size		20	.800		32.50	28		60.50	82	
	4400	3-1/2″ wall,										
	4412	12″ iron pipe size	Q-14	27	.593	L.F.	28.50	21		49.50	65.50	
	4414	14″ iron pipe size	″	25	.640	″	30.50	22.50		53	70	
	4430	4″ wall,										
	4446	16″ iron pipe size	Q-14	22	.727	L.F.	31	25.50		56.50	75.50	
	4448	18″ iron pipe size		20	.800		40	28		68	90	
	4450	20″ iron pipe size		18	.889		41.50	31		72.50	96.50	
	4480	Fittings, average with fabric and mastic										
	4484	1″ wall,										
	4486	1/2″ iron pipe size	1 Asbe	40	.200	Ea.	2.93	7.80		10.73	15.95	
	4500	1-1/2″ wall,										
	4502	1″ iron pipe size	1 Asbe	38	.211	Ea.	5.05	8.20		13.25	18.95	
	4504	2-1/2″ iron pipe size		32	.250		6.25	9.75		16	23	
	4506	3″ iron pipe size		30	.267		6.90	10.40		17.30	24.50	
	4508	4″ iron pipe size		28	.286		10.60	11.15		21.75	30	
	4510	5″ iron pipe size		24	.333		11.70	13		24.70	34	
	4520	2″ wall,										

MECHANICAL 15

15080 | Mechanical Insulation

			CREW	DAILY OUTPUT	LABOR-HOURS	UNIT	2006 BARE COSTS MAT.	LABOR	EQUIP.	TOTAL	TOTAL INCL O&P	
600	4522	1" iron pipe size	1 Asbe	36	.222	Ea.	5.85	8.70		14.55	20.50	600
	4524	2-1/2" iron pipe size		30	.267		6.90	10.40		17.30	24.50	
	4526	3" iron pipe size		28	.286		8.30	11.15		19.45	27.50	
	4528	4" iron pipe size		24	.333		11.15	13		24.15	33.50	
	4530	5" iron pipe size		22	.364		14.05	14.20		28.25	38.50	
	4532	6" iron pipe size		20	.400		16.45	15.60		32.05	43.50	
	4536	8" iron pipe size		12	.667		25	26		51	70	
	4538	10" iron pipe size		8	1		31	39		70	97.50	
	4550	2-1/2" wall,										
	4560	12" iron pipe size	1 Asbe	6	1.333	Ea.	53	52		105	144	
	4562	14" iron pipe size	"	4	2	"	61.50	78		139.50	195	
	4570	3" wall,										
	4578	6" iron pipe size	1 Asbe	16	.500	Ea.	19.10	19.55		38.65	53	
	4580	8" iron pipe size		10	.800		28	31		59	82	
	4582	10" iron pipe size		6	1.333		40	52		92	129	
	4584	16" iron pipe size		2	4		95.50	156		251.50	360	
	4586	18" iron pipe size		2	4		123	156		279	390	
	4588	20" iron pipe size		2	4		145	156		301	415	
	4600	3-1/2" wall,										
	4612	12" iron pipe size	1 Asbe	4	2	Ea.	69	78		147	203	
	4614	14" iron pipe size	"	2	4	"	76.50	156		232.50	340	
	4630	4" wall,										
	4646	16" iron pipe size	1 Asbe	2	4	Ea.	115	156		271	380	
	4648	18" iron pipe size		2	4		140	156		296	410	
	4650	20" iron pipe size		2	4		163	156		319	435	
	4900	Calcium silicate, with cover										
	5100	1" wall, 1/2" iron pipe size	Q-14	170	.094	L.F.	1.92	3.31		5.23	7.50	
	5130	3/4" iron pipe size		170	.094		1.94	3.31		5.25	7.55	
	5140	1" iron pipe size		170	.094		1.88	3.31		5.19	7.45	
	5150	1-1/4" iron pipe size		165	.097		1.92	3.41		5.33	7.65	
	5160	1-1/2" iron pipe size		165	.097		1.95	3.41		5.36	7.70	
	5170	2" iron pipe size		160	.100		2.24	3.52		5.76	8.20	
	5180	2-1/2" iron pipe size		160	.100		2.38	3.52		5.90	8.35	
	5190	3" iron pipe size		150	.107		2.57	3.75		6.32	8.95	
	5200	4" iron pipe size		140	.114		3.19	4.02		7.21	10.05	
	5210	5" iron pipe size		135	.119		3.36	4.17		7.53	10.50	
	5220	6" iron pipe size		130	.123		3.57	4.33		7.90	11	
	5280	1-1/2" wall, 1/2" iron pipe size		150	.107		2.08	3.75		5.83	8.40	
	5310	3/4" iron pipe size		150	.107		2.12	3.75		5.87	8.45	
	5320	1" iron pipe size		150	.107		2.30	3.75		6.05	8.65	
	5330	1-1/4" iron pipe size		145	.110		2.48	3.88		6.36	9.10	
	5340	1-1/2" iron pipe size		145	.110		2.66	3.88		6.54	9.30	
	5350	2" iron pipe size		140	.114		2.93	4.02		6.95	9.75	
	5360	2-1/2" iron pipe size		140	.114		3.19	4.02		7.21	10.05	
	5370	3" iron pipe size		135	.119		3.35	4.17		7.52	10.50	
	5380	4" iron pipe size		125	.128		3.88	4.50		8.38	11.60	
	5390	5" iron pipe size		120	.133		4.37	4.69		9.06	12.45	
	5400	6" iron pipe size		110	.145		4.52	5.10		9.62	13.30	
	5460	2" wall, 1/2" iron pipe size		135	.119		3.20	4.17		7.37	10.30	
	5490	3/4" iron pipe size		135	.119		3.37	4.17		7.54	10.50	
	5500	1" iron pipe size		135	.119		3.55	4.17		7.72	10.70	
	5510	1-1/4" iron pipe size		130	.123		3.80	4.33		8.13	11.25	
	5520	1-1/2" iron pipe size		130	.123		3.97	4.33		8.30	11.40	
	5530	2" iron pipe size		125	.128		4.17	4.50		8.67	11.95	
	5540	2-1/2" iron pipe size		125	.128		4.97	4.50		9.47	12.80	
600	5550	3" iron pipe size		120	.133		5	4.69		9.69	13.15	

	15080 | Mechanical Insulation	CREW	DAILY OUTPUT	LABOR-HOURS	UNIT	2006 BARE COSTS MAT.	LABOR	EQUIP.	TOTAL	TOTAL INCL O&P
600	5560 4" iron pipe size	Q-14	115	.139	L.F.	5.80	4.89		10.69	14.40
	5570 5" iron pipe size		110	.145		6.60	5.10		11.70	15.60
	5580 6" iron pipe size		105	.152		7.20	5.35		12.55	16.70
	5600 Calcium silicate, no cover									
	5720 1" wall, 1/2" iron pipe size	Q-14	180	.089	L.F.	1.75	3.12		4.87	7.05
	5740 3/4" iron pipe size		180	.089		1.75	3.12		4.87	7.05
	5750 1" iron pipe size		180	.089		1.67	3.12		4.79	6.95
	5760 1-1/4" iron pipe size		175	.091		1.70	3.21		4.91	7.10
	5770 1-1/2" iron pipe size		175	.091		1.72	3.21		4.93	7.15
	5780 2" iron pipe size		170	.094		1.98	3.31		5.29	7.60
	5790 2-1/2" iron pipe size		170	.094		2.09	3.31		5.40	7.70
	5800 3" iron pipe size		160	.100		2.25	3.52		5.77	8.25
	5810 4" iron pipe size		150	.107		2.81	3.75		6.56	9.20
	5820 5" iron pipe size		145	.110		2.93	3.88		6.81	9.55
	5830 6" iron pipe size		140	.114		3.08	4.02		7.10	9.95
	5900 1-1/2" wall, 1/2" iron pipe size		160	.100		1.85	3.52		5.37	7.80
	5920 3/4" iron pipe size		160	.100		1.88	3.52		5.40	7.80
	5930 1" iron pipe size		160	.100		2.05	3.52		5.57	8
	5940 1-1/4" iron pipe size		155	.103		2.21	3.63		5.84	8.35
	5950 1-1/2" iron pipe size		155	.103		2.37	3.63		6	8.50
	5960 2" iron pipe size		150	.107		2.61	3.75		6.36	8.95
	5970 2-1/2" iron pipe size		150	.107		2.85	3.75		6.60	9.25
	5980 3" iron pipe size		145	.110		2.98	3.88		6.86	9.65
	5990 4" iron pipe size		135	.119		3.45	4.17		7.62	10.60
	6000 5" iron pipe size		130	.123		3.89	4.33		8.22	11.35
	6010 6" iron pipe size		120	.133		3.97	4.69		8.66	12
	6020 7" iron pipe size		115	.139		4.66	4.89		9.55	13.15
	6030 8" iron pipe size		105	.152		5.25	5.35		10.60	14.55
	6040 9" iron pipe size		100	.160		6.30	5.60		11.90	16.10
	6050 10" iron pipe size		95	.168		6.95	5.90		12.85	17.30
	6060 12" iron pipe size		90	.178		8.25	6.25		14.50	19.30
	6070 14" iron pipe size		85	.188		9.35	6.60		15.95	21
	6080 16" iron pipe size		80	.200		10.45	7.05		17.50	23
	6090 18" iron pipe size		75	.213		11.55	7.50		19.05	25
	6120 2" wall, 1/2" iron pipe size		145	.110		2.92	3.88		6.80	9.55
	6140 3/4" iron pipe size		145	.110		3.08	3.88		6.96	9.75
	6150 1" iron pipe size		145	.110		3.24	3.88		7.12	9.90
	6160 1-1/4" iron pipe size		140	.114		3.47	4.02		7.49	10.35
	6170 1-1/2" iron pipe size		140	.114		3.62	4.02		7.64	10.55
	6180 2" iron pipe size		135	.119		3.80	4.17		7.97	11
	6190 2-1/2" iron pipe size		135	.119		4.58	4.17		8.75	11.85
	6200 3" iron pipe size		130	.123		4.59	4.33		8.92	12.10
	6210 4" iron pipe size		125	.128		5.30	4.50		9.80	13.20
	6220 5" iron pipe size		120	.133		6.05	4.69		10.74	14.30
	6230 6" iron pipe size		115	.139		6.60	4.89		11.49	15.30
	6240 7" iron pipe size		110	.145		7.20	5.10		12.30	16.25
	6250 8" iron pipe size		105	.152		7.90	5.35		13.25	17.45
	6260 9" iron pipe size		100	.160		8.95	5.60		14.55	18.95
	6270 10" iron pipe size		95	.168		9.90	5.90		15.80	20.50
	6280 12" iron pipe size		90	.178		11	6.25		17.25	22.50
	6290 14" iron pipe size		85	.188		12.10	6.60		18.70	24
	6300 16" iron pipe size		80	.200		13.35	7.05		20.40	26
	6310 18" iron pipe size		75	.213		14.55	7.50		22.05	28.50
	6320 20" iron pipe size		65	.246		17.95	8.65		26.60	34
	6330 22" iron pipe size		60	.267		19.85	9.35		29.20	37.50
	6340 24" iron pipe size		55	.291		20.50	10.25		30.75	39

15080 | Mechanical Insulation

		15080 Mechanical Insulation	CREW	DAILY OUTPUT	LABOR-HOURS	UNIT	2006 BARE COSTS MAT.	LABOR	EQUIP.	TOTAL	TOTAL INCL O&P	
600	6360	3" wall, 1/2" iron pipe size	Q-14	115	.139	L.F.	5.30	4.89		10.19	13.80	600
	6380	3/4" iron pipe size		115	.139		5.35	4.89		10.24	13.85	
	6390	1" iron pipe size		115	.139		5.35	4.89		10.24	13.90	
	6400	1-1/4" iron pipe size		110	.145		5.45	5.10		10.55	14.35	
	6410	1-1/2" iron pipe size		110	.145		5.50	5.10		10.60	14.40	
	6420	2" iron pipe size		105	.152		5.65	5.35		11	15	
	6430	2-1/2" iron pipe size		105	.152		6.65	5.35		12	16.05	
	6440	3" iron pipe size		100	.160		6.70	5.60		12.30	16.55	
	6450	4" iron pipe size		95	.168		8.55	5.90		14.45	19.10	
	6460	5" iron pipe size		90	.178		9.30	6.25		15.55	20.50	
	6470	6" iron pipe size		90	.178		10.45	6.25		16.70	21.50	
	6480	7" iron pipe size		85	.188		11.60	6.60		18.20	23.50	
	6490	8" iron pipe size		85	.188		12.45	6.60		19.05	24.50	
	6500	9" iron pipe size		80	.200		14.10	7.05		21.15	27	
	6510	10" iron pipe size		75	.213		15	7.50		22.50	29	
	6520	12" iron pipe size		70	.229		16.60	8.05		24.65	31.50	
	6530	14" iron pipe size		65	.246		18.65	8.65		27.30	34.50	
	6540	16" iron pipe size		60	.267		20.50	9.35		29.85	38	
	6550	18" iron pipe size		55	.291		22.50	10.25		32.75	41.50	
	6560	20" iron pipe size		50	.320		27	11.25		38.25	48	
	6570	22" iron pipe size		45	.356		29	12.50		41.50	52.50	
	6580	24" iron pipe size		40	.400		31.50	14.05		45.55	57.50	
	6600	Fiberglass, with all service jacket										
	6640	1/2" wall, 1/2" iron pipe size	Q-14	250	.064	L.F.	.73	2.25		2.98	4.47	
	6660	3/4" iron pipe size		240	.067		.83	2.34		3.17	4.73	
	6670	1" iron pipe size		230	.070		.86	2.45		3.31	4.94	
	6680	1-1/4" iron pipe size		220	.073		.91	2.56		3.47	5.15	
	6690	1-1/2" iron pipe size		220	.073		1.05	2.56		3.61	5.35	
	6700	2" iron pipe size		210	.076		1.12	2.68		3.80	5.60	
	6710	2-1/2" iron pipe size		200	.080		1.17	2.81		3.98	5.90	
	6840	1" wall, 1/2" iron pipe size		240	.067		.89	2.34		3.23	4.80	
	6860	3/4" iron pipe size		230	.070		.97	2.45		3.42	5.05	
	6870	1" iron pipe size		220	.073		1.04	2.56		3.60	5.30	
	6880	1-1/4" iron pipe size		210	.076		1.12	2.68		3.80	5.60	
	6890	1-1/2" iron pipe size		210	.076		1.21	2.68		3.89	5.70	
	6900	2" iron pipe size		200	.080		1.31	2.81		4.12	6.05	
	6910	2-1/2" iron pipe size		190	.084		1.49	2.96		4.45	6.45	
	6920	3" iron pipe size		180	.089		1.59	3.12		4.71	6.85	
	6930	3-1/2" iron pipe size		170	.094		1.73	3.31		5.04	7.30	
	6940	4" iron pipe size		150	.107		2.11	3.75		5.86	8.40	
	6950	5" iron pipe size		140	.114		2.38	4.02		6.40	9.15	
	6960	6" iron pipe size		120	.133		2.52	4.69		7.21	10.40	
	6970	7" iron pipe size		110	.145		3	5.10		8.10	11.65	
	6980	8" iron pipe size		100	.160		4.12	5.60		9.72	13.70	
	6990	9" iron pipe size		90	.178		4.34	6.25		10.59	14.95	
	7000	10" iron pipe size		90	.178		4.37	6.25		10.62	15	
	7010	12" iron pipe size		80	.200		4.76	7.05		11.81	16.70	
	7020	14" iron pipe size		80	.200		5.50	7.05		12.55	17.50	
	7030	16" iron pipe size		70	.229		7	8.05		15.05	21	
	7040	18" iron pipe size		70	.229		7.75	8.05		15.80	21.50	
	7050	20" iron pipe size		60	.267		8.65	9.35		18	25	
	7060	24" iron pipe size		60	.267		10.55	9.35		19.90	27	
	7080	1-1/2" wall, 1/2" iron pipe size		230	.070		1.67	2.45		4.12	5.85	
	7100	3/4" iron pipe size		220	.073		1.68	2.56		4.24	6	
	7110	1" iron pipe size		210	.076		1.80	2.68		4.48	6.35	
600	7120	1-1/4" iron pipe size		200	.080		1.96	2.81		4.77	6.75	

15080 | Mechanical Insulation		CREW	DAILY OUTPUT	LABOR-HOURS	UNIT	2006 BARE COSTS MAT.	LABOR	EQUIP.	TOTAL	TOTAL INCL O&P
600 7130	1-1/2" iron pipe size	Q-14	200	.080	L.F.	2.05	2.81		4.86	6.85 600
7140	2" iron pipe size		190	.084		2.26	2.96		5.22	7.30
7150	2-1/2" iron pipe size		180	.089		2.42	3.12		5.54	7.75
7160	3" iron pipe size		170	.094		2.54	3.31		5.85	8.20
7170	3-1/2" iron pipe size		160	.100		2.78	3.52		6.30	8.80
7180	4" iron pipe size		140	.114		2.88	4.02		6.90	9.70
7190	5" iron pipe size		130	.123		3.23	4.33		7.56	10.60
7200	6" iron pipe size		110	.145		3.41	5.10		8.51	12.10
7210	7" iron pipe size		100	.160		3.79	5.60		9.39	13.30
7220	8" iron pipe size		90	.178		4.78	6.25		11.03	15.45
7230	9" iron pipe size		85	.188		4.98	6.60		11.58	16.30
7240	10" iron pipe size		80	.200		5.15	7.05		12.20	17.10
7250	12" iron pipe size		75	.213		5.85	7.50		13.35	18.70
7260	14" iron pipe size		70	.229		6.70	8.05		14.75	20.50
7270	16" iron pipe size		65	.246		8.75	8.65		17.40	24
7280	18" iron pipe size		60	.267		9.85	9.35		19.20	26
7290	20" iron pipe size		55	.291		10.05	10.25		20.30	28
7300	24" iron pipe size		50	.320		12.30	11.25		23.55	32
7320	2" wall, 1/2" iron pipe size		220	.073		2.60	2.56		5.16	7.05
7340	3/4" iron pipe size		210	.076		2.68	2.68		5.36	7.30
7350	1" iron pipe size		200	.080		2.86	2.81		5.67	7.75
7360	1-1/4" iron pipe size		190	.084		3.02	2.96		5.98	8.15
7370	1-1/2" iron pipe size		190	.084		3.14	2.96		6.10	8.30
7380	2" iron pipe size		180	.089		3.32	3.12		6.44	8.75
7390	2-1/2" iron pipe size		170	.094		3.57	3.31		6.88	9.35
7400	3" iron pipe size		160	.100		3.79	3.52		7.31	9.90
7410	3-1/2" iron pipe size		150	.107		4.10	3.75		7.85	10.60
7420	4" iron pipe size		130	.123		4.42	4.33		8.75	11.90
7430	5" iron pipe size		120	.133		5.05	4.69		9.74	13.20
7440	6" iron pipe size		100	.160		5.20	5.60		10.80	14.90
7450	7" iron pipe size		90	.178		5.95	6.25		12.20	16.70
7460	8" iron pipe size		80	.200		6.35	7.05		13.40	18.40
7470	9" iron pipe size		75	.213		6.95	7.50		14.45	19.90
7480	10" iron pipe size		70	.229		7.60	8.05		15.65	21.50
7490	12" iron pipe size		65	.246		8.50	8.65		17.15	23.50
7500	14" iron pipe size		60	.267		10.55	9.35		19.90	27
7510	16" iron pipe size		55	.291		11.60	10.25		21.85	29.50
7520	18" iron pipe size		50	.320		12.95	11.25		24.20	32.50
7530	20" iron pipe size		45	.356		14.65	12.50		27.15	36.50
7540	24" iron pipe size	↓	40	.400	↓	15.75	14.05		29.80	40.50
7600	For fiberglass with standard canvas jacket, deduct					5%				
7660	For fittings, add 3 L.F. for each fitting									
7680	plus 4 L.F. for each flange of the fitting									
7700	For equipment or congested areas, add						20%			
7720	Finishes, for .010" aluminum jacket, add	Q-14	200	.080	S.F.	.59	2.81		3.40	5.25
7740	For .016" aluminum jacket, add		200	.080		.86	2.81		3.67	5.55
7760	For .010" stainless steel, add	↓	160	.100	↓	2.59	3.52		6.11	8.60
7780	For single layer of felt, add					10%	10%			
7879	Rubber tubing, flexible closed cell foam									
7880	3/8" wall, 1/4" iron pipe size	1 Asbe	120	.067	L.F.	.24	2.60		2.84	4.51
7900	3/8" iron pipe size		120	.067		.27	2.60		2.87	4.55
7910	1/2" iron pipe size		115	.070		.30	2.72		3.02	4.76
7920	3/4" iron pipe size		115	.070		.34	2.72		3.06	4.80
7930	1" iron pipe size		110	.073		.39	2.84		3.23	5.05
7940	1-1/4" iron pipe size		110	.073		.48	2.84		3.32	5.15
7950	1-1/2" iron pipe size	↓	110	.073	↓	.53	2.84		3.37	5.20

MECHANICAL 15

15050 | Basic Materials & Methods

	15080 | Mechanical Insulation		CREW	DAILY OUTPUT	LABOR-HOURS	UNIT	2006 BARE COSTS MAT.	LABOR	EQUIP.	TOTAL	TOTAL INCL O&P	
600	8100	1/2" wall, 1/4" iron pipe size	1 Asbe	90	.089	L.F.	.36	3.47		3.83	6.05	600
	8120	3/8" iron pipe size		90	.089		.39	3.47		3.86	6.10	
	8130	1/2" iron pipe size		89	.090		.44	3.51		3.95	6.25	
	8140	3/4" iron pipe size		89	.090		.49	3.51		4	6.30	
	8150	1" iron pipe size		88	.091		.54	3.55		4.09	6.40	
	8160	1-1/4" iron pipe size		87	.092		.63	3.59		4.22	6.55	
	8170	1-1/2" iron pipe size		87	.092		.76	3.59		4.35	6.70	
	8180	2" iron pipe size		86	.093		.97	3.63		4.60	7	
	8190	2-1/2" iron pipe size		86	.093		1.09	3.63		4.72	7.15	
	8200	3" iron pipe size		85	.094		1.37	3.68		5.05	7.50	
	8210	3-1/2" iron pipe size		85	.094		1.90	3.68		5.58	8.10	
	8220	4" iron pipe size		80	.100		2.03	3.91		5.94	8.60	
	8230	5" iron pipe size		80	.100		2.79	3.91		6.70	9.40	
	8300	3/4" wall, 1/4" iron pipe size		90	.089		.56	3.47		4.03	6.25	
	8320	3/8" iron pipe size		90	.089		.61	3.47		4.08	6.30	
	8330	1/2" iron pipe size		89	.090		.73	3.51		4.24	6.55	
	8340	3/4" iron pipe size		89	.090		.89	3.51		4.40	6.75	
	8350	1" iron pipe size		88	.091		1.01	3.55		4.56	6.90	
	8360	1-1/4" iron pipe size		87	.092		1.36	3.59		4.95	7.35	
	8370	1-1/2" iron pipe size		87	.092		1.54	3.59		5.13	7.55	
	8380	2" iron pipe size		86	.093		1.81	3.63		5.44	7.95	
	8390	2-1/2" iron pipe size		86	.093		2.42	3.63		6.05	8.60	
	8400	3" iron pipe size		85	.094		2.76	3.68		6.44	9.05	
	8410	3-1/2" iron pipe size		85	.094		3.28	3.68		6.96	9.60	
	8420	4" iron pipe size		80	.100		3.49	3.91		7.40	10.20	
	8430	5" iron pipe size		80	.100		4.21	3.91		8.12	11	
	8440	6" iron pipe size		80	.100		5.10	3.91		9.01	11.95	
	8444	1" wall, 1/2" iron pipe size		86	.093		1.41	3.63		5.04	7.50	
	8445	3/4" iron pipe size		84	.095		1.71	3.72		5.43	7.95	
	8446	1" iron pipe size		84	.095		1.99	3.72		5.71	8.25	
	8447	1-1/4" iron pipe size		82	.098		2.25	3.81		6.06	8.70	
	8448	1-1/2" iron pipe size		82	.098		2.61	3.81		6.42	9.05	
	8449	2" iron pipe size		80	.100		3.49	3.91		7.40	10.20	
	8450	2-1/2" iron pipe size	↓	80	.100	↓	4.55	3.91		8.46	11.35	
	8456	Rubber insulation tape, 1/8" x 2" x 30'				Ea.	11.40			11.40	12.50	
	9600	Minimum labor/equipment charge	1 Plum	4	2	Job		85.50		85.50	133	

15100 | Building Services Piping

	15106 | Glass Pipe & Fittings		CREW	DAILY OUTPUT	LABOR-HOURS	UNIT	2006 BARE COSTS MAT.	LABOR	EQUIP.	TOTAL	TOTAL INCL O&P	
120	0010	**PIPE, GLASS** Borosilicate, couplings & hangers 10' O.C.										120
	0020	Drainage										
	1100	1-1/2" diameter	Q-1	52	.308	L.F.	9.30	11.80		21.10	28.50	
	1120	2" diameter		44	.364		12.40	13.95		26.35	35	
	1140	3" diameter		39	.410		16.60	15.75		32.35	43	
	1160	4" diameter		30	.533		30	20.50		50.50	65	
	1180	6" diameter	↓	26	.615	↓	54	23.50		77.50	96.50	
	1870	To delete coupling & hanger, subtract										
	1880	1-1/2" diam. to 2" diam.					19%	22%				
	1890	3" diam. to 6" diam.					20%	17%				

15106 | Glass Pipe & Fittings

		15106 Glass Pipe & Fittings	CREW	DAILY OUTPUT	LABOR-HOURS	UNIT	2006 BARE COSTS MAT.	LABOR	EQUIP.	TOTAL	TOTAL INCL O&P	
120	2000	Process supply (pressure), beaded joints										120
	2040	1/2″ diameter	1 Plum	36	.222	L.F.	3.36	9.50		12.86	18.45	
	2060	3/4″ diameter		31	.258		3.78	11		14.78	21.50	
	2080	1″ diameter		27	.296		13.45	12.65		26.10	34.50	
	2100	1-1/2″ diameter	Q-1	47	.340		16.20	13.10		29.30	38.50	
	2120	2″ diameter		39	.410		21.50	15.75		37.25	48	
	2140	3″ diameter		34	.471		31	18.10		49.10	62.50	
	2160	4″ diameter		25	.640		43.50	24.50		68	86	
	2180	6″ diameter		21	.762		86	29.50		115.50	140	
	2860	To delete coupling & hanger, subtract										
	2870	1/2″ diam. to 1″ diam.					25%	33%				
	2880	1-1/2″ diam. to 3″ diam.					22%	21%				
	2890	4″ diam. to 6″ diam.					23%	15%				
	3800	Conical joint, transparent										
	3980	6″ diameter	Q-1	21	.762	L.F.	570	29.50		599.50	675	
	4500	To delete couplings & hangers, subtract										
	4530	6″ diam.					22%	26%				
	9000	Minimum labor/equipment charge	1 Plum	4	2	Job		85.50		85.50	133	
160	0010	**PIPE, GLASS, FITTINGS**										160
	0020	Drainage, beaded ends										
	0040	Coupling & labor required at joints not incl. in fitting										
	0050	price. Add 1 per joint for installed price										
	0070	90° Bend or sweep, 1-1/2″				Ea.	20.50			20.50	22.50	
	0090	2″					26			26	28.50	
	0100	3″					42.50			42.50	47	
	0110	4″					68.50			68.50	75	
	0120	6″ (sweep only)					208			208	229	
	0200	45° Bend or sweep same as 90°										
	0350	Tee, single sanitary, 1-1/2″				Ea.	33			33	36.50	
	0370	2″					33			33	36.50	
	0380	3″					49.50			49.50	54.50	
	0390	4″					88.50			88.50	97.50	
	0400	6″					238			238	262	
	0410	Tee, straight, 1-1/2″					41			41	45	
	0430	2″					41			41	45	
	0440	3″					59			59	65	
	0450	4″					104			104	115	
	0460	6″					256			256	282	
	0500	Coupling, stainless steel, TFE seal ring										
	0520	1-1/2″	Q-1	32	.500	Ea.	14.30	19.20		33.50	46	
	0530	2″		30	.533		18.10	20.50		38.60	52	
	0540	3″		25	.640		24.50	24.50		49	65	
	0550	4″		23	.696		42	26.50		68.50	87.50	
	0560	6″		20	.800		94.50	30.50		125	152	
	0600	Coupling, stainless steel, bead to plain end										
	0610	1-1/2″	Q-1	36	.444	Ea.	18.45	17.10		35.55	47	
	0620	2″		34	.471		23	18.10		41.10	53.50	
	0630	3″		29	.552		39	21		60	76	
	0640	4″		27	.593		58.50	23		81.50	100	
	0650	6″		24	.667		198	25.50		223.50	258	
	2000	Process supply (pressure), beaded ends										
	2050	90° Sweep elbow, 1/2″				Ea.	20.50			20.50	22.50	
	2070	3/4″					24.50			24.50	27	
	2080	1″					32.50			32.50	35.50	
	2090	1-1/2″					34			34	37.50	
	2100	2″					45.50			45.50	50	

MECHANICAL 15

15106 | Glass Pipe & Fittings

			CREW	DAILY OUTPUT	LABOR-HOURS	UNIT	2006 BARE COSTS MAT.	LABOR	EQUIP.	TOTAL	TOTAL INCL O&P	
160	2120	3″				Ea.	72			72	79	160
	2130	4″					101			101	111	
	2140	6″				↓	253			253	278	
	2150	45° Sweep elbow, same as 90°										
	2250	Tee, straight, 1/2″				Ea.	35			35	38.50	
	2270	3/4″					35			35	38.50	
	2280	1″					39			39	43	
	2290	1-1/2″					58			58	63.50	
	2300	2″					68.50			68.50	75.50	
	2310	3″					83.50			83.50	92	
	2320	4″					140			140	154	
	2330	6″				↓	261			261	287	
	2350	Coupling, Viton liner, for temperatures to 400°F										
	2370	1/2″	Q-1	40	.400	Ea.	29.50	15.35		44.85	56.50	
	2380	3/4″		37	.432		33.50	16.60		50.10	63	
	2390	1″		35	.457		40	17.55		57.55	71.50	
	2400	1-1/2″		32	.500		45	19.20		64.20	80	
	2410	2″		30	.533		54.50	20.50		75	92	
	2420	3″		25	.640		100	24.50		124.50	148	
	2430	4″		23	.696		120	26.50		146.50	175	
	2440	6″	↓	20	.800	↓	255	30.50		285.50	330	
	2550	For beaded joint armored fittings, add					200%					
	2600	Conical ends. Flange set, gasket & labor not incl. in fitting										
	2620	price. Add 1 per joint for installed price.										
	2650	90° Sweep elbow, 1″				Ea.	53			53	58.50	
	2670	1-1/2″					77			77	84.50	
	2680	2″					89			89	98	
	2690	3″					178			178	196	
	2700	4″					340			340	375	
	2710	6″					645			645	710	
	2750	Cross (straight), add					55%					
	2850	Tee, add				↓	20%					
	9000	Minimum labor/equipment charge	1 Plum	4	2	Job		85.50		85.50	133	

15107 | Metal Pipe & Fittings

			CREW	DAILY OUTPUT	LABOR-HOURS	UNIT	MAT.	LABOR	EQUIP.	TOTAL	TOTAL INCL O&P	
320	0010	**PIPE, CAST IRON** Soil, on hangers 5′ O.C. R221113-50										320
	0020	Single hub, service wt., lead & oakum joints 10′ O.C.										
	2120	2″ diameter	Q-1	63	.254	L.F.	5	9.75		14.75	20.50	
	2140	3″ diameter		60	.267		6.95	10.25		17.20	23.50	
	2160	4″ diameter	↓	55	.291		8.90	11.20		20.10	27	
	2180	5″ diameter 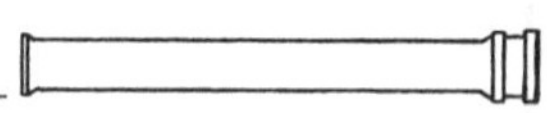	Q-2	76	.316		12.25	12.60		24.85	33	
	2200	6″ diameter	″	73	.329		15.05	13.10		28.15	37	
	2220	8″ diameter	Q-3	59	.542		23	22		45	60	
	2240	10″ diameter		54	.593		42	24		66	84	
	2260	12″ diameter		48	.667		59	27		86	107	
	2261	15″ diameter	↓	40	.800		91.50	32.50		124	152	
	2320	For service weight, double hub, add					10%					
	2340	For extra heavy, single hub, add					48%	4%				
	2360	For extra heavy, double hub, add				↓	71%	4%				
	2400	Lead for caulking, (1#/dia in)	Q-1	160	.100	Lb.	1.04	3.84		4.88	7.15	
	2420	Oakum for caulking, (1/8#/dia in)	″	40	.400	″	1.80	15.35		17.15	26	
	2960	To delete hangers, subtract										
	2970	2″ diam. to 4″ diam.					16%	19%				
	2980	5″ diam. to 8″ diam.					14%	14%				
	2990	10″ diam. to 15″ diam.					13%	19%				

15 MECHANICAL

15107 | Metal Pipe & Fittings

		15107 Metal Pipe & Fittings	CREW	DAILY OUTPUT	LABOR-HOURS	UNIT	2006 BARE COSTS MAT.	LABOR	EQUIP.	TOTAL	TOTAL INCL O&P	
320	3000	Single hub, service wt, push-on gasket joints 10′ O.C. R221113-50										320
	3010	2″ diameter	Q-1	66	.242	L.F.	5.35	9.30		14.65	20.50	
	3020	3″ diameter		63	.254		7.40	9.75		17.15	23.50	
	3030	4″ diameter		57	.281		9.50	10.80		20.30	27	
	3040	5″ diameter	Q-2	79	.304		13.10	12.10		25.20	33.50	
	3050	6″ diameter	"	75	.320		15.95	12.75		28.70	37.50	
	3060	8″ diameter	Q-3	62	.516		25	21		46	60	
	3070	10″ diameter		56	.571		45	23.50		68.50	85.50	
	3080	12″ diameter		49	.653		63	26.50		89.50	111	
	3082	15″ diameter		40	.800		96.50	32.50		129	157	
	3100	For service weight, double hub, add					65%					
	3110	For extra heavy, single hub, add					48%	4%				
	3120	For extra heavy, double hub, add					29%	4%				
	3130	To delete hangers, subtract										
	3140	2″ diam. to 4″ diam.					12%	21%				
	3150	5″ diam. to 8″ diam.					10%	16%				
	3160	10″ diam. to 15″ diam.					9%	21%				
	4000	No hub, couplings 10′ O.C.										
	4100	1-1/2″ diameter	Q-1	71	.225	L.F.	5.25	8.65		13.90	19.20	
	4120	2″ diameter		67	.239		5.40	9.20		14.60	20	
	4140	3″ diameter 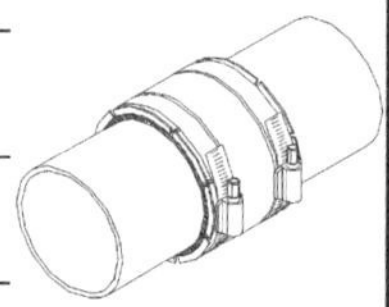		64	.250		7.20	9.60		16.80	23	
	4160	4″ diameter		58	.276		9.15	10.60		19.75	26.50	
	4180	5″ diameter	Q-2	83	.289		13	11.50		24.50	32	
	4200	6″ diameter	"	79	.304		16.25	12.10		28.35	37	
	4220	8″ diameter	Q-3	69	.464		25.50	18.85		44.35	57.50	
	4240	10″ diameter	"	61	.525		45.50	21.50		67	83	
	4280	To delete hangers, subtract										
	4290	1-1/2″ diam. to 6″ diam.					22%	47%				
	4300	8″ diam. to 10″ diam.					21%	44%				
	9000	Minimum labor/equipment charge	1 Plum	4	2	Job		85.50		85.50	133	
360	0010	**PIPE, CAST IRON, FITTINGS** Soil										360
	0040	Hub and spigot, service weight, lead & oakum joints										
	0080	1/4 bend, 2″	Q-1	16	1	Ea.	10.35	38.50		48.85	71.50	
	0120	3″		14	1.143		13.80	44		57.80	83.50	
	0140	4″ 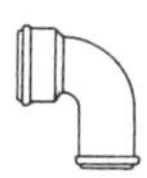		13	1.231		21.50	47.50		69	97	
	0160	5″	Q-2	18	1.333		30	53		83	116	
	0180	6″	"	17	1.412		37.50	56.50		94	129	
	0200	8″	Q-3	11	2.909		113	118		231	310	
	0220	10″		10	3.200		165	130		295	385	
	0224	12″		9	3.556		224	145		369	470	
	0266	Closet bend, 3″ diameter with flange 10″ x 16″	Q-1	14	1.143		67	44		111	142	
	0268	16″x16″		12	1.333		92.50	51		143.50	181	
	0270	Closet bend, 4″ diameter, 1″ x 4″ ring, 6″ x 16″		13	1.231		74	47.50		121.50	155	
	0280	8″ x 16″		13	1.231		66	47.50		113.50	147	
	0290	10″ x 12″		12	1.333		62.50	51		113.50	148	
	0300	10″ x 18″		11	1.455		87	56		143	183	
	0310	12″ x 16″		11	1.455		75	56		131	170	
	0330	16″ x 16″		10	1.600		94.50	61.50		156	200	
	0340	1/8 bend, 2″		16	1		7.35	38.50		45.85	68	
	0350	3″		14	1.143		11.55	44		55.55	81	
	0360	4″		13	1.231		16.90	47.50		64.40	92	
	0380	5″ 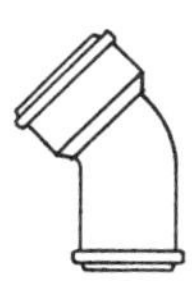	Q-2	18	1.333		23.50	53		76.50	108	
	0400	6″	"	17	1.412		28.50	56.50		85	119	
	0420	8″	Q-3	11	2.909		85	118		203	278	
	0440	10″		10	3.200		123	130		253	340	
	0460	12″		9	3.556		233	145		378	480	

MECHANICAL 15

15107 | Metal Pipe & Fittings

360		CREW	DAILY OUTPUT	LABOR-HOURS	UNIT	2006 BARE COSTS MAT.	LABOR	EQUIP.	TOTAL	TOTAL INCL O&P
0500	Sanitary tee, 2″	Q-1	10	1.600	Ea.	14.50	61.50		76	111
0540	3″		9	1.778		23.50	68.50		92	132
0620	4″		8	2		28.50	77		105.50	152
0700	5″	Q-2	12	2		57	79.50		136.50	187
0800	6″	"	11	2.182		64.50	87		151.50	206
0880	8″	Q-3	7	4.571		171	186		357	475
1000	Tee, 2″	Q-1	10	1.600		19.15	61.50		80.65	117
1060	3″		9	1.778		31	68.50		99.50	140
1120	4″		8	2		40	77		117	164
1200	5″	Q-2	12	2		61	79.50		140.50	192
1300	6″	"	11	2.182		83	87		170	227
1380	8″	Q-3	7	4.571		183	186		369	490
1400	Combination Y and 1/8 bend									
1420	2″	Q-1	10	1.600	Ea.	18	61.50		79.50	115
1460	3″		9	1.778		27.50	68.50		96	136
1520	4″		8	2		38	77		115	162
1540	5″	Q-2	12	2		72	79.50		151.50	203
1560	6″		11	2.182		91	87		178	235
1580	8″		7	3.429		224	137		361	460
1582	12″	Q-3	6	5.333		505	217		722	890
1600	Double Y, 2″	Q-1	8	2		23	77		100	146
1610	3″		7	2.286		40	88		128	181
1620	4″		6.50	2.462		52.50	94.50		147	205
1630	5″	Q-2	9	2.667		89.50	106		195.50	264
1640	6″	"	8	3		134	120		254	335
1650	8″	Q-3	5.50	5.818		330	237		567	730
1660	10″		5	6.400		465	260		725	915
1670	12″		4.50	7.111		665	289		954	1,175
1740	Reducer, 3″ x 2″	Q-1	15	1.067		10.10	41		51.10	74.50
1750	4″ x 2″		14.50	1.103		10.60	42.50		53.10	77.50
1760	4″ x 3″		14	1.143		13.15	44		57.15	83
1770	5″ x 2″		14	1.143		18.40	44		62.40	88.50
1780	5″ x 3″		13.50	1.185		17.90	45.50		63.40	90.50
1790	5″ x 4″		13	1.231		16.95	47.50		64.45	92
1800	6″ x 2″		13.50	1.185		23.50	45.50		69	97
1810	6″ x 3″		13	1.231		23.50	47.50		71	99.50
1840	6″ x 5″		11	1.455		28.50	56		84.50	119
1880	8″ x 3″	Q-2	13.50	1.778		48.50	71		119.50	163
1900	8″ x 4″		13	1.846		44	73.50		117.50	163
1920	8″ x 5″		12	2		46	79.50		125.50	175
1940	8″ x 6″		12	2		45.50	79.50		125	174
1960	Increaser, 2″ x 3″	Q-1	15	1.067		24	41		65	90
1980	2″ x 4″		14	1.143		24	44		68	95
2000	2″ x 5″		13	1.231		29	47.50		76.50	106
2020	3″ x 4″		13	1.231		26.50	47.50		74	103
2040	3″ x 5″		13	1.231		29	47.50		76.50	106
2060	3″ x 6″		12	1.333		36	51		87	119
2070	4″ x 5″		13	1.231		29	47.50		76.50	106
2080	4″ x 6″		12	1.333		36	51		87	119
2090	4″ x 8″	Q-2	13	1.846		73.50	73.50		147	195
2100	5″ x 6″	Q-1	11	1.455		53.50	56		109.50	146
2110	5″ x 8″	Q-2	12	2		86.50	79.50		166	219
2120	6″ x 8″		12	2		86.50	79.50		166	219
2130	6″ x 10″		8	3		156	120		276	360
2140	8″ x 10″		6.50	3.692		156	147		303	400
2150	10″ x 12″		5.50	4.364		279	174		453	575

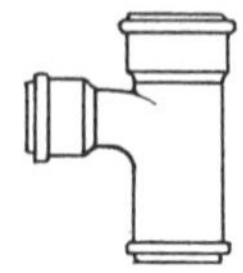
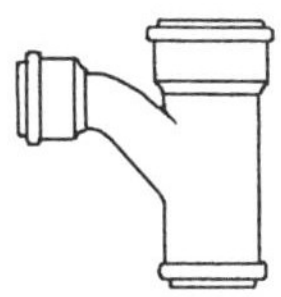
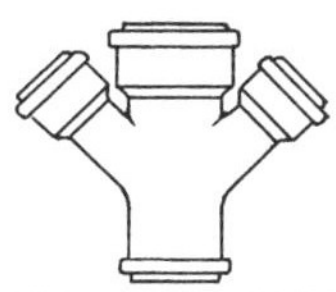
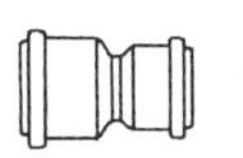
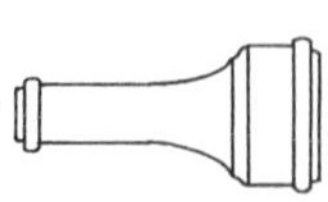

15107 | Metal Pipe & Fittings

360

		CREW	DAILY OUTPUT	LABOR-HOURS	UNIT	2006 BARE COSTS MAT.	LABOR	EQUIP.	TOTAL	TOTAL INCL O&P
2500	Y, 2″	Q-1	10	1.600	Ea.	13.15	61.50		74.65	110
2510	3″		9	1.778		24.50	68.50		93	133
2520	4″		8	2		32.50	77		109.50	156
2530	5″	Q-2	12	2		55	79.50		134.50	185
2540	6″	″	11	2.182		75	87		162	218
2550	8″	Q-3	7	4.571		184	186		370	490
2560	10″		6	5.333		297	217		514	660
2570	12″		5	6.400		510	260		770	965
2580	15″		4	8		1,075	325		1,400	1,700
3000	For extra heavy, add					44%	4%			
3600	Hub and spigot, service weight gasket joint									
3605	Note: gaskets and joint labor have									
3606	been included with all listed fittings.									
3610	1/4 bend, 2″	Q-1	20	.800	Ea.	13.70	30.50		44.20	63
3620	3″		17	.941		18.25	36		54.25	76
3630	4″		15	1.067		27	41		68	93.50
3640	5″	Q-2	21	1.143		39	45.50		84.50	114
3650	6″	″	19	1.263		46.50	50.50		97	130
3660	8″	Q-3	12	2.667		133	109		242	315
3670	10″		11	2.909		196	118		314	400
3680	12″		10	3.200		263	130		393	490
3700	Closet bend, 3″ diameter with ring 10″ x 16″	Q-1	17	.941		71	36		107	135
3710	16″ x 16″		15	1.067		96.50	41		137.50	170
3730	Closet bend, 4″ diameter, 1″ x 4″ ring, 6″ x 16″		15	1.067		79.50	41		120.50	151
3740	8″ x 16″		15	1.067		72	41		113	143
3750	10″ x 12″		14	1.143		68	44		112	144
3760	10″ x 18″		13	1.231		92.50	47.50		140	176
3770	12″ x 16″		13	1.231		80.50	47.50		128	162
3780	16″ x 16″		12	1.333		100	51		151	190
3800	1/8 bend, 2″		20	.800		10.75	30.50		41.25	60
3810	3″		17	.941		15.95	36		51.95	73.50
3820	4″		15	1.067		22.50	41		63.50	88.50
3830	5″	Q-2	21	1.143		32	45.50		77.50	106
3840	6″	″	19	1.263		37.50	50.50		88	120
3850	8″	Q-3	12	2.667		105	109		214	285
3860	10″		11	2.909		153	118		271	350
3870	12″		10	3.200		271	130		401	500
3900	Sanitary Tee, 2″	Q-1	12	1.333		21.50	51		72.50	103
3910	3″		10	1.600		32	61.50		93.50	131
3920	4″		9	1.778		40	68.50		108.50	150
3930	5″	Q-2	13	1.846		74	73.50		147.50	196
3940	6″	″	11	2.182		82.50	87		169.50	226
3950	8″	Q-3	8	4		211	163		374	485
3980	Tee, 2″	Q-1	12	1.333		26	51		77	108
3990	3″		10	1.600		40	61.50		101.50	140
4000	4″		9	1.778		51	68.50		119.50	162
4010	5″	Q-2	13	1.846		78.50	73.50		152	201
4020	6″	″	11	2.182		101	87		188	246
4030	8″	Q-3	8	4		222	163		385	500
4060	Combination Y and 1/8 bend									
4070	2″	Q-1	12	1.333	Ea.	25	51		76	107
4080	3″		10	1.600		36.50	61.50		98	136
4090	4″		9	1.778		49	68.50		117.50	160
4100	5″	Q-2	13	1.846		89	73.50		162.50	212
4110	6″	″	11	2.182		109	87		196	255
4120	8″	Q-3	8	4		264	163		427	545

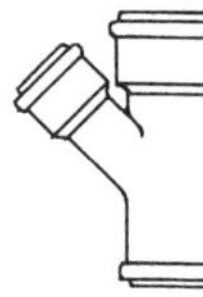

360

MECHANICAL 15

15107 | Metal Pipe & Fittings

		CREW	DAILY OUTPUT	LABOR-HOURS	UNIT	2006 BARE COSTS MAT.	LABOR	EQUIP.	TOTAL	TOTAL INCL O&P	
360 4160	Double Y, 2"	Q-1	10	1.600	Ea.	33	61.50		94.50	132	360
4170	3"		8	2		53	77		130	179	
4180	4"		7	2.286		69	88		157	213	
4190	5"	Q-2	10	2.400		116	95.50		211.50	276	
4200	6"	"	9	2.667		161	106		267	345	
4210	8"	Q-3	6	5.333		390	217		607	765	
4220	10"		5	6.400		555	260		815	1,025	
4230	12"		4.50	7.111		785	289		1,074	1,300	
4260	Reducer, 3" x 2"	Q-1	17	.941		17.90	36		53.90	75.50	
4270	4" x 2"		16.50	.970		19.60	37.50		57.10	79.50	
4280	4" x 3"		16	1		23	38.50		61.50	85.50	
4290	5" x 2"		16	1		30.50	38.50		69	93.50	
4300	5" x 3"		15.50	1.032		31	39.50		70.50	95.50	
4310	5" x 4"		15	1.067		31.50	41		72.50	98	
4320	6" x 2"		15.50	1.032		36	39.50		75.50	101	
4330	6" x 3"		15	1.067		37	41		78	104	
4340	6" x 5"		13	1.231		46.50	47.50		94	125	
4360	8" x 3"	Q-2	15	1.600		72.50	64		136.50	179	
4370	8" x 4"		15	1.600		70	64		134	176	
4380	8" x 5"		14	1.714		74.50	68.50		143	188	
4390	8" x 6"		14	1.714		74.50	68.50		143	188	
4430	Increaser, 2" x 3"	Q-1	17	.941		28.50	36		64.50	87.50	
4440	2" x 4"		16	1		30	38.50		68.50	93	
4450	2" x 5"		15	1.067		37.50	41		78.50	105	
4460	3" x 4"		15	1.067		32	41		73	99	
4470	3" x 5"		15	1.067		37.50	41		78.50	105	
4480	3" x 6"		14	1.143		45	44		89	118	
4490	4" x 5"		15	1.067		37.50	41		78.50	105	
4500	4" x 6"		14	1.143		45	44		89	118	
4510	4" x 8"	Q-2	15	1.600		93.50	64		157.50	202	
4520	5" x 6"	Q-1	13	1.231		62.50	47.50		110	142	
4530	5" x 8"	Q-2	14	1.714		106	68.50		174.50	223	
4540	6" x 8"		14	1.714		106	68.50		174.50	223	
4550	6" x 10"		10	2.400		187	95.50		282.50	355	
4560	8" x 10"		8.50	2.824		187	113		300	380	
4570	10" x 12"		7.50	3.200		320	128		448	550	
4600	Y, 2"	Q-1	12	1.333		19.95	51		70.95	102	
4610	3"		10	1.600		33	61.50		94.50	132	
4620	4"		9	1.778		44	68.50		112.50	155	
4630	5"	Q-2	13	1.846		72.50	73.50		146	194	
4640	6"	"	11	2.182		93	87		180	237	
4650	8"	Q-3	8	4		223	163		386	500	
4660	10"		7	4.571		355	186		541	685	
4670	12"		6	5.333		585	217		802	980	
4900	For extra heavy, add					44%	4%				
4940	Gasket and making push-on joint										
4950	2"	Q-1	40	.400	Ea.	3.38	15.35		18.73	27.50	
4960	3"		35	.457		4.43	17.55		21.98	32.50	
4970	4"		32	.500		5.65	19.20		24.85	36	
4980	5"	Q-2	43	.558		8.70	22		30.70	44	
4990	6"	"	40	.600		9	24		33	47	
5000	8"	Q-3	32	1		19.95	40.50		60.45	85.50	
5010	10"		29	1.103		30.50	45		75.50	104	
5020	12"		25	1.280		39	52		91	124	
5022	15"		21	1.524		46.50	62		108.50	148	
5030	Note: gaskets and joint labor have										

15107 | Metal Pipe & Fittings

360

	15107 Metal Pipe & Fittings	CREW	DAILY OUTPUT	LABOR-HOURS	UNIT	2006 BARE COSTS MAT.	LABOR	EQUIP.	TOTAL	TOTAL INCL O&P
5040	Been included with all listed fittings.									
5990	No hub									
6000	Cplg. & labor required at joints not incl. in fitting									
6010	price. Add 1 coupling per joint for installed price									
6020	1/4 Bend, 1-1/2"				Ea.	5.80			5.80	6.40
6060	2"					6.35			6.35	7
6080	3" 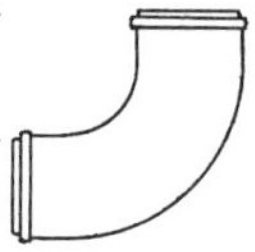					8.75			8.75	9.65
6120	4"					12.65			12.65	13.95
6140	5"					29			29	32
6160	6"					32			32	35
6180	8"					88			88	96.50
6184	1/4 Bend, long sweep, 1-1/2"					13.65			13.65	15.05
6186	2"					13.65			13.65	15.05
6188	3"					16.20			16.20	17.80
6189	4"					26			26	28.50
6190	5"					47.50			47.50	52.50
6191	6"					58			58	64
6192	8"					141			141	155
6193	10"					253			253	279
6200	1/8 Bend, 1-1/2"					4.85			4.85	5.35
6210	2"					5.40			5.40	5.95
6212	3" 					7.30			7.30	8.05
6214	4"					9.25			9.25	10.20
6216	5"					19.70			19.70	21.50
6218	6"					21.50			21.50	23.50
6220	8"					61.50			61.50	68
6222	10"					115			115	127
6380	Sanitary Tee, tapped, 1-1/2"					10.70			10.70	11.75
6382	2" x 1-1/2"					9.65			9.65	10.60
6384	2"					10.75			10.75	11.85
6386	3" x 2"					15			15	16.50
6388	3"					28			28	30.50
6390	4" x 1-1/2"					13.35			13.35	14.70
6392	4" x 2"					15.05			15.05	16.55
6394	6" x 1-1/2"					31			31	34
6396	6" x 2"					32			32	35
6459	Sanitary Tee, 1-1/2"					8.05			8.05	8.85
6460	2"					8.05			8.05	8.85
6470	3" 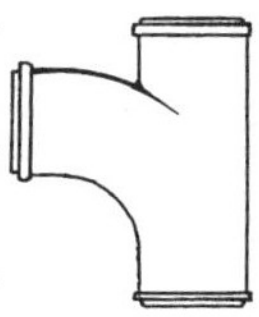					10.65			10.65	11.70
6472	4"					16.50			16.50	18.15
6474	5"					47			47	52
6476	6"					48			48	53
6478	8"					195			195	214
6730	Y, 1-1/2"					8.20			8.20	9
6740	2" 					8.20			8.20	9
6750	3"					11.60			11.60	12.75
6760	4"					18.95			18.95	21
6762	5"					44			44	48.50
6764	6"					50.50			50.50	55.50
6768	8"					119			119	131
6791	Y, reducing, 3" x 2"					8.75			8.75	9.65
6792	4" x 2"					12.15			12.15	13.40
6793	5" x 2"					27.50			27.50	30
6794	6" x 2"					30.50			30.50	33.50
6795	6" x 4"					39.50			39.50	43.50
6796	8" x 4"				↓	68			68	74.50

360

MECHANICAL 15

15107 | Metal Pipe & Fittings

			CREW	DAILY OUTPUT	LABOR-HOURS	UNIT	2006 BARE COSTS MAT.	LABOR	EQUIP.	TOTAL	TOTAL INCL O&P	
360	6797	8″ x 6″				Ea.	83.50			83.50	92	360
	6798	10″ x 6″					188			188	207	
	6799	10″ x 8″					226			226	248	
	6800	Double Y, 2″					12.65			12.65	13.95	
	6920	3″					23.50			23.50	25.50	
	7000	4″					50			50	54.50	
	7100	6″					88			88	96.50	
	7120	8″					251			251	276	
	7200	Combination Y and 1/8 Bend										
	7220	1-1/2″				Ea.	9.10			9.10	10	
	7260	2″					8.70			8.70	9.60	
	7320	3″					13.25			13.25	14.60	
	7400	4″					24.50			24.50	26.50	
	7480	5″					53.50			53.50	58.50	
	7500	6″					67.50			67.50	74.50	
	7520	8″					167			167	184	
	7800	Reducer, 3″ x 2″					4.62			4.62	5.10	
	7820	4″ x 2″					7.20			7.20	7.90	
	7840	4″ x 3″					7.20			7.20	7.90	
	8000	Coupling, standard (by CISPI Mfrs.)										
	8020	1-1/2″	Q-1	48	.333	Ea.	6	12.80		18.80	26.50	
	8040	2″		44	.364		6	13.95		19.95	28	
	8080	3″		38	.421		7.15	16.20		23.35	33	
	8120	4″		33	.485		8.45	18.65		27.10	38.50	
	8160	5″	Q-2	44	.545		20.50	21.50		42	56.50	
	8180	6″	″	40	.600		21.50	24		45.50	60.50	
	8200	8″	Q-3	33	.970		40.50	39.50		80	106	
	8220	10″	″	26	1.231		56	50		106	140	
	8300	Coupling, cast iron clamp & neoprene gasket (by MG)										
	8310	1-1/2″	Q-1	48	.333	Ea.	7.75	12.80		20.55	28.50	
	8320	2″		44	.364		8	13.95		21.95	30.50	
	8330	3″		38	.421		8.35	16.20		24.55	34	
	8340	4″		33	.485		9.80	18.65		28.45	40	
	8350	5″	Q-2	44	.545		18.55	21.50		40.05	54.50	
	8360	6″	″	40	.600		25.50	24		49.50	65	
	8380	8″	Q-3	33	.970		44.50	39.50		84	110	
	8400	10″	″	26	1.231		68	50		118	153	
	8600	Coupling, Stainless steel, heavy duty										
	8620	1-1/2″	Q-1	48	.333	Ea.	8.55	12.80		21.35	29.50	
	8630	2″		44	.364		8.90	13.95		22.85	31.50	
	8640	2″ x 1-1/2″		44	.364		13.70	13.95		27.65	36.50	
	8650	3″		38	.421		9.65	16.20		25.85	35.50	
	8660	4″		33	.485		10.90	18.65		29.55	41	
	8670	4″ x 3″		33	.485		23.50	18.65		42.15	55	
	8680	5″	Q-2	44	.545		23.50	21.50		45	60	
	8690	6″	″	40	.600		26.50	24		50.50	66	
	8700	8″	Q-3	33	.970		44.50	39.50		84	111	
	8710	10″	″	26	1.231		56	50		106	140	
	9000	Minimum labor/equipment charge	1 Plum	4	2	Job		85.50		85.50	133	
420	0010	**PIPE, COPPER** Solder joints R221113-50										420
	1000	Type K tubing, couplings & clevis hangers 10′ O.C.										
	1100	1/4″ diameter	1 Plum	84	.095	L.F.	1.66	4.07		5.73	8.15	
	1120	3/8″ diameter		82	.098		1.98	4.17		6.15	8.65	
	1140	1/2″ diameter		78	.103		2.27	4.38		6.65	9.30	
	1160	5/8″ diameter		77	.104		2.84	4.44		7.28	10	

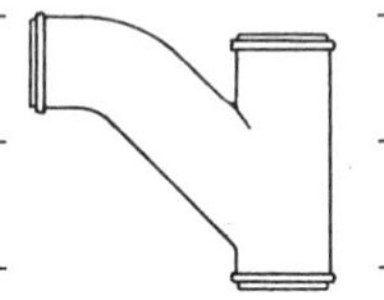

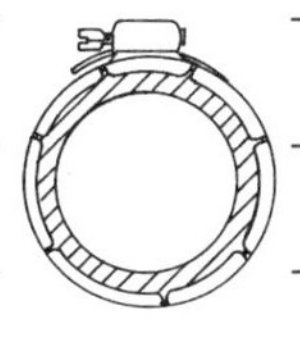

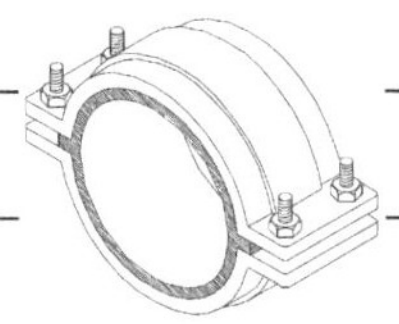

15 MECHANICAL

15107 | Metal Pipe & Fittings

R221113-50

		CREW	DAILY OUTPUT	LABOR-HOURS	UNIT	2006 BARE COSTS MAT.	LABOR	EQUIP.	TOTAL	TOTAL INCL O&P
420 1180	3/4" diameter	1 Plum	74	.108	L.F.	3.59	4.62		8.21	11.15
1200	1" diameter		66	.121		5.05	5.20		10.25	13.60
1220	1-1/4" diameter		56	.143		6.25	6.10		12.35	16.35
1240	1-1/2" diameter		50	.160		7.75	6.85		14.60	19.15
1260	2" diameter		40	.200		11.90	8.55		20.45	26.50
1280	2-1/2" diameter	Q-1	60	.267		18.10	10.25		28.35	36
1300	3" diameter		54	.296		25	11.40		36.40	45.50
1320	3-1/2" diameter		42	.381		34	14.65		48.65	60.50
1330	4" diameter		38	.421		42.50	16.20		58.70	72
1340	5" diameter		32	.500		111	19.20		130.20	152
1360	6" diameter	Q-2	38	.632		161	25		186	216
1380	8" diameter	"	34	.706		285	28		313	360
1390	For other than full hard temper, add					13%				
1440	For silver solder, add						15%			
1800	For medical clean, (oxygen class), add					12%				
1950	To delete cplgs. & hngrs., 1/4"-1" pipe, subtract					27%	60%			
1960	1-1/4" -3" pipe, subtract					14%	52%			
1970	3-1/2"-5" pipe, subtract					10%	60%			
1980	6"-8" pipe, subtract					19%	53%			
2000	Type L tubing, couplings & hangers 10' O.C.									
2100	1/4" diameter	1 Plum	88	.091	L.F.	1.10	3.88		4.98	7.25
2120	3/8" diameter		84	.095		1.53	4.07		5.60	8
2140	1/2" diameter		81	.099		1.77	4.22		5.99	8.50
2160	5/8" diameter		79	.101		2.48	4.32		6.80	9.40
2180	3/4" diameter		76	.105		2.60	4.49		7.09	9.85
2200	1" diameter		68	.118		3.65	5		8.65	11.80
2220	1-1/4" diameter		58	.138		5.05	5.90		10.95	14.75
2240	1-1/2" diameter		52	.154		6.50	6.55		13.05	17.30
2260	2" diameter		42	.190		10.05	8.15		18.20	23.50
2280	2-1/2" diameter	Q-1	62	.258		15.55	9.90		25.45	32.50
2300	3" diameter		56	.286		21	11		32	40
2320	3-1/2" diameter		43	.372		28.50	14.30		42.80	53.50
2340	4" diameter		39	.410		35.50	15.75		51.25	63.50
2360	5" diameter		34	.471		100	18.10		118.10	138
2380	6" diameter	Q-2	40	.600		124	24		148	173
2400	8" diameter	"	36	.667		219	26.50		245.50	283
2410	For other than full hard temper, add					21%				
2590	For silver solder, add						15%			
2900	For medical clean, (oxygen class), add					12%				
2940	To delete cplgs. & hngrs., 1/4"-1" pipe, subtract					37%	63%			
2960	1-1/4"-3" pipe, subtract					12%	53%			
2970	3-1/2"-5" pipe, subtract					12%	63%			
2980	6"-8" pipe, subtract					24%	55%			
3000	Type M tubing, couplings & hangers 10' O.C.									
3100	1/4" diameter	1 Plum	90	.089	L.F.	1.23	3.80		5.03	7.25
3120	3/8" diameter		87	.092		1.25	3.93		5.18	7.45
3140	1/2" diameter		84	.095		1.39	4.07		5.46	7.85
3160	5/8" diameter		81	.099		2.13	4.22		6.35	8.90
3180	3/4" diameter		78	.103		2.14	4.38		6.52	9.15
3200	1" diameter		70	.114		2.80	4.88		7.68	10.70
3220	1-1/4" diameter		60	.133		4.15	5.70		9.85	13.40
3240	1-1/2" diameter		54	.148		5.60	6.35		11.95	16
3260	2" diameter		44	.182		8.75	7.75		16.50	21.50
3280	2-1/2" diameter	Q-1	64	.250		13.30	9.60		22.90	29.50
3300	3" diameter		58	.276		17.75	10.60		28.35	36
3320	3-1/2" diameter		45	.356		25	13.65		38.65	48.50

MECHANICAL 15

15107 | Metal Pipe & Fittings

	Line	Description	Crew	Daily Output	Labor-Hours	Unit	2006 Bare Costs Mat.	Labor	Equip.	Total	Total Incl O&P	
420	3340	4″ diameter (R221113-50)	Q-1	40	.400	L.F.	32	15.35		47.35	59	420
	3360	5″ diameter		36	.444		96.50	17.10		113.60	133	
	3370	6″ diameter	Q-2	42	.571		124	23		147	172	
	3380	8″ diameter	″	38	.632		206	25		231	265	
	3440	For silver solder, add						15%				
	3960	To delete cplgs. & hngrs., 1/4″-1″ pipe, subtract					35%	65%				
	3970	1-1/4″-3″ pipe, subtract					19%	56%				
	3980	3-1/2″-5″ pipe, subtract					13%	65%				
	3990	6″-8″ pipe, subtract					28%	58%				
	4000	Type DWV tubing, couplings & hangers 10′ O.C.										
	4100	1-1/4″ diameter	1 Plum	60	.133	L.F.	4.45	5.70		10.15	13.75	
	4120	1-1/2″ diameter		54	.148		5.55	6.35		11.90	15.95	
	4140	2″ diameter		44	.182		7.40	7.75		15.15	20	
	4160	3″ diameter	Q-1	58	.276		13.15	10.60		23.75	31	
	4180	4″ diameter		40	.400		23	15.35		38.35	49.50	
	4200	5″ diameter		36	.444		68.50	17.10		85.60	102	
	4220	6″ diameter	Q-2	42	.571		98	23		121	144	
	4240	8″ diameter	″	38	.632		235	25		260	298	
	4730	To delete cplgs. & hngrs., 1-1/4″-2″ pipe, subtract					16%	53%				
	4740	2″-4″ pipe, subtract					13%	60%				
	4750	5″-8″ pipe, subtract					23%	58%				
	5200	ACR tubing, type L, hard temper, cleaned and										
	5220	capped, no couplings or hangers										
	5240	3/8″ OD				L.F.	.82			.82	.90	
	5250	1/2″ OD					1.31			1.31	1.44	
	5260	5/8″ OD					1.59			1.59	1.75	
	5270	3/4″ OD					2.25			2.25	2.48	
	5280	7/8″ OD					2.50			2.50	2.75	
	5290	1-1/8″ OD					3.61			3.61	3.97	
	5300	1-3/8″ OD					4.86			4.86	5.35	
	5310	1-5/8″ OD					6.20			6.20	6.85	
	5320	2-1/8″ OD					9.80			9.80	10.75	
	5330	2-5/8″ OD					14.60			14.60	16.05	
	5340	3-1/8″ OD					19.60			19.60	21.50	
	5350	3-5/8″ OD					25.50			25.50	28	
	5360	4-1/8″ OD					32.50			32.50	36	
	5380	ACR tubing, type L, hard, cleaned and capped										
	5381	No couplings or hangers										
	5384	3/8″	1 Stpi	160	.050	L.F.	.82	2.15		2.97	4.25	
	5385	1/2″		160	.050		1.31	2.15		3.46	4.79	
	5386	5/8″		160	.050		1.59	2.15		3.74	5.10	
	5387	3/4″		130	.062		2.25	2.65		4.90	6.60	
	5388	7/8″		130	.062		2.50	2.65		5.15	6.85	
	5389	1-1/8″		115	.070		3.61	3		6.61	8.65	
	5390	1-3/8″		100	.080		4.86	3.44		8.30	10.70	
	5391	1-5/8″		90	.089		6.20	3.83		10.03	12.80	
	5392	2-1/8″		80	.100		9.80	4.31		14.11	17.45	
	5393	2-5/8″	Q-5	125	.128		14.60	4.96		19.56	24	
	5394	3-1/8″		105	.152		19.60	5.90		25.50	30.50	
	5395	4-1/8″		95	.168		32.50	6.55		39.05	46	
	5800	Refrigeration tubing, dryseal, 50′ coils										
	5840	1/8″ OD				Coil	21.50			21.50	23.50	
	5850	3/16″ OD					23.50			23.50	26	
	5860	1/4″ OD					27			27	29.50	
	5870	5/16″ OD					34.50			34.50	38	
420	5880	3/8″ OD					39			39	43	

15 MECHANICAL

15107 | Metal Pipe & Fittings

				DAILY	LABOR-		2006 BARE COSTS				TOTAL	
			CREW	OUTPUT	HOURS	UNIT	MAT.	LABOR	EQUIP.	TOTAL	INCL O&P	
420	5890	1/2" OD (R221113 -50)				Coil	50			50	55	420
	5900	5/8" OD					67.50			67.50	74	
	5910	3/4" OD					79			79	87	
	5920	7/8" OD					117			117	129	
	5930	1-1/8" OD					171			171	188	
	5940	1-3/8" OD					264			264	290	
	5950	1-5/8" OD					340			340	375	
	9000	Minimum labor/equipment charge	1 Plum	4	2	Job		85.50		85.50	133	
460	0010	**PIPE, COPPER, FITTINGS** Wrought unless otherwise noted										460
	0040	Solder joints, copper x copper										
	0070	90° elbow, 1/4"	1 Plum	22	.364	Ea.	1.70	15.55		17.25	26	
	0090	3/8"		22	.364		1.61	15.55		17.16	26	
	0100	1/2"		20	.400		.51	17.10		17.61	27	
	0110	5/8"		19	.421		3.28	18		21.28	31.50	
	0120	3/4"		19	.421		1.14	18		19.14	29.50	
	0130	1"		16	.500		2.82	21.50		24.32	36	
	0140	1-1/4" 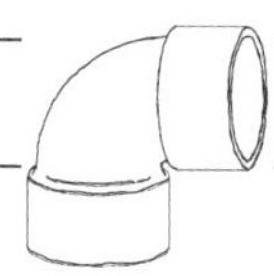		15	.533		4.26	23		27.26	40	
	0150	1-1/2"		13	.615		6.65	26.50		33.15	48.50	
	0160	2"		11	.727		12.10	31		43.10	62	
	0170	2-1/2"	Q-1	13	1.231		25.50	47.50		73	102	
	0180	3"		11	1.455		34	56		90	125	
	0190	3-1/2"		10	1.600		119	61.50		180.50	227	
	0200	4"		9	1.778		87	68.50		155.50	202	
	0210	5"		6	2.667		365	102		467	560	
	0220	6"	Q-2	9	2.667		485	106		591	700	
	0230	8"	"	8	3		1,800	120		1,920	2,150	
	0250	45° elbow, 1/4"	1 Plum	22	.364		3.18	15.55		18.73	27.50	
	0270	3/8"		22	.364		2.59	15.55		18.14	27	
	0280	1/2"		20	.400		.94	17.10		18.04	27.50	
	0290	5/8"		19	.421		4.93	18		22.93	33.50	
	0300	3/4" 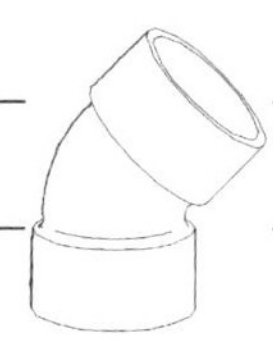		19	.421		1.64	18		19.64	30	
	0310	1"		16	.500		4.13	21.50		25.63	37.50	
	0320	1-1/4"		15	.533		5.90	23		28.90	42	
	0330	1-1/2"		13	.615		7.10	26.50		33.60	49	
	0340	2"		11	.727		11.80	31		42.80	61.50	
	0350	2-1/2"	Q-1	13	1.231		25	47.50		72.50	101	
	0360	3"		13	1.231		37.50	47.50		85	115	
	0370	3-1/2"		10	1.600		66	61.50		127.50	168	
	0380	4"		9	1.778		79.50	68.50		148	194	
	0390	5"		6	2.667		310	102		412	500	
	0400	6"	Q-2	9	2.667		485	106		591	700	
	0410	8"	"	8	3		1,650	120		1,770	2,000	
	0450	Tee, 1/4"	1 Plum	14	.571		3.57	24.50		28.07	42	
	0470	3/8"		14	.571		2.73	24.50		27.23	41	
	0480	1/2" 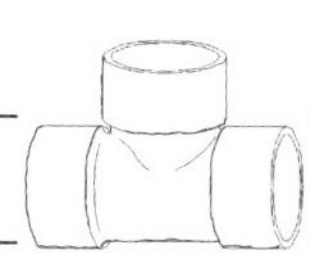		13	.615		.87	26.50		27.37	42	
	0490	5/8"		12	.667		5.95	28.50		34.45	51	
	0500	3/4"		12	.667		2.11	28.50		30.61	47	
	0510	1"		10	.800		6.50	34		40.50	60	
	0520	1-1/4"		9	.889		9.35	38		47.35	69.50	
	0530	1-1/2"		8	1		14.45	42.50		56.95	82.50	
	0540	2"		7	1.143		22.50	49		71.50	101	
	0550	2-1/2"	Q-1	8	2		45.50	77		122.50	170	
	0560	3"		7	2.286		69.50	88		157.50	214	
	0570	3-1/2"		6	2.667		201	102		303	380	
	0580	4"		5	3.200		168	123		291	375	
	0590	5"		4	4		550	154		704	845	

15107 | Metal Pipe & Fittings

460	15107 Metal Pipe & Fittings	CREW	DAILY OUTPUT	LABOR-HOURS	UNIT	2006 BARE COSTS MAT.	LABOR	EQUIP.	TOTAL	TOTAL INCL O&P
0600	6″	Q-2	6	4	Ea.	750	159		909	1,075
0610	8″	″	5	4.800		2,900	191		3,091	3,500
0612	Tee, reducing on the outlet, 1/4″	1 Plum	15	.533		6.15	23		29.15	42.50
0613	3/8″		15	.533		5.40	23		28.40	41.50
0614	1/2″		14	.571		4.74	24.50		29.24	43
0615	5/8″		13	.615		9.55	26.50		36.05	51.50
0616	3/4″		12	.667		2.97	28.50		31.47	48
0617	1″		11	.727		6.90	31		37.90	56
0618	1-1/4″		10	.800		10.10	34		44.10	64
0619	1-1/2″		9	.889		10.75	38		48.75	71
0620	2″		8	1		17	42.50		59.50	85
0621	2-1/2″	Q-1	9	1.778		53	68.50		121.50	165
0622	3″		8	2		58.50	77		135.50	184
0623	4″		6	2.667		113	102		215	284
0624	5″		5	3.200		520	123		643	765
0625	6″	Q-2	7	3.429		715	137		852	995
0626	8″	″	6	4		2,900	159		3,059	3,450
0630	Tee, reducing on the run, 1/4″	1 Plum	15	.533		7.20	23		30.20	43.50
0631	3/8″		15	.533		9.65	23		32.65	46
0632	1/2″		14	.571		6.50	24.50		31	45
0633	5/8″		13	.615		10.15	26.50		36.65	52
0634	3/4″		12	.667		5.25	28.50		33.75	50.50
0635	1″		11	.727		8.20	31		39.20	57.50
0636	1-1/4″		10	.800		13.30	34		47.30	67.50
0637	1-1/2″		9	.889		23	38		61	84
0638	2″		8	1		29.50	42.50		72	98.50
0639	2-1/2″	Q-1	9	1.778		68.50	68.50		137	182
0640	3″		8	2		101	77		178	231
0641	4″		6	2.667		224	102		326	405
0642	5″		5	3.200		520	123		643	765
0643	6″	Q-2	7	3.429		790	137		927	1,075
0644	8″	″	6	4		3,050	159		3,209	3,600
0650	Coupling, 1/4″	1 Plum	24	.333		.40	14.25		14.65	22.50
0670	3/8″		24	.333		.52	14.25		14.77	22.50
0680	1/2″		22	.364		.40	15.55		15.95	24.50
0690	5/8″		21	.381		1.22	16.25		17.47	27
0700	3/4″		21	.381		.76	16.25		17.01	26.50
0710	1″		18	.444		1.56	19		20.56	31
0715	1-1/4″		17	.471		2.90	20		22.90	34.50
0716	1-1/2″		15	.533		3.84	23		26.84	39.50
0718	2″		13	.615		6.40	26.50		32.90	48
0721	2-1/2″	Q-1	15	1.067		13.60	41		54.60	78.50
0722	3″		13	1.231		20.50	47.50		68	96
0724	3-1/2″		8	2		39	77		116	163
0726	4″		7	2.286		43	88		131	185
0728	5″		6	2.667		106	102		208	275
0731	6″	Q-2	8	3		175	120		295	380
0732	8″	″	7	3.429		565	137		702	835
0741	Coupling, reducing, concentric									
0743	1/2″	1 Plum	23	.348	Ea.	1.14	14.85		15.99	24.50
0745	3/4″		21.50	.372		1.84	15.90		17.74	26.50
0747	1″		19.50	.410		3.18	17.50		20.68	30.50
0748	1-1/4″		18	.444		3.92	19		22.92	34
0749	1-1/2″		16	.500		6.25	21.50		27.75	40
0751	2″		14	.571		10.40	24.50		34.90	49.50
0752	2-1/2″		13	.615		24.50	26.50		51	68

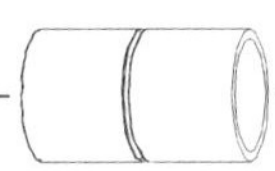

460	15107 | Metal Pipe & Fittings	CREW	DAILY OUTPUT	LABOR-HOURS	UNIT	2006 BARE COSTS MAT.	LABOR	EQUIP.	TOTAL	TOTAL INCL O&P
0753	3"	Q-1	14	1.143	Ea.	31	44		75	103
0755	4"	"	8	2		65.50	77		142.50	193
0757	5"	Q-2	7.50	3.200		216	128		344	435
0759	6"		7	3.429		355	137		492	600
0761	8"		6.50	3.692		585	147		732	870
0771	Cap, sweat									
0773	1/2"	1 Plum	40	.200	Ea.	.37	8.55		8.92	13.70
0775	3/4"		38	.211		.68	9		9.68	14.75
0777	1"		32	.250		1.56	10.70		12.26	18.30
0778	1-1/4"		29	.276		2.32	11.80		14.12	21
0779	1-1/2"		26	.308		3.39	13.15		16.54	24
0781	2"		22	.364		6.25	15.55		21.80	31
0791	Flange, sweat									
0793	3"	Q-1	22	.727	Ea.	50.50	28		78.50	99
0795	4"		18	.889		70	34		104	131
0797	5"		12	1.333		131	51		182	224
0799	6"	Q-2	18	1.333		137	53		190	234
0801	8"	"	16	1.500		565	60		625	720
0850	Unions, 1/4"	1 Plum	21	.381		10.30	16.25		26.55	37
0870	3/8"		21	.381		10.35	16.25		26.60	37
0880	1/2"		19	.421		5.50	18		23.50	34
0890	5/8"		18	.444		23.50	19		42.50	55.50
0900	3/4"		18	.444		6.90	19		25.90	37
0910	1"		15	.533		11.90	23		34.90	48.50
0920	1-1/4"		14	.571		20.50	24.50		45	60.50
0930	1-1/2"		12	.667		27	28.50		55.50	74.50
0940	2"		10	.800		46.50	34		80.50	104
0950	2-1/2"	Q-1	12	1.333		101	51		152	191
0960	3"	"	10	1.600		262	61.50		323.50	385
0980	Adapter, copper x male IPS, 1/4"	1 Plum	20	.400		5.60	17.10		22.70	32.50
0990	3/8"		20	.400		2.79	17.10		19.89	29.50
1000	1/2"		18	.444		1.07	19		20.07	30.50
1010	3/4"		17	.471		1.79	20		21.79	33.50
1020	1"		15	.533		4.64	23		27.64	40.50
1030	1-1/4" 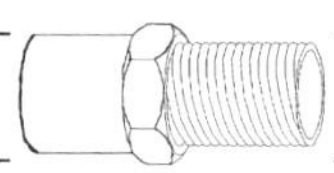		13	.615		7.25	26.50		33.75	49
1040	1-1/2"		12	.667		8.30	28.50		36.80	53.50
1050	2"		11	.727		14.05	31		45.05	64
1060	2-1/2"	Q-1	10.50	1.524		54	58.50		112.50	151
1070	3"		10	1.600		71.50	61.50		133	174
1080	3-1/2"		9	1.778		104	68.50		172.50	220
1090	4"		8	2		104	77		181	234
1200	5"		6	2.667		490	102		592	700
1210	6"	Q-2	8.50	2.824		535	113		648	765
1250	Cross, 1/2"	1 Plum	10	.800		7.45	34		41.45	61
1260	3/4"		9.50	.842		14.50	36		50.50	72
1270	1"		8	1		24.50	42.50		67	93.50
1280	1-1/4"		7.50	1.067		35.50	45.50		81	110
1290	1-1/2"		6.50	1.231		50.50	52.50		103	137
1300	2"		5.50	1.455		95.50	62		157.50	202
1310	2-1/2"	Q-1	6.50	2.462		220	94.50		314.50	390
1320	3"	"	5.50	2.909		276	112		388	480
1500	Tee fitting, mechanically formed, (Type 1).									
1520	1/2" run size, 3/8" to 1/2" branch size	1 Plum	80	.100	Ea.		4.27		4.27	6.65
1530	3/4" run size, 3/8" to 3/4" branch size		60	.133			5.70		5.70	8.85
1540	1" run size, 3/8" to 1" branch size		54	.148			6.35		6.35	9.85
1550	1-1/4" run size, 3/8" to 1-1/4" branch size		48	.167			7.10		7.10	11.05

15 MECHANICAL

15107 | Metal Pipe & Fittings

			Description	Crew	Daily Output	Labor-Hours	Unit	2006 Bare Costs Mat.	Labor	Equip.	Total	Total Incl O&P	
460	1560		1-1/2" run size, 3/8" to 1-1/2" branch size	1 Plum	40	.200	Ea.		8.55		8.55	13.30	460
	1570		2" run size, 3/8" to 2" branch size		35	.229			9.75		9.75	15.20	
	1580		2-1/2" run size, 1/2" to 2" branch size		32	.250			10.70		10.70	16.60	
	1590		3" run size, 1" to 2" branch size		26	.308			13.15		13.15	20.50	
	1600		4" run size, 1" to 2" branch size	↓	24	.333	↓		14.25		14.25	22	
	1640		Tee fitting, mechanically formed, (Type 2)										
	1650		2-1/2" run size, 2-1/2" branch size	1 Plum	12.50	.640	Ea.		27.50		27.50	42.50	
	1660		3" run size, 2-1/2" to 3" branch size		12	.667			28.50		28.50	44.50	
	1670		3-1/2" run size, 2-1/2" to 3-1/2" branch size		11	.727			31		31	48.50	
	1680		4" run size, 2-1/2" to 4" branch size		10.50	.762			32.50		32.50	50.50	
	1698		5" run size, 2" to 4" branch size		9.50	.842			36		36	56	
	1700		6" run size, 2" to 4" branch size		8.50	.941			40		40	62.50	
	1710		8" run size, 2" to 4" branch size	↓	7	1.143	↓		49		49	76	
	1800		ACR fittings, OD size										
	1802		Tee, straight										
	1808		5/8"	1 Stpi	12	.667	Ea.	.87	28.50		29.37	45.50	
	1810		3/4"		12	.667		5.95	28.50		34.45	51	
	1812		7/8"		10	.800		2.11	34.50		36.61	56	
	1813		1"		10	.800		16.10	34.50		50.60	71	
	1814		1-1/8"		10	.800		6.50	34.50		41	60.50	
	1816		1-3/8"		9	.889		9.35	38.50		47.85	70	
	1818		1-5/8"		8	1		14.45	43		57.45	83	
	1820		2-1/8"	↓	7	1.143		22.50	49		71.50	102	
	1822		2-5/8"	Q-5	8	2		45.50	77.50		123	171	
	1824		3-1/8"		7	2.286		69.50	88.50		158	215	
	1826		4-1/8"	↓	5	3.200	↓	168	124		292	380	
	1830		90° elbow										
	1836		5/8"	1 Stpi	19	.421	Ea.	1.80	18.15		19.95	30	
	1838		3/4"		19	.421		3.28	18.15		21.43	31.50	
	1840		7/8"		16	.500		3.25	21.50		24.75	37	
	1842		1-1/8"		16	.500		4.59	21.50		26.09	38.50	
	1844		1-3/8"		15	.533		4.48	23		27.48	40.50	
	1846		1-5/8"		13	.615		7	26.50		33.50	48.50	
	1848		2-1/8"	↓	11	.727		12.70	31.50		44.20	62.50	
	1850		2-5/8"	Q-5	13	1.231		27	47.50		74.50	104	
	1852		3-1/8"		11	1.455		36	56.50		92.50	127	
	1854		4-1/8"	↓	9	1.778	↓	92	69		161	208	
	1860		Coupling										
	1866		5/8"	1 Stpi	21	.381	Ea.	.42	16.40		16.82	26	
	1868		3/4"		21	.381		1.29	16.40		17.69	27	
	1870		7/8"		18	.444		.81	19.15		19.96	31	
	1871		1"		18	.444		2.93	19.15		22.08	33	
	1872		1-1/8"		18	.444		1.74	19.15		20.89	32	
	1874		1-3/8"		17	.471		3.05	20.50		23.55	35	
	1876		1-5/8"		15	.533		4.04	23		27.04	40	
	1878		2-1/8"	↓	13	.615		6.75	26.50		33.25	48.50	
	1880		2-5/8"	Q-5	15	1.067		15.20	41.50		56.70	81	
	1882		3-1/8"		13	1.231		22	47.50		69.50	98	
	1884		4-1/8"	↓	7	2.286	↓	50	88.50		138.50	193	
	2000		DWV, solder joints, copper x copper										
	2030		90° Elbow, 1-1/4"	1 Plum	13	.615	Ea.	5.20	26.50		31.70	47	
	2050		1-1/2"		12	.667		7.80	28.50		36.30	53	
	2070		2"	↓	10	.800		10.20	34		44.20	64	
	2090		3"	Q-1	10	1.600		25	61.50		86.50	123	
	2100		4"	"	9	1.778		120	68.50		188.50	238	
	2150		45° Elbow, 1-1/4"	1 Plum	13	.615	↓	4.80	26.50		31.30	46.50	

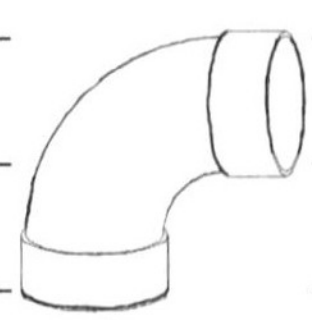

15 MECHANICAL

15107 | Metal Pipe & Fittings

		15107 | Metal Pipe & Fittings	CREW	DAILY OUTPUT	LABOR-HOURS	UNIT	2006 BARE COSTS MAT.	LABOR	EQUIP.	TOTAL	TOTAL INCL O&P	
460	2170	1-1/2"	1 Plum	12	.667	Ea.	3.97	28.50		32.47	49	460
	2180	2" 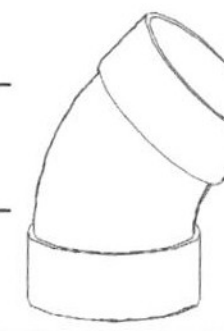	↓	10	.800		9.10	34		43.10	63	
	2190	3"	Q-1	10	1.600		19.35	61.50		80.85	117	
	2200	4"	"	9	1.778		92	68.50		160.50	207	
	2250	Tee, Sanitary, 1-1/4"	1 Plum	9	.889		9.50	38		47.50	69.50	
	2270	1-1/2"		8	1		12.80	42.50		55.30	80.50	
	2290	2" 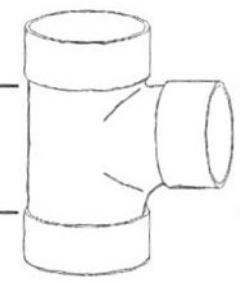	↓	7	1.143		14.95	49		63.95	92.50	
	2310	3"	Q-1	7	2.286		54	88		142	197	
	2330	4"	"	6	2.667		138	102		240	310	
	2400	Coupling, 1-1/4"	1 Plum	14	.571		2.43	24.50		26.93	40.50	
	2420	1-1/2"		13	.615		3.04	26.50		29.54	44.50	
	2440	2"	↓	11	.727		4.20	31		35.20	53	
	2460	3"	Q-1	11	1.455		8.15	56		64.15	96	
	2480	4"	"	10	1.600	↓	26	61.50		87.50	124	
	2600	Traps, see division 15150-800										
	6990	Polybutylene/polyethylene pipe, See 15108 for plastic ftng.										
	7000	Insert type Brass/copper, 100 psi @ 180°F, CTS										
	7010	Adapter MPT 3/8" x 3/8" CTS	1 Plum	29	.276	Ea.	.97	11.80		12.77	19.35	
	7020	1/2" x 1/2" 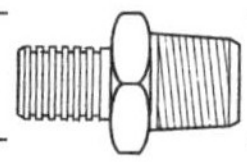		26	.308		.97	13.15		14.12	21.50	
	7030	3/4" x 1/2"		26	.308		1.44	13.15		14.59	22	
	7040	3/4" x 3/4"		25	.320		1.42	13.65		15.07	23	
	7050	Adapter CTS 1/2" x 1/2" sweat 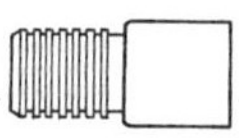		24	.333		.49	14.25		14.74	22.50	
	7060	3/4" x 3/4" sweat		22	.364		.88	15.55		16.43	25	
	7070	Coupler center set 3/8" CTS		25	.320		.91	13.65		14.56	22.50	
	7080	1/2" CTS 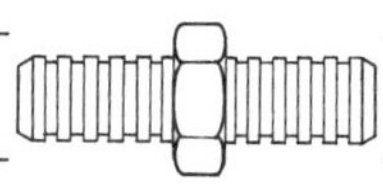		23	.348		.49	14.85		15.34	23.50	
	7090	3/4" CTS		22	.364		.88	15.55		16.43	25	
	7100	Elbow 90°, copper 3/8"		25	.320		1.08	13.65		14.73	22.50	
	7110	1/2" CTS 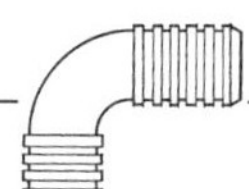		23	.348		.69	14.85		15.54	24	
	7120	3/4" CTS		22	.364		.84	15.55		16.39	25	
	7130	Tee copper 3/8" CTS		17	.471		1.28	20		21.28	33	
	7140	1/2" CTS		15	.533		.88	23		23.88	36.50	
	7150	3/4" CTS		14	.571		1.35	24.50		25.85	39.50	
	7160	3/8" x 3/8" x 1/2" 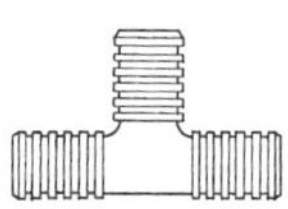		16	.500		.88	21.50		22.38	34	
	7170	1/2" x 3/8" x 1/2"		15	.533		.88	23		23.88	36.50	
	7180	3/4" x 1/2" x 3/4"		14	.571	↓	1.35	24.50		25.85	39.50	
	9000	Minimum labor/equipment charge	↓	4	2	Job		85.50		85.50	133	
500	0010	**PIPE, CORROSION RESISTANT** No couplings or hangers										500
	0020	Iron alloy, drain, mechanical joint										
	1000	1-1/2" diameter	Q-1	70	.229	L.F.	22.50	8.80		31.30	38.50	
	1100	2" diameter		66	.242		26	9.30		35.30	43	
	1120	3" diameter		60	.267		36	10.25		46.25	56	
	1140	4" diameter	↓	52	.308	↓	46.50	11.80		58.30	69.50	
	1980	Iron alloy, drain, B&S joint										
	2000	2" diameter	Q-1	54	.296	L.F.	28	11.40		39.40	48.50	
	2100	3" diameter		52	.308		40.50	11.80		52.30	63	
	2120	4" diameter	↓	48	.333		56	12.80		68.80	81.50	
	2140	6" diameter	Q-2	59	.407		93.50	16.20		109.70	128	
	2160	8" diameter	"	54	.444	↓	169	17.70		186.70	214	
	2980	Plastic, epoxy, fiberglass filament wound, B&S joint										
	3000	2" diameter	Q-1	62	.258	L.F.	7.20	9.90		17.10	23.50	
	3100	3" diameter		51	.314		8.30	12.05		20.35	28	
	3120	4" diameter		45	.356		12.15	13.65		25.80	34.50	
	3140	6" diameter	↓	32	.500		17.15	19.20		36.35	49	
	3160	8" diameter	Q-2	38	.632		30.50	25		55.50	72.50	
	3180	10" diameter		32	.750		42	30		72	93	
	3200	12" diameter	↓	28	.857	↓	50	34		84	108	

15100 | Building Services Piping

	15107 | Metal Pipe & Fittings		CREW	DAILY OUTPUT	LABOR-HOURS	UNIT	2006 BARE COSTS MAT.	LABOR	EQUIP.	TOTAL	TOTAL INCL O&P
500	3980	Polyester, fiberglass filament wound, B&S joint									
	4000	2″ diameter	Q-1	62	.258	L.F.	9.95	9.90		19.85	26.50
	4100	3″ diameter		51	.314		13.75	12.05		25.80	34
	4120	4″ diameter		45	.356		21	13.65		34.65	44.50
	4140	6″ diameter		32	.500		35	19.20		54.20	68.50
	4160	8″ diameter	Q-2	38	.632		35.50	25		60.50	78
	4180	10″ diameter		32	.750		54.50	30		84.50	107
	4200	12″ diameter		28	.857		66	34		100	126
	4980	Polypropylene, acid resistant, fire retardant, schedule 40									
	5000	1-1/2″ diameter	Q-1	68	.235	L.F.	1.13	9.05		10.18	15.30
	5100	2″ diameter		62	.258		1.53	9.90		11.43	17.10
	5120	3″ diameter		51	.314		4.28	12.05		16.33	23.50
	5140	4″ diameter		45	.356		4.55	13.65		18.20	26
	5160	6″ diameter		32	.500		7.65	19.20		26.85	38.50
	5980	Proxylene, fire retardant, Schedule 40									
	6000	1-1/2″ diameter	Q-1	68	.235	L.F.	5.05	9.05		14.10	19.60
	6100	2″ diameter		62	.258		6.90	9.90		16.80	23
	6120	3″ diameter		51	.314		12.50	12.05		24.55	32.50
	6140	4″ diameter		45	.356		17.75	13.65		31.40	40.50
	6160	6″ diameter		32	.500		32	19.20		51.20	65
	6820	For Schedule 80, add					35%	2%			
	9800	Minimum labor/equipment charge	1 Plum	4	2	Job		85.50		85.50	133
560	0010	**PIPE, CORROSION RESISTANT, FITTINGS**									
	0030	Iron alloy									
	0050	Mechanical joint									
	0060	1/4 Bend, 1-1/2″	Q-1	12	1.333	Ea.	47	51		98	131
	0080	2″		10	1.600		76.50	61.50		138	180
	0090	3″		9	1.778		92	68.50		160.50	207
	0100	4″		8	2		105	77		182	236
	0110	1/8 Bend, 1-1/2″		12	1.333		30	51		81	112
	0130	2″		10	1.600		51	61.50		112.50	152
	0140	3″		9	1.778		68	68.50		136.50	181
	0150	4″		8	2		91	77		168	220
	0160	Tee and Y, sanitary, straight									
	0170	1-1/2″	Q-1	8	2	Ea.	51	77		128	176
	0180	2″		7	2.286		68	88		156	212
	0190	3″		6	2.667		105	102		207	275
	0200	4″		5	3.200		194	123		317	405
	0360	Coupling, 1-1/2″		14	1.143		35.50	44		79.50	108
	0380	2″		12	1.333		41	51		92	125
	0390	3″		11	1.455		42.50	56		98.50	134
	0400	4″		10	1.600		48.50	61.50		110	149
	0500	Bell & Spigot									
	0510	1/4 and 1/16 bend, 2″	Q-1	16	1	Ea.	53.50	38.50		92	119
	0520	3″		14	1.143		125	44		169	206
	0530	4″		13	1.231		128	47.50		175.50	214
	0540	6″	Q-2	17	1.412		253	56.50		309.50	365
	0550	8″	″	12	2		1,050	79.50		1,129.50	1,275
	0620	1/8 bend, 2″	Q-1	16	1		60	38.50		98.50	126
	0640	3″		14	1.143		111	44		155	191
	0650	4″		13	1.231		117	47.50		164.50	203
	0660	6″	Q-2	17	1.412		212	56.50		268.50	320
	0680	8″	″	12	2		825	79.50		904.50	1,025
	0700	Tee, sanitary, 2″	Q-1	10	1.600		114	61.50		175.50	221
	0710	3″		9	1.778		370	68.50		438.50	510
	0720	4″		8	2		305	77		382	455

15107 | Metal Pipe & Fittings

			CREW	DAILY OUTPUT	LABOR-HOURS	UNIT	2006 BARE COSTS MAT.	LABOR	EQUIP.	TOTAL	TOTAL INCL O&P
560	0730	6"	Q-2	11	2.182	Ea.	385	87		472	555
	0740	8"	"	8	3		1,050	120		1,170	1,325
	1800	Y, sanitary, 2"	Q-1	10	1.600		120	61.50		181.50	228
	1820	3"		9	1.778		215	68.50		283.50	340
	1830	4"	↓	8	2		193	77		270	330
	1840	6"	Q-2	11	2.182		640	87		727	840
	1850	8"	"	8	3	↓	1,775	120		1,895	2,125
	3000	Epoxy, filament wound									
	3030	Quick-lock joint									
	3040	90° Elbow, 2"	Q-1	28	.571	Ea.	67.50	22		89.50	109
	3060	3"		16	1		78	38.50		116.50	146
	3070	4"		13	1.231		106	47.50		153.50	191
	3080	6"	↓	8	2		155	77		232	290
	3090	8"	Q-2	9	2.667		286	106		392	480
	3100	10"		7	3.429		360	137		497	605
	3110	12"	↓	6	4		515	159		674	815
	3120	45° Elbow, 2"	Q-1	28	.571		52.50	22		74.50	91.50
	3130	3"		16	1		74.50	38.50		113	142
	3140	4"		13	1.231		103	47.50		150.50	187
	3150	6"	↓	8	2		155	77		232	290
	3160	8"	Q-2	9	2.667		286	106		392	480
	3170	10"		7	3.429		360	137		497	605
	3180	12"	↓	6	4		515	159		674	815
	3190	Tee, 2"	Q-1	19	.842		161	32.50		193.50	228
	3200	3"		11	1.455		195	56		251	300
	3210	4"		9	1.778		234	68.50		302.50	365
	3220	6"	↓	5	3.200		360	123		483	585
	3230	8"	Q-2	6	4		445	159		604	740
	3240	10"		5	4.800		600	191		791	955
	3250	12"	↓	4	6	↓	875	239		1,114	1,325
	4000	Polypropylene, acid resistant									
	4020	Non-pressure, electrofusion joints									
	4050	1/4 bend, 1-1/2"	1 Plum	16	.500	Ea.	17	21.50		38.50	51.50
	4060	2"	Q-1	28	.571		21	22		43	57
	4080	3"		17	.941		36.50	36		72.50	96
	4090	4"		14	1.143		58.50	44		102.50	133
	4110	6"	↓	8	2	↓	140	77		217	274
	4150	1/4 Bend, long sweep									
	4170	1-1/2"	1 Plum	16	.500	Ea.	18.35	21.50		39.85	53
	4180	2"	Q-1	28	.571		24.50	22		46.50	61
	4200	3"		17	.941		41	36		77	101
	4210	4"	↓	14	1.143		59.50	44		103.50	134
	4250	1/8 bend, 1-1/2"	1 Plum	16	.500		17	21.50		38.50	51.50
	4260	2"	Q-1	28	.571		20.50	22		42.50	56.50
	4280	3"		17	.941		37.50	36		73.50	97.50
	4290	4"		14	1.143		42.50	44		86.50	115
	4310	6"	↓	8	2	↓	117	77		194	249
	4400	Tee, sanitary									
	4420	1-1/2"	1 Plum	10	.800	Ea.	21	34		55	76
	4430	2"	Q-1	17	.941		25	36		61	83.50
	4450	3"		11	1.455		50.50	56		106.50	143
	4460	4"		9	1.778		75.50	68.50		144	189
	4480	6"		5	3.200		128	123		251	330
	4490	Tee, sanitary reducing, 2" x 2" x 1-1/2"		17	.941		25	36		61	83.50
	4492	3" x 3" x 2"		11	1.455		49	56		105	141
	4494	4" x 4" x 3"	↓	9	1.778	↓	71	68.50		139.50	184

15107 | Metal Pipe & Fittings

	Line	Description	Crew	Daily Output	Labor-Hours	Unit	2006 Bare Costs Mat.	Labor	Equip.	Total	Total Incl O&P
560	4496	6" x 6" x 4"	Q-1	5	3.200	Ea.	244	123		367	460
	4500	Tee/wye, long turn									
	4520	1-1/2"	1 Plum	10	.800	Ea.	28.50	34		62.50	84
	4530	2"	Q-1	17	.941		35.50	36		71.50	95
	4550	3"		11	1.455		60.50	56		116.50	154
	4570	4"		9	1.778		85	68.50		153.50	200
	4650	Wye 45°, 1-1/2"	1 Plum	10	.800		21.50	34		55.50	77
	4652	2"	Q-1	17	.941		31	36		67	90
	4653	3"		11	1.455		53.50	56		109.50	146
	4654	4"		9	1.778		78.50	68.50		147	192
	4656	6"		5	3.200		200	123		323	410
	9000	Minimum labor/equipment charge	1 Plum	4	2	Job		85.50		85.50	133
620	0010	PIPE, STEEL (R221113-50)									
	0020	All pipe sizes are to Spec. A-53 unless noted otherwise									
	0030	Schedule 10, see 15107-690-0500									
	0050	Schedule 40, threaded, with couplings, and clevis type									
	0060	hangers sized for covering, 10' O.C.									
	0540	Black, 1/4" diameter	1 Plum	66	.121	L.F.	2.04	5.20		7.24	10.30
	0550	3/8" diameter		65	.123		2.03	5.25		7.28	10.40
	0560	1/2" diameter		63	.127		2.05	5.40		7.45	10.70
	0570	3/4" diameter		61	.131		2.40	5.60		8	11.35
	0580	1" diameter		53	.151		3.47	6.45		9.92	13.80
	0590	1-1/4" diameter	Q-1	89	.180		4.56	6.90		11.46	15.75
	0600	1-1/2" diameter		80	.200		5.35	7.70		13.05	17.85
	0610	2" diameter		64	.250		7.10	9.60		16.70	23
	0620	2-1/2" diameter		50	.320		11	12.30		23.30	31
	0630	3" diameter		43	.372		14.30	14.30		28.60	37.50
	0640	3-1/2" diameter		40	.400		19.35	15.35		34.70	45.50
	0650	4" diameter		36	.444		21	17.10		38.10	49.50
	0660	5" diameter		26	.615		31.50	23.50		55	71.50
	0670	6" diameter	Q-2	31	.774		46	31		77	99
	0680	8" diameter		27	.889		71	35.50		106.50	133
	0690	10" diameter		23	1.043		99.50	41.50		141	174
	0700	12" diameter		18	1.333		110	53		163	205
	0809	A-106, gr. A/B, seamless w/cplgs. & hangers									
	0811	1/4" diameter	1 Plum	66	.121	L.F.	6.65	5.20		11.85	15.40
	0812	3/8" diameter		65	.123		6.10	5.25		11.35	14.90
	0813	1/2" diameter		63	.127		3.89	5.40		9.29	12.75
	0814	3/4" diameter		61	.131		5.40	5.60		11	14.60
	0815	1" diameter		53	.151		6.65	6.45		13.10	17.30
	0816	1-1/4" diameter	Q-1	89	.180		8.40	6.90		15.30	20
	0817	1-1/2" diameter		80	.200		10.20	7.70		17.90	23
	0819	A-53, 2" diameter		64	.250		11.50	9.60		21.10	27.50
	0821	2-1/2" diameter		50	.320		14.25	12.30		26.55	35
	0822	3" diameter		43	.372		18.55	14.30		32.85	42.50
	0823	4" diameter		36	.444		28.50	17.10		45.60	58
	1220	To delete coupling & hanger, subtract									
	1230	1/4" diam. to 3/4" diam.					31%	56%			
	1240	1" diam. to 1-1/2" diam.					23%	51%			
	1250	2" diam. to 4" diam.					23%	41%			
	1260	5" diam. to 12" diam.					21%	45%			
	1280	All pipe sizes are to Spec. A-53 unless noted otherwise									
	1281	Schedule 40, threaded, with couplings and clevis type									
	1282	hangers sized for covering, 10' O. C.									
	1290	Galvanized, 1/4" diameter	1 Plum	66	.121	L.F.	2.42	5.20		7.62	10.70
	1300	3/8" diameter		65	.123		2.68	5.25		7.93	11.10

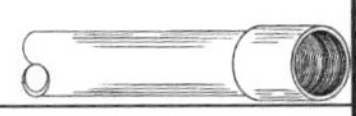

15 MECHANICAL

15107 | Metal Pipe & Fittings

620

	15107 Metal Pipe & Fittings	CREW	DAILY OUTPUT	LABOR-HOURS	UNIT	2006 BARE COSTS MAT.	LABOR	EQUIP.	TOTAL	TOTAL INCL O&P
1310	1/2″ diameter (R221113-50)	1 Plum	63	.127	L.F.	2.78	5.40		8.18	11.50
1320	3/4″ diameter		61	.131		3.32	5.60		8.92	12.35
1330	1″ diameter		53	.151		4.51	6.45		10.96	14.95
1340	1-1/4″ diameter	Q-1	89	.180		6	6.90		12.90	17.35
1350	1-1/2″ diameter		80	.200		7.10	7.70		14.80	19.75
1360	2″ diameter		64	.250		9.45	9.60		19.05	25.50
1370	2-1/2″ diameter		50	.320		15.60	12.30		27.90	36.50
1380	3″ diameter		43	.372		19.90	14.30		34.20	44
1390	3-1/2″ diameter		40	.400		25	15.35		40.35	52
1400	4″ diameter		36	.444		29	17.10		46.10	58
1410	5″ diameter		26	.615		39.50	23.50		63	80.50
1420	6″ diameter	Q-2	31	.774		53	31		84	106
1430	8″ diameter		27	.889		81.50	35.50		117	145
1440	10″ diameter		23	1.043		91	41.50		132.50	165
1450	12″ diameter		18	1.333		127	53		180	222
1750	To delete coupling & hanger, subtract									
1760	1/4″ diam. to 3/4″ diam.					31%	56%			
1770	1″ diam. to 1-1/2″ diam.					23%	51%			
1780	2″ diam. to 4″ diam.					23%	41%			
1790	5″ diam. to 12″ diam.					21%	45%			
2000	Welded, sch. 40, on yoke & roll hangers, sized for covering,									
2010	10′ O.C. (no hangers incl. for lines 2160 thru 2260)									
2040	Black, 1″ diameter	Q-15	93	.172	L.F.	5.05	6.60	.74	12.39	16.65
2050	1-1/4″ diameter		84	.190		6.70	7.30	.82	14.82	19.65
2060	1-1/2″ diameter		76	.211		7.35	8.10	.91	16.36	21.50
2070	2″ diameter		61	.262		8.90	10.10	1.13	20.13	26.50
2080	2-1/2″ diameter		47	.340		11.60	13.10	1.47	26.17	35
2090	3″ diameter		43	.372		14	14.30	1.61	29.91	39
2100	3-1/2″ diameter		39	.410		18.35	15.75	1.77	35.87	46.50
2110	4″ diameter		37	.432		19.25	16.60	1.87	37.72	49
2120	5″ diameter		32	.500		25.50	19.20	2.16	46.86	60.50
2130	6″ diameter	Q-16	36	.667		34.50	26.50	1.92	62.92	81.50
2140	8″ diameter		29	.828		51.50	33	2.38	86.88	111
2150	10″ diameter		24	1		93	40	2.88	135.88	167
2160	12″ diameter		19	1.263		109	50.50	3.64	163.14	203
2170	14″ diameter		15	1.600		66	64	4.61	134.61	177
2180	16″ diameter		13	1.846		83	73.50	5.30	161.80	211
2190	18″ diameter		11	2.182		93	87	6.30	186.30	244
2200	20″ diameter		9	2.667		103	106	7.70	216.70	286
2220	24″ diameter		8	3		111	120	8.65	239.65	320
2230	26″ diameter		7.50	3.200		261	128	9.20	398.20	495
2240	30″ diameter		6	4		285	159	11.50	455.50	575
2260	36″ diameter		4.50	5.333		330	213	15.35	558.35	710
2345	Sch. 40, A-53, gr. A/B, ERW, welded w/hngrs.									
2346	2″ diameter	Q-15	61	.262	L.F.	8.20	10.10	1.13	19.43	26
2347	2-1/2″ diameter		47	.340		12.95	13.10	1.47	27.52	36.50
2348	3″ diameter		43	.372		16.90	14.30	1.61	32.81	42.50
2349	4″ diameter		38	.421		23.50	16.20	1.82	41.52	53
2350	6″ diameter	Q-16	37	.649		33.50	26	1.87	61.37	79
2351	8″ diameter		29	.828		50.50	33	2.38	85.88	110
2352	10″ diameter		24	1		46	40	2.88	88.88	116
2363	.375″ wall, A-53, gr. A/B, ERW, welded w/hngrs.									
2364	12″ diameter	Q-16	20	1.200	L.F.	76	48	3.46	127.46	162
2365	14″ diameter		17	1.412		91	56.50	4.07	151.57	192
2366	16″ diameter		14	1.714		91.50	68.50	4.94	164.94	212
2560	To delete hanger, subtract									

MECHANICAL 15

15107 | Metal Pipe & Fittings		CREW	DAILY OUTPUT	LABOR-HOURS	UNIT	2006 BARE COSTS MAT.	LABOR	EQUIP.	TOTAL	TOTAL INCL O&P
620 2570	1" diam. to 1-1/2" diam. R221113-50					15%	34%			
2580	2" diam. to 3-1/2" diam.					9%	21%			
2590	4" diam. to 12" diam.					5%	12%			
3250	Flanged, 150 lb. weld neck, on yoke & roll hangers									
3260	sized for covering, 10' O.C.									
3290	Black, 1" diameter	Q-15	70	.229	L.F.	8.55	8.80	.99	18.34	24
3300	1-1/4" diameter		64	.250		10.20	9.60	1.08	20.88	27.50
3310	1-1/2" diameter		58	.276		10.85	10.60	1.19	22.64	29.50
3320	2" diameter		45	.356		12.75	13.65	1.54	27.94	36.50
3330	2-1/2" diameter		36	.444		16.50	17.10	1.92	35.52	47
3340	3" diameter		32	.500		19.20	19.20	2.16	40.56	53.50
3350	3-1/2" diameter		29	.552		25	21	2.38	48.38	63
3360	4" diameter		26	.615		26	23.50	2.66	52.16	68.50
3370	5" diameter		21	.762		34.50	29.50	3.29	67.29	87
3380	6" diameter	Q-16	25	.960		44.50	38.50	2.76	85.76	112
3390	8" diameter		19	1.263		67.50	50.50	3.64	121.64	157
3400	10" diameter		16	1.500		118	60	4.32	182.32	228
3410	12" diameter		14	1.714		153	68.50	4.94	226.44	279
3470	For 300 lb. flanges, add					63%				
3480	For 600 lb. flanges, add					310%				
3960	To delete flanges & hanger, subtract									
3970	1" diam. to 2" diam.					76%	65%			
3980	2-1/2" diam. to 4" diam.					62%	59%			
3990	5" diam. to 12" diam.					60%	46%			
4750	Schedule 80, threaded, with couplings, and clevis type hangers									
4760	sized for covering, 10' O.C.									
4790	Black, 1/4" diameter	1 Plum	54	.148	L.F.	2.61	6.35		8.96	12.70
4800	3/8" diameter		53	.151		2.92	6.45		9.37	13.20
4810	1/2" diameter		52	.154		3.03	6.55		9.58	13.55
4820	3/4" diameter		50	.160		3.63	6.85		10.48	14.60
4830	1" diameter		45	.178		4.94	7.60		12.54	17.25
4840	1-1/4" diameter	Q-1	75	.213		6.80	8.20		15	20
4850	1-1/2" diameter		69	.232		7.95	8.90		16.85	22.50
4860	2" diameter		56	.286		10.80	11		21.80	29
4870	2-1/2" diameter		44	.364		16.60	13.95		30.55	40
4880	3" diameter		38	.421		22	16.20		38.20	49
4890	3-1/2" diameter		35	.457		27	17.55		44.55	57
4900	4" diameter		32	.500		31.50	19.20		50.70	64.50
4910	5" diameter		23	.696		42.50	26.50		69	88.50
4920	6" diameter	Q-2	28	.857		55	34		89	114
4930	8" diameter		23	1.043		98	41.50		139.50	173
4940	10" diameter		20	1.200		109	48		157	195
4950	12" diameter		15	1.600		129	64		193	241
5061	A-106, gr. A/B seamless with cplgs. & hangers, 1/4" dia.	1 Plum	63	.127		3.96	5.40		9.36	12.80
5062	3/8" diameter		62	.129		3.24	5.50		8.74	12.10
5063	1/2" diameter		61	.131		3.36	5.60		8.96	12.40
5064	3/4" diameter		57	.140		7.50	6		13.50	17.55
5065	1" diameter		51	.157		11.75	6.70		18.45	23.50
5066	1-1/4" diameter	Q-1	85	.188		14	7.25		21.25	26.50
5067	1-1/2" diameter		77	.208		16.15	8		24.15	30
5071	A-53 2" diameter		61	.262		10.55	10.10		20.65	27.50
5072	2-1/2" diameter		48	.333		16.30	12.80		29.10	38
5073	3" diameter		41	.390		21.50	15		36.50	47.50
5074	4" diameter		35	.457		31	17.55		48.55	61.50
5430	To delete coupling & hanger, subtract									
5440	1/4" diam. to 1/2" diam.					31%	54%			620

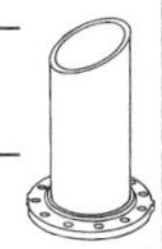

15 MECHANICAL

15107 | Metal Pipe & Fittings

				DAILY OUTPUT	LABOR-HOURS	UNIT	2006 BARE COSTS MAT.	LABOR	EQUIP.	TOTAL	TOTAL INCL O&P	
			CREW									
620	5450	3/4" diam. to 1-1/2" diam. R221113-50					28%	49%				620
	5460	2" diam. to 4" diam.					21%	40%				
	5470	5" diam. to 12" diam.					17%	43%				
	5510	Galvanized, 1/4" diameter	1 Plum	54	.148	L.F.	3.73	6.35		10.08	13.95	
	5520	3/8" diameter		53	.151		3.76	6.45		10.21	14.15	
	5530	1/2" diameter		52	.154		3.85	6.55		10.40	14.45	
	5540	3/4" diameter		50	.160		4.64	6.85		11.49	15.70	
	5550	1" diameter		45	.178		6.45	7.60		14.05	18.90	
	5560	1-1/4" diameter	Q-1	75	.213		9	8.20		17.20	22.50	
	5570	1-1/2" diameter		69	.232		10.60	8.90		19.50	25.50	
	5580	2" diameter		56	.286		14.60	11		25.60	33	
	5590	2-1/2" diameter		44	.364		22.50	13.95		36.45	46.50	
	5600	3" diameter		38	.421		30	16.20		46.20	58	
	5610	3-1/2" diameter		35	.457		35.50	17.55		53.05	66.50	
	5620	4" diameter		32	.500		42.50	19.20		61.70	76.50	
	5630	5" diameter		23	.696		56	26.50		82.50	103	
	5640	6" diameter	Q-2	28	.857		71	34		105	132	
	5650	8" diameter		23	1.043		128	41.50		169.50	205	
	5660	10" diameter		20	1.200		140	48		188	229	
	5670	12" diameter		15	1.600		165	64		229	280	
	5930	To delete coupling & hanger, subtract										
	5940	1/4" diam. to 1/2" diam.					31%	54%				
	5950	3/4" diam. to 1-1/2" diam.					28%	49%				
	5960	2" diam. to 4" diam.					21%	40%				
	5970	5" diam. to 12" diam.					17%	43%				
	6000	Welded, on yoke & roller hangers										
	6010	sized for covering, 10' O.C.										
	6040	Black, 1" diameter	Q-15	85	.188	L.F.	6.50	7.25	.81	14.56	19.35	
	6050	1-1/4" diameter		79	.203		8.85	7.80	.87	17.52	23	
	6060	1-1/2" diameter		72	.222		10	8.55	.96	19.51	25.50	
	6070	2" diameter		57	.281		12.50	10.80	1.21	24.51	32	
	6080	2-1/2" diameter		44	.364		18.25	13.95	1.57	33.77	43	
	6090	3" diameter		40	.400		23.50	15.35	1.73	40.58	51.50	
	6100	3-1/2" diameter		34	.471		28.50	18.10	2.03	48.63	61.50	
	6110	4" diameter		33	.485		32.50	18.65	2.09	53.24	67.50	
	6120	5" diameter		26	.615		44.50	23.50	2.66	70.66	89	
	6130	6" diameter	Q-16	30	.800		55.50	32	2.30	89.80	113	
	6140	8" diameter		25	.960		97.50	38.50	2.76	138.76	171	
	6150	10" diameter		20	1.200		120	48	3.46	171.46	210	
	6160	12" diameter		15	1.600		142	64	4.61	210.61	260	
	6540	To delete hanger, subtract										
	6550	1" diam. to 1-1/2" diam.					30%	14%				
	6560	2" diam. to 3" diam.					23%	9%				
	6570	3-1/2" diam. to 5" diam.					12%	6%				
	6580	6" diam. to 12" diam.					10%	4%				
	7250	Flanged, 300 lb. weld neck, on yoke & roll hangers										
	7260	sized for covering, 10' O.C.										
	7290	Black, 1" diameter	Q-15	66	.242	L.F.	10.80	9.30	1.05	21.15	27.50	
	7300	1-1/4" diameter		61	.262		13.15	10.10	1.13	24.38	31.50	
	7310	1-1/2" diameter		54	.296		14.30	11.40	1.28	26.98	35	
	7320	2" diameter		42	.381		17.55	14.65	1.65	33.85	44	
	7330	2-1/2" diameter		33	.485		24.50	18.65	2.09	45.24	58	
	7340	3" diameter		29	.552		30	21	2.38	53.38	68.50	
	7350	3-1/2" diameter		24	.667		39	25.50	2.88	67.38	86	
	7360	4" diameter		23	.696		43	26.50	3.01	72.51	92.50	
	7370	5" diameter		19	.842		60	32.50	3.64	96.14	121	

MECHANICAL 15

For expanded coverage of these items see *Means Mechanical or Plumbing Cost Data 2006*

15107 | Metal Pipe & Fittings

				DAILY	LABOR-		2006 BARE COSTS				TOTAL	
			CREW	OUTPUT	HOURS	UNIT	MAT.	LABOR	EQUIP.	TOTAL	INCL O&P	
620	7380	6" diameter (R221113-50)	Q-16	23	1.043	L.F.	72	41.50	3.01	116.51	147	620
	7390	8" diameter		17	1.412		124	56.50	4.07	184.57	228	
	7400	10" diameter		14	1.714		167	68.50	4.94	240.44	295	
	7410	12" diameter		12	2		201	79.50	5.75	286.25	350	
	7470	For 600 lb. flanges, add					100%					
	7940	To delete flanges & hanger, subtract										
	7950	1" diam. to 1-1/2" diam.					75%	66%				
	7960	2" diam. to 3" diam.					62%	60%				
	7970	3-1/2" diam. to 5" diam.					54%	66%				
	7980	6" diam. to 12" diam.					55%	62%				
	8040	Galvanized, 1" diameter	Q-15	66	.242	L.F.	12.35	9.30	1.05	22.70	29.50	
	8050	1-1/4" diameter		61	.262		15.35	10.10	1.13	26.58	34	
	8060	1-1/2" diameter		54	.296		16.95	11.40	1.28	29.63	38	
	8070	2" diameter		42	.381		21.50	14.65	1.65	37.80	48.50	
	8080	2-1/2" diameter		33	.485		30.50	18.65	2.09	51.24	65	
	8090	3" diameter		29	.552		38	21	2.38	61.38	77	
	8100	3-1/2" diameter		24	.667		48	25.50	2.88	76.38	95.50	
	8110	4" diameter		23	.696		54	26.50	3.01	83.51	104	
	8120	5" diameter		19	.842		73.50	32.50	3.64	109.64	135	
	8130	6" diameter	Q-16	23	1.043		88	41.50	3.01	132.51	165	
	8140	8" diameter		17	1.412		154	56.50	4.07	214.57	261	
	8150	10" diameter		14	1.714		198	68.50	4.94	271.44	330	
	8160	12" diameter		12	2		237	79.50	5.75	322.25	390	
	8240	For 600 lb. flanges, add					100%					
	8900	To delete flanges & hangers, subtract										
	8910	1" diam. to 1-1/2" diam.					72%	66%				
	8920	2" diam. to 3" diam.					59%	60%				
	8930	3-1/2" diam. to 5" diam.					51%	66%				
	8940	6" diam. to 12" diam.					49%	62%				
	9000	Threading pipe labor, one end, all schedules through 80										
	9010	1/4" through 3/4" pipe size	1 Plum	80	.100	Ea.		4.27		4.27	6.65	
	9020	1" through 2" pipe size		73	.110			4.68		4.68	7.30	
	9030	2-1/2" pipe size		53	.151			6.45		6.45	10	
	9040	3" pipe size		50	.160			6.85		6.85	10.60	
	9050	3-1/2" pipe size	Q-1	89	.180			6.90		6.90	10.75	
	9060	4" pipe size		73	.219			8.40		8.40	13.10	
	9070	5" pipe size		53	.302			11.60		11.60	18.05	
	9080	6" pipe size		46	.348			13.35		13.35	21	
	9090	8" pipe size		29	.552			21		21	33	
	9100	10" pipe size		21	.762			29.50		29.50	45.50	
	9110	12" pipe size		13	1.231			47.50		47.50	73.50	
	9200	Welding labor per joint										
	9210	Schedule 40,										
	9230	1/2" pipe size	Q-15	32	.500	Ea.		19.20	2.16	21.36	32.50	
	9240	3/4" pipe size		27	.593			23	2.56	25.56	38.50	
	9250	1" pipe size		23	.696			26.50	3.01	29.51	45	
	9260	1-1/4" pipe size		20	.800			30.50	3.46	33.96	52	
	9270	1-1/2" pipe size		19	.842			32.50	3.64	36.14	54.50	
	9280	2" pipe size		16	1			38.50	4.32	42.82	65	
	9290	2-1/2" pipe size		13	1.231			47.50	5.30	52.80	79.50	
	9300	3" pipe size		12	1.333			51	5.75	56.75	86	
	9310	4" pipe size		10	1.600			61.50	6.90	68.40	103	
	9320	5" pipe size		9	1.778			68.50	7.70	76.20	114	
	9330	6" pipe size		8	2			77	8.65	85.65	130	
	9340	8" pipe size		5	3.200			123	13.80	136.80	206	
	9350	10" pipe size		4	4			154	17.30	171.30	258	

15107 | Metal Pipe & Fittings

		15107 | Metal Pipe & Fittings	CREW	DAILY OUTPUT	LABOR-HOURS	UNIT	2006 BARE COSTS MAT.	LABOR	EQUIP.	TOTAL	TOTAL INCL O&P
620	9360	12″ pipe size (R221113 -50)	Q-15	3	5.333	Ea.		205	23	228	345
	9370	14″ pipe size		2.60	6.154			236	26.50	262.50	400
	9380	16″ pipe size		2.20	7.273			279	31.50	310.50	470
	9390	18″ pipe size		2	8			305	34.50	339.50	520
	9400	20″ pipe size		1.80	8.889			340	38.50	378.50	570
	9410	22″ pipe size		1.70	9.412			360	40.50	400.50	605
	9420	24″ pipe size		1.50	10.667			410	46	456	685
	9450	Schedule 80,									
	9460	1/2″ pipe size	Q-15	27	.593	Ea.		23	2.56	25.56	38.50
	9470	3/4″ pipe size		23	.696			26.50	3.01	29.51	45
	9480	1″ pipe size		20	.800			30.50	3.46	33.96	52
	9490	1-1/4″ pipe size		19	.842			32.50	3.64	36.14	54.50
	9500	1-1/2″ pipe size		18	.889			34	3.84	37.84	57
	9510	2″ pipe size		15	1.067			41	4.61	45.61	68.50
	9520	2-1/2″ pipe size		12	1.333			51	5.75	56.75	86
	9530	3″ pipe size		11	1.455			56	6.30	62.30	94
	9540	4″ pipe size		8	2			77	8.65	85.65	130
	9550	5″ pipe size		6	2.667			102	11.50	113.50	172
	9560	6″ pipe size		5	3.200			123	13.80	136.80	206
	9570	8″ pipe size		4	4			154	17.30	171.30	258
	9580	10″ pipe size		3	5.333			205	23	228	345
	9590	12″ pipe size		2	8			305	34.50	339.50	520
	9600	14″ pipe size	Q-16	2.60	9.231			370	26.50	396.50	600
	9610	16″ pipe size		2.30	10.435			415	30	445	680
	9620	18″ pipe size		2	12			480	34.50	514.50	785
	9630	20″ pipe size		1.80	13.333			530	38.50	568.50	870
	9640	22″ pipe size		1.60	15			600	43	643	980
	9650	24″ pipe size		1.50	16			640	46	686	1,050
	9990	Minimum labor/equipment charge	1 Plum	3	2.667	Job		114		114	177
640	0010	**PIPE, STEEL, FITTINGS** Threaded (R221113 -50)									
	0020	Cast Iron									
	0040	Standard weight, black									
	0060	90° Elbow, straight 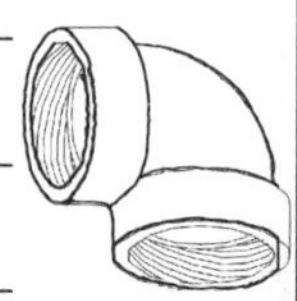									
	0070	1/4″	1 Plum	16	.500	Ea.	4.47	21.50		25.97	38
	0080	3/8″		16	.500		6.40	21.50		27.90	40
	0090	1/2″		15	.533		2.84	23		25.84	38.50
	0100	3/4″		14	.571		2.96	24.50		27.46	41.50
	0110	1″		13	.615		3.50	26.50		30	45
	0120	1-1/4″	Q-1	22	.727		4.97	28		32.97	49
	0130	1-1/2″		20	.800		6.85	30.50		37.35	55.50
	0140	2″		18	.889		10.70	34		44.70	65
	0150	2-1/2″		14	1.143		25.50	44		69.50	97
	0160	3″		10	1.600		42	61.50		103.50	142
	0170	3-1/2″		8	2		114	77		191	246
	0180	4″		6	2.667		78	102		180	245
	0190	5″		5	3.200		166	123		289	375
	0200	6″	Q-2	7	3.429		215	137		352	450
	0210	8″	″	6	4		525	159		684	830
	0250	45° Elbow, straight 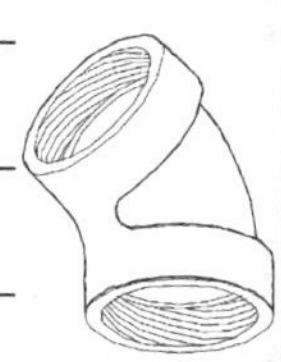									
	0260	1/4″	1 Plum	16	.500	Ea.	5.35	21.50		26.85	39
	0270	3/8″		16	.500		5.75	21.50		27.25	39.50
	0280	1/2″		15	.533		4.33	23		27.33	40.50
	0300	3/4″		14	.571		4.36	24.50		28.86	43
	0320	1″		13	.615		5.10	26.50		31.60	46.50
	0330	1-1/4″	Q-1	22	.727		6.90	28		34.90	51

MECHANICAL 15

15107 | Metal Pipe & Fittings

			CREW	DAILY OUTPUT	LABOR-HOURS	UNIT	2006 BARE COSTS MAT.	LABOR	EQUIP.	TOTAL	TOTAL INCL O&P	
640	0340	1-1/2" R221113-50	Q-1	20	.800	Ea.	11.40	30.50		41.90	60.50	640
	0350	2"		18	.889		13.15	34		47.15	67.50	
	0360	2-1/2"		14	1.143		34	44		78	106	
	0370	3"		10	1.600		54	61.50		115.50	155	
	0380	3-1/2"		8	2		120	77		197	252	
	0400	4"		6	2.667		113	102		215	283	
	0420	5"		5	3.200		247	123		370	465	
	0440	6"	Q-2	7	3.429		315	137		452	560	
	0460	8"	"	6	4		635	159		794	945	
	0500	Tee, straight										
	0510	1/4"	1 Plum	10	.800	Ea.	7	34		41	60.50	
	0520	3/8"		10	.800		6.80	34		40.80	60.50	
	0530	1/2"		9	.889		4.41	38		42.41	64	
	0540	3/4"		9	.889		5.15	38		43.15	64.50	
	0550	1"		8	1		4.58	42.50		47.08	71.50	
	0560	1-1/4"	Q-1	14	1.143		8.35	44		52.35	77.50	
	0570	1-1/2"		13	1.231		10.85	47.50		58.35	85.50	
	0580	2"		11	1.455		15.15	56		71.15	104	
	0590	2-1/2"		9	1.778		39.50	68.50		108	150	
	0600	3"		6	2.667		60.50	102		162.50	226	
	0610	3-1/2"		5	3.200		122	123		245	325	
	0620	4"		4	4		118	154		272	370	
	0630	5"		3	5.333		315	205		520	665	
	0640	6"	Q-2	4	6		355	239		594	760	
	0650	8"	"	3	8		805	320		1,125	1,375	
	0700	Standard weight, galvanized cast iron										
	0720	90° Elbow, straight										
	0730	1/4"	1 Plum	16	.500	Ea.	10.50	21.50		32	44.50	
	0740	3/8"		16	.500		10.50	21.50		32	44.50	
	0750	1/2"		15	.533		8.40	23		31.40	45	
	0760	3/4"		14	.571		11.75	24.50		36.25	51	
	0770	1"		13	.615		13.55	26.50		40.05	56	
	0780	1-1/4"	Q-1	22	.727		21	28		49	66.50	
	0790	1-1/2"		20	.800		29	30.50		59.50	80	
	0800	2"		18	.889		43	34		77	100	
	0810	2-1/2"		14	1.143		88	44		132	165	
	0820	3"		10	1.600		134	61.50		195.50	243	
	0830	3-1/2"		8	2		213	77		290	355	
	0840	4"		6	2.667		245	102		347	430	
	0850	5"		5	3.200		490	123		613	730	
	0860	6"	Q-2	7	3.429		640	137		777	910	
	0870	8"	"	6	4		1,550	159		1,709	1,975	
	0900	45° Elbow, straight										
	0910	1/4"	1 Plum	16	.500	Ea.	15.85	21.50		37.35	50.50	
	0920	3/8"		16	.500		17.05	21.50		38.55	52	
	0930	1/2"		15	.533		11.95	23		34.95	48.50	
	0940	3/4"		14	.571		13.70	24.50		38.20	53	
	0950	1"		13	.615		15.15	26.50		41.65	57.50	
	0960	1-1/4"	Q-1	22	.727		20.50	28		48.50	66	
	0970	1-1/2"		20	.800		35.50	30.50		66	87	
	0980	2"		18	.889		50	34		84	108	
	0990	2-1/2"		14	1.143		98	44		142	177	
	1000	3"		10	1.600		157	61.50		218.50	268	
	1010	3-1/2"		8	2		287	77		364	435	
	1020	4"		6	2.667		287	102		389	475	
640	1030	5"		5	3.200		730	123		853	995	

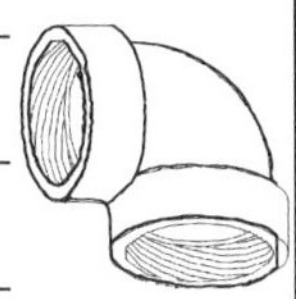

15 MECHANICAL

15107 | Metal Pipe & Fittings

	15107 | Metal Pipe & Fittings	CREW	DAILY OUTPUT	LABOR-HOURS	UNIT	2006 BARE COSTS MAT.	LABOR	EQUIP.	TOTAL	TOTAL INCL O&P
640 1040	6″ (R221113-50)	Q-2	7	3.429	Ea.	940	137		1,077	1,225
1050	8″	″	6	4		1,875	159		2,034	2,325
1100	Tee, straight									
1110	1/4″	1 Plum	10	.800	Ea.	20.50	34		54.50	76
1120	3/8″		10	.800		13.40	34		47.40	67.50
1130	1/2″		9	.889		13.10	38		51.10	73.50
1140	3/4″		9	.889		16.85	38		54.85	77.50
1150	1″		8	1		18.30	42.50		60.80	86.50
1160	1-1/4″	Q-1	14	1.143		32	44		76	104
1170	1-1/2″		13	1.231		42.50	47.50		90	120
1180	2″		11	1.455		53	56		109	145
1190	2-1/2″		9	1.778		114	68.50		182.50	231
1200	3″		6	2.667		295	102		397	485
1210	3-1/2″		5	3.200		305	123		428	525
1220	4″		4	4		340	154		494	615
1230	5″		3	5.333		930	205		1,135	1,350
1240	6″	Q-2	4	6		1,050	239		1,289	1,525
1250	8″	″	3	8		2,375	320		2,695	3,125
1300	Extra heavy weight, black									
1310	Couplings, steel straight									
1320	1/4″	1 Plum	19	.421	Ea.	1.86	18		19.86	30
1330	3/8″		19	.421		2.04	18		20.04	30
1340	1/2″		19	.421		2.75	18		20.75	31
1350	3/4″		18	.444		2.94	19		21.94	32.50
1360	1″		15	.533		3.74	23		26.74	39.50
1370	1-1/4″	Q-1	26	.615		6	23.50		29.50	43.50
1380	1-1/2″		24	.667		6	25.50		31.50	46.50
1390	2″		21	.762		9.15	29.50		38.65	55.50
1400	2-1/2″		18	.889		13.60	34		47.60	68
1410	3″		14	1.143		16.15	44		60.15	86.50
1420	3-1/2″		12	1.333		21.50	51		72.50	104
1430	4″		10	1.600		25.50	61.50		87	124
1440	5″		6	2.667		36.50	102		138.50	199
1450	6″	Q-2	8	3		44.50	120		164.50	235
1460	8″		7	3.429		81	137		218	300
1470	10″		6	4		128	159		287	390
1480	12″		4	6		171	239		410	560
1510	90° Elbow, straight									
1520	1/2″	1 Plum	15	.533	Ea.	14.05	23		37.05	51
1530	3/4″		14	.571		14.85	24.50		39.35	54.50
1540	1″		13	.615		17.95	26.50		44.45	61
1550	1-1/4″	Q-1	22	.727		26.50	28		54.50	73
1560	1-1/2″		20	.800		33	30.50		63.50	84.50
1580	2″		18	.889		41	34		75	98
1590	2-1/2″		14	1.143		95.50	44		139.50	174
1600	3″		10	1.600		126	61.50		187.50	235
1610	4″		6	2.667		265	102		367	450
1620	6″	Q-2	7	3.429		680	137		817	960
1650	45° Elbow, straight									
1660	1/2″	1 Plum	15	.533	Ea.	20	23		43	57.50
1670	3/4″		14	.571		19.35	24.50		43.85	59.50
1680	1″		13	.615		23	26.50		49.50	66.50
1690	1-1/4″	Q-1	22	.727		42	28		70	90
1700	1-1/2″		20	.800		60	30.50		90.50	114
1710	2″		18	.889		60	34		94	119
1720	2-1/2″		14	1.143		103	44		147	183

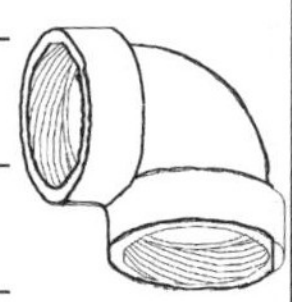

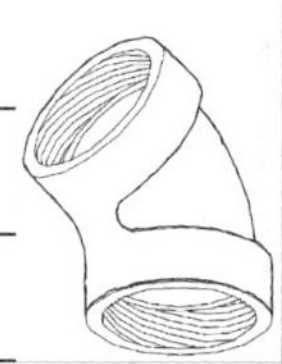

15107 | Metal Pipe & Fittings

R221113 -50

640		CREW	DAILY OUTPUT	LABOR-HOURS	UNIT	2006 BARE COSTS MAT.	LABOR	EQUIP.	TOTAL	TOTAL INCL O&P
1800	Tee, straight									
1810	1/2"	1 Plum	9	.889	Ea.	22	38		60	83.50
1820	3/4"		9	.889		22.50	38		60.50	84
1830	1"		8	1		27	42.50		69.50	96
1840	1-1/4"	Q-1	14	1.143		40	44		84	113
1850	1-1/2"		13	1.231		51.50	47.50		99	130
1860	2"		11	1.455		63.50	56		119.50	157
1870	2-1/2"		9	1.778		136	68.50		204.50	255
1880	3"		6	2.667		185	102		287	365
1890	4"		4	4		360	154		514	635
1900	6"	Q-2	4	6		785	239		1,024	1,225
4000	Standard weight, black									
4010	Couplings, steel straight, merchants									
4030	1/4"	1 Plum	19	.421	Ea.	.50	18		18.50	28.50
4040	3/8"		19	.421		.60	18		18.60	28.50
4050	1/2"		19	.421		.64	18		18.64	28.50
4060	3/4"		18	.444		.82	19		19.82	30.50
4070	1"		15	.533		1.15	23		24.15	37
4080	1-1/4"	Q-1	26	.615		1.46	23.50		24.96	38.50
4090	1-1/2"		24	.667		1.85	25.50		27.35	42
4100	2"		21	.762		2.65	29.50		32.15	48.50
4110	2-1/2"		18	.889		7.55	34		41.55	61.50
4120	3"		14	1.143		10.60	44		54.60	80
4130	3-1/2"		12	1.333		18.80	51		69.80	100
4140	4"		10	1.600		20	61.50		81.50	118
4150	5"		6	2.667		34.50	102		136.50	197
4160	6"	Q-2	8	3		41	120		161	232
4166	Plug, 1/4"	1 Plum	38	.211		1.47	9		10.47	15.60
4167	3/8"		38	.211		1.47	9		10.47	15.60
4168	1/2"		38	.211		1.47	9		10.47	15.60
4169	3/4"		32	.250		3.81	10.70		14.51	21
4170	1"		30	.267		4.06	11.40		15.46	22
4171	1-1/4"	Q-1	52	.308		4.66	11.80		16.46	23.50
4172	1-1/2"		48	.333		6.60	12.80		19.40	27
4173	2"		42	.381		8.55	14.65		23.20	32.50
4176	2-1/2"		36	.444		8.55	17.10		25.65	36
4180	4"		20	.800		21.50	30.50		52	71.50
4182	6"	Q-2	16	1.500		53.50	60		113.50	152
5000	Malleable iron, 150 lb.									
5020	Black									
5040	90° elbow, straight									
5060	1/4"	1 Plum	16	.500	Ea.	2.17	21.50		23.67	35.50
5070	3/8"		16	.500		2.17	21.50		23.67	35.50
5080	1/2"		15	.533		1.52	23		24.52	37
5090	3/4"		14	.571		1.82	24.50		26.32	40
5100	1"		13	.615		3.16	26.50		29.66	44.50
5110	1-1/4"	Q-1	22	.727		5.20	28		33.20	49
5120	1-1/2"		20	.800		6.85	30.50		37.35	55.50
5130	2"		18	.889		11.80	34		45.80	66
5140	2-1/2"		14	1.143		26.50	44		70.50	97.50
5150	3"		10	1.600		38.50	61.50		100	138
5160	3-1/2"		8	2		82.50	77		159.50	211
5170	4"		6	2.667		82.50	102		184.50	250
5180	5"		5	3.200		235	123		358	450
5190	6"	Q-2	7	3.429		279	137		416	515
5250	45° elbow, straight									

15107 | Metal Pipe & Fittings

			CREW	DAILY OUTPUT	LABOR-HOURS	UNIT	2006 BARE COSTS MAT.	LABOR	EQUIP.	TOTAL	TOTAL INCL O&P	
640	5270	1/4" R221113-50	1 Plum	16	.500	Ea.	3.27	21.50		24.77	36.50	640
	5280	3/8"		16	.500		3.34	21.50		24.84	36.50	
	5290	1/2"		15	.533		2.48	23		25.48	38	
	5300	3/4"		14	.571		3.07	24.50		27.57	41.50	
	5310	1"		13	.615		3.87	26.50		30.37	45.50	
	5320	1-1/4"	Q-1	22	.727		6.85	28		34.85	51	
	5330	1-1/2"		20	.800		8.50	30.50		39	57.50	
	5340	2"		18	.889		12.80	34		46.80	67	
	5350	2-1/2"		14	1.143		37	44		81	110	
	5360	3"		10	1.600		48	61.50		109.50	149	
	5370	3-1/2"		8	2		91.50	77		168.50	221	
	5380	4"		6	2.667		94.50	102		196.50	263	
	5390	5"		5	3.200		370	123		493	600	
	5400	6"	Q-2	7	3.429		370	137		507	620	
	5450	Tee, straight										
	5470	1/4"	1 Plum	10	.800	Ea.	3.16	34		37.16	56.50	
	5480	3/8"		10	.800		3.16	34		37.16	56.50	
	5490	1/2"		9	.889		2.01	38		40.01	61	
	5500	3/4"		9	.889		2.90	38		40.90	62	
	5510	1"		8	1		4.94	42.50		47.44	72	
	5520	1-1/4"	Q-1	14	1.143		8	44		52	77.50	
	5530	1-1/2"		13	1.231		9.95	47.50		57.45	84.50	
	5540	2"		11	1.455		17	56		73	106	
	5550	2-1/2"		9	1.778		36.50	68.50		105	147	
	5560	3"		6	2.667		54	102		156	219	
	5570	3-1/2"		5	3.200		130	123		253	335	
	5580	4"		4	4		130	154		284	380	
	5590	5"		3	5.333		375	205		580	730	
	5600	6"	Q-2	4	6		410	239		649	825	
	5650	Coupling										
	5670	1/4"	1 Plum	19	.421	Ea.	2.69	18		20.69	31	
	5680	3/8"		19	.421		2.69	18		20.69	31	
	5690	1/2"		19	.421		2.07	18		20.07	30.50	
	5700	3/4"		18	.444		2.44	19		21.44	32	
	5710	1"		15	.533		3.69	23		26.69	39.50	
	5720	1-1/4"	Q-1	26	.615		4.73	23.50		28.23	42	
	5730	1-1/2"		24	.667		6.40	25.50		31.90	47	
	5740	2"		21	.762		9.45	29.50		38.95	56	
	5750	2-1/2"		18	.889		26	34		60	82	
	5760	3"		14	1.143		35.50	44		79.50	108	
	5770	3-1/2"		12	1.333		71	51		122	158	
	5780	4"		10	1.600		71	61.50		132.50	174	
	5790	5"		6	2.667		204	102		306	385	
	5800	6"	Q-2	8	3		204	120		324	410	
	5840	Reducer, concentric, 1/4"	1 Plum	19	.421		2.84	18		20.84	31	
	5850	3/8"		19	.421		3.65	18		21.65	32	
	5860	1/2"		19	.421		2.88	18		20.88	31	
	5870	3/4"		16	.500		3.27	21.50		24.77	36.50	
	5880	1"		15	.533		5.35	23		28.35	41.50	
	5890	1-1/4"	Q-1	26	.615		6.45	23.50		29.95	44	
	5900	1-1/2"		24	.667		9.20	25.50		34.70	50	
	5910	2"		21	.762		13.30	29.50		42.80	60	
	5911	2-1/2"		18	.889		29.50	34		63.50	85.50	
	5981	Bushing, 1/4"	1 Plum	19	.421		2.83	18		20.83	31	
	5982	3/8"		19	.421		2.83	18		20.83	31	
	5983	1/2"		19	.421		2.83	18		20.83	31	640

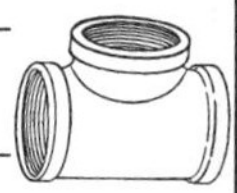

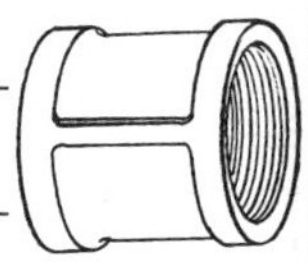

15107 | Metal Pipe & Fittings

		Description		Crew	Daily Output	Labor-Hours	Unit	2006 Bare Costs Mat.	Labor	Equip.	Total	Total Incl O&P
640	5984	3/4"	R221113-50	1 Plum	16	.500	Ea.	2.71	21.50		24.21	36
	5985	1"			15	.533		3.97	23		26.97	40
	5986	1-1/4"		Q-1	26	.615		4.93	23.50		28.43	42.50
	5987	1-1/2"			24	.667		6.60	25.50		32.10	47.50
	5988	2"			21	.762		7	29.50		36.50	53
	5989	Cap, 1/4"		1 Plum	38	.211		2	9		11	16.20
	5991	3/8"			38	.211		1.68	9		10.68	15.85
	5992	1/2"			38	.211		1.52	9		10.52	15.65
	5993	3/4"			32	.250		2.06	10.70		12.76	18.85
	5994	1"			30	.267		2.50	11.40		13.90	20.50
	5995	1-1/4"		Q-1	52	.308		3.29	11.80		15.09	22
	5996	1-1/2"			48	.333		4.52	12.80		17.32	25
	5997	2"			42	.381		6.60	14.65		21.25	30.50
	6000	For galvanized elbows, tees, and couplings add						20%				
	6058	For galvanized reducers, caps and plugs add						20%				
	7000	Union, with brass seat										
	7010	1/4"		1 Plum	15	.533	Ea.	11.70	23		34.70	48.50
	7020	3/8"			15	.533		8.05	23		31.05	44.50
	7030	1/2"			14	.571		7.30	24.50		31.80	46
	7040	3/4"			13	.615		7.65	26.50		34.15	49.50
	7050	1"			12	.667		10.95	28.50		39.45	56.50
	7060	1-1/4"		Q-1	21	.762		15	29.50		44.50	62
	7070	1-1/2"			19	.842		19.55	32.50		52.05	72
	7080	2"			17	.941		22.50	36		58.50	81
	7090	2-1/2"			13	1.231		67.50	47.50		115	148
	7100	3"			9	1.778		81.50	68.50		150	196
	7250	For galvanized unions, add						15%				
	9757	Forged steel, 3000 lb.										
	9758	Black										
	9760	90° Elbow, 1/4"		1 Plum	16	.500	Ea.	12.95	21.50		34.45	47.50
	9761	3/8"			16	.500		12.95	21.50		34.45	47.50
	9762	1/2"			15	.533		10.05	23		33.05	46.50
	9763	3/4"			14	.571		12.40	24.50		36.90	51.50
	9764	1"			13	.615		18	26.50		44.50	61
	9765	1-1/4"		Q-1	22	.727		36	28		64	83
	9766	1-1/2"			20	.800		46	30.50		76.50	98.50
	9767	2"			18	.889		55.50	34		89.50	114
	9780	45° Elbow 1/4"		1 Plum	16	.500		16.45	21.50		37.95	51
	9781	3/8"			16	.500		16.45	21.50		37.95	51
	9782	1/2"			15	.533		16.45	23		39.45	53.50
	9783	3/4"			14	.571		19	24.50		43.50	59
	9784	1"			13	.615		26	26.50		52.50	69.50
	9785	1-1/4"		Q-1	22	.727		36	28		64	83
	9786	1-1/2"			20	.800		51.50	30.50		82	105
	9787	2"			18	.889		71	34		105	131
	9800	Tee, 1/4"		1 Plum	10	.800		15.95	34		49.95	70.50
	9801	3/8"			10	.800		15.95	34		49.95	70.50
	9802	1/2"			9	.889		14.15	38		52.15	74.50
	9803	3/4"			9	.889		19.40	38		57.40	80.50
	9804	1"			8	1		24.50	42.50		67	93.50
	9805	1-1/4"		Q-1	14	1.143		49	44		93	123
	9806	1-1/2"			13	1.231		58	47.50		105.50	137
	9807	2"			11	1.455		75	56		131	170
	9820	Reducer, concentric, 1/4"		1 Plum	19	.421		9.25	18		27.25	38
	9821	3/8"			19	.421		9.25	18		27.25	38
	9822	1/2"			17	.471		9.25	20		29.25	41.50

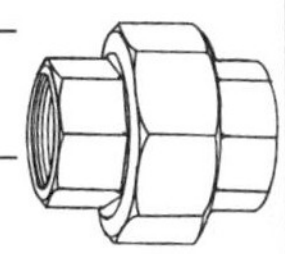

15 MECHANICAL

15107 | Metal Pipe & Fittings

			CREW	DAILY OUTPUT	LABOR-HOURS	UNIT	2006 BARE COSTS MAT.	LABOR	EQUIP.	TOTAL	TOTAL INCL O&P	
640	9823	3/4" R221113-50	1 Plum	16	.500	Ea.	10.40	21.50		31.90	44.50	640
	9824	1"		15	.533		14.25	23		37.25	51	
	9825	1-1/4"	Q-1	26	.615		22.50	23.50		46	61.50	
	9826	1-1/2"		24	.667		25.50	25.50		51	68	
	9827	2"		21	.762		37	29.50		66.50	86.50	
	9840	Cap, 1/4"	1 Plum	38	.211		4.91	9		13.91	19.40	
	9841	3/8"		38	.211		4.91	9		13.91	19.40	
	9842	1/2"		34	.235		4.77	10.05		14.82	21	
	9843	3/4"		32	.250		6.65	10.70		17.35	24	
	9844	1"		30	.267		10.20	11.40		21.60	29	
	9845	1-1/4"	Q-1	52	.308		15.95	11.80		27.75	36	
	9846	1-1/2"		48	.333		18.35	12.80		31.15	40	
	9847	2"		42	.381		28	14.65		42.65	53.50	
	9860	Plug, 1/4"	1 Plum	38	.211		2.36	9		11.36	16.60	
	9861	3/8"		38	.211		2.48	9		11.48	16.75	
	9862	1/2"		34	.235		2.48	10.05		12.53	18.35	
	9863	3/4"		32	.250		3.31	10.70		14.01	20	
	9864	1"		30	.267		4.73	11.40		16.13	23	
	9865	1-1/4"	Q-1	52	.308		8.95	11.80		20.75	28.50	
	9866	1-1/2"		48	.333		10.40	12.80		23.20	31.50	
	9867	2"		42	.381		17.40	14.65		32.05	42	
	9880	Union, bronze seat, 1/4"	1 Plum	15	.533		34	23		57	72.50	
	9881	3/8"		15	.533		34	23		57	72.50	
	9882	1/2"		14	.571		32.50	24.50		57	73.50	
	9883	3/4"		13	.615		43.50	26.50		70	88.50	
	9884	1"		12	.667		49	28.50		77.50	98.50	
	9885	1-1/4"	Q-1	21	.762		95	29.50		124.50	151	
	9886	1-1/2"		19	.842		98	32.50		130.50	159	
	9887	2"		17	.941		116	36		152	184	
	9900	Coupling, 1/4"	1 Plum	19	.421		4.75	18		22.75	33.50	
	9901	3/8"		19	.421		4.75	18		22.75	33.50	
	9902	1/2"		17	.471		4.13	20		24.13	36	
	9903	3/4"		16	.500		5.25	21.50		26.75	39	
	9904	1"		15	.533		9.30	23		32.30	46	
	9905	1-1/4"	Q-1	26	.615		15.50	23.50		39	54	
	9906	1-1/2"		24	.667		20	25.50		45.50	62	
	9907	2"		21	.762		24	29.50		53.50	72	
	9990	Minimum labor/equipment charge	1 Plum	4	2	Job		85.50		85.50	133	
660	0010	**PIPE, STEEL, FITTINGS** Flanged, welded and special type										660
	0020	flanged joints, C.I., standard weight, black. One gasket & bolt										
	0040	set, mat'l only, required at each joint, not included (see line 0620)										
	0060	90° Elbow, straight, 1-1/2" pipe size	Q-1	14	1.143	Ea.	244	44		288	340	
	0080	2" pipe size		13	1.231		149	47.50		196.50	238	
	0090	2-1/2" pipe size		12	1.333		161	51		212	257	
	0100	3" pipe size		11	1.455		134	56		190	235	
	0110	4" pipe size		8	2		166	77		243	305	
	0120	5" pipe size		7	2.286		395	88		483	570	
	0130	6" pipe size	Q-2	9	2.667		260	106		366	450	
	0140	8" pipe size		8	3		450	120		570	680	
	0150	10" pipe size		7	3.429		985	137		1,122	1,275	
	0160	12" pipe size		6	4		1,975	159		2,134	2,425	
	0200	45° Elbow, straight, 1-1/2" pipe size	Q-1	14	1.143		295	44		339	395	
	0220	2" pipe size		13	1.231		216	47.50		263.50	310	
	0230	2-1/2" pipe size		12	1.333		230	51		281	335	
	0240	3" pipe size		11	1.455		224	56		280	335	
	0250	4" pipe size		8	2		252	77		329	395	

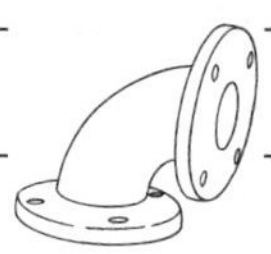

MECHANICAL 15

15107 | Metal Pipe & Fittings

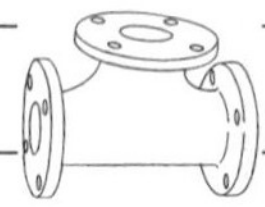

		15107 Metal Pipe & Fittings	CREW	DAILY OUTPUT	LABOR-HOURS	UNIT	2006 BARE COSTS MAT.	LABOR	EQUIP.	TOTAL	TOTAL INCL O&P	
660	0260	5″ pipe size	Q-1	7	2.286	Ea.	605	88		693	800	660
	0270	6″ pipe size	Q-2	9	2.667		410	106		516	620	
	0280	8″ pipe size		8	3		600	120		720	850	
	0290	10″ pipe size		7	3.429		1,275	137		1,412	1,600	
	0300	12″ pipe size		6	4		1,950	159		2,109	2,400	
	0350	Tee, straight, 1-1/2″ pipe size	Q-1	10	1.600		281	61.50		342.50	405	
	0370	2″ pipe size		9	1.778		163	68.50		231.50	285	
	0380	2-1/2″ pipe size		8	2		238	77		315	380	
	0390	3″ pipe size		7	2.286		167	88		255	320	
	0400	4″ pipe size		5	3.200		253	123		376	470	
	0410	5″ pipe size		4	4		680	154		834	985	
	0420	6″ pipe size	Q-2	6	4		370	159		529	655	
	0430	8″ pipe size		5	4.800		630	191		821	985	
	0440	10″ pipe size		4	6		2,375	239		2,614	2,975	
	0450	12″ pipe size		3	8		2,650	320		2,970	3,425	
	0500	For galvanized elbows and tees, add					100%					
	0520	For extra heavy weight elbows and tees, add					140%					
	0620	Gasket and bolt set, 150#, 1/2″ pipe size	1 Plum	20	.400		2.36	17.10		19.46	29	
	0622	3/4″ pipe size		19	.421		2.36	18		20.36	30.50	
	0624	1″ pipe size		18	.444		2.42	19		21.42	32	
	0626	1-1/4″ pipe size		17	.471		2.54	20		22.54	34.50	
	0628	1-1/2″ pipe size		15	.533		2.71	23		25.71	38.50	
	0630	2″ pipe size		13	.615		4.33	26.50		30.83	46	
	0640	2-1/2″ pipe size		12	.667		4.58	28.50		33.08	49.50	
	0650	3″ pipe size		11	.727		4.80	31		35.80	54	
	0660	3-1/2″ pipe size		9	.889		8.70	38		46.70	68.50	
	0670	4″ pipe size		8	1		8.70	42.50		51.20	76	
	0680	5″ pipe size		7	1.143		12.90	49		61.90	90	
	0690	6″ pipe size		6	1.333		13.55	57		70.55	103	
	0700	8″ pipe size		5	1.600		14.45	68.50		82.95	122	
	0710	10″ pipe size		4.50	1.778		30	76		106	151	
	0720	12″ pipe size		4.20	1.905		32	81.50		113.50	162	
	0730	14″ pipe size		4	2		45.50	85.50		131	183	
	0740	16″ pipe size		3	2.667		59.50	114		173.50	243	
	0750	18″ pipe size		2.70	2.963		83	127		210	289	
	0760	20″ pipe size		2.30	3.478		111	149		260	355	
	0780	24″ pipe size		1.90	4.211		149	180		329	445	
	0790	26″ pipe size		1.60	5		168	214		382	515	
	0810	30″ pipe size		1.40	5.714		208	244		452	610	
	0830	36″ pipe size		1.10	7.273		256	310		566	765	
	0850	For 300 lb gasket set, add					40%					
	2000	Flanged unions, 125 lb., black, 1/2″ pipe size	1 Plum	17	.471	Ea.	38	20		58	73.50	
	2040	3/4″ pipe size		17	.471		50	20		70	86.50	
	2050	1″ pipe size		16	.500		48.50	21.50		70	86.50	
	2060	1-1/4″ pipe size	Q-1	28	.571		57.50	22		79.50	97	
	2070	1-1/2″ pipe size		27	.593		53	23		76	94	
	2080	2″ pipe size		26	.615		62.50	23.50		86	106	
	2090	2-1/2″ pipe size		24	.667		84.50	25.50		110	133	
	2100	3″ pipe size		22	.727		95.50	28		123.50	149	
	2110	3-1/2″ pipe size		18	.889		162	34		196	231	
	2120	4″ pipe size		16	1		130	38.50		168.50	203	
	2130	5″ pipe size		14	1.143		305	44		349	405	
	2140	6″ pipe size	Q-2	19	1.263		284	50.50		334.50	395	
	2150	8″ pipe size	″	16	1.500		630	60		690	790	
	2200	For galvanized unions, add					150%					
	2290	Threaded flange										

15 MECHANICAL

15107 | Metal Pipe & Fittings

			CREW	DAILY OUTPUT	LABOR-HOURS	UNIT	2006 BARE COSTS MAT.	LABOR	EQUIP.	TOTAL	TOTAL INCL O&P	
660	2300	Cast iron										660
	2310	Black, 125 lb., per flange										
	2320	1" pipe size	1 Plum	27	.296	Ea.	21.50	12.65		34.15	43.50	
	2330	1-1/4" pipe size	Q-1	44	.364		26	13.95		39.95	50	
	2340	1-1/2" pipe size		40	.400		24	15.35		39.35	50.50	
	2350	2" pipe size		36	.444		24	17.10		41.10	52.50	
	2360	2-1/2" pipe size		28	.571		28	22		50	65	
	2370	3" pipe size		20	.800		36	30.50		66.50	87.50	
	2380	3-1/2" pipe size		16	1		48.50	38.50		87	114	
	2390	4" pipe size		12	1.333		48.50	51		99.50	133	
	2400	5" pipe size		10	1.600		69	61.50		130.50	171	
	2410	6" pipe size	Q-2	14	1.714		78	68.50		146.50	192	
	2420	8" pipe size		12	2		123	79.50		202.50	259	
	2430	10" pipe size		10	2.400		218	95.50		313.50	390	
	2440	12" pipe size		8	3		465	120		585	700	
	2460	For galvanized flanges, add					95%					
	2490	Blind flange										
	2492	Cast iron										
	2494	Black, 125 lb., per flange										
	2496	1" pipe size	1 Plum	27	.296	Ea.	34.50	12.65		47.15	57.50	
	2500	1-1/2" pipe size	Q-1	40	.400		39	15.35		54.35	66.50	
	2502	2" pipe size		36	.444		44	17.10		61.10	74.50	
	2504	2-1/2" pipe size		28	.571		48	22		70	87	
	2506	3" pipe size		20	.800		58	30.50		88.50	112	
	2508	4" pipe size		12	1.333		74	51		125	161	
	2510	5" pipe size		10	1.600		120	61.50		181.50	228	
	2512	6" pipe size	Q-2	14	1.714		131	68.50		199.50	250	
	2514	8" pipe size		12	2		205	79.50		284.50	350	
	2516	10" pipe size		10	2.400		305	95.50		400.50	485	
	2518	12" pipe size		8	3		555	120		675	795	
	2520	For galvanized flanges, add					80%					
	2570	Threaded flange										
	2580	Forged steel,										
	2590	Black 150 lb., per flange										
	2600	1/2" pipe size	1 Plum	30	.267	Ea.	17.30	11.40		28.70	37	
	2610	3/4" pipe size		28	.286		17.30	12.20		29.50	38	
	2620	1" pipe size		27	.296		17.30	12.65		29.95	38.50	
	2630	1-1/4" pipe size	Q-1	44	.364		17.30	13.95		31.25	40.50	
	2640	1-1/2" pipe size		40	.400		17.30	15.35		32.65	43	
	2650	2" pipe size		36	.444		17.30	17.10		34.40	45.50	
	2660	2-1/2" pipe size		28	.571		21.50	22		43.50	57.50	
	2670	3" pipe size		20	.800		21.50	30.50		52	71.50	
	2690	4" pipe size		12	1.333		26	51		77	108	
	2700	5" pipe size		10	1.600		41	61.50		102.50	141	
	2710	6" pipe size	Q-2	14	1.714		44.50	68.50		113	155	
	2720	8" pipe size		12	2		74.50	79.50		154	206	
	2730	10" pipe size		10	2.400		135	95.50		230.50	298	
	2860	Black 300 lb., per flange										
	2870	1/2" pipe size	1 Plum	30	.267	Ea.	16.50	11.40		27.90	36	
	2880	3/4" pipe size		28	.286		16.50	12.20		28.70	37	
	2890	1" pipe size		27	.296		16.50	12.65		29.15	38	
	2900	1-1/4" pipe size	Q-1	44	.364		16.50	13.95		30.45	39.50	
	2910	1-1/2" pipe size		40	.400		16.50	15.35		31.85	42	
	2920	2" pipe size		36	.444		17.20	17.10		34.30	45.50	
	2930	2-1/2" pipe size		28	.571		25	22		47	61.50	
	2940	3" pipe size		20	.800		27	30.50		57.50	77.50	

15107 | Metal Pipe & Fittings

			CREW	DAILY OUTPUT	LABOR-HOURS	UNIT	2006 BARE COSTS MAT.	LABOR	EQUIP.	TOTAL	TOTAL INCL O&P
660	2960	4" pipe size	Q-1	12	1.333	Ea.	39	51		90	122
	2970	6" pipe size	Q-2	14	1.714	↓	72	68.50		140.50	185
	3000	Weld joint, butt, carbon steel, standard weight									
	3040	90° elbow, long radius									
	3050	1/2" pipe size	Q-15	16	1	Ea.	25	38.50	4.32	67.82	92.50
	3060	3/4" pipe size		16	1		25	38.50	4.32	67.82	92.50
	3070	1" pipe size		16	1		12.15	38.50	4.32	54.97	78
	3080	1-1/4" pipe size		14	1.143		12.15	44	4.94	61.09	87.50
	3090	1-1/2" pipe size		13	1.231		12.15	47.50	5.30	64.95	93
	3100	2" pipe size		10	1.600		13.10	61.50	6.90	81.50	118
	3110	2-1/2" pipe size		8	2		15.70	77	8.65	101.35	147
	3120	3" pipe size		7	2.286		19.20	88	9.85	117.05	169
	3130	4" pipe size		5	3.200		31.50	123	13.80	168.30	241
	3136	5" pipe size	↓	4	4		66	154	17.30	237.30	330
	3140	6" pipe size	Q-16	5	4.800		69	191	13.80	273.80	390
	3150	8" pipe size		3.75	6.400		130	255	18.45	403.45	560
	3160	10" pipe size		3	8		261	320	23	604	810
	3170	12" pipe size		2.50	9.600		385	385	27.50	797.50	1,050
	3180	14" pipe size		2	12		510	480	34.50	1,024.50	1,350
	3190	16" pipe size		1.50	16		735	640	46	1,421	1,850
	3191	18" pipe size		1.25	19.200		970	765	55.50	1,790.50	2,325
	3192	20" pipe size		1.15	20.870		1,375	830	60	2,265	2,900
	3194	24" pipe size	↓	1.02	23.529	↓	1,850	940	68	2,858	3,550
	3200	45° Elbow, long									
	3210	1/2" pipe size	Q-15	16	1	Ea.	34.50	38.50	4.32	77.32	103
	3220	3/4" pipe size		16	1		34.50	38.50	4.32	77.32	103
	3230	1" pipe size		16	1		12.50	38.50	4.32	55.32	78.50
	3240	1-1/4" pipe size		14	1.143		12.50	44	4.94	61.44	87.50
	3250	1-1/2" pipe size		13	1.231		12.50	47.50	5.30	65.30	93
	3260	2" pipe size		10	1.600		12.50	61.50	6.90	80.90	117
	3270	2-1/2" pipe size		8	2		14.40	77	8.65	100.05	145
	3280	3" pipe size		7	2.286		15.70	88	9.85	113.55	165
	3290	4" pipe size		5	3.200		28	123	13.80	164.80	237
	3296	5" pipe size	↓	4	4		42	154	17.30	213.30	305
	3300	6" pipe size	Q-16	5	4.800		55.50	191	13.80	260.30	375
	3310	8" pipe size		3.75	6.400		93	255	18.45	366.45	520
	3320	10" pipe size		3	8		185	320	23	528	725
	3330	12" pipe size		2.50	9.600		261	385	27.50	673.50	915
	3340	14" pipe size		2	12		350	480	34.50	864.50	1,175
	3341	16" pipe size		1.50	16		620	640	46	1,306	1,725
	3342	18" pipe size		1.25	19.200		885	765	55.50	1,705.50	2,225
	3343	20" pipe size		1.15	20.870		915	830	60	1,805	2,375
	3345	24" pipe size		1.05	22.857		1,400	910	66	2,376	3,025
	3346	26" pipe size		.85	28.235		1,525	1,125	81.50	2,731.50	3,550
	3347	30" pipe size		.45	53.333		1,700	2,125	154	3,979	5,325
	3349	36" pipe size	↓	.38	63.158	↓	1,850	2,525	182	4,557	6,175
	3350	Tee, straight									
	3360	1/2" pipe size	Q-15	10	1.600	Ea.	63	61.50	6.90	131.40	173
	3370	3/4" pipe size		10	1.600		63	61.50	6.90	131.40	173
	3380	1" pipe size		10	1.600		32	61.50	6.90	100.40	138
	3390	1-1/4" pipe size		9	1.778		39	68.50	7.70	115.20	157
	3400	1-1/2" pipe size		8	2		39	77	8.65	124.65	173
	3410	2" pipe size		6	2.667		31.50	102	11.50	145	207
	3420	2-1/2" pipe size		5	3.200		43	123	13.80	179.80	254
	3430	3" pipe size		4	4		47.50	154	17.30	218.80	310
	3440	4" pipe size	↓	3	5.333	↓	67	205	23	295	420

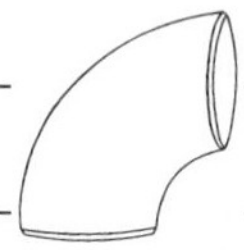

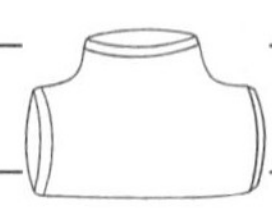

15 MECHANICAL

15107 | Metal Pipe & Fittings

			CREW	DAILY OUTPUT	LABOR-HOURS	UNIT	2006 BARE COSTS MAT.	LABOR	EQUIP.	TOTAL	TOTAL INCL O&P	
660	3446	5″ pipe size	Q-15	2.50	6.400	Ea.	111	246	27.50	384.50	535	660
	3450	6″ pipe size	Q-16	3	8		115	320	23	458	650	
	3460	8″ pipe size		2.50	9.600		203	385	27.50	615.50	850	
	3470	10″ pipe size		2	12		400	480	34.50	914.50	1,225	
	3480	12″ pipe size		1.60	15		560	600	43	1,203	1,600	
	3481	14″ pipe size		1.30	18.462		910	735	53	1,698	2,200	
	3482	16″ pipe size		1	24		1,100	955	69	2,124	2,750	
	3483	18″ pipe size		.80	30		1,750	1,200	86.50	3,036.50	3,875	
	3484	20″ pipe size		.75	32		2,700	1,275	92	4,067	5,050	
	3486	24″ pipe size		.70	34.286		3,475	1,375	98.50	4,948.50	6,050	
	3487	26″ pipe size		.55	43.636		4,125	1,750	126	6,001	7,400	
	3488	30″ pipe size		.30	80		4,550	3,200	230	7,980	10,200	
	3490	36″ pipe size		.20	120		5,000	4,775	345	10,120	13,300	
	3491	Eccentric reducer, 1-1/2″ pipe size	Q-15	14	1.143		76.50	44	4.94	125.44	158	
	3492	2″ pipe size		11	1.455		114	56	6.30	176.30	219	
	3493	2-1/2″ pipe size		9	1.778		137	68.50	7.70	213.20	264	
	3494	3″ pipe size		8	2		154	77	8.65	239.65	300	
	3495	4″ pipe size		6	2.667		168	102	11.50	281.50	355	
	3496	6″ pipe size	Q-16	5	4.800		293	191	13.80	497.80	630	
	3497	8″ pipe size		4	6		335	239	17.30	591.30	760	
	3498	10″ pipe size		3	8		390	320	23	733	950	
	3499	12″ pipe size		2.50	9.600		555	385	27.50	967.50	1,250	
	3501	Cap, 1-1/2″ pipe size	Q-15	28	.571		13.75	22	2.47	38.22	52	
	3502	2″ pipe size		22	.727		13.45	28	3.14	44.59	62	
	3503	2-1/2″ pipe size		18	.889		14.10	34	3.84	51.94	72.50	
	3504	3″ pipe size		16	1		14.10	38.50	4.32	56.92	80.50	
	3505	4″ pipe size		12	1.333		20.50	51	5.75	77.25	108	
	3506	6″ pipe size	Q-16	10	2.400		35	95.50	6.90	137.40	195	
	3507	8″ pipe size		8	3		53.50	120	8.65	182.15	255	
	3508	10″ pipe size		6	4		96.50	159	11.50	267	365	
	3509	12″ pipe size		5	4.800		146	191	13.80	350.80	475	
	3511	14″ pipe size		4	6		195	239	17.30	451.30	605	
	3512	16″ pipe size		4	6		226	239	17.30	482.30	640	
	3513	18″ pipe size		3	8		310	320	23	653	860	
	3517	Weld joint, butt, carbon steel, extra strong										
	3519	90° elbow, long										
	3520	1/2″ pipe size	Q-15	13	1.231	Ea.	31.50	47.50	5.30	84.30	114	
	3530	3/4″ pipe size		12	1.333		31.50	51	5.75	88.25	120	
	3540	1″ pipe size		11	1.455		15.70	56	6.30	78	111	
	3550	1-1/4″ pipe size		10	1.600		15.70	61.50	6.90	84.10	120	
	3560	1-1/2″ pipe size		9	1.778		15.70	68.50	7.70	91.90	132	
	3570	2″ pipe size		8	2		15.70	77	8.65	101.35	147	
	3580	2-1/2″ pipe size		7	2.286		22	88	9.85	119.85	172	
	3590	3″ pipe size		6	2.667		28	102	11.50	141.50	202	
	3600	4″ pipe size		4	4		46	154	17.30	217.30	310	
	3606	5″ pipe size		3.50	4.571		99	176	19.75	294.75	405	
	3610	6″ pipe size	Q-16	4.50	5.333		118	213	15.35	346.35	475	
	3620	8″ pipe size		3.50	6.857		220	273	19.75	512.75	690	
	3630	10″ pipe size		2.50	9.600		405	385	27.50	817.50	1,075	
	3640	12″ pipe size		2.25	10.667		575	425	30.50	1,030.50	1,325	
	3650	45° Elbow, long										
	3660	1/2″ pipe size	Q-15	13	1.231	Ea.	34.50	47.50	5.30	87.30	117	
	3670	3/4″ pipe size		12	1.333		34.50	51	5.75	91.25	124	
	3680	1″ pipe size		11	1.455		16	56	6.30	78.30	112	
	3690	1-1/4″ pipe size		10	1.600		16	61.50	6.90	84.40	121	
	3700	1-1/2″ pipe size		9	1.778		16	68.50	7.70	92.20	132	

15107	Metal Pipe & Fittings		CREW	DAILY OUTPUT	LABOR-HOURS	UNIT	2006 BARE COSTS MAT.	LABOR	EQUIP.	TOTAL	TOTAL INCL O&P
660	3710	2″ pipe size	Q-15	8	2	Ea.	16	77	8.65	101.65	147
	3720	2-1/2″ pipe size		7	2.286		35.50	88	9.85	133.35	187
	3730	3″ pipe size		6	2.667		21	102	11.50	134.50	195
	3740	4″ pipe size		4	4		32.50	154	17.30	203.80	294
	3746	5″ pipe size		3.50	4.571		78.50	176	19.75	274.25	380
	3750	6″ pipe size	Q-16	4.50	5.333		91.50	213	15.35	319.85	450
	3760	8″ pipe size		3.50	6.857		156	273	19.75	448.75	620
	3770	10″ pipe size		2.50	9.600		300	385	27.50	712.50	955
	3780	12″ pipe size		2.25	10.667		450	425	30.50	905.50	1,200
	3800	Tee, straight									
	3810	1/2″ pipe size	Q-15	9	1.778	Ea.	74.50	68.50	7.70	150.70	196
	3820	3/4″ pipe size		8.50	1.882		74.50	72.50	8.15	155.15	203
	3830	1″ pipe size		8	2		41.50	77	8.65	127.15	176
	3840	1-1/4″ pipe size		7	2.286		39.50	88	9.85	137.35	191
	3850	1-1/2″ pipe size		6	2.667		46.50	102	11.50	160	223
	3860	2″ pipe size		5	3.200		39.50	123	13.80	176.30	250
	3870	2-1/2″ pipe size		4	4		65.50	154	17.30	236.80	330
	3880	3″ pipe size		3.50	4.571		60.50	176	19.75	256.25	360
	3890	4″ pipe size		2.50	6.400		98	246	27.50	371.50	520
	3896	5″ pipe size		2.25	7.111		211	273	30.50	514.50	690
	3900	6″ pipe size	Q-16	2.25	10.667		165	425	30.50	620.50	875
	3910	8″ pipe size		2	12		310	480	34.50	824.50	1,125
	3920	10″ pipe size		1.75	13.714		555	545	39.50	1,139.50	1,500
	3930	12″ pipe size		1.50	16		810	640	46	1,496	1,925
	4000	Eccentric reducer, 1-1/2″ pipe size	Q-15	10	1.600		74	61.50	6.90	142.40	185
	4010	2″ pipe size		9	1.778		212	68.50	7.70	288.20	350
	4020	2-1/2″ pipe size		8	2		146	77	8.65	231.65	290
	4030	3″ pipe size		7	2.286		187	88	9.85	284.85	355
	4040	4″ pipe size		5	3.200		259	123	13.80	395.80	490
	4046	5″ pipe size		4.70	3.404		575	131	14.70	720.70	850
	4050	6″ pipe size	Q-16	4.50	5.333		365	213	15.35	593.35	745
	4060	8″ pipe size		3.50	6.857		420	273	19.75	712.75	910
	4070	10″ pipe size		2.50	9.600		535	385	27.50	947.50	1,225
	4080	12″ pipe size		2.25	10.667		560	425	30.50	1,015.50	1,300
	4090	14″ pipe size		2.10	11.429		1,075	455	33	1,563	1,950
	4100	16″ pipe size		1.90	12.632		1,150	505	36.50	1,691.50	2,100
	4151	Cap, 1-1/2″ pipe size	Q-15	24	.667		16.95	25.50	2.88	45.33	62
	4152	2″ pipe size		18	.889		13.45	34	3.84	51.29	72
	4153	2-1/2″ pipe size		16	1		18.55	38.50	4.32	61.37	85.50
	4154	3″ pipe size		14	1.143		20.50	44	4.94	69.44	96.50
	4155	4″ pipe size		10	1.600		28	61.50	6.90	96.40	134
	4156	6″ pipe size	Q-16	9	2.667		57.50	106	7.70	171.20	236
	4157	8″ pipe size		7	3.429		85.50	137	9.85	232.35	315
	4158	10″ pipe size		5	4.800		132	191	13.80	336.80	455
	4159	12″ pipe size		4	6		176	239	17.30	432.30	580
	4200	Welding ring w/spacer pins, 2″ pipe size					2.15			2.15	2.37
	4210	2-1/2″ pipe size					2.59			2.59	2.85
	4220	3″ pipe size					2.68			2.68	2.95
	4230	4″ pipe size					2.79			2.79	3.07
	4236	5″ pipe size					3.29			3.29	3.62
	4240	6″ pipe size					3.29			3.29	3.62
	4250	8″ pipe size					3.88			3.88	4.27
	4260	10″ pipe size					4.75			4.75	5.25
	4270	12″ pipe size					5.45			5.45	6
	4280	14″ pipe size					6.35			6.35	6.95
	4290	16″ pipe size					7.15			7.15	7.85

15107 | Metal Pipe & Fittings

			CREW	DAILY OUTPUT	LABOR-HOURS	UNIT	2006 BARE COSTS MAT.	LABOR	EQUIP.	TOTAL	TOTAL INCL O&P	
660	4300	18″ pipe size				Ea.	8.25			8.25	9.10	660
	4310	20″ pipe size					8.75			8.75	9.65	
	4330	24″ pipe size					10.20			10.20	11.20	
	4340	26″ pipe size					13.60			13.60	14.95	
	4350	30″ pipe size					13.60			13.60	14.95	
	4370	36″ pipe size				↓	19.55			19.55	21.50	
	5000	Weld joint, socket, forged steel, 3000 lb., schedule 40 pipe										
	5010	90° elbow, straight										
	5020	1/4″ pipe size	Q-15	22	.727	Ea.	13.60	28	3.14	44.74	62	
	5030	3/8″ pipe size		22	.727		13.75	28	3.14	44.89	62	
	5040	1/2″ pipe size		20	.800		8.95	30.50	3.46	42.91	61.50	
	5050	3/4″ pipe size		20	.800		9.50	30.50	3.46	43.46	62.50	
	5060	1″ pipe size		20	.800		11.75	30.50	3.46	45.71	65	
	5070	1-1/4″ pipe size		18	.889		22.50	34	3.84	60.34	82	
	5080	1-1/2″ pipe size		16	1		25.50	38.50	4.32	68.32	93	
	5090	2″ pipe size		12	1.333		39.50	51	5.75	96.25	129	
	5100	2-1/2″ pipe size		10	1.600		112	61.50	6.90	180.40	226	
	5110	3″ pipe size		8	2		197	77	8.65	282.65	345	
	5120	4″ pipe size	↓	6	2.667	↓	510	102	11.50	623.50	730	
	5130	45° Elbow, straight										
	5134	1/4″ pipe size	Q-15	22	.727	Ea.	15.90	28	3.14	47.04	64.50	
	5135	3/8″ pipe size		22	.727		15.90	28	3.14	47.04	64.50	
	5136	1/2″ pipe size		20	.800		11.95	30.50	3.46	45.91	65	
	5137	3/4″ pipe size		20	.800		13.85	30.50	3.46	47.81	67	
	5140	1″ pipe size		20	.800		18.15	30.50	3.46	52.11	72	
	5150	1-1/4″ pipe size		18	.889		24.50	34	3.84	62.34	84	
	5160	1-1/2″ pipe size		16	1		30.50	38.50	4.32	73.32	98.50	
	5170	2″ pipe size		12	1.333		49	51	5.75	105.75	140	
	5180	2-1/2″ pipe size		10	1.600		127	61.50	6.90	195.40	243	
	5190	3″ pipe size		8	2		211	77	8.65	296.65	360	
	5200	4″ pipe size	↓	6	2.667	↓	420	102	11.50	533.50	635	
	5250	Tee, straight										
	5254	1/4″ pipe size	Q-15	15	1.067	Ea.	16.85	41	4.61	62.46	87	
	5255	3/8″ pipe size		15	1.067		16.85	41	4.61	62.46	87	
	5256	1/2″ pipe size		13	1.231		11.10	47.50	5.30	63.90	91.50	
	5257	3/4″ pipe size		13	1.231		13.45	47.50	5.30	66.25	94	
	5260	1″ pipe size		13	1.231		18.45	47.50	5.30	71.25	100	
	5270	1-1/4″ pipe size		12	1.333		28	51	5.75	84.75	116	
	5280	1-1/2″ pipe size		11	1.455		37.50	56	6.30	99.80	135	
	5290	2″ pipe size		8	2		54	77	8.65	139.65	189	
	5300	2-1/2″ pipe size		6	2.667		156	102	11.50	269.50	345	
	5310	3″ pipe size		5	3.200		370	123	13.80	506.80	615	
	5320	4″ pipe size	↓	4	4	↓	600	154	17.30	771.30	920	
	5350	For reducing sizes, add					60%					
	5450	Couplings										
	5451	1/4″ pipe size	Q-15	23	.696	Ea.	10	26.50	3.01	39.51	56	
	5452	3/8″ pipe size		23	.696		10	26.50	3.01	39.51	56	
	5453	1/2″ pipe size		21	.762		4.60	29.50	3.29	37.39	54	
	5454	3/4″ pipe size		21	.762		5.95	29.50	3.29	38.74	55.50	
	5460	1″ pipe size		20	.800		6.55	30.50	3.46	40.51	59	
	5470	1-1/4″ pipe size		20	.800		11.65	30.50	3.46	45.61	64.50	
	5480	1-1/2″ pipe size		18	.889		12.90	34	3.84	50.74	71.50	
	5490	2″ pipe size		14	1.143		20.50	44	4.94	69.44	96.50	
	5500	2-1/2″ pipe size		12	1.333		46	51	5.75	102.75	136	
	5510	3″ pipe size		9	1.778		104	68.50	7.70	180.20	228	
	5520	4″ pipe size	↓	7	2.286	↓	156	88	9.85	253.85	320	

	15107 | Metal Pipe & Fittings	CREW	DAILY OUTPUT	LABOR-HOURS	UNIT	2006 BARE COSTS MAT.	LABOR	EQUIP.	TOTAL	TOTAL INCL O&P	
660	5570 Union, 1/4" pipe size	Q-15	21	.762	Ea.	20.50	29.50	3.29	53.29	71.50	660
	5571 3/8" pipe size		21	.762		20.50	29.50	3.29	53.29	71.50	
	5572 1/2" pipe size		19	.842		18.40	32.50	3.64	54.54	75	
	5573 3/4" pipe size		19	.842		21	32.50	3.64	57.14	77.50	
	5574 1" pipe size		19	.842		27	32.50	3.64	63.14	84	
	5575 1-1/4" pipe size		17	.941		44.50	36	4.07	84.57	109	
	5576 1-1/2" pipe size		15	1.067		47	41	4.61	92.61	121	
	5577 2" pipe size		11	1.455		65	56	6.30	127.30	165	
	5600 Reducer, 1/4" pipe size		23	.696		22	26.50	3.01	51.51	69.50	
	5601 3/8" pipe size		23	.696		23	26.50	3.01	52.51	70	
	5602 1/2" pipe size		21	.762		15.55	29.50	3.29	48.34	66	
	5603 3/4" pipe size		21	.762		16.75	29.50	3.29	49.54	67.50	
	5604 1" pipe size		21	.762		18.15	29.50	3.29	50.94	69	
	5605 1-1/4" pipe size		19	.842		33	32.50	3.64	69.14	90.50	
	5607 1-1/2" pipe size		17	.941		35.50	36	4.07	75.57	99.50	
	5608 2" pipe size		13	1.231		39	47.50	5.30	91.80	122	
	5612 Cap, 1/4" pipe size		46	.348		8.65	13.35	1.50	23.50	32	
	5613 3/8" pipe size		46	.348		8.65	13.35	1.50	23.50	32	
	5614 1/2" pipe size		42	.381		5.90	14.65	1.65	22.20	31.50	
	5615 3/4" pipe size		42	.381		6.90	14.65	1.65	23.20	32.50	
	5616 1" pipe size		42	.381		10.45	14.65	1.65	26.75	36.50	
	5617 1-1/4" pipe size		38	.421		12.30	16.20	1.82	30.32	40.50	
	5618 1-1/2" pipe size		34	.471		16.80	18.10	2.03	36.93	48.50	
	5619 2" pipe size		26	.615		26	23.50	2.66	52.16	68.50	
	5630 T-O-L, 1/4" pipe size, nozzle		23	.696		5	26.50	3.01	34.51	50.50	
	5631 3/8" pipe size, nozzle		23	.696		5	26.50	3.01	34.51	50.50	
	5632 1/2" pipe size, nozzle		22	.727		5	28	3.14	36.14	52.50	
	5633 3/4" pipe size, nozzle		21	.762		5.75	29.50	3.29	38.54	55.50	
	5634 1" pipe size, nozzle		20	.800		6.70	30.50	3.46	40.66	59	
	5635 1-1/4" pipe size, nozzle		18	.889		10.65	34	3.84	48.49	69	
	5636 1-1/2" pipe size, nozzle		16	1		10.65	38.50	4.32	53.47	76.50	
	5637 2" pipe size, nozzle		12	1.333		12.10	51	5.75	68.85	99	
	5638 2-1/2" pipe size, nozzle		10	1.600		40.50	61.50	6.90	108.90	148	
	5639 4" pipe size, nozzle		6	2.667		91.50	102	11.50	205	273	
	5640 W-O-L, 1/4" pipe size, nozzle		23	.696		12.10	26.50	3.01	41.61	58	
	5641 3/8" pipe size, nozzle		23	.696		12.10	26.50	3.01	41.61	58	
	5642 1/2" pipe size, nozzle		22	.727		12.10	28	3.14	43.24	60.50	
	5643 3/4" pipe size, nozzle		21	.762		12.40	29.50	3.29	45.19	63	
	5644 1" pipe size, nozzle		20	.800		12.95	30.50	3.46	46.91	66	
	5645 1-1/4" pipe size, nozzle		18	.889		15.65	34	3.84	53.49	74.50	
	5646 1-1/2" pipe size, nozzle		16	1		15.65	38.50	4.32	58.47	82	
	5647 2" pipe size, nozzle		12	1.333		15.75	51	5.75	72.50	103	
	5648 2-1/2" pipe size, nozzle		10	1.600		34.50	61.50	6.90	102.90	141	
	5649 3" pipe size, nozzle		8	2		37.50	77	8.65	123.15	171	
	5650 4" pipe size, nozzle		6	2.667		48	102	11.50	161.50	224	
	5651 5" pipe size, nozzle		5	3.200		108	123	13.80	244.80	325	
	5652 6" pipe size, nozzle		4	4		197	154	17.30	368.30	475	
	5653 8" pipe size, nozzle		3	5.333		360	205	23	588	740	
	5654 10" pipe size, nozzle		2.60	6.154		640	236	26.50	902.50	1,100	
	5655 12" pipe size, nozzle		2.20	7.273		1,175	279	31.50	1,485.50	1,750	
	5656 14" pipe size, nozzle		2	8		2,150	305	34.50	2,489.50	2,875	
	5674 S-O-L, 1/4" pipe size, outlet		23	.696		6.20	26.50	3.01	35.71	51.50	
	5675 3/8" pipe size, outlet		23	.696		6.20	26.50	3.01	35.71	51.50	
	5676 1/2" pipe size, outlet		22	.727		6.20	28	3.14	37.34	54	
	5677 3/4" pipe size, outlet		21	.762		6.30	29.50	3.29	39.09	56	
660	5678 1" pipe size, outlet	↓	20	.800	↓	7	30.50	3.46	40.96	59.50	

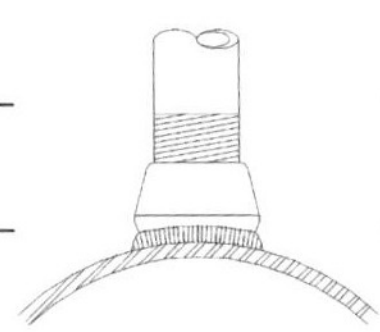

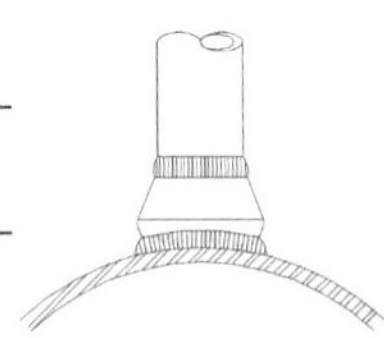

15107 | Metal Pipe & Fittings

		CREW	DAILY OUTPUT	LABOR-HOURS	UNIT	2006 BARE COSTS MAT.	LABOR	EQUIP.	TOTAL	TOTAL INCL O&P
660	5679 1-1/4" pipe size, outlet	Q-15	18	.889	Ea.	11.65	34	3.84	49.49	70
	5680 1-1/2" pipe size, outlet		16	1		11.65	38.50	4.32	54.47	77.50
	5681 2" pipe size, outlet		12	1.333		13.25	51	5.75	70	100
	6000 Weld-on flange, forged steel									
	6020 Slip-on, 150 lb. flange (welded front and back)									
	6050 1/2" pipe size	Q-15	18	.889	Ea.	11.60	34	3.84	49.44	70
	6060 3/4" pipe size		18	.889		11.60	34	3.84	49.44	70
	6070 1" pipe size		17	.941		11.60	36	4.07	51.67	73.50
	6080 1-1/4" pipe size 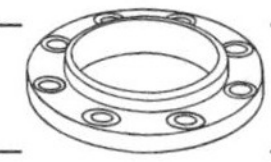		16	1		11.60	38.50	4.32	54.42	77.50
	6090 1-1/2" pipe size		15	1.067		11.60	41	4.61	57.21	81.50
	6100 2" pipe size		12	1.333		11.60	51	5.75	68.35	98.50
	6110 2-1/2" pipe size		10	1.600		20	61.50	6.90	88.40	125
	6120 3" pipe size		9	1.778		16.80	68.50	7.70	93	133
	6130 3-1/2" pipe size		7	2.286		22	88	9.85	119.85	172
	6140 4" pipe size		6	2.667		22	102	11.50	135.50	196
	6150 5" pipe size		5	3.200		32	123	13.80	168.80	242
	6160 6" pipe size	Q-16	6	4		33.50	159	11.50	204	298
	6170 8" pipe size		5	4.800		52	191	13.80	256.80	370
	6180 10" pipe size		4	6		88.50	239	17.30	344.80	485
	6190 12" pipe size		3	8		131	320	23	474	665
	6191 14" pipe size		2.50	9.600		175	385	27.50	587.50	820
	6192 16" pipe size		1.80	13.333		264	530	38.50	832.50	1,150
	6200 300 lb. flange									
	6210 1/2" pipe size	Q-15	17	.941	Ea.	15.45	36	4.07	55.52	77.50
	6220 3/4" pipe size		17	.941		15.45	36	4.07	55.52	77.50
	6230 1" pipe size		16	1		15.45	38.50	4.32	58.27	82
	6240 1-1/4" pipe size		13	1.231		15.45	47.50	5.30	68.25	96.50
	6250 1-1/2" pipe size		12	1.333		15.45	51	5.75	72.20	103
	6260 2" pipe size		11	1.455		18.45	56	6.30	80.75	114
	6270 2-1/2" pipe size		9	1.778		22	68.50	7.70	98.20	138
	6280 3" pipe size		7	2.286		27	88	9.85	124.85	178
	6290 4" pipe size		6	2.667		39.50	102	11.50	153	215
	6300 5" pipe size		4	4		57.50	154	17.30	228.80	320
	6310 6" pipe size	Q-16	5	4.800		66.50	191	13.80	271.30	385
	6320 8" pipe size		4	6		105	239	17.30	361.30	505
	6330 10" pipe size		3.40	7.059		173	281	20.50	474.50	650
	6340 12" pipe size		2.80	8.571		212	340	24.50	576.50	790
	6400 Welding neck, 150 lb. flange									
	6410 1/2" pipe size	Q-15	40	.400	Ea.	16.45	15.35	1.73	33.53	44
	6420 3/4" pipe size 		36	.444		16.45	17.10	1.92	35.47	46.50
	6430 1" pipe size		32	.500		16.45	19.20	2.16	37.81	50.50
	6440 1-1/4" pipe size		29	.552		16.45	21	2.38	39.83	53.50
	6450 1-1/2" pipe size		26	.615		16.45	23.50	2.66	42.61	58
	6460 2" pipe size		20	.800		17.30	30.50	3.46	51.26	71
	6470 2-1/2" pipe size		16	1		19.95	38.50	4.32	62.77	87
	6480 3" pipe size		14	1.143		21.50	44	4.94	70.44	97.50
	6500 4" pipe size		10	1.600		29.50	61.50	6.90	97.90	136
	6510 5" pipe size		8	2		39.50	77	8.65	125.15	173
	6520 6" pipe size	Q-16	10	2.400		44	95.50	6.90	146.40	205
	6530 8" pipe size		7	3.429		74.50	137	9.85	221.35	305
	6540 10" pipe size		6	4		112	159	11.50	282.50	385
	6550 12" pipe size		5	4.800		162	191	13.80	366.80	490
	6551 14" pipe size		4.50	5.333		237	213	15.35	465.35	610
	6552 16" pipe size		3	8		370	320	23	713	930
	6553 18" pipe size		2.50	9.600		485	385	27.50	897.50	1,150
	6554 20" pipe size		2.30	10.435		585	415	30	1,030	1,325

15107	Metal Pipe & Fittings	CREW	DAILY OUTPUT	LABOR-HOURS	UNIT	2006 BARE COSTS MAT.	LABOR	EQUIP.	TOTAL	TOTAL INCL O&P
660 6556	24" pipe size	Q-16	2	12	Ea.	795	480	34.50	1,309.50	1,650
6557	26" pipe size		1.70	14.118		905	565	40.50	1,510.50	1,925
6558	30" pipe size		.90	26.667		1,050	1,075	77	2,202	2,875
6559	36" pipe size		.75	32		1,200	1,275	92	2,567	3,400
6560	300 lb. flange									
6570	1/2" pipe size	Q-15	36	.444	Ea.	20.50	17.10	1.92	39.52	51
6580	3/4" pipe size		34	.471		20.50	18.10	2.03	40.63	52.50
6590	1" pipe size		30	.533		20.50	20.50	2.30	43.30	57
6600	1-1/4" pipe size		28	.571		20.50	22	2.47	44.97	59
6610	1-1/2" pipe size		24	.667		20.50	25.50	2.88	48.88	65.50
6620	2" pipe size		18	.889		23	34	3.84	60.84	82.50
6630	2-1/2" pipe size		14	1.143		28	44	4.94	76.94	104
6640	3" pipe size		12	1.333		31	51	5.75	87.75	120
6650	4" pipe size		8	2		49	77	8.65	134.65	184
6660	5" pipe size		7	2.286		70	88	9.85	167.85	225
6670	6" pipe size	Q-16	9	2.667		79.50	106	7.70	193.20	261
6680	8" pipe size		6	4		128	159	11.50	298.50	400
6690	10" pipe size		5	4.800		220	191	13.80	424.80	555
6700	12" pipe size		4	6		282	239	17.30	538.30	700
6710	14" pipe size		3.50	6.857		595	273	19.75	887.75	1,100
6720	16" pipe size		2	12		740	480	34.50	1,254.50	1,600
7740	Plain ends for plain end pipe, mechanically coupled									
7750	Cplg. & labor required at joints not included add 1 per									
7760	joint for installed price, see line 9180									
7770	Malleable iron, painted, unless noted otherwise									
7800	90° Elbow 1"				Ea.	46.50			46.50	51.50
7810	1-1/2"					55.50			55.50	61
7820	2"					83			83	91.50
7830	2-1/2"					97			97	107
7840	3"					101			101	111
7860	4"					114			114	126
7870	5" welded steel					131			131	144
7880	6"					166			166	183
7890	8" welded steel					315			315	345
7900	10" welded steel					415			415	455
7910	12" welded steel					460			460	505
7970	45° Elbow 1"					27.50			27.50	30.50
7980	1-1/2"					41.50			41.50	45.50
7990	2"					87.50			87.50	96.50
8000	2-1/2"					87.50			87.50	96.50
8010	3"					101			101	111
8030	4"					105			105	115
8040	5" welded steel					131			131	144
8050	6"					151			151	166
8060	8"					177			177	194
8070	10" welded steel					201			201	222
8080	12" welded steel					345			345	375
8140	Tee, straight 1"					52.50			52.50	57.50
8150	1-1/2"					67.50			67.50	74
8160	2"					67.50			67.50	74
8170	2-1/2"					88			88	97
8180	3"					136			136	149
8200	4"					195			195	214
8210	5" welded steel					267			267	294
8220	6"					310			310	345
8230	8" welded steel					335			335	370

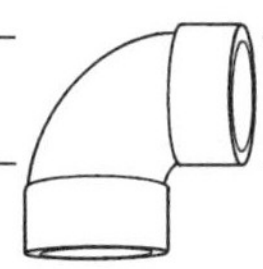

15 MECHANICAL

15107 | Metal Pipe & Fittings		CREW	DAILY OUTPUT	LABOR-HOURS	UNIT	2006 BARE COSTS MAT.	LABOR	EQUIP.	TOTAL	TOTAL INCL O&P
660 8240	10" welded steel				Ea.	520			520	570
8250	12" welded steel				↓	635			635	700
8340	Segmentally welded steel, painted									
8390	Wye 2"				Ea.	85			85	93.50
8400	2-1/2"					85			85	93.50
8410	3"					96			96	106
8430	4" 					142			142	157
8440	5"					173			173	191
8450	6"					238			238	262
8460	8"					310			310	340
8470	10"					455			455	500
8480	12"					700			700	775
8540	Wye, lateral 2"					91.50			91.50	101
8550	2-1/2"					105			105	116
8560	3"					125			125	138
8580	4" 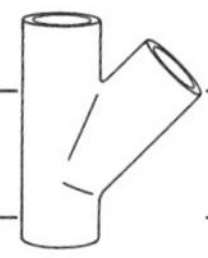					172			172	190
8590	5"					293			293	320
8600	6"					300			300	335
8610	8"					505			505	560
8620	10"					525			525	580
8630	12"					970			970	1,075
8690	Cross, 2"					97			97	107
8700	2-1/2"					97			97	107
8710	3"					115			115	127
8730	4" 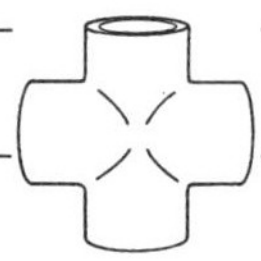					158			158	174
8740	5"					225			225	247
8750	6"					297			297	325
8760	8"					380			380	420
8770	10"					555			555	610
8780	12"					800			800	880
8800	Tees, reducing 2" x 1"					57.50			57.50	63.50
8810	2" x 1-1/2"					57.50			57.50	63.50
8820	3" x 1"					57.50			57.50	63.50
8830	3" x 1-1/2"					57.50			57.50	63.50
8840	3" x 2"					57.50			57.50	63.50
8850	4" x 1"					84			84	92.50
8860	4" x 1-1/2"					84			84	92.50
8870	4" x 2" 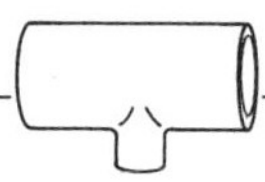					84			84	92.50
8880	4" x 2-1/2"					86.50			86.50	95
8890	4" x 3"					86.50			86.50	95
8900	6" x 2"					140			140	154
8910	6" x 3"					147			147	162
8920	6" x 4"					147			147	162
8930	8" x 2"					179			179	197
8940	8" x 3"					191			191	211
8950	8" x 4"					191			191	211
8960	8" x 5"					197			197	216
8970	8" x 6"					197			197	216
8980	10" x 4"					282			282	310
8990	10" x 6"					299			299	330
9000	10" x 8"					299			299	330
9010	12" x 6"					435			435	480
9020	12" x 8"					435			435	480
9030	12" x 10" 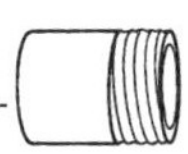				↓	450			450	495
9080	Adapter nipples 3" long									
9090	1"				Ea.	7.55			7.55	8.30

MECHANICAL 15

	15107 | Metal Pipe & Fittings		CREW	DAILY OUTPUT	LABOR-HOURS	UNIT	2006 BARE COSTS MAT.	LABOR	EQUIP.	TOTAL	TOTAL INCL O&P	
660	9100	1-1/2"				Ea.	7.55			7.55	8.30	660
	9110	2"					7.55			7.55	8.30	
	9120	2-1/2" 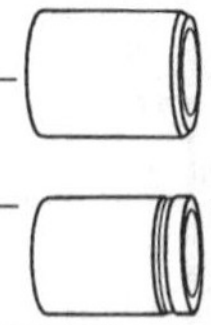					8.85			8.85	9.70	
	9130	3"					10.95			10.95	12.05	
	9140	4"					18.20			18.20	20	
	9150	6"					49			49	54	
	9180	Coupling, mechanical, plain end pipe to plain end pipe or fitting										
	9190	1"	Q-1	29	.552	Ea.	28	21		49	63.50	
	9200	1-1/2"		28	.571		28	22		50	64.50	
	9210	2"		27	.593		28	23		51	66	
	9220	2-1/2"		26	.615		28	23.50		51.50	67.50	
	9230	3" 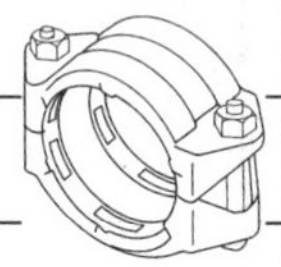		25	.640		41	24.50		65.50	83.50	
	9240	3-1/2"		24	.667		47.50	25.50		73	92.50	
	9250	4"		22	.727		47.50	28		75.50	96	
	9260	5"	Q-2	28	.857		68	34		102	128	
	9270	6"		24	1		83	40		123	154	
	9280	8"		19	1.263		147	50.50		197.50	241	
	9290	10"		16	1.500		192	60		252	305	
	9300	12"		12	2		240	79.50		319.50	390	
	9400	Mechanical joint ends for plain end pipe										
	9410	Malleable iron, black										
	9420	90° Elbows, 1-1/4"	Q-1	29	.552	Ea.	8.90	21		29.90	43	
	9430	1-1/2" 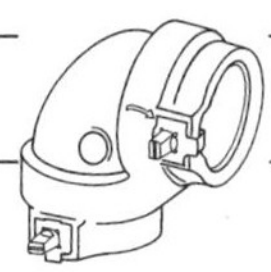		27	.593		9.85	23		32.85	46.50	
	9440	2"		24	.667		11.60	25.50		37.10	53	
	9490	Tee, reducing outlet										
	9510	1-1/4" x 1/2"	Q-1	18	.889	Ea.	7.20	34		41.20	61	
	9520	1-1/4" x 3/4"		18	.889		7.20	34		41.20	61	
	9530	1-1/4" x 1"		18	.889		7.20	34		41.20	61	
	9540	1-1/2" x 1/2"		17	.941		7.60	36		43.60	64.50	
	9550	1-1/2" x 3/4"		17	.941		7.60	36		43.60	64.50	
	9560	1-1/2" x 1" 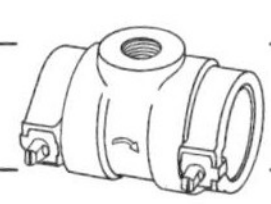		17	.941		7.60	36		43.60	64.50	
	9570	2" x 1/2"		15	1.067		9.30	41		50.30	74	
	9580	2" x 3/4"		15	1.067		9.30	41		50.30	74	
	9590	2" x 1"		15	1.067		9.30	41		50.30	74	
	9640	Tee, reducing run and outlet										
	9660	1-1/4" x 1" x 1/2"	Q-1	18	.889	Ea.	7.65	34		41.65	61.50	
	9670	1-1/4" x 1" x 3/4"		18	.889		7.65	34		41.65	61.50	
	9680	1-1/4" x 1" x 1"		18	.889		7.65	34		41.65	61.50	
	9690	1-1/2" x 1-1/4" x 1/2"		17	.941		8.15	36		44.15	65	
	9700	1-1/2" x 1-1/4" x 3/4" 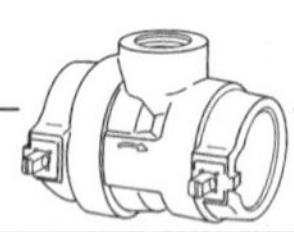		17	.941		8.15	36		44.15	65	
	9710	1-1/2" x 1-1/4" x 1"		17	.941		8.15	36		44.15	65	
	9720	2" x 1-1/2" x 1/2"		15	1.067		9.85	41		50.85	74.50	
	9730	2" x 1-1/2" x 3/4"		15	1.067		9.85	41		50.85	74.50	
	9740	2" x 1-1/2" x 1"		15	1.067		9.85	41		50.85	74.50	
	9790	Tee, outlet										
	9810	3" x 1-1/4"	Q-1	29	.552	Ea.	17.30	21		38.30	52	
	9820	3" x 1-1/2"		28	.571		17.30	22		39.30	53	
	9830	3" x 2" 		26	.615		19.35	23.50		42.85	58.50	
	9840	4" x 1-1/4"		28	.571		21.50	22		43.50	57.50	
	9850	4" x 1-1/2"		26	.615		22.50	23.50		46	62	
	9860	4" x 2"		24	.667		22.50	25.50		48	65	
	9940	For galvanized fittings for plain end pipe, add					20%					
	9990	Minimum labor/equipment charge	Q-15	3	5.333	Job		205	23	228	345	
690	0010	**PIPE, GROOVED-JOINT STEEL FITTINGS & VALVES**										690
	0012	Fittings are ductile iron. Steel fittings noted.										

15107 | Metal Pipe & Fittings

			CREW	DAILY OUTPUT	LABOR-HOURS	UNIT	2006 BARE COSTS MAT.	LABOR	EQUIP.	TOTAL	TOTAL INCL O&P	
690	0020	Pipe includes coupling & clevis type hanger 10′ O.C.										690
	0500	Schedule 10, black										
	0550	2″ diameter	1 Plum	43	.186	L.F.	7	7.95		14.95	20	
	0560	2-1/2″ diameter	Q-1	61	.262		8.60	10.10		18.70	25	
	0570	3″ diameter		55	.291		9.90	11.20		21.10	28.50	
	0580	3-1/2″ diameter		53	.302		12.10	11.60		23.70	31.50	
	0590	4″ diameter		49	.327		12.15	12.55		24.70	33	
	0600	5″ diameter		40	.400		20.50	15.35		35.85	46.50	
	0610	6″ diameter	Q-2	46	.522		25	21		46	60	
	0620	8″ diameter	″	41	.585		46	23.50		69.50	87	
	0700	To delete couplings & hangers, subtract										
	0710	2″ diam. to 5″ diam.					25%	20%				
	0720	6″ diam. to 8″ diam.					27%	15%				
	1000	Schedule 40, black										
	1040	3/4″ diameter	1 Plum	71	.113	L.F.	2.98	4.81		7.79	10.80	
	1050	1″ diameter		63	.127		3.62	5.40		9.02	12.45	
	1060	1-1/4″ diameter		58	.138		4.78	5.90		10.68	14.40	
	1070	1-1/2″ diameter		51	.157		5.55	6.70		12.25	16.50	
	1080	2″ diameter 		40	.200		7.15	8.55		15.70	21	
	1090	2-1/2″ diameter	Q-1	57	.281		10.15	10.80		20.95	28	
	1100	3″ diameter		50	.320		12.75	12.30		25.05	33	
	1110	4″ diameter		45	.356		17.30	13.65		30.95	40	
	1120	5″ diameter		37	.432		22.50	16.60		39.10	50.50	
	1130	6″ diameter	Q-2	42	.571		31	23		54	70	
	1140	8″ diameter		37	.649		47	26		73	92	
	1150	10″ diameter		31	.774		76	31		107	132	
	1160	12″ diameter		27	.889		94	35.50		129.50	158	
	1170	14″ diameter		20	1.200		121	48		169	208	
	1180	16″ diameter		17	1.412		131	56.50		187.50	233	
	1190	18″ diameter		14	1.714		156	68.50		224.50	277	
	1200	20″ diameter		12	2		235	79.50		314.50	385	
	1210	24″ diameter		10	2.400		276	95.50		371.50	455	
	1740	To delete coupling & hanger, subtract										
	1750	3/4″ diam. to 2″ diam.					65%	27%				
	1760	2-1/2″ diam. to 5″ diam.					41%	18%				
	1770	6″ diam. to 12″ diam.					31%	13%				
	1780	14″ diam. to 24″ diam.					35%	10%				
	1800	Galvanized										
	1840	3/4″ diameter	1 Plum	71	.113	L.F.	3.89	4.81		8.70	11.80	
	1850	1″ diameter		63	.127		4.51	5.40		9.91	13.40	
	1860	1-1/4″ diameter		58	.138		6	5.90		11.90	15.75	
	1870	1-1/2″ diameter		51	.157		7	6.70		13.70	18.10	
	1880	2″ diameter		40	.200		9	8.55		17.55	23	
	1890	2-1/2″ diameter	Q-1	57	.281		13.65	10.80		24.45	32	
	1900	3″ diameter		50	.320		17.50	12.30		29.80	38.50	
	1910	4″ diameter		45	.356		24.50	13.65		38.15	48	
	1920	5″ diameter		37	.432		29	16.60		45.60	58	
	1930	6″ diameter	Q-2	42	.571		39	23		62	78.50	
	1940	8″ diameter		37	.649		61	26		87	108	
	1950	10″ diameter		31	.774		98	31		129	156	
	1960	12″ diameter		27	.889		121	35.50		156.50	188	
	2540	To delete coupling & hanger, subtract										
	2550	3/4″ diam. to 2″ diam.					36%	27%				
	2560	2-1/2″ diam. to 5″ diam.					19%	18%				
	2570	6″ diam. to 12″ diam.					14%	13%				
	2600	Schedule 80, black										

	15107 | Metal Pipe & Fittings	CREW	DAILY OUTPUT	LABOR-HOURS	UNIT	2006 BARE COSTS MAT.	LABOR	EQUIP.	TOTAL	TOTAL INCL O&P	
690	2610 3/4" diameter	1 Plum	65	.123	L.F.	4	5.25		9.25	12.55	690
	2650 1" diameter		61	.131		5.15	5.60		10.75	14.35	
	2660 1-1/4" diameter		55	.145		7	6.20		13.20	17.35	
	2670 1-1/2" diameter		49	.163		8.25	6.95		15.20	19.90	
	2680 2" diameter		38	.211		10.80	9		19.80	26	
	2690 2-1/2" diameter	Q-1	54	.296		16.35	11.40		27.75	35.50	
	2700 3" diameter		48	.333		21.50	12.80		34.30	44	
	2710 4" diameter		44	.364		30.50	13.95		44.45	55.50	
	2720 5" diameter		35	.457		42	17.55		59.55	73.50	
	2730 6" diameter	Q-2	40	.600		51.50	24		75.50	93.50	
	2740 8" diameter		35	.686		92.50	27.50		120	145	
	2750 10" diameter		29	.828		103	33		136	165	
	2760 12" diameter		24	1		118	40		158	192	
	3240 To delete coupling & hanger, subtract										
	3250 3/4" diam. to 2" diam.					30%	25%				
	3260 2-1/2" diam. to 5" diam.					14%	17%				
	3270 6" diam. to 12" diam.					12%	12%				
	3300 Galvanized										
	3310 3/4" diameter	1 Plum	65	.123	L.F.	5	5.25		10.25	13.65	
	3350 1" diameter		61	.131		6.65	5.60		12.25	16.05	
	3360 1-1/4" diameter		55	.145		9.20	6.20		15.40	19.75	
	3370 1-1/2" diameter		46	.174		10.90	7.45		18.35	23.50	
	3380 2" diameter		38	.211		14.55	9		23.55	30	
	3390 2-1/2" diameter	Q-1	54	.296		22.50	11.40		33.90	42	
	3400 3" diameter		48	.333		29.50	12.80		42.30	52.50	
	3410 4" diameter		44	.364		41.50	13.95		55.45	67	
	3420 5" diameter		35	.457		54.50	17.55		72.05	87.50	
	3430 6" diameter	Q-2	40	.600		67.50	24		91.50	111	
	3440 8" diameter		35	.686		122	27.50		149.50	177	
	3450 10" diameter		29	.828		134	33		167	199	
	3460 12" diameter		24	1		153	40		193	230	
	3920 To delete coupling & hanger, subtract										
	3930 3/4" diam. to 2" diam.					30%	25%				
	3940 2-1/2" diam. to 5" diam.					15%	17%				
	3950 6" diam. to 12" diam.					11%	12%				
	3990 Fittings: cplg. & labor required at joints not incl. in fitting										
	3994 price. Add 1 per joint for installed price.										
	4000 Elbow, 90° or 45°, painted										
	4030 3/4" diameter	1 Plum	50	.160	Ea.	24.50	6.85		31.35	37.50	
	4040 1" diameter		50	.160		13.15	6.85		20	25	
	4050 1-1/4" diameter		40	.200		13.15	8.55		21.70	28	
	4060 1-1/2" diameter		33	.242		13.15	10.35		23.50	30.50	
	4070 2" diameter		25	.320		13.15	13.65		26.80	36	
	4080 2-1/2" diameter	Q-1	40	.400		13.15	15.35		28.50	38.50	
	4090 3" diameter		33	.485		23.50	18.65		42.15	54.50	
	4100 4" diameter		25	.640		25.50	24.50		50	66	
	4110 5" diameter		20	.800		61	30.50		91.50	116	
	4120 6" diameter	Q-2	25	.960		72	38.50		110.50	139	
	4130 8" diameter		21	1.143		150	45.50		195.50	236	
	4140 10" diameter		18	1.333		273	53		326	385	
	4150 12" diameter		15	1.600		440	64		504	580	
	4170 14" diameter		12	2		785	79.50		864.50	985	
	4180 16" diameter		11	2.182		1,025	87		1,112	1,250	
	4190 18" diameter	Q-3	15	2.133		1,300	87		1,387	1,550	
	4200 20" diameter		13	2.462		1,700	100		1,800	2,025	
	4210 24" diameter		11	2.909		2,475	118		2,593	2,900	

	15107 | Metal Pipe & Fittings	CREW	DAILY OUTPUT	LABOR-HOURS	UNIT	2006 BARE COSTS MAT.	LABOR	EQUIP.	TOTAL	TOTAL INCL O&P
690	4250 For galvanized elbows, add				Ea.	26%				
	4690 Tee, painted									
	4700 3/4" diameter	1 Plum	38	.211	Ea.	26.50	9		35.50	43
	4740 1" diameter		33	.242		20	10.35		30.35	38
	4750 1-1/4" diameter		27	.296		20	12.65		32.65	41.50
	4760 1-1/2" diameter		22	.364		20	15.55		35.55	46
	4770 2" diameter		17	.471		20	20		40	53.50
	4780 2-1/2" diameter	Q-1	27	.593		20	23		43	57.50
	4790 3" diameter		22	.727		28	28		56	74.50
	4800 4" diameter		17	.941		43	36		79	104
	4810 5" diameter		13	1.231		101	47.50		148.50	185
	4820 6" diameter	Q-2	17	1.412		117	56.50		173.50	216
	4830 8" diameter		14	1.714		256	68.50		324.50	390
	4840 10" diameter		12	2		535	79.50		614.50	710
	4850 12" diameter		10	2.400		745	95.50		840.50	965
	4851 14" diameter		9	2.667		785	106		891	1,025
	4852 16" diameter		8	3		890	120		1,010	1,175
	4853 18" diameter	Q-3	11	2.909		1,100	118		1,218	1,400
	4854 20" diameter		10	3.200		1,600	130		1,730	1,950
	4855 24" diameter		8	4		2,425	163		2,588	2,925
	4900 For galvanized tees, add					24%				
	4939 Couplings									
	4940 Flexible, standard, painted									
	4950 3/4" diameter	1 Plum	100	.080	Ea.	7	3.42		10.42	13
	4960 1" diameter		100	.080		7	3.42		10.42	13
	4970 1-1/4" diameter		80	.100		9.25	4.27		13.52	16.85
	4980 1-1/2" diameter		67	.119		10.15	5.10		15.25	19.10
	4990 2" diameter		50	.160		10.70	6.85		17.55	22.50
	5000 2-1/2" diameter	Q-1	80	.200		12.75	7.70		20.45	26
	5010 3" diameter		67	.239		14.10	9.20		23.30	30
	5020 3-1/2" diameter		57	.281		20.50	10.80		31.30	39.50
	5030 4" diameter		50	.320		20.50	12.30		32.80	41.50
	5040 5" diameter		40	.400		31.50	15.35		46.85	58.50
	5050 6" diameter	Q-2	50	.480		37	19.15		56.15	71
	5070 8" diameter		42	.571		60.50	23		83.50	102
	5090 10" diameter		35	.686		100	27.50		127.50	154
	5110 12" diameter		32	.750		114	30		144	173
	5120 14" diameter		24	1		139	40		179	215
	5130 16" diameter		20	1.200		182	48		230	275
	5140 18" diameter		18	1.333		213	53		266	320
	5150 20" diameter		16	1.500		335	60		395	465
	5160 24" diameter		13	1.846		370	73.50		443.50	520
	5200 For galvanized couplings, add					33%				
	5750 Flange, w/groove gasket, black steel									
	5752 See 15107-660-0620 for gasket & bolt set									
	5760 ANSI class 125 and 150, painted									
	5780 2" pipe size	1 Plum	23	.348	Ea.	44	14.85		58.85	71.50
	5790 2-1/2" pipe size	Q-1	37	.432		55	16.60		71.60	86.50
	5800 3" pipe size		31	.516		59	19.85		78.85	96
	5820 4" pipe size		23	.696		79.50	26.50		106	129
	5830 5" pipe size		19	.842		92	32.50		124.50	152
	5840 6" pipe size	Q-2	23	1.043		100	41.50		141.50	176
	5850 8" pipe size		17	1.412		113	56.50		169.50	213
	5860 10" pipe size		14	1.714		180	68.50		248.50	305
	5870 12" pipe size		12	2		235	79.50		314.50	385
	5880 14" pipe size		10	2.400		540	95.50		635.50	745

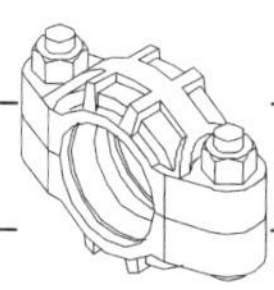

MECHANICAL 15

15107 | Metal Pipe & Fittings

			CREW	DAILY OUTPUT	LABOR-HOURS	UNIT	2006 BARE COSTS MAT.	LABOR	EQUIP.	TOTAL	TOTAL INCL O&P	
690	5890	16″ pipe size	Q-2	9	2.667	Ea.	625	106		731	855	690
	5900	18″ pipe size		6	4		770	159		929	1,100	
	5910	20″ pipe size		5	4.800		930	191		1,121	1,325	
	5920	24″ pipe size		4.50	5.333		1,200	213		1,413	1,625	
	8000	Butterfly valve, 2 position handle, with standard trim										
	8010	1-1/2″ pipe size	1 Plum	50	.160	Ea.	124	6.85		130.85	148	
	8020	2″ pipe size	"	38	.211		124	9		133	151	
	8030	3″ pipe size	Q-1	50	.320		178	12.30		190.30	215	
	8050	4″ pipe size	"	38	.421		196	16.20		212.20	241	
	8070	6″ pipe size	Q-2	38	.632		395	25		420	475	
	8080	8″ pipe size		27	.889		570	35.50		605.50	680	
	8090	10″ pipe size		20	1.200		1,000	48		1,048	1,175	
	8200	With stainless steel trim										
	8240	1-1/2″ pipe size	1 Plum	50	.160	Ea.	158	6.85		164.85	184	
	8250	2″ pipe size	"	38	.211		158	9		167	187	
	8270	3″ pipe size	Q-1	50	.320		212	12.30		224.30	252	
	8280	4″ pipe size	"	38	.421		230	16.20		246.20	278	
	8300	6″ pipe size	Q-2	38	.632		450	25		475	535	
	8310	8″ pipe size		27	.889		690	35.50		725.50	815	
	8320	10″ pipe size		20	1.200		985	48		1,033	1,150	
	9000	Cut one groove, labor										
	9010	3/4″ pipe size	Q-1	152	.105	Ea.		4.05		4.05	6.30	
	9020	1″ pipe size		140	.114			4.39		4.39	6.85	
	9030	1-1/4″ pipe size		124	.129			4.96		4.96	7.70	
	9040	1-1/2″ pipe size		114	.140			5.40		5.40	8.40	
	9050	2″ pipe size		104	.154			5.90		5.90	9.20	
	9060	2-1/2″ pipe size		96	.167			6.40		6.40	9.95	
	9070	3″ pipe size		88	.182			7		7	10.85	
	9080	3-1/2″ pipe size		83	.193			7.40		7.40	11.50	
	9090	4″ pipe size		78	.205			7.90		7.90	12.25	
	9100	5″ pipe size		72	.222			8.55		8.55	13.30	
	9110	6″ pipe size		70	.229			8.80		8.80	13.65	
	9120	8″ pipe size		54	.296			11.40		11.40	17.70	
	9130	10″ pipe size		38	.421			16.20		16.20	25	
	9140	12″ pipe size		30	.533			20.50		20.50	32	
	9150	14″ pipe size		20	.800			30.50		30.50	48	
	9160	16″ pipe size		19	.842			32.50		32.50	50.50	
	9170	18″ pipe size		18	.889			34		34	53	
	9180	20″ pipe size		17	.941			36		36	56	
	9190	24″ pipe size		15	1.067			41		41	63.50	
	9210	Roll one groove										
	9220	3/4″ pipe size	Q-1	266	.060	Ea.		2.31		2.31	3.59	
	9230	1″ pipe size		228	.070			2.70		2.70	4.19	
	9240	1-1/4″ pipe size		200	.080			3.07		3.07	4.78	
	9250	1-1/2″ pipe size		178	.090			3.45		3.45	5.35	
	9260	2″ pipe size		116	.138			5.30		5.30	8.25	
	9270	2-1/2″ pipe size		110	.145			5.60		5.60	8.70	
	9280	3″ pipe size		100	.160			6.15		6.15	9.55	
	9290	3-1/2″ pipe size		94	.170			6.55		6.55	10.15	
	9300	4″ pipe size		86	.186			7.15		7.15	11.10	
	9310	5″ pipe size		84	.190			7.30		7.30	11.40	
	9320	6″ pipe size		80	.200			7.70		7.70	11.95	
	9330	8″ pipe size		66	.242			9.30		9.30	14.50	
	9340	10″ pipe size		58	.276			10.60		10.60	16.50	
	9350	12″ pipe size		46	.348			13.35		13.35	21	
	9360	14″ pipe size		30	.533			20.50		20.50	32	

	15107 | Metal Pipe & Fittings		CREW	DAILY OUTPUT	LABOR-HOURS	UNIT	2006 BARE COSTS MAT.	LABOR	EQUIP.	TOTAL	TOTAL INCL O&P	
690	9370	16″ pipe size	Q-1	28	.571	Ea.		22		22	34	690
	9380	18″ pipe size		27	.593			23		23	35.50	
	9390	20″ pipe size		25	.640			24.50		24.50	38	
	9400	24″ pipe size		23	.696			26.50		26.50	41.50	
	9990	Minimum labor/equipment charge	1 Plum	4	2	Job		85.50		85.50	133	
900	5000	**TUBING, STAINLESS STEEL**										900
	5010	Type 304, no joints, no hangers										
	5020	.035 wall										
	5021	1/4″	1 Plum	160	.050	L.F.	1.54	2.14		3.68	5	
	5022	3/8″		160	.050		1.98	2.14		4.12	5.50	
	5023	1/2″		160	.050		1.44	2.14		3.58	4.90	
	5024	5/8″		160	.050		3.54	2.14		5.68	7.20	
	5025	3/4″		133	.060		2.46	2.57		5.03	6.70	
	5026	7/8″		133	.060		6.90	2.57		9.47	11.60	
	5027	1″		114	.070		6	3		9	11.25	
	5040	.049 wall										
	5041	1/4″	1 Plum	160	.050	L.F.	3.22	2.14		5.36	6.85	
	5042	3/8″		160	.050		2.18	2.14		4.32	5.70	
	5043	1/2″		160	.050		1.58	2.14		3.72	5.05	
	5044	5/8″		160	.050		2.09	2.14		4.23	5.60	
	5045	3/4″		133	.060		4.93	2.57		7.50	9.40	
	5046	7/8″		133	.060		5.50	2.57		8.07	10.05	
	5047	1″		114	.070		5.20	3		8.20	10.35	
	5060	.065″ wall										
	5061	1/4″	1 Plum	160	.050	L.F.	2.04	2.14		4.18	5.55	
	5062	3/8″		160	.050		2.43	2.14		4.57	6	
	5063	1/2″		160	.050		5	2.14		7.14	8.80	
	5064	5/8″		160	.050		5.05	2.14		7.19	8.85	
	5065	3/4″		133	.060		1.99	2.57		4.56	6.20	
	5066	7/8″		133	.060		8.20	2.57		10.77	13.05	
	5067	1″		114	.070		5.60	3		8.60	10.80	
	5210	Type 316										
	5220	.035 wall										
	5221	1/4″	1 Plum	160	.050	L.F.	1.72	2.14		3.86	5.20	
	5222	3/8″		160	.050		2.09	2.14		4.23	5.60	
	5223	1/2″		160	.050		2.02	2.14		4.16	5.55	
	5224	5/8″		160	.050		2.92	2.14		5.06	6.55	
	5225	3/4″		133	.060		3.69	2.57		6.26	8.05	
	5226	7/8″		133	.060		7.45	2.57		10.02	12.15	
	5227	1″		114	.070		4.84	3		7.84	9.95	
	5240	.049 wall										
	5241	1/4″	1 Plum	160	.050	L.F.	2.07	2.14		4.21	5.60	
	5242	3/8″		160	.050		2.38	2.14		4.52	5.95	
	5243	1/2″		160	.050		2.59	2.14		4.73	6.15	
	5244	5/8″		160	.050		3.88	2.14		6.02	7.60	
	5245	3/4″		133	.060		3.56	2.57		6.13	7.90	
	5246	7/8″		133	.060		5.45	2.57		8.02	10	
	5247	1″		114	.070		6.15	3		9.15	11.45	
	5260	.065 wall										
	5261	1/4″	1 Plum	160	.050	L.F.	3.19	2.14		5.33	6.85	
	5262	3/8″		160	.050		2.70	2.14		4.84	6.30	
	5263	1/2″		160	.050		3.27	2.14		5.41	6.90	
	5264	5/8″		160	.050		3.41	2.14		5.55	7.05	
	5265	3/4″		133	.060		4.46	2.57		7.03	8.90	
	5266	7/8″		133	.060		6	2.57		8.57	10.60	

MECHANICAL 15

15107 | Metal Pipe & Fittings

	Line	15107 Metal Pipe & Fittings	CREW	DAILY OUTPUT	LABOR-HOURS	UNIT	2006 BARE COSTS MAT.	LABOR	EQUIP.	TOTAL	TOTAL INCL O&P	
900	5267	1"	1 Plum	114	.070	L.F.	8	3		11	13.45	900
920	0010	PIPE, STAINLESS STEEL										920
	0020	Welded, with clevis type hangers 10' O.C.										
	0500	Schedule 5, type 304										
	0540	1/2" diameter	Q-15	128	.125	L.F.	5.40	4.80	.54	10.74	14	
	0550	3/4" diameter		116	.138		6.15	5.30	.60	12.05	15.70	
	0560	1" diameter		103	.155		7.65	5.95	.67	14.27	18.45	
	0570	1-1/4" diameter		93	.172		8.80	6.60	.74	16.14	21	
	0580	1-1/2" diameter		85	.188		10.15	7.25	.81	18.21	23.50	
	0590	2" diameter		69	.232		12.25	8.90	1	22.15	28.50	
	0600	2-1/2" diameter		53	.302		17.25	11.60	1.30	30.15	38.50	
	0610	3" diameter		48	.333		21	12.80	1.44	35.24	44.50	
	0620	4" diameter		44	.364		26.50	13.95	1.57	42.02	52	
	0630	5" diameter		36	.444		54	17.10	1.92	73.02	88	
	0640	6" diameter	Q-16	42	.571		50	23	1.65	74.65	92.50	
	0650	8" diameter		34	.706		77	28	2.03	107.03	130	
	0660	10" diameter		26	.923		108	37	2.66	147.66	179	
	0670	12" diameter		21	1.143		142	45.50	3.29	190.79	232	
	0700	To delete hangers, subtract										
	0710	1/2" diam. to 1-1/2" diam.					8%	19%				
	0720	2" diam. to 5" diam.					4%	9%				
	0730	6" diam. to 12" diam.					3%	4%				
	0750	For small quantities, add				L.F.	10%					
	1250	Schedule 5, type 316										
	1290	1/2" diameter	Q-15	128	.125	L.F.	6.90	4.80	.54	12.24	15.65	
	1300	3/4" diameter		116	.138		8.05	5.30	.60	13.95	17.80	
	1310	1" diameter		103	.155		9.25	5.95	.67	15.87	20	
	1320	1-1/4" diameter		93	.172		11.15	6.60	.74	18.49	23.50	
	1330	1-1/2" diameter		85	.188		12.70	7.25	.81	20.76	26	
	1340	2" diameter		69	.232		15.50	8.90	1	25.40	32	
	1350	2-1/2" diameter		53	.302		22.50	11.60	1.30	35.40	44	
	1360	3" diameter		48	.333		28.50	12.80	1.44	42.74	53	
	1370	4" diameter		44	.364		35.50	13.95	1.57	51.02	62	
	1380	5" diameter		36	.444		67.50	17.10	1.92	86.52	103	
	1390	6" diameter	Q-16	42	.571		66	23	1.65	90.65	110	
	1400	8" diameter		34	.706		99	28	2.03	129.03	155	
	1410	10" diameter		26	.923		144	37	2.66	183.66	218	
	1420	12" diameter		21	1.143		185	45.50	3.29	233.79	279	
	1490	For small quantities, add					10%					
	1940	To delete hanger, subtract										
	1950	1/2" diam. to 1-1/2" diam.					5%	19%				
	1960	2" diam. to 5" diam.					3%	9%				
	1970	6" diam. to 12" diam.					2%	4%				
	2000	Schedule 10, type 304										
	2040	1/4" diameter	Q-15	131	.122	L.F.	4.17	4.69	.53	9.39	12.45	
	2050	3/8" diameter		128	.125		4.43	4.80	.54	9.77	12.90	
	2060	1/2" diameter		125	.128		5.55	4.92	.55	11.02	14.35	
	2070	3/4" diameter		113	.142		6.50	5.45	.61	12.56	16.25	
	2080	1" diameter		100	.160		9.50	6.15	.69	16.34	21	
	2090	1-1/4" diameter		91	.176		11.55	6.75	.76	19.06	24	
	2100	1-1/2" diameter		83	.193		12.70	7.40	.83	20.93	26.50	
	2110	2" diameter		67	.239		15.85	9.20	1.03	26.08	33	
	2120	2-1/2" diameter		51	.314		19.75	12.05	1.36	33.16	42	
	2130	3" diameter		46	.348		24.50	13.35	1.50	39.35	49.50	
	2140	4" diameter		42	.381		32	14.65	1.65	48.30	60	

	15107 | Metal Pipe & Fittings		CREW	DAILY OUTPUT	LABOR-HOURS	UNIT	2006 BARE COSTS MAT.	LABOR	EQUIP.	TOTAL	TOTAL INCL O&P	
920	2150	5″ diameter	Q-15	35	.457	L.F.	43	17.55	1.97	62.52	76.50	920
	2160	6″ diameter	Q-16	40	.600		50	24	1.73	75.73	94	
	2170	8″ diameter		33	.727		77	29	2.09	108.09	132	
	2180	10″ diameter		25	.960		108	38.50	2.76	149.26	182	
	2190	12″ diameter		21	1.143		136	45.50	3.29	184.79	224	
	2250	For small quantities, add					10%					
	2650	To delete hanger, subtract										
	2660	1/4″ diam. to 3/4″ diam.					9%	22%				
	2670	1″ diam. to 2″ diam.					4%	15%				
	2680	2-1/2″ diam. to 5″ diam.					3%	8%				
	2690	6″ diam. to 12″ diam.					3%	4%				
	2750	Schedule 10, type 316										
	2790	1/4″ diameter	Q-15	131	.122	L.F.	4.75	4.69	.53	9.97	13.15	
	2800	3/8″ diameter		128	.125		5.35	4.80	.54	10.69	13.90	
	2810	1/2″ diameter		125	.128		6.90	4.92	.55	12.37	15.80	
	2820	3/4″ diameter		113	.142		8.10	5.45	.61	14.16	18	
	2830	1″ diameter		100	.160		10.75	6.15	.69	17.59	22	
	2840	1-1/4″ diameter		91	.176		14.60	6.75	.76	22.11	27.50	
	2850	1-1/2″ diameter		83	.193		16.35	7.40	.83	24.58	30.50	
	2860	2″ diameter		67	.239		20.50	9.20	1.03	30.73	38	
	2870	2-1/2″ diameter		51	.314		26.50	12.05	1.36	39.91	49	
	2880	3″ diameter		46	.348		33.50	13.35	1.50	48.35	59.50	
	2890	4″ diameter		42	.381		41	14.65	1.65	57.30	70	
	2900	5″ diameter		35	.457		61	17.55	1.97	80.52	96.50	
	2910	6″ diameter	Q-16	40	.600		65.50	24	1.73	91.23	111	
	2920	8″ diameter		33	.727		108	29	2.09	139.09	166	
	2930	10″ diameter		25	.960		144	38.50	2.76	185.26	222	
	2940	12″ diameter		21	1.143		180	45.50	3.29	228.79	273	
	2990	For small quantities, add					10%					
	3430	To delete hanger, subtract										
	3440	1/4″ diam. to 3/4″ diam.					6%	22%				
	3450	1″ diam. to 2″ diam.					3%	15%				
	3460	2-1/2″ diam. to 5″ diam.					2%	8%				
	3470	6″ diam. to 12″ diam.					2%	4%				
	3500	Threaded, couplings and hangers 10′ O.C.										
	3520	Schedule 40, type 304										
	3540	1/4″ diameter	1 Plum	54	.148	L.F.	5.75	6.35		12.10	16.15	
	3550	3/8″ diameter		53	.151		6.70	6.45		13.15	17.35	
	3560	1/2″ diameter		52	.154		7.90	6.55		14.45	18.90	
	3570	3/4″ diameter 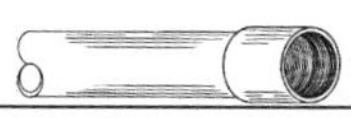		51	.157		10.55	6.70		17.25	22	
	3580	1″ diameter		45	.178		13.25	7.60		20.85	26.50	
	3590	1-1/4″ diameter	Q-1	76	.211		17.30	8.10		25.40	31.50	
	3600	1-1/2″ diameter		69	.232		20	8.90		28.90	36	
	3610	2″ diameter		57	.281		27	10.80		37.80	46.50	
	3620	2-1/2″ diameter		44	.364		46	13.95		59.95	72	
	3630	3″ diameter		38	.421		59.50	16.20		75.70	90	
	3640	4″ diameter	Q-2	51	.471		83.50	18.75		102.25	121	
	3740	For small quantities, add					10%					
	4200	To delete couplings & hangers, subtract										
	4210	1/4″ diam. to 3/4″ diam.					15%	56%				
	4220	1″ diam. to 2″ diam.					18%	49%				
	4230	2-1/2″ diam. to 4″ diam.					34%	40%				
	4250	Schedule 40, type 316										
	4290	1/4″ diameter	1 Plum	54	.148	L.F.	6.65	6.35		13	17.15	
	4300	3/8″ diameter		53	.151		8.10	6.45		14.55	18.90	
	4310	1/2″ diameter		52	.154		10.45	6.55		17	21.50	

MECHANICAL 15

15107 | Metal Pipe & Fittings

		Description	CREW	DAILY OUTPUT	LABOR-HOURS	UNIT	MAT. (2006 BARE COSTS)	LABOR (2006 BARE COSTS)	EQUIP. (2006 BARE COSTS)	TOTAL (2006 BARE COSTS)	TOTAL INCL O&P	
920	4320	3/4" diameter	1 Plum	51	.157	L.F.	12.95	6.70		19.65	24.50	920
	4330	1" diameter		45	.178		16.95	7.60		24.55	30.50	
	4340	1-1/4" diameter	Q-1	76	.211		22	8.10		30.10	36.50	
	4350	1-1/2" diameter		69	.232		25.50	8.90		34.40	42	
	4360	2" diameter		57	.281		34.50	10.80		45.30	55	
	4370	2-1/2" diameter		44	.364		59	13.95		72.95	86.50	
	4380	3" diameter		38	.421		79.50	16.20		95.70	113	
	4390	4" diameter	Q-2	51	.471		109	18.75		127.75	149	
	4490	For small quantities, add					10%					
	4900	To delete couplings & hangers, subtract										
	4910	1/4" diam. to 3/4" diam.					12%	56%				
	4920	1" diam. to 2" diam.					14%	49%				
	4930	2-1/2" diam. to 4" diam.					27%	40%				
	5000	Schedule 80, type 304										
	5040	1/4" diameter	1 Plum	53	.151	L.F.	7.80	6.45		14.25	18.55	
	5050	3/8" diameter		52	.154		8.90	6.55		15.45	20	
	5060	1/2" diameter		51	.157		11.20	6.70		17.90	22.50	
	5070	3/4" diameter		48	.167		14.55	7.10		21.65	27	
	5080	1" diameter		43	.186		20	7.95		27.95	34.50	
	5090	1-1/4" diameter	Q-1	73	.219		29.50	8.40		37.90	45.50	
	5100	1-1/2" diameter		67	.239		34	9.20		43.20	52	
	5110	2" diameter		54	.296		43.50	11.40		54.90	65	
	5190	For small quantities, add					10%					
	5700	To delete couplings & hangers, subtract										
	5710	1/4" diam. to 3/4" diam.					10%	53%				
	5720	1" diam. to 2" diam.					14%	47%				
	5750	Schedule 80, type 316										
	5790	1/4" diameter	1 Plum	53	.151	L.F.	9.70	6.45		16.15	20.50	
	5800	3/8" diameter		52	.154		11.05	6.55		17.60	22.50	
	5810	1/2" diameter		51	.157		15.95	6.70		22.65	28	
	5820	3/4" diameter		48	.167		18.25	7.10		25.35	31	
	5830	1" diameter		43	.186		25.50	7.95		33.45	40.50	
	5840	1-1/4" diameter	Q-1	73	.219		38.50	8.40		46.90	55	
	5850	1-1/2" diameter		67	.239		43.50	9.20		52.70	62.50	
	5860	2" diameter		54	.296		56	11.40		67.40	79	
	5950	For small quantities, add					10%					
	7000	To delete couplings & hangers, subtract										
	7010	1/4" diam. to 3/4" diam.					9%	53%				
	7020	1" diam. to 2" diam.					14%	47%				
	8000	Weld joints with clevis type hangers 10' O.C.										
	8010	Schedule 40, type 304										
	8050	1/8" pipe size	Q-15	126	.127	L.F.	4.80	4.88	.55	10.23	13.50	
	8060	1/4" pipe size		125	.128		4.92	4.92	.55	10.39	13.65	
	8070	3/8" pipe size		122	.131		5.65	5.05	.57	11.27	14.65	
	8080	1/2" pipe size 		118	.136		6.90	5.20	.59	12.69	16.30	
	8090	3/4" pipe size		109	.147		8.70	5.65	.63	14.98	19.05	
	8100	1" pipe size		95	.168		11	6.45	.73	18.18	23	
	8110	1-1/4" pipe size		86	.186		13.70	7.15	.80	21.65	27	
	8120	1-1/2" pipe size		78	.205		16	7.90	.89	24.79	31	
	8130	2" pipe size		62	.258		20.50	9.90	1.11	31.51	39	
	8140	2-1/2" pipe size		49	.327		31	12.55	1.41	44.96	55.50	
	8150	3" pipe size		44	.364		40	13.95	1.57	55.52	67	
	8160	3-1/2" pipe size		44	.364		47.50	13.95	1.57	63.02	75	
	8170	4" pipe size		39	.410		54	15.75	1.77	71.52	86	
	8180	5" pipe size		32	.500		78.50	19.20	2.16	99.86	119	
	8190	6" pipe size	Q-16	37	.649		94.50	26	1.87	122.37	146	

15 MECHANICAL

15107 | Metal Pipe & Fittings

		Description	CREW	DAILY OUTPUT	LABOR-HOURS	UNIT	2006 BARE COSTS MAT.	LABOR	EQUIP.	TOTAL	TOTAL INCL O&P	
920	8200	8″ pipe size	Q-16	29	.828	L.F.	153	33	2.38	188.38	222	920
	8210	10″ pipe size		24	1		260	40	2.88	302.88	350	
	8220	12″ pipe size		20	1.200		294	48	3.46	345.46	405	
	8300	Schedule 40, type 316										
	8310	1/8″ pipe size	Q-15	126	.127	L.F.	4.76	4.88	.55	10.19	13.45	
	8320	1/4″ pipe size		125	.128		5.70	4.92	.55	11.17	14.50	
	8330	3/8″ pipe size		122	.131		6.85	5.05	.57	12.47	16	
	8340	1/2″ pipe size		118	.136		9.25	5.20	.59	15.04	18.90	
	8350	3/4″ pipe size		109	.147		10.75	5.65	.63	17.03	21.50	
	8360	1″ pipe size		95	.168		14.25	6.45	.73	21.43	26.50	
	8370	1-1/4″ pipe size		86	.186		17.70	7.15	.80	25.65	31.50	
	8380	1-1/2″ pipe size		78	.205		20.50	7.90	.89	29.29	36	
	8390	2″ pipe size		62	.258		27	9.90	1.11	38.01	46	
	8400	2-1/2″ pipe size		49	.327		41.50	12.55	1.41	55.46	67	
	8410	3″ pipe size		44	.364		56	13.95	1.57	71.52	85	
	8420	3-1/2″ pipe size		44	.364		65	13.95	1.57	80.52	94.50	
	8430	4″ pipe size		39	.410		74	15.75	1.77	91.52	108	
	8440	5″ pipe size		32	.500		112	19.20	2.16	133.36	155	
	8450	6″ pipe size	Q-16	37	.649		126	26	1.87	153.87	180	
	8460	8″ pipe size		29	.828		235	33	2.38	270.38	315	
	8470	10″ pipe size		24	1		315	40	2.88	357.88	410	
	8480	12″ pipe size		20	1.200		430	48	3.46	481.46	550	
	8500	Schedule 80, type 304										
	8510	1/4″ pipe size	Q-15	110	.145	L.F.	6.80	5.60	.63	13.03	16.85	
	8520	3/8″ pipe size		109	.147		7.75	5.65	.63	14.03	17.95	
	8530	1/2″ pipe size		106	.151		9.80	5.80	.65	16.25	20.50	
	8540	3/4″ pipe size		96	.167		12.65	6.40	.72	19.77	24.50	
	8550	1″ pipe size		87	.184		17	7.05	.79	24.84	30.50	
	8560	1-1/4″ pipe size		81	.198		22	7.60	.85	30.45	36.50	
	8570	1-1/2″ pipe size		74	.216		25	8.30	.93	34.23	41.50	
	8580	2″ pipe size		58	.276		31.50	10.60	1.19	43.29	52.50	
	8590	2-1/2″ pipe size		46	.348		65	13.35	1.50	79.85	94	
	8600	3″ pipe size		41	.390		87.50	15	1.69	104.19	121	
	8610	4″ pipe size		33	.485		137	18.65	2.09	157.74	182	
	8630	6″ pipe size	Q-16	30	.800		300	32	2.30	334.30	380	
	8640	Schedule 80, type 316										
	8650	1/4″ pipe size	Q-15	110	.145	L.F.	8.55	5.60	.63	14.78	18.85	
	8660	3/8″ pipe size		109	.147		9.85	5.65	.63	16.13	20.50	
	8670	1/2″ pipe size		106	.151		14.25	5.80	.65	20.70	25.50	
	8680	3/4″ pipe size		96	.167		15.80	6.40	.72	22.92	28	
	8690	1″ pipe size		87	.184		21.50	7.05	.79	29.34	35.50	
	8700	1-1/4″ pipe size		81	.198		29	7.60	.85	37.45	44.50	
	8710	1-1/2″ pipe size		74	.216		33	8.30	.93	42.23	50	
	8720	2″ pipe size		58	.276		41	10.60	1.19	52.79	63	
	8730	2-1/2″ pipe size		46	.348		107	13.35	1.50	121.85	141	
	8740	3″ pipe size		41	.390		125	15	1.69	141.69	163	
	8760	4″ pipe size		33	.485		174	18.65	2.09	194.74	223	
	8770	6″ pipe size	Q-16	30	.800		470	32	2.30	504.30	565	
	8790	Schedule 160, type 304										
	8800	1/2″ pipe size	Q-15	96	.167	L.F.	14	6.40	.72	21.12	26	
	8810	3/4″ pipe size		88	.182		16.50	7	.79	24.29	30	
	8820	1″ pipe size		80	.200		28	7.70	.86	36.56	44	
	8840	1-1/2″ pipe size		67	.239		39	9.20	1.03	49.23	58	
	8850	2″ pipe size		53	.302		56	11.60	1.30	68.90	81	
	8870	3″ pipe size		37	.432		92	16.60	1.87	110.47	129	
	8900	Schedule 160, type 316										

MECHANICAL 15

15107 | Metal Pipe & Fittings

920

	15107 Metal Pipe & Fittings	CREW	DAILY OUTPUT	LABOR-HOURS	UNIT	2006 BARE COSTS MAT.	LABOR	EQUIP.	TOTAL	TOTAL INCL O&P
8910	1/2" pipe size	Q-15	96	.167	L.F.	21	6.40	.72	28.12	33.50
8920	3/4" pipe size		88	.182		22	7	.79	29.79	35.50
8930	1" pipe size		80	.200		31.50	7.70	.86	40.06	47.50
8940	1-1/4" pipe size		73	.219		45	8.40	.95	54.35	63.50
8950	1-1/2" pipe size		67	.239		51.50	9.20	1.03	61.73	72
8960	2" pipe size		53	.302		71.50	11.60	1.30	84.40	98.50
8990	4" pipe size		30	.533		181	20.50	2.30	203.80	234
9100	Threading pipe labor, sst, one end, schedules 40 & 80									
9110	1/4" through 3/4" pipe size	1 Plum	61.50	.130	Ea.		5.55		5.55	8.65
9120	1" through 2" pipe size		55.90	.143			6.10		6.10	9.50
9130	2-1/2" pipe size		41.50	.193			8.25		8.25	12.80
9140	3" pipe size		38.50	.208			8.85		8.85	13.80
9150	3-1/2" pipe size	Q-1	68.40	.234			9		9	14
9160	4" pipe size		73	.219			8.40		8.40	13.10
9170	5" pipe size		40.70	.393			15.10		15.10	23.50
9180	6" pipe size		35.40	.452			17.35		17.35	27
9190	8" pipe size		22.30	.717			27.50		27.50	43
9200	10" pipe size		16.10	.994			38		38	59.50
9210	12" pipe size		12.30	1.301			50		50	77.50
9250	Welding labor per joint for stainless steel									
9260	Schedule 5 and 10									
9270	1/4" pipe size	Q-15	36	.444	Ea.		17.10	1.92	19.02	28.50
9280	3/8" pipe size		35	.457			17.55	1.97	19.52	29.50
9290	1/2" pipe size		35	.457			17.55	1.97	19.52	29.50
9300	3/4" pipe size		28	.571			22	2.47	24.47	36.50
9310	1" pipe size		25	.640			24.50	2.76	27.26	41
9320	1-1/4" pipe size		22	.727			28	3.14	31.14	47
9330	1-1/2" pipe size		21	.762			29.50	3.29	32.79	49
9340	2" pipe size		18	.889			34	3.84	37.84	57
9350	2-1/2" pipe size		12	1.333			51	5.75	56.75	86
9360	3" pipe size		9.73	1.644			63	7.10	70.10	106
9370	4" pipe size		7.37	2.171			83.50	9.40	92.90	140
9380	5" pipe size		6.15	2.602			100	11.25	111.25	167
9390	6" pipe size		5.71	2.802			108	12.10	120.10	180
9400	8" pipe size		3.69	4.336			167	18.75	185.75	280
9410	10" pipe size		2.91	5.498			211	24	235	355
9420	12" pipe size		2.31	6.926			266	30	296	450
9500	Schedule 40									
9510	1/4" pipe size	Q-15	28	.571	Ea.		22	2.47	24.47	36.50
9520	3/8" pipe size		27	.593			23	2.56	25.56	38.50
9530	1/2" pipe size		25.40	.630			24	2.72	26.72	40.50
9540	3/4" pipe size		22.22	.720			27.50	3.11	30.61	46.50
9550	1" pipe size		20.25	.790			30.50	3.41	33.91	51
9560	1-1/4" pipe size		18.82	.850			32.50	3.67	36.17	55
9570	1-1/2" pipe size		17.78	.900			34.50	3.89	38.39	58.50
9580	2" pipe size		15.09	1.060			41	4.58	45.58	68.50
9590	2-1/2" pipe size		7.96	2.010			77.50	8.70	86.20	130
9600	3" pipe size		6.43	2.488			95.50	10.75	106.25	161
9610	4" pipe size		4.88	3.279			126	14.15	140.15	212
9620	5" pipe size		4.26	3.756			144	16.25	160.25	242
9630	6" pipe size		3.77	4.244			163	18.35	181.35	274
9640	8" pipe size		2.44	6.557			252	28.50	280.50	420
9650	10" pipe size		1.92	8.333			320	36	356	540
9660	12" pipe size		1.52	10.526			405	45.50	450.50	680
9750	Schedule 80									
9760	1/4" pipe size	Q-15	21.55	.742	Ea.		28.50	3.21	31.71	48

920

15107 | Metal Pipe & Fittings

				DAILY	LABOR-		2006 BARE COSTS				TOTAL	
			CREW	OUTPUT	HOURS	UNIT	MAT.	LABOR	EQUIP.	TOTAL	INCL O&P	
920	9770	3/8" pipe size	Q-15	20.75	.771	Ea.		29.50	3.33	32.83	49.50	920
	9780	1/2" pipe size		19.54	.819			31.50	3.54	35.04	53	
	9790	3/4" pipe size		17.09	.936			36	4.04	40.04	60.50	
	9800	1" pipe size		15.58	1.027			39.50	4.44	43.94	66.50	
	9810	1-1/4" pipe size		14.48	1.105			42.50	4.77	47.27	71.50	
	9820	1-1/2" pipe size		13.68	1.170			45	5.05	50.05	75.50	
	9830	2" pipe size		11.61	1.378			53	5.95	58.95	89	
	9840	2-1/2" pipe size		6.12	2.614			100	11.30	111.30	168	
	9850	3" pipe size		4.94	3.239			124	14	138	209	
	9860	4" pipe size		3.75	4.267			164	18.45	182.45	276	
	9870	5" pipe size		3.27	4.893			188	21	209	315	
	9880	6" pipe size		2.90	5.517			212	24	236	355	
	9890	8" pipe size		1.87	8.556			330	37	367	550	
	9900	10" pipe size		1.48	10.811			415	46.50	461.50	695	
	9910	12" pipe size		1.17	13.675			525	59	584	880	
	9920	Schedule 160, 1/2" pipe size		17	.941			36	4.07	40.07	60.50	
	9930	3/4" pipe size		14.81	1.080			41.50	4.67	46.17	69.50	
	9940	1" pipe size		13.50	1.185			45.50	5.10	50.60	76.50	
	9950	1-1/4" pipe size		12.55	1.275			49	5.50	54.50	82	
	9960	1-1/2" pipe size		11.85	1.350			52	5.85	57.85	87	
	9970	2" pipe size		10	1.600			61.50	6.90	68.40	103	
	9980	3" pipe size		4.28	3.738			144	16.15	160.15	241	
	9990	4" pipe size		3.25	4.923			189	21.50	210.50	320	
960	0010	**PIPE, STAINLESS STEEL, FITTINGS**										960
	0100	Butt weld joint, schedule 5, type 304										
	0120	90° Elbow, long										
	0140	1/2"	Q-15	17.50	.914	Ea.	12.15	35	3.95	51.10	72	
	0150	3/4"		14	1.143		19.10	44	4.94	68.04	95	
	0160	1"		12.50	1.280		13.95	49	5.55	68.50	98	
	0170	1-1/4"		11	1.455		22.50	56	6.30	84.80	118	
	0180	1-1/2"		10.50	1.524		17.50	58.50	6.60	82.60	118	
	0190	2"		9	1.778		23	68.50	7.70	99.20	140	
	0200	2-1/2"		6	2.667		33.50	102	11.50	147	209	
	0210	3"		4.86	3.292		42	127	14.20	183.20	259	
	0220	3-1/2"		4.27	3.747		223	144	16.20	383.20	485	
	0230	4"		3.69	4.336		78	167	18.75	263.75	365	
	0240	5"		3.08	5.195		560	200	22.50	782.50	950	
	0250	6"	Q-16	4.29	5.594		223	223	16.10	462.10	610	
	0260	8"		2.76	8.696		380	345	25	750	985	
	0270	10"		2.18	11.009		780	440	31.50	1,251.50	1,575	
	0280	12"		1.73	13.873		1,225	555	40	1,820	2,250	
	0320	For schedule 5, type 316, add					30%					
	0600	45° Elbow, long										
	0620	1/2"	Q-15	17.50	.914	Ea.	17.50	35	3.95	56.45	78	
	0630	3/4"		14	1.143		17.50	44	4.94	66.44	93	
	0640	1"		12.50	1.280		19.75	49	5.55	74.30	104	
	0650	1-1/4"		11	1.455		33.50	56	6.30	95.80	131	
	0660	1-1/2"		10.50	1.524		17.95	58.50	6.60	83.05	118	
	0670	2"		9	1.778		20.50	68.50	7.70	96.70	137	
	0680	2-1/2"		6	2.667		34	102	11.50	147.50	209	
	0690	3"		4.86	3.292		35	127	14.20	176.20	251	
	0700	3-1/2"		4.27	3.747		223	144	16.20	383.20	485	
	0710	4"		3.69	4.336		62.50	167	18.75	248.25	350	
	0720	5"		3.08	5.195		390	200	22.50	612.50	765	
	0730	6"	Q-16	4.29	5.594		134	223	16.10	373.10	510	

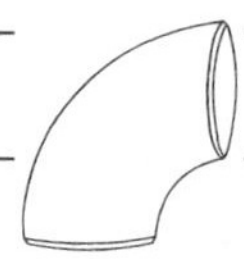

15107 | Metal Pipe & Fittings

			CREW	DAILY OUTPUT	LABOR-HOURS	UNIT	2006 BARE COSTS MAT.	LABOR	EQUIP.	TOTAL	TOTAL INCL O&P	
960	0740	8"	Q-16	2.76	8.696	Ea.	234	345	25	604	825	960
	0750	10"		2.18	11.009		525	440	31.50	996.50	1,300	
	0760	12"		1.73	13.873		835	555	40	1,430	1,825	
	0800	For schedule 5, type 316, add					25%					
	1100	Tee, straight										
	1130	1/2"	Q-15	11.66	1.372	Ea.	49	52.50	5.95	107.45	143	
	1140	3/4"		9.33	1.715		49	66	7.40	122.40	164	
	1150	1"		8.33	1.921		47	74	8.30	129.30	176	
	1160	1-1/4"		7.33	2.183		80	84	9.45	173.45	228	
	1170	1-1/2"		7	2.286		47	88	9.85	144.85	199	
	1180	2"		6	2.667		49	102	11.50	162.50	226	
	1190	2-1/2"		4	4		140	154	17.30	311.30	410	
	1200	3"		3.24	4.938		91.50	190	21.50	303	420	
	1210	3-1/2"		2.85	5.614		244	216	24.50	484.50	630	
	1220	4"		2.46	6.504		130	250	28	408	565	
	1230	5"		2	8		525	305	34.50	864.50	1,100	
	1240	6"	Q-16	2.85	8.421		390	335	24.50	749.50	975	
	1250	8"		1.84	13.043		890	520	37.50	1,447.50	1,825	
	1260	10"		1.45	16.552		1,400	660	47.50	2,107.50	2,625	
	1270	12"		1.15	20.870		1,775	830	60	2,665	3,325	
	1320	For schedule 5, type 316, add					25%					
	2000	Butt weld joint, schedule 10, type 304										
	2020	90° elbow, long										
	2040	1/2"	Q-15	17	.941	Ea.	11.45	36	4.07	51.52	73	
	2050	3/4"		14	1.143		11.45	44	4.94	60.39	86.50	
	2060	1"		12.50	1.280		11.45	49	5.55	66	95	
	2070	1-1/4"		11	1.455		21.50	56	6.30	83.80	117	
	2080	1-1/2"		10.50	1.524		16.70	58.50	6.60	81.80	117	
	2090	2"		9	1.778		22.50	68.50	7.70	98.70	139	
	2100	2-1/2"		6	2.667		17.80	102	11.50	131.30	191	
	2110	3"		4.86	3.292		23	127	14.20	164.20	238	
	2120	3-1/2"		4.27	3.747		93	144	16.20	253.20	345	
	2130	4"		3.69	4.336		36.50	167	18.75	222.25	320	
	2140	5"		3.08	5.195		190	200	22.50	412.50	545	
	2150	6"	Q-16	4.29	5.594		97	223	16.10	336.10	470	
	2160	8"		2.76	8.696		180	345	25	550	765	
	2170	10"		2.18	11.009		350	440	31.50	821.50	1,100	
	2180	12"		1.73	13.873		545	555	40	1,140	1,500	
	2500	45° elbow, long										
	2520	1/2"	Q-15	17.50	.914	Ea.	7.35	35	3.95	46.30	67	
	2530	3/4"		14	1.143		7.35	44	4.94	56.29	82	
	2540	1"		12.50	1.280		9.40	49	5.55	63.95	93	
	2550	1-1/4"		11	1.455		14.60	56	6.30	76.90	110	
	2560	1-1/2"		10.50	1.524		16.70	58.50	6.60	81.80	117	
	2570	2"		9	1.778		19.15	68.50	7.70	95.35	135	
	2580	2-1/2"		6	2.667		16.75	102	11.50	130.25	190	
	2590	3"		4.86	3.292		18.80	127	14.20	160	233	
	2600	3-1/2"		4.27	3.747		93	144	16.20	253.20	345	
	2610	4"		3.69	4.336		29	167	18.75	214.75	310	
	2620	5"		3.08	5.195		145	200	22.50	367.50	495	
	2630	6"	Q-16	4.29	5.594		64.50	223	16.10	303.60	435	
	2640	8"		2.76	8.696		144	345	25	514	725	
	2650	10"		2.18	11.009		350	440	31.50	821.50	1,100	
	2660	12"		1.73	13.873		380	555	40	975	1,325	
	3000	Tee, straight										
	3030	1/2"	Q-15	11.66	1.372	Ea.	23.50	52.50	5.95	81.95	115	

15107 | Metal Pipe & Fittings

Line	Description	CREW	DAILY OUTPUT	LABOR-HOURS	UNIT	2006 BARE COSTS MAT.	LABOR	EQUIP.	TOTAL	TOTAL INCL O&P
960 3040	3/4″	Q-15	9.33	1.715	Ea.	23.50	66	7.40	96.90	136
3050	1″		8.33	1.921		40	74	8.30	122.30	168
3060	1-1/4″		7.33	2.183		29	84	9.45	122.45	172
3070	1-1/2″		7	2.286		44	88	9.85	141.85	196
3080	2″		6	2.667		44.50	102	11.50	158	221
3090	2-1/2″		4	4		53	154	17.30	224.30	315
3100	3″		3.24	4.938		37.50	190	21.50	249	360
3110	3-1/2″		2.85	5.614		55.50	216	24.50	296	425
3120	4″		2.46	6.504		122	250	28	400	555
3130	5″		2	8		223	305	34.50	562.50	765
3140	6″	Q-16	2.85	8.421		148	335	24.50	507.50	710
3150	8″		1.84	13.043		325	520	37.50	882.50	1,200
3151	10″		1.45	16.552		535	660	47.50	1,242.50	1,675
3152	12″		1.15	20.870		740	830	60	1,630	2,175
3154	For schedule 10, type 316, add					25%				
3281	Butt weld joint, schedule 40, type 304									
3284	90° Elbow, long, 1/2″	Q-15	12.70	1.260	Ea.	7.35	48.50	5.45	61.30	89.50
3288	3/4″		11.10	1.441		7.35	55.50	6.25	69.10	101
3289	1″		10.13	1.579		8.40	60.50	6.80	75.70	111
3290	1-1/4″		9.40	1.702		11.55	65.50	7.35	84.40	123
3300	1-1/2″		8.89	1.800		9.40	69	7.80	86.20	127
3310	2″		7.55	2.119		13.60	81.50	9.15	104.25	152
3320	2-1/2″		3.98	4.020		27	154	17.35	198.35	289
3330	3″		3.21	4.984		34.50	192	21.50	248	360
3340	3-1/2″		2.83	5.654		145	217	24.50	386.50	525
3350	4″		2.44	6.557		64.50	252	28.50	345	490
3360	5″		2.13	7.512		227	289	32.50	548.50	735
3370	6″	Q-16	2.83	8.481		201	340	24.50	565.50	775
3380	8″		1.83	13.115		340	525	38	903	1,225
3390	10″		1.44	16.667		705	665	48	1,418	1,850
3400	12″		1.14	21.053		1,000	840	60.50	1,900.50	2,475
3410	For schedule 40, type 316, add					25%				
3460	45° Elbow, long, 1/2″	Q-15	12.70	1.260	Ea.	9.40	48.50	5.45	63.35	92
3470	3/4″		11.10	1.441		9.40	55.50	6.25	71.15	103
3480	1″		10.13	1.579		9.40	60.50	6.80	76.70	112
3490	1-1/4″		9.40	1.702		14.55	65.50	7.35	87.40	126
3500	1-1/2″		8.89	1.800		11.55	69	7.80	88.35	129
3510	2″		7.55	2.119		13.60	81.50	9.15	104.25	152
3520	2-1/2″		3.98	4.020		27	154	17.35	198.35	289
3530	3″		3.21	4.984		25	192	21.50	238.50	350
3540	3-1/2″		2.83	5.654		147	217	24.50	388.50	530
3550	4″		2.44	6.557		46	252	28.50	326.50	470
3560	5″		2.13	7.512		159	289	32.50	480.50	660
3570	6″	Q-16	2.83	8.481		119	340	24.50	483.50	685
3580	8″		1.83	13.115		228	525	38	791	1,100
3590	10″		1.44	16.667		420	665	48	1,133	1,550
3600	12″		1.14	21.053		580	840	60.50	1,480.50	2,000
3610	For schedule 40, type 316, add					25%				
3660	Tee, straight 1/2″	Q-15	8.46	1.891	Ea.	26.50	72.50	8.15	107.15	151
3670	3/4″		7.40	2.162		26.50	83	9.35	118.85	168
3680	1″		6.74	2.374		27.50	91	10.25	128.75	184
3690	1-1/4″		6.27	2.552		33.50	98	11	142.50	201
3700	1-1/2″		5.92	2.703		19.85	104	11.70	135.55	196
3710	2″		5.03	3.181		22	122	13.75	157.75	229
3720	2-1/2″		2.65	6.038		55.50	232	26	313.50	450
3730	3″		2.14	7.477		45	287	32.50	364.50	530

MECHANICAL 15

15107 | Metal Pipe & Fittings

		Description	Crew	Daily Output	Labor-Hours	Unit	2006 Bare Costs Mat.	Labor	Equip.	Total	Total Incl O&P
960	3740	3-1/2"	Q-15	1.88	8.511	Ea.	75	325	37	437	635
	3750	4"		1.62	9.877		75	380	42.50	497.50	720
	3760	5"		1.42	11.268		264	435	48.50	747.50	1,025
	3770	6"	Q-16	1.88	12.766		234	510	37	781	1,100
	3780	8"		1.22	19.672		475	785	56.50	1,316.50	1,800
	3790	10"		.96	25		995	995	72	2,062	2,725
	3800	12"		.76	31.579		1,225	1,250	91	2,566	3,400
	3810	For schedule 40, type 316, add					25%				
	3820	Tee, reducing on outlet, 3/4" x 1/2"	Q-15	7.73	2.070	Ea.	61.50	79.50	8.95	149.95	201
	3822	1" x 1/2"		7.24	2.210		85	85	9.55	179.55	236
	3824	1" x 3/4"		6.96	2.299		70.50	88.50	9.95	168.95	226
	3826	1-1/4" x 1"		6.43	2.488		224	95.50	10.75	330.25	405
	3828	1-1/2" x 1/2"		6.58	2.432		118	93.50	10.50	222	287
	3830	1-1/2" x 3/4"		6.35	2.520		118	97	10.90	225.90	293
	3832	1-1/2" x 1"		6.18	2.589		85	99.50	11.20	195.70	261
	3834	2" x 1"		5.50	2.909		82.50	112	12.55	207.05	278
	3836	2" x 1-1/2"		5.30	3.019		78	116	13.05	207.05	280
	3838	2-1/2" x 2"		3.15	5.079		200	195	22	417	550
	3840	3" x 1-1/2"		2.72	5.882		171	226	25.50	422.50	565
	3842	3" x 2"		2.65	6.038		144	232	26	402	545
	3844	4" x 2"		2.10	7.619		260	293	33	586	775
	3846	4" x 3"		1.77	9.040		212	345	39	596	815
	3848	5" x 4"		1.48	10.811		670	415	46.50	1,131.50	1,425
	3850	6" x 3"	Q-16	2.19	10.959		535	435	31.50	1,001.50	1,300
	3852	6" x 4"		2.04	11.765		485	470	34	989	1,300
	3854	8" x 4"		1.46	16.438		890	655	47.50	1,592.50	2,050
	3856	10" x 8"		.69	34.783		2,475	1,375	100	3,950	4,975
	3858	12" x 10"		.55	43.636		3,050	1,750	126	4,926	6,200
	3950	Reducer, concentric, 3/4" x 1/2"	Q-15	11.85	1.350		28.50	52	5.85	86.35	118
	3952	1" x 3/4"		10.60	1.509		33	58	6.50	97.50	134
	3954	1-1/4" x 3/4"		10.19	1.570		94	60.50	6.80	161.30	205
	3956	1-1/4" x 1"		9.76	1.639		50.50	63	7.10	120.60	162
	3958	1-1/2" x 3/4"		9.88	1.619		58.50	62	7	127.50	169
	3960	1-1/2" x 1"		9.47	1.690		19.65	65	7.30	91.95	131
	3962	2" x 1"		8.65	1.850		23.50	71	8	102.50	146
	3964	2" x 1-1/2"		8.16	1.961		19.65	75.50	8.45	103.60	148
	3966	2-1/2" x 1"		5.71	2.802		86.50	108	12.10	206.60	275
	3968	2-1/2" x 2"		5.21	3.071		70.50	118	13.25	201.75	276
	3970	3" x 1"		4.88	3.279		94	126	14.15	234.15	315
	3972	3" x 1-1/2"		4.72	3.390		43.50	130	14.65	188.15	267
	3974	3" x 2"		4.51	3.548		35.50	136	15.35	186.85	268
	3976	4" x 2"		3.69	4.336		48.50	167	18.75	234.25	335
	3978	4" x 3"		2.77	5.776		42.50	222	25	289.50	420
	3980	5" x 3"		2.56	6.250		355	240	27	622	795
	3982	5" x 4"		2.27	7.048		283	271	30.50	584.50	765
	3984	6" x 3"	Q-16	3.57	6.723		134	268	19.35	421.35	585
	3986	6" x 4"		3.19	7.524		112	300	21.50	433.50	610
	3988	8" x 4"		2.44	9.836		425	390	28.50	843.50	1,100
	3990	8" x 6"		2.22	10.811		271	430	31	732	1,000
	3992	10" x 6"		1.91	12.565		565	500	36	1,101	1,450
	3994	10" x 8"		1.61	14.907		335	595	43	973	1,350
	3995	12" x 6"		1.63	14.724		850	585	42.50	1,477.50	1,900
	3996	12" x 8"		1.41	17.021		775	680	49	1,504	1,950
	3997	12" x 10"		1.27	18.898		465	755	54.50	1,274.50	1,750
	4000	Socket weld joint, 3000 lb., type 304									
	4100	90° Elbow									

15 MECHANICAL

	15107 | Metal Pipe & Fittings		CREW	DAILY OUTPUT	LABOR-HOURS	UNIT	2006 BARE COSTS MAT.	LABOR	EQUIP.	TOTAL	TOTAL INCL O&P	
960	4140	1/4"	Q-15	13.47	1.188	Ea.	31.50	45.50	5.15	82.15	111	960
	4150	3/8"		12.97	1.234		41	47.50	5.35	93.85	124	
	4160	1/2"		12.21	1.310		42.50	50.50	5.65	98.65	132	
	4170	3/4"		10.68	1.498		51.50	57.50	6.45	115.45	153	
	4180	1"		9.74	1.643		77.50	63	7.10	147.60	191	
	4190	1-1/4"		9.05	1.768		135	68	7.65	210.65	262	
	4200	1-1/2"		8.55	1.871		164	72	8.10	244.10	300	
	4210	2"		7.26	2.204		264	84.50	9.50	358	435	
	4300	45° Elbow										
	4340	1/4"	Q-15	13.47	1.188	Ea.	59	45.50	5.15	109.65	142	
	4350	3/8"		12.97	1.234		59	47.50	5.35	111.85	144	
	4360	1/2"		12.21	1.310		59	50.50	5.65	115.15	150	
	4370	3/4"		10.68	1.498		67	57.50	6.45	130.95	171	
	4380	1"		9.74	1.643		97	63	7.10	167.10	213	
	4390	1-1/4"		9.05	1.768		160	68	7.65	235.65	290	
	4400	1-1/2"		8.55	1.871		161	72	8.10	241.10	298	
	4410	2"		7.26	2.204		292	84.50	9.50	386	460	
	4500	Tee										
	4540	1/4"	Q-15	8.97	1.784	Ea.	41.50	68.50	7.70	117.70	161	
	4550	3/8"		8.64	1.852		49.50	71	8	128.50	174	
	4560	1/2"		8.13	1.968		60.50	75.50	8.50	144.50	194	
	4570	3/4"		7.12	2.247		70	86.50	9.70	166.20	222	
	4580	1"		6.48	2.469		94	95	10.65	199.65	264	
	4590	1-1/4"		6.03	2.653		167	102	11.45	280.45	355	
	4600	1-1/2"		5.69	2.812		239	108	12.15	359.15	445	
	4610	2"		4.83	3.313		365	127	14.30	506.30	615	
	5000	Socket weld joint, 3000 lb., type 316										
	5100	90° Elbow										
	5140	1/4"	Q-15	13.47	1.188	Ea.	37	45.50	5.15	87.65	117	
	5150	3/8"		12.97	1.234		43	47.50	5.35	95.85	127	
	5160	1/2"		12.21	1.310		52	50.50	5.65	108.15	142	
	5170	3/4"		10.68	1.498		68	57.50	6.45	131.95	172	
	5180	1"		9.74	1.643		96.50	63	7.10	166.60	212	
	5190	1-1/4"		9.05	1.768		173	68	7.65	248.65	305	
	5200	1-1/2"		8.55	1.871		195	72	8.10	275.10	335	
	5210	2"		7.26	2.204		330	84.50	9.50	424	505	
	5300	45° Elbow										
	5340	1/4"	Q-15	13.47	1.188	Ea.	74.50	45.50	5.15	125.15	158	
	5350	3/8"		12.97	1.234		74.50	47.50	5.35	127.35	161	
	5360	1/2"		12.21	1.310		74.50	50.50	5.65	130.65	166	
	5370	3/4"		10.68	1.498		82.50	57.50	6.45	146.45	187	
	5380	1"		9.74	1.643		124	63	7.10	194.10	243	
	5390	1-1/4"		9.05	1.768		178	68	7.65	253.65	310	
	5400	1-1/2"		8.55	1.871		199	72	8.10	279.10	340	
	5410	2"		7.26	2.204		298	84.50	9.50	392	465	
	5500	Tee										
	5540	1/4"	Q-15	8.97	1.784	Ea.	50	68.50	7.70	126.20	170	
	5550	3/8"		8.64	1.852		61	71	8	140	187	
	5560	1/2"		8.13	1.968		68	75.50	8.50	152	202	
	5570	3/4"		7.12	2.247		84	86.50	9.70	180.20	237	
	5580	1"		6.48	2.469		128	95	10.65	233.65	300	
	5590	1-1/4"		6.03	2.653		204	102	11.45	317.45	395	
	5600	1-1/2"		5.69	2.812		288	108	12.15	408.15	495	
	5610	2"		4.83	3.313		460	127	14.30	601.30	720	
	5700	For socket weld joint, 6000 lb., type 304 and 316, add					100%					
	6000	Threaded companion flange										

15107 | Metal Pipe & Fittings

			CREW	DAILY OUTPUT	LABOR-HOURS	UNIT	2006 BARE COSTS MAT.	LABOR	EQUIP.	TOTAL	TOTAL INCL O&P	
960	6010	Stainless steel, 150 lb., type 304										960
	6020	1/2" diam.	1 Plum	30	.267	Ea.	68	11.40		79.40	92.50	
	6030	3/4" diam.		28	.286		76	12.20		88.20	102	
	6040	1" diam.		27	.296		80	12.65		92.65	108	
	6050	1-1/4" diam.	Q-1	44	.364		108	13.95		121.95	141	
	6060	1-1/2" diam.		40	.400		108	15.35		123.35	143	
	6070	2" diam.		36	.444		139	17.10		156.10	180	
	6080	2-1/2" diam.		28	.571		194	22		216	247	
	6090	3" diam.		20	.800		204	30.50		234.50	272	
	6110	4" diam.		12	1.333		279	51		330	385	
	6130	6" diam.	Q-2	14	1.714		505	68.50		573.50	660	
	6140	8" diam.	"	12	2		965	79.50		1,044.50	1,200	
	6150	For type 316 add					40%					
	6260	Weld flanges, stainless steel, type 304										
	6270	Slip on, 150 lb. (welded, front and back)										
	6280	1/2" diam.	Q-15	12.70	1.260	Ea.	59	48.50	5.45	112.95	147	
	6290	3/4" diam.		11.11	1.440		63	55.50	6.20	124.70	162	
	6300	1" diam.		10.13	1.579		68	60.50	6.80	135.30	177	
	6310	1-1/4" diam.		9.41	1.700		92	65.50	7.35	164.85	211	
	6320	1-1/2" diam.		8.89	1.800		92	69	7.80	168.80	218	
	6330	2" diam.		7.55	2.119		118	81.50	9.15	208.65	267	
	6340	2-1/2" diam.		3.98	4.020		160	154	17.35	331.35	435	
	6350	3" diam.		3.21	4.984		170	192	21.50	383.50	510	
	6370	4" diam.		2.44	6.557		232	252	28.50	512.50	675	
	6390	6" diam.	Q-16	1.89	12.698		320	505	36.50	861.50	1,175	
	6400	8" diam.	"	1.22	19.672		700	785	56.50	1,541.50	2,050	
	6410	For type 316, add					40%					
	6530	Weld neck 150 lb.										
	6540	1/2" diam.	Q-15	25.40	.630	Ea.	41.50	24	2.72	68.22	86	
	6550	3/4" diam.		22.22	.720		47	27.50	3.11	77.61	98	
	6560	1" diam.		20.25	.790		55.50	30.50	3.41	89.41	112	
	6570	1-1/4" diam.		18.82	.850		67	32.50	3.67	103.17	129	
	6580	1-1/2" diam.		17.78	.900		78	34.50	3.89	116.39	144	
	6590	2" diam.		15.09	1.060		100	41	4.58	145.58	179	
	6600	2-1/2" diam.		7.96	2.010		167	77.50	8.70	253.20	315	
	6610	3" diam.		6.43	2.488		190	95.50	10.75	296.25	370	
	6630	4" diam.		4.88	3.279		245	126	14.15	385.15	480	
	6640	5" diam.		4.26	3.756		325	144	16.25	485.25	600	
	6650	6" diam.	Q-16	5.66	4.240		380	169	12.20	561.20	690	
	6652	8" diam.		3.65	6.575		530	262	18.95	810.95	1,000	
	6654	10" diam.		2.88	8.333		730	330	24	1,084	1,350	
	6656	12" diam.		2.28	10.526		1,150	420	30.50	1,600.50	1,950	
	6670	For type 316 add					23%					
	7000	Threaded joint, 150 lb., type 304										
	7030	90° elbow										
	7040	1/8"	1 Plum	13	.615	Ea.	18.10	26.50		44.60	61	
	7050	1/4"		13	.615		18.10	26.50		44.60	61	
	7070	3/8"		13	.615		19.90	26.50		46.40	63	
	7080	1/2"		12	.667		13.30	28.50		41.80	59	
	7090	3/4"		11	.727		15.15	31		46.15	65	
	7100	1"		10	.800		20	34		54	75	
	7110	1-1/4"	Q-1	17	.941		41.50	36		77.50	102	
	7120	1-1/2"		16	1		34.50	38.50		73	98	
	7130	2"		14	1.143		49.50	44		93.50	123	
	7140	2-1/2"		11	1.455		182	56		238	288	
960	7150	3"		8	2		253	77		330	400	

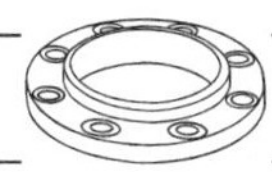

15107 | Metal Pipe & Fittings

		Description	CREW	DAILY OUTPUT	LABOR-HOURS	UNIT	2006 BARE COSTS MAT.	LABOR	EQUIP.	TOTAL	TOTAL INCL O&P	
960	7160	4"	Q-2	11	2.182	Ea.	340	87		427	505	960
	7180	45° elbow										
	7190	1/8"	1 Plum	13	.615	Ea.	26.50	26.50		53	70	
	7200	1/4"		13	.615		26.50	26.50		53	70	
	7210	3/8"		13	.615		25	26.50		51.50	68.50	
	7220	1/2"		12	.667		19.10	28.50		47.60	65.50	
	7230	3/4"		11	.727		20.50	31		51.50	71.50	
	7240	1"		10	.800		23	34		57	78	
	7250	1-1/4"	Q-1	17	.941		39	36		75	99	
	7260	1-1/2"		16	1		38	38.50		76.50	102	
	7270	2"		14	1.143		54	44		98	128	
	7280	2-1/2"		11	1.455		257	56		313	370	
	7290	3"		8	2		375	77		452	535	
	7300	4"	Q-2	11	2.182		510	87		597	695	
	7320	Tee, straight										
	7330	1/8"	1 Plum	9	.889	Ea.	28	38		66	89.50	
	7340	1/4"		9	.889		28	38		66	89.50	
	7350	3/8"		9	.889		28	38		66	90	
	7360	1/2"		8	1		20	42.50		62.50	88.50	
	7370	3/4"		7	1.143		21.50	49		70.50	99.50	
	7380	1"		6.50	1.231		25	52.50		77.50	109	
	7390	1-1/4"	Q-1	11	1.455		56.50	56		112.50	149	
	7400	1-1/2"		10	1.600		52	61.50		113.50	153	
	7410	2"		9	1.778		64.50	68.50		133	177	
	7420	2-1/2"		7	2.286		262	88		350	425	
	7430	3"		5	3.200		400	123		523	630	
	7440	4"	Q-2	7	3.429		725	137		862	1,000	
	7460	Coupling, straight										
	7470	1/8"	1 Plum	19	.421	Ea.	7.30	18		25.30	36	
	7480	1/4"		19	.421		8.05	18		26.05	37	
	7490	3/8"		19	.421		10.35	18		28.35	39.50	
	7500	1/2"		19	.421		10.35	18		28.35	39.50	
	7510	3/4"		18	.444		18.35	19		37.35	49.50	
	7520	1"		15	.533		22	23		45	60	
	7530	1-1/4"	Q-1	26	.615		35.50	23.50		59	76.50	
	7540	1-1/2"		24	.667		40	25.50		65.50	84	
	7550	2"		21	.762		63	29.50		92.50	115	
	7560	2-1/2"		18	.889		147	34		181	214	
	7570	3"		14	1.143		193	44		237	281	
	7580	4"	Q-2	16	1.500		293	60		353	420	
	7600	Reducer, concentric, 1/2"	1 Plum	12	.667		11.10	28.50		39.60	56.50	
	7610	3/4"		11	.727		14.30	31		45.30	64.50	
	7612	1"		10	.800		24	34		58	79	
	7614	1-1/4"	Q-1	17	.941		50.50	36		86.50	112	
	7616	1-1/2"		16	1		55.50	38.50		94	121	
	7618	2"		14	1.143		86	44		130	163	
	7620	2-1/2"		11	1.455		214	56		270	320	
	7622	3"		8	2		238	77		315	380	
	7624	4"	Q-2	11	2.182		395	87		482	570	
	7710	Union										
	7720	1/8"	1 Plum	12	.667	Ea.	34	28.50		62.50	82	
	7730	1/4"		12	.667		34	28.50		62.50	82	
	7740	3/8"		12	.667		39.50	28.50		68	88	
	7750	1/2"		11	.727		37	31		68	89.50	
	7760	3/4"		10	.800		51	34		85	109	
	7770	1"		9	.889		74	38		112	141	

15107 | Metal Pipe & Fittings

			CREW	DAILY OUTPUT	LABOR-HOURS	UNIT	2006 BARE COSTS MAT.	LABOR	EQUIP.	TOTAL	TOTAL INCL O&P	
960	7780	1-1/4"	Q-1	16	1	Ea.	201	38.50		239.50	281	960
	7790	1-1/2"		15	1.067		187	41		228	270	
	7800	2"		13	1.231		236	47.50		283.50	335	
	7810	2-1/2"		10	1.600		595	61.50		656.50	750	
	7820	3"		7	2.286		790	88		878	1,000	
	7830	4"	Q-2	10	2.400		1,100	95.50		1,195.50	1,350	
	7850	For 150 lb., type 316, add					25%					
	8750	Threaded joint, 2000 lb., type 304										
	8770	90° Elbow										
	8780	1/8"	1 Plum	13	.615	Ea.	23	26.50		49.50	66.50	
	8790	1/4"		13	.615		23	26.50		49.50	66.50	
	8800	3/8"		13	.615		28.50	26.50		55	72.50	
	8810	1/2"		12	.667		39	28.50		67.50	87	
	8820	3/4"		11	.727		45	31		76	98	
	8830	1"		10	.800		59.50	34		93.50	119	
	8840	1-1/4"	Q-1	17	.941		97.50	36		133.50	163	
	8850	1-1/2"		16	1		156	38.50		194.50	232	
	8860	2"		14	1.143		219	44		263	310	
	8880	45° Elbow										
	8890	1/8"	1 Plum	13	.615	Ea.	48	26.50		74.50	94	
	8900	1/4"		13	.615		48	26.50		74.50	94	
	8910	3/8"		13	.615		60.50	26.50		87	108	
	8920	1/2"		12	.667		60.50	28.50		89	111	
	8930	3/4"		11	.727		66.50	31		97.50	122	
	8940	1"		10	.800		79.50	34		113.50	140	
	8950	1-1/4"	Q-1	17	.941		139	36		175	208	
	8960	1-1/2"		16	1		186	38.50		224.50	264	
	8970	2"		14	1.143		216	44		260	305	
	8990	Tee, straight										
	9000	1/8"	1 Plum	9	.889	Ea.	29	38		67	91	
	9010	1/4"		9	.889		29	38		67	91	
	9020	3/8"		9	.889		38.50	38		76.50	102	
	9030	1/2"		8	1		49.50	42.50		92	121	
	9040	3/4"		7	1.143		56	49		105	138	
	9050	1"		6.50	1.231		77	52.50		129.50	167	
	9060	1-1/4"	Q-1	11	1.455		133	56		189	234	
	9070	1-1/2"		10	1.600		217	61.50		278.50	335	
	9080	2"		9	1.778		300	68.50		368.50	435	
	9100	For couplings and unions use 3000 lb., type 304										
	9120	2000 lb., type 316										
	9130	90° Elbow										
	9140	1/8"	1 Plum	13	.615	Ea.	27.50	26.50		54	71.50	
	9150	1/4"		13	.615		27.50	26.50		54	71.50	
	9160	3/8"		13	.615		32	26.50		58.50	76.50	
	9170	1/2"		12	.667		42	28.50		70.50	90.50	
	9180	3/4"		11	.727		51	31		82	105	
	9190	1"		10	.800		76	34		110	137	
	9200	1-1/4"	Q-1	17	.941		138	36		174	208	
	9210	1-1/2"		16	1		159	38.50		197.50	235	
	9220	2"		14	1.143		270	44		314	365	
	9240	45° Elbow										
	9250	1/8"	1 Plum	13	.615	Ea.	58.50	26.50		85	105	
	9260	1/4"		13	.615		58.50	26.50		85	105	
	9270	3/8"		13	.615		58.50	26.50		85	105	
	9280	1/2"		12	.667		58.50	28.50		87	109	
	9300	3/4"		11	.727		66	31		97	121	
960												

15107 | Metal Pipe & Fittings

			Crew	Daily Output	Labor-Hours	Unit	2006 Bare Costs Mat.	Labor	Equip.	Total	Total Incl O&P
960	9310	1"	1 Plum	10	.800	Ea.	99.50	34		133.50	162
	9320	1-1/4"	Q-1	17	.941		135	36		171	205
	9330	1-1/2"		16	1		195	38.50		233.50	274
	9340	2"		14	1.143		238	44		282	330
	9360	Tee, straight									
	9370	1/8"	1 Plum	9	.889	Ea.	35	38		73	97.50
	9380	1/4"		9	.889		35	38		73	97.50
	9390	3/8"		9	.889		42.50	38		80.50	106
	9400	1/2"		8	1		53.50	42.50		96	125
	9410	3/4"		7	1.143		66	49		115	149
	9420	1"		6.50	1.231		101	52.50		153.50	193
	9430	1-1/4"	Q-1	11	1.455		163	56		219	267
	9440	1-1/2"		10	1.600		234	61.50		295.50	355
	9450	2"		9	1.778		375	68.50		443.50	515
	9470	For couplings and unions use 3000 lb., type 316									
	9490	3000 lb., type 304									
	9510	Coupling									
	9520	1/8"	1 Plum	19	.421	Ea.	9.15	18		27.15	38
	9530	1/4"		19	.421		9.90	18		27.90	39
	9540	3/8"		19	.421		11.55	18		29.55	40.50
	9550	1/2"		19	.421		13.95	18		31.95	43.50
	9560	3/4"		18	.444		18.60	19		37.60	50
	9570	1"		15	.533		31.50	23		54.50	70
	9580	1-1/4"	Q-1	26	.615		76.50	23.50		100	121
	9590	1-1/2"		24	.667		88.50	25.50		114	138
	9600	2"		21	.762		118	29.50		147.50	175
	9620	Union									
	9630	1/8"	1 Plum	12	.667	Ea.	70	28.50		98.50	122
	9640	1/4"		12	.667		70	28.50		98.50	122
	9650	3/8"		12	.667		75	28.50		103.50	127
	9660	1/2"		11	.727		75	31		106	131
	9670	3/4"		10	.800		91.50	34		125.50	154
	9680	1"		9	.889		141	38		179	214
	9690	1-1/4"	Q-1	16	1		261	38.50		299.50	345
	9700	1-1/2"		15	1.067		294	41		335	390
	9710	2"		13	1.231		395	47.50		442.50	505
	9730	3000 lb., type 316									
	9750	Coupling									
	9770	1/8"	1 Plum	19	.421	Ea.	9.85	18		27.85	39
	9780	1/4"		19	.421		11.10	18		29.10	40
	9790	3/8"		19	.421		11.90	18		29.90	41
	9800	1/2"		19	.421		16.85	18		34.85	46.50
	9810	3/4"		18	.444		25	19		44	57
	9820	1"		15	.533		41.50	23		64.50	81
	9830	1-1/4"	Q-1	26	.615		93	23.50		116.50	139
	9840	1-1/2"		24	.667		108	25.50		133.50	159
	9850	2"		21	.762		150	29.50		179.50	211
	9870	Union									
	9880	1/8"	1 Plum	12	.667	Ea.	73.50	28.50		102	126
	9890	1/4"		12	.667		73.50	28.50		102	126
	9900	3/8"		12	.667		85	28.50		113.50	138
	9910	1/2"		11	.727		85	31		116	142
	9920	3/4"		10	.800		111	34		145	175
	9930	1"		9	.889		175	38		213	252
	9940	1-1/4"	Q-1	16	1		305	38.50		343.50	395
	9950	1-1/2"		15	1.067		380	41		421	480

MECHANICAL 15

	15107 | Metal Pipe & Fittings		Crew	Daily Output	Labor-Hours	Unit	2006 Bare Costs Mat.	Labor	Equip.	Total	Total Incl O&P	
960	9960	2″	Q-1	13	1.231	Ea.	480	47.50		527.50	605	960

	15108 | Plastic Pipe & Fittings		Crew	Daily Output	Labor-Hours	Unit	2006 Bare Costs Mat.	Labor	Equip.	Total	Total Incl O&P	
520	0010	**PIPE, PLASTIC**										520
	0020	Fiberglass reinforced, couplings 10′ O.C., hangers 3 per 10′										
	0080	General service										
	0120	2″ diameter	Q-1	59	.271	L.F.	5.10	10.40		15.50	22	
	0140	3″ diameter		52	.308		6.95	11.80		18.75	26	
	0150	4″ diameter		48	.333		9.30	12.80		22.10	30	
	0160	6″ diameter		39	.410		17.10	15.75		32.85	43.50	
	0170	8″ diameter	Q-2	49	.490		28	19.50		47.50	61.50	
	0180	10″ diameter		41	.585		48.50	23.50		72	89.50	
	0190	12″ diameter		36	.667		62.50	26.50		89	110	
	0200	High strength										
	0240	2″ diameter	Q-1	58	.276	L.F.	9.10	10.60		19.70	26.50	
	0260	3″ diameter		51	.314		13.10	12.05		25.15	33	
	0280	4″ diameter		47	.340		18.55	13.10		31.65	41	
	0300	6″ diameter		38	.421		30.50	16.20		46.70	59	
	0320	8″ diameter	Q-2	48	.500		44.50	19.95		64.45	79.50	
	0340	10″ diameter		40	.600		73.50	24		97.50	118	
	0360	12″ diameter		36	.667		85	26.50		111.50	135	
	0550	To delete coupling & hangers, subtract										
	0560	2″ diam. to 6″ diam.					33%	56%				
	0570	8″ diam. to 12″ diam.					31%	52%				
	0600	PVC, high impact/pressure, cplgs. 10′ O.C., hangers 3 per 10′										
	1020	Schedule 80										
	1040	1/4″ diameter	1 Plum	58	.138	L.F.	1.56	5.90		7.46	10.85	
	1060	3/8″ diameter		55	.145		1.75	6.20		7.95	11.55	
	1070	1/2″ diameter		50	.160		1.44	6.85		8.29	12.20	
	1080	3/4″ diameter		47	.170		1.63	7.25		8.88	13.10	
	1090	1″ diameter		43	.186		1.97	7.95		9.92	14.50	
	1100	1-1/4″ diameter		39	.205		2.40	8.75		11.15	16.25	
	1110	1-1/2″ diameter		34	.235		2.68	10.05		12.73	18.55	
	1120	2″ diameter	Q-1	55	.291		3.28	11.20		14.48	21	
	1130	2-1/2″ diameter		52	.308		4.67	11.80		16.47	23.50	
	1140	3″ diameter		50	.320		6.30	12.30		18.60	26	
	1150	4″ diameter		46	.348		8.55	13.35		21.90	30.50	
	1160	5″ diameter		42	.381		11.80	14.65		26.45	36	
	1170	6″ diameter		38	.421		15.85	16.20		32.05	42.50	
	1180	8″ diameter	Q-2	47	.511		24	20.50		44.50	58	
	1190	10″ diameter		42	.571		40.50	23		63.50	80	
	1200	12″ diameter		38	.632		52	25		77	96.50	
	1730	To delete coupling & hangers, subtract										
	1740	1/4″ diam. to 1/2″ diam.					62%	80%				
	1750	3/4″ diam. to 1-1/4″ diam.					58%	73%				
	1760	1-1/2″ diam. to 6″ diam.					40%	57%				
	1770	8″ diam. to 12″ diam.					34%	50%				
	1800	PVC, couplings 10′ O.C., hangers 3 per 10′										
	1820	Schedule 40										
	1860	1/2″ diameter	1 Plum	54	.148	L.F.	1.14	6.35		7.49	11.10	
	1870	3/4″ diameter		51	.157		1.23	6.70		7.93	11.75	
	1880	1″ diameter		46	.174		1.38	7.45		8.83	13.05	
	1890	1-1/4″ diameter		42	.190		1.57	8.15		9.72	14.40	
	1900	1-1/2″ diameter		36	.222		1.69	9.50		11.19	16.60	
	1910	2″ diameter	Q-1	59	.271		1.98	10.40		12.38	18.40	

15108 | Plastic Pipe & Fittings

520

		CREW	DAILY OUTPUT	LABOR-HOURS	UNIT	2006 BARE COSTS MAT.	LABOR	EQUIP.	TOTAL	TOTAL INCL O&P
1920	2-1/2″ diameter	Q-1	56	.286	L.F.	2.70	11		13.70	20
1930	3″ diameter		53	.302		3.47	11.60		15.07	22
1940	4″ diameter		48	.333		4.39	12.80		17.19	24.50
1950	5″ diameter		43	.372		5.75	14.30		20.05	28.50
1960	6″ diameter		39	.410		7.65	15.75		23.40	33
1970	8″ diameter	Q-2	48	.500		11.40	19.95		31.35	43.50
1980	10″ diameter		43	.558		27.50	22		49.50	65
1990	12″ diameter		42	.571		35.50	23		58.50	75
2000	14″ diameter		31	.774		57.50	31		88.50	112
2010	16″ diameter		23	1.043		78.50	41.50		120	151
2340	To delete coupling & hangers, subtract									
2360	1/2″ diam. to 1-1/4″ diam.					65%	74%			
2370	1-1/2″ diam. to 6″ diam.					44%	57%			
2380	8″ diam. to 12″ diam.					41%	53%			
2390	14″ diam. to 16″ diam.					48%	45%			
2420	Schedule 80									
2440	1/4″ diameter	1 Plum	58	.138	L.F.	1.01	5.90		6.91	10.25
2450	3/8″ diameter		55	.145		1.01	6.20		7.21	10.75
2460	1/2″ diameter		50	.160		1.18	6.85		8.03	11.90
2470	3/4″ diameter		47	.170		1.29	7.25		8.54	12.70
2480	1″ diameter		43	.186		1.47	7.95		9.42	13.95
2490	1-1/4″ diameter		39	.205		1.70	8.75		10.45	15.45
2500	1-1/2″ diameter		34	.235		1.85	10.05		11.90	17.65
2510	2″ diameter	Q-1	55	.291		2.14	11.20		13.34	19.75
2520	2-1/2″ diameter		52	.308		2.83	11.80		14.63	21.50
2530	3″ diameter		50	.320		3.95	12.30		16.25	23.50
2540	4″ diameter		46	.348		5.15	13.35		18.50	26.50
2550	5″ diameter		42	.381		8	14.65		22.65	32
2560	6″ diameter		38	.421		9.30	16.20		25.50	35
2570	8″ diameter	Q-2	47	.511		13.95	20.50		34.45	47
2580	10″ diameter		42	.571		33	23		56	72
2590	12″ diameter		38	.632		42.50	25		67.50	85.50
2830	To delete coupling & hangers, subtract									
2840	1/4″ diam. to 1/2″ diam.					66%	80%			
2850	3/4″ diam. to 1-1/4″ diam.					61%	73%			
2860	1-1/2″ diam. to 6″ diam.					41%	57%			
2870	8″ diam. to 12″ diam.					31%	50%			
2900	Schedule 120									
2910	1/2″ diameter	1 Plum	50	.160	L.F.	1.57	6.85		8.42	12.30
2950	3/4″ diameter		47	.170		1.79	7.25		9.04	13.25
2960	1″ diameter		43	.186		2.21	7.95		10.16	14.80
2970	1-1/4″ diameter		39	.205		2.74	8.75		11.49	16.60
2980	1-1/2″ diameter		33	.242		3.10	10.35		13.45	19.50
2990	2″ diameter	Q-1	54	.296		3.91	11.40		15.31	22
3000	2-1/2″ diameter		52	.308		5.90	11.80		17.70	25
3010	3″ diameter		49	.327		7.55	12.55		20.10	28
3020	4″ diameter		45	.356		11.20	13.65		24.85	33.50
3030	6″ diameter		37	.432		20.50	16.60		37.10	49
3240	To delete coupling & hangers, subtract									
3250	1/2″ diam. to 1-1/4″ diam.					52%	74%			
3260	1-1/2″ diam. to 4″ diam.					30%	57%			
3270	6″ diam.					17%	50%			
3300	PVC, pressure, couplings 10′ O.C., hangers 3 per 10′									
3310	SDR 26, 160 psi									
3350	1-1/4″ diameter	1 Plum	42	.190	L.F.	1.40	8.15		9.55	14.20
3360	1-1/2″ diameter	″	36	.222		1.50	9.50		11	16.40

	15108	Plastic Pipe & Fittings	CREW	DAILY OUTPUT	LABOR-HOURS	UNIT	2006 BARE COSTS MAT.	LABOR	EQUIP.	TOTAL	TOTAL INCL O&P	
520	3370	2" diameter	Q-1	59	.271	L.F.	1.67	10.40		12.07	18.05	520
	3380	2-1/2" diameter		56	.286		2.33	11		13.33	19.60	
	3390	3" diameter		53	.302		3.07	11.60		14.67	21.50	
	3400	4" diameter		48	.333		3.99	12.80		16.79	24.50	
	3420	6" diameter		39	.410		7.40	15.75		23.15	32.50	
	3430	8" diameter	Q-2	48	.500		11.40	19.95		31.35	43.50	
	3660	To delete coupling & hangers, subtract										
	3670	1-1/4" diam.					63%	68%				
	3680	1-1/2" diam. to 4" diam.					48%	57%				
	3690	6" diam. to 8" diam.					60%	54%				
	3720	SDR 21, 200 psi, 1/2" diameter	1 Plum	54	.148	L.F.	1.10	6.35		7.45	11.05	
	3740	3/4" diameter		51	.157		1.15	6.70		7.85	11.65	
	3750	1" diameter		46	.174		1.26	7.45		8.71	12.95	
	3760	1-1/4" diameter		42	.190		1.43	8.15		9.58	14.20	
	3770	1-1/2" diameter		36	.222		1.54	9.50		11.04	16.45	
	3780	2" diameter	Q-1	59	.271		1.75	10.40		12.15	18.10	
	3790	2-1/2" diameter		56	.286		2.85	11		13.85	20	
	3800	3" diameter		53	.302		3.23	11.60		14.83	21.50	
	3810	4" diameter		48	.333		4.25	12.80		17.05	24.50	
	3830	6" diameter		39	.410		7.95	15.75		23.70	33.50	
	3840	8" diameter	Q-2	48	.500		12.55	19.95		32.50	45	
	4000	To delete coupling & hangers, subtract										
	4010	1/2" diam. to 3/4" diam.					71%	77%				
	4020	1" diam. to 1-1/4" diam.					63%	70%				
	4030	1-1/2" diam. to 6" diam.					44%	57%				
	4040	8" diam.					46%	54%				
	4100	DWV type, schedule 40, couplings 10' O.C., hangers 3 per 10'										
	4120	ABS										
	4140	1-1/4" diameter	1 Plum	42	.190	L.F.	1.26	8.15		9.41	14.05	
	4150	1-1/2" diameter	"	36	.222		1.29	9.50		10.79	16.15	
	4160	2" diameter	Q-1	59	.271		1.42	10.40		11.82	17.75	
	4170	3" diameter		53	.302		2.64	11.60		14.24	21	
	4180	4" diameter		48	.333		3.49	12.80		16.29	23.50	
	4190	6" diameter		39	.410		7.95	15.75		23.70	33.50	
	4360	To delete coupling & hangers, subtract										
	4370	1-1/4" diam.					64%	68%				
	4380	1-1/2" diam. to 6" diam.					54%	57%				
	4400	PVC										
	4410	1-1/4" diameter	1 Plum	42	.190	L.F.	1.37	8.15		9.52	14.15	
	4420	1-1/2" diameter	"	36	.222		1.40	9.50		10.90	16.30	
	4460	2" diameter	Q-1	59	.271		1.58	10.40		11.98	17.95	
	4470	3" diameter		53	.302		2.96	11.60		14.56	21.50	
	4480	4" diameter		48	.333		3.84	12.80		16.64	24	
	4490	6" diameter		39	.410		7.10	15.75		22.85	32.50	
	4500	8" diameter	Q-2	48	.500		16.75	19.95		36.70	49.50	
	4750	To delete coupling & hangers, subtract										
	4760	1-1/4" diam. to 1-1/2" diam.					71%	64%				
	4770	2" diam. to 8" diam.					60%	57%				
	4800	PVC, clear pipe, cplgs. 10' O.C., hangers 3 per 10', Sched. 40										
	4840	1/4" diameter	1 Plum	59	.136	L.F.	1.50	5.80		7.30	10.65	
	4850	3/8" diameter		56	.143		1.70	6.10		7.80	11.35	
	4860	1/2" diameter		54	.148		2.10	6.35		8.45	12.15	
	4870	3/4" diameter		51	.157		2.52	6.70		9.22	13.15	
	4880	1" diameter		46	.174		3.58	7.45		11.03	15.50	
	4890	1-1/4" diameter		42	.190		4.25	8.15		12.40	17.35	
	4900	1-1/2" diameter		36	.222		4.89	9.50		14.39	20	

15108 | Plastic Pipe & Fittings

520

Line	Description	Crew	Daily Output	Labor-Hours	Unit	2006 Bare Costs Mat.	2006 Bare Costs Labor	2006 Bare Costs Equip.	2006 Bare Costs Total	Total Incl O&P
4910	2″ diameter	Q-1	59	.271	L.F.	6.30	10.40		16.70	23
4920	2-1/2″ diameter		56	.286		9.75	11		20.75	28
4930	3″ diameter		53	.302		12.50	11.60		24.10	32
4940	3-1/2″ diameter		50	.320		16.85	12.30		29.15	37.50
4950	4″ diameter		48	.333		19.70	12.80		32.50	41.50
5250	To delete coupling & hangers, subtract									
5260	1/4″ diam. to 3/8″ diam.					60%	81%			
5270	1/2″ diam. to 3/4″ diam.					41%	77%			
5280	1″ diam. to 1-1/2″ diam.					26%	67%			
5290	2″ diam. to 4″ diam.					16%	58%			
5360	CPVC, couplings 10′ O.C., hangers 3 per 10′									
5380	Schedule 40									
5460	1/2″ diameter	1 Plum	54	.148	L.F.	2.53	6.35		8.88	12.65
5470	3/4″ diameter		51	.157		3.32	6.70		10.02	14.05
5480	1″ diameter		46	.174		4.07	7.45		11.52	16.05
5490	1-1/4″ diameter		42	.190		4.71	8.15		12.86	17.85
5500	1-1/2″ diameter		36	.222		5.25	9.50		14.75	20.50
5510	2″ diameter	Q-1	59	.271		6.50	10.40		16.90	23.50
5520	2-1/2″ diameter		56	.286		10.45	11		21.45	28.50
5530	3″ diameter		53	.302		12.75	11.60		24.35	32
5540	4″ diameter		48	.333		20.50	12.80		33.30	42.50
5550	6″ diameter		43	.372		33	14.30		47.30	58
5730	To delete coupling & hangers, subtract									
5740	1/2″ diam. to 3/4″ diam.					37%	77%			
5750	1″ diam. to 1-1/4″ diam.					27%	70%			
5760	1-1/2″ diam. to 3″ diam.					21%	57%			
5770	4″ diam. to 6″ diam.					16%	57%			
5800	Schedule 80									
5860	1/2″ diameter	1 Plum	50	.160	L.F.	2.97	6.85		9.82	13.85
5870	3/4″ diameter		47	.170		3.53	7.25		10.78	15.20
5880	1″ diameter		43	.186		4.37	7.95		12.32	17.15
5890	1-1/4″ diameter		39	.205		5.15	8.75		13.90	19.30
5900	1-1/2″ diameter		34	.235		5.90	10.05		15.95	22
5910	2″ diameter	Q-1	55	.291		7.40	11.20		18.60	25.50
5920	2-1/2″ diameter		52	.308		11.70	11.80		23.50	31.50
5930	3″ diameter		50	.320		14.60	12.30		26.90	35
5940	4″ diameter		46	.348		23.50	13.35		36.85	47
5950	6″ diameter		38	.421		39.50	16.20		55.70	68.50
5960	8″ diameter	Q-2	47	.511		72.50	20.50		93	111
6060	To delete couplings & hangers, subtract									
6070	1/2″ diam. to 3/4″ diam.					44%	77%			
6080	1″ diam. to 1-1/4″ diam.					32%	71%			
6090	1-1/2″ diam. to 4″ diam.					25%	58%			
6100	6″ diam. to 8″ diam.					20%	53%			
6240	CTS, 1/2″ diameter	1 Plum	54	.148	L.F.	1.46	6.35		7.81	11.45
6250	3/4″ diameter		51	.157		1.98	6.70		8.68	12.60
6260	1″ diameter		46	.174		3.15	7.45		10.60	15
6270	1 1/4″		42	.190		4.23	8.15		12.38	17.30
6280	1 1/2″ diameter		36	.222		5.35	9.50		14.85	20.50
6290	2″ diameter	Q-1	59	.271		8.45	10.40		18.85	25.50
6370	To delete coupling & hangers, subtract									
6380	1/2″ diam.					51%	79%			
6390	3/4″ diam.					40%	76%			
6392	1″ thru 2″ diam.					72%	68%			
6500	Residential installation, plastic pipe									
6510	Couplings 10′ O.C., strap hangers 3 per 10′									

15108 | Plastic Pipe & Fittings

			CREW	DAILY OUTPUT	LABOR-HOURS	UNIT	2006 BARE COSTS				TOTAL INCL O&P
							MAT.	LABOR	EQUIP.	TOTAL	
520	6520	PVC, Schedule 40									
	6530	1/2″ diameter	1 Plum	138	.058	L.F.	.30	2.48		2.78	4.18
	6540	3/4″ diameter		128	.063		.36	2.67		3.03	4.55
	6550	1″ diameter		119	.067		.47	2.87		3.34	4.97
	6560	1-1/4″ diameter		111	.072		.59	3.08		3.67	5.45
	6570	1-1/2″ diameter		104	.077		.69	3.28		3.97	5.85
	6580	2″ diameter	Q-1	197	.081		.88	3.12		4	5.80
	6590	2-1/2″ diameter		162	.099		1.46	3.80		5.26	7.50
	6600	4″ diameter		123	.130		7.75	5		12.75	16.30
	6700	PVC, DWV, Schedule 40									
	6720	1-1/4″ diameter	1 Plum	100	.080	L.F.	.84	3.42		4.26	6.20
	6730	1-1/2″ diameter	"	94	.085		.69	3.63		4.32	6.40
	6740	2″ diameter	Q-1	178	.090		.87	3.45		4.32	6.30
	6760	4″ diameter	"	110	.145		7.75	5.60		13.35	17.20
	7280	Polyethylene, flexible, no couplings or hangers									
	7282	Note: For labor costs add 25% to the couplings and fittings labor total.									
	7300	SDR 15, 100 psi									
	7310	3/4″ diameter				L.F.	.24			.24	.26
	7350	1″ diameter					.32			.32	.35
	7360	1-1/4″ diameter					.57			.57	.63
	7370	1-1/2″ diameter					.69			.69	.76
	7380	2″ diameter					1.13			1.13	1.24
	7700	SDR 9, 160 psi									
	7710	1/2″ diameter				L.F.	.30			.30	.33
	7750	3/4″ diameter					.31			.31	.34
	7760	1″ diameter					.48			.48	.53
	7770	1-1/4″ diameter					.84			.84	.92
	7780	1-1/2″ diameter					1.17			1.17	1.29
	7790	2″ diameter					1.94			1.94	2.13
	8120	SDR 9, 200 psi									
	8150	3/4″ diameter				L.F.	.28			.28	.31
	8160	1″ diameter					.46			.46	.51
	8170	1-1/4″ diameter					.69			.69	.76
	8420	SDR 7, 250 psi									
	8440	3/4″ diameter				L.F.	.40			.40	.44
	8450	1″ diameter					.64			.64	.70
	8460	1-1/4″ diameter					1.13			1.13	1.24
	9900	Minimum labor/equipment charge	1 Plum	4	2	Job		85.50		85.50	133
560	0010	**PIPE, PLASTIC, FITTINGS**									
	0030	Epoxy resin, fiberglass reinforced, general service									
	0090	Elbow, 90°, 2″	Q-1	33.10	.483	Ea.	67.50	18.60		86.10	104
	0100	3″		20.80	.769		78	29.50		107.50	132
	0110	4″		16.50	.970		106	37.50		143.50	175
	0120	6″		10.10	1.584		155	61		216	265
	0130	8″	Q-2	9.30	2.581		286	103		389	475
	0140	10″		8.50	2.824		360	113		473	570
	0150	12″		7.60	3.158		515	126		641	760
	0160	45° Elbow, same as 90°									
	0290	Tee, 2″	Q-1	20	.800	Ea.	37	30.50		67.50	88.50
	0300	3″		13.90	1.151		44	44		88	118
	0310	4″		11	1.455		61.50	56		117.50	155
	0320	6″		6.70	2.388		172	92		264	330
	0330	8″	Q-2	6.20	3.871		1,100	154		1,254	1,450
	0340	10″		5.70	4.211		1,250	168		1,418	1,625
	0350	12″		5.10	4.706		1,525	188		1,713	1,975
	0380	Couplings									

15 MECHANICAL

15108 | Plastic Pipe & Fittings

			Crew	Daily Output	Labor-Hours	Unit	2006 Bare Costs Mat.	Labor	Equip.	Total	Total Incl O&P	
560	0410	2"	Q-1	33.10	.483	Ea.	11.85	18.60		30.45	42	560
	0420	3"		20.80	.769		13.30	29.50		42.80	60.50	
	0430	4"		16.50	.970		18.35	37.50		55.85	78	
	0440	6"		10.10	1.584		43.50	61		104.50	142	
	0450	8"	Q-2	9.30	2.581		74	103		177	242	
	0460	10"		8.50	2.824		108	113		221	294	
	0470	12"		7.60	3.158		146	126		272	355	
	0473	High corrosion resistant couplings, add					30%					
	2100	PVC schedule 80, socket joint										
	2110	90° elbow, 1/2"	1 Plum	30.30	.264	Ea.	.71	11.25		11.96	18.35	
	2130	3/4"		26	.308		.90	13.15		14.05	21.50	
	2140	1"		22.70	.352		1.46	15.05		16.51	25	
	2150	1-1/4"		20.20	.396		1.94	16.90		18.84	28.50	
	2160	1-1/2"		18.20	.440		2.08	18.75		20.83	31.50	
	2170	2"	Q-1	33.10	.483		2.50	18.60		21.10	32	
	2180	3"		20.80	.769		6.65	29.50		36.15	53.50	
	2190	4"		16.50	.970		7.50	37.50		45	66.50	
	2200	6"		10.10	1.584		67	61		128	168	
	2210	8"	Q-2	9.30	2.581		185	103		288	365	
	2250	45° elbow, 1/2"	1 Plum	30.30	.264		1.34	11.25		12.59	19	
	2270	3/4"		26	.308		2.03	13.15		15.18	22.50	
	2280	1"		22.70	.352		3.01	15.05		18.06	27	
	2290	1-1/4"		20.20	.396		3.87	16.90		20.77	31	
	2300	1-1/2"		18.20	.440		4.57	18.75		23.32	34	
	2310	2"	Q-1	33.10	.483		5.95	18.60		24.55	35.50	
	2320	3"		20.80	.769		15.10	29.50		44.60	62.50	
	2330	4"		16.50	.970		27	37.50		64.50	88	
	2340	6"		10.10	1.584		34.50	61		95.50	133	
	2350	8"	Q-2	9.30	2.581		175	103		278	350	
	2400	Tee, 1/2"	1 Plum	20.20	.396		1.98	16.90		18.88	28.50	
	2420	3/4"		17.30	.462		2.08	19.75		21.83	33	
	2430	1"		15.20	.526		2.59	22.50		25.09	38	
	2440	1-1/4"		13.50	.593		7.15	25.50		32.65	47.50	
	2450	1-1/2"		12.10	.661		7.15	28		35.15	52	
	2460	2"	Q-1	20	.800		8.90	30.50		39.40	58	
	2470	3"		13.90	1.151		12.15	44		56.15	82.50	
	2480	4"		11	1.455		14.05	56		70.05	102	
	2490	6"		6.70	2.388		47.50	92		139.50	196	
	2500	8"	Q-2	6.20	3.871		215	154		369	475	
	2510	Flange, socket, 150 lb., 1/2"	1 Plum	55.60	.144		3.86	6.15		10.01	13.80	
	2514	3/4"		47.60	.168		4.13	7.20		11.33	15.70	
	2518	1"		41.70	.192		4.59	8.20		12.79	17.80	
	2522	1-1/2"		33.30	.240		4.82	10.25		15.07	21.50	
	2526	2"	Q-1	60.60	.264		6.45	10.15		16.60	23	
	2530	4"		30.30	.528		13.85	20.50		34.35	46.50	
	2534	6"		18.50	.865		21.50	33		54.50	75.50	
	2538	8"	Q-2	17.10	1.404		107	56		163	204	
	2550	Coupling, 1/2"	1 Plum	30.30	.264		1.29	11.25		12.54	18.95	
	2570	3/4"		26	.308		1.73	13.15		14.88	22.50	
	2580	1"		22.70	.352		1.77	15.05		16.82	25.50	
	2590	1-1/4"		20.20	.396		2.69	16.90		19.59	29.50	
	2600	1-1/2"		18.20	.440		2.89	18.75		21.64	32	
	2610	2"	Q-1	33.10	.483		3.10	18.60		21.70	32.50	
	2620	3"		20.80	.769		8.80	29.50		38.30	55.50	
	2630	4"		16.50	.970		11.10	37.50		48.60	70	
	2640	6"		10.10	1.584		23.50	61		84.50	121	

15108 | Plastic Pipe & Fittings

560

		CREW	DAILY OUTPUT	LABOR-HOURS	UNIT	MAT.	LABOR	EQUIP.	TOTAL	TOTAL INCL O&P
						2006 BARE COSTS				
2650	8"	Q-2	9.30	2.581	Ea.	37.50	103		140.50	201
2660	10"		8.50	2.824		40.50	113		153.50	220
2670	12"		7.60	3.158		47	126		173	248
2700	PVC (white), schedule 40, socket joints									
2760	90° elbow, 1/2"	1 Plum	33.30	.240	Ea.	.33	10.25		10.58	16.30
2770	3/4"		28.60	.280		.37	11.95		12.32	18.95
2780	1"		25	.320		.66	13.65		14.31	22
2790	1-1/4"		22.20	.360		1.17	15.40		16.57	25.50
2800	1-1/2"		20	.400		1.25	17.10		18.35	28
2810	2"	Q-1	36.40	.440		1.97	16.90		18.87	28.50
2820	2-1/2"		26.70	.599		5.95	23		28.95	42.50
2830	3"		22.90	.699		7.15	27		34.15	50
2840	4"		18.20	.879		12.80	34		46.80	66.50
2850	5"		12.10	1.322		33	51		84	116
2860	6"		11.10	1.441		40.50	55.50		96	131
2870	8"	Q-2	10.30	2.330		89	93		182	242
2980	45° elbow, 1/2"	1 Plum	33.30	.240		.54	10.25		10.79	16.55
2990	3/4"		28.60	.280		.84	11.95		12.79	19.45
3000	1"		25	.320		1.01	13.65		14.66	22.50
3010	1-1/4"		22.20	.360		1.40	15.40		16.80	25.50
3020	1-1/2"		20	.400		1.76	17.10		18.86	28.50
3030	2"	Q-1	36.40	.440		2.30	16.90		19.20	29
3040	2-1/2"		26.70	.599		6	23		29	42.50
3050	3"		22.90	.699		8.65	27		35.65	51.50
3060	4"		18.20	.879		16.65	34		50.65	71
3070	5"		12.10	1.322		33	51		84	116
3080	6"		11.10	1.441		42	55.50		97.50	133
3090	8"	Q-2	10.30	2.330		132	93		225	289
3180	Tee, 1/2"	1 Plum	22.20	.360		.41	15.40		15.81	24.50
3190	3/4"		19	.421		.47	18		18.47	28.50
3200	1"		16.70	.479		.88	20.50		21.38	33
3210	1-1/4"		14.80	.541		1.37	23		24.37	37.50
3220	1-1/2"		13.30	.602		1.67	25.50		27.17	42
3230	2"	Q-1	24.20	.661		2.42	25.50		27.92	42
3240	2-1/2"		17.80	.899		8	34.50		42.50	62.50
3250	3"		15.20	1.053		10.50	40.50		51	74.50
3260	4"		12.10	1.322		19	51		70	100
3270	5"		8.10	1.975		46	76		122	169
3280	6"		7.40	2.162		64	83		147	200
3290	8"	Q-2	6.80	3.529		148	141		289	380
3380	Coupling, 1/2"	1 Plum	33.30	.240		.22	10.25		10.47	16.20
3390	3/4"		28.60	.280		.30	11.95		12.25	18.90
3400	1"		25	.320		.52	13.65		14.17	22
3410	1-1/4"		22.20	.360		.71	15.40		16.11	25
3420	1-1/2"		20	.400		.77	17.10		17.87	27.50
3430	2"	Q-1	36.40	.440		1.18	16.90		18.08	28
3440	2-1/2"		26.70	.599		2.60	23		25.60	39
3450	3"		22.90	.699		4.06	27		31.06	46.50
3460	4"		18.20	.879		5.85	34		39.85	59
3470	5"		12.10	1.322		14.35	51		65.35	95
3480	6"		11.10	1.441		18.55	55.50		74.05	107
3490	8"	Q-2	10.30	2.330		46	93		139	195
4500	DWV, ABS, non pressure, socket joints									
4540	1/4 Bend, 1-1/4"	1 Plum	20.20	.396	Ea.	3.15	16.90		20.05	30
4560	1-1/2"	"	18.20	.440		2.41	18.75		21.16	31.50
4570	2"	Q-1	33.10	.483		3.73	18.60		22.33	33

15108 | Plastic Pipe & Fittings

			CREW	DAILY OUTPUT	LABOR-HOURS	UNIT	2006 BARE COSTS MAT.	LABOR	EQUIP.	TOTAL	TOTAL INCL O&P
560	4580	3"	Q-1	20.80	.769	Ea.	8.95	29.50		38.45	56
	4590	4"		16.50	.970		15.10	37.50		52.60	74.50
	4600	6"		10.10	1.584		71.50	61		132.50	174
	4650	1/8 Bend, same as 1/4 Bend									
	4800	Tee, sanitary									
	4820	1-1/4"	1 Plum	13.50	.593	Ea.	3.57	25.50		29.07	43.50
	4830	1-1/2"	"	12.10	.661		3.29	28		31.29	47.50
	4840	2"	Q-1	20	.800		4.81	30.50		35.31	53.50
	4850	3"		13.90	1.151		12.15	44		56.15	82.50
	4860	4"		11	1.455		22.50	56		78.50	112
	4862	Tee, sanitary, reducing, 2" x 1-1/2"		22	.727		4.64	28		32.64	48.50
	4864	3" x 2"		15.30	1.046		8.85	40		48.85	72
	4868	4" x 3"		12.10	1.322		22	51		73	104
	4870	Combination Y and 1/8 bend									
	4872	1-1/2"	1 Plum	12.10	.661	Ea.	6.70	28		34.70	51.50
	4874	2"	Q-1	20	.800		7.75	30.50		38.25	56.50
	4876	3"		13.90	1.151		19.05	44		63.05	90
	4878	4"		11	1.455		38.50	56		94.50	129
	4880	3" x 1-1/2"		15.50	1.032		16.70	39.50		56.20	80
	4882	4" x 3"		12.10	1.322		28	51		79	110
	4900	Wye, 1-1/4"	1 Plum	13.50	.593		3.93	25.50		29.43	44
	4902	1-1/2"	"	12.10	.661		4.21	28		32.21	48.50
	4904	2"	Q-1	20	.800		5.95	30.50		36.45	54.50
	4906	3"		13.90	1.151		14.70	44		58.70	85
	4908	4"		11	1.455		29	56		85	119
	4910	6"		6.70	2.388		89	92		181	241
	4918	3" x 1-1/2"		15.50	1.032		9.80	39.50		49.30	72.50
	4920	4" x 3"		12.10	1.322		21	51		72	102
	4922	6" x 4"		6.90	2.319		73.50	89		162.50	220
	4930	Double Wye, 1-1/2"	1 Plum	9.10	.879		9.60	37.50		47.10	69
	4932	2"	Q-1	16.60	.964		11.45	37		48.45	70
	4934	3"		10.40	1.538		29.50	59		88.50	125
	4936	4"		8.25	1.939		60	74.50		134.50	182
	4940	2" x 1-1/2"		16.80	.952		11.45	36.50		47.95	69.50
	4942	3" x 2"		10.60	1.509		22	58		80	114
	4944	4" x 3"		8.45	1.893		47.50	73		120.50	166
	4946	6" x 4"		7.25	2.207		94.50	85		179.50	236
	4950	Reducer bushing, 2" x 1-1/2"		36.40	.440		1.58	16.90		18.48	28
	4952	3" x 1-1/2"		27.30	.586		5.95	22.50		28.45	41.50
	4954	4" x 2"		18.20	.879		12.20	34		46.20	66
	4956	6" x 4"		11.10	1.441		32.50	55.50		88	122
	4960	Couplings, 1-1/2"	1 Plum	18.20	.440		1.01	18.75		19.76	30
	4962	2"	Q-1	33.10	.483		1.50	18.60		20.10	30.50
	4963	3"		20.80	.769		4.35	29.50		33.85	51
	4964	4"		16.50	.970		6.80	37.50		44.30	65.50
	4966	6"		10.10	1.584		26.50	61		87.50	124
	4970	2" x 1-1/2"		33.30	.480		2.48	18.45		20.93	31
	4972	3" x 1-1/2"		21	.762		6.95	29.50		36.45	53
	4974	4" x 3"		16.70	.958		12.35	37		49.35	71
	4978	Closet flange, 4"	1 Plum	32	.250		6.40	10.70		17.10	23.50
	4980	4" x 3"	"	34	.235		6.60	10.05		16.65	23
	5000	DWV, PVC, schedule 40, socket joints									
	5040	1/4 bend, 1-1/4"	1 Plum	20.20	.396	Ea.	4.50	16.90		21.40	31.50
	5060	1-1/2"	"	18.20	.440		1.71	18.75		20.46	31
	5070	2"	Q-1	33.10	.483		2.52	18.60		21.12	32
	5080	3"		20.80	.769		7.20	29.50		36.70	54

15108 | Plastic Pipe & Fittings

				DAILY	LABOR-		2006 BARE COSTS				TOTAL
			CREW	OUTPUT	HOURS	UNIT	MAT.	LABOR	EQUIP.	TOTAL	INCL O&P
560	5090	4″	Q-1	16.50	.970	Ea.	12.50	37.50		50	72
	5100	6″		10.10	1.584		47.50	61		108.50	147
	5105	8″	Q-2	9.30	2.581		96.50	103		199.50	266
	5110	1/4 bend, long sweep, 1-1/2″	1 Plum	18.20	.440		3.74	18.75		22.49	33
	5112	2″	Q-1	33.10	.483		4.03	18.60		22.63	33.50
	5114	3″		20.80	.769		9.50	29.50		39	56.50
	5116	4″		16.50	.970		17.70	37.50		55.20	77.50
	5150	1/8 bend, 1-1/4″	1 Plum	20.20	.396		2.69	16.90		19.59	29.50
	5170	1-1/2″	″	18.20	.440		1.57	18.75		20.32	30.50
	5180	2″	Q-1	33.10	.483		2.43	18.60		21.03	31.50
	5190	3″		20.80	.769		6.65	29.50		36.15	53.50
	5200	4″		16.50	.970		11.10	37.50		48.60	70.50
	5210	6″		10.10	1.584		46	61		107	145
	5215	8″	Q-2	9.30	2.581		96	103		199	266
	5250	Tee, sanitary 1-1/4″	1 Plum	13.50	.593		4.12	25.50		29.62	44
	5254	1-1/2″	″	12.10	.661		2.95	28		30.95	47.50
	5255	2″	Q-1	20	.800		4.14	30.50		34.64	52.50
	5256	3″		13.90	1.151		9.75	44		53.75	80
	5257	4″		11	1.455		16.95	56		72.95	106
	5259	6″		6.70	2.388		91	92		183	243
	5261	8″	Q-2	6.20	3.871		190	154		344	450
	5264	2″ x 1-1/2″	Q-1	22	.727		4.67	28		32.67	48.50
	5266	3″ x 1-1/2″		15.50	1.032		5.90	39.50		45.40	68
	5268	4″ x 3″		12.10	1.322		22.50	51		73.50	104
	5271	6″ x 4″		6.90	2.319		76	89		165	223
	5314	Combination Y & 1/8 bend, 1-1/2″	1 Plum	12.10	.661		5.35	28		33.35	50
	5315	2″	Q-1	20	.800		7.15	30.50		37.65	56
	5317	3″		13.90	1.151		13	44		57	83.50
	5318	4″		11	1.455		22	56		78	112
	5324	Combination Y & 1/8 bend, reducing									
	5325	2″ x 2″ x 1-1/2″	Q-1	22	.727	Ea.	8.05	28		36.05	52.50
	5327	3″ x 3″ x 1-1/2″		15.50	1.032		14	39.50		53.50	77
	5328	3″ x 3″ x 2″		15.30	1.046		11.65	40		51.65	75.50
	5329	4″ x 4″ x 2″		12.20	1.311		19.50	50.50		70	100
	5331	Wye, 1-1/4″	1 Plum	13.50	.593		4	25.50		29.50	44
	5332	1-1/2″	″	12.10	.661		3.95	28		31.95	48.50
	5333	2″	Q-1	20	.800		5.05	30.50		35.55	53.50
	5334	3″		13.90	1.151		14.45	44		58.45	85
	5335	4″		11	1.455		21	56		77	110
	5336	6″		6.70	2.388		44	92		136	192
	5337	8″	Q-2	6.20	3.871		73.50	154		227.50	320
	5341	2″ x 1-1/2″	Q-1	22	.727		5.90	28		33.90	50
	5342	3″ x 1-1/2″		15.50	1.032		8.30	39.50		47.80	70.50
	5343	4″ x 3″		12.10	1.322		16.75	51		67.75	97.50
	5344	6″ x 4″		6.90	2.319		52	89		141	197
	5345	8″ x 6″	Q-2	6.40	3.750		56	149		205	294
	5347	Double wye, 1-1/2″	1 Plum	9.10	.879		8	37.50		45.50	67.50
	5348	2″	Q-1	16.60	.964		9.15	37		46.15	67.50
	5349	3″		10.40	1.538		23.50	59		82.50	118
	5350	4″		8.25	1.939		48	74.50		122.50	169
	5354	2″ x 1-1/2″		16.80	.952		8.40	36.50		44.90	66
	5355	3″ x 2″		10.60	1.509		17.60	58		75.60	109
	5356	4″ x 3″		8.45	1.893		38	73		111	155
	5357	6″ x 4″		7.25	2.207		79	85		164	219
	5374	Coupling, 1-1/4″	1 Plum	20.20	.396		2.50	16.90		19.40	29.50
	5376	1-1/2″	″	18.20	.440		.76	18.75		19.51	30

15 MECHANICAL

15108 | Plastic Pipe & Fittings

			CREW	DAILY OUTPUT	LABOR-HOURS	UNIT	2006 BARE COSTS MAT.	LABOR	EQUIP.	TOTAL	TOTAL INCL O&P	
560	5378	2″	Q-1	33.10	.483	Ea.	1.03	18.60		19.63	30	560
	5380	3″		20.80	.769		3.75	29.50		33.25	50	
	5390	4″		16.50	.970		6.15	37.50		43.65	65	
	5400	6″		10.10	1.584		20	61		81	117	
	5402	8″	Q-2	9.30	2.581		43	103		146	207	
	5404	2″ x 1-1/2″	Q-1	33.30	.480		1.97	18.45		20.42	30.50	
	5406	3″ x 1-1/2″		21	.762		6.25	29.50		35.75	52.50	
	5408	4″ x 3″		16.70	.958		11.05	37		48.05	69.50	
	5410	Reducer bushing, 2″ x 1-1/4″		36.50	.438		2.56	16.85		19.41	29	
	5412	3″ x 1-1/2″		27.30	.586		5.85	22.50		28.35	41.50	
	5414	4″ x 2″		18.20	.879		10.80	34		44.80	64.50	
	5416	6″ x 4″		11.10	1.441		31	55.50		86.50	120	
	5418	8″ x 6″	Q-2	10.20	2.353		62	94		156	215	
	5425	Closet flange 4″	Q-1	32	.500		6.30	19.20		25.50	37	
	5426	4″ x 3″	″	34	.471		10.85	18.10		28.95	40	
	5450	Solvent cement for PVC, industrial grade, per quart				Qt.	13.05			13.05	14.35	
	5500	CPVC, Schedule 80, threaded joints										
	5540	90° Elbow, 1/4″	1 Plum	32	.250	Ea.	8.25	10.70		18.95	25.50	
	5560	1/2″		30.30	.264		4.79	11.25		16.04	23	
	5570	3/4″		26	.308		7.15	13.15		20.30	28.50	
	5580	1″		22.70	.352		10.05	15.05		25.10	34.50	
	5590	1-1/4″		20.20	.396		19.35	16.90		36.25	48	
	5600	1-1/2″		18.20	.440		21	18.75		39.75	52	
	5610	2″	Q-1	33.10	.483		28	18.60		46.60	60	
	5620	2-1/2″		24.20	.661		86.50	25.50		112	135	
	5630	3″		20.80	.769		93	29.50		122.50	148	
	5640	4″		16.50	.970		146	37.50		183.50	218	
	5650	6″		10.10	1.584		172	61		233	284	
	5700	45° Elbow same as 90° Elbow										
	5850	Tee, 1/4″	1 Plum	22	.364	Ea.	16.45	15.55		32	42	
	5870	1/2″		20.20	.396		16.45	16.90		33.35	44.50	
	5880	3/4″		17.30	.462		23	19.75		42.75	56	
	5890	1″		15.20	.526		25	22.50		47.50	62.50	
	5900	1-1/4″		13.50	.593		25	25.50		50.50	67	
	5910	1-1/2″		12.10	.661		26.50	28		54.50	73	
	5920	2″	Q-1	20	.800		29	30.50		59.50	80	
	5930	2-1/2″		16.20	.988		144	38		182	217	
	5940	3″		13.90	1.151		167	44		211	252	
	5950	4″		11	1.455		395	56		451	520	
	5960	6″		6.70	2.388		214	92		306	380	
	6000	Coupling, 1/4″	1 Plum	32	.250		10.50	10.70		21.20	28	
	6020	1/2″		30.30	.264		8.65	11.25		19.90	27	
	6030	3/4″		26	.308		14	13.15		27.15	36	
	6040	1″		22.70	.352		15.85	15.05		30.90	41	
	6050	1-1/4″		20.20	.396		16.85	16.90		33.75	45	
	6060	1-1/2″		18.20	.440		18.10	18.75		36.85	49	
	6070	2″	Q-1	33.10	.483		21.50	18.60		40.10	52.50	
	6080	2-1/2″		24.20	.661		38	25.50		63.50	81.50	
	6090	3″		20.80	.769		44.50	29.50		74	95	
	6100	4″		16.50	.970		91	37.50		128.50	158	
	6110	6″		10.10	1.584		124	61		185	231	
	6120	8″	Q-2	9.30	2.581		265	103		368	450	
	6200	CTS, 100 psi at 180°F, hot and cold water										
	6230	90° Elbow, 1/2″	1 Plum	20	.400	Ea.	.08	17.10		17.18	26.50	
	6250	3/4″		19	.421		.17	18		18.17	28	
	6251	1″		16	.500		.58	21.50		22.08	33.50	

15108 | Plastic Pipe & Fittings

			CREW	DAILY OUTPUT	LABOR-HOURS	UNIT	2006 BARE COSTS MAT.	LABOR	EQUIP.	TOTAL	TOTAL INCL O&P	
560	6252	1-1/4"	1 Plum	15	.533	Ea.	.99	23		23.99	36.50	560
	6253	1-1/2"	↓	14	.571		1.59	24.50		26.09	40	
	6254	2"	Q-1	23	.696		3.43	26.50		29.93	45.50	
	6260	45° Elbow, 1/2"	1 Plum	20	.400		.19	17.10		17.29	26.50	
	6280	3/4"		19	.421		.23	18		18.23	28.50	
	6281	1"		16	.500		.53	21.50		22.03	33.50	
	6282	1-1/4"		15	.533		1.04	23		24.04	36.50	
	6283	1-1/2"	↓	14	.571		1.65	24.50		26.15	40	
	6284	2"	Q-1	23	.696		3.48	26.50		29.98	45.50	
	6290	Tee, 1/2"	1 Plum	13	.615		.14	26.50		26.64	41	
	6310	3/4"		12	.667		.26	28.50		28.76	45	
	6311	1"		11	.727		1.20	31		32.20	50	
	6312	1-1/4"		10	.800		1.60	34		35.60	55	
	6313	1-1/2"	↓	10	.800		2.60	34		36.60	56	
	6314	2"	Q-1	17	.941		4.22	36		40.22	60.50	
	6320	Coupling, 1/2"	1 Plum	22	.364		.10	15.55		15.65	24	
	6340	3/4"		21	.381		.14	16.25		16.39	25.50	
	6341	1"		18	.444		.50	19		19.50	30	
	6342	1-1/4"		17	.471		.62	20		20.62	32	
	6343	1-1/2"	↓	16	.500		.88	21.50		22.38	34	
	6344	2"	Q-1	28	.571	↓	1.75	22		23.75	36	
	6360	Solvent cement for CPVC, commercial grade, per quart				Qt.	17.85			17.85	19.60	
	7340	PVC flange, slip-on, Sch 80 std., 1/2"	1 Plum	22	.364	Ea.	8.20	15.55		23.75	33	
	7350	3/4"		21	.381		8.75	16.25		25	35	
	7360	1"		18	.444		9.75	19		28.75	40	
	7370	1-1/4"		17	.471		10.05	20		30.05	42.50	
	7380	1-1/2"	↓	16	.500		10.25	21.50		31.75	44.50	
	7390	2"	Q-1	26	.615		13.65	23.50		37.15	52	
	7400	2-1/2"		24	.667		21	25.50		46.50	63	
	7410	3"		18	.889		23.50	34		57.50	78.50	
	7420	4"		15	1.067		29.50	41		70.50	96	
	7430	6"	↓	10	1.600		49.50	61.50		111	150	
	7440	8"	Q-2	11	2.182		93.50	87		180.50	238	
	7550	Union, schedule 40, socket joints, 1/2"	1 Plum	19	.421		3	18		21	31.50	
	7560	3/4"		18	.444		3.25	19		22.25	33	
	7570	1"		15	.533		3.34	23		26.34	39	
	7580	1-1/4"		14	.571		10.75	24.50		35.25	50	
	7590	1-1/2"	↓	13	.615		11.20	26.50		37.70	53.50	
	7600	2"	Q-1	20	.800	↓	15.10	30.50		45.60	64.50	
	7990	Polybutyl/polyethyl pipe, for copper fittings see 15107-460-7000										
	8000	Compression type, PVC, 160 psi cold water										
	8010	Coupling, 3/4" CTS	1 Plum	21	.381	Ea.	2.43	16.25		18.68	28	
	8020	1" CTS		18	.444		3.02	19		22.02	33	
	8030	1-1/4" CTS		17	.471		4.20	20		24.20	36	
	8040	1-1/2" CTS		16	.500		5.80	21.50		27.30	39.50	
	8050	2" CTS		15	.533		8.05	23		31.05	44.50	
	8060	Female adapter, 3/4" FPT x 3/4" CTS		23	.348		4.18	14.85		19.03	27.50	
	8070	3/4" FPT x 1" CTS		21	.381		6.25	16.25		22.50	32.50	
	8080	1" FPT x 1" CTS		20	.400		6.30	17.10		23.40	33.50	
	8090	1-1/4" FPT x 1-1/4" CTS		18	.444		8.25	19		27.25	38.50	
	8100	1-1/2" FPT x 1-1/2" CTS		16	.500		9.65	21.50		31.15	43.50	
	8110	2" FPT x 2" CTS		13	.615		14.25	26.50		40.75	56.50	
	8130	Male adapter, 3/4" MPT x 3/4" CTS		23	.348		3.77	14.85		18.62	27	
	8140	3/4" MPT x 1" CTS		21	.381		4.63	16.25		20.88	30.50	
	8150	1" MPT x 1" CTS		20	.400		4.74	17.10		21.84	31.50	
560	8160	1-1/4" MPT x 1-1/4" CTS	↓	18	.444	↓	7.25	19		26.25	37.50	560

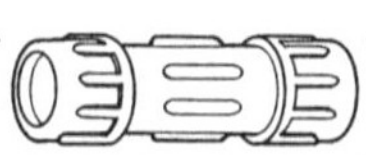

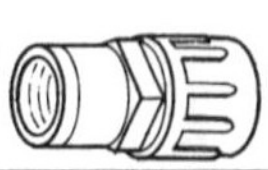

15 MECHANICAL

15108 | Plastic Pipe & Fittings

		15108 Plastic Pipe & Fittings	CREW	DAILY OUTPUT	LABOR-HOURS	UNIT	2006 BARE COSTS MAT.	LABOR	EQUIP.	TOTAL	TOTAL INCL O&P	
560	8170	1-1/2" MPT x 1-1/2" CTS	1 Plum	16	.500	Ea.	8.40	21.50		29.90	42.50	560
	8180	2" MPT x 2" CTS		13	.615		13.45	26.50		39.95	56	
	8200	Spigot adapter, 3/4" IPS x 3/4" CTS		23	.348		3.75	14.85		18.60	27	
	8210	3/4" IPS x 1" CTS		21	.381		3.77	16.25		20.02	29.50	
	8220	1" IPS x 1" CTS 		20	.400		4.06	17.10		21.16	31	
	8230	1-1/4" IPS x 1-1/4" CTS		18	.444		7.25	19		26.25	37.50	
	8240	1-1/2" IPS x 1-1/2" CTS		16	.500		8.40	21.50		29.90	42.50	
	8250	2" IPS x 2" CTS		13	.615		13.45	26.50		39.95	56	
	8270	Price includes insert stiffeners										
	8280	250 psi is same price as 160 psi										
	8300	Insert type, nylon, 160 & 250 psi, cold water										
	8310	Clamp ring stainless steel, 3/4" IPS	1 Plum	115	.070	Ea.	1.71	2.97		4.68	6.50	
	8320	1" IPS		107	.075		1.74	3.19		4.93	6.85	
	8330	1-1/4" IPS 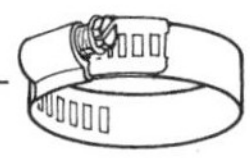		101	.079		1.76	3.38		5.14	7.20	
	8340	1-1/2" IPS		95	.084		2.40	3.60		6	8.25	
	8350	2" IPS		85	.094		2.72	4.02		6.74	9.25	
	8370	Coupling, 3/4" IPS		22	.364		.71	15.55		16.26	25	
	8380	1" IPS		19	.421		.94	18		18.94	29	
	8390	1-1/4" IPS 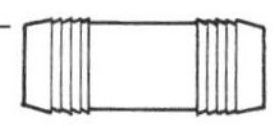		18	.444		1.68	19		20.68	31.50	
	8400	1-1/2" IPS		17	.471		2.19	20		22.19	34	
	8410	2" IPS		16	.500		2.63	21.50		24.13	36	
	8430	Elbow, 90°, 3/4" IPS		22	.364		1.26	15.55		16.81	25.50	
	8440	1" IPS 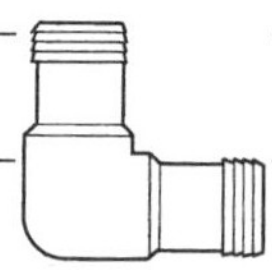		19	.421		1.34	18		19.34	29.50	
	8450	1-1/4" IPS		18	.444		2.27	19		21.27	32	
	8460	1-1/2" IPS		17	.471		3.34	20		23.34	35	
	8470	2" IPS		16	.500		5.60	21.50		27.10	39	
	8490	Male adapter, 3/4" IPS x 3/4" MPT		25	.320		.71	13.65		14.36	22.50	
	8500	1" IPS x 1" MPT 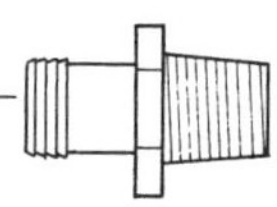		21	.381		.94	16.25		17.19	26.50	
	8510	1-1/4" IPS x 1-1/4" MPT		20	.400		1.55	17.10		18.65	28	
	8520	1-1/2" IPS x 1-1/2" MPT		18	.444		2.05	19		21.05	32	
	8530	2" IPS x 2" MPT		15	.533		2.63	23		25.63	38.50	
	8550	Tee, 3/4" IPS 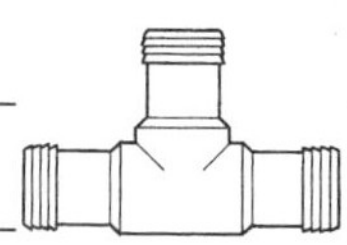		14	.571		1.34	24.50		25.84	39.50	
	8560	1" IPS		13	.615		2.21	26.50		28.71	43.50	
	8570	1-1/4" IPS		12	.667		3.96	28.50		32.46	49	
	8580	1-1/2" IPS		11	.727		4.97	31		35.97	54	
	8590	2" IPS		10	.800		6.70	34		40.70	60.50	
	8610	Insert type, PVC, 100 psi @ 180°F, hot & cold water										
	8620	Coupler, male, 3/8" CTS x 3/8" MPT	1 Plum	29	.276	Ea.	.50	11.80		12.30	18.85	
	8630	3/8" CTS x 1/2" MPT		28	.286		.50	12.20		12.70	19.50	
	8640	1/2" CTS x 1/2" MPT		27	.296		.50	12.65		13.15	20	
	8650	1/2" CTS x 3/4" MPT 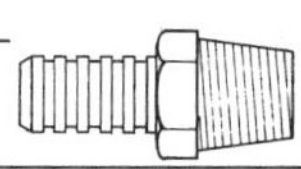		26	.308		1.56	13.15		14.71	22	
	8660	3/4" CTS x 1/2" MPT		25	.320		1.40	13.65		15.05	23	
	8670	3/4" CTS x 3/4" MPT		25	.320		.60	13.65		14.25	22	
	8700	Coupling, 3/8" CTS x 1/2" CTS		25	.320		2.15	13.65		15.80	24	
	8710	1/2" CTS 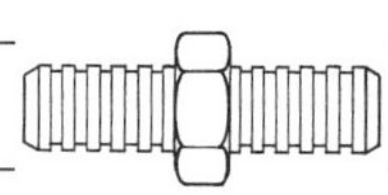		23	.348		2.50	14.85		17.35	26	
	8720	1/2" CTS x stub		23	.348		2.52	14.85		17.37	26	
	8730	3/4" CTS		22	.364		3.96	15.55		19.51	28.50	
	8750	Elbow 90°, 3/8" CTS		25	.320		2.55	13.65		16.20	24.50	
	8760	1/2" CTS 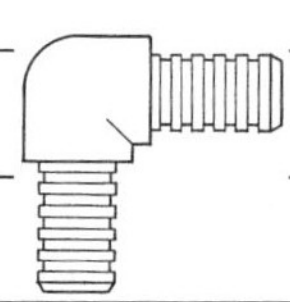		23	.348		3.01	14.85		17.86	26.50	
	8770	3/4" CTS		22	.364		4.61	15.55		20.16	29	
	8800	Rings, crimp, copper, 3/8" CTS		120	.067		.11	2.85		2.96	4.55	
	8810	1/2" CTS		117	.068		.12	2.92		3.04	4.67	
	8820	3/4" CTS		115	.070		.15	2.97		3.12	4.79	
	8850	Reducer tee, 3/8" x 3/8" x 1/2" CTS		17	.471		1.46	20		21.46	33	
	8860	1/2" x 3/8" x 1/2" CTS		15	.533		1.25	23		24.25	37	
560	8870	3/4" x 1/2" x 1/2" CTS		14	.571		1.31	24.50		25.81	39.50	

	15108 | Plastic Pipe & Fittings		CREW	DAILY OUTPUT	LABOR-HOURS	UNIT	2006 BARE COSTS MAT.	LABOR	EQUIP.	TOTAL	TOTAL INCL O&P
560	8890	3/4" x 3/4" x 1/2" CTS	1 Plum	14	.571	Ea.	1.24	24.50		25.74	39.50
	8900	3/4" x 1/2" x 3/8" CTS		14	.571		1.09	24.50		25.59	39
	8930	Tee, 3/8" CTS		17	.471		1.05	20		21.05	32.50
	8940	1/2" CTS		15	.533		1.19	23		24.19	37
	8950	3/4" CTS		14	.571		1.29	24.50		25.79	39.50
	8960	Copper rings included in fitting price 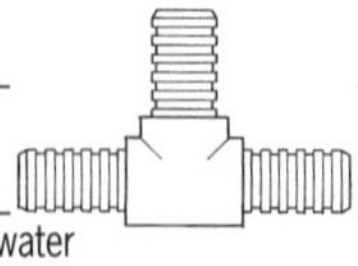									
	9000	Flare type, assembled, acetal, hot & cold water									
	9010	Coupling, 1/4" & 3/8" CTS 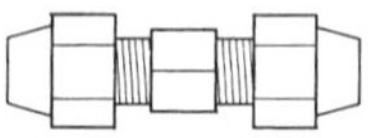	1 Plum	24	.333	Ea.	2.14	14.25		16.39	24.50
	9020	1/2" CTS		22	.364		2.45	15.55		18	26.50
	9030	3/4" CTS		21	.381		3.96	16.25		20.21	30
	9040	1" CTS		18	.444		3.96	19		22.96	34
	9050	Elbow 90°, 1/4" CTS 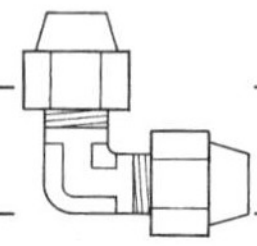		26	.308		2.38	13.15		15.53	23
	9060	3/8" CTS		24	.333		2.55	14.25		16.80	25
	9070	1/2" CTS		22	.364		3.01	15.55		18.56	27.50
	9080	3/4" CTS		21	.381		4.61	16.25		20.86	30.50
	9090	1" CTS		18	.444		5.80	19		24.80	36
	9110	Tee ,1/4" & 3/8" CTS 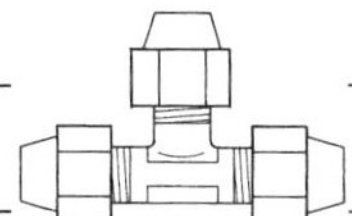		15	.533		2.62	23		25.62	38.50
	9120	1/2" CTS		14	.571		3.44	24.50		27.94	42
	9130	3/4" CTS		13	.615		5.25	26.50		31.75	47
	9140	1" CTS		12	.667		7	28.50		35.50	52
	9550	For plastic hangers see 15060-300-8000									
	9560	For copper/brass fittings see 15107-460-7000									
	9900	Minimum labor/equipment charge	1 Plum	4	2	Job		85.50		85.50	133
	15110 | Valves										
100	0010	**VALVES, BRASS**									
	0030	For motorized valves, see Division 13838-200									
	0500	Gas cocks, threaded									
	0510	1/4"	1 Plum	26	.308	Ea.	5.95	13.15		19.10	27
	0520	3/8"		24	.333		5.95	14.25		20.20	28.50
	0530	1/2"		24	.333		5.75	14.25		20	28.50
	0540	3/4"		22	.364		14.25	15.55		29.80	39.50
	0550	1"		19	.421		18.20	18		36.20	48
	0560	1-1/4"		15	.533		22.50	23		45.50	60.50
	0570	1-1/2"		13	.615		32.50	26.50		59	76.50
	0580	2"		11	.727		48	31		79	101
	0670	For larger sizes use lubricated plug valve, Section 15110-600									
160	0010	**VALVES, BRONZE** R220523-90									
	1020	Angle, 150 lb., rising stem, threaded									
	1030	1/8"	1 Plum	24	.333	Ea.	57	14.25		71.25	84.50
	1040	1/4"		24	.333		57	14.25		71.25	84.50
	1050	3/8"		24	.333		57	14.25		71.25	84.50
	1060	1/2" 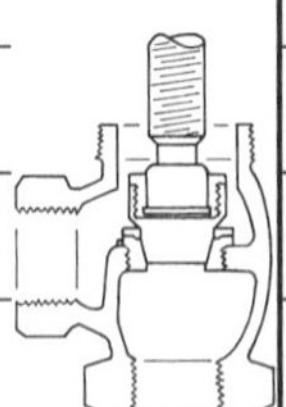		22	.364		57	15.55		72.55	86.50
	1070	3/4"		20	.400		77.50	17.10		94.60	112
	1080	1"		19	.421		110	18		128	149
	1090	1-1/4"		15	.533		143	23		166	193
	1100	1-1/2"		13	.615		218	26.50		244.50	281
	1110	2"		11	.727		345	31		376	430
	1380	Ball, 150 psi, threaded									
	1400	1/4"	1 Plum	24	.333	Ea.	8.10	14.25		22.35	31
	1430	3/8"		24	.333		8.10	14.25		22.35	31
	1450	1/2" 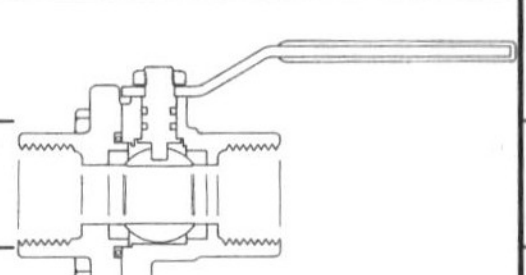		22	.364		8.10	15.55		23.65	33
	1460	3/4"		20	.400		13.35	17.10		30.45	41
	1470	1"		19	.421		16.85	18		34.85	46.50
	1480	1-1/4"		15	.533		19.05	23		42.05	56.50

15 MECHANICAL

15110 | Valves

		15110 Valves	CREW	DAILY OUTPUT	LABOR-HOURS	UNIT	2006 BARE COSTS MAT.	LABOR	EQUIP.	TOTAL	TOTAL INCL O&P	
160	1490	1-1/2" R220523-90	1 Plum	13	.615	Ea.	25	26.50		51.50	68.50	160
	1500	2"		11	.727		30	31		61	81.50	
	1510	2-1/2"		9	.889		100	38		138	170	
	1520	3"	↓	8	1	↓	153	42.50		195.50	235	
	1600	Butterfly, 175 psi, full port, solder or threaded ends										
	1610	Stainless steel disc and stem										
	1620	1/4"	1 Plum	24	.333	Ea.	7	14.25		21.25	29.50	
	1630	3/8"		24	.333		7	14.25		21.25	29.50	
	1640	1/2"		22	.364		7.40	15.55		22.95	32	
	1650	3/4"		20	.400		11.95	17.10		29.05	39.50	
	1660	1"		19	.421		14.60	18		32.60	44	
	1670	1-1/4"		15	.533		23.50	23		46.50	61.50	
	1680	1-1/2"		13	.615		30.50	26.50		57	74.50	
	1690	2"	↓	11	.727	↓	38	31		69	90	
	1750	Check, swing, class 150, regrinding disc, threaded										
	1800	1/8"	1 Plum	24	.333	Ea.	29	14.25		43.25	54	
	1830	1/4"		24	.333		29	14.25		43.25	54	
	1840	3/8" 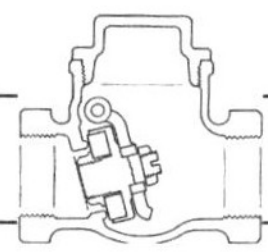		24	.333		30	14.25		44.25	55	
	1850	1/2"		24	.333		30	14.25		44.25	55	
	1860	3/4"		20	.400		40.50	17.10		57.60	71	
	1870	1"		19	.421		60	18		78	94	
	1880	1-1/4"		15	.533		84	23		107	128	
	1890	1-1/2"		13	.615		99	26.50		125.50	150	
	1900	2"	↓	11	.727		146	31		177	209	
	1910	2-1/2"	Q-1	15	1.067		242	41		283	330	
	1920	3"	"	13	1.231	↓	272	47.50		319.50	375	
	2000	For 200 lb, add					5%	10%				
	2040	For 300 lb, add					15%	15%				
	2060	Check swing, 300#, sweat, 3/8" size	1 Plum	24	.333	Ea.	15.90	14.25		30.15	39.50	
	2070	1/2"		24	.333		15.90	14.25		30.15	39.50	
	2080	3/4"		20	.400		20	17.10		37.10	49	
	2090	1"		19	.421		28	18		46	59	
	2100	1-1/4"		15	.533		36.50	23		59.50	75.50	
	2110	1-1/2 "		13	.615		48.50	26.50		75	94	
	2120	2"	↓	11	.727		73	31		104	129	
	2130	2-1/2"	Q-1	15	1.067		173	41		214	255	
	2140	3"	"	13	1.231	↓	231	47.50		278.50	330	
	2350	Check, lift, class 150, horizontal composition disc, threaded										
	2430	1/4"	1 Plum	24	.333	Ea.	80	14.25		94.25	110	
	2440	3/8"		24	.333		101	14.25		115.25	133	
	2450	1/2"		24	.333		88	14.25		102.25	119	
	2460	3/4" 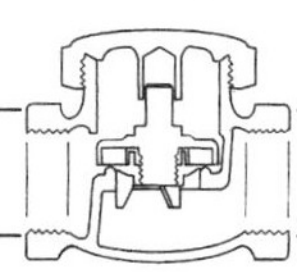		20	.400		108	17.10		125.10	145	
	2470	1"		19	.421		153	18		171	196	
	2480	1-1/4"		15	.533		206	23		229	262	
	2490	1-1/2"		13	.615		245	26.50		271.50	310	
	2500	2"	↓	11	.727	↓	410	31		441	500	
	2850	Gate, N.R.S., soldered, 125 psi										
	2900	3/8"	1 Plum	24	.333	Ea.	23	14.25		37.25	47	
	2920	1/2"		24	.333		19.50	14.25		33.75	43.50	
	2940	3/4"		20	.400		22.50	17.10		39.60	51	
	2950	1"		19	.421		32	18		50	63	
	2960	1-1/4"		15	.533		50	23		73	90.50	
	2970	1-1/2"		13	.615		56.50	26.50		83	103	
	2980	2"	↓	11	.727		76.50	31		107.50	133	
	2990	2-1/2"	Q-1	15	1.067		191	41		232	274	
	3000	3"	"	13	1.231	↓	250	47.50		297.50	350	

15110 | Valves

160 … 160

R220523-90

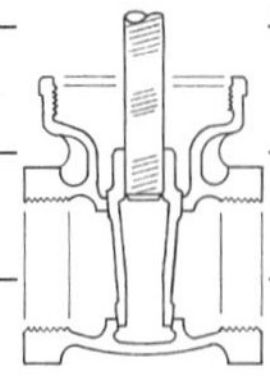

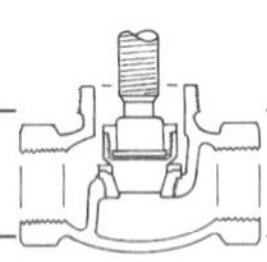

	15110 Valves	CREW	DAILY OUTPUT	LABOR-HOURS	UNIT	2006 BARE COSTS MAT.	LABOR	EQUIP.	TOTAL	TOTAL INCL O&P
3350	Threaded, class 150									
3410	1/4"	1 Plum	24	.333	Ea.	22.50	14.25		36.75	46.50
3420	3/8"		24	.333		22.50	14.25		36.75	46.50
3430	1/2"		24	.333		20	14.25		34.25	44
3440	3/4"		20	.400		24	17.10		41.10	53
3450	1"		19	.421		30.50	18		48.50	61.50
3460	1-1/4" size		15	.533		47	23		70	87
3470	1-1/2"		13	.615		54	26.50		80.50	101
3480	2"		11	.727		65	31		96	120
3490	2-1/2"	Q-1	15	1.067		184	41		225	267
3500	3" size	"	13	1.231		226	47.50		273.50	325
3850	Rising stem, soldered, 300 psi									
3900	3/8"	1 Plum	24	.333	Ea.	53	14.25		67.25	80
3920	1/2"		24	.333		53	14.25		67.25	80
3940	3/4"		20	.400		57.50	17.10		74.60	90
3950	1"		19	.421		83.50	18		101.50	120
3960	1-1/4"		15	.533		116	23		139	164
3970	1-1/2"		13	.615		137	26.50		163.50	191
3980	2"		11	.727		222	31		253	293
3990	2-1/2"	Q-1	15	1.067		500	41		541	615
4000	3"	"	13	1.231		700	47.50		747.50	845
4250	Threaded, class 150									
4310	1/4"	1 Plum	24	.333	Ea.	29	14.25		43.25	54
4320	3/8"		24	.333		29	14.25		43.25	54
4330	1/2"		24	.333		27	14.25		41.25	52
4340	3/4"		20	.400		32	17.10		49.10	62
4350	1"		19	.421		43	18		61	75.50
4360	1-1/4"		15	.533		57.50	23		80.50	99
4370	1-1/2"		13	.615		72.50	26.50		99	121
4380	2"		11	.727		98.50	31		129.50	158
4390	2-1/2"	Q-1	15	1.067		230	41		271	315
4400	3"	"	13	1.231		320	47.50		367.50	430
4500	For 300 psi, threaded, add					100%	15%			
4540	For chain operated type, add					15%				
4850	Globe, class 150, rising stem, threaded									
4920	1/4"	1 Plum	24	.333	Ea.	42	14.25		56.25	68
4940	3/8" size		24	.333		42	14.25		56.25	68
4950	1/2"		24	.333		42	14.25		56.25	68
4960	3/4"		20	.400		57	17.10		74.10	89
4970	1"		19	.421		89.50	18		107.50	126
4980	1-1/4"		15	.533		141	23		164	191
4990	1-1/2"		13	.615		172	26.50		198.50	230
5000	2"		11	.727		257	31		288	330
5010	2-1/2"	Q-1	15	1.067		515	41		556	635
5020	3"	"	13	1.231		735	47.50		782.50	885
5120	For 300 lb threaded, add					50%	15%			
5130	Globe, 300 lb., sweat, 3/8" size	1 Plum	24	.333	Ea.	40.50	14.25		54.75	66.50
5140	1/2"		24	.333		40.50	14.25		54.75	66.50
5150	3/4"		20	.400		54.50	17.10		71.60	86.50
5160	1"		19	.421		85.50	18		103.50	123
5170	1-1/4"		15	.533		135	23		158	185
5180	1-1/2"		13	.615		165	26.50		191.50	222
5190	2"		11	.727		247	31		278	320
5200	2-1/2"	Q-1	15	1.067		495	41		536	610
5210	3"	"	13	1.231		705	47.50		752.50	850
5600	Relief, pressure & temperature, self-closing, ASME, threaded									

15 MECHANICAL

15110 | Valves

			CREW	DAILY OUTPUT	LABOR-HOURS	UNIT	2006 BARE COSTS MAT.	LABOR	EQUIP.	TOTAL	TOTAL INCL O&P
160	5640	3/4" R220523-90	1 Plum	28	.286	Ea.	85.50	12.20		97.70	113
	5650	1"		24	.333		124	14.25		138.25	159
	5660	1-1/4"		20	.400		249	17.10		266.10	300
	5670	1-1/2"		18	.444		475	19		494	555
	5680	2"	↓	16	.500	↓	520	21.50		541.50	605
	5950	Pressure, poppet type, threaded									
	6000	1/2"	1 Plum	30	.267	Ea.	15.35	11.40		26.75	34.50
	6040	3/4"	"	28	.286	"	42	12.20		54.20	65
	6400	Pressure, water, ASME, threaded									
	6440	3/4"	1 Plum	28	.286	Ea.	49	12.20		61.20	73
	6450	1"		24	.333		53.50	14.25		67.75	81
	6460	1-1/4"		20	.400		157	17.10		174.10	200
	6470	1-1/2"		18	.444		232	19		251	286
	6480	2"		16	.500		360	21.50		381.50	430
	6490	2-1/2"	↓	15	.533	↓	1,175	23		1,198	1,325
	6900	Reducing, water pressure									
	6920	300 psi to 25-75 psi, threaded or sweat									
	6940	1/2"	1 Plum	24	.333	Ea.	145	14.25		159.25	182
	6950	3/4"		20	.400		145	17.10		162.10	187
	6960	1"		19	.421		225	18		243	275
	6970	1-1/4"		15	.533		400	23		423	480
	6980	1-1/2"		13	.615		605	26.50		631.50	710
	6990	2"	↓	11	.727		875	31		906	1,025
	7100	For built-in by-pass or 10-35 psi, add				↓	7.30			7.30	8
	7700	High capacity, 250 psi to 25-75 psi, threaded									
	7740	1/2"	1 Plum	24	.333	Ea.	145	14.25		159.25	182
	7780	3/4"		20	.400		145	17.10		162.10	187
	7790	1"		19	.421		202	18		220	251
	7800	1-1/4"		15	.533		400	23		423	480
	7810	1-1/2"		13	.615		605	26.50		631.50	710
	7820	2"		11	.727		875	31		906	1,025
	7830	2-1/2"		9	.889		1,750	38		1,788	1,975
	7840	3"	↓	8	1		1,375	42.50		1,417.50	1,575
	7850	3" flanged (iron body)	Q-1	10	1.600		1,800	61.50		1,861.50	2,075
	7860	4" flanged (iron body)	"	8	2	↓	1,825	77		1,902	2,125
	7920	For higher pressure, add					25%				
	8350	Tempering, water, sweat connections									
	8400	1/2"	1 Plum	24	.333	Ea.	52	14.25		66.25	79
	8440	3/4"	"	20	.400	"	63.50	17.10		80.60	96
	8650	Threaded connections									
	8700	1/2"	1 Plum	24	.333	Ea.	63.50	14.25		77.75	91.50
	8740	3/4"		20	.400		242	17.10		259.10	293
	8750	1"		19	.421		267	18		285	320
	8760	1-1/4"		15	.533		425	23		448	500
	8770	1-1/2"		13	.615		460	26.50		486.50	550
	8780	2"		11	.727	↓	690	31		721	810
	9000	Minimum labor/equipment charge	↓	4	2	Job		85.50		85.50	133
200	0010	**VALVES, IRON BODY** R220523-90									
	0020	For grooved joint, see Division 15107-690									
	0560	Butterfly, lug type, pneumatic operator, 2" size	1 Stpi	14	.571	Ea.	244	24.50		268.50	305
	0570	3"	Q-1	8	2		253	77		330	400
	0580	4"	"	5	3.200		276	123		399	495
	0590	6"	Q-2	5	4.800		370	191		561	700
	0600	8"		4.50	5.333		445	213		658	820
	0610	10"	↓	4	6	↓	560	239		799	990

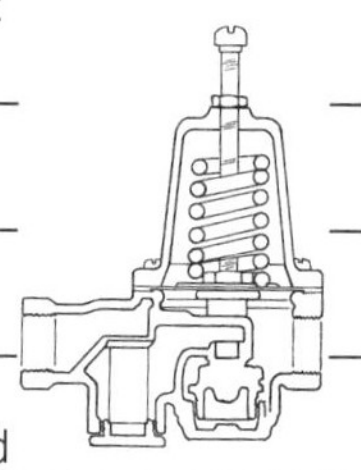

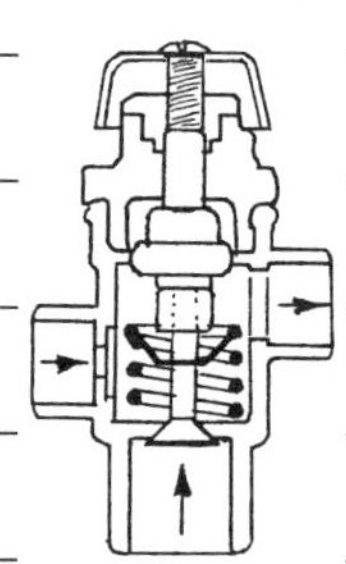

15110 | Valves

			CREW	DAILY OUTPUT	LABOR-HOURS	UNIT	2006 BARE COSTS MAT.	LABOR	EQUIP.	TOTAL	TOTAL INCL O&P	
200	0620	12" R220523-90	Q-2	3	8	Ea.	755	320		1,075	1,325	200
	0630	14"		2.30	10.435		1,075	415		1,490	1,850	
	0640	18"		1.50	16		1,875	640		2,515	3,050	
	0650	20"		1	24		2,250	955		3,205	3,950	
	0790	Butterfly, lug type, gear operated, 2" size, 200 lb. except noted	1 Plum	14	.571		200	24.50		224.50	258	
	0800	2-1/2"	Q-1	9	1.778		202	68.50		270.50	330	
	0810	3"		8	2		210	77		287	350	
	0820	4"		5	3.200		239	123		362	455	
	0830	5"	Q-2	5	4.800		298	191		489	625	
	0840	6"		5	4.800		330	191		521	660	
	0850	8"		4.50	5.333		425	213		638	800	
	0860	10"		4	6		565	239		804	990	
	0870	12"		3	8		820	320		1,140	1,400	
	0880	14", 150 lb.		2.30	10.435		895	415		1,310	1,625	
	0890	16", 150 lb.		1.75	13.714		1,275	545		1,820	2,250	
	0900	18", 150 lb.		1.50	16		1,650	640		2,290	2,800	
	0910	20", 150 lb.		1	24		1,950	955		2,905	3,625	
	0930	24", 150 lb.		.75	32		2,725	1,275		4,000	4,975	
	1020	Butterfly, wafer type, gear actuator, 200 lb.										
	1030	2"	1 Plum	14	.571	Ea.	120	24.50		144.50	170	
	1040	2-1/2"	Q-1	9	1.778		123	68.50		191.50	242	
	1050	3"		8	2		128	77		205	261	
	1060	4"		5	3.200		160	123		283	365	
	1070	5"	Q-2	5	4.800		193	191		384	510	
	1080	6"		5	4.800		218	191		409	535	
	1090	8"		4.50	5.333		290	213		503	650	
	1100	10"		4	6		365	239		604	770	
	1110	12"		3	8		505	320		825	1,050	
	1200	Wafer type, lever actuator, 200 lb.										
	1220	2"	1 Plum	14	.571	Ea.	88.50	24.50		113	136	
	1230	2-1/2"	Q-1	9	1.778		91.50	68.50		160	207	
	1240	3"		8	2		96	77		173	226	
	1250	4"		5	3.200		118	123		241	320	
	1260	5"	Q-2	5	4.800		164	191		355	475	
	1270	6"		5	4.800		197	191		388	515	
	1280	8"		4.50	5.333		291	213		504	650	
	1290	10"		4	6		525	239		764	945	
	1300	12"		3	8		740	320		1,060	1,300	
	1650	Gate, 125 lb., N.R.S.										
	2150	Flanged										
	2200	2"	1 Plum	5	1.600	Ea.	345	68.50		413.50	485	
	2240	2-1/2"	Q-1	5	3.200		355	123		478	580	
	2260	3"		4.50	3.556		400	137		537	650	
	2280	4"		3	5.333		570	205		775	950	
	2290	5"	Q-2	3.40	7.059		970	281		1,251	1,500	
	2300	6"		3	8		970	320		1,290	1,575	
	2320	8"		2.50	9.600		1,675	385		2,060	2,425	
	2340	10"		2.20	10.909		2,925	435		3,360	3,900	
	2360	12"		1.70	14.118		4,025	565		4,590	5,300	
	2370	14"		1.30	18.462		6,600	735		7,335	8,400	
	2380	16"		1	24		9,175	955		10,130	11,600	
	2420	For 250 lb flanged, add					200%	10%				
	3550	OS&Y, 125 lb., flanged										
	3600	2"	1 Plum	5	1.600	Ea.	251	68.50		319.50	380	
	3640	2-1/2"	Q-1	5	3.200		259	123		382	475	
200	3660	3"		4.50	3.556		290	137		427	530	

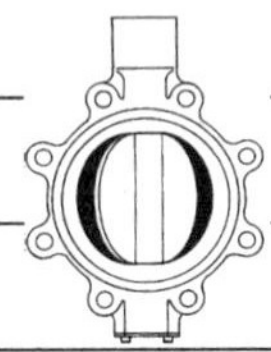

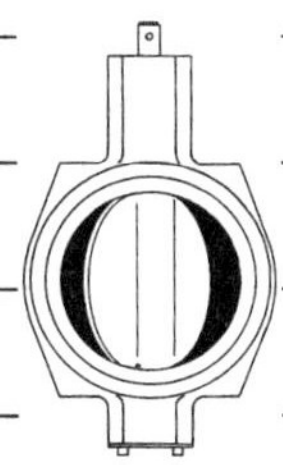

15 MECHANICAL

15110 | Valves

		Description	CREW	DAILY OUTPUT	LABOR-HOURS	UNIT	2006 BARE COSTS MAT.	LABOR	EQUIP.	TOTAL	TOTAL INCL O&P	
200	3670	3-1/2″ R220523-90	Q-1	3	5.333	Ea.	460	205		665	825	200
	3680	4″		3	5.333		415	205		620	775	
	3690	5″	Q-2	3.40	7.059		685	281		966	1,175	
	3700	6″		3	8		685	320		1,005	1,250	
	3720	8″		2.50	9.600		1,225	385		1,610	1,925	
	3740	10″		2.20	10.909		2,225	435		2,660	3,125	
	3760	12″		1.70	14.118		2,975	565		3,540	4,125	
	3770	14″		1.30	18.462		5,650	735		6,385	7,375	
	3780	16″		1	24		8,475	955		9,430	10,800	
	3790	18″		.80	30		11,400	1,200		12,600	14,400	
	3800	20″		.60	40		15,900	1,600		17,500	20,000	
	3830	24″		.50	48		23,600	1,925		25,525	28,900	
	3900	For 175 lb, flanged, add					200%	10%				
	4350	Globe, OS&Y										
	4540	Class 125, flanged										
	4550	2″	1 Plum	5	1.600	Ea.	515	68.50		583.50	670	
	4560	2-1/2″	Q-1	5	3.200		490	123		613	730	
	4570	3″		4.50	3.556		630	137		767	900	
	4580	4″		3	5.333		900	205		1,105	1,300	
	4590	5″	Q-2	3.40	7.059		1,650	281		1,931	2,225	
	4600	6″		3	8		1,650	320		1,970	2,300	
	4610	8″		2.50	9.600		3,200	385		3,585	4,125	
	5040	Class 250, flanged										
	5050	2″	1 Plum	4.50	1.778	Ea.	700	76		776	890	
	5060	2-1/2″	Q-1	4.50	3.556		910	137		1,047	1,200	
	5070	3″		4	4		940	154		1,094	1,275	
	5080	4″		2.70	5.926		1,375	228		1,603	1,875	
	5090	5″	Q-2	3	8		2,475	320		2,795	3,225	
	5100	6″		2.70	8.889		2,475	355		2,830	3,275	
	5110	8″		2.20	10.909		4,200	435		4,635	5,275	
	5120	10″		2	12		6,500	480		6,980	7,900	
	5130	12″		1.60	15		9,775	600		10,375	11,600	
	5450	Swing check, 125 lb., threaded										
	5470	1″	1 Plum	13	.615	Ea.	355	26.50		381.50	430	
	5500	2″	″	11	.727		445	31		476	540	
	5540	2-1/2″	Q-1	15	1.067		685	41		726	815	
	5550	3″		13	1.231		765	47.50		812.50	915	
	5560	4″		10	1.600		1,225	61.50		1,286.50	1,450	
	5950	Flanged										
	6000	2″	1 Plum	5	1.600	Ea.	172	68.50		240.50	295	
	6040	2-1/2″	Q-1	5	3.200		210	123		333	420	
	6050	3″		4.50	3.556		225	137		362	460	
	6060	4″		3	5.333		350	205		555	705	
	6070	6″	Q-2	3	8		605	320		925	1,150	
	6080	8″		2.50	9.600		1,125	385		1,510	1,850	
	6090	10″		2.20	10.909		1,925	435		2,360	2,800	
	6100	12″		1.70	14.118		3,000	565		3,565	4,200	
	6160	For 250 lb flanged, add					200%	20%				
	6600	Silent check, bronze trim										
	6610	Compact wafer type, for 125 or 150 lb. flanges										
	6630	1-1/2″	1 Plum	11	.727	Ea.	97	31		128	156	
	6640	2″	″	9	.889		115	38		153	186	
	6650	2-1/2″	Q-1	9	1.778		126	68.50		194.50	245	
	6660	3″		8	2		136	77		213	269	
	6670	4″		5	3.200		173	123		296	380	
	6680	5″	Q-2	6	4		226	159		385	495	

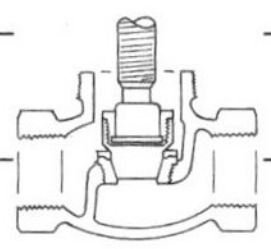

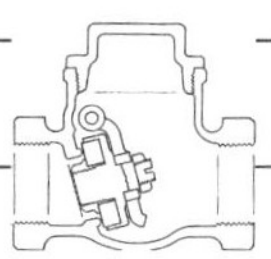

	15110 | Valves		CREW	DAILY OUTPUT	LABOR-HOURS	UNIT	2006 BARE COSTS MAT.	LABOR	EQUIP.	TOTAL	TOTAL INCL O&P	
200	6690	6″ (R220523-90)	Q-2	6	4	Ea.	305	159		464	585	200
	6700	8″		4.50	5.333		540	213		753	925	
	6710	10″		4	6		935	239		1,174	1,400	
	6720	12″		3	8		1,775	320		2,095	2,475	
	9000	Minimum labor/equipment charge	1 Plum	3	2.667	Job		114		114	177	
300	0010	**VALVES, LINED, CORROSION RESISTANT/HIGH PURITY** (R220523-90)										300
	2000	Butterfly 150 lb. ductile iron,										
	2010	Wafer type										
	2030	TFE lined 3″, lever handle	Q-1	8	2	Ea.	1,025	77		1,102	1,250	
	2050	4″, lever handle	"	5	3.200		1,425	123		1,548	1,775	
	2070	6″, gear operated	Q-2	4.50	5.333		2,025	213		2,238	2,550	
	2080	8″, gear operated		4	6		2,875	239		3,114	3,550	
	2090	10″, gear operated		3.50	6.857		3,700	273		3,973	4,500	
	2100	12″, gear operated		2.50	9.600		4,425	385		4,810	5,450	
	3500	Check lift, 125 lb., cast iron flanged										
	3510	Horizontal PPL or SL lined										
	3530	1″	1 Plum	14	.571	Ea.	475	24.50		499.50	565	
	3540	1-1/2″		11	.727		575	31		606	685	
	3550	2″		8	1		670	42.50		712.50	800	
	3560	2-1/2″	Q-1	5	3.200		875	123		998	1,150	
	3570	3″		4.50	3.556		1,100	137		1,237	1,400	
	3590	4″		3	5.333		1,450	205		1,655	1,900	
	3610	6″	Q-2	3	8		2,450	320		2,770	3,200	
	3620	8″	"	2.50	9.600		5,400	385		5,785	6,550	
	4250	Vertical PPL or SL lined										
	4270	1″	1 Plum	14	.571	Ea.	425	24.50		449.50	510	
	4290	1-1/2″		11	.727		525	31		556	630	
	4300	2″		8	1		615	42.50		657.50	740	
	4310	2-1/2″	Q-1	5	3.200		1,100	123		1,223	1,425	
	4320	3″		4.50	3.556		1,200	137		1,337	1,525	
	4340	4″		3	5.333		1,225	205		1,430	1,675	
	4360	6″	Q-2	3	8		2,050	320		2,370	2,775	
	4370	8″	"	2.50	9.600		3,725	385		4,110	4,675	
	5000	Clamp type, ductile iron, 150 lb. flanged										
	5010	TFE lined										
	5030	1″size, lever handle	1 Plum	9	.889	Ea.	885	38		923	1,025	
	5050	1-1/2″ size, lever handle		6	1.333		1,150	57		1,207	1,375	
	5060	2″ size, lever handle		5	1.600		1,425	68.50		1,493.50	1,650	
	5080	3″ size, lever handle	Q-1	4.50	3.556		1,775	137		1,912	2,150	
	5100	4″ size, gear operated	"	3	5.333		2,550	205		2,755	3,125	
	5120	6″ size, gear operated	Q-2	3	8		8,900	320		9,220	10,300	
	5130	8″ size, gear operated	"	2.50	9.600		9,900	385		10,285	11,500	
	6000	Diaphragm type, cast iron, 125 lb. flanged										
	6010	PTFE or VITON, lined										
	6030	1″ size, handwheel operated	1 Plum	9	.889	Ea.	189	38		227	267	
	6050	1-1/2″ size, handwheel operated		6	1.333		240	57		297	355	
	6060	2″ size, handwheel operated		5	1.600		305	68.50		373.50	440	
	6080	3″ size, handwheel operated	Q-1	4.50	3.556		520	137		657	780	
	6100	4″ size, handwheel operated	"	3	5.333		850	205		1,055	1,250	
	6120	6″ size, handwheel operated	Q-2	3	8		1,625	320		1,945	2,275	
	6130	8″ size, handwheel operated	"	2.50	9.600		3,425	385		3,810	4,375	
500	0010	**VALVES, PLASTIC** (R220523-90)										500
	1100	Angle, PVC, threaded										
	1110	1/4″	1 Plum	26	.308	Ea.	49.50	13.15		62.65	75	
	1120	1/2″		26	.308		49.50	13.15		62.65	75	

15110 | Valves

		Description	Crew	Daily Output	Labor-Hours	Unit	2006 Bare Costs Mat.	Labor	Equip.	Total	Total Incl O&P
500	1130	3/4" (R220523-90)	1 Plum	25	.320	Ea.	58.50	13.65		72.15	86
	1140	1"		23	.348		70	14.85		84.85	100
	1150	Ball, PVC, socket or threaded, single union									
	1230	1/2"	1 Plum	26	.308	Ea.	12.65	13.15		25.80	34.50
	1240	3/4"		25	.320		15.40	13.65		29.05	38.50
	1250	1"		23	.348		18.75	14.85		33.60	43.50
	1260	1-1/4"		21	.381		27.50	16.25		43.75	56
	1270	1-1/2"		20	.400		32	17.10		49.10	62
	1280	2"		17	.471		42	20		62	78
	1290	2-1/2"	Q-1	26	.615		196	23.50		219.50	252
	1300	3"		24	.667		238	25.50		263.50	300
	1310	4"		20	.800		400	30.50		430.50	490
	1360	For PVC, flanged, add					100%	15%			
	1450	Double union 1/2"	1 Plum	26	.308		13.80	13.15		26.95	35.50
	1460	3/4"		25	.320		16.20	13.65		29.85	39.50
	1470	1"		23	.348		17.75	14.85		32.60	42.50
	1480	1-1/4"		21	.381		27	16.25		43.25	55
	1490	1-1/2"		20	.400		30	17.10		47.10	59.50
	1500	2"		17	.471		42.50	20		62.50	78
	1650	CPVC, socket or threaded, single union									
	1700	1/2"	1 Plum	26	.308	Ea.	35	13.15		48.15	59
	1720	3/4"		25	.320		44	13.65		57.65	70
	1730	1"		23	.348		52.50	14.85		67.35	81
	1750	1-1/4"		21	.381		88	16.25		104.25	123
	1760	1-1/2"		20	.400		88	17.10		105.10	123
	1770	2"		17	.471		122	20		142	166
	1780	3"	Q-1	24	.667		365	25.50		390.50	440
	1840	For CPVC, flanged, add					65%	15%			
	1880	For true union, socket or threaded, add					50%	5%			
	2050	Polypropylene, threaded									
	2100	1/4"	1 Plum	26	.308	Ea.	36	13.15		49.15	60
	2120	3/8"		26	.308		36	13.15		49.15	60
	2130	1/2"		26	.308		36	13.15		49.15	60
	2140	3/4"		25	.320		45.50	13.65		59.15	71.50
	2150	1"		23	.348		53.50	14.85		68.35	82
	2160	1-1/4"		21	.381		77.50	16.25		93.75	111
	2170	1-1/2"		20	.400		89.50	17.10		106.60	125
	2180	2"		17	.471		122	20		142	167
	2190	3"	Q-1	24	.667		325	25.50		350.50	395
	2200	4"	"	20	.800		540	30.50		570.50	640
	2550	PVC, three way, socket or threaded									
	2600	1/2"	1 Plum	26	.308	Ea.	44.50	13.15		57.65	69.50
	2640	3/4"		25	.320		50.50	13.65		64.15	77
	2650	1"		23	.348		54.50	14.85		69.35	83
	2660	1-1/2"		20	.400		111	17.10		128.10	149
	2670	2"		17	.471		148	20		168	195
	2680	3"	Q-1	24	.667		320	25.50		345.50	395
	2740	For flanged, add					60%	15%			
	3150	Ball check, PVC, socket or threaded									
	3200	1/4"	1 Plum	26	.308	Ea.	30.50	13.15		43.65	54
	3220	3/8"		26	.308		30.50	13.15		43.65	54
	3240	1/2"		26	.308		30.50	13.15		43.65	54
	3250	3/4"		25	.320		34	13.65		47.65	59
	3260	1"		23	.348		43	14.85		57.85	70
	3270	1-1/4"		21	.381		71.50	16.25		87.75	105
	3280	1-1/2"		20	.400		71.50	17.10		88.60	106

15110 | Valves

500			CREW	DAILY OUTPUT	LABOR-HOURS	UNIT	2006 BARE COSTS MAT.	LABOR	EQUIP.	TOTAL	TOTAL INCL O&P
	3290	2″ (R220523-90)	1 Plum	17	.471	Ea.	98	20		118	140
	3310	3″	Q-1	24	.667		271	25.50		296.50	340
	3320	4″	″	20	.800		385	30.50		415.50	470
	3360	For PVC, flanged, add					50%	15%			
	3750	CPVC, socket or threaded									
	3800	1/2″	1 Plum	26	.308	Ea.	44.50	13.15		57.65	69.50
	3840	3/4″		25	.320		53.50	13.65		67.15	80
	3850	1″		23	.348		63	14.85		77.85	92
	3860	1-1/2″		20	.400		108	17.10		125.10	146
	3870	2″		17	.471		146	20		166	192
	3880	3″	Q-1	24	.667		395	25.50		420.50	475
	3920	4″	″	20	.800		530	30.50		560.50	635
	3930	For CPVC, flanged, add					40%	15%			
	4340	Polypropylene, threaded									
	4360	1/2″	1 Plum	26	.308	Ea.	29.50	13.15		42.65	53
	4400	3/4″		25	.320		34.50	13.65		48.15	59.50
	4440	1″		23	.348		44	14.85		58.85	71.50
	4450	1-1/2″		20	.400		85	17.10		102.10	120
	4460	2″		17	.471		106	20		126	149
	4500	For polypropylene flanged, add					200%	15%			
	4850	Foot valve, PVC, socket or threaded									
	4900	1/2″	1 Plum	34	.235	Ea.	49.50	10.05		59.55	70
	4930	3/4″		32	.250		56	10.70		66.70	78
	4940	1″		28	.286		73	12.20		85.20	99
	4950	1-1/4″		27	.296		140	12.65		152.65	174
	4960	1-1/2″		26	.308		140	13.15		153.15	175
	4970	2″		24	.333		162	14.25		176.25	200
	4980	3″		20	.400		385	17.10		402.10	450
	4990	4″		18	.444		680	19		699	775
	5000	For flanged, add					25%	10%			
	5050	CPVC, socket or threaded									
	5060	1/2″	1 Plum	34	.235	Ea.	48.50	10.05		58.55	68.50
	5070	3/4″		32	.250		55.50	10.70		66.20	78
	5080	1″		28	.286		69	12.20		81.20	94.50
	5090	1-1/4″		27	.296		110	12.65		122.65	141
	5100	1-1/2″		26	.308		110	13.15		123.15	142
	5110	2″		24	.333		141	14.25		155.25	177
	5120	3″		20	.400		288	17.10		305.10	340
	5130	4″		18	.444		525	19		544	605
	5140	For flanged, add					25%	10%			
	5280	Needle valve, PVC, threaded									
	5300	1/4″	1 Plum	26	.308	Ea.	35.50	13.15		48.65	60
	5340	3/8″		26	.308		41.50	13.15		54.65	66
	5360	1/2″		26	.308		41.50	13.15		54.65	66
	5380	For polypropylene, add					10%				
	5800	Y check, PVC, socket or threaded									
	5820	1/2″	1 Plum	26	.308	Ea.	55	13.15		68.15	81
	5840	3/4″		25	.320		59.50	13.65		73.15	87
	5850	1″		23	.348		65	14.85		79.85	94.50
	5860	1-1/4″		21	.381		102	16.25		118.25	138
	5870	1-1/2″		20	.400		111	17.10		128.10	149
	5880	2″		17	.471		138	20		158	183
	5890	2-1/2″		15	.533		291	23		314	355
	5900	3″	Q-1	24	.667		365	25.50		390.50	440
	5910	4″	″	20	.800		475	30.50		505.50	575
	5960	For PVC flanged, add					45%	15%			

15 MECHANICAL

	15110 | Valves		Crew	Daily Output	Labor-Hours	Unit	2006 Bare Costs Mat.	Labor	Equip.	Total	Total Incl O&P
500	6350	Y sediment strainer, PVC, socket or threaded (R220523-90)									
	6400	1/2″	1 Plum	26	.308	Ea.	36	13.15		49.15	60
	6440	3/4″		24	.333		39	14.25		53.25	65
	6450	1″		23	.348		47	14.85		61.85	75
	6460	1-1/4″		21	.381		77.50	16.25		93.75	111
	6470	1-1/2″		20	.400		77.50	17.10		94.60	112
	6480	2″		17	.471		95	20		115	137
	6490	2-1/2″		15	.533		233	23		256	292
	6500	3″	Q-1	24	.667		233	25.50		258.50	296
	6510	4″	″	20	.800		390	30.50		420.50	475
	6560	For PVC, flanged, add					55%	15%			
	9000	Minimum labor/equipment charge	1 Plum	3.50	2.286	Job		97.50		97.50	152
600	0010	**VALVES, SEMI-STEEL**									
	1020	Lubricated plug valve, threaded, 200 psi									
	1030	1/2″	1 Plum	18	.444	Ea.	67	19		86	103
	1040	3/4″		16	.500		67	21.50		88.50	107
	1050	1″		14	.571		85.50	24.50		110	132
	1060	1-1/4″		12	.667		101	28.50		129.50	156
	1070	1-1/2″		11	.727		110	31		141	169
	1080	2″		8	1		129	42.50		171.50	209
	1090	2-1/2″	Q-1	5	3.200		198	123		321	410
	1100	3″	″	4.50	3.556		245	137		382	480
	6990	Flanged, 200 psi									
	7000	2″	1 Plum	8	1	Ea.	160	42.50		202.50	243
	7010	2-1/2″	Q-1	5	3.200		239	123		362	455
	7020	3″		4.50	3.556		278	137		415	515
	7030	4″		3	5.333		350	205		555	705
	7036	5″		2.50	6.400		500	246		746	930
	7040	6″	Q-2	3	8		650	320		970	1,200
	7050	8″		2.50	9.600		1,175	385		1,560	1,900
	7060	10″		2.20	10.909		1,950	435		2,385	2,825
	7070	12″		1.70	14.118		2,975	565		3,540	4,125
700	0010	**VALVES, STEEL** (R220523-90)									
	0800	Cast									
	1350	Check valve, swing type, 150 lb., flanged									
	1370	1″	1 Plum	10	.800	Ea.	380	34		414	470
	1400	2″	″	8	1		610	42.50		652.50	735
	1440	2-1/2″	Q-1	5	3.200		550	123		673	795
	1450	3″		4.50	3.556		720	137		857	1,000
	1460	4″		3	5.333		1,025	205		1,230	1,450
	1470	6″	Q-2	3	8		1,600	320		1,920	2,275
	1480	8″		2.50	9.600		2,625	385		3,010	3,500
	1490	10″		2.20	10.909		3,900	435		4,335	4,950
	1500	12″		1.70	14.118		5,575	565		6,140	7,000
	1510	14″		1.30	18.462		8,025	735		8,760	10,000
	1520	16″		1	24		9,500	955		10,455	11,900
	1548	For 600 lb., flanged, add					110%	20%			
	1571	300 lb., 2″	1 Plum	7.40	1.081		645	46		691	780
	1572	2-1/2″	Q-1	4.20	3.810		970	146		1,116	1,300
	1573	3″		4	4		970	154		1,124	1,325
	1574	4″		2.80	5.714		1,150	220		1,370	1,600
	1575	6″	Q-2	2.90	8.276		1,950	330		2,280	2,675
	1576	8″		2.40	10		3,000	400		3,400	3,925
	1577	10″		2.10	11.429		4,450	455		4,905	5,600
	1578	12″		1.60	15		6,225	600		6,825	7,775
	1579	14″		1.20	20		11,100	795		11,895	13,500

15110 | Valves

			CREW	DAILY OUTPUT	LABOR-HOURS	UNIT	2006 BARE COSTS MAT.	LABOR	EQUIP.	TOTAL	TOTAL INCL O&P	
700	1581	16″	Q-2	.90	26.667	Ea.	14,300	1,075		15,375	17,400	700
	1950	Gate valve, 150 lb., flanged (R220523-90)										
	2000	2″	1 Plum	8	1	Ea.	650	42.50		692.50	780	
	2040	2-1/2″	Q-1	5	3.200		920	123		1,043	1,225	
	2050	3″		4.50	3.556		920	137		1,057	1,225	
	2060	4″		3	5.333		1,150	205		1,355	1,575	
	2070	6″	Q-2	3	8		1,775	320		2,095	2,475	
	2080	8″		2.50	9.600		2,725	385		3,110	3,600	
	2090	10″		2.20	10.909		4,025	435		4,460	5,125	
	2100	12″		1.70	14.118		5,500	565		6,065	6,925	
	2110	14″		1.30	18.462		8,775	735		9,510	10,800	
	2120	16″		1	24		11,300	955		12,255	14,000	
	2130	18″		.80	30		13,800	1,200		15,000	17,000	
	2140	20″		.60	40		16,500	1,600		18,100	20,600	
	2650	300 lb., flanged										
	2700	2″	1 Plum	7.40	1.081	Ea.	890	46		936	1,050	
	2740	2-1/2″	Q-1	4.20	3.810		1,200	146		1,346	1,550	
	2750	3″		4	4		1,200	154		1,354	1,575	
	2760	4″		2.80	5.714		1,675	220		1,895	2,200	
	2770	6″	Q-2	2.90	8.276		2,825	330		3,155	3,650	
	2780	8″		2.40	10		4,425	400		4,825	5,500	
	2790	10″		2.10	11.429		5,950	455		6,405	7,250	
	2800	12″		1.60	15		8,025	600		8,625	9,775	
	2810	14″		1.20	20		16,300	795		17,095	19,200	
	2820	16″		.90	26.667		20,400	1,075		21,475	24,100	
	2830	18″		.70	34.286		29,100	1,375		30,475	34,100	
	2840	20″		.50	48		32,400	1,925		34,325	38,600	
	3650	Globe valve, 150 lb., flanged										
	3700	2″	1 Plum	8	1	Ea.	820	42.50		862.50	965	
	3740	2-1/2″	Q-1	5	3.200		1,050	123		1,173	1,350	
	3750	3″		4.50	3.556		1,050	137		1,187	1,350	
	3760	4″		3	5.333		1,525	205		1,730	2,000	
	3770	6″	Q-2	3	8		2,400	320		2,720	3,125	
	3780	8″		2.50	9.600		4,400	385		4,785	5,425	
	3790	10″		2.20	10.909		7,775	435		8,210	9,225	
	3800	12″		1.70	14.118		10,500	565		11,065	12,400	
	4080	300 lb., flanged										
	4100	2″	1 Plum	7.40	1.081	Ea.	1,100	46		1,146	1,275	
	4140	2-1/2″	Q-1	4.20	3.810		1,500	146		1,646	1,875	
	4150	3″		4	4		1,500	154		1,654	1,900	
	4160	4″		2.80	5.714		2,075	220		2,295	2,625	
	4170	6″	Q-2	2.90	8.276		3,725	330		4,055	4,625	
	4180	8″		2.40	10		6,100	400		6,500	7,350	
	4190	10″		2.10	11.429		11,500	455		11,955	13,400	
	4200	12″		1.60	15		13,300	600		13,900	15,600	
	4680	600 lb., flanged										
	4700	2″	1 Plum	7	1.143	Ea.	1,525	49		1,574	1,750	
	4740	2-1/2″	Q-1	4	4		2,300	154		2,454	2,775	
	4750	3″		3.60	4.444		2,300	171		2,471	2,800	
	4760	4″		2.50	6.400		3,700	246		3,946	4,425	
	4770	6″	Q-2	2.60	9.231		7,675	370		8,045	9,025	
	4780	8″	″	2.10	11.429		12,500	455		12,955	14,400	
	5150	Forged										
	5340	Ball valve, 800 psi, threaded, 1/4″ size	1 Plum	24	.333	Ea.	42.50	14.25		56.75	68.50	
	5350	3/8″		24	.333		42.50	14.25		56.75	68.50	
	5360	1/2″		24	.333		52.50	14.25		66.75	79.50	

15 MECHANICAL

15110 | Valves

		15110 Valves	CREW	DAILY OUTPUT	LABOR-HOURS	UNIT	2006 BARE COSTS MAT.	LABOR	EQUIP.	TOTAL	TOTAL INCL O&P
700	5370	3/4" (R220523-90)	1 Plum	20	.400	Ea.	69.50	17.10		86.60	103
	5380	1"		19	.421		86.50	18		104.50	124
	5390	1-1/4"		15	.533		116	23		139	163
	5400	1-1/2"		13	.615		151	26.50		177.50	207
	5410	2"		11	.727		200	31		231	268
	5460	Ball valve, 800 lb., socket weld, 1/4" size	Q-15	19	.842		52.50	32.50	3.64	88.64	112
	5470	3/8"		19	.842		52	32.50	3.64	88.14	112
	5480	1/2"		19	.842		52.50	32.50	3.64	88.64	112
	5490	3/4"		19	.842		69.50	32.50	3.64	105.64	131
	5500	1"		15	1.067		86.50	41	4.61	132.11	164
	5510	1-1/4"		13	1.231		116	47.50	5.30	168.80	206
	5520	1-1/2"		11	1.455		151	56	6.30	213.30	260
	5530	2"		8.50	1.882		200	72.50	8.15	280.65	340
	5650	Check valve, class 800, horizontal, socket									
	5698	Threaded									
	5700	1/4"	1 Plum	24	.333	Ea.	67.50	14.25		81.75	96.50
	5720	3/8"		24	.333		67.50	14.25		81.75	96.50
	5730	1/2"		24	.333		67.50	14.25		81.75	96.50
	5740	3/4"		20	.400		73.50	17.10		90.60	108
	5750	1"		19	.421		85.50	18		103.50	123
	5760	1-1/4"		15	.533		168	23		191	221
	5770	1-1/2"		13	.615		168	26.50		194.50	226
	5780	2"		11	.727		235	31		266	310
	5840	For class 150, flanged, add					100%	15%			
	5860	For class 300, flanged, add					120%	20%			
	6100	Gate, class 800, OS&Y, socket									
	6102	3/8"	Q-15	19	.842	Ea.	47.50	32.50	3.64	83.64	107
	6103	1/2"		19	.842		47.50	32.50	3.64	83.64	107
	6104	3/4"		19	.842		52.50	32.50	3.64	88.64	113
	6105	1"		15	1.067		63.50	41	4.61	109.11	138
	6106	1-1/4"		13	1.231		120	47.50	5.30	172.80	211
	6107	1-1/2"		11	1.455		120	56	6.30	182.30	226
	6108	2"		8.50	1.882		156	72.50	8.15	236.65	293
	6118	Threaded									
	6120	3/8"	1 Plum	24	.333	Ea.	47.50	14.25		61.75	74.50
	6130	1/2"		24	.333		47.50	14.25		61.75	74.50
	6140	3/4"		20	.400		52.50	17.10		69.60	84.50
	6150	1"		19	.421		63.50	18		81.50	97.50
	6160	1-1/4"		15	.533		120	23		143	168
	6170	1-1/2"		13	.615		120	26.50		146.50	173
	6180	2"		11	.727		156	31		187	221
	6260	For OS&Y, flanged, add					100%	20%			
	6700	Globe, OS&Y, class 800, socket									
	6710	1/4"	Q-15	19	.842	Ea.	73.50	32.50	3.64	109.64	136
	6720	3/8"		19	.842		73.50	32.50	3.64	109.64	136
	6730	1/2"		19	.842		73.50	32.50	3.64	109.64	136
	6740	3/4"		19	.842		83.50	32.50	3.64	119.64	147
	6750	1"		15	1.067		109	41	4.61	154.61	189
	6760	1-1/4"		13	1.231		213	47.50	5.30	265.80	315
	6770	1-1/2"		11	1.455		213	56	6.30	275.30	330
	6780	2"		8.50	1.882		274	72.50	8.15	354.65	420
800	0010	**VALVES, STAINLESS STEEL** (R220523-90)									
	1610	Ball, threaded 1/4"	1 Stpi	24	.333	Ea.	35.50	14.35		49.85	61.50
	1620	3/8"		24	.333		35.50	14.35		49.85	61.50
	1630	1/2"		22	.364		35.50	15.65		51.15	63.50

15110 | Valves

	Line	Description	Crew	Daily Output	Labor-Hours	Unit	Mat. (2006 Bare Costs)	Labor (2006 Bare Costs)	Equip. (2006 Bare Costs)	Total (2006 Bare Costs)	Total Incl O&P	
800	1640	3/4″ R220523-90	1 Stpi	20	.400	Ea.	58.50	17.20		75.70	91.50	800
	1650	1″		19	.421		71	18.15		89.15	107	
	1660	1-1/4		15	.533		136	23		159	185	
	1670	1-1/2″		13	.615		142	26.50		168.50	197	
	1680	2″		11	.727		184	31.50		215.50	251	
	1700	Check, 200 lb., threaded										
	1710	1/4″	1 Plum	24	.333	Ea.	107	14.25		121.25	140	
	1720	1/2″		22	.364		107	15.55		122.55	142	
	1730	3/4″		20	.400		117	17.10		134.10	155	
	1750	1″		19	.421		150	18		168	193	
	1760	1-1/2″		13	.615		287	26.50		313.50	355	
	1770	2″		11	.727		505	31		536	605	
	1800	150 lb., flanged										
	1810	2-1/2″	Q-1	5	3.200	Ea.	1,150	123		1,273	1,475	
	1820	3″		4.50	3.556		1,150	137		1,287	1,475	
	1830	4″		3	5.333		1,725	205		1,930	2,225	
	1840	6″	Q-2	3	8		3,075	320		3,395	3,875	
	1850	8″	″	2.50	9.600		5,125	385		5,510	6,250	
	2100	Gate, OS&Y, 150 lb., flanged										
	2120	1/2″	1 Plum	18	.444	Ea.	340	19		359	405	
	2140	3/4″		16	.500		360	21.50		381.50	435	
	2150	1″		14	.571		445	24.50		469.50	530	
	2160	1-1/2″		11	.727		620	31		651	730	
	2170	2″		8	1		695	42.50		737.50	830	
	2180	2-1/2″	Q-1	5	3.200		1,150	123		1,273	1,450	
	2190	3″		4.50	3.556		1,150	137		1,287	1,450	
	2200	4″		3	5.333		1,700	205		1,905	2,175	
	2205	5″		2.80	5.714		2,925	220		3,145	3,575	
	2210	6″	Q-2	3	8		2,925	320		3,245	3,725	
	2220	8″		2.50	9.600		4,850	385		5,235	5,950	
	2230	10″		2.30	10.435		7,750	415		8,165	9,175	
	2240	12″		1.90	12.632		9,875	505		10,380	11,700	
	2260	For 300 lb., flanged, add					120%	15%				
	2600	600 lb., flanged										
	2620	1/2″	1 Plum	16	.500	Ea.	154	21.50		175.50	202	
	2640	3/4″		14	.571		166	24.50		190.50	221	
	2650	1″		12	.667		200	28.50		228.50	265	
	2660	1-1/2″		10	.800		320	34		354	405	
	2670	2″		7	1.143		440	49		489	560	
	2680	2-1/2″	Q-1	4	4		4,075	154		4,229	4,725	
	2690	3″	″	3.60	4.444		4,075	171		4,246	4,750	
	3100	Globe, OS&Y, 150 lb., flanged										
	3120	1/2″	1 Plum	18	.444	Ea.	340	19		359	405	
	3140	3/4″		16	.500		380	21.50		401.50	450	
	3150	1″		14	.571		470	24.50		494.50	555	
	3160	1-1/2″		11	.727		695	31		726	815	
	3170	2″		8	1		850	42.50		892.50	1,000	
	3180	2-1/2″	Q-1	5	3.200		1,825	123		1,948	2,200	
	3190	3″		4.50	3.556		1,825	137		1,962	2,200	
	3200	4″		3	5.333		2,150	205		2,355	2,700	
	3210	6″	Q-2	3	8		3,775	320		4,095	4,650	

15120 | Piping Specialties

	Line	Description	Crew	Daily Output	Labor-Hours	Unit	Mat.	Labor	Equip.	Total	Total Incl O&P	
120	0010	**AIR CONTROL**										120
	0030	Air separator, with strainer										

15 MECHANICAL

		15120 | Piping Specialties	Crew	Daily Output	Labor-Hours	Unit	2006 Bare Costs Mat.	Labor	Equip.	Total	Total Incl O&P	
120	0040	2″ diameter	Q-5	6	2.667	Ea.	655	103		758	880	120
	0080	2-1/2″ diameter		5	3.200		745	124		869	1,025	
	0100	3″ diameter		4	4		1,125	155		1,280	1,500	
	0120	4″ diameter		3	5.333		1,650	207		1,857	2,125	
	0130	5″ diameter	Q-6	3.60	6.667		2,100	268		2,368	2,725	
	0140	6″ diameter		3.40	7.059		2,500	284		2,784	3,200	
	0160	8″ diameter		3	8		3,750	320		4,070	4,600	
	0180	10″ diameter		2.20	10.909		5,800	440		6,240	7,050	
	0200	12″ diameter		1.70	14.118		9,725	565		10,290	11,600	
160	0010	**AUTOMATIC AIR VENT**										160
	0020	Cast iron body, stainless steel internals, float type										
	0060	1/2″ NPT inlet, 300 psi	1 Stpi	12	.667	Ea.	76.50	28.50		105	129	
	0140	3/4″ NPT inlet, 300 psi		12	.667		76.50	28.50		105	129	
	0180	1/2″ NPT inlet, 250 psi		10	.800		246	34.50		280.50	325	
	0220	3/4″ NPT inlet, 250 psi		10	.800		246	34.50		280.50	325	
	0260	1″ NPT inlet, 250 psi		10	.800		360	34.50		394.50	450	
	0340	1-1/2″ NPT inlet, 250 psi	Q-5	12	1.333		765	51.50		816.50	920	
	0380	2″ NPT inlet, 250 psi	″	12	1.333		765	51.50		816.50	920	
	0600	Forged steel body, stainless steel internals, float type										
	0640	1/2″ NPT inlet, 750 psi	1 Stpi	12	.667	Ea.	760	28.50		788.50	880	
	0680	3/4″ NPT inlet, 750 psi		12	.667		760	28.50		788.50	880	
	0760	3/4″ NPT inlet, 1000 psi		10	.800		1,150	34.50		1,184.50	1,300	
	0800	1″ NPT inlet, 1000 psi	Q-5	12	1.333		1,150	51.50		1,201.50	1,325	
	0880	1-1/2″ NPT inlet, 1000 psi		10	1.600		3,200	62		3,262	3,625	
	0920	2″ NPT inlet, 1000 psi		10	1.600		3,200	62		3,262	3,625	
	1100	Formed steel body, non corrosive										
	1110	1/8″ NPT inlet 150 psi	1 Stpi	32	.250	Ea.	7.70	10.75		18.45	25.50	
	1120	1/4″ NPT inlet 150 psi		32	.250		26.50	10.75		37.25	46.50	
	1130	3/4″ NPT inlet 150 psi		32	.250		26.50	10.75		37.25	46.50	
	1300	Chrome plated brass, automatic/manual, for radiators										
	1310	1/8″ NPT inlet, nickel plated brass	1 Stpi	32	.250	Ea.	4.59	10.75		15.34	22	
180	0010	**CIRCUIT SENSOR** Flow meter										180
	0020	Metering stations										
	0040	Wafer orifice insert type										
	0060	2-1/2″ pipe size	Q-5	12	1.333	Ea.	140	51.50		191.50	235	
	0100	3″ pipe size		11	1.455		154	56.50		210.50	257	
	0140	4″ pipe size		8	2		175	77.50		252.50	315	
	0180	5″ pipe size		7.30	2.192		224	85		309	380	
	0220	6″ pipe size		6.40	2.500		266	97		363	445	
	0260	8″ pipe size	Q-6	5.30	4.528		370	182		552	695	
	0280	10″ pipe size		4.60	5.217		430	210		640	800	
	0360	12″ pipe size		4.20	5.714		715	230		945	1,150	
200	0010	**CIRCUIT SETTER** Balance valve										200
	0020	3/4″ pipe size	1 Stpi	20	.400	Ea.	52.50	17.20		69.70	84.50	
	0040	1″ pipe size		18	.444		67.50	19.15		86.65	104	
	0060	1-1/2″ pipe size		12	.667		117	28.50		145.50	174	
	0080	2″ pipe size		10	.800		168	34.50		202.50	239	
	0100	2-1/2″ pipe size	Q-5	15	1.067		380	41.50		421.50	485	
	0120	3″ pipe size		10	1.600		535	62		597	685	
	0136	3″ pipe size, flanged		4	4		575	155		730	875	
	0140	4″ pipe size		3	5.333		800	207		1,007	1,200	
	0200	For differential meter, accurate to 1%, add					650			650	715	
250	0010	**DIELECTRIC UNIONS** Standard gaskets for water and air										250
	0020	250 psi maximum pressure										

MECHANICAL 15

15120 | Piping Specialties		CREW	DAILY OUTPUT	LABOR-HOURS	UNIT	2006 BARE COSTS MAT.	LABOR	EQUIP.	TOTAL	TOTAL INCL O&P
250 0280	Female IPT to sweat, straight									
0300	1/2" pipe size	1 Plum	24	.333	Ea.	3.63	14.25		17.88	26
0340	3/4" pipe size		20	.400		3.63	17.10		20.73	30.50
0360	1" pipe size		19	.421		7.05	18		25.05	36
0380	1-1/4" pipe size		15	.533		10.50	23		33.50	47
0400	1-1/2" pipe size		13	.615		18.15	26.50		44.65	61
0420	2" pipe size	↓	11	.727	↓	25.50	31		56.50	76.50
0580	Female IPT to brass pipe thread, straight									
0600	1/2" pipe size	1 Plum	24	.333	Ea.	6.85	14.25		21.10	29.50
0640	3/4" pipe size		20	.400		8.75	17.10		25.85	36
0660	1" pipe size		19	.421		13.15	18		31.15	42.50
0680	1-1/4" pipe size		15	.533		20.50	23		43.50	58
0700	1-1/2" pipe size		13	.615		28.50	26.50		55	72
0720	2" pipe size	↓	11	.727	↓	37	31		68	89.50
0780	Female IPT to female IPT, straight									
0800	1/2" pipe size	1 Plum	24	.333	Ea.	5.35	14.25		19.60	28
0840	3/4" pipe size		20	.400		5.45	17.10		22.55	32.50
0860	1" pipe size		19	.421		8.05	18		26.05	37
0880	1-1/4" pipe size		15	.533		11.80	23		34.80	48.50
0900	1-1/2" pipe size		13	.615		18.05	26.50		44.55	61
0920	2" pipe size	↓	11	.727	↓	26	31		57	77
2000	175 psi maximum pressure									
2180	Female IPT to sweat									
2200	1-1/2" pipe size	1 Plum	11	.727	Ea.	46.50	31		77.50	99.50
2240	2" pipe size	"	9	.889		53.50	38		91.50	118
2260	2-1/2" pipe size	Q-1	15	1.067		58	41		99	127
2280	3" pipe size		14	1.143		79	44		123	156
2300	4" pipe size	↓	11	1.455		210	56		266	320
2320	5" pipe size	Q-2	14	1.714		345	68.50		413.50	480
2340	6" pipe size		12	2		395	79.50		474.50	560
2360	8" pipe size	↓	10	2.400	↓	1,100	95.50		1,195.50	1,350
2480	Female IPT to brass pipe									
2500	1-1/2" pipe size	1 Plum	11	.727	Ea.	66.50	31		97.50	122
2540	2" pipe size	"	9	.889		79.50	38		117.50	147
2560	2-1/2" pipe size	Q-1	15	1.067		115	41		156	190
2580	3" pipe size		14	1.143		134	44		178	216
2600	4" pipe size	↓	11	1.455		240	56		296	350
2620	5" pipe size	Q-2	14	1.714		270	68.50		338.50	405
2640	6" pipe size		12	2		325	79.50		404.50	480
2680	8" pipe size	↓	10	2.400	↓	525	95.50		620.50	730
2880	Female IPT to female IPT									
2900	1-1/2" pipe size	1 Plum	11	.727	Ea.	41.50	31		72.50	94
2940	2" pipe size	"	9	.889		44.50	38		82.50	108
2960	2-1/2" pipe size	Q-1	15	1.067		55.50	41		96.50	125
2980	3" pipe size		14	1.143		69	44		113	145
3000	4" pipe size	↓	11	1.455		110	56		166	208
3020	5" pipe size	Q-2	14	1.714		127	68.50		195.50	246
3040	6" pipe size		12	2		170	79.50		249.50	310
3060	8" pipe size	↓	10	2.400	↓	203	95.50		298.50	370
3380	Copper to copper									
3400	1-1/2" pipe size	1 Plum	11	.727	Ea.	57.50	31		88.50	112
3440	2" pipe size	"	9	.889		65.50	38		103.50	131
3460	2-1/2" pipe size	Q-1	15	1.067		70	41		111	141
3480	3" pipe size		14	1.143		87.50	44		131.50	165
3500	4" pipe size	↓	11	1.455		277	56		333	390
3520	5" pipe size	Q-2	14	1.714	↓	445	68.50		513.50	595

15120 | Piping Specialties

		15120 Piping Specialties	CREW	DAILY OUTPUT	LABOR-HOURS	UNIT	2006 BARE COSTS MAT.	LABOR	EQUIP.	TOTAL	TOTAL INCL O&P	
250	3540	6" pipe size	Q-2	12	2	Ea.	545	79.50		624.50	725	250
	3560	8" pipe size	↓	10	2.400	↓	1,450	95.50		1,545.50	1,725	
	9000	Minimum labor/equipment charge	1 Plum	4	2	Job		85.50		85.50	133	
280	0010	**EXPANSION COUPLINGS** Hydronic										280
	0100	Copper to copper, sweat										
	0120	3/4" diameter	1 Stpi	20	.400	Ea.	26	17.20		43.20	56	
	0140	1" diameter		19	.421		33.50	18.15		51.65	65	
	0160	1-1/4" diameter	↓	15	.533	↓	45.50	23		68.50	85.50	
	1000	Baseboard riser fitting, 5" stub by coupling 12" long										
	1020	1/2" diameter	1 Stpi	24	.333	Ea.	12.75	14.35		27.10	36.50	
	1040	3/4" diameter		20	.400		18.10	17.20		35.30	47	
	1060	1" diameter		19	.421		25	18.15		43.15	55.50	
	1080	1-1/4" diameter	↓	15	.533	↓	33	23		56	71.50	
	1180	9" Stub by tubing 8" long										
	1200	1/2" diameter	1 Stpi	24	.333	Ea.	11.40	14.35		25.75	35	
	1220	3/4" diameter		20	.400		15.15	17.20		32.35	43.50	
	1240	1" diameter		19	.421		21	18.15		39.15	51	
	1260	1-1/4" diameter	↓	15	.533	↓	29.50	23		52.50	68	
300	0010	**EXPANSION JOINTS**										300
	0100	Bellows type, neoprene cover, flanged spool										
	0140	6" face to face, 1-1/4" diameter	1 Stpi	11	.727	Ea.	210	31.50		241.50	280	
	0160	1-1/2" diameter	"	10.60	.755		210	32.50		242.50	282	
	0180	2" diameter	Q-5	13.30	1.203		214	46.50		260.50	310	
	0190	2-1/2" diameter		12.40	1.290		222	50		272	320	
	0200	3" diameter		11.40	1.404		252	54.50		306.50	360	
	0480	10" face to face, 2" diameter		13	1.231		315	47.50		362.50	420	
	0500	2-1/2" diameter		12	1.333		330	51.50		381.50	445	
	0520	3" diameter		11	1.455		335	56.50		391.50	460	
	0540	4" diameter		8	2		375	77.50		452.50	535	
	0560	5" diameter		7	2.286		455	88.50		543.50	640	
	0580	6" diameter		6	2.667		465	103		568	675	
	0600	8" diameter		5	3.200		530	124		654	775	
	0620	10" diameter		4.60	3.478		590	135		725	860	
	0640	12" diameter		4	4		675	155		830	985	
	0660	14" diameter	↓	3.80	4.211	↓	850	163		1,013	1,200	
320	0010	**EXPANSION TANKS**										320
	1400	Plastic, corrosion resistant, see Plumbing Cost Data, Div. 15220										
	1505	Fiberglass and steel single / double wall storage, see Div 13201										
	1510	Tank leak detection systems, see Div 13851-350										
	2000	Steel, liquid expansion, ASME, painted, 15 gallon capacity	Q-5	17	.941	Ea.	365	36.50		401.50	455	
	2020	24 gallon capacity		14	1.143		380	44.50		424.50	485	
	2040	30 gallon capacity		12	1.333		420	51.50		471.50	540	
	2060	40 gallon capacity		10	1.600		460	62		522	600	
	2080	60 gallon capacity		8	2		550	77.50		627.50	725	
	2100	80 gallon capacity		7	2.286		615	88.50		703.50	815	
	2120	100 gallon capacity		6	2.667		805	103		908	1,050	
	2130	120 gallon capacity		5	3.200		915	124		1,039	1,200	
	2140	135 gallon capacity		4.50	3.556		970	138		1,108	1,300	
	2150	175 gallon capacity		4	4		1,375	155		1,530	1,750	
	2160	220 gallon capacity		3.60	4.444		1,625	172		1,797	2,050	
	2170	240 gallon capacity		3.30	4.848		1,850	188		2,038	2,325	
	2180	305 gallon capacity		3	5.333		2,675	207		2,882	3,275	
	2190	400 gallon capacity	↓	2.80	5.714	↓	3,125	221		3,346	3,775	

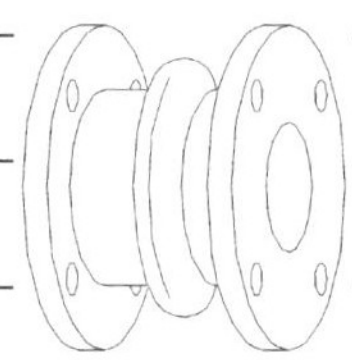

15120 | Piping Specialties

		15120 Piping Specialties	CREW	DAILY OUTPUT	LABOR-HOURS	UNIT	2006 BARE COSTS MAT.	LABOR	EQUIP.	TOTAL	TOTAL INCL O&P
320	2360	Galvanized									
	2370	15 gallon capacity	Q-5	17	.941	Ea.	630	36.50		666.50	750
	2380	24 gallon capacity		14	1.143		670	44.50		714.50	810
	2390	30 gallon capacity		12	1.333		755	51.50		806.50	910
	2400	40 gallon capacity		10	1.600		885	62		947	1,075
	2410	60 gallon capacity		8	2		1,025	77.50		1,102.50	1,250
	2420	80 gallon capacity		7	2.286		1,175	88.50		1,263.50	1,425
	2430	100 gallon capacity		6	2.667		1,525	103		1,628	1,825
	2440	120 gallon capacity		5	3.200		1,625	124		1,749	2,000
	2450	135 gallon capacity		4.50	3.556		1,725	138		1,863	2,125
	2460	175 gallon capacity		4	4		2,800	155		2,955	3,325
	2470	220 gallon capacity		3.60	4.444		3,200	172		3,372	3,800
	2480	240 gallon capacity		3.30	4.848		3,675	188		3,863	4,350
	2490	305 gallon capacity		3	5.333		4,850	207		5,057	5,650
	2500	400 gallon capacity		2.80	5.714		5,975	221		6,196	6,900
	3000	Steel ASME expansion, rubber diaphragm, 19 gal. cap. accept.		12	1.333		1,700	51.50		1,751.50	1,950
	3020	31 gallon capacity		8	2		1,875	77.50		1,952.50	2,200
	3040	61 gallon capacity		6	2.667		1,975	103		2,078	2,325
	3060	79 gallon capacity		5	3.200		2,700	124		2,824	3,150
	3080	119 gallon capacity		4	4		2,850	155		3,005	3,375
	3100	158 gallon capacity		3.80	4.211		3,950	163		4,113	4,600
	3120	211 gallon capacity		3.30	4.848		4,575	188		4,763	5,325
	3140	317 gallon capacity		2.80	5.714		5,975	221		6,196	6,925
	3160	422 gallon capacity		2.60	6.154		8,850	238		9,088	10,100
	3180	528 gallon capacity		2.40	6.667	↓	9,700	258		9,958	11,100
	9000	Minimum labor/equipment charge	↓	4	4	Job		155		155	241
350	0010	**FLEXIBLE CONNECTORS**, Corrugated, 7/8" O.D., 1/2" I.D.									
	0050	Gas, seamless brass, steel fittings									
	0200	12" long	1 Plum	36	.222	Ea.	12.35	9.50		21.85	28.50
	0220	18" long		36	.222		15.30	9.50		24.80	31.50
	0240	24" long		34	.235		18.05	10.05		28.10	35.50
	0260	30" long		34	.235		19.55	10.05		29.60	37
	0280	36" long		32	.250		21.50	10.70		32.20	40.50
	0320	48" long		30	.267		27.50	11.40		38.90	47.50
	0340	60" long		30	.267		32.50	11.40		43.90	53.50
	0360	72" long	↓	30	.267	↓	37.50	11.40		48.90	59
	2000	Water, copper tubing, dielectric separators									
	2100	12" long	1 Plum	36	.222	Ea.	9.70	9.50		19.20	25.50
	2220	15" long		36	.222		10.80	9.50		20.30	26.50
	2240	18" long		36	.222		11.75	9.50		21.25	27.50
	2260	24" long		34	.235	↓	14.40	10.05		24.45	31.50
	9000	Minimum labor/equipment charge	↓	4	2	Job		85.50		85.50	133
370	0010	**FLEXIBLE METAL HOSE** Connectors, standard lengths									
	0100	Bronze braided, bronze ends									
	0120	3/8" diameter x 12"	1 Stpi	26	.308	Ea.	16.95	13.25		30.20	39
	0140	1/2" diameter x 12"		24	.333		16.15	14.35		30.50	40.50
	0160	3/4" diameter x 12"		20	.400		23.50	17.20		40.70	53
	0180	1" diameter x 18"		19	.421		30.50	18.15		48.65	61.50
	0200	1-1/2" diameter x 18"		13	.615		48.50	26.50		75	94
	0220	2" diameter x 18"	↓	11	.727	↓	58	31.50		89.50	113
	1000	Carbon steel ends									
	1020	1/4" diameter x 12"	1 Stpi	28	.286	Ea.	13.90	12.30		26.20	34.50
	1040	3/8" diameter x 12"		26	.308		14.30	13.25		27.55	36.50
	1060	1/2" diameter x 12"		24	.333		13.90	14.35		28.25	38
	1080	1/2" diameter x 24"		24	.333		17.85	14.35		32.20	42
	1100	1/2" diameter x 36"	↓	24	.333	↓	22	14.35		36.35	46.50

15 MECHANICAL

	15120 | Piping Specialties		CREW	DAILY OUTPUT	LABOR-HOURS	UNIT	2006 BARE COSTS MAT.	LABOR	EQUIP.	TOTAL	TOTAL INCL O&P
370	1120	3/4" diameter x 12"	1 Stpi	20	.400	Ea.	18.45	17.20		35.65	47.50
	1140	3/4" diameter x 24"		20	.400		22.50	17.20		39.70	52
	1160	3/4" diameter x 36"		20	.400		27	17.20		44.20	56.50
	1180	1" diameter x 18"		19	.421		24.50	18.15		42.65	54.50
	1200	1" diameter x 30"		19	.421		28.50	18.15		46.65	59.50
	1220	1" diameter x 36"		19	.421		31	18.15		49.15	62
	1240	1-1/4" diameter x 18"		15	.533		25	23		48	63
	1260	1-1/4" diameter x 36"		15	.533		33	23		56	72
	1280	1-1/2" diameter x 18"		13	.615		26.50	26.50		53	70.50
	1300	1-1/2" diameter x 36"		13	.615		35.50	26.50		62	80
	1320	2" diameter x 24"		11	.727		35	31.50		66.50	87
	1340	2" diameter x 36"		11	.727		43.50	31.50		75	96
	1360	2-1/2" diameter x 24"		9	.889		49.50	38.50		88	114
	1380	2-1/2" diameter x 36"		9	.889		58.50	38.50		97	124
	1400	3" diameter x 24"		7	1.143		59	49		108	141
	1420	3" diameter x 36"		7	1.143		68.50	49		117.50	152
	2000	Carbon steel braid, carbon steel solid ends									
	2100	1/2" diameter x 12"	1 Stpi	24	.333	Ea.	13.90	14.35		28.25	38
	2120	3/4" diameter x 12"		20	.400		18.45	17.20		35.65	47.50
	2140	1" diameter x 12"		19	.421		22	18.15		40.15	52
	2160	1-1/4" diameter x 12"		15	.533		22	23		45	59.50
	2180	1-1/2" diameter x 12"		13	.615		24	26.50		50.50	67
	3000	Stainless steel braid, welded on carbon steel ends									
	3100	1/2" diameter x 12"	1 Stpi	24	.333	Ea.	20.50	14.35		34.85	45
	3120	3/4" diameter x 12"		20	.400		29	17.20		46.20	59
	3140	3/4" diameter x 24"		20	.400		35	17.20		52.20	65.50
	3160	3/4" diameter x 36"		20	.400		41	17.20		58.20	72
	3180	1" diameter x 12"		19	.421		33	18.15		51.15	64
	3200	1" diameter x 24"		19	.421		40	18.15		58.15	72
	3220	1" diameter x 36"		19	.421		47	18.15		65.15	79.50
	3240	1-1/4" diameter x 12"		15	.533		36	23		59	75
	3260	1-1/4" diameter x 24"		15	.533		45.50	23		68.50	85.50
	3280	1-1/4" diameter x 36"		15	.533		55	23		78	96
	3300	1-1/2" diameter x 12"		13	.615		39.50	26.50		66	84.50
	3320	1-1/2" diameter x 24"		13	.615		50.50	26.50		77	96.50
	3340	1-1/2" diameter x 36"		13	.615		61.50	26.50		88	109
520	0010	HYDRONIC HEATING CONTROL VALVES									
	0050	Hot water, nonelectric, thermostatic									
	0100	Radiator supply, 1/2" diameter	1 Stpi	24	.333	Ea.	41.50	14.35		55.85	68.50
	0120	3/4" diameter		20	.400		42.50	17.20		59.70	73.50
	0140	1" diameter		19	.421		53	18.15		71.15	86
	0160	1-1/4" diameter		15	.533		74	23		97	117
	0500	For low pressure steam, add					25%				
	1000	Manual, radiator supply									
	1010	1/2" pipe size, angle union	1 Stpi	24	.333	Ea.	32.50	14.35		46.85	58
	1020	3/4" pipe size, angle union		20	.400		30.50	17.20		47.70	60.50
	1030	1" pipe size, angle union		19	.421		39.50	18.15		57.65	71.50
	1100	Radiator, balancing, straight, sweat connections									
	1110	1/2" pipe size	1 Stpi	24	.333	Ea.	9.25	14.35		23.60	32.50
	1120	3/4" pipe size		20	.400		12.95	17.20		30.15	41
	1130	1" pipe size		19	.421		21	18.15		39.15	51
	1140	Balance and stop valve 1/2" size		22	.364		10.90	15.65		26.55	36.50
	1150	3/4" size		20	.400		13	17.20		30.20	41.50
	1160	1" size		19	.421		15.55	18.15		33.70	45
	1170	1-1/4" size		15	.533		19.75	23		42.75	57
	1200	Steam, radiator, supply									

15120 | Piping Specialties

			CREW	DAILY OUTPUT	LABOR-HOURS	UNIT	2006 BARE COSTS MAT.	LABOR	EQUIP.	TOTAL	TOTAL INCL O&P	
520	1210	1/2" pipe size, angle union	1 Stpi	24	.333	Ea.	23.50	14.35		37.85	48.50	520
	1220	3/4" pipe size, angle union		20	.400		26	17.20		43.20	55.50	
	1230	1" pipe size, angle union		19	.421		27.50	18.15		45.65	58.50	
	1240	1-1/4" pipe size, angle union		15	.533		38	23		61	77.50	
	8000	System balancing and shut-off										
	8020	Butterfly, quarter turn, calibrated, threaded or solder										
	8040	Bronze, -30° F to +350° F, pressure to 175 psi										
	8060	1/2" size	1 Stpi	22	.364	Ea.	6.65	15.65		22.30	32	
	8070	3/4" size		20	.400		10.80	17.20		28	39	
	8080	1" size		19	.421		13.15	18.15		31.30	42.50	
	8090	1-1/4" size		15	.533		21	23		44	59	
	8100	1-1/2" size		13	.615		27.50	26.50		54	71	
	8110	2" size		11	.727		34.50	31.50		66	86	
580	0010	**MIXING VALVE** Automatic water tempering										580
	0040	1/2" size	1 Stpi	19	.421	Ea.	415	18.15		433.15	485	
	0050	3/4" size		18	.444		415	19.15		434.15	485	
	0100	1" size		16	.500		595	21.50		616.50	690	
	0120	1-1/4" size		13	.615		860	26.50		886.50	985	
	0140	1-1/2" size		10	.800		1,025	34.50		1,059.50	1,175	
	0160	2" size		8	1		1,300	43		1,343	1,500	
	0170	2-1/2" size		6	1.333		1,300	57.50		1,357.50	1,525	
	0180	3" size		4	2		2,975	86		3,061	3,400	
	0190	4" size		3	2.667		2,975	115		3,090	3,450	
	9000	Minimum labor/equipment charge		5	1.600	Job		69		69	107	
640	0010	**PRESSURE REDUCING VALVE** Steam, pilot operated										640
	0100	Threaded, iron body										
	0200	1-1/2" size	1 Stpi	8	1	Ea.	860	43		903	1,000	
	0220	2" size	"	5	1.600	"	1,125	69		1,194	1,350	
	1000	Flanged, iron body, 125 lb. flanges										
	1020	2" size	1 Stpi	8	1	Ea.	1,250	43		1,293	1,450	
	1040	2-1/2" size	"	4	2		1,500	86		1,586	1,775	
	1060	3" size	Q-5	4.50	3.556		1,525	138		1,663	1,900	
	1080	4" size	"	3	5.333		1,925	207		2,132	2,450	
	1500	For 250 lb. flanges, add					5%					
670	0010	**PRESSURE REGULATOR**										670
	0200	Oil, light, hot water, ordinary steam, threaded										
	0220	Bronze body, 1/4" size	1 Stpi	24	.333	Ea.	150	14.35		164.35	188	
	0230	3/8" size		24	.333		162	14.35		176.35	201	
	0240	1/2" size		24	.333		201	14.35		215.35	244	
	0250	3/4" size		20	.400		233	17.20		250.20	283	
	0260	1" size		19	.421		365	18.15		383.15	430	
	0270	1-1/4" size		15	.533		490	23		513	575	
	0280	1-1/2" size		13	.615		530	26.50		556.50	620	
	0290	2" size		11	.727		960	31.50		991.50	1,100	
	0320	Iron body, 1/4" size		24	.333		117	14.35		131.35	152	
	0330	3/8" size		24	.333		138	14.35		152.35	174	
	0340	1/2" size		24	.333		146	14.35		160.35	183	
	0350	3/4" size		20	.400		177	17.20		194.20	222	
	0360	1" size		19	.421		231	18.15		249.15	282	
	0370	1-1/4" size		15	.533		320	23		343	390	
	0380	1-1/2" size		13	.615		355	26.50		381.50	430	
	0390	2" size		11	.727		515	31.50		546.50	620	
	0500	Oil, heavy, viscous fluids, threaded										
	0520	Bronze body, 3/8" size	1 Stpi	24	.333	Ea.	257	14.35		271.35	305	

	15120 | Piping Specialties	CREW	DAILY OUTPUT	LABOR-HOURS	UNIT	2006 BARE COSTS MAT.	LABOR	EQUIP.	TOTAL	TOTAL INCL O&P	
670	0530 1/2" size	1 Stpi	24	.333	Ea.	330	14.35		344.35	390	670
	0540 3/4" size		20	.400		370	17.20		387.20	430	
	0550 1" size		19	.421		460	18.15		478.15	535	
	0560 1-1/4" size		15	.533		650	23		673	750	
	0570 1-1/2" size		13	.615		745	26.50		771.50	855	
	0600 Iron body, 3/8" size		24	.333		205	14.35		219.35	248	
	0620 1/2" size		24	.333		248	14.35		262.35	296	
	0630 3/4" size		20	.400		281	17.20		298.20	335	
	0640 1" size		19	.421		335	18.15		353.15	395	
	0650 1-1/4" size		15	.533		480	23		503	565	
	0660 1-1/2" size		13	.615		515	26.50		541.50	605	
	0800 Process steam, wet or super heated, monel trim, threaded										
	0820 Bronze body, 1/4" size	1 Stpi	24	.333	Ea.	540	14.35		554.35	620	
	0830 3/8" size		24	.333		540	14.35		554.35	620	
	0840 1/2" size		24	.333		620	14.35		634.35	710	
	0850 3/4" size		20	.400		730	17.20		747.20	830	
	0860 1" size		19	.421		920	18.15		938.15	1,050	
	0870 1-1/4" size		15	.533		1,150	23		1,173	1,275	
	0880 1-1/2" size		13	.615		1,375	26.50		1,401.50	1,550	
	0890 2" size		11	.727		2,000	31.50		2,031.50	2,250	
	0920 Iron body, max 125 PSIG press out, 1/4" size		24	.333		117	14.35		131.35	152	
	0930 3/8" size		24	.333		103	14.35		117.35	137	
	0940 1/2" size		24	.333		109	14.35		123.35	143	
	0950 3/4" size		20	.400		133	17.20		150.20	173	
	0960 1" size		19	.421		173	18.15		191.15	219	
	0970 1-1/4" size		15	.533		240	23		263	300	
	0980 1-1/2" size		13	.615		267	26.50		293.50	335	
	0990 2" size		11	.727		390	31.50		421.50	475	
	3000 Steam, high capacity, bronze body, stainless steel trim										
	3020 Threaded, 1/2" diameter	1 Stpi	24	.333	Ea.	985	14.35		999.35	1,100	
	3030 3/4" diameter		24	.333		985	14.35		999.35	1,100	
	3040 1" diameter		19	.421		1,100	18.15		1,118.15	1,225	
	3060 1-1/4" diameter		15	.533		1,225	23		1,248	1,350	
	3080 1-1/2" diameter		13	.615		1,400	26.50		1,426.50	1,575	
	3100 2" diameter		11	.727		1,700	31.50		1,731.50	1,925	
	3120 2-1/2" diameter	Q-5	12	1.333		2,125	51.50		2,176.50	2,425	
	3140 3" diameter	"	11	1.455		2,425	56.50		2,481.50	2,775	
	3500 Flanged connection, iron body, 125 lb. W.S.P.										
	3520 3" diameter	Q-5	11	1.455	Ea.	2,675	56.50		2,731.50	3,025	
	3540 4" diameter	"	5	3.200	"	3,350	124		3,474	3,900	
	9000 For water pressure regulators, see Div. 15110-160										
730	0010 **SLEEVES AND ESCUTCHEONS**										730
	0100 Pipe sleeve										
	0110 Steel, w/water stop, 12" long, with link seal										
	0120 2" diam. for 1/2" carrier pipe	1 Plum	8.40	.952	Ea.	39.50	40.50		80	107	
	0130 2-1/2" diam. for 3/4" carrier pipe		8	1		45	42.50		87.50	116	
	0140 2-1/2" diam. for 1" carrier pipe		8	1		42.50	42.50		85	113	
	0150 3" diam. for 1-1/4" carrier pipe		7.20	1.111		54.50	47.50		102	134	
	0160 3-1/2" diam. for 1-1/2" carrier pipe		6.80	1.176		55.50	50		105.50	139	
	0170 4" diam. for 2" carrier pipe		6	1.333		60.50	57		117.50	155	
	0180 4" diam. for 2-1/2" carrier pipe		6	1.333		60	57		117	154	
	0190 5" diam. for 3" carrier pipe		5.40	1.481		72	63.50		135.50	178	
	0200 6" diam. for 4" carrier pipe		4.80	1.667		81.50	71		152.50	201	
	0210 10" diam. for 6" carrier pipe	Q-1	8	2		140	77		217	274	
	0220 12" diam. for 8" carrier pipe		7.20	2.222		196	85.50		281.50	350	

MECHANICAL 15

15120 | Piping Specialties

	Line	Description	Crew	Daily Output	Labor-Hours	Unit	2006 Bare Costs Mat.	Labor	Equip.	Total	Total Incl O&P
730	0230	14" diam. for 10" carrier pipe	Q-1	6.40	2.500	Ea.	225	96		321	395
	0240	16" diam. for 12" carrier pipe		5.80	2.759		258	106		364	450
	0250	18" diam. for 14" carrier pipe		5.20	3.077		282	118		400	495
	0260	24" diam. for 18" carrier pipe		4	4		460	154		614	745
	0270	24" diam. for 20" carrier pipe		4	4		380	154		534	660
	0280	30" diam. for 24" carrier pipe		3.20	5		870	192		1,062	1,250
	0500	Wall sleeve									
	0510	Ductile iron with rubber gasket seal									
	0520	3"	1 Plum	8.40	.952	Ea.	155	40.50		195.50	234
	0530	4"		7.20	1.111		210	47.50		257.50	305
	0540	6"		6	1.333		320	57		377	440
	0550	8"		4	2		380	85.50		465.50	555
	0560	10"		3	2.667		455	114		569	675
	0570	12"		2.40	3.333		575	142		717	850
	5000	Escutcheon									
	5100	Split ring, pipe									
	5110	Chrome plated									
	5120	1/2"	1 Plum	160	.050	Ea.	1.05	2.14		3.19	4.48
	5130	3/4"		160	.050		1.09	2.14		3.23	4.52
	5140	1"		135	.059		1.17	2.53		3.70	5.20
	5150	1-1/2"		115	.070		1.76	2.97		4.73	6.55
	5160	2"		100	.080		1.85	3.42		5.27	7.35
	5170	4"		80	.100		4.11	4.27		8.38	11.15
	5180	6"		68	.118		6.85	5		11.85	15.35
760	0010	**STEAM TRAP**									
	0030	Cast iron body, threaded									
	0040	Inverted bucket									
	0050	1/2" pipe size	1 Stpi	12	.667	Ea.	112	28.50		140.50	168
	0070	3/4" pipe size		10	.800		194	34.50		228.50	268
	0100	1" pipe size		9	.889		300	38.50		338.50	390
	0120	1-1/4" pipe size		8	1		450	43		493	560
	1000	Float & thermostatic, 15 psi									
	1010	3/4" pipe size	1 Stpi	16	.500	Ea.	89.50	21.50		111	132
	1020	1" pipe size		15	.533		107	23		130	154
	1030	1-1/4" pipe size		13	.615		131	26.50		157.50	185
	1040	1-1/2" pipe size		9	.889		191	38.50		229.50	270
	1060	2" pipe size		6	1.333		350	57.50		407.50	475
	1290	Brass body, threaded									
	1300	Thermostatic, angle union, 25 psi									
	1310	1/2" pipe size	1 Stpi	24	.333	Ea.	42	14.35		56.35	68.50
	1320	3/4" pipe size		20	.400		79	17.20		96.20	114
	1330	1" pipe size		19	.421		116	18.15		134.15	155
	9000	Minimum labor/equipment charge		4	2	Job		86		86	134
790	0010	**STRAINERS, BASKET TYPE** Perforated stainless steel basket									
	0100	Brass or monel available									
	2000	Simplex style									
	2300	Bronze body									
	2320	Screwed, 3/8" pipe size	1 Stpi	22	.364	Ea.	109	15.65		124.65	145
	2340	1/2" pipe size		20	.400		112	17.20		129.20	150
	2360	3/4" pipe size		17	.471		114	20.50		134.50	158
	2380	1" pipe size		15	.533		149	23		172	200
	2400	1-1/4" pipe size		13	.615		225	26.50		251.50	288
	2420	1-1/2" pipe size		12	.667		227	28.50		255.50	295
	2440	2" pipe size		10	.800		345	34.50		379.50	435
	2460	2-1/2" pipe size	Q-5	15	1.067		490	41.50		531.50	605

15120 | Piping Specialties

15120	Piping Specialties	CREW	DAILY OUTPUT	LABOR-HOURS	UNIT	2006 BARE COSTS MAT.	LABOR	EQUIP.	TOTAL	TOTAL INCL O&P
790 2480	3″ pipe size	Q-5	14	1.143	Ea.	710	44.50		754.50	855
2600	Flanged, 2″ pipe size	1 Stpi	6	1.333		520	57.50		577.50	665
2620	2-1/2″ pipe size	Q-5	4.50	3.556		815	138		953	1,125
2640	3″ pipe size		3.50	4.571		920	177		1,097	1,275
2660	4″ pipe size		3	5.333		1,550	207		1,757	2,050
2680	5″ pipe size	Q-6	3.40	7.059		2,350	284		2,634	3,025
2700	6″ pipe size		3	8		2,875	320		3,195	3,675
2710	8″ pipe size		2.50	9.600		4,925	385		5,310	6,025
3600	Iron body									
3700	Screwed, 3/8″ pipe size	1 Stpi	22	.364	Ea.	78	15.65		93.65	110
3720	1/2″ pipe size		20	.400		80.50	17.20		97.70	116
3740	3/4″ pipe size		17	.471		102	20.50		122.50	144
3760	1″ pipe size		15	.533		105	23		128	151
3780	1-1/4″ pipe size		13	.615		137	26.50		163.50	191
3800	1-1/2″ pipe size		12	.667		150	28.50		178.50	210
3820	2″ pipe size		10	.800		180	34.50		214.50	251
3840	2-1/2″ pipe size	Q-5	15	1.067		240	41.50		281.50	330
3860	3″ pipe size	″	14	1.143		291	44.50		335.50	390
4000	Flanged, 2″ pipe size	1 Stpi	6	1.333		242	57.50		299.50	355
4020	2-1/2″ pipe size	Q-5	4.50	3.556		325	138		463	575
4040	3″ pipe size		3.50	4.571		340	177		517	650
4060	4″ pipe size		3	5.333		525	207		732	895
4080	5″ pipe size	Q-6	3.40	7.059		795	284		1,079	1,325
4100	6″ pipe size		3	8		1,025	320		1,345	1,625
4120	8″ pipe size		2.50	9.600		1,825	385		2,210	2,625
4140	10″ pipe size		2.20	10.909		4,025	440		4,465	5,100
6000	Cast steel body									
6400	Screwed, 1″ pipe size	1 Stpi	15	.533	Ea.	166	23		189	219
6410	1-1/4″ pipe size		13	.615		252	26.50		278.50	320
6420	1-1/2″ pipe size		12	.667		252	28.50		280.50	325
6440	2″ pipe size		10	.800		345	34.50		379.50	435
6460	2-1/2″ pipe size	Q-5	15	1.067		490	41.50		531.50	605
6480	3″ pipe size	″	14	1.143		640	44.50		684.50	775
6560	Flanged, 2″ pipe size	1 Stpi	6	1.333		510	57.50		567.50	650
6580	2-1/2″ pipe size	Q-5	4.50	3.556		765	138		903	1,050
6600	3″ pipe size		3.50	4.571		815	177		992	1,175
6620	4″ pipe size		3	5.333		1,150	207		1,357	1,600
6640	6″ pipe size	Q-6	3	8		2,125	320		2,445	2,825
6660	8″ pipe size	″	2.50	9.600		3,425	385		3,810	4,375
7000	Stainless steel body									
7200	Screwed, 1″ pipe size	1 Stpi	15	.533	Ea.	243	23		266	305
7210	1-1/4″ pipe size		13	.615		375	26.50		401.50	455
7220	1-1/2″ pipe size		12	.667		375	28.50		403.50	460
7240	2″ pipe size		10	.800		560	34.50		594.50	675
7260	2-1/2″ pipe size	Q-5	15	1.067		795	41.50		836.50	940
7280	3″ pipe size	″	14	1.143		1,100	44.50		1,144.50	1,275
7400	Flanged, 2″ pipe size	1 Stpi	6	1.333		805	57.50		862.50	975
7420	2-1/2″ pipe size	Q-5	4.50	3.556		1,475	138		1,613	1,850
7440	3″ pipe size		3.50	4.571		1,500	177		1,677	1,925
7460	4″ pipe size		3	5.333		2,350	207		2,557	2,925
7480	6″ pipe size	Q-6	3	8		4,300	320		4,620	5,250
7500	8″ pipe size	″	2.50	9.600		5,975	385		6,360	7,175
8100	Duplex style									
8200	Bronze body									
8240	Screwed, 3/4″ pipe size	1 Stpi	16	.500	Ea.	770	21.50		791.50	880
8260	1″ pipe size		14	.571		770	24.50		794.50	885

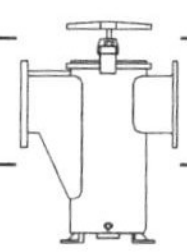

15120 | Piping Specialties

			CREW	DAILY OUTPUT	LABOR-HOURS	UNIT	2006 BARE COSTS MAT.	LABOR	EQUIP.	TOTAL	TOTAL INCL O&P	
790	8280	1-1/4" pipe size	1 Stpi	12	.667	Ea.	1,500	28.50		1,528.50	1,700	790
	8300	1-1/2" pipe size		11	.727		1,500	31.50		1,531.50	1,700	
	8320	2" pipe size		9	.889		2,450	38.50		2,488.50	2,725	
	8340	2-1/2" pipe size	Q-5	14	1.143		3,150	44.50		3,194.50	3,550	
	8420	Flanged, 2" pipe size	1 Stpi	6	1.333		3,450	57.50		3,507.50	3,875	
	8440	2-1/2" pipe size	Q-5	4.50	3.556		3,625	138		3,763	4,225	
	8460	3" pipe size		3.50	4.571		3,950	177		4,127	4,625	
	8480	4" pipe size		3	5.333		5,750	207		5,957	6,650	
	8500	5" pipe size	Q-6	3.40	7.059		12,600	284		12,884	14,300	
	8520	6" pipe size	"	3	8		12,600	320		12,920	14,400	
	8700	Iron body										
	8740	Screwed, 3/4" pipe size	1 Stpi	16	.500	Ea.	460	21.50		481.50	540	
	8760	1" pipe size		14	.571		460	24.50		484.50	545	
	8780	1-1/4" pipe size		12	.667		805	28.50		833.50	935	
	8800	1-1/2" pipe size		11	.727		805	31.50		836.50	940	
	8820	2" pipe size		9	.889		1,350	38.50		1,388.50	1,550	
	8840	2-1/2" pipe size	Q-5	14	1.143		1,500	44.50		1,544.50	1,725	
	9000	Flanged, 2" pipe size	1 Stpi	6	1.333		1,450	57.50		1,507.50	1,700	
	9020	2-1/2" pipe size	Q-5	4.50	3.556		1,550	138		1,688	1,925	
	9040	3" pipe size		3.50	4.571		1,700	177		1,877	2,150	
	9060	4" pipe size		3	5.333		2,850	207		3,057	3,475	
	9080	5" pipe size	Q-6	3.40	7.059		5,675	284		5,959	6,700	
	9100	6" pipe size		3	8		5,675	320		5,995	6,750	
	9120	8" pipe size		2.50	9.600		9,900	385		10,285	11,500	
	9140	10" pipe size		2.20	10.909		13,800	440		14,240	15,900	
	9160	12" pipe size		1.70	14.118		15,300	565		15,865	17,700	
	9170	14" pipe size		1.40	17.143		18,500	690		19,190	21,500	
	9180	16" pipe size		1	24		23,200	965		24,165	27,100	
	9300	Cast steel body										
	9340	Screwed, 1" pipe size	1 Stpi	14	.571	Ea.	1,000	24.50		1,024.50	1,150	
	9360	1-1/2" pipe size		11	.727		1,625	31.50		1,656.50	1,825	
	9380	2" pipe size		9	.889		1,625	38.50		1,663.50	1,825	
	9460	Flanged, 2" pipe size		6	1.333		2,150	57.50		2,207.50	2,475	
	9480	2-1/2" pipe size	Q-5	4.50	3.556		3,725	138		3,863	4,325	
	9500	3" pipe size		3.50	4.571		4,050	177		4,227	4,725	
	9520	4" pipe size		3	5.333		5,000	207		5,207	5,825	
	9540	6" pipe size	Q-6	3	8		9,375	320		9,695	10,800	
	9560	8" pipe size	"	2.50	9.600		19,800	385		20,185	22,400	
	9700	Stainless steel body										
	9740	Screwed, 1" pipe size	1 Stpi	14	.571	Ea.	1,275	24.50		1,299.50	1,450	
	9760	1-1/2" pipe size		11	.727		1,925	31.50		1,956.50	2,175	
	9780	2" pipe size		9	.889		2,700	38.50		2,738.50	3,000	
	9860	Flanged, 2" pipe size		6	1.333		2,925	57.50		2,982.50	3,325	
	9880	2-1/2" pipe size	Q-5	4.50	3.556		4,925	138		5,063	5,650	
	9900	3" pipe size		3.50	4.571		5,375	177		5,552	6,175	
	9920	4" pipe size		3	5.333		6,975	207		7,182	8,000	
	9940	6" pipe size	Q-6	3	8		10,700	320		11,020	12,200	
	9960	8" pipe size	"	2.50	9.600		29,800	385		30,185	33,400	
820	0010	**STRAINERS, Y TYPE** Bronze body										820
	0050	Screwed, 150 lb., 1/4" pipe size	1 Stpi	24	.333	Ea.	12.80	14.35		27.15	36.50	
	0070	3/8" pipe size		24	.333		16.85	14.35		31.20	41	
	0100	1/2" pipe size		20	.400		16.85	17.20		34.05	45.50	
	0120	3/4" pipe size		19	.421		18.45	18.15		36.60	48.50	
	0140	1" pipe size		17	.471		19.85	20.50		40.35	53.50	
	0150	1-1/4" pipe size		15	.533		44.50	23		67.50	84.50	
	0160	1-1/2" pipe size		14	.571		42.50	24.50		67	85.50	

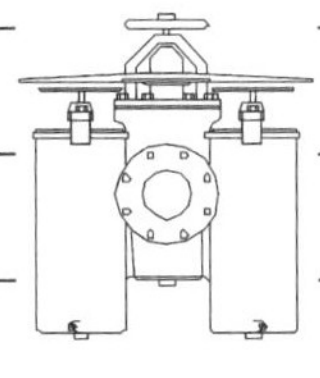

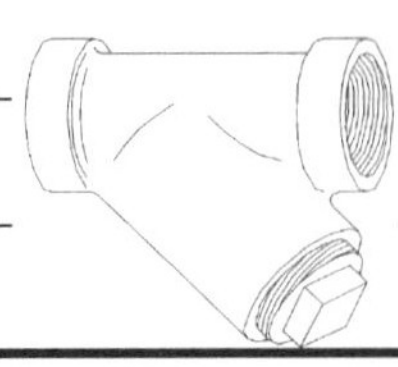

15 MECHANICAL

	15120 | Piping Specialties		CREW	DAILY OUTPUT	LABOR-HOURS	UNIT	2006 BARE COSTS MAT.	LABOR	EQUIP.	TOTAL	TOTAL INCL O&P	
820	0180	2" pipe size	1 Stpi	13	.615	Ea.	57	26.50		83.50	104	820
	0182	3" pipe size		12	.667		360	28.50		388.50	440	
	0200	300 lb., 2-1/2" pipe size	Q-5	17	.941		305	36.50		341.50	390	
	0220	3" pipe size		16	1		600	39		639	720	
	0240	4" pipe size		15	1.067		1,375	41.50		1,416.50	1,575	
	0500	For 300 lb rating 1/4" thru 2", add					15%					
	1000	Flanged, 150 lb., 1-1/2" pipe size	1 Stpi	11	.727	Ea.	315	31.50		346.50	395	
	1020	2" pipe size	"	8	1		365	43		408	470	
	1030	2-1/2" pipe size	Q-5	5	3.200		560	124		684	815	
	1040	3" pipe size		4.50	3.556		695	138		833	975	
	1060	4" pipe size		3	5.333		1,050	207		1,257	1,475	
	1080	5" pipe size	Q-6	3.40	7.059		2,150	284		2,434	2,825	
	1100	6" pipe size		3	8		2,800	320		3,120	3,575	
	1106	8" pipe size		2.60	9.231		3,425	370		3,795	4,350	
	1500	For 300 lb. rating, add					40%					
	9000	Minimum labor/equipment charge	1 Stpi	3.75	2.133	Job		92		92	143	
840	0010	**STRAINERS, Y TYPE** Iron body										840
	0050	Screwed, 250 lb., 1/4" pipe size	1 Stpi	20	.400	Ea.	7.05	17.20		24.25	35	
	0070	3/8" pipe size		20	.400		7.05	17.20		24.25	35	
	0100	1/2" pipe size		20	.400		7.05	17.20		24.25	35	
	0120	3/4" pipe size		18	.444		8.45	19.15		27.60	39.50	
	0140	1" pipe size		16	.500		11.40	21.50		32.90	46	
	0150	1-1/4" pipe size		15	.533		15.60	23		38.60	52.50	
	0160	1-1/2" pipe size		12	.667		18.80	28.50		47.30	65	
	0180	2" pipe size		8	1		29	43		72	99	
	0200	2-1/2" pipe size	Q-5	12	1.333		148	51.50		199.50	244	
	0220	3" pipe size		11	1.455		160	56.50		216.50	263	
	0240	4" pipe size		5	3.200		270	124		394	490	
	0500	For galvanized body, add					50%					
	1000	Flanged, 125 lb., 1-1/2" pipe size	1 Stpi	11	.727	Ea.	91	31.50		122.50	149	
	1020	2" pipe size	"	8	1		68	43		111	142	
	1030	2-1/2" pipe size	Q-5	5	3.200		77	124		201	278	
	1040	3" pipe size		4.50	3.556		90	138		228	315	
	1060	4" pipe size		3	5.333		164	207		371	500	
	1080	5" pipe size	Q-6	3.40	7.059		256	284		540	720	
	1100	6" pipe size		3	8		315	320		635	845	
	1120	8" pipe size		2.50	9.600		525	385		910	1,175	
	1140	10" pipe size		2	12		1,000	480		1,480	1,850	
	1160	12" pipe size		1.70	14.118		1,500	565		2,065	2,525	
	1170	14" pipe size		1.30	18.462		2,775	740		3,515	4,225	
	1180	16" pipe size		1	24		3,950	965		4,915	5,825	
	1500	For 250 lb rating, add					20%					
	2000	For galvanized body, add					50%					
	2500	For steel body, add					40%					
900	0010	**THERMOFLO INDICATOR** For balancing										900
	1000	Sweat connections, 1-1/4" pipe size	1 Stpi	12	.667	Ea.	450	28.50		478.50	540	
	1020	1-1/2" pipe size		10	.800		460	34.50		494.50	565	
	1040	2" pipe size		8	1		485	43		528	600	
	1060	2-1/2" pipe size		7	1.143		740	49		789	885	
	2000	Flange connections, 3" pipe size	Q-5	5	3.200		880	124		1,004	1,175	
	2020	4" pipe size		4	4		1,050	155		1,205	1,425	
	2030	5" pipe size		3.50	4.571		1,325	177		1,502	1,750	
	2040	6" pipe size		3	5.333		1,425	207		1,632	1,900	
	2060	8" pipe size		2	8		1,725	310		2,035	2,375	
920	0010	**VENTURI FLOW** Measuring device										920
	0050	1/2" diameter	1 Stpi	24	.333	Ea.	175	14.35		189.35	216	

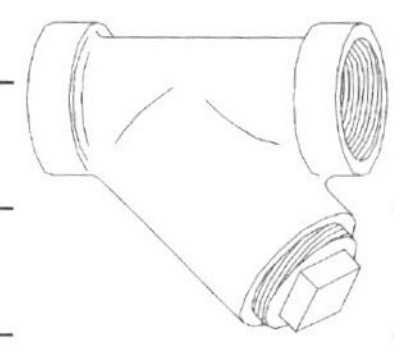

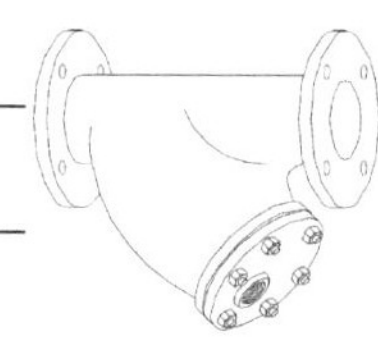

15120 | Piping Specialties

			CREW	DAILY OUTPUT	LABOR-HOURS	UNIT	2006 BARE COSTS MAT.	LABOR	EQUIP.	TOTAL	TOTAL INCL O&P	
920	0100	3/4″ diameter	1 Stpi	20	.400	Ea.	163	17.20		180.20	206	920
	0120	1″ diameter		19	.421		169	18.15		187.15	214	
	0140	1-1/4″ diameter		15	.533		203	23		226	259	
	0160	1-1/2″ diameter		13	.615		206	26.50		232.50	268	
	0180	2″ diameter		11	.727		222	31.50		253.50	293	
	0200	2-1/2″ diameter	Q-5	16	1		330	39		369	425	
	0220	3″ diameter		14	1.143		380	44.50		424.50	490	
	0240	4″ diameter		11	1.455		595	56.50		651.50	745	
	0260	5″ diameter	Q-6	4	6		720	241		961	1,175	
	0280	6″ diameter		3.50	6.857		780	276		1,056	1,275	
	0300	8″ diameter		3	8		1,075	320		1,395	1,675	
	0320	10″ diameter		2	12		2,050	480		2,530	3,025	
	0500	For meter, add					1,175			1,175	1,300	
940	0010	**WATER SUPPLY METERS**										940
	1000	Detector, serves dual systems such as fire and domestic or										
	1020	process water, wide range cap., UL and FM approved										
	1100	3″ mainline x 2″ by-pass, 400 GPM	Q-1	3.60	4.444	Ea.	5,150	171		5,321	5,925	
	1140	4″ mainline x 2″ by-pass, 700 GPM	″	2.50	6.400		5,150	246		5,396	6,025	
	1180	6″ mainline x 3″ by-pass, 1600 GPM	Q-2	2.60	9.231		7,775	370		8,145	9,125	
	1220	8″ mainline x 4″ by-pass, 2800 GPM		2.10	11.429		11,500	455		11,955	13,300	
	1260	10″ mainline x 6″ by-pass, 4400 GPM		2	12		15,700	480		16,180	18,000	
	1300	10″x12″ mainlines x 6″ by-pass,5400 GPM		1.70	14.118		20,700	565		21,265	23,700	
	2000	Domestic/commercial, bronze										
	2020	Threaded										
	2060	5/8″ diameter, to 20 GPM	1 Plum	16	.500	Ea.	40	21.50		61.50	77	
	2080	3/4″ diameter, to 30 GPM		14	.571		67.50	24.50		92	113	
	2100	1″ diameter, to 50 GPM		12	.667		94	28.50		122.50	148	
	2300	Threaded/flanged										
	2340	1-1/2″ diameter, to 100 GPM	1 Plum	8	1	Ea.	330	42.50		372.50	430	
	2360	2″ diameter, to 160 GPM	″	6	1.333	″	415	57		472	545	
	2600	Flanged, compound										
	2640	3″ diameter, 320 GPM	Q-1	3	5.333	Ea.	1,950	205		2,155	2,450	
	2660	4″ diameter, to 500 GPM		1.50	10.667		3,025	410		3,435	3,950	
	2680	6″ diameter, to 1,000 GPM		1	16		4,350	615		4,965	5,725	
	2700	8″ diameter, to 1,800 GPM		.80	20		8,600	770		9,370	10,700	
	7000	Turbine										
	7260	Flanged 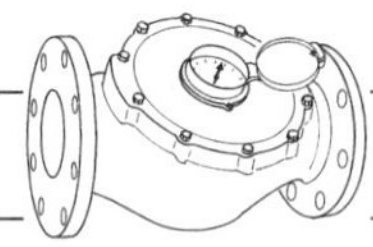										
	7300	2″ diameter, to 160 GPM	1 Plum	7	1.143	Ea.	455	49		504	575	
	7320	3″ diameter, to 450 GPM	Q-1	3.60	4.444		745	171		916	1,075	
	7340	4″ diameter, to 650 GPM	″	2.50	6.400		1,350	246		1,596	1,850	
	7360	6″ diameter, to 1800 GPM	Q-2	2.60	9.231		2,550	370		2,920	3,375	
	7380	8″ diameter, to 2500 GPM		2.10	11.429		4,050	455		4,505	5,150	
	7400	10″ diameter, to 5500 GPM		1.70	14.118		5,400	565		5,965	6,825	
	9000	Minimum labor/equipment charge	1 Plum	3.25	2.462	Job		105		105	163	
960	0010	**WELD END BALL JOINTS** Steel										960
	0050	2-1/2″ diameter	Q-17	13	1.231	Ea.	515	47.50	5.30	567.80	645	
	0100	3″ diameter		12	1.333		595	51.50	5.75	652.25	735	
	0120	4″ diameter		11	1.455		915	56.50	6.30	977.80	1,100	
	0140	5″ diameter	Q-18	14	1.714		1,200	69	4.94	1,273.94	1,425	
	0160	6″ diameter		12	2		1,600	80.50	5.75	1,686.25	1,900	
	0180	8″ diameter		9	2.667		2,175	107	7.70	2,289.70	2,550	
	0200	10″ diameter		8	3		3,075	121	8.65	3,204.65	3,600	
	0220	12″ diameter		6	4		4,050	161	11.50	4,222.50	4,725	
	0240	14″ diameter		4	6		4,825	241	17.30	5,083.30	5,725	

15140 | Domestic Water Piping

		15140 | Domestic Water Piping	CREW	DAILY OUTPUT	LABOR-HOURS	UNIT	2006 BARE COSTS MAT.	LABOR	EQUIP.	TOTAL	TOTAL INCL O&P	
100	0010	**BACKFLOW PREVENTER** Includes valves										100
	0020	and four test cocks, corrosion resistant, automatic operation										
	1000	Double check principle										
	1010	Threaded, with ball valves										
	1020	3/4" pipe size	1 Plum	16	.500	Ea.	148	21.50		169.50	196	
	1030	1" pipe size		14	.571		164	24.50		188.50	218	
	1040	1-1/2" pipe size		10	.800		295	34		329	380	
	1050	2" pipe size	↓	7	1.143	↓	365	49		414	475	
	1080	Threaded, with gate valves										
	1100	3/4" pipe size	1 Plum	16	.500	Ea.	580	21.50		601.50	675	
	1120	1" pipe size		14	.571		620	24.50		644.50	720	
	1140	1-1/2" pipe size		10	.800		755	34		789	885	
	1160	2" pipe size 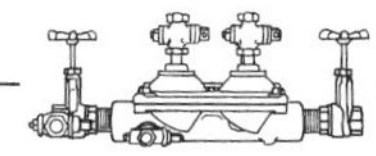	↓	7	1.143	↓	925	49		974	1,100	
	1300	Flanged, valves are OS&Y										
	1380	3" pipe size	Q-1	4.50	3.556	Ea.	2,300	137		2,437	2,725	
	1400	4" pipe size	"	3	5.333		3,750	205		3,955	4,450	
	1420	6" pipe size	Q-2	3	8	↓	5,975	320		6,295	7,050	
	4000	Reduced pressure principle										
	4100	Threaded, bronze, valves are ball										
	4120	3/4" pipe size	1 Plum	16	.500	Ea.	220	21.50		241.50	275	
	4140	1" pipe size		14	.571		236	24.50		260.50	298	
	4150	1-1/4" pipe size		12	.667		405	28.50		433.50	490	
	4160	1-1/2" pipe size		10	.800		445	34		479	545	
	4180	2" pipe size		7	1.143	↓	500	49		549	625	
	4500	Minimum labor/equipment charge	↓	4	2	Job		85.50		85.50	133	
	5000	Flanged, valves are OS&Y										
	5060	2-1/2" pipe size	Q-1	5	3.200	Ea.	2,550	123		2,673	3,000	
	5080	3" pipe size		4.50	3.556		2,675	137		2,812	3,150	
	5100	4" pipe size 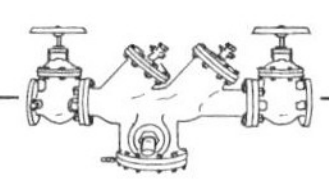	↓	3	5.333		3,375	205		3,580	4,025	
	5120	6" pipe size	Q-2	3	8	↓	4,875	320		5,195	5,875	
	5600	Flanged, iron, valves are OS&Y										
	5660	2-1/2" pipe size	Q-1	5	3.200	Ea.	2,050	123		2,173	2,450	
	5680	3" pipe size		4.50	3.556		2,150	137		2,287	2,575	
	5700	4" pipe size	↓	3	5.333		2,700	205		2,905	3,300	
	5720	6" pipe size	Q-2	3	8		3,900	320		4,220	4,800	
	5740	8" pipe size		2	12		6,875	480		7,355	8,300	
	5760	10" pipe size	↓	1	24	↓	9,200	955		10,155	11,600	
	9010	Minimum labor/equipment charge	1 Plum	2	4	Job		171		171	266	
600	0010	**VACUUM BREAKERS** Hot or cold water										600
	1030	Anti-siphon, brass										
	1040	1/4" size	1 Plum	24	.333	Ea.	24	14.25		38.25	48	
	1050	3/8" size		24	.333		24	14.25		38.25	48	
	1060	1/2" size		24	.333		27	14.25		41.25	51.50	
	1080	3/4" size		20	.400		32	17.10		49.10	61.50	
	1100	1" size		19	.421		50	18		68	82.50	
	1120	1-1/4" size		15	.533		87	23		110	132	
	1140	1-1/2" size		13	.615		103	26.50		129.50	154	
	1160	2" size		11	.727		160	31		191	225	
	1180	2-1/2" size		9	.889		460	38		498	565	
	1200	3" size	↓	7	1.143	↓	610	49		659	745	
	1300	For polished chrome, (1/4" thru 1"), add					50%					
	1900	Vacuum relief, water service, bronze										
	2000	1/2" size	1 Plum	30	.267	Ea.	23.50	11.40		34.90	43	
	2040	3/4" size		28	.286	"	29	12.20		41.20	51	
	9000	Minimum labor/equipment charge	↓	4	2	Job		85.50		85.50	133	

15140 | Domestic Water Piping

		15140 Domestic Water Piping	CREW	DAILY OUTPUT	LABOR-HOURS	UNIT	MAT. (2006 BARE COSTS)	LABOR (2006 BARE COSTS)	EQUIP. (2006 BARE COSTS)	TOTAL (2006 BARE COSTS)	TOTAL INCL O&P	
800	0010	**WATER HAMMER ARRESTORS / SHOCK ABSORBERS**										800
	0490	Copper										
	0500	3/4" male I.P.S. For 1 to 11 fixtures	1 Plum	12	.667	Ea.	15.55	28.50		44.05	61.50	
	0600	1" male I.P.S., For 12 to 32 fixtures		8	1		39.50	42.50		82	110	
	0700	1-1/4" male I.P.S. For 33 to 60 fixtures		8	1		46	42.50		88.50	117	
	0800	1-1/2" male I.P.S. For 61 to 113 fixtures		8	1		62.50	42.50		105	136	
	0900	2" male I.P.S.For 114 to 154 fixtures		8	1		100	42.50		142.50	177	
	1000	2-1/2" male I.P.S. For 155 to 330 fixtures		4	2		285	85.50		370.50	450	
	9000	Minimum labor/equipment charge		3.50	2.286	Job		97.50		97.50	152	

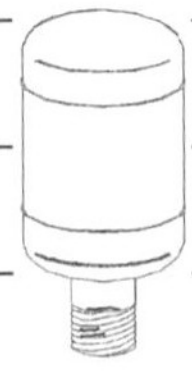

15150 | Sanitary Waste and Vent Piping

		15150 Sanitary Waste and Vent Piping	CREW	DAILY OUTPUT	LABOR-HOURS	UNIT	MAT.	LABOR	EQUIP.	TOTAL	TOTAL INCL O&P	
100	0010	**BACKWATER VALVES**, C.I. body										100
	6980	Bronze gate and automatic flapper valves										
	7000	3" and 4" pipe size	Q-1	13	1.231	Ea.	1,050	47.50		1,097.50	1,250	
	7100	5" and 6" pipe size	"	13	1.231	"	1,625	47.50		1,672.50	1,850	
	7240	Bronze flapper valve, bolted cover										
	7260	2" pipe size	Q-1	16	1	Ea.	310	38.50		348.50	400	
	7280	3" pipe size		14.50	1.103		485	42.50		527.50	600	
	7300	4" pipe size		13	1.231		595	47.50		642.50	730	
	7320	5" pipe size	Q-2	18	1.333		715	53		768	875	
	7340	6" pipe size	"	17	1.412		855	56.50		911.50	1,025	
	7360	8" pipe size	Q-3	10	3.200		1,200	130		1,330	1,525	
	7380	10" pipe size	"	9	3.556		1,975	145		2,120	2,400	
	7500	For threaded cover, same cost										
	7540	Revolving disk type, same cost as flapper type										
200	0010	**CLEANOUTS**										200
	0060	Floor type										
	0080	Round or square, scoriated nickel bronze top										
	0100	2" pipe size	1 Plum	10	.800	Ea.	102	34		136	165	
	0120	3" pipe size		8	1		153	42.50		195.50	235	
	0140	4" pipe size		6	1.333		153	57		210	257	
	0160	5" pipe size		4	2		194	85.50		279.50	345	
	0180	6" pipe size	Q-1	6	2.667		194	102		296	370	
	0200	8" pipe size	"	4	4		345	154		499	615	
	0340	Recessed for tile, same price										
	0980	Round top, recessed for terrazzo										
	1000	2" pipe size	1 Plum	9	.889	Ea.	102	38		140	171	
	1080	3" pipe size		6	1.333		153	57		210	257	
	1100	4" pipe size		4	2		153	85.50		238.50	300	
	1120	5" pipe size	Q-1	6	2.667		194	102		296	370	
	1140	6" pipe size		5	3.200		194	123		317	405	
	1160	8" pipe size		4	4		345	154		499	615	
	2000	Round scoriated nickel bronze top, extra heavy duty										
	2060	2" pipe size	1 Plum	9	.889	Ea.	139	38		177	211	
	2080	3" pipe size		6	1.333		190	57		247	298	
	2100	4" pipe size		4	2		190	85.50		275.50	340	
	2120	5" pipe size	Q-1	6	2.667		231	102		333	415	
	2140	6" pipe size		5	3.200		231	123		354	445	
	2160	8" pipe size		4	4		380	154		534	660	
	4000	Wall type, square smooth cover, over wall frame										
	4060	2" pipe size	1 Plum	14	.571	Ea.	167	24.50		191.50	221	
	4080	3" pipe size		12	.667		179	28.50		207.50	242	
	4100	4" pipe size		10	.800		193	34		227	265	
	4120	5" pipe size		9	.889		279	38		317	365	
	4140	6" pipe size		8	1		310	42.50		352.50	405	

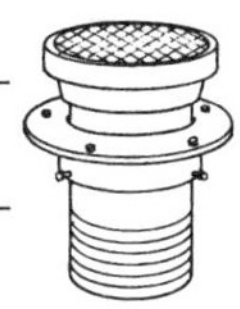

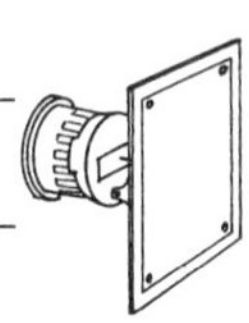

		15150 | Sanitary Waste and Vent Piping	CREW	DAILY OUTPUT	LABOR-HOURS	UNIT	2006 BARE COSTS MAT.	LABOR	EQUIP.	TOTAL	TOTAL INCL O&P	
200	4160	8″ pipe size	Q-1	11	1.455	Ea.	415	56		471	540	200
	5000	Extension, C.I.;bronze countersunk plug, 8″ long										
	5040	2″ pipe size	1 Plum	16	.500	Ea.	87	21.50		108.50	129	
	5060	3″ pipe size		14	.571		101	24.50		125.50	149	
	5080	4″ pipe size		13	.615		104	26.50		130.50	155	
	5100	5″ pipe size		12	.667		162	28.50		190.50	223	
	5120	6″ pipe size		11	.727		196	31		227	265	
	9000	Minimum labor/equipment charge		3	2.667	Job		114		114	177	

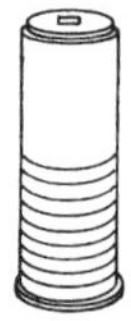

		15150 | Sanitary Waste and Vent Piping	CREW	DAILY OUTPUT	LABOR-HOURS	UNIT	2006 BARE COSTS MAT.	LABOR	EQUIP.	TOTAL	TOTAL INCL O&P	
250	0010	CLEANOUT TEE										250
	0100	Cast iron, B&S, with countersunk plug										
	0200	2″ pipe size	1 Plum	4	2	Ea.	139	85.50		224.50	285	
	0220	3″ pipe size		3.60	2.222		151	95		246	315	
	0240	4″ pipe size		3.30	2.424		188	104		292	370	
	0260	5″ pipe size	Q-1	5.50	2.909		410	112		522	625	
	0280	6″ pipe size	″	5	3.200		510	123		633	750	
	0300	8″ pipe size	Q-3	5	6.400		585	260		845	1,050	
	0500	For round smooth access cover, same price										
	0600	For round scoriated access cover, same price										
	0700	For square smooth access cover, add					60%					
	4000	Plastic, tees and adapters. Add plugs										
	4010	ABS, DWV										
	4020	Cleanout tee, 1-1/2″ pipe size	1 Plum	15	.533	Ea.	6.05	23		29.05	42	
	4030	2″ pipe size	Q-1	27	.593		7.15	23		30.15	43.50	
	4040	3″ pipe size		21	.762		14.30	29.50		43.80	61	
	4050	4″ pipe size		16	1		26.50	38.50		65	89	
	4100	Cleanout plug, 1-1/2″ pipe size	1 Plum	32	.250		1.11	10.70		11.81	17.80	
	4110	2″ pipe size	Q-1	56	.286		1.46	11		12.46	18.65	
	4120	3″ pipe size		36	.444		2.22	17.10		19.32	29	
	4130	4″ pipe size		30	.533		4.57	20.50		25.07	37	
	4180	Cleanout adapter fitting, 1-1/2″ pipe size	1 Plum	32	.250		1.75	10.70		12.45	18.55	
	4190	2″ pipe size	Q-1	56	.286		2.46	11		13.46	19.75	
	4200	3″ pipe size		36	.444		6.95	17.10		24.05	34	
	4210	4″ pipe size		30	.533		11.10	20.50		31.60	44.50	
	5000	PVC, DWV										
	5010	Cleanout tee, 1-1/2″ pipe size	1 Plum	15	.533	Ea.	6.05	23		29.05	42	
	5020	2″ pipe size	Q-1	27	.593		7.15	23		30.15	43.50	
	5030	3″ pipe size		21	.762		14.30	29.50		43.80	61	
	5040	4″ pipe size		16	1		26.50	38.50		65	89	
	5090	Cleanout plug, 1-1/2″ pipe size	1 Plum	32	.250		1.11	10.70		11.81	17.80	
	5100	2″ pipe size	Q-1	56	.286		1.46	11		12.46	18.65	
	5110	3″ pipe size		36	.444		2.22	17.10		19.32	29	
	5120	4″ pipe size		30	.533		4.58	20.50		25.08	37	
	5130	6″ pipe size		24	.667		13.95	25.50		39.45	55.50	
	5170	Cleanout adapter fitting, 1-1/2″ pipe size	1 Plum	32	.250		1.75	10.70		12.45	18.55	
	5180	2″ pipe size	Q-1	56	.286		2.46	11		13.46	19.75	
	5190	3″ pipe size		36	.444		6.95	17.10		24.05	34	
	5200	4″ pipe size		30	.533		11.10	20.50		31.60	44.50	
	5210	6″ pipe size		24	.667		25.50	25.50		51	68	
	9000	Minimum labor/equipment charge	1 Plum	2.75	2.909	Job		124		124	193	

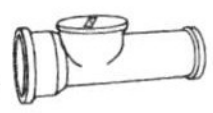

		15150 | Sanitary Waste and Vent Piping	CREW	DAILY OUTPUT	LABOR-HOURS	UNIT	2006 BARE COSTS MAT.	LABOR	EQUIP.	TOTAL	TOTAL INCL O&P	
300	0010	FLOOR AND AREA DRAINS										300
	0400	Deck, auto park, C.I., 13″ top										
	0440	3″, 4″, 5″, and 6″ pipe size	Q-1	8	2	Ea.	745	77		822	940	
	0480	For galvanized body, add				″	355			355	390	
	0800	Promenade, heelproof grate, C.I., 14″ top										
	0840	2″, 3″, and 4″ pipe size	Q-1	10	1.600	Ea.	286	61.50		347.50	410	

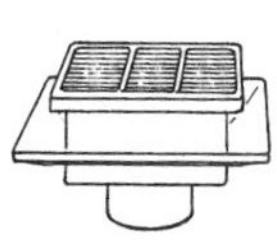

15150 | Sanitary Waste and Vent Piping

	15150 Sanitary Waste and Vent Piping	CREW	DAILY OUTPUT	LABOR-HOURS	UNIT	2006 BARE COSTS MAT.	LABOR	EQUIP.	TOTAL	TOTAL INCL O&P
300	0860 5″ and 6″ pipe size	Q-1	9	1.778	Ea.	355	68.50		423.50	495
	0880 8″ pipe size		8	2		420	77		497	585
	0940 For galvanized body, add					186			186	204
	0960 For polished bronze top, add					194			194	213
	1200 Promenade, heelproof grate, C.I., lateral, 14″ top									
	1240 2″, 3″ and 4″ pipe size	Q-1	10	1.600	Ea.	370	61.50		431.50	500
	1260 5″ and 6″ pipe size		9	1.778		440	68.50		508.50	590
	1280 8″ pipe size		8	2		510	77		587	680
	1340 For galvanized body, add					273			273	300
	1360 For polished bronze top, add					189			189	208
	1500 Promenade, slotted grate, C.I., 11″ top									
	1540 2″, 3″, 4″, 5″, and 6″ pipe size	Q-1	12	1.333	Ea.	255	51		306	360
	1600 For galvanized body, add					104			104	115
	1640 For polished bronze top, add					153			153	168
	2000 Floor, medium duty, C.I., deep flange, 7″ dia top									
	2040 2″ and 3″ pipe size	Q-1	12	1.333	Ea.	104	51		155	195
	2080 For galvanized body, add					44			44	48.50
	2120 For polished bronze top, add					58			58	64
	2160 Heavy duty, C.I., 12″ dia anti-tilt grate									
	2180 2″, 3″, 4″, 5″ and 6″ pipe size	Q-1	10	1.600	Ea.	296	61.50		357.50	420
	2220 For galvanized body, add					135			135	149
	2240 For polished bronze top, add					152			152	167
	2300 X-Heavy duty, C.I., 15″ antitilt grate									
	2320 4″, 5″, 6″, and 8″ pipe size	Q-1	8	2	Ea.	650	77		727	830
	2360 For galvanized body, add					220			220	242
	2380 For polished bronze top, add					257			257	283
	2400 Heavy duty, with sediment bucket, C.I., 12″ dia. loose grate									
	2420 2″, 3″, 4″, 5″, and 6″ pipe size	Q-1	9	1.778	Ea.	360	68.50		428.50	500
	2440 For galvanized body, add					200			200	220
	2460 For polished bronze top, add					148			148	162
	2500 Heavy duty, cleanout & trap w/bucket, C.I., 15″ top									
	2540 2″, 3″, and 4″ pipe size	Q-1	6	2.667	Ea.	3,325	102		3,427	3,800
	2560 For galvanized body, add					770			770	845
	2580 For polished bronze top, add					270			270	297
	2600 Medium duty, with perforated SS basket, C.I., body,									
	2610 18″ top for refuse container washing area									
	2620 2″ thru 6″ pipe size	Q-1	4	4	Ea.	2,025	154		2,179	2,475
	2630 Acid resistant									
	2638 PVC									
	2640 2″, 3″ and 4″ pipe size	Q-1	16	1	Ea.	250	38.50		288.50	335
	2644 Cast iron, epoxy coated									
	2646 2″, 3″ and 4″ pipe size	Q-1	14	1.143	Ea.	215	44		259	305
	2650 PVC or ABS thermoplastic									
	2660 3″ and 4″ pipe size	Q-1	16	1	Ea.	250	38.50		288.50	335
	2680 Extra heavy duty, oil intercepting, gas seal cone,									
	2690 with cleanout, loose grate, C.I., body 16″ top									
	2700 3″ and 4″ diameter outlet, 4″ slab depth	Q-1	4	4	Ea.	3,600	154		3,754	4,200
	2720 3″ and 4″ diameter outlet, 8″ slab depth		3	5.333		3,900	205		4,105	4,600
	2740 3″ & 4″ dia. outlet, 10″-12″ slab depth, 16″top		2	8		4,025	305		4,330	4,925
	2910 Prison cell, vandal-proof, 1-1/2″, and 2″ diam. pipe		12	1.333		230	51		281	335
	2920 3″ pipe size		10	1.600		258	61.50		319.50	380
	2930 Trap drain, light duty, backwater valve C.I. top									
	2950 8″ diameter top, 2″ pipe size	Q-1	12	1.333	Ea.	200	51		251	300
	2960 10″ diameter top, 3″ pipe size		10	1.600		277	61.50		338.50	400
	2970 12″ diameter top, 4″ pipe size		8	2		385	77		462	545

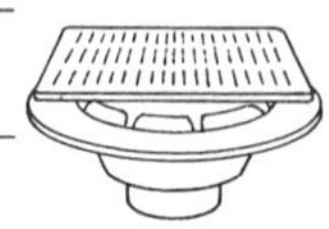

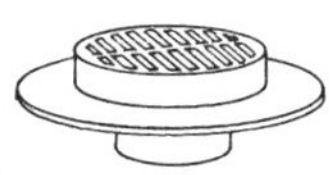

15 MECHANICAL

15150 | Sanitary Waste and Vent Piping

			CREW	DAILY OUTPUT	LABOR-HOURS	UNIT	2006 BARE COSTS MAT.	LABOR	EQUIP.	TOTAL	TOTAL INCL O&P	
350	0010	**FLOOR RECEPTORS** For connection to 2″, 3″ & 4″ diameter pipe										350
	0200	12-1/2″ square top, 25 sq in open area	Q-1	10	1.600	Ea.	505	61.50		566.50	650	
	0300	For grate with 4″ diam. x 3-3/4″ high funnel, add					89.50			89.50	98.50	
	0400	For grate with 6″ diameter x 6″ high funnel, add					116			116	127	
	0500	For full hinged grate with open center, add					84			84	92.50	
	0600	For aluminum bucket, add					75			75	82.50	
	0700	For acid-resisting bucket, add					142			142	156	
	0800	For galvanized bucket, add					92.50			92.50	102	
	0900	For stainless steel mesh bucket liner, add					112			112	123	
	1000	For bronze antisplash dome strainer, add					54			54	59.50	
	1100	For partial solid cover, add					26.50			26.50	29.50	
	1200	For trap primer connection, add					45.50			45.50	50	
	2000	12-5/8″ diameter top, 40 sq. in. open area	Q-1	10	1.600		400	61.50		461.50	535	
	2100	For options, add same prices as square top										
	3000	8″ x 4″ rectangular top, 7.5 sq. in. open area	Q-1	14	1.143	Ea.	385	44		429	495	
	3100	For trap primer connections, add					45.50			45.50	50	
	4000	24″ x 16″ rectangular top, 70 sq. in. open area	Q-1	4	4		2,250	154		2,404	2,725	
	4100	For trap primer connection, add					89.50			89.50	98.50	
	9000	Minimum labor/equipment charge	Q-1	3	5.333	Job		205		205	320	

			CREW	DAILY OUTPUT	LABOR-HOURS	UNIT	2006 BARE COSTS MAT.	LABOR	EQUIP.	TOTAL	TOTAL INCL O&P	
400	0010	**INTERCEPTORS**										400
	0150	Grease, cast iron, 4 GPM, 8 lb. fat capacity	1 Plum	4	2	Ea.	615	85.50		700.50	815	
	0200	7 GPM, 14 lb. fat capacity		4	2		855	85.50		940.50	1,075	
	1000	10 GPM, 20 lb. fat capacity		4	2		1,000	85.50		1,085.50	1,225	
	1040	15 GPM, 30 lb. fat capacity		4	2		1,500	85.50		1,585.50	1,775	
	1060	20 GPM, 40 lb. fat capacity		3	2.667		1,825	114		1,939	2,175	
	1080	25 GPM, 50 lb. fat capacity	Q-1	3.50	4.571		2,050	176		2,226	2,525	
	1100	35 GPM, 70 lb. fat capacity		3	5.333		2,525	205		2,730	3,100	
	1110	50 GPM, 100 lb. fat capacity		2	8		2,525	305		2,830	3,250	
	1120	Fabricated steel, 50 GPM, 100 lb. fat capacity		2	8		3,350	305		3,655	4,150	
	1140	75 GPM, 150 lb. fat capacity		2	8		6,750	305		7,055	7,900	
	1160	100 GPM, 200 lb. fat capacity		2	8		7,600	305		7,905	8,825	
	1180	150 GPM, 300 lb. fat capacity		2	8		8,450	305		8,755	9,750	
	1200	200 GPM, 400 lb. fat capacity		1.50	10.667		12,200	410		12,610	14,100	
	1220	250 GPM, 500 lb. fat capacity		1.30	12.308		14,300	475		14,775	16,400	
	1240	300 GPM, 600 lb. fat capacity		1	16		16,800	615		17,415	19,500	
	1260	400 GPM, 800 lb. fat capacity	Q-2	1.20	20		20,600	795		21,395	23,900	
	1280	500 GPM, 1000 lb. fat capacity	″	1	24		24,300	955		25,255	28,300	
	1580	For seepage pan, add					7%					
	3000	Hair, cast iron, 1-1/4″ and 1-1/2″ pipe connection	1 Plum	8	1	Ea.	231	42.50		273.50	320	
	3100	For chrome-plated cast iron, add					142			142	156	
	3200	For polished bronze, add					193			193	213	
	3400	Lint interceptor, fabricated steel										
	3410	Size based on 10 GPM per machine										
	3420	30 GPM, 2″ pipe size	Q-1	3	5.333	Ea.	2,700	205		2,905	3,300	
	3430	70 GPM, 3″ pipe size		2.50	6.400		3,150	246		3,396	3,850	
	3440	100 GPM, 4″ pipe size		2	8		3,700	305		4,005	4,550	
	3450	200 GPM, 4″ pipe size		1.50	10.667		4,500	410		4,910	5,575	
	3460	300 GPM, 6″ pipe size		1	16		5,225	615		5,840	6,700	
	3470	400 GPM, 6″ pipe size	Q-2	1.20	20		5,950	795		6,745	7,800	
	3480	500 GPM, 6″ pipe size	″	1	24		6,675	955		7,630	8,825	
	4000	Oil, fabricated steel, 10 GPM, 2″ pipe size	1 Plum	4	2		1,375	85.50		1,460.50	1,625	
	4100	15 GPM, 2″ or 3″ pipe size		4	2		1,875	85.50		1,960.50	2,200	
	4120	20 GPM, 2″ or 3″ pipe size		3	2.667		2,275	114		2,389	2,675	
	4140	25 GPM, 2″ or 3″ pipe size	Q-1	3.50	4.571		2,475	176		2,651	3,000	
	4160	35 GPM, 2″, 3″, or 4″ pipe size		3	5.333		3,000	205		3,205	3,625	

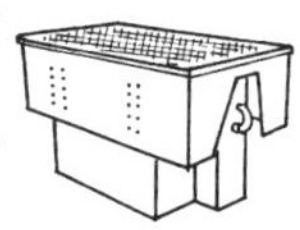

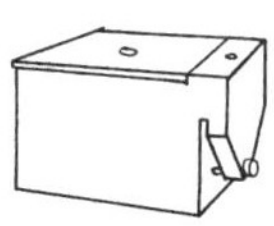

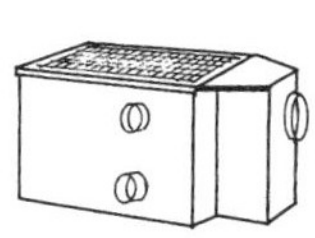

MECHANICAL 15

15150 | Sanitary Waste and Vent Piping

		15150 Sanitary Waste and Vent Piping	CREW	DAILY OUTPUT	LABOR-HOURS	UNIT	2006 BARE COSTS MAT.	LABOR	EQUIP.	TOTAL	TOTAL INCL O&P	
400	4180	50 GPM, 2″, 3″, or 4″ pipe size	Q-1	2	8	Ea.	4,050	305		4,355	4,925	400
	4200	75 GPM, 3″ pipe size		2	8		7,000	305		7,305	8,175	
	4220	100 GPM, 3″ pipe size		2	8		7,600	305		7,905	8,825	
	4240	150 GPM, 4″ pipe size		2	8		9,425	305		9,730	10,900	
	4260	200 GPM, 4″ pipe size		1.50	10.667		13,300	410		13,710	15,200	
	4280	250 GPM, 5″ pipe size		1.30	12.308		15,400	475		15,875	17,600	
	4300	300 GPM, 5″ pipe size		1	16		17,400	615		18,015	20,200	
	4320	400 GPM, 6″ pipe size	Q-2	1.20	20		22,400	795		23,195	26,000	
	4340	500 GPM, 6″ pipe size	″	1	24		28,300	955		29,255	32,600	
	5000	Sand interceptor, fabricated steel										
	5020	20 GPM, 4″ pipe size	Q-1	3	5.333	Ea.	3,400	205		3,605	4,050	
	5030	50 GPM, 4″ pipe size		2.50	6.400		5,750	246		5,996	6,700	
	5040	150 GPM, 4″ pipe size		2	8		17,400	305		17,705	19,600	
	5050	250 GPM, 6″ pipe size		1.30	12.308		18,100	475		18,575	20,600	
	5060	500 GPM, 6″ pipe size	Q-2	1	24		23,300	955		24,255	27,100	
	6000	Solids, precious metals recovery, C.I., 1-1/4″ to 2″ pipe	1 Plum	4	2		350	85.50		435.50	520	
	6100	Dental Lab., large, C.I., 1-1/2″ to 2″ pipe		3	2.667		1,225	114		1,339	1,525	
	9000	Minimum labor/equipment charge		3	2.667	Job		114		114	177	
500	0010	**SEPARATORS** Entrainment eliminator, steel body, 150 PSIG										500
	0100	1/4″ size	1 Stpi	24	.333	Ea.	179	14.35		193.35	220	
	0120	1/2″ size		24	.333		188	14.35		202.35	229	
	0140	3/4″ size		20	.400		200	17.20		217.20	247	
	0160	1″ size		19	.421		272	18.15		290.15	330	
	0180	1-1/4″ size		15	.533		310	23		333	375	
	0200	1-1/2″ size		13	.615		340	26.50		366.50	415	
	0220	2″ size		11	.727		380	31.50		411.50	465	
	0240	2-1/2″ size	Q-5	15	1.067		1,525	41.50		1,566.50	1,750	
	0260	3″ size		13	1.231		1,825	47.50		1,872.50	2,075	
	0280	4″ size		10	1.600		2,050	62		2,112	2,350	
	0300	5″ size		6	2.667		2,400	103		2,503	2,800	
	0320	6″ size		3	5.333		2,650	207		2,857	3,250	
	0340	8″ size	Q-6	4.40	5.455		3,300	219		3,519	3,975	
	0360	10″ size	″	4	6		4,800	241		5,041	5,650	
	1000	For 300 PSIG, add					15%					
600	0010	**SHOWER DRAINS**										600
	2780	Shower, with strainer, uniform diam. trap, bronze top										
	2800	2″ and 3″ pipe size	Q-1	8	2	Ea.	220	77		297	360	
	2820	4″ pipe size	″	7	2.286		243	88		331	405	
	2840	For galvanized body, add					86			86	94.50	
	2860	With strainer, backwater valve, drum trap										
	2880	1-1/2″, 2″,& 3″ pipe size	Q-1	8	2	Ea.	197	77		274	335	
	2890	4″ pipe size	″	7	2.286		275	88		363	440	
	2900	For galvanized body, add					86			86	94.50	
700	0010	**SINK WASTE TREATMENT** System for commercial kitchens										700
	0100	includes clock timer, & fittings										
	0200	System less chemical, wall mounted cabinet	1 Plum	16	.500	Ea.	430	21.50		451.50	510	
	2000	Chemical, 1 gallon, add					50			50	55	
	2100	6 gallons, add					485			485	535	
	2200	15 gallons, add					610			610	675	
	2300	30 gallons, add					1,150			1,150	1,250	
	2400	55 gallons, add					1,950			1,950	2,150	
800	0010	**TRAPS**										800
	0030	Cast iron, service weight										
	0050	Running P trap, without vent										
	1100	2″	Q-1	16	1	Ea.	28.50	38.50		67	91.50	

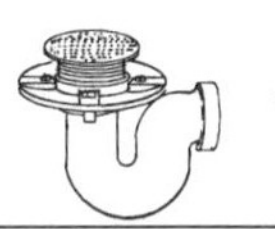

15150 | Sanitary Waste and Vent Piping

		Description	CREW	DAILY OUTPUT	LABOR-HOURS	UNIT	2006 BARE COSTS MAT.	LABOR	EQUIP.	TOTAL	TOTAL INCL O&P	
800	1140	3"	Q-1	14	1.143	Ea.	47.50	44		91.50	121	800
	1150	4"		13	1.231		82.50	47.50		130	165	
	1160	6"	Q-2	17	1.412		360	56.50		416.50	485	
	1180	Running trap, single hub, with vent										
	2080	3" pipe size, 3" vent	Q-1	14	1.143	Ea.	65.50	44		109.50	141	
	2120	4" pipe size, 4" vent 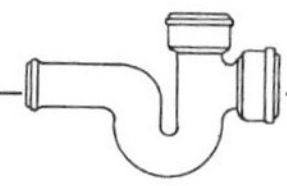	"	13	1.231		89	47.50		136.50	172	
	2140	5" pipe size, 4" vent	Q-2	11	2.182		142	87		229	291	
	2160	6" pipe size, 4" vent		10	2.400		415	95.50		510.50	610	
	2180	6" pipe size, 6" vent		8	3		415	120		535	645	
	2200	8" pipe size, 4" vent	Q-3	10	3.200		1,250	130		1,380	1,575	
	2220	8" pipe size, 6" vent	"	8	4		1,375	163		1,538	1,750	
	2300	For double hub, vent, add					10%	20%				
	2800	S trap, 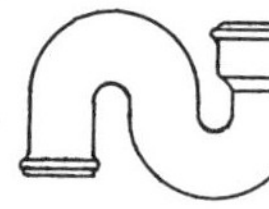										
	2850	4" pipe size	Q-1	13	1.231	Ea.	88	47.50		135.50	171	
	3000	P trap, B&S, 2" pipe size		16	1		19.60	38.50		58.10	81.50	
	3040	3" pipe size		14	1.143		29.50	44		73.50	101	
	3060	4" pipe size		13	1.231		42	47.50		89.50	120	
	3080	5" pipe size 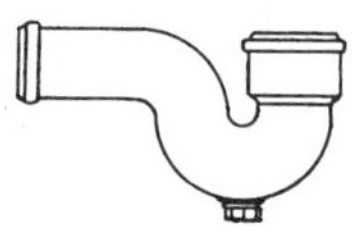	Q-2	18	1.333		89	53		142	181	
	3100	6" pipe size	"	17	1.412		130	56.50		186.50	231	
	3120	8" pipe size	Q-3	11	2.909		390	118		508	615	
	3130	10" pipe size	"	10	3.200		730	130		860	1,000	
	3150	P trap, no hub, 1-1/2" pipe size	Q-1	17	.941		11.40	36		47.40	68.50	
	3160	2" pipe size		16	1		11.05	38.50		49.55	72	
	3170	3" pipe size		14	1.143		24.50	44		68.50	95	
	3180	4" pipe size		13	1.231		42	47.50		89.50	120	
	3190	6" pipe size	Q-2	17	1.412		102	56.50		158.50	200	
	3350	Deep seal trap, B&S										
	3400	1-1/4" pipe size 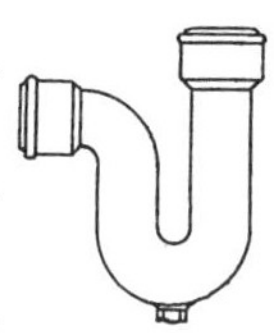	Q-1	14	1.143	Ea.	34	44		78	106	
	3410	1-1/2" pipe size		14	1.143		34	44		78	106	
	3420	2" pipe size		14	1.143		22	44		66	93	
	3440	3" pipe size		12	1.333		36	51		87	119	
	3460	4" pipe size		11	1.455		57	56		113	150	
	3500	For trap primer connection, add					57.50			57.50	63	
	3540	For trap with floor cleanout, add					70%	5%				
	3580	For trap with adjustable cleanout, add	Q-1	10	1.600	Ea.	158	61.50		219.50	270	
	4700	Copper, drainage, drum trap										
	4800	3" x 5" solid, 1-1/2" pipe size	1 Plum	16	.500	Ea.	43.50	21.50		65	80.50	
	4840	3" x 6" swivel, 1-1/2" pipe size	"	16	.500	"	69	21.50		90.50	109	
	5100	P trap, standard pattern										
	5200	1-1/4" pipe size	1 Plum	18	.444	Ea.	32	19		51	64.50	
	5240	1-1/2" pipe size		17	.471		31	20		51	65.50	
	5260	2" pipe size		15	.533		48	23		71	88	
	5280	3" pipe size		11	.727		115	31		146	175	
	5340	With cleanout and slip joint										
	5360	1-1/4" pipe size	1 Plum	18	.444	Ea.	30.50	19		49.50	63.50	
	5400	1-1/2" pipe size		17	.471		47	20		67	83.50	
	5420	2" pipe size		15	.533		76	23		99	119	
	5460	For swivel, 1-1/2"					43			43	47	
	5750	Chromed brass, tubular, P trap, without cleanout, 20 Ga.										
	5800	1-1/4" pipe size	1 Plum	18	.444	Ea.	15.45	19		34.45	46.50	
	5840	1-1/2" pipe size	"	17	.471	"	18.45	20		38.45	52	
	5900	With cleanout, 20 Ga.										
	5940	1-1/4" pipe size	1 Plum	18	.444	Ea.	34	19		53	67	
	6000	1-1/2" pipe size	"	17	.471	"	37	20		57	72.50	
	6350	S trap, without cleanout, 20 Ga.										
800	6400	1-1/4" pipe size	1 Plum	18	.444	Ea.	32.50	19		51.50	65.50	

MECHANICAL 15

15150 | Sanitary Waste and Vent Piping

		15150 Sanitary Waste and Vent Piping	CREW	DAILY OUTPUT	LABOR-HOURS	UNIT	2006 BARE COSTS MAT.	LABOR	EQUIP.	TOTAL	TOTAL INCL O&P	
800	6440	1-1/2″ pipe size	1 Plum	17	.471	Ea.	37	20		57	72.50	800
	6550	With cleanout, 20 Ga.										
	6600	1-1/4″ pipe size	1 Plum	18	.444	Ea.	59	19		78	94.50	
	6640	1-1/2″ pipe size	″	17	.471		64	20		84	102	
	6660	Corrosion resistant, glass, P trap, 1-1/2″ pipe size	Q-1	17	.941		46.50	36		82.50	107	
	6670	2″ pipe size		16	1		61	38.50		99.50	127	
	6680	3″ pipe size		14	1.143		125	44		169	207	
	6690	4″ pipe size		13	1.231		187	47.50		234.50	280	
	6700	6″ pipe size	Q-2	17	1.412		700	56.50		756.50	860	
	6710	ABS DWV P trap, solvent weld joint										
	6720	1-1/2″ pipe size	1 Plum	18	.444	Ea.	6.55	19		25.55	36.50	
	6722	2″ pipe size		17	.471		9.90	20		29.90	42.50	
	6724	3″ pipe size		15	.533		34	23		57	73	
	6726	4″ pipe size		14	.571		69.50	24.50		94	115	
	6760	PP DWV, dilution trap, 1-1/2″ pipe size		16	.500		234	21.50		255.50	290	
	6770	P trap, 1-1/2″ pipe size		17	.471		69.50	20		89.50	108	
	6780	2″ pipe size		16	.500		98.50	21.50		120	141	
	6790	3″ pipe size		14	.571		182	24.50		206.50	238	
	6800	4″ pipe size		13	.615		320	26.50		346.50	395	
	6810	Running trap, 1-1/2″ pipe size		16	.500		44.50	21.50		66	82	
	6820	2″ pipe size		15	.533		62.50	23		85.50	105	
	6830	S trap, 1-1/2″ pipe size		16	.500		45.50	21.50		67	83	
	6840	2″ pipe size		15	.533		57	23		80	98.50	
	6850	Universal trap, 1-1/2″ pipe size		14	.571		129	24.50		153.50	180	
	6860	PVC DWV hub x hub, basin trap, 1-1/4″ pipe size		18	.444		8.40	19		27.40	39	
	6870	Sink P trap, 1-1/2″ pipe size		18	.444		8.40	19		27.40	39	
	6880	Tubular S trap, 1-1/2″ pipe size		17	.471		15.60	20		35.60	48.50	
	6890	PVC sch. 40 DWV, drum trap										
	6900	1-1/2″ pipe size	1 Plum	16	.500	Ea.	22.50	21.50		44	57.50	
	6910	P trap, 1-1/2″ pipe size		18	.444		6	19		25	36	
	6920	2″ pipe size		17	.471		8.65	20		28.65	41	
	6930	3″ pipe size		15	.533		27.50	23		50.50	66	
	6940	4″ pipe size		14	.571		62.50	24.50		87	107	
	6950	P trap w/clean out, 1-1/2″ pipe size		18	.444		9.45	19		28.45	40	
	6960	2″ pipe size		17	.471		15.45	20		35.45	48.50	
	6970	P trap adjustable, 1-1/2″ pipe size		17	.471		7.70	20		27.70	40	
	6980	P trap adj. w/union & cleanout, 1-1/2″ pipe size		16	.500		17.30	21.50		38.80	52	
	7000	Trap primer, flow through type, 1/2″ diameter		24	.333		39.50	14.25		53.75	65.50	
	7100	With sediment strainer		22	.364		43.50	15.55		59.05	71.50	
	7450	Trap primer distribution unit										
	7500	2 openings	1 Plum	18	.444	Ea.	22	19		41	53.50	
	7540	3 opening		17	.471		22	20		42	55.50	
	7560	4 opening		16	.500		22	21.50		43.50	57	
	7850	Trap primer manifold										
	7900	2 outlet	1 Plum	18	.444	Ea.	50.50	19		69.50	85	
	7940	4 outlet		16	.500		78.50	21.50		100	119	
	7960	6 outlet		15	.533		98.50	23		121.50	144	
	7980	8 outlet		13	.615		123	26.50		149.50	177	
	9000	Minimum labor/equipment charge		3	2.667	Job		114		114	177	
900	0010	**VENT FLASHING, CAPS**										900
	0120	Vent caps										
	0140	Cast iron										
	0180	2-1/2″ - 3-5/8″ pipe	1 Plum	21	.381	Ea.	35	16.25		51.25	63.50	
	0190	4″ - 4-1/8″ pipe	″	19	.421	″	41.50	18		59.50	74	
	0900	Vent flashing										

	15150	Sanitary Waste and Vent Piping	CREW	DAILY OUTPUT	LABOR-HOURS	UNIT	2006 BARE COSTS MAT.	LABOR	EQUIP.	TOTAL	TOTAL INCL O&P	
900	1000	Aluminum with lead ring										900
	1020	1-1/4" pipe	1 Plum	20	.400	Ea.	6.65	17.10		23.75	34	
	1030	1-1/2" pipe		20	.400		7.15	17.10		24.25	34.50	
	1040	2" pipe		18	.444		7.15	19		26.15	37.50	
	1050	3" pipe		17	.471		7.90	20		27.90	40	
	1060	4" pipe		16	.500		9.55	21.50		31.05	43.50	
	1350	Copper with neoprene ring										
	1400	1-1/4" pipe	1 Plum	20	.400	Ea.	14.05	17.10		31.15	42	
	1430	1-1/2" pipe		20	.400		14.05	17.10		31.15	42	
	1440	2" pipe		18	.444		14.85	19		33.85	46	
	1450	3" pipe		17	.471		17.40	20		37.40	50.50	
	1460	4" pipe		16	.500		19.20	21.50		40.70	54	
	2000	Galvanized with neoprene ring										
	2020	1-1/4" pipe	1 Plum	20	.400	Ea.	5.30	17.10		22.40	32.50	
	2030	1-1/2" pipe		20	.400		5.30	17.10		22.40	32.50	
	2040	2" pipe		18	.444		5.60	19		24.60	35.50	
	2050	3" pipe		17	.471		6.10	20		26.10	38	
	2060	4" pipe		16	.500		7.90	21.50		29.40	41.50	
	2980	Neoprene, one piece										
	3000	1-1/4" pipe	1 Plum	24	.333	Ea.	5.15	14.25		19.40	27.50	
	3030	1-1/2" pipe		24	.333		5.40	14.25		19.65	28	
	3040	2" pipe		23	.348		5.40	14.85		20.25	29	
	3050	3" pipe		21	.381		6.55	16.25		22.80	33	
	3060	4" pipe		20	.400		8.90	17.10		26	36.50	
	9000	Minimum labor/equipment charge		4	2	Job		85.50		85.50	133	

	15160	Storm Drainage Piping	CREW	DAILY OUTPUT	LABOR-HOURS	UNIT	MAT.	LABOR	EQUIP.	TOTAL	TOTAL INCL O&P	
500	0010	**STORM AREA DRAINS**										500
	0140	Cornice, C.I., 45° or 90° outlet										
	0200	3" and 4" pipe size	Q-1	12	1.333	Ea.	171	51		222	268	
	0260	For galvanized body, add					34.50			34.50	37.50	
	0280	For polished bronze dome, add					29.50			29.50	32.50	
	3860	Roof, flat metal deck, C.I. body, 12" C.I. dome										
	3880	2" pipe size	Q-1	15	1.067	Ea.	149	41		190	228	
	3890	3" pipe size		14	1.143		204	44		248	294	
	3900	4" pipe size		13	1.231		204	47.50		251.50	299	
	3910	5" pipe size		12	1.333		355	51		406	470	
	3920	6" pipe size		10	1.600		355	61.50		416.50	485	
	4280	Integral expansion joint, C.I. body, 12" C.I. dome										
	4300	2" pipe size	Q-1	8	2	Ea.	305	77		382	455	
	4320	3" pipe size		7	2.286		370	88		458	545	
	4340	4" pipe size		6	2.667		395	102		497	595	
	4360	5" pipe size		4	4		425	154		579	710	
	4380	6" pipe size		3	5.333		615	205		820	995	
	4400	8" pipe size		3	5.333		800	205		1,005	1,200	
	4440	For galvanized body, add					119			119	131	
	4480	Flexible neoprene bellows, no-hub connection										
	4500	3" and 4" pipe size	Q-1	5	3.200	Ea.	88	123		211	288	
	4520	6" and 8" pipe size	"	4	4	"	195	154		349	455	
	4620	Main, all aluminum, 12" low profile dome										
	4640	2", 3" and 4" pipe size	Q-1	14	1.143	Ea.	226	44		270	320	
	4660	5" and 6" pipe size		13	1.231		298	47.50		345.50	405	
	4680	8" pipe size		10	1.600		370	61.50		431.50	505	
	4690	Main, CI body, 12" poly. dome, 2", 3", & 4" pipe		8	2		168	77		245	305	
	4710	5" and 6" pipe size		6	2.667		240	102		342	425	

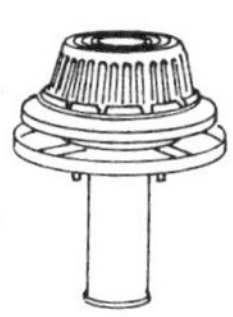

15160 | Storm Drainage Piping

		15160 Storm Drainage Piping	CREW	DAILY OUTPUT	LABOR-HOURS	UNIT	2006 BARE COSTS MAT.	LABOR	EQUIP.	TOTAL	TOTAL INCL O&P	
500	4720	8" pipe size	Q-1	4	4	Ea.	310	154		464	585	500
	4730	For underdeck clamp, add		22	.727		116	28		144	172	
	4740	For vandalproof dome, add					29.50			29.50	32.50	
	4750	For galvanized body, add					200			200	220	
	4760	Main, PVC body and dome, 2" pipe size	Q-1	14	1.143		89	44		133	166	
	4780	3" pipe size		14	1.143		89	44		133	166	
	4800	4" pipe size		14	1.143		89	44		133	166	
	4820	For underdeck clamp, add		24	.667		19.30	25.50		44.80	61	
	4900	Terrace planting area, with perforated overflow, C.I.										
	4920	2", 3" and 4" pipe size	Q-1	8	2	Ea.	320	77		397	470	
	4980	Scupper floor, oblique strainer, C.I.										
	5000	6" x 7" top, 2", 3" and 4" pipe size	Q-1	16	1	Ea.	147	38.50		185.50	222	
	5100	8" x 12" top, 5" and 6" pipe size	"	14	1.143		286	44		330	385	
	5160	For galvanized body, add					40%					
	5200	For polished bronze strainer, add					85%					
	5980	Trench, floor, hvy duty, modular, C.I., 12" x 12" top										
	6000	2", 3", 4", 5", & 6" pipe size	Q-1	8	2	Ea.	465	77		542	630	
	6100	For polished bronze top, add					228			228	250	
	6200	For 12" extension section, C.I. top, add					465			465	510	
	6240	For 12" extension section, polished bronze top, add					217			217	239	
	9000	Minimum labor/equipment charge	1 Plum	4	2	Job		85.50		85.50	133	

15170 | Swimming Pool and Fountain Piping

		15170 Swimming Pool and Fountain Piping	CREW	DAILY OUTPUT	LABOR-HOURS	UNIT	MAT.	LABOR	EQUIP.	TOTAL	TOTAL INCL O&P	
500	0010	**SWIMMING POOL DRAINS**										500
	2990	Pool main discharge, polished bronze top, C.I. body										
	3000	8" diam. grate, 2", 3" & 4" pipe size	Q-1	8	2	Ea.	375	77		452	530	
	3100	12" diam. grate, 4", 5" & 6" pipe size	"	8	2		630	77		707	815	
	3160	For galvanized body, add					151			151	166	
	3200	Gutter, bronze body, polished bronze top										
	3220	9" grate, 1-1/2" & 2" pipe size	Q-1	8	2	Ea.	108	77		185	239	
	3240	10" grate, 2", 2-1/2" & 3" pipe size	"	7	2.286	"	112	88		200	261	
	3380	Overflow, 5" dome & 6" high pipe polished bronze,										
	3400	cast iron body, 2", 3", & 4" pipe size	Q-1	10	1.600	Ea.	475	61.50		536.50	615	
	3580	Recirculating inlet, 6" sq face polished bronze,										
	3600	C.I. body, 2", 2-1/2" & 3" pipe size	Q-1	10	1.600	Ea.	68	61.50		129.50	170	
	3620	For galvanized body, add				"	29.50			29.50	32	

15180 | Heating and Cooling Piping

		15180 Heating and Cooling Piping	CREW	DAILY OUTPUT	LABOR-HOURS	UNIT	MAT.	LABOR	EQUIP.	TOTAL	TOTAL INCL O&P	
100	0010	**ANTI-FREEZE** Inhibited										100
	0900	Ethylene glycol concentrated										
	1000	55 gallon drums, small quantities				Gal.	9.70			9.70	10.70	
	1200	Large quantities					9.15			9.15	10.05	
	2000	Propylene glycol, for solar heat, small quantities					10.75			10.75	11.85	
	2100	Large quantities					10.55			10.55	11.60	
200	0010	**PUMPS, CIRCULATING** Heated or chilled water application										200
	0600	Bronze, sweat connections, 1/40 HP, in line										
	0640	3/4" size	Q-1	16	1	Ea.	134	38.50		172.50	208	
	1000	Flange connection, 3/4" to 1-1/2" size										
	1040	1/12 HP	Q-1	6	2.667	Ea.	365	102		467	565	
	1060	1/8 HP		6	2.667		630	102		732	855	
	1100	1/3 HP		6	2.667		705	102		807	935	
	1140	2" size, 1/6 HP		5	3.200		905	123		1,028	1,175	
	1180	2-1/2" size, 1/4 HP		5	3.200		1,150	123		1,273	1,475	
	1220	3" size, 1/4 HP		4	4		1,200	154		1,354	1,575	

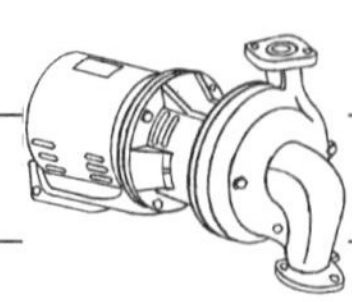

	15180 | Heating and Cooling Piping		CREW	DAILY OUTPUT	LABOR-HOURS	UNIT	2006 BARE COSTS MAT.	LABOR	EQUIP.	TOTAL	TOTAL INCL O&P	
200	1260	1/3 HP	Q-1	4	4	Ea.	1,475	154		1,629	1,875	200
	1300	1/2 HP	↓	4	4	↓	1,500	154		1,654	1,900	
	1340	3/4 HP		4	4		1,650	154		1,804	2,050	
	1380	1 HP		4	4		2,650	154		2,804	3,150	
	2000	Cast iron, flange connection										
	2040	3/4" to 1-1/2" size, in line, 1/12 HP	Q-1	6	2.667	Ea.	239	102		341	420	
	2060	1/8 HP	↓	6	2.667	↓	400	102		502	600	
	2100	1/3 HP		6	2.667		445	102		547	650	
	2140	2" size, 1/6 HP		5	3.200		490	123		613	725	
	2180	2-1/2" size, 1/4 HP		5	3.200		625	123		748	880	
	2220	3" size, 1/4 HP		4	4		640	154		794	940	
	2260	1/3 HP		4	4		860	154		1,014	1,175	
	2300	1/2 HP		4	4		895	154		1,049	1,225	
	2340	3/4 HP		4	4		1,025	154		1,179	1,375	
	2380	1 HP		4	4		1,500	154		1,654	1,900	
	2600	For non-ferrous impeller, add					3%					
	3000	High head, bronze impeller										
	3030	1-1/2" size 1/2 HP	Q-1	5	3.200	Ea.	740	123		863	1,000	
	3040	1-1/2" size 3/4 HP	↓	5	3.200	↓	790	123		913	1,050	
	3050	2" size 1 HP		4	4		950	154		1,104	1,300	
	3090	2" size 1-1/2 HP		4	4		1,100	154		1,254	1,475	
	4000	Close coupled, end suction, bronze impeller										
	4040	1-1/2" size, 1-1/2 HP, to 40 GPM	Q-1	3	5.333	Ea.	1,400	205		1,605	1,850	
	4090	2" size, 2 HP, to 50 GPM	↓	3	5.333	↓	1,650	205		1,855	2,125	
	4100	2" size, 3 HP, to 90 GPM		2.30	6.957		1,725	267		1,992	2,300	
	4190	2-1/2" size, 3 HP, to 150 GPM		2	8		1,850	305		2,155	2,525	
	4300	3" size, 5 HP, to 225 GPM		1.80	8.889		2,125	340		2,465	2,850	
	4410	3" size, 10 HP, to 350 GPM		1.60	10		2,850	385		3,235	3,725	
	4420	4" size, 7-1/2 HP, to 350 GPM		1.60	10		2,775	385		3,160	3,650	
	4520	4" size, 10 HP, to 600 GPM	Q-2	1.70	14.118		3,150	565		3,715	4,350	
	4530	5" size, 15 HP, to 1000 GPM	↓	1.70	14.118		3,175	565		3,740	4,375	
	4610	5" size, 20 HP, to 1350 GPM		1.50	16		3,375	640		4,015	4,700	
	4620	5" size, 25 HP, to 1550 GPM		1.50	16		4,600	640		5,240	6,050	
	5000	Base mounted, bronze impeller, coupling guard										
	5040	1-1/2" size, 1-1/2 HP, to 40 GPM	Q-1	2.30	6.957	Ea.	2,650	267		2,917	3,350	
	5090	2" size, 2 HP, to 50 GPM	↓	2.30	6.957	↓	2,900	267		3,167	3,625	
	5100	2" size, 3 HP, to 90 GPM		2	8		2,925	305		3,230	3,700	
	5190	2-1/2" size, 3 HP, to 150 GPM		1.80	8.889		3,150	340		3,490	4,000	
	5300	3" size, 5 HP, to 225 GPM		1.60	10		2,825	385		3,210	3,725	
	5410	4" size, 5 HP, to 350 GPM		1.50	10.667		2,900	410		3,310	3,825	
	5420	4" size, 7-1/2 HP, to 350 GPM		1.50	10.667		3,025	410		3,435	3,950	
	5520	5" size, 10 HP, to 600 GPM	Q-2	1.60	15		3,650	600		4,250	4,950	
	5530	5" size, 15 HP, to 1000 GPM	↓	1.60	15		3,800	600		4,400	5,100	
	5610	6" size, 20 HP, to 1350 GPM		1.40	17.143		4,925	685		5,610	6,450	
	5620	6" size, 25 HP, to 1550 GPM		1.40	17.143		5,050	685		5,735	6,600	
	9000	Minimum labor/equipment charge	Q-1	3.25	4.923	Job		189		189	294	
300	0010	PUMPS, CONDENSATE RETURN SYSTEM										300
	0200	Simplex, 3/4 H.P. mtr, float switch, controls, 10 Gal. C.I. rcvr, 6-15GPM	Q-1	1	16	Ea.	1,850	615		2,465	2,975	
	1000	Duplex, 2 pumps, 3/4 H.P. motors, float switch,										
	1060	alternator asssembly, 15 Gal. C.I. receiver	Q-1	.50	32	Ea.	5,050	1,225		6,275	7,450	
	4400	Refrigerant										
	4420	Refrigerant, R-22, 50 lb. disposable cylinder				Lb.	2.54			2.54	2.79	
	4440	Refrigerant, R-507, 25 lb. disposable cylinder				"	11.50			11.50	12.65	
800	0010	STEAM CONDENSATE METER										800
	0100	500 lb. per hour	1 Stpi	14	.571	Ea.	2,300	24.50		2,324.50	2,575	

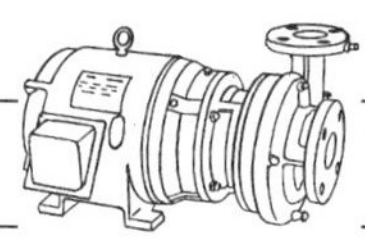

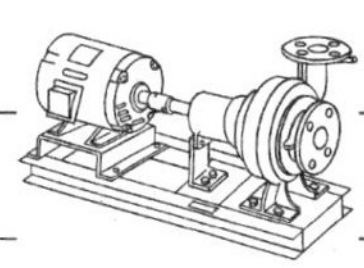

15180 | Heating and Cooling Piping

		15180 Heating and Cooling Piping	Crew	Daily Output	Labor-Hours	Unit	2006 Bare Costs Mat.	Labor	Equip.	Total	Total Incl O&P	
800	0140	1500 lb. per hour	1 Stpi	7	1.143	Ea.	2,475	49		2,524	2,800	800
	0160	3000 lb. per hour		5	1.600		3,325	69		3,394	3,750	
	0200	12,000 lb. per hour	Q-5	3.50	4.571		4,450	177		4,627	5,150	

15190 | Fuel Piping

		15190 Fuel Piping	Crew	Daily Output	Labor-Hours	Unit	2006 Bare Costs Mat.	Labor	Equip.	Total	Total Incl O&P	
490	0010	FUEL OIL SPECIALTIES										490
	0020	Foot valve, single poppet, metal to metal construction										
	0040	Bevel seat, 1/2" diameter	1 Stpi	20	.400	Ea.	36	17.20		53.20	66.50	
	0060	3/4" diameter		18	.444		37.50	19.15		56.65	71.50	
	0080	1" diameter		16	.500		28.50	21.50		50	65	
	0100	1-1/4" diameter		15	.533		45	23		68	85	
	0120	1-1/2" diameter		13	.615		49.50	26.50		76	95.50	
	0140	2" diameter		11	.727		54	31.50		85.50	108	
	0400	Fuel fill box, flush type										
	0408	Nonlocking, watertight										
	0410	1-1/2" diameter	1 Stpi	12	.667	Ea.	15.70	28.50		44.20	62	
	0440	2" diameter		10	.800		16.90	34.50		51.40	72	
	0450	2-1/2" diameter		9	.889		56.50	38.50		95	122	
	0460	3" diameter		7	1.143		66.50	49		115.50	150	
	0470	4" diameter		5	1.600		75.50	69		144.50	190	
	0500	Locking inner cover										
	0510	2" diameter	1 Stpi	8	1	Ea.	82	43		125	158	
	0520	2-1/2" diameter		7	1.143		110	49		159	198	
	0530	3" diameter		5	1.600		118	69		187	237	
	0540	4" diameter		4	2		152	86		238	300	
	0600	Fuel system components										
	0620	Spill container	1 Stpi	4	2	Ea.	485	86		571	670	
	0640	Fill adapter, 4", straight drop		8	1		33	43		76	103	
	0680	Fill cap, 4"		30	.267		28.50	11.50		40	49.50	
	0700	Extractor fitting, 4" x 1-1/2"		8	1		340	43		383	440	
	0740	Vapor hose adapter, 4"		8	1		70.50	43		113.50	145	
	0760	Wood gage stick, 10'					10.75			10.75	11.85	
	1000	Oil filters, 3/8" IPT., 20 gal. per hour	1 Stpi	20	.400		23	17.20		40.20	52	
	1020	32 gal. per hour		18	.444		34.50	19.15		53.65	68	
	1040	40 gal. per hour		16	.500		39.50	21.50		61	77	
	1060	50 gal. per hour		14	.571		41	24.50		65.50	83.50	
	1800	Pump and motor sets										
	1810	Light fuel and diesel oils, 100 PSI										
	1820	25 GPH 1/4 HP	Q-5	6	2.667	Ea.	795	103		898	1,025	
	1830	45 GPH, 1/4 HP		6	2.667		800	103		903	1,050	
	1840	90 GPH, 1/4 HP		5	3.200		860	124		984	1,150	
	1850	160 GPH, 1/3 HP		4	4		885	155		1,040	1,200	
	1860	325 GPH, 3/4 HP		4	4		970	155		1,125	1,325	
	1870	700 GPH, 1-1/2 HP		3	5.333		1,750	207		1,957	2,250	
	1880	1000 GPH, 2 HP		3	5.333		1,800	207		2,007	2,300	
	1890	1800 GPH, 5 HP		1.80	8.889		2,750	345		3,095	3,575	
	2000	Remote tank gauging system, self contained										
	2100	Single tank kit/8 sensors inputs w/printer	1 Stpi	2.50	3.200	Ea.	2,725	138		2,863	3,225	
	2120	Two tank kit/8 sensors inputs w/printer		2	4		3,675	172		3,847	4,325	
	3000	Valve, ball check, globe type, 3/8" diameter		24	.333		9.50	14.35		23.85	33	
	3500	Fusible, 3/8" diameter		24	.333		7.55	14.35		21.90	31	
	3600	1/2" diameter		24	.333		19.55	14.35		33.90	44	
	3610	3/4" diameter		20	.400		41.50	17.20		58.70	73	
	3620	1" diameter		19	.421		123	18.15		141.15	163	
	4000	Nonfusible, 3/8" diameter		24	.333		15.20	14.35		29.55	39	

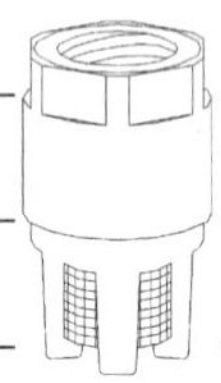

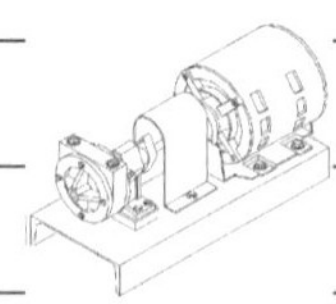

15 MECHANICAL

15190 | Fuel Piping

			CREW	DAILY OUTPUT	LABOR-HOURS	UNIT	2006 BARE COSTS MAT.	LABOR	EQUIP.	TOTAL	TOTAL INCL O&P	
490	4500	Shutoff, gate type, lever handle, spring-fusible kit										490
	4520	1/4" diameter	1 Stpi	14	.571	Ea.	19.30	24.50		43.80	59.50	
	4540	3/8" diameter		12	.667		19.10	28.50		47.60	65.50	
	4560	1/2" diameter		10	.800		29	34.50		63.50	85.50	
	4570	3/4" diameter		8	1		50	43		93	122	
	4580	Lever handle, requires weight and fusible kit										
	4600	1" diameter	1 Stpi	9	.889	Ea.	284	38.50		322.50	370	
	4620	1-1/4" diameter		8	1		294	43		337	390	
	4640	1-1/2" diameter		7	1.143		400	49		449	515	
	4660	2" diameter		6	1.333		420	57.50		477.50	550	
	4680	For fusible link, weight and braided wire, add					5%					
	5000	Vent alarm, whistling signal					26			26	28.50	
	5500	Vent protector/breather, 1-1/4" diameter	1 Stpi	32	.250		7.20	10.75		17.95	24.50	
	5520	1-1/2" diameter		32	.250		8	10.75		18.75	25.50	
	5540	2" diameter		32	.250		16	10.75		26.75	34.50	
	5560	3" diameter		28	.286		40	12.30		52.30	63	
	5580	4" diameter		24	.333		47	14.35		61.35	74.50	
	5600	Dust cap, breather, 2"		40	.200		11	8.60		19.60	25.50	
	8000	Fuel oil and tank heaters										
	8020	Electric, capacity rated at 230 volts										
	8040	Immersion element in steel manifold										
	8060	96 GPH at 50°F rise	Q-5	6.40	2.500	Ea.	630	97		727	840	
	8070	128 GPH at 50°F rise		6.20	2.581		665	100		765	885	
	8080	160 GPH at 50°F rise		5.90	2.712		695	105		800	930	
	8090	192 GPH at 50°F rise		5.50	2.909		740	113		853	990	
	8100	240 GPH at 50°F rise		5.10	3.137		890	122		1,012	1,175	
	8110	288 GPH at 50°F rise		4.60	3.478		1,050	135		1,185	1,375	
	8120	384 GPH at 50°F rise		3.10	5.161		1,200	200		1,400	1,625	
	8130	480 GPH at 50°F rise		2.30	6.957		1,500	270		1,770	2,075	
	8140	576 GPH at 50°F rise		2.10	7.619		1,625	295		1,920	2,250	
	8300	Suction stub, immersion type										
	8320	75" long, 750 watts	1 Stpi	14	.571	Ea.	530	24.50		554.50	625	
	8330	99" long, 2000 watts		12	.667		745	28.50		773.50	865	
	8340	123" long, 3000 watts		10	.800		805	34.50		839.50	940	
	8660	Steam, cross flow, rated at 5 PSIG										
	8680	42 GPH	Q-5	7	2.286	Ea.	1,225	88.50		1,313.50	1,500	
	8690	73 GPH		6.70	2.388		1,425	92.50		1,517.50	1,725	
	8700	112 GPH		6.20	2.581		1,475	100		1,575	1,775	
	8710	158 GPH		5.80	2.759		1,625	107		1,732	1,950	
	8720	187 GPH		4	4		2,225	155		2,380	2,700	
	8730	270 GPH		3.60	4.444		2,400	172		2,572	2,900	
	8740	365 GPH	Q-6	4.90	4.898		2,800	197		2,997	3,400	
	8750	635 GPH		3.70	6.486		3,550	261		3,811	4,300	
	8760	845 GPH		2.50	9.600		5,200	385		5,585	6,325	
	8770	1420 GPH		1.60	15		8,475	605		9,080	10,300	
	8780	2100 GPH		1.10	21.818		11,600	875		12,475	14,200	

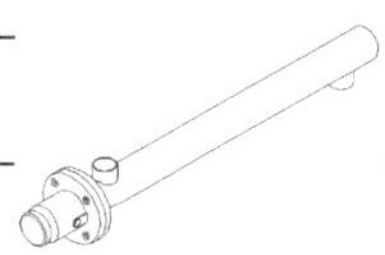

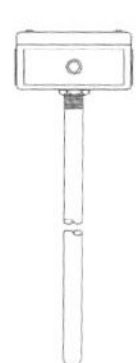

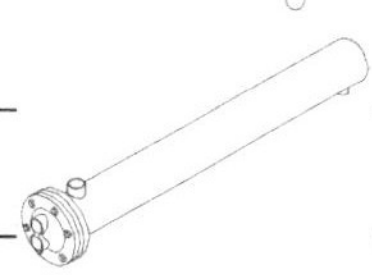

15210 | Process Air/Gas Piping

		15210 Process Air/Gas Piping	CREW	DAILY OUTPUT	LABOR-HOURS	UNIT	2006 BARE COSTS MAT.	LABOR	EQUIP.	TOTAL	TOTAL INCL O&P	
100	0010	**AIR COMPRESSORS**										100
	5250	Air, reciprocating air cooled, splash lubricated, tank mounted										
	5300	Single stage, 1 phase, 140 psi										
	5303	1/2 HP, 30 gal tank	1 Stpi	3	2.667	Ea.	890	115		1,005	1,150	
	5305	3/4 HP, 30 gal tank		2.60	3.077		920	132		1,052	1,200	
	5307	1 HP, 30 gal tank		2.20	3.636		1,150	157		1,307	1,500	
	5309	2 HP, 30 gal tank	Q-5	4	4		1,525	155		1,680	1,925	
	5310	3 HP, 30 gal tank		3.60	4.444		2,150	172		2,322	2,650	
	5314	3 HP, 60 gal tank		3.50	4.571		2,225	177		2,402	2,725	
	5320	5 HP, 60 gal tank		3.20	5		2,875	194		3,069	3,450	
	5330	5 HP, 80 gal tank		3	5.333		2,950	207		3,157	3,575	
	5340	7.5 HP, 80 gal tank		2.60	6.154		4,200	238		4,438	5,000	
	5600	2 stage pkg., 3 phase										
	5650	6 CFM at 125 psi 1-1/2 HP, 60 gal tank	Q-5	3	5.333	Ea.	1,950	207		2,157	2,475	
	5670	10.9 CFM at 125 psi, 3 HP, 80 gal tank		1.50	10.667		2,325	415		2,740	3,225	
	5680	38.7 CFM at 125 psi, 10 HP, 120 gal tank		.60	26.667		4,775	1,025		5,800	6,850	
	5690	105 CFM at 125 psi, 25 HP, 250 gal tank	Q-6	.60	40		9,575	1,600		11,175	13,000	
	5800	With single stage pump										
	5850	8.3 CFM at 125 psi, 2 HP, 80 gal tank	Q-6	3.50	6.857	Ea.	2,750	276		3,026	3,450	
	5860	38.7 CFM at 125 psi, 10 HP, 120 gal tank	"	.90	26.667	"	5,275	1,075		6,350	7,475	
	6000	Reciprocating, 2 stage, tank mtd, 3 Ph., Cap. rated @175 PSIG										
	6050	Pressure lubcatd, hvy duty, 9.7 CFM, 3 HP, 120 Gal. tank	Q-5	1.30	12.308	Ea.	4,325	475		4,800	5,500	
	6054	5 CFM, 1-1/2 HP, 80 gal tank		2.80	5.714		2,525	221		2,746	3,125	
	6056	6.4 CFM, 2 HP, 80 gal tank		2	8		2,750	310		3,060	3,500	
	6058	8.1 CFM, 3 HP, 80 gal tank		1.70	9.412		2,900	365		3,265	3,750	
	6059	14.8 CFM, 5 HP, 80 gal tank		1	16		2,950	620		3,570	4,225	
	6060	16.5 CFM, 5 HP, 120 gal. tank		1	16		4,400	620		5,020	5,800	
	6063	13 CFM, 6 HP, 80 gal tank		.90	17.778		4,950	690		5,640	6,500	
	6066	19.8 CFM, 7.5 HP, 80 gal tank		.80	20		5,200	775		5,975	6,925	
	6070	25.8 CFM, 7-1/2 HP, 120 gal. tank		.80	20		6,550	775		7,325	8,400	
	6078	34.8 CFM, 10 HP, 80 gal. tank		.70	22.857		6,175	885		7,060	8,175	
	6080	34.8 CFM, 10 HP, 120 gal. tank		.60	26.667		6,500	1,025		7,525	8,750	
	6090	53.7 CFM, 15 HP, 120 gal. tank	Q-6	.80	30		8,150	1,200		9,350	10,900	
	6100	76.7 CFM, 20 HP, 120 gal. tank		.70	34.286		10,100	1,375		11,475	13,300	
	6104	76.7 CFM, 20 HP, 240 gal tank		.68	35.294		11,000	1,425		12,425	14,300	
	6110	90.1 CFM, 25 HP, 120 gal. tank		.63	38.095		10,600	1,525		12,125	14,100	
	6120	101 CFM, 30 HP, 120 gal. tank		.57	42.105		11,400	1,700		13,100	15,100	
	6130	101 CFM, 30 HP, 250 gal. tank		.52	46.154		12,300	1,850		14,150	16,400	
	6200	Oil-less, 13.6 CFM, 5 HP, 120 gal. tank	Q-5	.88	18.182		13,700	705		14,405	16,100	
	6210	13.6 CFM, 5 HP, 250 gal. tank		.80	20		14,700	775		15,475	17,300	
	6220	18.2 CFM, 7.5 HP, 120 gal. tank		.73	21.918		13,900	850		14,750	16,500	
	6230	18.2 CFM, 7.5 HP, 250 gal. tank		.67	23.881		14,900	925		15,825	17,900	
	6250	30.5 CFM, 10 HP, 120 gal. tank		.57	28.070		16,100	1,100		17,200	19,400	
	6260	30.5 CFM, 10 HP, 250 gal. tank		.53	30.189		17,100	1,175		18,275	20,700	
	6270	41.3 CFM, 15 HP, 120 gal. tank	Q-6	.70	34.286		17,500	1,375		18,875	21,500	
	6280	41.3 CFM, 15 HP, 250 gal. tank	"	.67	35.821		18,500	1,450		19,950	22,700	
200	0010	**COMPRESSOR ACCESSORIES**										200
	1700	Refrigerated air dryers with ambient air filters										
	1710	10 CFM	Q-5	8	2	Ea.	570	77.50		647.50	745	
	1720	25 CFM		6.60	2.424		870	94		964	1,100	
	1730	50 CFM		6.20	2.581		1,325	100		1,425	1,625	
	1740	75 CFM		5.80	2.759		1,925	107		2,032	2,300	
	1750	100 CFM		5.60	2.857		2,450	111		2,561	2,875	
	4000	Couplers, air line, sleeve type										
	4010	Female, connection size NPT										
	4020	1/4"	1 Stpi	38	.211	Ea.	4.37	9.05		13.42	18.90	

15210 | Process Air/Gas Piping

	Line	Description	Crew	Daily Output	Labor-Hours	Unit	2006 Bare Costs Mat.	Labor	Equip.	Total	Total Incl O&P	
200	4030	3/8"	1 Stpi	36	.222	Ea.	5.45	9.55		15	21	200
	4040	1/2"		35	.229		11.40	9.85		21.25	28	
	4050	3/4"	↓	34	.235	↓	13.05	10.15		23.20	30	
	4100	Male										
	4110	1/4"	1 Stpi	38	.211	Ea.	4.32	9.05		13.37	18.85	
	4120	3/8"		36	.222		5.45	9.55		15	21	
	4130	1/2"		35	.229		10.80	9.85		20.65	27	
	4140	3/4"	↓	34	.235	↓	12.90	10.15		23.05	30	
	4150	Coupler, combined male and female halves										
	4160	1/2"	1 Stpi	17	.471	Ea.	22	20.50		42.50	56	
	4170	3/4"	"	15	.533	"	26	23		49	64	
500	0010	**MEDICAL GAS SYSTEM SPECIALTIES**										500
	0100	Vacuum system										
	0110	Vacuum outlet alarm panel	1 Plum	3.20	2.500	Ea.	850	107		957	1,100	
	1000	Nitrogen or oxygen system										
	1010	Cylinder manifold										
	1020	5 cylinder	1 Plum	.80	10	Ea.	4,575	425		5,000	5,700	
	1026	10 cylinder	"	.40	20	"	5,550	855		6,405	7,425	
	3000	Outlets and valves										
	3010	Recessed, wall mounted										
	3012	Single outlet	1 Plum	3.20	2.500	Ea.	53.50	107		160.50	225	
	3100	Ceiling outlet										
	3200	Zone valve with box										
	3210	2"	1 Plum	3.20	2.500	Ea.	380	107		487	585	
	4000	Alarm panel, medical gases and vacuum										
	4010	Alarm panel	1 Plum	3.20	2.500	Ea.	850	107		957	1,100	

15220 | Process Water/Waste Piping

	Line	Description	Crew	Daily Output	Labor-Hours	Unit	Mat.	Labor	Equip.	Total	Total Incl O&P	
100	0010	**LABORATORY EQUIPMENT** Corrosion resistant										100
	7000	Tanks, covers included										
	7800	Polyethylene										
	7810	Continuous service to 220°F										
	7830	2 gallon, 8" x 8" x 8"	Q-1	20	.800	Ea.	104	30.50		134.50	163	
	7850	7 gallon, 12" x 12" x 12"		20	.800		140	30.50		170.50	202	
	7870	15 gallon, 24" x 12" x 12"		17	.941		262	36		298	345	
	8010	30 gallon, 24" x 18" x 18"		12	1.333		320	51		371	430	
	8070	45 gallon, 24" x 18" x 24"		10	1.600		241	61.50		302.50	360	
	8080	90 gallon, 36" x 24" x 24"	↓	8	2	↓	415	77		492	580	
	8150	Polyethylene, heavy duty walls										
	8160	Continuous service to 180°F										
	8180	5 gallon, 11" I.D. x 15" deep	Q-1	20	.800	Ea.	75	30.50		105.50	131	
	8210	15 gallon, 14" I.D. x 27" deep		17	.941		115	36		151	182	
	8230	55 gallon, 22" I.D. x 36" deep		10	1.600		218	61.50		279.50	335	
	8250	100 gallon 28" I.D. x 42" deep		8	2		370	77		447	525	
	8270	200 gallon 36" I.D. x 48" deep		6	2.667		530	102		632	745	
	8290	360 gallon 48" I.D. x 48" deep	↓	5	3.200	↓	740	123		863	1,000	

15230 | Industrial Process Piping

	Line	Description	Crew	Daily Output	Labor-Hours	Unit	Mat.	Labor	Equip.	Total	Total Incl O&P	
500	0010	**PUMPS, GENERAL UTILITY** With motor										500
	2000	Single stage										
	3000	Double suction,										
	3190	75 HP, to 2500 GPM	Q-3	.28	114	Ea.	12,100	4,650		16,750	20,500	
	3220	100 HP, to 3000 GPM		.26	123		15,400	5,000		20,400	24,800	
	3240	150 HP, to 4000 GPM	↓	.24	133	↓	20,700	5,425		26,125	31,200	

15410 | Plumbing Fixtures

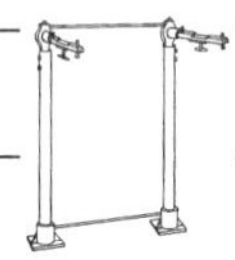

		15410 Plumbing Fixtures	Crew	Daily Output	Labor-Hours	Unit	2006 Bare Costs Mat.	Labor	Equip.	Total	Total Incl O&P	
040	0010	**FIXTURES** Includes trim fittings unless otherwise noted										040
	0080	For rough-in, supply, waste, and vent, see add for each type										
	0120	For electric water coolers, see division 15413										
	0160	For color, unless otherwise noted, add				Ea.	20%					
200	0010	**CARRIERS/SUPPORTS** For plumbing fixtures										200
	0500	Drinking fountain, wall mounted										
	0600	Plate type with studs, top back plate	1 Plum	7	1.143	Ea.	46	49		95	127	
	0700	Top front and back plate		7	1.143		57.50	49		106.50	139	
	0800	Top & bottom, front & back plates, w/bearing jacks		7	1.143		83.50	49		132.50	168	
	3000	Lavatory, concealed arm										
	3050	Floor mounted, single										
	3100	High back fixture	1 Plum	6	1.333	Ea.	177	57		234	284	
	3200	Flat slab fixture		6	1.333		223	57		280	335	
	3220	Paraplegic		6	1.333		290	57		347	410	
	3250	Floor mounted, back to back										
	3300	High back fixtures	1 Plum	5	1.600	Ea.	370	68.50		438.50	510	
	3400	Flat slab fixtures		5	1.600		455	68.50		523.50	605	
	3430	Paraplegic		5	1.600		320	68.50		388.50	455	
	3500	Wall mounted, in stud or masonry										
	3600	High back fixture	1 Plum	6	1.333	Ea.	153	57		210	257	
	3700	Flat slab fixture	″	6	1.333	″	196	57		253	305	
	4000	Exposed arm type, floor mounted										
	4100	Single high back or flat slab fixture	1 Plum	6	1.333	Ea.	370	57		427	495	
	4200	Back to back, high back or flat slab fixtures		5	1.600		580	68.50		648.50	740	
	4300	Wall mounted, High back or flat slab lavatory		6	1.333		255	57		312	370	
	4600	Sink, floor mounted										
	4650	Exposed arm system										
	4700	Single heavy fixture	1 Plum	5	1.600	Ea.	435	68.50		503.50	580	
	4750	Single heavy sink with slab		5	1.600		350	68.50		418.50	490	
	4800	Back to back, standard fixtures		5	1.600		320	68.50		388.50	455	
	4850	Back to back, heavy fixtures		5	1.600		350	68.50		418.50	490	
	4900	Back to back, heavy sink with slab		5	1.600		350	68.50		418.50	490	
	4950	Exposed offset arm system										
	5000	Single heavy deep fixture	1 Plum	5	1.600	Ea.	410	68.50		478.50	555	
	5100	Plate type system										
	5200	With bearing jacks, single fixture	1 Plum	5	1.600	Ea.	475	68.50		543.50	625	
	5300	With exposed arms, single heavy fixture		5	1.600		605	68.50		673.50	775	
	5400	Wall mounted, exposed arms, single heavy fixture		5	1.600		221	68.50		289.50	350	
	6000	Urinal, floor mounted, 2″ or 3″ coupling, blowout type		6	1.333		275	57		332	390	
	6100	With fixture or hanger bolts, blowout or washout		6	1.333		196	57		253	305	
	6200	With bearing plate		6	1.333		196	57		253	305	
	6300	Wall mounted, plate type system		6	1.333		170	57		227	276	
	6980	Water closet, siphon jet										
	7000	Horizontal, adjustable, caulk										
	7040	Single, 4″ pipe size	1 Plum	5.33	1.501	Ea.	360	64		424	495	
	7050	4″ pipe size, paraplegic		5.33	1.501		360	64		424	495	
	7060	5″ pipe size		5.33	1.501		475	64		539	620	
	7100	Double, 4″ pipe size		5	1.600		675	68.50		743.50	850	
	7110	4″ pipe size, paraplegic		5	1.600		675	68.50		743.50	850	
	7120	5″ pipe size		5	1.600		830	68.50		898.50	1,025	
	7160	Horizontal, adjustable, extended, caulk										
	7180	Single, 4″ pipe size	1 Plum	5.33	1.501	Ea.	360	64		424	495	
	7200	5″ pipe size		5.33	1.501		475	64		539	620	
	7240	Double, 4″ pipe size		5	1.600		670	68.50		738.50	840	
	7260	5″ pipe size		5	1.600		830	68.50		898.50	1,025	
	7400	Vertical, adjustable, caulk or thread										

15 MECHANICAL

15410 | Plumbing Fixtures

		15410 Plumbing Fixtures	CREW	DAILY OUTPUT	LABOR-HOURS	UNIT	2006 BARE COSTS MAT.	LABOR	EQUIP.	TOTAL	TOTAL INCL O&P	
200	7440	Single, 4″ pipe size	1 Plum	5.33	1.501	Ea.	495	64		559	640	200
	7460	5″ pipe size		5.33	1.501		605	64		669	770	
	7480	6″ pipe size		5	1.600		695	68.50		763.50	870	
	7520	Double, 4″ pipe size		5	1.600		735	68.50		803.50	915	
	7540	5″ pipe size		5	1.600		850	68.50		918.50	1,050	
	7560	6″ pipe size		4	2		940	85.50		1,025.50	1,150	
	7600	Vertical, adjustable, extended, caulk										
	7620	Single, 4″ pipe size	1 Plum	5.33	1.501	Ea.	495	64		559	640	
	7640	5″ pipe size		5.33	1.501		605	64		669	770	
	7680	6″ pipe size		5	1.600		695	68.50		763.50	870	
	7720	Double, 4″ pipe size		5	1.600		735	68.50		803.50	915	
	7740	5″ pipe size		5	1.600		850	68.50		918.50	1,050	
	7760	6″ pipe size		4	2		940	85.50		1,025.50	1,150	
	7780	Water closet, blow out										
	7800	Vertical offset, caulk or thread										
	7820	Single, 4″ pipe size	1 Plum	5.33	1.501	Ea.	370	64		434	510	
	7840	Double, 4″ pipe size	″	5	1.600	″	635	68.50		703.50	805	
	7880	Vertical offset, extended, caulk										
	7900	Single, 4″ pipe size	1 Plum	5.33	1.501	Ea.	465	64		529	615	
	7920	Double, 4″ pipe size	″	5	1.600	″	730	68.50		798.50	910	
	7960	Vertical, for floor mounted back-outlet										
	7980	Single, 4″ thread, 2″ vent	1 Plum	5.33	1.501	Ea.	330	64		394	465	
	8000	Double, 4″ thread, 2″ vent	″	6	1.333	″	625	57		682	775	
	8040	Vertical, for floor mounted back-outlet, extended										
	8060	Single, 4″ caulk, 2″ vent	1 Plum	6	1.333	Ea.	330	57		387	455	
	8080	Double, 4″ caulk, 2″ vent	″	6	1.333	″	625	57		682	775	
	8200	Water closet, residential										
	8220	Vertical centerline, floor mount										
	8240	Single, 3″ caulk, 2″ or 3″ vent	1 Plum	6	1.333	Ea.	297	57		354	415	
	8260	4″ caulk, 2″ or 4″ vent		6	1.333		385	57		442	510	
	8280	3″ copper sweat, 3″ vent		6	1.333		267	57		324	385	
	8300	4″ copper sweat, 4″ vent		6	1.333		325	57		382	445	
	8400	Vertical offset, floor mount										
	8420	Single, 3″ or 4″ caulk, vent	1 Plum	4	2	Ea.	370	85.50		455.50	545	
	8440	3″ or 4″ copper sweat, vent		5	1.600		370	68.50		438.50	515	
	8460	Double, 3″ or 4″ caulk, vent		4	2		635	85.50		720.50	835	
	8480	3″ or 4″ copper sweat, vent		5	1.600		635	68.50		703.50	805	
	9000	Water cooler (electric), floor mounted										
	9100	Plate type with bearing plate, single	1 Plum	6	1.333	Ea.	188	57		245	296	
	9120	Plate type with bearing plate, double		4	2		440	85.50		525.50	615	
	9140	Plate type with bearing plate, back to back		4	2		263	85.50		348.50	425	
	9990	Minimum labor/equipment charge		3.50	2.286	Job		97.50		97.50	152	
300	0010	**FAUCETS/FITTINGS**										300
	0150	Bath, faucets, diverter spout combination, sweat	1 Plum	8	1	Ea.	69.50	42.50		112	143	
	0200	For integral stops, IPS unions, add					73			73	80.50	
	0300	Three valve combinations, spout, head, arm, flange, sweat	1 Plum	6	1.333		69.50	57		126.50	165	
	0400	For integral stops, IPS unions, add				Pr.	86			86	94.50	
	0420	Bath, press-bal mix valve w/diverter, spout, shower hd, arm/flange	1 Plum	8	1	Ea.	116	42.50		158.50	195	
	0500	Drain, central lift, 1-1/2″ IPS male		20	.400		38	17.10		55.10	68.50	
	0600	Trip lever, 1-1/2″ IPS male		20	.400		38.50	17.10		55.60	69	
	0700	Pop up, 1-1/2″ IPS male		18	.444		45.50	19		64.50	79.50	
	0800	Chain and stopper, 1-1/2″ IPS male		24	.333		30.50	14.25		44.75	55.50	
	0810	Bidet										
	0812	Fitting, over the rim, swivel spray/pop-up drain	1 Plum	8	1	Ea.	138	42.50		180.50	219	
	0840	Flush valves, with vacuum breaker										
	0850	Water closet										

MECHANICAL 15

15410 | Plumbing Fixtures		CREW	DAILY OUTPUT	LABOR-HOURS	UNIT	2006 BARE COSTS MAT.	LABOR	EQUIP.	TOTAL	TOTAL INCL O&P
300 0860	Exposed, rear spud	1 Plum	8	1	Ea.	118	42.50		160.50	196
0870	Top spud		8	1		109	42.50		151.50	187
0880	Concealed, rear spud		8	1		154	42.50		196.50	237
0890	Top spud		8	1		127	42.50		169.50	207
0900	Wall hung		8	1		136	42.50		178.50	217
0920	Urinal									
0930	Exposed, stall	1 Plum	8	1	Ea.	108	42.50		150.50	186
0940	Wall, (washout)		8	1		108	42.50		150.50	186
0950	Pedestal, top spud		8	1		110	42.50		152.50	188
0960	Concealed, stall		8	1		124	42.50		166.50	203
0970	Wall (washout)		8	1		123	42.50		165.50	203
0971	Automatic flush sensor and operator for									
0972	urinals or water closets	1 Plum	5.33	1.501	Ea.	355	64		419	490
1000	Kitchen sink faucets, top mount, cast spout		10	.800		51.50	34		85.50	110
1100	For spray, add		24	.333		11.05	14.25		25.30	34
1110	For basket strainer w/tail piece, add		24	.333		12.15	14.25		26.40	35.50
1200	Wall type, swing tube spout		10	.800		64	34		98	124
1240	For soap dish, add					4.60			4.60	5.05
1250	For basket strainer w/tail piece, add					12.15			12.15	13.35
1300	Single control lever handle									
1310	With pull out spray									
1320	Polished chrome	1 Plum	10	.800	Ea.	167	34		201	236
1330	Polished brass		10	.800		200	34		234	273
1340	White		10	.800		184	34		218	255
1348	With spray thru escutcheon									
1350	Polished chrome	1 Plum	10	.800	Ea.	117	34		151	181
1360	Polished brass		10	.800		140	34		174	207
1370	White		10	.800		128	34		162	194
2000	Laundry faucets, shelf type, IPS or copper unions		12	.667		42.50	28.50		71	91
2100	Lavatory faucet, centerset, without drain		10	.800		37.50	34		71.50	94
2120	With pop-up drain		6.66	1.201		52	51.50		103.50	138
2130	For acrylic handles, add					4.56			4.56	5
2150	Concealed, 12" centers	1 Plum	10	.800		81.50	34		115.50	143
2160	With pop-up drain	"	6.66	1.201		96.50	51.50		148	186
2210	Porcelain cross handles and pop-up drain									
2220	Polished chrome	1 Plum	6.66	1.201	Ea.	110	51.50		161.50	201
2230	Polished brass	"	6.66	1.201	"	165	51.50		216.50	262
2260	Single lever handle and pop-up drain									
2270	Black nickel	1 Plum	6.66	1.201	Ea.	212	51.50		263.50	315
2280	Polished brass		6.66	1.201		212	51.50		263.50	315
2290	Polished chrome		6.66	1.201		141	51.50		192.50	235
2600	Shelfback, 4" to 6" centers, 17 Ga. tailpiece		10	.800		66	34		100	126
2650	With pop-up drain		6.66	1.201		81	51.50		132.50	169
2700	Shampoo faucet with supply tube		24	.333		40.50	14.25		54.75	67
2800	Self-closing, center set		10	.800		179	34		213	250
2810	Automatic sensor and operator, with faucet head		6.15	1.301		300	55.50		355.50	415
2850	Medical, bedpan cleanser, with pedal valve,		12	.667		470	28.50		498.50	560
2860	With screwdriver stop valve		12	.667		244	28.50		272.50	315
2870	With self-closing spray valve		12	.667		182	28.50		210.50	245
2900	Faucet, gooseneck spout, wrist handles, grid drain		10	.800		109	34		143	173
2940	Mixing valve, knee action, screwdriver stops		4	2		365	85.50		450.50	535
3000	Service sink faucet, cast spout, pail hook, hose end		14	.571		72	24.50		96.50	117
4000	Shower by-pass valve with union		18	.444		50.50	19		69.50	85
4100	Shower arm with flange and head		22	.364		79	15.55		94.55	111
4140	Shower, hand held, pin mount, massage action, chrome		22	.364		51.50	15.55		67.05	81
4142	Polished brass		22	.364		98	15.55		113.55	132

15 MECHANICAL

15410 | Plumbing Fixtures

		15410 Plumbing Fixtures	CREW	DAILY OUTPUT	LABOR-HOURS	UNIT	2006 BARE COSTS MAT.	LABOR	EQUIP.	TOTAL	TOTAL INCL O&P
300	4144	Shower, hand held, wall mtd, adj. spray, 2 wall mounts, chrome	1 Plum	20	.400	Ea.	110	17.10		127.10	148
	4146	Polished brass		20	.400		220	17.10		237.10	269
	4148	Shower, hand held head, bar mounted 24″, adj. spray, chrome		20	.400		144	17.10		161.10	185
	4150	Polished brass		20	.400		330	17.10		347.10	390
	4200	Shower thermostatic mixing valve, concealed		8	1		250	42.50		292.50	340
	4204	For inlet strainer, check, and stops, add					33			33	36
	4220	Shower pressure balancing mixing valve,									
	4230	With shower head, arm, flange and diverter tub spout									
	4240	Chrome	1 Plum	6.14	1.303	Ea.	138	55.50		193.50	238
	4250	Polished brass		6.14	1.303		193	55.50		248.50	299
	4260	Satin		6.14	1.303		193	55.50		248.50	299
	4270	Polished chrome/brass		6.14	1.303		158	55.50		213.50	261
	5000	Sillcock, compact, brass, IPS or copper to hose		24	.333		5.50	14.25		19.75	28
	6000	Stop and waste valves, bronze									
	6100	Angle, solder end 1/2″	1 Plum	24	.333	Ea.	4.87	14.25		19.12	27.50
	6110	3/4″		20	.400		5.35	17.10		22.45	32.50
	6300	Straightway, solder end 3/8″		24	.333		4.23	14.25		18.48	26.50
	6310	1/2″		24	.333		2.71	14.25		16.96	25
	6320	3/4″		20	.400		3.11	17.10		20.21	30
	6330	1″		19	.421		4.74	18		22.74	33
	6400	Straightway, threaded 3/8″		24	.333		3.87	14.25		18.12	26.50
	6410	1/2″		24	.333		3.67	14.25		17.92	26
	6420	3/4″		20	.400		4.38	17.10		21.48	31.50
	6430	1″		19	.421		8.90	18		26.90	38
	7800	Water closet, wax gasket		96	.083		.99	3.56		4.55	6.65
	7820	Gasket toilet tank to bowl		32	.250		1.24	10.70		11.94	17.95
	8000	Water supply stops, polished chrome plate									
	8200	Angle, 3/8″	1 Plum	24	.333	Ea.	4.73	14.25		18.98	27
	8300	1/2″		22	.364		4.73	15.55		20.28	29
	8400	Straight, 3/8″		26	.308		7.05	13.15		20.20	28.50
	8500	1/2″		24	.333		7.05	14.25		21.30	30
	8600	Water closet, angle, w/flex riser, 3/8″		24	.333		11.05	14.25		25.30	34
	9000	Minimum labor/equipment charge		4	2	Job		85.50		85.50	133
800	0010	TOILET SEATS									
	0100	Molded composition, white									
	0150	Industrial, w/o cover, open front, regular bowl	1 Plum	24	.333	Ea.	15.25	14.25		29.50	39
	0200	With self-sustaining hinge		24	.333		16.20	14.25		30.45	40
	0220	With self-sustaining check hinge		24	.333		16.20	14.25		30.45	40
	0240	Extra heavy, with check hinge		24	.333		15	14.25		29.25	38.50
	0260	Elongated bowl, same price									
	0300	Junior size, w/o cover, open front	1 Plum	24	.333	Ea.	31	14.25		45.25	56
	0320	Regular primary bowl, open front		24	.333		31	14.25		45.25	56
	0340	Regular baby bowl, open front, check hinge		24	.333		29	14.25		43.25	54
	0380	Open back & front, w/o cover, reg. or elongated bowl		24	.333		25	14.25		39.25	49.50
	0400	Residential									
	0420	Regular bowl, w/cover, closed front	1 Plum	24	.333	Ea.	21.50	14.25		35.75	45.50
	0440	Open front	″	24	.333	″	28.50	14.25		42.75	53.50
	0460	Elongated bowl, add					25%				
	0500	Self-raising hinge, w/o cover, open front									
	0520	Regular bowl	1 Plum	24	.333	Ea.	93	14.25		107.25	125
	0540	Elongated bowl	″	24	.333	″	80	14.25		94.25	110
	0700	Molded wood, white, with cover									
	0720	Closed front, regular bowl, square back	1 Plum	24	.333	Ea.	9.85	14.25		24.10	33
	0740	Extended back		24	.333		13.90	14.25		28.15	37.50
	0780	Elongated bowl, square back		24	.333		12.70	14.25		26.95	36

		15410 | Plumbing Fixtures	CREW	DAILY OUTPUT	LABOR-HOURS	UNIT	2006 BARE COSTS MAT.	LABOR	EQUIP.	TOTAL	TOTAL INCL O&P	
800	0800	Open front	1 Plum	24	.333	Ea.	13.90	14.25		28.15	37.50	800
	0850	Decorator styles										
	0870	Cane top	1 Plum	24	.333	Ea.	23	14.25		37.25	47.50	
	0890	Vinyl top, patterned		24	.333		14.70	14.25		28.95	38	
	0900	Vinyl padded, plain colors, regular bowl		24	.333		14.70	14.25		28.95	38	
	0930	Elongated bowl		24	.333		19.80	14.25		34.05	44	
	1000	Solid plastic, white										
	1030	Industrial, w/o cover, open front, regular bowl	1 Plum	24	.333	Ea.	28.50	14.25		42.75	53.50	
	1080	Extra heavy, concealed check hinge		24	.333		15.10	14.25		29.35	38.50	
	1100	Self-sustaining hinge		24	.333		17.30	14.25		31.55	41	
	1150	Elongated bowl		24	.333		24	14.25		38.25	48.50	
	1170	Concealed check		24	.333		15.20	14.25		29.45	38.50	
	1190	Self-sustaining hinge, concealed check		24	.333		40.50	14.25		54.75	66.50	
	1220	Residential, with cover, closed front, regular bowl		24	.333		24	14.25		38.25	48.50	
	1240	Elongated bowl		24	.333		32	14.25		46.25	57.50	
	1260	Open front, regular bowl		24	.333		30.50	14.25		44.75	56	
	1280	Elongated bowl		24	.333		38.50	14.25		52.75	64	
		15411 | Commercial/Indust Fixtures										
400	0010	HOT WATER DISPENSERS										400
	0160	Commercial, 100 cup, 11.3 amp	1 Plum	14	.571	Ea.	330	24.50		354.50	405	
	3180	Household, 60 cup	"	14	.571	"	166	24.50		190.50	221	
500	0010	HYDRANTS										500
	0050	Wall type, moderate climate, bronze, encased										
	0200	3/4" IPS connection	1 Plum	16	.500	Ea.	395	21.50		416.50	470	
	0300	1" IPS connection	"	14	.571		395	24.50		419.50	475	
	0500	Anti-siphon type					340			340	375	
	1000	Non-freeze, bronze, exposed										
	1100	3/4" IPS connection, 4" to 9" thick wall	1 Plum	14	.571	Ea.	268	24.50		292.50	335	
	1120	10" to 14" thick wall		12	.667		290	28.50		318.50	365	
	1140	15" to 19" thick wall		12	.667		325	28.50		353.50	400	
	1160	20" to 24" thick wall		10	.800		350	34		384	440	
	1200	For 1" IPS connection, add					15%	10%				
	1240	For 3/4" adapter type vacuum breaker, add				Ea.	33.50			33.50	37	
	1280	For anti-siphon type, add				"	70			70	77	
	2000	Non-freeze bronze, encased, anti-siphon type										
	2100	3/4" IPS connection, 5" to 9" thick wall	1 Plum	14	.571	Ea.	460	24.50		484.50	550	
	2120	10" to 14" thick wall		12	.667		485	28.50		513.50	580	
	2140	15" to 19" thick wall		12	.667		520	28.50		548.50	615	
	2160	20" to 24" thick wall		10	.800		545	34		579	650	
	2200	For 1" IPS connection, add					10%	10%				
	3000	Ground box type, bronze frame, 3/4" IPS connection										
	3080	Non-freeze, all bronze, polished face, set flush										
	3100	2 feet depth of bury	1 Plum	8	1	Ea.	495	42.50		537.50	605	
	3120	3 feet depth of bury		8	1		530	42.50		572.50	645	
	3140	4 feet depth of bury		8	1		565	42.50		607.50	690	
	3160	5 feet depth of bury		7	1.143		605	49		654	745	
	3180	6 feet depth of bury		7	1.143		645	49		694	785	
	3200	7 feet depth of bury		6	1.333		680	57		737	840	
	3220	8 feet depth of bury		5	1.600		720	68.50		788.50	895	
	3240	9 feet depth of bury		4	2		760	85.50		845.50	970	
	3260	10 feet depth of bury		4	2		795	85.50		880.50	1,000	
	3400	For 1" IPS connection, add					15%	10%				
	3450	For 1-1/4" IPS connection, add					325%	14%				

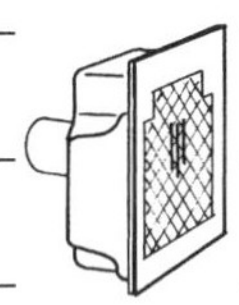

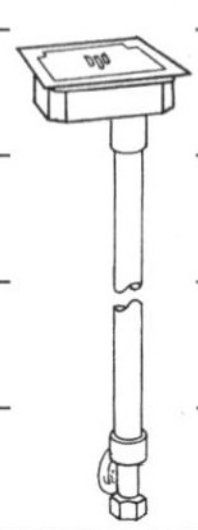

15411 | Commercial/Indust Fixtures

	15411	Commercial/Indust Fixtures	CREW	DAILY OUTPUT	LABOR-HOURS	UNIT	2006 BARE COSTS MAT.	LABOR	EQUIP.	TOTAL	TOTAL INCL O&P	
500	3500	For 1-1/2" connection, add				Ea.	370%	18%				500
	3550	For 2" connection, add					445%	24%				
	3600	For tapped drain port in box, add					45.50			45.50	50	
	4000	Non-freeze, CI body, bronze frame & scoriated cover										
	4010	with hose storage										
	4100	2 feet depth of bury	1 Plum	7	1.143	Ea.	930	49		979	1,100	
	4120	3 feet depth of bury 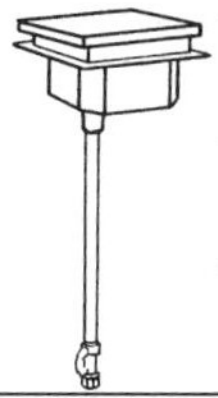		7	1.143		970	49		1,019	1,150	
	4140	4 feet depth of bury		7	1.143		1,000	49		1,049	1,175	
	4160	5 feet depth of bury		6.50	1.231		1,025	52.50		1,077.50	1,200	
	4180	6 feet depth of bury		6	1.333		1,050	57		1,107	1,250	
	4200	7 feet depth of bury		5.50	1.455		1,075	62		1,137	1,275	
	4220	8 feet depth of bury		5	1.600		1,125	68.50		1,193.50	1,325	
	4240	9 feet depth of bury		4.50	1.778		1,150	76		1,226	1,400	
	4260	10 feet depth of bury		4	2		1,200	85.50		1,285.50	1,425	
	4280	For 1" IPS connection, add					205			205	226	
	4300	For tapped drain port in box, add					45.50			45.50	50	
	5000	Moderate climate, all bronze, polished face										
	5020	and scoriated cover, set flush										
	5100	3/4" IPS connection	1 Plum	16	.500	Ea.	350	21.50		371.50	420	
	5120	1" IPS connection	"	14	.571		350	24.50		374.50	425	
	5200	For tapped drain port in box, add					45.50			45.50	50	
	6000	Ground post type, all non-freeze, all bronze, aluminum casing										
	6010	guard, exposed head, 3/4" IPS connection										
	6100	2 feet depth of bury	1 Plum	8	1	Ea.	495	42.50		537.50	610	
	6120	3 feet depth of bury		8	1		535	42.50		577.50	655	
	6140	4 feet depth of bury		8	1		575	42.50		617.50	700	
	6160	5 feet depth of bury 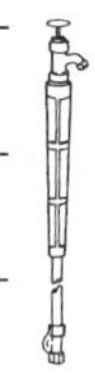		7	1.143		615	49		664	755	
	6180	6 feet depth of bury		7	1.143		660	49		709	800	
	6200	7 feet depth of bury		6	1.333		705	57		762	865	
	6220	8 feet depth of bury		5	1.600		740	68.50		808.50	920	
	6240	9 feet depth of bury		4	2		795	85.50		880.50	1,000	
	6260	10 feet depth of bury		4	2		840	85.50		925.50	1,050	
	6300	For 1" IPS connection, add					40%	10%				
	6350	For 1-1/4" IPS connection, add					140%	14%				
	6400	For 1-1/2" IPS connection, add					225%	18%				
	6450	For 2" IPS connection, add					315%	24%				
	9000	Minimum labor/equipment charge	1 Plum	3	2.667	Job		114		114	177	
700	0010	**URINALS** R224000-30										700
	0100	For automatic flush see 15410-300-0972										
	3000	Wall hung, vitreous china, with hanger & self-closing valve										
	3100	Siphon jet type	Q-1	3	5.333	Ea.	315	205		520	665	
	3120	Blowout type 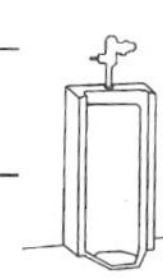		3	5.333		350	205		555	705	
	3300	Rough-in, supply, waste & vent		2.83	5.654		136	217		353	490	
	5000	Stall type, vitreous china, includes valve		2.50	6.400		520	246		766	955	
	5100	3" seam cover, add		12	1.333		154	51		205	249	
	5200	6" seam cover, add		12	1.333		170	51		221	267	
	6980	Rough-in, supply, waste and vent		1.99	8.040		184	310		494	685	
	8000	Waterless (no flush) urinal										
	8010	Wall hung										
	8020	Standard unit	Q-1	21.30	.751	Ea.	400	29		429	485	
	8030	ADA compliant unit	"	21.30	.751		385	29		414	465	
	8070	For solid color, add					60			60	66	
	8080	For 2" brass flange, (new const.), add	Q-1	96	.167		19.20	6.40		25.60	31	
	8090	Rough-in, supply, waste & vent	"	3.86	4.145		119	159		278	380	
	8100	Trap liquid										

MECHANICAL 15

15411 | Commercial/Indust Fixtures

		15411 Commercial/Indust Fixtures	CREW	DAILY OUTPUT	LABOR-HOURS	UNIT	2006 BARE COSTS MAT.	LABOR	EQUIP.	TOTAL	TOTAL INCL O&P	
700	8110	1 quart (R224000-30)				Ea.	14.10			14.10	15.50	700
	8120	1 gallon				"	55			55	60.50	
	9000	Minimum labor/equipment charge	Q-1	4	4	Job		154		154	239	
810	0010	**HANDWASHER-DRYER MODULE**										810
	0100	Wall mounted										
	0120	With electric dryer										
	0140	Sensor operated	Q-1	8	2	Ea.	2,300	77		2,377	2,650	
	0160	Sensor operated (ADA)		8	2		2,175	77		2,252	2,525	
	0180	Sensor operated (ADA), surface mounted	↓	8	2	↓	2,825	77		2,902	3,225	
	0200	With paper towels										
	0220	Sensor operated	Q-1	8	2	Ea.	2,250	77		2,327	2,600	
840	0010	**WASH FOUNTAINS**										840
	1900	Group, foot control										
	2000	Precast terrazzo, circular, 36" diam., 5 or 6 persons	Q-2	3	8	Ea.	2,975	320		3,295	3,750	
	2100	54" diameter for 8 or 10 persons		2.50	9.600		3,600	385		3,985	4,550	
	2400	Semi-circular, 36" diam. for 3 persons		3	8		2,650	320		2,970	3,425	
	2500	54" diam. for 4 or 5 persons		2.50	9.600		3,275	385		3,660	4,200	
	2700	Quarter circle (corner), 54" for 3 persons		3.50	6.857		3,275	273		3,548	4,025	
	3000	Stainless steel, circular, 36" diameter		3.50	6.857		3,400	273		3,673	4,150	
	3100	54" diameter		2.80	8.571		4,475	340		4,815	5,450	
	3400	Semi-circular, 36" diameter		3.50	6.857		2,925	273		3,198	3,650	
	3500	54" diameter		2.80	8.571		3,850	340		4,190	4,775	
	5000	Thermoplastic, pre-assembled, circular, 36" diameter		6	4		2,575	159		2,734	3,075	
	5100	54" diameter		4	6		3,025	239		3,264	3,700	
	5400	Semi-circular, 36" diameter		6	4		2,425	159		2,584	2,925	
	5600	54" diameter	↓	4	6	↓	2,875	239		3,114	3,550	
	5610	Group, infrared control, barrier free										
	5614	Precast terrazzo										
	5620	Semi-circular 36" diam. for 3 persons	Q-2	3	8	Ea.	3,900	320		4,220	4,800	
	5630	46" diam. for 4 persons		2.80	8.571		4,375	340		4,715	5,325	
	5640	Circular, 54" diam. for 8 persons, button control	↓	2.50	9.600		6,400	385		6,785	7,625	
	5700	Rough-in, supply, waste and vent for above wash fountains	Q-1	1.82	8.791		176	340		516	720	
	6200	Duo for small washrooms, stainless steel		2	8		2,000	305		2,305	2,675	
	6400	Bowl with backsplash		2	8		1,700	305		2,005	2,350	
	6500	Rough-in, supply, waste & vent for duo fountains	↓	2.02	7.921	↓	98	305		403	585	
	9000	Minimum labor/equipment charge	Q-2	3	8	Job		320		320	495	
900	0010	**WATER CHILLERS REMOTE** 80° F inlet										900
	0100	Air cooled, 50° F outlet, 115V, 4.1 GPH	1 Plum	6	1.333	Ea.	555	57		612	700	
	0200	5.7 GPH		5.50	1.455		780	62		842	950	
	0300	8.0 GPH		5	1.600		620	68.50		688.50	785	
	0400	10.0 GPH		4.50	1.778		735	76		811	930	
	0500	13.4 GPH	↓	4	2		900	85.50		985.50	1,125	
	0700	29 GPH	Q-1	5	3.200		1,350	123		1,473	1,700	
	1000	230V, 32 GPH	"	5	3.200		1,500	123		1,623	1,850	
	1200	For remote grill, add	1 Plum	16	.500	↓	144	21.50		165.50	192	

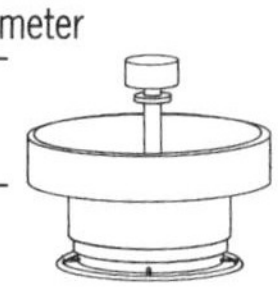

15412 | Drinking Fountains

		15412 Drinking Fountains	CREW	DAILY OUTPUT	LABOR-HOURS	UNIT	MAT.	LABOR	EQUIP.	TOTAL	TOTAL INCL O&P	
200	0010	**DRINKING FOUNTAIN** For connection to cold water supply (R224000-30)										200
	0800	For remote water chiller, see division 15411										
	1000	Wall mounted, non-recessed										
	1200	Aluminum,										
	1280	Dual bubbler type	1 Plum	3.20	2.500	Ea.	1,050	107		1,157	1,325	
	1400	Bronze, with no back	↓	4	2	↓	1,075	85.50		1,160.50	1,300	

	15412 | Drinking Fountains	CREW	DAILY OUTPUT	LABOR-HOURS	UNIT	2006 BARE COSTS MAT.	LABOR	EQUIP.	TOTAL	TOTAL INCL O&P
200 1600	Cast iron, enameled, low back, single bubbler (R224000-30)	1 Plum	4	2	Ea.	475	85.50		560.50	655
1640	Dual bubbler type		3.20	2.500		1,450	107		1,557	1,750
1680	Triple bubbler type		3.20	2.500		1,200	107		1,307	1,500
1800	Cast aluminum, enameled, for correctional institutions		4	2		660	85.50		745.50	865
2000	Fiberglass, 12″ back, single bubbler unit		4	2		590	85.50		675.50	785
2040	Dual bubbler		3.20	2.500		855	107		962	1,100
2080	Triple bubbler		3.20	2.500		1,100	107		1,207	1,400
2200	Polymarble, no back, single bubbler		4	2		530	85.50		615.50	715
2240	Dual bubbler		3.20	2.500		910	107		1,017	1,175
2280	Triple bubbler		3.20	2.500		1,200	107		1,307	1,500
2400	Precast stone, no back		4	2		535	85.50		620.50	725
2700	Stainless steel, single bubbler, no back		4	2		855	85.50		940.50	1,075
2740	With back		4	2		975	85.50		1,060.50	1,200
2780	Dual handle & wheelchair projection type		4	2		540	85.50		625.50	730
2820	Dual level for handicapped type		3.20	2.500		1,025	107		1,132	1,300
2840	Vandal resistant type		4	2		430	85.50		515.50	610
3300	Vitreous china									
3340	7″ back	1 Plum	4	2	Ea.	420	85.50		505.50	595
3940	For vandal-resistant bottom plate, add					52.50			52.50	57.50
3960	For freeze-proof valve system, add	1 Plum	2	4		410	171		581	715
3980	For rough-in, supply and waste, add	″	2.21	3.620		75.50	155		230.50	325
4000	Wall mounted, semi-recessed									
4200	Poly-marble, single bubbler	1 Plum	4	2	Ea.	680	85.50		765.50	880
4600	Stainless steel, satin finish, single bubbler		4	2		870	85.50		955.50	1,100
4900	Vitreous china, single bubbler		4	2		565	85.50		650.50	755
5980	For rough-in, supply and waste, add		1.83	4.372		75.50	187		262.50	375
6000	Wall mounted, fully recessed									
6400	Poly-marble, single bubbler	1 Plum	4	2	Ea.	765	85.50		850.50	975
6440	For water glass filler, add					129			129	142
6800	Stainless steel, single bubbler	1 Plum	4	2		700	85.50		785.50	900
6900	Fountain and cuspidor combination		2	4		1,450	171		1,621	1,850
7560	For freeze-proof valve system, add		2	4		500	171		671	820
7580	For rough-in, supply and waste, add		1.83	4.372		75.50	187		262.50	375
7600	Floor mounted, pedestal type									
7700	Aluminum, architectural style, C.I. base	1 Plum	2	4	Ea.	460	171		631	770
7780	Wheelchair handicap unit		2	4		1,050	171		1,221	1,450
8000	Bronze, architectural style		2	4		1,200	171		1,371	1,575
8040	Enameled steel cylindrical column style		2	4		1,000	171		1,171	1,375
8200	Precast stone/concrete, cylindrical column		1	8		770	340		1,110	1,375
8240	Wheelchair handicap unit		1	8		1,125	340		1,465	1,750
8400	Stainless steel, architectural style		2	4		1,200	171		1,371	1,575
8600	Enameled iron, heavy duty service, 2 bubblers		2	4		1,100	171		1,271	1,475
8660	4 bubblers		2	4		1,175	171		1,346	1,575
8880	For freeze-proof valve system, add		2	4		335	171		506	635
8900	For rough-in, supply and waste, add		1.83	4.372		75.50	187		262.50	375
9000	Minimum labor/equipment charge		2	4	Job		171		171	266
9100	Deck mounted									
9500	Stainless steel, circular receptor	1 Plum	4	2	Ea.	280	85.50		365.50	445
9540	14″ x 9″ receptor		4	2		252	85.50		337.50	410
9580	25″ x 17″ deep receptor, with water glass filler		3	2.667		208	114		322	405
9760	White enameled steel, 14″ x 9″ receptor		4	2		253	85.50		338.50	410
9860	White enameled cast iron, 24″ x 16″ receptor		3	2.667		242	114		356	445
9980	For rough-in, supply and waste, add		1.83	4.372		75.50	187		262.50	375

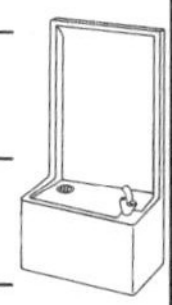

	15413 | Electric Water Coolers		CREW	DAILY OUTPUT	LABOR-HOURS	UNIT	2006 BARE COSTS MAT.	LABOR	EQUIP.	TOTAL	TOTAL INCL O&P	
900	0010	**WATER COOLER** R224000-30										900
	0030	See line 15413-900-9800 for rough-in, waste & vent										
	0040	for all water coolers										
	0100	Wall mounted, non-recessed										
	0140	4 GPH	Q-1	4	4	Ea.	500	154		654	790	
	0160	8 GPH, barrier free, sensor operated		4	4		630	154		784	935	
	0180	8.2 GPH		4	4		560	154		714	855	
	0220	14.3 GPH		4	4		645	154		799	950	
	0600	For hot and cold water, add					144			144	158	
	0640	For stainless steel cabinet, add					73.50			73.50	81	
	1000	Dual height, 8.2 GPH	Q-1	3.80	4.211		805	162		967	1,125	
	1040	14.3 GPH	"	3.80	4.211		855	162		1,017	1,200	
	1240	For stainless steel cabinet, add					112			112	123	
	2600	Wheelchair type, 8 GPH	Q-1	4	4		1,425	154		1,579	1,825	
	3000	Simulated recessed, 8 GPH		4	4		735	154		889	1,050	
	3040	11.5 GPH		4	4		775	154		929	1,100	
	3200	For glass filler, add					73.50			73.50	81	
	3240	For stainless steel cabinet, add					51			51	56.50	
	3300	Semi-recessed, 8.1 GPH	Q-1	4	4		840	154		994	1,150	
	3320	12 GPH	"	4	4		885	154		1,039	1,225	
	3340	For glass filler, add					73.50			73.50	81	
	3360	For stainless steel cabinet, add					51			51	56.50	
	3400	Full recessed, stainless steel, 8 GPH	Q-1	3.50	4.571		1,325	176		1,501	1,725	
	3420	11.5 GPH	"	3.50	4.571		1,375	176		1,551	1,775	
	3460	For glass filler, add					134			134	148	
	3600	For mounting can only					147			147	162	
	4600	Floor mounted, flush-to-wall										
	4640	4 GPH	1 Plum	3	2.667	Ea.	520	114		634	750	
	4680	8.2 GPH		3	2.667		575	114		689	805	
	4720	14.3 GPH		3	2.667		620	114		734	860	
	4960	14 GPH hot and cold water		3	2.667		865	114		979	1,125	
	4980	For stainless steel cabinet, add					112			112	123	
	5000	Dual height, 8.2 GPH	1 Plum	2	4		870	171		1,041	1,225	
	5040	14.3 GPH	"	2	4		900	171		1,071	1,250	
	5120	For stainless steel cabinet, add					147			147	162	
	5600	Explosion Proof, 16 GPH	1 Plum	3	2.667		1,350	114		1,464	1,650	
	5640	For stainless steel cabinet, add					96			96	106	
	6000	Refrigerator Compartment Type, 4.5 GPH	1 Plum	3	2.667		1,100	114		1,214	1,400	
	6600	Bottle Supply Type, 1.0 GPH		4	2		294	85.50		379.50	460	
	6640	Hot and cold, 1.0 GPH		4	2		400	85.50		485.50	575	
	9000	Minimum labor/equipment charge		2	4	Job		171		171	266	
	9800	For supply, waste & vent, all coolers		2.21	3.620	Ea.	75.50	155		230.50	325	

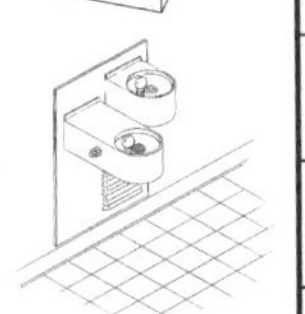

	15414 | Emergency Fixtures		CREW	DAILY OUTPUT	LABOR-HOURS	UNIT	MAT.	LABOR	EQUIP.	TOTAL	TOTAL INCL O&P	
200	0010	**INDUSTRIAL SAFETY FIXTURES** Rough-in not included										200
	1000	Eye wash fountain										
	1400	Plastic bowl, pedestal mounted	Q-1	4	4	Ea.	196	154		350	455	
	1600	Unmounted		4	4		138	154		292	390	
	1800	Wall mounted		4	4		141	154		295	395	
	2000	Stainless steel, pedestal mounted		4	4		293	154		447	560	
	2200	Unmounted		4	4		192	154		346	450	
	2400	Wall mounted		4	4		185	154		339	440	
	3000	Eye wash, portable, self-contained					395			395	435	
	4000	Eye and face wash, combination fountain										
	4200	Stainless steel, pedestal mounted	Q-1	4	4	Ea.	415	154		569	695	
	4400	Unmounted		4	4		315	154		469	590	

		15414 | Emergency Fixtures	CREW	DAILY OUTPUT	LABOR-HOURS	UNIT	2006 BARE COSTS MAT.	LABOR	EQUIP.	TOTAL	TOTAL INCL O&P	
200	4600	Wall mounted	Q-1	4	4	Ea.	320	154		474	595	200
	5000	Shower, single head, drench, ball valve, pull, freestanding		4	4		273	154		427	540	
	5200	Horizontal or vertical supply		4	4		249	154		403	515	
	6000	Multi-nozzle, eye/face wash combination		4	4		505	154		659	800	
	6400	Multi-nozzle, 12 spray, shower only		4	4		1,300	154		1,454	1,700	
	6600	For freeze-proof, add		6	2.667		287	102		389	475	
	8000	Walk-thru decontamination with eye-face wash		2	8		2,225	305		2,530	2,925	
	8200	For freeze proof, add		4	4		465	154		619	750	
	9000	Minimum labor/equipment charge		3	5.333	Job		205		205	320	

		15417 | Institutional Fixtures	CREW	DAILY OUTPUT	LABOR-HOURS	UNIT	MAT.	LABOR	EQUIP.	TOTAL	TOTAL INCL O&P	
700	0010	**PRISON/INSTITUTION FIXTURES,** Stainless steel										700
	1000	Lavatory, wall hung, push button filler valve										
	1100	Rectangular bowl	Q-1	8	2	Ea.	935	77		1,012	1,150	
	1200	Oval bowl		8	2		905	77		982	1,125	
	1240	Oval bowl, corner mount		8	2		1,000	77		1,077	1,225	
	1300	For lavatory rough-in, supply, waste and vent		1.50	10.667		135	410		545	785	
	1700	Service sink, with soap dish										
	1740	24″ x 19″ size	Q-1	3	5.333	Ea.	1,550	205		1,755	2,025	
	1790	For sink rough-in, supply, waste and vent	″	.89	17.978	″	305	690		995	1,400	
	1800	Shower cabinet, unitized										
	1840	36″ x 36″ x 88″	Q-1	2.20	7.273	Ea.	4,150	279		4,429	4,975	
	1900	Shower package for built-in										
	1940	Hot & cold valves, recessed soap dish	Q-1	6	2.667	Ea.	525	102		627	735	
	2000	Urinal, back supply and flush										
	2200	Wall hung	Q-1	4	4	Ea.	1,125	154		1,279	1,475	
	2240	Stall		2.50	6.400		1,275	246		1,521	1,775	
	2300	For urinal rough-in, supply, waste and vent		1.49	10.738		173	415		588	830	
	3000	Water closet, integral seat, back supply and flush										
	3300	Wall hung, wall outlet	Q-1	5.80	2.759	Ea.	885	106		991	1,125	
	3400	Floor mount, wall outlet		5.80	2.759		1,025	106		1,131	1,300	
	3440	Floor mount, floor outlet		5.80	2.759		1,100	106		1,206	1,400	
	3480	For recessed tissue holder, add					85			85	93.50	
	3500	For water closet rough-in, supply, waste and vent	Q-1	1.19	13.445		197	515		712	1,025	
	5000	Water closet and lavatory units, push button filler valves,										
	5010	soap & paper holders, seat										
	5300	Wall hung	Q-1	5	3.200	Ea.	2,075	123		2,198	2,475	
	5400	Floor mount		5	3.200		1,975	123		2,098	2,375	
	6300	For unit rough-in, supply, waste and vent		1	16		213	615		828	1,200	

		15418 | Resi/Comm/Industrial Fixtures	CREW	DAILY OUTPUT	LABOR-HOURS	UNIT	MAT.	LABOR	EQUIP.	TOTAL	TOTAL INCL O&P	
100	0010	**BATHS** R224000-30										100
	0100	Tubs, recessed porcelain enamel on cast iron, with trim										
	0180	48″ x 42″	Q-1	4	4	Ea.	1,625	154		1,779	2,025	
	0220	72″ x 36″	″	3	5.333	″	1,675	205		1,880	2,150	
	0300	Mat bottom										
	0340	4′-6″ long	Q-1	5	3.200	Ea.	820	123		943	1,100	
	0380	5′ long		4.40	3.636		730	140		870	1,025	
	0420	5′-6″ long		4	4		1,125	154		1,279	1,500	
	0480	Above floor drain, 5′ long		4	4		685	154		839	995	
	0560	Corner 48″ x 44″		4.40	3.636		1,625	140		1,765	2,000	
	0750	For color, add					30%					
	2000	Enameled formed steel, 4′-6″ long	Q-1	5.80	2.759	Ea.	335	106		441	530	
	2300	Above floor drain, 5′ long	″	5.50	2.909	″	305	112		417	510	
	2350	For color, add					10%					

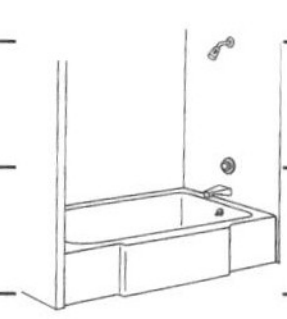

MECHANICAL 15

15418 | Resi/Comm/Industrial Fixtures

		15418 Resi/Comm/Industrial Fixtures	CREW	DAILY OUTPUT	LABOR-HOURS	UNIT	2006 BARE COSTS MAT.	LABOR	EQUIP.	TOTAL	TOTAL INCL O&P
100	4000	Soaking, acrylic with pop-up drain, 60" x 32" x 21" deep (R224000-30)	Q-1	5.50	2.909	Ea.	830	112		942	1,100
	4100	60" x 48" x 18-1/2" deep		5	3.200		635	123		758	885
	4200	72" x 45" x 21-1/2" deep		4.80	3.333		795	128		923	1,075
	4600	Module tub & showerwall surround, molded fiberglass									
	4610	5' long x 34" wide x 76" high	Q-1	4	4	Ea.	495	154		649	785
	4750	Handicap with 1-1/2" OD grab bar, anti-skid bottom									
	4760	60" x 32-3/4" x 72" high	Q-1	4	4	Ea.	670	154		824	980
	4770	60" x 30" x 71" high with molded seat	"	3.50	4.571	"	835	176		1,011	1,200
	5000	Hospital type, with trim (see 15410-300)									
	5050	Bathing pool, porcelain enamel on cast iron, grab bars									
	5060	pop-up drain, 72" x 36"	Q-1	3	5.333	Ea.	2,800	205		3,005	3,425
	5100	Perineal (sitz), vitreous china		3	5.333		895	205		1,100	1,300
	5120	Pedestal, vitreous china		8	2		135	77		212	269
	5180	Pier tub, porcelain enamel on cast iron, 66-3/4" x 30"		3	5.333		2,225	205		2,430	2,775
	5200	Base, porcelain enamel on cast iron		8	2		1,050	77		1,127	1,275
	5300	Whirlpool, porcelain enamel on cast iron, 72" x 36"		1	16		2,950	615		3,565	4,200
	6000	Whirlpool, bath with vented overflow, molded fiberglass									
	6100	66" x 48" x 24"	Q-1	1	16	Ea.	2,700	615		3,315	3,925
	6400	72" x 36" x 24"		1	16		2,675	615		3,290	3,875
	6500	60" x 30" x 21"		1	16		2,300	615		2,915	3,500
	6600	72" x 42" x 22"		1	16		3,475	615		4,090	4,775
	6700	83" x 65"		.30	53.333		4,700	2,050		6,750	8,350
	7000	Redwood hot tub system									
	7050	4' diameter x 4' deep	Q-1	1	16	Ea.	1,750	615		2,365	2,875
	7100	5' diameter x 4' deep		1	16		2,075	615		2,690	3,225
	7150	6' diameter x 4' deep		.80	20		2,575	770		3,345	4,025
	7200	8' diameter x 4' deep		.80	20		3,950	770		4,720	5,550
	9600	Rough-in, supply, waste and vent, for all above tubs, add		2.07	7.729		166	297		463	640
	9900	Minimum labor/equipment charge		3	5.333	Job		205		205	320
200	0010	**BIDET**									
	0180	Vitreous china, with trim on fixture	Q-1	5	3.200	Ea.	590	123		713	835
	0200	With trim for wall mounting		5	3.200		590	123		713	835
	9600	For rough-in, supply, waste and vent, add		1.78	8.989		143	345		488	690
400	0010	**LAUNDRY SINKS** With trim									
	0020	Porcelain enamel on cast iron, black iron frame									
	0050	24" x 21", single compartment	Q-1	6	2.667	Ea.	345	102		447	540
	0100	26" x 21", single compartment	"	6	2.667	"	340	102		442	535
	2000	Molded stone, on wall hanger or legs									
	2020	22" x 23", single compartment	Q-1	6	2.667	Ea.	124	102		226	295
	2100	45" x 21", double compartment	"	5	3.200	"	257	123		380	475
	3000	Plastic, on wall hanger or legs									
	3020	18" x 23", single compartment	Q-1	6.50	2.462	Ea.	94	94.50		188.50	251
	3100	20" x 24", single compartment		6.50	2.462		126	94.50		220.50	285
	3200	36" x 23", double compartment		5.50	2.909		151	112		263	340
	3300	40" x 24", double compartment		5.50	2.909		224	112		336	420
	5000	Stainless steel, counter top, 22" x 17" single compartment		6	2.667		54.50	102		156.50	219
	5200	33" x 22", double compartment		5	3.200		66.50	123		189.50	265
	9600	Rough-in, supply, waste and vent, for all laundry sinks		2.14	7.477		130	287		417	590
	9810	Minimum labor/equipment charge	1 Plum	3	2.667	Job		114		114	177
450	0010	**LAVATORIES** With trim, white unless noted otherwise (R224000-30)									
	0500	Vanity top, porcelain enamel on cast iron									
	0600	20" x 18"	Q-1	6.40	2.500	Ea.	212	96		308	385
	0640	33" x 19" oval		6.40	2.500		445	96		541	635
	0680	20" x 17" oval		6.40	2.500		138	96		234	300
	0720	19" round		6.40	2.500		199	96		295	370

	15418 | Resi/Comm/Industrial Fixtures		CREW	DAILY OUTPUT	LABOR-HOURS	UNIT	2006 BARE COSTS MAT.	LABOR	EQUIP.	TOTAL	TOTAL INCL O&P	
450	0760	20″ x 12″ triangular bowl R224000-30	Q-1	6.40	2.500	Ea.	215	96		311	385	450
	0860	For color, add					25%					
	1000	Cultured marble, 19″ x 17″, single bowl	Q-1	6.40	2.500	Ea.	163	96		259	330	
	1040	25″ x 19″, single bowl		6.40	2.500		175	96		271	340	
	1080	31″ x 19″, single bowl		6.40	2.500		194	96		290	360	
	1120	25″ x 22″, single bowl		6.40	2.500		175	96		271	340	
	1160	37″ x 22″, single bowl		6.40	2.500		208	96		304	380	
	1200	49″ x 22″, single bowl	↓	6.40	2.500	↓	244	96		340	415	
	1580	For color, same price										
	1900	Stainless steel, self-rimming, 25″ x 22″, single bowl, ledge	Q-1	6.40	2.500	Ea.	247	96		343	420	
	1960	17″ x 22″, single bowl		6.40	2.500		240	96		336	415	
	2040	18-3/4″ round		6.40	2.500		480	96		576	680	
	2600	Steel, enameled, 20″ x 17″, single bowl		5.80	2.759		138	106		244	315	
	2660	19″ round		5.80	2.759		135	106		241	315	
	2720	18″ round	↓	5.80	2.759	↓	112	106		218	289	
	2860	For color, add					10%					
	2900	Vitreous china, 20″ x 16″, single bowl	Q-1	5.40	2.963	Ea.	222	114		336	420	
	2960	20″ x 17″, single bowl		5.40	2.963		147	114		261	340	
	3020	19″ round, single bowl		5.40	2.963		145	114		259	335	
	3080	19″ x 16″, single bowl		5.40	2.963		224	114		338	425	
	3140	17″ x 14″, single bowl		5.40	2.963		183	114		297	380	
	3200	22″ x 13″, single bowl	↓	5.40	2.963	↓	234	114		348	435	
	3560	For color, add					50%					
	3580	Rough-in, supply, waste and vent for all above lavatories	Q-1	2.30	6.957	Ea.	113	267		380	540	
	4000	Wall hung										
	4040	Porcelain enamel on cast iron, 16″ x 14″, single bowl	Q-1	8	2	Ea.	345	77		422	500	
	4060	18″ x 15″ single bowl		8	2		288	77		365	435	
	4120	19″ x 17″, single bowl		8	2		269	77		346	415	
	4180	20″ x 18″, single bowl		8	2		241	77		318	385	
	4240	22″ x 19″, single bowl	↓	8	2	↓	410	77		487	575	
	4580	For color, add					30%					
	6000	Vitreous china, 18″ x 15″, single bowl with backsplash	Q-1	7	2.286	Ea.	204	88		292	360	
	6060	19″ x 17″, single bowl		7	2.286		170	88		258	325	
	6120	24″ x 20″, single bowl		7	2.286		320	88		408	485	
	6180	19″ x 19″, corner style		8	2		435	77		512	600	
	6210	27″ x 20″, wheelchair type	↓	7	2.286	↓	300	88		388	465	
	6500	For color, add					30%					
	6700	Hospital type, without trim (see 15410-300)										
	6710	20″ x 18″, contoured splash shield	Q-1	8	2	Ea.	114	77		191	245	
	6720	26″ x 20″, patient,corner,side deck & back		7	2.286		375	88		463	545	
	6730	28″ x 20″, surgeon, side decks		8	2		450	77		527	615	
	6740	28″ x 22″, surgeon scrub-up, deep bowl		8	2		645	77		722	830	
	6750	20″ x 27″, patient, wheelchair		7	2.286		305	88		393	470	
	6760	30″ x 22″, all purpose		7	2.286		570	88		658	760	
	6770	30″ x 22″, plaster work		7	2.286		585	88		673	780	
	6820	20″ x 24″ clinic service, liquid/solid waste		6	2.667		735	102		837	970	
	6960	Rough-in, supply, waste and vent for above lavatories	↓	1.66	9.639	↓	310	370		680	915	
	9000	Minimum labor/equipment charge	1 Plum	3	2.667	Job		114		114	177	
500	0010	**SHOWERS** R224000-30										500
	1500	Stall, with drain only. Add for valve and door/curtain										
	1510	Baked enamel, molded stone receptor, 30″ square	Q-1	5.20	3.077	Ea.	360	118		478	585	
	1520	32″ square		5	3.200		345	123		468	570	
	1530	36″ square		4.80	3.333		425	128		553	665	
	1540	Terrazzo receptor, 32″ square		5	3.200		730	123		853	995	
	1560	36″ square		4.80	3.333		875	128		1,003	1,150	
	1580	36″ corner angle	↓	4.80	3.333	↓	800	128		928	1,075	

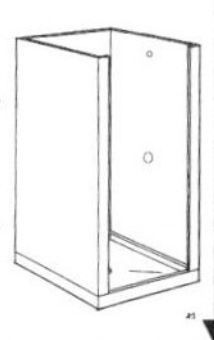

MECHANICAL 15

	15418 | Resi/Comm/Industrial Fixtures	CREW	DAILY OUTPUT	LABOR-HOURS	UNIT	2006 BARE COSTS MAT.	LABOR	EQUIP.	TOTAL	TOTAL INCL O&P	
500	1600 For color, add (R224000-30)					10%					500
	3000 Fiberglass, one piece, with 3 walls, 32″ x 32″ square	Q-1	5.50	2.909	Ea.	420	112		532	635	
	3100 36″ x 36″ square	″	5.50	2.909	″	465	112		577	685	
	3200 Handicap, 1-1/2″ O.D. grab bars, nonskid floor										
	3210 48″ x 34-1/2″ x 72″ corner seat	Q-1	5	3.200	Ea.	455	123		578	690	
	3220 60″ x 34-1/2″ x 72″ corner seat		4	4		570	154		724	865	
	3230 48″ x 34-1/2″ x 72″ fold up seat		5	3.200		630	123		753	885	
	3250 64″ x 65-3/4″ x 81-1/2″ fold. seat, whlchr.		3.80	4.211		2,900	162		3,062	3,450	
	4000 Polypropylene, stall only, w/ molded-stone floor, 30″ x 30″		2	8		330	305		635	840	
	4100 32″ x 32″		2	8		420	305		725	940	
	4200 Rough-in, supply, waste and vent for above showers		2.05	7.805		244	300		544	735	
	4520 Showers, fiberglass receptor only	1 Plum	8	1		187	42.50		229.50	272	
	5000 Built-in, head, arm, 4 GPM valve		4	2		123	85.50		208.50	269	
	5200 Head, arm, by-pass, integral stops, handles		3.60	2.222		196	95		291	365	
	5500 Head, water economizer		24	.333		65	14.25		79.25	93.50	
	5800 Mixing valve, built-in		6	1.333		106	57		163	206	
	5900 Exposed		6	1.333		505	57		562	645	
	5950 Module, handicap, SS panel, fixed & hand held head, control										
	5960 valves, grab bar, curtain & rod, folding seat	1 Plum	4	2	Ea.	1,725	85.50		1,810.50	2,025	
	6000 Group, w/pressure balancing valve, rough-in and rigging not included										
	6800 Column, 6 heads, no receptors, less partitions	Q-1	3	5.333	Ea.	2,125	205		2,330	2,675	
	6900 With stainless steel partitions		1	16		4,550	615		5,165	5,950	
	7600 5 heads, no receptors, less partitions		3	5.333		1,625	205		1,830	2,100	
	7620 4 heads (1 handicap) no receptors, less partitions		3	5.333		2,275	205		2,480	2,825	
	7700 With stainless steel partitions		1	16		1,625	615		2,240	2,725	
	8000 Wall, 2 heads, no receptors, less partitions		4	4		935	154		1,089	1,275	
	8100 With stainless steel partitions		2	8		2,275	305		2,580	2,975	
	9000 Minimum labor/equipment charge		4	4	Job		154		154	239	
600	0010 **SINKS** With faucets and drain (R224000-30)										600
	0050 Laboratory sinks, corrosion resistant										
	1000 Polyethylene, single sink, bench mounted, with										
	1020 plug & waste fitting with 1-1/2″ straight threads										
	1030 2 drainboards, backnut & strainer										
	1050 18-1/2″ x 15-1/2″ x 12-1/2″ sink, 54″ x 24″ O.D.	Q-1	3	5.333	Ea.	855	205		1,060	1,250	
	1100 Single drainboard, backnut & strainer										
	1130 18-1/2″ x 15-1/2″ x 12-1/2″ sink, 47″ x 24″ O.D.	Q-1	3	5.333	Ea.	970	205		1,175	1,400	
	1150 18-1/2″ x 15-1/2″ x 12-1/2″ sink, 70″ x 24″ O.D.	″	3	5.333	″	990	205		1,195	1,425	
	1290 Flanged 1-1/4″ wide, rectangular with strainer										
	1300 plug & waste fitting, 1-1/2″ straight threads										
	1320 12″ x 12″ x 8″ sink, 14-1/2″ x 14-1/2″ O.D.	Q-1	4	4	Ea.	163	154		317	420	
	1340 16″ x 16″ x 8″ sink, 18-1/2″ x 18-1/2″ O.D.		4	4		242	154		396	505	
	1360 21″ x 18″ x 10″ sink, 23-1/2″ x 20-1/2″ O.D.		4	4		291	154		445	560	
	1490 For rough-in, supply, waste & vent, add		2.02	7.921		128	305		433	615	
	1600 Polypropylene										
	1620 Cup sink, oval, integral strainers										
	1640 6″ x 3″ I.D., 7″ x 4″ O.D.	Q-1	6	2.667	Ea.	114	102		216	284	
	1660 9″ x 3″ I.D., 10″ x 4-1/2″ O.D.	″	6	2.667	″	108	102		210	278	
	1720 Overflow standpipe										
	1740 1-1/2″ diam. x 11″ long				Ea.	33.50			33.50	36.50	
	1980 For rough-in, supply, waste & vent, add	Q-1	1.70	9.412		128	360		488	700	
	2000 Kitchen, countertop, P.E. on C.I., 24″ x 21″ single bowl		5.60	2.857		232	110		342	425	
	2100 31″ x 22″ single bowl		5.60	2.857		217	110		327	410	
	2200 32″ x 21″ double bowl		4.80	3.333		315	128		443	545	
	2300 42″ x 21″ double bowl		4.80	3.333		480	128		608	730	
	3000 Stainless steel, self rimming, 19″ x 18″ single bowl		5.60	2.857		380	110		490	585	
	3100 25″ x 22″ single bowl		5.60	2.857		420	110		530	635	

	15418 | Resi/Comm/Industrial Fixtures		CREW	DAILY OUTPUT	LABOR-HOURS	UNIT	2006 BARE COSTS MAT.	LABOR	EQUIP.	TOTAL	TOTAL INCL O&P	
600	3200	33" x 22" double bowl R224000-30	Q-1	4.80	3.333	Ea.	610	128		738	870	600
	3300	43" x 22" double bowl		4.80	3.333		705	128		833	975	
	3400	22" x 43" triple bowl		4.40	3.636		810	140		950	1,100	
	3500	Corner double bowl each 14" x 16"		4.80	3.333		415	128		543	660	
	4000	Steel, enameled, with ledge, 24" x 21" single bowl		5.60	2.857		114	110		224	296	
	4100	32" x 21" double bowl		4.80	3.333		150	128		278	365	
	4960	For color sinks except stainless steel, add					10%					
	4980	For rough-in, supply, waste and vent, counter top sinks	Q-1	2.14	7.477		130	287		417	590	
	5000	Kitchen, raised deck, P.E. on C.I.										
	5100	32" x 21", dual level, double bowl	Q-1	2.60	6.154	Ea.	265	236		501	660	
	5200	42" x 21", double bowl & disposer well	"	2.20	7.273		660	279		939	1,150	
	5700	For color, add					20%					
	5790	For rough-in, supply, waste & vent, sinks	Q-1	1.85	8.649		130	330		460	660	
	6650	Service, floor, corner, P.E. on C.I., 28" x 28"	"	4.40	3.636		585	140		725	860	
	6750	Vinyl coated rim guard, add					67.50			67.50	74	
	6760	Mop sink, molded stone, 24" x 36"	1 Plum	3.33	2.402		194	103		297	375	
	6770	Mop sink, molded stone, 24" x 36", w/rim 3 sides	"	3.33	2.402		218	103		321	400	
	6790	For rough-in, supply, waste & vent, floor service sinks	Q-1	1.64	9.756		305	375		680	920	
	7000	Service, wall, P.E. on C.I., roll rim, 22" x 18"		4	4		480	154		634	770	
	7100	24" x 20"		4	4		530	154		684	825	
	7600	For stainless steel rim guard, add					95			95	104	
	7800	For stainless steel rim guard, front only, add					37			37	40.50	
	8600	Vitreous china, 22" x 20"	Q-1	4	4		420	154		574	705	
	8960	For stainless steel rim guard, front or side, add					37			37	40.50	
	8980	For rough-in, supply, waste & vent, wall service sinks	Q-1	1.30	12.308		500	475		975	1,275	
	9000	Minimum labor/equipment charge	"	4	4	Job		154		154	239	
900	0010	**WATER CLOSETS** R224000-30										900
	0020	For seats, see 15410-800										
	0030	For automatic flush, see 15410-300-0972										
	0150	Tank type, vitreous china, incl. seat, supply pipe w/stop										
	0200	Wall hung, one piece	Q-1	5.30	3.019	Ea.	390	116		506	610	
	0400	Two piece, close coupled		5.30	3.019		480	116		596	705	
	0960	For rough-in, supply, waste, vent and carrier		2.73	5.861		445	225		670	835	
	1000	Floor mounted, one piece		5.30	3.019		495	116		611	725	
	1020	One piece, low profile		5.30	3.019		395	116		511	615	
	1050	One piece combination		5.30	3.019		705	116		821	955	
	1100	Two piece, close coupled		5.30	3.019		169	116		285	365	
	1150	With wall outlet		5.30	3.019		540	116		656	770	
	1200	With 18" high bowl		5.30	3.019		385	116		501	600	
	1960	For color, add					30%					
	1980	For rough-in, supply, waste and vent	Q-1	3.05	5.246	Ea.	191	202		393	525	
	3000	Bowl only, with flush valve, seat										
	3100	Wall hung	Q-1	5.80	2.759	Ea.	350	106		456	550	
	3150	Hospital type, slotted rim for bed pan										
	3160	Elongated bowl	Q-1	5.80	2.759	Ea.	380	106		486	585	
	3200	For rough-in, supply, waste and vent, single WC		2.56	6.250		460	240		700	880	
	3300	Floor mounted		5.80	2.759		291	106		397	485	
	3350	With wall outlet		5.80	2.759		510	106		616	725	
	3360	Hospital type, slotted rim for bed pan										
	3370	Elongated bowl	Q-1	5	3.200	Ea.	310	123		433	530	
	3380	Elongated bowl, 18" high		5	3.200		380	123		503	610	
	3400	For rough-in, supply, waste and vent, single WC		2.84	5.634		209	217		426	565	
	3500	Gang side by side carrier system, rough-in, supply, waste & vent										
	3510	For single hook-up	Q-1	1.97	8.122	Ea.	520	310		830	1,050	
	3520	For each additional hook-up, add	"	2.14	7.477	"	490	287		777	985	
	3550	Gang back to back carrier system, rough-in, supply, waste & vent										

	15418 | Resi/Comm/Industrial Fixtures		Crew	Daily Output	Labor-Hours	Unit	2006 Bare Costs Mat.	Labor	Equip.	Total	Total Incl O&P	
900	3560	For pair hook-up (R224000-30)	Q-1	1.76	9.091	Pr.	855	350		1,205	1,475	900
	3570	For each additional pair hook-up, add	″	1.81	8.840	″	820	340		1,160	1,425	
	4000	Water conserving systems										
	4900	2 quart flush, residential	Q-1	5.40	2.963	Ea.	605	114		719	840	
	4980	For rough-in, supply, waste and vent		1.94	8.247		209	315		524	725	
	5100	2 quart flush		4.60	3.478		610	134		744	885	
	5200	For remote valve, add		24	.667		194	25.50		219.50	253	
	5300	For residential air compressor		6	2.667		705	102		807	935	
	5400	For light industrial air compressor		4	4		1,125	154		1,279	1,500	
	5500	For heavy duty industrial air compressor		1	16		1,675	615		2,290	2,800	
	5600	For dual compressor alternator		20	.800		565	30.50		595.50	670	
	9000	Minimum labor/equipment charge		4	4	Job		154		154	239	

	15440 | Plumbing Pumps		Crew	Daily Output	Labor-Hours	Unit	Mat.	Labor	Equip.	Total	Total Incl O&P	
400	0010	**PUMPS, GRINDER SYSTEM** Complete, incl. check valve, tank, std.										400
	0020	controls incl. alarm/disconnect panel w/wire. Excavation not included										
	0260	Simplex, 9 GPM at 60 PSIG, 70 gal. tank				Ea.	2,525			2,525	2,775	
	0300	For manway, 26″ I.D., 18″ high, add					320			320	355	
	0340	26″ I.D., 36″ high, add					445			445	485	
	0380	43″ I.D., 4′ high, add					495			495	545	
	0600	Simplex, 9 GPM at 60 PSIG, 150 gal. tank					2,725			2,725	3,000	
	0660	For manway, add										
	0700	26″ I.D., 36″ high, add				Ea.	590			590	650	
	0740	26″ I.D., 4′ high, add					670			670	740	
	2000	Duplex, 18 GPM at 60 PSIG, 150 gal. tank					1,800			1,800	1,975	
	2060	For manway 43″ I.D., 4′ high, add					1,475			1,475	1,625	
	2400	For core only					1,350			1,350	1,475	
800	0010	**PUMPS, SEWAGE EJECTOR** With operating and level controls										800
	0100	Simplex system incl. tank, cover, pump 15′ head										
	0500	37 gal PE tank, 12 GPM, 1/2 HP, 2″ discharge	Q-1	3.20	5	Ea.	385	192		577	725	
	0510	3″ discharge		3.10	5.161		415	198		613	765	
	0530	87 GPM, .7 HP, 2″ discharge		3.20	5		585	192		777	940	
	0540	3″ discharge		3.10	5.161		635	198		833	1,000	
	0600	45 gal. coated stl tank, 12 GPM, 1/2 HP, 2″ discharge		3	5.333		680	205		885	1,075	
	0610	3″ discharge		2.90	5.517		710	212		922	1,100	
	0630	87 GPM, .7 HP, 2″ discharge		3	5.333		870	205		1,075	1,275	
	0640	3″ discharge		2.90	5.517		920	212		1,132	1,350	
	0660	134 GPM, 1 HP, 2″ discharge		2.80	5.714		945	220		1,165	1,400	
	0680	3″ discharge		2.70	5.926		995	228		1,223	1,450	
	0700	70 gal. PE tank, 12 GPM, 1/2 HP, 2″ discharge		2.60	6.154		750	236		986	1,200	
	0710	3″ discharge		2.40	6.667		800	256		1,056	1,275	
	0730	87 GPM, 0.7 HP, 2″ discharge		2.50	6.400		960	246		1,206	1,425	
	0740	3″ discharge		2.30	6.957		1,025	267		1,292	1,550	
	0760	134 GPM, 1 HP, 2″ discharge		2.20	7.273		1,050	279		1,329	1,575	
	0770	3″ discharge		2	8		1,125	305		1,430	1,700	
	0800	75 gal. coated stl. tank, 12 GPM, 1/2 HP, 2 ″ discharge		2.40	6.667		775	256		1,031	1,250	
	0810	3″ discharge		2.20	7.273		805	279		1,084	1,325	
	0830	87 GPM, .7 HP, 2″ discharge		2.30	6.957		970	267		1,237	1,500	
	0840	3″ discharge		2.10	7.619		1,025	293		1,318	1,575	
	0860	134 GPM, 1 HP, 2″ discharge		2	8		1,050	305		1,355	1,625	
	0880	3″ discharge		1.80	8.889		1,100	340		1,440	1,725	
	1040	Duplex system incl. tank, covers, pumps										
	1060	110 gal. fiberglass tank, 24 GPM, 1/2 HP, 2″ discharge	Q-1	1.60	10	Ea.	1,475	385		1,860	2,200	
	1080	3″ discharge		1.40	11.429		1,525	440		1,965	2,375	
	1100	174 GPM, .7 HP, 2″ discharge		1.50	10.667		1,900	410		2,310	2,725	

15400 | Plumbing Fixtures & Equipment

	15440 | Plumbing Pumps		CREW	DAILY OUTPUT	LABOR-HOURS	UNIT	2006 BARE COSTS MAT.	LABOR	EQUIP.	TOTAL	TOTAL INCL O&P	
800	1120	3″ discharge	Q-1	1.30	12.308	Ea.	2,000	475		2,475	2,925	800
	1140	268 GPM, 1 HP, 2″ discharge		1.20	13.333		2,075	510		2,585	3,100	
	1160	3″ discharge	↓	1	16		2,150	615		2,765	3,325	
	1260	135 gal. coated stl. tank, 24 GPM, 1/2 HP, 2″ discharge	Q-2	1.70	14.118		1,525	565		2,090	2,550	
	2000	3″ discharge		1.60	15		1,600	600		2,200	2,700	
	2640	174 GPM, .7 HP, 2″ discharge		1.60	15		1,975	600		2,575	3,100	
	2660	3″ discharge		1.50	16		2,100	640		2,740	3,300	
	2700	268 GPM, 1 HP, 2″ discharge		1.30	18.462		2,150	735		2,885	3,525	
	3040	3″ discharge		1.10	21.818		2,300	870		3,170	3,875	
	3060	275 gal. coated stl. tank, 24 GPM, 1/2 HP, 2″ discharge		1.50	16		1,900	640		2,540	3,075	
	3080	3″ discharge		1.40	17.143		1,925	685		2,610	3,175	
	3100	174 GPM, .7 HP, 2″ discharge		1.40	17.143		2,450	685		3,135	3,750	
	3120	3″ discharge		1.30	18.462		2,475	735		3,210	3,875	
	3140	268 GPM, 1 HP, 2″ discharge		1.10	21.818		2,675	870		3,545	4,300	
	3160	3″ discharge	↓	.90	26.667	↓	2,825	1,075		3,900	4,750	
	3260	Pump system accessories, add										
	3300	Alarm horn and lights, 115V mercury switch	Q-1	8	2	Ea.	86.50	77		163.50	215	
	3340	Switch, mag. contactor, alarm bell, light, 3 level control		5	3.200		395	123		518	625	
	3380	Alternator, mercury switch activated		4	4	↓	750	154		904	1,075	
	9000	Minimum labor/equipment charge	↓	2.50	6.400	Job		246		246	380	
900	0010	**PUMPS, PEDESTAL SUMP** With float control										900
	0400	Molded PVC base, 21 GPM at 15′ head, 1/3 HP	1 Plum	5	1.600	Ea.	88.50	68.50		157	204	
	0800	Iron base, 21 GPM at 15′ head, 1/3 HP		5	1.600		113	68.50		181.50	231	
	1200	Solid brass, 21 GPM at 15′ head, 1/3 HP	↓	5	1.600	↓	185	68.50		253.50	310	
940	0010	**PUMPS, SUBMERSIBLE** Sump										940
	7000	Sump pump, automatic										
	7100	Plastic, 1-1/4″ discharge, 1/4 HP	1 Plum	6	1.333	Ea.	114	57		171	214	
	7140	1/3 HP		5	1.600		138	68.50		206.50	257	
	7160	1/2 HP		5	1.600		168	68.50		236.50	291	
	7180	1-1/2″ discharge, 1/2 HP		4	2		188	85.50		273.50	340	
	7500	Cast iron, 1-1/4″ discharge, 1/4 HP		6	1.333		131	57		188	233	
	7540	1/3 HP		6	1.333		155	57		212	259	
	7560	1/2 HP		5	1.600	↓	187	68.50		255.50	310	
	9000	Minimum labor/equipment charge	↓	4	2	Job		85.50		85.50	133	

	15450 | Potable Water Storage Tanks		CREW	DAILY OUTPUT	LABOR-HOURS	UNIT	MAT.	LABOR	EQUIP.	TOTAL	TOTAL INCL O&P	
900	0010	**WATER HEATER STORAGE TANKS** 125 psi ASME										900
	2000	Galvanized steel, 15 gal., 14″ diam., 26″ LOA	1 Plum	12	.667	Ea.	630	28.50		658.50	740	
	2060	30 gal., 14″ diam. x 49″ LOA		11	.727		755	31		786	880	
	2080	80 gal., 20″ diam. x 64″ LOA		9	.889		1,175	38		1,213	1,325	
	2100	135 gal., 24″ diam. x 75″ LOA		6	1.333		1,725	57		1,782	2,000	
	2120	240 gal., 30″ diam. x 86″ LOA		4	2		2,975	85.50		3,060.50	3,400	
	2140	300 gal., 36″ diam. x 76″ LOA	↓	3	2.667		4,850	114		4,964	5,500	
	2160	400 gal., 36″ diam. x 100″ LOA	Q-1	4	4		5,975	154		6,129	6,800	
	2180	500 gal., 36″ diam., x 126″ LOA	″	3	5.333		7,375	205		7,580	8,425	
	3000	Glass lined, P.E., 80 gal., 20″ diam. x 60″ LOA	1 Plum	9	.889		1,900	38		1,938	2,125	
	3060	140 gal., 24″ diam. x 80″ LOA		6	1.333		2,625	57		2,682	2,975	
	3080	225 gal., 30″ diam. x 78″ LOA		4	2		3,025	85.50		3,110.50	3,450	
	3100	325 gal., 36″ diam. x 81″ LOA	↓	3	2.667		3,875	114		3,989	4,425	
	3120	460 gal., 42″ diam. x 84″ LOA	Q-1	4	4		4,075	154		4,229	4,725	
	3140	605 gal., 48″ diam. x 87″ LOA		3	5.333		7,800	205		8,005	8,900	
	3160	740 gal., 54″ diam. x 91″ LOA		3	5.333		8,775	205		8,980	9,975	
	3180	940 gal., 60″ diam. x 93″ LOA		2.50	6.400		10,100	246		10,346	11,500	
	3200	1330 gal., 66″ diam. x 107″ LOA		2	8		13,900	305		14,205	15,800	
	3220	1615 gal., 72″ diam. x 110″ LOA		1.50	10.667		15,800	410		16,210	18,000	
	3240	2285 gal., 84″ diam. x 128″ LOA	↓	1	16	↓	19,500	615		20,115	22,500	

For expanded coverage of these items see *Means Mechanical or Plumbing Cost Data 2006*

	15450 | Potable Water Storage Tanks		CREW	DAILY OUTPUT	LABOR-HOURS	UNIT	2006 BARE COSTS MAT.	LABOR	EQUIP.	TOTAL	TOTAL INCL O&P	
900	3260	3440 gal., 96" diam. x 157" LOA	Q-2	1.50	16	Ea.	28,900	640		29,540	32,800	900

	15460 | Domestic Water Cond Equipment		CREW	DAILY OUTPUT	LABOR-HOURS	UNIT	MAT.	LABOR	EQUIP.	TOTAL	TOTAL INCL O&P	
900	0010	WATER SOFTENER										900
	5800	Softener systems, automatic, intermediate sizes										
	5820	available, may be used in multiples.										
	6000	Hardness capacity between regenerations and flow										
	6100	150,000 grains, 37 GPM cont., 51 GPM peak	Q-1	1.20	13.333	Ea.	4,250	510		4,760	5,475	
	6200	300,000 grains, 81 GPM cont., 113 GPM peak		1	16		6,300	615		6,915	7,875	
	6300	750,000 grains, 160 GPM cont., 230 GPM peak		.80	20		8,200	770		8,970	10,200	
	6400	900,000 grains, 185 GPM cont., 270 GPM peak	↓	.70	22.857	↓	13,200	880		14,080	15,900	

	15470 | Domestic Water Filtration Equip.		CREW	DAILY OUTPUT	LABOR-HOURS	UNIT	MAT.	LABOR	EQUIP.	TOTAL	TOTAL INCL O&P	
400	0010	WATER FILTERS Purification and treatment										400
	1000	Cartridge style, dirt and rust type	1 Plum	12	.667	Ea.	30.50	28.50		59	78	
	1200	Replacement cartridge		32	.250		11.60	10.70		22.30	29.50	
	1600	Taste and odor type		12	.667		31	28.50		59.50	79	
	1700	Replacement cartridge		32	.250		26.50	10.70		37.20	45.50	
	3000	Central unit, dirt/rust/odor/taste/scale		4	2		705	85.50		790.50	910	
	3100	Replacement cartridge, standard		20	.400		138	17.10		155.10	178	
	3600	Replacement cartridge, heavy duty	↓	20	.400	↓	154	17.10		171.10	197	
	8000	Commercial, fully automatic or push button automatic										
	8200	Iron removal, 660 GPH, 1" pipe size	Q-1	1.50	10.667	Ea.	1,900	410		2,310	2,700	
	8240	1500 GPH, 1-1/4" pipe size		1	16		3,175	615		3,790	4,450	
	8280	2340 GPH, 1-1/2" pipe size		.80	20		3,500	770		4,270	5,050	
	8320	3420 GPH, 2" pipe size		.60	26.667		6,450	1,025		7,475	8,675	
	8360	4620 GPH, 2-1/2" pipe size		.50	32		10,200	1,225		11,425	13,200	
	8500	Neutralizer for acid water, 780 GPH, 1" pipe size		1.50	10.667		1,825	410		2,235	2,625	
	8540	1140 GPH, 1-1/4" pipe size		1	16		2,050	615		2,665	3,200	
	8580	1740 GPH, 1-1/2" pipe size		.80	20		2,975	770		3,745	4,475	
	8620	2520 GPH, 2" pipe size		.60	26.667		3,875	1,025		4,900	5,850	
	8660	3480 GPH, 2-1/2" pipe size		.50	32		6,425	1,225		7,650	8,975	
	8800	Sediment removal, 780 GPH, 1" pipe size		1.50	10.667		1,750	410		2,160	2,550	
	8840	1140 GPH, 1-1/4" pipe size		1	16		2,075	615		2,690	3,250	
	8880	1740 GPH, 1-1/2" pipe size		.80	20		2,750	770		3,520	4,225	
	8920	2520 GPH, 2" pipe size		.60	26.667		4,000	1,025		5,025	6,000	
	8960	3480 GPH, 2-1/2" pipe size		.50	32		6,275	1,225		7,500	8,800	
	9200	Taste and odor removal, 660 GPH, 1" pipe size		1.50	10.667		2,425	410		2,835	3,300	
	9240	1500 GPH, 1-1/4" pipe size		1	16		4,175	615		4,790	5,550	
	9280	2340 GPH, 1-1/2" pipe size		.80	20		4,750	770		5,520	6,425	
	9320	3420 GPH, 2" pipe size		.60	26.667		7,325	1,025		8,350	9,650	
	9360	4620 GPH, 2-1/2" pipe size	↓	.50	32	↓	11,300	1,225		12,525	14,300	

	15480 | Domestic Water Heaters		CREW	DAILY OUTPUT	LABOR-HOURS	UNIT	MAT.	LABOR	EQUIP.	TOTAL	TOTAL INCL O&P	
200	0010	WATER HEATERS										200
	0020	For solar, see division 13600										
	1000	Residential, electric, glass lined tank, 5 yr, 10 gal., single element	1 Plum	2.30	3.478	Ea.	227	149		376	480	
	1040	20 gallon, single element		2.20	3.636		259	155		414	525	
	1060	30 gallon, double element		2.20	3.636		290	155		445	560	
	1080	40 gallon, double element		2	4		310	171		481	610	
	1100	52 gallon, double element		2	4		370	171		541	670	
	1120	66 gallon, double element		1.80	4.444		430	190		620	770	
	1140	80 gallon, double element		1.60	5		555	214		769	940	
	1180	120 gallon, double element	↓	1.40	5.714	↓	800	244		1,044	1,250	

15 MECHANICAL

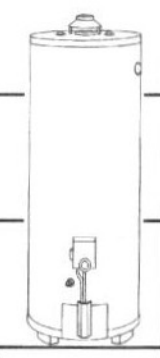

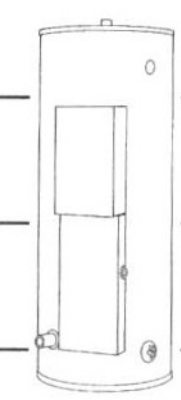

	15480 | Domestic Water Heaters	CREW	DAILY OUTPUT	LABOR-HOURS	UNIT	2006 BARE COSTS MAT.	LABOR	EQUIP.	TOTAL	TOTAL INCL O&P
200 2000	Gas fired, foam lined tank, 10 yr, vent not incl.,									
2040	30 gallon	1 Plum	2	4	Ea.	630	171		801	955
2060	40 gallon		1.90	4.211		635	180		815	975
2080	50 gallon		1.80	4.444		665	190		855	1,025
2100	75 gallon		1.50	5.333		820	228		1,048	1,250
2120	100 gallon		1.30	6.154		1,250	263		1,513	1,775
3000	Oil fired, glass lined tank, 5 yr, vent not included, 30 gallon		2	4		795	171		966	1,150
3040	50 gallon		1.80	4.444		1,375	190		1,565	1,800
3060	70 gallon		1.50	5.333		1,525	228		1,753	2,025
3080	85 gallon		1.40	5.714		4,750	244		4,994	5,600
4000	Commercial, 100° rise. NOTE: for each size tank, a range of									
4010	heaters between the ones shown are available									
4020	Electric									
4100	5 gal., 3 kW, 12 GPH, 208V	1 Plum	2	4	Ea.	1,800	171		1,971	2,250
4120	10 gal., 6 kW, 25 GPH, 208V		2	4		1,975	171		2,146	2,450
4140	50 gal., 9 kW, 37 GPH, 208V		1.80	4.444		2,725	190		2,915	3,300
4160	50 gal., 36 kW, 148 GPH, 208V		1.80	4.444		4,250	190		4,440	4,975
4180	80 gal., 12 kW, 49 GPH, 208V		1.50	5.333		3,350	228		3,578	4,050
4200	80 gal., 36 kW, 148 GPH, 208V		1.50	5.333		4,725	228		4,953	5,550
4220	100 gal., 36 kW, 148 GPH, 208V		1.20	6.667		4,800	285		5,085	5,750
4240	120 gal., 36 kW, 148 GPH, 208V		1.20	6.667		5,550	285		5,835	6,550
4260	150 gal., 15 kW , 61 GPH, 480V		1	8		12,000	340		12,340	13,700
4280	150 gal., 120 kW, 490 GPH, 480V		1	8		16,800	340		17,140	19,000
4300	200 gal., 15 kW, 61 GPH, 480V	Q-1	1.70	9.412		13,100	360		13,460	15,000
4320	200 gal., 120 kW , 490 GPH, 480V		1.70	9.412		17,900	360		18,260	20,300
4340	250 gal., 15 kW, 61 GPH, 480V		1.50	10.667		13,500	410		13,910	15,400
4360	250 gal., 150 kW, 615 GPH, 480V		1.50	10.667		19,600	410		20,010	22,200
4380	300 gal., 30 kW, 123 GPH, 480V		1.30	12.308		14,600	475		15,075	16,700
4400	300 gal., 180 kW, 738 GPH, 480V		1.30	12.308		21,100	475		21,575	23,900
4420	350 gal., 30 kW, 123 GPH, 480V		1.10	14.545		16,000	560		16,560	18,500
4440	350 gal, 180 kW, 738 GPH, 480V		1.10	14.545		21,700	560		22,260	24,800
4460	400 gal., 30 kW, 123 GPH, 480V		1	16		18,000	615		18,615	20,800
4480	400 gal., 210 kW, 860 GPH, 480V		1	16		25,100	615		25,715	28,600
4500	500 gal., 30 kW, 123 GPH, 480V		.80	20		20,800	770		21,570	24,000
4520	500 gal., 240 kW, 984 GPH, 480V		.80	20		29,900	770		30,670	34,100
4540	600 gal., 30 kW, 123 GPH, 480V	Q-2	1.20	20		24,000	795		24,795	27,700
4560	600 gal., 300 kW, 1230 GPH, 480V		1.20	20		36,000	795		36,795	40,900
4580	700 gal., 30 kW, 123 GPH, 480V		1	24		24,900	955		25,855	28,900
4600	700 gal., 300 kW, 1230 GPH, 480V		1	24		37,500	955		38,455	42,700
4620	800 gal., 60 kW, 245 GPH, 480V		.90	26.667		27,000	1,075		28,075	31,400
4640	800 gal., 300 kW, 1230 GPH, 480V		.90	26.667		38,300	1,075		39,375	43,900
4660	1000 gal., 60 kW, 245 GPH, 480V		.70	34.286		29,600	1,375		30,975	34,600
4680	1000 gal., 480 kW, 1970 GPH, 480V		.70	34.286		37,200	1,375		38,575	43,000
4700	1250 gal., 60 kW, 245 GPH, 480V		.60	40		33,400	1,600		35,000	39,200
4720	1250 gal., 480 kW, 1970 GPH, 480V		.60	40		50,500	1,600		52,100	58,500
4740	1500 gal., 60 kW, 245 GPH, 480V		.50	48		44,000	1,925		45,925	51,500
4760	1500 gal., 480 kW, 1970 GPH, 480V		.50	48		61,000	1,925		62,925	70,000
5400	Modulating step control, 2-5 steps	1 Elec	5.30	1.509		1,675	63.50		1,738.50	1,950
5440	6-10 steps		3.20	2.500		2,175	105		2,280	2,550
5460	11-15 steps		2.70	2.963		2,600	124		2,724	3,050
5480	16-20 steps		1.60	5		3,050	210		3,260	3,700
6000	Gas fired, flush jacket, std. controls, vent not incl.									
6040	75 MBH input, 73 GPH	1 Plum	1.40	5.714	Ea.	1,575	244		1,819	2,100
6060	98 MBH input, 95 GPH		1.40	5.714		2,775	244		3,019	3,425
6080	120 MBH input, 110 GPH		1.20	6.667		3,000	285		3,285	3,750
6100	120 MBH input, 115 GPH		1.10	7.273		3,375	310		3,685	4,175

MECHANICAL 15

15480 | Domestic Water Heaters

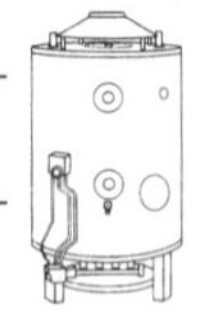

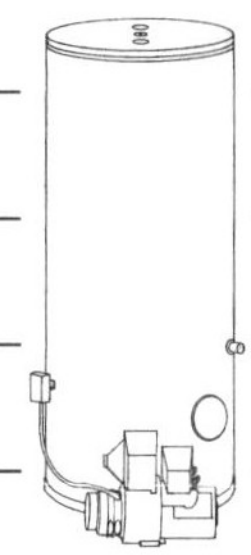

		15480 Domestic Water Heaters	CREW	DAILY OUTPUT	LABOR-HOURS	UNIT	2006 BARE COSTS MAT.	LABOR	EQUIP.	TOTAL	TOTAL INCL O&P	
200	6120	140 MBH input, 130 GPH	1 Plum	1	8	Ea.	3,775	340		4,115	4,675	200
	6140	155 MBH input, 150 GPH		.80	10		4,100	425		4,525	5,200	
	6160	180 MBH input, 170 GPH		.70	11.429		4,275	490		4,765	5,450	
	6180	200 MBH input, 192 GPH		.60	13.333		4,300	570		4,870	5,625	
	6200	250 MBH input, 245 GPH		.50	16		4,675	685		5,360	6,200	
	6220	260 MBH input, 250 GPH	Q-1	.80	20		5,900	770		6,670	7,675	
	6240	360 MBH input, 360 GPH		.80	20		6,550	770		7,320	8,400	
	6260	500 MBH input, 480 GPH		.70	22.857		8,200	880		9,080	10,400	
	6280	725 MBH input, 690 GPH		.60	26.667		9,425	1,025		10,450	12,000	
	6900	For low water cutoff, add	1 Plum	8	1		242	42.50		284.50	335	
	6960	For bronze body hot water circulator, add	"	4	2		1,150	85.50		1,235.50	1,400	
	8000	Oil fired, glass lined, UL listed, std. controls, vent not incl.										
	8060	140 gal., 140 MBH input, 134 GPH	Q-1	2.13	7.512	Ea.	1,150	289		1,439	1,700	
	8080	140 gal., 199 MBH input, 191 GPH		2	8		11,800	305		12,105	13,500	
	8100	140 gal., 255 MBH input, 247 GPH		1.60	10		12,100	385		12,485	13,900	
	8120	140 gal., 270 MBH input, 259 GPH		1.20	13.333		15,000	510		15,510	17,300	
	8140	140 gal., 400 MBH input, 384 GPH		1	16		15,400	615		16,015	17,900	
	8160	140 gal., 540 MBH input, 519 GPH		.96	16.667		16,100	640		16,740	18,700	
	8180	140 gal., 720 MBH input, 691 GPH		.92	17.391		16,400	670		17,070	19,100	
	8200	221 gal., 300 MBH input, 288 GPH		.88	18.182		21,700	700		22,400	24,900	
	8220	221 gal., 600 MBH input, 576 GPH		.86	18.605		24,200	715		24,915	27,700	
	8240	221 gal., 800 MBH input, 768 GPH		.82	19.512		24,300	750		25,050	28,000	
	8260	201 gal., 1000 MBH input, 960 GPH	Q-2	1.26	19.048		24,900	760		25,660	28,600	
	8280	201 gal., 1250 MBH input, 1200 GPH		1.22	19.672		25,100	785		25,885	28,900	
	8300	201 gal., 1500 MBH input, 1441 GPH		1.16	20.690		27,300	825		28,125	31,400	
	8320	411 gal., 600 MBH input, 576 GPH		1.12	21.429		27,600	855		28,455	31,700	
	8340	411 gal., 800 MBH input, 768 GPH		1.08	22.222		27,700	885		28,585	31,800	
	8360	411 gal., 1000 MBH input, 960 GPH		1.04	23.077		29,200	920		30,120	33,500	
	8380	411 gal., 1250 MBH input, 1200 GPH		.98	24.490		29,900	975		30,875	34,400	
	8400	397 gal., 1500 MBH input, 1441 GPH		.92	26.087		31,900	1,050		32,950	36,700	
	8420	397 gal., 1750 MBH input, 1681 GPH		.86	27.907		32,700	1,100		33,800	37,600	
	8430	397 gal., 2000 MBH input, 1921 GPH		.82	29.268		35,400	1,175		36,575	40,700	
	8440	375 gal., 2250 MBH input, 2161 GPH		.76	31.579		36,300	1,250		37,550	42,000	
	8450	375 gal., 2500 MBH input, 2401 GPH		.82	29.268		37,700	1,175		38,875	43,300	
	8900	For low water cutoff, add	1 Plum	8	1		242	42.50		284.50	335	
	8960	For bronze body hot water circulator, add		4	2		905	85.50		990.50	1,125	
	9000	Minimum labor/equipment charge		1.75	4.571	Job		195		195	305	
700	0010	**TEMPERATURE MAINTENANCE CABLE**										700
	0040	Components										
	0080	Heating cable										
	0100	208 V										
	0120	105 °F	Q-1	1060.80	.015	L.F.	8.55	.58		9.13	10.30	
	0130	115 °F		1060.80	.015		8.55	.58		9.13	10.30	
	0140	125 °F		1060.80	.015		8.55	.58		9.13	10.30	
	0150	140 °F		1060.80	.015		8.55	.58		9.13	10.30	
	0200	120 V										
	0220	125 °F	Q-1	1060.80	.015	L.F.	5.50	.58		6.08	6.95	
	0300	Power kit w/1 end seal	1 Elec	48.80	.164	Ea.	48	6.90		54.90	63.50	
	0310	Splice kit		35.50	.225		57	9.45		66.45	77	
	0320	End seal		160	.050		7.65	2.10		9.75	11.65	
	0330	Tee kit w/1 end seal		26.80	.299		60.50	12.55		73.05	86	
	0340	Powered splice w/2 end seals		20	.400		57.50	16.80		74.30	89	
	0350	Powered tee kit w/3 end seals		18	.444		73.50	18.65		92.15	110	
	0360	Cross kit w/2 end seals		18.60	.430		76	18.05		94.05	112	
	0500	Recommended thickness of fiberglass insulation										

15480 | Domestic Water Heaters

		15480 Domestic Water Heaters	CREW	DAILY OUTPUT	LABOR-HOURS	UNIT	2006 BARE COSTS MAT.	LABOR	EQUIP.	TOTAL	TOTAL INCL O&P
700	0510	Pipe size									
	0520	1/2" - 1" use 1" insulation									
	0530	1-1/4" - 2" use 1-1/2" insulation									
	0540	2-1/2" - 6" use 2" insulation									
	0560	NOTE: For pipe sizes 1-1/4" and smaller use 1/4" larger diameter									
	0570	insulation to allow room for installation over cable.									
900	0010	**HEAT TRANSFER PACKAGES** Complete, controls,									
	0020	expansion tank, converter, air separator									
	1000	Hot water, 180° F enter, 200° F leaving, 15# steam									
	1010	One pump system, 28 GPM	Q-6	.75	32	Ea.	13,400	1,275		14,675	16,700
	1020	35 GPM		.70	34.286		14,700	1,375		16,075	18,400
	1040	55 GPM		.65	36.923		16,700	1,475		18,175	20,700
	1060	130 GPM		.55	43.636		20,700	1,750		22,450	25,500
	1080	255 GPM		.40	60		28,200	2,400		30,600	34,800
	1100	550 GPM		.30	80		35,600	3,225		38,825	44,100
	1120	800 GPM		.25	96		42,300	3,850		46,150	52,500
	1220	Two pump system, 28 GPM		.70	34.286		18,200	1,375		19,575	22,200
	1240	35 GPM		.65	36.923		21,600	1,475		23,075	26,100
	1260	55 GPM		.60	40		22,100	1,600		23,700	26,800
	1280	130 GPM		.50	48		28,300	1,925		30,225	34,100
	1300	255 GPM		.35	68.571		38,200	2,750		40,950	46,300
	1320	550 GPM		.25	96		47,800	3,850		51,650	58,500
	1340	800 GPM		.20	120		55,500	4,825		60,325	68,500
940	0010	**WATER HEATER PACKAGED SYSTEMS**									
	1000	Car wash package, continuous duty, high recovery, gas fired									
	1040	100° rise, 180 MBH input, 174 GPH	1 Plum	3	2.667	Ea.	3,275	114		3,389	3,775
	1060	280 MBH input, 270 GPH		2.50	3.200		3,500	137		3,637	4,050
	1080	400 MBH input, 386 GPH		2.50	3.200		4,025	137		4,162	4,625
	1100	480 MBH input, 464 GPH		2	4		4,250	171		4,421	4,950
	1120	605 MBH input, 584 GPH		1.50	5.333		4,850	228		5,078	5,675
	1140	700 MBH input, 676 GPH		1	8		5,050	340		5,390	6,075
	1160	1000 MBH input, 966 GPH	Q-1	1.60	10		5,625	385		6,010	6,775
	1180	1200 MBH input, 1159 GPH		1.40	11.429		7,625	440		8,065	9,075
	1200	1400 MBH input, 1353 GPH		1.20	13.333		8,325	510		8,835	9,950
	3000	Combination dishwasher & general purpose, 2 temp. gas fired									
	3040	124 GPH @ 140° rise; 434 GPH @ 40° rise	1 Plum	2	4	Ea.	3,350	171		3,521	3,950
	3060	193 GPH @ 140° rise; 677 GPH @ 40° rise		1.60	5		3,825	214		4,039	4,525
	3080	276 GPH @ 140° rise; 969 GPH @ 40° rise		1.40	5.714		4,350	244		4,594	5,150
	3100	330 GPH @ 140° rise; 1160 GPH @ 40°rise		1.20	6.667		4,600	285		4,885	5,525
	3120	417 GPH @ 140° rise; 1460 GPH @ 40°rise		1	8		5,200	340		5,540	6,250
	3140	481 GPH @ 140° rise; 1685 GPH @ 40°rise	Q-1	1.40	11.429		6,000	440		6,440	7,275
	3160	688 GPH @ 140° rise; 2410 GPH @ 40°rise		1.20	13.333		6,575	510		7,085	8,050
	3180	828 GPH @ 140° rise; 2790 GPH @ 40°rise		1	16		8,625	615		9,240	10,400
	3200	965 GPH @ 140° rise; 3370 GPH @ 40°rise		.80	20		9,300	770		10,070	11,400
	3960	For unit base, add					257			257	283
	5000	Coin laundry units, gas fired, 100° rise									
	5020	Single heater,									
	5040	280 MBH input, 270 GPH	1 Plum	1.60	5	Ea.	3,825	214		4,039	4,525
	5060	400 MBH input, 386 GPH		1.30	6.154		5,025	263		5,288	5,950
	5080	480 MBH input, 464 GPH		1	8		5,300	340		5,640	6,350
	5100	605 MBH input, 584 GPH	Q-1	1.40	11.429		6,250	440		6,690	7,550
	5120	700 MBH input, 676 GPH		1.20	13.333		6,575	510		7,085	8,050
	5140	1000 MBH input, 966 GPH		1	16		7,150	615		7,765	8,825
	5160	1200 MBH input, 1159 GPH		.90	17.778		9,300	685		9,985	11,300
	5180	1400 MBH input, 1353 GPH		.70	22.857		10,300	880		11,180	12,800

15400 | Plumbing Fixtures & Equipment

15480 | Domestic Water Heaters

			CREW	DAILY OUTPUT	LABOR-HOURS	UNIT	2006 BARE COSTS MAT.	LABOR	EQUIP.	TOTAL	TOTAL INCL O&P	
940	6000	Multiple heater										940
	6040	560 MBH input, 540 GPH	1 Plum	1	8	Ea.	6,050	340		6,390	7,175	
	6060	800 MBH input, 772 GPH	Q-1	1.30	12.308		7,850	475		8,325	9,350	
	6080	960 MBH input, 928 GPH		1.20	13.333		8,850	510		9,360	10,500	
	6100	1210 MBH input, 1168 GPH		1	16		9,950	615		10,565	11,900	
	6120	1400 MBH input, 1352 GPH		1	16		10,400	615		11,015	12,400	
	6140	1700 MBH input, 1642 GPH		.80	20		11,900	770		12,670	14,300	
	6160	2000 MBH input, 1932 GPH		.70	22.857		14,400	880		15,280	17,200	
	6180	2400 MBH input, 2318 GPH	Q-2	.90	26.667		16,100	1,075		17,175	19,400	
	6200	2600 MBH input, 2512 GPH		.80	30		18,000	1,200		19,200	21,700	
	6220	2800 MBH input, 2706 GPH		.70	34.286		20,300	1,375		21,675	24,400	
	6800	Standard system sizing is based on										
	6820	80% of the washers operating at one time.										

15490 | Pool and Fountain Equipment

			CREW	DAILY OUTPUT	LABOR-HOURS	UNIT	MAT.	LABOR	EQUIP.	TOTAL	TOTAL INCL O&P	
100	0010	FOUNTAINS/AERATORS										100
	0100	Pump w/controls										
	0200	Single phase, 100′ cord, 1/2 H.P. pump	2 Skwk	4.40	3.636	Ea.	2,900	133		3,033	3,425	
	0300	3/4 H.P. pump		4.30	3.721		3,300	136		3,436	3,875	
	0400	1 H.P. pump		4.20	3.810		3,350	139		3,489	3,900	
	0500	1-1/2 H.P. pump		4.10	3.902		3,450	142		3,592	4,025	
	0600	2 H.P. pump		4	4		3,500	146		3,646	4,100	
	0700	Three phase, 200′ cord, 5 H.P. pump		3.90	4.103		8,500	150		8,650	9,600	
	0800	7-1/2 H.P. pump		3.80	4.211		9,575	154		9,729	10,800	
	0900	10 H.P. pump		3.70	4.324		10,700	158		10,858	12,100	
	1000	15 H.P. pump		3.60	4.444		12,700	162		12,862	14,200	
	1100	Nozzles, minimum		8	2		212	73		285	355	
	1200	Maximum		8	2		279	73		352	425	
	1300	Lights w/mounting kits, 200 watt		18	.889		350	32.50		382.50	440	
	1400	300 watt		18	.889		390	32.50		422.50	480	
	1500	500 watt		18	.889		425	32.50		457.50	525	
	1600	Color blender		12	1.333		330	48.50		378.50	445	

15 MECHANICAL

15500 | Heat Generation Equipment

15510 | Heating Boilers and Accessories

			CREW	DAILY OUTPUT	LABOR-HOURS	UNIT	2006 BARE COSTS MAT.	LABOR	EQUIP.	TOTAL	TOTAL INCL O&P	
110	0010	BOILER BLOWDOWN SYSTEMS										110
	1010	Boiler blowdown, auto/manual to 2000 MBH	Q-5	3.75	4.267	Ea.	3,275	165		3,440	3,875	
	1020	7300 MBH	"	3	5.333	"	3,775	207		3,982	4,475	
120	0010	BURNERS										120
	0990	Residential, conversion, gas fired, LP or natural										
	1000	Gun type, atmospheric input 72 to 200 MBH	Q-1	2.50	6.400	Ea.	655	246		901	1,100	
	1020	120 to 360 MBH		2	8		725	305		1,030	1,275	
	1040	280 to 800 MBH		1.70	9.412		1,400	360		1,760	2,100	
	2000	Commercial and industrial, gas/oil, input										
	2050	400 MBH	Q-1	1.50	10.667	Ea.	6,025	410		6,435	7,250	
	2080	For mounting plate, add		35	.457		166	17.55		183.55	210	

15510 | Heating Boilers and Accessories

			CREW	DAILY OUTPUT	LABOR-HOURS	UNIT	2006 BARE COSTS MAT.	LABOR	EQUIP.	TOTAL	TOTAL INCL O&P	
120	2090	670 MBH	Q-1	1.40	11.429	Ea.	8,900	440		9,340	10,500	120
	2110	For mounting plate, add		35	.457		166	17.55		183.55	210	
	2140	1155 MBH		1.30	12.308		10,800	475		11,275	12,500	
	2170	For mounting plate, add		35	.457		166	17.55		183.55	210	
	2200	1800 MBH		1.20	13.333		15,300	510		15,810	17,700	
	2230	For mounting plate, add		35	.457		166	17.55		183.55	210	
	2260	3000 MBH		1.10	14.545		21,000	560		21,560	24,100	
	2290	For mounting plate, add		35	.457		166	17.55		183.55	210	
	2320	4100 MBH		1	16		21,700	615		22,315	24,900	
	2350	For mounting plate, add		35	.457		166	17.55		183.55	210	
	2500	Impinged jet, rectangular, burner only, 3-1/2" WC,										
	2520	Gas fired, input, 420 to 640 MBH	Q-1	2	8	Ea.	690	305		995	1,250	
	2540	560 to 860 MBH		2	8		965	305		1,270	1,550	
	2560	630 to 950 MBH		1.90	8.421		1,025	325		1,350	1,650	
	2580	950 to 1200 MBH		1.80	8.889		1,450	340		1,790	2,125	
	2600	1120 to 1700 MBH		1.80	8.889		1,925	340		2,265	2,650	
	2620	1700 to 1900 MBH		1.70	9.412		2,175	360		2,535	2,950	
	2640	1400 to 2100 MBH		1.60	10		2,425	385		2,810	3,250	
	2660	1600 to 2400 MBH		1.60	10		2,600	385		2,985	3,450	
	2680	1700 to 2500 MBH		1.50	10.667		2,900	410		3,310	3,825	
	2700	1880 to 2880 MBH		1.40	11.429		3,100	440		3,540	4,100	
	2720	2260 to 3680 MBH		1.40	11.429		3,625	440		4,065	4,675	
	2740	2590 to 4320 MBH		1.30	12.308		4,350	475		4,825	5,500	
	2760	3020 to 5040 MBH		1.20	13.333		5,075	510		5,585	6,375	
	2780	3450 to 5760 MBH		1.20	13.333		5,800	510		6,310	7,175	
	2850	Burner pilot		50	.320		16.65	12.30		28.95	37.50	
	2860	Thermocouple		50	.320		17.20	12.30		29.50	38	
	2870	Thermocouple & pilot bracket		45	.356		45.50	13.65		59.15	71	
	2900	For 7" W.C. pressure, increase MBH 40%										
	3000	Flame retention oil fired assembly, input										
	3020	.50 to 2.25 GPH	Q-1	2.40	6.667	Ea.	291	256		547	720	
	3040	2.0 to 5.0 GPH		2	8		525	305		830	1,050	
	3060	3.0 to 7.0 GPH		1.80	8.889		535	340		875	1,125	
	3080	6.0 to 12.0 GPH		1.60	10		835	385		1,220	1,525	
	4600	Gas safety, shut off valve, 3/4" threaded	1 Stpi	20	.400		165	17.20		182.20	209	
	4610	1" threaded		19	.421		166	18.15		184.15	211	
	4620	1-1/4" threaded		15	.533		181	23		204	235	
	4630	1-1/2" threaded		13	.615		196	26.50		222.50	256	
	4640	2" threaded		11	.727		219	31.50		250.50	290	
	4650	2-1/2" threaded	Q-1	15	1.067		251	41		292	340	
	4660	3" threaded		13	1.231		345	47.50		392.50	455	
	4670	4" flanged		3	5.333		2,375	205		2,580	2,925	
	4680	6" flanged	Q-2	3	8		5,275	320		5,595	6,300	
300	0010	**BOILERS, ELECTRIC, ASME** Standard controls and trim R235000-50										300
	1000	Steam, 6 KW, 20.5 MBH	Q-19	1.20	20	Ea.	2,900	795		3,695	4,425	
	1040	9 KW, 30.7 MBH		1.20	20		2,975	795		3,770	4,500	
	1080	24 KW, 81.8 MBH		1.10	21.818		3,550	870		4,420	5,250	
	1160	60 KW, 205 MBH		1	24		4,925	955		5,880	6,875	
	1240	148 KW, 505 MBH		.65	36.923		7,100	1,475		8,575	10,100	
	1280	222 KW, 758 MBH		.55	43.636		11,100	1,750		12,850	15,000	
	1320	300 KW, 1023 MBH		.40	60		14,500	2,400		16,900	19,600	
	1360	444 KW, 1515 MBH		.30	80		16,600	3,175		19,775	23,300	
	1400	592 KW, 2020 MBH	Q-21	.34	94.118		21,300	3,825		25,125	29,300	
	1480	814 KW, 2778 MBH		.25	128		26,300	5,200		31,500	37,100	
	1520	1036 KW, 3536 MBH		.20	160		28,400	6,500		34,900	41,300	

	15510 | Heating Boilers and Accessories		CREW	DAILY OUTPUT	LABOR-HOURS	UNIT	2006 BARE COSTS MAT.	LABOR	EQUIP.	TOTAL	TOTAL INCL O&P	
300	1560	2070 KW, 7063 MBH R235000-50	Q-21	.18	177	Ea.	41,700	7,225		48,925	57,000	300
	1600	2,340 KW, 7984 MBH		.16	200		54,000	8,125		62,125	72,000	
	2000	Hot water, 7.5 KW, 25.6 MBH	Q-19	1.30	18.462		3,050	735		3,785	4,525	
	2040	30 KW, 102 MBH		1.20	20		3,250	795		4,045	4,800	
	2070	60 KW, 205 MBH		1.20	20		3,850	795		4,645	5,475	
	2100	90 KW, 307 MBH		1.10	21.818		4,475	870		5,345	6,275	
	2140	120 KW, 410 MBH 		.90	26.667		5,025	1,050		6,075	7,175	
	2180	150 KW, 512 MBH		.65	36.923		5,650	1,475		7,125	8,500	
	2220	296 KW, 1010 MBH		.55	43.636		9,850	1,750		11,600	13,500	
	2300	444 KW, 1515 MBH		.35	68.571		13,200	2,725		15,925	18,700	
	2340	518 KW, 1768 MBH	Q-21	.44	72.727		14,600	2,950		17,550	20,700	
	2420	740 KW, 2526 MBH		.39	82.051		17,700	3,325		21,025	24,600	
	2460	888 KW, 3031 MBH		.37	86.486		19,200	3,525		22,725	26,600	
	2500	1,036 KW, 3536 MBH		.34	94.118		22,900	3,825		26,725	31,100	
	2540	1440 KW, 4915 MBH		.32	100		29,000	4,075		33,075	38,200	
	2580	1680 KW, 5733 MBH		.30	106		34,800	4,325		39,125	45,000	
	2620	1980 KW, 6757 MBH		.28	114		39,200	4,650		43,850	50,500	
	2660	2220 KW, 7576 MBH		.26	123		45,200	5,000		50,200	57,500	
	2700	2610 KW, 8905 MBH		.24	133		51,000	5,425		56,425	65,000	
	2740	2970 KW, 10133 MBH		.21	152		55,000	6,200		61,200	70,000	
	2780	3240 KW, 11055 MBH		.18	177		61,000	7,225		68,225	78,500	
	2820	3,600 KW, 12,283 MBH		.16	200		67,000	8,125		75,125	86,500	
	9000	Minimum labor/equipment charge	Q-20	1	20	Job		775		775	1,225	
400	0010	**BOILERS, GAS FIRED** Natural or propane, standard controls R235000-50										400
	1000	Cast iron, with insulated jacket										
	2000	Steam, gross output, 81 MBH	Q-7	1.40	22.857	Ea.	1,575	940		2,515	3,175	
	2020	102 MBH 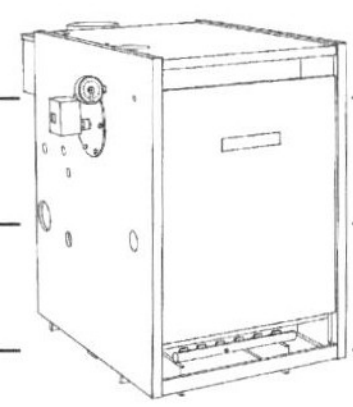		1.30	24.615		1,750	1,000		2,750	3,500	
	2040	122 MBH		1	32		1,925	1,325		3,250	4,175	
	2060	163 MBH		.90	35.556		2,325	1,450		3,775	4,825	
	2080	203 MBH		.90	35.556		2,550	1,450		4,000	5,075	
	2100	240 MBH		.85	37.647		2,600	1,550		4,150	5,250	
	2120	280 MBH		.80	40		3,000	1,650		4,650	5,850	
	2140	320 MBH		.70	45.714		3,275	1,875		5,150	6,525	
	2160	360 MBH		.63	51.200		3,700	2,100		5,800	7,325	
	2180	400 MBH		.56	56.838		3,925	2,325		6,250	7,950	
	2200	440 MBH		.51	62.500		4,200	2,575		6,775	8,600	
	2220	544 MBH		.45	71.588		7,050	2,925		9,975	12,300	
	2240	765 MBH		.43	74.419		8,525	3,050		11,575	14,100	
	2260	892 MBH		.38	84.211		9,875	3,450		13,325	16,300	
	2280	1275 MBH		.34	94.118		12,100	3,850		15,950	19,300	
	2300	1530 MBH		.32	100		12,600	4,100		16,700	20,200	
	2320	1,875 MBH		.30	106		12,900	4,375		17,275	21,000	
	2340	2170 MBH		.26	122		14,700	5,000		19,700	24,000	
	2360	2675 MBH		.20	163		16,900	6,700		23,600	29,000	
	2380	3060 MBH		.19	172		18,300	7,050		25,350	31,100	
	2400	3570 MBH		.18	181		20,400	7,450		27,850	34,000	
	2420	4207 MBH		.16	205		22,400	8,425		30,825	37,800	
	2440	4,720 MBH		.15	207		24,700	8,525		33,225	40,500	
	2460	5660 MBH		.15	220		53,500	9,050		62,550	72,500	
	2480	6,100 MBH		.13	246		61,000	10,100		71,100	82,500	
	2500	6390 MBH		.12	266		64,500	10,900		75,400	88,000	
	2520	6680 MBH		.11	290		65,500	11,900		77,400	90,500	
	2540	6,970 MBH		.10	320		68,500	13,100		81,600	96,000	
	3000	Hot water, gross output, 80 MBH		1.46	21.918		1,500	900		2,400	3,050	
	3020	100 MBH		1.35	23.704		1,700	975		2,675	3,375	

	15510 | Heating Boilers and Accessories		CREW	DAILY OUTPUT	LABOR-HOURS	UNIT	2006 BARE COSTS MAT.	LABOR	EQUIP.	TOTAL	TOTAL INCL O&P	
400	3040	122 MBH R235000-50	Q-7	1.10	29.091	Ea.	1,850	1,200		3,050	3,875	400
	3060	163 MBH		1	32		2,225	1,325		3,550	4,500	
	3080	203 MBH		1	32		2,475	1,325		3,800	4,750	
	3100	240 MBH		.95	33.684		2,600	1,375		3,975	5,000	
	3120	280 MBH		.90	35.556		2,650	1,450		4,100	5,175	
	3140	320 MBH		.80	40		3,050	1,650		4,700	5,900	
	3160	360 MBH		.71	45.070		3,475	1,850		5,325	6,700	
	3180	400 MBH		.64	50		3,700	2,050		5,750	7,275	
	3200	440 MBH		.58	54.983		3,975	2,250		6,225	7,875	
	3220	544 MBH		.51	62.992		6,825	2,575		9,400	11,600	
	3240	765 MBH		.46	70.022		8,275	2,875		11,150	13,600	
	3260	1,088 MBH		.40	80		10,300	3,275		13,575	16,400	
	3280	1,275 MBH		.36	89.888		11,800	3,700		15,500	18,700	
	3300	1,530 MBH		.31	104		12,400	4,300		16,700	20,300	
	3320	2,000 MBH		.26	125		13,400	5,125		18,525	22,800	
	3340	2,312 MBH		.22	148		15,500	6,075		21,575	26,600	
	3360	2,856 MBH		.20	160		18,300	6,575		24,875	30,300	
	3380	3,264 MBH		.18	179		19,400	7,375		26,775	32,800	
	3400	3,808 MBH		.16	195		21,400	8,000		29,400	35,900	
	3420	4,488 MBH		.15	210		23,800	8,650		32,450	39,500	
	3440	4,720 MBH		.15	220		54,000	9,050		63,050	73,500	
	3460	5,520 MBH		.14	228		61,000	9,375		70,375	81,500	
	3480	6,100 MBH		.13	250		61,500	10,300		71,800	83,500	
	3500	6,390 MBH		.11	285		63,000	11,700		74,700	87,000	
	3520	6,680 MBH		.10	310		66,500	12,700		79,200	93,000	
	3540	6,970 MBH		.09	359		69,000	14,800		83,800	99,000	
	7000	For tankless water heater, add					10%					
	7050	For additional zone valves up to 312 MBH add					124			124	136	
	7990	Special feature gas fired boilers										
	8000	Pulse combustion, standard controls / trim										
	8050	88,000 BTU	Q-5	1.40	11.429	Ea.	4,000	445		4,445	5,100	
	8080	134,000 BTU	″	1.20	13.333	″	4,775	515		5,290	6,075	
	9900	Minimum labor/equipment charge	Q-6	1	24	Job		965		965	1,500	
460	0010	**BOILERS, GAS/OIL** Combination with burners and controls R235000-50										460
	1000	Cast iron with insulated jacket										
	2000	Steam, gross output, 720 MBH	Q-7	.43	74.074	Ea.	7,600	3,050		10,650	13,100	
	2020	810 MBH		.38	83.990		8,125	3,450		11,575	14,300	
	2040	1,084 MBH		.34	93.023		9,350	3,825		13,175	16,200	
	2060	1,360 MBH		.33	98.160		11,100	4,025		15,125	18,500	
	2080	1,600 MBH		.30	107		12,500	4,400		16,900	20,500	
	2100	2,040 MBH		.25	130		15,300	5,325		20,625	25,200	
	2120	2,450 MBH		.21	156		16,800	6,400		23,200	28,500	
	2140	2,700 MBH		.19	165		17,800	6,800		24,600	30,100	
	2160	3,000 MBH		.18	175		19,200	7,225		26,425	32,300	
	2180	3,270 MBH		.17	183		20,600	7,550		28,150	34,400	
	2200	3,770 MBH		.17	191		50,500	7,850		58,350	67,500	
	2220	4,070 MBH		.16	200		53,000	8,200		61,200	71,500	
	2240	4,650 MBH		.15	210		56,000	8,650		64,650	75,000	
	2260	5,230 MBH		.14	223		62,000	9,175		71,175	82,500	
	2280	5,520 MBH		.14	235		67,500	9,650		77,150	89,000	
	2300	5,810 MBH		.13	248		68,000	10,200		78,200	91,000	
	2320	6,100 MBH		.12	260		68,500	10,700		79,200	92,000	
	2340	6,390 MBH		.11	296		72,000	12,200		84,200	98,500	
	2360	6,680 MBH		.10	320		76,000	13,100		89,100	104,000	
	2380	6,970 MBH		.09	372		76,500	15,300		91,800	107,500	

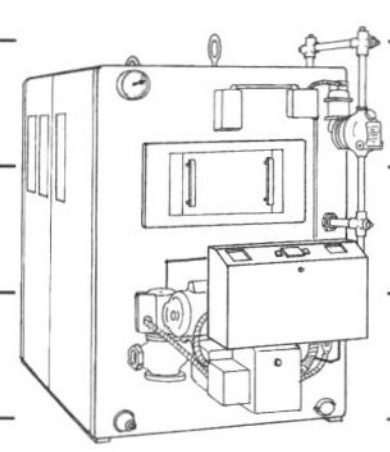

15510 | Heating Boilers and Accessories

		15510 Heating Boilers and Accessories	CREW	DAILY OUTPUT	LABOR-HOURS	UNIT	2006 BARE COSTS MAT.	LABOR	EQUIP.	TOTAL	TOTAL INCL O&P	
460	2900	Hot water, gross output (R235000-50)										460
	2910	200 MBH	Q-6	.62	39.024	Ea.	6,150	1,575		7,725	9,225	
	2920	300 MBH		.49	49.080		6,150	1,975		8,125	9,850	
	2930	400 MBH		.41	57.971		7,200	2,325		9,525	11,600	
	2940	500 MBH		.36	67.039		7,775	2,700		10,475	12,800	
	3000	584 MBH	Q-7	.44	72.072		10,400	2,950		13,350	16,000	
	3020	876 MBH		.41	79.012		11,300	3,250		14,550	17,600	
	3040	1,168 MBH		.31	103		13,200	4,275		17,475	21,100	
	3060	1,460 MBH		.28	113		15,300	4,650		19,950	24,000	
	3080	2,044 MBH		.26	122		19,100	5,000		24,100	28,900	
	3100	2,628 MBH		.21	150		23,000	6,175		29,175	34,900	
	3120	3,210 MBH		.18	174		28,300	7,175		35,475	42,300	
	3140	3,796 MBH		.17	186		31,700	7,625		39,325	46,700	
	3160	4,088 MBH		.16	195		36,000	8,000		44,000	52,000	
	3180	4,672 MBH		.16	203		39,300	8,375		47,675	56,000	
	3200	5,256 MBH		.15	217		44,000	8,925		52,925	62,500	
	3220	6,000 MBH, 179 BHP		.13	256		63,000	10,500		73,500	86,000	
	3240	7,130 MBH, 213 BHP		.08	385		66,000	15,800		81,800	97,000	
	3260	9,800 MBH, 286 BHP		.06	533		76,000	21,900		97,900	117,500	
	3280	10,900 MBH, 325.6 BHP		.05	592		92,000	24,300		116,300	139,500	
	3290	12,200 MBH, 364.5 BHP		.05	666		100,500	27,400		127,900	153,000	
	3300	13,500 MBH, 403.3 BHP		.04	727		111,500	29,800		141,300	169,000	
500	0010	**BOILERS, OIL FIRED** Standard controls, flame retention burner (R235000-50)										500
	1000	Cast iron, with insulated flush jacket										
	2000	Steam, gross output, 109 MBH	Q-7	1.20	26.667	Ea.	2,250	1,100		3,350	4,175	
	2020	144 MBH		1.10	29.091		2,425	1,200		3,625	4,525	
	2040	173 MBH		1	32		2,875	1,325		4,200	5,200	
	2060	207 MBH		.90	35.556		3,050	1,450		4,500	5,650	
	2080	236 MBH		.85	37.647		3,125	1,550		4,675	5,850	
	2100	300 MBH		.70	45.714		3,250	1,875		5,125	6,500	
	2120	480 MBH		.50	64		4,125	2,625		6,750	8,625	
	2140	665 MBH		.45	71.111		5,625	2,925		8,550	10,700	
	2160	794 MBH		.41	78.049		6,725	3,200		9,925	12,400	
	2180	1,084 MBH		.38	85.106		7,725	3,500		11,225	13,900	
	2200	1,360 MBH		.33	98.160		9,050	4,025		13,075	16,200	
	2220	1,600 MBH		.26	122		10,200	5,000		15,200	19,000	
	2240	2,175 MBH		.24	133		12,300	5,500		17,800	22,100	
	2260	2,480 MBH		.21	156		15,100	6,400		21,500	26,600	
	2280	3,000 MBH		.19	170		17,300	6,975		24,275	29,900	
	2300	3,550 MBH		.17	187		19,900	7,675		27,575	33,800	
	2320	3,820 MBH		.16	200		20,900	8,200		29,100	35,800	
	2340	4,360 MBH		.15	214		23,000	8,800		31,800	39,000	
	2360	4,940 MBH		.14	225		48,400	9,250		57,650	68,000	
	2380	5,520 MBH		.14	235		57,000	9,650		66,650	78,000	
	2400	6,100 MBH		.13	256		65,500	10,500		76,000	88,500	
	2420	6,390 MBH		.11	290		69,000	11,900		80,900	94,000	
	2440	6,680 MBH		.10	313		70,500	12,900		83,400	97,500	
	2460	6,970 MBH		.09	363		73,000	14,900		87,900	103,500	
	3000	Hot water, same price as steam										
	4000	For tankless coil in smaller sizes, add				Ea.	15%					
	5000	Steel, insulated jacket, burner										
	6000	Steam, full water leg construction, gross output										
	6020	144 MBH	Q-6	1.33	18.045	Ea.	4,250	725		4,975	5,800	
	6040	198 MBH		1.20	20		4,625	805		5,430	6,325	
	6060	252 MBH		1.13	21.295		5,750	855		6,605	7,650	
	6080	324 MBH		1.04	23.011		6,575	925		7,500	8,675	

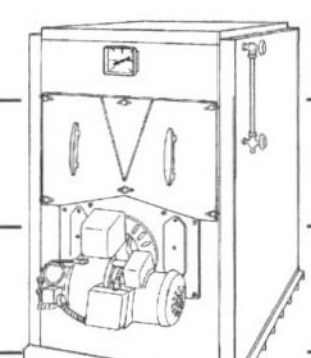

15 MECHANICAL

	15510	Heating Boilers and Accessories	Crew	Daily Output	Labor-Hours	Unit	2006 Bare Costs Mat.	Labor	Equip.	Total	Total Incl O&P	
500	6100	396 MBH (R235000-50)	Q-6	.78	30.691	Ea.	7,125	1,225		8,350	9,775	500
	6120	468 MBH		.70	34.483		7,825	1,375		9,200	10,800	
	6140	648 MBH		.52	45.977		8,325	1,850		10,175	12,100	
	6160	792 MBH		.44	55.046		9,800	2,200		12,000	14,300	
	6180	1,008 MBH		.39	61.381		12,100	2,475		14,575	17,200	
	6200	1,260 MBH		.35	68.966		13,400	2,775		16,175	19,100	
	6220	1,512 MBH		.31	78.689		16,100	3,150		19,250	22,600	
	6240	1,800 MBH		.29	83.624		17,600	3,350		20,950	24,600	
	6260	2,100 MBH		.23	106		20,800	4,275		25,075	29,500	
	6280	2,400 MBH		.19	125		23,200	5,025		28,225	33,300	
	6400	Larger sizes are same as steel, gas fired										
	7000	Hot water, gross output, 103 MBH	Q-6	1.60	15	Ea.	1,400	605		2,005	2,450	
	7020	122 MBH		1.45	16.506		2,625	665		3,290	3,900	
	7040	137 MBH		1.36	17.595		2,850	705		3,555	4,250	
	7060	168 MBH		1.30	18.405		3,300	740		4,040	4,775	
	7080	225 MBH		1.22	19.704		3,400	790		4,190	4,950	
	7100	315 MBH		.96	25.105		4,075	1,000		5,075	6,050	
	7120	420 MBH		.70	34.483		5,225	1,375		6,600	7,875	
	7140	525 MBH		.57	42.403		5,675	1,700		7,375	8,900	
	7160	630 MBH		.52	45.977		6,825	1,850		8,675	10,400	
	7180	735 MBH		.48	50.104		7,600	2,025		9,625	11,500	
	7200	840 MBH		.44	55.172		10,600	2,225		12,825	15,100	
	7220	1,050 MBH		.37	65.753		10,900	2,650		13,550	16,000	
	7240	1,365 MBH		.32	74.534		12,500	3,000		15,500	18,500	
	7260	1,680 MBH		.29	83.624		14,500	3,350		17,850	21,200	
	7280	2,310 MBH		.21	114		17,200	4,625		21,825	26,100	
	7320	3,150 MBH		.13	184		21,200	7,425		28,625	34,800	
	7340	For tankless coil in steam or hot water, add					7%					
	9000	Minimum labor/equipment charge	Q-6	1.75	13.714	Job		550		550	855	
700	0010	**BOILERS, PACKAGED SCOTCH MARINE** Steam or hot water										700
	1000	Packaged fire tube, #2 oil, gross output										
	1020	3348 MBH, 100 HP	Q-7	.21	152	Ea.	41,700	6,250		47,950	55,500	
	1040	6696 MBH, 200 HP		.14	223		60,000	9,175		69,175	80,500	
	1060	10,044 MBH, 300 HP		.13	251		86,500	10,300		96,800	111,500	
	1080	16,740 MBH, 500 HP		.08	380		124,000	15,600		139,600	161,000	
	1100	23,435 MBH, 700 HP		.07	484		152,000	19,900		171,900	198,000	
	1120	To fire #6, add		.83	38.554		9,200	1,575		10,775	12,600	
	1140	To fire #6, and gas, add		.42	76.190		13,500	3,125		16,625	19,800	
	1160	For high pressure, add					15,000			15,000	16,500	
	1180	For duplex package feed system										
	1200	To 3348 MBH boiler, add	Q-7	.54	59.259	Ea.	7,275	2,425		9,700	11,800	
	1220	To 6696 MBH boiler, add		.41	78.049		8,625	3,200		11,825	14,500	
	1240	To 10,044 MBH boiler, add		.38	84.211		9,900	3,450		13,350	16,300	
	1260	To 16,740 MBH boiler, add		.28	114		11,300	4,700		16,000	19,700	
	1280	To 23,435 MBH boiler, add		.25	128		13,300	5,250		18,550	22,900	
	2000	Packaged water tube, #2 oil, steam or hot water, gross output										
	2040	1200 MBH	Q-7	.50	64	Ea.	15,000	2,625		17,625	20,600	
	2060	1600 MBH		.40	80		20,200	3,275		23,475	27,300	
	2080	2400 MBH		.30	106		27,200	4,375		31,575	36,700	
	2100	3200 MBH		.25	128		31,100	5,250		36,350	42,400	
	2120	4800 MBH		.20	160		40,600	6,575		47,175	55,000	
	2140	For gas fired, add		.40	80		2,725	3,275		6,000	8,100	
760	0010	**BOILERS, SOLID FUEL** (R235000-50)										760
	0020	Wood, coal, wood/paper waste, peat. Steel, 3" diam. horiz.										

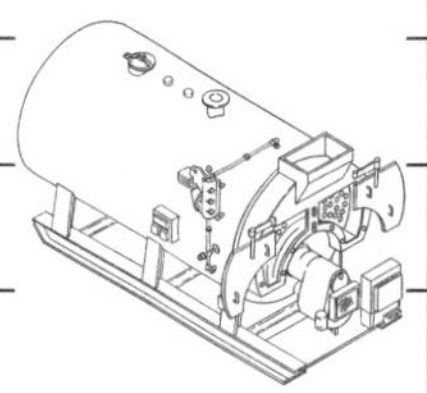

MECHANICAL 15

15510 | Heating Boilers and Accessories

		15510 Heating Boilers and Accessories	CREW	DAILY OUTPUT	LABOR-HOURS	UNIT	2006 BARE COSTS MAT.	LABOR	EQUIP.	TOTAL	TOTAL INCL O&P	
760	0040	fire tubes, low pressure steam or hot water, natural or (R235000-50)										760
	0060	mechanical draft. Not including dry base.										
	1000	Natural draft, gross output, 1445 MBH	Q-6	.45	53.333	Ea.	15,300	2,150		17,450	20,200	
	1050	1635 MBH		.43	55.814		16,000	2,250		18,250	21,100	
	1100	1805 MBH		.41	58.537		16,700	2,350		19,050	22,100	
	1150	2160 MBH		.38	63.158		17,900	2,550		20,450	23,700	
	1200	2520 MBH		.32	75		21,700	3,025		24,725	28,500	
	1250	3065 MBH	Q-7	.40	80		23,300	3,275		26,575	30,700	
	1300	3600 MBH		.38	84.211		24,900	3,450		28,350	32,800	
	1350	4500 MBH		.29	110		32,300	4,525		36,825	42,600	
	1400	5400 MBH		.26	123		34,700	5,050		39,750	46,100	
	1450	6295 MBH		.23	139		40,300	5,700		46,000	53,000	
	1500	7200 MBH		.21	152		45,600	6,250		51,850	59,500	
	1550	9000 MBH		.19	168		49,500	6,900		56,400	65,000	
	1600	10,800 MBH		.17	188		56,500	7,725		64,225	74,500	
	1650	12,600 MBH		.15	213		62,000	8,750		70,750	81,500	
	1700	14,390 MBH		.13	246		74,500	10,100		84,600	97,500	
	1750	16,190 MBH		.12	266		81,500	10,900		92,400	106,500	
	1800	17,990 MBH		.10	320		90,500	13,100		103,600	120,000	
	3000	Stoker fired (coal) cast iron with flush jacket and										
	3080	insulation, steam or water, gross output, 148 MBH	Q-6	1.50	16	Ea.	4,375	645		5,020	5,800	
	3100	200 MBH		1.30	18.462		5,000	740		5,740	6,650	
	3120	247 MBH		1.20	20		5,575	805		6,380	7,400	
	3140	300 MBH		1	24		6,250	965		7,215	8,375	
	3160	410 MBH		.96	25		7,175	1,000		8,175	9,425	
	3180	495 MBH		.88	27.273		8,000	1,100		9,100	10,500	
	3200	580 MBH		.80	30		8,850	1,200		10,050	11,600	
	3400	1280 MBH		.36	66.667		21,800	2,675		24,475	28,200	
	3420	1460 MBH		.30	80		24,000	3,225		27,225	31,400	
	3440	1640 MBH		.28	85.714		25,600	3,450		29,050	33,500	
	3460	1820 MBH		.26	92.308		27,400	3,700		31,100	35,900	
	3480	2000 MBH		.25	96		29,200	3,850		33,050	38,100	
	3500	2360 MBH		.23	104		33,000	4,200		37,200	42,800	
	3540	2725 MBH		.20	120		36,700	4,825		41,525	47,800	
	3800	2950 MBH	Q-7	.16	200		61,500	8,200		69,700	80,500	
	3820	3210 MBH		.15	213		64,500	8,750		73,250	84,500	
	3840	3480 MBH		.14	228		67,500	9,375		76,875	88,500	
	3860	3745 MBH		.14	228		70,500	9,375		79,875	92,000	
	3880	4000 MBH		.13	246		73,500	10,100		83,600	96,000	
	3900	4200 MBH		.13	246		76,500	10,100		86,600	100,000	
	3920	4400 MBH		.12	266		79,500	10,900		90,400	104,500	
	3940	4600 MBH		.12	266		82,500	10,900		93,400	108,000	
880	0010	**SWIMMING POOL HEATERS** Not including wiring, external										880
	0020	piping, base or pad,										
	0160	Gas fired, input, 155 MBH	Q-6	1.50	16	Ea.	1,925	645		2,570	3,125	
	0200	199 MBH		1	24		2,050	965		3,015	3,775	
	0220	250 MBH		.70	34.286		2,225	1,375		3,600	4,600	
	0240	300 MBH		.60	40		2,350	1,600		3,950	5,075	
	0260	399 MBH		.50	48		2,625	1,925		4,550	5,900	
	0280	500 MBH		.40	60		8,325	2,400		10,725	12,900	
	0300	650 MBH		.35	68.571		8,850	2,750		11,600	14,000	
	0320	750 MBH		.33	72.727		9,675	2,925		12,600	15,200	
	0360	990 MBH		.22	109		13,000	4,375		17,375	21,100	
	0370	1,260 MBH		.21	114		15,300	4,600		19,900	24,000	
	0380	1,440 MBH		.19	126		16,300	5,075		21,375	25,800	
	0400	1,800 MBH		.14	171		18,100	6,900		25,000	30,600	

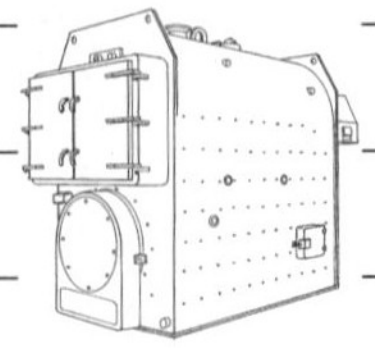

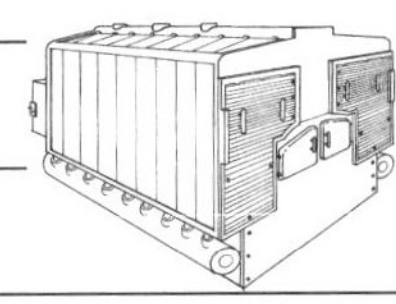

15 MECHANICAL

	15510 | Heating Boilers and Accessories		CREW	DAILY OUTPUT	LABOR-HOURS	UNIT	2006 BARE COSTS MAT.	LABOR	EQUIP.	TOTAL	TOTAL INCL O&P	
880	0410	2,070 MBH	Q-6	.13	184	Ea.	21,300	7,425		28,725	35,000	880
	1120	Oil fired, 315 MBH		.96	25		3,000	1,000		4,000	4,850	
	2000	Electric, 12 KW, 4,800 gallon pool	Q-19	3	8		1,925	320		2,245	2,625	
	2020	15 KW, 7,200 gallon pool		2.80	8.571		1,950	340		2,290	2,675	
	2040	24 KW, 9,600 gallon pool		2.40	10		2,625	400		3,025	3,525	
	2060	30 KW, 12,000 gallon pool		2	12		2,725	480		3,205	3,750	
	2080	35 KW, 14,400 gallon pool		1.60	15		2,725	595		3,320	3,925	
	2100	55 KW, 24,000 gallon pool		1.20	20		3,725	795		4,520	5,325	
	9000	To select pool heater: 12 BTUH x S.F. pool area										
	9010	X temperature differential =required output										
	9050	For electric, KW = gallons x 2.5 divided by 1000										
	9100	For family home type pool, double the										
	9110	Rated gallon capacity = 1/2°F rise per hour										

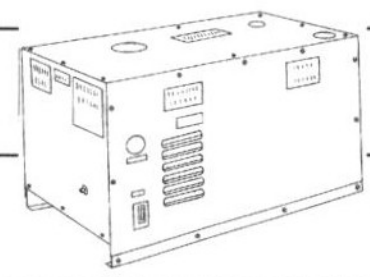

	15530 | Furnaces		CREW	DAILY OUTPUT	LABOR-HOURS	UNIT	MAT.	LABOR	EQUIP.	TOTAL	TOTAL INCL O&P	
200	0010	FURNACE COMPONENTS AND COMBINATIONS										200
	0080	Coils, A/C evaporator, for gas or oil furnaces										
	0090	Add-on, with holding charge										
	0100	Upflow										
	0120	1-1/2 ton cooling	Q-5	4	4	Ea.	114	155		269	365	
	0130	2 ton cooling		3.70	4.324		136	168		304	410	
	0140	3 ton cooling		3.30	4.848		172	188		360	480	
	0150	4 ton cooling		3	5.333		237	207		444	580	
	0160	5 ton cooling		2.70	5.926		280	230		510	665	
	0300	Downflow										
	0330	2-1/2 ton cooling	Q-5	3	5.333	Ea.	167	207		374	505	
	0340	3-1/2 ton cooling		2.60	6.154		199	238		437	590	
	0350	5 ton cooling		2.20	7.273		280	282		562	750	
	0600	Horizontal										
	0630	2 ton cooling	Q-5	3.90	4.103	Ea.	145	159		304	405	
	0640	3 ton cooling		3.50	4.571		164	177		341	455	
	0650	4 ton cooling		3.20	5		207	194		401	530	
	0660	5 ton cooling		2.90	5.517		275	214		489	630	
	2000	Cased evaporator coils for air handlers										
	2100	1-1/2 ton cooling	Q-5	4.40	3.636	Ea.	202	141		343	440	
	2110	2 ton cooling		4.10	3.902		205	151		356	460	
	2120	2-1/2 ton cooling		3.90	4.103		214	159		373	485	
	2130	3 ton cooling		3.70	4.324		237	168		405	520	
	2140	3-1/2 ton cooling		3.50	4.571		248	177		425	545	
	2150	4 ton cooling		3.20	5		296	194		490	625	
	2160	5 ton cooling		2.90	5.517		340	214		554	705	
	3010	Air handler, modular										
	3100	With cased evaporator cooling coil										
	3120	1-1/2 ton cooling	Q-5	3.80	4.211	Ea.	510	163		673	815	
	3130	2 ton cooling		3.50	4.571		535	177		712	865	
	3140	2-1/2 ton cooling		3.30	4.848		580	188		768	925	
	3150	3 ton cooling		3.10	5.161		625	200		825	1,000	
	3160	3-1/2 ton cooling		2.90	5.517		775	214		989	1,175	
	3170	4 ton cooling		2.50	6.400		830	248		1,078	1,300	
	3180	5 ton cooling		2.10	7.619		885	295		1,180	1,425	
	3500	With no cooling coil										
	3520	1-1/2 ton coil size	Q-5	12	1.333	Ea.	360	51.50		411.50	475	
	3530	2 ton coil size		10	1.600		385	62		447	520	
	3540	2-1/2 ton coil size		10	1.600		425	62		487	565	
	3554	3 ton coil size		9	1.778		475	69		544	625	

15530 | Furnaces

		15530 Furnaces	Crew	Daily Output	Labor-Hours	Unit	2006 Bare Costs Mat.	Labor	Equip.	Total	Total Incl O&P	
200	3560	3-1/2 ton coil size	Q-5	9	1.778	Ea.	530	69		599	685	200
	3570	4 ton coil size		8.50	1.882		680	73		753	860	
	3580	5 ton coil size		8	2		730	77.50		807.50	925	
	4000	With heater										
	4120	5 kW, 17.1 MBH	Q-5	16	1	Ea.	400	39		439	500	
	4130	7.5 kW, 25.6 MBH		15.60	1.026		440	39.50		479.50	540	
	4140	10 kW, 34.2 MBH		15.20	1.053		530	41		571	650	
	4150	12.5 KW, 42.7 MBH		14.80	1.081		650	42		692	780	
	4160	15 KW, 51.2 MBH		14.40	1.111		800	43		843	945	
	4170	25 KW, 85.4 MBH		14	1.143		905	44.50		949.50	1,075	
	4180	30 KW, 102 MBH		13	1.231		1,075	47.50		1,122.50	1,250	
400	0010	**FURNACES** Hot air heating, blowers, standard controls (R235000-50)										400
	0020	not including gas, oil or flue piping										
	1000	Electric, UL listed										
	1020	10.2 MBH	Q-20	5	4	Ea.	325	155		480	605	
	1040	17.1 MBH		4.80	4.167		340	161		501	630	
	1060	27.3 MBH		4.60	4.348		410	168		578	720	
	1100	34.1 MBH		4.40	4.545		425	176		601	745	
	1120	51.6 MBH		4.20	4.762		525	184		709	865	
	1140	68.3 MBH		4	5		635	194		829	1,000	
	1160	85.3 MBH		3.80	5.263		695	204		899	1,075	
	3000	Gas, AGA certified, upflow, direct drive models										
	3020	45 MBH input	Q-9	4	4	Ea.	455	152		607	740	
	3040	60 MBH input		3.80	4.211		605	160		765	920	
	3060	75 MBH input		3.60	4.444		640	169		809	975	
	3100	100 MBH input		3.20	5		680	190		870	1,050	
	3120	125 MBH input		3	5.333		785	202		987	1,175	
	3130	150 MBH input		2.80	5.714		910	217		1,127	1,350	
	3140	200 MBH input		2.60	6.154		2,100	233		2,333	2,675	
	3160	300 MBH input		2.30	6.957		2,200	264		2,464	2,825	
	3180	400 MBH input		2	8		3,025	305		3,330	3,800	
	6000	Oil, UL listed, atomizing gun type burner										
	6020	56 MBH output	Q-9	3.60	4.444	Ea.	745	169		914	1,100	
	6030	84 MBH output		3.50	4.571		775	173		948	1,125	
	6040	95 MBH output		3.40	4.706		790	178		968	1,150	
	6060	134 MBH output		3.20	5		1,100	190		1,290	1,500	
	6080	151 MBH output		3	5.333		1,225	202		1,427	1,650	
	6100	200 MBH input		2.60	6.154		2,225	233		2,458	2,825	
	6120	300 MBH input		2.30	6.957		2,450	264		2,714	3,125	
	6140	400 MBH input		2	8		2,850	305		3,155	3,600	
	9000	Minimum labor/equipment charge		2.75	5.818	Job		221		221	350	

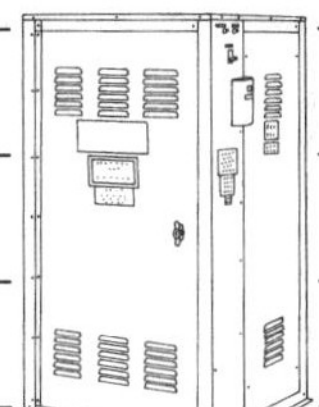

15540 | Fuel-Fired Heaters

		15540 Fuel-Fired Heaters	Crew	Daily Output	Labor-Hours	Unit	Mat.	Labor	Equip.	Total	Total Incl O&P	
300	0010	**DUCT FURNACES** Includes burner, controls, stainless steel (R235000-50)										300
	0020	heat exchanger. Gas fired, electric ignition										
	0030	Indoor installation										
	0080	100 MBH output	Q-5	5	3.200	Ea.	1,350	124		1,474	1,700	
	0100	120 MBH output		4	4		1,525	155		1,680	1,925	
	0130	200 MBH output		2.70	5.926		2,050	230		2,280	2,600	
	0140	240 MBH output		2.30	6.957		2,150	270		2,420	2,775	
	0160	280 MBH output		2	8		2,375	310		2,685	3,075	
	0180	320 MBH output		1.60	10		2,575	390		2,965	3,425	
	0300	For powered venter and adapter, add					355			355	390	
	0500	For required flue pipe, see 15550										
	1000	Outdoor installation, with vent cap										

15540 | Fuel-Fired Heaters

		15540 Fuel-Fired Heaters	CREW	DAILY OUTPUT	LABOR-HOURS	UNIT	2006 BARE COSTS MAT.	LABOR	EQUIP.	TOTAL	TOTAL INCL O&P	
300	1020	75 MBH output (R235000-50)	Q-5	4	4	Ea.	2,000	155		2,155	2,450	300
	1040	94 MBH output		4	4		2,275	155		2,430	2,750	
	1060	120 MBH output		4	4		2,550	155		2,705	3,075	
	1080	157 MBH output		3.50	4.571		2,975	177		3,152	3,525	
	1100	187 MBH output		3	5.333		3,650	207		3,857	4,350	
	1120	225 MBH output		2.50	6.400		3,800	248		4,048	4,550	
	1140	300 MBH output		1.80	8.889		4,750	345		5,095	5,750	
	1160	375 MBH output		1.60	10		5,050	390		5,440	6,150	
	1180	450 MBH output		1.40	11.429		5,275	445		5,720	6,500	
	1200	600 MBH output		1	16		6,425	620		7,045	8,050	
	1300	Aluminized exchanger, subtract					15%					
	1400	For powered venter, add					10%					
	1500	For two stage gas valve, add					194			194	213	
900	0010	**SPACE HEATERS** Cabinet, grilles, fan, controls, burner, (R235000-50)										900
	0020	thermostat, no piping. For flue see 15550										
	1000	Gas fired, floor mounted										
	1100	60 MBH output	Q-5	10	1.600	Ea.	620	62		682	780	
	1120	80 MBH output		9	1.778		630	69		699	800	
	1140	100 MBH output		8	2		680	77.50		757.50	865	
	1160	120 MBH output		7	2.286		765	88.50		853.50	980	
	1180	180 MBH output		6	2.667		925	103		1,028	1,175	
	1500	Rooftop mounted, gravity vent, stainless steel exchanger										
	1520	75 MBH output	Q-6	4	6	Ea.	4,025	241		4,266	4,800	
	1540	95 MBH output		3.60	6.667		4,075	268		4,343	4,925	
	1560	120 MBH output		3.30	7.273		4,375	292		4,667	5,250	
	1580	159 MBH output		3	8		4,500	320		4,820	5,450	
	1600	190 MBH output		2.60	9.231		5,075	370		5,445	6,175	
	1620	225 MBH output		2.30	10.435		5,175	420		5,595	6,325	
	1640	300 MBH output		1.90	12.632		6,150	510		6,660	7,575	
	1660	375 MBH output		1.40	17.143		6,975	690		7,665	8,750	
	1680	450 MBH output		1.20	20		7,125	805		7,930	9,075	
	1700	600 MBH output		1	24		8,600	965		9,565	11,000	
	1720	750 MBH output		.80	30		14,700	1,200		15,900	18,000	
	1740	900 MBH output		.60	40		16,500	1,600		18,100	20,600	
	1760	1200 MBH output		.30	80		20,000	3,225		23,225	27,000	
	1800	For power vent, add					10%					
	1900	For aluminized steel exchanger, subtract					10%					
	2000	Suspension mounted, propeller fan, 20 MBH output	Q-5	8.50	1.882		480	73		553	640	
	2020	40 MBH output		7.50	2.133		525	82.50		607.50	710	
	2040	60 MBH output		7	2.286		590	88.50		678.50	790	
	2060	80 MBH output		6	2.667		675	103		778	900	
	2080	100 MBH output		5.50	2.909		785	113		898	1,050	
	2100	130 MBH output		5	3.200		875	124		999	1,150	
	2120	140 MBH output		4.50	3.556		985	138		1,123	1,300	
	2140	160 MBH output		4	4		970	155		1,125	1,325	
	2160	180 MBH output		3.50	4.571		1,125	177		1,302	1,500	
	2180	200 MBH output		3	5.333		1,125	207		1,332	1,575	
	2200	240 MBH output		2.70	5.926		1,350	230		1,580	1,825	
	2220	280 MBH output		2.30	6.957		1,675	270		1,945	2,250	
	2240	320 MBH output		2	8		1,800	310		2,110	2,450	
	2500	For powered venter and adapter, add					305			305	335	
	3000	Suspension mounted, blower type, 40 MBH output	Q-5	6.80	2.353		520	91		611	715	
	3020	60 MBH output		6.60	2.424		625	94		719	835	
	3040	84 MBH output		5.80	2.759		695	107		802	930	
	3060	104 MBH output		5.20	3.077		765	119		884	1,025	

15540 | Fuel-Fired Heaters

		15540 Fuel-Fired Heaters	CREW	DAILY OUTPUT	LABOR-HOURS	UNIT	2006 BARE COSTS MAT.	LABOR	EQUIP.	TOTAL	TOTAL INCL O&P	
900	3080	140 MBH output (R235000-50)	Q-5	4.30	3.721	Ea.	920	144		1,064	1,225	900
	3100	180 MBH output		3.30	4.848		1,000	188		1,188	1,400	
	3120	240 MBH output		2.50	6.400		1,325	248		1,573	1,825	
	3140	280 MBH output		2	8		1,575	310		1,885	2,200	
	4000	Suspension mounted, sealed combustion system,										
	4020	Aluminized steel exchanger, powered vent										
	4040	100 MBH output	Q-5	5	3.200	Ea.	1,550	124		1,674	1,900	
	4060	120 MBH output		4.70	3.404		1,725	132		1,857	2,100	
	4080	160 MBH output		3.70	4.324		1,875	168		2,043	2,325	
	4100	200 MBH output		2.90	5.517		2,250	214		2,464	2,800	
	4120	240 MBH output		2.50	6.400		2,425	248		2,673	3,025	
	4140	320 MBH output		1.70	9.412		2,875	365		3,240	3,750	
	5000	Wall furnace, 17.5 MBH output		6	2.667		545	103		648	760	
	5020	24 MBH output		5	3.200		575	124		699	825	
	5040	35 MBH output		4	4		770	155		925	1,100	
	6000	Oil fired, suspension mounted, 94 MBH output		4	4		2,325	155		2,480	2,825	
	6040	140 MBH output		3	5.333		2,500	207		2,707	3,075	
	6060	184 MBH output		3	5.333		2,675	207		2,882	3,275	
	9000	Minimum labor/equipment charge		3.50	4.571	Job		177		177	275	

15550 | Breechings, Chimneys & Stacks

		15550 Breechings, Chimneys & Stacks	CREW	DAILY OUTPUT	LABOR-HOURS	UNIT	MAT.	LABOR	EQUIP.	TOTAL	TOTAL INCL O&P	
200	0010	**DRAFT CONTROL DEVICES**										200
	1000	Barometric,gas fired system only, 6" size for 5" and 6" pipes	1 Shee	20	.400	Ea.	45	16.85		61.85	76.50	
	1020	7" size, for 6" and 7" pipes		19	.421		49	17.75		66.75	82	
	1040	8" size, for 7" and 8" pipes		18	.444		62	18.75		80.75	98.50	
	1060	9" size, for 8" and 9" pipes		16	.500		72.50	21		93.50	114	
	2000	All fuel, oil, oil/gas, coal										
	2020	10" for 9" and 10" pipes	1 Shee	15	.533	Ea.	107	22.50		129.50	154	
	2040	12" for 11" and 12" pipes		15	.533		138	22.50		160.50	188	
	2060	14" for 13" and 14" pipes		14	.571		176	24		200	233	
	2080	16" for 15" and 16" pipes		13	.615		247	26		273	315	
	2100	18" for 17" and 18" pipes		12	.667		330	28		358	405	
	2120	20" for 19" and 21" pipes		10	.800		395	33.50		428.50	485	
	2140	24" for 22" and 25" pipes	Q-9	12	1.333		480	50.50		530.50	610	
	2160	28" for 26" and 30" pipes		10	1.600		595	60.50		655.50	750	
	2180	32" for 31" and 34" pipes		8	2		765	76		841	960	
	3260	For thermal switch for above, add	1 Shee	24	.333		44.50	14.05		58.55	71.50	
	5000	Vent damper, bi-metal, gas, 3" diameter	Q-9	24	.667		25	25.50		50.50	67.50	
	5010	4" diameter		24	.667		25	25.50		50.50	67.50	
	5020	5" diameter		23	.696		25	26.50		51.50	69.50	
	5030	6" diameter		22	.727		25	27.50		52.50	71.50	
	5040	7" diameter		21	.762		26.50	29		55.50	75.50	
	5050	8" diameter		20	.800		29	30.50		59.50	80	
	9000	Minimum labor/equipment charge	1 Shee	4	2	Job		84.50		84.50	134	
440	0010	**VENT CHIMNEY** Prefab metal, U.L. listed										440
	0020	Gas, double wall, galvanized steel										
	0080	3" diameter	Q-9	72	.222	V.L.F.	3.96	8.45		12.41	17.75	
	0100	4" diameter		68	.235		4.95	8.90		13.85	19.65	
	0120	5" diameter		64	.250		5.75	9.50		15.25	21.50	
	0140	6" diameter		60	.267		6.70	10.10		16.80	23.50	
	0160	7" diameter		56	.286		9.85	10.85		20.70	28	
	0180	8" diameter		52	.308		10.95	11.65		22.60	30.50	
	0200	10" diameter		48	.333		23	12.65		35.65	45.50	
	0220	12" diameter		44	.364		31	13.80		44.80	56	

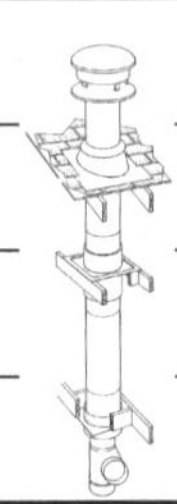

15550 | Breechings, Chimneys & Stacks

			CREW	DAILY OUTPUT	LABOR-HOURS	UNIT	2006 BARE COSTS MAT.	LABOR	EQUIP.	TOTAL	TOTAL INCL O&P	
440	0240	14″ diameter	Q-9	42	.381	V.L.F.	52	14.45		66.45	80	440
	0260	16″ diameter		40	.400		70.50	15.15		85.65	102	
	0280	18″ diameter	↓	38	.421		91	15.95		106.95	126	
	0300	20″ diameter	Q-10	36	.667		107	26		133	160	
	0320	22″ diameter		34	.706		136	28		164	193	
	0340	24″ diameter		32	.750		167	29.50		196.50	231	
	0360	26″ diameter		31	.774		201	30.50		231.50	270	
	0380	28″ diameter		30	.800		212	31.50		243.50	283	
	0400	30″ diameter		28	.857		224	33.50		257.50	300	
	0420	32″ diameter		27	.889		260	35		295	340	
	0440	34″ diameter		26	.923		296	36.50		332.50	385	
	0460	36″ diameter		25	.960		315	38		353	405	
	0480	38″ diameter		24	1		340	39.50		379.50	440	
	0500	40″ diameter		23	1.043		380	41		421	485	
	0520	42″ diameter		22	1.091		400	43		443	510	
	0540	44″ diameter		21	1.143		440	45		485	555	
	0560	46″ diameter		20	1.200		485	47		532	610	
	0580	48″ diameter	↓	19	1.263	↓	540	49.50		589.50	670	
	0600	For 4″, 5″ and 6″ oval, add					50%					
	0650	Gas, double wall, galvanized steel, fittings										
	0660	Elbow 45°, 3″ diameter	Q-9	36	.444	Ea.	9.80	16.85		26.65	38	
	0670	4″ diameter		34	.471		11.60	17.85		29.45	41.50	
	0680	5″ diameter		32	.500		13.60	18.95		32.55	45	
	0690	6″ diameter		30	.533		16.30	20		36.30	50	
	0700	7″ diameter		28	.571		23.50	21.50		45	60.50	
	0710	8″ diameter		26	.615		27	23.50		50.50	67	
	0720	10″ diameter		24	.667		67	25.50		92.50	114	
	0730	12″ diameter		22	.727		85	27.50		112.50	138	
	0740	14″ diameter		21	.762		119	29		148	177	
	0750	16″ diameter		20	.800		149	30.50		179.50	213	
	0760	18″ diameter	↓	19	.842		201	32		233	272	
	0770	20″ diameter	Q-10	18	1.333		227	52.50		279.50	335	
	0780	22″ diameter		17	1.412		340	55.50		395.50	465	
	0790	24″ diameter		16	1.500		435	59		494	570	
	0800	26″ diameter		16	1.500		855	59		914	1,025	
	0810	28″ diameter		15	1.600		875	63		938	1,075	
	0820	30″ diameter		14	1.714		925	67.50		992.50	1,125	
	0830	32″ diameter		14	1.714		1,050	67.50		1,117.50	1,250	
	0840	34″ diameter		13	1.846		1,150	72.50		1,222.50	1,375	
	0850	36″ diameter		12	2		1,175	78.50		1,253.50	1,425	
	0860	38″ diameter		12	2		1,275	78.50		1,353.50	1,525	
	0870	40″ diameter		11	2.182		1,325	86		1,411	1,600	
	0880	42″ diameter		11	2.182		1,425	86		1,511	1,700	
	0890	44″ diameter		10	2.400		1,500	94.50		1,594.50	1,800	
	0900	46″ diameter		10	2.400		1,600	94.50		1,694.50	1,900	
	0910	48″ diameter	↓	10	2.400		1,675	94.50		1,769.50	1,975	
	0950	Elbow 90°, adjustable, 3″ diameter	Q-9	36	.444		16.35	16.85		33.20	45	
	0960	4″ diameter		34	.471		19.90	17.85		37.75	50.50	
	0970	5″ diameter		32	.500		23	18.95		41.95	55.50	
	0980	6″ diameter		30	.533		28	20		48	62.50	
	0990	7″ diameter		28	.571		44	21.50		65.50	82.50	
	1010	8″ diameter		26	.615		46.50	23.50		70	88.50	
	1020	Wall thimble, 4 to 7″ adjustable, 3″ diameter		36	.444		10.35	16.85		27.20	38.50	
	1022	4″ diameter		34	.471		11.55	17.85		29.40	41	
	1024	5″ diameter		32	.500		12.95	18.95		31.90	44.50	
	1026	6″ diameter	↓	30	.533	↓	15.85	20		35.85	49.50	

15550 | Breechings, Chimneys & Stacks

		Description	CREW	DAILY OUTPUT	LABOR-HOURS	UNIT	2006 BARE COSTS MAT.	LABOR	EQUIP.	TOTAL	TOTAL INCL O&P	
440	1028	7″ diameter	Q-9	28	.571	Ea.	18.45	21.50		39.95	55	440
	1030	8″ diameter		26	.615		19.10	23.50		42.60	58	
	1040	Roof flashing, 3″ diameter		36	.444		5.90	16.85		22.75	33.50	
	1050	4″ diameter		34	.471		6.65	17.85		24.50	36	
	1060	5″ diameter		32	.500		10.35	18.95		29.30	41.50	
	1070	6″ diameter		30	.533		14.50	20		34.50	48	
	1080	7″ diameter		28	.571		18.30	21.50		39.80	54.50	
	1090	8″ diameter		26	.615		19.85	23.50		43.35	59	
	1100	10″ diameter		24	.667		54	25.50		79.50	99.50	
	1110	12″ diameter		22	.727		69	27.50		96.50	120	
	1120	14″ diameter		20	.800		85	30.50		115.50	142	
	1130	16″ diameter		18	.889		105	33.50		138.50	170	
	1140	18″ diameter	↓	16	1		142	38		180	218	
	1150	20″ diameter	Q-10	18	1.333		185	52.50		237.50	287	
	1160	22″ diameter		14	1.714		234	67.50		301.50	365	
	1170	24″ diameter	↓	12	2		281	78.50		359.50	435	
	1200	Tee, 3″ diameter	Q-9	27	.593		24	22.50		46.50	62.50	
	1210	4″ diameter		26	.615		25	23.50		48.50	64.50	
	1220	5″ diameter		25	.640		27.50	24.50		52	68.50	
	1230	6″ diameter		24	.667		30	25.50		55.50	73.50	
	1240	7″ diameter		23	.696		32.50	26.50		59	77.50	
	1250	8″ diameter		22	.727		44.50	27.50		72	93	
	1260	10″ diameter		21	.762		121	29		150	179	
	1270	12″ diameter		20	.800		145	30.50		175.50	208	
	1280	14″ diameter		18	.889		236	33.50		269.50	315	
	1290	16″ diameter		16	1		335	38		373	430	
	1300	18″ diameter	↓	14	1.143		420	43.50		463.50	530	
	1310	20″ diameter	Q-10	17	1.412		570	55.50		625.50	715	
	1320	22″ diameter		13	1.846		705	72.50		777.50	890	
	1330	24″ diameter		12	2		825	78.50		903.50	1,025	
	1340	26″ diameter		12	2		1,175	78.50		1,253.50	1,425	
	1350	28″ diameter		11	2.182		1,325	86		1,411	1,575	
	1360	30″ diameter		10	2.400		1,425	94.50		1,519.50	1,725	
	1370	32″ diameter		10	2.400		1,675	94.50		1,769.50	1,975	
	1380	34″ diameter		9	2.667		1,675	105		1,780	2,025	
	1390	36″ diameter		9	2.667		2,000	105		2,105	2,375	
	1400	38″ diameter		8	3		2,000	118		2,118	2,400	
	1410	40″ diameter		7	3.429		2,275	135		2,410	2,725	
	1420	42″ diameter		6	4		2,400	157		2,557	2,875	
	1430	44″ diameter		5	4.800		2,600	189		2,789	3,175	
	1440	46″ diameter		5	4.800		2,850	189		3,039	3,450	
	1450	48″ diameter	↓	5	4.800		3,050	189		3,239	3,650	
	1460	Tee cap, 3″ diameter	Q-9	45	.356		1.71	13.50		15.21	23.50	
	1470	4″ diameter		42	.381		1.94	14.45		16.39	25	
	1490	6″ diameter		37	.432		3	16.40		19.40	29.50	
	1510	8″ diameter		34	.471		5.70	17.85		23.55	35	
	1530	12″ diameter		30	.533		14.55	20		34.55	48	
	1550	16″ diameter	↓	25	.640		34.50	24.50		59	76.50	
	1570	20″ diameter	Q-10	27	.889		42	35		77	102	
	1590	24″ diameter		21	1.143		54.50	45		99.50	131	
	1610	28″ diameter		19	1.263		111	49.50		160.50	201	
	1630	32″ diameter		17	1.412		149	55.50		204.50	253	
	1650	36″ diameter		15	1.600		185	63		248	305	
	1670	40″ diameter		14	1.714		219	67.50		286.50	350	
	1690	44″ diameter		12	2		252	78.50		330.50	405	
	1710	48″ diameter	↓	11	2.182	↓	300	86		386	465	

	15550 | Breechings, Chimneys & Stacks		CREW	DAILY OUTPUT	LABOR-HOURS	UNIT	2006 BARE COSTS MAT.	LABOR	EQUIP.	TOTAL	TOTAL INCL O&P	
440	1750	Top, 3″ diameter	Q-9	46	.348	Ea.	9.10	13.20		22.30	31	440
	1760	4″ diameter		44	.364		9.50	13.80		23.30	32.50	
	1780	6″ diameter		40	.400		16.60	15.15		31.75	42.50	
	1800	8″ diameter		36	.444		32	16.85		48.85	62	
	1820	12″ diameter		32	.500		113	18.95		131.95	154	
	1840	16″ diameter		28	.571		232	21.50		253.50	290	
	1860	20″ diameter	Q-10	28	.857		455	33.50		488.50	555	
	1880	24″ diameter		20	1.200		770	47		817	920	
	1882	26″ diameter		19	1.263		1,175	49.50		1,224.50	1,350	
	1884	28″ diameter		18	1.333		1,325	52.50		1,377.50	1,525	
	1886	30″ diameter		17	1.412		1,500	55.50		1,555.50	1,725	
	1888	32″ diameter		16	1.500		1,575	59		1,634	1,825	
	1890	34″ diameter		15	1.600		1,675	63		1,738	1,950	
	1892	36″ diameter		14	1.714		1,800	67.50		1,867.50	2,075	
	3200	All fuel, pressure tight, double wall, U.L. listed, 1400° F.										
	3210	304 stainless steel liner, aluminized steel outer jacket										
	3220	6″ diameter	Q-9	60	.267	L.F.	40	10.10		50.10	60	
	3221	8″ diameter		52	.308		49	11.65		60.65	72.50	
	3222	10″ diameter		48	.333		54.50	12.65		67.15	80	
	3223	12″ diameter		44	.364		62.50	13.80		76.30	90.50	
	3224	14″ diameter		42	.381		71.50	14.45		85.95	102	
	3225	16″ diameter		40	.400		75.50	15.15		90.65	108	
	3226	18″ diameter		38	.421		85.50	15.95		101.45	120	
	3227	20″ diameter	Q-10	36	.667		98	26		124	150	
	3228	24″ diameter		32	.750		126	29.50		155.50	185	
	3229	28″ diameter		30	.800		144	31.50		175.50	209	
	3230	32″ diameter		27	.889		157	35		192	229	
	3231	36″ diameter		25	.960		185	38		223	263	
	3232	42″ diameter		22	1.091		212	43		255	300	
	3233	48″ diameter		19	1.263		248	49.50		297.50	350	
	3260	For 316 stainless steel liner add					6%					
	3280	All fuel, pressure tight, double wall fittings										
	3284	304 stainless steel inner, aluminized steel jacket										
	3288	Adjustable 18″/30″ section										
	3291	5″ diameter	Q-9	32	.500	Ea.	103	18.95		121.95	143	
	3292	6″ diameter		30	.533		130	20		150	175	
	3293	8″ diameter		26	.615		149	23.50		172.50	201	
	3294	10″ diameter		24	.667		167	25.50		192.50	224	
	3295	12″ diameter		22	.727		191	27.50		218.50	254	
	3296	14″ diameter		21	.762		221	29		250	289	
	3297	16″ diameter		20	.800		229	30.50		259.50	300	
	3298	18″ diameter		19	.842		283	32		315	360	
	3299	20″ diameter	Q-10	18	1.333		325	52.50		377.50	445	
	3300	24″ diameter		16	1.500		420	59		479	560	
	3301	28″ diameter		15	1.600		470	63		533	615	
	3302	32″ diameter		14	1.714		550	67.50		617.50	710	
	3303	36″ diameter		12	2		550	78.50		628.50	730	
	3304	42″ diameter		11	2.182		630	86		716	830	
	3305	48″ diameter		10	2.400		740	94.50		834.50	965	
	3326	For 316 stainless steel liner, add					15%					
	3350	Elbow 90° fixed										
	3353	5″ diameter	Q-9	32	.500	Ea.	192	18.95		210.95	241	
	3354	6″ diameter		30	.533		226	20		246	280	
	3355	8″ diameter		26	.615		249	23.50		272.50	310	
	3356	10″ diameter		24	.667		286	25.50		311.50	355	
	3357	12″ diameter		22	.727		325	27.50		352.50	400	

15550 | Breechings, Chimneys & Stacks		Crew	Daily Output	Labor-Hours	Unit	2006 Bare Costs Mat.	Labor	Equip.	Total	Total Incl O&P
440 3358	14" diameter	Q-9	21	.762	Ea.	380	29		409	460
3359	16" diameter		20	.800		425	30.50		455.50	515
3360	18" diameter		19	.842		495	32		527	590
3361	20" diameter	Q-10	18	1.333		550	52.50		602.50	690
3362	24" diameter		16	1.500		680	59		739	845
3363	28" diameter		15	1.600		850	63		913	1,025
3364	32" diameter		14	1.714		1,125	67.50		1,192.50	1,350
3365	36" diameter		12	2		1,350	78.50		1,428.50	1,600
3366	42" diameter		11	2.182		2,025	86		2,111	2,350
3367	48" diameter		10	2.400		2,550	94.50		2,644.50	2,975
3380	For 316 stainless steel liner, add					20%				
3400	Elbow 45°									
3404	6" diameter	Q-9	30	.533	Ea.	142	20		162	188
3405	8" diameter		26	.615		161	23.50		184.50	215
3406	10" diameter		24	.667		181	25.50		206.50	239
3407	12" diameter		22	.727		206	27.50		233.50	271
3408	14" diameter		21	.762		237	29		266	305
3409	16" diameter		20	.800		267	30.50		297.50	345
3410	18" diameter		19	.842		300	32		332	385
3411	20" diameter	Q-10	18	1.333		345	52.50		397.50	465
3412	24" diameter		16	1.500		445	59		504	585
3413	28" diameter		15	1.600		505	63		568	655
3414	32" diameter		14	1.714		675	67.50		742.50	845
3415	36" diameter		12	2		950	78.50		1,028.50	1,175
3416	42" diameter		11	2.182		1,300	86		1,386	1,550
3417	48" diameter		10	2.400		1,750	94.50		1,844.50	2,075
3430	For 316 stainless steel liner, add					20%				
3450	Tee 90°									
3454	6" diameter	Q-9	24	.667	Ea.	149	25.50		174.50	204
3455	8" diameter		22	.727		171	27.50		198.50	232
3456	10" diameter		21	.762		200	29		229	266
3457	12" diameter		20	.800		233	30.50		263.50	305
3458	14" diameter		18	.889		276	33.50		309.50	360
3459	16" diameter		16	1		310	38		348	400
3460	18" diameter		14	1.143		365	43.50		408.50	470
3461	20" diameter	Q-10	17	1.412		410	55.50		465.50	540
3462	24" diameter		12	2		505	78.50		583.50	685
3463	28" diameter		11	2.182		635	86		721	830
3464	32" diameter		10	2.400		870	94.50		964.50	1,100
3465	36" diameter		9	2.667		1,000	105		1,105	1,300
3466	42" diameter		6	4		1,625	157		1,782	2,025
3467	48" diameter		5	4.800		2,050	189		2,239	2,550
3480	For Tee Cap, add					35%	20%			
3500	For 316 stainless steel liner, add					22%				
3520	Plate support, galvanized									
3524	6" diameter	Q-9	26	.615	Ea.	81	23.50		104.50	126
3525	8" diameter		22	.727		92	27.50		119.50	145
3526	10" diameter		20	.800		100	30.50		130.50	159
3527	12" diameter		18	.889		104	33.50		137.50	168
3528	14" diameter		17	.941		127	35.50		162.50	196
3529	16" diameter		16	1		133	38		171	208
3530	18" diameter		15	1.067		140	40.50		180.50	219
3531	20" diameter	Q-10	16	1.500		149	59		208	258
3532	24" diameter		14	1.714		157	67.50		224.50	279
3533	28" diameter		13	1.846		192	72.50		264.50	325
3534	32" diameter		12	2		250	78.50		328.50	400

		15550 | Breechings, Chimneys & Stacks	CREW	DAILY OUTPUT	LABOR-HOURS	UNIT	2006 BARE COSTS MAT.	LABOR	EQUIP.	TOTAL	TOTAL INCL O&P	
440	3535	36″ diameter	Q-10	10	2.400	Ea.	310	94.50		404.50	495	440
	3536	42″ diameter		9	2.667		550	105		655	770	
	3537	48″ diameter		8	3		665	118		783	920	
	3570	Bellows, lined, 316 stainless steel only										
	3574	6″ diameter	Q-9	30	.533	Ea.	340	20		360	405	
	3575	8″ diameter		26	.615		370	23.50		393.50	445	
	3576	10″ diameter		24	.667		410	25.50		435.50	490	
	3577	12″ diameter		22	.727		450	27.50		477.50	545	
	3578	14″ diameter		21	.762		510	29		539	605	
	3579	16″ diameter		20	.800		565	30.50		595.50	670	
	3580	18″ diameter		19	.842		630	32		662	740	
	3581	20″ diameter	Q-10	18	1.333		885	52.50		937.50	1,050	
	3582	24″ diameter		16	1.500		890	59		949	1,075	
	3583	28″ diameter		15	1.600		1,000	63		1,063	1,200	
	3584	32″ diameter		14	1.714		1,150	67.50		1,217.50	1,350	
	3585	36″ diameter		12	2		1,325	78.50		1,403.50	1,575	
	3586	42″ diameter		11	2.182		1,975	86		2,061	2,300	
	3587	48″ diameter		10	2.400		2,050	94.50		2,144.50	2,400	
	3600	Ventilated roof thimble, 304 stainless steel										
	3620	6″ diameter	Q-9	26	.615	Ea.	200	23.50		223.50	257	
	3624	8″ diameter		22	.727		214	27.50		241.50	280	
	3625	10″ diameter		20	.800		232	30.50		262.50	305	
	3626	12″ diameter		18	.889		250	33.50		283.50	330	
	3627	14″ diameter		17	.941		269	35.50		304.50	355	
	3628	16″ diameter		16	1		286	38		324	375	
	3629	18″ diameter		15	1.067		310	40.50		350.50	405	
	3630	20″ diameter	Q-10	16	1.500		330	59		389	460	
	3631	24″ diameter		14	1.714		370	67.50		437.50	515	
	3632	28″ diameter		13	1.846		425	72.50		497.50	585	
	3633	32″ diameter		12	2		480	78.50		558.50	650	
	3634	36″ diameter		10	2.400		550	94.50		644.50	755	
	3635	42″ diameter		9	2.667		730	105		835	970	
	3636	48″ diameter		8	3		920	118		1,038	1,225	
	3650	For 316 stainless steel, add					10%					
	3670	Exit cone, 316 stainless steel only										
	3674	6″ diameter	Q-9	46	.348	Ea.	141	13.20		154.20	176	
	3675	8″ diameter		42	.381		148	14.45		162.45	186	
	3676	10″ diameter		40	.400		160	15.15		175.15	200	
	3677	12″ diameter		38	.421		164	15.95		179.95	206	
	3678	14″ diameter		37	.432		186	16.40		202.40	231	
	3679	16″ diameter		36	.444		205	16.85		221.85	253	
	3680	18″ diameter		35	.457		252	17.35		269.35	305	
	3681	20″ diameter	Q-10	28	.857		300	33.50		333.50	385	
	3682	24″ diameter		26	.923		390	36.50		426.50	485	
	3683	28″ diameter		25	.960		470	38		508	575	
	3684	32″ diameter		24	1		585	39.50		624.50	705	
	3685	36″ diameter		22	1.091		700	43		743	840	
	3686	42″ diameter		21	1.143		895	45		940	1,050	
	3687	48″ diameter		20	1.200		1,150	47		1,197	1,350	
	3720	Roof support assy., incl. 30″ pipe sect, 304 stainless steel										
	3724	6″ diameter	Q-9	25	.640	Ea.	230	24.50		254.50	292	
	3725	8″ diameter		21	.762		251	29		280	325	
	3726	10″ diameter		19	.842		275	32		307	355	
	3727	12″ diameter		17	.941		292	35.50		327.50	375	
	3728	14″ diameter		16	1		330	38		368	420	
	3729	16″ diameter		15	1.067		350	40.50		390.50	450	

MECHANICAL 15

440	15550 | Breechings, Chimneys & Stacks	CREW	DAILY OUTPUT	LABOR-HOURS	UNIT	2006 BARE COSTS MAT.	LABOR	EQUIP.	TOTAL	TOTAL INCL O&P
3730	18" diameter	Q-9	14	1.143	Ea.	375	43.50		418.50	485
3731	20" diameter	Q-10	15	1.600		400	63		463	540
3732	24" diameter		13	1.846		445	72.50		517.50	605
3733	28" diameter		12	2		525	78.50		603.50	700
3734	32" diameter		11	2.182		630	86		716	825
3735	36" diameter		9	2.667		760	105		865	1,000
3736	42" diameter		8	3		1,200	118		1,318	1,500
3737	48" diameter		7	3.429		1,475	135		1,610	1,850
3750	For 316 stainless steel, add					20%				
3770	Stack cap, 316 stainless steel only									
3773	5" diameter	Q-9	48	.333	Ea.	134	12.65		146.65	168
3774	6" diameter		46	.348		158	13.20		171.20	195
3775	8" diameter		42	.381		182	14.45		196.45	223
3776	10" diameter		40	.400		214	15.15		229.15	260
3777	12" diameter		38	.421		248	15.95		263.95	299
3778	14" diameter		37	.432		293	16.40		309.40	345
3779	16" diameter		36	.444		335	16.85		351.85	395
3780	18" diameter		35	.457		375	17.35		392.35	445
3781	20" diameter	Q-10	28	.857		430	33.50		463.50	530
3782	24" diameter		26	.923		520	36.50		556.50	630
3783	28" diameter		25	.960		565	38		603	680
3784	32" diameter		24	1		660	39.50		699.50	790
3785	36" diameter		22	1.091		795	43		838	945
3786	42" diameter		21	1.143		980	45		1,025	1,150
3787	48" diameter		20	1.200		1,150	47		1,197	1,350
3800	Round storm collar									
3806	5" diameter	Q-9	32	.500	Ea.	23.50	18.95		42.45	55.50
7800	All fuel, double wall, stainless steel, 6" diameter		60	.267	V.L.F.	30.50	10.10		40.60	49.50
7802	7" diameter		56	.286		39.50	10.85		50.35	61
7804	8" diameter		52	.308		46	11.65		57.65	69.50
7806	10" diameter		48	.333		67.50	12.65		80.15	94.50
7808	12" diameter		44	.364		90.50	13.80		104.30	122
7810	14" diameter		42	.381		119	14.45		133.45	154
8000	All fuel, double wall, stainless steel fittings									
8010	Roof support 6" diameter	Q-9	30	.533	Ea.	89	20		109	130
8020	7" diameter		28	.571		100	21.50		121.50	145
8030	8" diameter		26	.615		112	23.50		135.50	160
8040	10" diameter		24	.667		129	25.50		154.50	182
8050	12" diameter		22	.727		155	27.50		182.50	215
8060	14" diameter		21	.762		197	29		226	263
8100	Elbow 15°, 6" diameter		30	.533		142	20		162	188
8120	7" diameter		28	.571		160	21.50		181.50	211
8140	8" diameter		26	.615		182	23.50		205.50	237
8160	10" diameter		24	.667		238	25.50		263.50	300
8180	12" diameter		22	.727		288	27.50		315.50	360
8200	14" diameter		21	.762		345	29		374	425
8300	Insulated tee with insulated tee cap, 6" diameter		30	.533		134	20		154	180
8340	7" diameter		28	.571		176	21.50		197.50	229
8360	8" diameter		26	.615		199	23.50		222.50	255
8380	10" diameter		24	.667		279	25.50		304.50	345
8400	12" diameter		22	.727		390	27.50		417.50	475
8420	14" diameter		21	.762		515	29		544	610
8500	Joist shield, 6" diameter		30	.533		41	20		61	77
8510	7" diameter		28	.571		44.50	21.50		66	83.50
8520	8" diameter		26	.615		54.50	23.50		78	97
8530	10" diameter		24	.667		74	25.50		99.50	121

15550 | Breechings, Chimneys & Stacks			CREW	DAILY OUTPUT	LABOR-HOURS	UNIT	2006 BARE COSTS MAT.	LABOR	EQUIP.	TOTAL	TOTAL INCL O&P	
440	8540	12″ diameter	Q-9	22	.727	Ea.	92	27.50		119.50	145	440
	8550	14″ diameter		21	.762		114	29		143	172	
	8600	Round top, 6″ diameter		30	.533		45.50	20		65.50	82	
	8620	7″ diameter		28	.571		62	21.50		83.50	103	
	8640	8″ diameter		26	.615		83.50	23.50		107	129	
	8660	10″ diameter		24	.667		150	25.50		175.50	205	
	8680	12″ diameter		22	.727		214	27.50		241.50	279	
	8700	14″ diameter		21	.762		285	29		314	360	
	8800	Adjustable roof flashing, 6″ diameter		30	.533		54	20		74	91.50	
	8820	7″ diameter		28	.571		62	21.50		83.50	103	
	8840	8″ diameter		26	.615		67	23.50		90.50	111	
	8860	10″ diameter		24	.667		86.50	25.50		112	135	
	8880	12″ diameter		22	.727		111	27.50		138.50	166	
	8900	14″ diameter		21	.762		139	29		168	199	
	9000	High temp. (2000° F), steel jacket, acid resist. refractory lining										
	9010	11 ga. galvanized jacket, U.L. listed										
	9020	Straight section, 48″ long, 10″ diameter	Q-10	13.30	1.805	Ea.	365	71		436	520	
	9030	12″ diameter		11.20	2.143		400	84.50		484.50	575	
	9040	18″ diameter		7.40	3.243		560	128		688	820	
	9050	24″ diameter		4.60	5.217		765	205		970	1,175	
	9060	30″ diameter		3.70	6.486		975	255		1,230	1,475	
	9070	36″ diameter		2.70	8.889		1,200	350		1,550	1,875	
	9080	42″ diameter	Q-11	3.10	10.323		1,475	415		1,890	2,275	
	9090	48″ diameter		2.70	11.852		1,750	475		2,225	2,675	
	9100	54″ diameter		2.20	14.545		2,125	585		2,710	3,250	
	9110	60″ diameter		2	16		2,775	645		3,420	4,075	
	9120	Tee section, 10″ diameter	Q-10	4.40	5.455		775	215		990	1,200	
	9130	12″ diameter		3.70	6.486		820	255		1,075	1,300	
	9140	18″ diameter		2.40	10		1,100	395		1,495	1,850	
	9150	24″ diameter		1.50	16		1,525	630		2,155	2,675	
	9160	30″ diameter		1.20	20		2,150	785		2,935	3,625	
	9170	36″ diameter		.80	30		3,625	1,175		4,800	5,875	
	9180	42″ diameter	Q-11	1	32		4,775	1,275		6,050	7,300	
	9190	48″ diameter		.90	35.556		6,175	1,425		7,600	9,075	
	9200	54″ diameter		.70	45.714		9,000	1,825		10,825	12,800	
	9210	60″ diameter		.60	53.333		10,900	2,150		13,050	15,400	
	9220	Cleanout pier section, 10″ diameter	Q-10	3.50	6.857		600	270		870	1,100	
	9230	12″ diameter		2.50	9.600		630	380		1,010	1,300	
	9240	18″ diameter		1.90	12.632		805	495		1,300	1,675	
	9250	24″ diameter		1.30	18.462		1,025	725		1,750	2,275	
	9260	30″ diameter		1	24		1,250	945		2,195	2,875	
	9270	36″ diameter		.75	32		1,475	1,250		2,725	3,625	
	9280	42″ diameter	Q-11	.90	35.556		1,875	1,425		3,300	4,350	
	9290	48″ diameter		.75	42.667		2,250	1,725		3,975	5,200	
	9300	54″ diameter		.60	53.333		2,700	2,150		4,850	6,375	
	9310	60″ diameter		.50	64		3,425	2,575		6,000	7,875	
	9320	For drain, add					53%					
	9330	Elbow, 30° and 45°, 10″ diameter	Q-10	6.60	3.636		510	143		653	790	
	9340	12″ diameter		5.60	4.286		545	169		714	870	
	9350	18″ diameter		3.70	6.486		820	255		1,075	1,300	
	9360	24″ diameter		2.30	10.435		1,225	410		1,635	2,000	
	9370	30″ diameter		1.85	12.973		1,675	510		2,185	2,650	
	9380	36″ diameter		1.30	18.462		2,025	725		2,750	3,400	
	9390	42″ diameter	Q-11	1.60	20		2,775	805		3,580	4,350	
	9400	48″ diameter		1.35	23.704		3,550	950		4,500	5,425	
	9410	54″ diameter		1.10	29.091		4,600	1,175		5,775	6,900	

MECHANICAL 15

15550 | Breechings, Chimneys & Stacks

440	15550 Breechings, Chimneys & Stacks	CREW	DAILY OUTPUT	LABOR-HOURS	UNIT	2006 BARE COSTS MAT.	LABOR	EQUIP.	TOTAL	TOTAL INCL O&P
9420	60″ diameter	Q-11	1	32	Ea.	5,600	1,275		6,875	8,225
9430	For 60° and 90° elbow, add					112%				
9440	End cap, 10″ diameter	Q-10	26	.923		870	36.50		906.50	1,025
9450	12″ diameter		22	1.091		995	43		1,038	1,175
9460	18″ diameter		15	1.600		1,225	63		1,288	1,450
9470	24″ diameter		9	2.667		1,500	105		1,605	1,825
9480	30″ diameter		7.30	3.288		1,875	129		2,004	2,250
9490	36″ diameter		5	4.800		2,150	189		2,339	2,675
9500	42″ diameter	Q-11	6.30	5.079		2,675	204		2,879	3,275
9510	48″ diameter		5.30	6.038		3,150	242		3,392	3,850
9520	54″ diameter		4.40	7.273		3,600	292		3,892	4,450
9530	60″ diameter		3.90	8.205		4,025	330		4,355	4,950
9540	Increaser (1 diameter), 10″ diameter	Q-10	6.60	3.636		430	143		573	705
9550	12″ diameter		5.60	4.286		460	169		629	780
9560	18″ diameter		3.70	6.486		675	255		930	1,150
9570	24″ diameter		2.30	10.435		970	410		1,380	1,725
9580	30″ diameter		1.85	12.973		1,200	510		1,710	2,125
9590	36″ diameter		1.30	18.462		1,325	725		2,050	2,600
9592	42″ diameter	Q-11	1.60	20		1,450	805		2,255	2,875
9594	48″ diameter		1.35	23.704		1,750	950		2,700	3,450
9596	54″ diameter		1.10	29.091		2,125	1,175		3,300	4,175
9600	For expansion joints, add to straight section					7%				
9610	For 1/4″ hot rolled steel jacket, add					157%				
9620	For 2950°F very high temperature, add					89%				
9630	26 ga. aluminized jacket, straight section, 48″ long									
9640	10″ diameter	Q-10	15.30	1.569	V.L.F.	217	61.50		278.50	335
9650	12″ diameter		12.90	1.860		229	73		302	370
9660	18″ diameter		8.50	2.824		335	111		446	540
9670	24″ diameter		5.30	4.528		475	178		653	805
9680	30″ diameter		4.30	5.581		605	220		825	1,025
9690	36″ diameter		3.10	7.742		770	305		1,075	1,325
9700	Accessories (all models)									
9710	Guy band, 10″ diameter	Q-10	32	.750	Ea.	67	29.50		96.50	121
9720	12″ diameter		30	.800		71	31.50		102.50	129
9730	18″ diameter		26	.923		92	36.50		128.50	159
9740	24″ diameter		24	1		136	39.50		175.50	213
9750	30″ diameter		20	1.200		157	47		204	248
9760	36″ diameter		18	1.333		174	52.50		226.50	275
9770	42″ diameter	Q-11	22	1.455		315	58.50		373.50	445
9780	48″ diameter		20	1.600		345	64.50		409.50	480
9790	54″ diameter		16	2		385	80.50		465.50	550
9800	60″ diameter		12	2.667		415	107		522	625
9810	Draw band, galv. stl., 11 gauge, 10″ diameter	Q-10	32	.750		58.50	29.50		88	112
9820	12″ diameter		30	.800		63	31.50		94.50	119
9830	18″ diameter		26	.923		78.50	36.50		115	145
9840	24″ diameter		24	1		99.50	39.50		139	172
9850	30″ diameter		20	1.200		116	47		163	203
9860	36″ diameter		18	1.333		133	52.50		185.50	230
9870	42″ diameter	Q-11	22	1.455		178	58.50		236.50	289
9880	48″ diameter		20	1.600		228	64.50		292.50	355
9890	54″ diameter		16	2		266	80.50		346.50	420
9900	60″ diameter		12	2.667		335	107		442	540
9910	Draw band, aluminized stl., 26 gauge, 10″ diameter	Q-10	32	.750		16.75	29.50		46.25	65.50
9920	12″ diameter		30	.800		16.75	31.50		48.25	68.50
9930	18″ diameter		26	.923		21	36.50		57.50	81
9940	24″ diameter		24	1		25	39.50		64.50	90

	15550 | Breechings, Chimneys & Stacks		CREW	DAILY OUTPUT	LABOR-HOURS	UNIT	2006 BARE COSTS MAT.	LABOR	EQUIP.	TOTAL	TOTAL INCL O&P	
440	9950	30″ diameter	Q-10	20	1.200	Ea.	29.50	47		76.50	107	440
	9960	36″ diameter	↓	18	1.333		33.50	52.50		86	121	
	9962	42″ diameter	Q-11	22	1.455		37.50	58.50		96	135	
	9964	48″ diameter		20	1.600		43	64.50		107.50	149	
	9966	54″ diameter		16	2		46	80.50		126.50	179	
	9968	60″ diameter	↓	12	2.667	↓	50	107		157	226	
	9990	Minimum labor/equipment charge	Q-9	3	5.333	Job		202		202	320	
600	0010	**INDUCED DRAFT FANS**										600
	1000	Breeching installation										
	1800	Hot gas, 600°F, variable pitch pulley and motor										
	1840	6″ diam. inlet, 1/4 H.P., 1 phase, 400 CFM	Q-9	6	2.667	Ea.	1,025	101		1,126	1,275	
	1850	7″ diam. inlet, 1/4 H.P., 1 phase, 800 CFM		5	3.200		1,100	121		1,221	1,400	
	1860	8″ diam. inlet, 1/4 H.P., 1phase, 1120 CFM		4	4		1,600	152		1,752	2,000	
	1870	9″ diam. inlet, 3/4 H.P., 1 phase, 1440 CFM		3.60	4.444		1,875	169		2,044	2,350	
	1880	10″ diam. inlet, 3/4 H.P., 1 phase, 2000 CFM		3.30	4.848		2,125	184		2,309	2,625	
	1900	12″ diam. inlet, 3/4 H.P., 3 phase, 2960 CFM		3	5.333		2,175	202		2,377	2,725	
	1910	14″ diam. inlet, 1 H.P., 3 phase, 4160 CFM		2.60	6.154		2,575	233		2,808	3,200	
	1920	16″ diam. inlet, 2 H.P., 3 phase, 6720 CFM		2.30	6.957		2,800	264		3,064	3,500	
	1940	18″ diam. inlet, 3 H.P., 3 phase, 9120 CFM		2	8		3,375	305		3,680	4,200	
	1950	20″ diam. inlet, 3 H.P., 3 phase, 9760 CFM		1.50	10.667		3,675	405		4,080	4,700	
	1960	22″ diam. inlet, 5 H.P., 3 phase, 13,360 CFM		1	16		4,325	605		4,930	5,725	
	1980	24″ diam. inlet, 7-1/2 H.P., 3 phase, 17,760 CFM	↓	.80	20		5,300	760		6,060	7,050	
	2300	For multi-blade damper at fan inlet, add				↓	20%					
	3600	Chimneytop installation										
	3700	6″ size	1 Shee	8	1	Ea.	555	42		597	675	
	3740	8″ size		7	1.143		570	48		618	700	
	3780	13″ size	↓	6	1.333		765	56		821	935	
	3880	For speed control switch, add					45			45	49.50	
	3920	For thermal fan control, add				↓	45			45	49.50	
	5500	Flue shutter damper for draft control,										
	5510	parallel blades										
	5550	8″ size	Q-9	8	2	Ea.	360	76		436	515	
	5560	9″ size		7.50	2.133		390	81		471	555	
	5570	10″ size		7	2.286		400	86.50		486.50	580	
	5580	12″ size		6.50	2.462		440	93.50		533.50	630	
	5590	14″ size		6	2.667		500	101		601	710	
	5600	16″ size		5.50	2.909		540	110		650	770	
	5610	18″ size		5	3.200		595	121		716	850	
	5620	20″ size		4.50	3.556		655	135		790	940	
	5630	22″ size		4	4		725	152		877	1,025	
	5640	24″ size		3.50	4.571		790	173		963	1,150	
	5650	27″ size		3	5.333		810	202		1,012	1,200	
	5660	30″ size		2.50	6.400		840	243		1,083	1,300	
	5670	32″ size		2	8		880	305		1,185	1,450	
	5680	36″ size	↓	1.50	10.667	↓	950	405		1,355	1,700	

15600 | Refrigeration Equipment

		15610 | Refrigeration Compressors	Crew	Daily Output	Labor-Hours	Unit	2006 Bare Costs Mat.	Labor	Equip.	Total	Total Incl O&P	
400	0010	**ROTARY / RECIPROCATING REFRIGERANT COMPRESSOR**										400
	0100	Refrigeration, hermetic, switches and protective devices										
	0210	1.25 ton	1 Stpi	3	2.667	Ea.	360	115		475	575	
	0220	1.42 ton		3	2.667		370	115		485	585	
	0230	1.68 ton		2.80	2.857		420	123		543	650	
	0240	2.00 ton		2.60	3.077		515	132		647	775	
	0250	2.37 ton		2.50	3.200		625	138		763	900	
	0260	2.67 ton		2.40	3.333		370	144		514	630	
	0270	3.53 ton		2.30	3.478		455	150		605	735	
	0280	4.43 ton		2.20	3.636		645	157		802	955	
	0290	5.08 ton		2.10	3.810		795	164		959	1,125	
	0300	5.22 ton		2	4		795	172		967	1,150	
	0310	7.31 ton	Q-5	3	5.333		1,575	207		1,782	2,050	
	0320	9.95 ton		2.60	6.154		1,725	238		1,963	2,275	
	0330	11.6 ton		2.50	6.400		1,550	248		1,798	2,075	
	0340	15.25 ton		2.30	6.957		2,225	270		2,495	2,875	
	0350	17.7 ton		2.10	7.619		2,225	295		2,520	2,875	
	0990	Refrigeration, recip. hermetic, switches & protective devices										
	1000	10 ton	Q-5	1	16	Ea.	7,325	620		7,945	9,025	
	1100	20 ton	Q-6	.72	33.333		11,000	1,350		12,350	14,200	
	1200	30 ton		.64	37.500		13,200	1,500		14,700	16,900	
	1300	40 ton		.44	54.545		14,300	2,200		16,500	19,100	
	1400	50 ton		.20	120		15,100	4,825		19,925	24,100	
	1500	75 ton	Q-7	.27	118		16,700	4,875		21,575	25,900	
	1600	130 ton	″	.21	152		19,000	6,250		25,250	30,600	
800	0010	**SCROLL REFRIGERANT COMPRESSORS**										800
	1800	Refrigeration, scroll type										
	1810	1.9 ton	1 Stpi	2.70	2.963	Ea.	440	128		568	685	
	1820	2.35 ton		2.50	3.200		505	138		643	770	
	1830	2.82 ton		2.35	3.404		540	147		687	825	
	1840	3.3 ton		2.30	3.478		570	150		720	860	
	1850	3.83 ton		2.25	3.556		645	153		798	945	
	1860	4.1 ton		2.20	3.636		730	157		887	1,050	
	1870	4.8 ton		2.10	3.810		770	164		934	1,100	
	1880	5 ton		2.07	3.865		785	166		951	1,125	
		15620 | Packaged Water Chillers										
100	0010	**ABSORPTION WATER CHILLERS** R262213-27										100
	0020	Steam or hot water, water cooled										
	0050	100 ton	Q-7	.13	240	Ea.	122,500	9,850		132,350	150,500	
	0100	148 ton		.12	258		127,000	10,600		137,600	156,000	
	0200	200 ton		.12	275		141,000	11,300		152,300	173,000	
	0300	354 ton		.11	304		194,000	12,500		206,500	233,000	
	0400	420 ton		.10	323		212,000	13,300		225,300	254,000	
	0500	665 ton		.09	340		297,500	14,000		311,500	349,000	
	0600	750 ton		.09	363		348,000	14,900		362,900	405,500	
	0700	850 ton		.08	385		371,500	15,800		387,300	433,000	
	0800	955 ton		.08	410		397,000	16,800		413,800	462,500	
	0900	1125 ton		.08	421		458,500	17,300		475,800	531,000	
	1000	1250 ton		.07	444		493,000	18,200		511,200	570,500	
	1100	1465 ton		.07	463		584,500	19,000		603,500	672,500	
	1200	1660 ton		.07	477		687,500	19,600		707,100	786,500	
	2000	For two stage unit, add					80%	25%				
	3000	Gas fired, air cooled										
	3180	3 ton	Q-5	1.30	12.308	Ea.	6,625	475		7,100	8,050	

15620 | Packaged Water Chillers

		15620 Packaged Water Chillers	CREW	DAILY OUTPUT	LABOR-HOURS	UNIT	2006 BARE COSTS MAT.	LABOR	EQUIP.	TOTAL	TOTAL INCL O&P	
100	3220	5 ton (R262213-27)	Q-5	.60	26.667	Ea.	8,850	1,025		9,875	11,300	100
	3270	10 ton		.40	40		24,700	1,550		26,250	29,500	
	4000	Water cooled, duplex										
	4130	100 ton	Q-7	.13	246	Ea.	133,000	10,100		143,100	162,000	
	4140	200 ton		.11	283		184,000	11,600		195,600	220,500	
	4150	300 ton		.11	299		213,000	12,300		225,300	253,500	
	4160	400 ton		.10	316		277,000	13,000		290,000	324,500	
	4170	500 ton		.10	329		349,000	13,500		362,500	404,500	
	4180	600 ton		.09	340		412,000	14,000		426,000	474,500	
	4190	700 ton		.09	359		457,000	14,800		471,800	525,500	
	4200	800 ton		.08	380		551,500	15,600		567,100	631,000	
	4210	900 ton		.08	400		620,500	16,400		636,900	708,000	
	4220	1000 ton		.08	421		689,500	17,300		706,800	785,500	
	9000	Minimum labor/equipment charge	Q-5	1	16	Job		620		620	965	
600	0010	**CENTRIFUGAL/SCREW/RECIP. WATER CHILLERS,** W/ standard controls										600
	0020	Centrifugal liquid chiller, water cooled										
	0030	not including water tower (R262213-27)										
	0100	Open drive, 2000 ton	Q-7	.07	477	Ea.	617,500	19,600		637,100	709,500	
	0110	Screw, liquid chiller, air cooled, insulated evaporator										
	0120	130 ton	Q-7	.14	228	Ea.	63,500	9,375		72,875	84,000	
	0124	160 ton		.13	246		78,000	10,100		88,100	101,000	
	0128	180 ton		.13	250		87,500	10,300		97,800	112,500	
	0132	210 ton		.12	258		96,500	10,600		107,100	122,500	
	0136	270 ton		.12	266		110,500	10,900		121,400	138,500	
	0140	320 ton		.12	275		138,500	11,300		149,800	170,000	
	0240	200 ton		.13	251		67,000	10,300		77,300	90,000	
	0274	Centrifugal, packaged unit, water cooled, not incl. tower										
	0280	400 ton	Q-7	.11	283	Ea.	135,500	11,600		147,100	167,000	
	0320	1000 ton		.09	372		352,500	15,300		367,800	411,500	
	0340	1300 ton		.08	410		453,000	16,800		469,800	524,500	
	0360	1500 ton		.08	426		520,000	17,500		537,500	599,000	
	0490	Packaged w/integral air cooled condenser, 15 ton cool		.37	86.486		15,300	3,550		18,850	22,300	
	0500	20 ton cooling		.34	94.955		20,300	3,900		24,200	28,400	
	0510	25 ton cooling		.33	97.859		21,200	4,025		25,225	29,600	
	0515	30 ton cooling		.31	101		22,700	4,175		26,875	31,500	
	0517	35 ton cooling		.31	104		25,400	4,300		29,700	34,600	
	0520	40 ton cooling		.30	108		27,500	4,425		31,925	37,100	
	0528	45 ton cooling		.29	109		29,200	4,500		33,700	39,100	
	0536	50 ton cooling		.28	113		32,000	4,650		36,650	42,400	
	0546	70 ton cooling		.27	119		45,600	4,925		50,525	57,500	
	0554	90 ton cooling		.25	125		53,000	5,175		58,175	66,000	
	0600	100 ton cooling		.25	129		59,000	5,300		64,300	73,000	
	0620	110 ton cooling		.24	132		64,500	5,425		69,925	79,500	
	0630	130 ton cooling		.24	135		71,000	5,550		76,550	87,000	
	0640	150 ton cooling		.23	137		82,500	5,650		88,150	99,500	
	0650	175 ton cooling		.23	140		87,500	5,775		93,275	105,500	
	0654	190 ton cooling		.22	144		99,500	5,925		105,425	118,500	
	0660	210 ton cooling		.22	148		108,500	6,075		114,575	129,000	
	0662	240 ton cooling		.21	151		130,500	6,225		136,725	153,000	
	0664	270 ton cooling		.21	156		147,000	6,400		153,400	171,500	
	0666	300 ton cooling		.20	160		158,500	6,575		165,075	184,500	
	0668	330 ton cooling		.20	164		175,500	6,725		182,225	204,000	
	0670	360 ton cooling		.19	168		199,000	6,900		205,900	229,000	
	0672	390 ton cooling		.19	172		208,500	7,100		215,600	240,500	
	0674	420 ton cooling		.18	177		218,000	7,300		225,300	251,500	
	0680	Scroll water cooled, single compressor, hermetic, tower not incl.										

15620 | Packaged Water Chillers

			CREW	DAILY OUTPUT	LABOR-HOURS	UNIT	2006 BARE COSTS MAT.	LABOR	EQUIP.	TOTAL	TOTAL INCL O&P	
600	0700	2 to 5 ton cooling R262213-27	Q-5	.57	28.021	Ea.	3,175	1,075		4,250	5,175	600
	0720	6 ton cooling		.42	38.005		4,500	1,475		5,975	7,250	
	0740	8 ton cooling		.31	52.117		4,675	2,025		6,700	8,275	
	0760	10 ton cooling	Q-6	.36	67.039		6,300	2,700		9,000	11,100	
	0780	15 ton cooling	"	.33	72.727		7,625	2,925		10,550	12,900	
	0800	20 ton cooling	Q-7	.38	83.990		9,275	3,450		12,725	15,600	
	0820	30 ton cooling		.33	96.096		10,700	3,950		14,650	17,800	
	0840	35 ton cooling		.33	97.859		12,200	4,025		16,225	19,700	
	0980	Water cooled, multiple compr., semi-hermetic, tower not incl.										
	1000	15 ton cooling	Q-6	.36	65.934	Ea.	12,600	2,650		15,250	18,000	
	1020	20 ton cooling	Q-7	.41	78.049		13,800	3,200		17,000	20,100	
	1040	25 ton cooling		.36	89.888		15,500	3,700		19,200	22,800	
	1060	30 ton cooling		.31	101		16,300	4,175		20,475	24,500	
	1080	40 ton cooling		.30	108		18,900	4,425		23,325	27,700	
	1100	50 ton cooling		.28	113		29,200	4,675		33,875	39,500	
	1120	60 ton cooling		.25	125		32,200	5,175		37,375	43,400	
	1130	70 ton cooling		.23	139		43,500	5,700		49,200	57,000	
	1140	80 ton cooling		.21	151		45,500	6,225		51,725	59,500	
	1150	90 ton cooling		.19	164		48,700	6,775		55,475	64,000	
	1160	100 ton cooling		.18	179		52,000	7,375		59,375	69,000	
	1170	110 ton cooling		.17	190		55,000	7,825		62,825	72,500	
	1180	120 ton cooling		.16	196		59,000	8,050		67,050	77,500	
	1200	140 ton cooling		.16	202		69,500	8,300		77,800	89,500	
	1210	160 ton cooling		.15	210		73,500	8,650		82,150	94,000	
	1300	Water cooled, single compressor, semi-hermetic, tower not incl.										
	1320	40 ton cooling,	Q-7	.30	108	Ea.	13,400	4,425		17,825	21,600	
	1340	50 ton cooling		.28	113		17,300	4,675		21,975	26,300	
	1360	60 ton cooling		.25	125		19,500	5,175		24,675	29,500	
	1380	80 ton cooling		.21	151		23,500	6,225		29,725	35,600	
	1450	Water cooled, dual compressors, semi-hermetic, tower not incl.										
	1500	80 ton cooling	Q-7	.14	222	Ea.	25,600	9,125		34,725	42,400	
	1520	100 ton cooling		.14	228		32,000	9,375		41,375	49,800	
	1540	120 ton cooling		.14	231		36,100	9,525		45,625	54,500	
	1560	135 ton cooling, screw compressors		.14	235		48,400	9,650		58,050	68,000	
	1580	150 ton cooling, screw compressors		.13	240		55,500	9,875		65,375	76,500	
	1600	175 ton cooling, screw compressors		.13	244		61,500	10,000		71,500	83,000	
	1620	200 ton cooling, screw compressors		.13	250		67,500	10,300		77,800	90,500	
	1640	225 ton cooling, screw compressors		.13	256		72,500	10,500		83,000	96,500	
	1660	291 ton cooling, screw compressors		.12	260		76,500	10,700		87,200	100,500	
	4000	Packaged chiller, remote air cooled condensers not incl.										
	4020	15 ton cooling	Q-7	.30	108	Ea.	12,600	4,425		17,025	20,800	
	4030	20 ton cooling		.28	115		13,800	4,750		18,550	22,500	
	4040	25 ton cooling		.25	125		15,500	5,175		20,675	25,100	
	4050	30 ton cooling		.24	133		16,300	5,500		21,800	26,600	
	4060	40 ton cooling		.22	144		18,900	5,925		24,825	30,000	
	4070	50 ton cooling		.21	153		23,200	6,300		29,500	35,300	
	4080	60 ton cooling		.20	164		27,000	6,725		33,725	40,200	
	4090	70 ton cooling		.18	173		43,500	7,125		50,625	59,000	
	4100	80 ton cooling		.17	183		45,500	7,550		53,050	61,500	
	4110	90 ton cooling		.17	193		48,700	7,950		56,650	66,000	
	4120	100 ton cooling		.16	203		52,000	8,375		60,375	70,500	
	4130	110 ton cooling		.15	213		55,000	8,750		63,750	74,000	
	4140	120 ton cooling		.14	223		59,000	9,175		68,175	79,500	
	4150	140 ton cooling		.14	233		69,500	9,575		79,075	91,500	
	9000	Minimum labor/equipment charge	Q-6	1	24	Job		965		965	1,500	
600												600

 Important: See the Reference Section for supporting data - Crews, Rental Equipment, City Cost Indexes and Reference Data

15640 | Packaged Cooling Towers

	Line	Description	Crew	Daily Output	Labor-Hours	Unit	2006 Bare Costs Mat.	Labor	Equip.	Total	Total Incl O&P	
400	0010	**COOLING TOWERS** Packaged units (R262213-27)										400
	0070	Galvanized steel										
	0080	Induced draft, crossflow										
	0100	Vertical, belt drive, 61 tons	Q-6	90	.267	TonAC	87	10.70		97.70	112	
	0150	100 ton		100	.240		69	9.65		78.65	91	
	0200	115 ton		109	.220		67	8.85		75.85	88	
	0250	131 ton		120	.200		59.50	8.05		67.55	78	
	0260	162 ton		132	.182		50	7.30		57.30	66.50	
	1000	For higher capacities, use multiples										
	1500	Induced air, double flow										
	1900	Vertical, gear drive, 167 ton	Q-6	126	.190	TonAC	92.50	7.65		100.15	114	
	2000	297 ton		129	.186		66	7.50		73.50	84	
	2100	582 ton		132	.182		54.50	7.30		61.80	71.50	
	2150	849 ton		142	.169		54	6.80		60.80	70	
	2200	1016 ton		150	.160		51.50	6.45		57.95	66.50	
	3000	For higher capacities, use multiples										
	3500	For pumps and piping, add	Q-6	38	.632	TonAC	43.50	25.50		69	87.50	
	4000	For absorption systems, add				"	75%	75%				
	5000	Fiberglass										
	5010	Draw thru										
	5100	60 ton	Q-6	1.50	16	Ea.	3,275	645		3,920	4,600	
	5120	125 ton		.99	24.242		6,700	975		7,675	8,900	
	5140	300 ton		.43	55.814		15,700	2,250		17,950	20,800	
	5160	600 ton		.22	109		28,200	4,375		32,575	37,800	
	5180	1000 ton		.15	160		48,400	6,425		54,825	63,000	
	6000	Stainless steel										
	6010	Induced draft, crossflow, horizontal, belt drive										
	6100	57 ton	Q-6	1.50	16	Ea.	8,775	645		9,420	10,700	
	6120	91 ton		.99	24.242		12,600	975		13,575	15,400	
	6140	111 ton		.43	55.814		14,900	2,250		17,150	19,900	
	6160	126 ton		.22	109		16,100	4,375		20,475	24,500	
	6170	Induced draft, crossflow, vertical, gear drive										
	6172	167 ton	Q-6	.75	32	Ea.	25,900	1,275		27,175	30,500	
	6174	297 ton		.43	55.814		33,300	2,250		35,550	40,100	
	6176	582 ton		.23	104		52,500	4,200		56,700	64,000	
	6178	849 ton		.17	141		84,000	5,675		89,675	101,500	
	6180	1016 ton		.15	160		91,500	6,425		97,925	111,000	
	9000	Minimum labor/equipment charge		1	24	Job		965		965	1,500	

15660 | Liquid Coolers/Evap Condensers

	Line	Description	Crew	Daily Output	Labor-Hours	Unit	2006 Bare Costs Mat.	Labor	Equip.	Total	Total Incl O&P	
100	0010	**CONDENSERS** Ratings are for 30° F TD, R-22 (R262213-27)										100
	0080	Air cooled, belt drive, propeller fan										
	0220	45 ton	Q-6	.70	34.286	Ea.	10,200	1,375		11,575	13,400	
	0240	50 ton		.69	34.985		10,500	1,400		11,900	13,800	
	0260	54 ton		.64	37.795		10,900	1,525		12,425	14,400	
	0280	59 ton		.58	41.308		11,700	1,650		13,350	15,500	
	0300	65 ton		.53	45.541		12,200	1,825		14,025	16,300	
	0320	73 ton		.47	51.173		13,800	2,050		15,850	18,400	
	0340	81 ton		.42	56.738		15,700	2,275		17,975	20,800	
	0360	86 ton		.40	60.302		16,400	2,425		18,825	21,800	
	0380	88 ton		.39	61.697		17,500	2,475		19,975	23,200	
	0400	101 ton	Q-7	.45	70.640		20,600	2,900		23,500	27,200	
	0500	159 ton		.31	102		30,800	4,225		35,025	40,400	
	0600	228 ton		.22	148		44,500	6,075		50,575	58,500	
	0700	314 ton		.16	203		66,500	8,375		74,875	86,500	
	0800	471 ton		.11	283		80,000	11,600		91,600	106,000	

	15660	Liquid Coolers/Evap Condensers	CREW	DAILY OUTPUT	LABOR-HOURS	UNIT	2006 BARE COSTS MAT.	LABOR	EQUIP.	TOTAL	TOTAL INCL O&P	
100	1500	May be specified single or multi-circuit R262213-27										100
	1550	Air cooled, direct drive, propeller fan										
	1590	1 ton	Q-5	3.80	4.211	Ea.	560	163		723	870	
	1600	1-1/2 ton		3.60	4.444		665	172		837	1,000	
	1620	2 ton		3.20	5		735	194		929	1,100	
	1630	3 ton		2.40	6.667		820	258		1,078	1,300	
	1640	5 ton		2	8		1,500	310		1,810	2,125	
	1650	8 ton		1.80	8.889		1,875	345		2,220	2,600	
	1660	10 ton		1.40	11.429		2,300	445		2,745	3,225	
	1670	12 ton		1.30	12.308		2,575	475		3,050	3,600	
	1680	14 ton		1.20	13.333		2,950	515		3,465	4,050	
	1690	16 ton		1.10	14.545		3,250	565		3,815	4,450	
	1700	21 ton		1	16		3,750	620		4,370	5,100	
	1720	26 ton		.84	19.002		4,200	735		4,935	5,775	
	1740	30 ton		.70	22.792		6,975	885		7,860	9,050	
	1760	41 ton	Q-6	.77	31.008		7,700	1,250		8,950	10,400	
	1780	52 ton		.66	36.419		11,000	1,475		12,475	14,400	
	1800	63 ton		.55	44.037		12,100	1,775		13,875	16,200	
	1820	76 ton		.45	52.980		13,800	2,125		15,925	18,500	
	1840	86 ton		.40	60		17,000	2,400		19,400	22,500	
	1860	97 ton		.35	67.989		19,400	2,725		22,125	25,600	
	1880	105 ton	Q-7	.44	73.563		22,300	3,025		25,325	29,300	
	1890	118 ton		.39	82.687		24,100	3,400		27,500	31,800	
	1900	126 ton		.36	88.154		28,000	3,625		31,625	36,400	
	1910	136 ton		.34	95.238		31,300	3,900		35,200	40,500	
	1920	142 ton		.32	99.379		32,100	4,075		36,175	41,700	
	3400	Evaporative, copper coil, pump, fan motor										
	3440	10 ton	Q-5	.54	29.630	Ea.	4,525	1,150		5,675	6,750	
	3460	15 ton		.50	32		4,700	1,250		5,950	7,100	
	3480	20 ton		.47	34.043		4,900	1,325		6,225	7,425	
	3500	25 ton		.45	35.556		5,225	1,375		6,600	7,900	
	3520	30 ton		.42	38.095		5,575	1,475		7,050	8,425	
	3540	40 ton	Q-6	.49	48.980		7,375	1,975		9,350	11,200	
	3560	50 ton		.39	61.538		9,225	2,475		11,700	14,100	
	3580	65 ton		.35	68.571		10,200	2,750		12,950	15,600	
	3600	80 ton		.33	72.727		11,100	2,925		14,025	16,800	
	3620	90 ton		.29	82.759		12,200	3,325		15,525	18,600	
	3640	100 ton	Q-7	.36	88.889		13,300	3,650		16,950	20,400	
	3660	110 ton		.33	96.970		14,700	3,975		18,675	22,400	
	3680	125 ton		.30	106		16,000	4,375		20,375	24,400	
	3700	135 ton		.28	114		17,100	4,700		21,800	26,200	
	3720	150 ton		.25	128		18,900	5,250		24,150	29,000	
	3740	165 ton		.23	139		20,800	5,700		26,500	31,700	
	3760	185 ton		.22	145		21,900	5,975		27,875	33,400	
	3860	For fan damper control, add	Q-5	2	8		470	310		780	995	
	8000	Oils										
	8100	Lubricating										
	8120	Oil, lubricating				Oz.	.17			.17	.19	
	8500	Refrigeration										
	8520	Oil, refrigeration				Gal.	14.50			14.50	15.95	
	8525	Oil, refrigeration				Qt.	3.62			3.62	3.98	

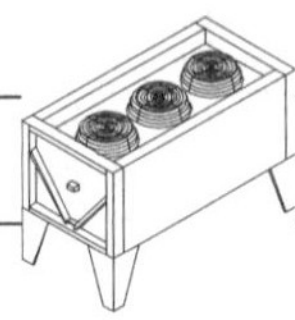

	15670	Refrigerant Condensing Units										
300	0010	CONDENSING UNITS										300
	0030	Air cooled, compressor, standard controls										

15 MECHANICAL

	15670 | Refrigerant Condensing Units		CREW	DAILY OUTPUT	LABOR-HOURS	UNIT	2006 BARE COSTS MAT.	LABOR	EQUIP.	TOTAL	TOTAL INCL O&P	
300	0050	1.5 ton	Q-5	2.50	6.400	Ea.	635	248		883	1,075	300
	0100	2 ton		2.10	7.619		670	295		965	1,200	
	0200	2.5 ton		1.70	9.412		770	365		1,135	1,400	
	0300	3 ton		1.30	12.308		910	475		1,385	1,750	
	0350	3.5 ton		1.10	14.545		1,025	565		1,590	2,000	
	0400	4 ton		.90	17.778		1,175	690		1,865	2,375	
	0500	5 ton		.60	26.667		1,425	1,025		2,450	3,175	
	0550	7.5 ton		.55	29.091		2,800	1,125		3,925	4,825	
	0560	8.5 ton		.53	30.189		3,550	1,175		4,725	5,725	
	0600	10 ton		.50	32		3,850	1,250		5,100	6,175	
	0620	11 ton		.48	33.333		5,950	1,300		7,250	8,550	
	0650	15 ton		.40	40		6,150	1,550		7,700	9,175	
	0700	20 ton	Q-6	.40	60		8,875	2,400		11,275	13,500	
	0720	25 ton		.35	68.571		13,300	2,750		16,050	18,900	
	0750	30 ton		.30	80		15,200	3,225		18,425	21,700	
	0800	40 ton		.20	120		18,400	4,825		23,225	27,700	
	0840	50 ton		.18	133		21,900	5,350		27,250	32,400	
	0860	60 ton		.16	150		25,200	6,025		31,225	37,100	
	0900	75 ton		.14	171		28,800	6,900		35,700	42,300	
	1000	80 ton		.12	200		32,000	8,025		40,025	47,700	
	1100	100 ton		.09	266		44,900	10,700		55,600	66,000	
	2000	Water cooled, compressor, heat exchanger, controls										
	2100	5 ton	Q-5	.70	22.857	Ea.	9,775	885		10,660	12,200	
	2200	15 ton	"	.50	32		15,400	1,250		16,650	18,900	
	2300	20 ton	Q-6	.40	60		17,200	2,400		19,600	22,800	
	2400	40 ton		.20	120		23,400	4,825		28,225	33,200	
	2500	100 ton		.11	218		50,000	8,775		58,775	68,500	
	9000	Minimum labor/equipment charge	Q-5	1.50	10.667	Job		415		415	645	

	15680 | Evaporators, DX Coils		CREW	DAILY OUTPUT	LABOR-HOURS	UNIT	MAT.	LABOR	EQUIP.	TOTAL	TOTAL INCL O&P	
300	0010	**EVAPORATORS** DX coils, remote compressors not included										300
	1000	Coolers, reach-in type, above freezing temperatures										
	1300	Shallow depth, wall mount, 7 fins per inch, air defrost										
	1310	600 BTUH, 8″ fan, rust-proof core	Q-5	3.80	4.211	Ea.	299	163		462	585	
	1320	900 BTUH, 8″ fan, rust-proof core		3.30	4.848		370	188		558	695	
	1330	1200 BTUH, 8″ fan, rust-proof core		3	5.333		435	207		642	800	
	1340	1800 BTUH, 8″ fan, rust-proof core		2.40	6.667		505	258		763	960	
	1350	2500 BTUH, 10″ fan, rust-proof core		2.20	7.273		550	282		832	1,050	
	1360	3500 BTUH, 10″ fan		2	8		645	310		955	1,200	
	1370	4500 BTUH, 10″ fan		1.90	8.421		680	325		1,005	1,250	
	1390	For pan drain, add					39.50			39.50	43.50	
	1600	Undercounter refrigerators, ceiling or wall mount,										
	1610	8 fins per inch, air defrost, rust-proof core										
	1630	800 BTUH, one 6″ fan	Q-5	5	3.200	Ea.	201	124		325	415	
	1640	1300 BTUH, two 6″ fans		4	4		244	155		399	510	
	1650	1700 BTUH, two 6″ fans		3.60	4.444		273	172		445	570	
	2000	Coolers, reach-in and walk-in types, above freezing temperatures										
	2600	Two-way discharge, ceiling mount, 150-4100 CFM,										
	2610	above 34°F applications, air defrost										
	2630	900 BTUH, 7 fins per inch, 8″ fan	Q-5	4	4	Ea.	225	155		380	490	
	2660	2500 BTUH, 7 fins per inch, 10″ fan		2.60	6.154		420	238		658	835	
	2690	5500 BTUH, 7 fins per inch, 12″ fan		1.70	9.412		835	365		1,200	1,475	
	2720	8500 BTUH, 8 fins per inch, 16″ fan		1.20	13.333		1,025	515		1,540	1,950	
	2750	15,000 BTUH, 7 fins per inch, 18″ fan		1.10	14.545		1,450	565		2,015	2,450	
	2770	24,000 BTUH, 7 fins per inch, two 16″ fans		1	16		2,175	620		2,795	3,350	
	2790	30,000 BTUH, 7 fins per inch, two 18″ fans		.90	17.778		2,675	690		3,365	4,025	

15680 | Evaporators, DX Coils

			DAILY	LABOR-		2006 BARE COSTS				TOTAL
		CREW	OUTPUT	HOURS	UNIT	MAT.	LABOR	EQUIP.	TOTAL	INCL O&P
300 2850	Two-way discharge, low profile, ceiling mount,									
2860	8 fins per inch, 200-570 CFM, air defrost									
2880	800 BTUH, one 6" fan	Q-5	4	4	Ea.	228	155		383	490
2890	1300 BTUH, two 6" fans		3.50	4.571		270	177		447	570
2900	1800 BTUH, two 6" fans		3.30	4.848		284	188		472	605
2910	2700 BTUH, three 6" fans		2.70	5.926		335	230		565	720
3000	Coolers, walk-in type, above freezing temperatures									
3300	General use, ceiling mount, 108-2080 CFM									
3320	600 BTUH, 7 fins per inch, 6" fan	Q-5	5.20	3.077	Ea.	191	119		310	395
3340	1200 BTUH, 7 fins per inch, 8" fan		3.70	4.324		263	168		431	550
3360	1800 BTUH, 7 fins per inch, 10" fan		2.70	5.926		370	230		600	760
3380	3500 BTUH, 7 fins per inch, 12" fan		2.40	6.667		545	258		803	995
3400	5500 BTUH, 8 fins per inch, 12" fan		2	8		605	310		915	1,150
3430	8500 BTUH, 8 fins per inch, 16" fan		1.40	11.429		900	445		1,345	1,675
3460	15,000 BTUH, 7 fins per inch, two 16" fans		1.10	14.545		1,450	565		2,015	2,475
3640	Low velocity, high latent load, ceiling mount,									
3650	1050-2420 CFM, air defrost, 6 fins per inch									
3670	6700 BTUH, two 10" fans	Q-5	1.30	12.308	Ea.	1,200	475		1,675	2,050
3680	10,000 BTUH, three 10" fans		1	16		1,375	620		1,995	2,500
3690	13,500 BTUH, three 10" fans		1	16		1,650	620		2,270	2,800
3700	18,000 BTUH, four 10" fans		1	16		2,200	620		2,820	3,400
3710	26,500 BTUH, four 10" fans		.80	20		2,725	775		3,500	4,200
3730	For electric defrost, add					17%				
5000	Freezers and coolers, reach-in type, above 34° F									
5030	to sub-freezing temperature range, low latent load,									
5050	air defrost, 7 fins per inch									
5070	1200 BTUH, 8" fan, rustproof core	Q-5	4.40	3.636	Ea.	249	141		390	495
5080	1500 BTUH, 8" fan, rustproof core		4.20	3.810		256	148		404	510
5090	1800 BTUH, 8" fan, rustproof core		3.70	4.324		305	168		473	595
5100	2500 BTUH, 10" fan, rustproof core		2.90	5.517		345	214		559	710
5110	3500 BTUH, 10" fan, rustproof core		2.50	6.400		420	248		668	845
5120	4500 BTUH, 12" fan, rustproof core		2.10	7.619		560	295		855	1,075
6000	Freezers and coolers, walk-in type									
6050	1960-18,000 CFM, medium profile									
6060	Standard motor, 6 fins per inch, to -30°F, all aluminum									
6080	10,500 BTUH, air defrost, one 18" fan	Q-5	1.40	11.429	Ea.	1,075	445		1,520	1,875
6090	12,500 BTUH, air defrost, one 18" fan		1.40	11.429		1,150	445		1,595	1,975
6100	16,400 BTUH, air defrost, one 18" fan		1.30	12.308		1,175	475		1,650	2,050
6110	20,900 BTUH, air defrost, one 18" fans		1	16		1,450	620		2,070	2,575
6120	27,000 BTUH, air defrost, two 18" fans		1	16		2,050	620		2,670	3,225
6130	32,900 BTUH, air defrost, two 20" fans	Q-6	1.30	18.462		2,300	740		3,040	3,675
6140	39,000 BTUH, air defrost, two 20" fans		1.25	19.200		2,675	770		3,445	4,125
6150	44,100 BTUH, air defrost, three 20" fans		1	24		3,025	965		3,990	4,825
6155	53,000 BTUH, air defrost, three 20" fans		1	24		3,125	965		4,090	4,925
6160	66,200 BTUH, air defrost, four 20" fans		1	24		3,450	965		4,415	5,275
6170	78,000 BTUH, air defrost, four 20" fans		.80	30		4,275	1,200		5,475	6,575
6180	88,200 BTUH, air defrost, four 24" fans		.60	40		4,425	1,600		6,025	7,350
6190	110,000 BTUH, air defrost, four 24" fans		.50	48		5,150	1,925		7,075	8,675
6330	Hot gas defrost, standard motor units, add					35%				
6370	Electric defrost, 230V, standard motor, add					16%				
6410	460V, standard or high capacity motor, add					25%				
6800	Eight fins per inch, increases BTUH 75%									
6810	Standard capacity, add				Ea.	5%				
6830	Hot gas & electric defrost not recommended									
7000	800-4150 CFM, 12" fans									
7010	Four fins per inch									

15600 | Refrigeration Equipment

15680	Evaporators, DX Coils	CREW	DAILY OUTPUT	LABOR-HOURS	UNIT	2006 BARE COSTS MAT.	LABOR	EQUIP.	TOTAL	TOTAL INCL O&P
300 7030	3400 BTUH, 1 fan	Q-5	2.80	5.714	Ea.	375	221		596	760
7040	4200 BTUH, 1 fan		2.40	6.667		390	258		648	830
7050	5300 BTUH, 1 fan		1.90	8.421		440	325		765	990
7060	6800 BTUH, 1 fan		1.70	9.412		495	365		860	1,100
7070	8400 BTUH, 2 fans		1.60	10		645	390		1,035	1,300
7080	10,500 BTUH, 2 fans		1.40	11.429		720	445		1,165	1,475
7090	13,000 BTUH, 3 fans		1.30	12.308		900	475		1,375	1,725
7100	17,000 BTUH, 4 fans		1.20	13.333		1,150	515		1,665	2,050
7110	21,500 BTUH, 5 fans		1	16		1,400	620		2,020	2,525
7160	Six fins per inch increases BTUH 35%, add					5%				
7180	Eight fins per inch increases BTUH 53%,									
7190	not recommended for below 35° F, add				Ea.	8%				
7210	For 12° F temperature differential, add					20%				
7230	For adjustable thermostat control, add					62.50			62.50	68.50
7240	For hot gas defrost, except 8 fin models									
7250	on applications below 35° F, add				Ea.	58%				
7260	For electric defrost, except 8 fin models									
7270	on applications below 35° F, add				Ea.	36%				
8000	Freezers, pass-thru door uprights, walk-in storage									
8300	Low temperature, thin profile, electric defrost									
8310	138-1800 CFM, 5 fins per inch									
8320	900 BTUH, one 6″ fan	Q-5	3.30	4.848	Ea.	282	188		470	600
8340	1500 BTUH, two 6″ fans		2.20	7.273		420	282		702	905
8360	2600 BTUH, three 6″ fans		1.90	8.421		610	325		935	1,175
8370	3300 BTUH, four 6″ fans		1.70	9.412		680	365		1,045	1,300
8380	4400 BTUH, five 6″ fans		1.40	11.429		950	445		1,395	1,750
8400	7000 BTUH, three 10″ fans		1.30	12.308		1,150	475		1,625	2,025
8410	8700 BTUH, four 10″ fans		1.10	14.545		1,300	565		1,865	2,300
8450	For air defrost, deduct					140			140	154

MECHANICAL 15

15700 | Heating/Ventilating/Air Conditioning Equipment

15710	Heat Exchangers	CREW	DAILY OUTPUT	LABOR-HOURS	UNIT	2006 BARE COSTS MAT.	LABOR	EQUIP.	TOTAL	TOTAL INCL O&P
900 0010	HEAT EXCHANGERS									
0016	Shell & tube type, 2 or 4 pass, 3/4″ O.D. copper tubes,									
0020	C.I. heads, C.I. tube sheet, steel shell									
0100	Hot water 40° F to 180° F, by steam at 10 PSI									
0120	8 GPM	Q-5	6	2.667	Ea.	1,175	103		1,278	1,425
0140	10 GPM		5	3.200		1,750	124		1,874	2,125
0160	40 GPM		4	4		2,725	155		2,880	3,250
0180	64 GPM		2	8		4,175	310		4,485	5,075
0200	96 GPM		1	16		5,575	620		6,195	7,125
0220	120 GPM	Q-6	1.50	16		7,325	645		7,970	9,050
0240	168 GPM		1	24		9,000	965		9,965	11,400
0260	240 GPM		.80	30		14,000	1,200		15,200	17,300
0300	600 GPM		.70	34.286		30,100	1,375		31,475	35,300
0500	For bronze head and tube sheet, add					50%				
1000	Hot water 40° F to 140° F, by water at 200° F									
1020	7 GPM	Q-5	6	2.667	Ea.	1,425	103		1,528	1,725

15710 | Heat Exchangers

	Line	Description	Crew	Daily Output	Labor-Hours	Unit	Mat. (2006 Bare Costs)	Labor (2006 Bare Costs)	Equip. (2006 Bare Costs)	Total (2006 Bare Costs)	Total Incl O&P	
900	1040	16 GPM	Q-5	5	3.200	Ea.	2,025	124		2,149	2,425	900
	1060	34 GPM		4	4		3,075	155		3,230	3,625	
	1080	55 GPM		3	5.333		4,450	207		4,657	5,225	
	1100	74 GPM		1.50	10.667		5,575	415		5,990	6,775	
	1120	86 GPM		1.40	11.429		7,450	445		7,895	8,900	
	1140	112 GPM	Q-6	2	12		9,275	480		9,755	11,000	
	1160	126 GPM		1.80	13.333		11,600	535		12,135	13,500	
	1180	152 GPM		1	24		15,100	965		16,065	18,100	
	3000	Plate type,										
	3100	400 GPM	Q-6	.80	30	Ea.	25,400	1,200		26,600	29,800	
	3120	800 GPM	"	.50	48		43,700	1,925		45,625	51,000	
	3140	1200 GPM	Q-7	.34	94.118		65,000	3,850		68,850	77,500	
	3160	1800 GPM	"	.24	133		86,500	5,475		91,975	103,500	

15720 | Air Handling Units

	Line	Description	Crew	Daily Output	Labor-Hours	Unit	Mat. (2006 Bare Costs)	Labor (2006 Bare Costs)	Equip. (2006 Bare Costs)	Total (2006 Bare Costs)	Total Incl O&P	
200	0010	**CENTRAL STATION AIR-HANDLING UNIT** Chilled water (R262213-27)										200
	3000	Packaged, 3000 CFM, 7.5 ton	Q-6	1.70	14.118	Ea.	2,175	565		2,740	3,275	
	3050	3200 CFM, 8 ton		1.60	15		2,450	605		3,055	3,625	
	3100	4000 CFM, 10 ton		1.40	17.143		2,575	690		3,265	3,925	
	3150	4400 CFM, 11 ton		1.35	17.778		3,075	715		3,790	4,475	
	3200	6000 CFM, 15 ton		1	24		3,325	965		4,290	5,150	
	3250	7000 CFM, 17.5 ton		.90	26.667		4,025	1,075		5,100	6,100	
	3300	10,000 CFM, 25 ton		.65	36.923		6,125	1,475		7,600	9,050	
	3400	12,000 CFM, 30 ton		.50	48		6,600	1,925		8,525	10,300	
	3500	For hot water heat coil, add					30%	5%				
300	0010	**EVAPORATIVE COOLERS** Ducted, not incl. duct.										300
	0100	Side discharge style, capacities at .25" S.P.										
	0120	1785 CFM, 1/3 HP, 115 V	Q-9	5	3.200	Ea.	395	121		516	630	
	0140	2740 CFM, 1/3 HP, 115 V		4.50	3.556		485	135		620	745	
	0160	3235 CFM, 1/2 HP, 115 V		4	4		495	152		647	785	
	0180	3615 CFM, 1/2 HP, 230 V		3.60	4.444		505	169		674	825	
	0200	4215 CFM, 3/4 HP, 230 V		3.20	5		640	190		830	1,000	
	0220	5255 CFM. 1 HP, 115/230 V		3	5.333		1,050	202		1,252	1,475	
	0240	6090 CFM, 1 HP, 230/460 V		2.80	5.714		1,475	217		1,692	1,975	
	0260	8300 CFM, 1-1/2 HP, 230/460 V		2.60	6.154		1,550	233		1,783	2,075	
	0280	8360 CFM, 1-1/2 HP, 230/460 V		2.20	7.273		1,975	276		2,251	2,625	
	0300	9725 CFM, 2 HP, 230/460 V		1.80	8.889		2,000	335		2,335	2,725	
	0320	11,715 CFM, 3 HP, 230/460 V		1.40	11.429		2,050	435		2,485	2,950	
	0340	14,410 CFM, 5 HP, 230/460 V		1	16		2,175	605		2,780	3,375	
	0400	For two-speed motor, add					5%					
	0500	For down discharge style, add					10%					
	1000	High capacities at .4" S.P.										
	1040	15,000 CFM, 5 HP, 230/460 V	Q-9	.54	29.630	Ea.	5,975	1,125		7,100	8,375	
	1080	20,000 CFM, 10HP, 230/460 V	"	.48	33.333		7,675	1,275		8,950	10,500	
	1100	25,000 CFM, 20 HP, 230/460 V	Q-10	.54	44.444		8,475	1,750		10,225	12,100	
	1140	30,000 CFM, 20 HP, 230/460 V		.46	52.174		9,775	2,050		11,825	14,000	
	1160	35,000 CFM, 20 HP, 230/460 V		.44	54.545		10,700	2,150		12,850	15,100	
	1180	40,000 CFM, 30 HP, 230/460 V		.42	57.143		11,200	2,250		13,450	15,900	
500	0010	**MAKE-UP AIR UNIT** (R233400-10)										500
	0020	Indoor suspension, natural/LP gas, direct fired,										
	0030	standard control. For flue see division 15550-440 (R235000-50)										
	0040	70°F temperature rise, MBH is input										
	0100	2000 CFM, 168 MBH	Q-6	3	8	Ea.	6,950	320		7,270	8,125	
	0120	3000 CFM, 252 MBH		2	12		7,150	480		7,630	8,600	

15720 | Air Handling Units

	Line	Description	Crew	Daily Output	Labor-Hours	Unit	2006 Bare Costs Mat.	Labor	Equip.	Total	Total Incl O&P	
500	0140	4000 CFM, 336 MBH (R233400-10)	Q-6	1.80	13.333	Ea.	7,250	535		7,785	8,800	500
	0160	6000 CFM, 502 MBH		1.50	16		8,750	645		9,395	10,600	
	0180	8000 CFM, 670 MBH (R235000-50)		1.40	17.143		9,575	690		10,265	11,600	
	0200	10,000 CFM, 838 MBH		1.20	20		10,800	805		11,605	13,200	
	0220	12,000 CFM, 1005 MBH		1	24		10,900	965		11,865	13,500	
	0240	14,000 CFM, 1180 MBH		.94	25.532		11,000	1,025		12,025	13,700	
	0260	18,000 CFM, 1340 MBH		.86	27.907		11,100	1,125		12,225	14,000	
	0280	20,000 CFM, 1675 MBH		.76	31.579		13,700	1,275		14,975	17,100	
	0300	24,000 CFM, 2007 MBH	Q-7	1	32		14,200	1,325		15,525	17,700	
	0320	30,000 CFM, 2510 MBH		.96	33.333		14,600	1,375		15,975	18,100	
	0340	35,000 CFM, 2930 MBH		.92	34.783		18,300	1,425		19,725	22,400	
	0360	40,000 CFM, 3350 MBH		.88	36.364		18,800	1,500		20,300	23,000	
	0380	45,000 CFM, 3770 MBH		.84	38.095		19,200	1,575		20,775	23,600	
	0400	50,000 CFM, 4180 MBH		.80	40		19,500	1,650		21,150	24,000	
	0420	55,000 CFM, 4600 MBH		.75	42.667		24,900	1,750		26,650	30,000	
	0440	60,000 CFM, 5020 MBH		.70	45.714		25,800	1,875		27,675	31,300	
	0460	65,000 CFM, 5435 MBH		.60	53.333		26,000	2,200		28,200	32,000	
	0480	75,000 CFM, 6275 MBH		.50	64		27,400	2,625		30,025	34,200	
	0600	For discharge louver assembly, add					10%					
	0700	For filters, add					20%					
	0800	For air shut-off damper section, add					10%					
	0900	For vertical unit, add					10%					
	1000	Rooftop unit, natural gas, gravity vent, S.S. exchanger										
	1010	70°F temperature rise, MBH is input										
	1020	995 CFM, 75 MBH	Q-6	4	6	Ea.	4,675	241		4,916	5,500	
	1040	1245 CFM, 95 MBH		3.60	6.667		4,725	268		4,993	5,600	
	1060	1590 CFM, 120 MBH		3.30	7.273		5,075	292		5,367	6,025	
	1080	2090 CFM, 159 MBH		3	8		5,225	320		5,545	6,250	
	1100	2490 CFM, 190 MBH		2.60	9.231		5,975	370		6,345	7,150	
	1120	2985 CFM, 225 MBH		2.30	10.435		6,075	420		6,495	7,325	
	1140	3975 CFM, 300 MBH		1.90	12.632		7,200	510		7,710	8,725	
	1160	4970 CFM, 375 MBH		1.40	17.143		8,325	690		9,015	10,300	
	1180	5950 CFM, 450 MBH		1.20	20		8,800	805		9,605	10,900	
	1200	7940 CFM, 600 MBH		1	24		10,200	965		11,165	12,700	
	1220	9930 CFM, 750 MBH		.80	30		17,400	1,200		18,600	21,100	
	1240	11,900 CFM, 900 MBH		.60	40		19,500	1,600		21,100	23,900	
	1260	15,880 CFM, 1200 MBH		.30	80		23,400	3,225		26,625	30,800	
	1500	For power vent, add					8%					
	1600	For cleanable filters, add				Ea.	136			136	149	
	1700	For electric modulating gas control, add				"	1,450			1,450	1,600	
	1800	For aluminized steel exchanger, deduct					10%					
	9000	Minimum labor/equipment charge	Q-6	2.75	8.727	Job		350		350	545	

15730 | Unitary Air Conditioning Equip

	Line	Description	Crew	Daily Output	Labor-Hours	Unit	2006 Bare Costs Mat.	Labor	Equip.	Total	Total Incl O&P	
200	0010	**COMPUTER ROOM UNITS** (R235000-50)										200
	1000	Air cooled, includes remote condenser but not										
	1020	interconnecting tubing or refrigerant (R236000-30)										
	1080	3 ton	Q-5	.50	32	Ea.	9,000	1,250		10,250	11,800	
	1120	5 ton (R262213-27)		.45	35.556		12,000	1,375		13,375	15,400	
	1160	6 ton		.30	53.333		22,200	2,075		24,275	27,600	
	1200	8 ton		.27	59.259		25,100	2,300		27,400	31,200	
	1240	10 ton		.25	64		26,100	2,475		28,575	32,600	
	1280	15 ton		.22	72.727		29,200	2,825		32,025	36,500	
	1290	18 ton		.20	80		33,200	3,100		36,300	41,300	
	1320	20 ton	Q-6	.29	82.759		34,800	3,325		38,125	43,500	
	1360	23 ton	"	.28	85.714		36,600	3,450		40,050	45,600	

15730 | Unitary Air Conditioning Equip

	Line	Description	Ref.	Crew	Daily Output	Labor-Hours	Unit	2006 Bare Costs Mat.	Labor	Equip.	Total	Total Incl O&P	
200	2200	Chilled water, for connection to	R235000-50										200
	2220	existing chiller system of adequate capacity											
	2260	5 ton	R236000-30	Q-5	.74	21.622	Ea.	10,200	840		11,040	12,500	
	2280	6 ton			.52	30.769		13,500	1,200		14,700	16,700	
	2300	8 ton	R262213-27		.50	32		14,500	1,250		15,750	17,900	
	2320	10 ton			.49	32.653		16,100	1,275		17,375	19,700	
	2330	12 ton			.49	32.990		16,700	1,275		17,975	20,400	
	2360	15 ton			.48	33.333		17,300	1,300		18,600	21,000	
	2400	20 ton			.46	34.783		17,700	1,350		19,050	21,500	
	2440	23 ton			.42	38.095		18,600	1,475		20,075	22,700	
	4000	Glycol system, complete except for interconnecting tubing											
	4060	3 ton		Q-5	.40	40	Ea.	11,000	1,550		12,550	14,500	
	4100	5 ton			.38	42.105		14,400	1,625		16,025	18,300	
	4120	6 ton			.25	64		28,700	2,475		31,175	35,500	
	4140	8 ton			.23	69.565		31,000	2,700		33,700	38,300	
	4160	10 ton			.21	76.190		33,600	2,950		36,550	41,600	
	4200	15 ton		Q-6	.26	92.308		40,800	3,700		44,500	50,500	
	4240	20 ton			.24	100		43,700	4,025		47,725	54,500	
	4280	23 ton			.22	109		47,600	4,375		51,975	59,500	
	8000	Water cooled system, not including condenser,											
	8020	water supply or cooling tower											
	8060	3 ton		Q-5	.62	25.806	Ea.	8,125	1,000		9,125	10,500	
	8100	5 ton			.54	29.630		11,500	1,150		12,650	14,400	
	8120	6 ton			.35	45.714		21,500	1,775		23,275	26,400	
	8140	8 ton			.33	48.485		23,600	1,875		25,475	28,900	
	8160	10 ton			.31	51.613		25,100	2,000		27,100	30,700	
	8200	15 ton			.27	59.259		28,900	2,300		31,200	35,400	
	8240	20 ton		Q-6	.38	63.158		34,100	2,550		36,650	41,500	
	8280	23 ton		″	.34	70.588		36,100	2,825		38,925	44,100	
500	0010	**PACKAGED TERMINAL AIR CONDITIONER** Cabinet, wall sleeve,											500
	0100	louver, electric heat, thermostat, manual changeover, 208 V											
	0200	6,000 BTUH cooling, 8800 BTU heat	R235000-50	Q-5	6	2.667	Ea.	955	103		1,058	1,200	
	0220	9,000 BTUH cooling, 13,900 BTU heat			5	3.200		1,000	124		1,124	1,300	
	0240	12,000 BTUH cooling, 13,900 BTU heat	R262213-27		4	4		1,100	155		1,255	1,475	
	0260	15,000 BTUH cooling, 13,900 BTU heat			3	5.333		1,300	207		1,507	1,775	
	0280	18,000 BTUH cooling, 10 KW heat			2.40	6.667		1,400	258		1,658	1,950	
	0300	24,000 BTUH cooling, 10 KW heat			1.90	8.421		1,425	325		1,750	2,075	
	0320	30,000 BTUH cooling, 10 KW heat			1.40	11.429		1,625	445		2,070	2,500	
	0340	36,000 BTUH cooling, 10 KW heat			1.25	12.800		1,700	495		2,195	2,625	
	0360	42,000 BTUH cooling, 10 KW heat			1	16		2,125	620		2,745	3,300	
	0380	48,000 BTUH cooling, 10 KW heat			.90	17.778		2,300	690		2,990	3,600	
	0500	For hot water coil, increase heat by 10%, add						5%	10%				
	1000	For steam, increase heat output by 30%, add						8%	10%				
600	0010	**ROOF TOP AIR CONDITIONERS** Standard controls, curb, economizer											600
	1000	Single zone, electric cool, gas heat											
	1090	2 ton cooling, 55 MBH heating	R235000-50	Q-5	.93	17.204	Ea.	2,900	665		3,565	4,225	
	1100	3 ton cooling, 60 MBH heating			.70	22.857		2,925	885		3,810	4,600	
	1120	4 ton cooling, 95 MBH heating	R262213-27		.61	26.403		4,250	1,025		5,275	6,275	
	1140	5 ton cooling, 112 MBH heating			.56	28.521		4,650	1,100		5,750	6,850	
	1145	6 ton cooling, 140 MBH heating			.52	30.769		5,450	1,200		6,650	7,850	
	1150	7.5 ton cooling, 170 MBH heating			.50	32.258		6,800	1,250		8,050	9,425	
	1160	10 ton cooling, 200 MBH heating		Q-6	.67	35.982		8,725	1,450		10,175	11,900	
	1170	12.5 ton cooling, 230 MBH heating			.63	37.975		9,950	1,525		11,475	13,300	
	1180	15 ton cooling, 270 MBH heating			.57	42.032		13,700	1,700		15,400	17,700	
	1190	18 ton cooling, 330 MBH heating			.52	45.889		16,300	1,850		18,150	20,800	

15730 | Unitary Air Conditioning Equip

				DAILY	LABOR-		2006 BARE COSTS				TOTAL
			CREW	OUTPUT	HOURS	UNIT	MAT.	LABOR	EQUIP.	TOTAL	INCL O&P
600	1200	20 ton cooling, 360 MBH heating (R235000-50)	Q-7	.67	47.976	Ea.	18,200	1,975		20,175	23,100
	1210	25 ton cooling, 450 MBH heating		.56	57.554		22,400	2,350		24,750	28,400
	1220	30 ton cooling, 540 MBH heating (R262213-27)		.47	68.376		28,300	2,800		31,100	35,500
	1240	40 ton cooling, 675 MBH heating		.35	91.168		36,900	3,750		40,650	46,400
	2000	Multizone, electric cool, gas heat, economizer									
	2100	15 ton cooling, 360 MBH heating	Q-7	.61	52.545	Ea.	76,000	2,150		78,150	87,500
	2120	20 ton cooling, 360 MBH heating		.53	60.038		87,500	2,475		89,975	100,000
	2140	25 ton cooling, 450 MBH heating		.45	71.910		99,500	2,950		102,450	114,000
	2160	28 ton cooling, 450 MBH heating		.41	79.012		104,000	3,250		107,250	119,500
	2180	30 ton cooling, 540 MBH heating		.37	85.562		113,000	3,500		116,500	129,500
	2200	40 ton cooling, 540 MBH heating		.28	113		118,000	4,675		122,675	137,500
	2210	50 ton cooling, 540 MBH heating		.23	142		129,500	5,825		135,325	151,000
	2220	70 ton cooling, 1500 MBH heating		.16	198		140,000	8,150		148,150	166,500
	2240	80 ton cooling, 1500 MBH heating		.14	228		160,000	9,375		169,375	190,500
	2260	90 ton cooling, 1500 MBH heating		.13	256		164,000	10,500		174,500	196,500
	2280	105 ton cooling, 1500 MBH heating		.11	290		185,500	11,900		197,400	222,500
	2400	For hot water heat coil, deduct					5%				
	2500	For steam heat coil, deduct					2%				
	2600	For electric heat, deduct					3%	5%			
	5000	Single zone, electric cool only									
	5050	3 ton cooling	Q-5	.88	18.203	Ea.	2,900	705		3,605	4,300
	5060	4 ton cooling		.76	21.108		3,250	820		4,070	4,850
	5070	5 ton cooling		.70	22.792		3,700	885		4,585	5,450
	5080	6 ton cooling		.65	24.502		4,225	950		5,175	6,100
	5090	7.5 ton cooling		.62	25.806		5,500	1,000		6,500	7,600
	5100	8.5 ton cooling		.59	26.981		6,075	1,050		7,125	8,300
	5110	10 ton cooling	Q-6	.83	28.812		6,850	1,150		8,000	9,350
	5120	12 ton cooling		.80	30		8,525	1,200		9,725	11,300
	5130	15 ton cooling		.71	33.613		11,700	1,350		13,050	15,000
	5140	18 ton cooling		.65	36.697		14,000	1,475		15,475	17,700
	5150	20 ton cooling	Q-7	.83	38.415		16,200	1,575		17,775	20,400
	5160	25 ton cooling		.70	45.977		19,900	1,875		21,775	24,800
	5170	30 ton cooling		.59	54.701		24,900	2,250		27,150	30,900
	5180	40 ton cooling		.44	73.059		32,600	3,000		35,600	40,600
	5400	For low heat, add					7%				
	5410	For high heat, add					10%				
	6000	Single zone electric cooling with variable volume distribution									
	6020	20 ton cooling	Q-7	.72	44.199	Ea.	20,100	1,825		21,925	24,900
	6030	25 ton cooling		.61	52.893		23,800	2,175		25,975	29,600
	6040	30 ton cooling		.51	62.868		28,700	2,575		31,275	35,600
	6050	40 ton cooling		.38	83.990		36,300	3,450		39,750	45,400
	6060	50 ton cooling		.31	104		40,400	4,300		44,700	51,000
	6070	60 ton cooling		.25	125		45,200	5,175		50,375	57,500
	6200	For low heat, add					7%				
	6210	For high heat, add					10%				
	7000	Multizone, cool/heat, variable volume distribution									
	7100	50 ton cooling	Q-7	.20	164	Ea.	82,500	6,725		89,225	101,000
	7110	70 ton cooling		.14	228		115,000	9,375		124,375	141,000
	7120	90 ton cooling		.11	296		132,500	12,200		144,700	164,500
	7130	105 ton cooling		.10	333		148,000	13,700		161,700	184,500
	7140	120 ton cooling		.08	380		169,000	15,600		184,600	210,500
	7150	140 ton cooling		.07	444		181,500	18,200		199,700	228,000
	7300	Penthouse unit, cool/heat, variable volume distribution									
	7400	150 ton cooling	Q-7	.10	310	Ea.	265,000	12,700		277,700	311,000
	7410	170 ton cooling		.10	320		288,000	13,100		301,100	337,000
	7420	200 ton cooling		.10	326		338,500	13,400		351,900	393,500

		15730 Unitary Air Conditioning Equip	CREW	DAILY OUTPUT	LABOR-HOURS	UNIT	2006 BARE COSTS MAT.	LABOR	EQUIP.	TOTAL	TOTAL INCL O&P	
600	7430	225 ton cooling R235000-50	Q-7	.09	340	Ea.	370,500	14,000		384,500	429,000	600
	7440	250 ton cooling		.09	359		382,500	14,800		397,300	443,500	
	7450	270 ton cooling R262213-27		.09	363		413,000	14,900		427,900	477,000	
	7460	300 ton cooling		.09	372		458,500	15,300		473,800	528,000	
	9000	Minimum labor/equipment charge	Q-5	1.50	10.667	Job		415		415	645	
800	0010	WINDOW UNIT AIR CONDITIONERS										800
	4000	Portable/window, 15 amp 125V grounded receptacle required										
	4060	5000 BTUH	1 Carp	8	1	Ea.	194	35.50		229.50	272	
	4340	6000 BTUH		8	1		204	35.50		239.50	283	
	4480	8000 BTUH		6	1.333		290	47.50		337.50	400	
	4500	10,000 BTUH		6	1.333		350	47.50		397.50	465	
	4520	12,000 BTUH	L-2	8	2		630	62		692	795	
	4600	Window/thru-the-wall, 15 amp 230V grounded receptacle required										
	4780	17,000 BTUH	L-2	6	2.667	Ea.	615	82.50		697.50	810	
	4940	25,000 BTUH		4	4		790	124		914	1,075	
	4960	29,000 BTUH		4	4		850	124		974	1,150	
	9000	Minimum labor/equipment charge	1 Carp	2	4	Job		142		142	236	
840	0010	SELF-CONTAINED SINGLE PACKAGE R262213-27										840
	0100	Air cooled, for free blow or duct, not incl. remote condenser										
	0200	3 ton cooling	Q-5	1	16	Ea.	2,975	620		3,595	4,250	
	0210	4 ton cooling	"	.80	20		3,225	775		4,000	4,750	
	0220	5 ton cooling	Q-6	1.20	20		3,525	805		4,330	5,125	
	0230	7.5 ton cooling	"	.90	26.667		4,550	1,075		5,625	6,675	
	0240	10 ton cooling	Q-7	1	32		6,025	1,325		7,350	8,675	
	0250	15 ton cooling		.95	33.684		8,375	1,375		9,750	11,400	
	0260	20 ton cooling		.90	35.556		14,100	1,450		15,550	17,800	
	0270	25 ton cooling		.85	37.647		14,300	1,550		15,850	18,100	
	0280	30 ton cooling		.80	40		14,700	1,650		16,350	18,800	
	0300	40 ton cooling		.60	53.333		20,400	2,200		22,600	25,900	
	0320	50 ton cooling		.50	64		25,100	2,625		27,725	31,700	
	0340	60 ton cooling	Q-8	.40	80		29,300	3,275	173	32,748	37,500	
	0490	For duct mounting no price change										
	0500	For steam heating coils, add				Ea.	10%	10%				
	1000	Water cooled for free blow or duct, not including tower										
	1010	Constant volume										
	1100	3 ton cooling	Q-6	1	24	Ea.	2,825	965		3,790	4,600	
	1120	5 ton cooling		1	24		3,650	965		4,615	5,525	
	1130	7.5 ton cooling		.80	30		5,125	1,200		6,325	7,525	
	1140	10 ton cooling	Q-7	.90	35.556		7,225	1,450		8,675	10,200	
	1150	15 ton cooling		.85	37.647		10,400	1,550		11,950	13,900	
	1160	20 ton cooling		.80	40		24,200	1,650		25,850	29,300	
	1170	25 ton cooling		.75	42.667		28,600	1,750		30,350	34,200	
	1180	30 ton cooling		.70	45.714		32,100	1,875		33,975	38,300	
	1200	40 ton cooling		.40	80		40,700	3,275		43,975	49,900	
	1220	50 ton cooling		.30	106		50,500	4,375		54,875	62,500	
	1240	60 ton cooling	Q-8	.30	106		59,000	4,375	230	63,605	72,000	
	1300	For hot water or steam heat coils, add					12%	10%				
	9000	Minimum labor/equipment charge	Q-5	1	16	Job		620		620	965	
900	0010	SPLIT DUCTLESS SYSTEM R262213-27										900
	0100	Cooling only, single zone										
	0110	Wall mount										
	0120	3/4 ton cooling	Q-5	2	8	Ea.	1,000	310		1,310	1,575	
	0130	1 ton cooling		1.80	8.889		1,150	345		1,495	1,775	
	0140	1-1/2 ton cooling		1.60	10		1,400	390		1,790	2,150	

15 MECHANICAL

	15730 | Unitary Air Conditioning Equip		CREW	DAILY OUTPUT	LABOR-HOURS	UNIT	2006 BARE COSTS MAT.	LABOR	EQUIP.	TOTAL	TOTAL INCL O&P	
900	0150	2 ton cooling (R262213-27)	Q-5	1.40	11.429	Ea.	1,800	445		2,245	2,675	900
	1000	Ceiling mount										
	1020	2 ton cooling	Q-5	1.40	11.429	Ea.	1,225	445		1,670	2,050	
	1030	3 ton cooling	"	1.20	13.333	"	3,300	515		3,815	4,425	
	2000	T-Bar mount										
	2010	2 ton cooling	Q-5	1.40	11.429	Ea.	2,450	445		2,895	3,400	
	2020	3 ton cooling	↓	1.20	13.333	↓	2,950	515		3,465	4,050	
	2030	3-1/2 ton cooling	↓	1.10	14.545	↓	3,550	565		4,115	4,775	
	3000	Multizone										
	3010	Wall mount										
	3020	2 @ 3/4 ton cooling	Q-5	1.80	8.889	Ea.	1,200	345		1,545	1,850	
	5000	Cooling / Heating										
	5010	Wall mount										
	5110	1 ton cooling	Q-5	1.70	9.412	Ea.	890	365		1,255	1,550	
	5120	1-1/2 ton cooling	"	1.50	10.667	"	1,425	415		1,840	2,225	
	5300	Ceiling mount										
	5310	3 ton cooling	Q-5	1	16	Ea.	3,825	620		4,445	5,175	
	7000	Accessories for all split ductless systems										
	7010	Add for ambient frost control	Q-5	8	2	Ea.	120	77.50		197.50	253	
	7020	Add for tube / wiring kit										
	7030	15' kit	Q-5	32	.500	Ea.	31.50	19.40		50.90	65	
	7040	35' kit	"	24	.667	"	102	26		128	152	

	15740 | Heat Pumps		CREW	DAILY OUTPUT	LABOR-HOURS	UNIT	MAT.	LABOR	EQUIP.	TOTAL	TOTAL INCL O&P	
100	0010	**AIR-SOURCE HEAT PUMPS** (Not including interconnecting tubing)										100
	1000	Air to air, split system, not including curbs, pads, or ductwork										
	1015	1.5 ton cooling, 7 MBH heat @ 0° F (R235000-50)	Q-5	1.22	13.115	Ea.	1,575	510		2,085	2,525	
	1020	2 ton cooling, 8.5 MBH heat @ 0° F		1.20	13.333		1,625	515		2,140	2,600	
	1030	2.5 ton cooling, 10 MBH heat @ 0° F (R262213-27)		1	16		1,825	620		2,445	2,975	
	1040	3 ton cooling, 13 MBH heat @ 0° F		.80	20		1,975	775		2,750	3,375	
	1050	3.5 ton cooling, 18 MBH heat @ 0° F		.75	21.333		2,275	825		3,100	3,775	
	1054	4 ton cooling, 24 MBH heat @ 0° F		.60	26.667		2,475	1,025		3,500	4,325	
	1060	5 ton cooling, 27 MBH heat @ 0° F		.50	32		2,825	1,250		4,075	5,050	
	1080	7.5 ton cooling, 33 MBH heat @ 0° F	↓	.30	53.333		6,250	2,075		8,325	10,100	
	1100	10 ton cooling, 50 MBH heat @ 0° F	Q-6	.38	63.158		8,375	2,550		10,925	13,200	
	1120	15 ton cooling, 64 MBH heat @ 0° F		.26	92.308		11,900	3,700		15,600	18,900	
	1130	20 ton cooling, 85 MBH heat @ 0° F		.20	120		15,600	4,825		20,425	24,700	
	1140	25 ton cooling, 119 MBH heat @ 0° F	↓	.20	120	↓	18,800	4,825		23,625	28,200	
	1300	Supplementary electric heat coil, included										
	1500	Single package, not including curbs, pads, or plenums										
	1502	1/2 ton cooling, supplementary heat not incl.	Q-5	8	2	Ea.	1,025	77.50		1,102.50	1,250	
	1504	3/4 ton cooling, supplementary heat not incl.		6	2.667		1,100	103		1,203	1,350	
	1506	1 ton cooling, supplementary heat not incl.		4	4		1,275	155		1,430	1,650	
	1510	1.5 ton cooling, 5 MBH heat @ 0° F		1.55	10.323		2,225	400		2,625	3,075	
	1520	2 ton cooling, 6.5 MBH heat @ 0° F		1.50	10.667		2,525	415		2,940	3,425	
	1540	2.5 ton cooling, 8 MBH heat @ 0° F		1.40	11.429		2,750	445		3,195	3,725	
	1560	3 ton cooling, 10 MBH heat @ 0° F		1.20	13.333		3,025	515		3,540	4,125	
	1570	3.5 ton cooling, 11 MBH heat @ 0° F		1	16		3,300	620		3,920	4,600	
	1580	4 ton cooling, 13 MBH heat @ 0° F		.96	16.667		3,550	645		4,195	4,900	
	1620	5 ton cooling, 27 MBH heat @ 0° F		.65	24.615		3,825	955		4,780	5,675	
	1640	7.5 ton cooling, 35 MBH heat @ 0° F	↓	.40	40		5,850	1,550		7,400	8,825	
	1648	10 ton cooling, 45 MBH heat @ 0° F	Q-6	.40	60		7,725	2,400		10,125	12,300	
	1652	12 ton cooling, 50 MBH heat @ 0° F	"	.36	66.667	↓	10,000	2,675		12,675	15,200	
	1696	Supplementary electric heat coil incl., except as noted										
800	0010	**WATER-SOURCE HEAT PUMPS** (Not including interconnecting tubing)										800
	2000	Water source to air, single package										

MECHANICAL 15

15740 | Heat Pumps

			CREW	DAILY OUTPUT	LABOR-HOURS	UNIT	2006 BARE COSTS MAT.	LABOR	EQUIP.	TOTAL	TOTAL INCL O&P	
800	2100	1 ton cooling, 13 MBH heat @ 75° F	Q-5	2	8	Ea.	1,125	310		1,435	1,725	800
	2120	1.5 ton cooling, 17 MBH heat @ 75° F		1.80	8.889		1,250	345		1,595	1,900	
	2140	2 ton cooling, 19 MBH heat @ 75° F		1.70	9.412		1,300	365		1,665	2,000	
	2160	2.5 ton cooling, 25 MBH heat @ 75° F		1.60	10		1,375	390		1,765	2,125	
	2180	3 ton cooling, 27 MBH heat @ 75° F		1.40	11.429		1,475	445		1,920	2,325	
	2190	3.5 ton cooling, 29 MBH heat @ 75° F		1.30	12.308		1,525	475		2,000	2,450	
	2200	4 ton cooling, 31 MBH heat @ 75° F		1.20	13.333		1,700	515		2,215	2,675	
	2220	5 ton cooling, 29 MBH heat @ 75° F		.90	17.778		2,000	690		2,690	3,250	
	2240	7.5 ton cooling, 35 MBH heat @ 75° F		.60	26.667		5,750	1,025		6,775	7,925	
	2250	8.5 ton cooling, 40 MBH heat @ 75° F		.58	27.586		6,200	1,075		7,275	8,475	
	2260	10 ton cooling, 50 MBH heat @ 75° F		.53	30.189		6,275	1,175		7,450	8,725	
	2280	15 ton cooling, 64 MBH heat @ 75° F	Q-6	.47	51.064		10,200	2,050		12,250	14,500	
	2300	20 ton cooling, 100 MBH heat @ 75° F		.41	58.537		11,200	2,350		13,550	16,000	
	2310	25 ton cooling, 100 MBH heat @ 75° F		.32	75		15,200	3,025		18,225	21,400	
	2320	30 ton cooling		.24	102		16,700	4,100		20,800	24,800	
	2340	40 ton cooling		.21	117		23,600	4,700		28,300	33,200	
	2360	50 ton cooling		.15	160		26,600	6,425		33,025	39,300	
	3960	For supplementary heat coil, add					10%					
	4000	For increase in capacity thru use										
	4020	of solar collector, size boiler at 60%										
	9000	Minimum labor/equipment charge	Q-5	1.75	9.143	Job		355		355	550	

15750 | Humidity Control Equipment

			CREW	DAILY OUTPUT	LABOR-HOURS	UNIT	2006 BARE COSTS MAT.	LABOR	EQUIP.	TOTAL	TOTAL INCL O&P	
300	0010	**DEHUMIDIFIERS** R236000-30										300
	6000	Self contained with filters and standard controls										
	6040	1.5 lb/hr, 50 cfm	1 Plum	8	1	Ea.	3,400	42.50		3,442.50	3,825	
	6060	3 lb/hr, 150 cfm	Q-1	12	1.333		4,000	51		4,051	4,475	
	6065	6 lb/hr, 150 cfm		9	1.778		7,200	68.50		7,268.50	8,000	
	6070	16 to 20 lb/hr, 600 cfm		5	3.200		14,300	123		14,423	15,900	
	6080	30 to 40 lb/hr, 1125 cfm		4	4		23,700	154		23,854	26,200	
	6090	60 to 75 lb/hr, 2250 cfm		3	5.333		31,100	205		31,305	34,500	
	6100	120 to 155 lb/hr, 4500 cfm		2	8		56,500	305		56,805	63,000	
	6110	240 to 310 lb/hr, 9000 cfm		1.50	10.667		80,000	410		80,410	88,500	
	6120	400 to 515 lb/hr, 15,000 cfm	Q-2	1.60	15		108,500	600		109,100	120,500	
	6130	530 to 690 lb/hr, 20,000 cfm		1.40	17.143		117,500	685		118,185	130,500	
	6140	800 to 1030 lb/hr, 30,000 cfm		1.20	20		134,000	795		134,795	149,000	
	6150	1060 to 1375 lb/hr, 40,000 cfm		1	24		187,000	955		187,955	207,500	
500	0010	**HUMIDIFIERS**										500
	0520	Steam, room or duct, filter, regulators, auto. controls, 220 V										
	0540	11 lb. per hour R236000-30	Q-5	6	2.667	Ea.	2,175	103		2,278	2,550	
	0560	22 lb. per hour		5	3.200		2,400	124		2,524	2,825	
	0580	33 lb. per hour		4	4		2,450	155		2,605	2,950	
	0600	50 lb. per hour		4	4		3,025	155		3,180	3,575	
	0620	100 lb. per hour		3	5.333		3,600	207		3,807	4,300	
	0640	150 lb. per hour		2.50	6.400		4,775	248		5,023	5,625	
	0660	200 lb. per hour		2	8		5,925	310		6,235	6,975	
	0700	With blower										
	0720	11 lb. per hour	Q-5	5.50	2.909	Ea.	3,025	113		3,138	3,500	
	0740	22 lb. per hour		4.75	3.368		3,250	131		3,381	3,775	
	0760	33 lb. per hour		3.75	4.267		3,325	165		3,490	3,900	
	0780	50 lb. per hour		3.50	4.571		4,000	177		4,177	4,675	
	0800	100 lb. per hour		2.75	5.818		4,500	225		4,725	5,300	
	0820	150 lb. per hour		2	8		6,675	310		6,985	7,825	
	0840	200 lb. per hour		1.50	10.667		7,825	415		8,240	9,250	
	5000	Furnace type, wheel bypass										

15700 | Heating/Ventilating/Air Conditioning Equipment

		15750 | Humidity Control Equipment	Crew	Daily Output	Labor-Hours	Unit	2006 Bare Costs Mat.	Labor	Equip.	Total	Total Incl O&P	
500	5020	10 GPD (R236000-30)	1 Stpi	4	2	Ea.	144	86		230	292	500
	5040	14 GPD		3.80	2.105		175	90.50		265.50	335	
	5060	19 GPD		3.60	2.222		204	95.50		299.50	375	
	9000	Minimum labor/equipment charge	Q-5	3.50	4.571	Job		177		177	275	

		15760 | Terminal Heating & Cooling Units	Crew	Daily Output	Labor-Hours	Unit	Mat.	Labor	Equip.	Total	Total Incl O&P	
100	0010	**COILS, FLANGED**										100
	0500	Chilled water cooling, 6 rows, 24" x 48"	Q-5	3.20	5	Ea.	2,875	194		3,069	3,450	
	1000	Direct expansion cooling, 6 rows, 24" x 48"		2.80	5.714		3,125	221		3,346	3,775	
	1500	Hot water heating, 1 row, 24" x 48"		4	4		1,150	155		1,305	1,500	
	2000	Steam heating, 1 row, 24" x 48"		3.06	5.229		1,625	203		1,828	2,100	
	9000	Minimum labor/equipment charge	1 Plum	1	8	Job		340		340	530	
200	0010	**DUCT HEATERS** Electric, 480 V, 3 Ph										200
	0020	Finned tubular insert, 500°F										
	0100	8" wide x 6" high, 4.0 kW	Q-20	16	1.250	Ea.	570	48.50		618.50	700	
	0120	12" high, 8.0 kW		15	1.333		940	51.50		991.50	1,100	
	0140	18" high, 12.0 kW		14	1.429		1,325	55.50		1,380.50	1,550	
	0160	24" high, 16.0 kW		13	1.538		1,700	59.50		1,759.50	1,975	
	0180	30" high, 20.0 kW		12	1.667		2,075	64.50		2,139.50	2,375	
	0300	12" wide x 6" high, 6.7 kW		15	1.333		600	51.50		651.50	745	
	0320	12" high, 13.3 kW		14	1.429		975	55.50		1,030.50	1,175	
	0340	18" high, 20.0 kW		13	1.538		1,375	59.50		1,434.50	1,600	
	0360	24" high, 26.7 kW		12	1.667		1,750	64.50		1,814.50	2,025	
	0380	30" high, 33.3 kW		11	1.818		2,150	70.50		2,220.50	2,450	
	0500	18" wide x 6" high, 13.3 kW		14	1.429		640	55.50		695.50	790	
	0520	12" high, 26.7 kW		13	1.538		1,125	59.50		1,184.50	1,325	
	0540	18" high, 40.0 kW		12	1.667		1,475	64.50		1,539.50	1,725	
	0560	24" high, 53.3 kW		11	1.818		1,975	70.50		2,045.50	2,275	
	0580	30" high, 66.7 kW		10	2		2,475	77.50		2,552.50	2,825	
	0700	24" wide x 6" high, 17.8 kW		13	1.538		705	59.50		764.50	870	
	0720	12" high, 35.6 kW		12	1.667		1,200	64.50		1,264.50	1,425	
	0740	18" high, 53.3 kW		11	1.818		1,675	70.50		1,745.50	1,950	
	0760	24" high, 71.1 kW		10	2		2,200	77.50		2,277.50	2,525	
	0780	30" high, 88.9 kW		9	2.222		2,675	86		2,761	3,075	
	0900	30" wide x 6" high, 22.2 kW		12	1.667		745	64.50		809.50	920	
	0920	12" high, 44.4 kW		11	1.818		1,275	70.50		1,345.50	1,500	
	0940	18" high, 66.7 kW		10	2		1,775	77.50		1,852.50	2,075	
	0960	24" high, 88.9 kW		9	2.222		2,275	86		2,361	2,650	
	0980	30" high, 111.0 kW		8	2.500		2,850	97		2,947	3,275	
	1400	Note decreased kW available for										
	1410	each duct size at same cost										
	1420	See line 5000 for modifications and accessories										
	2000	Finned tubular flange with insulated										
	2020	terminal box, 500°F										
	2100	12" wide x 36" high, 54 kW	Q-20	10	2	Ea.	2,800	77.50		2,877.50	3,200	
	2120	40" high, 60 kW		9	2.222		3,225	86		3,311	3,650	
	2200	24" wide x 36" high, 118.8 kW		9	2.222		3,075	86		3,161	3,500	
	2220	40" high, 132 kW		8	2.500		3,500	97		3,597	4,025	
	2400	36" wide x 8" high, 40 kW		11	1.818		1,400	70.50		1,470.50	1,650	
	2420	16" high, 80 kW		10	2		1,900	77.50		1,977.50	2,200	
	2440	24" high, 120 kW		9	2.222		2,475	86		2,561	2,850	
	2460	32" high, 160 kW		8	2.500		3,175	97		3,272	3,650	
	2480	36" high, 180 kW		7	2.857		3,925	111		4,036	4,500	
	2500	40" high, 200 kW		6	3.333		4,350	129		4,479	4,975	
	2600	40" wide x 8" high, 45 kW		11	1.818		1,475	70.50		1,545.50	1,725	
	2620	16" high, 90 kW		10	2		2,050	77.50		2,127.50	2,375	

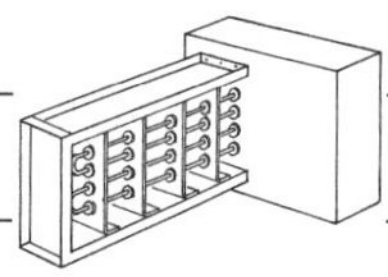

	15760 | Terminal Heating & Cooling Units		CREW	DAILY OUTPUT	LABOR-HOURS	UNIT	2006 BARE COSTS MAT.	LABOR	EQUIP.	TOTAL	TOTAL INCL O&P	
200	2640	24" high, 135 kW	Q-20	9	2.222	Ea.	2,600	86		2,686	2,975	200
	2660	32" high, 180 kW		8	2.500		3,450	97		3,547	3,950	
	2680	36" high, 202.5 kW		7	2.857		3,975	111		4,086	4,550	
	2700	40" high, 225 kW		6	3.333		4,525	129		4,654	5,175	
	2800	48" wide x 8" high, 54.8 kW		10	2		1,550	77.50		1,627.50	1,825	
	2820	16" high, 109.8 kW		9	2.222		2,150	86		2,236	2,500	
	2840	24" high, 164.4 kW		8	2.500		2,750	97		2,847	3,175	
	2860	32" high, 219.2 kW		7	2.857		3,675	111		3,786	4,200	
	2880	36" high, 246.6 kW		6	3.333		4,225	129		4,354	4,825	
	2900	40" high, 274 kW		5	4		4,775	155		4,930	5,500	
	3000	56" wide x 8" high, 64 kW		9	2.222		1,775	86		1,861	2,075	
	3020	16" high, 128 kW		8	2.500		2,450	97		2,547	2,850	
	3040	24" high, 192 kW		7	2.857		2,975	111		3,086	3,450	
	3060	32" high, 256 kW		6	3.333		4,150	129		4,279	4,775	
	3080	36" high, 288kW		5	4		4,775	155		4,930	5,500	
	3100	40" high, 320kW		4	5		5,275	194		5,469	6,100	
	3200	64" wide x 8" high, 74kW		8	2.500		1,825	97		1,922	2,150	
	3220	16" high, 148kW		7	2.857		2,525	111		2,636	2,950	
	3240	24" high, 222kW		6	3.333		3,225	129		3,354	3,725	
	3260	32" high, 296kW		5	4		4,300	155		4,455	5,000	
	3280	36" high, 333kW		4	5		5,150	194		5,344	5,950	
	3300	40" high, 370kW	↓	3	6.667	↓	5,675	258		5,933	6,650	
	3800	Note decreased kW available for										
	3820	each duct size at same cost										
	5000	Duct heater modifications and accessories										
	5120	T.C.O. limit auto or manual reset	Q-20	42	.476	Ea.	93	18.45		111.45	131	
	5140	Thermostat		28	.714		395	27.50		422.50	480	
	5160	Overheat thermocouple (removable)		7	2.857		560	111		671	795	
	5180	Fan interlock relay		18	1.111		136	43		179	217	
	5200	Air flow switch		20	1		117	38.50		155.50	190	
	5220	Split terminal box cover	↓	100	.200	↓	39	7.75		46.75	55	
	8000	To obtain BTU multiply kW by 3413										
	9900	Minimum labor/equipment charge	1 Shee	4	2	Job		84.50		84.50	134	
250	0010	**ELECTRIC HEATING**, not incl. conduit or feed wiring R235000-50										250
	1100	Rule of thumb: Baseboard units, including control	1 Elec	4.40	1.818	kW	79	76.50		155.50	205	
	1300	Baseboard heaters, 2' long, 375 watt		8	1	Ea.	33	42		75	101	
	1400	3' long, 500 watt		8	1		39	42		81	108	
	1600	4' long, 750 watt		6.70	1.194		46	50		96	129	
	1800	5' long, 935 watt 		5.70	1.404		55	59		114	152	
	2000	6' long, 1125 watt		5	1.600		61	67		128	171	
	2200	7' long, 1310 watt		4.40	1.818		66.50	76.50		143	191	
	2400	8' long, 1500 watt		4	2		76.50	84		160.50	214	
	2600	9' long, 1680 watt		3.60	2.222		85.50	93.50		179	238	
	2800	10' long, 1875 watt	↓	3.30	2.424	↓	95	102		197	262	
	2950	Wall heaters with fan, 120 to 277 volt										
	3160	Recessed, residential, 750 watt	1 Elec	6	1.333	Ea.	89.50	56		145.50	185	
	3170	1000 watt		6	1.333		92	56		148	188	
	3180	1250 watt		5	1.600		101	67		168	215	
	3190	1500 watt		4	2		101	84		185	240	
	3210	2000 watt		4	2		106	84		190	246	
	3230	2500 watt		3.50	2.286		229	96		325	400	
	3240	3000 watt		3	2.667		345	112		457	555	
	3250	4000 watt		2.70	2.963		350	124		474	575	
	3260	Commercial, 750 watt		6	1.333		146	56		202	247	
	3270	1000 watt	↓	6	1.333	↓	146	56		202	247	

15 MECHANICAL

15760 | Terminal Heating & Cooling Units

250

	15760 Terminal Heating & Cooling Units	CREW	DAILY OUTPUT	LABOR-HOURS	UNIT	2006 BARE COSTS MAT.	LABOR	EQUIP.	TOTAL	TOTAL INCL O&P
3280	1250 watt (R235000-50)	1 Elec	5	1.600	Ea.	146	67		213	264
3290	1500 watt		4	2		146	84		230	289
3300	2000 watt		4	2		146	84		230	289
3310	2500 watt		3.50	2.286		146	96		242	310
3320	3000 watt		3	2.667		257	112		369	455
3330	4000 watt		2.70	2.963		257	124		381	475
3600	Thermostats, integral		16	.500		18	21		39	52.50
3800	Line voltage, 1 pole		8	1		22.50	42		64.50	89
3810	2 pole		8	1		29	42		71	96.50
3820	Low voltage, 1 pole		8	1		25	42		67	92
4000	Heat trace system, 400 degree									
4020	115V, 2.5 watts per L.F.	1 Elec	530	.015	L.F.	6.70	.63		7.33	8.35
4030	5 watts per L.F.		530	.015		6.70	.63		7.33	8.35
4050	10 watts per L.F.		530	.015		6.70	.63		7.33	8.35
4060	208V, 5 watts per L.F.		530	.015		6.70	.63		7.33	8.35
4080	480V, 8 watts per L.F.		530	.015		6.70	.63		7.33	8.35
4200	Heater raceway									
4260	Heat transfer cement									
4280	1 gallon				Ea.	57.50			57.50	63
4300	5 gallon				"	235			235	258
4320	Snap band, clamp									
4340	3/4" pipe size	1 Elec	470	.017	Ea.		.71		.71	1.10
4360	1" pipe size		444	.018			.76		.76	1.17
4380	1-1/4" pipe size		400	.020			.84		.84	1.29
4400	1-1/2" pipe size		355	.023			.95		.95	1.46
4420	2" pipe size		320	.025			1.05		1.05	1.62
4440	3" pipe size		160	.050			2.10		2.10	3.24
4460	4" pipe size		100	.080			3.36		3.36	5.20
4480	Thermostat NEMA 3R, 22 amp, 0-150 Deg, 10' cap.		8	1		196	42		238	280
4500	Thermostat NEMA 4X, 25 amp, 40 Deg, 5-1/2' cap.		7	1.143		196	48		244	289
4520	Thermostat NEMA 4X, 22 amp, 25-325 Deg, 10' cap.		7	1.143		530	48		578	655
4540	Thermostat NEMA 4X, 22 amp, 15-140 Deg,		6	1.333		445	56		501	575
4580	Thermostat NEMA 4,7,9, 22 amp, 25-325 Deg, 10' cap.		3.60	2.222		650	93.50		743.50	860
4600	Thermostat NEMA 4,7,9, 22 amp, 15-140 Deg,		3	2.667		645	112		757	885
4720	Fiberglass application tape, 36 yard roll		11	.727		59.50	30.50		90	113
5000	Radiant heating ceiling panels, 2' x 4', 500 watt		16	.500		203	21		224	256
5050	750 watt		16	.500		223	21		244	279
5200	For recessed plaster frame, add		32	.250		42	10.50		52.50	62.50
5300	Infrared quartz heaters, 120 volts, 1000 watts		6.70	1.194		135	50		185	227
5350	1500 watt		5	1.600		135	67		202	253
5400	240 volts, 1500 watt		5	1.600		135	67		202	253
5450	2000 watt		4	2		135	84		219	278
5500	3000 watt		3	2.667		155	112		267	345
5550	4000 watt		2.60	3.077		155	129		284	370
5570	Modulating control		.80	10		78	420		498	730
5600	Unit heaters, heavy duty, with fan & mounting bracket									
5650	Single phase, 208-240-277 volt, 3 kW	1 Elec	3.20	2.500	Ea.	335	105		440	525
5750	5 kW		2.40	3.333		345	140		485	595
5800	7 kW		1.90	4.211		530	177		707	855
5850	10 kW		1.30	6.154		605	258		863	1,075
5950	15 kW		.90	8.889		975	375		1,350	1,650
6000	480 volt, 3 kW		3.30	2.424		345	102		447	535
6020	4 kW		3	2.667		375	112		487	585
6040	5 kW		2.60	3.077		385	129		514	620
6060	7 kW		2	4		570	168		738	885
6080	10 kW		1.40	5.714		625	240		865	1,050

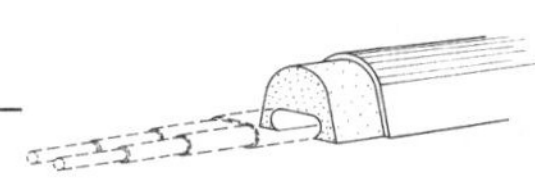

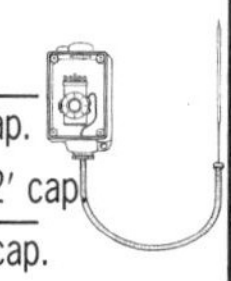

MECHANICAL 15

15760 | Terminal Heating & Cooling Units

		15760 Terminal Heating & Cooling Units	CREW	DAILY OUTPUT	LABOR-HOURS	UNIT	2006 BARE COSTS MAT.	LABOR	EQUIP.	TOTAL	TOTAL INCL O&P	
250	6100	13 kW (R235000-50)	1 Elec	1.10	7.273	Ea.	975	305		1,280	1,550	250
	6120	15 kW		1	8		975	335		1,310	1,600	
	6140	20 kW		.90	8.889		1,275	375		1,650	1,975	
	6300	3 phase, 208-240 volt, 5 kW		2.40	3.333		325	140		465	570	
	6320	7 kW		1.90	4.211		500	177		677	820	
	6340	10 kW		1.30	6.154		540	258		798	995	
	6360	15 kW		.90	8.889		905	375		1,280	1,575	
	6380	20 kW		.70	11.429		1,300	480		1,780	2,175	
	6400	25 kW		.50	16		1,525	670		2,195	2,700	
	6500	480 volt, 5 kW		2.60	3.077		455	129		584	700	
	6520	7 kW		2	4		570	168		738	890	
	6540	10 kW		1.40	5.714		605	240		845	1,025	
	6560	13 kW		1.10	7.273		975	305		1,280	1,550	
	6580	15 kW		1	8		975	335		1,310	1,600	
	6600	20 kW		.90	8.889		1,075	375		1,450	1,775	
	6620	25 kW		.60	13.333		1,250	560		1,810	2,250	
	6630	30 kW		.70	11.429		1,350	480		1,830	2,225	
	6640	40 kW		.60	13.333		2,250	560		2,810	3,350	
	6650	50 kW	↓	.50	16	↓	2,700	670		3,370	4,000	
	6800	Vertical discharge heaters, with fan										
	6820	Single phase, 208-240-277 volt, 10 kW	1 Elec	1.30	6.154	Ea.	560	258		818	1,025	
	6840	15 kW		.90	8.889		940	375		1,315	1,600	
	6900	3 phase, 208-240 volt, 10 kW		1.30	6.154		540	258		798	995	
	6920	15 kW		.90	8.889		905	375		1,280	1,575	
	6940	20 kW		.70	11.429		1,300	480		1,780	2,175	
	6960	25 kW		.50	16		1,525	670		2,195	2,700	
	6980	30 kW		.40	20		1,750	840		2,590	3,225	
	7000	40 kW		.36	22.222		2,250	935		3,185	3,925	
	7020	50 kW		.32	25		2,700	1,050		3,750	4,600	
	7100	480 volt, 10 kW		1.40	5.714		605	240		845	1,025	
	7120	15 kW		1	8		975	335		1,310	1,600	
	7140	20 kW		.90	8.889		1,275	375		1,650	1,975	
	7160	25 kW		.60	13.333		1,525	560		2,085	2,550	
	7180	30 kW		.50	16		1,750	670		2,420	2,950	
	7200	40 kW		.40	20		2,250	840		3,090	3,775	
	7220	50 kW		.35	22.857		2,700	960		3,660	4,450	
	7410	Sill height convector heaters, 5″ high x 2′ long, 500 watt		6.70	1.194		236	50		286	340	
	7420	3′ long, 750 watt		6.50	1.231		263	51.50		314.50	370	
	7430	4′ long, 1000 watt		6.20	1.290		315	54		369	435	
	7440	5′ long, 1250 watt		5.50	1.455		360	61		421	490	
	7450	6′ long, 1500 watt		4.80	1.667		405	70		475	560	
	7460	8′ long, 2000 watt		3.60	2.222		610	93.50		703.50	815	
	7470	10′ long, 2500 watt	↓	3	2.667	↓	685	112		797	930	
	7900	Cabinet convector heaters, 240 volt										
	7920	3′ long, 2000 watt	1 Elec	5.30	1.509	Ea.	1,625	63.50		1,688.50	1,900	
	7940	3000 watt		5.30	1.509		1,700	63.50		1,763.50	1,975	
	7960	4000 watt		5.30	1.509		1,750	63.50		1,813.50	2,025	
	7980	6000 watt		4.60	1.739		1,825	73		1,898	2,125	
	8000	8000 watt		4.60	1.739		1,875	73		1,948	2,200	
	8020	4′ long, 4000 watt		4.60	1.739		1,775	73		1,848	2,075	
	8040	6000 watt		4	2		1,850	84		1,934	2,150	
	8060	8000 watt		4	2		1,900	84		1,984	2,225	
	8080	10,000 watt	↓	4	2	↓	1,925	84		2,009	2,225	
	8100	Available also in 208 or 277 volt										
	8200	Cabinet unit heaters, 120 to 277 volt, 1 pole,										
	8220	wall mounted, 2 kW	1 Elec	4.60	1.739	Ea.	1,500	73		1,573	1,775	

	15760	Terminal Heating & Cooling Units	CREW	DAILY OUTPUT	LABOR-HOURS	UNIT	2006 BARE COSTS MAT.	LABOR	EQUIP.	TOTAL	TOTAL INCL O&P	
250	8230	3 kW R235000-50	1 Elec	4.60	1.739	Ea.	1,550	73		1,623	1,850	250
	8240	4 kW		4.40	1.818		1,600	76.50		1,676.50	1,900	
	8250	5 kW		4.40	1.818		1,675	76.50		1,751.50	1,950	
	8260	6 kW		4.20	1.905		1,675	80		1,755	1,975	
	8270	8 kW		4	2		1,750	84		1,834	2,050	
	8280	10 kW		3.80	2.105		1,775	88.50		1,863.50	2,100	
	8290	12 kW		3.50	2.286		1,825	96		1,921	2,150	
	8300	13.5 kW		2.90	2.759		1,850	116		1,966	2,225	
	8310	16 kW		2.70	2.963		1,900	124		2,024	2,300	
	8320	20 kW		2.30	3.478		2,850	146		2,996	3,375	
	8330	24 kW		1.90	4.211		2,900	177		3,077	3,475	
	8350	Recessed, 2 kW		4.40	1.818		1,500	76.50		1,576.50	1,750	
	8370	3 kW		4.40	1.818		1,550	76.50		1,626.50	1,850	
	8380	4 kW		4.20	1.905		1,600	80		1,680	1,900	
	8390	5 kW		4.20	1.905		1,675	80		1,755	1,950	
	8400	6 kW		4	2		1,700	84		1,784	2,000	
	8410	8 kW		3.80	2.105		1,775	88.50		1,863.50	2,075	
	8420	10 kW		3.50	2.286		2,200	96		2,296	2,575	
	8430	12 kW		2.90	2.759		2,200	116		2,316	2,600	
	8440	13.5 kW		2.70	2.963		2,225	124		2,349	2,650	
	8450	16 kW		2.30	3.478		2,250	146		2,396	2,700	
	8460	20 kW		1.90	4.211		2,700	177		2,877	3,250	
	8470	24 kW		1.60	5		2,775	210		2,985	3,375	
	8490	Ceiling mounted, 2 kW		3.20	2.500		1,500	105		1,605	1,775	
	8510	3 kW		3.20	2.500		1,550	105		1,655	1,875	
	8520	4 kW		3	2.667		1,625	112		1,737	1,950	
	8530	5 kW		3	2.667		1,675	112		1,787	2,000	
	8540	6 kW		2.80	2.857		1,675	120		1,795	2,025	
	8550	8 kW		2.40	3.333		1,750	140		1,890	2,150	
	8560	10 kW		2.20	3.636		1,775	153		1,928	2,200	
	8570	12 kW		2	4		1,775	168		1,943	2,225	
	8580	13.5 kW		1.50	5.333		1,850	224		2,074	2,400	
	8590	16 kW		1.30	6.154		1,900	258		2,158	2,500	
	8600	20 kW		.90	8.889		2,700	375		3,075	3,550	
	8610	24 kW		.60	13.333		2,750	560		3,310	3,900	
	8630	208 to 480 V, 3 pole										
	8650	Wall mounted, 2 kW	1 Elec	4.60	1.739	Ea.	1,500	73		1,573	1,750	
	8670	3 kW		4.60	1.739		1,550	73		1,623	1,850	
	8680	4 kW		4.40	1.818		1,600	76.50		1,676.50	1,900	
	8690	5 kW		4.40	1.818		1,675	76.50		1,751.50	1,950	
	8700	6 kW		4.20	1.905		1,675	80		1,755	1,975	
	8710	8 kW		4	2		1,750	84		1,834	2,050	
	8720	10 kW		3.80	2.105		1,775	88.50		1,863.50	2,100	
	8730	12 kW		3.50	2.286		1,775	96		1,871	2,125	
	8740	13.5 kW		2.90	2.759		1,850	116		1,966	2,225	
	8750	16 kW		2.70	2.963		2,050	124		2,174	2,450	
	8760	20 kW		2.30	3.478		2,700	146		2,846	3,200	
	8770	24 kW		1.90	4.211		2,750	177		2,927	3,300	
	8790	Recessed, 2 kW		4.40	1.818		1,500	76.50		1,576.50	1,750	
	8810	3 kW		4.40	1.818		1,550	76.50		1,626.50	1,850	
	8820	4 kW		4.20	1.905		1,600	80		1,680	1,900	
	8830	5 kW		4.20	1.905		1,675	80		1,755	1,950	
	8840	6 kW		4	2		1,675	84		1,759	1,975	
	8850	8 kW		3.80	2.105		1,750	88.50		1,838.50	2,050	
	8860	10 kW		3.50	2.286		1,775	96		1,871	2,100	
250	8870	12 kW		2.90	2.759		1,775	116		1,891	2,125	

15760 | Terminal Heating & Cooling Units

		15760 Terminal Heating & Cooling Units	CREW	DAILY OUTPUT	LABOR-HOURS	UNIT	2006 BARE COSTS MAT.	LABOR	EQUIP.	TOTAL	TOTAL INCL O&P	
250	8880	13.5 kW (R235000-50)	1 Elec	2.70	2.963	Ea.	1,900	124		2,024	2,300	250
	8890	16 kW		2.30	3.478		1,900	146		2,046	2,325	
	8900	20 kW		1.90	4.211		2,700	177		2,877	3,250	
	8920	24 kW		1.60	5		2,750	210		2,960	3,350	
	8940	Ceiling mount, 2 kW		3.20	2.500		1,500	105		1,605	1,775	
	8950	3 kW		3.20	2.500		1,550	105		1,655	1,875	
	8960	4 kW		3	2.667		1,600	112		1,712	1,950	
	8970	5 kW		3	2.667		1,675	112		1,787	2,000	
	8980	6 kW		2.80	2.857		1,675	120		1,795	2,025	
	8990	8 kW		2.40	3.333		1,750	140		1,890	2,150	
	9000	10 kW		2.20	3.636		1,775	153		1,928	2,200	
	9020	13.5 kW		1.50	5.333		1,850	224		2,074	2,400	
	9030	16 kW		1.30	6.154		1,900	258		2,158	2,500	
	9040	20 kW		.90	8.889		2,700	375		3,075	3,550	
	9060	24 kW		.60	13.333		2,750	560		3,310	3,900	
	9990	Minimum labor/equipment charge		4	2	Job		84		84	129	
300	0010	**FAN COIL AIR CONDITIONING** Cabinet mtd., filters, controls (R262213-27)										300
	0100	Chilled water, 1/2 ton cooling	Q-5	8	2	Ea.	600	77.50		677.50	780	
	0110	3/4 ton cooling		7	2.286		635	88.50		723.50	835	
	0120	1 ton cooling		6	2.667		720	103		823	955	
	0140	1.5 ton cooling		5.50	2.909		825	113		938	1,075	
	0150	2 ton cooling		5.25	3.048		1,050	118		1,168	1,325	
	0160	2.5 ton cooling		5	3.200		1,550	124		1,674	1,900	
	0180	3 ton cooling		4	4		1,700	155		1,855	2,125	
	0190	7.5 ton cooling		2.70	5.926		1,825	230		2,055	2,350	
	0262	For hot water coil, add					40%	10%				
	0940	Direct expansion, for use w/air cooled condensing, 1.5 ton cooling	Q-5	5	3.200	Ea.	470	124		594	715	
	0950	2 ton cooling		4.80	3.333		500	129		629	750	
	0960	2.5 ton cooling		4.40	3.636		575	141		716	855	
	0970	3 ton cooling		3.80	4.211		600	163		763	915	
	0980	3.5 ton cooling		3.60	4.444		735	172		907	1,075	
	0990	4 ton cooling		3.40	4.706		775	182		957	1,150	
	1000	5 ton cooling		3	5.333		875	207		1,082	1,275	
	1020	7.5 ton cooling		3	5.333		2,025	207		2,232	2,575	
	1040	10 ton cooling	Q-6	2.60	9.231		2,450	370		2,820	3,275	
	1042	11 ton cooling		2.30	10.435		3,150	420		3,570	4,125	
	1050	15 ton cooling		1.60	15		3,250	605		3,855	4,500	
	1060	20 ton cooling		.70	34.286		4,300	1,375		5,675	6,900	
	1070	25 ton cooling		.65	36.923		5,400	1,475		6,875	8,250	
	1080	30 ton cooling		.60	40		5,975	1,600		7,575	9,050	
	1500	For hot water coil, add					40%	10%				
	1510	For condensing unit add see division 15670.										
	9000	Minimum labor/equipment charge	Q-5	3.75	4.267	Job		165		165	257	
500	0010	**HEATING & VENTILATING UNITS** Classroom (R233400-10)										500
	0020	Includes filter, heating/cooling coils, standard controls										
	0080	750 CFM, 2 tons cooling (R235000-50)	Q-6	2	12	Ea.	3,275	480		3,755	4,350	
	0100	1000 CFM, 2-1/2 tons cooling		1.60	15		3,850	605		4,455	5,175	
	0120	1250 CFM, 3 tons cooling (R262213-27)		1.40	17.143		3,975	690		4,665	5,450	
	0140	1500 CFM, 4 tons cooling		.80	30		4,275	1,200		5,475	6,575	
	0160	2000 CFM 5 tons cooling		.50	48		5,000	1,925		6,925	8,500	
	0500	For electric heat, add					35%					
	1000	For no cooling, deduct					25%	10%				
600	0010	**HYDRONIC HEATING** Terminal units, not incl. main supply pipe (R235000-50)										600
	1000	Radiation										

15 MECHANICAL

		15760 | Terminal Heating & Cooling Units	CREW	DAILY OUTPUT	LABOR-HOURS	UNIT	2006 BARE COSTS MAT.	LABOR	EQUIP.	TOTAL	TOTAL INCL O&P	
600	1100	Panel, baseboard, C.I., including supports, no covers (R235000-50)	Q-5	46	.348	L.F.	27	13.50		40.50	50.50	600
	1150	Fin tube, wall hung, 14" slope top cover, with damper										
	1200	1-1/4" copper tube, 4-1/4" alum. fin	Q-5	38	.421	L.F.	27.50	16.30		43.80	56	
	1250	1-1/4" steel tube, 4-1/4" steel fin		36	.444		25	17.20		42.20	54.50	
	1255	2" steel tube, 4-1/4" steel fin		32	.500		27.50	19.40		46.90	60.50	
	1310	Baseboard, pkgd, 1/2" copper tube, alum. fin, 7" high		60	.267		6.25	10.35		16.60	23	
	1320	3/4" copper tube, alum. fin, 7" high		58	.276		6.70	10.70		17.40	24	
	1340	1" copper tube, alum. fin, 8-7/8" high		56	.286		13.80	11.05		24.85	32.50	
	1360	1-1/4" copper tube, alum. fin, 8-7/8" high		54	.296		20.50	11.50		32	40.50	
	1380	1-1/4" IPS steel tube with steel fins		52	.308		20.50	11.90		32.40	41	
	1500	Note: fin tube may also require corners, caps, etc.										
	1990	Convector unit, floor recessed, flush, with trim										
	2000	for under large glass wall areas, no damper	Q-5	20	.800	L.F.	27.50	31		58.50	78.50	
	2100	For unit with damper				"	39.50			39.50	43.50	
	2210	17" H x 24" L	Q-5	10	1.600	Ea.	66.50	62		128.50	170	
	2214	17" H x 36" L		8.60	1.860		99.50	72		171.50	222	
	2218	17" H x 48" L		7.40	2.162		133	84		217	276	
	2222	21" H x 24" L		9	1.778		66.50	69		135.50	180	
	2226	21" H x 36" L		8.20	1.951		100	75.50		175.50	228	
	2228	21" H x 48" L		6.80	2.353		133	91		224	288	
	2240	For knob operated damper, add					140%					
	2241	For metal trim strips, add	Q-5	64	.250	Ea.	4.12	9.70		13.82	19.60	
	2243	For snap-on inlet grille, add					10%	10%				
	2245	For hinged access door, add	Q-5	64	.250	Ea.	14.70	9.70		24.40	31	
	2246	For air chamber, auto-venting, add	"	58	.276	"	4.59	10.70		15.29	21.50	
	3000	Radiators, cast iron										
	3100	Free standing or wall hung, 6 tube, 25" high	Q-5	96	.167	Section	24	6.45		30.45	36.50	
	3150	4 tube 25" high		96	.167		17.90	6.45		24.35	30	
	3200	4 tube, 19" high		96	.167		16.65	6.45		23.10	28.50	
	3250	Adj. brackets, 2 per wall radiator up to 30 sections	1 Stpi	32	.250	Ea.	19.25	10.75		30	38	
	3500	Recessed, 20" high x 5" deep, without grille	Q-5	60	.267	Section	20	10.35		30.35	38	
	3600	For inlet grille, add				"	1.40			1.40	1.54	
	9000	Minimum labor/equipment charge	Q-5	3.20	5	Job		194		194	300	
	9500	To convert SFR to BTU rating: Hot water, 150 x SFR										
	9510	Forced hot water, 180 x SFR; steam, 240 x SFR										
700	0010	**INFRARED UNIT** (R235000-50)										700
	0020	Gas fired, unvented, electric ignition, 100% shutoff.										
	0030	Piping and wiring not included										
	0060	Input, 15 MBH	Q-5	7	2.286	Ea.	380	88.50		468.50	560	
	0100	30 MBH		6	2.667		390	103		493	585	
	0120	45 MBH		5	3.200		430	124		554	670	
	0140	50 MBH		4.50	3.556		450	138		588	710	
	0160	60 MBH		4	4		465	155		620	750	
	0180	75 MBH		3	5.333		540	207		747	915	
	0200	90 MBH		2.50	6.400		575	248		823	1,025	
	0220	105 MBH		2	8		575	310		885	1,125	
	0240	120 MBH		2	8		895	310		1,205	1,475	
	2000	Electric, single or three phase										
	2050	6 kW, 20,478 BTU	1 Elec	2.30	3.478	Ea.	370	146		516	635	
	2100	13.5 KW, 40,956 BTU		2.20	3.636		580	153		733	875	
	2150	24 KW, 81,912 BTU		2	4		1,250	168		1,418	1,625	
	3000	Oil fired, pump, controls, fusible valve, oil supply tank										
	3050	91,000 BTU	Q-5	2.50	6.400	Ea.	5,775	248		6,023	6,725	
	3080	105.000 BTU		2.25	7.111		6,000	276		6,276	7,025	
	3110	119,000 BTU		2	8		6,600	310		6,910	7,725	

15760 | Terminal Heating & Cooling Units

		15760 Terminal Heating & Cooling Units	CREW	DAILY OUTPUT	LABOR-HOURS	UNIT	2006 BARE COSTS MAT.	LABOR	EQUIP.	TOTAL	TOTAL INCL O&P	
800	0010	UNIT HEATERS										800
	3950	Unit heaters, propeller, 115 V 2 psi steam, 60° F entering air										
	4000	Horizontal, 12 MBH	Q-5	12	1.333	Ea.	278	51.50		329.50	385	
	4020	28.2 MBH		10	1.600		370	62		432	500	
	4040	36.5 MBH		8	2		425	77.50		502.50	590	
	4060	43.9 MBH		8	2		425	77.50		502.50	590	
	4080	56.5 MBH		7.50	2.133		435	82.50		517.50	610	
	4100	65.6 MBH		7	2.286		470	88.50		558.50	655	
	4120	87.6 MBH		6.50	2.462		505	95.50		600.50	705	
	4140	96.8 MBH		6	2.667		620	103		723	840	
	4160	133.3 MBH		5	3.200		665	124		789	925	
	4180	157.6 MBH		4	4		830	155		985	1,150	
	4200	197.7 MBH		3	5.333		940	207		1,147	1,350	
	4220	257.2 MBH		2.50	6.400		1,275	248		1,523	1,775	
	4240	286.9 MBH		2	8		1,325	310		1,635	1,925	
	4250	326.0 MBH		1.90	8.421		1,625	325		1,950	2,275	
	4260	364 MBH		1.80	8.889		1,700	345		2,045	2,400	
	4270	404 MBH	↓	1.60	10		2,200	390		2,590	3,025	
	4300	For vertical diffuser, add					154			154	169	
	4310	Vertical flow, 40 MBH	Q-5	11	1.455		415	56.50		471.50	550	
	4314	58.5 MBH		8	2		470	77.50		547.50	635	
	4318	92 MBH		7	2.286		565	88.50		653.50	760	
	4322	109.7 MBH		6	2.667		655	103		758	880	
	4326	131.0 MBH		4	4		655	155		810	960	
	4330	160.0 MBH		3	5.333		780	207		987	1,175	
	4334	194.0 MBH		2.20	7.273		845	282		1,127	1,375	
	4338	212.0 MBH		2.10	7.619		955	295		1,250	1,500	
	4342	247.0 MBH		1.96	8.163		1,075	315		1,390	1,700	
	4346	297.0 MBH	↓	1.80	8.889		1,125	345		1,470	1,750	
	4350	333.0 MBH	Q-6	1.90	12.632		1,550	510		2,060	2,500	
	4354	420 MBH,(460 V)		1.80	13.333		1,575	535		2,110	2,550	
	4358	500 MBH, (460 V)		1.71	14.035		2,225	565		2,790	3,325	
	4362	570 MBH, (460 V)		1.40	17.143		3,275	690		3,965	4,675	
	4366	620 MBH, (460 V)		1.30	18.462		3,750	740		4,490	5,275	
	4370	960 MBH, (460 V)	↓	1.10	21.818	↓	6,500	875		7,375	8,525	
	6000	Valance units, complete with 1/2″ cooling coil, enclosure										
	6020	2 tube	Q-5	18	.889	L.F.	29.50	34.50		64	86	
	6040	3 tube		16	1		35	39		74	99	
	6060	4 tube		16	1		40	39		79	105	
	6080	5 tube		15	1.067		45.50	41.50		87	115	
	6100	6 tube		15	1.067		50.50	41.50		92	120	
	6120	8 tube	↓	14	1.143		73	44.50		117.50	150	
	6200	For 3/4″ cooling coil, add				↓	10%					

15770 | Floor-Heating & Snow-Melting Eq.

		15770 Floor-Heating & Snow-Melting Eq.	CREW	DAILY OUTPUT	LABOR-HOURS	UNIT	MAT.	LABOR	EQUIP.	TOTAL	TOTAL INCL O&P	
500	0010	RADIANT FLOOR HEATING										500
	0100	Tubing, PEX (cross-linked polyethylene)										
	0110	Oxygen barrier type for systems with ferrous materials										
	0120	1/2″	Q-5	800	.020	L.F.	.44	.78		1.22	1.69	
	0130	3/4″		535	.030		.56	1.16		1.72	2.42	
	0140	1″	↓	400	.040	↓	1.27	1.55		2.82	3.81	
	0200	Non barrier type for ferrous free systems										
	0210	1/2″	Q-5	800	.020	L.F.	.24	.78		1.02	1.47	
	0220	3/4″		535	.030		.39	1.16		1.55	2.23	
	0230	1″	↓	400	.040	↓	.68	1.55		2.23	3.16	

15770 | Floor-Heating & Snow-Melting Eq.

		Description	Crew	Daily Output	Labor-Hours	Unit	2006 Bare Costs Mat.	Labor	Equip.	Total	Total Incl O&P	
500	1000	Manifolds										500
	1110	Brass										
	1120	Valved										
	1130	1", 2 circuit	Q-5	13	1.231	Ea.	40.50	47.50		88	119	
	1140	1", 3 circuit		11	1.455		55	56.50		111.50	148	
	1150	1", 4 circuit		9	1.778		76	69		145	191	
	1160	1-1/4", 3 circuit		10	1.600		63	62		125	166	
	1170	1-1/4", 4 circuit		8	2		76.50	77.50		154	205	
	1300	Valveless										
	1310	1", 2 circuit	Q-5	14	1.143	Ea.	21	44.50		65.50	92	
	1320	1", 3 circuit		12	1.333		28	51.50		79.50	111	
	1330	1", 4 circuit		10	1.600		36	62		98	136	
	1340	1-1/4", 3 circuit		11	1.455		34.50	56.50		91	126	
	1350	1-1/4", 4 circuit		9	1.778		39	69		108	150	
	1400	Manifold connector kit										
	1410	Mounting brackets, couplings, end cap, air vent, valve, plug										
	1420	1"	Q-5	8	2	Ea.	70.50	77.50		148	199	
	1430	1-1/4"	"	7	2.286	"	83.50	88.50		172	230	
	1610	Copper, (cut to size)										
	1620	1" x 1/2" x 72" Lg, 2" sweat drops, 24 circuit	Q-5	3.33	4.805	Ea.	73.50	186		259.50	370	
	1630	1-1/4" x 1/2" x 72" Lg, 2" sweat drops, 24 circuit		3.20	5		85.50	194		279.50	395	
	1640	1-1/2" x 1/2" x 72" Lg, 2" sweat drops, 24 circuit		3	5.333		105	207		312	435	
	1650	1-1/4" x 3/4" x 72" Lg, 2" sweat drops, 24 circuit		3.10	5.161		91.50	200		291.50	410	
	1660	1-1/2" x 3/4" x 72" Lg, 2" sweat drops, 24 circuit		2.90	5.517		111	214		325	450	
	1710	Copper, (modular)										
	1720	1" x 1/2", 2" sweat drops, 2 circuit	Q-5	17.50	.914	Ea.	7.50	35.50		43	63.50	
	1730	1" x 1/2", 2" sweat drops, 3 circuit		14.50	1.103		10.50	43		53.50	78	
	1740	1" x 1/2", 2" sweat drops, 4 circuit		12.30	1.301		12	50.50		62.50	91.50	
	1750	1-1/4" x 1/2", 2" sweat drops, 2 circuit		17	.941		8.25	36.50		44.75	65.50	
	1760	1-1/4" x 1/2", 2" sweat drops, 3 circuit		14	1.143		12	44.50		56.50	82	
	1770	1-1/4" x 1/2", 2" sweat drops, 4 circuit		11.80	1.356		15.75	52.50		68.25	99	
	3000	Valves										
	3110	Motorized zone valve operator, 24V	Q-5	30	.533	Ea.	43	20.50		63.50	79	
	3120	Motorized zone valve with operator complete, 24V										
	3130	3/4"	Q-5	35	.457	Ea.	74.50	17.70		92.20	109	
	3140	1"		32	.500		84	19.40		103.40	123	
	3150	1-1/4"		29.60	.541		104	21		125	147	
	3200	Flow control/isolation valve, 1/2"		32	.500		24	19.40		43.40	56.50	
	3300	Thermostatic mixing valves										
	3310	1/2"	Q-5	26.60	.602	Ea.	77.50	23.50		101	121	
	3320	3/4"		25	.640		94	25		119	142	
	3330	1"		21.40	.748		101	29		130	156	
	3400	3 Way mixing / diverting valve, manual, brass										
	3410	1/2"	Q-5	21.30	.751	Ea.	102	29		131	158	
	3420	3/4"		20	.800		105	31		136	164	
	3430	1"		17.80	.899		122	35		157	188	
	3500	4 Way mixing / diverting valve, manual, brass										
	3510	1/2"	Q-5	16	1	Ea.	107	39		146	178	
	3520	3/4"		15	1.067		119	41.50		160.50	195	
	3530	1"		13.30	1.203		133	46.50		179.50	219	
	3540	1-1/4"		11.40	1.404		149	54.50		203.50	249	
	3550	1-1/2"		10.60	1.509		196	58.50		254.50	305	
	3560	2"		10	1.600		211	62		273	330	
	3570	2-1/2"		8	2		580	77.50		657.50	760	
	3800	Motor control, 3 or 4 way, for valves up to 1-1/2"		30	.533		147	20.50		167.50	194	
	5000	Radiant floor heating, zone control panel										

MECHANICAL 15

15700 | Heating/Ventilating/Air Conditioning Equipment

15770 | Floor-Heating & Snow-Melting Eq.

		15770 Floor-Heating & Snow-Melting Eq.	CREW	DAILY OUTPUT	LABOR-HOURS	UNIT	2006 BARE COSTS MAT.	LABOR	EQUIP.	TOTAL	TOTAL INCL O&P	
500	5110	6 Zone control										500
	5120	For zone valves	Q-5	16	1	Ea.	138	39		177	213	
	5130	For circulators	"	16	1	"	197	39		236	277	
	7000	PEX tubing fittings, compression type										
	7110	PEX tubing to brass valve conn., 1/2"	Q-5	54	.296	Ea.	2.37	11.50		13.87	20.50	
	7200	PEX x male NPT										
	7210	1/2" x 1/2"	Q-5	54	.296	Ea.	4.62	11.50		16.12	23	
	7220	3/4" x 3/4"		46	.348		7.90	13.50		21.40	29.50	
	7230	1" x 1"	↓	40	.400	↓	18.75	15.50		34.25	44.50	
	7300	PEX coupling										
	7310	1/2" x 1/2"	Q-5	50	.320	Ea.	5.65	12.40		18.05	25.50	
	7320	3/4" x 3/4"		42	.381		16.15	14.75		30.90	41	
	7330	1" x 1"	↓	36	.444	↓	22	17.20		39.20	51	
	7400	PEX x copper sweat										
	7410	1/2" x 1/2"	Q-5	38	.421	Ea.	4.13	16.30		20.43	30	
	7420	3/4" x 3/4"		34	.471		8.65	18.25		26.90	38	
	7430	1" x 1"	↓	29	.552	↓	18.75	21.50		40.25	53.50	

15780 | Energy Recovery Equipment

		15780 Energy Recovery Equipment	CREW	DAILY OUTPUT	LABOR-HOURS	UNIT	MAT.	LABOR	EQUIP.	TOTAL	TOTAL INCL O&P	
100	0010	HEAT RECOVERY PACKAGES										100
	0100	Air to air										
	4000	Enthalpy recovery wheel										
	4010	1000 max CFM	Q-9	1.20	13.333	Ea.	4,675	505		5,180	5,950	
	4020	2000 max CFM		1	16		5,475	605		6,080	7,000	
	4030	4000 max CFM		.80	20		6,350	760		7,110	8,175	
	4040	6000 max CFM	↓	.70	22.857		7,400	865		8,265	9,525	
	4050	8000 max CFM	Q-10	1	24		8,200	945		9,145	10,500	
	4060	10,000 max CFM		.90	26.667		9,800	1,050		10,850	12,500	
	4070	20,000 max CFM		.80	30		17,800	1,175		18,975	21,400	
	4080	25,000 max CFM		.70	34.286		21,700	1,350		23,050	26,100	
	4090	30,000 max CFM		.50	48		24,100	1,900		26,000	29,500	
	4100	40,000 max CFM		.45	53.333		33,200	2,100		35,300	39,800	
	4110	50,000 max CFM	↓	.40	60	↓	38,600	2,350		40,950	46,200	
	9000	Minimum labor/equipment charge	1 Shee	4	2	Job		84.50		84.50	134	

15800 | Air Distribution

15810 | Ducts

		15810 Ducts	CREW	DAILY OUTPUT	LABOR-HOURS	UNIT	2006 BARE COSTS MAT.	LABOR	EQUIP.	TOTAL	TOTAL INCL O&P	
100	0010	METAL DUCTWORK (R233100-40)										100
	0020	Fabricated rectangular, includes fittings, joints, supports,										
	0030	allowance for flexible connections, no insulation										
	0031	NOTE: Fabrication and installation are combined										
	0040	as LABOR cost. Approx. 25% fittings assumed.										
	0042	Fabrication/Inst. is to commercial quality standards										
	0043	(SMACNA or equiv.) for structure, sealing, leak testing, etc.										
	0050	Add to labor for elevated installation										
	0051	of fabricated ductwork										
	0052	10' to 15' high						6%				

	15810 | Ducts	CREW	DAILY OUTPUT	LABOR-HOURS	UNIT	2006 BARE COSTS MAT.	LABOR	EQUIP.	TOTAL	TOTAL INCL O&P
100	0053 15' to 20' high R233100-40						12%			
	0054 20' to 25' high						15%			
	0055 25' to 30' high						21%			
	0056 30' to 35' high						24%			
	0057 35' to 40' high						30%			
	0058 Over 40' high						33%			
	0070 For duct insulation and lining see 15080-200-3000									
	0100 Aluminum, alloy 3003-H14, under 100 lb.	Q-10	75	.320	Lb.	2.50	12.60		15.10	23
	0110 100 to 500 lb.		80	.300		2	11.80		13.80	21
	0120 500 to 1,000 lb.		95	.253		1.65	9.95		11.60	17.60
	0140 1,000 to 2,000 lb.		120	.200		1.60	7.85		9.45	14.25
	0150 2,000 to 5,000 lb.		130	.185		1.58	7.25		8.83	13.30
	0160 Over 5,000 lb.		145	.166		1.55	6.50		8.05	12.05
	0500 Galvanized steel, under 200 lb.		235	.102		.95	4.02		4.97	7.45
	0520 200 to 500 lb.		245	.098		.81	3.85		4.66	7.05
	0540 500 to 1,000 lb.		255	.094		.71	3.70		4.41	6.70
	0560 1,000 to 2,000 lb.		265	.091		.54	3.56		4.10	6.25
	0570 2,000 to 5,000 lb.		275	.087		.50	3.43		3.93	6
	0580 Over 5,000 lb.	↓	285	.084	↓	.48	3.31		3.79	5.80
	0600 For large quantities special prices available from supplier									
	1000 Stainless steel, type 304, under 100 lb.	Q-10	165	.145	Lb.	2.50	5.70		8.20	11.85
	1020 100 to 500 lb.		175	.137		2.30	5.40		7.70	11.15
	1030 500 to 1,000 lb.		190	.126		2.20	4.97		7.17	10.30
	1040 1,000 to 2,000 lb.		200	.120		1.90	4.72		6.62	9.60
	1050 2,000 to 5,000 lb.		225	.107		1.90	4.20		6.10	8.75
	1060 Over 5,000 lb.	↓	235	.102	↓	1.90	4.02		5.92	8.50
	1080 Note: Minimum order cost exceeds per lb. cost for min. wt.									
	1100 For medium pressure ductwork, add				Ea.		15%			
	1200 For high pressure ductwork, add				"		40%			
	1280 Add to labor for elevated installation									
	1282 of prefabricated (purchased) ductwork									
	1283 10' to 15' high						10%			
	1284 15' to 20' high						20%			
	1285 20' to 25' high						25%			
	1286 25' to 30' high						35%			
	1287 30' to 35' high						40%			
	1288 35' to 40' high						50%			
	1289 Over 40' high						55%			
	5400 Spiral preformed, steel, galv., straight lengths, Max 10" spwg.									
	5410 4" diameter, 26 ga.	Q-9	360	.044	L.F.	1.21	1.69		2.90	4.01
	5416 5" diameter, 26 ga.		320	.050		1.54	1.90		3.44	4.71
	5420 6" diameter, 26 ga.		280	.057		1.76	2.17		3.93	5.40
	5425 7" diameter, 26 ga.		240	.067		2.09	2.53		4.62	6.30
	5430 8" diameter, 26 ga.		200	.080		2.31	3.03		5.34	7.35
	5440 10" diameter, 26 ga.		160	.100		2.86	3.79		6.65	9.20
	5450 12" diameter, 26 ga.		120	.133		3.52	5.05		8.57	11.90
	5460 14" diameter, 26 ga.		80	.200		4.07	7.60		11.67	16.55
	5480 16" diameter, 24 ga.		60	.267		6.05	10.10		16.15	23
	5490 18" diameter, 24 ga.	↓	50	.320		6.70	12.15		18.85	26.50
	5500 20" diameter, 24 ga.	Q-10	65	.369		7.50	14.50		22	31.50
	5510 22" diameter, 24 ga.		60	.400		8.25	15.75		24	34
	5520 24" diameter, 24 ga.		55	.436		9	17.15		26.15	37.50
	5540 30" diameter, 22 ga.		45	.533		13.65	21		34.65	48.50
	5600 36" diameter, 22 ga.	↓	40	.600	↓	16.40	23.50		39.90	55.50
	5800 Connector, 4" diameter	Q-9	100	.160	Ea.	2.26	6.05		8.31	12.15
	5810 5" diameter	↓	94	.170	↓	2.42	6.45		8.87	12.90

MECHANICAL 15

		15810 | Ducts	CREW	DAILY OUTPUT	LABOR-HOURS	UNIT	2006 BARE COSTS MAT.	LABOR	EQUIP.	TOTAL	TOTAL INCL O&P	
100	5820	6″ diameter (R233100-40)	Q-9	88	.182	Ea.	2.60	6.90		9.50	13.80	100
	5840	8″ diameter		78	.205		3.36	7.80		11.16	16.10	
	5860	10″ diameter		70	.229		5.30	8.65		13.95	19.65	
	5880	12″ diameter		50	.320		6.40	12.15		18.55	26.50	
	5900	14″ diameter		44	.364		7.40	13.80		21.20	30	
	5920	16″ diameter		40	.400		8.30	15.15		23.45	33	
	5930	18″ diameter		37	.432		9.30	16.40		25.70	36.50	
	5940	20″ diameter		34	.471		10.05	17.85		27.90	39.50	
	5950	22″ diameter		31	.516		11	19.60		30.60	43	
	5960	24″ diameter		28	.571		11.80	21.50		33.30	47.50	
	5980	30″ diameter		22	.727		15.35	27.50		42.85	61	
	6000	36″ diameter		18	.889		23	33.50		56.50	79	
	6300	Elbow, 45°, 4″ diameter		60	.267		2.55	10.10		12.65	18.90	
	6310	5″ diameter		52	.308		3.28	11.65		14.93	22	
	6320	6″ diameter		44	.364		3.58	13.80		17.38	26	
	6340	8″ diameter		28	.571		4.73	21.50		26.23	39.50	
	6360	10″ diameter		18	.889		7.20	33.50		40.70	61.50	
	6380	12″ diameter		13	1.231		9.30	46.50		55.80	85	
	6400	14″ diameter		11	1.455		17.45	55		72.45	107	
	6420	16″ diameter		10	1.600		28.50	60.50		89	128	
	6430	18″ diameter		9.60	1.667		35	63		98	140	
	6440	20″ diameter	Q-10	14	1.714		44.50	67.50		112	156	
	6450	22″ diameter		13	1.846		49.50	72.50		122	171	
	6460	24″ diameter		12	2		59.50	78.50		138	191	
	6480	30″ diameter		9	2.667		91.50	105		196.50	268	
	6500	36″ diameter		7	3.429		111	135		246	335	
	6600	Elbow, 90°, 4″ diameter	Q-9	60	.267		4.36	10.10		14.46	21	
	6610	5″ diameter		52	.308		5	11.65		16.65	24	
	6620	6″ diameter		44	.364		5.75	13.80		19.55	28.50	
	6625	7″ diameter		36	.444		6.05	16.85		22.90	33.50	
	6630	8″ diameter		28	.571		7.05	21.50		28.55	42.50	
	6640	10″ diameter		18	.889		10.05	33.50		43.55	64.50	
	6650	12″ diameter		13	1.231		13.30	46.50		59.80	89	
	6660	14″ diameter		11	1.455		25	55		80	116	
	6670	16″ diameter		10	1.600		39.50	60.50		100	140	
	6676	18″ diameter		9.60	1.667		51.50	63		114.50	158	
	6680	20″ diameter	Q-10	14	1.714		68.50	67.50		136	182	
	6684	22″ diameter		13	1.846		106	72.50		178.50	233	
	6690	24″ diameter		12	2		119	78.50		197.50	256	
	6700	30″ diameter		9	2.667		183	105		288	370	
	6710	36″ diameter		7	3.429		223	135		358	460	
	6800	Reducing coupling, 6″ x 4″	Q-9	46	.348		6.50	13.20		19.70	28	
	6820	8″ x 6″		40	.400		8	15.15		23.15	33	
	6840	10″ x 8″		32	.500		10.10	18.95		29.05	41	
	6860	12″ x 10″		24	.667		14.45	25.50		39.95	56	
	6880	14″ x 12″		20	.800		24	30.50		54.50	74.50	
	6900	16″ x 14″		18	.889		34.50	33.50		68	91.50	
	6920	18″ x 16″		16	1		43	38		81	108	
	6940	20″ x 18″	Q-10	24	1		52.50	39.50		92	120	
	6950	22″ x 20″		23	1.043		61.50	41		102.50	133	
	6960	24″ x 22″		22	1.091		70	43		113	146	
	6980	30″ x 28″		18	1.333		104	52.50		156.50	199	
	7000	36″ x 34″		16	1.500		140	59		199	248	
	7100	Tee, 90°, 4″ diameter	Q-9	40	.400		5.70	15.15		20.85	30.50	
	7110	5″ diameter		34.60	.462		7.60	17.55		25.15	36.50	
	7120	6″ diameter		30	.533		8.20	20		28.20	41	100

	15810 | Ducts		CREW	DAILY OUTPUT	LABOR-HOURS	UNIT	2006 BARE COSTS MAT.	LABOR	EQUIP.	TOTAL	TOTAL INCL O&P	
100	7130	8″ diameter R233100-40	Q-9	19	.842	Ea.	10.70	32		42.70	63	100
	7140	10″ diameter		12	1.333		15.25	50.50		65.75	97.50	
	7150	12″ diameter		9	1.778		19.50	67.50		87	129	
	7160	14″ diameter		7	2.286		37	86.50		123.50	179	
	7170	16″ diameter		6.70	2.388		52	90.50		142.50	202	
	7176	18″ diameter		6	2.667		72	101		173	241	
	7180	20″ diameter	Q-10	8.70	2.759		79.50	109		188.50	261	
	7190	24″ diameter		7	3.429		95	135		230	320	
	7200	30″ diameter		6	4		117	157		274	380	
	7210	36″ diameter		5	4.800		148	189		337	465	
	7400	Steel, PVC coated both sides, straight lengths										
	7410	4″ diameter, 26 ga.	Q-9	360	.044	L.F.	1.70	1.69		3.39	4.55	
	7416	5″ diameter, 26 ga.		240	.067		2.12	2.53		4.65	6.35	
	7420	6″ diameter, 26 ga.		280	.057		2.36	2.17		4.53	6.05	
	7440	8″ diameter, 26 ga.		200	.080		3.26	3.03		6.29	8.40	
	7460	10″ diameter, 26 ga.		160	.100		4.38	3.79		8.17	10.85	
	7480	12″ diameter, 26 ga.		120	.133		5.45	5.05		10.50	14	
	7500	14″ diameter, 26 ga.		80	.200		7.45	7.60		15.05	20	
	7520	16″ diameter, 24 ga.		60	.267		10.50	10.10		20.60	27.50	
	7540	18″ diameter, 24 ga.		45	.356		11.35	13.50		24.85	34	
	7560	20″ diameter, 24 ga.	Q-10	65	.369		13.40	14.50		27.90	37.50	
	7580	24″ diameter, 24 ga.		55	.436		17	17.15		34.15	46	
	7600	30″ diameter, 22 ga.		45	.533		24.50	21		45.50	60.50	
	7620	36″ diameter, 22 ga.		40	.600		30	23.50		53.50	70.50	
	7890	Connector, 4″ diameter	Q-9	100	.160	Ea.	3.56	6.05		9.61	13.55	
	7896	5″ diameter		83	.193		3.80	7.30		11.10	15.85	
	7900	6″ diameter		88	.182		4	6.90		10.90	15.35	
	7920	8″ diameter		78	.205		5.05	7.80		12.85	17.95	
	7940	10″ diameter		70	.229		6.95	8.65		15.60	21.50	
	7960	12″ diameter		50	.320		8.90	12.15		21.05	29	
	7980	14″ diameter		44	.364		11.45	13.80		25.25	34.50	
	8000	16″ diameter		40	.400		12.70	15.15		27.85	38	
	8020	18″ diameter		37	.432		14.30	16.40		30.70	42	
	8040	20″ diameter		34	.471		15.40	17.85		33.25	45.50	
	8060	24″ diameter		28	.571		17.20	21.50		38.70	53.50	
	8080	30″ diameter		22	.727		22.50	27.50		50	68.50	
	8100	36″ diameter		18	.889		30.50	33.50		64	87	
	8390	Elbow, 45°, 4″ diameter		60	.267		4.48	10.10		14.58	21	
	8396	5″ diameter		36	.444		5.40	16.85		22.25	33	
	8400	6″ diameter		44	.364		5.90	13.80		19.70	28.50	
	8420	8″ diameter		28	.571		7.80	21.50		29.30	43	
	8440	10″ diameter		18	.889		11.30	33.50		44.80	66	
	8460	12″ diameter		13	1.231		16.65	46.50		63.15	93	
	8480	14″ diameter		11	1.455		30	55		85	121	
	8500	16″ diameter		10	1.600		43	60.50		103.50	144	
	8520	18″ diameter		9	1.778		49.50	67.50		117	162	
	8540	20″ diameter	Q-10	13	1.846		66.50	72.50		139	189	
	8560	24″ diameter		11	2.182		84	86		170	230	
	8580	30″ diameter		9	2.667		129	105		234	310	
	8600	36″ diameter		7	3.429		157	135		292	390	
	8800	Elbow, 90°, 4″ diameter	Q-9	60	.267		7.45	10.10		17.55	24.50	
	8804	5″ diameter		52	.308		8.50	11.65		20.15	28	
	8806	6″ diameter		44	.364		9.30	13.80		23.10	32.50	
	8808	8″ diameter		28	.571		11.60	21.50		33.10	47.50	
	8810	10″ diameter		18	.889		15.65	33.50		49.15	70.50	
	8812	12″ diameter		13	1.231		21.50	46.50		68	98	

15810 | Ducts

15810	Ducts		CREW	DAILY OUTPUT	LABOR-HOURS	UNIT	2006 BARE COSTS MAT.	LABOR	EQUIP.	TOTAL	TOTAL INCL O&P
100	8814	14" diameter R233100-40	Q-9	11	1.455	Ea.	38.50	55		93.50	131
	8816	16" diameter		10	1.600		60	60.50		120.50	163
	8818	18" diameter		9.60	1.667		73.50	63		136.50	182
	8820	20" diameter	Q-10	14	1.714		93.50	67.50		161	210
	8822	24" diameter		12	2		168	78.50		246.50	310
	8824	30" diameter		9	2.667		258	105		363	450
	8826	36" diameter		7	3.429		315	135		450	560
	8850	Tee, 90°, 4" diameter	Q-9	40	.400		8.35	15.15		23.50	33
	8854	5" diameter		34.60	.462		11	17.55		28.55	40
	8856	6" diameter		30	.533		12.20	20		32.20	45.50
	8858	8" diameter		19	.842		15.65	32		47.65	68.50
	8860	10" diameter		12	1.333		20.50	50.50		71	103
	8862	12" diameter		9	1.778		29	67.50		96.50	139
	8864	14" diameter		7	2.286		50.50	86.50		137	194
	8866	16" diameter		6.70	2.388		73.50	90.50		164	225
	8868	18" diameter		6	2.667		107	101		208	278
	8870	20" diameter	Q-10	8.70	2.759		115	109		224	300
	8872	24" diameter		7	3.429		158	135		293	390
	8874	30" diameter		6	4		198	157		355	470
	8876	36" diameter		5	4.800		251	189		440	575
	9000	Reducing coupling, tapered offset									
	9020	Diameter x any smaller diameter									
	9040	6" x	Q-9	46	.348	Ea.	9.25	13.20		22.45	31
	9060	8" x		40	.400		11.45	15.15		26.60	36.50
	9080	10" x		32	.500		13.80	18.95		32.75	45
	9100	12" x		24	.667		19.20	25.50		44.70	61
	9120	14" x		20	.800		31	30.50		61.50	82.50
	9140	16" x		18	.889		47	33.50		80.50	105
	9160	18" x		16	1		57	38		95	123
	9180	20" x	Q-10	24	1		70	39.50		109.50	140
	9200	24" x		22	1.091		95	43		138	173
	9220	30" x		18	1.333		134	52.50		186.50	231
	9240	36" x		16	1.500		175	59		234	286
	9400	Lateral 45°, straight barrel									
	9410	4" size	Q-9	40	.400	Ea.	9.60	15.15		24.75	34.50
	9416	5" size		34.60	.462		12.65	17.55		30.20	42
	9420	6" size		30	.533		14.05	20		34.05	47.50
	9440	8" size		19	.842		18	32		50	71
	9460	10" size		12	1.333		23.50	50.50		74	107
	9480	12" size		9	1.778		29	67.50		96.50	139
	9500	14" size		7	2.286		39	86.50		125.50	181
	9520	16" size		6.70	2.388		73	90.50		163.50	225
	9540	18" size		6	2.667		103	101		204	274
	9560	20" size	Q-10	8.70	2.759		122	109		231	305
	9580	24" size		7	3.429		170	135		305	400
	9600	30" size		6	4		198	157		355	470
	9620	36" size		5	4.800		259	189		448	585
	9640	For reducing lateral, add					40%				
	9700	Register boot, elbow or perimeter									
	9710	6" duct diameter	Q-9	60	.267	Ea.	34.50	10.10		44.60	54
	9720	8" duct diameter		56	.286		41.50	10.85		52.35	63
	9730	10" duct diameter		45	.356		57	13.50		70.50	84
	9740	12" duct diameter		36	.444		57	16.85		73.85	89.50
	9800	Steel, stainless, straight lengths									
	9810	3" diameter, 24 Ga.	Q-9	400	.040	L.F.	4.28	1.52		5.80	7.10
	9820	4" diameter, 24 Ga.		360	.044		5.40	1.69		7.09	8.65

15810 | Ducts

		15810 Ducts	Crew	Daily Output	Labor-Hours	Unit	2006 Bare Costs Mat.	Labor	Equip.	Total	Total Incl O&P	
100	9830	5″ diameter, 24 Ga. (R233100-40)	Q-9	320	.050	L.F.	6.65	1.90		8.55	10.30	100
	9840	6″ diameter, 24 Ga.		280	.057		7.80	2.17		9.97	12.05	
	9850	7″ diameter, 24 Ga.		240	.067		8.95	2.53		11.48	13.85	
	9860	8″ diameter, 24 Ga.		200	.080		10.15	3.03		13.18	16	
	9870	9″ diameter, 24 Ga.		180	.089		11.30	3.37		14.67	17.80	
	9880	10″ diameter, 24 Ga.		160	.100		12.45	3.79		16.24	19.75	
	9890	12″ diameter, 24 Ga.		120	.133		14.85	5.05		19.90	24.50	
	9900	14″ diameter, 24 Ga.		80	.200		17.15	7.60		24.75	31	
	9910	16″ diameter, 24 Ga.		60	.267		19.50	10.10		29.60	37.50	
	9920	18″ diameter, 24 Ga.		45	.356		22	13.50		35.50	45.50	
	9930	20″ diameter, 24 Ga.	Q-10	65	.369		24	14.50		38.50	49.50	
	9940	24″ diameter, 24 Ga.		55	.436		29	17.15		46.15	59.50	
	9950	30″ diameter, 22 Ga.		45	.533		44	21		65	82	
	9960	36″ diameter, 22 Ga.		40	.600		53.50	23.50		77	96.50	
	9970	40″ diameter, 20 Ga.		35	.686		71	27		98	121	
	9990	Minimum labor/equipment charge	1 Shee	3	2.667	Job		112		112	179	
300	0010	**FIBEROUS GLASS DUCTWORK** (R233100-40)										300
	1280	Add to labor for elevated installation										
	1282	of prefabricated (purchased) ductwork										
	1283	10′ to 15′ high						10%				
	1284	15′ to 20′ high						20%				
	1285	20′ to 25′ high						25%				
	1286	25′ to 30′ high						35%				
	1287	30′ to 35′ high						40%				
	1288	35′ to 40′ high						50%				
	1289	Over 40′ high						55%				
	3490	Rigid fiberglass duct board, foil reinf. kraft facing										
	3500	Rectangular, 1″ thick, alum. faced, (FRK), std. weight	Q-10	350	.069	SF Surf	.62	2.70		3.32	4.97	
	9990	Minimum labor/equipment charge	1 Shee	3	2.667	Job		112		112	179	
500	0010	**FLEXIBLE DUCTS** (R233100-40)										500
	1280	Add to labor for elevated installation										
	1282	of prefabricated (purchased) ductwork										
	1283	10′ to 15′ high						10%				
	1284	15′ to 20′ high						20%				
	1285	20′ to 25′ high						25%				
	1286	25′ to 30′ high						35%				
	1287	30′ to 35′ high						40%				
	1288	35′ to 40′ high						50%				
	1289	Over 40′ high						55%				
	1300	Flexible, coated fiberglass fabric on corr. resist. metal helix										
	1400	pressure to 12″ (WG) UL-181										
	1500	Non-insulated, 3″ diameter	Q-9	400	.040	L.F.	1.08	1.52		2.60	3.60	
	1520	4″ diameter		360	.044		1.08	1.69		2.77	3.87	
	1540	5″ diameter		320	.050		1.35	1.90		3.25	4.51	
	1560	6″ diameter		280	.057		1.60	2.17		3.77	5.20	
	1580	7″ diameter		240	.067		1.90	2.53		4.43	6.10	
	1600	8″ diameter		200	.080		2.20	3.03		5.23	7.25	
	1620	9″ diameter		180	.089		2.60	3.37		5.97	8.20	
	1640	10″ diameter		160	.100		2.76	3.79		6.55	9.10	
	1660	12″ diameter		120	.133		3.34	5.05		8.39	11.70	
	1680	14″ diameter		80	.200		4.06	7.60		11.66	16.50	
	1700	16″ diameter		60	.267		6	10.10		16.10	22.50	
	1800	For adjustable clamps, add				Ea.	.78			.78	.86	
	1900	Insulated, 1″ thick, PE jacket, 3″ diameter	Q-9	380	.042	L.F.	2.22	1.60		3.82	4.98	
	1910	4″ diameter		340	.047		2.22	1.79		4.01	5.30	

For expanded coverage of these items see *Means Mechanical or Plumbing Cost Data 2006*

	15810 | Ducts		CREW	DAILY OUTPUT	LABOR-HOURS	UNIT	2006 BARE COSTS MAT.	LABOR	EQUIP.	TOTAL	TOTAL INCL O&P	
500	1920	5″ diameter R233100-40	Q-9	300	.053	L.F.	2.39	2.02		4.41	5.85	500
	1940	6″ diameter		260	.062		2.54	2.33		4.87	6.50	
	1960	7″ diameter		220	.073		2.95	2.76		5.71	7.65	
	1980	8″ diameter		180	.089		3.15	3.37		6.52	8.80	
	2000	9″ diameter		160	.100		3.58	3.79		7.37	10	
	2020	10″ diameter		140	.114		3.71	4.34		8.05	11	
	2040	12″ diameter		100	.160		4.62	6.05		10.67	14.75	
	2060	14″ diameter		80	.200		5.60	7.60		13.20	18.20	
	2080	16″ diameter		60	.267		7	10.10		17.10	24	
	2100	18″ diameter		45	.356		8.80	13.50		22.30	31	
	2120	20″ diameter	Q-10	65	.369		9.70	14.50		24.20	33.50	
	2500	Insulated, heavy duty, coated fiberglass fabric										
	2520	4″ diameter	Q-9	340	.047	L.F.	3.63	1.79		5.42	6.85	
	2540	5″ diameter		300	.053		3.89	2.02		5.91	7.50	
	2560	6″ diameter		260	.062		4.34	2.33		6.67	8.50	
	2580	7″ diameter		220	.073		4.88	2.76		7.64	9.75	
	2600	8″ diameter		180	.089		5.55	3.37		8.92	11.45	
	2620	9″ diameter		160	.100		5.90	3.79		9.69	12.55	
	2640	10″ diameter		140	.114		7.10	4.34		11.44	14.70	
	2660	12″ diameter		100	.160		7.60	6.05		13.65	18	
	2680	14″ diameter		80	.200		9.95	7.60		17.55	23	
	2700	16″ diameter		60	.267		14.95	10.10		25.05	32.50	
	2720	18″ diameter		45	.356		18.30	13.50		31.80	41.50	
	9990	Minimum labor/equipment charge	1 Shee	3	2.667	Job		112		112	179	
700	0010	**GLASS FIBER-REINFORCED PLASTIC DUCTS** R233100-40										700
	1280	Add to labor for elevated installation										
	1282	of prefabricated (purchased) ductwork										
	1283	10′ to 15′ high						10%				
	1284	15′ to 20′ high						20%				
	1285	20′ to 25′ high						25%				
	1286	25′ to 30′ high						35%				
	1287	30′ to 35′ high						40%				
	1288	35′ to 40′ high						50%				
	1289	Over 40′ high						55%				
	4000	Rigid plastic, corrosive fume resistant PVC										
	4020	Straight, 6″ diameter	Q-9	220	.073	L.F.	6.80	2.76		9.56	11.90	
	4030	7″ diameter		160	.100		10.10	3.79		13.89	17.15	
	4040	8″ diameter		160	.100		10.10	3.79		13.89	17.15	
	4050	9″ diameter		140	.114		11.40	4.34		15.74	19.45	
	4060	10″ diameter		120	.133		12.75	5.05		17.80	22	
	4070	12″ diameter		100	.160		14.85	6.05		20.90	26	
	4080	14″ diameter		70	.229		19.25	8.65		27.90	35	
	4090	16″ diameter		62	.258		22.50	9.80		32.30	40	
	4100	18″ diameter		58	.276		34.50	10.45		44.95	54.50	
	4110	20″ diameter	Q-10	75	.320		44.50	12.60		57.10	69	
	4120	22″ diameter		60	.400		51	15.75		66.75	81	
	4130	24″ diameter		55	.436		57.50	17.15		74.65	90.50	
	4140	26″ diameter		42	.571		65	22.50		87.50	108	
	4150	28″ diameter		35	.686		69.50	27		96.50	120	
	4160	30″ diameter		28	.857		75.50	33.50		109	137	
	4170	36″ diameter		23	1.043		90	41		131	165	
	4180	42″ diameter		20	1.200		105	47		152	191	
	4200	48″ diameter		18	1.333		120	52.50		172.50	216	
	4250	Coupling, 6″ diameter	Q-9	88	.182	Ea.	7.10	6.90		14	18.75	
	4260	7″ diameter		82	.195		7.60	7.40		15	20	
	4270	8″ diameter		78	.205		7.95	7.80		15.75	21	

15800 | Air Distribution

15810 | Ducts

			CREW	DAILY OUTPUT	LABOR-HOURS	UNIT	2006 BARE COSTS MAT.	LABOR	EQUIP.	TOTAL	TOTAL INCL O&P	
700	4280	9″ diameter R233100-40	Q-9	74	.216	Ea.	8.65	8.20		16.85	22.50	700
	4290	10″ diameter		70	.229		8.65	8.65		17.30	23.50	
	4300	12″ diameter		55	.291		9.25	11.05		20.30	28	
	4310	14″ diameter		44	.364		11.25	13.80		25.05	34.50	
	4320	16″ diameter		40	.400		13.35	15.15		28.50	38.50	
	4330	18″ diameter		38	.421		15.50	15.95		31.45	42.50	
	4340	20″ diameter	Q-10	55	.436		27	17.15		44.15	57	
	4350	22″ diameter		50	.480		31	18.90		49.90	64	
	4360	24″ diameter		45	.533		35	21		56	72	
	4370	26″ diameter		39	.615		37	24		61	79	
	4380	28″ diameter		33	.727		38.50	28.50		67	88	
	4390	30″ diameter		27	.889		42	35		77	102	
	4400	36″ diameter		25	.960		66.50	38		104.50	133	
	4410	42″ diameter		23	1.043		74.50	41		115.50	148	
	4420	48″ diameter		21	1.143		87	45		132	168	
	4470	Elbow, 90°, 6″ diameter	Q-9	44	.364		66.50	13.80		80.30	95	
	4480	7″ diameter		36	.444		71.50	16.85		88.35	106	
	4490	8″ diameter		28	.571		71.50	21.50		93	113	
	4500	9″ diameter		22	.727		98	27.50		125.50	152	
	4510	10″ diameter		18	.889		116	33.50		149.50	181	
	4520	12″ diameter		15	1.067		158	40.50		198.50	238	
	4530	14″ diameter		11	1.455		216	55		271	325	
	4540	16″ diameter		10	1.600		260	60.50		320.50	385	
	4550	18″ diameter	Q-10	15	1.600		282	63		345	410	
	4560	20″ diameter		14	1.714		299	67.50		366.50	435	
	4570	22″ diameter		13	1.846		380	72.50		452.50	535	
	4580	24″ diameter		12	2		430	78.50		508.50	595	
	4590	26″ diameter		11	2.182		490	86		576	675	
	4600	28″ diameter		10	2.400		580	94.50		674.50	790	
	4610	30″ diameter		9	2.667		685	105		790	920	
	4620	36″ diameter		7	3.429		940	135		1,075	1,250	
	4630	42″ diameter		5	4.800		1,300	189		1,489	1,725	
	4640	48″ diameter		4	6		1,700	236		1,936	2,250	
	4750	Elbow 45°, use 90° and deduct					45%					
	9990	Minimum labor/equipment charge	1 Shee	3	2.667	Job		112		112	179	

15820 | Duct Accessories

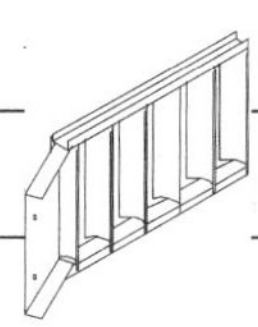

			CREW	DAILY OUTPUT	LABOR-HOURS	UNIT	MAT.	LABOR	EQUIP.	TOTAL	TOTAL INCL O&P	
300	0010	DUCT ACCESSORIES										300
	0050	Air extractors, 12″ x 4″	1 Shee	24	.333	Ea.	16.50	14.05		30.55	40.50	
	0100	8″ x 6″		22	.364		16.50	15.35		31.85	42.50	
	0120	12″ x 6″		21	.381		23	16.05		39.05	50.50	
	0140	16″ x 6″		20	.400		28.50	16.85		45.35	58.50	
	0160	24″ x 6″		18	.444		48	18.75		66.75	82.50	
	0180	12″ x 8″		20	.400		25.50	16.85		42.35	55	
	0200	20″ x 8″		16	.500		37.50	21		58.50	74.50	
	0240	18″ x 10″		14	.571		36.50	24		60.50	78.50	
	0260	24″ x 10″		12	.667		47	28		75	96	
	0280	24″ x 12″		10	.800		50.50	33.50		84	109	
	0300	30″ x 12″		8	1		63.50	42		105.50	137	
	1000	Duct access door, insulated, 6″ x 6″		14	.571		12.90	24		36.90	52.50	
	1020	10″ x 10″		11	.727		19.80	30.50		50.30	71	
	1040	12″ x 12″		10	.800		21	33.50		54.50	77	
	1050	12″ x 18″		9	.889		33	37.50		70.50	95.50	
	1060	16″ x 12″		9	.889		30	37.50		67.50	92.50	
	1070	18″ x 18″		8	1		37.50	42		79.50	109	

MECHANICAL 15

15820 | Duct Accessories		CREW	DAILY OUTPUT	LABOR-HOURS	UNIT	2006 BARE COSTS MAT.	LABOR	EQUIP.	TOTAL	TOTAL INCL O&P
300 1074	24" x 18"	1 Shee	8	1	Ea.	45	42		87	117
1080	24" x 24"		8	1	↓	48	42		90	120
2000	Fabrics for flexible connections, with metal edge		100	.080	L.F.	1.84	3.37		5.21	7.35
2100	Without metal edge		160	.050	"	1.33	2.11		3.44	4.81
3000	Fire damper, curtain type, 1-1/2 hr rated, vertical, 6" x 6"		24	.333	Ea.	14.75	14.05		28.80	39
3020	8" x 6"		22	.364		15.10	15.35		30.45	41
3040	12" x 6"		22	.364		23.50	15.35		38.85	50
3060	20" x 6"		18	.444		27.50	18.75		46.25	60
3080	12" x 8"		22	.364		23.50	15.35		38.85	50
3100	24" x 8"		16	.500		32.50	21		53.50	69.50
3120	12" x 10"		21	.381		23.50	16.05		39.55	51
3140	24" x 10"		15	.533		36.50	22.50		59	76
3160	36" x 10"		12	.667		45	28		73	94
3180	16" x 12"		20	.400		27.50	16.85		44.35	57
3200	24" x 12"		13	.615		24.50	26		50.50	68.50
3220	48" x 12"		10	.800		56	33.50		89.50	115
3240	16" x 14"		18	.444		32.50	18.75		51.25	66
3260	24" x 14"		12	.667		43.50	28		71.50	92.50
3280	30" x 14"		11	.727		34	30.50		64.50	86
3300	18" x 16"		17	.471		25.50	19.85		45.35	59.50
3320	24" x 16"		11	.727		42.50	30.50		73	95.50
3340	36" x 16"		10	.800		37	33.50		70.50	94
3360	24" x 18"		10	.800		41.50	33.50		75	99.50
3380	48" x 18"		8	1		71.50	42		113.50	146
3400	24" x 20"		8	1		45.50	42		87.50	117
3420	36" x 20"		7	1.143		60	48		108	143
3440	24" x 22"		7	1.143		47	48		95	128
3460	30" x 22"		6	1.333		54.50	56		110.50	150
3480	26" x 24"	↓	7	1.143		51.50	48		99.50	133
3500	48" x 24"	Q-9	12	1.333		78.50	50.50		129	167
3520	28" x 26"		13	1.231		57	46.50		103.50	138
3540	30" x 28"		12	1.333		63.50	50.50		114	151
3560	48" x 30"		11	1.455		93	55		148	190
3580	48" x 36"		10	1.600		101	60.50		161.50	209
3600	44" x 44"		10	1.600		106	60.50		166.50	214
3620	48" x 48"	↓	8	2	↓	95	76		171	226
3700	U.L. label included in above									
3800	For horizontal operation, add				Ea.	20%				
3900	For cap for blades out of air stream, add					20%				
4000	For 10" 22 ga., U.L. approved sleeve, add					35%				
4100	For 10" 22 ga., U.L. sleeve, 100% free area, add				↓	65%				
4200	For oversize openings group dampers									
4500	Fire/smoke combination damper, louver type, UL									
4510	8" x 8"	1 Shee	22	.364	Ea.	136	15.35		151.35	174
4520	16" x 8"		20	.400		148	16.85		164.85	190
4540	18" x 8"		18	.444		148	18.75		166.75	193
4560	20" x 8"		16	.500		159	21		180	209
4580	10" x 10"		21	.381		153	16.05		169.05	195
4600	24" x 10"		15	.533		180	22.50		202.50	234
4620	30" x 10"		12	.667		186	28		214	250
4640	12" x 12"		20	.400		160	16.85		176.85	203
4660	18" x 12"		18	.444		163	18.75		181.75	209
4680	24" x 12"		13	.615		172	26		198	231
4700	30" x 12"		11	.727		179	30.50		209.50	246
4720	14" x 14"		17	.471		172	19.85		191.85	222
4740	16" x 14"	↓	18	.444	↓	177	18.75		195.75	225

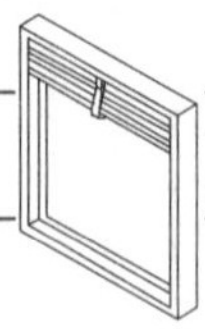

	15820 | Duct Accessories		CREW	DAILY OUTPUT	LABOR-HOURS	UNIT	2006 BARE COSTS MAT.	LABOR	EQUIP.	TOTAL	TOTAL INCL O&P	
300	4760	20″ x 14″	1 Shee	14	.571	Ea.	180	24		204	237	300
	4780	24″ x 14″		12	.667		185	28		213	248	
	4800	30″ x 14″		10	.800		196	33.50		229.50	269	
	4820	16″ x 16″		16	.500		191	21		212	244	
	4840	20″ x 16″		14	.571		196	24		220	254	
	4860	24″ x 16″		11	.727		202	30.50		232.50	271	
	4880	30″ x 16″		8	1		211	42		253	299	
	4900	18″ x 18″		15	.533		189	22.50		211.50	244	
	5000	24″ x 18″		10	.800		202	33.50		235.50	276	
	5020	36″ x 18″		7	1.143		221	48		269	320	
	5040	20″ x 20″		13	.615		211	26		237	274	
	5060	24″ x 20″		8	1		216	42		258	305	
	5080	30″ x 20″		8	1		227	42		269	315	
	5100	36″ x 20″		7	1.143		239	48		287	340	
	5120	24″ x 24″		8	1		233	42		275	325	
	5130	30″ x 24″		7.60	1.053		187	44.50		231.50	277	
	5140	36″ x 24″	Q-9	12	1.333		255	50.50		305.50	360	
	5141	36″ x 30″		10	1.600		310	60.50		370.50	435	
	5142	40″ x 36″		8	2		480	76		556	645	
	5143	48″ x 48″		4	4		935	152		1,087	1,275	
	5150	Damper operator motor, 24 or 120 volt	1 Shee	16	.500		279	21		300	340	
	5990	Multi-blade dampers, opposed blade, 8″ x 6″		24	.333		20	14.05		34.05	44.50	
	5994	8″ x 8″		22	.364		21	15.35		36.35	47.50	
	5996	10″ x 10″		21	.381		24	16.05		40.05	52	
	6000	12″ x 12″		21	.381		27	16.05		43.05	55.50	
	6020	12″ x 18″		18	.444		37	18.75		55.75	70.50	
	6030	14″ x 10″		20	.400		26.50	16.85		43.35	56	
	6031	14″ x 14″		17	.471		33	19.85		52.85	67.50	
	6033	16″ x 12″		17	.471		33	19.85		52.85	67.50	
	6035	16″ x 16″		16	.500		41	21		62	78.50	
	6037	18″ x 16″		15	.533		45	22.50		67.50	85.50	
	6038	18″ x 18″		15	.533		49	22.50		71.50	89.50	
	6040	18″ x 24″		12	.667		62.50	28		90.50	113	
	6060	18″ x 28″		10	.800		72	33.50		105.50	133	
	6070	20″ x 16″		14	.571		49	24		73	92	
	6072	20″ x 20″		13	.615		58.50	26		84.50	106	
	6074	22″ x 18″		14	.571		58.50	24		82.50	103	
	6076	24″ x 16″		11	.727		57.50	30.50		88	113	
	6078	24″ x 20″		8	1		68	42		110	142	
	6080	24″ x 24″		8	1		79	42		121	154	
	6100	24″ x 28″		6	1.333		92	56		148	191	
	6110	26″ x 26″		6	1.333		85.50	56		141.50	184	
	6120	28″ x 28″	Q-9	11	1.455		98.50	55		153.50	196	
	6130	30″ x 18″		10	1.600		83	60.50		143.50	188	
	6132	30″ x 24″		7	2.286		104	86.50		190.50	252	
	6133	30″ x 30″		6.60	2.424		124	92		216	282	
	6135	32″ x 32″		6.40	2.500		144	95		239	310	
	6151	36″ x 12″		10	1.600		65.50	60.50		126	169	
	6152	36″ x 16″		8	2		89	76		165	219	
	6158	36″ x 36″		6	2.667		176	101		277	355	
	6160	44″ x 28″		5.80	2.759		181	105		286	365	
	6180	48″ x 36″		5.60	2.857		236	108		344	430	
	6200	56″ x 36″		5.40	2.963		280	112		392	490	
	6220	60″ x 36″		5.20	3.077		296	117		413	510	
	6240	60″ x 44″		5	3.200		360	121		481	590	
	7000	Splitter damper assembly, self-locking, 1′ rod	1 Shee	24	.333		17.40	14.05		31.45	41.50	

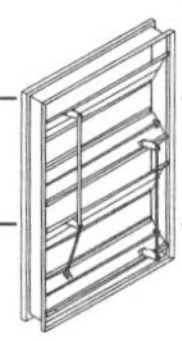

		15820 | Duct Accessories	CREW	DAILY OUTPUT	LABOR-HOURS	UNIT	2006 BARE COSTS MAT.	LABOR	EQUIP.	TOTAL	TOTAL INCL O&P	
300	7020	3′ rod	1 Shee	22	.364	Ea.	22.50	15.35		37.85	49.50	300
	7040	4′ rod		20	.400		25	16.85		41.85	54.50	
	7060	6′ rod	↓	18	.444	↓	30.50	18.75		49.25	63.50	
	7500	Variable volume modulating motorized damper, incl. elect. mtr.										
	7504	8″ x 6″	1 Shee	15	.533	Ea.	111	22.50		133.50	158	
	7506	10″ x 6″		14	.571		111	24		135	161	
	7510	10″ x 10″		13	.615		114	26		140	168	
	7520	12″ x 12″		12	.667		118	28		146	174	
	7522	12″ x 16″		11	.727		121	30.50		151.50	182	
	7524	16″ x 10″		12	.667		118	28		146	175	
	7526	16″ x 14″		10	.800		123	33.50		156.50	190	
	7528	16″ x 18″		9	.889		131	37.50		168.50	204	
	7540	18″ x 12″		10	.800		126	33.50		159.50	192	
	7542	18″ x 18″		8	1		136	42		178	217	
	7544	20″ x 14″		8	1		138	42		180	219	
	7546	20″ x 18″		7	1.143		147	48		195	239	
	7560	24″ x 12″		8	1		140	42		182	221	
	7562	24″ x 18″		7	1.143		156	48		204	249	
	7564	24″ x 24″		6	1.333		162	56		218	268	
	7568	28″ x 10″		7	1.143		234	48		282	335	
	7580	28″ x 16″		6	1.333		246	56		302	360	
	7590	30″ x 14″		5	1.600		246	67.50		313.50	380	
	7600	30″ x 18″		4	2		262	84.50		346.50	425	
	7610	30″ x 24″		3.80	2.105		299	88.50		387.50	470	
	7700	For thermostat, add		8	1		40	42		82	111	
	7800	For transformer 40 VA capacity, add	↓	16	.500	↓	20	21		41	55.50	
	8000	Multi-blade dampers, parallel blade										
	8100	8″ x 8″	1 Shee	24	.333	Ea.	59	14.05		73.05	87.50	
	8120	12″ x 8″		22	.364		59	15.35		74.35	89.50	
	8140	16″ x 10″		20	.400		78	16.85		94.85	113	
	8160	18″ x 12″		18	.444		85	18.75		103.75	124	
	8180	22″ x 12″		15	.533		87.50	22.50		110	133	
	8200	24″ x 16″		11	.727		100	30.50		130.50	159	
	8220	28″ x 16″		10	.800		110	33.50		143.50	175	
	8240	30″ x 16″		8	1		115	42		157	194	
	8260	30″ x 18″	↓	7	1.143	↓	127	48		175	217	
	9000	Silencers, noise control for air flow, duct				MCFM	49			49	54	
	9900	Minimum labor/equipment charge	1 Shee	4	2	Job		84.50		84.50	134	

		15830 | Fans	CREW	DAILY OUTPUT	LABOR-HOURS	UNIT	MAT.	LABOR	EQUIP.	TOTAL	TOTAL INCL O&P	
100	0010	**FANS** R233400-10										100
	0020	Air conditioning and process air handling										
	0030	Axial flow, compact, low sound, 2.5″ S.P.										
	0050	3,800 CFM, 5 HP	Q-20	3.40	5.882	Ea.	3,925	228		4,153	4,650	
	0080	6,400 CFM, 5 HP		2.80	7.143		4,375	277		4,652	5,250	
	0100	10,500 CFM, 7-1/2 HP		2.40	8.333		5,450	325		5,775	6,500	
	0120	15,600 CFM, 10 HP		1.60	12.500		6,875	485		7,360	8,325	
	0140	23,000 CFM, 15 HP		.70	28.571		10,500	1,100		11,600	13,400	
	0160	28,000 CFM, 20 HP	↓	.40	50	↓	11,800	1,925		13,725	16,000	
	0200	In-line centrifugal, supply/exhaust booster										
	0220	aluminum wheel/hub, disconnect switch, 1/4″ S.P.										
	0240	500 CFM, 10″ diameter connection	Q-20	3	6.667	Ea.	940	258		1,198	1,425	
	0260	1,380 CFM, 12″ diameter connection		2	10		995	385		1,380	1,700	
	0280	1,520 CFM, 16″ diameter connection		2	10		1,075	385		1,460	1,800	
	0300	2,560 CFM, 18″ diameter connection		1	20		1,175	775		1,950	2,525	
	0320	3,480 CFM, 20″ diameter connection	↓	.80	25	↓	1,400	970		2,370	3,075	

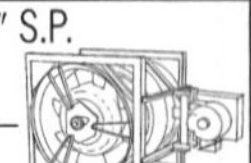

15830 | Fans

	Line	Description	Crew	Daily Output	Labor-Hours	Unit	2006 Bare Costs Mat.	2006 Bare Costs Labor	2006 Bare Costs Equip.	2006 Bare Costs Total	Total Incl O&P	
100	0326	5,080 CFM, 20″ diameter connection (R233400-10)	Q-20	.75	26.667	Ea.	1,525	1,025		2,550	3,300	100
	1500	Vaneaxial, low pressure, 2000 CFM, 1/2 HP		3.60	5.556		1,350	215		1,565	1,850	
	1520	4,000 CFM, 1 HP		3.20	6.250		1,475	242		1,717	2,000	
	1540	8,000 CFM, 2 HP		2.80	7.143		2,000	277		2,277	2,600	
	1560	16,000 CFM, 5 HP		2.40	8.333		2,800	325		3,125	3,575	
	2000	Blowers, direct drive with motor, complete										
	2020	1045 CFM @ .5″ S.P., 1/5 HP	Q-20	18	1.111	Ea.	156	43		199	240	
	2040	1385 CFM @ .5″ S.P., 1/4 HP		18	1.111		159	43		202	243	
	2060	1640 CFM @ .5″ S.P., 1/3 HP		18	1.111		189	43		232	275	
	2080	1760 CFM @ .5″ S.P., 1/2 HP		18	1.111		204	43		247	293	
	2090	4 speed										
	2100	1164 to 1739 CFM @ .5″ S.P., 1/3 HP	Q-20	16	1.250	Ea.	251	48.50		299.50	355	
	2120	1467 to 2218 CFM @ 1.0″ S.P., 3/4 HP	″	14	1.429	″	325	55.50		380.50	450	
	2500	Ceiling fan, right angle, extra quiet, 0.10″ S.P.										
	2520	95 CFM	Q-20	20	1	Ea.	160	38.50		198.50	237	
	2540	210 CFM		19	1.053		189	41		230	273	
	2560	385 CFM		18	1.111		240	43		283	330	
	2580	885 CFM		16	1.250		470	48.50		518.50	595	
	2600	1,650 CFM		13	1.538		655	59.50		714.50	815	
	2620	2,960 CFM		11	1.818		870	70.50		940.50	1,075	
	2640	For wall or roof cap, add	1 Shee	16	.500		160	21		181	210	
	2660	For straight thru fan, add					10%					
	2680	For speed control switch, add	1 Elec	16	.500		87	21		108	129	
	3000	Paddle blade air circulator, 3 speed switch										
	3020	42″, 5,000 CFM high, 3000 CFM low	1 Elec	2.40	3.333	Ea.	73.50	140		213.50	297	
	3040	52″, 6,500 CFM high, 4000 CFM low	″	2.20	3.636	″	74.50	153		227.50	315	
	3100	For antique white motor, same cost										
	3200	For brass plated motor, same cost										
	3300	For light adaptor kit, add				Ea.	27.50			27.50	30	
	3500	Centrifugal, airfoil, motor and drive, complete										
	3520	1000 CFM, 1/2 HP	Q-20	2.50	8	Ea.	1,100	310		1,410	1,700	
	3540	2,000 CFM, 1 HP		2	10		1,175	385		1,560	1,900	
	3560	4,000 CFM, 3 HP		1.80	11.111		1,475	430		1,905	2,300	
	3580	8,000 CFM, 7-1/2 HP		1.40	14.286		2,375	555		2,930	3,475	
	3600	12,000 CFM, 10 HP		1	20		2,925	775		3,700	4,450	
	4000	Single width, belt drive, not incl. motor, capacities										
	4020	at 2000 FPM, 2.5″ S.P. for indicated motor										
	4040	6900 CFM, 5 HP	Q-9	2.40	6.667	Ea.	2,875	253		3,128	3,550	
	4060	10,340 CFM, 7-1/2″ HP		2.20	7.273		3,975	276		4,251	4,825	
	4080	15,320 CFM, 10 HP		2	8		4,700	305		5,005	5,650	
	4100	22,780 CFM, 15 HP		1.80	8.889		6,975	335		7,310	8,200	
	4120	33,840 CFM, 20 HP		1.60	10		9,275	380		9,655	10,800	
	4140	41,400 CFM, 25 HP		1.40	11.429		12,300	435		12,735	14,200	
	4160	50,100 CFM, 30 HP		.80	20		15,600	760		16,360	18,400	
	4200	Double width wheel, 12,420 CFM, 7.5 HP		2.20	7.273		4,100	276		4,376	4,950	
	4220	18,620 CFM, 15 HP		2	8		5,850	305		6,155	6,925	
	4240	27,580 CFM, 20 HP		1.80	8.889		7,175	335		7,510	8,400	
	4260	40,980 CFM, 25 HP		1.50	10.667		11,200	405		11,605	12,900	
	4280	60,920 CFM, 40 HP		1	16		15,100	605		15,705	17,600	
	4300	74,520 CFM, 50 HP		.80	20		17,900	760		18,660	20,900	
	4320	90,160 CFM, 50 HP		.70	22.857		24,500	865		25,365	28,300	
	4340	110,300 CFM, 60 HP		.50	32		33,200	1,225		34,425	38,400	
	4360	134,960 CFM, 75 HP		.40	40		44,100	1,525		45,625	51,000	
	4500	Corrosive fume resistant, plastic										
	4600	roof ventilators, centrifugal, V belt drive, motor										
	4620	1/4″ S.P., 250 CFM, 1/4 HP	Q-20	6	3.333	Ea.	2,525	129		2,654	3,000	

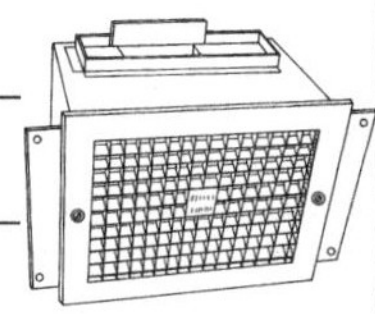

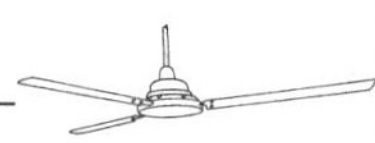

15830 | Fans		Crew	Daily Output	Labor-Hours	Unit	2006 Bare Costs Mat.	Labor	Equip.	Total	Total Incl O&P
100 4640	895 CFM, 1/3 HP (R233400-10)	Q-20	5	4	Ea.	2,750	155		2,905	3,275 100
4660	1630 CFM, 1/2 HP		4	5		3,250	194		3,444	3,875
4680	2240 CFM, 1 HP		3	6.667		3,400	258		3,658	4,125
4700	3810 CFM, 2 HP		2	10		3,775	385		4,160	4,750
4710	5000 CFM, 2 HP		1.80	11.111		7,925	430		8,355	9,400
4715	8000 CFM, 5 HP		1.40	14.286		8,825	555		9,380	10,600
4720	11760 CFM, 5 HP		1	20		8,975	775		9,750	11,100
4740	18810 CFM, 10 HP	↓	.70	28.571	↓	9,300	1,100		10,400	12,000
4800	For intermediate capacity, motors may be varied									
4810	For explosion proof motor, add				Ea.	15%				
5000	Utility set, centrifugal, V belt drive, motor									
5020	1/4" S.P., 1200 CFM, 1/4 HP	Q-20	6	3.333	Ea.	2,850	129		2,979	3,350
5040	1520 CFM, 1/3 HP		5	4		2,850	155		3,005	3,400
5060	1850 CFM, 1/2 HP		4	5		2,875	194		3,069	3,450
5080	2180 CFM, 3/4 HP		3	6.667		2,900	258		3,158	3,600
5100	1/2" S.P., 3600 CFM, 1 HP		2	10		4,200	385		4,585	5,225
5120	4250 CFM, 1-1/2 HP		1.60	12.500		4,250	485		4,735	5,450
5140	4800 CFM, 2 HP		1.40	14.286		4,325	555		4,880	5,625
5160	6920 CFM, 5 HP		1.30	15.385		4,450	595		5,045	5,850
5180	7700 CFM, 7-1/2 HP	↓	1.20	16.667	↓	4,575	645		5,220	6,075
5200	For explosion proof motor, add					15%				
5500	Fans, industrial exhauster, for air which may contain granular matl.									
5520	1000 CFM, 1-1/2 HP	Q-20	2.50	8	Ea.	1,575	310		1,885	2,225
5540	2000 CFM, 3 HP		2	10		2,000	385		2,385	2,800
5560	4000 CFM, 7-1/2 HP		1.80	11.111		2,150	430		2,580	3,050
5580	8000 CFM, 15 HP		1.40	14.286		3,825	555		4,380	5,075
5600	12,000 CFM, 30 HP	↓	1	20	↓	5,975	775		6,750	7,775
6000	Propeller exhaust, wall shutter, 1/4" S.P.									
6020	Direct drive, two speed									
6100	375 CFM, 1/10 HP	Q-20	10	2	Ea.	287	77.50		364.50	435
6120	730 CFM, 1/7 HP		9	2.222		320	86		406	485
6140	1000 CFM, 1/8 HP		8	2.500		445	97		542	645
6160	1890 CFM, 1/4 HP		7	2.857		455	111		566	675
6180	3275 CFM, 1/2 HP		6	3.333		460	129		589	710
6200	4720 CFM, 1 HP	↓	5	4	↓	710	155		865	1,025
6300	V-belt drive, 3 phase									
6320	6175 CFM, 3/4 HP	Q-20	5	4	Ea.	595	155		750	900
6340	7500 CFM, 3/4 HP		5	4		630	155		785	935
6360	10,100 CFM, 1 HP		4.50	4.444		770	172		942	1,125
6380	14,300 CFM, 1-1/2 HP		4	5		905	194		1,099	1,300
6400	19,800 CFM, 2 HP		3	6.667		1,025	258		1,283	1,550
6420	26,250 CFM, 3 HP		2.60	7.692		1,250	298		1,548	1,850
6440	38,500 CFM, 5 HP		2.20	9.091		1,475	350		1,825	2,175
6460	46,000 CFM, 7-1/2 HP		2	10		1,575	385		1,960	2,325
6480	51,500 CFM, 10 HP	↓	1.80	11.111	↓	1,625	430		2,055	2,475
6490	V-belt drive, 115V., residential, whole house									
6500	Ceiling-wall, 5200 CFM, 1/4 HP, 30" x 30"	1 Shee	6	1.333	Ea.	310	56		366	435
6530	13,200 CFM, 1/3 HP, 48" x 48"		4	2		380	84.50		464.50	550
6540	15,445 CFM, 1/2 HP, 48" x 48"		4	2		390	84.50		474.50	565
6550	17,025 CFM, 1/2 HP, 54" x 54"	↓	4	2	↓	755	84.50		839.50	970
6560	For two speed motor, add					20%				
6570	Shutter, automatic, ceiling/wall									
6580	30" x 30"	1 Shee	8	1	Ea.	102	42		144	179
6590	36" x 36"		8	1		116	42		158	195
6600	42" x 42"		8	1		145	42		187	227
6610	48" x 48"	↓	7	1.143	↓	158	48		206	250

15 MECHANICAL

15830 | Fans

		15830 Fans	CREW	DAILY OUTPUT	LABOR-HOURS	UNIT	2006 BARE COSTS MAT.	LABOR	EQUIP.	TOTAL	TOTAL INCL O&P	
100	6620	54″ x 54″ R233400-10	1 Shee	6	1.333	Ea.	208	56		264	320	100
	6630	Timer, shut off, to 12 Hr.	↓	20	.400	↓	33.50	16.85		50.35	64	
	6650	Residential, bath exhaust, grille, back draft damper										
	6660	50 CFM	Q-20	24	.833	Ea.	30	32.50		62.50	84	
	6670	110 CFM		22	.909		50	35		85	111	
	6680	Light combination, squirrel cage, 100 watt, 70 CFM	↓	24	.833	↓	63	32.50		95.50	120	
	6700	Light/heater combination, ceiling mounted										
	6710	70 CFM, 1450 watt	Q-20	24	.833	Ea.	74.50	32.50		107	133	
	6800	Heater combination, recessed, 70 CFM		24	.833		36	32.50		68.50	90.50	
	6820	With 2 infrared bulbs		23	.870		53.50	33.50		87	112	
	6900	Kitchen exhaust, grille, complete, 160 CFM		22	.909		63.50	35		98.50	126	
	6910	180 CFM		20	1		54	38.50		92.50	121	
	6920	270 CFM		18	1.111		97.50	43		140.50	175	
	6930	350 CFM	↓	16	1.250	↓	74.50	48.50		123	159	
	6940	Residential roof jacks and wall caps										
	6944	Wall cap with back draft damper										
	6946	3″ & 4″ dia. round duct	1 Shee	11	.727	Ea.	13.05	30.50		43.55	63.50	
	6948	6″ dia. round duct	″	11	.727	″	31	30.50		61.50	83	
	6958	Roof jack with bird screen and back draft damper										
	6960	3″ & 4″ dia. round duct	1 Shee	11	.727	Ea.	12.55	30.50		43.05	63	
	6962	3-1/4″ x 10″ rectangular duct	″	10	.800	″	22.50	33.50		56	78.50	
	6980	Transition										
	6982	3-1/4″ x 10″ to 6″ dia. round	1 Shee	20	.400	Ea.	13.85	16.85		30.70	42.50	
	7000	Roof exhauster, centrifugal, aluminum housing, 12″ galvanized										
	7020	curb, bird screen, back draft damper, 1/4″ S.P.										
	7100	Direct drive, 320 CFM, 11″ sq. damper	Q-20	7	2.857	Ea.	360	111		471	570	
	7120	600 CFM, 11″ sq. damper		6	3.333		365	129		494	605	
	7140	815 CFM, 13″ sq. damper		5	4		365	155		520	645	
	7160	1450 CFM, 13″ sq. damper		4.20	4.762		465	184		649	800	
	7180	2050 CFM, 16″ sq. damper		4	5		465	194		659	815	
	7200	V-belt drive, 1650 CFM, 12″ sq. damper		6	3.333		790	129		919	1,075	
	7220	2750 CFM, 21″ sq. damper		5	4		890	155		1,045	1,225	
	7230	3500 CFM, 21″ sq. damper		4.50	4.444		985	172		1,157	1,350	
	7240	4910 CFM, 23″ sq. damper		4	5		1,225	194		1,419	1,650	
	7260	8525 CFM, 28″ sq. damper		3	6.667		1,525	258		1,783	2,075	
	7280	13,760 CFM, 35″ sq. damper		2	10		2,100	385		2,485	2,925	
	7300	20,558 CFM, 43″ sq. damper	↓	1	20		4,475	775		5,250	6,150	
	7320	For 2 speed winding, add					15%					
	7340	For explosion-proof motor, add					330			330	360	
	7360	For belt driven, top discharge, add				↓	15%					
	7500	Utility set, steel construction, pedestal, 1/4″ S.P.										
	7520	Direct drive, 150 CFM, 1/8 HP	Q-20	6.40	3.125	Ea.	660	121		781	915	
	7540	485 CFM, 1/6 HP		5.80	3.448		830	134		964	1,125	
	7560	1950 CFM, 1/2 HP		4.80	4.167		970	161		1,131	1,325	
	7580	2410 CFM, 3/4 HP		4.40	4.545		1,800	176		1,976	2,250	
	7600	3328 CFM, 1-1/2 HP	↓	3	6.667	↓	2,000	258		2,258	2,600	
	7680	V-belt drive, drive cover, 3 phase										
	7700	800 CFM, 1/4 HP	Q-20	6	3.333	Ea.	520	129		649	775	
	7720	1,300 CFM, 1/3 HP		5	4		545	155		700	840	
	7740	2,000 CFM, 1 HP		4.60	4.348		640	168		808	970	
	7760	2,900 CFM, 3/4 HP		4.20	4.762		865	184		1,049	1,250	
	7780	3,600 CFM, 3/4 HP		4	5		1,075	194		1,269	1,475	
	7800	4,800 CFM, 1 HP		3.50	5.714		1,250	221		1,471	1,725	
	7820	6,700 CFM, 1-1/2 HP		3	6.667		1,550	258		1,808	2,100	
	7830	7,500 CFM, 2 HP		2.50	8		2,100	310		2,410	2,825	
	7840	11,000 CFM, 3 HP	↓	2	10	↓	2,825	385		3,210	3,700	100

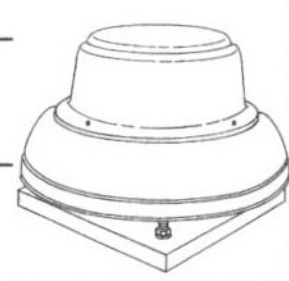

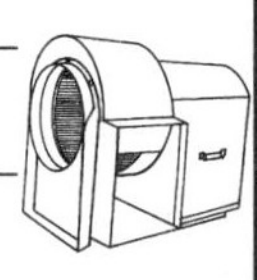

15800 | Air Distribution

15830 | Fans

			CREW	DAILY OUTPUT	LABOR-HOURS	UNIT	2006 BARE COSTS MAT.	LABOR	EQUIP.	TOTAL	TOTAL INCL O&P	
100	7860	13,000 CFM, 3 HP R233400-10	Q-20	1.60	12.500	Ea.	2,850	485		3,335	3,925	100
	7880	15,000 CFM, 5 HP		1	20		2,950	775		3,725	4,475	
	7900	17,000 CFM, 7-1/2 HP		.80	25		3,175	970		4,145	5,000	
	7920	20,000 CFM, 7-1/2 HP		.80	25		3,775	970		4,745	5,675	
	8000	Ventilation, residential										
	8020	Attic, roof type										
	8030	Aluminum dome, damper & curb										
	8040	6″ diameter, 300 CFM	1 Elec	16	.500	Ea.	266	21		287	325	
	8050	7″ diameter, 450 CFM		15	.533		291	22.50		313.50	355	
	8060	9″ diameter, 900 CFM		14	.571		465	24		489	550	
	8080	12″ diameter, 1000 CFM (gravity)		10	.800		330	33.50		363.50	415	
	8090	16″ diameter, 1500 CFM (gravity)		9	.889		400	37.50		437.50	500	
	8100	20″ diameter, 2500 CFM (gravity)		8	1		490	42		532	600	
	8110	26″ diameter, 4000 CFM (gravity)		7	1.143		590	48		638	725	
	8120	32″ diameter, 6500 CFM (gravity)		6	1.333		810	56		866	980	
	8130	38″ diameter, 8000 CFM (gravity)		5	1.600		1,200	67		1,267	1,425	
	8140	50″ diameter, 13,000 CFM (gravity)		4	2		1,750	84		1,834	2,050	
	8160	Plastic, ABS dome										
	8180	1050 CFM	1 Elec	14	.571	Ea.	97	24		121	144	
	8200	1600 CFM	″	12	.667	″	145	28		173	203	
	8240	Attic, wall type, with shutter, one speed										
	8250	12″ diameter, 1000 CFM	1 Elec	14	.571	Ea.	210	24		234	268	
	8260	14″ diameter, 1500 CFM		12	.667		227	28		255	293	
	8270	16″ diameter, 2000 CFM		9	.889		257	37.50		294.50	340	
	8290	Whole house, wall type, with shutter, one speed										
	8300	30″ diameter, 4800 CFM	1 Elec	7	1.143	Ea.	550	48		598	680	
	8310	36″ diameter, 7000 CFM		6	1.333		600	56		656	745	
	8320	42″ diameter, 10,000 CFM		5	1.600		670	67		737	845	
	8330	48″ diameter, 16,000 CFM		4	2		835	84		919	1,050	
	8340	For two speed, add					50			50	55	
	8350	Whole house, lay-down type, with shutter, one speed										
	8360	30″ diameter, 4500 CFM	1 Elec	8	1	Ea.	585	42		627	710	
	8370	36″ diameter, 6500 CFM		7	1.143		630	48		678	770	
	8380	42″ diameter, 9000 CFM		6	1.333		690	56		746	845	
	8390	48″ diameter, 12,000 CFM		5	1.600		785	67		852	970	
	8440	For two speed, add					37.50			37.50	41.50	
	8450	For 12 hour timer switch, add	1 Elec	32	.250		37.50	10.50		48	57.50	
	8500	Wall exhausters, centrifugal, auto damper, 1/8″ S.P.										
	8520	Direct drive, 610 CFM, 1/20 HP	Q-20	14	1.429	Ea.	232	55.50		287.50	345	
	8540	796 CFM, 1/12 HP		13	1.538		240	59.50		299.50	360	
	8560	822 CFM, 1/6 HP		12	1.667		380	64.50		444.50	515	
	8580	1,320 CFM, 1/4 HP		12	1.667		385	64.50		449.50	520	
	8600	1756 CFM, 1/4 HP		11	1.818		445	70.50		515.50	600	
	8620	1983 CFM, 1/4 HP		10	2		535	77.50		612.50	705	
	8640	2900 CFM, 1/2 HP		9	2.222		645	86		731	845	
	8660	3307 CFM, 3/4 HP		8	2.500		710	97		807	935	
	9500	V-belt drive, 3 phase										
	9520	2,800 CFM, 1/4 HP	Q-20	9	2.222	Ea.	1,050	86		1,136	1,275	
	9540	3,740 CFM, 1/2 HP		8	2.500		1,100	97		1,197	1,350	
	9560	4400 CFM, 3/4 HP		7	2.857		1,100	111		1,211	1,400	
	9580	5700 CFM, 1-1/2 HP		6	3.333		1,150	129		1,279	1,475	
	9900	Minimum labor/equipment charge	1 Elec	4	2	Job		84		84	129	

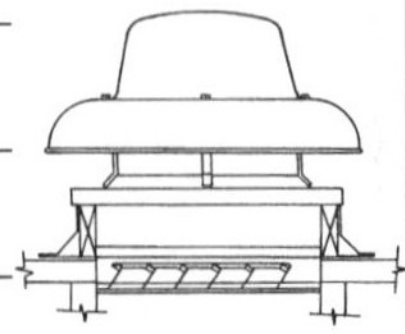

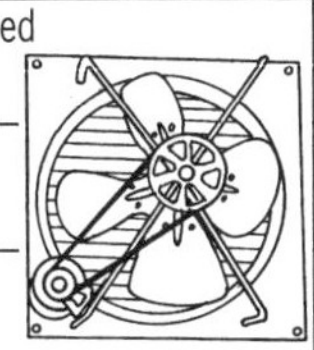

15840 | Air Terminal Units

			CREW	DAILY OUTPUT	LABOR-HOURS	UNIT	MAT.	LABOR	EQUIP.	TOTAL	TOTAL INCL O&P	
200	0010	**AIR CURTAINS** Incl. motor starters, transformers,										200
	0050	door switches & temperature controls										

15 MECHANICAL

	15840 | Air Terminal Units		Crew	Daily Output	Labor-Hours	Unit	2006 Bare Costs Mat.	Labor	Equip.	Total	Total Incl O&P
200	0100	Shipping and receiving doors, unheated, minimal wind stoppage									
	0150	8′ high, multiples of 3′ wide	2 Shee	6	2.667	L.F.	320	112		432	530
	0160	5′ wide		10	1.600		310	67.50		377.50	450
	0210	10′ high, multiples of 4′ wide		8	2		305	84.50		389.50	470
	0250	12′ high, 3′-6″ wide		7	2.286		335	96.50		431.50	520
	0260	12′ wide		6	2.667		305	112		417	515
	0350	16′ high, 3′-6″ wide		7	2.286		320	96.50		416.50	505
	0360	12′ wide		6	2.667		355	112		467	570
	0500	Maximum wind stoppage									
	0550	10′ high, multiples of 4′ wide	2 Shee	8	2	L.F.	246	84.50		330.50	405
	0650	14′ high, multiples of 4′ wide		8	2		246	84.50		330.50	405
	0750	20′ high, multiples of 8′ wide		6	2.667		855	112		967	1,125
	1100	Heated, maximum wind stoppage, steam heat									
	1150	10′ high, multiples of 4′ wide	2 Shee	6	2.667	L.F.	425	112		537	645
	1250	14′ high, multiples of 4′ wide		6	2.667		460	112		572	685
	1350	20′ high, multiples of 8′ wide		4	4		1,275	169		1,444	1,675
	1500	Customer entrance doors, unheated, minimal wind stoppage									
	1550	10′ high, multiples of 3′ wide	2 Shee	6	2.667	L.F.	282	112		394	490
	1560	5′ wide		10	1.600		305	67.50		372.50	440
	1650	Maximum wind stoppage, 12′ high, multiples of 4′ wide		8	2		225	84.50		309.50	380
	1700	Heated, minimal wind stoppage, electric heat									
	1750	8′ high, multiples of 3′ wide	2 Shee	6	2.667	L.F.	560	112		672	800
	1760	Multiples of 5′ wide		10	1.600		400	67.50		467.50	545
	1850	10′ high, multiples of 3′ wide		6	2.667		480	112		592	710
	1860	Multiples of 5′ wide		10	1.600		540	67.50		607.50	700
	1950	Maximum wind stoppage, steam heat									
	1960	12′ high, multiples of 4′ wide	2 Shee	8	2	L.F.	510	84.50		594.50	695
	2000	Walk-in coolers and freezers, ambient air, minimal wind stoppage									
	2050	8′ high, multiples of 3′ wide	2 Shee	6	2.667	L.F.	210	112		322	410
	2060	Multiples of 5′ wide		10	1.600		229	67.50		296.50	360
	2250	Maximum wind stoppage, 12′ high, multiples of 3′ wide		6	2.667		241	112		353	445
	2450	Conveyor openings or service windows, unheated, 5′ high		5	3.200		268	135		403	510
	2460	Heated, electric, 5′ high, 2′-6″ wide		5	3.200		253	135		388	495
500	0010	**CONSTANT VOLUME MIXING BOXES**									
	5180	Mixing box, includes electric or pneumatic motor									
	5190	Recommend use with attenuator, see 15820-300-9000									
	5200	Constant volume, 150 to 270 CFM	Q-9	12	1.333	Ea.	545	50.50		595.50	680
	5210	270 to 600 CFM		11	1.455		560	55		615	705
	5230	550 to 1000 CFM		9	1.778		560	67.50		627.50	720
	5240	1000 to 1600 CFM		8	2		575	76		651	750
	5250	1300 to 1900 CFM		6	2.667		585	101		686	805
	5260	550 to 2640 CFM		5.60	2.857		645	108		753	880
	5270	650 to 3120 CFM		5.20	3.077		685	117		802	940
700	0010	**VARIABLE VOLUME MIXING BOXES**									
	5500	VAV Cool only, pneum. pressure indep. 300 to 600 CFM	Q-9	11	1.455	Ea.	335	55		390	455
	5510	500 to 1000 CFM		9	1.778		345	67.50		412.50	485
	5520	800 to 1600 CFM		9	1.778		360	67.50		427.50	500
	5530	1100 to 2000 CFM		8	2		370	76		446	530
	5540	1500 to 3000 CFM		7	2.286		400	86.50		486.50	580
	5550	2000 to 4000 CFM		6	2.667		410	101		511	610
	5560	For electric, w/thermostat, pressure dependent, add					20			20	22

	15850 | Air Outlets & Inlets		Crew	Daily Output	Labor-Hours	Unit	Mat.	Labor	Equip.	Total	Total Incl O&P
300	0010	**DIFFUSERS** Aluminum, opposed blade damper unless noted									
	0100	Ceiling, linear, also for sidewall									

15850 | Air Outlets & Inlets

	Line	Description	CREW	DAILY OUTPUT	LABOR-HOURS	UNIT	2006 BARE COSTS MAT.	LABOR	EQUIP.	TOTAL	TOTAL INCL O&P	
300	0120	2″ wide	1 Shee	32	.250	L.F.	30.50	10.55		41.05	50.50	300
	0140	3″ wide		30	.267		35.50	11.25		46.75	57	
	0160	4″ wide		26	.308		39.50	12.95		52.45	64	
	0180	6″ wide		24	.333		47.50	14.05		61.55	75	
	0200	8″ wide		22	.364		54.50	15.35		69.85	84.50	
	0220	10″ wide		20	.400		61	16.85		77.85	94	
	0240	12″ wide	↓	18	.444	↓	75	18.75		93.75	113	
	0260	For floor or sill application, add					15%					
	0500	Perforated, 24″ x 24″ lay-in panel size, 6″ x 6″	1 Shee	16	.500	Ea.	72	21		93	113	
	0520	8″ x 8″		15	.533		74	22.50		96.50	117	
	0530	9″ x 9″		14	.571		77	24		101	124	
	0540	10″ x 10″		14	.571		79.50	24		103.50	126	
	0560	12″ x 12″		12	.667		82	28		110	135	
	0580	15″ x 15″		11	.727		102	30.50		132.50	161	
	0590	16″ x 16″		11	.727		112	30.50		142.50	173	
	0600	18″ x 18″		10	.800		120	33.50		153.50	186	
	0610	20″ x 20″		10	.800		137	33.50		170.50	205	
	0620	24″ x 24″		9	.889		159	37.50		196.50	235	
	1000	Rectangular, 1 to 4 way blow, 6″ x 6″		16	.500		44.50	21		65.50	82.50	
	1010	8″ x 8″		15	.533		53	22.50		75.50	94.50	
	1014	9″ x 9″		15	.533		57	22.50		79.50	99	
	1016	10″ x 10″		15	.533		67.50	22.50		90	110	
	1020	12″ x 6″		15	.533		77.50	22.50		100	121	
	1040	12″ x 9″		14	.571		84.50	24		108.50	132	
	1060	12″ x 12″		12	.667		78.50	28		106.50	131	
	1070	14″ x 6″		13	.615		84	26		110	134	
	1074	14″ x 14″		12	.667		103	28		131	158	
	1080	18″ x 12″		11	.727		149	30.50		179.50	213	
	1120	15″ x 15″		10	.800		146	33.50		179.50	214	
	1140	18″ x 15″		9	.889		175	37.50		212.50	252	
	1150	18″ x 18″		9	.889		145	37.50		182.50	219	
	1160	21″ x 21″		8	1		183	42		225	268	
	1170	24″ x 12″		10	.800		136	33.50		169.50	204	
	1180	24″ x 24″		7	1.143		300	48		348	405	
	1500	Round, butterfly damper, steel, 6″ diameter		18	.444		17.95	18.75		36.70	49.50	
	1520	8″ diameter		16	.500		19.35	21		40.35	55	
	1540	10″ diameter		14	.571		24	24		48	65	
	1560	12″ diameter		12	.667		31.50	28		59.50	79	
	1580	14″ diameter		10	.800		41	33.50		74.50	98.50	
	1600	18″ diameter		9	.889		97	37.50		134.50	167	
	1610	20″ diameter		9	.889		114	37.50		151.50	185	
	1620	22″ diameter		8	1		128	42		170	207	
	1630	24″ diameter		8	1		141	42		183	222	
	1640	28″ diameter		7	1.143		157	48		205	250	
	1650	32″ diameter		6	1.333		172	56		228	279	
	2000	T bar mounting, 24″ x 24″ lay-in frame, 6″ x 6″		16	.500		94.50	21		115.50	138	
	2020	9″ x 9″		14	.571		105	24		129	154	
	2040	12″ x 12″		12	.667		135	28		163	193	
	2060	15″ x 15″		11	.727		173	30.50		203.50	239	
	2080	18″ x 18″	↓	10	.800	↓	190	33.50		223.50	264	
	2500	Combination supply and return										
	2520	21″ x 21″ supply, 15″ x 15″ return	Q-9	10	1.600	Ea.	208	60.50		268.50	325	
	2540	24″ x 24″ supply, 18″ x 18″ return		9.50	1.684		262	64		326	390	
	2560	27″ x 27″ supply, 18″ x 18″ return		9	1.778		320	67.50		387.50	455	
	2580	30″ x 30″ supply, 21″ x 21″ return		8.50	1.882		405	71.50		476.50	560	
	2600	33″ x 33″ supply, 24″ x 24″ return	↓	8	2	↓	475	76		551	640	

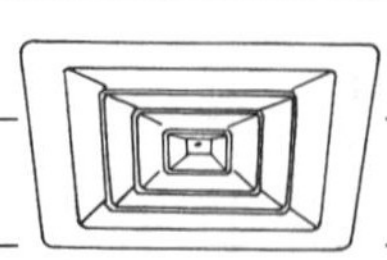

15 MECHANICAL

	15850 | Air Outlets & Inlets		CREW	DAILY OUTPUT	LABOR-HOURS	UNIT	2006 BARE COSTS MAT.	LABOR	EQUIP.	TOTAL	TOTAL INCL O&P	
300	2620	36″ x 36″ supply, 24″ x 24″ return	Q-9	7.50	2.133	Ea.	555	81		636	740	300
	3000	Baseboard, white enameled steel										
	3100	18″ long	1 Shee	20	.400	Ea.	9.15	16.85		26	37	
	3120	24″ long		18	.444		17.85	18.75		36.60	49.50	
	3140	48″ long		16	.500		32.50	21		53.50	69.50	
	3400	For matching return, deduct					10%					
	4000	Floor, steel, adjustable pattern										
	4100	2″ x 10″	1 Shee	34	.235	Ea.	12.35	9.90		22.25	29.50	
	4120	2″ x 12″		32	.250		13.55	10.55		24.10	31.50	
	4140	2″ x 14″		30	.267		15.35	11.25		26.60	35	
	4200	4″ x 10″		28	.286		14.25	12.05		26.30	35	
	4220	4″ x 12″		26	.308		15.80	12.95		28.75	38	
	4240	4″ x 14″		25	.320		17.85	13.50		31.35	41	
	4260	6″ x 10″		26	.308		21.50	12.95		34.45	44	
	4280	6″ x 12″		24	.333		22.50	14.05		36.55	47.50	
	4300	6″ x 14″		22	.364		23.50	15.35		38.85	50.50	
	5000	Sidewall, aluminum, 3 way dispersion										
	5100	8″ x 4″	1 Shee	24	.333	Ea.	6.75	14.05		20.80	30	
	5110	8″ x 6″		24	.333		7.25	14.05		21.30	30.50	
	5120	10″ x 4″		22	.364		7.10	15.35		22.45	32.50	
	5130	10″ x 6″		22	.364		7.65	15.35		23	33	
	5140	10″ x 8″		20	.400		10.40	16.85		27.25	38.50	
	5160	12″ x 4″		18	.444		7.65	18.75		26.40	38.50	
	5170	12″ x 6″		17	.471		8.45	19.85		28.30	41	
	5180	12″ x 8″		16	.500		11	21		32	45.50	
	5200	14″ x 4″		15	.533		8.55	22.50		31.05	45.50	
	5220	14″ x 6″		14	.571		9.40	24		33.40	49	
	5240	14″ x 8″		13	.615		12.20	26		38.20	55	
	5260	16″ x 6″		12.50	.640		12.40	27		39.40	56.50	
	6000	For steel diffusers instead of aluminum, deduct					10%					
	9000	Minimum labor/equipment charge	1 Shee	4	2	Job		84.50		84.50	134	
500	0010	**GRILLES**										500
	0020	Aluminum										
	0100	Air supply, single deflection, adjustable										
	0120	8″ x 4″	1 Shee	30	.267	Ea.	12.60	11.25		23.85	32	
	0140	8″ x 8″		28	.286		15.25	12.05		27.30	36	
	0160	10″ x 4″		24	.333		13.90	14.05		27.95	38	
	0180	10″ x 10″		23	.348		17.65	14.65		32.30	43	
	0200	12″ x 6″		23	.348		16.40	14.65		31.05	41.50	
	0220	12″ x 12″		22	.364		21.50	15.35		36.85	48	
	0230	14″ x 6″		23	.348		18.35	14.65		33	43.50	
	0240	14″ x 8″		23	.348		19.55	14.65		34.20	45	
	0260	14″ x 14″		22	.364		26.50	15.35		41.85	53.50	
	0270	18″ x 8″		23	.348		22.50	14.65		37.15	48.50	
	0280	18″ x 10″		23	.348		26	14.65		40.65	52	
	0300	18″ x 18″		21	.381		40.50	16.05		56.55	70	
	0320	20″ x 12″		22	.364		29.50	15.35		44.85	57	
	0340	20″ x 20″		21	.381		49	16.05		65.05	79.50	
	0350	24″ x 8″		20	.400		28.50	16.85		45.35	58	
	0360	24″ x 14″		18	.444		39.50	18.75		58.25	73.50	
	0380	24″ x 24″		15	.533		69	22.50		91.50	112	
	0400	30″ x 8″		20	.400		36.50	16.85		53.35	67	
	0420	30″ x 10″		19	.421		40	17.75		57.75	72	
	0440	30″ x 12″		18	.444		44	18.75		62.75	78.50	
	0460	30″ x 16″		17	.471		57	19.85		76.85	94.50	

	15850 | Air Outlets & Inlets		Crew	Daily Output	Labor-Hours	Unit	2006 Bare Costs Mat.	Labor	Equip.	Total	Total Incl O&P	
500	0480	30″ x 18″	1 Shee	17	.471	Ea.	64	19.85		83.85	102	500
	0500	30″ x 30″		14	.571		114	24		138	165	
	0520	36″ x 12″		17	.471		54.50	19.85		74.35	91.50	
	0540	36″ x 16″		16	.500		70	21		91	110	
	0560	36″ x 18″		15	.533		77.50	22.50		100	121	
	0570	36″ x 20″		15	.533		84.50	22.50		107	129	
	0580	36″ x 24″		14	.571		110	24		134	160	
	0600	36″ x 28″		13	.615		134	26		160	190	
	0620	36″ x 30″		12	.667		148	28		176	208	
	0640	36″ x 32″		12	.667		163	28		191	224	
	0660	36″ x 34″		11	.727		177	30.50		207.50	244	
	0680	36″ x 36″	↓	11	.727		192	30.50		222.50	260	
	0700	For double deflecting, add					70%					
	1000	Air return, 6″ x 6″	1 Shee	26	.308		14.30	12.95		27.25	36.50	
	1020	10″ x 6″		24	.333		16.90	14.05		30.95	41	
	1040	14″ x 6″		23	.348		19.50	14.65		34.15	45	
	1060	10″ x 8″		23	.348		16.90	14.65		31.55	42	
	1080	16″ x 8″		22	.364		25.50	15.35		40.85	52.50	
	1100	12″ x 12″		22	.364		25.50	15.35		40.85	52.50	
	1120	24″ x 12″		18	.444		45	18.75		63.75	79.50	
	1140	30″ x 12″		16	.500		53.50	21		74.50	92	
	1160	14″ x 14″		22	.364		31	15.35		46.35	59	
	1180	16″ x 16″		22	.364		36.50	15.35		51.85	64.50	
	1200	18″ x 18″		21	.381		42.50	16.05		58.55	72	
	1220	24″ x 18″		16	.500		49.50	21		70.50	88	
	1240	36″ x 18″		15	.533		83	22.50		105.50	128	
	1260	24″ x 24″		15	.533		65	22.50		87.50	108	
	1280	36″ x 24″		14	.571		99	24		123	148	
	1300	48″ x 24″		12	.667		174	28		202	236	
	1320	48″ x 30″		11	.727		224	30.50		254.50	295	
	1340	36″ x 36″		13	.615		179	26		205	239	
	1360	48″ x 36″		11	.727		252	30.50		282.50	325	
	1380	48″ x 48″		8	1		335	42		377	435	
	2000	Door grilles, 12″ x 12″		22	.364		36.50	15.35		51.85	64.50	
	2020	18″ x 12″		22	.364		45	15.35		60.35	74	
	2040	24″ x 12″		18	.444		53.50	18.75		72.25	88.50	
	2060	18″ x 18″		18	.444		67	18.75		85.75	104	
	2080	24″ x 18″		16	.500		84	21		105	126	
	2100	24″ x 24″		15	.533		112	22.50		134.50	159	
	3000	Filter grille with filter, 12″ x 12″		24	.333		42.50	14.05		56.55	69	
	3020	18″ x 12″		20	.400		62	16.85		78.85	95	
	3040	24″ x 18″		18	.444		75.50	18.75		94.25	113	
	3060	24″ x 24″		16	.500		118	21		139	163	
	3080	30″ x 24″		14	.571		151	24		175	205	
	3100	30″ x 30″		13	.615		190	26		216	252	
	3950	Eggcrate, framed, 6″ x 6″ opening		26	.308		15.55	12.95		28.50	37.50	
	3954	8″ x 8″ opening		24	.333		16.05	14.05		30.10	40	
	3960	10″ x 10″ opening		23	.348		18.55	14.65		33.20	44	
	3970	12″ x 12″ opening		22	.364		22	15.35		37.35	48.50	
	3980	14″ x 14″ opening		22	.364		27	15.35		42.35	54.50	
	3984	16″ x 16″ opening		21	.381		34	16.05		50.05	63	
	3990	18″ x 18″ opening		21	.381		40.50	16.05		56.55	70	
	4020	22″ x 22″ opening		17	.471		64.50	19.85		84.35	103	
	4040	24″ x 24″ opening		15	.533		67.50	22.50		90	110	
	4044	28″ x 28″ opening		14	.571		96	24		120	145	
	4048	36″ x 36″ opening	↓	12	.667	↓	115	28		143	171	

		15850 | Air Outlets & Inlets	CREW	DAILY OUTPUT	LABOR-HOURS	UNIT	2006 BARE COSTS MAT.	LABOR	EQUIP.	TOTAL	TOTAL INCL O&P	
500	4050	48″ x 24″ opening	1 Shee	12	.667	Ea.	202	28		230	267	500
	4060	Eggcrate, lay-in, T-bar system										
	4070	48″ x 24″ sheet	1 Shee	40	.200	Ea.	97	8.45		105.45	120	
	5000	Transfer grille, vision proof, 8″ x 4″		30	.267		23	11.25		34.25	43.50	
	5020	8″ x 6″		28	.286		23	12.05		35.05	44.50	
	5040	8″ x 8″		26	.308		26	12.95		38.95	49	
	5060	10″ x 6″		24	.333		24	14.05		38.05	49	
	5080	10″ x 10″		23	.348		30.50	14.65		45.15	57	
	5090	12″ x 6″		23	.348		32.50	14.65		47.15	59	
	5100	12″ x 10″		22	.364		33.50	15.35		48.85	61	
	5110	12″ x 12″		22	.364		36	15.35		51.35	64.50	
	5120	14″ x 10″		22	.364		36	15.35		51.35	64.50	
	5140	16″ x 8″		21	.381		36	16.05		52.05	65.50	
	5160	18″ x 12″		20	.400		47.50	16.85		64.35	79	
	5170	18″ x 18″		21	.381		66.50	16.05		82.55	99	
	5180	20″ x 12″		20	.400		51	16.85		67.85	83	
	5200	20″ x 20″		19	.421		80.50	17.75		98.25	117	
	5220	24″ x 12″		17	.471		57.50	19.85		77.35	94.50	
	5240	24″ x 24″		16	.500		110	21		131	155	
	5260	30″ x 6″		15	.533		51	22.50		73.50	92	
	5280	30″ x 8″		15	.533		59.50	22.50		82	101	
	5300	30″ x 12″		14	.571		74	24		98	120	
	5320	30″ x 16″		13	.615		88	26		114	139	
	5340	30″ x 20″		12	.667		111	28		139	167	
	5360	30″ x 24″		11	.727		137	30.50		167.50	200	
	5380	30″ x 30″		10	.800		181	33.50		214.50	253	
	6000	For steel grilles instead of aluminum in above, deduct					10%					
	6200	Plastic, eggcrate, lay-in, T-bar system										
	6210	48″ x 24″ sheet	1 Shee	50	.160	Ea.	23.50	6.75		30.25	36.50	
	6250	Steel door louver										
	6270	With fire link, steel only										
	6350	12″ x 18″	1 Shee	22	.364	Ea.	148	15.35		163.35	188	
	6410	12″ x 24″		21	.381		165	16.05		181.05	208	
	6470	18″ x 18″		20	.400		165	16.85		181.85	209	
	6530	18″ x 24″		19	.421		189	17.75		206.75	236	
	6600	18″ x 36″		17	.471		215	19.85		234.85	269	
	6660	24″ x 24″		15	.533		199	22.50		221.50	255	
	6720	24″ x 30″		15	.533		215	22.50		237.50	273	
	6780	30″ x 30″		13	.615		252	26		278	320	
	6840	30″ x 36″		12	.667		292	28		320	365	
	6900	36″ x 36″		10	.800		335	33.50		368.50	425	
	9000	Minimum labor/equipment charge		4	2	Job		84.50		84.50	134	
600	0010	**LOUVERS**										600
	0100	Aluminum, extruded, with screen, mill finish										
	1000	Brick vent, (see also division 04090-860)										
	1100	Standard, 4″ deep, 8″ wide, 5″ high	1 Shee	24	.333	Ea.	25	14.05		39.05	50	
	1200	Modular, 4″ deep, 7-3/4″ wide, 5″ high		24	.333		26	14.05		40.05	51	
	1300	Speed brick, 4″ deep, 11-5/8″ wide, 3-7/8″ high		24	.333		26	14.05		40.05	51	
	1400	Fuel oil brick, 4″ deep, 8″ wide, 5″ high		24	.333		44.50	14.05		58.55	71.50	
	2000	Cooling tower and mechanical equip., screens, light weight		40	.200	S.F.	11.15	8.45		19.60	25.50	
	2020	Standard weight		35	.229		29.50	9.65		39.15	48	
	2500	Dual combination, automatic, intake or exhaust		20	.400		40.50	16.85		57.35	71.50	
	2520	Manual operation		20	.400		30	16.85		46.85	60	
	2540	Electric or pneumatic operation		20	.400		30	16.85		46.85	60	
	2560	Motor, for electric or pneumatic		14	.571	Ea.	350	24		374	425	
	3000	Fixed blade, continuous line										

15850 | Air Outlets & Inlets

				DAILY	LABOR-		2006 BARE COSTS				TOTAL
			CREW	OUTPUT	HOURS	UNIT	MAT.	LABOR	EQUIP.	TOTAL	INCL O&P
600	3100	Mullion type, stormproof	1 Shee	28	.286	S.F.	30	12.05		42.05	52
	3200	Stormproof		28	.286		30	12.05		42.05	52
	3300	Vertical line		28	.286		40	12.05		52.05	63
	3500	For damper to use with above, add					50%	30%			
	3520	Motor, for damper, electric or pneumatic	1 Shee	14	.571	Ea.	350	24		374	425
	4000	Operating, 45°, manual, electric or pneumatic		24	.333	S.F.	30	14.05		44.05	55.50
	4100	Motor, for electric or pneumatic		14	.571	Ea.	350	24		374	425
	4200	Penthouse, roof		56	.143	S.F.	17.85	6		23.85	29
	4300	Walls		40	.200		42	8.45		50.45	59.50
	5000	Thinline, under 4″ thick, fixed blade		40	.200		17.40	8.45		25.85	32.50
	5010	Finishes, applied by mfr. at additional cost, available in colors									
	5020	Prime coat only, add				S.F.	2.42			2.42	2.66
	5040	Baked enamel finish coating, add					4.44			4.44	4.88
	5060	Anodized finish, add					4.82			4.82	5.30
	5080	Duranodic finish, add					8.75			8.75	9.60
	5100	Fluoropolymer finish coating, add					13.65			13.65	15.05
	9980	For small orders (under 10 pieces), add					25%				
700	0010	**REGISTERS**									
	0980	Air supply									
	1000	Ceiling/wall, O.B. damper, anodized aluminum									
	1010	One or two way deflection, adj. curved face bars									
	1014	6″ x 6″	1 Shee	24	.333	Ea.	21.50	14.05		35.55	46.50
	1020	8″ x 4″		26	.308		21.50	12.95		34.45	44.50
	1040	8″ x 8″		24	.333		27	14.05		41.05	52.50
	1060	10″ x 6″		20	.400		25	16.85		41.85	54.50
	1080	10″ x 10″		19	.421		33.50	17.75		51.25	65
	1100	12″ x 6″		19	.421		24.50	17.75		42.25	55
	1120	12″ x 12″		18	.444		35.50	18.75		54.25	69
	1140	14″ x 8″		17	.471		34.50	19.85		54.35	69.50
	1160	14″ x 14″		18	.444		41.50	18.75		60.25	75.50
	1170	16″ x 16″		17	.471		46	19.85		65.85	82
	1180	18″ x 8″		18	.444		56	18.75		74.75	91.50
	1200	18″ x 18″		17	.471		70.50	19.85		90.35	109
	1220	20″ x 4″		19	.421		35.50	17.75		53.25	67
	1240	20″ x 6″		18	.444		35.50	18.75		54.25	69
	1260	20″ x 8″		18	.444		39	18.75		57.75	73
	1280	20″ x 20″		17	.471		86	19.85		105.85	126
	1290	22″ x 22″		15	.533		98.50	22.50		121	144
	1300	24″ x 4″		17	.471		44	19.85		63.85	79.50
	1320	24″ x 6″		16	.500		44	21		65	81.50
	1340	24″ x 8″		13	.615		48	26		74	94
	1350	24″ x 18″		12	.667		89	28		117	143
	1360	24″ x 24″		11	.727		121	30.50		151.50	182
	1380	30″ x 4″		16	.500		55	21		76	94
	1400	30″ x 6″		15	.533		55	22.50		77.50	96.50
	1420	30″ x 8″		14	.571		63	24		87	108
	1440	30″ x 24″		12	.667		156	28		184	217
	1460	30″ x 30″		10	.800		200	33.50		233.50	274
	1504	4 way deflection, adjustable curved face bars									
	1510	6″ x 6″	1 Shee	26	.308	Ea.	26	12.95		38.95	49
	1514	8″ x 8″		24	.333		32.50	14.05		46.55	58.50
	1518	10″ x 10″		19	.421		40.50	17.75		58.25	72.50
	1522	12″ x 6″		19	.421		29.50	17.75		47.25	60.50
	1526	12″ x 12″		18	.444		42.50	18.75		61.25	77
	1530	14″ x 14″		18	.444		50	18.75		68.75	85

15850 | Air Outlets & Inlets

			CREW	DAILY OUTPUT	LABOR-HOURS	UNIT	2006 BARE COSTS MAT.	LABOR	EQUIP.	TOTAL	TOTAL INCL O&P
700	1534	16" x 16"	1 Shee	17	.471	Ea.	55.50	19.85		75.35	92.50
	1538	18" x 18"		17	.471		84.50	19.85		104.35	125
	1542	22" x 22"		16	.500		118	21		139	164
	1980	One way deflection, adj. vert. or horiz. face bars									
	1990	6" x 6"	1 Shee	26	.308	Ea.	30.50	12.95		43.45	54
	2000	8" x 4"		26	.308		32.50	12.95		45.45	56
	2020	8" x 8"		24	.333		35	14.05		49.05	61
	2040	10" x 6"		20	.400		35	16.85		51.85	65.50
	2060	10" x 10"		19	.421		41	17.75		58.75	73
	2080	12" x 6"		19	.421		37.50	17.75		55.25	69.50
	2100	12" x 12"		18	.444		49.50	18.75		68.25	84.50
	2120	14" x 8"		17	.471		48.50	19.85		68.35	84.50
	2140	14" x 12"		18	.444		54	18.75		72.75	89.50
	2160	14" x 14"		18	.444		61.50	18.75		80.25	97.50
	2180	16" x 6"		18	.444		45	18.75		63.75	79.50
	2200	16" x 12"		18	.444		60	18.75		78.75	96
	2220	16" x 16"		17	.471		76	19.85		95.85	115
	2240	18" x 8"		18	.444		53	18.75		71.75	88.50
	2260	18" x 12"		17	.471		64	19.85		83.85	102
	2280	18" x 18"		16	.500		93.50	21		114.50	137
	2300	20" x 10"		18	.444		63	18.75		81.75	99
	2320	20" x 16"		16	.500		90.50	21		111.50	133
	2340	20" x 20"		16	.500		114	21		135	159
	2360	24" x 12"		15	.533		78.50	22.50		101	123
	2380	24" x 16"		14	.571		105	24		129	154
	2400	24" x 20"		12	.667		131	28		159	189
	2420	24" x 24"		10	.800		160	33.50		193.50	230
	2440	30" x 12"		13	.615		102	26		128	154
	2460	30" x 16"		13	.615		132	26		158	187
	2480	30" x 24"		11	.727		206	30.50		236.50	276
	2500	30" x 30"		9	.889		265	37.50		302.50	350
	2520	36" x 12"		11	.727		124	30.50		154.50	185
	2540	36" x 24"		10	.800		255	33.50		288.50	335
	2560	36" x 36"		8	1		445	42		487	555
	2600	For 2 way deflect., adj. vert. or horiz. face bars, add					40%				
	2700	Above registers in steel instead of aluminum, deduct					10%				
	3000	Baseboard, hand adj. damper, enameled steel									
	3012	8" x 6"	1 Shee	26	.308	Ea.	13.05	12.95		26	35
	3020	10" x 6"		24	.333		14.70	14.05		28.75	38.50
	3040	12" x 5"		23	.348		16.80	14.65		31.45	42
	3060	12" x 6"		23	.348		15.50	14.65		30.15	40.50
	3080	12" x 8"		22	.364		22.50	15.35		37.85	49
	3100	14" x 6"		20	.400		16.80	16.85		33.65	45.50
	4000	Floor, toe operated damper, enameled steel									
	4020	4" x 8"	1 Shee	32	.250	Ea.	18.65	10.55		29.20	37.50
	4040	4" x 12"		26	.308		22	12.95		34.95	44.50
	4060	6" x 8"		28	.286		19	12.05		31.05	40
	4080	6" x 14"		22	.364		25.50	15.35		40.85	52.50
	4100	8" x 10"		22	.364		23	15.35		38.35	49.50
	4120	8" x 16"		20	.400		38	16.85		54.85	69
	4140	10" x 10"		20	.400		27	16.85		43.85	56.50
	4160	10" x 16"		18	.444		41.50	18.75		60.25	75.50
	4180	12" x 12"		18	.444		33.50	18.75		52.25	67
	4200	12" x 24"		16	.500		115	21		136	161
	4220	14" x 14"		16	.500		95	21		116	139
	4240	14" x 20"		15	.533		127	22.50		149.50	176

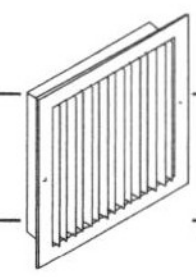

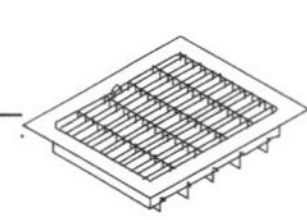

MECHANICAL 15

15850 | Air Outlets & Inlets

		15850 Air Outlets & Inlets	CREW	DAILY OUTPUT	LABOR-HOURS	UNIT	2006 BARE COSTS MAT.	LABOR	EQUIP.	TOTAL	TOTAL INCL O&P	
700	4300	Spiral pipe supply register										700
	4310	Aluminum, double deflection, w/damper extractor										
	4320	4" x 12", for 6" thru 10" diameter duct	1 Shee	25	.320	Ea.	62	13.50		75.50	89.50	
	4330	4" x 18", for 6" thru 10" diameter duct		18	.444		76.50	18.75		95.25	115	
	4340	6" x 12", for 8" thru 12" diameter duct		19	.421		67.50	17.75		85.25	103	
	4350	6" x 18", for 8" thru 12" diameter duct		18	.444		88	18.75		106.75	127	
	4360	6" x 24", for 8" thru 12" diameter duct		16	.500		109	21		130	153	
	4370	6" x 30", for 8" thru 12" diameter duct		15	.533		138	22.50		160.50	188	
	4380	8" x 18", for 10" thru 14" diameter duct		18	.444		93.50	18.75		112.25	133	
	4390	8" x 24", for 10" thru 14" diameter duct		15	.533		117	22.50		139.50	165	
	4400	8" x 30", for 10" thru 14" diameter duct		14	.571		155	24		179	210	
	4410	10" x 24", for 12" thru 18" diameter duct		13	.615		129	26		155	184	
	4420	10" x 30", for 12" thru 18" diameter duct		12	.667		170	28		198	232	
	4430	10" x 36", for 12" thru 18" diameter duct	↓	11	.727	↓	211	30.50		241.50	281	
	4980	Air return										
	5000	Ceiling or wall, fixed 45° face blades										
	5010	Adjustable O.B. damper, anodized aluminum										
	5020	4" x 8"	1 Shee	26	.308	Ea.	25.50	12.95		38.45	48.50	
	5040	6" x 8"		24	.333		25.50	14.05		39.55	50.50	
	5060	6" x 10"		19	.421		28	17.75		45.75	59	
	5080	6" x 16"		18	.444		36.50	18.75		55.25	70	
	5100	8" x 10"		19	.421		31	17.75		48.75	62.50	
	5120	8" x 12"		16	.500		34	21		55	70.50	
	5140	10" x 10"		18	.444		34	18.75		52.75	67	
	5160	10" x 16"		17	.471		45	19.85		64.85	81	
	5180	12" x 18"		18	.444		59	18.75		77.75	95	
	5200	12" x 30"		12	.667		87	28		115	141	
	5220	16" x 16"		17	.471		62	19.85		81.85	99.50	
	5240	18" x 18"		16	.500		73	21		94	114	
	5260	18" x 36"		10	.800		146	33.50		179.50	214	
	5280	24" x 24"		11	.727		120	30.50		150.50	181	
	5300	24" x 36"		8	1		187	42		229	273	
	5320	24" x 48"	↓	6	1.333	↓	274	56		330	390	
	6000	For steel construction instead of aluminum, deduct					10%					
	9000	Minimum labor/equipment charge	1 Shee	4	2	Job		84.50		84.50	134	
800	0010	**VENTILATORS** Base & damper										800
	1280	Rotary ventilators, wind driven, galvanized										
	1300	4" neck diameter	Q-9	20	.800	Ea.	52	30.50		82.50	106	
	1320	5" neck diameter		18	.889		58	33.50		91.50	118	
	1340	6" neck diameter		16	1		58	38		96	125	
	1360	8" neck diameter		14	1.143		60.50	43.50		104	136	
	1380	10" neck diameter		12	1.333		77.50	50.50		128	166	
	1400	12" neck diameter		10	1.600		84.50	60.50		145	189	
	1420	14" neck diameter		10	1.600		117	60.50		177.50	226	
	1440	16" neck diameter		9	1.778		141	67.50		208.50	263	
	1460	18" neck diameter, 1700 CFM		9	1.778		164	67.50		231.50	288	
	1480	20" neck diameter, 2100 CFM		8	2		182	76		258	320	
	1500	24" neck diameter, 3,100 CFM		8	2		245	76		321	390	
	1520	30" neck diameter, 4500 CFM		7	2.286		510	86.50		596.50	700	
	1540	36" neck diameter, 5,500 CFM	↓	6	2.667		720	101		821	950	
	1600	For aluminum, add					300%					
	1620	For stainless steel, add				↓	600%					
	2000	Stationary, gravity, syphon, galvanized										
	2100	3" neck diameter, 40 CFM	Q-9	24	.667	Ea.	47	25.50		72.50	91.50	
	2120	4" neck diameter, 50 CFM	↓	20	.800	↓	51.50	30.50		82	106	

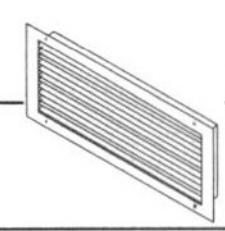

15 MECHANICAL

	15850 | Air Outlets & Inlets	CREW	DAILY OUTPUT	LABOR-HOURS	UNIT	2006 BARE COSTS MAT.	LABOR	EQUIP.	TOTAL	TOTAL INCL O&P
800 2140	5″ neck diameter, 58 CFM	Q-9	18	.889	Ea.	52	33.50		85.50	111 800
2160	6″ neck diameter, 66 CFM 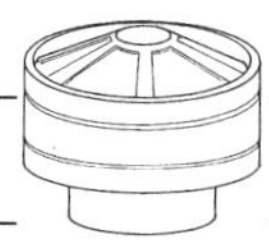		16	1		45	38		83	110
2180	7″ neck diameter, 86 CFM		15	1.067		56.50	40.50		97	127
2200	8″ neck diameter, 110 CFM		14	1.143		57	43.50		100.50	132
2220	10″ neck diameter, 140 CFM		12	1.333		73.50	50.50		124	162
2240	12″ neck diameter, 160 CFM		10	1.600		94.50	60.50		155	201
2260	14″ neck diameter, 250 CFM		10	1.600		119	60.50		179.50	227
2280	16″ neck diameter, 380 CFM		9	1.778		158	67.50		225.50	281
2300	18″ neck diameter, 500 CFM		9	1.778		211	67.50		278.50	340
2320	20″ neck diameter, 625 CFM		8	2		223	76		299	365
2340	24″ neck diameter, 900 CFM		8	2		257	76		333	405
2360	30″ neck diameter, 1375 CFM		7	2.286		585	86.50		671.50	785
2380	36″ neck diameter, 2,000 CFM		6	2.667		805	101		906	1,050
2400	42″ neck diameter, 3000 CFM	↓	4	4		1,125	152		1,277	1,500
2500	For aluminum, add					300%				
2520	For stainless steel, add					600%				
3000	Rotating chimney cap, galvanized, 4″ neck diameter	Q-9	20	.800		40.50	30.50		71	93
3020	5″ neck diameter		18	.889		41.50	33.50		75	99
3040	6″ neck diameter		16	1		42	38		80	107
3060	7″ neck diameter		15	1.067		45.50	40.50		86	115
3080	8″ neck diameter		14	1.143		47	43.50		90.50	121
3100	10″ neck diameter		12	1.333		61	50.50		111.50	148
3600	Stationary chimney rain cap, galvanized, 3″ neck diameter		24	.667		35.50	25.50		61	79
3620	4″ neck diameter		20	.800		36	30.50		66.50	88
3640	6″ neck diameter		16	1		37.50	38		75.50	102
3680	8″ neck diameter		14	1.143		40.50	43.50		84	114
3700	10″ neck diameter		12	1.333		51.50	50.50		102	137
3720	12″ neck diameter		10	1.600		85.50	60.50		146	191
3740	14″ neck diameter		10	1.600		91.50	60.50		152	198
3760	16″ neck diameter		9	1.778		112	67.50		179.50	230
4200	Stationary mushroom, aluminum, 16″ orifice diameter		10	1.600		360	60.50		420.50	490
4220	26″ orifice diameter		6.15	2.602		530	98.50		628.50	735
4230	30″ orifice diameter		5.71	2.802		775	106		881	1,025
4240	38″ orifice diameter		5	3.200		1,100	121		1,221	1,425
4250	42″ orifice diameter		4.70	3.404		1,475	129		1,604	1,800
4260	50″ orifice diameter	↓	4.44	3.604	↓	1,750	137		1,887	2,150
5000	Relief vent									
5500	Rectangular, aluminum, galvanized curb									
5510	intake/exhaust, 0.05″ SP									
5600	500 CFM, 12″ x 16″	Q-9	8	2	Ea.	410	76		486	570
5620	750 CFM, 12″ x 20″		7.20	2.222		450	84.50		534.50	630
5640	1000 CFM, 12″ x 24″ 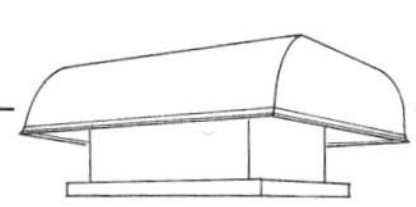		6.60	2.424		485	92		577	680
5660	1500 CFM, 12″ x 36″		5.80	2.759		590	105		695	810
5680	3000 CFM, 20″ x 42″		4	4		845	152		997	1,175
5700	6000 CFM, 20″ x 84″		2.60	6.154		1,275	233		1,508	1,800
5720	8000 CFM, 24″ x 96″		2.30	6.957		1,475	264		1,739	2,050
5740	10,000 CFM, 48″ x 60″		1.80	8.889		1,800	335		2,135	2,500
5760	12,500 CFM, 48″ x 72″		1.60	10		2,125	380		2,505	2,925
5780	15,000 CFM, 48″ x 96″		1.30	12.308		2,450	465		2,915	3,450
5800	20,000 CFM, 48″ x 120″		1.20	13.333		2,825	505		3,330	3,925
5820	25,000 CFM, 60″ x 120″		.90	17.778		3,750	675		4,425	5,200
5840	30,000 CFM, 72″ x 120″		.70	22.857		4,925	865		5,790	6,800
5860	40,000 CFM, 96″ x 120″		.60	26.667		6,325	1,000		7,325	8,575
5880	50,000 CFM, 96″ x 144″	↓	.50	32	↓	6,725	1,225		7,950	9,325
7000	Note: sizes based on exhaust. Intake, with 0.125″ SP									
7100	loss, approximately twice listed capacity.									

MECHANICAL 15

15850 | Air Outlets & Inlets

		15850 Air Outlets & Inlets	CREW	DAILY OUTPUT	LABOR-HOURS	UNIT	2006 BARE COSTS MAT.	LABOR	EQUIP.	TOTAL	TOTAL INCL O&P	
800	7500	For power open, spring return damper										800
	7600	motor for each S.F. of throat area	Q-9	100	.160	S.F.	49.50	6.05		55.55	64	
	9000	Minimum labor/equipment charge	1 Plum	2	4	Job		171		171	266	

15860 | Air Cleaning Devices

		15860 Air Cleaning Devices	CREW	DAILY OUTPUT	LABOR-HOURS	UNIT	MAT.	LABOR	EQUIP.	TOTAL	TOTAL INCL O&P	
100	0010	AIR FILTERS										100
	0050	Activated charcoal type, full flow				MCFM	600			600	660	
	0060	Activated charcoal type, full flow, impregnated media 12" deep					175			175	193	
	0070	Activated charcoal type, HEPA filter & frame for field erection					175			175	193	
	0080	Activated charcoal type, HEPA filter-diffuser, ceiling install.					250			250	275	
	0500	Chemical media filtration type										
	1100	Industrial air fume & odor scrubber unit w/pump & motor										
	1110	corrosion resistant PVC construction										
	1120	Single pack filter, horizontal type										
	1130	500 CFM	Q-9	14	1.143	Ea.	4,125	43.50		4,168.50	4,600	
	1140	1000 CFM		11	1.455		4,825	55		4,880	5,400	
	1150	2000 CFM		8	2		5,850	76		5,926	6,550	
	1160	3000 CFM		7	2.286		6,775	86.50		6,861.50	7,600	
	1170	5000 CFM		5	3.200		8,825	121		8,946	9,900	
	1180	8000 CFM		4	4		12,300	152		12,452	13,800	
	1190	12,000 CFM		3	5.333		16,400	202		16,602	18,300	
	1200	16,000 CFM		2.50	6.400		20,200	243		20,443	22,600	
	1210	20,000 CFM		2	8		25,300	305		25,605	28,400	
	1220	26,000 CFM		1.50	10.667		29,200	405		29,605	32,700	
	1230	30,000 CFM	Q-10	2	12		33,200	470		33,670	37,300	
	1240	40,000 CFM	"	1.50	16		43,400	630		44,030	48,800	
	1250	50,000 CFM	Q-11	2	16		53,000	645		53,645	59,000	
	1260	55,000 CFM		1.80	17.778		57,500	715		58,215	64,500	
	1270	60,000 CFM		1.50	21.333		62,000	855		62,855	70,000	
	1300	Double pack filter, horizontal type										
	1310	500 CFM	Q-9	10	1.600	Ea.	5,350	60.50		5,410.50	6,000	
	1320	1000 CFM		8	2		6,425	76		6,501	7,175	
	1330	2000 CFM		6	2.667		7,725	101		7,826	8,650	
	1340	3000 CFM		5	3.200		9,250	121		9,371	10,400	
	1350	5000 CFM		4	4		11,700	152		11,852	13,100	
	1360	8000 CFM		3	5.333		15,500	202		15,702	17,400	
	1370	12,000 CFM		2.50	6.400		21,100	243		21,343	23,600	
	1380	16,000 CFM		2	8		27,500	305		27,805	30,700	
	1390	20,000 CFM		1.50	10.667		33,300	405		33,705	37,200	
	1400	26,000 CFM		1	16		38,600	605		39,205	43,400	
	1410	30,000 CFM	Q-10	1.50	16		45,300	630		45,930	51,000	
	1420	40,000 CFM	"	1.30	18.462		58,000	725		58,725	64,500	
	1430	50,000 CFM	Q-11	1.50	21.333		72,500	855		73,355	81,000	
	1440	55,000 CFM		1.30	24.615		79,000	990		79,990	88,500	
	1450	60,000 CFM		1	32		86,000	1,275		87,275	96,500	
	1500	Single pack filter, vertical type										
	1510	500 CFM	Q-9	24	.667	Ea.	3,925	25.50		3,950.50	4,375	
	1520	1000 CFM		18	.889		4,650	33.50		4,683.50	5,150	
	1530	2000 CFM		12	1.333		5,575	50.50		5,625.50	6,200	
	1540	3000 CFM		9	1.778		6,500	67.50		6,567.50	7,250	
	1550	5000 CFM		6	2.667		8,675	101		8,776	9,675	
	1560	8000 CFM		4	4		12,300	152		12,452	13,800	
	1570	12,000 CFM		3	5.333		16,800	202		17,002	18,800	
	1580	16,000 CFM		2	8		19,600	305		19,905	22,100	
	1590	20,000 CFM		1.80	8.889		24,800	335		25,135	27,700	

15860 | Air Cleaning Devices

			CREW	DAILY OUTPUT	LABOR-HOURS	UNIT	2006 BARE COSTS MAT.	LABOR	EQUIP.	TOTAL	TOTAL INCL O&P
100	1600	24,000 CFM	Q-9	1.60	10	Ea.	27,700	380		28,080	31,100
	1650	Double pack filter, vertical type									
	1660	500 CFM	Q-9	22	.727	Ea.	5,100	27.50		5,127.50	5,650
	1670	1000 CFM		16	1		6,075	38		6,113	6,725
	1680	2000 CFM		10	1.600		7,300	60.50		7,360.50	8,125
	1690	3000 CFM		7	2.286		8,575	86.50		8,661.50	9,600
	1700	5000 CFM		5	3.200		11,200	121		11,321	12,600
	1710	8000 CFM		3	5.333		15,400	202		15,602	17,200
	1720	12,000 CFM		2.50	6.400		20,800	243		21,043	23,300
	1730	16,000 CFM		2	8		26,100	305		26,405	29,200
	1740	20,000 CFM		1.50	10.667		32,600	405		33,005	36,500
	1750	24,000 CFM	↓	1	16	↓	36,400	605		37,005	41,000
	1800	Inlet or outlet transition, horizontal									
	1810	Single pack to 12,000 CFM, add					4%				
	1820	Single pack to 30,000 CFM, add					6%				
	1830	Single pack to 60,000 CFM, add					8%				
	1840	Double pack to 12,000 CFM, add					3%				
	1850	Double pack to 30,000 CFM, add					5%				
	1860	Double pack to 60,000 CFM, add					6%				
	1870	Inlet or outlet transition, vertical									
	1880	Single pack to 5000 CFM, add					2%				
	1890	Single pack to 24,000 CFM, add					3%				
	1900	Double pack to 24,000 CFM, add					2%				
	2000	Electronic air cleaner, duct mounted									
	2150	400 - 1000 CFM	1 Shee	2.30	3.478	Ea.	695	147		842	1,000
	2200	1000 - 1400 CFM		2.20	3.636		725	153		878	1,050
	2250	1400 - 2000 CFM	↓	2.10	3.810	↓	800	161		961	1,125
	2950	Mechanical media filtration units									
	3000	High efficiency type, with frame, non-supported				MCFM	45			45	49.50
	3100	Supported type					55			55	60.50
	4000	Medium efficiency, extended surface					5.50			5.50	6.05
	4500	Permanent washable					20			20	22
	5000	Renewable disposable roll				↓	120			120	132
	5500	Throwaway glass or paper media type				Ea.	4.60			4.60	5.05
	9000	Minimum labor/equipment charge	1 Shee	2.25	3.556	Job		150		150	238
500	0010	**EXHAUST SYSTEMS**									
	0500	Engine exhaust, garage, in-floor system									
	0510	Single tube outlet assemblies									
	0520	For transite pipe ducting, self-storing tube 									
	0530	3" tubing adapter plate	1 Shee	16	.500	Ea.	180	21		201	232
	0540	4" tubing adapter plate		16	.500		180	21		201	232
	0550	5" tubing adapter plate	↓	16	.500	↓	180	21		201	232
	0600	For vitrified tile ducting 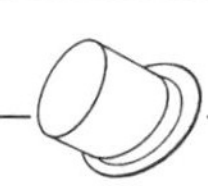									
	0610	3" tubing adapter plate, self-storing tube	1 Shee	16	.500	Ea.	180	21		201	232
	0620	4" tubing adapter plate, self-storing tube		16	.500		180	21		201	232
	0660	5" tubing adapter plate, self-storing tube	↓	16	.500	↓	180	21		201	232
	0800	Two tube outlet assemblies									
	0810	For transite pipe ducting, self-storing tube									
	0820	3" tubing, dual exhaust adapter plate 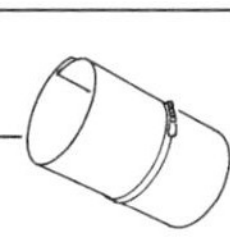	1 Shee	16	.500	Ea.	180	21		201	232
	0850	For vitrified tile ducting									
	0860	3" tubing, dual exhaust, self-storing tube	1 Shee	16	.500	Ea.	180	21		201	232
	0870	3" tubing, double outlet, non-storing tubes	"	16	.500	"	180	21		201	232
	0900	Accessories for metal tubing, (overhead systems also)									
	0910	Adapters, for metal tubing end									
	0920	3" tail pipe type				Ea.	40			40	44

15860	Air Cleaning Devices	CREW	DAILY OUTPUT	LABOR-HOURS	UNIT	2006 BARE COSTS MAT.	LABOR	EQUIP.	TOTAL	TOTAL INCL O&P
500 0930	4" tail pipe type				Ea.	40			40	44
0940	5" tail pipe type					40			40	44
0990	5" diesel stack type					280			280	310
1000	6" diesel stack type					280			280	310
1100	Bullnose (guide) required for in-floor assemblies									
1110	3" tubing size				Ea.	20			20	22
1120	4" tubing size					20			20	22
1130	5" tubing size					20			20	22
1150	Plain rings, for tubing end									
1160	3" tubing size				Ea.	9.50			9.50	10.45
1170	4" tubing size				"	12			12	13.20
1200	Tubing, galvanized, flexible, (for overhead systems also)									
1210	3" ID				L.F.	5.35			5.35	5.90
1220	4" ID					6.45			6.45	7.10
1230	5" ID					7.50			7.50	8.25
1240	6" ID					8.60			8.60	9.45
1250	Stainless steel, flexible, (for overhead systems also)									
1260	3" ID				L.F.	11.10			11.10	12.20
1270	4" ID					14.45			14.45	15.90
1280	5" ID					17.45			17.45	19.20
1290	6" ID					21			21	23
1500	Engine exhaust, garage, ovrhd. components, for neoprene tubing									
1510	Alternate metal tubing & accessories see above									
1550	Adapters, for neoprene tubing end									
1560	3" tail pipe, adjustable, neoprene				Ea.	35			35	38.50
1570	3" tail pipe, heavy wall neoprene					45			45	49.50
1580	4" tail pipe, heavy wall neoprene					48			48	53
1590	5" tail pipe, heavy wall neoprene					65			65	71.50
1650	Connectors, tubing									
1660	3" interior, aluminum				Ea.	9.50			9.50	10.45
1670	4" interior, aluminum					12			12	13.20
1710	5" interior, neoprene					30			30	33
1750	3" spiralock, neoprene					9.50			9.50	10.45
1760	4" spiralock, neoprene					12			12	13.20
1780	Y for 3" ID tubing, neoprene, dual exhaust					25			25	27.50
1790	Y for 4" ID tubing, aluminum, dual exhaust					45			45	49.50
1850	Elbows, aluminum, splice into tubing for strap									
1860	3" neoprene tubing size				Ea.	30			30	33
1870	4" neoprene tubing size					35			35	38.50
1900	Flange assemblies, connect tubing to overhead duct					33			33	36.50
2000	Hardware and accessories									
2020	Cable, galvanized, 1/8" diameter				L.F.	.35			.35	.39
2040	Cleat, tie down cable or rope				Ea.	4			4	4.40
2060	Pulley					5.50			5.50	6.05
2080	Pulley hook, universal					4			4	4.40
2100	Rope, nylon, 1/4" diameter				L.F.	.30			.30	.33
2120	Winch, 1" diameter				Ea.	85			85	93.50
2150	Lifting strap, mounts on neoprene									
2160	3" tubing size				Ea.	20			20	22
2170	4" tubing size					20			20	22
2180	5" tubing size					20			20	22
2190	6" tubing size					20			20	22
2200	Tubing, neoprene, 11' lengths									
2210	3" ID				L.F.	6			6	6.60
2220	4" ID					7.50			7.50	8.25
2230	5" ID					13.50			13.50	14.85 500

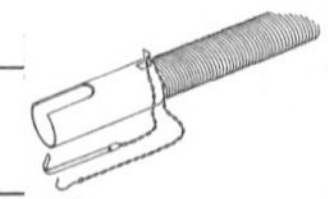

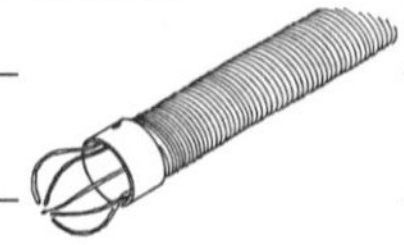

15 MECHANICAL

15800 | Air Distribution

	15860 | Air Cleaning Devices		CREW	DAILY OUTPUT	LABOR-HOURS	UNIT	2006 BARE COSTS MAT.	LABOR	EQUIP.	TOTAL	TOTAL INCL O&P	
500	2500	Engine exhaust, thru-door outlet										500
	2510	3″ tube size	1 Carp	16	.500	Ea.	35	17.80		52.80	68	
	2530	4″ tube size	″	16	.500	″	35	17.80		52.80	68	
	7500	Welding fume elimination accessories for garage exhaust systems										
	7600	Cut off (blast gate)										
	7610	3″ tubing size, 3″ x 6″ opening	1 Shee	24	.333	Ea.	16	14.05		30.05	40	
	7620	4″ tubing size, 4″ x 8″ opening		24	.333		18	14.05		32.05	42.50	
	7630	5″ tubing size, 5″ x 10″ opening		24	.333		24	14.05		38.05	49	
	7640	6″ tubing size		24	.333		28	14.05		42.05	53.50	
	7650	8″ tubing size	↓	24	.333	↓	40	14.05		54.05	66.50	
	7700	Hoods, magnetic, with handle & screen										
	7710	3″ tubing size, 3″ x 6″ opening	1 Shee	24	.333	Ea.	75	14.05		89.05	105	
	7720	4″ tubing size, 4″ x 8″ opening		24	.333		75	14.05		89.05	105	
	7730	5″ tubing size, 5″ x 10″ opening	↓	24	.333	↓	75	14.05		89.05	105	

15900 | HVAC Instrumentation & Controls

	15905 | HVAC Instrumentation		CREW	DAILY OUTPUT	LABOR-HOURS	UNIT	2006 BARE COSTS MAT.	LABOR	EQUIP.	TOTAL	TOTAL INCL O&P	
960	0010	WATER LEVEL CONTROLS										960
	1000	Electric water feeder	1 Stpi	12	.667	Ea.	214	28.50		242.50	280	
	2000	Feeder cut-off combination										
	2100	Steam system up to 5000 sq. ft.	1 Stpi	12	.667	Ea.	435	28.50		463.50	520	
	2200	Steam system above 5000 sq. ft.		12	.667		535	28.50		563.50	630	
	2300	Steam and hot water, high pressure	↓	10	.800	↓	625	34.50		659.50	745	
	3000	Low water cut-off for hot water boiler, 50 psi maximum										
	3100	1″ top & bottom equalizing pipes, manual reset	1 Stpi	14	.571	Ea.	242	24.50		266.50	305	
	3200	1″ top & bottom equalizing pipes		14	.571		242	24.50		266.50	305	
	3300	2-1/2″ side connection for nipple-to-boiler	↓	14	.571	↓	218	24.50		242.50	279	
	4000	Low water cut-off for low pressure steam with quick hook-up ftgs.										
	4100	For installation in gauge glass tappings	1 Stpi	16	.500	Ea.	174	21.50		195.50	225	
	4200	Built-in type, 2-1/2″ tap - 3-1/8″ insertion		16	.500		146	21.50		167.50	195	
	4300	Built-in type, 2-1/2″ tap - 1-3/4″ insertion		16	.500		155	21.50		176.50	205	
	4400	Side connection to 2-1/2″ tapping		16	.500		166	21.50		187.50	216	
	5000	Pump control, low water cut-off and alarm switch	↓	14	.571	↓	455	24.50		479.50	540	

15950 | Testing/Adjusting/Balancing

	15955 | HVAC Test/Adjust/Balance		CREW	DAILY OUTPUT	LABOR-HOURS	UNIT	2006 BARE COSTS MAT.	LABOR	EQUIP.	TOTAL	TOTAL INCL O&P	
100	0010	**BALANCING, AIR** (Subcontractor's quote incl. material & labor)										100
	0900	Heating and ventilating equipment										
	1000	Centrifugal fans, utility sets				Ea.					321.84	
	1100	Heating and ventilating unit									482.76	
	1200	In-line fan									482.76	
	1300	Propeller and wall fan				↓					91.19	

MECHANICAL 15

	15955 | HVAC Test/Adjust/Balance		Crew	Daily Output	Labor-Hours	Unit	2006 Bare Costs Mat.	Labor	Equip.	Total	Total Incl O&P	
100	1400	Roof exhaust fan				Ea.					214.56	100
	2000	Air conditioning equipment, central station									697.32	
	2100	Built-up low pressure unit									643.68	
	2200	Built-up high pressure unit									750.96	
	2300	Built-up high pressure dual duct									1,180.08	
	2400	Built-up variable volume									1,394.64	
	2500	Multi-zone A.C. and heating unit									482.76	
	2600	For each zone over one, add									107.28	
	2700	Package A.C. unit									268.20	
	2800	Rooftop heating and cooling unit									375.48	
	3000	Supply, return, exhaust, registers & diffusers, avg. height ceiling									64.37	
	3100	High ceiling									96.55	
	3200	Floor height									53.64	
	3300	Off mixing box									42.91	
	3500	Induction unit									69.73	
	3600	Lab fume hood									321.84	
	3700	Linear supply									160.92	
	3800	Linear supply high									187.74	
	4000	Linear return									53.64	
	4100	Light troffers									64.37	
	4200	Moduline - master									64.37	
	4300	Moduline - slaves									32.18	
	4400	Regenerators									429.12	
	4500	Taps into ceiling plenums									80.46	
	4600	Variable volume boxes									64.37	
700	0010	**PIPING, TESTING**										700
	0100	Nondestructive testing										
	0110	Nondestructive hydraulic pressure test, isolate & 1 hr. hold										
	0120	1" - 4" pipe										
	0140	0 - 250 L.F.	1 Stpi	1.33	6.015	Ea.		259		259	405	
	0160	250 - 500 L.F.	"	.80	10			430		430	670	
	0180	500 - 1000 L.F.	Q-5	1.14	14.035			545		545	845	
	0200	1000 - 2000 L.F.	"	.80	20			775		775	1,200	
	0300	6" - 10" pipe										
	0320	0 - 250 L.F.	Q-5	1	16	Ea.		620		620	965	
	0340	250 - 500 L.F.		.73	21.918			850		850	1,325	
	0360	500 - 1000 L.F.		.53	30.189			1,175		1,175	1,825	
	0380	1000 - 2000 L.F.		.38	42.105			1,625		1,625	2,525	
	2000	X-Ray of welds										
	2110	2" diam.	1 Stpi	8	1	Ea.	8.85	43		51.85	77	
	2120	3" diam.		8	1		8.85	43		51.85	77	
	2130	4" diam.		8	1		13.30	43		56.30	81.50	
	2140	6" diam.		8	1		13.30	43		56.30	81.50	
	2150	8" diam.		6.60	1.212		13.30	52		65.30	95.50	
	2160	10" diam.		6	1.333		17.70	57.50		75.20	109	
900	0010	**BALANCING, WATER** (Subcontractor's quote incl. material & labor)										900
	0050	Air cooled condenser				Ea.					187.46	
	0100	Cabinet unit heater									64.27	
	0200	Chiller									455.26	
	0300	Convector									53.56	
	0400	Converter									267.80	
	0500	Cooling tower									348.14	
	0600	Fan coil unit, unit ventilator									96.41	
	0700	Fin tube and radiant panels									107.12	
	0800	Main and duct re-heat coils									99.09	

15950 | Testing/Adjusting/Balancing

	15955 | HVAC Test/Adjust/Balance		CREW	DAILY OUTPUT	LABOR-HOURS	UNIT	2006 BARE COSTS MAT.	LABOR	EQUIP.	TOTAL	TOTAL INCL O&P	
900	0810	Heat exchanger				Ea.					99.09	900
	0900	Main balancing cocks									80.34	
	1000	Pumps									235.66	
	1100	Unit heater				↓					74.98	

For information about Means Estimating Seminars, see yellow pages 12 and 13 in back of book

Division Notes

		CREW	DAILY OUTPUT	LABOR-HOURS	UNIT	2006 BARE COSTS MAT.	LABOR	EQUIP.	TOTAL	TOTAL INCL O&P

Division 16 Electrical

Estimating Tips

16060 Grounding & Bonding

- When taking off grounding system, identify separately the type and size of wire and list each unique type of ground connection.

16100 Wiring Methods

- Conduit should be taken off in three main categories: power distribution, branch power, and branch lighting, so the estimator can concentrate on systems and components, therefore making it easier to ensure all items have been accounted for.
- For cost modifications for elevated conduit installation, add the percentages to labor according to the height of installation and only to the quantities exceeding the different height levels, not to the total conduit quantities.
- Remember that aluminum wiring of equal ampacity is larger in diameter than copper and may require larger conduit.
- If more than three wires at a time are being pulled, deduct percentages from the labor hours of that grouping of wires.
- The estimator should take the weights of materials into consideration when completing a takeoff. Topics to consider include: How will the materials be supported? What methods of support are available? How high will the support structure have to reach? Will the final support structure be able to withstand the total burden? Is the support material included or separate from the fixture, equipment, and material specified?

16200 Electrical Power

- Do not overlook the costs for equipment used in the installation. If scaffolding or highlifts are available in the field, contractors may use them in lieu of the proposed ladders and rolling staging.

16400 Low-Voltage Distribution

- Supports and concrete pads may be shown on drawings for the larger equipment, or the support system may be only a piece of plywood for the back of a panelboard. In either case, it must be included in the costs.

16500 Lighting

- Fixtures should be taken off room by room, using the fixture schedule, specifications, and the ceiling plan. For large concentrations of lighting fixtures in the same area, deduct the percentages from labor hours.

16700 Communications
16800 Sound & Video

- When estimating material costs for special systems, it is always prudent to obtain manufacturers' quotations for equipment prices and special installation requirements which will affect the total costs.

Reference Numbers

Reference numbers are shown in bold squares at the beginning of some major classifications. These numbers refer to related items in the Reference Section. The reference information may be an estimating procedure, an alternate pricing method, or technical information.

Note: Not all subdivisions listed here necessarily appear in this publication.

Note: **i2 Trade Service,** *in part, has been used as a reference source for some of the material prices used in Division 16.*

16055 | Selective Demolition

	Line	Description	Crew	Daily Output	Labor-Hours	Unit	2006 Bare Costs Mat.	Labor	Equip.	Total	Total Incl O&P	
300	0010	**ELECTRICAL DEMOLITION** R260105-30										300
	0020	Conduit to 15′ high, including fittings & hangers										
	0100	Rigid galvanized steel, 1/2″ to 1″ diameter R024119-10	1 Elec	242	.033	L.F.		1.39		1.39	2.14	
	0120	1-1/4″ to 2″	″	200	.040			1.68		1.68	2.59	
	0140	2-1/2″ to 3-1/2″	2 Elec	302	.053			2.23		2.23	3.43	
	0160	4″ to 6″	″	160	.100			4.20		4.20	6.45	
	0170	PVC #40, 1/2″ to 1″	1 Elec	410	.020			.82		.82	1.26	
	0172	1-1/4″ to 2″		350	.023			.96		.96	1.48	
	0174	2-1/2″		250	.032			1.34		1.34	2.07	
	0176	3″ to 3-1/2″	2 Elec	340	.047			1.98		1.98	3.04	
	0178	4″ to 6″	″	230	.070			2.92		2.92	4.50	
	0200	Electric metallic tubing (EMT), 1/2″ to 1″	1 Elec	394	.020			.85		.85	1.31	
	0220	1-1/4″ to 1-1/2″		326	.025			1.03		1.03	1.59	
	0240	2″ to 3″		236	.034			1.42		1.42	2.19	
	0260	3-1/2″ to 4″	2 Elec	310	.052			2.17		2.17	3.34	
	0270	Armored cable, (BX) avg. 50′ runs										
	0280	#14, 2 wire	1 Elec	690	.012	L.F.		.49		.49	.75	
	0290	#14, 3 wire		571	.014			.59		.59	.91	
	0300	#12, 2 wire		605	.013			.56		.56	.86	
	0310	#12, 3 wire		514	.016			.65		.65	1.01	
	0320	#10, 2 wire		514	.016			.65		.65	1.01	
	0330	#10, 3 wire		425	.019			.79		.79	1.22	
	0340	#8, 3 wire		342	.023			.98		.98	1.51	
	0350	Non metallic sheathed cable (Romex)										
	0360	#14, 2 wire	1 Elec	720	.011	L.F.		.47		.47	.72	
	0370	#14, 3 wire		657	.012			.51		.51	.79	
	0380	#12, 2 wire		629	.013			.53		.53	.82	
	0390	#10, 3 wire		450	.018			.75		.75	1.15	
	0400	Wiremold raceway, including fittings & hangers										
	0420	No. 3000	1 Elec	250	.032	L.F.		1.34		1.34	2.07	
	0440	No. 4000		217	.037			1.55		1.55	2.39	
	0460	No. 6000		166	.048			2.02		2.02	3.12	
	0465	Telephone/power pole		12	.667	Ea.		28		28	43	
	0470	Non-metallic, straight section		480	.017	L.F.		.70		.70	1.08	
	0500	Channels, steel, including fittings & hangers										
	0520	3/4″ x 1-1/2″	1 Elec	308	.026	L.F.		1.09		1.09	1.68	
	0540	1-1/2″ x 1-1/2″		269	.030			1.25		1.25	1.92	
	0560	1-1/2″ x 1-7/8″		229	.035			1.47		1.47	2.26	
	0600	Copper bus duct, indoor, 3 phase										
	0610	Including hangers & supports										
	0620	225 amp	2 Elec	135	.119	L.F.		4.98		4.98	7.65	
	0640	400 amp		106	.151			6.35		6.35	9.75	
	0660	600 amp		86	.186			7.80		7.80	12.05	
	0680	1000 amp		60	.267			11.20		11.20	17.25	
	0700	1600 amp		40	.400			16.80		16.80	26	
	0720	3000 amp		10	1.600			67		67	104	
	0800	Plug-in switches, 600V 3 ph, incl. disconnecting										
	0820	wire, conduit terminations, 30 amp	1 Elec	15.50	.516	Ea.		21.50		21.50	33.50	
	0840	60 amp		13.90	.576			24		24	37	
	0850	100 amp		10.40	.769			32.50		32.50	50	
	0860	200 amp		6.20	1.290			54		54	83.50	
	0880	400 amp	2 Elec	5.40	2.963			124		124	192	
	0900	600 amp		3.40	4.706			198		198	305	
	0920	800 amp		2.60	6.154			258		258	400	
	0940	1200 amp		2	8			335		335	520	
	0960	1600 amp		1.70	9.412			395		395	610	

16055 | Selective Demolition

		16055 Selective Demolition	Crew	Daily Output	Labor-Hours	Unit	Mat. (2006 Bare Costs)	Labor (2006 Bare Costs)	Equip. (2006 Bare Costs)	Total (2006 Bare Costs)	Total Incl O&P
300	1010	Safety switches, 250 or 600V, incl. disconnection [R260105-30]									
	1050	of wire & conduit terminations									
	1100	30 amp [R024119-10]	1 Elec	12.30	.650	Ea.		27.50		27.50	42
	1120	60 amp		8.80	.909			38		38	59
	1140	100 amp		7.30	1.096			46		46	71
	1160	200 amp		5	1.600			67		67	104
	1180	400 amp	2 Elec	6.80	2.353			99		99	152
	1200	600 amp	″	4.60	3.478			146		146	225
	1210	Panel boards, incl. removal of all breakers,									
	1220	conduit terminations & wire connections									
	1230	3 wire, 120/240V, 100A, to 20 circuits	1 Elec	2.60	3.077	Ea.		129		129	199
	1240	200 amps, to 42 circuits	2 Elec	2.60	6.154			258		258	400
	1250	400 amps, to 42 circuits	″	2.20	7.273			305		305	470
	1260	4 wire, 120/208V, 125A, to 20 circuits	1 Elec	2.40	3.333			140		140	216
	1270	200 amps, to 42 circuits	2 Elec	2.40	6.667			280		280	430
	1280	400 amps, to 42 circuits	″	1.92	8.333			350		350	540
	1300	Transformer, dry type, 1 ph, incl. removal of									
	1320	supports, wire & conduit terminations									
	1340	1 kVA	1 Elec	7.70	1.039	Ea.		43.50		43.50	67
	1360	5 kVA		4.70	1.702			71.50		71.50	110
	1380	10 kVA		3.60	2.222			93.50		93.50	144
	1400	37.5 kVA	2 Elec	3	5.333			224		224	345
	1420	75 kVA	″	2.50	6.400			269		269	415
	1440	3 Phase to 600V, primary									
	1460	3 kVA	1 Elec	3.85	2.078	Ea.		87.50		87.50	134
	1480	15 kVA	2 Elec	4.20	3.810			160		160	246
	1500	30 kVA		3.50	4.571			192		192	296
	1510	45 kVA		3.10	5.161			217		217	335
	1520	75 kVA		2.70	5.926			249		249	385
	1530	112.5 kVA	R-3	2.90	6.897			286	52.50	338.50	500
	1540	150 kVA		2.70	7.407			305	56.50	361.50	535
	1560	500 kVA		1.40	14.286			590	109	699	1,025
	1570	750 kVA		1.10	18.182			755	139	894	1,325
	1600	Pull boxes & cabinets, sheet metal, incl. removal									
	1620	of supports and conduit terminations									
	1640	6″ x 6″ x 4″	1 Elec	31.10	.257	Ea.		10.80		10.80	16.65
	1660	12″ x 12″ x 4″		23.30	.343			14.40		14.40	22
	1680	24″ x 24″ x 6″		12.30	.650			27.50		27.50	42
	1700	36″ x 36″ x 8″		7.70	1.039			43.50		43.50	67
	1720	Junction boxes, 4″ sq. & oct.		80	.100			4.20		4.20	6.45
	1740	Handy box		107	.075			3.14		3.14	4.84
	1760	Switch box		107	.075			3.14		3.14	4.84
	1780	Receptacle & switch plates		257	.031			1.31		1.31	2.01
	1790	Receptacles & switches, 15 to 30 amp		135	.059			2.49		2.49	3.83
	1800	Wire, THW-THWN-THHN, removed from									
	1810	in place conduit, to 15′ high									
	1830	#14	1 Elec	65	.123	C.L.F.		5.15		5.15	7.95
	1840	#12		55	.145			6.10		6.10	9.40
	1850	#10		45.50	.176			7.40		7.40	11.40
	1860	#8		40.40	.198			8.30		8.30	12.80
	1870	#6		32.60	.245			10.30		10.30	15.90
	1880	#4	2 Elec	53	.302			12.70		12.70	19.55
	1890	#3		50	.320			13.45		13.45	20.50
	1900	#2		44.60	.359			15.05		15.05	23
	1910	1/0		33.20	.482			20		20	31
	1920	2/0		29.20	.548			23		23	35.50

		16055 | Selective Demolition	CREW	DAILY OUTPUT	LABOR-HOURS	UNIT	2006 BARE COSTS MAT.	LABOR	EQUIP.	TOTAL	TOTAL INCL O&P	
300	1930	3/0 (R260105-30)	2 Elec	25	.640	C.L.F.		27		27	41.50	300
	1940	4/0		22	.727			30.50		30.50	47	
	1950	250 kcmil (R024119-10)		20	.800			33.50		33.50	52	
	1960	300 kcmil		19	.842			35.50		35.50	54.50	
	1970	350 kcmil		18	.889			37.50		37.50	57.50	
	1980	400 kcmil		17	.941			39.50		39.50	61	
	1990	500 kcmil		16.20	.988			41.50		41.50	64	
	2000	Interior fluorescent fixtures, incl. supports										
	2010	& whips, to 15′ high										
	2100	Recessed drop-in 2′ x 2′, 2 lamp	2 Elec	35	.457	Ea.		19.20		19.20	29.50	
	2120	2′ x 4′, 2 lamp		33	.485			20.50		20.50	31.50	
	2140	2′ x 4′, 4 lamp		30	.533			22.50		22.50	34.50	
	2160	4′ x 4′, 4 lamp		20	.800			33.50		33.50	52	
	2180	Surface mount, acrylic lens & hinged frame										
	2200	1′ x 4′, 2 lamp	2 Elec	44	.364	Ea.		15.25		15.25	23.50	
	2220	2′ x 2′, 2 lamp		44	.364			15.25		15.25	23.50	
	2260	2′ x 4′, 4 lamp		33	.485			20.50		20.50	31.50	
	2280	4′ x 4′, 4 lamp		23	.696			29		29	45	
	2300	Strip fixtures, surface mount										
	2320	4′ long, 1 lamp	2 Elec	53	.302	Ea.		12.70		12.70	19.55	
	2340	4′ long, 2 lamp		50	.320			13.45		13.45	20.50	
	2360	8′ long, 1 lamp		42	.381			16		16	24.50	
	2380	8′ long, 2 lamp		40	.400			16.80		16.80	26	
	2400	Pendant mount, industrial, incl. removal										
	2410	of chain or rod hangers, to 15′ high										
	2420	4′ long, 2 lamp	2 Elec	35	.457	Ea.		19.20		19.20	29.50	
	2440	8′ long, 2 lamp	″	27	.593	″		25		25	38.50	
	2460	Interior incandescent, surface, ceiling										
	2470	or wall mount, to 12′ high										
	2480	Metal cylinder type, 75 Watt	2 Elec	62	.258	Ea.		10.85		10.85	16.70	
	2500	150 Watt	″	62	.258	″		10.85		10.85	16.70	
	2520	Metal halide, high bay										
	2540	400 Watt	2 Elec	15	1.067	Ea.		45		45	69	
	2560	1000 Watt		12	1.333			56		56	86.50	
	2580	150 Watt, low bay		20	.800			33.50		33.50	52	
	2600	Exterior fixtures, incandescent, wall mount										
	2620	100 Watt	2 Elec	50	.320	Ea.		13.45		13.45	20.50	
	2640	Quartz, 500 Watt		33	.485			20.50		20.50	31.50	
	2660	1500 Watt		27	.593			25		25	38.50	
	2680	Wall pack, mercury vapor										
	2700	175 Watt	2 Elec	25	.640	Ea.		27		27	41.50	
	2720	250 Watt	″	25	.640			27		27	41.50	
	7000	Weatherhead/mast, 2″	1 Elec	16	.500			21		21	32.50	
	7002	3″		10	.800			33.50		33.50	52	
	7004	3-1/2″		8.50	.941			39.50		39.50	61	
	7006	4″		8	1			42		42	64.50	
	7100	Service entry cable, #6, +#6 neutral		420	.019	L.F.		.80		.80	1.23	
	7102	#4, +#4 neutral		360	.022			.93		.93	1.44	
	7104	#2, +#4 neutral		345	.023			.97		.97	1.50	
	7106	#2, +#2 neutral		330	.024			1.02		1.02	1.57	
	9000	Minimum labor/equipment charge		4	2	Job		84		84	129	
320	0010	**ELECTRICAL DEMOLITION, GROUNDING** Addition										320
	0100	Ground clamp, bronze	1 Elec	64	.125	Ea.		5.25		5.25	8.10	
	0140	Water pipe ground clamp, bronze, heavy duty	″	24	.333	″		14		14	21.50	
	0200	Ground wire, bare armored	2 Elec	900	.018	L.F.		.75		.75	1.15	

		16055 | Selective Demolition	CREW	DAILY OUTPUT	LABOR-HOURS	UNIT	2006 BARE COSTS MAT.	LABOR	EQUIP.	TOTAL	TOTAL INCL O&P	
320	0240	Bare copper or aluminum	2 Elec	2800	.006	L.F.		.24		.24	.37	320
340	0010	**ELECTRICAL DEMOLITION, WIRING METHODS** Addition R024119-10										340
	0100	Armored cable, w/ PVC jacket, in cable tray										
	0110	#6	1 Elec	9.30	.860	C.L.F.		36		36	55.50	
	0120	#4	2 Elec	16.20	.988			41.50		41.50	64	
	0130	#2		13.80	1.159			48.50		48.50	75	
	0140	#1		12	1.333			56		56	86.50	
	0150	1/0		10.80	1.481			62		62	96	
	0160	2/0		10.20	1.569			66		66	101	
	0170	3/0		9.60	1.667			70		70	108	
	0180	4/0		9	1.778			74.50		74.50	115	
	0190	250 kcmil	3 Elec	10.80	2.222			93.50		93.50	144	
	0210	350 kcmil		9.90	2.424			102		102	157	
	0230	500 kcmil		9	2.667			112		112	173	
	0240	750 kcmil		8.10	2.963			124		124	192	
	1100	Control cable, 600 V or less										
	1110	3 wires	1 Elec	24	.333	C.L.F.		14		14	21.50	
	1120	5 wires		20	.400			16.80		16.80	26	
	1130	7 wires		16.50	.485			20.50		20.50	31.50	
	1140	9 wires		15	.533			22.50		22.50	34.50	
	1150	12 wires	2 Elec	26	.615			26		26	40	
	1160	15 wires		22	.727			30.50		30.50	47	
	1170	19 wires		19	.842			35.50		35.50	54.50	
	1180	25 wires		16	1			42		42	64.50	
	1500	Mineral insulated (MI) cable, 600 V										
	1510	#10	1 Elec	4.80	1.667	C.L.F.		70		70	108	
	1520	#8		4.50	1.778			74.50		74.50	115	
	1530	#6		4.20	1.905			80		80	123	
	1540	#4	2 Elec	7.20	2.222			93.50		93.50	144	
	1550	#2		6.60	2.424			102		102	157	
	1560	#1		6.30	2.540			107		107	164	
	1570	1/0		6	2.667			112		112	173	
	1580	2/0		5.70	2.807			118		118	182	
	1590	3/0		5.40	2.963			124		124	192	
	1600	4/0		4.80	3.333			140		140	216	
	1610	250 kcmil	3 Elec	7.20	3.333			140		140	216	
	1620	500 kcmil	"	5.90	4.068			171		171	263	
	2000	Shielded cable, XLP shielding, to 35 kV										
	2010	#4	2 Elec	13.20	1.212	C.L.F.		51		51	78.50	
	2020	#1		12	1.333			56		56	86.50	
	2030	1/0		11.40	1.404			59		59	91	
	2040	2/0		10.80	1.481			62		62	96	
	2050	4/0		9.60	1.667			70		70	108	
	2060	250 kcmil	3 Elec	13.50	1.778			74.50		74.50	115	
	2070	350 kcmil		11.70	2.051			86		86	133	
	2080	500 kcmil		11.20	2.143			90		90	139	
	2090	750 kcmil		10.80	2.222			93.50		93.50	144	
	3000	Modular flexible wiring										
	3010	Cable set	1 Elec	120	.067	Ea.		2.80		2.80	4.31	
	3020	Conversion module		48	.167			7		7	10.80	
	3030	Switching assembly		96	.083			3.50		3.50	5.40	
	3200	Undercarpet										
	3210	Power or telephone, flat cable	1 Elec	1320	.006	L.F.		.25		.25	.39	
	3220	Transition block assemblies w/ fitting		75	.107	Ea.		4.48		4.48	6.90	
	3230	Floor box with fitting		60	.133			5.60		5.60	8.65	

For expanded coverage of these items see *Means Electrical Cost Data 2006*

	16055 | Selective Demolition		CREW	DAILY OUTPUT	LABOR-HOURS	UNIT	2006 BARE COSTS MAT.	LABOR	EQUIP.	TOTAL	TOTAL INCL O&P	
340	3300	Data system, cable with connection R024119-10	1 Elec	50	.160	Ea.		6.70		6.70	10.35	340
	4000	Cable tray, including fitting & support										
	4010	Galvanized steel, 6″ wide	2 Elec	310	.052	L.F.		2.17		2.17	3.34	
	4020	9″ wide		295	.054			2.28		2.28	3.51	
	4030	12″ wide		285	.056			2.36		2.36	3.63	
	4040	18″ wide		270	.059			2.49		2.49	3.83	
	4050	24″ wide		260	.062			2.58		2.58	3.98	
	4060	30″ wide		240	.067			2.80		2.80	4.31	
	4070	36″ wide		220	.073			3.05		3.05	4.71	
	4110	Aluminum, 6″ wide		420	.038			1.60		1.60	2.47	
	4120	9″ wide		415	.039			1.62		1.62	2.49	
	4130	12″ wide		390	.041			1.72		1.72	2.65	
	4140	18″ wide		370	.043			1.82		1.82	2.80	
	4150	24″ wide		350	.046			1.92		1.92	2.96	
	4160	30″ wide		325	.049			2.07		2.07	3.19	
	4170	36″ wide		300	.053			2.24		2.24	3.45	
	4310	Cable channel, aluminum, 4″ wide straight	1 Elec	240	.033			1.40		1.40	2.16	
	5000	Conduit nipples, with locknuts and bushings										
	5020	1/2″	1 Elec	108	.074	Ea.		3.11		3.11	4.79	
	5040	3/4″		96	.083			3.50		3.50	5.40	
	5060	1″		81	.099			4.15		4.15	6.40	
	5080	1-1/4″		69	.116			4.87		4.87	7.50	
	5100	1-1/2″		60	.133			5.60		5.60	8.65	
	5120	2″		54	.148			6.20		6.20	9.60	
	5140	2-1/2″		45	.178			7.45		7.45	11.50	
	5160	3″		36	.222			9.35		9.35	14.40	
	5180	3-1/2″		33	.242			10.20		10.20	15.70	
	5200	4″		27	.296			12.45		12.45	19.15	
	5220	5″		21	.381			16		16	24.50	
	5240	6″		18	.444			18.65		18.65	29	
	5500	Electric nonmetallic tubing (ENT), flexible, 1/2″ - 1″ diameter		690	.012	L.F.		.49		.49	.75	
	5510	1-1/4″ - 2″ diameter		300	.027			1.12		1.12	1.73	
	5600	Flexible metallic tubing, steel, 3/8″ - 3/4″ diameter		600	.013			.56		.56	.86	
	5610	1″ - 1-1/4″ diameter		260	.031			1.29		1.29	1.99	
	5620	1-1/2″ - 2″ diameter		140	.057			2.40		2.40	3.70	
	5630	2-1/2″ diameter		100	.080			3.36		3.36	5.20	
	5640	3″ - 3-1/2″ diameter	2 Elec	150	.107			4.48		4.48	6.90	
	5650	4″ diameter	″	100	.160			6.70		6.70	10.35	
	5700	Sealtite flexible conduit, 3/8″ - 3/4″ diameter	1 Elec	420	.019			.80		.80	1.23	
	5710	1″ - 1-1/4″ diameter		180	.044			1.87		1.87	2.88	
	5720	1-1/2″ - 2″ diameter		120	.067			2.80		2.80	4.31	
	5730	2-1/2″ diameter		80	.100			4.20		4.20	6.45	
	5740	3″ diameter	2 Elec	150	.107			4.48		4.48	6.90	
	5750	4″ diameter	″	100	.160			6.70		6.70	10.35	
	5800	Wiring duct, plastic, 1-1/2″ - 2-1/2″ wide	1 Elec	360	.022			.93		.93	1.44	
	5810	3″ wide		330	.024			1.02		1.02	1.57	
	5820	4″ wide		300	.027			1.12		1.12	1.73	
	6300	Wireway, with fittings and supports, to 15′ high										
	6310	2-1/2″ x 2-1/2″	1 Elec	180	.044	L.F.		1.87		1.87	2.88	
	6320	4″ x 4″	″	160	.050			2.10		2.10	3.24	
	6330	6″ x 6″	2 Elec	240	.067			2.80		2.80	4.31	
	6340	8″ x 8″		160	.100			4.20		4.20	6.45	
	6350	10″ x 10″		120	.133			5.60		5.60	8.65	
	6360	12″ x 12″		80	.200			8.40		8.40	12.95	
350	0010	**ELECTRICAL DEMOLITION, ELECTRICAL POWER** R024119-10										350
	0100	Meter centers and sockets										

16 ELECTRICAL

16055 | Selective Demolition

			CREW	DAILY OUTPUT	LABOR-HOURS	UNIT	2006 BARE COSTS MAT.	LABOR	EQUIP.	TOTAL	TOTAL INCL O&P
350	0120	Meter socket, 4 terminal R024119-10	1 Elec	8.40	.952	Ea.		40		40	61.50
	0140	Trans-socket, 13 terminal, 400 A	"	3.60	2.222			93.50		93.50	144
	0160	800 A R260105-30	2 Elec	4.40	3.636			153		153	235
	0200	Meter center, 400 A		5.80	2.759			116		116	178
	0210	600 A		4	4			168		168	259
	0220	800 A		3.30	4.848			204		204	315
	0230	1200 A		2.80	5.714			240		240	370
	0240	1600 A		2.50	6.400			269		269	415
	0300	Base meter devices, 3 meter		3.60	4.444			187		187	288
	0310	4 meter		3.30	4.848			204		204	315
	0320	5 meter		2.90	5.517			232		232	355
	0330	6 meter		2.20	7.273			305		305	470
	0340	7 meter		2	8			335		335	520
	0350	8 meter		1.90	8.421			355		355	545
	0400	Branch meter devices									
	0410	Socket w/ circuit breaker 200 A, 2 meter	2 Elec	3.30	4.848	Ea.		204		204	315
	0420	3 meter		2.90	5.517			232		232	355
	0430	4 meter		2.60	6.154			258		258	400
	0450	Main circuit breaker, 400 A		5.80	2.759			116		116	178
	0460	600 A		4	4			168		168	259
	0470	800 A		3.30	4.848			204		204	315
	0480	1200 A		2.80	5.714			240		240	370
	0490	1600 A		2.50	6.400			269		269	415
	0500	Main lug terminal box, 800 A		3.40	4.706			198		198	305
	0510	1200 A		2.60	6.154			258		258	400
	1000	Motors, 230/460 V, 60 Hz, 3/4 HP	1 Elec	10.70	.748			31.50		31.50	48.50
	1010	5 HP		9	.889			37.50		37.50	57.50
	1020	10 HP		8	1			42		42	64.50
	1030	15 HP		6.40	1.250			52.50		52.50	81
	1040	20 HP	2 Elec	10.40	1.538			64.50		64.50	99.50
	1050	50 HP		9.60	1.667			70		70	108
	1060	75 HP		5.60	2.857			120		120	185
	1070	100 HP	3 Elec	5.40	4.444			187		187	288
	1080	150 HP		3.60	6.667			280		280	430
	1090	200 HP		3	8			335		335	520
	1200	Variable frequency drive, 460 V, for 5 HP motor size	1 Elec	3.20	2.500			105		105	162
	1210	10 HP motor size	"	2.70	2.963			124		124	192
	1220	20 HP motor size	2 Elec	3.60	4.444			187		187	288
	1230	50 HP motor size	"	2.10	7.619			320		320	495
	1240	75 HP motor size	R-3	2.20	9.091			375	69.50	444.50	660
	1250	100 HP motor size		2	10			415	76.50	491.50	725
	1260	150 HP motor size		2	10			415	76.50	491.50	725
	1270	200 HP motor size		1.70	11.765			485	90	575	855
	2000	Generator set w/ accessories, 3 ph 4 wire, 277/480 V									
	2010	7.5 kW	3 Elec	2	12	Ea.		505		505	775
	2020	20 kW		1.70	14.118			595		595	915
	2040	30 kW		1.33	18.045			760		760	1,175
	2060	50 kW		1	24			1,000		1,000	1,550
	2080	100 kW		.75	32			1,350		1,350	2,075
	2100	150 kW		.63	38.095			1,600		1,600	2,475
	2120	250 kW		.59	40.678			1,700		1,700	2,625
	2140	400 kW		.50	48			2,025		2,025	3,100
	2160	500 kW		.44	54.545			2,300		2,300	3,525
	2180	750 kW	4 Elec	.56	57.143			2,400		2,400	3,700
	2200	1000 kW	"	.46	69.565			2,925		2,925	4,500
	3000	Uninterruptible power supply system (UPS)									

ELECTRICAL 16

16055 | Selective Demolition		CREW	DAILY OUTPUT	LABOR-HOURS	UNIT	2006 BARE COSTS MAT.	LABOR	EQUIP.	TOTAL	TOTAL INCL O&P
350 3010	Single phase, 120 V, 1 kVA (R024119-10)	1 Elec	3.20	2.500	Ea.		105		105	162
3020	2 kVA	2 Elec	3.60	4.444			187		187	288
3030	5 kVA (R260105-30)	3 Elec	2.50	9.600			405		405	620
3040	10 kVA		2.30	10.435			440		440	675
3050	15 kVA		1.80	13.333			560		560	865
4000	Transformer, incl support, wire & conduit termination									
4010	Buck-boost, single phase, 120/240 V, 0.1 kVA	1 Elec	25	.320	Ea.		13.45		13.45	20.50
4020	0.5 kVA		12.50	.640			27		27	41.50
4030	1 kVA		6.30	1.270			53.50		53.50	82
4040	5 kVA		3.80	2.105			88.50		88.50	136
4800	5 kV or 15 kV primary, 277/480 V second, 112.5 kVA	R-3	3.50	5.714			237	43.50	280.50	415
4810	150 kVA		2.70	7.407			305	56.50	361.50	535
4820	225 kVA		2.30	8.696			360	66.50	426.50	630
4830	500 kVA		1.45	13.793			570	105	675	1,000
4840	750 kVA		1.33	15.038			625	115	740	1,100
4850	1000 kVA		1.25	16			665	122	787	1,150
4860	2000 kVA		1.05	19.048			790	145	935	1,375
4870	3000 kVA		.75	26.667			1,100	203	1,303	1,925
6010	Isolation panel, 3 kVA	1 Elec	2.30	3.478			146		146	225
6020	5 kVA		2.20	3.636			153		153	235
6030	7.5 kVA		2.10	3.810			160		160	246
6040	10 kVA		1.80	4.444			187		187	288
6050	15 kVA		1.40	5.714			240		240	370
7000	Power filters & conditioners									
7100	Automatic voltage regulator	2 Elec	6	2.667	Ea.		112		112	173
7200	Capacitor, 1 kVAR	1 Elec	8.60	.930			39		39	60
7210	5 kVAR		5.80	1.379			58		58	89
7220	10 kVAR		4.80	1.667			70		70	108
7230	15 kVAR		4.20	1.905			80		80	123
7240	20 kVAR		3.50	2.286			96		96	148
7250	30 kVAR		3.40	2.353			99		99	152
7260	50 kVAR		3.20	2.500			105		105	162
7400	Computer isolator transformer									
7410	Single phase, 120/240 V, 0.5 kVAR	1 Elec	12.80	.625	Ea.		26.50		26.50	40.50
7420	1 kVAR		8.50	.941			39.50		39.50	61
7430	2.5 kVAR		6.40	1.250			52.50		52.50	81
7440	5 kVAR		3.70	2.162			91		91	140
7500	Computer regulator transformer									
7510	Single phase, 240 V, 0.5 kVAR	1 Elec	8.50	.941	Ea.		39.50		39.50	61
7520	1 kVAR		6.40	1.250			52.50		52.50	81
7530	2 kVAR		3.20	2.500			105		105	162
7540	Single phase, plug-in unit 120 V, 0.5 kVAR		26	.308			12.90		12.90	19.90
7550	1 kVAR		17	.471			19.75		19.75	30.50
7600	Power conditioner transformer									
7610	Single phase 115 V - 240 V, 3 kVA	2 Elec	5.10	3.137	Ea.		132		132	203
7620	5 kVA	"	3.70	4.324			182		182	280
7630	7.5 kVA	3 Elec	4.80	5			210		210	325
7640	10 kVA	"	4.30	5.581			234		234	360
7700	Transient suppressor/voltage regulator									
7710	Single phase 115 or 220 V, 1 kVA	1 Elec	8.50	.941	Ea.		39.50		39.50	61
7720	2 kVA		7.30	1.096			46		46	71
7730	4 kVA		6.80	1.176			49.50		49.50	76
7800	Transient voltage suppressor transformer									
7810	Single phase, 115 or 220 V, 3.6 kVA	1 Elec	12.80	.625	Ea.		26.50		26.50	40.50
7820	7.2 kVA		11.50	.696			29		29	45
7830	14.4 kVA		10.20	.784			33		33	50.50

		16055 | Selective Demolition	CREW	DAILY OUTPUT	LABOR-HOURS	UNIT	2006 BARE COSTS MAT.	LABOR	EQUIP.	TOTAL	TOTAL INCL O&P	
350	7840	Single phase, plug-in, 120 V, 1.8 kVA (R024119-10)	1 Elec	26	.308	Ea.		12.90		12.90	19.90	350
	8000	Power measurement & control										
	8010	Switchboard instruments, 3 phase 4 wire, indicating unit (R260105-30)	1 Elec	25	.320	Ea.		13.45		13.45	20.50	
	8020	Recording unit		12	.667			28		28	43	
	8100	3 current transformers, 3 phase 4 wire, 5 to 800 A		6.40	1.250			52.50		52.50	81	
	8110	1000 to 1500 A		4.20	1.905			80		80	123	
	8120	2000 to 4000 A		3.20	2.500			105		105	162	
360	0010	**ELECTRICAL DEMOLITION, TRANSMISSION & DISTRIBUTION**										360
	0100	Load interrupter switch, 600 A, NEMA 1, 4.8 kV	R-3	1.33	15.038	Ea.		625	115	740	1,100	
	0120	13.8 kV (R024119-10)	"	1.27	15.748			650	120	770	1,125	
	0200	Lightning arrester, 4.8 kV	1 Elec	9	.889			37.50		37.50	57.50	
	0220	13.8 kV (R260105-30)		6.67	1.199			50.50		50.50	77.50	
	0300	Alarm or option items		3.33	2.402			101		101	155	
400	0010	**ELECTRICAL DEMOLITION, L.V. DISTRIBUTION** Addition (R024119-10)										400
	0100	Circuit breakers in enclosure										
	0120	Enclosed (NEMA 1), 600 V, 3 pole, 30 A (R260105-30)	1 Elec	12.30	.650	Ea.		27.50		27.50	42	
	0140	60 A		8.80	.909			38		38	59	
	0160	100 A		7.30	1.096			46		46	71	
	0180	225 A		5	1.600			67		67	104	
	0200	400 A	2 Elec	6.80	2.353			99		99	152	
	0220	600 A		4.60	3.478			146		146	225	
	0240	800 A		3.60	4.444			187		187	288	
	0260	1200 A		3.10	5.161			217		217	335	
	0280	1600 A		2.80	5.714			240		240	370	
	0300	2000 A		2.50	6.400			269		269	415	
	0400	Enclosed (NEMA 7), 600 V, 3 pole, 50 A	1 Elec	8.80	.909			38		38	59	
	0410	100 A		5.80	1.379			58		58	89	
	0420	150 A		3.80	2.105			88.50		88.50	136	
	0430	250 A	2 Elec	6.20	2.581			108		108	167	
	0440	400 A	"	4.60	3.478			146		146	225	
	0460	Manual motor starter, NEMA 1	1 Elec	24	.333			14		14	21.50	
	0480	NEMA 4 or NEMA 7		15	.533			22.50		22.50	34.50	
	0500	Time switches, single pole single throw		15.40	.519			22		22	33.50	
	0520	Photo cell		30	.267			11.20		11.20	17.25	
	0540	Load management device, 4 loads		7.70	1.039			43.50		43.50	67	
	0550	8 loads		3.80	2.105			88.50		88.50	136	
	0600	Transfer switches, enclosed, 30 A		12	.667			28		28	43	
	0610	60 A		9.50	.842			35.50		35.50	54.50	
	0620	100 A		6.50	1.231			51.50		51.50	79.50	
	0630	150 A	2 Elec	12	1.333			56		56	86.50	
	0640	260 A		10	1.600			67		67	104	
	0660	400 A		8	2			84		84	129	
	0670	600 A		5	3.200			134		134	207	
	0680	800 A		4	4			168		168	259	
	0690	1200 A		3.50	4.571			192		192	296	
	0700	1600 A		3	5.333			224		224	345	
	0710	2000 A		2.50	6.400			269		269	415	
	1000	Enclosed controller, NEMA 1, 30 A	1 Elec	13.80	.580			24.50		24.50	37.50	
	1020	60 A		11.50	.696			29		29	45	
	1040	100 A		9.60	.833			35		35	54	
	1060	150 A		7.70	1.039			43.50		43.50	67	
	1080	200 A		5.40	1.481			62		62	96	
	1100	400 A	2 Elec	6.90	2.319			97.50		97.50	150	
	1120	600 A		4.60	3.478			146		146	225	
	1140	800 A		3.80	4.211			177		177	272	

16055 | Selective Demolition

		CREW	DAILY OUTPUT	LABOR-HOURS	UNIT	MAT.	LABOR	EQUIP.	TOTAL	TOTAL INCL O&P
400	1160 1200 A (R024119-10)	2 Elec	3.10	5.161	Ea.		217		217	335
	1200 Control station, NEMA 1	1 Elec	30	.267			11.20		11.20	17.25
	1220 NEMA 7 (R260105-30)		23	.348			14.60		14.60	22.50
	1230 Control switches, push button		69	.116			4.87		4.87	7.50
	1240 Indicating light unit		120	.067			2.80		2.80	4.31
	1250 Relay		15	.533			22.50		22.50	34.50
	2000 Motor control center components									
	2020 Starter, NEMA 1, size 1	1 Elec	9	.889	Ea.		37.50		37.50	57.50
	2040 Size 2	2 Elec	13.30	1.203			50.50		50.50	78
	2060 Size 3		6.70	2.388			100		100	155
	2080 Size 4		5.30	3.019			127		127	195
	2100 Size 5		3.30	4.848			204		204	315
	2120 NEMA 7, size 1	1 Elec	8.70	.920			38.50		38.50	59.50
	2140 Size 2	2 Elec	12.70	1.260			53		53	81.50
	2160 Size 3		6.30	2.540			107		107	164
	2180 Size 4		5	3.200			134		134	207
	2200 Size 5		3.20	5			210		210	325
	2300 Fuse, light contactor, NEMA 1, 30 A	1 Elec	9	.889			37.50		37.50	57.50
	2310 60 A		6.70	1.194			50		50	77.50
	2320 100 A		3.30	2.424			102		102	157
	2330 200 A		2.70	2.963			124		124	192
	2350 Motor control center, incoming section	2 Elec	4	4			168		168	259
	2400 Starter & structure, 10 HP	1 Elec	9	.889			37.50		37.50	57.50
	2410 25 HP	2 Elec	13.30	1.203			50.50		50.50	78
	2420 50 HP		6.70	2.388			100		100	155
	2430 75 HP		5.30	3.019			127		127	195
	2440 100 HP		4.70	3.404			143		143	220
	2450 200 HP		3.30	4.848			204		204	315
	2460 400 HP		2.70	5.926			249		249	385
	2500 Motor starter & control									
	2510 Motor starter, NEMA 1, 5 HP	1 Elec	8.90	.899	Ea.		38		38	58
	2520 10 HP	"	6.20	1.290			54		54	83.50
	2530 25 HP	2 Elec	8.40	1.905			80		80	123
	2540 50 HP		6.90	2.319			97.50		97.50	150
	2550 100 HP		4.60	3.478			146		146	225
	2560 200 HP		3.50	4.571			192		192	296
	2570 400 HP		3.10	5.161			217		217	335
	2610 Motor starter, NEMA 7, 5 HP	1 Elec	6.20	1.290			54		54	83.50
	2620 10 HP	"	4.20	1.905			80		80	123
	2630 25 HP	2 Elec	6.90	2.319			97.50		97.50	150
	2640 50 HP		4.60	3.478			146		146	225
	2650 100 HP		3.50	4.571			192		192	296
	2660 200 HP		1.90	8.421			355		355	545
	2710 Combination control unit, NEMA 1, 5 HP	1 Elec	6.90	1.159			48.50		48.50	75
	2720 10 HP	"	5	1.600			67		67	104
	2730 25 HP	2 Elec	7.70	2.078			87.50		87.50	134
	2740 50 HP		5.10	3.137			132		132	203
	2750 100 HP		3.10	5.161			217		217	335
	2810 NEMA 7, 5 HP	1 Elec	5	1.600			67		67	104
	2820 10 HP	"	3.90	2.051			86		86	133
	2830 25 HP	2 Elec	5.10	3.137			132		132	203
	2840 50 HP		3.10	5.161			217		217	335
	2850 100 HP		2.30	6.957			292		292	450
	2860 200 HP		1.50	10.667			450		450	690
	3000 Panelboard or load center circuit breaker									
	3010 Bolt-on or plug in, 15 A - 50 A	1 Elec	20	.400	Ea.		16.80		16.80	26

16 ELECTRICAL

		16055 | Selective Demolition	CREW	DAILY OUTPUT	LABOR-HOURS	UNIT	2006 BARE COSTS MAT.	LABOR	EQUIP.	TOTAL	TOTAL INCL O&P	
400	3020	60 A - 70 A (R024119-10)	1 Elec	16	.500	Ea.		21		21	32.50	400
	3030	Bolt-on, 80 A - 100 A		14	.571			24		24	37	
	3040	Up to 250 A (R260105-30)		7.30	1.096			46		46	71	
	3050	Motor operated, 30 A		13	.615			26		26	40	
	3060	60 A		10	.800			33.50		33.50	52	
	3070	100 A	↓	8	1	↓		42		42	64.50	
	3200	Switchboard circuit breaker										
	3210	15 A - 60 A	1 Elec	19	.421	Ea.		17.70		17.70	27	
	3220	70 A - 100 A		14	.571			24		24	37	
	3230	125 A - 400 A		10	.800			33.50		33.50	52	
	3240	450 A - 600 A		5.30	1.509			63.50		63.50	97.50	
	3250	700 A - 800 A		4.30	1.860			78		78	120	
	3260	1000 A		3.30	2.424			102		102	157	
	3270	1200 A	↓	2.70	2.963			124		124	192	
	3500	Switchboard, incoming section, 400 A	2 Elec	3.70	4.324			182		182	280	
	3510	600 A		3.30	4.848			204		204	315	
	3520	800 A		2.90	5.517			232		232	355	
	3530	1200 A		2.40	6.667			280		280	430	
	3540	1600 A		2.20	7.273			305		305	470	
	3550	2000 A		2.10	7.619			320		320	495	
	3560	3000 A		1.90	8.421			355		355	545	
	3570	4000 A		1.70	9.412			395		395	610	
	3610	Distribution section, 600 A		4	4			168		168	259	
	3620	800 A		3.60	4.444			187		187	288	
	3630	1200 A		3.10	5.161			217		217	335	
	3640	1600 A		2.90	5.517			232		232	355	
	3650	2000 A		2.70	5.926			249		249	385	
	3710	Transition section, 600 A		3.80	4.211			177		177	272	
	3720	800 A		3.30	4.848			204		204	315	
	3730	1200 A		2.70	5.926			249		249	385	
	3740	1600 A	2 Elec	2.40	6.667			280		280	430	
	3750	2000 A	2 Elec	2.20	7.273	Ea.		305		305	470	
	3760	2500 A	2 Elec	2.10	7.619			320		320	495	
	3770	3000 A	"	1.90	8.421	↓		355		355	545	
	4000	Bus duct, aluminum or copper, 30 A	1 Elec	200	.040	L.F.		1.68		1.68	2.59	
	4020	60 A		160	.050			2.10		2.10	3.24	
	4040	100 A		140	.057			2.40		2.40	3.70	
	5000	Feedrail, trolley busway, up to 60 A		160	.050			2.10		2.10	3.24	
	5020	100 A		120	.067			2.80		2.80	4.31	
	5040	Busway, 50 A		200	.040			1.68		1.68	2.59	
	6000	Fuse, 30 A		133	.060			2.53		2.53	3.89	
	6010	60 A		133	.060			2.53		2.53	3.89	
	6020	100 A		105	.076			3.20		3.20	4.93	
	6030	200 A		95	.084			3.54		3.54	5.45	
	6040	400 A		80	.100			4.20		4.20	6.45	
	6050	600 A		53	.151			6.35		6.35	9.75	
	6060	601 A - 1200 A		42	.190			8		8	12.30	
	6070	1500 A - 1600 A		34	.235			9.90		9.90	15.20	
	6080	Up to 2500 A		26	.308			12.90		12.90	19.90	
	6090	4000 A		21	.381			16		16	24.50	
	6100	4500 A - 5000 A		18	.444			18.65		18.65	29	
	6120	6000 A		15	.533			22.50		22.50	34.50	
	6200	Fuse plug or fustat	↓	100	.080	↓		3.36		3.36	5.20	
500	0010	**ELECTRICAL DEMOLITION, LIGHTING** Addition (R024119-10)										500
	0100	Fixture hanger, flexible, 1/2" diameter	1 Elec	36	.222	Ea.		9.35		9.35	14.40	

	16055 | Selective Demolition	CREW	DAILY OUTPUT	LABOR-HOURS	UNIT	2006 BARE COSTS MAT.	LABOR	EQUIP.	TOTAL	TOTAL INCL O&P	
500	0120 3/4" diameter (R024119-10)	1 Elec	30	.267	Ea.		11.20		11.20	17.25	500
	3000 Light pole, anchor base, excl. concrete bases										
	3010 Metal light pole, 10' (R260105-30)	2 Elec	24	.667	Ea.		28		28	43	
	3020 16'	"	18	.889			37.50		37.50	57.50	
	3030 20'	R-3	8.70	2.299			95	17.55	112.55	166	
	3040 40'	"	6	3.333			138	25.50	163.50	242	
	3100 Wood light pole, 10'	2 Elec	36	.444			18.65		18.65	29	
	3120 20'	"	24	.667			28		28	43	
	3140 Bollard light, 42"	1 Elec	9	.889			37.50		37.50	57.50	
	3160 Walkway luminaire	"	8.10	.988			41.50		41.50	64	
	4000 Explosionproof										
	4010 Metal halide, 175 W	1 Elec	8.70	.920	Ea.		38.50		38.50	59.50	
	4020 250 W	"	8.10	.988			41.50		41.50	64	
	4030 400 W	2 Elec	14.40	1.111			46.50		46.50	72	
	4050 High pressure sodium, 70 W	1 Elec	9	.889			37.50		37.50	57.50	
	4060 100 W		9	.889			37.50		37.50	57.50	
	4070 150 W		8.10	.988			41.50		41.50	64	
	4100 Incandescent		8.70	.920			38.50		38.50	59.50	
	4200 Flurorescent		8.10	.988			41.50		41.50	64	
	5000 Ballast, flurorescent fixture		24	.333			14		14	21.50	
	5040 High intensity discharge fixture		24	.333			14		14	21.50	
	5300 Exit and emergency lighting										
	5310 Exit light	1 Elec	24	.333	Ea.		14		14	21.50	
	5320 Emergency battery pack lighting unit		12	.667			28		28	43	
	5330 Remote lamp only		80	.100			4.20		4.20	6.45	
	5340 Self-contained fluorescent lamp pack		30	.267			11.20		11.20	17.25	
	5800 Energy saving devices										
	5810 Occupancy sensor	1 Elec	21	.381	Ea.		16		16	24.50	
	5820 Automatic wall switch		72	.111			4.67		4.67	7.20	
	5830 Remote power pack		30	.267			11.20		11.20	17.25	
	5840 Photoelectric control		24	.333			14		14	21.50	
	6000 Lamps										
	6010 Fluorescent	1 Elec	200	.040	Ea.		1.68		1.68	2.59	
	6030 High intensity discharge lamp, up to 400 W		68	.118			4.94		4.94	7.60	
	6040 Up to 1000 W		45	.178			7.45		7.45	11.50	
	6050 Quartz		90	.089			3.73		3.73	5.75	
	6070 Incandescent		360	.022			.93		.93	1.44	
	6080 Exterior, PAR		290	.028			1.16		1.16	1.79	
	6090 Guards for fluorescent lamp		24	.333			14		14	21.50	
700	0010 **ELECTRICAL DEMOLITION, COMMUNICATIONS** (R024119-10)										700
	0100 Fiber optics										
	0120 Fiber optic cable (R260105-30)	1 Elec	2400	.003	L.F.		.14		.14	.22	
	0160 Multi-channel rack enclosure		6	1.333	Ea.		56		56	86.50	
	0180 Fiber optic patch panel		18	.444	"		18.65		18.65	29	
	0200 Communication cables & fittings										
	0220 Voice/data outlet	1 Elec	140	.057	Ea.		2.40		2.40	3.70	
	0240 Telephone cable		2800	.003	L.F.		.12		.12	.19	
	0260 Phone jack		135	.059	Ea.		2.49		2.49	3.83	
	0300 High performance cable, 2 pair		3000	.003	L.F.		.11		.11	.17	
	0320 4 pair		2100	.004			.16		.16	.25	
	0340 25 pair		900	.009			.37		.37	.58	
	1000 Nurse call system										
	1020 Nurse call station	1 Elec	24	.333	Ea.		14		14	21.50	
	1040 Nurse call standard call button		24	.333			14		14	21.50	
	1060 Corridor, dome light or zone indictor		24	.333			14		14	21.50	

16050 | Basic Electrical Materials & Methods

16055 | Selective Demolition

			CREW	DAILY OUTPUT	LABOR-HOURS	UNIT	2006 BARE COSTS MAT.	LABOR	EQUIP.	TOTAL	TOTAL INCL O&P	
700	1080	Master control station (R024119-10)	2 Elec	2	8	Ea.		335		335	520	700
800	0010	ELECTRICAL DEMOLITION, SOUND & VIDEO (R024119-10)										800
	0100	Sound & video cables (R260105-30)										
	0120	TV antenna lead-in cable	1 Elec	2100	.004	L.F.		.16		.16	.25	
	0140	Sound cable		2400	.003			.14		.14	.22	
	0160	Microphone cable		2400	.003			.14		.14	.22	
	0180	Coaxial cable		2400	.003			.14		.14	.22	
	0200	Doorbell system		16	.500	Ea.		21		21	32.50	
	0220	Door chime or devices		36	.222	"		9.35		9.35	14.40	
	0300	Public address system, not including rough-in										
	0320	Conventional office	1 Elec	16	.500	Speaker		21		21	32.50	
	0340	Conventional industrial	"	8	1	"		42		42	64.50	
	0400	Sound system, not including rough-in										
	0410	Sound components	1 Elec	24	.333	Ea.		14		14	21.50	
	0420	Intercom, master station		6	1.333			56		56	86.50	
	0440	Remote station		24	.333			14		14	21.50	
	0460	Intercom outlets		24	.333			14		14	21.50	
	0480	Handset		12	.667			28		28	43	
	0500	Emergency call system, not including rough-in										
	0520	Annunciator	1 Elec	4	2	Ea.		84		84	129	
	0540	Devices	"	16	.500			21		21	32.50	
	0600	Master door, buzzer type unit	2 Elec	1.60	10			420		420	645	
	0800	TV System, not including rough-in										
	0820	Master TV antenna system, per outlet	1 Elec	39	.205	Outlet		8.60		8.60	13.25	
	0840	School & deluxe, per outlet		16	.500	"		21		21	32.50	
	0860	Amplifier		12	.667	Ea.		28		28	43	
	0880	Antenna		6	1.333	"		56		56	86.50	
	0900	One camera & one monitor	2 Elec	7.80	2.051	Total		86		86	133	
	0920	One camera	1 Elec	8	1	Ea.		42		42	64.50	

16060 | Grounding & Bonding

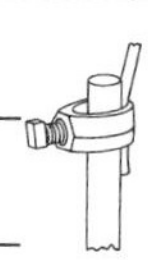

			CREW	DAILY OUTPUT	LABOR-HOURS	UNIT	MAT.	LABOR	EQUIP.	TOTAL	TOTAL INCL O&P	
800	0010	GROUNDING										800
	0030	Rod, copper clad, 8′ long, 1/2″ diameter	1 Elec	5.50	1.455	Ea.	13.70	61		74.70	109	
	0040	5/8″ diameter		5.50	1.455		15	61		76	111	
	0050	3/4″ diameter		5.30	1.509		28.50	63.50		92	129	
	0080	10′ long, 1/2″ diameter		4.80	1.667		17.70	70		87.70	127	
	0090	5/8″ diameter		4.60	1.739		22	73		95	137	
	0100	3/4″ diameter		4.40	1.818		31.50	76.50		108	153	
	0130	15′ long, 3/4″ diameter		4	2		87	84		171	225	
	0150	Coupling, bronze, 1/2″ diameter					2.95			2.95	3.25	
	0160	5/8″ diameter					4.30			4.30	4.73	
	0170	3/4″ diameter					9.75			9.75	10.75	
	0190	Drive studs, 1/2″ diameter					5.80			5.80	6.40	
	0210	5/8″ diameter					7.10			7.10	7.80	
	0220	3/4″ diameter					8			8	8.80	
	0230	Clamp, bronze, 1/2″ diameter	1 Elec	32	.250		3.85	10.50		14.35	20.50	
	0240	5/8″ diameter		32	.250		4.33	10.50		14.83	21	
	0250	3/4″ diameter		32	.250		5.75	10.50		16.25	22.50	
	0260	Wire ground bare armored, #8-1 conductor		2	4	C.L.F.	80.50	168		248.50	350	
	0270	#6-1 conductor		1.80	4.444		102	187		289	400	
	0280	#4-1 conductor		1.60	5		149	210		359	490	
	0320	Bare copper wire, #14 solid		14	.571		3.62	24		27.62	41	
	0330	#12		13	.615		5.60	26		31.60	46	
	0340	#10		12	.667		8.35	28		36.35	52	
	0350	#8		11	.727		14.65	30.50		45.15	63	

ELECTRICAL 16

16060 | Grounding & Bonding

		Description	CREW	DAILY OUTPUT	LABOR-HOURS	UNIT	MAT.	LABOR	EQUIP.	TOTAL	TOTAL INCL O&P	
							2006 BARE COSTS					
800	0360	#6	1 Elec	10	.800	C.L.F.	28	33.50		61.50	83	800
	0370	#4		8	1		38.50	42		80.50	107	
	0380	#2		5	1.600		63	67		130	174	
	0390	Bare copper wire, #8 stranded		11	.727		15.90	30.50		46.40	64.50	
	0400	#6		10	.800		28.50	33.50		62	83.50	
	0450	#4	2 Elec	16	1		44.50	42		86.50	114	
	0600	#2		10	1.600		66	67		133	177	
	0650	#1		9	1.778		88	74.50		162.50	212	
	0700	1/0		8	2		98	84		182	237	
	0750	2/0		7.20	2.222		116	93.50		209.50	272	
	0800	3/0		6.60	2.424		147	102		249	320	
	1000	4/0		5.70	2.807		184	118		302	385	
	1200	250 kcmil	3 Elec	7.20	3.333		218	140		358	455	
	1210	300 kcmil		6.60	3.636		257	153		410	515	
	1220	350 kcmil		6	4		305	168		473	595	
	1230	400 kcmil		5.70	4.211		365	177		542	670	
	1240	500 kcmil		5.10	4.706		435	198		633	780	
	1250	600 kcmil		3.90	6.154		470	258		728	915	
	1260	750 kcmil		3.60	6.667		705	280		985	1,200	
	1270	1000 kcmil		3	8		905	335		1,240	1,525	
	1350	Bare aluminum, #8 stranded	1 Elec	10	.800		10.75	33.50		44.25	64	
	1360	#6	″	9	.889		17.40	37.50		54.90	76.50	
	1370	#4	2 Elec	16	1		22.50	42		64.50	89.50	
	1380	#2		13	1.231		34	51.50		85.50	117	
	1390	#1		10.60	1.509		40	63.50		103.50	142	
	1400	1/0		9	1.778		46	74.50		120.50	166	
	1410	2/0		8	2		56.50	84		140.50	192	
	1420	3/0		7.20	2.222		72.50	93.50		166	224	
	1430	4/0		6.60	2.424		91.50	102		193.50	258	
	1440	250 kcmil	3 Elec	9.30	2.581		131	108		239	310	
	1450	300 kcmil		8.70	2.759		162	116		278	355	
	1460	400 kcmil		7.50	3.200		194	134		328	420	
	1470	500 kcmil		6.90	3.478		249	146		395	500	
	1480	600 kcmil		6	4		287	168		455	575	
	1490	700 kcmil		5.70	4.211		320	177		497	625	
	1500	750 kcmil		5.10	4.706		340	198		538	680	
	1510	1000 kcmil		4.80	5		485	210		695	855	
	1800	Water pipe ground clamps, heavy duty										
	2000	Bronze, 1/2″ to 1″ diameter	1 Elec	8	1	Ea.	14.40	42		56.40	80.50	
	2100	1-1/4″ to 2″ diameter		8	1		18.95	42		60.95	85.50	
	2200	2-1/2″ to 3″ diameter		6	1.333		41	56		97	132	
	2730	Exothermic weld, 4/0 wire to 1″ ground rod		7	1.143		9.15	48		57.15	84	
	2740	4/0 wire to building steel		7	1.143		7.30	48		55.30	82	
	2750	4/0 wire to motor frame		7	1.143		7.30	48		55.30	82	
	2760	4/0 wire to 4/0 wire		7	1.143		6.40	48		54.40	81	
	2770	4/0 wire to #4 wire		7	1.143		6.40	48		54.40	81	
	2780	4/0 wire to #8 wire		7	1.143		6.40	48		54.40	81	
	2790	Mold, reusable, for above					105			105	115	
	2800	Brazed connections, #6 wire	1 Elec	12	.667		12.35	28		40.35	56.50	
	3000	#2 wire		10	.800		16.55	33.50		50.05	70	
	3100	3/0 wire		8	1		25	42		67	92	
	3200	4/0 wire		7	1.143		28.50	48		76.50	106	
	3400	250 kcmil wire		5	1.600		33	67		100	141	
	3600	500 kcmil wire		4	2		41	84		125	174	
	3700	Insulated ground wire, copper #14		13	.615	C.L.F.	4.85	26		30.85	45.50	
800	3710	#12		11	.727		7.30	30.50		37.80	55	

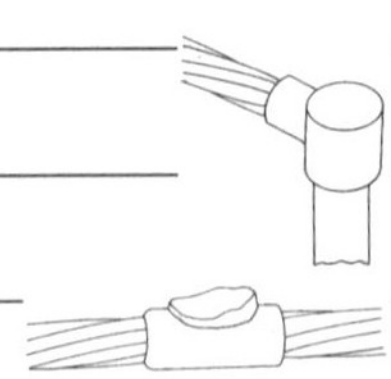

16 ELECTRICAL

	16060 | Grounding & Bonding	CREW	DAILY OUTPUT	LABOR-HOURS	UNIT	2006 BARE COSTS MAT.	LABOR	EQUIP.	TOTAL	TOTAL INCL O&P	
800	3720 #10	1 Elec	10	.800	C.L.F.	10.95	33.50		44.45	64	800
	3730 #8		8	1		19.70	42		61.70	86	
	3740 #6		6.50	1.231		32.50	51.50		84	116	
	3750 #4	2 Elec	10.60	1.509		50	63.50		113.50	153	
	3770 #2		9	1.778		79	74.50		153.50	202	
	3780 #1		8	2		100	84		184	239	
	3790 1/0		6.60	2.424		111	102		213	279	
	3800 2/0		5.80	2.759		130	116		246	320	
	3810 3/0		5	3.200		166	134		300	390	
	3820 4/0		4.40	3.636		206	153		359	460	
	3830 250 kcmil	3 Elec	6	4		245	168		413	530	
	3840 300 kcmil		5.70	4.211		291	177		468	590	
	3850 350 kcmil		5.40	4.444		340	187		527	665	
	3860 400 kcmil		5.10	4.706		390	198		588	735	
	3870 500 kcmil		4.80	5		485	210		695	855	
	3880 600 kcmil		3.90	6.154		580	258		838	1,025	
	3890 750 kcmil		3.30	7.273		770	305		1,075	1,325	
	3900 1000 kcmil		2.70	8.889		1,000	375		1,375	1,700	
	3950 Insulated ground wire, aluminum #8	1 Elec	9	.889		17.90	37.50		55.40	77	
	3960 #6	"	8	1		24.50	42		66.50	91.50	
	3970 #4	2 Elec	13	1.231		30.50	51.50		82	113	
	3980 #2		10.60	1.509		41	63.50		104.50	143	
	3990 #1		9	1.778		60.50	74.50		135	182	
	4000 1/0		8	2		72.50	84		156.50	209	
	4010 2/0		7.20	2.222		85.50	93.50		179	238	
	4020 3/0		6.60	2.424		106	102		208	274	
	4030 4/0		6.20	2.581		118	108		226	297	
	4040 250 kcmil	3 Elec	8.70	2.759		144	116		260	335	
	4050 300 kcmil		8.10	2.963		199	124		323	410	
	4060 350 kcmil		7.50	3.200		202	134		336	430	
	4070 400 kcmil		6.90	3.478		237	146		383	485	
	4080 500 kcmil		6	4		261	168		429	545	
	4090 600 kcmil		5.70	4.211		330	177		507	635	
	4100 700 kcmil		5.10	4.706		380	198		578	725	
	4110 750 kcmil		4.80	5		385	210		595	750	
	5000 Copper Electrolytic ground rod system										
	5010 Includes augering hole, mixing bentonite clay,										
	5020 Installing rod, and terminating ground wire										
	5100 Straight Vertical type, 2″ Dia.										
	5120 8.5′ long, Clamp Connection	1 Elec	2.67	2.996	Ea.	585	126		711	840	
	5130 With exothermic weld Connection		1.95	4.103		585	172		757	910	
	5140 10′ long		2.35	3.404		650	143		793	935	
	5150 With exothermic weld Connection		1.78	4.494		650	189		839	1,000	
	5160 12′ long		2.16	3.704		785	156		941	1,100	
	5170 With exothermic weld Connection		1.67	4.790		785	201		986	1,175	
	5180 20′ long		1.74	4.598		1,200	193		1,393	1,625	
	5190 With exothermic weld Connection		1.40	5.714		1,225	240		1,465	1,725	
	5195 40′ long with exothermic weld connection	2 Elec	2	8		2,350	335		2,685	3,100	
	5200 L-Shaped, 2″ Dia.										
	5220 4′ Vert. x 10′ Horz., Clamp Connection	1 Elec	5.33	1.501	Ea.	900	63		963	1,075	
	5230 With exothermic weld Connection	"	3.08	2.597	"	930	109		1,039	1,200	
	5300 Protective Box at grade level, with breather slots										
	5320 Round 12″ long, Plastic	1 Elec	32	.250	Ea.	40	10.50		50.50	60	
	5330 Concrete	"	16	.500		50	21		71	87.50	
	5400 Bentonite Clay, 50# bag, 1 per 10′ of rod					35			35	38.50	
	9000 Minimum labor/equipment charge	1 Elec	4	2	Job		84		84	129	

16070 | Hangers & Supports

			CREW	DAILY OUTPUT	LABOR-HOURS	UNIT	2006 BARE COSTS MAT.	LABOR	EQUIP.	TOTAL	TOTAL INCL O&P	
310	0010	**FASTENERS** see division 05090 and 15060										310
320	0010	**HANGERS** Steel R260533-21										320
	0030	Conduit supports										
	0050	Strap w/2 holes, rigid steel conduit										
	0100	1/2" diameter	1 Elec	470	.017	Ea.	.20	.71		.91	1.32	
	0150	3/4" diameter		440	.018		.24	.76		1	1.44	
	0200	1" diameter		400	.020		.39	.84		1.23	1.72	
	0300	1-1/4" diameter		355	.023		.62	.95		1.57	2.14	
	0350	1-1/2" diameter		320	.025		.73	1.05		1.78	2.42	
	0400	2" diameter 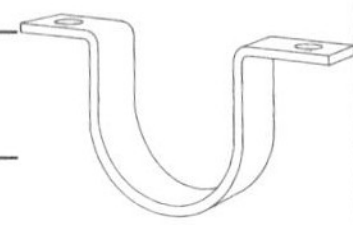		266	.030		.85	1.26		2.11	2.89	
	0500	2-1/2" diameter		160	.050		1.75	2.10		3.85	5.15	
	0550	3" diameter		133	.060		2.17	2.53		4.70	6.30	
	0600	3-1/2" diameter		100	.080		2.84	3.36		6.20	8.30	
	0650	4" diameter		80	.100		3.07	4.20		7.27	9.85	
	0700	EMT, 1/2" diameter		470	.017		.13	.71		.84	1.24	
	0800	3/4" diameter		440	.018		.19	.76		.95	1.39	
	0850	1" diameter		400	.020		.28	.84		1.12	1.60	
	0900	1-1/4" diameter		355	.023		.42	.95		1.37	1.92	
	0950	1-1/2" diameter		320	.025		.50	1.05		1.55	2.17	
	1000	2" diameter 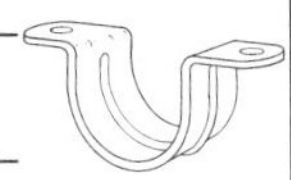		266	.030		.75	1.26		2.01	2.78	
	1100	2-1/2" diameter		160	.050		1.05	2.10		3.15	4.40	
	1150	3" diameter		133	.060		1.32	2.53		3.85	5.35	
	1200	3-1/2" diameter		100	.080		1.64	3.36		5	7	
	1250	4" diameter		80	.100		1.85	4.20		6.05	8.50	
	1400	Hanger, with bolt, 1/2" diameter		200	.040		.50	1.68		2.18	3.14	
	1450	3/4" diameter		190	.042		.51	1.77		2.28	3.28	
	1500	1" diameter		176	.045		.84	1.91		2.75	3.86	
	1550	1-1/4" diameter 		160	.050		1.20	2.10		3.30	4.56	
	1600	1-1/2" diameter		140	.057		1.41	2.40		3.81	5.25	
	1650	2" diameter		130	.062		1.55	2.58		4.13	5.70	
	1700	2-1/2" diameter		100	.080		2.06	3.36		5.42	7.45	
	1750	3" diameter		64	.125		2.42	5.25		7.67	10.75	
	1800	3-1/2" diameter		50	.160		3.56	6.70		10.26	14.25	
	1850	4" diameter		40	.200		8.10	8.40		16.50	22	
	1900	Riser clamps, conduit, 1/2" diameter		40	.200		8.30	8.40		16.70	22	
	1950	3/4" diameter		36	.222		8.30	9.35		17.65	23.50	
	2000	1" diameter		30	.267		8.35	11.20		19.55	26.50	
	2100	1-1/4" diameter		27	.296		11.15	12.45		23.60	31.50	
	2150	1-1/2" diameter		27	.296		11.75	12.45		24.20	32	
	2200	2" diameter 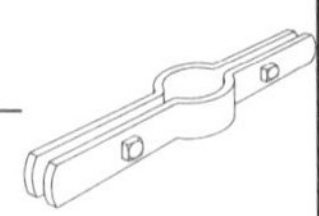		20	.400		12.30	16.80		29.10	39.50	
	2250	2-1/2" diameter		20	.400		12.90	16.80		29.70	40	
	2300	3" diameter		18	.444		13.90	18.65		32.55	44.50	
	2350	3-1/2" diameter		18	.444		17.70	18.65		36.35	48.50	
	2400	4" diameter		14	.571	↓	18.35	24		42.35	57	
	2500	Threaded rod, painted, 1/4" diameter		260	.031	L.F.	1.75	1.29		3.04	3.92	
	2600	3/8" diameter		200	.040		2.02	1.68		3.70	4.81	
	2700	1/2" diameter 		140	.057		3.05	2.40		5.45	7.05	
	2800	5/8" diameter		100	.080		4.20	3.36		7.56	9.80	
	2900	3/4" diameter	↓	60	.133	↓	5.30	5.60		10.90	14.45	
	2940	Couplings painted, 1/4" diameter				C	173			173	190	
	2960	3/8" diameter					235			235	259	
	2970	1/2" diameter 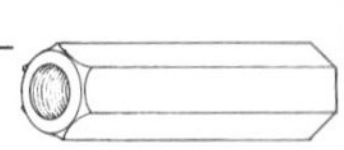					360			360	395	
	2980	5/8" diameter					510			510	560	
	2990	3/4" diameter					755			755	830	
	3000	Nuts, galvanized, 1/4" diameter				↓	11.95			11.95	13.10	

16 ELECTRICAL

16070 | Hangers & Supports

		16070 Hangers & Supports	CREW	DAILY OUTPUT	LABOR-HOURS	UNIT	MAT.	LABOR	EQUIP.	TOTAL	TOTAL INCL O&P
320	3050	3/8″ diameter (R260533-21)				C	17.95			17.95	19.75
	3100	1/2″ diameter					40			40	44
	3150	5/8″ diameter					60.50			60.50	66.50
	3200	3/4″ diameter					90			90	99
	3250	Washers, galvanized, 1/4″ diameter					6.40			6.40	7.05
	3300	3/8″ diameter					13.40			13.40	14.75
	3350	1/2″ diameter					16.25			16.25	17.85
	3400	5/8″ diameter					50			50	55
	3450	3/4″ diameter					72.50			72.50	80
	3500	Lock washers, galvanized, 1/4″ diameter					5.40			5.40	5.95
	3550	3/8″ diameter					11.70			11.70	12.85
	3600	1/2″ diameter					16.10			16.10	17.70
	3650	5/8″ diameter					29.50			29.50	32.50
	3700	3/4″ diameter					46			46	50.50
	3800	Channels, steel, 3/4″ x 1-1/2″	1 Elec	80	.100	L.F.	4.31	4.20		8.51	11.20
	3900	1-1/2″ x 1-1/2″		70	.114		5.75	4.80		10.55	13.75
	4000	1-7/8″ x 1-1/2″		60	.133		13.90	5.60		19.50	24
	4100	3″ x 1-1/2″		50	.160		16.10	6.70		22.80	28
	4200	Spring nuts, long, 1/4″		120	.067	Ea.	1.42	2.80		4.22	5.85
	4250	3/8″		100	.080		2.48	3.36		5.84	7.95
	4300	1/2″		80	.100		2.66	4.20		6.86	9.40
	4350	Spring nuts, short, 1/4″		120	.067		1.58	2.80		4.38	6.05
	4400	3/8″		100	.080		1.74	3.36		5.10	7.10
	4450	1/2″		80	.100		1.93	4.20		6.13	8.55
	4500	Closure strip		200	.040	L.F.	2.63	1.68		4.31	5.50
	4550	End cap		60	.133	Ea.	1.07	5.60		6.67	9.85
	4600	End connector 3/4″ conduit		40	.200		3.99	8.40		12.39	17.35
	4650	Junction box, 1 channel		16	.500		24.50	21		45.50	59.50
	4700	2 channel		14	.571		30	24		54	70
	4750	3 channel		12	.667		34	28		62	80.50
	4800	4 channel		10	.800		38	33.50		71.50	93.50
	4850	Spliceplate		40	.200		9.30	8.40		17.70	23
	4900	Continuous concrete insert, 1-1/2″ deep, 1′ long		16	.500		12.85	21		33.85	46.50
	4950	2′ long		14	.571		16.05	24		40.05	54.50
	5000	3′ long		12	.667		19.25	28		47.25	64
	5050	4′ long		10	.800		22.50	33.50		56	76.50
	5100	6′ long		8	1		33.50	42		75.50	101
	5150	3/4″ deep, 1′ long		16	.500		10.40	21		31.40	44
	5200	2′ long		14	.571		12.85	24		36.85	51
	5250	3′ long		12	.667		15.45	28		43.45	60
	5300	4′ long		10	.800		17.80	33.50		51.30	71.50
	5350	6′ long		8	1		26.50	42		68.50	93.50
	5400	90° angle fitting 2-1/8″ x 2-1/8″		60	.133		2.20	5.60		7.80	11.05
	5450	Supports, suspension rod type, small		60	.133		9.30	5.60		14.90	18.90
	5500	Large		40	.200		10.25	8.40		18.65	24.50
	5550	Beam clamp, small		60	.133		4.85	5.60		10.45	14
	5600	Large		40	.200		5.55	8.40		13.95	19.05
	5650	U-support, small		60	.133		2.46	5.60		8.06	11.35
	5700	Large		40	.200		8.60	8.40		17	22.50
	5750	Concrete insert, cast, for up to 1/2″ threaded rod		16	.500		2.88	21		23.88	35.50
	5800	Beam clamp, 1/4″ clamp, for 1/4″ threaded drop rod		32	.250		2.58	10.50		13.08	19.05
	5900	3/8″ clamp, for 3/8″ threaded drop rod		32	.250		5.60	10.50		16.10	22.50
	6000	Strap, rigid conduit, 1/2″ diameter		540	.015		1.12	.62		1.74	2.19
	6050	3/4″ diameter		440	.018		1.25	.76		2.01	2.56
	6100	1″ diameter		420	.019		1.34	.80		2.14	2.70
	6150	1-1/4″ diameter		400	.020		1.55	.84		2.39	3

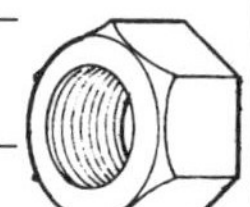
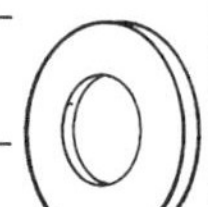
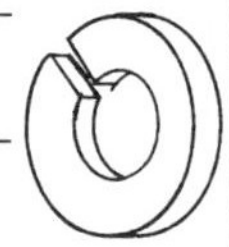
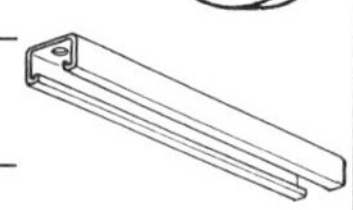
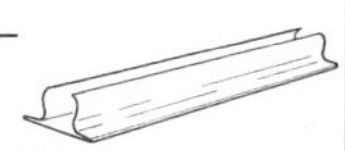
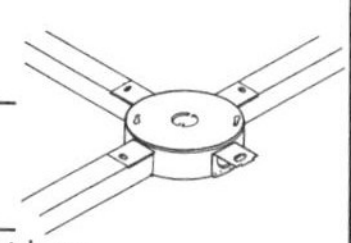
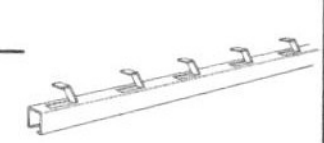

ELECTRICAL 16

16070 | Hangers & Supports		CREW	DAILY OUTPUT	LABOR-HOURS	UNIT	2006 BARE COSTS MAT.	LABOR	EQUIP.	TOTAL	TOTAL INCL O&P
320 6200	1-1/2" diameter R260533-21	1 Elec	400	.020	Ea.	1.76	.84		2.60	3.23
6250	2" diameter		267	.030		1.99	1.26		3.25	4.13
6300	2-1/2" diameter		267	.030		2.16	1.26		3.42	4.32
6350	3" diameter		160	.050		2.39	2.10		4.49	5.85
6400	3-1/2" diameter		133	.060		2.92	2.53		5.45	7.10
6450	4" diameter		100	.080		3.25	3.36		6.61	8.80
6500	5" diameter		80	.100		7.65	4.20		11.85	14.85
6550	6" diameter		60	.133		8.65	5.60		14.25	18.15
6600	EMT, 1/2" diameter		540	.015		1.12	.62		1.74	2.19
6650	3/4" diameter 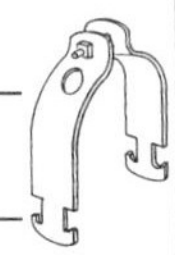		440	.018		1.19	.76		1.95	2.49
6700	1" diameter		420	.019		1.34	.80		2.14	2.70
6750	1-1/4" diameter		400	.020		1.50	.84		2.34	2.94
6800	1-1/2" diameter		400	.020		1.82	.84		2.66	3.29
6850	2" diameter		267	.030		1.96	1.26		3.22	4.10
6900	2-1/2" diameter		267	.030		2.48	1.26		3.74	4.67
6950	3" diameter		160	.050		2.61	2.10		4.71	6.10
6970	3-1/2" diameter		133	.060		2.98	2.53		5.51	7.15
6990	4" diameter		100	.080		3.42	3.36		6.78	8.95
7000	Clip, 1 hole for rigid conduit, 1/2" diameter		500	.016		.61	.67		1.28	1.71
7050	3/4" diameter		470	.017		.90	.71		1.61	2.09
7100	1" diameter 		440	.018		1.12	.76		1.88	2.41
7150	1-1/4" diameter		400	.020		2.40	.84		3.24	3.93
7200	1-1/2" diameter		355	.023		2.80	.95		3.75	4.54
7250	2" diameter		320	.025		5.40	1.05		6.45	7.55
7300	2-1/2" diameter		266	.030		12.10	1.26		13.36	15.25
7350	3" diameter		160	.050		17.30	2.10		19.40	22.50
7400	3-1/2" diameter		133	.060		24.50	2.53		27.03	31
7450	4" diameter		100	.080		55	3.36		58.36	65.50
7500	5" diameter		80	.100		173	4.20		177.20	196
7550	6" diameter		60	.133		185	5.60		190.60	213
7820	Conduit hangers, with bolt & 12" rod, 1/2" diameter		150	.053		2.52	2.24		4.76	6.20
7830	3/4" diameter		145	.055		2.53	2.32		4.85	6.35
7840	1" diameter 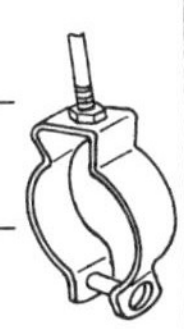		135	.059		2.86	2.49		5.35	7
7850	1-1/4" diameter		120	.067		3.22	2.80		6.02	7.85
7860	1-1/2" diameter		110	.073		3.43	3.05		6.48	8.50
7870	2" diameter		100	.080		4.60	3.36		7.96	10.25
7880	2-1/2" diameter		80	.100		5.10	4.20		9.30	12.05
7890	3" diameter		60	.133		5.45	5.60		11.05	14.65
7900	3-1/2" diameter		45	.178		6.60	7.45		14.05	18.75
7910	4" diameter		35	.229		11.15	9.60		20.75	27
7920	5" diameter		30	.267		12.30	11.20		23.50	31
7930	6" diameter		25	.320		23	13.45		36.45	46
7950	Jay clamp, 1/2" diameter		32	.250		4.59	10.50		15.09	21.50
7960	3/4" diameter		32	.250		4.59	10.50		15.09	21.50
7970	1" diameter 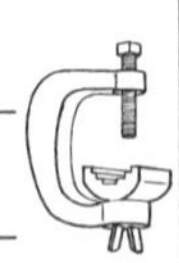		32	.250		4.59	10.50		15.09	21.50
7980	1-1/4" diameter		30	.267		6.65	11.20		17.85	24.50
7990	1-1/2" diameter		30	.267		6.65	11.20		17.85	24.50
8000	2" diameter		30	.267		8.25	11.20		19.45	26.50
8010	2-1/2" diameter		28	.286		11.85	12		23.85	31.50
8020	3" diameter		28	.286		16.10	12		28.10	36
8030	3-1/2" diameter		25	.320		20.50	13.45		33.95	43
8040	4" diameter		25	.320		36	13.45		49.45	60.50
8050	5" diameter		20	.400		91	16.80		107.80	126
8060	6" diameter		16	.500		172	21		193	222
8070	Channels, 3/4" x 1-1/2" w/12" rods for 1/2" to 1" conduit		30	.267		8.50	11.20		19.70	26.50
8080	1-1/2" x 1-1/2" w/12" rods for 1-1/4" to 2" conduit		28	.286		9.20	12		21.20	28.50

16070 | Hangers & Supports

	Line	Description	Crew	Daily Output	Labor-Hours	Unit	2006 Bare Costs Mat.	Labor	Equip.	Total	Total Incl O&P	
320	8090	1-1/2" x 1-1/2" w/12" rods for 2-1/2" to 4" conduit (R260533-21)	1 Elec	26	.308	Ea.	11.40	12.90		24.30	32.50	320
	8100	1-1/2" x 1-7/8" w/12" rods for 5" to 6" conduit		24	.333		33	14		47	58	
	8110	Beam clamp, conduit, plastic coated steel, 1/2" diam.		30	.267		14.80	11.20		26	33.50	
	8120	3/4" diameter		30	.267		15.55	11.20		26.75	34.50	
	8130	1" diameter		30	.267		15.80	11.20		27	34.50	
	8140	1-1/4" diameter		28	.286		21.50	12		33.50	42	
	8150	1-1/2" diameter 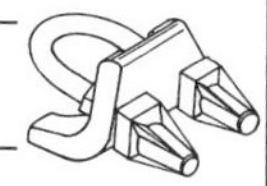		28	.286		25.50	12		37.50	47	
	8160	2" diameter		28	.286		34	12		46	56	
	8170	2-1/2" diameter		26	.308		37.50	12.90		50.40	61.50	
	8180	3" diameter		26	.308		42	12.90		54.90	66.50	
	8190	3-1/2" diameter		23	.348		43.50	14.60		58.10	70.50	
	8200	4" diameter		23	.348		47.50	14.60		62.10	75	
	8210	5" diameter	↓	18	.444	↓	140	18.65		158.65	182	
	8220	Channels, plastic coated										
	8250	3/4" x 1-1/2", w/12" rods for 1/2" to 1" conduit	1 Elec	28	.286	Ea.	22.50	12		34.50	43	
	8260	1-1/2" x 1-1/2", w/12" rods for 1-1/4" to 2" conduit		26	.308		25	12.90		37.90	47.50	
	8270	1-1/2" x 1-1/2", w/12" rods for 2-1/2" to 3-1/2" conduit		24	.333		27.50	14		41.50	51.50	
	8280	1-1/2" x 1-7/8", w/12" rods for 4" to 5" conduit		22	.364		45.50	15.25		60.75	73.50	
	8290	1-1/2" x 1-7/8", w/12" rods for 6" conduit		20	.400		50	16.80		66.80	81	
	8320	Conduit hangers, plastic coated steel, with bolt & 12" rod, 1/2" diam.		140	.057		12.90	2.40		15.30	17.90	
	8330	3/4" diameter		135	.059		13.20	2.49		15.69	18.35	
	8340	1" diameter		125	.064		13.50	2.69		16.19	19	
	8350	1-1/4" diameter		110	.073		14.30	3.05		17.35	20.50	
	8360	1-1/2" diameter		100	.080		16.10	3.36		19.46	23	
	8370	2" diameter		90	.089		17.70	3.73		21.43	25	
	8380	2-1/2" diameter		70	.114		21	4.80		25.80	30.50	
	8390	3" diameter 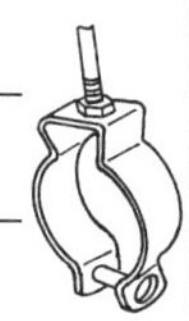		50	.160		25.50	6.70		32.20	38.50	
	8400	3-1/2" diameter		35	.229		27	9.60		36.60	45	
	8410	4" diameter		25	.320		38	13.45		51.45	62	
	8420	5" diameter		20	.400		41.50	16.80		58.30	71.50	
	9000	Parallel type, conduit beam clamp, 1/2"		32	.250		3.95	10.50		14.45	20.50	
	9010	3/4"		32	.250		4.35	10.50		14.85	21	
	9020	1"		32	.250		4.65	10.50		15.15	21.50	
	9030	1-1/4"		30	.267		6.40	11.20		17.60	24.50	
	9040	1-1/2" 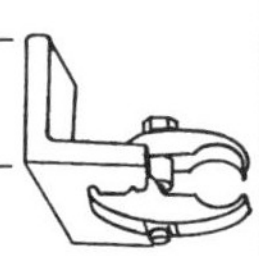		30	.267		7.25	11.20		18.45	25.50	
	9050	2"		30	.267		9.40	11.20		20.60	27.50	
	9060	2-1/2"		28	.286		12.50	12		24.50	32.50	
	9070	3"		28	.286		15.90	12		27.90	36	
	9090	4"		25	.320		20.50	13.45		33.95	43	
	9110	Right angle, conduit beam clamp, 1/2"		32	.250		2.60	10.50		13.10	19.05	
	9120	3/4"		32	.250		2.70	10.50		13.20	19.15	
	9130	1"		32	.250		3	10.50		13.50	19.50	
	9140	1-1/4"		30	.267		3.70	11.20		14.90	21.50	
	9150	1-1/2" 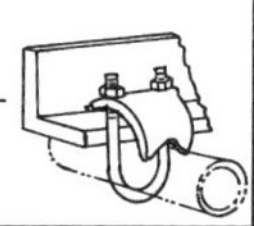		30	.267		4.15	11.20		15.35	22	
	9160	2"		30	.267		6	11.20		17.20	24	
	9170	2-1/2"		28	.286		7.50	12		19.50	27	
	9180	3"		28	.286		8.60	12		20.60	28	
	9190	3-1/2"		25	.320		10.80	13.45		24.25	32.50	
	9200	4"		25	.320		11.70	13.45		25.15	33.50	
	9230	Adjustable, conduit hanger, 1/2"		32	.250		3.12	10.50		13.62	19.65	
	9240	3/4" 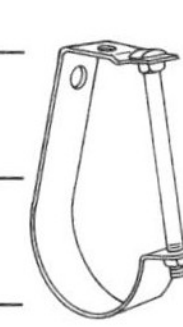		32	.250		2.94	10.50		13.44	19.45	
	9250	1"		32	.250		3.60	10.50		14.10	20	
	9260	1-1/4"		30	.267		4.26	11.20		15.46	22	
	9270	1-1/2"		30	.267		4.92	11.20		16.12	22.50	
	9280	2"		30	.267		5.30	11.20		16.50	23	
	9290	2-1/2"	↓	28	.286	↓	6.60	12		18.60	26	

16050 | Basic Electrical Materials & Methods

		16070 | Hangers & Supports	CREW	DAILY OUTPUT	LABOR-HOURS	UNIT	2006 BARE COSTS MAT.	LABOR	EQUIP.	TOTAL	TOTAL INCL O&P	
320	9300	3″ R260533-21	1 Elec	28	.286	Ea.	6.95	12		18.95	26	320
	9310	3-1/2″		25	.320		9.05	13.45		22.50	30.50	
	9320	4″		25	.320		15.85	13.45		29.30	38	
	9330	5″		20	.400		20.50	16.80		37.30	48.50	
	9340	6″		16	.500		25.50	21		46.50	60.50	
	9350	Combination conduit hanger, 3/8″		32	.250		7.20	10.50		17.70	24	
	9360	Adjustable flange 3/8″		32	.250		8.40	10.50		18.90	25.50	

16100 | Wiring Methods

		16120 | Conductors & Cables	CREW	DAILY OUTPUT	LABOR-HOURS	UNIT	2006 BARE COSTS MAT.	LABOR	EQUIP.	TOTAL	TOTAL INCL O&P	
120	0010	**ARMORED CABLE**										120
	0050	600 volt, copper (BX), #14, 2 conductor, solid	1 Elec	2.40	3.333	C.L.F.	58.50	140		198.50	281	
	0100	3 conductor, solid		2.20	3.636		93	153		246	335	
	0120	4 conductor, solid		2	4		134	168		302	405	
	0152	#12, 2 conductor, solid		2.10	3.810		59.50	160		219.50	310	
	0202	3 conductor, solid		1.80	4.444		95	187		282	390	
	0220	4 conductor, solid		1.80	4.444		137	187		324	440	
	0240	#12, 19 conductor, stranded		1.10	7.273		705	305		1,010	1,250	
	0252	#10, 2 conductor, solid		1.80	4.444		108	187		295	405	
	0302	3 conductor, solid		1.50	5.333		149	224		373	510	
	0320	4 conductor, solid		1.40	5.714		229	240		469	620	
	0352	#8, 3 conductor, solid		1.20	6.667		268	280		548	725	
	0370	4 conductor, stranded		1.10	7.273		385	305		690	890	
	0380	#6, 2 conductor, stranded		1.30	6.154		281	258		539	710	
	0400	3 conductor with PVC jacket, in cable tray, #6		3.10	2.581		320	108		428	520	
	0450	#4	2 Elec	5.40	2.963		415	124		539	645	
	0500	#2		4.60	3.478		550	146		696	830	
	0550	#1		4	4		745	168		913	1,075	
	0600	1/0		3.60	4.444		875	187		1,062	1,250	
	0650	2/0		3.40	4.706		1,075	198		1,273	1,475	
	0700	3/0		3.20	5		1,250	210		1,460	1,700	
	0750	4/0		3	5.333		1,425	224		1,649	1,925	
	0800	250 kcmil	3 Elec	3.60	6.667		1,600	280		1,880	2,175	
	0850	350 kcmil		3.30	7.273		2,150	305		2,455	2,825	
	0900	500 kcmil		3	8		2,775	335		3,110	3,575	
	0910	4 conductor with PVC jacket, in cable tray, #6	1 Elec	2.70	2.963		410	124		534	640	
	0920	#4	2 Elec	4.60	3.478		525	146		671	800	
	0930	#2		4	4		650	168		818	975	
	0940	#1		3.60	4.444		945	187		1,132	1,350	
	0950	1/0		3.40	4.706		1,075	198		1,273	1,500	
	0960	2/0		3.20	5		1,225	210		1,435	1,675	
	0970	3/0		3	5.333		1,475	224		1,699	1,975	
	0980	4/0		2.40	6.667		1,875	280		2,155	2,500	
	0990	250 kcmil	3 Elec	3.30	7.273		2,225	305		2,530	2,925	
	1000	350 kcmil		3	8		2,650	335		2,985	3,425	
	1010	500 kcmil		2.70	8.889		3,625	375		4,000	4,550	
	1050	5 kV, copper, 3 conductor with PVC jacket,										
	1060	non-shielded, in cable tray, #4	2 Elec	380	.042	L.F.	6.35	1.77		8.12	9.65	

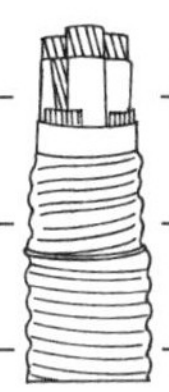

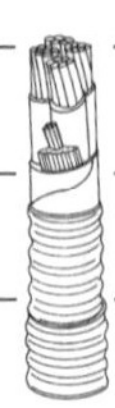

16 ELECTRICAL

16120 | Conductors & Cables

		16120 Conductors & Cables	CREW	DAILY OUTPUT	LABOR-HOURS	UNIT	2006 BARE COSTS MAT.	LABOR	EQUIP.	TOTAL	TOTAL INCL O&P	
120	1100	#2	2 Elec	360	.044	L.F.	8.20	1.87		10.07	11.95	120
	1200	#1 		300	.053		10.45	2.24		12.69	14.95	
	1400	1/0		290	.055		12.10	2.32		14.42	16.85	
	1600	2/0		260	.062		13.95	2.58		16.53	19.35	
	2000	4/0	↓	240	.067		18.65	2.80		21.45	25	
	2100	250 kcmil	3 Elec	330	.073		25.50	3.05		28.55	32.50	
	2150	350 kcmil		315	.076		31.50	3.20		34.70	39.50	
	2200	500 kcmil	↓	270	.089	↓	38.50	3.73		42.23	48.50	
	2400	15 kV, copper, 3 conductor with PVC jacket galv. steel armored										
	2500	grounded neutral, in cable tray, #2	2 Elec	300	.053	L.F.	13.30	2.24		15.54	18.05	
	2600	#1		280	.057		14.15	2.40		16.55	19.25	
	2800	1/0		260	.062		16.35	2.58		18.93	22	
	2900	2/0		220	.073		21.50	3.05		24.55	28	
	3000	4/0	↓	190	.084		24.50	3.54		28.04	32.50	
	3100	250 kcmil	3 Elec	270	.089		27	3.73		30.73	36	
	3150	350 kcmil		240	.100		32	4.20		36.20	41.50	
	3200	500 kcmil	↓	210	.114	↓	43	4.80		47.80	55	
	3400	15 kV, copper, 3 conductor with PVC jacket,										
	3450	ungrounded neutral, in cable tray, #2	2 Elec	260	.062	L.F.	14.30	2.58		16.88	19.75	
	3500	#1 		230	.070		15.85	2.92		18.77	22	
	3600	1/0		200	.080		18.15	3.36		21.51	25	
	3700	2/0		190	.084		22.50	3.54		26.04	30	
	3800	4/0	↓	160	.100		27	4.20		31.20	36	
	4000	250 kcmil	3 Elec	210	.114		31.50	4.80		36.30	42	
	4050	350 kcmil		195	.123		41.50	5.15		46.65	53.50	
	4100	500 kcmil	↓	180	.133	↓	50.50	5.60		56.10	64	
	4200	600 volt, aluminum, 3 conductor in cable tray with PVC jacket										
	4300	#2	2 Elec	540	.030	L.F.	3.16	1.24		4.40	5.40	
	4400	#1		460	.035		3.51	1.46		4.97	6.10	
	4500	#1/0 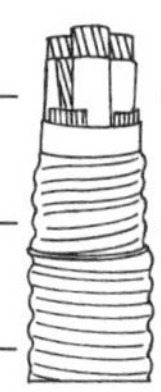		400	.040		4.37	1.68		6.05	7.40	
	4600	#2/0		360	.044		4.43	1.87		6.30	7.75	
	4700	#3/0		340	.047		5.20	1.98		7.18	8.75	
	4800	#4/0	↓	320	.050		6.25	2.10		8.35	10.15	
	4900	250 kcmil	3 Elec	450	.053		7.55	2.24		9.79	11.75	
	5000	350 kcmil		360	.067		8.95	2.80		11.75	14.15	
	5200	500 kcmil		330	.073		11.15	3.05		14.20	17	
	5300	750 kcmil	↓	285	.084	↓	14.40	3.54		17.94	21.50	
	5400	600 volt, aluminum, 4 conductor in cable tray with PVC jacket										
	5410	#2	2 Elec	520	.031	L.F.	3.62	1.29		4.91	5.95	
	5430	#1		440	.036		4.43	1.53		5.96	7.20	
	5450	1/0 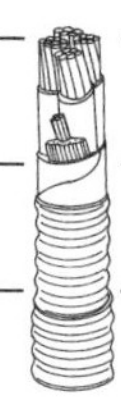		380	.042		5.20	1.77		6.97	8.40	
	5470	2/0		340	.047		5.30	1.98		7.28	8.85	
	5480	3/0		320	.050		6.20	2.10		8.30	10.10	
	5500	4/0	↓	300	.053		7.30	2.24		9.54	11.50	
	5520	250 kcmil	3 Elec	420	.057		7.80	2.40		10.20	12.30	
	5540	350 kcmil		330	.073		10.10	3.05		13.15	15.85	
	5560	500 kcmil		300	.080		12.65	3.36		16.01	19.10	
	5580	750 kcmil	↓	270	.089		17.40	3.73		21.13	25	
	5600	5 kV, aluminum, unshielded in cable tray, #2 with PVC jacket	2 Elec	380	.042		4.43	1.77		6.20	7.60	
	5700	#1 with PVC jacket 		360	.044		4.95	1.87		6.82	8.35	
	5800	1/0 with PVC jacket		300	.053		5.05	2.24		7.29	9	
	6000	2/0 with PVC jacket		290	.055		5.20	2.32		7.52	9.25	
	6200	3/0 with PVC jacket		260	.062		6.20	2.58		8.78	10.85	
	6300	4/0 with PVC jacket	↓	240	.067		7.30	2.80		10.10	12.35	
	6400	250 kcmil with PVC jacket	3 Elec	330	.073		8	3.05		11.05	13.50	
120	6500	350 kcmil with PVC jacket	↓	315	.076	↓	9.35	3.20		12.55	15.25	

		16120 | Conductors & Cables	Crew	Daily Output	Labor-Hours	Unit	2006 Bare Costs Mat.	Labor	Equip.	Total	Total Incl O&P	
120	6600	500 kcmil with PVC jacket	3 Elec	300	.080	L.F.	11.10	3.36		14.46	17.40	120
	6800	750 kcmil with PVC jacket		270	.089		13.70	3.73		17.43	21	
	6900	15 KV, aluminum, shielded-grounded, #2 with PVC jacket	2 Elec	320	.050		10.05	2.10		12.15	14.30	
	7000	#1 with PVC jacket		300	.053		10.35	2.24		12.59	14.85	
	7200	1/0 with PVC jacket		280	.057		11.15	2.40		13.55	16	
	7300	2/0 with PVC jacket		260	.062		11.35	2.58		13.93	16.45	
	7400	3/0 with PVC jacket		240	.067		12.70	2.80		15.50	18.30	
	7500	4/0 with PVC jacket		220	.073		13.05	3.05		16.10	19.05	
	7600	250 kcmil with PVC jacket	3 Elec	300	.080		14.30	3.36		17.66	21	
	7700	350 kcmil with PVC jacket		270	.089		16.80	3.73		20.53	24	
	7800	500 kcmil with PVC jacket		240	.100		20	4.20		24.20	28.50	
	8000	750 kcmil with PVC jacket		204	.118		24	4.94		28.94	34	
	8200	15 kV, aluminum, shielded-ungrounded, #1 with PVC jacket	2 Elec	250	.064		12.60	2.69		15.29	18	
	8300	1/0 with PVC jacket		230	.070		13.05	2.92		15.97	18.85	
	8400	2/0 with PVC jacket		210	.076		14.30	3.20		17.50	20.50	
	8500	3/0 with PVC jacket		200	.080		14.55	3.36		17.91	21	
	8600	4/0 with PVC jacket		190	.084		15.80	3.54		19.34	23	
	8700	250 kcmil with PVC jacket	3 Elec	270	.089		16.95	3.73		20.68	24.50	
	8800	350 kcmil with PVC jacket		240	.100		19.40	4.20		23.60	28	
	8900	500 kcmil with PVC jacket		210	.114		23.50	4.80		28.30	33	
	8950	750 kcmil with PVC jacket		174	.138		29	5.80		34.80	41	
	9010	600 volt, copper (MC) steel clad, #14, 2 wire	1 Elec	2.40	3.333	C.L.F.	61.50	140		201.50	284	
	9020	3 wire		2.20	3.636		95.50	153		248.50	340	
	9030	4 wire		2	4		128	168		296	400	
	9040	#12, 2 wire		2.30	3.478		62.50	146		208.50	294	
	9050	3 wire		2	4		96.50	168		264.50	365	
	9060	4 wire		1.80	4.444		131	187		318	430	
	9070	#10, 2 wire		2	4		111	168		279	380	
	9080	3 wire		1.60	5		171	210		381	515	
	9090	4 wire		1.40	5.714		271	240		511	670	
	9091	For Health Care Facilities cable,add,minimum					5%					
	9092	Maximum					20%					
	9100	#8, 2 wire, stranded	1 Elec	1.80	4.444	C.L.F.	210	187		397	520	
	9110	3 wire, stranded		1.30	6.154		325	258		583	755	
	9120	4 wire, stranded		1.10	7.273		475	305		780	990	
	9130	#6, 2 wire, stranded		1.30	6.154		325	258		583	755	
	9200	600 volt, copper (MC) aluminum clad, #14, 2 wire		2.65	3.019		61.50	127		188.50	263	
	9210	3 wire		2.45	3.265		95.50	137		232.50	315	
	9220	4 wire		2.20	3.636		128	153		281	375	
	9230	#12, 2 wire		2.55	3.137		62.50	132		194.50	272	
	9240	3 wire		2.20	3.636		96.50	153		249.50	340	
	9250	4 wire		2	4		131	168		299	405	
	9260	#10, 2 wire		2.20	3.636		113	153		266	360	
	9270	3 wire		1.80	4.444		176	187		363	480	
	9280	4 wire		1.55	5.161		277	217		494	640	
	9600	Alum. (MC) aluminum clad, #6, 3 conductor w/#6 grnd.		1.67	4.790		112	201		313	435	
	9610	4 conductor w/#6 grnd.		1.64	4.878		121	205		326	450	
	9620	#4, 3 conductor w/#6 grnd.	2 Elec	2.86	5.594		120	235		355	490	
	9630	4 conductor w/#6 grnd.		2.82	5.674		150	238		388	530	
	9640	#2, 3 conductor w/#4 grnd.		2.50	6.400		164	269		433	595	
	9650	4 conductor w/#4 grnd.		2.47	6.478		184	272		456	620	
	9660	#1, 3 conductor w/#4 grnd.		2	8		192	335		527	730	
	9670	4 conductor w/#4 grnd.		1.98	8.081		232	340		572	780	
	9680	1/0, 3 conductor w/#4 grnd.		1.82	8.791		226	370		596	820	
	9690	4 conductor w/#4 grnd.		1.79	8.939		286	375		661	895	
	9700	2/0, 3 conductor w/#4 grnd.		1.75	9.143		248	385		633	865	

	16120 | Conductors & Cables		CREW	DAILY OUTPUT	LABOR-HOURS	UNIT	2006 BARE COSTS MAT.	LABOR	EQUIP.	TOTAL	TOTAL INCL O&P	
120	9710	4 conductor w/#4 grnd.	2 Elec	1.71	9.357	C.L.F.	315	395		710	955	120
	9720	3/0, 3 conductor w/#4 grnd.		1.71	9.357		310	395		705	950	
	9730	4 conductor w/#4 grnd.		1.67	9.581		365	400		765	1,025	
	9740	4/0, 3 conductor w/#2 grnd.		1.67	9.581		330	400		730	985	
	9750	4 conductor w/#2 grnd.	↓	1.61	9.938		440	415		855	1,125	
	9760	250 kcmil, 3 conductor w/#1 grnd.	3 Elec	2.42	9.917		390	415		805	1,075	
	9770	4 conductor w/#1 grnd.		2.36	10.169		570	425		995	1,300	
	9775	300 kcmil, 4 conductor w/ #1 grnd.		2.22	10.811		605	455		1,060	1,375	
	9780	350 kcmil, 3 conductor w/ 1/0 grnd.		2.19	10.959		600	460		1,060	1,375	
	9790	4 conductor w/ 1/0 grnd.		2.10	11.429		760	480		1,240	1,575	
	9800	500 kcmil, 3 conductor w/#1 grnd.		2.10	11.429		750	480		1,230	1,575	
	9810	4 conductor w/ 2/0 grnd.		2	12		900	505		1,405	1,775	
	9840	750 kcmil, 3 conductor w/ 1/0 grnd.		1.95	12.308		1,075	515		1,590	2,000	
	9850	4 conductor w/ 3/0 grnd.	↓	1.89	12.698	↓	1,275	535		1,810	2,225	
	9900	Minimum labor/equipment charge	1 Elec	4	2	Job		84		84	129	
210	0010	CABLE CONNECTORS										210
	0100	600 volt, nonmetallic, #14-2 wire	1 Elec	160	.050	Ea.	.49	2.10		2.59	3.78	
	0200	#14-3 wire to #12-2 wire		133	.060		.49	2.53		3.02	4.43	
	0300	#12-3 wire to #10-2 wire		114	.070		.49	2.95		3.44	5.10	
	0400	#10-3 wire to #14-4 and #12-4 wire		100	.080		.49	3.36		3.85	5.75	
	0500	#8-3 wire to #10-4 wire		80	.100		1.35	4.20		5.55	7.95	
	0600	#6-3 wire		40	.200		2.18	8.40		10.58	15.35	
	0800	SER aluminum, 3 #8 insulated + 1 #8 ground		32	.250		2.30	10.50		12.80	18.75	
	0900	3 #6 + 1 #6 ground		24	.333		2.30	14		16.30	24	
	1000	3 #4 + 1 #6 ground		22	.364		3.34	15.25		18.59	27	
	1100	3 #2 + 1 #4 ground		20	.400		6	16.80		22.80	32.50	
	1200	3 1/0 + 1 #2 ground		18	.444		15.50	18.65		34.15	46	
	1400	3 2/0 + 1 #1 ground		16	.500		15.50	21		36.50	49.50	
	1600	3 4/0 + 1 #2/0 ground		14	.571		18.65	24		42.65	57.50	
	1800	600 volt, armored, #14-2 wire		80	.100		.59	4.20		4.79	7.10	
	2200	#14-4, #12-3 and #10-2 wire		40	.200		.59	8.40		8.99	13.60	
	2400	#12-4, #10-3 and #8-2 wire		32	.250		1.39	10.50		11.89	17.75	
	2600	#8-3 and #10-4 wire		26	.308		2.38	12.90		15.28	22.50	
	2650	#8-4 wire		22	.364		4.17	15.25		19.42	28	
	2700	PVC jacket connector, #6-3 wire, #6-4 wire		16	.500		10.55	21		31.55	44	
	2800	#4-3 wire, #4-4 wire		16	.500		10.55	21		31.55	44	
	2900	#2-3 wire		12	.667		10.55	28		38.55	54.50	
	3000	#1-3 wire, #2-4 wire		12	.667		14.80	28		42.80	59.50	
	3200	1/0-3 wire		11	.727		14.80	30.50		45.30	63.50	
	3400	2/0-3 wire, 1/0-4 wire		10	.800		14.80	33.50		48.30	68.50	
	3500	3/0-3 wire, 2/0-4 wire		9	.889		30.50	37.50		68	91	
	3600	4/0-3 wire, 3/0-4 wire		7	1.143		30.50	48		78.50	108	
	3800	250 kcmil-3 wire, 4/0-4 wire		6	1.333		42.50	56		98.50	133	
	4000	350 kcmil-3 wire, 250 kcmil-4 wire		5	1.600		42.50	67		109.50	151	
	4100	350 kcmil-4 wire		4	2		164	84		248	310	
	4200	500 kcmil-3 wire		4	2		164	84		248	310	
	4250	500 kcmil-4 wire, 750 kcmil-3 wire		3.50	2.286		215	96		311	385	
	4300	750 kcmil-4 wire		3	2.667		215	112		327	410	
	4400	5 kV, armored, #4		8	1		57.50	42		99.50	128	
	4600	#2		8	1		57.50	42		99.50	128	
	4800	#1		8	1		57.50	42		99.50	128	
	5000	1/0		6.40	1.250		71	52.50		123.50	160	
	5200	2/0		5.30	1.509		71	63.50		134.50	176	
	5500	4/0		4	2		94	84		178	232	
	5600	250 kcmil	↓	3.60	2.222	↓	115	93.50		208.50	271	

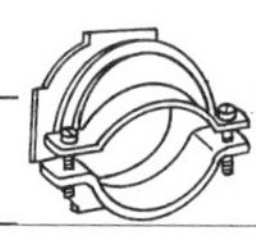

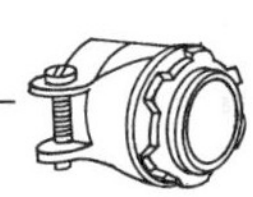

ELECTRICAL 16

	16120 | Conductors & Cables		Crew	Daily Output	Labor-Hours	Unit	2006 Bare Costs Mat.	Labor	Equip.	Total	Total Incl O&P
210	5650	350 kcmil	1 Elec	3.20	2.500	Ea.	136	105		241	310
	5700	500 kcmil		2.50	3.200		136	134		270	355
	5720	750 kcmil		2.20	3.636		198	153		351	455
	5750	1000 kcmil		2	4		264	168		432	550
	5800	15 kV, armored, #1		4	2		90.50	84		174.50	229
	5900	1/0		4	2		90.50	84		174.50	229
	6000	3/0		3.60	2.222		116	93.50		209.50	272
	6100	4/0		3.40	2.353		116	99		215	280
	6200	250 kcmil		3.20	2.500		129	105		234	305
	6300	350 kcmil		2.70	2.963		149	124		273	355
	6400	500 kcmil		2	4		149	168		317	425
220	0010	**CABLE SPLICING** URD or similar, ideal conditions									
	0100	#6 stranded to #1 stranded, 5 kV	1 Elec	4	2	Ea.	107	84		191	247
	0120	15 kV		3.60	2.222		107	93.50		200.50	262
	0140	25 kV		3.20	2.500		107	105		212	280
	0200	#1 stranded to 4/0 stranded, 5 kV		3.60	2.222		112	93.50		205.50	267
	0210	15 kV		3.20	2.500		112	105		217	285
	0220	25 kV		2.80	2.857		112	120		232	310
	0300	4/0 stranded to 500 kcmil stranded, 5 kV		3.30	2.424		252	102		354	435
	0310	15 kV		2.90	2.759		252	116		368	455
	0320	25 kV		2.50	3.200		252	134		386	485
	0400	500 kcmil, 5 kV		3.20	2.500		252	105		357	440
	0410	15 kV		2.80	2.857		252	120		372	460
	0420	25 kV		2.30	3.478		252	146		398	500
	0500	600 kcmil, 5 kV		2.90	2.759		252	116		368	455
	0510	15 kV		2.40	3.333		252	140		392	495
	0520	25 kV		2	4		252	168		420	535
	0600	750 kcmil, 5 kV		2.60	3.077		297	129		426	525
	0610	15 kV		2.20	3.636		297	153		450	560
	0620	25 kV		1.90	4.211		297	177		474	595
	0700	1000 kcmil, 5 kV		2.30	3.478		320	146		466	580
	0710	15 kV		1.90	4.211		320	177		497	625
	0720	25 kV		1.60	5		320	210		530	680
230	0010	**CABLE TERMINATIONS**									
	0015	Wire connectors, screw type, #22 to #14	1 Elec	260	.031	Ea.	.07	1.29		1.36	2.07
	0020	#18 to #12		240	.033		.08	1.40		1.48	2.25
	0025	#18 to #10		240	.033		.13	1.40		1.53	2.30
	0030	Screw-on connectors, insulated, #18 to #12		240	.033		.09	1.40		1.49	2.26
	0035	#16 to #10		230	.035		.12	1.46		1.58	2.38
	0040	#14 to #8		210	.038		.25	1.60		1.85	2.75
	0045	#12 to #6		180	.044		.42	1.87		2.29	3.34
	0050	Terminal lugs, solderless, #16 to #10		50	.160		.46	6.70		7.16	10.85
	0100	#8 to #4		30	.267		.64	11.20		11.84	17.95
	0150	#2 to #1		22	.364		.86	15.25		16.11	24.50
	0200	1/0 to 2/0		16	.500		2.22	21		23.22	35
	0250	3/0		12	.667		2.95	28		30.95	46.50
	0300	4/0		11	.727		2.95	30.50		33.45	50.50
	0350	250 kcmil		9	.889		2.95	37.50		40.45	61
	0400	350 kcmil		7	1.143		3.80	48		51.80	78
	0450	500 kcmil		6	1.333		7.40	56		63.40	94.50
	0500	600 kcmil		5.80	1.379		7.95	58		65.95	98
	0550	750 kcmil		5.20	1.538		8.70	64.50		73.20	109
	0600	Split bolt connectors, tapped, #6		16	.500		3.30	21		24.30	36
	0650	#4		14	.571		4	24		28	41.50
	0700	#2		12	.667		6.05	28		34.05	49.50

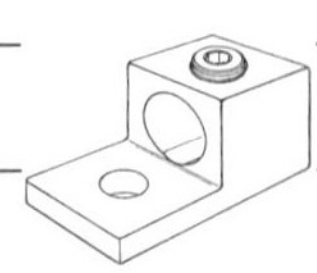

16120 | Conductors & Cables

		CREW	DAILY OUTPUT	LABOR-HOURS	UNIT	2006 BARE COSTS MAT.	LABOR	EQUIP.	TOTAL	TOTAL INCL O&P
230 0750	#1	1 Elec	11	.727	Ea.	7.65	30.50		38.15	55.50
0800	1/0		10	.800		7.65	33.50		41.15	60.50
0850	2/0		9	.889		12.40	37.50		49.90	71
0900	3/0		7.20	1.111		17.30	46.50		63.80	91
1000	4/0		6.40	1.250		21	52.50		73.50	104
1100	250 kcmil		5.70	1.404		21	59		80	114
1200	300 kcmil		5.30	1.509		38.50	63.50		102	140
1400	350 kcmil		4.60	1.739		38.50	73		111.50	156
1500	500 kcmil	↓	4	2	↓	50.50	84		134.50	185
1600	Crimp 1 hole lugs, copper or aluminum, 600 volt									
1620	#14	1 Elec	60	.133	Ea.	.32	5.60		5.92	9
1630	#12		50	.160		.45	6.70		7.15	10.85
1640	#10		45	.178		.45	7.45		7.90	12
1780	#8		36	.222		1.72	9.35		11.07	16.30
1800	#6		30	.267		2.78	11.20		13.98	20.50
2000	#4		27	.296		3	12.45		15.45	22.50
2200	#2		24	.333		4.07	14		18.07	26
2400	#1		20	.400		4.90	16.80		21.70	31.50
2500	1/0		17.50	.457		5.50	19.20		24.70	35.50
2600	2/0		15	.533		6.60	22.50		29.10	42
2800	3/0		12	.667		8.20	28		36.20	52
3000	4/0		11	.727		8.90	30.50		39.40	57
3200	250 kcmil		9	.889		10.60	37.50		48.10	69
3400	300 kcmil		8	1		11.90	42		53.90	77.50
3500	350 kcmil		7	1.143		12.30	48		60.30	87.50
3600	400 kcmil		6.50	1.231		15.30	51.50		66.80	96.50
3800	500 kcmil		6	1.333		18.60	56		74.60	107
4000	600 kcmil		5.80	1.379		29.50	58		87.50	122
4200	700 kcmil		5.50	1.455		32	61		93	129
4400	750 kcmil	↓	5.20	1.538	↓	32	64.50		96.50	135
4500	Crimp 2-way connectors, copper or alum., 600 volt,									
4510	#14	1 Elec	60	.133	Ea.	.51	5.60		6.11	9.20
4520	#12		50	.160		.81	6.70		7.51	11.25
4530	#10		45	.178		.81	7.45		8.26	12.40
4540	#8		27	.296		1.90	12.45		14.35	21
4600	#6		25	.320		3.90	13.45		17.35	25
4800	#4		23	.348		4.20	14.60		18.80	27
5000	#2		20	.400		6.05	16.80		22.85	32.50
5200	#1		16	.500		8.70	21		29.70	42
5400	1/0		13	.615		9.60	26		35.60	50.50
5420	2/0		12	.667		9.80	28		37.80	54
5440	3/0		11	.727		11.80	30.50		42.30	60
5460	4/0		10	.800		12.30	33.50		45.80	65.50
5480	250 kcmil		9	.889		14	37.50		51.50	73
5500	300 kcmil		8.50	.941		15.70	39.50		55.20	78.50
5520	350 kcmil		8	1		16.30	42		58.30	82.50
5540	400 kcmil		7.30	1.096		21	46		67	94
5560	500 kcmil		6.20	1.290		26	54		80	112
5580	600 kcmil		5.50	1.455		38.50	61		99.50	137
5600	700 kcmil		4.50	1.778		39.50	74.50		114	158
5620	750 kcmil		4	2		39.50	84		123.50	172
7000	Compression equipment adapter, aluminum wire, #6		30	.267		7.40	11.20		18.60	25.50
7020	#4		27	.296		7.80	12.45		20.25	28
7040	#2		24	.333		8.20	14		22.20	30.50
7060	#1		20	.400		9.45	16.80		26.25	36.50
7080	1/0	↓	18	.444	↓	9.90	18.65		28.55	40

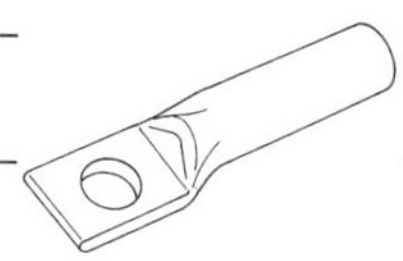

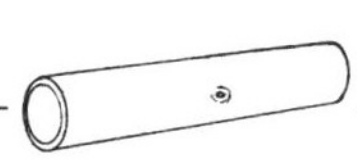

		16120 | Conductors & Cables	Crew	Daily Output	Labor-Hours	Unit	2006 Bare Costs Mat.	Labor	Equip.	Total	Total Incl O&P
230	7100	2/0	1 Elec	15	.533	Ea.	18.70	22.50		41.20	55
	7140	4/0		11	.727		18.70	30.50		49.20	67.50
	7160	250 kcmil		9	.889		18.70	37.50		56.20	78
	7180	300 kcmil		8	1		22	42		64	89
	7200	350 kcmil		7	1.143		22	48		70	98.50
	7220	400 kcmil		6.50	1.231		29	51.50		80.50	111
	7240	500 kcmil		6	1.333		29	56		85	118
	7260	600 kcmil		5.80	1.379		39.50	58		97.50	133
	7280	750 kcmil		5.20	1.538		39.50	64.50		104	143
	8000	Compression tool, hand					980			980	1,075
	8100	Hydraulic					2,525			2,525	2,775
	8500	Hydraulic dies					380			380	420
240	0010	**CABLE TERMINATIONS,** 5 kV to 35 kV									
	0100	Indoor, insulation diameter range .525" to 1.025"									
	0300	Padmount, 5 kV	1 Elec	8	1	Ea.	57	42		99	128
	0400	15 kV		6.40	1.250		103	52.50		155.50	194
	0500	25 kV		6	1.333		120	56		176	219
	0600	35 kV		5.60	1.429		153	60		213	261
	0700	insulation diameter range .975" to 1.570"									
	0800	Padmount, 5 kV	1 Elec	8	1	Ea.	69.50	42		111.50	141
	0900	15 kV		6	1.333		137	56		193	238
	1000	25 kV		5.60	1.429		158	60		218	267
	1100	35 kV		5.30	1.509		189	63.50		252.50	305
	1200	insulation diameter range 1.540" to 1.900"									
	1300	Padmount, 5 KV	1 Elec	7.40	1.081	Ea.	105	45.50		150.50	186
	1400	15 kV		5.60	1.429		174	60		234	284
	1500	25 kV		5.30	1.509		222	63.50		285.50	340
	1600	35 kV		5	1.600		248	67		315	375
	1700	Outdoor systems, #4 stranded to 1/0 stranded									
	1800	5 kV	1 Elec	7.40	1.081	Ea.	113	45.50		158.50	195
	1900	15 kV		5.30	1.509		140	63.50		203.50	252
	2000	25 kV		5	1.600		158	67		225	278
	2100	35 kV		4.80	1.667		184	70		254	310
	2200	#1 solid to 4/0 stranded, 5 kV		6.90	1.159		127	48.50		175.50	215
	2300	15 kV		5	1.600		140	67		207	258
	2400	25 kV		4.80	1.667		189	70		259	315
	2500	35 kV		4.60	1.739		189	73		262	320
	2600	3/0 solid to 350 kcmil stranded, 5 kV		6.40	1.250		158	52.50		210.50	255
	2700	15 kV		4.80	1.667		158	70		228	282
	2800	25 kV		4.60	1.739		222	73		295	355
	2900	35 kV		4.40	1.818		222	76.50		298.50	360
	3000	400 kcmil compact to 750 kcmil stranded, 5 kV		6	1.333		205	56		261	310
	3100	15 kV		4.60	1.739		205	73		278	340
	3200	25 kV		4.40	1.818		248	76.50		324.50	390
	3300	35 kV		4.20	1.905		248	80		328	395
	3400	1000 kcmil, 5 kV		5.60	1.429		222	60		282	335
	3500	15 kV		4.40	1.818		237	76.50		313.50	380
	3600	25 kV		4.20	1.905		248	80		328	395
	3700	35 kV		4	2		248	84		332	400
280	0010	**CONTROL CABLE**									
	0020	600 volt, copper, #14 THWN wire with PVC jacket, 2 wires	1 Elec	9	.889	C.L.F.	15.55	37.50		53.05	74.50
	0030	3 wires		8	1		21	42		63	87.50
	0100	4 wires		7	1.143		25.50	48		73.50	102
	0150	5 wires		6.50	1.231		31.50	51.50		83	115
	0200	6 wires		6	1.333		45.50	56		101.50	137

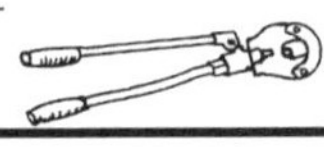

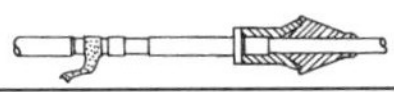

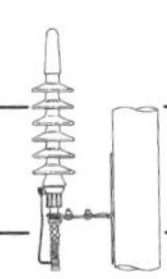

	16120	Conductors & Cables	Crew	Daily Output	Labor-Hours	Unit	2006 Bare Costs Mat.	Labor	Equip.	Total	Total Incl O&P	
280	0300	8 wires	1 Elec	5.30	1.509	C.L.F.	56	63.50		119.50	160	280
	0400	10 wires		4.80	1.667		66	70		136	181	
	0500	12 wires		4.30	1.860		75	78		153	203	
	0600	14 wires		3.80	2.105		92	88.50		180.50	237	
	0700	16 wires		3.50	2.286		94.50	96		190.50	252	
	0800	18 wires		3.30	2.424		112	102		214	280	
	0810	19 wires		3.10	2.581		115	108		223	293	
	0900	20 wires		3	2.667		130	112		242	315	
	1000	22 wires	↓	2.80	2.857	↓	133	120		253	330	
500	0010	**MINERAL INSULATED CABLE** 600 volt										500
	0100	1 conductor, #12	1 Elec	1.60	5	C.L.F.	236	210		446	585	
	0200	#10		1.60	5		305	210		515	665	
	0400	#8		1.50	5.333		360	224		584	740	
	0500	#6	↓	1.40	5.714		405	240		645	815	
	0600	#4	2 Elec	2.40	6.667		560	280		840	1,050	
	0800	#2		2.20	7.273		750	305		1,055	1,300	
	0900	#1		2.10	7.619		855	320		1,175	1,425	
	1000	1/0		2	8		985	335		1,320	1,600	
	1100	2/0		1.90	8.421		1,175	355		1,530	1,850	
	1200	3/0		1.80	8.889		1,400	375		1,775	2,100	
	1400	4/0	↓	1.60	10		1,600	420		2,020	2,400	
	1410	250 kcmil	3 Elec	2.40	10		1,825	420		2,245	2,650	
	1420	350 kcmil		1.95	12.308		2,200	515		2,715	3,225	
	1430	500 kcmil	↓	1.95	12.308		2,775	515		3,290	3,850	
	1500	2 conductor, #12	1 Elec	1.40	5.714		485	240		725	905	
	1600	#10		1.20	6.667		590	280		870	1,075	
	1800	#8		1.10	7.273		735	305		1,040	1,275	
	2000	#6	↓	1.05	7.619		940	320		1,260	1,525	
	2100	#4	2 Elec	2	8		1,225	335		1,560	1,875	
	2200	3 conductor, #12	1 Elec	1.20	6.667		615	280		895	1,100	
	2400	#10		1.10	7.273		710	305		1,015	1,250	
	2600	#8		1.05	7.619		855	320		1,175	1,450	
	2800	#6	↓	1	8		1,100	335		1,435	1,750	
	3000	#4	2 Elec	1.80	8.889		1,475	375		1,850	2,175	
	3100	4 conductor, #12	1 Elec	1.20	6.667		650	280		930	1,150	
	3200	#10		1.10	7.273		775	305		1,080	1,325	
	3400	#8		1	8		1,000	335		1,335	1,625	
	3600	#6		.90	8.889		1,275	375		1,650	1,975	
	3620	7 conductor, #12		1.10	7.273		825	305		1,130	1,375	
	3640	#10		1	8	↓	1,025	335		1,360	1,650	
	3800	M.I. terminations, 600 volt, 1 conductor, #12		8	1	Ea.	9.60	42		51.60	75	
	4000	#10		7.60	1.053		9.60	44		53.60	78.50	
	4100	#8		7.30	1.096		9.60	46		55.60	81.50	
	4200	#6		6.70	1.194		9.60	50		59.60	88	
	4400	#4		6.20	1.290		9.60	54		63.60	94	
	4600	#2		5.70	1.404		14.40	59		73.40	107	
	4800	#1		5.30	1.509		14.40	63.50		77.90	113	
	5000	1/0		5	1.600		14.40	67		81.40	120	
	5100	2/0		4.70	1.702		14.40	71.50		85.90	126	
	5200	3/0		4.30	1.860		14.40	78		92.40	136	
	5400	4/0		4	2		31	84		115	163	
	5410	250 kcmil		4	2		32	84		116	164	
	5420	350 kcmil		4	2		57	84		141	192	
	5430	500 kcmil		4	2		57	84		141	192	
	5500	2 conductor, #12	↓	6.70	1.194	↓	9.60	50		59.60	88	

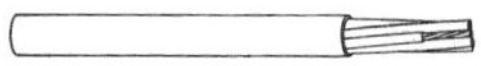
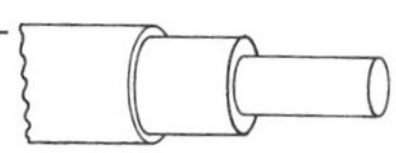
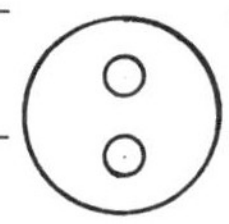

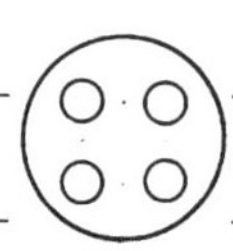
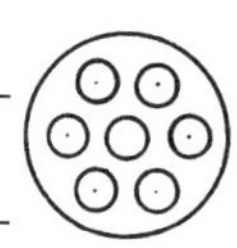

16120 | Conductors & Cables

		16120 Conductors & Cables	CREW	DAILY OUTPUT	LABOR-HOURS	UNIT	2006 BARE COSTS MAT.	LABOR	EQUIP.	TOTAL	TOTAL INCL O&P	
500	5600	#10	1 Elec	6.40	1.250	Ea.	14.40	52.50		66.90	97	500
	5800	#8		6.20	1.290		14.40	54		68.40	99.50	
	6000	#6 		5.70	1.404		14.40	59		73.40	107	
	6200	#4		5.30	1.509		32	63.50		95.50	133	
	6400	3 conductor, #12		5.70	1.404		14.40	59		73.40	107	
	6500	#10		5.50	1.455		14.40	61		75.40	110	
	6600	#8		5.20	1.538		14.40	64.50		78.90	115	
	6800	#6		4.80	1.667		14.40	70		84.40	124	
	7200	#4		4.60	1.739		32	73		105	148	
	7400	4 conductor, #12		4.60	1.739		15.95	73		88.95	131	
	7500	#10		4.40	1.818		15.95	76.50		92.45	136	
	7600	#8		4.20	1.905		15.95	80		95.95	141	
	8400	#6 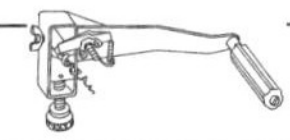		4	2		33.50	84		117.50	166	
	8500	7 conductor, #12		3.50	2.286		15.95	96		111.95	166	
	8600	#10		3	2.667		33.50	112		145.50	210	
	8800	Crimping tool, plier type					45.50			45.50	50	
	9000	Stripping tool 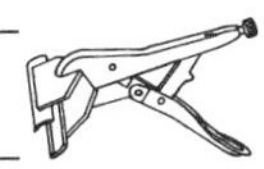					105			105	116	
	9200	Hand vise					51			51	56	
	9500	Minimum labor/equipment charge	1 Elec	4	2	Job		84		84	129	
550	0010	**NON-METALLIC SHEATHED CABLE** 600 volt										550
	0100	Copper with ground wire, (Romex)										
	0152	#14, 2 wire	1 Elec	2.50	3.200	C.L.F.	18.25	134		152.25	227	
	0202	3 wire		2.30	3.478		29	146		175	257	
	0220	4 conductor		2.20	3.636		49	153		202	289	
	0252	#12, 2 wire		2.20	3.636		27	153		180	265	
	0302	3 wire		2	4		48.50	168		216.50	315	
	0320	4 conductor		2	4		69.50	168		237.50	335	
	0352	#10, 2 wire		2	4		48.50	168		216.50	310	
	0402	3 wire		1.40	5.714		60.50	240		300.50	435	
	0420	4 conductor		1.60	5		108	210		318	445	
	0452	#8, 3 conductor		1.30	6.154		138	258		396	550	
	0480	4 conductor		1.40	5.714		208	240		448	600	
	0502	#6, 3 wire		1.20	6.667		214	280		494	665	
	0520	#4, 3 conductor	2 Elec	2.40	6.667		340	280		620	800	
	0540	#2, 3 conductor	"	2.20	7.273		505	305		810	1,025	
	0550	SE type SER aluminum cable, 3 RHW and										
	0602	1 bare neutral, 3 #8 & 1 #8	1 Elec	1.50	5.333	C.L.F.	133	224		357	490	
	0652	3 #6 & 1 #6	"	1.30	6.154		150	258		408	565	
	0702	3 #4 & 1 #6	2 Elec	2.20	7.273		168	305		473	655	
	0752	3 #2 & 1 #4		2	8		248	335		583	790	
	0802	3 #1/0 & 1 #2		1.80	8.889		375	375		750	985	
	0852	3 #2/0 & 1 #1		1.60	10		440	420		860	1,125	
	0902	3 #4/0 & 1 #2/0		1.40	11.429		630	480		1,110	1,425	
	1450	UF underground feeder cable, copper with ground, #14, 2 conductor	1 Elec	4	2		24.50	84		108.50	156	
	1500	#12, 2 conductor		3.50	2.286		35.50	96		131.50	187	
	1550	#10, 2 conductor		3	2.667		56.50	112		168.50	235	
	1600	#14, 3 conductor 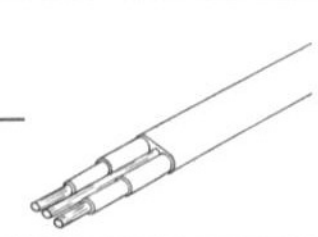		3.50	2.286		34	96		130	186	
	1650	#12, 3 conductor		3	2.667		49	112		161	227	
	1700	#10, 3 conductor		2.50	3.200		76.50	134		210.50	291	
	2400	SEU service entrance cable, copper 2 conductors, #8 + #8 neut.		1.50	5.333		111	224		335	465	
	2600	#6 + #8 neutral		1.30	6.154		167	258		425	585	
	2800	#6 + #6 neutral 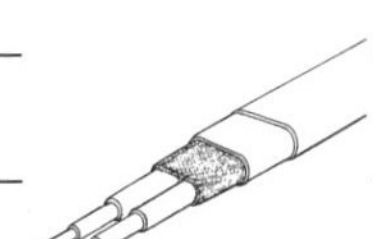		1.30	6.154		186	258		444	605	
	3000	#4 + #6 neutral	2 Elec	2.20	7.273		249	305		554	745	
	3200	#4 + #4 neutral		2.20	7.273		275	305		580	775	
	3400	#3 + #5 neutral		2.10	7.619		293	320		613	815	

16 ELECTRICAL

16120 | Conductors & Cables

		16120 Conductors & Cables	CREW	DAILY OUTPUT	LABOR-HOURS	UNIT	MAT.	LABOR	EQUIP.	TOTAL	TOTAL INCL O&P	
550	3600	#3 + #3 neutral	2 Elec	2.10	7.619	C.L.F.	320	320		640	850	550
	3800	#2 + #4 neutral		2	8		370	335		705	930	
	4000	#1 + #1 neutral		1.90	8.421		635	355		990	1,250	
	4200	1/0 + 1/0 neutral		1.80	8.889		720	375		1,095	1,375	
	4400	2/0 + 2/0 neutral		1.70	9.412		880	395		1,275	1,575	
	4600	3/0 + 3/0 neutral		1.60	10		1,100	420		1,520	1,850	
	4800	Aluminum 2 conductors, #8 + #8 neutral	1 Elec	1.60	5		109	210		319	445	
	5000	#6 + #6 neutral	"	1.40	5.714		110	240		350	490	
	5100	#4 + #6 neutral	2 Elec	2.50	6.400		141	269		410	570	
	5200	#4 + #4 neutral		2.40	6.667		141	280		421	585	
	5300	#2 + #4 neutral		2.30	6.957		188	292		480	655	
	5400	#2 + #2 neutral		2.20	7.273		188	305		493	675	
	5450	1/0 + #2 neutral		2.10	7.619		287	320		607	810	
	5500	1/0 + 1/0 neutral		2	8		287	335		622	835	
	5550	2/0 + #1 neutral		1.90	8.421		330	355		685	905	
	5600	2/0 + 2/0 neutral		1.80	8.889		330	375		705	935	
	5800	3/0 + 1/0 neutral		1.70	9.412		420	395		815	1,075	
	6000	3/0 + 3/0 neutral		1.70	9.412		420	395		815	1,075	
	6200	4/0 + 2/0 neutral		1.60	10		460	420		880	1,150	
	6400	4/0 + 4/0 neutral		1.60	10		460	420		880	1,150	
	6500	Service entrance cap for copper SEU										
	6600	100 amp	1 Elec	12	.667	Ea.	9.50	28		37.50	53.50	
	6700	150 amp		10	.800		13.95	33.50		47.45	67.50	
	6800	200 amp		8	1		29	42		71	96.50	
	9000	Minimum labor/equipment charge		4	2	Job		84		84	129	
700	0010	**SHIELDED CABLE** Splicing & terminations not included										700
	0040	Copper, XLP shielding, 5 kV, #6	2 Elec	4.40	3.636	C.L.F.	113	153		266	360	
	0050	#4		4.40	3.636		147	153		300	395	
	0100	#2		4	4		174	168		342	450	
	0200	#1		4	4		200	168		368	480	
	0400	1/0		3.80	4.211		220	177		397	515	
	0600	2/0		3.60	4.444		277	187		464	595	
	0800	4/0		3.20	5		360	210		570	720	
	1000	250 kcmil	3 Elec	4.50	5.333		445	224		669	835	
	1200	350 kcmil		3.90	6.154		575	258		833	1,025	
	1400	500 kcmil		3.60	6.667		700	280		980	1,200	
	1600	15 kV, ungrounded neutral, #1	2 Elec	4	4		243	168		411	525	
	1800	1/0		3.80	4.211		290	177		467	590	
	2000	2/0		3.60	4.444		330	187		517	655	
	2200	4/0		3.20	5		440	210		650	810	
	2400	250 kcmil	3 Elec	4.50	5.333		485	224		709	880	
	2600	350 kcmil		3.90	6.154		615	258		873	1,075	
	2800	500 kcmil		3.60	6.667		755	280		1,035	1,250	
	3000	25 kV, grounded neutral, #1/0	2 Elec	3.60	4.444		400	187		587	730	
	3200	2/0		3.40	4.706		440	198		638	790	
	3400	4/0		3	5.333		555	224		779	955	
	3600	250 kcmil	3 Elec	4.20	5.714		690	240		930	1,125	
	3800	350 kcmil		3.60	6.667		805	280		1,085	1,325	
	3900	500 kcmil		3.30	7.273		945	305		1,250	1,525	
	4000	35 kV, grounded neutral, #1/0	2 Elec	3.40	4.706		425	198		623	775	
	4200	2/0		3.20	5		500	210		710	875	
	4400	4/0		2.80	5.714		630	240		870	1,050	
	4600	250 kcmil	3 Elec	3.90	6.154		735	258		993	1,200	
	4800	350 kcmil		3.30	7.273		885	305		1,190	1,450	
	5000	500 kcmil		3	8		1,050	335		1,385	1,675	

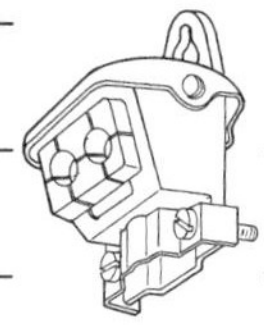

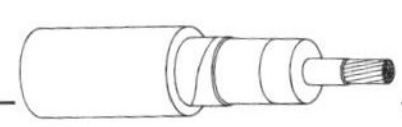

ELECTRICAL 16

	16120 | Conductors & Cables		CREW	DAILY OUTPUT	LABOR-HOURS	UNIT	2006 BARE COSTS MAT.	LABOR	EQUIP.	TOTAL	TOTAL INCL O&P	
700	5050	Aluminum, XLP shielding, 5 kV, #2	2 Elec	5	3.200	C.L.F.	140	134		274	360	700
	5070	#1		4.40	3.636		145	153		298	395	
	5090	1/0		4	4		168	168		336	445	
	5100	2/0		3.80	4.211		188	177		365	480	
	5150	4/0		3.60	4.444		223	187		410	535	
	5200	250 kcmil	3 Elec	4.80	5		270	210		480	620	
	5220	350 kcmil		4.50	5.333		320	224		544	695	
	5240	500 kcmil		3.90	6.154		410	258		668	850	
	5260	750 kcmil		3.60	6.667		540	280		820	1,025	
	5300	15 kV aluminum, XLP, #1	2 Elec	4.40	3.636		180	153		333	435	
	5320	1/0		4	4		186	168		354	465	
	5340	2/0 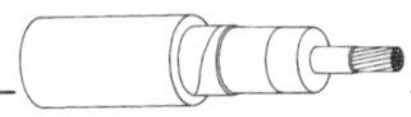		3.80	4.211		222	177		399	515	
	5360	4/0		3.60	4.444		246	187		433	560	
	5380	250 kcmil	3 Elec	4.80	5		294	210		504	650	
	5400	350 kcmil		4.50	5.333		330	224		554	710	
	5420	500 kcmil		3.90	6.154		455	258		713	900	
	5440	750 kcmil		3.60	6.667		680	280		960	1,175	
750	0010	**SPECIAL WIRES & FITTINGS**										750
	0100	Fixture TFFN 600 volt 90°C stranded, #18	1 Elec	13	.615	C.L.F.	5.40	26		31.40	46	
	0150	#16		13	.615		7.30	26		33.30	48	
	0500	Thermostat, no jacket, twisted, #18-2 conductor		8	1		8.90	42		50.90	74.50	
	0550	#18-3 conductor		7	1.143		12.35	48		60.35	87.50	
	0600	#18-4 conductor		6.50	1.231		17	51.50		68.50	98	
	0650	#18-5 conductor 		6	1.333		20.50	56		76.50	109	
	0700	#18-6 conductor		5.50	1.455		25.50	61		86.50	122	
	0750	#18-7 conductor		5	1.600		39	67		106	147	
	0800	#18-8 conductor		4.80	1.667		39	70		109	151	
	1500	Fire alarm FEP teflon 150 volt to 200°C										
	1550	#22, 1 pair	1 Elec	10	.800	C.L.F.	68.50	33.50		102	127	
	1600	2 pair		8	1		114	42		156	190	
	1650	4 pair		7	1.143		174	48		222	266	
	1700	6 pair		6	1.333		226	56		282	335	
	1750	8 pair		5.50	1.455		285	61		346	410	
	1800	10 pair		5	1.600		345	67		412	485	
	1850	#18, 1 pair		8	1		78	42		120	151	
	1900	2 pair		6.50	1.231		150	51.50		201.50	244	
	1950	4 pair		4.80	1.667		226	70		296	355	
	2000	6 pair		4	2		298	84		382	455	
	2050	8 pair		3.50	2.286		390	96		486	580	
	2100	10 pair		3	2.667		445	112		557	665	
	2460	Tray cable, type TC, copper, #16-2 conductor		9.40	.851		11.75	35.50		47.25	68	
	2464	#16-3 conductor		8.40	.952		15.05	40		55.05	78	
	2468	#16-4 conductor		7.30	1.096		18.80	46		64.80	91.50	
	2472	#16-5 conductor		6.70	1.194		22	50		72	102	
	2476	#16-7 conductor		5.50	1.455		30	61		91	127	
	2480	#16-9 conductor		5	1.600		39.50	67		106.50	147	
	2484	#16-12 conductor	2 Elec	8.80	1.818		51	76.50		127.50	174	
	2488	#16-15 conductor		7.20	2.222		65.50	93.50		159	216	
	2492	#16-19 conductor		6.60	2.424		76	102		178	241	
	2496	#16-25 conductor		6.40	2.500		106	105		211	278	
	2500	#14-2 conductor	1 Elec	9	.889		15.55	37.50		53.05	74.50	
	2520	#14-3 conductor		8	1		21	42		63	87.50	
	2540	#14-4 conductor		7	1.143		25.50	48		73.50	102	
	2560	#14-5 conductor		6.50	1.231		31.50	51.50		83	115	
	2564	#14-7 conductor		5.30	1.509		45.50	63.50		109	148	

		16120 | Conductors & Cables	CREW	DAILY OUTPUT	LABOR-HOURS	UNIT	2006 BARE COSTS MAT.	LABOR	EQUIP.	TOTAL	TOTAL INCL O&P	
750	2568	#14-9 conductor	2 Elec	9.60	1.667	C.L.F.	56	70		126	170	750
	2572	#14-12 conductor		8.60	1.860		75	78		153	203	
	2576	#14-15 conductor		7.20	2.222		92.50	93.50		186	246	
	2578	#14-19 conductor		6.20	2.581		113	108		221	291	
	2582	#14-25 conductor		5.30	3.019		151	127		278	360	
	2590	#12-2 conductor	1 Elec	8.40	.952		21.50	40		61.50	85	
	2592	#12-3 conductor		7.60	1.053		30.50	44		74.50	102	
	2594	#12-4 conductor		6.60	1.212		38	51		89	120	
	2596	#12-5 conductor		6.20	1.290		45.50	54		99.50	134	
	2598	#12-7 conductor	2 Elec	10.40	1.538		64	64.50		128.50	170	
	2602	#12-9 conductor		8.80	1.818		84	76.50		160.50	211	
	2604	#12-12 conductor		7.80	2.051		108	86		194	252	
	2606	#12-15 conductor		6.80	2.353		133	99		232	299	
	2608	#12-19 conductor		6	2.667		165	112		277	355	
	2610	#12-25 conductor	3 Elec	7.70	3.117		225	131		356	450	
	2618	#10-2 conductor	1 Elec	8	1		32	42		74	100	
	2622	#10-3 conductor	"	7.30	1.096		47	46		93	123	
	2624	#10-4 conductor	2 Elec	12.80	1.250		59	52.50		111.50	146	
	2626	#10-5 conductor		11.80	1.356		74	57		131	169	
	2628	#10-7 conductor		9.40	1.702		102	71.50		173.50	222	
	2630	#10-9 conductor		8.40	1.905		129	80		209	265	
	2632	#10-12 conductor		7.20	2.222		167	93.50		260.50	330	
	2640	300 V, copper braided shield, PVC jacket										
	2650	2 conductor #18 stranded	1 Elec	7	1.143	C.L.F.	47	48		95	126	
	2660	3 conductor #18	"	6	1.333	"	66	56		122	159	
	3000	Strain relief grip for cable										
	3050	Cord, top, #12-3	1 Elec	40	.200	Ea.	13.45	8.40		21.85	28	
	3060	#12-4		40	.200		13.45	8.40		21.85	28	
	3070	#12-5		39	.205		15.65	8.60		24.25	30.50	
	3100	#10-3		39	.205		15.65	8.60		24.25	30.50	
	3110	#10-4		38	.211		15.65	8.85		24.50	31	
	3120	#10-5		38	.211		17.05	8.85		25.90	32.50	
	3200	Bottom, #12-3		40	.200		34.50	8.40		42.90	51	
	3210	#12-4		40	.200		34.50	8.40		42.90	51	
	3220	#12-5		39	.205		35	8.60		43.60	52	
	3230	#10-3		39	.205		37	8.60		45.60	54.50	
	3300	#10-4		38	.211		40	8.85		48.85	57.50	
	3310	#10-5		38	.211		44	8.85		52.85	62	
	3400	Cable ties, standard, 4" length		190	.042		.18	1.77		1.95	2.92	
	3410	7" length		160	.050		.28	2.10		2.38	3.55	
	3420	14.5" length		90	.089		.61	3.73		4.34	6.40	
	3430	Heavy, 14.5" length		80	.100		.89	4.20		5.09	7.45	
	9000	Minimum labor/equipment charge		4	2	Job		84		84	129	
755	0010	**MODULAR FLEXIBLE WIRING SYSTEM**										755
	0050	Standard selector cable, 3 conductor, 15' long	1 Elec	40	.200	Ea.	44.50	8.40		52.90	61.50	
	0100	Conversion Module		16	.500		16.10	21		37.10	50	
	0150	Cable set, 3 conductor, 15' long		24	.333		17.25	14		31.25	40.50	
	0200	Switching assembly, 3 conductor, 10' long		32	.250		34	10.50		44.50	53.50	
800	0010	**UNDERCARPET** R260519-80										800
	0020	Power System										
	0100	Cable flat, 3 conductor, #12, w/attached bottom shield	1 Elec	982	.008	L.F.	4.42	.34		4.76	5.40	
	0200	Shield, top, steel 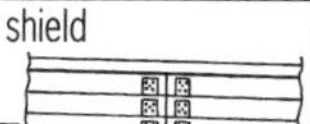		1768	.005	"	4.80	.19		4.99	5.60	
	0250	Splice, 3 conductor		48	.167	Ea.	14.65	7		21.65	27	
	0300	Top shield		96	.083		1.30	3.50		4.80	6.85	

ELECTRICAL 16

16120 | Conductors & Cables

800	Description	Crew	Daily Output	Labor-Hours	Unit	2006 Bare Costs Mat.	Labor	Equip.	Total	Total Incl O&P
0350	Tap R260519-80	1 Elec	40	.200	Ea.	18.80	8.40		27.20	33.50
0400	Insulating patch, splice, tap, & end		48	.167		46.50	7		53.50	62
0450	Fold		230	.035			1.46		1.46	2.25
0500	Top shield, tap & fold		96	.083		1.30	3.50		4.80	6.85
0700	Transition, block assembly 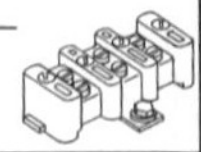		77	.104		67.50	4.36		71.86	81
0750	Receptacle frame & base		32	.250		37	10.50		47.50	56.50
0800	Cover receptacle		120	.067		3.15	2.80		5.95	7.80
0850	Cover blank		160	.050		3.70	2.10		5.80	7.30
0860	Receptacle, direct connected, single 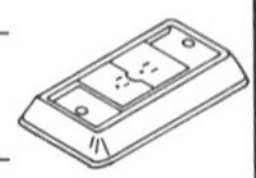		25	.320		79.50	13.45		92.95	108
0870	Dual		16	.500		130	21		151	176
0880	Combination Hi & Lo, tension		21	.381		96	16		112	130
0900	Box, floor with cover		20	.400		80.50	16.80		97.30	115
0920	Floor service w/barrier 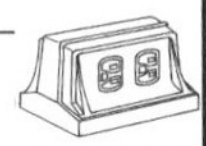		4	2		227	84		311	380
1000	Wall, surface, with cover		20	.400		53	16.80		69.80	84
1100	Wall, flush, with cover		20	.400		37	16.80		53.80	66.50
1450	Cable flat, 5 conductor #12, w/attached bottom shield		800	.010	L.F.	7.25	.42		7.67	8.60
1550	Shield, top, steel		1768	.005	"	7.20	.19		7.39	8.20
1600	Splice, 5 conductor		48	.167	Ea.	23.50	7		30.50	37
1650	Top shield		96	.083		1.30	3.50		4.80	6.85
1700	Tap		48	.167		31	7		38	45
1750	Insulating patch, splice tap, & end		83	.096		45.50	4.05		49.55	56.50
1800	Transition, block assembly		77	.104		48.50	4.36		52.86	60
1850	Box, wall, flush with cover		20	.400		49.50	16.80		66.30	80
1900	Cable flat, 4 conductor, #12		933	.009	L.F.	5.95	.36		6.31	7.10
1950	3 conductor #10		982	.008		5.15	.34		5.49	6.20
1960	4 conductor #10		933	.009		6.75	.36		7.11	8
1970	5 conductor #10		884	.009		8.25	.38		8.63	9.70
2500	Telephone System									
2510	Transition fitting wall box, surface 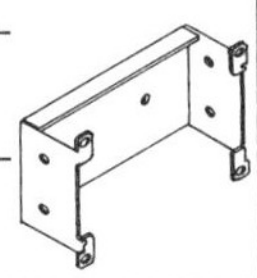	1 Elec	24	.333	Ea.	37.50	14		51.50	62.50
2520	Flush		24	.333		37.50	14		51.50	62.50
2530	Flush, for PC board		24	.333		37.50	14		51.50	62.50
2540	Floor service box		4	2		206	84		290	355
2550	Cover, surface 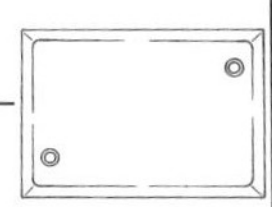					12.65			12.65	13.90
2560	Flush					12.65			12.65	13.90
2570	Flush for PC board					12.65			12.65	13.90
2700	Floor fitting w/duplex jack & cover	1 Elec	21	.381		39	16		55	67.50
2720	Low profile		53	.151		13.60	6.35		19.95	24.50
2740	Miniature w/duplex jack 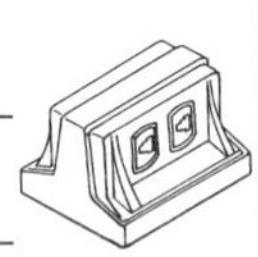		53	.151		21	6.35		27.35	33
2760	25 pair kit		21	.381		41	16		57	69.50
2780	Low profile		53	.151		13.90	6.35		20.25	25
2800	Call director kit for 5 cable		19	.421		65.50	17.70		83.20	99
2820	4 pair kit		19	.421		78	17.70		95.70	113
2840	3 pair kit		19	.421		82.50	17.70		100.20	118
2860	Comb. 25 pair & 3 cond., power		21	.381		65.50	16		81.50	96.50
2880	5 cond., power		21	.381		74.50	16		90.50	107
2900	PC board, 8-3 pair		161	.050		56.50	2.09		58.59	65.50
2920	6-4 pair		161	.050		56.50	2.09		58.59	65.50
2940	3 pair adapter		161	.050		51.50	2.09		53.59	59.50
2950	Plug		77	.104		2.40	4.36		6.76	9.35
2960	Couplers		321	.025		6.75	1.05		7.80	9.05
3000	Bottom shield for 25 pr. cable		4420	.002	L.F.	.67	.08		.75	.86
3020	4 pair		4420	.002		.32	.08		.40	.47
3040	Top shield for 25 pr. cable		4420	.002		.67	.08		.75	.86
3100	Cable assembly, double-end, 50', 25 pr.		11.80	.678	Ea.	199	28.50		227.50	263
3110	3 pair 		23.60	.339		58.50	14.25		72.75	86.50
3120	4 pair		23.60	.339		65.50	14.25		79.75	94

	16120 | Conductors & Cables		CREW	DAILY OUTPUT	LABOR-HOURS	UNIT	2006 BARE COSTS MAT.	LABOR	EQUIP.	TOTAL	TOTAL INCL O&P	
800	3140	Bulk 3 pair R260519-80	1 Elec	1473	.005	L.F.	1	.23		1.23	1.45	800
	3160	4 pair		1473	.005	″	1.25	.23		1.48	1.73	
	3500	Data System										
	3520	Cable 25 conductor w/conn. 40′, 75 ohm	1 Elec	14.50	.552	Ea.	59	23		82	101	
	3530	Single lead 		22	.364		161	15.25		176.25	201	
	3540	Dual lead		22	.364		200	15.25		215.25	244	
	3560	Shields same for 25 cond. as 25 pair tele. 										
	3570	Single & dual, none req'd.										
	3590	BNC coax connectors, Plug	1 Elec	40	.200	Ea.	8.55	8.40		16.95	22.50	
	3600	TNC coax connectors, Plug 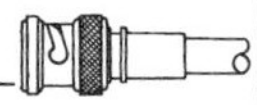	″	40	.200	″	11	8.40		19.40	25	
	3700	Cable-bulk										
	3710	Single lead	1 Elec	1473	.005	L.F.	2.15	.23		2.38	2.72	
	3720	Dual lead	″	1473	.005	″	3.10	.23		3.33	3.76	
	3730	Hand tool crimp 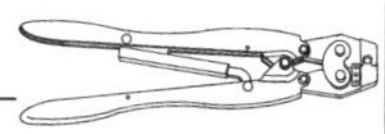				Ea.	315			315	345	
	3740	Hand tool notch				″	15.35			15.35	16.90	
	3750	Boxes & floor fitting same as telephone										
	3790	Data cable notching, 90° 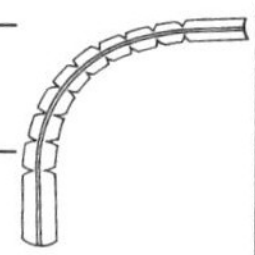	1 Elec	97	.082	Ea.		3.46		3.46	5.35	
	3800	180°		60	.133			5.60		5.60	8.65	
	8100	Drill floor		160	.050		1.70	2.10		3.80	5.10	
	8200	Marking floor		1600	.005	L.F.		.21		.21	.32	
	8300	Tape, hold down		6400	.001	″	.13	.05		.18	.22	
	8350	Tape primer, 500 ft. per can		96	.083	Ea.	26	3.50		29.50	34	
	8400	Tool, splicing				″	207			207	228	
900	0010	**WIRE** R260519-90										900
	0020	600 volt type THW, copper, solid, #14	1 Elec	13	.615	C.L.F.	4.25	26		30.25	44.50	
	0030	#12		11	.727		6.40	30.50		36.90	54	
	0040	#10 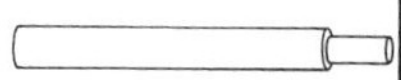		10	.800		9.70	33.50		43.20	62.50	
	0050	Stranded, #14 R260519-92		13	.615		4.85	26		30.85	45.50	
	0100	#12		11	.727		7.30	30.50		37.80	55	
	0120	#10 R260533-22		10	.800		10.95	33.50		44.45	64	
	0140	#8		8	1		19.70	42		61.70	86	
	0160	#6		6.50	1.231		32.50	51.50		84	116	
	0180	#4	2 Elec	10.60	1.509		50	63.50		113.50	153	
	0200	#3		10	1.600		62.50	67		129.50	173	
	0220	#2		9	1.778		79	74.50		153.50	202	
	0240	#1		8	2		100	84		184	239	
	0260	1/0		6.60	2.424		111	102		213	279	
	0280	2/0		5.80	2.759		130	116		246	320	
	0300	3/0		5	3.200		166	134		300	390	
	0350	4/0		4.40	3.636		206	153		359	460	
	0400	250 kcmil	3 Elec	6	4		245	168		413	530	
	0420	300 kcmil		5.70	4.211		291	177		468	590	
	0450	350 kcmil		5.40	4.444		340	187		527	665	
	0480	400 kcmil		5.10	4.706		390	198		588	735	
	0490	500 kcmil		4.80	5		480	210		690	855	
	0500	600 kcmil		3.90	6.154		580	258		838	1,025	
	0510	750 kcmil		3.30	7.273		770	305		1,075	1,325	
	0520	1000 kcmil		2.70	8.889		1,000	375		1,375	1,700	
	0530	Aluminum, stranded, #8	1 Elec	9	.889		17.90	37.50		55.40	77	
	0540	#6	″	8	1		24.50	42		66.50	91.50	
	0560	#4	2 Elec	13	1.231		30.50	51.50		82	113	
	0580	#2		10.60	1.509		41	63.50		104.50	143	
	0600	#1 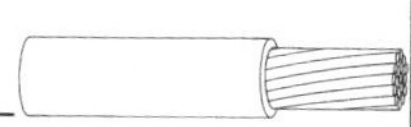		9	1.778		60.50	74.50		135	182	
	0620	1/0		8	2		72.50	84		156.50	209	
	0640	2/0		7.20	2.222		85.50	93.50		179	238	

16120 | Conductors & Cables

			CREW	DAILY OUTPUT	LABOR-HOURS	UNIT	2006 BARE COSTS MAT.	LABOR	EQUIP.	TOTAL	TOTAL INCL O&P	
900	0680	3/0 (R260519-90)	2 Elec	6.60	2.424	C.L.F.	106	102		208	274	900
	0700	4/0	↓	6.20	2.581		118	108		226	297	
	0720	250 kcmil (R260519-92)	3 Elec	8.70	2.759		144	116		260	335	
	0740	300 kcmil		8.10	2.963		199	124		323	410	
	0760	350 kcmil (R260533-22)		7.50	3.200		202	134		336	430	
	0780	400 kcmil		6.90	3.478		237	146		383	485	
	0800	500 kcmil		6	4		277	168		445	565	
	0850	600 kcmil		5.70	4.211		330	177		507	635	
	0880	700 kcmil		5.10	4.706		380	198		578	725	
	0900	750 kcmil	↓	4.80	5		385	210		595	750	
	0920	Type THWN-THHN, copper, solid, #14	1 Elec	13	.615		4.25	26		30.25	44.50	
	0940	#12		11	.727		6.40	30.50		36.90	54	
	0960	#10		10	.800		9.70	33.50		43.20	62.50	
	1000	Stranded, #14		13	.615		4.85	26		30.85	45.50	
	1200	#12		11	.727		7.30	30.50		37.80	55	
	1250	#10		10	.800		10.95	33.50		44.45	64	
	1300	#8		8	1		19.70	42		61.70	86	
	1350	#6	↓	6.50	1.231		32.50	51.50		84	116	
	1400	#4	2 Elec	10.60	1.509		50	63.50		113.50	153	
	1450	#3		10	1.600		62.50	67		129.50	173	
	1500	#2		9	1.778		79	74.50		153.50	202	
	1550	#1		8	2		100	84		184	239	
	1600	1/0 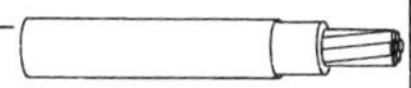		6.60	2.424		111	102		213	279	
	1650	2/0		5.80	2.759		130	116		246	320	
	1700	3/0		5	3.200		166	134		300	390	
	2000	4/0	↓	4.40	3.636		206	153		359	460	
	2200	250 kcmil	3 Elec	6	4		245	168		413	530	
	2400	300 kcmil		5.70	4.211		291	177		468	590	
	2600	350 kcmil		5.40	4.444		340	187		527	665	
	2700	400 kcmil		5.10	4.706		390	198		588	735	
	2800	500 kcmil	↓	4.80	5		485	210		695	855	
	2900	600 volt, copper type XHHW, solid, #14	1 Elec	13	.615		6.10	26		32.10	47	
	2920	#12		11	.727		9.25	30.50		39.75	57	
	2940	#10		10	.800		16	33.50		49.50	69.50	
	3000	Stranded, #14		13	.615		8.50	26		34.50	49.50	
	3020	#12		11	.727		11.50	30.50		42	59.50	
	3040	#10		10	.800		16	33.50		49.50	69.50	
	3060	#8		8	1		25	42		67	92	
	3080	#6	↓	6.50	1.231		39	51.50		90.50	123	
	3100	#4	2 Elec	10.60	1.509		60.50	63.50		124	164	
	3120	#2		9	1.778		93.50	74.50		168	218	
	3140	#1		8	2		123	84		207	264	
	3160	1/0		6.60	2.424		137	102		239	310	
	3180	2/0		5.80	2.759		155	116		271	350	
	3200	3/0 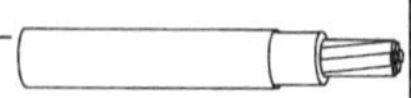		5	3.200		192	134		326	420	
	3220	4/0	↓	4.40	3.636		241	153		394	500	
	3240	250 kcmil	3 Elec	6	4		286	168		454	575	
	3260	300 kcmil		5.70	4.211		335	177		512	635	
	3280	350 kcmil		5.40	4.444		390	187		577	715	
	3300	400 kcmil		5.10	4.706		445	198		643	795	
	3320	500 kcmil		4.80	5		555	210		765	935	
	3340	600 kcmil		3.90	6.154		665	258		923	1,125	
	3360	750 kcmil		3.30	7.273		820	305		1,125	1,375	
	3380	1000 kcmil	↓	2.40	10		1,075	420		1,495	1,850	
	5000	600 volt, aluminum type XHHW, stranded, #8	1 Elec	9	.889		17.90	37.50		55.40	77	
	5020	#6	"	8	1	↓	24.50	42		66.50	91.50	

16 ELECTRICAL

16120 | Conductors & Cables		CREW	DAILY OUTPUT	LABOR-HOURS	UNIT	2006 BARE COSTS MAT.	LABOR	EQUIP.	TOTAL	TOTAL INCL O&P	
900	5040 #4 R260519-90	2 Elec	13	1.231	C.L.F.	30.50	51.50		82	113	900
	5060 #2		10.60	1.509		41	63.50		104.50	143	
	5080 #1 R260519-92		9	1.778		60	74.50		134.50	181	
	5100 1/0		8	2		72.50	84		156.50	209	
	5120 2/0 R260533-22		7.20	2.222		85.50	93.50		179	238	
	5140 3/0		6.60	2.424		106	102		208	274	
	5160 4/0		6.20	2.581		118	108		226	297	
	5180 250 kcmil 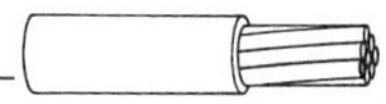	3 Elec	8.70	2.759		144	116		260	335	
	5200 300 kcmil		8.10	2.963		199	124		323	410	
	5220 350 kcmil		7.50	3.200		202	134		336	430	
	5240 400 kcmil		6.90	3.478		237	146		383	485	
	5260 500 kcmil		6	4		261	168		429	545	
	5280 600 kcmil		5.70	4.211		330	177		507	635	
	5300 700 kcmil		5.40	4.444		380	187		567	710	
	5320 750 kcmil		5.10	4.706		385	198		583	730	
	5340 1000 kcmil		3.60	6.667		575	280		855	1,050	
	5400 600 volt, copper type XLPE-USE(RHW), solid, #12	1 Elec	11	.727		13.75	30.50		44.25	62	
	5420 #10		10	.800		20	33.50		53.50	74	
	5440 Stranded, #14		13	.615		11.60	26		37.60	53	
	5460 #12		11	.727		16.10	30.50		46.60	64.50	
	5480 #10		10	.800		21.50	33.50		55	75.50	
	5500 #8		8	1		30	42		72	97.50	
	5520 #6		6.50	1.231		45.50	51.50		97	130	
	5540 #4	2 Elec	10.60	1.509		67	63.50		130.50	172	
	5560 #2		9	1.778		100	74.50		174.50	225	
	5580 #1		8	2		133	84		217	276	
	5600 1/0		6.60	2.424		153	102		255	325	
	5620 2/0		5.80	2.759		187	116		303	385	
	5640 3/0		5	3.200		233	134		367	465	
	5660 4/0 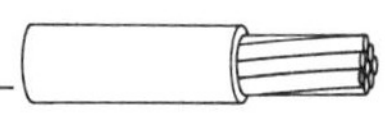		4.40	3.636		290	153		443	555	
	5680 250 kcmil	3 Elec	6	4		345	168		513	640	
	5700 300 kcmil		5.70	4.211		410	177		587	720	
	5720 350 kcmil		5.40	4.444		465	187		652	800	
	5740 400 kcmil		5.10	4.706		540	198		738	900	
	5760 500 kcmil		4.80	5		660	210		870	1,050	
	5780 600 kcmil		3.90	6.154		715	258		973	1,175	
	5800 750 kcmil		3.30	7.273		840	305		1,145	1,400	
	5820 1000 kcmil		2.70	8.889		1,075	375		1,450	1,775	
	5840 600 volt, type XLPE-USE (RHW) aluminum, stranded, #6	1 Elec	8	1		30	42		72	97.50	
	5860 #4	2 Elec	13	1.231		36	51.50		87.50	119	
	5880 #2		10.60	1.509		47.50	63.50		111	150	
	5900 #1		9	1.778		69.50	74.50		144	192	
	5920 1/0		8	2		81.50	84		165.50	219	
	5940 2/0		7.20	2.222		95.50	93.50		189	249	
	5960 3/0		6.60	2.424		118	102		220	287	
	5980 4/0		6.20	2.581		132	108		240	310	
	6000 250 kcmil	3 Elec	8.70	2.759		170	116		286	365	
	6020 300 kcmil 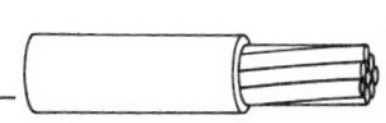		8.10	2.963		232	124		356	445	
	6040 350 kcmil		7.50	3.200		237	134		371	465	
	6060 400 kcmil		6.90	3.478		275	146		421	530	
	6080 500 kcmil		6	4		305	168		473	595	
	6100 600 kcmil		5.70	4.211		390	177		567	695	
	6110 700 kcmil		5.40	4.444		445	187		632	780	
	6120 750 kcmil		5.10	4.706		450	198		648	800	
	9000 Minimum labor/equipment charge	1 Elec	4	2	Job		84		84	129	

ELECTRICAL 16

105	16131 | Cable Trays		CREW	DAILY OUTPUT	LABOR-HOURS	UNIT	2006 BARE COSTS MAT.	LABOR	EQUIP.	TOTAL	TOTAL INCL O&P	
105	0010	**CABLE TRAY LADDER TYPE** w/ ftngs. & supports, 4″ dp., to 15′ elev.										105
	0160	Galvanized steel tray										
	0170	4″ rung spacing, 6″ wide	2 Elec	98	.163	L.F.	10.60	6.85		17.45	22	
	0180	9″ wide		92	.174		11.55	7.30		18.85	24	
	0200	12″ wide		86	.186		12.75	7.80		20.55	26	
	0400	18″ wide		82	.195		14.80	8.20		23	29	
	0600	24″ wide		78	.205		17.05	8.60		25.65	32	
	0650	30″ wide		68	.235		21.50	9.90		31.40	38.50	
	0700	36″ wide		60	.267		24	11.20		35.20	43.50	
	0800	6″ rung spacing, 6″ wide		100	.160		9.65	6.70		16.35	21	
	0850	9″ wide		94	.170		10.60	7.15		17.75	22.50	
	0860	12″ wide		88	.182		11.10	7.65		18.75	24	
	0870	18″ wide		84	.190		12.70	8		20.70	26.50	
	0880	24″ wide		80	.200		14.20	8.40		22.60	28.50	
	0890	30″ wide		70	.229		16.90	9.60		26.50	33.50	
	0900	36″ wide		64	.250		18.40	10.50		28.90	36.50	
	0910	9″ rung spacing, 6″ wide		102	.157		9.05	6.60		15.65	20	
	0920	9″ wide		98	.163		9.30	6.85		16.15	21	
	0930	12″ wide		94	.170		10.70	7.15		17.85	23	
	0940	18″ wide		90	.178		12.05	7.45		19.50	25	
	0950	24″ wide		86	.186		13.75	7.80		21.55	27	
	0960	30″ wide		80	.200		15	8.40		23.40	29.50	
	0970	36″ wide		74	.216		16	9.10		25.10	31.50	
	0980	12″ rung spacing, 6″ wide		106	.151		8.95	6.35		15.30	19.55	
	0990	9″ wide		104	.154		9.15	6.45		15.60	20	
	1000	12″ wide		100	.160		9.65	6.70		16.35	21	
	1010	18″ wide		96	.167		10.35	7		17.35	22	
	1020	24″ wide		94	.170		11.10	7.15		18.25	23.50	
	1030	30″ wide		88	.182		12.65	7.65		20.30	25.50	
	1040	36″ wide		84	.190		13.35	8		21.35	27	
	1041	18″ rung spacing, 6″ wide		108	.148		8.85	6.20		15.05	19.35	
	1042	9″ wide		106	.151		9	6.35		15.35	19.65	
	1043	12″ wide		102	.157		9.60	6.60		16.20	20.50	
	1044	18″ wide		98	.163		10.10	6.85		16.95	21.50	
	1045	24″ wide		96	.167		10.40	7		17.40	22.50	
	1046	30″ wide		90	.178		10.95	7.45		18.40	23.50	
	1047	36″ wide		86	.186		11.25	7.80		19.05	24.50	
	1050	Elbows horiz. 9″ rung spacing, 90°, 12″ radius, 6″ wide		9.60	1.667	Ea.	41.50	70		111.50	154	
	1060	9″ wide		8.40	1.905		44	80		124	171	
	1070	12″ wide		7.60	2.105		47.50	88.50		136	189	
	1080	18″ wide		6.20	2.581		62.50	108		170.50	236	
	1090	24″ wide		5.40	2.963		70	124		194	269	
	1100	30″ wide		4.80	3.333		92.50	140		232.50	320	
	1110	36″ wide		4.20	3.810		108	160		268	365	
	1120	90° 24″ radius, 6″ wide		9.20	1.739		85	73		158	207	
	1130	9″ wide		8	2		87.50	84		171.50	226	
	1140	12″ wide		7.20	2.222		90	93.50		183.50	243	
	1150	18″ wide		5.80	2.759		97.50	116		213.50	285	
	1160	24″ wide		5	3.200		105	134		239	325	
	1170	30″ wide		4.40	3.636		114	153		267	360	
	1180	36″ wide		3.80	4.211		135	177		312	420	
	1190	90° 36″ radius, 6″ wide		8.80	1.818		94	76.50		170.50	221	
	1200	9″ wide		7.60	2.105		99	88.50		187.50	245	
	1210	12″ wide		6.80	2.353		104	99		203	266	
	1220	18″ wide		5.40	2.963		119	124		243	325	
	1230	24″ wide		4.60	3.478		130	146		276	370	

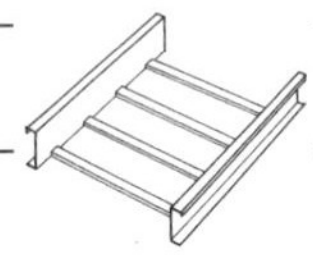

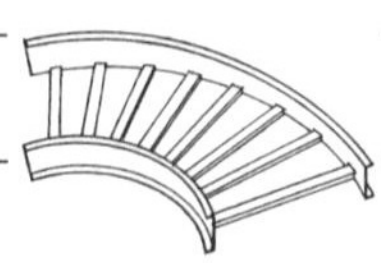

16131 | Cable Trays

	Line	Description	Crew	Daily Output	Labor-Hours	Unit	2006 Bare Costs Mat.	Labor	Equip.	Total	Total Incl O&P
105	1240	30″ wide	2 Elec	4	4	Ea.	156	168		324	430
	1250	36″ wide		3.40	4.706		170	198		368	490
	1260	45° 12″ radius, 6″ wide		13.20	1.212		32	51		83	114
	1270	9″ wide		11	1.455		33	61		94	131
	1280	12″ wide		9.60	1.667		34.50	70		104.50	146
	1290	18″ wide		7.60	2.105		36.50	88.50		125	176
	1300	24″ wide		6.20	2.581		47	108		155	219
	1310	30″ wide		5.40	2.963		60	124		184	258
	1320	36″ wide		4.60	3.478		62.50	146		208.50	294
	1330	45° 24″ radius, 6″ wide		12.80	1.250		40.50	52.50		93	126
	1340	9″ wide 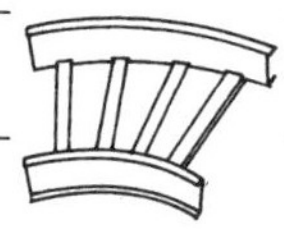		10.60	1.509		43	63.50		106.50	145
	1350	12″ wide		9.20	1.739		48	73		121	166
	1360	18″ wide		7.20	2.222		49.50	93.50		143	199
	1370	24″ wide		5.80	2.759		60	116		176	244
	1380	30″ wide		5	3.200		71.50	134		205.50	286
	1390	36″ wide		4.20	3.810		82	160		242	335
	1400	45° 36″ radius, 6″ wide		12.40	1.290		58	54		112	147
	1410	9″ wide		10.20	1.569		61.50	66		127.50	169
	1420	12″ wide		8.80	1.818		62.50	76.50		139	187
	1430	18″ wide		6.80	2.353		67.50	99		166.50	227
	1440	24″ wide		5.40	2.963		71.50	124		195.50	271
	1450	30″ wide		4.60	3.478		92.50	146		238.50	325
	1460	36″ wide	↓	3.80	4.211	↓	103	177		280	385
	1470	Elbows horizontal, 4″ rung spacing, use 9″ rung x 1.50									
	1480	6″ rung spacing use 9″ rung x 1.20									
	1490	12″ rung spacing use 9″ rung x .93									
	1500	Elbows vert. 9″ rung spacing, 90°, 12″ radius, 6″ wide	2 Elec	9.60	1.667	Ea.	56.50	70		126.50	170
	1510	9″ wide		8.40	1.905		58	80		138	187
	1520	12″ wide		7.60	2.105		60	88.50		148.50	202
	1530	18″ wide		6.20	2.581		62.50	108		170.50	236
	1540	24″ wide 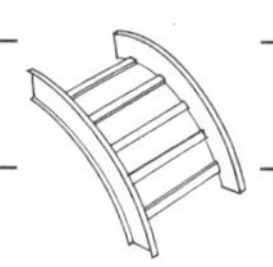		5.40	2.963		65	124		189	264
	1550	30″ wide		4.80	3.333		70	140		210	293
	1560	36″ wide		4.20	3.810		72.50	160		232.50	325
	1570	24″ radius, 6″ wide		9.20	1.739		80	73		153	201
	1580	9″ wide		8	2		82.50	84		166.50	220
	1590	12″ wide		7.20	2.222		85	93.50		178.50	238
	1600	18″ wide		5.80	2.759		90	116		206	277
	1610	24″ wide		5	3.200		95	134		229	310
	1620	30″ wide		4.40	3.636		99	153		252	345
	1630	36″ wide		3.80	4.211		111	177		288	395
	1640	36″ radius, 6″ wide		8.80	1.818		108	76.50		184.50	236
	1650	9″ wide		7.60	2.105		111	88.50		199.50	258
	1660	12″ wide		6.80	2.353		113	99		212	276
	1670	18″ wide		5.40	2.963		123	124		247	325
	1680	24″ wide 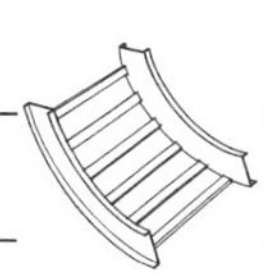		4.60	3.478		124	146		270	360
	1690	30″ wide		4	4		140	168		308	415
	1700	36″ wide	↓	3.40	4.706	↓	150	198		348	470
	1710	Elbows vertical, 4″ rung spacing, use 9″ rung x 1.25									
	1720	6″ rung spacing, use 9″ rung x 1.15									
	1730	12″ rung spacing, use 9″ rung x .90									
	1740	Tee horizontal, 9″ rung spacing, 12″ radius, 6″ wide	2 Elec	5	3.200	Ea.	87.50	134		221.50	305
	1750	9″ wide		4.60	3.478		94	146		240	330
	1760	12″ wide		4.40	3.636		99	153		252	345
	1770	18″ wide 		4	4		113	168		281	385
	1780	24″ wide		3.60	4.444		133	187		320	435
105	1790	30″ wide	↓	3.40	4.706	↓	159	198		357	480

ELECTRICAL 16

16131 | Cable Trays

105		CREW	DAILY OUTPUT	LABOR-HOURS	UNIT	2006 BARE COSTS MAT.	LABOR	EQUIP.	TOTAL	TOTAL INCL O&P
1800	36″ wide	2 Elec	3	5.333	Ea.	180	224		404	545
1810	24″ radius, 6″ wide		4.60	3.478		136	146		282	375
1820	9″ wide		4.20	3.810		143	160		303	405
1830	12″ wide		4	4		148	168		316	420
1840	18″ wide		3.60	4.444		160	187		347	465
1850	24″ wide		3.20	5		174	210		384	515
1860	30″ wide		3	5.333		189	224		413	555
1870	36″ wide		2.60	6.154		243	258		501	665
1880	36″ radius, 6″ wide		4.20	3.810		204	160		364	470
1890	9″ wide		3.80	4.211		213	177		390	505
1900	12″ wide		3.60	4.444		226	187		413	535
1910	18″ wide		3.20	5		243	210		453	590
1920	24″ wide		2.80	5.714		274	240		514	670
1930	30″ wide		2.60	6.154		299	258		557	730
1940	36″ wide		2.20	7.273		320	305		625	825
1980	Tee vertical, 9″ rung spacing, 12″ radius, 6″ wide		5.40	2.963		149	124		273	355
1990	9″ wide		5.20	3.077		150	129		279	365
2000	12″ wide		5	3.200		151	134		285	375
2010	18″ wide		4.60	3.478		153	146		299	395
2020	24″ wide		4.40	3.636		156	153		309	405
2030	30″ wide		4	4		166	168		334	440
2040	36″ wide		3.60	4.444		173	187		360	480
2050	24″ radius, 6″ wide		5	3.200		268	134		402	500
2060	9″ wide		4.80	3.333		271	140		411	515
2070	12″ wide		4.60	3.478		275	146		421	530
2080	18″ wide		4.20	3.810		284	160		444	555
2090	24″ wide		4	4		291	168		459	580
2100	30″ wide		3.60	4.444		299	187		486	620
2110	36″ wide		3.20	5		315	210		525	670
2120	36″ radius, 6″ wide		4.60	3.478		500	146		646	775
2130	9″ wide		4.40	3.636		510	153		663	795
2140	12″ wide		4.20	3.810		515	160		675	810
2150	18″ wide		3.80	4.211		520	177		697	840
2160	24″ wide		3.60	4.444		525	187		712	870
2170	30″ wide		3.20	5		545	210		755	925
2180	36″ wide		2.80	5.714		555	240		795	980
2190	Tee, 4″ rung spacing, use 9″ rung x 1.30									
2200	6″ rung spacing, use 9″ rung x 1.20									
2210	12″ rung spacing, use 9″ rung x .90									
2220	Cross horizontal, 9″ rung spacing, 12″ radius, 6″ wide	2 Elec	4	4	Ea.	108	168		276	380
2230	9″ wide		3.80	4.211		113	177		290	395
2240	12″ wide		3.60	4.444		124	187		311	425
2250	18″ wide		3.40	4.706		135	198		333	455
2260	24″ wide		3	5.333		150	224		374	510
2270	30″ wide		2.80	5.714		181	240		421	570
2280	36″ wide		2.60	6.154		201	258		459	620
2290	24″ radius, 6″ wide		3.60	4.444		210	187		397	520
2300	9″ wide		3.40	4.706		216	198		414	545
2310	12″ wide		3.20	5		223	210		433	570
2320	18″ wide		3	5.333		238	224		462	605
2330	24″ wide		2.60	6.154		253	258		511	680
2340	30″ wide		2.40	6.667		269	280		549	725
2350	36″ wide		2.20	7.273		315	305		620	815
2360	36″ radius, 6″ wide		3.20	5		293	210		503	645
2370	9″ wide		3	5.333		305	224		529	680
2380	12″ wide		2.80	5.714		320	240		560	720

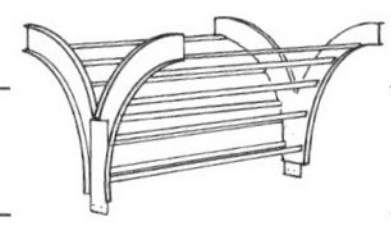

16131 | Cable Trays

105

	16131 Cable Trays	CREW	DAILY OUTPUT	LABOR-HOURS	UNIT	2006 BARE COSTS MAT.	LABOR	EQUIP.	TOTAL	TOTAL INCL O&P
2390	18" wide	2 Elec	2.60	6.154	Ea.	350	258		608	785
2400	24" wide		2.20	7.273		390	305		695	900
2410	30" wide		2	8		425	335		760	985
2420	36" wide	↓	1.80	8.889	↓	455	375		830	1,075
2430	Cross horizontal, 4" rung spacing, use 9" rung x 1.30									
2440	6" rung spacing, use 9" rung x 1.20									
2450	12" rung spacing, use 9" rung x .90									
2460	Reducer, 9" to 6" wide tray	2 Elec	13	1.231	Ea.	60	51.50		111.50	146
2470	12" to 9" wide tray		12	1.333		61.50	56		117.50	154
2480	18" to 12" wide tray		10.40	1.538		61.50	64.50		126	167
2490	24" to 18" wide tray		9	1.778		62.50	74.50		137	184
2500	30" to 24" wide tray		8	2		64	84		148	199
2510	36" to 30" wide tray		7	2.286		71.50	96		167.50	227
2511	Reducer, 18" to 6" wide tray		10.40	1.538		64	64.50		128.50	170
2512	24" to 12" wide tray		9	1.778		65	74.50		139.50	187
2513	30" to 18" wide tray		8	2		71.50	84		155.50	208
2514	30" to 12" wide tray		8	2		69	84		153	205
2515	36" to 24" wide tray		7	2.286		72.50	96		168.50	228
2516	36" to 18" wide tray		7	2.286		72.50	96		168.50	228
2517	36" to 12" wide tray		7	2.286		72.50	96		168.50	228
2520	Dropout or end plate, 6" wide		32	.500		6.70	21		27.70	40
2530	9" wide		28	.571		7.90	24		31.90	45.50
2540	12" wide		26	.615		8.65	26		34.65	49.50
2550	18" wide		22	.727		10.15	30.50		40.65	58
2560	24" wide		20	.800		12.25	33.50		45.75	65.50
2570	30" wide		18	.889		13.40	37.50		50.90	72
2580	36" wide		16	1		15.40	42		57.40	81.50
2590	Tray connector		48	.333	↓	11.90	14		25.90	34.50
3200	Aluminum tray, 4" deep, 6" rung spacing, 6" wide		134	.119	L.F.	12.80	5		17.80	22
3210	9" wide		128	.125		13.55	5.25		18.80	23
3220	12" wide		124	.129		14.30	5.40		19.70	24
3230	18" wide		114	.140		15.95	5.90		21.85	26.50
3240	24" wide		106	.151		18.55	6.35		24.90	30.50
3250	30" wide		100	.160		20.50	6.70		27.20	33
3260	36" wide		94	.170		22	7.15		29.15	35.50
3270	9" rung spacing, 6" wide		140	.114		10.40	4.80		15.20	18.85
3280	9" wide		134	.119		10.90	5		15.90	19.70
3290	12" wide		130	.123		11.10	5.15		16.25	20
3300	18" wide		122	.131		12.95	5.50		18.45	23
3310	24" wide		116	.138		15	5.80		20.80	25.50
3320	30" wide		108	.148		16.65	6.20		22.85	28
3330	36" wide		100	.160		17.60	6.70		24.30	29.50
3340	12" rung spacing, 6" wide		146	.110		11	4.60		15.60	19.20
3350	9" wide		140	.114		11.15	4.80		15.95	19.70
3360	12" wide		134	.119		11.50	5		16.50	20.50
3370	18" wide		128	.125		12.20	5.25		17.45	21.50
3380	24" wide		124	.129		13.55	5.40		18.95	23.50
3390	30" wide		114	.140		14.50	5.90		20.40	25
3400	36" wide		106	.151		15.25	6.35		21.60	26.50
3401	18" rung spacing, 6" wide		150	.107		10.95	4.48		15.43	18.95
3402	9" wide tray		144	.111		11.10	4.67		15.77	19.40
3403	12" wide tray		140	.114		11.50	4.80		16.30	20
3404	18" wide tray		134	.119		12.10	5		17.10	21
3405	24" wide tray		130	.123		13.05	5.15		18.20	22.50
3406	30" wide tray		120	.133		13.50	5.60		19.10	23.50
3407	36" wide tray	↓	110	.145	↓	14.15	6.10		20.25	25

105

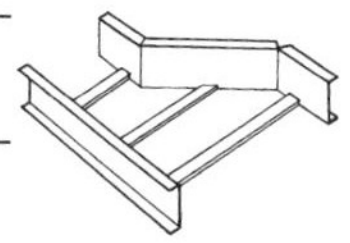

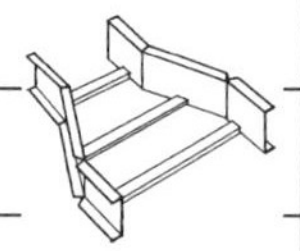

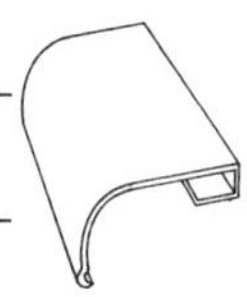

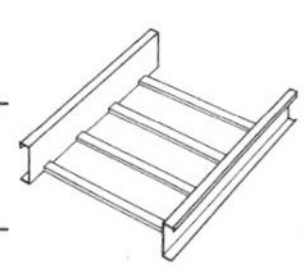

ELECTRICAL 16

16131 | Cable Trays		CREW	DAILY OUTPUT	LABOR-HOURS	UNIT	2006 BARE COSTS MAT.	LABOR	EQUIP.	TOTAL	TOTAL INCL O&P
105 3410	Elbows horiz. 9″ rung spacing, 90° 12″ radius, 6″ wide	2 Elec	9.60	1.667	Ea.	49	70		119	162
3420	9″ wide		8.40	1.905		50.50	80		130.50	179
3430	12″ wide		7.60	2.105		60	88.50		148.50	202
3440	18″ wide		6.20	2.581		67.50	108		175.50	241
3450	24″ wide		5.40	2.963		83.50	124		207.50	284
3460	30″ wide		4.80	3.333		88.50	140		228.50	315
3470	36″ wide		4.20	3.810		107	160		267	365
3480	24″ radius, 6″ wide		9.20	1.739		87.50	73		160.50	210
3490	9″ wide		8	2		91	84		175	229
3500	12″ wide		7.20	2.222		95.50	93.50		189	249
3510	18″ wide		5.80	2.759		104	116		220	292
3520	24″ wide		5	3.200		114	134		248	330
3530	30″ wide		4.40	3.636		124	153		277	370
3540	36″ wide		3.80	4.211		140	177		317	425
3550	90°, 36″ radius, 6″ wide		8.80	1.818		96	76.50		172.50	224
3560	9″ wide		7.60	2.105		98.50	88.50		187	244
3570	12″ wide		6.80	2.353		110	99		209	273
3580	18″ wide		5.40	2.963		124	124		248	330
3590	24″ wide		4.60	3.478		139	146		285	380
3600	30″ wide		4	4		154	168		322	430
3610	36″ wide		3.40	4.706		176	198		374	500
3620	45° 12″ radius, 6″ wide		13.20	1.212		32.50	51		83.50	114
3630	9″ wide		11	1.455		33.50	61		94.50	131
3640	12″ wide		9.60	1.667		35	70		105	147
3650	18″ wide		7.60	2.105		41	88.50		129.50	182
3660	24″ wide		6.20	2.581		51	108		159	223
3670	30″ wide		5.40	2.963		59	124		183	257
3680	36″ wide		4.60	3.478		61	146		207	293
3690	45° 24″ radius, 6″ wide		12.80	1.250		43	52.50		95.50	129
3700	9″ wide		10.60	1.509		49	63.50		112.50	152
3710	12″ wide		9.20	1.739		50.50	73		123.50	169
3720	18″ wide		7.20	2.222		54	93.50		147.50	204
3730	24″ wide		5.80	2.759		62.50	116		178.50	247
3740	30″ wide		5	3.200		74.50	134		208.50	289
3750	36″ wide		4.20	3.810		79	160		239	335
3760	45° 36″ radius, 6″ wide		12.40	1.290		59	54		113	148
3770	9″ wide		10.20	1.569		60	66		126	167
3780	12″ wide		8.80	1.818		61	76.50		137.50	186
3790	18″ wide		6.80	2.353		67	99		166	226
3800	24″ wide		5.40	2.963		83	124		207	283
3810	30″ wide		4.60	3.478		87.50	146		233.50	320
3820	36″ wide		3.80	4.211		96	177		273	380
3830	Elbows horizontal, 4″ rung spacing, use 9″ rung x 1.50									
3840	6″ rung spacing, use 9″ rung x 1.20									
3850	12″ rung spacing, use 9″ rung x .93									
3860	Elbows vertical 9″ rung spacing, 90° 12″ radius, 6″ wide	2 Elec	9.60	1.667	Ea.	68.50	70		138.50	183
3870	9″ wide		8.40	1.905		72	80		152	202
3880	12″ wide		7.60	2.105		73	88.50		161.50	217
3890	18″ wide		6.20	2.581		75.50	108		183.50	250
3900	24″ wide		5.40	2.963		80.50	124		204.50	281
3910	30″ wide		4.80	3.333		84	140		224	310
3920	36″ wide		4.20	3.810		86.50	160		246.50	340
3930	24″ radius, 6″ wide		9.20	1.739		74.50	73		147.50	195
3940	9″ wide		8	2		77	84		161	214
3950	12″ wide		7.20	2.222		79	93.50		172.50	231
3960	18″ wide		5.80	2.759		84	116		200	271 105

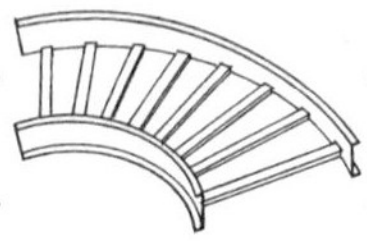

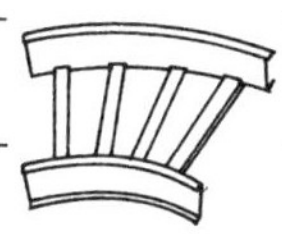

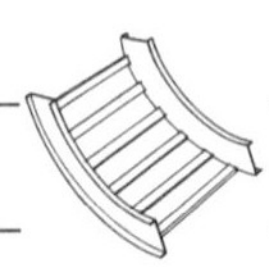

16 ELECTRICAL

16131 | Cable Trays

	Line	Description	Crew	Daily Output	Labor-Hours	Unit	Mat. (2006 Bare Costs)	Labor (2006 Bare Costs)	Equip. (2006 Bare Costs)	Total (2006 Bare Costs)	Total Incl O&P
105	3970	24″ wide	2 Elec	5	3.200	Ea.	89	134		223	305
	3980	30″ wide		4.40	3.636		93.50	153		246.50	340
	3990	36″ wide		3.80	4.211		99.50	177		276.50	380
	4000	36″ radius, 6″ wide		8.80	1.818		115	76.50		191.50	245
	4010	9″ wide		7.60	2.105		118	88.50		206.50	265
	4020	12″ wide		6.80	2.353		125	99		224	289
	4030	18″ wide		5.40	2.963		130	124		254	335
	4040	24″ wide		4.60	3.478		134	146		280	375
	4050	30″ wide		4	4		145	168		313	420
	4060	36″ wide		3.40	4.706		151	198		349	470
	4070	Elbows vertical, 4″ rung spacing, use 9″ rung x 1.25									
	4080	6″ rung spacing, use 9″ rung x 1.15									
	4090	12″ rung spacing, use 9″ rung x .90									
	4100	Tee horizontal 9″ rung spacing, 12″ radius, 6″ wide	2 Elec	5	3.200	Ea.	76	134		210	291
	4110	9″ wide		4.60	3.478		81.50	146		227.50	315
	4120	12″ wide		4.40	3.636		83	153		236	325
	4130	18″ wide		4.20	3.810		98.50	160		258.50	355
	4140	24″ wide		4	4		108	168		276	380
	4150	30″ wide		3.60	4.444		126	187		313	425
	4160	36″ wide		3.40	4.706		151	198		349	470
	4170	24″ radius, 6″ wide		4.60	3.478		130	146		276	370
	4180	9″ wide		4.20	3.810		137	160		297	395
	4190	12″ wide		4	4		140	168		308	415
	4200	18″ wide		3.80	4.211		154	177		331	440
	4210	24″ wide		3.60	4.444		168	187		355	475
	4220	30″ wide		3.20	5		182	210		392	525
	4230	36″ wide		3	5.333		281	224		505	655
	4240	36″ radius, 6″ wide		4.20	3.810		184	160		344	450
	4250	9″ wide		3.80	4.211		188	177		365	480
	4260	12″ wide		3.60	4.444		206	187		393	515
	4270	18″ wide		3.40	4.706		217	198		415	545
	4280	24″ wide		3.20	5		240	210		450	590
	4290	30″ wide		2.80	5.714		305	240		545	705
	4300	36″ wide		2.60	6.154		335	258		593	770
	4310	Tee vertical 9″ rung spacing, 12″ radius, 6″ wide		5.40	2.963		170	124		294	380
	4320	9″ wide		5.20	3.077		176	129		305	395
	4330	12″ wide		5	3.200		178	134		312	400
	4340	18″ wide		4.60	3.478		181	146		327	425
	4350	24″ wide		4.40	3.636		196	153		349	450
	4360	30″ wide		4.20	3.810		199	160		359	465
	4370	36″ wide		4	4		206	168		374	485
	4380	24″ radius, 6″ wide		5	3.200		289	134		423	525
	4390	9″ wide		4.80	3.333		293	140		433	535
	4400	12″ wide		4.60	3.478		298	146		444	550
	4410	18″ wide		4.20	3.810		305	160		465	580
	4420	24″ wide		4	4		315	168		483	605
	4430	30″ wide		3.80	4.211		320	177		497	625
	4440	36″ wide		3.60	4.444		350	187		537	675
	4450	36″ radius, 6″ wide		4.60	3.478		635	146		781	925
	4460	9″ wide		4.40	3.636		650	153		803	950
	4470	12″ wide		4.20	3.810		675	160		835	985
	4480	18″ wide		3.80	4.211		680	177		857	1,025
	4490	24″ wide		3.60	4.444		700	187		887	1,050
	4500	30″ wide		3.40	4.706		705	198		903	1,075
	4510	36″ wide		3.20	5		720	210		930	1,125
	4520	Tees, 4″ rung spacing, use 9″ rung x 1.30									

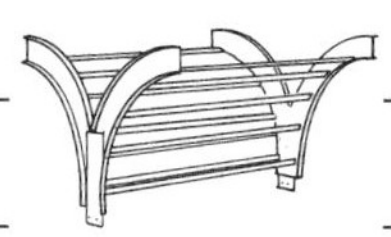

16131 | Cable Trays

			CREW	DAILY OUTPUT	LABOR-HOURS	UNIT	2006 BARE COSTS MAT.	LABOR	EQUIP.	TOTAL	TOTAL INCL O&P
105	4530	6″ rung spacing, use 9″ rung x 1.20									
	4540	12″ rung spacing, use 9″ rung x .90									
	4550	Cross horizontal 9″ rung spacing, 12″ radius, 6″ wide	2 Elec	4.40	3.636	Ea.	106	153		259	350
	4560	9″ wide		4.20	3.810		118	160		278	375
	4570	12″ wide		4	4		119	168		287	390
	4580	18″ wide		3.60	4.444		144	187		331	445
	4590	24″ wide		3.40	4.706		149	198		347	470
	4600	30″ wide		3	5.333		175	224		399	540
	4610	36″ wide		2.80	5.714		218	240		458	610
	4620	24″ radius, 6″ wide		4	4		203	168		371	480
	4630	9″ wide		3.80	4.211		210	177		387	505
	4640	12″ wide		3.60	4.444		216	187		403	525
	4650	18″ wide		3.20	5		227	210		437	575
	4660	24″ wide		3	5.333		247	224		471	615
	4670	30″ wide		2.60	6.154		264	258		522	690
	4680	36″ wide		2.40	6.667		299	280		579	760
	4690	36″ radius, 6″ wide		3.60	4.444		240	187		427	550
	4700	9″ wide		3.40	4.706		250	198		448	580
	4710	12″ wide		3.20	5		264	210		474	615
	4720	18″ wide		2.80	5.714		282	240		522	680
	4730	24″ wide		2.60	6.154		350	258		608	785
	4740	30″ wide		2.20	7.273		385	305		690	895
	4750	36″ wide		2	8		430	335		765	995
	4760	Cross horizontal, 4″ rung spacing, use 9″ rung x 1.30									
	4770	6″ rung spacing, use 9″ rung x 1.20									
	4780	12″ rung spacing, use 9″ rung x .90									
	4790	Reducer, 9″ to 6″ wide tray	2 Elec	16	1	Ea.	56.50	42		98.50	127
	4800	12″ to 9″ wide tray		14	1.143		59	48		107	139
	4810	18″ to 12″ wide tray		12.40	1.290		59	54		113	148
	4820	24″ to 18″ wide tray		10.60	1.509		60	63.50		123.50	164
	4830	30″ to 24″ wide tray		9.20	1.739		61	73		134	181
	4840	36″ to 30″ wide tray		8	2		65	84		149	201
	4841	Reducer, 18″ to 6″ wide tray		12.40	1.290		59	54		113	148
	4842	24″ to 12″ wide tray		10.60	1.509		60	63.50		123.50	164
	4843	30″ to 18″ wide tray		9.20	1.739		61	73		134	181
	4844	30″ to 12″ wide tray		9.20	1.739		61	73		134	181
	4845	36″ to 24″ wide tray		8	2		65	84		149	201
	4846	36″ to 18″ wide tray		8	2		65	84		149	201
	4847	36″ to 12″ wide tray		8	2		65	84		149	201
	4850	Dropout or end plate, 6″ wide		32	.500		7.45	21		28.45	40.50
	4860	9″ wide tray		28	.571		7.90	24		31.90	45.50
	4870	12″ wide tray		26	.615		8.65	26		34.65	49.50
	4880	18″ wide tray		22	.727		11.50	30.50		42	59.50
	4890	24″ wide tray		20	.800		13.55	33.50		47.05	67
	4900	30″ wide tray		18	.889		16.10	37.50		53.60	75
	4910	36″ wide tray		16	1		17.90	42		59.90	84
	4920	Tray connector		48	.333		10.80	14		24.80	33.50
	9200	Splice plate	1 Elec	48	.167	Pr.	11.30	7		18.30	23
	9210	Expansion joint		48	.167		13.55	7		20.55	25.50
	9220	Horizontal hinged		48	.167		11.30	7		18.30	23
	9230	Vertical hinged		48	.167		13.55	7		20.55	25.50
	9240	Ladder hanger, vertical		28	.286	Ea.	3.36	12		15.36	22
	9250	Ladder to channel connector		24	.333		39.50	14		53.50	65
	9260	Ladder to box connector 30″ wide		19	.421		39.50	17.70		57.20	70.50
	9270	24″ wide		20	.400		39.50	16.80		56.30	69.50
	9280	18″ wide		21	.381		38	16		54	66.50

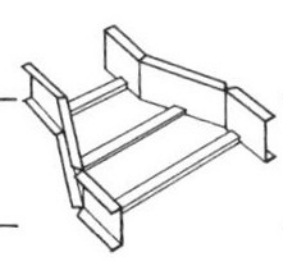

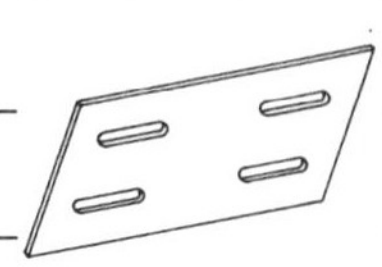

	16131 | Cable Trays		CREW	DAILY OUTPUT	LABOR-HOURS	UNIT	2006 BARE COSTS MAT.	LABOR	EQUIP.	TOTAL	TOTAL INCL O&P	
105	9290	12" wide	1 Elec	22	.364	Ea.	35.50	15.25		50.75	62.50	105
	9300	9" wide		23	.348		32	14.60		46.60	57.50	
	9310	6" wide		24	.333		32	14		46	56.50	
	9320	Ladder floor flange		24	.333		25.50	14		39.50	49.50	
	9330	Cable roller for tray 30" wide		10	.800		206	33.50		239.50	278	
	9340	24" wide		11	.727		175	30.50		205.50	239	
	9350	18" wide		12	.667		167	28		195	226	
	9360	12" wide		13	.615		131	26		157	184	
	9370	9" wide		14	.571		116	24		140	165	
	9380	6" wide		15	.533		94.50	22.50		117	139	
	9390	Pulley, single wheel		12	.667		215	28		243	280	
	9400	Triple wheel		10	.800		430	33.50		463.50	520	
	9440	Nylon cable tie, 14" long		80	.100		.44	4.20		4.64	6.95	
	9450	Ladder, hold down clamp		60	.133		7.20	5.60		12.80	16.55	
	9460	Cable clamp		60	.133		7.30	5.60		12.90	16.70	
	9470	Wall bracket, 30" wide tray		19	.421		38	17.70		55.70	69	
	9480	24" wide tray		20	.400		29.50	16.80		46.30	58.50	
	9490	18" wide tray		21	.381		26.50	16		42.50	53.50	
	9500	12" wide tray 		22	.364		24.50	15.25		39.75	50.50	
	9510	9" wide tray		23	.348		21.50	14.60		36.10	46.50	
	9520	6" wide tray		24	.333		21	14		35	45	
110	0010	**CABLE TRAY SOLID BOTTOM** w/ ftngs. & supports, 3" dp, to 15' high										110
	0220	Galvanized steel, tray, 6" wide	2 Elec	120	.133	L.F.	8.55	5.60		14.15	18.05	
	0240	12" wide		100	.160		10.95	6.70		17.65	22.50	
	0260	18" wide		70	.229		13.35	9.60		22.95	29.50	
	0280	24" wide 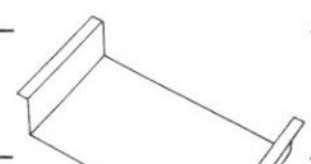		60	.267		15.65	11.20		26.85	34.50	
	0300	30" wide		50	.320		18.60	13.45		32.05	41	
	0320	36" wide		44	.364		26.50	15.25		41.75	53	
	0340	Elbow horizontal 90°, 12" radius, 6" wide		9.60	1.667	Ea.	65	70		135	180	
	0360	12" wide		6.80	2.353		73	99		172	232	
	0370	18" wide		5.40	2.963		83.50	124		207.50	284	
	0380	24" wide		4.40	3.636		101	153		254	345	
	0390	30" wide		3.80	4.211		121	177		298	405	
	0400	36" wide		3.40	4.706		138	198		336	455	
	0420	24" radius, 6" wide		9.20	1.739		92.50	73		165.50	215	
	0440	12" wide 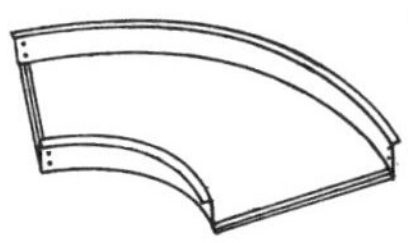		6.40	2.500		108	105		213	280	
	0450	18" wide		5	3.200		125	134		259	345	
	0460	24" wide		4	4		145	168		313	420	
	0470	30" wide		3.40	4.706		170	198		368	490	
	0480	36" wide		3	5.333		191	224		415	555	
	0500	36" radius, 6" wide		8.80	1.818		135	76.50		211.50	267	
	0520	12" wide		6	2.667		153	112		265	340	
	0530	18" wide		4.60	3.478		186	146		332	430	
	0540	24" wide		3.60	4.444		201	187		388	510	
	0550	30" wide		3	5.333		244	224		468	615	
	0560	36" wide		2.60	6.154		274	258		532	700	
	0580	Elbow vertical 90°, 12" radius, 6" wide		9.60	1.667		79	70		149	195	
	0600	12" wide		6.80	2.353		84	99		183	244	
	0610	18" wide 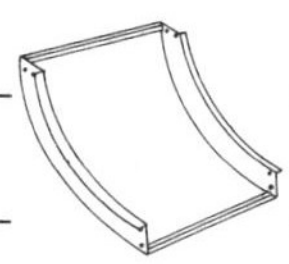		5.40	2.963		85	124		209	286	
	0620	24" wide		4.40	3.636		99	153		252	345	
	0630	30" wide		3.80	4.211		105	177		282	390	
	0640	36" wide		3.40	4.706		109	198		307	425	
	0670	24" radius, 6" wide		9.20	1.739		111	73		184	235	
	0690	12" wide		6.40	2.500		123	105		228	297	
	0700	18" wide		5	3.200		133	134		267	355	

16131 | Cable Trays

		Description	CREW	DAILY OUTPUT	LABOR-HOURS	UNIT	2006 BARE COSTS MAT.	LABOR	EQUIP.	TOTAL	TOTAL INCL O&P	
110	0710	24" wide	2 Elec	4	4	Ea.	144	168		312	415	110
	0720	30" wide		3.40	4.706		153	198		351	475	
	0730	36" wide		3	5.333		166	224		390	530	
	0750	36" radius, 6" wide		8.80	1.818		153	76.50		229.50	286	
	0770	12" wide		6.60	2.424		170	102		272	345	
	0780	18" wide		4.60	3.478		186	146		332	430	
	0790	24" wide		3.60	4.444		204	187		391	510	
	0800	30" wide		3	5.333		223	224		447	590	
	0810	36" wide		2.60	6.154		241	258		499	665	
	0840	Tee horizontal, 12" radius, 6" wide		5	3.200		91.50	134		225.50	305	
	0860	12" wide		4	4		99	168		267	370	
	0870	18" wide		3.40	4.706		118	198		316	435	
	0880	24" wide		2.80	5.714		131	240		371	515	
	0890	30" wide		2.60	6.154		150	258		408	565	
	0900	36" wide		2.20	7.273		171	305		476	660	
	0940	24" radius, 6" wide		4.60	3.478		141	146		287	380	
	0960	12" wide		3.60	4.444		165	187		352	470	
	0970	18" wide		3	5.333		184	224		408	545	
	0980	24" wide		2.40	6.667		249	280		529	705	
	0990	30" wide		2.20	7.273		270	305		575	765	
	1000	36" wide		1.80	8.889		293	375		668	895	
	1020	36" radius, 6" wide		4.20	3.810		223	160		383	490	
	1040	12" wide		3.20	5		249	210		459	600	
	1050	18" wide		2.60	6.154		270	258		528	695	
	1060	24" wide		2.20	7.273		350	305		655	850	
	1070	30" wide		2	8		380	335		715	935	
	1080	36" wide		1.60	10		390	420		810	1,075	
	1100	Tee vertical, 12" radius, 6" wide		5	3.200		150	134		284	370	
	1120	12" wide		4	4		156	168		324	430	
	1130	18" wide		3.60	4.444		159	187		346	465	
	1140	24" wide		3.40	4.706		174	198		372	495	
	1150	30" wide		3	5.333		186	224		410	550	
	1160	36" wide		2.60	6.154		193	258		451	610	
	1180	24" radius, 6" wide		4.60	3.478		221	146		367	470	
	1200	12" wide		3.60	4.444		231	187		418	540	
	1210	18" wide		3.20	5		243	210		453	590	
	1220	24" wide		3	5.333		255	224		479	625	
	1230	30" wide		2.60	6.154		280	258		538	710	
	1240	36" wide		2.20	7.273		293	305		598	790	
	1260	36" radius, 6" wide		4.20	3.810		345	160		505	625	
	1280	12" wide		3.20	5		350	210		560	710	
	1290	18" wide		2.80	5.714		380	240		620	785	
	1300	24" wide		2.60	6.154		390	258		648	830	
	1310	30" wide		2.20	7.273		455	305		760	970	
	1320	36" wide		2	8		480	335		815	1,050	
	1340	Cross horizontal, 12" radius, 6" wide		4	4		109	168		277	380	
	1360	12" wide		3.40	4.706		119	198		317	435	
	1370	18" wide		2.80	5.714		139	240		379	525	
	1380	24" wide		2.40	6.667		156	280		436	600	
	1390	30" wide		2	8		174	335		509	710	
	1400	36" wide		1.80	8.889		191	375		566	785	
	1420	24" radius, 6" wide		3.60	4.444		199	187		386	505	
	1440	12" wide		3	5.333		223	224		447	590	
	1450	18" wide		2.40	6.667		249	280		529	705	
	1460	24" wide		2	8		320	335		655	870	
110	1470	30" wide		1.80	8.889		345	375		720	955	110

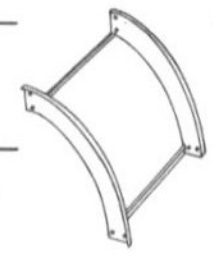

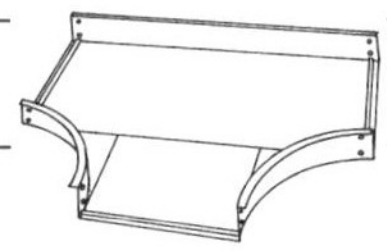

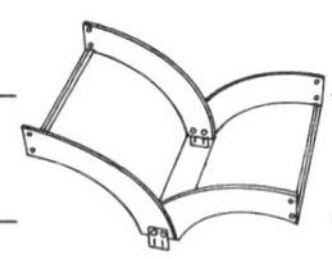

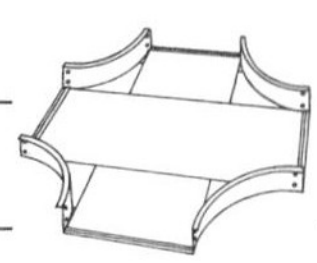

16131 | Cable Trays

110		CREW	DAILY OUTPUT	LABOR-HOURS	UNIT	2006 BARE COSTS MAT.	LABOR	EQUIP.	TOTAL	TOTAL INCL O&P
1480	36" wide	2 Elec	1.60	10	Ea.	365	420		785	1,050
1500	36" radius, 6" wide		3.20	5		325	210		535	685
1520	12" wide		2.60	6.154		360	258		618	795
1530	18" wide		2	8		390	335		725	950
1540	24" wide		1.80	8.889		460	375		835	1,075
1550	30" wide		1.60	10		500	420		920	1,200
1560	36" wide		1.40	11.429		535	480		1,015	1,325
1580	Drop out or end plate, 6" wide		32	.500		13.25	21		34.25	47
1600	12" wide		26	.615		16.50	26		42.50	58
1610	18" wide		22	.727		17.70	30.50		48.20	66.50
1620	24" wide		20	.800		21	33.50		54.50	75
1630	30" wide		18	.889		23.50	37.50		61	83
1640	36" wide		16	1		25	42		67	92
1660	Reducer, 12" to 6" wide		12	1.333		64	56		120	157
1680	18" to 12" wide		10.60	1.509		65	63.50		128.50	169
1700	18" to 6" wide		10.60	1.509		65	63.50		128.50	169
1720	24" to 18" wide		9.20	1.739		67.50	73		140.50	188
1740	24" to 12" wide		9.20	1.739		67.50	73		140.50	188
1760	30" to 24" wide		8	2		72.50	84		156.50	209
1780	30" to 18" wide		8	2		72.50	84		156.50	209
1800	30" to 12" wide		8	2		72.50	84		156.50	209
1820	36" to 30" wide		7.20	2.222		75	93.50		168.50	227
1840	36" to 24" wide		7.20	2.222		75	93.50		168.50	227
1860	36" to 18" wide		7.20	2.222		75	93.50		168.50	227
1880	36" to 12" wide		7.20	2.222		75	93.50		168.50	227
2000	Aluminum tray, 6" wide		150	.107	L.F.	9.95	4.48		14.43	17.85
2020	12" wide		130	.123		13.05	5.15		18.20	22.50
2030	18" wide		100	.160		16.35	6.70		23.05	28.50
2040	24" wide		90	.178		20.50	7.45		27.95	34
2050	30" wide		70	.229		24.50	9.60		34.10	41.50
2060	36" wide		64	.250		30.50	10.50		41	49.50
2080	Elbow horizontal 90°, 12" radius, 6" wide		9.60	1.667	Ea.	80	70		150	196
2100	12" wide		7.60	2.105		90	88.50		178.50	235
2110	18" wide		6.80	2.353		110	99		209	273
2120	24" wide		5.80	2.759		126	116		242	315
2130	30" wide		5	3.200		154	134		288	375
2140	36" wide		4.40	3.636		173	153		326	425
2160	24" radius, 6" wide		9.20	1.739		114	73		187	238
2180	12" wide		7.20	2.222		134	93.50		227.50	292
2190	18" wide		6.40	2.500		154	105		259	330
2200	24" wide		5.40	2.963		190	124		314	400
2210	30" wide		4.60	3.478		217	146		363	465
2220	36" wide		4	4		250	168		418	535
2240	36" radius, 6" wide		8.80	1.818		168	76.50		244.50	305
2260	12" wide		6.80	2.353		202	99		301	375
2270	18" wide		6	2.667		240	112		352	435
2280	24" wide		5	3.200		257	134		391	490
2290	30" wide		4.20	3.810		290	160		450	565
2300	36" wide		3.60	4.444		340	187		527	665
2320	Elbow vertical 90°, 12" radius, 6" wide		9.60	1.667		95	70		165	212
2340	12" wide		7.60	2.105		98.50	88.50		187	244
2350	18" wide		6.80	2.353		112	99		211	275
2360	24" wide		5.80	2.759		118	116		234	305
2370	30" wide		5	3.200		125	134		259	345
2380	36" wide		4.40	3.636		128	153		281	375
2400	24" radius, 6" wide		9.20	1.739		134	73		207	261

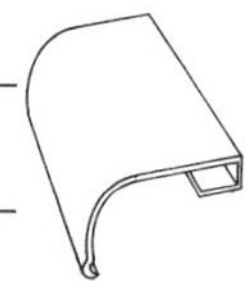

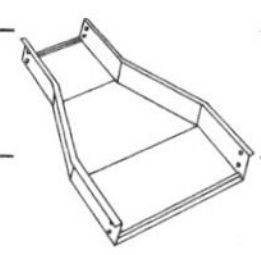

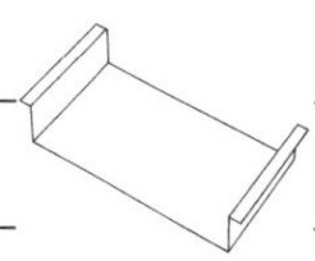

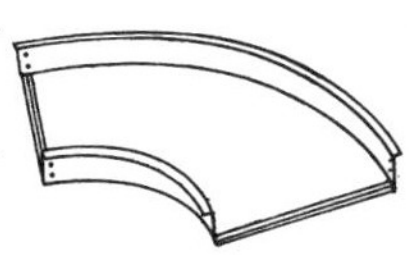

16131 | Cable Trays

			CREW	DAILY OUTPUT	LABOR-HOURS	UNIT	2006 BARE COSTS MAT.	LABOR	EQUIP.	TOTAL	TOTAL INCL O&P	
110	2420	12″ wide	2 Elec	7.20	2.222	Ea.	146	93.50		239.50	305	110
	2430	18″ wide		6.40	2.500		157	105		262	335	
	2440	24″ wide		5.40	2.963		170	124		294	380	
	2450	30″ wide		4.60	3.478		181	146		327	425	
	2460	36″ wide		4	4		199	168		367	480	
	2480	36″ radius, 6″ wide		8.80	1.818		178	76.50		254.50	315	
	2500	12″ wide 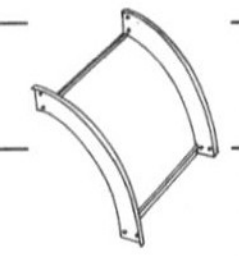		6.80	2.353		199	99		298	370	
	2510	18″ wide		6	2.667		217	112		329	410	
	2520	24″ wide		5	3.200		226	134		360	455	
	2530	30″ wide		4.20	3.810		250	160		410	520	
	2540	36″ wide		3.60	4.444		259	187		446	575	
	2560	Tee horizontal, 12″ radius, 6″ wide		5	3.200		124	134		258	345	
	2580	12″ wide		4.40	3.636		150	153		303	400	
	2590	18″ wide		4	4		168	168		336	445	
	2600	24″ wide		3.60	4.444		188	187		375	495	
	2610	30″ wide		3	5.333		217	224		441	585	
	2620	36″ wide		2.40	6.667		250	280		530	705	
	2640	24″ radius, 6″ wide		4.60	3.478		204	146		350	450	
	2660	12″ wide 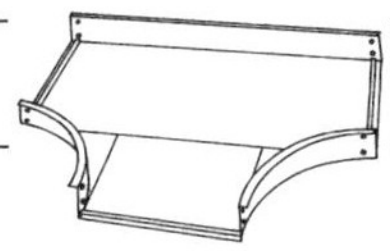		4	4		229	168		397	510	
	2670	18″ wide		3.60	4.444		259	187		446	575	
	2680	24″ wide		3	5.333		320	224		544	695	
	2690	30″ wide		2.40	6.667		360	280		640	830	
	2700	36″ wide		2.20	7.273		405	305		710	915	
	2720	36″ radius, 6″ wide		4.20	3.810		330	160		490	610	
	2740	12″ wide		3.60	4.444		360	187		547	690	
	2750	18″ wide		3.20	5		420	210		630	785	
	2760	24″ wide		2.60	6.154		485	258		743	935	
	2770	30″ wide		2	8		535	335		870	1,100	
	2780	36″ wide		1.80	8.889		600	375		975	1,225	
	2800	Tee vertical, 12″ radius, 6″ wide		5	3.200		176	134		310	400	
	2820	12″ wide		4.40	3.636		180	153		333	435	
	2830	18″ wide		4.20	3.810		187	160		347	450	
	2840	24″ wide		4	4		199	168		367	480	
	2850	30″ wide		3.60	4.444		217	187		404	525	
	2860	36″ wide		3	5.333		229	224		453	595	
	2880	24″ radius, 6″ wide		4.60	3.478		259	146		405	510	
	2900	12″ wide 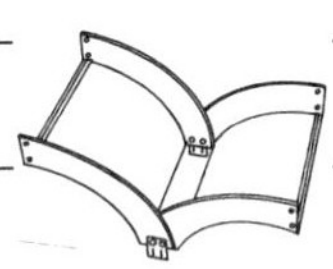		4	4		275	168		443	560	
	2910	18″ wide		3.80	4.211		293	177		470	590	
	2920	24″ wide		3.60	4.444		320	187		507	640	
	2930	30″ wide		3.20	5		350	210		560	710	
	2940	36″ wide		2.60	6.154		370	258		628	805	
	2960	36″ radius, 6″ wide		4.20	3.810		405	160		565	690	
	2980	12″ wide		3.40	4.706		425	198		623	770	
	2990	18″ wide		3.40	4.706		450	198		648	800	
	3000	24″ wide		3.20	5		460	210		670	835	
	3010	30″ wide		2.80	5.714		500	240		740	920	
	3020	36″ wide		2.20	7.273		550	305		855	1,075	
	3040	Cross horizontal, 12″ radius, 6″ wide		4.40	3.636		157	153		310	410	
	3060	12″ wide		4	4		178	168		346	455	
	3070	18″ wide		3.40	4.706		202	198		400	525	
	3080	24″ wide		2.80	5.714		229	240		469	620	
	3090	30″ wide		2.60	6.154		259	258		517	685	
	3100	36″ wide 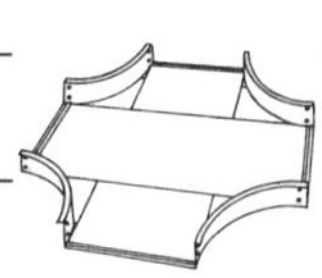		2.20	7.273		287	305		592	785	
	3120	24″ radius, 6″ wide		4	4		275	168		443	560	
	3140	12″ wide		3.60	4.444		320	187		507	640	
	3150	18″ wide	↓	3	5.333	↓	345	224		569	725	

16131 | Cable Trays

		16131 Cable Trays	CREW	DAILY OUTPUT	LABOR-HOURS	UNIT	2006 BARE COSTS MAT.	LABOR	EQUIP.	TOTAL	TOTAL INCL O&P	
110	3160	24″ wide	2 Elec	2.40	6.667	Ea.	420	280		700	890	110
	3170	30″ wide		2.20	7.273		460	305		765	980	
	3180	36″ wide		1.80	8.889		480	375		855	1,100	
	3200	36″ radius, 6″ wide		3.60	4.444		480	187		667	820	
	3220	12″ wide		3.20	5		505	210		715	880	
	3230	18″ wide		2.60	6.154		560	258		818	1,025	
	3240	24″ wide		2	8		610	335		945	1,200	
	3250	30″ wide 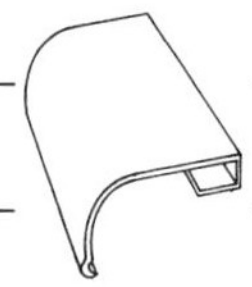		1.80	8.889		660	375		1,035	1,300	
	3260	36″ wide		1.60	10		775	420		1,195	1,500	
	3280	Dropout, or end plate, 6″ wide		32	.500		15.85	21		36.85	50	
	3300	12″ wide		26	.615		17.90	26		43.90	59.50	
	3310	18″ wide		22	.727		21.50	30.50		52	70.50	
	3320	24″ wide		20	.800		23.50	33.50		57	78	
	3330	30″ wide		18	.889		27.50	37.50		65	87.50	
	3340	36″ wide		16	1		32	42		74	99.50	
	3380	Reducer, 12″ to 6″ wide		14	1.143		77.50	48		125.50	159	
	3400	18″ to 12″ wide		12	1.333		80	56		136	175	
	3420	18″ to 6″ wide		12	1.333		80	56		136	175	
	3440	24″ to 18″ wide 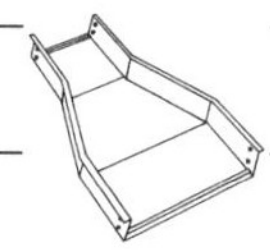		10.60	1.509		85	63.50		148.50	191	
	3460	24″ to 12″ wide		10.60	1.509		85	63.50		148.50	191	
	3480	30″ to 24″ wide		9.20	1.739		92.50	73		165.50	215	
	3500	30″ to 18″ wide		9.20	1.739		92.50	73		165.50	215	
	3520	30″ to 12″ wide		9.20	1.739		92.50	73		165.50	215	
	3540	36″ to 30″ wide		8	2		97	84		181	236	
	3560	36″ to 24″ wide		8	2		98.50	84		182.50	237	
	3580	36″ to 18″ wide		8	2		98.50	84		182.50	237	
	3600	36″ to 12″ wide	↓	8	2	↓	101	84		185	240	
120	0010	**CABLE TRAY TROUGH** vented, w/ ftngs & supports, 6″ dp, to 15′ high										120
	0200	Galvanized steel, tray, 6″ wide	2 Elec	90	.178	L.F.	11.25	7.45		18.70	24	
	0240	12″ wide		80	.200		13.75	8.40		22.15	28	
	0260	18″ wide		70	.229		16.60	9.60		26.20	33	
	0280	24″ wide 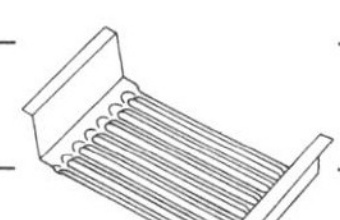		60	.267		19	11.20		30.20	38.50	
	0300	30″ wide		50	.320		27	13.45		40.45	50.50	
	0320	36″ wide		40	.400	↓	30.50	16.80		47.30	59.50	
	0340	Elbow horizontal 90°, 12″ radius, 6″ wide		7.60	2.105	Ea.	69.50	88.50		158	213	
	0360	12″ wide		5.60	2.857		82.50	120		202.50	276	
	0370	18″ wide		4.40	3.636		95	153		248	340	
	0380	24″ wide		3.60	4.444		118	187		305	415	
	0390	30″ wide		3.20	5		133	210		343	470	
	0400	36″ wide		2.80	5.714		154	240		394	540	
	0420	24″ radius, 6″ wide		7.20	2.222		103	93.50		196.50	257	
	0440	12″ wide 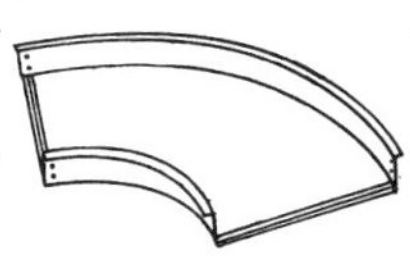		5.20	3.077		119	129		248	330	
	0450	18″ wide		4	4		143	168		311	415	
	0460	24″ wide		3.20	5		161	210		371	500	
	0470	30″ wide		2.80	5.714		186	240		426	575	
	0480	36″ wide		2.40	6.667		216	280		496	670	
	0500	36″ radius, 6″ wide		6.80	2.353		154	99		253	320	
	0520	12″ wide		4.80	3.333		175	140		315	410	
	0530	18″ wide		3.60	4.444		210	187		397	520	
	0540	24″ wide		2.80	5.714		223	240		463	615	
	0550	30″ wide		2.40	6.667		250	280		530	705	
	0560	36″ wide		2	8		279	335		614	825	
	0580	Elbow vertical 90°, 12″ radius, 6″ wide		7.60	2.105		90	88.50		178.50	235	
	0600	12″ wide		5.60	2.857		100	120		220	295	
	0610	18″ wide	↓	4.40	3.636	↓	101	153		254	345	

16131 | Cable Trays

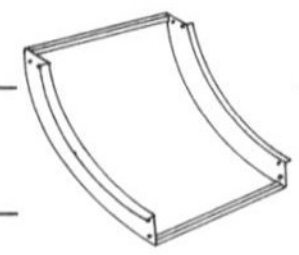

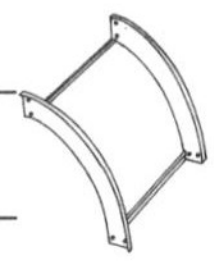

	16131 Cable Trays	CREW	DAILY OUTPUT	LABOR-HOURS	UNIT	2006 BARE COSTS MAT.	LABOR	EQUIP.	TOTAL	TOTAL INCL O&P
120	0620 24″ wide	2 Elec	3.60	4.444	Ea.	116	187		303	415
	0630 30″ wide		3.20	5		119	210		329	455
	0640 36″ wide		2.80	5.714		125	240		365	510
	0660 24″ radius, 6″ wide		7.20	2.222		129	93.50		222.50	286
	0680 12″ wide		5.20	3.077		136	129		265	350
	0690 18″ wide		4	4		151	168		319	425
	0700 24″ wide		3.20	5		156	210		366	495
	0710 30″ wide		2.80	5.714		175	240		415	565
	0720 36″ wide		2.40	6.667		184	280		464	630
	0740 36″ radius, 6″ wide		6.80	2.353		174	99		273	345
	0760 12″ wide		4.80	3.333		185	140		325	420
	0770 18″ wide		3.60	4.444		205	187		392	515
	0780 24″ wide		2.80	5.714		220	240		460	610
	0790 30″ wide		2.40	6.667		225	280		505	680
	0800 36″ wide		2	8		250	335		585	795
	0820 Tee horizontal, 12″ radius, 6″ wide		4	4		105	168		273	375
	0840 12″ wide		3.20	5		118	210		328	455
	0850 18″ wide		2.80	5.714		133	240		373	515
	0860 24″ wide		2.40	6.667		154	280		434	600
	0870 30″ wide		2.20	7.273		175	305		480	665
	0880 36″ wide		2	8		193	335		528	730
	0900 24″ radius, 6″ wide		3.60	4.444		174	187		361	480
	0920 12″ wide		2.80	5.714		190	240		430	580
	0930 18″ wide		2.40	6.667		213	280		493	665
	0940 24″ wide		2	8		275	335		610	825
	0950 30″ wide		1.80	8.889		299	375		674	905
	0960 36″ wide		1.60	10		330	420		750	1,000
	0980 36″ radius, 6″ wide		3.20	5		250	210		460	600
	1000 12″ wide		2.40	6.667		285	280		565	745
	1010 18″ wide		2	8		310	335		645	860
	1020 24″ wide		1.60	10		375	420		795	1,050
	1030 30″ wide		1.40	11.429		400	480		880	1,175
	1040 36″ wide		1.20	13.333		470	560		1,030	1,375
	1060 Tee vertical, 12″ radius, 6″ wide		4	4		175	168		343	450
	1080 12″ wide		3.20	5		178	210		388	520
	1090 18″ wide		3	5.333		183	224		407	545
	1100 24″ wide		2.80	5.714		193	240		433	580
	1110 30″ wide		2.60	6.154		204	258		462	625
	1120 36″ wide		2.20	7.273		208	305		513	700
	1140 24″ radius, 6″ wide		3.60	4.444		235	187		422	545
	1160 12″ wide		2.80	5.714		248	240		488	640
	1170 18″ wide		2.60	6.154		259	258		517	685
	1180 24″ wide		2.40	6.667		279	280		559	735
	1190 30″ wide		2.20	7.273		293	305		598	790
	1200 36″ wide		1.80	8.889		320	375		695	925
	1220 36″ radius, 6″ wide		3.20	5		375	210		585	740
	1240 12″ wide		2.40	6.667		385	280		665	850
	1250 18″ wide		2.20	7.273		400	305		705	905
	1260 24″ wide		2	8		420	335		755	980
	1270 30″ wide		1.80	8.889		490	375		865	1,125
	1280 36″ wide		1.40	11.429		495	480		975	1,275
	1300 Cross horizontal, 12″ radius, 6″ wide		3.20	5		135	210		345	475
	1320 12″ wide		2.80	5.714		136	240		376	520
	1330 18″ wide		2.40	6.667		153	280		433	600
	1340 24″ wide		2	8		160	335		495	695
	1350 30″ wide	↓	1.80	8.889	↓	180	375		555	775

16131 | Cable Trays

			CREW	DAILY OUTPUT	LABOR-HOURS	UNIT	2006 BARE COSTS MAT.	LABOR	EQUIP.	TOTAL	TOTAL INCL O&P	
120	1360	36" wide	2 Elec	1.60	10	Ea.	201	420		621	865	120
	1380	24" radius, 6" wide		2.80	5.714		213	240		453	605	
	1400	12" wide		2.40	6.667		223	280		503	675	
	1410	18" wide		2	8		245	335		580	790	
	1420	24" wide		1.60	10		310	420		730	985	
	1430	30" wide		1.40	11.429		345	480		825	1,125	
	1440	36" wide		1.20	13.333		365	560		925	1,275	
	1460	36" radius, 6" wide		2.40	6.667		365	280		645	830	
	1480	12" wide		2	8		375	335		710	935	
	1490	18" wide		1.60	10		385	420		805	1,075	
	1500	24" wide		1.20	13.333		460	560		1,020	1,375	
	1510	30" wide		1	16		510	670		1,180	1,575	
	1520	36" wide		.80	20		545	840		1,385	1,900	
	1540	Dropout or end plate, 6" wide		26	.615		15.65	26		41.65	57	
	1560	12" wide		22	.727		19.15	30.50		49.65	68	
	1580	18" wide		20	.800		21	33.50		54.50	75	
	1600	24" wide		18	.889		24	37.50		61.50	84	
	1620	30" wide		16	1		26	42		68	93	
	1640	36" wide		13.40	1.194		28.50	50		78.50	109	
	1660	Reducer, 12" to 6" wide		9.40	1.702		65.50	71.50		137	182	
	1680	18" to 12" wide		8.40	1.905		68.50	80		148.50	198	
	1700	18" to 6" wide		8.40	1.905		68.50	80		148.50	198	
	1720	24" to 18" wide		7.20	2.222		73	93.50		166.50	224	
	1740	24" to 12" wide		7.20	2.222		73	93.50		166.50	224	
	1760	30" to 24" wide		6.40	2.500		75.50	105		180.50	245	
	1780	30" to 18" wide		6.40	2.500		75.50	105		180.50	245	
	1800	30" to 12" wide		6.40	2.500		75.50	105		180.50	245	
	1820	36" to 30" wide		5.80	2.759		80.50	116		196.50	267	
	1840	36" to 24" wide		5.80	2.759		80.50	116		196.50	267	
	1860	36" to 18" wide		5.80	2.759		82	116		198	268	
	1880	36" to 12" wide		5.80	2.759		82	116		198	268	
	2000	Aluminum, tray, vented, 6" wide		120	.133	L.F.	13.05	5.60		18.65	23	
	2010	9" wide		110	.145		14.75	6.10		20.85	25.50	
	2020	12" wide		100	.160		16.45	6.70		23.15	28.50	
	2030	18" wide		90	.178		19.70	7.45		27.15	33	
	2040	24" wide		80	.200		23	8.40		31.40	38.50	
	2050	30" wide		70	.229		31	9.60		40.60	49	
	2060	36" wide		60	.267		34	11.20		45.20	55	
	2080	Elbow horiz. 90°, 12" radius, 6" wide		7.60	2.105	Ea.	86	88.50		174.50	231	
	2090	9" wide		7	2.286		93.50	96		189.50	251	
	2100	12" wide		6.20	2.581		99.50	108		207.50	277	
	2110	18" wide		5.60	2.857		116	120		236	315	
	2120	24" wide		4.60	3.478		134	146		280	375	
	2130	30" wide		4	4		166	168		334	440	
	2140	36" wide		3.60	4.444		181	187		368	485	
	2160	24" radius, 6" wide		7.20	2.222		125	93.50		218.50	281	
	2180	12" wide		5.80	2.759		146	116		262	340	
	2190	18" wide		5.20	3.077		168	129		297	385	
	2200	24" wide		4.20	3.810		192	160		352	455	
	2210	30" wide		3.60	4.444		220	187		407	530	
	2220	36" wide		3.20	5		238	210		448	585	
	2240	36" radius, 6" wide		6.80	2.353		176	99		275	345	
	2260	12" wide		5.40	2.963		200	124		324	410	
	2270	18" wide		4.80	3.333		228	140		368	465	
	2280	24" wide		3.80	4.211		250	177		427	545	
	2290	30" wide		3.40	4.706		293	198		491	625	

ELECTRICAL 16

16131 | Cable Trays

			CREW	DAILY OUTPUT	LABOR-HOURS	UNIT	2006 BARE COSTS MAT.	LABOR	EQUIP.	TOTAL	TOTAL INCL O&P
120	2300	36″ wide	2 Elec	2.80	5.714	Ea.	320	240		560	720
	2320	Elbow vertical 90°, 12″ radius, 6″ wide		7.60	2.105		104	88.50		192.50	251
	2330	9″ wide		7	2.286		112	96		208	271
	2340	12″ wide		6.20	2.581		113	108		221	291
	2350	18″ wide		5.60	2.857		116	120		236	315
	2360	24″ wide		4.60	3.478		128	146		274	365
	2370	30″ wide		4	4		134	168		302	405
	2380	36″ wide		3.60	4.444		136	187		323	435
	2400	24″ radius, 6″ wide		7.20	2.222		142	93.50		235.50	300
	2420	12″ wide		5.80	2.759		154	116		270	345
	2430	18″ wide		5.20	3.077		167	129		296	380
	2440	24″ wide		4.20	3.810		169	160		329	430
	2450	30″ wide		3.60	4.444		179	187		366	485
	2460	36″ wide		3.20	5		187	210		397	530
	2480	36″ radius, 6″ wide		6.80	2.353		175	99		274	345
	2500	12″ wide		5.40	2.963		187	124		311	400
	2510	18″ wide		4.80	3.333		210	140		350	445
	2520	24″ wide		3.80	4.211		227	177		404	520
	2530	30″ wide		3.40	4.706		250	198		448	580
	2540	36″ wide		2.80	5.714		257	240		497	650
	2560	Tee horizontal, 12″ radius, 6″ wide		4	4		131	168		299	405
	2570	9″ wide		3.80	4.211		134	177		311	420
	2580	12″ wide		3.60	4.444		146	187		333	450
	2590	18″ wide		3.20	5		169	210		379	510
	2600	24″ wide		2.80	5.714		194	240		434	585
	2610	30″ wide		2.40	6.667		209	280		489	660
	2620	36″ wide		2.20	7.273		238	305		543	730
	2640	24″ radius, 6″ wide		3.60	4.444		199	187		386	505
	2660	12″ wide		3.20	5		227	210		437	575
	2670	18″ wide		2.80	5.714		250	240		490	645
	2680	24″ wide		2.40	6.667		310	280		590	775
	2690	30″ wide		2	8		335	335		670	890
	2700	36″ wide		1.80	8.889		395	375		770	1,000
	2720	36″ radius, 6″ wide		3.20	5		320	210		530	675
	2740	12″ wide		2.80	5.714		370	240		610	775
	2750	18″ wide		2.40	6.667		420	280		700	890
	2760	24″ wide		2	8		485	335		820	1,050
	2770	30″ wide		1.60	10		520	420		940	1,225
	2780	36″ wide		1.40	11.429		600	480		1,080	1,400
	2800	Tee vertical, 12″ radius, 6″ wide		4	4		180	168		348	455
	2810	9″ wide		3.80	4.211		181	177		358	470
	2820	12″ wide		3.60	4.444		194	187		381	500
	2830	18″ wide		3.40	4.706		196	198		394	520
	2840	24″ wide		3.20	5		200	210		410	545
	2850	30″ wide		3	5.333		210	224		434	575
	2860	36″ wide		2.60	6.154		218	258		476	640
	2880	24″ radius, 6″ wide		3.60	4.444		254	187		441	570
	2900	12″ wide		3.20	5		275	210		485	625
	2910	18″ wide		3	5.333		293	224		517	665
	2920	24″ wide		2.80	5.714		310	240		550	715
	2930	30″ wide		2.60	6.154		335	258		593	770
	2940	36″ wide		2.20	7.273		345	305		650	850
	2960	36″ radius, 6″ wide		3.20	5		395	210		605	760
	2980	12″ wide		2.80	5.714		420	240		660	830
	2990	18″ wide		2.60	6.154		430	258		688	875
120	3000	24″ wide	↓	2.40	6.667	↓	460	280		740	940

	16131 | Cable Trays		CREW	DAILY OUTPUT	LABOR-HOURS	UNIT	2006 BARE COSTS MAT.	LABOR	EQUIP.	TOTAL	TOTAL INCL O&P	
120	3010	30″ wide	2 Elec	2.20	7.273	Ea.	500	305		805	1,025	120
	3020	36″ wide		1.80	8.889		525	375		900	1,150	
	3040	Cross horizontal, 12″ radius, 6″ wide		3.60	4.444		163	187		350	470	
	3050	9″ wide		3.40	4.706		175	198		373	500	
	3060	12″ wide		3.20	5		180	210		390	525	
	3070	18″ wide		2.80	5.714		188	240		428	575	
	3080	24″ wide		2.40	6.667		210	280		490	660	
	3090	30″ wide		2.20	7.273		254	305		559	750	
	3100	36″ wide		1.80	8.889		305	375		680	915	
	3120	24″ radius, 6″ wide		3.20	5		293	210		503	645	
	3140	12″ wide		2.80	5.714		320	240		560	720	
	3150	18″ wide		2.40	6.667		345	280		625	810	
	3160	24″ wide		2	8		400	335		735	960	
	3170	30″ wide		1.80	8.889		430	375		805	1,050	
	3180	36″ wide		1.40	11.429		485	480		965	1,275	
	3200	36″ radius, 6″ wide		2.80	5.714		485	240		725	905	
	3220	12″ wide		2.40	6.667		510	280		790	990	
	3230	18″ wide		2	8		555	335		890	1,125	
	3240	24″ wide		1.60	10		645	420		1,065	1,350	
	3250	30″ wide		1.40	11.429		700	480		1,180	1,500	
	3260	36″ wide		1.20	13.333		810	560		1,370	1,750	
	3280	Dropout, or end plate, 6″ wide		26	.615		16.30	26		42.30	58	
	3300	12″ wide		22	.727		19.30	30.50		49.80	68.50	
	3310	18″ wide		20	.800		23.50	33.50		57	77.50	
	3320	24″ wide		18	.889		28	37.50		65.50	88.50	
	3330	30″ wide		16	1		30	42		72	97.50	
	3340	36″ wide		14	1.143		34.50	48		82.50	112	
	3370	Reducer, 9″ to 6″ wide		12	1.333		80	56		136	175	
	3380	12″ to 6″ wide		11.40	1.404		84	59		143	184	
	3390	12″ to 9″ wide		11.40	1.404		84	59		143	184	
	3400	18″ to 12″ wide		9.60	1.667		90	70		160	207	
	3420	18″ to 6″ wide		9.60	1.667		90	70		160	207	
	3430	18″ to 9″ wide		9.60	1.667		90	70		160	207	
	3440	24″ to 18″ wide		8.40	1.905		96	80		176	229	
	3460	24″ to 12″ wide		8.40	1.905		96	80		176	229	
	3470	24″ to 9″ wide		8.40	1.905		97	80		177	230	
	3475	24″ to 6″ wide		8.40	1.905		97	80		177	230	
	3480	30″ to 24″ wide		7.20	2.222		98.50	93.50		192	252	
	3500	30″ to 18″ wide		7.20	2.222		98.50	93.50		192	252	
	3520	30″ to 12″ wide		7.20	2.222		101	93.50		194.50	255	
	3540	36″ to 30″ wide		6.40	2.500		102	105		207	274	
	3560	36″ to 24″ wide		6.40	2.500		102	105		207	274	
	3580	36″ to 18″ wide		6.40	2.500		102	105		207	274	
	3600	36″ to 12″ wide		6.40	2.500		102	105		207	274	
	3610	Elbow horizontal 60°, 12″ radius, 6″ wide		7.80	2.051		69.50	86		155.50	210	
	3620	9″ wide		7.20	2.222		76	93.50		169.50	228	
	3630	12″ wide		6.40	2.500		84	105		189	255	
	3640	18″ wide		5.80	2.759		90	116		206	277	
	3650	24″ wide		4.80	3.333		110	140		250	335	
	3680	Elbow horizontal 45°, 12″ radius, 6″ wide		8	2		60	84		144	195	
	3690	9″ wide		7.40	2.162		63.50	91		154.50	210	
	3700	12″ wide		6.60	2.424		67	102		169	231	
	3710	18″ wide		6	2.667		75.50	112		187.50	256	
	3720	24″ wide		5	3.200		87.50	134		221.50	305	
	3750	Elbow horizontal, 30° 12″ radius, 6″ wide		8.20	1.951		52.50	82		134.50	184	
	3760	9″ wide	↓	7.60	2.105	↓	55	88.50		143.50	197	

	16131 | Cable Trays		CREW	DAILY OUTPUT	LABOR-HOURS	UNIT	2006 BARE COSTS MAT.	LABOR	EQUIP.	TOTAL	TOTAL INCL O&P	
120	3770	12″ wide	2 Elec	6.80	2.353	Ea.	59	99		158	217	120
	3780	18″ wide		6.20	2.581		63.50	108		171.50	237	
	3790	24″ wide		5.20	3.077		69.50	129		198.50	276	
	3820	Elbow vertical 60° in/outside, 12″ radius, 6″ wide		7.80	2.051		86.50	86		172.50	228	
	3830	9″ wide		7.20	2.222		87.50	93.50		181	241	
	3840	12″ wide		6.40	2.500		89	105		194	260	
	3850	18″ wide		5.80	2.759		92.50	116		208.50	280	
	3860	24″ wide		4.80	3.333		96	140		236	320	
	3890	Elbow vertical 45° in/outside, 12″ radius, 6″ wide		8	2		69.50	84		153.50	206	
	3900	9″ wide		7.40	2.162		73.50	91		164.50	221	
	3910	12″ wide		6.60	2.424		76	102		178	241	
	3920	18″ wide		6	2.667		78.50	112		190.50	260	
	3930	24″ wide		5	3.200		86	134		220	300	
	3960	Elbow vertical 30° in/outside, 12″ radius, 6″ wide		8.20	1.951		60	82		142	192	
	3970	9″ wide		7.60	2.105		63.50	88.50		152	206	
	3980	12″ wide		6.80	2.353		65	99		164	224	
	3990	18″ wide		6.20	2.581		67.50	108		175.50	241	
	4000	24″ wide		5.20	3.077		68.50	129		197.50	274	
	4250	Reducer, left or right hand, 24″ to 18″ wide		8.40	1.905		91	80		171	223	
	4260	24″ to 12″ wide		8.40	1.905		91	80		171	223	
	4270	24″ to 9″ wide		8.40	1.905		91	80		171	223	
	4280	24″ to 6″ wide		8.40	1.905		92.50	80		172.50	225	
	4290	18″ to 12″ wide		9.60	1.667		85	70		155	202	
	4300	18″ to 9″ wide		9.60	1.667		85	70		155	202	
	4310	18″ to 6″ wide		9.60	1.667		85	70		155	202	
	4320	12″ to 9″ wide		11.40	1.404		81.50	59		140.50	181	
	4330	12″ to 6″ wide		11.40	1.404		81.50	59		140.50	181	
	4340	9″ to 6″ wide		12	1.333		79	56		135	174	
	4350	Splice plate	1 Elec	48	.167		7.55	7		14.55	19.10	
	4360	Splice plate, expansion joint		48	.167		7.75	7		14.75	19.30	
	4370	Splice plate, hinged, horizontal		48	.167		6.25	7		13.25	17.65	
	4380	Vertical		48	.167		8.90	7		15.90	20.50	
	4390	Trough, hanger, vertical		28	.286		25	12		37	46	
	4400	Box connector, 24″ wide		20	.400		33.50	16.80		50.30	63	
	4410	18″ wide		21	.381		28	16		44	55.50	
	4420	12″ wide		22	.364		27	15.25		42.25	53	
	4430	9″ wide		23	.348		25.50	14.60		40.10	50.50	
	4440	6″ wide		24	.333		24.50	14		38.50	48.50	
	4450	Floor flange		24	.333		25.50	14		39.50	49.50	
	4460	Hold down clamp		60	.133		2.94	5.60		8.54	11.90	
	4520	Wall bracket, 24″ wide tray		20	.400		24.50	16.80		41.30	53	
	4530	18″ wide tray		21	.381		24	16		40	51	
	4540	12″ wide tray		22	.364		12.90	15.25		28.15	37.50	
	4550	9″ wide tray		23	.348		11.50	14.60		26.10	35	
	4560	6″ wide tray		24	.333		10.45	14		24.45	33	
	5000	Cable channel aluminum, vented, 1-1/4″ deep, 4″ wide, straight		80	.100	L.F.	10	4.20		14.20	17.45	
	5010	Elbow horizontal, 36″ radius, 90°		5	1.600	Ea.	180	67		247	300	
	5020	60°		5.50	1.455		138	61		199	245	
	5030	45°		6	1.333		112	56		168	210	
	5040	30°		6.50	1.231		96.50	51.50		148	186	
	5050	Adjustable		6	1.333		92	56		148	188	
	5060	Elbow vertical, 36″ radius, 90°		5	1.600		193	67		260	315	
	5070	60°		5.50	1.455		151	61		212	260	
	5080	45°		6	1.333		124	56		180	224	
	5090	30°		6.50	1.231		110	51.50		161.50	201	
120	5100	Adjustable		6	1.333		92	56		148	188	120

	16131 | Cable Trays		CREW	DAILY OUTPUT	LABOR-HOURS	UNIT	2006 BARE COSTS MAT.	LABOR	EQUIP.	TOTAL	TOTAL INCL O&P	
120	5110	Splice plate, hinged, horizontal	1 Elec	48	.167	Ea.	6.90	7		13.90	18.35	120
	5120	Splice plate, hinged, vertical		48	.167		9.75	7		16.75	21.50	
	5130	Hanger, vertical		28	.286		10.75	12		22.75	30.50	
	5140	Single		28	.286		16.35	12		28.35	36.50	
	5150	Double		20	.400		16.75	16.80		33.55	44.50	
	5160	Channel to box connector		24	.333		22	14		36	45.50	
	5170	Hold down clip		80	.100		3.11	4.20		7.31	9.85	
	5180	Wall bracket, single		28	.286		8.35	12		20.35	27.50	
	5190	Double		20	.400		10.70	16.80		27.50	38	
	5200	Cable roller		16	.500		131	21		152	177	
	5210	Splice plate		48	.167		4.03	7		11.03	15.25	
130	0010	**CABLE TRAY, COVERS AND DIVIDERS** To 15′ high										130
	0100	Covers, ventilated galv. steel, straight, 6″ wide tray size	2 Elec	520	.031	L.F.	4.13	1.29		5.42	6.55	
	0200	9″ wide tray size		460	.035		5.05	1.46		6.51	7.80	
	0300	12″ wide tray size		400	.040		6.05	1.68		7.73	9.25	
	0400	18″ wide tray size		300	.053		8.25	2.24		10.49	12.55	
	0500	24″ wide tray size		220	.073		10.15	3.05		13.20	15.85	
	0600	30″ wide tray size		180	.089		12.25	3.73		15.98	19.25	
	0700	36″ wide tray size		160	.100		14	4.20		18.20	22	
	1000	Elbow horizontal 90°, 12″ radius, 6″ wide tray size		150	.107	Ea.	30.50	4.48		34.98	40.50	
	1020	9″ wide tray size		128	.125		34	5.25		39.25	45	
	1040	12″ wide tray size		108	.148		36	6.20		42.20	49	
	1060	18″ wide tray size		84	.190		49.50	8		57.50	67	
	1080	24″ wide tray size		66	.242		60	10.20		70.20	81	
	1100	30″ wide tray size		60	.267		74	11.20		85.20	99	
	1120	36″ wide tray size		50	.320		90	13.45		103.45	119	
	1160	24″ radius, 6″ wide tray size		136	.118		51	4.94		55.94	63.50	
	1180	9″ wide tray size		116	.138		52	5.80		57.80	66	
	1200	12″ wide tray size		96	.167		58	7		65	74.50	
	1220	18″ wide tray size		76	.211		71.50	8.85		80.35	92	
	1240	24″ wide tray size		60	.267		87	11.20		98.20	113	
	1260	30″ wide tray size		52	.308		114	12.90		126.90	146	
	1280	36″ wide tray size		44	.364		133	15.25		148.25	170	
	1320	36″ radius, 6″ wide tray size		120	.133		74	5.60		79.60	90	
	1340	9″ wide tray size		104	.154		82	6.45		88.45	100	
	1360	12″ wide tray size		84	.190		87.50	8		95.50	109	
	1380	18″ wide tray size		72	.222		111	9.35		120.35	136	
	1400	24″ wide tray size		52	.308		133	12.90		145.90	166	
	1420	30″ wide tray size		46	.348		158	14.60		172.60	196	
	1440	36″ wide tray size		40	.400		185	16.80		201.80	230	
	1480	Elbow horizontal 45°, 12″ radius, 6″ wide tray size		150	.107		22	4.48		26.48	31	
	1500	9″ wide tray size		128	.125		26	5.25		31.25	36.50	
	1520	12″ wide tray size		108	.148		28.50	6.20		34.70	40.50	
	1540	18″ wide tray size		88	.182		34.50	7.65		42.15	50	
	1560	24″ wide tray size		76	.211		40.50	8.85		49.35	58	
	1580	30″ wide tray size		66	.242		47.50	10.20		57.70	68	
	1600	36″ wide tray size		60	.267		53.50	11.20		64.70	76	
	1640	24″ radius, 6″ wide tray size		136	.118		32	4.94		36.94	42.50	
	1660	9″ wide tray size		116	.138		36	5.80		41.80	48.50	
	1680	12″ wide tray size		96	.167		40.50	7		47.50	55.50	
	1700	18″ wide tray size		80	.200		46	8.40		54.40	63.50	
	1720	24″ wide tray size		70	.229		53.50	9.60		63.10	73.50	
	1740	30″ wide tray size		60	.267		67.50	11.20		78.70	92	
	1760	36″ wide tray size		52	.308		74	12.90		86.90	101	
	1800	36″ radius, 6″ wide tray size		120	.133		46	5.60		51.60	59	

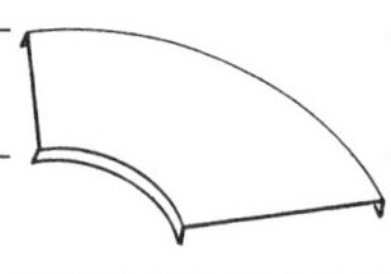

16131 | Cable Trays			CREW	DAILY OUTPUT	LABOR-HOURS	UNIT	2006 BARE COSTS MAT.	LABOR	EQUIP.	TOTAL	TOTAL INCL O&P	
130	1820	9" wide tray size	2 Elec	104	.154	Ea.	52	6.45		58.45	67	130
	1840	12" wide tray size		84	.190		57.50	8		65.50	75.50	
	1860	18" wide tray size		76	.211		65.50	8.85		74.35	85.50	
	1880	24" wide tray size		62	.258		80	10.85		90.85	105	
	1900	30" wide tray size		52	.308		87	12.90		99.90	116	
	1920	36" wide tray size		48	.333		105	14		119	138	
	1960	Elbow vertical 90°, 12" radius, 6" wide tray size		150	.107		25.50	4.48		29.98	35	
	1980	9" wide tray size		128	.125		26	5.25		31.25	36.50	
	2000	12" wide tray size		108	.148		28	6.20		34.20	40.50	
	2020	18" wide tray size		88	.182		32	7.65		39.65	47	
	2040	24" wide tray size		68	.235		32.50	9.90		42.40	51	
	2060	30" wide tray size		60	.267		36	11.20		47.20	57	
	2080	36" wide tray size		50	.320		46	13.45		59.45	71	
	2120	24" radius, 6" wide tray size		136	.118		31.50	4.94		36.44	42	
	2140	9" wide tray size		116	.138		34.50	5.80		40.30	47	
	2160	12" wide tray size		96	.167		36	7		43	50.50	
	2180	18" wide tray size		80	.200		47.50	8.40		55.90	65.50	
	2200	24" wide tray size		62	.258		53.50	10.85		64.35	75	
	2220	30" wide tray size		52	.308		60.50	12.90		73.40	86.50	
	2240	36" wide tray size		44	.364		69	15.25		84.25	99.50	
	2280	36" radius, 6" wide tray size		120	.133		36.50	5.60		42.10	48.50	
	2300	9" wide tray size		104	.154		46	6.45		52.45	60.50	
	2320	12" wide tray size		84	.190		52	8		60	69.50	
	2340	18" wide tray size		76	.211		65.50	8.85		74.35	85.50	
	2350	24" wide tray size		54	.296		73	12.45		85.45	99	
	2360	30" wide tray size		46	.348		87	14.60		101.60	119	
	2370	36" wide tray size		40	.400		103	16.80		119.80	139	
	2400	Tee horizontal, 12" radius, 6" wide tray size		92	.174		46.50	7.30		53.80	62.50	
	2410	9" wide tray size		80	.200		47.50	8.40		55.90	65.50	
	2420	12" wide tray size		68	.235		54	9.90		63.90	74	
	2430	18" wide tray size		60	.267		65	11.20		76.20	89	
	2440	24" wide tray size		52	.308		82	12.90		94.90	110	
	2460	30" wide tray size		36	.444		96.50	18.65		115.15	135	
	2470	36" wide tray size		30	.533		116	22.50		138.50	163	
	2500	24" radius, 6" wide tray size		88	.182		72.50	7.65		80.15	92	
	2510	9" wide tray size		76	.211		83.50	8.85		92.35	105	
	2520	12" wide tray size		64	.250		86	10.50		96.50	111	
	2530	18" wide tray size		56	.286		110	12		122	140	
	2540	24" wide tray size		48	.333		169	14		183	208	
	2560	30" wide tray size		32	.500		193	21		214	245	
	2570	36" wide tray size		26	.615		213	26		239	274	
	2600	36" radius, 6" wide tray size		84	.190		131	8		139	156	
	2610	9" wide tray size		72	.222		134	9.35		143.35	161	
	2620	12" wide tray size		60	.267		149	11.20		160.20	181	
	2630	18" wide tray size		52	.308		174	12.90		186.90	211	
	2640	24" wide tray size		44	.364		229	15.25		244.25	276	
	2660	30" wide tray size		28	.571		249	24		273	310	
	2670	36" wide tray size		22	.727		283	30.50		313.50	355	
	2700	Cross horizontal, 12" radius, 6" wide tray size		68	.235		69	9.90		78.90	91	
	2710	9" wide tray size		64	.250		73	10.50		83.50	96	
	2720	12" wide tray size		60	.267		81.50	11.20		92.70	107	
	2730	18" wide tray size		52	.308		96.50	12.90		109.40	126	
	2740	24" wide tray size		36	.444		118	18.65		136.65	158	
	2760	30" wide tray size		30	.533		134	22.50		156.50	182	
	2770	36" wide tray size		28	.571		156	24		180	209	
	2800	24" radius, 6" wide tray size	↓	64	.250	↓	131	10.50		141.50	160	130

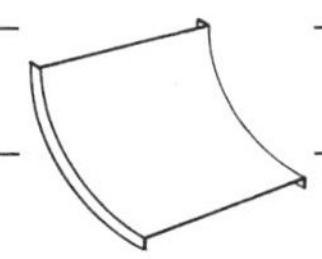

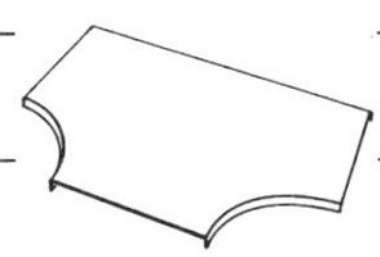

16 ELECTRICAL

16131 | Cable Trays

		Description	CREW	DAILY OUTPUT	LABOR-HOURS	UNIT	2006 BARE COSTS MAT.	LABOR	EQUIP.	TOTAL	TOTAL INCL O&P	
130	2810	9″ wide tray size	2 Elec	60	.267	Ea.	143	11.20		154.20	174	130
	2820	12″ wide tray size		56	.286		154	12		166	188	
	2830	18″ wide tray size		48	.333		184	14		198	224	
	2840	24″ wide tray size		32	.500		224	21		245	279	
	2860	30″ wide tray size		26	.615		253	26		279	320	
	2870	36″ wide tray size		24	.667		284	28		312	355	
	2900	36″ radius, 6″ wide tray size		60	.267		220	11.20		231.20	259	
	2910	9″ wide tray size		56	.286		228	12		240	269	
	2920	12″ wide tray size		52	.308		239	12.90		251.90	283	
	2930	18″ wide tray size		44	.364		275	15.25		290.25	330	
	2940	24″ wide tray size		28	.571		350	24		374	420	
	2960	30″ wide tray size		22	.727		380	30.50		410.50	460	
	2970	36″ wide tray size		20	.800		405	33.50		438.50	495	
	3000	Reducer, 9″ to 6″ wide tray size 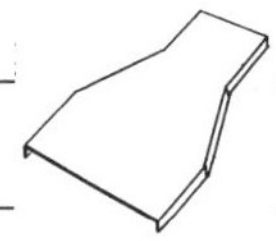		128	.125		29.50	5.25		34.75	40	
	3010	12″ to 6″ wide tray size		108	.148		30.50	6.20		36.70	43	
	3020	12″ to 9″ wide tray size		108	.148		30.50	6.20		36.70	43	
	3030	18″ to 12″ wide tray size		88	.182		32.50	7.65		40.15	48	
	3050	18″ to 6″ wide tray size		88	.182		32.50	7.65		40.15	48	
	3060	24″ to 18″ wide tray size		80	.200		45.50	8.40		53.90	63	
	3070	24″ to 12″ wide tray size		80	.200		41.50	8.40		49.90	59	
	3090	30″ to 24″ wide tray size		70	.229		48	9.60		57.60	68	
	3100	30″ to 18″ wide tray size 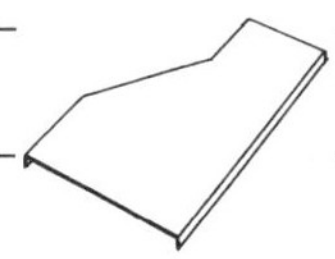		70	.229		48	9.60		57.60	68	
	3110	30″ to 12″ wide tray size		70	.229		41.50	9.60		51.10	61	
	3140	36″ to 30″ wide tray size		64	.250		52.50	10.50		63	74	
	3150	36″ to 24″ wide tray size		64	.250		52.50	10.50		63	74	
	3160	36″ to 18″ wide tray size		64	.250		52.50	10.50		63	74	
	3170	36″ to 12″ wide tray size		64	.250		52.50	10.50		63	74	
	3250	Covers, aluminum, straight, 6″ wide tray size 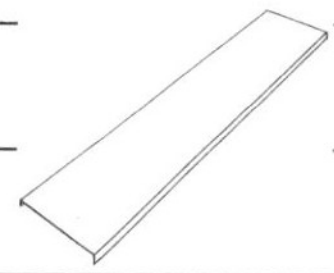		520	.031	L.F.	3.84	1.29		5.13	6.20	
	3270	9″ wide tray size		460	.035		4.62	1.46		6.08	7.35	
	3290	12″ wide tray size		400	.040		5.50	1.68		7.18	8.65	
	3310	18″ wide tray size		320	.050		7.30	2.10		9.40	11.30	
	3330	24″ wide tray size		260	.062		9.10	2.58		11.68	14.05	
	3350	30″ wide tray size		200	.080		10.10	3.36		13.46	16.30	
	3370	36″ wide tray size		180	.089		10.70	3.73		14.43	17.50	
	3400	Elbow horizontal 90°, 12″ radius, 6″ wide tray size		150	.107	Ea.	28.50	4.48		32.98	38.50	
	3410	9″ wide tray size		128	.125		30	5.25		35.25	41	
	3420	12″ wide tray size		108	.148		32.50	6.20		38.70	45	
	3430	18″ wide tray size		88	.182		42.50	7.65		50.15	58.50	
	3440	24″ wide tray size		70	.229		53.50	9.60		63.10	74	
	3460	30″ wide tray size		64	.250		64.50	10.50		75	87	
	3470	36″ wide tray size		54	.296		78.50	12.45		90.95	106	
	3500	24″ radius, 6″ wide tray size		136	.118		39.50	4.94		44.44	51	
	3510	9″ wide tray size		116	.138		48.50	5.80		54.30	62.50	
	3520	12″ wide tray size		96	.167		53.50	7		60.50	70	
	3530	18″ wide tray size 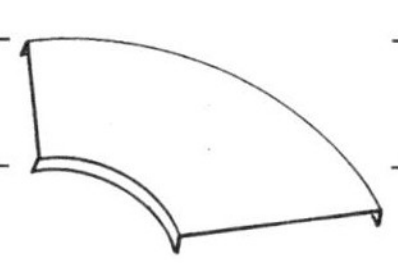		80	.200		63.50	8.40		71.90	83	
	3540	24″ wide tray size		64	.250		78.50	10.50		89	103	
	3560	30″ wide tray size		56	.286		96	12		108	125	
	3570	36″ wide tray size		48	.333		119	14		133	153	
	3600	36″ radius, 6″ wide tray size		120	.133		67.50	5.60		73.10	82.50	
	3610	9″ wide tray size		104	.154		73.50	6.45		79.95	91	
	3620	12″ wide tray size		84	.190		83.50	8		91.50	104	
	3630	18″ wide tray size		76	.211		99.50	8.85		108.35	124	
	3640	24″ wide tray size		56	.286		121	12		133	152	
	3660	30″ wide tray size		50	.320		139	13.45		152.45	174	
	3670	36″ wide tray size		44	.364		163	15.25		178.25	204	
	3700	Elbow horizontal 45°, 12″ radius, 6″ wide tray size		150	.107		21	4.48		25.48	30.50	

16131 | Cable Trays

16131	Cable Trays	CREW	DAILY OUTPUT	LABOR-HOURS	UNIT	2006 BARE COSTS MAT.	LABOR	EQUIP.	TOTAL	TOTAL INCL O&P
130 3710	9" wide tray size	2 Elec	128	.125	Ea.	22	5.25		27.25	32
3720	12" wide tray size		108	.148		24.50	6.20		30.70	36.50
3730	18" wide tray size		88	.182		28	7.65		35.65	43
3740	24" wide tray size		80	.200		31	8.40		39.40	47.50
3760	30" wide tray size		70	.229		39.50	9.60		49.10	58.50
3770	36" wide tray size		64	.250		47.50	10.50		58	68
3800	24" radius, 6" wide tray size		136	.118		24.50	4.94		29.44	34.50
3810	9" wide tray size		116	.138		31	5.80		36.80	43.50
3820	12" wide tray size		96	.167		32.50	7		39.50	46.50
3830	18" wide tray size		80	.200		39.50	8.40		47.90	56.50
3840	24" wide tray size		72	.222		48.50	9.35		57.85	68
3860	30" wide tray size		64	.250		56	10.50		66.50	78
3870	36" wide tray size		56	.286		65	12		77	90
3900	36" radius, 6" wide tray size		120	.133		42.50	5.60		48.10	55
3910	9" wide tray size		104	.154		46	6.45		52.45	61
3920	12" wide tray size		84	.190		48.50	8		56.50	66
3930	18" wide tray size		76	.211		58.50	8.85		67.35	78
3940	24" wide tray size		64	.250		71	10.50		81.50	94.50
3960	30" wide tray size		56	.286		81	12		93	108
3970	36" wide tray size		50	.320		92.50	13.45		105.95	123
4000	Elbow vertical 90°, 12" radius, 6" wide tray size		150	.107		24.50	4.48		28.98	34
4010	9" wide tray size		128	.125		24.50	5.25		29.75	35
4020	12" wide tray size		108	.148		27	6.20		33.20	39
4030	18" wide tray size		88	.182		31	7.65		38.65	46.50
4040	24" wide tray size		70	.229		32.50	9.60		42.10	50.50
4060	30" wide tray size		64	.250		33	10.50		43.50	52.50
4070	36" wide tray size		54	.296		40	12.45		52.45	63
4100	24" radius, 6" wide tray size		136	.118		28	4.94		32.94	38.50
4110	9" wide tray size		116	.138		30.50	5.80		36.30	42.50
4120	12" wide tray size		96	.167		33	7		40	47.50
4130	18" wide tray size		80	.200		40	8.40		48.40	57
4140	24" wide tray size		64	.250		43.50	10.50		54	64
4160	30" wide tray size		56	.286		56	12		68	80.50
4170	36" wide tray size		48	.333		61	14		75	89
4200	36" radius, 6" wide tray size		120	.133		32.50	5.60		38.10	44
4210	9" wide tray size		104	.154		40	6.45		46.45	54
4220	12" wide tray size		84	.190		46	8		54	63.50
4230	18" wide tray size		76	.211		56	8.85		64.85	75.50
4240	24" wide tray size		56	.286		66	12		78	91
4260	30" wide tray size		50	.320		83.50	13.45		96.95	113
4270	36" wide tray size		44	.364		90	15.25		105.25	123
4300	Tee horizontal, 12" radius, 6" wide tray size		108	.148		40	6.20		46.20	53.50
4310	9" wide tray size		88	.182		42.50	7.65		50.15	58.50
4320	12" wide tray size		80	.200		47.50	8.40		55.90	65
4330	18" wide tray size		68	.235		56	9.90		65.90	77
4340	24" wide tray size		56	.286		71	12		83	97
4360	30" wide tray size		44	.364		83.50	15.25		98.75	116
4370	36" wide tray size		36	.444		101	18.65		119.65	140
4400	24" radius, 6" wide tray size		96	.167		65	7		72	82.50
4410	9" wide tray size		80	.200		72.50	8.40		80.90	92.50
4420	12" wide tray size		72	.222		81	9.35		90.35	103
4430	18" wide tray size		60	.267		95	11.20		106.20	121
4440	24" wide tray size		48	.333		151	14		165	188
4460	30" wide tray size		40	.400		167	16.80		183.80	209
4470	36" wide tray size		32	.500		186	21		207	238
4500	36" radius, 6" wide tray size		88	.182		119	7.65		126.65	143

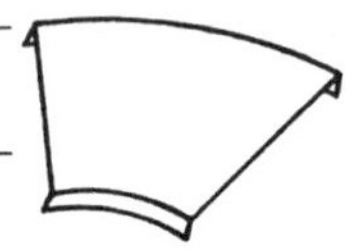

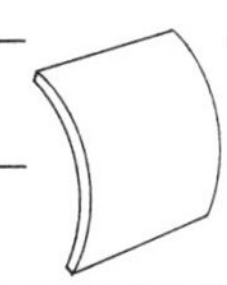

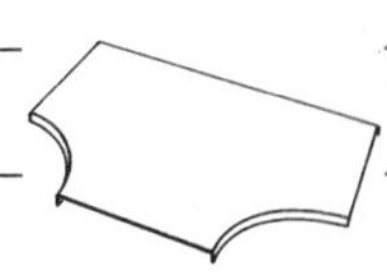

16 ELECTRICAL

	16131 | Cable Trays		CREW	DAILY OUTPUT	LABOR-HOURS	UNIT	2006 BARE COSTS MAT.	LABOR	EQUIP.	TOTAL	TOTAL INCL O&P
130	4510	9" wide tray size	2 Elec	72	.222	Ea.	121	9.35		130.35	147
	4520	12" wide tray size		64	.250		133	10.50		143.50	163
	4530	18" wide tray size		56	.286		151	12		163	185
	4540	24" wide tray size		44	.364		192	15.25		207.25	235
	4560	30" wide tray size		36	.444		216	18.65		234.65	267
	4570	36" wide tray size		28	.571		244	24		268	305
	4600	Cross horizontal, 12" radius, 6" wide tray size		80	.200		61	8.40		69.40	80.50
	4610	9" wide tray size		72	.222		65	9.35		74.35	86
	4620	12" wide tray size		64	.250		72	10.50		82.50	95
	4630	18" wide tray size		56	.286		87.50	12		99.50	115
	4640	24" wide tray size		48	.333		101	14		115	133
	4660	30" wide tray size		40	.400		120	16.80		136.80	158
	4670	36" wide tray size		32	.500		138	21		159	185
	4700	24" radius, 6" wide tray size		72	.222		120	9.35		129.35	146
	4710	9" wide tray size		64	.250		128	10.50		138.50	157
	4720	12" wide tray size		56	.286		138	12		150	171
	4730	18" wide tray size		48	.333		161	14		175	199
	4740	24" wide tray size		40	.400		192	16.80		208.80	237
	4760	30" wide tray size		32	.500		224	21		245	280
	4770	36" wide tray size		24	.667		244	28		272	310
	4800	36" radius, 6" wide tray size		64	.250		192	10.50		202.50	227
	4810	9" wide tray size		56	.286		200	12		212	239
	4820	12" wide tray size		50	.320		216	13.45		229.45	259
	4830	18" wide tray size		44	.364		238	15.25		253.25	285
	4840	24" wide tray size		36	.444		293	18.65		311.65	350
	4860	30" wide tray size		28	.571		335	24		359	405
	4870	36" wide tray size		22	.727		370	30.50		400.50	450
	4900	Reducer, 9" to 6" wide tray size		128	.125		30.50	5.25		35.75	41.50
	4910	12" to 6" wide tray size		108	.148		32	6.20		38.20	44.50
	4920	12" to 9" wide tray size		108	.148		32	6.20		38.20	44.50
	4930	18" to 12" wide tray size		88	.182		35	7.65		42.65	50.50
	4950	18" to 6" wide tray size		88	.182		35	7.65		42.65	50.50
	4960	24" to 18" wide tray size		80	.200		45	8.40		53.40	62.50
	4970	24" to 12" wide tray size		80	.200		38.50	8.40		46.90	55.50
	4990	30" to 24" wide tray size		70	.229		47	9.60		56.60	66.50
	5000	30" to 18" wide tray size		70	.229		47	9.60		56.60	66.50
	5010	30" to 12" wide tray size		70	.229		47	9.60		56.60	66.50
	5040	36" to 30" wide tray size		64	.250		52	10.50		62.50	73
	5050	36" to 24" wide tray size		64	.250		52	10.50		62.50	73
	5060	36" to 18" wide tray size		64	.250		52	10.50		62.50	73
	5070	36" to 12" wide tray size		64	.250		52	10.50		62.50	73
	5710	Tray cover hold down clamp	1 Elec	60	.133		8.30	5.60		13.90	17.75
	8000	Divider strip, straight, galvanized, 3" deep		200	.040	L.F.	3.63	1.68		5.31	6.60
	8020	4" deep		180	.044		4.44	1.87		6.31	7.75
	8040	6" deep		160	.050		5.80	2.10		7.90	9.65
	8060	Aluminum, straight, 3" deep		210	.038		3.48	1.60		5.08	6.30
	8080	4" deep		190	.042		4.32	1.77		6.09	7.45
	8100	6" deep		170	.047		5.50	1.98		7.48	9.10
	8110	Divider strip vertical fitting 3" deep									
	8120	12" radius, galvanized, 30°	1 Elec	28	.286	Ea.	17.55	12		29.55	38
	8140	45°		27	.296		22	12.45		34.45	43
	8160	60°		26	.308		23.50	12.90		36.40	45.50
	8180	90°		25	.320		29.50	13.45		42.95	52.50
	8200	Aluminum, 30°		29	.276		14.30	11.60		25.90	33.50
	8220	45°		28	.286		16.85	12		28.85	37
	8240	60°		27	.296		19.70	12.45		32.15	40.50

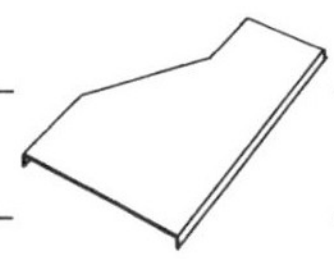

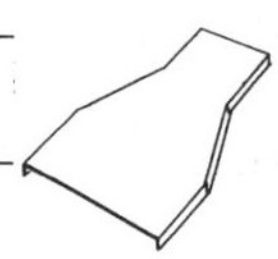

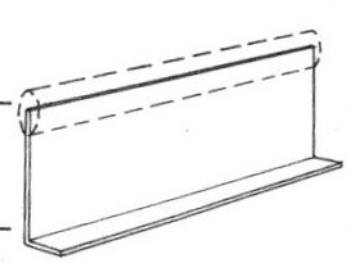

16131 | Cable Trays

		Description	Crew	Daily Output	Labor-Hours	Unit	2006 Bare Costs Mat.	Labor	Equip.	Total	Total Incl O&P
130	8260	90°	1 Elec	26	.308	Ea.	25	12.90		37.90	47.50
	8280	24" radius, galvanized, 30°		25	.320		26.50	13.45		39.95	50
	8300	45°		24	.333		30	14		44	54.50
	8320	60°		23	.348		38	14.60		52.60	64
	8340	90°		22	.364		51.50	15.25		66.75	80
	8360	Aluminum, 30°		26	.308		24	12.90		36.90	46
	8380	45°		25	.320		27.50	13.45		40.95	50.50
	8400	60°		24	.333		34.50	14		48.50	59.50
	8420	90°		23	.348		47	14.60		61.60	74
	8440	36" radius, galvanized, 30°		22	.364		36	15.25		51.25	63
	8460	45°		21	.381		41.50	16		57.50	70.50
	8480	60°		20	.400		49	16.80		65.80	79.50
	8500	90°		19	.421		67.50	17.70		85.20	102
	8520	Aluminum, 30°		23	.348		36	14.60		50.60	62.50
	8540	45°		22	.364		47	15.25		62.25	75
	8560	60°		21	.381		60.50	16		76.50	91
	8570	90°	↓	20	.400	↓	79.50	16.80		96.30	113
	8590	Divider strip vertical fitting 4" deep									
	8600	12" radius, galvanized, 30°	1 Elec	27	.296	Ea.	22.50	12.45		34.95	43.50
	8610	45°		26	.308		26	12.90		38.90	48.50
	8620	60°		25	.320		29.50	13.45		42.95	52.50
	8630	90°		24	.333		36	14		50	61
	8640	Aluminum, 30°		28	.286		20	12		32	40.50
	8650	45°		27	.296		23.50	12.45		35.95	44.50
	8660	60°		26	.308		27	12.90		39.90	49.50
	8670	90°		25	.320		32	13.45		45.45	55.50
	8680	24" radius, galvanized, 30°		24	.333		36	14		50	61
	8690	45°		23	.348		45	14.60		59.60	72
	8700	60°		22	.364		51	15.25		66.25	79.50
	8710	90°		21	.381		67.50	16		83.50	99
	8720	Aluminum, 30°		25	.320		32	13.45		45.45	55.50
	8730	45°		24	.333		38.50	14		52.50	64
	8740	60°		23	.348		45.50	14.60		60.10	72.50
	8750	90°		22	.364		62.50	15.25		77.75	92
	8760	36" radius, galvanized, 30°		23	.348		41.50	14.60		56.10	68.50
	8770	45°		22	.364		47.50	15.25		62.75	76
	8780	60°		21	.381		60	16		76	90
	8790	90°		20	.400		81.50	16.80		98.30	116
	8800	Aluminum, 30°		24	.333		48.50	14		62.50	75
	8810	45°		23	.348		62.50	14.60		77.10	91
	8820	60°		22	.364		76	15.25		91.25	107
	8830	90°	↓	21	.381	↓	96	16		112	131
	8840	Divider strip vertical fitting 6" deep									
	8850	12" radius, galvanized, 30°	1 Elec	24	.333	Ea.	24.50	14		38.50	48.50
	8860	45°		23	.348		27.50	14.60		42.10	52.50
	8870	60°		22	.364		31.50	15.25		46.75	58
	8880	90°		21	.381		39	16		55	67.50
	8890	Aluminum, 30°		25	.320		22	13.45		35.45	45
	8900	45°		24	.333		26	14		40	50.50
	8910	60°		23	.348		27.50	14.60		42.10	52.50
	8920	90°		22	.364		32.50	15.25		47.75	59
	8930	24" radius, galvanized, 30°		23	.348		36	14.60		50.60	62
	8940	45°		22	.364		45	15.25		60.25	73
	8950	60°		21	.381		51.50	16		67.50	81
	8960	90°		20	.400		67.50	16.80		84.30	101
130	8970	Aluminum, 30°	↓	24	.333	↓	32.50	14		46.50	57

16131 | Cable Trays

		16131 Cable Trays	Crew	Daily Output	Labor-Hours	Unit	2006 Bare Costs Mat.	Labor	Equip.	Total	Total Incl O&P
130	8980	45°	1 Elec	23	.348	Ea.	44.50	14.60		59.10	71
	8990	60°		22	.364		48.50	15.25		63.75	77
	9000	90°		21	.381		66	16		82	97
	9010	36″ radius, galvanized, 30°		22	.364		41.50	15.25		56.75	69.50
	9020	45°		21	.381		51.50	16		67.50	81
	9030	60°		20	.400		67.50	16.80		84.30	101
	9040	90°		19	.421		87	17.70		104.70	123
	9050	Aluminum, 30°		23	.348		49.50	14.60		64.10	77
	9060	45°		22	.364		66	15.25		81.25	96
	9070	60°		21	.381		78	16		94	111
	9080	90°		20	.400		93.50	16.80		110.30	129
	9120	Divider strip, horizontal fitting, galvanized, 3″ deep		33	.242		25	10.20		35.20	43
	9130	4″ deep		30	.267		28	11.20		39.20	48.50
	9140	6″ deep		27	.296		36	12.45		48.45	58.50
	9150	Aluminum, 3″ deep		35	.229		22.50	9.60		32.10	40
	9160	4″ deep 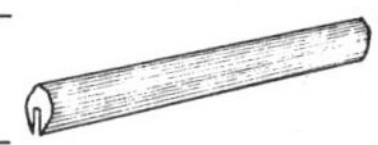		32	.250		25	10.50		35.50	43.50
	9170	6″ deep		29	.276		33	11.60		44.60	54.50
	9300	Divider strip protector		300	.027	L.F.	2.56	1.12		3.68	4.55
	9310	Fastener, ladder tray				Ea.	.46			.46	.51
	9320	Trough or solid bottom tray				″	.31			.31	.34
	9900	Add to labor for higher elevated installation									
	9910	15′ to 20′ high add						10%			
	9920	20′ to 25′ high add						20%			
	9930	25′ to 30′ high add						25%			
	9940	30′ to 35′ high add						30%			
	9960	Over 40′ high add						40%			

16132 | Conduit & Tubing

		16132 Conduit & Tubing	Crew	Daily Output	Labor-Hours	Unit	2006 Bare Costs Mat.	Labor	Equip.	Total	Total Incl O&P
205	0010	**CONDUIT** To 15′ high, includes 2 terminations, 2 elbows and R260533 -20									
	0020	11 beam clamps & couplings per 100 L.F.									
	0300	Aluminum, 1/2″ diameter	1 Elec	100	.080	L.F.	1.93	3.36		5.29	7.30
	0500	3/4″ diameter		90	.089		2.59	3.73		6.32	8.60
	0700	1″ diameter		80	.100		3.43	4.20		7.63	10.20
	1000	1-1/4″ diameter		70	.114		4.61	4.80		9.41	12.45
	1030	1-1/2″ diameter		65	.123		5.55	5.15		10.70	14.05
	1050	2″ diameter		60	.133		7.50	5.60		13.10	16.90
	1070	2-1/2″ diameter 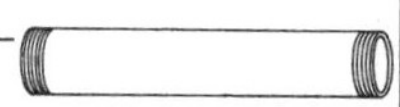		50	.160		12	6.70		18.70	23.50
	1100	3″ diameter	2 Elec	90	.178		16.30	7.45		23.75	29.50
	1130	3-1/2″ diameter		80	.200		21.50	8.40		29.90	36.50
	1140	4″ diameter		70	.229		25.50	9.60		35.10	43
	1150	5″ diameter		50	.320		52.50	13.45		65.95	78
	1160	6″ diameter		40	.400		77	16.80		93.80	111
	1161	Field bends, 45° to 90°, 1/2″ diameter	1 Elec	53	.151	Ea.		6.35		6.35	9.75
	1162	3/4″ diameter		47	.170			7.15		7.15	11
	1163	1″ diameter		44	.182			7.65		7.65	11.75
	1164	1-1/4″ diameter		23	.348			14.60		14.60	22.50
	1165	1-1/2″ diameter		21	.381			16		16	24.50
	1166	2″ diameter		16	.500			21		21	32.50
	1170	Elbows, 1/2″ diameter		40	.200		8.55	8.40		16.95	22.50
	1200	3/4″ diameter		32	.250		11.55	10.50		22.05	29
	1230	1″ diameter		28	.286		16.15	12		28.15	36.50
	1250	1-1/4″ diameter 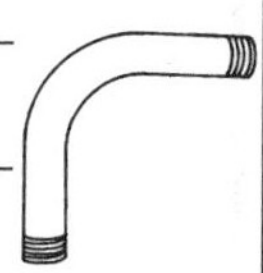		24	.333		25.50	14		39.50	49.50
	1270	1-1/2″ diameter		20	.400		34	16.80		50.80	63.50
	1300	2″ diameter		16	.500		50	21		71	88
	1330	2-1/2″ diameter		12	.667		84.50	28		112.50	136
	1350	3″ diameter		8	1		131	42		173	209

For expanded coverage of these items see *Means Electrical Cost Data 2006*

16132 | Conduit & Tubing

205		CREW	DAILY OUTPUT	LABOR-HOURS	UNIT	2006 BARE COSTS MAT.	LABOR	EQUIP.	TOTAL	TOTAL INCL O&P
1370	3-1/2" diameter R260533-20	1 Elec	6	1.333	Ea.	204	56		260	310
1400	4" diameter		5	1.600		242	67		309	370
1410	5" diameter		4	2		715	84		799	915
1420	6" diameter		2.50	3.200		990	134		1,124	1,300
1430	Couplings, 1/2" diameter					2.80			2.80	3.08
1450	3/4" diameter					4.24			4.24	4.66
1470	1" diameter					5.60			5.60	6.15
1500	1-1/4" diameter					6.85			6.85	7.55
1530	1-1/2" diameter					7.90			7.90	8.70
1550	2" diameter					11.25			11.25	12.35
1570	2-1/2" diameter					25.50			25.50	28
1600	3" diameter					33			33	36.50
1630	3-1/2" diameter					45.50			45.50	50
1650	4" diameter					54.50			54.50	60
1670	5" diameter					161			161	177
1690	6" diameter					251			251	276
1750	Rigid galvanized steel, 1/2" diameter	1 Elec	90	.089	L.F.	2.44	3.73		6.17	8.45
1770	3/4" diameter		80	.100		2.78	4.20		6.98	9.50
1800	1" diameter		65	.123		4.09	5.15		9.24	12.45
1830	1-1/4" diameter		60	.133		5.60	5.60		11.20	14.80
1850	1-1/2" diameter		55	.145		6.40	6.10		12.50	16.45
1870	2" diameter		45	.178		8.25	7.45		15.70	20.50
1900	2-1/2" diameter		35	.229		15.50	9.60		25.10	32
1930	3" diameter	2 Elec	50	.320		18.85	13.45		32.30	41
1950	3-1/2" diameter		44	.364		23	15.25		38.25	49
1970	4" diameter		40	.400		26.50	16.80		43.30	55
1980	5" diameter		30	.533		55.50	22.50		78	95.50
1990	6" diameter		20	.800		83	33.50		116.50	143
1991	Field bends, 45° to 90°, 1/2" diameter	1 Elec	44	.182	Ea.		7.65		7.65	11.75
1992	3/4" diameter		40	.200			8.40		8.40	12.95
1993	1" diameter		36	.222			9.35		9.35	14.40
1994	1-1/4" diameter		19	.421			17.70		17.70	27
1995	1-1/2" diameter		18	.444			18.65		18.65	29
1996	2" diameter		13	.615			26		26	40
2000	Elbows, 1/2" diameter		32	.250		8.35	10.50		18.85	25.50
2030	3/4" diameter		28	.286		8.80	12		20.80	28
2050	1" diameter		24	.333		13.45	14		27.45	36.50
2070	1-1/4" diameter		18	.444		18.45	18.65		37.10	49.50
2100	1-1/2" diameter		16	.500		23	21		44	57.50
2130	2" diameter		12	.667		33	28		61	79.50
2150	2-1/2" diameter		8	1		61.50	42		103.50	132
2170	3" diameter		6	1.333		85.50	56		141.50	181
2200	3-1/2" diameter		4.20	1.905		135	80		215	272
2220	4" diameter		4	2		153	84		237	297
2230	5" diameter		3.50	2.286		425	96		521	615
2240	6" diameter		2	4		635	168		803	960
2250	Couplings, 1/2" diameter					1.91			1.91	2.10
2270	3/4" diameter					2.34			2.34	2.57
2300	1" diameter					3.49			3.49	3.84
2330	1-1/4" diameter					4.34			4.34	4.77
2350	1-1/2" diameter					5.50			5.50	6.05
2370	2" diameter					7.25			7.25	7.95
2400	2-1/2" diameter					17.95			17.95	19.75
2430	3" diameter					23			23	25.50
2450	3-1/2" diameter					31			31	34
2470	4" diameter					31.50			31.50	34.50

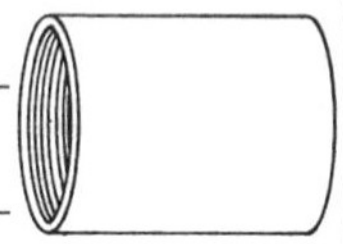
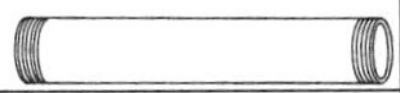
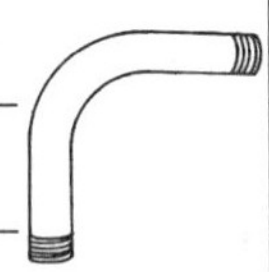
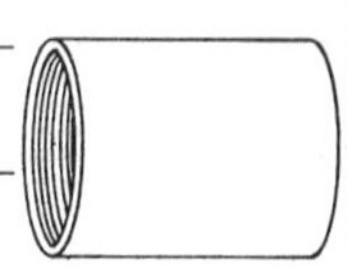

16132 | Conduit & Tubing

205

R260533-20

	16132 Conduit & Tubing	CREW	DAILY OUTPUT	LABOR-HOURS	UNIT	2006 BARE COSTS MAT.	LABOR	EQUIP.	TOTAL	TOTAL INCL O&P
2480	5″ diameter				Ea.	72.50			72.50	79.50
2490	6″ diameter					100			100	110
2500	Steel, intermediate conduit (IMC), 1/2″ diameter	1 Elec	100	.080	L.F.	2.14	3.36		5.50	7.55
2530	3/4″ diameter		90	.089		2.53	3.73		6.26	8.55
2550	1″ diameter		70	.114		3.65	4.80		8.45	11.40
2570	1-1/4″ diameter		65	.123		4.80	5.15		9.95	13.25
2600	1-1/2″ diameter 		60	.133		5.60	5.60		11.20	14.80
2630	2″ diameter		50	.160		7.15	6.70		13.85	18.20
2650	2-1/2″ diameter		40	.200		14.65	8.40		23.05	29
2670	3″ diameter	2 Elec	60	.267		19.10	11.20		30.30	38.50
2700	3-1/2″ diameter		54	.296		23.50	12.45		35.95	45
2730	4″ diameter		50	.320		27	13.45		40.45	50
2731	Field bends, 45° to 90°, 1/2″ diameter	1 Elec	44	.182	Ea.		7.65		7.65	11.75
2732	3/4″ diameter		40	.200			8.40		8.40	12.95
2733	1″ diameter		36	.222			9.35		9.35	14.40
2734	1-1/4″ diameter		19	.421			17.70		17.70	27
2735	1-1/2″ diameter		18	.444			18.65		18.65	29
2736	2″ diameter		13	.615			26		26	40
2750	Elbows, 1/2″ diameter		32	.250		8.35	10.50		18.85	25.50
2770	3/4″ diameter		28	.286		10.20	12		22.20	29.50
2800	1″ diameter		24	.333		14.70	14		28.70	37.50
2830	1-1/4″ diameter 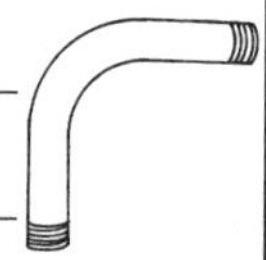		18	.444		24	18.65		42.65	55
2850	1-1/2″ diameter		16	.500		26.50	21		47.50	62
2870	2″ diameter		12	.667		38.50	28		66.50	85.50
2900	2-1/2″ diameter		8	1		68.50	42		110.50	140
2930	3″ diameter		6	1.333		104	56		160	202
2950	3-1/2″ diameter		4.20	1.905		159	80		239	298
2970	4″ diameter		4	2		185	84		269	335
3000	Couplings, 1/2″ diameter					1.91			1.91	2.10
3030	3/4″ diameter					2.34			2.34	2.57
3050	1″ diameter					3.49			3.49	3.84
3070	1-1/4″ diameter					4.34			4.34	4.77
3100	1-1/2″ diameter					5.50			5.50	6.05
3130	2″ diameter 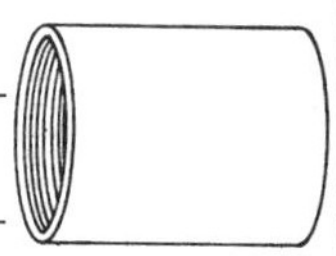					7.25			7.25	7.95
3150	2-1/2″ diameter					17.95			17.95	19.75
3170	3″ diameter					23			23	25.50
3200	3-1/2″ diameter					31			31	34
3230	4″ diameter					31.50			31.50	34.50
4100	Rigid steel, plastic coated, 40 mil. thick									
4130	1/2″ diameter	1 Elec	80	.100	L.F.	4.55	4.20		8.75	11.45
4150	3/4″ diameter		70	.114		5.15	4.80		9.95	13.05
4170	1″ diameter		55	.145		6.40	6.10		12.50	16.45
4200	1-1/4″ diameter 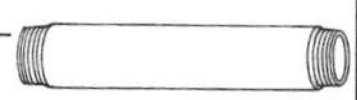		50	.160		8.15	6.70		14.85	19.30
4230	1-1/2″ diameter		45	.178		9.75	7.45		17.20	22.50
4250	2″ diameter		35	.229		12.90	9.60		22.50	29
4270	2-1/2″ diameter		25	.320		21	13.45		34.45	43.50
4300	3″ diameter	2 Elec	44	.364		26.50	15.25		41.75	52.50
4330	3-1/2″ diameter		40	.400		36	16.80		52.80	65.50
4350	4″ diameter		36	.444		38.50	18.65		57.15	71
4370	5″ diameter		30	.533		76	22.50		98.50	119
4400	Elbows, 1/2″ diameter	1 Elec	28	.286	Ea.	13.60	12		25.60	33.50
4430	3/4″ diameter 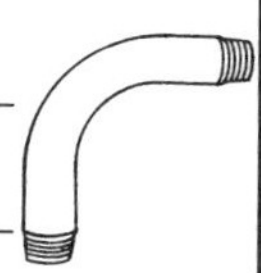		24	.333		14.20	14		28.20	37
4450	1″ diameter		18	.444		16.20	18.65		34.85	47
4470	1-1/4″ diameter		16	.500		19.95	21		40.95	54.50
4500	1-1/2″ diameter		12	.667		24.50	28		52.50	70
4530	2″ diameter		8	1		34	42		76	102

205

16132 | Conduit & Tubing

205

	16132 Conduit & Tubing	CREW	DAILY OUTPUT	LABOR-HOURS	UNIT	2006 BARE COSTS MAT.	LABOR	EQUIP.	TOTAL	TOTAL INCL O&P
4550	2-1/2" diameter R260533-20	1 Elec	6	1.333	Ea.	64.50	56		120.50	157
4570	3" diameter		4.20	1.905		104	80		184	237
4600	3-1/2" diameter		4	2		135	84		219	278
4630	4" diameter		3.80	2.105		147	88.50		235.50	298
4650	5" diameter		3.50	2.286		355	96		451	540
4680	Couplings, 1/2" diameter					3.90			3.90	4.29
4700	3/4" diameter					4.08			4.08	4.49
4730	1" diameter					5.30			5.30	5.85
4750	1-1/4" diameter					6.20			6.20	6.80
4770	1-1/2" diameter					7.35			7.35	8.10
4800	2" diameter					10.70			10.70	11.75
4830	2-1/2" diameter					26.50			26.50	29.50
4850	3" diameter					32.50			32.50	35.50
4870	3-1/2" diameter					41.50			41.50	45.50
4900	4" diameter					48.50			48.50	53.50
4950	5" diameter					158			158	174
5000	Electric metallic tubing (EMT), 1/2" diameter	1 Elec	170	.047	L.F.	.60	1.98		2.58	3.70
5020	3/4" diameter		130	.062		.98	2.58		3.56	5.05
5040	1" diameter		115	.070		1.74	2.92		4.66	6.40
5060	1-1/4" diameter		100	.080		2.67	3.36		6.03	8.15
5080	1-1/2" diameter		90	.089		3.37	3.73		7.10	9.45
5100	2" diameter		80	.100		4.27	4.20		8.47	11.15
5120	2-1/2" diameter		60	.133		8.15	5.60		13.75	17.60
5140	3" diameter	2 Elec	100	.160		10.95	6.70		17.65	22.50
5160	3-1/2" diameter		90	.178		14.05	7.45		21.50	27
5180	4" diameter		80	.200		15.95	8.40		24.35	30.50
5200	Field bends, 45° to 90°, 1/2" diameter	1 Elec	89	.090	Ea.		3.78		3.78	5.80
5220	3/4" diameter		80	.100			4.20		4.20	6.45
5240	1" diameter		73	.110			4.60		4.60	7.10
5260	1-1/4" diameter		38	.211			8.85		8.85	13.60
5280	1-1/2" diameter		36	.222			9.35		9.35	14.40
5300	2" diameter		26	.308			12.90		12.90	19.90
5320	Offsets, 1/2" diameter		65	.123			5.15		5.15	7.95
5340	3/4" diameter		62	.129			5.40		5.40	8.35
5360	1" diameter		53	.151			6.35		6.35	9.75
5380	1-1/4" diameter		30	.267			11.20		11.20	17.25
5400	1-1/2" diameter		28	.286			12		12	18.50
5420	2" diameter		20	.400			16.80		16.80	26
5700	Elbows, 1" diameter		40	.200		5.90	8.40		14.30	19.45
5720	1-1/4" diameter		32	.250		7.80	10.50		18.30	25
5740	1-1/2" diameter		24	.333		9.10	14		23.10	31.50
5760	2" diameter		20	.400		13.30	16.80		30.10	40.50
5780	2-1/2" diameter		12	.667		32.50	28		60.50	78.50
5800	3" diameter		9	.889		48.50	37.50		86	111
5820	3-1/2" diameter		7	1.143		65	48		113	146
5840	4" diameter		6	1.333		76	56		132	170
6200	Couplings, set screw, steel, 1/2" diameter					1.10			1.10	1.21
6220	3/4" diameter					1.67			1.67	1.84
6240	1" diameter					2.78			2.78	3.06
6260	1-1/4" diameter					5.85			5.85	6.40
6280	1-1/2" diameter					8.10			8.10	8.90
6300	2" diameter					10.90			10.90	11.95
6320	2-1/2" diameter					26.50			26.50	29
6340	3" diameter					37.50			37.50	41
6360	3-1/2" diameter					41.50			41.50	45.50
6380	4" diameter					44.50			44.50	49

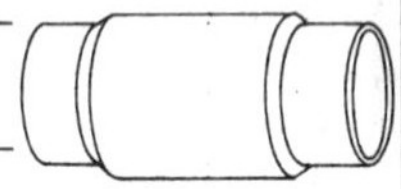
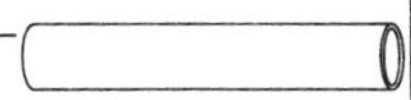
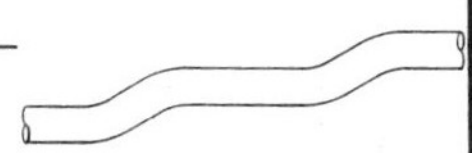
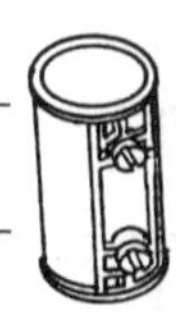

16 ELECTRICAL

16132 | Conduit & Tubing

205			CREW	DAILY OUTPUT	LABOR-HOURS	UNIT	2006 BARE COSTS MAT.	LABOR	EQUIP.	TOTAL	TOTAL INCL O&P
	6500	Box connectors, set screw, steel, 1/2" diameter (R260533-20)	1 Elec	120	.067	Ea.	.81	2.80		3.61	5.20
	6520	3/4" diameter		110	.073		1.32	3.05		4.37	6.15
	6540	1" diameter		90	.089		2.51	3.73		6.24	8.50
	6560	1-1/4" diameter		70	.114		4.88	4.80		9.68	12.75
	6580	1-1/2" diameter		60	.133		7.05	5.60		12.65	16.40
	6600	2" diameter		50	.160		9.75	6.70		16.45	21
	6620	2-1/2" diameter		36	.222		34	9.35		43.35	52
	6640	3" diameter		27	.296		41	12.45		53.45	64
	6680	3-1/2" diameter		21	.381		57	16		73	87
	6700	4" diameter		16	.500		63.50	21		84.50	103
	6740	Insulated box connectors, set screw, steel, 1/2" diameter		120	.067		.98	2.80		3.78	5.40
	6760	3/4" diameter		110	.073		1.55	3.05		4.60	6.40
	6780	1" diameter		90	.089		2.84	3.73		6.57	8.85
	6800	1-1/4" diameter		70	.114		5.20	4.80		10	13.10
	6820	1-1/2" diameter		60	.133		7.30	5.60		12.90	16.70
	6840	2" diameter		50	.160		10.60	6.70		17.30	22
	6860	2-1/2" diameter		36	.222		50.50	9.35		59.85	70
	6880	3" diameter		27	.296		59.50	12.45		71.95	84.50
	6900	3-1/2" diameter		21	.381		83	16		99	116
	6920	4" diameter		16	.500		90.50	21		111.50	132
	7000	EMT to conduit adapters, 1/2" diameter (compression)		70	.114		4.05	4.80		8.85	11.85
	7020	3/4" diameter		60	.133		5.75	5.60		11.35	15
	7040	1" diameter		50	.160		8.80	6.70		15.50	20
	7060	1-1/4" diameter		40	.200		15.20	8.40		23.60	29.50
	7080	1-1/2" diameter		30	.267		18.80	11.20		30	38
	7100	2" diameter		25	.320		27	13.45		40.45	50.50
	7200	EMT to Greenfield adapters, 1/2" to 3/8" diameter (compression)		90	.089		2.97	3.73		6.70	9
	7220	1/2" diameter		90	.089		5.55	3.73		9.28	11.85
	7240	3/4" diameter		80	.100		7.10	4.20		11.30	14.25
	7260	1" diameter		70	.114		18.80	4.80		23.60	28
	7270	1-1/4" diameter		60	.133		23.50	5.60		29.10	34.50
	7280	1-1/2" diameter		50	.160		26.50	6.70		33.20	39.50
	7290	2" diameter		40	.200		39.50	8.40		47.90	56.50
	7400	EMT, LB, LR or LL fittings with covers, 1/2" dia., set screw		24	.333		10.40	14		24.40	33
	7420	3/4" diameter		20	.400		12.80	16.80		29.60	40
	7440	1" diameter		16	.500		19.70	21		40.70	54
	7450	1-1/4" diameter		13	.615		29	26		55	72
	7460	1-1/2" diameter		11	.727		34.50	30.50		65	85
	7470	2" diameter		9	.889		55	37.50		92.50	118
	7600	EMT, "T" fittings with covers, 1/2" diameter, set screw		16	.500		13.40	21		34.40	47.50
	7620	3/4" diameter		15	.533		17	22.50		39.50	53
	7640	1" diameter		12	.667		24.50	28		52.50	70
	7650	1-1/4" diameter		11	.727		35	30.50		65.50	85.50
	7660	1-1/2" diameter		10	.800		43.50	33.50		77	100
	7670	2" diameter		8	1		57.50	42		99.50	128
	8000	EMT, expansion fittings, no jumper, 1/2" diameter		24	.333		41	14		55	66.50
	8020	3/4" diameter		20	.400		47.50	16.80		64.30	78.50
	8040	1" diameter		16	.500		56	21		77	94
	8060	1-1/4" diameter		13	.615		76	26		102	124
	8080	1-1/2" diameter		11	.727		104	30.50		134.50	161
	8100	2" diameter		9	.889		155	37.50		192.50	228
	8110	2-1/2" diameter		7	1.143		244	48		292	340
	8120	3" diameter		6	1.333		300	56		356	415
	8140	4" diameter		5	1.600		525	67		592	685
	8200	Split adapter, 1/2" diameter		110	.073		3.10	3.05		6.15	8.10
	8210	3/4" diameter		90	.089		2.80	3.73		6.53	8.85

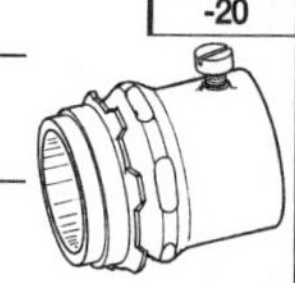

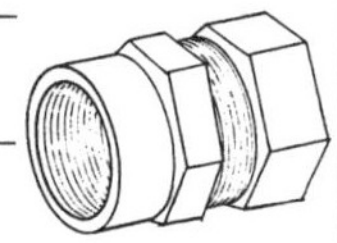

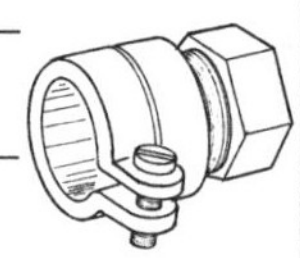

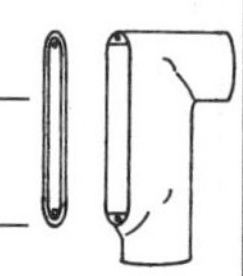

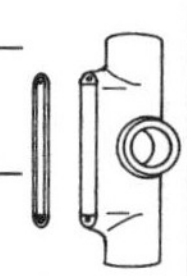

ELECTRICAL 16

16132 | Conduit & Tubing

R260533-20

16132	Conduit & Tubing	CREW	DAILY OUTPUT	LABOR-HOURS	UNIT	2006 BARE COSTS MAT.	LABOR	EQUIP.	TOTAL	TOTAL INCL O&P
205 8220	1" diameter	1 Elec	70	.114	Ea.	3.80	4.80		8.60	11.60
8230	1-1/4" diameter 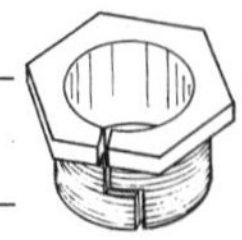		60	.133		6.20	5.60		11.80	15.45
8240	1-1/2" diameter		50	.160		9.50	6.70		16.20	21
8250	2" diameter		36	.222		28	9.35		37.35	45
8300	1 hole clips, 1/2" diameter		500	.016		.28	.67		.95	1.35
8320	3/4" diameter		470	.017		.37	.71		1.08	1.51
8340	1" diameter		444	.018		.68	.76		1.44	1.92
8360	1-1/4" diameter 		400	.020		.90	.84		1.74	2.28
8380	1-1/2" diameter		355	.023		1.38	.95		2.33	2.98
8400	2" diameter		320	.025		2.17	1.05		3.22	4.01
8420	2-1/2" diameter		266	.030		4.03	1.26		5.29	6.40
8440	3" diameter		160	.050		5.05	2.10		7.15	8.80
8460	3-1/2" diameter		133	.060		8	2.53		10.53	12.70
8480	4" diameter		100	.080		10.30	3.36		13.66	16.55
8500	Clamp back spacers, 1/2" diameter		500	.016		.84	.67		1.51	1.96
8510	3/4" diameter		470	.017		1	.71		1.71	2.20
8520	1" diameter		444	.018		1.76	.76		2.52	3.11
8530	1-1/4" diameter		400	.020		3.14	.84		3.98	4.74
8540	1-1/2" diameter 		355	.023		3.55	.95		4.50	5.35
8550	2" diameter		320	.025		5.95	1.05		7	8.15
8560	2-1/2" diameter		266	.030		11.40	1.26		12.66	14.50
8570	3" diameter		160	.050		16.30	2.10		18.40	21
8580	3-1/2" diameter		133	.060		22	2.53		24.53	28.50
8590	4" diameter		100	.080		48	3.36		51.36	58
8600	Offset connectors, 1/2" diameter 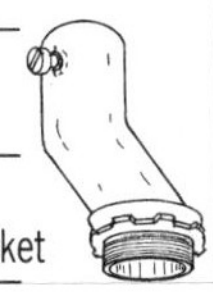		40	.200		3.03	8.40		11.43	16.30
8610	3/4" diameter		32	.250		4.40	10.50		14.90	21
8620	1" diameter		24	.333		8.15	14		22.15	30.50
8650	90° pulling elbows, female, 1/2" diameter, with gasket		24	.333		4.70	14		18.70	26.50
8660	3/4" diameter		20	.400		7.35	16.80		24.15	34
8700	Couplings, compression, 1/2" diameter, steel					2.24			2.24	2.46
8710	3/4" diameter					3.10			3.10	3.41
8720	1" diameter					5.35			5.35	5.90
8730	1-1/4" diameter					10.15			10.15	11.15
8740	1-1/2" diameter 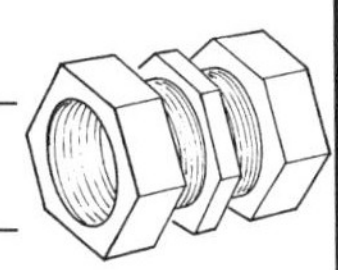					14.80			14.80	16.30
8750	2" diameter					20			20	22
8760	2-1/2" diameter					90			90	99
8770	3" diameter					113			113	124
8780	3-1/2" diameter					183			183	201
8790	4" diameter					187			187	206
8800	Box connectors, compression, 1/2" diam., steel	1 Elec	120	.067		2.60	2.80		5.40	7.15
8810	3/4" diameter		110	.073		3.80	3.05		6.85	8.90
8820	1" diameter		90	.089		6.20	3.73		9.93	12.55
8830	1-1/4" diameter		70	.114		12.25	4.80		17.05	21
8840	1-1/2" diameter		60	.133		17.90	5.60		23.50	28.50
8850	2" diameter		50	.160		25.50	6.70		32.20	38.50
8860	2-1/2" diameter 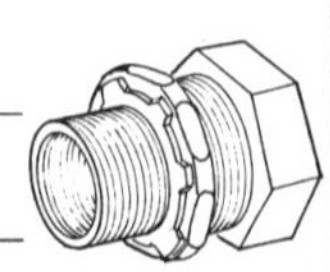		36	.222		80	9.35		89.35	102
8870	3" diameter		27	.296		101	12.45		113.45	130
8880	3-1/2" diameter		21	.381		153	16		169	193
8890	4" diameter		16	.500		169	21		190	219
8900	Box connectors, insulated compression, 1/2" diam., steel		120	.067		3.35	2.80		6.15	8
8910	3/4" diameter		110	.073		4.60	3.05		7.65	9.75
8920	1" diameter 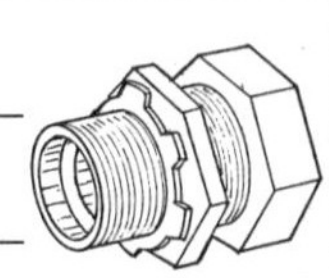		90	.089		7.65	3.73		11.38	14.15
8930	1-1/4" diameter		70	.114		15.90	4.80		20.70	25
8940	1-1/2" diameter		60	.133		24	5.60		29.60	35
8950	2" diameter		50	.160		33.50	6.70		40.20	47
8960	2-1/2" diameter		36	.222		99.50	9.35		108.85	123

16132 | Conduit & Tubing

205	Description	Crew	Daily Output	Labor-Hours	Unit	2006 Bare Costs Mat.	Labor	Equip.	Total	Total Incl O&P
8970	3″ diameter (R260533-20)	1 Elec	27	.296	Ea.	135	12.45		147.45	168
8980	3-1/2″ diameter		21	.381		224	16		240	271
8990	4″ diameter		16	.500		201	21		222	254
9100	PVC, #40, 1/2″ diameter		190	.042	L.F.	.86	1.77		2.63	3.67
9110	3/4″ diameter		145	.055		1.02	2.32		3.34	4.70
9120	1″ diameter		125	.064		1.59	2.69		4.28	5.90
9130	1-1/4″ diameter		110	.073		2.06	3.05		5.11	6.95
9140	1-1/2″ diameter		100	.080		2.34	3.36		5.70	7.75
9150	2″ diameter		90	.089		2.98	3.73		6.71	9.05
9160	2-1/2″ diameter		65	.123		4.99	5.15		10.14	13.45
9170	3″ diameter	2 Elec	110	.145		5.55	6.10		11.65	15.50
9180	3-1/2″ diameter		100	.160		7.55	6.70		14.25	18.65
9190	4″ diameter		90	.178		8.15	7.45		15.60	20.50
9200	5″ diameter		70	.229		11.45	9.60		21.05	27.50
9210	6″ diameter		60	.267		15.90	11.20		27.10	35
9220	Elbows, 1/2″ diameter	1 Elec	50	.160	Ea.	1.58	6.70		8.28	12.10
9230	3/4″ diameter		42	.190		1.59	8		9.59	14.05
9240	1″ diameter		35	.229		2.69	9.60		12.29	17.75
9250	1-1/4″ diameter		28	.286		3.83	12		15.83	22.50
9260	1-1/2″ diameter		20	.400		5.20	16.80		22	31.50
9270	2″ diameter		16	.500		7.55	21		28.55	41
9280	2-1/2″ diameter		11	.727		13.15	30.50		43.65	61.50
9290	3″ diameter		9	.889		23	37.50		60.50	83
9300	3-1/2″ diameter		7	1.143		32	48		80	109
9310	4″ diameter		6	1.333		40	56		96	131
9320	5″ diameter		4	2		70	84		154	206
9330	6″ diameter		3	2.667		119	112		231	305
9340	Field bends, 45° & 90°, 1/2″ diameter		45	.178			7.45		7.45	11.50
9350	3/4″ diameter		40	.200			8.40		8.40	12.95
9360	1″ diameter		35	.229			9.60		9.60	14.80
9370	1-1/4″ diameter		32	.250			10.50		10.50	16.20
9380	1-1/2″ diameter		27	.296			12.45		12.45	19.15
9390	2″ diameter		20	.400			16.80		16.80	26
9400	2-1/2″ diameter		16	.500			21		21	32.50
9410	3″ diameter		13	.615			26		26	40
9420	3-1/2″ diameter		12	.667			28		28	43
9430	4″ diameter		10	.800			33.50		33.50	52
9440	5″ diameter		9	.889			37.50		37.50	57.50
9450	6″ diameter		8	1			42		42	64.50
9460	PVC adapters, 1/2″ diameter		50	.160		.51	6.70		7.21	10.90
9470	3/4″ diameter		42	.190		.94	8		8.94	13.35
9480	1″ diameter		38	.211		1.19	8.85		10.04	14.90
9490	1-1/4″ diameter		35	.229		1.59	9.60		11.19	16.55
9500	1-1/2″ diameter		32	.250		1.85	10.50		12.35	18.25
9510	2″ diameter		27	.296		2.66	12.45		15.11	22
9520	2-1/2″ diameter		23	.348		4.57	14.60		19.17	27.50
9530	3″ diameter		18	.444		6.75	18.65		25.40	36.50
9540	3-1/2″ diameter		13	.615		8.70	26		34.70	49.50
9550	4″ diameter		11	.727		11.80	30.50		42.30	60
9560	5″ diameter		8	1		23	42		65	90
9570	6″ diameter		6	1.333		28	56		84	118
9580	PVC-LB, LR or LL fittings & covers									
9590	1/2″ diameter	1 Elec	20	.400	Ea.	5.50	16.80		22.30	32
9600	3/4″ diameter		16	.500		7.10	21		28.10	40.50
9610	1″ diameter		12	.667		7.80	28		35.80	51.50
9620	1-1/4″ diameter		9	.889		11.90	37.50		49.40	70.50

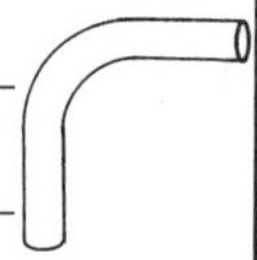
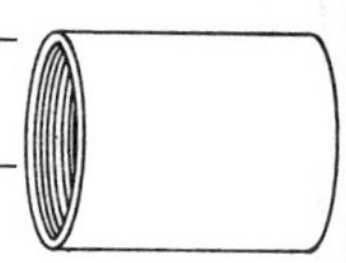
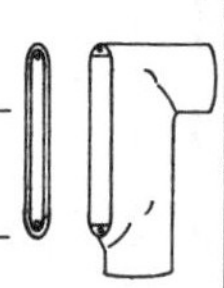

ELECTRICAL 16

	16132 | Conduit & Tubing		CREW	DAILY OUTPUT	LABOR-HOURS	UNIT	2006 BARE COSTS MAT.	LABOR	EQUIP.	TOTAL	TOTAL INCL O&P	
205	9630	1-1/2" diameter	1 Elec	7	1.143	Ea.	14.30	48		62.30	90	205
	9640	2" diameter		6	1.333		25.50	56		81.50	115	
	9650	2-1/2" diameter		6	1.333		92	56		148	188	
	9660	3" diameter		5	1.600		95	67		162	208	
	9670	3-1/2" diameter		4	2		102	84		186	241	
	9680	4" diameter		3	2.667		104	112		216	287	
	9690	PVC-tee fitting & cover										
	9700	1/2"	1 Elec	14	.571	Ea.	7.15	24		31.15	45	
	9710	3/4"		13	.615		8.70	26		34.70	49.50	
	9720	1"		10	.800		9.10	33.50		42.60	62	
	9730	1-1/4"		9	.889		14.80	37.50		52.30	74	
	9740	1-1/2"		8	1		19.50	42		61.50	86	
	9750	2"		7	1.143		28	48		76	105	
	9760	PVC-reducers, 3/4" x 1/2" diameter					1.30			1.30	1.43	
	9770	1" x 1/2" diameter					2.90			2.90	3.19	
	9780	1" x 3/4" diameter					3.10			3.10	3.41	
	9790	1-1/4" x 3/4" diameter					4.05			4.05	4.46	
	9800	1-1/4" x 1" diameter					4.05			4.05	4.46	
	9810	1-1/2" x 1-1/4" diameter					4.20			4.20	4.62	
	9820	2" x 1-1/4" diameter					5.10			5.10	5.60	
	9830	2-1/2" x 2" diameter					16.45			16.45	18.10	
	9840	3" x 2" diameter					17.55			17.55	19.30	
	9850	4" x 3" diameter					22			22	24	
	9860	Cement, quart					14.70			14.70	16.15	
	9870	Gallon					62.50			62.50	68.50	
	9880	Heat bender, to 6" diameter					1,225			1,225	1,350	
	9900	Add to labor for higher elevated installation										
	9910	15' to 20' high, add						10%				
	9920	20' to 25' high, add						20%				
	9930	25' to 30' high, add						25%				
	9940	30' to 35' high, add						30%				
	9950	35' to 40' high, add						35%				
	9960	Over 40' high, add						40%				
	9990	Minimum labor/equipment charge	1 Elec	4	2	Job		84		84	129	
210	0010	**CONDUIT** To 15' high, includes couplings only										210
	0200	Electric metallic tubing, 1/2"diameter	1 Elec	435	.018	L.F.	.44	.77		1.21	1.68	
	0220	3/4" diameter		253	.032		.80	1.33		2.13	2.93	
	0240	1" diameter		207	.039		1.41	1.62		3.03	4.05	
	0260	1-1/4"diameter		173	.046		2.25	1.94		4.19	5.45	
	0280	1-1/2" diameter		153	.052		2.88	2.20		5.08	6.55	
	0300	2" diameter		130	.062		3.61	2.58		6.19	7.95	
	0320	2-1/2" diameter		92	.087		6.40	3.65		10.05	12.70	
	0340	3" diameter	2 Elec	148	.108		8.75	4.54		13.29	16.60	
	0360	3-1/2" diameter		134	.119		11.15	5		16.15	20	
	0380	4" diameter		114	.140		12.40	5.90		18.30	23	
	0500	Steel rigid galvanized, 1/2" diameter	1 Elec	146	.055		1.76	2.30		4.06	5.50	
	0520	3/4" diameter		125	.064		2.08	2.69		4.77	6.45	
	0540	1" diameter		93	.086		3.28	3.61		6.89	9.15	
	0560	1-1/4" diameter		88	.091		4.44	3.82		8.26	10.80	
	0580	1-1/2" diameter		80	.100		5.15	4.20		9.35	12.10	
	0600	2" diameter		65	.123		6.60	5.15		11.75	15.20	
	0620	2-1/2" diameter		48	.167		12.75	7		19.75	25	
	0640	3" diameter	2 Elec	64	.250		15.55	10.50		26.05	33.50	
	0660	3-1/2" diameter		60	.267		19.05	11.20		30.25	38.50	
	0680	4" diameter		52	.308		21	12.90		33.90	43	
	0700	5" diameter		50	.320		43.50	13.45		56.95	68.50	

R260533 -20

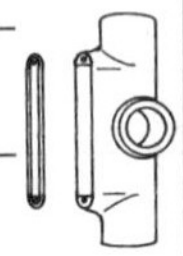

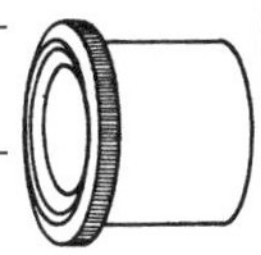

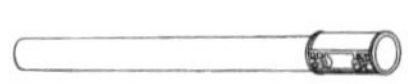

	16132 | Conduit & Tubing		CREW	DAILY OUTPUT	LABOR-HOURS	UNIT	2006 BARE COSTS MAT.	LABOR	EQUIP.	TOTAL	TOTAL INCL O&P	
210	0720	6″ diameter	2 Elec	48	.333	L.F.	63	14		77	90.50	210
	1000	Steel intermediate conduit (IMC), 1/2 diameter	1 Elec	155	.052		1.56	2.17		3.73	5.05	
	1010	3/4″ diameter		130	.062		1.83	2.58		4.41	6	
	1020	1″ diameter		100	.080		2.84	3.36		6.20	8.35	
	1030	1-1/4″ diameter		93	.086		3.60	3.61		7.21	9.50	
	1040	1-1/2″ diameter		85	.094		4.32	3.95		8.27	10.85	
	1050	2″ diameter		70	.114		5.60	4.80		10.40	13.55	
	1060	2-1/2″ diameter		53	.151		12.45	6.35		18.80	23.50	
	1070	3″ diameter	2 Elec	80	.200		15.75	8.40		24.15	30.50	
	1080	3-1/2″ diameter		70	.229		18.75	9.60		28.35	35.50	
	1090	4″ diameter		60	.267		21	11.20		32.20	40.50	
220	0010	**CONDUIT NIPPLES** With locknuts and bushings										220
	0100	Aluminum, 1/2″ diameter, close	1 Elec	36	.222	Ea.	4.52	9.35		13.87	19.35	
	0120	1-1/2″ long		36	.222		4.60	9.35		13.95	19.45	
	0140	2″ long		36	.222		4.82	9.35		14.17	19.70	
	0160	2-1/2″ long		36	.222		5.40	9.35		14.75	20.50	
	0180	3″ long		36	.222		5.60	9.35		14.95	20.50	
	0200	3-1/2″ long		36	.222		5.90	9.35		15.25	21	
	0220	4″ long		36	.222		6.20	9.35		15.55	21.50	
	0240	5″ long		36	.222		6.90	9.35		16.25	22	
	0260	6″ long		36	.222		7.15	9.35		16.50	22.50	
	0280	8″ long		36	.222		9	9.35		18.35	24.50	
	0300	10″ long		36	.222		10.60	9.35		19.95	26	
	0320	12″ long		36	.222		11.95	9.35		21.30	27.50	
	0340	3/4″ diameter, close		32	.250		6.40	10.50		16.90	23.50	
	0360	1-1/2″ long		32	.250		6.55	10.50		17.05	23.50	
	0380	2″ long		32	.250		6.55	10.50		17.05	23.50	
	0400	2-1/2″ long		32	.250		6.70	10.50		17.20	23.50	
	0420	3″ long		32	.250		6.95	10.50		17.45	24	
	0440	3-1/2″ long		32	.250		7.30	10.50		17.80	24.50	
	0460	4″ long		32	.250		7.50	10.50		18	24.50	
	0480	5″ long		32	.250		7.75	10.50		18.25	25	
	0500	6″ long		32	.250		8.75	10.50		19.25	26	
	0520	8″ long		32	.250		9.60	10.50		20.10	27	
	0540	10″ long		32	.250		11.80	10.50		22.30	29	
	0560	12″ long		32	.250		13.40	10.50		23.90	31	
	0580	1″ diameter, close		27	.296		9.20	12.45		21.65	29.50	
	0600	2″ long		27	.296		9.70	12.45		22.15	30	
	0620	2-1/2″ long		27	.296		10.20	12.45		22.65	30.50	
	0640	3″ long		27	.296		10.60	12.45		23.05	31	
	0660	3-1/2″ long		27	.296		11.40	12.45		23.85	31.50	
	0680	4″ long		27	.296		12	12.45		24.45	32.50	
	0700	5″ long		27	.296		13.40	12.45		25.85	34	
	0720	6″ long		27	.296		14.95	12.45		27.40	35.50	
	0740	8″ long		27	.296		17.45	12.45		29.90	38.50	
	0760	10″ long		27	.296		21	12.45		33.45	42	
	0780	12″ long		27	.296		24	12.45		36.45	45	
	0800	1-1/4″ diameter, close		23	.348		12.65	14.60		27.25	36.50	
	0820	2″ long		23	.348		12.80	14.60		27.40	36.50	
	0840	2-1/2″ long		23	.348		13.45	14.60		28.05	37.50	
	0860	3″ long		23	.348		14.35	14.60		28.95	38.50	
	0880	3-1/2″ long		23	.348		15.45	14.60		30.05	39.50	
	0900	4″ long		23	.348		16.35	14.60		30.95	40.50	
	0920	5″ long		23	.348		18.10	14.60		32.70	42.50	
	0940	6″ long		23	.348		19.95	14.60		34.55	44.50	

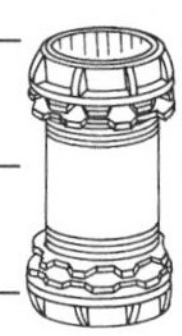

16132 | Conduit & Tubing

		Description	Crew	Daily Output	Labor-Hours	Unit	2006 Bare Costs Mat.	2006 Bare Costs Labor	2006 Bare Costs Equip.	2006 Bare Costs Total	Total Incl O&P
220	0960	8″ long	1 Elec	23	.348	Ea.	23.50	14.60		38.10	48.50
	0980	10″ long		23	.348		27.50	14.60		42.10	52.50
	1000	12″ long		23	.348		31	14.60		45.60	56.50
	1020	1-1/2″ diameter, close		20	.400		17.45	16.80		34.25	45
	1040	2″ long		20	.400		17.70	16.80		34.50	45.50
	1060	2-1/2″ long		20	.400		18.20	16.80		35	46
	1080	3″ long		20	.400		19.35	16.80		36.15	47.50
	1100	3-1/2″ long		20	.400		22	16.80		38.80	50
	1120	4″ long		20	.400		22	16.80		38.80	50
	1140	5″ long		20	.400		23.50	16.80		40.30	52
	1160	6″ long		20	.400		26	16.80		42.80	54.50
	1180	8″ long		20	.400		30.50	16.80		47.30	59.50
	1200	10″ long		20	.400		35	16.80		51.80	64.50
	1220	12″ long		20	.400		39	16.80		55.80	69
	1240	2″ diameter, close		18	.444		24.50	18.65		43.15	56
	1260	2-1/2″ long		18	.444		25	18.65		43.65	56.50
	1280	3″ long		18	.444		26.50	18.65		45.15	58
	1300	3-1/2″ long		18	.444		28	18.65		46.65	60
	1320	4″ long		18	.444		29	18.65		47.65	61
	1340	5″ long		18	.444		32	18.65		50.65	64
	1360	6″ long		18	.444		34.50	18.65		53.15	67
	1380	8″ long		18	.444		40	18.65		58.65	73.50
	1400	10″ long		18	.444		46	18.65		64.65	79.50
	1420	12″ long		18	.444		51.50	18.65		70.15	85.50
	1440	2-1/2″ diameter, close		15	.533		51.50	22.50		74	91
	1460	3″ long		15	.533		52	22.50		74.50	92
	1480	3-1/2″ long		15	.533		54.50	22.50		77	94.50
	1500	4″ long		15	.533		56	22.50		78.50	96
	1520	5″ long		15	.533		59	22.50		81.50	99.50
	1540	6″ long		15	.533		61	22.50		83.50	102
	1560	8″ long		15	.533		70	22.50		92.50	112
	1580	10″ long		15	.533		78.50	22.50		101	121
	1600	12″ long		15	.533		86	22.50		108.50	130
	1620	3″ diameter, close		12	.667		59.50	28		87.50	109
	1640	3″ long		12	.667		62.50	28		90.50	112
	1660	3-1/2″ long		12	.667		67	28		95	117
	1680	4″ long		12	.667		68.50	28		96.50	118
	1700	5″ long		12	.667		72.50	28		100.50	123
	1720	6″ long		12	.667		77.50	28		105.50	129
	1740	8″ long		12	.667		88.50	28		116.50	141
	1760	10″ long		12	.667		100	28		128	153
	1780	12″ long		12	.667		111	28		139	165
	1800	3-1/2″ diameter, close		11	.727		110	30.50		140.50	168
	1820	4″ long		11	.727		120	30.50		150.50	178
	1840	5″ long		11	.727		124	30.50		154.50	184
	1860	6″ long		11	.727		131	30.50		161.50	191
	1880	8″ long		11	.727		143	30.50		173.50	204
	1900	10″ long		11	.727		157	30.50		187.50	220
	1920	12″ long		11	.727		169	30.50		199.50	233
	1940	4″ diameter, close		9	.889		132	37.50		169.50	203
	1960	4″ long		9	.889		140	37.50		177.50	213
	1980	5″ long		9	.889		147	37.50		184.50	220
	2000	6″ long		9	.889		155	37.50		192.50	228
	2020	8″ long		9	.889		170	37.50		207.50	246
	2040	10″ long		9	.889		186	37.50		223.50	263
220	2060	12″ long	↓	9	.889	↓	202	37.50		239.50	280

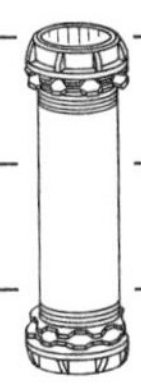

16132 | Conduit & Tubing

				DAILY OUTPUT	LABOR-HOURS		2006 BARE COSTS				TOTAL	
			CREW			UNIT	MAT.	LABOR	EQUIP.	TOTAL	INCL O&P	
220	2080	5″ diameter, close	1 Elec	7	1.143	Ea.	290	48		338	395	220
	2100	5″ long		7	1.143		305	48		353	410	
	2120	6″ long		7	1.143		310	48		358	415	
	2140	8″ long		7	1.143		335	48		383	445	
	2160	10″ long		7	1.143		360	48		408	470	
	2180	12″ long		7	1.143		385	48		433	495	
	2200	6″ diameter, close		6	1.333		480	56		536	610	
	2220	5″ long		6	1.333		505	56		561	640	
	2240	6″ long		6	1.333		515	56		571	650	
	2260	8″ long		6	1.333		550	56		606	690	
	2280	10″ long		6	1.333		585	56		641	730	
	2300	12″ long		6	1.333		605	56		661	750	
	2320	Rigid galvanized steel, 1/2″ diameter, close		32	.250		2.80	10.50		13.30	19.30	
	2340	1-1/2″ long		32	.250		3.08	10.50		13.58	19.60	
	2360	2″ long		32	.250		3.32	10.50		13.82	19.85	
	2380	2-1/2″ long		32	.250		3.47	10.50		13.97	20	
	2400	3″ long		32	.250		3.64	10.50		14.14	20	
	2420	3-1/2″ long		32	.250		3.86	10.50		14.36	20.50	
	2440	4″ long		32	.250		4.08	10.50		14.58	20.50	
	2460	5″ long		32	.250		4.40	10.50		14.90	21	
	2480	6″ long		32	.250		4.99	10.50		15.49	21.50	
	2500	8″ long		32	.250		7.70	10.50		18.20	24.50	
	2520	10″ long		32	.250		8.65	10.50		19.15	25.50	
	2540	12″ long		32	.250		9.75	10.50		20.25	27	
	2560	3/4″ diameter, close		27	.296		3.82	12.45		16.27	23.50	
	2580	2″ long		27	.296		4.15	12.45		16.60	23.50	
	2600	2-1/2″ long		27	.296		4.41	12.45		16.86	24	
	2620	3″ long		27	.296		4.63	12.45		17.08	24.50	
	2640	3-1/2″ long		27	.296		4.75	12.45		17.20	24.50	
	2660	4″ long		27	.296		5.15	12.45		17.60	25	
	2680	5″ long		27	.296		5.60	12.45		18.05	25.50	
	2700	6″ long		27	.296		6.20	12.45		18.65	26	
	2720	8″ long		27	.296		9	12.45		21.45	29	
	2740	10″ long		27	.296		10.40	12.45		22.85	30.50	
	2760	12″ long		27	.296		11.45	12.45		23.90	32	
	2780	1″ diameter, close		23	.348		5.80	14.60		20.40	29	
	2800	2″ long		23	.348		6.10	14.60		20.70	29	
	2820	2-1/2″ long		23	.348		6.35	14.60		20.95	29.50	
	2840	3″ long		23	.348		6.75	14.60		21.35	30	
	2860	3-1/2″ long		23	.348		9.95	14.60		24.55	33.50	
	2880	4″ long		23	.348		7.55	14.60		22.15	31	
	2900	5″ long		23	.348		8.20	14.60		22.80	31.50	
	2920	6″ long		23	.348		8.65	14.60		23.25	32	
	2940	8″ long		23	.348		11.90	14.60		26.50	35.50	
	2960	10″ long		23	.348		14.50	14.60		29.10	38.50	
	2980	12″ long		23	.348		16.10	14.60		30.70	40	
	3000	1-1/4″ diameter, close		20	.400		7.80	16.80		24.60	34.50	
	3020	2″ long		20	.400		8.05	16.80		24.85	35	
	3040	3″ long		20	.400		8.75	16.80		25.55	35.50	
	3060	3-1/2″ long		20	.400		9.40	16.80		26.20	36.50	
	3080	4″ long		20	.400		9.70	16.80		26.50	36.50	
	3100	5″ long		20	.400		10.60	16.80		27.40	37.50	
	3120	6″ long		20	.400		11.40	16.80		28.20	38.50	
	3140	8″ long		20	.400		16.05	16.80		32.85	43.50	
	3160	10″ long		20	.400		19.15	16.80		35.95	47	
	3180	12″ long	↓	20	.400	↓	21.50	16.80		38.30	50	

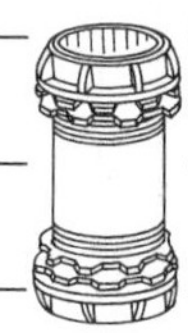

16132 | Conduit & Tubing

		CREW	DAILY OUTPUT	LABOR-HOURS	UNIT	2006 BARE COSTS MAT.	LABOR	EQUIP.	TOTAL	TOTAL INCL O&P
220 3200	1-1/2" diameter, close	1 Elec	18	.444	Ea.	10.65	18.65		29.30	41
3220	2" long		18	.444		10.90	18.65		29.55	41
3240	2-1/2" long		18	.444		11.50	18.65		30.15	41.50
3260	3" long		18	.444		11.90	18.65		30.55	42
3280	3-1/2" long		18	.444		12.75	18.65		31.40	43
3300	4" long		18	.444		13.35	18.65		32	43.50
3320	5" long		18	.444		14.20	18.65		32.85	44.50
3340	6" long		18	.444		16	18.65		34.65	46.50
3360	8" long		18	.444		21	18.65		39.65	52.50
3380	10" long		18	.444		24	18.65		42.65	55.50
3400	12" long		18	.444		26	18.65		44.65	57.50
3420	2" diameter, close		16	.500		14.90	21		35.90	49
3440	2-1/2" long		16	.500		15.80	21		36.80	50
3460	3" long		16	.500		16.80	21		37.80	51
3480	3-1/2" long		16	.500		17.85	21		38.85	52
3500	4" long		16	.500		18.70	21		39.70	53
3520	5" long		16	.500		20.50	21		41.50	55
3540	6" long		16	.500		22	21		43	56.50
3560	8" long		16	.500		27.50	21		48.50	62.50
3580	10" long		16	.500		31	21		52	66.50
3600	12" long		16	.500		34	21		55	69.50
3620	2-1/2" diameter, close		13	.615		37	26		63	80.50
3640	3" long		13	.615		37	26		63	81
3660	3-1/2" long		13	.615		40	26		66	84
3680	4" long		13	.615		40.50	26		66.50	85
3700	5" long		13	.615		44	26		70	88.50
3720	6" long		13	.615		47	26		73	92
3740	8" long		13	.615		55	26		81	101
3760	10" long		13	.615		60.50	26		86.50	107
3780	12" long		13	.615		66.50	26		92.50	114
3800	3" diameter, close		12	.667		45.50	28		73.50	93
3820	3" long		12	.667		48.50	28		76.50	96.50
3900	3-1/2" long		12	.667		48.50	28		76.50	96.50
3920	4" long		12	.667		50.50	28		78.50	98.50
3940	5" long		12	.667		54	28		82	103
3960	6" long		12	.667		58	28		86	107
3980	8" long		12	.667		67	28		95	117
4000	10" long		12	.667		74.50	28		102.50	125
4020	12" long		12	.667		84.50	28		112.50	136
4040	3-1/2" diameter, close		10	.800		73.50	33.50		107	133
4060	4" long		10	.800		79.50	33.50		113	140
4080	5" long		10	.800		83.50	33.50		117	144
4100	6" long		10	.800		88	33.50		121.50	149
4120	8" long		10	.800		97	33.50		130.50	159
4140	10" long		10	.800		106	33.50		139.50	169
4160	12" long		10	.800		116	33.50		149.50	179
4180	4" diameter, close		8	1		90	42		132	163
4200	4" long		8	1		95.50	42		137.50	170
4220	5" long		8	1		101	42		143	176
4240	6" long		8	1		105	42		147	181
4260	8" long		8	1		115	42		157	192
4280	10" long		8	1		127	42		169	205
4300	12" long		8	1		139	42		181	218
4320	5" diameter, close		6	1.333		180	56		236	286
4340	5" long		6	1.333		203	56		259	310
4360	6" long	↓	6	1.333	↓	212	56		268	320

16132 | Conduit & Tubing

		Crew	Daily Output	Labor-Hours	Unit	2006 Bare Costs Mat.	2006 Bare Costs Labor	2006 Bare Costs Equip.	2006 Bare Costs Total	Total Incl O&P
220	4380	8″ long	1 Elec	6	1.333	Ea.	223	56		279	335
	4400	10″ long		6	1.333		240	56		296	350
	4420	12″ long		6	1.333		262	56		318	375
	4440	6″ diameter, close		5	1.600		271	67		338	400
	4460	5″ long		5	1.600		295	67		362	430
	4480	6″ long		5	1.600		305	67		372	440
	4500	8″ long		5	1.600		320	67		387	460
	4520	10″ long		5	1.600		350	67		417	490
	4540	12″ long		5	1.600		365	67		432	505
	4560	Plastic coated, 40 mil thick, 1/2″ diameter, 2″ long		32	.250		11.70	10.50		22.20	29
	4580	2-1/2″ long		32	.250		13	10.50		23.50	30.50
	4600	3″ long		32	.250		13.15	10.50		23.65	30.50
	4680	3-1/2″ long		32	.250		14.50	10.50		25	32
	4700	4″ long		32	.250		14.55	10.50		25.05	32.50
	4720	5″ long		32	.250		14.75	10.50		25.25	32.50
	4740	6″ long		32	.250		15.10	10.50		25.60	33
	4760	8″ long		32	.250		15.65	10.50		26.15	33.50
	4780	10″ long		32	.250		16.30	10.50		26.80	34
	4800	12″ long		32	.250		16.80	10.50		27.30	34.50
	4820	3/4″ diameter, 2″ long		26	.308		12.55	12.90		25.45	33.50
	4840	2-1/2″ long		26	.308		13.90	12.90		26.80	35
	4860	3″ long		26	.308		14	12.90		26.90	35.50
	4880	3-1/2″ long		26	.308		15.35	12.90		28.25	37
	4900	4″ long		26	.308		15.60	12.90		28.50	37
	4920	5″ long		26	.308		15.80	12.90		28.70	37.50
	4940	6″ long		26	.308		16.05	12.90		28.95	37.50
	4960	8″ long		26	.308		16.65	12.90		29.55	38.50
	4980	10″ long		26	.308		17.25	12.90		30.15	39
	5000	12″ long		26	.308		17.85	12.90		30.75	39.50
	5020	1″ diameter, 2″ long		22	.364		14.20	15.25		29.45	39
	5040	2-1/2″ long		22	.364		15.55	15.25		30.80	40.50
	5060	3″ long		22	.364		15.80	15.25		31.05	41
	5080	3-1/2″ long		22	.364		17.20	15.25		32.45	42.50
	5100	4″ long		22	.364		17.40	15.25		32.65	42.50
	5120	5″ long		22	.364		17.65	15.25		32.90	43
	5140	6″ long		22	.364		18.05	15.25		33.30	43.50
	5160	8″ long		22	.364		18.85	15.25		34.10	44
	5180	10″ long		22	.364		19.65	15.25		34.90	45
	5200	12″ long		22	.364		22	15.25		37.25	47.50
	5220	1-1/4″ diameter, 2″ long		18	.444		17.90	18.65		36.55	48.50
	5240	2-1/2″ long		18	.444		19.35	18.65		38	50.50
	5260	3″ long		18	.444		19.50	18.65		38.15	50.50
	5280	3-1/2″ long		18	.444		19.40	18.65		38.05	50.50
	5300	4″ long		18	.444		21	18.65		39.65	52.50
	5320	5″ long		18	.444		21.50	18.65		40.15	52.50
	5340	6″ long		18	.444		22.50	18.65		41.15	53.50
	5360	8″ long		18	.444		22.50	18.65		41.15	53.50
	5380	10″ long		18	.444		25	18.65		43.65	56.50
	5400	12″ long		18	.444		29	18.65		47.65	61
	5420	1-1/2″ diameter, 2″ long		16	.500		20.50	21		41.50	55
	5440	2-1/2″ long		16	.500		22	21		43	56.50
	5460	3″ long		16	.500		22.50	21		43.50	57
	5480	3-1/2″ long		16	.500		23	21		44	58
	5500	4″ long		16	.500		24	21		45	59
	5520	5″ long		16	.500		24.50	21		45.50	59.50
	5540	6″ long		16	.500		26.50	21		47.50	61.50

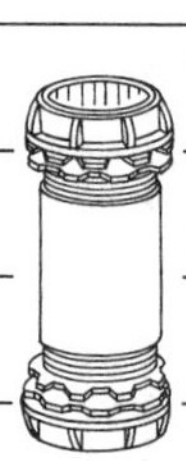

16132 | Conduit & Tubing

		16132 Conduit & Tubing	CREW	DAILY OUTPUT	LABOR-HOURS	UNIT	2006 BARE COSTS MAT.	LABOR	EQUIP.	TOTAL	TOTAL INCL O&P	
220	5560	8″ long	1 Elec	16	.500	Ea.	28	21		49	63.50	220
	5580	10″ long		16	.500		33.50	21		54.50	69.50	
	5600	12″ long		16	.500		38	21		59	74.50	
	5620	2″ diameter, 2-1/2″ long		14	.571		27.50	24		51.50	67	
	5640	3″ long		14	.571		27.50	24		51.50	67.50	
	5660	3-1/2″ long		14	.571		29.50	24		53.50	69.50	
	5680	4″ long		14	.571		32.50	24		56.50	73	
	5700	5″ long		14	.571		32.50	24		56.50	72.50	
	5720	6″ long		14	.571		33.50	24		57.50	74	
	5740	8″ long		14	.571		37.50	24		61.50	78	
	5760	10″ long		14	.571		44.50	24		68.50	86	
	5780	12″ long		14	.571		49.50	24		73.50	91.50	
	5800	2-1/2″ diameter, 3-1/2″ long		12	.667		53.50	28		81.50	102	
	5820	4″ long		12	.667		54.50	28		82.50	103	
	5840	5″ long		12	.667		63	28		91	112	
	5860	6″ long		12	.667		64	28		92	114	
	5880	8″ long		12	.667		72	28		100	123	
	5900	10″ long		12	.667		78.50	28		106.50	130	
	5920	12″ long		12	.667		88	28		116	140	
	5940	3″ diameter, 3-1/2″ long		11	.727		63.50	30.50		94	117	
	5960	4″ long		11	.727		67	30.50		97.50	121	
	5980	5″ long		11	.727		71.50	30.50		102	126	
	6000	6″ long		11	.727		77.50	30.50		108	132	
	6020	8″ long		11	.727		85.50	30.50		116	141	
	6040	10″ long		11	.727		97.50	30.50		128	154	
	6060	12″ long		11	.727		112	30.50		142.50	170	
	6080	3-1/2″ diameter, 4″ long		9	.889		101	37.50		138.50	169	
	6100	5″ long		9	.889		102	37.50		139.50	170	
	6120	6″ long		9	.889		107	37.50		144.50	176	
	6140	8″ long		9	.889		121	37.50		158.50	191	
	6160	10″ long		9	.889		135	37.50		172.50	206	
	6180	12″ long		9	.889		144	37.50		181.50	216	
	6200	4″ diameter, 4″ long		7.50	1.067		116	45		161	197	
	6220	5″ long		7.50	1.067		119	45		164	200	
	6240	6″ long		7.50	1.067		128	45		173	210	
	6260	8″ long		7.50	1.067		142	45		187	225	
	6280	10″ long		7.50	1.067		166	45		211	251	
	6300	12″ long		7.50	1.067		173	45		218	259	
	6320	5″ diameter, 5″ long		5.50	1.455		217	61		278	330	
	6340	6″ long		5.50	1.455		230	61		291	345	
	6360	8″ long		5.50	1.455		240	61		301	355	
	6380	10″ long		5.50	1.455		258	61		319	380	
	6400	12″ long		5.50	1.455		261	61		322	380	
	6420	6″ diameter, 5″ long		4.50	1.778		315	74.50		389.50	465	
	6440	6″ long		4.50	1.778		330	74.50		404.50	475	
	6460	8″ long		4.50	1.778		340	74.50		414.50	490	
	6480	10″ long		4.50	1.778		360	74.50		434.50	510	
	6500	12″ long	↓	4.50	1.778	↓	380	74.50		454.50	535	
230	0010	**CONDUIT IN CONCRETE SLAB** Including terminations,										230
	0020	fittings and supports										
	3230	PVC, schedule 40, 1/2″ diameter	1 Elec	270	.030	L.F.	.57	1.24		1.81	2.55	
	3250	3/4″ diameter		230	.035		.67	1.46		2.13	2.99	
	3270	1″ diameter		200	.040		.88	1.68		2.56	3.56	
	3300	1-1/4″ diameter		170	.047		1.26	1.98		3.24	4.43	
	3330	1-1/2″ diameter		140	.057		1.52	2.40		3.92	5.35	
	3350	2″ diameter	↓	120	.067	↓	1.95	2.80		4.75	6.45	

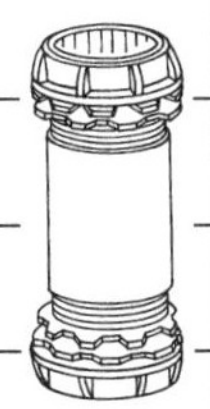

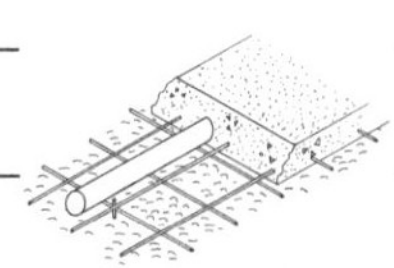

	16132 | Conduit & Tubing		Crew	Daily Output	Labor-Hours	Unit	2006 Bare Costs Mat.	Labor	Equip.	Total	Total Incl O&P
230	3370	2-1/2″ diameter	1 Elec	90	.089	L.F.	3.25	3.73		6.98	9.35
	3400	3″ diameter	2 Elec	160	.100		4.12	4.20		8.32	11
	3430	3-1/2″ diameter		120	.133		5.50	5.60		11.10	14.70
	3440	4″ diameter		100	.160		6.10	6.70		12.80	17.10
	3450	5″ diameter		80	.200		9.15	8.40		17.55	23
	3460	6″ diameter		60	.267		13.10	11.20		24.30	31.50
	3530	Sweeps, 1″ diameter, 30″ radius	1 Elec	32	.250	Ea.	15.70	10.50		26.20	33.50
	3550	1-1/4″ diameter		24	.333		19.85	14		33.85	43.50
	3570	1-1/2″ diameter		21	.381		20.50	16		36.50	47
	3600	2″ diameter		18	.444		21.50	18.65		40.15	52.50
	3630	2-1/2″ diameter		14	.571		36.50	24		60.50	77.50
	3650	3″ diameter		10	.800		42	33.50		75.50	98
	3670	3-1/2″ diameter		8	1		62.50	42		104.50	134
	3700	4″ diameter		7	1.143		60	48		108	140
	3710	5″ diameter		6	1.333		91	56		147	187
	3730	Couplings, 1/2″ diameter					.41			.41	.45
	3750	3/4″ diameter					.50			.50	.55
	3770	1″ diameter					.77			.77	.85
	3800	1-1/4″ diameter					1.02			1.02	1.12
	3830	1-1/2″ diameter					1.40			1.40	1.54
	3850	2″ diameter					1.86			1.86	2.05
	3870	2-1/2″ diameter					3.30			3.30	3.63
	3900	3″ diameter					5.40			5.40	5.95
	3930	3-1/2″ diameter					6			6	6.60
	3950	4″ diameter					8.35			8.35	9.20
	3960	5″ diameter					21			21	23.50
	3970	6″ diameter					27			27	30
	4030	End bells 1″ diameter, PVC	1 Elec	60	.133		4	5.60		9.60	13.05
	4050	1-1/4″ diameter		53	.151		4.95	6.35		11.30	15.20
	4100	1-1/2″ diameter		48	.167		4.95	7		11.95	16.25
	4150	2″ diameter		34	.235		7.30	9.90		17.20	23.50
	4170	2-1/2″ diameter		27	.296		8.10	12.45		20.55	28
	4200	3″ diameter		20	.400		8.60	16.80		25.40	35.50
	4250	3-1/2″ diameter		16	.500		9.40	21		30.40	43
	4300	4″ diameter		14	.571		10.20	24		34.20	48
	4310	5″ diameter		12	.667		16.05	28		44.05	60.50
	4320	6″ diameter		9	.889		17.60	37.50		55.10	77
	4350	Rigid galvanized steel, 1/2″ diameter		200	.040	L.F.	2.23	1.68		3.91	5.05
	4400	3/4″ diameter		170	.047		2.56	1.98		4.54	5.85
	4450	1″ diameter		130	.062		3.88	2.58		6.46	8.25
	4500	1-1/4″ diameter		110	.073		5.15	3.05		8.20	10.35
	4600	1-1/2″ diameter		100	.080		5.95	3.36		9.31	11.75
	4800	2″ diameter		90	.089		7.65	3.73		11.38	14.15
	9000	Minimum labor/equipment charge		4	2	Job		84		84	129
240	0010	**CONDUIT IN TRENCH** Includes terminations and fittings									
	0020	Does not include excavation or backfill, see div. 02315									
	0200	Rigid galvanized steel, 2″ diameter	1 Elec	150	.053	L.F.	7.35	2.24		9.59	11.55
	0400	2-1/2″ diameter	″	100	.080		14.30	3.36		17.66	21
	0600	3″ diameter	2 Elec	160	.100		17.60	4.20		21.80	26
	0800	3-1/2″ diameter		140	.114		22.50	4.80		27.30	32
	1000	4″ diameter		100	.160		24.50	6.70		31.20	37.50
	1200	5″ diameter		80	.200		53.50	8.40		61.90	72
	1400	6″ diameter		60	.267		78.50	11.20		89.70	103
	9000	Minimum labor/equipment charge	1 Elec	4	2	Job		84		84	129
250	0010	**CONDUIT FITTINGS FOR RIGID GALVANIZED STEEL**									
	0050	Standard, locknuts, 1/2″ diameter				Ea.	.20			.20	.22

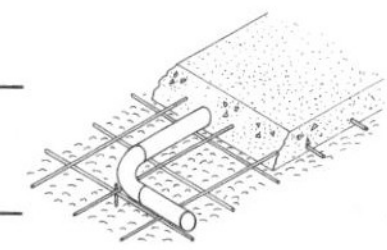

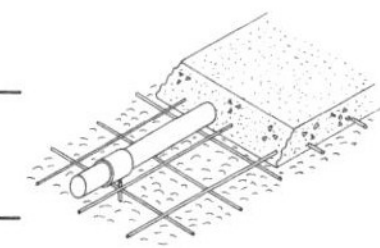

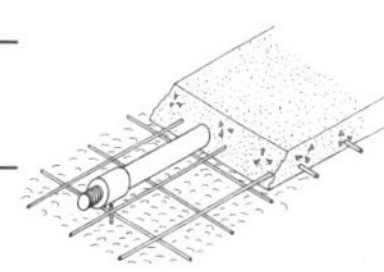

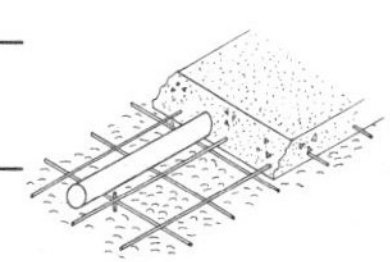

ELECTRICAL 16

16132 | Conduit & Tubing

		CREW	DAILY OUTPUT	LABOR-HOURS	UNIT	2006 BARE COSTS MAT.	LABOR	EQUIP.	TOTAL	TOTAL INCL O&P
250	0100 3/4" diameter				Ea.	.36			.36	.40
	0300 1" diameter					.57			.57	.63
	0500 1-1/4" diameter					.79			.79	.87
	0700 1-1/2" diameter					1.29			1.29	1.42
	1000 2" diameter					1.88			1.88	2.07
	1030 2-1/2" diameter					5.25			5.25	5.80
	1050 3" diameter					6.75			6.75	7.45
	1070 3-1/2" diameter					11.40			11.40	12.55
	1100 4" diameter					14.20			14.20	15.60
	1110 5" diameter					30.50			30.50	33.50
	1120 6" diameter					52			52	57
	1130 Bushings, plastic, 1/2" diameter	1 Elec	40	.200		.20	8.40		8.60	13.15
	1150 3/4" diameter		32	.250		.34	10.50		10.84	16.55
	1170 1" diameter		28	.286		.55	12		12.55	19.10
	1200 1-1/4" diameter		24	.333		.84	14		14.84	22.50
	1230 1-1/2" diameter		18	.444		1.14	18.65		19.79	30.50
	1250 2" diameter		15	.533		2.12	22.50		24.62	37
	1270 2-1/2" diameter		13	.615		4.90	26		30.90	45.50
	1300 3" diameter		12	.667		5.10	28		33.10	48.50
	1330 3-1/2" diameter		11	.727		6.70	30.50		37.20	54.50
	1350 4" diameter		9	.889		8.25	37.50		45.75	66.50
	1360 5" diameter		7	1.143		16.80	48		64.80	92.50
	1370 6" diameter		5	1.600		32	67		99	139
	1390 Steel, 1/2" diameter		40	.200		.47	8.40		8.87	13.45
	1400 3/4" diameter		32	.250		.61	10.50		11.11	16.85
	1430 1" diameter		28	.286		.93	12		12.93	19.50
	1450 Steel insulated, 1-1/4" diameter		24	.333		5.95	14		19.95	28
	1470 1-1/2" diameter		18	.444		7.50	18.65		26.15	37.50
	1500 2" diameter		15	.533		10.85	22.50		33.35	46.50
	1530 2-1/2" diameter		13	.615		20	26		46	62
	1550 3" diameter		12	.667		27	28		55	72.50
	1570 3-1/2" diameter		11	.727		34.50	30.50		65	85
	1600 4" diameter		9	.889		44	37.50		81.50	106
	1610 5" diameter		7	1.143		95	48		143	179
	1620 6" diameter		5	1.600		146	67		213	265
	1630 Sealing locknuts, 1/2" diameter		40	.200		1.55	8.40		9.95	14.65
	1650 3/4" diameter		32	.250		1.80	10.50		12.30	18.20
	1670 1" diameter		28	.286		2.65	12		14.65	21.50
	1700 1-1/4" diameter		24	.333		4.50	14		18.50	26.50
	1730 1-1/2" diameter		18	.444		5.70	18.65		24.35	35.50
	1750 2" diameter		15	.533		7.20	22.50		29.70	42.50
	1760 Grounding bushing, insulated, 1/2" diameter		32	.250		4.30	10.50		14.80	21
	1770 3/4" diameter		28	.286		5.45	12		17.45	24.50
	1780 1" diameter		20	.400		6.15	16.80		22.95	33
	1800 1-1/4" diameter		18	.444		7.75	18.65		26.40	37.50
	1830 1-1/2" diameter		16	.500		8.35	21		29.35	41.50
	1850 2" diameter		13	.615		10.15	26		36.15	51
	1870 2-1/2" diameter		12	.667		18.40	28		46.40	63
	1900 3" diameter		11	.727		23.50	30.50		54	73
	1930 3-1/2" diameter		9	.889		31	37.50		68.50	91.50
	1950 4" diameter		8	1		33	42		75	101
	1960 5" diameter		6	1.333		89	56		145	185
	1970 6" diameter		4	2		136	84		220	278
	1990 Coupling, with set screw, 1/2" diameter		50	.160		4.60	6.70		11.30	15.40
	2000 3/4" diameter		40	.200		5.80	8.40		14.20	19.35
	2030 1" diameter	↓	35	.229	↓	9.50	9.60		19.10	25.50

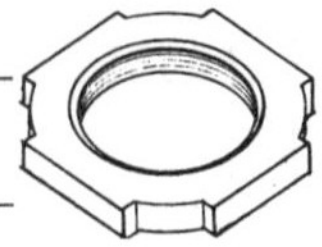

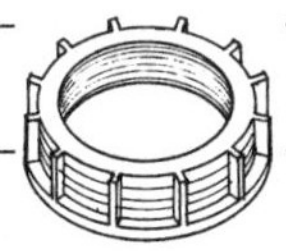

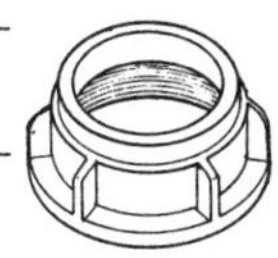

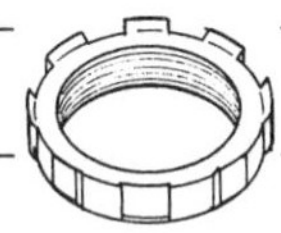

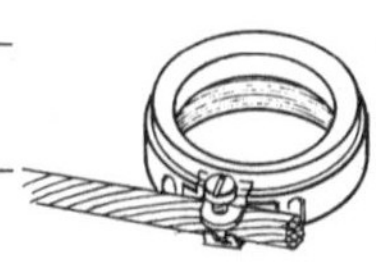

16 ELECTRICAL

	16132 | Conduit & Tubing		CREW	DAILY OUTPUT	LABOR-HOURS	UNIT	2006 BARE COSTS MAT.	LABOR	EQUIP.	TOTAL	TOTAL INCL O&P	
250	2050	1-1/4" diameter	1 Elec	28	.286	Ea.	16.20	12		28.20	36.50	250
	2070	1-1/2" diameter 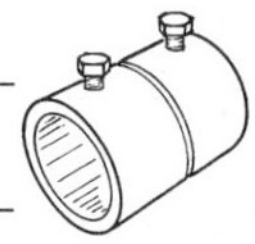		23	.348		21	14.60		35.60	45.50	
	2090	2" diameter		20	.400		46.50	16.80		63.30	77	
	2100	2-1/2" diameter		18	.444		99.50	18.65		118.15	138	
	2110	3" diameter		15	.533		118	22.50		140.50	165	
	2120	3-1/2" diameter		12	.667		169	28		197	229	
	2130	4" diameter		10	.800		221	33.50		254.50	295	
	2140	5" diameter		9	.889		355	37.50		392.50	455	
	2150	6" diameter		8	1		465	42		507	575	
	2160	Box connector with set screw, plain, 1/2" diameter		70	.114		3	4.80		7.80	10.70	
	2170	3/4" diameter		60	.133		4.25	5.60		9.85	13.35	
	2180	1" diameter		50	.160		6.70	6.70		13.40	17.70	
	2190	Insulated, 1-1/4" diameter		40	.200		13.90	8.40		22.30	28.50	
	2200	1-1/2" diameter 		30	.267		19.90	11.20		31.10	39.50	
	2210	2" diameter		20	.400		39.50	16.80		56.30	69.50	
	2220	2-1/2" diameter		18	.444		96	18.65		114.65	135	
	2230	3" diameter		15	.533		128	22.50		150.50	175	
	2240	3-1/2" diameter		12	.667		179	28		207	239	
	2250	4" diameter		10	.800		235	33.50		268.50	310	
	2260	5" diameter		9	.889		285	37.50		322.50	375	
	2270	6" diameter		8	1		365	42		407	465	
	2280	LB, LR or LL fittings & covers, 1/2" diameter		16	.500		9.70	21		30.70	43	
	2290	3/4" diameter		13	.615		11.70	26		37.70	53	
	2300	1" diameter 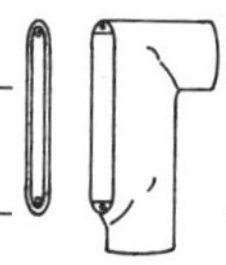		11	.727		17.30	30.50		47.80	66	
	2330	1-1/4" diameter		8	1		26	42		68	93	
	2350	1-1/2" diameter		6	1.333		32.50	56		88.50	122	
	2370	2" diameter		5	1.600		54	67		121	163	
	2380	2-1/2" diameter		4	2		112	84		196	252	
	2390	3" diameter		3.50	2.286		145	96		241	310	
	2400	3-1/2" diameter		3	2.667		234	112		346	430	
	2410	4" diameter		2.50	3.200		261	134		395	495	
	2420	T fittings, with cover, 1/2" diameter		12	.667		11.70	28		39.70	56	
	2430	3/4" diameter		11	.727		14	30.50		44.50	62.50	
	2440	1" diameter		9	.889		21	37.50		58.50	80.50	
	2450	1-1/4" diameter		6	1.333		29.50	56		85.50	119	
	2470	1-1/2" diameter		5	1.600		37.50	67		104.50	146	
	2500	2" diameter 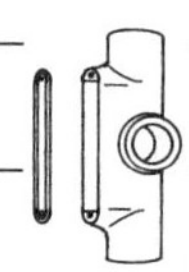		4	2		59	84		143	194	
	2510	2-1/2" diameter		3.50	2.286		117	96		213	277	
	2520	3" diameter		3	2.667		165	112		277	355	
	2530	3-1/2" diameter		2.50	3.200		266	134		400	500	
	2540	4" diameter		2	4		280	168		448	570	
	2550	Nipples chase, plain, 1/2" diameter		40	.200		.90	8.40		9.30	13.95	
	2560	3/4" diameter		32	.250		1.12	10.50		11.62	17.45	
	2570	1" diameter		28	.286		2.30	12		14.30	21	
	2600	Insulated, 1-1/4" diameter 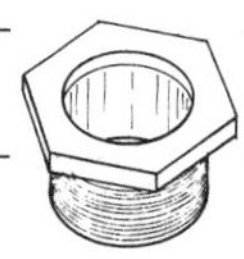		24	.333		8.50	14		22.50	31	
	2630	1-1/2" diameter		18	.444		11.20	18.65		29.85	41.50	
	2650	2" diameter		15	.533		16.90	22.50		39.40	53	
	2660	2-1/2" diameter		12	.667		40	28		68	87	
	2670	3" diameter		10	.800		43.50	33.50		77	100	
	2680	3-1/2" diameter		9	.889		61	37.50		98.50	125	
	2690	4" diameter 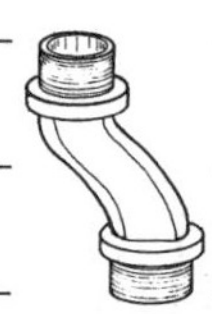		8	1		93	42		135	167	
	2700	5" diameter		7	1.143		274	48		322	375	
	2710	6" diameter		6	1.333		425	56		481	550	
	2720	Nipples offset, plain, 1/2" diameter		40	.200		4.48	8.40		12.88	17.90	
	2730	3/4" diameter		32	.250		4.68	10.50		15.18	21.50	
	2740	1" diameter	↓	24	.333	↓	6	14		20	28	

ELECTRICAL 16

16132 | Conduit & Tubing

250		CREW	DAILY OUTPUT	LABOR-HOURS	UNIT	2006 BARE COSTS MAT.	LABOR	EQUIP.	TOTAL	TOTAL INCL O&P
2750	Insulated, 1-1/4″ diameter	1 Elec	20	.400	Ea.	32	16.80		48.80	61
2760	1-1/2″ diameter		18	.444		39	18.65		57.65	72
2770	2″ diameter		16	.500		62	21		83	101
2780	3″ diameter		14	.571		124	24		148	173
2850	Coupling, expansion, 1/2″ diameter		12	.667		41	28		69	88
2880	3/4″ diameter		10	.800		47.50	33.50		81	105
2900	1″ diameter		8	1		56	42		98	126
2920	1-1/4″ diameter		6.40	1.250		76	52.50		128.50	165
2940	1-1/2″ diameter		5.30	1.509		104	63.50		167.50	212
2960	2″ diameter		4.60	1.739		155	73		228	284
2980	2-1/2″ diameter		3.60	2.222		244	93.50		337.50	410
3000	3″ diameter		3	2.667		300	112		412	505
3020	3-1/2″ diameter		2.80	2.857		445	120		565	670
3040	4″ diameter		2.40	3.333		525	140		665	795
3060	5″ diameter		2	4		1,100	168		1,268	1,450
3080	6″ diameter		1.80	4.444		1,775	187		1,962	2,250
3100	Expansion deflection, 1/2″ diameter		12	.667		145	28		173	203
3120	3/4″ diameter		12	.667		155	28		183	214
3140	1″ diameter		10	.800		196	33.50		229.50	268
3160	1-1/4″ diameter		6.40	1.250		244	52.50		296.50	350
3180	1-1/2″ diameter		5.30	1.509		258	63.50		321.50	380
3200	2″ diameter		4.60	1.739		335	73		408	480
3220	2-1/2″ diameter		3.60	2.222		470	93.50		563.50	665
3240	3″ diameter		3	2.667		595	112		707	830
3260	3-1/2″ diameter		2.80	2.857		710	120		830	965
3280	4″ diameter		2.40	3.333		790	140		930	1,075
3300	5″ diameter		2	4		1,275	168		1,443	1,650
3320	6″ diameter		1.80	4.444		2,175	187		2,362	2,675
3340	Ericson, 1/2″ diameter		16	.500		3.55	21		24.55	36.50
3360	3/4″ diameter		14	.571		4.55	24		28.55	42
3380	1″ diameter		11	.727		8.40	30.50		38.90	56.50
3400	1-1/4″ diameter		8	1		17.95	42		59.95	84.50
3420	1-1/2″ diameter		7	1.143		21.50	48		69.50	97.50
3440	2″ diameter		5	1.600		43	67		110	152
3460	2-1/2″ diameter		4	2		105	84		189	245
3480	3″ diameter		3.50	2.286		159	96		255	325
3500	3-1/2″ diameter		3	2.667		265	112		377	465
3520	4″ diameter		2.70	2.963		310	124		434	530
3540	5″ diameter		2.50	3.200		575	134		709	840
3560	6″ diameter		2.30	3.478		765	146		911	1,075
3580	Split, 1/2″ diameter		32	.250		4.30	10.50		14.80	21
3600	3/4″ diameter		27	.296		5.30	12.45		17.75	25
3620	1″ diameter		20	.400		9.85	16.80		26.65	37
3640	1-1/4″ diameter		16	.500		14.20	21		35.20	48
3660	1-1/2″ diameter		14	.571		17.85	24		41.85	56.50
3680	2″ diameter		12	.667		35.50	28		63.50	82.50
3700	2-1/2″ diameter		10	.800		82.50	33.50		116	143
3720	3″ diameter		9	.889		124	37.50		161.50	194
3740	3-1/2″ diameter		8	1		200	42		242	285
3760	4″ diameter		7	1.143		236	48		284	335
3780	5″ diameter		6	1.333		405	56		461	535
3800	6″ diameter		5	1.600		550	67		617	710
4600	Reducing bushings, 3/4″ to 1/2″ diameter		54	.148		1.05	6.20		7.25	10.75
4620	1″ to 3/4″ diameter		46	.174		1.61	7.30		8.91	13
4640	1-1/4″ to 1″ diameter		40	.200		3.75	8.40		12.15	17.10
4660	1-1/2″ to 1-1/4″ diameter		36	.222		5.40	9.35		14.75	20.50

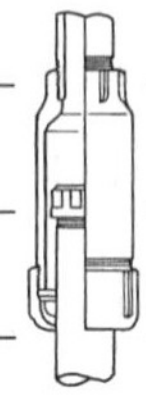

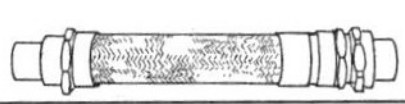

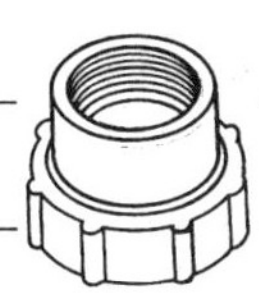

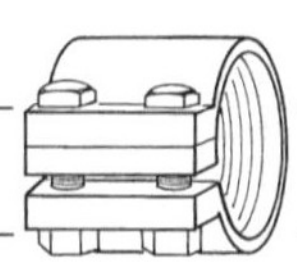

 Important: See the Reference Section for supporting data - Crews, Rental Equipment, City Cost Indexes and Reference Data

	16132 | Conduit & Tubing		CREW	DAILY OUTPUT	LABOR-HOURS	UNIT	2006 BARE COSTS MAT.	LABOR	EQUIP.	TOTAL	TOTAL INCL O&P	
250	4680	2″ to 1-1/2″ diameter	1 Elec	32	.250	Ea.	12.20	10.50		22.70	29.50	250
	4740	2-1/2″ to 2″ diameter		30	.267		16.50	11.20		27.70	35.50	
	4760	3″ to 2-1/2″ diameter		28	.286		21.50	12		33.50	42	
	4800	Through-wall seal, 1/2″ diameter		8	1		154	42		196	234	
	4820	3/4″ diameter		7.50	1.067		154	45		199	238	
	4840	1″ diameter		6.50	1.231		154	51.50		205.50	249	
	4860	1-1/4″ diameter		5.50	1.455		213	61		274	330	
	4880	1-1/2″ diameter		5	1.600		213	67		280	340	
	4900	2″ diameter		4.20	1.905		231	80		311	375	
	4920	2-1/2″ diameter		3.50	2.286		247	96		343	420	
	4940	3″ diameter 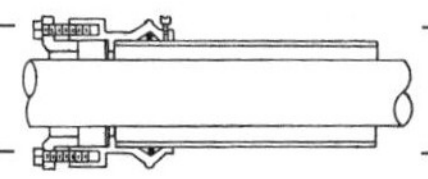		3	2.667		247	112		359	445	
	4960	3-1/2″ diameter		2.50	3.200		385	134		519	630	
	4980	4″ diameter		2	4		445	168		613	745	
	5000	5″ diameter		1.50	5.333		565	224		789	965	
	5020	6″ diameter	↓	1	8	↓	565	335		900	1,150	
	5100	Cable supports, 2 or more wires										
	5120	1-1/2″ diameter	1 Elec	8	1	Ea.	61	42		103	132	
	5140	2″ diameter 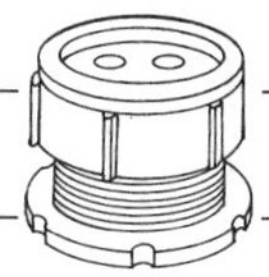		6	1.333		90	56		146	186	
	5160	2-1/2″ diameter		4	2		101	84		185	240	
	5180	3″ diameter		3.50	2.286		133	96		229	294	
	5200	3-1/2″ diameter		2.60	3.077		180	129		309	395	
	5220	4″ diameter		2	4		224	168		392	505	
	5240	5″ diameter		1.50	5.333		420	224		644	805	
	5260	6″ diameter		1	8		705	335		1,040	1,300	
	5280	Service entrance cap, 1/2″ diameter		16	.500		7.05	21		28.05	40.50	
	5300	3/4″ diameter		13	.615		7.85	26		33.85	48.50	
	5320	1″ diameter		10	.800		8.55	33.50		42.05	61.50	
	5340	1-1/4″ diameter		8	1		14.25	42		56.25	80	
	5360	1-1/2″ diameter		6.50	1.231		26	51.50		77.50	108	
	5380	2″ diameter		5.50	1.455		39	61		100	137	
	5400	2-1/2″ diameter 		4	2		122	84		206	263	
	5420	3″ diameter		3.40	2.353		186	99		285	355	
	5440	3-1/2″ diameter		3	2.667		227	112		339	425	
	5460	4″ diameter		2.70	2.963		345	124		469	570	
	5600	Fire stop fittings, to 3/4″ diameter		24	.333		107	14		121	140	
	5610	1″ diameter		22	.364		131	15.25		146.25	168	
	5620	1-1/2″ diameter		20	.400		165	16.80		181.80	208	
	5640	2″ diameter		16	.500		255	21		276	315	
	5660	3″ diameter		12	.667		320	28		348	395	
	5680	4″ diameter		10	.800		425	33.50		458.50	515	
	5700	6″ diameter		8	1		735	42		777	870	
	5750	90° pull elbows steel, female, 1/2″ diameter		16	.500		8.50	21		29.50	42	
	5760	3/4″ diameter		13	.615		9.90	26		35.90	51	
	5780	1″ diameter		11	.727		16.75	30.50		47.25	65.50	
	5800	1-1/4″ diameter 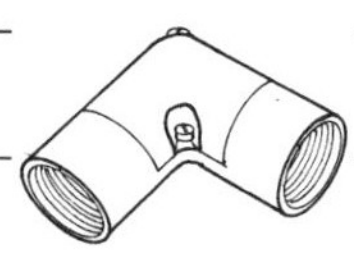		8	1		25	42		67	92	
	5820	1-1/2″ diameter		6	1.333		35.50	56		91.50	126	
	5840	2″ diameter	↓	5	1.600	↓	57	67		124	167	
	6000	Explosion proof, flexible coupling										
	6010	1/2″ diameter, 4″ long	1 Elec	12	.667	Ea.	95	28		123	147	
	6020	6″ long		12	.667		106	28		134	159	
	6050	12″ long		12	.667		135	28		163	192	
	6070	18″ long		12	.667		165	28		193	225	
	6090	24″ long		12	.667		195	28		223	258	
	6110	30″ long		12	.667		225	28		253	291	
	6130	36″ long 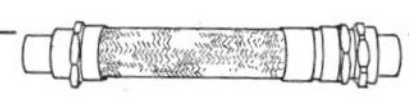		12	.667		256	28		284	325	
	6140	3/4″ diameter, 4″ long	↓	10	.800	↓	114	33.50		147.50	177	

16132 | Conduit & Tubing

		16132 Conduit & Tubing	CREW	DAILY OUTPUT	LABOR-HOURS	UNIT	2006 BARE COSTS MAT.	LABOR	EQUIP.	TOTAL	TOTAL INCL O&P	
250	6150	6" long	1 Elec	10	.800	Ea.	129	33.50		162.50	194	250
	6180	12" long		10	.800		170	33.50		203.50	238	
	6200	18" long		10	.800		212	33.50		245.50	285	
	6220	24" long		10	.800		256	33.50		289.50	335	
	6240	30" long		10	.800		296	33.50		329.50	375	
	6260	36" long		10	.800		340	33.50		373.50	425	
	6270	1" diameter, 6" long		8	1		230	42		272	320	
	6300	12" long		8	1		291	42		333	385	
	6320	18" long		8	1		350	42		392	450	
	6340	24" long		8	1		410	42		452	515	
	6360	30" long		8	1		560	42		602	680	
	6380	36" long		8	1		630	42		672	760	
	6390	1-1/4" diameter, 12" long		6.40	1.250		445	52.50		497.50	565	
	6410	18" long		6.40	1.250		605	52.50		657.50	745	
	6430	24" long		6.40	1.250		720	52.50		772.50	870	
	6450	30" long		6.40	1.250		815	52.50		867.50	975	
	6470	36" long		6.40	1.250		1,025	52.50		1,077.50	1,200	
	6480	1-1/2" diameter, 12" long		5.30	1.509		590	63.50		653.50	750	
	6500	18" long		5.30	1.509		690	63.50		753.50	855	
	6520	24" long		5.30	1.509		780	63.50		843.50	960	
	6540	30" long		5.30	1.509		1,050	63.50		1,113.50	1,250	
	6560	36" long		5.30	1.509		1,200	63.50		1,263.50	1,425	
	6570	2" diameter, 12" long		4.60	1.739		900	73		973	1,100	
	6590	18" long		4.60	1.739		1,125	73		1,198	1,350	
	6610	24" long		4.60	1.739		1,225	73		1,298	1,475	
	6630	30" long		4.60	1.739		1,325	73		1,398	1,600	
	6650	36" long		4.60	1.739		1,625	73		1,698	1,925	
	7000	Close up plug, 1/2" diameter, explosion proof		40	.200		1.99	8.40		10.39	15.15	
	7010	3/4" diameter		32	.250		2.23	10.50		12.73	18.65	
	7020	1" diameter		28	.286		2.72	12		14.72	21.50	
	7030	1-1/4" diameter		24	.333		2.97	14		16.97	25	
	7040	1-1/2" diameter		18	.444		4.22	18.65		22.87	33.50	
	7050	2" diameter		15	.533		7.20	22.50		29.70	42.50	
	7060	2-1/2" diameter		13	.615		11.65	26		37.65	53	
	7070	3" diameter		12	.667		17	28		45	61.50	
	7080	3-1/2" diameter		11	.727		21	30.50		51.50	70.50	
	7090	4" diameter		9	.889		28.50	37.50		66	88.50	
	7091	Elbow female, 45°, 1/2"		16	.500		8.40	21		29.40	42	
	7092	3/4"		13	.615		9.10	26		35.10	50	
	7093	1"		11	.727		12.75	30.50		43.25	61	
	7094	1-1/4"		8	1		17.90	42		59.90	84	
	7095	1-1/2"		6	1.333		18.50	56		74.50	107	
	7096	2"		5	1.600		23	67		90	129	
	7097	2-1/2"		4.50	1.778		62	74.50		136.50	183	
	7098	3"		4.20	1.905		67	80		147	197	
	7099	3-1/2"		4	2		101	84		185	240	
	7100	4"		3.80	2.105		124	88.50		212.50	272	
	7101	90°, 1/2"		16	.500		7.95	21		28.95	41.50	
	7102	3/4"		13	.615		8.65	26		34.65	49.50	
	7103	1"		11	.727		11.70	30.50		42.20	60	
	7104	1-1/4"		8	1		19.10	42		61.10	85.50	
	7105	1-1/2"		6	1.333		27.50	56		83.50	117	
	7106	2"		5	1.600		44	67		111	152	
	7107	2-1/2"		4.50	1.778		79	74.50		153.50	202	
	7110	Elbows 90°, long male & female, 1/2" diameter, explosion proof		16	.500		11.70	21		32.70	45.50	
	7120	3/4" diameter	↓	13	.615	↓	13.10	26		39.10	54.50	

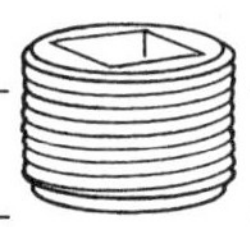

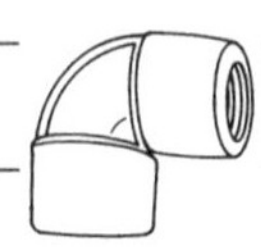

16 ELECTRICAL

16132 | Conduit & Tubing

			CREW	DAILY OUTPUT	LABOR-HOURS	UNIT	2006 BARE COSTS MAT.	LABOR	EQUIP.	TOTAL	TOTAL INCL O&P	
250	7130	1" diameter	1 Elec	11	.727	Ea.	19	30.50		49.50	68	250
	7140	1-1/4" diameter		8	1		22	42		64	88.50	
	7150	1-1/2" diameter		6	1.333		33	56		89	123	
	7160	2" diameter		5	1.600		49.50	67		116.50	158	
	7170	Capped elbow, 1/2" diameter, explosion proof		11	.727		11.35	30.50		41.85	59.50	
	7180	3/4" diameter		8	1		12.70	42		54.70	78.50	
	7190	1" diameter		6	1.333		15.70	56		71.70	104	
	7200	1-1/4" diameter		5	1.600		37	67		104	145	
	7210	Pulling elbow, 1/2" diameter, explosion proof		11	.727		55	30.50		85.50	108	
	7220	3/4" diameter		8	1		55	42		97	125	
	7230	1" diameter		6	1.333		150	56		206	252	
	7240	1-1/4" diameter		5	1.600		155	67		222	275	
	7250	1-1/2" diameter		5	1.600		228	67		295	355	
	7260	2" diameter		4	2		237	84		321	390	
	7270	2-1/2" diameter		3.50	2.286		520	96		616	725	
	7280	3" diameter		3	2.667		505	112		617	730	
	7290	3-1/2" diameter		2.50	3.200		940	134		1,074	1,225	
	7300	4" diameter		2.20	3.636		920	153		1,073	1,225	
	7310	LB conduit body, 1/2" diameter		11	.727		34	30.50		64.50	84.50	
	7320	3/4" diameter		8	1		38.50	42		80.50	107	
	7330	T conduit body, 1/2" diameter		9	.889		35.50	37.50		73	96.50	
	7340	3/4" diameter		6	1.333		41.50	56		97.50	133	
	7350	Explosion proof, round box w/cover, 3 threaded hubs, 1/2" diameter		8	1		33	42		75	101	
	7351	3/4" diameter		8	1		37	42		79	106	
	7352	1" diameter		7.50	1.067		46	45		91	120	
	7353	1-1/4" diameter		7	1.143		84.50	48		132.50	167	
	7354	1-1/2" diameter		7	1.143		153	48		201	242	
	7355	2" diameter		6	1.333		159	56		215	262	
	7356	Round box w/cover & mtng flange, 3 threaded hubs, 1/2" diameter		8	1		51.50	42		93.50	121	
	7357	3/4" diameter		8	1		53.50	42		95.50	124	
	7358	4 threaded hubs, 1" diameter		7	1.143		52	48		100	132	
	7400	Unions, 1/2" diameter		20	.400		8.75	16.80		25.55	35.50	
	7410	3/4" - 1/2" diameter		16	.500		12.50	21		33.50	46.50	
	7420	3/4" diameter		16	.500		12	21		33	45.50	
	7430	1" diameter		14	.571		22	24		46	61	
	7440	1-1/4" diameter		12	.667		33	28		61	79	
	7450	1-1/2" diameter		10	.800		42	33.50		75.50	98.50	
	7460	2" diameter		8.50	.941		54	39.50		93.50	121	
	7480	2-1/2" diameter		8	1		81.50	42		123.50	154	
	7490	3" diameter		7	1.143		117	48		165	203	
	7500	3-1/2" diameter		6	1.333		183	56		239	288	
	7510	4" diameter		5	1.600		203	67		270	325	
	7680	Reducer, 3/4" to 1/2"		54	.148		1.90	6.20		8.10	11.70	
	7690	1" to 1/2"		46	.174		2.65	7.30		9.95	14.15	
	7700	1" to 3/4"		46	.174		2.65	7.30		9.95	14.15	
	7710	1-1/4" to 3/4"		40	.200		3.80	8.40		12.20	17.15	
	7720	1-1/4" to 1"		40	.200		3.80	8.40		12.20	17.15	
	7730	1-1/2" to 1"		36	.222		6.25	9.35		15.60	21.50	
	7740	1-1/2" to 1-1/4"		36	.222		6.25	9.35		15.60	21.50	
	7750	2" to 3/4"		32	.250		12.15	10.50		22.65	29.50	
	7760	2" to 1-1/4"		32	.250		8.50	10.50		19	25.50	
	7770	2" to 1-1/2"		32	.250		8.50	10.50		19	25.50	
	7780	2-1/2" to 1-1/2"		30	.267		14.45	11.20		25.65	33	
	7790	3" to 2"		30	.267		17.05	11.20		28.25	36	
	7800	3-1/2" to 2-1/2"		28	.286		33.50	12		45.50	55	
	7810	4" to 3"	↓	28	.286	↓	39	12		51	61.50	

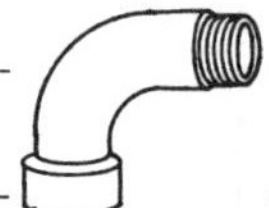

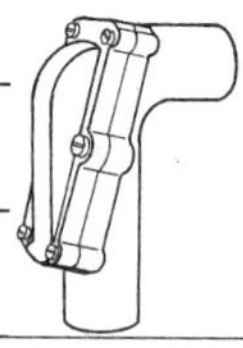

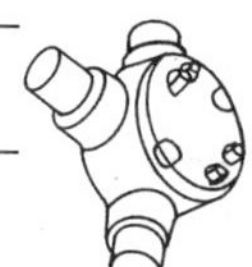

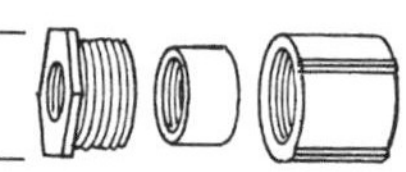

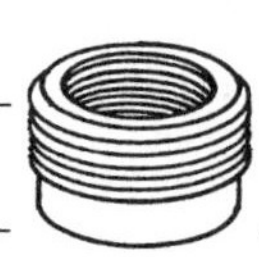

ELECTRICAL 16

	16132 | Conduit & Tubing	CREW	DAILY OUTPUT	LABOR-HOURS	UNIT	2006 BARE COSTS MAT.	LABOR	EQUIP.	TOTAL	TOTAL INCL O&P
250 7820	Sealing fitting, vertical/horizontal, 1/2" diameter	1 Elec	14.50	.552	Ea.	13.30	23		36.30	50
7830	3/4" diameter		13.30	.602		15.60	25.50		41.10	56
7840	1" diameter 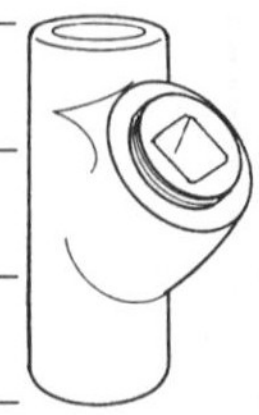		11.40	.702		20	29.50		49.50	67.50
7850	1-1/4" diameter		10	.800		24.50	33.50		58	78.50
7860	1-1/2" diameter		8.80	.909		37	38		75	99.50
7870	2" diameter		8	1		48	42		90	117
7880	2-1/2" diameter		6.70	1.194		66.50	50		116.50	151
7890	3" diameter		5.70	1.404		82	59		141	182
7900	3-1/2" diameter		4.70	1.702		219	71.50		290.50	350
7910	4" diameter		4	2		345	84		429	505
7920	Sealing hubs, 1" by 1-1/2"		12	.667		25	28		53	70.50
7930	1-1/4" by 2"		10	.800		28	33.50		61.50	83
7940	1-1/2" by 2"		9	.889		53.50	37.50		91	116
7950	2" by 2-1/2"		8	1		71	42		113	143
7960	3" by 4"		7	1.143		122	48		170	208
7970	4" by 5"		6	1.333		253	56		309	365
7980	Drain, 1/2"		32	.250		41.50	10.50		52	61.50
7990	Breather, 1/2"	↓	32	.250	↓	41.50	10.50		52	61.50
8000	Plastic coated 40 mil thick									
8010	LB, LR or LL conduit body w/cover, 1/2" diameter	1 Elec	13	.615	Ea.	35	26		61	78.50
8020	3/4" diameter		11	.727		42	30.50		72.50	93
8030	1" diameter 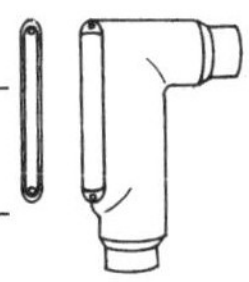		8	1		56	42		98	127
8040	1-1/4" diameter		6	1.333		81	56		137	176
8050	1-1/2" diameter		5	1.600		99.50	67		166.50	213
8060	2" diameter		4.50	1.778		146	74.50		220.50	276
8070	2-1/2" diameter		4	2		269	84		353	425
8080	3" diameter		3.50	2.286		340	96		436	525
8090	3-1/2" diameter		3	2.667		495	112		607	715
8100	4" diameter		2.50	3.200		550	134		684	810
8150	T conduit body with cover, 1/2" diameter		11	.727		43.50	30.50		74	95
8160	3/4" diameter		9	.889		55	37.50		92.50	118
8170	1" diameter		6	1.333		65.50	56		121.50	159
8180	1-1/4" diameter 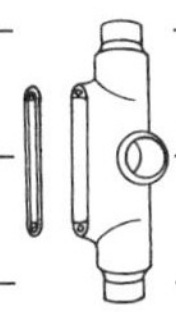		5	1.600		91.50	67		158.50	205
8190	1-1/2" diameter		4.50	1.778		115	74.50		189.50	242
8200	2" diameter		4	2		168	84		252	315
8210	2-1/2" diameter		3.50	2.286		282	96		378	460
8220	3" diameter		3	2.667		375	112		487	590
8230	3-1/2" diameter		2.50	3.200		500	134		634	755
8240	4" diameter 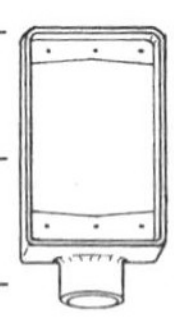		2	4		545	168		713	855
8300	FS conduit body, 1 gang, 3/4" diameter		11	.727		43	30.50		73.50	94.50
8310	1" diameter		10	.800		47	33.50		80.50	104
8350	2 gang, 3/4" diameter		9	.889		74.50	37.50		112	140
8360	1" diameter		8	1		89	42		131	163
8400	Duplex receptacle cover		64	.125		31.50	5.25		36.75	43
8410	Switch cover		64	.125		43	5.25		48.25	55.50
8420	Switch, vaportight cover		53	.151		99.50	6.35		105.85	119
8430	Blank, cover		64	.125		29	5.25		34.25	39.50
8520	FSC conduit body, 1 gang, 3/4" diameter 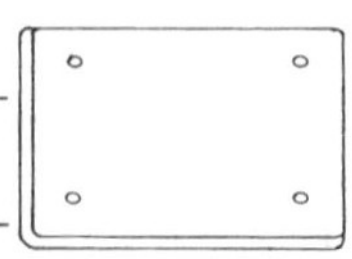		10	.800		51	33.50		84.50	108
8530	1" diameter		9	.889		57.50	37.50		95	121
8550	2 gang, 3/4" diameter		8	1		86.50	42		128.50	160
8560	1" diameter		7	1.143		95.50	48		143.50	179
8590	Conduit hubs, 1/2" diameter		18	.444		28.50	18.65		47.15	60.50
8600	3/4" diameter		16	.500		32.50	21		53.50	68.50
8610	1" diameter		14	.571		40.50	24		64.50	81.50
8620	1-1/4" diameter		12	.667		46.50	28		74.50	94
8630	1-1/2" diameter	↓	10	.800	↓	53.50	33.50		87	111

16132 | Conduit & Tubing

		16132 Conduit & Tubing	CREW	DAILY OUTPUT	LABOR-HOURS	UNIT	2006 BARE COSTS MAT.	LABOR	EQUIP.	TOTAL	TOTAL INCL O&P	
250	8640	2" diameter	1 Elec	8.80	.909	Ea.	76.50	38		114.50	144	250
	8650	2-1/2" diameter		8.50	.941		120	39.50		159.50	193	
	8660	3" diameter		8	1		165	42		207	247	
	8670	3-1/2" diameter		7.50	1.067		214	45		259	305	
	8680	4" diameter		7	1.143		262	48		310	360	
	8690	5" diameter		6	1.333		315	56		371	430	
	8700	Plastic coated 40 mil thick										
	8710	Pipe strap, stamped 1 hole, 1/2" diameter	1 Elec	470	.017	Ea.	7.20	.71		7.91	9	
	8720	3/4" diameter		440	.018		7.25	.76		8.01	9.15	
	8730	1" diameter		400	.020		9.30	.84		10.14	11.55	
	8740	1-1/4" diameter		355	.023		11.30	.95		12.25	13.90	
	8750	1-1/2" diameter		320	.025		13.95	1.05		15	16.95	
	8760	2" diameter		266	.030		14.50	1.26		15.76	17.90	
	8770	2-1/2" diameter		200	.040		17.40	1.68		19.08	21.50	
	8780	3" diameter		133	.060		18.30	2.53		20.83	24	
	8790	3-1/2" diameter		110	.073		19.60	3.05		22.65	26	
	8800	4" diameter		90	.089		25.50	3.73		29.23	34	
	8810	5" diameter		70	.114		99.50	4.80		104.30	116	
	8840	Clamp back spacers, 3/4" diameter		440	.018		10.25	.76		11.01	12.50	
	8850	1" diameter		400	.020		12.20	.84		13.04	14.70	
	8860	1-1/4" diameter		355	.023		16.15	.95		17.10	19.20	
	8870	1-1/2" diameter		320	.025		22	1.05		23.05	26	
	8880	2" diameter		266	.030		35	1.26		36.26	40.50	
	8900	3" diameter		133	.060		78.50	2.53		81.03	90	
	8920	4" diameter		90	.089		114	3.73		117.73	132	
	8950	Touch-up plastic coating, spray, 12 oz.					32.50			32.50	35.50	
	8960	Sealing fittings, 1/2" diameter	1 Elec	11	.727		47	30.50		77.50	98.50	
	8970	3/4" diameter		9	.889		48	37.50		85.50	111	
	8980	1" diameter		7.50	1.067		58	45		103	133	
	8990	1-1/4" diameter		6.50	1.231		69.50	51.50		121	156	
	9000	1-1/2" diameter		5.50	1.455		96	61		157	199	
	9010	2" diameter		4.80	1.667		116	70		186	236	
	9020	2-1/2" diameter		4	2		168	84		252	315	
	9030	3" diameter		3.50	2.286		211	96		307	380	
	9040	3-1/2" diameter		3	2.667		530	112		642	760	
	9050	4" diameter		2.50	3.200		860	134		994	1,150	
	9060	5" diameter		1.70	4.706		1,125	198		1,323	1,525	
	9070	Unions, 1/2" diameter		18	.444		39.50	18.65		58.15	72	
	9080	3/4" diameter		15	.533		40	22.50		62.50	78.50	
	9090	1" diameter		13	.615		53.50	26		79.50	99	
	9100	1-1/4" diameter		11	.727		86.50	30.50		117	142	
	9110	1-1/2" diameter		9.50	.842		104	35.50		139.50	170	
	9120	2" diameter		8	1		139	42		181	218	
	9130	2-1/2" diameter		7.50	1.067		210	45		255	300	
	9140	3" diameter		6.80	1.176		290	49.50		339.50	395	
	9150	3-1/2" diameter		5.80	1.379		360	58		418	485	
	9160	4" diameter		4.80	1.667		460	70		530	615	
	9170	5" diameter		4	2		680	84		764	880	
260	0010	CUTTING AND DRILLING										260
	0100	Hole drilling to 10' high, concrete wall										
	0110	8" thick, 1/2" pipe size	R-31	12	.667	Ea.	3.51	28	5.40	36.91	53	
	0120	3/4" pipe size		12	.667		3.51	28	5.40	36.91	53	
	0130	1" pipe size		9.50	.842		7.25	35.50	6.80	49.55	70	
	0140	1-1/4" pipe size		9.50	.842		7.25	35.50	6.80	49.55	70	
	0150	1-1/2" pipe size		9.50	.842		7.25	35.50	6.80	49.55	70	
	0160	2" pipe size		4.40	1.818		7.85	76.50	14.70	99.05	143	

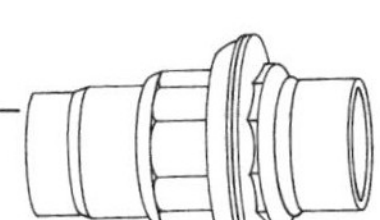

16132 | Conduit & Tubing		CREW	DAILY OUTPUT	LABOR-HOURS	UNIT	2006 BARE COSTS MAT.	LABOR	EQUIP.	TOTAL	TOTAL INCL O&P
260 0170	2-1/2" pipe size	R-31	4.40	1.818	Ea.	7.85	76.50	14.70	99.05	143
0180	3" pipe size		4.40	1.818		7.85	76.50	14.70	99.05	143
0190	3-1/2" pipe size		3.30	2.424		8.40	102	19.60	130	188
0200	4" pipe size		3.30	2.424		8.40	102	19.60	130	188
0500	12" thick, 1/2" pipe size		9.40	.851		5.35	35.50	6.90	47.75	68.50
0520	3/4" pipe size		9.40	.851		5.35	35.50	6.90	47.75	68.50
0540	1" pipe size		7.30	1.096		10.25	46	8.85	65.10	92
0560	1-1/4" pipe size		7.30	1.096		10.25	46	8.85	65.10	92
0570	1-1/2" pipe size		7.30	1.096		10.25	46	8.85	65.10	92
0580	2" pipe size		3.60	2.222		12	93.50	17.95	123.45	177
0590	2-1/2" pipe size		3.60	2.222		12	93.50	17.95	123.45	177
0600	3" pipe size		3.60	2.222		12	93.50	17.95	123.45	177
0610	3-1/2" pipe size		2.80	2.857		13.60	120	23	156.60	226
0630	4" pipe size		2.50	3.200		13.60	134	26	173.60	251
0650	16" thick, 1/2" pipe size		7.60	1.053		7.20	44	8.50	59.70	85.50
0670	3/4" pipe size		7	1.143		7.20	48	9.25	64.45	92
0690	1" pipe size		6	1.333		13.25	56	10.75	80	113
0710	1-1/4" pipe size		5.50	1.455		13.25	61	11.75	86	122
0730	1-1/2" pipe size		5.50	1.455		13.25	61	11.75	86	122
0750	2" pipe size		3	2.667		16.15	112	21.50	149.65	214
0770	2-1/2" pipe size		2.70	2.963		16.15	124	24	164.15	236
0790	3" pipe size		2.50	3.200		16.15	134	26	176.15	253
0810	3-1/2" pipe size		2.30	3.478		18.85	146	28	192.85	277
0830	4" pipe size		2	4		18.85	168	32.50	219.35	315
0850	20" thick, 1/2" pipe size		6.40	1.250		9.05	52.50	10.10	71.65	102
0870	3/4" pipe size		6	1.333		9.05	56	10.75	75.80	108
0890	1" pipe size		5	1.600		16.25	67	12.95	96.20	136
0910	1-1/4" pipe size		4.80	1.667		16.25	70	13.45	99.70	141
0930	1-1/2" pipe size		4.60	1.739		16.25	73	14.05	103.30	146
0950	2" pipe size		2.70	2.963		20.50	124	24	168.50	241
0970	2-1/2" pipe size		2.40	3.333		20.50	140	27	187.50	268
0990	3" pipe size		2.20	3.636		20.50	153	29.50	203	290
1010	3-1/2" pipe size		2	4		24	168	32.50	224.50	320
1030	4" pipe size		1.70	4.706		24	198	38	260	375
1050	24" thick, 1/2" pipe size		5.50	1.455		10.85	61	11.75	83.60	119
1070	3/4" pipe size		5.10	1.569		10.85	66	12.65	89.50	127
1090	1" pipe size		4.30	1.860		19.25	78	15.05	112.30	158
1110	1-1/4" pipe size		4	2		19.25	84	16.15	119.40	168
1130	1-1/2" pipe size		4	2		19.25	84	16.15	119.40	168
1150	2" pipe size		2.40	3.333		24.50	140	27	191.50	273
1170	2-1/2" pipe size		2.20	3.636		24.50	153	29.50	207	295
1190	3" pipe size		2	4		24.50	168	32.50	225	320
1210	3-1/2" pipe size		1.80	4.444		29.50	187	36	252.50	360
1230	4" pipe size		1.50	5.333		29.50	224	43	296.50	425
1500	Brick wall, 8" thick, 1/2" pipe size		18	.444		3.51	18.65	3.59	25.75	37
1520	3/4" pipe size		18	.444		3.51	18.65	3.59	25.75	37
1540	1" pipe size		13.30	.602		7.25	25.50	4.86	37.61	52.50
1560	1-1/4" pipe size		13.30	.602		7.25	25.50	4.86	37.61	52.50
1580	1-1/2" pipe size		13.30	.602		7.25	25.50	4.86	37.61	52.50
1600	2" pipe size		5.70	1.404		7.85	59	11.35	78.20	112
1620	2-1/2" pipe size		5.70	1.404		7.85	59	11.35	78.20	112
1640	3" pipe size		5.70	1.404		7.85	59	11.35	78.20	112
1660	3-1/2" pipe size		4.40	1.818		8.40	76.50	14.70	99.60	143
1680	4" pipe size		4	2		8.40	84	16.15	108.55	156
1700	12" thick, 1/2" pipe size		14.50	.552		5.35	23	4.46	32.81	46.50
1720	3/4" pipe size	↓	14.50	.552	↓	5.35	23	4.46	32.81	46.50 260

16132 | Conduit & Tubing

260		CREW	DAILY OUTPUT	LABOR-HOURS	UNIT	2006 BARE COSTS MAT.	LABOR	EQUIP.	TOTAL	TOTAL INCL O&P	260
1740	1" pipe size	R-31	11	.727	Ea.	10.25	30.50	5.90	46.65	65	
1760	1-1/4" pipe size		11	.727		10.25	30.50	5.90	46.65	65	
1780	1-1/2" pipe size		11	.727		10.25	30.50	5.90	46.65	65	
1800	2" pipe size		5	1.600		12	67	12.95	91.95	131	
1820	2-1/2" pipe size		5	1.600		12	67	12.95	91.95	131	
1840	3" pipe size		5	1.600		12	67	12.95	91.95	131	
1860	3-1/2" pipe size		3.80	2.105		13.60	88.50	17	119.10	170	
1880	4" pipe size		3.30	2.424		13.60	102	19.60	135.20	194	
1900	16" thick, 1/2" pipe size		12.30	.650		7.20	27.50	5.25	39.95	55.50	
1920	3/4" pipe size		12.30	.650		7.20	27.50	5.25	39.95	55.50	
1940	1" pipe size		9.30	.860		13.25	36	6.95	56.20	78	
1960	1-1/4" pipe size		9.30	.860		13.25	36	6.95	56.20	78	
1980	1-1/2" pipe size		9.30	.860		13.25	36	6.95	56.20	78	
2000	2" pipe size		4.40	1.818		16.15	76.50	14.70	107.35	152	
2010	2-1/2" pipe size		4.40	1.818		16.15	76.50	14.70	107.35	152	
2030	3" pipe size		4.40	1.818		16.15	76.50	14.70	107.35	152	
2050	3-1/2" pipe size		3.30	2.424		18.85	102	19.60	140.45	200	
2070	4" pipe size		3	2.667		18.85	112	21.50	152.35	218	
2090	20" thick, 1/2" pipe size		10.70	.748		9.05	31.50	6.05	46.60	65	
2110	3/4" pipe size		10.70	.748		9.05	31.50	6.05	46.60	65	
2130	1" pipe size		8	1		16.25	42	8.10	66.35	91.50	
2150	1-1/4" pipe size		8	1		16.25	42	8.10	66.35	91.50	
2170	1-1/2" pipe size		8	1		16.25	42	8.10	66.35	91.50	
2190	2" pipe size		4	2		20.50	84	16.15	120.65	169	
2210	2-1/2" pipe size		4	2		20.50	84	16.15	120.65	169	
2230	3" pipe size		4	2		20.50	84	16.15	120.65	169	
2250	3-1/2" pipe size		3	2.667		24	112	21.50	157.50	223	
2270	4" pipe size		2.70	2.963		24	124	24	172	245	
2290	24" thick, 1/2" pipe size		9.40	.851		10.85	35.50	6.90	53.25	74.50	
2310	3/4" pipe size		9.40	.851		10.85	35.50	6.90	53.25	74.50	
2330	1" pipe size		7.10	1.127		19.25	47.50	9.10	75.85	104	
2350	1-1/4" pipe size		7.10	1.127		19.25	47.50	9.10	75.85	104	
2370	1-1/2" pipe size		7.10	1.127		19.25	47.50	9.10	75.85	104	
2390	2" pipe size		3.60	2.222		24.50	93.50	17.95	135.95	191	
2410	2-1/2" pipe size		3.60	2.222		24.50	93.50	17.95	135.95	191	
2430	3" pipe size		3.60	2.222		24.50	93.50	17.95	135.95	191	
2450	3-1/2" pipe size		2.80	2.857		29.50	120	23	172.50	243	
2470	4" pipe size		2.50	3.200		29.50	134	26	189.50	268	
3000	Knockouts to 8' high, metal boxes & enclosures										
3020	With hole saw, 1/2" pipe size	1 Elec	53	.151	Ea.		6.35		6.35	9.75	
3040	3/4" pipe size		47	.170			7.15		7.15	11	
3050	1" pipe size		40	.200			8.40		8.40	12.95	
3060	1-1/4" pipe size		36	.222			9.35		9.35	14.40	
3070	1-1/2" pipe size		32	.250			10.50		10.50	16.20	
3080	2" pipe size		27	.296			12.45		12.45	19.15	
3090	2-1/2" pipe size		20	.400			16.80		16.80	26	
4010	3" pipe size		16	.500			21		21	32.50	
4030	3-1/2" pipe size		13	.615			26		26	40	
4050	4" pipe size		11	.727			30.50		30.50	47	
4070	With hand punch set, 1/2" pipe size		40	.200			8.40		8.40	12.95	
4090	3/4" pipe size		32	.250			10.50		10.50	16.20	
4110	1" pipe size		30	.267			11.20		11.20	17.25	
4130	1-1/4" pipe size		28	.286			12		12	18.50	
4150	1-1/2" pipe size		26	.308			12.90		12.90	19.90	
4170	2" pipe size		20	.400			16.80		16.80	26	
4190	2-1/2" pipe size		17	.471			19.75		19.75	30.50	

16132 | Conduit & Tubing

			CREW	DAILY OUTPUT	LABOR-HOURS	UNIT	2006 BARE COSTS MAT.	LABOR	EQUIP.	TOTAL	TOTAL INCL O&P
260	4200	3" pipe size	1 Elec	15	.533	Ea.		22.50		22.50	34.50
	4220	3-1/2" pipe size		12	.667			28		28	43
	4240	4" pipe size		10	.800			33.50		33.50	52
	4260	With hydraulic punch, 1/2" pipe size		44	.182			7.65		7.65	11.75
	4280	3/4" pipe size		38	.211			8.85		8.85	13.60
	4300	1" pipe size		38	.211			8.85		8.85	13.60
	4320	1-1/4" pipe size		38	.211			8.85		8.85	13.60
	4340	1-1/2" pipe size		38	.211			8.85		8.85	13.60
	4360	2" pipe size		32	.250			10.50		10.50	16.20
	4380	2-1/2" pipe size		27	.296			12.45		12.45	19.15
	4400	3" pipe size		23	.348			14.60		14.60	22.50
	4420	3-1/2" pipe size		20	.400			16.80		16.80	26
	4440	4" pipe size		18	.444			18.65		18.65	29
300	0010	**ELECTRICAL NONMETALLIC TUBING (ENT)**									
	0050	Flexible, 1/2" diameter	1 Elec	270	.030	L.F.	.46	1.24		1.70	2.43
	0100	3/4" diameter		230	.035		.65	1.46		2.11	2.97
	0200	1" diameter		145	.055		1.02	2.32		3.34	4.69
	0210	1-1/4" diameter		125	.064		1.41	2.69		4.10	5.70
	0220	1-1/2" diameter		100	.080		1.92	3.36		5.28	7.30
	0230	2" diameter		75	.107		2.24	4.48		6.72	9.35
	0300	Connectors, to outlet box, 1/2" diameter		230	.035	Ea.	.82	1.46		2.28	3.15
	0310	3/4" diameter		210	.038		1.50	1.60		3.10	4.12
	0320	1" diameter		200	.040		2.40	1.68		4.08	5.25
	0400	Couplings, to conduit, 1/2" diameter		145	.055		1.26	2.32		3.58	4.96
	0410	3/4" diameter		130	.062		2.40	2.58		4.98	6.60
	0420	1" diameter		125	.064		3.65	2.69		6.34	8.15
320	0010	**FLEXIBLE METALLIC CONDUIT**									
	0050	Steel, 3/8" diameter	1 Elec	200	.040	L.F.	.34	1.68		2.02	2.96
	0100	1/2" diameter		200	.040		.37	1.68		2.05	3
	0200	3/4" diameter		160	.050		.51	2.10		2.61	3.80
	0250	1" diameter		100	.080		.94	3.36		4.30	6.25
	0300	1-1/4" diameter		70	.114		1.23	4.80		6.03	8.75
	0350	1-1/2" diameter		50	.160		2.16	6.70		8.86	12.75
	0370	2" diameter		40	.200		2.65	8.40		11.05	15.85
	0380	2-1/2" diameter		30	.267		3.20	11.20		14.40	21
	0390	3" diameter	2 Elec	50	.320		4.73	13.45		18.18	25.50
	0400	3-1/2" diameter		40	.400		5.95	16.80		22.75	32.50
	0410	4" diameter		30	.533		8.10	22.50		30.60	43.50
	0420	Connectors, plain, 3/8" diameter	1 Elec	100	.080	Ea.	1.10	3.36		4.46	6.40
	0430	1/2" diameter		80	.100		1.65	4.20		5.85	8.25
	0440	3/4" diameter		70	.114		1.85	4.80		6.65	9.45
	0450	1" diameter		50	.160		4.50	6.70		11.20	15.30
	0452	1-1/4" diameter		45	.178		7.65	7.45		15.10	19.90
	0454	1-1/2" diameter		40	.200		11.25	8.40		19.65	25.50
	0456	2" diameter		28	.286		15.30	12		27.30	35.50
	0458	2-1/2" diameter		25	.320		31	13.45		44.45	54.50
	0460	3" diameter		20	.400		43	16.80		59.80	73.50
	0462	3-1/2" diameter		16	.500		125	21		146	171
	0464	4" diameter		13	.615		159	26		185	215
	0490	Insulated, 1" diameter		40	.200		7.75	8.40		16.15	21.50
	0500	1-1/4" diameter		40	.200		13.90	8.40		22.30	28.50
	0550	1-1/2" diameter		32	.250		19.90	10.50		30.40	38
	0600	2" diameter		23	.348		35.50	14.60		50.10	62
	0610	2-1/2" diameter		20	.400		90.50	16.80		107.30	126

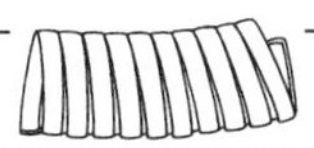

 Important: See the Reference Section for supporting data - Crews, Rental Equipment, City Cost Indexes and Reference Data

16132	Conduit & Tubing	CREW	DAILY OUTPUT	LABOR-HOURS	UNIT	2006 BARE COSTS MAT.	LABOR	EQUIP.	TOTAL	TOTAL INCL O&P
320 0620	3″ diameter	1 Elec	17	.471	Ea.	120	19.75		139.75	163 320
0630	3-1/2″ diameter		13	.615		210	26		236	271
0640	4″ diameter		10	.800		277	33.50		310.50	355
0650	Connectors 90°, plain, 3/8″ diameter		80	.100		1.87	4.20		6.07	8.50
0660	1/2″ diameter 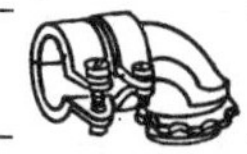		60	.133		3.14	5.60		8.74	12.10
0700	3/4″ diameter		50	.160		5.70	6.70		12.40	16.60
0750	1″ diameter		40	.200		9.90	8.40		18.30	24
0790	Insulated, 1″ diameter		40	.200		14.55	8.40		22.95	29
0800	1-1/4″ diameter		30	.267		31	11.20		42.20	51.50
0850	1-1/2″ diameter		23	.348		54	14.60		68.60	82
0900	2″ diameter		18	.444		72	18.65		90.65	108
0910	2-1/2″ diameter		16	.500		154	21		175	202
0920	3″ diameter		14	.571		192	24		216	248
0930	3-1/2″ diameter		11	.727		560	30.50		590.50	660
0940	4″ diameter		8	1		840	42		882	990
0960	Couplings, to flexible conduit, 1/2″ diameter		50	.160		.81	6.70		7.51	11.25
0970	3/4″ diameter		40	.200		1.43	8.40		9.83	14.50
0980	1″ diameter		35	.229		3.23	9.60		12.83	18.35
0990	1-1/4″ diameter		28	.286		4.75	12		16.75	24
1000	1-1/2″ diameter		23	.348		6.35	14.60		20.95	29.50
1010	2″ diameter		20	.400		11.95	16.80		28.75	39
1020	2-1/2″ diameter		18	.444		16.35	18.65		35	47
1030	3″ diameter		15	.533		41	22.50		63.50	79.50
1070	Sealtite, 3/8″ diameter		140	.057	L.F.	1.67	2.40		4.07	5.55
1080	1/2″ diameter		140	.057		1.72	2.40		4.12	5.60
1090	3/4″ diameter		100	.080		2.41	3.36		5.77	7.85
1100	1″ diameter		70	.114		3.64	4.80		8.44	11.40
1200	1-1/4″ diameter		50	.160		4.98	6.70		11.68	15.85
1300	1-1/2″ diameter		40	.200		5.80	8.40		14.20	19.30
1400	2″ diameter		30	.267		7.20	11.20		18.40	25
1410	2-1/2″ diameter		27	.296		13	12.45		25.45	33.50
1420	3″ diameter	2 Elec	50	.320		17.95	13.45		31.40	40.50
1440	4″ diameter	″	30	.533		28	22.50		50.50	65
1490	Connectors, plain, 3/8″ diameter 	1 Elec	70	.114	Ea.	3.29	4.80		8.09	11
1500	1/2″ diameter		70	.114		3.24	4.80		8.04	10.95
1700	3/4″ diameter		50	.160		4.74	6.70		11.44	15.55
1900	1″ diameter		40	.200		8.10	8.40		16.50	22
1910	Insulated, 1″ diameter		40	.200		10.50	8.40		18.90	24.50
2000	1-1/4″ diameter 		32	.250		16.40	10.50		26.90	34.50
2100	1-1/2″ diameter		27	.296		21.50	12.45		33.95	42.50
2200	2″ diameter		20	.400		38.50	16.80		55.30	68
2210	2-1/2″ diameter		15	.533		226	22.50		248.50	284
2220	3″ diameter		12	.667		252	28		280	320
2240	4″ diameter		8	1		315	42		357	410
2290	Connectors, 90°, 3/8″ diameter		70	.114		5.50	4.80		10.30	13.45
2300	1/2″ diameter		70	.114		5.30	4.80		10.10	13.25
2400	3/4″ diameter 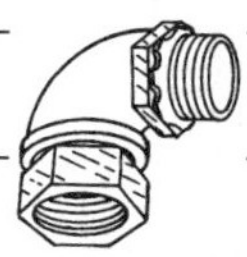		50	.160		8.20	6.70		14.90	19.35
2600	1″ diameter		40	.200		16.85	8.40		25.25	31.50
2790	Insulated, 1″ diameter		40	.200		22.50	8.40		30.90	37.50
2800	1-1/4″ diameter		32	.250		36.50	10.50		47	56.50
3000	1-1/2″ diameter		27	.296		40.50	12.45		52.95	63.50
3100	2″ diameter 		20	.400		57	16.80		73.80	88.50
3110	2-1/2″ diameter		14	.571		266	24		290	330
3120	3″ diameter		11	.727		310	30.50		340.50	385
3140	4″ diameter		7	1.143		400	48		448	515
4300	Coupling sealtite to rigid, 1/2″ diameter		20	.400		4.70	16.80		21.50	31 320

ELECTRICAL 16

16132 | Conduit & Tubing

	Line	Description	Crew	Daily Output	Labor-Hours	Unit	2006 Bare Costs Mat.	Labor	Equip.	Total	Total Incl O&P	
320	4500	3/4" diameter	1 Elec	18	.444	Ea.	6.65	18.65		25.30	36.50	320
	4800	1" diameter		14	.571		9.10	24		33.10	47	
	4900	1-1/4" diameter		12	.667		15.30	28		43.30	60	
	5000	1-1/2" diameter		11	.727		27.50	30.50		58	77.50	
	5100	2" diameter		10	.800		47.50	33.50		81	104	
	5110	2-1/2" diameter		9.50	.842		182	35.50		217.50	255	
	5120	3" diameter		9	.889		195	37.50		232.50	273	
	5130	3-1/2" diameter		9	.889		240	37.50		277.50	320	
	5140	4" diameter	↓	8.50	.941	↓	240	39.50		279.50	325	

16133 | Multi-outlet Assemblies

	Line	Description	Crew	Daily Output	Labor-Hours	Unit	2006 Bare Costs Mat.	Labor	Equip.	Total	Total Incl O&P	
540	0010	**TRENCH DUCT** Steel with cover										540
	0020	Standard adjustable, depths to 4"										
	0100	Straight, single compartment, 9" wide	2 Elec	40	.400	L.F.	80	16.80		96.80	114	
	0200	12" wide		32	.500		91	21		112	133	
	0400	18" wide		26	.615		121	26		147	173	
	0600	24" wide		22	.727		155	30.50		185.50	218	
	0700	27" wide		21	.762		167	32		199	234	
	0800	30" wide		20	.800		187	33.50		220.50	258	
	1000	36" wide		16	1		225	42		267	315	
	1020	Two compartment, 9" wide		38	.421		91	17.70		108.70	127	
	1030	12" wide		30	.533		103	22.50		125.50	148	
	1040	18" wide		24	.667		127	28		155	183	
	1050	24" wide		20	.800		166	33.50		199.50	235	
	1060	30" wide		18	.889		202	37.50		239.50	280	
	1070	36" wide		14	1.143		235	48		283	335	
	1090	Three compartment, 9" wide		36	.444		103	18.65		121.65	142	
	1100	12" wide		28	.571		114	24		138	162	
	1110	18" wide		22	.727		143	30.50		173.50	204	
	1120	24" wide		18	.889		180	37.50		217.50	256	
	1130	30" wide		16	1		218	42		260	305	
	1140	36" wide		12	1.333	↓	255	56		311	370	
	1200	Horizontal elbow, 9" wide		5.40	2.963	Ea.	296	124		420	515	
	1400	12" wide		4.60	3.478		340	146		486	600	
	1600	18" wide		4	4		440	168		608	740	
	1800	24" wide		3.20	5		615	210		825	1,000	
	1900	27" wide		3	5.333		700	224		924	1,125	
	2000	30" wide		2.60	6.154		820	258		1,078	1,300	
	2200	36" wide		2.40	6.667		1,075	280		1,355	1,600	
	2220	Two compartment, 9" wide		3.80	4.211		480	177		657	795	
	2230	12" wide		3	5.333		530	224		754	925	
	2240	18" wide		2.40	6.667		635	280		915	1,125	
	2250	24" wide		2	8		805	335		1,140	1,400	
	2260	30" wide		1.80	8.889		1,075	375		1,450	1,775	
	2270	36" wide		1.60	10		1,300	420		1,720	2,100	
	2290	Three compartment, 9" wide		3.60	4.444		500	187		687	840	
	2300	12" wide		2.80	5.714		560	240		800	985	
	2310	18" wide		2.20	7.273		685	305		990	1,225	
	2320	24" wide		1.80	8.889		860	375		1,235	1,525	
	2330	30" wide		1.60	10		1,125	420		1,545	1,875	
	2350	36" wide		1.40	11.429		1,375	480		1,855	2,250	
	2400	Vertical elbow, 9" wide		5.40	2.963		103	124		227	305	
	2600	12" wide		4.60	3.478		112	146		258	350	
	2800	18" wide		4	4		129	168		297	400	
	3000	24" wide	↓	3.20	5	↓	160	210		370	500	

16133 | Multi-outlet Assemblies		CREW	DAILY OUTPUT	LABOR-HOURS	UNIT	2006 BARE COSTS MAT.	LABOR	EQUIP.	TOTAL	TOTAL INCL O&P
540 3100	27" wide	2 Elec	3	5.333	Ea.	167	224		391	530
3200	30" wide		2.60	6.154		177	258		435	595
3400	36" wide		2.40	6.667		194	280		474	645
3600	Cross, 9" wide		4	4		485	168		653	795
3800	12" wide		3.20	5		515	210		725	890
4000	18" wide		2.60	6.154		615	258		873	1,075
4200	24" wide		2.20	7.273		780	305		1,085	1,325
4300	27" wide		2.20	7.273		915	305		1,220	1,475
4400	30" wide		2	8		1,000	335		1,335	1,650
4600	36" wide		1.80	8.889		1,275	375		1,650	1,975
4620	Two compartment, 9" wide		3.80	4.211		505	177		682	825
4630	12" wide		3	5.333		535	224		759	930
4640	18" wide		2.40	6.667		645	280		925	1,125
4650	24" wide		2	8		820	335		1,155	1,425
4660	30" wide		1.80	8.889		1,075	375		1,450	1,775
4670	36" wide		1.60	10		1,300	420		1,720	2,075
4690	Three compartment, 9" wide		3.60	4.444		515	187		702	855
4700	12" wide		2.80	5.714		585	240		825	1,025
4710	18" wide		2.20	7.273		695	305		1,000	1,225
4720	24" wide		1.80	8.889		870	375		1,245	1,525
4730	30" wide		1.60	10		1,125	420		1,545	1,900
4740	36" wide		1.40	11.429		1,400	480		1,880	2,300
4800	End closure, 9" wide		14.40	1.111		30.50	46.50		77	106
5000	12" wide		12	1.333		34.50	56		90.50	125
5200	18" wide		10	1.600		53	67		120	163
5400	24" wide		8	2		70	84		154	206
5500	27" wide		7	2.286		81	96		177	237
5600	30" wide		6.60	2.424		87.50	102		189.50	254
5800	36" wide		5.80	2.759		104	116		220	292
6000	Tees, 9" wide		4	4		295	168		463	585
6200	12" wide		3.60	4.444		340	187		527	665
6400	18" wide		3.20	5		440	210		650	805
6600	24" wide		3	5.333		630	224		854	1,025
6700	27" wide		2.80	5.714		695	240		935	1,125
6800	30" wide		2.60	6.154		820	258		1,078	1,300
7000	36" wide		2	8		1,075	335		1,410	1,700
7020	Two compartment, 9" wide		3.80	4.211		340	177		517	645
7030	12" wide		3.40	4.706		370	198		568	710
7040	18" wide		3	5.333		490	224		714	885
7050	24" wide		2.80	5.714		665	240		905	1,100
7060	30" wide		2.40	6.667		905	280		1,185	1,425
7070	36" wide		1.90	8.421		1,125	355		1,480	1,800
7090	Three compartment, 9" wide		3.60	4.444		390	187		577	720
7100	12" wide		3.20	5		410	210		620	775
7110	18" wide		2.80	5.714		520	240		760	940
7120	24" wide		2.60	6.154		715	258		973	1,200
7130	30" wide		2.20	7.273		935	305		1,240	1,500
7140	36" wide		1.80	8.889		1,200	375		1,575	1,900
7200	Riser, and cabinet connector, 9" wide		5.40	2.963		129	124		253	335
7400	12" wide		4.60	3.478		150	146		296	390
7600	18" wide		4	4		184	168		352	460
7800	24" wide		3.20	5		223	210		433	570
7900	27" wide		3	5.333		230	224		454	600
8000	30" wide		2.60	6.154		257	258		515	685
8200	36" wide	↓	2	8		299	335		634	850
8400	Insert assembly, cell to conduit adapter, 1-1/4"	1 Elec	16	.500	↓	51	21		72	88.50

540

ELECTRICAL 16

16133 | Multi-outlet Assemblies

		16133 Multi-outlet Assemblies	CREW	DAILY OUTPUT	LABOR-HOURS	UNIT	2006 BARE COSTS MAT.	LABOR	EQUIP.	TOTAL	TOTAL INCL O&P	
540	8500	Adjustable partition	1 Elec	320	.025	L.F.	18.10	1.05		19.15	21.50	540
	8600	Depth of duct over 4", per 1", add					7.90			7.90	8.70	
	8700	Support post	1 Elec	240	.033		18.10	1.40		19.50	22	
	8800	Cover double tile trim, 2 sides					28			28	31	
	8900	4 sides					83			83	91.50	
	9160	Trench duct 3-1/2" x 4-1/2", add					7.50			7.50	8.25	
	9170	Trench duct 4" x 5", add					7.50			7.50	8.25	
	9200	For carpet trim, add					25.50			25.50	28	
	9210	For double carpet trim, add					77.50			77.50	85.50	
560	0010	**UNDERFLOOR DUCT**										560
	0100	Duct, 1-3/8" x 3-1/8" blank, standard	2 Elec	160	.100	L.F.	9.75	4.20		13.95	17.20	
	0200	1-3/8" x 7-1/4" blank, super duct		120	.133		19.50	5.60		25.10	30	
	0400	7/8" or 1-3/8" insert type, 24" O.C., 1-3/8" x 3-1/8", std.		140	.114		13.10	4.80		17.90	22	
	0600	1-3/8" x 7-1/4", super duct		100	.160		23	6.70		29.70	35.50	
	0800	Junction box, single duct, 1 level, 3-1/8"	1 Elec	4	2	Ea.	294	84		378	455	
	0820	3-1/8" x 7-1/4"		4	2		340	84		424	505	
	0840	2 level, 3-1/8" upper & lower		3.20	2.500		340	105		445	535	
	0860	3-1/8" upper, 7-1/4" lower		2.70	2.963		340	124		464	565	
	0880	Carpet pan for above		80	.100		243	4.20		247.20	273	
	0900	Terrazzo pan for above		67	.119		570	5		575	635	
	1000	Junction box, single duct, 1 level, 7-1/4"		2.70	2.963		340	124		464	565	
	1020	2 level, 7-1/4" upper & lower		2.70	2.963		390	124		514	620	
	1040	2 duct, two 3-1/8" upper & lower		3.20	2.500		520	105		625	730	
	1200	1 level, 2 duct, 3-1/8"		3.20	2.500		390	105		495	590	
	1220	Carpet pan for above boxes		80	.100		243	4.20		247.20	273	
	1240	Terrazzo pan for above boxes		67	.119		570	5		575	635	
	1260	Junction box, 1 level, two 3-1/8" x one 3-1/8" + one 7-1/4"		2.30	3.478		665	146		811	955	
	1280	2 level, two 3-1/8" upper, one 3-1/8" + one 7-1/4" lower		2	4		730	168		898	1,075	
	1300	Carpet pan for above boxes		80	.100		243	4.20		247.20	273	
	1320	Terrazzo pan for above boxes		67	.119		570	5		575	635	
	1400	Junction box, 1 level, 2 duct, 7-1/4"		2.30	3.478		1,000	146		1,146	1,325	
	1420	Two 3-1/8" + one 7-1/4"		2	4		1,000	168		1,168	1,350	
	1440	Carpet pan for above		80	.100		243	4.20		247.20	273	
	1460	Terrazzo pan for above		67	.119		570	5		575	635	
	1580	Junction box, 1 level, one 3-1/8" + one 7-1/4" x same		2.30	3.478		665	146		811	955	
	1600	Triple duct, 3-1/8"		2.30	3.478		665	146		811	955	
	1700	Junction box, 1 level, one 3-1/8" + two 7-1/4"		2	4		1,125	168		1,293	1,475	
	1720	Carpet pan for above		80	.100		243	4.20		247.20	273	
	1740	Terrazzo pan for above		67	.119		570	5		575	635	
	1800	Insert to conduit adapter, 3/4" & 1"		32	.250		24	10.50		34.50	42.50	
	2000	Support, single cell		27	.296		36	12.45		48.45	58.50	
	2200	Super duct		16	.500		36	21		57	72	
	2400	Double cell		16	.500		36	21		57	72	
	2600	Triple cell		11	.727		36	30.50		66.50	86.50	
	2800	Vertical elbow, standard duct		10	.800		64.50	33.50		98	123	
	3000	Super duct		8	1		64.50	42		106.50	136	
	3200	Cabinet connector, standard duct		32	.250		49	10.50		59.50	69.50	
	3400	Super duct		27	.296		49	12.45		61.45	72.50	
	3600	Conduit adapter, 1" to 1-1/4"		32	.250		49	10.50		59.50	69.50	
	3800	2" to 1-1/4"		27	.296		58.50	12.45		70.95	83	
	4000	Outlet, low tension (telephone, computer, etc.)		8	1		68.50	42		110.50	140	
	4200	High tension, receptacle (120 volt)		8	1		68.50	42		110.50	140	
	4300	End closure, standard duct		160	.050		2.70	2.10		4.80	6.20	
	4310	Super duct		160	.050		5.20	2.10		7.30	8.95	
	4350	Elbow, horiz., standard duct		26	.308		163	12.90		175.90	199	

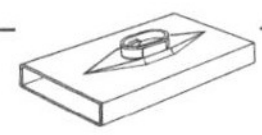

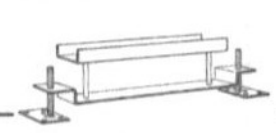

16 ELECTRICAL

16133 | Multi-outlet Assemblies

			CREW	DAILY OUTPUT	LABOR-HOURS	UNIT	2006 BARE COSTS MAT.	LABOR	EQUIP.	TOTAL	TOTAL INCL O&P	
560	4360	Super duct	1 Elec	26	.308	Ea.	163	12.90		175.90	199	560
	4380	Elbow, offset, standard duct		26	.308		64.50	12.90		77.40	91	
	4390	Super duct		26	.308		64.50	12.90		77.40	91	
	4400	Marker screw assembly for inserts		50	.160		11.40	6.70		18.10	23	
	4410	Y take off, standard duct		26	.308		97.50	12.90		110.40	127	
	4420	Super duct		26	.308		97.50	12.90		110.40	127	
	4430	Box opening plug, standard duct		160	.050		9.10	2.10		11.20	13.25	
	4440	Super duct		160	.050		9.10	2.10		11.20	13.25	
	4450	Sleeve coupling, standard duct		160	.050		24.50	2.10		26.60	30	
	4460	Super duct		160	.050		30	2.10		32.10	36	
	4470	Conduit adapter, standard duct, 3/4"		32	.250		49	10.50		59.50	69.50	
	4480	1" or 1-1/4"		32	.250		49	10.50		59.50	69.50	
	4500	1-1/2"	↓	32	.250	↓	49	10.50		59.50	69.50	
580	0010	**WIRING DUCT** Plastic										580
	1250	PVC, snap-in slots, adhesive backed										
	1270	1-1/2"W x 2"H	2 Elec	120	.133	L.F.	5.65	5.60		11.25	14.85	
	1280	1-1/2"W x 3"H		120	.133		6.90	5.60		12.50	16.25	
	1290	1-1/2"W x 4"H		120	.133		8.10	5.60		13.70	17.55	
	1300	2"W x 1"H		120	.133		5.35	5.60		10.95	14.55	
	1310	2"W x 1-1/2"H		120	.133		5.65	5.60		11.25	14.85	
	1320	2"W x 2"H		120	.133		5.75	5.60		11.35	15	
	1330	2"W x 2-1/2"H		120	.133		7.05	5.60		12.65	16.40	
	1340	2"W x 3"H		120	.133		7.30	5.60		12.90	16.70	
	1350	2"W x 4"H		120	.133		8.60	5.60		14.20	18.10	
	1360	2-1/2"W x 3"H		120	.133		7.65	5.60		13.25	17.05	
	1370	3"W x 1"H		110	.145		6	6.10		12.10	16	
	1380	3"W x 1-1/4"H		110	.145		7	6.10		13.10	17.10	
	1390	3"W x 2"H		110	.145		7.05	6.10		13.15	17.15	
	1400	3"W x 3"H		110	.145		8.30	6.10		14.40	18.55	
	1410	3"W x 4"H		110	.145		10.20	6.10		16.30	20.50	
	1420	3"W x 5"H		110	.145		13.10	6.10		19.20	24	
	1430	4"W x 1-1/2"H		100	.160		7.05	6.70		13.75	18.10	
	1440	4"W x 2"H		100	.160		7.90	6.70		14.60	19.05	
	1450	4"W x 3"H		100	.160		9.10	6.70		15.80	20.50	
	1460	4"W x 4"H		100	.160		11.10	6.70		17.80	22.50	
	1470	4"W x 5"H		100	.160		15.35	6.70		22.05	27.50	
	1550	Cover, 1-1/2"W		200	.080		1.04	3.36		4.40	6.35	
	1560	2"W		200	.080		1.27	3.36		4.63	6.60	
	1570	2-1/2"W		200	.080		1.62	3.36		4.98	7	
	1580	3"W	↓	200	.080	↓	1.93	3.36		5.29	7.30	
	1590	4"W	↓	200	.080	L.F.	2.34	3.36		5.70	7.75	
800	0010	**SURFACE RACEWAY**										800
	0090	Metal, straight section										
	0100	No. 500	1 Elec	100	.080	L.F.	.81	3.36		4.17	6.10	
	0110	No. 700		100	.080		.91	3.36		4.27	6.20	
	0400	No. 1500, small pancake		90	.089		1.65	3.73		5.38	7.55	
	0600	No. 2000, base & cover, blank		90	.089		1.61	3.73		5.34	7.50	
	0610	Receptacle, 6" O.C.		40	.200		9	8.40		17.40	23	
	0620	12" O.C.		44	.182		5.85	7.65		13.50	18.20	
	0630	18" O.C.		46	.174		6.20	7.30		13.50	18.05	
	0650	30" O.C.		50	.160		4.25	6.70		10.95	15.05	
	0660	60" O.C.		50	.160		3.95	6.70		10.65	14.70	
	0670	No. 2200, base & cover, blank		80	.100		2.35	4.20		6.55	9.05	
	0700	Receptacle, 18" O.C.		36	.222		5.80	9.35		15.15	21	
	0720	30" O.C.	↓	40	.200	↓	5.50	8.40		13.90	19	

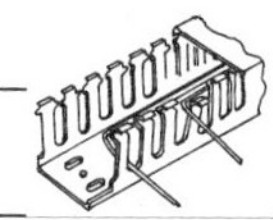

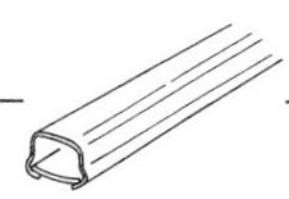

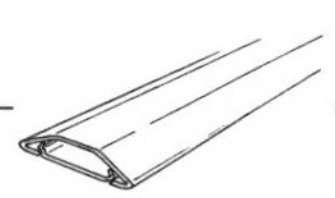

ELECTRICAL 16

16133 | Multi-outlet Assemblies

800

		CREW	DAILY OUTPUT	LABOR-HOURS	UNIT	2006 BARE COSTS MAT.	LABOR	EQUIP.	TOTAL	TOTAL INCL O&P
0730	60″ O.C.	1 Elec	40	.200	L.F.	4.85	8.40		13.25	18.30
0800	No. 3000, base & cover, blank		75	.107		3.25	4.48		7.73	10.50
0810	Receptacle, 6″ O.C.		45	.178		27.50	7.45		34.95	42
0820	12″ O.C.		62	.129		15.65	5.40		21.05	25.50
0830	18″ O.C.		64	.125		12.60	5.25		17.85	22
0840	24″ O.C.		66	.121		9.45	5.10		14.55	18.25
0850	30″ O.C.		68	.118		8.50	4.94		13.44	16.95
0860	60″ O.C.		70	.114		6.25	4.80		11.05	14.30
1000	No. 4000, base & cover, blank		65	.123		5.25	5.15		10.40	13.75
1010	Receptacle, 6″ O.C.		41	.195		38.50	8.20		46.70	54.50
1020	12″ O.C.		52	.154		23	6.45		29.45	35.50
1030	18″ O.C.		54	.148		19.10	6.20		25.30	30.50
1040	24″ O.C.		56	.143		15.65	6		21.65	26.50
1050	30″ O.C.		58	.138		13.90	5.80		19.70	24
1060	60″ O.C.		60	.133		10.90	5.60		16.50	20.50
1200	No. 6000, base & cover, blank		50	.160		8.80	6.70		15.50	20
1210	Receptacle, 6″ O.C.		30	.267		47	11.20		58.20	69
1220	12″ O.C. 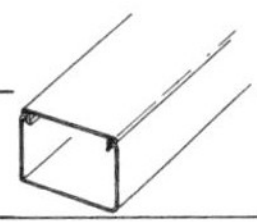		37	.216		30	9.10		39.10	47
1230	18″ O.C.		39	.205		25.50	8.60		34.10	42
1240	24″ O.C.		41	.195		21	8.20		29.20	35.50
1250	30″ O.C.		43	.186		20	7.80		27.80	34
1260	60″ O.C.		45	.178		15.50	7.45		22.95	28.50
2400	Fittings, elbows, No. 500		40	.200	Ea.	1.47	8.40		9.87	14.55
2800	Elbow cover, No. 2000		40	.200		2.80	8.40		11.20	16.05
2880	Tee, No. 500		42	.190		2.82	8		10.82	15.40
2900	No. 2000		27	.296		8.80	12.45		21.25	29
3000	Switch box, No. 500		16	.500		9.55	21		30.55	43
3400	Telephone outlet, No. 1500		16	.500		10.60	21		31.60	44
3600	Junction box, No. 1500		16	.500		7.40	21		28.40	40.50
3800	Plugmold wired sections, No. 2000									
4000	1 circuit, 6 outlets, 3 ft. long	1 Elec	8	1	Ea.	27	42		69	94
4100	2 circuits, 8 outlets, 6 ft. long		5.30	1.509		45	63.50		108.50	147
4110	Tele-power pole, alum., w/ 2 recept, 10′		4	2		139	84		223	282
4120	12′		3.85	2.078		167	87.50		254.50	320
4130	15′		3.70	2.162		219	91		310	380
4140	Steel, w/ 2 recept, 10′		4	2		96	84		180	235
4150	One phone fitting, 10′		4	2		132	84		216	274
4160	Alum., 4 outlets, 10′		3.70	2.162		176	91		267	335
4300	Overhead distribution systems, 125 volt									
4800	No. 2000, entrance end fitting	1 Elec	20	.400	Ea.	3.70	16.80		20.50	30
5000	Blank end fitting		40	.200		1.64	8.40		10.04	14.75
5200	Supporting clip		40	.200		.92	8.40		9.32	13.95
5800	No. 3000, entrance end fitting		20	.400		6.85	16.80		23.65	33.50
6000	Blank end fitting		40	.200		2.02	8.40		10.42	15.15
6020	Internal elbow		20	.400		9.20	16.80		26	36
6030	External elbow		20	.400		12.65	16.80		29.45	40
6040	Device bracket		53	.151		3.15	6.35		9.50	13.20
6400	Hanger clamp 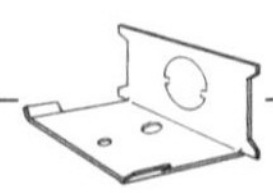		32	.250		4.45	10.50		14.95	21
7000	Base		90	.089	L.F.	3.35	3.73		7.08	9.45
7200	Divider		100	.080	″	.65	3.36		4.01	5.90
7400	Entrance end fitting		16	.500	Ea.	16.20	21		37.20	50.50
7600	Blank end fitting 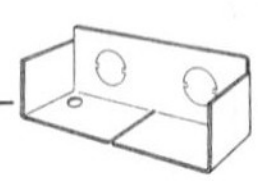		40	.200		4.80	8.40		13.20	18.25
7610	Recp. & tele. cover		53	.151		8.10	6.35		14.45	18.65
7620	External elbow		16	.500		25.50	21		46.50	60.50
7630	Coupling		53	.151		3.85	6.35		10.20	14
7640	Divider clip & coupling		80	.100		.76	4.20		4.96	7.30

800

16 ELECTRICAL

16133 | Multi-outlet Assemblies

			CREW	DAILY OUTPUT	LABOR-HOURS	UNIT	2006 BARE COSTS MAT.	LABOR	EQUIP.	TOTAL	TOTAL INCL O&P	
800	7650	Panel connector	1 Elec	16	.500	Ea.	16.15	21		37.15	50.50	800
	7800	Take off connector		16	.500		51	21		72	88.50	
	8000	No. 6000, take off connector		16	.500		61.50	21		82.50	100	
	8100	Take off fitting		16	.500		46	21		67	83	
	8200	Hanger clamp		32	.250		10.40	10.50		20.90	27.50	
	8230	Coupling					4.80			4.80	5.30	
	8240	One gang device plate	1 Elec	53	.151		7.50	6.35		13.85	18	
	8250	Two gang device plate		40	.200		7.50	8.40		15.90	21	
	8260	Blank end fitting		40	.200		6.80	8.40		15.20	20.50	
	8270	Combination elbow		14	.571		26.50	24		50.50	66	
	8300	Panel connector		16	.500		11.65	21		32.65	45.50	
	8500	Chan-L-Wire system installed in 1-5/8″ x 1-5/8″ strut. Strut										
	8600	not incl., 30 amp, 4 wire, 3 phase	1 Elec	200	.040	L.F.	3.60	1.68		5.28	6.55	
	8700	Junction box		8	1	Ea.	24.50	42		66.50	91	
	8800	Insulating end cap		40	.200		6.45	8.40		14.85	20	
	8900	Strut splice plate		40	.200		8.50	8.40		16.90	22.50	
	9000	Tap		40	.200		16.90	8.40		25.30	31.50	
	9100	Fixture hanger		60	.133		7.30	5.60		12.90	16.70	
	9200	Pulling tool					65.50			65.50	72	
	9300	Non-metallic, straight section										
	9310	7/16″ x 7/8″, base & cover, blank	1 Elec	160	.050	L.F.	1.25	2.10		3.35	4.62	
	9320	Base & cover w/ adhesive		160	.050		1.38	2.10		3.48	4.76	
	9340	7/16″ x 1-5/16″, base & cover, blank		145	.055		1.38	2.32		3.70	5.10	
	9350	Base & cover w/ adhesive		145	.055		1.60	2.32		3.92	5.35	
	9370	11/16″ x 2-1/4″, base & cover, blank		130	.062		1.94	2.58		4.52	6.10	
	9380	Base & cover w/ adhesive		130	.062		2.24	2.58		4.82	6.45	
	9400	Fittings, elbows, 7/16″ x 7/8″		50	.160	Ea.	1.38	6.70		8.08	11.85	
	9410	7/16″ x 1-5/16″		45	.178		1.44	7.45		8.89	13.10	
	9420	11/16″ x 2-1/4″		40	.200		1.56	8.40		9.96	14.65	
	9430	Tees, 7/16″ x 7/8″		35	.229		1.82	9.60		11.42	16.80	
	9440	7/16″ x 1-5/16″		32	.250		1.87	10.50		12.37	18.25	
	9450	11/16″ x 2-1/4″		30	.267		1.89	11.20		13.09	19.35	
	9460	Cover clip, 7/16″ x 7/8″		80	.100		.36	4.20		4.56	6.85	
	9470	7/16″ x 1-5/16″		72	.111		.33	4.67		5	7.55	
	9480	11/16″ x 2-1/4″		64	.125		.55	5.25		5.80	8.70	
	9490	Blank end, 7/16″ x 7/8″		50	.160		.52	6.70		7.22	10.90	
	9510	11/16″ x 2-1/4″		40	.200		.88	8.40		9.28	13.90	
	9520	Round fixture box 5.5″ dia x 1″		25	.320		8.70	13.45		22.15	30	
	9530	Device box, 1 gang		30	.267		3.85	11.20		15.05	21.50	
	9540	2 gang		25	.320		5.75	13.45		19.20	27	
	9990	Minimum labor/equipment charge		5	1.600	Job		67		67	104	

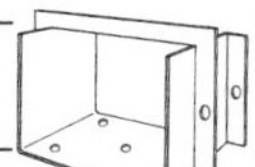

16134 | Wireway & Aux Gutters

			CREW	DAILY OUTPUT	LABOR-HOURS	UNIT	MAT.	LABOR	EQUIP.	TOTAL	TOTAL INCL O&P	
150	0010	**WIREWAY** to 15′ high										150
	0100	Screw cover, NEMA 1 w/ fittings and supports, 2-1/2″ x 2-1/2″	1 Elec	45	.178	L.F.	10.95	7.45		18.40	23.50	
	0200	4″ x 4″	″	40	.200		11.60	8.40		20	26	
	0400	6″ x 6″	2 Elec	60	.267		19.70	11.20		30.90	39	
	0600	8″ x 8″		40	.400		33	16.80		49.80	62.50	
	0620	10″ x 10″		30	.533		42	22.50		64.50	80.50	
	0640	12″ x 12″		20	.800		60	33.50		93.50	118	
	0800	Elbows, 90°, 2-1/2″	1 Elec	24	.333	Ea.	34	14		48	58.50	
	1000	4″		20	.400		38.50	16.80		55.30	68.50	
	1200	6″		18	.444		44	18.65		62.65	77	
	1400	8″		16	.500		70.50	21		91.50	110	
	1420	10″		12	.667		78.50	28		106.50	130	

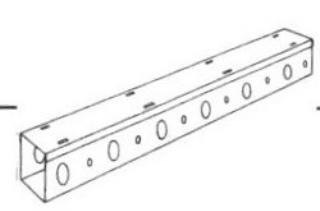

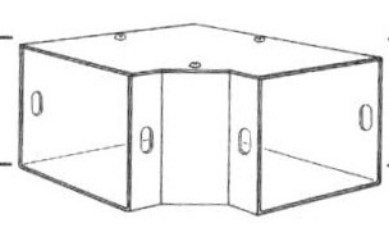

ELECTRICAL 16

16134 | Wireway & Aux Gutters

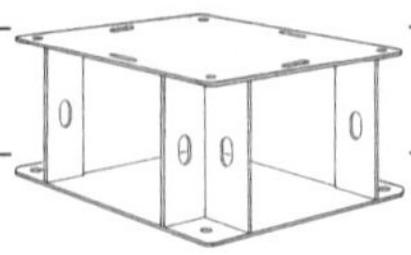
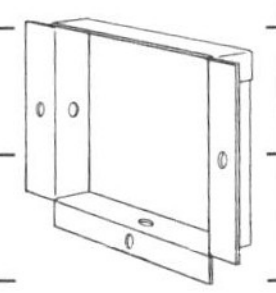
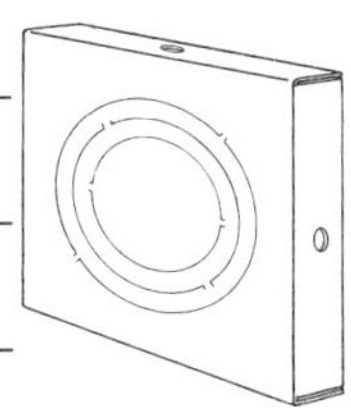
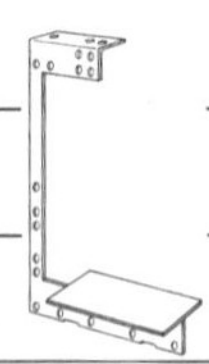
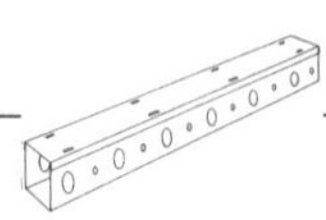

150		CREW	DAILY OUTPUT	LABOR-HOURS	UNIT	2006 BARE COSTS MAT.	LABOR	EQUIP.	TOTAL	TOTAL INCL O&P
1440	12″	1 Elec	10	.800	Ea.	127	33.50		160.50	192
1500	Elbows, 45°, 2-1/2″		24	.333		32.50	14		46.50	57
1510	4″		20	.400		42	16.80		58.80	72
1520	6″		18	.444		44	18.65		62.65	77
1530	8″		16	.500		70.50	21		91.50	110
1540	10″		12	.667		78	28		106	129
1550	12″		10	.800		178	33.50		211.50	248
1600	"T" box, 2-1/2″		18	.444		39	18.65		57.65	72
1800	4″		16	.500		47	21		68	84.50
2000	6″		14	.571		53	24		77	95.50
2200	8″		12	.667		98.50	28		126.50	151
2220	10″		10	.800		113	33.50		146.50	176
2240	12″		8	1		188	42		230	272
2300	Cross, 2-1/2″		16	.500		44	21		65	80.50
2310	4″		14	.571		53	24		77	95.50
2320	6″		12	.667		65.50	28		93.50	115
2400	Panel adapter, 2-1/2″		24	.333		14.50	14		28.50	37.50
2600	4″		20	.400		16.40	16.80		33.20	44
2800	6″		18	.444		21.50	18.65		40.15	52.50
3000	8″		16	.500		30.50	21		51.50	66
3020	10″		14	.571		36.50	24		60.50	77.50
3040	12″		12	.667		49	28		77	97
3200	Reducer, 4″ to 2-1/2″		24	.333		23	14		37	46.50
3400	6″ to 4″		20	.400		44	16.80		60.80	74
3600	8″ to 6″		18	.444		50.50	18.65		69.15	84.50
3620	10″ to 8″		16	.500		58.50	21		79.50	96.50
3640	12″ to 10″		14	.571		73	24		97	117
3780	End cap, 2-1/2″		24	.333		4.90	14		18.90	27
3800	4″		20	.400		6	16.80		22.80	32.50
4000	6″		18	.444		7.20	18.65		25.85	37
4200	8″		16	.500		9.80	21		30.80	43.50
4220	10″		14	.571		12.65	24		36.65	51
4240	12″		12	.667		22	28		50	67
4300	U-connector, 2-1/2″		200	.040		4.90	1.68		6.58	8
4320	4″		200	.040		6	1.68		7.68	9.20
4340	6″		180	.044		7.20	1.87		9.07	10.80
4360	8″		170	.047		12.20	1.98		14.18	16.45
4380	10″		150	.053		13.50	2.24		15.74	18.30
4400	12″		130	.062		29.50	2.58		32.08	36
4420	Hanger, 2-1/2″		100	.080		8.50	3.36		11.86	14.55
4430	4″		100	.080		10.80	3.36		14.16	17.10
4440	6″		80	.100		19.50	4.20		23.70	28
4450	8″		65	.123		25.50	5.15		30.65	36
4460	10″		50	.160		34	6.70		40.70	48
4470	12″		40	.200		68.50	8.40		76.90	88.50
4475	Screw cover, NEMA 3R w/ fittings and supports, 4″ x 4″		36	.222	L.F.	24.50	9.35		33.85	41.50
4480	6″ x 6″	2 Elec	55	.291		33	12.20		45.20	55
4485	8″ x 8″		36	.444		53	18.65		71.65	87
4490	12″ x 12″		18	.889		73.50	37.50		111	139
4500	Hinged cover, with fittings and supports, 2-1/2″ x 2-1/2″	1 Elec	60	.133		10.95	5.60		16.55	20.50
4520	4″ x 4″	″	45	.178		11.60	7.45		19.05	24.50
4540	6″ x 6″	2 Elec	80	.200		19.65	8.40		28.05	34.50
4560	8″ x 8″		60	.267		33	11.20		44.20	54
4580	10″ x 10″		50	.320		42	13.45		55.45	66.50
4600	12″ x 12″		24	.667		60	28		88	109
4700	Elbows 90°, hinged cover, 2-1/2″ x 2-1/2″	1 Elec	32	.250	Ea.	34	10.50		44.50	53

16 ELECTRICAL

16134 | Wireway & Aux Gutters

			CREW	DAILY OUTPUT	LABOR-HOURS	UNIT	2006 BARE COSTS MAT.	LABOR	EQUIP.	TOTAL	TOTAL INCL O&P	
150	4720	4"	1 Elec	27	.296	Ea.	38.50	12.45		50.95	61.50	150
	4730	6"		23	.348		44	14.60		58.60	70.50	
	4740	8"		18	.444		70.50	18.65		89.15	107	
	4750	10"		14	.571		78.50	24		102.50	124	
	4760	12"		12	.667		127	28		155	183	
	4800	Tee box, hinged cover, 2-1/2" x 2-1/2"		23	.348		39	14.60		53.60	65.50	
	4810	4"		20	.400		47	16.80		63.80	78	
	4820	6"		18	.444		53	18.65		71.65	87.50	
	4830	8"		16	.500		98.50	21		119.50	141	
	4840	10"		12	.667		113	28		141	167	
	4860	12"		10	.800		188	33.50		221.50	259	
	4880	Cross box, hinged cover, 2-1/2" x 2-1/2"		18	.444		44	18.65		62.65	77	
	4900	4"		16	.500		53	21		74	91	
	4920	6"		13	.615		65.50	26		91.50	112	
	4940	8"		11	.727		97.50	30.50		128	154	
	4960	10"		10	.800		167	33.50		200.50	236	
	4980	12"		9	.889		183	37.50		220.50	259	
	5000	Flanged, oil tite w/screw cover, 2-1/2" x 2-1/2"		40	.200	L.F.	25.50	8.40		33.90	41	
	5020	4" x 4"		35	.229		29.50	9.60		39.10	47.50	
	5040	6" x 6"	2 Elec	60	.267		42.50	11.20		53.70	64.50	
	5060	8" x 8"	"	50	.320		58.50	13.45		71.95	85	
	5120	Elbows 90°, flanged, 2-1/2" x 2-1/2"	1 Elec	23	.348	Ea.	54	14.60		68.60	82	
	5140	4"		20	.400		69	16.80		85.80	102	
	5160	6"		18	.444		85	18.65		103.65	123	
	5180	8"		15	.533		126	22.50		148.50	174	
	5240	Tee box, flanged, 2-1/2" x 2-1/2"		18	.444		74.50	18.65		93.15	111	
	5260	4"		16	.500		85.50	21		106.50	127	
	5280	6"		15	.533		115	22.50		137.50	162	
	5300	8"		13	.615		166	26		192	223	
	5360	Cross box, flanged, 2-1/2" x 2-1/2"		15	.533		92	22.50		114.50	136	
	5380	4"		13	.615		120	26		146	172	
	5400	6"		12	.667		156	28		184	214	
	5420	8"		10	.800		183	33.50		216.50	253	
	5480	Flange gasket, 2-1/2"		160	.050		2.71	2.10		4.81	6.20	
	5500	4"		80	.100		3.69	4.20		7.89	10.50	
	5520	6"		53	.151		4.91	6.35		11.26	15.15	
	5530	8"		40	.200		6.55	8.40		14.95	20	

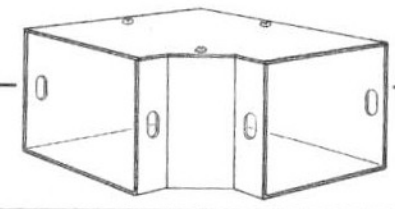
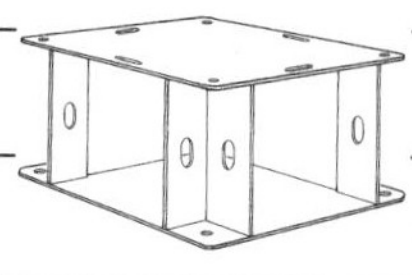
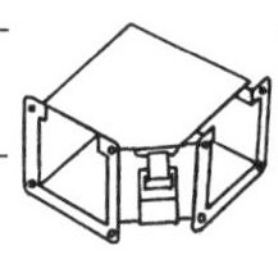
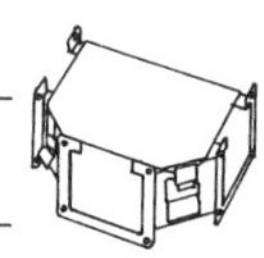
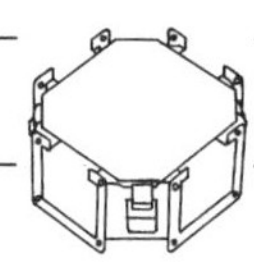

16136 | Boxes

			CREW	DAILY OUTPUT	LABOR-HOURS	UNIT	MAT.	LABOR	EQUIP.	TOTAL	TOTAL INCL O&P	
600	0010	OUTLET BOXES										600
	0020	Pressed steel, octagon, 4"	1 Elec	20	.400	Ea.	2.21	16.80		19.01	28.50	
	0040	For Romex or BX		20	.400		3.33	16.80		20.13	29.50	
	0050	For Romex or BX, with bracket		20	.400		5.20	16.80		22	32	
	0060	Covers, blank		64	.125		.92	5.25		6.17	9.10	
	0100	Extension rings		40	.200		3.64	8.40		12.04	16.95	
	0152	Square 4"		18	.444		2.24	18.65		20.89	31.50	
	0160	For Romex or BX		20	.400		5.35	16.80		22.15	32	
	0170	For Romex or BX, with bracket		20	.400		5.75	16.80		22.55	32.50	
	0200	Extension rings		40	.200		3.71	8.40		12.11	17.05	
	0220	2-1/8" deep, 1" KO		20	.400		4.94	16.80		21.74	31.50	
	0250	Covers, blank		64	.125		1.04	5.25		6.29	9.25	
	0260	Raised device		64	.125		1.96	5.25		7.21	10.25	
	0300	Plaster rings		64	.125		2.03	5.25		7.28	10.35	
	0350	Square, 4-11/16"		20	.400		5.35	16.80		22.15	32	
	0370	2-1/8" deep, 3/4" to 1-1/4" KO		20	.400		6.25	16.80		23.05	33	

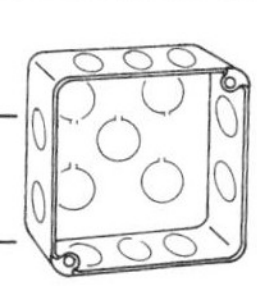

16136 | Boxes		CREW	DAILY OUTPUT	LABOR-HOURS	UNIT	2006 BARE COSTS MAT.	LABOR	EQUIP.	TOTAL	TOTAL INCL O&P
600	0400 Extension rings	1 Elec	40	.200	Ea.	8.35	8.40		16.75	22
	0450 Covers, blank		53	.151		1.86	6.35		8.21	11.80
	0460 Raised device		53	.151		5.60	6.35		11.95	15.90
	0500 Plaster rings		53	.151		5.55	6.35		11.90	15.85
	0550 Handy box		27	.296		2.80	12.45		15.25	22
	0560 Covers, device		64	.125		.86	5.25		6.11	9.05
	0600 Extension rings		54	.148		2.92	6.20		9.12	12.80
	0652 Switchbox		24	.333		3.54	14		17.54	25.50
	0660 Romex or BX		27	.296		5.10	12.45		17.55	25
	0670 with bracket		27	.296		5.85	12.45		18.30	25.50
	0680 Partition, metal		27	.296		2.22	12.45		14.67	21.50
	0700 Masonry, 1 gang, 2-1/2" deep		27	.296		6.75	12.45		19.20	26.50
	0710 3-1/2" deep		27	.296		6.75	12.45		19.20	26.50
	0750 2 gang, 2-1/2" deep		20	.400		12.65	16.80		29.45	40
	0760 3-1/2" deep		20	.400		10.20	16.80		27	37.50
	0800 3 gang, 2-1/2" deep		13	.615		15.20	26		41.20	56.50
	0850 4 gang, 2-1/2" deep		10	.800		17.55	33.50		51.05	71.50
	0860 5 gang, 2-1/2" deep		9	.889		18.80	37.50		56.30	78
	0870 6 gang, 2-1/2" deep		8	1		39	42		81	108
	0880 Masonry thru-the-wall, 1 gang, 4" block		16	.500		15.55	21		36.55	49.50
	0890 6" block		16	.500		17.80	21		38.80	52
	0900 8" block		16	.500		19.95	21		40.95	54.50
	0920 2 gang, 6" block		16	.500		18.05	21		39.05	52.50
	0940 Bar hanger with 3/8" stud, for wood and masonry boxes		53	.151		3.43	6.35		9.78	13.50
	0950 Concrete, set flush, 4" deep		20	.400		10.90	16.80		27.70	38
	1000 Plate with 3/8" stud		80	.100		4.45	4.20		8.65	11.35
	1102 Concrete, floor, 1 gang		4.80	1.667		67.50	70		137.50	182
	1150 2 gang		4	2		106	84		190	246
	1200 3 gang		2.70	2.963		151	124		275	360
	1250 For duplex receptacle, pedestal mounted, add		24	.333		65.50	14		79.50	93.50
	1270 Flush mounted, add		27	.296		17.80	12.45		30.25	39
	1300 For telephone, pedestal mounted, add		30	.267		70	11.20		81.20	94.50
	1350 Carpet flange, 1 gang		53	.151		34.50	6.35		40.85	48
	1400 Cast, 1 gang, FS (2" deep), 1/2" hub		12	.667		14.10	28		42.10	58.50
	1410 3/4" hub		12	.667		14.90	28		42.90	59.50
	1420 FD (2-11/16" deep), 1/2" hub		12	.667		16.35	28		44.35	61
	1430 3/4" hub		12	.667		17.80	28		45.80	62.50
	1450 2 gang, FS, 1/2" hub		10	.800		26.50	33.50		60	81
	1460 3/4" hub		10	.800		27	33.50		60.50	82
	1470 FD, 1/2" hub		10	.800		30.50	33.50		64	86
	1480 3/4" hub		10	.800		31.50	33.50		65	86.50
	1500 3 gang, FS, 3/4" hub		9	.889		39.50	37.50		77	101
	1510 Switch cover, 1 gang, FS		64	.125		3.75	5.25		9	12.25
	1520 2 gang		53	.151		6.15	6.35		12.50	16.50
	1530 Duplex receptacle cover, 1 gang, FS		64	.125		3.75	5.25		9	12.25
	1540 2 gang, FS		53	.151		6.15	6.35		12.50	16.50
	1550 Weatherproof switch cover		64	.125		5.35	5.25		10.60	14
	1600 Weatherproof receptacle cover		64	.125		5.35	5.25		10.60	14
	1750 FSC, 1 gang, 1/2" hub		11	.727		15.60	30.50		46.10	64
	1760 3/4" hub		11	.727		17.05	30.50		47.55	66
	1770 2 gang, 1/2" hub		9	.889		28	37.50		65.50	88
	1780 3/4" hub		9	.889		29.50	37.50		67	90
	1790 FDC, 1 gang, 1/2" hub		11	.727		18.60	30.50		49.10	67.50
	1800 3/4" hub		11	.727		20	30.50		50.50	69
	1810 2 gang, 1/2" hub		9	.889		34	37.50		71.50	95
	1820 3/4" hub		9	.889		33.50	37.50		71	94

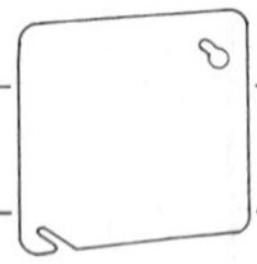

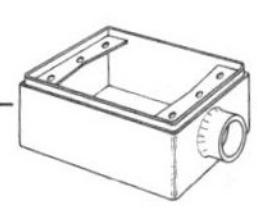

	16136	Boxes	CREW	DAILY OUTPUT	LABOR-HOURS	UNIT	2006 BARE COSTS MAT.	LABOR	EQUIP.	TOTAL	TOTAL INCL O&P	
600	2000	Poke-thru fitting, fire rated, for 3-3/4" floor	1 Elec	6.80	1.176	Ea.	100	49.50		149.50	186	600
	2040	For 7" floor		6.80	1.176		100	49.50		149.50	186	
	2100	Pedestal, 15 amp, duplex receptacle & blank plate		5.25	1.524		103	64		167	213	
	2120	Duplex receptacle and telephone plate		5.25	1.524		104	64		168	213	
	2140	Pedestal, 20 amp, duplex recept. & phone plate		5	1.600		105	67		172	219	
	2160	Telephone plate, both sides		5.25	1.524		99	64		163	208	
	2200	Abandonment plate		32	.250		30.50	10.50		41	49.50	
	9000	Minimum labor/equipment charge		4	2	Job		84		84	129	
620	0010	**OUTLET BOXES, PLASTIC**										620
	0050	4" diameter, round with 2 mounting nails	1 Elec	25	.320	Ea.	2.48	13.45		15.93	23	
	0102	Bar hanger mounted		23	.348		4.52	14.60		19.12	27.50	
	0202	Square with 2 mounting nails		23	.348		3.72	14.60		18.32	26.50	
	0300	Plaster ring		64	.125		1.55	5.25		6.80	9.80	
	0402	Switch box with 2 mounting nails, 1 gang 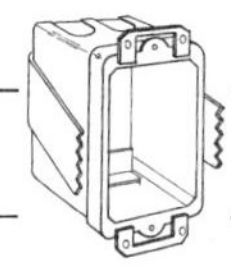		27	.296		1.69	12.45		14.14	21	
	0502	2 gang		23	.348		3.45	14.60		18.05	26.50	
	0602	3 gang		18	.444		5.45	18.65		24.10	35	
	0702	Old work box		27	.296		3.32	12.45		15.77	23	
	1400	PVC, FSS, 1 gang, 1/2" hub		14	.571		13.90	24		37.90	52.50	
	1410	3/4" hub		14	.571		14.90	24		38.90	53.50	
	1420	FD, 1 gang for variable terminations		14	.571		13	24		37	51.50	
	1450	FS, 2 gang for variable terminations		12	.667		17.45	28		45.45	62	
	1480	Blank cover, FS, 1 gang		64	.125		3.70	5.25		8.95	12.15	
	1500	2 gang		53	.151		4.20	6.35		10.55	14.35	
	1510	Switch cover, FS, 1 gang		64	.125		9.15	5.25		14.40	18.15	
	1520	2 gang		53	.151		14.50	6.35		20.85	25.50	
	1530	Duplex receptacle cover, FS, 1 gang		64	.125		13	5.25		18.25	22.50	
	1540	2 gang		53	.151		14.20	6.35		20.55	25.50	
	1750	FSC, 1 gang, 1/2" hub		13	.615		13.65	26		39.65	55	
	1760	3/4" hub		13	.615		14.90	26		40.90	56.50	
	1770	FSC, 2 gang, 1/2" hub		11	.727		17.80	30.50		48.30	66.50	
	1780	3/4" hub		11	.727		17.80	30.50		48.30	66.50	
	1790	FDC, 1 gang, 1/2" hub		13	.615		13.05	26		39.05	54.50	
	1800	3/4" hub		13	.615		14.20	26		40.20	55.50	
	1810	Weatherproof, T box w/ 3 holes		14	.571		13.40	24		37.40	52	
	1820	4" diameter round w/ 5 holes		14	.571		19.55	24		43.55	58.50	
	9000	Minimum labor/equipment charge		4	2	Job		84		84	129	
700	0010	**PULL BOXES & CABINETS**										700
	0100	Sheet metal, pull box, NEMA 1, type SC, 6" W x 6" H x 4" D	1 Elec	8	1	Ea.	10.80	42		52.80	76.50	
	0180	6" W x 8" H x 4" D		8	1		12.70	42		54.70	78.50	
	0200	8" W x 8" H x 4" D		8	1		14.80	42		56.80	81	
	0210	10" W x 10" H x 4" D		7	1.143		19.55	48		67.55	95.50	
	0220	12" W x 12" H x 4" D		6.50	1.231		25	51.50		76.50	107	
	0230	15" W x 15" H x 4" D		5.20	1.538		34	64.50		98.50	137	
	0240	18" W x 18" H x 4" D 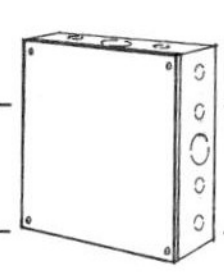		4.40	1.818		45	76.50		121.50	168	
	0250	6" W x 6" H x 6" D		8	1		13.10	42		55.10	79	
	0260	8" W x 8" H x 6" D		7.50	1.067		17.80	45		62.80	88.50	
	0270	10" W x 10" H x 6" D		5.50	1.455		23	61		84	120	
	0300	10" W x 12" H x 6" D		5.30	1.509		26	63.50		89.50	127	
	0310	12" W x 12" H x 6" D		5.20	1.538		29.50	64.50		94	132	
	0320	15" W x 15" H x 6" D		4.60	1.739		42	73		115	159	
	0330	18" W x 18" H x 6" D		4.20	1.905		52.50	80		132.50	181	
	0340	24" W x 24" H x 6" D		3.20	2.500		106	105		211	279	
	0350	12" W x 12" H x 8" D		5	1.600		33	67		100	140	
	0360	15" W x 15" H x 8" D		4.50	1.778		67.50	74.50		142	190	
	0370	18" W x 18" H x 8" D		4	2		104	84		188	243	
	0380	24" W x 18" H x 6" D		3.70	2.162		82.50	91		173.50	231	

16136 | Boxes

700

	16136 Boxes	Crew	Daily Output	Labor-Hours	Unit	Mat.	Labor	Equip.	Total	Total Incl O&P
0400	16" W x 20" H x 8" D	1 Elec	4	2	Ea.	98.50	84		182.50	237
0500	20" W x 24" H x 8" D		3.20	2.500		115	105		220	289
0510	24" W x 24" H x 8" D		3	2.667		118	112		230	300
0600	24" W x 36" H x 8" D		2.70	2.963		163	124		287	370
0610	30" W x 30" H x 8" D		2.70	2.963		218	124		342	430
0620	36" W x 36" H x 8" D		2	4		265	168		433	550
0630	24" W x 24" H x 10" D		2.50	3.200		134	134		268	355
0650	Hinged cabinets, NEMA 1, 6" W x 6" H x 4" D		8	1		11	42		53	76.50
0660	8" W x 8" H x 4" D		8	1		15.10	42		57.10	81
0670	10" W x 10" H x 4" D		7	1.143		19.90	48		67.90	96
0680	12" W x 12" H x 4" D		6	1.333		25.50	56		81.50	115
0690	15" W x 15" H x 4" D		5.20	1.538		39.50	64.50		104	143
0700	18" W x 18" H x 4" D		4.40	1.818		45.50	76.50		122	168
0710	6" W x 6" H x 6" D		8	1		13.40	42		55.40	79.50
0720	8" W x 8" H x 6" D		7.50	1.067		18.10	45		63.10	89
0730	10" W x 10" H x 6" D		5.50	1.455		23.50	61		84.50	120
0740	12" W x 12" H x 6" D		5.20	1.538		30	64.50		94.50	133
0802	12" W x 16" H x 6" D		4	2		36	84		120	169
0810	15" W x 15" H x 6" D		4.60	1.739		42	73		115	160
0820	18" W x 18" H x 6" D		4.20	1.905		53.50	80		133.50	182
1000	20" W x 20" H x 6" D		3.60	2.222		72.50	93.50		166	224
1010	24" W x 24" H x 6" D		3.20	2.500		108	105		213	281
1020	12" W x 12" H x 8" D		5	1.600		62	67		129	173
1030	15" W x 15" H x 8" D		4.50	1.778		88	74.50		162.50	212
1040	18" W x 18" H x 8" D		4	2		124	84		208	266
1200	20" W x 20" H x 8" D		3.20	2.500		145	105		250	320
1210	24" W x 24" H x 8" D		3	2.667		167	112		279	355
1220	30" W x 30" H x 8" D		2.70	2.963		236	124		360	450
1400	24" W x 36" H x 8" D		2.70	2.963		233	124		357	450
1600	24" W x 42" H x 8" D		2	4		355	168		523	650
1610	36" W x 36" H x 8" D	↓	2	4	↓	325	168		493	620
2100	Pull box, NEMA 3R, type SC, raintight & weatherproof									
2150	6" L x 6" W x 6" D	1 Elec	10	.800	Ea.	18.50	33.50		52	72.50
2200	8" L x 6" W x 6" D		8	1		23	42		65	90
2250	10" L x 6" W x 6" D		7	1.143		31	48		79	108
2300	12" L x 12" W x 6" D		5	1.600		43.50	67		110.50	152
2350	16" L x 16" W x 6" D		4.50	1.778		86.50	74.50		161	211
2400	20" L x 20" W x 6" D		4	2		119	84		203	260
2450	24" L x 18" W x 8" D		3	2.667		130	112		242	315
2500	24" L x 24" W x 10" D		2.50	3.200		174	134		308	400
2550	30" L x 24" W x 12" D		2	4		315	168		483	610
2600	36" L x 36" W x 12" D	↓	1.50	5.333	↓	415	224		639	805
2800	Cast iron, pull boxes for surface mounting									
3000	NEMA 4, watertight & dust tight									
3050	6" L x 6" W x 6" D	1 Elec	4	2	Ea.	164	84		248	310
3100	8" L x 6" W x 6" D		3.20	2.500		222	105		327	405
3150	10" L x 6" W x 6" D		2.50	3.200		277	134		411	510
3200	12" L x 12" W x 6" D		2.30	3.478		465	146		611	740
3250	16" L x 16" W x 6" D		1.30	6.154		955	258		1,213	1,450
3300	20" L x 20" W x 6" D		.80	10		1,775	420		2,195	2,600
3350	24" L x 18" W x 8" D		.70	11.429		1,875	480		2,355	2,800
3400	24" L x 24" W x 10" D		.50	16		2,875	670		3,545	4,175
3450	30" L x 24" W x 12" D		.40	20		4,800	840		5,640	6,600
3500	36" L x 36" W x 12" D		.20	40		5,950	1,675		7,625	9,150
3510	NEMA 4 clamp cover, 6" L x 6" W x 4" D		4	2		195	84		279	345
3520	8" L x 6" W x 4" D	↓	4	2	↓	230	84		314	380

700

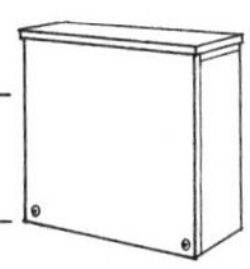

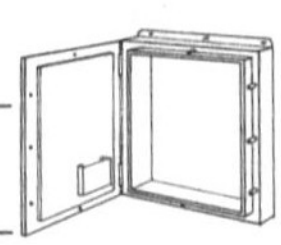

16 ELECTRICAL

16136 | Boxes

			CREW	DAILY OUTPUT	LABOR-HOURS	UNIT	2006 BARE COSTS MAT.	LABOR	EQUIP.	TOTAL	TOTAL INCL O&P
700	4000	NEMA 7, explosionproof									
	4050	6″ L x 6″ W x 6″ D	1 Elec	2	4	Ea.	500	168		668	810
	4100	8″ L x 6″ W x 6″ D		1.80	4.444		605	187		792	960
	4150	10″ L x 6″ W x 6″ D		1.60	5		745	210		955	1,150
	4200	12″ L x 12″ W x 6″ D		1	8		1,225	335		1,560	1,875
	4250	16″ L x 14″ W x 6″ D		.60	13.333		1,700	560		2,260	2,725
	4300	18″ L x 18″ W x 8″ D		.50	16		3,000	670		3,670	4,300
	4350	24″ L x 18″ W x 8″ D		.40	20		4,000	840		4,840	5,700
	4400	24″ L x 24″ W x 10″ D		.30	26.667		5,175	1,125		6,300	7,425
	4450	30″ L x 24″ W x 12″ D		.20	40		7,925	1,675		9,600	11,300
	5000	NEMA 9, dust tight 6″ L x 6″ W x 6″ D		3.20	2.500		151	105		256	330
	5050	8″ L x 6″ W x 6″ D		2.70	2.963		181	124		305	390
	5100	10″ L x 6″ W x 6″ D		2	4		243	168		411	525
	5150	12″ L x 12″ W x 6″ D		1.60	5		480	210		690	855
	5200	16″ L x 16″ W x 6″ D		1	8		830	335		1,165	1,425
	5250	18″ L x 18″ W x 8″ D		.70	11.429		1,300	480		1,780	2,200
	5300	24″ L x 18″ W x 8″ D		.60	13.333		1,825	560		2,385	2,875
	5350	24″ L x 24″ W x 10″ D		.40	20		2,350	840		3,190	3,900
	5400	30″ L x 24″ W x 12″ D	↓	.30	26.667	↓	3,725	1,125		4,850	5,825
	6000	J.I.C. wiring boxes, NEMA 12, dust tight & drip tight									
	6050	6″ L x 8″ W x 4″ D	1 Elec	10	.800	Ea.	43	33.50		76.50	99.50
	6100	8″ L x 10″ W x 4″ D		8	1		54	42		96	124
	6150	12″ L x 14″ W x 6″ D		5.30	1.509		88	63.50		151.50	194
	6200	14″ L x 16″ W x 6″ D		4.70	1.702		105	71.50		176.50	225
	6250	16″ L x 20″ W x 6″ D		4.40	1.818		223	76.50		299.50	365
	6300	24″ L x 30″ W x 6″ D		3.20	2.500		330	105		435	520
	6350	24″ L x 30″ W x 8″ D		2.90	2.759		350	116		466	565
	6400	24″ L x 36″ W x 8″ D		2.70	2.963		390	124		514	620
	6450	24″ L x 42″ W x 8″ D		2.30	3.478		425	146		571	695
	6500	24″ L x 48″ W x 8″ D	↓	2	4	↓	465	168		633	770
	7000	Cabinets, current transformer									
	7050	Single door, 24″ H x 24″ W x 10″ D	1 Elec	1.60	5	Ea.	139	210		349	480
	7100	30″ H x 24″ W x 10″ D		1.30	6.154		153	258		411	570
	7150	36″ H x 24″ W x 10″ D		1.10	7.273		161	305		466	645
	7200	30″ H x 30″ W x 10″ D		1	8		167	335		502	705
	7250	36″ H x 30″ W x 10″ D		.90	8.889		230	375		605	830
	7300	36″ H x 36″ W x 10″ D		.80	10		244	420		664	915
	7500	Double door, 48″ H x 36″ W x 10″ D		.60	13.333		440	560		1,000	1,350
	7550	24″ H x 24″ W x 12″ D	↓	1	8	↓	230	335		565	775
	7600	Telephone with wood backboard									
	7620	Single door, 12″ H x 12″ W x 4″ D	1 Elec	5.30	1.509	Ea.	101	63.50		164.50	210
	7650	18″ H x 12″ W x 4″ D		4.70	1.702		140	71.50		211.50	263
	7700	24″ H x 12″ W x 4″ D		4.20	1.905		158	80		238	297
	7720	18″ H x 18″ W x 4″ D		4.20	1.905		171	80		251	310
	7750	24″ H x 18″ W x 4″ D		4	2		233	84		317	385
	7780	36″ H x 36″ W x 4″ D		3.60	2.222		320	93.50		413.50	500
	7800	24″ H x 24″ W x 6″ D		3.60	2.222		305	93.50		398.50	485
	7820	30″ H x 24″ W x 6″ D		3.20	2.500		395	105		500	595
	7850	30″ H x 30″ W x 6″ D		2.70	2.963		460	124		584	695
	7880	36″ H x 30″ W x 6″ D		2.50	3.200		490	134		624	745
	7900	48″ H x 36″ W x 6″ D		2.20	3.636		865	153		1,018	1,175
	7920	Double door, 48″ H x 36″ W x 6″ D	↓	2	4	↓	895	168		1,063	1,250
	8000	NEMA 12, double door, floor mounted									
	8020	54″ H x 42″ W x 8″ D	2 Elec	6	2.667	Ea.	910	112		1,022	1,175
	8040	60″ H x 48″ W x 8″ D		5.40	2.963		1,225	124		1,349	1,550
	8060	60″ H x 48″ W x 10″ D	↓	5.40	2.963	↓	1,250	124		1,374	1,575

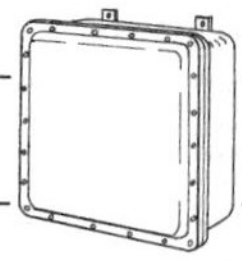

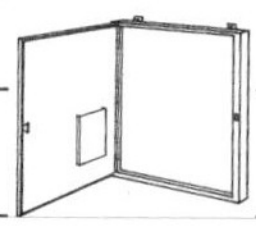

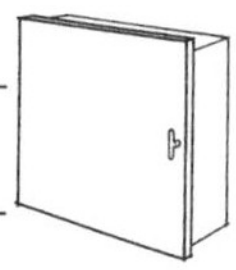

16136 | Boxes

			CREW	DAILY OUTPUT	LABOR-HOURS	UNIT	2006 BARE COSTS MAT.	LABOR	EQUIP.	TOTAL	TOTAL INCL O&P
700	8080	60″ H x 60″ W x 10″ D	2 Elec	5	3.200	Ea.	1,425	134		1,559	1,775
	8100	72″ H x 60″ W x 10″ D		4	4		1,625	168		1,793	2,050
	8120	72″ H x 72″ W x 10″ D		3.40	4.706		1,825	198		2,023	2,300
	8140	60″ H x 48″ W x 12″ D		3.40	4.706		1,275	198		1,473	1,700
	8160	60″ H x 60″ W x 12″ D		3.20	5		1,425	210		1,635	1,900
	8180	72″ H x 60″ W x 12″ D		3	5.333		1,450	224		1,674	1,950
	8200	72″ H x 72″ W x 12″ D		3	5.333		1,650	224		1,874	2,150
	8220	60″ H x 48″ W x 16″ D		3.20	5		1,350	210		1,560	1,825
	8240	72″ H x 72″ W x 16″ D		2.60	6.154		1,925	258		2,183	2,525
	8260	60″ H x 48″ W x 20″ D		3	5.333		1,475	224		1,699	1,975
	8280	72″ H x 72″ W x 20″ D		2.20	7.273		2,100	305		2,405	2,775
	8300	60″ H x 48″ W x 24″ D		2.60	6.154		1,550	258		1,808	2,100
	8320	72″ H x 72″ W x 24″ D	↓	2	8	↓	2,200	335		2,535	2,950
	8340	Pushbutton enclosure, oiltight									
	8360	3-1/2″ H x 3-1/4″ W x 2-3/4″ D, for 1 P.B.	1 Elec	12	.667	Ea.	41.50	28		69.50	88.50
	8380	5-3/4″ H x 3-1/4″ W x 2-3/4″ D, for 2 P.B.		11	.727		41.50	30.50		72	92.50
	8400	8″ H x 3-1/4″ W x 2-3/4″ D, for 3 P.B.		10.50	.762		56.50	32		88.50	112
	8420	10-1/4″ H x 3-1/4″ W x 2-3/4″ D, for 4 P.B.		10.50	.762		58.50	32		90.50	114
	8440	7-1/4″ H x 6-1/4″ W x 3″ D, for 4 P.B.		10	.800		61.50	33.50		95	120
	8460	12-1/2″ H x 3-1/4″ W x 3″ D, for 5 P.B.		9	.889		65	37.50		102.50	129
	8480	9-1/2″ H x 6-1/4″ W x 3″ D, for 6 P.B.		8.50	.941		74.50	39.50		114	143
	8500	9-1/2″ H x 8-1/2″ W x 3″ D, for 9 P.B.		8	1		80	42		122	153
	8510	11-3/4″ H x 8-1/2″ W x 3″ D, for 12 P.B.		7	1.143		88.50	48		136.50	172
	8520	11-3/4″ H x 10-3/4″ W x 3″ D, for 16 P.B.		6.50	1.231		115	51.50		166.50	207
	8540	14″ H x 10-3/4″ W x 3″ D, for 20 P.B.		5	1.600		130	67		197	247
	8560	14″ H x 13″ W x 3″ D, for 25 P.B.	↓	4.50	1.778	↓	148	74.50		222.50	277
	8580	Sloping front pushbutton enclosures									
	8600	3-1/2″ H x 7-3/4″ W x 4-7/8″ D, for 3 P.B	1 Elec	10	.800	Ea.	68	33.50		101.50	127
	8620	7-1/4″ H x 8-1/2″ W x 6-3/4″ D, for 6 P.B.		8	1		100	42		142	175
	8640	9-1/2″ H x 8-1/2″ W x 7-7/8″ D, for 9 P.B.		7	1.143		111	48		159	196
	8660	11-1/4″ H x 8-1/2″ W x 9″ D, for 12 P.B.		5	1.600		130	67		197	247
	8680	11-3/4″ H x 10″ W x 9″ D, for 16 P.B.		5	1.600		145	67		212	264
	8700	11-3/4″ H x 13″ W x 9″ D, for 20 P.B.		5	1.600		170	67		237	291
	8720	14″ H x 13″ W x 10-1/8″ D, for 25 P.B.	↓	4.50	1.778	↓	191	74.50		265.50	325
	8740	Pedestals, not including P.B. enclosure or base									
	8760	Straight column 4″ x 4″	1 Elec	4.50	1.778	Ea.	173	74.50		247.50	305
	8780	6″ x 6″		4	2		260	84		344	415
	8800	Angled column 4″ x 4″		4.50	1.778		203	74.50		277.50	340
	8820	6″ x 6″		4	2		305	84		389	470
	8840	Pedestal, base 18″ x 18″		10	.800		105	33.50		138.50	168
	8860	24″ x 24″	↓	9	.889	↓	238	37.50		275.50	320
	8900	Electronic rack enclosures									
	8920	72″ H x 25″ W x 24″ D	1 Elec	1.50	5.333	Ea.	1,775	224		1,999	2,300
	8940	72″ H x 30″ W x 24″ D		1.50	5.333		1,900	224		2,124	2,425
	8960	72″ H x 25″ W x 30″ D		1.30	6.154		1,925	258		2,183	2,500
	8980	72″ H x 25″ W x 36″ D		1.20	6.667		2,050	280		2,330	2,700
	9000	72″ H x 30″ W x 36″ D	↓	1.20	6.667	↓	2,175	280		2,455	2,825
	9020	NEMA 12 & 4 enclosure panels									
	9040	12″ x 24″	1 Elec	20	.400	Ea.	24.50	16.80		41.30	53
	9060	16″ x 12″		20	.400		17.40	16.80		34.20	45
	9080	20″ x 16″		20	.400		25.50	16.80		42.30	54
	9100	20″ x 20″		19	.421		31	17.70		48.70	61
	9120	24″ x 20″		18	.444		40	18.65		58.65	73
	9140	24″ x 24″		17	.471		46	19.75		65.75	81
	9160	30″ x 20″		16	.500		48	21		69	85.50
	9180	30″ x 24″	↓	16	.500	↓	55.50	21		76.50	94

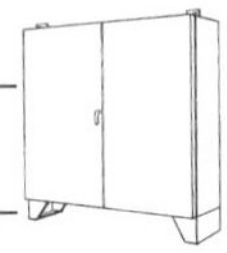

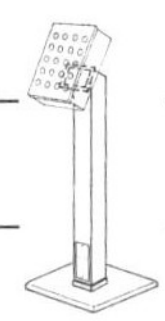
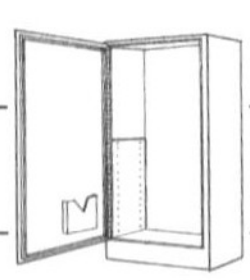
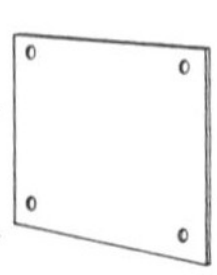

16 ELECTRICAL

	16136 | Boxes		CREW	DAILY OUTPUT	LABOR-HOURS	UNIT	2006 BARE COSTS MAT.	LABOR	EQUIP.	TOTAL	TOTAL INCL O&P	
700	9200	36″ x 24″	1 Elec	15	.533	Ea.	66.50	22.50		89	108	700
	9220	36″ x 30″		15	.533		89	22.50		111.50	133	
	9240	42″ x 24″		15	.533		76.50	22.50		99	119	
	9260	42″ x 30″		14	.571		102	24		126	149	
	9280	42″ x 36″		14	.571		120	24		144	169	
	9300	48″ x 24″		14	.571		86.50	24		110.50	133	
	9320	48″ x 30″		14	.571		114	24		138	163	
	9340	48″ x 36″		13	.615		135	26		161	189	
	9360	60″ x 36″	↓	12	.667	↓	166	28		194	226	
	9400	Wiring trough steel JIC, clamp cover										
	9490	4″ x 4″, 12″ long	1 Elec	12	.667	Ea.	60	28		88	109	
	9510	24″ long		10	.800		77	33.50		110.50	137	
	9530	36″ long		8	1		94	42		136	168	
	9540	48″ long		7	1.143		110	48		158	196	
	9550	60″ long		6	1.333		127	56		183	226	
	9560	6″ x 6″, 12″ long		11	.727		79	30.50		109.50	134	
	9580	24″ long		9	.889		103	37.50		140.50	172	
	9600	36″ long		7	1.143		130	48		178	217	
	9610	48″ long		6	1.333		156	56		212	258	
	9620	60″ long		5	1.600	↓	183	67		250	305	
	9990	Minimum labor/equipment charge	↓	2	4	Job		168		168	259	
720	0010	**PULL BOXES & CABINETS** Nonmetallic										720
	0080	Enclosures fiberglass NEMA 4X										
	0100	Wall mount, quick release latch door, 20″H x 16″W x 6″D	1 Elec	4.80	1.667	Ea.	430	70		500	585	
	0110	20″H x 20″W x 6″D		4.50	1.778		510	74.50		584.50	675	
	0120	24″H x 20″W x 6″D		4.20	1.905		545	80		625	725	
	0130	20″H x 16″W x 8″D		4.50	1.778		485	74.50		559.50	650	
	0140	20″H x 20″W x 8″D		4.20	1.905		540	80		620	720	
	0150	24″H x 24″W x 8″D		3.80	2.105		605	88.50		693.50	805	
	0160	30″H x 24″W x 8″D		3.20	2.500		675	105		780	905	
	0170	36″H x 30″W x 8″D		3	2.667		940	112		1,052	1,200	
	0180	20″H x 16″W x 10″D		3.50	2.286		555	96		651	760	
	0190	20″H x 20″W x 10″D		3.20	2.500		600	105		705	820	
	0200	24″H x 20″W x 10″D		3	2.667		645	112		757	880	
	0210	30″H x 24″W x 10″D		2.80	2.857		730	120		850	985	
	0220	20″H x 16″W x 12″D		3	2.667		615	112		727	850	
	0230	20″H x 20″W x 12″D		2.80	2.857		645	120		765	895	
	0240	24″H x 24″W x 12″D		2.60	3.077		705	129		834	975	
	0250	30″H x 24″W x 12″D		2.40	3.333		780	140		920	1,075	
	0260	36″H x 30″W x 12″D		2.20	3.636		1,075	153		1,228	1,400	
	0270	36″H x 36″W x 12″D		2.10	3.810		1,175	160		1,335	1,550	
	0280	48″H x 36″W x 12″D		2	4		1,400	168		1,568	1,775	
	0290	60″H x 36″W x 12″D		1.80	4.444		1,550	187		1,737	2,000	
	0300	30″H x 24″W x 16″D		1.40	5.714		925	240		1,165	1,400	
	0310	48″H x 36″W x 16″D		1.20	6.667		1,550	280		1,830	2,125	
	0320	60″H x 36″W x 16″D		1	8		1,750	335		2,085	2,450	
	0480	Freestanding, one door, 72″H x 25″W x 25″D		.80	10		3,000	420		3,420	3,950	
	0490	Two doors with two panels, 72″H x 49″W x 24″D		.50	16		7,100	670		7,770	8,850	
	0500	Floor stand kits, for NEMA 4 & 12, 20″W or more		24	.333		88	14		102	118	
	0510	6″H x 10″D		24	.333		96.50	14		110.50	128	
	0520	6″H x 12″D		24	.333		110	14		124	143	
	0530	6″H x 18″D		24	.333		130	14		144	165	
	0540	12″H x 8″D		22	.364		113	15.25		128.25	149	
	0550	12″H x 10″D		22	.364		122	15.25		137.25	158	
	0560	12″H x 12″D	↓	22	.364	↓	131	15.25		146.25	168	

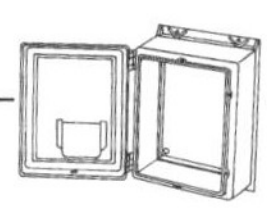

16136 | Boxes		CREW	DAILY OUTPUT	LABOR-HOURS	UNIT	2006 BARE COSTS MAT.	LABOR	EQUIP.	TOTAL	TOTAL INCL O&P
720 0570	12"H x 16"D	1 Elec	22	.364	Ea.	148	15.25		163.25	186 720
0580	12"H x 18"D		22	.364		155	15.25		170.25	195
0590	12"H x 20"D		22	.364		164	15.25		179.25	205
0600	18"H x 8"D		20	.400		139	16.80		155.80	179
0610	18"H x 10"D		20	.400		148	16.80		164.80	188
0620	18"H x 12"D		20	.400		155	16.80		171.80	197
0630	18"H x 16"D		20	.400		174	16.80		190.80	218
0640	24"H x 8"D		16	.500		164	21		185	214
0650	24"H x 10"D		16	.500		174	21		195	225
0660	24"H x 12"D		16	.500		182	21		203	233
0670	24"H x 16"D		16	.500		200	21		221	253
0680	Small, screw cover, 5-1/2"H x 4"W x 4-15/16"D		12	.667		53	28		81	102
0690	7-1/2"H x 4"W x 4-15/16"D		12	.667		55.50	28		83.50	104
0700	7-1/2"H x 6"W x 5-3/16"D		10	.800		64	33.50		97.50	123
0710	9-1/2"H x 6"W x 5-11/16"D		10	.800		66.50	33.50		100	126
0720	11-1/2"H x 8"W x 6-11/16"D		8	1		97.50	42		139.50	172
0730	13-1/2"H x 10"W x 7-3/16"D		7	1.143		120	48		168	206
0740	15-1/2"H x 12"W x 8-3/16"D		6	1.333		159	56		215	262
0750	17-1/2"H x 14"W x 8-11/16"D		5	1.600		182	67		249	305
0760	Screw cover with window, 6"H x 4"W x 5"D		12	.667		86	28		114	138
0770	8"H x 4"W x 5"D		11	.727		92	30.50		122.50	148
0780	8"H x 6"W x 5"D		11	.727		130	30.50		160.50	190
0790	10"H x 6"W x 6"D		10	.800		137	33.50		170.50	202
0800	12"H x 8"W x 7"D		8	1		177	42		219	259
0810	14"H x 10"W x 7"D		7	1.143		210	48		258	305
0820	16"H x 12"W x 8"D		6	1.333		253	56		309	365
0830	18"H x 14"W x 9"D		5	1.600		300	67		367	435
0840	Quick-release latch cover, 5-1/2"H x 4"W x 5"D		12	.667		68.50	28		96.50	119
0850	7-1/2"H x 4"W x 5"D		12	.667		71.50	28		99.50	122
0860	7-1/2"H x 6"W x 5-1/4"D		10	.800		82.50	33.50		116	143
0870	9-1/2"H x 6"W x 5-3/4"D		10	.800		85.50	33.50		119	146
0880	11-1/2"H x 8"W x 6-3/4"D		8	1		125	42		167	203
0890	13-1/2"H x 10"W x 7-1/4"D		7	1.143		154	48		202	244
0900	15-1/2"H x 12"W x 8-1/4"D		6	1.333		199	56		255	305
0910	17-1/2"H x 14"W x 8-3/4"D		5	1.600		235	67		302	365
0920	Pushbutton, 1 hole 5-1/2"H x 4"W x 4-15/16"D		12	.667		50.50	28		78.50	98.50
0930	2 hole 7-1/2"H x 4"W x 4-15/16"D		11	.727		57	30.50		87.50	110
0940	4 hole 7-1/2"H x 6"W x 5-3/16"D		10.50	.762		68	32		100	125
0950	6 hole 9-1/2"H x 6"W x 5-11/16"D		9	.889		83	37.50		120.50	149
0960	8 hole 11-1/2"H x 8"W x 6-11/16"D		8.50	.941		105	39.50		144.50	176
0970	12 hole 13-1/2"H x 10"W x 7-3/16"D		8	1		137	42		179	215
0980	20 hole 15-1/2"H x 12"W x 8-3/16"D		5	1.600		187	67		254	310
0990	30 hole 17-1/2"H x 14"W x 8-11/16"D	↓	4.50	1.778	↓	207	74.50		281.50	340
1450	Enclosures polyester NEMA 4X									
1460	Small, screw cover,									
1500	3-15/16"H x 3-15/16"W x 3-1/16"D	1 Elec	12	.667	Ea.	42.50	28		70.50	90
1510	5-3/16"H x 3-5/16"W x 3-1/16"D		12	.667		41	28		69	88.50
1520	5-7/8"H x 3-7/8"W x 4-3/16"D		12	.667		44	28		72	91
1530	5-7/8"H x 5-7/8"W x 4-3/16"D		12	.667		50	28		78	98
1540	7-5/8"H x 3-5/16"W x 3-1/16"D		12	.667		45.50	28		73.50	93
1550	10-3/16"H x 3-5/16"W x 3-1/16"D		10	.800		54.50	33.50		88	112
1560	Clear cover, 3-15/16"H x 3-15/16"W x 2-7/8"D		12	.667		48	28		76	95.50
1570	5-3/16"H x 3-5/16"W x 2-7/8"D		12	.667		51.50	28		79.50	99.50
1580	5-7/8"H x 3-7/8"W x 4"D		12	.667		63.50	28		91.50	113
1590	5-7/8"H x 5-7/8"W x 4"D		12	.667		78.50	28		106.50	129
1600	7-5/8"H x 3-5/16"W x 2-7/8"D	↓	12	.667	↓	57	28		85	106

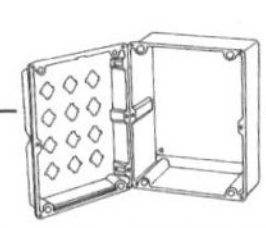

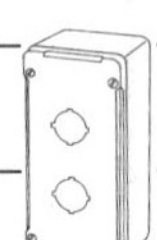

16 ELECTRICAL

16136 | Boxes

		16136 Boxes	CREW	DAILY OUTPUT	LABOR-HOURS	UNIT	2006 BARE COSTS MAT.	LABOR	EQUIP.	TOTAL	TOTAL INCL O&P	
720	1610	10-3/16"H x 3-5/16"W x 2-7/8"D	1 Elec	10	.800	Ea.	71.50	33.50		105	131	720
	1620	Pushbutton, 1 hole, 5-5/16"H x 3-5/16"W x 3-1/16"D		12	.667		37.50	28		65.50	84.50	
	1630	2 hole, 7-5/8"H x 3-5/16"W x 3-1/8"D		11	.727		42	30.50		72.50	93	
	1640	3 hole, 10-3/16"H x 3-5/16"W x 3-1/16"D		10.50	.762		50	32		82	105	
	8000	Wireway fiberglass, straight sect. screwcover, 12" L, 4" W x 4" D		40	.200		128	8.40		136.40	154	
	8010	6" W x 6" D		30	.267		172	11.20		183.20	206	
	8020	24" L, 4" W x 4" D		20	.400		163	16.80		179.80	205	
	8030	6" W x 6" D		15	.533		251	22.50		273.50	310	
	8040	36" L, 4" W x 4" D		13.30	.602		199	25.50		224.50	258	
	8050	6" W x 6" D		10	.800		315	33.50		348.50	395	
	8060	48" L, 4" W x 4" D		10	.800		233	33.50		266.50	310	
	8070	6" W x 6" D		7.50	1.067		380	45		425	490	
	8080	60" L, 4" W x 4" D		8	1		269	42		311	360	
	8090	6" W x 6" D		6	1.333		445	56		501	575	
	8100	Elbow, 90°, 4" W x 4" D		20	.400		127	16.80		143.80	165	
	8110	6" W x 6" D		18	.444		254	18.65		272.65	310	
	8120	Elbow, 45°, 4" W x 4" D		20	.400		127	16.80		143.80	165	
	8130	6" W x 6" D		18	.444		250	18.65		268.65	305	
	8140	Tee, 4" W x 4" D		16	.500		163	21		184	212	
	8150	6" W x 6" D		14	.571		294	24		318	360	
	8160	Cross, 4" W x 4" D		14	.571		229	24		253	289	
	8170	6" W x 6" D		12	.667		430	28		458	515	
	8180	Cut-off fitting, w/flange & adhesive, 4" W x 4" D		18	.444		98.50	18.65		117.15	137	
	8190	6" W x 6" D		16	.500		199	21		220	252	
	8200	Flexible ftng., heavy neoprene coated nylon, 4" W x 4" D		20	.400		163	16.80		179.80	205	
	8210	6" W x 6" D		18	.444		243	18.65		261.65	296	
	8220	Closure plate, fiberglass, 4" W x 4" D		20	.400		31.50	16.80		48.30	60.50	
	8230	6" W x 6" D		18	.444		35.50	18.65		54.15	68	
	8240	Box connector, stainless steel type 304, 4" W x 4" D		20	.400		55	16.80		71.80	86.50	
	8250	6" W x 6" D		18	.444		60.50	18.65		79.15	95.50	
	8260	Hanger, 4" W x 4" D		100	.080		13.70	3.36		17.06	20.50	
	8270	6" W x 6" D		80	.100		16.40	4.20		20.60	24.50	
	8280	Straight tube section fiberglass, 4"W x 4"D, 12" long		40	.200		111	8.40		119.40	135	
	8290	24" long		20	.400		129	16.80		145.80	168	
	8300	36" long		13.30	.602		160	25.50		185.50	215	
	8310	48" long		10	.800		178	33.50		211.50	247	
	8320	60" long		8	1		201	42		243	286	
	8330	120" long	↓	4	2	↓	310	84		394	470	

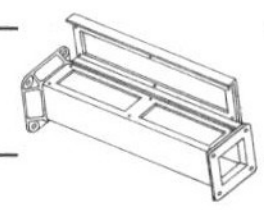

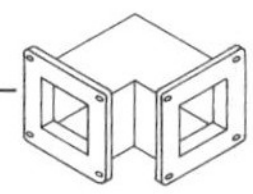

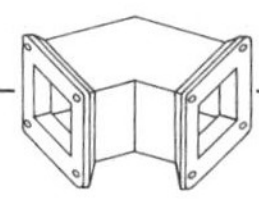

16139 | Residential Wiring

		16139 Residential Wiring	CREW	DAILY OUTPUT	LABOR-HOURS	UNIT	MAT.	LABOR	EQUIP.	TOTAL	TOTAL INCL O&P	
700	0010	**RESIDENTIAL WIRING**										700
	0020	20' avg. runs and #14/2 wiring incl. unless otherwise noted										
	1000	Service & panel, includes 24' SE-AL cable, service eye, meter,										
	1010	Socket, panel board, main bkr., ground rod, 15 or 20 amp										
	1020	1-pole circuit breakers, and misc. hardware										
	1100	100 amp, with 10 branch breakers	1 Elec	1.19	6.723	Ea.	490	282		772	975	
	1110	With PVC conduit and wire		.92	8.696		535	365		900	1,150	
	1120	With RGS conduit and wire		.73	10.959		710	460		1,170	1,500	
	1150	150 amp, with 14 branch breakers		1.03	7.767		760	325		1,085	1,350	
	1170	With PVC conduit and wire		.82	9.756		850	410		1,260	1,575	
	1180	With RGS conduit and wire	↓	.67	11.940		1,200	500		1,700	2,075	
	1200	200 amp, with 18 branch breakers	2 Elec	1.80	8.889		995	375		1,370	1,675	
	1220	With PVC conduit and wire		1.46	10.959		1,100	460		1,560	1,900	
	1230	With RGS conduit and wire	↓	1.24	12.903		1,525	540		2,065	2,525	
	1800	Lightning surge suppressor for above services, add	1 Elec	32	.250	↓	43.50	10.50		54	64	
	2000	Switch devices										

ELECTRICAL 16

16139 | Residential Wiring

		Description	Crew	Daily Output	Labor-Hours	Unit	2006 Bare Costs Mat.	Labor	Equip.	Total	Total Incl O&P
700	2100	Single pole, 15 amp, Ivory, with a 1-gang box, cover plate,									
	2110	Type NM (Romex) cable	1 Elec	17.10	.468	Ea.	8.45	19.65		28.10	40
	2120	Type MC (BX) cable		14.30	.559		22	23.50		45.50	60
	2130	EMT & wire		5.71	1.401		28	59		87	122
	2150	3-way, #14/3, type NM cable		14.55	.550		12.10	23		35.10	49
	2170	Type MC cable		12.31	.650		30	27.50		57.50	75.50
	2180	EMT & wire		5	1.600		30.50	67		97.50	138
	2200	4-way, #14/3, type NM cable		14.55	.550		25.50	23		48.50	64
	2220	Type MC cable		12.31	.650		44	27.50		71.50	90
	2230	EMT & wire		5	1.600		44	67		111	153
	2250	S.P., 20 amp, #12/2, type NM cable		13.33	.600		14.50	25		39.50	55
	2270	Type MC cable		11.43	.700		26.50	29.50		56	74.50
	2280	EMT & wire		4.85	1.649		34.50	69.50		104	145
	2290	S.P. rotary dimmer, 600W, no wiring		17	.471		16.90	19.75		36.65	49
	2300	S.P. rotary dimmer, 600W, type NM cable		14.55	.550		20.50	23		43.50	58
	2320	Type MC cable		12.31	.650		34	27.50		61.50	79.50
	2330	EMT & wire		5	1.600		41	67		108	149
	2350	3-way rotary dimmer, type NM cable		13.33	.600		18.30	25		43.30	59
	2370	Type MC cable		11.43	.700		31.50	29.50		61	80.50
	2380	EMT & wire		4.85	1.649		39	69.50		108.50	150
	2400	Interval timer wall switch, 20 amp, 1-30 min., #12/2									
	2410	Type NM cable	1 Elec	14.55	.550	Ea.	40	23		63	79
	2420	Type MC cable		12.31	.650		47.50	27.50		75	94
	2430	EMT & wire		5	1.600		60	67		127	170
	2500	Decorator style									
	2510	S.P., 15 amp, type NM cable	1 Elec	17.10	.468	Ea.	12	19.65		31.65	43.50
	2520	Type MC cable		14.30	.559		25.50	23.50		49	64
	2530	EMT & wire		5.71	1.401		31.50	59		90.50	126
	2550	3-way, #14/3, type NM cable		14.55	.550		15.65	23		38.65	53
	2570	Type MC cable		12.31	.650		34	27.50		61.50	79
	2580	EMT & wire		5	1.600		34	67		101	142
	2600	4-way, #14/3, type NM cable		14.55	.550		29.50	23		52.50	67.50
	2620	Type MC cable		12.31	.650		47.50	27.50		75	94
	2630	EMT & wire		5	1.600		47.50	67		114.50	157
	2650	S.P., 20 amp, #12/2, type NM cable		13.33	.600		18.05	25		43.05	59
	2670	Type MC cable		11.43	.700		30	29.50		59.50	78.50
	2680	EMT & wire		4.85	1.649		38	69.50		107.50	149
	2700	S.P., slide dimmer, type NM cable		17.10	.468		27	19.65		46.65	60.50
	2720	Type MC cable		14.30	.559		40.50	23.50		64	80.50
	2730	EMT & wire		5.71	1.401		47.50	59		106.50	143
	2750	S.P., touch dimmer, type NM cable		17.10	.468		24	19.65		43.65	57
	2770	Type MC cable		14.30	.559		37.50	23.50		61	77
	2780	EMT & wire		5.71	1.401		44.50	59		103.50	140
	2800	3-way touch dimmer, type NM cable		13.33	.600		44	25		69	87.50
	2820	Type MC cable		11.43	.700		57.50	29.50		87	109
	2830	EMT & wire		4.85	1.649		64.50	69.50		134	178
	3000	Combination devices									
	3100	S.P. switch/15 amp recpt., Ivory, 1-gang box, plate									
	3110	Type NM cable	1 Elec	11.43	.700	Ea.	17.50	29.50		47	65
	3120	Type MC cable		10	.800		31	33.50		64.50	86
	3130	EMT & wire		4.40	1.818		38	76.50		114.50	160
	3150	S.P. switch/pilot light, type NM cable		11.43	.700		18.20	29.50		47.70	65.50
	3170	Type MC cable		10	.800		31.50	33.50		65	87
	3180	EMT & wire		4.43	1.806		38.50	76		114.50	160
	3190	2-S.P. switches, 2-#14/2, no wiring		14	.571		7	24		31	44.50
	3200	2-S.P. switches, 2-#14/2, type NM cables		10	.800		20.50	33.50		54	74.50

16139 | Residential Wiring		CREW	DAILY OUTPUT	LABOR-HOURS	UNIT	2006 BARE COSTS MAT.	LABOR	EQUIP.	TOTAL	TOTAL INCL O&P
700 3220	Type MC cable	1 Elec	8.89	.900	Ea.	42	38		80	104
3230	EMT & wire		4.10	1.951		40	82		122	170
3250	3-way switch/15 amp recpt., #14/3, type NM cable		10	.800		24.50	33.50		58	79
3270	Type MC cable		8.89	.900		42.50	38		80.50	105
3280	EMT & wire		4.10	1.951		43	82		125	173
3300	2-3 way switches, 2-#14/3, type NM cables		8.89	.900		32.50	38		70.50	94
3320	Type MC cable		8	1		63.50	42		105.50	135
3330	EMT & wire		4	2		47.50	84		131.50	182
3350	S.P. switch/20 amp recpt., #12/2, type NM cable		10	.800		28	33.50		61.50	82.50
3370	Type MC cable		8.89	.900		35.50	38		73.50	97
3380	EMT & wire		4.10	1.951		48	82		130	179
3400	Decorator style									
3410	S.P. switch/15 amp recpt., type NM cable	1 Elec	11.43	.700	Ea.	21	29.50		50.50	68.50
3420	Type MC cable		10	.800		34.50	33.50		68	90
3430	EMT & wire		4.40	1.818		41.50	76.50		118	164
3450	S.P. switch/pilot light, type NM cable		11.43	.700		21.50	29.50		51	69.50
3470	Type MC cable		10	.800		35	33.50		68.50	90.50
3480	EMT & wire		4.40	1.818		42	76.50		118.50	165
3500	2-S.P. switches, 2-#14/2, type NM cables		10	.800		24	33.50		57.50	78.50
3520	Type MC cable		8.89	.900		45.50	38		83.50	108
3530	EMT & wire		4.10	1.951		43.50	82		125.50	174
3550	3-way/15 amp recpt., #14/3, type NM cable		10	.800		28	33.50		61.50	83
3570	Type MC cable		8.89	.900		46	38		84	109
3580	EMT & wire		4.10	1.951		46.50	82		128.50	177
3650	2-3 way switches, 2-#14/3, type NM cables		8.89	.900		36	38		74	98
3670	Type MC cable		8	1		67	42		109	139
3680	EMT & wire		4	2		51.50	84		135.50	186
3700	S.P. switch/20 amp recpt., #12/2, type NM cable		10	.800		31.50	33.50		65	86.50
3720	Type MC cable		8.89	.900		39	38		77	101
3730	EMT & wire		4.10	1.951		51.50	82		133.50	183
4000	Receptacle devices									
4010	Duplex outlet, 15 amp recpt., Ivory, 1-gang box, plate									
4015	Type NM cable	1 Elec	14.55	.550	Ea.	6.80	23		29.80	43
4020	Type MC cable		12.31	.650		20	27.50		47.50	64.50
4030	EMT & wire		5.33	1.501		26.50	63		89.50	126
4050	With #12/2, type NM cable		12.31	.650		8.55	27.50		36.05	51.50
4070	Type MC cable		10.67	.750		20.50	31.50		52	71
4080	EMT & wire		4.71	1.699		28.50	71.50		100	142
4100	20 amp recpt., #12/2, type NM cable		12.31	.650		16.20	27.50		43.70	60
4120	Type MC cable		10.67	.750		28	31.50		59.50	79.50
4130	EMT & wire		4.71	1.699		36.50	71.50		108	150
4140	For GFI see line 4300 below									
4150	Decorator style, 15 amp recpt., type NM cable	1 Elec	14.55	.550	Ea.	10.35	23		33.35	47
4170	Type MC cable		12.31	.650		24	27.50		51.50	68
4180	EMT & wire		5.33	1.501		30	63		93	130
4200	With #12/2, type NM cable		12.31	.650		12.10	27.50		39.60	55.50
4220	Type MC cable		10.67	.750		24	31.50		55.50	75
4230	EMT & wire		4.71	1.699		32	71.50		103.50	146
4250	20 amp recpt. #12/2, type NM cable		12.31	.650		19.75	27.50		47.25	63.50
4270	Type MC cable		10.67	.750		31.50	31.50		63	83.50
4280	EMT & wire		4.71	1.699		40	71.50		111.50	154
4300	GFI, 15 amp recpt., type NM cable		12.31	.650		35.50	27.50		63	81
4320	Type MC cable		10.67	.750		48.50	31.50		80	102
4330	EMT & wire		4.71	1.699		55	71.50		126.50	171
4350	GFI with #12/2, type NM cable		10.67	.750		37	31.50		68.50	89
4370	Type MC cable		9.20	.870		49	36.50		85.50	111

16139 | Residential Wiring		CREW	DAILY OUTPUT	LABOR-HOURS	UNIT	2006 BARE COSTS MAT.	LABOR	EQUIP.	TOTAL	TOTAL INCL O&P
700 4380	EMT & wire	1 Elec	4.21	1.900	Ea.	57	80		137	186 700
4400	20 amp recpt., #12/2 type NM cable		10.67	.750		38.50	31.50		70	91
4420	Type MC cable		9.20	.870		50.50	36.50		87	112
4430	EMT & wire		4.21	1.900		59	80		139	188
4500	Weather-proof cover for above receptacles, add		32	.250		4.40	10.50		14.90	21
4550	Air conditioner outlet, 20 amp-240 volt recpt.									
4560	30′ of #12/2, 2 pole circuit breaker									
4570	Type NM cable	1 Elec	10	.800	Ea.	48.50	33.50		82	106
4580	Type MC cable		9	.889		63.50	37.50		101	128
4590	EMT & wire		4	2		67.50	84		151.50	204
4600	Decorator style, type NM cable		10	.800		52.50	33.50		86	110
4620	Type MC cable		9	.889		67.50	37.50		105	132
4630	EMT & wire		4	2		72	84		156	208
4650	Dryer outlet, 30 amp-240 volt recpt., 20′ of #10/3									
4660	2 pole circuit breaker									
4670	Type NM cable	1 Elec	6.41	1.248	Ea.	57.50	52.50		110	145
4680	Type MC cable		5.71	1.401		70.50	59		129.50	168
4690	EMT & wire		3.48	2.299		71.50	96.50		168	228
4700	Range outlet, 50 amp-240 volt recpt., 30′ of #8/3									
4710	Type NM cable	1 Elec	4.21	1.900	Ea.	89	80		169	221
4720	Type MC cable		4	2		133	84		217	275
4730	EMT & wire		2.96	2.703		91	114		205	275
4750	Central vacuum outlet, Type NM cable		6.40	1.250		53.50	52.50		106	140
4770	Type MC cable		5.71	1.401		73.50	59		132.50	171
4780	EMT & wire		3.48	2.299		75	96.50		171.50	232
4800	30 amp-110 volt locking recpt., #10/2 circ. bkr.									
4810	Type NM cable	1 Elec	6.20	1.290	Ea.	60	54		114	150
4820	Type MC cable		5.40	1.481		85.50	62		147.50	190
4830	EMT & wire		3.20	2.500		82.50	105		187.50	253
4900	Low voltage outlets									
4910	Telephone recpt., 20′ of 4/C phone wire	1 Elec	26	.308	Ea.	7.95	12.90		20.85	28.50
4920	TV recpt., 20′ of RG59U coax wire, F type connector	"	16	.500	"	13.20	21		34.20	47
4950	Door bell chime, transformer, 2 buttons, 60′ of bellwire									
4970	Economy model	1 Elec	11.50	.696	Ea.	55	29		84	106
4980	Custom model		11.50	.696		86	29		115	140
4990	Luxury model, 3 buttons		9.50	.842		232	35.50		267.50	310
6000	Lighting outlets									
6050	Wire only (for fixture), type NM cable	1 Elec	32	.250	Ea.	4.71	10.50		15.21	21.50
6070	Type MC cable		24	.333		14.30	14		28.30	37
6080	EMT & wire		10	.800		19.35	33.50		52.85	73.50
6100	Box (4″), and wire (for fixture), type NM cable		25	.320		10.45	13.45		23.90	32
6120	Type MC cable		20	.400		20	16.80		36.80	48
6130	EMT & wire		11	.727		25	30.50		55.50	74.50
6200	Fixtures (use with lines 6050 or 6100 above)									
6210	Canopy style, economy grade	1 Elec	40	.200	Ea.	26	8.40		34.40	41.50
6220	Custom grade		40	.200		48	8.40		56.40	66
6250	Dining room chandelier, economy grade		19	.421		78	17.70		95.70	113
6260	Custom grade		19	.421		230	17.70		247.70	280
6270	Luxury grade		15	.533		510	22.50		532.50	595
6310	Kitchen fixture (fluorescent), economy grade		30	.267		57	11.20		68.20	80
6320	Custom grade		25	.320		165	13.45		178.45	203
6350	Outdoor, wall mounted, economy grade		30	.267		27.50	11.20		38.70	48
6360	Custom grade		30	.267		104	11.20		115.20	131
6370	Luxury grade		25	.320		235	13.45		248.45	280
6410	Outdoor PAR floodlights, 1 lamp, 150 watt		20	.400		26.50	16.80		43.30	55
6420	2 lamp, 150 watt each		20	.400		43	16.80		59.80	73.50

16139 | Residential Wiring

		Description	Crew	Daily Output	Labor-Hours	Unit	2006 Bare Costs Mat.	Labor	Equip.	Total	Total Incl O&P
700	6430	For infrared security sensor, add	1 Elec	32	.250	Ea.	90	10.50		100.50	115
	6450	Outdoor, quartz-halogen, 300 watt flood		20	.400		39	16.80		55.80	69
	6600	Recessed downlight, round, pre-wired, 50 or 75 watt trim		30	.267		36	11.20		47.20	57
	6610	With shower light trim		30	.267		45	11.20		56.20	67
	6620	With wall washer trim		28	.286		54	12		66	78
	6630	With eye-ball trim		28	.286		54	12		66	78
	6640	For direct contact with insulation, add					1.80			1.80	1.98
	6700	Porcelain lamp holder	1 Elec	40	.200		3.80	8.40		12.20	17.15
	6710	With pull switch		40	.200		4.13	8.40		12.53	17.50
	6750	Fluorescent strip, 1-20 watt tube, wrap around diffuser, 24″		24	.333		53	14		67	80
	6760	1-40 watt tube, 48″		24	.333		65	14		79	93
	6770	2-40 watt tubes, 48″		20	.400		79	16.80		95.80	113
	6780	With residential ballast		20	.400		89.50	16.80		106.30	125
	6800	Bathroom heat lamp, 1-250 watt		28	.286		38.50	12		50.50	61
	6810	2-250 watt lamps		28	.286		63	12		75	88
	6820	For timer switch, see line 2400									
	6900	Outdoor post lamp, incl. post, fixture, 35′ of #14/2									
	6910	Type NMC cable	1 Elec	3.50	2.286	Ea.	186	96		282	350
	6920	Photo-eye, add		27	.296		30	12.45		42.45	52
	6950	Clock dial time switch, 24 hr., w/enclosure, type NM cable		11.43	.700		56.50	29.50		86	108
	6970	Type MC cable		11	.727		70	30.50		100.50	124
	6980	EMT & wire		4.85	1.649		76	69.50		145.50	191
	7000	Alarm systems									
	7050	Smoke detectors, box, #14/3, type NM cable	1 Elec	14.55	.550	Ea.	30	23		53	68.50
	7070	Type MC cable		12.31	.650		44	27.50		71.50	90.50
	7080	EMT & wire		5	1.600		44	67		111	153
	7090	For relay output to security system, add					12.15			12.15	13.35
	8000	Residential equipment									
	8050	Disposal hook-up, incl. switch, outlet box, 3′ of flex									
	8060	20 amp-1 pole circ. bkr., and 25′ of #12/2									
	8070	Type NM cable	1 Elec	10	.800	Ea.	24.50	33.50		58	79
	8080	Type MC cable		8	1		38	42		80	107
	8090	EMT & wire		5	1.600		47.50	67		114.50	156
	8100	Trash compactor or dishwasher hook-up, incl. outlet box,									
	8110	3′ of flex, 15 amp-1 pole circ. bkr., and 25′ of #14/2									
	8120	Type NM cable	1 Elec	10	.800	Ea.	17.85	33.50		51.35	71.50
	8130	Type MC cable		8	1		33.50	42		75.50	102
	8140	EMT & wire		5	1.600		41	67		108	150
	8150	Hot water sink dispensor hook-up, use line 8100									
	8200	Vent/exhaust fan hook-up, type NM cable	1 Elec	32	.250	Ea.	4.71	10.50		15.21	21.50
	8220	Type MC cable		24	.333		14.30	14		28.30	37
	8230	EMT & wire		10	.800		19.35	33.50		52.85	73.50
	8250	Bathroom vent fan, 50 CFM (use with above hook-up)									
	8260	Economy model	1 Elec	15	.533	Ea.	22.50	22.50		45	59.50
	8270	Low noise model		15	.533		32	22.50		54.50	69.50
	8280	Custom model		12	.667		117	28		145	172
	8300	Bathroom or kitchen vent fan, 110 CFM									
	8310	Economy model	1 Elec	15	.533	Ea.	60.50	22.50		83	101
	8320	Low noise model	″	15	.533	″	79	22.50		101.50	122
	8350	Paddle fan, variable speed (w/o lights)									
	8360	Economy model (AC motor)	1 Elec	10	.800	Ea.	105	33.50		138.50	168
	8370	Custom model (AC motor)		10	.800		182	33.50		215.50	252
	8380	Luxury model (DC motor)		8	1		360	42		402	460
	8390	Remote speed switch for above, add		12	.667		26	28		54	71.50
	8500	Whole house exhaust fan, ceiling mount, 36″, variable speed									
	8510	Remote switch, incl. shutters, 20 amp-1 pole circ. bkr.									

16139 | Residential Wiring

			CREW	DAILY OUTPUT	LABOR-HOURS	UNIT	2006 BARE COSTS MAT.	LABOR	EQUIP.	TOTAL	TOTAL INCL O&P	
700	8520	30′ of #12/2, type NM cable	1 Elec	4	2	Ea.	685	84		769	885	700
	8530	Type MC cable		3.50	2.286		705	96		801	925	
	8540	EMT & wire		3	2.667		715	112		827	960	
	8600	Whirlpool tub hook-up, incl. timer switch, outlet box										
	8610	3′ of flex, 20 amp-1 pole GFI circ. bkr.										
	8620	30′ of #12/2, type NM cable	1 Elec	5	1.600	Ea.	100	67		167	214	
	8630	Type MC cable		4.20	1.905		109	80		189	243	
	8640	EMT & wire		3.40	2.353		118	99		217	281	
	8650	Hot water heater hook-up, incl. 1-2 pole circ. bkr., box;										
	8660	3′ of flex, 20′ of #10/2, type NM cable	1 Elec	5	1.600	Ea.	26.50	67		93.50	133	
	8670	Type MC cable		4.20	1.905		44	80		124	171	
	8680	EMT & wire		3.40	2.353		42.50	99		141.50	199	
	9000	Heating/air conditioning										
	9050	Furnace/boiler hook-up, incl. firestat, local on-off switch										
	9060	Emergency switch, and 40′ of type NM cable	1 Elec	4	2	Ea.	46	84		130	180	
	9070	Type MC cable		3.50	2.286		67.50	96		163.50	223	
	9080	EMT & wire		1.50	5.333		79.50	224		303.50	435	
	9100	Air conditioner hook-up, incl. local 60 amp disc. switch										
	9110	3′ sealtite, 40 amp, 2 pole circuit breaker										
	9130	40′ of #8/2, type NM cable	1 Elec	3.50	2.286	Ea.	169	96		265	335	
	9140	Type MC cable		3	2.667		240	112		352	435	
	9150	EMT & wire		1.30	6.154		190	258		448	610	
	9200	Heat pump hook-up, 1-40 & 1-100 amp 2 pole circ. bkr.										
	9210	Local disconnect switch, 3′ sealtite										
	9220	40′ of #8/2 & 30′ of #3/2										
	9230	Type NM cable	1 Elec	1.30	6.154	Ea.	425	258		683	870	
	9240	Type MC cable		1.08	7.407		585	310		895	1,125	
	9250	EMT & wire		.94	8.511		460	355		815	1,050	
	9500	Thermostat hook-up, using low voltage wire										
	9520	Heating only	1 Elec	24	.333	Ea.	6	14		20	28	
	9530	Heating/cooling	"	20	.400	"	7.15	16.80		23.95	34	

16140 | Wiring Devices

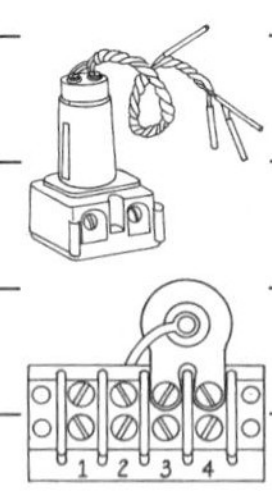

			CREW	DAILY OUTPUT	LABOR-HOURS	UNIT	MAT.	LABOR	EQUIP.	TOTAL	TOTAL INCL O&P	
500	0010	LOW VOLTAGE SWITCHING										500
	3600	Relays, 120 V or 277 V standard	1 Elec	12	.667	Ea.	26	28		54	71.50	
	3800	Flush switch, standard		40	.200		9.05	8.40		17.45	23	
	4000	Interchangeable		40	.200		11.80	8.40		20.20	26	
	4100	Surface switch, standard		40	.200		6.60	8.40		15	20.50	
	4200	Transformer 115 V to 25 V		12	.667		93	28		121	145	
	4400	Master control, 12 circuit, manual		4	2		94	84		178	232	
	4500	25 circuit, motorized		4	2		102	84		186	241	
	4600	Rectifier, silicon		12	.667		30.50	28		58.50	76.50	
	4800	Switchplates, 1 gang, 1, 2 or 3 switch, plastic		80	.100		3	4.20		7.20	9.75	
	5000	Stainless steel		80	.100		8.10	4.20		12.30	15.35	
	5400	2 gang, 3 switch, stainless steel		53	.151		15.65	6.35		22	27	
	5500	4 switch, plastic		53	.151		6.70	6.35		13.05	17.10	
	5600	2 gang, 4 switch, stainless steel		53	.151		16.35	6.35		22.70	28	
	5700	6 switch, stainless steel		53	.151		36	6.35		42.35	49.50	
	5800	3 gang, 9 switch, stainless steel		32	.250		50	10.50		60.50	71	
	5900	Receptacle, triple, 1 return, 1 feed		26	.308		32	12.90		44.90	55	
	6000	2 feed		20	.400		32	16.80		48.80	61	
	6100	Relay gang boxes, flush or surface, 6 gang		5.30	1.509		55	63.50		118.50	158	
	6200	12 gang		4.70	1.702		58	71.50		129.50	174	
	6400	18 gang		4	2		72	84		156	208	
	6500	Frame, to hold up to 6 relays		12	.667		48	28		76	96	

16 ELECTRICAL

16140 | Wiring Devices

			CREW	DAILY OUTPUT	LABOR-HOURS	UNIT	2006 BARE COSTS MAT.	LABOR	EQUIP.	TOTAL	TOTAL INCL O&P	
500	7200	Control wire, 2 conductor	1 Elec	6.30	1.270	C.L.F.	14.95	53.50		68.45	98.50	500
	7400	3 conductor		5	1.600		20.50	67		87.50	127	
	7600	19 conductor		2.50	3.200		150	134		284	370	
	7800	26 conductor		2	4		230	168		398	510	
	8000	Weatherproof, 3 conductor		5	1.600		44	67		111	153	
910	0010	WIRING DEVICES R262726-90										910
	0200	Toggle switch, quiet type, single pole, 15 amp	1 Elec	40	.200	Ea.	4.88	8.40		13.28	18.30	
	0500	20 amp		27	.296		7.15	12.45		19.60	27	
	0510	30 amp		23	.348		19.55	14.60		34.15	44	
	0530	Lock handle, 20 amp		27	.296		22	12.45		34.45	43	
	0540	Security key, 20 amp		26	.308		55	12.90		67.90	80.50	
	0550	Rocker, 15 amp		40	.200		5.40	8.40		13.80	18.90	
	0560	20 amp		27	.296		12.70	12.45		25.15	33	
	0600	3 way, 15 amp		23	.348		7	14.60		21.60	30	
	0800	20 amp		18	.444		8	18.65		26.65	38	
	0810	30 amp		9	.889		28	37.50		65.50	88	
	0830	Lock handle, 20 amp		18	.444		24.50	18.65		43.15	55.50	
	0840	Security key, 20 amp		17	.471		60	19.75		79.75	96.50	
	0850	Rocker, 15 amp		23	.348		7.60	14.60		22.20	31	
	0860	20 amp		18	.444		18.40	18.65		37.05	49	
	0900	4 way, 15 amp		15	.533		21	22.50		43.50	57.50	
	1000	20 amp		11	.727		34.50	30.50		65	85	
	1020	Lock handle, 20 amp		11	.727		54	30.50		84.50	107	
	1030	Rocker, 15 amp		15	.533		29.50	22.50		52	67	
	1040	20 amp		11	.727		45	30.50		75.50	96.50	
	1100	Toggle switch, quiet type, double pole, 15 amp		15	.533		14.10	22.50		36.60	50	
	1200	20 amp		11	.727		13.75	30.50		44.25	62	
	1210	30 amp		9	.889		28.50	37.50		66	88.50	
	1230	Lock handle, 20 amp		11	.727		25	30.50		55.50	74.50	
	1250	Security key, 20 amp		10	.800		60	33.50		93.50	118	
	1420	Toggle switch quiet type, 1 pole, 2 throw center off, 15 amp		23	.348		47	14.60		61.60	74.50	
	1440	20 amp		18	.444		51.50	18.65		70.15	85.50	
	1460	Lock handle, 20 amp		18	.444		52	18.65		70.65	86.50	
	1480	Momentary contact, 15 amp		23	.348		20.50	14.60		35.10	45	
	1500	20 amp		18	.444		28	18.65		46.65	60	
	1520	Momentary contact, lock handle, 20 amp		18	.444		35.50	18.65		54.15	68	
	1650	Dimmer switch, 120 volt, incandescent, 600 watt, 1 pole		16	.500		11.35	21		32.35	45	
	1700	600 watt, 3 way		12	.667		9.10	28		37.10	53	
	1750	1000 watt, 1 pole		16	.500		46.50	21		67.50	84	
	1800	1000 watt, 3 way		12	.667		64	28		92	114	
	2000	1500 watt, 1 pole		11	.727		87	30.50		117.50	143	
	2100	2000 watt, 1 pole		8	1		130	42		172	208	
	2110	Fluorescent, 600 watt		15	.533		71.50	22.50		94	114	
	2120	1000 watt		15	.533		89.50	22.50		112	133	
	2130	1500 watt		10	.800		168	33.50		201.50	237	
	2160	Explosionproof, toggle switch, wall, single pole 20 amp		5.30	1.509		148	63.50		211.50	261	
	2180	Receptacle, single outlet, 20 amp		5.30	1.509		236	63.50		299.50	360	
	2190	30 amp		4	2		470	84		554	645	
	2290	60 amp		2.50	3.200		515	134		649	775	
	2360	Plug, 20 amp		16	.500		109	21		130	153	
	2370	30 amp		12	.667		194	28		222	256	
	2380	60 amp		8	1		254	42		296	345	
	2410	Furnace, thermal cutoff switch with plate		26	.308		12.65	12.90		25.55	34	
	2460	Receptacle, duplex, 120 volt, grounded, 15 amp		40	.200		1.17	8.40		9.57	14.25	
	2470	20 amp		27	.296		8.85	12.45		21.30	29	

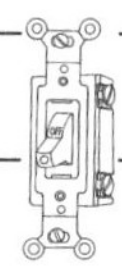

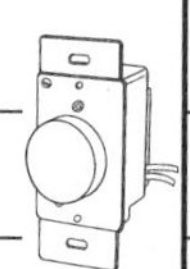

ELECTRICAL 16

16140 | Wiring Devices

16140 Wiring Devices		CREW	DAILY OUTPUT	LABOR-HOURS	UNIT	2006 BARE COSTS MAT.	LABOR	EQUIP.	TOTAL	TOTAL INCL O&P
910 2480	Ground fault interrupting, 15 amp (R262726-90)	1 Elec	27	.296	Ea.	29.50	12.45		41.95	51.50 910
2482	20 amp		27	.296		31.50	12.45		43.95	53.50
2490	Dryer, 30 amp		15	.533		10.80	22.50		33.30	46.50
2500	Range, 50 amp		11	.727		11.40	30.50		41.90	59.50
2540	Isolated ground receptacle, duplex, 20 amp		27	.296		20.50	12.45		32.95	41.50
2550	Simplex, 20 amp		27	.296		23	12.45		35.45	44.50
2560	Simplex, 30 amp		15	.533		27	22.50		49.50	64.50
2600	Wall plates, stainless steel, 1 gang		80	.100		1.80	4.20		6	8.45
2800	2 gang		53	.151		4.10	6.35		10.45	14.25
3000	3 gang		32	.250		5.50	10.50		16	22.50
3100	4 gang		27	.296		8.45	12.45		20.90	28.50
3110	Brown plastic, 1 gang		80	.100		.31	4.20		4.51	6.80
3120	2 gang		53	.151		.77	6.35		7.12	10.60
3130	3 gang		32	.250		1.14	10.50		11.64	17.45
3140	4 gang		27	.296		1.78	12.45		14.23	21
3150	Brushed brass, 1 gang		80	.100		3.85	4.20		8.05	10.70
3160	Anodized aluminum, 1 gang		80	.100		2.45	4.20		6.65	9.15
3170	Switch cover, weatherproof, 1 gang		60	.133		7.25	5.60		12.85	16.65
3180	Vandal proof lock, 1 gang		60	.133		6.55	5.60		12.15	15.85
3200	Lampholder, keyless		26	.308		9.70	12.90		22.60	30.50
3400	Pullchain with receptacle		22	.364		9.20	15.25		24.45	33.50
3500	Pilot light, neon with jewel		27	.296		12.20	12.45		24.65	32.50
3600	Receptacle, 20 amp, 250 volt, NEMA 6		27	.296		15.45	12.45		27.90	36
3620	277 volt NEMA 7		27	.296		17.45	12.45		29.90	38.50
3640	125/250 volt NEMA 10		27	.296		18.25	12.45		30.70	39
3680	125/250 volt NEMA 14		25	.320		23	13.45		36.45	46
3700	3 pole, 250 volt NEMA 15		25	.320		24	13.45		37.45	47
3720	120/208 volt NEMA 18		25	.320		25.50	13.45		38.95	48.50
3740	30 amp, 125 volt NEMA 5		15	.533		16.30	22.50		38.80	52.50
3760	250 volt NEMA 6		15	.533		18.60	22.50		41.10	55
3780	277 volt NEMA 7		15	.533		24	22.50		46.50	60.50
3820	125/250 volt NEMA 14		14	.571		51.50	24		75.50	93.50
3840	3 pole, 250 volt NEMA 15		14	.571		48	24		72	89.50
3880	50 amp, 125 volt NEMA 5		11	.727		21.50	30.50		52	70.50
3900	250 volt NEMA 6		11	.727		21.50	30.50		52	70.50
3920	277 volt NEMA 7		11	.727		28	30.50		58.50	77.50
3960	125/250 volt NEMA 14		10	.800		66	33.50		99.50	125
3980	3 pole 250 volt NEMA 15		10	.800		65.50	33.50		99	124
4020	60 amp, 125/250 volt, NEMA 14		8	1		73.50	42		115.50	146
4040	3 pole, 250 volt NEMA 15		8	1		77	42		119	149
4060	120/208 volt NEMA 18		8	1		66	42		108	137
4100	Receptacle locking, 20 amp, 125 volt NEMA L5		27	.296		16.45	12.45		28.90	37.50
4120	250 volt NEMA L6		27	.296		16.45	12.45		28.90	37.50
4140	277 volt NEMA L7		27	.296		16.45	12.45		28.90	37.50
4150	3 pole, 250 volt, NEMA L11		27	.296		26	12.45		38.45	47.50
4160	20 amp, 480 volt NEMA L8		27	.296		21	12.45		33.45	42
4180	600 volt NEMA L9		27	.296		31.50	12.45		43.95	53.50
4200	125/250 volt NEMA L10		27	.296		26	12.45		38.45	47.50
4230	125/250 volt NEMA L14		25	.320		23	13.45		36.45	46
4280	250 volt NEMA L15		25	.320		23	13.45		36.45	46
4300	480 volt NEMA L16		25	.320		27.50	13.45		40.95	50.50
4320	3 phase, 120/208 volt NEMA L18		25	.320		35.50	13.45		48.95	59.50
4340	277/480 volt NEMA L19		25	.320		35.50	13.45		48.95	59.50
4360	347/600 volt NEMA L20		25	.320		35.50	13.45		48.95	59.50
4380	120/208 volt NEMA L21		23	.348		28	14.60		42.60	53.50
4400	277/480 volt NEMA L22		23	.348		33.50	14.60		48.10	59.50

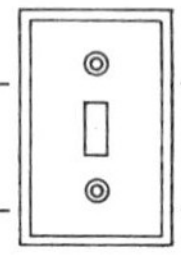

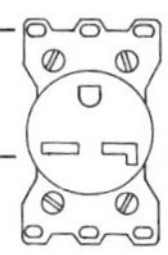

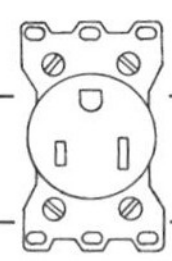

16 ELECTRICAL

16140 | Wiring Devices			CREW	DAILY OUTPUT	LABOR-HOURS	UNIT	2006 BARE COSTS MAT.	LABOR	EQUIP.	TOTAL	TOTAL INCL O&P	
910	4420	347/600 volt NEMA L23 R262726-90	1 Elec	23	.348	Ea.	33.50	14.60		48.10	59.50	910
	4440	30 amp, 125 volt NEMA L5		15	.533		23	22.50		45.50	60	
	4460	250 volt NEMA L6		15	.533		25	22.50		47.50	62	
	4480	277 volt NEMA L7		15	.533		25	22.50		47.50	62	
	4500	480 volt NEMA L8		15	.533		30	22.50		52.50	67.50	
	4520	600 volt NEMA L9		15	.533		30	22.50		52.50	67.50	
	4540	125/250 volt NEMA L10		15	.533		32	22.50		54.50	69.50	
	4560	3 phase, 250 volt NEMA L11		15	.533		32	22.50		54.50	69.50	
	4620	125/250 volt NEMA L14		14	.571		36.50	24		60.50	77	
	4640	250 volt NEMA L15		14	.571		36.50	24		60.50	77	
	4660	480 volt NEMA L16		14	.571		43	24		67	84.50	
	4680	600 volt NEMA L17		14	.571		41.50	24		65.50	83	
	4700	120/208 volt NEMA L18		14	.571		52	24		76	94.50	
	4720	277/480 volt NEMA L19		14	.571		52	24		76	94.50	
	4740	347/600 volt NEMA L20		14	.571		52.50	24		76.50	94.50	
	4760	120/208 volt NEMA L21		13	.615		39	26		65	82.50	
	4780	277/480 volt NEMA L22		13	.615		46.50	26		72.50	91	
	4800	347/600 volt NEMA L23		13	.615		58	26		84	104	
	4840	Receptacle, corrosion resistant, 15 or 20 amp, 125 volt NEMA L5		27	.296		23.50	12.45		35.95	44.50	
	4860	250 volt NEMA L6		27	.296		23.50	12.45		35.95	44.50	
	4900	Receptacle, cover plate, phenolic plastic, NEMA 5 & 6		80	.100		.46	4.20		4.66	6.95	
	4910	NEMA 7-23		80	.100		.53	4.20		4.73	7.05	
	4920	Stainless steel, NEMA 5 & 6		80	.100		1.83	4.20		6.03	8.45	
	4930	NEMA 7-23		80	.100		2.85	4.20		7.05	9.60	
	4940	Brushed brass NEMA 5 & 6		80	.100		4.50	4.20		8.70	11.40	
	4950	NEMA 7-23		80	.100		4.95	4.20		9.15	11.90	
	4960	Anodized aluminum, NEMA 5 & 6		80	.100		2.50	4.20		6.70	9.20	
	4970	NEMA 7-23		80	.100		4	4.20		8.20	10.85	
	4980	Weatherproof NEMA 7-23		60	.133		27	5.60		32.60	38	
	5100	Plug, 20 amp, 250 volt, NEMA 6		30	.267		13.95	11.20		25.15	32.50	
	5110	277 volt NEMA 7		30	.267		15.25	11.20		26.45	34	
	5120	3 pole, 120/250 volt, NEMA 10		26	.308		19.50	12.90		32.40	41.50	
	5130	125/250 volt NEMA 14		26	.308		38.50	12.90		51.40	62.50	
	5140	250 volt NEMA 15		26	.308		40.50	12.90		53.40	64.50	
	5150	120/208 volt NEMA 8		26	.308		41.50	12.90		54.40	66	
	5160	30 amp, 125 volt NEMA 5		13	.615		39.50	26		65.50	83.50	
	5170	250 volt NEMA 6		13	.615		41	26		67	85	
	5180	277 volt NEMA 7		13	.615		50.50	26		76.50	95.50	
	5190	125/250 volt NEMA 14		13	.615		45.50	26		71.50	90	
	5200	3 pole, 250 volt NEMA 15		12	.667		47.50	28		75.50	95.50	
	5210	50 amp, 125 volt NEMA 5		9	.889		49.50	37.50		87	112	
	5220	250 volt NEMA 6		9	.889		51.50	37.50		89	115	
	5230	277 volt NEMA 7		9	.889		55	37.50		92.50	118	
	5240	125/250 volt NEMA 14		9	.889		55	37.50		92.50	118	
	5250	3 pole, 250 volt NEMA 15		8	1		57.50	42		99.50	128	
	5260	60 amp, 125/250 volt NEMA 14		7	1.143		63.50	48		111.50	144	
	5270	3 pole, 250 volt NEMA 15		7	1.143		68	48		116	149	
	5280	120/208 volt NEMA 18		7	1.143		66.50	48		114.50	147	
	5300	Plug angle, 20 amp, 250 volt NEMA 6		30	.267		19.20	11.20		30.40	38.50	
	5310	30 amp, 125 volt NEMA 5		13	.615		39.50	26		65.50	83.50	
	5320	250 volt NEMA 6		13	.615		41	26		67	85	
	5330	277 volt NEMA 7		13	.615		50.50	26		76.50	95.50	
	5340	125/250 volt NEMA 14		13	.615		50.50	26		76.50	95.50	
	5350	3 pole, 250 volt NEMA 15		12	.667		53	28		81	102	
	5360	50 amp, 125 volt NEMA 5		9	.889		43.50	37.50		81	105	
910	5370	250 volt NEMA 6		9	.889		42.50	37.50		80	105	910

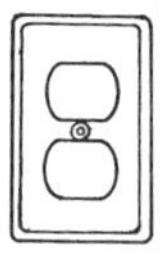

ELECTRICAL 16

16140 | Wiring Devices

		16140 Wiring Devices	CREW	DAILY OUTPUT	LABOR-HOURS	UNIT	2006 BARE COSTS MAT.	LABOR	EQUIP.	TOTAL	TOTAL INCL O&P	
910	5380	277 volt NEMA 7 R262726-90	1 Elec	9	.889	Ea.	55.50	37.50		93	119	910
	5390	125/250 volt NEMA 14		9	.889		61	37.50		98.50	125	
	5400	3 pole, 250 volt NEMA 15		8	1		64	42		106	135	
	5410	60 amp, 125/250 volt NEMA 14		7	1.143		71	48		119	152	
	5420	3 pole, 250 volt NEMA 15		7	1.143		74	48		122	155	
	5430	120/208 volt NEMA 18		7	1.143		74	48		122	156	
	5500	Plug, locking, 20 amp, 125 volt NEMA L5		30	.267		12.90	11.20		24.10	31.50	
	5510	250 volt NEMA L6		30	.267		12.90	11.20		24.10	31.50	
	5520	277 volt NEMA L7		30	.267		12.90	11.20		24.10	31.50	
	5530	480 volt NEMA L8		30	.267		16.20	11.20		27.40	35	
	5540	600 volt NEMA L9		30	.267		19.20	11.20		30.40	38.50	
	5550	3 pole, 125/250 volt NEMA L10		26	.308		17.55	12.90		30.45	39	
	5560	250 volt NEMA L11		26	.308		17.55	12.90		30.45	39	
	5570	480 volt NEMA L12		26	.308		20.50	12.90		33.40	42.50	
	5580	125/250 volt NEMA L14		26	.308		19.55	12.90		32.45	41.50	
	5590	250 volt NEMA L15		26	.308		21	12.90		33.90	43	
	5600	480 volt NEMA L16		26	.308		24.50	12.90		37.40	47	
	5610	4 pole, 120/208 volt NEMA L18		24	.333		31.50	14		45.50	56	
	5620	277/480 volt NEMA L19		24	.333		31.50	14		45.50	56	
	5630	347/600 volt NEMA L20		24	.333		31.50	14		45.50	56	
	5640	120/208 volt NEMA L21		24	.333		25	14		39	49	
	5650	277/480 volt NEMA L22		24	.333		30	14		44	54.50	
	5660	347/600 volt NEMA L23		24	.333		30	14		44	54.50	
	5670	30 amp, 125 volt NEMA L5		13	.615		20	26		46	62	
	5680	250 volt NEMA L6		13	.615		20	26		46	62	
	5690	277 volt NEMA L7		13	.615		19.75	26		45.75	61.50	
	5700	480 volt NEMA L8		13	.615		24	26		50	66	
	5710	600 volt NEMA L9		13	.615		23	26		49	65.50	
	5720	3 pole, 125/250 volt NEMA L10		11	.727		19.05	30.50		49.55	68	
	5730	250 volt NEMA L11		11	.727		19.05	30.50		49.55	68	
	5760	125/250 volt NEMA L14		11	.727		27	30.50		57.50	76.50	
	5770	250 volt NEMA L15		11	.727		27	30.50		57.50	76.50	
	5780	480 volt NEMA L16		11	.727		31.50	30.50		62	81.50	
	5790	600 volt NEMA L17		11	.727		32	30.50		62.50	82	
	5800	4 pole, 120/208 volt NEMA L18		10	.800		39.50	33.50		73	95	
	5810	120/208 volt NEMA L19		10	.800		39.50	33.50		73	95	
	5820	347/600 volt NEMA L20		10	.800		39.50	33.50		73	95	
	5830	120/208 volt NEMA L21		10	.800		30	33.50		63.50	84.50	
	5840	277/480 volt NEMA L22		10	.800		35.50	33.50		69	91.50	
	5850	347/600 volt NEMA L23		10	.800		44.50	33.50		78	101	
	6000	Connector, 20 amp, 250 volt NEMA 6		30	.267		22	11.20		33.20	42	
	6010	277 volt NEMA 7		30	.267		25	11.20		36.20	45	
	6020	3 pole, 120/250 volt NEMA 10		26	.308		28	12.90		40.90	51	
	6030	125/250 volt NEMA 14		26	.308		28	12.90		40.90	51	
	6040	250 volt NEMA 15		26	.308		28	12.90		40.90	51	
	6050	120/208 volt NEMA 18		26	.308		31	12.90		43.90	54	
	6060	30 amp, 125 volt NEMA 5		13	.615		48.50	26		74.50	93.50	
	6070	250 volt NEMA 6		13	.615		48.50	26		74.50	93.50	
	6080	277 volt NEMA 7		13	.615		48.50	26		74.50	93.50	
	6110	50 amp, 125 volt NEMA 5		9	.889		67	37.50		104.50	132	
	6120	250 volt NEMA 6		9	.889		67	37.50		104.50	132	
	6130	277 volt NEMA 7		9	.889		67	37.50		104.50	132	
	6200	Connector, locking, 20 amp, 125 volt NEMA L5		30	.267		20	11.20		31.20	39.50	
	6210	250 volt NEMA L6		30	.267		20	11.20		31.20	39.50	
	6220	277 volt NEMA L7		30	.267		20	11.20		31.20	39.50	
	6230	480 volt NEMA L8		30	.267		30	11.20		41.20	50.50	

		16140 | Wiring Devices	CREW	DAILY OUTPUT	LABOR-HOURS	UNIT	MAT.	LABOR	EQUIP.	TOTAL	TOTAL INCL O&P	
910	6240	600 volt NEMA L9 (R262726-90)	1 Elec	30	.267	Ea.	30	11.20		41.20	50.50	910
	6250	3 pole, 125/250 volt NEMA L10		26	.308		27.50	12.90		40.40	50	
	6260	250 volt NEMA L11		26	.308		27.50	12.90		40.40	50	
	6280	125/250 volt NEMA L14		26	.308		27.50	12.90		40.40	50	
	6290	250 volt NEMA L15		26	.308		27.50	12.90		40.40	50	
	6300	480 volt NEMA L16		26	.308		33	12.90		45.90	56.50	
	6310	4 pole, 120/208 volt NEMA L18		24	.333		43.50	14		57.50	69	
	6320	277/480 volt NEMA L19		24	.333		43.50	14		57.50	69	
	6330	347/600 volt NEMA L20		24	.333		43.50	14		57.50	69.50	
	6340	120/208 volt NEMA L21		24	.333		43	14		57	69	
	6350	277/480 volt NEMA L22		24	.333		52	14		66	78.50	
	6360	347/600 volt NEMA L23		24	.333		52	14		66	79	
	6370	30 amp, 125 volt NEMA L5		13	.615		40	26		66	84	
	6380	250 volt NEMA L6		13	.615		40	26		66	84	
	6390	277 volt NEMA L7		13	.615		40	26		66	84	
	6400	480 volt NEMA L8		13	.615		47.50	26		73.50	92.50	
	6410	600 volt NEMA L9		13	.615		48	26		74	92.50	
	6420	3 pole, 125/250 volt NEMA L10		11	.727		53	30.50		83.50	106	
	6430	250 volt NEMA L11		11	.727		53	30.50		83.50	106	
	6460	125/250 volt NEMA L14		11	.727		55.50	30.50		86	108	
	6470	250 volt NEMA L15		11	.727		55.50	30.50		86	108	
	6480	480 volt NEMA L16		11	.727		66.50	30.50		97	120	
	6490	600 volt NEMA L17		11	.727		66.50	30.50		97	121	
	6500	4 pole, 120/208 volt NEMA L18		10	.800		83	33.50		116.50	144	
	6510	120/208 volt NEMA L19		10	.800		83	33.50		116.50	144	
	6520	347/600 volt NEMA L20		10	.800		83	33.50		116.50	144	
	6530	120/208 volt NEMA L21		10	.800		55	33.50		88.50	113	
	6540	277/480 volt NEMA L22		10	.800		65.50	33.50		99	124	
	6550	347/600 volt NEMA L23		10	.800		81.50	33.50		115	142	
	7000	Receptacle computer, 250 volt, 15 amp, 3 pole 4 wire		8	1		63	42		105	134	
	7010	20 amp, 2 pole 3 wire		8	1		63	42		105	134	
	7020	30 amp, 2 pole 3 wire		6.50	1.231		99	51.50		150.50	189	
	7030	30 amp, 3 pole 4 wire		6.50	1.231		105	51.50		156.50	196	
	7040	60 amp, 3 pole 4 wire		4.50	1.778		177	74.50		251.50	310	
	7050	100 amp, 3 pole 4 wire		3	2.667		224	112		336	420	
	7100	Connector computer, 250 volt, 15 amp, 3 pole 4 wire		27	.296		87.50	12.45		99.95	115	
	7110	20 amp, 2 pole 3 wire		27	.296		87.50	12.45		99.95	115	
	7120	30 amp, 2 pole 3 wire		15	.533		139	22.50		161.50	187	
	7130	30 amp, 3 pole 4 wire		15	.533		131	22.50		153.50	179	
	7140	60 amp, 3 pole 4 wire		8	1		223	42		265	310	
	7150	100 amp, 3 pole 4 wire		4	2		310	84		394	475	
	7200	Plug, computer, 250 volt, 15 amp, 3 pole 4 wire		27	.296		79	12.45		91.45	106	
	7210	20 amp, 2 pole, 3 wire		27	.296		79	12.45		91.45	106	
	7220	30 amp, 2 pole, 3 wire		15	.533		116	22.50		138.50	163	
	7230	30 amp, 3 pole, 4 wire		15	.533		125	22.50		147.50	173	
	7240	60 amp, 3 pole, 4 wire		8	1		187	42		229	271	
	7250	100 amp, 3 pole, 4 wire		4	2		243	84		327	395	
	7300	Connector adapter to flexible conduit, 1/2"		60	.133		3.80	5.60		9.40	12.85	
	7310	3/4"		50	.160		5.50	6.70		12.20	16.40	
	7320	1-1/4"		30	.267		13.85	11.20		25.05	32.50	
	7330	1-1/2"		23	.348	↓	19.80	14.60		34.40	44.50	
	9000	Minimum labor/equipment charge	↓	4	2	Job		84		84	129	

		16150 | Wiring Connections	CREW	DAILY OUTPUT	LABOR-HOURS	UNIT	MAT.	LABOR	EQUIP.	TOTAL	TOTAL INCL O&P	
275	0010	MOTOR CONNECTIONS										275
	0020	Flexible conduit and fittings, 115 volt, 1 phase, up to 1 HP motor	1 Elec	8	1	Ea.	4.97	42		46.97	70	

ELECTRICAL 16

16100 | Wiring Methods

16150	Wiring Connections		Crew	Daily Output	Labor-Hours	Unit	2006 Bare Costs Mat.	Labor	Equip.	Total	Total Incl O&P
275	0050	2 HP motor	1 Elec	6.50	1.231	Ea.	5.10	51.50		56.60	85
	0100	3 HP motor		5.50	1.455		8.50	61		69.50	103
	0120	230 volt, 10 HP motor, 3 phase		4.20	1.905		9.15	80		89.15	133
	0150	15 HP motor		3.30	2.424		17.55	102		119.55	176
	0200	25 HP motor		2.70	2.963		19.80	124		143.80	214
	0400	50 HP motor		2.20	3.636		55	153		208	296
	0600	100 HP motor		1.50	5.333		141	224		365	500
	1500	460 volt, 5 HP motor, 3 phase		8	1		5.55	42		47.55	70.50
	1520	10 HP motor		8	1		5.55	42		47.55	70.50
	1530	25 HP motor		6	1.333		10.50	56		66.50	98
	1540	30 HP motor		6	1.333		10.50	56		66.50	98
	1550	40 HP motor		5	1.600		16.45	67		83.45	122
	1560	50 HP motor		5	1.600		18.50	67		85.50	125
	1570	60 HP motor		3.80	2.105		25	88.50		113.50	164
	1580	75 HP motor		3.50	2.286		33	96		129	185
	1590	100 HP motor		2.50	3.200		48	134		182	260
	1600	125 HP motor		2	4		61.50	168		229.50	325
	1610	150 HP motor		1.80	4.444		64	187		251	360
	1620	200 HP motor		1.50	5.333		111	224		335	465
	2005	460 Volt, 5 HP motor, 3 Phase, w/sealtite		8	1		10.75	42		52.75	76.50
	2010	10 HP motor		8	1		10.75	42		52.75	76.50
	2015	25 HP motor		6	1.333		19.15	56		75.15	108
	2020	30 HP motor		6	1.333		19.15	56		75.15	108
	2025	40 HP motor		5	1.600		36.50	67		103.50	144
	2030	50 HP motor		5	1.600		34.50	67		101.50	142
	2035	60 HP motor		3.80	2.105		52.50	88.50		141	194
	2040	75 HP motor		3.50	2.286		56.50	96		152.50	210
	2045	100 HP motor		2.50	3.200		74	134		208	289
	2055	150 HP motor		1.80	4.444		78.50	187		265.50	375
	2060	200 HP motor		1.50	5.333	↓	515	224		739	915
	9000	Minimum labor/equipment charge	↓	4	2	Job		84		84	129

16200 | Electrical Power

16210	Electrical Utility Services		Crew	Daily Output	Labor-Hours	Unit	2006 Bare Costs Mat.	Labor	Equip.	Total	Total Incl O&P
600	0010	**METER CENTERS AND SOCKETS**									
	0100	Sockets, single position, 4 terminal, 100 amp	1 Elec	3.20	2.500	Ea.	36	105		141	202
	0200	150 amp		2.30	3.478		40	146		186	270
	0300	200 amp		1.90	4.211		53.50	177		230.50	330
	0400	20 amp		3.20	2.500		83	105		188	253
	0500	Double position, 4 terminal, 100 amp		2.80	2.857		141	120		261	340
	0600	150 amp		2.10	3.810		160	160		320	420
	0700	200 amp		1.70	4.706		345	198		543	685
	0800	Trans-socket, 13 terminal, 3 CT mounts, 400 amp	↓	1	8		860	335		1,195	1,475
	0900	800 amp	2 Elec	1.20	13.333	↓	895	560		1,455	1,850
	2000	Meter center, main fusible switch, 1P 3W 120/240 volt									
	2030	400 amp	2 Elec	1.60	10	Ea.	1,400	420		1,820	2,175
	2040	600 amp		1.10	14.545		2,425	610		3,035	3,625
	2050	800 amp	↓	.90	17.778	↓	3,800	745		4,545	5,350

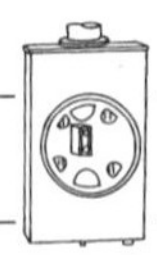

16210 | Electrical Utility Services

				DAILY	LABOR-		2006 BARE COSTS				TOTAL	
			CREW	OUTPUT	HOURS	UNIT	MAT.	LABOR	EQUIP.	TOTAL	INCL O&P	
600	2060	Rainproof 1P 3W 120/240 volt, 400 amp	2 Elec	1.60	10	Ea.	1,400	420		1,820	2,175	600
	2070	600 amp		1.10	14.545		2,425	610		3,035	3,625	
	2080	800 amp		.90	17.778		3,800	745		4,545	5,350	
	2100	3P 4W 120/208V, 400 amp		1.60	10		1,600	420		2,020	2,400	
	2110	600 amp		1.10	14.545		2,975	610		3,585	4,225	
	2120	800 amp		.90	17.778		5,550	745		6,295	7,250	
	2130	Rainproof 3P 4W 120/208V, 400 amp		1.60	10		1,600	420		2,020	2,400	
	2140	600 amp		1.10	14.545		2,975	610		3,585	4,225	
	2150	800 amp		.90	17.778		5,550	745		6,295	7,250	
	2170	Main circuit breaker, 1P 3W 120/240V										
	2180	400 amp	2 Elec	1.60	10	Ea.	2,500	420		2,920	3,400	
	2190	600 amp		1.10	14.545		3,375	610		3,985	4,650	
	2200	800 amp		.90	17.778		3,925	745		4,670	5,475	
	2210	1000 amp		.80	20		5,400	840		6,240	7,250	
	2220	1200 amp		.76	21.053		7,300	885		8,185	9,375	
	2230	1600 amp		.68	23.529		12,500	990		13,490	15,200	
	2240	Rainproof 1P 3W 120/240V, 400 amp		1.60	10		2,500	420		2,920	3,400	
	2250	600 amp		1.10	14.545		3,375	610		3,985	4,650	
	2260	800 amp		.90	17.778		3,925	745		4,670	5,475	
	2270	1000 amp		.80	20		5,400	840		6,240	7,250	
	2280	1200 amp		.76	21.053		7,300	885		8,185	9,375	
	2300	3P 4W 120/208V, 400 amp		1.60	10		2,850	420		3,270	3,775	
	2310	600 amp		1.10	14.545		4,000	610		4,610	5,350	
	2320	800 amp		.90	17.778		4,775	745		5,520	6,400	
	2330	1000 amp		.80	20		6,250	840		7,090	8,175	
	2340	1200 amp		.76	21.053		8,000	885		8,885	10,200	
	2350	1600 amp		.68	23.529		12,500	990		13,490	15,200	
	2360	Rainproof 3P 4W 120/208V, 400 amp		1.60	10		2,850	420		3,270	3,775	
	2370	600 amp		1.10	14.545		4,000	610		4,610	5,350	
	2380	800 amp		.90	17.778		4,775	745		5,520	6,400	
	2390	1000 amp		.76	21.053		6,250	885		7,135	8,225	
	2400	1200 amp		.68	23.529		8,000	990		8,990	10,300	
	2420	Main lugs terminal box, 1P 3W 120/240V										
	2430	800 amp	2 Elec	.94	17.021	Ea.	470	715		1,185	1,625	
	2440	1200 amp		.72	22.222		940	935		1,875	2,475	
	2450	Rainproof 1P 3W 120/240V, 225 amp		2.40	6.667		320	280		600	780	
	2460	800 amp		.94	17.021		470	715		1,185	1,625	
	2470	1200 amp		.72	22.222		940	935		1,875	2,475	
	2500	3P 4W 120/208V, 800 amp		.94	17.021		520	715		1,235	1,675	
	2510	1200 amp		.72	22.222		1,050	935		1,985	2,600	
	2520	Rainproof 3P 4W 120/208V, 225 amp		2.40	6.667		320	280		600	780	
	2530	800 amp		.94	17.021		520	715		1,235	1,675	
	2540	1200 amp		.72	22.222		1,050	935		1,985	2,600	
	2590	Basic meter device										
	2600	1P 3W 120/240V 4 jaw 125A sockets, 3 meter	2 Elec	1	16	Ea.	535	670		1,205	1,625	
	2610	4 meter		.90	17.778		645	745		1,390	1,850	
	2620	5 meter		.80	20		805	840		1,645	2,175	
	2630	6 meter		.60	26.667		925	1,125		2,050	2,750	
	2640	7 meter		.56	28.571		1,175	1,200		2,375	3,150	
	2650	8 meter		.52	30.769		1,275	1,300		2,575	3,425	
	2660	10 meter		.48	33.333		1,600	1,400		3,000	3,925	
	2680	Rainproof 1P 3W 120/240V 4 jaw 125A sockets										
	2690	3 meter	2 Elec	1	16	Ea.	535	670		1,205	1,625	
	2700	4 meter		.90	17.778		645	745		1,390	1,850	
	2710	6 meter		.60	26.667		925	1,125		2,050	2,750	
	2720	7 meter		.56	28.571		1,175	1,200		2,375	3,150	

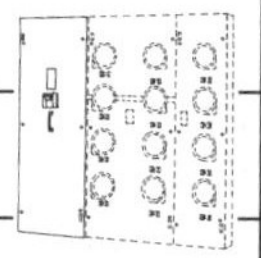

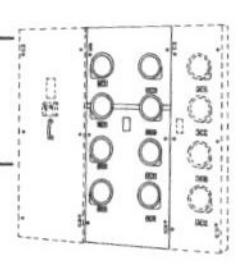

16210 | Electrical Utility Services

	16210 Electrical Utility Services	CREW	DAILY OUTPUT	LABOR-HOURS	UNIT	2006 BARE COSTS MAT.	LABOR	EQUIP.	TOTAL	TOTAL INCL O&P
600 2730	8 meter	2 Elec	.52	30.769	Ea.	1,275	1,300		2,575	3,425 600
2750	1P 3W 120/240V 4 jaw sockets									
2760	with 125A circuit breaker, 3 meter	2 Elec	1	16	Ea.	1,000	670		1,670	2,125
2770	4 meter		.90	17.778		1,275	745		2,020	2,550
2780	5 meter		.80	20		1,575	840		2,415	3,050
2790	6 meter		.60	26.667		1,875	1,125		3,000	3,775
2800	7 meter		.56	28.571		2,275	1,200		3,475	4,350
2810	8 meter		.52	30.769		2,525	1,300		3,825	4,800
2820	10 meter		.48	33.333		3,175	1,400		4,575	5,650
2830	Rainproof 1P 3W 120/240V 4 jaw sockets									
2840	with 125A circuit breaker, 3 meter	2 Elec	1	16	Ea.	1,000	670		1,670	2,125
2850	4 meter		.90	17.778		1,275	745		2,020	2,550
2870	6 meter		.60	26.667		1,875	1,125		3,000	3,775
2880	7 meter		.56	28.571		2,275	1,200		3,475	4,350
2890	8 meter		.52	30.769		2,525	1,300		3,825	4,800
2920	1P 3W on 3P 4W 120/208V system 5 jaw									
2930	125A sockets, 3 meter	2 Elec	1	16	Ea.	535	670		1,205	1,625
2940	4 meter		.90	17.778		645	745		1,390	1,850
2950	5 meter		.80	20		805	840		1,645	2,175
2960	6 meter		.60	26.667		925	1,125		2,050	2,750
2970	7 meter		.56	28.571		1,175	1,200		2,375	3,150
2980	8 meter		.52	30.769		1,275	1,300		2,575	3,425
2990	10 meter		.48	33.333		1,600	1,400		3,000	3,925
3000	Rainproof 1P 3W on 3P 4W 120/208V system									
3020	5 jaw 125A sockets, 3 meter	2 Elec	1	16	Ea.	535	670		1,205	1,625
3030	4 meter		.90	17.778		645	745		1,390	1,850
3050	6 meter		.60	26.667		925	1,125		2,050	2,750
3060	7 meter		.56	28.571		1,175	1,200		2,375	3,150
3070	8 meter		.52	30.769		1,275	1,300		2,575	3,425
3090	1P 3W on 3P 4W 120/208V system 5 jaw sockets									
3100	With 125A circuit breaker, 3 meter	2 Elec	1	16	Ea.	1,000	670		1,670	2,125
3110	4 meter		.90	17.778		1,275	745		2,020	2,550
3120	5 meter		.80	20		1,575	840		2,415	3,050
3130	6 meter		.60	26.667		1,875	1,125		3,000	3,775
3140	7 meter		.56	28.571		2,275	1,200		3,475	4,350
3150	8 meter		.52	30.769		2,525	1,300		3,825	4,800
3160	10 meter		.48	33.333		3,175	1,400		4,575	5,650
3170	Rainproof 1P 3W on 3P 4W 120/208V system									
3180	5 jaw sockets w/125A circuit breaker, 3 meter	2 Elec	1	16	Ea.	1,000	670		1,670	2,125
3190	4 meter		.90	17.778		1,275	745		2,020	2,550
3210	6 meter		.60	26.667		1,875	1,125		3,000	3,775
3220	7 meter		.56	28.571		2,275	1,200		3,475	4,350
3230	8 meter		.52	30.769		2,525	1,300		3,825	4,800
3250	1P 3W 120/240V 4 jaw sockets									
3260	with 200A circuit breaker, 3 meter	2 Elec	1	16	Ea.	1,500	670		2,170	2,675
3270	4 meter		.90	17.778		2,050	745		2,795	3,400
3290	6 meter		.60	26.667		3,025	1,125		4,150	5,050
3300	7 meter		.56	28.571		3,550	1,200		4,750	5,750
3310	8 meter		.56	28.571		4,075	1,200		5,275	6,325
3330	Rainproof 1P 3W 120/240V 4 jaw sockets									
3350	with 200A circuit breaker, 3 meter	2 Elec	1	16	Ea.	1,500	670		2,170	2,675
3360	4 meter		.90	17.778		2,050	745		2,795	3,400
3380	6 meter		.60	26.667		3,025	1,125		4,150	5,050
3390	7 meter		.56	28.571		3,550	1,200		4,750	5,750
3400	8 meter		.52	30.769		4,075	1,300		5,375	6,475
3420	1P 3W on 3P 4W 120/208V 5 jaw sockets									

16210 | Electrical Utility Services

		16210 Electrical Utility Services	CREW	DAILY OUTPUT	LABOR-HOURS	UNIT	2006 BARE COSTS MAT.	LABOR	EQUIP.	TOTAL	TOTAL INCL O&P	
600	3430	with 200A circuit breaker, 3 meter	2 Elec	1	16	Ea.	1,500	670		2,170	2,675	600
	3440	4 meter		.90	17.778		2,050	745		2,795	3,400	
	3460	6 meter		.60	26.667		3,025	1,125		4,150	5,050	
	3470	7 meter		.56	28.571		3,550	1,200		4,750	5,750	
	3480	8 meter		.52	30.769		4,075	1,300		5,375	6,475	
	3500	Rainproof 1P 3W on 3P 4W 120/208V 5 jaw socket										
	3510	with 200A circuit breaker, 3 meter	2 Elec	1	16	Ea.	1,500	670		2,170	2,675	
	3520	4 meter		.90	17.778		2,050	745		2,795	3,400	
	3540	6 meter		.60	26.667		3,025	1,125		4,150	5,050	
	3550	7 meter		.56	28.571		3,550	1,200		4,750	5,750	
	3560	8 meter		.52	30.769		4,075	1,300		5,375	6,475	
	3600	Automatic circuit closing, add					58.50			58.50	64	
	3610	Manual circuit closing, add					66.50			66.50	73	
	3650	Branch meter device										
	3660	3P 4W 208/120 or 240/120 V 7 jaw sockets										
	3670	with 200A circuit breaker, 2 meter	2 Elec	.90	17.778	Ea.	2,525	745		3,270	3,925	
	3680	3 meter		.80	20		3,775	840		4,615	5,450	
	3690	4 meter		.70	22.857		5,050	960		6,010	7,025	
	3700	Main circuit breaker 42,000 rms, 400 amp		1.60	10		2,200	420		2,620	3,075	
	3710	600 amp		1.10	14.545		3,550	610		4,160	4,850	
	3720	800 amp		.90	17.778		4,775	745		5,520	6,400	
	3730	Rainproof main circ. breaker 42,000 rms, 400 amp		1.60	10		2,200	420		2,620	3,075	
	3740	600 amp		1.10	14.545		3,550	610		4,160	4,850	
	3750	800 amp		.90	17.778		4,775	745		5,520	6,400	
	3760	Main circuit breaker 65,000 rms, 400 amp		1.60	10		3,000	420		3,420	3,950	
	3770	600 amp		1.10	14.545		4,200	610		4,810	5,575	
	3780	800 amp		.90	17.778		4,775	745		5,520	6,400	
	3790	1000 amp		.80	20		6,250	840		7,090	8,175	
	3800	1200 amp		.76	21.053		8,000	885		8,885	10,200	
	3810	1600 amp		.68	23.529		12,500	990		13,490	15,200	
	3820	Rainproof main circ. breaker 65,000 rms, 400 amp		1.60	10		3,000	420		3,420	3,950	
	3830	600 amp		1.10	14.545		4,200	610		4,810	5,575	
	3840	800 amp		.90	17.778		4,750	745		5,495	6,375	
	3850	1000 amp		.80	20		6,250	840		7,090	8,175	
	3860	1200 amp		.76	21.053		8,000	885		8,885	10,200	
	3880	Main circuit breaker 100,000 rms, 400 amp		1.60	10		3,000	420		3,420	3,950	
	3890	600 amp		1.10	14.545		4,200	610		4,810	5,575	
	3900	800 amp		.90	17.778		4,950	745		5,695	6,575	
	3910	Rainproof main circ. breaker 100,000 rms, 400 amp		1.60	10		3,000	420		3,420	3,950	
	3920	600 amp		1.10	14.545		4,200	610		4,810	5,575	
	3930	800 amp		.90	17.778		4,950	745		5,695	6,575	
	3940	Main lugs terminal box, 800 amp		.94	17.021		520	715		1,235	1,675	
	3950	1600 amp		.72	22.222		1,675	935		2,610	3,300	
	3960	Rainproof, 800 amp		.94	17.021		520	715		1,235	1,675	
	3970	1600 amp		.72	22.222		1,675	935		2,610	3,300	
	9000	Minimum labor/equipment charge	1 Elec	3	2.667	Job		112		112	173	

16220 | Motors & Generators

		16220 Motors & Generators	CREW	DAILY OUTPUT	LABOR-HOURS	UNIT	MAT.	LABOR	EQUIP.	TOTAL	TOTAL INCL O&P	
600	0010	**HANDLING** Add to normal labor cost for restricted areas										600
	5000	Motors										
	5100	1/2 HP, 23 pounds	1 Elec	4	2	Ea.		84		84	129	
	5110	3/4 HP, 28 pounds		4	2			84		84	129	
	5120	1 HP, 33 pounds		4	2			84		84	129	
	5130	1-1/2 HP, 44 pounds		3.20	2.500			105		105	162	
	5140	2 HP, 56 pounds	1 Elec	3	2.667			112		112	173	
	5150	3 HP, 71 pounds		2.30	3.478			146		146	225	

For expanded coverage of these items see *Means Electrical Cost Data 2006*

	16220 | Motors & Generators		Crew	Daily Output	Labor-Hours	Unit	2006 Bare Costs Mat.	Labor	Equip.	Total	Total Incl O&P	
600	5160	5 HP, 82 pounds	1 Elec	1.90	4.211	Ea.		177		177	272	600
	5170	7-1/2 HP, 124 pounds		1.50	5.333			224		224	345	
	5180	10 HP, 144 pounds		1.20	6.667			280		280	430	
	5190	15 HP, 185 pounds		1	8			335		335	520	
	5200	20 HP, 214 pounds	2 Elec	1.50	10.667			450		450	690	
	5210	25 HP, 266 pounds		1.40	11.429			480		480	740	
	5220	30 HP, 310 pounds		1.20	13.333			560		560	865	
	5230	40 HP, 400 pounds		1	16			670		670	1,025	
	5240	50 HP, 450 pounds		.90	17.778			745		745	1,150	
	5250	75 HP, 680 pounds		.80	20			840		840	1,300	
	5260	100 HP, 870 pounds	3 Elec	1	24			1,000		1,000	1,550	
	5270	125 HP, 940 pounds		.80	30			1,250		1,250	1,950	
	5280	150 HP, 1200 pounds		.70	34.286			1,450		1,450	2,225	
	5290	175 HP, 1300 pounds		.60	40			1,675		1,675	2,600	
	5300	200 HP, 1400 pounds		.50	48			2,025		2,025	3,100	
610	0010	**MOTORS** 230/460 volts, 60 HZ										610
	0050	Dripproof, premium efficiency, 1.15 service factor										
	0060	1800 RPM, 1/4 HP	1 Elec	5.33	1.501	Ea.	134	63		197	244	
	0070	1/3 HP		5.33	1.501		146	63		209	258	
	0080	1/2 HP		5.33	1.501		172	63		235	286	
	0090	3/4 HP		5.33	1.501		192	63		255	310	
	0100	1 HP		4.50	1.778		201	74.50		275.50	335	
	0150	2 HP		4.50	1.778		238	74.50		312.50	375	
	0200	3 HP		4.50	1.778		238	74.50		312.50	375	
	0250	5 HP		4.50	1.778		305	74.50		379.50	455	
	0300	7.5 HP		4.20	1.905		445	80		525	615	
	0350	10 HP		4	2		550	84		634	735	
	0400	15 HP		3.20	2.500		700	105		805	930	
	0450	20 HP	2 Elec	5.20	3.077		875	129		1,004	1,150	
	0500	25 HP		5	3.200		1,025	134		1,159	1,325	
	0550	30 HP		4.80	3.333		1,175	140		1,315	1,525	
	0600	40 HP		4	4		1,700	168		1,868	2,125	
	0650	50 HP		3.20	5		1,925	210		2,135	2,450	
	0700	60 HP		2.80	5.714		2,525	240		2,765	3,150	
	0750	75 HP		2.40	6.667		2,875	280		3,155	3,575	
	0800	100 HP	3 Elec	2.70	8.889		3,775	375		4,150	4,725	
	0850	125 HP		2.10	11.429		4,375	480		4,855	5,575	
	0900	150 HP		1.80	13.333		5,700	560		6,260	7,150	
	0950	200 HP		1.50	16		7,350	670		8,020	9,125	
	1000	1200 RPM, 1 HP	1 Elec	4.50	1.778		235	74.50		309.50	375	
	1050	2 HP		4.50	1.778		330	74.50		404.50	480	
	1100	3 HP		4.50	1.778		445	74.50		519.50	600	
	1150	5 HP		4.50	1.778		530	74.50		604.50	700	
	1200	3600 RPM, 2 HP		4.50	1.778		249	74.50		323.50	390	
	1250	3 HP		4.50	1.778		285	74.50		359.50	430	
	1300	5 HP		4.50	1.778		305	74.50		379.50	455	
	1350	Totally enclosed, premium efficiency 1.15 service factor										
	1360	1800 RPM, 1/4 HP	1 Elec	5.33	1.501	Ea.	135	63		198	246	
	1370	1/3 HP		5.33	1.501		151	63		214	263	
	1380	1/2 HP		5.33	1.501		178	63		241	293	
	1390	3/4 HP		5.33	1.501		192	63		255	310	
	1400	1 HP		4.50	1.778		233	74.50		307.50	370	
	1450	2 HP		4.50	1.778		281	74.50		355.50	425	
	1500	3 HP		4.50	1.778		310	74.50		384.50	455	
	1550	5 HP		4.50	1.778		355	74.50		429.50	505	

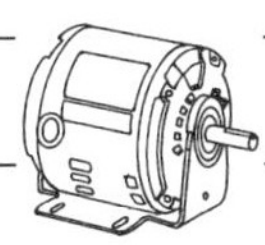

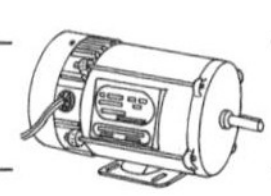

16 ELECTRICAL

16220 | Motors & Generators

			CREW	DAILY OUTPUT	LABOR-HOURS	UNIT	2006 BARE COSTS MAT.	LABOR	EQUIP.	TOTAL	TOTAL INCL O&P
610	1600	7.5 HP	1 Elec	4.20	1.905	Ea.	510	80		590	685
	1650	10 HP		4	2		615	84		699	805
	1700	15 HP		3.20	2.500		820	105		925	1,075
	1750	20 HP	2 Elec	5.20	3.077		1,000	129		1,129	1,300
	1800	25 HP		5	3.200		1,225	134		1,359	1,550
	1850	30 HP		4.80	3.333		1,425	140		1,565	1,800
	1900	40 HP		4	4		1,850	168		2,018	2,275
	1950	50 HP		3.20	5		2,275	210		2,485	2,825
	2000	60 HP		2.80	5.714		3,400	240		3,640	4,125
	2050	75 HP		2.40	6.667		4,350	280		4,630	5,200
	2100	100 HP	3 Elec	2.70	8.889		5,175	375		5,550	6,250
	2150	125 HP		2.10	11.429		7,500	480		7,980	9,000
	2200	150 HP		1.80	13.333		8,500	560		9,060	10,200
	2250	200 HP		1.50	16		10,200	670		10,870	12,200
	2300	1200 RPM, 1 HP	1 Elec	4.50	1.778		272	74.50		346.50	415
	2350	2 HP		4.50	1.778		330	74.50		404.50	480
	2400	3 HP		4.50	1.778		440	74.50		514.50	600
	2450	5 HP		4.50	1.778		650	74.50		724.50	830
	2500	3600 RPM, 2 HP		4.50	1.778		269	74.50		343.50	410
	2550	3 HP		4.50	1.778		315	74.50		389.50	460
	2600	5 HP		4.50	1.778		390	74.50		464.50	545
900	0010	**VARIABLE FREQUENCY DRIVES/ADJUSTABLE FREQUENCY DRIVES**									
	0100	Enclosed (NEMA 1), 460 volt, for 3 HP motor size	1 Elec	.80	10	Ea.	1,500	420		1,920	2,300
	0110	5 HP motor size		.80	10		1,625	420		2,045	2,450
	0120	7.5 HP motor size		.67	11.940		1,925	500		2,425	2,900
	0130	10 HP motor size		.67	11.940		1,925	500		2,425	2,900
	0140	15 HP motor size	2 Elec	.89	17.978		2,225	755		2,980	3,625
	0150	20 HP motor size		.89	17.978		3,300	755		4,055	4,800
	0160	25 HP motor size		.67	23.881		3,825	1,000		4,825	5,775
	0170	30 HP motor size		.67	23.881		4,675	1,000		5,675	6,700
	0180	40 HP motor size		.67	23.881		6,875	1,000		7,875	9,100
	0190	50 HP motor size		.53	30.189		7,475	1,275		8,750	10,200
	0200	60 HP motor size	R-3	.56	35.714		8,175	1,475	273	9,923	11,600
	0210	75 HP motor size		.56	35.714		11,200	1,475	273	12,948	14,900
	0220	100 HP motor size		.50	40		11,400	1,650	305	13,355	15,500
	0230	125 HP motor size		.50	40		12,700	1,650	305	14,655	16,800
	0240	150 HP motor size		.50	40		16,200	1,650	305	18,155	20,700
	0250	200 HP motor size		.42	47.619		20,000	1,975	365	22,340	25,500
	1100	Custom-engineered, 460 volt, for 3 HP motor size	1 Elec	.56	14.286		2,475	600		3,075	3,650
	1110	5 HP motor size		.56	14.286		2,600	600		3,200	3,775
	1120	7.5 HP motor size		.47	17.021		3,125	715		3,840	4,550
	1130	10 HP motor size		.47	17.021		3,125	715		3,840	4,550
	1140	15 HP motor size	2 Elec	.62	25.806		3,425	1,075		4,500	5,450
	1150	20 HP motor size		.62	25.806		4,675	1,075		5,750	6,825
	1160	25 HP motor size		.47	34.043		5,075	1,425		6,500	7,775
	1170	30 HP motor size		.47	34.043		6,925	1,425		8,350	9,800
	1180	40 HP motor size		.47	34.043		7,725	1,425		9,150	10,700
	1190	50 HP motor size		.37	43.243		7,725	1,825		9,550	11,300
	1200	60 HP motor size	R-3	.39	51.282		11,500	2,125	390	14,015	16,400
	1210	75 HP motor size		.39	51.282		13,300	2,125	390	15,815	18,400
	1220	100 HP motor size		.35	57.143		13,900	2,375	435	16,710	19,500
	1230	125 HP motor size		.35	57.143		14,700	2,375	435	17,510	20,300
	1240	150 HP motor size		.35	57.143		16,400	2,375	435	19,210	22,300
	1250	200 HP motor size		.29	68.966		22,700	2,850	525	26,075	30,000
	2000	For complex & special design systems to meet specific									

ELECTRICAL 16

	16220 | Motors & Generators		CREW	DAILY OUTPUT	LABOR-HOURS	UNIT	2006 BARE COSTS MAT.	LABOR	EQUIP.	TOTAL	TOTAL INCL O&P	
900	2010	requirements, obtain quote from vendor.										900
	16230 | Generator Assemblies											
450	0010	**GENERATOR SET**										450
	0020	Gas or gasoline operated, includes battery,										
	0050	charger, muffler & transfer switch										
	0200	3 phase 4 wire, 277/480 volt, 7.5 kW	R-3	.83	24.096	Ea.	6,000	1,000	184	7,184	8,350	
	0300	11.5 kW		.71	28.169		8,500	1,175	215	9,890	11,400	
	0400	20 kW		.63	31.746		10,000	1,325	242	11,567	13,300	
	0500	35 kW		.55	36.364		12,000	1,500	277	13,777	15,800	
	0520	60 kW		.50	40		16,500	1,650	305	18,455	21,100	
	0600	80 kW		.40	50		20,500	2,075	380	22,955	26,200	
	0700	100 kW		.33	60.606		22,500	2,500	460	25,460	29,200	
	0800	125 kW		.28	71.429		46,000	2,950	545	49,495	55,500	
	0900	185 kW	↓	.25	80	↓	61,000	3,325	610	64,935	73,000	
	2000	Diesel engine, including battery, charger,										
	2010	muffler, automatic transfer switch & day tank, 30 kW	R-3	.55	36.364	Ea.	16,000	1,500	277	17,777	20,200	
	2100	50 kW		.42	47.619		19,700	1,975	365	22,040	25,200	
	2110	60 kW		.39	51.282		21,300	2,125	390	23,815	27,100	
	2200	75 kW		.35	57.143		25,700	2,375	435	28,510	32,500	
	2300	100 kW		.31	64.516		28,500	2,675	490	31,665	36,100	
	2400	125 kW		.29	68.966		30,000	2,850	525	33,375	38,000	
	2500	150 kW		.26	76.923		34,400	3,175	585	38,160	43,400	
	2600	175 kW		.25	80		37,500	3,325	610	41,435	47,100	
	2700	200 kW		.24	83.333		38,700	3,450	635	42,785	48,700	
	2800	250 kW		.23	86.957		45,600	3,600	665	49,865	56,500	
	2850	275 kW		.22	90.909		45,600	3,775	695	50,070	56,500	
	2900	300 kW		.22	90.909		49,300	3,775	695	53,770	60,500	
	3000	350 kW		.20	100		55,500	4,150	765	60,415	68,500	
	3100	400 kW		.19	105		69,000	4,350	805	74,155	83,000	
	3200	500 kW		.18	111		86,500	4,600	850	91,950	103,000	
	3220	600 kW	↓	.17	117		113,500	4,875	900	119,275	133,500	
	3230	650 kW	R-13	.38	110		135,000	4,425	445	139,870	156,000	
	3240	750 kW		.38	110		142,500	4,425	445	147,370	164,500	
	3250	800 kW		.36	116		148,000	4,675	470	153,145	171,000	
	3260	900 kW		.31	135		171,000	5,425	545	176,970	197,000	
	3270	1000 kW	↓	.27	155	↓	176,000	6,225	625	182,850	204,500	
	16260 | Static Power Converters											
800	0010	**UNINTERRUPTIBLE POWER SUPPLY/CONDITIONER TRANSFORMERS**										800
	0100	Volt. regulating, isolating trans., w/invert. & 10 min. battery pack										
	0110	Single-phase, 120 V, 0.35 kVA	1 Elec	2.29	3.493	Ea.	920	147		1,067	1,250	
	0120	0.5 kVA		2	4		965	168		1,133	1,300	
	0130	For additional 55 min. battery, add to .35 kVA		2.29	3.493		545	147		692	820	
	0140	Add to 0.5 kVA		1.14	7.018		570	295		865	1,075	
	0150	Single-phase, 120 V, 0.75 kVA		.80	10		1,225	420		1,645	2,000	
	0160	1.0 kVA	↓	.80	10		1,750	420		2,170	2,575	
	0170	1.5 kVA	2 Elec	1.14	14.035		3,000	590		3,590	4,200	
	0180	2 kVA	"	.89	17.978		3,250	755		4,005	4,775	
	0190	3 kVA	R-3	.63	31.746		3,950	1,325	242	5,517	6,650	
	0200	5 kVA		.42	47.619		5,725	1,975	365	8,065	9,750	
	0210	7.5 kVA		.33	60.606		7,325	2,500	460	10,285	12,400	
	0220	10 kVA		.28	71.429		7,775	2,950	545	11,270	13,700	
	0230	15 kVA	↓	.22	90.909	↓	10,800	3,775	695	15,270	18,500	
	0500	For options & accessories add to above, minimum									10%	

16260 | Static Power Converters

	Line	Description	Crew	Daily Output	Labor-Hours	Unit	Mat.	Labor	Equip.	Total	Total Incl O&P	
800	0520	Maximum									35%	800
	0600	For complex & special design systems to meet specific										
	0610	requirements, obtain quote from vendor										

16270 | Transformers

	Line	Description	Crew	Daily Output	Labor-Hours	Unit	Mat.	Labor	Equip.	Total	Total Incl O&P	
100	0010	BUCK-BOOST TRANSFORMER										100
	0100	Single phase, 120/240 V primary, 12/24 V secondary										
	0200	0.10 kVA	1 Elec	8	1	Ea.	73	42		115	145	
	0400	0.25 kVA		5.70	1.404		107	59		166	209	
	0600	0.50 kVA		4	2		146	84		230	290	
	0800	0.75 kVA		3.10	2.581		188	108		296	375	
	1000	1.0 kVA		2	4		234	168		402	515	
	1200	1.5 kVA		1.80	4.444		287	187		474	605	
	1400	2.0 kVA		1.60	5		345	210		555	705	
	1600	3.0 kVA		1.40	5.714		465	240		705	885	
	1800	5.0 kVA		1.20	6.667		615	280		895	1,100	
	2000	3 phase, 240 V primary, 208/120 V secondary, 15 kVA	2 Elec	2.40	6.667		1,900	280		2,180	2,525	
	2200	30 kVA		1.60	10		2,125	420		2,545	2,975	
	2400	45 kVA		1.40	11.429		2,550	480		3,030	3,550	
	2600	75 kVA		1.20	13.333		3,075	560		3,635	4,275	
	2800	112.5 kVA	R-3	1.40	14.286		3,825	590	109	4,524	5,250	
	3000	150 kVA		1.10	18.182		5,100	755	139	5,994	6,925	
	3200	225 kVA		1	20		6,650	830	153	7,633	8,775	
	3400	300 kVA		.90	22.222		9,025	920	170	10,115	11,500	
200	0010	DRY TYPE TRANSFORMER										200
	0050	Single phase, 240/480 volt primary, 120/240 volt secondary										
	0100	1 kVA	1 Elec	2	4	Ea.	219	168		387	500	
	0300	2 kVA		1.60	5		330	210		540	685	
	0500	3 kVA		1.40	5.714		405	240		645	820	
	0700	5 kVA		1.20	6.667		560	280		840	1,050	
	0900	7.5 kVA	2 Elec	2.20	7.273		780	305		1,085	1,325	
	1100	10 kVA		1.60	10		955	420		1,375	1,700	
	1300	15 kVA		1.20	13.333		1,300	560		1,860	2,300	
	1500	25 kVA		1	16		1,775	670		2,445	2,975	
	1700	37.5 kVA		.80	20		2,300	840		3,140	3,825	
	1900	50 kVA		.70	22.857		2,750	960		3,710	4,475	
	2100	75 kVA		.65	24.615		3,625	1,025		4,650	5,575	
	2110	100 kVA	R-3	.90	22.222		4,725	920	170	5,815	6,800	
	2120	167 kVA	"	.80	25		7,825	1,025	191	9,041	10,400	
	2190	480V primary 120/240V secondary, nonvent., 15 kVA	2 Elec	1.20	13.333		1,375	560		1,935	2,400	
	2200	25 kVA		.90	17.778		2,025	745		2,770	3,375	
	2210	37 kVA		.75	21.333		2,400	895		3,295	4,025	
	2220	50 kVA		.65	24.615		2,850	1,025		3,875	4,750	
	2230	75 kVA		.60	26.667		3,775	1,125		4,900	5,900	
	2240	100 kVA		.50	32		4,925	1,350		6,275	7,500	
	2250	Low operating temperature (80° C), 25 kVA		1	16		3,675	670		4,345	5,075	
	2260	37 kVA		.80	20		3,975	840		4,815	5,675	
	2270	50 kVA		.70	22.857		5,175	960		6,135	7,175	
	2280	75 kVA		.65	24.615		8,250	1,025		9,275	10,700	
	2290	100 kVA		.55	29.091		8,600	1,225		9,825	11,300	
	2300	3 phase, 480 volt primary 120/208 volt secondary										
	2310	Ventilated, 3 kVA	1 Elec	1	8	Ea.	725	335		1,060	1,325	
	2700	6 kVA		.80	10		1,000	420		1,420	1,750	
	2900	9 kVA		.70	11.429		1,125	480		1,605	2,000	

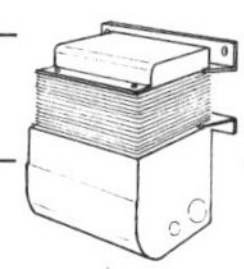

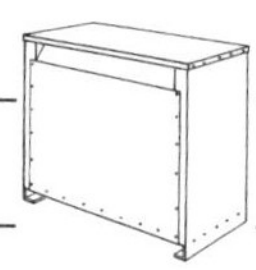

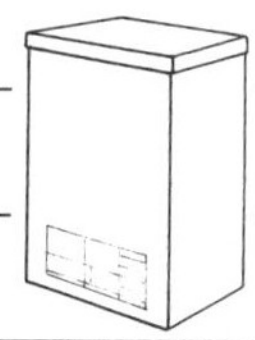

ELECTRICAL 16

16270 | Transformers

		16270 Transformers	Crew	Daily Output	Labor-Hours	Unit	2006 Bare Costs Mat.	Labor	Equip.	Total	Total Incl O&P	
200	3100	15 kVA	2 Elec	1.10	14.545	Ea.	1,500	610		2,110	2,600	200
	3300	30 kVA		.90	17.778		1,775	745		2,520	3,100	
	3500	45 kVA		.80	20		2,125	840		2,965	3,625	
	3700	75 kVA		.70	22.857		3,200	960		4,160	5,000	
	3900	112.5 kVA	R-3	.90	22.222		4,250	920	170	5,340	6,275	
	4100	150 kVA		.85	23.529		5,550	975	180	6,705	7,800	
	4300	225 kVA		.65	30.769		7,525	1,275	235	9,035	10,500	
	4500	300 kVA		.55	36.364		9,500	1,500	277	11,277	13,000	
	4700	500 kVA		.45	44.444		15,700	1,850	340	17,890	20,500	
	4800	750 kVA		.35	57.143		27,600	2,375	435	30,410	34,500	
	4820	1000 kVA		.32	62.500		31,600	2,600	475	34,675	39,200	
	4850	K-4 rated, 15 kVA	2 Elec	1.10	14.545		1,650	610		2,260	2,775	
	4855	30 kVA		.90	17.778		2,550	745		3,295	3,950	
	4860	45 kVA		.80	20		2,925	840		3,765	4,525	
	4865	75 kVA		.70	22.857		4,000	960		4,960	5,875	
	4870	112.5 kVA	R-3	.90	22.222		6,675	920	170	7,765	8,925	
	4875	150 kVA		.85	23.529		8,925	975	180	10,080	11,500	
	4880	225 kVA		.65	30.769		13,600	1,275	235	15,110	17,100	
	4885	300 kVA		.55	36.364		18,800	1,500	277	20,577	23,300	
	4890	500 kVA		.45	44.444		26,000	1,850	340	28,190	31,800	
	4900	K-13 rated, 15 kVA	2 Elec	1.10	14.545		2,175	610		2,785	3,350	
	4905	30 kVA		.90	17.778		2,975	745		3,720	4,425	
	4910	45 kVA		.80	20		3,500	840		4,340	5,150	
	4915	75 kVA		.70	22.857		4,975	960		5,935	6,925	
	4920	112.5 kVA	R-3	.90	22.222		10,200	920	170	11,290	12,900	
	4925	150 kVA		.85	23.529		12,400	975	180	13,555	15,400	
	4930	225 kVA		.65	30.769		18,100	1,275	235	19,610	22,100	
	4935	300 kVA		.55	36.364		25,500	1,500	277	27,277	30,600	
	4940	500 kVA		.45	44.444		38,900	1,850	340	41,090	46,000	
	5020	480 volt primary 120/208 volt secondary										
	5030	Nonventilated, 15 kVA	2 Elec	1.10	14.545	Ea.	3,150	610		3,760	4,425	
	5040	30 kVA		.80	20		3,550	840		4,390	5,200	
	5050	45 kVA		.70	22.857		4,850	960		5,810	6,800	
	5060	75 kVA		.65	24.615		7,050	1,025		8,075	9,350	
	5070	112.5 kVA	R-3	.85	23.529		9,575	975	180	10,730	12,200	
	5081	150 kVA		.85	23.529		11,600	975	180	12,755	14,400	
	5090	225 kVA		.60	33.333		14,200	1,375	254	15,829	18,000	
	5100	300 kVA		.50	40		17,000	1,650	305	18,955	21,600	
	5200	Low operating temperature (80°C), 30 kVA	2 Elec	.90	17.778		2,900	745		3,645	4,325	
	5210	45 kVA		.80	20		4,350	840		5,190	6,075	
	5220	75 kVA		.70	22.857		5,800	960		6,760	7,850	
	5230	112.5 kVA	R-3	.90	22.222		10,200	920	170	11,290	12,900	
	5240	150 kVA		.85	23.529		12,900	975	180	14,055	15,900	
	5250	225 kVA		.65	30.769		17,900	1,275	235	19,410	21,900	
	5260	300 kVA		.55	36.364		21,400	1,500	277	23,177	26,100	
	5270	500 kVA		.45	44.444		32,200	1,850	340	34,390	38,600	
	5380	3 phase, 5 kV primary 277/480 volt secondary										
	5400	High voltage, 112.5 kVA	R-3	.85	23.529	Ea.	12,500	975	180	13,655	15,400	
	5410	150 kVA		.65	30.769		13,600	1,275	235	15,110	17,200	
	5420	225 kVA		.55	36.364		16,100	1,500	277	17,877	20,300	
	5430	300 kVA		.45	44.444		20,500	1,850	340	22,690	25,800	
	5440	500 kVA		.35	57.143		26,400	2,375	435	29,210	33,300	
	5450	750 kVA		.32	62.500		37,700	2,600	475	40,775	45,900	
	5460	1000 kVA		.30	66.667		44,200	2,750	510	47,460	53,500	
	5470	1500 kVA		.27	74.074		51,500	3,075	565	55,140	62,000	
	5480	2000 kVA		.25	80		60,500	3,325	610	64,435	72,500	

16270 | Transformers

		16270 Transformers	Crew	Daily Output	Labor-Hours	Unit	2006 Bare Costs Mat.	Labor	Equip.	Total	Total Incl O&P	
200	5490	2500 kVA	R-3	.20	100	Ea.	68,000	4,150	765	72,915	81,500	200
	5500	3000 kVA		.18	111		89,000	4,600	850	94,450	106,000	
	5590	15 kV primary 277/480 volt secondary										
	5600	High voltage, 112.5 kVA	R-3	.85	23.529	Ea.	19,100	975	180	20,255	22,700	
	5610	150 kVA		.65	30.769		22,200	1,275	235	23,710	26,700	
	5620	225 kVA		.55	36.364		24,600	1,500	277	26,377	29,600	
	5630	300 kVA		.45	44.444		28,900	1,850	340	31,090	35,000	
	5640	500 kVA		.35	57.143		36,700	2,375	435	39,510	44,500	
	5650	750 kVA		.32	62.500		47,800	2,600	475	50,875	57,000	
	5660	1000 kVA		.30	66.667		54,500	2,750	510	57,760	65,000	
	5670	1500 kVA		.27	74.074		63,000	3,075	565	66,640	74,500	
	5680	2000 kVA		.25	80		70,000	3,325	610	73,935	83,000	
	5690	2500 kVA		.20	100		81,000	4,150	765	85,915	96,000	
	5700	3000 kVA		.18	111		96,500	4,600	850	101,950	114,000	
	6000	2400V primary, 480V secondary, 300 KVA		.45	44.444		20,500	1,850	340	22,690	25,800	
	6010	500 KVA		.35	57.143		26,400	2,375	435	29,210	33,300	
	6020	750 KVA		.32	62.500		34,800	2,600	475	37,875	42,800	
	9000	Minimum labor/equipment charge	1 Elec	1	8	Job		335		335	520	
300	0010	**ISOLATING PANELS** used with isolating transformers										300
	0100	Critical care area, 8 circuit, 3 kVA	1 Elec	.58	13.793	Ea.	6,325	580		6,905	7,850	
	0200	5 kVA		.54	14.815		6,500	620		7,120	8,100	
	0400	7.5 kVA		.52	15.385		6,625	645		7,270	8,300	
	0600	10 kVA		.44	18.182		6,850	765		7,615	8,725	
	0800	Operating room power & lighting, 8 circuit, 3 kVA		.58	13.793		5,300	580		5,880	6,725	
	1000	5 kVA		.54	14.815		5,575	620		6,195	7,075	
	1200	7.5 kVA		.52	15.385		6,100	645		6,745	7,700	
	1400	10 kVA		.44	18.182		6,500	765		7,265	8,325	
	1600	X-ray systems, 15 kVA, 90 amp		.44	18.182		13,500	765		14,265	16,100	
	1800	25 kVA, 125 amp		.36	22.222		13,900	935		14,835	16,700	
310	0010	**ISOLATING TRANSFORMER**										310
	0100	Single phase, 120/240 volt primary, 120/240 volt secondary										
	0200	0.50 kVA	1 Elec	4	2	Ea.	288	84		372	445	
	0400	1 kVA		2	4		360	168		528	655	
	0600	2 kVA		1.60	5		540	210		750	920	
	0800	3 kVA		1.40	5.714		615	240		855	1,050	
	1000	5 kVA		1.20	6.667		815	280		1,095	1,325	
	1200	7.5 kVA		1.10	7.273		1,025	305		1,330	1,600	
	1400	10 kVA		.80	10		1,325	420		1,745	2,100	
	1600	15 kVA		.60	13.333		1,675	560		2,235	2,725	
	1800	25 kVA		.50	16		2,425	670		3,095	3,675	
	1810	37.5 kVA	2 Elec	.80	20		4,000	840		4,840	5,700	
	1820	75 kVA	″	.65	24.615		6,000	1,025		7,025	8,200	
	1830	3 phase, 120/240 primary, 120/208V secondary, 112.5 kVA	R-3	.90	22.222		7,575	920	170	8,665	9,950	
	1840	150 kVA		.85	23.529		9,650	975	180	10,805	12,300	
	1850	225 kVA		.65	30.769		13,400	1,275	235	14,910	17,000	
	1860	300 kVA		.55	36.364		17,900	1,500	277	19,677	22,200	
	1870	500 kVA		.45	44.444		29,700	1,850	340	31,890	35,900	
	1880	750 kVA		.35	57.143		25,700	2,375	435	28,510	32,500	
600	0010	**OIL-FILLED TRANSFORMER** Pad mounted, primary delta or Y,										600
	0050	5 kV or 15 kV, with taps, 277/480 V secondary, 3 phase										
	0100	150 kVA	R-3	.65	30.769	Ea.	7,275	1,275	235	8,785	10,200	
	0110	225 kVA		.55	36.364		8,250	1,500	277	10,027	11,700	
	0200	300 kVA		.45	44.444		10,300	1,850	340	12,490	14,600	
	0300	500 kVA		.40	50		14,500	2,075	380	16,955	19,600	

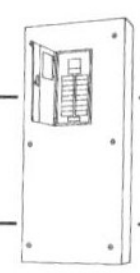

16270 | Transformers

		16270 Transformers	CREW	DAILY OUTPUT	LABOR-HOURS	UNIT	2006 BARE COSTS MAT.	LABOR	EQUIP.	TOTAL	TOTAL INCL O&P	
600	0400	750 kVA	R-3	.38	52.632	Ea.	18,600	2,175	400	21,175	24,200	600
	0500	1000 kVA		.26	76.923		22,000	3,175	585	25,760	29,800	
	0600	1500 kVA		.23	86.957		26,100	3,600	665	30,365	35,100	
	0700	2000 kVA		.20	100		33,000	4,150	765	37,915	43,500	
	0710	2500 kVA		.19	105		39,900	4,350	805	45,055	51,500	
	0720	3000 kVA		.17	117		48,200	4,875	900	53,975	61,500	
	0800	3750 kVA		.16	125		62,000	5,175	955	68,130	77,000	
610	0010	**TRANSFORMER, LIQUID-FILLED** Pad mounted										610
	0020	5 kV or 15 kV primary, 277/480 volt secondary, 3 phase										
	0050	225 kVA	R-3	.55	36.364	Ea.	11,200	1,500	277	12,977	14,900	
	0100	300 kVA		.45	44.444		13,300	1,850	340	15,490	17,800	
	0200	500 kVA		.40	50		16,800	2,075	380	19,255	22,100	
	0250	750 kVA		.38	52.632		21,700	2,175	400	24,275	27,600	
	0300	1000 kVA		.26	76.923		25,200	3,175	585	28,960	33,300	
	0350	1500 kVA		.23	86.957		29,400	3,600	665	33,665	38,600	
	0400	2000 kVA		.20	100		36,400	4,150	765	41,315	47,200	
	0450	2500 kVA		.19	105		43,400	4,350	805	48,555	55,500	
620	0010	**TRANSFORMER HANDLING** Add to normal labor cost in restricted areas										620
	5000	Transformers										
	5150	15 kVA, approximately 200 pounds	2 Elec	2.70	5.926	Ea.		249		249	385	
	5160	25 kVA, approximately 300 pounds		2.50	6.400			269		269	415	
	5170	37.5 kVA, approximately 400 pounds		2.30	6.957			292		292	450	
	5180	50 kVA, approximately 500 pounds		2	8			335		335	520	
	5190	75 kVA, approximately 600 pounds		1.80	8.889			375		375	575	
	5200	100 kVA, approximately 700 pounds		1.60	10			420		420	645	
	5210	112.5 kVA, approximately 800 pounds	3 Elec	2.20	10.909			460		460	705	
	5220	125 kVA, approximately 900 pounds		2	12			505		505	775	
	5230	150 kVA, approximately 1000 pounds		1.80	13.333			560		560	865	
	5240	167 kVA, approximately 1200 pounds		1.60	15			630		630	970	
	5250	200 kVA, approximately 1400 pounds		1.40	17.143			720		720	1,100	
	5260	225 kVA, approximately 1600 pounds		1.30	18.462			775		775	1,200	
	5270	250 kVA, approximately 1800 pounds		1.10	21.818			915		915	1,400	
	5280	300 kVA, approximately 2000 pounds		1	24			1,000		1,000	1,550	
	5290	500 kVA, approximately 3000 pounds		.75	32			1,350		1,350	2,075	
	5300	600 kVA, approximately 3500 pounds		.67	35.821			1,500		1,500	2,325	
	5310	750 kVA, approximately 4000 pounds		.60	40			1,675		1,675	2,600	
	5320	1000 kVA, approximately 5000 pounds		.50	48			2,025		2,025	3,100	

16280 | Power Filters & Conditioners

		16280 Power Filters & Conditioners	CREW	DAILY OUTPUT	LABOR-HOURS	UNIT	MAT.	LABOR	EQUIP.	TOTAL	TOTAL INCL O&P	
100	0010	**AUTOMATIC VOLTAGE REGULATORS**										100
	0100	Computer grade, solid state, variable trans. volt. regulator										
	0110	Single-phase, 120 V, 8.6 kVA	2 Elec	1.33	12.030	Ea.	5,050	505		5,555	6,350	
	0120	17.3 kVA		1.14	14.035		5,975	590		6,565	7,475	
	0130	208/240 V, 7.5/8.6 kVA		1.33	12.030		5,050	505		5,555	6,350	
	0140	13.5/15.6 kVA		1.33	12.030		5,975	505		6,480	7,350	
	0150	27.0/31.2 kVA		1.14	14.035		7,550	590		8,140	9,200	
	0210	Two-phase, single control, 208/240 V, 15.0/17.3 kVA		1.14	14.035		5,975	590		6,565	7,475	
	0220	Individual phase control, 15.0/17.3 kVA		1.14	14.035		5,975	590		6,565	7,475	
	0230	30.0/34.6 kVA	3 Elec	1.33	18.045		7,550	760		8,310	9,475	
	0310	Three-phase single control, 208/240 V, 26/30 kVA	2 Elec	1	16		5,975	670		6,645	7,600	
	0320	380/480 V, 24/30 kVA	"	1	16		5,975	670		6,645	7,600	
	0330	43/54 kVA	3 Elec	1.33	18.045		10,900	760		11,660	13,100	
	0340	Individual phase control, 208 V, 26 kVA	"	1.33	18.045		5,975	760		6,735	7,750	
	0350	52 kVA	R-3	.91	21.978		7,550	910	168	8,628	9,875	
	0360	340/480 V, 24/30 kVA	2 Elec	1	16		5,975	670		6,645	7,600	

16280 | Power Filters & Conditioners

	Line	Description	Crew	Daily Output	Labor-Hours	Unit	2006 Bare Costs Mat.	Labor	Equip.	Total	Total Incl O&P	
100	0370	43/54 kVA	2 Elec	1	16	Ea.	7,550	670		8,220	9,325	100
	0380	48/60 kVA	3 Elec	1.33	18.045		10,900	760		11,660	13,200	
	0390	86/108 kVA	R-3	.91	21.978		12,100	910	168	13,178	15,000	
	0500	Standard grade, solid state, variable transformer volt. regulator										
	0510	Single-phase, 115 V, 2.3 kVA	1 Elec	2.29	3.493	Ea.	1,650	147		1,797	2,050	
	0520	4.2 kVA		2	4		2,650	168		2,818	3,175	
	0530	6.6 kVA		1.14	7.018		3,250	295		3,545	4,025	
	0540	13.0 kVA		1.14	7.018		4,775	295		5,070	5,725	
	0550	16.6 kVA	2 Elec	1.23	13.008		5,650	545		6,195	7,075	
	0610	230 V, 8.3 kVA		1.33	12.030		4,775	505		5,280	6,050	
	0620	21.4 kVA		1.23	13.008		5,650	545		6,195	7,075	
	0630	29.9 kVA		1.23	13.008		5,650	545		6,195	7,075	
	0710	460 V, 9.2 kVA		1.33	12.030		4,775	505		5,280	6,050	
	0720	20.7 kVA		1.23	13.008		5,650	545		6,195	7,075	
	0810	Three-phase, 230 V, 13.1 kVA	3 Elec	1.41	17.021		4,775	715		5,490	6,375	
	0820	19.1 kVA		1.41	17.021		5,650	715		6,365	7,325	
	0830	25.1 kVA		1.23	19.512		5,650	820		6,470	7,475	
	0840	57.8 kVA	R-3	.95	21.053		10,300	870	161	11,331	12,800	
	0850	74.9 kVA	"	.91	21.978		10,300	910	168	11,378	12,900	
	0910	460 V, 14.3 kVA	3 Elec	1.41	17.021		4,775	715		5,490	6,375	
	0920	19.1 kVA		1.41	17.021		5,650	715		6,365	7,325	
	0930	27.9 kVA		1.23	19.512		5,650	820		6,470	7,475	
	0940	59.8 kVA	R-3	1	20		10,300	830	153	11,283	12,700	
	0950	79.7 kVA		.95	21.053		11,500	870	161	12,531	14,100	
	0960	118 kVA		.95	21.053		12,100	870	161	13,131	14,800	
	1000	Laboratory grade, precision, electronic voltage regulator										
	1110	Single-phase, 115 V, 0.5 kVA	1 Elec	2.29	3.493	Ea.	1,375	147		1,522	1,750	
	1120	1.0 kVA		2	4		1,475	168		1,643	1,875	
	1130	3.0 kVA		.80	10		2,050	420		2,470	2,900	
	1140	6.0 kVA	2 Elec	1.46	10.959		3,800	460		4,260	4,875	
	1150	10.0 kVA	3 Elec	1	24		4,925	1,000		5,925	6,975	
	1160	15.0 kVA	"	1.50	16		5,625	670		6,295	7,225	
	1210	230 V, 3.0 kVA	1 Elec	.80	10		2,325	420		2,745	3,200	
	1220	6.0 kVA	2 Elec	1.46	10.959		3,875	460		4,335	4,975	
	1230	10.0 kVA	3 Elec	1.71	14.035		5,150	590		5,740	6,575	
	1240	15.0 kVA	"	1.60	15		5,825	630		6,455	7,375	
300	0010	**CAPACITORS** Indoor										300
	0020	240 volts, single & 3 phase, 0.5 kVAR	1 Elec	2.70	2.963	Ea.	300	124		424	520	
	0100	1.0 kVAR		2.70	2.963		365	124		489	590	
	0150	2.5 kVAR		2	4		410	168		578	710	
	0200	5.0 kVAR		1.80	4.444		490	187		677	830	
	0250	7.5 kVAR		1.60	5		570	210		780	950	
	0300	10 kVAR		1.50	5.333		685	224		909	1,100	
	0350	15 kVAR		1.30	6.154		945	258		1,203	1,450	
	0400	20 kVAR		1.10	7.273		1,175	305		1,480	1,775	
	0450	25 kVAR		1	8		1,375	335		1,710	2,050	
	1000	480 volts, single & 3 phase, 1 kVAR		2.70	2.963		275	124		399	490	
	1050	2 kVAR		2.70	2.963		315	124		439	540	
	1100	5 kVAR		2	4		400	168		568	700	
	1150	7.5 kVAR		2	4		430	168		598	735	
	1200	10 kVAR		2	4		480	168		648	790	
	1250	15 kVAR		2	4		595	168		763	915	
	1300	20 kVAR		1.60	5		660	210		870	1,050	
	1350	30 kVAR		1.50	5.333		825	224		1,049	1,250	
	1400	40 kVAR		1.20	6.667		1,050	280		1,330	1,575	
	1450	50 kVAR		1.10	7.273		1,200	305		1,505	1,800	

	16280 | Power Filters & Conditioners		CREW	DAILY OUTPUT	LABOR-HOURS	UNIT	2006 BARE COSTS MAT.	LABOR	EQUIP.	TOTAL	TOTAL INCL O&P	
300	2000	600 volts, single & 3 phase, 1 kVAR	1 Elec	2.70	2.963	Ea.	283	124		407	500	300
	2050	2 kVAR		2.70	2.963		325	124		449	550	
	2100	5 kVAR		2	4		400	168		568	700	
	2150	7.5 kVAR		2	4		430	168		598	735	
	2200	10 kVAR		2	4		480	168		648	790	
	2250	15 kVAR		1.60	5		595	210		805	980	
	2300	20 kVAR		1.60	5		660	210		870	1,050	
	2350	25 kVAR		1.50	5.333		740	224		964	1,150	
	2400	35 kVAR		1.40	5.714		940	240		1,180	1,400	
	2450	50 kVAR		1.30	6.154		1,200	258		1,458	1,725	
340	0010	**COMPUTER ISOLATION TRANSFORMER**										340
	0100	Computer grade										
	0110	Single-phase, 120/240 V, 0.5 kVA	1 Elec	4	2	Ea.	370	84		454	535	
	0120	1.0 kVA		2.67	2.996		525	126		651	770	
	0130	2.5 kVA		2	4		795	168		963	1,125	
	0140	5 kVA		1.14	7.018		910	295		1,205	1,450	
360	0010	**COMPUTER REGULATOR TRANSFORMER**										360
	0100	Ferro-resonant, constant voltage, variable transformer										
	0110	Single-phase, 240 V, 0.5 kVA	1 Elec	2.67	2.996	Ea.	460	126		586	705	
	0120	1.0 kVA		2	4		635	168		803	960	
	0130	2.0 kVA		1	8		1,075	335		1,410	1,725	
	0210	Plug-in unit 120 V, 0.14 kVA		8	1		267	42		309	360	
	0220	0.25 kVA		8	1		310	42		352	410	
	0230	0.5 kVA		8	1		460	42		502	575	
	0240	1.0 kVA		5.33	1.501		635	63		698	795	
	0250	2.0 kVA		4	2		1,075	84		1,159	1,325	
600	0010	**POWER CONDITIONER TRANSFORMER**										600
	0100	Electronic solid state, buck-boost, transformer, w/tap switch										
	0110	Single-phase, 115 V, 3.0 kVA, + or - 3% accuracy	2 Elec	1.60	10	Ea.	2,500	420		2,920	3,400	
	0120	208, 220, 230, or 240 V, 5.0 kVA, + or - 1.5% accuracy	3 Elec	1.60	15		3,225	630		3,855	4,525	
	0130	5.0 kVA, + or - 6% accuracy	2 Elec	1.14	14.035		2,900	590		3,490	4,075	
	0140	7.5 kVA, + or - 1.5% accuracy	3 Elec	1.50	16		4,100	670		4,770	5,525	
	0150	7.5 kVA, + or - 6% accuracy		1.60	15		3,425	630		4,055	4,725	
	0160	10.0 kVA, + or - 1.5% accuracy		1.33	18.045		5,450	760		6,210	7,175	
	0170	10.0 kVA, + or - 6% accuracy		1.41	17.021		4,650	715		5,365	6,200	
820	0010	**TRANSIENT SUPPRESSOR/VOLTAGE REGULATOR** (without isolation)										820
	0110	Single-phase, 115 V, 1.0 kVA	1 Elec	2.67	2.996	Ea.	925	126		1,051	1,225	
	0120	2.0 kVA		2.29	3.493		1,250	147		1,397	1,600	
	0130	4.0 kVA		2.13	3.756		1,550	158		1,708	1,950	
	0140	220 V, 1.0 kVA		2.67	2.996		925	126		1,051	1,225	
	0150	2.0 kVA		2.29	3.493		1,250	147		1,397	1,600	
	0160	4.0 kVA		2.13	3.756		1,650	158		1,808	2,050	
	0210	Plug-in unit, 120 V, 1.0 kVA		8	1		880	42		922	1,025	
	0220	2.0 kVA		8	1		1,250	42		1,292	1,450	
840	0010	**TRANSIENT VOLTAGE SUPPRESSOR TRANSFORMER**										840
	0110	Single-phase, 120 V, 1.8 kVA	1 Elec	4	2	Ea.	530	84		614	715	
	0120	3.6 kVA		4	2		1,125	84		1,209	1,350	
	0130	7.2 kVA		3.20	2.500		1,450	105		1,555	1,750	
	0150	240 V, 3.6 kVA		4	2		1,125	84		1,209	1,350	
	0160	7.2 kVA		4	2		1,450	84		1,534	1,725	
	0170	14.4 kVA		3.20	2.500		1,850	105		1,955	2,200	
	0210	Plug-in unit, 120 V, 1.8 kVA		8	1		510	42		552	625	

16200 | Electrical Power

		16290 | Power Measurement & Control	CREW	DAILY OUTPUT	LABOR-HOURS	UNIT	2006 BARE COSTS MAT.	LABOR	EQUIP.	TOTAL	TOTAL INCL O&P	
800	0010	**SWITCHBOARD INSTRUMENTS** 3 phase, 4 wire										800
	0100	AC indicating, ammeter & switch	1 Elec	8	1	Ea.	1,675	42		1,717	1,900	
	0200	Voltmeter & switch		8	1		1,675	42		1,717	1,900	
	0300	Wattmeter		8	1		3,300	42		3,342	3,700	
	0400	AC recording, ammeter		4	2		5,875	84		5,959	6,600	
	0500	Voltmeter		4	2		5,875	84		5,959	6,600	
	0600	Ground fault protection, zero sequence		2.70	2.963		5,200	124		5,324	5,900	
	0700	Ground return path		2.70	2.963		5,200	124		5,324	5,900	
	0800	3 current transformers, 5 to 800 amp		2	4		2,425	168		2,593	2,900	
	0900	1000 to 1500 amp		1.30	6.154		3,475	258		3,733	4,225	
	1200	2000 to 4000 amp		1	8		4,100	335		4,435	5,025	
	1300	Fused potential transformer, maximum 600 volt		8	1		910	42		952	1,075	
860	0010	**VOLTAGE MONITOR SYSTEMS** (test equipment)										860
	0100	AC voltage monitor system, 120/240 V, one-channel				Ea.	3,050			3,050	3,375	
	0110	Modem adapter					385			385	420	
	0120	Add-on detector only					1,600			1,600	1,775	
	0150	AC voltage remote monitor sys., 3 channel, 120, 230, or 480 V					5,550			5,550	6,125	
	0160	With internal modem					5,875			5,875	6,450	
	0170	Combination temperature and humidity probe					860			860	950	
	0180	Add-on detector only					4,025			4,025	4,425	
	0190	With internal modem					4,375			4,375	4,825	

16300 | Transmission & Distribution

		16330 | Med-Voltage Switching & Protection	CREW	DAILY OUTPUT	LABOR-HOURS	UNIT	2006 BARE COSTS MAT.	LABOR	EQUIP.	TOTAL	TOTAL INCL O&P	
760	0010	**SWITCHGEAR**, Incorporate switch with cable connections, transformer,										760
	0100	& Low Voltage section										
	0200	Load interrupter switch, 600 amp, 2 position										
	0300	NEMA 1, 4.8 KV, 300 kVA & below w/CLF fuses 	R-3	.40	50	Ea.	15,900	2,075	380	18,355	21,100	
	0400	400 kVA & above w/CLF fuses		.38	52.632		17,800	2,175	400	20,375	23,300	
	0500	Non fusible		.41	48.780		12,900	2,025	370	15,295	17,700	
	0600	13.8 kV, 300 kVA & below		.38	52.632		20,000	2,175	400	22,575	25,800	
	0700	400 kVA & above		.36	55.556		20,000	2,300	425	22,725	26,000	
	0800	Non fusible		.40	50		15,200	2,075	380	17,655	20,300	
	0900	Cable lugs for 2 feeders 4.8 kV or 13.8 kV	1 Elec	8	1		485	42		527	600	
	1000	Pothead, one 3 conductor or three 1 conductor		4	2		2,325	84		2,409	2,675	
	1100	Two 3 conductor or six 1 conductor		2	4		4,600	168		4,768	5,300	
	1200	Key interlocks		8	1		540	42		582	655	
	1300	Lightning arresters, Distribution class (no charge)										
	1400	Intermediate class or line type 4.8 kV	1 Elec	2.70	2.963	Ea.	2,625	124		2,749	3,075	
	1500	13.8 kV		2	4		3,475	168		3,643	4,050	
	1600	Station class, 4.8 kV		2.70	2.963		4,475	124		4,599	5,125	
	1700	13.8 kV		2	4		7,700	168		7,868	8,725	
	1800	Transformers, 4800 volts to 480/277 volts, 75 kVA	R-3	.68	29.412		13,600	1,225	224	15,049	17,100	
	1900	112.5 kVA		.65	30.769		16,600	1,275	235	18,110	20,500	
	2000	150 kVA		.57	35.088		18,900	1,450	268	20,618	23,300	
	2100	225 kVA		.48	41.667		21,700	1,725	320	23,745	26,900	
	2200	300 kVA		.41	48.780		24,300	2,025	370	26,695	30,200	
	2300	500 kVA		.36	55.556		32,000	2,300	425	34,725	39,200	

16300 | Transmission & Distribution

16330 | Med-Voltage Switching & Protection

			CREW	DAILY OUTPUT	LABOR-HOURS	UNIT	2006 BARE COSTS MAT.	LABOR	EQUIP.	TOTAL	TOTAL INCL O&P	
760	2400	750 kVA	R-3	.29	68.966	Ea.	36,300	2,850	525	39,675	44,900	760
	2500	13,800 volts to 480/277 volts, 75 kVA		.61	32.787		19,200	1,350	250	20,800	23,500	
	2600	112.5 kVA		.55	36.364		25,500	1,500	277	27,277	30,600	
	2700	150 kVA		.49	40.816		25,700	1,700	310	27,710	31,300	
	2800	225 kVA		.41	48.780		29,700	2,025	370	32,095	36,100	
	2900	300 kVA		.37	54.054		30,300	2,250	410	32,960	37,200	
	3000	500 kVA		.31	64.516		33,500	2,675	490	36,665	41,600	
	3100	750 kVA		.42	47.733		36,900	1,975	365	39,240	44,100	
	3200	Forced air cooling & temperature alarm	1 Elec	1	8		2,975	335		3,310	3,775	
	3300	Low voltage components										
	3400	Maximum panel height 49-1/2", single or twin row										
	3500	Breaker heights, type FA or FH, 6"										
	3600	type KA or KH, 8"										
	3700	type LA, 11"										
	3800	type MA, 14"										
	3900	Breakers, 2 pole, 15 to 60 amp, type FA	1 Elec	5.60	1.429	Ea.	305	60		365	430	
	4000	70 to 100 amp, type FA		4.20	1.905		375	80		455	540	
	4100	15 to 60 amp, type FH		5.60	1.429		485	60		545	630	
	4200	70 to 100 amp, type FH		4.20	1.905		570	80		650	750	
	4300	125 to 225 amp, type KA		3.40	2.353		870	99		969	1,100	
	4400	125 to 225 amp, type KH		3.40	2.353		2,025	99		2,124	2,375	
	4500	125 to 400 amp, type LA		2.50	3.200		1,600	134		1,734	1,950	
	4600	125 to 600 amp, type MA		1.80	4.444		2,575	187		2,762	3,125	
	4700	700 & 800 amp, type MA		1.50	5.333		3,250	224		3,474	3,925	
	4800	3 pole, 15 to 60 amp, type FA		5.30	1.509		380	63.50		443.50	520	
	4900	70 to 100 amp, type FA		4	2		465	84		549	640	
	5000	15 to 60 amp, type FH		5.30	1.509		570	63.50		633.50	725	
	5100	70 to 100 amp, type FH		4	2		645	84		729	840	
	5200	125 to 225 amp, type KA		3.20	2.500		1,100	105		1,205	1,350	
	5300	125 to 225 amp, type KH		3.20	2.500		2,450	105		2,555	2,850	
	5400	125 to 400 amp, type LA		2.30	3.478		1,925	146		2,071	2,350	
	5500	125 to 600 amp, type MA		1.60	5		3,175	210		3,385	3,825	
	5600	700 & 800 amp, type MA		1.30	6.154		4,150	258		4,408	4,975	

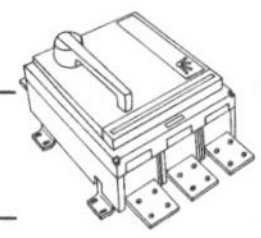

16400 | Low-Voltage Distribution

16410 | Encl. Switches & Circuit Breakers

			CREW	DAILY OUTPUT	LABOR-HOURS	UNIT	2006 BARE COSTS MAT.	LABOR	EQUIP.	TOTAL	TOTAL INCL O&P	
200	0010	**CIRCUIT BREAKERS** (in enclosure)										200
	0100	Enclosed (NEMA 1), 600 volt, 3 pole, 30 amp	1 Elec	3.20	2.500	Ea.	500	105		605	710	
	0200	60 amp		2.80	2.857		500	120		620	735	
	0400	100 amp		2.30	3.478		570	146		716	850	
	0600	225 amp		1.50	5.333		1,325	224		1,549	1,800	
	0700	400 amp	2 Elec	1.60	10		2,250	420		2,670	3,125	
	0800	600 amp		1.20	13.333		3,250	560		3,810	4,450	
	1000	800 amp		.94	17.021		4,250	715		4,965	5,775	
	1200	1000 amp		.84	19.048		5,350	800		6,150	7,125	
	1220	1200 amp		.80	20		6,850	840		7,690	8,850	
	1240	1600 amp		.72	22.222		14,000	935		14,935	16,900	
	1260	2000 amp		.64	25		15,100	1,050		16,150	18,200	

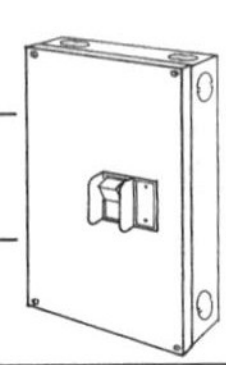

16410 | Encl. Switches & Circuit Breakers

			CREW	DAILY OUTPUT	LABOR-HOURS	UNIT	2006 BARE COSTS MAT.	LABOR	EQUIP.	TOTAL	TOTAL INCL O&P	
200	1400	1200 amp with ground fault	2 Elec	.80	20	Ea.	13,900	840		14,740	16,600	200
	1600	1600 amp with ground fault		.72	22.222		15,800	935		16,735	18,900	
	1800	2000 amp with ground fault		.64	25		17,000	1,050		18,050	20,300	
	2000	Disconnect, 240 volt 3 pole, 5 HP motor	1 Elec	3.20	2.500		350	105		455	545	
	2020	10 HP motor		3.20	2.500		350	105		455	545	
	2040	15 HP motor		2.80	2.857		350	120		470	570	
	2060	20 HP motor		2.30	3.478		425	146		571	690	
	2080	25 HP motor		2.30	3.478		425	146		571	690	
	2100	30 HP motor		2.30	3.478		425	146		571	690	
	2120	40 HP motor		2	4		725	168		893	1,050	
	2140	50 HP motor		1.50	5.333		725	224		949	1,150	
	2160	60 HP motor		1.50	5.333		1,675	224		1,899	2,175	
	2180	75 HP motor	2 Elec	2	8		1,675	335		2,010	2,350	
	2200	100 HP motor		1.60	10		1,675	420		2,095	2,475	
	2220	125 HP motor		1.60	10		1,675	420		2,095	2,475	
	2240	150 HP motor		1.20	13.333		3,250	560		3,810	4,450	
	2260	200 HP motor		1.20	13.333		4,250	560		4,810	5,550	
	2300	Enclosed (NEMA 7), explosion proof, 600 volt 3 pole, 50 amp	1 Elec	2.30	3.478		1,050	146		1,196	1,375	
	2350	100 amp		1.50	5.333		1,100	224		1,324	1,550	
	2400	150 amp		1	8		2,625	335		2,960	3,425	
	2450	250 amp	2 Elec	1.60	10		3,300	420		3,720	4,275	
	2500	400 amp	"	1.20	13.333		3,650	560		4,210	4,900	
	9000	Minimum labor/equipment charge	1 Elec	4	2	Job		84		84	129	
800	0010	SAFETY SWITCHES										800
	0100	General duty 240 volt, 3 pole NEMA 1, fusible, 30 amp	1 Elec	3.20	2.500	Ea.	82	105		187	252	
	0200	60 amp		2.30	3.478		138	146		284	375	
	0300	100 amp		1.90	4.211		238	177		415	535	
	0400	200 amp		1.30	6.154		510	258		768	965	
	0500	400 amp	2 Elec	1.80	8.889		1,300	375		1,675	2,000	
	0600	600 amp	"	1.20	13.333		2,400	560		2,960	3,525	
	0610	Nonfusible, 30 amp	1 Elec	3.20	2.500		66.50	105		171.50	235	
	0650	60 amp		2.30	3.478		88	146		234	320	
	0700	100 amp		1.90	4.211		203	177		380	495	
	0750	200 amp		1.30	6.154		375	258		633	815	
	0800	400 amp	2 Elec	1.80	8.889		910	375		1,285	1,575	
	0850	600 amp	"	1.20	13.333		1,725	560		2,285	2,775	
	1100	Heavy duty, 600 volt, 3 pole NEMA 1 nonfused										
	1110	30 amp	1 Elec	3.20	2.500	Ea.	116	105		221	290	
	1500	60 amp		2.30	3.478		211	146		357	455	
	1700	100 amp		1.90	4.211		330	177		507	635	
	1900	200 amp		1.30	6.154		505	258		763	955	
	2100	400 amp	2 Elec	1.80	8.889		1,125	375		1,500	1,825	
	2300	600 amp		1.20	13.333		2,000	560		2,560	3,075	
	2500	800 amp		.94	17.021		4,100	715		4,815	5,600	
	2700	1200 amp		.80	20		5,500	840		6,340	7,350	
	2900	Heavy duty, 240 volt, 3 pole NEMA 1 fusible										
	2910	30 amp	1 Elec	3.20	2.500	Ea.	132	105		237	305	
	3000	60 amp		2.30	3.478		224	146		370	470	
	3300	100 amp		1.90	4.211		350	177		527	660	
	3500	200 amp		1.30	6.154		605	258		863	1,075	
	3700	400 amp	2 Elec	1.80	8.889		1,550	375		1,925	2,275	
	3900	600 amp		1.20	13.333		2,675	560		3,235	3,825	
	4100	800 amp		.94	17.021		6,275	715		6,990	8,000	
	4300	1200 amp		.80	20		8,250	840		9,090	10,400	
	4340	2 pole fused, 30 amp	1 Elec	3.50	2.286		99.50	96		195.50	257	

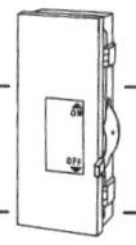

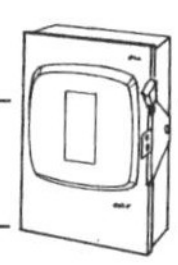

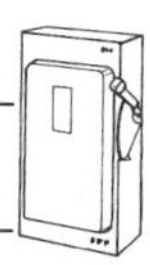

16410 | Encl. Switches & Circuit Breakers

			CREW	DAILY OUTPUT	LABOR-HOURS	UNIT	2006 BARE COSTS MAT.	LABOR	EQUIP.	TOTAL	TOTAL INCL O&P	
800	4350	600 volt, 3 pole, fusible, 30 amp	1 Elec	3.20	2.500	Ea.	224	105		329	410	800
	4380	60 amp		2.30	3.478		271	146		417	525	
	4400	100 amp		1.90	4.211		495	177		672	815	
	4420	200 amp		1.30	6.154		715	258		973	1,200	
	4440	400 amp	2 Elec	1.80	8.889		1,875	375		2,250	2,625	
	4450	600 amp		1.20	13.333		3,125	560		3,685	4,325	
	4460	800 amp		.94	17.021		6,275	715		6,990	8,000	
	4480	1200 amp		.80	20		8,250	840		9,090	10,400	
	4500	240 volt 3 pole NEMA 3R (no hubs), fusible										
	4510	30 amp	1 Elec	3.10	2.581	Ea.	237	108		345	425	
	4700	60 amp		2.20	3.636		370	153		523	645	
	4900	100 amp		1.80	4.444		535	187		722	875	
	5100	200 amp		1.20	6.667		735	280		1,015	1,250	
	5300	400 amp	2 Elec	1.60	10		1,625	420		2,045	2,450	
	5500	600 amp	"	1	16		3,400	670		4,070	4,775	
	5510	Heavy duty, 600 volt, 3 pole 3ph. NEMA 3R fusible, 30 amp	1 Elec	3.10	2.581		375	108		483	580	
	5520	60 amp		2.20	3.636		440	153		593	720	
	5530	100 amp		1.80	4.444		690	187		877	1,050	
	5540	200 amp		1.20	6.667		950	280		1,230	1,475	
	5550	400 amp	2 Elec	1.60	10		2,225	420		2,645	3,100	
	5700	600 volt, 3 pole NEMA 3R nonfused										
	5710	30 amp	1 Elec	3.10	2.581	Ea.	209	108		317	395	
	5900	60 amp		2.20	3.636		365	153		518	640	
	6100	100 amp		1.80	4.444		510	187		697	855	
	6300	200 amp		1.20	6.667		620	280		900	1,125	
	6500	400 amp	2 Elec	1.60	10		1,550	420		1,970	2,350	
	6700	600 amp	"	1	16		3,100	670		3,770	4,425	
	6900	600 volt, 6 pole NEMA 3R nonfused, 30 amp	1 Elec	2.70	2.963		1,325	124		1,449	1,675	
	7100	60 amp		2	4		1,550	168		1,718	1,950	
	7300	100 amp		1.50	5.333		1,900	224		2,124	2,425	
	7500	200 amp		1.20	6.667		4,275	280		4,555	5,125	
	7600	600 volt, 3 pole NEMA 7 explosion proof nonfused										
	7610	30 amp	1 Elec	2.20	3.636	Ea.	1,050	153		1,203	1,375	
	7620	60 amp		1.80	4.444		1,250	187		1,437	1,675	
	7630	100 amp		1.20	6.667		1,525	280		1,805	2,100	
	7640	200 amp		.80	10		3,175	420		3,595	4,125	
	7710	600 volt 6 pole, NEMA 3R fusible, 30 amp		2.70	2.963		1,500	124		1,624	1,850	
	7900	60 amp		2	4		1,800	168		1,968	2,225	
	8100	100 amp		1.50	5.333		2,175	224		2,399	2,750	
	8110	240 volt 3 pole, NEMA 12 fusible, 30 amp		3.10	2.581		289	108		397	485	
	8120	60 amp		2.20	3.636		390	153		543	665	
	8130	100 amp		1.80	4.444		575	187		762	920	
	8140	200 amp		1.20	6.667		820	280		1,100	1,325	
	8150	400 amp	2 Elec	1.60	10		1,875	420		2,295	2,725	
	8160	600 amp	"	1	16		3,050	670		3,720	4,375	
	8180	600 volt 3 pole, NEMA 12 fused, 30 amp	1 Elec	3.10	2.581		380	108		488	585	
	8190	60 amp		2.20	3.636		390	153		543	665	
	8200	100 amp		1.80	4.444		630	187		817	980	
	8210	200 amp		1.20	6.667		980	280		1,260	1,500	
	8220	400 amp	2 Elec	1.60	10		2,150	420		2,570	3,025	
	8230	600 amp	"	1	16		3,625	670		4,295	5,025	
	8240	600 volt 3 pole, NEMA 12 nonfused, 30 amp	1 Elec	3.10	2.581		268	108		376	460	
	8250	60 amp		2.20	3.636		345	153		498	615	
	8260	100 amp		1.80	4.444		495	187		682	830	
	8270	200 amp		1.20	6.667		655	280		935	1,150	
	8280	400 amp	2 Elec	1.60	10		1,625	420		2,045	2,425	

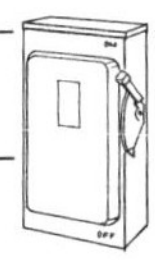

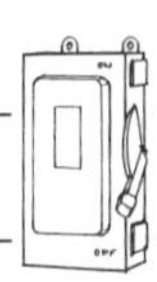

Important: See the Reference Section for supporting data - Crews, Rental Equipment, City Cost Indexes and Reference Data

16410 | Encl. Switches & Circuit Breakers

			CREW	DAILY OUTPUT	LABOR-HOURS	UNIT	2006 BARE COSTS MAT.	LABOR	EQUIP.	TOTAL	TOTAL INCL O&P	
800	8290	600 amp	2 Elec	1	16	Ea.	2,650	670		3,320	3,925	800
	8310	600 volt, 3 pole NEMA 4 fusible, 30 amp	1 Elec	3	2.667		1,050	112		1,162	1,325	
	8320	60 amp		2.20	3.636		1,150	153		1,303	1,500	
	8330	100 amp		1.80	4.444		2,575	187		2,762	3,125	
	8340	200 amp		1.20	6.667		3,250	280		3,530	4,000	
	8350	400 amp	2 Elec	1.60	10		6,425	420		6,845	7,700	
	8360	600 volt 3 pole NEMA 4 nonfused, 30 amp	1 Elec	3	2.667		860	112		972	1,125	
	8370	60 amp		2.20	3.636		1,025	153		1,178	1,350	
	8380	100 amp		1.80	4.444		2,100	187		2,287	2,600	
	8390	200 amp		1.20	6.667		2,850	280		3,130	3,575	
	8400	400 amp	2 Elec	1.60	10		5,775	420		6,195	7,025	
	8490	Motor starters, manual, single phase, NEMA 1	1 Elec	6.40	1.250		49	52.50		101.50	135	
	8500	NEMA 4		4	2		144	84		228	287	
	8700	NEMA 7		4	2		158	84		242	305	
	8900	NEMA 1 with pilot		6.40	1.250		66	52.50		118.50	154	
	8920	3 pole, NEMA 1, 230/460 volt, 5 HP, size 0		3.50	2.286		153	96		249	315	
	8940	10 HP, size 1		2	4		181	168		349	460	
	9010	Disc. switch, 600V 3 pole fusible, 30 amp, to 10 HP motor		3.20	2.500		239	105		344	425	
	9050	60 amp, to 30 HP motor		2.30	3.478		550	146		696	830	
	9070	100 amp, to 60 HP motor		1.90	4.211		550	177		727	875	
	9100	200 amp, to 125 HP motor		1.30	6.154		825	258		1,083	1,300	
	9110	400 amp, to 200 HP motor	2 Elec	1.80	8.889		2,075	375		2,450	2,850	
	9990	Minimum labor/equipment charge	1 Elec	3	2.667	Job		112		112	173	
840	0010	**TIME SWITCHES**										840
	0100	Single pole, single throw, 24 hour dial	1 Elec	4	2	Ea.	87	84		171	225	
	0200	24 hour dial with reserve power		3.60	2.222		370	93.50		463.50	550	
	0300	Astronomic dial		3.60	2.222		149	93.50		242.50	310	
	0400	Astronomic dial with reserve power		3.30	2.424		480	102		582	685	
	0500	7 day calendar dial		3.30	2.424		134	102		236	305	
	0600	7 day calendar dial with reserve power		3.20	2.500		520	105		625	730	
	0700	Photo cell 2000 watt		8	1		16.05	42		58.05	82	
	1080	Load management device, 4 loads		2	4		845	168		1,013	1,200	
	1100	Load management device, 8 loads		1	8		1,375	335		1,710	2,050	
	9000	Minimum labor/equipment charge		3.50	2.286	Job		96		96	148	

16415 | Transfer Switches

			CREW	DAILY OUTPUT	LABOR-HOURS	UNIT	MAT.	LABOR	EQUIP.	TOTAL	TOTAL INCL O&P	
100	0010	**NON-AUTOMATIC TRANSFER SWITCHES** enclosed										100
	0100	Manual operated, 480 volt 3 pole, 30 amp	1 Elec	2.30	3.478	Ea.	1,425	146		1,571	1,800	
	0150	60 amp		1.90	4.211		1,425	177		1,602	1,850	
	0200	100 amp		1.30	6.154		1,425	258		1,683	1,975	
	0250	200 amp	2 Elec	2	8		2,875	335		3,210	3,675	
	0300	400 amp		1.60	10		4,175	420		4,595	5,250	
	0350	600 amp		1	16		6,325	670		6,995	7,975	
	1000	250 volt 3 pole, 30 amp	1 Elec	2.30	3.478		1,425	146		1,571	1,800	
	1100	60 amp		1.90	4.211		1,425	177		1,602	1,850	
	1150	100 amp		1.30	6.154		1,425	258		1,683	1,975	
	1200	200 amp	2 Elec	2	8		2,875	335		3,210	3,675	
	1300	600 amp	″	1	16		6,325	670		6,995	7,975	
	1500	Electrically operated, 480 volt 3 pole, 60 amp	1 Elec	1.90	4.211		2,750	177		2,927	3,300	
	1600	100 amp	″	1.30	6.154		2,750	258		3,008	3,425	
	1650	200 amp	2 Elec	2	8		4,525	335		4,860	5,525	
	1700	400 amp		1.60	10		6,350	420		6,770	7,625	
	1750	600 amp		1	16		9,150	670		9,820	11,100	
	2000	250 volt 3 pole, 30 amp	1 Elec	2.30	3.478		2,750	146		2,896	3,250	

For expanded coverage of these items see *Means Electrical Cost Data 2006*

	16415 Transfer Switches		CREW	DAILY OUTPUT	LABOR-HOURS	UNIT	2006 BARE COSTS MAT.	LABOR	EQUIP.	TOTAL	TOTAL INCL O&P	
100	2050	60 amp	1 Elec	1.90	4.211	Ea.	2,750	177		2,927	3,300	100
	2150	200 amp	2 Elec	2	8		4,525	335		4,860	5,525	
	2200	400 amp		1.60	10		6,350	420		6,770	7,625	
	2250	600 amp		1	16		9,150	670		9,820	11,100	
	2500	NEMA 3R, 480 volt 3 pole, 60 amp	1 Elec	1.80	4.444		3,250	187		3,437	3,850	
	2550	100 amp	″	1.20	6.667		3,250	280		3,530	3,975	
	2600	200 amp	2 Elec	1.80	8.889		5,125	375		5,500	6,225	
	2650	400 amp	″	1.40	11.429		6,950	480		7,430	8,375	
	2800	NEMA 3R, 250 volt 3 pole solid state, 100 amp	1 Elec	1.20	6.667		3,250	280		3,530	3,975	
	2850	150 amp	2 Elec	1.80	8.889		4,300	375		4,675	5,300	
	2900	250 volt 2 pole solid state, 100 amp	1 Elec	1.30	6.154		3,150	258		3,408	3,875	
	2950	150 amp	2 Elec	2	8		4,175	335		4,510	5,125	
600	0010	**AUTOMATIC TRANSFER SWITCHES**										600
	0015	Switches, enclosed 120/240 volt, 2 pole, 30 amp	1 Elec	2.40	3.333	Ea.	2,750	140		2,890	3,225	
	0020	70 amp		2	4		2,750	168		2,918	3,250	
	0030	100 amp		1.35	5.926		2,750	249		2,999	3,375	
	0040	225 amp	2 Elec	2.10	7.619		4,050	320		4,370	4,950	
	0050	400 amp		1.70	9.412		6,150	395		6,545	7,350	
	0060	600 amp		1.06	15.094		9,025	635		9,660	10,900	
	0070	800 amp		.84	19.048		10,600	800		11,400	12,900	
	0100	Switches, enclosed 480 volt, 3 pole, 30 amp	1 Elec	2.30	3.478		3,050	146		3,196	3,575	
	0200	60 amp		1.90	4.211		3,050	177		3,227	3,625	
	0300	100 amp		1.30	6.154		3,050	258		3,308	3,750	
	0400	150 amp	2 Elec	2.40	6.667		3,750	280		4,030	4,550	
	0500	225 amp		2	8		4,800	335		5,135	5,800	
	0600	260 amp		2	8		5,550	335		5,885	6,625	
	0700	400 amp		1.60	10		6,575	420		6,995	7,875	
	0800	600 amp		1	16		10,000	670		10,670	12,000	
	0900	800 amp		.80	20		11,800	840		12,640	14,200	
	1000	1000 amp		.76	21.053		15,500	885		16,385	18,500	
	1100	1200 amp		.70	22.857		21,300	960		22,260	24,900	
	1200	1600 amp		.60	26.667		24,200	1,125		25,325	28,400	
	1300	2000 amp		.50	32		27,100	1,350		28,450	31,900	
	1600	Accessories, time delay on engine starting					238			238	261	
	1700	Adjustable time delay on retransfer					238			238	261	
	1800	Shunt trips for customer connections					425			425	465	
	1900	Maintenance select switch					96.50			96.50	106	
	2000	Auxiliary contact when normal fails					112			112	123	
	2100	Pilot light-emergency					96.50			96.50	106	
	2200	Pilot light-normal					96.50			96.50	106	
	2300	Auxiliary contact-closed on normal					112			112	123	
	2400	Auxiliary contact-closed on emergency					112			112	123	
	2500	Emergency source sensing, frequency relay					490			490	540	

	16420 Enclosed Controllers		CREW	DAILY OUTPUT	LABOR-HOURS	UNIT	MAT.	LABOR	EQUIP.	TOTAL	TOTAL INCL O&P	
200	0010	**CONTACTORS, AC** Enclosed (NEMA 1)										200
	0050	Lighting, 600 volt 3 pole, electrically held										
	0100	20 amp	1 Elec	4	2	Ea.	234	84		318	385	
	0200	30 amp		3.60	2.222		248	93.50		341.50	415	
	0300	60 amp		3	2.667		495	112		607	720	
	0400	100 amp		2.50	3.200		820	134		954	1,100	
	0500	200 amp		1.40	5.714		1,925	240		2,165	2,500	
	0600	300 amp	2 Elec	1.60	10		4,100	420		4,520	5,150	
	0800	600 volt 3 pole, mechanically held, 30 amp	1 Elec	3.60	2.222		375	93.50		468.50	555	
	0900	60 amp		3	2.667		740	112		852	990	

16420 | Enclosed Controllers

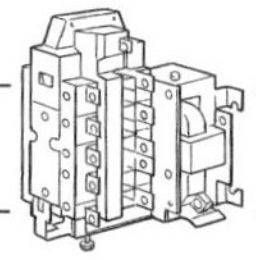

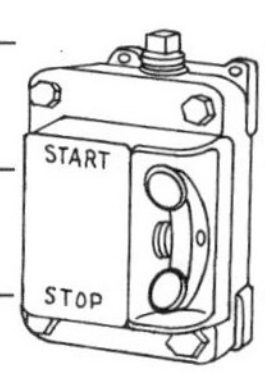

	16420 Enclosed Controllers		CREW	DAILY OUTPUT	LABOR-HOURS	UNIT	2006 BARE COSTS MAT.	LABOR	EQUIP.	TOTAL	TOTAL INCL O&P	
200	1000	75 amp	1 Elec	2.80	2.857	Ea.	1,050	120		1,170	1,325	200
	1100	100 amp		2.50	3.200		1,050	134		1,184	1,350	
	1200	150 amp		2	4		2,875	168		3,043	3,425	
	1300	200 amp		1.40	5.714		2,875	240		3,115	3,550	
	1500	Magnetic with auxiliary contact, size 00, 9 amp		4	2		171	84		255	315	
	1600	Size 0, 18 amp		4	2		204	84		288	355	
	1700	Size 1, 27 amp		3.60	2.222		232	93.50		325.50	400	
	1800	Size 2, 45 amp		3	2.667		430	112		542	650	
	1900	Size 3, 90 amp		2.50	3.200		695	134		829	970	
	2000	Size 4, 135 amp		2.30	3.478		1,600	146		1,746	1,975	
	2100	Size 5, 270 amp	2 Elec	1.80	8.889		3,350	375		3,725	4,250	
	2200	Size 6, 540 amp		1.20	13.333		9,775	560		10,335	11,700	
	2300	Size 7, 810 amp		1	16		13,200	670		13,870	15,500	
	2310	Size 8, 1215 amp		.80	20		20,500	840		21,340	23,800	
	2500	Magnetic, 240 volt, 1-2 pole, .75 HP motor	1 Elec	4	2		138	84		222	281	
	2520	2 HP motor		3.60	2.222		155	93.50		248.50	315	
	2540	5 HP motor		2.50	3.200		375	134		509	620	
	2560	10 HP motor		1.40	5.714		620	240		860	1,050	
	2600	240 volt or less, 3 pole, .75 HP motor		4	2		138	84		222	281	
	2620	5 HP motor		3.60	2.222		171	93.50		264.50	330	
	2640	10 HP motor		3.60	2.222		199	93.50		292.50	365	
	2660	15 HP motor		2.50	3.200		395	134		529	640	
	2700	25 HP motor		2.50	3.200		395	134		529	640	
	2720	30 HP motor	2 Elec	2.80	5.714		660	240		900	1,100	
	2740	40 HP motor		2.80	5.714		660	240		900	1,100	
	2760	50 HP motor		1.60	10		660	420		1,080	1,375	
	2800	75 HP motor		1.60	10		1,550	420		1,970	2,350	
	2820	100 HP motor		1	16		1,550	670		2,220	2,725	
	2860	150 HP motor		1	16		3,300	670		3,970	4,675	
	2880	200 HP motor		1	16		3,300	670		3,970	4,675	
	3000	600 volt, 3 pole, 5 HP motor	1 Elec	4	2		171	84		255	315	
	3020	10 HP motor		3.60	2.222		199	93.50		292.50	365	
	3040	25 HP motor		3	2.667		395	112		507	610	
	3100	50 HP motor		2.50	3.200		660	134		794	935	
	3160	100 HP motor	2 Elec	2.80	5.714		1,550	240		1,790	2,075	
	3220	200 HP motor	″	1.60	10		3,300	420		3,720	4,300	
220	0010	**CONTROL STATIONS**										220
	0050	NEMA 1, heavy duty, stop/start	1 Elec	8	1	Ea.	127	42		169	205	
	0100	Stop/start, pilot light		6.20	1.290		173	54		227	274	
	0200	Hand/off/automatic		6.20	1.290		94	54		148	187	
	0400	Stop/start/reverse		5.30	1.509		171	63.50		234.50	286	
	0500	NEMA 7, heavy duty, stop/start		6	1.333		216	56		272	325	
	0600	Stop/start, pilot light		4	2		265	84		349	420	
	0700	NEMA 7 or 9, 1 element		6	1.333		192	56		248	298	
	0800	2 element		6	1.333		229	56		285	340	
	0900	3 element		4	2		515	84		599	695	
	0910	Selector switch, 2 position		6	1.333		192	56		248	298	
	0920	3 position		4	2		198	84		282	345	
	0930	Oiltight, 1 element		8	1		88.50	42		130.50	162	
	0940	2 element		6.20	1.290		127	54		181	224	
	0950	3 element		5.30	1.509		171	63.50		234.50	286	
	0960	Selector switch, 2 position		6.20	1.290		94	54		148	187	
	0970	3 position		5.30	1.509		94	63.50		157.50	201	
240	0010	**CONTROL SWITCHES** Field installed										240
	6000	Push button 600V 10A, momentary contact										

ELECTRICAL 16

16420 | Enclosed Controllers

			Crew	Daily Output	Labor-Hours	Unit	2006 Bare Costs Mat.	Labor	Equip.	Total	Total Incl O&P	
240	6150	Standard operator with colored button	1 Elec	34	.235	Ea.	14.90	9.90		24.80	31.50	240
	6160	With single block 1NO 1NC		18	.444		31.50	18.65		50.15	63.50	
	6170	With double block 2NO 2NC		15	.533		48	22.50		70.50	87.50	
	6180	Stnd operator w/mushroom button 1-9/16" diam.		34	.235		31.50	9.90		41.40	49.50	
	6190	Stnd operator w/mushroom button 2-1/4" diam.										
	6200	With single block 1NO 1NC	1 Elec	18	.444	Ea.	48	18.65		66.65	82	
	6210	With double block 2NO 2NC		15	.533		64.50	22.50		87	106	
	6500	Maintained contact, selector operator		34	.235		48	9.90		57.90	68	
	6510	With single block 1NO 1NC		18	.444		64.50	18.65		83.15	100	
	6520	With double block 2NO 2NC		15	.533		81	22.50		103.50	124	
	6560	Spring-return selector operator		34	.235		48	9.90		57.90	68	
	6570	With single block 1NO 1NC		18	.444		64.50	18.65		83.15	100	
	6580	With double block 2NO 2NC		15	.533		81	22.50		103.50	124	
	6620	Transformer operator w/illuminated										
	6630	button 6V #12 lamp	1 Elec	32	.250	Ea.	81	10.50		91.50	106	
	6640	With single block 1NO 1NC w/guard		16	.500		97.50	21		118.50	140	
	6650	With double block 2NO 2NC w/guard		13	.615		114	26		140	166	
	6690	Combination operator		34	.235		48	9.90		57.90	68	
	6700	With single block 1NO 1NC		18	.444		64.50	18.65		83.15	100	
	6710	With double block 2NO 2NC		15	.533		81	22.50		103.50	124	
	9000	Indicating light unit, full voltage										
	9010	110-125V front mount	1 Elec	32	.250	Ea.	64.50	10.50		75	87	
	9020	130V resistor type		32	.250		48	10.50		58.50	69	
	9030	6V transformer type		32	.250		59	10.50		69.50	81	
800	0010	**RELAYS** Enclosed (NEMA 1)										800
	0050	600 volt AC, 1 pole, 12 amp	1 Elec	5.30	1.509	Ea.	83	63.50		146.50	189	
	0100	2 pole, 12 amp		5	1.600		83	67		150	195	
	0200	4 pole, 10 amp		4.50	1.778		110	74.50		184.50	236	
	0500	250 volt DC, 1 pole, 15 amp		5.30	1.509		109	63.50		172.50	217	
	0600	2 pole, 10 amp		5	1.600		104	67		171	218	
	0700	4 pole, 4 amp		4.50	1.778		137	74.50		211.50	266	

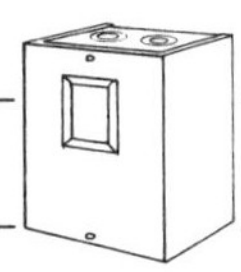

16440 | Swbds, Panels & Control Centers

			Crew	Daily Output	Labor-Hours	Unit	Mat.	Labor	Equip.	Total	Total Incl O&P	
500	0010	**LOAD CENTERS** (residential type) R262416-50										500
	0100	3 wire, 120/240V, 1 phase, including 1 pole plug-in breakers										
	0200	100 amp main lugs, indoor, 8 circuits	1 Elec	1.40	5.714	Ea.	155	240		395	540	
	0300	12 circuits		1.20	6.667		216	280		496	670	
	0400	Rainproof, 8 circuits		1.40	5.714		186	240		426	575	
	0500	12 circuits		1.20	6.667		262	280		542	720	
	0600	200 amp main lugs, indoor, 16 circuits	R-1A	1.80	8.889		300	305		605	810	
	0700	20 circuits		1.50	10.667		380	365		745	995	
	0800	24 circuits		1.30	12.308		495	420		915	1,200	
	0900	30 circuits		1.20	13.333		570	455		1,025	1,350	
	1000	42 circuits		.80	20		790	685		1,475	1,950	
	1200	Rainproof, 16 circuits		1.80	8.889		360	305		665	875	
	1300	20 circuits		1.50	10.667		430	365		795	1,050	
	1400	24 circuits		1.30	12.308		645	420		1,065	1,375	
	1500	30 circuits		1.20	13.333		720	455		1,175	1,500	
	1600	42 circuits		.80	20		940	685		1,625	2,100	
	1800	400 amp main lugs, indoor, 42 circuits		.72	22.222		1,100	760		1,860	2,425	
	1900	Rainproof, 42 circuit		.72	22.222		1,300	760		2,060	2,650	
	2200	Plug in breakers, 20 amp, 1 pole, 4 wire, 120/208 volts										
	2210	125 amp main lugs, indoor, 12 circuits	1 Elec	1.20	6.667	Ea.	283	280		563	740	
	2300	18 circuits		.80	10		400	420		820	1,075	
	2400	Rainproof, 12 circuits		1.20	6.667		325	280		605	790	

 Important: See the Reference Section for supporting data - Crews, Rental Equipment, City Cost Indexes and Reference Data

16440 | Swbds, Panels & Control Centers

R262416-50

			CREW	DAILY OUTPUT	LABOR-HOURS	UNIT	2006 BARE COSTS MAT.	LABOR	EQUIP.	TOTAL	TOTAL INCL O&P	
500	2500	18 circuits	1 Elec	.80	10	Ea.	440	420		860	1,125	500
	2600	200 amp main lugs, indoor, 24 circuits	R-1A	1.30	12.308		540	420		960	1,250	
	2700	30 circuits		1.20	13.333		615	455		1,070	1,400	
	2800	36 circuits		1	16		745	545		1,290	1,675	
	2900	42 circuits		.80	20		815	685		1,500	1,975	
	3000	Rainproof, 24 circuits		1.30	12.308		600	420		1,020	1,325	
	3100	30 circuits		1.20	13.333		675	455		1,130	1,475	
	3200	36 circuits		1	16		935	545		1,480	1,900	
	3300	42 circuits		.80	20		1,000	685		1,685	2,175	
	3500	400 amp main lugs, indoor, 42 circuits		.72	22.222		1,225	760		1,985	2,550	
	3600	Rainproof, 42 circuits		.72	22.222		1,475	760		2,235	2,825	
	3700	Plug-in breakers, 20 amp, 1 pole, 3 wire, 120/240 volts										
	3800	100 amp main breaker, indoor, 12 circuits	1 Elec	1.20	6.667	Ea.	305	280		585	765	
	3900	18 circuits	"	.80	10		410	420		830	1,100	
	4000	200 amp main breaker, indoor, 20 circuits	R-1A	1.50	10.667		550	365		915	1,175	
	4200	24 circuits		1.30	12.308		665	420		1,085	1,400	
	4300	30 circuits		1.20	13.333		740	455		1,195	1,525	
	4400	40 circuits		.90	17.778		920	610		1,530	2,000	
	4500	Rainproof, 20 circuits		1.50	10.667		600	365		965	1,250	
	4600	24 circuits		1.30	12.308		715	420		1,135	1,450	
	4700	30 circuits		1.20	13.333		820	455		1,275	1,625	
	4800	40 circuits		.90	17.778		1,000	610		1,610	2,075	
	5000	400 amp main breaker, indoor, 42 circuits		.72	22.222		2,650	760		3,410	4,100	
	5100	Rainproof, 42 circuits		.72	22.222		2,950	760		3,710	4,450	
	5300	Plug in breakers, 20 amp, 1 pole, 4 wire, 120/208 volts										
	5400	200 amp main breaker, indoor, 30 circuits	R-1A	1.20	13.333	Ea.	1,125	455		1,580	1,950	
	5500	42 circuits		.80	20		1,375	685		2,060	2,575	
	5600	Rainproof, 30 circuits		1.20	13.333		1,200	455		1,655	2,050	
	5700	42 circuits		.80	20		1,450	685		2,135	2,675	
620	0010	**MOTOR CONTROL CENTER COMPONENTS**										620
	0100	Starter, size 1, FVNR, NEMA 1, type A, fusible	1 Elec	2.70	2.963	Ea.	1,275	124		1,399	1,600	
	0120	Circuit breaker		2.70	2.963		1,375	124		1,499	1,725	
	0140	Type B, fusible		2.70	2.963		1,400	124		1,524	1,725	
	0160	Circuit breaker		2.70	2.963		1,525	124		1,649	1,875	
	0180	NEMA 12, type A, fusible		2.60	3.077		1,275	129		1,404	1,625	
	0200	Circuit breaker		2.60	3.077		1,400	129		1,529	1,725	
	0220	Type B, fusible		2.60	3.077		1,425	129		1,554	1,750	
	0240	Circuit breaker		2.60	3.077		1,525	129		1,654	1,900	
	0300	Starter, size 1, FVR, NEMA 1, type A, fusible		2	4		1,825	168		1,993	2,250	
	0320	Circuit breaker		2	4		1,825	168		1,993	2,250	
	0340	Type B, fusible		2	4		2,000	168		2,168	2,450	
	0360	Circuit breaker		2	4		2,000	168		2,168	2,450	
	0380	NEMA 12, type A, fusible		1.90	4.211		1,825	177		2,002	2,300	
	0400	Circuit breaker		1.90	4.211		1,825	177		2,002	2,300	
	0420	Type B, fusible		1.90	4.211		2,025	177		2,202	2,500	
	0440	Circuit breaker		1.90	4.211		2,025	177		2,202	2,500	
	0490	Starter size 1, 2 speed, separate winding										
	0500	NEMA 1, type A, fusible	1 Elec	2.60	3.077	Ea.	2,400	129		2,529	2,825	
	0520	Circuit breaker		2.60	3.077		2,400	129		2,529	2,825	
	0540	Type B, fusible		2.60	3.077		2,625	129		2,754	3,100	
	0560	Circuit breaker		2.60	3.077		2,625	129		2,754	3,100	
	0580	NEMA 12, type A, fusible		2.50	3.200		2,425	134		2,559	2,875	
	0600	Circuit breaker		2.50	3.200		2,425	134		2,559	2,875	
	0620	Type B, fusible		2.50	3.200		2,675	134		2,809	3,150	
	0640	Circuit breaker		2.50	3.200		2,675	134		2,809	3,150	

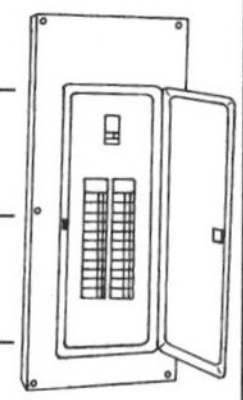

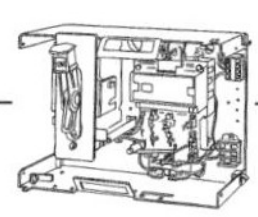

16440 | Swbds, Panels & Control Centers		Crew	Daily Output	Labor-Hours	Unit	2006 Bare Costs Mat.	Labor	Equip.	Total	Total Incl O&P
620 0650	Starter size 1, 2 speed, consequent pole									
0660	NEMA 1, type A, fusible	1 Elec	2.60	3.077	Ea.	2,400	129		2,529	2,825
0680	Circuit breaker		2.60	3.077		2,400	129		2,529	2,825
0700	Type B, fusible		2.60	3.077		2,625	129		2,754	3,100
0720	Circuit breaker		2.60	3.077		2,625	129		2,754	3,100
0740	NEMA 12, type A, fusible		2.50	3.200		2,425	134		2,559	2,875
0760	Circuit breaker		2.50	3.200		2,425	134		2,559	2,875
0780	Type B, fusible		2.50	3.200		2,650	134		2,784	3,125
0800	Circuit breaker		2.50	3.200		2,675	134		2,809	3,150
0810	Starter size 1, 2 speed, space only									
0820	NEMA 1, type A, fusible	1 Elec	16	.500	Ea.	535	21		556	625
0840	Circuit breaker		16	.500		535	21		556	625
0860	Type B, fusible		16	.500		535	21		556	625
0880	Circuit breaker		16	.500		535	21		556	625
0900	NEMA 12, type A, fusible		15	.533		560	22.50		582.50	650
0920	Circuit breaker		15	.533		560	22.50		582.50	650
0940	Type B, fusible		15	.533		560	22.50		582.50	650
0960	Circuit breaker		15	.533		560	22.50		582.50	650
1100	Starter size 2, FVNR, NEMA 1, type A, fusible	2 Elec	4	4		1,425	168		1,593	1,825
1120	Circuit breaker		4	4		1,575	168		1,743	1,975
1140	Type B, fusible		4	4		1,575	168		1,743	1,975
1160	Circuit breaker		4	4		1,725	168		1,893	2,125
1180	NEMA 12, type A, fusible		3.80	4.211		1,450	177		1,627	1,875
1200	Circuit breaker		3.80	4.211		1,600	177		1,777	2,025
1220	Type B, fusible		3.80	4.211		1,600	177		1,777	2,025
1240	Circuit breaker		3.80	4.211		1,725	177		1,902	2,175
1300	FVR, NEMA 1, type A, fusible		3.20	5		2,425	210		2,635	2,975
1320	Circuit breaker		3.20	5		2,425	210		2,635	2,975
1340	Type B, fusible		3.20	5		2,650	210		2,860	3,250
1360	Circuit breaker		3.20	5		2,650	210		2,860	3,250
1380	NEMA type 12, type A, fusible		3	5.333		2,475	224		2,699	3,050
1400	Circuit breaker		3	5.333		2,475	224		2,699	3,050
1420	Type B, fusible		3	5.333		2,700	224		2,924	3,325
1440	Circuit breaker		3	5.333		2,700	224		2,924	3,325
1490	Starter size 2, 2 speed, separate winding									
1500	NEMA 1, type A, fusible	2 Elec	3.80	4.211	Ea.	2,700	177		2,877	3,250
1520	Circuit breaker		3.80	4.211		2,700	177		2,877	3,250
1540	Type B, fusible		3.80	4.211		2,975	177		3,152	3,550
1560	Circuit breaker		3.80	4.211		2,975	177		3,152	3,550
1570	NEMA 12, type A, fusible		3.60	4.444		2,750	187		2,937	3,325
1580	Circuit breaker		3.60	4.444		2,750	187		2,937	3,325
1600	Type B, fusible		3.60	4.444		3,025	187		3,212	3,625
1620	Circuit breaker		3.60	4.444		3,025	187		3,212	3,625
1630	Starter size 2, 2 speed, consequent pole									
1640	NEMA 1, type A, fusible	2 Elec	3.80	4.211	Ea.	3,075	177		3,252	3,650
1660	Circuit breaker		3.80	4.211		3,075	177		3,252	3,650
1680	Type B, fusible		3.80	4.211		3,250	177		3,427	3,850
1700	Circuit breaker		3.80	4.211		3,250	177		3,427	3,850
1720	NEMA 12, type A, fusible		3.80	4.211		3,075	177		3,252	3,650
1740	Circuit breaker		3.60	4.444		3,100	187		3,287	3,725
1760	Type B, fusible		3.60	4.444		3,300	187		3,487	3,925
1780	Circuit breaker		3.60	4.444		3,300	187		3,487	3,925
1830	Starter size 2, autotransformer									
1840	NEMA 1, type A, fusible	2 Elec	3.40	4.706	Ea.	4,775	198		4,973	5,575
1860	Circuit breaker		3.40	4.706		4,900	198		5,098	5,675
1880	Type B, fusible		3.40	4.706		5,225	198		5,423	6,050 620

	16440 | Swbds, Panels & Control Centers	CREW	DAILY OUTPUT	LABOR-HOURS	UNIT	2006 BARE COSTS MAT.	LABOR	EQUIP.	TOTAL	TOTAL INCL O&P	
620 1900	Circuit breaker	2 Elec	3.40	4.706	Ea.	5,225	198		5,423	6,050	620
1920	NEMA 12, type A, fusible		3.20	5		4,875	210		5,085	5,675	
1940	Circuit breaker		3.20	5		4,875	210		5,085	5,675	
1960	Type B, fusible		3.20	5		5,325	210		5,535	6,175	
1980	Circuit breaker		3.20	5		5,325	210		5,535	6,175	
2030	Starter size 2, space only										
2040	NEMA 1, type A, fusible	1 Elec	16	.500	Ea.	535	21		556	625	
2060	Circuit breaker		16	.500		535	21		556	625	
2080	Type B, fusible		16	.500		535	21		556	625	
2100	Circuit breaker		16	.500		535	21		556	625	
2120	NEMA 12, type A, fusible		15	.533		560	22.50		582.50	650	
2140	Circuit breaker		15	.533		560	22.50		582.50	650	
2160	Type B, fusible		15	.533		560	22.50		582.50	650	
2180	Circuit breaker		15	.533		560	22.50		582.50	650	
2300	Starter size 3, FVNR, NEMA 1, type A, fusible	2 Elec	2	8		2,725	335		3,060	3,525	
2320	Circuit breaker		2	8		2,425	335		2,760	3,175	
2340	Type B, fusible		2	8		3,000	335		3,335	3,825	
2360	Circuit breaker		2	8		2,675	335		3,010	3,450	
2380	NEMA 12, type A, fusible		1.90	8.421		2,775	355		3,130	3,600	
2400	Circuit breaker		1.90	8.421		2,475	355		2,830	3,250	
2420	Type B, fusible		1.90	8.421		3,100	355		3,455	3,950	
2440	Circuit breaker		1.90	8.421		2,700	355		3,055	3,525	
2500	Starter size 3, FVR, NEMA 1, type A, fusible		1.60	10		3,725	420		4,145	4,725	
2520	Circuit breaker		1.60	10		3,550	420		3,970	4,550	
2540	Type B, fusible		1.60	10		4,050	420		4,470	5,100	
2560	Circuit breaker		1.60	10		3,875	420		4,295	4,925	
2580	NEMA 12, type A, fusible		1.50	10.667		3,800	450		4,250	4,875	
2600	Circuit breaker		1.50	10.667		3,625	450		4,075	4,675	
2620	Type B, fusible		1.50	10.667		4,125	450		4,575	5,225	
2640	Circuit breaker		1.50	10.667		4,275	450		4,725	5,400	
2690	Starter size 3, 2 speed, separate winding										
2700	NEMA 1, type A, fusible	2 Elec	2	8	Ea.	4,275	335		4,610	5,225	
2720	Circuit breaker		2	8		3,775	335		4,110	4,675	
2740	Type B, fusible		2	8		4,675	335		5,010	5,675	
2760	Circuit breaker		2	8		4,125	335		4,460	5,075	
2780	NEMA 12, type A, fusible		1.90	8.421		4,350	355		4,705	5,350	
2800	Circuit breaker		1.90	8.421		3,850	355		4,205	4,775	
2820	Type B, fusible		1.90	8.421		4,775	355		5,130	5,800	
2840	Circuit breaker		1.90	8.421		4,200	355		4,555	5,175	
2850	Starter size 3, 2 speed, consequent pole										
2860	NEMA 1, type A, fusible	2 Elec	2	8	Ea.	4,775	335		5,110	5,775	
2880	Circuit breaker		2	8		4,275	335		4,610	5,225	
2900	Type B, fusible		2	8		5,175	335		5,510	6,200	
2920	Circuit breaker		2	8		4,625	335		4,960	5,600	
2940	NEMA 12, type A, fusible		1.90	8.421		4,850	355		5,205	5,900	
2960	Circuit breaker		1.90	8.421		4,350	355		4,705	5,325	
2980	Type B, fusible		1.90	8.421		5,250	355		5,605	6,325	
3000	Circuit breaker		1.90	8.421		4,700	355		5,055	5,725	
3100	Starter size 3, autotransformer, NEMA 1, type A, fusible		1.60	10		5,925	420		6,345	7,175	
3120	Circuit breaker		1.60	10		5,925	420		6,345	7,175	
3140	Type B, fusible		1.60	10		6,475	420		6,895	7,775	
3160	Circuit breaker		1.60	10		6,475	420		6,895	7,775	
3180	NEMA 12, type A, fusible		1.50	10.667		6,050	450		6,500	7,350	
3200	Circuit breaker		1.50	10.667		6,050	450		6,500	7,350	
3220	Type B, fusible		1.50	10.667		6,600	450		7,050	7,950	
3240	Circuit breaker		1.50	10.667		6,600	450		7,050	7,950	

ELECTRICAL 16

For expanded coverage of these items see *Means Electrical Cost Data 2006*

16440 | Swbds, Panels & Control Centers

			CREW	DAILY OUTPUT	LABOR-HOURS	UNIT	2006 BARE COSTS MAT.	LABOR	EQUIP.	TOTAL	TOTAL INCL O&P	
620	3260	Starter size 3, space only, NEMA 1, type A, fusible	1 Elec	15	.533	Ea.	920	22.50		942.50	1,050	620
	3280	Circuit breaker		15	.533		715	22.50		737.50	820	
	3300	Type B, fusible		15	.533		920	22.50		942.50	1,050	
	3320	Circuit breaker		15	.533		715	22.50		737.50	820	
	3340	NEMA 12, type A, fusible		14	.571		975	24		999	1,100	
	3360	Circuit breaker		14	.571		760	24		784	870	
	3380	Type B, fusible		14	.571		975	24		999	1,100	
	3400	Circuit breaker		14	.571		760	24		784	870	
	3500	Starter size 4, FVNR, NEMA 1, type A, fusible	2 Elec	1.60	10		3,550	420		3,970	4,550	
	3520	Circuit breaker		1.60	10		3,225	420		3,645	4,200	
	3540	Type B, fusible		1.60	10		3,875	420		4,295	4,925	
	3560	Circuit breaker		1.60	10		3,550	420		3,970	4,550	
	3580	NEMA 12, type A, fusible		1.50	10.667		3,625	450		4,075	4,700	
	3600	Circuit breaker		1.50	10.667		3,275	450		3,725	4,300	
	3620	Type B, fusible		1.50	10.667		3,975	450		4,425	5,050	
	3640	Circuit breaker		1.50	10.667		3,600	450		4,050	4,675	
	3700	Starter size 4, FVR, NEMA 1, type A, fusible		1.20	13.333		4,850	560		5,410	6,200	
	3720	Circuit breaker		1.20	13.333		4,375	560		4,935	5,675	
	3740	Type B, fusible		1.20	13.333		5,300	560		5,860	6,700	
	3760	Circuit breaker		1.20	13.333		4,825	560		5,385	6,175	
	3780	NEMA 12, type A, fusible		1.16	13.793		4,950	580		5,530	6,350	
	3800	Circuit breaker		1.16	13.793		4,450	580		5,030	5,775	
	3820	Type B, fusible		1.16	13.793		4,475	580		5,055	5,825	
	3840	Circuit breaker		1.16	13.793		4,875	580		5,455	6,275	
	3890	Starter size 4, 2 speed, separate windings										
	3900	NEMA 1, type A, fusible	2 Elec	1.60	10	Ea.	6,125	420		6,545	7,375	
	3920	Circuit breaker		1.60	10		4,600	420		5,020	5,700	
	3940	Type B, fusible		1.60	10		6,725	420		7,145	8,050	
	3960	Circuit breaker		1.60	10		5,050	420		5,470	6,200	
	3980	NEMA 12, type A, fusible		1.50	10.667		6,225	450		6,675	7,550	
	4000	Circuit breaker		1.50	10.667		4,675	450		5,125	5,850	
	4020	Type B, fusible		1.50	10.667		6,825	450		7,275	8,225	
	4040	Circuit breaker		1.50	10.667		5,125	450		5,575	6,325	
	4050	Starter size 4, 2 speed, consequent pole										
	4060	NEMA 1, type A, fusible	2 Elec	1.60	10	Ea.	7,150	420		7,570	8,525	
	4080	Circuit breaker		1.60	10		5,325	420		5,745	6,500	
	4100	Type B, fusible		1.60	10		7,875	420		8,295	9,300	
	4120	Circuit breaker		1.60	10		5,825	420		6,245	7,075	
	4140	NEMA 12, type A, fusible		1.50	10.667		7,250	450		7,700	8,675	
	4160	Circuit breaker		1.50	10.667		5,400	450		5,850	6,625	
	4180	Type B, fusible		1.50	10.667		7,975	450		8,425	9,475	
	4200	Circuit breaker		1.50	10.667		5,900	450		6,350	7,200	
	4300	Starter size 4, autotransformer, NEMA 1, type A, fusible		1.30	12.308		7,050	515		7,565	8,550	
	4320	Circuit breaker		1.30	12.308		7,100	515		7,615	8,600	
	4340	Type B, fusible		1.30	12.308		7,750	515		8,265	9,325	
	4360	Circuit breaker		1.30	12.308		7,750	515		8,265	9,325	
	4380	NEMA 12, type A, fusible		1.24	12.903		7,150	540		7,690	8,700	
	4400	Circuit breaker		1.24	12.903		7,200	540		7,740	8,750	
	4420	Type B, fusible		1.24	12.903		7,875	540		8,415	9,500	
	4440	Circuit breaker		1.24	12.903		7,850	540		8,390	9,475	
	4500	Starter size 4, space only, NEMA 1, type A, fusible	1 Elec	14	.571		1,275	24		1,299	1,425	
	4520	Circuit breaker		14	.571		920	24		944	1,050	
	4540	Type B, fusible		14	.571		1,275	24		1,299	1,425	
	4560	Circuit breaker		14	.571		920	24		944	1,050	
	4580	NEMA 12, type A, fusible		13	.615		1,350	26		1,376	1,550	
	4600	Circuit breaker		13	.615		975	26		1,001	1,125	

16440 | Swbds, Panels & Control Centers

	Line	Description	Crew	Daily Output	Labor-Hours	Unit	Mat. (2006 Bare Costs)	Labor (2006 Bare Costs)	Equip. (2006 Bare Costs)	Total (2006 Bare Costs)	Total Incl O&P	
620	4620	Type B, fusible	1 Elec	13	.615	Ea.	1,350	26		1,376	1,550	620
	4640	Circuit breaker		13	.615		975	26		1,001	1,125	
	4800	Starter size 5, FVNR, NEMA 1, type A, fusible	2 Elec	1	16		7,325	670		7,995	9,075	
	4820	Circuit breaker		1	16		5,200	670		5,870	6,750	
	4840	Type B, fusible		1	16		8,050	670		8,720	9,875	
	4860	Circuit breaker		1	16		5,725	670		6,395	7,325	
	4880	NEMA 12, type A, fusible		.96	16.667		7,425	700		8,125	9,250	
	4900	Circuit breaker		.96	16.667		5,300	700		6,000	6,900	
	4920	Type B, fusible		.96	16.667		8,125	700		8,825	10,000	
	4940	Circuit breaker		.96	16.667		5,825	700		6,525	7,475	
	5000	Starter size 5, FVR, NEMA 1, type A, fusible		.80	20		10,800	840		11,640	13,200	
	5020	Circuit breaker		.80	20		8,475	840		9,315	10,600	
	5040	Type B, fusible		.80	20		11,800	840		12,640	14,300	
	5060	Circuit breaker		.80	20		9,300	840		10,140	11,500	
	5080	NEMA 12, type A, fusible		.76	21.053		11,000	885		11,885	13,500	
	5100	Circuit breaker		.76	21.053		8,600	885		9,485	10,800	
	5120	Type B, fusible		.76	21.053		12,000	885		12,885	14,600	
	5140	Circuit breaker		.76	21.053		9,425	885		10,310	11,800	
	5190	Starter size 5, 2 speed, separate windings										
	5200	NEMA 1, type A, fusible	2 Elec	1	16	Ea.	14,900	670		15,570	17,400	
	5220	Circuit breaker		1	16		11,100	670		11,770	13,200	
	5240	Type B, fusible		1	16		16,300	670		16,970	19,000	
	5260	Circuit breaker		1	16		12,200	670		12,870	14,400	
	5280	NEMA 12, type A, fusible		.96	16.667		15,100	700		15,800	17,700	
	5300	Circuit breaker		.96	16.667		11,200	700		11,900	13,500	
	5320	Type B, fusible		.96	16.667		16,500	700		17,200	19,300	
	5340	Circuit breaker		.96	16.667		12,300	700		13,000	14,700	
	5400	Starter size 5, autotransformer, NEMA 1, type A, fusible		.70	22.857		12,200	960		13,160	14,900	
	5420	Circuit breaker		.70	22.857		10,100	960		11,060	12,600	
	5440	Type B, fusible		.70	22.857		13,400	960		14,360	16,200	
	5460	Circuit breaker		.70	22.857		11,100	960		12,060	13,700	
	5480	NEMA 12, type A, fusible		.68	23.529		12,300	990		13,290	15,000	
	5500	Circuit breaker		.68	23.529		10,200	990		11,190	12,700	
	5520	Type B, fusible		.68	23.529		13,500	990		14,490	16,300	
	5540	Circuit breakers		.68	23.529		11,200	990		12,190	13,800	
	5600	Starter size 5, space only, NEMA 1, type A, fusible	1 Elec	12	.667		1,625	28		1,653	1,850	
	5620	Circuit breaker		12	.667		1,625	28		1,653	1,850	
	5640	Type B, fusible		12	.667		1,625	28		1,653	1,850	
	5660	Circuit breaker		12	.667		1,075	28		1,103	1,225	
	5680	NEMA 12, type A, fusible		11	.727		1,725	30.50		1,755.50	1,950	
	5700	Circuit breaker		11	.727		1,150	30.50		1,180.50	1,300	
	5720	Type B, fusible		11	.727		1,725	30.50		1,755.50	1,950	
	5740	Circuit breaker		11	.727		1,150	30.50		1,180.50	1,300	
	5800	Fuse, light contactor NEMA 1, type A, 30 amp		2.70	2.963		1,275	124		1,399	1,600	
	5820	60 amp		2	4		1,425	168		1,593	1,825	
	5840	100 amp		1	8		2,725	335		3,060	3,525	
	5860	200 amp		.80	10		6,325	420		6,745	7,600	
	5880	Type B, 30 amp		2.70	2.963		1,375	124		1,499	1,725	
	5900	60 amp		2	4		1,550	168		1,718	1,950	
	5920	100 amp		1	8		3,000	335		3,335	3,825	
	5940	200 amp	2 Elec	1.60	10		6,925	420		7,345	8,275	
	5960	NEMA 12, type A, 30 amp	1 Elec	2.60	3.077		1,275	129		1,404	1,625	
	5980	60 amp		1.90	4.211		1,450	177		1,627	1,875	
	6000	100 amp		.95	8.421		2,775	355		3,130	3,600	
	6020	200 amp	2 Elec	1.50	10.667		6,425	450		6,875	7,775	
	6040	Type B, 30 amp	1 Elec	2.60	3.077		1,400	129		1,529	1,725	

620	16440 | Swbds, Panels & Control Centers	CREW	DAILY OUTPUT	LABOR-HOURS	UNIT	2006 BARE COSTS MAT.	LABOR	EQUIP.	TOTAL	TOTAL INCL O&P
6060	60 amp	1 Elec	1.90	4.211	Ea.	1,550	177		1,727	2,000
6080	100 amp		.95	8.421		3,050	355		3,405	3,900
6100	200 amp	2 Elec	1.50	10.667		7,025	450		7,475	8,425
6200	Circuit breaker, light contactor NEMA 1, type A, 30 amp	1 Elec	2.70	2.963		1,375	124		1,499	1,725
6220	60 amp		2	4		1,575	168		1,743	1,975
6240	100 amp		1	8		2,425	335		2,760	3,175
6260	200 amp	2 Elec	1.60	10		5,250	420		5,670	6,425
6280	Type B, 30 amp	1 Elec	2.70	2.963		1,525	124		1,649	1,875
6300	60 amp		2	4		1,725	168		1,893	2,150
6320	100 amp		1	8		2,650	335		2,985	3,425
6340	200 amp	2 Elec	1.60	10		5,725	420		6,145	6,950
6360	NEMA 12, type A, 30 amp	1 Elec	2.60	3.077		1,400	129		1,529	1,725
6380	60 amp		1.90	4.211		1,600	177		1,777	2,025
6400	100 amp		.95	8.421		2,475	355		2,830	3,250
6420	200 amp	2 Elec	1.50	10.667		5,275	450		5,725	6,500
6440	Type B, 30 amp	1 Elec	2.60	3.077		1,650	129		1,779	2,000
6460	60 amp		1.90	4.211		1,750	177		1,927	2,200
6480	100 amp		.95	8.421		2,675	355		3,030	3,500
6500	200 amp	2 Elec	1.50	10.667		5,775	450		6,225	7,075
6600	Fusible switch, NEMA 1, type A, 30 amp	1 Elec	5.30	1.509		840	63.50		903.50	1,025
6620	60 amp		5	1.600		895	67		962	1,100
6640	100 amp		4	2		990	84		1,074	1,225
6660	200 amp		3.20	2.500		1,700	105		1,805	2,025
6680	400 amp	2 Elec	4.60	3.478		4,025	146		4,171	4,650
6700	600 amp		3.20	5		4,350	210		4,560	5,100
6720	800 amp		2.60	6.154		11,900	258		12,158	13,500
6740	NEMA 12, type A, 30 amp	1 Elec	5.20	1.538		860	64.50		924.50	1,050
6760	60 amp		4.90	1.633		915	68.50		983.50	1,100
6780	100 amp		3.90	2.051		1,000	86		1,086	1,250
6800	200 amp		3.10	2.581		1,750	108		1,858	2,100
6820	400 amp	2 Elec	4.40	3.636		4,100	153		4,253	4,750
6840	600 amp		3	5.333		4,475	224		4,699	5,275
6860	800 amp		2.40	6.667		12,000	280		12,280	13,700
6900	Circuit breaker, NEMA 1, type A, 30 amp	1 Elec	5.30	1.509		770	63.50		833.50	945
6920	60 amp		5	1.600		770	67		837	950
6940	100 amp		4	2		770	84		854	975
6960	225 amp		3.20	2.500		1,375	105		1,480	1,675
6980	400 amp	2 Elec	4.60	3.478		2,625	146		2,771	3,100
7000	600 amp		3.20	5		3,050	210		3,260	3,675
7020	800 amp		2.60	6.154		6,375	258		6,633	7,425
7040	NEMA 12, type A, 30 amp	1 Elec	5.20	1.538		790	64.50		854.50	970
7060	60 amp		4.90	1.633		790	68.50		858.50	975
7080	100 amp		3.90	2.051		790	86		876	1,000
7100	225 amp		3.10	2.581		1,400	108		1,508	1,700
7120	400 amp	2 Elec	4.40	3.636		2,650	153		2,803	3,150
7140	600 amp		3	5.333		3,100	224		3,324	3,750
7160	800 amp		2.40	6.667		6,500	280		6,780	7,575
7300	Incoming line, main lug only, 600 amp, alum., NEMA 1		1.60	10		975	420		1,395	1,725
7320	NEMA 12		1.50	10.667		995	450		1,445	1,800
7340	Copper, NEMA 1		1.60	10		1,025	420		1,445	1,775
7360	800 amp, alum., NEMA 1		1.50	10.667		2,450	450		2,900	3,400
7380	NEMA 12		1.40	11.429		2,475	480		2,955	3,475
7400	Copper, NEMA 1		1.50	10.667		2,550	450		3,000	3,500
7420	1200 amp, copper, NEMA 1		1.40	11.429		2,625	480		3,105	3,650
7440	Incoming line, fusible switch, 400 amp, alum., NEMA 1		1.20	13.333		3,350	560		3,910	4,575
7460	NEMA 12		1.10	14.545		3,425	610		4,035	4,725

		16440 | Swbds, Panels & Control Centers	CREW	DAILY OUTPUT	LABOR-HOURS	UNIT	2006 BARE COSTS MAT.	LABOR	EQUIP.	TOTAL	TOTAL INCL O&P	
620	7480	Copper, NEMA 1	2 Elec	1.20	13.333	Ea.	3,400	560		3,960	4,625	620
	7500	600 amp, alum., NEMA 1		1.10	14.545		4,100	610		4,710	5,475	
	7520	NEMA 12		1	16		4,175	670		4,845	5,600	
	7540	Copper, NEMA 1		1.10	14.545		4,150	610		4,760	5,525	
	7560	Incoming line, circuit breaker, 225 amp, alum., NEMA 1		1.20	13.333		1,750	560		2,310	2,800	
	7580	NEMA 12		1.10	14.545		1,775	610		2,385	2,900	
	7600	Copper, NEMA 1		1.20	13.333		1,800	560		2,360	2,850	
	7620	400 amp, alum., NEMA 1		1.20	13.333		2,625	560		3,185	3,750	
	7640	NEMA 12		1.10	14.545		2,650	610		3,260	3,875	
	7660	Copper, NEMA 1		1.20	13.333		2,675	560		3,235	3,800	
	7680	600 amp, alum., NEMA 1		1.10	14.545		3,050	610		3,660	4,300	
	7700	NEMA 12		1	16		3,100	670		3,770	4,425	
	7720	Copper, NEMA 1		1.10	14.545		3,100	610		3,710	4,375	
	7740	800 amp, copper, NEMA 1		.90	17.778		6,375	745		7,120	8,175	
	7760	Incoming line, for copper bus, add					96.50			96.50	106	
	7780	For 65000 amp bus bracing, add					143			143	157	
	7800	For NEMA 3R enclosure, add					4,500			4,500	4,950	
	7820	For NEMA 12 enclosure, add					124			124	136	
	7840	For 1/4" x 1" ground bus, add	1 Elec	16	.500		80	21		101	120	
	7860	For 1/4" x 2" ground bus, add	"	12	.667		80	28		108	131	
	7880	Main rating basic section, alum., NEMA 1, 600 amp	2 Elec	1.60	10			420		420	645	
	7900	800 amp		1.40	11.429		261	480		741	1,025	
	7920	1200 amp		1.20	13.333		515	560		1,075	1,425	
	7940	For copper bus, add					385			385	425	
	7960	For 65000 amp bus bracing, add					260			260	286	
	7980	For NEMA 3R enclosure, add					4,500			4,500	4,950	
	8000	For NEMA 12, enclosure, add					124			124	136	
	8020	For 1/4" x 1" ground bus, add	1 Elec	16	.500		80	21		101	120	
	8040	For 1/4" x 2" ground bus, add		12	.667		80	28		108	131	
	8060	Unit devices, pilot light, standard		16	.500		69	21		90	108	
	8080	Pilot light, push to test		16	.500		96.50	21		117.50	139	
	8100	Pilot light, standard, and push button		12	.667		165	28		193	225	
	8120	Pilot light, push to test, and push button		12	.667		193	28		221	255	
	8140	Pilot light, standard, and select switch		12	.667		165	28		193	225	
	8160	Pilot light, push to test, and select switch		12	.667		193	28		221	255	
640	0010	**MOTOR CONTROL CENTER** Consists of starters & structures										640
	0050	Starters, class 1, type B, comb. MCP, FVNR, with										
	0100	control transformer, 10 HP, size 1, 12" high	1 Elec	2.70	2.963	Ea.	1,375	124		1,499	1,725	
	0200	25 HP, size 2, 18" high	2 Elec	4	4		1,575	168		1,743	1,975	
	0300	50 HP, size 3, 24" high		2	8		2,425	335		2,760	3,175	
	0350	75 HP, size 4, 24" high		1.60	10		3,225	420		3,645	4,200	
	0400	100 HP, size 4, 30" high		1.40	11.429		4,250	480		4,730	5,425	
	0500	200 HP, size 5, 48" high		1	16		6,325	670		6,995	8,000	
	0600	400 HP, size 6, 72" high		.80	20		13,000	840		13,840	15,600	
	0800	Structures, 600 amp, 22,000 rms, takes any										
	0900	combination of starters up to 72" high	2 Elec	1.60	10	Ea.	1,800	420		2,220	2,625	
	1000	Back to back, 72" front & 66" back	"	1.20	13.333		2,425	560		2,985	3,550	
	1100	For copper bus add per structure					215			215	236	
	1200	For NEMA 12, add per structure					124			124	136	
	1300	For 42,000 rms, add per structure					166			166	183	
	1400	For 100,000 rms, size 1 & 2, add					550			550	605	
	1500	Size 3, add					885			885	975	
	1600	Size 4, add					715			715	785	
	1700	For pilot lights, add per starter	1 Elec	16	.500		96.50	21		117.50	139	
	1800	For push button, add per starter		16	.500		96.50	21		117.50	139	

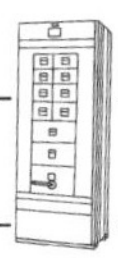

16440 | Swbds, Panels & Control Centers

			CREW	DAILY OUTPUT	LABOR-HOURS	UNIT	2006 BARE COSTS MAT.	LABOR	EQUIP.	TOTAL	TOTAL INCL O&P	
640	1900	For auxilliary contacts, add per starter	1 Elec	16	.500	Ea.	201	21		222	254	640
660	0010	**MOTOR STARTERS & CONTROLS**										660
	0050	Magnetic, FVNR, with enclosure and heaters, 480 volt										
	0080	2 HP, size 00	1 Elec	3.50	2.286	Ea.	185	96		281	350	
	0100	5 HP, size 0		2.30	3.478		224	146		370	470	
	0200	10 HP, size 1		1.60	5		251	210		461	600	
	0300	25 HP, size 2	2 Elec	2.20	7.273		470	305		775	990	
	0400	50 HP, size 3		1.80	8.889		770	375		1,145	1,425	
	0500	100 HP, size 4		1.20	13.333		1,700	560		2,260	2,750	
	0600	200 HP, size 5		.90	17.778		4,000	745		4,745	5,550	
	0610	400 HP, size 6		.80	20		11,300	840		12,140	13,800	
	0620	NEMA 7, 5 HP, size 0	1 Elec	1.60	5		930	210		1,140	1,350	
	0630	10 HP, size 1	"	1.10	7.273		970	305		1,275	1,550	
	0640	25 HP, size 2	2 Elec	1.80	8.889		1,550	375		1,925	2,275	
	0650	50 HP, size 3		1.20	13.333		2,325	560		2,885	3,425	
	0660	100 HP, size 4		.90	17.778		3,800	745		4,545	5,325	
	0670	200 HP, size 5		.50	32		8,850	1,350		10,200	11,800	
	0700	Combination, with motor circuit protectors, 5 HP, size 0	1 Elec	1.80	4.444		725	187		912	1,100	
	0800	10 HP, size 1	"	1.30	6.154		755	258		1,013	1,225	
	0900	25 HP, size 2	2 Elec	2	8		1,050	335		1,385	1,700	
	1000	50 HP, size 3		1.32	12.121		1,525	510		2,035	2,450	
	1200	100 HP, size 4		.80	20		3,325	840		4,165	4,950	
	1220	NEMA 7, 5 HP, size 0	1 Elec	1.30	6.154		1,650	258		1,908	2,200	
	1230	10 HP, size 1	"	1	8		1,675	335		2,010	2,375	
	1240	25 HP, size 2	2 Elec	1.32	12.121		2,250	510		2,760	3,250	
	1250	50 HP, size 3		.80	20		3,675	840		4,515	5,350	
	1260	100 HP, size 4		.60	26.667		5,725	1,125		6,850	8,025	
	1270	200 HP, size 5		.40	40		12,400	1,675		14,075	16,200	
	1400	Combination, with fused switch, 5 HP, size 0	1 Elec	1.80	4.444		555	187		742	900	
	1600	10 HP, size 1	"	1.30	6.154		595	258		853	1,050	
	1800	25 HP, size 2	2 Elec	2	8		965	335		1,300	1,575	
	2000	50 HP, size 3		1.32	12.121		1,625	510		2,135	2,575	
	2200	100 HP, size 4		.80	20		2,850	840		3,690	4,425	
	2610	NEMA 4, with start-stop pushbutton size 1	1 Elec	1.30	6.154		1,500	258		1,758	2,050	
	2620	Size 2	2 Elec	2	8		2,000	335		2,335	2,725	
	2630	Size 3		1.32	12.121		3,150	510		3,660	4,250	
	2640	Size 4		.80	20		4,850	840		5,690	6,650	
	2650	NEMA 4, FVNR, including control transformer										
	2660	Size 1	2 Elec	2.60	6.154	Ea.	1,375	258		1,633	1,900	
	2670	Size 2		2	8		1,950	335		2,285	2,675	
	2680	Size 3		1.32	12.121		3,125	510		3,635	4,200	
	2690	Size 4		.80	20		4,800	840		5,640	6,600	
	2710	Magnetic, FVR, control circuit transformer, NEMA 1, size 1	1 Elec	1.30	6.154		810	258		1,068	1,300	
	2720	Size 2	2 Elec	2	8		1,300	335		1,635	1,950	
	2730	Size 3		1.32	12.121		1,950	510		2,460	2,925	
	2740	Size 4		.80	20		4,275	840		5,115	6,000	
	2760	NEMA 4, size 1	1 Elec	1.10	7.273		1,150	305		1,455	1,750	
	2770	Size 2	2 Elec	1.60	10		1,875	420		2,295	2,700	
	2780	Size 3		1.20	13.333		2,800	560		3,360	3,975	
	2790	Size 4		.70	22.857		5,775	960		6,735	7,825	
	2820	NEMA 12, size 1	1 Elec	1.10	7.273		950	305		1,255	1,525	
	2830	Size 2	2 Elec	1.60	10		1,500	420		1,920	2,300	
	2840	Size 3		1.20	13.333		2,375	560		2,935	3,475	
	2850	Size 4		.70	22.857		4,875	960		5,835	6,850	
	2870	Combination FVR, fused, w/control XFMR & PB, NEMA 1, size 1	1 Elec	1	8		1,400	335		1,735	2,050	

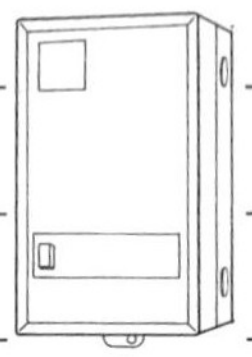

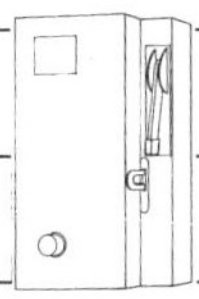

16440 | Swbds, Panels & Control Centers

			Crew	Daily Output	Labor-Hours	Unit	2006 Bare Costs Mat.	Labor	Equip.	Total	Total Incl O&P	
660	2880	Size 2	2 Elec	1.50	10.667	Ea.	1,975	450		2,425	2,875	660
	2890	Size 3		1.10	14.545		2,950	610		3,560	4,200	
	2900	Size 4		.70	22.857		5,975	960		6,935	8,050	
	2910	NEMA 4, size 1	1 Elec	.90	8.889		2,075	375		2,450	2,850	
	2920	Size 2	2 Elec	1.40	11.429		3,000	480		3,480	4,050	
	2930	Size 3		1	16		4,750	670		5,420	6,250	
	2940	Size 4		.60	26.667		7,900	1,125		9,025	10,400	
	2950	NEMA 12, size 1	1 Elec	1	8		1,575	335		1,910	2,275	
	2960	Size 2	2 Elec	1.40	11.429		2,225	480		2,705	3,200	
	2970	Size 3		1	16		3,275	670		3,945	4,625	
	2980	Size 4		.60	26.667		6,500	1,125		7,625	8,875	
	3010	Manual, single phase, w/pilot, 1 pole 120V NEMA 1	1 Elec	6.40	1.250		58	52.50		110.50	145	
	3020	NEMA 4		4	2		325	84		409	490	
	3030	2 pole, 120/240 V, NEMA 1		6.40	1.250		63.50	52.50		116	151	
	3040	NEMA 4		4	2		330	84		414	495	
	3041	3 phase, 3 pole 600V, NEMA 1		5.50	1.455		226	61		287	345	
	3042	NEMA 4		3.50	2.286		440	96		536	635	
	3043	NEMA 12		3.50	2.286		265	96		361	440	
	3070	Auxiliary contact, normally open					66			66	73	
	3500	Magnetic FVNR with NEMA 12, enclosure & heaters, 480 volt										
	3600	5 HP, size 0	1 Elec	2.20	3.636	Ea.	211	153		364	465	
	3700	10 HP, size 1	″	1.50	5.333		315	224		539	695	
	3800	25 HP, size 2	2 Elec	2	8		595	335		930	1,175	
	3900	50 HP, size 3		1.60	10		915	420		1,335	1,650	
	4000	100 HP, size 4		1	16		2,175	670		2,845	3,425	
	4100	200 HP, size 5		.80	20		5,250	840		6,090	7,075	
	4200	Combination, with motor circuit protectors, 5 HP, size 0	1 Elec	1.70	4.706		700	198		898	1,075	
	4300	10 HP, size 1	″	1.20	6.667		725	280		1,005	1,225	
	4400	25 HP, size 2	2 Elec	1.80	8.889		1,075	375		1,450	1,775	
	4500	50 HP, size 3		1.20	13.333		1,775	560		2,335	2,825	
	4600	100 HP, size 4		.74	21.622		4,000	910		4,910	5,800	
	4700	Combination, with fused switch, 5 HP, size 0	1 Elec	1.70	4.706		675	198		873	1,050	
	4800	10 HP, size 1	″	1.20	6.667		705	280		985	1,200	
	4900	25 HP, size 2	2 Elec	1.80	8.889		1,075	375		1,450	1,750	
	5000	50 HP, size 3		1.20	13.333		1,725	560		2,285	2,775	
	5100	100 HP, size 4		.74	21.622		3,475	910		4,385	5,225	
	5200	Factory installed controls, adders to size 0 thru 5										
	5300	Start-stop push button	1 Elec	32	.250	Ea.	44	10.50		54.50	64.50	
	5400	Hand-off-auto-selector switch		32	.250		44	10.50		54.50	64.50	
	5500	Pilot light		32	.250		83	10.50		93.50	107	
	5600	Start-stop-pilot		32	.250		127	10.50		137.50	156	
	5700	Auxiliary contact, NO or NC		32	.250		60.50	10.50		71	83	
	5800	NO-NC		32	.250		121	10.50		131.50	150	
	5810	Magnetic FVR, NEMA 7, w/heaters, size 1		.66	12.121		1,700	510		2,210	2,650	
	5830	Size 2	2 Elec	1.10	14.545		2,825	610		3,435	4,050	
	5840	Size 3		.70	22.857		4,525	960		5,485	6,450	
	5850	Size 4		.60	26.667		6,700	1,125		7,825	9,100	
	5860	Combination w/circuit breakers, heaters, control XFMR PB, size 1	1 Elec	.60	13.333		1,500	560		2,060	2,550	
	5870	Size 2	2 Elec	.80	20		1,950	840		2,790	3,450	
	5880	Size 3		.50	32		2,600	1,350		3,950	4,925	
	5890	Size 4		.40	40		5,125	1,675		6,800	8,250	
	5900	Manual, 240 volt, .75 HP motor	1 Elec	4	2		47	84		131	181	
	5910	2 HP motor		4	2		127	84		211	269	
	6000	Magnetic, 240 volt, 1 or 2 pole, .75 HP motor		4	2		185	84		269	330	
	6020	2 HP motor		4	2		204	84		288	355	
	6040	5 HP motor		3	2.667		293	112		405	495	

16440 | Swbds, Panels & Control Centers		CREW	DAILY OUTPUT	LABOR-HOURS	UNIT	2006 BARE COSTS MAT.	LABOR	EQUIP.	TOTAL	TOTAL INCL O&P
660 6060	10 HP motor	1 Elec	2.30	3.478	Ea.	725	146		871	1,025
6100	3 pole, .75 HP motor		3	2.667		185	112		297	375
6120	5 HP motor		2.30	3.478		251	146		397	500
6140	10 HP motor		1.60	5		470	210		680	845
6160	15 HP motor		1.60	5		470	210		680	845
6180	20 HP motor		1.10	7.273		770	305		1,075	1,325
6200	25 HP motor	2 Elec	2.20	7.273		770	305		1,075	1,325
6210	30 HP motor		1.80	8.889		770	375		1,145	1,425
6220	40 HP motor		1.80	8.889		1,700	375		2,075	2,450
6230	50 HP motor		1.80	8.889		1,700	375		2,075	2,450
6240	60 HP motor		1.20	13.333		4,000	560		4,560	5,275
6250	75 HP motor		1.20	13.333		4,000	560		4,560	5,275
6260	100 HP motor		1.20	13.333		4,000	560		4,560	5,275
6270	125 HP motor		.90	17.778		11,200	745		11,945	13,500
6280	150 HP motor		.90	17.778		11,200	745		11,945	13,500
6290	200 HP motor		.90	17.778		11,200	745		11,945	13,500
6400	Starter & nonfused disconnect, 240 volt, 1-2 pole, .75 HP motor	1 Elec	2	4		251	168		419	535
6410	2 HP motor		2	4		271	168		439	555
6420	5 HP motor		1.80	4.444		360	187		547	685
6430	10 HP motor		1.40	5.714		815	240		1,055	1,275
6440	3 pole, .75 HP motor		1.60	5		251	210		461	600
6450	5 HP motor		1.40	5.714		315	240		555	720
6460	10 HP motor		1.10	7.273		560	305		865	1,075
6470	15 HP motor		1	8		560	335		895	1,125
6480	20 HP motor	2 Elec	1.50	10.667		975	450		1,425	1,775
6490	25 HP motor		1.50	10.667		975	450		1,425	1,775
6500	30 HP motor		1.30	12.308		975	515		1,490	1,875
6510	40 HP motor		1.24	12.903		2,075	540		2,615	3,125
6520	50 HP motor		1.12	14.286		2,075	600		2,675	3,225
6530	60 HP motor		.90	17.778		4,375	745		5,120	5,950
6540	75 HP motor		.76	21.053		4,375	885		5,260	6,150
6550	100 HP motor		.70	22.857		4,375	960		5,335	6,275
6560	125 HP motor		.60	26.667		12,100	1,125		13,225	15,000
6570	150 HP motor		.52	30.769		12,100	1,300		13,400	15,300
6580	200 HP motor		.50	32		12,100	1,350		13,450	15,400
6600	Starter & fused disconnect, 240 volt, 1-2 pole, .75 HP motor	1 Elec	2	4		267	168		435	555
6610	2 HP motor		2	4		286	168		454	575
6620	5 HP motor		1.80	4.444		375	187		562	700
6630	10 HP motor		1.40	5.714		865	240		1,105	1,325
6640	3 pole, .75 HP motor		1.60	5		267	210		477	620
6650	5 HP motor		1.40	5.714		335	240		575	735
6660	10 HP motor		1.10	7.273		610	305		915	1,150
6690	15 HP motor		1	8		610	335		945	1,200
6700	20 HP motor	2 Elec	1.60	10		1,000	420		1,420	1,750
6710	25 HP motor		1.60	10		1,000	420		1,420	1,750
6720	30 HP motor		1.40	11.429		1,000	480		1,480	1,850
6730	40 HP motor		1.20	13.333		2,225	560		2,785	3,325
6740	50 HP motor	2 Elec	1.20	13.333		2,225	560		2,785	3,325
6750	60 HP motor		.90	17.778		4,500	745		5,245	6,100
6760	75 HP motor		.90	17.778		4,500	745		5,245	6,100
6770	100 HP motor		.70	22.857		4,500	960		5,460	6,425
6780	125 HP motor		.54	29.630		12,500	1,250		13,750	15,700
6790	Combination starter & nonfusible disconnect									
6800	240 volt, 1-2 pole, .75 HP motor	1 Elec	2	4	Ea.	525	168		693	840
6810	2 HP motor		2	4		525	168		693	840
6820	5 HP motor		1.50	5.333		555	224		779	955

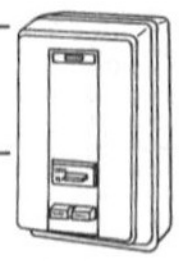

16440 | Swbds, Panels & Control Centers

			CREW	DAILY OUTPUT	LABOR-HOURS	UNIT	2006 BARE COSTS MAT.	LABOR	EQUIP.	TOTAL	TOTAL INCL O&P	
660	6830	10 HP motor	1 Elec	1.20	6.667	Ea.	865	280		1,145	1,375	660
	6840	3 pole, .75 HP motor		1.80	4.444		525	187		712	870	
	6850	5 HP motor		1.30	6.154		555	258		813	1,000	
	6860	10 HP motor		1	8		865	335		1,200	1,475	
	6870	15 HP motor		1	8		865	335		1,200	1,475	
	6880	20 HP motor	2 Elec	1.32	12.121		1,425	510		1,935	2,350	
	6890	25 HP motor		1.32	12.121		1,425	510		1,935	2,350	
	6900	30 HP motor		1.32	12.121		1,425	510		1,935	2,350	
	6910	40 HP motor		.80	20		2,725	840		3,565	4,300	
	6920	50 HP motor		.80	20		2,725	840		3,565	4,300	
	6930	60 HP motor		.70	22.857		6,075	960		7,035	8,150	
	6940	75 HP motor		.70	22.857		6,075	960		7,035	8,150	
	6950	100 HP motor		.70	22.857		6,075	960		7,035	8,150	
	6960	125 HP motor		.60	26.667		16,000	1,125		17,125	19,300	
	6970	150 HP motor		.60	26.667		16,000	1,125		17,125	19,300	
	6980	200 HP motor		.60	26.667		16,000	1,125		17,125	19,300	
	6990	Combination starter and fused disconnect										
	7000	240 volt, 1-2 pole, .75 HP motor	1 Elec	2	4	Ea.	545	168		713	860	
	7010	2 HP motor		2	4		545	168		713	860	
	7020	5 HP motor		1.50	5.333		570	224		794	975	
	7030	10 HP motor		1.20	6.667		885	280		1,165	1,400	
	7040	3 pole, .75 HP motor		1.80	4.444		545	187		732	890	
	7050	5 HP motor		1.30	6.154		570	258		828	1,025	
	7060	10 HP motor		1	8		885	335		1,220	1,500	
	7070	15 HP motor		1	8		885	335		1,220	1,500	
	7080	20 HP motor	2 Elec	1.32	12.121		1,475	510		1,985	2,400	
	7090	25 HP motor		1.32	12.121		1,475	510		1,985	2,400	
	7100	30 HP motor		1.32	12.121		1,475	510		1,985	2,400	
	7110	40 HP motor		.80	20		2,800	840		3,640	4,400	
	7120	50 HP motor		.80	20		2,800	840		3,640	4,400	
	7130	60 HP motor		.80	20		6,275	840		7,115	8,200	
	7140	75 HP motor		.70	22.857		6,275	960		7,235	8,375	
	7150	100 HP motor		.70	22.857		6,275	960		7,235	8,375	
	7160	125 HP motor		.70	22.857		16,600	960		17,560	19,700	
	7170	150 HP motor		.60	26.667		16,600	1,125		17,725	19,900	
	7180	200 HP motor		.60	26.667		16,600	1,125		17,725	19,900	
	7190	Combination starter & circuit breaker disconnect										
	7200	240 volt, 1-2 pole, .75 HP motor	1 Elec	2	4	Ea.	550	168		718	865	
	7210	2 HP motor		2	4		550	168		718	865	
	7220	5 HP motor		1.50	5.333		575	224		799	980	
	7230	10 HP motor		1.20	6.667		880	280		1,160	1,400	
	7240	3 pole, .75 HP motor		1.80	4.444		565	187		752	910	
	7250	5 HP motor		1.30	6.154		595	258		853	1,050	
	7260	10 HP motor		1	8		895	335		1,230	1,500	
	7270	15 HP motor		1	8		895	335		1,230	1,500	
	7280	20 HP motor	2 Elec	1.32	12.121		1,525	510		2,035	2,450	
	7290	25 HP motor		1.32	12.121		1,525	510		2,035	2,450	
	7300	30 HP motor		1.32	12.121		1,525	510		2,035	2,450	
	7310	40 HP motor		.80	20		3,325	840		4,165	4,950	
	7320	50 HP motor		.80	20		3,325	840		4,165	4,950	
	7330	60 HP motor		.80	20		7,675	840		8,515	9,725	
	7340	75 HP motor		.70	22.857		7,675	960		8,635	9,900	
	7350	100 HP motor		.70	22.857		7,675	960		8,635	9,900	
	7360	125 HP motor		.70	22.857		16,600	960		17,560	19,800	
	7370	150 HP motor		.60	26.667		16,600	1,125		17,725	20,000	
	7380	200 HP motor		.60	26.667		16,600	1,125		17,725	20,000	

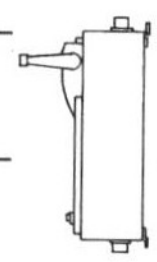

16440	Swbds, Panels & Control Centers	CREW	DAILY OUTPUT	LABOR-HOURS	UNIT	2006 BARE COSTS MAT.	LABOR	EQUIP.	TOTAL	TOTAL INCL O&P
660 7400	Magnetic FVNR with enclosure & heaters, 2 pole,									
7410	230 volt, 1 HP size 00	1 Elec	4	2	Ea.	168	84		252	315
7420	2 HP, size 0		4	2		188	84		272	335
7430	3 HP, size 1		3	2.667		215	112		327	410
7440	5 HP, size 1p		3	2.667		276	112		388	480
7450	115 volt, 1/3 HP, size 00		4	2		168	84		252	315
7460	1 HP, size 0		4	2		188	84		272	335
7470	2 HP, size 1		3	2.667		215	112		327	410
7480	3 HP, size 1P		3	2.667		268	112		380	465
7500	3 pole, 480 volt, 600 HP, size 7	2 Elec	.70	22.857		14,900	960		15,860	17,900
7590	Magnetic FVNR with heater, NEMA 1									
7600	600 volt, 3 pole, 5 HP motor	1 Elec	2.30	3.478	Ea.	224	146		370	470
7610	10 HP motor	"	1.60	5		251	210		461	600
7620	25 HP motor	2 Elec	2.20	7.273		470	305		775	990
7630	30 HP motor		1.80	8.889		770	375		1,145	1,425
7640	40 HP motor		1.80	8.889		770	375		1,145	1,425
7650	50 HP motor		1.80	8.889		770	375		1,145	1,425
7660	60 HP motor		1.20	13.333		1,700	560		2,260	2,750
7670	75 HP motor		1.20	13.333		1,700	560		2,260	2,750
7680	100 HP motor		1.20	13.333		1,700	560		2,260	2,750
7690	125 HP motor		.90	17.778		4,000	745		4,745	5,550
7700	150 HP motor		.90	17.778		4,000	745		4,745	5,550
7710	200 HP motor		.90	17.778		4,000	745		4,745	5,550
7750	Starter & nonfused disconnect, 600 volt, 3 pole, 5 HP motor	1 Elec	1.40	5.714		340	240		580	745
7760	10 HP motor	"	1.10	7.273		370	305		675	875
7770	25 HP motor	2 Elec	1.50	10.667		590	450		1,040	1,325
7780	30 HP motor		1.30	12.308		980	515		1,495	1,875
7790	40 HP motor		1.30	12.308		980	515		1,495	1,875
7800	50 HP motor		1.30	12.308		980	515		1,495	1,875
7810	60 HP motor		.92	17.391		2,050	730		2,780	3,375
7820	75 HP motor		.92	17.391		2,050	730		2,780	3,375
7830	100 HP motor		.84	19.048		2,050	800		2,850	3,475
7840	125 HP motor		.70	22.857		4,500	960		5,460	6,425
7850	150 HP motor		.70	22.857		4,500	960		5,460	6,425
7860	200 HP motor		.60	26.667		4,500	1,125		5,625	6,675
7870	Starter & fused disconnect, 600 volt, 3 pole, 5 HP motor	1 Elec	1.40	5.714		445	240		685	860
7880	10 HP motor	"	1.10	7.273		475	305		780	990
7890	25 HP motor	2 Elec	1.50	10.667		695	450		1,145	1,450
7900	30 HP motor		1.30	12.308		1,050	515		1,565	1,950
7910	40 HP motor		1.30	12.308		1,050	515		1,565	1,950
7920	50 HP motor		1.30	12.308		1,050	515		1,565	1,950
7930	60 HP motor		.92	17.391		2,200	730		2,930	3,550
7940	75 HP motor		.92	17.391		2,200	730		2,930	3,550
7950	100 HP motor		.84	19.048		2,200	800		3,000	3,650
7960	125 HP motor		.70	22.857		4,725	960		5,685	6,650
7970	150 HP motor		.70	22.857		4,725	960		5,685	6,650
7980	200 HP motor		.60	26.667		4,725	1,125		5,850	6,900
7990	Combination starter and nonfusible disconnect									
8000	600 volt, 3 pole, 5 HP motor	1 Elec	1.80	4.444	Ea.	525	187		712	870
8010	10 HP motor	"	1.30	6.154		555	258		813	1,000
8020	25 HP motor	2 Elec	2	8		865	335		1,200	1,475
8030	30 HP motor		1.32	12.121		1,475	510		1,985	2,400
8040	40 HP motor		1.32	12.121		1,475	510		1,985	2,400
8050	50 HP motor		1.32	12.121		1,425	510		1,935	2,350
8060	60 HP motor		.80	20		2,800	840		3,640	4,400
8070	75 HP motor		.80	20		2,800	840		3,640	4,400

16440 | Swbds, Panels & Control Centers		CREW	DAILY OUTPUT	LABOR-HOURS	UNIT	2006 BARE COSTS MAT.	LABOR	EQUIP.	TOTAL	TOTAL INCL O&P
660 8080	100 HP motor	2 Elec	.80	20	Ea.	2,725	840		3,565	4,300
8090	125 HP motor		.70	22.857		6,275	960		7,235	8,375
8100	150 HP motor		.70	22.857		6,275	960		7,235	8,375
8110	200 HP motor		.70	22.857		6,275	960		7,235	8,375
8140	Combination starter and fused disconnect									
8150	600 volt, 3 pole, 5 HP motor	1 Elec	1.80	4.444	Ea.	545	187		732	890
8160	10 HP motor	″	1.30	6.154		595	258		853	1,050
8170	25 HP motor	2 Elec	2	8		965	335		1,300	1,575
8180	30 HP motor		1.32	12.121		1,500	510		2,010	2,425
8190	40 HP motor		1.32	12.121		1,625	510		2,135	2,550
8200	50 HP motor		1.32	12.121		1,625	510		2,135	2,550
8210	60 HP motor		.80	20		2,825	840		3,665	4,425
8220	75 HP motor		.80	20		2,825	840		3,665	4,425
8230	100 HP motor		.80	20		2,825	840		3,665	4,425
8240	125 HP motor		.70	22.857		6,275	960		7,235	8,375
8250	150 HP motor		.70	22.857		6,275	960		7,235	8,375
8260	200 HP motor		.70	22.857		6,275	960		7,235	8,375
8290	Combination starter & circuit breaker disconnect									
8300	600 volt, 3 pole, 5 HP motor	1 Elec	1.80	4.444	Ea.	725	187		912	1,100
8310	10 HP motor	″	1.30	6.154		755	258		1,013	1,225
8320	25 HP motor	2 Elec	2	8		1,050	335		1,385	1,700
8330	30 HP motor		1.32	12.121		1,525	510		2,035	2,450
8340	40 HP motor		1.32	12.121		1,525	510		2,035	2,450
8350	50 HP motor		1.32	12.121		1,525	510		2,035	2,450
8360	60 HP motor		.80	20		3,325	840		4,165	4,950
8370	75 HP motor		.80	20		3,325	840		4,165	4,950
8380	100 HP motor		.80	20		3,325	840		4,165	4,950
8390	125 HP motor		.70	22.857		7,675	960		8,635	9,900
8400	150 HP motor		.70	22.857		7,175	960		8,135	9,350
8410	200 HP motor		.70	22.857		7,675	960		8,635	9,900
8430	Starter & circuit breaker disconnect									
8440	600 volt, 3 pole, 5 HP motor	1 Elec	1.40	5.714	Ea.	720	240		960	1,175
8450	10 HP motor	″	1.10	7.273		750	305		1,055	1,300
8460	25 HP motor	2 Elec	1.50	10.667		970	450		1,420	1,775
8470	30 HP motor		1.30	12.308		1,275	515		1,790	2,200
8480	40 HP motor		1.30	12.308		1,275	515		1,790	2,200
8490	50 HP motor		1.30	12.308		1,275	515		1,790	2,200
8500	60 HP motor		.92	17.391		3,025	730		3,755	4,450
8510	75 HP motor		.92	17.391		3,025	730		3,755	4,450
8520	100 HP motor		.84	19.048		3,025	800		3,825	4,550
8530	125 HP motor		.70	22.857		5,325	960		6,285	7,325
8540	150 HP motor		.70	22.857		5,325	960		6,285	7,325
8550	200 HP motor		.60	26.667		6,250	1,125		7,375	8,600
8900	240 volt, 1-2 pole, .75 HP motor	1 Elec	2	4		685	168		853	1,000
8910	2 HP motor		2	4		700	168		868	1,025
8920	5 HP motor		1.80	4.444		790	187		977	1,150
8930	10 HP motor		1.40	5.714		1,225	240		1,465	1,725
8950	3 pole, .75 HP motor		1.60	5		685	210		895	1,075
8970	5 HP motor		1.40	5.714		750	240		990	1,200
8980	10 HP motor		1.10	7.273		970	305		1,275	1,550
8990	15 HP motor		1	8		970	335		1,305	1,600
9100	20 HP motor	2 Elec	1.50	10.667		1,350	450		1,800	2,175
9110	25 HP motor		1.50	10.667		1,350	450		1,800	2,175
9120	30 HP motor		1.30	12.308		1,350	515		1,865	2,275
9130	40 HP motor		1.24	12.903		3,025	540		3,565	4,150
9140	50 HP motor		1.12	14.286		3,025	600		3,625	4,250

ELECTRICAL 16

For expanded coverage of these items see *Means Electrical Cost Data 2006*

16400 | Low-Voltage Distribution

	16440 | Swbds, Panels & Control Centers		CREW	DAILY OUTPUT	LABOR-HOURS	UNIT	2006 BARE COSTS MAT.	LABOR	EQUIP.	TOTAL	TOTAL INCL O&P
660	9150	60 HP motor	2 Elec	.90	17.778	Ea.	6,250	745		6,995	8,025
	9160	75 HP motor		.76	21.053		6,250	885		7,135	8,225
	9170	100 HP motor		.70	22.857		6,250	960		7,210	8,350
	9180	125 HP motor		.60	26.667		14,500	1,125		15,625	17,600
	9190	150 HP motor		.52	30.769		14,500	1,300		15,800	17,900
	9200	200 HP motor		.50	32		14,500	1,350		15,850	18,000
700	0010	PANELBOARD & LOAD CENTER CIRCUIT BREAKERS									
	0050	Bolt-on, 10,000 amp IC, 120 volt, 1 pole									
	0100	15 to 50 amp	1 Elec	10	.800	Ea.	13	33.50		46.50	66.50
	0200	60 amp		8	1		13	42		55	79
	0300	70 amp		8	1		24.50	42		66.50	91.50
	0350	240 volt, 2 pole									
	0400	15 to 50 amp	1 Elec	8	1	Ea.	28.50	42		70.50	96
	0500	60 amp		7.50	1.067		28.50	45		73.50	101
	0600	80 to 100 amp		5	1.600		73.50	67		140.50	185
	0700	3 pole, 15 to 60 amp		6.20	1.290		90.50	54		144.50	183
	0800	70 amp		5	1.600		114	67		181	229
	0900	80 to 100 amp		3.60	2.222		129	93.50		222.50	286
	1000	22,000 amp I.C., 240 volt, 2 pole, 70 - 225 amp		2.70	2.963		555	124		679	800
	1100	3 pole, 70 - 225 amp		2.30	3.478		620	146		766	905
	1200	14,000 amp I.C., 277 volts, 1 pole, 15 - 30 amp		8	1		34.50	42		76.50	103
	1300	22,000 amp I.C., 480 volts, 2 pole, 70 - 225 amp		2.70	2.963		555	124		679	800
	1400	3 pole, 70 - 225 amp		2.30	3.478		685	146		831	980
	2000	Plug-in panel or load center, 120/240 volt, to 60 amp, 1 pole		12	.667		10.10	28		38.10	54
	2010	2 pole		9	.889		23	37.50		60.50	82.50
	2020	3 pole		7.50	1.067		78.50	45		123.50	156
	2030	100 amp, 2 pole		6	1.333		105	56		161	203
	2040	3 pole		4.50	1.778		117	74.50		191.50	244
	2050	150 to 200 amp, 2 pole		3	2.667		192	112		304	385
	2060	Plug-in tandem, 120/240 V, 2-15 A, 1 pole		11	.727		24	30.50		54.50	73.50
	2070	1-15 A & 1-20 A		11	.727		24	30.50		54.50	73.50
	2080	2-20 A		11	.727		24	30.50		54.50	73.50
	2100	High interrupting capacity, 120/240 volt, plug-in, 30 amp, 1 pole		12	.667		19.85	28		47.85	65
	2110	60 amp, 2 pole		9	.889		46	37.50		83.50	109
	2120	3 pole		7.50	1.067		120	45		165	201
	2130	100 amp, 2 pole		6	1.333		133	56		189	233
	2140	3 pole		4.50	1.778		172	74.50		246.50	305
	2150	125 amp, 2 pole		3	2.667		355	112		467	565
	2200	Bolt-on, 30 amp, 1 pole		10	.800		25.50	33.50		59	80
	2210	60 amp, 2 pole		7.50	1.067		53.50	45		98.50	128
	2220	3 pole		6.20	1.290		138	54		192	236
	2230	100 amp, 2 pole		5	1.600		153	67		220	272
	2240	3 pole		3.60	2.222		200	93.50		293.50	365
	2300	Ground fault, 240 volt, 30 amp, 1 pole		7	1.143		78	48		126	160
	2310	2 pole		6	1.333		140	56		196	241
	2350	Key operated, 240 volt, 1 pole, 30 amp		7	1.143		68.50	48		116.50	149
	2360	Switched neutral, 240 volt, 30 amp, 2 pole		6	1.333		32.50	56		88.50	123
	2370	3 pole		5.50	1.455		49	61		110	148
	2400	Shunt trip, for 240 volt breaker, 60 amp, 1 pole		4	2		63.50	84		147.50	199
	2410	2 pole		3.50	2.286		63.50	96		159.50	218
	2420	3 pole		3	2.667		63.50	112		175.50	243
	2430	100 amp, 2 pole		3	2.667		63.50	112		175.50	243
	2440	3 pole		2.50	3.200		63.50	134		197.50	277
	2450	150 amp, 2 pole		2	4		151	168		319	425
	2500	Auxiliary switch, for 240 volt breaker, 60 amp, 1 pole		4	2		58	84		142	193
	2510	2 pole		3.50	2.286		58	96		154	212

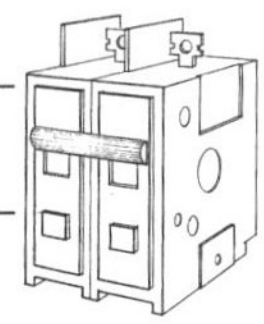

16440 | Swbds, Panels & Control Centers

		16440 Swbds, Panels & Control Centers	CREW	DAILY OUTPUT	LABOR-HOURS	UNIT	2006 BARE COSTS MAT.	LABOR	EQUIP.	TOTAL	TOTAL INCL O&P	
700	2520	3 pole	1 Elec	3	2.667	Ea.	58	112		170	237	700
	2530	100 amp, 2 pole		3	2.667		58	112		170	237	
	2540	3 pole		2.50	3.200		58	134		192	271	
	2550	150 amp, 2 pole		2	4		96	168		264	365	
	2600	Panel or load center, 277/480 volt, plug-in, 30 amp, 1 pole		12	.667		47	28		75	94.50	
	2610	60 amp, 2 pole		9	.889		119	37.50		156.50	189	
	2620	3 pole		7.50	1.067		205	45		250	295	
	2650	Bolt-on, 60 amp, 2 pole		7.50	1.067		140	45		185	223	
	2660	3 pole		6.20	1.290		205	54		259	310	
	2700	I-line, 277/480 volt, 30 amp, 1 pole		8	1		53.50	42		95.50	124	
	2710	60 amp, 2 pole		7.50	1.067		189	45		234	276	
	2720	3 pole		6.20	1.290		245	54		299	355	
	2730	100 amp, 1 pole		7.50	1.067		103	45		148	183	
	2740	2 pole		5	1.600		245	67		312	375	
	2750	3 pole		3.50	2.286		289	96		385	470	
	2800	High interrupting capacity, 277/480 volt, plug-in, 30 amp, 1 pole		12	.667		270	28		298	340	
	2810	60 amp, 2 pole		9	.889		415	37.50		452.50	520	
	2820	3 pole		7	1.143		485	48		533	610	
	2830	Bolt-on, 30 amp, 1 pole		8	1		270	42		312	360	
	2840	60 amp, 2 pole		7.50	1.067		415	45		460	530	
	2850	3 pole		6.20	1.290		485	54		539	620	
	2900	I-line, 30 amp, 1 pole		8	1		270	42		312	360	
	2910	60 amp, 2 pole		7.50	1.067		385	45		430	495	
	2920	3 pole		6.20	1.290		425	54		479	555	
	2930	100 amp, 1 pole		7.50	1.067		435	45		480	550	
	2940	2 pole		5	1.600		435	67		502	585	
	2950	3 pole		3.60	2.222		480	93.50		573.50	675	
	2960	Shunt trip, 277/480V breaker, remote oper., 30 amp, 1 pole		4	2		290	84		374	450	
	2970	60 amp, 2 pole		3.50	2.286		425	96		521	615	
	2980	3 pole		3	2.667		480	112		592	705	
	2990	100 amp, 1 pole		3.50	2.286		340	96		436	525	
	3000	2 pole		3	2.667		480	112		592	705	
	3010	3 pole		2.50	3.200		525	134		659	785	
	3050	Under voltage trip, 277/480 volt breaker, 30 amp, 1 pole		4	2		290	84		374	450	
	3060	60 amp, 2 pole		3.50	2.286		425	96		521	615	
	3070	3 pole		3	2.667		480	112		592	705	
	3080	100 amp, 1 pole		3.50	2.286		340	96		436	525	
	3090	2 pole		3	2.667		480	112		592	705	
	3100	3 pole		2.50	3.200		525	134		659	785	
	3150	Motor operated, 277/480 volt breaker, 30 amp, 1 pole		4	2		515	84		599	695	
	3160	60 amp, 2 pole		3.50	2.286		650	96		746	865	
	3170	3 pole		3	2.667		705	112		817	950	
	3180	100 amp, 1 pole		3.50	2.286		565	96		661	770	
	3190	2 pole		3	2.667		705	112		817	950	
	3200	3 pole		2.50	3.200		750	134		884	1,025	
	3250	Panelboard spacers, per pole		40	.200	↓	3.90	8.40		12.30	17.25	
	9000	Minimum labor/equipment charge	↓	3	2.667	Job		112		112	173	
720	0010	**PANELBOARDS** (Commercial use) R262416-50										720
	0050	NQOD, w/20 amp 1 pole bolt-on circuit breakers										
	0100	3 wire, 120/240 volts, 100 amp main lugs										
	0150	10 circuits	1 Elec	1	8	Ea.	420	335		755	985	
	0200	14 circuits		.88	9.091		495	380		875	1,125	
	0250	18 circuits		.75	10.667		540	450		990	1,275	
	0300	20 circuits	↓	.65	12.308		610	515		1,125	1,475	
	0350	225 amp main lugs, 24 circuits	2 Elec	1.20	13.333	↓	690	560		1,250	1,625	

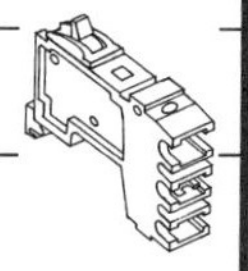

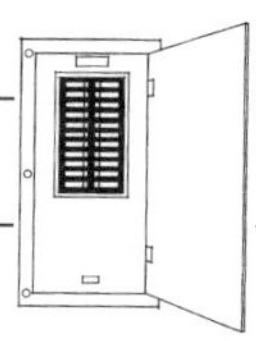

ELECTRICAL 16

	16440 | Swbds, Panels & Control Centers	CREW	DAILY OUTPUT	LABOR-HOURS	UNIT	2006 BARE COSTS MAT.	LABOR	EQUIP.	TOTAL	TOTAL INCL O&P	
720	0400 30 circuits (R262416-50)	2 Elec	.90	17.778	Ea.	805	745		1,550	2,025	720
	0450 36 circuits		.80	20		920	840		1,760	2,300	
	0500 38 circuits		.72	22.222		985	935		1,920	2,525	
	0550 42 circuits		.66	24.242		1,025	1,025		2,050	2,700	
	0600 4 wire, 120/208 volts, 100 amp main lugs, 12 circuits	1 Elec	1	8		475	335		810	1,050	
	0650 16 circuits		.75	10.667		550	450		1,000	1,300	
	0700 20 circuits		.65	12.308		640	515		1,155	1,500	
	0750 24 circuits		.60	13.333		695	560		1,255	1,625	
	0800 30 circuits		.53	15.094		800	635		1,435	1,850	
	0850 225 amp main lugs, 32 circuits	2 Elec	.90	17.778		900	745		1,645	2,150	
	0900 34 circuits		.84	19.048		925	800		1,725	2,250	
	0950 36 circuits		.80	20		945	840		1,785	2,350	
	1000 42 circuits		.68	23.529		1,050	990		2,040	2,700	
	1040 225 amp main lugs, NEMA 7, 12 circuits		1	16		2,550	670		3,220	3,825	
	1100 24 circuits		.40	40		3,175	1,675		4,850	6,075	
	1200 NEHB,w/20 amp, 1 pole bolt-on circuit breakers										
	1250 4 wire, 277/480 volts, 100 amp main lugs, 12 circuits	1 Elec	.88	9.091	Ea.	915	380		1,295	1,600	
	1300 20 circuits	"	.60	13.333		1,350	560		1,910	2,350	
	1350 225 amp main lugs, 24 circuits	2 Elec	.90	17.778		1,550	745		2,295	2,875	
	1400 30 circuits		.80	20		1,875	840		2,715	3,350	
	1450 36 circuits		.72	22.222		2,175	935		3,110	3,850	
	1500 42 circuits		.60	26.667		2,500	1,125		3,625	4,475	
	1510 225 amp main lugs, NEMA 7, 12 circuits		.90	17.778		3,700	745		4,445	5,225	
	1590 24 circuits		.30	53.333		4,475	2,250		6,725	8,375	
	1600 NQOD panel, w/20 amp, 1 pole, circuit breakers										
	1650 3 wire, 120/240 volt with main circuit breaker										
	1700 100 amp main, 12 circuits	1 Elec	.80	10	Ea.	585	420		1,005	1,300	
	1750 20 circuits	"	.60	13.333		750	560		1,310	1,700	
	1800 225 amp main, 30 circuits	2 Elec	.68	23.529		1,425	990		2,415	3,100	
	1850 42 circuits		.52	30.769		1,675	1,300		2,975	3,825	
	1900 400 amp main, 30 circuits		.54	29.630		2,000	1,250		3,250	4,125	
	1950 42 circuits		.50	32		2,225	1,350		3,575	4,525	
	2000 4 wire, 120/208 volts with main circuit breaker										
	2050 100 amp main, 24 circuits	1 Elec	.47	17.021	Ea.	875	715		1,590	2,050	
	2100 30 circuits	"	.40	20		990	840		1,830	2,400	
	2200 225 amp main, 32 circuits	2 Elec	.72	22.222		1,675	935		2,610	3,275	
	2250 42 circuits		.56	28.571		1,825	1,200		3,025	3,850	
	2300 400 amp main, 42 circuits		.48	33.333		2,475	1,400		3,875	4,875	
	2350 600 amp main, 42 circuits		.40	40		3,675	1,675		5,350	6,625	
	2400 NEHB, with 20 amp, 1 pole circuit breaker										
	2450 4 wire, 277/480 volts with main circuit breaker										
	2500 100 amp main, 24 circuits	1 Elec	.42	19.048	Ea.	1,800	800		2,600	3,200	
	2550 30 circuits	"	.38	21.053		2,100	885		2,985	3,675	
	2600 225 amp main, 30 circuits	2 Elec	.72	22.222		2,650	935		3,585	4,375	
	2650 42 circuits		.56	28.571		3,275	1,200		4,475	5,450	
	2700 400 amp main, 42 circuits		.46	34.783		3,950	1,450		5,400	6,600	
	2750 600 amp main, 42 circuits		.38	42.105		5,425	1,775		7,200	8,675	
	2900 Note: the following line items don't include branch circuit breakers										
	2910 For branch circuit breakers information, refer to										
	2920 Division 16440-700										
	3010 Main lug, no main breaker, 240 volt, 1 pole, 3 wire, 100 amp	1 Elec	2.30	3.478	Ea.	380	146		526	645	
	3020 225 amp	2 Elec	2.40	6.667		465	280		745	940	
	3030 400 amp	"	1.80	8.889		690	375		1,065	1,325	
	3060 3 pole, 3 wire, 100 amp	1 Elec	2.30	3.478		415	146		561	680	
	3070 225 amp	2 Elec	2.40	6.667		490	280		770	970	
	3080 400 amp		1.80	8.889		750	375		1,125	1,400	

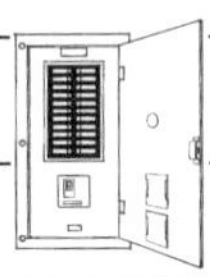

16 ELECTRICAL

	16440 | Swbds, Panels & Control Centers		CREW	DAILY OUTPUT	LABOR-HOURS	UNIT	2006 BARE COSTS MAT.	LABOR	EQUIP.	TOTAL	TOTAL INCL O&P	
720	3090	600 amp R262416-50	2 Elec	1.60	10	Ea.	875	420		1,295	1,600	720
	3110	3 pole, 4 wire, 100 amp	1 Elec	2.30	3.478		460	146		606	730	
	3120	225 amp	2 Elec	2.40	6.667		580	280		860	1,075	
	3130	400 amp		1.80	8.889		750	375		1,125	1,400	
	3140	600 amp		1.60	10		875	420		1,295	1,600	
	3160	480 volt, 3 pole, 3 wire, 100 amp	1 Elec	2.30	3.478		555	146		701	835	
	3170	225 amp	2 Elec	2.40	6.667		675	280		955	1,175	
	3180	400 amp		1.80	8.889		955	375		1,330	1,625	
	3190	600 amp		1.60	10		1,075	420		1,495	1,825	
	3210	277/480 volt, 3 pole, 4 wire, 100 amp	1 Elec	2.30	3.478		540	146		686	820	
	3220	225 amp	2 Elec	2.40	6.667		660	280		940	1,150	
	3230	400 amp		1.80	8.889		930	375		1,305	1,600	
	3240	600 amp		1.60	10		1,050	420		1,470	1,800	
	3260	Main circuit breaker, 240 volt, 1 pole, 3 wire, 100 amp	1 Elec	2	4		520	168		688	835	
	3270	225 amp	2 Elec	2	8		1,100	335		1,435	1,725	
	3280	400 amp	"	1.60	10		1,750	420		2,170	2,550	
	3310	3 pole, 3 wire, 100 amp	1 Elec	2	4		600	168		768	920	
	3320	225 amp	2 Elec	2	8		1,250	335		1,585	1,900	
	3330	400 amp		1.60	10		2,000	420		2,420	2,850	
	3360	120/208 volt, 3 pole, 4 wire, 100 amp		4	4		600	168		768	920	
	3370	225 amp		2	8		1,250	335		1,585	1,900	
	3380	400 amp		1.60	10		2,000	420		2,420	2,850	
	3410	480 volt, 3 pole, 3 wire, 100 amp	1 Elec	2	4		850	168		1,018	1,200	
	3420	225 amp	2 Elec	2	8		1,475	335		1,810	2,150	
	3430	400 amp		1.60	10		2,250	420		2,670	3,125	
	3460	277/480 volt, 3 pole, 4 wire, 100 amp		4	4		830	168		998	1,175	
	3470	225 amp		2	8		1,450	335		1,785	2,125	
	3480	400 amp		1.60	10		2,250	420		2,670	3,125	
	3510	Main circuit breaker, HIC, 240 volt, 1 pole, 3 wire, 100 amp	1 Elec	2	4		810	168		978	1,150	
	3520	225 amp	2 Elec	2	8		2,175	335		2,510	2,925	
	3530	400 amp	"	1.60	10		2,925	420		3,345	3,850	
	3560	3 pole, 3 wire, 100 amp	1 Elec	2	4		905	168		1,073	1,250	
	3570	225 amp	2 Elec	2	8		2,450	335		2,785	3,225	
	3580	400 amp	"	1.60	10		3,250	420		3,670	4,225	
	3610	120/208 volt, 3 pole, 4 wire, 100 amp	1 Elec	2	4		905	168		1,073	1,250	
	3620	225 amp	2 Elec	2	8		2,450	335		2,785	3,225	
	3630	400 amp	"	1.60	10		3,250	420		3,670	4,225	
	3660	480 volt, 3 pole, 3 wire, 100 amp	1 Elec	2	4		1,375	168		1,543	1,750	
	3670	225 amp	2 Elec	2	8		2,725	335		3,060	3,525	
	3680	400 amp	"	1.60	10		3,475	420		3,895	4,475	
	3710	277/480 volt, 3 pole, 4 wire, 100 amp	1 Elec	2	4		1,300	168		1,468	1,675	
	3720	225 amp	2 Elec	2	8		2,600	335		2,935	3,375	
	3730	400 amp	"	1.60	10		3,450	420		3,870	4,450	
	3760	Main circuit breaker, shunt trip, 100 amp	1 Elec	1.20	6.667		675	280		955	1,175	
	3770	225 amp	2 Elec	1.60	10		1,600	420		2,020	2,400	
	3780	400 amp	"	1.40	11.429		2,325	480		2,805	3,325	
	9000	Minimum labor/equipment charge	1 Elec	1	8	Job		335		335	520	
800	0010	**SWITCHBOARDS** distribution section										800
	0100	Aluminum bus bars, not including breakers										
	0160	Subfeed lug-rated at 60 amp	2 Elec	1.30	12.308	Ea.	1,425	515		1,940	2,375	
	0170	100 amp		1.26	12.698		1,425	535		1,960	2,400	
	0180	200 amp		1.20	13.333		1,425	560		1,985	2,450	
	0190	400 amp		1.10	14.545		1,425	610		2,035	2,525	
	0200	120/208 or 277/480 volt, 4 wire, 600 amp		1	16		1,475	670		2,145	2,650	
	0300	800 amp		.88	18.182		1,900	765		2,665	3,250	

		16440 | Swbds, Panels & Control Centers	CREW	DAILY OUTPUT	LABOR-HOURS	UNIT	2006 BARE COSTS MAT.	LABOR	EQUIP.	TOTAL	TOTAL INCL O&P	
800	0400	1000 amp	2 Elec	.80	20	Ea.	2,000	840		2,840	3,500	800
	0500	1200 amp		.72	22.222		2,700	935		3,635	4,425	
	0600	1600 amp		.66	24.242		3,150	1,025		4,175	5,050	
	0700	2000 amp		.62	25.806		3,700	1,075		4,775	5,750	
	0800	2500 amp		.60	26.667		4,175	1,125		5,300	6,325	
	0900	3000 amp		.56	28.571		5,050	1,200		6,250	7,425	
	0950	4000 amp	↓	.52	30.769	↓	7,375	1,300		8,675	10,100	
820	0010	**SWITCHBOARDS** feeder section group mounted devices										820
	0030	Circuit breakers										
	0160	FA frame, 15 to 60 amp, 240 volt, 1 pole	1 Elec	8	1	Ea.	83	42		125	156	
	0170	2 pole		7	1.143		166	48		214	257	
	0180	3 pole		5.30	1.509		239	63.50		302.50	360	
	0210	480 volt, 1 pole		8	1		105	42		147	180	
	0220	2 pole		7	1.143		272	48		320	375	
	0230	3 pole		5.30	1.509		350	63.50		413.50	485	
	0260	600 volt, 2 pole		7	1.143		325	48		373	435	
	0270	3 pole		5.30	1.509		405	63.50		468.50	545	
	0280	FA frame, 70 to 100 amp, 240 volt, 1 pole		7	1.143		109	48		157	194	
	0310	2 pole		5	1.600		258	67		325	385	
	0320	3 pole		4	2		294	84		378	455	
	0330	480 volt, 1 pole		7	1.143		130	48		178	217	
	0360	2 pole		5	1.600		350	67		417	490	
	0370	3 pole		4	2		415	84		499	590	
	0380	600 volt, 2 pole		5	1.600		395	67		462	540	
	0410	3 pole		4	2		485	84		569	665	
	0420	KA frame, 70 to 225 amp		3.20	2.500		915	105		1,020	1,150	
	0430	LA frame, 125 to 400 amp		2.30	3.478		2,075	146		2,221	2,500	
	0460	MA frame, 450 to 600 amp		1.60	5		3,425	210		3,635	4,075	
	0470	700 to 800 amp		1.30	6.154		4,425	258		4,683	5,275	
	0480	MAL frame, 1000 amp		1	8		4,600	335		4,935	5,600	
	0490	PA frame, 1200 amp		.80	10		9,350	420		9,770	10,900	
	0500	Branch circuit, fusible switch, 600 volt, double 30/30 amp		4	2		935	84		1,019	1,150	
	0550	60/60 amp		3.20	2.500		950	105		1,055	1,200	
	0600	100/100 amp		2.70	2.963		1,325	124		1,449	1,650	
	0650	Single, 30 amp		5.30	1.509		630	63.50		693.50	795	
	0700	60 amp		4.70	1.702		630	71.50		701.50	805	
	0750	100 amp		4	2		885	84		969	1,100	
	0800	200 amp		2.70	2.963		1,325	124		1,449	1,650	
	0850	400 amp		2.30	3.478		2,500	146		2,646	2,975	
	0900	600 amp		1.80	4.444		2,950	187		3,137	3,550	
	0950	800 amp		1.30	6.154		4,850	258		5,108	5,725	
	1000	1200 amp	↓	.80	10	↓	5,775	420		6,195	7,000	
	1080	Branch circuit, circuit breakers, high interrupting capacity										
	1100	60 amp, 240, 480 or 600 volt, 1 pole	1 Elec	8	1	Ea.	212	42		254	298	
	1120	2 pole		7	1.143		505	48		553	635	
	1140	3 pole		5.30	1.509		605	63.50		668.50	765	
	1150	100 amp, 240, 480 or 600 volt, 1 pole		7	1.143		235	48		283	335	
	1160	2 pole		5	1.600		605	67		672	775	
	1180	3 pole		4	2		680	84		764	880	
	1200	225 amp, 240, 480 or 600 volt, 2 pole		3.50	2.286		2,125	96		2,221	2,475	
	1220	3 pole		3.20	2.500		2,575	105		2,680	2,975	
	1240	400 amp, 240, 480 or 600 volt, 2 pole		2.50	3.200		2,825	134		2,959	3,300	
	1260	3 pole		2.30	3.478		3,400	146		3,546	3,975	
	1280	600 amp, 240, 480 or 600 volt, 2 pole		1.80	4.444		3,450	187		3,637	4,100	
	1300	3 pole	↓	1.60	5	↓	4,150	210		4,360	4,900	

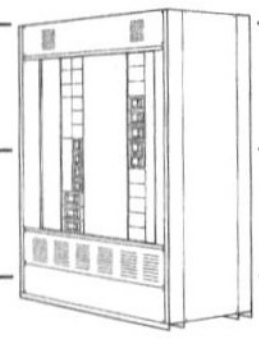

 Important: See the Reference Section for supporting data - Crews, Rental Equipment, City Cost Indexes and Reference Data

16400 | Low-Voltage Distribution

16440 | Swbds, Panels & Control Centers

				DAILY	LABOR-		2006 BARE COSTS				TOTAL	
			CREW	OUTPUT	HOURS	UNIT	MAT.	LABOR	EQUIP.	TOTAL	INCL O&P	
820	1320	800 amp, 240, 480 or 600 volt, 2 pole	1 Elec	1.50	5.333	Ea.	4,225	224		4,449	5,000	820
	1340	3 pole		1.30	6.154		5,275	258		5,533	6,200	
	1360	1000 amp, 240, 480 or 600 volt, 2 pole		1.10	7.273		7,650	305		7,955	8,900	
	1380	3 pole		1	8		8,350	335		8,685	9,725	
	1400	1200 amp, 240, 480 or 600 volt, 2 pole		.90	8.889		5,950	375		6,325	7,125	
	1420	3 pole		.80	10		8,975	420		9,395	10,500	
	1700	Fusible switch, 240 V, 60 amp, 2 pole		3.20	2.500		390	105		495	585	
	1720	3 pole		3	2.667		535	112		647	760	
	1740	100 amp, 2 pole		2.70	2.963		395	124		519	625	
	1760	3 pole		2.50	3.200		600	134		734	865	
	1780	200 amp, 2 pole		2	4		825	168		993	1,175	
	1820	400 amp, 2 pole		1.50	5.333		1,550	224		1,774	2,050	
	1840	3 pole		1.30	6.154		1,925	258		2,183	2,525	
	1860	600 amp, 2 pole		1	8		2,200	335		2,535	2,950	
	1880	3 pole		.90	8.889		2,700	375		3,075	3,550	
	1900	240-600 V, 800 amp, 2 pole		.70	11.429		4,900	480		5,380	6,125	
	1920	3 pole		.60	13.333		6,000	560		6,560	7,475	
	2000	600 V, 60 amp, 2 pole		3.20	2.500		590	105		695	805	
	2040	100 amp, 2 pole		2.70	2.963		605	124		729	855	
	2080	200 amp, 2 pole		2	4		1,050	168		1,218	1,400	
	2120	400 amp, 2 pole		1.50	5.333		2,075	224		2,299	2,650	
	2160	600 amp, 2 pole		1	8		2,525	335		2,860	3,325	
	2500	Branch circuit, circuit breakers, 60 amp, 600 volt, 3 pole		5.30	1.509		560	63.50		623.50	720	
	2520	240, 480 or 600 volt, 1 pole		8	1		91	42		133	165	
	2540	240 volt, 2 pole		7	1.143		156	48		204	246	
	2560	480 or 600 volt, 2 pole		7	1.143		330	48		378	440	
	2580	240 volt, 3 pole		5.30	1.509		254	63.50		317.50	375	
	2600	480 volt, 3 pole		5.30	1.509		440	63.50		503.50	585	
	2620	100 amp, 600 volt, 2 pole		5	1.600		425	67		492	570	
	2640	3 pole		4	2		535	84		619	715	
	2660	480 volt, 2 pole		5	1.600		365	67		432	505	
	2680	240 volt, 2 pole		5	1.600		208	67		275	335	
	2700	3 pole		4	2		335	84		419	500	
	2720	480 volt, 3 pole		4	2		490	84		574	670	
	2740	225 amp, 240, 480 or 600 volt, 2 pole		3.50	2.286		555	96		651	760	
	2760	3 pole		3.20	2.500		635	105		740	860	
	2780	400 amp, 240, 480 or 600 volt, 2 pole		2.50	3.200		1,275	134		1,409	1,600	
	2800	3 pole		2.30	3.478		1,475	146		1,621	1,850	
	2820	600 amp, 240 or 480 volt, 2 pole		1.80	4.444		2,075	187		2,262	2,575	
	2840	3 pole		1.60	5		2,550	210		2,760	3,125	
	2860	800 amp, 240, 480 volt or 600 volt, 2 pole		1.50	5.333		3,125	224		3,349	3,800	
	2880	3 pole		1.30	6.154		3,650	258		3,908	4,425	
	2900	1000 amp, 240, 480 or 600 volt, 2 pole		1.10	7.273		3,850	305		4,155	4,700	
	2920	480 volt, 600 volt, 3 pole		1	8		4,400	335		4,735	5,375	
	2940	1200 amp, 240, 480 or 600 volt, 2 pole		.90	8.889		5,400	375		5,775	6,525	
	2960	3 pole		.80	10		5,900	420		6,320	7,150	
	2980	600 volt, 3 pole		.80	10		5,900	420		6,320	7,150	
840	0010	**SWITCHBOARDS** Incoming main service section										840
	0100	Aluminum bus bars, not including CT's or PT's										
	0200	No main disconnect, includes CT compartment										
	0300	120/208 volt, 4 wire, 600 amp	2 Elec	1	16	Ea.	3,450	670		4,120	4,825	
	0400	800 amp		.88	18.182		3,450	765		4,215	4,975	
	0500	1000 amp		.80	20		4,150	840		4,990	5,850	
	0600	1200 amp		.72	22.222		4,150	935		5,085	6,000	
	0700	1600 amp		.66	24.242		4,150	1,025		5,175	6,125	

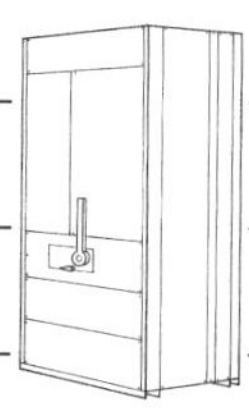

		16440 | Swbds, Panels & Control Centers	CREW	DAILY OUTPUT	LABOR-HOURS	UNIT	2006 BARE COSTS MAT.	LABOR	EQUIP.	TOTAL	TOTAL INCL O&P	
840	0800	2000 amp	2 Elec	.62	25.806	Ea.	4,450	1,075		5,525	6,575	840
	1000	3000 amp		.56	28.571		5,900	1,200		7,100	8,325	
	1200	277/480 volt, 4 wire, 600 amp		1	16		3,450	670		4,120	4,825	
	1300	800 amp		.88	18.182		3,450	765		4,215	4,975	
	1400	1000 amp		.80	20		4,375	840		5,215	6,125	
	1500	1200 amp		.72	22.222		4,375	935		5,310	6,275	
	1600	1600 amp		.66	24.242		4,375	1,025		5,400	6,400	
	1700	2000 amp		.62	25.806		4,450	1,075		5,525	6,575	
	1800	3000 amp		.56	28.571		5,900	1,200		7,100	8,325	
	1900	4000 amp		.52	30.769		7,300	1,300		8,600	10,000	
	2000	Fused switch & CT compartment										
	2100	120/208 volt, 4 wire, 400 amp	2 Elec	1.12	14.286	Ea.	3,775	600		4,375	5,075	
	2200	600 amp		.94	17.021		4,475	715		5,190	6,000	
	2300	800 amp		.84	19.048		9,325	800		10,125	11,500	
	2400	1200 amp		.68	23.529		12,100	990		13,090	14,800	
	2500	277/480 volt, 4 wire, 400 amp		1.14	14.035		3,950	590		4,540	5,250	
	2600	600 amp		.94	17.021		4,525	715		5,240	6,075	
	2700	800 amp		.84	19.048		9,325	800		10,125	11,500	
	2800	1200 amp		.68	23.529		12,100	990		13,090	14,800	
	2900	Pressure switch & CT compartment										
	3000	120/208 volt, 4 wire, 800 amp	2 Elec	.80	20	Ea.	8,375	840		9,215	10,500	
	3100	1200 amp		.66	24.242		16,200	1,025		17,225	19,400	
	3200	1600 amp		.62	25.806		17,200	1,075		18,275	20,700	
	3300	2000 amp		.56	28.571		18,300	1,200		19,500	22,000	
	3310	2500 amp		.50	32		22,500	1,350		23,850	26,900	
	3320	3000 amp		.44	36.364		30,400	1,525		31,925	35,800	
	3330	4000 amp		.40	40		39,000	1,675		40,675	45,500	
	3400	277/480 volt, 4 wire, 800 amp, with ground fault		.80	20		13,800	840		14,640	16,500	
	3600	1200 amp, with ground fault		.66	24.242		17,900	1,025		18,925	21,300	
	4000	1600 amp, with ground fault		.62	25.806		19,400	1,075		20,475	23,100	
	4200	2000 amp, with ground fault		.56	28.571		21,000	1,200		22,200	25,000	
	4400	Circuit breaker, molded case & CT compartment										
	4600	3 pole, 4 wire, 600 amp	2 Elec	.94	17.021	Ea.	7,225	715		7,940	9,025	
	4800	800 amp		.84	19.048		8,625	800		9,425	10,700	
	5000	1200 amp		.68	23.529		11,800	990		12,790	14,400	
	5100	Copper bus bars, not incl. CT's or PT's, add, minimum					15%					
860	0010	**SWITCHBOARDS** (in plant distribution)										860
	0100	Main lugs only, to 600 volt, 3 pole, 3 wire, 200 amp	2 Elec	1.20	13.333	Ea.	1,525	560		2,085	2,550	
	0110	400 amp		1.20	13.333		1,525	560		2,085	2,550	
	0120	600 amp		1.20	13.333		1,575	560		2,135	2,600	
	0130	800 amp		1.08	14.815		1,700	620		2,320	2,825	
	0140	1200 amp		.92	17.391		2,075	730		2,805	3,400	
	0150	1600 amp		.86	18.605		2,725	780		3,505	4,200	
	0160	2000 amp		.82	19.512		3,025	820		3,845	4,575	
	0250	To 480 volt, 3 pole, 4 wire, 200 amp		1.20	13.333		1,325	560		1,885	2,325	
	0260	400 amp		1.20	13.333		1,525	560		2,085	2,550	
	0270	600 amp		1.20	13.333		1,675	560		2,235	2,700	
	0280	800 amp		1.08	14.815		1,825	620		2,445	2,950	
	0290	1200 amp		.92	17.391		2,250	730		2,980	3,600	
	0300	1600 amp		.86	18.605		2,575	780		3,355	4,025	
	0310	2000 amp		.82	19.512		3,000	820		3,820	4,550	
	0400	Main circuit breaker, to 600 volt, 3 pole, 3 wire, 200 amp		1.20	13.333		2,625	560		3,185	3,775	
	0410	400 amp		1.14	14.035		2,625	590		3,215	3,800	
	0420	600 amp		1.10	14.545		3,350	610		3,960	4,625	
	0430	800 amp		1.04	15.385		5,575	645		6,220	7,150	
	0440	1200 amp		.88	18.182		7,300	765		8,065	9,225	

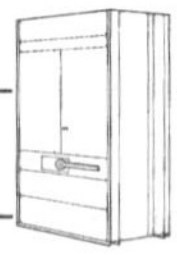

	16440 | Swbds, Panels & Control Centers	CREW	DAILY OUTPUT	LABOR-HOURS	UNIT	2006 BARE COSTS MAT.	LABOR	EQUIP.	TOTAL	TOTAL INCL O&P	
860	0450 1600 amp	2 Elec	.84	19.048	Ea.	11,700	800		12,500	14,100	860
	0460 2000 amp		.80	20		12,500	840		13,340	15,100	
	0550 277/480 volt, 3 pole, 4 wire, 200 amp		1.20	13.333		2,775	560		3,335	3,925	
	0560 400 amp		1.14	14.035		2,775	590		3,365	3,950	
	0570 600 amp		1.10	14.545		3,475	610		4,085	4,775	
	0580 800 amp		1.04	15.385		5,850	645		6,495	7,450	
	0590 1200 amp		.88	18.182		7,550	765		8,315	9,500	
	0600 1600 amp		.84	19.048		11,700	800		12,500	14,100	
	0610 2000 amp		.80	20		12,500	840		13,340	15,100	
	0700 Main fusible switch w/fuse, 208/240 volt, 3 pole, 3 wire, 200 amp		1.20	13.333		2,900	560		3,460	4,075	
	0710 400 amp		1.14	14.035		2,900	590		3,490	4,100	
	0720 600 amp		1.10	14.545		3,525	610		4,135	4,850	
	0730 800 amp		1.04	15.385		7,225	645		7,870	8,950	
	0740 1200 amp		.88	18.182		8,425	765		9,190	10,400	
	0800 120/208, 120/240 volt, 3 pole, 4 wire, 200 amp		1.20	13.333		2,575	560		3,135	3,725	
	0810 400 amp		1.14	14.035		2,575	590		3,165	3,750	
	0820 600 amp		1.10	14.545		3,375	610		3,985	4,650	
	0830 800 amp		1.04	15.385		5,375	645		6,020	6,900	
	0840 1200 amp		.88	18.182		6,225	765		6,990	8,025	
	0900 480 or 600 volt, 3 pole, 3 wire, 200 amp		1.20	13.333		2,800	560		3,360	3,950	
	0910 400 amp		1.14	14.035		2,800	590		3,390	3,975	
	0920 600 amp		1.10	14.545		3,350	610		3,960	4,625	
	0930 800 amp		1.04	15.385		5,175	645		5,820	6,675	
	0940 1200 amp		.88	18.182		6,000	765		6,765	7,775	
	1000 277 or 480 volt, 3 pole, 4 wire, 200 amp		1.20	13.333		2,900	560		3,460	4,075	
	1010 400 amp		1.14	14.035		2,900	590		3,490	4,100	
	1020 600 amp		1.10	14.545		3,500	610		4,110	4,800	
	1030 800 amp		1.04	15.385		5,375	645		6,020	6,900	
	1040 1200 amp		.88	18.182		6,225	765		6,990	8,025	
	1120 1600 amp		.76	21.053		11,400	885		12,285	13,900	
	1130 2000 amp		.68	23.529		15,000	990		15,990	18,000	
	1150 Pressure switch, bolted, 3 pole, 208/240 volt, 3 wire, 800 amp		.96	16.667		8,425	700		9,125	10,400	
	1160 1200 amp		.80	20		10,800	840		11,640	13,100	
	1170 1600 amp		.76	21.053		12,300	885		13,185	14,900	
	1180 2000 amp		.68	23.529		14,000	990		14,990	16,900	
	1200 120/208 or 120/240 volt, 3 pole, 4 wire, 800 amp		.96	16.667		6,650	700		7,350	8,400	
	1210 1200 amp		.80	20		7,775	840		8,615	9,875	
	1220 1600 amp		.76	21.053		12,300	885		13,185	14,900	
	1230 2000 amp		.68	23.529		14,000	990		14,990	16,900	
	1300 480 or 600 volt, 3 wire, 800 amp		.96	16.667		8,425	700		9,125	10,400	
	1310 1200 amp		.80	20		11,800	840		12,640	14,300	
	1320 1600 amp		.76	21.053		13,300	885		14,185	16,000	
	1330 2000 amp		.68	23.529		15,000	990		15,990	18,000	
	1400 277-480 volt, 4 wire, 800 amp		.96	16.667		8,425	700		9,125	10,400	
	1410 1200 amp		.80	20		11,800	840		12,640	14,300	
	1420 1600 amp		.76	21.053		13,300	885		14,185	16,000	
	1430 2000 amp		.68	23.529		15,000	990		15,990	18,000	
	1500 Main ground fault protector, 1200-2000 amp		5.40	2.963		4,550	124		4,674	5,200	
	1600 Busway connection, 200 amp		5.40	2.963		365	124		489	595	
	1610 400 amp		4.60	3.478		365	146		511	630	
	1620 600 amp		4	4		365	168		533	665	
	1630 800 amp		3.20	5		365	210		575	730	
	1640 1200 amp		2.60	6.154		365	258		623	805	
	1650 1600 amp		2.40	6.667		760	280		1,040	1,275	
	1660 2000 amp		2	8		760	335		1,095	1,350	
	1700 Shunt trip for remote operation 200 amp	↓	8	2	↓	905	84		989	1,125	

ELECTRICAL 16

	16440	Swbds, Panels & Control Centers	CREW	DAILY OUTPUT	LABOR-HOURS	UNIT	2006 BARE COSTS MAT.	LABOR	EQUIP.	TOTAL	TOTAL INCL O&P	
860	1710	400 amp	2 Elec	8	2	Ea.	1,450	84		1,534	1,725	860
	1720	600 amp		8	2		1,700	84		1,784	2,000	
	1730	800 amp		8	2		2,200	84		2,284	2,550	
	1740	1200-2000 amp		8	2		4,650	84		4,734	5,225	
	1800	Motor operated main breaker 200 amp		8	2		2,775	84		2,859	3,175	
	1810	400 amp		8	2		2,775	84		2,859	3,175	
	1820	600 amp		8	2		2,375	84		2,459	2,750	
	1830	800 amp		8	2		2,825	84		2,909	3,225	
	1840	1200-2000 amp		8	2		4,725	84		4,809	5,300	
	1900	Current/potential transformer metering compartment 200-800 amp		5.40	2.963		1,800	124		1,924	2,200	
	1940	1200 amp		5.40	2.963		2,500	124		2,624	2,950	
	1950	1600-2000 amp		5.40	2.963		2,500	124		2,624	2,950	
	2000	With watt meter 200-800 amp		4	4		7,125	168		7,293	8,075	
	2040	1200 amp		4	4		8,800	168		8,968	9,925	
	2050	1600-2000 amp		4	4		8,800	168		8,968	9,925	
	2100	Split bus 60-200 amp	1 Elec	5.30	1.509		273	63.50		336.50	400	
	2130	400 amp	2 Elec	4.60	3.478		475	146		621	745	
	2140	600 amp		3.60	4.444		580	187		767	925	
	2150	800 amp		2.60	6.154		735	258		993	1,200	
	2170	1200 amp		2	8		840	335		1,175	1,450	
	2250	Contactor control 60 amp	1 Elec	2	4		1,800	168		1,968	2,225	
	2260	100 amp		1.50	5.333		2,000	224		2,224	2,550	
	2270	200 amp		1	8		3,025	335		3,360	3,850	
	2280	400 amp	2 Elec	1	16		9,200	670		9,870	11,100	
	2290	600 amp		.84	19.048		10,300	800		11,100	12,500	
	2300	800 amp		.72	22.222		12,200	935		13,135	14,900	
	2500	Modifier, two distribution sections, add		.80	20		2,775	840		3,615	4,350	
	2520	Three distribution sections, add		.40	40		6,200	1,675		7,875	9,400	
	2560	Auxiliary pull section, 20", add		2	8		1,225	335		1,560	1,875	
	2580	24", add		1.80	8.889		1,225	375		1,600	1,925	
	2600	30", add		1.60	10		1,225	420		1,645	2,000	
	2620	36", add		1.40	11.429		1,475	480		1,955	2,350	
	2640	Dog house, 12", add		2.40	6.667		315	280		595	775	
	2660	18", add		2	8		630	335		965	1,200	
	3000	Transition section between switchboard and transformer										
	3050	or motor control center, 4 wire alum. bus, 600 amp	2 Elec	1.14	14.035	Ea.	2,075	590		2,665	3,175	
	3100	800 amp		1	16		2,325	670		2,995	3,575	
	3150	1000 amp		.88	18.182		2,600	765		3,365	4,050	
	3200	1200 amp		.80	20		2,875	840		3,715	4,450	
	3250	1600 amp		.72	22.222		3,375	935		4,310	5,175	
	3300	2000 amp		.66	24.242		3,900	1,025		4,925	5,850	
	3350	2500 amp		.62	25.806		4,525	1,075		5,600	6,675	
	3400	3000 amp		.56	28.571		5,225	1,200		6,425	7,600	
	4000	Weatherproof construction, per vertical section		1.76	9.091		2,875	380		3,255	3,750	

	16450	Enclosed Bus Assemblies	CREW	DAILY OUTPUT	LABOR-HOURS	UNIT	MAT.	LABOR	EQUIP.	TOTAL	TOTAL INCL O&P	
100	0010	**ALUMINUM BUS DUCT** 10 ft. long										100
	0050	3 pole 4 wire, plug-in/indoor, straight section, 225 amp	2 Elec	44	.364	L.F.	132	15.25		147.25	169	
	0100	400 amp		36	.444		132	18.65		150.65	174	
	0150	600 amp		32	.500		132	21		153	178	
	0200	800 amp		26	.615		152	26		178	207	
	0250	1000 amp		24	.667		172	28		200	232	
	0300	1350 amp		22	.727		263	30.50		293.50	335	
	0310	1600 amp		18	.889		305	37.50		342.50	395	
	0320	2000 amp		16	1		355	42		397	455	
	0330	2500 amp		14	1.143		455	48		503	575	

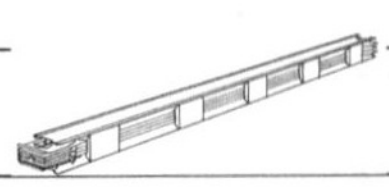

16450 | Enclosed Bus Assemblies

		16450 Enclosed Bus Assemblies	CREW	DAILY OUTPUT	LABOR-HOURS	UNIT	2006 BARE COSTS MAT.	LABOR	EQUIP.	TOTAL	TOTAL INCL O&P	
100	0340	3000 amp	2 Elec	12	1.333	L.F.	515	56		571	655	100
	0350	Feeder, 600 amp		34	.471		121	19.75		140.75	165	
	0400	800 amp		28	.571		142	24		166	193	
	0450	1000 amp 		26	.615		162	26		188	218	
	0455	1200 amp		25	.640		162	27		189	220	
	0500	1350 amp		24	.667		253	28		281	320	
	0550	1600 amp		20	.800		293	33.50		326.50	375	
	0600	2000 amp		18	.889		345	37.50		382.50	440	
	0620	2500 amp		14	1.143		445	48		493	565	
	0630	3000 amp		12	1.333		505	56		561	640	
	0640	4000 amp		10	1.600		730	67		797	905	
	0650	Elbow, 225 amp 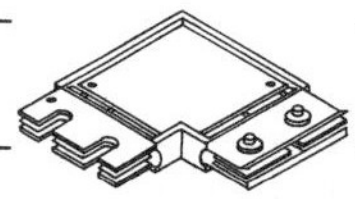		4.40	3.636	Ea.	715	153		868	1,025	
	0700	400 amp		3.80	4.211		715	177		892	1,050	
	0750	600 amp		3.40	4.706		715	198		913	1,100	
	0800	800 amp		3	5.333		745	224		969	1,175	
	0850	1000 amp		2.80	5.714		970	240		1,210	1,450	
	0870	1200 amp		2.70	5.926		1,200	249		1,449	1,700	
	0900	1350 amp		2.60	6.154		1,275	258		1,533	1,825	
	0950	1600 amp		2.40	6.667		1,525	280		1,805	2,100	
	1000	2000 amp		2	8		1,675	335		2,010	2,375	
	1020	2500 amp		1.80	8.889		1,975	375		2,350	2,750	
	1030	3000 amp		1.60	10		2,300	420		2,720	3,175	
	1040	4000 amp		1.40	11.429		3,700	480		4,180	4,825	
	1100	Cable tap box end, 225 amp		3.60	4.444		1,000	187		1,187	1,400	
	1150	400 amp		3.20	5		1,000	210		1,210	1,425	
	1200	600 amp		2.60	6.154		1,000	258		1,258	1,500	
	1250	800 amp		2.20	7.273		1,100	305		1,405	1,675	
	1300	1000 amp		2	8		1,200	335		1,535	1,825	
	1320	1200 amp		2	8		1,425	335		1,760	2,100	
	1350	1350 amp		1.60	10		1,550	420		1,970	2,350	
	1400	1600 amp		1.40	11.429		1,750	480		2,230	2,675	
	1450	2000 amp 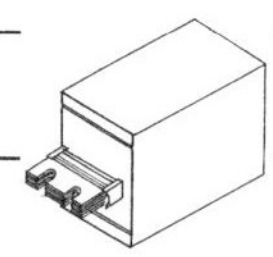		1.20	13.333		2,000	560		2,560	3,075	
	1460	2500 amp		1	16		2,425	670		3,095	3,700	
	1470	3000 amp		.80	20		2,700	840		3,540	4,250	
	1480	4000 amp		.60	26.667		3,575	1,125		4,700	5,675	
	1500	Switchboard stub, 225 amp		5.80	2.759		885	116		1,001	1,150	
	1550	400 amp		5.40	2.963		885	124		1,009	1,150	
	1600	600 amp		4.60	3.478		885	146		1,031	1,200	
	1650	800 amp		4	4		1,000	168		1,168	1,350	
	1700	1000 amp		3.20	5		1,125	210		1,335	1,550	
	1720	1200 amp		3.10	5.161		1,250	217		1,467	1,675	
	1750	1350 amp		3	5.333		1,375	224		1,599	1,875	
	1800	1600 amp		2.60	6.154		1,525	258		1,783	2,100	
	1850	2000 amp		2.40	6.667		1,775	280		2,055	2,375	
	1860	2500 amp		2.20	7.273		2,100	305		2,405	2,775	
	1870	3000 amp		2	8		2,450	335		2,785	3,225	
	1880	4000 amp		1.80	8.889		3,050	375		3,425	3,950	
	1890	Tee fittings, 225 amp		3.20	5		960	210		1,170	1,375	
	1900	400 amp		2.80	5.714		960	240		1,200	1,425	
	1950	600 amp		2.60	6.154		960	258		1,218	1,450	
	2000	800 amp		2.40	6.667		1,025	280		1,305	1,550	
	2050	1000 amp		2.20	7.273		1,075	305		1,380	1,675	
	2070	1200 amp		2.10	7.619		1,275	320		1,595	1,900	
	2100	1350 amp 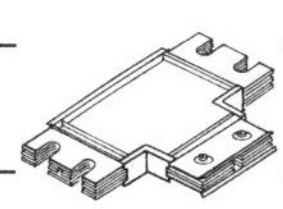		2	8		1,750	335		2,085	2,450	
	2150	1600 amp		1.60	10		2,100	420		2,520	2,975	
	2200	2000 amp		1.20	13.333		2,325	560		2,885	3,450	

16450 | Enclosed Bus Assemblies

		CREW	DAILY OUTPUT	LABOR-HOURS	UNIT	2006 BARE COSTS MAT.	LABOR	EQUIP.	TOTAL	TOTAL INCL O&P	
100	2220 2500 amp	2 Elec	1	16	Ea.	2,800	670		3,470	4,100	100
	2230 3000 amp		.80	20		3,175	840		4,015	4,800	
	2240 4000 amp		.60	26.667		5,300	1,125		6,425	7,550	
	2300 Wall flange, 600 amp		20	.800		119	33.50		152.50	183	
	2310 800 amp		16	1		119	42		161	196	
	2320 1000 amp		13	1.231		119	51.50		170.50	211	
	2325 1200 amp		12	1.333		119	56		175	218	
	2330 1350 amp		10.80	1.481		119	62		181	227	
	2340 1600 amp		9	1.778		119	74.50		193.50	246	
	2350 2000 amp		8	2		119	84		203	260	
	2360 2500 amp		6.60	2.424		119	102		221	288	
	2370 3000 amp		5.40	2.963		119	124		243	325	
	2380 4000 amp		4	4		119	168		287	390	
	2390 5000 amp		3	5.333		119	224		343	475	
	2400 Vapor barrier		8	2		300	84		384	460	
	2420 Roof flange kit		4	4		585	168		753	905	
	2600 Expansion fitting, 225 amp		10	1.600		1,175	67		1,242	1,400	
	2610 400 amp		8	2		1,175	84		1,259	1,425	
	2620 600 amp		6	2.667		1,175	112		1,287	1,475	
	2630 800 amp		4.60	3.478		1,375	146		1,521	1,725	
	2640 1000 amp		4	4		1,525	168		1,693	1,925	
	2650 1350 amp		3.60	4.444		2,225	187		2,412	2,750	
	2660 1600 amp		3.20	5		2,675	210		2,885	3,275	
	2670 2000 amp		2.80	5.714		2,975	240		3,215	3,650	
	2680 2500 amp		2.40	6.667		3,600	280		3,880	4,375	
	2690 3000 amp		2	8		4,150	335		4,485	5,075	
	2700 4000 amp		1.60	10		5,450	420		5,870	6,650	
	2800 Reducer nonfused, 400 amp		8	2		940	84		1,024	1,150	
	2810 600 amp		6	2.667		940	112		1,052	1,200	
	2820 800 amp		4.60	3.478		1,125	146		1,271	1,475	
	2830 1000 amp		4	4		1,325	168		1,493	1,700	
	2840 1350 amp		3.60	4.444		1,725	187		1,912	2,200	
	2850 1600 amp		3.20	5		2,325	210		2,535	2,900	
	2860 2000 amp		2.80	5.714		2,675	240		2,915	3,325	
	2870 2500 amp		2.40	6.667		3,375	280		3,655	4,125	
	2880 3000 amp		2	8		3,900	335		4,235	4,825	
	2890 4000 amp		1.60	10		5,225	420		5,645	6,400	
	2950 Reducer fuse included, 225 amp		4.40	3.636		2,750	153		2,903	3,250	
	2960 400 amp		4.20	3.810		2,800	160		2,960	3,325	
	2970 600 amp		3.60	4.444		3,300	187		3,487	3,950	
	2980 800 amp		3.20	5		5,275	210		5,485	6,125	
	2990 1000 amp		3	5.333		6,025	224		6,249	6,975	
	3000 1200 amp		2.80	5.714		6,025	240		6,265	7,000	
	3010 1600 amp		2.20	7.273		13,700	305		14,005	15,600	
	3020 2000 amp		1.80	8.889		15,300	375		15,675	17,400	
	3100 Reducer circuit breaker, 225 amp		4.40	3.636		2,725	153		2,878	3,200	
	3110 400 amp		4.20	3.810		3,300	160		3,460	3,900	
	3120 600 amp		3.60	4.444		4,700	187		4,887	5,475	
	3130 800 amp		3.20	5		5,500	210		5,710	6,375	
	3140 1000 amp		3	5.333		6,250	224		6,474	7,225	
	3150 1200 amp		2.80	5.714		7,500	240		7,740	8,650	
	3160 1600 amp		2.20	7.273		11,100	305		11,405	12,700	
	3170 2000 amp		1.80	8.889		12,200	375		12,575	14,000	
	3250 Reducer circuit breaker, 75,000 AIC, 225 amp		4.40	3.636		4,250	153		4,403	4,900	
	3260 400 amp		4.20	3.810		4,250	160		4,410	4,925	
	3270 600 amp	↓	3.60	4.444	↓	5,675	187		5,862	6,550	

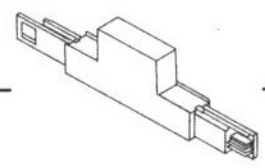

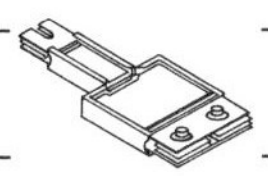

16450 | Enclosed Bus Assemblies

			CREW	DAILY OUTPUT	LABOR-HOURS	UNIT	2006 BARE COSTS MAT.	LABOR	EQUIP.	TOTAL	TOTAL INCL O&P
100	3280	800 amp	2 Elec	3.20	5	Ea.	6,225	210		6,435	7,175
	3290	1000 amp		3	5.333		10,000	224		10,224	11,300
	3300	1200 amp		2.80	5.714		10,000	240		10,240	11,400
	3310	1600 amp		2.20	7.273		11,100	305		11,405	12,700
	3320	2000 amp		1.80	8.889		12,200	375		12,575	14,000
	3400	Reducer circuit breaker CLF 225 amp		4.40	3.636		4,375	153		4,528	5,025
	3410	400 amp		4.20	3.810		5,200	160		5,360	5,950
	3420	600 amp		3.60	4.444		7,775	187		7,962	8,850
	3430	800 amp		3.20	5		8,100	210		8,310	9,250
	3440	1000 amp		3	5.333		8,450	224		8,674	9,650
	3450	1200 amp		2.80	5.714		11,000	240		11,240	12,600
	3460	1600 amp		2.20	7.273		11,100	305		11,405	12,700
	3470	2000 amp		1.80	8.889		12,200	375		12,575	14,000
	3550	Ground bus added to bus duct, 225 amp		320	.050	L.F.	36	2.10		38.10	42.50
	3560	400 amp		320	.050		36	2.10		38.10	42.50
	3570	600 amp		280	.057		36	2.40		38.40	43
	3580	800 amp		240	.067		36	2.80		38.80	44
	3590	1000 amp		200	.080		36	3.36		39.36	44.50
	3600	1350 amp		180	.089		36	3.73		39.73	45.50
	3610	1600 amp		160	.100		36	4.20		40.20	46
	3620	2000 amp		160	.100		36	4.20		40.20	46
	3630	2500 amp		140	.114		36	4.80		40.80	47
	3640	3000 amp		120	.133		36	5.60		41.60	48
	3650	4000 amp		100	.160		36	6.70		42.70	50
	3810	High short circuit, 400 amp		36	.444		121	18.65		139.65	163
	3820	600 amp		32	.500		121	21		142	167
	3830	800 amp		26	.615		142	26		168	196
	3840	1000 amp		24	.667		152	28		180	210
	3850	1350 amp		22	.727		202	30.50		232.50	270
	3860	1600 amp		18	.889		233	37.50		270.50	315
	3870	2000 amp		16	1		273	42		315	365
	3880	2500 amp		14	1.143		455	48		503	575
	3890	3000 amp		12	1.333		515	56		571	655
	3920	Cross, 225 amp		5.60	2.857	Ea.	1,450	120		1,570	1,750
	3930	400 amp		4.60	3.478		1,450	146		1,596	1,800
	3940	600 amp		4	4		1,450	168		1,618	1,825
	3950	800 amp		3.40	4.706		1,525	198		1,723	1,975
	3960	1000 amp		3	5.333		1,600	224		1,824	2,100
	3970	1350 amp		2.80	5.714		2,575	240		2,815	3,200
	3980	1600 amp		2.20	7.273		3,050	305		3,355	3,850
	3990	2000 amp		1.80	8.889		3,350	375		3,725	4,275
	4000	2500 amp		1.60	10		3,975	420		4,395	5,025
	4010	3000 amp		1.20	13.333		4,575	560		5,135	5,900
	4020	4000 amp		1	16		7,000	670		7,670	8,725
	4040	Cable tap box center, 225 amp		3.60	4.444		1,000	187		1,187	1,400
	4050	400 amp		3.20	5		1,000	210		1,210	1,425
	4060	600 amp		2.60	6.154		1,000	258		1,258	1,500
	4070	800 amp		2.20	7.273		1,100	305		1,405	1,675
	4080	1000 amp		2	8		1,200	335		1,535	1,825
	4090	1350 amp		1.60	10		1,550	420		1,970	2,350
	4100	1600 amp		1.40	11.429		1,750	480		2,230	2,675
	4110	2000 amp		1.20	13.333		2,000	560		2,560	3,075
	4120	2500 amp		1	16		2,425	670		3,095	3,700
	4130	3000 amp		.80	20		2,700	840		3,540	4,250
	4140	4000 amp		.60	26.667		3,575	1,125		4,700	5,675
	4500	Weatherproof, feeder, 600 amp		30	.533	L.F.	145	22.50		167.50	195

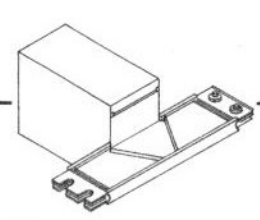

ELECTRICAL 16

For expanded coverage of these items see *Means Electrical Cost Data 2006*

16450 | Enclosed Bus Assemblies

		16450 Enclosed Bus Assemblies	CREW	DAILY OUTPUT	LABOR-HOURS	UNIT	2006 BARE COSTS MAT.	LABOR	EQUIP.	TOTAL	TOTAL INCL O&P	
100	4520	800 amp	2 Elec	24	.667	L.F.	170	28		198	230	100
	4540	1000 amp		22	.727		194	30.50		224.50	261	
	4550	1200 amp		21	.762		269	32		301	345	
	4560	1350 amp		20	.800		305	33.50		338.50	385	
	4580	1600 amp		17	.941		350	39.50		389.50	450	
	4600	2000 amp		16	1		415	42		457	520	
	4620	2500 amp		12	1.333		535	56		591	675	
	4640	3000 amp		10	1.600		605	67		672	775	
	4660	4000 amp		8	2		875	84		959	1,100	
	5000	3 pole, 3 wire, feeder, 600 amp		40	.400		111	16.80		127.80	148	
	5010	800 amp		32	.500		132	21		153	178	
	5020	1000 amp		30	.533		142	22.50		164.50	191	
	5025	1200 amp		29	.552		142	23		165	192	
	5030	1350 amp		28	.571		192	24		216	249	
	5040	1600 amp		24	.667		223	28		251	288	
	5050	2000 amp		20	.800		263	33.50		296.50	340	
	5060	2500 amp		16	1		365	42		407	465	
	5070	3000 amp		14	1.143		435	48		483	555	
	5080	4000 amp		12	1.333		525	56		581	665	
	5200	Plug-in type, 225 amp		50	.320		121	13.45		134.45	155	
	5210	400 amp		42	.381		121	16		137	159	
	5220	600 amp		36	.444		121	18.65		139.65	163	
	5230	800 amp		30	.533		142	22.50		164.50	191	
	5240	1000 amp		28	.571		152	24		176	204	
	5245	1200 amp		27	.593		172	25		197	228	
	5250	1350 amp		26	.615		202	26		228	263	
	5260	1600 amp		20	.800		233	33.50		266.50	310	
	5270	2000 amp		18	.889		273	37.50		310.50	360	
	5280	2500 amp		16	1		375	42		417	475	
	5290	3000 amp		14	1.143		445	48		493	565	
	5300	4000 amp		12	1.333		535	56		591	675	
	5330	High short circuit, 400 amp		42	.381		121	16		137	159	
	5340	600 amp		36	.444		121	18.65		139.65	163	
	5350	800 amp		30	.533		142	22.50		164.50	191	
	5360	1000 amp		28	.571		152	24		176	204	
	5370	1350 amp		26	.615		202	26		228	263	
	5380	1600 amp		20	.800		233	33.50		266.50	310	
	5390	2000 amp		18	.889		273	37.50		310.50	360	
	5400	2500 amp		16	1		430	42		472	540	
	5410	3000 amp		14	1.143		445	48		493	565	
	5440	Elbow, 225 amp		5	3.200	Ea.	620	134		754	885	
	5450	400 amp		4.40	3.636		620	153		773	915	
	5460	600 amp		4	4		620	168		788	940	
	5470	800 amp		3.40	4.706		660	198		858	1,025	
	5480	1000 amp		3.20	5		675	210		885	1,075	
	5485	1200 amp		3.10	5.161		720	217		937	1,125	
	5490	1350 amp		3	5.333		780	224		1,004	1,200	
	5500	1600 amp		2.80	5.714		1,200	240		1,440	1,700	
	5510	2000 amp		2.40	6.667		1,325	280		1,605	1,875	
	5520	2500 amp		2	8		1,625	335		1,960	2,300	
	5530	3000 amp		1.80	8.889		1,925	375		2,300	2,700	
	5540	4000 amp		1.60	10		2,575	420		2,995	3,475	
	5560	Tee fittings, 225 amp		3.60	4.444		860	187		1,047	1,225	
	5570	400 amp		3.20	5		860	210		1,070	1,275	
	5580	600 amp		3	5.333		860	224		1,084	1,300	
	5590	800 amp		2.80	5.714		920	240		1,160	1,375	

16450 | Enclosed Bus Assemblies

		16450 Enclosed Bus Assemblies	CREW	DAILY OUTPUT	LABOR-HOURS	UNIT	2006 BARE COSTS MAT.	LABOR	EQUIP.	TOTAL	TOTAL INCL O&P	
100	5600	1000 amp	2 Elec	2.60	6.154	Ea.	950	258		1,208	1,450	100
	5605	1200 amp		2.50	6.400		1,025	269		1,294	1,550	
	5610	1350 amp		2.40	6.667		1,400	280		1,680	1,975	
	5620	1600 amp		1.80	8.889		1,650	375		2,025	2,400	
	5630	2000 amp		1.40	11.429		1,850	480		2,330	2,775	
	5640	2500 amp		1.20	13.333		2,300	560		2,860	3,400	
	5650	3000 amp		1	16		2,700	670		3,370	4,000	
	5660	4000 amp		.70	22.857		3,950	960		4,910	5,825	
	5680	Cross, 225 amp		6.40	2.500		1,400	105		1,505	1,675	
	5690	400 amp		5.40	2.963		1,400	124		1,524	1,725	
	5700	600 amp		4.60	3.478		1,400	146		1,546	1,750	
	5710	800 amp		4	4		1,475	168		1,643	1,875	
	5720	1000 amp		3.60	4.444		1,525	187		1,712	1,975	
	5730	1350 amp		3.20	5		2,250	210		2,460	2,800	
	5740	1600 amp		2.60	6.154		2,625	258		2,883	3,300	
	5750	2000 amp		2.20	7.273		2,875	305		3,180	3,650	
	5760	2500 amp		1.80	8.889		3,475	375		3,850	4,400	
	5770	3000 amp		1.40	11.429		4,150	480		4,630	5,325	
	5780	4000 amp		1.20	13.333		5,775	560		6,335	7,225	
	5800	Expansion fitting, 225 amp		11.60	1.379		1,000	58		1,058	1,200	
	5810	400 amp		9.20	1.739		1,000	73		1,073	1,225	
	5820	600 amp		7	2.286		1,000	96		1,096	1,250	
	5830	800 amp		5.20	3.077		1,200	129		1,329	1,525	
	5840	1000 amp		4.60	3.478		1,300	146		1,446	1,650	
	5850	1350 amp		4.20	3.810		1,575	160		1,735	2,000	
	5860	1600 amp		3.60	4.444		1,975	187		2,162	2,475	
	5870	2000 amp		3.20	5		2,325	210		2,535	2,875	
	5880	2500 amp		2.80	5.714		2,625	240		2,865	3,275	
	5890	3000 amp		2.40	6.667		3,075	280		3,355	3,800	
	5900	4000 amp		1.80	8.889		4,225	375		4,600	5,225	
	5940	Reducer, nonfused, 400 amp		9.20	1.739		810	73		883	1,000	
	5950	600 amp		7	2.286		810	96		906	1,050	
	5960	800 amp		5.20	3.077		850	129		979	1,125	
	5970	1000 amp		4.60	3.478		1,050	146		1,196	1,375	
	5980	1350 amp		4.20	3.810		1,550	160		1,710	1,950	
	5990	1600 amp		3.60	4.444		1,750	187		1,937	2,250	
	6000	2000 amp		3.20	5		2,050	210		2,260	2,600	
	6010	2500 amp		2.80	5.714		2,625	240		2,865	3,275	
	6020	3000 amp		2.20	7.273		3,075	305		3,380	3,850	
	6030	4000 amp		1.80	8.889		4,225	375		4,600	5,225	
	6050	Reducer, fuse included, 225 amp		5	3.200		2,125	134		2,259	2,525	
	6060	400 amp		4.80	3.333		2,800	140		2,940	3,300	
	6070	600 amp		4.20	3.810		3,600	160		3,760	4,200	
	6080	800 amp		3.60	4.444		5,575	187		5,762	6,425	
	6090	1000 amp		3.40	4.706		6,075	198		6,273	6,975	
	6100	1350 amp		3.20	5		11,100	210		11,310	12,500	
	6110	1600 amp		2.60	6.154		13,100	258		13,358	14,900	
	6120	2000 amp		2	8		15,300	335		15,635	17,300	
	6160	Reducer, circuit breaker, 225 amp		5	3.200		2,600	134		2,734	3,050	
	6170	400 amp		4.80	3.333		3,175	140		3,315	3,725	
	6180	600 amp		4.20	3.810		4,575	160		4,735	5,275	
	6190	800 amp		3.60	4.444		5,350	187		5,537	6,175	
	6200	1000 amp		3.40	4.706		6,075	198		6,273	6,975	
	6210	1350 amp		3.20	5		7,350	210		7,560	8,400	
	6220	1600 amp		2.60	6.154		10,900	258		11,158	12,400	
100	6230	2000 amp		2	8		12,000	335		12,335	13,700	

ELECTRICAL 16

For expanded coverage of these items see *Means Electrical Cost Data 2006*

16450 | Enclosed Bus Assemblies

	16450 Enclosed Bus Assemblies		CREW	DAILY OUTPUT	LABOR-HOURS	UNIT	2006 BARE COSTS MAT.	LABOR	EQUIP.	TOTAL	TOTAL INCL O&P	
100	6270	Cable tap box center, 225 amp	2 Elec	4.20	3.810	Ea.	970	160		1,130	1,325	100
	6280	400 amp		3.60	4.444		970	187		1,157	1,375	
	6290	600 amp		3	5.333		970	224		1,194	1,425	
	6300	800 amp		2.60	6.154		1,050	258		1,308	1,575	
	6310	1000 amp		2.40	6.667		1,125	280		1,405	1,650	
	6320	1350 amp		1.80	8.889		1,350	375		1,725	2,075	
	6330	1600 amp		1.60	10		1,525	420		1,945	2,325	
	6340	2000 amp		1.40	11.429		1,750	480		2,230	2,650	
	6350	2500 amp		1.20	13.333		2,175	560		2,735	3,250	
	6360	3000 amp		1	16		2,450	670		3,120	3,725	
	6370	4000 amp		.70	22.857		2,900	960		3,860	4,675	
	6390	Cable tap box end, 225 amp		4.20	3.810		590	160		750	895	
	6400	400 amp		3.60	4.444		590	187		777	940	
	6410	600 amp		3	5.333		590	224		814	995	
	6420	800 amp		2.60	6.154		650	258		908	1,100	
	6430	1000 amp		2.40	6.667		700	280		980	1,200	
	6435	1200 amp		2.10	7.619		760	320		1,080	1,325	
	6440	1350 amp		1.80	8.889		845	375		1,220	1,500	
	6450	1600 amp		1.60	10		955	420		1,375	1,700	
	6460	2000 amp		1.40	11.429		1,075	480		1,555	1,950	
	6470	2500 amp		1.20	13.333		1,250	560		1,810	2,250	
	6480	3000 amp		1	16		1,475	670		2,145	2,650	
	6490	4000 amp		.70	22.857		1,775	960		2,735	3,425	
	7000	Weatherproof, feeder, 600 amp		34	.471	L.F.	144	19.75		163.75	189	
	7020	800 amp		28	.571		158	24		182	211	
	7040	1000 amp		26	.615		170	26		196	227	
	7050	1200 amp		25	.640		199	27		226	261	
	7060	1350 amp		24	.667		231	28		259	297	
	7080	1600 amp		20	.800		267	33.50		300.50	345	
	7100	2000 amp		18	.889		315	37.50		352.50	405	
	7120	2500 amp		14	1.143		435	48		483	555	
	7140	3000 amp		12	1.333		525	56		581	660	
	7160	4000 amp		10	1.600		630	67		697	800	
200	0010	**BUS DUCT** 100 amp and less, aluminum or copper, plug-in										200
	0080	Bus duct, 3 pole 3 wire, 100 amp	1 Elec	42	.190	L.F.	22	8		30	37	
	0110	Elbow		4	2	Ea.	57	84		141	192	
	0120	Tee		2	4		83	168		251	350	
	0130	Wall flange		8	1		7.35	42		49.35	72.50	
	0140	Ground kit		16	.500		15.65	21		36.65	49.50	
	0180	3 pole 4 wire, 100 amp		40	.200	L.F.	25.50	8.40		33.90	41	
	0200	Cable tap box		3.10	2.581	Ea.	69	108		177	243	
	0300	End closure		16	.500		9.20	21		30.20	42.50	
	0400	Elbow		4	2		57	84		141	192	
	0500	Tee		2	4		83	168		251	350	
	0600	Hangers		10	.800		5.50	33.50		39	58	
	0700	Circuit breakers, 15 to 50 amp, 1 pole		8	1		147	42		189	227	
	0800	15 to 60 amp, 2 pole		6.70	1.194		575	50		625	715	
	0900	3 pole		5.30	1.509		575	63.50		638.50	735	
	1000	60 to 100 amp, 1 pole		6.70	1.194		575	50		625	715	
	1100	70 to 100 amp, 2 pole		5.30	1.509		575	63.50		638.50	735	
	1200	3 pole		4.50	1.778		575	74.50		649.50	750	
	1220	Switch, nonfused, 3 pole, 4 wire		8	1		78	42		120	151	
	1240	Fused, 3 fuses, 4 wire, 30 amp		8	1		283	42		325	375	
	1260	60 amp		5.30	1.509		297	63.50		360.50	425	
	1280	100 amp		4.50	1.778		495	74.50		569.50	660	

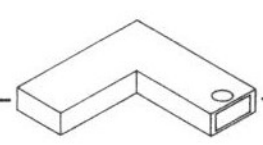

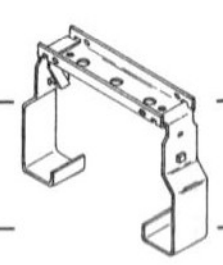

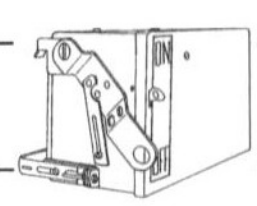

16 ELECTRICAL

16450 | Enclosed Bus Assemblies

		Description	CREW	DAILY OUTPUT	LABOR-HOURS	UNIT	2006 BARE COSTS MAT.	LABOR	EQUIP.	TOTAL	TOTAL INCL O&P	
200	1300	Plug, fusible, 3 pole 250 volt, 30 amp	1 Elec	5.30	1.509	Ea.	283	63.50		346.50	410	200
	1310	60 amp		5.30	1.509		335	63.50		398.50	470	
	1320	100 amp		4.50	1.778		495	74.50		569.50	660	
	1330	3 pole 480 volt, 30 amp		5.30	1.509		325	63.50		388.50	455	
	1340	60 amp		5.30	1.509		350	63.50		413.50	485	
	1350	100 amp		4.50	1.778		505	74.50		579.50	670	
	1360	Circuit breaker, 3 pole 250 volt, 60 amp		5.30	1.509		575	63.50		638.50	735	
	1370	3 pole 480 volt, 100 amp		4.50	1.778		575	74.50		649.50	750	
	2000	Bus duct, 2 wire, 250 volt, 30 amp		60	.133	L.F.	2.45	5.60		8.05	11.35	
	2100	60 amp		50	.160		2.45	6.70		9.15	13.05	
	2200	300 volt, 30 amp		60	.133		2.45	5.60		8.05	11.35	
	2300	60 amp		50	.160		2.45	6.70		9.15	13.05	
	2400	3 wire, 250 volt, 30 amp		60	.133		2.45	5.60		8.05	11.35	
	2500	60 amp		50	.160		2.45	6.70		9.15	13.05	
	2600	480/277 volt, 30 amp		60	.133		2.45	5.60		8.05	11.35	
	2700	60 amp		50	.160		2.45	6.70		9.15	13.05	
	2750	End feed, 300 volt 2 wire max. 30 amp		6	1.333	Ea.	48	56		104	139	
	2800	60 amp		5.50	1.455		48	61		109	147	
	2850	30 amp miniature		6	1.333		48	56		104	139	
	2900	3 wire, 30 amp		6	1.333		60.50	56		116.50	153	
	2950	60 amp		5.50	1.455		60.50	61		121.50	161	
	3000	30 amp miniature		6	1.333		60.50	56		116.50	153	
	3050	Center feed, 300 volt 2 wire, 30 amp		6	1.333		66	56		122	160	
	3100	60 amp		5.50	1.455		66	61		127	167	
	3150	3 wire, 30 amp		6	1.333		75.50	56		131.50	170	
	3200	60 amp		5.50	1.455		75.50	61		136.50	177	
	3220	Elbow, 30 amp		6	1.333		46	56		102	137	
	3240	60 amp		5.50	1.455		46	61		107	145	
	3260	End cap		40	.200		9.20	8.40		17.60	23	
	3280	Strength beam, 10 ft.		15	.533		26	22.50		48.50	63	
	3300	Hanger		24	.333		5.50	14		19.50	27.50	
	3320	Tap box, nonfusible		6.30	1.270		78	53.50		131.50	168	
	3340	Fusible switch 30 amp, 1 fuse		6	1.333		340	56		396	460	
	3360	2 fuse		6	1.333		345	56		401	465	
	3380	3 fuse		6	1.333		350	56		406	470	
	3400	Circuit breaker handle on cover, 1 pole		6	1.333		55	56		111	147	
	3420	2 pole		6	1.333		60	56		116	153	
	3440	3 pole		6	1.333		78	56		134	173	
	3460	Circuit breaker external operhandle, 1 pole		6	1.333		55	56		111	147	
	3480	2 pole		6	1.333		60	56		116	153	
	3500	3 pole		6	1.333		78	56		134	173	
	3520	Terminal plug only		16	.500		10.10	21		31.10	43.50	
	3540	Terminal with receptacle		16	.500		13.10	21		34.10	47	
	3560	Fixture plug		16	.500		9	21		30	42.50	
	4000	Copper bus duct, lighting, 2 wire 300 volt, 20 amp		70	.114	L.F.	2.53	4.80		7.33	10.20	
	4020	35 amp		60	.133		2.53	5.60		8.13	11.45	
	4040	50 amp		55	.145		2.53	6.10		8.63	12.20	
	4060	60 amp		50	.160		2.53	6.70		9.23	13.15	
	4080	3 wire 300 volt, 20 amp		70	.114		2.76	4.80		7.56	10.45	
	4100	35 amp		60	.133		2.76	5.60		8.36	11.70	
	4120	50 amp		55	.145		2.76	6.10		8.86	12.45	
	4140	60 amp		50	.160		2.76	6.70		9.46	13.40	
	4160	Feeder in box, end, 1 circuit		6	1.333	Ea.	55	56		111	147	
	4180	2 circuit		5.50	1.455		57	61		118	157	
	4200	Center, 1 circuit		6	1.333		75.50	56		131.50	170	
	4220	2 circuit		5.50	1.455		78	61		139	180	

		16450 | Enclosed Bus Assemblies	CREW	DAILY OUTPUT	LABOR-HOURS	UNIT	2006 BARE COSTS MAT.	LABOR	EQUIP.	TOTAL	TOTAL INCL O&P	
200	4240	End cap	1 Elec	40	.200	Ea.	9.20	8.40		17.60	23	200
	4260	Hanger, surface mount		24	.333		5.50	14		19.50	27.50	
	4280	Coupling		40	.200		7	8.40		15.40	20.50	
300	0010	COPPER BUS DUCT										300
	0100	3 pole 4 wire, weatherproof, feeder duct, 600 amp	2 Elec	24	.667	L.F.	155	28		183	213	
	0110	800 amp		18	.889		188	37.50		225.50	264	
	0120	1000 amp		17	.941		210	39.50		249.50	292	
	0125	1200 amp		16.50	.970		256	40.50		296.50	345	
	0130	1350 amp		16	1		310	42		352	405	
	0140	1600 amp		12	1.333		355	56		411	475	
	0150	2000 amp		10	1.600		455	67		522	605	
	0160	2500 amp		7	2.286		565	96		661	770	
	0170	3000 amp		5	3.200		685	134		819	960	
	0180	4000 amp		3.60	4.444		905	187		1,092	1,275	
	0200	Plug-in/indoor, bus duct high short circuit, 400 amp		32	.500		147	21		168	195	
	0210	600 amp		26	.615		147	26		173	202	
	0220	800 amp		20	.800		175	33.50		208.50	244	
	0230	1000 amp		18	.889		193	37.50		230.50	271	
	0240	1350 amp		16	1		276	42		318	370	
	0250	1600 amp		12	1.333		315	56		371	430	
	0260	2000 amp		10	1.600		395	67		462	540	
	0270	2500 amp		8	2		490	84		574	665	
	0280	3000 amp		6	2.667		590	112		702	825	
	0310	Cross, 225 amp		3	5.333	Ea.	1,750	224		1,974	2,275	
	0320	400 amp		2.80	5.714		1,750	240		1,990	2,300	
	0330	600 amp		2.60	6.154		1,750	258		2,008	2,325	
	0340	800 amp		2.20	7.273		1,900	305		2,205	2,550	
	0350	1000 amp		2	8		2,100	335		2,435	2,850	
	0360	1350 amp		1.80	8.889		2,400	375		2,775	3,200	
	0370	1600 amp		1.70	9.412		2,650	395		3,045	3,525	
	0380	2000 amp		1.60	10		4,350	420		4,770	5,450	
	0390	2500 amp		1.40	11.429		5,275	480		5,755	6,550	
	0400	3000 amp		1.20	13.333		6,100	560		6,660	7,575	
	0410	4000 amp		1	16		7,950	670		8,620	9,750	
	0430	Expansion fitting, 225 amp		5.40	2.963		915	124		1,039	1,200	
	0440	400 amp		4.60	3.478		1,025	146		1,171	1,375	
	0450	600 amp		4	4		1,275	168		1,443	1,650	
	0460	800 amp		3.40	4.706		1,500	198		1,698	1,950	
	0470	1000 amp		3	5.333		1,725	224		1,949	2,250	
	0480	1350 amp		2.80	5.714		2,175	240		2,415	2,775	
	0490	1600 amp		2.60	6.154		3,025	258		3,283	3,725	
	0500	2000 amp		2.20	7.273		3,500	305		3,805	4,325	
	0510	2500 amp		1.80	8.889		4,250	375		4,625	5,250	
	0520	3000 amp		1.60	10		5,025	420		5,445	6,175	
	0530	4000 amp		1.20	13.333		6,475	560		7,035	8,000	
	0550	Reducer nonfused, 225 amp		5.40	2.963		1,050	124		1,174	1,350	
	0560	400 amp		4.60	3.478		1,050	146		1,196	1,375	
	0570	600 amp		4	4		1,050	168		1,218	1,400	
	0580	800 amp		3.40	4.706		1,275	198		1,473	1,700	
	0590	1000 amp		3	5.333		1,525	224		1,749	2,025	
	0600	1350 amp		2.80	5.714		2,300	240		2,540	2,900	
	0610	1600 amp		2.60	6.154		2,650	258		2,908	3,325	
	0620	2000 amp		2.20	7.273		3,200	305		3,505	3,975	
	0630	2500 amp		1.80	8.889		4,025	375		4,400	5,000	
	0640	3000 amp		1.60	10		4,800	420		5,220	5,925	

16450 | Enclosed Bus Assemblies

			CREW	DAILY OUTPUT	LABOR-HOURS	UNIT	2006 BARE COSTS MAT.	LABOR	EQUIP.	TOTAL	TOTAL INCL O&P	
300	0650	4000 amp	2 Elec	1.20	13.333	Ea.	6,225	560		6,785	7,725	300
	0670	Reducer fuse included, 225 amp		4.40	3.636		2,250	153		2,403	2,700	
	0680	400 amp		4.20	3.810		2,800	160		2,960	3,325	
	0690	600 amp		3.60	4.444		3,450	187		3,637	4,100	
	0700	800 amp		3.20	5		4,875	210		5,085	5,700	
	0710	1000 amp		3	5.333		6,100	224		6,324	7,075	
	0720	1350 amp		2.80	5.714		6,200	240		6,440	7,200	
	0730	1600 amp		2.20	7.273		13,600	305		13,905	15,400	
	0740	2000 amp		1.80	8.889		15,300	375		15,675	17,400	
	0790	Reducer, circuit breaker, 225 amp		4.40	3.636		2,825	153		2,978	3,350	
	0800	400 amp		4.20	3.810		3,300	160		3,460	3,900	
	0810	600 amp		3.60	4.444		4,700	187		4,887	5,475	
	0820	800 amp		3.20	5		5,500	210		5,710	6,375	
	0830	1000 amp		3	5.333		6,250	224		6,474	7,225	
	0840	1350 amp		2.80	5.714		7,500	240		7,740	8,650	
	0850	1600 amp		2.20	7.273		11,100	305		11,405	12,700	
	0860	2000 amp		1.80	8.889		12,200	375		12,575	14,000	
	0910	Cable tap box, center, 225 amp		3.20	5		1,050	210		1,260	1,475	
	0920	400 amp		2.60	6.154		1,050	258		1,308	1,550	
	0930	600 amp		2.20	7.273		1,050	305		1,355	1,625	
	0940	800 amp		2	8		1,175	335		1,510	1,800	
	0950	1000 amp		1.60	10		1,250	420		1,670	2,025	
	0960	1350 amp		1.40	11.429		1,600	480		2,080	2,500	
	0970	1600 amp		1.20	13.333		1,775	560		2,335	2,825	
	0980	2000 amp		1	16		2,150	670		2,820	3,375	
	1040	2500 amp		.80	20		2,550	840		3,390	4,100	
	1060	3000 amp		.60	26.667		2,925	1,125		4,050	4,950	
	1080	4000 amp		.40	40		3,700	1,675		5,375	6,675	
	1800	3 pole 3 wire, feeder duct, weatherproof, 600 amp		28	.571	L.F.	155	24		179	207	
	1820	800 amp		22	.727		188	30.50		218.50	253	
	1840	1000 amp		20	.800		210	33.50		243.50	283	
	1850	1200 amp		19	.842		217	35.50		252.50	294	
	1860	1350 amp		18	.889		310	37.50		347.50	400	
	1880	1600 amp		14	1.143		355	48		403	465	
	1900	2000 amp		12	1.333		455	56		511	585	
	1920	2500 amp		8	2		565	84		649	750	
	1940	3000 amp		6	2.667		685	112		797	930	
	1960	4000 amp		4	4		905	168		1,073	1,250	
	2000	Feeder duct/indoor, 600 amp		32	.500		129	21		150	175	
	2010	800 amp		26	.615		156	26		182	212	
	2020	1000 amp		24	.667		175	28		203	235	
	2025	1200 amp		22	.727		230	30.50		260.50	300	
	2030	1350 amp		20	.800		258	33.50		291.50	335	
	2040	1600 amp		16	1		294	42		336	390	
	2050	2000 amp		14	1.143		375	48		423	490	
	2060	2500 amp		10	1.600		470	67		537	620	
	2070	3000 amp		8	2		570	84		654	755	
	2080	4000 amp		6	2.667		755	112		867	1,000	
	2090	5000 amp		5	3.200		920	134		1,054	1,200	
	2200	Bus duct plug-in/indoor, 225 amp		46	.348		147	14.60		161.60	185	
	2210	400 amp		36	.444		147	18.65		165.65	191	
	2220	600 amp		30	.533		147	22.50		169.50	197	
	2230	800 amp		24	.667		175	28		203	235	
	2240	1000 amp		20	.800		193	33.50		226.50	265	
	2250	1350 amp		18	.889		276	37.50		313.50	365	
	2260	1600 amp		14	1.143		315	48		363	420	

16450 | Enclosed Bus Assemblies

			CREW	DAILY OUTPUT	LABOR-HOURS	UNIT	2006 BARE COSTS MAT.	LABOR	EQUIP.	TOTAL	TOTAL INCL O&P	
300	2270	2000 amp	2 Elec	12	1.333	L.F.	395	56		451	520	300
	2280	2500 amp		10	1.600		490	67		557	640	
	2290	3000 amp		8	2		590	84		674	780	
	2330	High short circuit, 400 amp		36	.444		147	18.65		165.65	191	
	2340	600 amp		30	.533		147	22.50		169.50	197	
	2350	800 amp		24	.667		175	28		203	235	
	2360	1000 amp		20	.800		193	33.50		226.50	265	
	2370	1350 amp		18	.889		276	37.50		313.50	365	
	2380	1600 amp		14	1.143		315	48		363	420	
	2390	2000 amp		12	1.333		395	56		451	520	
	2400	2500 amp		10	1.600		490	67		557	640	
	2410	3000 amp		8	2		590	84		674	780	
	2440	Elbows, 225 amp		4.60	3.478	Ea.	735	146		881	1,025	
	2450	400 amp		4.20	3.810		735	160		895	1,050	
	2460	600 amp		3.60	4.444		735	187		922	1,100	
	2470	800 amp		3.20	5		790	210		1,000	1,200	
	2480	1000 amp		3	5.333		830	224		1,054	1,250	
	2485	1200 amp		2.90	5.517		890	232		1,122	1,325	
	2490	1350 amp		2.80	5.714		995	240		1,235	1,475	
	2500	1600 amp		2.60	6.154		1,075	258		1,333	1,575	
	2510	2000 amp		2	8		1,300	335		1,635	1,950	
	2520	2500 amp		1.80	8.889		1,975	375		2,350	2,750	
	2530	3000 amp		1.60	10		2,275	420		2,695	3,150	
	2540	4000 amp		1.40	11.429		2,900	480		3,380	3,950	
	2560	Tee fittings, 225 amp		2.80	5.714		870	240		1,110	1,325	
	2570	400 amp		2.40	6.667		870	280		1,150	1,375	
	2580	600 amp		2	8		870	335		1,205	1,475	
	2590	800 amp		1.80	8.889		945	375		1,320	1,625	
	2600	1000 amp		1.60	10		1,050	420		1,470	1,800	
	2605	1200 amp		1.50	10.667		1,125	450		1,575	1,950	
	2610	1350 amp		1.40	11.429		1,250	480		1,730	2,125	
	2620	1600 amp		1.20	13.333		1,450	560		2,010	2,450	
	2630	2000 amp		1	16		2,400	670		3,070	3,650	
	2640	2500 amp		.70	22.857		2,800	960		3,760	4,575	
	2650	3000 amp		.60	26.667		3,275	1,125		4,400	5,325	
	2660	4000 amp		.50	32		4,225	1,350		5,575	6,725	
	2680	Cross, 225 amp		3.60	4.444		1,325	187		1,512	1,750	
	2690	400 amp		3.20	5		1,325	210		1,535	1,775	
	2700	600 amp		3	5.333		1,325	224		1,549	1,800	
	2710	800 amp		2.60	6.154		1,575	258		1,833	2,150	
	2720	1000 amp		2.40	6.667		1,650	280		1,930	2,250	
	2730	1350 amp		2.20	7.273		1,975	305		2,280	2,650	
	2740	1600 amp		2	8		2,125	335		2,460	2,875	
	2750	2000 amp		1.80	8.889		3,375	375		3,750	4,300	
	2760	2500 amp		1.60	10		3,925	420		4,345	4,975	
	2770	3000 amp		1.40	11.429		4,550	480		5,030	5,750	
	2780	4000 amp		1	16		5,800	670		6,470	7,425	
	2800	Expansion fitting, 225 amp		6.40	2.500		1,050	105		1,155	1,325	
	2810	400 amp		5.40	2.963		1,050	124		1,174	1,375	
	2820	600 amp		4.60	3.478		1,050	146		1,196	1,400	
	2830	800 amp		4	4		1,275	168		1,443	1,650	
	2840	1000 amp		3.60	4.444		1,400	187		1,587	1,825	
	2850	1350 amp		3.20	5		1,725	210		1,935	2,225	
	2860	1600 amp		3	5.333		1,900	224		2,124	2,450	
	2870	2000 amp		2.60	6.154		2,300	258		2,558	2,925	
	2880	2500 amp	↓	2.20	7.273	↓	3,200	305		3,505	4,000	

		16450 | Enclosed Bus Assemblies	CREW	DAILY OUTPUT	LABOR-HOURS	UNIT	2006 BARE COSTS MAT.	LABOR	EQUIP.	TOTAL	TOTAL INCL O&P	
300	2890	3000 amp	2 Elec	1.80	8.889	Ea.	3,775	375		4,150	4,725	300
	2900	4000 amp		1.40	11.429		4,825	480		5,305	6,050	
	2920	Reducer nonfused, 225 amp		6.40	2.500		860	105		965	1,100	
	2930	400 amp		5.40	2.963		860	124		984	1,125	
	2940	600 amp		4.60	3.478		860	146		1,006	1,175	
	2950	800 amp		4	4		1,000	168		1,168	1,350	
	2960	1000 amp		3.60	4.444		1,125	187		1,312	1,550	
	2970	1350 amp		3.20	5		1,500	210		1,710	1,975	
	2980	1600 amp		3	5.333		1,700	224		1,924	2,200	
	2990	2000 amp		2.60	6.154		2,025	258		2,283	2,625	
	3000	2500 amp		2.20	7.273		2,975	305		3,280	3,750	
	3010	3000 amp		1.80	8.889		3,500	375		3,875	4,425	
	3020	4000 amp		1.40	11.429		4,550	480		5,030	5,775	
	3040	Reducer fuse included, 225 amp		5	3.200		2,025	134		2,159	2,450	
	3050	400 amp		4.80	3.333		2,700	140		2,840	3,200	
	3060	600 amp		4.20	3.810		3,225	160		3,385	3,800	
	3070	800 amp		3.60	4.444		4,600	187		4,787	5,350	
	3080	1000 amp		3.40	4.706		5,450	198		5,648	6,300	
	3090	1350 amp		3.20	5		5,550	210		5,760	6,425	
	3100	1600 amp		2.60	6.154		12,900	258		13,158	14,600	
	3110	2000 amp		2	8		14,600	335		14,935	16,600	
	3160	Reducer circuit breaker, 225 amp		5	3.200		2,600	134		2,734	3,050	
	3170	400 amp		4.80	3.333		3,175	140		3,315	3,725	
	3180	600 amp		4.20	3.810		4,575	160		4,735	5,275	
	3190	800 amp		3.60	4.444		5,350	187		5,537	6,175	
	3200	1000 amp		3.40	4.706		6,075	198		6,273	6,975	
	3210	1350 amp		3.20	5		7,350	210		7,560	8,400	
	3220	1600 amp		2.60	6.154		10,900	258		11,158	12,400	
	3230	2000 amp		2	8		12,000	335		12,335	13,700	
	3280	3 pole, 3 wire, cable tap box center, 225 amp		3.60	4.444		1,200	187		1,387	1,625	
	3290	400 amp		3	5.333		1,200	224		1,424	1,675	
	3300	600 amp		2.60	6.154		1,200	258		1,458	1,725	
	3310	800 amp		2.40	6.667		1,350	280		1,630	1,900	
	3320	1000 amp		1.80	8.889		1,450	375		1,825	2,175	
	3330	1350 amp		1.60	10		1,875	420		2,295	2,700	
	3340	1600 amp		1.40	11.429		2,100	480		2,580	3,050	
	3350	2000 amp		1.20	13.333		2,550	560		3,110	3,675	
	3360	2500 amp		1	16		3,025	670		3,695	4,350	
	3370	3000 amp		.70	22.857		3,525	960		4,485	5,350	
	3380	4000 amp		.50	32		4,475	1,350		5,825	7,000	
	3400	Cable tap box end, 225 amp		3.60	4.444		650	187		837	1,000	
	3410	400 amp		3	5.333		650	224		874	1,050	
	3420	600 amp		2.60	6.154		730	258		988	1,200	
	3430	800 amp		2.40	6.667		730	280		1,010	1,225	
	3440	1000 amp		1.80	8.889		805	375		1,180	1,450	
	3445	1200 amp		1.70	9.412		905	395		1,300	1,600	
	3450	1350 amp		1.60	10		1,000	420		1,420	1,750	
	3460	1600 amp		1.40	11.429		1,150	480		1,630	2,025	
	3470	2000 amp		1.20	13.333		1,325	560		1,885	2,325	
	3480	2500 amp		1	16		1,575	670		2,245	2,750	
	3490	3000 amp		.70	22.857		1,825	960		2,785	3,475	
	3500	4000 amp		.50	32		2,300	1,350		3,650	4,625	
	4600	Plug-in, fusible switch w/3 fuses, 3 pole, 250 volt, 30 amp	1 Elec	4	2		255	84		339	410	
	4610	60 amp		3.60	2.222		330	93.50		423.50	510	
	4620	100 amp		2.70	2.963		475	124		599	715	
	4630	200 amp	2 Elec	3.20	5		800	210		1,010	1,200	

16450 | Enclosed Bus Assemblies

16450	Enclosed Bus Assemblies	CREW	DAILY OUTPUT	LABOR-HOURS	UNIT	2006 BARE COSTS MAT.	LABOR	EQUIP.	TOTAL	TOTAL INCL O&P
300 4640	400 amp	2 Elec	1.40	11.429	Ea.	2,075	480		2,555	3,050 300
4650	600 amp		.90	17.778		2,875	745		3,620	4,325
4700	4 pole, 120/208 volt, 30 amp	1 Elec	3.90	2.051		350	86		436	520
4710	60 amp		3.50	2.286		375	96		471	565
4720	100 amp		2.60	3.077		525	129		654	775
4730	200 amp	2 Elec	3	5.333		880	224		1,104	1,300
4740	400 amp		1.30	12.308		2,050	515		2,565	3,075
4750	600 amp		.80	20		2,875	840		3,715	4,475
4800	3 pole, 480 volt, 30 amp	1 Elec	4	2		261	84		345	415
4810	60 amp		3.60	2.222		276	93.50		369.50	450
4820	100 amp		2.70	2.963		465	124		589	700
4830	200 amp	2 Elec	3.20	5		800	210		1,010	1,200
4840	400 amp		1.40	11.429		1,875	480		2,355	2,800
4850	600 amp		.90	17.778		2,650	745		3,395	4,075
4860	800 amp		.66	24.242		8,175	1,025		9,200	10,600
4870	1000 amp		.60	26.667		9,100	1,125		10,225	11,700
4880	1200 amp		.50	32		9,125	1,350		10,475	12,100
4890	1600 amp		.44	36.364		10,600	1,525		12,125	14,000
4900	4 pole, 277/480 volt, 30 amp	1 Elec	3.90	2.051		375	86		461	550
4910	60 amp		3.50	2.286		400	96		496	590
4920	100 amp		2.60	3.077		585	129		714	845
4930	200 amp	2 Elec	3	5.333		1,175	224		1,399	1,625
4940	400 amp		1.30	12.308		2,200	515		2,715	3,225
4950	600 amp		.80	20		3,000	840		3,840	4,600
5050	800 amp		.60	26.667		9,925	1,125		11,050	12,600
5060	1000 amp		.56	28.571		11,400	1,200		12,600	14,500
5070	1200 amp		.48	33.333		11,500	1,400		12,900	14,900
5080	1600 amp		.42	38.095		13,400	1,600		15,000	17,200
5150	Fusible with starter, 3 pole 250 volt, 30 amp	1 Elec	3.50	2.286		1,350	96		1,446	1,625
5160	60 amp		3.20	2.500		1,425	105		1,530	1,725
5170	100 amp		2.50	3.200		1,625	134		1,759	1,975
5180	200 amp	2 Elec	2.80	5.714		2,675	240		2,915	3,300
5200	3 pole 480 volt, 30 amp	1 Elec	3.50	2.286		1,350	96		1,446	1,625
5210	60 amp		3.20	2.500		1,425	105		1,530	1,725
5220	100 amp		2.50	3.200		1,625	134		1,759	1,975
5230	200 amp	2 Elec	2.80	5.714		2,675	240		2,915	3,300
5300	Fusible with contactor, 3 pole 250 volt, 30 amp	1 Elec	3.50	2.286		1,325	96		1,421	1,600
5310	60 amp		3.20	2.500		1,675	105		1,780	2,000
5320	100 amp		2.50	3.200		2,350	134		2,484	2,775
5330	200 amp	2 Elec	2.80	5.714		2,675	240		2,915	3,325
5400	3 pole 480 volt, 30 amp	1 Elec	3.50	2.286		1,425	96		1,521	1,700
5410	60 amp		3.20	2.500		2,000	105		2,105	2,350
5420	100 amp		2.50	3.200		2,750	134		2,884	3,225
5430	200 amp	2 Elec	2.80	5.714		2,800	240		3,040	3,475
5450	Fusible with capacitor, 3 pole 250 volt, 30 amp	1 Elec	3	2.667		3,425	112		3,537	3,925
5460	60 amp		2	4		3,975	168		4,143	4,625
5500	3 pole 480 volt, 30 amp		3	2.667		2,875	112		2,987	3,325
5510	60 amp		2	4		3,575	168		3,743	4,200
5600	Circuit breaker, 3 pole, 250 volt, 60 amp		4.50	1.778		360	74.50		434.50	510
5610	100 amp		3.20	2.500		440	105		545	645
5650	4 pole, 120/208 volt, 60 amp		4.40	1.818		405	76.50		481.50	565
5660	100 amp		3.10	2.581		480	108		588	690
5700	3 pole, 4 wire 277/480 volt, 60 amp		4.30	1.860		545	78		623	720
5710	100 amp		3	2.667		605	112		717	840
5720	225 amp	2 Elec	3.20	5		1,350	210		1,560	1,800
5730	400 amp		1.20	13.333		2,775	560		3,335	3,950

		16450 | Enclosed Bus Assemblies	CREW	DAILY OUTPUT	LABOR-HOURS	UNIT	2006 BARE COSTS MAT.	LABOR	EQUIP.	TOTAL	TOTAL INCL O&P	
300	5740	600 amp	2 Elec	.96	16.667	Ea.	3,750	700		4,450	5,200	300
	5750	700 amp		.60	26.667		4,750	1,125		5,875	6,950	
	5760	800 amp		.60	26.667		4,750	1,125		5,875	6,950	
	5770	900 amp		.54	29.630		6,250	1,250		7,500	8,800	
	5780	1000 amp		.54	29.630		6,250	1,250		7,500	8,800	
	5790	1200 amp		.42	38.095		7,500	1,600		9,100	10,800	
	5810	Circuit breaker w/HIC fuses, 3 pole 480 volt, 60 amp	1 Elec	4.40	1.818		685	76.50		761.50	875	
	5820	100 amp	"	3.10	2.581		750	108		858	990	
	5830	225 amp	2 Elec	3.40	4.706		2,400	198		2,598	2,925	
	5840	400 amp		1.40	11.429		3,800	480		4,280	4,925	
	5850	600 amp		1	16		3,850	670		4,520	5,275	
	5860	700 amp		.64	25		5,100	1,050		6,150	7,250	
	5870	800 amp		.64	25		5,100	1,050		6,150	7,250	
	5880	900 amp		.56	28.571		11,000	1,200		12,200	14,100	
	5890	1000 amp		.56	28.571		11,000	1,200		12,200	14,100	
	5950	3 pole 4 wire 277/480 volt, 60 amp	1 Elec	4.30	1.860		685	78		763	875	
	5960	100 amp	"	3	2.667		750	112		862	1,000	
	5970	225 amp	2 Elec	3	5.333		2,400	224		2,624	2,975	
	5980	400 amp		1.10	14.545		3,800	610		4,410	5,125	
	5990	600 amp		.94	17.021		3,850	715		4,565	5,350	
	6000	700 amp		.58	27.586		5,100	1,150		6,250	7,400	
	6010	800 amp		.58	27.586		5,100	1,150		6,250	7,400	
	6020	900 amp		.52	30.769		11,000	1,300		12,300	14,200	
	6030	1000 amp		.52	30.769		11,000	1,300		12,300	14,200	
	6040	1200 amp		.40	40		11,000	1,675		12,675	14,800	
	6100	Circuit breaker with starter, 3 pole 250 volt, 60 amp	1 Elec	3.20	2.500		1,025	105		1,130	1,275	
	6110	100 amp	"	2.50	3.200		1,425	134		1,559	1,750	
	6120	225 amp	2 Elec	3	5.333		1,825	224		2,049	2,350	
	6130	3 pole 480 volt, 60 amp	1 Elec	3.20	2.500		1,025	105		1,130	1,275	
	6140	100 amp	"	2.50	3.200		1,425	134		1,559	1,750	
	6150	225 amp	2 Elec	3	5.333		1,825	224		2,049	2,350	
	6200	Circuit breaker with contactor, 3 pole 250 volt, 60 amp	1 Elec	3.20	2.500		965	105		1,070	1,200	
	6210	100 amp	"	2.50	3.200		1,300	134		1,434	1,650	
	6220	225 amp	2 Elec	3	5.333		1,700	224		1,924	2,200	
	6250	3 pole 480 volt, 60 amp	1 Elec	3.20	2.500		965	105		1,070	1,200	
	6260	100 amp	"	2.50	3.200		1,300	134		1,434	1,650	
	6270	225 amp	2 Elec	3	5.333		1,700	224		1,924	2,200	
	6300	Circuit breaker with capacitor, 3 pole 250 volt, 60 amp	1 Elec	2	4		4,125	168		4,293	4,800	
	6310	3 pole 480 volt, 60 amp		2	4		3,825	168		3,993	4,475	
	6400	Add control transformer with pilot light to above starter		16	.500		294	21		315	360	
	6410	Switch, fusible, mechanically held contactor optional		16	.500		775	21		796	890	
	6430	Circuit breaker, mechanically held contactor optional		16	.500		775	21		796	890	
	6450	Ground neutralizer, 3 pole		16	.500		36	21		57	72	
320	0010	**COPPER BUS DUCT** 10 ft. long										320
	0050	3 pole 4 wire, plug-in/indoor, straight section, 225 amp	2 Elec	40	.400	L.F.	147	16.80		163.80	188	
	1000	400 amp		32	.500		147	21		168	195	
	1500	600 amp		26	.615		147	26		173	202	
	2400	800 amp		20	.800		175	33.50		208.50	244	
	2450	1000 amp		18	.889		193	37.50		230.50	271	
	2470	1200 amp		17	.941		248	39.50		287.50	335	
	2500	1350 amp		16	1		276	42		318	370	
	2510	1600 amp		12	1.333		315	56		371	430	
	2520	2000 amp		10	1.600		395	67		462	540	
	2530	2500 amp		8	2		490	84		574	665	
	2540	3000 amp		6	2.667		590	112		702	825	

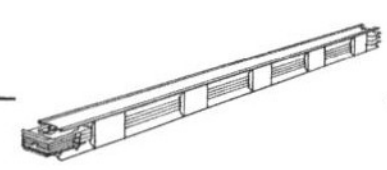

ELECTRICAL 16

16450 | Enclosed Bus Assemblies

	16450 Enclosed Bus Assemblies	CREW	DAILY OUTPUT	LABOR-HOURS	UNIT	2006 BARE COSTS MAT.	LABOR	EQUIP.	TOTAL	TOTAL INCL O&P
320	2550 Feeder, 600 amp	2 Elec	28	.571	L.F.	129	24		153	179
	2600 800 amp		22	.727		156	30.50		186.50	219
	2700 1000 amp		20	.800		175	33.50		208.50	244
	2750 1200 amp		19	.842		230	35.50		265.50	310
	2800 1350 amp		18	.889		258	37.50		295.50	340
	2900 1600 amp		14	1.143		294	48		342	400
	3000 2000 amp		12	1.333		375	56		431	500
	3010 2500 amp		8	2		470	84		554	645
	3020 3000 amp		6	2.667		570	112		682	800
	3030 4000 amp		4	4		755	168		923	1,100
	3040 5000 amp		2	8		920	335		1,255	1,525
	3100 Elbows, 225 amp		4	4	Ea.	870	168		1,038	1,225
	3200 400 amp		3.60	4.444		870	187		1,057	1,250
	3300 600 amp		3.20	5		870	210		1,080	1,275
	3400 800 amp		2.80	5.714		945	240		1,185	1,425
	3500 1000 amp		2.60	6.154		1,050	258		1,308	1,550
	3550 1200 amp		2.50	6.400		1,175	269		1,444	1,700
	3600 1350 amp		2.40	6.667		1,250	280		1,530	1,800
	3700 1600 amp		2.20	7.273		1,350	305		1,655	1,950
	3800 2000 amp		1.80	8.889		1,675	375		2,050	2,400
	3810 2500 amp		1.60	10		2,625	420		3,045	3,550
	3820 3000 amp		1.40	11.429		3,050	480		3,530	4,100
	3830 4000 amp		1.20	13.333		3,950	560		4,510	5,200
	3840 5000 amp		1	16		6,375	670		7,045	8,025
	4000 End box, 225 amp		34	.471		118	19.75		137.75	161
	4100 400 amp		32	.500		118	21		139	163
	4200 600 amp		28	.571		118	24		142	167
	4300 800 amp		26	.615		118	26		144	170
	4400 1000 amp		24	.667		118	28		146	173
	4410 1200 amp		23	.696		118	29		147	175
	4500 1350 amp		22	.727		118	30.50		148.50	177
	4600 1600 amp		20	.800		118	33.50		151.50	182
	4700 2000 amp		18	.889		144	37.50		181.50	217
	4710 2500 amp		16	1		144	42		186	224
	4720 3000 amp		14	1.143		144	48		192	233
	4730 4000 amp		12	1.333		175	56		231	279
	4740 5000 amp		10	1.600		175	67		242	296
	4800 Cable tap box end, 225 amp		3.20	5		875	210		1,085	1,300
	5000 400 amp		2.60	6.154		875	258		1,133	1,375
	5100 600 amp	2 Elec	2.20	7.273	Ea.	880	305		1,185	1,425
	5200 800 amp		2	8		985	335		1,320	1,600
	5300 1000 amp		1.60	10		1,075	420		1,495	1,850
	5350 1200 amp		1.50	10.667		1,225	450		1,675	2,050
	5400 1350 amp		1.40	11.429		1,300	480		1,780	2,175
	5500 1600 amp		1.20	13.333		1,450	560		2,010	2,475
	5600 2000 amp		1	16		1,675	670		2,345	2,850
	5610 2500 amp		.80	20		2,000	840		2,840	3,500
	5620 3000 amp		.60	26.667		2,250	1,125		3,375	4,200
	5630 4000 amp		.40	40		2,850	1,675		4,525	5,750
	5640 5000 amp		.20	80		3,325	3,350		6,675	8,850
	5700 Switchboard stub, 225 amp		5.40	2.963		885	124		1,009	1,150
	5800 400 amp		4.60	3.478		885	146		1,031	1,200
	5900 600 amp		4	4		885	168		1,053	1,225
	6000 800 amp		3.20	5		1,075	210		1,285	1,500
	6100 1000 amp		3	5.333		1,250	224		1,474	1,700
	6150 1200 amp		2.80	5.714		1,450	240		1,690	1,950

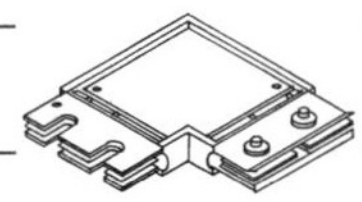

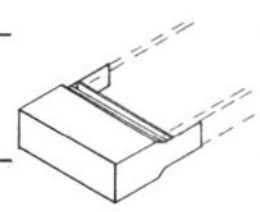

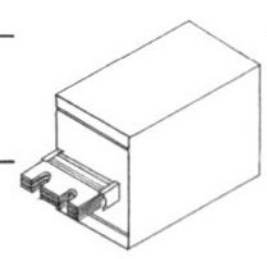

	16450 | Enclosed Bus Assemblies	CREW	DAILY OUTPUT	LABOR-HOURS	UNIT	2006 BARE COSTS MAT.	LABOR	EQUIP.	TOTAL	TOTAL INCL O&P	
320 6200	1350 amp	2 Elec	2.60	6.154	Ea.	1,600	258		1,858	2,150	320
6300	1600 amp		2.40	6.667		1,800	280		2,080	2,400	
6400	2000 amp		2	8		2,175	335		2,510	2,925	
6410	2500 amp		1.80	8.889		2,650	375		3,025	3,475	
6420	3000 amp		1.60	10		3,100	420		3,520	4,075	
6430	4000 amp		1.40	11.429		4,050	480		4,530	5,200	
6440	5000 amp		1.20	13.333		4,950	560		5,510	6,325	
6490	Tee fittings, 225 amp		2.40	6.667		1,200	280		1,480	1,750	
6500	400 amp		2	8		1,200	335		1,535	1,850	
6600	600 amp		1.80	8.889		1,200	375		1,575	1,900	
6700	800 amp		1.60	10		1,375	420		1,795	2,175	
6750	1000 amp 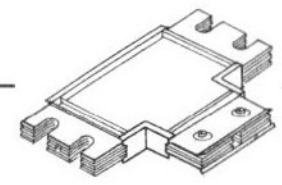		1.40	11.429		1,600	480		2,080	2,500	
6770	1200 amp		1.30	12.308		1,800	515		2,315	2,775	
6800	1350 amp		1.20	13.333		2,000	560		2,560	3,075	
7000	1600 amp		1	16		2,275	670		2,945	3,525	
7100	2000 amp		.80	20		2,700	840		3,540	4,250	
7110	2500 amp		.60	26.667		3,300	1,125		4,425	5,350	
7120	3000 amp		.50	32		3,850	1,350		5,200	6,300	
7130	4000 amp		.40	40		4,975	1,675		6,650	8,075	
7140	5000 amp		.20	80		5,875	3,350		9,225	11,700	
7200	Plug-in fusible switches w/3 fuses, 600 volt, 3 pole, 30 amp	1 Elec	4	2		525	84		609	705	
7300	60 amp		3.60	2.222		590	93.50		683.50	795	
7400	100 amp		2.70	2.963		900	124		1,024	1,175	
7500	200 amp	2 Elec	3.20	5		1,600	210		1,810	2,100	
7600	400 amp		1.40	11.429		4,700	480		5,180	5,925	
7700	600 amp		.90	17.778		5,325	745		6,070	7,025	
7800	800 amp		.66	24.242		7,525	1,025		8,550	9,850	
7900	1200 amp		.50	32		14,100	1,350		15,450	17,600	
7910	1600 amp		.44	36.364		14,100	1,525		15,625	17,900	
8000	Plug-in circuit breakers, molded case, 15 to 50 amp	1 Elec	4.40	1.818		495	76.50		571.50	665	
8100	70 to 100 amp	"	3.10	2.581		550	108		658	770	
8200	150 to 225 amp	2 Elec	3.40	4.706		1,425	198		1,623	1,875	
8300	250 to 400 amp		1.40	11.429		2,625	480		3,105	3,625	
8400	500 to 600 amp		1	16		3,550	670		4,220	4,925	
8500	700 to 800 amp		.64	25		4,375	1,050		5,425	6,425	
8600	900 to 1000 amp		.56	28.571		6,250	1,200		7,450	8,725	
8700	1200 amp		.44	36.364		7,500	1,525		9,025	10,600	
8720	1400 amp		.40	40		11,100	1,675		12,775	14,800	
8730	1600 amp		.40	40		12,200	1,675		13,875	16,000	
8750	Circuit breakers, with current limiting fuse, 15 to 50 amp	1 Elec	4.40	1.818		995	76.50		1,071.50	1,225	
8760	70 to 100 amp	"	3.10	2.581		1,175	108		1,283	1,475	
8770	150 to 225 amp	2 Elec	3.40	4.706		2,525	198		2,723	3,075	
8780	250 to 400 amp		1.40	11.429		3,900	480		4,380	5,050	
8790	500 to 600 amp		1	16		4,500	670		5,170	5,975	
8800	700 to 800 amp		.64	25		7,400	1,050		8,450	9,775	
8810	900 to 1000 amp		.56	28.571		8,450	1,200		9,650	11,200	
8850	Combination starter FVNR, fusible switch, NEMA size 0, 30 amp	1 Elec	2	4		1,350	168		1,518	1,725	
8860	NEMA size 1, 60 amp		1.80	4.444		1,425	187		1,612	1,875	
8870	NEMA size 2, 100 amp		1.30	6.154		1,800	258		2,058	2,375	
8880	NEMA size 3, 200 amp	2 Elec	2	8		2,850	335		3,185	3,650	
8900	Circuit breaker, NEMA size 0, 30 amp	1 Elec	2	4		1,375	168		1,543	1,775	
8910	NEMA size 1, 60 amp		1.80	4.444		1,450	187		1,637	1,875	
8920	NEMA size 2, 100 amp		1.30	6.154		2,075	258		2,333	2,675	
8930	NEMA size 3, 200 amp	2 Elec	2	8		2,625	335		2,960	3,425	
8950	Combination contactor, fusible switch, NEMA size 0, 30 amp	1 Elec	2	4		910	168		1,078	1,250	
8960	NEMA size 1, 60 amp		1.80	4.444		935	187		1,122	1,325	

16450 | Enclosed Bus Assemblies

		16450 Enclosed Bus Assemblies	CREW	DAILY OUTPUT	LABOR-HOURS	UNIT	2006 BARE COSTS MAT.	LABOR	EQUIP.	TOTAL	TOTAL INCL O&P	
320	8970	NEMA size 2, 100 amp	1 Elec	1.30	6.154	Ea.	1,400	258		1,658	1,925	320
	8980	NEMA size 3, 200 amp	2 Elec	2	8		1,600	335		1,935	2,300	
	9000	Circuit breaker, NEMA size 0, 30 amp	1 Elec	2	4		1,050	168		1,218	1,400	
	9010	NEMA size 1, 60 amp		1.80	4.444		1,075	187		1,262	1,475	
	9020	NEMA size 2, 100 amp		1.30	6.154		1,650	258		1,908	2,225	
	9030	NEMA size 3, 200 amp	2 Elec	2	8		2,000	335		2,335	2,725	
	9050	Control transformer for above, NEMA size 0, 30 amp	1 Elec	8	1		152	42		194	232	
	9060	NEMA size 1, 60 amp		8	1		152	42		194	232	
	9070	NEMA size 2, 100 amp		7	1.143		212	48		260	305	
	9080	NEMA size 3, 200 amp	2 Elec	14	1.143		294	48		342	400	
	9100	Comb. fusible switch & lighting control, electrically held, 30 amp	1 Elec	2	4		620	168		788	945	
	9110	60 amp		1.80	4.444		890	187		1,077	1,275	
	9120	100 amp		1.30	6.154		1,150	258		1,408	1,675	
	9130	200 amp	2 Elec	2	8		2,825	335		3,160	3,625	
	9150	Mechanically held, 30 amp	1 Elec	2	4		775	168		943	1,125	
	9160	60 amp		1.80	4.444		1,150	187		1,337	1,575	
	9170	100 amp		1.30	6.154		1,500	258		1,758	2,050	
	9180	200 amp	2 Elec	2	8		3,025	335		3,360	3,875	
	9200	Ground bus added to bus duct, 225 amp		320	.050	L.F.	27.50	2.10		29.60	33.50	
	9210	400 amp		240	.067		27.50	2.80		30.30	35	
	9220	600 amp		240	.067		27.50	2.80		30.30	35	
	9230	800 amp		160	.100		32	4.20		36.20	42	
	9240	1000 amp		160	.100		37	4.20		41.20	47	
	9250	1350 amp		140	.114		55	4.80		59.80	68	
	9260	1600 amp		120	.133		60	5.60		65.60	74.50	
	9270	2000 amp		110	.145		78	6.10		84.10	95.50	
	9280	2500 amp		100	.160		96.50	6.70		103.20	116	
	9290	3000 amp		90	.178		115	7.45		122.45	139	
	9300	4000 amp		80	.200		152	8.40		160.40	180	
	9310	5000 amp		70	.229		184	9.60		193.60	217	
	9320	High short circuit bracing, add					13.75			13.75	15.15	
360	0010	**COPPER OR ALUMINUM BUS DUCT FITTINGS**										360
	0100	Flange, wall, with vapor barrier, 225 amp	2 Elec	6.20	2.581	Ea.	555	108		663	775	
	0110	400 amp		6	2.667		555	112		667	785	
	0120	600 amp		5.80	2.759		555	116		671	790	
	0130	800 amp		5.40	2.963		555	124		679	800	
	0140	1000 amp		5	3.200		555	134		689	815	
	0145	1200 amp		4.80	3.333		555	140		695	825	
	0150	1350 amp		4.60	3.478		555	146		701	835	
	0160	1600 amp		4.20	3.810		555	160		715	855	
	0170	2000 amp		4	4		555	168		723	870	
	0180	2500 amp		3.60	4.444		555	187		742	900	
	0190	3000 amp		3.20	5		555	210		765	935	
	0200	4000 amp		2.60	6.154		645	258		903	1,100	
	0300	Roof, 225 amp		6.20	2.581		585	108		693	810	
	0310	400 amp		6	2.667		585	112		697	820	
	0320	600 amp		5.80	2.759		585	116		701	825	
	0330	800 amp		5.40	2.963		585	124		709	835	
	0340	1000 amp		5	3.200		585	134		719	850	
	0345	1200 amp		4.80	3.333		585	140		725	860	
	0350	1350 amp		4.60	3.478		585	146		731	870	
	0360	1600 amp		4.20	3.810		585	160		745	890	
	0370	2000 amp		4	4		585	168		753	905	
	0380	2500 amp		3.60	4.444		585	187		772	935	
	0390	3000 amp		3.20	5		585	210		795	970	

16450 | Enclosed Bus Assemblies			CREW	DAILY OUTPUT	LABOR-HOURS	UNIT	2006 BARE COSTS MAT.	LABOR	EQUIP.	TOTAL	TOTAL INCL O&P	
360	0400	4000 amp	2 Elec	2.60	6.154	Ea.	585	258		843	1,050	360
	0420	Support, floor mounted, 225 amp		20	.800		119	33.50		152.50	183	
	0430	400 amp		20	.800		119	33.50		152.50	183	
	0440	600 amp		18	.889		119	37.50		156.50	189	
	0450	800 amp		16	1		119	42		161	196	
	0460	1000 amp		13	1.231		119	51.50		170.50	211	
	0465	1200 amp		11.80	1.356		119	57		176	219	
	0470	1350 amp		10.60	1.509		119	63.50		182.50	229	
	0480	1600 amp		9.20	1.739		119	73		192	244	
	0490	2000 amp		8	2		119	84		203	260	
	0500	2500 amp		6.40	2.500		119	105		224	293	
	0510	3000 amp		5.40	2.963		119	124		243	325	
	0520	4000 amp		4	4		119	168		287	390	
	0540	Weather stop, 225 amp		12	1.333		360	56		416	480	
	0550	400 amp		10	1.600		360	67		427	500	
	0560	600 amp		9	1.778		360	74.50		434.50	510	
	0570	800 amp		8	2		360	84		444	525	
	0580	1000 amp		6.40	2.500		360	105		465	555	
	0585	1200 amp		5.90	2.712		360	114		474	570	
	0590	1350 amp		5.40	2.963		360	124		484	585	
	0600	1600 amp		4.60	3.478		360	146		506	620	
	0610	2000 amp		4	4		360	168		528	655	
	0620	2500 amp		3.20	5		360	210		570	720	
	0630	3000 amp		2.60	6.154		360	258		618	795	
	0640	4000 amp		2	8		360	335		695	915	
	0660	End closure, 225 amp		34	.471		118	19.75		137.75	161	
	0670	400 amp		32	.500		118	21		139	163	
	0680	600 amp		28	.571		118	24		142	167	
	0690	800 amp		26	.615		118	26		144	170	
	0700	1000 amp		24	.667		118	28		146	173	
	0705	1200 amp		23	.696		118	29		147	175	
	0710	1350 amp		22	.727		118	30.50		148.50	177	
	0720	1600 amp		20	.800		118	33.50		151.50	182	
	0730	2000 amp		18	.889		144	37.50		181.50	217	
	0740	2500 amp		16	1		144	42		186	224	
	0750	3000 amp		14	1.143		144	48		192	233	
	0760	4000 amp		12	1.333		175	56		231	279	
	0780	Switchboard stub, 3 pole 3 wire, 225 amp		6	2.667		710	112		822	955	
	0790	400 amp		5.20	3.077		710	129		839	980	
	0800	600 amp		4.60	3.478		710	146		856	1,000	
	0810	800 amp		3.60	4.444		855	187		1,042	1,225	
	0820	1000 amp		3.40	4.706		1,000	198		1,198	1,400	
	0825	1200 amp		3.20	5		1,000	210		1,210	1,425	
	0830	1350 amp		3	5.333		1,200	224		1,424	1,675	
	0840	1600 amp		2.80	5.714		1,425	240		1,665	1,925	
	0850	2000 amp		2.40	6.667		1,650	280		1,930	2,250	
	0860	2500 amp		2	8		2,000	335		2,335	2,725	
	0870	3000 amp		1.80	8.889		2,300	375		2,675	3,100	
	0880	4000 amp		1.60	10		2,975	420		3,395	3,925	
	0890	5000 amp		1.40	11.429		3,650	480		4,130	4,775	
	0900	3 pole 4 wire, 225 amp		5.40	2.963		885	124		1,009	1,150	
	0910	400 amp		4.60	3.478		885	146		1,031	1,200	
	0920	600 amp		4	4		885	168		1,053	1,225	
	0930	800 amp		3.20	5		1,075	210		1,285	1,500	
	0940	1000 amp		3	5.333		1,250	224		1,474	1,700	
360	0950	1350 amp	↓	2.60	6.154	↓	1,600	258		1,858	2,150	

16450 | Enclosed Bus Assemblies

		Description	Crew	Daily Output	Labor-Hours	Unit	2006 Bare Costs Mat.	2006 Bare Costs Labor	2006 Bare Costs Equip.	2006 Bare Costs Total	Total Incl O&P
360	0960	1600 amp	2 Elec	2.40	6.667	Ea.	1,800	280		2,080	2,400
	0970	2000 amp		2	8		2,175	335		2,510	2,925
	0980	2500 amp		1.80	8.889		2,650	375		3,025	3,500
	0990	3000 amp		1.60	10		3,100	420		3,520	4,075
	1000	4000 amp		1.40	11.429		4,025	480		4,505	5,175
	1050	Service head, weatherproof, 3 pole 3 wire, 225 amp		3	5.333		1,200	224		1,424	1,675
	1060	400 amp		2.80	5.714		1,200	240		1,440	1,700
	1070	600 amp		2.60	6.154		1,200	258		1,458	1,725
	1080	800 amp		2.40	6.667		1,350	280		1,630	1,900
	1090	1000 amp		2	8		1,450	335		1,785	2,125
	1100	1350 amp		1.80	8.889		1,900	375		2,275	2,650
	1110	1600 amp		1.60	10		2,125	420		2,545	2,975
	1120	2000 amp		1.40	11.429		2,575	480		3,055	3,600
	1130	2500 amp		1.20	13.333		3,075	560		3,635	4,275
	1140	3000 amp		.90	17.778		3,600	745		4,345	5,100
	1150	4000 amp		.70	22.857		4,575	960		5,535	6,500
	1200	3 pole 4 wire, 225 amp		2.60	6.154		1,350	258		1,608	1,875
	1210	400 amp		2.40	6.667		1,350	280		1,630	1,900
	1220	600 amp		2.20	7.273		1,350	305		1,655	1,950
	1230	800 amp		2	8		1,575	335		1,910	2,250
	1240	1000 amp		1.70	9.412		1,800	395		2,195	2,575
	1250	1350 amp		1.50	10.667		2,150	450		2,600	3,050
	1260	1600 amp		1.40	11.429		2,325	480		2,805	3,325
	1270	2000 amp		1.20	13.333		3,125	560		3,685	4,300
	1280	2500 amp		1	16		3,850	670		4,520	5,250
	1290	3000 amp		.80	20		4,500	840		5,340	6,250
	1300	4000 amp		.60	26.667		5,825	1,125		6,950	8,150
	1350	Flanged end, 3 pole 3 wire, 225 amp		6	2.667		655	112		767	895
	1360	400 amp		5.20	3.077		655	129		784	920
	1370	600 amp		4.60	3.478		655	146		801	945
	1380	800 amp		3.60	4.444		725	187		912	1,100
	1390	1000 amp		3.40	4.706		815	198		1,013	1,200
	1395	1200 amp		3.20	5		910	210		1,120	1,325
	1400	1350 amp		3	5.333		975	224		1,199	1,425
	1410	1600 amp		2.80	5.714		1,100	240		1,340	1,600
	1420	2000 amp		2.40	6.667		1,300	280		1,580	1,850
	1430	2500 amp		2	8		1,525	335		1,860	2,200
	1440	3000 amp		1.80	8.889		1,750	375		2,125	2,500
	1450	4000 amp		1.60	10		2,175	420		2,595	3,025
	1500	3 pole 4 wire, 225 amp		5.40	2.963		750	124		874	1,025
	1510	400 amp		4.60	3.478		750	146		896	1,050
	1520	600 amp		4	4		750	168		918	1,075
	1530	800 amp		3.20	5		885	210		1,095	1,300
	1540	1000 amp		3	5.333		990	224		1,214	1,450
	1545	1200 amp		2.80	5.714		1,125	240		1,365	1,600
	1550	1350 amp		2.60	6.154		1,200	258		1,458	1,725
	1560	1600 amp		2.40	6.667		1,375	280		1,655	1,950
	1570	2000 amp		2	8		1,625	335		1,960	2,325
	1580	2500 amp		1.80	8.889		1,950	375		2,325	2,725
	1590	3000 amp		1.60	10		2,225	420		2,645	3,100
	1600	4000 amp		1.40	11.429		2,875	480		3,355	3,900
	1650	Hanger, standard, 225 amp		64	.250		13.80	10.50		24.30	31.50
	1660	400 amp		48	.333		13.80	14		27.80	36.50
	1670	600 amp		40	.400		13.80	16.80		30.60	41
	1680	800 amp		32	.500		13.80	21		34.80	47.50
	1690	1000 amp		24	.667		13.80	28		41.80	58

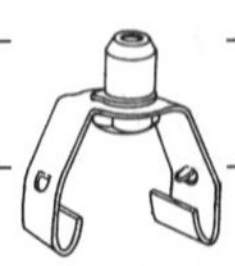

16 ELECTRICAL

16450 | Enclosed Bus Assemblies

				DAILY	LABOR-		2006 BARE COSTS				TOTAL	
			CREW	OUTPUT	HOURS	UNIT	MAT.	LABOR	EQUIP.	TOTAL	INCL O&P	
360	1695	1200 amp	2 Elec	22	.727	Ea.	13.80	30.50		44.30	62	360
	1700	1350 amp		20	.800		13.80	33.50		47.30	67	
	1710	1600 amp		20	.800		13.80	33.50		47.30	67	
	1720	2000 amp		18	.889		13.80	37.50		51.30	72.50	
	1730	2500 amp		16	1		13.80	42		55.80	79.50	
	1740	3000 amp		16	1		13.80	42		55.80	79.50	
	1750	4000 amp 		16	1		13.80	42		55.80	79.50	
	1800	Spring type, 225 amp		16	1		61.50	42		103.50	133	
	1810	400 amp		14	1.143		61.50	48		109.50	142	
	1820	600 amp		14	1.143		61.50	48		109.50	142	
	1830	800 amp		14	1.143		61.50	48		109.50	142	
	1840	1000 amp		14	1.143		61.50	48		109.50	142	
	1845	1200 amp 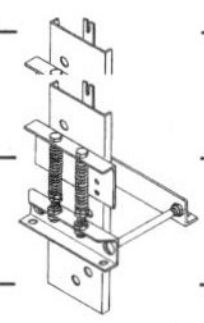		14	1.143		61.50	48		109.50	142	
	1850	1350 amp		14	1.143		61.50	48		109.50	142	
	1860	1600 amp		12	1.333		61.50	56		117.50	155	
	1870	2000 amp		12	1.333		61.50	56		117.50	155	
	1880	2500 amp		12	1.333		61.50	56		117.50	155	
	1890	3000 amp		10	1.600		61.50	67		128.50	172	
	1900	4000 amp		10	1.600		61.50	67		128.50	172	
500	0010	**FEEDRAIL**, 12 foot mounting										500
	0050	Trolley busway, 3 pole										
	0100	300 volt 60 amp, plain, 10 ft. lengths	1 Elec	50	.160	L.F.	21	6.70		27.70	33.50	
	0300	Door track		50	.160		23	6.70		29.70	36	
	0500	Curved track		30	.267		267	11.20		278.20	310	
	0700	Coupling				Ea.	11.95			11.95	13.15	
	0900	Center feed	1 Elec	5.30	1.509		32	63.50		95.50	133	
	1100	End feed		5.30	1.509		38.50	63.50		102	140	
	1300	Hanger set		24	.333		4.60	14		18.60	26.50	
	3000	600 volt 100 amp, plain, 10 ft. lengths		35	.229	L.F.	45	9.60		54.60	64.50	
	3300	Door track		35	.229	"	52.50	9.60		62.10	72.50	
	3700	Coupling				Ea.	37.50			37.50	41.50	
	4000	End cap	1 Elec	40	.200		32	8.40		40.40	48.50	
	4200	End feed		4	2		197	84		281	345	
	4500	Trolley, 600 volt, 20 amp		5.30	1.509		254	63.50		317.50	375	
	4700	30 amp		5.30	1.509		254	63.50		317.50	375	
	4900	Duplex, 40 amp		4	2		535	84		619	720	
	5000	60 amp		4	2		535	84		619	720	
	5300	Fusible, 20 amp		4	2		650	84		734	840	
	5500	30 amp		4	2		650	84		734	840	
	5900	300 volt, 20 amp		5.30	1.509		198	63.50		261.50	315	
	6000	30 amp		5.30	1.509		220	63.50		283.50	340	
	6300	Fusible, 20 amp		4.70	1.702		305	71.50		376.50	445	
	6500	30 amp		4.70	1.702		435	71.50		506.50	590	
	7300	Busway, 250 volt 50 amp, 2 wire		70	.114	L.F.	13.25	4.80		18.05	22	
	7330	Coupling				Ea.	26			26	28.50	
	7340	Center feed	1 Elec	6	1.333		295	56		351	410	
	7350	End feed		6	1.333		53.50	56		109.50	145	
	7360	End cap		40	.200		19.30	8.40		27.70	34.50	
	7370	Hanger set		24	.333		4.60	14		18.60	26.50	
	7400	125/250 volt 50 amp, 3 wire		60	.133	L.F.	13.90	5.60		19.50	24	
	7430	Coupling		6	1.333	Ea.	31.50	56		87.50	121	
	7440	Center feed		6	1.333		315	56		371	430	
	7450	End feed		6	1.333		57	56		113	149	
	7460	End cap		40	.200		21	8.40		29.40	36.50	
	7470	Hanger set		24	.333		4.60	14		18.60	26.50	

16450 | Enclosed Bus Assemblies

			CREW	DAILY OUTPUT	LABOR-HOURS	UNIT	2006 BARE COSTS MAT.	LABOR	EQUIP.	TOTAL	TOTAL INCL O&P	
500	7480	Trolley, 250 volt, 2 pole, 20 amp	1 Elec	6	1.333	Ea.	32	56		88	122	500
	7490	30 amp		6	1.333		32	56		88	122	
	7500	125/250 volt, 3 pole, 20 amp		6	1.333		34	56		90	124	
	7510	30 amp	↓	6	1.333		34	56		90	124	
	8000	Cleaning tools, 300 volt, dust remover					60			60	66	
	8100	Bus bar cleaner					105			105	115	
	8300	600 volt, dust remover, 60 amp					222			222	244	
	8400	100 amp					305			305	335	
	8600	Bus bar cleaner, 60 amp					270			270	297	
	8700	100 amp				↓	375			375	410	

16490 | Low V Dist Components/Accessories

			CREW	DAILY OUTPUT	LABOR-HOURS	UNIT	MAT.	LABOR	EQUIP.	TOTAL	TOTAL INCL O&P	
300	0010	**FUSES**										300
	0020	Cartridge, nonrenewable										
	0050	250 volt, 30 amp	1 Elec	50	.160	Ea.	1.41	6.70		8.11	11.90	
	0100	60 amp		50	.160		2.37	6.70		9.07	12.95	
	0150	100 amp		40	.200		9.95	8.40		18.35	24	
	0200	200 amp		36	.222		24	9.35		33.35	41	
	0250	400 amp		30	.267		56	11.20		67.20	79.50	
	0300	600 amp		24	.333		93	14		107	124	
	0400	600 volt, 30 amp		40	.200		7.90	8.40		16.30	21.50	
	0450	60 amp		40	.200		11.45	8.40		19.85	25.50	
	0500	100 amp		36	.222		24	9.35		33.35	41	
	0550	200 amp		30	.267		58.50	11.20		69.70	82	
	0600	400 amp		24	.333		120	14		134	154	
	0650	600 amp		20	.400		181	16.80		197.80	225	
	0800	Dual element, time delay, 250 volt, 30 amp		50	.160		3.60	6.70		10.30	14.30	
	0850	60 amp		50	.160		6.55	6.70		13.25	17.55	
	0900	100 amp		40	.200		14.70	8.40		23.10	29	
	0950	200 amp		36	.222		32.50	9.35		41.85	50.50	
	1000	400 amp		30	.267		59	11.20		70.20	82.50	
	1050	600 amp		24	.333		90	14		104	121	
	1300	600 volt, 15 to 30 amp		40	.200		7.85	8.40		16.25	21.50	
	1350	35 to 60 amp		40	.200		13.50	8.40		21.90	28	
	1400	70 to 100 amp		36	.222		28	9.35		37.35	45.50	
	1450	110 to 200 amp		30	.267		56	11.20		67.20	79	
	1500	225 to 400 amp		24	.333		112	14		126	145	
	1550	600 amp		20	.400		161	16.80		177.80	203	
	1800	Class RK1, high capacity, 250 volt, 30 amp		50	.160		4.90	6.70		11.60	15.75	
	1850	60 amp		50	.160		8.90	6.70		15.60	20	
	1900	100 amp		40	.200		20	8.40		28.40	35	
	1950	200 amp		36	.222		44	9.35		53.35	63	
	2000	400 amp		30	.267		79.50	11.20		90.70	104	
	2050	600 amp		24	.333		122	14		136	156	
	2200	600 volt, 30 amp		40	.200		11.10	8.40		19.50	25	
	2250	60 amp		40	.200		19.10	8.40		27.50	34	
	2300	100 amp		36	.222		39.50	9.35		48.85	58	
	2350	200 amp		30	.267		79	11.20		90.20	104	
	2400	400 amp		24	.333		158	14		172	196	
	2450	600 amp		20	.400		227	16.80		243.80	276	
	2700	Class J, CLF, 250 or 600 volt, 30 amp		40	.200		13.95	8.40		22.35	28.50	
	2750	60 amp		40	.200		23	8.40		31.40	38	
	2800	100 amp		36	.222		34.50	9.35		43.85	52.50	
	2850	200 amp		30	.267		65	11.20		76.20	89	
	2900	400 amp		24	.333		143	14		157	179	
	2950	600 amp	↓	20	.400	↓	202	16.80		218.80	248	

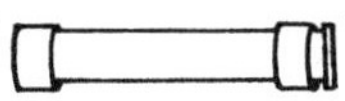

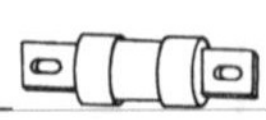

16 ELECTRICAL

16400 | Low-Voltage Distribution

16490 | Low V Dist Components/Accessories

			CREW	DAILY OUTPUT	LABOR-HOURS	UNIT	2006 BARE COSTS MAT.	LABOR	EQUIP.	TOTAL	TOTAL INCL O&P	
300	3100	Class L, 250 or 600 volt, 601 to 1200 amp	1 Elec	16	.500	Ea.	370	21		391	445	300
	3150	1500-1600 amp		13	.615		475	26		501	565	
	3200	1800-2000 amp		10	.800		640	33.50		673.50	750	
	3250	2500 amp		10	.800		845	33.50		878.50	980	
	3300	3000 amp		8	1		975	42		1,017	1,150	
	3350	3500-4000 amp		8	1		1,350	42		1,392	1,550	
	3400	4500-5000 amp		6.70	1.194		2,100	50		2,150	2,375	
	3450	6000 amp		5.70	1.404		3,475	59		3,534	3,925	
	3600	Plug, 120 volt, 1 to 10 amp		50	.160		1.14	6.70		7.84	11.60	
	3650	15 to 30 amp		50	.160		1.52	6.70		8.22	12	
	3700	Dual element 0.3 to 14 amp		50	.160		6.35	6.70		13.05	17.35	
	3750	15 to 30 amp		50	.160		1.90	6.70		8.60	12.45	
	3800	Fustat, 120 volt, 15 to 30 amp		50	.160		2.70	6.70		9.40	13.30	
	3850	0.3 to 14 amp		50	.160		6.55	6.70		13.25	17.55	
	3900	Adapters 0.3 to 10 amp		50	.160		3.45	6.70		10.15	14.15	
	3950	15 to 30 amp	↓	50	.160	↓	2.95	6.70		9.65	13.60	

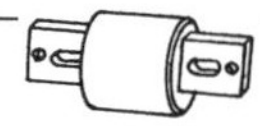

16500 | Lighting

16510 | Interior Luminaires

			CREW	DAILY OUTPUT	LABOR-HOURS	UNIT	2006 BARE COSTS MAT.	LABOR	EQUIP.	TOTAL	TOTAL INCL O&P	
300	0010	FIXTURE HANGERS										300
	0220	Box hub cover	1 Elec	32	.250	Ea.	3.75	10.50		14.25	20.50	
	0240	Canopy		12	.667		7.75	28		35.75	51.50	
	0260	Connecting block		40	.200		3.60	8.40		12	16.90	
	0280	Cushion hanger		16	.500		19.50	21		40.50	54	
	0300	Box hanger, with mounting strap		8	1		6	42		48	71	
	0320	Connecting block		40	.200		1.40	8.40		9.80	14.50	
	0340	Flexible, 1/2" diameter, 4" long		12	.667		9.20	28		37.20	53	
	0360	6" long		12	.667		9.90	28		37.90	54	
	0380	8" long		12	.667		11.05	28		39.05	55	
	0400	10" long		12	.667		11.70	28		39.70	56	
	0420	12" long		12	.667		12.70	28		40.70	57	
	0440	15" long		12	.667		13.40	28		41.40	58	
	0460	18" long		12	.667		15.30	28		43.30	60	
	0480	3/4" diameter, 4" long		10	.800		11.10	33.50		44.60	64	
	0500	6" long		10	.800		12.50	33.50		46	66	
	0520	8" long		10	.800		13.65	33.50		47.15	67	
	0540	10" long		10	.800		14.45	33.50		47.95	68	
	0560	12" long		10	.800		15.80	33.50		49.30	69.50	
	0580	15" long		10	.800		17.40	33.50		50.90	71	
	0600	18" long	↓	10	.800	↓	19.45	33.50		52.95	73.50	
430	0010	**INTERIOR H.I.D. FIXTURES** Incl. lamps, and mounting hardware										430
	0700	High pressure sodium, recessed, round, 70 watt	1 Elec	3.50	2.286	Ea.	380	96		476	570	
	0720	100 watt		3.50	2.286		400	96		496	590	
	0740	150 watt		3.20	2.500		450	105		555	655	
	0760	Square, 70 watt		3.60	2.222		380	93.50		473.50	560	
	0780	100 watt		3.60	2.222		405	93.50		498.50	590	
	0820	250 watt	↓	3	2.667		565	112		677	800	
	0840	1000 watt	2 Elec	4.80	3.333	↓	1,075	140		1,215	1,425	

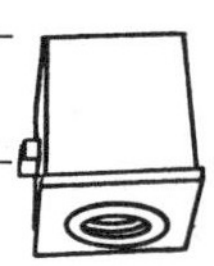

ELECTRICAL 16

16510 | Interior Luminaires

		16510 Interior Luminaires	CREW	DAILY OUTPUT	LABOR-HOURS	UNIT	MAT.	LABOR	EQUIP.	TOTAL	TOTAL INCL O&P	
							2006 BARE COSTS					
430	0860	Surface, round, 70 watt	1 Elec	3	2.667	Ea.	540	112		652	770	430
	0880	100 watt		3	2.667		555	112		667	785	
	0900	150 watt		2.70	2.963		580	124		704	825	
	0920	Square, 70 watt		3	2.667		525	112		637	750	
	0940	100 watt		3	2.667		555	112		667	785	
	0980	250 watt		2.50	3.200		570	134		704	830	
	1040	Pendent, round, 70 watt		3	2.667		575	112		687	805	
	1060	100 watt		3	2.667		550	112		662	780	
	1080	150 watt		2.70	2.963		560	124		684	810	
	1100	Square, 70 watt		3	2.667		590	112		702	820	
	1120	100 watt		3	2.667		595	112		707	830	
	1140	150 watt		2.70	2.963		615	124		739	865	
	1160	250 watt		2.50	3.200		825	134		959	1,125	
	1180	400 watt		2.40	3.333		870	140		1,010	1,175	
	1220	Wall, round, 70 watt		3	2.667		495	112		607	720	
	1240	100 watt		3	2.667		505	112		617	730	
	1260	150 watt		2.70	2.963		515	124		639	760	
	1300	Square, 70 watt		3	2.667		525	112		637	750	
	1320	100 watt		3	2.667		550	112		662	780	
	1340	150 watt		2.70	2.963		585	124		709	830	
	1360	250 watt		2.50	3.200		605	134		739	870	
	1380	400 watt	2 Elec	4.80	3.333		745	140		885	1,025	
	1400	1000 watt	"	3.60	4.444		1,050	187		1,237	1,450	
	1500	Metal halide, recessed, round, 175 watt	1 Elec	3.40	2.353		380	99		479	570	
	1520	250 watt	"	3.20	2.500		425	105		530	625	
	1540	400 watt	2 Elec	5.80	2.759		600	116		716	840	
	1580	Square, 175 watt	1 Elec	3.40	2.353		350	99		449	530	
	1640	Surface, round, 175 watt		2.90	2.759		470	116		586	695	
	1660	250 watt		2.70	2.963		715	124		839	980	
	1680	400 watt	2 Elec	4.80	3.333		830	140		970	1,125	
	1720	Square, 175 watt	1 Elec	2.90	2.759		510	116		626	740	
	1800	Pendent, round, 175 watt		2.90	2.759		555	116		671	790	
	1820	250 watt		2.70	2.963		775	124		899	1,050	
	1840	400 watt	2 Elec	4.80	3.333		860	140		1,000	1,175	
	1880	Square, 175 watt	1 Elec	2.90	2.759		535	116		651	770	
	1900	250 watt	"	2.70	2.963		545	124		669	785	
	1920	400 watt	2 Elec	4.80	3.333		875	140		1,015	1,175	
	1980	Wall, round, 175 watt	1 Elec	2.90	2.759		470	116		586	695	
	2000	250 watt	"	2.70	2.963		715	124		839	975	
	2020	400 watt	2 Elec	4.80	3.333		735	140		875	1,025	
	2060	Square, 175 watt	1 Elec	2.90	2.759		495	116		611	720	
	2080	250 watt	"	2.70	2.963		520	124		644	760	
	2100	400 watt	2 Elec	4.80	3.333		730	140		870	1,025	
	2800	High pressure sodium, recessed, 70 watt	1 Elec	3.50	2.286		490	96		586	685	
	2820	100 watt		3.50	2.286		500	96		596	700	
	2840	150 watt		3.20	2.500		515	105		620	725	
	2900	Surface, 70 watt		3	2.667		550	112		662	780	
	2920	100 watt		3	2.667		575	112		687	805	
	2940	150 watt		2.70	2.963		600	124		724	850	
	3000	Pendent, 70 watt		3	2.667		545	112		657	770	
	3020	100 watt		3	2.667		565	112		677	795	
	3040	150 watt		2.70	2.963		590	124		714	840	
	3100	Wall, 70 watt		3	2.667		585	112		697	815	
	3120	100 watt		3	2.667		610	112		722	845	
	3140	150 watt		2.70	2.963		635	124		759	885	
	3200	Metal halide, recessed, 175 watt		3.40	2.353		470	99		569	670	

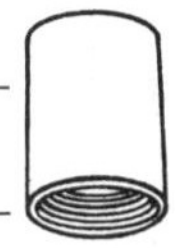
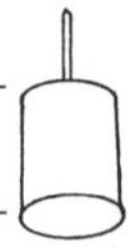
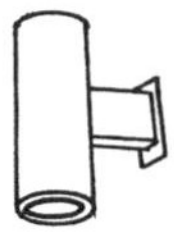
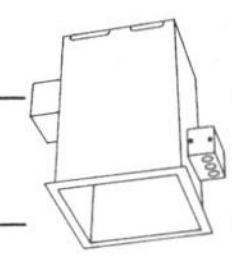
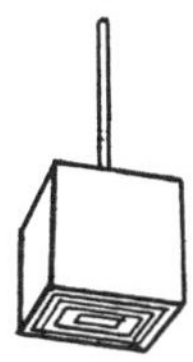
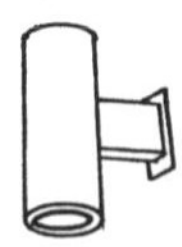
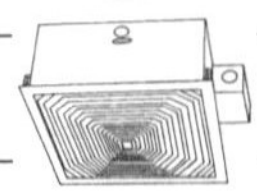
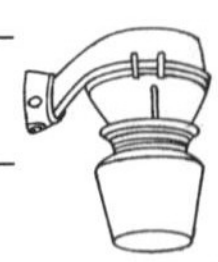

	16510 | Interior Luminaires		CREW	DAILY OUTPUT	LABOR-HOURS	UNIT	2006 BARE COSTS MAT.	LABOR	EQUIP.	TOTAL	TOTAL INCL O&P	
430	3220	250 watt	1 Elec	3.20	2.500	Ea.	505	105		610	715	430
	3240	400 watt	2 Elec	5.80	2.759		630	116		746	870	
	3260	1000 watt	"	4.80	3.333		1,150	140		1,290	1,500	
	3280	Surface, 175 watt 	1 Elec	2.90	2.759		480	116		596	705	
	3300	250 watt	"	2.70	2.963		735	124		859	1,000	
	3320	400 watt	2 Elec	4.80	3.333		885	140		1,025	1,175	
	3340	1000 watt	"	3.60	4.444		1,300	187		1,487	1,725	
	3360	Pendent, 175 watt 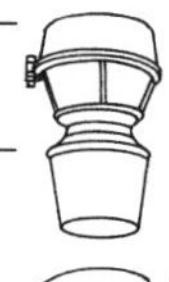	1 Elec	2.90	2.759		480	116		596	705	
	3380	250 watt	"	2.70	2.963		735	124		859	1,000	
	3400	400 watt	2 Elec	4.80	3.333		880	140		1,020	1,175	
	3420	1000 watt	"	3.60	4.444		1,425	187		1,612	1,875	
	3440	Wall, 175 watt 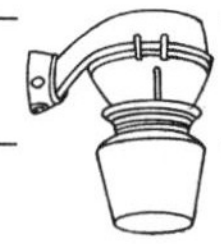	1 Elec	2.90	2.759		520	116		636	755	
	3460	250 watt	"	2.70	2.963		780	124		904	1,050	
	3480	400 watt	2 Elec	4.80	3.333		920	140		1,060	1,225	
	3500	1000 watt	"	3.60	4.444	↓	1,475	187		1,662	1,925	
440	0010	**INTERIOR LIGHTING FIXTURES** Including lamps, mounting (R265113-40)										440
	0030	hardware and connections										
	0100	Fluorescent, C.W. lamps, troffer, recess mounted in grid, RS (R265723-05)										
	0200	Acrylic lens, 1'W x 4'L, two 40 watt	1 Elec	5.70	1.404	Ea.	48.50	59		107.50	145	
	0210	1'W x 4'L, three 40 watt (R265723-20)		5.40	1.481		56.50	62		118.50	159	
	0300	2'W x 2'L, two U40 watt		5.70	1.404		52	59		111	148	
	0400	2'W x 4'L, two 40 watt		5.30	1.509		52	63.50		115.50	155	
	0500	2'W x 4'L, three 40 watt		5	1.600		55.50	67		122.50	165	
	0600	2'W x 4'L, four 40 watt	↓	4.70	1.702		59	71.50		130.50	175	
	0700	4'W x 4'L, four 40 watt 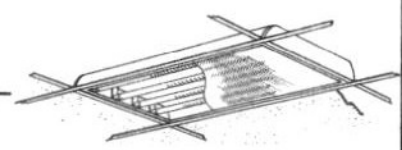	2 Elec	6.40	2.500		310	105		415	500	
	0800	4'W x 4'L, six 40 watt		6.20	2.581		320	108		428	520	
	0900	4'W x 4'L, eight 40 watt	↓	5.80	2.759		325	116		441	540	
	0910	Acrylic lens, 1'W x 4'L, two 32 watt	1 Elec	5.70	1.404		56	59		115	153	
	0930	2'W x 2'L, two U32 watt		5.70	1.404		80	59		139	179	
	0940	2'W x 4'L, two 32 watt		5.30	1.509		72	63.50		135.50	177	
	0950	2'W x 4'L, three 32 watt		5	1.600		76.50	67		143.50	188	
	0960	2'W x 4'L, four 32 watt	↓	4.70	1.702	↓	78.50	71.50		150	197	
	1000	Surface mounted, RS										
	1030	Acrylic lens with hinged & latched door frame										
	1100	1'W x 4'L, two 40 watt	1 Elec	7	1.143	Ea.	73.50	48		121.50	155	
	1110	1'W x 4'L, three 40 watt		6.70	1.194		75.50	50		125.50	161	
	1200	2'W x 2'L, two U40 watt 		7	1.143		79	48		127	161	
	1300	2'W x 4'L, two 40 watt		6.20	1.290		90.50	54		144.50	183	
	1400	2'W x 4'L, three 40 watt		5.70	1.404		91.50	59		150.50	191	
	1500	2'W x 4'L, four 40 watt	↓	5.30	1.509		93.50	63.50		157	201	
	1600	4'W x 4'L, four 40 watt	2 Elec	7.20	2.222		430	93.50		523.50	620	
	1700	4'W x 4'L, six 40 watt		6.60	2.424		470	102		572	670	
	1800	4'W x 4'L, eight 40 watt		6.20	2.581		485	108		593	700	
	1900	2'W x 8'L, four 40 watt		6.40	2.500		163	105		268	340	
	2000	2'W x 8'L, eight 40 watt	↓	6.20	2.581	↓	187	108		295	375	
	2010	Acrylic wrap around lens										
	2020	6"W x 4'L, one 40 watt	1 Elec	8	1	Ea.	69	42		111	141	
	2030	6"W x 8'L, two 40 watt	2 Elec	8	2		75.50	84		159.50	212	
	2040	11"W x 4'L, two 40 watt	1 Elec	7	1.143		45.50	48		93.50	124	
	2050	11"W x 8'L, four 40 watt 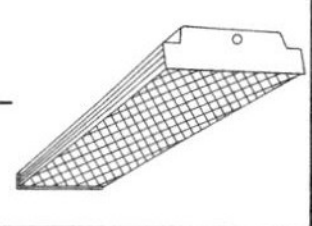	2 Elec	6.60	2.424		75.50	102		177.50	240	
	2060	16"W x 4'L, four 40 watt	1 Elec	5.30	1.509		75.50	63.50		139	181	
	2070	16"W x 8'L, eight 40 watt	2 Elec	6.40	2.500		164	105		269	345	
	2080	2'W x 2'L, two U40 watt	1 Elec	7	1.143	↓	93	48		141	176	
	2100	Strip fixture										
	2130	Surface mounted										

ELECTRICAL 16

16510 | Interior Luminaires

	16510 Interior Luminaires	CREW	DAILY OUTPUT	LABOR-HOURS	UNIT	2006 BARE COSTS MAT.	LABOR	EQUIP.	TOTAL	TOTAL INCL O&P
440 2200	4′ long, one 40 watt RS (R265113-40)	1 Elec	8.50	.941	Ea.	28.50	39.50		68	92
2300	4′ long, two 40 watt RS		8	1		30.50	42		72.50	98
2400	4′ long, one 40 watt, SL (R265723-05)		8	1		41.50	42		83.50	110
2500	4′ long, two 40 watt, SL		7	1.143		56.50	48		104.50	137
2600	8′ long, one 75 watt, SL (R265723-20)	2 Elec	13.40	1.194		42.50	50		92.50	125
2700	8′ long, two 75 watt, SL	"	12.40	1.290		51.50	54		105.50	140
2800	4′ long, two 60 watt, HO	1 Elec	6.70	1.194		83	50		133	169
2900	8′ long, two 110 watt, HO	2 Elec	10.60	1.509		87.50	63.50		151	194
2910	4′ long, two 115 watt, VHO	1 Elec	6.50	1.231		116	51.50		167.50	207
2920	8′ long, two 215 watt, VHO	2 Elec	10.40	1.538		109	64.50		173.50	220
3000	Pendent mounted, industrial, white porcelain enamel									
3100	4′ long, two 40 watt, RS	1 Elec	5.70	1.404	Ea.	47	59		106	143
3200	4′ long, two 60 watt, HO	"	5	1.600		74.50	67		141.50	186
3300	8′ long, two 75 watt, SL	2 Elec	8.80	1.818		88.50	76.50		165	216
3400	8′ long, two 110 watt, HO	"	8	2		113	84		197	254
3410	Acrylic finish, 4′ long, two 40 watt, RS	1 Elec	5.70	1.404		78	59		137	177
3420	4′ long, two 60 watt, HO		5	1.600		144	67		211	262
3430	4′ long, two 115 watt, VHO		4.80	1.667		186	70		256	310
3440	8′ long, two 75 watt, SL	2 Elec	8.80	1.818		152	76.50		228.50	286
3450	8′ long, two 110 watt, HO		8	2		170	84		254	315
3460	8′ long, two 215 watt, VHO		7.60	2.105		240	88.50		328.50	400
3470	Troffer, air handling, 2′W x 4′L with four 40 watt, RS	1 Elec	4	2		83	84		167	221
3480	2′W x 2′L with two U40 watt RS		5.50	1.455		72.50	61		133.50	174
3490	Air connector insulated, 5″ diameter		20	.400		56.50	16.80		73.30	88.50
3500	6″ diameter		20	.400		58	16.80		74.80	89.50
3510	Troffer parabolic lay-in, 1′W x 4′L with one F40		5.70	1.404		77	59		136	176
3520	1′W x 4′L with two F40		5.30	1.509		81	63.50		144.50	187
3530	2′W x 4′L with three F40		5	1.600		112	67		179	227
3535	Downlight, recess mounted		8	1		89.50	42		131.50	163
3540	Wall washer, recess mounted		8	1		89.50	42		131.50	163
3550	Direct/indirect, 4′ long, stl., pendent mtd.		5	1.600		124	67		191	241
3560	4′ long, alum., pendent mtd.		5	1.600		305	67		372	440
3565	Prefabricated cove, 4′ long, stl. continuous row		5	1.600		185	67		252	310
3570	4′ long, alum. continuous row		5	1.600		315	67		382	450
4220	Metal halide, integral ballast, ceiling, recess mounted									
4230	prismatic glass lens, floating door									
4240	2′W x 2′L, 250 watt	1 Elec	3.20	2.500	Ea.	291	105		396	480
4250	2′W x 2′L, 400 watt	2 Elec	5.80	2.759		330	116		446	540
4260	Surface mounted, 2′W x 2′L, 250 watt	1 Elec	2.70	2.963		288	124		412	505
4270	400 watt	2 Elec	4.80	3.333		340	140		480	590
4280	High bay, aluminum reflector,									
4290	Single unit, 400 watt	2 Elec	4.60	3.478	Ea.	345	146		491	605
4300	Single unit, 1000 watt		4	4		495	168		663	805
4310	Twin unit, 400 watt		3.20	5		690	210		900	1,075
4320	Low bay, aluminum reflector, 250W DX lamp	1 Elec	3.20	2.500		340	105		445	530
4330	400 watt lamp	2 Elec	5	3.200		490	134		624	745
4340	High pressure sodium integral ballast ceiling, recess mounted									
4350	prismatic glass lens, floating door									
4360	2′W x 2′L, 150 watt lamp	1 Elec	3.20	2.500	Ea.	350	105		455	545
4370	2′W x 2′L, 400 watt lamp	2 Elec	5.80	2.759		420	116		536	640
4380	Surface mounted, 2′W x 2′L, 150 watt lamp	1 Elec	2.70	2.963		395	124		519	625
4390	400 watt lamp	2 Elec	4.80	3.333		445	140		585	700
4400	High bay, aluminum reflector,									
4410	Single unit, 400 watt lamp	2 Elec	4.60	3.478	Ea.	320	146		466	575
4430	Single unit, 1000 watt lamp	"	4	4		460	168		628	765
4440	Low bay, aluminum reflector, 150 watt lamp	1 Elec	3.20	2.500		275	105		380	465

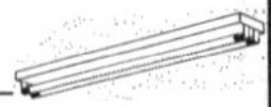

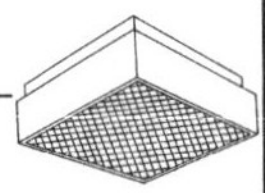

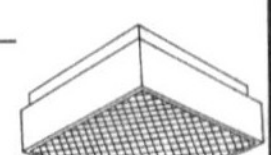

	16510 | Interior Luminaires	Crew	Daily Output	Labor-Hours	Unit	2006 Bare Costs Mat.	Labor	Equip.	Total	Total Incl O&P
440 4445	High bay H.I.D. quartz restrike (R265113-40)	1 Elec	16	.500	Ea.	137	21		158	184
4450	Incandescent, high hat can, round alzak reflector, prewired									
4470	100 watt (R265723-05)	1 Elec	8	1	Ea.	60.50	42		102.50	131
4480	150 watt		8	1		87.50	42		129.50	161
4500	300 watt (R265723-20)		6.70	1.194		204	50		254	305
4520	Round with reflector and baffles, 150 watt		8	1		42	42		84	111
4540	Round with concentric louver, 150 watt PAR		8	1		66	42		108	137
4600	Square glass lens with metal trim, prewired									
4630	100 watt	1 Elec	6.70	1.194	Ea.	47	50		97	129
4700	200 watt		6.70	1.194		82	50		132	168
4800	300 watt		5.70	1.404		122	59		181	225
4810	500 watt		5	1.600		239	67		306	365
4900	Ceiling/wall, surface mounted, metal cylinder, 75 watt		10	.800		48	33.50		81.50	105
4920	150 watt		10	.800		69	33.50		102.50	128
4930	300 watt		8	1		142	42		184	221
5000	500 watt		6.70	1.194		325	50		375	435
5010	Square, 100 watt		8	1		96	42		138	171
5020	150 watt		8	1		105	42		147	181
5030	300 watt		7	1.143		299	48		347	405
5040	500 watt		6	1.333		299	56		355	415
5200	Ceiling, surface mounted, opal glass drum									
5300	8″, one 60 watt lamp	1 Elec	10	.800	Ea.	37.50	33.50		71	93.50
5400	10″, two 60 watt lamps		8	1		42	42		84	111
5500	12″, four 60 watt lamps		6.70	1.194		60	50		110	144
5510	Pendent, round, 100 watt		8	1		96	42		138	171
5520	150 watt		8	1		103	42		145	178
5530	300 watt		6.70	1.194		143	50		193	235
5540	500 watt		5.50	1.455		272	61		333	395
5550	Square, 100 watt		6.70	1.194		127	50		177	218
5560	150 watt		6.70	1.194		133	50		183	224
5570	300 watt		5.70	1.404		196	59		255	305
5580	500 watt		5	1.600		265	67		332	395
5600	Wall, round, 100 watt		8	1		56	42		98	126
5620	300 watt		8	1		95.50	42		137.50	170
5630	500 watt		6.70	1.194		340	50		390	455
5640	Square, 100 watt		8	1		91	42		133	165
5650	150 watt		8	1		93.50	42		135.50	168
5660	300 watt		7	1.143		139	48		187	227
5670	500 watt		6	1.333		246	56		302	360
6010	Vapor tight, incandescent, ceiling mounted, 200 watt		6.20	1.290		53.50	54		107.50	142
6020	Recessed, 200 watt		6.70	1.194		94.50	50		144.50	182
6030	Pendent, 200 watt		6.70	1.194		53.50	50		103.50	136
6040	Wall, 200 watt		8	1		62.50	42		104.50	134
6100	Fluorescent, surface mounted, 2 lamps, 4′L, RS, 40 watt		3.20	2.500		97	105		202	269
6110	Industrial, 2 lamps 4′ long in tandem, 430 MA		2.20	3.636		178	153		331	430
6130	2 lamps 4′ long, 800 MA		1.90	4.211		152	177		329	440
6160	Pendent, indust, 2 lamps 4′L in tandem, 430 MA		1.90	4.211		208	177		385	500
6170	2 lamps 4′ long, 430 MA		2.30	3.478		139	146		285	380
6180	2 lamps 4′ long, 800 MA		1.70	4.706		176	198		374	500
6300	Explosionproof									
6310	Metal halide with ballast, ceiling, surface mounted, 175 W	1 Elec	2.90	2.759	Ea.	725	116		841	975
6320	250 watt	″	2.70	2.963		870	124		994	1,150
6330	400 watt	2 Elec	4.80	3.333		930	140		1,070	1,250
6340	Ceiling, pendent mounted, 175 watt	1 Elec	2.60	3.077		690	129		819	960
6350	250 watt	″	2.40	3.333		835	140		975	1,125
6360	400 watt	2 Elec	4.20	3.810		895	160		1,055	1,225

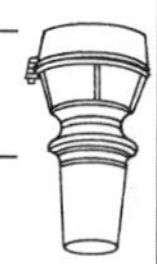

	16510 | Interior Luminaires	CREW	DAILY OUTPUT	LABOR-HOURS	UNIT	2006 BARE COSTS MAT.	LABOR	EQUIP.	TOTAL	TOTAL INCL O&P	
440 6370	Wall, surface mounted, 175 watt (R265113-40)	1 Elec	2.90	2.759	Ea.	780	116		896	1,025	440
6380	250 watt	"	2.70	2.963		925	124		1,049	1,225	
6390	400 watt (R265723-05)	2 Elec	4.80	3.333		985	140		1,125	1,300	
6400	High pressure sodium, ceiling surface mounted, 70 watt	1 Elec	3	2.667		785	112		897	1,025	
6410	100 watt (R265723-20)		3	2.667		810	112		922	1,075	
6420	150 watt		2.70	2.963		840	124		964	1,100	
6430	Pendent mounted, 70 watt		2.70	2.963		720	124		844	985	
6440	100 watt		2.70	2.963		775	124		899	1,050	
6450	150 watt		2.40	3.333		805	140		945	1,100	
6460	Wall mounted, 70 watt		3	2.667		840	112		952	1,100	
6470	100 watt		3	2.667		865	112		977	1,125	
6480	150 watt		2.70	2.963		890	124		1,014	1,175	
6510	Incandescent, ceiling mounted, 200 watt		4	2		700	84		784	900	
6520	Pendent mounted, 200 watt		3.50	2.286		600	96		696	810	
6530	Wall mounted, 200 watt		4	2		695	84		779	895	
6600	Fluorescent, RS, 4′ long, ceiling mounted, two 40 watt		2.70	2.963		1,775	124		1,899	2,150	
6610	Three 40 watt		2.20	3.636		2,550	153		2,703	3,025	
6620	Four 40 watt		1.90	4.211		3,275	177		3,452	3,900	
6630	Pendent mounted, two 40 watt		2.30	3.478		2,075	146		2,221	2,500	
6640	Three 40 watt		1.90	4.211		2,950	177		3,127	3,500	
6650	Four 40 watt		1.70	4.706		3,875	198		4,073	4,550	
6850	Vandalproof, surface mounted, fluorescent, two 40 watt		3.20	2.500		211	105		316	395	
6860	Incandescent, one 150 watt		8	1		54.50	42		96.50	125	
6900	Mirror light, fluorescent, RS, acrylic enclosure, two 40 watt		8	1		87.50	42		129.50	161	
6910	One 40 watt		8	1		69.50	42		111.50	141	
6920	One 20 watt		12	.667		55	28		83	104	
7000	Low bay, aluminum reflector. 70 watt, high pressure sodium		4	2		243	84		327	395	
7010	250 watt		3.20	2.500		325	105		430	520	
7020	400 watt	2 Elec	5	3.200		330	134		464	570	
7500	Ballast replacement, by weight of ballast, to 15′ high										
7520	Indoor fluorescent, less than 2 lbs.	1 Elec	10	.800	Ea.		33.50		33.50	52	
7540	Two 40W, watt reducer, 2 to 5 lbs.		9.40	.851		23	35.50		58.50	80.50	
7560	Two F96 slimline, over 5 lbs.		8	1		39	42		81	107	
7580	Vaportite ballast, less than 2 lbs.		9.40	.851			35.50		35.50	55	
7600	2 lbs. to 5 lbs.		8.90	.899			38		38	58	
7620	Over 5 lbs.		7.60	1.053			44		44	68	
7630	Electronic ballast for two tubes		8	1		34.50	42		76.50	103	
7640	Dimmable ballast one lamp		8	1		49.50	42		91.50	119	
7650	Dimmable ballast two-lamp		7.60	1.053		80	44		124	156	
7690	Emergency ballast (factory installed in fixture)					133			133	146	
7990	Decorator										
8000	Pendent RLM in colors, shallow dome, 12″ diam. 100 W	1 Elec	8	1	Ea.	68	42		110	140	
8010	Regular dome, 12″ diam., 100 watt		8	1		70	42		112	142	
8020	16″ diam., 200 watt		7	1.143		71.50	48		119.50	153	
8030	18″ diam., 500 watt		6	1.333		86.50	56		142.50	182	
8100	Picture framing light, minimum		16	.500		71.50	21		92.50	111	
8110	Maximum		16	.500		94.50	21		115.50	137	
8150	Miniature low voltage, recessed, pinhole		8	1		139	42		181	217	
8160	Star		8	1		114	42		156	191	
8170	Adjustable cone		8	1		155	42		197	236	
8180	Eyeball		8	1		127	42		169	204	
8190	Cone		8	1		111	42		153	187	
8200	Coilex baffle		8	1		109	42		151	185	
8210	Surface mounted, adjustable cylinder		8	1		112	42		154	188	
8250	Chandeliers, incandescent										
8260	24″ diam. x 42″ high, 6 light candle	1 Elec	6	1.333	Ea.	350	56		406	470	

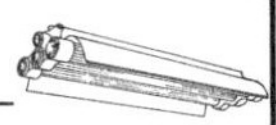

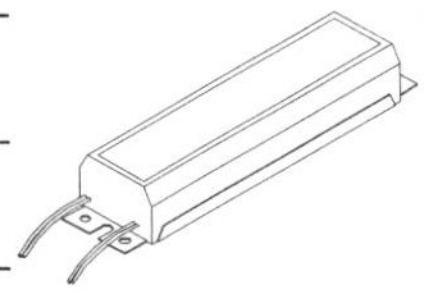

16 ELECTRICAL

16510 | Interior Luminaires

				DAILY	LABOR-		2006 BARE COSTS				TOTAL
			CREW	OUTPUT	HOURS	UNIT	MAT.	LABOR	EQUIP.	TOTAL	INCL O&P
440	8270	24" diam. x 42" high, 6 light candle w/glass shade (R265113-40)	1 Elec	6	1.333	Ea.	230	56		286	340
	8280	17" diam. x 12" high, 8 light w/glass panels		8	1		261	42		303	350
	8290	32" diam. x 48"H, 10 light bohemian lead crystal (R265723-05)		4	2		580	84		664	770
	8300	27" diam. x 29"H, 10 light bohemian lead crystal		4	2		540	84		624	725
	8310	21" diam. x 9" high 6 light sculptured ice crystal (R265723-20)		8	1		420	42		462	525
	8500	Accent lights, on floor or edge, 0.5W low volt incand.									
	8520	incl. transformer & fastenings, based on 100' lengths									
	8550	Lights in clear tubing, 12" on center	1 Elec	230	.035	L.F.	7.70	1.46		9.16	10.70
	8560	6" on center		160	.050		10	2.10		12.10	14.25
	8570	4" on center		130	.062		15.30	2.58		17.88	21
	8580	3" on center		125	.064		17	2.69		19.69	23
	8590	2" on center		100	.080		24.50	3.36		27.86	32
	8600	Carpet, lights both sides 6" OC, in alum. extrusion		270	.030		23	1.24		24.24	27.50
	8610	In bronze extrusion		270	.030		26	1.24		27.24	31
	8620	Carpet-bare floor, lights 18" OC, in alum. extrusion		270	.030		18.50	1.24		19.74	22.50
	8630	In bronze extrusion		270	.030		21.50	1.24		22.74	26
	8640	Carpet edge-wall, lights 6" OC in alum. extrusion		270	.030		23	1.24		24.24	27.50
	8650	In bronze extrusion		270	.030		26	1.24		27.24	31
	8660	Bare floor, lights 18" OC, in aluminum extrusion		300	.027		18.50	1.12		19.62	22
	8670	In bronze extrusion		300	.027		21.50	1.12		22.62	25.50
	8680	Bare floor conduit, aluminum extrusion		300	.027		6.10	1.12		7.22	8.50
	8690	In bronze extrusion		300	.027		12.25	1.12		13.37	15.20
	8700	Step edge to 36", lights 6" OC, in alum. extrusion		100	.080	Ea.	61.50	3.36		64.86	73
	8710	In bronze extrusion		100	.080		64	3.36		67.36	75.50
	8720	Step edge to 54", lights 6" OC, in alum. extrusion		100	.080		92.50	3.36		95.86	107
	8730	In bronze extrusion		100	.080		97.50	3.36		100.86	112
	8740	Step edge to 72", lights 6" OC, in alum. extrusion		100	.080		123	3.36		126.36	141
	8750	In bronze extrusion		100	.080		135	3.36		138.36	153
	8760	Connector, male		32	.250		2.28	10.50		12.78	18.70
	8770	Female with pigtail		32	.250		4.80	10.50		15.30	21.50
	8780	Clamps		400	.020		.46	.84		1.30	1.80
	8790	Transformers, 50 watt		8	1		65	42		107	136
	8800	250 watt		4	2		222	84		306	375
	8810	1000 watt		2.70	2.963		410	124		534	640
	9000	Minimum labor/equipment charge		3	2.667	Job		112		112	173
800	0010	**RESIDENTIAL FIXTURES**									
	0400	Fluorescent, interior, surface, circline, 32 watt & 40 watt	1 Elec	20	.400	Ea.	77.50	16.80		94.30	112
	0500	2' x 2', two U 40 watt		8	1		103	42		145	179
	0700	Shallow under cabinet, two 20 watt		16	.500		44	21		65	81
	0900	Wall mounted, 4'L, one 40 watt, with baffle		10	.800		119	33.50		152.50	183
	2000	Incandescent, exterior lantern, wall mounted, 60 watt		16	.500		31.50	21		52.50	67.50
	2100	Post light, 150W, with 7' post		4	2		110	84		194	250
	2500	Lamp holder, weatherproof with 150W PAR		16	.500		19.50	21		40.50	54
	2550	With reflector and guard		12	.667		54.50	28		82.50	103
	2600	Interior pendent, globe with shade, 150 watt		20	.400		128	16.80		144.80	167
	9000	Minimum labor/equipment charge		4	2	Job		84		84	129

16520 | Exterior Luminaires

			CREW	DAILY OUTPUT	LABOR-HOURS	UNIT	MAT.	LABOR	EQUIP.	TOTAL	TOTAL INCL O&P
300	0010	**EXTERIOR FIXTURES** With lamps									
	0200	Wall mounted, incandescent, 100 watt	1 Elec	8	1	Ea.	28	42		70	95.50
	0400	Quartz, 500 watt		5.30	1.509		54	63.50		117.50	157
	0420	1500 watt		4.20	1.905		102	80		182	236
	1100	Wall pack, low pressure sodium, 35 watt		4	2		214	84		298	365
	1150	55 watt		4	2		255	84		339	410

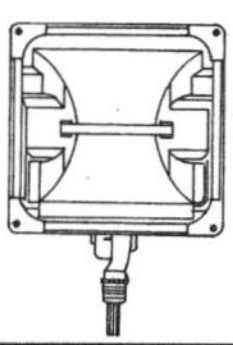

For expanded coverage of these items see *Means Electrical Cost Data 2006*

16520 | Exterior Luminaires

			CREW	DAILY OUTPUT	LABOR-HOURS	UNIT	2006 BARE COSTS MAT.	LABOR	EQUIP.	TOTAL	TOTAL INCL O&P	
300	1160	High pressure sodium, 70 watt	1 Elec	4	2	Ea.	200	84		284	350	300
	1170	150 watt		4	2		230	84		314	380	
	1180	Metal Halide, 175 watt		4	2		255	84		339	410	
	1190	250 watt		4	2		265	84		349	420	
	1200	Floodlights with ballast and lamp,										
	1400	pole mounted, pole not included										
	1950	Metal halide, 175 watt	1 Elec	2.70	2.963	Ea.	300	124		424	520	
	2000	400 watt	2 Elec	4.40	3.636		360	153		513	630	
	2200	1000 watt		4	4		505	168		673	815	
	2210	1500 watt		3.70	4.324		530	182		712	865	
	2250	Low pressure sodium, 55 watt	1 Elec	2.70	2.963		485	124		609	725	
	2270	90 watt		2	4		535	168		703	850	
	2290	180 watt		2	4		680	168		848	1,000	
	2340	High pressure sodium, 70 watt		2.70	2.963		207	124		331	420	
	2360	100 watt		2.70	2.963		212	124		336	425	
	2380	150 watt		2.70	2.963		249	124		373	465	
	2400	400 watt	2 Elec	4.40	3.636		365	153		518	635	
	2600	1000 watt	"	4	4		530	168		698	845	
	2610	Incandescent, 300 watt	1 Elec	4	2		76.50	84		160.50	213	
	2620	500 watt	"	4	2		122	84		206	263	
	2630	1000 watt	2 Elec	6	2.667		131	112		243	315	
	2640	1500 watt	"	6	2.667		144	112		256	330	
	2650	Roadway area luminaire, low pressure sodium, 135 watt	1 Elec	2	4		535	168		703	850	
	2700	180 watt	"	2	4		565	168		733	880	
	2750	Metal halide, 400 watt	2 Elec	4.40	3.636		460	153		613	740	
	2760	1000 watt		4	4		520	168		688	830	
	2780	High pressure sodium, 400 watt		4.40	3.636		475	153		628	760	
	2790	1000 watt		4	4		545	168		713	860	
	2800	Light poles, anchor base										
	2820	not including concrete bases										
	2840	Aluminum pole, 8′ high	1 Elec	4	2	Ea.	585	84		669	775	
	2850	10′ high		4	2		610	84		694	800	
	2860	12′ high		3.80	2.105		635	88.50		723.50	835	
	2870	14′ high		3.40	2.353		665	99		764	880	
	2880	16′ high		3	2.667		730	112		842	975	
	3000	20′ high	R-3	2.90	6.897		760	286	52.50	1,098.50	1,350	
	3200	30′ high		2.60	7.692		1,475	320	58.50	1,853.50	2,175	
	3400	35′ high		2.30	8.696		1,600	360	66.50	2,026.50	2,400	
	3600	40′ high		2	10		1,825	415	76.50	2,316.50	2,750	
	3800	Bracket arms, 1 arm	1 Elec	8	1		100	42		142	175	
	4000	2 arms		8	1		201	42		243	286	
	4200	3 arms		5.30	1.509		300	63.50		363.50	430	
	4400	4 arms		5.30	1.509		400	63.50		463.50	540	
	4500	Steel pole, galvanized, 8′ high		3.80	2.105		510	88.50		598.50	700	
	4510	10′ high		3.70	2.162		535	91		626	730	
	4520	12′ high		3.40	2.353		580	99		679	785	
	4530	14′ high		3.10	2.581		615	108		723	845	
	4540	16′ high		2.90	2.759		655	116		771	900	
	4550	18′ high		2.70	2.963		690	124		814	950	
	4600	20′ high	R-3	2.60	7.692		910	320	58.50	1,288.50	1,550	
	4800	30′ high		2.30	8.696		1,075	360	66.50	1,501.50	1,800	
	5000	35′ high		2.20	9.091		1,175	375	69.50	1,619.50	1,950	
	5200	40′ high		1.70	11.765		1,450	485	90	2,025	2,450	
	5400	Bracket arms, 1 arm	1 Elec	8	1		149	42		191	229	
	5600	2 arms		8	1		231	42		273	320	
	5800	3 arms		5.30	1.509		250	63.50		313.50	375	

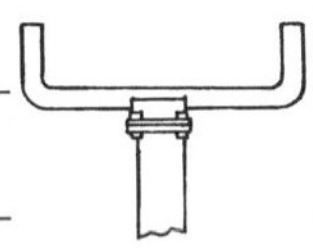

 Important: See the Reference Section for supporting data - Crews, Rental Equipment, City Cost Indexes and Reference Data

16520 | Exterior Luminaires

				DAILY	LABOR-		2006 BARE COSTS				TOTAL
			CREW	OUTPUT	HOURS	UNIT	MAT.	LABOR	EQUIP.	TOTAL	INCL O&P
300	6000	4 arms	1 Elec	5.30	1.509	Ea.	350	63.50		413.50	480
	6100	Fiberglass pole, 1 or 2 fixtures, 20′ high	R-3	4	5		465	207	38	710	870
	6200	30′ high		3.60	5.556		575	230	42.50	847.50	1,025
	6300	35′ high		3.20	6.250		900	259	47.50	1,206.50	1,450
	6400	40′ high		2.80	7.143		1,050	296	54.50	1,400.50	1,675
	6420	Wood pole, 4-1/2″ x 5-1/8″, 8′ high	1 Elec	6	1.333		269	56		325	385
	6430	10′ high		6	1.333		305	56		361	420
	6440	12′ high		5.70	1.404		385	59		444	515
	6450	15′ high		5	1.600		450	67		517	600
	6460	20′ high		4	2		545	84		629	730
	6500	Bollard light, lamp & ballast, 42″ high with polycarbonate lens									
	6800	Metal halide, 175 watt	1 Elec	3	2.667	Ea.	645	112		757	880
	6900	High pressure sodium, 70 watt		3	2.667		660	112		772	900
	7000	100 watt		3	2.667		660	112		772	900
	7100	150 watt		3	2.667		645	112		757	880
	7200	Incandescent, 150 watt		3	2.667		465	112		577	685
	7300	Transformer bases, not including concrete bases									
	7320	Maximum pole size, steel, 40′ high	1 Elec	2	4	Ea.	1,100	168		1,268	1,475
	7340	Cast aluminum, 30′ high		3	2.667		600	112		712	835
	7350	40′ high		2.50	3.200		905	134		1,039	1,200
	7380	Landscape recessed uplight, incl. housing, ballast, transformer									
	7390	& reflector									
	7420	Incandescent, 250 watt	1 Elec	5	1.600	Ea.	465	67		532	620
	7440	Quartz, 250 watt		5	1.600		445	67		512	590
	7460	500 watt		4	2		455	84		539	630
	7500	Replacement (H.I.D.) ballasts,									
	7510	Multi-tap 120/208/240/277 volt									
	7550	High pressure sodium, 70 watt	1 Elec	10	.800	Ea.	142	33.50		175.50	208
	7560	100 watt		9.40	.851		148	35.50		183.50	218
	7570	150 watt		9	.889		160	37.50		197.50	234
	7580	250 watt		8.50	.941		239	39.50		278.50	325
	7590	400 watt		7	1.143		269	48		317	370
	7600	1000 watt		6	1.333		370	56		426	495
	7610	Metal halide, 175 watt		8	1		87	42		129	161
	7620	250 watt		8	1		113	42		155	189
	7630	400 watt		7	1.143		141	48		189	229
	7640	1000 watt		6	1.333		241	56		297	350
	7650	1500 watt		5	1.600		299	67		366	435
	7810	Walkway luminaire, square 16″, metal halide 250 watt		2.70	2.963		520	124		644	760
	7820	High pressure sodium, 70 watt		3	2.667		595	112		707	830
	7830	100 watt		3	2.667		605	112		717	840
	7840	150 watt		3	2.667		605	112		717	840
	7850	200 watt		3	2.667		610	112		722	845
	7910	Round 19″, metal halide, 250 watt		2.70	2.963		755	124		879	1,025
	7920	High pressure sodium, 70 watt		3	2.667		830	112		942	1,100
	7930	100 watt		3	2.667		830	112		942	1,100
	7940	150 watt		3	2.667		835	112		947	1,100
	7950	250 watt		2.70	2.963		875	124		999	1,150
	8000	Sphere 14″ opal, incandescent, 200 watt		4	2		244	84		328	395
	8020	Sphere 18″ opal, incandescent, 300 watt		3.50	2.286		295	96		391	475
	8040	Sphere 16″ clear, high pressure sodium, 70 watt		3	2.667		510	112		622	735
	8050	100 watt		3	2.667		545	112		657	775
	8100	Cube 16″ opal, incandescent, 300 watt		3.50	2.286		325	96		421	510
	8120	High pressure sodium, 70 watt		3	2.667		475	112		587	700
	8130	100 watt		3	2.667		490	112		602	710
	8230	Lantern, high pressure sodium, 70 watt		3	2.667		420	112		532	640

16520 | Exterior Luminaires

			CREW	DAILY OUTPUT	LABOR-HOURS	UNIT	2006 BARE COSTS MAT.	LABOR	EQUIP.	TOTAL	TOTAL INCL O&P	
300	8240	100 watt	1 Elec	3	2.667	Ea.	455	112		567	675	300
	8250	150 watt		3	2.667		425	112		537	645	
	8260	250 watt		2.70	2.963		595	124		719	845	
	8270	Incandescent, 300 watt		3.50	2.286		315	96		411	495	
	8330	Reflector 22" w/globe, high pressure sodium, 70 watt		3	2.667		395	112		507	610	
	8340	100 watt		3	2.667		405	112		517	620	
	8350	150 watt		3	2.667		410	112		522	625	
	8360	250 watt		2.70	2.963		520	124		644	765	
	9000	Minimum labor/equipment charge		3.75	2.133	Job		89.50		89.50	138	

16525 | Aviation Lighting

			CREW	DAILY OUTPUT	LABOR-HOURS	UNIT	MAT.	LABOR	EQUIP.	TOTAL	TOTAL INCL O&P	
100	0010	AIRPORT LIGHTING										100
	0100	Runway centerline, bidir., semi-flush, 200 W, w/shallow insert base	R-22	12.40	3.006	Ea.	1,375	106		1,481	1,675	
	0120	Flush, 200 W, w/shallow insert base		12.40	3.006		1,375	106		1,481	1,675	
	0130	for mounting in base housing		18.64	2		720	71		791	900	
	0150	Touchdown zone light, unidirectional, 200 W, w/shallow insert base		12.40	3.006		1,125	106		1,231	1,400	
	0160	115 W		12.40	3.006		1,200	106		1,306	1,475	
	0180	Unidirectional, 200 W, for mounting in base housing		18.60	2.004		545	71		616	710	
	0190	115 W		18.60	2.004		590	71		661	760	
	0210	Runway edge & threshold light, bidir., 200 W, for base housing		9.36	3.983		775	141		916	1,075	
	0240	Threshold & approach light, unidir., 200 W, for base housing		9.36	3.983		450	141		591	715	
	0260	Runway edge, bi-directional, 2-115 W, for base housing		12.40	3.006		1,200	106		1,306	1,500	
	0280	Runway threshold & end, bidir., 2-115 W, for base housing		12.40	3.006		1,325	106		1,431	1,625	
	0370	45 W, flush, for mounting in base housing		18.64	2		650	71		721	825	
	0380	115 W		18.64	2		665	71		736	840	
	1200	Wind cone, 12' lighted assembly, rigid, w/obstruction light	R-21	1.36	24.118		7,600	1,025	51	8,676	9,950	
	1210	Without obstruction light		1.52	21.579		6,375	905	46	7,326	8,450	
	1220	Unlighted assembly, w/obstruction light		1.68	19.524		6,950	820	41.50	7,811.50	8,975	
	1230	Without obstruction light		1.84	17.826		6,375	750	38	7,163	8,200	
	1240	Wind cone slip fitter, 2-1/2" pipe		21.84	1.502		83.50	63	3.18	149.68	193	
	1250	Wind cone sock, 12' x 3', cotton		6.56	5		460	210	10.60	680.60	840	
	1260	Nylon		6.56	5		470	210	10.60	690.60	850	

16530 | Emergency Lighting

			CREW	DAILY OUTPUT	LABOR-HOURS	UNIT	MAT.	LABOR	EQUIP.	TOTAL	TOTAL INCL O&P	
320	0010	EXIT AND EMERGENCY LIGHTING										320
	0080	Exit light ceiling or wall mount, incandescent, single face	1 Elec	8	1	Ea.	42	42		84	111	
	0100	Double face		6.70	1.194		46	50		96	128	
	0120	Explosion proof		3.80	2.105		440	88.50		528.50	615	
	0150	Fluorescent, single face		8	1		64.50	42		106.50	136	
	0160	Double face		6.70	1.194		67.50	50		117.50	152	
	0200	L.E.D. standard, single face		8	1		64.50	42		106.50	136	
	0220	Double face		6.70	1.194		67.50	50		117.50	152	
	0240	L.E.D. w/battery unit, single face		4.40	1.818		125	76.50		201.50	256	
	0260	Double face		4	2		127	84		211	269	
	0300	Emergency light units, battery operated										
	0350	Twin sealed beam light, 25 watt, 6 volt each										
	0500	Lead battery operated	1 Elec	4	2	Ea.	122	84		206	263	
	0700	Nickel cadmium battery operated		4	2		580	84		664	770	
	0780	Additional remote mount, sealed beam, 25W 6V		26.70	.300		26.50	12.60		39.10	48.50	
	0790	Twin sealed beam light, 25W 6V each		26.70	.300		46	12.60		58.60	70	
	0900	Self-contained fluorescent lamp pack		10	.800		135	33.50		168.50	201	
	9000	Minimum labor/equipment charge		4	2	Job		84		84	129	

16550 | Special Purpose Lighting

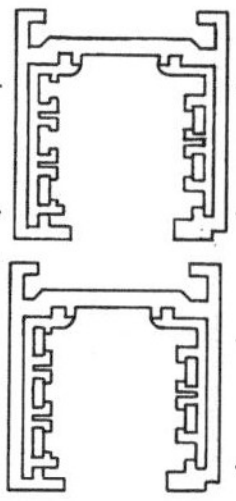

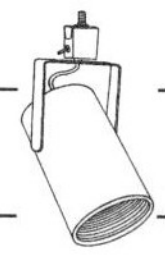

		16550 Special Purpose Lighting	CREW	DAILY OUTPUT	LABOR-HOURS	UNIT	2006 BARE COSTS MAT.	LABOR	EQUIP.	TOTAL	TOTAL INCL O&P	
820	0010	**TRACK LIGHTING**										820
	0080	Track, 1 circuit, 4′ section	1 Elec	6.70	1.194	Ea.	40.50	50		90.50	122	
	0100	8′ section		5.30	1.509		68	63.50		131.50	173	
	0200	12′ section		4.40	1.818		105	76.50		181.50	234	
	0300	3 circuits, 4′ section		6.70	1.194		54.50	50		104.50	138	
	0400	8′ section		5.30	1.509		84.50	63.50		148	191	
	0500	12′ section		4.40	1.818		163	76.50		239.50	297	
	1000	Feed kit, surface mounting		16	.500		10.05	21		31.05	43.50	
	1100	End cover		24	.333		3.50	14		17.50	25.50	
	1200	Feed kit, stem mounting, 1 circuit		16	.500		29.50	21		50.50	64.50	
	1300	3 circuit		16	.500		29.50	21		50.50	64.50	
	2000	Electrical joiner, for continuous runs, 1 circuit		32	.250		13.20	10.50		23.70	30.50	
	2100	3 circuit		32	.250		31	10.50		41.50	50	
	2200	Fixtures, spotlight, 75W PAR halogen		16	.500		87	21		108	128	
	2210	50W MR16 halogen		16	.500		106	21		127	150	
	3000	Wall washer, 250 watt tungsten halogen		16	.500		101	21		122	144	
	3100	Low voltage, 25/50 watt, 1 circuit		16	.500		102	21		123	145	
	3120	3 circuit		16	.500		105	21		126	149	
	9000	Minimum labor/equipment charge		3	2.667	Job		112		112	173	

16580 | Lighting Accessories

		16580 Lighting Accessories	CREW	DAILY OUTPUT	LABOR-HOURS	UNIT	MAT.	LABOR	EQUIP.	TOTAL	TOTAL INCL O&P	
200	0010	**ENERGY SAVING LIGHTING DEVICES**										200
	0100	Occupancy sensors infrared, ceiling mounted	1 Elec	7	1.143	Ea.	98	48		146	182	
	0150	Automatic wall switches		24	.333		60	14		74	87.50	
	0200	Remote power pack		10	.800		27	33.50		60.50	81.50	
	0250	Photoelectric control, S.P.S.T. 120 V		8	1		12.70	42		54.70	78.50	
	0300	S.P.S.T. 208 V/277 V		8	1		16.10	42		58.10	82	
	0350	D.P.S.T. 120 V		6	1.333		128	56		184	228	
	0400	D.P.S.T. 208 V/277 V		6	1.333		133	56		189	233	
	0450	S.P.D.T. 208 V/277 V		6	1.333		158	56		214	261	
300	0010	**FIXTURE WHIPS**										300
	0080	3/8″ Greenfield, 2 connectors, 6′ long										
	0100	TFFN wire, three #18	1 Elec	32	.250	Ea.	7.70	10.50		18.20	24.50	
	0150	Four #18		28	.286		7.65	12		19.65	27	
	0200	Three #16		32	.250		7.35	10.50		17.85	24.50	
	0250	Four #16		28	.286		8	12		20	27.50	
	0300	THHN wire, three #14		32	.250		10	10.50		20.50	27	
	0350	Four #14		28	.286		10.80	12		22.80	30.50	
	0360	Three #12		32	.250		12.85	10.50		23.35	30.50	

16585 | Lamps

		16585 Lamps	CREW	DAILY OUTPUT	LABOR-HOURS	UNIT	MAT.	LABOR	EQUIP.	TOTAL	TOTAL INCL O&P	
600	0010	**LAMPS**										600
	0080	Fluorescent, rapid start, cool white, 2′ long, 20 watt	1 Elec	1	8	C	310	335		645	860	
	0100	4′ long, 40 watt		.90	8.889		305	375		680	910	
	0120	3′ long, 30 watt		.90	8.889		390	375		765	1,000	
	0125	3′ long, 25 watt energy saver		.90	8.889		595	375		970	1,225	
	0150	U-40 watt		.80	10		955	420		1,375	1,700	
	0155	U-34 watt energy saver		.80	10		955	420		1,375	1,700	
	0170	4′ long, 34 watt energy saver		.90	8.889		296	375		671	900	
	0176	2′ long, T8, 17 W engergy saver		1	8		440	335		775	1,000	
	0178	3′ long, T8, 25 W energy saver		.90	8.889		450	375		825	1,075	
	0180	4′ long, T8, 32 watt energy saver		.90	8.889		273	375		648	875	
	0200	Slimline, 4′ long, 40 watt		.90	8.889		700	375		1,075	1,350	

16585 Lamps		CREW	DAILY OUTPUT	LABOR-HOURS	UNIT	2006 BARE COSTS MAT.	LABOR	EQUIP.	TOTAL	TOTAL INCL O&P
600 0210	4′ long, 30 watt energy saver	1 Elec	.90	8.889	C	700	375		1,075	1,350 600
0300	8′ long, 75 watt		.80	10		775	420		1,195	1,500
0350	8′ long, 60 watt energy saver		.80	10		490	420		910	1,175
0400	High output, 4′ long, 60 watt		.90	8.889		890	375		1,265	1,550
0410	8′ long, 95 watt energy saver		.80	10		855	420		1,275	1,575
0500	8′ long, 110 watt		.80	10		855	420		1,275	1,575
0512	2′ long, T5, 14 watt energy saver		1	8		1,225	335		1,560	1,875
0514	3′ long, T5, 21 watt energy saver		.90	8.889		1,225	375		1,600	1,925
0516	4′ long, T5, 28 watt energy saver		.90	8.889		1,050	375		1,425	1,725
0520	Very high output, 4′ long, 110 watt		.90	8.889		2,250	375		2,625	3,050
0525	8′ long, 195 watt energy saver		.70	11.429		2,350	480		2,830	3,325
0550	8′ long, 215 watt		.70	11.429		2,350	480		2,830	3,325
0554	Full spectrum, 4′ long, 60 watt		.90	8.889		1,650	375		2,025	2,400
0556	6′ long, 85 watt		.90	8.889		1,800	375		2,175	2,550
0558	8′ long, 110 watt		.80	10		1,550	420		1,970	2,375
0560	Twin tube compact lamp		.90	8.889		535	375		910	1,175
0570	Double twin tube compact lamp		.80	10		1,300	420		1,720	2,075
0600	Mercury vapor, mogul base, deluxe white, 100 watt		.30	26.667		2,725	1,125		3,850	4,700
0650	175 watt		.30	26.667		2,000	1,125		3,125	3,925
0700	250 watt		.30	26.667		3,600	1,125		4,725	5,675
0800	400 watt		.30	26.667		2,875	1,125		4,000	4,900
0900	1000 watt		.20	40		6,750	1,675		8,425	10,000
1000	Metal halide, mogul base, 175 watt		.30	26.667		3,400	1,125		4,525	5,450
1100	250 watt		.30	26.667		3,850	1,125		4,975	5,950
1200	400 watt		.30	26.667		3,650	1,125		4,775	5,725
1300	1000 watt		.20	40		9,100	1,675		10,775	12,600
1320	1000 watt, 125,000 initial lumens		.20	40		12,700	1,675		14,375	16,500
1330	1500 watt		.20	40		12,900	1,675		14,575	16,800
1350	High pressure sodium, 70 watt		.30	26.667		3,950	1,125		5,075	6,075
1360	100 watt		.30	26.667		4,275	1,125		5,400	6,425
1370	150 watt		.30	26.667		4,225	1,125		5,350	6,375
1380	250 watt		.30	26.667		4,500	1,125		5,625	6,675
1400	400 watt		.30	26.667		4,625	1,125		5,750	6,800
1450	1000 watt		.20	40		13,200	1,675		14,875	17,100
1500	Low pressure sodium, 35 watt		.30	26.667		5,975	1,125		7,100	8,300
1550	55 watt		.30	26.667		6,550	1,125		7,675	8,950
1600	90 watt		.30	26.667		7,600	1,125		8,725	10,100
1650	135 watt		.20	40		9,725	1,675		11,400	13,300
1700	180 watt		.20	40		10,700	1,675		12,375	14,300
1750	Quartz line, clear, 500 watt		1.10	7.273		930	305		1,235	1,500
1760	1500 watt		.20	40		3,650	1,675		5,325	6,600
1762	Spot, MR 16, 50 watt		1.30	6.154		785	258		1,043	1,275
1770	Tungsten halogen, T4, 400 watt		1.10	7.273		3,000	305		3,305	3,775
1775	T3, 1200 watt		.30	26.667		3,625	1,125		4,750	5,725
1778	PAR 30, 50 watt		1.30	6.154		840	258		1,098	1,325
1780	PAR 38, 90 watt		1.30	6.154		750	258		1,008	1,225
1800	Incandescent, interior, A21, 100 watt		1.60	5		140	210		350	480
1900	A21, 150 watt		1.60	5		143	210		353	485
2000	A23, 200 watt		1.60	5		201	210		411	545
2200	PS 30, 300 watt		1.60	5		485	210		695	860
2210	PS 35, 500 watt		1.60	5		825	210		1,035	1,225
2230	PS 52, 1000 watt		1.30	6.154		1,750	258		2,008	2,325
2240	PS 52, 1500 watt		1.30	6.154		4,725	258		4,983	5,575
2300	R30, 75 watt		1.30	6.154		535	258		793	990
2400	R40, 100 watt		1.30	6.154		580	258		838	1,050
2500	Exterior, PAR 38, 75 watt	↓	1.30	6.154	↓	1,100	258		1,358	1,600

16500 | Lighting

	16585 | Lamps		Crew	Daily Output	Labor-Hours	Unit	2006 Bare Costs Mat.	Labor	Equip.	Total	Total Incl O&P	
600	2600	PAR 38, 150 watt	1 Elec	1.30	6.154	C	1,200	258		1,458	1,725	600
	2700	PAR 46, 200 watt		1.10	7.273		2,400	305		2,705	3,125	
	2800	PAR 56, 300 watt		1.10	7.273		2,975	305		3,280	3,750	
	3000	Guards, fluorescent lamp, 4′ long		1	8		905	335		1,240	1,525	
	3200	8′ long		.90	8.889		1,800	375		2,175	2,575	
	9000	Minimum labor/equipment charge		4	2	Job		84		84	129	

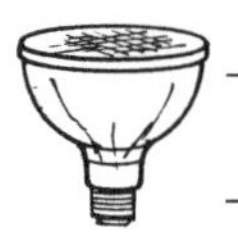

16700 | Communications

	16710 | Communication Circuits		Crew	Daily Output	Labor-Hours	Unit	2006 Bare Costs Mat.	Labor	Equip.	Total	Total Incl O&P	
400	0010	FIBER OPTICS										400
	0020	Fiber optics cable only. Added costs depend on the type of fiber										
	0030	special connectors, optical modems, and networking parts.										
	0040	Specialized tools & techniques cause installation costs to vary.										
	0070	Cable, minimum, bulk simplex (R271323-40)	1 Elec	8	1	C.L.F.	23	42		65	90	
	0080	Cable, maximum, bulk plenum quad	"	2.29	3.493	"	117	147		264	355	
	0150	Fiber optic jumper				Ea.	55.50			55.50	61	
	0200	Fiber optic pigtail					30			30	33	
	0300	Fiber optic connector	1 Elec	24	.333		16.90	14		30.90	40	
	0350	Fiber optic finger splice		32	.250		32	10.50		42.50	51	
	0400	Transceiver (low cost bi-directional)		8	1		425	42		467	535	
	0450	Multi-channel rack enclosure (10 modules)		2	4		475	168		643	785	
	0500	Fiber optic patch panel (12 ports)		6	1.333		178	56		234	283	
750	0010	COMMUNICATION CABLES & FITTINGS										750
	0100	Communication outlets, voice/data devises not included										
	0120	Voice/Data outlets, single opening	1 Elec	48	.167	Ea.	6.10	7		13.10	17.50	
	0140	Two jack openings		48	.167		2.70	7		9.70	13.75	
	0160	One jack & one 3/4″ round opening		48	.167		6.10	7		13.10	17.50	
	0180	One jack & one twinaxial opening		48	.167		6.10	7		13.10	17.50	
	0200	One jack & one connector cabling opening		48	.167		6.10	7		13.10	17.50	
	0220	Two 3/8″ coaxial openings		48	.167		6.10	7		13.10	17.50	
	0300	Data outlets, single opening		48	.167		6.10	7		13.10	17.50	
	0320	One 25-pin subminiature opening		48	.167		6.10	7		13.10	17.50	
	2200	Telephone twisted, PVC insulation, #22-2 conductor		10	.800	C.L.F.	7.20	33.50		40.70	60	
	2250	#22-3 conductor		9	.889		9.50	37.50		47	68	
	2300	#22-4 conductor		8	1		11.30	42		53.30	77	
	2350	#18-2 conductor		9	.889		8.75	37.50		46.25	67	
	2370	Telephone jack, eight pins		32	.250	Ea.	6.90	10.50		17.40	24	
	5000	High performance unshielded twisted pair (UTP)										
	5100	Category 3, #24, 2 pair solid, PVC jacket (R271513-75)	1 Elec	10	.800	C.L.F.	3	33.50		36.50	55.50	
	5200	4 pair solid		7	1.143		4.90	48		52.90	79.50	
	5300	25 pair solid		3	2.667		26.50	112		138.50	202	
	5400	2 pair solid, plenum		10	.800		6.50	33.50		40	59	
	5500	4 pair solid		7	1.143		8.70	48		56.70	83.50	
	5600	25 pair solid		3	2.667		53	112		165	232	
	5700	4 pair stranded, PVC jacket		7	1.143		14	48		62	89.50	
	7000	Category 5, #24, 4 pair solid, PVC jacket		7	1.143		7.80	48		55.80	82.50	
	7100	4 pair solid, plenum		7	1.143		24	48		72	101	
	7200	4 pair stranded, PVC jacket		7	1.143		11.70	48		59.70	87	

16700 | Communications

16710 | Communication Circuits

	16710	Communication Circuits	CREW	DAILY OUTPUT	LABOR-HOURS	UNIT	2006 BARE COSTS MAT.	LABOR	EQUIP.	TOTAL	TOTAL INCL O&P	
750	7210	Category 5e, #24, 4 pair solid, PVC jacket (R271513-75)	1 Elec	7	1.143	C.L.F.	6.90	48		54.90	81.50	750
	7212	4 pair solid, plenum		7	1.143		19.10	48		67.10	95	
	7214	4 pair stranded, PVC jacket		7	1.143		13.90	48		61.90	89.50	
	7240	Category 6, #24, 4 pair solid, PVC jacket		7	1.143		15.30	48		63.30	91	
	7242	4 pair solid, plenum		7	1.143		57	48		105	137	
	7244	4 pair stranded, PVC jacket		7	1.143		16.90	48		64.90	92.50	
	7300	Category 5, connector, UTP RJ-45		80	.100	Ea.	1.05	4.20		5.25	7.60	
	7302	shielded RJ-45		72	.111		3	4.67		7.67	10.50	
	7310	Category 3, jack, UTP RJ-45		72	.111		3.05	4.67		7.72	10.55	
	7312	Category 5		65	.123		3.95	5.15		9.10	12.30	
	7314	Category 5e		65	.123		3.95	5.15		9.10	12.30	
	7316	Category 6		65	.123		3.95	5.15		9.10	12.30	
	7322	Category 5, jack, shielded RJ-45		60	.133		4.95	5.60		10.55	14.10	
	7324	Category 5e		60	.133		4.95	5.60		10.55	14.10	
	7326	Category 6		60	.133		4.95	5.60		10.55	14.10	
	7400	Category 5e, voice/data expansion module		8	1		37	42		79	105	

16720 | Tel and Intercomm Equipment

	16720	Tel and Intercomm Equipment	CREW	DAILY OUTPUT	LABOR-HOURS	UNIT	MAT.	LABOR	EQUIP.	TOTAL	TOTAL INCL O&P	
600	0010	NURSE CALL SYSTEMS										600
	0100	Single bedside call station	1 Elec	8	1	Ea.	190	42		232	274	
	0200	Ceiling speaker station		8	1		55.50	42		97.50	126	
	0400	Emergency call station		8	1		94	42		136	169	
	0600	Pillow speaker		8	1		174	42		216	257	
	0800	Double bedside call station		4	2		180	84		264	325	
	1000	Duty station		4	2		143	84		227	286	
	1200	Standard call button		8	1		76	42		118	148	
	1400	Lights, corridor, dome or zone indicator		8	1		42.50	42		84.50	111	
	1600	Master control station for 20 stations	2 Elec	.65	24.615	Total	3,375	1,025		4,400	5,325	
900	0010	TELEPHONE EQUIPMENT										900
	1620	Telephone, TTY, compact, pocket type				Ea.	289			289	320	
	1630	Advanced, desk type	2 Elec	20	.800		570	33.50		603.50	675	
	1640	Full-featured public, wall type	"	4	4		970	168		1,138	1,325	

16800 | Sound & Video

16810 | Sound and Video Circuits

	16810	Sound and Video Circuits	CREW	DAILY OUTPUT	LABOR-HOURS	UNIT	2006 BARE COSTS MAT.	LABOR	EQUIP.	TOTAL	TOTAL INCL O&P	
750	0010	SOUND AND VIDEO CABLES & FITTINGS										750
	0900	TV antenna lead-in, 300 ohm, #20-2 conductor	1 Elec	7	1.143	C.L.F.	14.40	48		62.40	90	
	0950	Coaxial, feeder outlet		7	1.143		17.30	48		65.30	93	
	1000	Coaxial, main riser		6	1.333		25	56		81	114	
	1100	Sound, shielded with drain, #22-2 conductor		8	1		17.30	42		59.30	83.50	
	1150	#22-3 conductor		7.50	1.067		23.50	45		68.50	95	
	1200	#22-4 conductor		6.50	1.231		28.50	51.50		80	111	
	1250	Nonshielded, #22-2 conductor		10	.800		10.30	33.50		43.80	63.50	
	1300	#22-3 conductor		9	.889		14.05	37.50		51.55	73	
	1350	#22-4 conductor		8	1		17.75	42		59.75	84	
	1400	Microphone cable		8	1		61	42		103	132	
	3500	Coaxial connectors, 50 ohm impedance quick disconnect										
	3540	BNC plug, for RG A/U #58 cable	1 Elec	42	.190	Ea.	3.66	8		11.66	16.35	
	3550	RG A/U #59 cable 		42	.190		3.66	8		11.66	16.35	

16810 | Sound and Video Circuits

			CREW	DAILY OUTPUT	LABOR-HOURS	UNIT	2006 BARE COSTS MAT.	LABOR	EQUIP.	TOTAL	TOTAL INCL O&P	
750	3560	RG A/U #62 cable	1 Elec	42	.190	Ea.	3.66	8		11.66	16.35	750
	3600	BNC jack, for RG A/U #58 cable		42	.190		3.78	8		11.78	16.45	
	3610	RG A/U #59 cable		42	.190		3.78	8		11.78	16.45	
	3620	RG A/U #62 cable 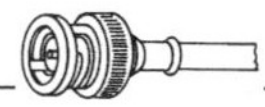		42	.190		3.78	8		11.78	16.45	
	3660	BNC panel jack, for RG A/U #58 cable		40	.200		5.90	8.40		14.30	19.40	
	3670	RG A/U #59 cable 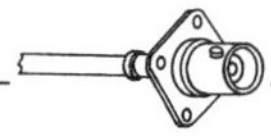		40	.200		5.90	8.40		14.30	19.40	
	3680	RG A/U #62 cable		40	.200		5.90	8.40		14.30	19.40	
	3720	BNC bulkhead jack, for RG A/U #58 cable		40	.200		6.85	8.40		15.25	20.50	
	3730	RG A/U #59 cable		40	.200		6.85	8.40		15.25	20.50	
	3740	RG A/U #62 cable		40	.200	↓	6.85	8.40		15.25	20.50	
	3850	Coaxial cable, RG A/U 58, 50 ohm 		8	1	C.L.F.	25	42		67	92	
	3860	RG A/U 59, 75 ohm		8	1		23.50	42		65.50	90	
	3870	RG A/U 62, 93 ohm		8	1		25	42		67	92	
	3875	RG 6/U, 75 ohm		8	1		21.50	42		63.50	88	
	3950	RG A/U 58, 50 ohm fire rated		8	1		54.50	42		96.50	125	
	3960	RG A/U 59, 75 ohm fire rated		8	1		95.50	42		137.50	170	
	3970	RG A/U 62, 93 ohm fire rated	↓	8	1	↓	77.50	42		119.50	150	

16820 | Sound Reinforcement

			CREW	DAILY OUTPUT	LABOR-HOURS	UNIT	MAT.	LABOR	EQUIP.	TOTAL	TOTAL INCL O&P	
300	0010	**DOORBELL SYSTEM** Incl. transformer, button & signal										300
	0100	6″ bell	1 Elec	4	2	Ea.	86	84		170	224	
	0200	Buzzer		4	2		68.50	84		152.50	204	
	1000	Door chimes, 2 notes, minimum		16	.500		22	21		43	57	
	1020	Maximum		12	.667		118	28		146	172	
	1100	Tube type, 3 tube system		12	.667		167	28		195	227	
	1180	4 tube system		10	.800		268	33.50		301.50	345	
	1900	For transformer & button, minimum add		5	1.600		12.70	67		79.70	118	
	1960	Maximum, add		4.50	1.778		38	74.50		112.50	157	
	3000	For push button only, minimum		24	.333		2.48	14		16.48	24	
	3100	Maximum		20	.400		20	16.80		36.80	48	
	3200	Bell transformer		16	.500	↓	17.65	21		38.65	52	
	9000	Minimum labor/equipment charge	↓	4	2	Job		84		84	129	
800	0010	**PUBLIC ADDRESS SYSTEM**										800
	0100	Conventional, office	1 Elec	5.33	1.501	Speaker	107	63		170	215	
	0200	Industrial	"	2.70	2.963	"	207	124		331	420	
	0400	Explosionproof system is 3 times cost of central control										
	0600	Installation costs run about 120% of material cost										
	9000	Minimum labor/equipment charge	1 Elec	3.50	2.286	Job		96		96	148	
840	0010	**SOUND SYSTEM** not including rough-in wires, cables & conduits										840
	0100	Components, outlet, projector	1 Elec	8	1	Ea.	49.50	42		91.50	119	
	0200	Microphone		4	2		55	84		139	190	
	0400	Speakers, ceiling or wall		8	1		93.50	42		135.50	168	
	0600	Trumpets		4	2		174	84		258	320	
	0800	Privacy switch		8	1		69.50	42		111.50	141	
	1000	Monitor panel		4	2		310	84		394	470	
	1200	Antenna, AM/FM		4	2		173	84		257	320	
	1400	Volume control		8	1		69.50	42		111.50	141	
	1600	Amplifier, 250 watts		1	8		1,375	335		1,710	2,050	
	1800	Cabinets	↓	1	8		675	335		1,010	1,250	
	2000	Intercom, 25 station capacity, master station	2 Elec	2	8		1,625	335		1,960	2,300	
	2020	11 station capacity	"	4	4		760	168		928	1,100	
	2200	Remote station	1 Elec	8	1		130	42		172	208	
	2400	Intercom outlets		8	1		76	42		118	149	
	2600	Handset	↓	4	2	↓	252	84		336	405	

16800 | Sound & Video

	16820 | Sound Reinforcement		Crew	Daily Output	Labor-Hours	Unit	2006 Bare Costs Mat.	Labor	Equip.	Total	Total Incl O&P	
840	2800	Emergency call system, 12 zones, annunciator	1 Elec	1.30	6.154	Ea.	760	258		1,018	1,225	840
	3000	Bell		5.30	1.509		78.50	63.50		142	184	
	3200	Light or relay		8	1		39.50	42		81.50	108	
	3400	Transformer		4	2		173	84		257	320	
	3600	House telephone, talking station		1.60	5		370	210		580	735	
	3800	Press to talk, release to listen		5.30	1.509		86.50	63.50		150	193	
	4000	System-on button					51.50			51.50	57	
	4200	Door release	1 Elec	4	2		92.50	84		176.50	231	
	4400	Combination speaker and microphone		8	1		157	42		199	238	
	4600	Termination box		3.20	2.500		49.50	105		154.50	217	
	4800	Amplifier or power supply		5.30	1.509		570	63.50		633.50	725	
	5000	Vestibule door unit		16	.500	Name	105	21		126	148	
	5200	Strip cabinet		27	.296	Ea.	197	12.45		209.45	236	
	5400	Directory		16	.500		93	21		114	135	
	6000	Master door, button buzzer type, 100 unit	2 Elec	.54	29.630		945	1,250		2,195	2,975	
	6020	200 unit		.30	53.333		1,775	2,250		4,025	5,400	
	6040	300 unit		.20	80		2,725	3,350		6,075	8,175	
	6060	Transformer	1 Elec	8	1		24	42		66	91	
	6080	Door opener		5.30	1.509		34.50	63.50		98	136	
	6100	Buzzer with door release and plate		4	2		34.50	84		118.50	167	
	6200	Intercom type, 100 unit	2 Elec	.54	29.630		1,175	1,250		2,425	3,225	
	6220	200 unit		.30	53.333		2,300	2,250		4,550	5,975	
	6240	300 unit		.20	80		3,475	3,350		6,825	8,975	
	6260	Amplifier	1 Elec	2	4		173	168		341	450	
	6280	Speaker with door release		4	2		51.50	84		135.50	186	
	9000	Minimum labor/equipment charge		3.50	2.286	Job		96		96	148	

	16850 | Television Equipment		Crew	Daily Output	Labor-Hours	Unit	Mat.	Labor	Equip.	Total	Total Incl O&P	
600	0010	**T.V. SYSTEMS** not including rough-in wires, cables & conduits										600
	0100	Master TV antenna system										
	0200	VHF reception & distribution, 12 outlets	1 Elec	6	1.333	Outlet	171	56		227	275	
	0400	30 outlets		10	.800		112	33.50		145.50	175	
	0600	100 outlets		13	.615		114	26		140	166	
	0800	VHF & UHF reception & distribution, 12 outlets		6	1.333		169	56		225	273	
	1000	30 outlets		10	.800		112	33.50		145.50	175	
	1200	100 outlets		13	.615		114	26		140	166	
	1400	School and deluxe systems, 12 outlets		2.40	3.333		223	140		363	460	
	1600	30 outlets		4	2		196	84		280	345	
	1800	80 outlets		5.30	1.509		188	63.50		251.50	305	
	1900	Amplifier		4	2	Ea.	560	84		644	745	
	1910	Antenna		2	4	"	270	168		438	555	
	2000	Closed circuit, surveillance, one station (camera & monitor)	2 Elec	2.60	6.154	Total	1,075	258		1,333	1,600	
	2200	For additional camera stations, add	1 Elec	2.70	2.963	Ea.	605	124		729	855	
	2400	Industrial quality, one station (camera & monitor)	2 Elec	2.60	6.154	Total	2,250	258		2,508	2,875	
	2600	For additional camera stations, add	1 Elec	2.70	2.963	Ea.	1,375	124		1,499	1,700	
	2610	For low light, add		2.70	2.963		1,100	124		1,224	1,425	
	2620	For very low light, add		2.70	2.963		8,125	124		8,249	9,150	
	2800	For weatherproof camera station, add		1.30	6.154		845	258		1,103	1,325	
	3000	For pan and tilt, add		1.30	6.154		2,200	258		2,458	2,800	
	3200	For zoom lens - remote control, add, minimum		2	4		2,025	168		2,193	2,475	
	3400	Maximum		2	4		7,375	168		7,543	8,375	
	3410	For automatic iris for low light, add		2	4		1,775	168		1,943	2,200	
	3600	Educational T.V. studio, basic 3 camera system, black & white,										
	3800	electrical & electronic equip. only, minimum	4 Elec	.80	40	Total	10,500	1,675		12,175	14,100	
	4000	Maximum (full console)		.28	114		44,700	4,800		49,500	56,500	
	4100	As above, but color system, minimum		.28	114		59,000	4,800		63,800	72,500	

16 ELECTRICAL

		16850 Television Equipment	CREW	DAILY OUTPUT	LABOR-HOURS	UNIT	2006 BARE COSTS MAT.	LABOR	EQUIP.	TOTAL	TOTAL INCL O&P	
600	4120	Maximum	4 Elec	.12	266	Total	257,000	11,200		268,200	300,000	600
	4200	For film chain, black & white, add	1 Elec	1	8	Ea.	12,000	335		12,335	13,700	
	4250	Color, add		.25	32		14,600	1,350		15,950	18,100	
	4400	For video tape recorders, add, minimum	↓	1	8		2,525	335		2,860	3,300	
	4600	Maximum	4 Elec	.40	80	↓	21,000	3,350		24,350	28,300	

For information about Means Estimating Seminars, see yellow pages 12 and 13 in back of book

Division Notes

		CREW	DAILY OUTPUT	LABOR-HOURS	UNIT	2006 BARE COSTS MAT.	LABOR	EQUIP.	TOTAL	TOTAL INCL O&P

Assemblies Section

Table of Contents

How to Use the Assemblies Cost Tables

The following is a detailed explanation of a sample Assemblies Cost Table. Most Assembly Tables are separated into three parts: 1) an illustration of the system to be estimated; 2) the components and related costs of a typical system; and 3) the costs for similar systems with dimensional and/or size variations. For costs of the components that comprise these systems, or assemblies, refer to the Unit Price Section. Next to each bold number below is the described item with the appropriate component of the sample entry following in parentheses. In most cases, if the work is to be subcontracted, the general contractor will need to add an additional markup (RSMeans suggests using 10%) to the "Total" figures.

1 System/Line Numbers (C1010 124 1250)

Each Assemblies Cost Line has been assigned a unique identification number based on the UNIFORMAT II classification system.

UNIFORMAT II Major Group

C1010 124 1250

UNIFORMAT II Level 3

Means Major Classification

Means Individual Line Number

C10 Interior Construction

C1010 Partitions

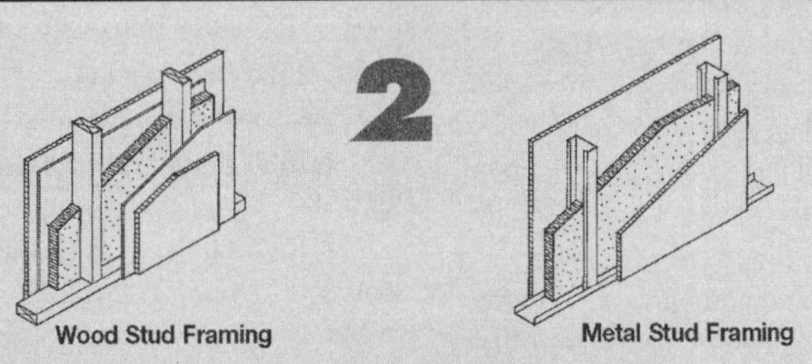

The Drywall Partitions/Stud Framing Systems are defined by type of drywall and number of layers, type and spacing of stud framing, and treatment on the opposite face. Components include taping and finishing.

Cost differences between regular and fire resistant drywall are negligible, and terminology is interchangeable. In some cases fiberglass insulation is included for additional sound deadening.

System Components	QUANTITY	UNIT	COST PER S.F. MAT.	INST.	TOTAL
SYSTEM C1010 124 1250					
DRYWALL PARTITION,5/8" F.R.1 SIDE,5/8" REG.1 SIDE,2"X4"STUDS,16" O.C.					
Gypsum plasterboard, nailed/screwed to studs, 5/8"F.R. fire resistant	1.000	S.F.	.33	.47	.80
Gypsum plasterboard, nailed/screwed to studs, 5/8" regular	1.000	S.F.	.33	.47	.80
Taping and finishing joints	2.000	S.F.	.08	.94	1.02
Framing, 2 x 4 studs @ 16" O.C., 10' high	1.000	S.F.	.43	.94	1.37
TOTAL			1.17	2.82	3.99

C1010 124	Drywall Partitions/Wood Stud Framing							
	FACE LAYER	BASE LAYER	FRAMING	OPPOSITE FACE	INSULATION	COST PER S.F. MAT.	INST.	TOTAL
1202	5/8" FR drywall	none	2 x 4, @ 16" O.C.	same	[illegible]	1.21	2.82	4.03
1250				5/8" reg. drywall	[illegible]	1.17	2.82	3.99
1302				nothing	[illegible]	.84	1.88	2.72
1402		1/4" SD gypsum	2 x 4 @ 16" O.C.	same	1-1/2" fiberglass	2.35	4.33	6.68
1452				5/8" FR drywall	1-1/2" fiberglass	2.05	[illegible]	[illegible]
1502				nothing	1-1/2" fiberglass	1.68	[illegible]	[illegible]
1602		resil. channels	2 x 4 @ 16", O.C.	same	1-1/2" fiberglass	2.06	5.50	7.56
1652				5/8" FR drywall	1-1/2" fiberglass	1.90	4.40	6.30
1702				nothing	1-1/2" fiberglass	1.53	3.46	4.99

2

Illustration

At the top of most assembly pages are an illustration, a brief description, and the design criteria used to develop the cost.

3

System Components

The components of a typical system are listed separately to show what has been included in the development of the total system price. The table below contains prices for other similar systems with dimensional and/or size variations.

4

Quantity

This is the number of line item units required for one system unit. For example, we assume that it will take 1 S.F. of 5/8″ fire-resistant gypsum board per square foot of wall area.

5

Unit of Measure for Each Item

The abbreviated designation indicates the unit of measure, as defined by industry standards, upon which the price of the component is based. For example, gypsum drywall is priced by S.F. (square foot).

6

Unit of Measure for Each System (Cost per S.F.)

Costs shown in the three right-hand columns have been adjusted by the component quantity and unit of measure for the entire system. In this example, "Cost per S.F." is the unit of measure for this system, or assembly.

7

Materials (1.17)

This column contains the Materials Cost of each component. These cost figures are bare costs plus 10% for profit.

8

Installation (2.82)

Installation includes labor and equipment plus the installing contractor's overhead and profit. Equipment costs are the bare rental costs plus 10% for profit. The labor overhead and profit are defined on the inside back cover of this book.

9

Total (3.99)

The figure in this column is the sum of the material and installation costs.

Material Cost	+	Installation Cost	=	Total
$1.17	+	$2.82	=	$3.99

A SUBSTRUCTURE

A1010 Standard Foundations

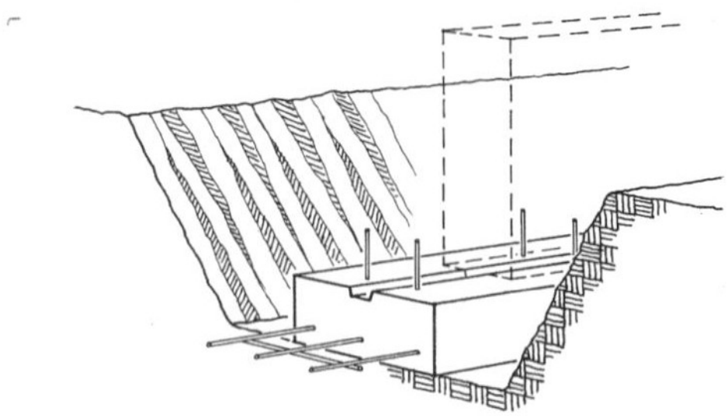

The Strip Footing System includes: excavation; hand trim; all forms needed for footing placement; forms for 2″ x 6″ keyway (four uses); dowels; and 3,000 p.s.i. concrete.

The footing size required varies for different soils. Soil bearing capacities are listed for 3 KSF and 6 KSF. Depths of the system range from 8″ to 24″. Widths range from 16″ to 96″. Smaller strip footings may not require reinforcement.

Please see the reference section for further design and cost information.

System Components	QUANTITY	UNIT	COST PER L.F. MAT.	INST.	TOTAL
SYSTEM A1010 110 2500					
STRIP FOOTING, LOAD 5.1KLF, SOIL CAP. 3 KSF, 24″WIDE X12″DEEP, REINF.					
Trench excavation	.148	C.Y.		1.06	1.06
Hand trim	2.000	S.F.		1.54	1.54
Compacted backfill	.074	C.Y.		.23	.23
Formwork, 4 uses	2.000	S.F.	1.88	7.32	9.20
Keyway form, 4 uses	1.000	L.F.	.33	.94	1.27
Reinforcing, fy = 60000 psi	3.000	Lb.	1.47	1.56	3.03
Dowels	2.000	Ea.	1.38	4.52	5.90
Concrete, f′c = 3000 psi	.074	C.Y.	7.07		7.07
Place concrete, direct chute	.074	C.Y.		1.43	1.43
Screed finish	2.000	S.F.		1.20	1.20
TOTAL			12.13	19.80	31.93

A1010 110	Strip Footings	COST PER L.F. MAT.	INST.	TOTAL
2100	Strip footing, load 2.6KLF, soil capacity 3KSF, 16″wide x 8″deep plain	5.90	11.90	17.80
2300	Load 3.9 KLF, soil capacity, 3 KSF, 24″wide x 8″deep, plain	7.55	13.20	20.75
2500	Load 5.1KLF, soil capacity 3 KSF, 24″wide x 12″deep, reinf.	12.15	19.80	31.95
2700	Load 11.1KLF, soil capacity 6 KSF, 24″wide x 12″deep, reinf.	12.15	19.80	31.95
2900	Load 6.8 KLF, soil capacity 3 KSF, 32″wide x 12″deep, reinf.	15.05	22	37.05
3100	Load 14.8 KLF, soil capacity 6 KSF, 32″wide x 12″deep, reinf.	15.05	22	37.05
3300	Load 9.3 KLF, soil capacity 3 KSF, 40″wide x 12″deep, reinf.	17.90	23.50	41.40
3500	Load 18.4 KLF, soil capacity 6 KSF, 40″wide x 12″deep, reinf.	18.10	24	42.10
3700	Load 10.1KLF, soil capacity 3 KSF, 48″wide x 12″deep, reinf.	20	25.50	45.50
3900	Load 22.1KLF, soil capacity 6 KSF, 48″wide x 12″deep, reinf.	21.50	27	48.50
4100	Load 11.8KLF, soil capacity 3 KSF, 56″wide x 12″deep, reinf.	24	28	52
4300	Load 25.8KLF, soil capacity 6 KSF, 56″wide x 12″deep, reinf.	26	30.50	56.50
4500	Load 10KLF, soil capacity 3 KSF, 48″wide x 16″deep, reinf.	25.50	29.50	55
4700	Load 22KLF, soil capacity 6 KSF, 48″wide, 16″deep, reinf.	26	30.50	56.50
4900	Load 11.6KLF, soil capacity 3 KSF, 56″wide x 16″deep, reinf.	29	41.50	70.50
5100	Load 25.6KLF, soil capacity 6 KSF, 56″wide x 16″deep, reinf.	31	43	74
5300	Load 13.3KLF, soil capacity 3 KSF, 64″wide x 16″deep, reinf.	34	35	69
5500	Load 29.3KLF, soil capacity 6 KSF, 64″wide x 16″deep, reinf.	36	37.50	73.50
5700	Load 15KLF, soil capacity 3 KSF, 72″wide x 20″deep, reinf.	45	42	87
5900	Load 33KLF, soil capacity 6 KSF, 72″wide x 20″deep, reinf.	48	45	93
6100	Load 18.3KLF, soil capacity 3 KSF, 88″wide x 24″deep, reinf.	64	53.50	117.50
6300	Load 40.3KLF, soil capacity 6 KSF, 88″wide x 24″deep, reinf.	69.50	59	128.50
6500	Load 20KLF, soil capacity 3 KSF, 96″wide x 24″deep, reinf.	70	56.50	126.50
6700	Load 44 KLF, soil capacity 6 KSF, 96″ wide x 24″ deep, reinf.	74	61	135

A1010 Standard Foundations

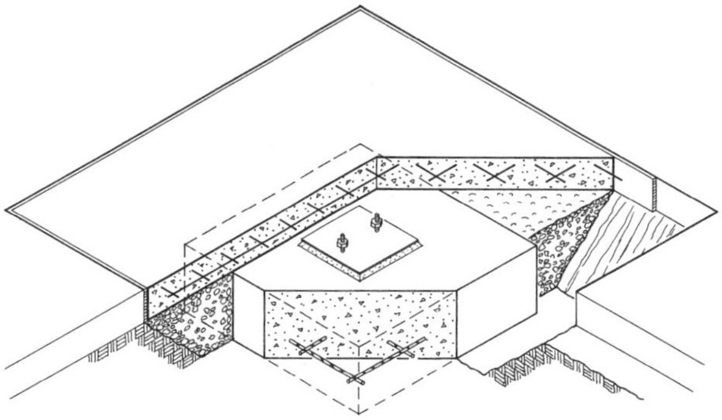

Column footing includes: Remove existing slab; excavation; forms for concrete; reinforcing steel; expansion joint; anchor bolts; grouting column base plate; and concrete placed and finished.

The footing size required varies for different soils.

Design assumptions:

Conc. slab f'c = 3.5 ksi
Conc. foundation f'c = 3.0 ksi
WWF 6 x 6 #10/#10
Reinforcement, fy = 60 ksi

System Components	QUANTITY	UNIT	COST EACH		
			MAT.	INST.	TOTAL
SYSTEM A1010 260 0200					
CONC. COLUMN FOOTING IN EXISTING BUILDING, LOAD 10K,					
SOIL CAPACITY 3KSF, 3′ SQ X 12″ DEEP, 3,000 PSI					
Saw cut slab	16.000	S.F.	40.32	110.40	150.72
Break up slab, 6″	16.000	S.F.		183.68	183.68
Remove concrete; hand	.755	C.Y.		23.03	23.03
Excavate; hand	.592	C.Y.		53.87	53.87
Forms, 1 use	12.000	SFCA	33.60	205.80	239.40
Reinforcing steel, #4 bar	15.000	Lb.	7.35	7.80	15.15
Concrete, 3,000 psi, for footing, incl. premium delv. chg.	.333	C.Y.	44.50		44.50
Place conc. footing	.333	C.Y.		19.74	19.74
Backfill by hand	.311	C.Y.		8.09	8.09
Compaction; hand tamp	.311	C.Y.		5.49	5.49
Welded wire fabric	.160	C.S.F.	2.11	4.96	7.07
Premolded, bituminous fiber, 1/2″ x 6″	16.000	L.F.	6.56	20.16	26.72
Concrete, 3500 psi, for slab, incl. premium delv. chg.	.166	C.Y.	22.74		22.74
Place conc. slab, 6″	.166	C.Y.		3.27	3.27
Finish slab; steel trowel	16.000	S.F.		12.64	12.64
Anchor bolts, 3/4″ x 12″	2.000	Ea.	4.22	13.50	17.72
Grout level plate	1.000	S.F.	4.75	12.40	17.15
TOTAL			166.15	684.83	850.98
COST per C.Y. of FOOTING			498.95	2,056.55	2,555.50

A1020 Special Foundations

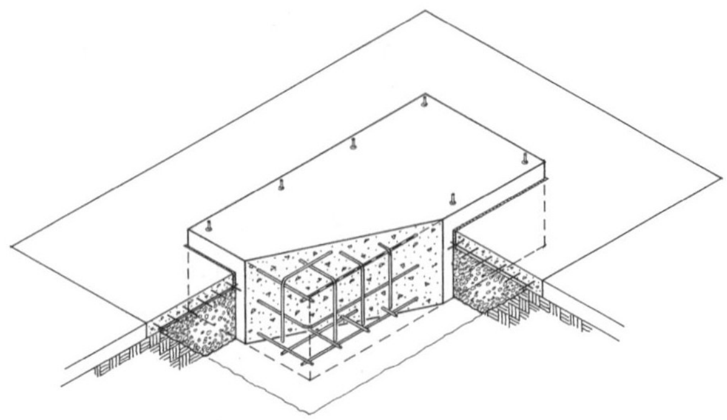

Equipment foundation includes: Remove existing slab; excavation; forms for concrete; reinforcing steel; expansion joint; anchor bolts; and concrete placed and screeded.

The foundation size required varies for different soils.

Design assumptions:

Conc. slab f'c = 3.5 ksi
Conc. foundation f'c = 3.0 ksi
WWF 6 x 6 #10/#10
Reinforcement, fy = 60 ksi

System Components	QUANTITY	UNIT	COST EACH MAT.	COST EACH INST.	COST EACH TOTAL
SYSTEM A1020 810 0170					
CONC. EQUIPMENT FOUNDATION, 4'X8'X30" DEEP, 3,000 PSI.					
Saw cut slab	45.000	S.F.	70.56	193.20	263.76
Break up slab; 6"	45.000	S.F.		516.60	516.60
Remove concrete; hand	3.133	C.Y.		95.56	95.56
Excavate; hand	2.500	C.Y.		227.50	227.50
Forms, 1 use	60.000	SFCA	168	1,029	1,197
Reinforcing steel, #6 bar	318.000	Lb.	155.82	165.36	321.18
Concrete, 3000 psi, for foundation, incl. premium delv. chg.	3.000	C.Y.	396.13		396.13
Anchor bolts, 1" diam., with sleeves	8.000	Ea.	38.72	83.60	122.32
Place concrete foundation	3.000	C.Y.		80.22	80.22
Backfill by hand	.866	C.Y.		22.52	22.52
Compaction in 6" layers, hand tamp	.866	C.Y.		15.28	15.28
Premolded, bituminous fiber, 1/2"x6"	24.000	L.F.	9.84	30.24	40.08
Welded wire fabric; 6x6-#10/10	.130	C.S.F.	1.72	4.03	5.75
Concrete, 3500 psi, for slab, incl. premium delv. chg.	.241	C.Y.	33.03		33.03
Place conc. slab	.241	C.Y.		4.75	4.75
Finish slab; steel trowel	13.000	S.F.		10.27	10.27
Finish equip. foundation; screed	32.000	S.F.		15.36	15.36
TOTAL			873.82	2,493.49	3,367.31
COST per C.Y. of FOUNDATION			291.27	831.16	1,122.43

For information about Means Estimating Seminars, see yellow pages 12 and 13 in back of book

B SHELL

B1010 Floor Construction

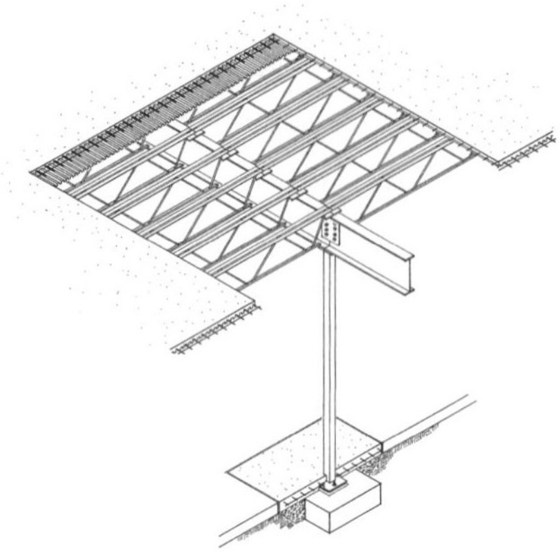

Mezzanine addition to existing building includes: Column footings; steel columns; structural steel; open web steel joists; uncoated 28 ga. steel slab forms; 2-1/2″ concrete slab reinforced with welded wire fabric; steel trowel finish.

Design assumptions:

Structural steel is A36, high strength bolted. Slab form is 28 gauge, galvanized.
WWF 6 x 6 #10/#10
Conc. slab f'c = 3 ksi

System Components	QUANTITY	UNIT	COST EACH		
			MAT.	INST.	TOTAL
SYSTEM B1010 310 0142					
MEZZANINE ADDITION TO EXISTING BUILDING; 100 PSF SUPERIMPOSED LOAD, SLAB FORM DECK, MTL. JOISTS, 2.5″ CONC. SLAB, 3,000 PSI, 3,000 S.F.					
Saw cut existing slab, 6″ thick	192.000	L.F.	483.84	1,324.80	1,808.64
Cut out existing slab, 4′ x 4′ openings	192.000	S.F.		2,204.16	2,204.16
Remove concrete debris	9.060	C.Y.		276.33	276.33
Excavate by hand	7.104	C.Y.		646.46	646.46
Backfill by hand	3.732	C.Y.		97.03	97.03
Compaction of backfill	3.732	C.Y.		65.87	65.87
Forms for footing, 3′ x 3′ x 1′	144.000	SFCA	403.20	2,469.60	2,872.80
Reinforcing bar for footing, 3#5 E.W. x 30″ long	180.000	Lb.	88.20	93.60	181.80
Anchor bolts, 12″, 2 per column	24.000	Ea.	50.64	162	212.64
Concrete, 3000 psi, footing	5.592	C.Y.	534.04		534.04
Place concrete in footing, direct chute	5.592	C.Y.		236.88	236.88
Premolded, bituminous fiber, 1/2″ x 6″	192.000	L.F.	78.72	241.92	320.64
Reinforcing mesh for slab cutouts	1.920	C.S.F.	25.34	59.52	84.86
Concrete, 3500 psi, slab cutouts	2.784	C.Y.	272.83		272.83
Place concrete slab cutouts, direct chute	2.784	C.Y.		39.23	39.23
Machine trowel slab cutouts	192.000	S.F.		151.68	151.68
Grout for leveling plates	12.000	S.F.	57	148.80	205.80
Column, 4″ x 4″ x 1/4″ x12′	12.000	Ea.	2,064	1,134	3,198
Structural steel W 21x50	80.000	L.F.	6,900	836.40	7,736.40
W 16x31	80.000	L.F.	4,260	729.60	4,989.60
Open web joists, H or K series	6.800	Ton	14,790	6,395.40	21,185.40
Slab form, steel 28 gauge, galvanized	30.000	C.S.F.	3,240	1,890	5,130
Concrete 3,000 psi, 2 1/2″ slab, incl. premium delv. chg.	23.500	C.Y.	3,366.38		3,366.38
Welded wire fabric 6x6 #10/#10 (w1.4/w1.4)	30.000	C.S.F.	396	930	1,326
Place concrete	23.500	C.Y.		981.71	981.71
Monolithic steel trowel finish	30.000	C.S.F.		2,370	2,370
Curing with sprayed membrane curing compound	30.000	C.S.F.	183	229.50	412.50
TOTAL			37,193.19	23,714.49	60,907.68
COST PER S.F.			12.40	7.90	20.30

B1010 Floor Construction

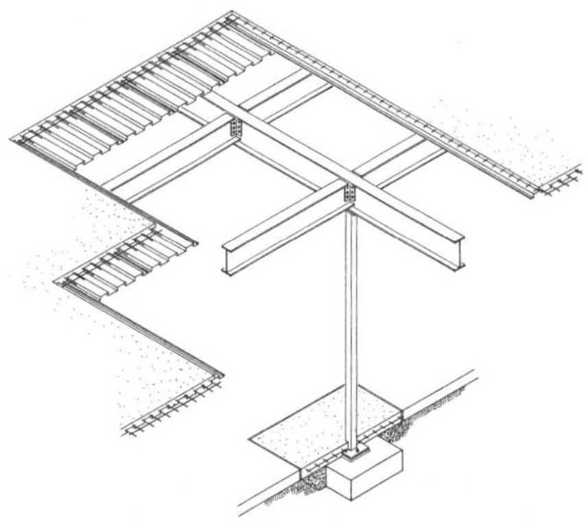

Mezzanine addition to existing building includes: Column footings; steel columns; structural steel; galvanized composite steel deck; 4″ concrete slab reinforced with welded wire fabric; steel trowel finish.

Design assumptions:

Structural steel is A36, high strength bolted.
Composite deck is 22 gauge, galvanized.
WWF 6 x 6 #10/#10
Conc. slab f'c = 3 ksi

System Components	QUANTITY	UNIT	COST EACH MAT.	COST EACH INST.	COST EACH TOTAL
SYSTEM B1010 320 0152					
MEZZANINE ADDITION TO EXISTING BUILDING; 100 PSF SUPERIMPOSED LOAD					
COMPOSITE DECK, BEAMS, 4″ CONC. SLAB, 3,000 PSI, 3,000 S.F.					
Saw cut existing slab, 6″ thick	192.000	L.F.	483.84	1,324.80	1,808.64
Remove existing slab, 4′ x 4′ openings	192.000	S.F.		2,204.16	2,204.16
Remove concrete debris	9.060	C.Y.		276.33	276.33
Excavate by hand	7.104	C.Y.		646.46	646.46
Backfill by hand, no compaction, light soil	3.732	C.Y.		97.03	97.03
Compaction of backfill	3.732	C.Y.		65.87	65.87
Forms in place, footing, 3′ x 3′ x 1′	144.000	SFCA	403.20	2,469.60	2,872.80
Footings, #4 to #7 bars, 3#5 E.W. x 30″ long	180.000	Lb.	88.20	93.60	181.80
Anchor bolts, 12″ long, 2 per column	24.000	Ea.	50.64	162	212.64
Concrete, 3000 psi, footing	5.592	C.Y.	534.04		534.04
Place concrete in footings, direct chute	5.592	C.Y.		236.88	236.88
Premolded, bituminous fiber, 1/2″ x 6″	192.000	L.F.	78.72	241.92	320.64
Structural steel W 12x19	225.000	L.F.	5,737.50	1,399.50	7,137
Reinforcing mesh for slab cutouts	1.920	C.S.F.	25.34	59.52	84.86
Concrete, 3500 psi, for slab cutouts	2.784	C.Y.	272.83		272.83
W 14x30	375.000	L.F.	19,406.25	3,420	22,826.25
Place concrete, slab cutouts, direct chute	2.784	C.Y.		39.23	39.23
W 16x31	80.000	L.F.	4,260	729.60	4,989.60
Machine trowel slab cutouts	192.000	S.F.		151.68	151.68
Grout for leveling plates	12.000	S.F.	57	148.80	205.80
Column, 4″ x 4″ x 1/4″ x 12′	12.000	Ea.	2,064	1,134	3,198
W 18x50	80.000	L.F.	6,900	975.60	7,875.60
Metal deck, galvanized, 2″ deep, 22 gauge.	30.000	C.S.F.	4,620	1,980	6,600
Concrete, 3,000 p.s.i., 4″ slab, incl. premium delv. chg.	37.000	C.Y.	5,300.25		5,300.25
Welded wire fabric 6x6 #10/#10 (w1.4/w1.4)	30.000	C.S.F.	396	930	1,326
Place concrete	37.000	C.Y.		1,545.68	1,545.68
Monolithic steel trowel finish	30.000	C.S.F.		2,370	2,370
Curing with sprayed membrane curing compound	30.000	C.S.F.	183	229.50	412.50
TOTAL			50,860.81	22,931.76	73,792.57
COST PER S.F.			16.95	7.64	24.59

B1010 Floor Construction

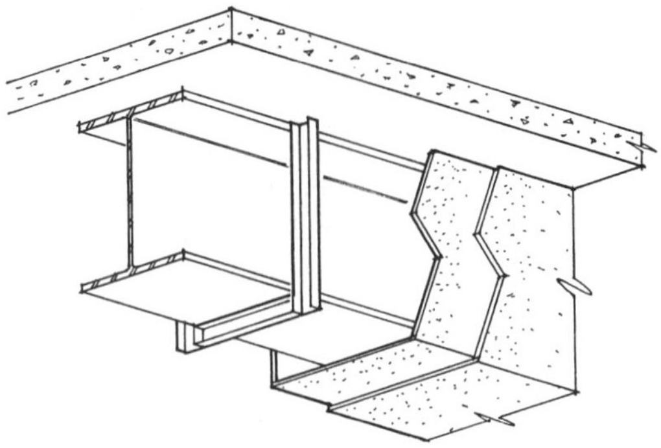

The table below lists fireproofing costs for steel beams by type, beam size, thickness and fire rating. Weights listed are for the fireproofing material only.

System Components	QUANTITY	UNIT	COST PER L.F. MAT.	COST PER L.F. INST.	COST PER L.F. TOTAL
SYSTEM B1010 710 1300					
FIREPROOFING, 5/8″ F.R. GYP. BOARD, 12″X 4″ BEAM, 2″ THICK, 2 HR. F.R.					
Corner bead for drywall, 1-1/4″ x 1-1/4″, galvanized	.020	C.L.F.	.45	2.70	3.15
L bead for drywall, galvanized	.020	C.L.F.	.35	3.14	3.49
Furring, beams & columns, 3/4″ galv. channels, 24″ O.C.	2.330	S.F.	.35	5.24	5.59
Drywall on beam, no finish, 2 layers at 5/8″ thick	3.000	S.F.	2.22	9.42	11.64
Drywall, taping and finishing joints, add	3.000	S.F.	.12	1.41	1.53
TOTAL			3.49	21.91	25.40

B1010 710 Steel Beam Fireproofing

	ENCASEMENT SYSTEM	BEAM SIZE (IN.)	THICKNESS (IN.)	FIRE RATING (HRS.)	WEIGHT (P.L.F.)	COST PER L.F. MAT.	COST PER L.F. INST.	COST PER L.F. TOTAL
0400	Concrete	12x4	1	1	77	5.50	23	28.50
0450	3000 PSI		1-1/2	2	93	6.40	25	31.40
0500			2	3	121	7.15	27	34.15
0550		14x5	1	1	100	7.15	28.50	35.65
0600			1-1/2	2	122	8.20	31.50	39.70
0650			2	3	142	9.20	34.50	43.70
0700		16x7	1	1	147	8.40	30	38.40
0750			1-1/2	2	169	9.10	31	40.10
0800			2	3	195	10.15	34	44.15
0850		18x7-1/2	1	1	172	9.70	34.50	44.20
0900			1-1/2	2	196	10.85	38	48.85
0950			2	3	225	12.05	42	54.05
1000		24x9	1	1	264	13.10	42	55.10
1050			1-1/2	2	295	14.20	44.50	58.70
1100			2	3	328	15.30	46.50	61.80
1150		30x10-1/2	1	1	366	17.10	53	70.10
1200			1-1/2	2	404	18.50	56.50	75
1250			2	3	449	19.50	59.50	79
1300	5/8″ fire rated	12x4	2	2	15	3.49	22	25.49
1350	Gypsum board		2-5/8	3	24	4.86	27	31.86
1400		14x5	2	2	17	3.74	23	26.74
1450			2-5/8	3	27	5.35	29	34.35
1500		16x7	2	2	20	4.19	25.50	29.69
1550			2-5/8	3	31	5.50	25	30.50
1600		18x7-1/2	2	2	22	4.49	27.50	31.99
1650			2-5/8	3	34	6.40	34.50	40.90

Important: See the Reference Section for critical supporting data - Reference Numbers and City Cost Indexes

B1010 Floor Construction

B1010 710 Steel Beam Fireproofing

	ENCASEMENT SYSTEM	BEAM SIZE (IN.)	THICKNESS (IN.)	FIRE RATING (HRS.)	WEIGHT (P.L.F.)	COST PER L.F.		
						MAT.	INST.	TOTAL
1700	5/8" fire rated	24x9	2	2	27	5.50	33.50	39
1750	Gypsum board		2-5/8	3	42	7.85	42.50	50.35
1800		30x10-1/2	2	2	33	6.50	39.50	46
1850			2-5/8	3	51	9.20	49.50	58.70
1900	Gypsum	12x4	1-1/8	3	18	4.92	16.50	21.42
1950	Plaster on	14x5	1-1/8	3	21	5.60	18.55	24.15
2000	Metal lath	16x7	1-1/8	3	25	6.45	21.50	27.95
2050		18x7-1/2	1-1/8	3	32	7.30	24	31.30
2100		24x9	1-1/8	3	35	9	28.50	37.50
2150		30x10-1/2	1-1/8	3	44	10.85	34.50	45.35
2200	Perlite plaster	12x4	1-1/8	2	16	4.47	18.35	22.82
2250	On metal lath		1-1/4	3	20	4.74	19.20	23.94
2300			1-1/2	4	22	5.05	20	25.05
2350		14x5	1-1/8	2	18	5.70	22.50	28.20
2400			1-1/4	3	23	6	23.50	29.50
2450			1-1/2	4	25	6.60	25.50	32.10
2500		16x7	1-1/8	2	21	5.45	22.50	27.95
2550			1-1/4	3	26	5.75	23.50	29.25
2600			1-1/2	4	29	6.05	24.50	30.55
2650		18x7-1/2	1-1/8	2	26	6.60	26.50	33.10
2700			1-1/4	3	33	6.95	27.50	34.45
2750			1-1/2	4	36	7.50	29	36.50
2800		24x9	1-1/8	2	30	7.05	30	37.05
2850			1-1/4	3	38	7.35	30.50	37.85
2900			1-1/2	4	41	7.65	31.50	39.15
2950		30x10-1/2	1-1/8	2	37	9.05	37	46.05
3000			1-1/4	3	46	9.35	38	47.35
3050			1-1/2	4	51	9.90	39.50	49.40
3100	Sprayed Fiber	12x4	5/8	1	12	1.02	1.85	2.87
3150	(non asbestos)		1-1/8	2	21	1.83	3.32	5.15
3200			1-1/4	3	22	2.04	3.69	5.73
3250		14x5	5/8	1	14	1.02	1.85	2.87
3300			1-1/8	2	26	1.83	3.32	5.15
3350			1-1/4	3	29	2.04	3.69	5.73
3400		16x7	5/8	1	18	1.29	2.34	3.63
3450			1-1/8	2	32	2.32	4.19	6.51
3500			1-1/4	3	36	2.58	4.66	7.24
3550		18x7-1/2	5/8	1	20	1.42	2.57	3.99
3600			1-1/8	2	35	2.56	4.64	7.20
3650			1-1/4	3	39	2.83	5.10	7.93
3700		24x9	5/8	1	25	1.80	3.25	5.05
3750			1-1/8	2	45	3.26	5.85	9.11
3800			1-1/4	3	50	3.62	6.55	10.17
3850		30x10-1/2	5/8	1	31	2.18	3.94	6.12
3900			1-1/8	2	55	3.92	7.10	11.02
3950			1-1/4	3	61	4.36	7.90	12.26
4000	On Decking	Flat	1		6	.47	.53	1
4050	Per S.F.	Corrugated	1		7	.70	1.01	1.71
4100		Fluted	1		7	.70	1.01	1.71

B1010 Floor Construction

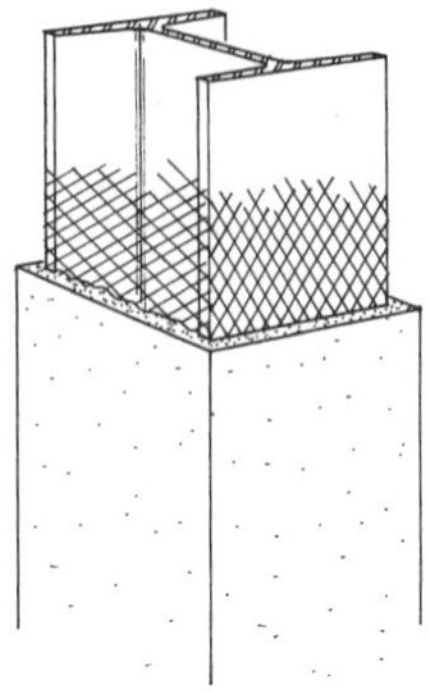

Listed below are costs per V.L.F. for fireproofing by material, column size, thickness and fire rating. Weights listed are for the fireproofing material only.

System Components	QUANTITY	UNIT	COST PER V.L.F. MAT.	INST.	TOTAL
SYSTEM B1010 720 3000					
CONCRETE FIREPROOFING, 8″ STEEL COLUMN, 1″ THICK, 1 HR. FIRE RATING					
Forms in place, columns, plywood, 4 uses	3.330	SFCA	2.70	23.64	26.34
Welded wire fabric, 2 x 2 #14 galv. 21 lb./C.S.F., column wrap	2.700	S.F.	.90	4.51	5.41
Concrete ready mix, regular weight, 3000 psi	.621	C.F.	2.20		2.20
Place and vibrate concrete, 12″ sq./round columns, pumped	.621	C.F.		1.49	1.49
TOTAL			5.80	29.64	35.44

B1010 720 Steel Column Fireproofing

	ENCASEMENT SYSTEM	COLUMN SIZE (IN.)	THICKNESS (IN.)	FIRE RATING (HRS.)	WEIGHT (P.L.F.)	COST PER V.L.F. MAT.	INST.	TOTAL
3000	Concrete	8	1	1	110	5.80	30	35.80
3050			1-1/2	2	133	6.65	33	39.65
3100			2	3	145	7.50	35.50	43
3150		10	1	1	145	7.55	36.50	44.05
3200			1-1/2	2	168	8.40	39.50	47.90
3250			2	3	196	9.30	42	51.30
3600	Gypsum board	8	1	3	14	4.42	23.50	27.92
3650	1/2″ fire rated	10	1	3	17	4.77	25	29.77
3700	2 layers	14	1	3	22	4.96	25.50	30.46
3750	Gypsum board	8	1-1/2	3	23	6.05	29.50	35.55
3800	1/2″ fire rated	10	1-1/2	3	27	6.85	32.50	39.35
3850	3 layers	14	1-1/2	3	35	7.60	35.50	43.10
3900	Sprayed fiber	8	1-1/2	2	6.3	3.19	5.75	8.94
3950	Direct application		2	3	8.3	4.39	7.95	12.34
4000			2-1/2	4	10.4	5.70	10.30	16
4050		10	1-1/2	2	7.9	3.85	6.95	10.80
4100			2	3	10.5	5.25	9.55	14.80
4150			2-1/2	4	13.1	6.80	12.25	19.05
4500	Perlite plaster	8	1	2	18	7	26	33
4550	On metal lath		1-3/8	3	23	7.65	28.50	36.15
4600			1-3/4	4	35	9	34.50	43.50
4650	Perlite plaster	10	1	2	21	8	30	38
4700			1-3/8	3	27	9	34.50	43.50
4750			1-3/4	4	41	10.15	39	49.15

Important: See the Reference Section for critical supporting data - Reference Numbers and City Cost Indexes

B1010 Floor Construction

B1010 720 Steel Column Fireproofing

	ENCASEMENT SYSTEM	COLUMN SIZE (IN.)	THICKNESS (IN.)	FIRE RATING (HRS.)	WEIGHT (P.L.F.)	COST PER V.L.F.		
						MAT.	INST.	TOTAL
5100	5/8" gypsum plaster	8	1	1-1/2	20	6.60	21	27.60
5150	On 3/8" gypsum lath	10	1	1-1/2	24	7.85	25.50	33.35
5200		14	1	1-1/2	33	9.65	30.50	40.15
5250	1" perlite plaster	8	1-3/8	2	23	6.15	22.50	28.65
5300	On 3/8" gypsum lath	10	1-3/8	2	28	7	26	33
5350		14	1-3/8	2	37	8.70	32.50	41.20
5400	1-3/8" perlite plaster	8	1-3/4	3	27	7.85	25	32.85
5450	On 3/8" gypsum lath	10	1-3/4	3	33	9	29	38
5500		14	1-3/4	3	43	11.35	36.50	47.85
5550	Concrete masonry	8	4-3/4	4	126	8.35	27.50	35.85
5600	Units 4" thick	10	4-3/4	4	166	10.50	34.50	45
5650	75% solid	14	4-3/4	4	262	12.60	41.50	54.10

B2010 Exterior Walls

This page illustrates masonry cleaning and restoration systems, including staging, cleaning and repointing.

System Components	QUANTITY	UNIT	COST PER S.F. MAT.	COST PER S.F. INST.	COST PER S.F. TOTAL
SYSTEM B2010 143 1300					
Repoint existing building, brick, running bond, high pressure cleaning, Water only, soft old mortar.					
Scaffold, building exterior	.010	C.S.F.	.28	.59	.87
Cleaning, high pressure water only	1.000	S.F.		1.47	1.47
Repoint, brick running bond	1.000	S.F.	.49	4.74	5.23
TOTAL			.77	6.80	7.57

B2010 143	Masonry Restoration - Cleaning	COST PER S.F. MAT.	COST PER S.F. INST.	COST PER S.F. TOTAL
1000	Repoint existing building, high pressure cleaning, water only,			
1100	soft old mortar			
1200	For alternate masonry surfaces:			
1300	Brick, common bond	.75	7	7.75
1400	Flemish bond	.78	7.30	8.08
1500	English bond	.78	7.85	8.63
1600	Add for wire cut face brick	.16		.16
1700	Stone, 2′ x 2′ blocks	.91	5.05	5.96
1800	2′ x 4′ blocks	.75	4.29	5.04
2000	Add to above prices for alternate cleaning systems:			
2100	Chemical brush and wash	.03	.52	.55
2200	High pressure chemical and water	.04	.50	.54
2300	Sandblasting, wet system	.20	1.36	1.56
2400	Dry system	.19	.80	.99
2500	Steam cleaning		.40	.40

B3010 Roof Coverings

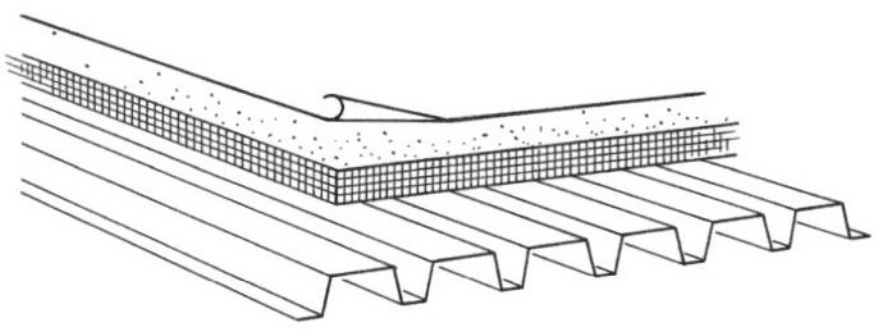

Fully Adhered

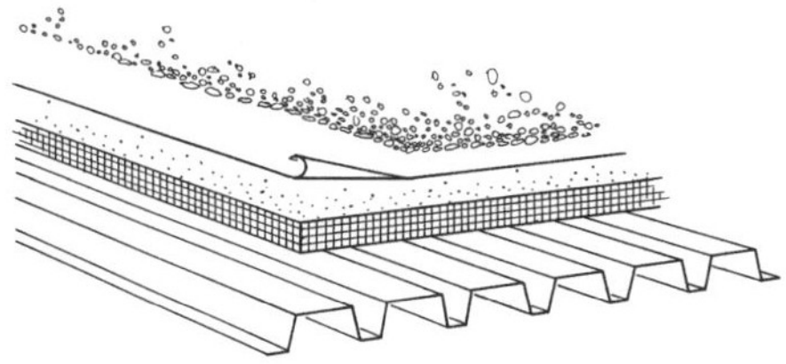

Ballasted

The systems listed below reflect only the cost for the single ply membrane.

B3010 120	Single Ply Membrane	COST PER S.F.		
		MAT.	INST.	TOTAL
1000	CSPE (Chlorosulfonated polyethylene), 35 mils, fully adhered	1.36	.83	2.19
1100	Loosely laid with stone ballast	1.43	.42	1.85
2000	EPDM (Ethylene propylene diene monomer), 45 mils, fully adhered	.85	.83	1.68
2100	Loosely laid with stone ballast	.67	.42	1.09
2200	Mechanically fastened with batten strips	.59	.62	1.21
3300	60 mils, fully adhered	1	.83	1.83
3400	Loosely laid with stone ballast	.83	.42	1.25
3500	Mechanically fastened with batten strips	.74	.62	1.36
4000	Modified bit., SBS modified, granule surface cap sheet, mopped, 150 mils	.54	1.63	2.17
4100	Smooth surface cap sheet, mopped, 145 mils	.50	1.55	2.05
4500	APP modified, granule surface cap sheet, torched, 180 mils	.61	1.08	1.69
4600	Smooth surface cap sheet, torched, 170 mils	.55	1.03	1.58
5000	PIB (Polyisobutylene), 100 mils, fully adhered with contact cement	1.82	.83	2.65
5100	Loosely laid with stone ballast	1.41	.42	1.83
5200	Partially adhered with adhesive	1.76	.62	2.38
5300	Hot asphalt attachment	1.68	.62	2.30
6000	Reinforced PVC, 48 mils, loose laid and ballasted with stone	1.07	.42	1.49
6100	Partially adhered with mechanical fasteners	.98	.62	1.60
6200	Fully adhered with adhesive	1.30	.83	2.13
6300	Reinforced PVC, 60 mils, loose laid and ballasted with stone	.99	.42	1.41
6400	Partially adhered with mechanical fasteners	.89	.62	1.51
6500	Fully adhered with adhesive	1.22	.83	2.05

C INTERIORS

C1010 Partitions

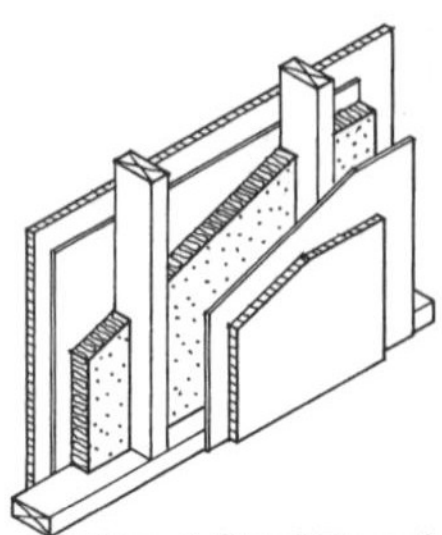

Wood Stud Framing

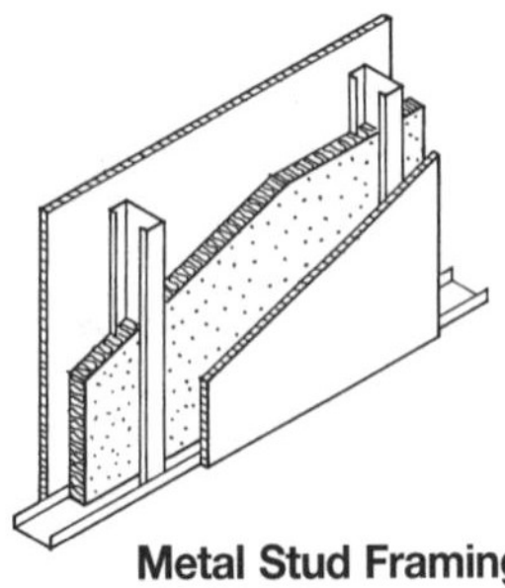

Metal Stud Framing

The Drywall Partitions/Stud Framing Systems are defined by type of drywall and number of layers, type and spacing of stud framing, and treatment on the opposite face. Components include taping and finishing.

Cost differences between regular and fire resistant drywall are negligible, and terminology is interchangeable. In some cases fiberglass insulation is included for additional sound deadening.

System Components	QUANTITY	UNIT	COST PER S.F. MAT.	COST PER S.F. INST.	COST PER S.F. TOTAL
SYSTEM C1010 124 1250					
DRYWALL PARTITION,5/8″ F.R.1 SIDE,5/8″ REG.1 SIDE,2″X4″STUDS,16″ O.C.					
Gypsum plasterboard, nailed/screwed to studs, 5/8″F.R. fire resistant	1.000	S.F.	.33	.47	.80
Gypsum plasterboard, nailed/screwed to studs, 5/8″ regular	1.000	S.F.	.33	.47	.80
Taping and finishing joints	2.000	S.F.	.08	.94	1.02
Framing, 2 x 4 studs @ 16″ O.C., 10′ high	1.000	S.F.	.43	.94	1.37
TOTAL			1.17	2.82	3.99

C1010 124 Drywall Partitions/Wood Stud Framing

	FACE LAYER	BASE LAYER	FRAMING	OPPOSITE FACE	INSULATION	COST PER S.F. MAT.	COST PER S.F. INST.	COST PER S.F. TOTAL
1202	5/8″ FR drywall	none	2 x 4, @ 16″ O.C.	same	0	1.21	2.82	4.03
1250				5/8″ reg. drywall	0	1.17	2.82	3.99
1302				nothing	0	.84	1.88	2.72
1402		1/4″ SD gypsum	2 x 4 @ 16″ O.C.	same	1-1/2″ fiberglass	2.35	4.33	6.68
1452				5/8″ FR drywall	1-1/2″ fiberglass	2.05	3.81	5.86
1502				nothing	1-1/2″ fiberglass	1.68	2.87	4.55
1602		resil. channels	2 x 4 @ 16″, O.C.	same	1-1/2″ fiberglass	2.06	5.50	7.56
1652				5/8″ FR drywall	1-1/2″ fiberglass	1.90	4.40	6.30
1702				nothing	1-1/2″ fiberglass	1.53	3.46	4.99
1802		5/8″ FR drywall	2 x 4 @ 24″ O.C.	same	0	1.76	3.58	5.34
1852				5/8″ FR drywall	0	1.43	3.11	4.54
1902				nothing	0	1.06	2.17	3.23
1952		5/8″ FR drywall	2 x 4, 16″ O.C.	same	0	1.87	3.76	5.63
1957				5/8″ FR drywall	0	1.54	3.29	4.83
2002				nothing	0	1.17	2.35	3.52
2012		5/8″ FR drywall	staggered, 6″ plate	same	0	2.36	4.72	7.08
2017				5/8″ FR drywall	0	2.03	4.25	6.28
2022				nothing	0	1.66	3.31	4.97
2202		5/8″ FR drywall	2 rows-2 x 4 16″O.C.	same	2″ fiberglass	3	5.15	8.15
2252				5/8″ FR drywall	2″ fiberglass	2.67	4.70	7.37
2302				nothing	2″ fiberglass	2.30	3.76	6.06
2402	5/8″ WR drywall	none	2 x 4, @ 16″ O.C.	same	0	1.15	2.82	3.97
2452				5/8″ FR drywall	0	1.18	2.82	4
2502				nothing	0	.81	1.88	2.69
2602		5/8″ FR drywall	2 x 4, @ 24″ O.C.	same	0	1.70	3.58	5.28
2652				5/8″ FR drywall	0	1.40	3.11	4.51
2702				nothing	0	1.03	2.17	3.20

Important: See the Reference Section for critical supporting data - Reference Numbers and City Cost Indexes

C1010 Partitions

C1010 128	Drywall Components	COST PER S.F. MAT.	INST.	TOTAL
0140	Metal studs, 24" O.C. including track, load bearing, 18 gage, 2-1/2"	.58	.88	1.46
0160	3-5/8"	.69	.90	1.59
0180	4"	.62	.92	1.54
0200	6"	.92	.94	1.86
0220	16 gage, 2-1/2"	.67	1.01	1.68
0240	3-5/8"	.80	1.03	1.83
0260	4"	.85	1.05	1.90
0280	6"	1.07	1.07	2.14
0300	Non load bearing, 25 gage, 1-5/8"	.18	.62	.80
0340	3-5/8"	.25	.64	.89
0360	4"	.27	.64	.91
0380	6"	.39	.65	1.04
0400	20 gage, 2-1/2"	.35	.79	1.14
0420	3-5/8"	.42	.80	1.22
0440	4"	.46	.80	1.26
0460	6"	.59	.81	1.40
0542	Wood studs including blocking, shoe and double top plate, 2"x4", 12"O.C.	.59	1.18	1.77
0562	16" O.C.	.47	.94	1.41
0582	24" O.C.	.36	.76	1.12
0602	2"x6", 12" O.C.	.93	1.35	2.28
0622	16" O.C.	.75	1.05	1.80
0641	24" O.C.	.58	.82	1.40
0642	Furring one side only, steel channels, 3/4", 12" O.C.	.25	1.77	2.02
0644	16" O.C.	.23	1.57	1.80
0646	24" O.C.	.15	1.19	1.34
0647	1-1/2" , 12" O.C.	.35	1.98	2.33
0648	16" O.C.	.32	1.73	2.05
0649	24" O.C.	.21	1.37	1.58
0656	Wood strips, 1" x 3", on wood, 12" O.C.	.31	.86	1.17
0657	16" O.C.	.23	.65	.88
0658	On masonry, 12" O.C.	.31	.95	1.26
0659	16" O.C.	.23	.71	.94
0660	On concrete, 12" O.C.	.31	1.81	2.12
0661	16" O.C.	.23	1.36	1.59
0662	Gypsum board, one face only, exterior sheathing, 1/2"	.68	.90	1.58
0680	Interior, fire resistant, 1/2"	.33	.47	.80
0700	5/8"	.33	.47	.80
0720	Sound deadening board 1/4"	.30	.52	.82
0740	Standard drywall 3/8"	.26	.47	.73
0760	1/2"	.30	.47	.77
0780	5/8"	.33	.47	.80
0800	Tongue & groove coreboard 1"	.53	1.96	2.49
0820	Water resistant, 1/2"	.28	.47	.75
0840	5/8"	.30	.47	.77
0860	Add for the following:, foil backing	.11		.11
0880	Fiberglass insulation, 3-1/2"	.52	.29	.81
0900	6"	.56	.35	.91
0920	Rigid insulation 1"	.47	.47	.94
0940	Resilient furring @ 16" O.C.	.20	1.48	1.68
0960	Taping and finishing	.04	.47	.51
0980	Texture spray	.06	.52	.58
1000	Thin coat plaster	.08	.58	.66
1042	2"x4" staggered studs 2"x6" plates & blocking	.05	.11	.16

C1020 Interior Doors

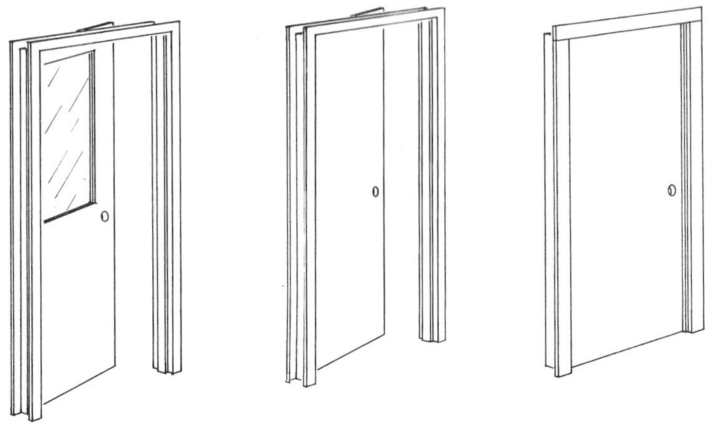

Steel Door, Half Glass Steel Frame

Steel Door, Flush Steel Frame

Wood Door, Flush Wood Frame

The Metal Door/Metal Frame Systems are defined as follows: door type, design and size, frame type and depth. Included in the components for each system is painting the door and frame. No hardware has been included in the systems.

System Components	QUANTITY	UNIT	COST EACH MAT.	COST EACH INST.	COST EACH TOTAL
SYSTEM C1020 114 1200					
STEEL DOOR, HOLLOW, 20 GA., HALF GLASS, 2′-8″X6′-8″, D.W.FRAME, 4-7/8″ DP					
Steel door, flush, hollow core, 1-3/4″ thick,half glass, 20 ga.,2′-8″x6′-8″	1.000	Ea.	420	47	467
Steel frame, KD, 16 ga., drywall, 4-7/8″ deep, 2′-8″x6′-8″, single	1.000	Ea.	114	63	177
Float glass, 3/16″ thick, clear, tempered	5.000	S.F.	30	34	64
Paint door and frame each side, primer	1.000	Ea.	4.84	81	85.84
Paint door and frame each side, 2 coats	1.000	Ea.	7.56	136	143.56
TOTAL			576.40	361	937.40

C1020 114 Metal Door/Metal Frame

	TYPE	DESIGN	SIZE	FRAME	DEPTH	COST EACH MAT.	COST EACH INST.	COST EACH TOTAL
1000	Flush-hollow	20 ga. full panel	2′-8″ x 6′-8″	drywall K.D.	4-7/8″	430	335	765
1020				butt welded	8-3/4″	535	375	910
1160			6′-0″ x 7′-0″	drywall K.D.	4-7/8″	850	535	1,385
1180				butt welded	8-3/4″	970	580	1,550
1200		20 ga. half glass	2′-8″ x 6′-8″	drywall K.D.	4-7/8″	575	360	935
1220				butt welded	8-3/4″	680	405	1,085
1360			6′-0″ x 7′-0″	drywall K.D.	4-7/8″	1,100	600	1,700
1380				butt welded	8-3/4″	1,200	640	1,840
1800		18 ga. full panel	2′-8″ x 6′-8″	drywall K.D.	4-7/8″	350	335	685
1820				butt welded	8-3/4″	455	380	835
1960			6′-0″ x 7′-0″	drywall K.D.	4-7/8″	920	535	1,455
1980				butt welded	8-3/4″	1,050	580	1,630
2000		18 ga. half glass	2′-8″ x 6′-8″	drywall K.D.	4-7/8″	625	365	990
2020				butt welded	8-3/4″	730	410	1,140
2160			6′-0″ x 7′-0″	drywall K.D.	4-7/8″	1,175	605	1,780
2180				butt welded	8-3/4″	1,300	650	1,950

Important: See the Reference Section for critical supporting data - Reference Numbers and City Cost Indexes

C1020 Interior Doors

C1020 310	Hardware	COST EACH		
		MAT.	INST.	TOTAL
0060	**HINGES**			
0080				
0100	Full mortise, low frequency, steel base, 4-1/2" x 4-1/2", USP	5.20		5.20
0120	5" x 5" USP	14.25		14.25
0140	6" x 6" USP	28.50		28.50
0160	Average frequency, steel base, 4-1/2" x 4-1/2" USP	11.75		11.75
0180	5" x 5" USP	19.50		19.50
0200	6" x 6" USP	41.50		41.50
0220	High frequency, steel base, 4-1/2" x 4-1/2", USP	28.50		28.50
0240	5" x 5" USP	26.50		26.50
0260	6" x 6" USP	64.50		64.50
0280				
0300	**LOCKSETS**			
0320				
0340	Heavy duty cylindrical, passage door			
0360	Non-keyed, passage	47	39.50	86.50
0380	Privacy	58.50	39.50	98
0400	Keyed, single cylinder function	82	47	129
0420	Hotel	117	59	176
0440				
0460	For re-core cylinder, add	34		34
0480				
0500	**CLOSERS**			
0520				
0540	Rack & pinion			
0560	Adjustable backcheck, 3 way mount, all sizes, regular arm	151	78.50	229.50
0580	Hold open arm	151	78.50	229.50
0600	Fusible link	134	72.50	206.50
0620	Non-sized, regular arm	150	78.50	228.50
0640	4 way mount, non-sized, regular arm	206	78.50	284.50
0660	Hold open arm	221	78.50	299.50
0680				
0700	**PUSH, PULL**			
0720				
0740	Push plate, aluminum	7.55	39.50	47.05
0760	Bronze	19.10	39.50	58.60
0780	Pull handle, push bar, aluminum	133	43	176
0800	Bronze	172	47	219
0810				
0820	**PANIC DEVICES**			
0840				
0860	Narrow stile, rim mounted, bar, exit only	635	78.50	713.50
0880	Outside key & pull	690	94	784
0900	Bar and vertical rod, exit only	655	94	749
0920	Outside key & pull	655	118	773
0940	Bar and concealed rod, exit only	765	157	922
0960	Mortise, bar, exit only	555	118	673
0980	Touch bar, exit only	810	118	928
1000				
1020	**WEATHERSTRIPPING**			
1040				
1060	Interlocking, 3' x 7', zinc	15.15	157	172.15
1080	Spring type, 3' x 7', bronze	33	157	190

C3010 Wall Finishes

C3010 230	Paint & Covering	COST PER S.F.		
		MAT.	INST.	TOTAL
0060	Painting, interior on plaster and drywall, brushwork, primer & 1 coat	.11	.63	.74
0080	Primer & 2 coats	.16	.83	.99
0100	Primer & 3 coats	.21	1.02	1.23
0120	Walls & ceilings, roller work, primer & 1 coat	.11	.42	.53
0140	Primer & 2 coats	.16	.54	.70
0160	Woodwork incl. puttying, brushwork, primer & 1 coat	.11	.91	1.02
0180	Primer & 2 coats	.16	1.20	1.36
0200	Primer & 3 coats	.21	1.63	1.84
0260	Cabinets and casework, enamel, primer & 1 coat	.10	1.02	1.12
0280	Primer & 2 coats	.15	1.25	1.40
0300	Masonry or concrete, latex, brushwork, primer & 1 coat	.19	.85	1.04
0320	Primer & 2 coats	.25	1.22	1.47
0340	Addition for block filler	.12	1.04	1.16
0380	Fireproof paints, intumescent, 1/8" thick 3/4 hour	1.87	.83	2.70
0420	7/16" thick 2 hour	5.40	2.91	8.31
0440	1-1/16" thick 3 hour	8.80	5.80	14.60
0500	Gratings, primer & 1 coat	.20	1.27	1.47
0600	Pipes over 12" diameter	.51	4.07	4.58
0720	Heavy framing 50-100 S.F./Ton	.06	.80	.86
0740	Spraywork, light framing 300-500 S.F./Ton	.07	.35	.42
0800	Varnish, interior wood trim, no sanding sealer & 1 coat	.08	1.02	1.10
0820	Hardwood floor, no sanding 2 coats	.14	.22	.36
0860	Maximum	.59	1.34	1.93
0880	Epoxy coatings, minimum	.36	.78	1.14
1080	Maximum	.96	2.81	3.77
1100	High build epoxy 50 mil, minimum	.62	1.04	1.66
1120	Maximum	1.06	4.29	5.35
1140	Laminated epoxy with fiberglass minimum	.67	1.38	2.05
1160	Maximum	1.19	2.81	4
1180	Sprayed perlite or vermiculite 1/16" thick, minimum	.24	.14	.38
1200	Maximum	.68	.64	1.32
1260	Wall coatings, vinyl plastic, minimum	.31	.55	.86
1280	Maximum	.75	1.70	2.45
1300	Urethane on smooth surface, 2 coats, minimum	.23	.36	.59
1320	Maximum	.51	.61	1.12
1340	3 coats, minimum	.31	.48	.79
1360	Maximum	.68	.87	1.55
1380	Ceramic-like glazed coating, cementitious, minimum	.44	.93	1.37
1420	Resin base, minimum	.31	.64	.95
1440	Maximum	.50	1.23	1.73
1460	Wall coverings, aluminum foil	.96	1.49	2.45
1500	Vinyl backing	5.15	1.70	6.85

C30 Interior Finishes

C3020 Floor Finishes

C3020 410	Tile & Covering	COST PER S.F.		
		MAT.	INST.	TOTAL
0060	Carpet tile, nylon, fusion bonded, 18" x 18" or 24" x 24", 24 oz.	3.72	.59	4.31
0080	35 oz.	3.94	.59	4.53
0100	42 oz.	5.15	.59	5.74
0140	Carpet, tufted, nylon, roll goods, 12' wide, 26 oz.	4.05	.68	4.73
0160	36 oz.	6.45	.68	7.13
0180	Woven, wool, 36 oz.	10.20	.68	10.88
0200	42 oz.	10.40	.68	11.08
0220	Padding, add to above, minimum	.48	.32	.80
0240	Maximum	.83	.32	1.15
0260	Composition flooring, acrylic, 1/4" thick	1.43	4.47	5.90
0280	3/8" thick	1.88	5.15	7.03
0300	Epoxy, minimum	2.53	3.44	5.97
0320	Maximum	3.07	4.75	7.82
0340	Epoxy terrazzo, minimum	6	5.15	11.15
0360	Maximum	9.20	6.90	16.10
0380	Mastic, hot laid, 1-1/2" thick, minimum	3.71	3.37	7.08
0400	Maximum	4.76	4.47	9.23
0420	Neoprene 1/4" thick, minimum	3.65	4.27	7.92
0440	Maximum	5.05	5.40	10.45
0460	Polyacrylate with ground granite 1/4", minimum	3.34	3.16	6.50
0480	Maximum	6.45	4.84	11.29
0500	Polyester with colored quart 2 chips 1/16", minimum	3.25	2.18	5.43
0520	Maximum	5.05	3.44	8.49
0540	Polyurethane with vinyl chips, minimum	7.10	2.18	9.28
0560	Maximum	10.25	2.70	12.95
0620	1" thick	.43	3.74	4.17
0640	2" thick	.85	4.30	5.15
0680	Maximum	.29	6.70	6.99
0700	For colors, add to above, minimum	.66	1.31	1.97
0740	Exposed aggregate finish, minimum	.18	.69	.87
0760	Maximum	.26	.93	1.19
0800	Maximum	.49	.51	1
0820	Dust on coloring, add, minimum	.66	.33	.99
0860	1/2" integral, minimum	5.25	3.65	8.90
0880	Maximum	5.70	3.65	9.35
1040	1/2" thick	.19	2.26	2.45
1060	3/4" thick	.29	2.53	2.82
1080	1" thick	.38	2.87	3.25
1100	Terrazzo, minimum	2.81	7.95	10.76
1120	Maximum	5.25	17.15	22.40
1280	Resilient, asphalt tile, 1/8" thick on concrete, minimum	1.17	1.07	2.24
1300	Maximum	1.28	1.07	2.35
1320	On wood, add for felt underlay	.21		.21

C3030 Ceiling Finishes

Plaster Partitions are defined as follows: type of plaster, type and spacing of framing, type of lath and treatment of opposite face.

Included in the system components are expansion joints. Metal studs are assumed to be nonload bearing.

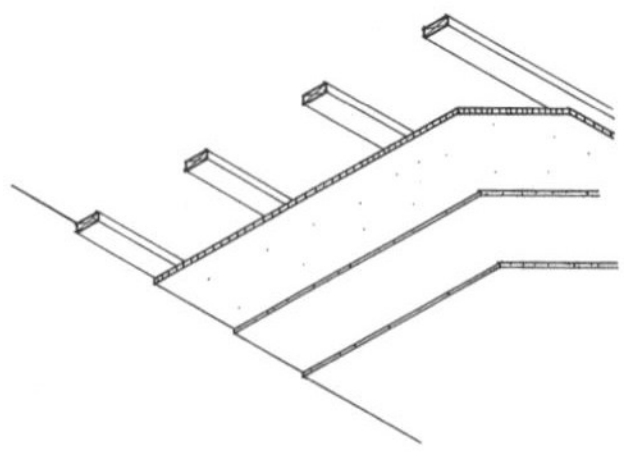

2 Coats of Plaster on Gypsum Lath on Wood Furring

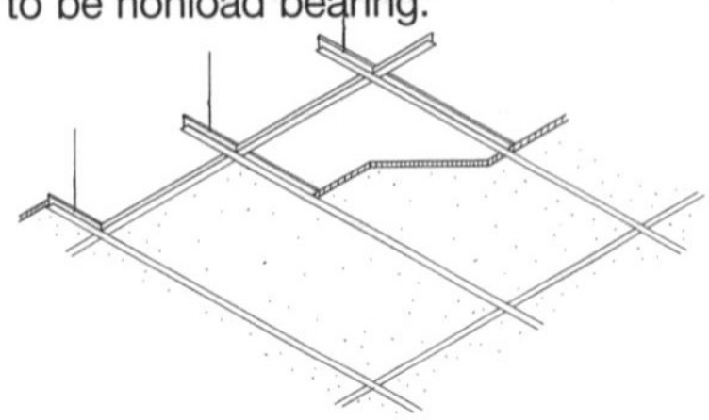

Fiberglass Board on Exposed Suspended Grid System

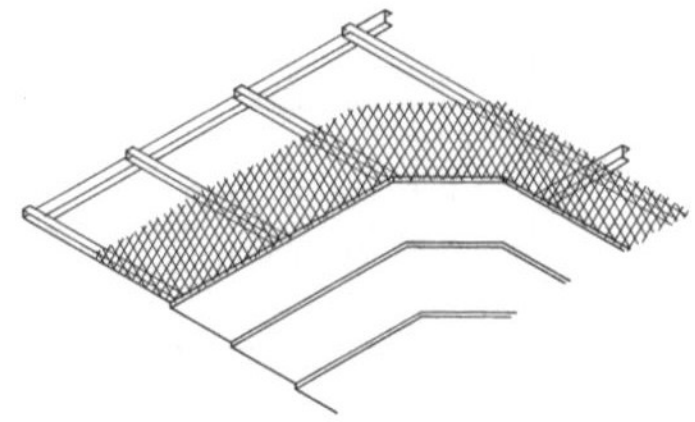

Plaster and Metal Lath on Metal Furring

System Components	QUANTITY	UNIT	COST PER S.F. MAT.	COST PER S.F. INST.	COST PER S.F. TOTAL
SYSTEM C3030 105 2400					
GYPSUM PLASTER, 2 COATS, 3/8"GYP. LATH, WOOD FURRING, FRAMING					
Gypsum plaster 2 coats no lath, on ceilings	.110	S.Y.	.40	2.53	2.93
Gypsum lath plain/perforated nailed, 3/8" thick	.110	S.Y.	.66	.54	1.20
Add for ceiling installation	.110	S.Y.		.21	.21
Furring, 1" x 3" wood strips on wood joists	.750	L.F.	.23	1.01	1.24
Paint, primer and one coat	1.000	S.F.	.11	.42	.53
TOTAL			1.40	4.71	6.11

C3030 105 Plaster Ceilings

	TYPE	LATH	FURRING	SUPPORT		COST PER S.F. MAT.	COST PER S.F. INST.	COST PER S.F. TOTAL
2400	2 coat gypsum	3/8" gypsum	1"x3" wood, 16" O.C.	wood		1.40	4.71	6.11
2500	Painted			masonry		1.40	4.80	6.20
2600				concrete		1.40	5.40	6.80
2700	3 coat gypsum	3.4# metal	1"x3" wood, 16" O.C.	wood		1.31	5.10	6.41
2800	Painted			masonry		1.31	5.15	6.46
2900				concrete		1.31	5.75	7.06
3000	2 coat perlite	3/8" gypsum	1"x3" wood, 16" O.C.	wood		1.40	4.91	6.31
3100	Painted			masonry		1.40	5	6.40
3200				concrete		1.40	5.55	6.95
3300	3 coat perlite	3.4# metal	1"x3" wood, 16" O.C.	wood		1.14	5.05	6.19
3400	Painted			masonry		1.14	5.10	6.24
3500				concrete		1.14	5.70	6.84
3600	2 coat gypsum	3/8" gypsum	3/4" CRC, 12" O.C.	1-1/2" CRC, 48"O.C.		1.42	5.70	7.12
3700	Painted		3/4" CRC, 16" O.C.	1-1/2" CRC, 48"O.C.		1.40	5.15	6.55
3800			3/4" CRC, 24" O.C.	1-1/2" CRC, 48"O.C.		1.32	4.69	6.01
3900	2 coat perlite	3/8" gypsum	3/4" CRC, 12" O.C.	1-1/2" CRC, 48"O.C		1.42	6.10	7.52
4000	Painted		3/4" CRC, 16" O.C.	1-1/2" CRC, 48"O.C.		1.40	5.55	6.95
4100			3/4" CRC, 24" O.C.	1-1/2" CRC, 48"O.C.		1.32	5.10	6.42
4200	3 coat gypsum	3.4# metal	3/4" CRC, 12" O.C.	1-1/2" CRC, 36" O.C.		1.54	7.45	8.99
4300	Painted		3/4" CRC, 16" O.C.	1-1/2" CRC, 36" O.C.		1.52	6.90	8.42
4400			3/4" CRC, 24" O.C.	1-1/2" CRC, 36" O.C.		1.44	6.45	7.89
4500	3 coat perlite	3.4# metal	3/4" CRC, 12" O.C.	1-1/2" CRC,36" O.C.		1.69	8.15	9.84
4600	Painted		3/4" CRC, 16" O.C.	1-1/2" CRC, 36" O.C.		1.67	7.60	9.27
4700			3/4" CRC, 24" O.C.	1-1/2" CRC, 36" O.C.		1.59	7.15	8.74

Important: See the Reference Section for critical supporting data - Reference Numbers and City Cost Indexes

D SERVICES

D1010 Elevators and Lifts

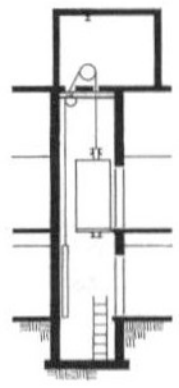

Geared traction elevators are the intermediate group both in cost and in operating areas. These are in buildings of four to fifteen floors and speed ranges from 200′ to 500′ per minute.

System Components	QUANTITY	UNIT	COST EACH MAT.	COST EACH INST.	COST EACH TOTAL
SYSTEM D1010 140 1600					
PASSENGER 2500 LB., 5 FLOORS, 200 FPM					
Passenger elevator, geared, 2000 lb capacity, 4 stop, 200 FPM	1.000	Ea.	75,500	24,600	100,100
Over 40′ travel height, passenger elevator electric, add	10.000	V.L.F.	1,180	1,700	2,880
Passenger elevator, electric, 2500 lb. cap., add	1.000	Ea.	3,075		3,075
Over 4 stops, passenger elevator, electric, add	1.000	Stop	2,625	4,575	7,200
Hall lantern	5.000	Ea.	1,900	770	2,670
Maintenance agreement for passenger elevator, 9 months	1.000	Ea.	3,425		3,425
Position indicator at lobby	1.000	Ea.	72.50	38.50	111
TOTAL			87,777.50	31,683.50	119,461

D1010 140	Traction Geared Elevators	COST EACH MAT.	COST EACH INST.	COST EACH TOTAL
1300	Passenger, 2000 Lb., 5 floors, 200 FPM	84,500	31,700	116,200
1500	15 floors, 350 FPM	133,000	96,000	229,000
1600	2500 Lb., 5 floors, 200 FPM	87,777.50	31,683.50	119,366
1800	15 floors, 350 FPM	136,000	96,000	232,000
1900	3000 Lb., 5 floors, 200 FPM	89,500	31,700	121,200
2100	15 floors, 350 FPM	137,500	96,000	233,500
2200	3500 Lb., 5 floors, 200 FPM	91,000	31,700	122,700
2400	15 floors, 350 FPM	139,500	96,000	235,500
2500	4000 Lb., 5 floors, 200 FPM	91,500	31,700	123,200
2700	15 floors, 350 FPM	140,000	96,000	236,000
2800	4500 Lb., 5 floors, 200 FPM	94,000	31,700	125,700
3000	15 floors, 350 FPM	142,000	96,000	238,000
3100	5000 Lb., 5 floors, 200 FPM	96,500	31,700	128,200
3300	15 floors, 350 FPM	144,500	96,000	240,500
4000	Hospital, 3500 Lb., 5 floors, 200 FPM	104,000	31,700	135,700
4200	15 floors, 350 FPM	179,500	96,000	275,500
4300	4000 Lb., 5 floors, 200 FPM	104,000	31,700	135,700
4500	15 floors, 350 FPM	179,500	96,000	275,500
4600	4500 Lb., 5 floors, 200 FPM	110,000	31,700	141,700
4800	15 floors, 350 FPM	185,000	96,000	281,000
4900	5000 Lb., 5 floors, 200 FPM	111,500	31,700	143,200
5100	15 floors, 350 FPM	187,000	96,000	283,000
6000	Freight, 4000 Lb., 5 floors, 50 FPM class 'B'	93,000	32,800	125,800
6200	15 floors, 200 FPM class 'B'	171,000	114,500	285,500
6300	8000 Lb., 5 floors, 50 FPM class 'B'	111,500	32,800	144,300
6500	15 floors, 200 FPM class 'B'	189,500	114,500	304,000
7000	10,000 Lb., 5 floors, 50 FPM class 'B'	129,000	32,800	161,800
7200	15 floors, 200 FPM class 'B'	347,000	114,500	461,500
8000	20,000 Lb., 5 floors, 50 FPM class 'B'	143,000	32,800	175,800
8200	15 floors, 200 FPM class 'B'	361,000	114,500	475,500

Important: See the Reference Section for critical supporting data - Reference Numbers and City Cost Indexes

D1020 Escalators and Moving Walks

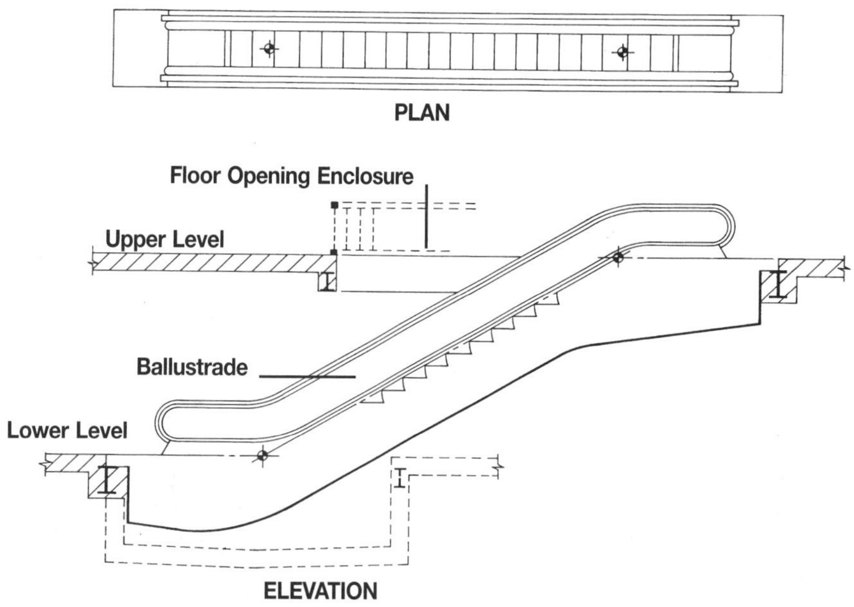

Moving stairs can be used for buildings where 600 or more people are to be carried to the second floor or beyond. Freight cannot be carried on escalators and at least one elevator must be available for this function.

Carrying capacity is 5000 to 8000 people per hour. Power requirement is 2 KW to 3 KW per hour and incline angle is 30°.

D1020 110 Moving Stairs

	TYPE	HEIGHT	WIDTH	BALUSTRADE MATERIAL		COST EACH		
						MAT.	INST.	TOTAL
0100	Escalator	10ft	32″	glass		70,500	34,300	104,800
0150				metal		72,000	34,300	106,300
0200			48″	glass		77,500	34,300	111,800
0250				metal		77,500	34,300	111,800
0300		15ft	32″	glass		74,000	40,000	114,000
0350				metal		74,000	40,000	114,000
0400			48″	glass		78,000	40,000	118,000
0450				metal		78,000	40,000	118,000
0500		20ft	32″	glass		76,000	49,000	125,000
0550				metal		76,000	49,000	125,000
0600			48″	glass		81,500	49,000	130,500
0650				metal		81,500	49,000	130,500
0700		25ft	32″	glass		82,000	60,000	142,000
0750				metal		82,000	60,000	142,000
0800			48″	glass		88,000	60,000	148,000
0850				metal		88,000	60,000	148,000

D1020 210 Moving Walks

	TYPE	STORY HEIGHT	DEGREE SLOPE	WIDTH	COST RANGE	COST PER L.F.		
						MAT.	INST.	TOTAL
1500	Ramp	10′-23′	12	3′-4″	minimum	2,125	670	2,795
1600					maximum	2,675	825	3,500
2000	Walk	0′	0	2′-0″	minimum	790	370	1,160
2500					maximum	1,100	545	1,645
3000				3′-4″	minimum	1,775	545	2,320
3500					maximum	2,075	630	2,705

D SERVICES

D2010 Plumbing Fixtures

Systems are complete with trim seat and rough-in (supply, waste and vent) for connection to supply branches and waste mains.

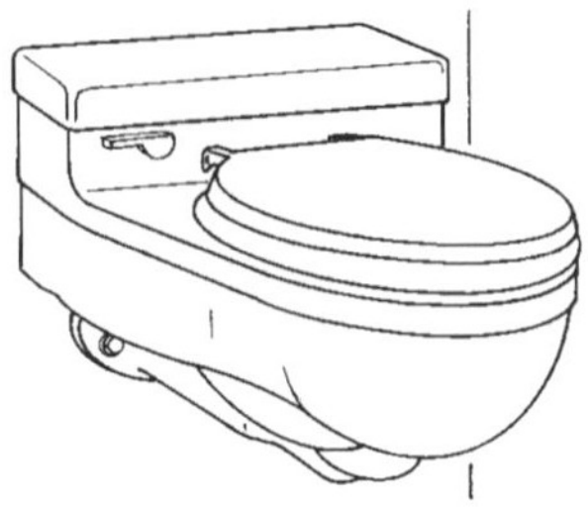

One Piece Wall Hung

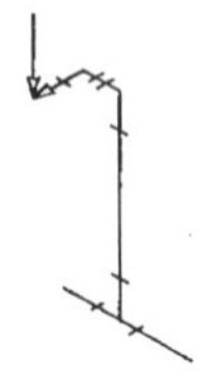

Supply

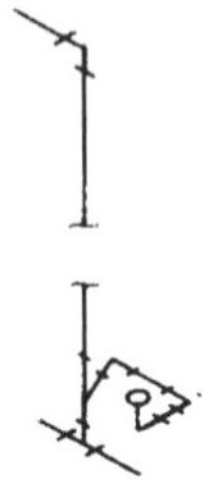

Waste/Vent

Floor Mount

System Components	QUANTITY	UNIT	COST EACH MAT.	COST EACH INST.	COST EACH TOTAL
SYSTEM D2010 110 1840					
WATER CLOSET, VITREOUS CHINA, ELONGATED					
TANK TYPE, WALL HUNG, ONE PIECE					
Wtr. closet tank type vit. china wall hung 1 pc. w/seat supply & stop	1.000	Ea.	430	180	610
Pipe steel galvanized, schedule 40, threaded, 2″ diam.	4.000	L.F.	41.60	59.80	101.40
Pipe, CI soil, no hub, cplg.10′ OC, hanger 5′ OC, 4″ diam.	2.000	L.F.	20.10	33	53.10
Pipe, coupling, standard coupling, CI soil, no hub, 4″ diam.	2.000	Ea.	18.60	58	76.60
Copper tubing type L, solder joint, hanger 10′OC, 1/2″ diam.	6.000	L.F.	11.70	39.30	51
Wrought copper 90° elbow for solder joints 1/2″ diam.	2.000	Ea.	1.12	53	54.12
Wrought copper Tee for solder joints 1/2″ diam.	1.000	Ea.	.96	41	41.96
Support/carrier, for water closet, siphon jet, horiz., single, 4″ waste	1.000	Ea.	395	99.50	494.50
TOTAL			919.08	563.60	1,482.68

D2010 110	Water Closet Systems		COST EACH MAT.	COST EACH INST.	COST EACH TOTAL
1800	Water closet, vitreous china, elongated				
1840	Tank type, wall hung, one piece		920	565	1,485
1880	Close coupled two piece	R224000-30	1,025	565	1,590
1920	Floor mount, one piece		775	600	1,375
1960	One piece low profile		665	600	1,265
2000	Two piece close coupled		415	600	1,015
2040	Bowl only with flush valve				
2080	Wall hung		910	640	1,550
2120	Floor mount		570	610	1,180

Important: See the Reference Section for critical supporting data - Reference Numbers and City Cost Indexes

D2010 Plumbing Fixtures

Systems are complete with trim, flush valve and rough-in (supply, waste and vent) for connection to supply branches and waste mains.

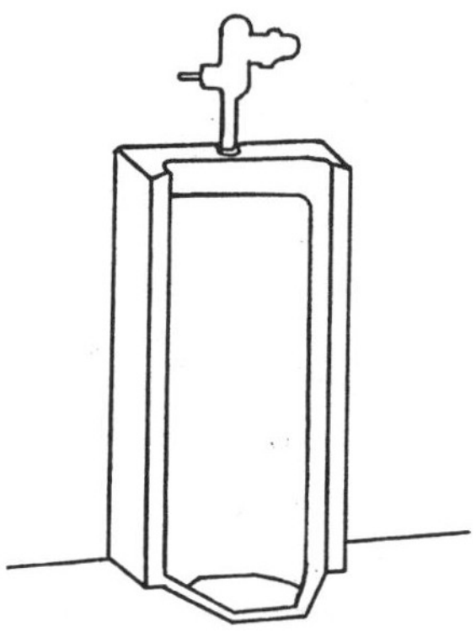

Stall Type

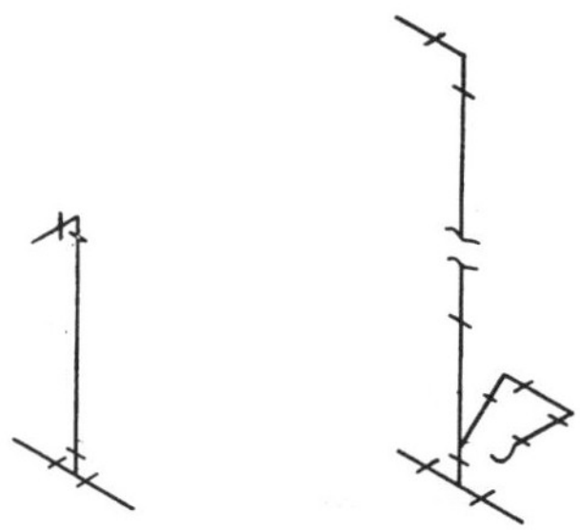

Supply Waste/Vent

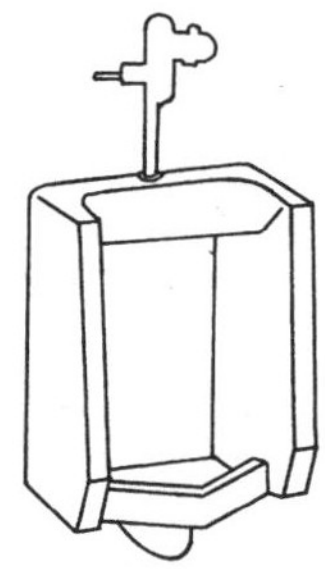

Wall Hung

System Components	QUANTITY	UNIT	COST EACH MAT.	COST EACH INST.	COST EACH TOTAL
SYSTEM D2010 210 2000					
URINAL, VITREOUS CHINA, WALL HUNG					
Urinal, wall hung, vitreous china, incl. hanger	1.000	Ea.	345	320	665
Pipe, steel, galvanized, schedule 40, threaded, 1-1/2" diam.	5.000	L.F.	39	59.75	98.75
Copper tubing type DWV, solder joint, hangers 10' OC, 2" diam.	3.000	L.F.	24.45	36.15	60.60
Combination Y & 1/8 bend for CI soil pipe, no hub, 3" diam.	1.000	Ea.	11.70		11.70
Pipe, CI, no hub, cplg. 10' OC, hanger 5' OC, 3" diam.	4.000	L.F.	31.80	59.80	91.60
Pipe coupling standard, CI soil, no hub, 3" diam.	3.000	Ea.	15.70	50	65.70
Copper tubing type L, solder joint, hanger 10' OC 3/4" diam.	5.000	L.F.	14.30	35	49.30
Wrought copper 90° elbow for solder joints 3/4" diam.	1.000	Ea.	1.25	28	29.25
Wrought copper Tee for solder joints, 3/4" diam.	1.000	Ea.	2.32	44.50	46.82
TOTAL			485.52	633.20	1,118.72

D2010 210	Urinal Systems		COST EACH MAT.	COST EACH INST.	COST EACH TOTAL
2000	Urinal, vitreous china, wall hung	R224000 -30	485	635	1,120
2040	Stall type		780	755	1,535

D SERVICES

D2010 Plumbing Fixtures

Systems are complete with trim and rough-in (supply, waste and vent) to connect to supply branches and waste mains.

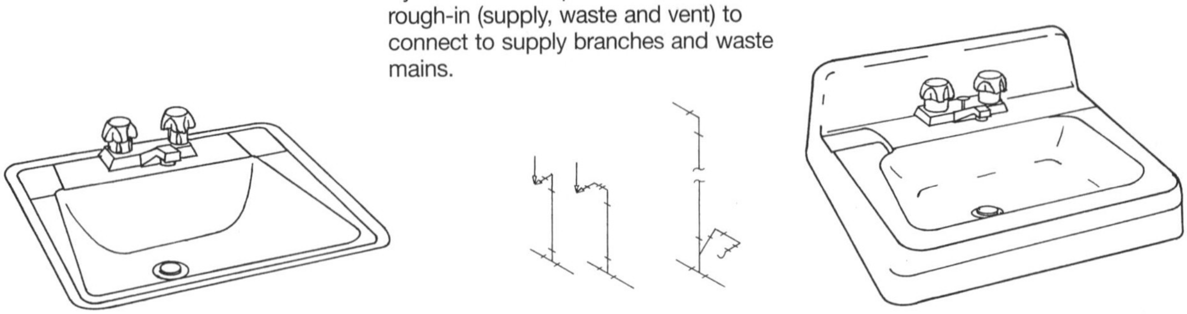

Vanity Top   Supply   Waste/Vent   Wall Hung

System Components	QUANTITY	UNIT	COST EACH MAT.	COST EACH INST.	COST EACH TOTAL
SYSTEM D2010 310 1560					
LAVATORY W/TRIM, VANITY TOP, P.E. ON C.I., 20″ X 18″					
Lavatory w/trim, PE on CI, white, vanity top, 20″ x 18″ oval	1.000	Ea.	234	149	383
Pipe, steel, galvanized, schedule 40, threaded, 1-1/4″ diam.	4.000	L.F.	26.40	43	69.40
Copper tubing type DWV, solder joint, hanger 10′ OC 1-1/4″ diam.	4.000	L.F.	19.60	35.40	55
Wrought copper DWV, Tee, sanitary, 1-1/4″ diam.	1.000	Ea.	10.45	59	69.45
P trap w/cleanout, 20 ga., 1-1/4″ diam.	1.000	Ea.	35	29.50	64.50
Copper tubing type L, solder joint, hanger 10′ OC 1/2″ diam.	10.000	L.F.	19.50	65.50	85
Wrought copper 90° elbow for solder joints 1/2″ diam.	2.000	Ea.	1.12	53	54.12
Wrought copper Tee for solder joints, 1/2″ diam.	2.000	Ea.	1.92	82	83.92
Stop, chrome, angle supply, 1/2″ diam.	2.000	Ea.	10.40	48	58.40
TOTAL			358.39	564.40	922.79

D2010 310	Lavatory Systems	COST EACH MAT.	COST EACH INST.	COST EACH TOTAL
1560	Lavatory w/trim, vanity top, PE on CI, 20″ x 18″, Vanity top by others.	360	565	925
1600	19″ x 16″ oval	276	565	841
1640	18″ round	345	565	910
1680	Cultured marble, 19″ x 17″	305	565	870
1720	25″ x 19″	315	565	880
1760	Stainless, self-rimming, 25″ x 22″	395	565	960
1800	17″ x 22″ R224000-30	390	565	955
1840	Steel enameled, 20″ x 17″	276	580	856
1880	19″ round	273	580	853
1920	Vitreous china, 20″ x 16″	370	590	960
1960	19″ x 16″	370	590	960
2000	22″ x 13″	380	590	970
2040	Wall hung, PE on CI, 18″ x 15″	660	625	1,285
2080	19″ x 17″	640	625	1,265
2120	20″ x 18″	610	625	1,235
2160	Vitreous china, 18″ x 15″	565	640	1,205
2200	19″ x 17″	530	640	1,170
2240	24″ x 20″	695	640	1,335

Important: See the Reference Section for critical supporting data - Reference Numbers and City Cost Indexes

D2010 Plumbing Fixtures

Systems are complete with trim and rough-in (supply, waste and vent) to connect to supply branches and waste mains.

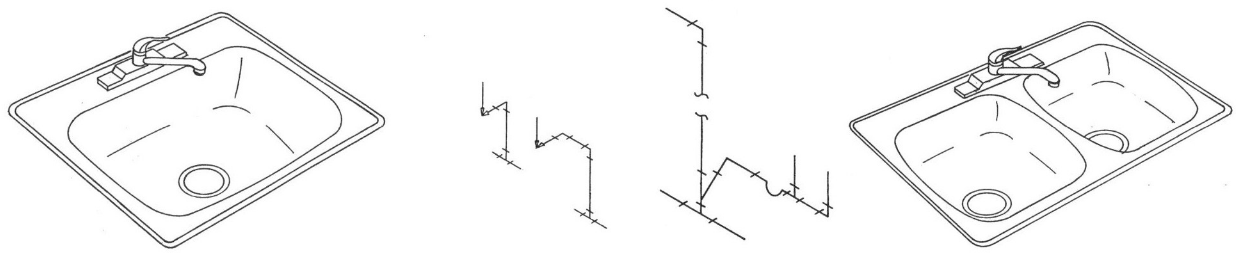

Countertop Single Bowl — Supply — Waste/Vent — Countertop Double Bowl

System Components	QUANTITY	UNIT	COST EACH MAT.	COST EACH INST.	COST EACH TOTAL
SYSTEM D2010 410 1720					
KITCHEN SINK W/TRIM, COUNTERTOP, P.E. ON C.I., 24″ X 21″, SINGLE BOWL					
Kitchen sink, countertop, PE on CI, 1 bowl, 24″ x 21″ OD	1.000	Ea.	255	171	426
Pipe, steel, galvanized, schedule 40, threaded, 1-1/4″ diam.	4.000	L.F.	26.40	43	69.40
Copper tubing, type DWV, solder, hangers 10′ OC 1-1/2″ diam.	6.000	L.F.	36.60	59.10	95.70
Wrought copper, DWV, Tee, sanitary, 1-1/2″ diam.	1.000	Ea.	14.10	66.50	80.60
P trap, standard, copper, 1-1/2″ diam.	1.000	Ea.	34	31.50	65.50
Copper tubing, type L, solder joints, hangers 10′ OC 1/2″ diam.	10.000	L.F.	19.50	65.50	85
Wrought copper 90° elbow for solder joints 1/2″ diam.	2.000	Ea.	1.12	53	54.12
Wrought copper Tee for solder joints, 1/2″ diam.	2.000	Ea.	1.92	82	83.92
Stop, angle supply, chrome, 1/2″ CTS	2.000	Ea.	10.40	48	58.40
TOTAL			399.04	619.60	1,018.64

D2010 410	Kitchen Sink Systems	COST EACH MAT.	COST EACH INST.	COST EACH TOTAL
1720	Kitchen sink w/trim, countertop, PE on CI, 24″x21″, single bowl	400	620	1,020
1760	30″ x 21″ single bowl	385	620	1,005
1800	32″ x 21″ double bowl	500	665	1,165
1840	42″ x 21″ double bowl	690	675	1,365
1880	Stainless steel, 19″ x 18″ single bowl	560	620	1,180
1920	25″ x 22″ single bowl	610	620	1,230
1960	33″ x 22″ double bowl	825	665	1,490
2000	43″ x 22″ double bowl	935	675	1,610
2040	44″ x 22″ triple bowl	1,050	705	1,755
2080	44″ x 24″ corner double bowl	620	675	1,295
2120	Steel, enameled, 24″ x 21″ single bowl	269	620	889
2160	32″ x 21″ double bowl	320	665	985
2240	Raised deck, PE on CI, 32″ x 21″, dual level, double bowl	455	850	1,305
2280	42″ x 21″ dual level, triple bowl	895	925	1,820

D2010 Plumbing Fixtures

Systems are complete with trim and rough-in (supply, waste and vent) to connect to supply branches and waste mains.

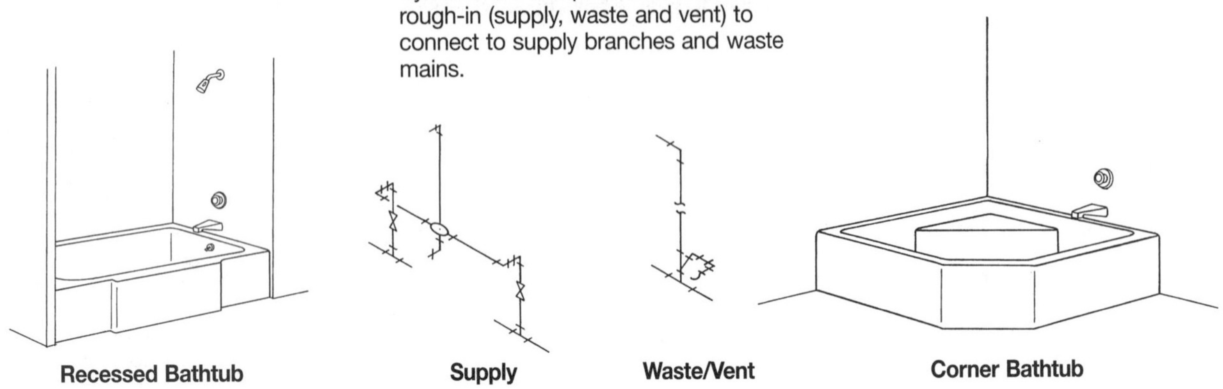

System Components	QUANTITY	UNIT	COST EACH MAT.	INST.	TOTAL
SYSTEM D2010 510 2000					
BATHTUB, RECESSED, PORCELAIN ENAMEL ON CAST IRON,, 48″ x 42″					
Bath tub, porcelain enamel on cast iron, w/fittings, 48″ x 42″	1.000	Ea.	1,775	239	2,014
Pipe, steel, galvanized, schedule 40, threaded, 1-1/4″ diam	4.000	L.F.	26.40	43	69.40
Pipe, CI no hub soil w/couplings 10′ OC, hangers 5′ OC, 4″ diam	3.000	L.F.	30.15	49.50	79.65
Combination Y and 1/8 bend for C.I. soil pipe, no hub, 4″ pipe size	1.000	Ea.	26.50		26.50
Drum trap, 3″ x 5″, copper, 1-1/2″ diam	1.000	Ea.	47.50	33	80.50
Copper tubing type L, solder joints, hangers 10′ OC 1/2″ diam	10.000	L.F.	19.50	65.50	85
Wrought copper 90° elbow, solder joints, 1/2″ diam	2.000	Ea.	1.12	53	54.12
Wrought copper Tee, solder joints, 1/2″ diam	2.000	Ea.	1.92	82	83.92
Stop, angle supply, 1/2″ diameter	2.000	Ea.	10.40	48	58.40
Copper tubing type DWV, solder joints, hanger 10′ OC 1-1/2″ diam	3.000	L.F.	18.30	29.55	47.85
Pipe coupling, standard, C.I. soil no hub, 4″ pipe size	2.000	Ea.	18.60	58	76.60
TOTAL			1,975.39	700.55	2,675.94

D2010 510	Bathtub Systems		COST EACH MAT.	INST.	TOTAL
2000	Bathtub, recessed, P.E. on CI., 48″ x 42″	R224000 -30	1,975	700	2,675
2040	72″ x 36″		2,025	780	2,805
2080	Mat bottom, 5′ long		1,000	680	1,680
2120	5′-6″ long		1,450	700	2,150
2160	Corner, 48″ x 42″		1,975	680	2,655
2200	Formed steel, enameled, 4′-6″ long		565	625	1,190

SERVICES D

D2010 Plumbing Fixtures

Systems are complete with trim, flush valve and rough-in (supply, waste and vent) for connection to supply branches and waste mains.

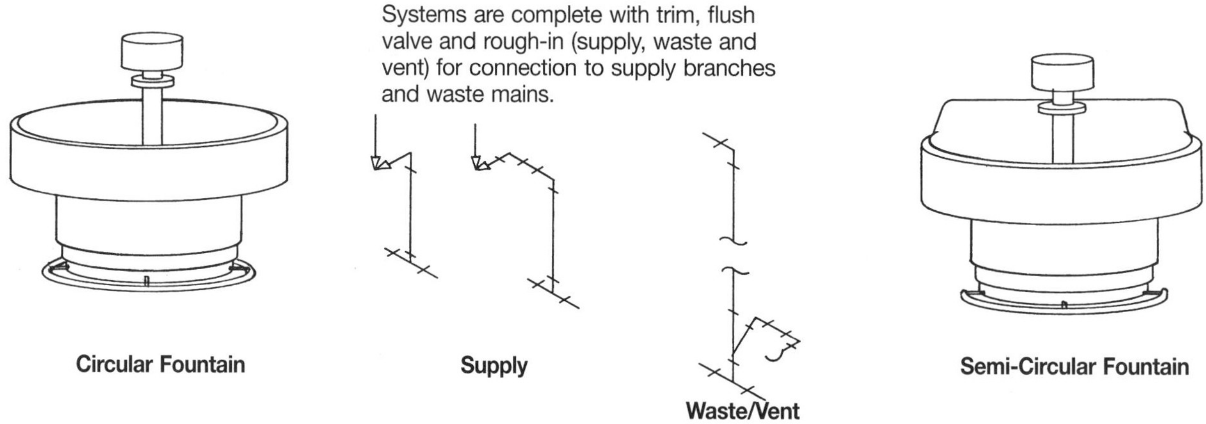

System Components	QUANTITY	UNIT	COST EACH MAT.	COST EACH INST.	COST EACH TOTAL
SYSTEM D2010 610 1760					
GROUP WASH FOUNTAIN, PRECAST TERRAZZO					
CIRCULAR, 36″ DIAMETER					
Wash fountain, group, precast terrazzo, foot control 36″ diam	1.000	Ea.	3,250	495	3,745
Copper tubing type DWV, solder joint, hanger 10′ OC, 2″ diam.	10.000	L.F.	81.50	120.50	202
P trap, standard, copper, 2″ diam.	1.000	Ea.	52.50	35.50	88
Wrought copper, Tee, sanitary, 2″ diam.	1.000	Ea.	16.45	76	92.45
Copper tubing type L, solder joint, hanger 10′ OC 1/2″ diam.	20.000	L.F.	39	131	170
Wrought copper 90° elbow for solder joints 1/2″ diam.	3.000	Ea.	1.68	79.50	81.18
Wrought copper Tee for solder joints, 1/2″ diam.	2.000	Ea.	1.92	82	83.92
TOTAL			3,443.05	1,019.50	4,462.55

D2010 610	Group Wash Fountain Systems	COST EACH MAT.	COST EACH INST.	COST EACH TOTAL
1740	Group wash fountain, precast terrazzo			
1760	Circular, 36″ diameter	3,450	1,025	4,475
1800	54″ diameter	4,150	1,125	5,275
1840	Semi-circular, 36″ diameter (R224000-30)	3,125	1,025	4,150
1880	54″ diameter	3,800	1,125	4,925
1960	Stainless steel, circular, 36″ diameter	3,925	950	4,875
2000	54″ diameter	5,125	1,050	6,175
2040	Semi-circular, 36″ diameter	3,425	950	4,375
2080	54″ diameter	4,450	1,050	5,500
2160	Thermoplastic, circular, 36″ diameter	3,025	775	3,800
2200	54″ diameter	3,525	895	4,420
2240	Semi-circular, 36″ diameter	2,875	775	3,650
2280	54″ diameter	3,375	895	4,270

D2010 Plumbing Fixtures

Systems are complete with trim and rough-in (supply, waste and vent) for connection to supply branches and waste mains.

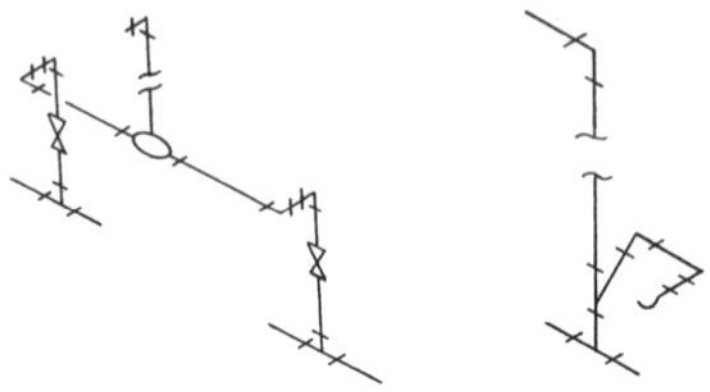

System Components	QUANTITY	UNIT	COST EACH MAT.	COST EACH INST.	COST EACH TOTAL
SYSTEM D2010 710 1560					
SHOWER, STALL, BAKED ENAMEL, MOLDED STONE RECEPTOR, 30″ SQUARE					
Shower stall, enameled steel, molded stone receptor, 30″ square	1.000	Ea.	400	184	584
Copper tubing type DWV, solder joints, hangers 10′ OC, 2″ diam.	6.000	L.F.	36.60	59.10	95.70
Wrought copper DWV, Tee, sanitary, 2″ diam.	1.000	Ea.	14.10	66.50	80.60
Trap, standard, copper, 2″ diam.	1.000	Ea.	34	31.50	65.50
Copper tubing type L, solder joint, hanger 10′ OC 1/2″ diam.	16.000	L.F.	31.20	104.80	136
Wrought copper 90° elbow for solder joints 1/2″ diam.	3.000	Ea.	1.68	79.50	81.18
Wrought copper Tee for solder joints, 1/2″ diam.	2.000	Ea.	1.92	82	83.92
Stop and waste, straightway, bronze, solder joint 1/2″ diam	2.000	Ea.	5.96	44	49.96
TOTAL			525.46	651.40	1,176.86

D2010 710	Shower Systems		COST EACH MAT.	COST EACH INST.	COST EACH TOTAL
1560	Shower, stall, baked enamel, molded stone receptor, 30″ square		525	650	1,175
1600	32″ square		505	660	1,165
1640	Terrazzo receptor, 32″ square	R224000 -30	930	660	1,590
1680	36″ square		1,075	665	1,740
1720	36″ corner angle		1,000	665	1,665
1800	Fiberglass one piece, three walls, 32″ square		585	640	1,225
1840	36″ square		635	640	1,275
1880	Polypropylene, molded stone receptor, 30″ square		485	945	1,430
1920	32″ square		585	945	1,530
1960	Built-in head, arm, bypass, stops and handles		147	245	392

SERVICES D

D2010 Plumbing Fixtures

Systems are complete with trim and rough-in (supply, waste and vent) for connection to supply branches and waste mains.

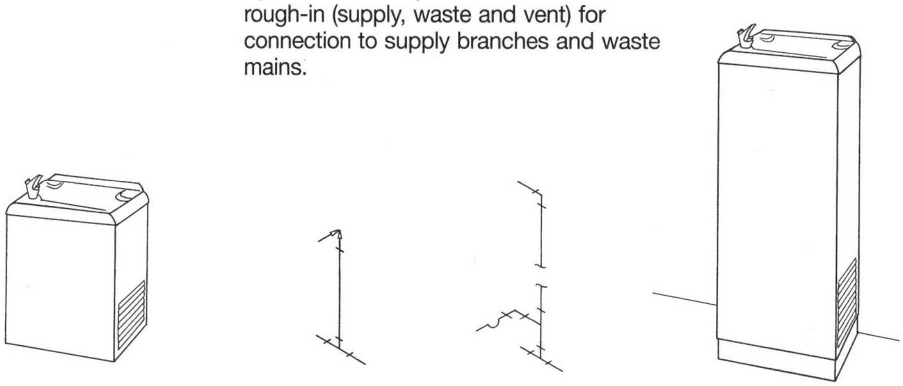

Wall Hung · Supply · Waste/Vent · Floor Mounted

System Components	QUANTITY	UNIT	COST EACH MAT.	COST EACH INST.	COST EACH TOTAL
SYSTEM D2010 820 1840					
WATER COOLER, ELECTRIC, SELF CONTAINED, WALL HUNG, 8.2 GPH					
Water cooler, wall mounted, 8.2 GPH	1.000	Ea.	615	239	854
Copper tubing type DWV, solder joint, hanger 10′ OC 1-1/4″ diam.	4.000	L.F.	19.60	35.40	55
Wrought copper DWV, Tee, sanitary 1-1/4″ diam.	1.000	Ea.	10.45	59	69.45
P trap, copper drainage, 1-1/4″ diam.	1.000	Ea.	35	29.50	64.50
Copper tubing type L, solder joint, hanger 10′ OC 3/8″ diam.	5.000	L.F.	8.40	31.50	39.90
Wrought copper 90° elbow for solder joints 3/8″ diam.	1.000	Ea.	1.77	24	25.77
Wrought copper Tee for solder joints, 3/8″ diam.	1.000	Ea.	3	38	41
Stop and waste, straightway, bronze, solder, 3/8″ diam.	1.000	Ea.	4.65	22	26.65
TOTAL			697.87	478.40	1,176.27

D2010 820	Water Cooler Systems		COST EACH MAT.	COST EACH INST.	COST EACH TOTAL
1840	Water cooler, electric, wall hung, 8.2 GPH	R224000 -30	700	480	1,180
1880	Dual height, 14.3 GPH		1,025	490	1,515
1920	Wheelchair type, 7.5 G.P.H.		1,650	480	2,130
1960	Semi recessed, 8.1 G.P.H.		1,000	480	1,480
2000	Full recessed, 8 G.P.H.		1,525	510	2,035
2040	Floor mounted, 14.3 G.P.H.		770	415	1,185
2080	Dual height, 14.3 G.P.H.		1,075	505	1,580
2120	Refrigerated compartment type, 1.5 G.P.H.		1,300	415	1,715

D2010 Plumbing Fixtures

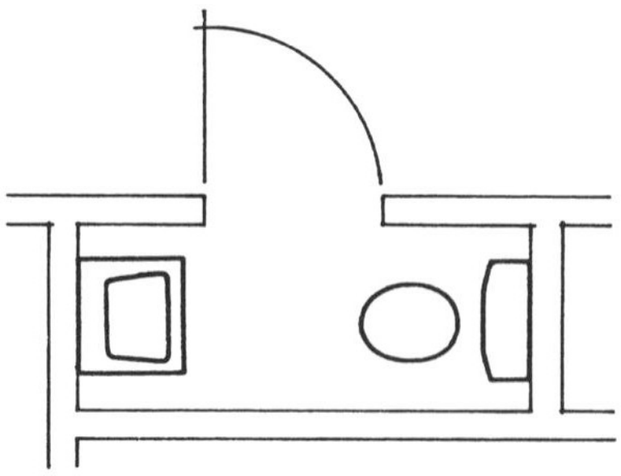

*Common wall is with an adjacent bathroom

Two Fixture Bathroom Systems consisting of a lavatory, water closet, and rough-in service piping.

- Prices for plumbing and fixtures only.

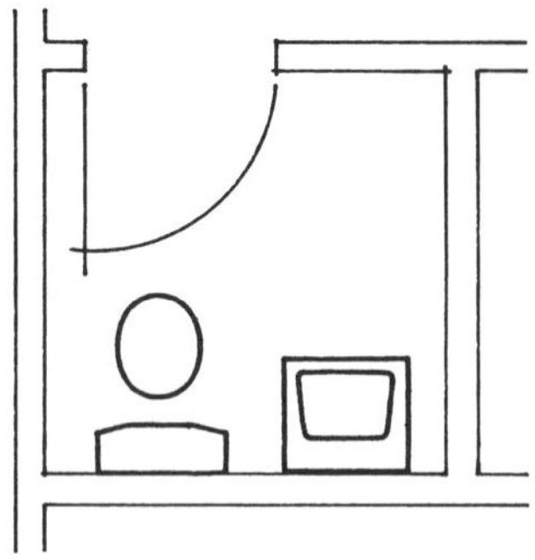

System Components	QUANTITY	UNIT	COST EACH MAT.	COST EACH INST.	COST EACH TOTAL
SYSTEM D2010 920 1180					
BATHROOM, LAVATORY & WATER CLOSET, 2 WALL PLUMBING, STAND ALONE					
Water closet, 2 Pc. close cpld. vit .china flr. mntd. w/seat, supply & stop	1.000	Ea.	186	180	366
Water closet, rough-in waste & vent	1.000	Set	210	315	525
Lavatory w/ftngs., wall hung, white, PE on CI, 20″ x 18″	1.000	Ea.	265	120	385
Lavatory, rough-in waste & vent	1.000	Set	340	575	915
Copper tubing type L, solder joint, hanger 10′ OC 1/2″ diam.	10.000	L.F.	19.50	65.50	85
Pipe, steel, galvanized, schedule 40, threaded, 2″ diam.	12.000	L.F.	124.80	179.40	304.20
Pipe, CI soil, no hub, coupling 10′ OC, hanger 5′ OC, 4″ diam.	7.000	L.F.	68.60	121.80	190.40
TOTAL			1,213.90	1,556.70	2,770.60

D2010 920	Two Fixture Bathroom, Two Wall Plumbing	COST EACH MAT.	COST EACH INST.	COST EACH TOTAL
1180	Bathroom, lavatory & water closet, 2 wall plumbing, stand alone	1,225	1,550	2,775
1200	Share common plumbing wall*	1,100	1,350	2,450

D2010 922	Two Fixture Bathroom, One Wall Plumbing	COST EACH MAT.	COST EACH INST.	COST EACH TOTAL
2220	Bathroom, lavatory & water closet, one wall plumbing, stand alone	1,125	1,400	2,525
2240	Share common plumbing wall*	1,000	1,200	2,200

D3030 Cooling Generating Systems

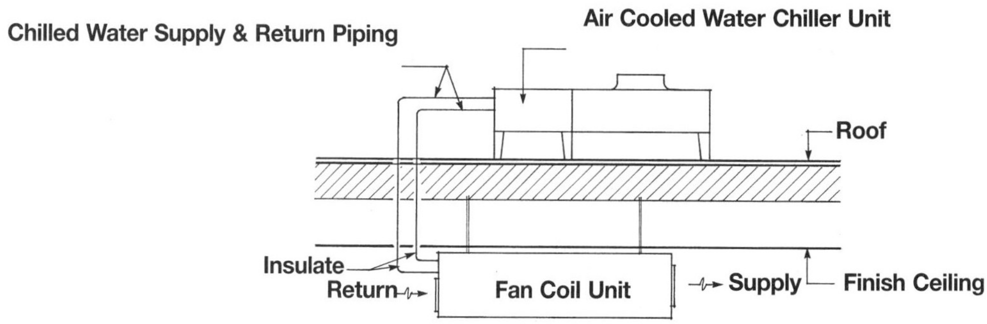

Design Assumptions: The chilled water, air cooled systems priced, utilize reciprocating hermetic compressors and propeller-type condenser fans. Piping with pumps and expansion tanks is included based on a two pipe system. No ducting is included and the fan-coil units are cooling only. Water treatment and balancing are not included. Chilled water piping is insulated. Area distribution is through the use of multiple fan coil units. Fewer but larger fan coil units with duct distribution would be approximately the same S.F. cost.

System Components	QUANTITY	UNIT	COST EACH		
			MAT.	INST.	TOTAL
SYSTEM D3030 110 1200					
PACKAGED CHILLER, AIR COOLED, WITH FAN COIL UNIT					
APARTMENT CORRIDORS, 3,000 S.F., 5.50 TON					
Fan coil air conditioning unit, cabinet mounted & filters chilled water	1.000	Ea.	3,436.88	441.75	3,878.63
Water chiller,air conditioning unit, reciprocating, air cooled,	1.000	Ea.	6,132.50	1,663.75	7,796.25
Chilled water unit coil connections	1.000	Ea.	790	1,200	1,990
Chilled water distribution piping	440.000	L.F.	9,240	16,280	25,520
TOTAL			19,599.38	19,585.50	39,184.88
COST PER S.F.			6.53	6.53	13.06

D3030 110	Chilled Water, Air Cooled Condenser Systems	COST PER S.F.		
		MAT.	INST.	TOTAL
1180	Packaged chiller, air cooled, with fan coil unit			
1200	Apartment corridors, 3,000 S.F., 5.50 ton	6.53	6.53	13.06
1360	40,000 S.F., 73.33 ton (R262213 -27)	3.89	2.91	6.80
1440	Banks and libraries, 3,000 S.F., 12.50 ton	9.65	7.70	17.35
1560	20,000 S.F., 83.33 ton	6.75	3.97	10.72
1680	Bars and taverns, 3,000 S.F., 33.25 ton	15.85	9.10	24.95
1760	10,000 S.F., 110.83 ton	11.30	2.27	13.57
1920	Bowling alleys, 3,000 S.F., 17.00 ton	12.05	8.65	20.70
2040	20,000 S.F., 113.33 ton	7.70	4.01	11.71
2160	Department stores, 3,000 S.F., 8.75 ton	8.85	7.25	16.10
2320	40,000 S.F., 116.66 ton	4.92	3	7.92
2400	Drug stores, 3,000 S.F., 20.00 ton	13.60	8.95	22.55
2520	20,000 S.F., 133.33 ton	9.15	4.23	13.38
2640	Factories, 2,000 S.F., 10.00 ton	8.45	7.40	15.85
2800	40,000 S.F., 133.33 ton	5.35	3.09	8.44
2880	Food supermarkets, 3,000 S.F., 8.50 ton	8.70	7.25	15.95
3040	40,000 S.F., 113.33 ton	4.81	3.01	7.82
3120	Medical centers, 3,000 S.F., 7.00 ton	7.80	7.05	14.85
3280	40,000 S.F., 93.33 ton	4.35	2.95	7.30
3360	Offices, 3,000 S.F., 9.50 ton	7.90	7.20	15.10
3520	40,000 S.F., 126.66 ton	5.25	3.12	8.37
3600	Restaurants, 3,000 S.F., 15.00 ton	10.80	8	18.80
3720	20,000 S.F., 100.00 ton	7.70	4.19	11.89
3840	Schools and colleges, 3,000 S.F., 11.50 ton	9.15	7.55	16.70
3960	20,000 S.F., 76.66 ton	6.55	4.01	10.56

D3030 Cooling Generating Systems

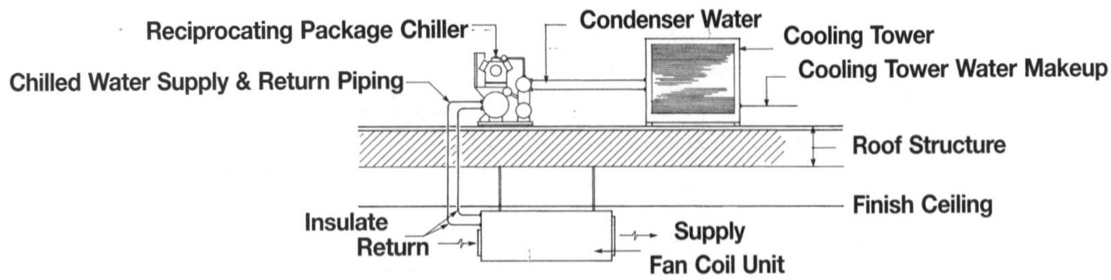

General: Water cooled chillers are available in the same sizes as air cooled units. They are also available in larger capacities.

Design Assumptions: The chilled water systems with water cooled condenser include reciprocating hermetic compressors, water cooling tower, pumps, piping and expansion tanks and are based on a two pipe system. Chilled water piping is insulated. No ducts are included and fan-coil units are cooling only. Area distribution is through use of multiple fan coil units. Fewer but larger fan coil units with duct distribution would be approximately the same S.F. cost. Water treatment and balancing are not included.

System Components	QUANTITY	UNIT	COST EACH		
			MAT.	INST.	TOTAL
SYSTEM D3030 115 1320					
PACKAGED CHILLER, WATER COOLED, WITH FAN COIL UNIT					
APARTMENT CORRIDORS, 4,000 S.F., 7.33 TON					
Fan coil air conditioner unit, cabinet mounted & filters, chilled water	2.000	Ea.	4,580.63	588.76	5,169.39
Water chiller, water cooled, 1 compressor, hermetic scroll,	1.000	Ea.	4,694.50	2,885.40	7,579.90
Cooling tower, draw thru single flow, belt drive	1.000	Ea.	700.02	122.04	822.06
Cooling tower pumps & piping	1.000	System	351.84	289.54	641.38
Chilled water unit coil connections	2.000	Ea.	1,580	2,400	3,980
Chilled water distribution piping	520.000	L.F.	10,920	19,240	30,160
TOTAL			22,826.99	25,525.74	48,352.73
COST PER S.F.			5.71	6.38	12.09

D3030 115	Chilled Water, Cooling Tower Systems	COST PER S.F.		
		MAT.	INST.	TOTAL
1300	Packaged chiller, water cooled, with fan coil unit			
1320	Apartment corridors, 4,000 S.F., 7.33 ton	5.70	6.38	12.08
1600	Banks and libraries, 4,000 S.F., 16.66 ton (R262213 -27)	8.95	7	15.95
1800	60,000 S.F., 250.00 ton	6.65	5.70	12.35
1880	Bars and taverns, 4,000 S.F., 44.33 ton	14.20	8.85	23.05
2000	20,000 S.F., 221.66 ton	12.85	7.75	20.60
2160	Bowling alleys, 4,000 S.F., 22.66 ton	10.35	7.70	18.05
2320	40,000 S.F., 226.66 ton	7.55	5.55	13.10
2440	Department stores, 4,000 S.F., 11.66 ton	6.55	6.95	13.50
2640	60,000 S.F., 175.00 ton	5.55	5.30	10.85
2720	Drug stores, 4,000 S.F., 26.66 ton	11.05	7.95	19
2880	40,000 S.F., 266.67 ton	8.75	6.10	14.85
3000	Factories, 4,000 S.F., 13.33 ton	7.80	6.70	14.50
3200	60,000 S.F., 200.00 ton	6	5.45	11.45
3280	Food supermarkets, 4,000 S.F., 11.33 ton	6.45	6.90	13.35
3480	60,000 S.F., 170.00 ton	5.50	5.30	10.80
3560	Medical centers, 4.000 S.F., 9.33 ton	5.70	6.30	12
3760	60,000 S.F., 140.00 ton	5.10	5.25	10.35
3840	Offices, 4,000 S.F., 12.66 ton	7.55	6.60	14.15
4040	60,000 S.F., 190.00 ton	5.80	5.40	11.20
4120	Restaurants, 4,000 S.F., 20.00 ton	9.35	7.15	16.50
4320	60,000 S.F., 300.00 ton	7.35	5.90	13.25
4400	Schools and colleges, 4,000 S.F., 15.33 ton	8.50	6.85	15.35
4600	60,000 S.F., 230.00 ton	6.20	5.45	11.65

D3050 Terminal & Package Units

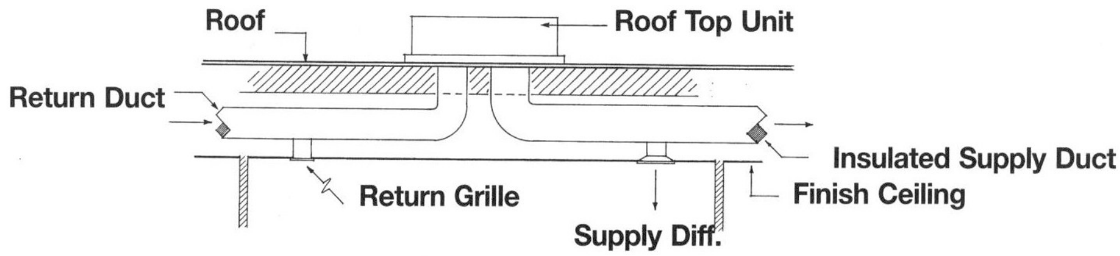

System Description: Rooftop single zone units are electric cooling and gas heat. Duct systems are low velocity, galvanized steel supply and return. Price variations between sizes are due to several factors. Jumps in the cost of the rooftop unit occur when the manufacturer shifts from the largest capacity unit on a small frame to the smallest capacity on the next larger frame, or changes from one compressor to two. As the unit capacity increases for larger areas the duct distribution grows in proportion. For most applications there is a tradeoff point where it is less expensive and more efficient to utilize smaller units with short simple distribution systems. Larger units also require larger initial supply and return ducts which can create a space problem. Supplemental heat may be desired in colder locations. The table below is based on one unit supplying the area listed. The 10,000 S.F. unit for bars and taverns is not listed because a nominal 110 ton unit would be required and this is above the normal single zone rooftop capacity.

System Components	QUANTITY	UNIT	COST EACH MAT.	COST EACH INST.	COST EACH TOTAL
SYSTEM D3050 150 1280					
ROOFTOP, SINGLE ZONE, AIR CONDITIONER					
APARTMENT CORRIDORS, 500 S.F., .92 TON					
Rooftop air-conditioner, 1 zone, electric cool, standard controls, curb	1.000	Ea.	1,483.50	632.50	2,116
Ductwork package for rooftop single zone units	1.000	System	299	920	1,219
TOTAL			1,782.50	1,552.50	3,335
COST PER S.F.			3.57	3.11	6.68

D3050 150	Rooftop Single Zone Unit Systems	COST PER S.F. MAT.	COST PER S.F. INST.	COST PER S.F. TOTAL
1260	Rooftop, single zone, air conditioner			
1280	Apartment corridors, 500 S.F., .92 ton	3.55	3.10	6.65
1480	10,000 S.F., 18.33 ton	2.43	2.11	4.54
1560	Banks or libraries, 500 S.F., 2.08 ton R262213 -27	8.05	7	15.05
1760	10,000 S.F., 41.67 ton	5.60	4.78	10.38
1840	Bars and taverns, 500 S.F. 5.54 ton	13.15	9.35	22.50
2000	5,000 S.F., 55.42 ton	11.75	7.15	18.90
2080	Bowling alleys, 500 S.F., 2.83 ton	8.45	7.95	16.40
2280	10,000 S.F., 56.67 ton	6.95	6.50	13.45
2360	Department stores, 500 S.F., 1.46 ton	5.65	4.92	10.57
2560	10,000 S.F., 29.17 ton	3.97	3.34	7.31
2640	Drug stores, 500 S.F., 3.33 ton	9.95	9.35	19.30
2840	10,000 S.F., 66.67 ton	8.15	7.65	15.80
2920	Factories, 500 S.F., 1.67 ton	6.45	5.60	12.05
3120	10,000 S.F., 33.33 ton	4.54	3.81	8.35
3200	Food supermarkets, 500 S.F., 1.42 ton	5.50	4.79	10.29
3400	10,000 S.F., 28.33 ton	3.86	3.24	7.10
3480	Medical centers, 500 S.F., 1.17 ton	4.52	3.94	8.46
3680	10,000 S.F., 23.33 ton	3.09	2.69	5.78
3760	Offices, 500 S.F., 1.58 ton	6.15	5.35	11.50
3960	10,000 S.F., 31.67 ton	4.31	3.63	7.94
4000	Restaurants, 500 S.F., 2.50 ton	9.70	8.45	18.15
4200	10,000 S.F., 50.00 ton	6.10	5.75	11.85
4240	Schools and colleges, 500 S.F., 1.92 ton	7.45	6.45	13.90
4440	10,000 S.F., 38.33 ton	5.15	4.39	9.54

D3050 Terminal & Package Units

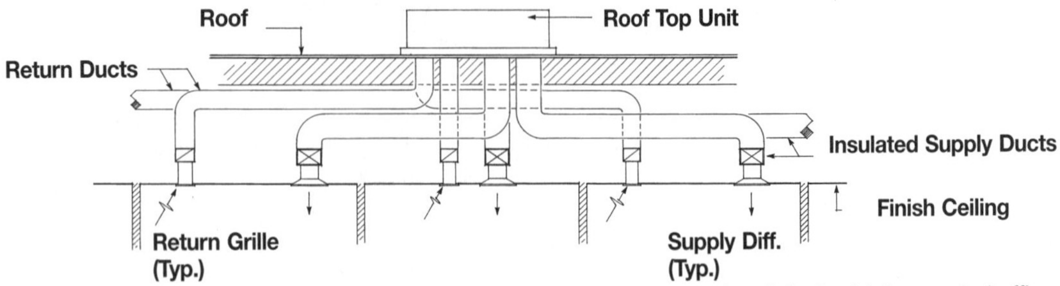

System Description: Rooftop units are multizone with up to 12 zones, and include electric cooling, gas heat, thermostats, filters, supply and return fans complete. Duct systems are low velocity, galvanized steel supply and return with insulated supplies.

Multizone units cost more per ton of cooling than single zone. However, they offer flexibility where load conditions are varied due to heat generating areas or exposure to radiational heating. For example, perimeter offices on the "sunny side" may require cooling at the same time "shady side" or central offices may require heating. It is possible to accomplish similar results using duct heaters in branches of the single zone unit. However, heater location could be a problem and total system operating energy efficiency could be lower.

System Components	QUANTITY	UNIT	COST EACH MAT.	COST EACH INST.	COST EACH TOTAL
SYSTEM D3050 155 1280					
ROOFTOP, MULTIZONE, AIR CONDITIONER					
APARTMENT CORRIDORS, 3,000 S.F., 5.50 TON					
Rooftop multizone unit, standard controls, curb	1.000	Ea.	36,960	1,474	38,434
Ductwork package for rooftop multizone units	1.000	System	2,502.50	10,587.50	13,090
TOTAL			39,462.50	12,061.50	51,524
COST PER S.F.			13.15	4.02	17.17
Note A: Small single zone unit recommended					

D3050 155	Rooftop Multizone Unit Systems		COST PER S.F. MAT.	COST PER S.F. INST.	COST PER S.F. TOTAL
1240	Rooftop, multizone, air conditioner				
1260	Apartment corridors, 1,500 S.F., 2.75 ton. See Note A.				
1280	3,000 S.F., 5.50 ton	R262213 -27	13.17	4.03	17.20
1440	25,000 S.F., 45.80 ton		7.25	3.89	11.14
1520	Banks or libraries, 1,500 S.F., 6.25 ton		30	9.15	39.15
1640	15,000 S.F., 62.50 ton		11.05	8.80	19.85
1720	25,000 S.F., 104.00 ton		9.95	8.75	18.70
1800	Bars and taverns, 1,500 S.F., 16.62 ton		64.50	13.15	77.65
1840	3,000 S.F., 33.24 ton		48.50	12.70	61.20
1880	10,000 S.F., 110.83 ton		24	12.65	36.65
2080	Bowling alleys, 1,500 S.F., 8.50 ton		40.50	12.45	52.95
2160	10,000 S.F., 56.70 ton		22.50	12	34.50
2240	20,000 S.F., 113.00 ton		13.55	11.90	25.45
2640	Drug stores, 1,500 S.F., 10.00 ton		48	14.60	62.60
2680	3,000 S.F., 20.00 ton		35	14.10	49.10
2760	15,000 S.F., 100.00 ton		16	14	30
3760	Offices, 1,500 S.F., 4.75 ton, See Note A				
3880	15,000 S.F., 47.50 ton		12.55	6.70	19.25
3960	25,000 S.F., 79.16 ton		8.40	6.70	15.10
4000	Restaurants, 1,500 S.F., 7.50 ton		36	10.95	46.95
4080	10,000 S.F., 50.00 ton		19.85	10.60	30.45
4160	20,000 S.F., 100.00 ton		12	10.50	22.50
4240	Schools and colleges, 1,500 S.F., 5.75 ton		27.50	8.40	35.90
4360	15,000 S.F., 57.50 ton		10.15	8.10	18.25
4441	25,000 S.F., 95.83 ton		9.40	8.05	17.45

SERVICES D

Important: See the Reference Section for critical supporting data - Reference Numbers and City Cost Indexes

D3050 Terminal & Package Units

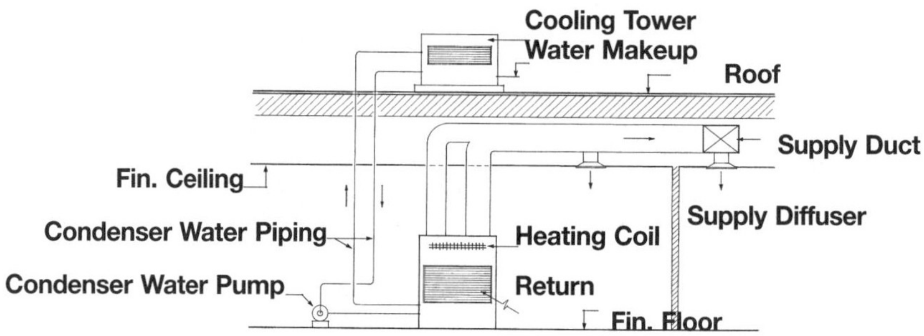

System Description: Self-contained, single package water cooled units include cooling tower, pump, piping allowance. Systems for 1000 S.F. and up include duct and diffusers to provide for even distribution of air. Smaller units distribute air through a supply air plenum, which is integral with the unit.

Returns are not ducted and supplies are not insulated.

Hot water or steam heating coils are included but piping to boiler and the boiler itself is not included.

Where local codes or conditions permit single pass cooling for the smaller units, deduct 10%.

System Components	QUANTITY	UNIT	COST EACH		
			MAT.	INST.	TOTAL
SYSTEM D3050 160 1300					
SELF-CONTAINED, WATER COOLED UNIT					
APARTMENT CORRIDORS, 500 S.F., .92 TON					
Self contained, water cooled, single package air conditioner unit	1.000	Ea.	1,041.60	504	1,545.60
Ductwork package for water or air cooled packaged units	1.000	System	88.32	703.80	792.12
Cooling tower, draw thru single flow, belt drive	1.000	Ea.	87.86	15.32	103.18
Cooling tower pumps & piping	1.000	System	44.16	36.34	80.50
TOTAL			1,261.94	1,259.46	2,521.40
COST PER S.F.			2.52	2.52	5.04

D3050 160	Self-contained, Water Cooled Unit Systems	COST PER S.F.		
		MAT.	INST.	TOTAL
1280	Self-contained, water cooled unit	2.53	2.52	5.05
1300	Apartment corridors, 500 S.F., .92 ton	2.50	2.50	5
1440	10,000 S.F., 18.33 ton R262213-27	3.14	1.76	4.90
1520	Banks or libraries, 500 S.F., 2.08 ton	5.30	2.52	7.82
1680	10,000 S.F., 41.66 ton	6.15	4	10.15
1760	Bars and taverns, 500 S.F., 5.54 ton	11.35	4.25	15.60
1920	10,000 S.F., 110.00 ton	14.95	6.20	21.15
2000	Bowling alleys, 500 S.F., 2.83 ton	7.25	3.42	10.67
2160	10,000 S.F., 56.66 ton	8.10	5.40	13.50
2200	Department stores, 500 S.F., 1.46 ton	3.74	1.78	5.52
2360	10,000 S.F., 29.17 ton	4.48	2.71	7.19
2440	Drug stores, 500 S.F., 3.33 ton	8.55	4.04	12.59
2600	10,000 S.F., 66.66 ton	9.55	6.35	15.90
2680	Factories, 500 S.F., 1.66 ton	4.26	2.02	6.28
2840	10,000 S.F., 33.33 ton	5.15	3.10	8.25
2920	Food supermarkets, 500 S.F., 1.42 ton	3.63	1.72	5.35
3080	10,000 S.F., 28.33 ton	4.36	2.63	6.99
3160	Medical centers, 500 S.F., 1.17 ton	2.98	1.41	4.39
3320	10,000 S.F., 23.33 ton	3.98	2.24	6.22
3400	Offices, 500 S.F., 1.58 ton	4.05	1.92	5.97
3560	10,000 S.F., 31.67 ton	4.86	2.94	7.80
3640	Restaurants, 500 S.F., 2.50 ton	6.40	3.03	9.43
3800	10,000 S.F., 50.00 ton	5.10	4.43	9.53
3880	Schools and colleges, 500 S.F., 1.92 ton	4.90	2.32	7.22
4040	10,000 S.F., 38.33 ton	5.65	3.68	9.33

D3050 Terminal & Package Units

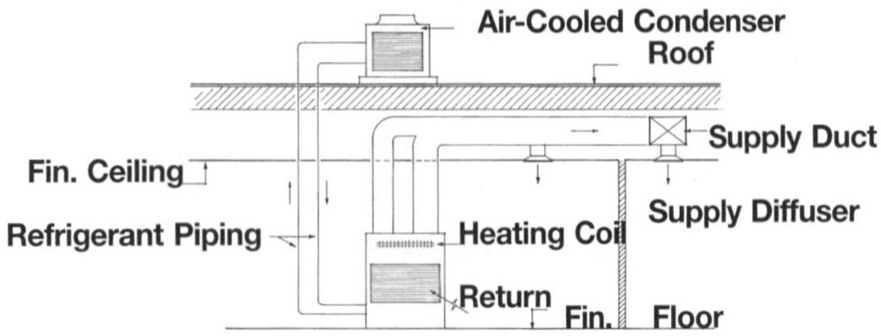

System Description: Self-contained air cooled units with remote air cooled condenser and interconnecting tubing. Systems for 1000 S.F. and up include duct and diffusers. Smaller units distribute air directly.

Returns are not ducted and supplies are not insulated.

Potential savings may be realized by using a single zone rooftop system or through-the-wall unit, especially in the smaller capacities, if the application permits.

Hot water or steam heating coils are included but piping to boiler and the boiler itself is not included.

Condenserless models are available for 15% less where remote refrigerant source is available.

System Components	QUANTITY	UNIT	COST EACH		
			MAT.	INST.	TOTAL
SYSTEM D3050 165 1320					
SELF-CONTAINED, AIR COOLED UNIT					
APARTMENT CORRIDORS, 500 S.F., .92 TON					
Air cooled, package unit	1.000	Ea.	1,113.50	328.10	1,441.60
Ductwork package for water or air cooled packaged units	1.000	System	88.32	703.80	792.12
Refrigerant piping	1.000	System	203.32	441.60	644.92
Air cooled condenser, direct drive, propeller fan	1.000	Ea.	279	124	403
TOTAL			1,684.14	1,597.50	3,281.64
COST PER S.F.			3.37	3.20	6.57

D3050 165	Self-contained, Air Cooled Unit Systems	COST PER S.F.		
		MAT.	INST.	TOTAL
1300	Self-contained, air cooled unit			
1320	Apartment corridors, 500 S.F., .92 ton	3.35	3.20	6.55
1480	10,000 S.F., 18.33 ton R262213 -27	2.52	2.59	5.11
1560	Banks or libraries, 500 S.F., 2.08 ton	7.10	4.01	11.11
1720	10,000 S.F., 41.66 ton	4.76	5.80	10.56
1800	Bars and taverns, 500 S.F., 5.54 ton	14.10	8.75	22.85
1960	10,000 S.F., 110.00 ton	12.05	11.05	23.10
2040	Bowling alleys, 500 S.F., 2.83 ton	9.75	5.50	15.25
2200	10,000 S.F., 56.66 ton	6.35	7.85	14.20
2240	Department stores, 500 S.F., 1.46 ton	5	2.82	7.82
2400	10,000 S.F., 29.17 ton	3.40	4.03	7.43
2480	Drug stores, 500 S.F., 3.33 ton	11.45	6.45	17.90
2640	10,000 S.F., 66.66 ton	7.55	9.25	16.80
2720	Factories, 500 S.F., 1.66 ton	5.80	3.25	9.05
2880	10,000 S.F., 33.33 ton	3.89	4.61	8.50
3200	Medical centers, 500 S.F., 1.17 ton	4.04	2.26	6.30
3360	10,000 S.F., 23.33 ton	3.20	3.30	6.50
3440	Offices, 500 S.F., 1.58 ton	5.45	3.06	8.51
3600	10,000 S.F., 31.66 ton	3.69	4.39	8.08
3680	Restaurants, 500 S.F., 2.50 ton	8.55	4.82	13.37
3840	10,000 S.F., 50.00 ton	5.80	6.90	12.70
3920	Schools and colleges, 500 S.F., 1.92 ton	6.60	3.70	10.30
4080	10,000 S.F., 38.33 ton	4.40	5.30	9.70

 Important: See the Reference Section for critical supporting data - Reference Numbers and City Cost Indexes

D3050 Terminal & Package Units

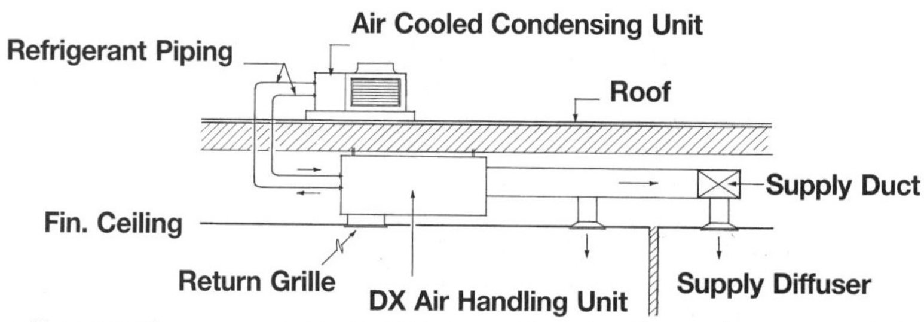

General: Split systems offer several important advantages which should be evaluated when a selection is to be made. They provide a greater degree of flexibility in component selection which permits an accurate match-up of the proper equipment size and type with the particular needs of the building. This allows for maximum use of modern energy saving concepts in heating and cooling. Outdoor installation of the air cooled condensing unit allows space savings in the building and also isolates the equipment operating sounds from building occupants.

Design Assumptions: The systems below are comprised of a direct expansion air handling unit and air cooled condensing unit with interconnecting copper tubing. Ducts and diffusers are also included for distribution of air. Systems are priced for cooling only. Heat can be added as desired either by putting hot water/steam coils into the air unit or into the duct supplying the particular area of need. Gas fired duct furnaces are also available. Refrigerant liquid line is insulated.

System Components	QUANTITY	UNIT	COST EACH MAT.	COST EACH INST.	COST EACH TOTAL
SYSTEM D3050 170 1280					
SPLIT SYSTEM, AIR COOLED CONDENSING UNIT					
APARTMENT CORRIDORS, 1,000 S.F., 1.80 TON					
Fan coil AC unit, cabinet mntd. & filters direct expansion air cool	1.000	Ea.	353.19	117.12	470.31
Ductwork package, for split system, remote condensing unit	1.000	System	82.81	667.95	750.76
Refrigeration piping	1.000	System	174.77	732	906.77
Condensing unit, air cooled, incls. compressor & standard controls	1.000	Ea.	854	469.70	1,323.70
TOTAL			1,464.77	1,986.77	3,451.54
COST PER S.F.			1.46	1.99	3.45

*Cooling requirements would lead to choosing a water cooled unit

D3050 170	Split Systems With Air Cooled Condensing Units	COST PER S.F. MAT.	COST PER S.F. INST.	COST PER S.F. TOTAL
1260	Split system, air cooled condensing unit			
1280	Apartment corridors, 1,000 S.F., 1.83 ton	1.48	1.98	3.46
1440	20,000 S.F., 36.66 ton (R262213-27)	1.88	2.81	4.69
1520	Banks and libraries, 1,000 S.F., 4.17 ton	2.74	4.58	7.32
1680	20,000 S.F., 83.32 ton	4.94	6.65	11.59
1760	Bars and taverns, 1,000 S.F., 11.08 ton	8.75	9.05	17.80
1880	10,000 S.F., 110.84 ton	11.95	10.55	22.50
2000	Bowling alleys, 1,000 S.F., 5.66 ton	3.92	8.60	12.52
2160	20,000 S.F., 113.32 ton	7.40	9.40	16.80
2320	Department stores, 1,000 S.F., 2.92 ton	1.94	3.14	5.08
2480	20,000 S.F., 58.33 ton	2.97	4.46	7.43
2560	Drug stores, 1,000 S.F., 6.66 ton	4.63	10.10	14.73
2720	20,000 S.F., 133.32 ton*			
2800	Factories, 1,000 S.F., 3.33 ton	2.22	3.57	5.79
2960	20,000 S.F., 66.66 ton	3.43	5.35	8.78
3040	Food supermarkets, 1,000 S.F., 2.83 ton	1.89	3.04	4.93
3200	20,000 S.F., 56.66 ton	2.89	4.33	7.22
3280	Medical centers, 1,000 S.F., 2.33 ton	1.57	2.46	4.03
3440	20,000 S.F., 46.66 ton	2.38	3.57	5.95
3520	Offices, 1,000 S.F., 3.17 ton	2.11	3.41	5.52
3680	20,000 S.F., 63.32 ton	3.25	5.05	8.30
3760	Restaurants, 1,000 S.F., 5.00 ton	3.48	7.55	11.03
3920	20,000 S.F., 100.00 ton	6.55	8.30	14.85
4000	Schools and colleges, 1,000 S.F., 3.83 ton	2.53	4.21	6.74
4160	20,000 S.F., 76.66 ton	3.96	6.10	10.06

D3050 Terminal & Package Units

Computer rooms impose special requirements on air conditioning systems. A prime requirement is reliability, due to the potential monetary loss that could be incurred by a system failure. A second basic requirement is the tolerance of control with which temperature and humidity are regulated, and dust eliminated. As the air conditioning system reliability is so vital, the additional cost of reserve capacity and redundant components is often justified.

System Descriptions: Computer areas may be environmentally controlled by one of three methods as follows:

1. **Self-contained Units**
   These are units built to higher standards of performance and reliability. They usually contain alarms and controls to indicate component operation failure, filter change, etc. It should be remembered that these units in the room will occupy space that is relatively expensive to build and that all alterations and service of the equipment will also have to be accomplished within the computer area.
2. **Decentralized Air Handling Units**
   In operation these are similar to the self-contained units except that their cooling capability comes from remotely located refrigeration equipment as refrigerant or chilled water. As no compressors or refrigerating equipment are required in the air units, they are smaller and require less service than self-contained units. An added plus for this type of system occurs if some of the computer components themselves also require chilled water for cooling.
3. **Central System Supply**
   Cooling is obtained from a central source which, since it is not located within the computer room, may have excess capacity and permit greater flexibility without interfering with the computer components. System performance criteria must still be met.

Note: The costs shown below do not include an allowance for ductwork or piping.

D3050 190	Computer Room Cooling Units	COST EACH		
		MAT.	INST.	TOTAL
0560	Computer room unit, air cooled, includes remote condenser			
0580	3 ton	9,900	1,925	11,825
0600	5 ton	13,200	2,150	15,350
0620	8 ton	27,600	3,575	31,175
0640	10 ton	28,700	3,850	32,550
0660	15 ton	32,100	4,375	36,475
0680	20 ton	38,300	5,175	43,475
0700	23 ton	40,200	5,350	45,550
0800	Chilled water, for connection to existing chiller system			
0820	5 ton	11,200	1,300	12,500
0840	8 ton	16,000	1,925	17,925
0860	10 ton	17,700	1,975	19,675
0880	15 ton	19,000	2,000	21,000
0900	20 ton	19,400	2,100	21,500
0920	23 ton	20,400	2,300	22,700
1000	Glycol system, complete except for interconnecting tubing			
1020	3 ton	12,100	2,400	14,500
1040	5 ton	15,800	2,525	18,325
1060	8 ton	34,100	4,200	38,300
1080	10 ton	37,000	4,600	41,600
1100	15 ton	44,800	5,775	50,575
1120	20 ton	48,000	6,250	54,250
1140	23 ton	52,500	6,825	59,325
1240	Water cooled, not including condenser water supply or cooling tower			
1260	3 ton	8,925	1,550	10,475
1280	5 ton	12,600	1,775	14,375
1300	8 ton	26,000	2,925	28,925
1320	15 ton	31,800	3,575	35,375
1340	20 ton	37,500	3,950	41,450
1360	23 ton	39,700	4,400	44,100

D4010 Sprinklers

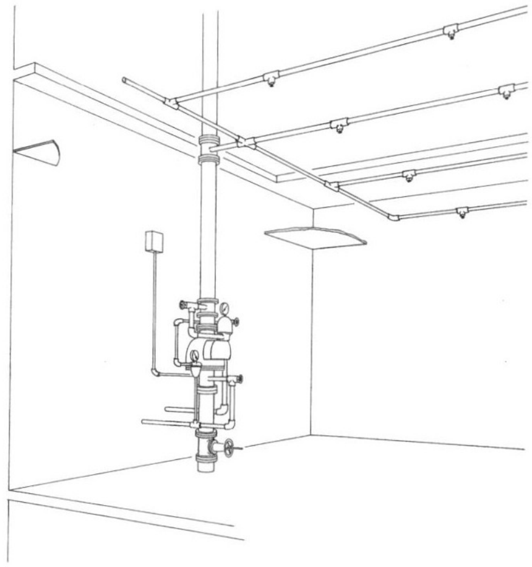

Dry Pipe System: A system employing automatic sprinklers attached to a piping system containing air under pressure, the release of which as from the opening of sprinklers permits the water pressure to open a valve known as a "dry pipe valve". The water then flows into the piping system and out the opened sprinklers.

All areas are assumed to be open.

System Components	QUANTITY	UNIT	COST EACH MAT.	COST EACH INST.	COST EACH TOTAL
SYSTEM D4010 310 0580					
DRY PIPE SPRINKLER, STEEL BLACK, SCH. 40 PIPE					
LIGHT HAZARD, ONE FLOOR, 2000 S.F.					
Valve, gate, iron body 125 lb., OS&Y, flanged, 4″ pipe size	1.000	Ea.	341.25	240	581.25
Valve, swing check, bronze, 125 lb., regrinding disc, 2-1/2″ pipe size	1.000	Ea.	199.50	47.63	247.13
Valve, angle, bronze, 150 lb., rising stem, threaded, 2″ pipe size	1.000	Ea.	285	36.38	321.38
*Alarm valve, 2-1/2″ pipe size	1.000	Ea.	701.25	236.25	937.50
Alarm, water motor, complete with gong,	1.000	Ea.	155.25	97.50	252.75
Fire alarm horn, electric	1.000	Ea.	30	58.13	88.13
Valve swing check w/balldrip CI with brass trim, 4″ pipe size	1.000	Ea.	135	236.25	371.25
Pipe, steel, black, schedule 40, 4″ diam.	10.000	L.F.	157.50	210.38	367.88
Dry pipe valve, trim & gauges, 4″ pipe size	1.000	Ea.	1,068.75	705	1,773.75
Pipe, steel, black, schedule 40, threaded, cplg & hngr 10′OC 2-1/2″ diam.	20.000	L.F.	181.50	286.50	468
Pipe, steel, black, schedule 40, threaded, cplg & hngr 10′OC 2″ diam.	12.500	L.F.	73.13	140.16	213.29
Pipe, steel, black, schedule 40, threaded, cplg & hngr 10′OC 1-1/4″ diam.	37.500	L.F.	140.63	302.34	442.97
Pipe, steel, black, schedule 40, threaded, cplg & hngr 10′OC 1″ diam.	112.000	L.F.	320.88	840	1,160.88
Pipe Tee, malleable iron black, 150 lb. threaded, 4″ pipe size	2.000	Ea.	214.50	358.50	573
Pipe Tee, malleable iron black, 150 lb. threaded, 2-1/2″ pipe size	2.000	Ea.	60.75	159	219.75
Pipe Tee, malleable iron black, 150 lb. threaded, 2″ pipe size	1.000	Ea.	14.03	65.25	79.28
Pipe Tee, malleable iron black, 150 lb. threaded, 1-1/4″ pipe size	5.000	Ea.	33	256.88	289.88
Pipe Tee, malleable iron black, 150 lb. threaded, 1″ pipe size	4.000	Ea.	16.35	199.50	215.85
Pipe 90° elbow malleable iron black, 150 lb. threaded, 1″ pipe size	6.000	Ea.	15.66	184.50	200.16
Sprinkler head dry 1/2″ orifice 1″ NPT, 3″ to 4-3/4″ length	12.000	Ea.	582	444	1,026
Air compressor, 200 Gal sprinkler system capacity, 1/3 HP	1.000	Ea.	615	300	915
*Standpipe connection, wall, flush, brs. w/plug & chain 2-1/2″x2-1/2″	1.000	Ea.	83.25	141	224.25
Valve gate bronze, 300 psi, NRS, class 150, threaded, 1″ pipe size	1.000	Ea.	25.13	21	46.13
TOTAL			5,449.31	5,566.15	11,015.46
COST PER S.F.			2.72	2.78	5.50

D4010 310	Dry Pipe Sprinkler Systems		COST PER S.F. MAT.	COST PER S.F. INST.	COST PER S.F. TOTAL
0520	Dry pipe sprinkler systems, steel, black, sch. 40 pipe				
0530	Light hazard, one floor, 500 S.F.		5.30	4.72	10.02
0560	1000 S.F.	R211313 -20	3.02	2.76	5.78
0580	2000 S.F.		2.73	2.79	5.52
0600	5000 S.F.		1.46	1.90	3.36
0620	10,000 S.F.		1.05	1.55	2.60
0640	50,000 S.F.		.84	1.39	2.23
0660	Each additional floor, 500 S.F.		1.22	2.30	3.52

D4010 Sprinklers

D4010 310	Dry Pipe Sprinkler Systems	COST PER S.F.		
		MAT.	INST.	TOTAL
0680	1000 S.F.	.97	1.89	2.86
0700	2000 S.F.	.97	1.75	2.72
0720	5000 S.F.	.77	1.50	2.27
0740	10,000 S.F.	.73	1.39	2.12
0760	50,000 S.F.	.64	1.25	1.89
1000	Ordinary hazard, one floor, 500 S.F.	5.35	4.77	10.12
1020	1000 S.F.	3.07	2.79	5.86
1040	2000 S.F.	2.79	2.90	5.69
1060	5000 S.F.	1.66	2.03	3.69
1080	10,000 S.F.	1.36	2.04	3.40
1100	50,000 S.F.	1.32	2	3.32
1140	Each additional floor, 500 S.F.	1.29	2.35	3.64
1160	1000 S.F.	1.13	2.13	3.26
1180	2000 S.F.	1.10	1.92	3.02
1200	5000 S.F.	1.02	1.66	2.68
1220	10,000 S.F.	.92	1.62	2.54
1240	50,000 S.F.	.92	1.48	2.40
1500	Extra hazard, one floor, 500 S.F.	7.20	5.95	13.15
1520	1000 S.F.	4.29	4.29	8.58
1540	2000 S.F.	3.11	3.73	6.84
1560	5000 S.F.	1.94	2.83	4.77
1580	10,000 S.F.	2.02	2.68	4.70
1600	50,000 S.F.	2.22	2.57	4.79
1660	Each additional floor, 500 S.F.	1.64	2.91	4.55
1680	1000 S.F.	1.61	2.76	4.37
1700	2000 S.F.	1.51	2.77	4.28
1720	5000 S.F.	1.29	2.42	3.71
1740	10,000 S.F.	1.47	2.21	3.68
1760	50,000 S.F.	1.47	2.12	3.59
2020	Grooved steel, black, sch. 40 pipe, light hazard, one floor, 2000 S.F.	2.63	2.39	5.02
2060	10,000 S.F.	1.02	1.36	2.38
2100	Each additional floor, 2000 S.F.	.94	1.41	2.35
2150	10,000 S.F.	.70	1.20	1.90
2200	Ordinary hazard, one floor, 2000 S.F.	2.74	2.51	5.25
2250	10,000 S.F.	1.25	1.72	2.97
2300	Each additional floor, 2000 S.F.	1.05	1.53	2.58
2350	10,000 S.F.	.94	1.56	2.50
2400	Extra hazard, one floor, 2000 S.F.	3.09	3.19	6.28
2450	10,000 S.F.	1.76	2.24	4
2500	Each additional floor, 2000 S.F.	1.50	2.28	3.78
2550	10,000 S.F.	1.33	1.96	3.29
3050	Grooved steel black sch. 10 pipe, light hazard, one floor, 2000 S.F.	2.60	2.37	4.97
3100	10,000 S.F.	1	1.32	2.32
3150	Each additional floor, 2000 S.F.	.91	1.39	2.30
3200	10,000 S.F.	.68	1.16	1.84
3250	Ordinary hazard, one floor, 2000 S.F.	2.70	2.50	5.20
3300	10,000 S.F.	1.23	1.67	2.90
3350	Each additional floor, 2000 S.F.	1.01	1.52	2.53
3400	10,000 S.F.	.92	1.51	2.43
3450	Extra hazard, one floor, 2000 S.F.	3.07	3.17	6.24
3500	10,000 S.F.	1.69	2.20	3.89
3550	Each additional floor, 2000 S.F.	1.48	2.26	3.74
3600	10,000 S.F.	1.28	1.93	3.21
4050	Copper tubing, type M, light hazard, one floor, 2000 S.F.	2.67	2.36	5.03
4100	10,000 S.F.	1.11	1.34	2.45
4150	Each additional floor, 2000 S.F.	.99	1.41	2.40
4200	10,000 S.F.	.79	1.19	1.98
4250	Ordinary hazard, one floor, 2000 S.F.	2.82	2.64	5.46

D4010 Sprinklers

D4010 310	Dry Pipe Sprinkler Systems	COST PER S.F.		
		MAT.	INST.	TOTAL
4300	10,000 S.F.	1.34	1.58	2.92
4350	Each additional floor, 2000 S.F.	1.31	1.63	2.94
4400	10,000 S.F.	1	1.39	2.39
4450	Extra hazard, one floor, 2000 S.F.	3.20	3.20	6.40
4500	10,000 S.F.	2.62	2.43	5.05
4550	Each additional floor, 2000 S.F.	1.61	2.29	3.90
4600	10,000 S.F.	1.79	2.13	3.92
5050	Copper tubing, type M, T-drill system, light hazard, one floor			
5060	2000 S.F.	2.67	2.21	4.88
5100	10,000 S.F.	1.06	1.12	2.18
5150	Each additional floor, 2000 S.F.	.99	1.26	2.25
5200	10,000 S.F.	.74	.97	1.71
5250	Ordinary hazard, one floor, 2000 S.F.	2.74	2.25	4.99
5300	10,000 S.F.	1.29	1.40	2.69
5350	Each additional floor, 2000 S.F.	1.05	1.27	2.32
5400	10,000 S.F.	.93	1.18	2.11
5450	Extra hazard, one floor, 2000 S.F.	2.96	2.65	5.61
5500	10,000 S.F.	2.20	1.77	3.97
5550	Each additional floor, 2000 S.F.	1.37	1.74	3.11
5600	10,000 S.F.	1.37	1.47	2.84

D4010 Sprinklers

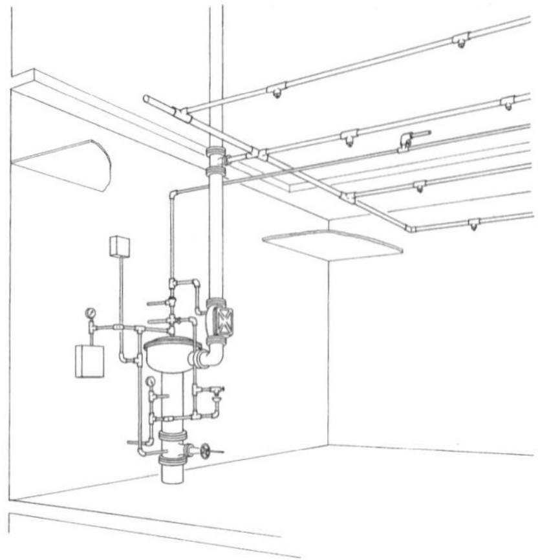

Pre-Action System: A system employing automatic sprinklers attached to a piping system containing air that may or may not be under pressure, with a supplemental heat responsive system of generally more sensitive characteristics than the automatic sprinklers themselves, installed in the same areas as the sprinklers. Actuation of the heat responsive system, as from a fire, opens a valve which permits water to flow into the sprinkler piping system and to be discharged from those sprinklers which were opened by heat from the fire.

All areas are assumed to be open.

System Components	QUANTITY	UNIT	COST EACH		
			MAT.	INST.	TOTAL
SYSTEM D4010 350 0580					
PREACTION SPRINKLER SYSTEM, STEEL BLACK SCH. 40 PIPE					
LIGHT HAZARD, 1 FLOOR, 2000 S.F.					
Valve, gate, iron body 125 lb., OS&Y, flanged, 4″ pipe size	1.000	Ea.	341.25	240	581.25
*Valve, swing check w/ball drip CI with brass trim 4″ pipe size	1.000	Ea.	135	236.25	371.25
Valve, swing check, bronze, 125 lb., regrinding disc, 2-1/2″ pipe size	1.000	Ea.	199.50	47.63	247.13
Valve, angle, bronze, 150 lb., rising stem, threaded, 2″ pipe size	1.000	Ea.	285	36.38	321.38
*Alarm valve, 2-1/2″ pipe size	1.000	Ea.	701.25	236.25	937.50
Alarm, water motor, complete with gong	1.000	Ea.	155.25	97.50	252.75
Fire alarm horn, electric	1.000	Ea.	30	58.13	88.13
Thermostatic release for release line	2.000	Ea.	495	39	534
Pipe, steel, black, schedule 40, 4″ diam	10.000	L.F.	157.50	210.38	367.88
Dry pipe valve, trim & gauges, 4″ pipe size	1.000	Ea.	1,068.75	705	1,773.75
Pipe, steel, black, schedule 40, threaded, cplg. & hngr. 10′OC 2-1/2″ diam.	20.000	L.F.	181.50	286.50	468
Pipe steel black, schedule 40, threaded, cplg. & hngr.10′OC 2″ diam.	12.500	L.F.	73.13	140.16	213.29
Pipe, steel, black, schedule 40, threaded, cplg. & hngr. 10′OC 1-1/4″ diam.	37.500	L.F.	140.63	302.34	442.97
Pipe, steel, black, schedule 40, threaded, cplg. & hngr. 10′OC 1″ diam.	112.000	L.F.	320.88	840	1,160.88
Pipe, Tee, malleable iron, black, 150 lb. threaded, 4″ diam	2.000	Ea.	214.50	358.50	573
Pipe, Tee, malleable iron, black, 150 lb. threaded, 2-1/2″ pipe size	2.000	Ea.	60.75	159	219.75
Pipe, Tee, malleable iron, black, 150 lb. threaded, 2″ pipe size	1.000	Ea.	14.03	65.25	79.28
Pipe, Tee, malleable iron, black, 150 lb. threaded, 1-1/4″ pipe size	5.000	Ea.	33	256.88	289.88
Pipe, Tee, malleable iron, black, 150 lb. threaded, 1″ pipe size	4.000	Ea.	16.35	199.50	215.85
Pipe, 90° elbow, malleable iron, blk., 150 lb. threaded, 1″ pipe size	6.000	Ea.	15.66	184.50	200.16
Sprinkler head, std. spray, brass 135°-286° F 1/2″ NPT, 3/8″ orifice	12.000	Ea.	88.80	390	478.80
Air compressor auto complete 200 Gal sprinkler sys. cap., 1/3 HP	1.000	Ea.	615	300	915
*Standpipe conn.,wall, flush, brass w/plug & chain 2-1/2″ x 2-1/2″	1.000	Ea.	83.25	141	224.25
Valve, gate, bronze, 300 psi, NRS, class 150, threaded, 1″ pipe size	1.000	Ea.	25.13	21	46.13
TOTAL			5,451.11	5,551.15	11,002.26
COST PER S.F.			2.73	2.78	5.51
*Not included in systems under 2000 S.F.					

D4010 350	Preaction Sprinkler Systems		COST PER S.F.		
			MAT.	INST.	TOTAL
0520	Preaction sprinkler systems, steel, black, sch. 40 pipe	R211313 -20			
0530	Light hazard, one floor, 500 S.F.		5.30	3.76	9.06
0560	1000 S.F.		3.09	2.83	5.92
0580	2000 S.F.		2.73	2.78	5.51

D4010 Sprinklers

D4010 350	Preaction Sprinkler Systems	COST PER S.F.		
		MAT.	INST.	TOTAL
0600	5000 S.F.	1.47	1.90	3.37
0620	10,000 S.F.	1.05	1.54	2.59
0640	50,000 S.F.	.84	1.39	2.23
0660	Each additional floor, 500 S.F.	1.29	2.05	3.34
0680	1000 S.F.	1.03	1.89	2.92
0700	2000 S.F.	1.03	1.75	2.78
0720	5000 S.F.	.78	1.50	2.28
0740	10,000 S.F.	.73	1.38	2.11
0760	50,000 S.F.	.72	1.29	2.01
1000	Ordinary hazard, one floor, 500 S.F.	5.55	4.07	9.62
1020	1000 S.F.	3.07	2.79	5.86
1040	2000 S.F.	2.91	2.90	5.81
1060	5000 S.F.	1.61	2.02	3.63
1080	10,000 S.F.	1.28	2.03	3.31
1100	50,000 S.F.	1.23	1.97	3.20
1140	Each additional floor, 500 S.F.	1.53	2.36	3.89
1160	1000 S.F.	1.02	1.92	2.94
1180	2000 S.F.	.97	1.90	2.87
1200	5000 S.F.	1.08	1.79	2.87
1220	10,000 S.F.	.97	1.87	2.84
1240	50,000 S.F.	.96	1.73	2.69
1500	Extra hazard, one floor, 500 S.F.	7.25	5.20	12.45
1520	1000 S.F.	4.13	3.91	8.04
1540	2000 S.F.	2.99	3.71	6.70
1560	5000 S.F.	2.02	3.06	5.08
1580	10,000 S.F.	1.97	2.95	4.92
1600	50,000 S.F.	2.14	2.86	5
1660	Each additional floor, 500 S.F.	1.76	2.91	4.67
1680	1000 S.F.	1.49	2.74	4.23
1700	2000 S.F.	1.39	2.75	4.14
1720	5000 S.F.	1.19	2.44	3.63
1740	10,000 S.F.	1.30	2.23	3.53
1760	50,000 S.F.	1.24	2.09	3.33
2020	Grooved steel, black, sch. 40 pipe, light hazard, one floor, 2000 S.F.	2.69	2.39	5.08
2060	10,000 S.F.	1.02	1.35	2.37
2100	Each additional floor of 2000 S.F.	1	1.41	2.41
2150	10,000 S.F.	.70	1.19	1.89
2200	Ordinary hazard, one floor, 2000 S.F.	2.74	2.50	5.24
2250	10,000 S.F.	1.17	1.71	2.88
2300	Each additional floor, 2000 S.F.	1.05	1.52	2.57
2350	10,000 S.F.	.86	1.55	2.41
2400	Extra hazard, one floor, 2000 S.F.	2.97	3.17	6.14
2450	10,000 S.F.	1.58	2.20	3.78
2500	Each additional floor, 2000 S.F.	1.38	2.26	3.64
2550	10,000 S.F.	1.13	1.92	3.05
3050	Grooved steel, black, sch. 10 pipe light hazard, one floor, 2000 S.F.	2.66	2.37	5.03
3100	10,000 S.F.	1	1.31	2.31
3150	Each additional floor, 2000 S.F.	.97	1.39	2.36
3200	10,000 S.F.	.68	1.15	1.83
3250	Ordinary hazard, one floor, 2000 S.F.	2.67	2.34	5.01
3300	10,000 S.F.	1.03	1.65	2.68
3350	Each additional floor, 2000 S.F.	1.01	1.51	2.52
3400	10,000 S.F.	.84	1.50	2.34
3450	Extra hazard, one floor, 2000 S.F.	2.95	3.15	6.10
3500	10,000 S.F.	1.49	2.16	3.65
3550	Each additional floor, 2000 S.F.	1.36	2.24	3.60
3600	10,000 S.F.	1.08	1.89	2.97
4050	Copper tubing, type M, light hazard, one floor, 2000 S.F.	2.73	2.36	5.09

D40 Fire Protection

D4010 Sprinklers

D4010 350	Preaction Sprinkler Systems	MAT.	[illegible]	[illegible]
4100	10,000 S.F.	1.11	[illegible]	[illegible]
4150	Each additional floor, 2000 S.F.	1.05	[illegible]	[illegible]
4200	10,000 S.F.	.67	[illegible]	[illegible]
4250	Ordinary hazard, one floor, 2000 S.F.	2.82	2.63	[illegible]
4300	10,000 S.F.	1.26	[illegible]	[illegible]
4350	Each additional floor, 2000 S.F.	1.05	1.44	[illegible]
4400	10,000 S.F.	.84	1.27	[illegible]
4450	Extra hazard, one floor, 2000 S.F.	3.08	3.18	[illegible]
4500	10,000 S.F.	2.40	2.39	[illegible]
4550	Each additional floor, 2000 S.F.	1.49	2.27	[illegible]
4600	10,000 S.F.	1.59	2.09	[illegible]
5050	Copper tubing, type M, T-drill system, light hazard, one floor			
5060	2000 S.F.	2.73	2.21	[illegible]
5100	10,000 S.F.	1.06	1.11	2.17
5150	Each additional floor, 2000 S.F.	1.05	1.26	2.31
5200	10,000 S.F.	.74	.96	1.70
5250	Ordinary hazard, one floor, 2000 S.F.	2.74	2.24	4.98
5300	10,000 S.F.	1.21	1.39	2.60
5350	Each additional floor, 2000 S.F.	1.05	1.27	2.32
5400	10,000 S.F.	.90	1.23	2.13
5450	Extra hazard, one floor, 2000 S.F.	2.84	2.63	5.47
5500	10,000 S.F.	1.98	1.73	3.71
5550	Each additional floor, 2000 S.F.	1.25	1.72	2.97
5600	10,000 S.F.	1.17	1.43	2.60

SERVICES D

D4010 Sprinklers

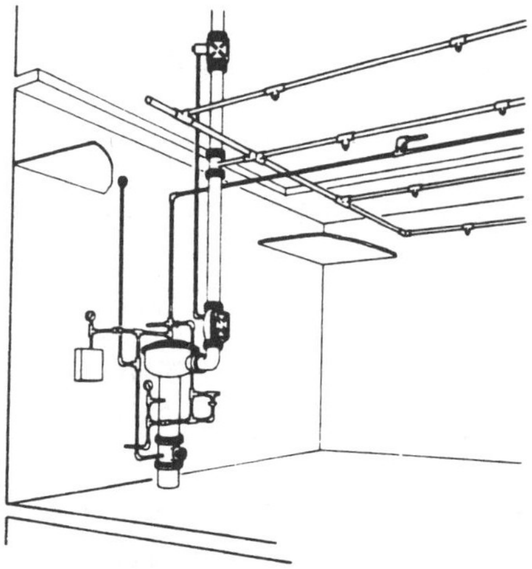

Deluge System: A system employing open sprinklers attached to a piping system connected to a water supply through a valve which is opened by the operation of a heat responsive system installed in the same areas as the sprinklers. When this valve opens, water flows into the piping system and discharges from all sprinklers attached thereto.

All areas are assumed to be open.

System Components	QUANTITY	UNIT	COST EACH MAT.	INST.	TOTAL
SYSTEM D4010 370 0580					
DELUGE SPRINKLER SYSTEM, STEEL BLACK SCH. 40 PIPE					
LIGHT HAZARD, 1 FLOOR, 2000 S.F.					
Valve, gate, iron body 125 lb., OS&Y, flanged, 4″ pipe size	1.000	Ea.	341.25	240	581.25
Valve, swing check w/ball drip, CI w/brass ftngs., 4″ pipe size	1.000	Ea.	135	236.25	371.25
Valve, swing check, bronze, 125 lb., regrinding disc, 2-1/2″ pipe size	1.000	Ea.	199.50	47.63	247.13
Valve, angle, bronze, 150 lb., rising stem, threaded, 2″ pipe size	1.000	Ea.	285	36.38	321.38
*Alarm valve, 2-1/2″ pipe size	1.000	Ea.	701.25	236.25	937.50
Alarm, water motor, complete with gong	1.000	Ea.	155.25	97.50	252.75
Fire alarm horn, electric	1.000	Ea.	30	58.13	88.13
Thermostatic release for release line	2.000	Ea.	495	39	534
Pipe, steel, black, schedule 40, 4″ diam.	10.000	L.F.	157.50	210.38	367.88
Deluge valve trim, pressure relief, emergency release, gauge, 4″ pipe size	1.000	Ea.	1,218.75	705	1,923.75
Deluge system pressured monitoring panel, 120V	1.000	Ea.	727.50	21.75	749.25
Pipe, steel, black, schedule 40, threaded, cplg & hngr 10′ OC 2-1/2″ diam.	20.000	L.F.	181.50	286.50	468
Pipe, steel, black, schedule 40, threaded, cplg & hngr 10′ OC 2″ diam.	12.500	L.F.	73.13	140.16	213.29
Pipe, steel, black, schedule 40, threaded, cplg & hngr 10′ OC 1-1/4″ diam.	37.500	L.F.	140.63	302.34	442.97
Pipe, steel, black, schedule 40, threaded, cplg & hngr 10′ OC 1″ diam.	112.000	L.F.	320.88	840	1,160.88
Pipe, Tee, malleable iron, black, 150 lb. threaded, 4″ pipe size	2.000	Ea.	214.50	358.50	573
Pipe, Tee, malleable iron, black, 150 lb. threaded, 2-1/2″ pipe size	2.000	Ea.	60.75	159	219.75
Pipe, Tee, malleable iron, black, 150 lb. threaded, 2″ pipe size	1.000	Ea.	14.03	65.25	79.28
Pipe, Tee, malleable iron, black, 150 lb. threaded, 1-1/4″ pipe size	5.000	Ea.	33	256.88	289.88
Pipe, Tee, malleable iron, black, 150 lb. threaded, 1″ pipe size	4.000	Ea.	16.35	199.50	215.85
Pipe, 90° elbow, malleable iron, black, 150 lb. threaded 1″ pipe size	6.000	Ea.	15.66	184.50	200.16
Sprinkler head, std spray, brass 135°-286° F 1/2″ NPT, 3/8″ orifice	9.720	Ea.	88.80	390	478.80
Air compressor, auto, complete, 200 Gal sprinkler sys. cap., 1/3 HP	1.000	Ea.	615	300	915
*Standpipe connection, wall, flush w/plug & chain 2-1/2″ x 2-1/2″	1.000	Ea.	83.25	141	224.25
Valve, gate, bronze, 300 psi, NRS, class 150, threaded, 1″ pipe size	1.000	Ea.	25.13	21	46.13
TOTAL			6,328.61	5,572.90	11,901.51
COST PER S.F.			3.16	2.79	5.95
*Not included in systems under 2000 S.F.					

D4010 370	Deluge Sprinkler Systems		COST PER S.F. MAT.	INST.	TOTAL
0520	Deluge sprinkler systems, steel, black, sch. 40 pipe	R211313 -20			
0530	Light hazard, one floor, 500 S.F.		6.35	3.80	10.15

D SERVICES

D4010 370	Deluge Sprinkler Systems	COST PER S.F.		
		MAT.	INST.	TOTAL
0560	1000 S.F.	3.60	2.72	6.32
0580	2000 S.F.	3.16	2.79	5.95
0600	5000 S.F.	1.65	1.90	3.55
0620	10,000 S.F.	1.13	1.54	2.67
0640	50,000 S.F.	.85	1.39	2.24
0660	Each additional floor, 500 S.F.	1.29	2.05	3.34
0680	1000 S.F.	1.03	1.89	2.92
0700	2000 S.F.	1.03	1.75	2.78
0720	5000 S.F.	.78	1.50	2.28
0740	10,000 S.F.	.73	1.38	2.11
0760	50,000 S.F.	.72	1.29	2.01
1000	Ordinary hazard, one floor, 500 S.F.	6.95	4.35	11.30
1020	1000 S.F.	3.61	2.81	6.42
1040	2000 S.F.	3.35	2.91	6.26
1060	5000 S.F.	1.79	2.02	3.81
1080	10,000 S.F.	1.36	2.03	3.39
1100	50,000 S.F.	1.27	2.02	3.29
1140	Each additional floor, 500 S.F.	1.53	2.36	3.89
1160	1000 S.F.	1.02	1.92	2.94
1180	2000 S.F.	.97	1.90	2.87
1200	5000 S.F.	.97	1.65	2.62
1220	10,000 S.F.	.90	1.65	2.55
1240	50,000 S.F.	.90	1.59	2.49
1500	Extra hazard, one floor, 500 S.F.	8.35	5.25	13.60
1520	1000 S.F.	4.89	4.06	8.95
1540	2000 S.F.	3.43	3.72	7.15
1560	5000 S.F.	1.98	2.81	4.79
1580	10,000 S.F.	1.93	2.70	4.63
1600	50,000 S.F.	2.10	2.63	4.73
1660	Each additional floor, 500 S.F.	1.76	2.91	4.67
1680	1000 S.F.	1.49	2.74	4.23
1700	2000 S.F.	1.39	2.75	4.14
1720	5000 S.F.	1.19	2.44	3.63
1740	10,000 S.F.	1.34	2.31	3.65
1760	50,000 S.F.	1.33	2.23	3.56
2000	Grooved steel, black, sch. 40 pipe, light hazard, one floor			
2020	2000 S.F.	3.14	2.40	5.54
2060	10,000 S.F.	1.11	1.37	2.48
2100	Each additional floor, 2,000 S.F.	1	1.41	2.41
2150	10,000 S.F.	.70	1.19	1.89
2200	Ordinary hazard, one floor, 2000 S.F.	2.74	2.50	5.24
2250	10,000 S.F.	1.25	1.71	2.96
2300	Each additional floor, 2000 S.F.	1.05	1.52	2.57
2350	10,000 S.F.	.86	1.55	2.41
2400	Extra hazard, one floor, 2000 S.F.	3.41	3.18	6.59
2450	10,000 S.F.	1.67	2.21	3.88
2500	Each additional floor, 2000 S.F.	1.38	2.26	3.64
2550	10,000 S.F.	1.13	1.92	3.05
3000	Grooved steel, black, sch. 10 pipe, light hazard, one floor			
3050	2000 S.F.	2.79	2.27	5.06
3100	10,000 S.F.	1.08	1.31	2.39
3150	Each additional floor, 2000 S.F.	.97	1.39	2.36
3200	10,000 S.F.	.68	1.15	1.83
3250	Ordinary hazard, one floor, 2000 S.F.	3.14	2.50	5.64
3300	10,000 S.F.	1.11	1.65	2.76
3350	Each additional floor, 2000 S.F.	1.01	1.51	2.52
3400	10,000 S.F.	.84	1.50	2.34
3450	Extra hazard, one floor, 2000 S.F.	3.39	3.16	6.55

D4010 Sprinklers

D4010 370	Deluge Sprinkler Systems	COST PER S.F.		
		MAT.	INST.	TOTAL
3500	10,000 S.F.	1.57	2.16	3.73
3550	Each additional floor, 2000 S.F.	1.36	2.24	3.60
3600	10,000 S.F.	1.08	1.89	2.97
4000	Copper tubing, type M, light hazard, one floor			
4050	2000 S.F.	3.17	2.37	5.54
4100	10,000 S.F.	1.19	1.33	2.52
4150	Each additional floor, 2000 S.F.	1.05	1.41	2.46
4200	10,000 S.F.	.67	1.17	1.84
4250	Ordinary hazard, one floor, 2000 S.F.	3.26	2.64	5.90
4300	10,000 S.F.	1.34	1.57	2.91
4350	Each additional floor, 2000 S.F.	1.05	1.44	2.49
4400	10,000 S.F.	.84	1.27	2.11
4450	Extra hazard, one floor, 2000 S.F.	3.52	3.19	6.71
4500	10,000 S.F.	2.51	2.40	4.91
4550	Each additional floor, 2000 S.F.	1.49	2.27	3.76
4600	10,000 S.F.	1.59	2.09	3.68
5000	Copper tubing, type M, T-drill system, light hazard, one floor			
5050	2000 S.F.	3.17	2.22	5.39
5100	10,000 S.F.	1.14	1.11	2.25
5150	Each additional floor, 2000 S.F.	1.05	1.28	2.33
5200	10,000 S.F.	.74	.96	1.70
5250	Ordinary hazard, one floor, 2000 S.F.	3.18	2.25	5.43
5300	10,000 S.F.	1.29	1.39	2.68
5350	Each additional floor, 2000 S.F.	1.05	1.26	2.31
5400	10,000 S.F.	.90	1.23	2.13
5450	Extra hazard, one floor, 2000 S.F.	3.28	2.64	5.92
5500	10,000 S.F.	2.06	1.73	3.79
5550	Each additional floor, 2000 S.F.	1.25	1.72	2.97
5600	10,000 S.F.	1.17	1.43	2.60

D4010 Sprinklers

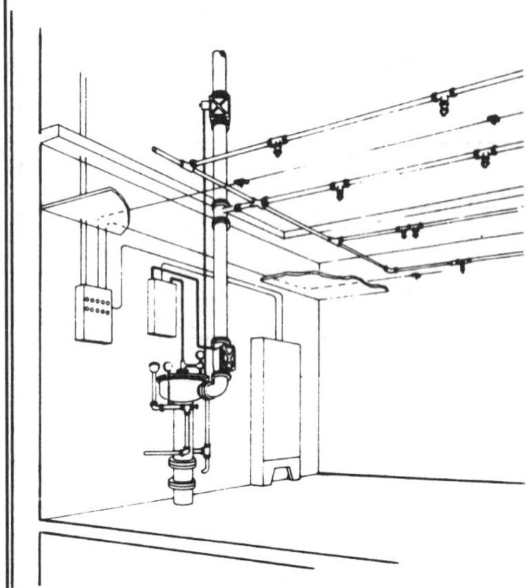

Firecycle is a fixed fire protection sprinkler system utilizing water as its extinguishing agent. It is a time delayed, recycling, preaction type which automatically shuts the water off when heat is reduced below the detector operating temperature and turns the water back on when that temperature is exceeded.

The system senses a fire condition through a closed circuit electrical detector system which controls water flow to the fire automatically. Batteries supply up to 90 hour emergency power supply for system operation. The piping system is dry (until water is required) and is monitored with pressurized air. Should any leak in the system piping occur, an alarm will sound, but water will not enter the system until heat is sensed by a Firecycle detector.

All areas are assumed to be open.

System Components	QUANTITY	UNIT	COST EACH MAT.	COST EACH INST.	COST EACH TOTAL
SYSTEM D4010 390 0580					
FIRECYCLE SPRINKLER SYSTEM, STEEL BLACK SCH. 40 PIPE					
LIGHT HAZARD, ONE FLOOR, 2000 S.F.					
Valve, gate, iron body 125 lb., OS&Y, flanged, 4″ pipe size	1.000	Ea.	341.25	240	581.25
Valve, angle, bronze, 150 lb., rising stem, threaded, 2″ pipe size	1.000	Ea.	285	36.38	321.38
Valve, swing check, bronze, 125 lb., regrinding disc, 2-1/2″ pipe size	1.000	Ea.	199.50	47.63	247.13
*Alarm valve, 2-1/2″ pipe size	1.000	Ea.	701.25	236.25	937.50
Alarm, water motor, complete with gong	1.000	Ea.	155.25	97.50	252.75
Pipe, steel, black, schedule 40, 4″ diam.	10.000	L.F.	157.50	210.38	367.88
Fire alarm, horn, electric	1.000	Ea.	30	58.13	88.13
Pipe, steel, black, schedule 40, threaded, cplg & hngr 10′ OC 2-1/2″ diam.	20.000	L.F.	181.50	286.50	468
Pipe, steel, black, schedule 40, threaded, cplg & hngr 10′ OC 2″ diam.	12.500	L.F.	73.13	140.16	213.29
Pipe, steel, black, schedule 40, threaded, cplg & hngr 10′ OC 1-1/4″ diam.	37.500	L.F.	140.63	302.34	442.97
Pipe, steel, black, schedule 40, threaded, cplg & hngr 10′ OC 1″ diam.	112.000	L.F.	320.88	840	1,160.88
Pipe, Tee, malleable iron, black, 150 lb. threaded, 4″ pipe size	2.000	Ea.	214.50	358.50	573
Pipe, Tee, malleable iron, black, 150 lb. threaded, 2-1/2″ pipe size	2.000	Ea.	60.75	159	219.75
Pipe, Tee, malleable iron, black, 150 lb. threaded, 2″ pipe size	1.000	Ea.	14.03	65.25	79.28
Pipe, Tee, malleable iron, black, 150 lb. threaded, 1-1/4″ pipe size	5.000	Ea.	33	256.88	289.88
Pipe, Tee, malleable iron, black, 150 lb. threaded, 1″ pipe size	4.000	Ea.	16.35	199.50	215.85
Pipe, 90° elbow, malleable iron, black, 150 lb. threaded, 1″ pipe size	6.000	Ea.	15.66	184.50	200.16
Sprinkler head std spray, brass 135°-286° F 1/2″ NPT, 3/8″ orifice	12.000	Ea.	88.80	390	478.80
Firecycle controls, incls panel, battery, solenoid valves, press switches	1.000	Ea.	7,256.25	1,481.25	8,737.50
Detector, firecycle system	2.000	Ea.	532.50	48.75	581.25
Firecycle pkg, swing check & flow control valves w/trim 4″ pipe size	1.000	Ea.	1,893.75	705	2,598.75
Air compressor, auto, complete, 200 Gal sprinkler sys. cap., 1/3 HP	1.000	Ea.	615	300	915
*Standpipe connection, wall, flush, brass w/plug & chain 2-1/2″x2-1/2″	1.000	Ea.	83.25	141	224.25
Valve, gate, bronze 300 psi, NRS, class 150, threaded, 1″ diam.	1.000	Ea.	25.13	21	46.13
TOTAL			13,434.86	6,805.90	20,240.76
COST PER S.F.			6.72	3.40	10.12
*Not included in systems under 2000 S.F.					

D4010 390	Firecycle Sprinkler Systems		COST PER S.F. MAT.	COST PER S.F. INST.	COST PER S.F. TOTAL
0520	Firecycle sprinkler systems, steel black sch. 40 pipe	R211313-20			
0530	Light hazard, one floor, 500 S.F.		21.50	7.40	28.90

Important: See the Reference Section for critical supporting data - Reference Numbers and City Cost Indexes

D4010 Sprinklers

D4010 390	Firecycle Sprinkler Systems	COST PER S.F. MAT.	COST PER S.F. INST.	COST PER S.F. TOTAL
0560	1000 S.F.	11.10	4.63	15.73
0580	2000 S.F.	6.70	3.41	10.11
0600	5000 S.F.	3.07	2.15	5.22
0620	10,000 S.F.	1.89	1.68	3.57
0640	50,000 S.F.	1.03	1.42	2.45
0660	Each additional floor of 500 S.F.	1.33	2.06	3.39
0680	1000 S.F.	1.05	1.89	2.94
0700	2000 S.F.	.91	1.74	2.65
0720	5000 S.F.	.79	1.50	2.29
0740	10,000 S.F.	.77	1.39	2.16
0760	50,000 S.F.	.74	1.29	2.03
1000	Ordinary hazard, one floor, 500 S.F.	21.50	7.70	29.20
1020	1000 S.F.	11.10	4.59	15.69
1040	2000 S.F.	6.80	3.51	10.31
1060	5000 S.F.	3.21	2.27	5.48
1080	10,000 S.F.	2.12	2.17	4.29
1100	50,000 S.F.	1.60	2.25	3.85
1140	Each additional floor, 500 S.F.	1.57	2.37	3.94
1160	1000 S.F.	1.04	1.92	2.96
1180	2000 S.F.	1.06	1.74	2.80
1200	5000 S.F.	.98	1.65	2.63
1220	10,000 S.F.	.88	1.62	2.50
1240	50,000 S.F.	.89	1.52	2.41
1500	Extra hazard, one floor, 500 S.F.	23.50	8.85	32.35
1520	1000 S.F.	12.05	5.70	17.75
1540	2000 S.F.	7	4.33	11.33
1560	5000 S.F.	3.40	3.06	6.46
1580	10,000 S.F.	2.78	3.07	5.85
1600	50,000 S.F.	2.61	3.40	6.01
1660	Each additional floor, 500 S.F.	1.80	2.92	4.72
1680	1000 S.F.	1.51	2.74	4.25
1700	2000 S.F.	1.41	2.75	4.16
1720	5000 S.F.	1.20	2.44	3.64
1740	10,000 S.F.	1.34	2.24	3.58
1760	50,000 S.F.	1.32	2.15	3.47
2020	Grooved steel, black, sch. 40 pipe, light hazard, one floor			
2030	2000 S.F.	6.70	3.01	9.71
2060	10,000 S.F.	2	2.05	4.05
2100	Each additional floor, 2000 S.F.	1.02	1.41	2.43
2150	10,000 S.F.	.74	1.20	1.94
2200	Ordinary hazard, one floor, 2000 S.F.	6.75	3.12	9.87
2250	10,000 S.F.	2.14	1.95	4.09
2300	Each additional floor, 2000 S.F.	1.07	1.52	2.59
2350	10,000 S.F.	.90	1.56	2.46
2400	Extra hazard, one floor, 2000 S.F.	6.95	3.79	10.74
2450	10,000 S.F.	2.39	2.32	4.71
2500	Each additional floor, 2000 S.F.	1.40	2.26	3.66
2550	10,000 S.F.	1.17	1.93	3.10
3050	Grooved steel, black, sch. 10 pipe light hazard, one floor,			
3060	2000 S.F.	6.65	2.99	9.64
3100	10,000 S.F.	1.84	1.45	3.29
3150	Each additional floor, 2000 S.F.	.99	1.39	2.38
3200	10,000 S.F.	.72	1.16	1.88
3250	Ordinary hazard, one floor, 2000 S.F.	6.70	3.11	9.81
3300	10,000 S.F.	1.99	1.80	3.79
3350	Each additional floor, 2000 S.F.	1.03	1.51	2.54
3400	10,000 S.F.	.88	1.51	2.39
3450	Extra hazard, one floor, 2000 S.F.	6.95	3.77	10.72

D4010 Sprinklers

D4010 390	Firecycle Sprinkler Systems	COST PER S.F.		
		MAT.	INST.	TOTAL
3500	10,000 S.F.	2.32	2.28	4.60
3550	Each additional floor, 2000 S.F.	1.38	2.24	3.62
3600	10,000 S.F.	1.12	1.90	3.02
4060	Copper tubing, type M, light hazard, one floor, 2000 S.F.	6.75	2.98	9.73
4100	10,000 S.F.	1.95	1.47	3.42
4150	Each additional floor, 2000 S.F.	1.07	1.41	2.48
4200	10,000 S.F.	.83	1.19	2.02
4250	Ordinary hazard, one floor, 2000 S.F.	6.80	3.25	10.05
4300	10,000 S.F.	2.10	1.71	3.81
4350	Each additional floor, 2000 S.F.	1.07	1.44	2.51
4400	10,000 S.F.	.86	1.26	2.12
4450	Extra hazard, one floor, 2000 S.F.	7.10	3.80	10.90
4500	10,000 S.F.	3.30	2.55	5.85
4550	Each additional floor, 2000 S.F.	1.51	2.27	3.78
4600	10,000 S.F.	1.63	2.10	3.73
5060	Copper tubing, type M, T-drill system, light hazard, one floor 2000 S.F.	6.75	2.83	9.58
5100	10,000 S.F.	1.90	1.25	3.15
5150	Each additional floor, 2000 S.F.	1.17	1.33	2.50
5200	10,000 S.F.	.78	.97	1.75
5250	Ordinary hazard, one floor, 2000 S.F.	6.75	2.86	9.61
5300	10,000 S.F.	2.05	1.53	3.58
5350	Each additional floor, 2000 S.F.	1.07	1.26	2.33
5400	10,000 S.F.	.94	1.24	2.18
5450	Extra hazard, one floor, 2000 S.F.	6.85	3.25	10.10
5500	10,000 S.F.	2.81	1.85	4.66
5550	Each additional floor, 2000 S.F.	1.27	1.72	2.99
5600	10,000 S.F.	1.21	1.44	2.65

D4010 Sprinklers

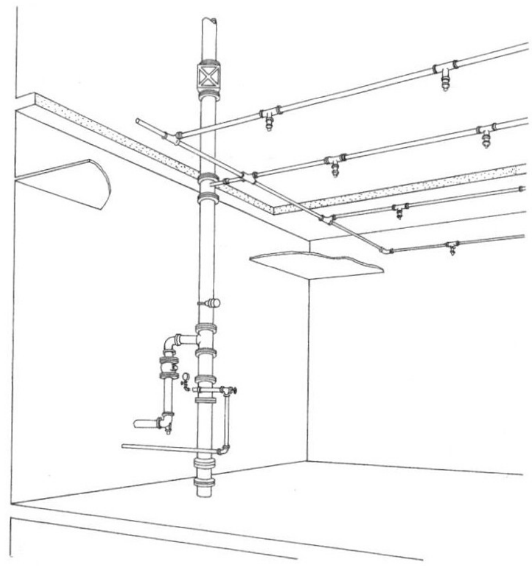

Wet Pipe System. A system employing automatic sprinklers attached to a piping system containing water and connected to a water supply so that water discharges immediately from sprinklers opened by heat from a fire.

All areas are assumed to be open.

System Components	QUANTITY	UNIT	COST EACH MAT.	COST EACH INST.	COST EACH TOTAL
SYSTEM D4010 410 0580					
WET PIPE SPRINKLER, STEEL, BLACK, SCH. 40 PIPE					
LIGHT HAZARD, ONE FLOOR, 2000 S.F.					
Valve, gate, iron body, 125 lb., OS&Y, flanged, 4″ diam.	1.000	Ea.	341.25	240	581.25
Valve, swing check, bronze, 125 lb., regrinding disc, 2-1/2″ pipe size	1.000	Ea.	199.50	47.63	247.13
Valve, angle, bronze, 150 lb., rising stem, threaded, 2″ diam.	1.000	Ea.	285	36.38	321.38
*Alarm valve, 2-1/2″ pipe size	1.000	Ea.	701.25	236.25	937.50
Alarm, water motor, complete with gong	1.000	Ea.	155.25	97.50	252.75
Valve, swing check, w/balldrip CI with brass trim 4″ pipe size	1.000	Ea.	135	236.25	371.25
Pipe, steel, black, schedule 40, 4″ diam.	10.000	L.F.	157.50	210.38	367.88
*Flow control valve, trim & gauges, 4″ pipe size	1.000	Set	1,350	532.50	1,882.50
Fire alarm horn, electric	1.000	Ea.	30	58.13	88.13
Pipe, steel, black, schedule 40, threaded, cplg & hngr 10′ OC, 2-1/2″ diam.	20.000	L.F.	181.50	286.50	468
Pipe, steel, black, schedule 40, threaded, cplg & hngr 10′ OC, 2″ diam.	12.500	L.F.	73.13	140.16	213.29
Pipe, steel, black, schedule 40, threaded, cplg & hngr 10′ OC, 1-1/4″ diam.	37.500	L.F.	140.63	302.34	442.97
Pipe steel, black, schedule 40, threaded cplg & hngr 10′ OC, 1″ diam.	112.000	L.F.	320.88	840	1,160.88
Pipe Tee, malleable iron black, 150 lb. threaded, 4″ pipe size	2.000	Ea.	214.50	358.50	573
Pipe Tee, malleable iron black, 150 lb. threaded, 2-1/2″ pipe size	2.000	Ea.	60.75	159	219.75
Pipe Tee, malleable iron black, 150 lb. threaded, 2″ pipe size	1.000	Ea.	14.03	65.25	79.28
Pipe Tee, malleable iron black, 150 lb. threaded, 1-1/4″ pipe size	5.000	Ea.	33	256.88	289.88
Pipe Tee, malleable iron black, 150 lb. threaded, 1″ pipe size	4.000	Ea.	16.35	199.50	215.85
Pipe 90° elbow, malleable iron black, 150 lb. threaded, 1″ pipe size	6.000	Ea.	15.66	184.50	200.16
Sprinkler head, standard spray, brass 135°-286° F 1/2″ NPT, 3/8″ orifice	12.000	Ea.	88.80	390	478.80
Valve, gate, bronze, NRS, class 150, threaded, 1″ pipe size	1.000	Ea.	25.13	21	46.13
*Standpipe connection, wall, single, flush w/plug & chain 2-1/2″x2-1/2″	1.000	Ea.	83.25	141	224.25
TOTAL			4,622.36	5,039.65	9,662.01
COST PER S.F.			2.31	2.52	4.83

*Not included in systems under 2000 S.F.

D4010 410	Wet Pipe Sprinkler Systems		COST PER S.F. MAT.	COST PER S.F. INST.	COST PER S.F. TOTAL
0520	Wet pipe sprinkler systems, steel, black, sch. 40 pipe				
0530	Light hazard, one floor, 500 S.F.		1.51	2.42	3.93
0560	1000 S.F.	R211313 -20	2.15	2.51	4.66
0580	2000 S.F.		2.31	2.53	4.84
0600	5000 S.F.		1.21	1.79	3
0620	10,000 S.F.		.85	1.48	2.33

D SERVICES

D4010 Sprinklers

D4010 410	Wet Pipe Sprinkler Systems	COST PER S.F.		
		MAT.	INST.	TOTAL
0640	50,000 S.F.	.68	1.37	2.05
0660	Each additional floor, 500 S.F.	.81	2.05	2.86
0680	1000 S.F.	.79	1.92	2.71
0700	2000 S.F.	.78	1.73	2.51
0720	5000 S.F.	.58	1.48	2.06
0740	10,000 S.F.	.56	1.37	1.93
0760	50,000 S.F.	.45	1.09	1.54
1000	Ordinary hazard, one floor, 500 S.F.	1.67	2.58	4.25
1020	1000 S.F.	2.13	2.47	4.60
1040	2000 S.F.	2.37	2.64	5.01
1060	5000 S.F.	1.35	1.91	3.26
1080	10,000 S.F.	1.08	1.97	3.05
1100	50,000 S.F.	1.04	1.93	2.97
1140	Each additional floor, 500 S.F.	1.04	2.32	3.36
1160	1000 S.F.	.77	1.90	2.67
1180	2000 S.F.	.85	1.89	2.74
1200	5000 S.F.	.90	1.81	2.71
1220	10,000 S.F.	.80	1.86	2.66
1240	50,000 S.F.	.80	1.72	2.52
1500	Extra hazard, one floor, 500 S.F.	3.97	3.99	7.96
1520	1000 S.F.	2.66	3.47	6.13
1540	2000 S.F.	2.58	3.57	6.15
1560	5000 S.F.	1.92	3.12	5.04
1580	10,000 S.F.	1.76	2.93	4.69
1600	50,000 S.F.	1.93	2.82	4.75
1660	Each additional floor, 500 S.F.	1.27	2.87	4.14
1680	1000 S.F.	1.24	2.72	3.96
1700	2000 S.F.	1.14	2.73	3.87
1720	5000 S.F.	.99	2.42	3.41
1740	10,000 S.F.	1.13	2.22	3.35
1760	50,000 S.F.	1.10	2.12	3.22
2020	Grooved steel, black sch. 40 pipe, light hazard, one floor, 2000 S.F.	2.27	2.14	4.41
2060	10,000 S.F.	.95	1.35	2.30
2100	Each additional floor, 2000 S.F.	.75	1.39	2.14
2150	10,000 S.F.	.53	1.18	1.71
2200	Ordinary hazard, one floor, 2000 S.F.	2.32	2.25	4.57
2250	10,000 S.F.	.97	1.65	2.62
2300	Each additional floor, 2000 S.F.	.80	1.50	2.30
2350	10,000 S.F.	.69	1.54	2.23
2400	Extra hazard, one floor, 2000 S.F.	2.55	2.92	5.47
2450	10,000 S.F.	1.36	2.14	3.50
2500	Each additional floor, 2000 S.F.	1.13	2.24	3.37
2550	10,000 S.F.	.96	1.91	2.87
3050	Grooved steel black sch. 10 pipe, light hazard, one floor, 2000 S.F.	2.24	2.12	4.36
3100	10,000 S.F.	.80	1.25	2.05
3150	Each additional floor, 2000 S.F.	.72	1.37	2.09
3200	10,000 S.F.	.51	1.14	1.65
3250	Ordinary hazard, one floor, 2000 S.F.	2.28	2.24	4.52
3300	10,000 S.F.	.95	1.60	2.55
3350	Each additional floor, 2000 S.F.	.76	1.49	2.25
3400	10,000 S.F.	.67	1.49	2.16
3450	Extra hazard, one floor, 2000 S.F.	2.53	2.90	5.43
3500	10,000 S.F.	1.29	2.10	3.39
3550	Each additional floor, 2000 S.F.	1.11	2.22	3.33
3600	10,000 S.F.	.91	1.88	2.79
4050	Copper tubing, type M, light hazard, one floor, 2000 S.F.	2.31	2.11	4.42
4100	10,000 S.F.	.91	1.27	2.18
4150	Each additional floor, 2000 S.F.	.80	1.39	2.19

Important: See the Reference Section for critical supporting data - Reference Numbers and City Cost Indexes

D4010 Sprinklers

D4010 410	Wet Pipe Sprinkler Systems	COST PER S.F.		
		MAT.	INST.	TOTAL
4200	10,000 S.F.	.62	1.17	1.79
4250	Ordinary hazard, one floor, 2000 S.F.	2.40	2.38	4.78
4300	10,000 S.F.	1.06	1.51	2.57
4350	Each additional floor, 2000 S.F.	.92	1.54	2.46
4400	10,000 S.F.	.75	1.37	2.12
4450	Extra hazard, one floor, 2000 S.F.	2.66	2.93	5.59
4500	10,000 S.F.	2.20	2.33	4.53
4550	Each additional floor, 2000 S.F.	1.24	2.25	3.49
4600	10,000 S.F.	1.42	2.08	3.50
5050	Copper tubing, type M, T-drill system, light hazard, one floor			
5060	2000 S.F.	2.31	1.96	4.27
5100	10,000 S.F.	.86	1.05	1.91
5150	Each additional floor, 2000 S.F.	.80	1.24	2.04
5200	10,000 S.F.	.57	.95	1.52
5250	Ordinary hazard, one floor, 2000 S.F.	2.32	1.99	4.31
5300	10,000 S.F.	1.01	1.33	2.34
5350	Each additional floor, 2000 S.F.	.80	1.24	2.04
5400	10,000 S.F.	.73	1.22	1.95
5450	Extra hazard, one floor, 2000 S.F.	2.42	2.38	4.80
5500	10,000 S.F.	1.78	1.67	3.45
5550	Each additional floor, 2000 S.F.	1.06	1.74	2.80
5600	10,000 S.F.	1	1.42	2.42

D4020 Standpipes

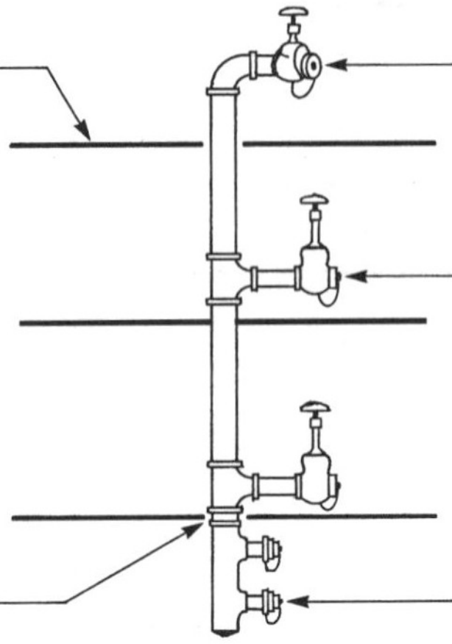

System Components	QUANTITY	UNIT	COST PER FLOOR MAT.	INST.	TOTAL
SYSTEM D4020 310 0560					
WET STANDPIPE RISER, CLASS I, STEEL, BLACK, SCH. 40 PIPE, 10′ HEIGHT					
4″ DIAMETER PIPE, ONE FLOOR					
Pipe, steel, black, schedule 40, threaded, 4″ diam.	20.000	L.F.	460	530	990
Pipe, Tee, malleable iron, black, 150 lb. threaded, 4″ pipe size	2.000	Ea.	286	478	764
Pipe, 90° elbow, malleable iron, black, 150 lb. threaded 4″ pipe size	1.000	Ea.	91	159	250
Pipe, nipple, steel, black, schedule 40, 2-1/2″ pipe size x 3″ long	2.000	Ea.	19.50	120	139.50
Fire valve, gate, 300 lb., brass w/handwheel, 2-1/2″ pipe size	1.000	Ea.	163	74.50	237.50
Fire valve, pressure restricting, adj., rgh. brs., 2-1/2″ pipe size	1.000	Ea.	218	149	367
Valve, swing check, w/ball drip, CI w/brs. ftngs., 4″ pipe size	1.000	Ea.	180	315	495
Standpipe conn wall dble. flush brs. w/plugs & chains 2-1/2″x2-1/2″x4″	1.000	Ea.	360	188	548
Valve, swing check, bronze, 125 lb., regrinding disc, 2-1/2″ pipe size	1.000	Ea.	266	63.50	329.50
Roof manifold, fire, w/valves & caps, horiz./vert. brs. 2-1/2″x2-1/2″x4″	1.000	Ea.	113	196	309
Fire, hydrolator, vent & drain, 2-1/2″ pipe size	1.000	Ea.	46	43.50	89.50
Valve, gate, iron body 125 lb., OS&Y, threaded, 4″ pipe size	1.000	Ea.	455	320	775
TOTAL			2,657.50	2,636.50	5,294

D4020 310	Wet Standpipe Risers, Class I	COST PER FLOOR MAT.	INST.	TOTAL
0550	Wet standpipe risers, Class I, steel black sch. 40, 10′ height			
0560	4″ diameter pipe, one floor	2,650	2,625	5,275
0580	Additional floors	710	815	1,525
0600	6″ diameter pipe, one floor	4,625	4,625	9,250
0620	Additional floors R211226-20	1,250	1,300	2,550
0640	8″ diameter pipe, one floor	7,100	5,600	12,700
0660	Additional floors	1,800	1,600	3,400
0680				

D4020 310	Wet Standpipe Risers, Class II	COST PER FLOOR MAT.	INST.	TOTAL
1030	Wet standpipe risers, Class II, steel black sch. 40, 10′ height			
1040	2″ diameter pipe, one floor	1,150	950	2,100
1060	Additional floors	390	365	755
1080	2-1/2″ diameter pipe, one floor	1,525	1,400	2,925
1100	Additional floors	440	430	870
1120				

D4020 Standpipes

D4020 310	Wet Standpipe Risers, Class III	COST PER FLOOR		
		MAT.	INST.	TOTAL
1530	Wet standpipe risers, Class III, steel black sch. 40, 10′ height			
1540	4″ diameter pipe, one floor	2,725	2,625	5,350
1560	Additional floors	570	680	1,250
1580	6″ diameter pipe, one floor	4,700	4,625	9,325
1600	Additional floors	1,275	1,300	2,575
1620	8″ diameter pipe, one floor	7,150	5,600	12,750
1640	Additional floors	1,825	1,600	3,425

D4020 Standpipes

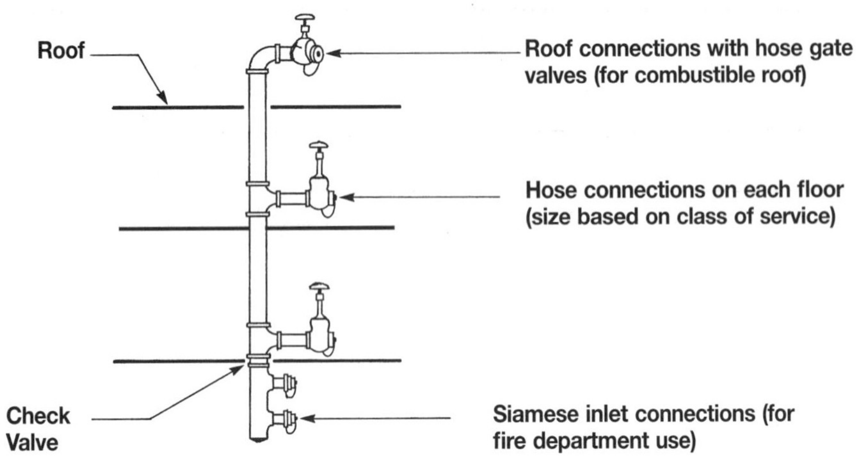

System Components	QUANTITY	UNIT	COST PER FLOOR MAT.	INST.	TOTAL
SYSTEM D4020 330 0540					
DRY STANDPIPE RISER, CLASS I, PIPE, STEEL, BLACK, SCH 40, 10′ HEIGHT					
4″ DIAMETER PIPE, ONE FLOOR					
Pipe, steel, black, schedule 40, threaded, 4″ diam.	20.000	L.F.	460	530	990
Pipe, Tee, malleable iron, black, 150 lb. threaded, 4″ pipe size	2.000	Ea.	286	478	764
Pipe, 90° elbow, malleable iron, black, 150 lb. threaded 4″ pipe size	1.000	Ea.	91	159	250
Pipe, nipple, steel, black, schedule 40, 2-1/2″ pipe size x 3″ long	2.000	Ea.	19.50	120	139.50
Fire valve gate NRS 300 lb., brass w/handwheel, 2-1/2″ pipe size	1.000	Ea.	163	74.50	237.50
Fire valve, pressure restricting, adj., rgh. brs., 2-1/2″ pipe size	1.000	Ea.	109	74.50	183.50
Standpipe conn wall dble. flush brs. w/plugs & chains 2-1/2″x2-1/2″x4″	1.000	Ea.	360	188	548
Valve swing check w/ball drip CI w/brs. ftngs., 4″pipe size	1.000	Ea.	180	315	495
Roof manifold, fire, w/valves & caps, horiz./vert. brs. 2-1/2″x2-1/2″x4″	1.000	Ea.	113	196	309
TOTAL			1,781.50	2,135	3,916.50

D4020 330	Dry Standpipe Risers, Class I		COST PER FLOOR MAT.	INST.	TOTAL
0530	Dry standpipe riser, Class I, steel black sch. 40, 10′ height				
0540	4″ diameter pipe, one floor		1,775	2,125	3,900
0560	Additional floors	R211226 -20	665	775	1,440
0580	6″ diameter pipe, one floor		3,525	3,675	7,200
0600	Additional floors		1,200	1,250	2,450
0620	8″ diameter pipe, one floor		5,400	4,475	9,875
0640	Additional floors		1,750	1,550	3,300
0660					

D4020 330	Dry Standpipe Risers, Class II	COST PER FLOOR MAT.	INST.	TOTAL
1030	Dry standpipe risers, Class II, steel black sch. 40, 10′ height			
1040	2″ diameter pipe, one floor	985	990	1,975
1060	Additional floor	345	325	670
1080	2-1/2″ diameter pipe, one floor	1,200	1,150	2,350
1100	Additional floors	395	385	780
1120				

D4020 Standpipes

D4020 330	Dry Standpipe Risers, Class III	COST PER FLOOR		
		MAT.	INST.	TOTAL
1530	Dry standpipe risers, Class III, steel black sch. 40, 10′ height			
1540	4″ diameter pipe, one floor	1,800	2,100	3,900
1560	Additional floors	535	700	1,235
1580	6″ diameter pipe, one floor	3,550	3,675	7,225
1600	Additional floors	1,250	1,250	2,500
1620	8″ diameter pipe, one floor	5,425	4,475	9,900
1640	Additional floor	1,775	1,550	3,325

D4020 Standpipes

D4020 410	Fire Hose Equipment	COST EACH		
		MAT.	INST.	TOTAL
0100	Adapters, reducing, 1 piece, FxM, hexagon, cast brass, 2-1/2" x 1-1/2"	39		39
0200	Pin lug, 1-1/2" x 1"	33		33
0250	3" x 2-1/2"	87		87
0300	For polished chrome, add 75% mat.			
0400	Cabinets, D.S. glass in door, recessed, steel box, not equipped			
0500	Single extinguisher, steel door & frame	77	117	194
0550	Stainless steel door & frame	164	117	281
0600	Valve, 2-1/2" angle, steel door & frame	94.50	78	172.50
0650	Aluminum door & frame	115	78	193
0700	Stainless steel door & frame	154	78	232
0750	Hose rack assy, 2-1/2" x 1-1/2" valve & 100' hose, steel door & frame	176	156	332
0800	Aluminum door & frame	260	156	416
0850	Stainless steel door & frame	345	156	501
0900	Hose rack assy,& extinguisher,2-1/2"x1-1/2" valve & hose,steel door & frame	193	188	381
0950	Aluminum	330	188	518
1000	Stainless steel	370	188	558
1550	Compressor, air, dry pipe system, automatic, 200 gal., 1/3 H.P.	820	400	1,220
1600	520 gal., 1 H.P.	970	400	1,370
1650	Alarm, electric pressure switch (circuit closer)	163	20	183
2500	Couplings, hose, rocker lug, cast brass, 1-1/2"	37		37
2550	2-1/2"	48		48
3000	Escutcheon plate, for angle valves, polished brass, 1-1/2"	12.15		12.15
3050	2-1/2"	29		29
3500	Fire pump, electric, w/controller, fittings, relief valve			
3550	4" pump, 30 H.P., 500 G.P.M.	17,100	2,925	20,025
3600	5" pump, 40 H.P., 1000 G.P.M.	25,100	3,300	28,400
3650	5" pump, 100 H.P., 1000 G.P.M.	28,100	3,675	31,775
3700	For jockey pump system, add	3,075	470	3,545
3800	Fire pump, diesel, w/controller, 6" pump, 140 HP, 1500 GPM, 100 psi	76,500	4,925	81,425
3900	Fire pump, diesel, w/controller, 10" pump, 300 HP, 3500 GPM, 100 psi	132,500	8,925	141,425
4100	Fire pump, electric, w/controller, 6" pump, 139 HP, 1500 GPM, 1770 RPM	37,300	4,925	42,225
4200	Fire pump, electric, w/controller, 10" pump, 300 HP, 3500 GPM, 1770 RPM	70,500	8,300	78,800
5000	Hose, per linear foot, synthetic jacket, lined,			
5100	300 lb. test, 1-1/2" diameter	2.04	.36	2.40
5150	2-1/2" diameter	3.38	.43	3.81
5200	500 lb. test, 1-1/2" diameter	2.09	.36	2.45
5250	2-1/2" diameter	3.63	.43	4.06
5500	Nozzle, plain stream, polished brass, 1-1/2" x 10"	33		33
5550	2-1/2" x 15" x 13/16" or 1-1/2"	60.50		60.50
5600	Heavy duty combination adjustable fog and straight stream w/handle 1-1/2"	279		279
5650	2-1/2" direct connection	420		420
6000	Rack, for 1-1/2" diameter hose 100 ft. long, steel	36.50	47	83.50
6050	Brass	56	47	103
6500	Reel, steel, for 50 ft. long 1-1/2" diameter hose	91.50	67	158.50
6550	For 75 ft. long 2-1/2" diameter hose	148	67	215
7050	Siamese, w/plugs & chains, polished brass, sidewalk, 4" x 2-1/2" x 2-1/2"	345	375	720
7100	6" x 2-1/2" x 2-1/2"	580	470	1,050
7200	Wall type, flush, 4" x 2-1/2" x 2-1/2"	360	188	548
7250	6" x 2-1/2" x 2-1/2"	505	204	709
7300	Projecting, 4" x 2-1/2" x 2-1/2"	300	188	488
7350	6" x 2-1/2" x 2-1/2"	495	204	699
7400	For chrome plate, add 15% mat.			
8000	Valves, angle, wheel handle, 300 Lb., rough brass, 1-1/2"	33	43.50	76.50
8050	2-1/2"	56.50	74.50	131
8100	Combination pressure restricting, 1-1/2"	59.50	43.50	103
8150	2-1/2"	121	74.50	195.50
8200	Pressure restricting, adjustable, satin brass, 1-1/2"	77.50	43.50	121
8250	2-1/2"	109	74.50	183.50

D4020 Standpipes

D4020 410	Fire Hose Equipment	COST EACH		
		MAT.	INST.	TOTAL
8300	Hydrolator, vent and drain, rough brass, 1-1/2"	46	43.50	89.50
8350	2-1/2"	46	43.50	89.50
8400	Cabinet assy, incls. adapter, rack, hose, and nozzle	510	283	793

D4090 Other Fire Protection Systems

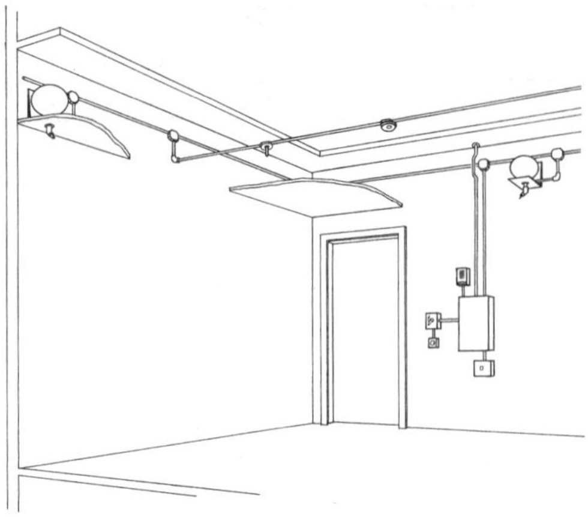

General: Automatic fire protection (suppression) systems other than water sprinklers may be desired for special environments, high risk areas, isolated locations or unusual hazards. Some typical applications would include:

Paint dip tanks
Securities vaults
Electronic data processing
Tape and data storage
Transformer rooms
Spray booths
Petroleum storage
High rack storage

Piping and wiring costs are dependent on the individual application and must be added to the component costs shown below.

All areas are assumed to be open.

D4090 910	Unit Components	COST EACH		
		MAT.	INST.	TOTAL
0020	Detectors with brackets			
0040	Fixed temperature heat detector	31	64.50	95.50
0060	Rate of temperature rise detector	37	64.50	101.50
0080	Ion detector (smoke) detector	82.50	83.50	166
0200	Extinguisher agent			
0240	200 lb FM200, container	6,375	239	6,614
0280	75 lb carbon dioxide cylinder	1,050	159	1,209
0320	Dispersion nozzle			
0340	FM200 1-1/2" dispersion nozzle	55	38	93
0380	Carbon dioxide 3" x 5" dispersion nozzle	55	29.50	84.50
0420	Control station			
0440	Single zone control station with batteries	1,400	520	1,920
0470	Multizone (4) control station with batteries	2,700	1,025	3,725
0490				
0500	Electric mechanical release	138	262	400
0520				
0550	Manual pull station	49.50	88.50	138
0570				
0640	Battery standby power 10" x 10" x 17"	760	129	889
0700				
0740	Bell signalling device	54.50	64.50	119

D4090 920	FM200 Systems	COST PER C.F.		
		MAT.	INST.	TOTAL
0820	Average FM200 system, minimum			1.38
0840	Maximum			2.75

Important: See the Reference Section for critical supporting data - Reference Numbers and City Cost Indexes

D5010 Electrical Service/Distribution

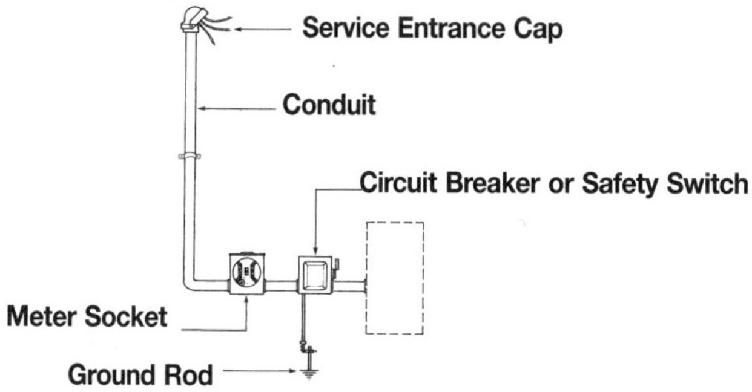

System Components	QUANTITY	UNIT	COST EACH MAT.	COST EACH INST.	COST EACH TOTAL
SYSTEM D5010 120 0220					
SERVICE INSTALLATION, INCLUDES BREAKERS, METERING, 20′ CONDUIT & WIRE					
3 PHASE, 4 WIRE, 60 A					
Circuit breaker, enclosed (NEMA 1), 600 volt, 3 pole, 60 A	1.000	Ea.	550	185	735
Meter socket, single position, 4 terminal, 100 A	1.000	Ea.	39.50	162	201.50
Rigid galvanized steel conduit, 3/4″, including fittings	20.000	L.F.	61.20	129	190.20
Wire, 600V type XHHW, copper stranded #6	.900	C.L.F.	38.70	71.55	110.25
Service entrance cap 3/4″ diameter	1.000	Ea.	8.65	40	48.65
Conduit LB fitting with cover, 3/4″ diameter	1.000	Ea.	12.85	40	52.85
Ground rod, copper clad, 8′ long, 3/4″ diameter	1.000	Ea.	31.50	97.50	129
Ground rod clamp, bronze, 3/4″ diameter	1.000	Ea.	6.35	16.20	22.55
Ground wire, bare armored, #6-1 conductor	.200	C.L.F.	22.40	57.60	80
TOTAL			771.15	798.85	1,570

D5010 120	Electric Service, 3 Phase - 4 Wire	COST EACH MAT.	COST EACH INST.	COST EACH TOTAL
0200	Service installation, includes breakers, metering, 20′ conduit & wire			
0220	3 phase, 4 wire, 120/208 volts, 60 A	770	800	1,570
0240	100 A	965	960	1,925
0280	200 A	1,300	1,475	2,775
0320	400 A	3,050	2,725	5,775
0360	600 A	5,900	3,675	9,575
0400	800 A	7,825	4,400	12,225
0440	1000 A	9,875	5,075	14,950
0480	1200 A	12,000	5,200	17,200
0520	1600 A	23,600	7,475	31,075
0560	2000 A	26,400	8,525	34,925
0570	Add 25% for 277/480 volt			
0580				
0610	1 phase, 3 wire, 120/240 volts, 100 A	460	870	1,330
0620	200 A	960	1,275	2,235

D5020 Lighting and Branch Wiring

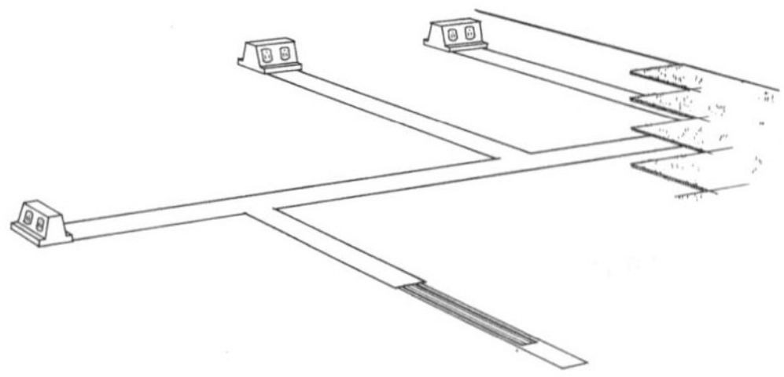

Underfloor Receptacle System

Description: Table D5020 115 includes installed costs of raceways and copper wire from panel to and including receptacle.

National Electrical Code prohibits use of undercarpet system in residential, school or hospital buildings. Can only be used with carpet squares.

Low density = (1) Outlet per 259 S.F. of floor area.

High density = (1) Outlet per 127 S.F. of floor area.

System Components	QUANTITY	UNIT	COST PER S.F. MAT.	INST.	TOTAL
SYSTEM D5020 115 0200					
RECEPTACLE SYSTEMS, UNDERFLOOR DUCT, 5′ ON CENTER, LOW DENSITY					
Underfloor duct 3-1/8″ x 7/8″ w/insert 24″ on center	.190	L.F.	2.74	1.41	4.15
Vertical elbow for underfloor duct, 3-1/8″, included					
Underfloor duct conduit adapter, 2″ x 1-1/4″, included					
Underfloor duct junction box, single duct, 3-1/8″	.003	Ea.	.98	.39	1.37
Underfloor junction box carpet pan	.003	Ea.	.80	.02	.82
Underfloor duct outlet, high tension receptacle	.004	Ea.	.30	.26	.56
Wire 600V type THWN-THHN copper solid #12	.010	C.L.F.	.07	.47	.54
TOTAL			4.89	2.55	7.44

D5020 115	Receptacles, Floor	COST PER S.F. MAT.	INST.	TOTAL
0200	Receptacle systems, underfloor duct, 5′ on center, low density	4.89	2.55	7.44
0240	High density	5.25	3.28	8.53
0280	7′ on center, low density	3.85	2.18	6.03
0320	High density	4.22	2.91	7.13
0400	Poke thru fittings, low density	.98	1.16	2.14
0440	High density	1.94	2.30	4.24
0520	Telepoles, using Romex, low density	.81	.77	1.58
0560	High density	1.61	1.53	3.14
0600	Using EMT, low density	.84	1	1.84
0640	High density	1.71	2.03	3.74
0720	Conduit system with floor boxes, low density	.87	.87	1.74
0760	High density	1.74	1.76	3.50
0840	Undercarpet power system, 3 conductor with 5 conductor feeder, low density	1.30	.31	1.61
0880	High density	2.53	.63	3.16

D5020 Lighting and Branch Wiring

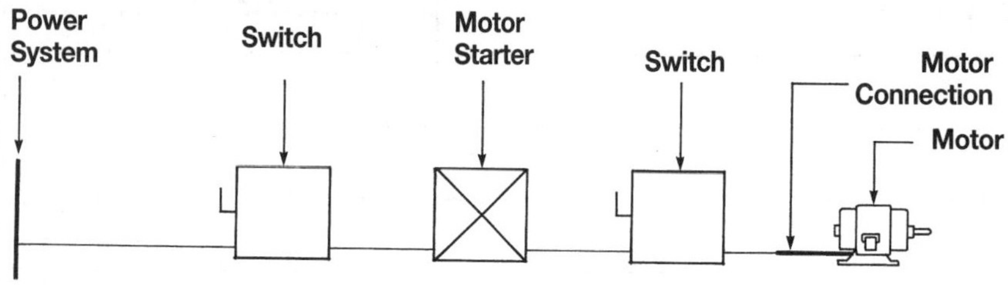

System D5020 145 installed cost of motor wiring using 50′ of rigid conduit and copper wire. **Cost and setting of motor not included.**

System Components	QUANTITY	UNIT	COST EACH MAT.	COST EACH INST.	COST EACH TOTAL
SYSTEM D5020 145 0200					
MOTOR INSTALLATION, SINGLE PHASE, 115V, TO AND INCLUDING 1/3 HP MOTOR SIZE					
Wire 600V type THWN-THHN, copper solid #12	1.250	C.L.F.	8.81	58.75	67.56
Steel intermediate conduit, (IMC) 1/2″ diam.	50.000	L.F.	118	260	378
Magnetic FVNR, 115V, 1/3 HP, size 00 starter	1.000	Ea.	185	129	314
Safety switch, fused, heavy duty, 240V 2P 30 amp	1.000	Ea.	109	148	257
Safety switch, non fused, heavy duty, 600V, 3 phase, 30 A	1.000	Ea.	128	162	290
Flexible metallic conduit, Greenfield 1/2″ diam.	1.500	L.F.	.62	3.89	4.51
Connectors for flexible metallic conduit Greenfield 1/2″ diam.	1.000	Ea.	1.82	6.45	8.27
Coupling for Greenfield to conduit 1/2″ diam flexible metalic conduit	1.000	Ea.	.89	10.35	11.24
Fuse cartridge nonrenewable, 250V 30 amp	1.000	Ea.	1.55	10.35	11.90
TOTAL			553.69	788.79	1,342.48

D5020 145	Motor Installation	COST EACH MAT.	COST EACH INST.	COST EACH TOTAL
0200	Motor installation, single phase, 115V, to and including 1/3 HP motor size	555	790	1,345
0240	To and incl. 1 HP motor size	575	790	1,365
0280	To and incl. 2 HP motor size	615	840	1,455
0320	To and incl. 3 HP motor size	685	855	1,540
0360	230V, to and including 1 HP motor size	555	800	1,355
0400	To and incl. 2 HP motor size	580	800	1,380
0440	To and incl. 3 HP motor size	650	860	1,510
0520	Three phase, 200V, to and including 1-1/2 HP motor size	645	880	1,525
0560	To and incl. 3 HP motor size	690	955	1,645
0600	To and incl. 5 HP motor size	735	1,075	1,810
0640	To and incl. 7-1/2 HP motor size	755	1,100	1,855
0680	To and incl. 10 HP motor size	1,200	1,350	2,550
0720	To and incl. 15 HP motor size	1,550	1,500	3,050
0760	To and incl. 20 HP motor size	1,950	1,750	3,700
0800	To and incl. 25 HP motor size	1,975	1,750	3,725
0840	To and incl. 30 HP motor size	3,075	2,050	5,125
0880	To and incl. 40 HP motor size	3,775	2,450	6,225
0920	To and incl. 50 HP motor size	6,450	2,850	9,300
0960	To and incl. 60 HP motor size	6,650	3,000	9,650
1000	To and incl. 75 HP motor size	8,650	3,425	12,075
1040	To and incl. 100 HP motor size	17,800	4,050	21,850
1080	To and incl. 125 HP motor size	18,200	4,450	22,650
1120	To and incl. 150 HP motor size	22,100	5,225	27,325
1160	To and incl. 200 HP motor size	26,500	6,350	32,850
1240	230V, to and including 1-1/2 HP motor size	615	870	1,485
1280	To and incl. 3 HP motor size	660	945	1,605
1320	To and incl. 5 HP motor size	700	1,050	1,750
1360	To and incl. 7-1/2 HP motor size	700	1,050	1,750
1400	To and incl. 10 HP motor size	1,075	1,275	2,350
1440	To and incl. 15 HP motor size	1,225	1,400	2,625
1480	To and incl. 20 HP motor size	1,825	1,700	3,525
1520	To and incl. 25 HP motor size	1,950	1,750	3,700

D5020 Lighting and Branch Wiring

D5020 145	Motor Installation	COST EACH		
		MAT.	INST.	TOTAL
1560	To and incl. 30 HP motor size	1,975	1,750	3,725
1600	To and incl. 40 HP motor size	3,700	2,375	6,075
1640	To and incl. 50 HP motor size	3,800	2,525	6,325
1680	To and incl. 60 HP motor size	6,450	2,850	9,300
1720	To and incl. 75 HP motor size	7,875	3,225	11,100
1760	To and incl. 100 HP motor size	9,150	3,600	12,750
1800	To and incl. 125 HP motor size	18,000	4,175	22,175
1840	To and incl. 150 HP motor size	19,500	4,750	24,250
1880	To and incl. 200 HP motor size	21,100	5,250	26,350
1960	460V, to and including 2 HP motor size	735	875	1,610
2000	To and incl. 5 HP motor size	780	955	1,735
2040	To and incl. 10 HP motor size	810	1,050	1,860
2080	To and incl. 15 HP motor size	1,050	1,200	2,250
2120	To and incl. 20 HP motor size	1,100	1,300	2,400
2160	To and incl. 25 HP motor size	1,200	1,350	2,550
2200	To and incl. 30 HP motor size	1,525	1,450	2,975
2240	To and incl. 40 HP motor size	1,925	1,575	3,500
2280	To and incl. 50 HP motor size	2,150	1,750	3,900
2320	To and incl. 60 HP motor size	3,225	2,050	5,275
2360	To and incl. 75 HP motor size	3,700	2,250	5,950
2400	To and incl. 100 HP motor size	4,000	2,525	6,525
2440	To and incl. 125 HP motor size	6,650	2,850	9,500
2480	To and incl. 150 HP motor size	8,225	3,200	11,425
2520	To and incl. 200 HP motor size	9,675	3,600	13,275
2600	575V, to and including 2 HP motor size	735	875	1,610
2640	To and incl. 5 HP motor size	780	955	1,735
2680	To and incl. 10 HP motor size	810	1,050	1,860
2720	To and incl. 20 HP motor size	1,050	1,200	2,250
2760	To and incl. 25 HP motor size	1,100	1,300	2,400
2800	To and incl. 30 HP motor size	1,525	1,450	2,975
2840	To and incl. 50 HP motor size	1,625	1,525	3,150
2880	To and incl. 60 HP motor size	3,200	2,025	5,225
2920	To and incl. 75 HP motor size	3,225	2,050	5,275
2960	To and incl. 100 HP motor size	3,700	2,250	5,950
3000	To and incl. 125 HP motor size	6,525	2,800	9,325
3040	To and incl. 150 HP motor size	6,650	2,850	9,500
3080	To and incl. 200 HP motor size	8,400	3,250	11,650

Important: See the Reference Section for critical supporting data - Reference Numbers and City Cost Indexes

D5020 Lighting and Branch Wiring

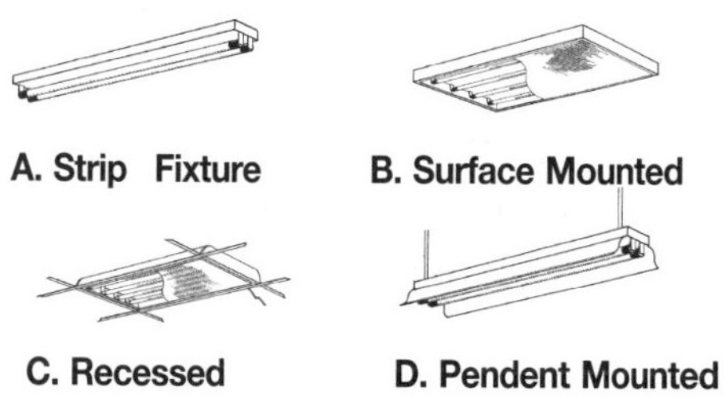

Design Assumptions:

1. A 100 footcandle average maintained level of illumination.
2. Ceiling heights range from 9′ to 11′.
3. Average reflectance values are assumed for ceilings, walls and floors.
4. Cool white (CW) fluorescent lamps with 3150 lumens for 40 watt lamps and 6300 lumens for 8′ slimline lamps.
5. Four 40 watt lamps per 4′ fixture and two 8′ lamps per 8′ fixture.
6. Average fixture efficiency values and spacing to mounting height ratios.
7. Installation labor is average U.S. rate as of January 1.

System Components	QUANTITY	UNIT	COST PER S.F.		
			MAT.	INST.	TOTAL
SYSTEM D5020 208 0520					
FLUORESCENT FIXTURES MOUNTED 9′-11″ ABOVE FLOOR, 100 FC					
TYPE A, 8 FIXTURES PER 400 S.F.					
Steel intermediate conduit, (IMC) 1/2″ diam.	.404	L.F.	.95	2.10	3.05
Wire, 600V, type THWN-THHN, copper, solid, #12	.008	C.L.F.	.06	.38	.44
Fluorescent strip fixture 8′ long, surface mounted, two 75W SL	.020	Ea.	1.13	1.67	2.80
Steel outlet box 4″ concrete	.020	Ea.	.24	.52	.76
Steel outlet box plate with stud, 4″ concrete	.020	Ea.	.10	.13	.23
TOTAL			2.48	4.80	7.28

D5020 208	Fluorescent Fixtures (by Type)	COST PER S.F.		
		MAT.	INST.	TOTAL
0520	Fluorescent fixtures, type A, 8 fixtures per 400 S.F.	2.48	4.80	7.28
0560	11 fixtures per 600 S.F.	2.38	4.65	7.03
0600	17 fixtures per 1000 S.F.	2.29	4.57	6.86
0640	23 fixtures per 1600 S.F.	2.13	4.33	6.46
0680	28 fixtures per 2000 S.F.	2.13	4.33	6.46
0720	41 fixtures per 3000 S.F.	2.10	4.32	6.42
0800	53 fixtures per 4000 S.F.	2.05	4.22	6.27
0840	64 fixtures per 5000 S.F.	2.05	4.22	6.27
0880	Type B, 11 fixtures per 400 S.F.	4.74	7	11.74
0920	15 fixtures per 600 S.F.	4.43	6.75	11.18
0960	24 fixtures per 1000 S.F.	4.33	6.70	11.03
1000	35 fixtures per 1600 S.F.	4.07	6.40	10.47
1040	42 fixtures per 2000 S.F.	4.01	6.40	10.41
1080	61 fixtures per 3000 S.F.	3.97	6.20	10.17
1160	80 fixtures per 4000 S.F.	3.91	6.35	10.26
1200	98 fixtures per 5000 S.F.	3.90	6.30	10.20
1240	Type C, 11 fixtures per 400 S.F.	3.69	7.40	11.09
1280	14 fixtures per 600 S.F.	3.36	6.90	10.26
1320	23 fixtures per 1000 S.F.	3.34	6.85	10.19
1360	34 fixtures per 1600 S.F.	3.26	6.80	10.06
1400	43 fixtures per 2000 S.F.	3.25	6.70	9.95
1440	63 fixtures per 3000 S.F.	3.19	6.65	9.84
1520	81 fixtures per 4000 S.F.	3.14	6.55	9.69
1560	101 fixtures per 5000 S.F.	3.14	6.55	9.69
1600	Type D, 8 fixtures per 400 S.F.	3.40	5.75	9.15
1640	12 fixtures per 600 S.F.	3.39	5.70	9.09
1680	19 fixtures per 1000 S.F.	3.27	5.55	8.82
1720	27 fixtures per 1600 S.F.	3.11	5.40	8.51
1760	34 fixtures per 2000 S.F.	3.09	5.35	8.44
1800	48 fixtures per 3000 S.F.	2.98	5.20	8.18
1880	64 fixtures per 4000 S.F.	2.98	5.20	8.18
1920	79 fixtures per 5000 S.F.	2.98	5.20	8.18

D5020 Lighting and Branch Wiring

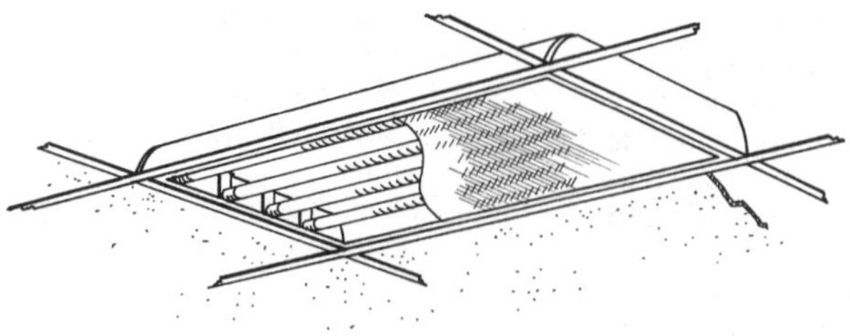

Type C. Recessed, mounted on grid ceiling suspension system, 2′ x 4′, four 40 watt lamps, acrylic prismatic diffusers.

5.3 watts per S.F. for 100 footcandles.

3 watts per S.F. for 57 footcandles.

System Components	QUANTITY	UNIT	COST PER S.F. MAT.	COST PER S.F. INST.	COST PER S.F. TOTAL
SYSTEM D5020 210 0200					
FLUORESCENT FIXTURES RECESS MOUNTED IN CEILING					
1 WATT PER S.F., 20 FC, 5 FIXTURES PER 1000 S.F.					
Steel intermediate conduit, (IMC) 1/2″ diam.	.128	L.F.	.30	.67	.97
Wire, 600 volt, type THW, copper, solid, #12	.003	C.L.F.	.02	.14	.16
Fluorescent fixture, recessed, 2′x 4′, four 40W, w/ lens, for grid ceiling	.005	Ea.	.33	.55	.88
Steel outlet box 4″ square	.005	Ea.	.06	.13	.19
Fixture whip, Greenfield w/#12 THHN wire	.005	Ea.	.02	.03	.05
TOTAL			.73	1.52	2.25

D5020 210	Fluorescent Fixtures (by Wattage)	COST PER S.F. MAT.	COST PER S.F. INST.	COST PER S.F. TOTAL
0190	Fluorescent fixtures recess mounted in ceiling			
0200	1 watt per S.F., 20 FC, 5 fixtures per 1000 S.F.	.73	1.52	2.25
0240	2 watts per S.F., 40 FC, 10 fixtures per 1000 S.F.	1.46	2.99	4.45
0280	3 watts per S.F., 60 FC, 15 fixtures per 1000 S.F	2.20	4.52	6.72
0320	4 watts per S.F., 80 FC, 20 fixtures per 1000 S.F.	2.92	6	8.92
0400	5 watts per S.F., 100 FC, 25 fixtures per 1000 S.F.	3.65	7.50	11.15

Important: See the Reference Section for critical supporting data - Reference Numbers and City Cost Indexes

D5030 Communications and Security

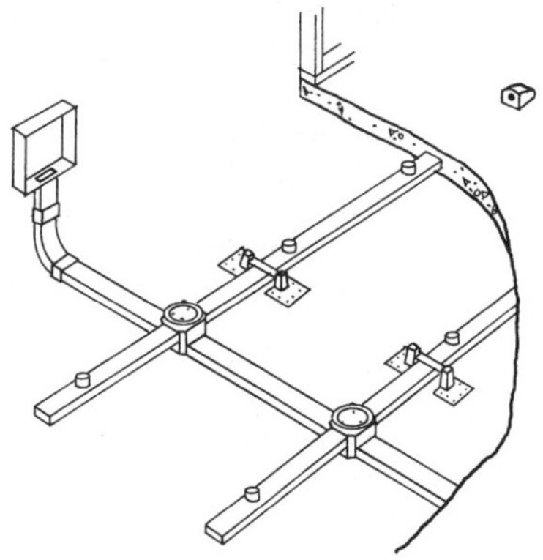

Description: System below includes telephone fitting installed. Does not include cable.

When poke thru fittings and telepoles are used for power, they can also be used for telephones at a negligible additional cost.

System Components	QUANTITY	UNIT	COST PER S.F. MAT.	COST PER S.F. INST.	COST PER S.F. TOTAL
SYSTEM D5030 310 0200					
TELEPHONE SYSTEMS, UNDERFLOOR DUCT, 5′ ON CENTER, LOW DENSITY					
Underfloor duct 7-1/4″ w/insert 2′ O.C. 1-3/8″ x 7-1/4″ super duct	.190	L.F.	4.75	1.97	6.72
Vertical elbow for underfloor superduct, 7-1/4″, included					
Underfloor duct conduit adapter, 2″ x 1-1/4″, included					
Underfloor duct junction box, single duct, 7-1/4″ x 3 1/8″	.003	Ea.	1.13	.39	1.52
Underfloor junction box carpet pan	.003	Ea.	.80	.02	.82
Underfloor duct outlet, low tension	.004	Ea.	.30	.26	.56
TOTAL			6.98	2.64	9.62

D5030 310	Telephone Systems	COST PER S.F. MAT.	COST PER S.F. INST.	COST PER S.F. TOTAL
0200	Telephone systems, underfloor duct, 5′ on center, low density	7	2.64	9.64
0240	5′ on center, high density	7.30	2.90	10.20
0280	7′ on center, low density	5.60	2.19	7.79
0320	7′ on center, high density	5.90	2.45	8.35
0400	Poke thru fittings, low density	.93	.86	1.79
0440	High density	1.84	1.73	3.57
0520	Telepoles, low density	.78	.56	1.34
0560	High density	1.55	1.12	2.67
0640	Conduit system with floor boxes, low density	1	.92	1.92
0680	High density	1.97	1.82	3.79

D5030 Communications and Security

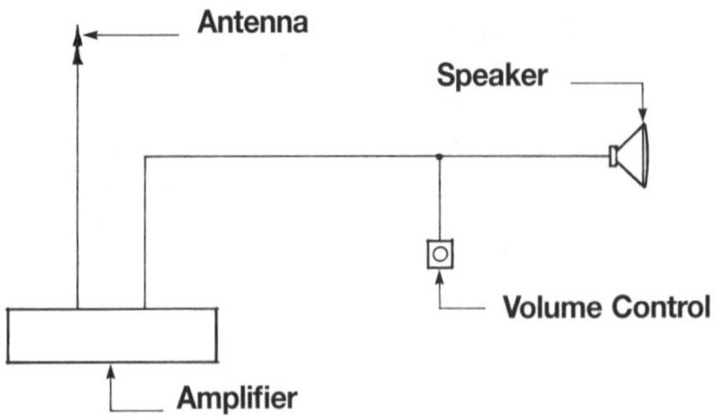

Sound System Includes AM–FM antenna, outlets, rigid conduit, and copper wire.
Fire Detection System Includes pull stations, signals, smoke and heat detectors, rigid conduit, and copper wire.
Intercom System Includes master and remote stations, rigid conduit, and copper wire.
Master Clock System Includes clocks, bells, rigid conduit, and copper wire.
Master TV Antenna Includes antenna, VHF–UHF reception and distribution, rigid conduit, and copper wire.

System Components	QUANTITY	UNIT	COST EACH		
			MAT.	INST.	TOTAL
SYSTEM D5030 910 0220					
SOUND SYSTEM, INCLUDES OUTLETS, BOXES, CONDUIT & WIRE					
Steel intermediate conduit, (IMC) 1/2″ diam	1200.000	L.F.	2,832	6,240	9,072
Wire sound shielded w/drain, #22-2 conductor	15.500	C.L.F.	294.50	999.75	1,294.25
Sound system speakers ceiling or wall	12.000	Ea.	1,236	774	2,010
Sound system volume control	12.000	Ea.	912	774	1,686
Sound system amplifier, 250 Watts	1.000	Ea.	1,525	520	2,045
Sound system antenna, AM FM	1.000	Ea.	190	129	319
Sound system monitor panel	1.000	Ea.	340	129	469
Sound system cabinet	1.000	Ea.	740	520	1,260
Steel outlet box 4″ square	12.000	Ea.	29.52	312	341.52
Steel outlet box 4″ plaster rings	12.000	Ea.	26.76	97.20	123.96
TOTAL			8,125.78	10,494.95	18,620.73

D5030 910	Communication & Alarm Systems	COST EACH		
		MAT.	INST.	TOTAL
0200	Communication & alarm systems, includes outlets, boxes, conduit & wire			
0210	Sound system, 6 outlets	5,650	6,550	12,200
0220	12 outlets	8,125	10,500	18,625
0240	30 outlets	14,200	19,800	34,000
0280	100 outlets	43,700	66,500	110,200
0320	Fire detection systems, 12 detectors	2,850	5,450	8,300
0360	25 detectors	4,925	9,200	14,125
0400	50 detectors	9,550	18,100	27,650
0440	100 detectors	17,200	32,600	49,800
0480	Intercom systems, 6 stations	3,300	4,475	7,775
0520	12 stations	6,675	8,950	15,625
0560	25 stations	11,700	17,200	28,900
0600	50 stations	22,400	32,300	54,700
0640	100 stations	44,100	63,000	107,100
0680	Master clock systems, 6 rooms	4,550	7,325	11,875
0720	12 rooms	7,050	12,500	19,550
0760	20 rooms	9,400	17,700	27,100
0800	30 rooms	16,100	32,600	48,700
0840	50 rooms	26,200	55,000	81,200
0880	100 rooms	50,500	109,000	159,500
0920	Master TV antenna systems, 6 outlets	2,925	4,650	7,575
0960	12 outlets	5,475	8,625	14,100
1000	30 outlets	11,300	20,000	31,300
1040	100 outlets	37,700	65,000	102,700

E EQUIPMENT & FURNISHINGS

E2020 Moveable Furnishings

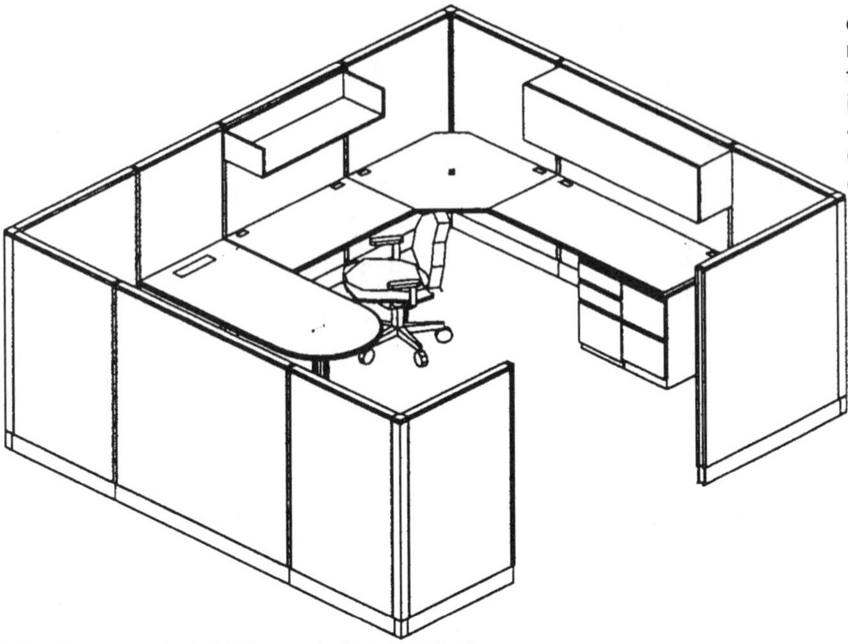

Note: 4′ raceway included in panels (not pre-wired)

The work station shown here represents one of many types and sizes on the market. Costs can vary considerably due to the various parameters involved. For instance, wood systems can cost up to 40% more than metal or fabric systems. Generally, the cost per work station will decrease for "quantity" purchases.

System Components	QUANTITY	UNIT	COST EACH MAT.	COST EACH INST.	COST EACH TOTAL
SYSTEM E2020 410 2600					
WORK STATION - MANAGER					
Acoustical panel, aluminum frame, fabric covered, 30″ wide, 64″ high	1.000	Ea.	315		315
36″ wide, 64″ high	6.000	Ea.	2,040		2,040
42″ wide, 64″ high	2.000	Ea.	790		790
48″ wide, 64″ high	1.000	Ea.	415		415
60″ wide, 64″ high	2.000	Ea.	920		920
Straight connector kit, 64″ high	7.000	Ea.	192.50		192.50
Ell connector kit, 64″ high, 90″ deep	4.000	Ea.	156		156
Panel end cover, 64″ high	2.000	Ea.	55		55
Worksurface, 24″ deep, 42″ wide	1.000	Ea.	160		160
Worksurface, 24″ deep, 60″ wide	1.000	Ea.	215		215
Corner worksurface, 24″ deep, 36″ wide	1.000	Ea.	315		315
Peninsula worksurface, 36″ wide, 66″ long	1.000	Ea.	410		410
Peninsula support column, 29-1/2″ high	1.000	Ea.	95.50		95.50
Cantilever bracket, 20″ deep, 24″ deep	3.000	Ea.	112.50		112.50
Pedestal spacer, 22″ deep, 15″ wide	2.000	Ea.	89		89
Pedestal, floorstanding, 2 box, 1 file	1.000	Ea.	315		315
Pedestal, floorstanding, 2 file, 22″ deep	1.000	Ea.	291		291
Overhead storage cabinet with door, 60″ wide	1.000	Ea.	390		390
Open bookshelf, 42″ wide	1.000	Ea.	116		116
Task light, recessed, 42″ - 48″ wide	1.000	Ea.	131		131
Task light, recessed, 60″ wide	1.000	Ea.	142		142
Electrical power harness, 36″ wide	2.000	Ea.	179		179
Electrical power harness, 42″ wide	1.000	Ea.	188		188
Electrical power harness, 60″ wide	1.000	Ea.	94		94
Base power in-feed cable	1.000	Ea.	99.50		99.50
Duplex receptacle, circuit 1	2.000	Ea.	27.30		27.30
Duplex receptacle, circuit 2	2.000	Ea.	27.30		27.30
Installation	1.000	Ea.		470	470
TOTAL			8,280.60	470	8,750.60

E2020 Moveable Furnishings

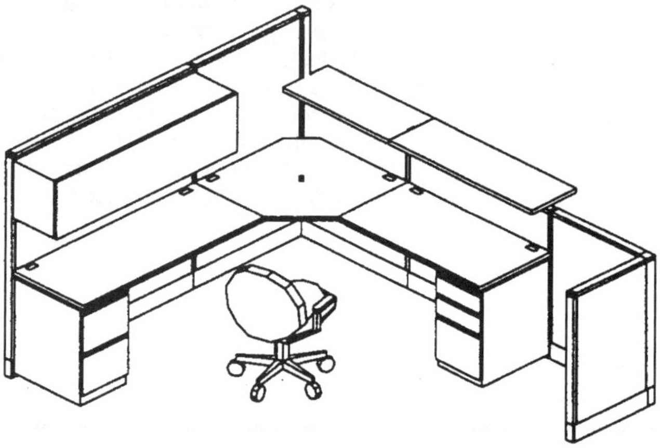

The work station shown here represents one of many types and sizes on the market. Costs can vary considerably due to the various parameters involved. For instance, wood systems can cost up to 40% more than metal or fabric systems. Generally, the cost per work station will decrease for "quantity" purchases.

System Components

System Components	QUANTITY	UNIT	COST EACH MAT.	INST.	TOTAL
SYSTEM E2020 420 2300					
WORK STATION - PROFESSIONAL/EXECUTIVE SECRETARY					
Acoustical panel, aluminum frame, fabric covered, 24″ wide x 43″ high	1.000	Ea.	239		239
30″ wide x 43″ high	1.000	Ea.	266		266
36″ wide x 43″ high	1.000	Ea.	294		294
48″ wide x 43″ high	1.000	Ea.	350		350
36″ wide x 64″ high	1.000	Ea.	340		340
60″ wide x 64″ high	1.000	Ea.	460		460
Straight connector kit, 43″ high	2.000	Ea.	55		55
Straight connector kit, 64″ high	1.000	Ea.	27.50		27.50
Ell connector kit, 43″ high, 90″ deep	1.000	Ea.	66.50		66.50
Ell connector kit, 64″ high	1.000	Ea.	39		39
Panel end cover, 43″ high	1.000	Ea.	27.50		27.50
Panel end cover, 64″ high	1.000	Ea.	27.50		27.50
Finish end cover, variable height, 2-way	1.000	Ea.	41		41
Corner worksurface, 24″ deep, 36″ wide	1.000	Ea.	315		315
Worksurface, 24″ deep, 48″ wide	1.000	Ea.	171		171
Worksurface, 24″ deep, 60″ wide	1.000	Ea.	215		215
Cantilever bracket, 20″ deep, 24″ deep	2.000	Ea.	75		75
Countertop brackets, 1 pair	2.000	Ea.	31		31
Countertop, 15″ deep, 36″ wide	1.000	Ea.	126		126
Countertop, 15″ deep, 48″ wide	1.000	Ea.	140		140
Pedestal spacer, 22″ deep, 15″ wide	2.000	Ea.	89		89
Pedestal, floorstanding, 2 box, 1 file, 22″ deep	1.000	Ea.	315		315
Pedestal, floorstanding, 2 file, 22″ deep	1.000	Ea.	291		291
Overhead storage cabinet with door, 60″ wide	1.000	Ea.	390		390
Task light, recessed, 60″ wide	1.000	Ea.	142		142
Electrical power harness, 36″ wide	2.000	Ea.	179		179
Electrical power harness, 60″ wide	1.000	Ea.	94		94
Base power in-feed cable	1.000	Ea.	99.50		99.50
Duplex receptacle, circuit 1	2.000	Ea.	27.30		27.30
Duplex receptacle, circuit 2	1.000	Ea.	13.65		13.65
Installation	1.000	Ea.		286	286
TOTAL			4,946.45	286	5,232.45

E2020 Moveable Furnishings

The work station shown here represents one of many types and sizes on the market. Costs can vary considerably due to the various parameters involved. For instance, wood systems can cost up to 40% more than metal or fabric systems. Generally, the cost per work station will decrease for "quantity" purchases.

System Components	QUANTITY	UNIT	COST EACH MAT.	COST EACH INST.	COST EACH TOTAL
SYSTEM E2020 430 1500					
WORK STATION - SECRETARY/CLERK					
Acoustical panel, aluminum frame, fabric covered, 30″ wide x 64″ high	2.000	Ea.	630		630
36″ wide x 64″ high	2.000	Ea.	680		680
42″ wide x 64″ high	1.000	Ea.	395		395
Straight connector kit, 64″ high	2.000	Ea.	55		55
Ell connector kit, 64″ high, 90″ deep	2.000	Ea.	78		78
Panel end cover, 64″ high	2.000	Ea.	55		55
Worksurface, 24″ deep, 42″ wide	1.000	Ea.	160		160
Worksurface, 72″ wide, 30″ deep	1.000	Ea.	273		273
Worksurface bracket kit, pair	2.000	Ea.	31		31
Flat bracket, 20″ deep, 24″ deep	1.000	Ea.	19.15		19.15
Pedestal spacer 22″ deep, 15″ wide	1.000	Ea.	44.50		44.50
Pedestal spacer, 28″ deep, 15″ wide	1.000	Ea.	54		54
Pedestal, 2 box, 1 file, 28″ deep	1.000	Ea.	325		325
Pedestal, 2 file, 22″ deep	1.000	Ea.	291		291
Overhead storage cabinet with door, 42″ wide	1.000	Ea.	259		259
Task light, recessed, 42″ - 48″ wide	1.000	Ea.	131		131
Installation	1.000	Ea.		188	188
TOTAL			3,480.65	188	3,668.65

Important: See the Reference Section for critical supporting data - Reference Numbers and City Cost Indexes

E2020 Moveable Furnishings

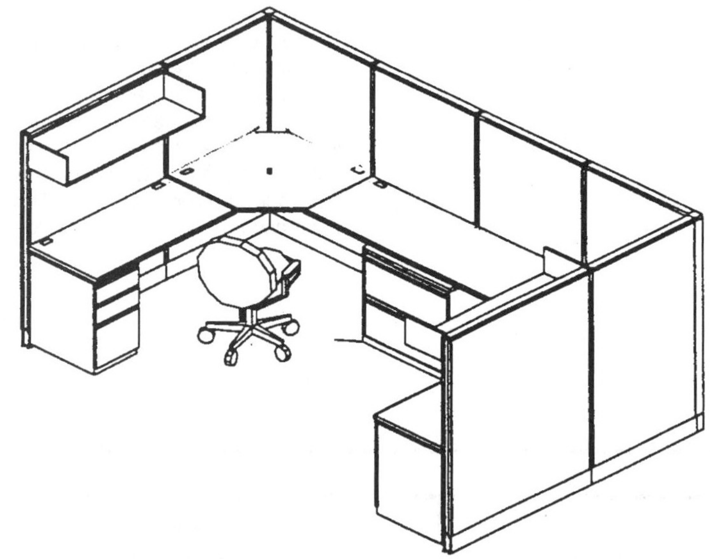

The work station shown here represents one of many types and sizes on the market. Costs can vary considerably due to the various parameters involved. For instance, wood systems can cost up to 40% more than metal or fabric systems. Generally, the cost per work station will decrease for "quantity" purchases.

System Components	QUANTITY	UNIT	COST EACH MAT.	COST EACH INST.	COST EACH TOTAL
SYSTEM E2020 440 1600					
WORK STATION - CLERICAL, TWO PERSON, UNIT					
Acoustical panel, aluminum frame, fabric covered, 36″ wide, 64″ high	6.000	Ea.	2,040		2,040
48″ wide x 64″ high	2.000	Ea.	830		830
Lateral file, 2 drawer, 30″ wide	1.000	Ea.	410		410
Ell connector kit, 64″ high, 90″ deep	2.000	Ea.	78		78
Panel end cover, 64″ high	2.000	Ea.	55		55
Corner worksurface, 36″ wide x 24″ deep	2.000	Ea.	630		630
Worksurface, 48″ wide x 24″ deep	2.000	Ea.	342		342
Worksurface, 72″ wide x 24″ deep	1.000	Ea.	243		243
Cantilever bracket, 20″ deep, 24″ deep	4.000	Ea.	150		150
Open bookshelf, 48″ wide	2.000	Ea.	238		238
Pedestal spacer, 22″ deep, 15″ wide	2.000	Ea.	89		89
Pedestal, floorstanding, 2 box, 1 file, 22″ deep	2.000	Ea.	630		630
Straight connector kit, 64″ high	5.000	Ea.	137.50		137.50
Task light, recessed, 42″ - 48″ wide	2.000	Ea.	262		262
Electrical power harness, 36″ wide	4.000	Ea.	358		358
Base power in-feed cable	1.000	Ea.	99.50		99.50
Duplex receptacle, circuit 1	2.000	Ea.	27.30		27.30
Duplex receptacle, circuit 2	2.000	Ea.	27.30		27.30
Installation	1.000	Ea.		375	375
TOTAL			6,646.60	375	7,021.60

E2020 Moveable Furnishings

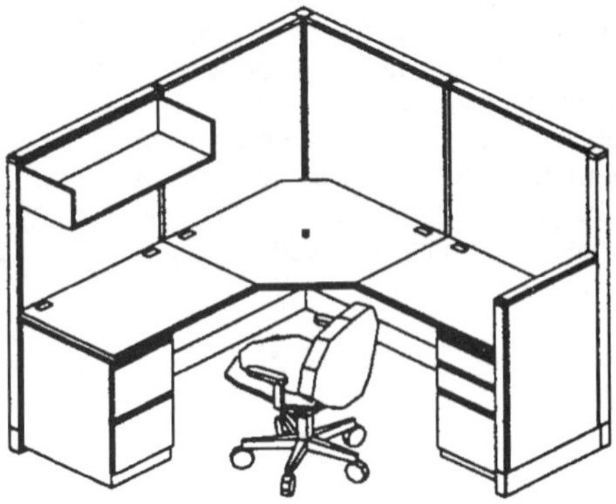

The work station shown here represents one of many types and sizes on the market. Costs can vary considerably due to the various parameters involved. For instance, wood systems can cost up to 40% more than metal or fabric systems. Generally, the cost per work station will decrease for "quantity" purchases.

System Components	QUANTITY	UNIT	COST EACH MAT.	COST EACH INST.	COST EACH TOTAL
SYSTEM E2020 450 1600					
WORK STATION - SECRETARY/WORD PROCESSING					
Acoustical panel, aluminum frame, fabric covered, 43″ high, 24″ wide	1.000	Ea.	239		239
Acoustical panel, 64″ high, 36″ wide	4.000	Ea.	1,360		1,360
Straight connector kit, 64″ high	2.000	Ea.	55		55
Ell connector kit, 64″ high, 90″ deep	2.000	Ea.	78		78
Panel end cover, 43″ high	1.000	Ea.	27.50		27.50
Panel end cover, 64″ high	1.000	Ea.	27.50		27.50
Variable height finish end cover, 2-way	1.000	Ea.	41		41
Corner worksurface, 24″ deep, 36″ wide	1.000	Ea.	315		315
Worksurface, 24″ deep, 36″ wide	2.000	Ea.	260		260
Cantilever bracket, 20″ deep, 24″ deep	2.000	Ea.	75		75
Pedestal spacer, 22″ deep, 15″ wide	2.000	Ea.	89		89
Pedestal, floorstanding, 2 box, 1 file, 22″ deep	1.000	Ea.	315		315
Pedestal, floorstanding, 2 file, 22″ deep	1.000	Ea.	291		291
Open bookshelf, 36″ wide	1.000	Ea.	110		110
Task light, recessed, 30″ - 36″ wide	1.000	Ea.	121		121
Electrical power harness, 36″ wide	3.000	Ea.	268.50		268.50
Base power in-feed cable	1.000	Ea.	99.50		99.50
Duplex receptacle, circuit 1	2.000	Ea.	27.30		27.30
Duplex receptacle, circuit 2	1.000	Ea.	13.65		13.65
Installation	1.000	Ea.		236	236
TOTAL			3,812.95	236	4,048.95

F SPECIAL CONSTRUCTION

F1020 Integrated Construction

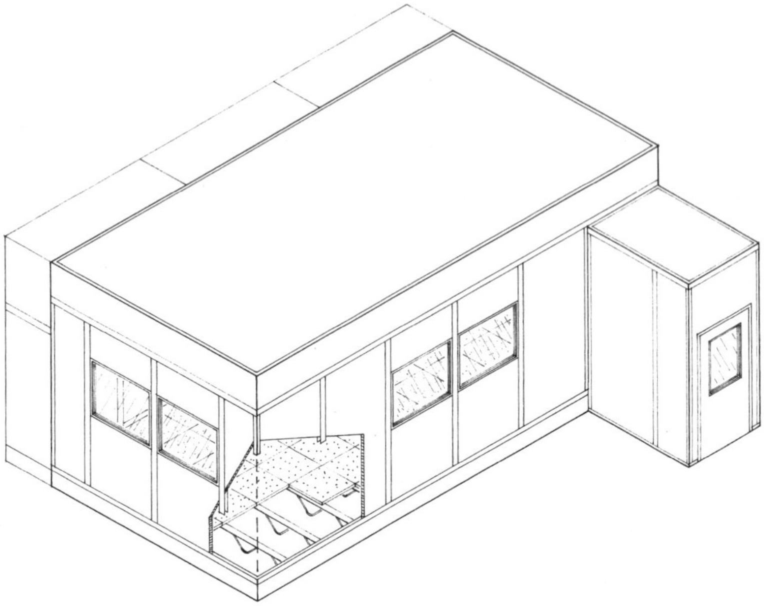

A clean room is a free standing enclosure designed to control particulate levels through air filtration and pressurization control, as well as temperature and humidity levels. These controlled environments are necessary for semi-conductor and electronics-related processes, precision machinery and metrology applications, laser and optic devices, pharmaceutical, biotechnology and medical device manufacture.

The class 10 and 100 prefab clean rooms include: 100% filtered ceiling coverage using air plenum modules, complete packaged HVAC with integrated electronic controls, fully insulated steel demountable modular wall partitions with integral utility chases, ante-room gown-up/airlock with interlocked doors and passthru.

All costs shown include factory pre-assembled and pre-tested components, field installed and tested, with only electrical power, water supply and drain connections required.

Costs are based on 288 S.F. net usable floor area.

System Components	QUANTITY	UNIT	COST EACH MAT.	COST EACH INST.	COST EACH TOTAL
CLASS 10 CLEAN ROOM SYSTEM					
Flooring, raised pedestal type w/steel perforated tiles, incl. steps	1.000	Ea.	19,550	5,950	25,500
Walls, modular baked enamel panels, insulated, all steel, incl. windows, doors, power outlets & light switches	1.000	Ea.	23,900	16,900	40,800
Ceiling, sealed plenum type, 99.9995% on .12 micron, HEPA, 100% coverage, 2″ gel grid tee bar incl. lights	1.000	Ea.	98,300	12,300	110,600
HVAC sys. incl. pressure, temp. and humidity control system and recirc. fan	1.000	Ea.	107,700	13,150	120,850
Gown-up, incl. floor, walls, ceiling, filters, lights, doors	1.000	Ea.	16,000	6,700	22,700
Test, balance and certification	1.000	Ea.	670	9,500	10,170
TOTAL			266,120	64,500	330,620

System Components	QUANTITY	UNIT	COST EACH MAT.	COST EACH INST.	COST EACH TOTAL
CLASS 100 CLEAN ROOM SYSTEM					
Flooring, seamless vinyl with coving	1.000	Ea.	4,575	1,100	5,675
Walls, modular baked enamel insul. panels, all steel incl. windows, doors, power outlets and light switches	1.000	Ea.	23,900	17,100	41,000
Ceiling, sealed plenum type, 99.99% on .3 micron HEPA, 100% coverage, 2″ gel grid tee bar incl. lights	1.000	Ea.	83,600	10,350	93,950
HVAC sys. incl. pressure, temp., and humidity control system, & recirc. fan	1.000	Ea.	97,750	10,900	108,650
Gown-up, incl. floor, walls, ceiling, filters, lights, doors with interlock	1.000	Ea.	16,000	6,700	22,700
Test, balance and certification	1.000	Ea.	670	9,500	10,170
TOTAL			226,495	55,650	282,145

F1020 Integrated Construction

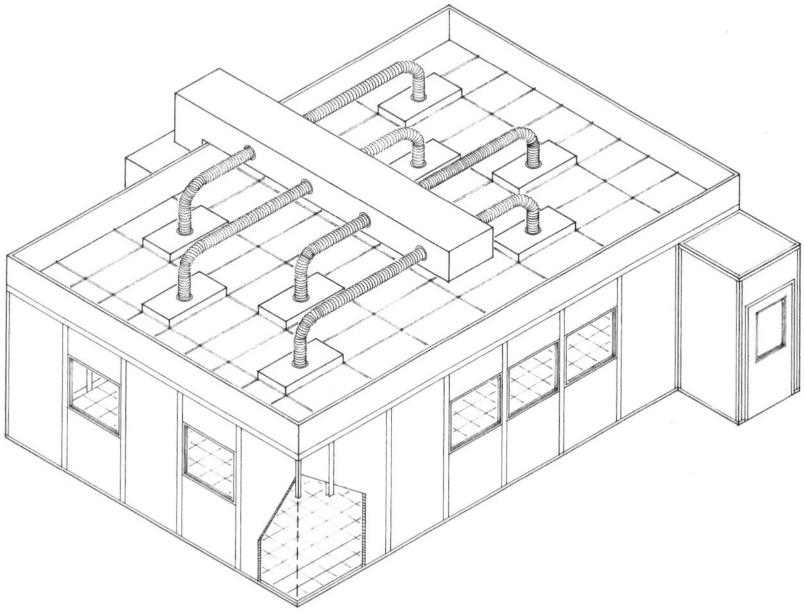

The Class 1000 through Class 100,000 prefab clean rooms use gasketed 2″ heavy duty tee-bar ceilings, ducted supply to and from the HVAC unit, sidewall returns, complete packaged HVAC and integrated electronic controls, fully insulated steel demountable modular wall partitions with integral utility chases, ante-room gown-up/airlock with interlocked doors and passthru.

All costs include factory pre-assembled and pre-tested components, field installation and certification, with the only connections required being electrical power, water supply, and drain.

Costs are based on 600 S.F. net usable floor area.

System Components	QUANTITY	UNIT	COST EACH MAT.	COST EACH INST.	COST EACH TOTAL
SYSTEM F1020 220 0310					
CLASS 1000 CLEAN ROOM SYSTEM					
Flooring, seamless vinyl with coving	1.000	Ea.	10,050	1,700	11,750
Walls, modular baked enamel insul panels, all steel incl. windows, doors, power outlets and light switch	1.000	Ea.	28,800	27,800	56,600
Ceiling, thermal ducted type, 99.99% on .3 micron HEPA, steel blank tiles, 2″ gasketed grid tee bars including lights	1.000	Ea.	39,800	36,300	76,100
HVAC system incl. pressure, temp., and humidity control sys. and recirc. fan	1.000	Ea.	63,200	6,600	69,800
Gown-up/air lock, incl. floor, walls, ceiling, filter, light, door	1.000	Ea.	11,900	5,100	17,000
Test, balance and certification	1.000	Ea.	230	5,100	5,330
TOTAL			153,980	82,600	236,580

F1020 220	Prefab Clean Room	COST EACH MAT.	COST EACH INST.	COST EACH TOTAL
0310	Prefab clean room, Class 1000	154,000	82,600	236,600
0320	Class 10,000	137,000	72,800	209,800
0330	Class 100,000	130,200	69,700	199,900

G BUILDING SITEWORK

G1030 Site Earthwork

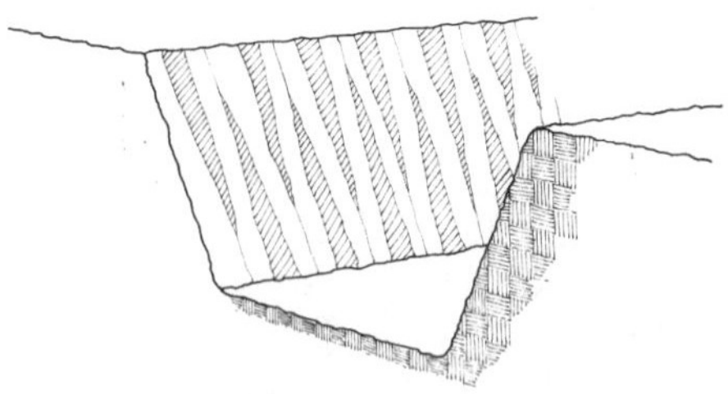

Trenching Systems are shown on a cost per linear foot basis. The systems include: excavation; backfill and removal of spoil; and compaction for various depths and trench bottom widths. The backfill has been reduced to accommodate a pipe of suitable diameter and bedding.

The slope for trench sides varies from 0:1 to 2:1.

The Expanded System Listing shows Trenching Systems that range from 2′ to 12′ in width. Depths range from 2′ to 25′.

System Components	QUANTITY	UNIT	COST PER L.F. EQUIP.	LABOR	TOTAL
SYSTEM G1030 805 1310					
TRENCHING, BACKHOE, 0 TO 1 SLOPE, 2′ WIDE, 2′ DP, 3/8 C.Y. BUCKET					
Excavation, trench, hyd. backhoe, track mtd., 3/8 C.Y. bucket	.174	C.Y.	.29	.96	1.25
Backfill and load spoil, from stockpile	.174	C.Y.	.09	.28	.37
Compaction by rammer tamper, 8″ lifts, 4 passes	.014	C.Y.		.04	.04
Remove excess spoil, 6 C.Y. dump truck, 2 mile roundtrip	.160	C.Y.	.57	.59	1.16
TOTAL			.95	1.87	2.82

G1030 805	Trenching	COST PER L.F. EQUIP.	LABOR	TOTAL
1310	Trenching, backhoe, 0 to 1 slope, 2′ wide, 2′ deep, 3/8 C.Y. bucket	.95	1.87	2.82
1320	3′ deep, 3/8 C.Y. bucket	1.17	2.73	3.90
1330	4′ deep, 3/8 C.Y. bucket	1.40	3.59	4.99
1340	6′ deep, 3/8 C.Y. bucket	1.76	4.59	6.35
1350	8′ deep, 1/2 C.Y. bucket	2.09	5.85	7.94
1360	10′ deep, 1 C.Y. bucket	3.05	6.90	9.95
1400	4′ wide, 2′ deep, 3/8 C.Y. bucket	2.02	3.75	5.77
1410	3′ deep, 3/8 C.Y. bucket	2.86	5.60	8.46
1420	4′ deep, 1/2 C.Y. bucket	2.89	5.60	8.49
1430	6′ deep, 1/2 C.Y. bucket	3.71	8.55	12.26
1440	8′ deep, 1/2 C.Y. bucket	5.65	11	16.65
1450	10′ deep, 1 C.Y. bucket	6.35	13.65	20
1460	12′ deep, 1 C.Y. bucket	8.05	17.45	25.50
1470	15′ deep, 1-1/2 C.Y. bucket	6.85	15	21.85
1480	18′ deep, 2-1/2 C.Y. bucket	9.40	21.50	30.90
1520	6′ wide, 6′ deep, 5/8 C.Y. bucket	7.30	11.85	19.15
1530	8′ deep, 3/4 C.Y. bucket	8.90	14.20	23.10
1540	10′ deep, 1 C.Y. bucket	9.55	17.10	26.65
1550	12′ deep, 1-1/4 C.Y. bucket	8.40	13.55	21.95
1560	16′ deep, 2 C.Y. bucket	11.05	15.45	26.50
1570	20′ deep, 3-1/2 C.Y. bucket	15	27.50	42.50
1580	24′ deep, 3-1/2 C.Y. bucket	24	46	70
1640	8′ wide, 12′ deep, 1-1/4 C.Y. bucket	12.45	18.50	30.95
1650	15′ deep, 1-1/2 C.Y. bucket	14.65	23	37.65
1660	18′ deep, 2-1/2 C.Y. bucket	17.10	23	40.10
1680	24′ deep, 3-1/2 C.Y. bucket	33.50	61.50	95
1730	10′ wide, 20′ deep, 3-1/2 C.Y. bucket	27	46.50	73.50
1740	24′ deep, 3-1/2 C.Y. bucket	42.50	77	119.50
1780	12′ wide, 20′ deep, 3-1/2 C.Y. bucket	33	56	89
1790	25′ deep, bucket	56.50	103	159.50
1800	1/2 to 1 slope, 2′ wide, 2′ deep, 3/8 C.Y. bucket	1.29	2.40	3.69
1810	3′ deep, 3/8 C.Y. bucket	1.79	3.95	5.74
1820	4′ deep, 3/8 C.Y. bucket	2.27	5.75	8.02
1840	6′ deep, 3/8 C.Y. bucket	3.38	8.90	12.28

G2020 Parking Lots

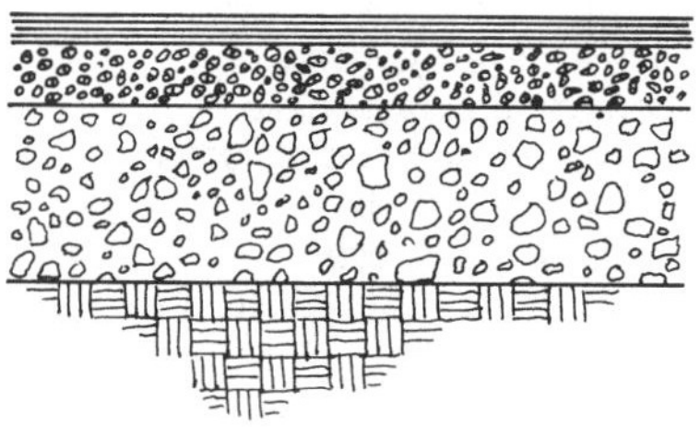

The Parking Lot System includes: compacted bank-run gravel; fine grading with a grader and roller; and bituminous concrete wearing course. All Parking Lot systems are on a cost per car basis. There are three basic types of systems: 90° angle, 60° angle, and 45° angle. The gravel base is compacted to 98%. Final stall design and lay-out of the parking lot with precast bumpers, sealcoating and white paint is also included.

The Expanded System Listing shows the three basic parking lot types with various depths of both gravel base and wearing course. The gravel base depths range from 6″ to 10″. The bituminous paving wearing course varies from a depth of 3″ to 6″.

System Components	QUANTITY	UNIT	COST PER CAR MAT.	COST PER CAR INST.	COST PER CAR TOTAL
SYSTEM G2020 220 1500					
PARKING LOT, 90° ANGLE PARKING, 3″ BITUMINOUS PAVING, 6″ GRAVEL BASE					
Surveying crew for layout, 4 man crew	.020	Day		40.28	40.28
Borrow, bank run gravel, haul 2 mi., spread w/dozer, no compaction	5.556	C.Y.	110.84	31.94	142.78
Grading, fine grade 3 passes with motor grader	33.333	S.Y.		56.61	56.61
Compact w/ vibrating plate, 8″ lifts, granular mat'l. to 98%	5.556	C.Y.		29.06	29.06
Bituminous paving, 3″ thick	33.333	S.Y.	225	50	275
Seal coating, petroleum resistant under 1,000 S.Y.	33.333	S.Y.	24.33	35.33	59.66
Painting lines on pavement, parking stall, white	1.000	Ea.	3.77	5.97	9.74
Precast concrete parking bar, 6″ x 10″ x 6′-0″	1.000	Ea.	44	15.35	59.35
TOTAL			407.94	264.54	672.48

G2020 220	Parking Lots	COST PER CAR MAT.	COST PER CAR INST.	COST PER CAR TOTAL
1500	Parking lot, 90° angle parking, 3″ bituminous paving, 6″ gravel base	410	265	675
1520	8″ gravel base	445	285	730
1540	10″ gravel base	480	305	785
1560	4″ bituminous paving, 6″ gravel base	495	263	758
1580	8″ gravel base	530	283	813
1600	10″ gravel base	570	305	875
1620	6″ bituminous paving, 6″ gravel base	640	275	915
1640	8″ gravel base	680	295	975
1661	10″ gravel base	715	315	1,030
1800	60° angle parking, 3″ bituminous paving, 6″ gravel base	470	289	759
1820	8″ gravel base	510	315	825
1840	10″ gravel base	555	335	890
1860	4″ bituminous paving, 6″ gravel base	570	288	858
1880	8″ gravel base	610	310	920
1900	10″ gravel base	655	335	990
1920	6″ bituminous paving, 6″ gravel base	740	300	1,040
1940	8″gravel base	785	325	1,110
1961	10″ gravel base	825	350	1,175
2200	45° angle parking, 3″ bituminous paving, 6″ gravel base	530	315	845
2220	8″ gravel base	575	340	915
2240	10″ gravel base	625	365	990
2260	4″ bituminous paving, 6″ gravel base	645	310	955
2280	8″ gravel base	695	340	1,035
2300	10″ gravel base	740	365	1,105
2320	6″ bituminous paving, 6″ gravel base	840	325	1,165
2340	8″ gravel base	890	355	1,245
2361	10″ gravel base	940	380	1,320

G2030 Pedestrian Paving

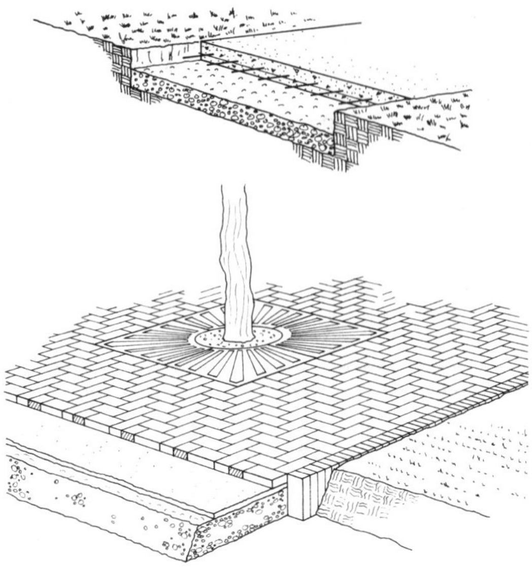

The Bituminous and Concrete Sidewalk Systems include excavation, hand grading and compacted gravel base. Pavements are shown for two conditions of each of the following variables; pavement thickness, gravel base thickness and pavement width. Costs are given on a linear foot basis.

The Plaza Systems listed include several brick and tile paving surfaces on two different bases: gravel and slab on grade. The type of bedding for the pavers depends on the base being used, and alternate bedding may be desirable. Also included in the paving costs are edging and precast grating costs and where concrete bases are involved, expansion joints. Costs are given on a square foot basis.

G2030 110	Bituminous Sidewalks	COST PER L.F.		
		MAT.	INST.	TOTAL
1580	Bituminous sidewalk, 1″ thick paving, 4″ gravel base, 3′ width	1.48	3.39	4.87
1600	4′ width	1.96	3.70	5.66
1640	6″ gravel base, 3′ width	1.86	3.56	5.42
1660	4′ width	2.46	3.91	6.37
2120	2″ thick paving, 4″ gravel base, 3′ width	2.21	4.12	6.33
2140	4′ width	2.94	4.60	7.54
2180	6″ gravel base, 3′ width	2.59	4.28	6.87
2201	4′ width	3.44	4.82	8.26

G2030 120	Concrete Sidewalks	COST PER L.F.		
		MAT.	INST.	TOTAL
1580	Concrete sidewalk, 4″ thick, 4″ gravel base, 3′ wide	5.55	9.40	14.95
1600	4′ wide	7.40	11.60	19
1640	6″ gravel base, 3′ wide	5.90	9.55	15.45
1660	4′ wide	7.90	11.85	19.75
2120	6″ thick concrete, 4″ gravel base, 3′ wide	8.20	10.75	18.95
2140	4′ wide	10.90	13.45	24.35
2180	6″ gravel base, 3′ wide	8.55	10.95	19.50
2201	4′ wide	11.40	14.15	25.55

G2030 150	Brick & Tile Plazas	COST PER S.F.		
		MAT.	INST.	TOTAL
2050	Brick pavers, 4″ x 8″ x 2-1/4″, aggregate base, course washed sand bedding	4.98	4.79	9.77
2100	Slab on grade, asphalt bedding	6.70	6.90	13.60
3550	Concrete paving stone, 4″ x 8″ x 2-1/2″, gravel base, sand bedding	3.70	3.31	7.01
3600	Slab on grade, asphalt bedding	5.10	5.05	10.15
4050	Concrete patio blocks, 8″ x 16″ x 2″, gravel base, sand bedding	3.46	4.35	7.81
4100	Slab on grade, asphalt bedding	5.05	6.40	11.45
6050	Granite pavers, 3-1/2″ x 3-1/2″ x 3-1/2″, gravel base, sand bedding	9.25	10.65	19.90
6101	Slab on grade, mortar bedding	11	14.30	25.30
7050	Limestone, 3″ thick, gravel base, sand bedding	11.10	13.20	24.30
7100	Slab on grade, mortar bedding	12.85	16.75	29.60
8050	Slate flagging, 3/4″ thick, gravel base, sand bedding	8.05	10.65	18.70
8100	Slab on grade, mastic bedding	16.80	15.25	32.05

G4020 Site Lighting

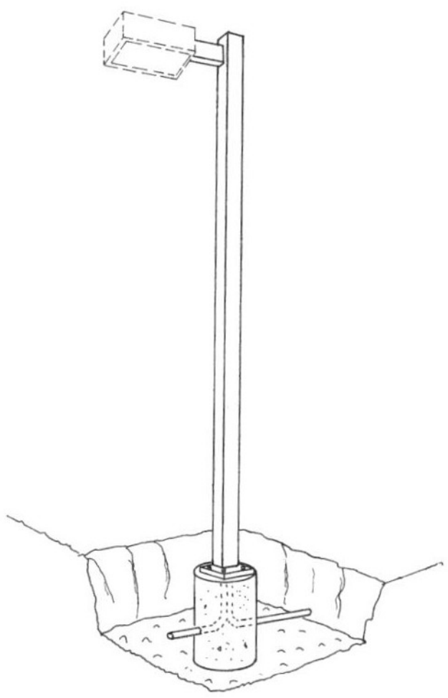

Table G4020 210 Procedure for Calculating Floodlights Required for Various Footcandles
Poles should not be spaced more than 4 times the fixture mounting height for good light distribution. To maintain 1 footcandle over a large area use these watts per square foot:

Incandescent	0.15
Metal Halide	0.032
Mercury Vapor	0.05
High Pressure Sodium	0.024

Estimating Chart
Select Lamp type.

Determine total square feet.

Chart will show quantity of fixtures to provide 1 footcandle initial, at intersection of lines. Multiply fixture quantity by desired footcandle level.

Chart based on use of wide beam luminaires in an area whose dimensions are large compared to mounting height and is approximate only.

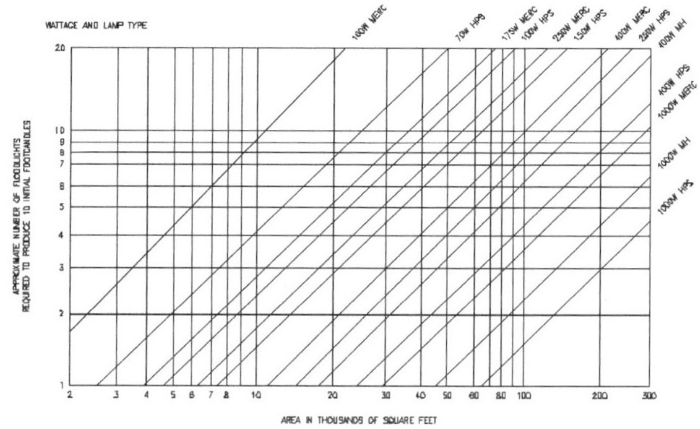

System Components	QUANTITY	UNIT	COST EACH MAT.	COST EACH INST.	COST EACH TOTAL
SYSTEM G4020 210 0200					
LIGHT POLES, ALUMINUM, 20′ HIGH, 1 ARM BRACKET					
Aluminum light pole, 20′, no concrete base	1.000	Ea.	840	498	1,338
Bracket arm for Aluminum light pole	1.000	Ea.	110	64.50	174.50
Excavation by hand, pits to 6′ deep, heavy soil or clay	2.368	C.Y.		215.49	215.49
Footing, concrete incl forms, reinforcing, spread, under 1 C.Y.	.465	C.Y.	63.71	76.22	139.93
Backfill by hand	1.903	C.Y.		62.80	62.80
Compaction vibrating plate	1.903	C.Y.		8.49	8.49
TOTAL			1,013.71	925.50	1,939.21

G4020 210	Light Pole (Installed)	COST EACH MAT.	COST EACH INST.	COST EACH TOTAL
0200	Light pole, aluminum, 20′ high, 1 arm bracket	1,025	925	1,950
0240	2 arm brackets	1,125	925	2,050
0280	3 arm brackets RD5020 -250	1,225	960	2,185
0320	4 arm brackets	1,350	960	2,310
0360	30′ high, 1 arm bracket	1,825	1,175	3,000
0400	2 arm brackets	1,925	1,175	3,100
0440	3 arm brackets	2,025	1,225	3,250
0480	4 arm brackets	2,150	1,225	3,375
0680	40′ high, 1 arm bracket	2,225	1,550	3,775
0720	2 arm brackets	2,350	1,550	3,900
0760	3 arm brackets	2,450	1,575	4,025
0800	4 arm brackets	2,550	1,575	4,125
0840	Steel, 20′ high, 1 arm bracket	1,225	985	2,210
0880	2 arm brackets	1,325	985	2,310
0920	3 arm brackets	1,350	1,025	2,375
0960	4 arm brackets	1,450	1,025	2,475
1000	30′ high, 1 arm bracket	1,425	1,250	2,675
1040	2 arm brackets	1,500	1,250	2,750
1080	3 arm brackets	1,525	1,275	2,800
1120	4 arm brackets	1,625	1,275	2,900
1320	40′ high, 1 arm bracket	1,850	1,700	3,550
1360	2 arm brackets	1,950	1,700	3,650
1400	3 arm brackets	1,975	1,725	3,700
1440	4 arm brackets	2,075	1,725	3,800

Reference Section

All the reference information is in one section, making it easy to find what you need to know . . . and easy to use the book on a daily basis. This section is visually identified by a vertical gray bar on the page edges.

In this Reference Section, we've included Equipment Rental Costs, a listing of rental and operating costs; Crew Listings, a full listing of all crews and equipment, and their costs; Historical Cost Indexes for cost comparisons over time; City Cost Indexes and Location Factors for adjusting costs to the region you are in; Reference Tables, where you will find explanations, estimating information and procedures, or technical data; Change Orders, information on pricing changes to contract documents; Square Foot Costs that allow you to make a rough estimate for the overall cost of a project; and an explanation of all the Abbreviations in the book.

Table of Contents

Reference Tables (cont.)

Construction Equipment Rental Costs

Estimating Tips

- This section contains the average costs to rent and operate hundreds of pieces of construction equipment. This is useful information when estimating the time and material requirements of any particular operation in order to establish a unit or total cost. Equipment costs include not only rental, but also operating costs for equipment under normal use.

Rental Costs

- Equipment rental rates are obtained from industry sources throughout North America - contractors, suppliers, dealers, manufacturers, and distributors.
- Rental rates vary throughout the country with larger cities generally having lower rates. Lease plans for new equipment are available for periods in excess of six months with a percentage of payments applying toward purchase.
- Monthly rental rates vary from 2% to 5% of the purchase price of the equipment depending on the anticipated life of the equipment and its wearing parts.
- Weekly rental rates are about 1/3 the monthly rates, and daily rental rates are about 1/3 the weekly rate.

Operating Costs

- The operating costs include parts and labor for routine servicing such as repair and replacement of pumps, filters and worn lines. Normal operating expendables such as fuel, lubricants, tires and electricity (where applicable) are also included.
- Extraordinary operating expendables with highly variable wear patterns such as diamond bits and blades are excluded. These costs can be found as material costs in the unit price section.
- The hourly operating costs listed do not include the operator's wages.

Crew Equipment Cost/Day

- Any power equipment required by a crew is shown in the Crew Listings with a daily cost.
- The daily cost of crew equipment is based on dividing the weekly rental rate by 5 (number of working days in the week), and then adding the hourly operating cost times 8 (the number of hours in a day). This "Crew Equipment Cost/Day" is shown in the far right column of the Equipment Rental pages.
- If equipment is needed for only one or two days, it is best to develop your own cost by including components for daily rent and hourly operating cost. This is important when the listed Crew for a task does not contain the equipment needed, such as a crane for lifting mechanical heating/cooling equipment up onto a roof.

Mobilization/ Demobilization

- The cost to move construction equipment from an equipment yard or rental company to the jobsite and back again is not included in equipment rental costs listed in the Reference section, nor in the bare equipment cost of any Unit Cost line item, nor in any equipment costs shown in the Crew listings.
- Mobilization (to the site) and demobilization (from the site) costs can be found in the Unit Cost section.
- If a piece of equipment is already at the jobsite, it is not appropriate to utilize mob/ demob costs again in an estimate.

01 54 33 | Rental Equipment

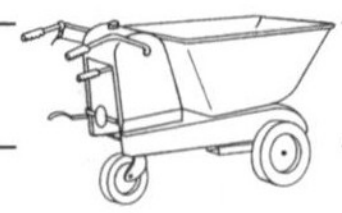
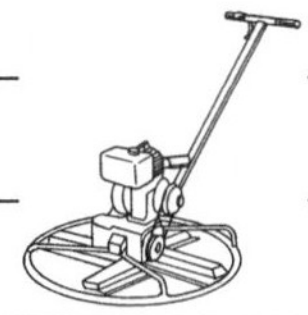

				UNIT	HOURLY OPER. COST	RENT PER DAY	RENT PER WEEK	RENT PER MONTH	CREW EQUIPMENT COST/DAY	
10	0010	**CONCRETE EQUIPMENT RENTAL**, without operators	R015433-10							10
	0150	For batch plant, see div. 01590-500								
	0200	Bucket, concrete lightweight, 1/2 C.Y.	R033105-70	Ea.	.55	16	48	144	14	
	0300	1 C.Y.			.60	19	57	171	16.20	
	0400	1-1/2 C.Y.			.75	26	78	234	21.60	
	0500	2 C.Y.			.85	30.50	92	276	25.20	
	0580	8 C.Y.			4.55	205	615	1,850	159.40	
	0600	Cart, concrete, self propelled, operator walking, 10 C.F.			2.15	58.50	175	525	52.20	
	0700	Operator riding, 18 C.F.			3.30	80	240	720	74.40	
	0800	Conveyer for concrete, portable, gas, 16" wide, 26' long			7.30	117	350	1,050	128.40	
	0900	46' long			7.65	142	425	1,275	146.20	
	1000	56' long			7.80	150	450	1,350	152.40	
	1100	Core drill, electric, 2-1/2 H.P., 1" to 8" bit diameter			2.00	81	243	730	64.60	
	1150	11 HP, 8" to 18" cores			6.61	104	312.80	940	115.45	
	1200	Finisher, concrete floor, gas, riding trowel, 48" diameter			4.80	86.50	260	780	90.40	
	1300	Gas, manual, 3 blade, 36" trowel			1.00	16.65	50	150	18	
	1400	4 blade, 48" trowel			1.45	21	63	189	24.20	
	1500	Float, hand-operated (Bull float) 48" wide			.08	13.35	40	120	8.65	
	1570	Curb builder, 14 H.P., gas, single screw			9.75	210	630	1,900	204	
	1590	Double screw			10.40	242	725	2,175	228.20	
	1600	Grinder, concrete and terrazzo, electric, floor			2.68	116	349	1,050	91.25	
	1700	Wall grinder			1.35	58.50	175	525	45.80	
	1800	Mixer, powered, mortar and concrete, gas, 6 C.F., 18 H.P.			5.20	107	320	960	105.60	
	1900	10 C.F., 25 H.P.			6.30	128	385	1,150	127.40	
	2000	16 C.F.			6.65	152	455	1,375	144.20	
	2100	Concrete, stationary, tilt drum, 2 C.Y.			5.35	200	600	1,800	162.80	
	2120	Pump, concrete, truck mounted 4" line 80' boom			21.70	880	2,635	7,900	700.60	
	2140	5" line, 110' boom			28.40	1,175	3,495	10,500	926.20	
	2160	Mud jack, 50 C.F. per hr.			5.79	118	355	1,075	117.30	
	2180	225 C.F. per hr.			7.56	158	473.20	1,425	155.10	
	2190	Shotcrete pump rig, 12 CY/hr			11.80	233	700	2,100	234.40	
	2600	Saw, concrete, manual, gas, 18 H.P.			3.50	35	105	315	49	
	2650	Self-propelled, gas, 30 H.P.			6.75	88.50	265	795	107	
	2700	Vibrators, concrete, electric, 60 cycle, 2 H.P.			.37	8.35	25	75	7.95	
	2800	3 H.P.			.56	12.35	37	111	11.90	
	2900	Gas engine, 5 H.P.			.90	15	45	135	16.20	
	3000	8 H.P.			1.30	18.35	55	165	21.40	
	3050	Vibrating screed, gas engine, 8 H.P.			2.27	66.50	199	595	57.95	
	3100	Concrete transit mixer, hydraulic drive								
	3120	6 x 4, 250 H.P., 8 C.Y., rear discharge			34.85	540	1,625	4,875	603.80	
	3200	Front discharge			40.95	675	2,020	6,050	731.60	
	3300	6 x 6, 285 H.P., 12 C.Y., rear discharge			40.05	635	1,900	5,700	700.40	
	3400	Front discharge		▼	41.85	680	2,045	6,125	743.80	
20	0010	**EARTHWORK EQUIPMENT RENTAL**, without operators	R015433-10							20
	0040	Aggregate spreader, push type 8' to 12' wide		Ea.	1.80	29.50	88	264	32	
	0045	Tailgate type, 8' wide	R312323-30	"	1.75	31.50	95	285	33	
	0050	Augers for vertical drilling								
	0055	Earth auger, truck-mounted, for fence & sign posts	R312316-40	Ea.	8.40	480	1,445	4,325	356.20	
	0060	For borings and monitoring wells			30.95	640	1,920	5,750	631.60	
	0070	Earth auger, portable, trailer mounted	R312316-45		1.75	23	69	207	27.80	
	0075	Earth auger, truck-mounted, for caissons, water wells, utility poles			160.90	3,475	10,460	31,400	3,379	
	0080	Auger, horizontal boring machine, 12" to 36" diameter, 45 H.P.			16.70	190	570	1,700	247.60	
	0090	12" to 48" diameter, 65 H.P.			23.50	340	1,020	3,050	392	
	0095	Auger, for fence posts, gas engine, hand held			.35	4.33	13	39	5.40	
	0100	Excavator, diesel hydraulic, crawler mounted, 1/2 C.Y. cap.			16.05	345	1,035	3,100	335.40	
	0120	5/8 C.Y. capacity			19.55	465	1,395	4,175	435.40	
	0140	3/4 C.Y. capacity		▼	22.85	490	1,475	4,425	477.80	

01 54 33 | Rental Equipment

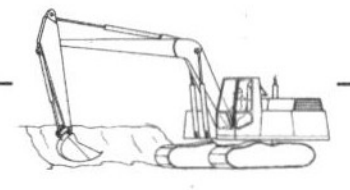
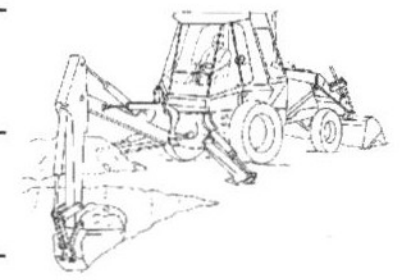
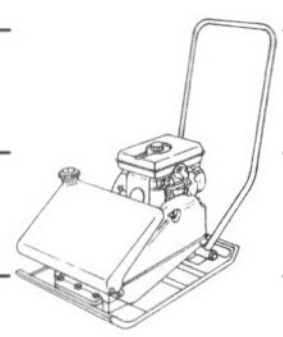
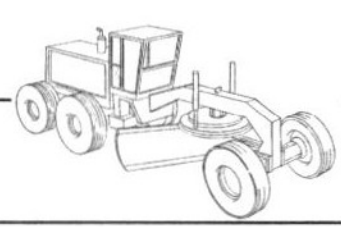

		01 54 33 Rental Equipment		UNIT	HOURLY OPER. COST	RENT PER DAY	RENT PER WEEK	RENT PER MONTH	CREW EQUIPMENT COST/DAY	
20	0150	1 C.Y. capacity	R015433-10	Ea.	27.40	570	1,705	5,125	560.20	20
	0200	1-1/2 C.Y. capacity			33.85	755	2,260	6,775	722.80	
	0300	2 C.Y. capacity	R312323-30		43.70	950	2,850	8,550	919.60	
	0320	2-1/2 C.Y. capacity			56.20	1,275	3,840	11,500	1,218	
	0340	3-1/2 C.Y. capacity	R312316-40		95.20	2,100	6,305	18,900	2,023	
	0341	Attachments								
	0342	Bucket thumbs	R312316-45		2.55	208	625	1,875	145.40	
	0345	Grapples			1.00	219	656	1,975	139.20	
	0350	Gradall type, truck mounted, 3 ton @ 15' radius, 5/8 C.Y.			38.75	885	2,655	7,975	841	
	0370	1 C.Y. capacity			44.65	1,025	3,085	9,250	974.20	
	0400	Backhoe-loader, 40 to 45 H.P., 5/8 C.Y. capacity			8.60	183	550	1,650	178.80	
	0450	45 H.P. to 60 H.P., 3/4 C.Y. capacity			10.90	232	695	2,075	226.20	
	0460	80 H.P., 1-1/4 C.Y. capacity			13.95	277	830	2,500	277.60	
	0470	112 H.P., 1-1/2 C.Y. capacity			18.55	400	1,200	3,600	388.40	
	0480	Attachments								
	0482	Compactor, 20,000 lb			4.35	118	355	1,075	105.80	
	0485	Hydraulic hammer, 750 ft-lbs			2.00	70	210	630	58	
	0486	Hydraulic hammer, 1200 ft-lbs			4.20	133	400	1,200	113.60	
	0500	Brush chipper, gas engine, 6" cutter head, 35 H.P.			6.60	102	305	915	113.80	
	0550	12" cutter head, 130 H.P.			10.25	152	455	1,375	173	
	0600	15" cutter head, 165 H.P.			14.30	162	485	1,450	211.40	
	0750	Bucket, clamshell, general purpose, 3/8 C.Y.			1.00	35	105	315	29	
	0800	1/2 C.Y.			1.10	41.50	125	375	33.80	
	0850	3/4 C.Y.			1.25	51.50	155	465	41	
	0900	1 C.Y.			1.30	55	165	495	43.40	
	0950	1-1/2 C.Y.			2.05	75	225	675	61.40	
	1000	2 C.Y.			2.20	85	255	765	68.60	
	1010	Bucket, dragline, medium duty, 1/2 C.Y.			.60	22.50	67	201	18.20	
	1020	3/4 C.Y.			.60	23.50	71	213	19	
	1030	1 C.Y.			.65	25.50	76	228	20.40	
	1040	1-1/2 C.Y.			1.00	38.50	115	345	31	
	1050	2 C.Y.			1.05	43.50	130	390	34.40	
	1070	3 C.Y.			1.60	58.50	175	525	47.80	
	1200	Compactor, manually guided 2-drum vibratory smooth roller, 7.5 H.P.			4.85	145	435	1,300	125.80	
	1250	Rammer compactor, gas, 1000 lb. blow			1.75	38.50	115	345	37	
	1300	Vibratory plate, gas, 18" plate, 3000 lb. blow			1.65	22	66	198	26.40	
	1350	21" plate, 5000 lb. blow			2.00	31	93	279	34.60	
	1370	Curb builder/extruder, 14 H.P., gas, single screw			9.75	212	635	1,900	205	
	1390	Double screw			10.40	248	745	2,225	232.20	
	1500	Disc harrow attachment, for tractor			.37	61.50	185	555	39.95	
	1750	Extractor, piling, see lines 2500 to 2750								
	1810	Feller buncher, shearing & accumulating trees, 100 H.P.		Ea.	22.60	520	1,555	4,675	491.80	
	1860	Grader, self-propelled, 25,000 lb.			18.25	400	1,200	3,600	386	
	1910	30,000 lb.			21.05	480	1,445	4,325	457.40	
	1920	40,000 lb.			30.55	745	2,230	6,700	690.40	
	1930	55,000 lb.			40.80	1,050	3,130	9,400	952.40	
	1950	Hammer, pavement demo., hyd., gas, self-prop., 1000 to 1250 lb.			18.30	305	915	2,750	329.40	
	2000	Diesel 1300 to 1500 lb.			28.20	610	1,830	5,500	591.60	
	2050	Pile driving hammer, steam or air, 4150 ft.-lb. @ 225 BPM			6.60	272	815	2,450	215.80	
	2100	8750 ft.-lb. @ 145 BPM			8.50	440	1,325	3,975	333	
	2150	15,000 ft.-lb. @ 60 BPM			8.85	475	1,425	4,275	355.80	
	2200	24,450 ft.-lb. @ 111 BPM			11.75	525	1,570	4,700	408	
	2250	Leads, 15,000 ft.-lb. hammers		L.F.	.03	1.45	4.35	13.05	1.10	
	2300	24,450 ft.-lb. hammers and heavier		"	.05	2.47	7.40	22	1.90	
	2350	Diesel type hammer, 22,400 ft.-lb.		Ea.	25.50	605	1,820	5,450	568	
	2400	41,300 ft.-lb.			34.75	655	1,970	5,900	672	
	2450	141,000 ft.-lb.			56.45	1,125	3,380	10,100	1,128	
	2500	Vib. elec. hammer/extractor, 200 KW diesel generator, 34 H.P.			28.70	635	1,910	5,725	611.60	

01 54 33 | Rental Equipment

	Line	Description	Ref.	Unit	Hourly Oper. Cost	Rent per Day	Rent per Week	Rent per Month	Crew Equipment Cost/Day	
20	2550	80 H.P.	R015433-10	Ea.	49.50	930	2,785	8,350	953	20
	2600	150 H.P.			91.90	1,825	5,440	16,300	1,823	
	2700	Extractor, steam or air, 700 ft.-lb.	R312323-30		14.85	435	1,310	3,925	380.80	
	2750	1000 ft.-lb.			16.95	540	1,620	4,850	459.60	
	2800	Log chipper, up to 22" diam, 600 H.P.	R312316-40		41.71	1,100	3,331.60	10,000	1,000	
	2850	Logger, for skidding & stacking logs, 150 H.P.			36.70	825	2,470	7,400	787.60	
	2900	Rake, spring tooth, with tractor	R312316-45		8.04	214	643	1,925	192.90	
	3000	Roller, vibratory, tandem, smooth drum, 20 H.P.			5.40	118	355	1,075	114.20	
	3050	35 H.P.			7.75	225	675	2,025	197	
	3100	Towed type vibratory compactor, smooth drum, 50 H.P.			24.70	435	1,300	3,900	457.60	
	3150	Sheepsfoot, 50 H.P.			25.50	455	1,370	4,100	478	
	3170	Landfill compactor, 220 HP			50.30	1,200	3,595	10,800	1,121	
	3200	Pneumatic tire roller, 80 H.P.			9.15	310	930	2,800	259.20	
	3250	120 H.P.			14.75	540	1,615	4,850	441	
	3300	Sheepsfoot vibratory roller, 200 H.P.			38.75	895	2,690	8,075	848	
	3320	340 H.P.			53.55	1,300	3,870	11,600	1,202	
	3350	Smooth drum vibratory roller, 75 H.P.			15.90	485	1,460	4,375	419.20	
	3400	125 H.P.			20.25	600	1,800	5,400	522	
	3410	Rotary mower, brush, 60", with tractor			11.50	232	695	2,075	231	
	3450	Scrapers, towed type, 9 to 12 C.Y. capacity			4.42	211	632	1,900	161.75	
	3500	12 to 17 C.Y. capacity			1.50	281	842	2,525	180.40	
	3550	Scrapers, self-propelled, 4 x 4 drive, 2 engine, 14 C.Y. capacity			84.60	1,500	4,500	13,500	1,577	
	3600	2 engine, 24 C.Y. capacity			121.45	2,300	6,890	20,700	2,350	
	3640	32 - 44 C.Y. capacity			143.20	2,675	8,060	24,200	2,758	
	3650	Self-loading, 11 C.Y. capacity			40.95	825	2,475	7,425	822.60	
	3700	22 C.Y. capacity			76.90	1,725	5,190	15,600	1,653	
	3710	Screening plant 110 H.P. w/ 5' x 10' screen			23.00	380	1,135	3,400	411	
	3720	5' x 16' screen		↓	25.05	475	1,430	4,300	486.40	
	3850	Shovels, see Cranes division 01590-600								
	3860	Shovel/backhoe bucket, 1/2 C.Y.		Ea.	1.85	56.50	170	510	48.80	
	3870	3/4 C.Y.			1.90	63.50	190	570	53.20	
	3880	1 C.Y.			2.00	71.50	215	645	59	
	3890	1-1/2 C.Y.			2.15	86.50	260	780	69.20	
	3910	3 C.Y.			2.45	120	360	1,075	91.60	
	3950	Stump chipper, 18" deep, 30 H.P.			4.56	45	135	405	63.50	
	4110	Tractor, crawler, with bulldozer, torque converter, diesel 80 H.P.			16.35	315	940	2,825	318.80	
	4150	105 H.P.			22.35	475	1,420	4,250	462.80	
	4200	140 H.P.			27.50	590	1,765	5,300	573	
	4260	200 H.P.			41.10	985	2,960	8,875	920.80	
	4310	300 H.P.			53.40	1,275	3,850	11,600	1,197	
	4360	410 H.P.			72.30	1,600	4,790	14,400	1,536	
	4370	500 H.P.			95.55	2,125	6,395	19,200	2,043	
	4380	700 H.P.			142.40	3,350	10,045	30,100	3,148	
	4400	Loader, crawler, torque conv., diesel, 1-1/2 C.Y., 80 H.P.			15.15	325	970	2,900	315.20	
	4450	1-1/2 to 1-3/4 C.Y., 95 H.P.			17.55	385	1,160	3,475	372.40	
	4510	1-3/4 to 2-1/4 C.Y., 130 H.P.			23.75	610	1,835	5,500	557	
	4530	2-1/2 to 3-1/4 C.Y., 190 H.P.			35.75	845	2,540	7,625	794	
	4560	3-1/2 to 5 C.Y., 275 H.P.			48.55	1,200	3,610	10,800	1,110	
	4610	Tractor loader, wheel, torque conv., 4 x 4, 1 to 1-1/4 C.Y., 65 H.P.			9.95	192	575	1,725	194.60	
	4620	1-1/2 to 1-3/4 C.Y., 80 H.P.			12.55	235	705	2,125	241.40	
	4650	1-3/4 to 2 C.Y., 100 H.P.			14.25	280	840	2,525	282	
	4710	2-1/2 to 3-1/2 C.Y., 130 H.P.			15.30	305	915	2,750	305.40	
	4730	3 to 4-1/2 C.Y., 170 H.P.			21.00	460	1,380	4,150	444	
	4760	5-1/4 to 5-3/4 C.Y., 270 H.P.			34.75	705	2,115	6,350	701	
	4810	7 to 8 C.Y., 375 H.P.			59.05	1,250	3,740	11,200	1,220	
	4870	12-1/2 C.Y., 690 H.P.			82.40	1,925	5,810	17,400	1,821	
	4880	Wheeled, skid steer, 10 C.F., 30 H.P. gas			8.75	140	420	1,250	154	
	4890	1 C.Y., 78 H.P., diesel		↓	10.50	197	590	1,775	202	

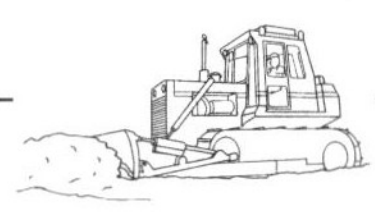

01 54 33 | Rental Equipment

		01 54 33 | Rental Equipment		UNIT	HOURLY OPER. COST	RENT PER DAY	RENT PER WEEK	RENT PER MONTH	CREW EQUIPMENT COST/DAY	
20	4891	Attachments for all skid steer loaders	R015433 -10							20
	4892	Auger		Ea.	.40	66.50	200	600	43.20	
	4893	Backhoe	R312323 -30		.67	112	336	1,000	72.55	
	4894	Broom			.61	101	303	910	65.50	
	4895	Forks	R312316 -40		.22	37	111	335	23.95	
	4896	Grapple			.50	83.50	250	750	54	
	4897	Concrete hammer	R312316 -45		.96	159	478	1,425	103.30	
	4898	Tree spade			.95	158	475	1,425	102.60	
	4899	Trencher			.64	106	318	955	68.70	
	4900	Trencher, chain, boom type, gas, operator walking, 12 H.P.			2.70	43.50	130	390	47.60	
	4910	Operator riding, 40 H.P.			8.85	243	730	2,200	216.80	
	5000	Wheel type, diesel, 4' deep, 12" wide			47.60	745	2,230	6,700	826.80	
	5100	Diesel, 6' deep, 20" wide			64.90	1,775	5,295	15,900	1,578	
	5150	Ladder type, diesel, 5' deep, 8" wide			23.25	705	2,115	6,350	609	
	5200	Diesel, 8' deep, 16" wide			57.25	1,675	5,010	15,000	1,460	
	5210	Tree spade, self-propelled			11.10	267	800	2,400	248.80	
	5250	Truck, dump, tandem, 12 ton payload			20.45	272	815	2,450	326.60	
	5300	Three axle dump, 16 ton payload			27.95	420	1,265	3,800	476.60	
	5350	Dump trailer only, rear dump, 16-1/2 C.Y.			4.25	115	345	1,025	103	
	5400	20 C.Y.			4.60	132	395	1,175	115.80	
	5450	Flatbed, single axle, 1-1/2 ton rating			11.85	56.50	170	510	128.80	
	5500	3 ton rating			15.15	91.50	275	825	176.20	
	5550	Off highway rear dump, 25 ton capacity			40.40	980	2,945	8,825	912.20	
	5600	35 ton capacity			41.30	1,000	3,015	9,050	933.40	
	5610	50 ton capacity			53.15	1,250	3,780	11,300	1,181	
	5620	65 ton capacity			56.85	1,375	4,105	12,300	1,276	
	5630	100 ton capacity			73.00	1,775	5,325	16,000	1,649	
	6000	Vibratory plow, 25 H.P., walking		↓	4.65	58.50	175	525	72.20	
40	0010	**GENERAL EQUIPMENT RENTAL**, without operators	R314116 -45							40
	0150	Aerial lift, scissor type, to 15' high, 1000 lb. cap., electric		Ea.	2.35	48.50	145	435	47.80	
	0160	To 25' high, 2000 lb. capacity	R314116 -40		2.85	70	210	630	64.80	
	0170	Telescoping boom to 40' high, 500 lb. capacity, gas			12.25	278	835	2,500	265	
	0180	To 45' high, 500 lb. capacity	R312323 -30		13.05	320	960	2,875	296.40	
	0190	To 60' high, 600 lb. capacity			15.00	425	1,280	3,850	376	
	0195	Air compressor, portable, 6.5 CFM, electric			.40	12.35	37	111	10.60	
	0196	Gasoline			.57	18.65	56	168	15.75	
	0200	Air compressor, portable, gas engine, 60 C.F.M.			4.90	33	99	297	59	
	0300	160 C.F.M.			8.70	46.50	140	420	97.60	
	0400	Diesel engine, rotary screw, 250 C.F.M.			8.60	85	255	765	119.80	
	0500	365 C.F.M.			11.50	112	335	1,000	159	
	0550	450 C.F.M.			13.90	130	390	1,175	189.20	
	0600	600 C.F.M.			23.65	218	655	1,975	320.20	
	0700	750 C.F.M.			25.55	228	685	2,050	341.40	
	0800	For silenced models, small sizes, add								
	0900	Large sizes, add		↓						
	0920	Air tools and accessories								
	0930	Breaker, pavement, 60 lb.		Ea.	.40	9.65	29	87	9	
	0940	80 lb.			.40	10.65	32	96	9.60	
	0950	Drills, hand (jackhammer) 65 lb.			.45	15.35	46	138	12.80	
	0960	Track or wagon, swing boom, 4" drifter			38.85	695	2,090	6,275	728.80	
	0970	5" drifter			45.75	760	2,280	6,850	822	
	0975	Track mounted quarry drill, 6" diameter drill			55.65	965	2,900	8,700	1,025	
	0980	Dust control per drill			.86	14.65	44	132	15.70	
	0990	Hammer, chipping, 12 lb.			.40	20	60	180	15.20	
	1000	Hose, air with couplings, 50' long, 3/4" diameter			.04	6	18	54	3.90	
	1100	1" diameter			.04	6	18	54	3.90	
	1200	1-1/2" diameter			.04	7.35	22	66	4.70	
	1300	2" diameter		↓	.10	17.35	52	156	11.20	

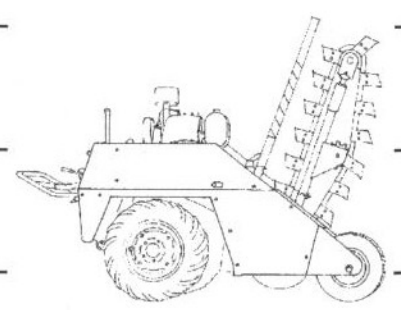

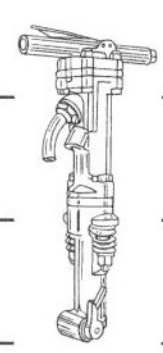

01 54 33 | Rental Equipment

					UNIT	HOURLY OPER. COST	RENT PER DAY	RENT PER WEEK	RENT PER MONTH	CREW EQUIPMENT COST/DAY	
40	1400	2-1/2" diameter	R314116 -45		Ea.	.12	19.35	58	174	12.55	40
	1410	3" diameter				.17	28.50	86	258	18.55	
	1450	Drill, steel, 7/8" x 2'	R314116 -40			.05	8.35	25	75	5.40	
	1460	7/8" x 6'				.06	9.35	28	84	6.10	
	1520	Moil points	R312323 -30			.03	4.33	13	39	2.85	
	1525	Pneumatic nailer w/accessories				.41	27.50	82	246	19.70	
	1530	Sheeting driver for 60 lb. breaker				.10	7.35	22	66	5.20	
	1540	For 90 lb. breaker				.15	10	30	90	7.20	
	1550	Spade, 25 lb.				.35	6.65	20	60	6.80	
	1560	Tamper, single, 35 lb.				.58	38.50	116	350	27.85	
	1570	Triple, 140 lb.				.87	58	174	520	41.75	
	1580	Wrenches, impact, air powered, up to 3/4" bolt				.25	7.65	23	69	6.60	
	1590	Up to 1-1/4" bolt				.35	16.35	49	147	12.60	
	1600	Barricades, barrels, reflectorized, 1 to 50 barrels				.02	3.10	9.30	28	2	
	1610	100 to 200 barrels				.01	2.33	7	21	1.50	
	1620	Barrels with flashers, 1 to 50 barrels				.02	3.77	11.30	34	2.40	
	1630	100 to 200 barrels				.02	3	9	27	1.95	
	1640	Barrels with steady burn type C lights				.03	5	15	45	3.25	
	1650	Illuminated board, trailer mounted, with generator				.65	115	345	1,025	74.20	
	1670	Portable barricade, stock, with flashers, 1 to 6 units				.02	3.77	11.30	34	2.40	
	1680	25 to 50 units				.02	3.50	10.50	31.50	2.25	
	1690	Butt fusion machine, electric				18.65	330	990	2,975	347.20	
	1695	Electro fusion machine				8.55	133	400	1,200	148.40	
	1700	Carts, brick, hand powered, 1000 lb. capacity				.30	49.50	148	445	32	
	1800	Gas engine, 1500 lb., 7-1/2' lift				3.09	94.50	284	850	81.50	
	1822	Dehumidifier, medium, 6 lb/hr, 150 CFM				.75	45.50	137	410	33.40	
	1824	Large, 18 lb/hr, 600 CFM				1.48	90.50	272	815	66.25	
	1830	Distributor, asphalt, trailer mtd, 2000 gal., 38 H.P. diesel		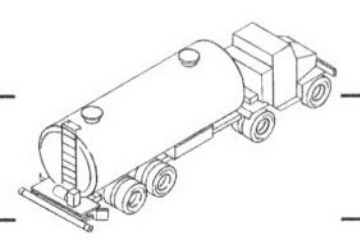		7.25	252	755	2,275	209	
	1840	3000 gal., 38 H.P. diesel				8.55	290	870	2,600	242.40	
	1850	Drill, rotary hammer, electric, 1-1/2" diameter				.40	24.50	74	222	18	
	1860	Carbide bit for above				.04	6	18	54	3.90	
	1865	Rotary, crawler, 250 H.P.				95.45	1,800	5,410	16,200	1,846	
	1870	Emulsion sprayer, 65 gal., 5 H.P. gas engine				2.00	75	225	675	61	
	1880	200 gal., 5 H.P. engine				5.15	125	375	1,125	116.20	
	1920	Floodlight, mercury vapor, or quartz, on tripod									
	1930	1000 watt			Ea.	.30	12	36	108	9.60	
	1940	2000 watt				.52	22	66	198	17.35	
	1950	Floodlights, trailer mounted with generator, 1 - 300 watt light				2.55	65	195	585	59.40	
	1960	2 - 1000 watt lights				3.60	110	330	990	94.80	
	2000	4 - 300 watt lights				2.95	76.50	230	690	69.60	
	2020	Forklift, wheeled, for brick, 18', 3000 lb., 2 wheel drive, gas				15.25	182	545	1,625	231	
	2040	28', 4000 lb., 4 wheel drive, diesel				12.55	237	710	2,125	242.40	
	2050	For rough terrain, 8000 lb., 16' lift, 68 H.P.				16.90	375	1,125	3,375	360.20	
	2060	For plant, 4 T. capacity, 80 H.P., 2 wheel drive, gas				8.85	88.50	265	795	123.80	
	2080	10 T. capacity, 120 H.P., 2 wheel drive, diesel				13.30	162	485	1,450	203.40	
	2100	Generator, electric, gas engine, 1.5 KW to 3 KW				1.90	12	36	108	22.40	
	2200	5 KW				2.60	17.35	52	156	31.20	
	2300	10 KW				4.30	33.50	100	300	54.40	
	2400	25 KW				8.35	61.50	185	555	103.80	
	2500	Diesel engine, 20 KW				6.40	66.50	200	600	91.20	
	2600	50 KW				12.25	71.50	215	645	141	
	2700	100 KW				18.15	80	240	720	193.20	
	2800	250 KW				51.65	148	445	1,325	502.20	
	2850	Hammer, hydraulic, for mounting on boom, to 500 ft.-lb.		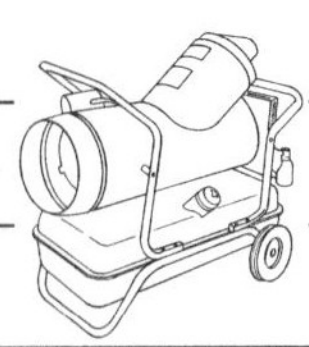		1.85	63.50	190	570	52.80	
	2860	1000 ft.-lb.				3.25	103	310	930	88	
	2900	Heaters, space, oil or electric, 50 MBH				1.25	7.35	22	66	14.40	
	3000	100 MBH				2.26	10.65	32	96	24.50	
	3100	300 MBH				7.24	33.50	100	300	77.90	

RENTAL EQUIPMENT

01 54 33 | Rental Equipment

40	01 54 33 Rental Equipment	Unit	Hourly Oper. Cost	Rent per Day	Rent per Week	Rent per Month	Crew Equipment Cost/Day
3150	500 MBH (R314116-45)	Ea.	14.60	50	150	450	146.80
3200	Hose, water, suction with coupling, 20' long, 2" diameter		.02	3.33	10	30	2.15
3210	3" diameter (R314116-40)		.03	5.65	17	51	3.65
3220	4" diameter		.03	6.35	19	57	4.05
3230	6" diameter (R312323-30)		.11	18.65	56	168	12.10
3240	8" diameter		.27	45.50	137	410	29.55
3250	Discharge hose with coupling, 50' long, 2" diameter		.01	2	6	18	1.30
3260	3" diameter		.02	3.33	10	30	2.15
3270	4" diameter		.02	4.67	14	42	2.95
3280	6" diameter		.06	11.65	35	105	7.50
3290	8" diameter		.33	54.50	164	490	35.45
3295	Insulation blower		.11	7.35	22	66	5.30
3300	Ladders, extension type, 16' to 36' long		.20	33	99	297	21.40
3400	40' to 60' long		.30	49.50	149	445	32.20
3405	Lance for cutting concrete		2.36	77	231	695	65.10
3407	Lawn mower, rotary, 22", 5HP		1.05	23.50	71	213	22.60
3408	48" self propelled		3.52	126	377	1,125	103.55
3410	Level, laser type, for pipe and sewer leveling		1.22	81	243	730	58.35
3430	Electronic		.81	54	162	485	38.90
3440	Laser type, rotating beam for grade control		1.20	79.50	239	715	57.40
3460	Builders level with tripod and rod		.08	12.65	38	114	8.25
3500	Light towers, towable, with diesel generator, 2000 watt		2.95	76.50	230	690	69.60
3600	4000 watt		3.60	110	330	990	94.80
3700	Mixer, powered, plaster and mortar, 6 C.F., 7 H.P.		1.30	21.50	65	195	23.40
3800	10 C.F., 9 H.P.		1.55	35	105	315	33.40
3850	Nailer, pneumatic		.41	27.50	82	246	19.70
3900	Paint sprayers complete, 8 CFM		.69	45.50	137	410	32.90
4000	17 CFM		1.12	74.50	223	670	53.55
4020	Pavers, bituminous, rubber tires, 8' wide, 50 H.P., diesel		28.05	800	2,405	7,225	705.40
4030	10' wide, 150 H.P.		62.65	1,525	4,595	13,800	1,420
4050	Crawler, 8' wide, 100 H.P., diesel		58.30	1,650	4,915	14,700	1,449
4060	10' wide, 150 H.P.		70.45	1,925	5,810	17,400	1,726
4070	Concrete paver, 12' to 24' wide, 250 H.P.		58.75	1,350	4,035	12,100	1,277
4080	Placer-spreader-trimmer, 24' wide, 300 H.P.		92.80	2,500	7,475	22,400	2,237
4100	Pump, centrifugal gas pump, 1-1/2", 4 MGPH		2.75	40	120	360	46
4200	2", 8 MGPH		3.55	43.50	130	390	54.40
4300	3", 15 MGPH		3.70	45	135	405	56.60
4400	6", 90 MGPH		17.75	163	490	1,475	240
4500	Submersible electric pump, 1-1/4", 55 GPM		.36	15	45	135	11.90
4600	1-1/2", 83 GPM		.39	17	51	153	13.30
4700	2", 120 GPM		.45	21.50	64	192	16.40
4800	3", 300 GPM		.95	36.50	110	330	29.60
4900	4", 560 GPM		6.34	145	435	1,300	137.70
5000	6", 1590 GPM		9.47	213	640	1,925	203.75
5100	Diaphragm pump, gas, single, 1-1/2" diameter		.88	40	120	360	31.05
5200	2" diameter		2.90	50	150	450	53.20
5300	3" diameter		2.90	50	150	450	53.20
5400	Double, 4" diameter		3.90	70	210	630	73.20
5500	Trash pump, self-priming, gas, 2" diameter		2.80	19.65	59	177	34.20
5600	Diesel, 4" diameter		6.85	55	165	495	87.80
5650	Diesel, 6" diameter		21.40	122	365	1,100	244.20
5655	Grout Pump		8.05	70	210	630	106.40
5660	Rollers, see division 01590-200						
5700	Salamanders, L.P. gas fired, 100,000 BTU	Ea.	2.27	11.35	34	102	24.95
5705	50,000 BTU		1.70	8	24	72	18.40
5720	Sandblaster, portable, open top, 3 C.F. capacity		.40	20.50	62	186	15.60
5730	6 C.F. capacity		.65	29.50	88	264	22.80
5740	Accessories for above		.11	18.65	56	168	12.10

01 54 33 | Rental Equipment

		Description	Ref.	Unit	Hourly Oper. Cost	Rent per Day	Rent per Week	Rent per Month	Crew Equipment Cost/Day	
40	5750	Sander, floor	R314116 -45	Ea.	.75	18.35	55	165	17	40
	5760	Edger			.55	17.65	53	159	15	
	5800	Saw, chain, gas engine, 18" long	R314116 -40		1.35	16	48	144	20.40	
	5900	36" long			.55	48.50	145	435	33.40	
	5950	60" long	R312323 -30		.55	50	150	450	34.40	
	6000	Masonry, table mounted, 14" diameter, 5 H.P.			1.30	56	168	505	44	
	6050	Portable cut-off, 8 H.P.			1.45	29.50	88	264	29.20	
	6100	Circular, hand held, electric, 7-1/4" diameter			.20	5	15	45	4.60	
	6200	12" diameter			.27	8.35	25	75	7.15	
	6250	Wall saw, w/hydraulic power, 10 H.P			2.04	108	323.20	970	80.95	
	6275	Shot blaster, walk behind, 20" wide			1.70	495	1,490	4,475	311.60	
	6300	Steam cleaner, 100 gallons per hour			2.40	63.50	190	570	57.20	
	6310	200 gallons per hour			3.25	76.50	230	690	72	
	6340	Tar Kettle/Pot, 400 gallon			3.29	51.50	155	465	57.30	
	6350	Torch, cutting, acetylene-oxygen, 150' hose			1.50	7	21	63	16.20	
	6360	Hourly operating cost includes tips and gas			8.10				64.80	
	6410	Toilet, portable chemical			.11	18	54	162	11.70	
	6420	Recycle flush type			.13	22	66	198	14.25	
	6430	Toilet, fresh water flush, garden hose,			.15	24.50	74	222	16	
	6440	Hoisted, non-flush, for high rise			.13	21.50	65	195	14.05	
	6450	Toilet, trailers, minimum			.22	37.50	112	335	24.15	
	6460	Maximum			.67	112	337	1,000	72.75	
	6465	Tractor, farm with attachment			10.30	223	670	2,000	216.40	
	6470	Trailer, office, see division 01520-500								
	6500	Trailers, platform, flush deck, 2 axle, 25 ton capacity		Ea.	4.25	90	270	810	88	
	6600	40 ton capacity			5.55	125	375	1,125	119.40	
	6700	3 axle, 50 ton capacity			6.00	138	415	1,250	131	
	6800	75 ton capacity			7.45	182	545	1,625	168.60	
	6810	Trailer mounted cable reel for H.V. line work			4.56	217	652	1,950	166.90	
	6820	Trailer mounted cable tensioning rig			8.96	425	1,280	3,850	327.70	
	6830	Cable pulling rig			58.52	2,425	7,260	21,800	1,920	
	6850	Trailer, storage, see division 01520-500								
	6900	Water tank, engine driven discharge, 5000 gallons		Ea.	5.60	120	360	1,075	116.80	
	6925	10,000 gallons			7.70	168	505	1,525	162.60	
	6950	Water truck, off highway, 6000 gallons			49.25	700	2,095	6,275	813	
	7010	Tram car for H.V. line work, powered, 2 conductor			5.78	118	354	1,050	117.05	
	7020	Transit (builder's level) with tripod			.08	12.65	38	114	8.25	
	7030	Trench box, 3000 lbs. 6'x8'			.62	104	311	935	67.15	
	7040	7200 lbs. 6'x20'			.71	118	354	1,050	76.50	
	7050	8000 lbs., 8' x 16'			.93	156	467	1,400	100.85	
	7060	9500 lbs., 8'x20'			1.20	200	599	1,800	129.40	
	7065	11,000 lbs., 8'x24'			1.35	226	677	2,025	146.20	
	7070	12,000 lbs., 10' x 20'			1.77	294	883	2,650	190.75	
	7100	Truck, pickup, 3/4 ton, 2 wheel drive			5.75	53.50	160	480	78	
	7200	4 wheel drive			5.90	63.50	190	570	85.20	
	7250	Crew carrier, 9 passenger			9.40	97.50	292.80	880	133.75	
	7290	Tool van, 24,000 G.V.W.			8.80	103	310	930	132.40	
	7300	Tractor, 4 x 2, 30 ton capacity, 195 H.P.			13.65	165	495	1,475	208.20	
	7410	250 H.P.			18.20	243	730	2,200	291.60	
	7500	6 x 2, 40 ton capacity, 240 H.P.			17.45	270	810	2,425	301.60	
	7600	6 x 4, 45 ton capacity, 240 H.P.			22.20	293	880	2,650	353.60	
	7620	Vacuum truck, hazardous material, 2500 gallon			7.38	310	926	2,775	244.25	
	7625	5,000 gallon			11.37	410	1,234.80	3,700	337.90	
	7640	Tractor, with A frame, boom and winch, 225 H.P.			15.75	223	670	2,000	260	
	7650	Vacuum, H.E.P.A., 16 gal., wet/dry		Ea.	.27	24	72	216	16.55	
	7655	55 gal, wet/dry			.60	36	108	325	26.40	
	7660	Water tank, portable			1.00	9.65	29	87	13.80	
	7690	Large production vacuum loader, 3150 CFM			16.25	615	1,850	5,550	500	

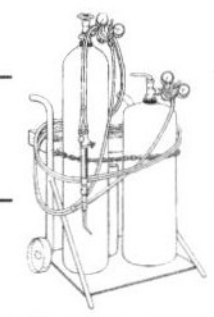

01 54 33 | Rental Equipment

		01 54 33 Rental Equipment		UNIT	HOURLY OPER. COST	RENT PER DAY	RENT PER WEEK	RENT PER MONTH	CREW EQUIPMENT COST/DAY	
40	7700	Welder, electric, 200 amp	R314116 -45	Ea.	3.52	42.50	127	380	53.55	40
	7800	300 amp			5.04	48	144	430	69.10	
	7900	Gas engine, 200 amp	R314116 -40		6.65	25.50	76	228	68.40	
	8000	300 amp			9.00	28.50	86	258	89.20	
	8100	Wheelbarrow, any size	R312323 -30		.06	10.65	32	96	6.90	
	8200	Wrecking ball, 4000 lb.			1.95	70	210	630	57.60	
50	0010	**HIGHWAY EQUIPMENT RENTAL**, without operators								50
	0050	Asphalt batch plant, portable drum mixer, 100 ton/hr.		Ea.	57.45	1,350	4,025	12,100	1,265	
	0060	200 ton/hr.			63.80	1,400	4,230	12,700	1,356	
	0070	300 ton/hr.			74.25	1,675	5,005	15,000	1,595	
	0100	Backhoe attachment, long stick, up to 185 HP, 10.5' long			.31	20.50	61	183	14.70	
	0140	Up to 250 HP, 12' long			.33	22	66	198	15.85	
	0180	Over 250 HP, 15' long			.44	29	87	261	20.90	
	0200	Special dipper arm, up to 100 HP, 32' long			.90	59.50	179	535	43	
	0240	Over 100 HP, 33' long			1.12	74.50	224	670	53.75	
	0300	Concrete batch plant, portable, electric, 200 CY/Hr			10.42	495	1,490	4,475	381.35	
	0500	Grader attachment, ripper/scarifier, rear mounted								
	0520	Up to 135 HP		Ea.	2.85	60	180	540	58.80	
	0540	Up to 180 HP			3.40	76.50	230	690	73.20	
	0580	Up to 250 HP			3.75	86.50	260	780	82	
	0700	Pvmt. removal bucket, for hyd. excavator, up to 90 HP			1.45	45	135	405	38.60	
	0740	Up to 200 HP			1.65	65	195	585	52.20	
	0780	Over 200 HP			1.75	78.50	235	705	61	
	0900	Aggregate spreader, self-propelled, 187 HP			37.80	790	2,375	7,125	777.40	
	1000	Chemical spreader, 3 C.Y.			2.25	51.50	155	465	49	
	1900	Hammermill, traveling, 250 HP			47.35	1,750	5,220	15,700	1,423	
	2000	Horizontal borer, 3" diam, 13 HP gas driven			4.25	53.50	160	480	66	
	2200	Hydromulchers, gas power, 3000 gal., for truck mounting			10.65	187	560	1,675	197.20	
	2400	Joint & crack cleaner, walk behind, 25 HP			2.10	48.50	145	435	45.80	
	2500	Filler, trailer mounted, 400 gal., 20 HP			6.05	182	545	1,625	157.40	
	3000	Paint striper, self propelled, double line, 30 HP			5.20	155	465	1,400	134.60	
	3200	Post drivers, 6" I-Beam frame, for truck mounting			8.25	415	1,240	3,725	314	
	3400	Road sweeper, self propelled, 8' wide, 90 HP			25.40	485	1,450	4,350	493.20	
	4000	Road mixer, self-propelled, 130 HP			30.65	595	1,780	5,350	601.20	
	4100	310 HP			58.55	2,050	6,150	18,500	1,698	
	4200	Cold mix paver, incl pug mill and bitumen tank,								
	4220	165 HP		Ea.	70.45	1,950	5,825	17,500	1,729	
	4250	Paver, asphalt, wheel or crawler, 130 H.P., diesel			69.15	1,850	5,540	16,600	1,661	
	4300	Paver, road widener, gas 1' to 6', 67 HP			31.70	680	2,035	6,100	660.60	
	4400	Diesel, 2' to 14', 88 HP			42.35	995	2,980	8,950	934.80	
	4600	Slipform pavers, curb and gutter, 2 track, 75 HP			25.20	655	1,960	5,875	593.60	
	4700	4 track, 165 HP			35.25	760	2,280	6,850	738	
	4800	Median barrier, 215 HP			35.85	790	2,370	7,100	760.80	
	4901	Trailer, low bed, 75 ton capacity			7.90	178	535	1,600	170.20	
	5000	Road planer, walk behind, 10" cutting width, 10 HP			2.05	27	81	243	32.60	
	5100	Self propelled, 12" cutting width, 64 HP			5.45	150	450	1,350	133.60	
	5200	Pavement profiler, 4' to 6' wide, 450 HP			161.05	2,925	8,755	26,300	3,039	
	5300	8' to 10' wide, 750 HP			256.35	4,350	13,040	39,100	4,659	
	5400	Roadway plate, steel, 1"x8'x20'			.06	10.35	31	93	6.70	
	5600	Stabilizer,self-propelled, 150 HP			29.65	565	1,690	5,075	575.20	
	5700	310 HP			48.65	1,200	3,570	10,700	1,103	
	5800	Striper, thermal, truck mounted 120 gal. paint, 150 H.P.			33.15	485	1,450	4,350	555.20	
	6000	Tar kettle, 330 gal., trailer mounted			2.97	36.50	110	330	45.75	
	7000	Tunnel locomotive, diesel, 8 to 12 ton			22.10	555	1,670	5,000	510.80	
	7005	Electric, 10 ton			21.20	630	1,895	5,675	548.60	
	7010	Muck cars, 1/2 C.Y. capacity			1.55	21	63	189	25	
	7020	1 C.Y. capacity			1.75	29	87	261	31.40	
	7030	2 C.Y. capacity			1.90	33.50	100	300	35.20	

01 54 | Construction Aids

01 54 33 | Rental Equipment

		01 54 33 Rental Equipment		UNIT	HOURLY OPER. COST	RENT PER DAY	RENT PER WEEK	RENT PER MONTH	CREW EQUIPMENT COST/DAY	
50	7040	Side dump, 2 C.Y. capacity		Ea.	2.10	41.50	125	375	41.80	50
	7050	3 C.Y. capacity			2.80	48.50	145	435	51.40	
	7060	5 C.Y. capacity			3.95	60	180	540	67.60	
	7100	Ventilating blower for tunnel, 7-1/2 H.P.			1.25	47	141	425	38.20	
	7110	10 H.P.			1.44	48.50	146	440	40.70	
	7120	20 H.P.			2.33	62	186	560	55.85	
	7140	40 H.P.			4.04	88.50	265	795	85.30	
	7160	60 H.P.			6.16	137	410	1,225	131.30	
	7175	75 H.P.			7.89	182	545	1,625	172.10	
	7180	200 H.P.		↓	17.60	270	810	2,425	302.80	
	7800	Windrow loader, elevating			35.00	905	2,715	8,150	823	
60	0010	**LIFTING** & HOISTING EQUIPMENT RENTAL, without operators	R015433 -10							60
	0120	Aerial lift truck, 2 person, to 80'		Ea.	19.35	600	1,805	5,425	515.80	
	0140	Boom work platform, 40' snorkel	R015433 -15		10.60	222	665	2,000	217.80	
	0150	Crane, flatbed mntd, 3 ton cap.			14.65	197	590	1,775	235.20	
	0200	Crane, climbing, 106' jib, 6000 lb. capacity, 410 FPM	R312316 -45		54.75	1,375	4,150	12,500	1,268	
	0300	101' jib, 10,250 lb. capacity, 270 FPM		↓	60.25	1,750	5,250	15,800	1,532	
	0400	Tower, static, 130' high, 106' jib,								
	0500	6200 lb. capacity at 400 FPM		Ea.	57.95	1,600	4,790	14,400	1,422	
	0600	Crawler mounted, lattice boom, 1/2 C.Y., 15 tons at 12' radius			22.08	595	1,790	5,375	534.65	
	0700	3/4 C.Y., 20 tons at 12' radius			29.44	715	2,140	6,425	663.50	
	0800	1 C.Y., 25 tons at 12' radius			39.25	945	2,830	8,500	880	
	0900	1-1/2 C.Y., 40 tons at 12' radius			43.65	975	2,925	8,775	934.20	
	1000	2 C.Y., 50 tons at 12' radius			54.25	1,375	4,155	12,500	1,265	
	1100	3 C.Y., 75 tons at 12' radius			51.75	1,350	4,060	12,200	1,226	
	1200	100 ton capacity, 60' boom			66.30	1,775	5,300	15,900	1,590	
	1300	165 ton capacity, 60' boom			96.50	2,300	6,890	20,700	2,150	
	1400	200 ton capacity, 70' boom			102.55	2,450	7,375	22,100	2,295	
	1500	350 ton capacity, 80' boom			149.45	3,475	10,450	31,400	3,286	
	1600	Truck mounted, lattice boom, 6 x 4, 20 tons at 10' radius			30.88	1,175	3,500	10,500	947.05	
	1700	25 tons at 10' radius			32.98	1,250	3,740	11,200	1,012	
	1800	8 x 4, 30 tons at 10' radius			42.67	1,325	3,970	11,900	1,135	
	1900	40 tons at 12' radius			36.20	1,400	4,200	12,600	1,130	
	2000	8 x 4, 60 tons at 15' radius			45.96	1,475	4,440	13,300	1,256	
	2050	82 tons at 15' radius			48.65	1,550	4,670	14,000	1,323	
	2100	90 tons at 15' radius			50.65	1,700	5,110	15,300	1,427	
	2200	115 tons at 15' radius			51.45	1,900	5,675	17,000	1,547	
	2300	150 tons at 18' radius			47.40	1,925	5,800	17,400	1,539	
	2350	165 tons at 18' radius			80.90	2,275	6,825	20,500	2,012	
	2400	Truck mounted, hydraulic, 12 ton capacity			35.20	590	1,775	5,325	636.60	
	2500	25 ton capacity			35.35	615	1,840	5,525	650.80	
	2550	33 ton capacity			36.20	645	1,935	5,800	676.60	
	2560	40 ton capacity			34.65	620	1,855	5,575	648.20	
	2600	55 ton capacity			51.45	900	2,700	8,100	951.60	
	2700	80 ton capacity			61.10	950	2,850	8,550	1,059	
	2720	100 ton capacity			83.70	2,425	7,250	21,800	2,120	
	2740	120 ton capacity			90.35	2,625	7,900	23,700	2,303	
	2760	150 ton capacity			110.25	3,325	9,990	30,000	2,880	
	2800	Self-propelled, 4 x 4, with telescoping boom, 5 ton			16.15	293	880	2,650	305.20	
	2900	12-1/2 ton capacity			27.05	535	1,600	4,800	536.40	
	3000	15 ton capacity			28.20	595	1,785	5,350	582.60	
	3050	20 ton capacity			29.05	620	1,855	5,575	603.40	
	3100	25 ton capacity			33.35	715	2,150	6,450	696.80	
	3150	40 ton capacity			48.00	885	2,660	7,975	916	
	3200	Derricks, guy, 20 ton capacity, 60' boom, 75' mast			15.61	340	1,016	3,050	328.10	
	3300	100' boom, 115' mast			24.93	580	1,740	5,225	547.45	
	3400	Stiffleg, 20 ton capacity, 70' boom, 37' mast		↓	17.67	435	1,310	3,925	403.35	

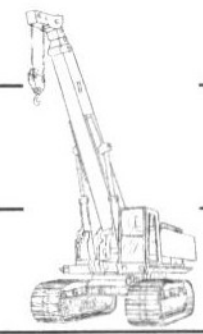

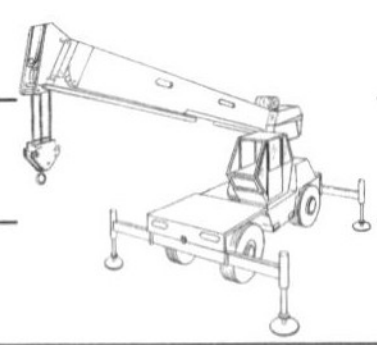

RENTAL EQUIPMENT

01 54 33 | Rental Equipment

			UNIT	HOURLY OPER. COST	RENT PER DAY	RENT PER WEEK	RENT PER MONTH	CREW EQUIPMENT COST/DAY
60	3500	100' boom, 47' mast R015433 -10	Ea.	27.59	705	2,120	6,350	644.70
	3550	Helicopter, small, lift to 1250 lbs. maximum, w/pilot		74.47	2,725	8,210	24,600	2,238
	3600	Hoists, chain type, overhead, manual, 3/4 ton R015433 -15		.10	1.33	4	12	1.60
	3900	10 ton		.60	10	30	90	10.80
	4000	Hoist and tower, 5000 lb. cap., portable electric, 40' high R312316 -45		4.25	195	585	1,750	151
	4100	For each added 10' section, add		.09	15.35	46	138	9.90
	4200	Hoist and single tubular tower, 5000 lb. electric, 100' high		5.74	272	817	2,450	209.30
	4300	For each added 6'-6" section, add		.15	25.50	77	231	16.60
	4400	Hoist and double tubular tower, 5000 lb., 100' high		6.15	300	900	2,700	229.20
	4500	For each added 6'-6" section, add		.17	29	87	261	18.75
	4550	Hoist and tower, mast type, 6000 lb., 100' high		6.65	310	933	2,800	239.80
	4570	For each added 10' section, add		.11	18.65	56	168	12.10
	4600	Hoist and tower, personnel, electric, 2000 lb., 100' @ 125 FPM		13.77	830	2,490	7,475	608.15
	4700	3000 lb., 100' @ 200 FPM		15.70	935	2,810	8,425	687.60
	4800	3000 lb., 150' @ 300 FPM		17.40	1,050	3,150	9,450	769.20
	4900	4000 lb., 100' @ 300 FPM		18.03	1,075	3,210	9,625	786.25
	5000	6000 lb., 100' @ 275 FPM		19.49	1,125	3,370	10,100	829.90
	5100	For added heights up to 500', add	L.F.	.01	1.67	5	15	1.10
	5200	Jacks, hydraulic, 20 ton	Ea.	.05	4.67	14	42	3.20
	5500	100 ton	"	.30	13.35	40	120	10.40
	6000	Jacks, hydraulic, climbing with 50' jackrods						
	6010	and control consoles, minimum 3 mo. rental						
	6100	30 ton capacity	Ea.	1.68	112	336	1,000	80.65
	6150	For each added 10' jackrod section, add		.05	3.33	10	30	2.40
	6300	50 ton capacity		2.71	180	541	1,625	129.90
	6350	For each added 10' jackrod section, add		.06	4	12	36	2.90
	6500	125 ton capacity		7.10	475	1,420	4,250	340.80
	6550	For each added 10' jackrod section, add		.49	32.50	97	291	23.30
	6600	Cable jack, 10 ton capacity with 200' cable		1.41	93.50	281	845	67.50
	6650	For each added 50' of cable, add		.16	10.35	31	93	7.50
70	0010	**WELLPOINT EQUIPMENT RENTAL**, without operators R312319 -90						
	0020	Based on 2 months rental						
	0100	Combination jetting & wellpoint pump, 60 H.P. diesel	Ea.	10.93	278	833	2,500	254.05
	0200	High pressure gas jet pump, 200 H.P., 300 psi	"	21.98	237	712	2,125	318.25
	0300	Discharge pipe, 8" diameter	L.F.	.01	.45	1.35	4.05	.35
	0350	12" diameter		.01	.67	2	6	.50
	0400	Header pipe, flows up to 150 G.P.M., 4" diameter		.01	.41	1.22	3.66	.30
	0500	400 G.P.M., 6" diameter		.01	.48	1.45	4.35	.35
	0600	800 G.P.M., 8" diameter		.01	.67	2	6	.50
	0700	1500 G.P.M., 10" diameter		.01	.70	2.10	6.30	.50
	0800	2500 G.P.M., 12" diameter		.02	1.32	3.97	11.90	.95
	0900	4500 G.P.M., 16" diameter		.03	1.69	5.08	15.25	1.25
	0950	For quick coupling aluminum and plastic pipe, add		.03	1.75	5.25	15.75	1.30
	1100	Wellpoint, 25' long, with fittings & riser pipe, 1-1/2" or 2" diameter	Ea.	.05	3.50	10.49	31.50	2.50
	1200	Wellpoint pump, diesel powered, 4" diameter, 20 H.P.		5.06	160	480	1,450	136.50
	1300	6" diameter, 30 H.P.		6.72	199	596	1,800	172.95
	1400	8" suction, 40 H.P.		9.12	272	817	2,450	236.35
	1500	10" suction, 75 H.P.		13.06	320	955	2,875	295.50
	1600	12" suction, 100 H.P.		19.21	510	1,530	4,600	459.70
	1700	12" suction, 175 H.P.		26.64	560	1,680	5,050	549.10
80	0010	**MARINE EQUIPMENT RENTAL**, without operators						
	0200	Barge, 400 Ton, 30' wide x 90' long	Ea.	17.00	240	720	2,150	280
	0240	800 Ton, 45' wide x 90' long		28.10	345	1,030	3,100	430.80
	2000	Tugboat, diesel, 100 HP		18.80	170	510	1,525	252.40
	2040	250 HP		35.50	315	945	2,825	473
	2080	380 HP	Ea.	80.55	935	2,805	8,425	1,205

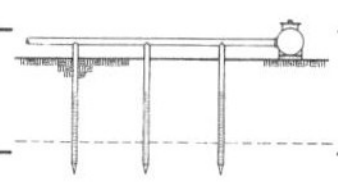

RENTAL EQUIPMENT

Crews

Crew No.	Bare Costs	Incl. Subs O & P	Cost Per Labor-Hour

Crew A-1	Hr.	Daily	Hr.	Daily	Bare Costs	Incl. O&P
1 Building Laborer	$27.40	$219.20	$45.40	$363.20	$27.40	$45.40
1 Concrete saw, gas manual		49.00		53.90	6.13	6.74
8 L.H., Daily Totals		$268.20		$417.10	$33.53	$52.14

Crew A-1A	Hr.	Daily	Hr.	Daily	Bare Costs	Incl. O&P
1 Skilled Worker	$36.50	$292.00	$59.70	$477.60	$36.50	$59.70
1 Shot Blaster, 20″		311.60		342.75	38.95	42.85
8 L.H., Daily Totals		$603.60		$820.35	$75.45	$102.55

Crew A-1B	Hr.	Daily	Hr.	Daily	Bare Costs	Incl. O&P
1 Building Laborer	$27.40	$219.20	$45.40	$363.20	$27.40	$45.40
1 Concr. saw, gas, self-prop.		107.00		117.70	13.38	14.71
8 L.H., Daily Totals		$326.20		$480.90	$40.78	$60.11

Crew A-1C	Hr.	Daily	Hr.	Daily	Bare Costs	Incl. O&P
1 Building Laborer	$27.40	$219.20	$45.40	$363.20	$27.40	$45.40
1 Brush saw		20.40		22.45	2.55	2.81
8 L.H., Daily Totals		$239.60		$385.65	$29.95	$48.21

Crew A-1D	Hr.	Daily	Hr.	Daily	Bare Costs	Incl. O&P
1 Building Laborer	$27.40	$219.20	$45.40	$363.20	$27.40	$45.40
1 Vibrating plate, gas, 18″		26.40		29.05	3.30	3.63
8 L.H., Daily Totals		$245.60		$392.25	$30.70	$49.03

Crew A-1E	Hr.	Daily	Hr.	Daily	Bare Costs	Incl. O&P
1 Building Laborer	$27.40	$219.20	$45.40	$363.20	$27.40	$45.40
1 Vibrating plate, gas, 21″		34.60		38.05	4.33	4.76
8 L.H., Daily Totals		$253.80		$401.25	$31.73	$50.16

Crew A-1F	Hr.	Daily	Hr.	Daily	Bare Costs	Incl. O&P
1 Building Laborer	$27.40	$219.20	$45.40	$363.20	$27.40	$45.40
1 Rammer/tamper, gas, 8″		37.00		40.70	4.63	5.09
8 L.H., Daily Totals		$256.20		$403.90	$32.03	$50.49

Crew A-1G	Hr.	Daily	Hr.	Daily	Bare Costs	Incl. O&P
1 Building Laborer	$27.40	$219.20	$45.40	$363.20	$27.40	$45.40
1 Rammer/tamper, gas, 8″		37.00		40.70	4.63	5.09
8 L.H., Daily Totals		$256.20		$403.90	$32.03	$50.49

Crew A-1H	Hr.	Daily	Hr.	Daily	Bare Costs	Incl. O&P
1 Building Laborer	$27.40	$219.20	$45.40	$363.20	$27.40	$45.40
1 Pressure washer		57.20		62.90	7.15	7.87
8 L.H., Daily Totals		$276.40		$426.10	$34.55	$53.27

Crew A-1J	Hr.	Daily	Hr.	Daily	Bare Costs	Incl. O&P
1 Building Laborer	$27.40	$219.20	$45.40	$363.20	$27.40	$45.40
1 Rototiller		103.55		113.90	12.94	14.24
8 L.H., Daily Totals		$322.75		$477.10	$40.34	$59.64

Crew A-1K	Hr.	Daily	Hr.	Daily	Bare Costs	Incl. O&P
1 Building Laborer	$27.40	$219.20	$45.40	$363.20	$27.40	$45.40
1 Lawn aerator		22.60		24.85	2.83	3.11
8 L.H., Daily Totals		$241.80		$388.05	$30.23	$48.51

Crew A-1L	Hr.	Daily	Hr.	Daily	Bare Costs	Incl. O&P
1 Building Laborer	$27.40	$219.20	$45.40	$363.20	$27.40	$45.40
1 Power blower/vacuum		22.60		24.85	2.83	3.11
8 L.H., Daily Totals		$241.80		$388.05	$30.23	$48.51

Crew A-1M	Hr.	Daily	Hr.	Daily	Bare Costs	Incl. O&P
1 Building Laborer	$27.40	$219.20	$45.40	$363.20	$27.40	$45.40
1 Snow blower		103.55		113.90	12.94	14.24
8 L.H., Daily Totals		$322.75		$477.10	$40.34	$59.64

Crew A-2	Hr.	Daily	Hr.	Daily	Bare Costs	Incl. O&P
2 Laborers	$27.40	$438.40	$45.40	$726.40	$27.47	$45.23
1 Truck Driver (light)	27.60	220.80	44.90	359.20		
1 Light Truck, 1.5 Ton		128.80		141.70	5.37	5.90
24 L.H., Daily Totals		$788.00		$1227.30	$32.84	$51.13

Crew A-2A	Hr.	Daily	Hr.	Daily	Bare Costs	Incl. O&P
2 Laborers	$27.40	$438.40	$45.40	$726.40	$27.47	$45.23
1 Truck Driver (light)	27.60	220.80	44.90	359.20		
1 Light Truck, 1.5 Ton		128.80		141.70		
1 Concrete Saw		107.00		117.70	9.83	10.81
24 L.H., Daily Totals		$895.00		$1345.00	$37.30	$56.04

Crew A-3	Hr.	Daily	Hr.	Daily	Bare Costs	Incl. O&P
1 Truck Driver (heavy)	$28.35	$226.80	$46.15	$369.20	$28.35	$46.15
1 Dump Truck, 12 Ton		326.60		359.25	40.83	44.91
8 L.H., Daily Totals		$553.40		$728.45	$69.18	$91.06

Crew A-3A	Hr.	Daily	Hr.	Daily	Bare Costs	Incl. O&P
1 Truck Driver (light)	$27.60	$220.80	$44.90	$359.20	$27.60	$44.90
1 Pickup Truck (4x4)		85.20		93.70	10.65	11.72
8 L.H., Daily Totals		$306.00		$452.90	$38.25	$56.62

Crew A-3B	Hr.	Daily	Hr.	Daily	Bare Costs	Incl. O&P
1 Equip. Oper. (medium)	$36.70	$293.60	$57.95	$463.60	$32.53	$52.05
1 Truck Driver (heavy)	28.35	226.80	46.15	369.20		
1 Dump Truck, 16 Ton		476.60		524.25		
1 F.E. Loader, 3 C.Y.		305.40		335.95	48.88	53.76
16 L.H., Daily Totals		$1302.40		$1693.00	$81.41	$105.81

Crew A-3C	Hr.	Daily	Hr.	Daily	Bare Costs	Incl. O&P
1 Equip. Oper. (light)	$35.20	$281.60	$55.60	$444.80	$35.20	$55.60
1 Wheeled Skid Steer Loader		202.00		222.20	25.25	27.78
8 L.H., Daily Totals		$483.60		$667.00	$60.45	$83.38

Crew A-3D	Hr.	Daily	Hr.	Daily	Bare Costs	Incl. O&P
1 Truck Driver, Light	$27.60	$220.80	$44.90	$359.20	$27.60	$44.90
1 Pickup Truck (4x4)		85.20		93.70		
1 Flatbed Trailer, 25 Ton		88.00		96.80	21.65	23.82
8 L.H., Daily Totals		$394.00		$549.70	$49.25	$68.72

Crew A-3E	Hr.	Daily	Hr.	Daily	Bare Costs	Incl. O&P
1 Equip. Oper. (crane)	$38.10	$304.80	$60.15	$481.20	$33.23	$53.15
1 Truck Driver (heavy)	28.35	226.80	46.15	369.20		
1 Pickup Truck (4x4)		85.20		93.70	5.33	5.86
16 L.H., Daily Totals		$616.80		$944.10	$38.56	$59.01

Crew A-3F	Hr.	Daily	Hr.	Daily	Bare Costs	Incl. O&P
1 Equip. Oper. (crane)	$38.10	$304.80	$60.15	$481.20	$33.23	$53.15
1 Truck Driver (heavy)	28.35	226.80	46.15	369.20		
1 Pickup Truck (4x4)		85.20		93.70		
1 Tractor, 6x2, 40 Ton Cap.		301.60		331.75		
1 Lowbed Trailer, 75 Ton		170.20		187.20	34.81	38.29
16 L.H., Daily Totals		$1088.60		$1463.05	$68.04	$91.44

Crews

Crew No.	Bare Costs		Incl. Subs O & P		Cost Per Labor-Hour	
Crew A-3G	**Hr.**	**Daily**	**Hr.**	**Daily**	**Bare Costs**	**Incl. O&P**
1 Equip. Oper. (crane)	$38.10	$304.80	$60.15	$481.20	$33.23	$53.15
1 Truck Driver (heavy)	28.35	226.80	46.15	369.20		
1 Pickup Truck (4x4)		85.20		93.70		
1 Tractor, 6x4, 45 Ton Cap.		353.60		388.95		
1 Lowbed Trailer, 75 Ton		170.20		187.20	38.06	41.87
16 L.H., Daily Totals		$1140.60		$1520.25	$71.29	$95.02

Crew A-4	Hr.	Daily	Hr.	Daily	Bare Costs	Incl. O&P
2 Carpenters	$35.55	$568.80	$58.90	$942.40	$34.27	$56.23
1 Painter, Ordinary	31.70	253.60	50.90	407.20		
24 L.H., Daily Totals		$822.40		$1349.60	$34.27	$56.23

Crew A-5	Hr.	Daily	Hr.	Daily	Bare Costs	Incl. O&P
2 Laborers	$27.40	$438.40	$45.40	$726.40	$27.42	$45.34
.25 Truck Driver (light)	27.60	55.20	44.90	89.80		
.25 Light Truck, 1.5 Ton		32.20		35.40	1.79	1.97
18 L.H., Daily Totals		$525.80		$851.60	$29.21	$47.31

Crew A-6	Hr.	Daily	Hr.	Daily	Bare Costs	Incl. O&P
1 Instrument Man	$36.50	$292.00	$59.70	$477.60	$35.38	$57.45
1 Rodman/Chainman	34.25	274.00	55.20	441.60		
1 Laser Transit/Level		58.35		64.20	3.64	4.01
16 L.H., Daily Totals		$624.35		$983.40	$39.02	$61.46

Crew A-7	Hr.	Daily	Hr.	Daily	Bare Costs	Incl. O&P
1 Chief Of Party	$45.80	$366.40	$73.60	$588.80	$38.85	$62.83
1 Instrument Man	36.50	292.00	59.70	477.60		
1 Rodman/Chainman	34.25	274.00	55.20	441.60		
1 Laser Transit/Level		58.35		64.20	2.43	2.67
24 L.H., Daily Totals		$990.75		$1572.20	$41.28	$65.50

Crew A-8	Hr.	Daily	Hr.	Daily	Bare Costs	Incl. O&P
1 Chief Of Party	$45.80	$366.40	$73.60	$588.80	$37.70	$60.93
1 Instrument Man	36.50	292.00	59.70	477.60		
2 Rodmen/Chainmen	34.25	548.00	55.20	883.20		
1 Laser Transit/Level		58.35		64.20	1.82	2.00
32 L.H., Daily Totals		$1264.75		$2013.80	$39.52	$62.93

Crew A-9	Hr.	Daily	Hr.	Daily	Bare Costs	Incl. O&P
1 Asbestos Foreman	$39.55	$316.40	$64.50	$516.00	$39.11	$63.80
7 Asbestos Workers	39.05	2186.80	63.70	3567.20		
64 L.H., Daily Totals		$2503.20		$4083.20	$39.11	$63.80

Crew A-10	Hr.	Daily	Hr.	Daily	Bare Costs	Incl. O&P
1 Asbestos Foreman	$39.55	$316.40	$64.50	$516.00	$39.11	$63.80
7 Asbestos Workers	39.05	2186.80	63.70	3567.20		
64 L.H., Daily Totals		$2503.20		$4083.20	$39.11	$63.80

Crew A-10A	Hr.	Daily	Hr.	Daily	Bare Costs	Incl. O&P
1 Asbestos Foreman	$39.55	$316.40	$64.50	$516.00	$39.22	$63.97
2 Asbestos Workers	39.05	624.80	63.70	1019.20		
24 L.H., Daily Totals		$941.20		$1535.20	$39.22	$63.97

Crew A-10B	Hr.	Daily	Hr.	Daily	Bare Costs	Incl. O&P
1 Asbestos Foreman	$39.55	$316.40	$64.50	$516.00	$39.18	$63.90
3 Asbestos Workers	39.05	937.20	63.70	1528.80		
32 L.H., Daily Totals		$1253.60		$2044.80	$39.18	$63.90

Crew No.	Bare Costs		Incl. Subs O & P		Cost Per Labor-Hour	
Crew A-10C	**Hr.**	**Daily**	**Hr.**	**Daily**	**Bare Costs**	**Incl. O&P**
3 Asbestos Workers	$39.05	$937.20	$63.70	$1528.80	$39.05	$63.70
1 Flatbed Truck		128.80		141.70	5.37	5.90
24 L.H., Daily Totals		$1066.00		$1670.50	$44.42	$69.60

Crew A-10D	Hr.	Daily	Hr.	Daily	Bare Costs	Incl. O&P
2 Asbestos Workers	$39.05	$624.80	$63.70	$1019.20	$37.16	$59.70
1 Equip. Oper. (crane)	38.10	304.80	60.15	481.20		
1 Equip. Oper. Oiler	32.45	259.60	51.25	410.00		
1 Hydraulic Crane, 33 Ton		676.60		744.25	21.14	23.26
32 L.H., Daily Totals		$1865.80		$2654.65	$58.30	$82.96

Crew A-11	Hr.	Daily	Hr.	Daily	Bare Costs	Incl. O&P
1 Asbestos Foreman	$39.55	$316.40	$64.50	$516.00	$39.11	$63.80
7 Asbestos Workers	39.05	2186.80	63.70	3567.20		
2 Chipping Hammers		30.40		33.45	.48	.52
64 L.H., Daily Totals		$2533.60		$4116.65	$39.59	$64.32

Crew A-12	Hr.	Daily	Hr.	Daily	Bare Costs	Incl. O&P
1 Asbestos Foreman	$39.55	$316.40	$64.50	$516.00	$39.11	$63.80
7 Asbestos Workers	39.05	2186.80	63.70	3567.20		
1 Large Prod. Vac. Loader		500.00		550.00	7.81	8.59
64 L.H., Daily Totals		$3003.20		$4633.20	$46.92	$72.39

Crew A-13	Hr.	Daily	Hr.	Daily	Bare Costs	Incl. O&P
1 Equip. Oper. (light)	$35.20	$281.60	$55.60	$444.80	$35.20	$55.60
1 Large Prod. Vac. Loader		500.00		550.00	62.50	68.75
8 L.H., Daily Totals		$781.60		$994.80	$97.70	$124.35

Crew B-1	Hr.	Daily	Hr.	Daily	Bare Costs	Incl. O&P
1 Laborer Foreman (outside)	$29.40	$235.20	$48.70	$389.60	$28.07	$46.50
2 Laborers	27.40	438.40	45.40	726.40		
24 L.H., Daily Totals		$673.60		$1116.00	$28.07	$46.50

Crew B-1A	Hr.	Daily	Hr.	Daily	Bare Costs	Incl. O&P
1 Laborer Foreman	$29.40	$235.20	$48.70	$389.60	$28.07	$46.50
2 Laborers	27.40	438.40	45.40	726.40		
2 Cutting Torches		32.40		35.65		
2 Gases		129.60		142.55	6.75	7.43
24 L.H., Daily Totals		$835.60		$1294.20	$34.82	$53.93

Crew B-1B	Hr.	Daily	Hr.	Daily	Bare Costs	Incl. O&P
1 Laborer Foreman	$29.40	$235.20	$48.70	$389.60	$30.58	$49.91
2 Laborers	27.40	438.40	45.40	726.40		
1 Equip. Oper. (crane)	38.10	304.80	60.15	481.20		
2 Cutting Torches		32.40		35.65		
2 Gases		129.60		142.55		
1 Hyd. Crane, 12 Ton		636.60		700.25	24.96	27.45
32 L.H., Daily Totals		$1777.00		$2475.65	$55.54	$77.36

Crew B-2	Hr.	Daily	Hr.	Daily	Bare Costs	Incl. O&P
1 Laborer Foreman (outside)	$29.40	$235.20	$48.70	$389.60	$27.80	$46.06
4 Laborers	27.40	876.80	45.40	1452.80		
40 L.H., Daily Totals		$1112.00		$1842.40	$27.80	$46.06

Crew No.	Bare Costs		Incl. Subs O & P		Cost Per Labor-Hour	
Crew B-3	**Hr.**	**Daily**	**Hr.**	**Daily**	**Bare Costs**	**Incl. O&P**
1 Laborer Foreman (outside)	$29.40	$235.20	$48.70	$389.60	$29.60	$48.29
2 Laborers	27.40	438.40	45.40	726.40		
1 Equip. Oper. (med.)	36.70	293.60	57.95	463.60		
2 Truck Drivers (heavy)	28.35	453.60	46.15	738.40		
1 F.E. Loader, T.M., 2.5 C.Y.		794.00		873.40		
2 Dump Trucks, 16 Ton		953.20		1048.50	36.40	40.04
48 L.H., Daily Totals		$3168.00		$4239.90	$66.00	$88.33

Crew B-3A	Hr.	Daily	Hr.	Daily	Bare Costs	Incl. O&P
4 Laborers	$27.40	$876.80	$45.40	$1452.80	$29.26	$47.91
1 Equip. Oper. (med.)	36.70	293.60	57.95	463.60		
1 Hyd. Excavator, 1.5 C.Y.		722.80		795.10	18.07	19.88
40 L.H., Daily Totals		$1893.20		$2711.50	$47.33	$67.79

Crew B-3B	Hr.	Daily	Hr.	Daily	Bare Costs	Incl. O&P
2 Laborers	$27.40	$438.40	$45.40	$726.40	$29.96	$48.73
1 Equip. Oper. (med.)	36.70	293.60	57.95	463.60		
1 Truck Driver (heavy)	28.35	226.80	46.15	369.20		
1 Backhoe Loader, 80 H.P.		277.60		305.35		
1 Dump Truck, 16 Ton		476.60		524.25	23.57	25.93
32 L.H., Daily Totals		$1713.00		$2388.80	$53.53	$74.66

Crew B-3C	Hr.	Daily	Hr.	Daily	Bare Costs	Incl. O&P
3 Laborers	$27.40	$657.60	$45.40	$1089.60	$29.73	$48.54
1 Equip. Oper. (med.)	36.70	293.60	57.95	463.60		
1 F.E. Crawler Ldr, 4 C.Y.		1110.00		1221.00	34.69	38.16
32 L.H., Daily Totals		$2061.20		$2774.20	$64.42	$86.70

Crew B-4	Hr.	Daily	Hr.	Daily	Bare Costs	Incl. O&P
1 Laborer Foreman (outside)	$29.40	$235.20	$48.70	$389.60	$27.89	$46.08
4 Laborers	27.40	876.80	45.40	1452.80		
1 Truck Driver (heavy)	28.35	226.80	46.15	369.20		
1 Tractor, 4 x 2, 195 H.P.		208.20		229.00		
1 Platform Trailer		119.40		131.35	6.83	7.51
48 L.H., Daily Totals		$1666.40		$2571.95	$34.72	$53.59

Crew B-5	Hr.	Daily	Hr.	Daily	Bare Costs	Incl. O&P
1 Laborer Foreman (outside)	$29.40	$235.20	$48.70	$389.60	$30.34	$49.46
4 Laborers	27.40	876.80	45.40	1452.80		
2 Equip. Oper. (med.)	36.70	587.20	57.95	927.20		
1 Air Compr., 250 C.F.M.		119.80		131.80		
2 Air Tools & Accessories		18.00		19.80		
2-50 Ft. Air Hoses, 1.5″ Dia.		9.40		10.35		
1 F.E. Loader, T.M., 2.5 C.Y.		794.00		873.40	16.81	18.49
56 L.H., Daily Totals		$2640.40		$3804.95	$47.15	$67.95

Crew B-6	Hr.	Daily	Hr.	Daily	Bare Costs	Incl. O&P
2 Laborers	$27.40	$438.40	$45.40	$726.40	$30.00	$48.80
1 Equip. Oper. (light)	35.20	281.60	55.60	444.80		
1 Backhoe Loader, 48 H.P.		226.20		248.80	9.43	10.37
24 L.H., Daily Totals		$946.20		$1420.00	$39.43	$59.17

Crew B-6B	Hr.	Daily	Hr.	Daily	Bare Costs	Incl. O&P
2 Laborer Foremen (out)	$29.40	$470.40	$48.70	$779.20	$28.07	$46.50
4 Laborers	27.40	876.80	45.40	1452.80		
1 Winch Truck		305.20		335.70		
1 Flatbed Truck		128.80		141.70		
1 Butt Fusion Machine		347.20		381.90	16.28	17.90
48 L.H., Daily Totals		$2128.40		$3091.30	$44.35	$64.40

Crew No.	Bare Costs		Incl. Subs O & P		Cost Per Labor-Hour	
Crew B-7	**Hr.**	**Daily**	**Hr.**	**Daily**	**Bare Costs**	**Incl. O&P**
1 Laborer Foreman (outside)	$29.40	$235.20	$48.70	$389.60	$29.28	$48.04
4 Laborers	27.40	876.80	45.40	1452.80		
1 Equip. Oper. (med.)	36.70	293.60	57.95	463.60		
1 Chipping Machine		173.00		190.30		
1 F.E. Loader, T.M., 2.5 C.Y.		794.00		873.40		
2 Chain Saws, 36″		66.80		73.50	21.54	23.69
48 L.H., Daily Totals		$2439.40		$3443.20	$50.82	$71.73

Crew B-7A	Hr.	Daily	Hr.	Daily	Bare Costs	Incl. O&P
2 Laborers	$27.40	$438.40	$45.40	$726.40	$30.00	$48.80
1 Equip. Oper. (light)	35.20	281.60	55.60	444.80		
1 Rake w/Tractor		192.90		212.20		
2 Chain Saws, 18″		40.80		44.90	9.74	10.71
24 L.H., Daily Totals		$953.70		$1428.30	$39.74	$59.51

Crew B-8	Hr.	Daily	Hr.	Daily	Bare Costs	Incl. O&P
1 Laborer Foreman (outside)	$29.40	$235.20	$48.70	$389.60	$30.84	$49.87
2 Laborers	27.40	438.40	45.40	726.40		
2 Equip. Oper. (med.)	36.70	587.20	57.95	927.20		
1 Equip. Oper. Oiler	32.45	259.60	51.25	410.00		
2 Truck Drivers (heavy)	28.35	453.60	46.15	738.40		
1 Hyd. Crane, 25 Ton		650.80		715.90		
1 F.E. Loader, T.M., 2.5 C.Y.		794.00		873.40		
2 Dump Trucks, 16 Ton		953.20		1048.50	37.47	41.22
64 L.H., Daily Totals		$4372.00		$5829.40	$68.31	$91.09

Crew B-9	Hr.	Daily	Hr.	Daily	Bare Costs	Incl. O&P
1 Laborer Foreman (outside)	$29.40	$235.20	$48.70	$389.60	$27.80	$46.06
4 Laborers	27.40	876.80	45.40	1452.80		
1 Air Compr., 250 C.F.M.		119.80		131.80		
2 Air Tools & Accessories		18.00		19.80		
2-50 Ft. Air Hoses, 1.5″ Dia.		9.40		10.35	3.68	4.05
40 L.H., Daily Totals		$1259.20		$2004.35	$31.48	$50.11

Crew B-9A	Hr.	Daily	Hr.	Daily	Bare Costs	Incl. O&P
2 Laborers	$27.40	$438.40	$45.40	$726.40	$27.72	$45.65
1 Truck Driver (heavy)	28.35	226.80	46.15	369.20		
1 Water Tanker		116.80		128.50		
1 Tractor		208.20		229.00		
2-50 Ft. Disch. Hoses		4.30		4.75	13.72	15.09
24 L.H., Daily Totals		$994.50		$1457.85	$41.44	$60.74

Crew B-9B	Hr.	Daily	Hr.	Daily	Bare Costs	Incl. O&P
2 Laborers	$27.40	$438.40	$45.40	$726.40	$27.72	$45.65
1 Truck Driver (heavy)	28.35	226.80	46.15	369.20		
2-50 Ft. Disch. Hoses		4.30		4.75		
1 Water Tanker		116.80		128.50		
1 Tractor		208.20		229.00		
1 Pressure Washer		50.60		55.65	15.83	17.41
24 L.H., Daily Totals		$1045.10		$1513.50	$43.55	$63.06

Crew B-9C	Hr.	Daily	Hr.	Daily	Bare Costs	Incl. O&P
1 Laborer Foreman (outside)	$29.40	$235.20	$48.70	$389.60	$27.80	$46.06
4 Laborers	27.40	876.80	45.40	1452.80		
1 Air Compr., 250 C.F.M.		119.80		131.80		
2-50 Ft. Air Hoses, 1.5″ Dia.		9.40		10.35		
2 Breaker, Pavement, 60 lb.		18.00		19.80	3.68	4.05
40 L.H., Daily Totals		$1259.20		$2004.35	$31.48	$50.11

Crew No.	Bare Costs		Incl. Subs O & P		Cost Per Labor-Hour	
Crew B-9D	**Hr.**	**Daily**	**Hr.**	**Daily**	**Bare Costs**	**Incl. O&P**
1 Laborer Foreman (Outside)	$29.40	$235.20	$48.70	$389.60	$27.80	$46.06
4 Common Laborers	27.40	876.80	45.40	1452.80		
1 Air Compressor, 250 CFM		119.80		131.80		
2 Air hoses, 1.5" x 50'		9.40		10.35		
2 Air tampers		55.70		61.25	4.63	5.09
40 L.H., Daily Totals		$1296.90		$2045.80	$32.43	$51.15
Crew B-10	**Hr.**	**Daily**	**Hr.**	**Daily**	**Bare Costs**	**Incl. O&P**
1 Equip. Oper. (med.)	$36.70	$293.60	$57.95	$463.60	$33.60	$53.77
.5 Laborer	27.40	109.60	45.40	181.60		
12 L.H., Daily Totals		$403.20		$645.20	$33.60	$53.77
Crew B-10A	**Hr.**	**Daily**	**Hr.**	**Daily**	**Bare Costs**	**Incl. O&P**
1 Equip. Oper. (med.)	$36.70	$293.60	$57.95	$463.60	$33.60	$53.77
.5 Laborer	27.40	109.60	45.40	181.60		
1 Walk behind compactor, 7.5 HP		125.80		138.40	10.48	11.53
12 L.H., Daily Totals		$529.00		$783.60	$44.08	$65.30
Crew B-10B	**Hr.**	**Daily**	**Hr.**	**Daily**	**Bare Costs**	**Incl. O&P**
1 Equip. Oper. (med.)	$36.70	$293.60	$57.95	$463.60	$33.60	$53.77
.5 Laborer	27.40	109.60	45.40	181.60		
1 Dozer, 200 H.P.		920.80		1012.90	76.73	84.41
12 L.H., Daily Totals		$1324.00		$1658.10	$110.33	$138.18
Crew B-10C	**Hr.**	**Daily**	**Hr.**	**Daily**	**Bare Costs**	**Incl. O&P**
1 Equip. Oper. (med.)	$36.70	$293.60	$57.95	$463.60	$33.60	$53.77
.5 Laborer	27.40	109.60	45.40	181.60		
1 Dozer, 200 H.P.		920.80		1012.90		
1 Vibratory Roller, Towed		457.60		503.35	114.87	126.35
12 L.H., Daily Totals		$1781.60		$2161.45	$148.47	$180.12
Crew B-10D	**Hr.**	**Daily**	**Hr.**	**Daily**	**Bare Costs**	**Incl. O&P**
1 Equip. Oper. (med.)	$36.70	$293.60	$57.95	$463.60	$33.60	$53.77
.5 Laborer	27.40	109.60	45.40	181.60		
1 Dozer, 200 H.P.		920.80		1012.90		
1 Sheepsft. Roller, Towed		478.00		525.80	116.57	128.22
12 L.H., Daily Totals		$1802.00		$2183.90	$150.17	$181.99
Crew B-10E	**Hr.**	**Daily**	**Hr.**	**Daily**	**Bare Costs**	**Incl. O&P**
1 Equip. Oper. (med.)	$36.70	$293.60	$57.95	$463.60	$33.60	$53.77
.5 Laborer	27.40	109.60	45.40	181.60		
1 Tandem Roller, 5 Ton		114.20		125.60	9.52	10.47
12 L.H., Daily Totals		$517.40		$770.80	$43.12	$64.24
Crew B-10F	**Hr.**	**Daily**	**Hr.**	**Daily**	**Bare Costs**	**Incl. O&P**
1 Equip. Oper. (med.)	$36.70	$293.60	$57.95	$463.60	$33.60	$53.77
.5 Laborer	27.40	109.60	45.40	181.60		
1 Tandem Roller, 10 Ton		197.00		216.70	16.42	18.06
12 L.H., Daily Totals		$600.20		$861.90	$50.02	$71.83
Crew B-10G	**Hr.**	**Daily**	**Hr.**	**Daily**	**Bare Costs**	**Incl. O&P**
1 Equip. Oper. (med.)	$36.70	$293.60	$57.95	$463.60	$33.60	$53.77
.5 Laborer	27.40	109.60	45.40	181.60		
1 Sheepsft. Roll., 130 H.P.		848.00		932.80	70.67	77.73
12 L.H., Daily Totals		$1251.20		$1578.00	$104.27	$131.50

Crew No.	Bare Costs		Incl. Subs O & P		Cost Per Labor-Hour	
Crew B-10H	**Hr.**	**Daily**	**Hr.**	**Daily**	**Bare Costs**	**Incl. O&P**
1 Equip. Oper. (med.)	$36.70	$293.60	$57.95	$463.60	$33.60	$53.77
.5 Laborer	27.40	109.60	45.40	181.60		
1 Diaphr. Water Pump, 2"		53.20		58.50		
1-20 Ft. Suction Hose, 2"		2.15		2.35		
2-50 Ft. Disch. Hoses, 2"		2.60		2.85	4.83	5.31
12 L.H., Daily Totals		$461.15		$708.90	$38.43	$59.08
Crew B-10I	**Hr.**	**Daily**	**Hr.**	**Daily**	**Bare Costs**	**Incl. O&P**
1 Equip. Oper. (med.)	$36.70	$293.60	$57.95	$463.60	$33.60	$53.77
.5 Laborer	27.40	109.60	45.40	181.60		
1 Diaphr. Water Pump, 4"		73.20		80.50		
1-20 Ft. Suction Hose, 4"		4.05		4.45		
2-50 Ft. Disch. Hoses, 4"		5.90		6.50	6.93	7.62
12 L.H., Daily Totals		$486.35		$736.65	$40.53	$61.39
Crew B-10J	**Hr.**	**Daily**	**Hr.**	**Daily**	**Bare Costs**	**Incl. O&P**
1 Equip. Oper. (med.)	$36.70	$293.60	$57.95	$463.60	$33.60	$53.77
.5 Laborer	27.40	109.60	45.40	181.60		
1 Centr. Water Pump, 3"		56.60		62.25		
1-20 Ft. Suction Hose, 3"		3.65		4.00		
2-50 Ft. Disch. Hoses, 3"		4.30		4.75	5.38	5.92
12 L.H., Daily Totals		$467.75		$716.20	$38.98	$59.69
Crew B-10K	**Hr.**	**Daily**	**Hr.**	**Daily**	**Bare Costs**	**Incl. O&P**
1 Equip. Oper. (med.)	$36.70	$293.60	$57.95	$463.60	$33.60	$53.77
.5 Laborer	27.40	109.60	45.40	181.60		
1 Centr. Water Pump, 6"		240.00		264.00		
1-20 Ft. Suction Hose, 6"		12.10		13.30		
2-50 Ft. Disch. Hoses, 6"		15.00		16.50	22.26	24.48
12 L.H., Daily Totals		$670.30		$939.00	$55.86	$78.25
Crew B-10L	**Hr.**	**Daily**	**Hr.**	**Daily**	**Bare Costs**	**Incl. O&P**
1 Equip. Oper. (med.)	$36.70	$293.60	$57.95	$463.60	$33.60	$53.77
.5 Laborer	27.40	109.60	45.40	181.60		
1 Dozer, 80 H.P.		318.80		350.70	26.57	29.22
12 L.H., Daily Totals		$722.00		$995.90	$60.17	$82.99
Crew B-10M	**Hr.**	**Daily**	**Hr.**	**Daily**	**Bare Costs**	**Incl. O&P**
1 Equip. Oper. (med.)	$36.70	$293.60	$57.95	$463.60	$33.60	$53.77
.5 Laborer	27.40	109.60	45.40	181.60		
1 Dozer, 300 H.P.		1197.00		1316.70	99.75	109.73
12 L.H., Daily Totals		$1600.20		$1961.90	$133.35	$163.50
Crew B-10N	**Hr.**	**Daily**	**Hr.**	**Daily**	**Bare Costs**	**Incl. O&P**
1 Equip. Oper. (med.)	$36.70	$293.60	$57.95	$463.60	$33.60	$53.77
.5 Laborer	27.40	109.60	45.40	181.60		
1 F.E. Loader, T.M., 1.5 C.Y		315.20		346.70	26.27	28.89
12 L.H., Daily Totals		$718.40		$991.90	$59.87	$82.66
Crew B-10O	**Hr.**	**Daily**	**Hr.**	**Daily**	**Bare Costs**	**Incl. O&P**
1 Equip. Oper. (med.)	$36.70	$293.60	$57.95	$463.60	$33.60	$53.77
.5 Laborer	27.40	109.60	45.40	181.60		
1 F.E. Loader, T.M., 2.25 C.Y.		557.00		612.70	46.42	51.06
12 L.H., Daily Totals		$960.20		$1257.90	$80.02	$104.83
Crew B-10P	**Hr.**	**Daily**	**Hr.**	**Daily**	**Bare Costs**	**Incl. O&P**
1 Equip. Oper. (med.)	$36.70	$293.60	$57.95	$463.60	$33.60	$53.77
.5 Laborer	27.40	109.60	45.40	181.60		
1 F.E. Loader, T.M., 2.5 C.Y.		794.00		873.40	66.17	72.78
12 L.H., Daily Totals		$1197.20		$1518.60	$99.77	$126.55

Crew No.	Bare Costs		Incl. Subs O & P		Cost Per Labor-Hour	
Crew B-10Q	**Hr.**	**Daily**	**Hr.**	**Daily**	**Bare Costs**	**Incl. O&P**
1 Equip. Oper. (med.)	$36.70	$293.60	$57.95	$463.60	$33.60	$53.77
.5 Laborer	27.40	109.60	45.40	181.60		
1 F.E. Loader, T.M., 5 C.Y.		1110.00		1221.00	92.50	101.75
12 L.H., Daily Totals		$1513.20		$1866.20	$126.10	$155.52

Crew B-10R	Hr.	Daily	Hr.	Daily	Bare Costs	Incl. O&P
1 Equip. Oper. (med.)	$36.70	$293.60	$57.95	$463.60	$33.60	$53.77
.5 Laborer	27.40	109.60	45.40	181.60		
1 F.E. Loader, W.M., 1 C.Y.		194.60		214.05	16.22	17.84
12 L.H., Daily Totals		$597.80		$859.25	$49.82	$71.61

Crew B-10S	Hr.	Daily	Hr.	Daily	Bare Costs	Incl. O&P
1 Equip. Oper. (med.)	$36.70	$293.60	$57.95	$463.60	$33.60	$53.77
.5 Laborer	27.40	109.60	45.40	181.60		
1 F.E. Loader, W.M., 1.5 C.Y.		241.40		265.55	20.12	22.13
12 L.H., Daily Totals		$644.60		$910.75	$53.72	$75.90

Crew B-10T	Hr.	Daily	Hr.	Daily	Bare Costs	Incl. O&P
1 Equip. Oper. (med.)	$36.70	$293.60	$57.95	$463.60	$33.60	$53.77
.5 Laborer	27.40	109.60	45.40	181.60		
1 F.E. Loader, W.M.,2.5 C.Y.		305.40		335.95	25.45	28.00
12 L.H., Daily Totals		$708.60		$981.15	$59.05	$81.77

Crew B-10U	Hr.	Daily	Hr.	Daily	Bare Costs	Incl. O&P
1 Equip. Oper. (med.)	$36.70	$293.60	$57.95	$463.60	$33.60	$53.77
.5 Laborer	27.40	109.60	45.40	181.60		
1 F.E. Loader, W.M., 5.5 C.Y.		701.00		771.10	58.42	64.26
12 L.H., Daily Totals		$1104.20		$1416.30	$92.02	$118.03

Crew B-10V	Hr.	Daily	Hr.	Daily	Bare Costs	Incl. O&P
1 Equip. Oper. (med.)	$36.70	$293.60	$57.95	$463.60	$33.60	$53.77
.5 Laborer	27.40	109.60	45.40	181.60		
1 Dozer, 700 H.P.		3148.00		3462.80	262.33	288.57
12 L.H., Daily Totals		$3551.20		$4108.00	$295.93	$342.34

Crew B-10W	Hr.	Daily	Hr.	Daily	Bare Costs	Incl. O&P
1 Equip. Oper. (med.)	$36.70	$293.60	$57.95	$463.60	$33.60	$53.77
.5 Laborer	27.40	109.60	45.40	181.60		
1 Dozer, 105 H.P.		462.80		509.10	38.57	42.42
12 L.H., Daily Totals		$866.00		$1154.30	$72.17	$96.19

Crew B-10X	Hr.	Daily	Hr.	Daily	Bare Costs	Incl. O&P
1 Equip. Oper. (med.)	$36.70	$293.60	$57.95	$463.60	$33.60	$53.77
.5 Laborer	27.40	109.60	45.40	181.60		
1 Dozer, 410 H.P.		1536.00		1689.60	128.00	140.80
12 L.H., Daily Totals		$1939.20		$2334.80	$161.60	$194.57

Crew B-10Y	Hr.	Daily	Hr.	Daily	Bare Costs	Incl. O&P
1 Equip. Oper. (med.)	$36.70	$293.60	$57.95	$463.60	$33.60	$53.77
.5 Laborer	27.40	109.60	45.40	181.60		
1 Vibratory Drum Roller		419.20		461.10	34.93	38.43
12 L.H., Daily Totals		$822.40		$1106.30	$68.53	$92.20

Crew B-11A	Hr.	Daily	Hr.	Daily	Bare Costs	Incl. O&P
1 Equipment Oper. (med.)	$36.70	$293.60	$57.95	$463.60	$32.05	$51.68
1 Laborer	27.40	219.20	45.40	363.20		
1 Dozer, 200 H.P.		920.80		1012.90	57.55	63.31
16 L.H., Daily Totals		$1433.60		$1839.70	$89.60	$114.99

Crew B-11B	Hr.	Daily	Hr.	Daily	Bare Costs	Incl. O&P
1 Equipment Oper. (light)	$35.20	$281.60	$55.60	$444.80	$31.30	$50.50
1 Laborer	27.40	219.20	45.40	363.20		
1 Air Powered Tamper		27.85		30.65		
1 Air Compr. 365 C.F.M.		159.00		174.90		
2-50 Ft. Air Hoses, 1.5" Dia.		9.40		10.35	12.27	13.50
16 L.H., Daily Totals		$697.05		$1023.90	$43.57	$64.00

Crew B-11C	Hr.	Daily	Hr.	Daily	Bare Costs	Incl. O&P
1 Equipment Oper. (med.)	$36.70	$293.60	$57.95	$463.60	$32.05	$51.68
1 Laborer	27.40	219.20	45.40	363.20		
1 Backhoe Loader, 48 H.P.		226.20		248.80	14.14	15.55
16 L.H., Daily Totals		$739.00		$1075.60	$46.19	$67.23

Crew B-11K	Hr.	Daily	Hr.	Daily	Bare Costs	Incl. O&P
1 Equipment Oper. (med.)	$36.70	$293.60	$57.95	$463.60	$32.05	$51.68
1 Laborer	27.40	219.20	45.40	363.20		
1 Trencher, 8' D., 16" W.		1460.00		1606.00	91.25	100.38
16 L.H., Daily Totals		$1972.80		$2432.80	$123.30	$152.06

Crew B-11L	Hr.	Daily	Hr.	Daily	Bare Costs	Incl. O&P
1 Equipment Oper. (med.)	$36.70	$293.60	$57.95	$463.60	$32.05	$51.68
1 Laborer	27.40	219.20	45.40	363.20		
1 Grader, 30,000 Lbs.		457.40		503.15	28.59	31.45
16 L.H., Daily Totals		$970.20		$1329.95	$60.64	$83.13

Crew B-11M	Hr.	Daily	Hr.	Daily	Bare Costs	Incl. O&P
1 Equipment Oper. (med.)	$36.70	$293.60	$57.95	$463.60	$32.05	$51.68
1 Laborer	27.40	219.20	45.40	363.20		
1 Backhoe Loader, 80 H.P.		277.60		305.35	17.35	19.09
16 L.H., Daily Totals		$790.40		$1132.15	$49.40	$70.77

Crew B-11S	Hr.	Daily	Hr.	Daily	Bare Costs	Incl. O&P
1 Equipment Operator (med.)	$36.70	$293.60	$57.95	$463.60	$33.60	$53.77
.5 Laborer	27.40	109.60	45.40	181.60		
1 Dozer, 300 H.P.		1197.00		1316.70		
1 Ripper, Beam & 1 Shank		73.20		80.50	105.85	116.44
12 L.H., Daily Totals		$1673.40		$2042.40	$139.45	$170.21

Crew B-11T	Hr.	Daily	Hr.	Daily	Bare Costs	Incl. O&P
1 Equipment Operator (med.)	$36.70	$293.60	$57.95	$463.60	$33.60	$53.77
.5 Laborer	27.40	109.60	45.40	181.60		
1 Dozer, 410 H.P.		1536.00		1689.60		
1 Ripper, Beam & 2 Shanks		82.00		90.20	134.83	148.32
12 L.H., Daily Totals		$2021.20		$2425.00	$168.43	$202.09

Crew B-11W	Hr.	Daily	Hr.	Daily	Bare Costs	Incl. O&P
1 Equipment Operator (med.)	$36.70	$293.60	$57.95	$463.60	$28.97	$47.07
1 Common Laborer	27.40	219.20	45.40	363.20		
10 Truck Drivers, Heavy	28.35	2268.00	46.15	3692.00		
1 Dozer, 200 H.P.		920.80		1012.90		
1 Vib. roller, smth, towed, 23 Ton		457.60		503.35		
10 Dump Truck, 10 Ton		3266.00		3592.60	48.38	53.22
96 L.H., Daily Totals		$7425.20		$9627.65	$77.35	$100.29

CREWS

Crews

Crew No.	Bare Costs		Incl. Subs O & P		Cost Per Labor-Hour	
Crew B-11Y	**Hr.**	**Daily**	**Hr.**	**Daily**	**Bare Costs**	**Incl. O&P**
1 Laborer Foreman (Outside)	$29.40	$235.20	$48.70	$389.60	$30.72	$49.95
5 Common Laborers	27.40	1096.00	45.40	1816.00		
3 Equipment Operators (med.)	36.70	880.80	57.95	1390.80		
1 Dozer, 80 H.P.		318.80		350.70		
2 Walk behind compactors, 7.5 HP		251.60		276.75		
4 Vibratory plates, gas, 21"		138.40		152.25	9.84	10.83
72 L.H., Daily Totals		$2920.80		$4376.10	$40.56	$60.78

Crew B-12A	Hr.	Daily	Hr.	Daily	Bare Costs	Incl. O&P
1 Equip. Oper. (crane)	$38.10	$304.80	$60.15	$481.20	$32.75	$52.78
1 Laborer	27.40	219.20	45.40	363.20		
1 Hyd. Excavator, 1 C.Y.		560.20		616.20	35.01	38.51
16 L.H., Daily Totals		$1084.20		$1460.60	$67.76	$91.29

Crew B-12B	Hr.	Daily	Hr.	Daily	Bare Costs	Incl. O&P
1 Equip. Oper. (crane)	$38.10	$304.80	$60.15	$481.20	$32.75	$52.78
1 Laborer	27.40	219.20	45.40	363.20		
1 Hyd. Excavator, 1.5 C.Y.		722.80		795.10	45.18	49.69
16 L.H., Daily Totals		$1246.80		$1639.50	$77.93	$102.47

Crew B-12C	Hr.	Daily	Hr.	Daily	Bare Costs	Incl. O&P
1 Equip. Oper. (crane)	$38.10	$304.80	$60.15	$481.20	$32.75	$52.78
1 Laborer	27.40	219.20	45.40	363.20		
1 Hyd. Excavator, 2 C.Y.		919.60		1011.55	57.48	63.22
16 L.H., Daily Totals		$1443.60		$1855.95	$90.23	$116.00

Crew B-12D	Hr.	Daily	Hr.	Daily	Bare Costs	Incl. O&P
1 Equip. Oper. (crane)	$38.10	$304.80	$60.15	$481.20	$32.75	$52.78
1 Laborer	27.40	219.20	45.40	363.20		
1 Hyd. Excavator, 3.5 C.Y.		2023.00		2225.30	126.44	139.08
16 L.H., Daily Totals		$2547.00		$3069.70	$159.19	$191.86

Crew B-12E	Hr.	Daily	Hr.	Daily	Bare Costs	Incl. O&P
1 Equip. Oper. (crane)	$38.10	$304.80	$60.15	$481.20	$32.75	$52.78
1 Laborer	27.40	219.20	45.40	363.20		
1 Hyd. Excavator, .5 C.Y.		335.40		368.95	20.96	23.06
16 L.H., Daily Totals		$859.40		$1213.35	$53.71	$75.84

Crew B-12F	Hr.	Daily	Hr.	Daily	Bare Costs	Incl. O&P
1 Equip. Oper. (crane)	$38.10	$304.80	$60.15	$481.20	$32.75	$52.78
1 Laborer	27.40	219.20	45.40	363.20		
1 Hyd. Excavator, .75 C.Y.		477.80		525.60	29.86	32.85
16 L.H., Daily Totals		$1001.80		$1370.00	$62.61	$85.63

Crew B-12G	Hr.	Daily	Hr.	Daily	Bare Costs	Incl. O&P
1 Equip. Oper. (crane)	$38.10	$304.80	$60.15	$481.20	$32.75	$52.78
1 Laborer	27.40	219.20	45.40	363.20		
1 Power Shovel, .5 C.Y.		534.65		588.10		
1 Clamshell Bucket, .5 C.Y.		33.80		37.20	35.53	39.08
16 L.H., Daily Totals		$1092.45		$1469.70	$68.28	$91.86

Crew B-12H	Hr.	Daily	Hr.	Daily	Bare Costs	Incl. O&P
1 Equip. Oper. (crane)	$38.10	$304.80	$60.15	$481.20	$32.75	$52.78
1 Laborer	27.40	219.20	45.40	363.20		
1 Power Shovel, 1 C.Y.		880.00		968.00		
1 Clamshell Bucket, 1 C.Y.		43.40		47.75	57.71	63.48
16 L.H., Daily Totals		$1447.40		$1860.15	$90.46	$116.26

Crew B-12I	Hr.	Daily	Hr.	Daily	Bare Costs	Incl. O&P
1 Equip. Oper. (crane)	$38.10	$304.80	$60.15	$481.20	$32.75	$52.78
1 Laborer	27.40	219.20	45.40	363.20		
1 Power Shovel, .75 C.Y.		663.50		729.85		
1 Dragline Bucket, .75 C.Y.		19.00		20.90	42.66	46.92
16 L.H., Daily Totals		$1206.50		$1595.15	$75.41	$99.70

Crew B-12J	Hr.	Daily	Hr.	Daily	Bare Costs	Incl. O&P
1 Equip. Oper. (crane)	$38.10	$304.80	$60.15	$481.20	$32.75	$52.78
1 Laborer	27.40	219.20	45.40	363.20		
1 Gradall, 3 Ton, .5 C.Y.		841.00		925.10	52.56	57.82
16 L.H., Daily Totals		$1365.00		$1769.50	$85.31	$110.60

Crew B-12K	Hr.	Daily	Hr.	Daily	Bare Costs	Incl. O&P
1 Equip. Oper. (crane)	$38.10	$304.80	$60.15	$481.20	$32.75	$52.78
1 Laborer	27.40	219.20	45.40	363.20		
1 Gradall, 3 Ton, 1 C.Y.		974.20		1071.60	60.89	66.98
16 L.H., Daily Totals		$1498.20		$1916.00	$93.64	$119.76

Crew B-12L	Hr.	Daily	Hr.	Daily	Bare Costs	Incl. O&P
1 Equip. Oper. (crane)	$38.10	$304.80	$60.15	$481.20	$32.75	$52.78
1 Laborer	27.40	219.20	45.40	363.20		
1 Power Shovel, .5 C.Y.		534.65		588.10		
1 F.E. Attachment, .5 C.Y.		48.80		53.70	36.46	40.11
16 L.H., Daily Totals		$1107.45		$1486.20	$69.21	$92.89

Crew B-12M	Hr.	Daily	Hr.	Daily	Bare Costs	Incl. O&P
1 Equip. Oper. (crane)	$38.10	$304.80	$60.15	$481.20	$32.75	$52.78
1 Laborer	27.40	219.20	45.40	363.20		
1 Power Shovel, .75 C.Y.		663.50		729.85		
1 F.E. Attachment, .75 C.Y.		53.20		58.50	44.79	49.27
16 L.H., Daily Totals		$1240.70		$1632.75	$77.54	$102.05

Crew B-12N	Hr.	Daily	Hr.	Daily	Bare Costs	Incl. O&P
1 Equip. Oper. (crane)	$38.10	$304.80	$60.15	$481.20	$32.75	$52.78
1 Laborer	27.40	219.20	45.40	363.20		
1 Power Shovel, 1 C.Y.		880.00		968.00		
1 F.E. Attachment, 1 C.Y.		59.00		64.90	58.69	64.56
16 L.H., Daily Totals		$1463.00		$1877.30	$91.44	$117.34

Crew B-12O	Hr.	Daily	Hr.	Daily	Bare Costs	Incl. O&P
1 Equip. Oper. (crane)	$38.10	$304.80	$60.15	$481.20	$32.75	$52.78
1 Laborer	27.40	219.20	45.40	363.20		
1 Power Shovel, 1.5 C.Y.		934.20		1027.60		
1 F.E. Attachment, 1.5 C.Y.		69.20		76.10	62.71	68.98
16 L.H., Daily Totals		$1527.40		$1948.10	$95.46	$121.76

Crew B-12P	Hr.	Daily	Hr.	Daily	Bare Costs	Incl. O&P
1 Equip. Oper. (crane)	$38.10	$304.80	$60.15	$481.20	$32.75	$52.78
1 Laborer	27.40	219.20	45.40	363.20		
1 Crawler Crane, 40 Ton		934.20		1027.60		
1 Dragline Bucket, 1.5 C.Y.		31.00		34.10	60.33	66.36
16 L.H., Daily Totals		$1489.20		$1906.10	$93.08	$119.14

Crew B-12Q	Hr.	Daily	Hr.	Daily	Bare Costs	Incl. O&P
1 Equip. Oper. (crane)	$38.10	$304.80	$60.15	$481.20	$32.75	$52.78
1 Laborer	27.40	219.20	45.40	363.20		
1 Hyd. Excavator, 5/8 C.Y.		435.40		478.95	27.21	29.93
16 L.H., Daily Totals		$959.40		$1323.35	$59.96	$82.71

Crews

Crew No.	Bare Costs		Incl. Subs O & P		Cost Per Labor-Hour	
Crew B-12R	**Hr.**	**Daily**	**Hr.**	**Daily**	**Bare Costs**	**Incl. O&P**
1 Equip. Oper. (crane)	$38.10	$304.80	$60.15	$481.20	$32.75	$52.78
1 Laborer	27.40	219.20	45.40	363.20		
1 Hyd. Excavator, 1.5 C.Y.		722.80		795.10	45.18	49.69
16 L.H., Daily Totals		$1246.80		$1639.50	$77.93	$102.47

Crew B-12S	Hr.	Daily	Hr.	Daily	Bare Costs	Incl. O&P
1 Equip. Oper. (crane)	$38.10	$304.80	$60.15	$481.20	$32.75	$52.78
1 Laborer	27.40	219.20	45.40	363.20		
1 Hyd. Excavator, 2.5 C.Y.		1218.00		1339.80	76.13	83.74
16 L.H., Daily Totals		$1742.00		$2184.20	$108.88	$136.52

Crew B-12T	Hr.	Daily	Hr.	Daily	Bare Costs	Incl. O&P
1 Equip. Oper. (crane)	$38.10	$304.80	$60.15	$481.20	$32.75	$52.78
1 Laborer	27.40	219.20	45.40	363.20		
1 Crawler Crane, 75 Ton		1226.00		1348.60		
1 F.E. Attachment, 3 C.Y.		91.60		100.75	82.35	90.59
16 L.H., Daily Totals		$1841.60		$2293.75	$115.10	$143.37

Crew B-12V	Hr.	Daily	Hr.	Daily	Bare Costs	Incl. O&P
1 Equip. Oper. (crane)	$38.10	$304.80	$60.15	$481.20	$32.75	$52.78
1 Laborer	27.40	219.20	45.40	363.20		
1 Crawler Crane, 75 Ton		1226.00		1348.60		
1 Dragline Bucket, 3 C.Y.		47.80		52.60	79.61	87.57
16 L.H., Daily Totals		$1797.80		$2245.60	$112.36	$140.35

Crew B-13	Hr.	Daily	Hr.	Daily	Bare Costs	Incl. O&P
1 Laborer Foreman (outside)	$29.40	$235.20	$48.70	$389.60	$29.94	$48.81
4 Laborers	27.40	876.80	45.40	1452.80		
1 Equip. Oper. (crane)	38.10	304.80	60.15	481.20		
1 Equip. Oper. Oiler	32.45	259.60	51.25	410.00		
1 Hyd. Crane, 25 Ton		650.80		715.90	11.62	12.78
56 L.H., Daily Totals		$2327.20		$3449.50	$41.56	$61.59

Crew B-13A	Hr.	Daily	Hr.	Daily	Bare Costs	Incl. O&P
1 Laborer Foreman	$29.40	$235.20	$48.70	$389.60	$30.61	$49.67
2 Laborers	27.40	438.40	45.40	726.40		
2 Equipment Operators	36.70	587.20	57.95	927.20		
2 Truck Drivers (heavy)	28.35	453.60	46.15	738.40		
1 Crane, 75 Ton		1226.00		1348.60		
1 F.E. Lder, 3.75 C.Y.		1110.00		1221.00		
2 Dump Trucks, 12 Ton		653.20		718.50	53.38	58.72
56 L.H., Daily Totals		$4703.60		$6069.70	$83.99	$108.39

Crew B-13B	Hr.	Daily	Hr.	Daily	Bare Costs	Incl. O&P
1 Laborer Foreman (outside)	$29.40	$235.20	$48.70	$389.60	$29.94	$48.81
4 Laborers	27.40	876.80	45.40	1452.80		
1 Equip. Oper. (crane)	38.10	304.80	60.15	481.20		
1 Equip. Oper. Oiler	32.45	259.60	51.25	410.00		
1 Hyd. Crane, 55 Ton		951.60		1046.75	16.99	18.69
56 L.H., Daily Totals		$2628.00		$3780.35	$46.93	$67.50

Crew B-13C	Hr.	Daily	Hr.	Daily	Bare Costs	Incl. O&P
1 Laborer Foreman (outside)	$29.40	$235.20	$48.70	$389.60	$29.94	$48.81
4 Laborers	27.40	876.80	45.40	1452.80		
1 Equip. Oper. (crane)	38.10	304.80	60.15	481.20		
1 Equip. Oper. Oiler	32.45	259.60	51.25	410.00		
1 Crawler Crane, 100 Ton		1590.00		1749.00	28.39	31.23
56 L.H., Daily Totals		$3266.40		$4482.60	$58.33	$80.04

Crew No.	Bare Costs		Incl. Subs O & P		Cost Per Labor-Hour	
Crew B-14	**Hr.**	**Daily**	**Hr.**	**Daily**	**Bare Costs**	**Incl. O&P**
1 Laborer Foreman (outside)	$29.40	$235.20	$48.70	$389.60	$29.03	$47.65
4 Laborers	27.40	876.80	45.40	1452.80		
1 Equip. Oper. (light)	35.20	281.60	55.60	444.80		
1 Backhoe Loader, 48 H.P.		226.20		248.80	4.71	5.18
48 L.H., Daily Totals		$1619.80		$2536.00	$33.74	$52.83

Crew B-15	Hr.	Daily	Hr.	Daily	Bare Costs	Incl. O&P
1 Equipment Oper. (med)	$36.70	$293.60	$57.95	$463.60	$30.60	$49.41
.5 Laborer	27.40	109.60	45.40	181.60		
2 Truck Drivers (heavy)	28.35	453.60	46.15	738.40		
2 Dump Trucks, 16 Ton		953.20		1048.50		
1 Dozer, 200 H.P.		920.80		1012.90	66.93	73.62
28 L.H., Daily Totals		$2730.80		$3445.00	$97.53	$123.03

Crew B-16	Hr.	Daily	Hr.	Daily	Bare Costs	Incl. O&P
1 Laborer Foreman (outside)	$29.40	$235.20	$48.70	$389.60	$28.14	$46.41
2 Laborers	27.40	438.40	45.40	726.40		
1 Truck Driver (heavy)	28.35	226.80	46.15	369.20		
1 Dump Truck, 16 Ton		476.60		524.25	14.89	16.38
32 L.H., Daily Totals		$1377.00		$2009.45	$43.03	$62.79

Crew B-17	Hr.	Daily	Hr.	Daily	Bare Costs	Incl. O&P
2 Laborers	$27.40	$438.40	$45.40	$726.40	$29.59	$48.14
1 Equip. Oper. (light)	35.20	281.60	55.60	444.80		
1 Truck Driver (heavy)	28.35	226.80	46.15	369.20		
1 Backhoe Loader, 48 H.P.		226.20		248.80		
1 Dump Truck, 12 Ton		326.60		359.25	17.28	19.00
32 L.H., Daily Totals		$1499.60		$2148.45	$46.87	$67.14

Crew B-17A	Hr.	Daily	Hr.	Daily	Bare Costs	Incl. O&P
2 Laborer Foremen	$29.40	$470.40	$48.70	$779.20	$29.82	$49.25
6 Laborers	27.40	1315.20	45.40	2179.20		
1 Skilled Worker Foreman	38.50	308.00	63.00	504.00		
1 Skilled Worker	36.50	292.00	59.70	477.60		
80 L.H., Daily Totals		$2385.60		$3940.00	$29.82	$49.25

Crew B-18	Hr.	Daily	Hr.	Daily	Bare Costs	Incl. O&P
1 Laborer Foreman (outside)	$29.40	$235.20	$48.70	$389.60	$28.07	$46.50
2 Laborers	27.40	438.40	45.40	726.40		
1 Vibrating Compactor		34.60		38.05	1.44	1.59
24 L.H., Daily Totals		$708.20		$1154.05	$29.51	$48.09

Crew B-19	Hr.	Daily	Hr.	Daily	Bare Costs	Incl. O&P
1 Pile Driver Foreman	$36.70	$293.60	$62.15	$497.20	$35.52	$58.61
4 Pile Drivers	34.70	1110.40	58.80	1881.60		
2 Equip. Oper. (crane)	38.10	609.60	60.15	962.40		
1 Equip. Oper. Oiler	32.45	259.60	51.25	410.00		
1 Crane, 40 Ton & Access.		934.20		1027.60		
60 L.F. Pile Leads		66.00		72.60		
1 Hammer, Diesel, 22k Ft-Lb		568.00		624.80	24.55	27.01
64 L.H., Daily Totals		$3841.40		$5476.20	$60.07	$85.62

Crew B-19A	Hr.	Daily	Hr.	Daily	Bare Costs	Incl. O&P
1 Pile Driver Foreman	$36.70	$293.60	$62.15	$497.20	$35.52	$58.61
4 Pile Drivers	34.70	1110.40	58.80	1881.60		
2 Equip. Oper. (crane)	38.10	609.60	60.15	962.40		
1 Equip. Oper. Oiler	32.45	259.60	51.25	410.00		
1 Crawler Crane, 75 Ton		1226.00		1348.60		
60 L.F. Leads, 25K Ft. Lbs.		114.00		125.40		
1 Hammer, Diesel, 41k Ft-Lb		672.00		739.20	31.39	34.53
64 L.H., Daily Totals		$4285.20		$5964.40	$66.91	$93.14

Crew No.	Bare Costs		Incl. Subs O & P		Cost Per Labor-Hour	
Crew B-20	**Hr.**	**Daily**	**Hr.**	**Daily**	**Bare Costs**	**Incl. O&P**
1 Laborer Foreman (out)	$29.40	$235.20	$48.70	$389.60	$31.10	$51.27
1 Skilled Worker	36.50	292.00	59.70	477.60		
1 Laborer	27.40	219.20	45.40	363.20		
24 L.H., Daily Totals		$746.40		$1230.40	$31.10	$51.27
Crew B-20A	**Hr.**	**Daily**	**Hr.**	**Daily**	**Bare Costs**	**Incl. O&P**
1 Laborer Foreman	$29.40	$235.20	$48.70	$389.60	$33.41	$53.40
1 Laborer	27.40	219.20	45.40	363.20		
1 Plumber	42.70	341.60	66.40	531.20		
1 Plumber Apprentice	34.15	273.20	53.10	424.80		
32 L.H., Daily Totals		$1069.20		$1708.80	$33.41	$53.40
Crew B-21	**Hr.**	**Daily**	**Hr.**	**Daily**	**Bare Costs**	**Incl. O&P**
1 Laborer Foreman (out)	$29.40	$235.20	$48.70	$389.60	$32.10	$52.54
1 Skilled Worker	36.50	292.00	59.70	477.60		
1 Laborer	27.40	219.20	45.40	363.20		
.5 Equip. Oper. (crane)	38.10	152.40	60.15	240.60		
.5 S.P. Crane, 5 Ton		152.60		167.85	5.45	6.00
28 L.H., Daily Totals		$1051.40		$1638.85	$37.55	$58.54
Crew B-21A	**Hr.**	**Daily**	**Hr.**	**Daily**	**Bare Costs**	**Incl. O&P**
1 Laborer Foreman	$29.40	$235.20	$48.70	$389.60	$34.35	$54.75
1 Laborer	27.40	219.20	45.40	363.20		
1 Plumber	42.70	341.60	66.40	531.20		
1 Plumber Apprentice	34.15	273.20	53.10	424.80		
1 Equip. Oper. (crane)	38.10	304.80	60.15	481.20		
1 S.P. Crane, 12 Ton		536.40		590.05	13.41	14.75
40 L.H., Daily Totals		$1910.40		$2780.05	$47.76	$69.50
Crew B-21B	**Hr.**	**Daily**	**Hr.**	**Daily**	**Bare Costs**	**Incl. O&P**
1 Laborer Foreman	$29.40	$235.20	$48.70	$389.60	$29.94	$49.01
3 Laborers	27.40	657.60	45.40	1089.60		
1 Equip. Oper. (crane)	38.10	304.80	60.15	481.20		
1 Hyd. Crane, 12 Ton		636.60		700.25	15.92	17.51
40 L.H., Daily Totals		$1834.20		$2660.65	$45.86	$66.52
Crew B-21C	**Hr.**	**Daily**	**Hr.**	**Daily**	**Bare Costs**	**Incl. O&P**
1 Laborer Foreman	$29.40	$235.20	$48.70	$389.60	$29.94	$48.81
4 Laborers	27.40	876.80	45.40	1452.80		
1 Equip. Oper. (crane)	38.10	304.80	60.15	481.20		
1 Equip. Oper. Oiler	32.45	259.60	51.25	410.00		
2 Cutting Torches		32.40		35.65		
2 Gases		129.60		142.55		
1 Crane, 90 Ton		1427.00		1569.70	28.38	31.21
56 L.H., Daily Totals		$3265.40		$4481.50	$58.32	$80.02
Crew B-22	**Hr.**	**Daily**	**Hr.**	**Daily**	**Bare Costs**	**Incl. O&P**
1 Laborer Foreman (out)	$29.40	$235.20	$48.70	$389.60	$32.50	$53.04
1 Skilled Worker	36.50	292.00	59.70	477.60		
1 Laborer	27.40	219.20	45.40	363.20		
.75 Equip. Oper. (crane)	38.10	228.60	60.15	360.90		
.75 S.P. Crane, 5 Ton		228.90		251.80	7.63	8.39
30 L.H., Daily Totals		$1203.90		$1843.10	$40.13	$61.43

Crew No.	Bare Costs		Incl. Subs O & P		Cost Per Labor-Hour	
Crew B-22A	**Hr.**	**Daily**	**Hr.**	**Daily**	**Bare Costs**	**Incl. O&P**
1 Laborer Foreman (out)	$29.40	$235.20	$48.70	$389.60	$31.43	$51.43
1 Skilled Worker	36.50	292.00	59.70	477.60		
2 Laborers	27.40	438.40	45.40	726.40		
.75 Equipment Oper. (crane)	38.10	228.60	60.15	360.90		
.75 Crane, 5 Ton		228.90		251.80		
1 Generator, 5 KW		31.20		34.30		
1 Butt Fusion Machine		347.20		381.90	15.98	17.58
38 L.H., Daily Totals		$1801.50		$2622.50	$47.41	$69.01
Crew B-22B	**Hr.**	**Daily**	**Hr.**	**Daily**	**Bare Costs**	**Incl. O&P**
1 Skilled Worker	$36.50	$292.00	$59.70	$477.60	$31.95	$52.55
1 Laborer	27.40	219.20	45.40	363.20		
1 Electro Fusion Machine		148.40		163.25	9.28	10.20
16 L.H., Daily Totals		$659.60		$1004.05	$41.23	$62.75
Crew B-23	**Hr.**	**Daily**	**Hr.**	**Daily**	**Bare Costs**	**Incl. O&P**
1 Laborer Foreman (outside)	$29.40	$235.20	$48.70	$389.60	$27.80	$46.06
4 Laborers	27.40	876.80	45.40	1452.80		
1 Drill Rig, Wells		3379.00		3716.90		
1 Light Truck, 3 Ton		176.20		193.80	88.88	97.77
40 L.H., Daily Totals		$4667.20		$5753.10	$116.68	$143.83
Crew B-23A	**Hr.**	**Daily**	**Hr.**	**Daily**	**Bare Costs**	**Incl. O&P**
1 Laborer Foreman (outside)	$29.40	$235.20	$48.70	$389.60	$31.17	$50.68
1 Laborer	27.40	219.20	45.40	363.20		
1 Equip. Operator (medium)	36.70	293.60	57.95	463.60		
1 Drill Rig, Wells		3379.00		3716.90		
1 Pickup Truck, 3/4 Ton		78.00		85.80	144.04	158.45
24 L.H., Daily Totals		$4205.00		$5019.10	$175.21	$209.13
Crew B-23B	**Hr.**	**Daily**	**Hr.**	**Daily**	**Bare Costs**	**Incl. O&P**
1 Laborer Foreman (outside)	$29.40	$235.20	$48.70	$389.60	$31.17	$50.68
1 Laborer	27.40	219.20	45.40	363.20		
1 Equip. Operator (medium)	36.70	293.60	57.95	463.60		
1 Drill Rig, Wells		3379.00		3716.90		
1 Pickup Truck, 3/4 Ton		78.00		85.80		
1 Pump, Cntfgl, 6"		240.00		264.00	154.04	169.45
24 L.H., Daily Totals		$4445.00		$5283.10	$185.21	$220.13
Crew B-24	**Hr.**	**Daily**	**Hr.**	**Daily**	**Bare Costs**	**Incl. O&P**
1 Cement Finisher	$34.40	$275.20	$54.20	$433.60	$32.45	$52.83
1 Laborer	27.40	219.20	45.40	363.20		
1 Carpenter	35.55	284.40	58.90	471.20		
24 L.H., Daily Totals		$778.80		$1268.00	$32.45	$52.83
Crew B-25	**Hr.**	**Daily**	**Hr.**	**Daily**	**Bare Costs**	**Incl. O&P**
1 Laborer Foreman	$29.40	$235.20	$48.70	$389.60	$30.12	$49.12
7 Laborers	27.40	1534.40	45.40	2542.40		
3 Equip. Oper. (med.)	36.70	880.80	57.95	1390.80		
1 Asphalt Paver, 130 H.P		1661.00		1827.10		
1 Tandem Roller, 10 Ton		197.00		216.70		
1 Roller, Pneumatic Wheel		259.20		285.10	24.06	26.47
88 L.H., Daily Totals		$4767.60		$6651.70	$54.18	$75.59

Crew No.	Bare Costs		Incl. Subs O & P		Cost Per Labor-Hour	
Crew B-25B	**Hr.**	**Daily**	**Hr.**	**Daily**	**Bare Costs**	**Incl. O&P**
1 Laborer Foreman	$29.40	$235.20	$48.70	$389.60	$30.67	$49.86
7 Laborers	27.40	1534.40	45.40	2542.40		
4 Equip. Oper. (medium)	36.70	1174.40	57.95	1854.40		
1 Asphalt Paver, 130 H.P.		1661.00		1827.10		
2 Rollers, Steel Wheel		394.00		433.40		
1 Roller, Pneumatic Wheel		259.20		285.10	24.11	26.52
96 L.H., Daily Totals		$5258.20		$7332.00	$54.78	$76.38

Crew B-25C	Hr.	Daily	Hr.	Daily	Bare Costs	Incl. O&P
1 Laborer Foreman	$29.40	$235.20	$48.70	$389.60	$30.83	$50.13
3 Laborers	27.40	657.60	45.40	1089.60		
2 Equip. Oper. (medium)	36.70	587.20	57.95	927.20		
1 Asphalt Paver, 130 H.P.		1661.00		1827.10		
1 Roller, Steel Wheel		197.00		216.70	38.71	42.58
48 L.H., Daily Totals		$3338.00		$4450.20	$69.54	$92.71

Crew B-26	Hr.	Daily	Hr.	Daily	Bare Costs	Incl. O&P
1 Laborer Foreman (outside)	$29.40	$235.20	$48.70	$389.60	$31.01	$50.82
6 Laborers	27.40	1315.20	45.40	2179.20		
2 Equip. Oper. (med.)	36.70	587.20	57.95	927.20		
1 Rodman (reinf.)	39.50	316.00	67.80	542.40		
1 Cement Finisher	34.40	275.20	54.20	433.60		
1 Grader, 30,000 Lbs.		457.40		503.15		
1 Paving Mach. & Equip.		2237.00		2460.70	30.62	33.68
88 L.H., Daily Totals		$5423.20		$7435.85	$61.63	$84.50

Crew B-27	Hr.	Daily	Hr.	Daily	Bare Costs	Incl. O&P
1 Laborer Foreman (outside)	$29.40	$235.20	$48.70	$389.60	$27.90	$46.23
3 Laborers	27.40	657.60	45.40	1089.60		
1 Berm Machine		232.20		255.40	7.26	7.98
32 L.H., Daily Totals		$1125.00		$1734.60	$35.16	$54.21

Crew B-28	Hr.	Daily	Hr.	Daily	Bare Costs	Incl. O&P
2 Carpenters	$35.55	$568.80	$58.90	$942.40	$32.83	$54.40
1 Laborer	27.40	219.20	45.40	363.20		
24 L.H., Daily Totals		$788.00		$1305.60	$32.83	$54.40

Crew B-29	Hr.	Daily	Hr.	Daily	Bare Costs	Incl. O&P
1 Laborer Foreman (outside)	$29.40	$235.20	$48.70	$389.60	$29.94	$48.81
4 Laborers	27.40	876.80	45.40	1452.80		
1 Equip. Oper. (crane)	38.10	304.80	60.15	481.20		
1 Equip. Oper. Oiler	32.45	259.60	51.25	410.00		
1 Gradall, 3 Ton, 1/2 C.Y.		841.00		925.10	15.02	16.52
56 L.H., Daily Totals		$2517.40		$3658.70	$44.96	$65.33

Crew B-30	Hr.	Daily	Hr.	Daily	Bare Costs	Incl. O&P
1 Equip. Oper. (med.)	$36.70	$293.60	$57.95	$463.60	$31.13	$50.08
2 Truck Drivers (heavy)	28.35	453.60	46.15	738.40		
1 Hyd. Excavator, 1.5 C.Y.		722.80		795.10		
2 Dump Trucks, 16 Ton		953.20		1048.50	69.83	76.82
24 L.H., Daily Totals		$2423.20		$3045.60	$100.96	$126.90

Crew B-31	Hr.	Daily	Hr.	Daily	Bare Costs	Incl. O&P
1 Laborer Foreman (outside)	$29.40	$235.20	$48.70	$389.60	$29.43	$48.76
3 Laborers	27.40	657.60	45.40	1089.60		
1 Carpenter	35.55	284.40	58.90	471.20		
1 Air Compr., 250 C.F.M.		119.80		131.80		
1 Sheeting Driver		7.20		7.90		
2-50 Ft. Air Hoses, 1.5" Dia.		9.40		10.35	3.41	3.75
40 L.H., Daily Totals		$1313.60		$2100.45	$32.84	$52.51

Crew B-32	Hr.	Daily	Hr.	Daily	Bare Costs	Incl. O&P
1 Laborer	$27.40	$219.20	$45.40	$363.20	$34.38	$54.81
3 Equip. Oper. (med.)	36.70	880.80	57.95	1390.80		
1 Grader, 30,000 Lbs.		457.40		503.15		
1 Tandem Roller, 10 Ton		197.00		216.70		
1 Dozer, 200 H.P.		920.80		1012.90	49.23	54.15
32 L.H., Daily Totals		$2675.20		$3486.75	$83.61	$108.96

Crew B-32A	Hr.	Daily	Hr.	Daily	Bare Costs	Incl. O&P
1 Laborer	$27.40	$219.20	$45.40	$363.20	$33.60	$53.77
2 Equip. Oper. (medium)	36.70	587.20	57.95	927.20		
1 Grader, 30,000 Lbs.		457.40		503.15		
1 Roller, Vibratory, 29,000 Lbs.		522.00		574.20	40.81	44.89
24 L.H., Daily Totals		$1785.80		$2367.75	$74.41	$98.66

Crew B-32B	Hr.	Daily	Hr.	Daily	Bare Costs	Incl. O&P
1 Laborer	$27.40	$219.20	$45.40	$363.20	$33.60	$53.77
2 Equip. Oper. (medium)	36.70	587.20	57.95	927.20		
1 Dozer, 200 H.P.		920.80		1012.90		
1 Roller, Vibratory, 29,000 Lbs.		522.00		574.20	60.12	66.13
24 L.H., Daily Totals		$2249.20		$2877.50	$93.72	$119.90

Crew B-32C	Hr.	Daily	Hr.	Daily	Bare Costs	Incl. O&P
1 Laborer Foreman	$29.40	$235.20	$48.70	$389.60	$32.38	$52.23
2 Laborers	27.40	438.40	45.40	726.40		
3 Equip. Oper. (medium)	36.70	880.80	57.95	1390.80		
1 Grader, 30,000 Lbs.		457.40		503.15		
1 Roller, Steel Wheel		197.00		216.70		
1 Dozer, 200 H.P.		920.80		1012.90	32.82	36.10
48 L.H., Daily Totals		$3129.60		$4239.55	$65.20	$88.33

Crew B-33A	Hr.	Daily	Hr.	Daily	Bare Costs	Incl. O&P
1 Equip. Oper. (med.)	$36.70	$293.60	$57.95	$463.60	$34.04	$54.36
.5 Laborer	27.40	109.60	45.40	181.60		
.25 Equip. Oper. (med.)	36.70	73.40	57.95	115.90		
1 Scraper, Towed, 7 C.Y.		161.75		177.95		
1.25 Dozer, 300 H.P.		1496.25		1645.90	118.43	130.28
14 L.H., Daily Totals		$2134.60		$2584.95	$152.47	$184.64

Crew B-33B	Hr.	Daily	Hr.	Daily	Bare Costs	Incl. O&P
1 Equip. Oper. (med.)	$36.70	$293.60	$57.95	$463.60	$34.04	$54.36
.5 Laborer	27.40	109.60	45.40	181.60		
.25 Equip. Oper. (med.)	36.70	73.40	57.95	115.90		
1 Scraper, Towed, 10 C.Y.		180.40		198.45		
1.25 Dozer, 300 H.P.		1496.25		1645.90	119.76	131.74
14 L.H., Daily Totals		$2153.25		$2605.45	$153.80	$186.10

Crew B-33C	Hr.	Daily	Hr.	Daily	Bare Costs	Incl. O&P
1 Equip. Oper. (med.)	$36.70	$293.60	$57.95	$463.60	$34.04	$54.36
.5 Laborer	27.40	109.60	45.40	181.60		
.25 Equip. Oper. (med.)	36.70	73.40	57.95	115.90		
1 Scraper, Towed, 12 C.Y.		180.40		198.45		
1.25 Dozer, 300 H.P.		1496.25		1645.90	119.76	131.74
14 L.H., Daily Totals		$2153.25		$2605.45	$153.80	$186.10

Crew B-33D	Hr.	Daily	Hr.	Daily	Bare Costs	Incl. O&P
1 Equip. Oper. (med.)	$36.70	$293.60	$57.95	$463.60	$34.04	$54.36
.5 Laborer	27.40	109.60	45.40	181.60		
.25 Equip. Oper. (med.)	36.70	73.40	57.95	115.90		
1 S.P. Scraper, 14 C.Y.		1577.00		1734.70		
.25 Dozer, 300 H.P.		299.25		329.20	134.02	147.42
14 L.H., Daily Totals		$2352.85		$2825.00	$168.06	$201.78

Crew No.	Bare Costs		Incl. Subs O & P		Cost Per Labor-Hour	
Crew B-33E	**Hr.**	**Daily**	**Hr.**	**Daily**	**Bare Costs**	**Incl. O&P**
1 Equip. Oper. (med.)	$36.70	$293.60	$57.95	$463.60	$34.04	$54.36
.5 Laborer	27.40	109.60	45.40	181.60		
.25 Equip. Oper. (med.)	36.70	73.40	57.95	115.90		
1 S.P. Scraper, 24 C.Y.		2350.00		2585.00		
.25 Dozer, 300 H.P.		299.25		329.20	189.23	208.16
14 L.H., Daily Totals		$3125.85		$3675.30	$223.27	$262.52

Crew B-33F	Hr.	Daily	Hr.	Daily	Bare Costs	Incl. O&P
1 Equip. Oper. (med.)	$36.70	$293.60	$57.95	$463.60	$34.04	$54.36
.5 Laborer	27.40	109.60	45.40	181.60		
.25 Equip. Oper. (med.)	36.70	73.40	57.95	115.90		
1 Elev. Scraper, 11 C.Y.		822.60		904.85		
.25 Dozer, 300 H.P.		299.25		329.20	80.13	88.15
14 L.H., Daily Totals		$1598.45		$1995.15	$114.17	$142.51

Crew B-33G	Hr.	Daily	Hr.	Daily	Bare Costs	Incl. O&P
1 Equip. Oper. (med.)	$36.70	$293.60	$57.95	$463.60	$34.04	$54.36
.5 Laborer	27.40	109.60	45.40	181.60		
.25 Equip. Oper. (med.)	36.70	73.40	57.95	115.90		
1 Elev. Scraper, 20 C.Y.		1653.00		1818.30		
.25 Dozer, 300 H.P.		299.25		329.20	139.45	153.39
14 L.H., Daily Totals		$2428.85		$2908.60	$173.49	$207.75

Crew B-34A	Hr.	Daily	Hr.	Daily	Bare Costs	Incl. O&P
1 Truck Driver (heavy)	$28.35	$226.80	$46.15	$369.20	$28.35	$46.15
1 Dump Truck, 12 Ton		326.60		359.25	40.83	44.91
8 L.H., Daily Totals		$553.40		$728.45	$69.18	$91.06

Crew B-34B	Hr.	Daily	Hr.	Daily	Bare Costs	Incl. O&P
1 Truck Driver (heavy)	$28.35	$226.80	$46.15	$369.20	$28.35	$46.15
1 Dump Truck, 16 Ton		476.60		524.25	59.58	65.53
8 L.H., Daily Totals		$703.40		$893.45	$87.93	$111.68

Crew B-34C	Hr.	Daily	Hr.	Daily	Bare Costs	Incl. O&P
1 Truck Driver (heavy)	$28.35	$226.80	$46.15	$369.20	$28.35	$46.15
1 Truck Tractor, 40 Ton		301.60		331.75		
1 Dump Trailer, 16.5 C.Y.		103.00		113.30	50.58	55.63
8 L.H., Daily Totals		$631.40		$814.25	$78.93	$101.78

Crew B-34D	Hr.	Daily	Hr.	Daily	Bare Costs	Incl. O&P
1 Truck Driver (heavy)	$28.35	$226.80	$46.15	$369.20	$28.35	$46.15
1 Truck Tractor, 40 Ton		301.60		331.75		
1 Dump Trailer, 20 C.Y.		115.80		127.40	52.18	57.39
8 L.H., Daily Totals		$644.20		$828.35	$80.53	$103.54

Crew B-34E	Hr.	Daily	Hr.	Daily	Bare Costs	Incl. O&P
1 Truck Driver (heavy)	$28.35	$226.80	$46.15	$369.20	$28.35	$46.15
1 Truck, Off Hwy., 25 Ton		912.20		1003.40	114.03	125.43
8 L.H., Daily Totals		$1139.00		$1372.60	$142.38	$171.58

Crew B-34F	Hr.	Daily	Hr.	Daily	Bare Costs	Incl. O&P
1 Truck Driver (heavy)	$28.35	$226.80	$46.15	$369.20	$28.35	$46.15
1 Truck, Off Hwy., 22 C.Y.		933.40		1026.75	116.68	128.34
8 L.H., Daily Totals		$1160.20		$1395.95	$145.03	$174.49

Crew B-34G	Hr.	Daily	Hr.	Daily	Bare Costs	Incl. O&P
1 Truck Driver (heavy)	$28.35	$226.80	$46.15	$369.20	$28.35	$46.15
1 Truck, Off Hwy., 34 C.Y.		1181.00		1299.10	147.63	162.39
8 L.H., Daily Totals		$1407.80		$1668.30	$175.98	$208.54

Crew No.	Bare Costs		Incl. Subs O & P		Cost Per Labor-Hour	
Crew B-34H	**Hr.**	**Daily**	**Hr.**	**Daily**	**Bare Costs**	**Incl. O&P**
1 Truck Driver (heavy)	$28.35	$226.80	$46.15	$369.20	$28.35	$46.15
1 Truck, Off Hwy., 42 C.Y.		1276.00		1403.60	159.50	175.45
8 L.H., Daily Totals		$1502.80		$1772.80	$187.85	$221.60

Crew B-34J	Hr.	Daily	Hr.	Daily	Bare Costs	Incl. O&P
1 Truck Driver (heavy)	$28.35	$226.80	$46.15	$369.20	$28.35	$46.15
1 Truck, Off Hwy., 60 C.Y.		1649.00		1813.90	206.13	226.74
8 L.H., Daily Totals		$1875.80		$2183.10	$234.48	$272.89

Crew B-34K	Hr.	Daily	Hr.	Daily	Bare Costs	Incl. O&P
1 Truck Driver (heavy)	$28.35	$226.80	$46.15	$369.20	$28.35	$46.15
1 Truck Tractor, 240 H.P.		353.60		388.95		
1 Low Bed Trailer		170.20		187.20	65.48	72.02
8 L.H., Daily Totals		$750.60		$945.35	$93.83	$118.17

Crew B-34N	Hr.	Daily	Hr.	Daily	Bare Costs	Incl. O&P
1 Truck Driver (heavy)	$28.35	$226.80	$46.15	$369.20	$28.35	$46.15
1 Dump Truck, 12 Ton		326.60		359.25		
1 Flatbed Trailer, 40 Ton		119.40		131.35	55.75	61.33
8 L.H., Daily Totals		$672.80		$859.80	$84.10	$107.48

Crew B-34P	Hr.	Daily	Hr.	Daily	Bare Costs	Incl. O&P
1 Pipe Fitter	$43.05	$344.40	$66.95	$535.60	$35.78	$56.60
1 Truck Driver (light)	27.60	220.80	44.90	359.20		
1 Equip. Oper. (medium)	36.70	293.60	57.95	463.60		
1 Flatbed Truck, 3 Ton		176.20		193.80		
1 Backhoe Loader, 48 H.P.		226.20		248.80	16.77	18.44
24 L.H., Daily Totals		$1261.20		$1801.00	$52.55	$75.04

Crew B-34Q	Hr.	Daily	Hr.	Daily	Bare Costs	Incl. O&P
1 Pipe Fitter	$43.05	$344.40	$66.95	$535.60	$36.25	$57.33
1 Truck Driver (light)	27.60	220.80	44.90	359.20		
1 Eqip. Oper. (crane)	38.10	304.80	60.15	481.20		
1 Flatbed Trailer, 25 Ton		88.00		96.80		
1 Dump Truck, 12 Ton		326.60		359.25		
1 Hyd. Crane, 25 Ton		650.80		715.90	44.39	48.83
24 L.H., Daily Totals		$1935.40		$2547.95	$80.64	$106.16

Crew B-34R	Hr.	Daily	Hr.	Daily	Bare Costs	Incl. O&P
1 Pipe Fitter	$43.05	$344.40	$66.95	$535.60	$36.25	$57.33
1 Truck Driver (light)	27.60	220.80	44.90	359.20		
1 Equip. Oper. (crane)	38.10	304.80	60.15	481.20		
1 Flatbed Trailer, 25 Ton		88.00		96.80		
1 Dump Truck, 12 Ton		326.60		359.25		
1 Hyd. Crane, 25 Ton		650.80		715.90		
1 Hyd. Excavator, 1 C.Y.		560.20		616.20	67.73	74.51
24 L.H., Daily Totals		$2495.60		$3164.15	$103.98	$131.84

Crew B-34S	Hr.	Daily	Hr.	Daily	Bare Costs	Incl. O&P
2 Pipe Fitters	$43.05	$688.80	$66.95	$1071.20	$38.14	$60.05
1 Truck Driver (heavy)	28.35	226.80	46.15	369.20		
1 Equip. Oper. (crane)	38.10	304.80	60.15	481.20		
1 Flatbed Trailer, 40 Ton		119.40		131.35		
1 Truck Tractor, 40 Ton		301.60		331.75		
1 Truck Crane, 80 Ton		1059.00		1164.90		
1 Hyd. Excavator, 2 C.Y.		919.60		1011.55	74.99	82.49
32 L.H., Daily Totals		$3620.00		$4561.15	$113.13	$142.54

Crew No.	Bare Costs		Incl. Subs O & P		Cost Per Labor-Hour	
Crew B-34T	**Hr.**	**Daily**	**Hr.**	**Daily**	**Bare Costs**	**Incl. O&P**
2 Pipe Fitters	$43.05	$688.80	$66.95	$1071.20	$38.14	$60.05
1 Truck Driver (heavy)	28.35	226.80	46.15	369.20		
1 Equip. Oper. (crane)	38.10	304.80	60.15	481.20		
1 Flatbed Trailer, 40 Ton		119.40		131.35		
1 Truck Tractor, 40 Ton		301.60		331.75		
1 Truck Crane, 80 Ton		1059.00		1164.90	46.25	50.88
32 L.H., Daily Totals		$2700.40		$3549.60	$84.39	$110.93

Crew B-35	Hr.	Daily	Hr.	Daily	Bare Costs	Incl. O&P
1 Laborer Foreman (out)	$29.40	$235.20	$48.70	$389.60	$34.43	$55.27
1 Skilled Worker	36.50	292.00	59.70	477.60		
1 Welder (plumber)	42.70	341.60	66.40	531.20		
1 Laborer	27.40	219.20	45.40	363.20		
1 Equip. Oper. (crane)	38.10	304.80	60.15	481.20		
1 Equip. Oper. Oiler	32.45	259.60	51.25	410.00		
1 Electric Welding Mach.		69.10		76.00		
1 Hyd. Excavator, .75 C.Y.		477.80		525.60	11.39	12.53
48 L.H., Daily Totals		$2199.30		$3254.40	$45.82	$67.80

Crew B-35A	Hr.	Daily	Hr.	Daily	Bare Costs	Incl. O&P
1 Laborer Foreman (out)	$29.40	$235.20	$48.70	$389.60	$33.42	$53.86
2 Laborers	27.40	438.40	45.40	726.40		
1 Skilled Worker	36.50	292.00	59.70	477.60		
1 Welder (plumber)	42.70	341.60	66.40	531.20		
1 Equip. Oper. (crane)	38.10	304.80	60.15	481.20		
1 Equip. Oper. Oiler	32.45	259.60	51.25	410.00		
1 Welder, 300 amp		89.20		98.10		
1 Crane, 75 Ton		1226.00		1348.60	23.49	25.83
56 L.H., Daily Totals		$3186.80		$4462.70	$56.91	$79.69

Crew B-36	Hr.	Daily	Hr.	Daily	Bare Costs	Incl. O&P
1 Laborer Foreman (outside)	$29.40	$235.20	$48.70	$389.60	$31.52	$51.08
2 Laborers	27.40	438.40	45.40	726.40		
2 Equip. Oper. (med.)	36.70	587.20	57.95	927.20		
1 Dozer, 200 H.P.		920.80		1012.90		
1 Aggregate Spreader		32.00		35.20		
1 Tandem Roller, 10 Ton		197.00		216.70	28.75	31.62
40 L.H., Daily Totals		$2410.60		$3308.00	$60.27	$82.70

Crew B-36A	Hr.	Daily	Hr.	Daily	Bare Costs	Incl. O&P
1 Laborer Foreman (outside)	$29.40	$235.20	$48.70	$389.60	$33.00	$53.04
2 Laborers	27.40	438.40	45.40	726.40		
4 Equip. Oper. (med.)	36.70	1174.40	57.95	1854.40		
1 Dozer, 200 H.P.		920.80		1012.90		
1 Aggregate Spreader		32.00		35.20		
1 Roller, Steel Wheel		197.00		216.70		
1 Roller, Pneumatic Wheel		259.20		285.10	25.16	27.68
56 L.H., Daily Totals		$3257.00		$4520.30	$58.16	$80.72

Crew B-36B	Hr.	Daily	Hr.	Daily	Bare Costs	Incl. O&P
1 Laborer Foreman (outside)	$29.40	$235.20	$48.70	$389.60	$32.42	$52.18
2 Laborers	27.40	438.40	45.40	726.40		
4 Equip. Oper. (medium)	36.70	1174.40	57.95	1854.40		
1 Truck Driver, Heavy	28.35	226.80	46.15	369.20		
1 Grader, 30,000 Lbs.		457.40		503.15		
1 F.E. Loader, crl, 1.5 C.Y.		372.40		409.65		
1 Dozer, 300 H.P.		1197.00		1316.70		
1 Roller, Vibratory		522.00		574.20		
1 Truck, Tractor, 240 H.P.		353.60		388.95		
1 Water Tanker, 5000 Gal.		116.80		128.50	47.18	51.89
64 L.H., Daily Totals		$5094.00		$6660.75	$79.60	$104.07

Crew B-36C	Hr.	Daily	Hr.	Daily	Bare Costs	Incl. O&P
1 Laborer Foreman (outside)	$29.40	$235.20	$48.70	$389.60	$33.57	$53.74
3 Equip. Oper. (medium)	36.70	880.80	57.95	1390.80		
1 Truck Driver, Heavy	28.35	226.80	46.15	369.20		
1 Grader, 30,000 Lbs.		457.40		503.15		
1 Dozer, 300 H.P.		1197.00		1316.70		
1 Roller, Vibratory		522.00		574.20		
1 Truck, Tractor, 240 H.P.		353.60		388.95		
1 Water Tanker, 5000 Gal.		116.80		128.50	66.17	72.79
40 L.H., Daily Totals		$3989.60		$5061.10	$99.74	$126.53

Crew B-37	Hr.	Daily	Hr.	Daily	Bare Costs	Incl. O&P
1 Laborer Foreman (outside)	$29.40	$235.20	$48.70	$389.60	$29.03	$47.65
4 Laborers	27.40	876.80	45.40	1452.80		
1 Equip. Oper. (light)	35.20	281.60	55.60	444.80		
1 Tandem Roller, 5 Ton		114.20		125.60	2.38	2.62
48 L.H., Daily Totals		$1507.80		$2412.80	$31.41	$50.27

Crew B-38	Hr.	Daily	Hr.	Daily	Bare Costs	Incl. O&P
1 Laborer Foreman (outside)	$29.40	$235.20	$48.70	$389.60	$31.22	$50.61
2 Laborers	27.40	438.40	45.40	726.40		
1 Equip. Oper. (light)	35.20	281.60	55.60	444.80		
1 Equip. Oper. (medium)	36.70	293.60	57.95	463.60		
1 Backhoe Loader, 48 H.P.		226.20		248.80		
1 Hyd.Hammer, (1200 lb)		113.60		124.95		
1 F.E. Loader (170 H.P.)		444.00		488.40		
1 Pavt. Rem. Bucket		52.20		57.40	20.90	22.99
40 L.H., Daily Totals		$2084.80		$2943.95	$52.12	$73.60

Crew B-39	Hr.	Daily	Hr.	Daily	Bare Costs	Incl. O&P
1 Laborer Foreman (outside)	$29.40	$235.20	$48.70	$389.60	$29.03	$47.65
4 Laborers	27.40	876.80	45.40	1452.80		
1 Equip. Oper. (light)	35.20	281.60	55.60	444.80		
1 Air Compr., 250 C.F.M.		119.80		131.80		
2 Air Tools & Accessories		18.00		19.80		
2-50 Ft. Air Hoses, 1.5" Dia.		9.40		10.35	3.07	3.38
48 L.H., Daily Totals		$1540.80		$2449.15	$32.10	$51.03

Crew B-40	Hr.	Daily	Hr.	Daily	Bare Costs	Incl. O&P
1 Pile Driver Foreman (out)	$36.70	$293.60	$62.15	$497.20	$35.52	$58.61
4 Pile Drivers	34.70	1110.40	58.80	1881.60		
2 Equip. Oper. (crane)	38.10	609.60	60.15	962.40		
1 Equip. Oper. Oiler	32.45	259.60	51.25	410.00		
1 Crane, 40 Ton		934.20		1027.60		
1 Vibratory Hammer & Gen.		1823.00		2005.30	43.08	47.39
64 L.H., Daily Totals		$5030.40		$6784.10	$78.60	$106.00

Crew B-40B	Hr.	Daily	Hr.	Daily	Bare Costs	Incl. O&P
1 Laborer Foreman	$29.40	$235.20	$48.70	$389.60	$30.36	$49.38
3 Laborers	27.40	657.60	45.40	1089.60		
1 Equip. Oper. (crane)	38.10	304.80	60.15	481.20		
1 Equip. Oper. Oiler	32.45	259.60	51.25	410.00		
1 Crane, 40 Ton		1130.00		1243.00	23.54	25.90
48 L.H., Daily Totals		$2587.20		$3613.40	$53.90	$75.28

Crew B-41	Hr.	Daily	Hr.	Daily	Bare Costs	Incl. O&P
1 Laborer Foreman (outside)	$29.40	$235.20	$48.70	$389.60	$28.48	$46.94
4 Laborers	27.40	876.80	45.40	1452.80		
.25 Equip. Oper. (crane)	38.10	76.20	60.15	120.30		
.25 Equip. Oper. Oiler	32.45	64.90	51.25	102.50		
.25 Crawler Crane, 40 Ton		233.55		256.90	5.31	5.84
44 L.H., Daily Totals		$1486.65		$2322.10	$33.79	$52.78

Crew No.	Bare Costs		Incl. Subs O & P		Cost Per Labor-Hour	
Crew B-42	**Hr.**	**Daily**	**Hr.**	**Daily**	**Bare Costs**	**Incl. O&P**
1 Laborer Foreman (outside)	$29.40	$235.20	$48.70	$389.60	$31.19	$52.05
4 Laborers	27.40	876.80	45.40	1452.80		
1 Equip. Oper. (crane)	38.10	304.80	60.15	481.20		
1 Equip. Oper. Oiler	32.45	259.60	51.25	410.00		
1 Welder	39.95	319.60	74.70	597.60		
1 Hyd. Crane, 25 Ton		650.80		715.90		
1 Gas Welding Machine		89.20		98.10		
1 Horiz. Boring Csg. Mch.		392.00		431.20	17.69	19.46
64 L.H., Daily Totals		$3128.00		$4576.40	$48.88	$71.51
Crew B-43	**Hr.**	**Daily**	**Hr.**	**Daily**	**Bare Costs**	**Incl. O&P**
1 Laborer Foreman (outside)	$29.40	$235.20	$48.70	$389.60	$30.36	$49.38
3 Laborers	27.40	657.60	45.40	1089.60		
1 Equip. Oper. (crane)	38.10	304.80	60.15	481.20		
1 Equip. Oper. Oiler	32.45	259.60	51.25	410.00		
1 Drill Rig & Augers		3379.00		3716.90	70.40	77.44
48 L.H., Daily Totals		$4836.20		$6087.30	$100.76	$126.82
Crew B-44	**Hr.**	**Daily**	**Hr.**	**Daily**	**Bare Costs**	**Incl. O&P**
1 Pile Driver Foreman	$36.70	$293.60	$62.15	$497.20	$34.89	$57.88
4 Pile Drivers	34.70	1110.40	58.80	1881.60		
2 Equip. Oper. (crane)	38.10	609.60	60.15	962.40		
1 Laborer	27.40	219.20	45.40	363.20		
1 Crane, 40 Ton, & Access.		934.20		1027.60		
45 L.F. Leads, 15K Ft. Lbs.		49.50		54.45	15.41	16.95
64 L.H., Daily Totals		$3216.50		$4786.45	$50.30	$74.83
Crew B-45	**Hr.**	**Daily**	**Hr.**	**Daily**	**Bare Costs**	**Incl. O&P**
1 Equip. Oper. (med.)	$36.70	$293.60	$57.95	$463.60	$32.53	$52.05
1 Truck Driver (heavy)	28.35	226.80	46.15	369.20		
1 Dist. Tank Truck, 3K Gal.		242.40		266.65		
1 Tractor, 4 x 2, 250 H.P.		291.60		320.75	33.38	36.71
16 L.H., Daily Totals		$1054.40		$1420.20	$65.91	$88.76
Crew B-46	**Hr.**	**Daily**	**Hr.**	**Daily**	**Bare Costs**	**Incl. O&P**
1 Pile Driver Foreman	$36.70	$293.60	$62.15	$497.20	$31.38	$52.66
2 Pile Drivers	34.70	555.20	58.80	940.80		
3 Laborers	27.40	657.60	45.40	1089.60		
1 Chain Saw, 36" Long		33.40		36.75	.70	.77
48 L.H., Daily Totals		$1539.80		$2564.35	$32.08	$53.43
Crew B-47	**Hr.**	**Daily**	**Hr.**	**Daily**	**Bare Costs**	**Incl. O&P**
1 Blast Foreman	$29.40	$235.20	$48.70	$389.60	$30.67	$49.90
1 Driller	27.40	219.20	45.40	363.20		
1 Equip. Oper. (light)	35.20	281.60	55.60	444.80		
1 Crawler Type Drill, 4"		728.80		801.70		
1 Air Compr., 600 C.F.M.		320.20		352.20		
2-50 Ft. Air Hoses, 3" Dia.		37.10		40.80	45.26	49.78
24 L.H., Daily Totals		$1822.10		$2392.30	$75.93	$99.68
Crew B-47A	**Hr.**	**Daily**	**Hr.**	**Daily**	**Bare Costs**	**Incl. O&P**
1 Drilling Foreman	$29.40	$235.20	$48.70	$389.60	$33.32	$53.37
1 Equip. Oper. (heavy)	38.10	304.80	60.15	481.20		
1 Oiler	32.45	259.60	51.25	410.00		
1 Quarry Drill		822.00		904.20	34.25	37.68
24 L.H., Daily Totals		$1621.60		$2185.00	$67.57	$91.05

Crew No.	Bare Costs		Incl. Subs O & P		Cost Per Labor-Hour	
Crew B-47C	**Hr.**	**Daily**	**Hr.**	**Daily**	**Bare Costs**	**Incl. O&P**
1 Laborer	$27.40	$219.20	$45.40	$363.20	$31.30	$50.50
1 Equip. Oper. (light)	35.20	281.60	55.60	444.80		
1 Air Compressor, 750 CFM		341.40		375.55		
2-50' Air Hoses, 3"		37.10		40.80		
1 Air Track Drill, 4"		728.80		801.70	69.21	76.13
16 L.H., Daily Totals		$1608.10		$2026.05	$100.51	$126.63
Crew B-47E	**Hr.**	**Daily**	**Hr.**	**Daily**	**Bare Costs**	**Incl. O&P**
1 Laborer Foreman	$29.40	$235.20	$48.70	$389.60	$27.90	$46.23
3 Laborers	27.40	657.60	45.40	1089.60		
1 Truck, Flatbed, 3 Ton		176.20		193.80	5.51	6.06
32 L.H., Daily Totals		$1069.00		$1673.00	$33.41	$52.29
Crew B-47G	**Hr.**	**Daily**	**Hr.**	**Daily**	**Bare Costs**	**Incl. O&P**
1 Laborer Foreman	$29.40	$235.20	$48.70	$389.60	$29.85	$48.78
2 Laborers	27.40	438.40	45.40	726.40		
1 Equip. Oper. (light)	35.20	281.60	55.60	444.80		
1 Air Track Drill, 4"		728.80		801.70		
1 Air Compr., 600 C.F.M.		320.20		352.20		
2-50 Ft. Air Hoses, 3" Dia.		37.10		40.80		
1 Grout Pump		106.40		117.05	37.27	41.00
32 L.H., Daily Totals		$2147.70		$2872.55	$67.12	$89.78
Crew B-48	**Hr.**	**Daily**	**Hr.**	**Daily**	**Bare Costs**	**Incl. O&P**
1 Laborer Foreman (outside)	$29.40	$235.20	$48.70	$389.60	$31.05	$50.27
3 Laborers	27.40	657.60	45.40	1089.60		
1 Equip. Oper. (crane)	38.10	304.80	60.15	481.20		
1 Equip. Oper. Oiler	32.45	259.60	51.25	410.00		
1 Equip. Oper. (light)	35.20	281.60	55.60	444.80		
1 Centr. Water Pump, 6"		240.00		264.00		
1-20 Ft. Suction Hose, 6"		12.10		13.30		
1-50 Ft. Disch. Hose, 6"		7.50		8.25		
1 Drill Rig & Augers		3379.00		3716.90	64.98	71.47
56 L.H., Daily Totals		$5377.40		$6817.65	$96.03	$121.74
Crew B-49	**Hr.**	**Daily**	**Hr.**	**Daily**	**Bare Costs**	**Incl. O&P**
1 Laborer Foreman (outside)	$29.40	$235.20	$48.70	$389.60	$32.48	$52.81
3 Laborers	27.40	657.60	45.40	1089.60		
2 Equip. Oper. (crane)	38.10	609.60	60.15	962.40		
2 Equip. Oper. Oilers	32.45	519.20	51.25	820.00		
1 Equip. Oper. (light)	35.20	281.60	55.60	444.80		
2 Pile Drivers	34.70	555.20	58.80	940.80		
1 Hyd. Crane, 25 Ton		650.80		715.90		
1 Centr. Water Pump, 6"		240.00		264.00		
1-20 Ft. Suction Hose, 6"		12.10		13.30		
1-50 Ft. Disch. Hose, 6"		7.50		8.25		
1 Drill Rig & Augers		3379.00		3716.90	48.74	53.62
88 L.H., Daily Totals		$7147.80		$9365.55	$81.22	$106.43
Crew B-50	**Hr.**	**Daily**	**Hr.**	**Daily**	**Bare Costs**	**Incl. O&P**
2 Pile Driver Foremen	$36.70	$587.20	$62.15	$994.40	$33.75	$56.06
6 Pile Drivers	34.70	1665.60	58.80	2822.40		
2 Equip. Oper. (crane)	38.10	609.60	60.15	962.40		
1 Equip. Oper. Oiler	32.45	259.60	51.25	410.00		
3 Laborers	27.40	657.60	45.40	1089.60		
1 Crane, 40 Ton		934.20		1027.60		
60 L.F. Leads, 15K Ft. Lbs.		66.00		72.60		
1 Hammer, 15K Ft. Lbs.		355.80		391.40		
1 Air Compr., 600 C.F.M.		320.20		352.20		
2-50 Ft. Air Hoses, 3" Dia.		37.10		40.80		
1 Chain Saw, 36" Long		33.40		36.75	15.62	17.19
112 L.H., Daily Totals		$5526.30		$8200.15	$49.37	$73.25

Crews

Crew No.	Bare Costs		Incl. Subs O & P		Cost Per Labor-Hour	

Crew B-51	Hr.	Daily	Hr.	Daily	Bare Costs	Incl. O&P
1 Laborer Foreman (outside)	$29.40	$235.20	$48.70	$389.60	$27.77	$45.87
4 Laborers	27.40	876.80	45.40	1452.80		
1 Truck Driver (light)	27.60	220.80	44.90	359.20		
1 Light Truck, 1.5 Ton		128.80		141.70	2.68	2.95
48 L.H., Daily Totals		$1461.60		$2343.30	$30.45	$48.82

Crew B-52	Hr.	Daily	Hr.	Daily	Bare Costs	Incl. O&P
1 Carpenter Foreman	$37.55	$300.40	$62.20	$497.60	$32.54	$53.48
1 Carpenter	35.55	284.40	58.90	471.20		
3 Laborers	27.40	657.60	45.40	1089.60		
1 Cement Finisher	34.40	275.20	54.20	433.60		
.5 Rodman (reinf.)	39.50	158.00	67.80	271.20		
.5 Equip. Oper. (med.)	36.70	146.80	57.95	231.80		
.5 F.E. Ldr., T.M., 2.5 C.Y.		397.00		436.70	7.09	7.80
56 L.H., Daily Totals		$2219.40		$3431.70	$39.63	$61.28

Crew B-53	Hr.	Daily	Hr.	Daily	Bare Costs	Incl. O&P
1 Equip. Oper. (light)	$35.20	$281.60	$55.60	$444.80	$35.20	$55.60
1 Trencher, Chain, 12 H.P.		47.60		52.35	5.95	6.55
8 L.H., Daily Totals		$329.20		$497.15	$41.15	$62.15

Crew B-54	Hr.	Daily	Hr.	Daily	Bare Costs	Incl. O&P
1 Equip. Oper. (light)	$35.20	$281.60	$55.60	$444.80	$35.20	$55.60
1 Trencher, Chain, 40 H.P.		216.80		238.50	27.10	29.81
8 L.H., Daily Totals		$498.40		$683.30	$62.30	$85.41

Crew B-54A	Hr.	Daily	Hr.	Daily	Bare Costs	Incl. O&P
.17 Labor Foreman (outside)	$29.40	$39.98	$48.70	$66.23	$35.64	$56.61
1 Equipment Operator (med.)	36.70	293.60	57.95	463.60		
1 Wheel Trencher, 67 H.P.		826.80		909.50	88.33	97.17
9.36 L.H., Daily Totals		$1160.38		$1439.33	$123.97	$153.78

Crew B-54B	Hr.	Daily	Hr.	Daily	Bare Costs	Incl. O&P
.25 Labor Foreman (outside)	$29.40	$58.80	$48.70	$97.40	$35.24	$56.10
1 Equipment Operator (med.)	36.70	293.60	57.95	463.60		
1 Wheel Trencher, 150 H.P.		1578.00		1735.80	157.80	173.58
10 L.H., Daily Totals		$1930.40		$2296.80	$193.04	$229.68

Crew B-55	Hr.	Daily	Hr.	Daily	Bare Costs	Incl. O&P
2 Laborers	$27.40	$438.40	$45.40	$726.40	$27.47	$45.23
1 Truck Driver (light)	27.60	220.80	44.90	359.20		
1 Auger, 4″ to 36″ Dia		631.60		694.75		
1 Flatbed 3 Ton Truck		176.20		193.80	33.66	37.02
24 L.H., Daily Totals		$1467.00		$1974.15	$61.13	$82.25

Crew B-56	Hr.	Daily	Hr.	Daily	Bare Costs	Incl. O&P
1 Laborer	$27.40	$219.20	$45.40	$363.20	$31.30	$50.50
1 Equip. Oper. (light)	35.20	281.60	55.60	444.80		
1 Crawler Type Drill, 4″		728.80		801.70		
1 Air Compr., 600 C.F.M.		320.20		352.20		
1-50 Ft. Air Hose, 3″ Dia.		18.55		20.40	66.73	73.40
16 L.H., Daily Totals		$1568.35		$1982.30	$98.03	$123.90

Crew No.	Bare Costs		Incl. Subs O & P		Cost Per Labor-Hour	

Crew B-57	Hr.	Daily	Hr.	Daily	Bare Costs	Incl. O&P
1 Laborer Foreman (outside)	$29.40	$235.20	$48.70	$389.60	$31.66	$51.08
2 Laborers	27.40	438.40	45.40	726.40		
1 Equip. Oper. (crane)	38.10	304.80	60.15	481.20		
1 Equip. Oper. (light)	35.20	281.60	55.60	444.80		
1 Equip. Oper. Oiler	32.45	259.60	51.25	410.00		
1 Power Shovel, 1 C.Y.		880.00		968.00		
1 Clamshell Bucket, 1 C.Y.		43.40		47.75		
1 Centr. Water Pump, 6″		240.00		264.00		
1-20 Ft. Suction Hose, 6″		12.10		13.30		
20-50 Ft. Disch. Hoses, 6″		150.00		165.00	27.61	30.38
48 L.H., Daily Totals		$2845.10		$3910.05	$59.27	$81.46

Crew B-58	Hr.	Daily	Hr.	Daily	Bare Costs	Incl. O&P
2 Laborers	$27.40	$438.40	$45.40	$726.40	$30.00	$48.80
1 Equip. Oper. (light)	35.20	281.60	55.60	444.80		
1 Backhoe Loader, 48 H.P.		226.20		248.80		
1 Small Helicopter, w/pilot		2238.00		2461.80	102.68	112.94
24 L.H., Daily Totals		$3184.20		$3881.80	$132.68	$161.74

Crew B-59	Hr.	Daily	Hr.	Daily	Bare Costs	Incl. O&P
1 Truck Driver (heavy)	$28.35	$226.80	$46.15	$369.20	$28.35	$46.15
1 Truck, 30 Ton		208.20		229.00		
1 Water tank, 6000 Gal.		116.80		128.50	40.63	44.69
8 L.H., Daily Totals		$551.80		$726.70	$68.98	$90.84

Crew B-59A	Hr.	Daily	Hr.	Daily	Bare Costs	Incl. O&P
2 Laborers	$27.40	$438.40	$45.40	$726.40	$27.72	$45.65
1 Truck Driver (heavy)	28.35	226.80	46.15	369.20		
1 Water tank, 5K w/pum		116.80		128.50		
1 Truck, 30 Ton		208.20		229.00	13.54	14.90
24 L.H., Daily Totals		$990.20		$1453.10	$41.26	$60.55

Crew B-60	Hr.	Daily	Hr.	Daily	Bare Costs	Incl. O&P
1 Laborer Foreman (outside)	$29.40	$235.20	$48.70	$389.60	$32.16	$51.73
2 Laborers	27.40	438.40	45.40	726.40		
1 Equip. Oper. (crane)	38.10	304.80	60.15	481.20		
2 Equip. Oper. (light)	35.20	563.20	55.60	889.60		
1 Equip. Oper. Oiler	32.45	259.60	51.25	410.00		
1 Crawler Crane, 40 Ton		934.20		1027.60		
45 L.F. Leads, 15K Ft. Lbs.		49.50		54.45		
1 Backhoe Loader, 48 H.P.		226.20		248.80	21.65	23.81
56 L.H., Daily Totals		$3011.10		$4227.65	$53.81	$75.54

Crew B-61	Hr.	Daily	Hr.	Daily	Bare Costs	Incl. O&P
1 Laborer Foreman (outside)	$29.40	$235.20	$48.70	$389.60	$29.36	$48.10
3 Laborers	27.40	657.60	45.40	1089.60		
1 Equip. Oper. (light)	35.20	281.60	55.60	444.80		
1 Cement Mixer, 2 C.Y.		162.80		179.10		
1 Air Compr., 160 C.F.M.		97.60		107.35	6.51	7.16
40 L.H., Daily Totals		$1434.80		$2210.45	$35.87	$55.26

Crew B-62	Hr.	Daily	Hr.	Daily	Bare Costs	Incl. O&P
2 Laborers	$27.40	$438.40	$45.40	$726.40	$30.00	$48.80
1 Equip. Oper. (light)	35.20	281.60	55.60	444.80		
1 Loader, Skid Steer		154.00		169.40	6.42	7.06
24 L.H., Daily Totals		$874.00		$1340.60	$36.42	$55.86

Crews

Crew No.	Bare Costs		Incl. Subs O & P		Cost Per Labor-Hour	
Crew B-63	**Hr.**	**Daily**	**Hr.**	**Daily**	**Bare Costs**	**Incl. O&P**
4 Laborers	$27.40	$876.80	$45.40	$1452.80	$28.96	$47.44
1 Equip. Oper. (light)	35.20	281.60	55.60	444.80		
1 Loader, Skid Steer		154.00		169.40	3.85	4.24
40 L.H., Daily Totals		$1312.40		$2067.00	$32.81	$51.68

Crew B-64	Hr.	Daily	Hr.	Daily	Bare Costs	Incl. O&P
1 Laborer	$27.40	$219.20	$45.40	$363.20	$27.50	$45.15
1 Truck Driver (light)	27.60	220.80	44.90	359.20		
1 Power Mulcher (small)		113.80		125.20		
1 Light Truck, 1.5 Ton		128.80		141.70	15.16	16.68
16 L.H., Daily Totals		$682.60		$989.30	$42.66	$61.83

Crew B-65	Hr.	Daily	Hr.	Daily	Bare Costs	Incl. O&P
1 Laborer	$27.40	$219.20	$45.40	$363.20	$27.50	$45.15
1 Truck Driver (light)	27.60	220.80	44.90	359.20		
1 Power Mulcher (large)		185.20		203.70		
1 Light Truck, 1.5 Ton		128.80		141.70	19.63	21.59
16 L.H., Daily Totals		$754.00		$1067.80	$47.13	$66.74

Crew B-66	Hr.	Daily	Hr.	Daily	Bare Costs	Incl. O&P
1 Equip. Oper. (light)	$35.20	$281.60	$55.60	$444.80	$35.20	$55.60
1 Backhoe Ldr. w/Attchmt.		178.80		196.70	22.35	24.59
8 L.H., Daily Totals		$460.40		$641.50	$57.55	$80.19

Crew B-67	Hr.	Daily	Hr.	Daily	Bare Costs	Incl. O&P
1 Millwright	$37.10	$296.80	$58.50	$468.00	$36.15	$57.05
1 Equip. Oper. (light)	35.20	281.60	55.60	444.80		
1 Forklift		242.40		266.65	15.15	16.67
16 L.H., Daily Totals		$820.80		$1179.45	$51.30	$73.72

Crew B-68	Hr.	Daily	Hr.	Daily	Bare Costs	Incl. O&P
2 Millwrights	$37.10	$593.60	$58.50	$936.00	$36.47	$57.53
1 Equip. Oper. (light)	35.20	281.60	55.60	444.80		
1 Forklift		242.40		266.65	10.10	11.11
24 L.H., Daily Totals		$1117.60		$1647.45	$46.57	$68.64

Crew B-69	Hr.	Daily	Hr.	Daily	Bare Costs	Incl. O&P
1 Laborer Foreman (outside)	$29.40	$235.20	$48.70	$389.60	$30.36	$49.38
3 Laborers	27.40	657.60	45.40	1089.60		
1 Equip Oper. (crane)	38.10	304.80	60.15	481.20		
1 Equip Oper. Oiler	32.45	259.60	51.25	410.00		
1 Truck Crane, 80 Ton		1059.00		1164.90	22.06	24.27
48 L.H., Daily Totals		$2516.20		$3535.30	$52.42	$73.65

Crew B-69A	Hr.	Daily	Hr.	Daily	Bare Costs	Incl. O&P
1 Laborer Foreman	$29.40	$235.20	$48.70	$389.60	$30.45	$49.51
3 Laborers	27.40	657.60	45.40	1089.60		
1 Equip. Oper. (medium)	36.70	293.60	57.95	463.60		
1 Concrete Finisher	34.40	275.20	54.20	433.60		
1 Curb Paver		593.60		652.95	12.37	13.60
48 L.H., Daily Totals		$2055.20		$3029.35	$42.82	$63.11

Crew B-69B	Hr.	Daily	Hr.	Daily	Bare Costs	Incl. O&P
1 Laborer Foreman	$29.40	$235.20	$48.70	$389.60	$30.45	$49.51
3 Laborers	27.40	657.60	45.40	1089.60		
1 Equip. Oper. (medium)	36.70	293.60	57.95	463.60		
1 Cement Finisher	34.40	275.20	54.20	433.60		
1 Curb/Gutter Paver		738.00		811.80	15.38	16.91
48 L.H., Daily Totals		$2199.60		$3188.20	$45.83	$66.42

Crew No.	Bare Costs		Incl. Subs O & P		Cost Per Labor-Hour	
Crew B-70	**Hr.**	**Daily**	**Hr.**	**Daily**	**Bare Costs**	**Incl. O&P**
1 Laborer Foreman (outside)	$29.40	$235.20	$48.70	$389.60	$31.67	$51.25
3 Laborers	27.40	657.60	45.40	1089.60		
3 Equip. Oper. (med.)	36.70	880.80	57.95	1390.80		
1 Motor Grader, 30,000 Lb.		457.40		503.15		
1 Grader Attach., Ripper		73.20		80.50		
1 Road Sweeper, S.P.		493.20		542.50		
1 F.E. Loader, 1-3/4 C.Y.		241.40		265.55	22.59	24.85
56 L.H., Daily Totals		$3038.80		$4261.70	$54.26	$76.10

Crew B-71	Hr.	Daily	Hr.	Daily	Bare Costs	Incl. O&P
1 Laborer Foreman (outside)	$29.40	$235.20	$48.70	$389.60	$31.67	$51.25
3 Laborers	27.40	657.60	45.40	1089.60		
3 Equip. Oper. (med.)	36.70	880.80	57.95	1390.80		
1 Pvmt. Profiler, 750 H.P.		4659.00		5124.90		
1 Road Sweeper, S.P.		493.20		542.50		
1 F.E. Loader, 1-3/4 C.Y.		241.40		265.55	96.31	105.95
56 L.H., Daily Totals		$7167.20		$8802.95	$127.98	$157.20

Crew B-72	Hr.	Daily	Hr.	Daily	Bare Costs	Incl. O&P
1 Laborer Foreman (outside)	$29.40	$235.20	$48.70	$389.60	$32.30	$52.09
3 Laborers	27.40	657.60	45.40	1089.60		
4 Equip. Oper. (med.)	36.70	1174.40	57.95	1854.40		
1 Pvmt. Profiler, 750 H.P.		4659.00		5124.90		
1 Hammermill, 250 H.P.		1423.00		1565.30		
1 Windrow Loader		823.00		905.30		
1 Mix Paver 165 H.P.		1729.00		1901.90		
1 Roller, Pneu. Tire, 12 T.		259.20		285.10	138.96	152.85
64 L.H., Daily Totals		$10960.40		$13116.10	$171.26	$204.94

Crew B-73	Hr.	Daily	Hr.	Daily	Bare Costs	Incl. O&P
1 Laborer Foreman (outside)	$29.40	$235.20	$48.70	$389.60	$33.46	$53.66
2 Laborers	27.40	438.40	45.40	726.40		
5 Equip. Oper. (med.)	36.70	1468.00	57.95	2318.00		
1 Road Mixer, 310 H.P.		1698.00		1867.80		
1 Roller, Tandem, 12 Ton		197.00		216.70		
1 Hammermill, 250 H.P.		1423.00		1565.30		
1 Motor Grader, 30,000 Lb.		457.40		503.15		
.5 F.E. Loader, 1-3/4 C.Y.		120.70		132.75		
.5 Truck, 30 Ton		104.10		114.50		
.5 Water Tank 5000 Gal.		58.40		64.25	63.42	69.76
64 L.H., Daily Totals		$6200.20		$7898.45	$96.88	$123.42

Crew B-74	Hr.	Daily	Hr.	Daily	Bare Costs	Incl. O&P
1 Laborer Foreman (outside)	$29.40	$235.20	$48.70	$389.60	$32.54	$52.28
1 Laborer	27.40	219.20	45.40	363.20		
4 Equip. Oper. (med.)	36.70	1174.40	57.95	1854.40		
2 Truck Drivers (heavy)	28.35	453.60	46.15	738.40		
1 Motor Grader, 30,000 Lb.		457.40		503.15		
1 Grader Attach., Ripper		73.20		80.50		
2 Stabilizers, 310 H.P.		2206.00		2426.60		
1 Flatbed Truck, 3 Ton		176.20		193.80		
1 Chem. Spreader, Towed		49.00		53.90		
1 Vibr. Roller, 29,000 Lb.		522.00		574.20		
1 Water Tank 5000 Gal.		116.80		128.50		
1 Truck, 30 Ton		208.20		229.00	59.51	65.46
64 L.H., Daily Totals		$5891.20		$7535.25	$92.05	$117.74

Crew No.	Bare Costs		Incl. Subs O & P		Cost Per Labor-Hour	
Crew B-75	**Hr.**	**Daily**	**Hr.**	**Daily**	**Bare Costs**	**Incl. O&P**
1 Laborer Foreman (outside)	$29.40	$235.20	$48.70	$389.60	$33.14	$53.15
1 Laborer	27.40	219.20	45.40	363.20		
4 Equip. Oper. (med.)	36.70	1174.40	57.95	1854.40		
1 Truck Driver (heavy)	28.35	226.80	46.15	369.20		
1 Motor Grader, 30,000 Lb.		457.40		503.15		
1 Grader Attach., Ripper		73.20		80.50		
2 Stabilizers, 310 H.P.		2206.00		2426.60		
1 Dist. Truck, 3000 Gal.		242.40		266.65		
1 Vibr. Roller, 29,000 Lb.		522.00		574.20	62.52	68.77
56 L.H., Daily Totals		$5356.60		$6827.50	$95.66	$121.92

Crew B-76	Hr.	Daily	Hr.	Daily	Bare Costs	Incl. O&P
1 Dock Builder Foreman	$36.70	$293.60	$62.15	$497.20	$35.43	$58.63
5 Dock Builders	34.70	1388.00	58.80	2352.00		
2 Equip. Oper. (crane)	38.10	609.60	60.15	962.40		
1 Equip. Oper. Oiler	32.45	259.60	51.25	410.00		
1 Crawler Crane, 50 Ton		1265.00		1391.50		
1 Barge, 400 Ton		280.00		308.00		
1 Hammer, 15K Ft. Lbs.		355.80		391.40		
60 L.F. Leads, 15K Ft. Lbs.		66.00		72.60		
1 Air Compr., 600 C.F.M.		320.20		352.20		
2-50 Ft. Air Hoses, 3" Dia.		37.10		40.80	32.32	35.55
72 L.H., Daily Totals		$4874.90		$6778.10	$67.75	$94.18

Crew B-76A	Hr.	Daily	Hr.	Daily	Bare Costs	Incl. O&P
1 Laborer Foreman	$29.40	$235.20	$48.70	$389.60	$29.62	$48.39
5 Laborers	27.40	1096.00	45.40	1816.00		
1 Equip. Oper. (crane)	38.10	304.80	60.15	481.20		
1 Equip. Oper. Oiler	32.45	259.60	51.25	410.00		
1 Crawler Crane, 50 Ton		1265.00		1391.50		
1 Barge, 400 Ton		280.00		308.00	24.14	26.55
64 L.H., Daily Totals		$3440.60		$4796.30	$53.76	$74.94

Crew B-77	Hr.	Daily	Hr.	Daily	Bare Costs	Incl. O&P
1 Laborer Foreman	$29.40	$235.20	$48.70	$389.60	$27.84	$45.96
3 Laborers	27.40	657.60	45.40	1089.60		
1 Truck Driver (light)	27.60	220.80	44.90	359.20		
1 Crack Cleaner, 25 H.P.		45.80		50.40		
1 Crack Filler, Trailer Mtd.		157.40		173.15		
1 Flatbed Truck, 3 Ton		176.20		193.80	9.49	10.43
40 L.H., Daily Totals		$1493.00		$2255.75	$37.33	$56.39

Crew B-78	Hr.	Daily	Hr.	Daily	Bare Costs	Incl. O&P
1 Laborer Foreman	$29.40	$235.20	$48.70	$389.60	$27.77	$45.87
4 Laborers	27.40	876.80	45.40	1452.80		
1 Truck Driver (light)	27.60	220.80	44.90	359.20		
1 Paint Striper, S.P.		134.60		148.05		
1 Flatbed Truck, 3 Ton		176.20		193.80		
1 Pickup Truck, 3/4 Ton		78.00		85.80	8.10	8.91
48 L.H., Daily Totals		$1721.60		$2629.25	$35.87	$54.78

Crew B-78A	Hr.	Daily	Hr.	Daily	Bare Costs	Incl. O&P
1 Equip. Oper. (light)	$35.20	$281.60	$55.60	$444.80	$35.20	$55.60
1 Line Remov. (metal balls) 115 HP		795.80		875.40	99.48	109.42
8 L.H., Daily Totals		$1077.40		$1320.20	$134.68	$165.02

Crew No.	Bare Costs		Incl. Subs O & P		Cost Per Labor-Hour	
Crew B-78B	**Hr.**	**Daily**	**Hr.**	**Daily**	**Bare Costs**	**Incl. O&P**
2 Laborers	$27.40	$438.40	$45.40	$726.40	$28.27	$46.53
.25 Equip. Oper. (light)	35.20	70.40	55.60	111.20		
1 Pickup Truck, 3/4 Ton		78.00		85.80		
1 Line Remov. (walk behind) 11 HP		44.80		49.30		
.25 Road Sweeper, SP		123.30		135.65	13.67	15.04
18 L.H., Daily Totals		$754.90		$1108.35	$41.94	$61.57

Crew B-79	Hr.	Daily	Hr.	Daily	Bare Costs	Incl. O&P
1 Laborer Foreman	$29.40	$235.20	$48.70	$389.60	$27.84	$45.96
3 Laborers	27.40	657.60	45.40	1089.60		
1 Truck Driver (light)	27.60	220.80	44.90	359.20		
1 Thermo. Striper, T.M.		555.20		610.70		
1 Flatbed Truck, 3 Ton		176.20		193.80		
2 Pickup Truck, 3/4 Ton		156.00		171.60	22.19	24.40
40 L.H., Daily Totals		$2001.00		$2814.50	$50.03	$70.36

Crew B-79A	Hr.	Daily	Hr.	Daily	Bare Costs	Incl. O&P
1.5 Equip. Oper. (light)	$35.20	$422.40	$55.60	$667.20	$35.20	$55.60
.5 Line Remov. (grinder) 115 HP		332.70		365.95		
1 Line Remov. (metal balls) 115 HP		795.80		875.40	94.04	103.45
12 L.H., Daily Totals		$1550.90		$1908.55	$129.24	$159.05

Crew B-80	Hr.	Daily	Hr.	Daily	Bare Costs	Incl. O&P
1 Laborer Foreman	$29.40	$235.20	$48.70	$389.60	$29.90	$48.65
1 Laborer	27.40	219.20	45.40	363.20		
1 Truck Driver (light)	27.60	220.80	44.90	359.20		
1 Equip. Oper. (light)	35.20	281.60	55.60	444.80		
1 Flatbed Truck, 3 Ton		176.20		193.80		
1 Fence Post Auger, T.M.		356.20		391.80	16.64	18.30
32 L.H., Daily Totals		$1489.20		$2142.40	$46.54	$66.95

Crew B-80A	Hr.	Daily	Hr.	Daily	Bare Costs	Incl. O&P
3 Laborers	$27.40	$657.60	$45.40	$1089.60	$27.40	$45.40
1 Flatbed Truck, 3 Ton		176.20		193.80	7.34	8.08
24 L.H., Daily Totals		$833.80		$1283.40	$34.74	$53.48

Crew B-80B	Hr.	Daily	Hr.	Daily	Bare Costs	Incl. O&P
3 Laborers	$27.40	$657.60	$45.40	$1089.60	$29.35	$47.95
1 Equip. Oper. (light)	35.20	281.60	55.60	444.80		
1 Crane, Flatbed Mnt.		235.20		258.70	7.35	8.09
32 L.H., Daily Totals		$1174.40		$1793.10	$36.70	$56.04

Crew B-80C	Hr.	Daily	Hr.	Daily	Bare Costs	Incl. O&P
2 Laborers	$27.40	$438.40	$45.40	$726.40	$27.47	$45.23
1 Truck Driver (light)	27.60	220.80	44.90	359.20		
1 Light Truck, 1.5 Ton		128.80		141.70		
1 Manual fence post auger, gas		5.40		5.95	5.59	6.15
24 L.H., Daily Totals		$793.40		$1233.25	$33.06	$51.38

Crew B-81	Hr.	Daily	Hr.	Daily	Bare Costs	Incl. O&P
1 Laborer	$27.40	$219.20	$45.40	$363.20	$30.82	$49.83
1 Equip. Oper. (med.)	36.70	293.60	57.95	463.60		
1 Truck Driver (heavy)	28.35	226.80	46.15	369.20		
1 Hydromulcher, T.M.		197.20		216.90		
1 Tractor Truck, 4x2		208.20		229.00	16.89	18.58
24 L.H., Daily Totals		$1145.00		$1641.90	$47.71	$68.41

Crew No.	Bare Costs Hr.	Bare Costs Daily	Incl. Subs O & P Hr.	Incl. Subs O & P Daily	Cost Per Labor-Hour Bare Costs	Cost Per Labor-Hour Incl. O&P
Crew B-82	**Hr.**	**Daily**	**Hr.**	**Daily**	**Bare Costs**	**Incl. O&P**
1 Laborer	$27.40	$219.20	$45.40	$363.20	$31.30	$50.50
1 Equip. Oper. (light)	35.20	281.60	55.60	444.80		
1 Horiz. Borer, 6 H.P.		66.00		72.60	4.13	4.54
16 L.H., Daily Totals		$566.80		$880.60	$35.43	$55.04

Crew B-83	Hr.	Daily	Hr.	Daily	Bare Costs	Incl. O&P
1 Tugboat Captain	$36.70	$293.60	$57.95	$463.60	$32.05	$51.68
1 Tugboat Hand	27.40	219.20	45.40	363.20		
1 Tugboat, 250 H.P.		473.00		520.30	29.56	32.52
16 L.H., Daily Totals		$985.80		$1347.10	$61.61	$84.20

Crew B-84	Hr.	Daily	Hr.	Daily	Bare Costs	Incl. O&P
1 Equip. Oper. (med.)	$36.70	$293.60	$57.95	$463.60	$36.70	$57.95
1 Rotary Mower/Tractor		231.00		254.10	28.88	31.76
8 L.H., Daily Totals		$524.60		$717.70	$65.58	$89.71

Crew B-85	Hr.	Daily	Hr.	Daily	Bare Costs	Incl. O&P
3 Laborers	$27.40	$657.60	$45.40	$1089.60	$29.45	$48.06
1 Equip. Oper. (med.)	36.70	293.60	57.95	463.60		
1 Truck Driver (heavy)	28.35	226.80	46.15	369.20		
1 Aerial Lift Truck, 80'		515.80		567.40		
1 Brush Chipper, 130 H.P.		173.00		190.30		
1 Pruning Saw, Rotary		7.15		7.85	17.40	19.14
40 L.H., Daily Totals		$1873.95		$2687.95	$46.85	$67.20

Crew B-86	Hr.	Daily	Hr.	Daily	Bare Costs	Incl. O&P
1 Equip. Oper. (med.)	$36.70	$293.60	$57.95	$463.60	$36.70	$57.95
1 Stump Chipper, S.P.		63.50		69.85	7.94	8.73
8 L.H., Daily Totals		$357.10		$533.45	$44.64	$66.68

Crew B-86A	Hr.	Daily	Hr.	Daily	Bare Costs	Incl. O&P
1 Equip. Oper. (medium)	$36.70	$293.60	$57.95	$463.60	$36.70	$57.95
1 Grader, 30,000 Lbs.		457.40		503.15	57.18	62.89
8 L.H., Daily Totals		$751.00		$966.75	$93.88	$120.84

Crew B-86B	Hr.	Daily	Hr.	Daily	Bare Costs	Incl. O&P
1 Equip. Oper. (medium)	$36.70	$293.60	$57.95	$463.60	$36.70	$57.95
1 Dozer, 200 H.P.		920.80		1012.90	115.10	126.61
8 L.H., Daily Totals		$1214.40		$1476.50	$151.80	$184.56

Crew B-87	Hr.	Daily	Hr.	Daily	Bare Costs	Incl. O&P
1 Laborer	$27.40	$219.20	$45.40	$363.20	$34.84	$55.44
4 Equip. Oper. (med.)	36.70	1174.40	57.95	1854.40		
2 Feller Bunchers, 50 H.P.		983.60		1081.95		
1 Log Chipper, 22" Tree		1000.00		1100.00		
1 Dozer, 105 H.P.		462.80		509.10		
1 Chainsaw, Gas, 36" Long		33.40		36.75	62.00	68.19
40 L.H., Daily Totals		$3873.40		$4945.40	$96.84	$123.63

Crew B-88	Hr.	Daily	Hr.	Daily	Bare Costs	Incl. O&P
1 Laborer	$27.40	$219.20	$45.40	$363.20	$35.37	$56.16
6 Equip. Oper. (med.)	36.70	1761.60	57.95	2781.60		
2 Feller Bunchers, 50 H.P.		983.60		1081.95		
1 Log Chipper, 22" Tree		1000.00		1100.00		
2 Log Skidders, 50 H.P.		1575.20		1732.70		
1 Dozer, 105 H.P.		462.80		509.10		
1 Chainsaw, Gas, 36" Long		33.40		36.75	72.41	79.65
56 L.H., Daily Totals		$6035.80		$7605.30	$107.78	$135.81

Crew B-89	Hr.	Daily	Hr.	Daily	Bare Costs	Incl. O&P
1 Equip. Oper. (light)	$35.20	$281.60	$55.60	$444.80	$31.40	$50.25
1 Truck Driver (light)	27.60	220.80	44.90	359.20		
1 Truck, Stake Body, 3 Ton		176.20		193.80		
1 Concrete Saw		107.00		117.70		
1 Water Tank, 65 Gal.		13.80		15.20	18.56	20.42
16 L.H., Daily Totals		$799.40		$1130.70	$49.96	$70.67

Crew B-89A	Hr.	Daily	Hr.	Daily	Bare Costs	Incl. O&P
1 Skilled Worker	$36.50	$292.00	$59.70	$477.60	$31.95	$52.55
1 Laborer	27.40	219.20	45.40	363.20		
1 Core Drill (large)		115.45		127.00	7.22	7.94
16 L.H., Daily Totals		$626.65		$967.80	$39.17	$60.49

Crew B-89B	Hr.	Daily	Hr.	Daily	Bare Costs	Incl. O&P
1 Equip. Oper. (light)	$35.20	$281.60	$55.60	$444.80	$31.40	$50.25
1 Truck Driver, Light	27.60	220.80	44.90	359.20		
1 Wall Saw, Hydraulic, 10 H.P.		80.95		89.05		
1 Generator, Diesel, 100 KW		193.20		212.50		
1 Water Tank, 65 Gal.		13.80		15.20		
1 Flatbed Truck, 3 Ton		176.20		193.80	29.01	31.91
16 L.H., Daily Totals		$966.55		$1314.55	$60.41	$82.16

Crew B-90	Hr.	Daily	Hr.	Daily	Bare Costs	Incl. O&P
1 Laborer Foreman (outside)	$29.40	$235.20	$48.70	$389.60	$29.84	$48.55
3 Laborers	27.40	657.60	45.40	1089.60		
2 Equip. Oper. (light)	35.20	563.20	55.60	889.60		
2 Truck Drivers (heavy)	28.35	453.60	46.15	738.40		
1 Road Mixer, 310 H.P.		1698.00		1867.80		
1 Dist. Truck, 2000 Gal.		209.00		229.90	29.80	32.78
64 L.H., Daily Totals		$3816.60		$5204.90	$59.64	$81.33

Crew B-90A	Hr.	Daily	Hr.	Daily	Bare Costs	Incl. O&P
1 Laborer Foreman	$29.40	$235.20	$48.70	$389.60	$33.00	$53.04
2 Laborers	27.40	438.40	45.40	726.40		
4 Equip. Oper. (medium)	36.70	1174.40	57.95	1854.40		
2 Graders, 30,000 Lbs.		914.80		1006.30		
1 Roller, Steel Wheel		197.00		216.70		
1 Roller, Pneumatic Wheel		259.20		285.10	24.48	26.93
56 L.H., Daily Totals		$3219.00		$4478.50	$57.48	$79.97

Crew B-90B	Hr.	Daily	Hr.	Daily	Bare Costs	Incl. O&P
1 Laborer Foreman	$29.40	$235.20	$48.70	$389.60	$32.38	$52.23
2 Laborers	27.40	438.40	45.40	726.40		
3 Equip. Oper. (medium)	36.70	880.80	57.95	1390.80		
1 Roller, Steel Wheel		197.00		216.70		
1 Roller, Pneumatic Wheel		259.20		285.10		
1 Road Mixer, 310 H.P.		1698.00		1867.80	44.88	49.37
48 L.H., Daily Totals		$3708.60		$4876.40	$77.26	$101.60

Crew B-91	Hr.	Daily	Hr.	Daily	Bare Costs	Incl. O&P
1 Laborer Foreman (outside)	$29.40	$235.20	$48.70	$389.60	$32.42	$52.18
2 Laborers	27.40	438.40	45.40	726.40		
4 Equip. Oper. (med.)	36.70	1174.40	57.95	1854.40		
1 Truck Driver (heavy)	28.35	226.80	46.15	369.20		
1 Dist. Truck, 3000 Gal.		242.40		266.65		
1 Aggreg. Spreader, S.P.		777.40		855.15		
1 Roller, Pneu. Tire, 12 Ton		259.20		285.10		
1 Roller, Steel, 10 Ton		197.00		216.70	23.06	25.37
64 L.H., Daily Totals		$3550.80		$4963.20	$55.48	$77.55

Crew No.	Bare Costs		Incl. Subs O & P		Cost Per Labor-Hour	

Crew B-92	Hr.	Daily	Hr.	Daily	Bare Costs	Incl. O&P
1 Laborer Foreman (outside)	$29.40	$235.20	$48.70	$389.60	$27.90	$46.23
3 Laborers	27.40	657.60	45.40	1089.60		
1 Crack Cleaner, 25 H.P.		45.80		50.40		
1 Air Compressor		59.00		64.90		
1 Tar Kettle, T.M.		45.75		50.35		
1 Flatbed Truck, 3 Ton		176.20		193.80	10.21	11.23
32 L.H., Daily Totals		$1219.55		$1838.65	$38.11	$57.46

Crew B-93	Hr.	Daily	Hr.	Daily	Bare Costs	Incl. O&P
1 Equip. Oper. (med.)	$36.70	$293.60	$57.95	$463.60	$36.70	$57.95
1 Feller Buncher, 50 H.P.		491.80		541.00	61.48	67.62
8 L.H., Daily Totals		$785.40		$1004.60	$98.18	$125.57

Crew B-94A	Hr.	Daily	Hr.	Daily	Bare Costs	Incl. O&P
1 Laborer	$27.40	$219.20	$45.40	$363.20	$27.40	$45.40
1 Diaph. Water Pump, 2″		53.20		58.50		
1-20 Ft. Suction Hose, 2″		2.15		2.35		
2-50 Ft. Disch. Hoses, 2″		2.60		2.85	7.24	7.97
8 L.H., Daily Totals		$277.15		$426.90	$34.64	$53.37

Crew B-94B	Hr.	Daily	Hr.	Daily	Bare Costs	Incl. O&P
1 Laborer	$27.40	$219.20	$45.40	$363.20	$27.40	$45.40
1 Diaph. Water Pump, 4″		73.20		80.50		
1-20 Ft. Suction Hose, 4″		4.05		4.45		
2-50 Ft. Disch. Hoses, 4″		5.90		6.50	10.39	11.43
8 L.H., Daily Totals		$302.35		$454.65	$37.79	$56.83

Crew B-94C	Hr.	Daily	Hr.	Daily	Bare Costs	Incl. O&P
1 Laborer	$27.40	$219.20	$45.40	$363.20	$27.40	$45.40
1 Centr. Water Pump, 3″		56.60		62.25		
1-20 Ft. Suction Hose, 3″		3.65		4.00		
2-50 Ft. Disch. Hoses, 3″		4.30		4.75	8.07	8.88
8 L.H., Daily Totals		$283.75		$434.20	$35.47	$54.28

Crew B-94D	Hr.	Daily	Hr.	Daily	Bare Costs	Incl. O&P
1 Laborer	$27.40	$219.20	$45.40	$363.20	$27.40	$45.40
1 Centr. Water Pump, 6″		240.00		264.00		
1-20 Ft. Suction Hose, 6″		12.10		13.30		
2-50 Ft. Disch. Hoses, 6″		15.00		16.50	33.39	36.73
8 L.H., Daily Totals		$486.30		$657.00	$60.79	$82.13

Crew B-95A	Hr.	Daily	Hr.	Daily	Bare Costs	Incl. O&P
1 Equip. Oper. (crane)	$38.10	$304.80	$60.15	$481.20	$32.75	$52.78
1 Laborer	27.40	219.20	45.40	363.20		
1 Hyd. Excavator, 5/8 C.Y.		435.40		478.95	27.21	29.93
16 L.H., Daily Totals		$959.40		$1323.35	$59.96	$82.71

Crew B-95B	Hr.	Daily	Hr.	Daily	Bare Costs	Incl. O&P
1 Equip. Oper. (crane)	$38.10	$304.80	$60.15	$481.20	$32.75	$52.78
1 Laborer	27.40	219.20	45.40	363.20		
1 Hyd. Excavator, 1.5 C.Y.		722.80		795.10	45.18	49.69
16 L.H., Daily Totals		$1246.80		$1639.50	$77.93	$102.47

Crew B-95C	Hr.	Daily	Hr.	Daily	Bare Costs	Incl. O&P
1 Equip. Oper. (crane)	$38.10	$304.80	$60.15	$481.20	$32.75	$52.78
1 Laborer	27.40	219.20	45.40	363.20		
1 Hyd. Excavator, 2.5 C.Y.		1218.00		1339.80	76.13	83.74
16 L.H., Daily Totals		$1742.00		$2184.20	$108.88	$136.52

Crew No.	Bare Costs		Incl. Subs O & P		Cost Per Labor-Hour	

Crew C-1	Hr.	Daily	Hr.	Daily	Bare Costs	Incl. O&P
3 Carpenters	$35.55	$853.20	$58.90	$1413.60	$33.51	$55.53
1 Laborer	27.40	219.20	45.40	363.20		
32 L.H., Daily Totals		$1072.40		$1776.80	$33.51	$55.53

Crew C-2	Hr.	Daily	Hr.	Daily	Bare Costs	Incl. O&P
1 Carpenter Foreman (out)	$37.55	$300.40	$62.20	$497.60	$34.53	$57.20
4 Carpenters	35.55	1137.60	58.90	1884.80		
1 Laborer	27.40	219.20	45.40	363.20		
48 L.H., Daily Totals		$1657.20		$2745.60	$34.53	$57.20

Crew C-2A	Hr.	Daily	Hr.	Daily	Bare Costs	Incl. O&P
1 Carpenter Foreman (out)	$37.55	$300.40	$62.20	$497.60	$34.33	$56.42
3 Carpenters	35.55	853.20	58.90	1413.60		
1 Cement Finisher	34.40	275.20	54.20	433.60		
1 Laborer	27.40	219.20	45.40	363.20		
48 L.H., Daily Totals		$1648.00		$2708.00	$34.33	$56.42

Crew C-3	Hr.	Daily	Hr.	Daily	Bare Costs	Incl. O&P
1 Rodman Foreman	$41.50	$332.00	$71.25	$570.00	$36.19	$61.11
4 Rodmen (reinf.)	39.50	1264.00	67.80	2169.60		
1 Equip. Oper. (light)	35.20	281.60	55.60	444.80		
2 Laborers	27.40	438.40	45.40	726.40		
3 Stressing Equipment		31.20		34.30		
.5 Grouting Equipment		77.55		85.30	1.70	1.87
64 L.H., Daily Totals		$2424.75		$4030.40	$37.89	$62.98

Crew C-4	Hr.	Daily	Hr.	Daily	Bare Costs	Incl. O&P
1 Rodman Foreman	$41.50	$332.00	$71.25	$570.00	$40.00	$68.66
3 Rodmen (reinf.)	39.50	948.00	67.80	1627.20		
3 Stressing Equipment		31.20		34.30	.98	1.07
32 L.H., Daily Totals		$1311.20		$2231.50	$40.98	$69.73

Crew C-5	Hr.	Daily	Hr.	Daily	Bare Costs	Incl. O&P
1 Rodman Foreman	$41.50	$332.00	$71.25	$570.00	$38.58	$64.84
4 Rodmen (reinf.)	39.50	1264.00	67.80	2169.60		
1 Equip. Oper. (crane)	38.10	304.80	60.15	481.20		
1 Equip. Oper. Oiler	32.45	259.60	51.25	410.00		
1 Hyd. Crane, 25 Ton		650.80		715.90	11.62	12.78
56 L.H., Daily Totals		$2811.20		$4346.70	$50.20	$77.62

Crew C-6	Hr.	Daily	Hr.	Daily	Bare Costs	Incl. O&P
1 Laborer Foreman (outside)	$29.40	$235.20	$48.70	$389.60	$28.90	$47.42
4 Laborers	27.40	876.80	45.40	1452.80		
1 Cement Finisher	34.40	275.20	54.20	433.60		
2 Gas Engine Vibrators		42.80		47.10	.89	.98
48 L.H., Daily Totals		$1430.00		$2323.10	$29.79	$48.40

Crew C-7	Hr.	Daily	Hr.	Daily	Bare Costs	Incl. O&P
1 Laborer Foreman (outside)	$29.40	$235.20	$48.70	$389.60	$29.99	$48.79
5 Laborers	27.40	1096.00	45.40	1816.00		
1 Cement Finisher	34.40	275.20	54.20	433.60		
1 Equip. Oper. (med.)	36.70	293.60	57.95	463.60		
1 Equip. Oper. Oiler	32.45	259.60	51.25	410.00		
2 Gas Engine Vibrators		42.80		47.10		
1 Concrete Bucket, 1 C.Y.		16.20		17.80		
1 Hyd. Crane, 55 Ton		951.60		1046.75	14.04	15.44
72 L.H., Daily Totals		$3170.20		$4624.45	$44.03	$64.23

Crews

Crew No.	Bare Costs		Incl. Subs O & P		Cost Per Labor-Hour	
Crew C-8	**Hr.**	**Daily**	**Hr.**	**Daily**	**Bare Costs**	**Incl. O&P**
1 Laborer Foreman (outside)	$29.40	$235.20	$48.70	$389.60	$31.01	$50.18
3 Laborers	27.40	657.60	45.40	1089.60		
2 Cement Finishers	34.40	550.40	54.20	867.20		
1 Equip. Oper. (med.)	36.70	293.60	57.95	463.60		
1 Concrete Pump (small)		700.60		770.65	12.51	13.76
56 L.H., Daily Totals		$2437.40		$3580.65	$43.52	$63.94

Crew C-8A	**Hr.**	**Daily**	**Hr.**	**Daily**	**Bare Costs**	**Incl. O&P**
1 Laborer Foreman (outside)	$29.40	$235.20	$48.70	$389.60	$30.07	$48.88
3 Laborers	27.40	657.60	45.40	1089.60		
2 Cement Finishers	34.40	550.40	54.20	867.20		
48 L.H., Daily Totals		$1443.20		$2346.40	$30.07	$48.88

Crew C-8B	**Hr.**	**Daily**	**Hr.**	**Daily**	**Bare Costs**	**Incl. O&P**
1 Laborer Foreman (outside)	$29.40	$235.20	$48.70	$389.60	$29.66	$48.57
3 Laborers	27.40	657.60	45.40	1089.60		
1 Equipment Operator	36.70	293.60	57.95	463.60		
1 Vibrating Screed		57.95		63.75		
1 Vibratory Roller		522.00		574.20		
1 Dozer, 200 H.P.		920.80		1012.90	37.52	41.27
40 L.H., Daily Totals		$2687.15		$3593.65	$67.18	$89.84

Crew C-8C	**Hr.**	**Daily**	**Hr.**	**Daily**	**Bare Costs**	**Incl. O&P**
1 Laborer Foreman (outside)	$29.40	$235.20	$48.70	$389.60	$30.45	$49.51
3 Laborers	27.40	657.60	45.40	1089.60		
1 Cement Finisher	34.40	275.20	54.20	433.60		
1 Equipment Operator (med.)	36.70	293.60	57.95	463.60		
1 Shotcrete Rig, 12 CY/hr		234.40		257.85	4.88	5.37
48 L.H., Daily Totals		$1696.00		$2634.25	$35.33	$54.88

Crew C-8D	**Hr.**	**Daily**	**Hr.**	**Daily**	**Bare Costs**	**Incl. O&P**
1 Laborer Foreman (outside)	$29.40	$235.20	$48.70	$389.60	$31.60	$50.98
1 Laborer	27.40	219.20	45.40	363.20		
1 Cement Finisher	34.40	275.20	54.20	433.60		
1 Equipment Operator (light)	35.20	281.60	55.60	444.80		
1 Compressor, 250 CFM		119.80		131.80		
2 Hoses, 1", 50'		7.80		8.60	3.99	4.39
32 L.H., Daily Totals		$1138.80		$1771.60	$35.59	$55.37

Crew C-8E	**Hr.**	**Daily**	**Hr.**	**Daily**	**Bare Costs**	**Incl. O&P**
1 Laborer Foreman (outside)	$29.40	$235.20	$48.70	$389.60	$31.60	$50.98
1 Laborer	27.40	219.20	45.40	363.20		
1 Cement Finisher	34.40	275.20	54.20	433.60		
1 Equipment Operator (light)	35.20	281.60	55.60	444.80		
1 Compressor, 250 CFM		119.80		131.80		
2 Hoses, 1", 50'		7.80		8.60		
1 Concrete Pump (small)		700.60		770.65	25.88	28.47
32 L.H., Daily Totals		$1839.40		$2542.25	$57.48	$79.45

Crew C-10	**Hr.**	**Daily**	**Hr.**	**Daily**	**Bare Costs**	**Incl. O&P**
1 Laborer	$27.40	$219.20	$45.40	$363.20	$32.07	$51.27
2 Cement Finishers	34.40	550.40	54.20	867.20		
24 L.H., Daily Totals		$769.60		$1230.40	$32.07	$51.27

Crew C-10B	**Hr.**	**Daily**	**Hr.**	**Daily**	**Bare Costs**	**Incl. O&P**
3 Laborers	$27.40	$657.60	$45.40	$1089.60	$30.20	$48.92
2 Cement Finishers	34.40	550.40	54.20	867.20		
1 Concrete mixer, 10 CF		127.40		140.15		
2 Concrete finishers, 48" dia		48.40		53.25	4.40	4.83
40 L.H., Daily Totals		$1383.80		$2150.20	$34.60	$53.75

Crew No.	Bare Costs		Incl. Subs O & P		Cost Per Labor-Hour	
Crew C-11	**Hr.**	**Daily**	**Hr.**	**Daily**	**Bare Costs**	**Incl. O&P**
1 Struc. Steel Foreman	$41.95	$335.60	$78.45	$627.60	$39.13	$70.89
6 Struc. Steel Workers	39.95	1917.60	74.70	3585.60		
1 Equip. Oper. (crane)	38.10	304.80	60.15	481.20		
1 Equip. Oper. Oiler	32.45	259.60	51.25	410.00		
1 Truck Crane, 150 Ton		1539.00		1692.90	21.38	23.51
72 L.H., Daily Totals		$4356.60		$6797.30	$60.51	$94.40

Crew C-12	**Hr.**	**Daily**	**Hr.**	**Daily**	**Bare Costs**	**Incl. O&P**
1 Carpenter Foreman (out)	$37.55	$300.40	$62.20	$497.60	$34.95	$57.41
3 Carpenters	35.55	853.20	58.90	1413.60		
1 Laborer	27.40	219.20	45.40	363.20		
1 Equip. Oper. (crane)	38.10	304.80	60.15	481.20		
1 Hyd. Crane, 12 Ton		636.60		700.25	13.26	14.59
48 L.H., Daily Totals		$2314.20		$3455.85	$48.21	$72.00

Crew C-13	**Hr.**	**Daily**	**Hr.**	**Daily**	**Bare Costs**	**Incl. O&P**
1 Struc. Steel Worker	$39.95	$319.60	$74.70	$597.60	$38.48	$69.43
1 Welder	39.95	319.60	74.70	597.60		
1 Carpenter	35.55	284.40	58.90	471.20		
1 Gas Welding Machine		89.20		98.10	3.72	4.09
24 L.H., Daily Totals		$1012.80		$1764.50	$42.20	$73.52

Crew C-14	**Hr.**	**Daily**	**Hr.**	**Daily**	**Bare Costs**	**Incl. O&P**
1 Carpenter Foreman (out)	$37.55	$300.40	$62.20	$497.60	$34.57	$57.18
5 Carpenters	35.55	1422.00	58.90	2356.00		
4 Laborers	27.40	876.80	45.40	1452.80		
4 Rodmen (reinf.)	39.50	1264.00	67.80	2169.60		
2 Cement Finishers	34.40	550.40	54.20	867.20		
1 Equip. Oper. (crane)	38.10	304.80	60.15	481.20		
1 Equip. Oper. Oiler	32.45	259.60	51.25	410.00		
1 Crane, 80 Ton, & Tools		1059.00		1164.90	7.35	8.09
144 L.H., Daily Totals		$6037.00		$9399.30	$41.92	$65.27

Crew C-14A	**Hr.**	**Daily**	**Hr.**	**Daily**	**Bare Costs**	**Incl. O&P**
1 Carpenter Foreman (out)	$37.55	$300.40	$62.20	$497.60	$35.61	$59.15
16 Carpenters	35.55	4550.40	58.90	7539.20		
4 Rodmen (reinf.)	39.50	1264.00	67.80	2169.60		
2 Laborers	27.40	438.40	45.40	726.40		
1 Cement Finisher	34.40	275.20	54.20	433.60		
1 Equip. Oper. (med.)	36.70	293.60	57.95	463.60		
1 Gas Engine Vibrator		21.40		23.55		
1 Concrete Pump (small)		700.60		770.65	3.61	3.97
200 L.H., Daily Totals		$7844.00		$12624.20	$39.22	$63.12

Crew C-14B	**Hr.**	**Daily**	**Hr.**	**Daily**	**Bare Costs**	**Incl. O&P**
1 Carpenter Foreman (out)	$37.55	$300.40	$62.20	$497.60	$35.56	$58.96
16 Carpenters	35.55	4550.40	58.90	7539.20		
4 Rodmen (reinf.)	39.50	1264.00	67.80	2169.60		
2 Laborers	27.40	438.40	45.40	726.40		
2 Cement Finishers	34.40	550.40	54.20	867.20		
1 Equip. Oper. (med.)	36.70	293.60	57.95	463.60		
1 Gas Engine Vibrator		21.40		23.55		
1 Concrete Pump (small)		700.60		770.65	3.47	3.82
208 L.H., Daily Totals		$8119.20		$13057.80	$39.03	$62.78

Crew No.	Bare Costs		Incl. Subs O & P		Cost Per Labor-Hour	
Crew C-14C	**Hr.**	**Daily**	**Hr.**	**Daily**	**Bare Costs**	**Incl. O&P**
1 Carpenter Foreman (out)	$37.55	$300.40	$62.20	$497.60	$33.85	$56.21
6 Carpenters	35.55	1706.40	58.90	2827.20		
2 Rodmen (reinf.)	39.50	632.00	67.80	1084.80		
4 Laborers	27.40	876.80	45.40	1452.80		
1 Cement Finisher	34.40	275.20	54.20	433.60		
1 Gas Engine Vibrator		21.40		23.55	.19	.21
112 L.H., Daily Totals		$3812.20		$6319.55	$34.04	$56.42

Crew C-14D	**Hr.**	**Daily**	**Hr.**	**Daily**	**Bare Costs**	**Incl. O&P**
1 Carpenter Foreman (out)	$37.55	$300.40	$62.20	$497.60	$35.29	$58.44
18 Carpenters	35.55	5119.20	58.90	8481.60		
2 Rodmen (reinf.)	39.50	632.00	67.80	1084.80		
2 Laborers	27.40	438.40	45.40	726.40		
1 Cement Finisher	34.40	275.20	54.20	433.60		
1 Equip. Oper. (med.)	36.70	293.60	57.95	463.60		
1 Gas Engine Vibrator		21.40		23.55		
1 Concrete Pump (small)		700.60		770.65	3.61	3.97
200 L.H., Daily Totals		$7780.80		$12481.80	$38.90	$62.41

Crew C-14E	**Hr.**	**Daily**	**Hr.**	**Daily**	**Bare Costs**	**Incl. O&P**
1 Carpenter Foreman (out)	$37.55	$300.40	$62.20	$497.60	$34.84	$58.33
2 Carpenters	35.55	568.80	58.90	942.40		
4 Rodmen (reinf.)	39.50	1264.00	67.80	2169.60		
3 Laborers	27.40	657.60	45.40	1089.60		
1 Cement Finisher	34.40	275.20	54.20	433.60		
1 Gas Engine Vibrator		21.40		23.55	.24	.27
88 L.H., Daily Totals		$3087.40		$5156.35	$35.08	$58.60

Crew C-14F	**Hr.**	**Daily**	**Hr.**	**Daily**	**Bare Costs**	**Incl. O&P**
1 Laborer Foreman (out)	$29.40	$235.20	$48.70	$389.60	$32.29	$51.63
2 Laborers	27.40	438.40	45.40	726.40		
6 Cement Finishers	34.40	1651.20	54.20	2601.60		
1 Gas Engine Vibrator		21.40		23.55	.30	.33
72 L.H., Daily Totals		$2346.20		$3741.15	$32.59	$51.96

Crew C-14G	**Hr.**	**Daily**	**Hr.**	**Daily**	**Bare Costs**	**Incl. O&P**
1 Laborer Foreman (out)	$29.40	$235.20	$48.70	$389.60	$31.69	$50.90
2 Laborers	27.40	438.40	45.40	726.40		
4 Cement Finishers	34.40	1100.80	54.20	1734.40		
1 Gas Engine Vibrator		21.40		23.55	.38	.42
56 L.H., Daily Totals		$1795.80		$2873.95	$32.07	$51.32

Crew C-14H	**Hr.**	**Daily**	**Hr.**	**Daily**	**Bare Costs**	**Incl. O&P**
1 Carpenter Foreman (out)	$37.55	$300.40	$62.20	$497.60	$34.99	$57.90
2 Carpenters	35.55	568.80	58.90	942.40		
1 Rodman (reinf.)	39.50	316.00	67.80	542.40		
1 Laborer	27.40	219.20	45.40	363.20		
1 Cement Finisher	34.40	275.20	54.20	433.60		
1 Gas Engine Vibrator		21.40		23.55	.45	.49
48 L.H., Daily Totals		$1701.00		$2802.75	$35.44	$58.39

Crew C-15	**Hr.**	**Daily**	**Hr.**	**Daily**	**Bare Costs**	**Incl. O&P**
1 Carpenter Foreman (out)	$37.55	$300.40	$62.20	$497.60	$33.24	$54.71
2 Carpenters	35.55	568.80	58.90	942.40		
3 Laborers	27.40	657.60	45.40	1089.60		
2 Cement Finishers	34.40	550.40	54.20	867.20		
1 Rodman (reinf.)	39.50	316.00	67.80	542.40		
72 L.H., Daily Totals		$2393.20		$3939.20	$33.24	$54.71

Crew No.	Bare Costs		Incl. Subs O & P		Cost Per Labor-Hour	
Crew C-16	**Hr.**	**Daily**	**Hr.**	**Daily**	**Bare Costs**	**Incl. O&P**
1 Laborer Foreman (outside)	$29.40	$235.20	$48.70	$389.60	$32.90	$54.09
3 Laborers	27.40	657.60	45.40	1089.60		
2 Cement Finishers	34.40	550.40	54.20	867.20		
1 Equip. Oper. (med.)	36.70	293.60	57.95	463.60		
2 Rodmen (reinf.)	39.50	632.00	67.80	1084.80		
1 Concrete Pump (small)		700.60		770.65	9.73	10.70
72 L.H., Daily Totals		$3069.40		$4665.45	$42.63	$64.79

Crew C-17	**Hr.**	**Daily**	**Hr.**	**Daily**	**Bare Costs**	**Incl. O&P**
2 Skilled Worker Foremen	$38.50	$616.00	$63.00	$1008.00	$36.90	$60.36
8 Skilled Workers	36.50	2336.00	59.70	3820.80		
80 L.H., Daily Totals		$2952.00		$4828.80	$36.90	$60.36

Crew C-17A	**Hr.**	**Daily**	**Hr.**	**Daily**	**Bare Costs**	**Incl. O&P**
2 Skilled Worker Foremen	$38.50	$616.00	$63.00	$1008.00	$36.91	$60.36
8 Skilled Workers	36.50	2336.00	59.70	3820.80		
.125 Equip. Oper. (crane)	38.10	38.10	60.15	60.15		
.125 Crane, 80 Ton, & Tools		132.38		145.60	1.63	1.80
81 L.H., Daily Totals		$3122.48		$5034.55	$38.54	$62.16

Crew C-17B	**Hr.**	**Daily**	**Hr.**	**Daily**	**Bare Costs**	**Incl. O&P**
2 Skilled Worker Foremen	$38.50	$616.00	$63.00	$1008.00	$36.93	$60.35
8 Skilled Workers	36.50	2336.00	59.70	3820.80		
.25 Equip. Oper. (crane)	38.10	76.20	60.15	120.30		
.25 Crane, 80 Ton, & Tools		264.75		291.25		
.25 Walk Behind Power Tool		6.05		6.65	3.30	3.63
82 L.H., Daily Totals		$3299.00		$5247.00	$40.23	$63.98

Crew C-17C	**Hr.**	**Daily**	**Hr.**	**Daily**	**Bare Costs**	**Incl. O&P**
2 Skilled Worker Foremen	$38.50	$616.00	$63.00	$1008.00	$36.94	$60.35
8 Skilled Workers	36.50	2336.00	59.70	3820.80		
.375 Equip. Oper. (crane)	38.10	114.30	60.15	180.45		
.375 Crane, 80 Ton & Tools		397.13		436.85	4.78	5.26
83 L.H., Daily Totals		$3463.43		$5446.10	$41.72	$65.61

Crew C-17D	**Hr.**	**Daily**	**Hr.**	**Daily**	**Bare Costs**	**Incl. O&P**
2 Skilled Worker Foremen	$38.50	$616.00	$63.00	$1008.00	$36.96	$60.35
8 Skilled Workers	36.50	2336.00	59.70	3820.80		
.5 Equip. Oper. (crane)	38.10	152.40	60.15	240.60		
.5 Crane, 80 Ton & Tools		529.50		582.45	6.30	6.93
84 L.H., Daily Totals		$3633.90		$5651.85	$43.26	$67.28

Crew C-17E	**Hr.**	**Daily**	**Hr.**	**Daily**	**Bare Costs**	**Incl. O&P**
2 Skilled Worker Foremen	$38.50	$616.00	$63.00	$1008.00	$36.90	$60.36
8 Skilled Workers	36.50	2336.00	59.70	3820.80		
1 Hyd. Jack with Rods		80.65		88.70	1.01	1.11
80 L.H., Daily Totals		$3032.65		$4917.50	$37.91	$61.47

Crew C-18	**Hr.**	**Daily**	**Hr.**	**Daily**	**Bare Costs**	**Incl. O&P**
.125 Laborer Foreman (out)	$29.40	$29.40	$48.70	$48.70	$27.62	$45.77
1 Laborer	27.40	219.20	45.40	363.20		
1 Concrete Cart, 10 C.F.		52.20		57.40	5.80	6.38
9 L.H., Daily Totals		$300.80		$469.30	$33.42	$52.15

Crew C-19	**Hr.**	**Daily**	**Hr.**	**Daily**	**Bare Costs**	**Incl. O&P**
.125 Laborer Foreman (out)	$29.40	$29.40	$48.70	$48.70	$27.62	$45.77
1 Laborer	27.40	219.20	45.40	363.20		
1 Concrete Cart, 18 C.F.		74.40		81.85	8.27	9.09
9 L.H., Daily Totals		$323.00		$493.75	$35.89	$54.86

Crew No.	Bare Costs		Incl. Subs O & P		Cost Per Labor-Hour	

Crew C-20	Hr.	Daily	Hr.	Daily	Bare Costs	Incl. O&P
1 Laborer Foreman (outside)	$29.40	$235.20	$48.70	$389.60	$29.69	$48.48
5 Laborers	27.40	1096.00	45.40	1816.00		
1 Cement Finisher	34.40	275.20	54.20	433.60		
1 Equip. Oper. (med.)	36.70	293.60	57.95	463.60		
2 Gas Engine Vibrators		42.80		47.10		
1 Concrete Pump (small)		700.60		770.65	11.62	12.78
64 L.H., Daily Totals		$2643.40		$3920.55	$41.31	$61.26

Crew C-21	Hr.	Daily	Hr.	Daily	Bare Costs	Incl. O&P
1 Laborer Foreman (outside)	$29.40	$235.20	$48.70	$389.60	$29.69	$48.48
5 Laborers	27.40	1096.00	45.40	1816.00		
1 Cement Finisher	34.40	275.20	54.20	433.60		
1 Equip. Oper. (med.)	36.70	293.60	57.95	463.60		
2 Gas Engine Vibrators		42.80		47.10		
1 Concrete Conveyer		152.40		167.65	3.05	3.36
64 L.H., Daily Totals		$2095.20		$3317.55	$32.74	$51.84

Crew C-22	Hr.	Daily	Hr.	Daily	Bare Costs	Incl. O&P
1 Rodman Foreman	$41.50	$332.00	$71.25	$570.00	$39.68	$67.88
4 Rodmen (reinf.)	39.50	1264.00	67.80	2169.60		
.125 Equip. Oper. (crane)	38.10	38.10	60.15	60.15		
.125 Equip. Oper. Oiler	32.45	32.45	51.25	51.25		
.125 Hyd. Crane, 25 Ton		81.35		89.50	1.94	2.13
42 L.H., Daily Totals		$1747.90		$2940.50	$41.62	$70.01

Crew C-23	Hr.	Daily	Hr.	Daily	Bare Costs	Incl. O&P
2 Skilled Worker Foremen	$38.50	$616.00	$63.00	$1008.00	$36.66	$59.56
6 Skilled Workers	36.50	1752.00	59.70	2865.60		
1 Equip. Oper. (crane)	38.10	304.80	60.15	481.20		
1 Equip. Oper. Oiler	32.45	259.60	51.25	410.00		
1 Crane, 90 Ton		1427.00		1569.70	17.84	19.62
80 L.H., Daily Totals		$4359.40		$6334.50	$54.50	$79.18

Crew C-24	Hr.	Daily	Hr.	Daily	Bare Costs	Incl. O&P
2 Skilled Worker Foremen	$38.50	$616.00	$63.00	$1008.00	$36.66	$59.56
6 Skilled Workers	36.50	1752.00	59.70	2865.60		
1 Equip. Oper. (crane)	38.10	304.80	60.15	481.20		
1 Equip. Oper. Oiler	32.45	259.60	51.25	410.00		
1 Truck Crane, 150 Ton		1539.00		1692.90	19.24	21.16
80 L.H., Daily Totals		$4471.40		$6457.70	$55.90	$80.72

Crew C-25	Hr.	Daily	Hr.	Daily	Bare Costs	Incl. O&P
2 Rodmen (reinf.)	$39.50	$632.00	$67.80	$1084.80	$31.03	$54.15
2 Rodman Helpers	22.55	360.80	40.50	648.00		
32 L.H., Daily Totals		$992.80		$1732.80	$31.03	$54.15

Crew C-27	Hr.	Daily	Hr.	Daily	Bare Costs	Incl. O&P
2 Cement Finishers	$34.40	$550.40	$54.20	$867.20	$34.40	$54.20
1 Concrete Saw		107.00		117.70	6.69	7.36
16 L.H., Daily Totals		$657.40		$984.90	$41.09	$61.56

Crew C-28	Hr.	Daily	Hr.	Daily	Bare Costs	Incl. O&P
1 Cement Finisher	$34.40	$275.20	$54.20	$433.60	$34.40	$54.20
1 Portable Air Compressor		15.75		17.35	1.96	2.16
8 L.H., Daily Totals		$290.95		$450.95	$36.36	$56.36

Crew D-1	Hr.	Daily	Hr.	Daily	Bare Costs	Incl. O&P
1 Bricklayer	$36.55	$292.40	$59.30	$474.40	$32.15	$52.15
1 Bricklayer Helper	27.75	222.00	45.00	360.00		
16 L.H., Daily Totals		$514.40		$834.40	$32.15	$52.15

Crew No.	Bare Costs		Incl. Subs O & P		Cost Per Labor-Hour	

Crew D-2	Hr.	Daily	Hr.	Daily	Bare Costs	Incl. O&P
3 Bricklayers	$36.55	$877.20	$59.30	$1423.20	$33.26	$54.06
2 Bricklayer Helpers	27.75	444.00	45.00	720.00		
.5 Carpenter	35.55	142.20	58.90	235.60		
44 L.H., Daily Totals		$1463.40		$2378.80	$33.26	$54.06

Crew D-3	Hr.	Daily	Hr.	Daily	Bare Costs	Incl. O&P
3 Bricklayers	$36.55	$877.20	$59.30	$1423.20	$33.15	$53.83
2 Bricklayer Helpers	27.75	444.00	45.00	720.00		
.25 Carpenter	35.55	71.10	58.90	117.80		
42 L.H., Daily Totals		$1392.30		$2261.00	$33.15	$53.83

Crew D-4	Hr.	Daily	Hr.	Daily	Bare Costs	Incl. O&P
1 Bricklayer	$36.55	$292.40	$59.30	$474.40	$31.81	$51.23
2 Bricklayer Helpers	27.75	444.00	45.00	720.00		
1 Equip. Oper. (light)	35.20	281.60	55.60	444.80		
1 Grout Pump, 50 C.F./hr		117.30		129.05		
1 Hose & Hopper		15.60		17.15		
1 Accessory		12.10		13.30	4.53	4.98
32 L.H., Daily Totals		$1163.00		$1798.70	$36.34	$56.21

Crew D-5	Hr.	Daily	Hr.	Daily	Bare Costs	Incl. O&P
1 Bricklayer	$36.55	$292.40	$59.30	$474.40	$36.55	$59.30
8 L.H., Daily Totals		$292.40		$474.40	$36.55	$59.30

Crew D-6	Hr.	Daily	Hr.	Daily	Bare Costs	Incl. O&P
3 Bricklayers	$36.55	$877.20	$59.30	$1423.20	$32.29	$52.42
3 Bricklayer Helpers	27.75	666.00	45.00	1080.00		
.25 Carpenter	35.55	71.10	58.90	117.80		
50 L.H., Daily Totals		$1614.30		$2621.00	$32.29	$52.42

Crew D-7	Hr.	Daily	Hr.	Daily	Bare Costs	Incl. O&P
1 Tile Layer	$34.25	$274.00	$53.70	$429.60	$30.38	$47.63
1 Tile Layer Helper	26.50	212.00	41.55	332.40		
16 L.H., Daily Totals		$486.00		$762.00	$30.38	$47.63

Crew D-8	Hr.	Daily	Hr.	Daily	Bare Costs	Incl. O&P
3 Bricklayers	$36.55	$877.20	$59.30	$1423.20	$33.03	$53.58
2 Bricklayer Helpers	27.75	444.00	45.00	720.00		
40 L.H., Daily Totals		$1321.20		$2143.20	$33.03	$53.58

Crew D-9	Hr.	Daily	Hr.	Daily	Bare Costs	Incl. O&P
3 Bricklayers	$36.55	$877.20	$59.30	$1423.20	$32.15	$52.15
3 Bricklayer Helpers	27.75	666.00	45.00	1080.00		
48 L.H., Daily Totals		$1543.20		$2503.20	$32.15	$52.15

Crew D-10	Hr.	Daily	Hr.	Daily	Bare Costs	Incl. O&P
1 Bricklayer Foreman	$38.55	$308.40	$62.55	$500.40	$33.74	$54.40
1 Bricklayer	36.55	292.40	59.30	474.40		
2 Bricklayer Helpers	27.75	444.00	45.00	720.00		
1 Equip. Oper. (crane)	38.10	304.80	60.15	481.20		
1 Truck Crane, 12.5 Ton		536.40		590.05	13.41	14.75
40 L.H., Daily Totals		$1886.00		$2766.05	$47.15	$69.15

Crew D-11	Hr.	Daily	Hr.	Daily	Bare Costs	Incl. O&P
1 Bricklayer Foreman	$38.55	$308.40	$62.55	$500.40	$34.28	$55.62
1 Bricklayer	36.55	292.40	59.30	474.40		
1 Bricklayer Helper	27.75	222.00	45.00	360.00		
24 L.H., Daily Totals		$822.80		$1334.80	$34.28	$55.62

Crew No.	Bare Costs		Incl. Subs O & P		Cost Per Labor-Hour	
Crew D-12	**Hr.**	**Daily**	**Hr.**	**Daily**	**Bare Costs**	**Incl. O&P**
1 Bricklayer Foreman	$38.55	$308.40	$62.55	$500.40	$32.65	$52.96
1 Bricklayer	36.55	292.40	59.30	474.40		
2 Bricklayer Helpers	27.75	444.00	45.00	720.00		
32 L.H., Daily Totals		$1044.80		$1694.80	$32.65	$52.96

Crew D-13	Hr.	Daily	Hr.	Daily	Bare Costs	Incl. O&P
1 Bricklayer Foreman	$38.55	$308.40	$62.55	$500.40	$34.04	$55.15
1 Bricklayer	36.55	292.40	59.30	474.40		
2 Bricklayer Helpers	27.75	444.00	45.00	720.00		
1 Carpenter	35.55	284.40	58.90	471.20		
1 Equip. Oper. (crane)	38.10	304.80	60.15	481.20		
1 Truck Crane, 12.5 Ton		536.40		590.05	11.18	12.29
48 L.H., Daily Totals		$2170.40		$3237.25	$45.22	$67.44

Crew E-1	Hr.	Daily	Hr.	Daily	Bare Costs	Incl. O&P
1 Welder Foreman	$41.95	$335.60	$78.45	$627.60	$39.03	$69.58
1 Welder	39.95	319.60	74.70	597.60		
1 Equip. Oper. (light)	35.20	281.60	55.60	444.80		
1 Gas Welding Machine		89.20		98.10	3.72	4.09
24 L.H., Daily Totals		$1026.00		$1768.10	$42.75	$73.67

Crew E-2	Hr.	Daily	Hr.	Daily	Bare Costs	Incl. O&P
1 Struc. Steel Foreman	$41.95	$335.60	$78.45	$627.60	$38.90	$69.81
4 Struc. Steel Workers	39.95	1278.40	74.70	2390.40		
1 Equip. Oper. (crane)	38.10	304.80	60.15	481.20		
1 Equip. Oper. Oiler	32.45	259.60	51.25	410.00		
1 Crane, 90 Ton		1427.00		1569.70	25.48	28.03
56 L.H., Daily Totals		$3605.40		$5478.90	$64.38	$97.84

Crew E-3	Hr.	Daily	Hr.	Daily	Bare Costs	Incl. O&P
1 Struc. Steel Foreman	$41.95	$335.60	$78.45	$627.60	$40.62	$75.95
1 Struc. Steel Worker	39.95	319.60	74.70	597.60		
1 Welder	39.95	319.60	74.70	597.60		
1 Gas Welding Machine		89.20		98.10	3.72	4.09
24 L.H., Daily Totals		$1064.00		$1920.90	$44.34	$80.04

Crew E-4	Hr.	Daily	Hr.	Daily	Bare Costs	Incl. O&P
1 Struc. Steel Foreman	$41.95	$335.60	$78.45	$627.60	$40.45	$75.64
3 Struc. Steel Workers	39.95	958.80	74.70	1792.80		
1 Gas Welding Machine		89.20		98.10	2.79	3.07
32 L.H., Daily Totals		$1383.60		$2518.50	$43.24	$78.71

Crew E-5	Hr.	Daily	Hr.	Daily	Bare Costs	Incl. O&P
2 Struc. Steel Foremen	$41.95	$671.20	$78.45	$1255.20	$39.42	$71.65
5 Struc. Steel Workers	39.95	1598.00	74.70	2988.00		
1 Equip. Oper. (crane)	38.10	304.80	60.15	481.20		
1 Welder	39.95	319.60	74.70	597.60		
1 Equip. Oper. Oiler	32.45	259.60	51.25	410.00		
1 Crane, 90 Ton		1427.00		1569.70		
1 Gas Welding Machine		89.20		98.10	18.95	20.85
80 L.H., Daily Totals		$4669.40		$7399.80	$58.37	$92.50

Crew No.	Bare Costs		Incl. Subs O & P		Cost Per Labor-Hour	
Crew E-6	**Hr.**	**Daily**	**Hr.**	**Daily**	**Bare Costs**	**Incl. O&P**
3 Struc. Steel Foremen	$41.95	$1006.80	$78.45	$1882.80	$39.44	$71.83
9 Struc. Steel Workers	39.95	2876.40	74.70	5378.40		
1 Equip. Oper. (crane)	38.10	304.80	60.15	481.20		
1 Welder	39.95	319.60	74.70	597.60		
1 Equip. Oper. Oiler	32.45	259.60	51.25	410.00		
1 Equip. Oper. (light)	35.20	281.60	55.60	444.80		
1 Crane, 90 Ton		1427.00		1569.70		
1 Gas Welding Machine		89.20		98.10		
1 Air Compr., 160 C.F.M.		97.60		107.35		
2 Impact Wrenches		25.20		27.70	12.80	14.09
128 L.H., Daily Totals		$6687.80		$10997.65	$52.24	$85.92

Crew E-7	Hr.	Daily	Hr.	Daily	Bare Costs	Incl. O&P
1 Struc. Steel Foreman	$41.95	$335.60	$78.45	$627.60	$39.42	$71.65
4 Struc. Steel Workers	39.95	1278.40	74.70	2390.40		
1 Equip. Oper. (crane)	38.10	304.80	60.15	481.20		
1 Equip. Oper. Oiler	32.45	259.60	51.25	410.00		
1 Welder Foreman	41.95	335.60	78.45	627.60		
2 Welders	39.95	639.20	74.70	1195.20		
1 Crane, 90 Ton		1427.00		1569.70		
2 Gas Welding Machines		178.40		196.25	20.07	22.07
80 L.H., Daily Totals		$4758.60		$7497.95	$59.49	$93.72

Crew E-8	Hr.	Daily	Hr.	Daily	Bare Costs	Incl. O&P
1 Struc. Steel Foreman	$41.95	$335.60	$78.45	$627.60	$39.17	$70.88
4 Struc. Steel Workers	39.95	1278.40	74.70	2390.40		
1 Welder Foreman	41.95	335.60	78.45	627.60		
4 Welders	39.95	1278.40	74.70	2390.40		
1 Equip. Oper. (crane)	38.10	304.80	60.15	481.20		
1 Equip. Oper. Oiler	32.45	259.60	51.25	410.00		
1 Equip. Oper. (light)	35.20	281.60	55.60	444.80		
1 Crane, 90 Ton		1427.00		1569.70		
4 Gas Welding Machines		356.80		392.50	17.15	18.87
104 L.H., Daily Totals		$5857.80		$9334.20	$56.32	$89.75

Crew E-9	Hr.	Daily	Hr.	Daily	Bare Costs	Incl. O&P
2 Struc. Steel Foremen	$41.95	$671.20	$78.45	$1255.20	$39.44	$71.83
5 Struc. Steel Workers	39.95	1598.00	74.70	2988.00		
1 Welder Foreman	41.95	335.60	78.45	627.60		
5 Welders	39.95	1598.00	74.70	2988.00		
1 Equip. Oper. (crane)	38.10	304.80	60.15	481.20		
1 Equip. Oper. Oiler	32.45	259.60	51.25	410.00		
1 Equip. Oper. (light)	35.20	281.60	55.60	444.80		
1 Crane, 90 Ton		1427.00		1569.70		
5 Gas Welding Machines		446.00		490.60	14.63	16.10
128 L.H., Daily Totals		$6921.80		$11255.10	$54.07	$87.93

Crew E-10	Hr.	Daily	Hr.	Daily	Bare Costs	Incl. O&P
1 Welder Foreman	$41.95	$335.60	$78.45	$627.60	$40.95	$76.58
1 Welder	39.95	319.60	74.70	597.60		
1 Gas Welding Machines		89.20		98.10		
1 Truck, 3 Ton		176.20		193.80	16.59	18.25
16 L.H., Daily Totals		$920.60		$1517.10	$57.54	$94.83

Crew E-11	Hr.	Daily	Hr.	Daily	Bare Costs	Incl. O&P
2 Painters, Struc. Steel	$32.45	$519.20	$63.70	$1019.20	$31.88	$57.10
1 Building Laborer	27.40	219.20	45.40	363.20		
1 Equip. Oper. (light)	35.20	281.60	55.60	444.80		
1 Air Compressor 250 C.F.M.		119.80		131.80		
1 Sand Blaster		15.60		17.15		
1 Sand Blasting Accessories		12.10		13.30	4.61	5.07
32 L.H., Daily Totals		$1167.50		$1989.45	$36.49	$62.17

Crew No.	Bare Costs		Incl. Subs O & P		Cost Per Labor-Hour	

Crew E-12	Hr.	Daily	Hr.	Daily	Bare Costs	Incl. O&P
1 Welder Foreman	$41.95	$335.60	$78.45	$627.60	$38.58	$67.03
1 Equip. Oper. (light)	35.20	281.60	55.60	444.80		
1 Gas Welding Machine		89.20		98.10	5.58	6.13
16 L.H., Daily Totals		$706.40		$1170.50	$44.16	$73.16

Crew E-13	Hr.	Daily	Hr.	Daily	Bare Costs	Incl. O&P
1 Welder Foreman	$41.95	$335.60	$78.45	$627.60	$39.70	$70.83
.5 Equip. Oper. (light)	35.20	140.80	55.60	222.40		
1 Gas Welding Machine		89.20		98.10	7.43	8.18
12 L.H., Daily Totals		$565.60		$948.10	$47.13	$79.01

Crew E-14	Hr.	Daily	Hr.	Daily	Bare Costs	Incl. O&P
1 Welder Foreman	$41.95	$335.60	$78.45	$627.60	$41.95	$78.45
1 Gas Welding Machine		89.20		98.10	11.15	12.27
8 L.H., Daily Totals		$424.80		$725.70	$53.10	$90.72

Crew E-16	Hr.	Daily	Hr.	Daily	Bare Costs	Incl. O&P
1 Welder Foreman	$41.95	$335.60	$78.45	$627.60	$40.95	$76.58
1 Welder	39.95	319.60	74.70	597.60		
1 Gas Welding Machine		89.20		98.10	5.58	6.13
16 L.H., Daily Totals		$744.40		$1323.30	$46.53	$82.71

Crew E-17	Hr.	Daily	Hr.	Daily	Bare Costs	Incl. O&P
1 Structural Steel Foreman	$41.95	$335.60	$78.45	$627.60	$40.95	$76.58
1 Structural Steel Worker	39.95	319.60	74.70	597.60		
1 Power Tool		5.40		5.95	.34	.37
16 L.H., Daily Totals		$660.60		$1231.15	$41.29	$76.95

Crew E-18	Hr.	Daily	Hr.	Daily	Bare Costs	Incl. O&P
1 Structural Steel Foreman	$41.95	$335.60	$78.45	$627.60	$39.70	$72.10
3 Structural Steel Workers	39.95	958.80	74.70	1792.80		
1 Equipment Operator (med.)	36.70	293.60	57.95	463.60		
1 Crane, 20 Ton		947.05		1041.75	23.68	26.04
40 L.H., Daily Totals		$2535.05		$3925.75	$63.38	$98.14

Crew E-19	Hr.	Daily	Hr.	Daily	Bare Costs	Incl. O&P
1 Structural Steel Worker	$39.95	$319.60	$74.70	$597.60	$39.03	$69.58
1 Structural Steel Foreman	41.95	335.60	78.45	627.60		
1 Equip. Oper. (light)	35.20	281.60	55.60	444.80		
1 Power Tool		5.40		5.95		
1 Crane, 20 Ton		947.05		1041.75	39.68	43.65
24 L.H., Daily Totals		$1889.25		$2717.70	$78.71	$113.23

Crew E-20	Hr.	Daily	Hr.	Daily	Bare Costs	Incl. O&P
1 Structural Steel Foreman	$41.95	$335.60	$78.45	$627.60	$39.03	$70.42
5 Structural Steel Workers	39.95	1598.00	74.70	2988.00		
1 Equip. Oper. (crane)	38.10	304.80	60.15	481.20		
1 Oiler	32.45	259.60	51.25	410.00		
1 Power Tool		5.40		5.95		
1 Crane, 40 Ton		1130.00		1243.00	17.74	19.51
64 L.H., Daily Totals		$3633.40		$5755.75	$56.77	$89.93

Crew E-22	Hr.	Daily	Hr.	Daily	Bare Costs	Incl. O&P
1 Skilled Worker Foreman	$38.50	$308.00	$63.00	$504.00	$37.17	$60.80
2 Skilled Workers	36.50	584.00	59.70	955.20		
24 L.H., Daily Totals		$892.00		$1459.20	$37.17	$60.80

Crew No.	Bare Costs		Incl. Subs O & P		Cost Per Labor-Hour	

Crew E-24	Hr.	Daily	Hr.	Daily	Bare Costs	Incl. O&P
3 Structural Steel Workers	$39.95	$958.80	$74.70	$1792.80	$39.14	$70.51
1 Equipment Operator (medium)	36.70	293.60	57.95	463.60		
1-25 Ton Crane		650.80		715.90	20.34	22.37
32 L.H., Daily Totals		$1903.20		$2972.30	$59.48	$92.88

Crew E-25	Hr.	Daily	Hr.	Daily	Bare Costs	Incl. O&P
1 Welder Foreman	$41.95	$335.60	$78.45	$627.60	$41.95	$78.45
1 Cutting Torch		16.20		17.80		
1 Gas		64.80		71.30	10.13	11.14
8 L.H., Daily Totals		$416.60		$716.70	$52.08	$89.59

Crew F-3	Hr.	Daily	Hr.	Daily	Bare Costs	Incl. O&P
4 Carpenters	$35.55	$1137.60	$58.90	$1884.80	$36.06	$59.15
1 Equip. Oper. (crane)	38.10	304.80	60.15	481.20		
1 Hyd. Crane, 12 Ton		636.60		700.25	15.92	17.51
40 L.H., Daily Totals		$2079.00		$3066.25	$51.98	$76.66

Crew F-4	Hr.	Daily	Hr.	Daily	Bare Costs	Incl. O&P
4 Carpenters	$35.55	$1137.60	$58.90	$1884.80	$35.46	$57.83
1 Equip. Oper. (crane)	38.10	304.80	60.15	481.20		
1 Equip. Oper. Oiler	32.45	259.60	51.25	410.00		
1 Hyd. Crane, 55 Ton		951.60		1046.75	19.83	21.81
48 L.H., Daily Totals		$2653.60		$3822.75	$55.29	$79.64

Crew F-5	Hr.	Daily	Hr.	Daily	Bare Costs	Incl. O&P
1 Carpenter Foreman	$37.55	$300.40	$62.20	$497.60	$36.05	$59.73
3 Carpenters	35.55	853.20	58.90	1413.60		
32 L.H., Daily Totals		$1153.60		$1911.20	$36.05	$59.73

Crew F-6	Hr.	Daily	Hr.	Daily	Bare Costs	Incl. O&P
2 Carpenters	$35.55	$568.80	$58.90	$942.40	$32.80	$53.75
2 Building Laborers	27.40	438.40	45.40	726.40		
1 Equip. Oper. (crane)	38.10	304.80	60.15	481.20		
1 Hyd. Crane, 12 Ton		636.60		700.25	15.92	17.51
40 L.H., Daily Totals		$1948.60		$2850.25	$48.72	$71.26

Crew F-7	Hr.	Daily	Hr.	Daily	Bare Costs	Incl. O&P
2 Carpenters	$35.55	$568.80	$58.90	$942.40	$31.48	$52.15
2 Building Laborers	27.40	438.40	45.40	726.40		
32 L.H., Daily Totals		$1007.20		$1668.80	$31.48	$52.15

Crew G-1	Hr.	Daily	Hr.	Daily	Bare Costs	Incl. O&P
1 Roofer Foreman	$32.60	$260.80	$58.60	$468.80	$28.59	$51.37
4 Roofers, Composition	30.60	979.20	55.00	1760.00		
2 Roofer Helpers	22.55	360.80	40.50	648.00		
1 Application Equipment		146.20		160.80		
1 Tar Kettle/Pot		57.30		63.05		
1 Crew Truck		133.75		147.15	6.02	6.62
56 L.H., Daily Totals		$1938.05		$3247.80	$34.61	$57.99

Crew G-2	Hr.	Daily	Hr.	Daily	Bare Costs	Incl. O&P
1 Plasterer	$32.45	$259.60	$52.35	$418.80	$29.25	$47.58
1 Plasterer Helper	27.90	223.20	45.00	360.00		
1 Building Laborer	27.40	219.20	45.40	363.20		
1 Grouting Equipment		117.30		129.05	4.89	5.38
24 L.H., Daily Totals		$819.30		$1271.05	$34.14	$52.96

Crew No.	Bare Costs		Incl. Subs O & P		Cost Per Labor-Hour	

Crew G-2A	Hr.	Daily	Hr.	Daily	Bare Costs	Incl. O&P
1 Roofer, composition	$30.60	$244.80	$55.00	$440.00	$26.85	$46.97
1 Roofer Helper	22.55	180.40	40.50	324.00		
1 Building Laborer	27.40	219.20	45.40	363.20		
1 Spray Equipment		117.30		129.05	4.89	5.38
24 L.H., Daily Totals		$761.70		$1256.25	$31.74	$52.35

Crew G-3	Hr.	Daily	Hr.	Daily	Bare Costs	Incl. O&P
2 Sheet Metal Workers	$42.15	$674.40	$67.05	$1072.80	$34.78	$56.23
2 Building Laborers	27.40	438.40	45.40	726.40		
32 L.H., Daily Totals		$1112.80		$1799.20	$34.78	$56.23

Crew G-4	Hr.	Daily	Hr.	Daily	Bare Costs	Incl. O&P
1 Laborer Foreman (outside)	$29.40	$235.20	$48.70	$389.60	$28.07	$46.50
2 Building Laborers	27.40	438.40	45.40	726.40		
1 Light Truck, 1.5 Ton		128.80		141.70		
1 Air Compr., 160 C.F.M.		97.60		107.35	9.43	10.38
24 L.H., Daily Totals		$900.00		$1365.05	$37.50	$56.88

Crew G-5	Hr.	Daily	Hr.	Daily	Bare Costs	Incl. O&P
1 Roofer Foreman	$32.60	$260.80	$58.60	$468.80	$27.78	$49.92
2 Roofers, Composition	30.60	489.60	55.00	880.00		
2 Roofer Helpers	22.55	360.80	40.50	648.00		
1 Application Equipment		146.20		160.80	3.66	4.02
40 L.H., Daily Totals		$1257.40		$2157.60	$31.44	$53.94

Crew G-6A	Hr.	Daily	Hr.	Daily	Bare Costs	Incl. O&P
2 Roofers, Composition	$30.60	$489.60	$55.00	$880.00	$30.60	$55.00
1 Small Compressor		10.60		11.65		
2 Pneumatic Nailers		39.40		43.35	3.12	3.43
16 L.H., Daily Totals		$539.60		$935.00	$33.72	$58.43

Crew G-7	Hr.	Daily	Hr.	Daily	Bare Costs	Incl. O&P
1 Carpenter	$35.55	$284.40	$58.90	$471.20	$35.55	$58.90
1 Small Compressor		10.60		11.65		
1 Pneumatic Nailer		19.70		21.65	3.78	4.16
8 L.H., Daily Totals		$314.70		$504.50	$39.33	$63.06

Crew H-1	Hr.	Daily	Hr.	Daily	Bare Costs	Incl. O&P
2 Glaziers	$34.25	$548.00	$55.20	$883.20	$37.10	$64.95
2 Struc. Steel Workers	39.95	639.20	74.70	1195.20		
32 L.H., Daily Totals		$1187.20		$2078.40	$37.10	$64.95

Crew H-2	Hr.	Daily	Hr.	Daily	Bare Costs	Incl. O&P
2 Glaziers	$34.25	$548.00	$55.20	$883.20	$31.97	$51.93
1 Building Laborer	27.40	219.20	45.40	363.20		
24 L.H., Daily Totals		$767.20		$1246.40	$31.97	$51.93

Crew H-3	Hr.	Daily	Hr.	Daily	Bare Costs	Incl. O&P
1 Glazier	$34.25	$274.00	$55.20	$441.60	$30.33	$49.40
1 Helper	26.40	211.20	43.60	348.80		
16 L.H., Daily Totals		$485.20		$790.40	$30.33	$49.40

Crew J-1	Hr.	Daily	Hr.	Daily	Bare Costs	Incl. O&P
3 Plasterers	$32.45	$778.80	$52.35	$1256.40	$30.63	$49.41
2 Plasterer Helpers	27.90	446.40	45.00	720.00		
1 Mixing Machine, 6 C.F.		105.60		116.15	2.64	2.90
40 L.H., Daily Totals		$1330.80		$2092.55	$33.27	$52.31

Crew J-2	Hr.	Daily	Hr.	Daily	Bare Costs	Incl. O&P
3 Plasterers	$32.45	$778.80	$52.35	$1256.40	$30.99	$49.85
2 Plasterer Helpers	27.90	446.40	45.00	720.00		
1 Lather	32.80	262.40	52.05	416.40		
1 Mixing Machine, 6 C.F.		105.60		116.15	2.20	2.42
48 L.H., Daily Totals		$1593.20		$2508.95	$33.19	$52.27

Crew J-3	Hr.	Daily	Hr.	Daily	Bare Costs	Incl. O&P
1 Terrazzo Worker	$34.10	$272.80	$53.45	$427.60	$30.80	$48.28
1 Terrazzo Helper	27.50	220.00	43.10	344.80		
1 Terrazzo Grinder, Electric		91.25		100.40		
1 Terrazzo Mixer		144.20		158.60	14.72	16.19
16 L.H., Daily Totals		$728.25		$1031.40	$45.52	$64.47

Crew J-4	Hr.	Daily	Hr.	Daily	Bare Costs	Incl. O&P
1 Tile Layer	$34.25	$274.00	$53.70	$429.60	$30.38	$47.63
1 Tile Layer Helper	26.50	212.00	41.55	332.40		
16 L.H., Daily Totals		$486.00		$762.00	$30.38	$47.63

Crew K-1	Hr.	Daily	Hr.	Daily	Bare Costs	Incl. O&P
1 Carpenter	$35.55	$284.40	$58.90	$471.20	$31.58	$51.90
1 Truck Driver (light)	27.60	220.80	44.90	359.20		
1 Truck w/Power Equip.		176.20		193.80	11.01	12.11
16 L.H., Daily Totals		$681.40		$1024.20	$42.59	$64.01

Crew K-2	Hr.	Daily	Hr.	Daily	Bare Costs	Incl. O&P
1 Struc. Steel Foreman	$41.95	$335.60	$78.45	$627.60	$36.50	$66.02
1 Struc. Steel Worker	39.95	319.60	74.70	597.60		
1 Truck Driver (light)	27.60	220.80	44.90	359.20		
1 Truck w/Power Equip.		176.20		193.80	7.34	8.08
24 L.H., Daily Totals		$1052.20		$1778.20	$43.84	$74.10

Crew L-1	Hr.	Daily	Hr.	Daily	Bare Costs	Incl. O&P
1 Electrician	$42.00	$336.00	$64.70	$517.60	$42.35	$65.55
1 Plumber	42.70	341.60	66.40	531.20		
16 L.H., Daily Totals		$677.60		$1048.80	$42.35	$65.55

Crew L-2	Hr.	Daily	Hr.	Daily	Bare Costs	Incl. O&P
1 Carpenter	$35.55	$284.40	$58.90	$471.20	$30.98	$51.25
1 Carpenter Helper	26.40	211.20	43.60	348.80		
16 L.H., Daily Totals		$495.60		$820.00	$30.98	$51.25

Crew L-3	Hr.	Daily	Hr.	Daily	Bare Costs	Incl. O&P
1 Carpenter	$35.55	$284.40	$58.90	$471.20	$38.81	$62.39
.5 Electrician	42.00	168.00	64.70	258.80		
.5 Sheet Metal Worker	42.15	168.60	67.05	268.20		
16 L.H., Daily Totals		$621.00		$998.20	$38.81	$62.39

Crew L-3A	Hr.	Daily	Hr.	Daily	Bare Costs	Incl. O&P
1 Carpenter Foreman (outside)	$37.55	$300.40	$62.20	$497.60	$39.08	$63.82
.5 Sheet Metal Worker	42.15	168.60	67.05	268.20		
12 L.H., Daily Totals		$469.00		$765.80	$39.08	$63.82

Crew L-4	Hr.	Daily	Hr.	Daily	Bare Costs	Incl. O&P
2 Skilled Workers	$36.50	$584.00	$59.70	$955.20	$33.13	$54.33
1 Helper	26.40	211.20	43.60	348.80		
24 L.H., Daily Totals		$795.20		$1304.00	$33.13	$54.33

Crews

Crew No.	Bare Costs		Incl. Subs O & P		Cost Per Labor-Hour	
Crew L-5	**Hr.**	**Daily**	**Hr.**	**Daily**	**Bare Costs**	**Incl. O&P**
1 Struc. Steel Foreman	$41.95	$335.60	$78.45	$627.60	$39.97	$73.16
5 Struc. Steel Workers	39.95	1598.00	74.70	2988.00		
1 Equip. Oper. (crane)	38.10	304.80	60.15	481.20		
1 Hyd. Crane, 25 Ton		650.80		715.90	11.62	12.78
56 L.H., Daily Totals		$2889.20		$4812.70	$51.59	$85.94

Crew L-5A	Hr.	Daily	Hr.	Daily	Bare Costs	Incl. O&P
1 Structural Steel Foreman	$41.95	$335.60	$78.45	$627.60	$39.99	$72.00
2 Structural Steel Workers	39.95	639.20	74.70	1195.20		
1 Equip. Oper. (crane)	38.10	304.80	60.15	481.20		
1 Crane,SP, 25 Ton		696.80		766.50	21.78	23.95
32 L.H., Daily Totals		$1976.40		$3070.50	$61.77	$95.95

Crew L-6	Hr.	Daily	Hr.	Daily	Bare Costs	Incl. O&P
1 Plumber	$42.70	$341.60	$66.40	$531.20	$42.47	$65.83
.5 Electrician	42.00	168.00	64.70	258.80		
12 L.H., Daily Totals		$509.60		$790.00	$42.47	$65.83

Crew L-7	Hr.	Daily	Hr.	Daily	Bare Costs	Incl. O&P
2 Carpenters	$35.55	$568.80	$58.90	$942.40	$34.14	$55.87
1 Building Laborer	27.40	219.20	45.40	363.20		
.5 Electrician	42.00	168.00	64.70	258.80		
28 L.H., Daily Totals		$956.00		$1564.40	$34.14	$55.87

Crew L-8	Hr.	Daily	Hr.	Daily	Bare Costs	Incl. O&P
2 Carpenters	$35.55	$568.80	$58.90	$942.40	$36.98	$60.40
.5 Plumber	42.70	170.80	66.40	265.60		
20 L.H., Daily Totals		$739.60		$1208.00	$36.98	$60.40

Crew L-9	Hr.	Daily	Hr.	Daily	Bare Costs	Incl. O&P
1 Laborer Foreman (inside)	$27.90	$223.20	$46.25	$370.00	$31.92	$54.24
2 Building Laborers	27.40	438.40	45.40	726.40		
1 Struc. Steel Worker	39.95	319.60	74.70	597.60		
.5 Electrician	42.00	168.00	64.70	258.80		
36 L.H., Daily Totals		$1149.20		$1952.80	$31.92	$54.24

Crew L-10	Hr.	Daily	Hr.	Daily	Bare Costs	Incl. O&P
1 Structural Steel Foreman	$41.95	$335.60	$78.45	$627.60	$40.00	$71.10
1 Structural Steel Worker	39.95	319.60	74.70	597.60		
1 Equip. Oper. (crane)	38.10	304.80	60.15	481.20		
1 Hyd. Crane, 12 Ton		636.60		700.25	26.53	29.18
24 L.H., Daily Totals		$1596.60		$2406.65	$66.53	$100.28

Crew L-11	Hr.	Daily	Hr.	Daily	Bare Costs	Incl. O&P
2 Wreckers	$27.40	$438.40	$51.35	$821.60	$32.03	$54.61
1 Equip. Oper. (crane)	38.10	304.80	60.15	481.20		
1 Equip. Oper. (light)	35.20	281.60	55.60	444.80		
1 Hyd. Excavator, 2.5 C.Y.		1218.00		1339.80		
1 Skid steer loader		202.00		222.20	44.38	48.81
32 L.H., Daily Totals		$2444.80		$3309.60	$76.41	$103.42

Crew M-1	Hr.	Daily	Hr.	Daily	Bare Costs	Incl. O&P
3 Elevator Constructors	$49.95	$1198.80	$77.00	$1848.00	$47.45	$73.15
1 Elevator Apprentice	39.95	319.60	61.60	492.80		
5 Hand Tools		54.00		59.40	1.69	1.86
32 L.H., Daily Totals		$1572.40		$2400.20	$49.14	$75.01

Crew M-3	Hr.	Daily	Hr.	Daily	Bare Costs	Incl. O&P
1 Electrician Foreman (out)	$44.00	$352.00	$67.80	$542.40	$40.11	$62.66
1 Common Laborer	27.40	219.20	45.40	363.20		
.25 Equipment Operator, Medium	36.70	73.40	57.95	115.90		
1 Elevator Constructor	49.95	399.60	77.00	616.00		
1 Elevator Apprentice	39.95	319.60	61.60	492.80		
.25 Crane, SP, 4 x 4, 20 ton		150.85		165.95	4.44	4.88
34 L.H., Daily Totals		$1514.65		$2296.25	$44.55	$67.54

Crew M-4	Hr.	Daily	Hr.	Daily	Bare Costs	Incl. O&P
1 Electrician Foreman (out)	$44.00	$352.00	$67.80	$542.40	$39.76	$62.14
1 Common Laborer	27.40	219.20	45.40	363.20		
.25 Equipment Operator, Crane	38.10	76.20	60.15	120.30		
.25 Equipment Operator, Oiler	32.45	64.90	51.25	102.50		
1 Elevator Constructor	49.95	399.60	77.00	616.00		
1 Elevator Apprentice	39.95	319.60	61.60	492.80		
.25 Crane, Hyd, SP, 4WD, 40 Ton		229.00		251.90	6.36	7.00
36 L.H., Daily Totals		$1660.50		$2489.10	$46.12	$69.14

Crew Q-1	Hr.	Daily	Hr.	Daily	Bare Costs	Incl. O&P
1 Plumber	$42.70	$341.60	$66.40	$531.20	$38.43	$59.75
1 Plumber Apprentice	34.15	273.20	53.10	424.80		
16 L.H., Daily Totals		$614.80		$956.00	$38.43	$59.75

Crew Q-1C	Hr.	Daily	Hr.	Daily	Bare Costs	Incl. O&P
1 Plumber	$42.70	$341.60	$66.40	$531.20	$37.85	$59.15
1 Plumber Apprentice	34.15	273.20	53.10	424.80		
1 Equip. Oper. (medium)	36.70	293.60	57.95	463.60		
1 Trencher, Chain		1460.00		1606.00	60.83	66.92
24 L.H., Daily Totals		$2368.40		$3025.60	$98.68	$126.07

Crew Q-2	Hr.	Daily	Hr.	Daily	Bare Costs	Incl. O&P
2 Plumbers	$42.70	$683.20	$66.40	$1062.40	$39.85	$61.97
1 Plumber Apprentice	34.15	273.20	53.10	424.80		
24 L.H., Daily Totals		$956.40		$1487.20	$39.85	$61.97

Crew Q-3	Hr.	Daily	Hr.	Daily	Bare Costs	Incl. O&P
1 Plumber Foreman (inside)	$43.20	$345.60	$67.20	$537.60	$40.69	$63.28
2 Plumbers	42.70	683.20	66.40	1062.40		
1 Plumber Apprentice	34.15	273.20	53.10	424.80		
32 L.H., Daily Totals		$1302.00		$2024.80	$40.69	$63.28

Crew Q-4	Hr.	Daily	Hr.	Daily	Bare Costs	Incl. O&P
1 Plumber Foreman (inside)	$43.20	$345.60	$67.20	$537.60	$40.69	$63.28
1 Plumber	42.70	341.60	66.40	531.20		
1 Welder (plumber)	42.70	341.60	66.40	531.20		
1 Plumber Apprentice	34.15	273.20	53.10	424.80		
1 Electric Welding Mach.		69.10		76.00	2.16	2.38
32 L.H., Daily Totals		$1371.10		$2100.80	$42.85	$65.66

Crew Q-5	Hr.	Daily	Hr.	Daily	Bare Costs	Incl. O&P
1 Steamfitter	$43.05	$344.40	$66.95	$535.60	$38.75	$60.25
1 Steamfitter Apprentice	34.45	275.60	53.55	428.40		
16 L.H., Daily Totals		$620.00		$964.00	$38.75	$60.25

Crew Q-6	Hr.	Daily	Hr.	Daily	Bare Costs	Incl. O&P
2 Steamfitters	$43.05	$688.80	$66.95	$1071.20	$40.18	$62.48
1 Steamfitter Apprentice	34.45	275.60	53.55	428.40		
24 L.H., Daily Totals		$964.40		$1499.60	$40.18	$62.48

Crew No.	Bare Costs		Incl. Subs O & P		Cost Per Labor-Hour	
Crew Q-7	**Hr.**	**Daily**	**Hr.**	**Daily**	**Bare Costs**	**Incl. O&P**
1 Steamfitter Foreman (inside)	$43.55	$348.40	$67.70	$541.60	$41.03	$63.79
2 Steamfitters	43.05	688.80	66.95	1071.20		
1 Steamfitter Apprentice	34.45	275.60	53.55	428.40		
32 L.H., Daily Totals		$1312.80		$2041.20	$41.03	$63.79

Crew Q-8	**Hr.**	**Daily**	**Hr.**	**Daily**	**Bare Costs**	**Incl. O&P**
1 Steamfitter Foreman (inside)	$43.55	$348.40	$67.70	$541.60	$41.03	$63.79
1 Steamfitter	43.05	344.40	66.95	535.60		
1 Welder (steamfitter)	43.05	344.40	66.95	535.60		
1 Steamfitter Apprentice	34.45	275.60	53.55	428.40		
1 Electric Welding Mach.		69.10		76.00	2.16	2.38
32 L.H., Daily Totals		$1381.90		$2117.20	$43.19	$66.17

Crew Q-9	**Hr.**	**Daily**	**Hr.**	**Daily**	**Bare Costs**	**Incl. O&P**
1 Sheet Metal Worker	$42.15	$337.20	$67.05	$536.40	$37.93	$60.33
1 Sheet Metal Apprentice	33.70	269.60	53.60	428.80		
16 L.H., Daily Totals		$606.80		$965.20	$37.93	$60.33

Crew Q-10	**Hr.**	**Daily**	**Hr.**	**Daily**	**Bare Costs**	**Incl. O&P**
2 Sheet Metal Workers	$42.15	$674.40	$67.05	$1072.80	$39.33	$62.57
1 Sheet Metal Apprentice	33.70	269.60	53.60	428.80		
24 L.H., Daily Totals		$944.00		$1501.60	$39.33	$62.57

Crew Q-11	**Hr.**	**Daily**	**Hr.**	**Daily**	**Bare Costs**	**Incl. O&P**
1 Sheet Metal Foreman (inside)	$42.65	$341.20	$67.85	$542.80	$40.16	$63.89
2 Sheet Metal Workers	42.15	674.40	67.05	1072.80		
1 Sheet Metal Apprentice	33.70	269.60	53.60	428.80		
32 L.H., Daily Totals		$1285.20		$2044.40	$40.16	$63.89

Crew Q-12	**Hr.**	**Daily**	**Hr.**	**Daily**	**Bare Costs**	**Incl. O&P**
1 Sprinkler Installer	$41.80	$334.40	$65.15	$521.20	$37.63	$58.65
1 Sprinkler Apprentice	33.45	267.60	52.15	417.20		
16 L.H., Daily Totals		$602.00		$938.40	$37.63	$58.65

Crew Q-13	**Hr.**	**Daily**	**Hr.**	**Daily**	**Bare Costs**	**Incl. O&P**
1 Sprinkler Foreman (inside)	$42.30	$338.40	$65.95	$527.60	$39.84	$62.10
2 Sprinkler Installers	41.80	668.80	65.15	1042.40		
1 Sprinkler Apprentice	33.45	267.60	52.15	417.20		
32 L.H., Daily Totals		$1274.80		$1987.20	$39.84	$62.10

Crew Q-14	**Hr.**	**Daily**	**Hr.**	**Daily**	**Bare Costs**	**Incl. O&P**
1 Asbestos Worker	$39.05	$312.40	$63.70	$509.60	$35.15	$57.33
1 Asbestos Apprentice	31.25	250.00	50.95	407.60		
16 L.H., Daily Totals		$562.40		$917.20	$35.15	$57.33

Crew Q-15	**Hr.**	**Daily**	**Hr.**	**Daily**	**Bare Costs**	**Incl. O&P**
1 Plumber	$42.70	$341.60	$66.40	$531.20	$38.43	$59.75
1 Plumber Apprentice	34.15	273.20	53.10	424.80		
1 Electric Welding Mach.		69.10		76.00	4.32	4.75
16 L.H., Daily Totals		$683.90		$1032.00	$42.75	$64.50

Crew Q-16	**Hr.**	**Daily**	**Hr.**	**Daily**	**Bare Costs**	**Incl. O&P**
2 Plumbers	$42.70	$683.20	$66.40	$1062.40	$39.85	$61.97
1 Plumber Apprentice	34.15	273.20	53.10	424.80		
1 Electric Welding Mach.		69.10		76.00	2.88	3.17
24 L.H., Daily Totals		$1025.50		$1563.20	$42.73	$65.14

Crew No.	Bare Costs		Incl. Subs O & P		Cost Per Labor-Hour	
Crew Q-17	**Hr.**	**Daily**	**Hr.**	**Daily**	**Bare Costs**	**Incl. O&P**
1 Steamfitter	$43.05	$344.40	$66.95	$535.60	$38.75	$60.25
1 Steamfitter Apprentice	34.45	275.60	53.55	428.40		
1 Electric Welding Mach.		69.10		76.00	4.32	4.75
16 L.H., Daily Totals		$689.10		$1040.00	$43.07	$65.00

Crew Q-17A	**Hr.**	**Daily**	**Hr.**	**Daily**	**Bare Costs**	**Incl. O&P**
1 Steamfitter	$43.05	$344.40	$66.95	$535.60	$38.53	$60.22
1 Steamfitter Apprentice	34.45	275.60	53.55	428.40		
1 Equip. Oper. (crane)	38.10	304.80	60.15	481.20		
1 Truck Crane, 12 Ton		636.60		700.25		
1 Electric Welding Mach.		69.10		76.00	29.40	32.34
24 L.H., Daily Totals		$1630.50		$2221.45	$67.93	$92.56

Crew Q-18	**Hr.**	**Daily**	**Hr.**	**Daily**	**Bare Costs**	**Incl. O&P**
2 Steamfitters	$43.05	$688.80	$66.95	$1071.20	$40.18	$62.48
1 Steamfitter Apprentice	34.45	275.60	53.55	428.40		
1 Electric Welding Mach.		69.10		76.00	2.88	3.17
24 L.H., Daily Totals		$1033.50		$1575.60	$43.06	$65.65

Crew Q-19	**Hr.**	**Daily**	**Hr.**	**Daily**	**Bare Costs**	**Incl. O&P**
1 Steamfitter	$43.05	$344.40	$66.95	$535.60	$39.83	$61.73
1 Steamfitter Apprentice	34.45	275.60	53.55	428.40		
1 Electrician	42.00	336.00	64.70	517.60		
24 L.H., Daily Totals		$956.00		$1481.60	$39.83	$61.73

Crew Q-20	**Hr.**	**Daily**	**Hr.**	**Daily**	**Bare Costs**	**Incl. O&P**
1 Sheet Metal Worker	$42.15	$337.20	$67.05	$536.40	$38.74	$61.20
1 Sheet Metal Apprentice	33.70	269.60	53.60	428.80		
.5 Electrician	42.00	168.00	64.70	258.80		
20 L.H., Daily Totals		$774.80		$1224.00	$38.74	$61.20

Crew Q-21	**Hr.**	**Daily**	**Hr.**	**Daily**	**Bare Costs**	**Incl. O&P**
2 Steamfitters	$43.05	$688.80	$66.95	$1071.20	$40.64	$63.04
1 Steamfitter Apprentice	34.45	275.60	53.55	428.40		
1 Electrician	42.00	336.00	64.70	517.60		
32 L.H., Daily Totals		$1300.40		$2017.20	$40.64	$63.04

Crew Q-22	**Hr.**	**Daily**	**Hr.**	**Daily**	**Bare Costs**	**Incl. O&P**
1 Plumber	$42.70	$341.60	$66.40	$531.20	$38.43	$59.75
1 Plumber Apprentice	34.15	273.20	53.10	424.80		
1 Truck Crane, 12 Ton		636.60		700.25	39.79	43.77
16 L.H., Daily Totals		$1251.40		$1656.25	$78.22	$103.52

Crew Q-22A	**Hr.**	**Daily**	**Hr.**	**Daily**	**Bare Costs**	**Incl. O&P**
1 Plumber	$42.70	$341.60	$66.40	$531.20	$35.59	$56.26
1 Plumber Apprentice	34.15	273.20	53.10	424.80		
1 Laborer	27.40	219.20	45.40	363.20		
1 Equip. Oper. (crane)	38.10	304.80	60.15	481.20		
1 Truck Crane, 12 Ton		636.60		700.25	19.89	21.88
32 L.H., Daily Totals		$1775.40		$2500.65	$55.48	$78.14

Crew Q-23	**Hr.**	**Daily**	**Hr.**	**Daily**	**Bare Costs**	**Incl. O&P**
1 Plumber Foreman	$44.70	$357.60	$69.50	$556.00	$41.37	$64.62
1 Plumber	42.70	341.60	66.40	531.20		
1 Equip. Oper. (medium)	36.70	293.60	57.95	463.60		
1 Power Tool		5.40		5.95		
1 Crane, 20 Ton		947.05		1041.75	39.68	43.65
24 L.H., Daily Totals		$1945.25		$2598.50	$81.05	$108.27

Crews

Crew No.	Bare Costs		Incl. Subs O & P		Cost Per Labor-Hour	
Crew R-1	**Hr.**	**Daily**	**Hr.**	**Daily**	**Bare Costs**	**Incl. O&P**
1 Electrician Foreman	$42.50	$340.00	$65.50	$524.00	$36.88	$57.80
3 Electricians	42.00	1008.00	64.70	1552.80		
2 Helpers	26.40	422.40	43.60	697.60		
48 L.H., Daily Totals		$1770.40		$2774.40	$36.88	$57.80

Crew R-1A	Hr.	Daily	Hr.	Daily	Bare Costs	Incl. O&P
1 Electrician	$42.00	$336.00	$64.70	$517.60	$34.20	$54.15
1 Helper	26.40	211.20	43.60	348.80		
16 L.H., Daily Totals		$547.20		$866.40	$34.20	$54.15

Crew R-2	Hr.	Daily	Hr.	Daily	Bare Costs	Incl. O&P
1 Electrician Foreman	$42.50	$340.00	$65.50	$524.00	$37.06	$58.14
3 Electricians	42.00	1008.00	64.70	1552.80		
2 Helpers	26.40	422.40	43.60	697.60		
1 Equip. Oper. (crane)	38.10	304.80	60.15	481.20		
1 S.P. Crane, 5 Ton		305.20		335.70	5.45	6.00
56 L.H., Daily Totals		$2380.40		$3591.30	$42.51	$64.14

Crew R-3	Hr.	Daily	Hr.	Daily	Bare Costs	Incl. O&P
1 Electrician Foreman	$42.50	$340.00	$65.50	$524.00	$41.42	$64.11
1 Electrician	42.00	336.00	64.70	517.60		
.5 Equip. Oper. (crane)	38.10	152.40	60.15	240.60		
.5 S.P. Crane, 5 Ton		152.60		167.85	7.63	8.39
20 L.H., Daily Totals		$981.00		$1450.05	$49.05	$72.50

Crew R-4	Hr.	Daily	Hr.	Daily	Bare Costs	Incl. O&P
1 Struc. Steel Foreman	$41.95	$335.60	$78.45	$627.60	$40.76	$73.45
3 Struc. Steel Workers	39.95	958.80	74.70	1792.80		
1 Electrician	42.00	336.00	64.70	517.60		
1 Gas Welding Machine		89.20		98.10	2.23	2.45
40 L.H., Daily Totals		$1719.60		$3036.10	$42.99	$75.90

Crew R-5	Hr.	Daily	Hr.	Daily	Bare Costs	Incl. O&P
1 Electrician Foreman	$42.50	$340.00	$65.50	$524.00	$36.37	$57.10
4 Electrician Linemen	42.00	1344.00	64.70	2070.40		
2 Electrician Operators	42.00	672.00	64.70	1035.20		
4 Electrician Groundmen	26.40	844.80	43.60	1395.20		
1 Crew Truck		133.75		147.15		
1 Tool Van		132.40		145.65		
1 Pickup Truck, 3/4 Ton		78.00		85.80		
.2 Crane, 55 Ton		190.32		209.35		
.2 Crane, 12 Ton		127.32		140.05		
.2 Auger, Truck Mtd.		675.80		743.40		
1 Tractor w/Winch		260.00		286.00	18.15	19.97
88 L.H., Daily Totals		$4798.39		$6782.20	$54.52	$77.07

Crew No.	Bare Costs		Incl. Subs O & P		Cost Per Labor-Hour	
Crew R-6	**Hr.**	**Daily**	**Hr.**	**Daily**	**Bare Costs**	**Incl. O&P**
1 Electrician Foreman	$42.50	$340.00	$65.50	$524.00	$36.37	$57.10
4 Electrician Linemen	42.00	1344.00	64.70	2070.40		
2 Electrician Operators	42.00	672.00	64.70	1035.20		
4 Electrician Groundmen	26.40	844.80	43.60	1395.20		
1 Crew Truck		133.75		147.15		
1 Tool Van		132.40		145.65		
1 Pickup Truck, 3/4 Ton		78.00		85.80		
.2 Crane, 55 Ton		190.32		209.35		
.2 Crane, 12 Ton		127.32		140.05		
.2 Auger, Truck Mtd.		675.80		743.40		
1 Tractor w/Winch		260.00		286.00		
3 Cable Trailers		500.70		550.75		
.5 Tensioning Rig		163.85		180.25		
.5 Cable Pulling Rig		960.00		1056.00	36.62	40.28
88 L.H., Daily Totals		$6422.94		$8569.20	$72.99	$97.38

Crew R-7	Hr.	Daily	Hr.	Daily	Bare Costs	Incl. O&P
1 Electrician Foreman	$42.50	$340.00	$65.50	$524.00	$29.08	$47.25
5 Electrician Groundmen	26.40	1056.00	43.60	1744.00		
1 Crew Truck		133.75		147.15	2.79	3.07
48 L.H., Daily Totals		$1529.75		$2415.15	$31.87	$50.32

Crew R-8	Hr.	Daily	Hr.	Daily	Bare Costs	Incl. O&P
1 Electrician Foreman	$42.50	$340.00	$65.50	$524.00	$36.88	$57.80
3 Electrician Linemen	42.00	1008.00	64.70	1552.80		
2 Electrician Groundmen	26.40	422.40	43.60	697.60		
1 Pickup Truck, 3/4 Ton		78.00		85.80		
1 Crew Truck		133.75		147.15	4.41	4.85
48 L.H., Daily Totals		$1982.15		$3007.35	$41.29	$62.65

Crew R-9	Hr.	Daily	Hr.	Daily	Bare Costs	Incl. O&P
1 Electrician Foreman	$42.50	$340.00	$65.50	$524.00	$34.26	$54.25
1 Electrician Lineman	42.00	336.00	64.70	517.60		
2 Electrician Operators	42.00	672.00	64.70	1035.20		
4 Electrician Groundmen	26.40	844.80	43.60	1395.20		
1 Pickup Truck, 3/4 Ton		78.00		85.80		
1 Crew Truck		133.75		147.15	3.31	3.64
64 L.H., Daily Totals		$2404.55		$3704.95	$37.57	$57.89

Crew R-10	Hr.	Daily	Hr.	Daily	Bare Costs	Incl. O&P
1 Electrician Foreman	$42.50	$340.00	$65.50	$524.00	$39.48	$61.32
4 Electrician Linemen	42.00	1344.00	64.70	2070.40		
1 Electrician Groundman	26.40	211.20	43.60	348.80		
1 Crew Truck		133.75		147.15		
3 Tram Cars		351.15		386.25	10.10	11.11
48 L.H., Daily Totals		$2380.10		$3476.60	$49.58	$72.43

Crew R-11	Hr.	Daily	Hr.	Daily	Bare Costs	Incl. O&P
1 Electrician Foreman	$42.50	$340.00	$65.50	$524.00	$39.43	$61.41
4 Electricians	42.00	1344.00	64.70	2070.40		
1 Equip. Oper. (crane)	38.10	304.80	60.15	481.20		
1 Common Laborer	27.40	219.20	45.40	363.20		
1 Crew Truck		133.75		147.15		
1 Crane, 12 Ton		636.60		700.25	13.76	15.13
56 L.H., Daily Totals		$2978.35		$4286.20	$53.19	$76.54

Crew No.	Bare Costs		Incl. Sub O & P		Cost Per Labor-Hour	
Crew R-12	**Hr.**	**Daily**	**Hr.**	**Daily**	**Bare Costs**	**Incl. O&P**
1 Carpenter Foreman	$36.05	$288.40	$59.75	$478.00	$33.14	$55.42
4 Carpenters	35.55	1137.60	58.90	1884.80		
4 Common Laborers	27.40	876.80	45.40	1452.80		
1 Equip. Oper. (med.)	36.70	293.60	57.95	463.60		
1 Steel Worker	39.95	319.60	74.70	597.60		
1 Dozer, 200 H.P.		920.80		1012.90		
1 Pickup Truck, 3/4 Ton		78.00		85.80	11.35	12.49
88 L.H., Daily Totals		$3914.80		$5975.50	$44.49	$67.91
Crew R-13	**Hr.**	**Daily**	**Hr.**	**Daily**	**Bare Costs**	**Incl. O&P**
1 Electrician Foreman	$42.50	$340.00	$65.50	$524.00	$40.09	$62.07
3 Electricians	42.00	1008.00	64.70	1552.80		
.25 Equip. Oper. (crane)	38.10	76.20	60.15	120.30		
1 Equipment Oiler	32.45	259.60	51.25	410.00		
.25-1 Hyd. Crane, 33 Ton		169.15		186.05	4.03	4.43
42 L.H., Daily Totals		$1852.95		$2793.15	$44.12	$66.50
Crew R-15	**Hr.**	**Daily**	**Hr.**	**Daily**	**Bare Costs**	**Incl. O&P**
1 Electrician Foreman	$42.50	$340.00	$65.50	$524.00	$40.95	$63.32
4 Electricians	42.00	1344.00	64.70	2070.40		
1 Equipment Operator	35.20	281.60	55.60	444.80		
1 Aerial Lift Truck		265.00		291.50	5.52	6.07
48 L.H., Daily Totals		$2230.60		$3330.70	$46.47	$69.39
Crew R-18	**Hr.**	**Daily**	**Hr.**	**Daily**	**Bare Costs**	**Incl. O&P**
.25 Electrician Foreman	$42.50	$85.00	$65.50	$131.00	$32.44	$51.78
1 Electrician	42.00	336.00	64.70	517.60		
2 Helpers	26.40	422.40	43.60	697.60		
26 L.H., Daily Totals		$843.40		$1346.20	$32.44	$51.78
Crew R-19	**Hr.**	**Daily**	**Hr.**	**Daily**	**Bare Costs**	**Incl. O&P**
.5 Electrician Foreman	$42.50	$170.00	$65.50	$262.00	$42.10	$64.86
2 Electricians	42.00	672.00	64.70	1035.20		
20 L.H., Daily Totals		$842.00		$1297.20	$42.10	$64.86
Crew R-21	**Hr.**	**Daily**	**Hr.**	**Daily**	**Bare Costs**	**Incl. O&P**
1 Electrician Foreman	$42.50	$340.00	$65.50	$524.00	$41.99	$64.73
3 Electricians	42.00	1008.00	64.70	1552.80		
.1 Equip. Oper. (med.)	36.70	29.36	57.95	46.36		
.1 Hyd. Crane 25 Ton		69.68		76.65	2.12	2.34
32. L.H., Daily Totals		$1447.04		$2199.81	$44.11	$67.07
Crew R-22	**Hr.**	**Daily**	**Hr.**	**Daily**	**Bare Costs**	**Incl. O&P**
.66 Electrician Foreman	$42.50	$224.40	$65.50	$345.84	$35.38	$55.76
2 Helpers	26.40	422.40	43.60	697.60		
2 Electricians	42.00	672.00	64.70	1035.20		
37.28 L.H., Daily Totals		$1318.80		$2078.64	$35.38	$55.76
Crew R-30	**Hr.**	**Daily**	**Hr.**	**Daily**	**Bare Costs**	**Incl. O&P**
.25 Electrician	$44.00	$88.00	$67.80	$135.60	$33.17	$53.06
1 Electrician	42.00	336.00	64.70	517.60		
2 Laborers (Semi-Skilled)	27.40	438.40	45.40	726.40		
26 L.H., Daily Totals		$862.40		$1379.60	$33.17	$53.06
Crew R-31	**Hr.**	**Daily**	**Hr.**	**Daily**	**Bare Costs**	**Incl. O&P**
1 Electrician	$42.00	$336.00	$64.70	$517.60	$42.00	$64.70
1 Core Drill, Elec, 2.5 HP		64.60		71.05	8.08	8.88
8 L.H., Daily Totals		$400.60		$588.65	$50.08	$73.58

Historical Cost Indexes

The table below lists both the Means Historical Cost Index based on Jan. 1, 1993 = 100 as well as the computed value of an index based on Jan. 1, 2006 costs. Since the Jan. 1, 2006 figure is estimated, space is left to write in the actual index figures as they become available through either the quarterly "Means Construction Cost Indexes" or as printed in the "Engineering News-Record." To compute the actual index based on Jan. 1, 2006 = 100, divide the Historical Cost Index for a particular year by the actual Jan. 1, 2006 Construction Cost Index. Space has been left to advance the index figures as the year progresses.

Year	Historical Cost Index Jan. 1, 1993 = 100		Current Index Based on Jan. 1, 2006 = 100	
	Est.	Actual	Est.	Actual
Oct 2006				
July 2006				
April 2006				
Jan 2006	155.9		100.0	100.0
July 2005		151.6	97.2	
2004		143.7	92.2	
2003		132.0	84.7	
2002		128.7	82.6	
2001		125.1	80.2	
2000		120.9	77.5	
1999		117.6	75.4	
1998		115.1	73.8	
1997		112.8	72.4	
1996		110.2	70.7	
1995		107.6	69.0	
1994		104.4	67.0	
1993		101.7	65.2	
▼ 1992		99.4	63.8	

Year	Historical Cost Index Jan. 1, 1993 = 100	Current Index Based on Jan. 1, 2006 = 100	
	Actual	Est.	Actual
July 1991	96.8	62.1	
1990	94.3	60.5	
1989	92.1	59.1	
1988	89.9	57.6	
1987	87.7	56.2	
1986	84.2	54.0	
1985	82.6	53.0	
1984	82.0	52.6	
1983	80.2	51.4	
1982	76.1	48.8	
1981	70.0	44.9	
1980	62.9	40.3	
1979	57.8	37.1	
1978	53.5	34.3	
1977	49.5	31.8	
1976	46.9	30.1	
1975	44.8	28.7	
▼ 1974	41.4	26.6	

Year	Historical Cost Index Jan. 1, 1993 = 100	Current Index Based on Jan. 1, 2006 = 100	
	Actual	Est.	Actual
July 1973	37.7	24.2	
1972	34.8	22.3	
1971	32.1	20.6	
1970	28.7	18.4	
1969	26.9	17.3	
1968	24.9	16.0	
1967	23.5	15.1	
1966	22.7	14.6	
1965	21.7	13.9	
1964	21.2	13.6	
1963	20.7	13.3	
1962	20.2	13.0	
1961	19.8	12.7	
1960	19.7	12.6	
1959	19.3	12.4	
1958	18.8	12.1	
1957	18.4	11.8	
▼ 1956	17.6	11.3	

Adjustments to Costs

The Historical Cost Index can be used to convert National Average building costs at a particular time to the approximate building costs for some other time.

Time Adjustment using the Historical Cost Indexes:

$$\frac{\text{Index for Year A}}{\text{Index for Year B}} \text{ x Cost in Year B = Cost in Year A}$$

Example:

Estimate and compare construction costs for different years in the same city.

To estimate the National Average construction cost of a building in 1970, knowing that it cost $900,000 in 2006:

INDEX in 1970 = 28.7

INDEX in 2006 = 155.9

$$\frac{\text{INDEX 1970}}{\text{INDEX 2006}} \text{ x Cost 2006 = Cost 1970}$$

$$\frac{28.7}{155.9} \text{ x \$900,000 = .184 x \$900,000 = \$165,600}$$

The construction cost of the building in 1970 is $165,600.

Note: The City Cost Indexes for Canada can be used to convert U.S. National averages to local costs in Canadian dollars.

How to Use the City Cost Indexes

What you should know before you begin

Means City Cost Indexes (CCI) are an extremely useful tool to use when you want to compare costs from city to city and region to region.

This publication contains average construction cost indexes for 719 U.S. and Canadian cities covering over 930 three-digit zip code locations, as listed directly under each city.

Keep in mind that a City Cost Index number is a *percentage ratio* of a specific city's cost to the national average cost of the same item at a stated time period.

In other words, these index figures represent relative construction *factors* (or, if you prefer, multipliers) for Material and Installation costs, as well as the weighted average for Total In Place costs for each CSI MasterFormat division. Installation costs include both labor and equipment rental costs. When estimating equipment rental rates only, for a specific location, use 01590 EQUIPMENT RENTAL index.

The 30 City Average Index is the average of 30 major U.S. cities and serves as a National Average.

Index figures for both material and installation are based on the 30 major city average of 100 and represent the cost relationship as of July 1, 2005. The index for each division is computed from representative material and labor quantities for that division. The weighted average for each city is a weighted total of the components listed above it, but does not include relative productivity between trades or cities.

As changes occur in local material prices, labor rates and equipment rental rates, the impact of these changes should be accurately measured by the change in the City Cost Index for each particular city (as compared to the 30 City Average).

Therefore, if you know (or have estimated) building costs in one city today, you can easily convert those costs to expected building costs in another city.

In addition, by using the Historical Cost Index, you can easily convert National Average building costs at a particular time to the approximate building costs for some other time. The City Cost Indexes can then be applied to calculate the costs for a particular city.

Quick Calculations

Location Adjustment Using the City Cost Indexes:

$$\frac{\text{Index for City A}}{\text{Index for City B}} \times \text{Cost in City B} = \text{Cost in City A}$$

Time Adjustment for the National Average Using the Historical Cost Index:

$$\frac{\text{Index for Year A}}{\text{Index for Year B}} \times \text{Cost in Year B} = \text{Cost in Year A}$$

Adjustment from the National Average:

$$\frac{\text{Index for City A}}{100} \times \text{National Average Cost} = \text{Cost in City A}$$

Since each of the other RSMeans publications contains many different items, any *one* item multiplied by the particular city index may give incorrect results. However, the larger the number of items compiled, the closer the results should be to actual costs for that particular city.

The City Cost Indexes for Canadian cities are calculated using Canadian material and equipment prices and labor rates, in Canadian dollars. Therefore, indexes for Canadian cities can be used to convert U.S. National Average prices to local costs in Canadian dollars.

How to use this section

1. Compare costs from city to city.

In using the Means Indexes, remember that an index number is not a fixed number but a *ratio:* It's a percentage ratio of a building component's cost at any stated time to the National Average cost of that same component at the same time period. Put in the form of an equation:

$$\frac{\text{Specific City Cost}}{\text{National Average Cost}} \times 100 = \text{City Index Number}$$

Therefore, when making cost comparisons between cities, do not subtract one city's index number from the index number of another city and read the result as a percentage difference. Instead, divide one city's index number by that of the other city. The resulting number may then be used as a multiplier to calculate cost differences from city to city.

The formula used to find cost differences between cities for the purpose of comparison is as follows:

$$\frac{\text{City A Index}}{\text{City B Index}} \times \text{City B Cost (Known)} = \text{City A Cost (Unknown)}$$

In addition, you can use *Means CCI* to calculate and compare costs division by division between cities using the same basic formula. (Just be sure that you're comparing similar divisions.)

2. Compare a specific city's construction costs with the National Average.

When you're studying construction location feasibility, it's advisable to compare a prospective project's cost index with an index of the National Average cost.

For example, divide the weighted average index of construction costs of a specific city by that of the 30 City Average, which = 100.

$$\frac{\text{City Index}}{100} = \text{\% of National Average}$$

As a result, you get a ratio that indicates the relative cost of construction in that city in comparison with the National Average.

3. Covert U.S. National Average to actual costs in Canadian City.

$$\frac{\text{Index for Canadian City}}{100} \times \text{National Average Cost} = \text{Cost in Canadian City in \\$ CAN}$$

4. Adjust construction cost data based on a National Average.

When you use a source of construction cost data which is based on a National Average (such as *Means cost data publications*), it is necessary to adjust those costs to a specific location.

$$\frac{\text{City Index}}{100} \times \begin{array}{c}\text{“Book” Cost Based on}\\ \text{National Average Costs}\end{array} = \begin{array}{c}\text{City Cost}\\ \text{(Unknown)}\end{array}$$

5. When applying the City Cost Indexes to demolition projects, use the appropriate division installation index. For example, for removal of existing doors and windows, use the Division 8 index.

What you might like to know about how we developed the Indexes

To create a reliable index, RSMeans researched the building type most often constructed in the United States and Canada. Because it was concluded that no one type of building completely represented the building construction industry, nine different types of buildings were combined to create a composite model.

The exact material, labor and equipment quantities are based on detailed analysis of these nine building types, then each quantity is weighted in proportion to expected usage. These various material items, labor hours, and equipment rental rates are thus combined to form a composite building representing as closely as possible the actual usage of materials, labor and equipment used in the North American Building Construction Industry.

The following structures were chosen to make up that composite model:

1. Factory, 1 story
2. Office, 2–4 story
3. Store, Retail
4. Town Hall, 2–3 story
5. High School, 2–3 story
6. Hospital, 4–8 story
7. Garage, Parking
8. Apartment, 1–3 story
9. Hotel/Motel, 2–3 story

For the purposes of ensuring the timeliness of the data, the components of the index for the composite model have been streamlined. They currently consist of:

- specific quantities of 66 commonly used construction materials;
- specific labor-hours for 21 building construction trades; and
- specific days of equipment rental for 6 types of construction equipment (normally used to install the 66 material items by the 21 trades.)

A sophisticated computer program handles the updating of all costs for each city on a quarterly basis. Material and equipment price quotations are gathered quarterly from 719 cities in the United States and Canada. These prices and the latest negotiated labor wage rates for 21 different building trades are used to compile the quarterly update of the City Cost Index.

The 30 major U.S. cities used to calculate the National Average are:

Atlanta, GA	Memphis, TN
Baltimore, MD	Milwaukee, WI
Boston, MA	Minneapolis, MN
Buffalo, NY	Nashville, TN
Chicago, IL	New Orleans, LA
Cincinnati, OH	New York, NY
Cleveland, OH	Philadelphia, PA
Columbus, OH	Phoenix, AZ
Dallas, TX	Pittsburgh, PA
Denver, CO	St. Louis, MO
Detroit, MI	San Antonio, TX
Houston, TX	San Diego, CA
Indianapolis, IN	San Francisco, CA
Kansas City, MO	Seattle, WA
Los Angeles, CA	Washington, DC

F.Y.I.: The CSI MasterFormat Divisions

1. General Requirements
2. Site Construction
3. Concrete
4. Masonry
5. Metals
6. Wood & Plastics
7. Thermal & Moisture Protection
8. Doors & Windows
9. Finishes
10. Specialties
11. Equipment
12. Furnishings
13. Special Construction
14. Conveying Systems
15. Mechanical
16. Electrical

The information presented in the CCI is organized according to the Construction Specifications Institute (CSI) MasterFormat.

What the CCI does not *indicate*

The weighted average for each city is a total of the components listed above weighted to reflect typical usage, but it does *not* include the productivity variations between trades or cities.

In addition, the CCI does not take into consideration factors such as the following:

- managerial efficiency
- competitive conditions
- automation
- restrictive union practices
- unique local requirements
- regional variations due to specific building codes

City Cost Indexes

DIVISION		UNITED STATES			ALABAMA														
		30 CITY			ANNISTON			BIRMINGHAM			BUTLER			DECATUR			DOTHAN		
		AVERAGE			362			350 - 352			369			356			363		
		MAT.	INST.	TOTAL	MAT.	INST.	TOTAL	MAT.	INST.	TOTAL	MAT.	INST.	TOTAL	MAT.	INST.	TOTAL	MAT.	INST.	TOTAL
01590	EQUIPMENT RENTAL	.0	100.0	100.0	.0	101.2	101.2	.0	101.3	101.3	.0	97.7	97.7	.0	101.2	101.2	.0	97.7	97.7
02	SITE CONSTRUCTION	100.0	100.0	100.0	88.9	91.5	90.8	83.2	92.9	90.3	101.8	85.3	89.7	81.6	91.3	88.7	99.4	85.3	89.1
03100	CONCRETE FORMS & ACCESSORIES	100.0	100.0	100.0	92.3	34.5	42.5	95.3	73.4	76.4	87.7	45.3	51.1	97.5	48.3	55.0	98.5	45.1	52.5
03200	CONCRETE REINFORCEMENT	100.0	100.0	100.0	92.1	86.2	89.1	92.1	87.9	90.0	97.3	55.3	76.0	92.1	53.0	72.3	97.3	55.3	76.0
03300	CAST-IN-PLACE CONCRETE	100.0	100.0	100.0	91.1	39.3	69.8	93.3	65.9	82.1	88.9	47.1	71.7	90.6	54.3	75.7	88.9	47.0	71.7
03	CONCRETE	100.0	100.0	100.0	94.8	47.9	71.5	91.7	74.6	83.2	95.8	49.5	72.8	90.5	52.9	71.8	95.3	49.4	72.5
04	MASONRY	100.0	100.0	100.0	86.0	34.2	53.9	88.5	76.2	80.9	90.4	35.5	56.3	86.9	50.4	64.3	91.2	35.5	56.6
05	METALS	100.0	100.0	100.0	92.4	89.3	91.4	92.8	94.2	93.3	91.3	76.5	86.6	94.2	77.7	88.9	91.3	76.3	86.5
06	WOOD & PLASTICS	100.0	100.0	100.0	90.9	33.9	61.2	96.3	73.7	84.5	85.5	46.8	65.3	96.3	47.5	70.9	97.7	46.8	71.2
07	THERMAL & MOISTURE PROTECTION	100.0	100.0	100.0	96.8	39.8	72.8	96.8	79.3	89.4	96.8	53.1	78.4	96.6	58.8	80.7	96.8	47.9	76.2
08	DOORS & WINDOWS	100.0	100.0	100.0	94.4	46.2	81.8	98.3	76.5	92.6	94.4	49.4	82.6	98.3	48.0	85.1	94.4	49.4	82.7
09200	PLASTER & GYPSUM BOARD	100.0	100.0	100.0	96.7	32.7	56.9	103.0	73.5	84.7	95.2	45.9	64.6	100.6	46.6	67.0	102.6	45.9	67.4
095,098	CEILINGS & ACOUSTICAL TREATMENT	100.0	100.0	100.0	97.3	32.7	58.7	104.3	73.5	85.9	97.3	45.9	66.6	102.4	46.6	69.1	97.3	45.9	66.6
09600	FLOORING	100.0	100.0	100.0	101.2	34.3	83.1	104.3	48.3	89.2	107.7	27.1	85.8	104.3	50.4	89.7	115.9	32.7	93.4
097,099	WALL FINISHES, PAINTS & COATINGS	100.0	100.0	100.0	94.8	31.9	56.4	94.8	70.9	80.2	94.8	54.0	69.9	94.8	50.5	67.8	94.8	54.0	69.9
09	FINISHES	100.0	100.0	100.0	97.8	33.6	64.3	101.1	68.1	83.9	101.0	42.5	70.6	100.1	48.5	73.2	104.4	43.5	72.7
10 - 14	TOTAL DIV. 10000 - 14000	100.0	100.0	100.0	100.0	45.7	88.4	100.0	85.3	96.9	100.0	48.6	89.0	100.0	50.5	89.5	100.0	48.6	89.0
15	MECHANICAL	100.0	100.0	100.0	99.9	37.1	73.8	99.9	67.0	86.3	97.2	39.7	73.3	99.9	48.1	78.4	97.2	39.6	73.3
16	ELECTRICAL	100.0	100.0	100.0	95.2	26.0	61.6	97.5	67.0	82.7	97.0	36.6	67.7	96.7	48.2	73.1	95.9	36.6	67.1
01 - 16	WEIGHTED AVERAGE	100.0	100.0	100.0	95.7	46.5	73.9	96.5	75.2	87.0	96.1	49.0	75.2	96.2	55.9	78.3	96.3	49.0	75.3

DIVISION		ALABAMA																	
		EVERGREEN			GADSDEN			HUNTSVILLE			JASPER			MOBILE			MONTGOMERY		
		364			359			357 - 358			355			365 - 366			360 - 361		
		MAT.	INST.	TOTAL	MAT.	INST.	TOTAL	MAT.	INST.	TOTAL	MAT.	INST.	TOTAL	MAT.	INST.	TOTAL	MAT.	INST.	TOTAL
01590	EQUIPMENT RENTAL	.0	97.7	97.7	.0	101.2	101.2	.0	101.2	101.2	.0	101.2	101.2	.0	97.7	97.7	.0	97.7	97.7
02	SITE CONSTRUCTION	102.3	85.6	90.0	87.6	91.9	90.7	81.3	92.9	89.8	87.3	91.7	90.5	94.2	85.8	88.0	92.5	86.4	88.0
03100	CONCRETE FORMS & ACCESSORIES	83.4	46.8	51.8	87.9	44.4	50.4	97.5	66.7	70.9	94.5	36.5	44.4	97.6	51.4	57.7	96.3	46.7	53.5
03200	CONCRETE REINFORCEMENT	97.4	55.3	76.0	97.5	86.6	92.0	92.1	74.8	83.3	92.1	86.3	89.2	95.1	51.8	73.1	95.1	86.3	90.6
03300	CAST-IN-PLACE CONCRETE	88.9	49.4	72.7	90.7	50.2	74.0	88.3	64.5	78.5	100.4	44.8	77.6	93.2	53.0	76.7	94.9	47.8	75.5
03	CONCRETE	95.8	50.9	73.5	94.6	56.1	75.5	89.4	68.7	79.1	97.9	50.7	74.4	92.3	53.6	73.0	93.0	56.2	74.7
04	MASONRY	90.4	39.8	59.0	85.4	42.2	58.6	88.1	68.4	75.9	82.1	36.9	54.1	88.8	50.8	65.3	87.9	35.2	55.3
05	METALS	91.3	76.3	86.5	92.4	91.4	92.1	94.2	89.8	92.8	92.4	90.1	91.7	93.0	77.4	88.0	92.8	90.8	92.1
06	WOOD & PLASTICS	80.5	46.8	62.9	86.2	42.7	63.6	96.3	66.3	80.7	93.3	34.7	62.8	96.5	51.1	72.9	94.9	46.8	69.8
07	THERMAL & MOISTURE PROTECTION	96.7	49.8	77.0	96.6	63.2	82.6	96.6	74.3	87.2	96.7	44.5	74.7	96.6	67.5	84.4	96.2	61.2	81.5
08	DOORS & WINDOWS	94.4	49.4	82.6	94.3	52.2	83.3	98.3	63.2	89.1	94.3	51.1	83.0	98.3	51.4	86.0	98.3	56.8	87.4
09200	PLASTER & GYPSUM BOARD	92.2	45.9	63.5	94.8	41.7	61.8	100.6	65.9	79.0	97.2	33.5	57.6	100.6	50.3	69.3	100.6	45.9	66.6
095,098	CEILINGS & ACOUSTICAL TREATMENT	97.3	45.9	66.6	97.3	41.7	64.1	104.3	65.9	81.4	96.4	33.5	58.9	104.3	50.3	72.1	104.3	45.9	69.5
09600	FLOORING	104.7	27.1	83.7	98.9	43.2	83.9	104.3	51.9	90.1	102.3	32.9	83.5	114.8	52.4	97.9	113.1	27.1	89.8
097,099	WALL FINISHES, PAINTS & COATINGS	94.8	54.0	69.9	94.8	61.5	74.5	94.8	61.9	74.8	94.8	39.1	60.8	99.6	54.9	72.3	94.8	54.0	69.9
09	FINISHES	99.7	43.6	70.5	96.8	44.6	69.6	100.6	62.8	80.9	97.8	35.1	65.1	105.3	51.3	77.2	104.3	42.4	72.1
10 - 14	TOTAL DIV. 10000 - 14000	100.0	50.3	89.4	100.0	76.5	95.0	100.0	83.4	96.5	100.0	46.5	88.6	100.0	64.3	92.4	100.0	76.5	95.0
15	MECHANICAL	97.2	42.0	74.3	101.9	42.6	77.3	99.9	67.2	86.3	101.9	58.1	83.7	99.9	68.2	86.7	99.9	40.2	75.1
16	ELECTRICAL	94.6	36.6	66.4	96.4	67.0	82.1	97.5	68.8	83.6	96.3	26.3	62.3	97.5	50.1	74.4	96.7	67.4	82.5
01 - 16	WEIGHTED AVERAGE	95.7	50.2	75.5	96.1	58.8	79.6	96.3	72.0	85.5	96.5	52.1	76.8	97.2	61.1	81.2	97.0	57.1	79.3

DIVISION		ALABAMA									ALASKA								
		PHENIX CITY			SELMA			TUSCALOOSA			ANCHORAGE			FAIRBANKS			JUNEAU		
		368			367			354			995 - 996			997			998		
		MAT.	INST.	TOTAL	MAT.	INST.	TOTAL	MAT.	INST.	TOTAL	MAT.	INST.	TOTAL	MAT.	INST.	TOTAL	MAT.	INST.	TOTAL
01590	EQUIPMENT RENTAL	.0	97.7	97.7	.0	97.7	97.7	.0	101.2	101.2	.0	118.4	118.4	.0	118.4	118.4	.0	118.4	118.4
02	SITE CONSTRUCTION	106.2	86.4	91.7	99.1	86.3	89.7	81.8	91.6	89.0	143.7	133.7	136.4	127.6	133.7	132.1	139.3	133.7	135.2
03100	CONCRETE FORMS & ACCESSORIES	92.1	41.1	48.1	89.2	46.0	51.9	97.4	38.5	46.6	133.2	115.6	118.0	136.3	119.9	122.1	134.7	115.6	118.2
03200	CONCRETE REINFORCEMENT	97.3	75.1	86.0	97.3	86.2	91.7	92.1	86.6	89.3	141.4	108.0	124.4	144.1	108.0	125.7	108.4	108.0	108.2
03300	CAST-IN-PLACE CONCRETE	88.9	48.3	72.2	88.9	47.7	71.9	91.9	45.7	72.9	197.4	115.2	163.6	164.6	115.7	144.5	198.3	115.2	164.1
03	CONCRETE	98.5	51.8	75.4	94.5	55.8	75.3	91.2	51.9	71.7	153.6	113.5	133.7	134.1	115.5	124.9	148.8	113.4	131.3
04	MASONRY	90.4	40.3	59.3	94.3	35.6	57.9	87.0	37.5	56.3	227.3	121.5	161.7	214.0	121.5	156.7	221.8	121.5	159.6
05	METALS	91.3	86.6	89.8	91.3	89.9	90.8	93.4	91.3	92.7	129.2	102.5	120.7	129.3	102.8	120.8	129.3	102.4	120.7
06	WOOD & PLASTICS	90.5	39.0	63.7	87.3	46.8	66.2	96.3	37.1	65.5	119.3	113.4	116.2	119.6	118.8	119.2	119.3	113.4	116.2
07	THERMAL & MOISTURE PROTECTION	97.1	61.9	82.3	96.6	53.1	78.3	96.6	60.0	81.2	172.8	115.6	148.7	170.2	118.7	148.6	170.8	115.6	147.5
08	DOORS & WINDOWS	94.4	49.9	82.7	94.4	56.8	84.6	98.3	56.7	87.4	125.9	111.0	122.0	123.1	113.4	120.6	123.1	111.0	119.9
09200	PLASTER & GYPSUM BOARD	97.8	37.9	60.6	96.3	45.9	65.0	100.6	36.0	60.5	132.8	113.6	120.9	142.5	119.1	128.0	132.8	113.6	120.9
095,098	CEILINGS & ACOUSTICAL TREATMENT	97.3	37.9	61.9	97.3	45.9	66.6	104.3	36.0	63.6	132.4	113.6	121.2	132.4	119.1	124.4	132.4	113.6	121.2
09600	FLOORING	110.7	27.1	88.1	108.5	27.1	86.5	104.3	39.7	86.8	165.0	125.3	154.2	165.0	125.3	154.2	165.0	125.3	154.2
097,099	WALL FINISHES, PAINTS & COATINGS	94.8	54.0	69.9	94.8	54.0	69.9	94.8	46.5	65.3	166.6	114.6	134.9	166.6	128.9	143.6	166.6	114.6	134.9
09	FINISHES	102.6	38.1	69.0	101.2	42.5	70.6	100.6	37.9	67.9	154.3	117.4	135.1	153.7	122.1	137.3	153.0	117.4	134.5
10 - 14	TOTAL DIV. 10000 - 14000	100.0	75.9	94.9	100.0	48.6	89.0	100.0	74.1	94.5	100.0	115.2	103.2	100.0	115.9	103.4	100.0	115.2	103.2
15	MECHANICAL	97.2	40.7	73.8	97.2	40.0	73.5	99.9	35.4	73.1	100.5	108.7	103.9	100.5	116.9	107.3	100.5	105.5	102.6
16	ELECTRICAL	96.5	61.3	79.4	95.7	36.6	67.0	97.0	67.0	82.4	148.4	114.2	131.8	150.2	114.2	132.7	150.2	114.2	132.7
01 - 16	WEIGHTED AVERAGE	96.6	54.9	78.1	95.9	51.6	76.2	96.3	55.4	78.1	133.7	114.7	125.3	130.3	117.6	124.6	132.5	114.0	124.3

City Cost Indexes

	DIVISION	ALASKA			ARIZONA														
		KETCHIKAN			CHAMBERS			FLAGSTAFF			GLOBE			KINGMAN			MESA/TEMPE		
		999			865			860			855			864			852		
		MAT.	INST.	TOTAL	MAT.	INST.	TOTAL	MAT.	INST.	TOTAL	MAT.	INST.	TOTAL	MAT.	INST.	TOTAL	MAT.	INST.	TOTAL
01590	**EQUIPMENT RENTAL**	.0	118.4	118.4	.0	95.0	95.0	.0	95.0	95.0	.0	98.3	98.3	.0	95.0	95.0	.0	98.3	98.3
02	**SITE CONSTRUCTION**	190.6	133.7	148.9	63.5	100.1	90.3	81.4	100.8	95.6	95.0	103.6	101.3	63.4	100.8	90.8	86.0	103.9	99.1
03100	CONCRETE FORMS & ACCESSORIES	126.1	115.5	117.0	96.5	54.9	60.6	102.7	67.6	72.4	93.9	54.6	60.0	94.1	66.4	70.2	97.3	63.0	67.7
03200	CONCRETE REINFORCEMENT	113.0	108.0	110.5	105.8	73.7	89.5	105.6	74.5	89.8	104.3	73.7	88.8	105.9	70.2	87.7	105.0	74.5	89.5
03300	CAST-IN-PLACE CONCRETE	320.0	115.2	235.8	90.0	62.4	78.7	90.2	76.2	84.4	97.0	62.7	82.9	89.8	62.5	78.6	97.7	69.5	86.1
03	**CONCRETE**	224.0	113.4	169.1	97.6	61.1	79.5	116.9	71.7	94.5	108.7	61.1	85.0	97.2	65.6	81.5	101.1	67.4	84.4
04	**MASONRY**	222.9	121.5	160.0	100.5	50.7	69.6	100.1	65.1	78.4	105.0	50.6	71.3	100.5	63.1	77.3	105.1	52.5	72.5
05	**METALS**	129.4	102.4	120.8	93.3	65.7	84.5	93.8	69.5	86.0	94.1	66.1	85.1	93.9	65.2	84.7	94.3	70.3	86.7
06	**WOOD & PLASTICS**	109.5	113.4	111.5	101.8	52.9	76.3	108.2	68.2	87.4	95.5	53.0	73.4	97.9	68.2	82.4	99.1	68.4	83.1
07	**THERMAL & MOISTURE PROTECTION**	173.4	115.6	149.0	97.8	61.1	82.4	99.7	69.1	86.8	99.8	58.0	82.2	97.8	66.9	84.8	99.1	63.4	84.1
08	**DOORS & WINDOWS**	123.5	111.0	120.2	102.0	56.7	90.1	102.1	68.3	93.3	98.8	56.8	87.8	102.2	66.9	92.9	98.8	68.1	90.8
09200	PLASTER & GYPSUM BOARD	127.7	113.6	118.9	89.5	51.6	66.0	92.3	67.4	76.8	94.9	51.6	68.0	85.6	67.4	74.3	96.8	67.4	78.5
095,098	CEILINGS & ACOUSTICAL TREATMENT	127.6	113.6	119.2	103.9	51.6	72.7	104.8	67.4	82.5	97.4	51.6	70.1	104.8	67.4	82.5	97.4	67.4	79.5
09600	FLOORING	165.0	125.3	154.2	92.0	49.4	80.4	94.5	49.6	82.4	95.0	49.4	82.7	91.0	67.0	84.5	96.5	61.8	87.1
097,099	WALL FINISHES, PAINTS & COATINGS	166.6	114.6	134.9	93.2	47.1	65.1	93.2	56.3	70.7	102.8	47.1	68.8	93.2	56.3	70.7	102.8	56.3	74.5
09	**FINISHES**	154.1	117.4	135.0	94.5	52.0	72.4	97.4	62.6	79.3	98.0	52.2	74.1	93.8	65.2	78.9	97.8	61.5	78.9
10 - 14	**TOTAL DIV. 10000 - 14000**	100.0	115.2	103.2	100.0	78.6	95.4	100.0	81.2	96.0	100.0	79.0	95.5	100.0	80.4	95.8	100.0	76.8	95.0
15	**MECHANICAL**	98.6	103.3	100.6	97.0	75.8	88.2	100.2	76.8	90.5	95.3	69.1	84.4	97.0	71.3	86.3	100.2	70.0	87.7
16	**ELECTRICAL**	150.2	114.2	132.7	100.5	77.9	89.5	99.5	64.1	82.3	94.4	65.8	80.5	100.5	45.1	73.6	91.4	64.1	78.2
01 - 16	**WEIGHTED AVERAGE**	141.9	113.6	129.3	96.8	68.3	84.1	100.4	72.3	87.9	98.1	65.5	83.6	96.7	67.2	83.6	97.8	68.9	85.0

	DIVISION	ARIZONA												ARKANSAS					
		PHOENIX			PRESCOTT			SHOW LOW			TUCSON			BATESVILLE			CAMDEN		
		850,853			863			859			856 - 857			725			717		
		MAT.	INST.	TOTAL	MAT.	INST.	TOTAL	MAT.	INST.	TOTAL	MAT.	INST.	TOTAL	MAT.	INST.	TOTAL	MAT.	INST.	TOTAL
01590	**EQUIPMENT RENTAL**	.0	98.9	98.9	.0	95.0	95.0	.0	98.3	98.3	.0	98.3	98.3	.0	86.0	86.0	.0	86.0	86.0
02	**SITE CONSTRUCTION**	86.5	104.6	99.8	69.6	100.0	91.9	97.1	103.6	101.9	82.9	104.4	98.6	72.8	83.5	80.6	73.6	83.1	80.6
03100	CONCRETE FORMS & ACCESSORIES	98.4	68.9	73.0	98.1	54.6	60.5	101.8	55.0	61.5	97.9	68.6	72.6	80.5	43.3	48.4	80.2	30.5	37.3
03200	CONCRETE REINFORCEMENT	103.3	74.6	88.7	105.6	70.8	87.9	105.0	73.8	89.2	86.8	74.5	80.5	96.2	47.0	71.2	97.5	25.9	61.1
03300	CAST-IN-PLACE CONCRETE	97.8	76.7	89.1	90.1	62.3	78.7	97.0	62.9	83.0	100.5	76.5	90.7	77.7	44.8	64.2	79.6	33.2	60.6
03	**CONCRETE**	100.7	72.5	86.7	102.7	60.4	81.7	110.8	61.4	86.3	99.3	72.3	85.9	79.3	45.3	62.4	80.8	31.8	56.5
04	**MASONRY**	93.3	66.2	76.5	100.5	57.1	73.6	105.0	52.5	72.5	95.0	65.1	76.5	99.9	41.2	63.5	108.6	35.6	63.3
05	**METALS**	95.8	71.2	87.9	93.8	65.1	84.6	93.9	66.7	85.2	95.0	70.1	87.1	94.5	58.4	82.9	94.5	49.5	80.1
06	**WOOD & PLASTICS**	100.1	69.7	84.3	103.3	52.9	77.1	103.8	53.0	77.3	99.3	69.7	83.9	82.2	44.5	62.6	82.0	31.5	55.7
07	**THERMAL & MOISTURE PROTECTION**	99.0	69.2	86.5	98.3	60.0	82.2	100.0	61.4	83.8	100.7	65.6	85.9	97.9	44.5	75.4	97.7	33.5	70.7
08	**DOORS & WINDOWS**	100.0	69.1	91.9	102.1	56.2	90.1	97.9	56.8	87.1	95.9	69.1	88.9	96.1	41.6	81.9	92.4	27.5	75.4
09200	PLASTER & GYPSUM BOARD	98.9	68.7	80.1	89.6	51.6	66.0	98.6	51.6	69.4	97.1	68.7	79.5	79.0	43.6	57.0	79.0	30.3	48.8
095,098	CEILINGS & ACOUSTICAL TREATMENT	104.2	68.7	83.0	103.1	51.6	72.4	97.4	51.6	70.1	97.4	68.7	80.3	84.5	43.6	60.1	84.5	30.3	52.2
09600	FLOORING	96.8	64.6	88.1	92.8	49.4	81.1	98.5	56.2	87.0	96.0	49.6	83.4	106.1	48.2	90.5	106.0	16.0	81.6
097,099	WALL FINISHES, PAINTS & COATINGS	102.8	58.3	75.7	93.2	47.1	65.1	102.8	47.1	68.8	100.5	56.3	73.5	97.0	43.0	64.1	97.0	33.6	58.3
09	**FINISHES**	100.0	66.5	82.6	95.0	52.0	72.6	99.8	53.4	75.6	97.5	63.5	79.8	90.6	44.4	66.6	90.6	28.7	58.4
10 - 14	**TOTAL DIV. 10000 - 14000**	100.0	81.7	96.1	100.0	78.6	95.4	100.0	79.0	95.5	100.0	81.7	96.1	100.0	44.5	88.2	100.0	40.7	87.4
15	**MECHANICAL**	100.2	76.9	90.6	100.2	75.4	89.9	95.3	75.8	87.2	100.2	70.5	87.9	95.1	40.7	72.6	95.1	31.4	68.7
16	**ELECTRICAL**	100.5	64.1	82.8	99.2	64.1	82.1	91.8	69.2	80.8	93.4	62.0	78.1	97.7	46.6	72.8	94.4	38.8	67.4
01 - 16	**WEIGHTED AVERAGE**	98.8	73.6	87.6	98.3	66.8	84.3	98.2	68.0	84.8	97.1	71.2	85.6	93.3	48.4	73.3	93.2	38.9	69.1

	DIVISION	ARKANSAS																	
		FAYETTEVILLE			FORT SMITH			HARRISON			HOT SPRINGS			JONESBORO			LITTLE ROCK		
		727			729			726			719			724			720 - 722		
		MAT.	INST.	TOTAL	MAT.	INST.	TOTAL	MAT.	INST.	TOTAL	MAT.	INST.	TOTAL	MAT.	INST.	TOTAL	MAT.	INST.	TOTAL
01590	**EQUIPMENT RENTAL**	.0	86.0	86.0	.0	86.0	86.0	.0	86.0	86.0	.0	86.0	86.0	.0	107.8	107.8	.0	86.0	86.0
02	**SITE CONSTRUCTION**	72.3	83.5	80.5	77.2	83.6	81.9	77.5	83.5	81.9	77.1	82.8	81.3	99.8	98.9	99.1	77.0	83.6	81.8
03100	CONCRETE FORMS & ACCESSORIES	75.6	39.5	44.4	97.5	41.0	48.7	85.5	43.3	49.1	77.2	28.6	35.3	84.2	46.3	51.5	91.7	57.6	62.3
03200	CONCRETE REINFORCEMENT	96.2	45.4	70.4	97.3	74.4	85.7	95.7	47.0	71.0	95.6	24.6	59.6	93.0	47.7	70.0	97.5	69.6	83.3
03300	CAST-IN-PLACE CONCRETE	77.7	44.3	64.0	88.9	66.5	79.7	86.2	44.8	69.2	81.5	32.3	61.3	84.7	55.3	72.6	88.9	66.7	79.8
03	**CONCRETE**	78.9	43.2	61.2	86.9	56.8	72.0	85.8	45.3	65.7	83.7	30.4	57.2	83.5	51.3	67.5	86.5	63.4	75.0
04	**MASONRY**	89.9	41.7	60.0	96.0	55.2	70.7	100.2	41.2	63.6	80.8	27.6	47.8	91.4	45.5	62.9	94.1	55.2	70.0
05	**METALS**	94.5	57.7	82.7	96.3	70.7	88.2	95.3	58.4	83.5	94.5	48.2	79.7	90.6	73.8	85.3	92.2	69.4	84.9
06	**WOOD & PLASTICS**	77.3	39.8	57.8	100.0	39.3	68.4	87.9	44.5	65.3	78.6	28.5	52.5	85.7	47.3	65.7	96.9	61.3	78.4
07	**THERMAL & MOISTURE PROTECTION**	99.1	46.3	76.9	99.6	47.8	77.8	98.2	44.5	75.6	98.0	34.4	71.2	101.5	50.1	79.9	98.0	50.1	77.8
08	**DOORS & WINDOWS**	96.1	39.8	81.4	96.9	44.8	83.3	96.9	41.6	82.5	92.4	27.6	75.4	98.5	46.2	84.9	96.9	56.3	86.3
09200	PLASTER & GYPSUM BOARD	76.1	38.8	52.9	83.8	38.4	55.6	82.7	43.6	58.4	77.2	27.3	46.2	92.6	46.2	63.7	83.8	60.9	69.6
095,098	CEILINGS & ACOUSTICAL TREATMENT	84.5	38.8	57.2	89.7	38.4	59.1	88.7	43.6	61.8	84.5	27.3	50.4	87.6	46.2	62.9	89.7	60.9	72.5
09600	FLOORING	102.5	48.2	87.8	115.3	73.1	103.9	109.3	48.2	92.8	104.4	48.2	89.2	77.1	43.6	68.0	116.6	73.1	104.8
097,099	WALL FINISHES, PAINTS & COATINGS	97.0	34.9	59.1	97.0	56.3	72.1	97.0	43.0	64.1	97.0	35.9	59.7	85.5	50.6	64.2	97.0	57.7	73.0
09	**FINISHES**	89.2	40.5	63.8	95.5	48.0	70.8	93.4	44.4	67.9	90.1	33.0	60.4	87.4	46.4	66.0	95.8	61.1	77.7
10 - 14	**TOTAL DIV. 10000 - 14000**	100.0	58.9	91.2	100.0	71.9	94.0	100.0	44.5	88.2	100.0	40.5	87.3	100.0	51.2	89.6	100.0	74.6	94.6
15	**MECHANICAL**	95.1	37.4	71.2	100.1	41.6	75.8	95.1	40.7	72.6	95.1	32.4	69.1	100.2	41.5	75.8	100.1	57.9	82.6
16	**ELECTRICAL**	92.0	43.0	68.2	95.2	66.6	81.3	96.3	46.6	72.1	96.4	38.0	68.0	102.1	46.6	75.1	94.9	68.4	82.0
01 - 16	**WEIGHTED AVERAGE**	91.9	46.7	71.8	95.9	56.9	78.6	94.5	48.4	74.0	92.4	38.4	68.4	95.2	53.4	76.6	95.1	63.8	81.2

	DIVISION	ARKANSAS												CALIFORNIA					
		PINE BLUFF			RUSSELLVILLE			TEXARKANA			WEST MEMPHIS			ALHAMBRA			ANAHEIM		
		716			728			718			723			917 - 918			928		
		MAT.	INST.	TOTAL	MAT.	INST.	TOTAL	MAT.	INST.	TOTAL	MAT.	INST.	TOTAL	MAT.	INST.	TOTAL	MAT.	INST.	TOTAL
01590	**EQUIPMENT RENTAL**	.0	86.0	86.0	.0	86.0	86.0	.0	86.7	86.7	.0	107.8	107.8	.0	97.0	97.0	.0	102.2	102.2
02	**SITE CONSTRUCTION**	79.2	83.6	82.4	74.2	83.5	81.0	94.6	84.4	87.1	107.2	98.9	101.1	98.8	107.3	105.1	103.2	109.0	107.5
03100	CONCRETE FORMS & ACCESSORIES	76.8	57.4	60.0	81.4	39.4	45.1	84.7	38.4	44.8	90.3	46.3	52.4	118.6	116.5	116.8	105.7	121.7	119.5
03200	CONCRETE REINFORCEMENT	97.4	69.6	83.3	96.8	48.9	72.5	97.0	46.2	71.2	93.0	47.7	70.0	107.3	112.7	110.0	95.4	112.6	104.1
03300	CAST-IN-PLACE CONCRETE	81.5	66.6	75.4	81.4	44.7	66.3	88.9	43.1	70.1	88.8	55.3	75.0	107.7	109.7	108.5	108.1	118.5	112.4
03	**CONCRETE**	84.3	63.2	73.9	82.1	43.9	63.1	83.8	42.5	63.3	89.7	51.3	70.6	111.9	112.4	112.2	107.3	118.0	112.6
04	**MASONRY**	115.9	55.2	78.3	95.4	41.2	61.8	96.0	31.9	56.2	78.3	45.5	58.0	125.6	111.8	117.0	93.2	111.0	104.2
05	**METALS**	95.0	69.1	86.8	94.5	59.0	83.1	87.6	57.7	78.1	90.0	73.8	84.8	91.0	96.4	92.8	110.5	99.8	107.1
06	**WOOD & PLASTICS**	78.1	61.3	69.4	83.7	39.3	60.6	87.6	41.0	63.3	91.9	47.3	68.7	103.0	116.5	110.0	96.3	121.9	109.6
07	**THERMAL & MOISTURE PROTECTION**	98.0	50.1	77.9	99.4	43.9	76.0	99.2	41.0	74.7	102.0	50.1	80.2	98.7	110.0	103.5	102.7	115.2	107.9
08	**DOORS & WINDOWS**	92.4	56.3	83.0	96.1	39.3	81.3	97.4	39.7	82.3	98.5	46.2	84.9	95.6	113.6	100.3	102.7	116.6	106.3
09200	PLASTER & GYPSUM BOARD	76.8	60.9	66.9	79.0	38.4	53.8	81.9	40.1	55.9	94.4	46.2	64.4	96.8	117.1	109.4	98.8	122.4	113.5
095,098	CEILINGS & ACOUSTICAL TREATMENT	85.4	60.9	70.8	84.5	38.4	57.0	92.3	40.1	61.1	86.6	46.2	62.5	97.4	117.1	109.1	120.4	122.4	121.6
09600	FLOORING	104.2	73.1	95.8	105.7	48.2	90.1	107.0	41.6	89.3	79.9	43.6	70.1	107.5	104.8	106.7	122.8	104.8	117.9
097,099	WALL FINISHES, PAINTS & COATINGS	97.0	57.7	73.0	97.0	43.0	64.1	97.0	35.5	59.4	85.5	50.6	64.2	111.1	109.0	109.8	110.9	109.0	109.8
09	**FINISHES**	90.3	61.1	75.1	90.7	41.4	65.0	94.1	38.8	65.3	88.9	46.4	66.7	104.5	113.8	109.3	114.3	117.5	116.0
10 - 14	**TOTAL DIV. 10000 - 14000**	100.0	74.6	94.6	100.0	43.9	88.0	100.0	36.6	86.5	100.0	51.2	89.6	100.0	98.0	99.6	100.0	113.0	102.8
15	**MECHANICAL**	100.1	43.9	76.8	95.1	35.6	70.4	100.1	36.7	73.8	95.3	41.5	73.0	95.3	106.0	99.7	100.2	107.3	103.1
16	**ELECTRICAL**	94.6	68.4	81.8	95.2	46.6	71.5	96.3	38.8	68.4	103.6	46.6	75.9	122.3	113.4	118.0	89.5	105.3	97.2
01 - 16	**WEIGHTED AVERAGE**	95.2	60.9	80.0	93.2	46.6	72.5	94.6	44.0	72.1	94.6	53.4	76.3	102.7	109.1	105.5	102.6	110.6	106.1

	DIVISION	CALIFORNIA																	
		BAKERSFIELD			BERKELEY			EUREKA			FRESNO			INGLEWOOD			LONG BEACH		
		932 - 933			947			955			936 - 938			903 - 905			906 - 908		
		MAT.	INST.	TOTAL	MAT.	INST.	TOTAL	MAT.	INST.	TOTAL	MAT.	INST.	TOTAL	MAT.	INST.	TOTAL	MAT.	INST.	TOTAL
01590	**EQUIPMENT RENTAL**	.0	99.6	99.6	.0	103.5	103.5	.0	99.2	99.2	.0	99.6	99.6	.0	96.7	96.7	.0	96.7	96.7
02	**SITE CONSTRUCTION**	108.0	106.1	106.6	145.4	104.6	115.5	118.6	103.1	107.2	109.5	106.2	107.1	93.4	104.6	101.6	100.8	104.6	103.6
03100	CONCRETE FORMS & ACCESSORIES	97.9	121.2	118.0	120.3	133.9	132.1	116.6	119.3	119.0	101.9	124.6	121.5	109.2	116.4	115.4	102.7	116.4	114.5
03200	CONCRETE REINFORCEMENT	108.6	112.5	110.6	100.6	113.6	107.2	105.3	112.5	109.0	91.9	112.8	102.6	112.8	112.7	112.7	111.8	112.7	112.2
03300	CAST-IN-PLACE CONCRETE	103.2	117.4	109.1	145.7	117.3	134.0	112.0	106.7	109.8	113.0	113.3	113.1	83.8	110.6	94.8	95.4	110.6	101.7
03	**CONCRETE**	106.5	117.3	111.9	125.8	123.3	124.6	119.1	112.7	115.9	108.9	117.5	113.2	97.7	112.8	105.2	107.3	112.8	110.0
04	**MASONRY**	109.1	110.3	109.9	145.8	124.0	132.3	115.4	103.8	108.2	112.7	113.5	113.2	80.1	111.9	99.8	89.0	111.9	103.2
05	**METALS**	105.1	98.9	103.1	110.3	105.1	108.6	110.1	95.8	105.5	110.7	100.4	107.4	104.9	97.6	102.6	104.8	97.6	102.5
06	**WOOD & PLASTICS**	86.8	122.0	105.1	118.0	135.5	127.1	110.0	125.8	118.2	100.2	125.8	113.5	96.3	116.4	106.8	89.2	116.4	103.4
07	**THERMAL & MOISTURE PROTECTION**	99.2	109.1	103.4	113.2	128.2	119.5	106.6	97.6	102.8	95.9	111.4	102.4	100.5	111.9	105.3	100.7	111.9	105.4
08	**DOORS & WINDOWS**	101.5	114.4	104.9	106.0	127.7	111.7	102.8	110.1	104.7	101.9	116.5	105.7	91.0	113.6	96.9	91.0	113.6	96.9
09200	PLASTER & GYPSUM BOARD	98.8	122.4	113.5	108.0	135.9	125.3	103.5	126.3	117.6	96.6	126.3	115.0	102.6	117.1	111.6	99.7	117.1	110.5
095,098	CEILINGS & ACOUSTICAL TREATMENT	120.4	122.4	121.6	113.1	135.9	126.7	122.5	126.3	124.7	120.4	126.3	123.9	110.8	117.1	114.5	110.8	117.1	114.5
09600	FLOORING	117.0	93.7	110.7	121.9	126.1	123.0	122.0	103.8	117.1	133.3	135.6	133.9	115.7	104.8	112.7	111.8	104.8	109.9
097,099	WALL FINISHES, PAINTS & COATINGS	110.5	95.5	101.4	115.4	137.8	129.0	109.9	59.0	78.8	133.7	96.7	111.1	100.3	109.0	105.6	100.3	109.0	105.6
09	**FINISHES**	114.5	114.1	114.3	118.4	133.7	126.3	118.0	112.0	114.9	120.7	124.5	122.7	108.9	113.7	111.4	107.7	113.7	110.8
10 - 14	**TOTAL DIV. 10000 - 14000**	100.0	111.9	102.5	100.0	128.3	106.0	100.0	121.8	104.6	100.0	125.3	105.4	100.0	97.8	99.5	100.0	97.8	99.5
15	**MECHANICAL**	100.2	96.2	98.6	95.4	135.4	112.0	95.3	95.4	95.4	100.2	109.1	103.9	95.2	105.9	99.7	95.2	105.9	99.7
16	**ELECTRICAL**	89.6	97.2	93.3	104.1	138.9	121.0	98.6	80.5	89.8	88.4	96.6	92.4	100.0	113.4	106.5	99.7	113.4	106.4
01 - 16	**WEIGHTED AVERAGE**	102.3	105.9	103.9	110.1	126.7	117.5	105.9	101.1	103.8	104.1	110.9	107.1	98.2	109.0	103.0	99.6	109.0	103.8

	DIVISION	CALIFORNIA																	
		LOS ANGELES			MARYSVILLE			MODESTO			MOJAVE			OAKLAND			OXNARD		
		900 - 902			959			953			935			946			930		
		MAT.	INST.	TOTAL	MAT.	INST.	TOTAL	MAT.	INST.	TOTAL	MAT.	INST.	TOTAL	MAT.	INST.	TOTAL	MAT.	INST.	TOTAL
01590	**EQUIPMENT RENTAL**	.0	98.6	98.6	.0	99.2	99.2	.0	99.2	99.2	.0	99.6	99.6	.0	103.5	103.5	.0	98.2	98.2
02	**SITE CONSTRUCTION**	100.5	107.8	105.8	114.3	105.0	107.5	107.8	105.6	106.2	101.6	106.0	104.8	154.4	104.7	117.9	109.5	103.9	105.4
03100	CONCRETE FORMS & ACCESSORIES	105.8	121.6	119.4	104.0	123.6	120.9	98.9	124.7	121.1	112.5	111.8	111.9	106.3	134.8	130.9	104.0	121.8	119.3
03200	CONCRETE REINFORCEMENT	112.8	112.9	112.9	105.3	112.8	109.1	109.2	112.8	111.1	110.6	112.4	111.5	102.9	113.6	108.3	108.6	112.5	110.6
03300	CAST-IN-PLACE CONCRETE	91.3	116.8	101.8	125.2	111.5	119.6	112.0	113.3	112.5	94.1	114.1	102.3	141.7	117.9	131.9	109.6	117.7	112.9
03	**CONCRETE**	104.0	117.3	110.6	120.9	116.4	118.7	111.0	117.5	114.2	100.9	112.0	106.4	126.7	123.9	125.3	110.0	117.7	113.8
04	**MASONRY**	95.6	117.3	109.1	116.4	110.8	112.9	113.1	113.5	113.3	109.4	108.0	108.6	154.1	125.1	136.1	113.7	109.6	111.1
05	**METALS**	111.7	98.7	107.6	109.6	99.7	106.4	106.6	100.6	104.7	106.7	98.6	104.1	104.6	105.1	104.8	104.9	99.5	103.1
06	**WOOD & PLASTICS**	92.3	121.6	107.6	95.6	125.8	111.3	90.6	125.8	108.9	101.7	110.7	106.4	102.8	136.2	120.2	95.1	122.0	109.1
07	**THERMAL & MOISTURE PROTECTION**	100.7	117.4	107.7	105.9	108.9	107.2	105.5	110.8	107.7	99.9	106.9	102.9	110.1	128.7	118.0	105.4	113.5	108.8
08	**DOORS & WINDOWS**	97.0	116.4	102.1	102.0	118.1	106.2	100.8	118.1	105.3	97.3	108.3	100.2	106.1	128.1	111.8	100.4	116.6	104.6
09200	PLASTER & GYPSUM BOARD	104.1	122.4	115.5	96.8	126.3	115.1	99.2	126.3	116.0	107.5	110.8	109.5	101.1	136.5	123.1	98.8	122.4	113.5
095,098	CEILINGS & ACOUSTICAL TREATMENT	123.5	122.4	122.9	120.8	126.3	124.1	115.7	126.3	122.0	120.8	110.8	114.8	116.7	136.5	128.5	120.4	122.4	121.6
09600	FLOORING	113.5	104.8	111.2	116.4	113.9	115.7	117.0	114.1	116.2	124.6	93.7	116.2	116.1	126.1	118.8	117.0	104.8	113.7
097,099	WALL FINISHES, PAINTS & COATINGS	100.3	109.0	105.6	109.9	115.8	113.5	109.9	110.4	110.2	109.9	87.8	96.4	115.4	137.8	129.0	109.9	102.7	105.5
09	**FINISHES**	112.0	117.3	114.7	114.4	122.1	118.4	113.2	122.3	118.0	117.8	106.1	111.7	117.0	134.3	126.0	114.3	116.9	115.7
10 - 14	**TOTAL DIV. 10000 - 14000**	100.0	112.3	102.6	100.0	124.4	105.2	100.0	125.3	105.4	100.0	109.6	102.0	100.0	128.8	106.1	100.0	113.3	102.8
15	**MECHANICAL**	100.1	107.2	103.1	95.3	105.6	99.6	100.2	109.1	103.9	95.3	94.9	95.1	100.3	136.0	115.1	100.2	107.3	103.2
16	**ELECTRICAL**	97.0	114.2	105.3	95.3	120.1	107.4	97.9	99.7	98.8	87.7	97.2	92.3	103.4	138.9	120.6	95.4	106.8	100.9
01 - 16	**WEIGHTED AVERAGE**	102.5	112.1	106.8	105.0	112.5	108.3	104.3	111.0	107.3	100.5	103.1	101.7	110.7	127.1	118.0	103.8	110.0	106.5

City Cost Indexes

	DIVISION	CALIFORNIA																	
		PALM SPRINGS			PALO ALTO			PASADENA			REDDING			RICHMOND			RIVERSIDE		
		922			943			910 - 912			960			948			925		
		MAT.	INST.	TOTAL	MAT.	INST.	TOTAL	MAT.	INST.	TOTAL	MAT.	INST.	TOTAL	MAT.	INST.	TOTAL	MAT.	INST.	TOTAL
01590	EQUIPMENT RENTAL	.0	100.7	100.7	.0	103.5	103.5	.0	97.0	97.0	.0	99.2	99.2	.0	103.5	103.5	.0	100.7	100.7
02	SITE CONSTRUCTION	93.7	106.4	103.1	139.9	104.6	114.0	95.5	107.3	104.2	114.3	105.0	107.5	153.2	104.7	117.6	100.8	106.7	105.2
03100	CONCRETE FORMS & ACCESSORIES	101.4	116.6	114.5	103.8	133.9	129.7	104.7	117.6	115.8	103.7	123.5	120.8	123.8	134.7	133.2	106.2	121.6	119.5
03200	CONCRETE REINFORCEMENT	110.8	112.5	111.6	100.6	113.6	107.2	108.2	112.7	110.5	105.3	112.8	109.1	100.6	113.6	107.2	107.6	112.5	110.1
03300	CAST-IN-PLACE CONCRETE	98.4	112.6	104.3	123.1	117.2	120.7	102.2	109.7	105.3	124.1	111.5	118.9	141.7	117.9	131.9	107.1	118.5	111.8
03	CONCRETE	102.3	113.6	107.9	112.6	123.2	117.9	106.7	112.9	109.8	119.7	116.4	118.1	127.6	123.8	125.7	108.8	117.9	113.3
04	MASONRY	85.9	108.3	99.8	120.7	124.0	122.7	107.6	111.8	110.2	116.3	110.8	112.9	145.5	125.1	132.8	86.9	110.6	101.6
05	METALS	111.3	99.2	107.5	102.7	105.0	103.4	91.1	96.5	92.8	109.9	99.6	106.6	102.7	105.0	103.5	110.8	99.7	107.2
06	WOOD & PLASTICS	90.5	116.7	104.1	99.8	135.5	118.4	88.2	118.0	103.7	95.3	125.8	111.1	122.3	136.2	129.5	96.3	121.9	109.6
07	THERMAL & MOISTURE PROTECTION	102.7	112.1	106.6	109.7	128.4	117.6	98.3	110.2	103.3	105.9	109.5	107.4	110.4	128.1	117.8	102.9	114.7	107.9
08	DOORS & WINDOWS	99.6	113.7	103.3	106.1	125.0	111.0	95.6	114.4	100.5	103.2	118.1	107.1	106.1	128.1	111.8	102.7	116.6	106.3
09200	PLASTER & GYPSUM BOARD	94.0	117.1	108.3	99.6	135.9	122.1	92.0	118.6	108.5	97.2	126.3	115.3	109.6	136.5	126.3	97.7	122.4	113.0
095,098	CEILINGS & ACOUSTICAL TREATMENT	111.4	117.1	114.8	114.8	135.9	127.4	97.4	118.6	110.0	127.2	126.3	126.7	114.8	136.5	127.8	116.2	122.4	119.9
09600	FLOORING	119.1	104.8	115.2	115.1	126.1	118.1	102.0	104.8	102.7	116.2	113.9	115.6	124.0	126.1	124.6	121.5	104.8	117.0
097,099	WALL FINISHES, PAINTS & COATINGS	107.4	109.0	108.4	115.4	137.8	129.0	111.1	109.0	109.8	109.9	115.8	113.5	115.4	137.8	129.0	107.4	109.0	108.4
09	FINISHES	109.4	113.9	111.7	115.3	133.7	124.9	102.0	114.6	108.5	116.0	122.1	119.2	120.2	134.3	127.5	112.2	117.5	115.0
10 - 14	TOTAL DIV. 10000 - 14000	100.0	98.5	99.7	100.0	128.3	106.0	100.0	98.2	99.6	100.0	124.4	105.2	100.0	128.8	106.1	100.0	113.0	102.8
15	MECHANICAL	95.3	106.0	99.7	95.4	137.5	112.8	95.3	106.0	99.7	100.2	105.6	102.5	95.4	133.4	111.2	100.1	107.3	103.1
16	ELECTRICAL	93.0	99.7	96.3	103.3	140.8	121.5	119.3	113.4	116.5	98.1	104.2	101.1	103.9	136.0	119.5	89.6	99.7	94.5
01 - 16	WEIGHTED AVERAGE	100.1	107.2	103.2	105.5	127.2	115.1	100.4	109.3	104.4	106.6	110.3	108.3	109.5	126.1	116.9	102.2	109.5	105.5

	DIVISION	CALIFORNIA																	
		SACRAMENTO			SALINAS			SAN BERNARDINO			SAN DIEGO			SAN FRANCISCO			SAN JOSE		
		942,956 - 958			939			923 - 924			919 - 921			940 - 941			951		
		MAT.	INST.	TOTAL	MAT.	INST.	TOTAL	MAT.	INST.	TOTAL	MAT.	INST.	TOTAL	MAT.	INST.	TOTAL	MAT.	INST.	TOTAL
01590	EQUIPMENT RENTAL	.0	103.0	103.0	.0	99.6	99.6	.0	100.7	100.7	.0	97.1	97.1	.0	108.8	108.8	.0	100.0	100.0
02	SITE CONSTRUCTION	118.7	111.3	113.3	125.0	106.3	111.3	78.3	106.5	99.0	101.2	100.7	100.8	156.7	111.4	123.5	147.9	99.8	112.6
03100	CONCRETE FORMS & ACCESSORIES	104.8	125.2	122.4	108.2	125.4	123.0	110.6	117.2	116.3	108.1	108.7	108.6	106.7	135.6	131.7	106.3	134.7	130.8
03200	CONCRETE REINFORCEMENT	95.2	112.9	104.2	109.2	113.3	111.3	107.6	112.5	110.1	104.9	112.4	108.7	117.2	113.9	115.5	95.9	113.6	104.9
03300	CAST-IN-PLACE CONCRETE	122.1	114.1	118.8	108.3	113.6	110.4	73.9	113.7	90.2	112.0	106.1	109.6	141.6	119.5	132.5	129.7	117.5	124.7
03	CONCRETE	115.9	118.0	116.9	120.4	118.0	119.2	81.4	114.3	97.7	112.9	107.9	110.4	129.0	124.9	127.0	117.9	123.8	120.8
04	MASONRY	123.3	113.6	117.3	108.9	120.9	116.3	93.4	110.2	103.8	99.4	108.6	105.1	154.4	128.5	138.3	149.4	125.2	134.4
05	METALS	100.0	100.0	100.0	109.8	101.7	107.2	110.7	99.2	107.0	108.4	98.8	105.3	110.7	107.1	109.5	104.2	106.9	105.1
06	WOOD & PLASTICS	97.5	126.0	112.4	100.4	125.8	113.6	100.7	116.7	109.0	102.2	105.8	104.1	102.8	136.4	120.3	102.0	135.9	119.7
07	THERMAL & MOISTURE PROTECTION	110.6	113.2	111.7	101.0	122.5	110.1	101.8	112.6	106.4	107.8	103.4	105.9	113.3	130.6	120.6	101.9	129.5	113.5
08	DOORS & WINDOWS	117.6	118.2	117.7	101.8	119.7	106.4	99.7	113.7	103.4	103.8	106.6	104.5	110.5	128.4	115.2	92.7	127.9	101.9
09200	PLASTER & GYPSUM BOARD	98.7	126.3	115.8	100.1	126.3	116.4	99.9	117.1	110.6	97.4	105.7	102.6	103.9	136.5	124.2	100.4	136.5	122.8
095,098	CEILINGS & ACOUSTICAL TREATMENT	116.7	126.3	122.4	120.8	126.3	124.1	115.7	117.1	116.5	104.2	105.7	105.1	125.8	136.5	132.2	108.7	136.5	125.3
09600	FLOORING	118.9	113.9	117.6	117.9	126.1	120.2	123.8	104.8	118.7	110.7	104.8	109.1	116.1	126.1	118.8	113.9	126.1	117.2
097,099	WALL FINISHES, PAINTS & COATINGS	112.8	115.8	114.7	110.8	137.8	127.3	107.4	109.0	108.4	108.2	109.0	108.7	115.4	149.6	136.3	111.5	137.8	127.5
09	FINISHES	115.6	123.1	119.5	116.5	127.5	122.2	111.7	114.2	113.0	107.7	107.8	107.8	119.9	136.1	128.3	112.5	134.1	123.7
10 - 14	TOTAL DIV. 10000 - 14000	100.0	126.1	105.6	100.0	125.4	105.4	100.0	99.2	99.8	100.0	110.9	102.3	100.0	129.4	106.3	100.0	128.1	106.0
15	MECHANICAL	100.2	110.6	104.5	95.3	109.1	101.1	95.3	104.5	99.1	100.3	104.0	101.8	100.3	148.6	120.3	100.2	138.0	115.9
16	ELECTRICAL	97.7	106.1	101.8	88.4	123.2	105.3	93.0	105.2	98.9	98.9	96.4	97.7	103.4	150.5	126.3	103.3	140.8	121.5
01 - 16	WEIGHTED AVERAGE	106.6	113.0	109.4	104.1	116.4	109.6	97.9	108.1	102.4	104.0	103.9	104.0	112.8	132.9	121.7	107.3	127.5	116.3

	DIVISION	CALIFORNIA																	
		SAN LUIS OBISPO			SAN MATEO			SAN RAFAEL			SANTA ANA			SANTA BARBARA			SANTA CRUZ		
		934			944			949			926 - 927			931			950		
		MAT.	INST.	TOTAL	MAT.	INST.	TOTAL	MAT.	INST.	TOTAL	MAT.	INST.	TOTAL	MAT.	INST.	TOTAL	MAT.	INST.	TOTAL
01590	EQUIPMENT RENTAL	.0	99.6	99.6	.0	103.5	103.5	.0	103.6	103.6	.0	100.7	100.7	.0	99.6	99.6	.0	100.0	100.0
02	SITE CONSTRUCTION	115.8	106.1	108.7	149.8	104.7	116.7	132.8	111.1	116.9	91.9	106.5	102.6	109.4	106.1	107.0	148.3	99.6	112.6
03100	CONCRETE FORMS & ACCESSORIES	114.8	117.4	117.0	111.4	134.8	131.6	118.2	134.1	131.9	111.4	116.6	115.9	104.7	121.7	119.4	106.4	125.5	122.9
03200	CONCRETE REINFORCEMENT	110.6	112.5	111.6	100.6	113.8	107.3	101.2	113.7	107.5	111.3	112.5	111.9	108.6	112.6	110.6	118.9	113.4	116.1
03300	CAST-IN-PLACE CONCRETE	116.4	115.3	115.9	137.2	117.9	129.2	159.7	116.4	141.9	94.4	115.1	102.9	109.2	117.6	112.6	129.4	115.7	123.8
03	CONCRETE	118.7	114.9	116.8	123.3	123.9	123.6	148.1	122.8	135.5	99.8	114.5	107.1	109.9	117.6	113.7	121.4	119.1	120.3
04	MASONRY	111.1	110.3	110.6	145.2	128.9	135.1	119.1	128.3	124.8	82.2	108.7	98.6	109.6	110.7	110.3	151.2	121.1	132.5
05	METALS	107.4	99.1	104.8	102.6	105.4	103.5	103.4	101.1	102.6	110.8	99.3	107.1	105.3	99.5	103.5	112.6	105.9	110.5
06	WOOD & PLASTICS	104.2	116.8	110.7	108.5	136.2	122.9	108.9	135.9	122.9	102.5	116.7	109.9	95.1	122.0	109.1	102.0	125.9	114.5
07	THERMAL & MOISTURE PROTECTION	100.9	110.2	104.8	110.1	130.3	118.6	115.9	126.5	120.4	103.1	112.2	106.9	100.3	112.5	105.4	101.9	123.8	111.1
08	DOORS & WINDOWS	99.6	111.6	102.7	106.0	125.4	111.1	116.4	127.9	119.4	98.9	113.7	102.8	101.5	116.6	105.5	94.8	119.8	101.4
09200	PLASTER & GYPSUM BOARD	108.6	117.1	113.8	103.7	136.5	124.1	107.0	136.5	125.3	100.3	117.1	110.7	98.8	122.4	113.5	104.3	126.3	118.0
095,098	CEILINGS & ACOUSTICAL TREATMENT	120.8	117.1	118.6	114.8	136.5	127.8	124.1	136.5	131.5	115.7	117.1	116.5	120.4	122.4	121.6	117.4	126.3	122.7
09600	FLOORING	125.9	96.1	117.8	118.2	126.1	120.4	129.1	126.1	128.3	124.5	104.8	119.2	117.0	96.1	111.3	113.9	126.1	117.2
097,099	WALL FINISHES, PAINTS & COATINGS	109.9	102.7	105.5	115.4	137.8	129.0	111.2	137.8	127.4	107.4	109.0	108.4	109.9	102.7	105.5	111.5	137.8	127.5
09	FINISHES	119.3	111.8	115.4	117.4	134.3	126.2	120.8	134.0	127.7	113.0	113.9	113.5	114.5	115.4	115.0	115.2	127.6	121.7
10 - 14	TOTAL DIV. 10000 - 14000	100.0	111.1	102.4	100.0	128.8	106.1	100.0	127.9	105.9	100.0	98.5	99.7	100.0	113.3	102.8	100.0	125.8	105.5
15	MECHANICAL	95.3	107.0	100.2	95.4	134.8	111.7	95.3	133.7	111.2	95.3	106.0	99.8	100.2	107.3	103.2	100.2	109.2	103.9
16	ELECTRICAL	87.7	100.5	93.9	103.3	142.8	122.5	96.9	110.0	103.2	93.0	105.0	98.8	86.7	110.0	98.0	102.2	123.2	112.4
01 - 16	WEIGHTED AVERAGE	103.4	107.9	105.4	108.4	127.8	117.0	110.2	122.8	115.8	99.9	108.2	103.5	102.6	110.5	106.1	109.4	116.4	112.5

	DIVISION	CALIFORNIA															COLORADO		
		SANTA ROSA			STOCKTON			SUSANVILLE			VALLEJO			VAN NUYS			ALAMOSA		
		954			952			961			945			913 - 916			811		
		MAT.	INST.	TOTAL	MAT.	INST.	TOTAL	MAT.	INST.	TOTAL	MAT.	INST.	TOTAL	MAT.	INST.	TOTAL	MAT.	INST.	TOTAL
01590	EQUIPMENT RENTAL	.0	99.9	99.9	.0	99.2	99.2	.0	99.2	99.2	.0	103.6	103.6	.0	97.0	97.0	.0	96.9	96.9
02	SITE CONSTRUCTION	107.4	105.6	106.1	107.5	105.6	106.1	122.1	105.0	109.6	115.5	111.4	112.5	113.6	107.3	109.0	128.9	94.6	103.7
03100	CONCRETE FORMS & ACCESSORIES	102.8	133.2	129.0	104.4	124.8	122.0	105.4	123.6	121.1	106.1	133.4	129.6	112.8	116.5	116.0	101.1	79.4	82.4
03200	CONCRETE REINFORCEMENT	106.3	113.7	110.0	109.2	112.9	111.1	105.3	112.8	109.1	102.5	113.7	108.2	108.2	112.7	110.5	114.5	80.9	97.4
03300	CAST-IN-PLACE CONCRETE	123.0	114.7	119.6	109.1	113.3	110.8	112.9	111.5	112.3	127.1	115.3	122.2	107.8	109.7	108.5	100.5	83.6	93.5
03	CONCRETE	120.6	122.0	121.3	109.9	117.6	113.7	121.0	116.4	118.7	119.6	122.2	120.9	121.6	112.4	117.1	116.4	81.2	98.9
04	MASONRY	113.9	126.0	121.4	113.3	113.5	113.4	115.4	110.8	112.5	87.4	125.9	111.3	125.6	111.8	117.0	135.9	79.7	101.0
05	METALS	111.1	102.9	108.5	106.8	100.7	104.8	108.9	99.7	106.0	103.3	102.1	102.9	90.2	96.4	92.2	96.0	82.9	91.8
06	WOOD & PLASTICS	91.9	135.6	114.7	97.1	125.8	112.0	97.5	125.8	112.2	95.9	135.9	116.7	97.2	116.5	107.3	100.2	81.6	90.5
07	THERMAL & MOISTURE PROTECTION	103.6	126.5	113.2	105.6	110.8	107.8	106.6	109.4	107.8	113.1	126.0	118.5	99.5	110.0	103.9	100.9	83.9	93.8
08	DOORS & WINDOWS	100.5	128.0	107.7	100.8	118.1	105.3	103.1	118.1	107.0	118.9	128.1	121.3	95.4	113.6	100.2	93.4	85.8	91.4
09200	PLASTER & GYPSUM BOARD	95.5	136.5	121.0	99.2	126.3	116.0	97.9	126.3	115.5	101.4	136.5	123.2	94.8	117.1	108.6	83.1	80.8	81.6
095,098	CEILINGS & ACOUSTICAL TREATMENT	115.7	136.5	128.1	120.4	126.3	123.9	120.8	126.3	124.1	126.1	136.5	132.3	94.0	117.1	107.7	100.5	80.8	88.7
09600	FLOORING	119.8	126.1	121.5	117.0	114.1	116.2	117.0	113.9	116.1	123.7	126.1	124.4	105.2	104.8	105.1	110.6	68.4	99.2
097,099	WALL FINISHES, PAINTS & COATINGS	107.4	137.8	125.9	109.9	116.6	114.0	109.9	131.2	122.9	112.0	137.8	127.7	111.1	109.0	109.8	115.2	30.1	63.3
09	FINISHES	111.8	133.6	123.1	114.4	123.0	118.9	115.5	123.9	119.9	117.4	133.7	125.9	103.5	113.8	108.9	103.1	72.0	86.9
10 - 14	TOTAL DIV. 10000 - 14000	100.0	126.3	105.6	100.0	125.4	105.4	100.0	124.3	105.2	100.0	127.0	105.8	100.0	98.0	99.6	100.0	89.8	97.8
15	MECHANICAL	95.3	133.5	111.2	100.2	104.4	101.9	95.3	106.5	100.0	100.2	122.9	109.6	95.3	106.0	99.7	95.2	71.0	85.2
16	ELECTRICAL	93.3	110.0	101.4	97.9	123.1	110.1	98.5	104.2	101.3	93.0	123.5	107.9	119.3	113.4	116.4	91.8	80.5	86.3
01 - 16	WEIGHTED AVERAGE	104.1	122.0	112.1	104.4	113.5	108.4	105.7	110.7	107.9	105.5	122.2	112.9	103.5	109.1	106.0	101.3	79.5	91.7

	DIVISION	COLORADO																	
		BOULDER			COLORADO SPRINGS			DENVER			DURANGO			FORT COLLINS			FORT MORGAN		
		803			808 - 809			800 - 802			813			805			807		
		MAT.	INST.	TOTAL	MAT.	INST.	TOTAL	MAT.	INST.	TOTAL	MAT.	INST.	TOTAL	MAT.	INST.	TOTAL	MAT.	INST.	TOTAL
01590	EQUIPMENT RENTAL	.0	96.7	96.7	.0	95.2	95.2	.0	100.3	100.3	.0	96.9	96.9	.0	96.7	96.7	.0	96.7	96.7
02	SITE CONSTRUCTION	91.2	98.7	96.7	93.3	96.8	95.8	91.9	106.0	102.3	122.6	94.6	102.0	102.2	99.2	100.0	92.9	99.3	97.6
03100	CONCRETE FORMS & ACCESSORIES	104.0	82.8	85.7	92.5	85.1	86.1	101.8	85.9	88.1	107.9	79.5	83.4	101.2	81.4	84.1	104.3	82.7	85.7
03200	CONCRETE REINFORCEMENT	113.6	80.9	97.0	112.8	88.5	100.5	112.8	88.6	100.5	114.5	80.9	97.4	113.7	79.9	96.5	113.9	80.9	97.1
03300	CAST-IN-PLACE CONCRETE	91.4	83.8	88.3	95.1	86.5	91.5	88.9	86.9	88.1	115.5	84.9	102.9	103.2	82.2	94.6	89.7	83.7	87.2
03	CONCRETE	103.0	82.8	93.0	106.7	86.3	96.6	100.9	86.8	93.9	119.0	81.7	100.5	113.1	81.4	97.4	101.6	82.7	92.2
04	MASONRY	100.7	79.9	87.8	106.0	81.1	90.6	105.7	83.3	91.8	122.2	79.7	95.9	118.2	59.1	81.6	117.2	79.9	94.1
05	METALS	95.4	82.8	91.4	98.0	89.3	95.2	100.3	89.5	96.9	96.0	83.1	91.9	96.2	81.3	91.5	95.2	82.9	91.2
06	WOOD & PLASTICS	104.5	85.4	94.6	93.3	87.8	90.4	102.9	87.7	95.0	108.6	81.6	94.5	102.7	85.4	93.7	104.9	85.4	94.8
07	THERMAL & MOISTURE PROTECTION	105.4	82.0	95.6	105.9	85.4	97.3	105.4	82.1	95.6	100.8	83.9	93.7	105.8	70.9	91.1	105.3	81.3	95.2
08	DOORS & WINDOWS	95.9	87.9	93.8	98.5	91.3	96.6	99.8	91.2	97.6	100.6	85.8	96.7	95.8	87.9	93.8	95.8	87.9	93.7
09200	PLASTER & GYPSUM BOARD	99.6	85.3	90.7	87.8	87.6	87.7	97.4	87.6	91.3	92.3	80.8	85.1	95.9	85.3	89.3	100.7	85.3	91.1
095,098	CEILINGS & ACOUSTICAL TREATMENT	95.9	85.3	89.6	101.7	87.6	93.3	101.8	87.6	93.3	100.5	80.8	88.7	95.9	85.3	89.6	95.9	85.3	89.6
09600	FLOORING	112.0	96.6	107.8	106.9	85.2	101.0	107.8	96.6	104.8	115.8	68.4	103.0	107.7	68.4	97.1	112.5	68.4	100.6
097,099	WALL FINISHES, PAINTS & COATINGS	105.6	75.0	86.9	105.6	51.7	72.7	105.6	78.4	89.0	115.2	30.1	63.3	105.6	51.3	72.5	105.6	68.3	82.8
09	FINISHES	100.2	84.2	91.9	98.4	81.6	89.7	100.0	87.2	93.3	105.3	72.0	88.0	98.8	76.0	86.9	100.4	78.6	89.0
10 - 14	TOTAL DIV. 10000 - 14000	100.0	89.0	97.7	100.0	89.8	97.8	100.0	89.8	97.8	100.0	89.8	97.8	100.0	88.0	97.4	100.0	89.0	97.7
15	MECHANICAL	95.1	84.6	90.8	100.1	79.9	91.7	100.1	87.4	94.8	95.2	79.6	88.8	100.1	83.1	93.0	95.1	84.6	90.8
16	ELECTRICAL	96.4	94.2	95.3	99.6	88.6	94.3	101.5	94.9	98.3	91.2	80.3	85.9	96.4	94.2	95.3	96.7	94.2	95.4
01 - 16	WEIGHTED AVERAGE	97.7	86.5	92.7	100.4	85.7	93.9	100.6	89.9	95.8	101.7	81.4	92.7	101.1	82.4	92.8	98.4	85.8	92.8

	DIVISION	COLORADO																	
		GLENWOOD SPRINGS			GOLDEN			GRAND JUNCTION			GREELEY			MONTROSE			PUEBLO		
		816			804			815			806			814			810		
		MAT.	INST.	TOTAL	MAT.	INST.	TOTAL	MAT.	INST.	TOTAL	MAT.	INST.	TOTAL	MAT.	INST.	TOTAL	MAT.	INST.	TOTAL
01590	EQUIPMENT RENTAL	.0	100.0	100.0	.0	96.7	96.7	.0	100.0	100.0	.0	96.7	96.7	.0	98.4	98.4	.0	96.9	96.9
02	SITE CONSTRUCTION	137.7	102.3	111.7	104.1	99.6	100.8	121.8	101.8	107.1	90.1	97.9	95.8	131.3	98.0	106.9	114.1	96.2	101.0
03100	CONCRETE FORMS & ACCESSORIES	97.6	82.6	84.7	95.7	82.6	84.4	106.6	81.8	85.2	98.6	48.1	55.1	97.0	82.4	84.4	103.4	85.4	87.8
03200	CONCRETE REINFORCEMENT	113.4	80.9	96.9	113.9	80.9	97.1	113.7	79.8	96.5	113.6	79.3	96.2	113.2	80.8	96.8	109.5	88.5	98.8
03300	CAST-IN-PLACE CONCRETE	100.4	84.1	93.7	89.8	83.7	87.3	111.2	83.2	99.7	86.3	58.9	75.0	100.5	84.1	93.7	99.7	87.7	94.8
03	CONCRETE	121.4	82.9	102.3	112.3	82.7	97.6	115.4	81.9	98.8	98.4	58.6	78.6	112.5	82.7	97.7	105.3	86.9	96.2
04	MASONRY	109.1	79.9	91.0	120.7	80.0	95.4	144.2	61.5	92.9	112.2	39.2	66.9	114.6	79.7	93.0	105.2	80.2	89.7
05	METALS	95.7	83.7	91.9	95.3	83.0	91.4	96.9	80.5	91.7	96.2	79.6	90.9	94.8	82.5	90.9	98.4	90.3	95.8
06	WOOD & PLASTICS	95.0	85.6	90.1	97.0	85.4	91.0	105.9	85.6	95.3	99.8	46.8	72.2	95.1	85.8	90.2	102.7	88.2	95.1
07	THERMAL & MOISTURE PROTECTION	100.7	80.5	92.2	106.3	78.6	94.7	99.7	67.1	86.0	105.2	59.0	85.7	100.9	84.4	93.9	99.2	84.9	93.2
08	DOORS & WINDOWS	99.8	88.0	96.7	95.9	87.3	93.6	100.5	88.0	97.2	95.8	66.9	88.3	100.8	88.1	97.5	95.2	91.5	94.2
09200	PLASTER & GYPSUM BOARD	101.0	85.3	91.2	94.0	85.3	88.6	109.1	85.3	94.3	94.8	45.6	64.2	79.7	85.3	83.2	87.4	87.6	87.5
095,098	CEILINGS & ACOUSTICAL TREATMENT	98.8	85.3	90.7	95.9	85.3	89.6	98.8	85.3	90.7	95.9	45.6	65.9	100.5	85.3	91.4	111.6	87.6	97.3
09600	FLOORING	109.6	63.0	97.0	105.1	68.4	95.1	115.2	68.4	102.5	106.3	68.4	96.1	113.0	56.6	97.7	111.5	96.6	107.5
097,099	WALL FINISHES, PAINTS & COATINGS	115.2	68.3	86.6	105.6	75.0	86.9	115.2	75.0	90.7	105.6	31.9	60.7	115.2	30.1	63.3	115.2	47.9	74.1
09	FINISHES	105.4	77.7	91.0	98.3	79.3	88.4	106.8	79.4	92.5	97.6	48.5	72.0	103.2	72.1	87.0	105.2	83.8	94.1
10 - 14	TOTAL DIV. 10000 - 14000	100.0	89.4	97.7	100.0	89.0	97.7	100.0	89.3	97.7	100.0	79.3	95.6	100.0	89.8	97.8	100.0	90.7	98.0
15	MECHANICAL	95.2	79.7	88.8	95.1	84.4	90.7	100.1	72.7	88.7	100.1	77.5	90.7	95.2	79.7	88.8	100.1	71.2	88.1
16	ELECTRICAL	88.9	81.9	85.5	96.7	94.2	95.4	90.9	63.9	77.8	96.4	94.1	95.3	90.9	63.9	77.8	91.8	80.6	86.3
01 - 16	WEIGHTED AVERAGE	101.1	83.3	93.2	99.9	85.8	93.6	103.6	76.8	91.7	98.7	70.4	86.1	100.3	79.6	91.1	100.0	83.1	92.5

City Cost Indexes

	DIVISION	COLORADO			CONNECTICUT															
		SALIDA			BRIDGEPORT			BRISTOL			HARTFORD			MERIDEN			NEW BRITAIN		
		812			066			060			061			064			060		
		MAT.	INST.	TOTAL	MAT.	INST.	TOTAL	MAT.	INST.	TOTAL	MAT.	INST.	TOTAL	MAT.	INST.	TOTAL	MAT.	INST.	TOTAL
01590	**EQUIPMENT RENTAL**	.0	98.4	98.4	.0	101.5	101.5	.0	101.5	101.5	.0	101.5	101.5	.0	102.1	102.1	.0	101.5	101.5
02	**SITE CONSTRUCTION**	122.2	98.4	104.7	103.3	104.9	104.5	102.4	104.9	104.2	102.8	104.9	104.4	100.3	105.8	104.3	102.6	104.9	104.3
03100	CONCRETE FORMS & ACCESSORIES	106.4	82.4	85.7	102.2	123.2	120.3	102.2	122.7	119.9	101.1	122.8	119.8	102.0	123.1	120.2	102.5	122.8	120.0
03200	CONCRETE REINFORCEMENT	113.0	80.8	96.6	107.5	128.1	117.9	107.5	128.0	117.9	107.5	128.0	117.9	107.5	128.0	117.9	107.5	128.0	117.9
03300	CAST-IN-PLACE CONCRETE	115.1	84.0	102.3	108.1	117.0	111.7	101.2	116.9	107.7	102.4	116.9	108.3	97.4	117.0	105.4	102.9	116.9	108.6
03	**CONCRETE**	114.3	82.6	98.6	109.3	121.7	115.5	106.0	121.4	113.7	106.5	121.4	113.9	104.1	121.6	112.8	106.8	121.4	114.1
04	**MASONRY**	146.1	79.7	104.9	109.2	127.8	120.7	101.0	127.8	117.6	101.1	127.8	117.6	100.7	127.8	117.5	101.1	127.8	117.6
05	**METALS**	94.5	82.5	90.7	100.9	124.6	108.5	100.9	124.1	108.3	106.0	124.1	111.8	97.9	124.4	106.4	97.3	124.1	105.9
06	**WOOD & PLASTICS**	104.3	85.8	94.7	100.4	122.3	111.8	100.4	122.3	111.8	100.4	122.3	111.8	100.4	122.3	111.8	100.4	122.3	111.8
07	**THERMAL & MOISTURE PROTECTION**	99.8	84.4	93.3	101.8	126.2	112.1	102.0	122.4	110.6	100.8	122.4	109.9	102.0	122.4	110.6	102.0	123.4	111.0
08	**DOORS & WINDOWS**	93.5	88.1	92.1	106.9	131.8	113.4	106.9	121.8	110.8	106.9	121.8	110.8	109.6	131.8	115.4	106.9	121.8	110.8
09200	PLASTER & GYPSUM BOARD	82.7	85.3	84.3	105.5	122.0	115.8	105.5	122.0	115.8	105.5	122.0	115.8	107.2	122.0	116.4	105.5	122.0	115.8
095,098	CEILINGS & ACOUSTICAL TREATMENT	100.5	85.3	91.4	98.5	122.0	112.5	98.5	122.0	112.5	98.5	122.0	112.5	103.4	122.0	114.5	98.5	122.0	112.5
09600	FLOORING	118.6	56.6	101.9	99.6	120.6	105.3	99.6	120.6	105.3	99.6	120.6	105.3	99.6	120.6	105.3	99.6	120.6	105.3
097,099	WALL FINISHES, PAINTS & COATINGS	115.2	30.1	63.3	91.5	118.5	108.0	91.5	114.1	105.3	91.5	114.1	105.3	91.5	121.8	110.0	91.5	114.1	105.3
09	**FINISHES**	104.5	72.1	87.6	102.1	122.0	112.5	102.1	121.5	112.2	102.1	121.5	112.2	103.6	122.4	113.4	102.1	121.5	112.2
10 - 14	**TOTAL DIV. 10000 - 14000**	100.0	89.9	97.8	100.0	115.4	103.3	100.0	115.3	103.3	100.0	115.4	103.3	100.0	115.4	103.3	100.0	115.3	103.3
15	**MECHANICAL**	95.2	70.9	85.1	100.1	111.8	104.9	100.1	111.8	104.9	100.1	111.8	104.9	95.1	111.8	102.1	100.1	111.8	104.9
16	**ELECTRICAL**	91.1	80.5	86.0	97.7	109.6	103.5	97.7	105.8	101.6	97.3	110.7	103.8	97.6	109.7	103.5	97.7	105.8	101.7
01 - 16	**WEIGHTED AVERAGE**	101.3	80.2	91.9	102.4	117.8	109.2	101.6	116.6	108.2	102.3	117.3	109.0	100.1	117.8	108.0	101.1	116.6	108.0

	DIVISION	CONNECTICUT																	
		NEW HAVEN			NEW LONDON			NORWALK			STAMFORD			WATERBURY			WILLIMANTIC		
		065			063			068			069			067			062		
		MAT.	INST.	TOTAL	MAT.	INST.	TOTAL	MAT.	INST.	TOTAL	MAT.	INST.	TOTAL	MAT.	INST.	TOTAL	MAT.	INST.	TOTAL
01590	**EQUIPMENT RENTAL**	.0	102.1	102.1	.0	102.1	102.1	.0	101.5	101.5	.0	101.5	101.5	.0	101.5	101.5	.0	101.5	101.5
02	**SITE CONSTRUCTION**	102.4	105.8	104.9	94.7	105.8	102.8	103.0	104.9	104.4	103.7	105.0	104.6	102.8	104.9	104.4	103.1	104.9	104.4
03100	CONCRETE FORMS & ACCESSORIES	102.0	123.1	120.2	102.0	122.8	119.9	102.2	123.4	120.5	102.2	123.6	120.7	102.2	123.1	120.2	102.2	122.8	120.0
03200	CONCRETE REINFORCEMENT	107.5	128.0	117.9	84.3	128.0	106.5	107.5	128.1	117.9	107.5	128.1	118.0	107.5	128.0	117.9	107.5	128.0	117.9
03300	CAST-IN-PLACE CONCRETE	104.7	117.0	109.7	89.1	116.9	100.5	106.3	125.5	114.2	108.1	125.6	115.3	108.1	117.0	111.7	100.9	105.5	102.8
03	**CONCRETE**	121.5	121.6	121.5	93.6	121.4	107.4	108.5	124.7	116.5	109.3	124.9	117.0	109.3	121.6	115.4	105.9	117.6	111.7
04	**MASONRY**	101.2	127.8	117.7	99.6	127.8	117.1	101.4	127.6	117.6	101.5	127.6	117.7	101.5	127.8	117.8	100.9	127.8	117.6
05	**METALS**	97.5	124.4	106.1	97.2	124.1	105.8	100.9	124.7	108.5	100.9	125.0	108.6	100.9	124.5	108.4	100.7	124.2	108.2
06	**WOOD & PLASTICS**	100.4	122.3	111.8	100.4	122.3	111.8	100.4	122.3	111.8	100.4	122.3	111.8	100.4	122.3	111.8	100.4	122.3	111.8
07	**THERMAL & MOISTURE PROTECTION**	102.1	123.4	111.0	101.9	122.4	110.6	102.0	127.4	112.7	102.0	127.4	112.7	102.0	123.4	111.0	102.2	120.7	110.0
08	**DOORS & WINDOWS**	106.9	131.8	113.4	110.3	121.2	113.1	106.9	131.8	113.4	106.9	131.8	113.4	106.9	131.8	113.4	110.3	131.8	115.9
09200	PLASTER & GYPSUM BOARD	105.5	122.0	115.8	105.5	122.0	115.8	105.5	122.0	115.8	105.5	122.0	115.8	105.5	122.0	115.8	105.5	122.0	115.8
095,098	CEILINGS & ACOUSTICAL TREATMENT	98.5	122.0	112.5	96.6	122.0	111.8	98.5	122.0	112.5	98.5	122.0	112.5	98.5	122.0	112.5	96.6	122.0	111.8
09600	FLOORING	99.6	120.6	105.3	99.6	120.6	105.3	99.6	120.6	105.3	99.6	120.6	105.3	99.6	120.6	105.3	99.6	120.6	105.3
097,099	WALL FINISHES, PAINTS & COATINGS	91.5	118.5	108.0	91.5	114.1	105.3	91.5	118.5	108.0	91.5	118.5	108.0	91.5	121.8	110.0	91.5	121.8	110.0
09	**FINISHES**	102.2	122.0	112.5	101.2	121.5	111.8	102.1	122.0	112.5	102.2	122.0	112.5	102.0	122.4	112.6	101.8	122.4	112.5
10 - 14	**TOTAL DIV. 10000 - 14000**	100.0	115.4	103.3	100.0	115.4	103.3	100.0	115.4	103.3	100.0	115.5	103.3	100.0	115.4	103.3	100.0	115.4	103.3
15	**MECHANICAL**	100.1	111.8	104.9	95.1	111.8	102.1	100.1	111.9	105.0	100.1	111.9	105.0	100.1	111.8	104.9	100.1	111.8	104.9
16	**ELECTRICAL**	97.6	109.7	103.5	94.8	110.7	102.5	97.7	109.6	103.5	97.7	150.4	123.3	97.2	109.6	103.2	97.7	110.6	104.0
01 - 16	**WEIGHTED AVERAGE**	102.8	117.8	109.4	98.1	117.3	106.7	101.9	118.2	109.1	102.0	124.0	111.8	101.9	117.7	108.9	101.8	117.2	108.7

	DIVISION	D.C.			DELAWARE									FLORIDA					
		WASHINGTON			DOVER			NEWARK			WILMINGTON			DAYTONA BEACH			FORT LAUDERDALE		
		200 - 205			199			197			198			321			333		
		MAT.	INST.	TOTAL	MAT.	INST.	TOTAL	MAT.	INST.	TOTAL	MAT.	INST.	TOTAL	MAT.	INST.	TOTAL	MAT.	INST.	TOTAL
01590	**EQUIPMENT RENTAL**	.0	102.3	102.3	.0	118.2	118.2	.0	118.2	118.2	.0	118.4	118.4	.0	97.7	97.7	.0	89.3	89.3
02	**SITE CONSTRUCTION**	102.8	89.7	93.2	101.7	110.8	108.4	101.7	110.8	108.4	87.4	111.1	104.8	113.4	86.5	93.7	99.5	73.6	80.5
03100	CONCRETE FORMS & ACCESSORIES	98.5	80.8	83.2	96.5	98.5	98.2	96.5	98.5	98.2	96.2	98.5	98.2	98.5	67.8	72.0	96.0	66.3	70.4
03200	CONCRETE REINFORCEMENT	105.5	92.8	99.0	101.4	95.6	98.5	102.3	95.6	98.9	102.3	95.6	98.9	95.1	84.3	89.6	95.1	70.1	82.4
03300	CAST-IN-PLACE CONCRETE	114.5	86.2	102.9	85.6	91.4	88.0	85.6	91.4	88.0	80.5	91.4	85.0	90.4	69.2	81.7	94.9	67.2	83.5
03	**CONCRETE**	111.9	86.3	99.2	100.9	96.5	98.7	101.1	96.5	98.8	98.6	96.5	97.5	91.1	72.6	81.9	93.1	68.6	80.9
04	**MASONRY**	95.0	81.4	86.6	107.0	91.8	97.6	105.0	91.8	96.8	111.1	91.8	99.1	87.6	63.7	72.8	87.8	65.6	74.1
05	**METALS**	104.9	110.0	106.5	105.5	112.6	107.8	105.5	112.6	107.8	102.4	112.6	105.6	94.6	94.3	94.5	94.5	87.4	92.2
06	**WOOD & PLASTICS**	96.4	81.1	88.4	94.5	100.0	97.4	94.5	100.0	97.4	94.5	100.0	97.4	97.5	70.3	83.3	92.9	63.8	77.7
07	**THERMAL & MOISTURE PROTECTION**	103.6	84.8	95.7	102.1	105.0	103.3	102.2	105.0	103.4	101.8	105.0	103.2	99.6	68.0	86.3	99.6	72.6	88.2
08	**DOORS & WINDOWS**	102.7	90.9	99.6	94.6	102.5	96.7	94.6	102.5	96.7	94.3	102.5	96.5	100.7	68.0	92.1	98.3	61.1	88.6
09200	PLASTER & GYPSUM BOARD	108.8	80.6	91.3	111.5	99.9	104.3	111.5	99.9	104.3	110.7	99.9	104.0	100.6	70.0	81.6	100.2	63.4	77.3
095,098	CEILINGS & ACOUSTICAL TREATMENT	103.7	80.6	90.0	104.3	99.9	101.6	104.3	99.9	101.6	99.9	99.9	99.9	104.3	70.0	83.9	104.3	63.4	79.9
09600	FLOORING	115.9	103.1	112.5	85.2	105.3	90.6	85.2	105.3	90.6	85.0	105.3	90.5	118.7	71.8	106.0	118.7	58.3	102.4
097,099	WALL FINISHES, PAINTS & COATINGS	117.2	84.9	97.4	92.8	93.6	93.3	92.8	93.6	93.3	92.8	93.6	93.3	110.6	69.5	85.6	106.7	48.1	71.0
09	**FINISHES**	104.6	85.0	94.4	102.8	99.2	100.9	102.8	99.2	100.9	101.6	99.2	100.3	108.7	68.8	87.9	106.7	62.3	83.6
10 - 14	**TOTAL DIV. 10000 - 14000**	100.0	96.1	99.2	100.0	93.2	98.6	100.0	93.2	98.6	100.0	93.2	98.6	100.0	81.6	96.1	100.0	87.9	97.4
15	**MECHANICAL**	99.9	91.4	96.4	100.3	113.6	105.8	100.3	113.6	105.8	100.3	113.6	105.8	99.9	70.3	87.6	99.9	67.2	86.3
16	**ELECTRICAL**	99.2	98.3	98.7	96.3	102.2	99.2	96.3	102.2	99.2	96.2	102.2	99.1	96.6	65.4	81.4	96.6	73.4	85.3
01 - 16	**WEIGHTED AVERAGE**	102.5	91.2	97.5	100.7	103.9	102.1	100.7	103.9	102.1	99.7	103.9	101.5	98.3	72.8	87.0	97.7	70.3	85.5

	DIVISION	FLORIDA																	
		FORT MYERS			GAINESVILLE			JACKSONVILLE			LAKELAND			MELBOURNE			MIAMI		
		339,341			326,344			320,322			338			329			330 - 332,340		
		MAT.	INST.	TOTAL	MAT.	INST.	TOTAL	MAT.	INST.	TOTAL	MAT.	INST.	TOTAL	MAT.	INST.	TOTAL	MAT.	INST.	TOTAL
01590	EQUIPMENT RENTAL	.0	97.7	97.7	.0	97.7	97.7	.0	97.7	97.7	.0	97.7	97.7	.0	97.7	97.7	.0	89.3	89.3
02	SITE CONSTRUCTION	110.7	86.4	92.9	123.4	86.0	96.0	113.5	87.0	94.1	112.8	86.0	93.2	121.5	87.1	96.2	98.8	73.5	80.2
03100	CONCRETE FORMS & ACCESSORIES	91.0	72.1	74.7	92.2	51.4	57.0	98.2	51.8	58.2	86.4	72.6	74.5	93.9	72.0	75.0	96.3	66.4	70.5
03200	CONCRETE REINFORCEMENT	96.2	85.9	90.9	100.9	48.8	74.4	95.1	49.2	71.8	98.5	87.1	92.7	96.2	84.4	90.2	95.1	70.1	82.4
03300	CAST-IN-PLACE CONCRETE	99.0	58.7	82.4	103.9	48.1	81.0	91.3	57.0	77.2	101.2	59.7	84.1	109.0	75.0	95.0	92.4	67.4	82.1
03	CONCRETE	93.9	71.2	82.7	102.0	51.5	76.9	91.5	54.8	73.3	95.7	71.9	83.9	102.4	76.5	89.5	91.9	68.7	80.4
04	MASONRY	81.5	50.9	62.5	101.4	42.4	64.8	87.2	49.9	64.1	98.0	73.5	82.8	86.6	73.5	78.5	86.2	66.0	73.7
05	METALS	97.1	91.8	95.4	93.8	77.3	88.5	94.1	78.2	89.1	97.0	93.0	95.7	103.2	95.0	100.6	99.6	87.7	95.8
06	WOOD & PLASTICS	89.6	75.0	82.0	90.8	50.5	69.8	97.5	50.5	73.0	84.7	75.0	79.7	92.7	70.3	81.0	92.9	63.8	77.7
07	THERMAL & MOISTURE PROTECTION	99.3	54.7	80.5	99.9	54.2	80.6	99.8	57.1	81.8	99.2	60.6	83.0	100.0	75.6	89.7	103.3	69.4	89.0
08	DOORS & WINDOWS	99.4	70.8	91.9	98.8	47.4	85.3	100.7	47.9	86.9	99.3	68.6	91.3	99.9	73.9	93.1	98.3	61.7	88.7
09200	PLASTER & GYPSUM BOARD	96.3	74.9	83.0	96.4	49.7	67.4	100.6	49.7	69.0	94.2	74.9	82.2	96.7	70.0	80.1	100.2	63.4	77.3
095,098	CEILINGS & ACOUSTICAL TREATMENT	97.3	74.9	83.9	96.4	49.7	68.6	104.3	49.7	71.7	96.4	74.9	83.6	99.2	70.0	81.8	104.3	63.4	79.9
09600	FLOORING	114.4	40.0	94.3	115.0	34.3	93.1	118.7	47.9	99.5	111.2	50.7	94.9	115.2	71.8	103.5	126.0	59.0	107.9
097,099	WALL FINISHES, PAINTS & COATINGS	110.6	44.9	70.5	110.6	43.2	69.5	110.6	46.4	71.4	110.6	44.9	70.5	110.6	94.7	100.9	106.7	48.1	71.0
09	FINISHES	104.8	63.9	83.5	105.8	46.7	75.0	108.7	49.8	78.0	103.5	66.1	84.0	106.3	74.0	89.5	108.8	62.4	84.7
10 - 14	TOTAL DIV. 10000 - 14000	100.0	82.1	96.2	100.0	79.3	95.6	100.0	76.7	95.0	100.0	82.1	96.2	100.0	85.3	96.9	100.0	87.9	97.4
15	MECHANICAL	97.2	51.7	78.3	98.7	66.9	85.5	99.9	49.5	79.0	97.2	71.3	86.5	99.9	75.9	89.9	99.9	73.8	89.1
16	ELECTRICAL	98.7	39.5	69.9	96.9	40.9	69.7	96.4	66.9	82.0	96.9	45.7	72.0	97.5	72.1	85.2	97.5	71.6	84.9
01 - 16	WEIGHTED AVERAGE	97.7	62.8	82.2	99.6	57.7	81.0	98.3	59.6	81.1	98.4	70.5	86.0	100.8	77.8	90.6	98.7	71.4	86.6

	DIVISION	FLORIDA																	
		ORLANDO			PANAMA CITY			PENSACOLA			SARASOTA			ST. PETERSBURG			TALLAHASSEE		
		327 - 328,347			324			325			342			337			323		
		MAT.	INST.	TOTAL	MAT.	INST.	TOTAL	MAT.	INST.	TOTAL	MAT.	INST.	TOTAL	MAT.	INST.	TOTAL	MAT.	INST.	TOTAL
01590	EQUIPMENT RENTAL	.0	97.7	97.7	.0	97.7	97.7	.0	97.7	97.7	.0	97.7	97.7	.0	97.7	97.7	.0	97.7	97.7
02	SITE CONSTRUCTION	114.1	86.2	93.6	127.7	83.6	95.4	125.4	86.1	96.6	114.9	86.1	93.8	114.4	85.5	93.2	114.8	85.6	93.4
03100	CONCRETE FORMS & ACCESSORIES	98.2	69.4	73.3	97.3	26.9	36.5	87.5	49.5	54.8	98.5	72.2	75.8	95.8	45.8	52.6	98.2	38.0	46.3
03200	CONCRETE REINFORCEMENT	95.1	80.8	87.8	99.3	48.1	73.3	101.8	48.5	74.7	95.1	87.1	91.0	98.5	55.4	76.6	95.1	48.7	71.5
03300	CAST-IN-PLACE CONCRETE	98.3	69.1	86.3	96.0	33.5	70.3	96.0	53.2	78.4	103.5	59.5	85.4	102.3	53.5	82.2	94.7	47.0	75.1
03	CONCRETE	92.7	72.6	82.8	99.7	34.9	67.5	98.7	52.3	75.7	97.4	71.7	84.6	97.2	52.0	74.8	93.1	45.1	69.3
04	MASONRY	91.3	63.7	74.2	92.0	26.8	51.6	89.1	48.4	63.9	88.5	73.5	79.2	136.3	47.8	81.4	90.2	37.7	57.6
05	METALS	100.5	92.7	98.0	94.4	63.8	84.6	94.3	77.5	89.0	97.4	92.6	95.9	97.7	78.3	91.5	87.8	76.6	84.2
06	WOOD & PLASTICS	97.5	72.8	84.7	96.3	26.9	60.2	85.8	49.8	67.1	97.5	75.0	85.8	94.6	45.7	69.2	95.9	36.2	64.8
07	THERMAL & MOISTURE PROTECTION	99.8	68.3	86.5	100.0	30.5	70.8	99.8	51.3	79.4	99.6	60.6	83.1	99.5	48.2	77.9	99.8	45.2	76.8
08	DOORS & WINDOWS	100.7	67.7	92.1	98.3	25.7	79.3	98.3	48.4	85.2	100.7	68.0	92.2	99.3	44.8	85.1	99.3	39.6	83.7
09200	PLASTER & GYPSUM BOARD	103.0	72.7	84.1	98.7	25.5	53.2	94.6	49.0	66.3	100.6	74.9	84.6	97.9	44.8	64.9	100.6	35.0	59.9
095,098	CEILINGS & ACOUSTICAL TREATMENT	104.3	72.7	85.4	98.3	25.5	54.9	98.3	49.0	68.9	102.4	74.9	86.0	98.3	44.8	66.4	104.3	35.0	63.0
09600	FLOORING	118.7	71.8	106.0	118.1	18.4	91.1	112.0	50.9	95.5	118.7	50.5	100.3	117.0	50.7	99.1	118.7	37.2	96.7
097,099	WALL FINISHES, PAINTS & COATINGS	110.6	53.3	75.6	110.6	23.7	57.6	110.6	53.9	76.0	110.6	44.9	70.5	110.6	44.9	70.5	110.6	38.2	66.4
09	FINISHES	109.0	68.4	87.9	107.8	24.5	64.4	105.2	50.1	76.5	108.2	66.0	86.2	106.3	46.4	75.1	108.8	36.9	71.4
10 - 14	TOTAL DIV. 10000 - 14000	100.0	81.8	96.1	100.0	42.9	87.8	100.0	48.0	88.9	100.0	82.1	96.2	100.0	50.2	89.4	100.0	72.9	94.2
15	MECHANICAL	99.9	65.8	85.8	99.9	24.3	68.5	99.9	48.2	78.5	99.8	52.1	80.0	99.9	48.0	78.4	99.9	38.5	74.4
16	ELECTRICAL	97.3	43.5	71.2	95.6	32.6	64.9	98.7	62.5	81.1	96.6	35.7	67.0	96.9	47.0	72.6	97.0	40.3	69.4
01 - 16	WEIGHTED AVERAGE	99.7	68.6	85.9	99.4	36.8	71.6	99.1	57.2	80.5	99.5	65.1	84.2	101.6	54.2	80.5	97.6	48.2	75.6

	DIVISION	FLORIDA						GEORGIA											
		TAMPA			WEST PALM BEACH			ALBANY			ATHENS			ATLANTA			AUGUSTA		
		335 - 336,346			334,349			317,398			306			300 - 303,399			308 - 309		
		MAT.	INST.	TOTAL	MAT.	INST.	TOTAL	MAT.	INST.	TOTAL	MAT.	INST.	TOTAL	MAT.	INST.	TOTAL	MAT.	INST.	TOTAL
01590	EQUIPMENT RENTAL	.0	97.7	97.7	.0	89.3	89.3	.0	90.1	90.1	.0	92.2	92.2	.0	92.8	92.8	.0	92.2	92.2
02	SITE CONSTRUCTION	114.6	86.1	93.7	96.1	73.8	79.7	99.7	76.0	82.3	103.2	92.6	95.4	99.4	95.7	96.7	96.0	92.4	93.4
03100	CONCRETE FORMS & ACCESSORIES	99.7	72.7	76.4	100.1	65.9	70.6	97.5	44.6	51.9	93.8	40.9	48.2	98.1	76.6	79.6	95.1	50.2	56.4
03200	CONCRETE REINFORCEMENT	95.1	87.1	91.1	97.8	69.7	83.5	94.7	90.8	92.7	99.6	87.4	93.4	98.9	92.6	95.7	100.0	74.6	87.1
03300	CAST-IN-PLACE CONCRETE	100.0	59.7	83.4	90.2	63.3	79.1	96.6	44.8	75.3	99.0	47.8	78.0	99.0	72.8	88.2	93.5	47.3	74.5
03	CONCRETE	95.8	72.0	84.0	90.4	67.0	78.8	93.8	54.9	74.5	103.2	52.7	78.1	100.6	78.3	89.5	96.1	54.3	75.4
04	MASONRY	88.7	73.5	79.2	87.4	59.7	70.2	90.6	38.7	58.4	75.3	53.7	61.9	90.7	68.6	77.0	90.9	39.1	58.8
05	METALS	97.8	93.0	96.3	93.8	86.5	91.5	93.8	87.3	91.7	90.9	72.6	85.1	91.8	80.4	88.1	90.6	67.8	83.3
06	WOOD & PLASTICS	99.0	75.0	86.5	97.8	63.8	80.1	96.3	41.2	67.6	92.5	35.5	62.8	97.0	78.4	87.3	93.9	51.7	71.9
07	THERMAL & MOISTURE PROTECTION	99.8	60.6	83.3	99.3	64.9	84.8	96.6	53.4	78.4	101.8	51.7	80.7	101.8	74.9	90.5	101.3	48.4	79.0
08	DOORS & WINDOWS	100.7	68.7	92.3	97.4	61.1	87.9	98.3	48.9	85.4	93.8	45.4	81.1	99.6	75.9	93.4	93.8	52.8	83.1
09200	PLASTER & GYPSUM BOARD	100.6	74.9	84.6	102.2	63.4	78.1	100.3	40.2	63.0	112.5	34.2	63.9	114.0	78.3	91.8	112.9	50.8	74.3
095,098	CEILINGS & ACOUSTICAL TREATMENT	104.3	74.9	86.8	97.3	63.4	77.1	103.4	40.2	65.7	115.5	34.2	67.0	115.5	78.3	93.3	116.4	50.8	77.3
09600	FLOORING	118.7	50.7	100.3	121.5	48.7	101.8	118.7	31.1	95.0	82.3	56.1	75.2	83.9	74.6	81.4	82.7	39.6	71.1
097,099	WALL FINISHES, PAINTS & COATINGS	110.6	44.9	70.5	106.7	44.5	68.7	106.7	39.9	65.9	90.7	33.3	55.6	90.7	81.5	85.1	90.7	37.9	58.5
09	FINISHES	108.7	66.1	86.5	105.9	59.9	82.0	106.5	40.2	71.9	98.2	41.5	68.6	98.5	76.7	87.2	98.1	46.6	71.3
10 - 14	TOTAL DIV. 10000 - 14000	100.0	84.0	96.6	100.0	87.8	97.4	100.0	76.8	95.0	100.0	50.0	89.3	100.0	83.3	96.4	100.0	63.0	92.1
15	MECHANICAL	99.9	71.3	88.0	97.2	71.7	86.7	99.9	47.9	78.3	95.1	76.4	87.4	100.1	78.1	91.0	100.1	63.8	85.0
16	ELECTRICAL	96.5	47.0	72.4	97.5	50.7	74.8	92.2	65.0	78.9	95.1	61.8	78.9	95.5	80.1	88.0	95.7	48.1	72.6
01 - 16	WEIGHTED AVERAGE	99.4	70.8	86.7	96.6	66.6	83.3	97.2	56.3	79.1	95.3	62.2	80.6	97.7	79.0	89.4	96.3	57.5	79.0

DIVISION		GEORGIA																	
		COLUMBUS			DALTON			GAINESVILLE			MACON			SAVANNAH			STATESBORO		
		318 - 319			307			305			310 - 312			313 - 314			304		
		MAT.	INST.	TOTAL	MAT.	INST.	TOTAL	MAT.	INST.	TOTAL	MAT.	INST.	TOTAL	MAT.	INST.	TOTAL	MAT.	INST.	TOTAL
01590	EQUIPMENT RENTAL	.0	90.1	90.1	.0	106.1	106.1	.0	92.2	92.2	.0	102.6	102.6	.0	91.2	91.2	.0	91.2	91.2
02	SITE CONSTRUCTION	99.6	76.8	82.9	104.0	95.6	97.8	102.4	91.9	94.7	100.4	93.9	95.6	100.2	76.9	83.1	104.5	74.4	82.5
03100	CONCRETE FORMS & ACCESSORIES	97.4	57.1	62.6	84.7	44.9	50.4	97.7	43.8	51.2	96.8	54.1	59.9	97.7	51.2	57.6	78.2	40.8	45.9
03200	CONCRETE REINFORCEMENT	95.1	91.9	93.5	99.1	53.2	75.8	99.4	65.9	82.4	96.3	91.1	93.7	96.0	74.8	85.3	98.7	32.0	64.8
03300	CAST-IN-PLACE CONCRETE	96.2	53.7	78.7	96.2	34.9	71.0	104.1	48.2	81.1	94.9	44.8	74.3	93.2	49.9	75.4	98.8	42.5	75.7
03	CONCRETE	93.7	63.8	78.9	102.3	44.4	73.5	104.7	50.1	77.6	93.2	59.3	76.4	92.4	56.6	74.6	102.3	41.6	72.1
04	MASONRY	89.8	55.6	68.6	78.5	29.2	47.9	85.9	31.7	52.3	103.8	37.4	62.6	92.6	52.3	67.6	77.9	39.4	54.0
05	METALS	93.4	92.0	92.9	92.4	65.6	83.8	90.2	61.7	81.1	89.2	90.6	89.7	89.5	82.1	87.2	95.6	67.1	86.5
06	WOOD & PLASTICS	96.3	56.8	75.7	75.9	48.7	61.7	96.7	41.1	67.8	103.8	55.8	78.8	109.5	48.7	77.8	68.2	42.7	54.9
07	THERMAL & MOISTURE PROTECTION	96.3	61.0	81.5	103.1	40.4	76.7	101.7	44.4	77.6	95.1	56.1	78.7	96.7	51.5	77.7	101.9	37.6	74.9
08	DOORS & WINDOWS	98.3	59.3	88.1	94.9	36.3	79.6	93.8	39.1	79.5	96.7	57.8	86.5	99.4	49.2	86.3	95.7	33.1	79.4
09200	PLASTER & GYPSUM BOARD	100.3	56.2	72.9	98.6	47.8	67.0	114.0	40.0	68.0	105.8	55.1	74.3	100.6	47.8	67.8	98.7	41.7	63.3
095,098	CEILINGS & ACOUSTICAL TREATMENT	103.4	56.2	75.3	120.3	47.8	77.0	115.5	40.0	70.4	99.5	55.1	73.0	104.3	47.8	70.6	121.3	41.7	73.8
09600	FLOORING	118.7	52.4	100.8	83.4	13.4	64.5	83.8	13.4	64.8	92.3	36.5	77.2	118.7	46.7	99.2	96.4	46.7	83.0
097,099	WALL FINISHES, PAINTS & COATINGS	106.7	47.4	70.5	81.8	39.9	56.2	90.7	34.9	56.6	108.5	46.7	70.8	107.5	47.4	70.8	89.2	41.4	60.0
09	FINISHES	106.4	54.4	79.3	104.0	39.2	70.2	98.7	36.2	66.2	94.6	49.4	71.0	107.0	49.6	77.1	107.0	42.6	73.4
10 - 14	TOTAL DIV. 10000 - 14000	100.0	78.6	95.4	100.0	24.9	84.0	100.0	50.7	89.5	100.0	76.8	95.1	100.0	65.9	92.7	100.0	47.4	88.8
15	MECHANICAL	99.9	48.3	78.5	95.2	32.8	69.3	95.1	67.5	83.7	99.9	40.4	75.2	99.9	46.1	77.6	95.7	35.9	70.9
16	ELECTRICAL	92.6	49.7	71.7	105.5	73.3	89.9	95.1	61.7	78.9	90.3	66.4	78.7	94.1	61.9	78.4	96.2	30.2	64.1
01 - 16	WEIGHTED AVERAGE	97.2	60.4	80.8	97.4	49.4	76.1	96.0	55.7	78.1	95.7	59.1	79.4	97.0	57.7	79.6	97.1	43.6	73.4

DIVISION		GEORGIA						HAWAII									IDAHO		
		VALDOSTA			WAYCROSS			HILO			HONOLULU			STATES & POSS., GUAM			BOISE		
		316			315			967			968			969			836 - 837		
		MAT.	INST.	TOTAL	MAT.	INST.	TOTAL	MAT.	INST.	TOTAL	MAT.	INST.	TOTAL	MAT.	INST.	TOTAL	MAT.	INST.	TOTAL
01590	EQUIPMENT RENTAL	.0	90.1	90.1	.0	90.1	90.1	.0	99.4	99.4	.0	99.4	99.4	.0	176.0	176.0	.0	101.7	101.7
02	SITE CONSTRUCTION	109.7	76.2	85.1	106.5	74.4	83.0	134.3	106.4	113.8	139.5	106.4	115.2	175.1	105.3	123.9	78.2	102.4	95.9
03100	CONCRETE FORMS & ACCESSORIES	84.3	44.3	49.8	86.7	41.5	47.7	107.0	142.9	138.0	107.5	142.9	138.1	109.1	48.7	57.0	97.6	81.3	83.5
03200	CONCRETE REINFORCEMENT	97.0	42.3	69.2	97.0	64.3	80.4	109.6	126.2	118.0	108.6	126.2	117.5	1386.8	26.7	695.8	108.2	76.7	92.2
03300	CAST-IN-PLACE CONCRETE	94.5	50.5	76.4	106.8	42.9	80.5	209.1	127.9	175.7	201.5	127.9	171.3	181.1	107.1	150.7	94.2	88.1	91.7
03	CONCRETE	97.6	47.7	72.9	101.6	48.0	75.0	158.5	133.3	146.0	154.7	133.3	144.1	350.5	65.5	209.0	104.3	82.7	93.6
04	MASONRY	95.9	47.6	65.9	96.8	39.4	61.2	134.1	128.9	130.9	133.8	128.9	130.8	185.1	33.0	90.8	133.5	72.3	95.6
05	METALS	93.0	70.5	85.8	92.4	77.2	87.5	118.7	110.9	116.2	144.4	110.9	133.7	155.2	77.5	130.4	99.6	76.1	92.1
06	WOOD & PLASTICS	81.8	40.2	60.1	84.9	42.6	62.9	99.1	146.9	124.0	98.9	146.9	123.9	107.8	48.5	76.9	95.4	81.3	88.0
07	THERMAL & MOISTURE PROTECTION	96.5	56.7	79.7	96.3	39.7	72.5	105.6	128.9	115.4	107.5	128.9	116.5	115.3	58.7	91.5	92.7	78.0	86.5
08	DOORS & WINDOWS	93.7	37.3	79.0	93.7	42.5	80.3	100.6	137.9	110.3	105.8	137.9	114.2	100.4	41.6	85.0	94.7	76.6	89.9
09200	PLASTER & GYPSUM BOARD	92.0	39.1	59.2	94.6	41.7	61.7	99.2	148.0	129.5	120.6	148.0	137.6	3174.6	34.0	1223.5	85.5	80.6	82.5
095,098	CEILINGS & ACOUSTICAL TREATMENT	98.3	39.1	63.0	96.4	41.7	63.8	117.4	148.0	135.6	120.4	148.0	136.9	258.8	34.0	124.7	110.7	80.6	92.7
09600	FLOORING	110.3	37.3	90.5	111.8	31.1	89.9	145.4	132.6	141.9	168.2	132.6	158.6	117.8	32.7	94.8	99.0	53.8	86.7
097,099	WALL FINISHES, PAINTS & COATINGS	106.7	34.6	62.7	106.7	41.4	66.8	103.3	148.6	131.0	104.0	148.6	131.2	109.9	39.8	67.1	102.2	50.3	70.5
09	FINISHES	102.1	41.3	70.4	102.2	39.3	69.5	123.1	143.0	133.5	134.5	143.0	138.9	646.4	44.8	333.1	99.0	73.2	85.6
10 - 14	TOTAL DIV. 10000 - 14000	100.0	64.7	92.5	100.0	47.3	88.8	100.0	123.5	105.0	100.0	123.5	105.0	100.0	97.1	99.4	100.0	80.8	95.9
15	MECHANICAL	99.9	40.6	75.3	96.7	42.0	74.0	100.2	116.5	107.0	100.2	116.5	107.0	102.8	34.9	74.6	100.1	75.5	89.9
16	ELECTRICAL	90.8	29.5	61.0	94.7	53.7	74.8	106.1	123.3	114.5	108.9	123.3	115.9	118.8	31.9	76.5	96.0	77.0	86.8
01 - 16	WEIGHTED AVERAGE	96.9	47.6	75.0	97.0	50.0	76.1	115.0	124.8	119.3	120.6	124.8	122.5	196.4	52.7	132.5	100.2	78.9	90.7

DIVISION		IDAHO															ILLINOIS		
		COEUR D'ALENE			IDAHO FALLS			LEWISTON			POCATELLO			TWIN FALLS			BLOOMINGTON		
		838			834			835			832			833			617		
		MAT.	INST.	TOTAL	MAT.	INST.	TOTAL	MAT.	INST.	TOTAL	MAT.	INST.	TOTAL	MAT.	INST.	TOTAL	MAT.	INST.	TOTAL
01590	EQUIPMENT RENTAL	.0	94.6	94.6	.0	101.7	101.7	.0	94.6	94.6	.0	101.7	101.7	.0	101.7	101.7	.0	101.6	101.6
02	SITE CONSTRUCTION	76.7	96.5	91.2	76.1	100.9	94.3	82.1	96.7	92.8	79.3	102.4	96.2	85.8	100.3	96.5	95.0	95.0	95.0
03100	CONCRETE FORMS & ACCESSORIES	105.3	66.8	72.1	90.3	35.6	43.1	111.5	68.0	74.0	97.7	80.8	83.1	98.5	35.2	43.9	83.7	104.3	101.5
03200	CONCRETE REINFORCEMENT	115.7	97.1	106.3	110.1	75.5	92.5	115.7	97.3	106.4	108.5	76.2	92.1	110.4	75.9	92.9	100.5	96.4	98.4
03300	CAST-IN-PLACE CONCRETE	98.3	83.5	92.2	86.5	54.7	73.4	102.2	92.7	98.3	93.5	87.9	91.2	95.9	44.2	74.7	98.9	100.5	99.5
03	CONCRETE	108.3	78.5	93.5	93.8	50.7	72.4	112.1	82.3	97.3	100.9	82.3	91.7	108.5	46.9	78.0	97.2	101.9	99.5
04	MASONRY	134.4	85.0	103.8	127.6	31.0	67.7	134.7	85.0	103.9	129.6	64.5	89.2	132.8	37.8	73.9	111.5	110.3	110.8
05	METALS	93.0	85.9	90.7	107.6	73.0	96.6	92.6	87.3	90.9	107.7	75.0	97.3	107.7	72.6	96.5	96.4	105.9	99.4
06	WOOD & PLASTICS	97.1	61.8	78.7	87.7	33.0	59.2	103.3	61.8	81.7	95.4	81.3	88.0	96.0	34.1	63.7	83.1	101.0	92.4
07	THERMAL & MOISTURE PROTECTION	135.7	77.7	111.3	92.4	47.5	73.5	135.9	79.7	112.2	92.9	68.8	82.7	93.7	45.4	73.3	94.0	103.2	97.8
08	DOORS & WINDOWS	114.6	64.3	101.4	97.9	44.0	83.8	114.5	67.6	102.3	94.7	70.1	88.2	97.9	41.9	83.3	89.8	97.2	91.7
09200	PLASTER & GYPSUM BOARD	140.0	60.7	90.7	80.3	31.0	49.7	141.8	60.7	91.4	85.5	80.6	82.5	85.0	32.1	52.1	98.3	100.8	99.9
095,098	CEILINGS & ACOUSTICAL TREATMENT	131.5	60.7	89.3	101.4	31.0	59.4	131.5	60.7	89.3	111.6	80.6	93.1	103.9	32.1	61.1	88.6	100.8	95.9
09600	FLOORING	133.2	57.3	112.7	95.6	53.8	84.3	136.4	91.3	124.2	99.3	53.8	87.0	100.6	53.8	87.9	89.0	111.1	95.0
097,099	WALL FINISHES, PAINTS & COATINGS	127.5	74.7	95.3	102.2	41.4	65.1	127.5	74.7	95.3	102.2	52.4	71.8	102.2	30.9	58.7	90.8	96.6	94.3
09	FINISHES	156.2	64.9	108.6	94.5	38.3	65.3	157.5	71.9	112.9	99.4	73.4	85.9	98.3	37.2	66.5	90.7	105.1	98.2
10 - 14	TOTAL DIV. 10000 - 14000	100.0	65.9	92.7	100.0	46.3	88.5	100.0	82.8	96.3	100.0	80.8	95.9	100.0	45.5	88.4	100.0	88.3	97.5
15	MECHANICAL	100.0	62.1	84.3	101.0	62.3	85.0	101.2	89.1	96.2	100.1	75.4	89.8	100.1	62.1	84.3	95.1	96.4	95.6
16	ELECTRICAL	83.0	83.6	83.3	82.0	76.1	79.1	81.3	83.6	82.4	91.1	76.1	83.8	83.3	69.9	76.8	96.0	87.3	91.7
01 - 16	WEIGHTED AVERAGE	106.6	75.9	93.0	98.3	58.8	80.8	107.4	83.6	96.8	100.3	77.3	90.1	100.8	57.7	81.7	95.8	99.2	97.3

DIVISION		ILLINOIS																	
		CARBONDALE			CENTRALIA			CHAMPAIGN			CHICAGO			DECATUR			EAST ST. LOUIS		
		629			628			618 - 619			606 - 608			625			620 - 622		
		MAT.	INST.	TOTAL	MAT.	INST.	TOTAL	MAT.	INST.	TOTAL	MAT.	INST.	TOTAL	MAT.	INST.	TOTAL	MAT.	INST.	TOTAL
01590	EQUIPMENT RENTAL	.0	109.1	109.1	.0	109.1	109.1	.0	102.4	102.4	.0	91.5	91.5	.0	102.4	102.4	.0	109.1	109.1
02	SITE CONSTRUCTION	102.9	95.7	97.6	103.2	96.3	98.2	104.3	96.0	98.2	99.1	90.8	93.0	89.3	96.0	94.2	105.4	96.4	98.8
03100	CONCRETE FORMS & ACCESSORIES	94.0	99.8	99.0	96.4	112.7	110.4	91.1	102.6	101.1	99.7	137.6	132.4	97.3	102.2	101.5	91.0	110.2	107.6
03200	CONCRETE REINFORCEMENT	99.3	86.2	92.6	99.3	99.1	99.2	100.5	92.8	96.6	100.5	146.9	124.1	97.4	100.7	99.1	99.2	99.0	99.1
03300	CAST-IN-PLACE CONCRETE	92.5	85.3	89.5	92.9	104.4	97.6	114.7	105.5	110.9	113.0	137.9	123.2	101.0	102.6	101.7	94.5	105.3	99.0
03	CONCRETE	88.4	93.3	90.8	88.8	108.0	98.4	109.7	102.1	106.0	110.4	138.5	124.4	98.7	102.4	100.5	89.7	107.3	98.4
04	MASONRY	74.4	100.2	90.4	74.4	109.7	96.3	134.1	93.4	108.9	93.7	136.8	120.4	73.9	106.8	94.3	74.6	111.5	97.5
05	METALS	95.6	109.5	100.0	95.6	117.4	102.6	96.4	102.6	98.4	98.5	124.9	106.9	98.7	106.9	101.3	96.4	117.1	103.0
06	WOOD & PLASTICS	98.3	99.6	99.0	101.2	115.0	108.4	91.1	99.0	95.2	101.6	135.2	119.1	99.7	99.0	99.3	95.3	111.2	103.6
07	THERMAL & MOISTURE PROTECTION	89.4	90.1	89.7	89.5	106.1	96.5	94.6	103.2	98.2	101.2	129.7	113.2	94.7	99.9	96.9	89.5	101.9	94.7
08	DOORS & WINDOWS	86.0	102.1	90.2	86.0	113.8	93.3	90.4	98.2	92.5	103.2	139.4	112.6	96.3	100.1	97.3	86.1	111.7	92.8
09200	PLASTER & GYPSUM BOARD	102.7	99.5	100.7	103.8	115.2	110.9	101.3	98.8	99.7	101.3	136.0	122.9	105.6	98.8	101.4	102.0	111.3	107.8
095,098	CEILINGS & ACOUSTICAL TREATMENT	88.6	99.5	95.1	88.6	115.2	104.5	88.6	98.8	94.7	104.2	136.0	123.2	95.5	98.8	97.4	88.6	111.3	102.2
09600	FLOORING	106.5	97.2	103.9	107.8	118.0	110.5	92.5	106.5	96.3	83.5	127.9	95.6	96.7	105.6	99.1	105.1	118.0	108.6
097,099	WALL FINISHES, PAINTS & COATINGS	98.8	93.0	95.3	98.8	101.9	100.6	90.8	106.2	100.2	78.9	134.3	112.7	90.8	101.7	97.4	98.8	101.9	100.6
09	FINISHES	94.9	99.1	97.1	95.5	113.3	104.7	92.7	103.9	98.5	92.6	135.7	115.0	95.8	102.9	99.5	94.4	110.0	102.6
10 - 14	TOTAL DIV. 10000 - 14000	100.0	87.3	97.3	100.0	90.0	97.9	100.0	89.0	97.7	100.0	123.8	105.1	100.0	88.6	97.6	100.0	90.2	97.9
15	MECHANICAL	95.0	98.7	96.5	95.0	85.0	90.9	95.1	108.4	100.6	100.0	126.1	110.8	100.0	106.2	102.6	99.9	92.3	96.8
16	ELECTRICAL	95.0	90.6	92.8	96.4	89.5	93.0	99.2	95.4	97.3	98.8	117.1	107.7	98.7	89.1	94.0	95.9	89.5	92.8
01 - 16	WEIGHTED AVERAGE	92.9	97.3	94.9	93.3	101.1	96.8	99.2	100.8	99.9	100.1	126.3	111.7	97.0	101.1	98.8	94.5	102.0	97.8

DIVISION		ILLINOIS																	
		EFFINGHAM			GALESBURG			JOLIET			KANKAKEE			LA SALLE			NORTH SUBURBAN		
		624			614			604			609			613			600 - 603		
		MAT.	INST.	TOTAL	MAT.	INST.	TOTAL	MAT.	INST.	TOTAL	MAT.	INST.	TOTAL	MAT.	INST.	TOTAL	MAT.	INST.	TOTAL
01590	EQUIPMENT RENTAL	.0	102.4	102.4	.0	101.6	101.6	.0	89.3	89.3	.0	89.3	89.3	.0	101.6	101.6	.0	89.3	89.3
02	SITE CONSTRUCTION	96.2	95.8	95.9	97.3	95.0	95.6	99.0	89.6	92.1	92.7	88.8	89.9	96.8	95.6	95.9	98.2	90.1	92.3
03100	CONCRETE FORMS & ACCESSORIES	102.3	100.5	100.8	91.0	104.2	102.4	100.6	138.1	133.0	92.5	108.4	106.2	107.4	111.0	110.5	99.7	136.0	131.0
03200	CONCRETE REINFORCEMENT	100.5	95.8	98.1	100.0	96.4	98.1	100.5	123.7	112.3	101.4	112.5	107.1	100.1	104.9	102.5	100.5	146.9	124.1
03300	CAST-IN-PLACE CONCRETE	100.7	77.2	91.0	101.8	102.2	102.0	112.9	128.4	119.3	105.2	115.0	109.2	101.7	109.6	104.9	113.0	139.9	124.0
03	CONCRETE	99.7	92.1	95.9	100.1	102.4	101.2	110.4	131.0	120.6	103.9	111.1	107.5	101.2	109.7	105.4	110.4	138.4	124.3
04	MASONRY	78.2	86.5	83.3	111.6	110.2	110.8	94.7	127.6	115.1	93.2	112.5	105.2	111.6	109.0	110.0	93.7	134.1	118.8
05	METALS	96.3	103.1	98.5	96.4	105.7	99.4	96.2	112.5	101.4	96.2	106.0	99.4	96.5	112.8	101.7	97.5	122.7	105.5
06	WOOD & PLASTICS	102.3	99.0	100.6	90.9	101.2	96.3	103.2	141.5	123.1	94.5	108.2	101.6	107.6	111.6	109.7	101.6	134.7	118.9
07	THERMAL & MOISTURE PROTECTION	94.3	92.6	93.6	94.1	101.2	97.1	100.6	123.4	110.2	99.8	114.0	105.8	94.4	99.7	96.6	101.2	129.4	113.1
08	DOORS & WINDOWS	90.4	99.1	92.6	89.8	95.1	91.1	101.1	136.6	110.4	93.0	111.8	97.9	89.8	105.2	93.8	101.1	139.2	111.1
09200	PLASTER & GYPSUM BOARD	104.9	98.8	101.1	101.3	101.1	101.2	98.7	142.5	125.9	95.0	108.3	103.3	107.2	111.7	110.0	101.3	135.6	122.6
095,098	CEILINGS & ACOUSTICAL TREATMENT	88.6	98.8	94.7	88.6	101.1	96.1	104.2	142.5	127.1	104.2	108.3	106.7	88.6	111.7	102.4	104.2	135.6	122.9
09600	FLOORING	98.1	106.5	100.3	92.4	107.5	96.5	83.1	127.9	95.2	80.0	106.8	87.2	100.5	108.1	102.5	83.5	111.2	91.0
097,099	WALL FINISHES, PAINTS & COATINGS	90.8	102.0	97.6	90.8	97.9	95.1	76.7	132.0	110.5	76.7	79.9	78.7	90.8	96.6	94.3	78.9	132.0	111.3
09	FINISHES	94.4	102.4	98.6	92.3	104.6	98.7	91.9	136.9	115.3	90.1	103.8	97.3	95.5	109.4	102.8	92.5	131.0	112.6
10 - 14	TOTAL DIV. 10000 - 14000	100.0	87.5	97.3	100.0	86.4	97.1	100.0	120.9	104.5	100.0	99.5	99.9	100.0	86.0	97.0	100.0	117.9	103.8
15	MECHANICAL	95.1	106.4	99.8	95.1	98.5	96.5	100.1	113.4	105.6	95.2	101.3	97.7	95.1	92.1	93.9	100.0	121.1	108.7
16	ELECTRICAL	96.6	90.6	93.7	96.9	83.9	90.6	97.1	103.8	100.4	92.4	100.2	96.2	94.0	100.2	97.1	97.0	126.9	111.6
01 - 16	WEIGHTED AVERAGE	95.1	97.1	96.0	96.5	98.9	97.6	99.4	118.5	107.9	95.7	104.1	99.4	96.7	102.6	99.3	99.5	125.3	111.0

DIVISION		ILLINOIS																	
		PEORIA			QUINCY			ROCKFORD			ROCK ISLAND			SOUTH SUBURBAN			SPRINGFIELD		
		615 - 616			623			610 - 611			612			605			626 - 627		
		MAT.	INST.	TOTAL	MAT.	INST.	TOTAL	MAT.	INST.	TOTAL	MAT.	INST.	TOTAL	MAT.	INST.	TOTAL	MAT.	INST.	TOTAL
01590	EQUIPMENT RENTAL	.0	101.6	101.6	.0	102.4	102.4	.0	101.6	101.6	.0	101.6	101.6	.0	89.3	89.3	.0	102.4	102.4
02	SITE CONSTRUCTION	97.9	95.1	95.8	94.9	95.5	95.4	97.1	95.7	96.1	95.6	94.6	94.8	98.2	90.1	92.3	95.3	96.0	95.8
03100	CONCRETE FORMS & ACCESSORIES	95.6	105.0	103.7	99.6	98.8	98.9	99.9	117.1	114.8	93.1	104.6	103.0	99.7	136.0	131.0	99.2	102.5	102.0
03200	CONCRETE REINFORCEMENT	97.4	96.4	96.9	100.1	96.6	98.3	97.4	125.2	111.5	100.0	96.1	98.0	100.5	146.9	124.1	97.4	96.8	97.1
03300	CAST-IN-PLACE CONCRETE	98.7	103.8	100.8	100.8	99.1	100.1	101.0	96.1	99.0	99.6	99.7	99.7	113.0	139.9	124.0	93.8	103.0	97.6
03	CONCRETE	97.5	103.3	100.4	99.2	98.9	99.1	98.9	111.6	105.2	98.3	101.7	100.0	110.4	138.4	124.3	95.4	102.0	98.6
04	MASONRY	115.2	111.9	113.1	97.3	96.6	96.9	89.2	107.7	100.7	111.5	101.2	105.1	93.7	134.1	118.8	75.5	107.7	95.4
05	METALS	98.7	106.0	101.0	96.4	104.5	99.0	98.7	121.1	105.8	96.4	104.8	99.1	97.5	122.7	105.5	101.3	105.2	102.5
06	WOOD & PLASTICS	99.7	101.1	100.4	99.5	99.0	99.2	99.7	117.4	108.9	93.1	105.0	99.3	101.6	134.7	118.9	102.3	99.0	100.6
07	THERMAL & MOISTURE PROTECTION	94.9	103.9	98.7	94.3	94.4	94.4	97.9	112.6	104.1	94.1	97.3	95.5	101.2	129.4	113.1	94.4	100.9	97.1
08	DOORS & WINDOWS	96.3	99.1	97.0	91.2	99.1	93.3	96.3	115.7	101.4	89.8	97.6	91.8	101.1	139.2	111.1	96.3	99.1	97.0
09200	PLASTER & GYPSUM BOARD	105.6	100.9	102.7	103.8	98.8	100.7	105.6	117.7	113.1	102.0	104.9	103.8	101.3	135.6	122.6	105.6	98.8	101.4
095,098	CEILINGS & ACOUSTICAL TREATMENT	95.5	100.9	98.7	88.6	98.8	94.7	95.5	117.7	108.7	88.6	104.9	98.4	104.2	135.6	122.9	95.5	98.8	97.4
09600	FLOORING	96.7	107.5	99.6	96.7	101.1	97.9	96.7	89.3	94.7	93.8	91.6	93.2	83.5	111.2	91.0	96.9	105.6	99.3
097,099	WALL FINISHES, PAINTS & COATINGS	90.8	93.0	92.1	90.8	101.7	97.4	90.8	104.9	99.4	90.8	97.9	95.1	78.9	132.0	111.3	90.8	101.7	97.4
09	FINISHES	95.9	104.3	100.3	93.8	100.1	97.1	95.9	110.9	103.7	92.7	102.2	97.6	92.5	131.0	112.6	95.9	103.2	99.7
10 - 14	TOTAL DIV. 10000 - 14000	100.0	88.8	97.6	100.0	89.0	97.7	100.0	98.0	99.6	100.0	85.5	96.9	100.0	117.9	103.8	100.0	88.9	97.6
15	MECHANICAL	100.0	98.4	99.3	95.1	93.5	94.4	100.1	101.9	100.8	95.1	95.6	95.3	100.0	121.1	108.7	100.0	101.9	100.8
16	ELECTRICAL	98.0	86.6	92.4	94.1	77.8	86.2	98.1	107.4	102.6	89.3	85.0	87.2	97.0	126.9	111.6	98.6	86.9	92.9
01 - 16	WEIGHTED AVERAGE	99.1	99.9	99.4	95.7	94.6	95.2	98.1	107.9	102.4	95.4	97.0	96.1	99.5	125.3	111.0	97.3	99.8	98.4

City Cost Indexes

DIVISION		INDIANA																	
		ANDERSON			BLOOMINGTON			COLUMBUS			EVANSVILLE			FORT WAYNE			GARY		
		460			474			472			476 - 477			467 - 468			463 - 464		
		MAT.	INST.	TOTAL	MAT.	INST.	TOTAL	MAT.	INST.	TOTAL	MAT.	INST.	TOTAL	MAT.	INST.	TOTAL	MAT.	INST.	TOTAL
01590	**EQUIPMENT RENTAL**	.0	96.7	96.7	.0	85.9	85.9	.0	85.9	85.9	.0	121.3	121.3	.0	96.7	96.7	.0	96.7	96.7
02	**SITE CONSTRUCTION**	88.2	96.4	94.2	77.0	95.8	90.8	74.1	95.5	89.8	82.0	130.1	117.2	89.2	96.5	94.6	88.7	99.2	96.4
03100	CONCRETE FORMS & ACCESSORIES	96.4	79.9	82.2	100.5	80.7	83.4	94.1	77.9	80.2	93.1	82.2	83.7	95.0	77.8	80.2	96.5	106.3	105.0
03200	CONCRETE REINFORCEMENT	94.9	80.7	87.7	84.1	80.4	82.2	84.4	80.4	82.4	92.2	80.4	86.2	94.9	77.5	86.0	94.9	100.3	97.6
03300	CAST-IN-PLACE CONCRETE	101.8	82.0	93.6	100.8	79.5	92.0	100.3	75.6	90.2	96.3	92.0	94.5	108.1	80.4	96.8	106.3	111.1	108.3
03	**CONCRETE**	96.2	81.3	88.8	102.7	79.9	91.4	102.0	77.3	89.7	102.9	85.4	94.2	99.1	79.2	89.3	98.4	106.9	102.6
04	**MASONRY**	92.3	84.9	87.7	90.3	80.3	84.1	90.2	80.2	84.0	86.6	86.6	86.6	92.4	84.7	87.6	93.7	107.8	102.5
05	**METALS**	91.6	88.7	90.7	94.4	75.7	88.4	94.4	75.1	88.2	87.1	86.3	86.8	91.6	87.2	90.2	91.6	103.7	95.5
06	**WOOD & PLASTICS**	113.1	79.2	95.4	118.2	79.7	98.2	112.1	76.1	93.4	97.3	79.8	88.2	112.9	76.1	93.7	111.1	106.1	108.5
07	**THERMAL & MOISTURE PROTECTION**	100.3	82.3	92.7	91.4	84.3	88.4	91.0	84.1	88.1	95.5	88.1	92.4	100.1	86.1	94.2	99.6	105.8	102.2
08	**DOORS & WINDOWS**	98.9	80.9	94.2	103.1	81.2	97.4	98.6	79.2	93.5	95.6	80.4	91.7	98.9	76.0	92.9	98.9	104.4	100.4
09200	PLASTER & GYPSUM BOARD	96.6	79.2	85.8	97.0	79.9	86.4	93.3	76.2	82.7	91.6	78.5	83.4	95.8	76.1	83.6	91.7	106.9	101.1
095,098	CEILINGS & ACOUSTICAL TREATMENT	91.5	79.2	84.2	81.6	79.9	80.6	81.6	76.2	78.4	86.2	78.5	81.6	91.5	76.1	82.3	91.5	106.9	100.7
09600	FLOORING	88.2	92.7	89.4	104.5	92.7	101.3	98.6	92.7	97.0	97.5	88.7	95.1	88.2	80.8	86.2	88.2	119.9	96.8
097,099	WALL FINISHES, PAINTS & COATINGS	91.5	77.0	82.6	90.2	89.0	89.5	90.2	89.0	89.5	96.8	88.0	91.4	91.5	80.1	84.5	91.5	107.9	101.5
09	**FINISHES**	89.7	81.9	85.7	94.7	83.7	89.0	92.3	81.6	86.7	92.8	83.7	88.1	89.5	78.2	83.6	88.9	109.2	99.5
10 - 14	**TOTAL DIV. 10000 - 14000**	100.0	90.7	98.0	100.0	90.2	97.9	100.0	89.8	97.8	100.0	95.3	99.0	100.0	89.3	97.7	100.0	94.3	98.8
15	**MECHANICAL**	99.9	78.8	91.2	99.7	84.9	93.6	94.8	86.9	91.5	99.9	86.8	94.5	99.9	81.2	92.2	99.9	101.3	100.5
16	**ELECTRICAL**	85.1	90.6	87.8	101.6	88.0	95.0	100.9	90.6	95.8	95.0	87.6	91.4	86.2	82.9	84.6	94.4	104.1	99.1
01 - 16	**WEIGHTED AVERAGE**	94.9	84.8	90.4	98.2	84.1	91.9	96.1	84.0	90.7	95.3	90.0	93.0	95.4	83.1	89.9	96.2	104.3	99.8

DIVISION		INDIANA																	
		INDIANAPOLIS			KOKOMO			LAFAYETTE			LAWRENCEBURG			MUNCIE			NEW ALBANY		
		461 - 462			469			479			470			473			471		
		MAT.	INST.	TOTAL	MAT.	INST.	TOTAL	MAT.	INST.	TOTAL	MAT.	INST.	TOTAL	MAT.	INST.	TOTAL	MAT.	INST.	TOTAL
01590	**EQUIPMENT RENTAL**	.0	91.5	91.5	.0	96.7	96.7	.0	85.9	85.9	.0	108.1	108.1	.0	96.2	96.2	.0	96.0	96.0
02	**SITE CONSTRUCTION**	87.9	100.1	96.9	84.9	96.5	93.4	74.8	95.5	90.0	72.9	115.4	104.0	77.2	95.8	90.8	69.6	99.7	91.6
03100	CONCRETE FORMS & ACCESSORIES	97.0	84.4	86.1	99.7	78.7	81.6	91.4	81.9	83.2	89.9	76.5	78.3	90.9	79.7	81.2	88.6	74.3	76.3
03200	CONCRETE REINFORCEMENT	94.2	82.2	88.1	85.8	80.6	83.2	84.1	78.7	81.4	83.3	81.4	82.3	93.0	80.7	86.8	84.7	82.1	83.4
03300	CAST-IN-PLACE CONCRETE	97.9	85.0	92.6	100.7	83.0	93.5	100.9	82.5	93.3	94.3	77.2	87.3	106.0	81.4	95.9	97.4	77.1	89.0
03	**CONCRETE**	98.1	83.8	91.0	93.1	81.1	87.2	102.1	81.1	91.7	95.1	78.2	86.7	101.4	81.0	91.3	100.6	77.1	88.9
04	**MASONRY**	99.6	84.4	90.1	92.0	83.0	86.5	96.1	82.2	87.4	75.7	79.0	77.8	92.6	84.9	87.8	82.7	73.4	76.9
05	**METALS**	93.5	78.2	88.6	88.8	88.9	88.8	93.3	74.5	87.3	89.4	88.1	89.0	96.1	89.0	93.8	91.6	83.3	89.0
06	**WOOD & PLASTICS**	110.1	83.8	96.4	116.5	77.1	96.0	108.7	81.6	94.6	96.1	75.1	85.1	110.3	79.0	94.0	98.2	73.9	85.5
07	**THERMAL & MOISTURE PROTECTION**	98.5	86.4	93.4	100.0	81.3	92.1	91.0	83.0	87.6	90.8	81.5	86.9	93.7	82.3	88.9	85.9	70.1	79.3
08	**DOORS & WINDOWS**	106.7	83.8	100.8	93.2	79.8	89.7	97.2	81.7	93.1	97.5	78.0	92.4	98.0	80.8	93.5	95.3	77.9	90.7
09200	PLASTER & GYPSUM BOARD	94.8	83.7	87.9	100.8	77.1	86.1	88.9	81.8	84.5	80.2	75.0	77.0	91.3	79.2	83.8	90.5	73.6	80.0
095,098	CEILINGS & ACOUSTICAL TREATMENT	95.8	83.7	88.6	90.7	77.1	82.6	73.1	81.8	78.3	91.3	75.0	81.6	82.5	79.2	80.5	86.2	73.6	78.7
09600	FLOORING	88.0	92.7	89.3	92.8	88.1	91.5	97.3	94.4	96.5	72.6	92.7	78.0	96.6	92.7	95.5	95.6	68.8	88.3
097,099	WALL FINISHES, PAINTS & COATINGS	91.5	89.0	90.0	91.5	87.7	89.2	90.2	84.1	86.5	92.0	79.1	84.1	90.2	77.0	82.2	96.8	72.1	81.7
09	**FINISHES**	90.8	86.4	88.5	91.4	81.1	86.0	89.2	84.5	86.8	85.0	79.9	82.4	91.4	81.8	86.4	92.3	73.0	82.3
10 - 14	**TOTAL DIV. 10000 - 14000**	100.0	92.4	98.4	100.0	90.3	97.9	100.0	90.8	98.0	100.0	55.3	90.5	100.0	90.3	97.9	100.0	55.3	90.5
15	**MECHANICAL**	99.9	87.7	94.9	95.0	87.0	91.7	94.8	76.5	87.2	95.5	80.9	89.5	99.7	78.7	91.0	95.0	78.9	88.3
16	**ELECTRICAL**	104.4	90.6	97.7	89.4	88.2	88.8	100.4	85.7	93.3	92.4	71.8	82.4	89.7	74.6	82.4	93.7	71.6	83.0
01 - 16	**WEIGHTED AVERAGE**	98.8	87.2	93.7	93.0	85.7	89.8	95.7	82.4	89.8	92.0	81.8	87.5	96.2	82.4	90.1	93.6	77.5	86.5

DIVISION		INDIANA									IOWA								
		SOUTH BEND			TERRE HAUTE			WASHINGTON			BURLINGTON			CARROLL			CEDAR RAPIDS		
		465 - 466			478			475			526			514			522 - 524		
		MAT.	INST.	TOTAL	MAT.	INST.	TOTAL	MAT.	INST.	TOTAL	MAT.	INST.	TOTAL	MAT.	INST.	TOTAL	MAT.	INST.	TOTAL
01590	**EQUIPMENT RENTAL**	.0	106.8	106.8	.0	121.3	121.3	.0	121.3	121.3	.0	99.6	99.6	.0	99.6	99.6	.0	95.6	95.6
02	**SITE CONSTRUCTION**	88.2	96.2	94.1	83.3	130.0	117.6	83.3	128.5	116.5	87.6	95.5	93.4	78.2	93.3	89.3	89.7	94.6	93.3
03100	CONCRETE FORMS & ACCESSORIES	100.9	81.5	84.2	94.6	82.2	83.9	95.3	81.3	83.2	95.7	72.0	75.3	81.2	49.9	54.2	102.1	80.6	83.6
03200	CONCRETE REINFORCEMENT	94.9	76.7	85.6	92.2	80.9	86.4	85.1	63.7	74.2	92.6	81.3	86.9	93.3	51.9	72.3	93.3	84.5	88.8
03300	CAST-IN-PLACE CONCRETE	99.0	85.1	93.3	93.2	91.2	92.4	101.5	97.1	99.7	107.8	53.5	85.5	107.8	58.5	87.5	108.0	78.2	95.8
03	**CONCRETE**	91.1	83.2	87.2	105.7	85.2	95.5	111.3	83.4	97.4	100.1	68.3	84.3	98.7	54.4	76.7	100.3	81.1	90.8
04	**MASONRY**	90.6	84.6	86.9	93.4	85.7	88.6	86.7	88.4	87.7	100.0	62.0	76.4	102.7	46.5	67.9	106.7	79.6	89.9
05	**METALS**	91.6	101.2	94.7	87.8	87.0	87.6	83.2	76.0	80.9	85.7	92.2	87.8	85.8	72.4	81.5	87.8	92.8	89.4
06	**WOOD & PLASTICS**	113.0	80.3	96.0	99.6	81.2	90.0	100.0	79.6	89.4	102.1	77.5	89.3	86.1	51.5	68.1	109.0	80.0	93.9
07	**THERMAL & MOISTURE PROTECTION**	99.4	87.4	94.3	95.6	85.1	91.1	95.7	86.5	91.8	97.9	65.2	84.2	98.2	53.7	79.5	98.9	81.8	91.7
08	**DOORS & WINDOWS**	92.3	79.8	89.0	96.2	82.0	92.5	93.1	75.7	88.6	93.1	72.0	87.6	98.3	47.7	85.1	99.3	84.7	95.5
09200	PLASTER & GYPSUM BOARD	96.2	80.3	86.3	91.6	79.9	84.3	89.8	78.3	82.7	100.2	77.0	85.8	93.9	50.2	66.8	103.7	79.8	88.9
095,098	CEILINGS & ACOUSTICAL TREATMENT	91.5	80.3	84.9	86.2	79.9	82.5	75.1	78.3	77.0	109.6	77.0	90.1	109.6	50.2	74.2	114.7	79.8	93.9
09600	FLOORING	88.2	85.3	87.4	97.5	97.2	97.4	98.7	83.5	94.6	105.9	47.2	90.0	98.5	41.5	83.1	122.3	55.6	104.2
097,099	WALL FINISHES, PAINTS & COATINGS	91.5	83.8	86.8	96.8	93.3	94.7	96.8	91.9	93.8	101.4	84.7	91.2	101.4	39.7	63.7	105.5	76.5	87.8
09	**FINISHES**	89.7	82.4	85.9	92.8	86.1	89.4	90.4	83.4	86.8	105.1	69.0	86.3	100.9	46.9	72.8	111.9	74.6	92.5
10 - 14	**TOTAL DIV. 10000 - 14000**	100.0	90.2	97.9	100.0	94.4	98.8	100.0	95.3	99.0	100.0	85.1	96.8	100.0	80.9	95.9	100.0	90.4	98.0
15	**MECHANICAL**	99.9	82.8	92.8	99.9	76.9	90.4	95.0	86.2	91.4	95.4	74.2	86.6	95.4	51.7	77.3	100.3	85.9	94.3
16	**ELECTRICAL**	99.2	88.6	94.1	93.5	88.0	90.9	94.0	87.5	90.9	98.3	78.0	88.4	99.0	41.0	70.8	96.1	85.5	90.9
01 - 16	**WEIGHTED AVERAGE**	95.3	86.8	91.5	96.0	88.3	92.6	94.0	88.4	91.5	96.0	75.5	86.9	95.8	55.7	78.0	98.9	84.4	92.5

	DIVISION	IOWA																	
		COUNCIL BLUFFS			CRESTON			DAVENPORT			DECORAH			DES MOINES			DUBUQUE		
		515			508			527 - 528			521			500 - 503,509			520		
		MAT.	INST.	TOTAL	MAT.	INST.	TOTAL	MAT.	INST.	TOTAL	MAT.	INST.	TOTAL	MAT.	INST.	TOTAL	MAT.	INST.	TOTAL
01590	EQUIPMENT RENTAL	.0	95.3	95.3	.0	99.6	99.6	.0	99.6	99.6	.0	99.6	99.6	.0	101.6	101.6	.0	94.2	94.2
02	SITE CONSTRUCTION	95.4	91.4	92.4	78.2	94.4	90.0	87.8	99.1	96.1	86.3	93.9	91.8	80.9	99.9	94.9	87.7	91.6	90.5
03100	CONCRETE FORMS & ACCESSORIES	80.2	56.9	60.1	81.8	73.4	74.6	101.7	93.1	94.3	92.6	46.8	53.1	103.7	80.4	83.6	82.0	69.2	70.9
03200	CONCRETE REINFORCEMENT	95.2	78.9	86.9	92.8	78.4	85.4	93.3	95.5	94.4	92.6	52.0	72.0	93.3	79.6	86.3	92.0	84.2	88.1
03300	CAST-IN-PLACE CONCRETE	112.4	72.1	95.8	107.4	61.0	88.3	104.0	99.0	102.0	104.8	55.1	84.4	108.3	76.5	95.2	105.8	91.0	99.7
03	CONCRETE	102.1	67.5	84.9	98.5	70.8	84.7	98.3	96.0	97.2	97.9	51.9	75.1	99.4	79.6	89.6	96.6	80.4	88.6
04	MASONRY	106.4	76.0	87.6	102.7	50.5	70.3	102.7	91.8	96.0	121.1	43.3	72.8	100.5	77.9	86.5	107.6	73.9	86.7
05	METALS	93.0	89.3	91.8	85.7	85.0	85.4	87.8	101.9	92.3	85.8	71.9	81.4	87.9	91.9	89.2	86.4	91.9	88.2
06	WOOD & PLASTICS	84.8	52.8	68.1	86.7	80.3	83.4	109.0	91.5	99.9	98.8	47.8	72.2	110.5	80.3	94.8	86.8	66.9	76.5
07	THERMAL & MOISTURE PROTECTION	98.2	66.9	85.0	99.0	58.4	81.9	98.4	92.1	95.7	98.1	48.1	77.1	99.1	78.4	90.4	98.5	70.3	86.6
08	DOORS & WINDOWS	98.3	60.4	88.4	111.2	75.5	101.9	99.3	92.4	97.5	96.6	49.8	84.4	99.3	83.5	95.1	98.3	77.6	92.9
09200	PLASTER & GYPSUM BOARD	93.9	51.9	67.8	93.9	79.9	85.2	103.7	91.3	96.0	99.1	46.4	66.4	100.9	79.9	87.8	94.3	66.4	76.9
095,098	CEILINGS & ACOUSTICAL TREATMENT	109.6	51.9	75.1	109.6	79.9	91.9	114.7	91.3	100.8	109.6	46.4	71.9	113.0	79.9	93.2	109.6	66.4	83.8
09600	FLOORING	97.3	45.5	83.3	98.8	41.5	83.3	108.8	84.1	102.1	105.2	58.9	92.6	108.6	41.5	90.4	111.1	38.7	91.5
097,099	WALL FINISHES, PAINTS & COATINGS	97.2	69.5	80.3	101.4	49.0	69.4	101.4	93.3	96.5	101.4	42.0	65.1	101.4	78.7	87.6	104.4	76.5	87.4
09	FINISHES	101.2	54.5	76.9	100.9	65.9	82.7	107.7	91.1	99.1	104.6	48.7	75.5	105.9	72.4	88.4	105.7	63.0	83.5
10 - 14	TOTAL DIV. 10000 - 14000	100.0	69.3	93.4	100.0	86.0	97.0	100.0	95.1	98.9	100.0	51.1	89.6	100.0	90.9	98.1	100.0	87.6	97.4
15	MECHANICAL	100.3	79.0	91.5	95.4	53.8	78.1	100.3	95.3	98.2	95.4	44.8	74.4	100.3	80.2	92.0	100.3	78.9	91.4
16	ELECTRICAL	101.0	85.5	93.4	94.2	53.5	74.4	94.7	93.0	93.9	96.1	40.8	69.2	96.6	81.2	89.1	99.6	68.6	84.6
01 - 16	WEIGHTED AVERAGE	99.3	75.2	88.6	96.4	66.0	82.9	97.9	94.9	96.5	96.8	52.8	77.3	97.8	82.2	90.9	97.8	77.2	88.7

	DIVISION	IOWA																	
		FORT DODGE			MASON CITY			OTTUMWA			SHENANDOAH			SIBLEY			SIOUX CITY		
		505			504			525			516			512			510 - 511		
		MAT.	INST.	TOTAL	MAT.	INST.	TOTAL	MAT.	INST.	TOTAL	MAT.	INST.	TOTAL	MAT.	INST.	TOTAL	MAT.	INST.	TOTAL
01590	EQUIPMENT RENTAL	.0	99.6	99.6	.0	99.6	99.6	.0	94.2	94.2	.0	95.3	95.3	.0	99.6	99.6	.0	99.6	99.6
02	SITE CONSTRUCTION	83.7	93.0	90.5	83.7	93.8	91.1	88.0	90.0	89.4	93.3	89.0	90.2	95.8	93.3	94.0	97.3	95.5	95.9
03100	CONCRETE FORMS & ACCESSORIES	82.6	52.5	56.6	88.1	47.6	53.2	90.4	60.2	64.3	82.3	60.9	63.9	82.9	42.7	48.2	102.1	69.0	73.5
03200	CONCRETE REINFORCEMENT	92.8	51.5	71.8	92.6	66.7	79.5	92.6	75.6	84.0	95.2	77.1	86.0	95.2	66.1	80.4	93.3	66.8	79.9
03300	CAST-IN-PLACE CONCRETE	100.8	46.4	78.5	100.8	55.4	82.2	108.5	61.8	89.3	108.6	50.5	84.7	106.4	50.5	83.4	107.1	56.5	86.3
03	CONCRETE	94.0	51.4	72.8	94.3	55.2	74.9	99.5	64.8	82.3	99.7	61.4	80.7	98.7	51.2	75.1	99.8	65.2	82.6
04	MASONRY	100.6	27.3	55.2	114.4	49.0	73.8	104.2	51.9	71.8	106.1	35.2	62.2	127.1	39.6	72.9	99.7	59.4	74.7
05	METALS	85.8	71.0	81.0	85.8	79.5	83.8	85.7	86.8	86.0	92.3	82.0	89.0	85.9	77.9	83.3	87.8	82.9	86.2
06	WOOD & PLASTICS	87.4	58.5	72.3	93.3	47.8	69.6	96.0	61.9	78.2	86.8	68.6	77.3	87.8	42.5	64.2	109.0	69.3	88.3
07	THERMAL & MOISTURE PROTECTION	98.3	35.9	72.0	97.8	49.4	77.5	98.7	59.4	82.1	97.5	45.5	75.6	97.9	50.9	78.1	98.4	57.6	81.2
08	DOORS & WINDOWS	103.7	49.6	89.5	94.0	53.8	83.5	98.3	65.9	89.8	87.9	65.2	81.9	94.4	49.3	82.6	99.3	66.4	90.7
09200	PLASTER & GYPSUM BOARD	94.3	57.5	71.4	96.1	46.4	65.3	97.6	61.2	75.0	94.3	68.1	78.0	94.3	41.0	61.2	103.7	68.5	81.9
095,098	CEILINGS & ACOUSTICAL TREATMENT	109.6	57.5	78.5	109.6	46.4	71.9	109.6	61.2	80.7	109.6	68.1	84.8	109.6	41.0	68.7	114.7	68.5	87.1
09600	FLOORING	100.4	58.9	89.2	103.3	58.9	91.3	115.1	62.7	100.9	98.2	42.8	83.2	99.6	42.4	84.1	109.3	58.6	95.6
097,099	WALL FINISHES, PAINTS & COATINGS	101.4	18.0	50.5	101.4	38.4	62.9	104.4	82.4	90.9	97.2	27.1	54.5	101.4	39.7	63.7	102.6	66.0	80.3
09	FINISHES	102.4	50.8	75.5	103.5	48.3	74.7	107.5	62.4	84.0	101.5	55.0	77.3	103.3	41.9	71.3	109.0	66.8	87.0
10 - 14	TOTAL DIV. 10000 - 14000	100.0	66.4	92.8	100.0	52.0	89.8	100.0	82.4	96.2	100.0	81.2	96.0	100.0	63.0	92.1	100.0	87.6	97.4
15	MECHANICAL	95.4	67.9	84.0	95.4	69.6	84.7	95.4	72.3	85.8	95.4	34.1	70.0	95.4	40.7	72.7	100.3	82.4	92.9
16	ELECTRICAL	100.6	54.4	78.1	99.6	56.3	78.6	98.2	78.0	88.4	96.1	32.5	65.2	96.1	41.0	69.3	96.1	78.0	87.3
01 - 16	WEIGHTED AVERAGE	96.2	58.1	79.3	96.0	62.0	80.9	96.8	71.1	85.4	96.2	52.9	77.0	97.0	51.5	76.8	98.4	74.9	88.0

	DIVISION	IOWA						KANSAS											
		SPENCER			WATERLOO			BELLEVILLE			COLBY			DODGE CITY			EMPORIA		
		513			506 - 507			669			677			678			668		
		MAT.	INST.	TOTAL	MAT.	INST.	TOTAL	MAT.	INST.	TOTAL	MAT.	INST.	TOTAL	MAT.	INST.	TOTAL	MAT.	INST.	TOTAL
01590	EQUIPMENT RENTAL	.0	99.6	99.6	.0	99.6	99.6	.0	103.6	103.6	.0	103.6	103.6	.0	103.6	103.6	.0	101.6	101.6
02	SITE CONSTRUCTION	95.9	93.3	94.0	88.3	94.5	92.9	112.1	92.5	97.7	109.5	92.4	97.0	112.0	92.9	98.0	103.1	89.9	93.4
03100	CONCRETE FORMS & ACCESSORIES	90.2	42.5	49.1	102.6	48.1	55.6	93.9	47.7	54.1	101.6	42.0	50.2	93.4	67.9	71.4	84.1	38.6	44.8
03200	CONCRETE REINFORCEMENT	95.2	41.8	68.1	93.3	84.0	88.6	105.4	56.9	80.8	108.0	56.3	81.7	105.4	57.3	81.0	103.9	49.5	76.3
03300	CAST-IN-PLACE CONCRETE	106.4	50.5	83.4	108.0	47.1	83.0	119.8	55.9	93.5	115.2	55.5	90.7	117.3	56.4	92.2	115.8	42.1	85.5
03	CONCRETE	99.2	46.5	73.1	100.3	55.9	78.3	117.5	53.7	85.8	114.8	50.9	83.1	115.8	63.0	89.6	109.9	43.5	76.9
04	MASONRY	127.1	39.6	72.9	101.3	60.8	76.2	95.2	50.6	67.6	96.7	40.9	62.1	104.1	57.6	75.3	101.0	40.4	63.5
05	METALS	85.9	66.9	79.8	87.8	90.5	88.7	94.7	77.7	89.3	95.1	76.0	89.0	96.1	79.8	90.9	94.5	72.6	87.5
06	WOOD & PLASTICS	95.9	42.5	68.1	109.7	47.8	77.5	93.0	48.6	69.9	100.9	42.1	70.3	92.3	75.4	83.5	83.3	38.5	60.0
07	THERMAL & MOISTURE PROTECTION	98.8	44.7	76.1	98.2	52.9	79.1	95.5	57.7	79.6	95.5	47.8	75.4	95.4	62.2	81.4	94.1	40.7	71.6
08	DOORS & WINDOWS	107.8	42.2	90.7	95.0	60.8	86.0	96.1	49.0	83.8	96.2	45.4	82.9	96.1	63.5	87.6	93.7	41.1	80.0
09200	PLASTER & GYPSUM BOARD	97.6	41.0	62.4	103.7	46.4	68.1	103.0	47.1	68.3	106.7	40.3	65.5	102.3	74.6	85.1	99.0	36.7	60.3
095,098	CEILINGS & ACOUSTICAL TREATMENT	109.6	41.0	68.7	114.7	46.4	74.0	86.9	47.1	63.2	86.9	40.3	59.1	86.9	74.6	79.6	86.9	36.7	57.0
09600	FLOORING	103.0	42.4	86.6	110.5	58.9	96.5	96.2	48.9	83.4	100.4	48.9	86.4	96.0	48.9	83.3	90.3	45.5	78.1
097,099	WALL FINISHES, PAINTS & COATINGS	101.4	39.7	63.7	102.6	35.5	61.7	90.8	48.2	64.8	90.8	48.2	64.8	90.8	48.2	64.8	90.8	48.2	64.8
09	FINISHES	104.9	41.9	72.0	108.3	47.9	76.8	94.2	47.4	69.8	95.8	43.6	68.6	94.0	63.1	77.9	91.0	40.2	64.6
10 - 14	TOTAL DIV. 10000 - 14000	100.0	79.7	95.7	100.0	80.5	95.9	100.0	47.7	88.9	100.0	46.8	88.7	100.0	51.1	89.6	100.0	42.8	87.8
15	MECHANICAL	95.4	40.8	72.7	100.3	45.2	77.4	95.1	71.3	85.2	95.1	46.1	74.8	100.0	64.3	85.2	95.1	64.1	82.2
16	ELECTRICAL	97.7	41.0	70.1	96.1	67.0	82.0	102.5	69.6	86.5	102.5	49.1	76.6	99.4	74.8	87.4	99.7	73.2	86.9
01 - 16	WEIGHTED AVERAGE	98.8	49.9	77.1	97.8	62.0	81.9	99.2	63.7	83.5	99.2	53.1	78.8	100.4	68.4	86.2	97.4	57.7	79.8

DIVISION		KANSAS																	
		FORT SCOTT			HAYS			HUTCHINSON			INDEPENDENCE			KANSAS CITY			LIBERAL		
		667			676			675			673			660 - 662			679		
		MAT.	INST.	TOTAL	MAT.	INST.	TOTAL	MAT.	INST.	TOTAL	MAT.	INST.	TOTAL	MAT.	INST.	TOTAL	MAT.	INST.	TOTAL
01590	EQUIPMENT RENTAL	.0	102.8	102.8	.0	103.6	103.6	.0	103.6	103.6	.0	103.6	103.6	.0	100.1	100.1	.0	103.6	103.6
02	SITE CONSTRUCTION	99.8	91.9	94.0	114.8	93.2	98.9	92.1	92.8	92.6	112.8	93.3	98.5	93.1	91.3	91.8	115.2	92.8	98.7
03100	CONCRETE FORMS & ACCESSORIES	103.2	72.6	76.8	98.7	67.9	72.2	87.0	29.1	37.1	112.1	60.0	67.2	99.3	94.4	95.1	93.9	28.7	37.6
03200	CONCRETE REINFORCEMENT	103.3	71.9	87.3	105.4	57.0	80.8	105.4	46.9	75.7	104.8	64.5	84.3	100.1	87.4	93.7	106.9	46.8	76.4
03300	CAST-IN-PLACE CONCRETE	107.4	60.5	88.1	90.8	56.7	76.7	84.1	40.9	66.3	117.7	42.5	86.8	91.9	96.1	93.6	90.8	40.7	70.2
03	CONCRETE	105.3	69.2	87.4	105.8	63.1	84.6	88.9	38.4	63.8	117.4	56.1	86.9	97.0	94.0	95.5	107.9	38.1	73.2
04	MASONRY	102.0	59.9	75.9	104.3	50.9	71.2	96.3	39.3	60.9	93.7	51.5	67.5	104.3	97.0	99.8	102.1	39.3	63.2
05	METALS	94.5	85.8	91.7	94.7	80.4	90.1	94.5	72.2	87.4	94.5	82.7	90.7	100.9	96.4	99.5	95.0	71.7	87.5
06	WOOD & PLASTICS	103.5	78.5	90.4	97.8	75.4	86.1	86.3	27.3	55.6	111.9	66.1	88.0	99.4	93.4	96.3	93.0	27.3	58.8
07	THERMAL & MOISTURE PROTECTION	94.9	70.2	84.5	95.8	60.5	80.9	94.4	41.9	72.3	95.6	64.4	82.4	93.2	98.6	95.5	95.9	36.9	71.1
08	DOORS & WINDOWS	93.7	73.1	88.3	96.1	63.5	87.6	96.1	35.0	80.1	93.7	60.3	85.0	95.1	90.3	93.9	96.2	35.0	80.2
09200	PLASTER & GYPSUM BOARD	105.2	77.7	88.1	105.2	74.6	86.2	100.8	25.2	53.8	111.9	65.0	82.8	98.8	93.1	95.2	103.0	25.2	54.7
095,098	CEILINGS & ACOUSTICAL TREATMENT	86.9	77.7	81.4	86.9	74.6	79.6	86.9	25.2	50.1	86.9	65.0	73.9	87.8	93.1	91.0	86.9	25.2	50.1
09600	FLOORING	106.7	49.2	91.2	99.0	48.9	85.5	92.4	48.9	80.7	105.7	48.9	90.4	87.2	103.1	91.5	96.2	48.9	83.4
097,099	WALL FINISHES, PAINTS & COATINGS	93.1	48.2	65.7	90.8	48.2	64.8	90.8	48.2	64.8	90.8	48.2	64.8	98.8	86.4	91.2	90.8	48.2	64.8
09	FINISHES	97.0	66.3	81.0	95.6	63.1	78.7	91.5	33.9	61.5	98.3	56.9	76.7	92.3	94.0	93.2	94.7	33.9	63.0
10 - 14	TOTAL DIV. 10000 - 14000	100.0	58.4	91.1	100.0	51.1	89.6	100.0	43.4	87.9	100.0	48.7	89.1	100.0	80.7	95.9	100.0	43.3	87.9
15	MECHANICAL	95.1	67.2	83.5	95.1	64.3	82.3	95.1	37.7	71.3	95.1	69.7	84.6	99.9	95.9	98.2	95.1	31.8	68.8
16	ELECTRICAL	99.0	78.7	89.1	101.4	69.6	85.9	96.4	66.4	81.8	98.6	78.7	88.9	104.3	106.8	105.5	99.4	38.8	69.9
01 - 16	WEIGHTED AVERAGE	97.6	72.3	86.4	98.5	67.0	84.5	94.5	49.5	74.5	99.0	67.7	85.1	98.9	96.1	97.7	98.3	44.2	74.3

DIVISION		KANSAS									KENTUCKY								
		SALINA			TOPEKA			WICHITA			ASHLAND			BOWLING GREEN			CAMPTON		
		674			664 - 666			670 - 672			411 - 412			421 - 422			413 - 414		
		MAT.	INST.	TOTAL	MAT.	INST.	TOTAL	MAT.	INST.	TOTAL	MAT.	INST.	TOTAL	MAT.	INST.	TOTAL	MAT.	INST.	TOTAL
01590	EQUIPMENT RENTAL	.0	103.6	103.6	.0	101.6	101.6	.0	103.6	103.6	.0	98.8	98.8	.0	96.0	96.0	.0	103.5	103.5
02	SITE CONSTRUCTION	101.9	93.3	95.6	94.9	90.6	91.8	96.0	93.2	94.0	103.5	86.2	90.8	69.6	100.2	92.0	77.9	99.3	93.6
03100	CONCRETE FORMS & ACCESSORIES	89.2	47.9	53.6	98.6	44.2	51.7	95.4	56.4	61.7	85.7	102.6	100.3	84.1	83.2	83.3	87.3	33.0	40.4
03200	CONCRETE REINFORCEMENT	104.8	79.5	92.0	97.4	95.0	96.2	97.4	80.0	88.5	86.2	100.5	93.5	83.5	71.7	77.5	84.3	61.4	72.7
03300	CAST-IN-PLACE CONCRETE	101.7	48.2	79.7	93.0	52.4	76.3	88.6	57.9	76.0	88.6	102.2	94.2	87.9	81.4	85.2	98.1	39.0	73.8
03	CONCRETE	102.8	55.5	79.4	94.9	58.2	76.7	92.6	62.6	77.7	96.2	102.7	99.4	94.4	80.5	87.5	98.4	41.5	70.2
04	MASONRY	118.6	50.5	76.4	98.6	62.3	76.1	92.5	65.4	75.7	92.4	105.1	100.3	93.9	78.7	84.5	92.0	29.8	53.5
05	METALS	95.9	89.4	93.8	101.3	97.4	100.0	101.3	88.8	97.3	90.7	110.0	96.9	92.1	78.1	87.6	91.6	67.2	83.8
06	WOOD & PLASTICS	88.4	48.6	67.7	96.4	40.0	67.0	94.3	55.1	73.9	79.3	101.3	90.8	92.9	82.8	87.6	91.3	31.9	60.4
07	THERMAL & MOISTURE PROTECTION	95.0	57.2	79.1	94.9	69.3	84.1	94.6	65.2	82.2	86.7	99.6	92.1	85.8	79.1	83.0	95.8	40.5	72.5
08	DOORS & WINDOWS	96.1	52.9	84.8	96.3	59.1	86.6	96.3	61.7	87.3	93.6	96.9	94.4	95.3	75.5	90.1	96.8	48.2	84.1
09200	PLASTER & GYPSUM BOARD	101.6	47.1	67.7	103.6	38.3	63.0	103.6	53.7	72.6	63.6	101.4	87.1	87.5	82.6	84.5	87.5	29.3	51.4
095,098	CEILINGS & ACOUSTICAL TREATMENT	86.9	47.1	63.2	87.8	38.3	58.2	87.8	53.7	67.5	81.5	101.4	93.4	86.2	82.6	84.1	86.2	29.3	52.3
09600	FLOORING	93.9	35.7	78.2	97.5	45.5	83.4	96.7	78.7	91.8	77.9	101.8	84.3	93.2	60.3	84.3	94.6	41.9	80.4
097,099	WALL FINISHES, PAINTS & COATINGS	90.8	38.0	58.5	90.8	67.0	76.3	90.8	61.2	72.7	99.9	99.9	99.9	96.8	74.4	83.1	96.8	32.4	57.5
09	FINISHES	92.7	43.4	67.0	93.8	44.5	68.1	93.7	60.7	76.5	80.5	102.5	92.0	91.1	78.1	84.3	91.6	33.7	61.4
10 - 14	TOTAL DIV. 10000 - 14000	100.0	62.0	91.9	100.0	65.3	92.6	100.0	65.9	92.7	100.0	84.6	96.7	100.0	60.8	91.6	100.0	52.4	89.8
15	MECHANICAL	100.0	64.2	85.2	100.0	71.6	88.2	100.0	65.4	85.7	95.0	95.0	95.0	99.9	86.6	94.4	95.1	34.4	69.9
16	ELECTRICAL	99.1	70.8	85.4	103.0	78.7	91.2	101.3	70.8	86.5	90.6	82.8	86.8	93.9	85.5	89.8	91.5	27.0	60.1
01 - 16	WEIGHTED AVERAGE	99.2	63.9	83.5	98.7	69.3	85.6	97.9	69.5	85.3	92.5	96.9	94.5	94.6	82.7	89.3	94.1	43.7	71.7

DIVISION		KENTUCKY																	
		CORBIN			COVINGTON			ELIZABETHTOWN			FRANKFORT			HAZARD			HENDERSON		
		407 - 409			410			427			406			417 - 418			424		
		MAT.	INST.	TOTAL	MAT.	INST.	TOTAL	MAT.	INST.	TOTAL	MAT.	INST.	TOTAL	MAT.	INST.	TOTAL	MAT.	INST.	TOTAL
01590	EQUIPMENT RENTAL	.0	103.5	103.5	.0	108.1	108.1	.0	96.0	96.0	.0	103.5	103.5	.0	103.5	103.5	.0	121.3	121.3
02	SITE CONSTRUCTION	75.4	99.7	93.2	74.3	116.5	105.2	64.5	100.0	90.5	76.7	101.3	94.8	76.0	101.0	94.3	73.0	129.7	114.6
03100	CONCRETE FORMS & ACCESSORIES	84.0	32.6	39.7	82.9	86.9	86.4	78.7	82.6	82.0	94.8	57.4	62.5	84.0	32.8	39.8	91.0	82.7	83.8
03200	CONCRETE REINFORCEMENT	83.9	61.4	72.5	82.9	99.0	91.1	83.9	91.3	87.6	85.1	72.8	78.9	84.7	61.4	72.9	83.6	94.4	89.1
03300	CAST-IN-PLACE CONCRETE	92.8	39.4	70.9	93.9	85.0	90.2	79.5	79.7	79.6	92.8	66.2	81.9	94.3	39.5	71.8	77.7	94.4	84.6
03	CONCRETE	93.5	41.6	67.7	96.6	88.8	92.7	86.7	83.3	85.0	94.5	63.9	79.3	95.1	41.7	68.6	91.6	89.0	90.3
04	MASONRY	92.2	30.0	53.6	106.3	95.3	99.5	77.0	76.6	76.8	90.9	72.4	79.4	88.9	30.0	52.4	98.1	94.8	96.0
05	METALS	91.6	67.6	83.9	89.4	95.6	91.4	91.5	86.7	90.0	99.1	77.9	92.3	91.6	67.8	84.0	83.0	91.2	85.6
06	WOOD & PLASTICS	87.7	31.9	58.6	88.5	84.0	86.1	87.2	82.8	84.9	99.6	55.8	76.8	87.7	31.9	58.6	95.2	79.3	86.9
07	THERMAL & MOISTURE PROTECTION	95.6	40.5	72.4	91.0	93.0	91.8	85.4	79.5	82.9	95.7	66.8	83.5	95.7	40.5	72.5	95.1	94.3	94.8
08	DOORS & WINDOWS	96.1	37.8	80.9	98.5	87.0	95.5	95.3	84.4	92.5	95.4	63.6	87.1	97.2	37.8	81.7	93.6	83.7	91.0
09200	PLASTER & GYPSUM BOARD	86.0	29.3	50.8	76.6	84.1	81.3	85.6	82.6	83.8	91.6	53.9	68.2	86.0	29.3	50.8	87.6	78.0	81.6
095,098	CEILINGS & ACOUSTICAL TREATMENT	86.2	29.3	52.3	88.7	84.1	86.0	86.2	82.6	84.1	86.2	53.9	66.9	86.2	29.3	52.3	75.1	78.0	76.8
09600	FLOORING	92.6	41.9	78.9	69.8	97.0	77.2	90.1	73.3	85.6	98.3	51.2	85.6	92.6	41.9	78.9	96.6	91.4	95.2
097,099	WALL FINISHES, PAINTS & COATINGS	96.8	32.4	57.5	92.0	89.4	90.5	96.8	78.4	85.6	96.8	49.0	67.6	96.8	32.4	57.5	96.8	99.6	98.5
09	FINISHES	90.6	33.7	61.0	83.2	88.7	86.1	89.6	80.4	84.8	93.1	54.6	73.1	90.7	33.7	61.0	88.9	85.8	87.3
10 - 14	TOTAL DIV. 10000 - 14000	100.0	52.4	89.8	100.0	97.1	99.4	100.0	80.2	95.8	100.0	63.8	92.3	100.0	52.4	89.8	100.0	75.4	94.8
15	MECHANICAL	95.0	33.9	69.7	95.6	89.6	93.1	95.3	85.0	91.1	100.0	60.0	83.4	95.1	33.9	69.7	95.3	82.4	90.0
16	ELECTRICAL	91.7	27.0	60.2	94.2	81.4	88.0	91.3	85.5	88.4	94.0	59.5	77.2	91.5	27.0	60.1	93.4	84.1	88.9
01 - 16	WEIGHTED AVERAGE	93.3	43.3	71.1	93.8	91.9	93.0	91.1	84.5	88.2	96.3	66.7	83.1	93.5	43.4	71.3	92.0	90.4	91.3

	DIVISION	KENTUCKY																	
		LEXINGTON			LOUISVILLE			OWENSBORO			PADUCAH			PIKEVILLE			SOMERSET		
		403 - 405			400 - 402			423			420			415 - 416			425 - 426		
		MAT.	INST.	TOTAL	MAT.	INST.	TOTAL	MAT.	INST.	TOTAL	MAT.	INST.	TOTAL	MAT.	INST.	TOTAL	MAT.	INST.	TOTAL
01590	EQUIPMENT RENTAL	.0	103.5	103.5	.0	96.0	96.0	.0	121.3	121.3	.0	121.3	121.3	.0	98.8	98.8	.0	103.5	103.5
02	SITE CONSTRUCTION	77.3	102.2	95.6	67.1	100.0	91.2	81.9	129.3	116.6	75.4	129.7	115.3	113.3	84.2	91.9	69.4	99.3	91.3
03100	CONCRETE FORMS & ACCESSORIES	98.1	70.4	74.2	92.8	82.4	83.8	89.1	63.1	66.6	86.6	81.5	82.2	95.7	57.5	62.7	84.7	33.1	40.2
03200	CONCRETE REINFORCEMENT	92.2	95.4	93.8	92.2	96.7	94.5	83.6	92.6	88.2	84.1	84.7	84.4	86.7	53.8	70.0	83.9	61.5	72.5
03300	CAST-IN-PLACE CONCRETE	95.0	78.9	88.4	91.9	79.2	86.7	90.7	74.4	84.0	82.8	88.0	84.9	97.4	69.8	86.1	77.7	70.0	74.5
03	CONCRETE	96.9	78.4	87.7	95.0	84.1	89.6	103.6	73.0	88.4	96.0	84.4	90.3	109.8	62.7	86.4	81.7	52.2	67.1
04	MASONRY	90.6	46.7	63.4	92.2	81.4	85.5	90.7	50.5	65.8	93.6	93.8	93.7	90.7	54.0	67.9	84.8	29.8	50.7
05	METALS	93.6	88.1	91.8	101.0	89.3	97.3	84.0	86.2	84.7	81.9	86.3	83.3	90.6	82.7	88.1	91.6	67.3	83.8
06	WOOD & PLASTICS	104.0	73.9	88.3	102.5	82.8	92.2	93.1	61.7	76.7	90.6	79.5	84.8	89.7	54.8	71.5	88.7	31.9	59.1
07	THERMAL & MOISTURE PROTECTION	95.9	77.0	87.9	86.0	81.4	84.0	95.5	66.6	83.3	95.2	82.6	89.9	87.4	53.7	73.2	95.1	40.5	72.1
08	DOORS & WINDOWS	96.2	79.5	91.8	96.2	86.4	93.6	93.6	74.8	88.7	92.7	78.6	89.1	94.3	57.8	84.8	96.1	48.2	83.6
09200	PLASTER & GYPSUM BOARD	94.3	72.5	80.8	95.3	82.6	87.4	86.5	59.9	70.0	85.8	78.2	81.1	66.9	53.7	58.7	87.1	29.3	51.2
095,098	CEILINGS & ACOUSTICAL TREATMENT	91.3	72.5	80.1	91.3	82.6	86.1	75.1	59.9	66.0	75.1	78.2	77.0	81.5	53.7	64.9	86.2	29.3	52.3
09600	FLOORING	98.3	41.9	83.0	96.8	73.3	90.5	95.7	60.3	86.1	94.3	60.3	85.1	83.0	49.8	74.0	92.9	41.9	79.1
097,099	WALL FINISHES, PAINTS & COATINGS	96.8	57.6	72.9	96.8	81.5	87.5	96.8	99.6	98.5	96.8	77.3	84.9	99.9	64.2	78.1	96.8	32.4	57.5
09	FINISHES	94.9	63.3	78.4	94.3	80.5	87.1	88.8	65.6	76.7	88.1	76.7	82.2	83.1	56.3	69.2	90.4	33.7	60.9
10 - 14	TOTAL DIV. 10000 - 14000	100.0	76.7	95.0	100.0	79.9	95.7	100.0	94.6	98.9	100.0	70.1	93.6	100.0	56.2	90.7	100.0	52.4	89.8
15	MECHANICAL	99.9	47.0	78.0	99.9	84.6	93.6	99.9	51.7	79.9	95.3	86.2	91.5	95.0	54.7	78.3	95.3	34.0	69.9
16	ELECTRICAL	94.4	49.5	72.6	94.4	85.5	90.1	93.5	84.1	88.9	95.7	85.7	90.8	93.4	48.1	71.4	91.9	27.0	60.3
01 - 16	WEIGHTED AVERAGE	96.0	65.7	82.5	96.4	85.4	91.5	94.4	73.5	85.1	92.2	88.4	90.5	95.0	60.3	79.6	91.6	45.1	70.9

	DIVISION	LOUISIANA																	
		ALEXANDRIA			BATON ROUGE			HAMMOND			LAFAYETTE			LAKE CHARLES			MONROE		
		713 - 714			707 - 708			704			705			706			712		
		MAT.	INST.	TOTAL	MAT.	INST.	TOTAL	MAT.	INST.	TOTAL	MAT.	INST.	TOTAL	MAT.	INST.	TOTAL	MAT.	INST.	TOTAL
01590	EQUIPMENT RENTAL	.0	86.7	86.7	.0	86.2	86.2	.0	87.0	87.0	.0	87.0	87.0	.0	86.2	86.2	.0	86.7	86.7
02	SITE CONSTRUCTION	102.0	85.5	89.9	107.4	85.0	91.0	107.4	86.5	92.1	108.3	86.4	92.2	108.9	84.8	91.2	102.0	85.0	89.5
03100	CONCRETE FORMS & ACCESSORIES	79.6	43.5	48.5	98.1	59.7	65.0	77.9	58.8	61.4	97.5	53.5	59.5	98.6	59.2	64.6	78.9	43.7	48.5
03200	CONCRETE REINFORCEMENT	98.8	66.2	82.2	94.4	64.3	79.1	93.1	65.3	78.9	94.4	64.2	79.1	94.4	64.3	79.1	97.7	66.3	81.8
03300	CAST-IN-PLACE CONCRETE	92.9	43.0	72.4	83.4	51.0	70.1	84.3	49.9	70.2	83.9	50.3	70.1	88.3	52.3	73.5	92.9	46.6	73.8
03	CONCRETE	88.4	48.8	68.7	91.8	58.1	75.0	90.7	57.6	74.2	91.9	55.1	73.7	94.2	58.4	76.4	88.2	50.0	69.2
04	MASONRY	113.5	52.4	75.6	98.3	53.9	70.8	97.6	60.7	74.7	97.6	52.9	69.9	96.0	54.9	70.6	108.3	56.3	76.1
05	METALS	86.4	75.0	82.8	94.0	71.2	86.7	89.3	72.6	84.0	88.6	71.5	83.1	88.6	71.2	83.0	86.4	73.3	82.2
06	WOOD & PLASTICS	81.8	42.5	61.3	104.7	63.0	83.0	82.4	60.8	71.2	104.7	55.4	79.0	102.7	62.3	81.7	81.0	42.9	61.2
07	THERMAL & MOISTURE PROTECTION	99.6	53.6	80.2	98.9	58.9	82.1	101.0	64.2	85.5	101.7	57.0	82.9	101.1	58.6	83.2	99.6	56.3	81.4
08	DOORS & WINDOWS	99.0	49.6	86.1	103.8	60.0	92.4	99.1	63.8	89.9	103.8	55.9	91.3	103.8	61.2	92.7	99.0	54.5	87.4
09200	PLASTER & GYPSUM BOARD	78.1	41.6	55.4	97.9	62.6	75.9	89.1	60.3	71.2	97.9	54.7	71.1	97.9	61.8	75.5	77.7	42.1	55.6
095,098	CEILINGS & ACOUSTICAL TREATMENT	88.8	41.6	60.7	98.4	62.6	77.0	99.2	60.3	76.0	97.5	54.7	72.0	98.4	61.8	76.6	88.8	42.1	60.9
09600	FLOORING	104.4	65.9	94.0	108.0	64.6	96.2	96.5	57.8	86.0	108.2	43.3	90.6	108.2	50.5	92.6	104.0	41.7	87.1
097,099	WALL FINISHES, PAINTS & COATINGS	97.0	40.6	62.6	98.0	46.6	66.6	98.0	61.6	75.8	98.0	46.6	66.6	98.0	42.9	64.4	97.0	48.5	67.4
09	FINISHES	92.3	46.5	68.4	101.9	59.5	79.8	97.3	58.9	77.3	101.7	50.3	74.9	101.9	56.1	78.0	92.2	42.8	66.5
10 - 14	TOTAL DIV. 10000 - 14000	100.0	66.7	92.9	100.0	76.7	95.0	100.0	63.7	92.3	100.0	75.3	94.7	100.0	76.7	95.0	100.0	61.0	91.7
15	MECHANICAL	100.1	62.5	84.5	100.0	58.0	82.6	95.1	58.4	79.9	100.0	62.8	84.6	100.0	63.5	84.8	100.1	53.8	80.9
16	ELECTRICAL	92.0	60.4	76.6	94.0	63.7	79.3	93.2	56.9	75.5	94.4	66.2	80.7	94.0	66.2	80.4	93.7	61.3	78.0
01 - 16	WEIGHTED AVERAGE	95.4	59.5	79.5	98.1	62.9	82.5	95.0	62.7	80.7	97.4	62.3	81.8	97.5	64.1	82.7	95.3	57.9	78.7

	DIVISION	LOUISIANA									MAINE								
		NEW ORLEANS			SHREVEPORT			THIBODAUX			AUGUSTA			BANGOR			BATH		
		700 - 701			710 - 711			703			043			044			045		
		MAT.	INST.	TOTAL	MAT.	INST.	TOTAL	MAT.	INST.	TOTAL	MAT.	INST.	TOTAL	MAT.	INST.	TOTAL	MAT.	INST.	TOTAL
01590	EQUIPMENT RENTAL	.0	87.3	87.3	.0	86.7	86.7	.0	87.0	87.0	.0	101.5	101.5	.0	101.5	101.5	.0	101.5	101.5
02	SITE CONSTRUCTION	111.5	87.5	93.9	100.7	84.8	89.1	109.9	86.5	92.7	84.3	102.4	97.6	84.2	100.4	96.1	82.3	102.5	97.1
03100	CONCRETE FORMS & ACCESSORIES	98.2	67.7	71.9	98.5	46.5	53.7	91.5	68.2	71.4	100.9	63.4	68.5	95.0	89.9	90.6	90.1	63.5	67.2
03200	CONCRETE REINFORCEMENT	94.4	67.3	80.6	97.3	66.0	81.4	93.1	62.5	77.5	87.5	111.6	99.7	87.5	111.8	99.8	86.5	111.6	99.2
03300	CAST-IN-PLACE CONCRETE	87.5	75.5	82.6	91.6	50.2	74.6	90.6	53.5	75.4	83.7	61.8	74.7	83.7	63.1	75.3	83.7	61.8	74.7
03	CONCRETE	93.7	70.7	82.3	88.3	52.4	70.5	95.6	62.4	79.2	98.6	72.0	85.4	97.7	84.1	90.9	97.7	72.1	85.0
04	MASONRY	97.8	62.4	75.8	101.0	49.6	69.1	122.3	59.6	83.5	97.1	50.9	68.5	114.9	58.4	79.9	121.8	50.5	77.6
05	METALS	101.2	75.5	93.0	82.7	72.7	79.5	89.3	71.1	83.5	87.7	85.4	87.0	87.3	82.1	85.7	86.2	85.5	86.0
06	WOOD & PLASTICS	99.0	70.4	84.1	101.9	46.7	73.1	91.9	73.2	82.1	99.5	61.4	79.7	93.0	96.0	94.5	87.4	61.4	73.8
07	THERMAL & MOISTURE PROTECTION	102.2	63.6	85.9	98.7	55.0	80.3	102.3	62.2	85.4	102.2	52.9	81.5	102.1	59.4	84.1	102.1	57.2	83.2
08	DOORS & WINDOWS	104.5	70.3	95.6	96.9	52.0	85.2	104.8	69.7	95.6	103.8	60.6	92.5	103.7	81.0	97.8	103.7	62.3	92.9
09200	PLASTER & GYPSUM BOARD	96.8	70.2	80.3	83.8	45.9	60.3	94.2	73.0	81.1	104.0	59.5	76.3	100.4	95.0	97.0	97.5	59.5	73.9
095,098	CEILINGS & ACOUSTICAL TREATMENT	97.5	70.2	81.2	89.7	45.9	63.6	99.2	73.0	83.6	90.0	59.5	71.8	87.5	95.0	91.9	85.5	59.5	70.0
09600	FLOORING	108.6	49.5	92.6	115.1	61.8	100.7	104.4	55.1	91.1	99.6	42.0	84.0	97.1	52.4	85.0	94.9	42.0	80.6
097,099	WALL FINISHES, PAINTS & COATINGS	99.7	67.2	79.9	97.0	40.6	62.6	99.7	61.6	76.5	91.5	35.1	57.1	91.5	32.4	55.4	91.5	35.1	57.1
09	FINISHES	102.1	64.2	82.4	96.5	48.0	71.3	100.7	65.7	82.5	98.6	55.2	76.0	96.7	77.9	86.9	95.1	55.2	74.3
10 - 14	TOTAL DIV. 10000 - 14000	100.0	79.3	95.6	100.0	70.0	93.6	100.0	78.7	95.5	100.0	59.6	91.4	100.0	79.4	95.6	100.0	59.6	91.4
15	MECHANICAL	100.0	65.4	85.6	100.1	62.4	84.4	95.1	62.2	81.4	100.1	76.0	90.1	100.1	77.1	90.5	95.1	76.0	87.2
16	ELECTRICAL	95.1	69.0	82.4	92.6	70.2	81.7	91.9	69.0	80.7	97.6	87.8	92.9	95.9	85.0	90.6	94.2	87.8	91.1
01 - 16	WEIGHTED AVERAGE	99.8	69.7	86.4	94.6	61.3	79.8	97.7	67.2	84.2	97.4	73.2	86.7	97.7	79.8	89.8	96.3	73.4	86.1

	DIVISION	MAINE																	
		HOULTON			KITTERY			LEWISTON			MACHIAS			PORTLAND			ROCKLAND		
		047			039			042			046			040 - 041			048		
		MAT.	INST.	TOTAL	MAT.	INST.	TOTAL	MAT.	INST.	TOTAL	MAT.	INST.	TOTAL	MAT.	INST.	TOTAL	MAT.	INST.	TOTAL
01590	**EQUIPMENT RENTAL**	.0	101.5	101.5	.0	101.5	101.5	.0	101.5	101.5	.0	101.5	101.5	.0	101.5	101.5	.0	101.5	101.5
02	**SITE CONSTRUCTION**	83.8	102.5	97.5	78.7	102.5	96.1	82.9	100.4	95.7	83.2	102.4	97.3	81.6	100.4	95.4	79.9	102.4	96.4
03100	CONCRETE FORMS & ACCESSORIES	98.8	77.4	80.3	90.7	63.5	67.3	101.0	89.9	91.4	95.7	63.4	67.8	100.1	89.9	91.3	96.8	63.5	68.1
03200	CONCRETE REINFORCEMENT	87.5	111.6	99.7	85.9	111.6	99.0	107.5	111.8	109.7	87.5	111.5	99.7	107.5	111.8	109.7	87.5	111.6	99.7
03300	CAST-IN-PLACE CONCRETE	83.7	61.9	74.7	83.7	61.8	74.7	95.3	63.1	82.1	83.7	61.8	74.7	86.8	63.1	77.0	85.5	61.8	75.7
03	**CONCRETE**	98.6	78.3	88.5	93.3	72.1	82.8	103.1	84.1	93.6	98.2	72.0	85.2	98.9	84.1	91.5	95.5	72.1	83.9
04	**MASONRY**	97.6	50.5	68.4	112.3	50.5	74.0	98.9	58.4	73.8	97.6	50.5	68.4	96.5	58.4	72.8	91.2	51.8	66.8
05	**METALS**	86.4	85.6	86.1	86.1	85.5	85.9	90.4	82.1	87.8	86.4	85.3	86.1	91.8	82.1	88.7	86.3	85.5	86.1
06	**WOOD & PLASTICS**	97.3	80.0	88.3	87.9	61.4	74.1	99.5	96.0	97.7	94.0	61.4	77.0	99.5	96.0	97.7	95.0	61.4	77.5
07	**THERMAL & MOISTURE PROTECTION**	102.2	59.1	84.1	101.7	57.2	83.0	102.0	59.4	84.1	102.2	57.1	83.2	101.8	59.4	84.0	101.9	57.5	83.2
08	**DOORS & WINDOWS**	103.8	72.4	95.6	103.7	62.3	92.9	106.9	81.0	100.2	103.8	60.1	92.4	106.9	81.0	100.2	103.7	60.1	92.3
09200	PLASTER & GYPSUM BOARD	102.9	78.6	87.8	97.5	59.5	73.9	106.2	95.0	99.2	101.8	59.5	75.5	106.2	95.0	99.2	101.1	59.5	75.3
095,098	CEILINGS & ACOUSTICAL TREATMENT	88.1	78.6	82.4	85.5	59.5	70.0	98.5	95.0	96.4	88.1	59.5	71.0	98.5	95.0	96.4	85.5	59.5	70.0
09600	FLOORING	98.5	42.0	83.2	95.2	42.0	80.8	99.6	52.4	86.8	97.4	20.6	76.6	99.6	52.4	86.8	97.7	42.0	82.6
097,099	WALL FINISHES, PAINTS & COATINGS	91.5	35.1	57.1	91.5	35.1	57.1	91.5	32.4	55.4	91.5	35.1	57.1	91.5	32.4	55.4	91.5	35.1	57.1
09	**FINISHES**	97.7	66.1	81.2	94.9	55.2	74.2	100.8	77.9	88.9	97.2	51.5	73.4	100.8	77.9	88.9	96.2	55.2	74.9
10 - 14	**TOTAL DIV. 10000 - 14000**	100.0	61.9	91.9	100.0	59.6	91.4	100.0	79.4	95.6	100.0	59.6	91.4	100.0	79.4	95.6	100.0	59.6	91.4
15	**MECHANICAL**	95.1	76.0	87.2	95.1	76.0	87.2	100.1	77.1	90.5	95.1	76.0	87.2	100.1	77.1	90.5	95.1	76.0	87.2
16	**ELECTRICAL**	97.7	87.8	92.9	94.2	87.8	91.1	97.7	85.0	91.5	97.7	87.8	92.9	97.8	85.0	91.6	97.6	87.8	92.9
01 - 16	**WEIGHTED AVERAGE**	96.0	76.4	87.3	95.2	73.4	85.5	98.9	79.8	90.4	95.9	72.8	85.6	98.5	79.8	90.2	95.1	73.4	85.4

	DIVISION	MAINE			MARYLAND														
		WATERVILLE			ANNAPOLIS			BALTIMORE			COLLEGE PARK			CUMBERLAND			EASTON		
		049			214			210 - 212			207 - 208			215			216		
		MAT.	INST.	TOTAL	MAT.	INST.	TOTAL	MAT.	INST.	TOTAL	MAT.	INST.	TOTAL	MAT.	INST.	TOTAL	MAT.	INST.	TOTAL
01590	**EQUIPMENT RENTAL**	.0	101.5	101.5	.0	99.1	99.1	.0	103.3	103.3	.0	102.2	102.2	.0	99.1	99.1	.0	99.1	99.1
02	**SITE CONSTRUCTION**	84.0	102.4	97.5	96.1	88.6	90.6	97.1	93.1	94.1	100.0	87.9	91.2	88.4	88.6	88.5	94.9	85.3	87.9
03100	CONCRETE FORMS & ACCESSORIES	89.5	63.5	67.0	102.8	57.5	63.8	102.9	74.8	78.6	82.9	71.3	72.9	92.3	79.5	81.3	89.8	34.9	42.4
03200	CONCRETE REINFORCEMENT	87.5	111.6	99.7	93.3	88.0	90.6	93.3	88.1	90.6	105.5	81.7	93.4	82.4	76.0	79.1	81.7	20.9	50.8
03300	CAST-IN-PLACE CONCRETE	83.7	61.8	74.7	99.0	78.0	90.4	99.0	79.0	90.8	115.6	74.5	98.7	88.4	82.6	86.1	98.2	43.1	75.6
03	**CONCRETE**	99.1	72.1	85.7	100.5	71.8	86.3	100.5	79.8	90.3	111.5	75.6	93.7	89.7	80.8	85.3	97.2	37.0	67.3
04	**MASONRY**	107.7	50.9	72.5	93.2	72.5	80.4	92.7	72.5	80.2	108.3	69.2	84.0	90.9	77.0	82.3	104.2	37.4	62.8
05	**METALS**	86.4	85.4	86.1	100.3	97.8	99.5	101.3	98.7	100.5	91.1	95.6	92.5	97.4	91.0	95.4	97.6	61.0	85.9
06	**WOOD & PLASTICS**	86.6	61.4	73.5	100.0	53.0	75.5	100.0	76.0	87.5	79.7	74.4	77.0	88.9	78.9	83.7	85.9	35.1	59.5
07	**THERMAL & MOISTURE PROTECTION**	102.2	56.3	82.9	97.8	77.4	89.2	97.8	80.1	90.4	103.3	75.9	91.8	97.5	80.1	90.2	97.6	41.3	73.9
08	**DOORS & WINDOWS**	103.8	60.6	92.5	92.5	70.8	86.8	94.5	83.2	91.5	97.2	77.4	92.0	93.6	83.6	91.0	91.7	32.1	76.1
09200	PLASTER & GYPSUM BOARD	97.7	59.5	74.0	103.3	52.1	71.5	103.3	75.5	86.1	99.8	73.7	83.6	99.6	78.8	86.7	98.1	33.8	58.2
095,098	CEILINGS & ACOUSTICAL TREATMENT	88.1	59.5	71.0	103.1	52.1	72.7	103.1	75.5	86.7	94.4	73.7	82.1	103.1	78.8	88.6	103.1	33.8	61.8
09600	FLOORING	94.6	42.0	80.4	92.0	78.6	88.4	92.0	78.6	88.4	106.1	64.5	94.9	87.5	79.2	85.2	86.7	16.0	67.5
097,099	WALL FINISHES, PAINTS & COATINGS	91.5	35.1	57.1	95.9	87.4	90.7	95.9	87.4	90.7	117.2	81.7	95.5	95.9	61.8	75.1	95.9	25.9	53.2
09	**FINISHES**	95.9	55.2	74.7	96.0	62.8	78.7	96.1	76.3	85.8	98.0	71.0	83.9	93.8	77.4	85.3	93.6	31.3	61.2
10 - 14	**TOTAL DIV. 10000 - 14000**	100.0	59.6	91.4	100.0	82.5	96.3	100.0	85.8	97.0	100.0	67.2	93.0	100.0	89.0	97.7	100.0	49.0	89.1
15	**MECHANICAL**	95.1	76.0	87.2	100.0	86.0	94.2	100.0	86.1	94.2	95.0	82.7	89.9	95.1	74.8	86.7	95.1	27.9	67.2
16	**ELECTRICAL**	97.7	87.8	92.9	103.1	82.5	93.1	105.1	91.2	98.4	98.3	95.8	97.0	101.9	84.3	93.4	101.4	63.4	82.9
01 - 16	**WEIGHTED AVERAGE**	96.3	73.3	86.1	98.9	79.2	90.1	99.5	84.7	92.9	98.4	81.4	90.8	95.4	81.2	89.1	96.8	44.8	73.7

	DIVISION	MARYLAND															MASSACHUSETTS		
		ELKTON			HAGERSTOWN			SALISBURY			SILVER SPRING			WALDORF			BOSTON		
		219			217			218			209			206			020 - 022, 024		
		MAT.	INST.	TOTAL	MAT.	INST.	TOTAL	MAT.	INST.	TOTAL	MAT.	INST.	TOTAL	MAT.	INST.	TOTAL	MAT.	INST.	TOTAL
01590	**EQUIPMENT RENTAL**	.0	99.1	99.1	.0	99.1	99.1	.0	99.1	99.1	.0	93.7	93.7	.0	93.7	93.7	.0	108.4	108.4
02	**SITE CONSTRUCTION**	82.9	86.4	85.5	86.9	88.7	88.2	94.9	85.5	88.0	88.8	80.5	82.7	95.3	80.7	84.6	89.4	109.1	103.8
03100	CONCRETE FORMS & ACCESSORIES	97.1	64.6	69.1	91.2	76.4	78.4	106.8	33.0	43.1	92.1	61.2	65.5	100.0	54.2	60.5	106.5	137.9	133.6
03200	CONCRETE REINFORCEMENT	81.7	61.6	71.5	82.4	75.8	79.1	81.7	61.4	71.4	104.4	81.6	92.8	105.1	61.4	82.9	106.3	145.1	126.0
03300	CAST-IN-PLACE CONCRETE	79.5	50.4	67.5	84.2	60.4	74.4	98.2	42.4	75.3	118.3	67.0	97.2	132.6	60.5	103.0	108.1	145.5	123.5
03	**CONCRETE**	82.7	60.0	71.4	86.1	71.8	79.0	98.3	42.8	70.8	109.5	68.3	89.1	120.6	59.1	90.1	112.2	141.0	126.5
04	**MASONRY**	91.0	41.1	60.1	96.6	77.0	84.4	103.7	35.0	61.1	106.9	69.6	83.7	91.2	65.9	75.5	118.0	148.1	136.7
05	**METALS**	97.7	72.6	89.7	97.5	90.2	95.2	97.7	63.7	86.8	95.3	91.1	94.0	95.3	80.1	90.4	103.0	127.4	110.8
06	**WOOD & PLASTICS**	93.8	71.3	82.1	87.5	75.8	81.4	104.3	31.1	66.2	87.1	61.3	73.6	94.9	54.8	74.0	103.5	138.9	121.9
07	**THERMAL & MOISTURE PROTECTION**	97.2	50.2	77.4	97.2	71.2	86.3	97.9	59.0	81.5	102.6	80.1	93.1	103.1	75.1	91.3	102.4	142.5	119.3
08	**DOORS & WINDOWS**	91.7	58.5	83.1	91.7	76.8	87.8	92.0	26.8	75.0	90.6	70.3	85.3	91.5	53.6	81.6	102.9	136.8	111.7
09200	PLASTER & GYPSUM BOARD	101.1	71.0	82.4	98.5	75.5	84.2	105.2	29.7	58.2	106.6	61.2	78.4	109.2	54.6	75.3	108.6	139.1	127.6
095,098	CEILINGS & ACOUSTICAL TREATMENT	103.1	71.0	83.9	104.0	75.5	87.0	103.1	29.7	59.3	103.7	61.2	78.4	103.7	54.6	74.4	97.6	139.1	122.3
09600	FLOORING	89.4	61.1	81.8	87.2	79.2	85.0	93.7	61.1	84.9	113.6	64.5	100.3	118.6	61.1	103.1	100.3	163.7	117.4
097,099	WALL FINISHES, PAINTS & COATINGS	95.9	79.2	85.7	95.9	39.7	61.6	95.9	29.3	55.2	122.8	81.7	97.7	122.8	68.3	89.5	94.6	153.1	130.3
09	**FINISHES**	94.3	66.6	79.9	93.6	73.0	82.9	96.8	36.8	65.5	100.3	63.2	81.0	102.4	56.2	78.4	101.1	145.0	124.0
10 - 14	**TOTAL DIV. 10000 - 14000**	100.0	54.8	90.4	100.0	88.6	97.6	100.0	47.9	88.9	100.0	63.3	92.2	100.0	51.4	89.6	100.0	120.4	104.4
15	**MECHANICAL**	95.1	79.7	88.7	100.0	90.0	95.9	95.1	61.7	81.2	95.0	82.4	89.8	95.0	79.9	88.7	100.1	128.8	112.0
16	**ELECTRICAL**	103.1	64.2	84.2	101.7	84.3	93.2	100.1	63.4	82.3	94.1	95.7	94.9	92.0	95.7	93.8	96.8	120.4	108.3
01 - 16	**WEIGHTED AVERAGE**	94.6	66.7	82.2	96.2	81.9	89.8	97.3	53.5	77.8	97.6	77.9	88.9	98.3	72.6	86.9	102.5	132.0	115.6

	DIVISION	MASSACHUSETTS																	
		BROCKTON			BUZZARDS BAY			FALL RIVER			FITCHBURG			FRAMINGHAM			GREENFIELD		
		023			025			027			014			017			013		
		MAT.	INST.	TOTAL	MAT.	INST.	TOTAL	MAT.	INST.	TOTAL	MAT.	INST.	TOTAL	MAT.	INST.	TOTAL	MAT.	INST.	TOTAL
01590	**EQUIPMENT RENTAL**	.0	103.7	103.7	.0	103.7	103.7	.0	104.9	104.9	.0	101.5	101.5	.0	102.8	102.8	.0	101.5	101.5
02	**SITE CONSTRUCTION**	87.2	105.0	100.2	78.0	105.0	97.8	86.2	105.1	100.1	79.6	104.9	98.2	76.4	104.6	97.1	83.0	103.8	98.3
03100	CONCRETE FORMS & ACCESSORIES	106.1	124.2	121.7	103.2	122.8	120.1	106.1	123.1	120.8	96.6	122.8	119.2	104.4	122.5	120.0	94.6	105.1	103.6
03200	CONCRETE REINFORCEMENT	107.5	144.9	126.5	86.2	124.1	105.4	107.5	124.2	115.9	86.2	140.2	113.6	86.1	144.8	115.9	90.1	122.2	106.4
03300	CAST-IN-PLACE CONCRETE	102.9	141.7	118.8	85.4	141.5	108.5	99.5	142.2	117.1	89.6	141.3	110.9	89.6	106.3	96.5	92.1	119.4	103.3
03	**CONCRETE**	109.4	133.4	121.3	92.4	128.8	110.4	107.7	129.2	118.4	90.7	131.5	111.0	93.7	120.5	107.0	94.5	112.7	103.5
04	**MASONRY**	114.1	141.8	131.3	104.2	141.8	127.5	113.9	141.7	131.1	100.2	141.5	125.8	106.6	141.6	128.3	104.8	116.8	112.3
05	**METALS**	99.8	124.5	107.6	95.1	115.5	101.6	99.8	116.0	105.0	95.0	119.2	102.7	95.0	123.5	104.1	96.7	106.7	99.9
06	**WOOD & PLASTICS**	102.5	123.5	113.4	99.2	122.1	111.2	102.5	122.5	112.9	94.9	122.1	109.1	101.7	121.8	112.2	92.6	103.7	98.4
07	**THERMAL & MOISTURE PROTECTION**	102.3	135.2	116.1	101.6	131.7	114.2	102.2	130.7	114.2	101.6	129.2	113.2	101.7	131.9	114.4	101.7	112.5	106.2
08	**DOORS & WINDOWS**	100.8	127.7	107.8	96.3	120.0	102.5	100.8	120.1	105.8	105.8	126.5	111.2	96.8	127.6	104.8	105.7	109.8	106.7
09200	PLASTER & GYPSUM BOARD	103.8	123.2	115.8	99.1	121.8	113.2	103.8	121.8	115.0	100.4	121.8	113.7	102.6	121.8	114.6	102.1	102.9	102.6
095,098	CEILINGS & ACOUSTICAL TREATMENT	99.5	123.2	113.6	86.5	121.8	107.6	99.5	121.8	112.8	85.5	121.8	107.2	85.5	121.8	107.2	96.6	102.9	100.4
09600	FLOORING	100.7	163.7	117.8	99.2	163.7	116.6	100.5	163.7	117.6	97.1	163.7	115.1	99.1	163.7	116.6	96.2	128.6	104.9
097,099	WALL FINISHES, PAINTS & COATINGS	94.0	130.1	116.0	94.0	130.1	116.0	94.0	130.1	116.0	91.5	130.1	115.1	92.8	130.1	115.5	91.5	104.5	99.4
09	**FINISHES**	100.9	132.3	117.3	96.0	131.5	114.5	100.9	131.7	117.0	95.3	131.5	114.1	96.1	131.3	114.4	98.3	109.2	104.0
10 - 14	**TOTAL DIV. 10000 - 14000**	100.0	116.9	103.6	100.0	116.8	103.6	100.0	117.6	103.8	100.0	110.8	102.3	100.0	116.0	103.4	100.0	106.5	101.4
15	**MECHANICAL**	100.1	110.1	104.2	95.1	109.8	101.2	100.1	109.9	104.2	95.8	108.0	100.9	95.8	122.5	106.9	95.8	98.4	96.9
16	**ELECTRICAL**	95.8	103.4	99.5	93.4	103.4	98.3	95.4	103.4	99.3	98.1	106.8	102.3	95.1	120.4	107.4	98.1	93.7	96.0
01 - 16	**WEIGHTED AVERAGE**	101.1	121.0	110.0	95.5	119.0	105.9	100.9	119.1	109.0	96.6	119.8	106.9	96.1	123.8	108.4	97.9	105.3	101.2

	DIVISION	MASSACHUSETTS																	
		HYANNIS			LAWRENCE			LOWELL			NEW BEDFORD			PITTSFIELD			SPRINGFIELD		
		026			019			018			027			012			010 - 011		
		MAT.	INST.	TOTAL	MAT.	INST.	TOTAL	MAT.	INST.	TOTAL	MAT.	INST.	TOTAL	MAT.	INST.	TOTAL	MAT.	INST.	TOTAL
01590	**EQUIPMENT RENTAL**	.0	103.7	103.7	.0	103.7	103.7	.0	101.5	101.5	.0	104.9	104.9	.0	101.5	101.5	.0	101.5	101.5
02	**SITE CONSTRUCTION**	83.7	105.0	99.3	87.7	105.0	100.4	86.8	104.9	100.1	85.8	105.1	100.0	87.9	103.6	99.4	87.3	103.8	99.4
03100	CONCRETE FORMS & ACCESSORIES	96.5	122.8	119.2	105.9	123.2	120.9	102.6	123.4	120.6	106.1	123.1	120.8	102.6	99.2	99.7	102.8	107.7	107.0
03200	CONCRETE REINFORCEMENT	86.2	124.1	105.4	106.6	132.5	119.8	107.5	132.6	120.2	107.5	124.2	115.9	89.4	116.9	103.4	107.5	122.2	114.9
03300	CAST-IN-PLACE CONCRETE	93.9	141.5	113.5	103.7	141.7	119.3	94.3	141.8	113.8	96.6	142.2	115.3	102.9	117.8	109.0	98.1	119.4	106.9
03	**CONCRETE**	99.0	128.8	113.8	109.6	130.6	120.1	100.4	130.6	115.4	106.3	129.2	117.7	101.6	108.5	105.1	102.3	113.8	108.0
04	**MASONRY**	113.0	141.8	130.8	113.4	141.8	131.0	98.8	141.5	125.3	113.7	141.7	131.0	99.5	114.2	108.6	99.1	116.8	110.1
05	**METALS**	96.1	115.5	102.3	97.1	119.4	104.2	97.1	116.7	103.3	99.8	116.0	105.0	96.9	104.4	99.3	99.7	106.7	101.9
06	**WOOD & PLASTICS**	92.1	122.1	107.8	102.5	122.1	112.7	101.5	122.1	112.3	102.5	122.5	112.9	101.5	97.3	99.3	101.5	107.2	104.5
07	**THERMAL & MOISTURE PROTECTION**	101.8	131.7	114.4	102.2	135.0	116.0	102.0	134.2	115.6	102.2	130.7	114.2	102.0	110.7	105.7	102.0	112.9	106.6
08	**DOORS & WINDOWS**	97.0	120.0	103.0	100.8	124.5	107.0	106.9	124.5	111.5	100.8	120.1	105.8	106.9	104.9	106.4	106.9	111.7	108.2
09200	PLASTER & GYPSUM BOARD	96.1	121.8	112.1	106.2	121.8	115.9	106.2	121.8	115.9	103.8	121.8	115.0	106.2	96.3	100.0	106.2	106.5	106.4
095,098	CEILINGS & ACOUSTICAL TREATMENT	88.4	121.8	108.4	98.5	121.8	112.4	98.5	121.8	112.4	99.5	121.8	112.8	98.5	96.3	97.2	98.5	106.5	103.3
09600	FLOORING	96.6	163.7	114.7	99.6	163.7	117.0	99.6	163.7	117.0	100.5	163.7	117.6	99.8	128.6	107.6	99.6	128.6	107.4
097,099	WALL FINISHES, PAINTS & COATINGS	94.0	130.1	116.0	91.6	130.1	115.1	91.5	133.1	116.9	94.0	130.1	116.0	91.5	104.5	99.4	93.3	104.5	100.1
09	**FINISHES**	95.6	131.5	114.3	100.6	131.5	116.7	100.6	131.8	116.9	100.8	131.7	116.9	100.6	104.8	102.8	100.7	111.3	106.2
10 - 14	**TOTAL DIV. 10000 - 14000**	100.0	116.8	103.6	100.0	116.8	103.6	100.0	116.8	103.6	100.0	117.6	103.8	100.0	104.7	101.0	100.0	106.9	101.5
15	**MECHANICAL**	100.1	109.8	104.1	100.1	111.4	104.8	100.1	124.6	110.2	100.1	109.9	104.2	100.1	97.0	98.8	100.1	98.4	99.4
16	**ELECTRICAL**	94.0	103.4	98.6	97.3	120.4	108.5	97.7	120.4	108.7	96.7	103.4	100.0	97.7	93.7	95.8	97.7	93.7	95.8
01 - 16	**WEIGHTED AVERAGE**	98.1	119.0	107.4	100.9	122.6	110.5	99.7	125.0	111.0	100.8	119.1	109.0	99.9	103.0	101.3	100.4	105.9	102.8

	DIVISION	MASSACHUSETTS			MICHIGAN														
		WORCESTER			ANN ARBOR			BATTLE CREEK			BAY CITY			DEARBORN			DETROIT		
		015 - 016			481			490			487			481			482		
		MAT.	INST.	TOTAL	MAT.	INST.	TOTAL	MAT.	INST.	TOTAL	MAT.	INST.	TOTAL	MAT.	INST.	TOTAL	MAT.	INST.	TOTAL
01590	**EQUIPMENT RENTAL**	.0	101.5	101.5	.0	112.2	112.2	.0	105.1	105.1	.0	112.2	112.2	.0	112.2	112.2	.0	97.7	97.7
02	**SITE CONSTRUCTION**	86.8	104.9	100.1	77.2	95.3	90.5	82.3	88.0	86.5	69.6	93.8	87.3	76.9	95.7	90.7	90.4	97.5	95.6
03100	CONCRETE FORMS & ACCESSORIES	103.2	122.8	120.1	97.4	119.4	116.4	97.2	87.2	88.5	97.5	95.5	95.8	97.3	123.9	120.2	98.8	124.0	120.5
03200	CONCRETE REINFORCEMENT	107.5	143.9	126.0	95.2	122.2	108.9	92.2	85.0	88.5	95.2	121.3	108.5	95.2	123.1	109.4	94.6	123.1	109.1
03300	CAST-IN-PLACE CONCRETE	94.3	141.5	113.7	92.3	107.9	98.7	94.9	98.8	96.5	88.3	92.1	89.9	90.2	121.6	103.1	99.1	121.7	108.4
03	**CONCRETE**	100.4	132.3	116.3	94.0	116.4	105.1	96.8	90.1	93.4	92.0	100.1	96.1	92.9	123.2	108.0	97.2	121.9	109.5
04	**MASONRY**	98.8	141.7	125.4	98.8	103.1	101.5	97.4	84.4	89.4	98.5	86.1	90.8	98.7	122.7	113.6	97.8	122.7	113.2
05	**METALS**	99.7	120.7	106.4	102.4	125.5	109.8	92.4	82.7	89.3	103.1	121.4	108.9	102.5	127.8	110.6	107.1	106.2	106.8
06	**WOOD & PLASTICS**	102.0	122.1	112.5	100.1	122.6	111.8	99.6	85.5	92.3	100.1	96.3	98.1	100.1	124.0	112.5	101.1	124.0	113.0
07	**THERMAL & MOISTURE PROTECTION**	102.0	129.3	113.5	102.8	108.6	105.2	91.0	83.1	87.7	100.7	88.2	95.4	101.9	124.8	111.5	99.5	124.8	110.2
08	**DOORS & WINDOWS**	106.9	127.5	112.3	96.6	117.8	102.1	91.1	80.2	88.2	96.6	101.1	97.8	96.6	118.5	102.3	98.2	121.1	104.2
09200	PLASTER & GYPSUM BOARD	106.2	121.8	115.9	98.1	122.0	112.9	92.9	80.4	85.1	98.1	94.9	96.1	98.1	123.4	113.8	98.1	123.4	113.8
095,098	CEILINGS & ACOUSTICAL TREATMENT	98.5	121.8	112.4	91.3	122.0	109.6	91.3	80.4	84.8	92.3	94.9	93.8	91.3	123.4	110.4	92.3	123.4	110.8
09600	FLOORING	99.6	160.4	116.1	88.0	105.0	92.6	97.0	88.0	94.6	87.8	57.0	79.5	87.6	122.8	97.2	87.7	122.8	97.2
097,099	WALL FINISHES, PAINTS & COATINGS	91.5	130.1	115.1	86.6	99.7	94.6	96.8	83.6	88.7	86.6	78.2	81.5	86.6	113.8	103.2	88.4	113.8	103.9
09	**FINISHES**	100.6	130.8	116.3	92.4	115.4	104.4	94.2	86.7	90.3	92.4	86.1	89.1	92.3	123.0	108.3	93.3	123.0	108.8
10 - 14	**TOTAL DIV. 10000 - 14000**	100.0	110.8	102.3	100.0	110.1	102.2	100.0	97.8	99.5	100.0	91.7	98.2	100.0	112.7	102.7	100.0	112.7	102.7
15	**MECHANICAL**	100.1	108.1	103.4	100.0	99.6	99.8	99.9	89.0	95.4	100.0	92.4	96.8	100.0	117.5	107.3	100.0	120.0	108.3
16	**ELECTRICAL**	97.7	106.8	102.1	94.1	85.8	90.0	94.0	92.5	93.3	93.2	81.2	87.4	94.1	119.2	106.3	95.8	119.1	107.1
01 - 16	**WEIGHTED AVERAGE**	100.1	120.0	109.0	97.4	105.9	101.2	95.5	87.9	92.1	96.9	93.5	95.4	97.3	118.9	106.9	99.1	117.6	107.3

DIVISION		MICHIGAN																	
		FLINT			GAYLORD			GRAND RAPIDS			IRON MOUNTAIN			JACKSON			KALAMAZOO		
		484 - 485			497			493,495			498 - 499			492			491		
		MAT.	INST.	TOTAL	MAT.	INST.	TOTAL	MAT.	INST.	TOTAL	MAT.	INST.	TOTAL	MAT.	INST.	TOTAL	MAT.	INST.	TOTAL
01590	EQUIPMENT RENTAL	.0	112.2	112.2	.0	107.6	107.6	.0	105.1	105.1	.0	94.2	94.2	.0	107.6	107.6	.0	105.1	105.1
02	SITE CONSTRUCTION	68.2	94.3	87.3	78.8	85.1	83.4	82.1	87.6	86.1	84.9	95.4	92.6	97.7	87.2	90.0	82.6	88.0	86.5
03100	CONCRETE FORMS & ACCESSORIES	101.0	94.9	95.7	94.2	62.6	66.9	97.7	76.6	79.5	88.3	86.9	87.1	91.7	87.3	87.9	97.2	86.9	88.3
03200	CONCRETE REINFORCEMENT	95.2	121.9	108.8	85.9	114.2	100.2	92.2	83.3	87.7	85.8	101.6	93.8	83.3	121.6	102.8	92.2	85.0	88.5
03300	CAST-IN-PLACE CONCRETE	93.0	96.4	94.4	94.6	75.1	86.6	95.1	96.7	95.8	112.1	91.6	103.7	94.5	92.7	93.8	96.8	98.7	97.6
03	CONCRETE	94.5	101.4	97.9	93.0	77.8	85.5	96.9	84.3	90.7	102.7	91.3	97.1	87.8	96.5	92.1	100.0	89.9	95.0
04	MASONRY	98.9	89.8	93.3	108.9	74.6	87.6	94.8	54.4	69.7	94.4	91.4	92.6	88.6	89.7	89.3	97.6	84.4	89.4
05	METALS	102.5	123.0	109.0	93.9	108.6	98.6	93.9	78.4	89.0	93.5	92.9	93.3	94.0	119.9	102.3	92.4	82.3	89.2
06	WOOD & PLASTICS	103.6	96.3	99.8	92.5	59.6	75.3	98.0	77.2	87.2	88.6	86.3	87.4	91.3	84.9	88.0	99.6	85.5	92.3
07	THERMAL & MOISTURE PROTECTION	100.8	91.0	96.6	89.8	67.0	80.2	91.4	60.6	78.4	92.5	82.0	88.1	89.1	92.4	90.5	91.0	83.1	87.7
08	DOORS & WINDOWS	96.6	101.1	97.8	92.5	66.4	85.7	94.3	68.6	87.6	97.2	76.4	91.8	91.6	91.9	91.7	91.1	80.2	88.2
09200	PLASTER & GYPSUM BOARD	99.6	94.9	96.7	94.6	56.9	71.2	92.9	71.9	79.8	63.2	86.5	77.7	94.0	82.9	87.1	92.9	80.4	85.1
095,098	CEILINGS & ACOUSTICAL TREATMENT	91.3	94.9	93.4	88.7	56.9	69.8	91.3	71.9	79.7	89.4	86.5	87.6	90.5	82.9	86.0	91.3	80.4	84.8
09600	FLOORING	87.8	71.6	83.5	90.6	60.1	82.3	97.0	42.3	82.2	109.3	96.9	105.9	88.8	82.8	87.2	97.0	67.1	88.9
097,099	WALL FINISHES, PAINTS & COATINGS	86.6	85.0	85.6	92.0	53.9	68.8	96.8	41.2	62.8	117.4	45.0	73.2	92.0	91.6	91.8	96.8	83.6	88.7
09	FINISHES	92.1	89.0	90.5	94.3	59.9	76.4	94.2	66.3	79.7	95.8	84.3	89.8	95.0	86.2	90.5	94.2	82.6	88.1
10 - 14	TOTAL DIV. 10000 - 14000	100.0	90.3	97.9	100.0	83.7	96.5	100.0	94.5	98.8	100.0	93.3	98.6	100.0	90.8	98.0	100.0	97.8	99.5
15	MECHANICAL	100.0	99.9	99.9	95.3	85.7	91.3	99.9	55.0	81.3	95.2	85.3	91.1	95.3	89.9	93.1	99.9	84.6	93.6
16	ELECTRICAL	94.1	108.0	100.9	92.0	59.9	76.4	94.8	59.8	77.8	98.5	83.6	91.2	95.6	85.7	90.8	93.7	88.6	91.2
01 - 16	WEIGHTED AVERAGE	97.2	99.9	98.4	94.5	76.9	86.7	96.0	68.1	83.6	96.3	87.7	92.5	93.8	92.3	93.2	95.8	85.8	91.4

DIVISION		MICHIGAN															MINNESOTA		
		LANSING			MUSKEGON			ROYAL OAK			SAGINAW			TRAVERSE CITY			BEMIDJI		
		488 - 489			494			480,483			486			496			566		
		MAT.	INST.	TOTAL	MAT.	INST.	TOTAL	MAT.	INST.	TOTAL	MAT.	INST.	TOTAL	MAT.	INST.	TOTAL	MAT.	INST.	TOTAL
01590	EQUIPMENT RENTAL	.0	112.2	112.2	.0	105.1	105.1	.0	94.6	94.6	.0	112.2	112.2	.0	94.2	94.2	.0	98.3	98.3
02	SITE CONSTRUCTION	83.2	93.8	91.0	80.4	87.8	85.8	80.0	94.6	90.7	70.4	93.8	87.5	72.3	94.0	88.2	85.3	98.2	94.7
03100	CONCRETE FORMS & ACCESSORIES	100.9	95.9	96.6	97.9	83.3	85.3	93.3	110.0	107.7	97.4	95.4	95.7	88.3	60.3	64.1	84.5	96.2	94.6
03200	CONCRETE REINFORCEMENT	95.2	121.7	108.7	92.7	84.8	88.7	86.3	116.0	101.4	95.2	121.3	108.5	87.1	84.0	85.5	95.3	109.7	102.6
03300	CAST-IN-PLACE CONCRETE	92.3	91.8	92.1	94.6	95.1	94.8	80.8	113.8	94.4	91.2	92.1	91.6	87.5	65.7	78.5	102.3	102.8	102.5
03	CONCRETE	94.2	100.3	97.2	95.0	87.0	91.1	81.3	111.3	96.2	93.4	100.1	96.7	84.7	67.3	76.0	95.1	102.1	98.6
04	MASONRY	92.6	91.8	92.1	94.9	77.8	84.3	93.0	120.3	110.0	100.3	86.1	91.5	92.6	74.8	81.6	101.0	106.4	104.4
05	METALS	100.8	122.1	107.6	90.1	81.9	87.5	106.1	95.5	102.7	102.5	121.3	108.5	93.4	82.8	90.0	88.8	123.7	100.0
06	WOOD & PLASTICS	103.0	96.4	99.6	96.5	82.5	89.2	95.6	107.6	101.8	96.2	96.3	96.2	88.6	57.5	72.4	77.2	93.7	85.8
07	THERMAL & MOISTURE PROTECTION	100.8	90.9	96.6	90.2	76.4	84.4	98.9	114.0	105.3	101.2	88.2	95.7	91.6	67.3	81.3	99.9	95.1	97.9
08	DOORS & WINDOWS	96.6	101.2	97.8	90.3	79.2	87.4	96.7	104.9	98.8	96.1	101.1	97.4	97.2	58.6	87.1	99.1	112.4	102.6
09200	PLASTER & GYPSUM BOARD	100.5	95.0	97.1	82.1	77.3	79.1	95.8	106.6	102.5	98.1	94.9	96.1	63.2	56.9	59.3	105.6	93.8	98.3
095,098	CEILINGS & ACOUSTICAL TREATMENT	91.3	95.0	93.5	93.0	77.3	83.7	90.6	106.6	100.1	91.3	94.9	93.4	89.4	56.9	70.0	126.6	93.8	107.1
09600	FLOORING	96.3	79.4	91.7	96.2	77.9	91.3	85.1	114.6	93.1	88.0	57.0	79.6	109.2	60.1	95.9	102.0	131.2	109.9
097,099	WALL FINISHES, PAINTS & COATINGS	100.8	91.1	94.9	96.0	59.7	73.9	88.4	109.8	101.4	86.6	78.2	81.5	117.4	55.1	79.4	97.2	98.5	98.0
09	FINISHES	96.7	92.3	94.4	92.3	79.4	85.6	91.2	110.8	101.4	92.2	86.1	89.1	95.1	58.6	76.1	108.5	102.4	105.4
10 - 14	TOTAL DIV. 10000 - 14000	100.0	91.7	98.2	100.0	96.1	99.2	100.0	96.7	99.3	100.0	91.7	98.2	100.0	78.4	95.4	100.0	87.5	97.3
15	MECHANICAL	100.0	90.0	95.8	99.8	86.5	94.3	95.4	111.8	102.2	100.0	92.1	96.7	95.2	85.1	91.0	95.6	91.0	93.7
16	ELECTRICAL	92.3	86.1	89.3	94.2	82.5	88.5	96.0	105.8	100.7	93.4	81.2	87.5	94.1	59.2	77.1	103.3	99.1	101.3
01 - 16	WEIGHTED AVERAGE	97.2	95.2	96.3	94.4	83.6	89.6	95.3	107.9	100.9	97.0	93.4	95.4	93.3	73.1	84.3	97.3	101.2	99.0

DIVISION		MINNESOTA																	
		BRAINERD			DETROIT LAKES			DULUTH			MANKATO			MINNEAPOLIS			ROCHESTER		
		564			565			556 - 558			560			553 - 555			559		
		MAT.	INST.	TOTAL	MAT.	INST.	TOTAL	MAT.	INST.	TOTAL	MAT.	INST.	TOTAL	MAT.	INST.	TOTAL	MAT.	INST.	TOTAL
01590	EQUIPMENT RENTAL	.0	101.6	101.6	.0	98.3	98.3	.0	101.8	101.8	.0	101.6	101.6	.0	106.2	106.2	.0	101.8	101.8
02	SITE CONSTRUCTION	84.4	103.8	98.6	83.5	98.7	94.6	83.0	102.7	97.4	81.1	103.2	97.3	82.9	108.9	101.9	82.1	101.9	96.7
03100	CONCRETE FORMS & ACCESSORIES	86.5	99.6	97.8	81.1	104.6	101.3	99.7	122.4	119.3	95.9	104.0	102.9	100.4	138.3	133.1	100.2	111.1	109.6
03200	CONCRETE REINFORCEMENT	94.2	109.7	102.1	95.3	109.7	102.6	95.6	110.1	102.9	94.1	128.1	111.4	95.8	128.9	112.6	95.6	128.3	112.2
03300	CAST-IN-PLACE CONCRETE	111.3	108.7	110.2	99.3	107.7	102.8	109.7	108.8	109.3	102.4	113.2	106.8	108.0	126.4	115.6	105.9	103.4	104.9
03	CONCRETE	99.2	105.6	102.4	92.7	107.4	100.0	101.0	116.0	108.4	94.8	112.7	103.7	101.8	132.7	117.2	99.2	112.7	105.9
04	MASONRY	124.8	115.6	119.1	123.3	115.4	118.4	109.1	118.0	114.6	111.9	113.2	112.7	109.1	137.5	126.7	108.4	116.5	113.4
05	METALS	90.1	123.5	100.8	88.8	123.1	99.8	93.0	126.6	103.8	90.0	133.9	104.0	95.8	139.1	109.6	92.9	136.6	106.9
06	WOOD & PLASTICS	94.8	93.4	94.1	73.5	101.2	87.9	112.7	123.1	118.1	105.5	102.2	103.8	113.1	137.3	125.7	113.1	110.1	111.6
07	THERMAL & MOISTURE PROTECTION	98.2	110.1	103.2	99.7	87.1	94.4	100.4	121.0	109.0	98.5	100.6	99.4	100.3	136.5	115.5	100.2	107.6	103.4
08	DOORS & WINDOWS	86.7	112.3	93.4	99.1	116.5	103.6	96.7	124.4	104.0	91.6	123.7	100.0	99.9	142.8	111.1	96.7	128.1	104.9
09200	PLASTER & GYPSUM BOARD	88.4	93.8	91.8	103.7	101.5	102.4	105.9	124.2	117.3	93.2	102.8	99.2	106.1	138.6	126.3	105.7	110.9	108.9
095,098	CEILINGS & ACOUSTICAL TREATMENT	72.6	93.8	85.3	126.6	101.5	111.6	100.3	124.2	114.6	72.6	102.8	90.6	101.2	138.6	123.5	99.5	110.9	106.3
09600	FLOORING	100.7	131.2	109.0	100.5	131.2	108.8	105.0	122.6	109.8	103.3	120.9	108.1	102.4	131.2	110.2	104.8	78.5	97.7
097,099	WALL FINISHES, PAINTS & COATINGS	93.1	98.5	96.4	97.2	98.5	98.0	91.4	112.0	104.0	101.3	104.7	103.4	97.6	127.9	116.1	93.7	103.8	99.9
09	FINISHES	91.8	104.5	98.4	107.7	109.0	108.4	102.5	121.8	112.5	93.6	107.1	100.6	102.3	136.2	120.0	102.3	104.8	103.6
10 - 14	TOTAL DIV. 10000 - 14000	100.0	90.3	97.9	100.0	91.7	98.2	100.0	102.8	100.6	100.0	89.1	97.7	100.0	109.3	102.0	100.0	99.9	100.0
15	MECHANICAL	94.7	95.8	95.2	95.6	95.4	95.5	100.1	114.6	106.1	94.7	98.5	96.3	100.1	119.3	108.1	100.1	106.4	102.7
16	ELECTRICAL	101.4	100.6	101.0	103.2	70.1	87.1	103.5	100.6	102.1	107.5	87.2	97.6	105.3	114.3	109.7	103.5	87.2	95.6
01 - 16	WEIGHTED AVERAGE	96.0	105.1	100.0	97.9	100.6	99.1	99.5	114.5	106.2	96.3	105.9	100.6	100.5	126.7	112.2	99.2	108.5	103.4

COST INDEXES

	DIVISION	MINNESOTA															MISSISSIPPI		
		SAINT PAUL			ST. CLOUD			THIEF RIVER FALLS			WILLMAR			WINDOM			BILOXI		
		550 - 551			563			567			562			561			395		
		MAT.	INST.	TOTAL	MAT.	INST.	TOTAL	MAT.	INST.	TOTAL	MAT.	INST.	TOTAL	MAT.	INST.	TOTAL	MAT.	INST.	TOTAL
01590	EQUIPMENT RENTAL	.0	101.8	101.8	.0	101.6	101.6	.0	98.3	98.3	.0	101.6	101.6	.0	101.6	101.6	.0	98.3	98.3
02	SITE CONSTRUCTION	85.1	103.4	98.5	79.9	105.4	98.6	84.1	98.1	94.4	79.2	99.9	94.4	73.4	98.7	92.0	101.5	85.9	90.1
03100	CONCRETE FORMS & ACCESSORIES	91.1	134.3	128.4	83.1	133.0	126.2	85.4	95.5	94.2	82.8	59.0	62.3	88.1	55.0	59.5	95.9	41.2	48.8
03200	CONCRETE REINFORCEMENT	92.3	128.8	110.8	94.3	128.4	111.6	95.6	109.6	102.7	94.0	127.7	111.1	94.0	126.5	110.5	95.1	64.7	79.6
03300	CAST-IN-PLACE CONCRETE	109.0	125.6	115.8	98.0	123.8	108.6	101.4	93.9	98.3	99.7	81.4	92.2	86.2	58.2	74.7	105.1	44.8	80.3
03	CONCRETE	103.3	130.7	116.9	90.6	129.3	109.8	93.9	98.7	96.3	90.6	81.4	86.1	81.7	71.5	76.7	97.9	48.8	73.5
04	MASONRY	118.8	137.4	130.3	109.5	128.6	121.3	101.0	106.4	104.4	112.3	92.6	100.1	123.0	63.2	85.9	90.1	37.0	57.2
05	METALS	92.4	138.7	107.2	90.6	136.1	105.1	89.0	123.0	99.8	90.0	126.2	101.5	89.9	124.5	100.9	90.4	83.0	88.0
06	WOOD & PLASTICS	102.7	132.1	118.0	91.2	132.0	112.5	78.4	93.7	86.3	90.9	56.1	72.8	96.6	56.1	75.5	96.2	41.1	67.5
07	THERMAL & MOISTURE PROTECTION	100.2	135.8	115.2	98.2	121.3	108.0	101.1	81.2	92.7	98.1	62.0	82.9	98.1	51.2	78.4	96.4	45.9	75.2
08	DOORS & WINDOWS	94.1	140.0	106.1	91.6	139.9	104.2	99.1	112.4	102.6	89.1	81.4	87.1	93.4	81.4	90.3	98.3	48.0	85.1
09200	PLASTER & GYPSUM BOARD	100.4	133.5	121.0	87.3	133.5	116.0	105.2	93.8	98.2	87.3	55.5	67.6	89.1	55.5	68.3	103.5	40.0	64.1
095,098	CEILINGS & ACOUSTICAL TREATMENT	97.8	133.5	119.1	72.6	133.5	108.9	126.6	93.8	107.1	72.6	55.5	62.4	72.6	55.5	62.4	104.3	40.0	66.0
09600	FLOORING	97.8	131.2	106.8	96.7	131.2	106.0	101.6	131.2	109.6	98.5	120.9	104.5	101.4	120.9	106.7	118.7	39.6	97.3
097,099	WALL FINISHES, PAINTS & COATINGS	97.6	127.9	116.1	101.3	127.9	117.5	97.2	98.5	98.0	97.2	98.5	98.0	97.2	104.7	101.8	106.7	35.2	63.1
09	FINISHES	99.5	133.2	117.0	90.6	132.5	112.4	108.3	102.4	105.3	90.9	71.9	81.0	91.6	70.2	80.5	107.2	39.8	72.1
10 - 14	TOTAL DIV. 10000 - 14000	100.0	108.3	101.8	100.0	107.1	101.5	100.0	87.4	97.3	100.0	58.7	91.2	100.0	55.2	90.4	100.0	51.4	89.6
15	MECHANICAL	100.1	115.7	106.6	99.7	117.9	107.2	95.6	90.6	93.5	94.7	93.7	94.3	94.7	77.5	87.6	99.9	53.9	80.8
16	ELECTRICAL	104.1	114.2	109.0	101.4	114.2	107.7	100.5	70.1	85.7	101.4	76.7	89.4	107.5	87.1	97.6	97.0	61.0	79.5
01 - 16	WEIGHTED AVERAGE	99.7	124.5	110.7	95.7	123.3	108.0	96.8	96.1	96.5	94.4	87.4	91.3	95.1	80.3	88.5	97.8	55.5	79.0

	DIVISION	MISSISSIPPI																	
		CLARKSDALE			COLUMBUS			GREENVILLE			GREENWOOD			JACKSON			LAUREL		
		386			397			387			389			390 - 392			394		
		MAT.	INST.	TOTAL	MAT.	INST.	TOTAL	MAT.	INST.	TOTAL	MAT.	INST.	TOTAL	MAT.	INST.	TOTAL	MAT.	INST.	TOTAL
01590	EQUIPMENT RENTAL	.0	98.3	98.3	.0	98.3	98.3	.0	98.3	98.3	.0	98.3	98.3	.0	98.3	98.3	.0	98.3	98.3
02	SITE CONSTRUCTION	98.8	85.1	88.7	100.5	85.4	89.5	105.5	85.7	91.0	102.3	84.8	89.5	98.1	85.7	89.0	106.6	84.1	90.1
03100	CONCRETE FORMS & ACCESSORIES	84.5	18.3	27.4	82.2	29.5	36.8	80.4	34.0	40.4	95.8	20.1	30.5	95.6	38.8	46.6	82.5	23.3	31.4
03200	CONCRETE REINFORCEMENT	102.4	16.9	59.0	101.8	30.6	65.6	103.1	43.2	72.6	102.4	45.9	73.7	95.1	46.8	70.6	102.4	30.0	65.6
03300	CAST-IN-PLACE CONCRETE	102.5	26.7	71.3	107.2	36.9	78.3	105.5	39.2	78.3	110.2	26.6	75.8	103.1	41.7	77.8	104.8	29.7	73.9
03	CONCRETE	96.1	23.3	60.0	99.5	34.4	67.2	101.0	39.6	70.5	102.4	29.6	66.2	96.9	43.2	70.3	101.6	29.0	65.6
04	MASONRY	90.0	19.8	46.5	113.8	23.3	57.7	133.3	38.3	74.4	90.6	18.5	45.9	92.0	38.3	58.8	110.3	23.9	56.8
05	METALS	88.0	59.0	78.8	87.9	65.8	80.8	88.8	72.8	83.7	88.0	71.2	82.6	90.4	74.6	85.3	88.0	64.2	80.4
06	WOOD & PLASTICS	83.1	17.0	48.7	80.8	30.6	54.7	78.4	33.3	54.9	96.3	19.2	56.1	96.3	39.6	66.7	81.1	24.1	51.4
07	THERMAL & MOISTURE PROTECTION	96.2	27.6	67.3	96.3	27.5	67.4	96.5	41.2	73.2	96.6	23.2	65.7	96.3	42.4	73.6	96.5	26.2	66.9
08	DOORS & WINDOWS	97.6	21.1	77.6	97.6	28.7	79.6	97.7	37.8	82.0	97.6	30.4	80.1	98.7	42.3	83.9	94.2	25.0	76.1
09200	PLASTER & GYPSUM BOARD	94.6	15.3	45.4	94.2	29.3	53.9	92.0	32.1	54.8	102.0	17.6	49.5	103.5	38.5	63.1	92.0	22.6	48.9
095,098	CEILINGS & ACOUSTICAL TREATMENT	96.4	15.3	48.1	96.4	29.3	56.4	98.3	32.1	58.8	96.4	17.6	49.4	104.3	38.5	65.0	96.4	22.6	52.4
09600	FLOORING	111.3	23.0	87.4	109.9	19.9	85.6	108.9	33.8	88.6	118.7	20.7	92.2	118.7	39.1	97.2	107.7	18.7	83.6
097,099	WALL FINISHES, PAINTS & COATINGS	106.7	14.8	50.6	106.7	21.8	54.9	106.7	34.6	62.7	106.7	20.2	53.9	106.7	34.6	62.7	106.7	21.8	54.9
09	FINISHES	101.5	18.6	58.3	101.3	27.4	62.8	101.4	33.8	66.2	105.1	20.1	60.9	107.2	38.5	71.4	100.7	22.8	60.1
10 - 14	TOTAL DIV. 10000 - 14000	100.0	40.9	87.4	100.0	43.5	87.9	100.0	49.8	89.3	100.0	41.3	87.5	100.0	50.6	89.5	100.0	26.1	84.2
15	MECHANICAL	97.7	18.8	65.0	97.2	25.6	67.5	99.9	34.3	72.7	97.7	19.8	65.4	99.9	34.0	72.5	97.2	21.8	65.9
16	ELECTRICAL	96.3	24.6	61.4	94.6	27.0	61.7	96.3	37.1	67.6	96.3	25.1	61.7	97.8	37.1	68.3	96.0	25.0	61.5
01 - 16	WEIGHTED AVERAGE	95.9	30.8	67.0	97.1	36.7	70.3	99.3	44.6	75.0	97.2	33.3	68.8	97.8	46.1	74.8	97.1	33.4	68.8

	DIVISION	MISSISSIPPI									MISSOURI								
		MCCOMB			MERIDIAN			TUPELO			BOWLING GREEN			CAPE GIRARDEAU			CHILLICOTHE		
		396			393			388			633			637			646		
		MAT.	INST.	TOTAL	MAT.	INST.	TOTAL	MAT.	INST.	TOTAL	MAT.	INST.	TOTAL	MAT.	INST.	TOTAL	MAT.	INST.	TOTAL
01590	EQUIPMENT RENTAL	.0	98.3	98.3	.0	98.3	98.3	.0	98.3	98.3	.0	108.3	108.3	.0	108.3	108.3	.0	103.4	103.4
02	SITE CONSTRUCTION	92.6	85.4	87.3	97.0	85.7	88.7	97.0	85.0	88.2	93.0	92.7	92.8	94.3	93.3	93.5	104.0	91.4	94.7
03100	CONCRETE FORMS & ACCESSORIES	82.2	57.6	61.0	78.9	33.0	39.3	81.1	23.6	31.5	91.4	90.9	91.0	83.4	77.2	78.1	91.2	79.8	81.3
03200	CONCRETE REINFORCEMENT	103.1	39.5	70.8	101.8	46.3	73.6	100.3	46.5	73.0	93.4	98.4	95.9	94.5	84.5	89.4	108.4	94.9	101.6
03300	CAST-IN-PLACE CONCRETE	93.0	40.2	71.3	99.6	41.7	75.7	102.5	38.3	76.1	88.4	84.1	86.6	87.4	83.3	85.7	97.0	71.8	86.6
03	CONCRETE	89.6	49.7	69.8	93.7	40.6	67.3	95.5	35.3	65.6	88.8	91.3	90.0	87.8	82.4	85.1	101.1	80.6	90.9
04	MASONRY	116.2	38.5	68.0	89.8	28.6	51.8	123.2	30.5	65.7	117.3	97.6	105.1	113.8	81.7	93.9	104.5	90.7	96.0
05	METALS	88.1	70.8	82.6	88.7	73.7	83.9	87.9	72.8	83.1	92.0	116.2	99.7	92.7	109.1	98.0	96.5	96.8	96.6
06	WOOD & PLASTICS	80.8	64.8	72.5	77.1	32.0	53.6	79.0	22.5	49.6	87.0	91.7	89.5	78.5	74.5	76.5	93.9	81.5	87.5
07	THERMAL & MOISTURE PROTECTION	95.9	42.9	73.6	96.1	39.1	72.1	96.2	34.3	70.1	92.8	100.9	96.2	92.7	80.8	87.7	92.7	82.1	88.3
08	DOORS & WINDOWS	97.7	53.1	86.0	97.6	36.6	81.7	97.6	28.9	79.7	93.9	88.3	92.5	93.9	75.5	89.1	90.9	76.6	87.2
09200	PLASTER & GYPSUM BOARD	94.2	64.4	75.7	92.0	30.7	53.9	92.0	20.9	47.9	104.5	91.3	96.3	100.8	73.7	84.0	97.7	80.7	87.2
095,098	CEILINGS & ACOUSTICAL TREATMENT	96.4	64.4	77.3	98.3	30.7	58.0	96.4	20.9	51.4	86.1	91.3	89.2	86.1	73.7	78.7	82.0	80.7	81.3
09600	FLOORING	109.9	23.0	86.4	107.6	27.5	85.9	109.3	25.4	86.6	97.1	74.8	91.0	92.7	74.8	87.9	88.9	53.7	79.3
097,099	WALL FINISHES, PAINTS & COATINGS	106.7	32.6	61.5	106.7	34.2	62.5	106.7	34.6	62.7	101.0	104.3	103.0	101.0	75.7	85.6	93.1	57.2	71.2
09	FINISHES	100.8	50.2	74.4	100.5	31.5	64.5	100.5	24.8	61.1	92.9	89.0	90.9	91.0	75.5	82.9	91.5	73.5	82.1
10 - 14	TOTAL DIV. 10000 - 14000	100.0	72.6	94.2	100.0	49.8	89.3	100.0	42.1	87.7	100.0	69.2	93.4	100.0	66.8	92.9	100.0	65.4	92.6
15	MECHANICAL	97.2	30.3	69.4	99.9	34.0	72.6	98.0	24.5	67.5	95.2	101.4	97.8	100.1	100.9	100.5	95.1	51.2	76.9
16	ELECTRICAL	93.3	61.3	77.7	96.0	61.3	79.1	96.1	24.6	61.3	101.4	92.5	97.1	101.4	114.6	107.8	98.7	42.8	71.5
01 - 16	WEIGHTED AVERAGE	95.7	52.1	76.4	96.0	46.7	74.1	97.3	37.3	70.6	95.7	95.8	95.7	96.4	92.3	94.6	96.7	71.2	85.4

	DIVISION	MISSOURI																	
		COLUMBIA			FLAT RIVER			HANNIBAL			HARRISONVILLE			JEFFERSON CITY			JOPLIN		
		652			636			634			647			650 - 651			648		
		MAT.	INST.	TOTAL	MAT.	INST.	TOTAL	MAT.	INST.	TOTAL	MAT.	INST.	TOTAL	MAT.	INST.	TOTAL	MAT.	INST.	TOTAL
01590	EQUIPMENT RENTAL	.0	109.1	109.1	.0	108.3	108.3	.0	108.3	108.3	.0	103.4	103.4	.0	109.1	109.1	.0	107.1	107.1
02	SITE CONSTRUCTION	103.4	94.6	96.9	95.7	93.4	94.0	90.2	91.6	91.2	95.8	95.6	95.7	105.5	94.0	97.0	103.6	98.8	100.1
03100	CONCRETE FORMS & ACCESSORIES	85.0	72.0	73.8	98.3	86.6	88.2	89.4	70.7	73.2	87.6	95.5	94.4	99.6	63.6	68.5	105.0	73.5	77.8
03200	CONCRETE REINFORCEMENT	101.1	113.3	107.3	94.5	109.5	102.2	92.9	83.2	88.0	108.0	103.6	105.8	99.7	110.3	105.1	111.9	75.7	93.5
03300	CAST-IN-PLACE CONCRETE	90.4	84.3	87.9	91.2	84.6	88.5	83.7	78.1	81.4	99.3	99.1	99.3	94.5	71.1	84.9	105.1	72.6	91.7
03	CONCRETE	86.8	85.8	86.3	91.5	91.7	91.6	85.3	77.3	81.3	97.8	98.9	98.3	90.4	77.0	83.7	101.6	74.6	88.2
04	MASONRY	131.0	86.2	103.2	115.1	83.6	95.6	108.3	88.5	96.0	99.0	103.6	101.9	94.4	86.2	89.3	96.5	61.7	74.9
05	METALS	93.5	121.2	102.4	91.9	120.2	100.9	92.0	107.2	96.8	97.0	109.1	100.9	93.3	119.6	101.7	99.4	89.8	96.3
06	WOOD & PLASTICS	89.1	66.2	77.2	94.6	85.4	89.8	84.8	70.1	77.1	89.2	92.1	90.7	105.1	55.3	79.2	106.9	73.3	89.4
07	THERMAL & MOISTURE PROTECTION	89.6	84.8	87.6	93.0	96.7	94.6	92.6	89.1	91.1	91.9	103.6	96.9	89.7	81.9	86.4	93.5	69.6	83.4
08	DOORS & WINDOWS	90.8	91.5	91.0	93.9	88.2	92.4	93.9	72.6	88.4	91.3	98.8	93.3	87.7	84.8	86.9	92.2	76.2	88.0
09200	PLASTER & GYPSUM BOARD	100.1	65.2	78.4	108.2	84.9	93.7	103.0	69.1	82.0	90.7	91.6	91.3	106.8	54.0	74.0	103.4	72.3	84.1
095,098	CEILINGS & ACOUSTICAL TREATMENT	88.6	65.2	74.6	86.1	84.9	85.4	86.1	69.1	76.0	82.0	91.6	87.8	88.6	54.0	68.0	85.4	72.3	77.6
09600	FLOORING	101.9	91.1	98.9	100.8	74.8	93.7	95.9	74.8	90.2	85.0	98.3	88.6	110.3	91.1	105.1	111.5	50.3	95.0
097,099	WALL FINISHES, PAINTS & COATINGS	98.8	72.2	82.6	101.0	75.7	85.6	101.0	81.6	89.2	97.5	90.4	93.2	98.8	72.2	82.6	92.5	44.7	63.4
09	FINISHES	93.1	72.3	82.3	94.7	82.3	88.2	92.2	71.6	81.5	89.0	94.8	92.0	96.7	65.9	80.7	98.3	66.2	81.6
10 - 14	TOTAL DIV. 10000 - 14000	100.0	90.6	98.0	100.0	68.8	93.4	100.0	63.1	92.1	100.0	86.8	97.2	100.0	89.2	97.7	100.0	77.0	95.1
15	MECHANICAL	99.9	101.8	100.7	95.2	101.9	98.0	95.2	96.5	95.8	95.0	102.1	97.9	99.9	101.8	100.7	100.1	57.5	82.4
16	ELECTRICAL	99.0	92.5	95.8	105.9	114.6	110.1	100.2	84.8	92.7	105.2	109.4	107.3	100.8	92.5	96.8	96.8	62.1	80.0
01 - 16	WEIGHTED AVERAGE	97.1	92.4	95.0	96.7	97.0	96.8	94.5	86.3	90.9	96.4	101.3	98.6	96.1	89.7	93.3	98.6	70.5	86.1

	DIVISION	MISSOURI																	
		KANSAS CITY			KIRKSVILLE			POPLAR BLUFF			ROLLA			SEDALIA			SIKESTON		
		640 - 641			635			639			654 - 655			653			638		
		MAT.	INST.	TOTAL	MAT.	INST.	TOTAL	MAT.	INST.	TOTAL	MAT.	INST.	TOTAL	MAT.	INST.	TOTAL	MAT.	INST.	TOTAL
01590	EQUIPMENT RENTAL	.0	104.9	104.9	.0	99.5	99.5	.0	102.0	102.0	.0	109.1	109.1	.0	100.1	100.1	.0	102.0	102.0
02	SITE CONSTRUCTION	97.4	98.1	97.9	89.1	87.7	88.1	75.6	92.2	87.7	102.8	92.9	95.6	94.4	90.3	91.4	79.0	92.3	88.7
03100	CONCRETE FORMS & ACCESSORIES	103.9	106.6	106.3	81.2	60.3	63.2	81.2	73.2	74.3	93.8	86.9	87.8	91.1	66.0	69.4	82.4	73.4	74.6
03200	CONCRETE REINFORCEMENT	106.3	114.2	110.3	93.7	82.4	88.0	96.6	85.2	90.8	101.6	58.8	79.9	100.1	93.8	96.9	95.9	75.7	85.7
03300	CAST-IN-PLACE CONCRETE	97.6	107.7	101.8	91.2	67.5	81.4	70.1	65.6	68.3	92.5	77.5	86.3	96.6	70.8	86.0	75.0	74.9	74.9
03	CONCRETE	97.4	108.7	103.0	100.9	68.3	84.7	78.8	74.0	76.4	88.8	79.5	84.2	100.4	74.1	87.4	82.3	75.4	78.9
04	MASONRY	101.3	106.8	104.7	120.8	75.1	92.5	112.9	63.0	82.0	107.7	50.3	72.1	113.7	57.7	79.0	112.8	68.8	85.5
05	METALS	106.9	114.9	109.5	91.7	94.7	92.6	92.2	97.2	93.8	93.3	90.2	92.3	91.8	100.9	94.7	92.5	92.6	92.5
06	WOOD & PLASTICS	106.2	106.1	106.2	72.8	58.9	65.5	71.9	74.6	73.3	98.1	97.9	98.0	91.4	59.5	74.8	73.2	74.6	73.9
07	THERMAL & MOISTURE PROTECTION	92.3	108.0	98.9	97.0	81.3	90.4	97.0	78.8	89.4	89.7	64.0	78.9	93.6	80.9	88.2	97.2	72.2	86.6
08	DOORS & WINDOWS	98.3	109.1	101.1	97.8	66.7	89.7	98.7	78.5	93.4	90.8	75.2	86.7	94.2	76.4	89.5	98.7	67.9	90.7
09200	PLASTER & GYPSUM BOARD	100.1	106.0	103.7	97.2	57.6	72.6	97.9	73.7	82.8	102.7	97.7	99.6	96.0	58.3	72.5	99.0	73.7	83.3
095,098	CEILINGS & ACOUSTICAL TREATMENT	91.4	106.0	100.1	83.5	57.6	68.1	86.1	73.7	78.7	88.6	97.7	94.1	86.9	58.3	69.8	86.1	73.7	78.7
09600	FLOORING	91.6	98.3	93.4	75.1	55.1	69.7	88.0	62.3	81.1	106.4	46.9	90.3	83.4	98.3	87.4	88.6	50.3	78.2
097,099	WALL FINISHES, PAINTS & COATINGS	97.5	114.6	107.9	96.6	44.7	64.9	96.0	87.3	90.7	98.8	86.7	91.4	98.8	90.4	93.7	96.0	64.3	76.7
09	FINISHES	94.7	105.7	100.4	88.8	56.2	71.8	90.4	72.6	81.1	94.9	81.0	87.6	90.7	73.0	81.5	91.0	67.8	78.9
10 - 14	TOTAL DIV. 10000 - 14000	100.0	97.4	99.4	100.0	58.3	91.1	100.0	61.1	91.7	100.0	63.8	92.3	100.0	78.5	95.4	100.0	62.1	91.9
15	MECHANICAL	99.9	105.4	102.2	95.3	94.8	95.1	95.3	82.5	90.0	95.0	73.3	86.0	95.0	91.4	93.5	95.3	83.5	90.4
16	ELECTRICAL	108.3	109.4	108.8	99.5	84.7	92.3	100.2	114.6	107.2	97.5	92.4	95.0	97.5	109.4	103.3	99.3	114.6	106.7
01 - 16	WEIGHTED AVERAGE	100.9	106.8	103.5	96.8	79.2	89.0	94.0	83.8	89.5	95.0	78.6	87.7	96.1	84.9	91.2	94.5	83.2	89.5

	DIVISION	MISSOURI									MONTANA								
		SPRINGFIELD			ST. JOSEPH			ST. LOUIS			BILLINGS			BUTTE			GREAT FALLS		
		656 - 658			644 - 645			630 - 631			590 - 591			597			594		
		MAT.	INST.	TOTAL	MAT.	INST.	TOTAL	MAT.	INST.	TOTAL	MAT.	INST.	TOTAL	MAT.	INST.	TOTAL	MAT.	INST.	TOTAL
01590	EQUIPMENT RENTAL	.0	102.8	102.8	.0	103.4	103.4	.0	109.6	109.6	.0	98.6	98.6	.0	98.3	98.3	.0	98.3	98.3
02	SITE CONSTRUCTION	97.3	93.8	94.7	98.9	93.7	95.1	94.0	96.4	95.8	87.1	96.0	93.6	94.5	94.6	94.6	98.1	95.4	96.1
03100	CONCRETE FORMS & ACCESSORIES	100.4	73.7	77.4	103.8	85.8	88.2	98.3	105.4	104.4	96.8	63.6	68.2	84.1	61.2	64.3	98.0	64.5	69.1
03200	CONCRETE REINFORCEMENT	97.4	112.9	105.3	105.0	103.3	104.2	86.5	111.3	99.1	93.3	75.7	84.3	101.2	75.5	88.1	93.3	75.6	84.3
03300	CAST-IN-PLACE CONCRETE	103.6	66.9	88.5	97.6	102.8	99.7	87.4	107.4	95.6	119.3	69.6	98.9	122.8	71.1	101.5	129.8	56.8	99.8
03	CONCRETE	100.2	79.7	90.0	97.2	95.6	96.4	87.5	108.2	97.8	105.3	68.9	87.2	105.3	68.3	86.9	110.5	64.9	87.8
04	MASONRY	88.0	82.9	84.8	100.8	93.1	96.0	94.9	111.6	105.3	121.6	70.9	90.1	121.2	65.4	86.6	125.1	74.1	93.5
05	METALS	97.4	107.2	100.5	102.8	106.4	104.0	97.4	123.4	105.7	106.3	86.1	99.8	99.2	85.4	94.8	102.8	85.7	97.3
06	WOOD & PLASTICS	100.0	73.2	86.0	106.9	83.2	94.6	94.4	103.1	98.9	101.2	63.3	81.4	88.5	61.7	74.5	103.9	63.3	82.7
07	THERMAL & MOISTURE PROTECTION	94.1	78.7	87.6	92.7	94.5	93.5	92.6	106.7	98.5	98.8	68.2	85.9	98.3	67.3	85.2	98.9	67.8	85.8
08	DOORS & WINDOWS	96.3	82.2	92.6	97.0	94.0	96.2	93.0	110.7	97.6	99.0	62.7	89.5	95.9	61.7	86.9	99.3	62.7	89.7
09200	PLASTER & GYPSUM BOARD	103.8	72.3	84.3	104.7	82.5	90.9	109.1	103.0	105.3	103.7	62.6	78.2	95.0	61.0	73.8	103.7	62.6	78.2
095,098	CEILINGS & ACOUSTICAL TREATMENT	88.6	72.3	78.9	90.6	82.5	85.7	91.2	103.0	98.2	112.8	62.6	82.8	112.1	61.0	81.6	114.7	62.6	83.6
09600	FLOORING	105.4	50.3	90.5	93.9	80.6	90.3	100.5	97.1	99.6	107.3	49.0	91.5	98.4	39.0	82.3	107.3	52.9	92.6
097,099	WALL FINISHES, PAINTS & COATINGS	93.1	64.0	75.4	93.1	77.3	83.5	101.0	108.4	105.6	97.2	59.7	74.3	97.2	40.6	62.7	97.2	45.8	65.8
09	FINISHES	96.6	67.7	81.6	95.7	83.4	89.3	95.9	103.5	99.9	106.4	60.0	82.2	102.2	54.2	77.2	107.0	59.9	82.4
10 - 14	TOTAL DIV. 10000 - 14000	100.0	88.2	97.5	100.0	92.0	98.3	100.0	99.8	100.0	100.0	67.1	93.0	100.0	66.1	92.8	100.0	68.4	93.3
15	MECHANICAL	100.0	69.7	87.4	100.1	92.4	96.9	100.1	108.3	103.5	100.3	76.5	90.4	100.3	69.5	87.5	100.3	78.2	91.1
16	ELECTRICAL	102.4	67.1	85.2	105.3	93.2	99.4	104.3	114.6	109.3	94.2	78.7	86.6	100.9	75.6	88.6	94.2	73.3	84.0
01 - 16	WEIGHTED AVERAGE	98.4	78.7	89.6	99.9	93.3	97.0	97.0	109.0	102.3	102.1	74.5	89.8	101.2	70.9	87.7	102.7	73.8	89.9

	DIVISION	MONTANA																	
		HAVRE			HELENA			KALISPELL			MILES CITY			MISSOULA			WOLF POINT		
		595			596			599			593			598			592		
		MAT.	INST.	TOTAL	MAT.	INST.	TOTAL	MAT.	INST.	TOTAL	MAT.	INST.	TOTAL	MAT.	INST.	TOTAL	MAT.	INST.	TOTAL
01590	**EQUIPMENT RENTAL**	.0	98.3	98.3	.0	98.3	98.3	.0	98.3	98.3	.0	98.3	98.3	.0	98.3	98.3	.0	98.3	98.3
02	**SITE CONSTRUCTION**	102.1	95.1	97.0	100.1	95.4	96.6	84.7	94.5	91.9	91.2	95.1	94.1	77.3	94.5	89.9	108.9	95.1	98.8
03100	CONCRETE FORMS & ACCESSORIES	75.9	60.6	62.7	98.1	65.0	69.5	87.7	57.2	61.4	95.8	60.7	65.5	87.7	57.4	61.5	87.6	59.3	63.2
03200	CONCRETE REINFORCEMENT	102.1	63.6	82.5	96.6	63.3	79.7	104.0	80.8	92.2	101.6	63.8	82.4	103.0	81.1	91.9	103.1	63.6	83.0
03300	CAST-IN-PLACE CONCRETE	132.3	61.6	103.2	132.2	66.6	105.3	106.6	63.0	88.7	116.7	61.4	94.0	90.4	63.1	79.2	130.8	57.0	100.4
03	**CONCRETE**	113.2	62.5	88.0	112.2	66.1	89.3	95.3	64.7	80.1	102.4	62.5	82.6	84.1	64.9	74.5	116.9	60.3	88.8
04	**MASONRY**	122.1	67.6	88.3	120.3	66.7	87.1	120.3	53.3	78.8	127.2	72.1	93.0	146.1	58.7	91.9	128.2	67.6	90.6
05	**METALS**	95.7	79.0	90.3	102.0	78.9	94.6	95.5	86.4	92.6	95.0	79.6	90.0	95.9	86.9	93.0	95.0	79.0	89.9
06	**WOOD & PLASTICS**	79.5	59.8	69.2	104.0	63.3	82.8	92.9	58.0	74.7	100.6	59.8	79.3	92.9	58.0	74.7	91.8	58.0	74.2
07	**THERMAL & MOISTURE PROTECTION**	98.6	63.2	83.7	98.9	67.6	85.7	98.0	65.5	84.3	98.3	68.5	85.8	97.6	68.8	85.5	99.1	64.0	84.3
08	**DOORS & WINDOWS**	96.0	57.4	85.9	98.7	59.5	88.5	95.9	62.0	87.1	95.4	57.4	85.5	95.9	62.0	87.1	95.4	56.4	85.2
09200	PLASTER & GYPSUM BOARD	91.3	59.0	71.2	103.1	62.6	77.9	96.8	57.2	72.2	102.4	59.0	75.5	96.8	57.2	72.2	98.7	57.2	72.9
095,098	CEILINGS & ACOUSTICAL TREATMENT	112.1	59.0	80.4	112.1	62.6	82.6	112.1	57.2	79.3	109.6	59.0	79.4	112.1	57.2	79.3	109.6	57.2	78.3
09600	FLOORING	94.9	42.9	80.8	107.3	51.3	92.1	100.6	61.0	89.9	107.0	48.1	91.1	100.6	61.0	89.9	102.4	42.9	86.3
097,099	WALL FINISHES, PAINTS & COATINGS	97.2	45.2	65.5	97.2	50.8	68.9	97.2	45.8	65.8	97.2	42.7	63.9	97.2	45.8	65.8	97.2	42.7	63.9
09	**FINISHES**	101.0	55.3	77.2	106.3	60.7	82.5	102.7	56.2	78.5	105.1	56.0	79.5	102.3	56.2	78.3	104.1	53.9	78.0
10 - 14	**TOTAL DIV. 10000 - 14000**	100.0	54.3	90.3	100.0	56.4	90.7	100.0	51.5	89.7	100.0	54.3	90.3	100.0	51.5	89.7	100.0	54.1	90.2
15	**MECHANICAL**	95.4	77.1	87.8	100.3	73.9	89.4	95.4	69.3	84.6	95.4	77.0	87.8	100.3	69.3	87.4	95.4	77.1	87.8
16	**ELECTRICAL**	94.2	61.6	78.4	94.2	74.6	84.7	98.0	78.4	88.5	94.2	63.9	79.4	99.0	78.4	89.0	94.2	63.9	79.4
01 - 16	**WEIGHTED AVERAGE**	99.6	68.9	86.0	102.5	71.5	88.7	97.8	69.4	85.2	98.8	70.0	86.0	98.9	70.1	86.1	100.8	68.7	86.6

	DIVISION	NEBRASKA																	
		ALLIANCE			COLUMBUS			GRAND ISLAND			HASTINGS			LINCOLN			MCCOOK		
		693			686			688			689			683 - 685			690		
		MAT.	INST.	TOTAL	MAT.	INST.	TOTAL	MAT.	INST.	TOTAL	MAT.	INST.	TOTAL	MAT.	INST.	TOTAL	MAT.	INST.	TOTAL
01590	**EQUIPMENT RENTAL**	.0	96.6	96.6	.0	101.6	101.6	.0	101.6	101.6	.0	101.6	101.6	.0	101.6	101.6	.0	101.6	101.6
02	**SITE CONSTRUCTION**	97.1	95.3	95.8	95.6	89.0	90.7	100.3	91.5	93.9	99.1	89.6	92.2	91.6	91.5	91.5	99.8	89.0	91.9
03100	CONCRETE FORMS & ACCESSORIES	87.5	25.7	34.2	97.4	24.8	34.8	96.8	49.6	56.1	100.7	45.6	53.2	101.5	45.8	53.4	94.0	31.0	39.7
03200	CONCRETE REINFORCEMENT	116.3	44.5	79.8	106.6	49.7	77.7	106.0	74.7	90.1	106.0	49.6	77.3	97.4	75.6	86.3	108.0	74.3	90.9
03300	CAST-IN-PLACE CONCRETE	108.8	36.2	78.9	111.4	46.0	84.5	118.0	57.4	93.1	118.0	50.7	90.3	106.5	59.9	87.4	118.0	36.4	84.4
03	**CONCRETE**	119.3	34.0	76.9	105.2	38.7	72.2	109.7	58.4	84.2	110.0	49.5	80.0	101.7	57.7	79.8	109.9	42.7	76.5
04	**MASONRY**	108.1	29.0	59.1	113.6	30.0	61.7	105.9	48.6	70.4	114.9	39.3	68.0	96.1	64.9	76.7	103.7	28.9	57.4
05	**METALS**	100.5	55.4	86.1	92.3	70.7	85.4	93.6	83.9	90.5	95.0	70.9	87.3	98.7	85.4	94.5	95.2	79.4	90.1
06	**WOOD & PLASTICS**	85.4	24.6	53.7	97.0	22.0	57.9	96.3	44.1	69.1	100.5	44.1	71.1	101.0	38.5	68.5	93.9	30.6	60.9
07	**THERMAL & MOISTURE PROTECTION**	101.9	34.6	73.6	94.4	31.8	68.0	94.5	57.4	78.9	94.6	46.0	74.1	94.8	61.1	80.6	94.5	35.1	69.5
08	**DOORS & WINDOWS**	90.3	27.9	74.0	90.1	33.0	75.2	90.1	49.0	79.4	90.1	45.0	78.3	95.5	48.6	83.3	90.2	38.1	76.6
09200	PLASTER & GYPSUM BOARD	94.4	22.5	49.7	101.9	19.7	50.8	101.6	42.4	64.8	103.0	42.4	65.4	105.6	36.7	62.8	101.6	28.5	56.2
095,098	CEILINGS & ACOUSTICAL TREATMENT	89.2	22.5	49.4	86.9	19.7	46.8	86.9	42.4	60.4	86.9	42.4	60.4	95.5	36.7	60.4	86.9	28.5	52.1
09600	FLOORING	94.0	28.5	76.2	95.0	36.1	79.1	94.9	34.7	78.6	96.5	36.1	80.2	96.7	41.8	81.8	93.7	28.5	76.0
097,099	WALL FINISHES, PAINTS & COATINGS	145.1	21.8	69.9	90.8	50.0	65.9	90.8	41.3	60.6	90.8	50.0	65.9	90.8	41.7	60.8	90.8	23.7	49.9
09	**FINISHES**	93.8	25.3	58.1	93.0	28.0	59.2	93.1	44.2	67.6	93.8	43.8	67.7	96.2	42.5	68.2	92.8	28.8	59.5
10 - 14	**TOTAL DIV. 10000 - 14000**	100.0	46.4	88.6	100.0	59.2	91.3	100.0	68.9	93.4	100.0	66.2	92.8	100.0	68.2	93.2	100.0	45.5	88.4
15	**MECHANICAL**	95.3	26.2	66.7	95.1	64.5	82.4	100.0	74.8	89.6	95.1	70.4	84.9	100.0	74.9	89.6	95.1	67.9	83.8
16	**ELECTRICAL**	92.6	18.1	56.4	92.3	43.0	68.3	91.0	70.3	80.9	90.4	43.0	67.4	96.6	70.3	83.8	94.7	18.2	57.5
01 - 16	**WEIGHTED AVERAGE**	99.0	35.9	71.0	96.1	49.6	75.5	97.5	65.6	83.3	97.1	56.7	79.2	98.2	67.1	84.4	96.9	48.2	75.3

	DIVISION	NEBRASKA												NEVADA					
		NORFOLK			NORTH PLATTE			OMAHA			VALENTINE			CARSON CITY			ELKO		
		687			691			680 - 681			692			897			898		
		MAT.	INST.	TOTAL	MAT.	INST.	TOTAL	MAT.	INST.	TOTAL	MAT.	INST.	TOTAL	MAT.	INST.	TOTAL	MAT.	INST.	TOTAL
01590	**EQUIPMENT RENTAL**	.0	90.9	90.9	.0	101.6	101.6	.0	90.9	90.9	.0	94.0	94.0	.0	101.7	101.7	.0	101.7	101.7
02	**SITE CONSTRUCTION**	79.0	88.3	85.8	101.0	89.7	92.7	80.3	90.3	87.6	84.2	92.1	90.0	61.9	104.0	92.8	56.6	103.8	91.3
03100	CONCRETE FORMS & ACCESSORIES	82.6	45.6	50.7	96.6	45.9	52.8	96.5	73.6	76.7	83.8	22.9	31.3	94.2	100.7	99.8	106.4	89.1	91.5
03200	CONCRETE REINFORCEMENT	106.8	57.1	81.5	107.4	74.4	90.7	102.0	76.4	89.0	108.1	40.0	73.5	113.7	114.1	113.9	115.5	113.8	114.7
03300	CAST-IN-PLACE CONCRETE	112.1	56.0	89.0	118.0	50.8	90.4	110.6	74.1	95.6	103.8	42.5	78.6	109.7	91.4	102.2	98.4	80.9	91.2
03	**CONCRETE**	103.7	52.1	78.1	110.0	54.3	82.3	103.1	74.5	88.9	106.4	34.3	70.6	109.3	99.9	104.6	101.6	91.0	96.4
04	**MASONRY**	119.9	61.6	83.7	91.3	39.3	59.1	101.4	78.0	86.9	102.9	30.0	57.7	131.3	80.6	99.9	131.8	79.0	99.0
05	**METALS**	95.6	65.4	86.0	93.8	81.0	89.7	98.0	77.4	91.5	106.7	56.0	90.5	93.3	102.3	96.2	92.6	100.8	95.2
06	**WOOD & PLASTICS**	80.2	43.4	61.1	96.2	44.1	69.0	95.9	74.9	84.9	80.7	21.3	49.8	90.7	103.2	97.2	105.0	89.9	97.2
07	**THERMAL & MOISTURE PROTECTION**	92.4	54.7	76.5	94.5	46.1	74.1	90.2	71.5	82.3	93.0	31.1	66.9	99.8	88.9	95.2	99.7	80.9	91.8
08	**DOORS & WINDOWS**	92.1	46.1	80.0	89.4	51.3	79.5	98.7	67.5	90.6	92.2	26.7	75.1	94.3	108.8	98.1	96.9	94.0	96.1
09200	PLASTER & GYPSUM BOARD	101.0	42.4	64.6	102.0	42.4	65.0	107.9	74.7	87.3	104.6	19.7	51.9	83.6	103.0	95.7	93.2	89.5	90.9
095,098	CEILINGS & ACOUSTICAL TREATMENT	100.6	42.4	65.9	88.6	42.4	61.1	105.7	74.7	87.2	104.6	19.7	54.0	103.9	103.0	103.4	103.9	89.5	95.3
09600	FLOORING	117.0	43.4	97.1	94.8	36.1	78.9	122.7	46.3	102.1	119.9	43.4	99.2	100.3	63.8	90.5	106.6	63.8	95.0
097,099	WALL FINISHES, PAINTS & COATINGS	148.6	50.0	88.4	90.8	50.0	65.9	148.6	71.0	101.2	148.6	33.5	78.4	102.2	78.7	87.9	102.2	83.5	90.8
09	**FINISHES**	110.1	44.8	76.1	93.6	43.8	67.6	114.1	67.9	90.0	112.9	27.2	68.3	96.8	91.3	93.9	99.8	83.9	91.5
10 - 14	**TOTAL DIV. 10000 - 14000**	100.0	64.6	92.4	100.0	51.4	89.6	100.0	71.2	93.9	100.0	37.6	86.7	100.0	107.8	101.7	100.0	85.3	96.9
15	**MECHANICAL**	94.9	70.4	84.7	100.0	73.8	89.1	99.8	81.1	92.0	94.7	23.4	65.2	100.1	93.0	97.2	97.7	81.3	90.9
16	**ELECTRICAL**	91.4	54.0	73.2	92.9	43.0	68.7	90.6	86.3	88.5	90.5	42.9	67.4	91.0	110.9	100.7	90.5	68.4	79.7
01 - 16	**WEIGHTED AVERAGE**	97.7	60.6	81.3	97.1	58.8	80.1	99.3	78.2	89.9	99.2	38.5	72.3	98.6	97.9	98.3	97.6	85.4	92.2

DIVISION		NEVADA									NEW HAMPSHIRE								
		ELY			LAS VEGAS			RENO			CHARLESTON			CLAREMONT			CONCORD		
		893			889 - 891			894 - 895			036			037			032 - 033		
		MAT.	INST.	TOTAL	MAT.	INST.	TOTAL	MAT.	INST.	TOTAL	MAT.	INST.	TOTAL	MAT.	INST.	TOTAL	MAT.	INST.	TOTAL
01590	EQUIPMENT RENTAL	.0	101.7	101.7	.0	101.7	101.7	.0	101.7	101.7	.0	101.5	101.5	.0	101.5	101.5	.0	101.5	101.5
02	SITE CONSTRUCTION	61.5	103.4	92.2	61.5	106.2	94.3	61.6	104.0	92.7	80.9	98.2	93.6	75.2	98.2	92.0	87.8	99.9	96.6
03100	CONCRETE FORMS & ACCESSORIES	97.8	71.9	75.5	93.1	108.7	106.6	94.3	100.8	99.9	88.0	37.5	44.4	95.2	37.5	45.4	89.6	62.2	66.0
03200	CONCRETE REINFORCEMENT	114.2	105.1	109.6	108.1	119.3	113.8	108.1	118.5	113.4	85.9	57.9	71.7	85.9	57.9	71.7	85.9	88.6	87.3
03300	CAST-IN-PLACE CONCRETE	105.7	78.2	94.4	103.5	108.8	105.6	111.7	91.4	103.4	100.2	53.0	80.8	92.1	53.0	76.0	97.4	96.3	96.9
03	CONCRETE	109.9	80.9	95.5	105.4	110.4	107.8	109.4	100.7	105.1	103.0	47.6	75.5	94.9	47.6	71.4	99.8	79.2	89.6
04	MASONRY	137.0	70.6	95.8	122.2	100.4	108.7	130.7	80.6	99.6	93.3	41.4	61.1	93.5	41.4	61.2	96.6	92.2	93.9
05	METALS	92.5	96.9	93.9	93.9	107.4	98.2	93.7	104.2	97.1	94.5	66.0	85.4	94.5	66.0	85.4	95.1	85.2	91.9
06	WOOD & PLASTICS	94.7	71.3	82.5	90.4	107.7	99.4	90.7	103.2	97.2	85.3	36.8	60.1	93.3	36.8	63.9	86.7	54.2	69.8
07	THERMAL & MOISTURE PROTECTION	100.3	75.4	89.8	103.2	101.9	102.7	99.8	88.9	95.2	101.3	45.3	77.7	101.2	45.3	77.6	101.8	87.3	95.7
08	DOORS & WINDOWS	96.8	81.5	92.8	94.7	112.4	99.3	94.7	109.8	98.6	103.6	40.2	87.0	104.9	40.2	88.0	104.9	63.1	94.0
09200	PLASTER & GYPSUM BOARD	89.1	70.3	77.5	84.4	107.7	98.8	85.5	103.0	96.4	96.0	34.3	57.6	99.3	34.3	58.9	96.0	52.1	68.7
095,098	CEILINGS & ACOUSTICAL TREATMENT	103.9	70.3	83.9	109.9	107.7	108.6	111.6	103.0	106.5	85.5	34.3	54.9	85.5	34.3	54.9	85.5	52.1	65.6
09600	FLOORING	104.1	71.4	95.2	100.3	99.2	100.0	100.3	63.8	90.5	93.3	40.1	78.9	96.3	40.1	81.1	93.9	102.9	96.3
097,099	WALL FINISHES, PAINTS & COATINGS	102.2	78.7	87.9	102.2	117.1	111.3	102.2	78.7	87.9	91.5	32.1	55.3	91.5	32.1	55.3	91.5	94.9	93.6
09	FINISHES	98.9	71.9	84.8	98.3	107.7	103.2	99.0	91.3	95.0	93.4	36.7	63.9	94.4	36.7	64.4	94.4	71.7	82.6
10 - 14	TOTAL DIV. 10000 - 14000	100.0	70.9	93.8	100.0	103.3	100.7	100.0	107.8	101.7	100.0	45.7	88.4	100.0	45.7	88.4	100.0	74.6	94.6
15	MECHANICAL	97.7	76.8	89.0	100.1	107.1	103.0	100.1	93.1	97.2	95.1	39.5	72.1	95.1	39.5	72.1	100.1	90.3	96.0
16	ELECTRICAL	90.6	68.4	79.8	93.0	108.8	100.6	91.0	110.9	100.7	95.5	35.7	66.4	95.5	35.7	66.4	97.6	72.7	85.5
01 - 16	WEIGHTED AVERAGE	98.7	79.0	90.0	98.3	107.1	102.2	98.9	98.2	98.6	96.6	47.9	75.0	95.9	47.9	74.6	98.4	82.4	91.3

DIVISION		NEW HAMPSHIRE															NEW JERSEY		
		KEENE			LITTLETON			MANCHESTER			NASHUA			PORTSMOUTH			ATLANTIC CITY		
		034			035			031			030			038			082,084		
		MAT.	INST.	TOTAL	MAT.	INST.	TOTAL	MAT.	INST.	TOTAL	MAT.	INST.	TOTAL	MAT.	INST.	TOTAL	MAT.	INST.	TOTAL
01590	EQUIPMENT RENTAL	.0	101.5	101.5	.0	101.5	101.5	.0	101.5	101.5	.0	101.5	101.5	.0	101.5	101.5	.0	99.4	99.4
02	SITE CONSTRUCTION	87.9	98.2	95.5	75.1	97.8	91.8	87.7	99.9	96.6	89.0	99.9	97.0	83.2	98.9	94.7	91.5	105.1	101.5
03100	CONCRETE FORMS & ACCESSORIES	93.6	38.6	46.1	105.3	43.8	52.3	102.3	62.2	67.7	102.8	62.2	67.8	89.6	57.8	62.2	112.9	125.4	123.7
03200	CONCRETE REINFORCEMENT	85.9	69.0	77.3	86.7	41.4	63.7	107.5	88.6	97.9	107.5	88.6	97.9	85.9	88.5	87.2	81.9	110.4	96.4
03300	CAST-IN-PLACE CONCRETE	100.7	53.4	81.2	90.3	47.5	72.7	103.8	96.3	100.7	95.3	96.3	95.7	90.4	90.0	90.2	85.4	129.9	103.7
03	CONCRETE	102.9	50.3	76.8	94.4	45.4	70.1	107.3	79.2	93.3	103.2	79.2	91.3	94.1	75.0	84.6	96.1	122.9	109.4
04	MASONRY	97.4	41.4	62.7	106.9	40.0	65.4	99.1	92.2	94.8	99.4	92.2	94.9	94.7	81.9	86.8	101.7	121.3	113.9
05	METALS	95.0	71.3	87.5	95.1	58.8	83.5	99.7	85.2	95.1	99.7	85.2	95.0	96.1	83.1	92.0	94.8	100.1	96.5
06	WOOD & PLASTICS	91.4	36.8	63.0	103.6	42.2	71.6	101.5	54.2	76.9	101.5	54.2	76.9	86.7	54.2	69.8	113.2	126.0	119.9
07	THERMAL & MOISTURE PROTECTION	101.8	47.1	78.8	101.3	49.7	79.6	101.8	87.3	95.7	102.2	87.3	95.9	101.8	90.3	97.0	101.9	121.5	110.2
08	DOORS & WINDOWS	102.1	49.3	88.3	106.0	39.8	88.7	106.9	63.1	95.5	106.9	63.1	95.5	107.8	57.3	94.6	104.5	119.1	108.3
09200	PLASTER & GYPSUM BOARD	98.6	34.3	58.6	111.5	39.8	66.9	106.2	52.1	72.6	106.2	52.1	72.6	96.0	52.1	68.7	108.5	125.8	119.3
095,098	CEILINGS & ACOUSTICAL TREATMENT	85.5	34.3	54.9	85.5	39.8	58.2	98.5	52.1	70.8	98.5	52.1	70.8	87.5	52.1	66.3	85.5	125.8	109.6
09600	FLOORING	95.7	65.7	87.6	106.6	40.1	88.6	99.8	102.9	100.7	99.6	102.9	100.5	93.9	102.9	96.3	104.0	126.8	110.2
097,099	WALL FINISHES, PAINTS & COATINGS	91.5	32.1	55.3	91.5	65.6	75.7	91.5	94.9	93.6	91.5	94.9	93.6	91.5	38.9	59.4	91.5	115.5	106.1
09	FINISHES	95.5	42.0	67.6	99.3	44.8	70.9	100.8	71.7	85.6	101.0	71.7	85.7	94.6	62.8	78.1	99.8	124.9	112.9
10 - 14	TOTAL DIV. 10000 - 14000	100.0	64.9	92.5	100.0	48.2	89.0	100.0	74.6	94.6	100.0	74.6	94.6	100.0	70.7	93.7	100.0	116.4	103.5
15	MECHANICAL	95.1	42.3	73.2	95.1	80.5	89.1	100.1	90.3	96.0	100.1	90.3	96.0	100.1	84.8	93.7	99.7	116.3	106.6
16	ELECTRICAL	95.5	35.7	66.4	96.5	72.7	84.9	97.8	72.7	85.6	97.6	72.7	85.5	95.9	72.7	84.6	93.6	114.8	103.9
01 - 16	WEIGHTED AVERAGE	97.2	50.9	76.6	97.4	61.6	81.5	100.9	82.4	92.7	100.6	82.4	92.5	97.8	78.0	89.0	98.4	116.5	106.4

DIVISION		NEW JERSEY																	
		CAMDEN			DOVER			ELIZABETH			HACKENSACK			JERSEY CITY			LONG BRANCH		
		081			078			072			076			073			077		
		MAT.	INST.	TOTAL	MAT.	INST.	TOTAL	MAT.	INST.	TOTAL	MAT.	INST.	TOTAL	MAT.	INST.	TOTAL	MAT.	INST.	TOTAL
01590	EQUIPMENT RENTAL	.0	99.4	99.4	.0	101.5	101.5	.0	101.5	101.5	.0	101.5	101.5	.0	99.4	99.4	.0	98.9	98.9
02	SITE CONSTRUCTION	92.2	105.0	101.6	102.7	105.4	104.7	106.2	105.4	105.6	103.0	104.8	104.3	92.2	104.8	101.4	96.9	105.3	103.0
03100	CONCRETE FORMS & ACCESSORIES	102.8	125.6	122.4	99.0	130.2	125.9	113.1	130.1	127.8	99.0	130.0	125.7	102.8	130.2	126.4	103.4	133.3	129.2
03200	CONCRETE REINFORCEMENT	107.5	114.7	111.1	82.8	114.4	98.8	82.8	114.4	98.8	82.8	114.4	98.8	107.5	114.4	111.0	82.8	114.1	98.7
03300	CAST-IN-PLACE CONCRETE	82.8	125.3	100.3	106.3	131.2	116.5	91.3	131.1	107.7	103.9	131.1	115.1	82.8	131.1	102.7	92.1	130.1	107.7
03	CONCRETE	97.2	122.3	109.6	104.3	126.5	115.3	99.6	126.5	113.0	102.4	126.4	114.3	97.2	126.3	111.6	101.4	127.3	114.2
04	MASONRY	91.3	121.9	110.3	96.4	127.6	115.8	113.5	127.6	122.3	100.6	127.6	117.4	91.3	127.6	113.8	105.7	121.8	115.7
05	METALS	99.5	101.7	100.2	94.8	107.3	98.8	95.8	107.2	99.5	94.9	107.1	98.7	99.6	104.3	101.1	94.9	103.7	97.7
06	WOOD & PLASTICS	101.5	126.0	114.3	100.5	131.1	116.4	117.0	131.1	124.4	100.5	131.1	116.4	101.5	131.1	116.9	102.7	135.4	119.8
07	THERMAL & MOISTURE PROTECTION	101.7	122.5	110.5	102.2	133.5	115.4	102.5	133.5	115.6	102.0	125.9	112.1	101.7	133.5	115.1	101.9	131.5	114.3
08	DOORS & WINDOWS	106.9	120.2	110.4	110.3	124.1	113.9	108.1	124.1	112.3	107.3	124.1	111.7	106.9	124.1	111.4	102.9	126.4	109.1
09200	PLASTER & GYPSUM BOARD	106.2	125.8	118.4	102.3	131.0	120.1	109.6	131.0	122.9	102.3	131.0	120.1	106.2	131.0	121.6	104.1	135.6	123.7
095,098	CEILINGS & ACOUSTICAL TREATMENT	98.5	125.8	114.8	85.5	131.0	112.7	87.5	131.0	113.4	85.5	131.0	112.7	98.5	131.0	117.9	85.5	135.6	115.4
09600	FLOORING	99.6	126.8	107.0	98.4	126.8	106.1	104.5	126.8	110.5	98.4	126.8	106.1	99.6	126.8	107.0	99.9	126.8	107.2
097,099	WALL FINISHES, PAINTS & COATINGS	91.5	115.5	106.1	91.5	133.8	117.3	91.5	133.8	117.3	91.5	133.8	117.3	91.5	133.8	117.3	91.5	115.5	106.1
09	FINISHES	101.3	125.1	113.7	97.3	130.1	114.4	101.1	130.1	116.2	97.2	130.1	114.3	101.3	130.1	116.3	98.2	130.6	115.1
10 - 14	TOTAL DIV. 10000 - 14000	100.0	116.6	103.5	100.0	118.7	104.0	100.0	118.7	104.0	100.0	118.7	104.0	100.0	118.7	104.0	100.0	117.7	103.8
15	MECHANICAL	100.1	116.6	106.9	99.7	125.9	110.6	100.1	121.9	109.1	99.7	125.9	110.6	100.1	125.9	110.8	99.7	127.2	111.1
16	ELECTRICAL	98.0	114.8	106.2	95.3	135.0	114.6	95.9	135.0	114.9	95.3	137.6	115.9	99.5	137.6	118.0	95.0	144.2	118.9
01 - 16	WEIGHTED AVERAGE	99.6	116.8	107.2	99.7	124.5	110.7	100.8	123.6	110.9	99.5	124.5	110.6	99.8	124.5	110.8	99.1	125.3	110.8

DIVISION		NEW JERSEY																
		NEWARK			NEW BRUNSWICK			PATERSON			POINT PLEASANT			SUMMIT			TRENTON	
		070 - 071			088 - 089			074 - 075			087			079			085 - 086	
		MAT.	INST.	TOTAL	MAT.	INST.	TOTAL	MAT.	INST.	TOTAL	MAT.	INST.	TOTAL	MAT.	INST.	TOTAL	MAT.	INST.	TOTAL
01590	EQUIPMENT RENTAL	.0	101.5	101.5	.0	98.9	98.9	.0	101.5	101.5	.0	98.9	98.9	.0	101.5	101.5	.0	98.9	98.9
02	SITE CONSTRUCTION	110.6	105.4	106.8	104.6	105.3	105.1	104.7	104.8	104.8	105.9	105.3	105.4	103.8	105.4	105.0	94.0	105.3	102.2
03100	CONCRETE FORMS & ACCESSORIES	101.0	130.2	126.2	106.5	130.0	126.8	101.6	130.0	126.1	100.0	122.7	119.6	102.6	130.1	126.3	101.7	129.7	125.8
03200	CONCRETE REINFORCEMENT	107.5	114.4	111.0	82.8	114.1	98.7	107.5	114.4	111.0	82.8	114.0	98.7	82.8	114.4	98.8	107.5	115.0	111.3
03300	CAST-IN-PLACE CONCRETE	96.7	131.2	110.9	105.6	125.2	113.7	105.7	131.1	116.1	105.6	130.0	115.6	88.2	131.1	105.9	98.1	125.1	109.2
03	CONCRETE	103.8	126.5	115.1	113.6	124.2	118.9	108.1	126.4	117.2	113.2	122.5	117.8	96.2	126.5	111.2	104.5	124.1	114.2
04	MASONRY	101.1	127.6	117.6	99.5	121.8	113.3	97.0	127.6	116.0	91.4	121.8	110.3	99.3	127.6	116.9	90.9	121.8	110.1
05	METALS	99.6	107.3	102.0	94.9	103.6	97.7	94.4	107.0	98.4	94.9	103.4	97.6	94.8	107.2	98.8	94.4	102.2	96.9
06	WOOD & PLASTICS	103.3	131.1	117.8	106.1	131.0	119.1	103.3	131.1	117.8	98.8	121.4	110.6	104.6	131.1	118.4	101.5	131.0	116.9
07	THERMAL & MOISTURE PROTECTION	101.9	133.5	115.2	102.1	131.0	114.3	102.3	125.9	112.2	102.1	130.0	113.8	102.7	133.5	115.7	100.6	122.1	109.6
08	DOORS & WINDOWS	113.1	124.1	116.0	98.8	124.0	105.4	113.1	124.1	116.0	100.9	118.5	105.5	115.7	124.1	117.9	106.9	123.3	111.2
09200	PLASTER & GYPSUM BOARD	106.2	131.0	121.6	105.6	131.0	121.4	106.2	131.0	121.6	102.3	121.2	114.0	104.1	131.0	120.8	106.2	131.0	121.6
095,098	CEILINGS & ACOUSTICAL TREATMENT	98.5	131.0	117.9	85.5	131.0	112.7	98.5	131.0	117.9	85.5	121.2	106.8	85.5	131.0	112.7	98.5	131.0	117.9
09600	FLOORING	99.8	126.8	107.1	101.3	126.8	108.2	99.6	126.8	107.0	98.4	126.8	106.1	99.9	126.8	107.2	99.8	126.8	107.1
097,099	WALL FINISHES, PAINTS & COATINGS	91.5	133.8	117.3	91.5	133.8	117.3	91.5	133.8	117.3	91.5	115.5	106.1	91.5	133.8	117.3	91.5	115.5	106.2
09	FINISHES	101.8	130.1	116.5	99.6	130.1	115.4	101.4	130.1	116.4	98.2	122.4	110.8	98.3	130.1	114.9	101.3	128.0	115.2
10 - 14	TOTAL DIV. 10000 - 14000	100.0	118.7	104.0	100.0	118.5	104.0	100.0	118.7	104.0	100.0	114.3	103.0	100.0	118.7	104.0	100.0	117.1	103.6
15	MECHANICAL	100.1	124.7	110.3	99.7	127.3	111.1	100.1	125.9	110.8	99.7	123.9	109.7	99.7	121.9	108.9	100.1	127.2	111.3
16	ELECTRICAL	99.4	137.6	118.0	94.2	135.0	114.1	99.5	135.0	116.8	93.6	144.2	118.2	95.9	135.0	114.9	98.1	134.7	115.9
01 - 16	WEIGHTED AVERAGE	102.1	124.6	112.1	100.0	123.4	110.4	101.4	124.2	111.5	99.5	122.3	109.6	99.8	123.6	110.4	99.6	122.6	109.8

DIVISION		NEW JERSEY			NEW MEXICO													
		VINELAND			ALBUQUERQUE			CARRIZOZO			CLOVIS			FARMINGTON			GALLUP	
		080,083			870 - 872			883			881			874			873	
		MAT.	INST.	TOTAL	MAT.	INST.	TOTAL	MAT.	INST.	TOTAL	MAT.	INST.	TOTAL	MAT.	INST.	TOTAL	MAT.	INST.	TOTAL
01590	EQUIPMENT RENTAL	.0	99.4	99.4	.0	116.3	116.3	.0	116.3	116.3	.0	116.3	116.3	.0	116.3	116.3	.0	116.3	116.3
02	SITE CONSTRUCTION	96.0	105.0	102.6	78.5	111.2	102.5	98.1	111.2	107.7	86.1	111.2	104.5	83.8	111.2	103.9	91.9	111.2	106.1
03100	CONCRETE FORMS & ACCESSORIES	97.1	125.6	121.7	96.0	68.6	72.4	96.0	68.6	72.4	96.0	68.4	72.1	96.0	68.6	72.4	96.0	68.6	72.4
03200	CONCRETE REINFORCEMENT	81.9	110.4	96.4	108.7	68.1	88.1	116.6	68.1	91.9	117.9	57.0	87.0	118.8	68.1	93.0	113.7	68.1	90.5
03300	CAST-IN-PLACE CONCRETE	92.1	130.2	107.8	102.6	74.2	90.9	95.2	74.2	86.5	95.1	74.0	86.4	103.2	74.2	91.3	97.1	74.2	87.7
03	CONCRETE	100.8	123.1	111.9	105.3	71.5	88.5	119.9	71.5	95.8	108.5	69.2	89.0	109.1	71.5	90.4	115.1	71.5	93.4
04	MASONRY	88.7	121.9	109.3	119.3	64.4	85.3	112.9	64.4	82.9	113.0	64.4	82.9	123.6	64.4	86.9	113.0	64.4	82.9
05	METALS	94.7	100.1	96.5	100.1	87.7	96.2	97.2	87.7	94.2	96.9	82.1	92.2	97.8	87.7	94.6	97.2	87.7	94.2
06	WOOD & PLASTICS	95.6	126.0	111.5	95.4	70.1	82.2	95.4	70.1	82.2	95.4	70.1	82.2	95.4	70.1	82.2	95.4	70.1	82.2
07	THERMAL & MOISTURE PROTECTION	101.6	121.7	110.1	99.0	73.4	88.2	101.1	73.4	89.4	99.8	73.4	88.7	99.6	73.4	88.6	100.7	73.4	89.2
08	DOORS & WINDOWS	100.4	119.1	105.3	94.6	72.2	88.8	93.5	72.2	88.0	93.7	68.8	87.2	97.5	72.2	90.9	97.5	72.2	90.9
09200	PLASTER & GYPSUM BOARD	101.1	125.8	116.5	87.5	68.8	75.9	82.7	68.8	74.0	82.7	68.8	74.0	82.7	68.8	74.0	82.7	68.8	74.0
095,098	CEILINGS & ACOUSTICAL TREATMENT	85.5	125.8	109.6	109.0	68.8	85.0	100.5	68.8	81.6	100.5	68.8	81.6	100.5	68.8	81.6	100.5	68.8	81.6
09600	FLOORING	97.3	126.8	105.3	100.3	62.2	90.0	100.3	62.2	90.0	100.3	62.2	90.0	100.3	62.2	90.0	100.3	62.2	90.0
097,099	WALL FINISHES, PAINTS & COATINGS	91.5	115.5	106.1	102.2	58.6	75.6	102.2	58.6	75.6	102.2	58.6	75.6	102.2	58.6	75.6	102.2	58.6	75.6
09	FINISHES	97.0	125.1	111.6	99.1	66.2	82.0	97.9	66.2	81.4	96.8	66.2	80.9	96.5	66.2	80.7	97.4	66.2	81.2
10 - 14	TOTAL DIV. 10000 - 14000	100.0	116.6	103.5	100.0	71.5	93.9	100.0	71.5	93.9	100.0	71.5	93.9	100.0	71.5	93.9	100.0	71.5	93.9
15	MECHANICAL	99.7	119.6	108.0	100.1	72.2	88.5	97.2	72.2	86.8	97.2	71.7	86.6	100.1	72.2	88.5	97.1	72.2	86.7
16	ELECTRICAL	93.6	114.8	103.9	84.0	78.1	81.1	84.0	78.1	81.1	81.9	78.1	80.0	82.4	78.1	80.3	81.7	78.1	80.0
01 - 16	WEIGHTED AVERAGE	97.6	117.3	106.3	98.5	76.2	88.6	99.0	76.2	88.9	97.0	75.2	87.3	98.7	76.2	88.7	98.3	76.2	88.5

DIVISION		NEW MEXICO																
		LAS CRUCES			LAS VEGAS			ROSWELL			SANTA FE			SOCORRO			TRUTH/CONSEQUENCES	
		880			877			882			875			878			879	
		MAT.	INST.	TOTAL	MAT.	INST.	TOTAL	MAT.	INST.	TOTAL	MAT.	INST.	TOTAL	MAT.	INST.	TOTAL	MAT.	INST.	TOTAL
01590	EQUIPMENT RENTAL	.0	86.8	86.8	.0	116.3	116.3	.0	116.3	116.3	.0	116.3	116.3	.0	116.3	116.3	.0	86.8	86.8
02	SITE CONSTRUCTION	89.2	86.6	87.3	82.7	111.2	103.6	88.2	111.2	105.1	78.2	111.2	102.4	80.0	111.2	102.9	99.4	86.6	90.0
03100	CONCRETE FORMS & ACCESSORIES	93.4	66.9	70.5	96.0	68.6	72.4	96.0	68.5	72.3	96.0	68.6	72.4	96.0	68.6	72.4	93.4	67.0	70.7
03200	CONCRETE REINFORCEMENT	112.7	56.6	84.2	115.6	68.0	91.4	117.9	57.0	87.0	116.6	68.1	91.9	117.9	68.1	92.6	111.3	56.7	83.5
03300	CAST-IN-PLACE CONCRETE	89.8	64.7	79.5	100.4	74.1	89.6	95.1	74.1	86.5	97.0	74.2	87.6	98.4	74.2	88.4	107.8	64.8	90.1
03	CONCRETE	87.2	64.9	76.1	106.3	71.4	89.0	109.4	69.3	89.5	103.8	71.5	87.7	105.1	71.5	88.4	99.0	65.0	82.1
04	MASONRY	108.7	58.9	77.9	113.3	63.8	82.6	124.4	64.4	87.2	116.8	64.4	84.3	113.2	64.4	83.0	111.1	60.1	79.5
05	METALS	97.7	73.6	90.0	97.0	87.7	94.0	97.8	82.3	92.8	97.8	87.7	94.6	97.2	87.7	94.2	96.9	73.8	89.5
06	WOOD & PLASTICS	87.2	68.7	77.6	95.4	70.1	82.2	95.4	70.1	82.2	95.4	70.1	82.2	95.4	70.1	82.2	87.2	68.7	77.6
07	THERMAL & MOISTURE PROTECTION	84.3	66.6	76.8	99.3	73.3	88.3	99.9	73.4	88.7	99.2	73.4	88.3	99.2	73.4	88.4	84.8	66.9	77.3
08	DOORS & WINDOWS	87.1	68.0	82.1	93.7	72.2	88.1	93.5	68.8	87.1	93.7	72.2	88.1	93.6	72.2	88.0	87.0	68.0	82.0
09200	PLASTER & GYPSUM BOARD	85.1	68.8	75.0	82.7	68.8	74.0	82.7	68.8	74.0	82.7	68.8	74.0	82.7	68.8	74.0	85.1	68.8	75.0
095,098	CEILINGS & ACOUSTICAL TREATMENT	98.5	68.8	80.7	100.5	68.8	81.6	100.5	68.8	81.6	100.5	68.8	81.6	100.5	68.8	81.6	98.5	68.8	80.7
09600	FLOORING	132.4	62.4	113.5	100.3	62.2	90.0	100.3	62.2	90.0	100.3	62.2	90.0	100.3	62.2	90.0	132.4	64.5	114.1
097,099	WALL FINISHES, PAINTS & COATINGS	95.6	58.6	73.0	102.2	58.6	75.6	102.2	58.6	75.6	102.2	58.6	75.6	102.2	58.6	75.6	95.6	58.6	73.0
09	FINISHES	110.8	65.2	87.1	96.4	66.2	80.7	96.9	66.2	80.9	96.3	66.2	80.6	96.3	66.2	80.7	111.3	65.7	87.5
10 - 14	TOTAL DIV. 10000 - 14000	100.0	67.7	93.1	100.0	71.5	93.9	100.0	71.5	93.9	100.0	71.5	93.9	100.0	71.5	93.9	100.0	67.8	93.1
15	MECHANICAL	100.3	71.6	88.4	97.1	72.2	86.7	100.1	72.0	88.4	100.1	72.2	88.5	97.1	72.2	86.7	96.8	71.6	86.4
16	ELECTRICAL	84.7	58.9	72.2	84.0	78.1	81.1	83.1	78.1	80.7	84.0	78.1	81.1	82.2	78.1	80.2	87.3	78.1	82.8
01 - 16	WEIGHTED AVERAGE	95.8	67.9	83.4	96.9	76.2	87.7	98.7	75.3	88.3	97.5	76.2	88.0	96.5	76.2	87.5	96.9	70.8	85.3

	DIVISION	NEW MEXICO			NEW YORK															
		TUCUMCARI			ALBANY			BINGHAMTON			BRONX			BROOKLYN			BUFFALO		
		884			120 - 122			137 - 139			104			112			140 - 142		
		MAT.	INST.	TOTAL	MAT.	INST.	TOTAL	MAT.	INST.	TOTAL	MAT.	INST.	TOTAL	MAT.	INST.	TOTAL	MAT.	INST.	TOTAL
01590	EQUIPMENT RENTAL	.0	116.3	116.3	.0	115.6	115.6	.0	115.8	115.8	.0	117.2	117.2	.0	117.2	117.2	.0	93.1	93.1
02	SITE CONSTRUCTION	85.8	111.2	104.4	72.1	106.6	97.4	94.4	90.2	91.3	126.1	128.3	127.7	119.9	130.9	128.0	96.3	93.7	94.4
03100	CONCRETE FORMS & ACCESSORIES	96.0	68.4	72.1	97.4	93.1	93.7	102.5	80.5	83.5	105.9	169.5	160.7	109.7	168.9	160.7	98.9	115.2	113.0
03200	CONCRETE REINFORCEMENT	115.6	57.0	85.8	103.3	99.1	101.2	102.3	86.9	94.5	105.0	197.8	152.2	103.6	182.5	143.7	104.1	104.6	104.4
03300	CAST-IN-PLACE CONCRETE	95.1	74.0	86.4	87.7	100.8	93.1	104.8	95.5	101.0	105.1	160.6	127.9	106.7	158.6	128.0	115.7	119.1	117.1
03	CONCRETE	107.9	69.2	88.7	100.6	97.7	99.2	100.5	88.8	94.7	106.3	169.8	137.9	112.6	166.1	139.2	106.8	113.7	110.2
04	MASONRY	127.0	64.4	88.2	89.1	97.2	94.2	105.8	88.3	95.0	92.4	158.4	133.3	114.9	158.3	141.8	104.7	121.1	114.9
05	METALS	96.9	82.1	92.2	96.4	108.6	100.3	92.9	115.2	100.0	92.5	145.1	109.3	102.0	139.6	114.0	100.3	94.9	98.6
06	WOOD & PLASTICS	95.4	70.1	82.2	97.0	92.4	94.6	105.9	79.6	92.2	103.5	173.8	140.1	108.7	173.4	142.4	102.4	114.6	108.8
07	THERMAL & MOISTURE PROTECTION	99.7	73.4	88.7	93.8	92.8	93.4	102.5	87.3	96.1	113.9	156.6	131.9	110.3	156.1	129.6	97.7	110.5	103.1
08	DOORS & WINDOWS	93.5	68.8	87.0	95.6	88.7	93.8	90.8	77.6	87.3	92.1	169.8	112.4	89.9	166.2	109.8	92.9	103.7	95.7
09200	PLASTER & GYPSUM BOARD	82.7	68.8	74.0	109.8	91.9	98.7	114.6	78.5	92.2	104.7	175.3	148.6	108.5	175.3	150.0	99.6	114.7	109.0
095,098	CEILINGS & ACOUSTICAL TREATMENT	100.5	68.8	81.6	97.9	91.9	94.3	97.9	78.5	86.4	80.5	175.3	137.1	78.4	175.3	136.2	94.6	114.7	106.6
09600	FLOORING	100.3	62.2	90.0	87.1	96.5	89.7	97.7	91.2	95.9	95.1	173.9	116.4	104.4	173.9	123.2	97.9	119.6	103.8
097,099	WALL FINISHES, PAINTS & COATINGS	102.2	58.6	75.6	86.1	78.4	81.4	91.1	82.4	85.8	96.9	151.4	130.2	114.1	151.4	136.9	95.1	114.3	106.8
09	FINISHES	96.8	66.2	80.9	97.2	92.0	94.5	98.3	82.2	89.9	98.8	169.1	135.4	107.0	168.8	139.2	95.5	116.7	106.6
10 - 14	TOTAL DIV. 10000 - 14000	100.0	71.5	93.9	100.0	95.7	99.1	100.0	92.8	98.5	100.0	140.4	108.6	100.0	139.3	108.4	100.0	107.3	101.6
15	MECHANICAL	97.2	71.7	86.6	100.3	91.8	96.8	100.5	84.5	93.9	100.2	164.6	126.9	99.8	164.3	126.6	99.8	99.6	99.7
16	ELECTRICAL	84.0	78.1	81.1	103.6	94.1	99.0	101.6	87.4	94.7	102.4	166.1	133.4	102.3	167.9	134.2	100.2	99.2	99.7
01 - 16	WEIGHTED AVERAGE	97.9	75.2	87.8	97.8	96.4	97.2	98.5	88.8	94.2	99.8	159.9	126.5	103.3	159.1	128.1	99.8	105.7	102.4

	DIVISION	NEW YORK																	
		ELMIRA			FAR ROCKAWAY			FLUSHING			GLENS FALLS			HICKSVILLE			JAMAICA		
		148 - 149			116			113			128			115,117,118			114		
		MAT.	INST.	TOTAL	MAT.	INST.	TOTAL	MAT.	INST.	TOTAL	MAT.	INST.	TOTAL	MAT.	INST.	TOTAL	MAT.	INST.	TOTAL
01590	EQUIPMENT RENTAL	.0	116.0	116.0	.0	117.2	117.2	.0	117.2	117.2	.0	115.6	115.6	.0	117.2	117.2	.0	117.2	117.2
02	SITE CONSTRUCTION	91.5	89.0	89.7	123.9	130.9	129.0	123.9	130.9	129.0	62.2	106.2	94.4	112.7	131.1	126.2	117.7	130.9	127.4
03100	CONCRETE FORMS & ACCESSORIES	81.1	74.2	75.1	93.5	168.9	158.5	98.0	168.9	159.1	82.8	87.6	86.9	89.3	154.4	145.5	98.0	168.8	159.1
03200	CONCRETE REINFORCEMENT	105.4	84.7	94.9	103.6	182.5	143.7	105.5	182.5	144.6	103.0	94.3	98.6	103.6	182.3	143.6	103.6	182.5	143.7
03300	CAST-IN-PLACE CONCRETE	108.1	79.0	96.2	115.5	158.6	133.2	115.5	158.6	133.2	79.6	93.7	85.4	98.0	159.1	123.1	106.7	158.5	128.0
03	CONCRETE	100.4	80.3	90.4	118.8	166.1	142.3	119.4	166.1	142.6	90.1	91.9	91.0	103.9	159.8	131.7	111.8	166.0	138.7
04	MASONRY	101.2	71.5	82.8	119.2	158.3	143.5	113.1	158.3	141.1	91.2	87.7	89.0	109.0	160.5	140.9	117.1	158.3	142.7
05	METALS	95.8	117.5	102.7	102.1	139.5	114.0	102.1	139.5	114.0	93.0	106.1	97.2	103.1	138.8	114.5	102.1	139.5	114.0
06	WOOD & PLASTICS	84.5	72.2	78.1	91.2	173.4	134.0	96.4	173.4	136.5	81.0	88.6	84.9	87.2	154.5	122.3	96.4	173.4	136.5
07	THERMAL & MOISTURE PROTECTION	98.1	79.2	90.2	110.1	156.1	129.5	110.2	156.1	129.5	92.8	88.4	91.0	109.7	149.3	126.4	110.0	154.8	128.9
08	DOORS & WINDOWS	93.6	73.4	88.3	88.3	166.2	108.7	88.3	166.2	108.7	89.4	85.6	88.4	88.3	156.0	106.0	88.3	166.2	108.7
09200	PLASTER & GYPSUM BOARD	94.1	71.2	79.9	102.2	175.3	147.6	104.4	175.3	148.5	99.7	88.0	92.4	101.4	156.0	135.3	104.4	175.3	148.5
095,098	CEILINGS & ACOUSTICAL TREATMENT	91.2	71.2	79.3	78.4	175.3	136.2	78.4	175.3	136.2	91.2	88.0	89.3	77.5	156.0	124.3	78.4	175.3	136.2
09600	FLOORING	87.5	74.1	83.9	98.6	173.9	119.0	100.3	173.9	120.2	78.7	88.3	81.3	97.3	86.2	94.3	100.3	173.9	120.2
097,099	WALL FINISHES, PAINTS & COATINGS	93.3	80.7	85.6	114.1	151.4	136.9	114.1	151.4	136.9	86.1	78.4	81.4	114.1	151.4	136.9	114.1	151.4	136.9
09	FINISHES	90.5	74.0	81.9	104.7	168.8	138.1	105.5	168.8	138.5	90.7	86.7	88.6	103.2	140.1	122.4	105.1	168.8	138.3
10 - 14	TOTAL DIV. 10000 - 14000	100.0	91.0	98.1	100.0	139.3	108.4	100.0	139.3	108.4	100.0	64.9	92.5	100.0	137.8	108.1	100.0	139.3	108.4
15	MECHANICAL	95.0	83.9	90.4	94.9	164.3	123.7	94.9	164.3	123.7	95.5	86.0	91.6	99.8	152.0	121.5	94.9	164.3	123.7
16	ELECTRICAL	98.9	95.1	97.0	109.3	166.1	136.9	109.3	166.1	136.9	96.8	94.1	95.5	102.0	155.1	127.8	101.0	166.1	132.7
01 - 16	WEIGHTED AVERAGE	96.3	85.4	91.4	103.4	158.8	128.0	103.3	158.8	128.0	92.7	91.3	92.1	101.2	149.5	122.7	101.5	158.8	126.9

	DIVISION	NEW YORK																	
		JAMESTOWN			KINGSTON			LONG ISLAND CITY			MONTICELLO			MOUNT VERNON			NEW ROCHELLE		
		147			124			111			127			105			108		
		MAT.	INST.	TOTAL	MAT.	INST.	TOTAL	MAT.	INST.	TOTAL	MAT.	INST.	TOTAL	MAT.	INST.	TOTAL	MAT.	INST.	TOTAL
01590	EQUIPMENT RENTAL	.0	89.5	89.5	.0	117.2	117.2	.0	117.2	117.2	.0	117.2	117.2	.0	117.2	117.2	.0	117.2	117.2
02	SITE CONSTRUCTION	92.8	89.4	90.3	119.5	128.2	125.9	121.5	130.9	128.4	114.7	128.2	124.6	135.3	125.3	127.9	133.8	125.2	127.5
03100	CONCRETE FORMS & ACCESSORIES	81.2	85.0	84.5	84.5	112.0	108.2	103.2	168.8	159.8	92.9	112.1	109.5	93.9	139.7	133.4	112.0	139.5	135.7
03200	CONCRETE REINFORCEMENT	105.6	90.3	97.8	103.4	132.2	118.0	103.6	182.5	143.7	102.6	132.2	117.7	103.8	181.3	143.2	103.9	181.3	143.2
03300	CAST-IN-PLACE CONCRETE	112.2	110.5	111.5	107.5	122.9	113.9	110.2	158.5	130.0	100.7	123.0	109.8	117.6	132.3	123.6	117.6	132.2	123.6
03	CONCRETE	103.4	94.5	99.0	112.6	119.2	115.9	115.2	166.0	140.5	107.3	119.2	113.2	116.6	143.8	130.1	116.4	143.7	129.9
04	MASONRY	109.3	94.6	100.2	107.0	125.6	118.5	111.7	158.3	140.6	99.4	125.6	115.6	98.4	132.0	119.2	98.4	132.0	119.2
05	METALS	93.1	85.2	90.6	102.0	113.6	105.7	102.0	139.5	114.0	102.0	113.8	105.8	92.3	134.8	105.9	92.5	134.7	106.0
06	WOOD & PLASTICS	83.3	78.5	80.8	82.3	106.8	95.1	102.2	173.4	139.3	91.5	106.8	99.5	90.9	140.3	116.6	111.0	140.3	126.3
07	THERMAL & MOISTURE PROTECTION	97.5	92.9	95.6	112.1	133.4	121.1	110.2	154.8	129.0	111.8	133.4	120.9	115.0	140.0	125.5	115.1	140.0	125.6
08	DOORS & WINDOWS	93.4	80.6	90.1	93.4	119.2	100.2	88.3	166.2	108.7	88.4	119.2	96.4	92.1	148.2	106.8	92.2	148.2	106.8
09200	PLASTER & GYPSUM BOARD	89.5	77.7	82.2	97.3	107.0	103.3	107.0	175.3	149.5	100.6	107.0	104.6	98.9	140.9	125.0	110.3	140.9	129.3
095,098	CEILINGS & ACOUSTICAL TREATMENT	86.9	77.7	81.4	72.5	107.0	93.1	78.4	175.3	136.2	72.5	107.0	93.1	78.0	140.9	115.5	78.0	140.9	115.5
09600	FLOORING	90.4	88.7	89.9	91.1	90.5	90.9	102.0	173.9	121.5	94.2	90.5	93.2	87.2	168.7	109.2	94.7	168.7	114.7
097,099	WALL FINISHES, PAINTS & COATINGS	95.1	83.0	87.7	111.8	124.4	119.5	114.1	151.4	136.9	111.8	105.3	107.8	94.5	175.0	143.6	94.5	141.5	123.2
09	FINISHES	89.7	84.5	87.0	99.8	107.6	103.9	106.2	168.8	138.8	100.9	105.4	103.3	95.6	149.1	123.5	99.3	145.3	123.3
10 - 14	TOTAL DIV. 10000 - 14000	100.0	100.9	100.2	100.0	121.0	104.5	100.0	139.3	108.4	100.0	121.0	104.5	100.0	133.2	107.1	100.0	133.2	107.1
15	MECHANICAL	94.9	80.6	88.9	95.5	106.7	100.1	99.8	164.3	126.6	95.5	120.4	105.8	95.5	131.7	110.5	95.5	131.7	110.5
16	ELECTRICAL	97.7	90.6	94.3	98.5	125.7	111.7	101.5	166.1	132.9	98.5	125.7	111.7	99.0	147.1	122.4	99.0	147.1	122.4
01 - 16	WEIGHTED AVERAGE	96.3	87.9	92.6	100.9	117.3	108.2	103.1	158.8	127.8	99.5	119.9	108.5	99.5	138.5	116.8	100.0	138.0	116.9

City Cost Indexes

DIVISION		NEW YORK																	
		NEW YORK			NIAGARA FALLS			PLATTSBURGH			POUGHKEEPSIE			QUEENS			RIVERHEAD		
		100 - 102			143			129			125 - 126			110			119		
		MAT.	INST.	TOTAL	MAT.	INST.	TOTAL	MAT.	INST.	TOTAL	MAT.	INST.	TOTAL	MAT.	INST.	TOTAL	MAT.	INST.	TOTAL
01590	EQUIPMENT RENTAL	.0	117.5	117.5	.0	89.5	89.5	.0	97.8	97.8	.0	117.2	117.2	.0	117.2	117.2	.0	117.2	117.2
02	SITE CONSTRUCTION	136.2	128.7	130.7	95.1	90.7	91.9	86.5	102.9	98.5	115.8	128.3	125.0	116.0	130.9	126.9	113.4	131.1	126.3
03100	CONCRETE FORMS & ACCESSORIES	111.1	181.7	172.0	81.1	115.8	111.0	87.9	84.4	84.9	84.5	119.2	114.5	89.6	168.8	157.9	94.9	154.4	146.2
03200	CONCRETE REINFORCEMENT	111.1	180.5	146.4	104.1	98.5	101.3	107.5	98.6	103.0	103.4	132.4	118.1	105.5	182.5	144.6	105.6	182.3	144.6
03300	CAST-IN-PLACE CONCRETE	118.4	160.9	135.8	116.1	115.4	115.8	97.4	94.2	96.1	104.1	122.4	111.7	101.4	158.5	124.9	99.7	159.1	124.1
03	CONCRETE	117.7	172.4	144.9	105.9	111.4	108.6	104.0	90.0	97.1	109.6	122.3	115.9	107.3	166.0	136.5	104.8	159.8	132.1
04	MASONRY	102.8	158.4	137.3	117.7	120.2	119.3	87.8	87.9	87.9	100.3	123.8	114.9	106.1	158.3	138.5	115.1	160.5	143.2
05	METALS	104.4	145.4	117.5	95.9	89.6	93.8	97.8	84.9	93.7	102.1	115.2	106.3	102.0	139.5	114.0	103.5	138.8	114.8
06	WOOD & PLASTICS	108.6	189.9	150.9	83.2	113.9	99.2	89.1	81.9	85.4	82.3	116.3	100.0	87.4	173.4	132.2	93.1	154.5	125.1
07	THERMAL & MOISTURE PROTECTION	114.3	161.4	134.1	97.6	109.3	102.5	101.3	86.5	95.1	112.0	133.6	121.1	109.8	154.8	128.7	110.3	149.3	126.7
08	DOORS & WINDOWS	97.7	178.7	118.8	93.5	101.4	95.5	98.5	83.0	94.5	93.4	124.3	101.5	88.3	166.2	108.7	88.3	156.0	106.0
09200	PLASTER & GYPSUM BOARD	113.9	191.9	162.4	89.5	114.1	104.7	119.1	80.6	95.2	97.3	116.7	109.4	101.4	175.3	147.4	104.2	156.0	136.4
095,098	CEILINGS & ACOUSTICAL TREATMENT	107.8	191.9	158.0	86.9	114.1	103.1	97.9	80.6	87.5	72.5	116.7	98.9	78.4	175.3	136.2	82.6	156.0	126.4
09600	FLOORING	96.6	173.9	117.5	90.2	112.6	96.2	101.1	96.5	99.8	91.1	125.2	100.3	97.3	173.9	118.0	98.6	86.2	95.2
097,099	WALL FINISHES, PAINTS & COATINGS	96.9	151.6	130.3	95.1	107.7	102.8	105.3	74.7	86.6	111.8	105.3	107.8	114.1	151.4	136.9	114.1	151.4	136.9
09	FINISHES	108.0	178.6	144.8	89.8	115.1	103.0	99.8	85.3	92.2	99.7	117.9	109.1	103.6	168.8	137.5	105.2	140.1	123.4
10 - 14	TOTAL DIV. 10000 - 14000	100.0	142.4	109.1	100.0	108.6	101.8	100.0	60.8	91.6	100.0	121.5	104.6	100.0	139.3	108.4	100.0	137.8	108.1
15	MECHANICAL	100.2	164.6	126.9	94.9	102.0	97.8	95.4	88.3	92.5	95.5	106.9	100.2	99.8	164.3	126.6	99.9	152.0	121.5
16	ELECTRICAL	109.9	181.3	144.6	96.4	98.9	97.6	91.3	91.3	91.3	98.5	125.7	111.7	102.0	166.1	133.2	103.5	155.1	128.6
01 - 16	WEIGHTED AVERAGE	106.0	164.4	131.9	97.3	104.7	100.6	96.8	88.4	93.1	100.1	119.4	108.7	101.4	158.8	126.9	102.2	149.5	123.2

DIVISION		NEW YORK																	
		ROCHESTER			SCHENECTADY			STATEN ISLAND			SUFFERN			SYRACUSE			UTICA		
		144 - 146			123			103			109			130 - 132			133 - 135		
		MAT.	INST.	TOTAL	MAT.	INST.	TOTAL	MAT.	INST.	TOTAL	MAT.	INST.	TOTAL	MAT.	INST.	TOTAL	MAT.	INST.	TOTAL
01590	EQUIPMENT RENTAL	.0	116.5	116.5	.0	115.6	115.6	.0	117.2	117.2	.0	117.2	117.2	.0	115.6	115.6	.0	115.6	115.6
02	SITE CONSTRUCTION	74.3	107.4	98.6	71.8	106.6	97.3	140.6	128.3	131.6	130.2	125.3	126.6	93.2	107.0	103.3	69.9	105.4	96.0
03100	CONCRETE FORMS & ACCESSORIES	99.0	102.7	102.2	102.1	92.6	93.9	93.2	169.4	159.0	104.5	133.8	129.8	101.3	89.9	91.4	102.6	84.0	86.6
03200	CONCRETE REINFORCEMENT	105.1	85.9	95.3	102.0	99.1	100.5	105.0	197.8	152.2	103.9	132.7	118.5	103.3	96.1	99.6	103.3	84.5	93.8
03300	CAST-IN-PLACE CONCRETE	110.1	101.6	106.6	96.2	100.1	97.8	117.6	160.6	135.2	113.9	128.2	119.8	97.2	97.2	97.2	89.0	93.3	90.7
03	CONCRETE	112.1	100.0	106.1	104.8	97.2	101.0	118.4	169.8	143.9	113.0	130.7	121.8	103.6	94.4	99.1	101.5	88.4	95.0
04	MASONRY	102.7	100.2	101.2	90.0	95.9	93.7	105.7	158.4	138.4	98.2	126.6	115.8	96.4	99.3	98.2	88.7	91.2	90.2
05	METALS	96.7	107.1	100.0	96.4	108.6	100.3	90.5	145.0	107.9	90.5	114.9	98.3	96.3	106.6	99.6	94.4	102.1	96.9
06	WOOD & PLASTICS	99.3	103.7	101.6	102.5	92.4	97.2	89.4	173.8	133.4	102.8	138.2	121.2	102.5	87.5	94.7	102.5	83.6	92.7
07	THERMAL & MOISTURE PROTECTION	94.6	98.4	96.2	93.9	92.2	93.2	114.2	155.4	131.6	115.0	135.5	123.6	102.2	96.1	99.6	93.8	92.4	93.2
08	DOORS & WINDOWS	96.0	92.6	95.1	95.6	88.7	93.8	92.1	169.8	112.4	92.2	136.2	103.7	93.5	84.2	91.1	95.6	78.9	91.2
09200	PLASTER & GYPSUM BOARD	101.5	103.8	102.9	109.8	91.9	98.7	99.1	175.3	146.5	103.7	138.8	125.5	109.8	87.0	95.6	109.8	82.9	93.1
095,098	CEILINGS & ACOUSTICAL TREATMENT	99.0	103.8	101.9	97.9	91.9	94.3	80.5	175.3	137.1	78.0	138.8	114.3	97.9	87.0	91.4	97.9	82.9	89.0
09600	FLOORING	87.4	104.8	92.1	87.1	96.5	89.7	91.0	173.9	113.4	90.7	60.0	82.4	88.6	92.4	89.6	87.1	87.1	87.1
097,099	WALL FINISHES, PAINTS & COATINGS	93.0	103.4	99.4	86.1	78.4	81.4	96.9	151.4	130.2	94.5	109.7	103.8	91.6	86.2	88.3	86.1	84.0	84.8
09	FINISHES	96.7	103.8	100.4	96.9	91.7	94.2	97.8	169.1	134.9	96.9	116.8	107.3	98.2	89.6	93.7	96.9	84.3	90.4
10 - 14	TOTAL DIV. 10000 - 14000	100.0	101.1	100.2	100.0	95.2	99.0	100.0	140.4	108.6	100.0	130.8	106.6	100.0	96.3	99.2	100.0	92.7	98.4
15	MECHANICAL	99.8	94.7	97.7	100.3	91.1	96.5	100.2	164.6	126.9	95.5	123.0	106.9	100.3	89.0	95.6	100.3	87.2	94.9
16	ELECTRICAL	103.9	94.9	99.5	101.1	94.1	97.7	102.4	166.1	133.4	106.3	135.6	120.6	101.1	92.2	96.8	97.7	91.8	94.8
01 - 16	WEIGHTED AVERAGE	99.8	99.7	99.8	98.1	95.9	97.1	101.6	159.9	127.5	99.8	126.2	111.5	99.0	94.7	97.1	96.9	90.9	94.2

DIVISION		NEW YORK									NORTH CAROLINA								
		WATERTOWN			WHITE PLAINS			YONKERS			ASHEVILLE			CHARLOTTE			DURHAM		
		136			106			107			287 - 288			281 - 282			277		
		MAT.	INST.	TOTAL	MAT.	INST.	TOTAL	MAT.	INST.	TOTAL	MAT.	INST.	TOTAL	MAT.	INST.	TOTAL	MAT.	INST.	TOTAL
01590	EQUIPMENT RENTAL	.0	115.6	115.6	.0	117.2	117.2	.0	117.2	117.2	.0	93.3	93.3	.0	93.3	93.3	.0	99.6	99.6
02	SITE CONSTRUCTION	77.9	107.4	99.6	125.6	125.2	125.4	135.5	125.1	127.8	102.7	72.2	80.3	102.9	72.2	80.4	102.7	82.0	87.5
03100	CONCRETE FORMS & ACCESSORIES	84.7	90.5	89.7	110.7	139.5	135.6	110.9	139.6	135.7	95.1	42.7	49.9	102.7	42.8	51.0	97.0	43.0	50.4
03200	CONCRETE REINFORCEMENT	104.0	85.1	94.4	103.9	181.3	143.2	108.0	181.3	145.2	93.8	46.9	70.0	94.2	47.0	70.2	94.2	55.0	74.3
03300	CAST-IN-PLACE CONCRETE	103.6	83.1	95.2	104.3	132.3	115.8	116.8	132.3	123.1	96.5	49.3	77.1	99.0	49.3	78.5	96.8	49.4	77.3
03	CONCRETE	114.3	87.9	101.2	105.9	143.7	124.7	116.5	143.8	130.1	100.3	47.6	74.1	101.3	47.7	74.7	99.8	49.3	74.8
04	MASONRY	89.8	91.2	90.7	97.5	132.0	118.9	102.4	132.0	120.7	80.5	40.1	55.5	85.9	40.1	57.5	82.8	40.1	56.3
05	METALS	94.5	101.6	96.7	91.6	134.7	105.4	100.5	134.8	111.4	92.4	76.6	87.4	94.9	76.7	89.1	97.5	80.3	92.0
06	WOOD & PLASTICS	83.0	92.4	87.9	109.5	140.3	125.5	109.3	140.3	125.4	94.7	43.2	67.9	103.9	43.2	72.3	97.0	43.2	69.0
07	THERMAL & MOISTURE PROTECTION	94.0	92.1	93.2	114.8	140.0	125.4	115.1	140.0	125.6	101.3	45.7	77.9	101.3	47.3	78.6	101.8	46.2	78.4
08	DOORS & WINDOWS	95.6	81.0	91.8	92.2	148.2	106.8	95.8	148.3	109.5	92.6	43.4	79.7	96.6	43.4	82.7	96.6	46.1	83.4
09200	PLASTER & GYPSUM BOARD	102.0	92.0	95.8	106.6	140.9	127.9	113.5	140.9	130.5	103.4	41.5	64.9	109.6	41.5	67.3	109.6	41.5	67.3
095,098	CEILINGS & ACOUSTICAL TREATMENT	97.9	92.0	94.4	78.0	140.9	115.5	106.1	140.9	126.9	94.1	41.5	62.7	100.9	41.5	65.5	100.9	41.5	65.5
09600	FLOORING	79.3	87.1	81.4	92.8	168.7	113.3	92.4	168.7	113.1	101.8	48.1	87.3	105.8	48.1	90.2	106.0	48.1	90.3
097,099	WALL FINISHES, PAINTS & COATINGS	86.1	70.8	76.8	94.5	151.4	129.2	94.5	175.0	143.6	110.4	37.7	66.1	110.4	37.7	66.1	110.4	37.7	66.1
09	FINISHES	94.0	87.6	90.7	97.6	146.4	123.0	106.1	149.1	128.5	97.7	43.1	69.3	101.5	43.1	71.1	101.6	43.1	71.2
10 - 14	TOTAL DIV. 10000 - 14000	100.0	78.5	95.4	100.0	133.2	107.1	100.0	133.2	107.1	100.0	59.0	91.3	100.0	59.0	91.3	100.0	74.5	94.6
15	MECHANICAL	100.3	74.1	89.5	100.5	131.7	113.4	100.5	131.7	113.4	100.1	42.6	76.2	100.1	42.6	76.2	100.1	42.8	76.3
16	ELECTRICAL	101.1	91.4	96.4	99.0	147.1	122.4	106.4	147.1	126.2	97.8	39.4	69.4	99.6	38.5	69.9	97.3	37.7	68.3
01 - 16	WEIGHTED AVERAGE	98.5	88.5	94.1	99.5	138.2	116.7	104.5	138.5	119.6	96.8	48.9	75.5	98.6	48.9	76.5	98.4	50.7	77.2

DIVISION		NORTH CAROLINA																	
		ELIZABETH CITY			FAYETTEVILLE			GASTONIA			GREENSBORO			HICKORY			KINSTON		
		279			283			280			270,272 - 274			286			285		
		MAT.	INST.	TOTAL	MAT.	INST.	TOTAL	MAT.	INST.	TOTAL	MAT.	INST.	TOTAL	MAT.	INST.	TOTAL	MAT.	INST.	TOTAL
01590	EQUIPMENT RENTAL	.0	104.7	104.7	.0	99.6	99.6	.0	93.3	93.3	.0	99.6	99.6	.0	99.6	99.6	.0	99.6	99.6
02	SITE CONSTRUCTION	107.3	82.8	89.3	100.7	82.0	87.0	102.3	72.2	80.2	102.6	82.0	87.5	101.7	80.8	86.4	100.3	81.5	86.5
03100	CONCRETE FORMS & ACCESSORIES	81.0	25.1	32.8	92.7	43.1	49.9	102.9	42.8	51.1	97.1	42.9	50.3	89.9	24.1	33.2	85.7	24.4	32.8
03200	CONCRETE REINFORCEMENT	93.3	45.6	69.0	93.3	55.0	73.8	94.2	47.0	70.2	94.2	55.0	74.3	93.8	20.5	56.5	93.3	20.6	56.3
03300	CAST-IN-PLACE CONCRETE	96.8	33.0	70.6	93.4	49.4	75.3	94.3	49.3	75.8	95.9	49.3	76.8	96.5	35.1	71.2	93.1	36.2	69.7
03	CONCRETE	100.0	33.9	67.2	97.4	49.4	73.6	99.1	47.7	73.6	99.4	49.2	74.5	99.9	29.6	65.0	96.8	30.2	63.7
04	MASONRY	96.5	25.9	52.7	86.6	40.1	57.8	83.9	40.1	56.8	82.6	40.1	56.3	70.3	25.7	42.7	76.4	26.1	45.2
05	METALS	92.0	74.2	86.3	97.4	80.3	91.9	92.9	76.7	87.7	98.0	80.1	92.3	92.6	65.8	84.0	91.4	67.4	83.8
06	WOOD & PLASTICS	79.2	25.3	51.1	91.8	43.2	66.5	103.9	43.2	72.3	97.0	43.2	69.0	88.5	24.5	55.2	84.2	24.5	53.1
07	THERMAL & MOISTURE PROTECTION	101.2	26.0	69.5	101.5	46.9	78.5	101.5	47.3	78.7	101.8	46.2	78.4	101.5	28.6	70.8	101.4	29.1	71.0
08	DOORS & WINDOWS	93.5	27.9	76.4	92.7	46.1	80.5	96.6	43.4	82.7	96.6	46.1	83.4	92.6	24.0	74.7	92.7	24.4	74.9
09200	PLASTER & GYPSUM BOARD	100.3	22.3	51.8	103.6	41.5	65.0	109.6	41.5	67.3	109.6	41.5	67.3	103.4	22.3	53.0	101.9	22.3	52.4
095,098	CEILINGS & ACOUSTICAL TREATMENT	100.9	22.3	54.0	94.9	41.5	63.1	100.9	41.5	65.5	100.9	41.5	65.5	94.1	22.3	51.2	98.3	22.3	53.0
09600	FLOORING	95.1	15.5	73.5	101.9	48.1	87.4	106.0	48.1	90.3	106.0	48.1	90.3	101.7	14.1	78.0	98.3	21.2	77.4
097,099	WALL FINISHES, PAINTS & COATINGS	110.4	12.8	50.9	110.4	37.7	66.1	110.4	37.7	66.1	110.4	37.7	66.1	110.4	15.1	52.3	110.4	16.6	53.2
09	FINISHES	97.2	21.3	57.7	98.0	43.1	69.4	101.5	43.1	71.1	101.6	43.1	71.2	97.8	20.8	57.7	97.5	22.6	58.5
10 - 14	TOTAL DIV. 10000 - 14000	100.0	72.0	94.0	100.0	74.5	94.6	100.0	59.0	91.3	100.0	59.0	91.3	100.0	53.9	90.2	100.0	69.4	93.5
15	MECHANICAL	95.1	26.9	66.8	100.1	42.8	76.3	100.1	42.6	76.2	100.1	42.7	76.3	95.1	18.5	63.3	95.1	19.2	63.7
16	ELECTRICAL	98.2	29.1	64.6	94.7	37.7	67.0	97.3	38.5	68.7	97.9	37.7	68.6	95.5	29.1	63.2	95.3	29.1	63.1
01 - 16	WEIGHTED AVERAGE	96.4	37.8	70.4	97.1	50.8	76.5	97.7	48.9	76.0	98.5	50.3	77.1	94.8	33.9	67.8	94.4	35.1	68.1

DIVISION		NORTH CAROLINA															NORTH DAKOTA		
		MURPHY			RALEIGH			ROCKY MOUNT			WILMINGTON			WINSTON-SALEM			BISMARCK		
		289			275 - 276			278			284			271			585		
		MAT.	INST.	TOTAL	MAT.	INST.	TOTAL	MAT.	INST.	TOTAL	MAT.	INST.	TOTAL	MAT.	INST.	TOTAL	MAT.	INST.	TOTAL
01590	EQUIPMENT RENTAL	.0	93.3	93.3	.0	99.6	99.6	.0	99.6	99.6	.0	93.3	93.3	.0	99.6	99.6	.0	98.3	98.3
02	SITE CONSTRUCTION	104.0	70.9	79.7	103.7	82.0	87.8	105.5	80.8	87.4	103.6	72.2	80.6	102.9	82.0	87.6	90.5	96.5	94.9
03100	CONCRETE FORMS & ACCESSORIES	103.6	24.1	35.1	99.9	43.0	50.8	88.6	29.0	37.2	96.6	43.1	50.4	98.5	42.8	50.4	95.1	49.8	56.0
03200	CONCRETE REINFORCEMENT	93.3	20.4	56.3	94.2	55.0	74.3	93.3	20.5	56.3	94.5	55.0	74.4	94.2	47.0	70.2	101.2	78.4	89.7
03300	CAST-IN-PLACE CONCRETE	99.8	31.3	71.6	102.4	49.4	80.6	94.7	30.5	68.3	96.1	49.4	76.9	99.0	49.3	78.5	105.1	55.2	84.6
03	CONCRETE	103.4	28.3	66.1	102.8	49.3	76.2	101.1	30.2	65.9	100.2	49.4	75.0	101.0	47.7	74.5	99.6	58.2	79.1
04	MASONRY	72.5	23.4	42.1	87.4	40.1	58.1	76.6	27.7	46.3	70.4	40.1	51.6	82.8	40.1	56.3	104.5	65.1	80.1
05	METALS	90.5	65.4	82.5	95.3	80.3	90.5	91.4	65.8	83.2	91.9	80.3	88.2	95.3	76.7	89.3	92.6	83.0	89.5
06	WOOD & PLASTICS	104.7	24.5	62.9	100.3	43.2	70.6	87.2	29.9	57.4	96.5	43.2	68.8	97.0	43.2	69.0	88.3	44.7	65.6
07	THERMAL & MOISTURE PROTECTION	101.5	30.5	71.6	101.6	46.3	78.3	101.7	27.8	70.6	101.3	46.9	78.4	101.8	46.2	78.4	99.2	55.6	80.8
08	DOORS & WINDOWS	92.6	27.7	75.6	93.4	46.1	81.1	92.7	26.2	75.3	92.7	46.1	80.5	96.6	43.9	82.8	99.4	50.4	86.6
09200	PLASTER & GYPSUM BOARD	107.8	22.3	54.7	109.6	41.5	67.3	101.8	27.8	55.8	104.3	41.5	65.3	109.6	41.5	67.3	113.5	43.6	70.0
095,098	CEILINGS & ACOUSTICAL TREATMENT	94.1	22.3	51.2	100.9	41.5	65.5	94.9	27.8	54.9	94.9	41.5	63.1	100.9	41.5	65.5	138.5	43.6	81.9
09600	FLOORING	106.5	22.6	83.8	106.0	48.1	90.3	99.8	15.0	76.9	102.8	48.1	88.0	106.0	48.1	90.3	107.8	75.0	98.9
097,099	WALL FINISHES, PAINTS & COATINGS	110.4	17.2	53.5	110.4	37.7	66.1	110.4	28.7	60.5	110.4	37.7	66.1	110.4	37.7	66.1	97.2	37.5	60.8
09	FINISHES	99.9	22.5	59.6	101.7	43.1	71.2	97.5	26.6	60.5	98.3	43.1	69.6	101.6	43.1	71.2	114.7	52.0	82.0
10 - 14	TOTAL DIV. 10000 - 14000	100.0	53.9	90.2	100.0	74.5	94.6	100.0	70.8	93.8	100.0	74.5	94.6	100.0	59.0	91.3	100.0	69.1	93.4
15	MECHANICAL	95.1	18.6	63.4	100.1	42.8	76.3	95.1	18.6	63.4	100.1	42.8	76.3	100.1	42.7	76.2	100.5	65.3	85.9
16	ELECTRICAL	98.6	29.1	64.8	97.9	37.7	68.6	100.1	29.1	65.6	98.3	37.7	68.8	97.9	37.7	68.6	93.6	69.7	82.0
01 - 16	WEIGHTED AVERAGE	95.7	33.0	67.9	98.4	50.7	77.2	95.7	35.5	69.0	96.4	49.9	75.7	98.2	49.6	76.6	99.3	66.6	84.8

DIVISION		NORTH DAKOTA																	
		DEVILS LAKE			DICKINSON			FARGO			GRAND FORKS			JAMESTOWN			MINOT		
		583			586			580 - 581			582			584			587		
		MAT.	INST.	TOTAL	MAT.	INST.	TOTAL	MAT.	INST.	TOTAL	MAT.	INST.	TOTAL	MAT.	INST.	TOTAL	MAT.	INST.	TOTAL
01590	EQUIPMENT RENTAL	.0	98.3	98.3	.0	98.3	98.3	.0	98.3	98.3	.0	98.3	98.3	.0	98.3	98.3	.0	98.3	98.3
02	SITE CONSTRUCTION	94.9	95.0	95.0	102.8	94.0	96.4	89.9	96.5	94.8	98.7	94.0	95.3	94.0	94.0	94.0	96.2	96.5	96.4
03100	CONCRETE FORMS & ACCESSORIES	99.4	44.0	51.6	87.2	43.3	49.3	96.1	50.1	56.4	91.8	43.4	50.1	89.1	43.1	49.4	86.9	56.4	60.6
03200	CONCRETE REINFORCEMENT	101.2	78.7	89.8	102.2	51.3	76.4	93.3	77.6	85.3	99.7	78.6	89.0	101.8	61.7	81.5	103.1	78.5	90.6
03300	CAST-IN-PLACE CONCRETE	119.0	53.9	92.2	107.6	52.6	85.0	110.2	57.3	88.4	107.6	52.6	85.0	117.5	52.5	90.8	107.6	53.8	85.5
03	CONCRETE	109.3	55.2	82.5	108.4	49.2	79.0	110.0	58.9	84.7	105.5	54.3	80.1	107.7	51.0	79.6	104.2	60.7	82.6
04	MASONRY	112.2	66.9	84.1	115.5	59.0	80.5	106.9	48.7	70.8	106.1	66.6	81.6	125.9	42.6	74.3	107.1	66.4	81.9
05	METALS	92.6	82.7	89.4	92.5	68.4	84.8	95.0	81.6	90.7	92.6	79.0	88.2	92.6	71.1	85.7	92.9	83.1	89.8
06	WOOD & PLASTICS	92.7	41.1	65.9	80.1	41.1	59.8	88.3	45.4	66.0	84.7	41.1	62.0	82.2	41.1	60.8	79.8	53.7	66.2
07	THERMAL & MOISTURE PROTECTION	99.6	56.0	81.3	100.1	53.1	80.3	99.8	52.6	79.9	99.8	56.0	81.4	99.4	49.8	78.5	99.5	56.7	81.5
08	DOORS & WINDOWS	99.4	45.5	85.3	99.4	39.1	83.6	99.4	50.7	86.7	99.4	45.5	85.3	99.4	41.8	84.3	99.5	55.3	88.0
09200	PLASTER & GYPSUM BOARD	118.3	39.9	69.6	110.1	39.9	66.5	113.5	44.3	70.5	111.6	39.9	67.0	111.2	39.9	66.9	110.1	52.8	74.5
095,098	CEILINGS & ACOUSTICAL TREATMENT	138.5	39.9	79.7	138.5	39.9	79.7	138.5	44.3	82.3	138.5	39.9	79.7	138.5	39.9	79.7	138.5	52.8	87.4
09600	FLOORING	112.2	44.8	94.0	103.4	44.8	87.6	107.6	44.8	90.6	106.0	44.8	89.4	104.7	44.8	88.5	103.2	77.5	96.3
097,099	WALL FINISHES, PAINTS & COATINGS	97.2	28.7	55.4	97.2	28.7	55.4	97.2	71.0	81.2	97.2	28.7	55.4	97.2	28.7	55.4	97.2	31.4	57.1
09	FINISHES	117.0	41.2	77.5	113.8	41.2	76.0	114.5	49.9	80.9	114.4	41.2	76.3	113.6	41.2	75.9	113.2	57.0	84.0
10 - 14	TOTAL DIV. 10000 - 14000	100.0	44.9	88.2	100.0	45.0	88.3	100.0	69.2	93.4	100.0	45.0	88.3	100.0	66.4	92.8	100.0	70.3	93.7
15	MECHANICAL	95.6	66.0	83.3	95.6	63.2	82.1	100.5	69.3	87.6	100.5	45.2	77.5	95.6	45.0	74.6	100.5	63.4	85.1
16	ELECTRICAL	93.0	46.6	70.4	101.0	72.5	87.2	93.5	68.7	81.4	96.2	69.7	83.3	93.0	46.5	70.4	99.2	70.9	85.4
01 - 16	WEIGHTED AVERAGE	99.9	60.8	82.6	100.7	60.5	82.9	100.9	65.2	85.1	100.5	59.2	82.2	100.0	52.6	78.9	100.6	67.8	86.1

DIVISION		NORTH DAKOTA			OHIO														
		WILLISTON			AKRON			ATHENS			CANTON			CHILLICOTHE			CINCINNATI		
		588			442 - 443			457			446 - 447			456			451 - 452		
		MAT.	INST.	TOTAL	MAT.	INST.	TOTAL	MAT.	INST.	TOTAL	MAT.	INST.	TOTAL	MAT.	INST.	TOTAL	MAT.	INST.	TOTAL
01590	EQUIPMENT RENTAL	.0	98.3	98.3	.0	97.4	97.4	.0	90.5	90.5	.0	97.4	97.4	.0	102.8	102.8	.0	102.5	102.5
02	SITE CONSTRUCTION	96.7	94.0	94.7	101.7	106.7	105.4	87.0	94.0	92.2	101.8	106.8	105.5	76.5	108.6	100.1	74.4	108.5	99.4
03100	CONCRETE FORMS & ACCESSORIES	93.8	43.3	50.2	98.7	96.7	96.9	92.1	90.0	90.3	98.7	85.6	87.4	94.7	98.0	97.6	97.0	88.3	89.5
03200	CONCRETE REINFORCEMENT	104.2	51.3	77.3	94.8	96.8	95.9	96.8	88.2	92.4	94.8	83.6	89.1	93.3	83.1	88.1	98.9	86.0	92.4
03300	CAST-IN-PLACE CONCRETE	107.6	52.6	85.0	98.5	105.1	101.2	101.8	101.7	101.7	99.5	101.8	100.4	92.4	97.0	94.3	85.1	88.8	86.6
03	CONCRETE	105.6	49.2	77.6	98.6	98.6	98.6	102.8	92.8	97.9	99.1	90.1	94.6	97.5	94.6	96.1	92.4	88.1	90.3
04	MASONRY	101.9	59.0	75.3	89.7	102.5	97.7	74.9	98.7	89.6	90.5	92.7	91.9	82.7	97.9	92.1	82.4	94.2	89.7
05	METALS	92.7	68.4	85.0	89.8	81.8	87.2	93.7	74.4	87.5	89.8	75.5	85.2	86.0	85.7	86.0	88.0	88.0	88.0
06	WOOD & PLASTICS	86.3	41.1	62.8	92.3	94.8	93.6	82.7	88.0	85.5	92.6	84.0	88.1	94.3	97.1	95.8	96.8	86.7	91.5
07	THERMAL & MOISTURE PROTECTION	99.7	52.9	80.0	99.6	100.7	100.0	96.4	84.1	91.2	100.1	96.2	98.4	96.0	95.1	95.6	94.2	96.2	95.1
08	DOORS & WINDOWS	99.5	39.1	83.7	104.3	96.7	102.3	95.5	77.6	90.8	98.7	77.2	93.1	88.8	89.3	88.9	96.1	85.5	93.4
09200	PLASTER & GYPSUM BOARD	111.6	39.9	67.0	97.7	94.2	95.6	97.3	87.3	91.1	98.5	83.2	89.0	98.5	97.3	97.7	99.8	86.6	91.6
095,098	CEILINGS & ACOUSTICAL TREATMENT	138.5	39.9	79.7	90.4	94.2	92.7	103.5	87.3	93.8	90.4	83.2	86.1	98.0	97.3	97.6	98.8	86.6	91.6
09600	FLOORING	107.2	44.8	90.3	103.4	94.3	100.9	130.0	96.7	121.0	103.6	85.6	98.7	107.3	96.0	104.2	108.4	99.3	106.0
097,099	WALL FINISHES, PAINTS & COATINGS	97.2	28.7	55.4	107.9	118.5	114.4	111.3	53.2	75.9	107.9	90.4	97.2	107.8	98.5	102.1	107.8	92.8	98.6
09	FINISHES	114.7	41.2	76.4	99.4	98.5	98.9	102.9	88.1	95.2	99.6	85.1	92.1	98.7	98.2	98.5	99.3	90.6	94.8
10 - 14	TOTAL DIV. 10000 - 14000	100.0	45.0	88.3	100.0	86.8	97.2	100.0	70.6	93.7	100.0	75.0	94.7	100.0	81.4	96.0	100.0	92.7	98.4
15	MECHANICAL	95.6	61.9	81.6	99.9	102.0	100.8	94.9	69.8	84.5	99.9	86.3	94.2	95.4	93.0	94.4	99.8	88.5	95.1
16	ELECTRICAL	96.6	72.5	84.9	98.2	94.9	96.6	101.0	57.0	79.6	97.4	91.0	94.3	98.3	89.3	93.9	98.1	81.5	90.1
01 - 16	WEIGHTED AVERAGE	99.3	60.2	81.9	97.8	98.0	97.9	96.3	79.9	89.0	97.3	88.4	93.4	93.5	94.2	93.8	94.9	90.2	92.8

DIVISION		OHIO																	
		CLEVELAND			COLUMBUS			DAYTON			HAMILTON			LIMA			LORAIN		
		441			430 - 432			453 - 454			450			458			440		
		MAT.	INST.	TOTAL	MAT.	INST.	TOTAL	MAT.	INST.	TOTAL	MAT.	INST.	TOTAL	MAT.	INST.	TOTAL	MAT.	INST.	TOTAL
01590	EQUIPMENT RENTAL	.0	97.7	97.7	.0	96.4	96.4	.0	96.7	96.7	.0	102.8	102.8	.0	93.5	93.5	.0	97.4	97.4
02	SITE CONSTRUCTION	101.7	106.9	105.5	85.6	101.8	97.5	73.2	108.1	98.8	73.5	108.7	99.3	82.1	94.3	91.1	101.1	106.0	104.7
03100	CONCRETE FORMS & ACCESSORIES	98.8	102.7	102.1	99.2	86.3	88.1	96.9	79.2	81.6	97.0	88.3	89.5	92.1	86.9	87.6	98.7	103.7	103.0
03200	CONCRETE REINFORCEMENT	95.4	97.0	96.2	76.7	83.2	80.0	98.9	82.1	90.4	98.9	86.0	92.4	96.8	82.2	89.4	94.8	96.9	95.9
03300	CAST-IN-PLACE CONCRETE	96.7	112.5	103.2	88.9	90.9	89.7	77.5	89.1	82.3	84.8	87.4	85.9	93.5	99.2	95.8	93.8	107.5	99.4
03	CONCRETE	97.8	104.0	100.9	91.4	87.0	89.2	88.7	82.9	85.8	92.3	87.6	90.0	96.2	89.9	93.1	96.3	102.6	99.4
04	MASONRY	94.0	107.3	102.3	95.8	91.1	92.9	81.8	88.8	86.1	82.3	94.2	89.7	103.4	89.4	94.7	86.7	99.9	94.9
05	METALS	91.2	85.0	89.2	96.1	80.6	91.1	87.3	77.8	84.2	87.4	88.0	87.6	93.7	81.2	89.7	90.3	82.8	87.9
06	WOOD & PLASTICS	91.5	99.2	95.5	103.7	84.9	93.9	97.9	75.8	86.4	96.8	86.7	91.5	82.7	85.2	84.0	92.3	105.0	98.9
07	THERMAL & MOISTURE PROTECTION	98.3	112.4	104.2	98.1	93.0	96.0	99.4	89.7	95.3	96.1	96.0	96.1	96.0	100.2	97.8	100.0	107.2	103.1
08	DOORS & WINDOWS	95.2	99.1	96.2	99.5	83.5	95.3	96.3	78.6	91.7	94.0	85.5	91.7	95.5	82.6	92.1	98.7	102.2	99.6
09200	PLASTER & GYPSUM BOARD	96.9	98.8	98.1	95.2	84.3	88.5	99.8	75.4	84.7	99.8	86.6	91.6	97.3	84.3	89.2	97.7	104.7	102.0
095,098	CEILINGS & ACOUSTICAL TREATMENT	88.6	98.8	94.7	93.9	84.3	88.2	99.8	75.4	85.3	98.8	86.6	91.6	102.5	84.3	91.7	90.4	104.7	98.9
09600	FLOORING	103.2	108.0	104.5	97.5	91.3	95.8	111.2	86.3	104.4	108.4	99.3	106.0	129.1	93.3	119.5	103.6	106.7	104.4
097,099	WALL FINISHES, PAINTS & COATINGS	107.9	121.9	116.4	98.4	98.5	98.4	107.8	90.3	97.1	107.8	91.6	97.9	111.4	75.8	89.6	107.9	121.9	116.4
09	FINISHES	98.9	105.3	102.2	97.5	88.2	92.7	100.3	80.8	90.2	99.3	90.5	94.7	102.1	86.4	94.0	99.5	106.7	103.3
10 - 14	TOTAL DIV. 10000 - 14000	100.0	105.2	101.1	100.0	92.8	98.5	100.0	90.6	98.0	100.0	92.7	98.4	100.0	83.5	96.5	100.0	103.2	100.7
15	MECHANICAL	99.9	108.1	103.3	100.0	91.7	96.6	100.8	88.0	95.5	100.3	88.5	95.4	94.9	89.4	92.6	99.9	90.8	96.1
16	ELECTRICAL	98.2	107.1	102.6	98.1	88.4	93.4	96.1	88.3	92.3	96.3	77.7	87.3	101.2	84.4	93.1	97.5	88.6	93.2
01 - 16	WEIGHTED AVERAGE	97.2	104.3	100.4	97.3	89.6	93.9	94.6	87.0	91.2	94.5	89.6	92.3	96.8	87.9	92.8	96.9	97.1	97.0

DIVISION		OHIO																	
		MANSFIELD			MARION			SPRINGFIELD			STEUBENVILLE			TOLEDO			YOUNGSTOWN		
		448 - 449			433			455			439			434 - 436			444 - 445		
		MAT.	INST.	TOTAL	MAT.	INST.	TOTAL	MAT.	INST.	TOTAL	MAT.	INST.	TOTAL	MAT.	INST.	TOTAL	MAT.	INST.	TOTAL
01590	EQUIPMENT RENTAL	.0	97.4	97.4	.0	96.1	96.1	.0	96.7	96.7	.0	101.7	101.7	.0	99.5	99.5	.0	97.4	97.4
02	SITE CONSTRUCTION	96.9	106.8	104.1	81.1	102.0	96.4	73.7	107.2	98.3	121.5	112.2	114.6	84.9	102.9	98.1	101.6	107.1	105.6
03100	CONCRETE FORMS & ACCESSORIES	87.1	102.8	100.7	95.2	85.5	86.9	96.9	85.4	87.0	96.1	92.7	93.1	99.2	96.4	96.8	98.7	89.8	91.0
03200	CONCRETE REINFORCEMENT	86.3	83.9	85.1	70.8	83.4	77.2	98.9	82.1	90.4	68.8	93.4	81.3	76.7	95.3	86.2	94.8	96.7	95.8
03300	CAST-IN-PLACE CONCRETE	91.2	94.0	92.4	81.2	90.6	85.1	81.3	89.0	84.5	88.2	101.3	93.6	88.9	105.7	95.8	97.6	103.4	100.0
03	CONCRETE	90.8	95.2	93.0	83.8	86.4	85.1	90.6	85.6	88.1	87.6	94.9	91.3	91.4	98.9	95.1	98.2	95.0	96.6
04	MASONRY	88.9	101.2	96.5	98.3	97.4	97.8	82.1	88.6	86.2	84.3	95.9	91.5	105.7	102.8	103.9	89.9	95.2	93.2
05	METALS	90.5	77.2	86.2	95.0	77.8	89.5	87.3	77.6	84.2	91.2	81.3	88.0	95.9	89.2	93.8	89.8	82.1	87.3
06	WOOD & PLASTICS	79.8	105.0	92.9	99.5	84.9	91.9	99.1	84.5	91.5	95.3	90.4	92.8	103.7	94.2	98.7	92.3	87.4	89.7
07	THERMAL & MOISTURE PROTECTION	99.4	99.5	99.5	97.6	87.2	93.2	99.3	90.5	95.6	109.9	97.5	104.7	100.3	107.8	103.4	100.2	99.0	99.7
08	DOORS & WINDOWS	98.1	91.2	96.3	94.1	80.2	90.4	94.4	80.7	90.8	94.8	90.4	93.7	97.4	93.7	96.4	98.7	93.6	97.4
09200	PLASTER & GYPSUM BOARD	93.2	104.7	100.3	93.7	84.3	87.9	99.8	84.3	90.2	94.5	89.5	91.4	95.2	93.9	94.4	97.7	86.6	90.8
095,098	CEILINGS & ACOUSTICAL TREATMENT	92.9	104.7	99.9	93.9	84.3	88.2	99.8	84.3	90.6	90.9	89.5	90.0	93.9	93.9	93.9	90.4	86.6	88.1
09600	FLOORING	98.4	107.5	100.8	95.6	87.1	93.3	111.2	86.3	104.4	124.0	99.3	117.3	96.6	96.4	96.6	103.6	93.7	100.9
097,099	WALL FINISHES, PAINTS & COATINGS	107.9	70.7	85.2	98.4	60.6	75.3	107.8	90.3	97.1	115.6	97.4	104.5	98.4	106.5	103.4	107.9	100.0	103.1
09	FINISHES	97.7	101.3	99.6	96.5	83.2	89.5	100.3	85.9	92.8	113.2	94.0	103.2	97.2	97.0	97.1	99.6	90.9	95.0
10 - 14	TOTAL DIV. 10000 - 14000	100.0	82.7	96.3	100.0	71.3	93.9	100.0	91.6	98.2	100.0	99.8	100.0	100.0	87.6	97.3	100.0	85.7	97.0
15	MECHANICAL	95.0	90.8	93.2	95.1	91.7	93.7	100.8	88.0	95.4	95.4	86.7	91.8	100.0	102.9	101.2	99.9	92.5	96.8
16	ELECTRICAL	95.2	88.3	91.9	92.6	88.3	90.5	96.1	87.0	91.7	84.6	106.3	95.1	97.9	105.7	101.7	97.5	93.1	95.4
01 - 16	WEIGHTED AVERAGE	94.6	93.8	94.2	93.9	88.4	91.5	94.6	87.9	91.6	95.2	95.1	95.1	97.6	100.0	98.7	97.2	93.3	95.5

DIVISION		OHIO			OKLAHOMA															
		ZANESVILLE			ARDMORE			CLINTON			DURANT			ENID			GUYMON			
		437 - 438			734			736			747			737			739			
		MAT.	INST.	TOTAL	MAT.	INST.	TOTAL	MAT.	INST.	TOTAL	MAT.	INST.	TOTAL	MAT.	INST.	TOTAL	MAT.	INST.	TOTAL	
01590	**EQUIPMENT RENTAL**	.0	96.1	96.1	.0	78.2	78.2	.0	77.1	77.1	.0	77.1	77.1	.0	77.1	77.1	.0	77.1	77.1	
02	**SITE CONSTRUCTION**	83.6	102.5	97.5	104.8	90.5	94.3	106.1	88.9	93.5	99.0	88.6	91.4	108.0	88.9	94.0	111.4	87.7	94.0	
03100	CONCRETE FORMS & ACCESSORIES	91.6	85.5	86.3	90.8	46.9	52.9	89.5	40.2	47.0	81.3	46.9	51.6	93.5	39.4	46.8	97.9	24.2	34.3	
03200	CONCRETE REINFORCEMENT	70.3	92.9	81.8	97.2	82.4	89.6	97.7	82.4	89.9	98.1	66.9	82.2	97.0	82.4	89.6	97.7	33.7	65.2	
03300	CAST-IN-PLACE CONCRETE	85.6	90.1	87.4	94.8	47.1	75.2	91.6	47.1	73.3	88.6	47.0	71.5	91.6	50.0	74.5	91.6	30.9	66.7	
03	**CONCRETE**	87.3	88.0	87.6	90.2	53.7	72.1	89.4	50.7	70.2	85.2	50.8	68.1	90.0	51.4	70.8	92.4	28.8	60.8	
04	**MASONRY**	94.7	86.8	89.8	92.8	61.4	73.3	119.7	61.4	83.5	91.6	62.0	73.3	101.6	61.4	76.6	96.1	23.2	50.9	
05	**METALS**	96.6	81.1	91.6	88.8	67.1	81.9	88.9	67.0	81.9	88.9	60.4	79.8	89.9	67.1	82.6	89.3	39.1	73.3	
06	**WOOD & PLASTICS**	95.5	84.9	89.9	93.8	46.4	69.1	92.7	37.4	63.9	83.7	46.4	64.3	96.9	36.4	65.4	101.1	23.8	60.8	
07	**THERMAL & MOISTURE PROTECTION**	97.7	90.9	94.8	99.6	62.6	84.0	99.7	61.7	83.7	99.3	60.8	83.1	99.9	61.6	83.7	100.2	29.4	70.4	
08	**DOORS & WINDOWS**	94.1	85.0	91.7	95.4	56.1	85.1	95.4	51.3	83.9	95.4	52.2	84.1	95.4	50.5	83.7	95.5	25.9	77.3	
09200	PLASTER & GYPSUM BOARD	91.9	84.3	87.2	80.7	45.7	59.0	80.3	36.6	53.1	76.6	45.7	57.4	81.4	35.5	52.9	82.1	22.5	45.1	
095,098	CEILINGS & ACOUSTICAL TREATMENT	93.9	84.3	88.2	79.4	45.7	59.3	79.4	36.6	53.8	79.4	45.7	59.3	80.3	35.5	53.6	81.9	22.5	46.5	
09600	FLOORING	93.4	87.1	91.7	111.6	53.9	96.0	109.9	51.4	94.0	105.3	70.5	95.9	112.4	51.4	95.9	114.8	30.4	92.0	
097,099	WALL FINISHES, PAINTS & COATINGS	98.4	64.9	77.9	97.0	65.4	77.7	97.0	65.4	77.7	97.0	65.4	77.7	97.0	65.4	77.7	97.0	19.4	49.7	
09	**FINISHES**	95.7	83.4	89.3	93.0	48.9	70.0	92.6	43.2	66.9	90.2	52.5	70.6	93.8	42.5	67.1	95.3	24.7	58.6	
10 - 14	**TOTAL DIV. 10000 - 14000**	100.0	77.6	95.2	100.0	60.9	91.7	100.0	59.7	91.4	100.0	61.1	91.7	100.0	59.6	91.4	100.0	55.5	90.5	
15	**MECHANICAL**	95.1	83.8	90.4	95.1	65.3	82.8	95.1	65.3	82.8	95.1	65.5	82.8	100.1	65.3	85.6	95.1	27.2	67.0	
16	**ELECTRICAL**	92.5	83.3	88.0	93.9	71.5	83.0	94.9	71.5	83.5	96.4	63.3	80.3	94.9	71.5	83.5	96.4	20.0	59.3	
01 - 16	**WEIGHTED AVERAGE**	94.3	86.1	90.6	94.0	63.7	80.5	95.3	62.0	80.6	93.1	61.7	79.2	96.0	62.0	80.9	95.2	32.9	67.5	

DIVISION		OKLAHOMA																	
		LAWTON			MCALESTER			MIAMI			MUSKOGEE			OKLAHOMA CITY			PONCA CITY		
		735			745			743			744			730 - 731			746		
		MAT.	INST.	TOTAL	MAT.	INST.	TOTAL	MAT.	INST.	TOTAL	MAT.	INST.	TOTAL	MAT.	INST.	TOTAL	MAT.	INST.	TOTAL
01590	**EQUIPMENT RENTAL**	.0	78.2	78.2	.0	77.1	77.1	.0	86.7	86.7	.0	86.7	86.7	.0	78.5	78.5	.0	77.1	77.1
02	**SITE CONSTRUCTION**	103.4	90.5	94.0	91.3	88.2	89.0	93.4	85.9	87.9	94.1	84.9	87.4	104.7	91.1	94.7	99.9	88.9	91.8
03100	CONCRETE FORMS & ACCESSORIES	97.4	53.9	59.9	78.9	48.7	52.8	93.7	57.9	62.8	98.8	35.7	44.3	98.5	46.5	53.6	88.6	47.3	53.0
03200	CONCRETE REINFORCEMENT	97.3	82.4	89.7	97.7	50.2	73.6	96.2	82.4	89.2	97.0	38.5	67.3	97.3	82.4	89.7	97.0	82.3	89.6
03300	CAST-IN-PLACE CONCRETE	88.6	50.1	72.8	77.7	49.3	66.0	81.4	49.1	68.1	82.4	38.5	64.4	95.9	53.3	78.4	91.0	41.2	70.5
03	**CONCRETE**	86.8	57.9	72.4	76.9	49.1	63.1	81.1	60.2	70.7	82.6	38.3	60.6	90.4	55.7	73.1	87.2	51.8	69.6
04	**MASONRY**	96.0	61.4	74.5	109.7	61.8	80.0	94.6	62.8	74.8	111.2	54.9	76.3	97.8	63.5	76.5	86.3	62.0	71.2
05	**METALS**	93.3	67.2	84.9	88.8	50.0	76.4	88.8	81.9	86.6	89.8	58.6	79.9	95.2	67.1	86.2	88.8	66.8	81.8
06	**WOOD & PLASTICS**	100.0	55.8	77.0	81.0	50.4	65.1	96.8	60.4	77.9	102.1	36.0	67.7	101.9	44.8	72.1	92.0	47.1	68.6
07	**THERMAL & MOISTURE PROTECTION**	99.6	63.6	84.5	99.0	61.7	83.3	99.4	63.7	84.4	99.5	47.6	77.6	98.9	63.3	83.9	99.5	61.6	83.6
08	**DOORS & WINDOWS**	96.9	61.0	87.5	95.4	49.3	83.4	95.4	64.1	87.2	95.4	35.4	79.7	96.9	55.0	86.0	95.4	56.5	85.2
09200	PLASTER & GYPSUM BOARD	83.8	55.5	66.2	75.5	49.8	59.6	81.4	60.0	68.1	85.3	34.9	54.0	83.8	44.1	59.1	80.3	46.4	59.3
095,098	CEILINGS & ACOUSTICAL TREATMENT	89.7	55.5	69.3	79.4	49.8	61.8	79.4	60.0	67.8	89.7	34.9	57.0	89.7	44.1	62.5	79.4	46.4	59.7
09600	FLOORING	115.3	51.4	98.0	104.0	51.4	89.7	113.3	70.5	101.8	116.5	41.4	96.2	115.3	51.4	98.0	109.3	51.4	93.7
097,099	WALL FINISHES, PAINTS & COATINGS	97.0	65.4	77.7	97.0	48.2	67.2	97.0	80.4	86.9	97.0	35.0	59.2	97.0	65.4	77.7	97.0	65.4	77.7
09	**FINISHES**	96.9	53.9	74.5	89.1	48.9	68.2	92.7	62.6	77.0	96.9	37.0	65.7	97.0	48.0	71.4	92.0	49.0	69.6
10 - 14	**TOTAL DIV. 10000 - 14000**	100.0	62.0	91.9	100.0	61.6	91.8	100.0	63.8	92.3	100.0	59.0	91.3	100.0	61.5	91.8	100.0	61.2	91.7
15	**MECHANICAL**	100.1	65.3	85.6	95.1	38.3	71.6	95.1	65.1	82.7	100.1	29.2	70.6	100.1	66.4	86.1	95.1	64.6	82.5
16	**ELECTRICAL**	96.4	71.5	84.3	94.9	68.0	81.8	96.3	68.1	82.6	94.5	34.3	65.2	95.8	71.5	84.0	94.6	64.6	80.0
01 - 16	**WEIGHTED AVERAGE**	96.4	65.3	82.6	92.6	55.1	75.9	93.0	67.5	81.7	95.6	44.0	72.7	97.1	64.3	82.6	93.1	62.2	79.4

DIVISION		OKLAHOMA												OREGON					
		POTEAU			SHAWNEE			TULSA			WOODWARD			BEND			EUGENE		
		749			748			740 - 741			738			977			974		
		MAT.	INST.	TOTAL	MAT.	INST.	TOTAL	MAT.	INST.	TOTAL	MAT.	INST.	TOTAL	MAT.	INST.	TOTAL	MAT.	INST.	TOTAL
01590	**EQUIPMENT RENTAL**	.0	86.0	86.0	.0	77.1	77.1	.0	86.7	86.7	.0	77.1	77.1	.0	99.6	99.6	.0	99.6	99.6
02	**SITE CONSTRUCTION**	75.0	84.4	81.9	102.9	88.6	92.4	100.4	85.9	89.8	106.7	88.9	93.7	121.3	104.7	109.1	108.7	104.7	105.7
03100	CONCRETE FORMS & ACCESSORIES	85.9	47.7	52.9	81.1	45.7	50.6	98.5	46.8	53.9	89.6	40.1	46.9	108.2	105.8	106.1	104.0	105.7	105.5
03200	CONCRETE REINFORCEMENT	98.2	82.4	90.2	97.0	55.2	75.8	97.3	82.3	89.7	97.0	82.4	89.6	101.4	99.3	100.4	105.8	99.3	102.5
03300	CAST-IN-PLACE CONCRETE	81.4	48.6	68.0	93.9	46.2	74.3	89.8	48.8	72.9	91.6	47.1	73.3	110.4	105.4	108.4	106.9	105.4	106.3
03	**CONCRETE**	82.7	55.5	69.2	88.7	47.7	68.4	87.4	55.1	71.4	89.7	50.7	70.3	117.8	104.1	111.0	108.3	104.0	106.2
04	**MASONRY**	95.4	62.1	74.7	110.9	60.5	79.6	95.3	63.5	75.6	89.9	61.4	72.2	115.7	103.3	108.0	113.4	103.3	107.1
05	**METALS**	88.8	81.7	86.6	88.8	54.3	77.8	92.6	81.0	88.9	89.0	67.1	82.0	93.6	96.8	94.7	94.2	96.8	95.0
06	**WOOD & PLASTICS**	88.8	47.2	67.2	83.6	46.1	64.1	100.7	46.1	72.3	92.9	37.3	64.0	99.7	105.5	102.7	95.2	105.5	100.6
07	**THERMAL & MOISTURE PROTECTION**	99.5	61.7	83.6	99.5	62.3	83.9	99.5	60.2	82.9	99.8	61.7	83.7	106.0	97.6	102.5	105.2	95.3	101.0
08	**DOORS & WINDOWS**	95.4	56.9	85.3	95.4	45.9	82.4	96.9	55.3	86.0	95.4	51.2	83.9	98.5	106.2	100.5	98.9	106.2	100.8
09200	PLASTER & GYPSUM BOARD	79.6	46.4	59.0	76.6	45.5	57.3	83.8	45.3	59.9	81.0	36.4	53.3	96.9	105.5	102.2	95.1	105.5	101.5
095,098	CEILINGS & ACOUSTICAL TREATMENT	79.4	46.4	59.7	79.4	45.5	59.1	89.7	45.3	63.2	81.9	36.4	54.8	101.2	105.5	103.7	106.0	105.5	105.7
09600	FLOORING	108.5	70.5	98.2	105.2	40.9	87.8	115.1	53.5	98.4	109.9	53.9	94.8	117.5	98.6	112.3	115.4	98.6	110.9
097,099	WALL FINISHES, PAINTS & COATINGS	97.0	65.4	77.7	97.0	43.5	64.4	97.0	51.7	69.3	97.0	65.4	77.7	115.2	72.0	88.8	115.2	72.0	88.8
09	**FINISHES**	90.3	53.0	70.9	90.4	43.7	66.1	96.4	47.6	71.0	93.4	43.6	67.5	110.7	100.8	105.6	110.1	100.8	105.2
10 - 14	**TOTAL DIV. 10000 - 14000**	100.0	61.6	91.8	100.0	60.5	91.6	100.0	62.2	91.9	100.0	59.7	91.4	100.0	106.4	101.4	100.0	106.4	101.4
15	**MECHANICAL**	95.1	64.7	82.5	95.1	64.6	82.5	100.1	60.3	83.6	95.1	65.3	82.8	95.3	106.1	99.8	100.2	106.1	102.7
16	**ELECTRICAL**	94.7	68.1	81.7	96.5	71.5	84.3	96.4	44.7	71.3	96.3	71.5	84.2	99.5	100.7	100.0	98.2	100.7	99.4
01 - 16	**WEIGHTED AVERAGE**	92.3	64.7	80.1	94.6	60.2	79.3	96.2	59.9	80.1	94.2	62.1	79.9	102.2	102.9	102.5	101.7	102.8	102.2

		OREGON																	
DIVISION		KLAMATH FALLS 976			MEDFORD 975			PENDLETON 978			PORTLAND 970 - 972			SALEM 973			VALE 979		
		MAT.	INST.	TOTAL	MAT.	INST.	TOTAL	MAT.	INST.	TOTAL	MAT.	INST.	TOTAL	MAT.	INST.	TOTAL	MAT.	INST.	TOTAL
01590	**EQUIPMENT RENTAL**	.0	99.6	99.6	.0	99.6	99.6	.0	96.5	96.5	.0	99.6	99.6	.0	99.6	99.6	.0	96.5	96.5
02	**SITE CONSTRUCTION**	126.5	104.7	110.5	118.2	104.7	108.3	114.8	97.2	101.9	109.8	104.7	106.1	108.3	104.7	105.6	100.6	96.5	97.6
03100	CONCRETE FORMS & ACCESSORIES	99.9	105.5	104.7	98.8	105.5	104.6	101.6	105.8	105.3	105.6	106.0	105.9	105.4	105.8	105.8	109.3	104.7	105.3
03200	CONCRETE REINFORCEMENT	101.4	99.3	100.3	103.1	99.3	101.2	100.7	99.5	100.1	106.6	99.5	103.0	106.7	99.5	103.0	98.2	99.3	98.8
03300	CAST-IN-PLACE CONCRETE	110.5	105.3	108.3	110.4	105.3	108.3	111.2	106.8	109.4	114.8	105.5	111.0	110.9	105.4	108.7	87.5	101.3	93.2
03	**CONCRETE**	120.7	103.9	112.4	114.9	103.9	109.4	100.2	104.6	102.4	112.4	104.2	108.3	110.5	104.1	107.3	84.0	102.0	92.9
04	**MASONRY**	132.2	103.3	114.3	110.6	103.3	106.1	120.4	106.0	111.5	114.8	105.9	109.3	118.4	105.9	110.6	118.6	105.7	110.6
05	**METALS**	93.6	96.5	94.6	93.8	96.5	94.7	101.0	97.3	99.8	95.4	97.2	96.0	94.6	97.1	95.4	100.9	94.2	98.8
06	**WOOD & PLASTICS**	89.9	105.5	98.0	88.6	105.5	97.4	92.7	105.6	99.4	96.3	105.5	101.1	96.3	105.5	101.1	102.1	105.6	103.9
07	**THERMAL & MOISTURE PROTECTION**	106.2	92.6	100.5	105.8	92.6	100.3	93.9	97.4	95.4	105.1	101.5	103.6	105.1	98.2	102.2	93.3	95.9	94.4
08	**DOORS & WINDOWS**	98.5	106.2	100.5	101.5	106.2	102.7	94.5	106.3	97.6	96.4	106.2	99.0	98.3	106.2	100.3	94.4	99.4	95.7
09200	PLASTER & GYPSUM BOARD	94.0	105.5	101.1	93.3	105.5	100.8	77.7	105.5	94.9	92.9	105.5	100.7	92.9	105.5	100.7	83.2	105.5	97.0
095,098	CEILINGS & ACOUSTICAL TREATMENT	111.4	105.5	107.9	116.2	105.5	109.8	68.9	105.5	90.7	106.0	105.5	105.7	106.0	105.5	105.7	68.9	105.5	90.7
09600	FLOORING	113.7	98.6	109.6	112.9	98.6	109.1	80.1	98.6	85.1	115.4	98.6	110.9	115.4	98.6	110.9	82.3	98.6	86.7
097,099	WALL FINISHES, PAINTS & COATINGS	115.2	62.6	83.1	115.2	62.6	83.1	96.0	72.0	81.3	115.2	72.0	88.8	115.2	72.0	88.8	96.0	72.0	81.3
09	**FINISHES**	112.1	99.7	105.7	112.3	99.7	105.8	75.9	100.9	88.9	109.8	100.8	105.1	109.7	100.8	105.0	76.6	100.9	89.2
10 - 14	**TOTAL DIV. 10000 - 14000**	100.0	106.4	101.4	100.0	106.4	101.4	100.0	106.6	101.4	100.0	106.4	101.4	100.0	106.4	101.4	100.0	106.7	101.4
15	**MECHANICAL**	95.3	106.1	99.8	100.2	106.1	102.7	97.0	106.4	100.9	100.2	108.9	103.8	100.2	106.1	102.7	97.0	81.4	90.5
16	**ELECTRICAL**	98.3	90.2	94.3	101.5	90.2	96.0	89.1	99.8	94.3	98.6	105.4	101.9	98.3	100.7	99.5	89.1	77.1	83.2
01 - 16	**WEIGHTED AVERAGE**	103.4	101.1	102.4	103.3	101.1	102.3	96.5	102.6	99.2	102.2	104.6	103.3	102.2	103.2	102.6	94.4	93.2	93.9

		PENNSYLVANIA																	
DIVISION		ALLENTOWN 181			ALTOONA 166			BEDFORD 155			BRADFORD 167			BUTLER 160			CHAMBERSBURG 172		
		MAT.	INST.	TOTAL	MAT.	INST.	TOTAL	MAT.	INST.	TOTAL	MAT.	INST.	TOTAL	MAT.	INST.	TOTAL	MAT.	INST.	TOTAL
01590	**EQUIPMENT RENTAL**	.0	115.6	115.6	.0	115.6	115.6	.0	113.8	113.8	.0	115.6	115.6	.0	115.6	115.6	.0	114.8	114.8
02	**SITE CONSTRUCTION**	91.9	106.8	102.8	96.1	106.7	103.9	98.7	104.7	103.1	91.1	106.5	102.4	87.0	107.6	102.1	86.5	103.7	99.1
03100	CONCRETE FORMS & ACCESSORIES	100.5	111.9	110.3	82.7	85.8	85.4	83.7	84.5	84.4	85.6	86.8	86.6	84.6	98.8	96.8	84.7	83.6	83.8
03200	CONCRETE REINFORCEMENT	103.3	101.9	102.6	100.2	93.8	96.9	99.8	81.8	90.7	102.3	93.8	98.0	100.9	109.6	105.3	100.7	97.9	99.3
03300	CAST-IN-PLACE CONCRETE	88.1	101.9	93.7	98.2	85.6	93.0	102.9	78.9	93.0	93.8	91.4	92.8	86.7	95.7	90.4	95.2	78.3	88.3
03	**CONCRETE**	98.1	107.5	102.8	93.5	88.7	91.1	97.4	83.4	90.5	99.9	91.1	95.5	85.2	101.0	93.0	105.7	85.9	95.9
04	**MASONRY**	93.7	99.2	97.1	95.8	83.7	88.3	96.2	85.8	89.8	92.4	89.7	90.7	98.0	95.4	96.4	94.2	82.6	87.0
05	**METALS**	96.9	121.4	104.7	90.5	115.3	98.5	95.1	107.0	98.9	95.3	115.6	101.8	90.3	124.7	101.3	95.2	113.8	101.2
06	**WOOD & PLASTICS**	101.8	115.2	108.8	79.2	87.6	83.6	81.4	85.7	83.6	85.6	85.7	85.7	81.0	98.5	90.1	89.7	85.7	87.6
07	**THERMAL & MOISTURE PROTECTION**	102.2	116.2	108.1	101.1	93.0	97.7	101.7	89.1	96.4	102.1	95.2	99.2	100.8	99.4	100.2	101.1	76.8	90.8
08	**DOORS & WINDOWS**	93.5	110.8	98.0	88.0	93.1	89.4	90.9	89.0	90.4	93.9	96.6	94.6	88.0	108.4	93.3	90.4	86.5	89.4
09200	PLASTER & GYPSUM BOARD	107.4	115.3	112.3	100.3	87.0	92.1	91.4	85.0	87.4	101.2	85.0	91.2	100.7	98.2	99.2	105.2	85.0	92.7
095,098	CEILINGS & ACOUSTICAL TREATMENT	88.5	115.3	104.5	92.8	87.0	89.4	88.6	85.0	86.5	91.2	85.0	87.5	93.8	98.2	96.4	91.2	85.0	87.5
09600	FLOORING	88.6	96.4	90.7	82.6	57.0	75.6	88.2	57.8	79.9	81.8	102.6	87.5	83.5	101.1	88.3	86.0	57.6	78.3
097,099	WALL FINISHES, PAINTS & COATINGS	91.6	90.0	90.6	86.8	107.8	99.6	99.5	86.3	91.4	91.6	87.6	89.2	86.8	105.8	98.4	91.6	73.7	80.7
09	**FINISHES**	95.5	106.3	101.2	93.9	81.9	87.7	92.4	78.9	85.3	93.3	88.4	90.7	94.1	99.2	96.7	94.4	77.5	85.6
10 - 14	**TOTAL DIV. 10000 - 14000**	100.0	105.9	101.3	100.0	97.8	99.5	100.0	98.4	99.7	100.0	99.8	100.0	100.0	103.1	100.7	100.0	95.8	99.1
15	**MECHANICAL**	100.3	109.5	104.1	99.8	90.4	95.9	95.2	91.4	93.6	95.5	90.8	93.6	95.0	93.8	94.5	95.5	91.4	93.8
16	**ELECTRICAL**	100.7	99.5	100.1	89.8	105.0	97.2	93.4	105.0	99.0	94.9	105.0	99.8	90.3	105.1	97.5	93.2	86.8	90.1
01 - 16	**WEIGHTED AVERAGE**	98.0	107.4	102.2	94.4	94.5	94.5	95.2	92.6	94.0	95.7	96.6	96.1	92.3	102.3	96.7	95.9	89.9	93.3

		PENNSYLVANIA																	
DIVISION		DOYLESTOWN 189			DUBOIS 158			ERIE 164 - 165			GREENSBURG 156			HARRISBURG 170 - 171			HAZLETON 182		
		MAT.	INST.	TOTAL	MAT.	INST.	TOTAL	MAT.	INST.	TOTAL	MAT.	INST.	TOTAL	MAT.	INST.	TOTAL	MAT.	INST.	TOTAL
01590	**EQUIPMENT RENTAL**	.0	92.6	92.6	.0	113.8	113.8	.0	115.6	115.6	.0	113.8	113.8	.0	114.8	114.8	.0	115.6	115.6
02	**SITE CONSTRUCTION**	107.1	88.6	93.5	103.4	105.4	104.9	92.7	107.4	103.5	94.9	106.5	103.4	82.9	105.5	99.5	85.0	106.8	101.0
03100	CONCRETE FORMS & ACCESSORIES	81.6	131.9	125.0	82.9	87.0	86.4	99.8	95.6	96.1	91.3	98.8	97.8	93.2	88.2	88.9	79.1	89.3	87.9
03200	CONCRETE REINFORCEMENT	99.9	130.8	115.6	99.2	93.8	96.5	102.3	90.9	96.5	99.2	109.6	104.4	103.3	98.4	100.8	100.3	112.8	106.7
03300	CAST-IN-PLACE CONCRETE	83.2	96.4	88.7	99.2	92.4	96.4	96.5	84.2	91.5	95.4	95.4	95.4	97.2	93.4	95.6	83.2	91.5	86.6
03	**CONCRETE**	93.4	118.8	106.0	98.2	91.5	94.9	92.6	92.0	92.3	92.7	100.8	96.8	100.3	93.4	96.9	90.4	96.0	93.2
04	**MASONRY**	97.1	130.8	118.0	96.3	89.7	92.2	86.0	95.9	92.2	106.8	94.1	99.0	94.4	89.6	91.4	105.8	96.5	100.0
05	**METALS**	94.3	117.7	101.8	95.1	114.7	101.4	90.7	114.4	98.3	95.0	123.5	104.1	98.8	119.1	105.3	96.6	125.5	105.9
06	**WOOD & PLASTICS**	80.8	134.5	108.7	80.4	85.7	83.1	98.3	95.1	96.6	89.7	98.5	94.3	95.5	87.1	91.1	79.1	87.1	83.3
07	**THERMAL & MOISTURE PROTECTION**	100.8	130.8	113.4	102.0	95.5	99.2	100.8	98.2	99.7	101.6	98.7	100.4	105.2	108.6	106.6	101.6	106.0	103.5
08	**DOORS & WINDOWS**	95.3	137.4	106.3	90.9	96.6	92.4	88.2	93.2	89.5	90.9	108.3	95.5	93.5	94.2	93.7	94.2	98.7	95.4
09200	PLASTER & GYPSUM BOARD	98.8	135.2	121.4	90.6	85.0	87.2	107.4	94.7	99.5	94.1	98.2	96.7	107.4	86.5	94.4	99.3	86.5	91.3
095,098	CEILINGS & ACOUSTICAL TREATMENT	87.8	135.2	116.1	88.6	85.0	86.5	88.5	94.7	92.2	87.8	98.2	94.0	88.5	86.5	87.3	89.5	86.5	87.7
09600	FLOORING	72.6	103.5	81.0	87.8	102.6	91.8	90.8	84.7	89.1	92.0	84.9	90.1	88.8	87.5	88.4	78.9	94.8	83.2
097,099	WALL FINISHES, PAINTS & COATINGS	91.0	88.0	89.2	99.5	105.8	103.3	98.1	91.3	94.0	99.5	105.8	103.3	91.6	87.3	89.0	91.6	98.6	95.9
09	**FINISHES**	86.8	122.6	105.5	92.5	89.8	91.1	97.0	92.8	94.8	93.5	97.1	95.4	94.5	87.4	90.8	91.3	89.7	90.5
10 - 14	**TOTAL DIV. 10000 - 14000**	100.0	91.4	98.2	100.0	99.8	100.0	100.0	102.9	100.6	100.0	103.0	100.6	100.0	97.4	99.4	100.0	101.6	100.3
15	**MECHANICAL**	95.0	126.4	108.0	95.2	93.6	94.6	99.8	94.0	97.4	95.2	96.2	95.6	100.3	94.0	97.7	95.5	85.7	91.4
16	**ELECTRICAL**	93.9	117.7	105.5	93.9	105.0	99.3	91.3	90.6	91.0	93.9	105.0	99.3	99.7	86.8	93.4	95.6	91.7	93.7
01 - 16	**WEIGHTED AVERAGE**	94.7	119.7	105.8	95.4	97.3	96.2	94.4	96.7	95.4	95.3	102.2	98.3	98.2	95.3	96.9	95.3	96.6	95.9

	DIVISION	PENNSYLVANIA																	
		INDIANA			JOHNSTOWN			KITTANNING			LANCASTER			LEHIGH VALLEY			MONTROSE		
		157			159			162			175 - 176			180			188		
		MAT.	INST.	TOTAL	MAT.	INST.	TOTAL	MAT.	INST.	TOTAL	MAT.	INST.	TOTAL	MAT.	INST.	TOTAL	MAT.	INST.	TOTAL
01590	EQUIPMENT RENTAL	.0	113.8	113.8	.0	113.8	113.8	.0	115.6	115.6	.0	114.8	114.8	.0	115.6	115.6	.0	115.6	115.6
02	SITE CONSTRUCTION	93.2	106.0	102.6	99.3	106.0	104.2	89.6	107.7	102.9	78.3	105.5	98.3	89.2	106.8	102.1	87.9	105.4	100.7
03100	CONCRETE FORMS & ACCESSORIES	84.5	89.1	88.5	82.9	88.6	87.9	84.6	98.9	96.9	87.0	87.7	87.6	93.0	112.1	109.5	80.1	89.4	88.2
03200	CONCRETE REINFORCEMENT	98.4	109.7	104.1	99.8	109.4	104.7	100.9	109.7	105.4	100.3	98.4	99.3	100.3	102.3	101.3	105.1	112.7	109.0
03300	CAST-IN-PLACE CONCRETE	93.6	95.4	94.3	103.8	89.0	97.7	90.1	95.8	92.4	81.0	93.1	86.0	90.1	102.0	95.0	88.3	90.8	89.3
03	CONCRETE	90.5	96.6	93.5	98.1	94.1	96.1	87.8	101.1	94.4	93.1	93.0	93.1	97.1	107.7	102.4	95.6	95.6	95.6
04	MASONRY	92.9	97.1	95.5	93.7	90.1	91.5	101.6	97.1	98.9	99.6	89.6	93.4	93.8	99.2	97.2	93.8	97.2	95.9
05	METALS	95.2	123.7	104.3	95.1	121.9	103.7	90.4	125.1	101.5	95.2	118.5	102.7	96.6	121.7	104.6	95.4	121.7	103.8
06	WOOD & PLASTICS	82.3	85.7	84.1	80.4	87.6	84.2	81.0	98.5	90.1	92.5	87.1	89.7	93.7	115.2	104.9	80.0	87.9	84.1
07	THERMAL & MOISTURE PROTECTION	101.5	97.2	99.7	101.7	95.0	98.9	100.9	99.8	100.4	100.6	103.4	101.8	102.0	109.9	105.3	101.6	98.3	100.3
08	DOORS & WINDOWS	90.9	101.3	93.6	90.9	97.8	92.7	88.0	108.4	93.3	90.4	94.6	91.5	94.2	110.5	98.5	90.6	99.2	92.9
09200	PLASTER & GYPSUM BOARD	91.7	85.0	87.6	90.4	87.0	88.3	100.7	98.2	99.2	107.1	86.5	94.3	103.7	115.3	110.9	99.7	87.3	92.0
095,098	CEILINGS & ACOUSTICAL TREATMENT	88.6	85.0	86.5	87.8	87.0	87.3	93.8	98.2	96.4	91.2	86.5	88.4	89.5	115.3	104.9	91.2	87.3	88.9
09600	FLOORING	88.6	102.6	92.4	87.8	72.5	83.6	83.5	102.6	88.7	87.0	91.2	88.1	84.8	100.6	89.1	79.6	72.3	77.6
097,099	WALL FINISHES, PAINTS & COATINGS	99.5	105.8	103.3	99.5	107.8	104.6	86.8	105.8	98.4	91.6	72.9	80.2	91.6	84.8	87.4	91.6	98.6	95.9
09	FINISHES	92.3	92.0	92.1	91.9	86.7	89.2	94.2	99.6	97.0	94.6	84.9	89.6	94.0	106.5	100.5	92.2	86.9	89.4
10 - 14	TOTAL DIV. 10000 - 14000	100.0	101.4	100.3	100.0	100.2	100.0	100.0	103.1	100.7	100.0	97.3	99.4	100.0	106.1	101.3	100.0	102.0	100.4
15	MECHANICAL	95.2	96.2	95.6	95.2	94.1	94.7	95.0	98.5	96.4	95.5	93.9	94.9	95.5	109.5	101.3	95.5	85.8	91.5
16	ELECTRICAL	93.9	105.0	99.3	93.9	105.0	99.3	89.8	105.1	97.2	94.5	57.3	76.4	95.6	150.2	122.1	94.9	90.8	92.9
01 - 16	WEIGHTED AVERAGE	94.1	100.7	97.0	95.1	98.1	96.5	92.8	103.6	97.6	94.8	90.6	92.9	95.9	114.4	104.1	94.8	95.5	95.1

	DIVISION	PENNSYLVANIA																	
		NEW CASTLE			NORRISTOWN			OIL CITY			PHILADELPHIA			PITTSBURGH			POTTSVILLE		
		161			194			163			190 - 191			150 - 152			179		
		MAT.	INST.	TOTAL	MAT.	INST.	TOTAL	MAT.	INST.	TOTAL	MAT.	INST.	TOTAL	MAT.	INST.	TOTAL	MAT.	INST.	TOTAL
01590	EQUIPMENT RENTAL	.0	115.6	115.6	.0	95.3	95.3	.0	115.6	115.6	.0	94.7	94.7	.0	115.4	115.4	.0	114.8	114.8
02	SITE CONSTRUCTION	87.5	107.7	102.3	94.8	95.7	95.5	85.9	106.7	101.2	101.8	95.2	97.0	98.3	109.1	106.2	81.3	105.2	98.9
03100	CONCRETE FORMS & ACCESSORIES	84.6	98.4	96.5	80.9	129.5	122.8	84.6	85.2	85.1	99.3	136.4	131.3	100.6	99.0	99.2	77.4	88.0	86.6
03200	CONCRETE REINFORCEMENT	99.6	104.3	102.0	100.9	136.5	119.0	100.9	104.2	102.6	104.1	137.0	120.8	100.3	109.9	105.2	99.5	99.4	99.4
03300	CAST-IN-PLACE CONCRETE	87.5	95.9	91.0	80.8	127.1	99.8	85.0	96.5	89.7	99.2	129.3	111.6	99.1	95.5	97.6	86.3	85.0	85.8
03	CONCRETE	85.6	99.9	92.7	95.3	129.5	112.3	84.0	94.1	89.0	108.1	133.5	120.7	96.1	101.0	98.6	96.8	90.6	93.7
04	MASONRY	96.5	95.4	95.8	112.6	125.6	120.7	97.6	93.9	95.3	98.8	135.5	121.5	88.6	100.3	95.8	93.5	90.1	91.4
05	METALS	90.4	122.0	100.5	101.2	126.1	109.2	90.4	120.4	100.0	105.0	127.8	112.3	96.3	124.3	105.3	95.4	118.7	102.9
06	WOOD & PLASTICS	81.0	98.5	90.1	77.7	129.0	104.4	81.0	81.5	81.2	97.2	136.3	117.6	100.0	98.5	99.2	82.1	87.1	84.7
07	THERMAL & MOISTURE PROTECTION	100.8	97.4	99.4	102.1	133.0	115.1	100.7	96.7	99.0	102.9	136.8	117.1	101.9	100.3	101.2	100.7	105.1	102.5
08	DOORS & WINDOWS	88.0	106.5	92.8	88.7	136.1	101.1	88.0	87.6	87.9	96.7	140.2	108.1	93.5	108.3	97.4	90.5	94.5	91.5
09200	PLASTER & GYPSUM BOARD	100.7	98.2	99.2	96.3	129.7	117.0	100.7	80.7	88.3	104.1	137.2	124.6	99.1	98.2	98.6	102.3	86.5	92.4
095,098	CEILINGS & ACOUSTICAL TREATMENT	93.8	98.2	96.4	98.3	129.7	117.0	93.8	80.7	86.0	98.3	137.2	121.5	88.6	98.2	94.4	91.2	86.5	88.4
09600	FLOORING	83.5	71.8	80.3	78.4	131.8	92.9	83.5	102.6	88.7	87.0	138.6	101.0	96.6	104.3	98.7	82.7	91.2	85.0
097,099	WALL FINISHES, PAINTS & COATINGS	86.8	105.8	98.4	94.6	138.4	121.3	86.8	105.8	98.4	94.6	138.4	121.3	99.5	113.5	108.0	91.6	98.6	95.9
09	FINISHES	94.1	93.6	93.8	96.7	130.8	114.5	94.0	88.5	91.1	100.7	137.2	119.7	95.9	101.0	98.6	92.7	89.5	91.0
10 - 14	TOTAL DIV. 10000 - 14000	100.0	103.1	100.7	100.0	113.1	102.8	100.0	100.9	100.2	100.0	125.9	105.5	100.0	103.0	100.6	100.0	97.6	99.5
15	MECHANICAL	95.0	93.8	94.5	95.2	129.2	109.3	95.0	93.4	94.3	100.1	133.4	113.9	100.1	98.8	99.6	95.5	83.7	90.6
16	ELECTRICAL	90.3	104.5	97.2	92.0	130.9	110.9	91.9	105.0	98.3	96.1	138.2	116.6	95.9	105.1	100.3	92.8	93.4	93.1
01 - 16	WEIGHTED AVERAGE	92.3	101.0	96.2	96.6	126.0	109.7	92.3	98.1	94.9	101.0	131.1	114.4	97.0	104.2	100.2	94.5	94.0	94.3

	DIVISION	PENNSYLVANIA																	
		READING			SCRANTON			STATE COLLEGE			STROUDSBURG			SUNBURY			UNIONTOWN		
		195 - 196			184 - 185			168			183			178			154		
		MAT.	INST.	TOTAL	MAT.	INST.	TOTAL	MAT.	INST.	TOTAL	MAT.	INST.	TOTAL	MAT.	INST.	TOTAL	MAT.	INST.	TOTAL
01590	EQUIPMENT RENTAL	.0	118.2	118.2	.0	115.6	115.6	.0	114.8	114.8	.0	115.6	115.6	.0	115.6	115.6	.0	113.8	113.8
02	SITE CONSTRUCTION	99.2	111.1	107.9	92.4	106.8	103.0	83.4	105.5	99.6	86.8	105.3	100.4	90.9	106.5	102.3	93.7	106.5	103.1
03100	CONCRETE FORMS & ACCESSORIES	98.8	88.6	90.0	100.7	88.7	90.3	83.7	85.7	85.5	86.5	91.4	90.8	90.5	86.4	87.0	76.2	98.6	95.5
03200	CONCRETE REINFORCEMENT	102.3	99.6	100.9	103.3	112.7	108.1	101.5	93.6	97.5	103.7	102.5	103.1	102.3	98.4	100.3	99.2	109.2	104.3
03300	CAST-IN-PLACE CONCRETE	72.5	98.9	83.3	92.0	92.1	92.1	88.7	75.3	83.2	86.7	80.8	84.3	94.3	92.9	93.8	93.6	95.3	94.3
03	CONCRETE	94.9	95.7	95.3	100.0	95.9	98.0	99.9	85.1	92.6	94.4	91.1	92.8	100.7	92.4	96.5	90.0	100.6	95.3
04	MASONRY	100.4	92.1	95.3	94.1	94.5	94.4	99.4	81.4	88.2	90.7	95.6	93.8	92.4	89.7	90.7	108.4	88.2	95.9
05	METALS	101.5	120.2	107.5	98.8	125.3	107.3	95.1	115.1	101.5	96.7	116.0	102.9	95.2	118.0	102.5	94.9	122.8	103.8
06	WOOD & PLASTICS	97.3	86.2	91.6	101.8	85.6	93.3	88.5	87.6	88.1	87.0	91.1	89.1	90.8	85.3	87.9	73.0	98.5	86.3
07	THERMAL & MOISTURE PROTECTION	102.3	110.8	105.8	102.1	102.8	102.4	100.9	96.7	99.1	101.8	85.1	94.8	102.0	103.6	102.7	101.4	97.1	99.6
08	DOORS & WINDOWS	94.3	94.4	94.4	93.5	97.9	94.7	90.4	93.1	91.1	94.2	89.0	92.9	90.6	93.2	91.3	90.9	108.3	95.4
09200	PLASTER & GYPSUM BOARD	107.4	85.7	93.9	109.8	84.9	94.3	103.8	87.0	93.4	101.4	90.6	94.7	102.4	84.6	91.3	87.5	98.2	94.2
095,098	CEILINGS & ACOUSTICAL TREATMENT	87.1	85.7	86.3	97.9	84.9	90.2	88.6	87.0	87.7	87.8	90.6	89.5	88.6	84.6	86.2	87.8	98.2	94.0
09600	FLOORING	85.2	101.6	89.6	88.6	91.1	89.3	85.4	62.1	79.1	82.4	64.1	77.4	83.8	53.0	75.5	84.5	80.0	83.3
097,099	WALL FINISHES, PAINTS & COATINGS	92.8	99.8	97.1	91.6	98.6	95.9	91.6	107.8	101.5	91.6	82.4	86.0	91.6	98.6	95.9	99.5	105.8	103.3
09	FINISHES	97.8	91.1	94.3	98.2	89.4	93.6	93.2	83.0	87.9	92.3	85.4	88.7	93.4	81.3	87.1	90.2	95.3	92.8
10 - 14	TOTAL DIV. 10000 - 14000	100.0	100.2	100.0	100.0	101.8	100.4	100.0	95.3	99.0	100.0	80.4	95.8	100.0	97.2	99.4	100.0	103.0	100.6
15	MECHANICAL	100.3	107.9	103.5	100.3	86.2	94.5	95.5	90.7	93.5	95.5	86.2	91.6	95.5	93.9	94.8	95.2	96.2	95.6
16	ELECTRICAL	97.9	93.5	95.7	100.7	91.7	96.3	93.6	105.0	99.2	95.6	150.1	122.1	93.6	88.2	91.0	91.2	105.0	98.0
01 - 16	WEIGHTED AVERAGE	98.8	100.9	99.7	98.8	96.3	97.7	95.4	93.9	94.7	95.2	101.1	97.8	95.4	94.4	95.0	94.2	101.2	97.3

	DIVISION	PENNSYLVANIA																	
		WASHINGTON			WELLSBORO			WESTCHESTER			WILKES-BARRE			WILLIAMSPORT			YORK		
		153			169			193			186 - 187			177			173 - 174		
		MAT.	INST.	TOTAL	MAT.	INST.	TOTAL	MAT.	INST.	TOTAL	MAT.	INST.	TOTAL	MAT.	INST.	TOTAL	MAT.	INST.	TOTAL
01590	EQUIPMENT RENTAL	.0	113.8	113.8	.0	115.6	115.6	.0	95.3	95.3	.0	115.6	115.6	.0	115.6	115.6	.0	114.8	114.8
02	SITE CONSTRUCTION	93.7	106.5	103.1	95.1	105.2	102.5	101.1	94.4	96.2	84.7	106.6	100.7	82.0	105.1	98.9	82.2	105.5	99.3
03100	CONCRETE FORMS & ACCESSORIES	84.6	98.9	96.9	84.9	83.1	83.3	88.5	133.1	127.0	89.5	87.8	88.1	86.5	68.2	70.7	81.3	88.1	87.2
03200	CONCRETE REINFORCEMENT	99.2	109.7	104.5	101.5	112.6	107.1	99.9	105.9	103.0	102.3	102.3	102.3	101.5	83.8	92.5	102.3	98.4	100.3
03300	CAST-IN-PLACE CONCRETE	93.6	95.4	94.3	93.0	87.4	90.7	89.5	128.6	105.6	83.2	91.1	86.5	79.8	77.3	78.8	86.7	93.3	89.4
03	CONCRETE	90.6	100.9	95.7	102.3	91.6	97.0	103.6	126.1	114.7	91.5	93.1	92.3	88.2	76.2	82.2	98.1	93.3	95.8
04	MASONRY	93.2	97.1	95.6	98.7	88.5	92.4	107.2	131.4	122.2	106.7	92.9	98.2	84.8	79.6	81.5	93.4	89.6	91.0
05	METALS	94.9	123.8	104.1	95.2	122.3	103.9	101.2	116.1	106.0	95.3	120.0	103.2	95.3	110.5	100.1	96.4	119.0	103.6
06	WOOD & PLASTICS	82.4	98.5	90.8	85.0	82.5	83.7	85.8	134.4	111.1	89.8	85.4	87.5	87.0	66.2	76.2	86.2	87.1	86.6
07	THERMAL & MOISTURE PROTECTION	101.5	99.6	100.7	102.3	94.4	99.0	102.5	130.8	114.4	101.6	104.9	103.0	101.5	96.5	99.4	100.8	108.6	104.0
08	DOORS & WINDOWS	90.8	108.3	95.4	93.9	96.2	94.5	88.7	115.0	95.6	90.6	95.0	91.8	90.6	69.9	85.2	90.4	94.2	91.4
09200	PLASTER & GYPSUM BOARD	91.5	98.2	95.7	100.5	81.8	88.9	99.2	135.2	121.6	102.6	84.8	91.5	102.3	65.1	79.2	103.8	86.5	93.0
095,098	CEILINGS & ACOUSTICAL TREATMENT	87.8	98.2	94.0	88.6	81.8	84.6	98.3	135.2	120.3	91.2	84.8	87.4	91.2	65.1	75.6	90.2	86.5	88.0
09600	FLOORING	88.7	102.6	92.5	81.6	63.4	76.7	82.2	138.6	97.5	83.5	91.1	85.5	82.4	62.1	76.9	84.3	91.2	86.2
097,099	WALL FINISHES, PAINTS & COATINGS	99.5	105.8	103.3	91.6	98.6	95.9	94.6	138.4	121.3	91.6	97.1	95.0	91.6	97.1	95.0	91.6	87.3	89.0
09	FINISHES	92.0	99.7	96.0	92.7	80.2	86.2	98.7	134.3	117.2	93.6	88.7	91.0	93.1	68.8	80.4	93.2	87.4	90.2
10 - 14	TOTAL DIV. 10000 - 14000	100.0	103.0	100.6	100.0	96.2	99.2	100.0	123.6	105.0	100.0	101.2	100.3	100.0	91.7	98.2	100.0	97.4	99.4
15	MECHANICAL	95.2	96.2	95.6	95.5	92.7	94.4	95.2	130.2	109.7	95.5	85.3	91.3	95.5	89.4	93.0	100.3	94.0	97.7
16	ELECTRICAL	93.3	105.1	99.0	94.9	90.3	92.6	91.9	108.2	99.8	95.6	91.7	93.7	93.9	60.6	77.7	94.5	86.8	90.7
01 - 16	WEIGHTED AVERAGE	94.0	102.8	98.0	96.4	94.2	95.4	97.6	122.0	108.5	95.2	94.9	95.1	93.4	82.4	88.5	96.2	95.3	95.8

	DIVISION	PUERTO RICO			RHODE ISLAND						SOUTH CAROLINA								
		SAN JUAN			NEWPORT			PROVIDENCE			AIKEN			BEAUFORT			CHARLESTON		
		009			028			029			298			299			294		
		MAT.	INST.	TOTAL	MAT.	INST.	TOTAL	MAT.	INST.	TOTAL	MAT.	INST.	TOTAL	MAT.	INST.	TOTAL	MAT.	INST.	TOTAL
01590	EQUIPMENT RENTAL	.0	89.1	89.1	.0	103.4	103.4	.0	103.4	103.4	.0	99.3	99.3	.0	99.3	99.3	.0	99.3	99.3
02	SITE CONSTRUCTION	116.4	91.8	98.4	82.5	103.7	98.1	81.9	104.0	98.1	119.1	83.5	93.0	113.9	80.2	89.2	96.5	80.9	85.1
03100	CONCRETE FORMS & ACCESSORIES	94.9	19.5	29.9	106.1	117.6	116.0	104.9	117.7	115.9	100.0	72.1	75.9	98.2	28.8	38.3	97.1	37.3	45.5
03200	CONCRETE REINFORCEMENT	193.0	12.0	101.1	107.5	115.8	111.7	107.5	115.8	111.7	95.3	71.1	83.0	94.4	29.5	61.4	94.2	67.2	80.5
03300	CAST-IN-PLACE CONCRETE	108.0	30.4	76.1	83.7	111.0	94.9	88.5	111.0	97.7	72.8	73.0	72.8	72.7	39.9	59.2	85.3	47.8	69.9
03	CONCRETE	114.5	22.5	68.8	100.1	114.5	107.2	102.3	114.5	108.4	103.8	73.3	88.7	100.8	34.8	68.0	94.3	48.3	71.5
04	MASONRY	220.6	16.8	94.3	106.5	128.4	120.1	108.9	128.4	121.0	74.5	64.2	68.1	89.0	25.0	49.3	90.3	35.1	56.1
05	METALS	114.8	30.4	87.8	99.7	109.9	103.0	99.7	109.9	103.0	91.2	88.0	90.1	91.2	64.0	82.5	92.7	80.6	88.8
06	WOOD & PLASTICS	91.1	19.8	53.9	102.4	116.5	109.8	102.4	116.5	109.8	100.5	74.0	86.7	98.4	29.0	62.3	97.0	36.5	65.5
07	THERMAL & MOISTURE PROTECTION	148.2	23.2	95.6	102.1	116.5	108.2	101.0	116.5	107.5	102.6	71.2	89.3	102.2	33.1	73.1	101.4	42.6	76.7
08	DOORS & WINDOWS	153.5	16.5	117.7	100.8	115.8	104.7	100.8	115.8	104.7	92.6	70.2	86.8	92.6	29.0	76.0	96.6	42.1	82.4
09200	PLASTER & GYPSUM BOARD	258.0	17.4	108.5	103.3	116.1	111.2	103.3	116.1	111.2	106.3	73.1	85.7	110.3	26.9	58.5	109.6	34.6	63.0
095,098	CEILINGS & ACOUSTICAL TREATMENT	334.4	17.4	145.3	95.9	116.1	107.9	97.8	116.1	108.7	94.1	73.1	81.6	100.9	26.9	56.8	100.9	34.6	61.3
09600	FLOORING	199.8	16.4	150.1	100.5	129.7	108.4	100.5	129.7	108.4	104.8	16.8	81.0	106.6	30.7	86.0	106.0	39.2	87.9
097,099	WALL FINISHES, PAINTS & COATINGS	199.0	15.2	86.8	94.0	120.2	110.0	94.0	120.2	110.0	110.4	74.7	88.6	110.4	26.0	58.9	110.4	38.2	66.3
09	FINISHES	250.3	18.7	129.7	100.0	120.7	110.8	100.3	120.7	110.9	101.3	62.0	80.8	103.8	28.8	64.7	101.9	37.2	68.2
10 - 14	TOTAL DIV. 10000 - 14000	100.0	19.6	82.9	100.0	97.4	99.4	100.0	97.4	99.4	100.0	74.0	94.4	100.0	37.7	86.7	100.0	64.6	92.4
15	MECHANICAL	103.7	15.3	67.0	100.1	109.6	104.0	100.1	109.6	104.0	95.1	68.8	84.2	95.1	25.4	66.2	100.1	43.0	76.4
16	ELECTRICAL	129.8	15.3	74.2	96.7	105.4	101.0	96.5	105.4	100.8	95.6	72.4	84.3	98.7	25.9	63.3	97.3	32.4	65.8
01 - 16	WEIGHTED AVERAGE	135.2	25.5	86.5	99.6	112.7	105.4	99.9	112.7	105.6	96.1	72.0	85.4	97.0	36.2	70.0	97.3	47.9	75.3

	DIVISION	SOUTH CAROLINA															SOUTH DAKOTA		
		COLUMBIA			FLORENCE			GREENVILLE			ROCK HILL			SPARTANBURG			ABERDEEN		
		290 - 292			295			296			297			293			574		
		MAT.	INST.	TOTAL	MAT.	INST.	TOTAL	MAT.	INST.	TOTAL	MAT.	INST.	TOTAL	MAT.	INST.	TOTAL	MAT.	INST.	TOTAL
01590	EQUIPMENT RENTAL	.0	99.3	99.3	.0	99.3	99.3	.0	99.3	99.3	.0	99.3	99.3	.0	99.3	99.3	.0	98.3	98.3
02	SITE CONSTRUCTION	95.8	80.9	84.9	107.4	80.9	88.0	102.3	80.6	86.3	99.4	80.2	85.3	102.1	80.6	86.3	86.1	93.5	91.5
03100	CONCRETE FORMS & ACCESSORIES	101.9	39.5	48.1	82.4	39.4	45.3	96.9	39.2	47.2	95.1	27.8	37.0	100.8	39.2	47.7	95.6	38.8	46.6
03200	CONCRETE REINFORCEMENT	94.2	67.2	80.5	93.9	67.2	80.3	93.8	47.4	70.2	94.6	18.3	55.8	93.8	47.4	70.2	99.9	47.4	73.2
03300	CAST-IN-PLACE CONCRETE	80.5	47.4	66.9	72.7	47.7	62.4	72.7	47.6	62.4	72.7	35.2	57.3	72.7	47.6	62.4	102.6	49.5	80.8
03	CONCRETE	92.3	49.2	70.9	94.8	49.2	72.2	93.5	45.5	69.7	91.6	30.7	61.3	93.8	45.5	69.8	98.9	45.6	72.4
04	MASONRY	89.1	36.0	56.1	74.6	35.1	50.1	72.5	35.1	49.3	96.6	23.0	50.9	74.5	35.1	50.1	108.2	57.1	76.5
05	METALS	92.7	80.5	88.8	91.7	80.4	88.1	91.7	73.0	85.7	91.2	59.6	81.1	91.7	73.0	85.7	98.9	68.9	89.3
06	WOOD & PLASTICS	102.9	39.7	70.0	80.6	39.7	59.3	96.9	39.7	67.1	94.9	28.0	60.1	101.3	39.7	69.2	101.2	37.0	67.7
07	THERMAL & MOISTURE PROTECTION	101.4	42.4	76.6	101.6	42.9	76.9	101.6	42.9	76.9	101.5	28.8	70.9	101.7	42.9	77.0	98.1	50.9	78.3
08	DOORS & WINDOWS	96.6	43.8	82.8	92.7	43.8	79.9	92.6	39.2	78.6	92.6	25.9	75.2	92.6	39.2	78.7	95.1	39.7	80.6
09200	PLASTER & GYPSUM BOARD	109.6	37.9	65.0	98.8	37.9	60.9	104.9	37.9	63.2	104.1	25.9	55.5	106.7	37.9	63.9	105.5	35.6	62.1
095,098	CEILINGS & ACOUSTICAL TREATMENT	100.9	37.9	63.3	94.9	37.9	60.9	94.1	37.9	60.5	94.1	25.9	53.4	94.1	37.9	60.5	113.4	35.6	67.0
09600	FLOORING	105.8	39.2	87.8	95.7	39.2	80.4	103.3	40.0	86.1	102.1	30.3	82.7	105.0	40.0	87.4	108.1	56.1	94.1
097,099	WALL FINISHES, PAINTS & COATINGS	110.4	38.2	66.3	110.4	38.2	66.3	110.4	38.2	66.3	110.4	26.0	58.9	110.4	38.2	66.3	97.2	34.4	58.9
09	FINISHES	101.8	39.0	69.1	96.7	39.0	66.7	99.3	39.1	68.0	98.6	28.2	61.9	100.1	39.1	68.3	106.9	41.1	72.6
10 - 14	TOTAL DIV. 10000 - 14000	100.0	65.0	92.5	100.0	64.9	92.5	100.0	64.9	92.5	100.0	62.2	91.9	100.0	64.9	92.5	100.0	53.6	90.1
15	MECHANICAL	100.1	36.2	73.6	100.1	36.2	73.6	100.1	36.2	73.6	95.1	18.2	63.2	100.1	36.2	73.6	100.2	38.6	74.6
16	ELECTRICAL	97.9	34.2	66.9	95.5	19.5	58.6	97.4	31.1	65.2	97.4	19.2	59.4	97.4	31.1	65.2	96.7	59.3	78.5
01 - 16	WEIGHTED AVERAGE	97.1	47.3	75.0	95.4	45.2	73.1	95.7	45.4	73.4	95.3	33.0	67.6	95.9	45.4	73.5	99.5	53.1	78.9

DIVISION		SOUTH DAKOTA																	
		MITCHELL			MOBRIDGE			PIERRE			RAPID CITY			SIOUX FALLS			WATERTOWN		
		573			576			575			577			570 - 571			572		
		MAT.	INST.	TOTAL	MAT.	INST.	TOTAL	MAT.	INST.	TOTAL	MAT.	INST.	TOTAL	MAT.	INST.	TOTAL	MAT.	INST.	TOTAL
01590	EQUIPMENT RENTAL	.0	98.3	98.3	.0	98.3	98.3	.0	98.3	98.3	.0	98.3	98.3	.0	99.5	99.5	.0	98.3	98.3
02	SITE CONSTRUCTION	83.1	93.5	90.7	83.1	93.5	90.7	84.4	93.5	91.1	84.6	93.3	90.9	85.6	95.4	92.8	83.0	93.5	90.7
03100	CONCRETE FORMS & ACCESSORIES	94.6	40.2	47.6	84.3	38.7	45.0	94.0	40.5	47.8	103.1	36.0	45.2	94.1	41.5	48.7	80.3	38.8	44.5
03200	CONCRETE REINFORCEMENT	99.2	47.6	73.0	101.8	47.4	74.2	99.4	61.1	79.9	93.3	61.0	76.9	93.3	61.1	77.0	96.6	47.7	71.7
03300	CAST-IN-PLACE CONCRETE	99.6	48.0	78.4	99.6	49.5	79.0	99.6	45.4	77.4	98.9	43.0	75.9	103.0	48.3	80.5	99.6	49.5	79.0
03	CONCRETE	96.6	45.6	71.3	96.3	45.5	71.1	96.6	47.4	72.2	95.9	44.6	70.4	96.1	48.9	72.7	95.2	45.5	70.6
04	MASONRY	96.5	60.5	74.2	104.8	57.1	75.2	104.8	60.5	77.3	104.7	53.7	73.1	102.2	60.5	76.3	130.4	61.6	87.8
05	METALS	98.2	68.9	88.9	98.2	68.8	88.8	98.9	75.1	91.3	100.7	75.3	92.6	100.8	75.7	92.8	98.2	69.2	88.9
06	WOOD & PLASTICS	100.0	39.1	68.3	88.7	37.0	61.8	99.2	39.1	67.9	105.4	34.9	68.7	99.2	39.7	68.2	84.2	37.0	59.6
07	THERMAL & MOISTURE PROTECTION	98.0	49.9	77.7	98.0	50.9	78.2	98.2	50.1	78.0	98.6	47.5	77.1	98.2	51.3	78.4	97.7	52.1	78.5
08	DOORS & WINDOWS	93.5	40.3	79.6	96.5	38.8	81.4	98.3	44.6	84.3	99.3	42.4	84.4	99.3	45.0	85.1	93.5	38.8	79.2
09200	PLASTER & GYPSUM BOARD	104.1	37.8	62.9	99.6	35.6	59.8	103.7	37.8	62.8	105.5	33.5	60.8	105.5	38.4	63.8	96.7	35.6	58.8
095,098	CEILINGS & ACOUSTICAL TREATMENT	110.9	37.8	67.3	113.4	35.6	67.0	110.9	37.8	67.3	117.7	33.5	67.5	117.7	38.4	70.4	110.9	35.6	66.0
09600	FLOORING	107.6	56.1	93.7	101.8	56.1	89.4	107.3	40.0	89.1	107.3	75.8	98.8	107.3	79.7	99.8	100.0	56.1	88.1
097,099	WALL FINISHES, PAINTS & COATINGS	97.2	38.9	61.6	97.2	34.4	58.9	97.2	39.2	61.8	97.2	39.2	61.8	97.2	39.2	61.8	97.2	34.4	58.9
09	FINISHES	105.8	42.9	73.0	104.0	41.1	71.3	105.6	39.6	71.2	107.7	43.4	74.2	107.6	48.1	76.6	102.4	41.1	70.5
10 - 14	TOTAL DIV. 10000 - 14000	100.0	46.7	88.6	100.0	53.5	90.1	100.0	63.8	92.3	100.0	61.5	91.8	100.0	64.0	92.3	100.0	53.5	90.1
15	MECHANICAL	95.3	35.6	70.5	95.3	39.7	72.2	100.2	37.0	73.9	100.2	34.5	72.9	100.2	36.0	73.6	95.3	39.7	72.2
16	ELECTRICAL	95.2	44.4	70.5	96.7	39.6	68.9	93.0	54.7	74.4	93.5	54.8	74.7	92.9	71.9	82.7	94.4	39.6	67.8
01 - 16	WEIGHTED AVERAGE	96.9	50.8	76.4	97.5	50.4	76.6	98.8	53.5	78.7	99.4	52.2	78.4	99.2	57.3	80.6	97.9	51.0	77.1

DIVISION		TENNESSEE																	
		CHATTANOOGA			COLUMBIA			COOKEVILLE			JACKSON			JOHNSON CITY			KNOXVILLE		
		373 - 374			384			385			383			376			377 - 379		
		MAT.	INST.	TOTAL	MAT.	INST.	TOTAL	MAT.	INST.	TOTAL	MAT.	INST.	TOTAL	MAT.	INST.	TOTAL	MAT.	INST.	TOTAL
01590	EQUIPMENT RENTAL	.0	104.5	104.5	.0	97.7	97.7	.0	97.7	97.7	.0	104.9	104.9	.0	97.4	97.4	.0	97.4	97.4
02	SITE CONSTRUCTION	101.4	96.2	97.6	90.3	84.7	86.2	96.8	85.1	88.3	98.9	95.4	96.3	110.8	84.9	91.8	88.4	85.0	85.9
03100	CONCRETE FORMS & ACCESSORIES	98.3	45.6	52.8	81.0	50.6	54.8	81.2	39.8	45.5	89.0	37.2	44.3	82.9	46.4	51.4	97.2	46.4	53.4
03200	CONCRETE REINFORCEMENT	87.9	42.9	65.0	89.7	44.9	66.9	89.7	47.0	68.0	89.7	42.7	65.8	88.5	43.1	65.4	87.9	43.1	65.1
03300	CAST-IN-PLACE CONCRETE	102.0	47.8	79.7	91.4	51.4	75.0	103.6	49.0	81.1	101.1	41.3	76.5	82.0	52.4	69.8	95.8	48.0	76.1
03	CONCRETE	92.9	47.7	70.4	93.3	51.5	72.6	102.9	46.3	74.8	94.7	41.7	68.4	96.9	49.6	73.4	90.0	48.2	69.2
04	MASONRY	99.1	43.1	64.4	112.7	51.3	74.6	107.8	39.5	65.5	113.0	33.5	63.7	111.8	42.9	69.1	76.5	42.9	55.7
05	METALS	93.5	75.5	87.8	89.3	76.5	85.2	89.4	77.2	85.5	91.0	73.8	85.5	90.9	75.1	85.9	94.1	75.4	88.1
06	WOOD & PLASTICS	101.3	46.6	72.8	70.9	53.9	62.0	71.1	39.5	54.7	86.7	37.1	60.8	75.3	47.7	60.9	90.9	47.7	68.4
07	THERMAL & MOISTURE PROTECTION	97.9	50.2	77.8	93.0	56.4	77.6	93.5	46.9	73.9	95.1	39.5	71.7	92.2	47.9	73.5	90.3	48.1	72.6
08	DOORS & WINDOWS	101.2	46.5	86.9	92.6	48.2	81.0	92.7	43.0	79.7	101.8	39.2	85.5	97.0	47.6	84.1	93.7	47.6	81.6
09200	PLASTER & GYPSUM BOARD	84.6	45.7	60.4	89.5	53.2	66.9	89.5	38.4	57.7	90.1	35.9	56.4	90.0	46.9	63.2	97.3	46.9	66.0
095,098	CEILINGS & ACOUSTICAL TREATMENT	96.4	45.7	66.2	87.9	53.2	67.2	87.9	38.4	58.4	96.2	35.9	60.2	91.9	46.9	65.0	94.5	46.9	66.1
09600	FLOORING	104.4	49.4	89.5	91.3	25.6	73.6	91.5	23.9	73.2	92.8	23.9	74.2	97.7	50.8	85.0	103.5	50.8	89.2
097,099	WALL FINISHES, PAINTS & COATINGS	108.9	45.8	70.4	95.2	42.6	63.1	95.2	46.5	65.5	96.0	31.0	56.3	106.7	53.7	74.4	106.7	53.7	74.4
09	FINISHES	98.2	46.3	71.2	95.1	45.9	69.5	95.6	36.9	65.0	95.9	34.1	63.7	98.5	47.9	72.1	94.0	47.9	70.0
10 - 14	TOTAL DIV. 10000 - 14000	100.0	51.6	89.7	100.0	51.0	89.6	100.0	57.0	90.8	100.0	48.7	89.1	100.0	59.3	91.3	100.0	59.4	91.3
15	MECHANICAL	99.9	44.8	77.1	97.2	52.1	78.5	97.2	44.0	75.1	100.0	55.5	81.5	99.8	59.0	82.9	99.8	59.0	82.9
16	ELECTRICAL	104.2	52.8	79.2	93.4	44.8	69.8	94.9	43.7	70.0	100.9	51.1	76.7	92.5	51.1	72.4	97.9	59.7	79.3
01 - 16	WEIGHTED AVERAGE	98.6	54.1	78.8	94.9	55.2	77.3	96.2	49.9	75.6	98.3	51.7	77.6	97.3	56.4	79.1	94.7	57.5	78.2

DIVISION		TENNESSEE									TEXAS								
		MCKENZIE			MEMPHIS			NASHVILLE			ABILENE			AMARILLO			AUSTIN		
		382			375,380 - 381			370 - 372			795 - 796			790 - 791			786 - 787		
		MAT.	INST.	TOTAL	MAT.	INST.	TOTAL	MAT.	INST.	TOTAL	MAT.	INST.	TOTAL	MAT.	INST.	TOTAL	MAT.	INST.	TOTAL
01590	EQUIPMENT RENTAL	.0	97.7	97.7	.0	102.3	102.3	.0	105.4	105.4	.0	86.7	86.7	.0	86.7	86.7	.0	85.8	85.8
02	SITE CONSTRUCTION	96.1	85.7	88.5	93.8	91.7	92.3	93.5	98.7	97.3	100.7	85.2	89.3	101.1	86.0	90.1	89.4	84.9	86.1
03100	CONCRETE FORMS & ACCESSORIES	90.8	37.0	44.4	98.5	67.1	71.4	97.7	65.0	69.5	95.2	42.7	49.9	97.4	54.0	59.9	92.8	57.1	62.0
03200	CONCRETE REINFORCEMENT	89.9	35.3	62.1	102.3	69.3	85.5	86.9	68.2	77.4	94.5	52.6	73.2	94.5	54.0	73.9	98.8	52.0	75.0
03300	CAST-IN-PLACE CONCRETE	101.3	44.4	77.9	88.9	71.3	81.6	96.6	69.2	85.3	93.1	42.2	72.2	97.9	48.2	77.5	87.7	49.1	71.9
03	CONCRETE	101.7	41.5	71.8	89.2	70.5	79.9	90.6	68.5	79.6	88.4	45.4	67.0	90.8	52.8	71.9	80.5	54.0	67.3
04	MASONRY	110.8	53.6	75.3	89.1	76.4	81.2	84.4	65.7	72.8	98.8	53.8	70.9	100.7	51.3	70.1	92.3	54.7	69.0
05	METALS	89.4	76.6	85.3	94.8	94.4	94.7	95.1	91.5	94.0	96.3	66.6	86.8	96.3	67.1	87.0	95.4	65.0	85.7
06	WOOD & PLASTICS	81.7	34.8	57.3	96.7	67.9	81.7	97.6	66.2	81.2	96.6	42.0	68.2	98.0	57.1	76.7	93.4	60.9	76.5
07	THERMAL & MOISTURE PROTECTION	93.5	52.2	76.1	91.9	74.4	84.5	95.9	66.7	83.6	99.6	49.8	78.6	101.6	48.9	79.4	88.8	54.1	74.2
08	DOORS & WINDOWS	92.7	35.4	77.7	100.8	69.3	92.6	98.5	67.4	90.4	92.8	45.1	80.3	92.8	51.4	81.9	94.5	58.8	85.2
09200	PLASTER & GYPSUM BOARD	93.9	33.6	56.4	94.4	67.4	77.7	92.4	65.8	75.9	83.8	41.1	57.3	83.8	56.6	66.9	88.1	60.5	71.0
095,098	CEILINGS & ACOUSTICAL TREATMENT	87.9	33.6	55.5	104.2	67.4	82.3	94.8	65.8	77.5	89.7	41.1	60.7	89.7	56.6	70.0	87.7	60.5	71.5
09600	FLOORING	95.1	49.1	82.7	97.7	45.1	83.4	107.6	77.3	99.4	115.3	68.0	102.5	115.1	63.8	101.2	100.5	48.9	86.6
097,099	WALL FINISHES, PAINTS & COATINGS	95.2	54.2	70.2	99.8	59.4	75.2	110.5	68.1	84.6	95.8	54.6	70.7	95.8	37.8	60.4	92.5	42.5	62.0
09	FINISHES	97.3	39.9	67.4	96.5	61.6	78.3	103.4	67.7	84.8	96.4	48.3	71.4	96.4	54.4	74.5	91.4	54.1	72.0
10 - 14	TOTAL DIV. 10000 - 14000	100.0	34.3	86.0	100.0	80.4	95.8	100.0	79.4	95.6	100.0	73.3	94.3	100.0	61.9	91.9	100.0	62.1	91.9
15	MECHANICAL	97.2	72.7	87.0	99.9	75.5	89.8	99.8	71.0	87.9	100.1	45.1	77.3	100.1	51.5	79.9	99.8	59.5	83.1
16	ELECTRICAL	94.7	67.2	81.3	98.3	76.7	87.8	102.3	69.4	86.3	96.5	48.8	73.3	97.0	60.0	79.0	99.5	69.8	85.0
01 - 16	WEIGHTED AVERAGE	96.4	59.4	80.0	96.6	76.2	87.5	97.5	73.7	86.9	96.6	53.3	77.4	97.2	58.0	79.8	94.6	61.7	80.0

DIVISION		TEXAS																	
		BEAUMONT			BROWNWOOD			BRYAN			CHILDRESS			CORPUS CHRISTI			DALLAS		
		776 - 777			768			778			792			783 - 784			752 - 753		
		MAT.	INST.	TOTAL	MAT.	INST.	TOTAL	MAT.	INST.	TOTAL	MAT.	INST.	TOTAL	MAT.	INST.	TOTAL	MAT.	INST.	TOTAL
01590	**EQUIPMENT RENTAL**	.0	88.6	88.6	.0	86.7	86.7	.0	88.6	88.6	.0	86.7	86.7	.0	94.6	94.6	.0	98.1	98.1
02	**SITE CONSTRUCTION**	96.9	87.1	89.7	107.4	83.9	90.2	86.5	88.4	87.9	112.9	84.5	92.1	124.2	80.2	92.0	126.2	86.3	97.0
03100	CONCRETE FORMS & ACCESSORIES	102.3	53.0	59.8	93.3	28.8	37.7	79.0	43.2	48.1	93.5	52.4	58.0	96.3	38.7	46.6	93.7	59.8	64.5
03200	CONCRETE REINFORCEMENT	92.9	44.1	68.1	94.9	26.5	60.1	95.3	52.0	73.3	94.7	41.3	67.5	97.9	50.1	73.6	96.8	56.2	76.2
03300	CAST-IN-PLACE CONCRETE	87.4	54.5	73.9	100.5	37.4	74.6	70.1	67.8	69.1	95.6	46.5	75.4	104.0	45.5	79.9	96.6	56.6	80.2
03	**CONCRETE**	86.4	52.7	69.7	95.0	32.6	64.0	72.7	54.4	63.6	95.5	49.0	72.4	90.1	45.2	67.8	88.4	59.5	74.0
04	**MASONRY**	100.4	58.8	74.6	126.2	49.7	78.7	139.9	63.6	92.6	102.9	49.6	69.8	84.0	50.0	62.9	102.3	61.1	76.8
05	**METALS**	99.8	64.2	88.4	94.5	51.0	80.6	100.2	68.4	90.0	94.3	59.2	83.1	94.9	77.9	89.5	93.4	82.3	89.9
06	**WOOD & PLASTICS**	110.9	53.4	81.0	95.1	27.6	59.9	77.1	35.6	55.5	96.0	57.1	75.7	105.9	38.1	70.6	95.8	60.4	77.4
07	**THERMAL & MOISTURE PROTECTION**	105.1	58.5	85.5	99.7	41.2	75.1	99.8	60.1	83.1	100.2	46.1	77.4	91.3	46.9	72.6	99.5	63.6	84.4
08	**DOORS & WINDOWS**	97.4	49.2	84.8	88.9	28.5	73.1	99.3	46.3	85.5	89.9	48.8	79.2	101.5	39.9	85.4	105.5	55.9	92.6
09200	PLASTER & GYPSUM BOARD	93.6	52.8	68.3	82.7	26.3	47.7	83.3	34.5	53.0	82.7	56.6	66.5	88.1	36.9	56.3	87.6	59.8	70.3
095,098	CEILINGS & ACOUSTICAL TREATMENT	91.3	52.8	68.3	84.5	26.3	49.8	85.1	34.5	54.9	84.5	56.6	67.9	87.7	36.9	57.4	92.9	59.8	73.2
09600	FLOORING	112.1	72.0	101.2	113.9	40.6	94.1	83.6	65.2	78.6	112.8	42.5	93.8	113.6	48.4	95.9	104.3	59.3	92.1
097,099	WALL FINISHES, PAINTS & COATINGS	94.0	51.0	67.8	95.8	30.4	55.9	89.3	61.8	72.5	95.8	40.2	61.9	105.9	41.9	66.8	104.7	54.1	73.8
09	**FINISHES**	92.8	56.1	73.6	95.0	30.8	61.6	80.5	47.5	63.3	95.3	49.9	71.6	97.1	40.1	67.4	98.4	59.0	77.8
10 - 14	**TOTAL DIV. 10000 - 14000**	100.0	77.3	95.2	100.0	44.3	88.1	100.0	63.2	92.1	100.0	61.2	91.7	100.0	74.0	94.5	100.0	78.8	95.5
15	**MECHANICAL**	100.0	66.1	85.9	95.1	29.7	68.0	95.1	72.2	85.6	95.1	50.2	76.5	99.8	46.0	77.5	99.8	67.8	86.5
16	**ELECTRICAL**	94.4	72.5	83.7	92.8	28.7	61.6	94.0	70.9	82.8	96.5	39.1	68.6	97.6	54.3	76.6	96.1	65.5	81.2
01 - 16	**WEIGHTED AVERAGE**	97.1	64.1	82.4	96.5	39.5	71.2	94.7	64.5	81.3	96.1	52.5	76.7	97.3	53.1	77.7	98.3	67.1	84.4

DIVISION		TEXAS																	
		DEL RIO			DENTON			EASTLAND			EL PASO			FORT WORTH			GALVESTON		
		788			762			764			798 - 799,885			760 - 761			775		
		MAT.	INST.	TOTAL	MAT.	INST.	TOTAL	MAT.	INST.	TOTAL	MAT.	INST.	TOTAL	MAT.	INST.	TOTAL	MAT.	INST.	TOTAL
01590	**EQUIPMENT RENTAL**	.0	85.8	85.8	.0	93.9	93.9	.0	86.7	86.7	.0	86.7	86.7	.0	86.7	86.7	.0	99.1	99.1
02	**SITE CONSTRUCTION**	109.4	82.7	89.8	108.1	77.5	85.6	110.7	84.0	91.1	101.2	84.8	89.2	100.6	85.6	89.6	126.4	85.9	96.7
03100	CONCRETE FORMS & ACCESSORIES	90.5	25.3	34.3	101.1	38.1	46.7	94.4	33.9	42.2	94.7	47.2	53.7	96.2	59.0	64.1	88.4	68.1	70.9
03200	CONCRETE REINFORCEMENT	98.8	18.6	58.0	96.3	52.5	74.1	95.1	42.9	68.6	94.5	48.9	71.4	94.5	56.1	75.0	94.8	64.1	79.2
03300	CAST-IN-PLACE CONCRETE	111.9	32.3	79.2	77.7	49.4	66.1	106.2	38.2	78.2	95.4	37.0	71.4	91.6	51.6	75.2	93.0	69.1	83.2
03	**CONCRETE**	103.5	27.7	65.9	76.1	46.7	61.5	99.2	38.0	68.8	89.4	44.8	67.3	87.7	56.6	72.2	87.8	69.2	78.5
04	**MASONRY**	98.6	44.8	65.2	134.4	51.3	82.9	97.3	48.7	67.2	97.5	51.4	68.9	94.3	61.1	73.7	96.6	63.6	76.2
05	**METALS**	94.1	46.4	78.8	94.1	78.6	89.1	94.3	53.9	81.4	96.1	62.3	85.3	93.1	68.3	85.2	101.9	89.2	97.9
06	**WOOD & PLASTICS**	89.4	25.6	56.2	106.3	38.2	70.8	101.4	34.6	66.6	96.6	50.5	72.6	103.3	60.2	80.8	93.6	68.6	80.6
07	**THERMAL & MOISTURE PROTECTION**	88.8	34.2	65.8	99.6	47.8	77.8	100.1	41.5	75.4	99.2	53.9	80.1	99.9	54.8	80.9	99.5	66.1	85.4
08	**DOORS & WINDOWS**	95.3	21.9	76.1	104.7	42.9	88.6	77.5	31.9	65.5	92.8	45.8	80.5	86.8	55.8	78.7	104.8	67.4	95.0
09200	PLASTER & GYPSUM BOARD	84.3	24.3	47.0	87.2	37.1	56.1	82.7	33.5	52.1	83.8	49.8	62.7	83.8	59.8	68.9	89.7	68.3	76.4
095,098	CEILINGS & ACOUSTICAL TREATMENT	84.3	24.3	48.5	90.5	37.1	58.6	84.5	33.5	54.1	89.7	49.8	65.9	89.7	59.8	71.9	88.5	68.3	76.5
09600	FLOORING	97.2	19.5	76.2	105.9	47.4	90.1	148.1	29.3	115.9	115.3	69.5	102.9	149.4	47.4	121.8	98.4	65.2	89.4
097,099	WALL FINISHES, PAINTS & COATINGS	92.5	26.8	52.4	106.7	30.8	60.4	97.0	31.8	57.2	95.8	36.3	59.5	97.0	54.0	70.7	102.3	58.6	75.6
09	**FINISHES**	90.2	24.8	56.1	92.5	38.5	64.3	105.4	33.0	67.7	96.4	50.7	72.6	106.6	56.4	80.5	90.6	66.5	78.1
10 - 14	**TOTAL DIV. 10000 - 14000**	100.0	41.6	87.6	100.0	46.1	88.5	100.0	44.8	88.2	100.0	59.9	91.5	100.0	78.3	95.4	100.0	82.7	96.3
15	**MECHANICAL**	94.9	19.4	63.5	95.1	47.5	75.4	95.1	30.0	68.1	100.1	34.4	72.8	100.1	59.7	83.3	95.0	72.0	85.5
16	**ELECTRICAL**	99.2	20.4	60.9	95.7	46.5	71.8	92.8	35.7	65.0	95.7	56.2	76.5	95.7	63.8	80.2	95.2	56.4	76.3
01 - 16	**WEIGHTED AVERAGE**	96.5	33.2	68.4	96.5	51.6	76.6	95.6	42.0	71.8	96.6	51.6	76.6	96.2	62.8	81.4	97.3	70.6	85.4

DIVISION		TEXAS																	
		GIDDINGS			GREENVILLE			HOUSTON			HUNTSVILLE			LAREDO			LONGVIEW		
		789			754			770 - 772			773			780			756		
		MAT.	INST.	TOTAL	MAT.	INST.	TOTAL	MAT.	INST.	TOTAL	MAT.	INST.	TOTAL	MAT.	INST.	TOTAL	MAT.	INST.	TOTAL
01590	**EQUIPMENT RENTAL**	.0	85.8	85.8	.0	95.3	95.3	.0	98.8	98.8	.0	88.6	88.6	.0	85.8	85.8	.0	88.9	88.9
02	**SITE CONSTRUCTION**	95.9	83.2	86.6	117.2	81.2	90.8	124.2	85.6	95.9	106.3	85.9	91.4	89.9	84.7	86.1	109.7	89.3	94.7
03100	CONCRETE FORMS & ACCESSORIES	88.6	33.9	41.4	84.5	28.6	36.3	90.8	68.2	71.3	87.3	35.3	42.4	90.5	39.0	46.1	80.0	29.6	36.5
03200	CONCRETE REINFORCEMENT	99.5	43.7	71.1	97.2	26.6	61.3	94.6	64.2	79.2	95.6	29.1	61.8	98.8	50.3	74.2	96.2	24.3	59.7
03300	CAST-IN-PLACE CONCRETE	94.8	38.1	71.5	88.0	37.7	67.3	90.2	69.2	81.5	96.3	45.4	75.4	80.0	61.3	72.3	102.3	34.9	74.6
03	**CONCRETE**	86.0	38.1	62.2	81.9	33.4	57.8	85.8	69.3	77.6	92.8	38.8	66.0	80.2	49.8	65.1	94.5	31.6	63.3
04	**MASONRY**	105.8	50.7	71.6	157.1	50.0	90.7	96.4	65.5	77.2	138.5	34.0	73.7	92.6	53.8	68.5	153.0	46.6	87.0
05	**METALS**	93.6	54.1	81.0	91.2	64.5	82.7	104.4	89.5	99.6	100.1	53.3	85.2	96.3	63.9	85.9	84.5	50.1	73.5
06	**WOOD & PLASTICS**	88.5	34.6	60.4	86.4	27.7	55.8	96.1	68.6	81.8	86.9	36.9	60.9	89.4	37.9	62.6	79.3	30.0	53.6
07	**THERMAL & MOISTURE PROTECTION**	89.3	42.0	69.4	99.3	42.5	75.4	99.4	66.9	85.7	100.9	39.0	74.8	87.9	51.1	72.4	98.9	36.9	72.8
08	**DOORS & WINDOWS**	94.5	31.8	78.1	104.3	28.6	84.5	107.1	67.4	96.7	99.3	31.7	81.6	95.1	40.6	80.9	93.9	27.5	76.6
09200	PLASTER & GYPSUM BOARD	82.9	33.5	52.2	83.7	26.3	48.1	91.9	68.3	77.2	87.7	35.9	55.5	85.1	36.9	55.2	79.7	28.8	48.1
095,098	CEILINGS & ACOUSTICAL TREATMENT	83.4	33.5	53.6	89.5	26.3	51.8	92.8	68.3	78.2	85.1	35.9	55.8	87.7	36.9	57.4	85.2	28.8	51.6
09600	FLOORING	96.9	29.3	78.6	99.0	47.4	85.0	99.6	65.2	90.3	87.3	29.3	71.6	97.1	48.4	83.9	105.6	41.8	88.3
097,099	WALL FINISHES, PAINTS & COATINGS	92.5	31.8	55.4	104.7	26.2	56.8	102.3	61.8	77.6	89.3	32.6	54.7	92.5	55.5	69.9	96.2	34.9	58.8
09	**FINISHES**	89.0	32.9	59.8	94.9	31.5	61.9	97.3	66.9	81.5	83.2	34.2	57.7	90.1	41.5	64.8	99.0	32.3	64.2
10 - 14	**TOTAL DIV. 10000 - 14000**	100.0	43.1	87.9	100.0	44.2	88.1	100.0	82.8	96.3	100.0	43.2	87.9	100.0	71.4	93.9	100.0	24.3	83.9
15	**MECHANICAL**	94.9	30.0	68.0	94.9	29.2	67.6	99.9	72.4	88.5	95.1	28.3	67.4	99.8	41.5	75.6	94.8	28.8	67.4
16	**ELECTRICAL**	96.3	35.7	66.9	92.8	28.7	61.7	97.0	71.0	84.3	94.0	35.0	65.4	99.3	62.8	81.5	92.4	55.9	74.6
01 - 16	**WEIGHTED AVERAGE**	94.0	42.1	71.0	97.7	40.7	72.4	99.6	73.0	87.7	97.7	40.3	72.2	94.6	53.7	76.5	96.9	42.6	72.8

City Cost Indexes

	DIVISION	TEXAS																	
		LUBBOCK			LUFKIN			MCALLEN			MCKINNEY			MIDLAND			ODESSA		
		793 - 794			759			785			750			797			797		
		MAT.	INST.	TOTAL	MAT.	INST.	TOTAL	MAT.	INST.	TOTAL	MAT.	INST.	TOTAL	MAT.	INST.	TOTAL	MAT.	INST.	TOTAL
01590	EQUIPMENT RENTAL	.0	96.9	96.9	.0	88.9	88.9	.0	94.6	94.6	.0	95.3	95.3	.0	96.9	96.9	.0	86.7	86.7
02	SITE CONSTRUCTION	130.2	83.3	95.8	103.4	89.3	93.1	128.5	80.2	93.1	112.3	82.0	90.1	133.9	83.3	96.8	100.9	85.5	89.6
03100	CONCRETE FORMS & ACCESSORIES	94.8	41.9	49.2	83.6	41.4	47.2	95.1	36.5	44.6	83.3	43.7	49.1	99.1	40.0	48.1	95.1	39.9	47.5
03200	CONCRETE REINFORCEMENT	95.8	52.5	73.8	97.8	32.5	64.6	98.2	50.1	73.7	97.2	52.5	74.5	96.9	52.4	74.3	94.5	52.3	73.1
03300	CAST-IN-PLACE CONCRETE	93.3	48.3	74.8	91.6	40.4	70.5	113.0	43.8	84.6	82.5	51.1	69.6	99.2	45.5	77.1	93.1	44.3	73.1
03	CONCRETE	87.5	48.1	68.0	87.9	40.5	64.3	95.5	43.6	69.7	77.9	49.6	63.9	91.8	46.3	69.2	88.4	44.8	66.7
04	MASONRY	98.2	49.7	68.1	118.4	39.8	69.7	98.1	50.0	68.2	168.6	54.6	97.9	116.1	50.1	75.2	98.8	50.0	68.6
05	METALS	99.9	80.8	93.8	91.3	57.6	80.5	93.9	77.5	88.7	91.1	77.5	86.8	98.5	80.3	92.7	95.7	65.6	86.1
06	WOOD & PLASTICS	97.1	42.1	68.5	87.7	43.0	64.4	104.3	35.5	68.4	84.9	43.8	63.5	102.5	40.1	70.0	96.6	39.9	67.1
07	THERMAL & MOISTURE PROTECTION	91.9	50.8	74.6	98.7	42.1	74.9	91.3	45.1	71.8	99.1	51.7	79.2	92.2	47.1	73.2	99.6	46.2	77.1
08	DOORS & WINDOWS	104.1	43.3	88.2	75.7	39.7	66.3	100.1	39.0	84.1	104.3	46.0	89.0	103.0	41.8	87.0	92.8	41.7	79.4
09200	PLASTER & GYPSUM BOARD	84.5	41.1	57.5	80.0	42.1	56.5	88.1	34.3	54.6	82.6	42.8	57.9	86.6	39.0	57.0	83.8	39.0	56.0
095,098	CEILINGS & ACOUSTICAL TREATMENT	92.3	41.1	61.7	75.0	42.1	55.4	87.7	34.3	55.8	89.5	42.8	61.6	87.9	39.0	58.7	89.7	39.0	59.4
09600	FLOORING	106.6	40.2	88.6	142.2	39.0	114.3	112.9	64.5	99.8	98.5	47.4	84.7	108.4	39.7	89.8	115.3	39.7	94.9
097,099	WALL FINISHES, PAINTS & COATINGS	107.3	34.0	62.5	96.2	39.3	61.5	105.9	30.0	59.6	104.7	49.0	70.7	107.3	33.5	62.3	95.8	34.0	58.0
09	FINISHES	98.4	39.9	67.9	106.9	40.7	72.4	97.2	40.5	67.7	94.3	44.6	68.4	98.5	38.6	67.3	96.4	38.6	66.3
10 - 14	TOTAL DIV. 10000 - 14000	100.0	73.0	94.2	100.0	61.0	91.7	100.0	71.3	93.9	100.0	48.0	88.9	100.0	59.6	91.4	100.0	59.2	91.3
15	MECHANICAL	99.7	48.3	78.3	94.8	33.5	69.4	94.9	32.5	69.0	94.9	49.3	76.0	94.7	37.7	71.1	100.1	37.7	74.2
16	ELECTRICAL	95.1	45.9	71.2	93.6	49.0	71.9	96.9	36.3	67.4	92.9	55.9	74.9	95.1	42.8	69.6	96.5	42.7	70.4
01 - 16	WEIGHTED AVERAGE	98.6	53.5	78.6	94.5	46.7	73.3	97.2	47.4	75.1	97.6	55.5	78.9	98.7	49.9	77.0	96.6	48.5	75.2

	DIVISION	TEXAS																	
		PALESTINE			SAN ANGELO			SAN ANTONIO			TEMPLE			TEXARKANA			TYLER		
		758			769			781 - 782			765			755			757		
		MAT.	INST.	TOTAL	MAT.	INST.	TOTAL	MAT.	INST.	TOTAL	MAT.	INST.	TOTAL	MAT.	INST.	TOTAL	MAT.	INST.	TOTAL
01590	EQUIPMENT RENTAL	.0	88.9	88.9	.0	86.7	86.7	.0	88.7	88.7	.0	86.7	86.7	.0	88.9	88.9	.0	88.9	88.9
02	SITE CONSTRUCTION	112.3	88.8	95.1	103.2	85.9	90.5	89.6	89.5	89.5	90.5	84.6	86.2	96.8	89.8	91.6	109.3	89.8	95.0
03100	CONCRETE FORMS & ACCESSORIES	74.0	33.9	39.4	93.6	37.8	45.5	90.5	58.5	62.9	97.6	36.1	44.6	91.5	36.0	43.6	85.5	34.8	41.8
03200	CONCRETE REINFORCEMENT	95.5	43.6	69.1	94.7	52.4	73.2	105.5	52.5	78.6	94.9	51.2	72.7	95.4	55.9	75.3	96.2	56.0	75.8
03300	CAST-IN-PLACE CONCRETE	83.6	37.8	64.8	94.8	42.6	73.3	78.5	68.9	74.6	77.6	48.5	65.7	84.3	44.1	67.8	100.5	43.0	76.9
03	CONCRETE	87.3	37.9	62.8	90.7	43.2	67.1	80.6	61.6	71.1	78.3	44.3	61.4	82.2	43.7	63.1	94.1	42.8	68.7
04	MASONRY	113.1	48.1	72.8	122.9	49.6	77.5	92.4	64.3	75.0	133.6	52.7	83.4	172.8	50.2	96.8	161.8	51.2	93.3
05	METALS	91.1	53.0	78.9	94.6	63.3	84.6	97.0	68.0	87.7	94.3	63.0	84.3	84.4	66.8	78.8	91.0	67.2	83.4
06	WOOD & PLASTICS	77.0	34.6	54.9	95.4	38.2	65.6	89.4	57.3	72.7	104.6	34.6	68.2	91.7	34.1	61.7	89.4	32.2	59.6
07	THERMAL & MOISTURE PROTECTION	99.1	43.1	75.5	99.5	44.4	76.3	87.9	64.8	78.1	99.1	47.2	77.2	98.5	49.1	77.7	99.0	53.5	79.8
08	DOORS & WINDOWS	75.6	32.5	64.4	88.9	40.7	76.3	97.1	56.1	86.4	75.4	37.3	65.4	93.9	40.3	79.9	75.6	39.2	66.1
09200	PLASTER & GYPSUM BOARD	74.8	33.5	49.2	82.7	37.2	54.4	85.1	56.7	67.5	84.2	33.5	52.7	83.6	33.0	52.2	80.0	31.0	49.6
095,098	CEILINGS & ACOUSTICAL TREATMENT	75.0	33.5	50.3	84.5	37.2	56.3	87.7	56.7	69.2	84.5	33.5	54.1	81.8	33.0	52.7	75.0	31.0	48.8
09600	FLOORING	131.3	29.3	103.7	114.1	35.5	92.8	97.1	55.3	85.8	151.3	48.7	123.5	115.7	64.9	102.0	144.4	43.3	117.0
097,099	WALL FINISHES, PAINTS & COATINGS	96.2	31.8	56.9	95.8	35.0	58.7	92.5	55.5	69.9	97.0	35.0	59.2	96.2	41.0	62.5	96.2	41.0	62.5
09	FINISHES	103.6	32.7	66.7	94.8	36.4	64.4	90.1	56.9	72.8	105.5	37.4	70.0	101.0	41.0	69.8	107.9	35.4	70.1
10 - 14	TOTAL DIV. 10000 - 14000	100.0	44.8	88.2	100.0	58.1	91.1	100.0	77.9	95.3	100.0	44.7	88.2	100.0	71.6	93.9	100.0	71.3	93.9
15	MECHANICAL	94.8	30.0	67.9	95.1	43.5	73.7	99.8	72.4	88.4	95.1	35.7	70.5	94.8	39.4	71.8	94.8	63.0	81.6
16	ELECTRICAL	90.1	40.5	66.0	96.5	34.1	66.1	99.5	62.8	81.7	93.9	64.7	79.7	93.5	72.5	83.3	92.4	57.9	75.6
01 - 16	WEIGHTED AVERAGE	93.6	43.0	71.1	96.2	47.7	74.7	95.0	67.0	82.5	94.5	50.3	74.9	96.6	54.0	77.7	97.4	56.2	79.1

	DIVISION	TEXAS															UTAH		
		VICTORIA			WACO			WAXAHACKIE			WHARTON			WICHITA FALLS			LOGAN		
		779			766 - 767			751			774			763			843		
		MAT.	INST.	TOTAL	MAT.	INST.	TOTAL	MAT.	INST.	TOTAL	MAT.	INST.	TOTAL	MAT.	INST.	TOTAL	MAT.	INST.	TOTAL
01590	EQUIPMENT RENTAL	.0	97.6	97.6	.0	86.7	86.7	.0	95.3	95.3	.0	99.1	99.1	.0	86.7	86.7	.0	100.7	100.7
02	SITE CONSTRUCTION	132.5	81.5	95.1	99.6	85.5	89.3	115.4	81.8	90.8	140.7	83.4	98.7	100.3	85.2	89.2	89.7	100.6	97.7
03100	CONCRETE FORMS & ACCESSORIES	89.0	41.5	48.0	95.8	41.4	48.8	83.3	48.6	53.4	82.4	37.9	44.0	95.8	43.0	50.3	100.9	59.6	65.3
03200	CONCRETE REINFORCEMENT	91.0	32.5	61.3	94.5	51.8	72.8	97.2	52.5	74.5	94.7	63.4	78.8	94.5	53.0	73.4	108.5	77.6	92.8
03300	CAST-IN-PLACE CONCRETE	104.5	41.9	78.7	84.1	53.2	71.4	87.1	41.8	68.5	107.3	49.7	83.7	89.8	47.9	72.6	86.7	70.5	80.0
03	CONCRETE	94.4	42.2	68.5	84.0	48.5	66.4	81.1	48.7	65.0	97.8	48.8	73.5	86.8	47.6	67.3	108.6	67.2	88.0
04	MASONRY	113.1	39.9	67.7	95.7	59.7	73.4	157.7	52.5	92.5	97.6	34.9	58.7	96.1	59.6	73.4	118.6	60.4	82.6
05	METALS	100.1	75.4	92.2	96.2	65.9	86.5	91.2	77.4	86.8	101.9	81.8	95.5	96.2	67.4	87.0	99.4	75.4	91.7
06	WOOD & PLASTICS	97.4	43.1	69.2	102.3	37.1	68.3	84.9	51.6	67.6	87.0	39.1	62.0	102.3	42.0	70.9	87.9	57.9	72.3
07	THERMAL & MOISTURE PROTECTION	101.9	43.5	77.3	100.1	50.6	79.2	99.2	48.8	78.0	99.8	49.6	78.7	100.1	54.2	80.8	99.4	67.7	86.1
08	DOORS & WINDOWS	104.7	41.5	88.2	86.8	38.8	74.3	104.3	50.2	90.2	104.8	43.5	88.8	86.8	45.0	75.9	89.3	59.7	81.6
09200	PLASTER & GYPSUM BOARD	88.4	42.1	59.7	83.8	36.1	54.2	82.8	50.9	63.0	86.7	37.9	56.4	83.8	41.1	57.3	82.9	56.5	66.5
095,098	CEILINGS & ACOUSTICAL TREATMENT	89.4	42.1	61.2	89.7	36.1	57.7	90.4	50.9	66.8	88.5	37.9	58.4	89.7	41.1	60.7	101.4	56.5	74.6
09600	FLOORING	97.3	39.0	81.5	149.6	36.9	119.1	98.5	39.0	82.4	95.5	29.3	77.6	150.0	77.7	130.5	100.3	59.6	89.3
097,099	WALL FINISHES, PAINTS & COATINGS	102.5	39.3	63.9	97.0	35.0	59.2	104.7	39.3	64.8	102.3	32.6	59.8	98.7	55.5	72.4	102.2	47.9	69.0
09	FINISHES	89.0	40.9	63.9	106.6	38.5	71.2	94.8	46.2	69.4	90.1	35.6	61.7	106.9	50.4	77.5	97.6	57.6	76.7
10 - 14	TOTAL DIV. 10000 - 14000	100.0	47.6	88.8	100.0	75.4	94.8	100.0	49.2	89.2	100.0	46.0	88.5	100.0	60.1	91.5	100.0	70.7	93.7
15	MECHANICAL	95.1	33.5	69.6	100.1	55.5	81.6	94.9	48.2	75.5	95.0	27.8	67.2	100.1	51.3	79.8	100.1	71.2	88.1
16	ELECTRICAL	99.8	49.0	75.1	96.5	67.1	82.2	92.8	63.1	78.4	98.8	35.1	67.8	99.6	65.6	83.1	91.0	74.5	83.0
01 - 16	WEIGHTED AVERAGE	99.2	47.6	76.3	96.4	57.5	79.1	97.5	56.3	79.2	99.1	45.1	75.1	97.1	57.9	79.7	99.1	70.5	86.4

	DIVISION	UTAH												VERMONT					
		OGDEN			PRICE			PROVO			SALT LAKE CITY			BELLOWS FALLS			BENNINGTON		
		842,844			845			846 - 847			840 - 841			051			052		
		MAT.	INST.	TOTAL	MAT.	INST.	TOTAL	MAT.	INST.	TOTAL	MAT.	INST.	TOTAL	MAT.	INST.	TOTAL	MAT.	INST.	TOTAL
01590	EQUIPMENT RENTAL	.0	100.7	100.7	.0	99.5	99.5	.0	99.5	99.5	.0	100.7	100.7	.0	101.5	101.5	.0	101.5	101.5
02	SITE CONSTRUCTION	78.4	100.6	94.6	87.3	97.4	94.7	86.3	98.6	95.3	78.4	100.5	94.6	77.7	98.0	92.6	77.2	98.0	92.4
03100	CONCRETE FORMS & ACCESSORIES	100.9	59.6	65.3	103.5	32.8	42.5	102.5	59.7	65.6	99.4	59.7	65.2	102.6	39.6	48.3	99.7	39.4	47.7
03200	CONCRETE REINFORCEMENT	113.4	77.6	95.2	116.4	77.1	96.4	117.4	77.7	97.2	110.7	77.7	93.9	85.9	78.4	82.1	85.9	78.3	82.1
03300	CAST-IN-PLACE CONCRETE	88.0	70.5	80.8	86.8	46.5	70.2	86.8	70.5	80.1	95.5	70.5	85.2	97.2	54.6	79.7	97.2	54.5	79.6
03	CONCRETE	99.3	67.2	83.3	110.1	46.9	78.7	108.6	67.2	88.0	117.1	67.2	92.3	99.5	52.8	76.3	99.3	52.6	76.1
04	MASONRY	111.9	60.4	80.0	124.3	41.7	73.0	124.3	60.4	84.7	129.2	60.4	86.6	97.1	38.1	60.5	107.2	38.1	64.3
05	METALS	99.9	75.4	92.0	97.0	73.4	89.5	97.7	75.4	90.6	103.0	75.4	94.2	94.8	71.5	87.3	94.8	71.2	87.2
06	WOOD & PLASTICS	87.9	57.9	72.3	90.8	30.4	59.4	89.5	57.9	73.0	88.1	57.9	72.3	101.5	37.5	68.2	98.2	37.5	66.6
07	THERMAL & MOISTURE PROTECTION	98.2	67.7	85.4	101.2	54.5	81.6	101.2	67.7	87.1	100.9	67.7	86.9	101.3	56.2	82.3	101.3	47.8	78.8
08	DOORS & WINDOWS	89.3	59.7	81.6	93.5	43.5	80.4	93.5	59.7	84.7	89.3	59.7	81.6	103.7	45.5	88.5	103.7	43.0	87.8
09200	PLASTER & GYPSUM BOARD	82.9	56.5	66.5	84.4	28.4	49.6	83.3	56.5	66.7	82.9	56.5	66.5	103.4	34.9	60.9	101.5	34.9	60.2
095,098	CEILINGS & ACOUSTICAL TREATMENT	101.4	56.5	74.6	101.4	28.4	57.8	101.4	56.5	74.6	100.4	56.5	74.2	85.5	34.9	55.4	85.5	34.9	55.4
09600	FLOORING	100.3	59.6	89.3	101.5	44.3	86.0	101.1	59.6	89.9	99.9	59.6	89.0	99.6	38.1	83.0	98.2	38.1	82.0
097,099	WALL FINISHES, PAINTS & COATINGS	102.2	47.9	69.0	102.2	47.9	69.0	102.2	61.1	77.1	102.2	61.1	77.1	91.5	34.1	56.5	91.5	34.1	56.5
09	FINISHES	96.5	57.6	76.2	98.6	35.0	65.5	98.2	59.1	77.8	96.6	59.1	77.1	96.1	37.8	65.8	95.5	37.8	65.4
10 - 14	TOTAL DIV. 10000 - 14000	100.0	70.7	93.7	100.0	53.5	90.1	100.0	70.7	93.7	100.0	70.7	93.7	100.0	47.3	88.8	100.0	47.3	88.8
15	MECHANICAL	100.1	71.2	88.1	97.5	40.4	73.8	100.1	71.2	88.1	100.2	71.2	88.2	95.1	36.2	70.7	95.1	39.8	72.2
16	ELECTRICAL	91.5	74.5	83.3	95.6	44.2	70.6	91.2	74.5	83.1	95.0	74.5	85.0	97.6	36.0	67.6	97.6	36.0	67.6
01 - 16	WEIGHTED AVERAGE	97.4	70.5	85.5	99.6	50.1	77.6	99.6	70.5	86.7	101.3	70.7	87.7	97.1	48.8	75.6	97.4	49.1	76.0

	DIVISION	VERMONT																	
		BRATTLEBORO			BURLINGTON			GUILDHALL			MONTPELIER			RUTLAND			ST. JOHNSBURY		
		053			054			059			056			057			058		
		MAT.	INST.	TOTAL	MAT.	INST.	TOTAL	MAT.	INST.	TOTAL	MAT.	INST.	TOTAL	MAT.	INST.	TOTAL	MAT.	INST.	TOTAL
01590	EQUIPMENT RENTAL	.0	101.5	101.5	.0	101.5	101.5	.0	101.5	101.5	.0	101.5	101.5	.0	101.5	101.5	.0	101.5	101.5
02	SITE CONSTRUCTION	78.5	98.0	92.8	80.6	98.8	93.9	76.8	97.6	92.0	79.2	98.8	93.5	80.6	98.8	93.9	76.8	97.6	92.0
03100	CONCRETE FORMS & ACCESSORIES	102.9	39.5	48.2	92.4	50.5	56.2	99.2	35.4	44.2	96.5	50.5	56.9	103.2	50.5	57.8	96.5	35.5	43.9
03200	CONCRETE REINFORCEMENT	85.1	78.4	81.7	107.5	53.9	80.2	86.7	34.3	60.1	85.5	53.9	69.4	107.5	53.9	80.2	85.1	52.8	68.7
03300	CAST-IN-PLACE CONCRETE	100.2	54.6	81.5	99.8	62.9	84.6	94.2	50.2	76.1	100.2	62.9	84.9	95.2	62.9	81.9	94.2	50.2	76.1
03	CONCRETE	102.0	52.7	77.5	104.7	56.0	80.5	97.1	41.2	69.4	101.6	56.0	79.0	103.2	56.0	79.7	96.7	44.4	70.7
04	MASONRY	106.9	38.1	64.2	99.8	60.2	75.2	107.2	39.9	65.5	90.8	60.2	71.8	87.5	60.2	70.6	134.2	39.9	75.7
05	METALS	94.8	71.4	87.3	101.4	65.9	90.0	94.8	54.7	82.0	94.8	66.0	85.6	99.7	66.0	88.9	94.8	54.8	82.0
06	WOOD & PLASTICS	101.9	37.5	68.4	89.0	48.5	67.9	97.0	34.0	64.2	92.8	48.5	69.8	102.0	48.5	74.2	92.8	34.0	62.2
07	THERMAL & MOISTURE PROTECTION	101.4	48.8	79.3	101.7	55.2	82.1	101.1	42.8	76.6	101.2	56.4	82.3	101.5	56.4	82.5	101.0	42.8	76.5
08	DOORS & WINDOWS	103.7	45.5	88.5	106.9	45.2	90.8	103.7	32.6	85.1	103.7	45.2	88.4	106.9	45.2	90.8	103.7	32.6	85.1
09200	PLASTER & GYPSUM BOARD	103.4	34.9	60.9	104.7	46.3	68.4	108.2	31.3	60.4	112.6	46.3	71.4	104.7	46.3	68.4	112.6	31.3	62.1
095,098	CEILINGS & ACOUSTICAL TREATMENT	85.5	34.9	55.4	92.6	46.3	64.9	85.5	31.3	53.2	85.5	46.3	62.1	92.6	46.3	64.9	85.5	31.3	53.2
09600	FLOORING	99.8	38.1	83.1	99.6	68.8	91.3	103.4	38.1	85.7	107.7	68.8	97.2	99.6	68.8	91.3	107.7	38.1	88.8
097,099	WALL FINISHES, PAINTS & COATINGS	91.5	34.1	56.5	91.5	39.6	59.9	91.5	26.4	51.8	91.5	39.6	59.9	91.5	39.6	59.9	91.5	27.7	52.6
09	FINISHES	96.3	37.8	65.8	98.3	52.1	74.2	97.9	34.1	64.7	100.0	52.1	75.0	98.2	52.1	74.2	99.9	34.3	65.7
10 - 14	TOTAL DIV. 10000 - 14000	100.0	47.3	88.8	100.0	92.0	98.3	100.0	45.7	88.4	100.0	92.0	98.3	100.0	92.0	98.3	100.0	45.7	88.4
15	MECHANICAL	95.1	39.8	72.2	100.1	66.1	86.0	95.1	62.4	81.6	95.1	66.1	83.1	100.1	66.1	86.0	95.1	62.4	81.6
16	ELECTRICAL	97.6	36.0	67.6	98.3	68.7	83.9	97.6	28.0	63.7	97.6	68.7	83.5	97.6	68.7	83.6	97.6	28.0	63.7
01 - 16	WEIGHTED AVERAGE	97.9	49.3	76.3	100.5	64.9	84.7	97.4	48.6	75.7	97.3	64.9	82.9	99.5	64.9	84.1	98.8	49.1	76.7

	DIVISION	VERMONT			VIRGINIA														
		WHITE RIVER JCT.			ALEXANDRIA			ARLINGTON			BRISTOL			CHARLOTTESVILLE			CULPEPER		
		050			223			222			242			229			227		
		MAT.	INST.	TOTAL	MAT.	INST.	TOTAL	MAT.	INST.	TOTAL	MAT.	INST.	TOTAL	MAT.	INST.	TOTAL	MAT.	INST.	TOTAL
01590	EQUIPMENT RENTAL	.0	101.5	101.5	.0	101.2	101.2	.0	99.6	99.6	.0	99.6	99.6	.0	104.8	104.8	.0	99.6	99.6
02	SITE CONSTRUCTION	80.6	98.3	93.6	113.3	86.2	93.4	124.2	82.5	93.6	107.4	80.6	87.8	112.0	83.7	91.3	111.5	81.9	89.8
03100	CONCRETE FORMS & ACCESSORIES	96.6	40.7	48.4	92.4	75.2	77.5	91.1	71.7	74.4	85.8	33.1	40.4	83.9	40.6	46.5	80.9	66.2	68.2
03200	CONCRETE REINFORCEMENT	85.9	31.1	58.1	83.7	85.9	84.8	94.4	62.8	78.3	94.5	72.4	83.3	93.9	71.8	82.7	94.4	56.3	75.0
03300	CAST-IN-PLACE CONCRETE	100.2	51.5	80.2	99.9	80.8	92.0	97.2	76.4	88.6	96.8	41.3	74.0	100.7	52.8	81.0	99.5	42.1	75.9
03	CONCRETE	103.6	43.4	73.7	102.7	80.2	91.5	107.5	72.7	90.2	103.6	45.3	74.6	104.1	52.6	78.5	100.9	57.3	79.3
04	MASONRY	119.9	46.0	74.1	87.0	73.7	78.7	99.1	67.8	79.7	87.3	30.8	52.2	112.0	42.1	68.7	101.0	34.3	59.6
05	METALS	94.8	54.3	81.9	101.6	98.2	100.5	100.4	85.6	95.7	99.5	83.7	94.4	99.7	88.9	96.3	99.8	80.0	93.5
06	WOOD & PLASTICS	94.9	40.4	66.5	96.0	75.6	85.3	92.4	75.6	83.6	84.0	32.4	57.1	81.9	39.1	59.6	80.5	75.6	77.9
07	THERMAL & MOISTURE PROTECTION	101.4	40.2	75.7	100.6	81.5	92.6	102.2	72.3	89.6	101.7	39.9	75.7	101.4	49.8	79.7	101.5	47.3	78.6
08	DOORS & WINDOWS	103.7	38.0	86.5	96.6	78.0	91.7	94.6	71.7	88.6	97.9	43.2	83.6	95.9	45.4	82.7	96.3	67.0	88.6
09200	PLASTER & GYPSUM BOARD	100.4	37.9	61.6	109.6	74.7	87.9	105.8	74.7	86.5	101.5	30.4	57.3	100.4	36.4	60.7	101.0	74.7	84.7
095,098	CEILINGS & ACOUSTICAL TREATMENT	85.5	37.9	57.1	100.9	74.7	85.3	94.9	74.7	82.9	94.1	30.4	56.1	94.1	36.4	59.7	94.9	74.7	82.9
09600	FLOORING	97.1	38.1	81.1	106.0	91.4	102.0	104.2	67.5	94.2	99.6	33.7	81.8	97.9	39.8	82.2	97.9	66.7	89.5
097,099	WALL FINISHES, PAINTS & COATINGS	91.5	21.1	48.5	121.8	84.2	98.8	121.8	84.2	98.8	110.4	32.8	63.0	110.4	54.9	76.5	121.8	39.2	71.4
09	FINISHES	95.2	37.9	65.3	102.3	78.8	90.1	100.6	72.6	86.0	97.2	32.3	63.4	96.6	40.6	67.4	97.2	65.2	80.5
10 - 14	TOTAL DIV. 10000 - 14000	100.0	47.4	88.8	100.0	86.6	97.1	100.0	68.2	93.2	100.0	70.1	93.6	100.0	74.1	94.5	100.0	55.1	90.4
15	MECHANICAL	95.1	41.0	72.7	99.9	86.3	94.3	100.1	82.7	92.8	95.1	55.6	78.8	95.1	65.7	82.9	95.1	38.3	71.5
16	ELECTRICAL	97.6	32.1	65.7	98.3	95.1	96.8	95.8	95.2	95.5	97.7	29.9	64.7	97.7	71.5	84.9	100.0	48.1	74.8
01 - 16	WEIGHTED AVERAGE	98.6	46.5	75.4	99.9	84.9	93.3	100.5	79.3	91.1	97.9	49.0	76.2	99.1	61.5	82.4	98.5	55.3	79.3

DIVISION		VIRGINIA																	
		FAIRFAX			FARMVILLE			FREDERICKSBURG			GRUNDY			HARRISONBURG			LYNCHBURG		
		220 - 221			239			224 - 225			246			228			245		
		MAT.	INST.	TOTAL	MAT.	INST.	TOTAL	MAT.	INST.	TOTAL	MAT.	INST.	TOTAL	MAT.	INST.	TOTAL	MAT.	INST.	TOTAL
01590	EQUIPMENT RENTAL	.0	99.6	99.6	.0	104.8	104.8	.0	99.6	99.6	.0	99.6	99.6	.0	99.6	99.6	.0	99.6	99.6
02	SITE CONSTRUCTION	122.9	82.5	93.3	108.1	82.9	89.6	111.0	82.0	89.7	105.2	80.3	86.9	119.6	82.0	92.0	105.9	80.8	87.5
03100	CONCRETE FORMS & ACCESSORIES	84.2	71.8	73.5	98.2	27.2	36.9	84.2	45.6	50.9	89.4	29.0	37.3	79.6	28.0	35.1	85.8	36.3	43.1
03200	CONCRETE REINFORCEMENT	94.4	85.8	90.0	94.5	45.5	69.6	95.2	84.2	89.6	93.2	45.9	69.2	94.4	56.2	75.0	93.9	72.5	83.0
03300	CAST-IN-PLACE CONCRETE	97.2	76.3	88.6	100.6	45.8	78.1	98.7	47.4	77.6	96.8	40.4	73.6	97.1	58.9	81.4	96.8	46.4	76.1
03	CONCRETE	107.0	77.0	92.1	103.3	38.7	71.2	100.6	55.2	78.1	102.3	37.9	70.3	104.6	46.1	75.6	102.1	48.5	75.5
04	MASONRY	99.0	67.3	79.3	96.1	32.3	56.5	99.4	35.3	59.6	90.7	33.3	55.1	98.2	32.8	57.6	104.9	33.4	60.6
05	METALS	99.9	94.4	98.1	99.7	63.8	88.2	99.8	91.7	97.2	99.5	67.0	89.1	99.7	79.9	93.4	99.7	84.0	94.6
06	WOOD & PLASTICS	84.0	75.6	79.6	98.3	26.1	60.7	84.0	45.9	64.2	87.7	28.0	56.6	79.2	22.9	49.9	84.0	35.3	58.6
07	THERMAL & MOISTURE PROTECTION	102.0	72.1	89.4	101.5	33.9	73.0	101.5	45.0	77.7	101.7	36.6	74.3	101.8	51.7	80.7	101.5	41.9	76.4
08	DOORS & WINDOWS	94.6	77.2	90.0	96.2	26.0	77.9	95.9	57.6	85.9	97.9	27.9	79.6	96.2	38.5	81.1	96.2	44.7	82.8
09200	PLASTER & GYPSUM BOARD	101.8	74.7	85.0	107.8	23.1	55.2	101.8	44.3	66.0	102.6	25.8	54.9	100.4	20.7	50.9	101.5	33.4	59.2
095,098	CEILINGS & ACOUSTICAL TREATMENT	94.9	74.7	82.9	94.1	23.1	51.7	94.9	44.3	64.7	94.1	25.8	53.4	94.1	20.7	50.3	94.1	33.4	57.9
09600	FLOORING	99.9	58.2	88.6	106.0	54.4	92.0	99.9	66.7	90.9	101.4	30.4	82.1	97.4	66.7	89.1	99.6	37.0	82.6
097,099	WALL FINISHES, PAINTS & COATINGS	121.8	84.2	98.8	110.4	34.7	64.2	121.8	39.2	71.4	110.4	32.8	63.0	121.8	42.7	73.5	110.4	32.8	63.0
09	FINISHES	98.8	70.9	84.2	99.8	32.9	65.0	97.9	47.9	71.8	97.8	29.2	62.0	97.3	35.2	65.0	97.1	35.1	64.8
10 - 14	TOTAL DIV. 10000 - 14000	100.0	68.0	93.2	100.0	44.2	88.1	100.0	72.3	94.1	100.0	43.5	88.0	100.0	69.8	93.6	100.0	71.2	93.9
15	MECHANICAL	95.1	82.6	89.9	95.1	26.5	66.6	95.1	78.1	88.0	95.1	34.5	70.0	95.1	38.9	71.8	95.1	65.1	82.7
16	ELECTRICAL	98.7	95.2	97.0	95.2	44.1	70.3	96.0	95.2	95.6	97.7	58.0	78.4	97.9	36.9	68.3	98.7	29.2	64.9
01 - 16	WEIGHTED AVERAGE	99.3	80.6	91.0	98.3	41.1	72.9	98.0	68.4	84.9	98.0	44.4	74.2	98.8	47.1	75.8	98.6	52.2	78.0

DIVISION		VIRGINIA																	
		NEWPORT NEWS			NORFOLK			PETERSBURG			PORTSMOUTH			PULASKI			RICHMOND		
		236			233 - 235			238			237			243			230 - 232		
		MAT.	INST.	TOTAL	MAT.	INST.	TOTAL	MAT.	INST.	TOTAL	MAT.	INST.	TOTAL	MAT.	INST.	TOTAL	MAT.	INST.	TOTAL
01590	EQUIPMENT RENTAL	.0	104.8	104.8	.0	105.5	105.5	.0	104.8	104.8	.0	104.7	104.7	.0	99.6	99.6	.0	104.8	104.8
02	SITE CONSTRUCTION	106.7	84.5	90.4	106.0	85.6	91.1	110.4	84.9	91.7	105.3	83.9	89.6	104.7	80.1	86.7	107.3	84.9	90.9
03100	CONCRETE FORMS & ACCESSORIES	97.2	59.4	64.6	101.7	59.6	65.4	89.7	56.0	60.6	84.9	59.4	62.9	89.4	28.0	36.4	98.2	56.0	61.8
03200	CONCRETE REINFORCEMENT	94.2	74.6	84.2	94.2	74.6	84.2	93.9	72.1	82.8	93.9	74.6	84.1	93.2	45.9	69.1	94.2	72.1	83.0
03300	CAST-IN-PLACE CONCRETE	97.8	54.8	80.1	100.7	54.9	81.9	100.7	54.7	81.8	96.8	54.3	79.4	96.8	39.0	73.0	104.1	54.7	83.8
03	CONCRETE	100.4	62.1	81.4	102.1	62.2	82.3	104.2	60.1	82.3	99.0	61.9	80.6	102.3	37.0	69.8	103.4	60.1	81.9
04	MASONRY	91.7	53.8	68.2	98.0	53.8	70.6	103.9	49.4	70.1	96.5	53.6	69.9	84.6	30.8	51.3	90.3	49.4	65.0
05	METALS	101.7	90.5	98.1	100.7	90.7	97.5	99.8	89.8	96.6	100.7	89.0	97.0	99.5	66.9	89.1	103.6	89.8	99.2
06	WOOD & PLASTICS	97.0	63.3	79.4	102.6	63.3	82.1	88.2	57.7	72.3	83.4	63.3	72.9	87.7	28.0	56.6	98.2	57.7	77.1
07	THERMAL & MOISTURE PROTECTION	101.4	51.2	80.3	101.2	51.2	80.2	101.4	54.2	81.5	101.4	51.2	80.3	101.7	35.6	73.9	101.1	54.2	81.3
08	DOORS & WINDOWS	96.6	60.0	87.0	96.6	62.6	87.7	95.9	55.5	85.4	96.7	62.6	87.8	97.9	27.9	79.6	96.6	55.5	85.9
09200	PLASTER & GYPSUM BOARD	109.6	61.3	79.6	109.6	61.3	79.6	102.9	55.6	73.5	101.8	61.3	76.6	102.6	25.8	54.9	109.6	55.6	76.0
095,098	CEILINGS & ACOUSTICAL TREATMENT	100.9	61.3	77.3	100.9	61.3	77.3	94.9	55.6	71.4	100.9	61.3	77.3	94.1	25.8	53.4	100.9	55.6	73.8
09600	FLOORING	106.0	39.4	88.0	105.8	39.4	87.8	100.9	62.5	90.5	96.9	39.4	81.3	101.4	30.4	82.1	105.8	62.5	94.1
097,099	WALL FINISHES, PAINTS & COATINGS	110.4	42.0	68.7	110.4	63.3	81.7	110.4	54.9	76.5	110.4	63.3	81.7	110.4	32.8	63.0	110.4	54.9	76.5
09	FINISHES	101.7	53.8	76.8	101.6	56.3	78.0	98.0	56.9	76.6	97.8	56.3	76.2	97.8	28.5	61.7	101.5	56.9	78.3
10 - 14	TOTAL DIV. 10000 - 14000	100.0	78.1	95.3	100.0	78.1	95.3	100.0	77.5	95.2	100.0	78.4	95.4	100.0	42.6	87.8	100.0	77.5	95.2
15	MECHANICAL	100.1	63.4	84.8	100.1	63.6	84.9	95.1	67.4	83.6	100.1	63.6	84.9	95.1	27.1	66.9	100.1	67.4	86.5
16	ELECTRICAL	97.7	59.2	79.0	97.8	63.1	80.9	97.9	71.5	85.1	96.2	63.1	80.1	97.7	58.0	78.4	98.4	71.5	85.3
01 - 16	WEIGHTED AVERAGE	99.6	64.7	84.1	100.0	65.8	84.8	98.9	66.7	84.6	98.9	65.5	84.0	97.7	42.4	73.1	100.3	66.7	85.4

DIVISION		VIRGINIA									WASHINGTON								
		ROANOKE			STAUNTON			WINCHESTER			CLARKSTON			EVERETT			OLYMPIA		
		240 - 241			244			226			994			982			985		
		MAT.	INST.	TOTAL	MAT.	INST.	TOTAL	MAT.	INST.	TOTAL	MAT.	INST.	TOTAL	MAT.	INST.	TOTAL	MAT.	INST.	TOTAL
01590	EQUIPMENT RENTAL	.0	99.6	99.6	.0	104.8	104.8	.0	99.6	99.6	.0	90.0	90.0	.0	104.0	104.0	.0	104.0	104.0
02	SITE CONSTRUCTION	104.3	80.7	87.0	108.7	83.1	89.9	118.3	81.5	91.3	103.0	88.4	92.3	93.8	115.2	109.5	97.0	115.2	110.4
03100	CONCRETE FORMS & ACCESSORIES	97.0	36.0	44.4	89.1	35.3	42.7	82.4	37.7	43.9	113.2	79.0	83.7	108.0	101.8	102.7	98.3	102.1	101.6
03200	CONCRETE REINFORCEMENT	94.2	72.7	83.3	93.9	45.7	69.4	93.8	81.1	87.4	107.1	87.7	97.2	108.0	94.7	101.2	108.5	94.7	101.5
03300	CAST-IN-PLACE CONCRETE	109.9	46.3	83.7	100.7	46.6	78.4	97.1	57.9	81.0	118.2	85.0	104.5	94.5	106.3	99.3	99.2	108.3	102.9
03	CONCRETE	106.2	48.3	77.5	103.2	43.2	73.4	104.0	54.7	79.5	108.4	82.6	95.6	95.7	101.4	98.5	98.7	102.2	100.4
04	MASONRY	91.8	38.8	58.9	101.8	34.5	60.1	95.1	32.7	56.4	115.1	83.3	95.4	132.8	103.0	114.4	125.8	103.7	112.1
05	METALS	101.4	84.0	95.9	99.7	76.8	92.4	99.8	88.7	96.3	93.8	81.9	90.0	106.0	88.1	100.3	100.5	88.1	96.5
06	WOOD & PLASTICS	97.0	35.3	64.9	87.7	35.3	60.4	82.3	36.0	58.2	95.8	78.0	86.5	98.7	101.1	99.9	87.8	101.1	94.7
07	THERMAL & MOISTURE PROTECTION	101.4	44.6	77.5	101.3	40.3	75.6	102.0	53.0	81.3	138.8	78.6	113.4	104.6	98.4	102.0	104.2	97.3	101.3
08	DOORS & WINDOWS	96.6	44.7	83.1	96.2	38.6	81.2	98.0	51.5	85.8	111.1	80.8	103.2	98.6	98.8	98.7	98.6	98.8	98.6
09200	PLASTER & GYPSUM BOARD	109.6	33.4	62.2	102.6	32.6	59.1	101.8	34.1	59.7	124.7	77.2	95.2	104.7	101.1	102.4	98.9	101.1	100.3
095,098	CEILINGS & ACOUSTICAL TREATMENT	100.9	33.4	60.6	94.1	32.6	57.4	94.9	34.1	58.6	102.9	77.2	87.6	111.3	101.1	105.2	107.7	101.1	103.7
09600	FLOORING	106.0	35.3	86.9	100.6	35.6	83.0	99.1	66.7	90.4	109.9	58.9	96.1	119.1	105.1	115.3	111.6	105.1	109.9
097,099	WALL FINISHES, PAINTS & COATINGS	110.4	32.8	63.0	110.4	27.1	59.6	121.8	27.6	64.4	112.6	74.5	89.3	111.1	90.5	98.5	111.1	92.2	99.6
09	FINISHES	101.5	34.8	66.8	97.5	34.6	64.7	98.2	40.9	68.3	121.8	74.2	97.0	113.4	101.3	107.1	109.5	101.6	105.4
10 - 14	TOTAL DIV. 10000 - 14000	100.0	71.3	93.9	100.0	72.7	94.2	100.0	50.4	89.4	100.0	96.8	99.3	100.0	102.7	100.6	100.0	103.2	100.7
15	MECHANICAL	100.1	39.3	74.9	95.1	34.1	69.8	95.1	60.1	80.6	95.6	82.7	90.2	99.9	101.9	100.7	100.0	100.6	100.3
16	ELECTRICAL	97.7	34.2	66.8	96.7	37.4	67.9	96.4	33.4	65.8	98.8	97.6	98.2	103.6	97.7	100.7	103.4	100.6	102.0
01 - 16	WEIGHTED AVERAGE	100.2	48.1	77.1	98.4	45.6	75.0	98.6	53.8	78.7	103.9	84.3	95.2	103.6	101.0	102.4	102.3	101.3	101.9

City Cost Indexes

	DIVISION	WASHINGTON																	
		RICHLAND			SEATTLE			SPOKANE			TACOMA			VANCOUVER			WENATCHEE		
		993			980 - 981,987			990 - 992			983 - 984			986			988		
		MAT.	INST.	TOTAL	MAT.	INST.	TOTAL	MAT.	INST.	TOTAL	MAT.	INST.	TOTAL	MAT.	INST.	TOTAL	MAT.	INST.	TOTAL
01590	EQUIPMENT RENTAL	.0	90.0	90.0	.0	103.7	103.7	.0	90.0	90.0	.0	104.0	104.0	.0	97.1	97.1	.0	104.0	104.0
02	SITE CONSTRUCTION	104.4	89.4	93.3	97.8	113.0	109.0	105.2	89.4	93.6	96.6	115.2	110.3	108.7	100.7	102.8	108.1	114.2	112.6
03100	CONCRETE FORMS & ACCESSORIES	113.4	80.3	84.8	98.4	102.5	101.9	120.4	80.3	85.8	98.4	102.2	101.7	99.2	96.9	97.2	99.9	80.7	83.3
03200	CONCRETE REINFORCEMENT	102.9	87.7	95.1	106.8	94.8	100.7	103.5	87.7	95.5	106.8	94.7	100.6	107.6	94.5	100.9	107.5	90.1	98.7
03300	CAST-IN-PLACE CONCRETE	118.5	83.6	104.1	99.2	108.5	103.0	122.7	83.6	106.6	97.2	108.3	101.8	108.6	100.1	105.1	99.3	86.1	93.9
03	CONCRETE	108.0	82.7	95.4	98.4	102.5	100.5	110.7	82.7	96.8	97.4	102.3	99.8	107.2	97.2	102.2	105.2	84.3	94.8
04	MASONRY	115.0	83.3	95.3	126.8	103.7	112.5	116.5	83.3	95.9	126.6	103.7	112.4	127.2	100.1	110.4	129.8	84.9	101.9
05	METALS	94.1	82.0	90.2	107.7	90.1	102.1	96.7	82.0	92.0	107.7	88.2	101.4	104.8	89.6	100.0	105.4	84.0	98.6
06	WOOD & PLASTICS	96.0	79.6	87.5	88.8	101.1	95.2	105.8	79.6	92.2	87.8	101.1	94.7	82.3	96.6	89.7	88.9	78.7	83.6
07	THERMAL & MOISTURE PROTECTION	139.9	81.2	115.2	104.6	100.6	102.9	137.7	81.1	113.8	104.2	99.2	102.1	106.8	90.9	100.1	104.9	79.5	94.2
08	DOORS & WINDOWS	111.7	76.2	102.4	100.8	98.8	100.3	111.5	76.2	102.2	99.3	98.8	99.1	96.2	95.3	96.0	98.8	76.2	92.9
09200	PLASTER & GYPSUM BOARD	124.7	78.9	96.3	98.2	101.1	100.0	129.5	78.9	98.1	100.2	101.1	100.7	98.7	96.8	97.5	101.5	78.1	86.9
095,098	CEILINGS & ACOUSTICAL TREATMENT	107.7	78.9	90.5	114.5	101.1	106.5	107.7	78.9	90.5	114.7	101.1	106.6	109.7	96.8	102.0	107.7	78.1	90.0
09600	FLOORING	110.2	43.2	92.0	111.0	105.1	109.4	113.9	74.2	103.1	111.6	105.1	109.9	117.5	79.5	107.2	115.0	74.2	103.9
097,099	WALL FINISHES, PAINTS & COATINGS	112.6	74.5	89.3	111.1	92.2	99.6	112.6	74.5	89.3	111.1	92.2	99.6	119.0	68.1	87.9	111.1	74.5	88.8
09	FINISHES	123.2	71.9	96.5	111.0	101.6	106.1	125.0	78.3	100.7	111.5	101.6	106.3	109.5	90.3	99.5	111.8	78.1	94.2
10 - 14	TOTAL DIV. 10000 - 14000	100.0	83.1	96.4	100.0	103.2	100.7	100.0	83.0	96.4	100.0	103.2	100.7	100.0	80.6	95.9	100.0	97.1	99.4
15	MECHANICAL	100.5	98.5	99.7	99.9	112.2	105.0	100.5	87.7	95.2	100.0	100.6	100.3	100.1	104.0	101.7	95.1	86.0	91.3
16	ELECTRICAL	95.8	97.6	96.7	103.4	106.5	104.9	94.1	81.8	88.1	103.4	100.6	102.0	114.0	103.1	108.7	104.2	85.9	95.3
01 - 16	WEIGHTED AVERAGE	105.0	86.9	96.9	103.8	104.7	104.2	105.7	83.3	95.8	103.5	101.4	102.6	105.4	97.7	102.0	103.5	86.6	96.0

	DIVISION	WASHINGTON			WEST VIRGINIA														
		YAKIMA			BECKLEY			BLUEFIELD			BUCKHANNON			CHARLESTON			CLARKSBURG		
		989			258 - 259			247 - 248			262			250 - 253			263 - 264		
		MAT.	INST.	TOTAL	MAT.	INST.	TOTAL	MAT.	INST.	TOTAL	MAT.	INST.	TOTAL	MAT.	INST.	TOTAL	MAT.	INST.	TOTAL
01590	EQUIPMENT RENTAL	.0	104.0	104.0	.0	99.6	99.6	.0	99.6	99.6	.0	99.6	99.6	.0	99.6	99.6	.0	99.6	99.6
02	SITE CONSTRUCTION	99.8	113.8	110.1	99.1	84.9	88.7	98.9	84.8	88.6	105.1	84.6	90.0	102.1	86.0	90.3	105.6	84.6	90.2
03100	CONCRETE FORMS & ACCESSORIES	98.9	95.8	96.2	84.5	90.6	89.8	86.2	89.5	89.0	85.4	91.3	90.5	104.9	91.7	93.5	82.4	91.3	90.1
03200	CONCRETE REINFORCEMENT	107.2	93.4	100.2	92.8	87.1	89.9	92.8	69.6	81.0	93.4	89.0	91.2	94.2	87.3	90.7	93.4	91.6	92.4
03300	CAST-IN-PLACE CONCRETE	103.9	81.6	94.7	94.6	106.9	99.7	94.6	105.0	98.9	94.4	101.5	97.3	95.7	108.0	100.8	103.4	101.6	102.6
03	CONCRETE	102.1	89.9	96.0	97.7	96.2	97.0	97.8	91.8	94.8	101.2	95.0	98.1	99.9	97.1	98.5	104.9	95.5	100.2
04	MASONRY	118.8	71.1	89.2	90.6	92.1	91.6	86.2	92.1	89.9	97.0	90.3	92.9	89.0	93.4	91.8	98.5	90.3	93.4
05	METALS	105.9	82.8	98.5	99.7	98.4	99.3	99.7	91.1	96.9	99.9	99.8	99.8	101.7	99.4	100.9	99.9	100.8	100.2
06	WOOD & PLASTICS	88.2	101.1	95.0	83.7	90.9	87.5	85.6	90.9	88.4	84.7	90.6	87.8	105.6	90.5	97.7	81.3	90.6	86.1
07	THERMAL & MOISTURE PROTECTION	104.4	79.5	93.9	101.4	90.6	96.9	101.5	84.9	94.5	101.7	87.7	95.8	101.3	91.1	97.0	101.7	87.7	95.8
08	DOORS & WINDOWS	98.8	85.0	95.2	97.2	84.2	93.8	98.1	80.2	93.4	98.2	84.5	94.6	97.8	84.0	94.2	98.2	90.2	96.1
09200	PLASTER & GYPSUM BOARD	100.0	101.1	100.7	99.0	90.5	93.7	99.4	90.5	93.9	99.8	90.2	93.8	108.9	90.1	97.2	98.4	90.2	93.3
095,098	CEILINGS & ACOUSTICAL TREATMENT	109.6	101.1	104.5	91.5	90.5	90.9	91.5	90.5	90.9	93.2	90.2	91.4	98.3	90.1	93.4	93.2	90.2	91.4
09600	FLOORING	113.0	65.0	100.0	96.2	107.1	99.1	96.9	107.1	99.7	96.4	103.4	98.3	105.8	107.1	106.2	95.1	103.4	97.3
097,099	WALL FINISHES, PAINTS & COATINGS	111.1	74.5	88.8	110.4	44.7	70.3	110.4	44.7	70.3	110.4	101.3	104.9	110.4	95.9	101.6	110.4	101.3	104.9
09	FINISHES	110.7	88.3	99.0	94.7	89.3	91.9	95.0	89.3	92.0	95.7	94.8	95.3	100.9	95.2	97.9	95.1	94.8	94.9
10 - 14	TOTAL DIV. 10000 - 14000	100.0	98.5	99.7	100.0	71.2	93.9	100.0	71.1	93.8	100.0	94.6	98.8	100.0	95.0	98.9	100.0	94.6	98.8
15	MECHANICAL	100.0	97.6	99.0	95.1	83.4	90.3	95.1	67.1	83.5	95.1	94.7	95.0	100.1	84.9	93.8	95.1	92.7	94.1
16	ELECTRICAL	106.1	97.6	102.0	94.4	87.3	90.9	96.8	56.0	76.9	98.0	96.4	97.2	97.7	87.3	92.6	98.0	96.4	97.2
01 - 16	WEIGHTED AVERAGE	103.7	91.8	98.4	96.6	88.8	93.1	96.8	79.4	89.1	98.1	93.4	96.0	99.5	91.0	95.7	98.5	93.4	96.2

	DIVISION	WEST VIRGINIA																	
		GASSAWAY			HUNTINGTON			LEWISBURG			MARTINSBURG			MORGANTOWN			PARKERSBURG		
		266			255 - 257			249			254			265			261		
		MAT.	INST.	TOTAL	MAT.	INST.	TOTAL	MAT.	INST.	TOTAL	MAT.	INST.	TOTAL	MAT.	INST.	TOTAL	MAT.	INST.	TOTAL
01590	EQUIPMENT RENTAL	.0	99.6	99.6	.0	99.6	99.6	.0	99.6	99.6	.0	99.6	99.6	.0	99.6	99.6	.0	99.6	99.6
02	SITE CONSTRUCTION	102.4	85.1	89.7	103.5	86.8	91.2	115.1	84.9	92.9	102.7	82.9	88.2	99.7	84.9	88.8	108.3	85.9	91.9
03100	CONCRETE FORMS & ACCESSORIES	84.7	91.4	90.5	98.1	93.7	94.3	82.8	90.2	89.2	84.5	81.2	81.7	82.9	91.3	90.2	87.5	91.7	91.1
03200	CONCRETE REINFORCEMENT	93.4	87.3	90.3	94.2	85.5	89.8	93.4	69.7	81.4	92.8	76.0	84.3	93.4	91.6	92.4	92.8	89.0	90.9
03300	CAST-IN-PLACE CONCRETE	98.8	107.4	102.3	103.2	112.5	107.0	94.7	106.8	99.7	99.2	93.7	96.9	94.4	101.6	97.3	96.5	102.1	98.8
03	CONCRETE	101.2	96.8	99.0	103.0	99.2	101.1	108.0	92.7	100.4	101.3	85.2	93.3	97.6	95.5	96.5	103.1	95.4	99.2
04	MASONRY	100.1	92.3	95.3	91.2	94.4	93.2	90.2	92.1	91.4	93.1	80.0	85.0	118.1	90.3	100.9	77.0	91.0	85.7
05	METALS	99.8	99.4	99.7	101.7	98.8	100.8	99.8	91.8	97.2	100.0	88.1	96.2	99.9	100.8	100.2	100.3	99.8	100.2
06	WOOD & PLASTICS	83.9	90.6	87.4	97.0	91.8	94.3	81.6	90.9	86.5	83.7	83.1	83.4	81.6	90.6	86.3	85.8	90.6	88.3
07	THERMAL & MOISTURE PROTECTION	101.5	89.6	96.5	101.6	91.6	97.4	102.3	90.6	97.4	101.7	72.1	89.2	101.5	87.7	95.7	101.7	88.4	96.1
08	DOORS & WINDOWS	96.1	84.0	92.9	96.6	84.3	93.4	98.1	80.2	93.5	99.5	71.9	92.3	99.5	90.2	97.1	97.2	82.0	93.2
09200	PLASTER & GYPSUM BOARD	99.5	90.2	93.7	107.8	91.4	97.6	98.4	90.5	93.5	99.5	82.4	88.9	98.4	90.2	93.3	102.1	90.2	94.7
095,098	CEILINGS & ACOUSTICAL TREATMENT	93.2	90.2	91.4	94.1	91.4	92.5	93.2	90.5	91.6	93.2	82.4	86.8	93.2	90.2	91.4	93.2	90.2	91.4
09600	FLOORING	96.2	107.1	99.2	105.8	104.3	105.4	95.2	107.1	98.4	96.2	67.0	88.3	95.2	103.4	97.4	100.2	102.9	100.9
097,099	WALL FINISHES, PAINTS & COATINGS	110.4	95.9	101.6	110.4	93.1	99.9	110.4	44.7	70.3	110.4	52.8	75.2	110.4	101.3	104.9	110.4	96.6	102.0
09	FINISHES	95.4	95.0	95.2	99.5	95.7	97.5	95.9	89.3	92.5	95.4	76.3	85.4	94.7	94.8	94.8	97.2	94.4	95.8
10 - 14	TOTAL DIV. 10000 - 14000	100.0	94.6	98.8	100.0	96.2	99.2	100.0	71.1	93.8	100.0	68.5	93.3	100.0	84.5	96.7	100.0	95.0	98.9
15	MECHANICAL	95.1	92.4	94.0	100.1	88.4	95.2	95.1	83.3	90.2	95.1	80.3	89.0	95.1	94.5	94.9	100.1	89.7	95.8
16	ELECTRICAL	98.0	87.3	92.8	97.7	95.8	96.8	94.4	56.0	75.7	99.6	76.7	88.5	98.2	96.4	97.3	98.2	96.9	97.5
01 - 16	WEIGHTED AVERAGE	97.9	92.2	95.4	99.6	93.4	96.9	98.3	83.1	91.6	98.1	80.0	90.1	98.6	93.5	96.4	98.7	92.6	96.0

	DIVISION	WEST VIRGINIA									WISCONSIN								
		PETERSBURG			ROMNEY			WHEELING			BELOIT			EAU CLAIRE			GREEN BAY		
		268			267			260			535			547			541 - 543		
		MAT.	INST.	TOTAL	MAT.	INST.	TOTAL	MAT.	INST.	TOTAL	MAT.	INST.	TOTAL	MAT.	INST.	TOTAL	MAT.	INST.	TOTAL
01590	EQUIPMENT RENTAL	.0	99.6	99.6	.0	99.6	99.6	.0	99.6	99.6	.0	100.1	100.1	.0	100.6	100.6	.0	98.3	98.3
02	SITE CONSTRUCTION	99.1	85.3	89.0	102.0	85.3	89.7	108.9	85.6	91.8	91.4	105.3	101.6	86.5	102.9	98.5	89.2	98.8	96.2
03100	CONCRETE FORMS & ACCESSORIES	86.5	85.1	85.3	81.9	84.6	84.2	89.5	91.6	91.3	99.0	98.6	98.6	98.1	99.3	99.2	109.2	98.5	100.0
03200	CONCRETE REINFORCEMENT	92.8	80.7	86.7	93.4	72.5	82.8	92.2	91.6	91.9	89.2	107.0	98.3	91.8	105.6	98.8	90.1	93.3	91.7
03300	CAST-IN-PLACE CONCRETE	94.4	99.0	96.3	98.8	98.8	98.8	96.5	101.7	98.6	106.2	99.3	103.4	100.3	98.4	99.5	103.8	100.4	102.4
03	CONCRETE	97.7	89.8	93.8	101.0	88.0	94.6	103.1	95.7	99.4	98.5	100.5	99.5	96.9	100.4	98.7	98.2	98.4	98.3
04	MASONRY	93.1	90.3	91.4	90.5	75.7	81.3	100.1	90.3	94.1	102.8	101.1	101.7	92.8	101.5	98.2	124.3	100.5	109.5
05	METALS	99.9	96.2	98.8	100.0	92.0	97.4	100.4	101.1	100.6	94.7	104.0	97.7	91.9	105.2	96.2	94.3	99.5	96.0
06	WOOD & PLASTICS	85.8	83.1	84.4	80.5	83.1	81.8	87.7	90.6	89.2	110.1	98.3	104.0	112.2	98.4	105.0	118.1	98.4	107.8
07	THERMAL & MOISTURE PROTECTION	101.6	79.1	92.1	101.6	75.5	90.6	101.8	88.6	96.2	96.2	100.5	98.0	97.4	90.9	94.6	99.4	89.0	95.0
08	DOORS & WINDOWS	99.5	78.5	94.0	99.5	76.8	93.5	98.1	90.2	96.0	101.9	99.4	101.2	100.9	96.7	99.8	99.1	89.9	96.7
09200	PLASTER & GYPSUM BOARD	99.8	82.4	89.0	97.6	82.4	88.2	102.4	90.2	94.8	98.1	98.7	98.4	105.0	98.7	101.1	97.3	98.7	98.1
095,098	CEILINGS & ACOUSTICAL TREATMENT	93.2	82.4	86.8	93.2	82.4	86.8	93.2	90.2	91.4	86.1	98.7	93.6	98.5	98.7	98.6	91.8	98.7	95.9
09600	FLOORING	97.4	103.4	99.0	94.9	75.5	89.7	101.4	103.4	101.9	93.5	114.5	99.2	90.5	109.4	95.6	108.3	109.4	108.6
097,099	WALL FINISHES, PAINTS & COATINGS	110.4	80.6	92.2	110.4	80.6	92.2	110.4	101.3	104.9	91.5	96.3	94.4	87.7	94.2	91.7	97.2	77.1	84.9
09	FINISHES	95.6	88.1	91.7	94.7	82.3	88.3	97.6	94.8	96.2	93.9	101.6	97.9	97.3	100.9	99.2	100.7	98.7	99.7
10 - 14	TOTAL DIV. 10000 - 14000	100.0	83.6	96.5	100.0	93.6	98.6	100.0	102.5	100.5	100.0	92.1	98.3	100.0	94.8	98.9	100.0	93.1	98.5
15	MECHANICAL	95.1	95.5	95.3	95.1	95.2	95.2	100.1	94.6	97.8	99.8	92.6	96.8	100.2	90.9	96.3	100.5	90.1	96.2
16	ELECTRICAL	101.1	85.5	93.5	100.4	85.5	93.2	95.6	96.4	95.9	96.0	92.8	94.4	101.5	83.1	92.6	96.6	90.8	93.8
01 - 16	WEIGHTED AVERAGE	97.9	89.3	94.1	97.9	86.5	92.9	99.7	94.2	97.2	98.0	98.4	98.2	97.8	96.2	97.1	99.7	95.2	97.7

	DIVISION	WISCONSIN																	
		KENOSHA			LA CROSSE			LANCASTER			MADISON			MILWAUKEE			NEW RICHMOND		
		531			546			538			537			530,532			540		
		MAT.	INST.	TOTAL	MAT.	INST.	TOTAL	MAT.	INST.	TOTAL	MAT.	INST.	TOTAL	MAT.	INST.	TOTAL	MAT.	INST.	TOTAL
01590	EQUIPMENT RENTAL	.0	97.9	97.9	.0	100.6	100.6	.0	100.1	100.1	.0	100.1	100.1	.0	86.3	86.3	.0	101.6	101.6
02	SITE CONSTRUCTION	96.5	102.1	100.6	80.6	102.8	96.9	90.4	105.0	101.1	90.8	105.5	101.6	92.2	93.8	93.4	81.8	104.0	98.1
03100	CONCRETE FORMS & ACCESSORIES	109.2	108.7	108.7	82.6	98.8	96.6	98.0	96.9	97.1	99.8	99.4	99.5	102.1	114.2	112.6	91.6	109.6	107.2
03200	CONCRETE REINFORCEMENT	89.1	101.5	95.4	91.5	92.2	91.9	90.3	92.2	91.3	89.2	92.4	90.8	89.2	101.7	95.6	89.4	105.5	97.6
03300	CAST-IN-PLACE CONCRETE	115.6	102.6	110.2	90.2	97.8	93.3	105.5	98.6	102.7	104.2	100.9	102.8	104.2	108.5	105.9	104.4	95.0	100.5
03	CONCRETE	103.7	105.1	104.4	88.3	97.5	92.9	98.2	96.7	97.5	97.6	98.7	98.1	97.9	109.0	103.4	95.0	103.8	99.4
04	MASONRY	99.2	111.4	106.8	92.0	101.5	97.9	102.8	101.2	101.8	103.2	104.3	103.9	102.7	118.9	112.7	118.3	101.7	108.0
05	METALS	95.3	102.7	97.7	91.9	99.0	94.1	92.8	96.1	93.9	96.4	97.4	96.7	98.1	94.1	96.8	92.1	104.5	96.0
06	WOOD & PLASTICS	115.9	108.3	111.9	94.7	98.4	96.6	109.2	98.3	103.5	110.2	98.3	104.0	112.9	113.1	113.0	100.4	112.8	106.8
07	THERMAL & MOISTURE PROTECTION	96.4	103.7	99.5	96.8	90.0	93.9	96.0	84.2	91.1	95.7	97.3	96.4	94.9	111.9	102.1	98.3	102.3	100.0
08	DOORS & WINDOWS	96.9	108.0	99.8	100.9	86.7	97.2	97.3	87.7	94.8	101.9	96.3	100.4	104.2	110.6	105.8	88.8	104.5	92.9
09200	PLASTER & GYPSUM BOARD	90.0	108.9	101.8	98.3	98.7	98.5	96.4	98.7	97.8	98.1	98.7	98.4	99.0	113.7	108.1	90.0	113.7	104.7
095,098	CEILINGS & ACOUSTICAL TREATMENT	80.5	108.9	97.4	96.8	98.7	97.9	79.3	98.7	90.8	86.1	98.7	93.6	89.5	113.7	103.9	70.0	113.7	96.1
09600	FLOORING	109.7	110.7	109.9	82.9	109.5	90.1	92.8	114.6	98.7	93.5	107.2	97.2	96.2	121.1	102.9	100.8	109.4	103.1
097,099	WALL FINISHES, PAINTS & COATINGS	101.5	101.1	101.3	87.7	71.6	77.9	91.5	84.7	87.3	91.5	96.3	94.4	93.8	111.3	104.5	101.3	94.2	96.9
09	FINISHES	96.6	109.0	103.1	93.3	98.4	96.0	91.7	98.5	95.2	93.8	101.0	97.5	95.8	115.8	106.2	91.8	107.6	100.0
10 - 14	TOTAL DIV. 10000 - 14000	100.0	101.0	100.2	100.0	94.8	98.9	100.0	60.4	91.6	100.0	93.3	98.6	100.0	103.7	100.8	100.0	95.9	99.1
15	MECHANICAL	100.0	98.2	99.3	100.2	90.6	96.2	94.9	88.7	92.3	99.8	93.2	97.0	99.8	103.2	101.2	94.7	85.1	90.7
16	ELECTRICAL	96.0	96.0	96.0	101.8	91.7	96.9	95.5	91.6	93.6	97.5	95.3	96.4	97.1	103.1	100.0	99.6	91.7	95.7
01 - 16	WEIGHTED AVERAGE	98.5	103.1	100.5	96.1	95.7	95.9	95.8	94.0	95.0	98.3	98.1	98.2	99.0	106.2	102.2	95.5	98.5	96.8

	DIVISION	WISCONSIN																	
		OSHKOSH			PORTAGE			RACINE			RHINELANDER			SUPERIOR			WAUSAU		
		549			539			534			545			548			544		
		MAT.	INST.	TOTAL	MAT.	INST.	TOTAL	MAT.	INST.	TOTAL	MAT.	INST.	TOTAL	MAT.	INST.	TOTAL	MAT.	INST.	TOTAL
01590	EQUIPMENT RENTAL	.0	98.3	98.3	.0	100.1	100.1	.0	100.1	100.1	.0	98.3	98.3	.0	101.6	101.6	.0	98.3	98.3
02	SITE CONSTRUCTION	81.4	98.8	94.1	81.2	105.3	98.9	91.2	106.0	102.0	93.1	98.6	97.1	78.8	104.1	97.3	77.4	98.8	93.1
03100	CONCRETE FORMS & ACCESSORIES	89.8	98.2	97.1	89.7	99.1	97.8	99.3	108.6	107.3	86.9	97.9	96.4	89.8	99.5	98.1	89.1	98.1	96.9
03200	CONCRETE REINFORCEMENT	90.3	93.2	91.8	90.4	92.4	91.4	89.2	101.4	95.4	90.4	93.6	92.0	89.4	94.2	91.8	90.4	92.1	91.3
03300	CAST-IN-PLACE CONCRETE	96.2	100.3	97.9	90.6	103.1	95.7	104.2	102.2	103.3	109.1	100.2	105.4	98.4	100.9	99.4	89.7	100.3	94.0
03	CONCRETE	90.2	98.2	94.2	86.2	99.3	92.7	97.5	104.9	101.2	101.9	98.1	100.0	89.6	99.2	94.4	85.3	98.0	91.6
04	MASONRY	105.0	100.5	102.2	101.7	104.3	103.3	102.6	111.3	108.0	123.9	100.5	109.4	117.6	110.3	113.1	104.5	100.5	102.0
05	METALS	92.4	99.2	94.6	93.4	97.1	94.6	96.4	102.7	98.4	92.3	99.1	94.4	92.9	101.0	95.5	92.2	98.7	94.2
06	WOOD & PLASTICS	96.4	98.4	97.4	99.5	98.3	98.8	110.4	108.3	109.3	93.1	98.4	95.9	98.5	97.3	97.9	95.5	98.4	97.0
07	THERMAL & MOISTURE PROTECTION	98.5	90.6	95.1	95.4	97.1	96.1	96.3	103.2	99.2	99.2	85.3	93.4	98.0	103.0	100.1	98.3	84.2	92.3
08	DOORS & WINDOWS	94.4	89.9	93.2	97.5	96.3	97.2	101.9	108.0	103.5	94.4	90.0	93.2	88.7	95.2	90.4	94.6	89.6	93.3
09200	PLASTER & GYPSUM BOARD	87.7	98.7	94.5	92.2	98.7	96.2	98.1	108.9	104.8	86.6	98.7	94.1	89.8	97.8	94.8	87.0	98.7	94.2
095,098	CEILINGS & ACOUSTICAL TREATMENT	91.8	98.7	95.9	80.1	98.7	91.2	86.1	108.9	99.7	91.8	98.7	95.9	70.9	97.8	87.0	91.8	98.7	95.9
09600	FLOORING	98.5	109.4	101.4	88.8	114.5	95.8	93.5	110.7	98.1	97.5	109.4	100.7	102.1	123.2	107.8	98.3	109.4	101.3
097,099	WALL FINISHES, PAINTS & COATINGS	97.2	77.1	84.9	91.5	84.7	87.3	91.4	92.2	91.9	97.2	77.0	84.9	93.1	104.4	100.0	97.2	79.1	86.2
09	FINISHES	96.0	97.5	96.8	89.6	100.4	95.2	93.8	107.9	101.2	96.3	98.3	97.3	91.6	104.5	98.3	95.6	99.0	97.3
10 - 14	TOTAL DIV. 10000 - 14000	100.0	83.8	96.6	100.0	82.0	96.2	100.0	101.0	100.2	100.0	69.6	93.5	100.0	94.5	98.8	100.0	93.1	98.5
15	MECHANICAL	95.6	90.0	93.3	94.9	93.0	94.1	99.8	95.4	97.9	95.6	89.8	93.2	94.7	106.7	99.7	95.6	89.7	93.1
16	ELECTRICAL	100.1	93.4	96.8	99.1	95.3	97.2	95.8	98.0	96.9	99.5	86.4	93.1	104.1	98.5	101.4	101.2	86.4	94.0
01 - 16	WEIGHTED AVERAGE	95.5	95.2	95.4	94.4	97.7	95.9	98.1	102.9	100.2	98.0	93.7	96.1	95.4	102.8	98.7	94.9	94.3	94.6

DIVISION		WYOMING																
		CASPER			CHEYENNE			NEWCASTLE			RAWLINS			RIVERTON			ROCK SPRINGS		
		826			820			827			823			825			829 - 831		
		MAT.	INST.	TOTAL	MAT.	INST.	TOTAL	MAT.	INST.	TOTAL	MAT.	INST.	TOTAL	MAT.	INST.	TOTAL	MAT.	INST.	TOTAL
01590	**EQUIPMENT RENTAL**	.0	101.7	101.7	.0	101.7	101.7	.0	101.7	101.7	.0	101.7	101.7	.0	101.7	101.7	.0	101.7	101.7
02	**SITE CONSTRUCTION**	79.2	100.0	94.4	79.0	100.0	94.4	79.3	99.6	94.2	94.3	99.6	98.2	88.0	99.7	96.6	82.0	99.6	94.9
03100	CONCRETE FORMS & ACCESSORIES	97.7	42.6	50.1	98.5	42.8	50.5	92.4	30.4	38.9	97.4	30.4	39.6	91.0	35.1	42.8	99.2	30.4	39.8
03200	CONCRETE REINFORCEMENT	115.6	51.7	83.2	108.5	51.9	79.7	117.3	51.5	83.9	116.8	51.5	83.6	117.9	51.6	84.2	117.9	51.5	84.2
03300	CAST-IN-PLACE CONCRETE	97.6	75.8	88.7	97.6	75.9	88.7	98.1	42.5	75.3	98.2	42.5	75.3	98.2	43.6	75.7	98.1	42.5	75.3
03	**CONCRETE**	104.1	56.5	80.5	103.0	56.7	80.0	104.2	39.7	72.2	117.6	39.7	78.9	112.2	42.1	77.4	104.8	39.6	72.4
04	**MASONRY**	107.0	38.1	64.3	104.4	40.2	64.6	105.4	31.1	59.3	105.4	31.1	59.3	105.4	33.7	60.9	170.7	31.1	84.2
05	**METALS**	97.2	62.2	86.1	98.0	62.7	86.7	94.4	61.2	83.8	94.4	61.2	83.8	94.5	61.3	83.9	95.0	61.2	84.2
06	**WOOD & PLASTICS**	95.1	40.5	66.7	95.1	40.5	66.7	89.9	28.7	58.0	95.1	28.7	60.5	88.5	34.1	60.2	97.9	28.7	61.9
07	**THERMAL & MOISTURE PROTECTION**	99.6	53.5	80.2	99.6	54.0	80.4	99.9	44.1	76.4	101.7	44.1	77.4	101.0	47.1	78.3	100.0	44.1	76.5
08	**DOORS & WINDOWS**	93.5	43.2	80.3	94.2	43.2	80.9	100.1	36.8	83.5	99.7	36.8	83.3	99.9	39.7	84.2	100.4	36.8	83.8
09200	PLASTER & GYPSUM BOARD	82.7	38.7	55.4	82.7	38.7	55.4	80.5	26.6	47.0	82.7	26.6	47.9	80.1	32.2	50.3	88.2	26.6	50.0
095,098	CEILINGS & ACOUSTICAL TREATMENT	100.5	38.7	63.7	99.6	38.7	63.3	100.5	26.6	56.4	100.5	26.6	56.4	100.5	32.2	59.7	100.5	26.6	56.4
09600	FLOORING	100.3	41.2	84.3	100.3	66.2	91.1	97.4	40.4	82.0	100.3	40.4	84.1	96.7	40.4	81.5	102.6	40.4	85.8
097,099	WALL FINISHES, PAINTS & COATINGS	102.2	54.9	73.3	102.0	54.9	73.3	102.2	33.3	60.2	102.2	33.3	60.2	102.2	30.6	58.5	102.2	33.3	60.2
09	**FINISHES**	96.5	42.4	68.3	96.2	47.6	70.9	95.2	31.3	61.9	97.8	31.3	63.1	95.8	34.6	63.9	98.0	31.3	63.2
10 - 14	**TOTAL DIV. 10000 - 14000**	100.0	79.2	95.6	100.0	79.3	95.6	100.0	47.3	88.8	100.0	47.3	88.8	100.0	48.7	89.1	100.0	47.3	88.8
15	**MECHANICAL**	100.1	65.9	85.9	100.1	60.0	83.4	97.5	54.9	79.9	97.5	54.9	79.9	97.5	55.9	80.3	100.1	54.9	81.3
16	**ELECTRICAL**	90.6	63.3	77.3	90.6	74.4	82.7	91.5	74.4	83.2	91.5	74.4	83.2	91.5	63.3	77.8	89.9	62.7	76.7
01 - 16	**WEIGHTED AVERAGE**	97.7	59.9	80.9	97.6	61.2	81.4	97.2	53.1	77.6	99.4	53.1	78.8	98.4	53.1	78.3	101.5	51.4	79.2

DIVISION		WYOMING												CANADA					
		SHERIDAN			WHEATLAND			WORLAND			YELLOWSTONE NAT'L PA			BARRIE, ONTARIO			BATHURST, NEW BRUNSWICK		
		828			822			824			821								
		MAT.	INST.	TOTAL	MAT.	INST.	TOTAL	MAT.	INST.	TOTAL	MAT.	INST.	TOTAL	MAT.	INST.	TOTAL	MAT.	INST.	TOTAL
01590	**EQUIPMENT RENTAL**	.0	101.7	101.7	.0	101.7	101.7	.0	101.7	101.7	.0	101.7	101.7	.0	100.4	100.4	.0	98.6	98.6
02	**SITE CONSTRUCTION**	85.4	100.0	96.1	84.7	99.7	95.7	80.5	99.6	94.5	80.6	99.9	94.7	116.6	98.7	103.5	96.0	94.3	94.8
03100	CONCRETE FORMS & ACCESSORIES	91.4	42.5	49.2	94.7	37.9	45.7	94.8	30.4	39.2	94.9	34.9	43.1	122.6	80.2	86.1	97.6	57.0	62.6
03200	CONCRETE REINFORCEMENT	117.9	51.8	84.3	117.3	51.6	83.9	117.9	51.5	84.2	120.0	51.6	85.2	179.9	85.7	132.0	148.3	57.8	102.3
03300	CAST-IN-PLACE CONCRETE	98.2	75.8	89.0	107.0	44.0	81.1	98.1	42.5	75.3	98.1	55.3	80.5	174.7	79.0	135.4	144.0	55.3	107.5
03	**CONCRETE**	110.2	56.5	83.5	111.4	43.5	77.7	104.5	39.6	72.3	104.8	46.0	75.6	156.2	81.1	118.9	137.9	57.2	97.8
04	**MASONRY**	105.4	40.4	65.1	106.1	34.5	61.7	105.4	31.1	59.3	105.5	36.9	62.9	170.2	88.9	119.8	158.7	59.0	96.9
05	**METALS**	94.5	62.4	84.2	94.4	61.6	83.9	94.5	61.2	83.9	95.0	61.4	84.2	107.8	86.2	100.9	105.3	67.6	93.3
06	**WOOD & PLASTICS**	88.8	40.5	63.6	92.1	37.2	63.5	92.1	28.7	59.1	92.1	31.8	60.7	120.2	78.9	98.7	93.4	57.2	74.5
07	**THERMAL & MOISTURE PROTECTION**	100.7	54.1	81.1	100.3	48.3	78.4	100.0	44.1	76.4	99.6	48.9	78.2	109.5	82.7	98.2	103.8	58.3	84.6
08	**DOORS & WINDOWS**	100.3	44.3	85.6	98.7	41.4	83.7	100.3	36.8	83.7	93.7	38.5	79.2	91.1	79.0	87.9	86.1	50.9	76.9
09200	PLASTER & GYPSUM BOARD	80.1	38.7	54.4	80.9	35.3	52.5	80.9	26.6	47.2	81.3	29.7	49.3	158.6	78.4	108.8	171.1	56.1	99.7
095,098	CEILINGS & ACOUSTICAL TREATMENT	100.5	38.7	63.7	100.5	35.3	61.6	100.5	26.6	56.4	102.2	29.7	59.0	95.8	78.4	85.4	96.6	56.1	72.4
09600	FLOORING	96.9	45.4	82.9	98.8	40.4	83.0	98.8	40.4	83.0	98.8	49.9	85.6	132.9	87.9	120.7	113.7	42.0	94.3
097,099	WALL FINISHES, PAINTS & COATINGS	102.2	54.9	73.3	102.2	33.3	60.2	102.2	33.3	60.2	102.2	33.3	60.2	109.8	81.6	92.6	109.1	46.9	71.1
09	**FINISHES**	95.6	43.3	68.4	96.0	36.8	65.2	95.7	31.3	62.2	96.2	36.0	64.9	120.4	82.1	100.5	115.4	53.2	83.0
10 - 14	**TOTAL DIV. 10000 - 14000**	100.0	79.2	95.6	100.0	68.7	93.3	100.0	47.3	88.8	100.0	53.4	90.1	140.0	65.4	124.1	140.0	56.6	122.2
15	**MECHANICAL**	97.5	58.5	81.4	97.5	56.2	80.4	97.5	54.9	79.9	97.5	57.8	81.1	103.7	78.4	93.2	103.7	55.0	83.5
16	**ELECTRICAL**	91.5	62.7	77.5	91.5	74.4	83.2	91.5	62.7	77.5	90.5	62.7	77.0	124.7	89.2	107.4	126.1	61.0	94.4
01 - 16	**WEIGHTED AVERAGE**	98.1	58.7	80.6	98.1	56.0	79.4	97.4	51.4	77.0	96.8	54.6	78.0	119.9	84.1	104.0	115.1	60.9	91.1

DIVISION		CANADA																	
		BRANDON, MANITOBA			BRANTFORD, ONTARIO			BRIDGEWATER, NS			CALGARY, ALBERTA			CAP-DE-LA-MADELEINE, PQ			CHARLESBOURG, QUEBEC		
		MAT.	INST.	TOTAL	MAT.	INST.	TOTAL	MAT.	INST.	TOTAL	MAT.	INST.	TOTAL	MAT.	INST.	TOTAL	MAT.	INST.	TOTAL
01590	**EQUIPMENT RENTAL**	.0	102.9	102.9	.0	100.4	100.4	.0	98.6	98.6	.0	107.0	107.0	.0	99.2	99.2	.0	99.2	99.2
02	**SITE CONSTRUCTION**	111.8	98.1	101.8	114.3	99.2	103.2	98.7	95.5	96.4	116.6	104.8	107.9	93.7	96.5	95.8	93.7	96.5	95.8
03100	CONCRETE FORMS & ACCESSORIES	121.7	65.1	72.9	122.5	87.0	91.9	90.3	64.1	67.7	122.0	74.0	80.6	129.8	78.9	85.9	129.8	78.9	85.9
03200	CONCRETE REINFORCEMENT	161.5	54.3	107.0	174.7	84.3	128.8	148.3	47.6	97.1	161.5	64.2	112.0	148.3	80.6	113.9	148.3	80.6	113.9
03300	CAST-IN-PLACE CONCRETE	157.5	68.9	121.1	170.5	97.8	140.6	175.2	64.3	129.6	191.6	84.3	147.5	137.4	86.6	116.6	137.4	86.6	116.6
03	**CONCRETE**	138.2	65.1	101.9	153.3	90.3	122.0	150.4	61.8	106.4	161.3	76.4	119.1	133.6	82.1	108.0	133.6	82.1	108.0
04	**MASONRY**	164.6	61.4	100.6	166.7	93.0	121.0	160.9	65.2	101.6	181.1	71.9	113.4	161.1	79.1	110.2	161.1	79.1	110.2
05	**METALS**	107.9	73.5	96.9	107.2	87.4	100.8	106.3	70.3	94.8	135.6	83.2	118.9	105.6	83.1	98.4	105.6	83.1	98.4
06	**WOOD & PLASTICS**	116.1	65.8	89.9	118.4	85.7	101.4	84.5	63.5	73.6	116.2	73.6	94.0	131.5	78.7	104.0	131.5	78.7	104.0
07	**THERMAL & MOISTURE PROTECTION**	104.2	66.9	88.5	109.4	87.6	100.2	104.4	65.2	87.9	117.5	77.6	100.7	104.4	83.3	95.5	104.4	83.3	95.5
08	**DOORS & WINDOWS**	91.9	59.1	83.4	90.9	84.5	89.2	82.8	59.0	76.6	91.9	70.6	86.3	91.9	74.0	87.2	91.9	74.0	87.2
09200	PLASTER & GYPSUM BOARD	131.4	64.5	89.8	158.6	85.4	113.1	164.6	62.6	101.3	152.4	72.4	102.7	195.3	78.1	122.5	195.3	78.1	122.5
095,098	CEILINGS & ACOUSTICAL TREATMENT	95.8	64.5	77.1	95.8	85.4	89.6	95.8	62.6	76.0	110.2	72.4	87.7	95.8	78.1	85.2	95.8	78.1	85.2
09600	FLOORING	132.9	62.5	113.8	132.9	87.9	120.7	109.6	59.9	96.2	132.9	82.4	119.2	132.9	88.6	120.9	132.9	88.6	120.9
097,099	WALL FINISHES, PAINTS & COATINGS	109.2	53.3	75.1	109.1	89.6	97.2	109.1	58.2	78.1	109.1	73.8	87.5	109.1	83.3	93.3	109.1	83.3	93.3
09	**FINISHES**	115.3	63.6	88.4	119.3	87.8	102.9	112.8	62.9	86.8	122.3	75.4	97.9	124.2	81.5	102.0	124.2	81.5	102.0
10 - 14	**TOTAL DIV. 10000 - 14000**	140.0	59.3	122.8	140.0	67.6	124.6	140.0	58.1	122.5	140.0	79.8	127.2	140.0	73.8	125.9	140.0	73.8	125.9
15	**MECHANICAL**	103.7	65.9	88.0	103.7	81.0	94.3	103.7	66.4	88.2	101.4	74.7	90.3	104.1	71.8	90.7	104.1	71.8	90.7
16	**ELECTRICAL**	125.4	68.6	97.8	124.7	88.7	107.2	128.2	63.5	96.7	117.6	84.2	101.4	120.5	71.6	96.7	120.5	71.6	96.7
01 - 16	**WEIGHTED AVERAGE**	117.0	68.5	95.5	119.1	87.6	105.1	116.5	67.1	94.6	124.3	79.6	104.4	115.9	78.9	99.5	115.9	78.9	99.5

	DIVISION	CANADA																	
		CHARLOTTETOWN, PEI			CHICOUTIMI, QUEBEC			CORNER BROOK, NFLD			CORNWALL, ONTARIO			DALHOUSIE, NB			DARTMOUTH, NOVA SCOTIA		
		MAT.	INST.	TOTAL	MAT.	INST.	TOTAL	MAT.	INST.	TOTAL	MAT.	INST.	TOTAL	MAT.	INST.	TOTAL	MAT.	INST.	TOTAL
01590	EQUIPMENT RENTAL	.0	98.6	98.6	.0	99.2	99.2	.0	98.6	98.6	.0	100.4	100.4	.0	98.6	98.6	.0	98.6	98.6
02	SITE CONSTRUCTION	106.7	93.8	97.2	93.0	96.5	95.6	114.3	94.2	99.6	112.1	98.8	102.3	96.0	94.3	94.8	106.9	95.5	98.5
03100	CONCRETE FORMS & ACCESSORIES	90.6	52.1	57.4	129.8	78.7	85.7	99.5	55.1	61.2	119.8	80.3	85.8	97.6	57.0	62.6	90.4	64.1	67.8
03200	CONCRETE REINFORCEMENT	148.3	47.7	97.2	111.9	80.5	96.0	148.3	49.5	98.1	174.7	84.1	128.7	148.3	57.8	102.3	154.9	47.6	100.4
03300	CAST-IN-PLACE CONCRETE	166.1	54.6	120.3	135.6	86.6	115.5	179.6	62.6	131.5	153.3	89.3	127.0	144.0	55.3	107.5	177.1	64.3	130.7
03	CONCRETE	154.4	52.9	104.0	126.9	82.0	104.6	164.7	57.5	111.5	144.8	84.4	114.8	137.9	57.2	97.8	152.4	61.8	107.4
04	MASONRY	160.4	56.9	96.3	160.1	79.1	109.9	161.3	58.3	97.4	165.4	85.4	115.8	158.6	59.0	96.9	172.7	65.2	106.1
05	METALS	106.4	63.9	92.8	105.2	83.0	98.1	108.2	68.3	95.5	107.2	86.0	100.4	105.3	67.6	93.3	108.4	70.3	96.2
06	WOOD & PLASTICS	84.8	51.6	67.5	131.5	78.7	104.0	94.6	54.2	73.6	116.9	79.7	97.5	93.4	57.2	74.5	84.5	63.5	73.6
07	THERMAL & MOISTURE PROTECTION	104.8	56.0	84.2	104.3	83.3	95.5	106.5	57.7	86.0	109.2	81.7	97.6	106.1	58.3	85.9	104.4	65.2	87.9
08	DOORS & WINDOWS	94.1	45.5	81.4	91.5	68.2	85.4	98.0	51.9	86.0	91.9	78.7	88.5	86.1	50.9	76.9	82.8	59.0	76.6
09200	PLASTER & GYPSUM BOARD	168.9	50.4	95.3	195.3	78.1	122.5	171.4	53.1	97.9	230.5	79.2	136.5	171.1	56.1	99.7	167.9	62.6	102.5
095,098	CEILINGS & ACOUSTICAL TREATMENT	111.1	50.4	74.9	95.8	78.1	85.2	97.5	53.1	71.0	99.2	79.2	87.3	96.6	56.1	72.4	108.5	62.6	81.1
09600	FLOORING	109.7	56.0	95.2	132.9	88.6	120.9	114.6	50.9	97.4	132.9	86.7	120.4	113.7	42.0	94.3	109.6	59.9	96.2
097,099	WALL FINISHES, PAINTS & COATINGS	109.1	39.0	66.3	109.1	83.3	93.3	109.1	56.1	76.8	109.1	83.5	93.5	109.1	46.9	71.1	109.1	58.2	78.1
09	FINISHES	118.0	51.7	83.5	124.2	81.5	102.0	116.7	54.5	84.3	131.1	82.2	105.6	115.4	53.2	83.0	116.5	62.9	88.6
10 - 14	TOTAL DIV. 10000 - 14000	140.0	56.0	122.1	140.0	73.8	125.9	140.0	56.7	122.2	140.0	65.0	124.0	140.0	56.6	122.2	140.0	58.1	122.5
15	MECHANICAL	103.7	50.4	81.5	103.7	71.8	90.4	103.7	56.5	84.1	104.1	78.9	93.6	103.7	55.0	83.5	103.7	66.4	88.2
16	ELECTRICAL	120.0	51.6	86.8	120.5	71.6	96.7	120.4	57.4	89.8	123.0	89.6	106.8	125.4	61.0	94.1	125.1	63.5	95.2
01 - 16	WEIGHTED AVERAGE	117.7	56.9	90.7	115.0	78.6	98.8	119.8	60.9	93.7	119.1	84.3	103.6	115.1	60.9	91.1	117.8	67.1	95.3

	DIVISION	CANADA																	
		EDMONTON, ALBERTA			FORT MCMURRAY, ALBERTA			FREDERICTON, NB			GATINEAU, QUEBEC			GRANBY, QUEBEC			HALIFAX, NOVA SCOTIA		
		MAT.	INST.	TOTAL	MAT.	INST.	TOTAL	MAT.	INST.	TOTAL	MAT.	INST.	TOTAL	MAT.	INST.	TOTAL	MAT.	INST.	TOTAL
01590	EQUIPMENT RENTAL	.0	107.0	107.0	.0	103.7	103.7	.0	98.6	98.6	.0	99.2	99.2	.0	99.2	99.2	.0	98.6	98.6
02	SITE CONSTRUCTION	120.3	104.8	108.9	111.6	101.2	104.0	95.4	94.3	94.6	93.4	96.5	95.7	94.0	96.5	95.9	99.1	95.5	96.5
03100	CONCRETE FORMS & ACCESSORIES	119.7	74.0	80.3	119.2	73.9	80.1	120.1	57.3	65.9	129.8	78.7	85.8	129.8	78.7	85.8	90.5	64.2	67.8
03200	CONCRETE REINFORCEMENT	161.5	64.2	112.0	161.5	64.2	112.0	149.2	57.9	102.8	157.0	80.5	118.1	157.0	80.5	118.1	154.9	47.8	100.4
03300	CAST-IN-PLACE CONCRETE	207.0	84.3	156.5	157.5	83.3	127.0	137.1	55.3	103.4	135.6	86.6	115.5	140.1	86.6	118.1	178.5	66.1	132.3
03	CONCRETE	168.6	76.4	122.8	144.6	76.0	110.5	136.4	57.4	97.2	134.1	82.0	108.2	136.3	82.0	109.3	153.1	62.5	108.1
04	MASONRY	182.3	71.9	113.9	161.6	71.8	105.9	158.1	60.4	97.6	161.0	79.1	110.2	161.3	79.1	110.3	172.7	67.5	107.5
05	METALS	136.4	83.2	119.4	107.9	83.1	100.0	105.2	68.0	93.4	105.6	83.0	98.4	105.6	83.0	98.4	120.8	70.5	104.7
06	WOOD & PLASTICS	112.7	73.6	92.3	112.7	73.5	92.3	119.2	57.2	86.9	131.5	78.7	104.0	131.5	78.7	104.0	84.5	63.5	73.6
07	THERMAL & MOISTURE PROTECTION	117.7	77.6	100.8	114.9	76.8	98.8	106.4	59.2	86.5	104.4	83.3	95.5	104.3	83.3	95.5	104.8	65.8	88.3
08	DOORS & WINDOWS	91.9	70.6	86.3	91.9	70.5	86.3	86.0	49.9	76.6	91.9	69.8	86.1	91.9	69.8	86.1	82.8	59.0	76.6
09200	PLASTER & GYPSUM BOARD	154.5	72.4	103.5	137.2	72.4	96.9	189.2	56.1	106.5	146.7	78.1	104.1	149.2	78.1	105.0	167.9	62.6	102.5
095,098	CEILINGS & ACOUSTICAL TREATMENT	125.6	72.4	93.9	95.8	72.4	81.8	105.1	56.1	75.9	95.8	78.1	85.2	95.8	78.1	85.2	108.5	62.6	81.1
09600	FLOORING	132.9	82.4	119.2	132.9	82.4	119.2	128.0	42.0	104.7	132.9	88.6	120.9	132.9	88.6	120.9	109.6	59.9	96.2
097,099	WALL FINISHES, PAINTS & COATINGS	109.2	73.8	87.6	109.2	73.8	87.6	109.1	59.8	79.0	109.1	83.3	93.3	109.1	83.3	93.3	109.1	58.2	78.1
09	FINISHES	126.5	75.4	99.9	116.4	75.4	95.0	124.5	54.7	88.1	116.8	81.5	98.4	117.2	81.5	98.6	116.5	62.9	88.6
10 - 14	TOTAL DIV. 10000 - 14000	140.0	79.8	127.2	140.0	79.5	127.1	140.0	56.6	122.2	140.0	73.8	125.9	140.0	73.8	125.9	140.0	58.1	122.5
15	MECHANICAL	101.5	74.7	90.4	103.7	74.7	91.7	103.7	62.1	86.4	104.1	71.8	90.7	103.7	71.8	90.4	101.0	66.4	86.7
16	ELECTRICAL	118.3	84.2	101.8	114.4	84.2	99.7	133.6	76.0	105.6	120.5	71.6	96.7	121.7	71.6	97.3	127.4	68.3	98.7
01 - 16	WEIGHTED AVERAGE	125.9	79.6	105.3	116.6	79.1	100.0	117.0	64.9	93.8	115.3	78.7	99.1	115.7	78.7	99.3	119.3	68.2	96.6

	DIVISION	CANADA																	
		HAMILTON, ONTARIO			HULL, QUEBEC			JOLIETTE, QUEBEC			KAMLOOPS, BC			KINGSTON, ONTARIO			KITCHENER, ONTARIO		
		MAT.	INST.	TOTAL	MAT.	INST.	TOTAL	MAT.	INST.	TOTAL	MAT.	INST.	TOTAL	MAT.	INST.	TOTAL	MAT.	INST.	TOTAL
01590	EQUIPMENT RENTAL	.0	103.1	103.1	.0	99.2	99.2	.0	99.2	99.2	.0	103.5	103.5	.0	103.1	103.1	.0	103.0	103.0
02	SITE CONSTRUCTION	112.7	103.2	105.7	93.4	96.5	95.7	94.2	96.5	95.9	117.3	100.4	104.9	112.1	103.0	105.4	101.9	102.6	102.4
03100	CONCRETE FORMS & ACCESSORIES	121.4	84.1	89.2	129.8	78.7	85.8	129.8	78.9	85.9	117.3	81.5	86.4	119.9	80.5	85.9	115.6	77.1	82.4
03200	CONCRETE REINFORCEMENT	175.4	85.7	129.8	157.0	80.5	118.1	148.3	80.6	113.9	116.5	78.4	97.1	174.7	84.1	128.7	110.3	85.6	97.8
03300	CAST-IN-PLACE CONCRETE	157.2	91.8	130.3	135.6	86.6	115.5	141.2	86.6	118.8	122.2	90.7	109.2	153.3	89.3	127.0	146.7	74.0	116.8
03	CONCRETE	146.9	87.2	117.2	134.1	82.0	108.2	135.4	82.1	108.9	140.1	84.2	112.3	144.8	84.4	114.8	124.8	77.9	101.5
04	MASONRY	168.1	88.5	118.7	161.0	79.1	110.2	161.4	79.1	110.3	168.4	86.1	117.4	172.0	85.4	118.3	163.6	84.5	114.6
05	METALS	139.3	86.8	122.5	105.6	83.0	98.4	105.6	83.1	98.4	107.8	83.3	100.0	108.7	86.0	101.5	116.3	86.2	106.7
06	WOOD & PLASTICS	117.4	82.9	99.5	131.5	78.7	104.0	131.5	78.7	104.0	98.1	80.0	88.7	116.9	79.8	97.6	110.9	75.5	92.4
07	THERMAL & MOISTURE PROTECTION	110.0	85.8	99.8	104.3	83.3	95.5	104.3	83.3	95.5	121.5	80.5	104.2	109.2	82.8	98.1	108.6	81.5	97.2
08	DOORS & WINDOWS	91.9	82.7	89.5	91.9	69.8	86.1	91.9	74.0	87.2	87.8	78.1	85.3	91.9	78.5	88.4	83.4	77.1	81.8
09200	PLASTER & GYPSUM BOARD	191.0	82.5	123.6	146.7	78.1	104.1	195.3	78.1	122.5	140.7	79.1	102.4	234.6	79.3	138.1	157.6	74.9	106.2
095,098	CEILINGS & ACOUSTICAL TREATMENT	113.7	82.5	95.1	95.8	78.1	85.2	95.8	78.1	85.2	95.8	79.1	85.8	115.4	79.3	93.9	107.7	74.9	88.1
09600	FLOORING	132.9	87.9	120.7	132.9	88.6	120.9	132.9	88.6	120.9	130.0	49.5	108.2	132.9	86.7	120.4	128.4	87.9	117.5
097,099	WALL FINISHES, PAINTS & COATINGS	109.1	95.8	101.0	109.1	83.3	93.3	109.1	83.3	93.3	109.1	76.2	89.0	109.1	91.0	98.1	109.1	87.7	96.1
09	FINISHES	128.6	86.2	106.5	116.8	81.5	98.4	124.2	81.5	102.0	117.3	75.3	95.4	135.7	83.1	108.3	120.1	80.2	99.4
10 - 14	TOTAL DIV. 10000 - 14000	140.0	85.3	128.3	140.0	73.8	125.9	140.0	73.8	125.9	140.0	82.9	127.8	140.0	65.0	124.0	140.0	83.3	127.9
15	MECHANICAL	101.5	82.4	93.6	103.7	71.8	90.4	103.7	71.8	90.5	103.7	74.8	91.7	104.1	79.0	93.7	101.0	78.8	91.8
16	ELECTRICAL	130.4	91.7	111.6	124.2	71.6	98.6	121.7	71.6	97.3	126.8	80.8	104.4	123.0	88.4	106.1	125.2	88.5	107.4
01 - 16	WEIGHTED AVERAGE	124.5	87.9	108.2	115.6	78.7	99.2	116.2	78.9	99.7	117.8	81.8	101.8	120.1	84.7	104.4	115.6	83.7	101.4

	DIVISION	CANADA																	
		LAVAL, QUEBEC			LETHBRIDGE, ALBERTA			LLOYDMINSTER, ALBERTA			LONDON, ONTARIO			MEDICINE HAT, ALBERTA			MONCTON, NEW BRUNSWICK		
		MAT.	INST.	TOTAL	MAT.	INST.	TOTAL	MAT.	INST.	TOTAL	MAT.	INST.	TOTAL	MAT.	INST.	TOTAL	MAT.	INST.	TOTAL
01590	EQUIPMENT RENTAL	.0	99.2	99.2	.0	103.7	103.7	.0	103.7	103.7	.0	103.2	103.2	.0	103.7	103.7	.0	98.6	98.6
02	SITE CONSTRUCTION	94.0	96.5	95.9	112.2	101.2	104.1	111.6	101.2	104.0	112.6	103.0	105.5	110.3	101.2	103.6	95.3	94.3	94.6
03100	CONCRETE FORMS & ACCESSORIES	130.0	78.7	85.8	121.5	73.9	80.4	119.2	73.9	80.1	127.1	79.8	86.3	121.5	73.9	80.4	97.6	57.0	62.6
03200	CONCRETE REINFORCEMENT	157.0	80.5	118.1	161.5	64.2	112.0	161.5	64.2	112.0	135.2	84.2	109.3	161.5	64.2	112.0	148.3	57.8	102.3
03300	CAST-IN-PLACE CONCRETE	140.1	86.6	118.1	172.4	83.3	135.8	157.5	83.3	127.0	157.2	89.5	129.4	157.5	83.3	127.0	138.7	55.3	104.4
03	CONCRETE	136.3	82.0	109.3	152.0	75.9	114.2	144.6	76.0	110.5	140.9	84.2	112.7	144.8	75.9	110.6	135.3	57.2	96.5
04	MASONRY	161.3	79.1	110.3	162.7	71.8	106.4	161.6	71.8	105.9	168.0	85.8	117.0	161.6	71.8	105.9	158.3	59.0	96.7
05	METALS	105.6	83.0	98.4	107.9	83.0	100.0	107.9	83.1	100.0	124.5	86.3	112.3	107.9	83.0	100.0	105.3	67.6	93.3
06	WOOD & PLASTICS	131.5	78.7	104.0	116.2	73.5	94.0	112.7	73.5	92.3	117.4	78.1	97.0	116.2	73.5	94.0	93.4	57.2	74.5
07	THERMAL & MOISTURE PROTECTION	104.7	83.3	95.7	114.9	76.7	98.9	113.8	76.8	98.2	112.6	84.0	100.6	117.2	76.7	100.2	106.1	58.3	85.9
08	DOORS & WINDOWS	91.9	69.8	86.1	91.9	70.5	86.3	91.9	70.5	86.3	92.9	79.1	89.3	91.9	70.5	86.3	86.1	50.9	76.9
09200	PLASTER & GYPSUM BOARD	151.6	78.1	105.9	139.0	72.4	97.6	137.2	72.4	96.9	191.5	77.6	120.7	139.0	72.4	97.6	171.1	56.1	99.7
095,098	CEILINGS & ACOUSTICAL TREATMENT	95.8	78.1	85.2	95.8	72.4	81.8	95.8	72.4	81.8	115.4	77.6	92.8	95.8	72.4	81.8	96.6	56.1	72.4
09600	FLOORING	132.9	88.6	120.9	132.9	82.4	119.2	132.9	82.4	119.2	134.9	87.9	122.2	132.9	82.4	119.2	113.7	42.0	94.3
097,099	WALL FINISHES, PAINTS & COATINGS	109.1	83.3	93.3	109.1	73.8	87.5	109.2	73.8	87.6	109.1	95.8	101.0	109.1	73.8	87.5	109.1	46.9	71.1
09	FINISHES	117.5	81.5	98.8	116.6	75.4	95.1	116.4	75.4	95.0	129.8	83.0	105.4	116.6	75.4	95.1	115.4	53.2	83.0
10 - 14	TOTAL DIV. 10000 - 14000	140.0	73.8	125.9	140.0	79.5	127.1	140.0	79.5	127.1	140.0	84.1	128.1	140.0	79.5	127.1	140.0	56.6	122.2
15	MECHANICAL	101.0	71.8	88.9	103.7	72.2	90.6	104.1	74.7	91.9	101.6	79.0	92.2	103.7	72.2	90.6	103.7	55.0	83.5
16	ELECTRICAL	122.1	71.6	97.5	114.4	84.2	99.7	114.4	84.2	99.7	130.7	88.7	110.3	114.4	84.2	99.7	130.6	61.0	96.7
01 - 16	WEIGHTED AVERAGE	115.2	78.7	99.0	117.6	78.6	100.3	116.7	79.1	100.0	121.9	85.3	105.6	116.7	78.6	99.8	115.4	60.9	91.2

	DIVISION	CANADA																	
		MONTREAL, QUEBEC			MOOSE JAW, SASKATCHEWAN			NEWCASTLE, NB			NEW GLASGOW, NOVA SCOTIA			NORTH BAY, ONTARIO			OSHAWA, ONTARIO		
		MAT.	INST.	TOTAL	MAT.	INST.	TOTAL	MAT.	INST.	TOTAL	MAT.	INST.	TOTAL	MAT.	INST.	TOTAL	MAT.	INST.	TOTAL
01590	EQUIPMENT RENTAL	.0	101.0	101.0	.0	98.6	98.6	.0	98.6	98.6	.0	98.6	98.6	.0	100.4	100.4	.0	103.0	103.0
02	SITE CONSTRUCTION	93.7	96.8	96.0	111.4	94.5	99.0	96.0	94.3	94.8	99.0	95.5	96.4	114.3	98.5	102.7	114.2	102.8	105.8
03100	CONCRETE FORMS & ACCESSORIES	130.3	79.0	86.1	101.9	54.2	60.8	97.6	57.0	62.6	90.3	64.1	67.7	122.5	78.1	84.2	122.6	80.6	86.4
03200	CONCRETE REINFORCEMENT	159.6	80.6	119.4	113.9	63.3	88.2	148.3	57.8	102.3	148.3	47.6	97.1	174.7	83.6	128.4	174.7	86.2	129.8
03300	CAST-IN-PLACE CONCRETE	137.5	88.2	117.2	152.8	63.5	116.1	144.0	55.3	107.5	177.1	64.3	130.7	170.5	77.6	132.3	169.7	79.2	132.5
03	CONCRETE	135.5	82.7	109.3	127.2	59.9	93.8	137.9	57.2	97.8	151.3	61.8	106.9	153.3	79.3	116.5	152.9	81.4	117.4
04	MASONRY	164.5	79.1	111.5	160.1	58.5	97.1	158.7	59.0	96.9	161.1	65.2	101.7	166.7	81.8	114.0	166.7	87.8	117.7
05	METALS	125.0	83.5	111.7	104.7	70.8	93.9	105.3	67.6	93.3	106.3	70.3	94.8	107.2	85.8	100.3	107.2	86.5	100.6
06	WOOD & PLASTICS	131.5	79.0	104.2	95.2	52.8	73.2	93.4	57.2	74.5	84.5	63.5	73.6	118.4	78.4	97.6	118.4	78.9	97.9
07	THERMAL & MOISTURE PROTECTION	105.3	83.9	96.3	103.6	59.9	85.2	106.1	58.3	85.9	104.4	65.2	87.9	109.4	79.6	96.8	109.6	82.0	98.0
08	DOORS & WINDOWS	91.9	69.9	86.2	86.9	51.9	77.8	86.1	50.9	76.9	82.8	59.0	76.6	90.9	76.5	87.2	90.9	80.2	88.1
09200	PLASTER & GYPSUM BOARD	154.6	78.1	107.1	136.9	51.6	83.9	171.1	56.1	99.7	164.6	62.6	101.3	158.6	77.9	108.5	159.9	78.4	109.3
095,098	CEILINGS & ACOUSTICAL TREATMENT	107.7	78.1	90.0	95.8	51.6	69.4	96.6	56.1	72.4	95.8	62.6	76.0	95.8	77.9	85.1	100.9	78.4	87.5
09600	FLOORING	132.9	88.6	120.9	120.0	56.3	102.7	113.7	42.0	94.3	109.6	59.9	96.2	132.9	86.7	120.4	132.9	90.2	121.3
097,099	WALL FINISHES, PAINTS & COATINGS	109.1	83.3	93.3	109.1	60.7	79.6	109.1	46.9	71.1	109.1	58.2	78.1	109.1	82.8	93.1	109.1	100.6	103.9
09	FINISHES	121.0	81.7	100.5	112.5	55.1	82.6	115.4	53.2	83.0	112.8	62.9	86.8	119.3	80.6	99.1	120.7	84.7	102.0
10 - 14	TOTAL DIV. 10000 - 14000	140.0	74.6	126.0	140.0	56.0	122.1	140.0	56.6	122.2	140.0	58.1	122.5	140.0	63.8	123.7	140.0	84.5	128.2
15	MECHANICAL	101.6	71.9	89.3	104.1	59.7	85.7	103.7	55.0	83.5	103.7	66.4	88.2	103.7	77.1	92.6	101.0	80.4	92.5
16	ELECTRICAL	121.1	71.6	97.0	126.0	61.9	94.8	125.4	61.0	94.1	121.1	63.5	93.1	124.7	89.5	107.6	126.3	89.2	108.3
01 - 16	WEIGHTED AVERAGE	118.5	79.0	101.0	114.2	62.9	91.4	115.1	60.9	91.1	115.8	67.1	94.2	119.1	82.4	102.8	118.7	85.7	104.1

	DIVISION	CANADA																	
		OTTAWA, ONTARIO			OWEN SOUND, ONTARIO			PETERBOROUGH, ONTARIO			PORTAGE LA PRAIRIE, MB			PRINCE ALBERT, SS			PRINCE GEORGE, BC		
		MAT.	INST.	TOTAL	MAT.	INST.	TOTAL	MAT.	INST.	TOTAL	MAT.	INST.	TOTAL	MAT.	INST.	TOTAL	MAT.	INST.	TOTAL
01590	EQUIPMENT RENTAL	.0	103.0	103.0	.0	100.4	100.4	.0	100.4	100.4	.0	102.9	102.9	.0	98.6	98.6	.0	103.5	103.5
02	SITE CONSTRUCTION	112.9	102.8	105.5	116.6	98.6	103.4	114.3	98.8	102.9	112.2	98.1	101.9	107.6	94.6	98.1	121.3	100.4	106.0
03100	CONCRETE FORMS & ACCESSORIES	120.1	80.6	86.1	122.6	76.7	83.0	122.5	79.0	85.0	121.7	65.1	72.9	101.9	54.1	60.7	108.0	81.5	85.1
03200	CONCRETE REINFORCEMENT	173.8	84.1	128.2	179.9	85.6	132.0	174.7	84.1	128.7	161.5	54.3	107.0	119.0	63.2	90.7	116.5	78.4	97.1
03300	CAST-IN-PLACE CONCRETE	159.6	89.4	130.7	174.7	73.8	133.3	170.5	79.0	132.9	157.5	68.9	121.1	145.1	63.5	111.6	153.3	90.7	127.5
03	CONCRETE	147.7	84.6	116.3	156.2	77.7	117.2	153.3	80.3	117.0	138.2	65.1	101.9	124.3	59.8	92.3	154.4	84.2	119.6
04	MASONRY	168.1	85.4	116.8	170.2	86.8	118.5	166.7	87.8	117.8	164.6	61.4	100.6	159.7	58.5	97.0	170.7	86.1	118.3
05	METALS	116.4	86.3	106.8	107.8	86.0	100.9	107.2	86.2	100.5	107.9	73.5	96.9	104.8	70.7	93.9	107.8	83.3	100.0
06	WOOD & PLASTICS	116.9	79.7	97.5	120.2	75.5	96.9	118.4	77.2	97.0	116.1	65.8	89.9	95.2	52.8	73.2	98.1	80.0	88.7
07	THERMAL & MOISTURE PROTECTION	110.2	82.6	98.6	109.5	80.3	97.2	109.4	83.9	98.7	104.2	66.9	88.5	103.5	59.0	84.8	116.7	80.5	101.5
08	DOORS & WINDOWS	91.9	79.9	88.8	91.1	75.9	87.1	90.9	78.1	87.6	91.9	59.1	83.4	85.5	51.9	76.7	87.8	78.1	85.3
09200	PLASTER & GYPSUM BOARD	232.4	79.2	137.2	158.6	74.9	106.6	158.6	76.7	107.7	131.4	64.5	89.8	136.9	51.6	83.9	138.2	79.1	101.5
095,098	CEILINGS & ACOUSTICAL TREATMENT	106.8	79.2	90.3	95.8	74.9	83.3	95.8	76.7	84.4	95.8	64.5	77.1	95.8	51.6	69.4	95.8	79.1	85.8
09600	FLOORING	132.9	86.7	120.4	132.9	87.9	120.7	132.9	86.7	120.4	132.9	62.5	113.8	120.0	56.3	102.7	127.1	49.5	106.1
097,099	WALL FINISHES, PAINTS & COATINGS	109.1	89.6	97.2	109.8	81.6	92.6	109.1	84.9	94.3	109.2	53.3	75.1	109.1	51.8	74.1	109.1	76.2	89.0
09	FINISHES	133.3	82.9	107.0	120.4	79.5	99.1	119.3	81.2	99.4	115.4	63.6	88.4	112.5	54.1	82.0	116.1	75.3	94.8
10 - 14	TOTAL DIV. 10000 - 14000	140.0	82.3	127.7	140.0	64.2	123.8	140.0	65.2	124.1	140.0	59.3	122.8	140.0	56.0	122.1	140.0	82.9	127.8
15	MECHANICAL	101.6	79.3	92.4	103.7	77.2	92.7	103.7	80.2	93.9	103.7	65.9	88.0	104.1	59.7	85.7	103.7	74.8	91.7
16	ELECTRICAL	122.1	89.7	106.3	126.6	88.3	108.0	124.7	89.2	107.4	125.4	68.6	97.8	126.0	61.9	94.8	124.3	80.8	103.2
01 - 16	WEIGHTED AVERAGE	120.5	85.5	105.0	120.1	82.5	103.4	119.1	84.1	103.5	117.0	68.5	95.5	113.7	62.7	91.1	119.1	81.8	102.5

	DIVISION	CANADA																	
		QUEBEC, QUEBEC			RED DEER, ALBERTA			REGINA, SASKATCHEWAN			RIMOUSKI, QUEBEC			ROUYN-NORANDA, QUEBEC			SAINT HYACINTHE, QUEBEC		
		MAT.	INST.	TOTAL	MAT.	INST.	TOTAL	MAT.	INST.	TOTAL	MAT.	INST.	TOTAL	MAT.	INST.	TOTAL	MAT.	INST.	TOTAL
01590	EQUIPMENT RENTAL	.0	101.4	101.4	.0	103.7	103.7	.0	98.6	98.6	.0	99.2	99.2	.0	99.2	99.2	.0	99.2	99.2
02	SITE CONSTRUCTION	95.5	96.9	96.5	110.3	101.2	103.6	112.8	94.5	99.4	93.9	96.5	95.8	93.4	96.5	95.7	94.0	96.5	95.9
03100	CONCRETE FORMS & ACCESSORIES	130.2	79.3	86.3	135.1	73.9	82.3	102.0	54.2	60.8	129.8	78.7	85.7	129.8	78.7	85.8	129.8	78.7	85.8
03200	CONCRETE REINFORCEMENT	148.3	80.6	113.9	161.5	64.2	112.0	126.1	63.3	94.2	111.9	80.5	96.0	157.0	80.5	118.1	157.0	80.5	118.1
03300	CAST-IN-PLACE CONCRETE	151.5	88.6	125.7	157.5	83.3	127.0	162.8	63.5	122.0	142.7	86.6	119.6	135.6	86.6	115.5	140.1	86.6	118.1
03	CONCRETE	140.4	82.9	111.9	145.7	75.9	111.1	134.0	59.9	97.2	130.3	82.0	106.3	134.1	82.0	108.2	136.3	82.0	109.3
04	MASONRY	165.4	79.1	111.9	161.6	71.8	105.9	168.3	58.9	100.5	160.6	79.1	110.1	161.0	79.1	110.2	161.3	79.1	110.3
05	METALS	120.7	83.8	108.9	107.9	83.0	100.0	105.6	70.8	94.5	105.2	83.0	98.1	105.6	83.0	98.4	105.6	83.0	98.4
06	WOOD & PLASTICS	132.2	79.1	104.5	116.2	73.5	94.0	95.2	52.8	73.2	131.5	78.7	104.0	131.5	78.7	104.0	131.5	78.7	104.0
07	THERMAL & MOISTURE PROTECTION	104.9	84.1	96.1	124.6	76.7	104.5	105.5	60.0	86.4	104.3	83.3	95.5	104.3	83.3	95.5	104.7	83.3	95.7
08	DOORS & WINDOWS	91.9	76.8	88.0	91.9	70.5	86.3	86.9	51.9	77.8	91.5	68.2	85.4	91.9	69.8	86.1	91.9	69.8	86.1
09200	PLASTER & GYPSUM BOARD	197.2	78.1	123.2	139.0	72.4	97.6	168.0	51.6	95.7	194.7	78.1	122.2	146.1	78.1	103.8	150.9	78.1	105.7
095,098	CEILINGS & ACOUSTICAL TREATMENT	97.5	78.1	85.9	95.8	72.4	81.8	122.2	51.6	80.1	93.2	78.1	84.2	93.2	78.1	84.2	93.2	78.1	84.2
09600	FLOORING	132.9	88.6	120.9	134.9	82.4	120.7	120.0	56.3	102.7	132.9	88.6	120.9	132.9	88.6	120.9	132.9	88.6	120.9
097,099	WALL FINISHES, PAINTS & COATINGS	109.1	83.3	93.3	109.1	73.8	87.5	109.1	60.7	79.6	109.1	83.3	93.3	109.1	83.3	93.3	109.1	83.3	93.3
09	FINISHES	125.0	81.7	102.4	117.2	75.4	95.4	124.0	55.1	88.1	123.5	81.5	101.6	116.1	81.5	98.1	116.8	81.5	98.4
10 - 14	TOTAL DIV. 10000 - 14000	140.0	74.7	126.1	140.0	79.5	127.1	140.0	56.0	122.1	140.0	73.8	125.9	140.0	73.8	125.9	140.0	73.8	125.9
15	MECHANICAL	101.5	71.9	89.2	103.7	72.2	90.6	101.3	59.7	84.0	103.7	71.8	90.4	103.7	71.8	90.4	101.0	71.8	88.9
16	ELECTRICAL	125.9	71.6	99.5	114.4	84.2	99.7	126.4	61.9	95.0	121.7	71.6	97.3	121.7	71.6	97.3	122.1	71.6	97.5
01 - 16	WEIGHTED AVERAGE	119.4	79.4	101.6	117.1	78.6	100.0	116.1	62.9	92.5	115.5	78.6	99.1	115.3	78.7	99.0	115.1	78.7	98.9

	DIVISION	CANADA																	
		SAINT JOHN, N B			SARNIA, ONTARIO			SASKATOON, SASKATCHEWAN			SAULT STE MARIE, ON			SHERBROOKE, QUEBEC			SOREL, QUEBEC		
		MAT.	INST.	TOTAL	MAT.	INST.	TOTAL	MAT.	INST.	TOTAL	MAT.	INST.	TOTAL	MAT.	INST.	TOTAL	MAT.	INST.	TOTAL
01590	EQUIPMENT RENTAL	.0	98.6	98.6	.0	100.4	100.4	.0	98.6	98.6	.0	100.4	100.4	.0	99.2	99.2	.0	99.2	99.2
02	SITE CONSTRUCTION	96.1	94.3	94.8	112.6	98.8	102.5	107.9	94.6	98.1	102.3	98.5	99.5	94.0	96.5	95.9	94.2	96.5	95.9
03100	CONCRETE FORMS & ACCESSORIES	120.1	57.3	65.9	120.9	85.5	90.4	102.0	54.2	60.7	108.5	78.4	82.5	129.8	78.7	85.8	129.8	78.9	85.9
03200	CONCRETE REINFORCEMENT	148.3	57.9	102.3	123.5	85.5	104.2	119.0	63.3	90.7	112.2	83.7	97.7	157.0	80.5	118.1	148.3	80.6	113.9
03300	CAST-IN-PLACE CONCRETE	141.9	55.3	106.3	157.3	90.7	129.9	148.0	63.5	113.2	141.2	77.7	115.1	140.1	86.6	118.1	141.2	86.6	118.8
03	CONCRETE	138.6	57.4	98.3	138.6	87.4	113.1	125.7	59.8	93.0	122.0	79.5	100.9	136.3	82.0	109.3	135.4	82.1	108.9
04	MASONRY	176.5	60.4	104.5	177.4	90.3	123.4	168.0	58.9	100.3	163.2	81.8	112.7	161.3	79.1	110.3	161.4	79.1	110.3
05	METALS	105.2	68.0	93.4	107.1	86.5	100.5	105.7	70.7	94.5	106.5	86.1	100.0	105.6	83.0	98.4	105.6	83.1	98.4
06	WOOD & PLASTICS	119.2	57.2	86.9	117.4	84.8	100.4	93.8	52.8	72.5	103.4	78.4	90.4	131.5	78.7	104.0	131.5	78.7	104.0
07	THERMAL & MOISTURE PROTECTION	106.4	59.2	86.5	109.5	87.7	100.3	104.3	59.1	85.3	108.2	81.0	96.8	104.4	83.3	95.5	104.3	83.3	95.5
08	DOORS & WINDOWS	86.0	49.9	76.6	92.9	81.6	89.9	86.0	51.9	77.0	83.9	77.7	82.3	91.9	69.8	86.1	91.9	74.0	87.2
09200	PLASTER & GYPSUM BOARD	189.2	56.1	106.5	187.8	84.4	123.5	148.5	51.6	88.3	145.6	77.9	103.5	148.5	78.1	104.8	194.7	78.1	122.2
095,098	CEILINGS & ACOUSTICAL TREATMENT	105.1	56.1	75.9	100.9	84.4	91.1	122.2	51.6	80.1	95.8	77.9	85.1	93.2	78.1	84.2	93.2	78.1	84.2
09600	FLOORING	128.0	42.0	104.7	132.9	91.6	121.7	120.0	56.3	102.7	123.6	86.7	113.6	132.9	88.6	120.9	132.9	88.6	120.9
097,099	WALL FINISHES, PAINTS & COATINGS	109.1	59.8	79.0	109.1	96.0	101.1	109.1	51.8	74.1	109.1	89.0	96.8	109.1	83.3	93.3	109.1	83.3	93.3
09	FINISHES	124.5	54.7	88.1	125.0	88.2	105.9	120.8	54.1	86.1	113.9	81.3	96.9	116.4	81.5	98.2	123.5	81.5	101.6
10 - 14	TOTAL DIV. 10000 - 14000	140.0	56.6	122.2	140.0	66.7	124.4	140.0	56.0	122.1	140.0	82.8	127.8	140.0	73.8	125.9	140.0	73.8	125.9
15	MECHANICAL	103.7	62.1	86.4	103.7	84.9	95.9	101.2	59.7	84.0	103.7	74.6	91.6	104.1	71.8	90.7	103.7	71.8	90.5
16	ELECTRICAL	133.4	76.0	105.5	130.2	91.4	111.3	126.5	61.9	95.1	127.8	89.5	109.2	121.7	71.6	97.3	121.7	71.6	97.3
01 - 16	WEIGHTED AVERAGE	118.1	64.9	94.5	119.3	87.9	105.4	114.6	62.8	91.6	114.0	82.7	100.1	115.7	78.7	99.3	116.2	78.9	99.6

	DIVISION	CANADA																	
		ST CATHARINES, ONTARIO			ST JEROME, QUEBEC			ST JOHNS, NEWFOUNDLAND			SUDBURY, ONTARIO			SUMMERSIDE, PEI			SYDNEY, NOVA SCOTIA		
		MAT.	INST.	TOTAL	MAT.	INST.	TOTAL	MAT.	INST.	TOTAL	MAT.	INST.	TOTAL	MAT.	INST.	TOTAL	MAT.	INST.	TOTAL
01590	EQUIPMENT RENTAL	.0	100.4	100.4	.0	99.2	99.2	.0	98.6	98.6	.0	100.4	100.4	.0	98.6	98.6	.0	98.6	98.6
02	SITE CONSTRUCTION	102.2	98.6	99.5	93.4	96.5	95.7	114.8	94.2	99.7	102.3	98.5	99.5	106.9	93.8	97.3	93.8	95.5	95.1
03100	CONCRETE FORMS & ACCESSORIES	113.2	78.7	83.4	129.8	78.7	85.8	99.5	55.1	61.2	108.5	78.4	82.5	90.6	52.1	57.4	90.3	64.1	67.7
03200	CONCRETE REINFORCEMENT	111.3	85.6	98.3	157.0	80.5	118.1	164.7	49.5	106.2	112.2	83.7	97.7	148.3	47.7	97.2	148.3	47.6	97.1
03300	CAST-IN-PLACE CONCRETE	140.1	78.3	114.7	135.6	86.6	115.5	179.6	62.6	131.5	141.2	77.7	115.1	168.0	54.6	121.4	136.4	64.3	106.7
03	CONCRETE	121.6	80.1	101.0	134.1	82.0	108.2	167.3	57.5	112.8	122.0	79.5	100.9	155.4	52.9	104.5	131.7	61.8	97.0
04	MASONRY	163.1	82.0	112.8	161.0	79.1	110.2	163.0	58.3	98.1	163.2	81.8	112.7	160.4	56.9	96.3	158.1	65.2	100.5
05	METALS	106.6	86.3	100.1	105.6	83.0	98.4	108.2	68.3	95.5	106.5	86.1	100.0	106.4	63.9	92.8	106.3	70.3	94.8
06	WOOD & PLASTICS	108.3	79.5	93.3	131.5	78.7	104.0	94.6	54.2	73.6	103.4	78.4	90.4	84.8	51.6	67.5	84.5	63.5	73.6
07	THERMAL & MOISTURE PROTECTION	108.6	82.6	97.6	104.3	83.3	95.5	108.1	57.7	86.9	108.2	81.0	96.8	104.8	56.0	84.2	104.4	65.2	87.9
08	DOORS & WINDOWS	83.0	79.6	82.1	91.9	69.8	86.1	98.0	51.9	86.0	83.9	77.7	82.3	94.1	45.5	81.4	82.8	59.0	76.6
09200	PLASTER & GYPSUM BOARD	142.2	79.0	103.0	146.1	78.1	103.8	173.8	53.1	98.8	145.6	77.9	103.5	165.0	50.4	93.8	164.6	62.6	101.3
095,098	CEILINGS & ACOUSTICAL TREATMENT	100.9	79.0	87.8	93.2	78.1	84.2	106.8	53.1	74.8	95.8	77.9	85.1	95.8	50.4	68.7	95.8	62.6	76.0
09600	FLOORING	126.6	87.9	116.1	132.9	88.6	120.9	114.6	50.9	97.4	123.6	86.7	113.6	109.7	56.0	95.2	109.6	59.9	96.2
097,099	WALL FINISHES, PAINTS & COATINGS	109.1	89.6	97.2	109.1	83.3	93.3	109.1	56.1	76.8	109.1	89.0	96.8	109.1	39.0	66.3	109.1	58.2	78.1
09	FINISHES	115.6	81.9	98.0	116.1	81.5	98.1	119.4	54.5	85.6	113.9	81.3	96.9	113.6	51.7	81.4	112.8	62.9	86.8
10 - 14	TOTAL DIV. 10000 - 14000	140.0	63.4	123.7	140.0	73.8	125.9	140.0	56.7	122.2	140.0	82.8	127.8	140.0	56.0	122.1	140.0	58.1	122.5
15	MECHANICAL	101.0	77.3	91.2	103.7	71.8	90.4	101.4	56.5	82.8	103.7	74.6	91.7	103.7	50.4	81.5	103.7	66.4	88.2
16	ELECTRICAL	128.2	89.7	109.5	122.9	71.6	98.0	121.0	57.4	90.1	124.7	89.5	107.6	119.3	51.6	86.4	121.1	63.5	93.1
01 - 16	WEIGHTED AVERAGE	113.6	83.1	100.0	115.4	78.7	99.1	120.1	60.9	93.8	113.7	82.7	99.9	117.3	56.9	90.5	113.3	67.1	92.8

DIVISION		CANADA																	
		THUNDER BAY, ONTARIO			TIMMINS, ONTARIO			TORONTO, ONTARIO			TROIS RIVIERES, QUEBEC			TRURO, NOVA SCOTIA			VANCOUVER, B C		
		MAT.	INST.	TOTAL	MAT.	INST.	TOTAL	MAT.	INST.	TOTAL	MAT.	INST.	TOTAL	MAT.	INST.	TOTAL	MAT.	INST.	TOTAL
01590	**EQUIPMENT RENTAL**	.0	100.4	100.4	.0	100.4	100.4	.0	103.2	103.2	.0	99.2	99.2	.0	98.6	98.6	.0	111.3	111.3
02	**SITE CONSTRUCTION**	107.9	98.7	101.1	114.3	98.5	102.7	113.8	103.6	106.3	94.2	96.5	95.9	99.0	95.5	96.4	118.6	107.2	110.3
03100	CONCRETE FORMS & ACCESSORIES	122.6	79.7	85.6	122.5	78.1	84.2	123.0	90.4	94.8	129.8	78.9	85.9	90.3	64.1	67.7	114.5	82.2	86.7
03200	CONCRETE REINFORCEMENT	99.5	85.0	92.1	174.7	83.6	128.4	170.9	86.6	128.0	148.3	80.6	113.9	148.3	47.6	97.1	173.2	78.6	125.1
03300	CAST-IN-PLACE CONCRETE	154.2	89.6	127.6	170.5	77.6	132.3	163.3	98.8	136.7	141.2	86.6	118.8	177.1	64.3	130.7	153.2	91.5	127.9
03	**CONCRETE**	130.4	84.3	107.5	153.3	79.3	116.5	149.2	92.5	121.1	135.4	82.1	108.9	151.3	61.8	106.9	147.9	85.2	116.8
04	**MASONRY**	163.9	82.8	113.6	166.7	81.8	114.0	186.9	96.2	130.7	161.4	79.1	110.3	161.1	65.2	101.7	177.3	86.1	120.8
05	**METALS**	106.4	85.4	99.7	107.2	85.8	100.3	130.4	88.5	117.0	105.6	83.1	98.4	106.3	70.3	94.8	141.5	88.4	124.6
06	**WOOD & PLASTICS**	118.4	79.9	98.3	118.4	78.4	97.6	118.4	88.9	103.0	131.5	78.7	104.0	84.5	63.5	73.6	113.9	80.2	96.3
07	**THERMAL & MOISTURE PROTECTION**	108.8	81.9	97.5	109.4	79.6	96.8	110.2	92.3	102.7	104.3	83.3	95.5	104.4	65.2	87.9	116.6	86.9	104.1
08	**DOORS & WINDOWS**	82.0	79.2	81.2	90.9	76.5	87.2	90.9	87.7	90.1	91.9	74.0	87.2	82.8	59.0	76.6	94.4	78.4	90.2
09200	PLASTER & GYPSUM BOARD	170.8	79.4	114.0	158.6	77.9	108.5	174.9	88.6	121.3	194.7	78.1	122.2	164.6	62.6	101.3	149.2	79.1	105.7
095,098	CEILINGS & ACOUSTICAL TREATMENT	95.8	79.4	86.0	95.8	77.9	85.1	121.3	88.6	101.8	93.2	78.1	84.2	95.8	62.6	76.0	111.1	79.1	92.0
09600	FLOORING	132.9	50.6	110.6	132.9	86.7	120.4	132.9	93.2	122.1	132.9	88.6	120.9	109.6	59.9	96.2	132.9	85.3	120.0
097,099	WALL FINISHES, PAINTS & COATINGS	109.1	91.6	98.4	109.1	82.8	93.1	112.1	100.6	105.1	109.1	83.3	93.3	109.1	58.2	78.1	109.1	95.2	100.6
09	**FINISHES**	120.8	76.4	97.7	119.3	80.6	99.1	128.4	92.2	109.6	123.5	81.5	101.6	112.8	62.9	86.8	122.6	83.8	102.4
10 - 14	**TOTAL DIV. 10000 - 14000**	140.0	64.3	123.9	140.0	63.8	123.7	140.0	87.4	128.8	140.0	73.8	125.9	140.0	58.1	122.5	140.0	83.2	127.9
15	**MECHANICAL**	101.0	78.0	91.5	103.7	77.1	92.6	101.5	88.0	95.9	103.7	71.8	90.5	103.7	66.4	88.2	101.5	81.4	93.2
16	**ELECTRICAL**	125.2	87.9	107.1	127.8	89.5	109.2	129.8	92.3	111.6	121.7	71.6	97.3	121.1	63.5	93.1	132.9	80.8	107.5
01 - 16	**WEIGHTED AVERAGE**	114.8	82.8	100.6	119.4	82.4	103.0	124.2	92.1	109.9	116.2	78.9	99.6	115.8	67.1	94.2	125.7	85.6	107.9

DIVISION		CANADA																	
		VICTORIA, B C			WHITEHORSE, YUKON			WINDSOR, ONTARIO			WINNIPEG, MANITOBA			YARMOUTH, NOVA SCOTIA			YELLOWKNIFE, NWT		
		MAT.	INST.	TOTAL	MAT.	INST.	TOTAL	MAT.	INST.	TOTAL	MAT.	INST.	TOTAL	MAT.	INST.	TOTAL	MAT.	INST.	TOTAL
01590	**EQUIPMENT RENTAL**	.0	107.3	107.3	.0	99.2	99.2	.0	100.4	100.4	.0	105.3	105.3	.0	98.6	98.6	.0	98.7	98.7
02	**SITE CONSTRUCTION**	121.2	106.4	110.3	108.5	94.6	98.3	98.3	98.8	98.7	112.2	100.0	103.3	98.7	95.5	96.4	121.3	97.5	103.8
03100	CONCRETE FORMS & ACCESSORIES	108.0	81.5	85.1	102.0	53.4	60.0	122.6	80.3	86.1	122.2	65.2	73.0	90.3	64.1	67.7	102.0	73.2	77.2
03200	CONCRETE REINFORCEMENT	116.5	78.4	97.1	119.0	61.3	89.7	109.0	84.3	96.5	161.5	54.3	107.0	148.3	47.6	97.1	119.0	65.1	91.6
03300	CAST-IN-PLACE CONCRETE	153.3	90.7	127.5	152.8	62.8	115.8	143.5	90.7	121.8	157.5	70.2	121.6	175.2	64.3	129.6	253.7	79.2	182.0
03	**CONCRETE**	154.4	84.2	119.6	128.0	58.8	93.7	123.5	84.9	104.3	144.6	65.6	105.4	150.4	61.8	106.4	176.8	74.2	125.8
04	**MASONRY**	170.7	86.1	118.3	160.2	57.1	96.3	163.3	89.0	117.2	178.5	61.5	106.0	160.9	65.2	101.6	167.8	69.2	106.7
05	**METALS**	107.8	83.3	100.0	105.8	69.9	94.4	106.5	86.6	100.1	139.2	73.7	118.3	106.3	70.3	94.8	106.6	76.3	96.9
06	**WOOD & PLASTICS**	98.1	80.0	88.7	95.2	52.1	72.8	118.4	78.5	97.6	116.1	65.9	90.0	84.5	63.5	73.6	95.2	74.4	84.4
07	**THERMAL & MOISTURE PROTECTION**	116.7	80.5	101.5	103.5	58.1	84.4	108.6	85.5	98.9	105.3	68.2	89.6	104.4	65.2	87.9	104.7	73.9	91.7
08	**DOORS & WINDOWS**	87.8	78.1	85.3	85.5	50.7	76.4	81.7	79.0	81.0	91.9	59.2	83.4	82.8	59.0	76.6	85.5	64.3	80.0
09200	PLASTER & GYPSUM BOARD	138.2	79.1	101.5	119.9	50.8	77.0	163.5	78.0	110.4	153.9	64.5	98.4	164.6	62.6	101.3	122.3	73.8	92.2
095,098	CEILINGS & ACOUSTICAL TREATMENT	95.8	79.1	85.8	95.8	50.8	68.9	95.8	78.0	85.1	107.7	64.5	81.9	95.8	62.6	76.0	95.8	73.8	82.6
09600	FLOORING	127.1	49.5	106.1	120.0	54.6	102.3	132.9	88.6	120.9	132.9	62.5	113.8	109.6	59.9	96.2	120.0	83.8	110.2
097,099	WALL FINISHES, PAINTS & COATINGS	109.1	76.2	89.0	109.1	50.9	73.6	109.1	91.2	98.2	109.2	53.3	75.1	109.1	58.2	78.1	109.1	74.2	87.8
09	**FINISHES**	116.1	75.3	94.8	109.9	53.1	80.3	119.4	83.1	100.5	122.0	63.7	91.6	112.8	62.9	86.8	110.2	74.7	91.7
10 - 14	**TOTAL DIV. 10000 - 14000**	140.0	82.9	127.8	140.0	55.4	122.0	140.0	65.5	124.1	140.0	59.6	122.9	140.0	58.1	122.5	140.0	78.9	127.0
15	**MECHANICAL**	103.7	74.8	91.7	103.7	58.7	85.0	101.0	80.5	92.5	101.5	65.9	86.7	103.7	66.4	88.2	103.7	74.6	91.6
16	**ELECTRICAL**	124.9	80.8	103.5	126.0	61.3	94.6	133.4	89.7	112.1	131.9	68.6	101.1	121.1	63.5	93.1	127.2	85.5	106.9
01 - 16	**WEIGHTED AVERAGE**	119.1	82.3	102.8	114.0	61.9	90.8	114.6	85.4	101.6	124.1	68.9	99.6	115.6	67.1	94.1	120.4	77.4	101.3

Location Factors

Costs shown in *Means cost data publications* are based on National Averages for materials and installation. To adjust these costs to a specific location, simply multiply the base cost by the factor and divide by 100 for that city. The data is arranged alphabetically by state and postal zip code numbers. For a city not listed, use the factor for a nearby city with similar economic characteristics.

STATE/ZIP	CITY	MAT.	INST.	TOTAL
ALABAMA				
350-352	Birmingham	96.5	75.2	87.0
354	Tuscaloosa	96.3	55.4	78.1
355	Jasper	96.5	52.1	76.8
356	Decatur	96.2	55.9	78.3
357-358	Huntsville	96.3	72.0	85.5
359	Gadsden	96.1	58.8	79.6
360-361	Montgomery	97.0	57.1	79.3
362	Anniston	95.7	46.5	73.9
363	Dothan	96.3	49.0	75.3
364	Evergreen	95.7	50.2	75.5
365-366	Mobile	97.2	61.1	81.2
367	Selma	95.9	51.6	76.2
368	Phenix City	96.6	54.9	78.1
369	Butler	96.1	49.0	75.2
ALASKA				
995-996	Anchorage	133.7	114.7	125.3
997	Fairbanks	130.3	117.6	124.6
998	Juneau	132.5	114.0	124.3
999	Ketchikan	141.9	113.6	129.3
ARIZONA				
850,853	Phoenix	98.8	73.6	87.6
852	Mesa/Tempe	97.8	68.9	85.0
855	Globe	98.1	65.5	83.6
856-857	Tucson	97.1	71.2	85.6
859	Show Low	98.2	68.0	84.8
860	Flagstaff	100.4	72.3	87.9
863	Prescott	98.3	66.8	84.3
864	Kingman	96.7	67.2	83.6
865	Chambers	96.8	68.3	84.1
ARKANSAS				
716	Pine Bluff	95.2	60.9	80.0
717	Camden	93.2	38.9	69.1
718	Texarkana	94.6	44.0	72.1
719	Hot Springs	92.4	38.4	68.4
720-722	Little Rock	95.1	63.8	81.2
723	West Memphis	94.6	53.4	76.3
724	Jonesboro	95.2	53.4	76.6
725	Batesville	93.3	48.4	73.3
726	Harrison	94.5	48.4	74.0
727	Fayetteville	91.9	46.7	71.8
728	Russellville	93.2	46.6	72.5
729	Fort Smith	95.9	56.9	78.6
CALIFORNIA				
900-902	Los Angeles	102.5	112.1	106.8
903-905	Inglewood	98.2	109.0	103.0
906-908	Long Beach	99.6	109.0	103.8
910-912	Pasadena	100.4	109.3	104.4
913-916	Van Nuys	103.5	109.1	106.0
917-918	Alhambra	102.7	109.1	105.5
919-921	San Diego	104.0	103.9	104.0
922	Palm Springs	100.1	107.2	103.2
923-924	San Bernardino	97.9	108.1	102.4
925	Riverside	102.2	109.5	105.5
926-927	Santa Ana	99.9	108.2	103.5
928	Anaheim	102.6	110.6	106.1
930	Oxnard	103.8	110.0	106.5
931	Santa Barbara	102.6	110.5	106.1
932-933	Bakersfield	102.3	105.9	103.9
934	San Luis Obispo	103.4	107.9	105.4
935	Mojave	100.5	103.1	101.7
936-938	Fresno	104.1	110.9	107.1
939	Salinas	104.1	116.4	109.6
940-941	San Francisco	112.8	132.9	121.7
942,956-958	Sacramento	106.6	113.0	109.4
943	Palo Alto	105.5	127.2	115.1
944	San Mateo	108.4	127.8	117.0
945	Vallejo	105.5	122.2	112.9
946	Oakland	110.7	127.1	118.0
947	Berkeley	110.1	126.7	117.5
948	Richmond	109.5	126.1	116.9
949	San Rafael	110.2	122.8	115.8
950	Santa Cruz	109.4	116.4	112.5
CALIFORNIA (CONT'D)				
951	San Jose	107.3	127.5	116.3
952	Stockton	104.4	113.5	108.4
953	Modesto	104.3	111.0	107.3
954	Santa Rosa	104.1	122.0	112.1
955	Eureka	105.9	101.1	103.8
959	Marysville	105.0	112.5	108.3
960	Redding	106.6	110.3	108.3
961	Susanville	105.7	110.7	107.9
COLORADO				
800-802	Denver	100.6	89.9	95.8
803	Boulder	97.7	86.5	92.7
804	Golden	99.9	85.8	93.6
805	Fort Collins	101.1	82.4	92.8
806	Greeley	98.7	70.4	86.1
807	Fort Morgan	98.4	85.8	92.8
808-809	Colorado Springs	100.4	85.7	93.9
810	Pueblo	100.0	83.1	92.5
811	Alamosa	101.3	79.5	91.7
812	Salida	101.3	80.2	91.9
813	Durango	101.7	81.4	92.7
814	Montrose	100.3	79.6	91.1
815	Grand Junction	103.6	76.8	91.7
816	Glenwood Springs	101.1	83.3	93.2
CONNECTICUT				
060	New Britain	101.1	116.6	108.0
061	Hartford	102.3	117.3	109.0
062	Willimantic	101.8	117.2	108.7
063	New London	98.1	117.3	106.7
064	Meriden	100.1	117.8	108.0
065	New Haven	102.8	117.8	109.4
066	Bridgeport	102.4	117.8	109.2
067	Waterbury	101.9	117.7	108.9
068	Norwalk	101.9	118.2	109.1
069	Stamford	102.0	124.0	111.8
D.C.				
200-205	Washington	102.5	91.2	97.5
DELAWARE				
197	Newark	100.7	103.9	102.1
198	Wilmington	99.7	103.9	101.5
199	Dover	100.7	103.9	102.1
FLORIDA				
320,322	Jacksonville	98.3	59.6	81.1
321	Daytona Beach	98.3	72.8	87.0
323	Tallahassee	97.6	48.2	75.6
324	Panama City	99.4	36.8	71.6
325	Pensacola	99.1	57.2	80.5
326,344	Gainesville	99.6	57.7	81.0
327-328,347	Orlando	99.7	68.6	85.9
329	Melbourne	100.8	77.8	90.6
330-332,340	Miami	98.7	71.4	86.6
333	Fort Lauderdale	97.7	70.3	85.5
334,349	West Palm Beach	96.6	66.6	83.3
335-336,346	Tampa	99.4	70.8	86.7
337	St. Petersburg	101.6	54.2	80.5
338	Lakeland	98.4	70.5	86.0
339,341	Fort Myers	97.7	62.8	82.2
342	Sarasota	99.5	65.1	84.2
GEORGIA				
300-303,399	Atlanta	97.7	79.0	89.4
304	Statesboro	97.1	43.6	73.4
305	Gainesville	96.0	55.7	78.1
306	Athens	95.3	62.2	80.6
307	Dalton	97.4	49.4	76.1
308-309	Augusta	96.3	57.5	79.0
310-312	Macon	95.7	59.1	79.4
313-314	Savannah	97.0	57.7	79.6
315	Waycross	97.0	50.0	76.1
316	Valdosta	96.9	47.6	75.0
317,398	Albany	97.2	56.3	79.1
318-319	Columbus	97.2	60.4	80.8

STATE/ZIP	CITY	MAT.	INST.	TOTAL
HAWAII				
967	Hilo	115.0	124.8	119.3
968	Honolulu	120.6	124.8	122.5
STATES & POSS.				
969	Guam	196.4	52.7	132.5
IDAHO				
832	Pocatello	100.3	77.3	90.1
833	Twin Falls	100.8	57.7	81.7
834	Idaho Falls	98.3	58.8	80.8
835	Lewiston	107.4	83.6	96.8
836-837	Boise	100.2	78.9	90.7
838	Coeur d'Alene	106.6	75.9	93.0
ILLINOIS				
600-603	North Suburban	99.5	125.3	111.0
604	Joliet	99.4	118.5	107.9
605	South Suburban	99.5	125.3	111.0
606-608	Chicago	100.1	126.3	111.7
609	Kankakee	95.7	104.1	99.4
610-611	Rockford	98.1	107.9	102.4
612	Rock Island	95.4	97.0	96.1
613	La Salle	96.7	102.6	99.3
614	Galesburg	96.5	98.9	97.6
615-616	Peoria	99.1	99.9	99.4
617	Bloomington	95.8	99.2	97.3
618-619	Champaign	99.2	100.8	99.9
620-622	East St. Louis	94.5	102.0	97.8
623	Quincy	95.7	94.6	95.2
624	Effingham	95.1	97.1	96.0
625	Decatur	97.0	101.1	98.8
626-627	Springfield	97.3	99.8	98.4
628	Centralia	93.3	101.1	96.8
629	Carbondale	92.9	97.3	94.9
INDIANA				
460	Anderson	94.9	84.8	90.4
461-462	Indianapolis	98.8	87.2	93.7
463-464	Gary	96.2	104.3	99.8
465-466	South Bend	95.3	86.8	91.5
467-468	Fort Wayne	95.4	83.1	89.9
469	Kokomo	93.0	85.7	89.8
470	Lawrenceburg	92.0	81.8	87.5
471	New Albany	93.6	77.5	86.5
472	Columbus	96.1	84.0	90.7
473	Muncie	96.2	82.4	90.1
474	Bloomington	98.2	84.1	91.9
475	Washington	94.0	88.4	91.5
476-477	Evansville	95.3	90.0	93.0
478	Terre Haute	96.0	88.3	92.6
479	Lafayette	95.7	82.4	89.8
IOWA				
500-503,509	Des Moines	97.8	82.2	90.9
504	Mason City	96.0	62.0	80.9
505	Fort Dodge	96.2	58.1	79.3
506-507	Waterloo	97.8	62.0	81.9
508	Creston	96.4	66.0	82.9
510-511	Sioux City	98.4	74.9	88.0
512	Sibley	97.0	51.5	76.8
513	Spencer	98.8	49.9	77.1
514	Carroll	95.8	55.7	78.0
515	Council Bluffs	99.3	75.2	88.6
516	Shenandoah	96.2	52.9	77.0
520	Dubuque	97.8	77.2	88.7
521	Decorah	96.8	52.8	77.3
522-524	Cedar Rapids	98.9	84.4	92.5
525	Ottumwa	96.8	71.1	85.4
526	Burlington	96.0	75.5	86.9
527-528	Davenport	97.9	94.9	96.5
KANSAS				
660-662	Kansas City	98.9	96.1	97.7
664-666	Topeka	98.7	69.3	85.6
667	Fort Scott	97.6	72.3	86.4
668	Emporia	97.4	57.7	79.8
669	Belleville	99.2	63.7	83.5
670-672	Wichita	97.9	69.5	85.3
673	Independence	99.0	67.7	85.1
674	Salina	99.2	63.9	83.5
675	Hutchinson	94.5	49.5	74.5
676	Hays	98.5	67.0	84.5
677	Colby	99.2	53.1	78.8

STATE/ZIP	CITY	MAT.	INST.	TOTAL
KANSAS (CONT'D)				
678	Dodge City	100.4	68.4	86.2
679	Liberal	98.3	44.2	74.3
KENTUCKY				
400-402	Louisville	96.4	85.4	91.5
403-405	Lexington	96.0	65.7	82.5
406	Frankfort	96.3	66.7	83.1
407-409	Corbin	93.3	43.3	71.1
410	Covington	93.8	91.9	93.0
411-412	Ashland	92.5	96.9	94.5
413-414	Campton	94.1	43.7	71.7
415-416	Pikeville	95.0	60.3	79.6
417-418	Hazard	93.5	43.4	71.3
420	Paducah	92.2	88.4	90.5
421-422	Bowling Green	94.6	82.7	89.3
423	Owensboro	94.4	73.5	85.1
424	Henderson	92.0	90.4	91.3
425-426	Somerset	91.6	45.1	70.9
427	Elizabethtown	91.1	84.5	88.2
LOUISIANA				
700-701	New Orleans	99.8	69.7	86.4
703	Thibodaux	97.7	67.2	84.2
704	Hammond	95.0	62.7	80.7
705	Lafayette	97.4	62.3	81.8
706	Lake Charles	97.5	64.1	82.7
707-708	Baton Rouge	98.1	62.9	82.5
710-711	Shreveport	94.6	61.3	79.8
712	Monroe	95.3	57.9	78.7
713-714	Alexandria	95.4	59.5	79.5
MAINE				
039	Kittery	95.2	73.4	85.5
040-041	Portland	98.5	79.8	90.2
042	Lewiston	98.9	79.8	90.4
043	Augusta	97.4	73.2	86.7
044	Bangor	97.7	79.8	89.8
045	Bath	96.3	73.4	86.1
046	Machias	95.9	72.8	85.6
047	Houlton	96.0	76.4	87.3
048	Rockland	95.1	73.4	85.4
049	Waterville	96.3	73.3	86.1
MARYLAND				
206	Waldorf	98.3	72.6	86.9
207-208	College Park	98.4	81.4	90.8
209	Silver Spring	97.6	77.9	88.9
210-212	Baltimore	99.5	84.7	92.9
214	Annapolis	98.9	79.2	90.1
215	Cumberland	95.4	81.2	89.1
216	Easton	96.8	44.8	73.7
217	Hagerstown	96.2	81.9	89.8
218	Salisbury	97.3	53.5	77.8
219	Elkton	94.6	66.7	82.2
MASSACHUSETTS				
010-011	Springfield	100.4	105.9	102.8
012	Pittsfield	99.9	103.0	101.3
013	Greenfield	97.9	105.3	101.2
014	Fitchburg	96.6	119.8	106.9
015-016	Worcester	100.1	120.0	109.0
017	Framingham	96.1	123.8	108.4
018	Lowell	99.7	125.0	111.0
019	Lawrence	100.9	122.6	110.5
020-022, 024	Boston	102.5	132.0	115.6
023	Brockton	101.1	121.0	110.0
025	Buzzards Bay	95.5	119.0	105.9
026	Hyannis	98.1	119.0	107.4
027	New Bedford	100.8	119.1	109.0
MICHIGAN				
480,483	Royal Oak	95.3	107.9	100.9
481	Ann Arbor	97.4	105.9	101.2
482	Detroit	99.1	117.6	107.3
484-485	Flint	97.2	99.9	98.4
486	Saginaw	97.0	93.4	95.4
487	Bay City	96.9	93.5	95.4
488-489	Lansing	97.2	95.2	96.3
490	Battle Creek	95.5	87.9	92.1
491	Kalamazoo	95.8	85.8	91.4
492	Jackson	93.8	92.3	93.2
493,495	Grand Rapids	96.0	68.1	83.6
494	Muskegon	94.4	83.6	89.6

Location Factors

STATE/ZIP	CITY	MAT.	INST.	TOTAL
MICHIGAN (CONT'D)				
496	Traverse City	93.3	73.1	84.3
497	Gaylord	94.5	76.9	86.7
498-499	Iron Mountain	96.3	87.7	92.5
MINNESOTA				
550-551	Saint Paul	99.7	124.5	110.7
553-555	Minneapolis	100.5	126.7	112.2
556-558	Duluth	99.5	114.5	106.2
559	Rochester	99.2	108.5	103.4
560	Mankato	96.3	105.9	100.6
561	Windom	95.1	80.3	88.5
562	Willmar	94.4	87.4	91.3
563	St. Cloud	95.7	123.3	108.0
564	Brainerd	96.0	105.1	100.0
565	Detroit Lakes	97.9	100.6	99.1
566	Bemidji	97.3	101.2	99.0
567	Thief River Falls	96.8	96.1	96.5
MISSISSIPPI				
386	Clarksdale	95.9	30.8	67.0
387	Greenville	99.3	44.6	75.0
388	Tupelo	97.3	37.3	70.6
389	Greenwood	97.2	33.3	68.8
390-392	Jackson	97.8	46.1	74.8
393	Meridian	96.0	46.7	74.1
394	Laurel	97.1	33.4	68.8
395	Biloxi	97.8	55.5	79.0
396	McComb	95.7	52.1	76.4
397	Columbus	97.1	36.7	70.3
MISSOURI				
630-631	St. Louis	97.0	109.0	102.3
633	Bowling Green	95.7	95.8	95.7
634	Hannibal	94.5	86.3	90.9
635	Kirksville	96.8	79.2	89.0
636	Flat River	96.7	97.0	96.8
637	Cape Girardeau	96.4	92.3	94.6
638	Sikeston	94.5	83.2	89.5
639	Poplar Bluff	94.0	83.8	89.5
640-641	Kansas City	100.9	106.8	103.5
644-645	St. Joseph	99.9	93.3	97.0
646	Chillicothe	96.7	71.2	85.4
647	Harrisonville	96.4	101.3	98.6
648	Joplin	98.6	70.5	86.1
650-651	Jefferson City	96.1	89.7	93.3
652	Columbia	97.1	92.4	95.0
653	Sedalia	96.1	84.9	91.2
654-655	Rolla	95.0	78.6	87.7
656-658	Springfield	98.4	78.7	89.6
MONTANA				
590-591	Billings	102.1	74.5	89.8
592	Wolf Point	100.8	68.7	86.6
593	Miles City	98.8	70.0	86.0
594	Great Falls	102.7	73.8	89.9
595	Havre	99.6	68.9	86.0
596	Helena	102.5	71.5	88.7
597	Butte	101.2	70.9	87.7
598	Missoula	98.9	70.1	86.1
599	Kalispell	97.8	69.4	85.2
NEBRASKA				
680-681	Omaha	99.3	78.2	89.9
683-685	Lincoln	98.2	67.1	84.4
686	Columbus	96.1	49.6	75.5
687	Norfolk	97.7	60.6	81.3
688	Grand Island	97.5	65.6	83.3
689	Hastings	97.1	56.7	79.2
690	Mccook	96.9	48.2	75.3
691	North Platte	97.1	58.8	80.1
692	Valentine	99.2	38.5	72.3
693	Alliance	99.0	35.9	71.0
NEVADA				
889-891	Las Vegas	98.3	107.1	102.2
893	Ely	98.7	79.0	90.0
894-895	Reno	98.9	98.2	98.6
897	Carson City	98.6	97.9	98.3
898	Elko	97.6	85.4	92.2
NEW HAMPSHIRE				
030	Nashua	100.6	82.4	92.5
031	Manchester	100.9	82.4	92.7
NEW HAMPSHIRE (CONT'D)				
032-033	Concord	98.4	82.4	91.3
034	Keene	97.2	50.9	76.6
035	Littleton	97.4	61.6	81.5
036	Charleston	96.6	47.9	75.0
037	Claremont	95.9	47.9	74.6
038	Portsmouth	97.8	78.0	89.0
NEW JERSEY				
070-071	Newark	102.1	124.6	112.1
072	Elizabeth	100.8	123.6	110.9
073	Jersey City	99.8	124.5	110.8
074-075	Paterson	101.4	124.2	111.5
076	Hackensack	99.5	124.5	110.6
077	Long Branch	99.1	125.3	110.8
078	Dover	99.7	124.5	110.7
079	Summit	99.8	123.6	110.4
080,083	Vineland	97.6	117.3	106.3
081	Camden	99.6	116.8	107.2
082,084	Atlantic City	98.4	116.5	106.4
085-086	Trenton	99.6	122.6	109.8
087	Point Pleasant	99.5	122.3	109.6
088-089	New Brunswick	100.0	123.4	110.4
NEW MEXICO				
870-872	Albuquerque	98.5	76.2	88.6
873	Gallup	98.3	76.2	88.5
874	Farmington	98.7	76.2	88.7
875	Santa Fe	97.5	76.2	88.0
877	Las Vegas	96.9	76.2	87.7
878	Socorro	96.5	76.2	87.5
879	Truth/Consequences	96.9	70.8	85.3
880	Las Cruces	95.8	67.9	83.4
881	Clovis	97.0	75.2	87.3
882	Roswell	98.7	75.3	88.3
883	Carrizozo	99.0	76.2	88.9
884	Tucumcari	97.9	75.2	87.8
NEW YORK				
100-102	New York	106.0	164.4	131.9
103	Staten Island	101.6	159.9	127.5
104	Bronx	99.8	159.9	126.5
105	Mount Vernon	99.5	138.5	116.8
106	White Plains	99.5	138.2	116.7
107	Yonkers	104.5	138.5	119.6
108	New Rochelle	100.0	138.0	116.9
109	Suffern	99.8	126.2	111.5
110	Queens	101.4	158.8	126.9
111	Long Island City	103.1	158.8	127.8
112	Brooklyn	103.3	159.1	128.1
113	Flushing	103.3	158.8	128.0
114	Jamaica	101.5	158.8	126.9
115,117,118	Hicksville	101.2	149.5	122.7
116	Far Rockaway	103.4	158.8	128.0
119	Riverhead	102.2	149.5	123.2
120-122	Albany	97.8	96.4	97.2
123	Schenectady	98.1	95.9	97.1
124	Kingston	100.9	117.3	108.2
125-126	Poughkeepsie	100.1	119.4	108.7
127	Monticello	99.5	119.9	108.5
128	Glens Falls	92.7	91.3	92.1
129	Plattsburgh	96.8	88.4	93.1
130-132	Syracuse	99.0	94.7	97.1
133-135	Utica	96.9	90.9	94.2
136	Watertown	98.5	88.5	94.1
137-139	Binghamton	98.5	88.8	94.2
140-142	Buffalo	99.8	105.7	102.4
143	Niagara Falls	97.3	104.7	100.6
144-146	Rochester	99.8	99.7	99.8
147	Jamestown	96.3	87.9	92.6
148-149	Elmira	96.3	85.4	91.4
NORTH CAROLINA				
270,272-274	Greensboro	98.5	50.3	77.1
271	Winston-Salem	98.2	49.6	76.6
275-276	Raleigh	98.4	50.7	77.2
277	Durham	98.4	50.7	77.2
278	Rocky Mount	95.7	35.5	69.0
279	Elizabeth City	96.4	37.8	70.4
280	Gastonia	97.7	48.9	76.0
281-282	Charlotte	98.6	48.9	76.5
283	Fayetteville	97.1	50.8	76.5
284	Wilmington	96.4	49.9	75.7
285	Kinston	94.4	35.1	68.1

Location Factors

STATE/ZIP	CITY	MAT.	INST.	TOTAL
NORTH CAROLINA (CONT'D)				
286	Hickory	94.8	33.9	67.8
287-288	Asheville	96.8	48.9	75.5
289	Murphy	95.7	33.0	67.9
NORTH DAKOTA				
580-581	Fargo	100.9	65.2	85.1
582	Grand Forks	100.5	59.2	82.2
583	Devils Lake	99.9	60.8	82.6
584	Jamestown	100.0	52.6	78.9
585	Bismarck	99.3	66.6	84.8
586	Dickinson	100.7	60.5	82.9
587	Minot	100.6	67.8	86.1
588	Williston	99.3	60.2	81.9
OHIO				
430-432	Columbus	97.3	89.6	93.9
433	Marion	93.9	88.4	91.5
434-436	Toledo	97.6	100.0	98.7
437-438	Zanesville	94.3	86.1	90.6
439	Steubenville	95.2	95.1	95.1
440	Lorain	96.9	97.1	97.0
441	Cleveland	97.2	104.3	100.4
442-443	Akron	97.8	98.0	97.9
444-445	Youngstown	97.2	93.3	95.5
446-447	Canton	97.3	88.4	93.4
448-449	Mansfield	94.6	93.8	94.2
450	Hamilton	94.5	89.6	92.3
451-452	Cincinnati	94.9	90.2	92.8
453-454	Dayton	94.6	87.0	91.2
455	Springfield	94.6	87.9	91.6
456	Chillicothe	93.5	94.2	93.8
457	Athens	96.3	79.9	89.0
458	Lima	96.8	87.9	92.8
OKLAHOMA				
730-731	Oklahoma City	97.1	64.3	82.6
734	Ardmore	94.0	63.7	80.5
735	Lawton	96.4	65.3	82.6
736	Clinton	95.3	62.0	80.6
737	Enid	96.0	62.0	80.9
738	Woodward	94.2	62.1	79.9
739	Guymon	95.2	32.9	67.5
740-741	Tulsa	96.2	59.9	80.1
743	Miami	93.0	67.5	81.7
744	Muskogee	95.6	44.0	72.7
745	Mcalester	92.6	55.1	75.9
746	Ponca City	93.1	62.2	79.4
747	Durant	93.1	61.7	79.2
748	Shawnee	94.6	60.2	79.3
749	Poteau	92.3	64.7	80.1
OREGON				
970-972	Portland	102.2	104.6	103.3
973	Salem	102.2	103.2	102.6
974	Eugene	101.7	102.8	102.2
975	Medford	103.3	101.1	102.3
976	Klamath Falls	103.4	101.1	102.4
977	Bend	102.2	102.9	102.5
978	Pendleton	96.5	102.6	99.2
979	Vale	94.4	93.2	93.9
PENNSYLVANIA				
150-152	Pittsburgh	97.0	104.2	100.2
153	Washington	94.0	102.8	98.0
154	Uniontown	94.2	101.2	97.3
155	Bedford	95.2	92.6	94.0
156	Greensburg	95.3	102.2	98.3
157	Indiana	94.1	100.7	97.0
158	Dubois	95.4	97.3	96.2
159	Johnstown	95.1	98.1	96.5
160	Butler	92.3	102.3	96.7
161	New Castle	92.3	101.0	96.2
162	Kittanning	92.8	103.6	97.6
163	Oil City	92.3	98.1	94.9
164-165	Erie	94.4	96.7	95.4
166	Altoona	94.4	94.5	94.5
167	Bradford	95.7	96.6	96.1
168	State College	95.4	93.9	94.7
169	Wellsboro	96.4	94.2	95.4
170-171	Harrisburg	98.2	95.3	96.9
172	Chambersburg	95.9	89.9	93.3
173-174	York	96.2	95.3	95.8
175-176	Lancaster	94.8	90.6	92.9

STATE/ZIP	CITY	MAT.	INST.	TOTAL
PENNSYLVANIA (CONT'D)				
177	Williamsport	93.4	82.4	88.5
178	Sunbury	95.4	94.4	95.0
179	Pottsville	94.5	94.0	94.3
180	Lehigh Valley	95.9	114.4	104.1
181	Allentown	98.0	107.4	102.2
182	Hazleton	95.3	96.6	95.9
183	Stroudsburg	95.2	101.1	97.8
184-185	Scranton	98.8	96.3	97.7
186-187	Wilkes-Barre	95.2	94.9	95.1
188	Montrose	94.8	95.5	95.1
189	Doylestown	94.7	119.7	105.8
190-191	Philadelphia	101.0	131.1	114.4
193	Westchester	97.6	122.0	108.5
194	Norristown	96.6	126.0	109.7
195-196	Reading	98.8	100.9	99.7
PUERTO RICO				
009	San Juan	135.2	25.5	86.5
RHODE ISLAND				
028	Newport	99.6	112.7	105.4
029	Providence	99.9	112.7	105.6
SOUTH CAROLINA				
290-292	Columbia	97.1	47.3	75.0
293	Spartanburg	95.9	45.4	73.5
294	Charleston	97.3	47.9	75.3
295	Florence	95.4	45.2	73.1
296	Greenville	95.7	45.4	73.4
297	Rock Hill	95.3	33.0	67.6
298	Aiken	96.1	72.0	85.4
299	Beaufort	97.0	36.2	70.0
SOUTH DAKOTA				
570-571	Sioux Falls	99.2	57.3	80.6
572	Watertown	97.9	51.0	77.1
573	Mitchell	96.9	50.8	76.4
574	Aberdeen	99.5	53.1	78.9
575	Pierre	98.8	53.5	78.7
576	Mobridge	97.5	50.4	76.6
577	Rapid City	99.4	52.2	78.4
TENNESSEE				
370-372	Nashville	97.5	73.7	86.9
373-374	Chattanooga	98.6	54.1	78.8
375,380-381	Memphis	96.6	76.2	87.5
376	Johnson City	97.3	56.4	79.1
377-379	Knoxville	94.7	57.5	78.2
382	Mckenzie	96.4	59.4	80.0
383	Jackson	98.3	51.7	77.6
384	Columbia	94.9	55.2	77.3
385	Cookeville	96.2	49.9	75.6
TEXAS				
750	Mckinney	97.6	55.5	78.9
751	Waxahackie	97.5	56.3	79.2
752-753	Dallas	98.3	67.1	84.4
754	Greenville	97.7	40.7	72.4
755	Texarkana	96.6	54.0	77.7
756	Longview	96.9	42.6	72.8
757	Tyler	97.4	56.2	79.1
758	Palestine	93.6	43.0	71.1
759	Lufkin	94.5	46.7	73.3
760-761	Fort Worth	96.2	62.8	81.4
762	Denton	96.5	51.6	76.6
763	Wichita Falls	97.1	57.9	79.7
764	Eastland	95.6	42.0	71.8
765	Temple	94.5	50.3	74.9
766-767	Waco	96.4	57.5	79.1
768	Brownwood	96.5	39.5	71.2
769	San Angelo	96.2	47.7	74.7
770-772	Houston	99.6	73.0	87.7
773	Huntsville	97.7	40.3	72.2
774	Wharton	99.1	45.1	75.1
775	Galveston	97.3	70.6	85.4
776-777	Beaumont	97.1	64.1	82.4
778	Bryan	94.7	64.5	81.3
779	Victoria	99.2	47.6	76.3
780	Laredo	94.6	53.7	76.5
781-782	San Antonio	95.0	67.0	82.5
783-784	Corpus Christi	97.3	53.1	77.7
785	Mc Allen	97.2	47.4	75.1
786-787	Austin	94.6	61.7	80.0

Location Factors

STATE/ZIP	CITY	MAT.	INST.	TOTAL
TEXAS (CONT'D)				
788	Del Rio	96.5	33.2	68.4
789	Giddings	94.0	42.1	71.0
790-791	Amarillo	97.2	58.0	79.8
792	Childress	96.1	52.5	76.7
793-794	Lubbock	98.6	53.5	78.6
795-796	Abilene	96.6	53.3	77.4
797	Midland	98.7	49.9	77.0
798-799,885	El Paso	96.6	51.6	76.6
UTAH				
840-841	Salt Lake City	101.3	70.7	87.7
842,844	Ogden	97.4	70.5	85.5
843	Logan	99.1	70.5	86.4
845	Price	99.6	50.1	77.6
846-847	Provo	99.6	70.5	86.7
VERMONT				
050	White River Jct.	98.6	46.5	75.4
051	Bellows Falls	97.1	48.8	75.6
052	Bennington	97.4	49.1	76.0
053	Brattleboro	97.9	49.3	76.3
054	Burlington	100.5	64.9	84.7
056	Montpelier	97.3	64.9	82.9
057	Rutland	99.5	64.9	84.1
058	St. Johnsbury	98.8	49.1	76.7
059	Guildhall	97.4	48.6	75.7
VIRGINIA				
220-221	Fairfax	99.3	80.6	91.0
222	Arlington	100.5	79.3	91.1
223	Alexandria	99.9	84.9	93.3
224-225	Fredericksburg	98.0	68.4	84.9
226	Winchester	98.6	53.8	78.7
227	Culpeper	98.5	55.3	79.3
228	Harrisonburg	98.8	47.1	75.8
229	Charlottesville	99.1	61.5	82.4
230-232	Richmond	100.3	66.7	85.4
233-235	Norfolk	100.0	65.8	84.8
236	Newport News	99.6	64.7	84.1
237	Portsmouth	98.9	65.5	84.0
238	Petersburg	98.9	66.7	84.6
239	Farmville	98.3	41.1	72.9
240-241	Roanoke	100.2	48.1	77.1
242	Bristol	97.9	49.0	76.2
243	Pulaski	97.7	42.4	73.1
244	Staunton	98.4	45.6	75.0
245	Lynchburg	98.6	52.2	78.0
246	Grundy	98.0	44.4	74.2
WASHINGTON				
980-981,987	Seattle	103.8	104.7	104.2
982	Everett	103.6	101.0	102.4
983-984	Tacoma	103.5	101.4	102.6
985	Olympia	102.3	101.3	101.9
986	Vancouver	105.4	97.7	102.0
988	Wenatchee	103.5	86.6	96.0
989	Yakima	103.7	91.8	98.4
990-992	Spokane	105.7	83.3	95.8
993	Richland	105.0	86.9	96.9
994	Clarkston	103.9	84.3	95.2
WEST VIRGINIA				
247-248	Bluefield	96.8	79.4	89.1
249	Lewisburg	98.3	83.1	91.6
250-253	Charleston	99.5	91.0	95.7
254	Martinsburg	98.1	80.0	90.1
255-257	Huntington	99.6	93.4	96.9
258-259	Beckley	96.6	88.8	93.1
260	Wheeling	99.7	94.2	97.2
261	Parkersburg	98.7	92.6	96.0
262	Buckhannon	98.1	93.4	96.0
263-264	Clarksburg	98.5	93.4	96.2
265	Morgantown	98.6	93.5	96.4
266	Gassaway	97.9	92.2	95.4
267	Romney	97.9	86.5	92.9
268	Petersburg	97.9	89.3	94.1
WISCONSIN				
530,532	Milwaukee	99.0	106.2	102.2
531	Kenosha	98.5	103.1	100.5
534	Racine	98.1	102.9	100.2
535	Beloit	98.0	98.4	98.2
537	Madison	98.3	98.1	98.2

STATE/ZIP	CITY	MAT.	INST.	TOTAL
WISCONSIN (CONT'D)				
538	Lancaster	95.8	94.0	95.0
539	Portage	94.4	97.7	95.9
540	New Richmond	95.5	98.5	96.8
541-543	Green Bay	99.7	95.2	97.7
544	Wausau	94.9	94.3	94.6
545	Rhinelander	98.0	93.7	96.1
546	La Crosse	96.1	95.7	95.9
547	Eau Claire	97.8	96.2	97.1
548	Superior	95.4	102.8	98.7
549	Oshkosh	95.5	95.2	95.4
WYOMING				
820	Cheyenne	97.6	61.2	81.4
821	Yellowstone Nat'l Park	96.8	54.6	78.0
822	Wheatland	98.1	56.0	79.4
823	Rawlins	99.4	53.1	78.8
824	Worland	97.4	51.4	77.0
825	Riverton	98.4	53.1	78.3
826	Casper	97.7	59.9	80.9
827	Newcastle	97.2	53.1	77.6
828	Sheridan	98.1	58.7	80.6
829-831	Rock Springs	101.5	51.4	79.2
CANADIAN FACTORS (reflect Canadian currency)				
ALBERTA				
	Calgary	124.3	79.6	104.4
	Edmonton	125.9	79.6	105.3
	Fort McMurray	116.6	79.1	100.0
	Lethbridge	117.6	78.6	100.3
	Lloydminster	116.7	79.1	100.0
	Medicine Hat	116.7	78.6	99.8
	Red Deer	117.1	78.6	100.0
BRITISH COLUMBIA				
	Kamloops	117.8	81.8	101.8
	Prince George	119.1	81.8	102.5
	Vancouver	125.7	85.6	107.9
	Victoria	119.1	82.3	102.8
MANITOBA				
	Brandon	117.0	68.5	95.5
	Portage la Prairie	117.0	68.5	95.5
	Winnipeg	124.1	68.9	99.6
NEW BRUNSWICK				
	Bathurst	115.1	60.9	91.1
	Dalhousie	115.1	60.9	91.1
	Fredericton	117.0	64.9	93.8
	Moncton	115.4	60.9	91.2
	Newcastle	115.1	60.9	91.1
	Saint John	118.1	64.9	94.5
NEWFOUNDLAND				
	Corner Brook	119.8	60.9	93.7
	St. John's	120.1	60.9	93.8
NORTHWEST TERRITORIES				
	Yellowknife	120.4	77.4	101.3
NOVA SCOTIA				
	Dartmouth	117.8	67.1	95.3
	Halifax	119.3	68.2	96.6
	New Glasgow	115.8	67.1	94.2
	Sydney	113.3	67.1	92.8
	Yarmouth	115.6	67.1	94.1
ONTARIO				
	Barrie	119.9	84.1	104.0
	Brantford	119.1	87.6	105.1
	Cornwall	119.1	84.3	103.6
	Hamilton	124.5	87.9	108.2
	Kingston	120.1	84.7	104.4
	Kitchener	115.6	83.7	101.4
	London	121.9	85.3	105.6
	North Bay	119.1	82.4	102.8
	Oshawa	118.7	85.7	104.1
	Ottawa	120.5	85.5	105.0
	Owen Sound	120.1	82.5	103.4
	Peterborough	119.1	84.1	103.5
	Sarnia	119.3	87.9	105.4
	St. Catharines	113.6	83.1	100.0
	Sudbury	113.7	82.7	99.9

Location Factors

STATE/ZIP	CITY	MAT.	INST.	TOTAL
ONTARIO (CONT'D)	Thunder Bay	114.8	82.8	100.6
	Toronto	124.2	92.1	109.9
	Windsor	114.6	85.4	101.6
PRINCE EDWARD ISLAND	Charlottetown	117.7	56.9	90.7
	Summerside	117.3	56.9	90.5
QUEBEC				
	Cap-de-la-Madeleine	115.9	78.9	99.5
	Charlesbourg	115.9	78.9	99.5
	Chicoutimi	115.0	78.6	98.8
	Gatineau	115.3	78.7	99.1
	Laval	115.2	78.7	99.0
	Montreal	118.5	79.0	101.0
	Quebec	119.4	79.4	101.6
	Sherbrooke	115.7	78.7	99.3
	Trois Rivieres	116.2	78.9	99.6
SASKATCHEWAN				
	Moose Jaw	114.2	62.9	91.4
	Prince Albert	113.7	62.7	91.1
	Regina	116.1	62.9	92.5
	Saskatoon	114.6	62.8	91.6
YUKON				
	Whitehorse	114.0	61.9	90.8

R011105-05 Tips for Accurate Estimating

1. Use pre-printed or columnar forms for orderly sequence of dimensions and locations and for recording telephone quotations.
2. Use only the front side of each paper or form except for certain pre-printed summary forms.
3. Be consistent in listing dimensions: For example, length x width x height. This helps in rechecking to ensure that, the total length of partitions is appropriate for the building area.
4. Use printed (rather than measured) dimensions where given.
5. Add up multiple printed dimensions for a single entry where possible.
6. Measure all other dimensions carefully.
7. Use each set of dimensions to calculate multiple related quantities.
8. Convert foot and inch measurements to decimal feet when listing. Memorize decimal equivalents to .01 parts of a foot (1/8″ equals approximately .01′).
9. Do not "round off" quantities until the final summary.
10. Mark drawings with different colors as items are taken off.
11. Keep similar items together, different items separate.
12. Identify location and drawing numbers to aid in future checking for completeness.
13. Measure or list everything on the drawings or mentioned in the specifications.
14. It may be necessary to list items not called for to make the job complete.
15. Be alert for: Notes on plans such as N.T.S. (not to scale); changes in scale throughout the drawings; reduced size drawings; discrepancies between the specifications and the drawings.
16. Develop a consistent pattern of performing an estimate. For example:
    a. Start the quantity takeoff at the lower floor and move to the next higher floor.
    b. Proceed from the main section of the building to the wings.
    c. Proceed from south to north or vice versa, clockwise or counterclockwise.
    d. Take off floor plan quantities first, elevations next, then detail drawings.
17. List all gross dimensions that can be either used again for different quantities, or used as a rough check of other quantities for verification (exterior perimeter, gross floor area, individual floor areas, etc.).
18. Utilize design symmetry or repetition (repetitive floors, repetitive wings, symmetrical design around a center line, similar room layouts, etc.). Note: Extreme caution is needed here so as not to omit or duplicate an area.
19. Do not convert units until the final total is obtained. For instance, when estimating concrete work, keep all units to the nearest cubic foot, then summarize and convert to cubic yards.
20. When figuring alternatives, it is best to total all items involved in the basic system, then total all items involved in the alternates. Therefore you work with positive numbers in all cases. When adds and deducts are used, it is often confusing whether to add or subtract a portion of an item; especially on a complicated or involved alternate.

R011105-10 Unit Gross Area Requirements

The figures in the table below indicate typical ranges in square feet as a function of the "occupant" unit. This table is best used in the preliminary design stages to help determine the probable size requirement for the total project.

Building Type	Unit	Gross Area in S.F.		
		1/4	Median	3/4
Apartments	Unit	660	860	1,100
Auditorium & Play Theaters	Seat	18	25	38
Bowling Alleys	Lane		940	
Churches & Synagogues	Seat	20	28	39
Dormitories	Bed	200	230	275
Fraternity & Sorority Houses	Bed	220	315	370
Garages, Parking	Car	325	355	385
Hospitals	Bed	685	850	1,075
Hotels	Rental Unit	475	600	710
Housing for the elderly	Unit	515	635	755
Housing, Public	Unit	700	875	1,030
Ice Skating Rinks	Total	27,000	30,000	36,000
Motels	Rental Unit	360	465	620
Nursing Homes	Bed	290	350	450
Restaurants	Seat	23	29	39
Schools, Elementary	Pupil	65	77	90
Junior High & Middle	↓	85	110	129
Senior High	↓	102	130	145
Vocational	↓	110	135	195
Shooting Ranges	Point		450	
Theaters & Movies	Seat		15	

R011105-20 Floor Area Ratios

Table below lists commonly used gross to net area and net to gross area ratios expressed in % for various building types.

Building Type	Gross to Net Ratio	Net to Gross Ratio	Building Type	Gross to Net Ratio	Net to Gross Ratio
Apartment	156	64	School Buildings (campus type)		
Bank	140	72	Administrative	150	67
Church	142	70	Auditorium	142	70
Courthouse	162	61	Biology	161	62
Department Store	123	81	Chemistry	170	59
Garage	118	85	Classroom	152	66
Hospital	183	55	Dining Hall	138	72
Hotel	158	63	Dormitory	154	65
Laboratory	171	58	Engineering	164	61
Library	132	76	Fraternity	160	63
Office	135	75	Gymnasium	142	70
Restaurant	141	70	Science	167	60
Warehouse	108	93	Service	120	83
			Student Union	172	59

The gross area of a building is the total floor area based on outside dimensions.

The net area of a building is the usable floor area for the function intended and excludes such items as stairways, corridors and mechanical rooms. In the case of a commercial building, it might be considered as the "leasable area."

R011105-30 Occupancy Determinations

Description		S.F. Required per Person		
		BOCA	SBC	UBC
Assembly Areas	Fixed Seats	**	6	7
	Movable Seats		15	15
	Concentrated	7		
	Unconcentrated	15		
	Standing Space	3		
Educational	Unclassified			
	Classrooms	20	40	20
	Shop Areas	50	100	50
Institutional	Unclassified		125	
	In-Patient Areas	240		
	Sleeping Areas	120		
Mercantile	Basement	30	30	20
	Ground Floor	30	30	30
	Upper Floors	60	60	50
Office		100	100	100

BOCA=Building Officials & Code Administrators
SBC=Southern Building Code
UBC=Uniform Building Code

** The occupancy load for assembly area with fixed seats shall be determined by the number of fixed seats installed.

R011105-40 Weather Data and Design Conditions

City	Latitude (1)		Winter Temperatures (1)			Winter Degree Days (2)	Summer (Design Dry Bulb) Temperatures and Relative Humidity		
	0	1′	Med. of Annual Extremes	99%	97½%		1%	2½%	5%
UNITED STATES									
Albuquerque, NM	35	0	5.1	12	16	4,400	96/61	94/61	92/61
Atlanta, GA	33	4	11.9	17	22	3,000	94/74	92/74	90/73
Baltimore, MD	39	2	7	14	17	4,600	94/75	91/75	89/74
Birmingham, AL	33	3	13	17	21	2,600	96/74	94/75	92/74
Bismarck, ND	46	5	-32	-23	-19	8,800	95/68	91/68	88/67
Boise, ID	43	3	1	3	10	5,800	96/65	94/64	91/64
Boston, MA	42	2	-1	6	9	5,600	91/73	88/71	85/70
Burlington, VT	44	3	-17	-12	-7	8,200	88/72	85/70	82/69
Charleston, WV	38	2	3	7	11	4,400	92/74	90/73	87/72
Charlotte, NC	35	1	13	18	22	3,200	95/74	93/74	91/74
Casper, WY	42	5	-21	-11	-5	7,400	92/58	90/57	87/57
Chicago, IL	41	5	-8	-3	2	6,600	94/75	91/74	88/73
Cincinnati, OH	39	1	0	1	6	4,400	92/73	90/72	88/72
Cleveland, OH	41	2	-3	1	5	6,400	91/73	88/72	86/71
Columbia, SC	34	0	16	20	24	2,400	97/76	95/75	93/75
Dallas, TX	32	5	14	18	22	2,400	102/75	100/75	97/75
Denver, CO	39	5	-10	-5	1	6,200	93/59	91/59	89/59
Des Moines, IA	41	3	-14	-10	-5	6,600	94/75	91/74	88/73
Detroit, MI	42	2	-3	3	6	6,200	91/73	88/72	86/71
Great Falls, MT	47	3	-25	-21	-15	7,800	91/60	88/60	85/59
Hartford, CT	41	5	-4	3	7	6,200	91/74	88/73	85/72
Houston, TX	29	5	24	28	33	1,400	97/77	95/77	93/77
Indianapolis, IN	39	4	-7	-2	2	5,600	92/74	90/74	87/73
Jackson, MS	32	2	16	21	25	2,200	97/76	95/76	93/76
Kansas City, MO	39	1	-4	2	6	4,800	99/75	96/74	93/74
Las Vegas, NV	36	1	18	25	28	2,800	108/66	106/65	104/65
Lexington, KY	38	0	-1	3	8	4,600	93/73	91/73	88/72
Little Rock, AR	34	4	11	15	20	3,200	99/76	96/77	94/77
Los Angeles, CA	34	0	36	41	43	2,000	93/70	89/70	86/69
Memphis, TN	35	0	10	13	18	3,200	98/77	95/76	93/76
Miami, FL	25	5	39	44	47	200	91/77	90/77	89/77
Milwaukee, WI	43	0	-11	-8	-4	7,600	90/74	87/73	84/71
Minneapolis, MN	44	5	-22	-16	-12	8,400	92/75	89/73	86/71
New Orleans, LA	30	0	28	29	33	1,400	93/78	92/77	90/77
New York, NY	40	5	6	11	15	5,000	92/74	89/73	87/72
Norfolk, VA	36	5	15	20	22	3,400	93/77	91/76	89/76
Oklahoma City, OK	35	2	4	9	13	3,200	100/74	97/74	95/73
Omaha, NE	41	2	-13	-8	-3	6,600	94/76	91/75	88/74
Philadelphia, PA	39	5	6	10	14	4,400	93/75	90/74	87/72
Phoenix, AZ	33	3	27	31	34	1,800	109/71	107/71	105/71
Pittsburgh, PA	40	3	-1	3	7	6,000	91/72	88/71	86/70
Portland, ME	43	4	-10	-6	-1	7,600	87/72	84/71	81/69
Portland, OR	45	4	18	17	23	4,600	89/68	85/67	81/65
Portsmouth, NH	43	1	-8	-2	2	7,200	89/73	85/71	83/70
Providence, RI	41	4	-1	5	9	6,000	89/73	86/72	83/70
Rochester, NY	43	1	-5	1	5	6,800	91/73	88/71	85/70
Salt Lake City, UT	40	5	0	3	8	6,000	97/62	95/62	92/61
San Francisco, CA	37	5	36	38	40	3,000	74/63	71/62	69/61
Seattle, WA	47	4	22	22	27	5,200	85/68	82/66	78/65
Sioux Falls, SD	43	4	-21	-15	-11	7,800	94/73	91/72	88/71
St. Louis, MO	38	4	-3	3	8	5,000	98/75	94/75	91/75
Tampa, FL	28	0	32	36	40	680	92/77	91/77	90/76
Trenton, NJ	40	1	4	11	14	5,000	91/75	88/74	85/73
Washington, DC	38	5	7	14	17	4,200	93/75	91/74	89/74
Wichita, KS	37	4	-3	3	7	4,600	101/72	98/73	96/73
Wilmington, DE	39	4	5	10	14	5,000	92/74	89/74	87/73
ALASKA									
Anchorage	61	1	-29	-23	-18	10,800	71/59	68/58	66/56
Fairbanks	64	5	-59	-51	-47	14,280	82/62	78/60	75/59
CANADA									
Edmonton, Alta.	53	3	-30	-29	-25	11,000	85/66	82/65	79/63
Halifax, N.S.	44	4	-4	1	5	8,000	79/66	76/65	74/64
Montreal, Que.	45	3	-20	-16	-10	9,000	88/73	85/72	83/71
Saskatoon, Sask.	52	1	-35	-35	-31	11,000	89/68	86/66	83/65
St. John's, N.F.	47	4	1	3	7	8,600	77/66	75/65	73/64
Saint John, N.B.	45	2	-15	-12	-8	8,200	80/67	77/65	75/64
Toronto, Ont.	43	4	-10	-5	-1	7,000	90/73	87/72	85/71
Vancouver, B.C.	49	1	13	15	19	6,000	79/67	77/66	74/65
Winnipeg, Man.	49	5	-31	-30	-27	10,800	89/73	86/71	84/70

(1) Handbook of Fundamentals, ASHRAE, Inc., NY 1989
(2) Local Climatological Annual Survey, USDC Env. Science Services Administration, Asheville, NC

R011105-50 Metric Conversion Factors

Description: This table is primarily for converting customary U.S. units in the left hand column to SI metric units in the right hand column. In addition, conversion factors for some commonly encountered Canadian and non-SI metric units are included.

	If You Know		Multiply By		To Find
Length	Inches	x	25.4[a]	=	Millimeters
	Feet	x	0.3048[a]	=	Meters
	Yards	x	0.9144[a]	=	Meters
	Miles (statute)	x	1.609	=	Kilometers
Area	Square inches	x	645.2	=	Square millimeters
	Square feet	x	0.0929	=	Square meters
	Square yards	x	0.8361	=	Square meters
Volume (Capacity)	Cubic inches	x	16,387	=	Cubic millimeters
	Cubic feet	x	0.02832	=	Cubic meters
	Cubic yards	x	0.7646	=	Cubic meters
	Gallons (U.S. liquids)[b]	x	0.003785	=	Cubic meters[c]
	Gallons (Canadian liquid)[b]	x	0.004546	=	Cubic meters[c]
	Ounces (U.S. liquid)[b]	x	29.57	=	Milliliters[c, d]
	Quarts (U.S. liquid)[b]	x	0.9464	=	Liters[c, d]
	Gallons (U.S. liquid)[b]	x	3.785	=	Liters[c, d]
Force	Kilograms force[d]	x	9.807	=	Newtons
	Pounds force	x	4.448	=	Newtons
	Pounds force	x	0.4536	=	Kilograms force[d]
	Kips	x	4448	=	Newtons
	Kips	x	453.6	=	Kilograms force[d]
Pressure, Stress, Strength (Force per unit area)	Kilograms force per square centimeter[d]	x	0.09807	=	Megapascals
	Pounds force per square inch (psi)	x	0.006895	=	Megapascals
	Kips per square inch	x	6.895	=	Megapascals
	Pounds force per square inch (psi)	x	0.07031	=	Kilograms force per square centimeter[d]
	Pounds force per square foot	x	47.88	=	Pascals
	Pounds force per square foot	x	4.882	=	Kilograms force per square meter[d]
Bending Moment Or Torque	Inch-pounds force	x	0.01152	=	Meter-kilograms force[d]
	Inch-pounds force	x	0.1130	=	Newton-meters
	Foot-pounds force	x	0.1383	=	Meter-kilograms force[d]
	Foot-pounds force	x	1.356	=	Newton-meters
	Meter-kilograms force[d]	x	9.807	=	Newton-meters
Mass	Ounces (avoirdupois)	x	28.35	=	Grams
	Pounds (avoirdupois)	x	0.4536	=	Kilograms
	Tons (metric)	x	1000	=	Kilograms
	Tons, short (2000 pounds)	x	907.2	=	Kilograms
	Tons, short (2000 pounds)	x	0.9072	=	Megagrams[e]
Mass per Unit Volume	Pounds mass per cubic foot	x	16.02	=	Kilograms per cubic meter
	Pounds mass per cubic yard	x	0.5933	=	Kilograms per cubic meter
	Pounds mass per gallon (U.S. liquid)[b]	x	119.8	=	Kilograms per cubic meter
	Pounds mass per gallon (Canadian liquid)[b]	x	99.78	=	Kilograms per cubic meter
Temperature	Degrees Fahrenheit		(F-32)/1.8	=	Degrees Celsius
	Degrees Fahrenheit		(F+459.67)/1.8	=	Degrees Kelvin
	Degrees Celsius		C+273.15	=	Degrees Kelvin

[a]The factor given is exact
[b]One U.S. gallon = 0.8327 Canadian gallon
[c]1 liter = 1000 milliliters = 1000 cubic centimeters
1 cubic decimeter = 0.001 cubic meter
[d]Metric but not SI unit
[e]Called "tonne" in England and "metric ton" in other metric countries

R011105-60 Weights and Measures

Measures of Length
1 Mile = 1760 Yards = 5280 Feet
1 Yard = 3 Feet = 36 inches
1 Foot = 12 Inches
1 Mil = 0.001 Inch
1 Fathom = 2 Yards = 6 Feet
1 Rod = 5.5 Yards = 16.5 Feet
1 Hand = 4 Inches
1 Span = 9 Inches
1 Micro-inch = One Millionth Inch or 0.000001 Inch
1 Micron = One Millionth Meter + 0.00003937 Inch

Surveyor's Measure
1 Mile = 8 Furlongs = 80 Chains
1 Furlong = 10 Chains = 220 Yards
1 Chain = 4 Rods = 22 Yards = 66 Feet = 100 Links
1 Link = 7.92 Inches

Square Measure
1 Square Mile = 640 Acres = 6400 Square Chains
1 Acre = 10 Square Chains = 4840 Square Yards = 43,560 Sq. Ft.
1 Square Chain = 16 Square Rods = 484 Square Yards = 4356 Sq. Ft.
1 Square Rod = 30.25 Square Yards = 272.25 Square Feet = 625 Square Lines
1 Square Yard = 9 Square Feet
1 Square Foot = 144 Square Inches
An Acre equals a Square 208.7 Feet per Side

Cubic Measure
1 Cubic Yard = 27 Cubic Feet
1 Cubic Foot = 1728 Cubic Inches
1 Cord of Wood = 4 x 4 x 8 Feet = 128 Cubic Feet
1 Perch of Masonry = 16½ x 1½ x 1 Foot = 24.75 Cubic Feet

Avoirdupois or Commercial Weight
1 Gross or Long Ton = 2240 Pounds
1 Net or Short ton = 2000 Pounds
1 Pound = 16 Ounces = 7000 Grains
1 Ounce = 16 Drachms = 437.5 Grains
1 Stone = 14 Pounds

Shipping Measure
For Measuring Internal Capacity of a Vessel:
 1 Register Ton = 100 Cubic Feet

For Measurement of Cargo:
 Approximately 40 Cubic Feet of Merchandise is considered a Shipping Ton, unless that bulk would weigh more than 2000 Pounds, in which case Freight Charge may be based upon weight.

40 Cubic Feet = 32.143 U.S. Bushels = 31.16 Imp. Bushels

Liquid Measure
1 Imperial Gallon = 1.2009 U.S. Gallon = 277.42 Cu. In.
1 Cubic Foot = 7.48 U.S. Gallons

R011110-10 Architectural Fees

Tabulated below are typical percentage fees by project size, for good professional architectural service. Fees may vary from those listed depending upon degree of design difficulty and economic conditions in any particular area.

Rates can be interpolated horizontally and vertically. Various portions of the same project requiring different rates should be adjusted proportionately. For alterations, add 50% to the fee for the first $500,000 of project cost and add 25% to the fee for project cost over $500,000.

Architectural fees tabulated below include Structural, Mechanical and Electrical Engineering Fees. They do not include the fees for special consultants such as kitchen planning, security, acoustical, interior design, etc.

Civil Engineering fees are included in the Architectural fee for project sites requiring minimal design such as city sites. However, separate Civil Engineering fees must be added when utility connections require design, drainage calculations are needed, stepped foundations are required, or provisions are required to protect adjacent wetlands.

Building Types	Total Project Size in Thousands of Dollars						
	100	250	500	1,000	5,000	10,000	50,000
Factories, garages, warehouses, repetitive housing	9.0%	8.0%	7.0%	6.2%	5.3%	4.9%	4.5%
Apartments, banks, schools, libraries, offices, municipal buildings	12.2	12.3	9.2	8.0	7.0	6.6	6.2
Churches, hospitals, homes, laboratories, museums, research	15.0	13.6	12.7	11.9	9.5	8.8	8.0
Memorials, monumental work, decorative furnishings	—	16.0	14.5	13.1	10.0	9.0	8.3

General Requirements | R0111 | Summary of Work

R011110-30 Engineering Fees

Typical **Structural Engineering Fees** based on type of construction and total project size. These fees are included in Architectural Fees.

Type of Construction	Total Project Size (in thousands of dollars)			
	$500	$500-$1,000	$1,000-$5,000	Over $5000
Industrial buildings, factories & warehouses	Technical payroll times 2.0 to 2.5	1.60%	1.25%	1.00%
Hotels, apartments, offices, dormitories, hospitals, public buildings, food stores	↓	2.00%	1.70%	1.20%
Museums, banks, churches and cathedrals	↓	2.00%	1.75%	1.25%
Thin shells, prestressed concrete, earthquake resistive	↓	2.00%	1.75%	1.50%
Parking ramps, auditoriums, stadiums, convention halls, hangars & boiler houses	↓	2.50%	2.00%	1.75%
Special buildings, major alterations, underpinning & future expansion	↓	Add to above 0.5%	Add to above 0.5%	Add to above 0.5%

For complex reinforced concrete or unusually complicated structures, add 20% to 50%.

Typical **Mechanical and Electrical Engineering Fees** are based on the size of the subcontract. The fee structure for both are shown below. These fees are included in Architectural Fees.

Type of Construction	Subcontract Size							
	$25,000	$50,000	$100,000	$225,000	$350,000	$500,000	$750,000	$1,000,000
Simple structures	6.4%	5.7%	4.8%	4.5%	4.4%	4.3%	4.2%	4.1%
Intermediate structures	8.0	7.3	6.5	5.6	5.1	5.0	4.9	4.8
Complex structures	10.1	9.0	9.0	8.0	7.5	7.5	7.0	7.0

For renovations, add 15% to 25% to applicable fee.

General Requirements | R0121 | Allowances

R012153-10 Repair and Remodeling

Cost figures are based on new construction utilizing the most cost-effective combination of labor, equipment and material with the work scheduled in proper sequence to allow the various trades to accomplish their work in an efficient manner.

The costs for repair and remodeling work must be modified due to the following factors that may be present in any given repair and remodeling project.

1. Equipment usage curtailment due to the physical limitations of the project, with only hand-operated equipment being used.
2. Increased requirement for shoring and bracing to hold up the building while structural changes are being made and to allow for temporary storage of construction materials on above-grade floors.
3. Material handling becomes more costly due to having to move within the confines of an enclosed building. For multi-story construction, low capacity elevators and stairwells may be the only access to the upper floors.
4. Large amount of cutting and patching and attempting to match the existing construction is required. It is often more economical to remove entire walls rather than create many new door and window openings. This sort of trade-off has to be carefully analyzed.
5. Cost of protection of completed work is increased since the usual sequence of construction usually cannot be accomplished.
6. Economies of scale usually associated with new construction may not be present. If small quantities of components must be custom fabricated due to job requirements, unit costs will naturally increase. Also, if only small work areas are available at a given time, job scheduling between trades becomes difficult and subcontractor quotations may reflect the excessive start-up and shut-down phases of the job.
7. Work may have to be done on other than normal shifts and may have to be done around an existing production facility which has to stay in production during the course of the repair and remodeling.
8. Dust and noise protection of adjoining non-construction areas can involve substantial special protection and alter usual construction methods.
9. Job may be delayed due to unexpected conditions discovered during demolition or removal. These delays ultimately increase construction costs.
10. Piping and ductwork runs may not be as simple as for new construction. Wiring may have to be snaked through walls and floors.
11. Matching "existing construction" may be impossible because materials may no longer be manufactured. Substitutions may be expensive.
12. Weather protection of existing structure requires additional temporary structures to protect building at openings.
13. On small projects, because of local conditions, it may be necessary to pay a tradesman for a minimum of four hours for a task that is completed in one hour.

All of the above areas can contribute to increased costs for a repair and remodeling project. Each of the above factors should be considered in the planning, bidding and construction stage in order to minimize the increased costs associated with repair and remodeling jobs.

R012157-20 Construction Time Requirements

Table at left is average construction time in months for different types of building projects. Table at right is the construction time in months for different size projects. Design time runs 25% to 40% of construction time.

Type Building	Construction Time	Project Value	Construction Time
Industrial Buildings	12 Months	Under $1,400,000	10 Months
Commercial Buildings	15 Months	Up to $3,800,000	15 Months
Research & Development	18 Months	Up to $19,000,000	21 Months
Institutional Buildings	20 Months	over $19,000,000	28 Months

R012909-80 Sales Tax by State

State sales tax on materials is tabulated below (5 states have no sales tax). Many states allow local jurisdictions, such as a county or city, to levy additional sales tax.

Some projects may be sales tax exempt, particularly those constructed with public funds.

State	Tax (%)	State	Tax (%)	State	Tax (%)	State	Tax (%)
Alabama	4	Illinois	6.25	Montana	0	Rhode Island	7
Alaska	0	Indiana	6	Nebraska	5.5	South Carolina	5
Arizona	5.6	Iowa	5	Nevada	6.5	South Dakota	4
Arkansas	6	Kansas	5.3	New Hampshire	0	Tennessee	7
California	6.25	Kentucky	6	New Jersey	6	Texas	6.25
Colorado	2.9	Louisiana	4	New Mexico	5	Utah	4.75
Connecticut	6	Maine	5	New York	4.25	Vermont	6
Delaware	0	Maryland	5	North Carolina	4.5	Virginia	4
District of Columbia	5.75	Massachusetts	5	North Dakota	5	Washington	6.5
Florida	6	Michigan	6	Ohio	6	West Virginia	6
Georgia	4	Minnesota	6.5	Oklahoma	4.5	Wisconsin	5
Hawaii	4	Mississippi	7	Oregon	0	Wyoming	4
Idaho	6	Missouri	4.225	Pennsylvania	6	Average	4.83 %

Sales Tax by Province (Canada)

GST - a value-added tax, which the government imposes on most goods and services provided in or imported into Canada. PST - a retail sales tax, which five of the provinces impose on the price of most goods and some services. QST - a value-added tax, similar to the federal GST, which Quebec imposes. HST - Three provinces have combined their retail sales tax with the federal GST into one harmonized tax.

Province	PST (%)	QST (%)	GST(%)	HST(%)
Alberta	0	0	7	0
British Columbia	7.5	0	7	0
Manitoba	7	0	7	0
New Brunswick	0	0	0	15
Newfoundland	0	0	0	15
Northwest Territories	0	0	7	0
Nova Scotia	0	0	0	15
Ontario	8	0	7	0
Prince Edward Island	10	0	7	0
Quebec	0	7.5	7	0
Saskatchewan	6	0	7	0
Yukon	0	0	7	0

General Requirements | R0129 | Payment Procedures

R012909-85 Unemployment Taxes and Social Security Taxes

State Unemployment Tax rates vary not only from state to state, but also with the experience rating of the contractor. The Federal Unemployment Tax rate is 6.2% of the first $7,000 of wages. This is reduced by a credit of up to 5.4% for timely payment to the state. The minimum Federal Unemployment Tax is 0.8% after all credits.

Social Security (FICA) for 2006 is estimated at time of publication to be 7.65% of wages up to $90,000.

R012909-90 Overtime

One way to improve the completion date of a project or eliminate negative float from a schedule is to compress activity duration times. This can be achieved by increasing the crew size or working overtime with the proposed crew.

To determine the costs of working overtime to compress activity duration times, consider the following examples. Below is an overtime efficiency and cost chart based on a five, six, or seven day week with an eight through twelve hour day. Payroll percentage increases for time and one half and double time are shown for the various working days.

Days per Week	Hours per Day	Production Efficiency					Payroll Cost Factors	
		1 Week	2 Weeks	3 Weeks	4 Weeks	Average 4 Weeks	@ 1-1/2 Times	@ 2 Times
5	8	100%	100%	100%	100%	100 %	100 %	100 %
	9	100	100	95	90	96.25	105.6	111.1
	10	100	95	90	85	91.25	110.0	120.0
	11	95	90	75	65	81.25	113.6	127.3
	12	90	85	70	60	76.25	116.7	133.3
6	8	100	100	95	90	96.25	108.3	116.7
	9	100	95	90	85	92.50	113.0	125.9
	10	95	90	85	80	87.50	116.7	133.3
	11	95	85	70	65	78.75	119.7	139.4
	12	90	80	65	60	73.75	122.2	144.4
7	8	100	95	85	75	88.75	114.3	128.6
	9	95	90	80	70	83.75	118.3	136.5
	10	90	85	75	65	78.75	121.4	142.9
	11	85	80	65	60	72.50	124.0	148.1
	12	85	75	60	55	68.75	126.2	152.4

General Requirements | R0131 | Project Management & Coordination

R013113-40 Builder's Risk Insurance

Builder's Risk Insurance is insurance on a building during construction. Premiums are paid by the owner or the contractor. Blasting, collapse and underground insurance would raise total insurance costs above those listed. Floater policy for materials delivered to the job runs $.75 to $1.25 per $100 value. Contractor equipment insurance runs $.50 to $1.50 per $100 value. Insurance for miscellaneous tools to $1,500 value runs from $3.00 to $7.50 per $100 value.

Tabulated below are New England Builder's Risk insurance rates in dollars per $100 value for $1,000 deductible. For $25,000 deductible, rates can be reduced 13% to 34%. On contracts over $1,000,000, rates may be lower than those tabulated. Policies are written annually for the total completed value in place. For "all risk" insurance (excluding flood, earthquake and certain other perils) add $.025 to total rates below.

Coverage	Frame Construction (Class 1)				Brick Construction (Class 4)				Fire Resistive (Class 6)			
	Range			Average	Range			Average	Range			Average
Fire Insurance	$.350	to	$ .850	$.600	$.158	to	$.189	$.174	$.052	to	$.080	$.070
Extended Coverage	.115	to	.200	.158	.080	to	.105	.101	.081	to	.105	.100
Vandalism	.012	to	.016	.014	.008	to	.011	.011	.008	to	.011	.010
Total Annual Rate	$.477	to	$1.066	$.772	$.246	to	$.305	$.286	$.141	to	$.196	$.180

R013113-50 General Contractor's Overhead

There are two distinct types of overhead on a construction project: Project Overhead and Main Office Overhead. Project Overhead includes those costs at a construction site not directly associated with the installation of construction materials. Examples of Project Overhead costs include the following:

1. Superintendent
2. Construction office and storage trailers
3. Temporary sanitary facilities
4. Temporary utilities
5. Security fencing
6. Photographs
7. Clean up
8. Performance and payment bonds

The above Project Overhead items are also referred to as General Requirements and therefore are estimated in Division 1. Division 1 is the first division listed in the CSI MasterFormat but it is usually the last division estimated. The sum of the costs in Divisions 1 through 16 is referred to as the sum of the direct costs.

All construction projects also include indirect costs. The primary components of indirect costs are the contractor's Main Office Overhead and profit. The amount of the Main Office Overhead expense varies depending on the the following:

1. Owner's compensation
2. Project managers and estimator's wages
3. Clerical support wages
4. Office rent and utilities
5. Corporate legal and accounting costs
6. Advertising
7. Automobile expenses
8. Association dues
9. Travel and entertainment expenses

These costs are usually calculated as a percentage of annual sales volume. This percentage can range from 35% for a small contractor doing less than $500,000 to 5% for a large contractor with sales in excess of $100 million.

R013113-60 Workers' Compensation Insurance Rates by Trade

The table below tabulates the national averages for Workers' Compensation insurance rates by trade and type of building. The average "Insurance Rate" is multiplied by the "% of Building Cost" for each trade. This produces the "Workers' Compensation Cost" by % of total labor cost, to be added for each trade by building type to determine the weighted average Workers' Compensation rate for the building types analyzed.

Trade	Insurance Rate (% Labor Cost) Range	Insurance Rate (% Labor Cost) Average	% of Building Cost Office Bldgs.	% of Building Cost Schools & Apts.	% of Building Cost Mfg.	Workers' Compensation Office Bldgs.	Workers' Compensation Schools & Apts.	Workers' Compensation Mfg.
Excavation, Grading, etc.	4.2 % to 19.5%	10.6%	4.8%	4.9%	4.5%	.51%	.52%	.48%
Piles & Foundations	7.7 to 76.7	22.1	7.1	5.2	8.7	1.57	1.15	1.92
Concrete	5.2 to 35.8	15.8	5.0	14.8	3.7	.79	2.34	.58
Masonry	5.1 to 28.3	14.9	6.9	7.5	1.9	1.03	1.12	.28
Structural Steel	7.8 to 103.8	39.7	10.7	3.9	17.6	4.25	1.55	6.99
Miscellaneous & Ornamental Metals	4.9 to 24.8	12.2	2.8	4.0	3.6	.34	.49	.44
Carpentry & Millwork	7.0 to 53.2	18.4	3.7	4.0	0.5	.68	.74	.09
Metal or Composition Siding	5.3 to 32.1	17.3	2.3	0.3	4.3	.40	.05	.74
Roofing	7.8 to 78.9	32.4	2.3	2.6	3.1	.75	.84	1.00
Doors & Hardware	4.5 to 24.9	11.8	0.9	1.4	0.4	.11	.17	.05
Sash & Glazing	4.6 to 33.7	13.9	3.5	4.0	1.0	.49	.56	.14
Lath & Plaster	4.3 to 40.2	14.0	3.3	6.9	0.8	.46	.97	.11
Tile, Marble & Floors	2.4 to 31.8	9.5	2.6	3.0	0.5	.25	.29	.05
Acoustical Ceilings	2.4 to 29.7	11.4	2.4	0.2	0.3	.27	.02	.03
Painting	4.5 to 33.7	13.3	1.5	1.6	1.6	.20	.21	.21
Interior Partitions	7.0 to 53.2	18.4	3.9	4.3	4.4	.72	.79	.81
Miscellaneous Items	2.6 to 137.6	17.3	5.2	3.7	9.7	.90	.64	1.68
Elevators	2.8 to 42.8	6.9	2.1	1.1	2.2	.14	.08	.15
Sprinklers	3.2 to 22.1	8.6	0.5	—	2.0	.04	—	.17
Plumbing	2.7 to 12.4	8.2	4.9	7.2	5.2	.40	.59	.43
Heat., Vent., Air Conditioning	4.0 to 27.2	11.8	13.5	11.0	12.9	1.59	1.30	1.52
Electrical	2.6 to 12.7	6.8	10.1	8.4	11.1	.69	.57	.75
Total	2.4 % to 137.6%	—	100.0%	100.0%	100.0%	16.58%	14.99%	18.62%
		Overall Weighted Average	16.73%					

Workers' Compensation Insurance Rates by States

The table below lists the weighted average Workers' Compensation base rate for each state with a factor comparing this with the national average of 16.3%.

State	Weighted Average	Factor	State	Weighted Average	Factor	State	Weighted Average	Factor
Alabama	25.7%	158	Kentucky	20.5%	126	North Dakota	13.1%	80
Alaska	24.0	147	Louisiana	28.2	173	Ohio	14.5	89
Arizona	7.5	46	Maine	20.9	128	Oklahoma	14.7	90
Arkansas	13.7	84	Maryland	19.1	117	Oregon	13.0	80
California	18.8	115	Massachusetts	13.4	82	Pennsylvania	14.3	88
Colorado	10.9	67	Michigan	17.4	107	Rhode Island	21.1	129
Connecticut	22.4	137	Minnesota	25.5	156	South Carolina	16.8	103
Delaware	17.3	106	Mississippi	17.9	110	South Dakota	17.1	105
District of Columbia	17.1	105	Missouri	17.7	109	Tennessee	18.1	111
Florida	23.3	143	Montana	20.4	125	Texas	12.7	78
Georgia	22.9	140	Nebraska	22.9	140	Utah	12.3	75
Hawaii	16.9	104	Nevada	13.6	83	Vermont	25.0	153
Idaho	11.7	72	New Hampshire	22.6	139	Virginia	13.5	83
Illinois	18.8	115	New Jersey	12.2	75	Washington	10.9	67
Indiana	7.1	44	New Mexico	17.7	109	West Virginia	13.2	81
Iowa	12.3	75	New York	14.4	88	Wisconsin	14.7	90
Kansas	8.9	55	North Carolina	15.4	94	Wyoming	9.1	56
			Weighted Average for U.S. is	16.7% of payroll = 100%				

Rates in the following table are the base or manual costs per $100 of payroll for Workers' Compensation in each state. Rates are usually applied to straight time wages only and not to premium time wages and bonuses.

The weighted average skilled worker rate for 35 trades is 16.3%. For bidding purposes, apply the full value of Workers' Compensation directly to total labor costs, or if labor is 38%, materials 42% and overhead and profit 20% of total cost, carry 38/80 x 16.3% =7.7% of cost (before overhead and profit) into overhead. Rates vary not only from state to state but also with the experience rating of the contractor.

Rates are the most current available at the time of publication.

R013113-60 Workers' Compensation Insurance Rates by Trade and State (cont.)

State	Carpentry — 3 stories or less 5651	Carpentry — interior cab. work 5437	Carpentry — general 5403	Concrete Work — NOC 5213	Concrete Work — flat (flr., sdwk.) 5221	Electrical Wiring — inside 5190	Excavation — earth NOC 6217	Excavation — rock 6217	Glaziers 5462	Insulation Work 5479	Lathing 5443	Masonry 5022	Painting & Decorating 5474	Pile Driving 6003	Plastering 5480	Plumbing 5183	Roofing 5551	Sheet Metal Work (HVAC) 5538	Steel Erection — door & sash 5102	Steel Erection — inter., ornam. 5102	Steel Erection — structure 5040	Steel Erection — NOC 5057	Tile Work — (interior ceramic) 5348	Waterproofing 9014	Wrecking 5701
AL	29.57	16.77	31.62	11.24	10.07	8.55	12.90	12.90	26.88	17.64	12.34	23.46	32.15	38.02	32.13	11.79	65.04	21.69	19.37	19.37	56.15	31.21	12.21	6.54	56.15
AK	21.71	15.70	15.84	13.63	12.88	13.24	20.07	20.07	26.84	31.05	11.29	31.31	21.16	51.97	26.78	12.29	41.90	14.82	12.85	12.85	50.64	25.20	9.35	9.63	50.64
AZ	6.88	5.05	11.62	5.56	3.49	3.85	4.74	4.74	7.40	7.69	4.27	5.37	4.87	9.83	5.20	3.90	12.19	5.65	9.50	9.50	18.98	8.11	2.04	2.34	50.31
AR	11.87	9.14	15.06	11.65	6.81	5.31	9.16	9.16	11.83	30.45	7.61	10.44	10.67	16.93	12.24	5.61	22.08	12.99	7.43	7.43	36.81	16.98	6.36	4.67	38.24
CA	28.28	9.07	28.28	13.18	13.18	9.93	7.88	7.88	16.37	23.34	10.90	14.55	19.45	21.18	18.12	11.59	39.79	15.33	13.55	13.55	23.93	21.91	7.74	19.45	21.91
CO	14.58	6.77	10.47	11.38	6.17	4.52	8.41	8.41	7.33	10.76	4.73	10.57	8.54	13.61	8.00	7.15	19.17	10.23	6.98	6.98	27.39	12.94	7.07	4.70	27.39
CT	18.69	17.64	27.54	26.02	12.53	8.70	10.59	10.59	16.86	19.70	19.27	24.01	18.88	25.92	24.21	10.86	43.51	13.48	16.02	16.02	61.01	29.58	9.39	6.44	50.86
DE	17.74	17.74	14.29	14.46	11.19	7.77	11.13	11.13	13.41	14.29	15.42	14.16	18.57	23.83	15.42	9.28	31.57	11.76	14.60	14.60	33.94	14.60	11.81	14.16	33.94
DC	12.26	11.99	12.28	14.08	16.31	7.27	9.99	9.99	23.64	10.47	9.64	19.37	10.29	17.95	12.01	14.00	18.86	8.55	17.12	17.12	46.07	20.00	28.91	4.70	46.07
FL	24.98	19.24	28.32	24.68	12.25	10.16	12.77	12.77	18.92	17.94	13.27	20.39	19.51	47.98	24.53	10.82	37.58	16.40	13.31	13.31	56.10	32.45	10.28	8.90	56.10
GA	32.05	17.69	24.16	15.16	10.44	8.97	16.86	16.86	15.69	22.02	17.16	20.30	17.13	31.16	19.94	10.77	46.74	17.52	13.97	13.97	44.46	48.01	9.64	8.55	44.46
HI	17.38	11.62	28.20	13.74	12.27	7.10	7.73	7.73	20.55	21.19	10.94	17.87	11.33	20.15	15.61	6.06	33.24	8.02	11.11	11.11	31.71	21.70	9.78	11.54	31.71
ID	11.92	6.31	13.47	10.64	5.99	5.25	6.90	6.90	9.58	8.14	6.31	9.91	8.88	15.41	9.45	5.99	29.54	9.11	7.94	7.94	33.81	14.13	7.27	4.09	33.81
IL	17.63	14.46	17.30	27.55	11.47	19.13	9.77	9.77	17.62	12.10	9.88	16.21	10.52	28.78	11.82	9.62	27.47	13.74	15.18	15.18	52.69	23.84	13.78	4.43	52.69
IN	5.29	14.84	6.93	5.29	13.20	2.58	4.34	4.34	4.67	4.81	2.49	5.20	5.25	8.82	4.02	2.73	11.74	4.35	5.20	5.20	18.17	8.49	3.31	2.82	18.17
IA	11.94	5.77	9.18	13.10	7.10	4.40	5.19	5.19	12.22	8.84	5.48	9.22	8.30	10.69	8.48	5.61	17.71	6.88	8.97	8.97	42.71	42.83	4.81	3.70	31.77
KS	9.72	8.65	10.24	7.86	5.92	3.32	4.15	4.15	7.75	7.72	4.21	8.13	6.79	10.74	7.35	5.65	18.03	6.03	7.26	7.26	22.40	11.92	5.40	3.31	23.40
KY	19.74	12.89	18.65	24.60	7.70	8.10	15.00	15.00	21.87	24.34	11.81	10.91	14.25	22.06	15.53	8.42	54.39	17.05	14.51	14.51	50.29	28.09	12.52	5.04	50.26
LA	23.14	24.93	53.17	26.43	15.61	9.88	17.49	17.49	20.14	21.43	24.51	28.33	29.58	31.25	22.37	8.64	77.12	21.20	18.98	18.98	51.76	24.21	13.77	13.78	66.41
ME	13.43	10.42	43.59	24.29	11.12	3.72	12.56	12.56	13.35	15.39	14.03	17.26	16.07	31.44	17.95	9.07	32.07	10.07	15.08	15.08	37.84	61.81	11.66	5.92	37.84
MD	13.71	5.95	10.55	18.82	8.70	5.15	12.91	12.91	25.52	20.07	11.37	13.31	9.37	26.22	14.21	8.58	48.72	12.74	13.66	13.66	56.25	37.81	8.03	7.14	26.80
MA	9.93	6.07	16.09	17.95	8.10	3.69	6.49	6.49	7.55	13.48	5.76	13.04	7.54	14.31	5.34	4.55	38.25	7.47	12.35	12.35	35.13	27.39	9.24	3.32	28.82
MI	19.68	11.72	21.53	19.00	8.83	5.94	9.96	9.96	11.76	13.31	13.59	17.06	13.34	36.96	15.69	7.35	33.24	9.51	9.92	9.92	36.96	27.43	10.97	5.38	39.06
MN	19.53	19.80	41.48	14.65	15.73	7.53	14.95	14.95	15.68	11.90	22.05	19.18	16.95	24.63	22.05	10.88	70.18	14.23	10.58	10.58	104.13	33.30	14.55	6.68	132.92
MS	16.68	14.87	24.71	12.67	9.14	5.85	12.50	12.50	13.10	15.77	8.50	11.88	13.12	19.91	28.30	9.82	38.94	20.61	10.87	10.87	32.50	30.23	10.01	6.49	32.50
MO	24.01	11.40	15.92	15.49	10.23	6.90	9.30	9.30	11.03	17.88	11.24	14.55	13.46	16.69	15.51	9.66	31.18	13.82	11.26	11.26	53.28	36.77	8.63	6.32	53.28
MT	21.18	10.79	19.52	12.88	10.39	6.82	19.69	19.69	13.67	20.72	22.29	15.50	10.31	60.73	12.40	10.22	45.28	12.09	12.45	12.45	40.26	22.93	6.26	5.14	40.26
NE	23.85	14.32	22.60	27.52	12.77	9.52	18.07	18.07	21.82	33.02	12.77	23.10	17.70	23.25	17.17	11.13	38.42	17.27	13.90	13.90	60.90	30.50	10.52	7.63	50.40
NV	20.81	8.10	10.79	9.56	7.96	6.16	10.76	10.76	12.49	10.61	9.45	9.87	9.00	16.26	9.36	7.27	18.09	21.51	10.59	10.59	24.87	29.06	6.55	5.43	4.09
NH	27.93	14.04	25.82	31.58	12.51	7.80	18.33	18.33	13.55	38.86	9.24	22.82	12.51	20.88	15.76	11.83	49.65	16.44	13.62	13.62	52.78	22.94	14.95	6.55	52.78
NJ	12.63	9.48	12.63	10.77	8.25	4.50	8.33	8.33	7.06	12.93	11.17	12.87	10.95	13.78	11.17	6.14	32.47	6.17	10.72	10.72	22.56	12.00	4.75	4.81	26.39
NM	29.00	8.32	20.65	18.60	11.34	7.16	10.46	10.46	14.57	12.61	8.59	12.43	13.86	21.32	11.32	8.92	30.32	10.81	19.90	19.90	40.88	27.34	8.55	7.65	40.88
NY	17.66	6.98	15.91	17.16	13.57	6.32	10.02	10.02	12.02	9.09	5.11	19.20	13.75	16.81	8.43	8.53	33.44	17.47	9.63	9.63	18.80	19.14	9.43	6.09	11.29
NC	16.52	12.25	14.99	13.22	7.32	8.85	9.45	9.45	11.16	11.72	10.68	9.98	10.17	15.13	14.39	8.39	24.40	12.94	7.40	7.40	68.39	20.27	6.05	4.83	68.39
ND	10.49	10.49	10.49	5.94	5.94	4.15	5.93	5.93	10.49	10.49	9.25	8.19	6.57	21.54	9.25	5.46	20.98	5.46	21.45	21.45	21.54	21.54	10.49	20.98	14.48
OH	7.48	9.85	9.98	11.81	11.78	6.16	9.02	9.02	8.65	14.76	38.58	12.85	13.30	23.94	3.76	7.06	24.80	9.48	10.09	10.09	31.53	15.16	9.91	5.99	31.53
OK	13.44	9.11	12.70	12.65	6.63	5.82	10.25	10.25	16.10	19.50	8.46	10.98	9.09	20.29	12.21	7.24	22.32	9.49	13.78	13.78	39.78	24.01	6.95	5.88	39.78
OR	16.72	9.16	15.83	14.23	8.06	4.50	9.48	9.48	14.13	8.79	7.41	14.91	11.59	15.81	11.05	5.30	22.92	9.68	9.53	9.53	29.02	11.69	10.63	4.52	29.02
PA	13.48	13.48	12.50	14.67	10.14	6.58	8.34	8.34	9.68	12.50	11.87	12.38	14.03	17.07	11.87	7.29	27.70	8.07	15.29	15.29	24.67	15.29	8.67	12.38	24.67
RI	19.53	11.65	18.07	18.23	16.24	4.43	10.38	10.38	12.85	22.78	11.97	25.11	24.13	37.66	17.25	8.52	33.92	10.29	14.07	14.07	59.49	37.50	14.36	7.70	78.79
SC	22.26	17.21	19.93	13.51	7.81	9.42	11.95	11.95	14.32	11.08	8.53	13.27	16.47	18.32	18.15	8.83	43.19	10.97	11.47	11.47	24.55	27.29	8.90	5.98	24.55
SD	21.00	17.87	17.14	19.81	7.46	5.25	15.58	15.58	11.54	12.77	8.94	7.73	12.16	23.97	11.77	10.03	22.29	11.34	11.11	11.11	62.70	23.72	7.24	4.06	62.70
TN	30.13	13.71	22.70	18.06	9.32	8.15	13.14	13.14	11.29	13.81	12.27	15.91	13.61	16.19	16.22	10.57	36.23	13.55	12.52	12.52	42.02	23.33	9.07	6.20	42.02
TX	13.47	9.95	13.47	11.11	8.00	7.76	8.79	8.79	9.83	12.47	6.78	12.04	8.63	16.18	8.63	7.35	19.58	16.38	9.69	9.69	31.53	13.20	6.81	7.37	14.13
UT	9.45	6.23	10.52	7.68	7.99	7.05	6.38	6.38	9.55	10.95	12.32	13.75	15.29	17.24	10.60	6.41	26.27	6.30	8.99	8.99	23.51	23.51	6.06	5.83	26.53
VT	24.65	11.60	22.01	36.23	15.66	8.18	13.75	13.75	25.11	29.18	12.53	23.26	13.86	23.35	18.10	11.77	42.54	15.20	15.81	15.81	72.09	57.39	9.47	12.14	72.09
VA	13.36	8.95	11.28	12.51	5.79	5.62	7.58	7.58	9.32	10.10	17.52	9.64	10.83	15.93	9.09	6.69	21.06	8.52	11.01	11.01	49.54	21.40	6.14	3.14	49.54
WA	10.29	10.29	10.29	7.89	7.89	3.28	8.11	8.11	12.02	10.01	10.29	9.82	9.08	18.17	11.05	4.63	18.54	4.64	12.96	12.96	8.65	8.65	10.03	10.23	8.65
WV	12.30	12.30	12.30	27.37	27.37	5.86	8.90	8.90	6.99	13.15	13.15	11.94	12.35	13.34	12.35	5.48	14.63	6.99	10.89	10.89	10.89	10.89	13.15	4.46	48.62
WI	12.40	11.36	16.86	12.29	9.90	4.24	7.44	7.44	13.31	11.62	8.09	18.99	14.32	13.64	12.11	5.42	34.72	7.16	9.18	9.18	37.88	24.06	13.99	5.52	37.88
WY	8.27	8.27	8.27	8.27	8.27	8.27	8.27	8.27	8.27	8.27	8.27	8.27	8.27	8.27	8.27	8.27	8.27	8.27	8.27	8.27	8.27	8.27	8.27	8.27	8.27
AVG.	17.27	11.81	18.39	15.82	10.29	6.84	10.63	10.63	13.87	15.76	11.44	14.92	13.29	22.07	14.00	8.22	32.40	11.76	12.19	12.19	39.65	24.35	9.52	7.04	40.06

R013113-60 Workers' Compensation (cont.) (Canada in Canadian dollars)

Province		Alberta	British Columbia	Manitoba	Ontario	New Brunswick	Newfndld. & Labrador	Northwest Territories	Nova Scotia	Prince Edward Island	Quebec	Saskat-chewan	Yukon
Carpentry—3 stories or less	Rate	9.71	7.19	4.57	4.62	4.83	9.78	4.61	7.67	6.04	15.41	7.33	5.58
	Code	42143	721028	40102	723	422	4226	4-41	4226	401	80110	B1317	202
Carpentry—interior cab. work	Rate	1.60	6.16	4.57	4.62	4.31	6.37	4.61	5.74	4.05	15.41	3.68	5.58
	Code	42133	721021	40102	723	427	4270	4-41	4274	402	80110	B11-27	202
CARPENTRY—general	Rate	9.71	7.19	4.57	4.62	4.83	6.37	4.61	7.67	6.04	15.41	7.33	5.58
	Code	42143	721028	40102	723	422	4299	4-41	4226	401	80110	B1317	202
CONCRETE WORK—NOC	Rate	5.96	7.53	7.60	15.25	4.83	9.78	4.61	4.83	6.04	16.51	7.33	5.58
	Code	42104	721010	40110	748	422	4224	4-41	4224	401	80100	B13-14	203
CONCRETE WORK—flat	Rate	5.96	7.53	7.60	15.25	4.83	9.78	4.61	4.83	6.04	16.51	7.33	5.58
(flr. sidewalk)	Code	42104	721010	40110	748	422	4224	4-41	4224	401	80100	B13-14	203
ELECTRICAL Wiring—inside	Rate	2.50	2.17	2.52	2.94	2.35	3.25	4.13	2.23	4.05	7.64	3.68	5.58
	Code	42124	721019	40203	704	426	4261	4-46	4261	402	80170	B11-05	206
EXCAVATION—earth NOC	Rate	3.31	3.85	3.84	4.20	3.16	5.53	3.71	4.11	4.25	8.36	4.37	5.58
	Code	40604	721031	40706	711	421	4214	4-43	4214	404	80030	R11-06	207
EXCAVATION—rock	Rate	3.31	3.85	3.84	4.20	3.16	5.53	3.71	4.11	4.25	8.36	4.37	5.58
	Code	40604	721031	40706	711	421	4214	4-43	4214	404	80030	R11-06	207
GLAZIERS	Rate	4.17	4.07	4.57	8.12	5.58	7.07	4.61	7.67	4.05	14.36	7.33	5.58
	Code	42121	715020	40109	751	423	4233	4-41	4233	402	80150	B13-04	212
INSULATION WORK	Rate	3.33	5.88	4.57	8.12	5.58	7.07	4.61	7.67	6.04	15.41	6.15	5.58
	Code	42184	721029	40102	751	423	4234	4-41	4234	401	80110	B12-07	202
LATHING	Rate	6.71	8.99	4.57	4.62	4.31	6.37	4.61	5.74	4.05	15.41	7.33	5.58
	Code	42135	721033	40102	723	427	4279	4-41	4271	402	80110	B13-16	202
MASONRY	Rate	5.96	8.99	4.57	11.44	5.58	7.07	4.61	7.67	6.04	16.51	7.33	5.58
	Code	42102	721037	40102	741	423	4231	4-41	4231	401	80100	B13-18	202
PAINTING & DECORATING	Rate	5.20	5.88	3.45	6.41	4.31	6.37	4.61	5.74	4.05	15.41	6.15	5.58
	Code	42111	721041	40105	719	427	4275	4-41	4275	402	80110	B12-01	202
PILE DRIVING	Rate	5.96	5.63	3.84	5.84	4.83	9.78	3.71	4.83	6.04	8.36	7.33	5.58
	Code	42159	722004	40706	732	422	4221	4-43	4221	401	80030	B13-10	202
PLASTERING	Rate	6.71	8.99	5.13	6.41	4.31	6.37	4.61	5.74	4.05	15.41	6.15	5.58
	Code	42135	721042	40108	719	427	4271	4-41	4271	402	80110	B12-21	202
PLUMBING	Rate	2.50	4.67	2.88	3.67	3.09	3.87	4.13	2.23	4.05	7.61	3.68	5.58
	Code	42122	721043	40204	707	424	4241	4-46	4241	402	80160	B11-01	214
ROOFING	Rate	9.88	10.10	6.52	11.60	10.25	9.78	4.61	9.49	6.04	22.49	7.33	5.58
	Code	42118	721036	40403	728	430	4236	4-41	4236	401	80130	B13-20	202
SHEET METAL WORK (HVAC)	Rate	2.50	4.67	6.52	3.67	3.09	3.87	4.13	3.23	4.05	7.61	3.68	7.39
	Code	42117	721043	40402	707	424	4244	4-46	4244	402	80160	B11-07	208
STEEL ERECTION—door & sash	Rate	3.33	13.44	9.89	15.25	4.83	9.78	4.61	7.67	6.04	29.92	7.33	5.58
	Code	42106	722005	40502	748	422	4227	4-41	4227	401	80080	B13-22	202
STEEL ERECTION—inter., ornam.	Rate	3.33	13.44	9.89	15.25	4.83	9.78	4.61	7.67	6.04	29.92	7.33	5.58
	Code	42106	722005	40502	748	422	4227	4-41	4227	401	80080	B13-22	202
STEEL ERECTION—structure	Rate	3.33	13.44	9.89	15.25	4.83	9.78	4.61	7.67	6.04	29.92	7.33	5.58
	Code	42106	722005	40502	748	422	4227	4-41	4227	401	80080	B13-22	202
STEEL ERECTION—NOC	Rate	3.33	13.44	9.89	15.25	4.83	9.78	4.61	7.67	6.04	29.92	7.33	5.58
	Code	42106	722005	40502	748	422	4227	4-41	4227	401	80080	B13-22	202
TILE WORK—inter. (ceramic)	Rate	4.62	5.67	2.17	6.41	4.31	6.37	4.61	5.74	4.05	15.41	7.33	5.58
	Code	42113	721054	40103	719	427	4276	4-41	4276	402	80110	B13-01	202
WATERPROOFING	Rate	5.20	5.88	4.57	4.62	5.58	6.37	4.61	7.67	4.05	22.49	6.15	5.58
	Code	42139	721016	40102	723	423	4299	4-41	4239	402	80130	B12-17	202
WRECKING	Rate	3.31	5.91	6.46	15.25	3.16	5.53	3.71	4.11	6.04	15.41	7.33	5.58
	Code	40604	721005	40106	748	421	4211	4-43	4211	401	80110	B13-09	202

General Requirements | R0131 | Project Management & Coordination

R013113-80 Performance Bond

This table shows the cost of a Performance Bond for a construction job scheduled to be completed in 12 months. Add 1% of the premium cost per month for jobs requiring more than 12 months to complete. The rates are "standard" rates offered to contractors that the bonding company considers financially sound and capable of doing the work. Preferred rates are offered by some bonding companies based upon financial strength of the contractor. Actual rates vary from contractor to contractor and from bonding company to bonding company. Contractors should prequalify through a bonding agency before submitting a bid on a contract that requires a bond.

Contract Amount	Building Construction Class B Projects	Highways & Bridges	
		Class A New Construction	Class A-1 Highway Resurfacing
First $ 100,000 bid	$25.00 per M	$15.00 per M	$9.40 per M
Next 400,000 bid	$ 2,500 plus $15.00 per M	$ 1,500 plus $10.00 per M	$ 940 plus $7.20 per M
Next 2,000,000 bid	8,500 plus 10.00 per M	5,500 plus 7.00 per M	3,820 plus 5.00 per M
Next 2,500,000 bid	28,500 plus 7.50 per M	19,500 plus 5.50 per M	15,820 plus 4.50 per M
Next 2,500,000 bid	47,250 plus 7.00 per M	33,250 plus 5.00 per M	28,320 plus 4.50 per M
Over 7,500,000 bid	64,750 plus 6.00 per M	45,750 plus 4.50 per M	39,570 plus 4.00 per M

General Requirements | R0151 | Temporary Utilities

R015113-65 Temporary Power Equipment

Cost data for the temporary equipment was developed utilizing the following information.

1) Re-usable material-services, transformers, equipment and cords are based on new purchase and prorated to three projects.
2) PVC feeder includes trench and backfill.
3) Connections include disconnects and fuses.
4) Labor units include an allowance for removal.
5) No utility company charges or fees are included.
6) Concrete pads or vaults are not included.
7) Utility company conduits not included.

R015423-10 Steel Tubular Scaffolding

On new construction, tubular scaffolding is efficient up to 60′ high or five stories. Above this it is usually better to use a hung scaffolding if construction permits. Swing scaffolding operations may interfere with tenants. In this case, the tubular is more practical at all heights.

In repairing or cleaning the front of an existing building the cost of tubular scaffolding per S.F. of building front increases as the height increases above the first tier. The first tier cost is relatively high due to leveling and alignment.

The minimum efficient crew for erection is three workers. For heights over 50′, a crew of four is more efficient. Use two or more on top and two at the bottom for handing up or hoisting. Four workers can erect and dismantle about nine frames per hour up to five stories. From five to eight stories they will average six frames per hour. With 7′ horizontal spacing this will run about 400 S.F. and 265 S.F. of wall surface, respectively. Time for placing planks must be added to the above. On heights above 50′, five planks can be placed per labor-hour.

The table below shows the number of pieces required to erect tubular steel scaffolding for 1000 S.F. of building frontage. This area is made up of a scaffolding system that is 12 frames (11 bays) long by 2 frames high.

For jobs under twenty-five frames, add 50% to rental cost. Rental rates will be lower for jobs over three months duration. Large quantities for long periods can reduce rental rates by 20%.

Description of Component	CSI Line Item	Number of Pieces for 1000 S.F. of Building Front	Unit
5′ Wide Standard Frame, 6′-4″ High	01540-750-2200	24	Ea.
Leveling Jack & Plate	01540-750-2650	24	
Cross Brace	01540-750-2500	44	
Side Arm Bracket, 21″	01540-750-2700	12	
Guardrail Post	01540-750-2550	12	
Guardrail, 7′ section	01540-750-2600	22	
Stairway Section	01540-750-2900	2	
Stairway Starter Bar	01540-750-2910	1	
Stairway Inside Handrail	01540-750-2920	2	
Stairway Outside Handrail	01540-750-2930	2	
Walk-Thru Frame Guardrail	01540-750-2940	2	↓

Scaffolding is often used as falsework over 15′ high during construction of cast-in-place concrete beams and slabs. Two foot wide scaffolding is generally used for heavy beam construction. The span between frames depends upon the load to be carried with a maximum span of 5′.

Heavy duty shoring frames with a capacity of 10,000#/leg can be spaced up to 10′ O.C. depending upon form support design and loading.

Scaffolding used as horizontal shoring requires less than half the material required with conventional shoring.

On new construction, erection is done by carpenters.

Rolling towers supporting horizontal shores can reduce labor and speed the job. For maintenance work, catwalks with spans up to 70′ can be supported by the rolling towers.

R015423-20 Pump Staging

Pump staging is generally not available for rent. The table below shows the number of pieces required to erect pump staging for 2400 S.F. of building frontage. This area is made up of a pump jack system that is 3 poles (2 bays) wide by 2 poles high.

Item	CSI Line Item	Number of Pieces for 2400 S.F. of Building Front	Unit
Aluminum pole section, 24′ long	01540-550-0200	6	Ea.
Aluminum splice joint, 6′ long	01540-550-0600	3	
Aluminum foldable brace	01540-550-0900	3	
Aluminum pump jack	01540-550-0700	3	
Aluminum support for workbench/back safety rail	01540-550-1000	3	
Aluminum scaffold plank/workbench, 14″ wide x 24′ long	01540-550-1100	4	
Safety net, 22′ long	01540-550-1250	2	
Aluminum plank end safety rail	01540-550-1200	2	↓

The cost in place for this 2400 S.F. will depend on how many uses are realized during the life of the equipment.

R015433-10 Contractor Equipment

Rental Rates shown elsewhere in the book pertain to late model high quality machines in excellent working condition, rented from equipment dealers. Rental rates from contractors may be substantially lower than the rental rates from equipment dealers depending upon economic conditions; for older, less productive machines, reduce rates by a maximum of 15%. Any overtime must be added to the base rates. For shift work, rates are lower. Usual rule of thumb is 150% of one shift rate for two shifts; 200% for three shifts.

For periods of less than one week, operated equipment is usually more economical to rent than renting bare equipment and hiring an operator.

Costs to move equipment to a job site (mobilization) or from a job site (demobilization) are not included in rental rates, nor in any Equipment costs on any Unit Price line items or crew listings. These costs can be found elsewhere. If a piece of equipment is already at a job site, it is not appropriate to utilize mob/demob costs in an estimate again.

Rental rates vary throughout the country with larger cities generally having lower rates. Lease plans for new equipment are available for periods in excess of six months with a percentage of payments applying toward purchase.

Monthly rental rates vary from 2% to 5% of the cost of the equipment depending on the anticipated life of the equipment and its wearing parts. Weekly rates are about 1/3 the monthly rates and daily rental rates about 1/3 the weekly rate.

The hourly operating costs for each piece of equipment include costs to the user such as fuel, oil, lubrication, normal expendables for the equipment, and a percentage of mechanic's wages chargeable to maintenance. The hourly operating costs listed do not include the operator's wages.

The daily cost for equipment used in the standard crews is figured by dividing the weekly rate by five, then adding eight times the hourly operating cost to give the total daily equipment cost, not including the operator. This figure is in the right hand column of the Equipment listings under Crew Equipment Cost/Day.

Pile Driving rates shown for pile hammer and extractor do not include leads, crane, boiler or compressor. Vibratory pile driving requires an added field specialist during set-up and pile driving operation for the electric model. The hydraulic model requires a field specialist for set-up only. Up to 125 reuses of sheet piling are possible using vibratory drivers. For normal conditions, crane capacity for hammer type and size are as follows.

Crane Capacity	Hammer Type and Size		
	Air or Steam	Diesel	Vibratory
25 ton	to 8,750 ft.-lb.		70 H.P.
40 ton	15,000 ft.-lb.	to 32,000 ft.-lb.	170 H.P.
60 ton	25,000 ft.-lb.		300 H.P.
100 ton		112,000 ft.-lb.	

Cranes should be specified for the job by size, building and site characteristics, availability, performance characteristics, and duration of time required.

Backhoes & Shovels rent for about the same as equivalent size cranes but maintenance and operating expense is higher. Crane operators rate must be adjusted for high boom heights. Average adjustments: for 150′ boom add 2% per hour; over 185′, add 4% per hour; over 210′, add 6% per hour; over 250′, add 8% per hour and over 295′, add 12% per hour.

Tower Cranes of the climbing or static type have jibs from 50′ to 200′ and capacities at maximum reach range from 4,000 to 14,000 pounds. Lifting capacities increase up to maximum load as the hook radius decreases.

Typical rental rates, based on purchase price are about 2% to 3% per month.

Erection and dismantling runs between 500 and 2000 labor hours. Climbing operation takes 10 labor hours per 20′ climb. Crane dead time is about 5 hours per 40′ climb. If crane is bolted to side of the building add cost of ties and extra mast sections. Climbing cranes have from 80′ to 180′ of mast while static cranes have 80′ to 800′ of mast.

Truck Cranes can be converted to tower cranes by using tower attachments. Mast heights over 400′ have been used.

A single 100′ high material **Hoist and Tower** can be erected and dismantled in about 400 labor hours; a double 100′ high hoist and tower in about 600 labor hours. Erection times for additional heights are 3 and 4 labor hours per vertical foot respectively up to 150′, and 4 to 5 labor hours per vertical foot over 150′ high. A 40′ high portable Buck hoist takes about 160 labor hours to erect and dismantle. Additional heights take 2 labor hours per vertical foot to 80′ and 3 labor hours per vertical foot for the next 100′. Most material hoists do not meet local code requirements for carrying personnel.

A 150′ high **Personnel Hoist** requires about 500 to 800 labor hours to erect and dismantle. Budget erection time at 5 labor hours per vertical foot for all trades. Local code requirements or labor scarcity requiring overtime can add up to 50% to any of the above erection costs.

Earthmoving Equipment: The selection of earthmoving equipment depends upon the type and quantity of material, moisture content, haul distance, haul road, time available, and equipment available. Short haul cut and fill operations may require dozers only, while another operation may require excavators, a fleet of trucks, and spreading and compaction equipment. Stockpiled material and granular material are easily excavated with front end loaders. Scrapers are most economically used with hauls between 300′ and 1-1/2 miles if adequate haul roads can be maintained. Shovels are often used for blasted rock and any material where a vertical face of 8′ or more can be excavated. Special conditions may dictate the use of draglines, clamshells, or backhoes. Spreading and compaction equipment must be matched to the soil characteristics, the compaction required and the rate the fill is being supplied.

R015433-15 Heavy Lifting

Hydraulic Climbing Jacks

The use of hydraulic heavy lift systems is an alternative to conventional type crane equipment. The lifting, lowering, pushing, or pulling mechanism is a hydraulic climbing jack moving on a square steel jackrod from 1-5/8″ to 4″ square, or a steel cable. The jackrod or cable can be vertical or horizontal, stationary or movable, depending on the individual application. When the jackrod is stationary, the climbing jack will climb the rod and push or pull the load along with itself. When the climbing jack is stationary, the jackrod is movable with the load attached to the end and the climbing jack will lift or lower the jackrod with the attached load. The heavy lift system is normally operated by a single control lever located at the hydraulic pump.

The system is flexible in that one or more climbing jacks can be applied wherever a load support point is required, and the rate of lift synchronized.

Economic benefits have been demonstrated on projects such as: erection of ground assembled roofs and floors, complete bridge spans, girders and trusses, towers, chimney liners and steel vessels, storage tanks, and heavy machinery. Other uses are raising and lowering offshore work platforms, caissons, tunnel sections and pipelines.

R019313-10 Facility Maintenance - Frequency Table

The following table lists "average" frequency data for selected facility maintenance activities. The frequencies given are for a normal standard of maintenance under average conditions.

Activity	Average Frequency	Notes
Acoustical tile, cleaning		
Heavy smoking	2-3 years	
Non-smoking	10 years	
Carpet, cleaning		Frequency depends on the type of carpet and the occupancy in the building.
Heavy traffic	Weekly	
Light traffic	Every 6 weeks	
Carpet, vacuuming		
Heavy traffic	Daily	6-8 passes by machine in key areas.
Light traffic	Twice weekly	
Offices	Weekly	
Corridor and lobby, policing		Includes picking up loose trash, removing cigarette butts from in and around jardinieres and/or sand urns. Frequency depends on occupancy in the building.
Main	4 times daily	
Secondary	Daily	
Elevator, cleaning		The frequency of elevator cleaning is dependent upon the amount of traffic, size of the elevator car and the occupancy in the building.
Passenger	Daily	
Freight	Weekly	
Elevator lobby, cleaning	Daily	Includes sweeping, mopping and rinsing.
Escalator, cleaning	Daily	Frequency given is for a normal standard of cleaning under average conditions. Approximately 40% of escalator treads are exposed when escalator is stopped.
Floors, mopping		Frequency given is for a normal standard of cleaning under average conditions.
Main corridors	Daily	
Secondary	Weekly	
Floors, sweeping	Daily	Frequency given is for a normal standard of cleaning under average conditions.
Floors, waxing and polishing		Frequency given is for a normal standard of cleaning under average conditions.
Office area	Every 9 weeks	
Open area	Every 9 weeks	
Flower beds		Frequency depends on geographic location.
Fall clean-up	Every year	
Fertilize	2 times per year	
Mulch	Every year	
Police-up	30 times per year	
Weed		
With mulch	15 times per year	
No mulch	25 times per year	
Lawn, mowing	30 times per year	Frequency depends on geographic location.
Fertilize	2 times per year	
Sweep	3 times per year	
Weed control	2 times per year	
Edge-trim		
Walks	30 times per year	
Shrub	10 times per year	
Light fixtures		Frequency depends on occupancy in the building.
Dusting	Every month	
Washing	Every 3 months	
Shrub Areas		Frequency depends on geographic location.
Fertilize	Every year	
Mulch	2 times per year	
Police-up	30 times per year	
Prune	5 times per year	
Weed	10 times per year	
Stairway		Frequency is dependent upon weather conditions, type of building, type of occupancy, and amount of traffic.
Sweeping and dusting	Daily	
Mopping or scrubbing	Weekly	
Toilet, cleaning	Daily	Includes toilet cleaning, collect waste, cleaning wash basins, urinals, water closets, partitions, walls and floors.
Trash collection	Daily	
Trees		Frequency depends on geographic location.
Fertilize	Every year	
Prune	2 times per year	
Pest control		
Spray	3 times per year	
Systemic	Every year	
Urn and Jardiniere cleaning	Daily	Includes the removal of refuse and debris, cleaning and polishing.
Walks, sweeping	30 times per year	
Walls, periodic cleaning	Every 6 months	Damp wipe, spot removal.
Windows, washing	Every month	Frequency given is for a normal standard of cleaning under average conditions.

R019313-20 Facilities Maintenance Labor-Hours

This section lists minimum and maximum cleaning times per unit.
For more information, see *Means Facilities Maintenance Standards* book.

	Unit	
	Minimum	Maximum
Blast Cleaning		
White-metal	100 S.F./Hr	
Near-white	175 S.F./Hr.	
Commercial	370 S.F./Hr.	
Brush-off	870 S.F./Hr.	
Paint Application		
Brushing	125 S.F./Hr.	
Rolling	125 S.F./Hr	
Spraying	500 S.F./Hr.	
Plaster Cleaning		
Wall dusting	2 sec./S.F.	3 sec./S.F.
Vacuuming	4 sec./S.F.	5 sec./S.F.
Spot washing	125 S.F./Hr.	175 S.F./Hr.
Thorough cleaning	275 S.F./Hr.	
Plaster Repair		
Gypsum and lime repair	5 S.Y./Hr.	10 S.Y./Hr.
Ceramic Tile Repair		
General	7 S.F./Hr.	10 S.F./Hr.
Adhesive tile setting	9S.F./Hr	12 S.F./Hr.
Pointing tile joints	10 S.F./Hr.	15 S.F./Hr.
Floor Cleaning		
Manual sweeping	10 min./1000 S.F.	25 min./1000 S.F.
Dust mopping	5 min./1000 S.F.	20 min./1000 S.F.
Buffing	15 min./1000 S.F.	40 min./1000 S.F.
Spray buffin	20 min./1000 S.F.	50 min./1000 S.F.
Damp mopping	15 min./1000 S.F.	30 min./1000 S.F.
Wet mopping	30 min./1000 S.F.	50 min./1000 S.F.
Scrubbing	50 min./1000 S.F.	140 min./1000 S.F.
Scrubbing using electric floor machine		
General	15 min./1000 S.F.	30 min./1000 S.F.
Stripping	100 min./1000 S.F.	200 min./1000 S.F.
Waxing and Buffing using power maching		
Rewaxing	15 min./1000 S.F.	30 min./1000 S.F.
Stripping and rewaxing (two coats)	100 min./1000 S.F.	300 min./1000 S.F.
Waxing and buffing (one coat)	30 min./1000 S.F.	70 min./1000 S.F.
Carpets		
Dry vacuuming	15 min./1000 S.F.	40 min./1000 S.F.
Wet vacuuming	30 min./1000 S.F.	50 min./1000 S.F.
Carpet mopping	20 min./1000 S.F.	40 min./1000 S.F.
Shampooing	175 min./1000 S.F.	250 min./1000 S.F.
Resilient Floor Repair		
Grinding	50 S.F./Hr.	80 S.F./Hr.
Floor Replacement		
Removal (by hand)		
Tiles	100 S.F./Hr.	130 S.F./Hr.
Sheet	120 S.F./Hr	160 S.F./Hr.
Hardwood	40 S.F./Hr.	60 S.F./Hr.
Replacement		
Ceramic	10 S.F./Hr.	20 S.F./Hr.
Resilient	40 S.F./Hr.	70 S.F./Hr.
Hardwood	25 S.F./Hr.	35 S.F./Hr.
Add for related items:		
Replace wood subfloor	80 S.F./Hr.	100 S.F./Hr.
Replace underlayment	75 S.F./Hr.	90 S.F./Hr.
Replace floor moulding	10 S.F./Hr.	30 S.F./Hr.

Unit		
	Minimum	Maximum
Seamless Floor Repair	5 S.F./Hr.	15 S.F./Hr.
Wood Floors		
Sanding	40 S.F./Hr.	60 S.F./Hr.
Sealing	200 S.F/Hr.	300 S.F./Hr.
Waxing*		
Wood Floor Repair		
Loose boards or tiles	50 S.F./Hr.	250S.F./Hr.
Wood strip floor replacement	30 S.F./Hr.	60 S.F./Hr.
Window Washing	300 S.F./Hr.	450 S.F./Hr.
Venetian Blinds, cleaning	15 min./set of blinds	

This section lists average cleaning times per each, or square foot	
	Average
Floor Operations	
Sweeping	
Halls and Corridors	15min./1000 S.F.
General Rooms	30 min./100 S.F.
Dust Mop (unobstructed)	10 min./1000 S.F.
Dust Mop (obstructed)	15 min./1000 S.F.
Damp Mop (unobstructed)	20 min./1000 S.F.
Damp Mop (obstructed)	40 min./1000 S.F.
Wet Mop and Rinse	100 min./1000 S.F.
Hand Scrubbing 12″ brush	300 min./1000 S.F.
Deck Scrubbing	100 min./1000 S.F.
Machine Scrubbing	
12″ diameter	50 min./1000 S.F.
14″diameter	40 min./1000 S.F.
16″ diameter	35 min./1000 S.F.
18″ diameter	31 min./1000 S.F.
19″ diameter	28 min./1000 S.F.
21″ diameter	25 min./1000 S.F.
23″ diameter	23 min./1000 S.F.
24″ diameter	20 min./1000 S.F.
32″ diameter	18 min./1000 S.F.
36″ diameter	15 min./1000 S.F.
Automatic Scrub Machine (24″)	5 min./1000 S.F.
Vacuum (unobstructed)	20 min./1000 S.F.
Vacuum (obstructed)	30 min./1000 S.F.
Waxing	30 min./1000 S.F.
Machine Polish (19″ machine)	15 min./1000 S.F.
Rectangular Machine (48″ plate)	5 min./1000 S.F.
Buff with steel wool	20 min./1000 S.F.
Strip and Rewax	150 min./1000 S.F.
Dry Strip and Rewax	120 min./1000 S.F.
Spray Buffing (unobstructed)	30 min./1000 S.F.
Spray Buffing (obstructed)	45 min./1000 S.F.
Carpeting	
Vacuuming (unobstructed)	20 min./1000 S.F.
Vacuuming (obstructed)	30 min./1000 S.F.
Spot Vacuuming	15 min./1000 S.F.
Shampoo (dry foam)	60 min. 1000 S.F.
Pile Lift	30 min 1000 S.F.
Lockers	.20 min
Radiators	.30 min.
Tables (medium)	.50 min.
Telephones	.15 min.
Towel dispensers	.12 min.

*See waxing and buffing using electric power machine.

R019313-20 Facilities Maintenance Labor-Hours (cont.)

	Average
Towel Disposal Cans	.40 min.
Typewriter and Stand	.50 min
Wash Basin (office)	.60 min
Waste Basin (office)	.50 min.
Window Sill	.20 min.
Venetian Blinds, std. size	3.50 min.
Washrooms	
Cleaning commode	4 min.
Door (spot wash both sides)	1 min.
Mirrors	1 min.
Sanitary napkin dispenser	.50 min.
Urinals	3 min.
Wash basin-soap dispenser	3 min.
General cleaning	120 min./1000 S.F.
Wall Washing	
Painted walls (manual)	240 min./1000 S.F.
Painted walls (machine)	150 min./1000 S.F.
Marble walls (manual)	90 min./1000 S.F.
Ceiling Washing	
Ceiling washing (manual)	300 min./1000 S.F.
Ceiling washing (machine)	180 min./1000 S.F.
Window Washing	
Single pane	125 min./1000 S.F.
Multi-pane	170 min./1000 S.F.
Frosted single pane	190 min./1000 S.F.
Opaque glass	50 min./1000 S.F.
Plate glass	35 min./1000 S.F.
Office partitions (glass)	110 min./1000 S.F.
Dusting Lamps and Lighting Fixtures	
Wall fluorescent fixtures	.15 min.
Desk fluorescent lamp	.30 min.
Table lamp and shade	.60 min.
Floor lamp and shade	.60 min.
Washing Fluorescent Light Fixtures	
Ceiling fixtures (egg crate) 4′ ea.	9 min.
Ceiling fixtures (egg crate) 8′ ea.	12 min.
Dusting	
Air conditioners	.30 min.
Ash trays (desk)	.25 min.
Book cases (3-tier set)	.30 min.
Chairs	.30 min.
Cigarette stands	.40 min.

	Average
Couch	.25 min.
Desks	.80 min.
Desk Trays	.15 min.
File cabinets (4 drawer)	.40 min.
Fabric Upholstery Cleaning	
Whisk or vacuum armless chair	.50 min.
Armchair	1 min.
Couch	2 min.
Shampooing armless chair	4 min.
Armchair	7 min.
Couch	20 min.
Stairway Cleaning	
Sweep and dust, 1 flight, 15 steps	6 min.
Damp mop, 1 flight, 15 steps	5 min.
Scrubbing (hand)	20 min.
Carpentry	
Repair door surface closer	1-1/2 Hrs.
Repair concealed door closer	4 Hrs.
Repair door damage at shop	4 Hrs.
Repair and replace screens	1 Hr.
Repair broken glass	1-1/2 Hrs.
Repair and replace ceiling tile	2 Hrs.
Repair small drywall damage	4 Hrs.
Repair and replace sash balance	4 Hrs.
Change lock on door	1-1/2 Hrs.
Painting	
Repaint 20′ x 15′ room, 1 coat	34 Hrs.
Repaint small bathroom, 1 coat	11-1/2 Hrs.
Repaint exterior window	2-1/2 Hrs.
Plumbing and Steamfitting	
Clear stopped water closet	3 Hrs.
Clear stopped basin	1-1/2 Hrs.
Replace leaking radiator valve	6 Hrs.
Clear external sewer stoppage	9 Hrs.
Install replacement valve, faucet, or trap	10 Hrs.
Electrical	
Replace fluorescent lamp ballast	2 Hrs.
Replace fractional HP motor	10 Hrs.
Replace blown fuse or reset Circuit breaker	1-1/2 Hrs.
Repair exterior light damage	4 Hrs.
Control circuit problems	5 Hrs.

R019313-30 A Review of Major Building Materials–Advantages and Disadvantages

Material	Advantages	Disadvantages
Aluminum	Lightweight High resistance to corrosion Low electrical resistance Good conductor	Softness Limited strength for structural uses Low stiffness High rate of thermal expansion Low rate of fire resistance Relatively high cost
Concrete	High compressive resistance Durable Resistance to moisture, rot, insects, fire and wear Watertight (depending on water-cement ration)	Workability Lack of tensile strength Lack of resistance to many types of chemical exposure, such as salt Hard to remove stains
Copper	High resistance to corrosion Good electrical conductivity Workable Forms its own surface protection Thermal contraction and expansion not high	Costly Strength varies with treatment and mechanical working
Epoxy	Can have varied properties depending on composition Liquid use very applicable Controllable Excellent strength properties Small creep during curing Hard Tough Resistant to abrasion Resistant to corrosion, salts, acids, petroleum products, solvents and other chemicals Adhesion to surfaces good	Color Form
Glass	Considerably strong Non-corrosive	Brittle, subject to shattering under shock Transmits energy at a rapid rate Large sizes expensive
Lead	Good resistance to corrosion	Heavy High coefficient of thermal expansion Difficult to hold in place
Masonry	Available in small units Appearance (available in many textures, sizes and colors) Good insulator	Stains hard to remove Porous (absorption rate high) Shrinkage of mortar Thermal and expansion cracking Others — see concrete
Paper	Readily available Low in cost Used in conjunction with other materials	Susceptible to water damage Susceptible to rot Relatively weak Highly combustible
Plastics — general	Applicable to many uses Relatively low in cost Workable into many shapes High strength Lightweight Non-corrosive Others — see specific plastic entry	Lack of resistance to fire Low stiffness High rate of thermal expansion Low thermal conductivity Some cases of chemical or physical instability with time Non-salvageable Others — see specific plastic entry

R019313-30 A Review of Major Building Materials—Advantages and Disadvantages (cont.)

Material	Advantages	Disadvantages
Plastics — specific		
Acrylics	Transparent Hard Weather resistant Shatter resistant	Easily scratched
Alkyds	Water resistant Touch Good adhesive properties	
Melamines	Hard Durable Abrasive resistant Chemical and heat resistant	
Polyamides (nylon)	Hard Tough Wear resistant	Costly
Polyesters	Weather and chemical resistant Stiff Hard	
Polyethylene	Flexible Tough Translucent Low cost	Easily scratched
Polystyrene	Hard Clear Water and chemical resistant Low cost	Brittle
Vinyls	Tough Wear resistant Stain resistant	
Plywood	Stronger than standard lumber Durable High resistance to impact and load Not as affected by moisture changes as standard wood Workability See Wood for others	Dimensional stability not as good as that of standard structural lumber Poor quality glues are possible Thermal expansion and contraction more of a problem
Steel	Strong Most resistant to aging Most reliable in quality, non-combustible, non-rotting, dimensionally stable with time and moisture change Resistant to staining	Costly Resultant loss of strength when exposed to intense heat, heat, rapid heat gain and loss Corrosive when exposed to moisture and air or other corrosive conditions
Tin	Extremely workable Very resistant to corrosion	Heavy Loss of strength when exposed to heat
Wood	Readily available Relatively low in cost Simple to work with Available in many shapes and forms Good insulating properties	Paintability changes with moisture content Combustible Susceptible to rot and insect infestation Soft and easily damaged (porous) Dimensional changes due to changes in temperature and moisture Strength changes with changing moisture content
Zinc	Fairly high resistance to corrosion Forms its own protective surface Workable	Brittle

R024119-10 Demolition Defined

Whole Building Demolition - Demolition of the whole building with no concern for any particular building element, component, or material type being demolished. This type of demolition is accomplished with large pieces of construction equipment that break up the structure, load it into trucks and haul it to a disposal site, but disposal or dump fees are not included. Demolition of below-grade foundation elements, such as footings, foundation walls, grade beams, slabs on grade, etc., is not included. Certain mechanical equipment containing flammable liquids or ozone-depleting refrigerants, electric lighting elements, communication equipment components, and other building elements may contain hazardous waste, and must be removed, either selectively or carefully, as hazardous waste before the building can be demolished.

Foundation Demolition - Demolition of below-grade foundation footings, foundation walls, grade beams, and slabs on grade. This type of demolition is accomplished by hand or pneumatic hand tools, and does not include saw cutting, or handling, loading, hauling, or disposal of the debris.

Gutting - Removal of building interior finishes and electrical/mechanical systems down to the load-bearing and sub-floor elements of the rough building frame, with no concern for any particular building element, component, or material type being demolished. This type of demolition is accomplished by hand or pneumatic hand tools, and includes loading into trucks, but not hauling, disposal or dump fees, scaffolding, or shoring. Certain mechanical equipment containing flammable liquids or ozone-depleting refrigerants, electric lighting elements, communication equipment components, and other building elements may contain hazardous waste, and must be removed, either selectively or carefully, as hazardous waste, before the building is gutted.

Selective Demolition - Demolition of a selected building element, component, or finish, with some concern for surrounding or adjacent elements, components, or finishes (see the first Subdivision (s) at the beginning of appropriate Divisions). This type of demolition is accomplished by hand or pneumatic hand tools, and does not include handling, loading, storing, hauling, or disposal of the debris, scaffolding, or shoring. "Gutting" methods may be used in order to save time, but damage that is caused to surrounding or adjacent elements, components, or finishes may have to be repaired at a later time.

Careful Removal - Removal of a piece of service equipment, building element or component, or material type, with great concern for both the removed item and surrounding or adjacent elements, components or finishes. The purpose of careful removal may be to protect the removed item for later re-use, preserve a higher salvage value of the removed item, or replace an item while taking care to protect surrounding or adjacent elements, components, connections, or finishes from cosmetic and/or structural damage. An approximation of the time required to perform this type of removal is 1/3 to 1/2 the time it would take to install a new item of like kind (see Reference Numbers R220105-10 and R260105-30). This type of removal is accomplished by hand or pneumatic hand tools, and does not include loading, hauling, or storing the removed item, scaffolding, shoring, or lifting equipment.

Cutout Demolition - Demolition of a small quantity of floor, wall, roof, or other assembly, with concern for the appearance and structural integrity of the surrounding materials. This type of demolition is accomplished by hand or pneumatic hand tools, and does not include saw cutting, handling, loading, hauling, or disposal of debris, scaffolding, or shoring.

Rubbish Handling - Work activities that involve handling, loading or hauling of debris. Generally, the cost of rubbish handling must be added to the cost of all types of demolition, with the exception of whole building demolition.

Minor Site Demolition - Demolition of site elements outside the footprint of a building. This type of demolition is accomplished by hand or pneumatic hand tools, or with larger pieces of construction equipment, and may include loading a removed item onto a truck (check the Crew for equipment used). It does not include saw cutting, hauling or disposal of debris, and, sometimes, handling or loading.

Existing Conditions | R0265 | Underground Storage Tank Removal

R026510-20 Underground Storage Tank Removal

Underground Storage Tank Removal can be divided into two categories: Non-Leaking and Leaking. Prior to removing an underground storage tank, tests should be made, with the proper authorities present, to determine whether a tank has been leaking or the surrounding soil has been contaminated.

To safely remove Liquid Underground Storage Tanks:

1. Excavate to the top of the tank.
2. Disconnect all piping.
3. Open all tank vents and access ports.
4. Remove all liquids and/or sludge.
5. Purge the tank with an inert gas.
6. Provide access to the inside of the tank and clean out the interior using proper personal protective equipment (PPE).
7. Excavate soil surrounding the tank using proper PPE for on-site personnel.
8. Pull and properly dispose of the tank.
9. Clean up the site of all contaminated material.
10. Install new tanks or close the excavation.

Existing Conditions | R0282 | Asbestos Remediation

R028213-20 Asbestos Removal Process

Asbestos removal is accomplished by a specialty contractor who understands the federal and state regulations regarding the handling and disposal of the material. The process of asbestos removal is divided into many individual steps. An accurate estimate can be calculated only after all the steps have been priced.

The steps are generally as follows:

1. Obtain an asbestos abatement plan from an industrial hygienist.
2. Monitor the air quality in and around the removal area and along the path of travel between the removal area and transport area. This establishes the background contamination.
3. Construct a two part decontamination chamber at entrance to removal area.
4. Install a HEPA filter to create a negative pressure in the removal area.
5. Install wall, floor and ceiling protection as required by the plan, usually 2 layers of fireproof 6 mil polyethylene.
6. Industrial hygienist visually inspects work area to verify compliance with plan.
7. Provide temporary supports for conduit and piping affected by the removal process.
8. Proceed with asbestos removal and bagging process. Monitor air quality as described in Step #2. Discontinue operations when contaminate levels exceed applicable standards.
9. Document the legal disposal of materials in accordance with EPA standards.
10. Thoroughly clean removal area including all ledges, crevices and surfaces.
11. Post abatement inspection by industrial hygienist to verify plan compliance.
12. Provide a certificate from a licensed industrial hygienist attesting that contaminate levels are within acceptable standards before returning area to regular use.

Existing Conditions | R0283 | Lead Remediation

R028319-60 Lead Paint Remediation Methods

Lead paint remediation can be accomplished by the following methods.

1. Abrasive blast
2. Chemical stripping
3. Power tool cleaning with vacuum collection system
4. Encapsulation
5. Remove and replace
6. Enclosure

Each of these methods has strengths and weakness depending on the specific circumstances of the project. The following is an overview of each method.

1. **Abrasive blasting** is usually accomplished with sand or recyclable metallic blast. Before work can begin, the area must be contained to ensure the blast material with lead does not escape to the atmosphere. The use of vacuum blast greatly reduces the containment requirements. Lead abatement equipment that may be associated with this work includes a negative air machine. In addition, it is necessary to have an industrial hygienist monitor the project on a continual basis. When the work is complete, the spent blast sand with lead must be disposed of as a hazardous material. If metallic shot was used, the lead is separated from the shot and disposed of as hazardous material. Worker protection includes disposable clothing and respiratory protection.
2. **Chemical stripping** requires strong chemicals be applied to the surface to remove the lead paint. Before the work can begin, the area under/adjacent to the work area must be covered to catch the chemical and removed lead. After the chemical is applied to the painted surface it is usually covered with paper. The chemical is left in place for the specified period, then the paper with lead paint is pulled or scraped off. The process may require several chemical applications. The paper with chemicals and lead paint adhered to it, plus the containment and loose scrapings collected by a HEPA (High Efficiency Particulate Air Filter) vac, must be disposed of as a hazardous material. The chemical stripping process usually requires a neutralizing agent and several wash downs after the paint is removed. Worker protection includes a neoprene or other compatible protective clothing and respiratory protection with face shield. An industrial hygienist is required intermittently during the process.
3. **Power tool cleaning** is accomplished using shrouded needle blasting guns. The shrouding with different end configurations is held up against the surface to be cleaned. The area is blasted with hardened needles and the shroud captures the lead with a HEPA vac and deposits it in a holding tank. An industrial hygienist monitors the project, protective clothing and a respirator is required until air samples prove otherwise. When the work is complete the lead must be disposed of as a hazardous material.
4. **Encapsulation** is a method that leaves the well bonded lead paint in place after the peeling paint has been removed. Before the work can begin, the area under/adjacent to the work must be covered to catch the scrapings. The scraped surface is then washed with a detergent and rinsed. The prepared surface is covered with approximately 10 mils of paint. A reinforcing fabric can also be embedded in the paint covering. The scraped paint and containment must be disposed of as a hazardous material. Workers must wear protective clothing and respirators.
5. **Remove and replace** is an effective way to remove lead paint from windows, gypsum walls and concrete masonry surfaces. The painted materials are removed and new materials are installed. Workers should wear a respirator and tyvek suit. The demolished materials must be disposed of as hazardous waste if it fails the TCLP (Toxicity Characteristic Leachate Process) test.
6. **Enclosure** is the process that permanently seals lead painted materials in place. This process has many applications such as covering lead painted drywall with new drywall, covering exterior construction with tyvek paper then residing, or covering lead painted structural members with aluminum or plastic. The seams on all enclosing materials must be securely sealed. An industrial hygienist monitors the project, and protective clothing and a respirator is required until air samples prove otherwise.

All the processes require clearance monitoring and wipe testing as required by the hygienist.

R031113-10 Wall Form Materials

Aluminum Forms

Approximate weight is 3 lbs. per S.F.C.A. Standard widths are available from 4" to 36" with 36" most common. Standard lengths of 2', 4', 6' to 8' are available. Forms are lightweight and fewer ties are needed with the wider widths. The form face is either smooth or textured.

Metal Framed Plywood Forms

Manufacturers claim over 75 reuses of plywood and over 300 reuses of steel frames. Many specials such as corners, fillers, pilasters, etc. are available. Monthly rental is generally about 15% of purchase price for first month and 9% per month thereafter with 90% of rental applied to purchase for the first month and decreasing percentages thereafter. Aluminum framed forms cost 25% to 30% more than steel framed.

After the first month, extra days may be prorated from the monthly charge. Rental rates do not include ties, accessories, cleaning, loss of hardware or freight in and out. Approximate weight is 5 lbs. per S.F. for steel; 3 lbs. per S.F. for aluminum.

Forms can be rented with option to buy.

Plywood Forms, Job Fabricated

There are two types of plywood used for concrete forms.

1. Exterior plyform which is completely waterproof. This is face oiled to facilitate stripping. Ten reuses can be expected with this type with 25 reuses possible.
2. An overlaid type consists of a resin fiber fused to exterior plyform. No oiling is required except to facilitate cleaning. This is available in both high density (HDO) and medium density overlaid (MDO). Using HDO, 50 reuses can be expected with 200 possible.

Plyform is available in 5/8" and 3/4" thickness. High density overlaid is available in 3/8", 1/2", 5/8" and 3/4" thickness.

5/8" thick is sufficient for most building forms, while 3/4" is best on heavy construction.

Plywood Forms, Modular, Prefabricated

There are many plywood forming systems without frames. Most of these are manufactured from 1-1/8" (HDO) plywood and have some hardware attached. These are used principally for foundation walls 8' or less high. With care and maintenance, 100 reuses can be attained with decreasing quality of surface finish.

Steel Forms

Approximate weight is 6-1/2 lbs. per S.F.C.A. including accessories. Standard widths are available from 2" to 24", with 24" most common. Standard lengths are from 2' to 8', with 4' the most common. Forms are easily ganged into modular units.

Forms are usually leased for 15% of the purchase price per month prorated daily over 30 days.

Rental may be applied to sale price, and usually rental forms are bought. With careful handling and cleaning 200 to 400 reuses are possible.

Straight wall gang forms up to 12' x 20' or 8' x 30' can be fabricated. These crane handled forms usually lease for approx. 9% per month.

Individual job analysis is available from the manufacturer at no charge.

R031113-30 Slipforms

The slipform method of forming may be used for forming circular silo and multi-celled storage bin type structures over 30' high, and building core shear walls over eight stories high. The shear walls, usually enclose elevator shafts, stairwells, mechanical spaces, and toilet rooms. Reuse of the form on duplicate structures will reduce the height necessary and spread the cost of building the form. Slipform systems can be used to cast chimneys, towers, piers, dams, underground shafts or other structures capable of being extruded.

Slipforms are usually 4' high and are raised semi-continuously by jacks climbing on rods which are embedded in the concrete. The jacks are powered by a hydraulic, pneumatic, or electric source and are available in 3, 6, and 22 ton capacities. Interior work decks and exterior scaffolds must be provided for placing inserts, embedded items, reinforcing steel, and concrete. Scaffolds below the form for finishers may be required. The interior work decks are often used as roof slab forms on silos and bin work. Form raising rates will range from 6" to 20" per hour for silos; 6" to 30" per hour for buildings; and 6" to 48" per hour for shaft work.

Reinforcing bars and stressing strands are usually hoisted by crane or gin pole, and the concrete material can be hoisted by crane, winch-powered skip, or pumps. The slipform system is operated on a continuous 24-hour day when a monolithic structure is desired. For least cost, the system is operated only during normal working hours.

Placing concrete will range from 0.5 to 1.5 labor-hours per C.Y. Bucks, blockouts, keyways, weldplates, etc. are extra.

R031113-40 Forms for Reinforced Concrete

Design Economy

Avoid many sizes in proportioning beams and columns.

From story to story avoid changing column dimensions. Gain strength by adding steel or using a richer mix. If a change in size of column is necessary, vary one dimension only to minimize form alterations. Keep beams and columns the same width.

From floor to floor in a multi-story building vary beam depth, not width, as that will leave slab panel form unchanged. It is cheaper to vary the strength of a beam from floor to floor by means of steel area than by 2″ changes in either width or depth.

Cost Factors

Material includes the cost of lumber, cost of rent for metal pans or forms if used, nails, form ties, form oil, bolts and accessories.

Labor includes the cost of carpenters to make up, erect, remove and repair, plus common labor to clean and move. Having carpenters remove forms minimizes repairs.

Improper alignment and condition of forms will increase finishing cost. When forms are heavily oiled, concrete surfaces must be neutralized before finishing. Special curing compounds will cause spillages to spall off in first frost. Gang forming methods will reduce costs on large projects.

Materials Used

Boards are seldom used unless their architectural finish is required. Generally, steel, fiberglass and plywood are used for contact surfaces. Labor on plywood is 10% less than with boards. The plywood is backed up with 2 x 4′s at 12″ to 32″ O.C. Walers are generally 2 - 2 x 4′s. Column forms are held together with steel yokes or bands. Shoring is with adjustable shoring or scaffolding for high ceilings.

Reuse

Floor and column forms can be reused four or possibly five times without excessive repair. Remember to allow for 10% waste on each reuse.

When modular sized wall forms are made, up to twenty uses can be expected with exterior plyform.

When forms are reused, the cost to erect, strip, clean and move will not be affected. 10% replacement of lumber should be included and about one hour of carpenter time for repairs on each reuse per 100 S.F.

The reuse cost for certain accessory items normally rented on a monthly basis will be lower than the cost for the first use.

After fifth use, new material required plus time needed for repair prevent form cost from dropping further and it may go up. Much depends on care in stripping, the number of special bays, changes in beam or column sizes and other factors.

Costs for multiple use of formwork may be developed as follows:

2 Uses

$$\frac{\text{(1st Use + Reuse)}}{2} = \text{avg. cost/2 uses}$$

3 Uses

$$\frac{\text{(1st Use + 2 Reuse)}}{3} = \text{avg. cost/3 uses}$$

4 Uses

$$\frac{\text{(1st use + 3 Reuse)}}{4} = \text{avg. cost/4 uses}$$

R031113-60 Formwork Labor-Hours

Item	Hours Required				Total Hours	Multiple Use		
	Unit	Fabricate	Erect & Strip	Clean & Move	1 Use	2 Use	3 Use	4 Use
Beam and Girder, interior beams, 12″ wide	100 S.F.	6.4	8.3	1.3	16.0	13.3	12.4	12.0
Hung from steel beams		5.8	7.7	1.3	14.8	12.4	11.6	11.2
Beam sides only, 36″ high		5.8	7.2	1.3	14.3	11.9	11.1	10.7
Beam bottoms only, 24″ wide		6.6	13.0	1.3	20.9	18.1	17.2	16.7
Box out for openings		9.9	10.0	1.1	21.0	16.6	15.1	14.3
Buttress forms, to 8′ high		6.0	6.5	1.2	13.7	11.2	10.4	10.0
Centering, steel, 3/4″ rib lath			1.0		1.0			
3/8″ rib lath or slab form			0.9		0.9			
Chamfer strip or keyway	100 L.F.		1.5		1.5	1.5	1.5	1.5
Columns, fiber tube 8″ diameter			20.6		20.6			
12″			21.3		21.3			
16″			22.9		22.9			
20″			23.7		23.7			
24″			24.6		24.6			
30″			25.6		25.6			
Round Steel, 12″ diameter			22.0		22.0	22.0	22.0	22.0
16″			25.6		25.6	25.6	25.6	25.6
20″			30.5		30.5	30.5	30.5	30.5
24″			37.7		37.7	37.7	37.7	37.7
Plywood 8″ x 8″	100 S.F.	7.0	11.0	1.2	19.2	16.2	15.2	14.7
12″ x 12″		6.0	10.5	1.2	17.7	15.2	14.4	14.0
16″ x 16″		5.9	10.0	1.2	17.1	14.7	13.8	13.4
24″ x 24″		5.8	9.8	1.2	16.8	14.4	13.6	13.2
Steel framed plywood 8″ x 8″			10.0	1.0	11.0	11.0	11.0	11.0
12″ x 12″			9.3	1.0	10.3	10.3	10.3	10.3
16″ x 16″			8.5	1.0	9.5	9.5	9.5	9.5
24″ x 24″			7.8	1.0	8.8	8.8	8.8	8.8
Drop head forms, plywood		9.0	12.5	1.5	23.0	19.0	17.7	17.0
Coping forms		8.5	15.0	1.5	25.0	21.3	20.0	19.4
Culvert, box			14.5	4.3	18.8	18.8	18.8	18.8
Curb forms, 6″ to 12″ high, on grade		5.0	8.5	1.2	14.7	12.7	12.1	11.7
On elevated slabs		6.0	10.8	1.2	18.0	15.5	14.7	14.3
Edge forms to 6″ high, on grade	100 L.F.	2.0	3.5	0.6	6.1	5.6	5.4	5.3
7″ to 12″ high	100 S.F.	2.5	5.0	1.0	8.5	7.8	7.5	7.4
Equipment foundations		10.0	18.0	2.0	30.0	25.5	24.0	23.3
Flat slabs, including drops		3.5	6.0	1.2	10.7	9.5	9.0	8.8
Hung from steel		3.0	5.5	1.2	9.7	8.7	8.4	8.2
Closed deck for domes		3.0	5.8	1.2	10.0	9.0	8.7	8.5
Open deck for pans		2.2	5.3	1.0	8.5	7.9	7.7	7.6
Footings, continuous, 12″ high		3.5	3.5	1.5	8.5	7.3	6.8	6.6
Spread, 12″ high		4.7	4.2	1.6	10.5	8.7	8.0	7.7
Pile caps, square or rectangular		4.5	5.0	1.5	11.0	9.3	8.7	8.4
Grade beams, 24″ deep		2.5	5.3	1.2	9.0	8.3	8.0	7.9
Lintel or Sill forms		8.0	17.0	2.0	27.0	23.5	22.3	21.8
Spandrel beams, 12″ wide		9.0	11.2	1.3	21.5	17.5	16.2	15.5
Stairs			25.0	4.0	29.0	29.0	29.0	29.0
Trench forms in floor		4.5	14.0	1.5	20.0	18.3	17.7	17.4
Walls, Plywood, at grade, to 8′ high		5.0	6.5	1.5	13.0	11.0	9.7	9.5
8′ to 16′		7.5	8.0	1.5	17.0	13.8	12.7	12.1
16′ to 20′		9.0	10.0	1.5	20.5	16.5	15.2	14.5
Foundation walls, to 8′ high		4.5	6.5	1.0	12.0	10.3	9.7	9.4
8′ to 16′ high		5.5	7.5	1.0	14.0	11.8	11.0	10.6
Retaining wall to 12′ high, battered		6.0	8.5	1.5	16.0	13.5	12.7	12.3
Radial walls to 12′ high, smooth		8.0	9.5	2.0	19.5	16.0	14.8	14.3
2′ chords		7.0	8.0	1.5	16.5	13.5	12.5	12.0
Prefabricated modular, to 8′ high		—	4.3	1.0	5.3	5.3	5.3	5.3
Steel, to 8′ high		—	6.8	1.2	8.0	8.0	8.0	8.0
8′ to 16′ high		—	9.1	1.5	10.6	10.3	10.2	10.2
Steel framed plywood to 8′ high		—	6.8	1.2	8.0	7.5	7.3	7.2
8′ to 16′ high		—	9.3	1.2	10.5	9.5	9.2	9.0

R032110-10 Reinforcing Steel Weights and Measures

Bar Designation No.**	Nominal Weight Lb./Ft.	U.S. Customary Units			SI Units			
		Nominal Dimensions*			Nominal Dimensions*			
		Diameter in.	Cross Sectional Area, in.^{2}	Perimeter in.	Nominal Weight kg/m	Diameter mm	Cross Sectional Area, cm^2	Perimeter mm
3	.376	.375	.11	1.178	.560	9.52	.71	29.9
4	.668	.500	.20	1.571	.994	12.70	1.29	39.9
5	1.043	.625	.31	1.963	1.552	15.88	2.00	49.9
6	1.502	.750	.44	2.356	2.235	19.05	2.84	59.8
7	2.044	.875	.60	2.749	3.042	22.22	3.87	69.8
8	2.670	1.000	.79	3.142	3.973	25.40	5.10	79.8
9	3.400	1.128	1.00	3.544	5.059	28.65	6.45	90.0
10	4.303	1.270	1.27	3.990	6.403	32.26	8.19	101.4
11	5.313	1.410	1.56	4.430	7.906	35.81	10.06	112.5
14	7.650	1.693	2.25	5.320	11.384	43.00	14.52	135.1
18	13.600	2.257	4.00	7.090	20.238	57.33	25.81	180.1

* The nominal dimensions of a deformed bar are equivalent to those of a plain round bar having the same weight per foot as the deformed bar.
** Bar numbers are based on the number of eighths of an inch included in the nominal diameter of the bars.

R032110-20 Metric Rebar Specification - ASTM A615-81

Grade 300 (300 MPa* = 43,560 psi; +8.7% vs. Grade 40)				
Grade 400 (400 MPa* = 58,000 psi; −3.4% vs. Grade 60)				
Bar No.	Diameter mm	Area mm^2	Equivalent in.^{2}	Comparison with U.S. Customary Bars
10M	11.3	100	.16	Between #3 & #4
15M	16.0	200	.31	#5 (.31 in.^{2})
20M	19.5	300	.47	#6 (.44 in.^{2})
25M	25.2	500	.78	#8 (.79 in.^{2})
30M	29.9	700	1.09	#9 (1.00 in.^{2})
35M	35.7	1000	1.55	#11 (1.56 in.^{2})
45M	43.7	1500	2.33	#14 (2.25 in.^{2})
55M	56.4	2500	3.88	#18 (4.00 in.^{2})

* MPa = megapascals

R032110-25 Comparison of U.S. Customary Units and SI Units for Reinforcing Bars

U.S. Customary Units							
		Nominal Dimensions[a]			Deformation Requirements, in.		
Bar Designation No.[b]	Nominal Weight, lb/ft	Diameter in.	Cross Sectional Area, in.^{2}	Perimeter in.	Maximum Average Spacing	Minimum Average Height	Maximum Gap (Chord of 12-1/2% of Nominal Perimeter)
3	0.376	0.375	0.11	1.178	0.262	0.015	0.143
4	0.668	0.500	0.20	1.571	0.350	0.020	0.191
5	1.043	0.625	0.31	1.963	0.437	0.028	0.239
6	1.502	0.750	0.44	2.356	0.525	0.038	0.286
7	2.044	0.875	0.60	2.749	0.612	0.044	0.334
8	2.670	1.000	0.79	3.142	0.700	0.050	0.383
9	3.400	1.128	1.00	3.544	0.790	0.056	0.431
10	4.303	1.270	1.27	3.990	0.889	0.064	0.487
11	5.313	1.410	1.56	4.430	0.987	0.071	0.540
14	7.65	1.693	2.25	5.32	1.185	0.085	0.648
18	13.60	2.257	4.00	7.09	1.58	0.102	0.864
SI UNITS							
		Nominal Dimensions[a]			Deformation Requirements, mm		
Bar Designation No.[b]	Nominal Weight kg/m	Diameter, mm	Cross Sectional Area, cm^2	Perimeter, mm	Maximum Average Spacing	Minimum Average Height	Maximum Gap (Chord of 12-1/2% of Nominal Perimeter)
3	0.560	9.52	0.71	29.9	6.7	0.38	3.5
4	0.994	12.70	1.29	39.9	8.9	0.51	4.9
5	1.552	15.88	2.00	49.9	11.1	0.71	6.1
6	2.235	19.05	2.84	59.8	13.3	0.96	7.3
7	3.042	22.22	3.87	69.8	15.5	1.11	8.5
8	3.973	25.40	5.10	79.8	17.8	1.27	9.7
9	5.059	28.65	6.45	90.0	20.1	1.42	10.9
10	6.403	32.26	8.19	101.4	22.6	1.62	11.4
11	7.906	35.81	10.06	112.5	25.1	1.80	13.6
14	11.384	43.00	14.52	135.1	30.1	2.16	16.5
18	20.238	57.33	25.81	180.1	40.1	2.59	21.9

[a]Nominal dimensions of a deformed bar are equivalent to those of a plain round bar having the same weight per foot as the deformed bar.

[b]Bar numbers are based on the number of eighths of an inch included in the nominal diameter of the bars.

R032110-40 Weight of Steel Reinforcing Per Square Foot of Wall (PSF)

Reinforced Weights: The table below suggests the weights per square foot for reinforcing steel in walls. Weights are approximate and will be the same for all grades of steel bars. For bars in two directions, add weights for each size and spacing.

C/C Spacing in Inches	Bar Size								
	#3 Wt. (PSF)	#4 Wt. (PSF)	#5 Wt. (PSF)	#6 Wt. (PSF)	#7 Wt. (PSF)	#8 Wt. (PSF)	#9 Wt. (PSF)	#10 Wt. (PSF)	#11 Wt. (PSF)
2″	2.26	4.01	6.26	9.01	12.27				
3″	1.50	2.67	4.17	6.01	8.18	10.68	13.60	17.21	21.25
4″	1.13	2.01	3.13	4.51	6.13	8.10	10.20	12.91	15.94
5″	.90	1.60	2.50	3.60	4.91	6.41	8.16	10.33	12.75
6″	.752	1.34	2.09	3.00	4.09	5.34	6.80	8.61	10.63
8″	.564	1.00	1.57	2.25	3.07	4.01	5.10	6.46	7.97
10″	.451	.802	1.25	1.80	2.45	3.20	4.08	5.16	6.38
12″	.376	.668	1.04	1.50	2.04	2.67	3.40	4.30	5.31
18″	.251	.445	.695	1.00	1.32	1.78	2.27	2.86	3.54
24″	.188	.334	.522	.751	1.02	1.34	1.70	2.15	2.66
30″	.150	.267	.417	.600	.817	1.07	1.36	1.72	2.13
36″	.125	.223	.348	.501	.681	.890	1.13	1.43	1.77
42″	.107	.191	.298	.429	.584	.753	.97	1.17	1.52
48″	.094	.167	.261	.376	.511	.668	.85	1.08	1.33

R032110-50 Minimum Wall Reinforcement Weight (PSF)

This table lists the approximate minimum wall reinforcement weights per S.F. according to the specification of .12% of gross area for vertical bars and .20% of gross area for horizontal bars.

Location	Wall Thickness	Bar Size	Horizontal Steel Spacing C/C	Sq. In. Req'd per S.F.	Total Wt. per S.F.	Bar Size	Vertical Steel Spacing C/C	Sq. In. Req'd per S.F.	Total Wt. per S.F.	Horizontal & Vertical Steel Total Weight per S.F.
Both Faces	10″	#4	18″	.24	.89#	#3	18″	.14	.50#	1.39#
	12″	#4	16″	.29	1.00	#3	16″	.17	.60	1.60
	14″	#4	14″	.34	1.14	#3	13″	.20	.69	1.84
	16″	#4	12″	.38	1.34	#3	11″	.23	.82	2.16
	18″	#5	17″	.43	1.47	#4	18″	.26	.89	2.36
One Face	6″	#3	9″	.15	.50	#3	18″	.09	.25	.75
	8″	#4	12″	.19	.67	#3	11″	.12	.41	1.08
	10″	#5	15″	.24	.83	#4	16″	.14	.50	1.34

R032110-70 Bend, Place and Tie Reinforcing

Placing and tying by rodmen for footings and slabs runs from nine hrs. per ton for heavy bars to fifteen hrs. per ton for light bars. For beams, columns, and walls, production runs from eight hrs. per ton for heavy bars to twenty hrs. per ton for light bars. Overall average for typical reinforced concrete buildings is about fourteen hrs. per ton. These production figures include the time for placing of accessories and usual inserts, but not their material cost (allow 15% of the cost of delivered bent rods). Equipment handling is necessary for the larger-sized bars so that installation costs for the very heavy bars will not decrease proportionately.

Installation costs for splicing reinforcing bars include allowance for equipment to hold the bars in place while splicing as well as necessary scaffolding for iron workers.

R032110-80 Shop-Fabricated Reinforcing Steel

The material prices for reinforcing, shown in the unit cost sections of the book, are for 50 tons or more of shop-fabricated reinforcing steel and include:

1. Mill base price of reinforcing steel
2. Mill grade/size/length extras
3. Mill delivery to the fabrication shop
4. Shop storage and handling
5. Shop drafting/detailing
6. Shop shearing and bending
7. Shop listing
8. Shop delivery to the job site

Both material and installation costs can be considerably higher for small jobs consisting primarily of smaller bars, while material costs may be slightly lower for larger jobs.

R032205-30 Common Stock Styles of Welded Wire Fabric

This table provides some of the basic specifications, sizes, and weights of welded wire fabric used for reinforcing concrete.

New Designation		Old Designation	Steel Area per Foot				Approximate Weight per 100 S.F.	
Spacing — Cross Sectional Area (in.) — (Sq. in. 100)		Spacing — Wire Gauge (in.) — (AS & W)	Longitudinal		Transverse			
			in.	cm	in.	cm	lbs	kg
Rolls	6 x 6 — W1.4 x W1.4	6 x 6 — 10 x 10	.028	.071	.028	.071	21	9.53
	6 x 6 — W2.0 x W2.0	6 x 6 — 8 x 8 [1]	.040	.102	.040	.102	29	13.15
	6 x 6 — W2.9 x W2.9	6 x 6 — 6 x 6	.058	.147	.058	.147	42	19.05
	6 x 6 — W4.0 x W4.0	6 x 6 — 4 x 4	.080	.203	.080	.203	58	26.91
	4 x 4 — W1.4 x W1.4	4 x 4 — 10 x 10	.042	.107	.042	.107	31	14.06
	4 x 4 — W2.0 x W2.0	4 x 4 — 8 x 8 [1]	.060	.152	.060	.152	43	19.50
	4 x 4 — W2.9 x W2.9	4 x 4 — 6 x 6	.087	.227	.087	.227	62	28.12
	4 x 4 — W4.0 x W4.0	4 x 4 — 4 x 4	.120	.305	.120	.305	85	38.56
Sheets	6 x 6 — W2.9 x W2.9	6 x 6 — 6 x 6	.058	.147	.058	.147	42	19.05
	6 x 6 — W4.0 x W4.0	6 x 6 — 4 x 4	.080	.203	.080	.203	58	26.31
	6 x 6 — W5.5 x W5.5	6 x 6 — 2 x 2 [2]	.110	.279	.110	.279	80	36.29
	4 x 4 — W1.4 x W1.4	4 x 4 — 4 x 4	.120	.305	.120	.305	85	38.56

NOTES: 1. Exact W—number size for 8 gauge is W2.1
2. Exact W—number size for 2 gauge is W5.4

R033053-10 Spread Footings

General: A spread footing is used to convert a concentrated load (from one superstructure column, or substructure grade beams) into an allowable area load on supporting soil.

Because of punching action from the column load, a spread footing is usually thicker than strip footings which support wall loads. One or two story commercial or residential buildings should have no less than 1′ thick spread footings. Heavier loads require no less than 2′ thick. Spread footings may be square, rectangular or octagonal in plan.

Spread footings tend to minimize excavation and foundation materials, as well as labor and equipment. Another advantage is that footings and soil conditions can be readily examined. They are the most widely used type of footing, especially in mild climates and for buildings of four stories or under. This is because they are usually more economical than other types, if suitable soil and site conditions exist.

They are used when suitable supporting soil is located within several feet of the surface or line of subsurface excavation. Suitable soil types include sands and gravels, gravels with a small amount of clay or silt, hardpan, chalk, and rock. Pedestals may be used to bring the column base load down to the top of footing. Alternately, undesirable soil between underside of footing and top of bearing level can be removed and replaced with lean concrete mix or compacted granular material.

Depth of footing should be below topsoil, uncompacted fill, muck, etc. It must be lower than frost penetration but should be above the water table. It must not be at the ground surface because of potential surface erosion. If the ground slopes, approximately three horizontal feet of edge protection must remain. Differential footing elevations may overlap soil stresses or cause excavation problems if clear spacing between footings is less than the difference in depth.

Other footing types are usually used for the following reasons:

A. Bearing capacity of soil is low.
B. Very large footings are required, at a cost disadvantage.
C. Soil under footing (shallow or deep) is very compressible, with probability of causing excessive or differential settlement.
D. Good bearing soil is deep.
E. Potential for scour action exists.
F. Varying subsoil conditions within building perimeter.

Cost of spread footings for a building is determined by:

1. The soil bearing capacity.
2. Typical bay size.
3. Total load (live plus dead) per S.F. for roof and elevated floor levels.
4. The size and shape of the building.
5. Footing configuration. Does the building utilize outer spread footings or are there continuous perimeter footings only or a combination of spread footings plus continuous footings?

Soil Bearing Capacity in Kips per S.F.

Bearing Material	Typical Allowable Bearing Capacity
Hard sound rock	120 KSF
Medium hard rock	80
Hardpan overlaying rock	24
Compact gravel and boulder-gravel; very compact sandy gravel	20
Soft rock	16
Loose gravel; sandy gravel; compact sand; very compact sand-inorganic silt	12
Hard dry consolidated clay	10
Loose coarse to medium sand; medium compact fine sand	8
Compact sand-clay	6
Loose fine sand; medium compact sand-inorganic silts	4
Firm or stiff clay	3
Loose saturated sand-clay; medium soft clay	2

R033053-50 Industrial Chimneys

Foundation requirements in C.Y. of concrete for various sized chimneys.

Size Chimney	2 Ton Soil	3 Ton Soil	Size Chimney	2 Ton Soil	3 Ton Soil	Size Chimney	2 Ton Soil	3 Ton Soil
75′ x 3′-0″	13 C.Y.	11 C.Y.	160′ x 6′-6″	86 C.Y.	76 C.Y.	300′ x 10′-0″	325 C.Y.	245 C.Y.
85′ x 5′-6″	19	16	175′ x 7′-0″	108	95	350′ x 12′-0″	422	320
100′ x 5′-0″	24	20	200′ x 6′-0″	125	105	400′ x 14′-0″	520	400
125′ x 5′-6″	43	36	250′ x 8′-0″	230	175	500′ x 18′-0″	725	575

R033053-60 Maximum Depth of Frost Penetration in Inches

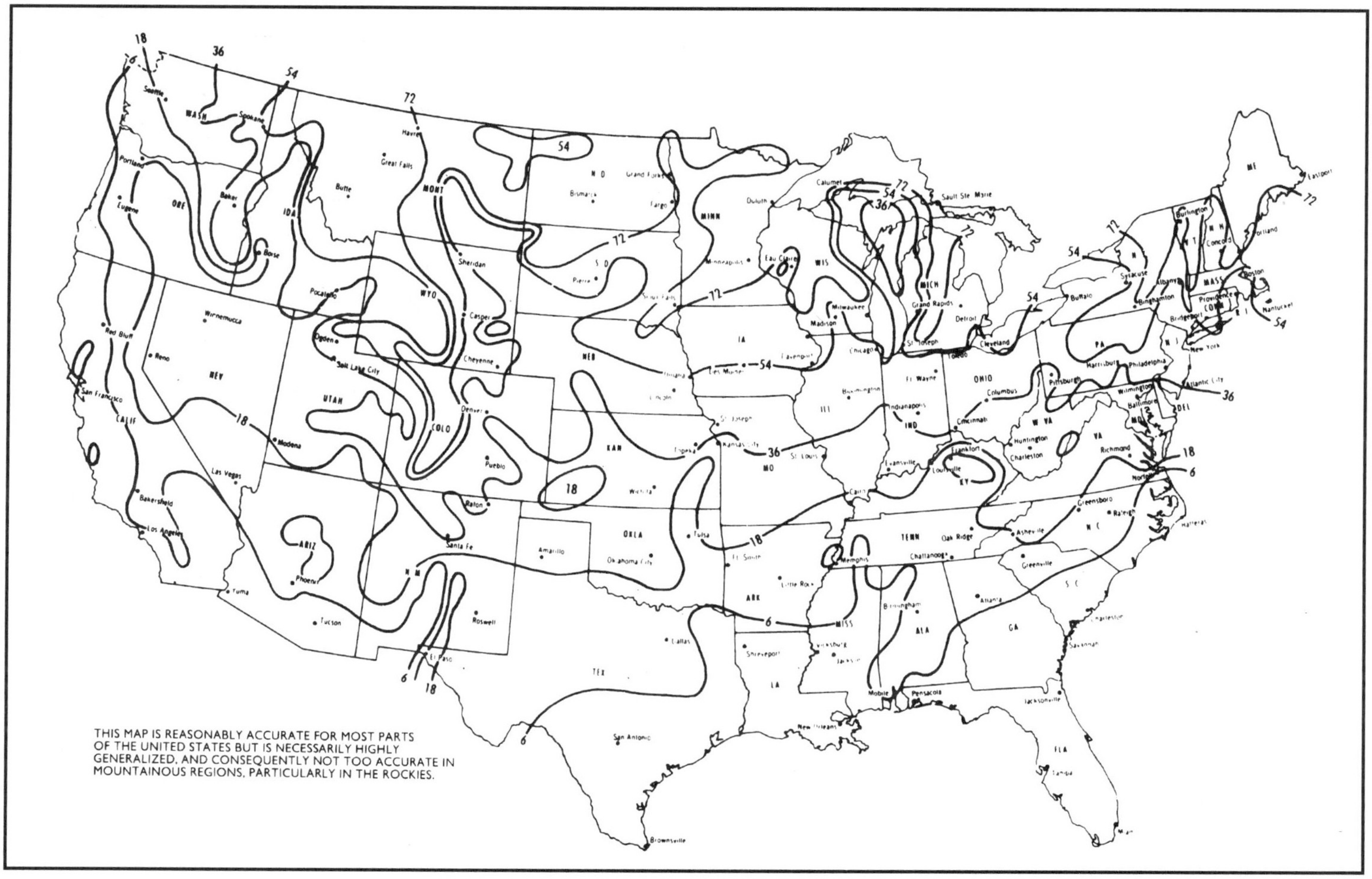

R033105-10 Proportionate Quantities

The tables below show both quantities per S.F. of floor areas as well as form and reinforcing quantities per C.Y. Unusual structural requirements would increase the ratios below. High strength reinforcing would reduce the steel weights. Figures are for 3000 psi concrete and 60,000 psi reinforcing unless specified otherwise.

Type of Construction	Live Load	Span	Per S.F. of Floor Area				Per C.Y. of Concrete		
			Concrete	Forms	Reinf.	Pans	Forms	Reinf.	Pans
Flat Plate	50 psf	15 Ft.	.46 C.F.	1.06 S.F.	1.71lb.		62 S.F.	101 lb.	
		20	.63	1.02	2.40		44	104	
		25	.79	1.02	3.03		35	104	
	100	15	.46	1.04	2.14		61	126	
		20	.71	1.02	2.72		39	104	
		25	.83	1.01	3.47		33	113	
Flat Plate (waffle construction) 20″ domes	50	20	.43	1.00	2.10	.84 S.F.	63	135	53 S.F.
		25	.52	1.00	2.90	.89	52	150	46
		30	.64	1.00	3.70	.87	42	155	37
	100	20	.51	1.00	2.30	.84	53	125	45
		25	.64	1.00	3.20	.83	42	135	35
		30	.76	1.00	4.40	.81	36	160	29
Waffle Construction 30″ domes	50	25	.69	1.06	1.83	.68	42	72	40
		30	.74	1.06	2.39	.69	39	87	39
		35	.86	1.05	2.71	.69	33	85	39
		40	.78	1.00	4.80	.68	35	165	40
Flat Slab (two way with drop panels)	50	20	.62	1.03	2.34		45	102	
		25	.77	1.03	2.99		36	105	
		30	.95	1.03	4.09		29	116	
	100	20	.64	1.03	2.83		43	119	
		25	.79	1.03	3.88		35	133	
		30	.96	1.03	4.66		29	131	
	200	20	.73	1.03	3.03		38	112	
		25	.86	1.03	4.23		32	133	
		30	1.06	1.03	5.30		26	135	
One Way Joists 20″ Pans	50	15	.36	1.04	1.40	.93	78	105	70
		20	.42	1.05	1.80	.94	67	120	60
		25	.47	1.05	2.60	.94	60	150	54
	100	15	.38	1.07	1.90	.93	77	140	66
		20	.44	1.08	2.40	.94	67	150	58
		25	.52	1.07	3.50	.94	55	185	49
One Way Joists 8″ x 16″ filler blocks	50	15	.34	1.06	1.80	.81 Ea.	84	145	64 Ea.
		20	.40	1.08	2.20	.82	73	145	55
		25	.46	1.07	3.20	.83	63	190	49
	100	15	.39	1.07	1.90	.81	74	130	56
		20	.46	1.09	2.80	.82	64	160	48
		25	.53	1.10	3.60	.83	56	190	42
One Way Beam & Slab	50	15	.42	1.30	1.73		84	111	
		20	.51	1.28	2.61		68	138	
		25	.64	1.25	2.78		53	117	
	100	15	.42	1.30	1.90		84	122	
		20	.54	1.35	2.69		68	154	
		25	.69	1.37	3.93		54	145	
	200	15	.44	1.31	2.24		80	137	
		20	.58	1.40	3.30		65	163	
		25	.69	1.42	4.89		53	183	
Two Way Beam & Slab	100	15	.47	1.20	2.26		69	130	
		20	.63	1.29	3.06		55	131	
		25	.83	1.33	3.79		43	123	
	200	15	.49	1.25	2.70		41	149	
		20	.66	1.32	4.04		54	165	
		25	.88	1.32	6.08		41	187	

R033105-10 Proportionate Quantities (cont.)

4000 psi Concrete and 60,000 psi Reinforcing—Form and Reinforcing Quantities per C.Y.					
Item	Size	Forms	Reinforcing	Minimum	Maximum
Columns (square tied)	10″ x 10″	130 S.F.C.A.	#5 to #11	220 lbs.	875 lbs.
	12″ x 12″	108	#6 to #14	200	955
	14″ x 14″	92	#7 to #14	190	900
	16″ x 16″	81	#6 to #14	187	1082
	18″ x 18″	72	#6 to #14	170	906
	20″ x 20″	65	#7 to #18	150	1080
	22″ x 22″	59	#8 to #18	153	902
	24″ x 24″	54	#8 to #18	164	884
	26″ x 26″	50	#9 to #18	169	994
	28″ x 28″	46	#9 to #18	147	864
	30″ x 30″	43	#10 to #18	146	983
	32″ x 32″	40	#10 to #18	175	866
	34″ x 34″	38	#10 to #18	157	772
	36″ x 36″	36	#10 to #18	175	852
	38″ x 38″	34	#10 to #18	158	765
	40″ x 40″	32 ▼	#10 to #18	143	692

Item	Size	Form	Spiral	Reinforcing	Minimum	Maximum
Columns (spirally reinforced)	12″ diameter	34.5 L.F.	190 lbs.	#4 to #11	165 lbs.	1505 lb.
		34.5	190	#14 & #18	—	1100
	14″	25	170	#4 to #11	150	970
		25	170	#14 & #18	800	1000
	16″	19	160	#4 to #11	160	950
		19	160	#14 & #18	605	1080
	18″	15	150	#4 to #11	160	915
		15	150	#14 & #18	480	1075
	20″	12	130	#4 to #11	155	865
		12	130	#14 & #18	385	1020
	22″	10	125	#4 to #11	165	775
		10	125	#14 & #18	320	995
	24″	9	120	#4 to #11	195	800
		9	120	#14 & #18	290	1150
	26″	7.3	100	#4 to #11	200	729
		7.3	100	#14 & #18	235	1035
	28″	6.3	95	#4 to #11	175	700
		6.3	95	#14 & #18	200	1075
	30″	5.5	90	#4 to #11	180	670
		5.5	90	#14 & #18	175	1015
	32″	4.8	85	#4 to #11	185	615
		4.8	85	#14 & #18	155	955
	34″	4.3	80	#4 to #11	180	600
		4.3	80	#14 & #18	170	855
	36″	3.8	75	#4 to #11	165	570
		3.8	75	#14 & #18	155	865
	40″	3.0	70	#4 to #11	165	500
		3.0	70	#14 & #18	145	765

R033105-10 Proportionate Quantities (cont.)

3000 psi Concrete and 60,000 psi Reinforcing—Form and Reinforcing Quantities per C.Y.						
Item	Type	Loading	Height	C.Y./L.F.	Forms/C.Y.	Reinf./C.Y.
Retaining Walls	Cantilever	Level Backfill	4 Ft.	0.2 C.Y.	49 S.F.	35 lbs.
			8	0.5	42	45
			12	0.8	35	70
			16	1.1	32	85
			20	1.6	28	105
		Highway Surcharge	4	0.3	41	35
			8	0.5	36	55
			12	0.8	33	90
			16	1.2	30	120
			20	1.7	27	155
		Railroad Surcharge	4	0.4	28	45
			8	0.8	25	65
			12	1.3	22	90
			16	1.9	20	100
			20	2.6	18	120
	Gravity, with Vertical Face	Level Backfill	4	0.4	37	None
			7	0.6	27	↓
			10	1.2	20	↓
		Sloping Backfill	4	0.3	31	↓
			7	0.8	21	↓
			10	1.6	15	↓

Item	Span	Live Load in Kips per Linear Foot							
		Under 1 Kip		2 to 3 Kips		4 to 5 Kips		6 to 7 Kips	
		Forms	Reinf.	Forms	Reinf.	Forms	Reinf.	Forms	Reinf.
Beams	10 Ft.	—	—	90 S.F.	170 #	85 S.F.	175 #	75 S.F.	185 #
	16	130 S.F.	165 #	85	180	75	180	65	225
	20	110	170	75	185	62	200	51	200
	26	90	170	65	215	62	215	—	—
	30	85	175	60	200	—	—	—	—

Item	Size	Type	Forms per C.Y.	Reinforcing per C.Y.
Spread Footings	Under 1 C.Y.	1,000 psf soil	24 S.F.	44 lbs.
		5,000	24	42
		10,000	24	52
	1 C.Y. to 5 C.Y.	1,000	14	49
		5,000	14	50
		10,000	14	50
	Over 5 C.Y.	1,000	9	54
		5,000	9	52
		10,000	9	56
Pile Caps (30 Ton Concrete Piles)	Under 5 C.Y.	shallow caps	20	65
		medium	20	50
		deep	20	40
	5 C.Y. to 10 C.Y.	shallow	14	55
		medium	15	45
		deep	15	40
	10 C.Y. to 20 C.Y.	shallow	11	60
		medium	11	45
		deep	12	35
	Over 20 C.Y.	shallow	9	60
		medium	9	45
		deep	10	40

R033105-10 Proportionate Quantities (cont.)

3000 psi Concrete and 60,000 psi Reinforcing — Form and Reinforcing Quantities per C.Y.						
Item	Size	Pile Spacing	50 T Pile	100 T Pile	50 T Pile	100 T Pile
Pile Caps (Steel H Piles)	Under 5 C.Y.	24″ O.C.	24 S.F.	24 S.F.	75 lbs.	90 lbs.
		30″	25	25	80	100
		36″	24	24	80	110
	5 C.Y. to 10 C.Y.	24″	15	15	80	110
		30″	15	15	85	110
		36″	15	15	75	90
	Over 10 C.Y.	24″	13	13	85	90
		30″	11	11	85	95
		36″	10	10	85	90

Item		8″ Thick		10″ Thick		12″ Thick		15″ Thick	
	Height	Forms	Reinf.	Forms	Reinf.	Forms	Reinf.	Forms	Reinf.
Basement Walls	7 Ft.	81 S.F.	44 lbs.	65 S.F.	45 lbs.	54 S.F.	44 lbs.	41 S.F.	43 lbs.
	8	↓	44	↓	45	↓	44	↓	43
	9	↓	46	↓	45	↓	44	↓	43
	10	↓	57	↓	45	↓	44	↓	43
	12	↓	83	↓	50	↓	52	↓	43
	14	↓	116	↓	65	↓	64	↓	51
	16	↓		↓	86	↓	90	↓	65
	18	↓		↓		↓	106	↓	70

R033105-20 Materials for One C.Y. of Concrete

This is an approximate method of figuring quantities of cement, sand and coarse aggregate for a field mix with waste allowance included.

With crushed gravel as coarse aggregate, to determine barrels of cement required, divide 10 by total mix; that is, for 1:2:4 mix, 10 divided by 7 = 1-3/7 barrels.

If the coarse aggregate is crushed stone, use 10-1/2 instead of 10 as given for gravel.

To determine tons of sand required, multiply barrels of cement by parts of sand and then by 0.2; that is, for the 1:2:4 mix, as above, 1-3/7 x 2 x .2 = .57 tons.

Tons of crushed gravel are in the same ratio to tons of sand as parts in the mix, or 4/2 x .57 = 1.14 tons.

1 bag cement = 94#
4 bags = 1 barrel

1 C.Y. sand or crushed gravel = 2700#
1 ton sand or crushed gravel = 20 C.F.

1 C.Y. crushed stone = 2575#
1 ton crushed stone = 21 C.F.

Average carload of cement is 692 bags; of sand or gravel is 56 tons.

Do not stack stored cement over 10 bags high.

R033105-30 Metric Equivalents of Cement Content for Concrete Mixes

94 Pound Bags per Cubic Yard	Kilograms per Cubic Meter	94 Pound Bags per Cubic Yard	Kilograms per Cubic Meter
1.0	55.77	7.0	390.4
1.5	83.65	7.5	418.3
2.0	111.5	8.0	446.2
2.5	139.4	8.5	474.0
3.0	167.3	9.0	501.9
3.5	195.2	9.5	529.8
4.0	223.1	10.0	557.7
4.5	251.0	10.5	585.6
5.0	278.8	11.0	613.5
5.5	306.7	11.5	641.3
6.0	334.6	12.0	669.2
6.5	362.5	12.5	697.1

a. If you know the cement content in pounds per cubic yard, multiply by .5933 to obtain kilograms per cubic meter.

b. If you know the cement content in 94 pound bags per cubic yard, multiply by 55.77 to obtain kilograms per cubic meter.

R033105-40 Metric Equivalents of Common Concrete Strengths (to convert other psi values to megapascals, multiply by 0.006895)

U.S. Values psi	SI Value Megapascals	Non-SI Metric Value kgf/cm²*
2000	14	140
2500	17	175
3000	21	210
3500	24	245
4000	28	280
4500	31	315
5000	34	350
6000	41	420
7000	48	490
8000	55	560
9000	62	630
10,000	69	705

* kilograms force per square centimeter

R033105-50 Quantities of Cement, Sand and Stone for One C.Y. of Concrete per Various Mixes

This table can be used to determine the quantities of the ingredients for smaller quantities of site mixed concrete.

Concrete (C.Y.)	Mix = 1:1:1-3/4			Mix = 1:2:2.25			Mix = 1:2.25:3			Mix = 1:3:4		
	Cement (sacks)	Sand (C.Y.)	Stone (C.Y.)	Cement (sacks)	Sand (C.Y.)	Stone (C.Y.)	Cement (sacks)	Sand (C.Y.)	Stone (C.Y.)	Cement (sacks)	Sand (C.Y.)	Stone (C.Y.)
1	10	.37	.63	7.75	.56	.65	6.25	.52	.70	5	.56	.74
2	20	.74	1.26	15.50	1.12	1.30	12.50	1.04	1.40	10	1.12	1.48
3	30	1.11	1.89	23.25	1.68	1.95	18.75	1.56	2.10	15	1.68	2.22
4	40	1.48	2.52	31.00	2.24	2.60	25.00	2.08	2.80	20	2.24	2.96
5	50	1.85	3.15	38.75	2.80	3.25	31.25	2.60	3.50	25	2.80	3.70
6	60	2.22	3.78	46.50	3.36	3.90	37.50	3.12	4.20	30	3.36	4.44
7	70	2.59	4.41	54.25	3.92	4.55	43.75	3.64	4.90	35	3.92	5.18
8	80	2.96	5.04	62.00	4.48	5.20	50.00	4.16	5.60	40	4.48	5.92
9	90	3.33	5.67	69.75	5.04	5.85	56.25	4.68	6.30	45	5.04	6.66
10	100	3.70	6.30	77.50	5.60	6.50	62.50	5.20	7.00	50	5.60	7.40
11	110	4.07	6.93	85.25	6.16	7.15	68.75	5.72	7.70	55	6.16	8.14
12	120	4.44	7.56	93.00	6.72	7.80	75.00	6.24	8.40	60	6.72	8.88
13	130	4.82	8.20	100.76	7.28	8.46	81.26	6.76	9.10	65	7.28	9.62
14	140	5.18	8.82	108.50	7.84	9.10	87.50	7.28	9.80	70	7.84	10.36
15	150	5.56	9.46	116.26	8.40	9.76	93.76	7.80	10.50	75	8.40	11.10
16	160	5.92	10.08	124.00	8.96	10.40	100.00	8.32	11.20	80	8.96	11.84
17	170	6.30	10.72	131.76	9.52	11.06	106.26	8.84	11.90	85	9.52	12.58
18	180	6.66	11.34	139.50	10.08	11.70	112.50	9.36	12.60	90	10.08	13.32
19	190	7.04	11.98	147.26	10.64	12.36	118.76	9.84	13.30	95	10.64	14.06
20	200	7.40	12.60	155.00	11.20	13.00	125.00	10.40	14.00	100	11.20	14.80
21	210	7.77	13.23	162.75	11.76	13.65	131.25	10.92	14.70	105	11.76	15.54
22	220	8.14	13.86	170.05	12.32	14.30	137.50	11.44	15.40	110	12.32	16.28
23	230	8.51	14.49	178.25	12.88	14.95	143.75	11.96	16.10	115	12.88	17.02
24	240	8.88	15.12	186.00	13.44	15.60	150.00	12.48	16.80	120	13.44	17.76
25	250	9.25	15.75	193.75	14.00	16.25	156.25	13.00	17.50	125	14.00	18.50
26	260	9.64	16.40	201.52	14.56	16.92	162.52	13.52	18.20	130	14.56	19.24
27	270	10.00	17.00	209.26	15.12	17.56	168.76	14.04	18.90	135	15.02	20.00
28	280	10.36	17.64	217.00	15.68	18.20	175.00	14.56	19.60	140	15.68	20.72
29	290	10.74	18.28	224.76	16.24	18.86	181.26	15.08	20.30	145	16.24	21.46

R033105-65 Field-Mix Concrete

Presently most building jobs are built with ready-mixed concrete except at isolated locations and some larger jobs requiring over 10,000 C.Y. where land is readily available for setting up a temporary batch plant.

The most economical mix is a controlled mix using local aggregate proportioned by trial to give the required strength with the least cost of material.

R033105-70 Placing Ready-Mixed Concrete

For ground pours allow for 5% waste when figuring quantities.

Prices in the front of the book assume normal deliveries. If deliveries are made before 8 A.M. or after 5 P.M. or on Saturday afternoons add 30%. Negotiated discounts for large volumes are not included in prices in front of book.

For the lower floors without truck access, concrete may be wheeled in rubber-tired buggies, conveyer handled, crane handled or pumped. Pumping is economical if there is top steel. Conveyers are more efficient for thick slabs.

At higher floors the rubber-tired buggies may be hoisted by a hoisting tower and wheeled to location. Placement by a conveyer is limited to three floors and is best for high-volume pours. Pumped concrete is best when building has no crane access. Concrete may be pumped directly as high as thirty-six stories using special pumping techniques. Normal maximum height is about fifteen stories.

Best pumping aggregate is screened and graded bank gravel rather than crushed stone.

Pumping downward is more difficult than pumping upward. Horizontal distance from pump to pour may increase preparation time prior to pour. Placing by cranes, either mobile, climbing or tower types, continues as the most efficient method for high-rise concrete buildings.

R033105-80 Slab on Grade

General: Ground slabs are classified on the basis of use. Thickness is generally controlled by the heaviest concentrated load supported. If load area is greater than 80 sq. in., soil bearing may be important. The base granular fill must be a uniformly compacted material of limited capillarity, such as gravel or crushed rock. Concrete is placed on this surface of the vapor barrier on top of base.

Ground slabs are either single or two course floors. Single course are widely used. Two course floors have a subsequent wear resistant topping.

Reinforcement is provided to maintain tightly closed cracks.

Control joints limit crack locations and provide for differential horizontal movement only. Isolation joints allow both horizontal and vertical differential movement.

Use of Table: Determine appropriate type of slab (A, B, C, or D) by considering type of use or amount of abrasive wear of traffic type.

Determine thickness by maximum allowable wheel load or uniform load, opposite 1st column, thickness. Increase the controlling thickness if details require, and select either plain or reinforced slab thickness and type.

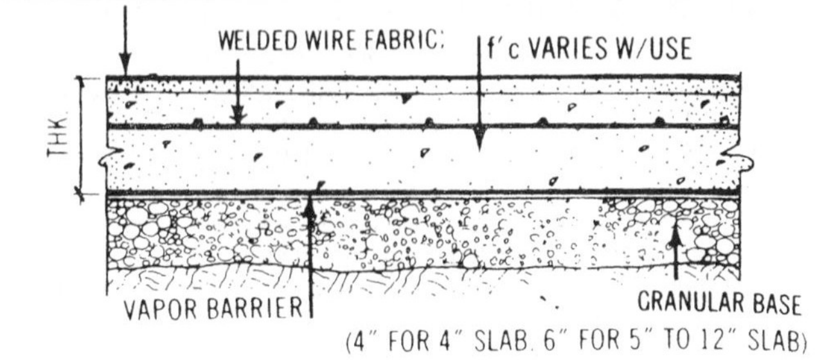

Slab on Grade

Thickness and Loading Assumptions by Type of Use

		A	B	C	D	◄ Slab I.D.
SLAB THICKNESS (IN.)	TYPE	Non	Light	Normal	Heavy	◄ Industrial
		Little	Light	Moderate	Severe	◄ Abrasion
		Foot Only	Pneumatic Wheels	Solid Rubber Wheels	Steel Tires	◄ Type of Traffic
		Load* (K)	Load* (K)	Load* (K)	Load* (K)	Max. Uniform Load to Slab ▼ (PSF)
4"	Reinf. Plain	4K				100
5"	Reinf. Plain	6K	4K			200
6"	Reinf. Plain		8K	6K	6K	500 to 800
7"	Reinf. Plain			9K	8K	1,500
8"	Reinf. Plain				11K	* Max. Wheel Load in Kips (incl. impact)
10"	Reinf. Plain				14K	
12"	Reinf. Plain					
DESIGN ASSUMPTIONS	Concrete, Chuted	f'c = 3.5 KSI	4 KSI	4.5 KSI	Slab @ 3.5 KSI	ASSUMPTIONS BY SLAB TYPE
	Toppings			1" Integral	1" Bonded	
	Finish	Steel Trowel	Steel Trowel	Steel Trowel	Screed & Steel Trowel	
	Compacted Granular Base	4" deep for 4" slab thickness 6" deep for 5" slab thickness & greater				ASSUMPTIONS FOR ALL SLAB TYPES
	Vapor Barrier	6 mil polyethylene				
	Forms & Joints	Allowances included				
	Rein-forcement	WWF as required ≥ 60,000 psi				

Concrete | R0331 | Structural Concrete

R033105-85 Lift Slabs

The cost advantage of the lift slab method is due to placing all concrete, reinforcing steel, inserts and electrical conduit at ground level and in reduction of formwork. Minimum economical project size is about 30,000 S.F. Slabs may be tilted for parking garage ramps.

It is now used in all types of buildings and has gone up to 22 stories high in apartment buildings. Current trend is to use post-tensioned flat plate slabs with spans from 22′ to 35′. Cylindrical void forms are used when deep slabs are required. One pound of prestressing steel is about equal to seven pounds of conventional reinforcing.

To be considered cured for stressing and lifting, a slab must have attained 75% of design strength. Seven days are usually sufficient with four to five days possible if high early strength cement is used. Slabs can be stacked using two coats of a non-bonding agent to insure that slabs do not stick to each other. Lifting is done by companies specializing in this work. Lift rate is 5′ to 15′ per hour with an average of 10′ per hour. Total areas up to 33,000 S.F. have been lifted at one time. 24 to 36 jacking columns are common. Most economical bay sizes are 24′ to 28′ with four to fourteen stories most efficient. Continuous design reduces reinforcing steel cost. Use of post-tensioned slabs allows larger bay sizes.

Concrete | R0341 | Precast Structural Concrete

R034105-30 Prestressed Precast Concrete Structural Units

Type	Location	Depth	Span in Ft.	Live Load Lb. per S.F.
Double Tee 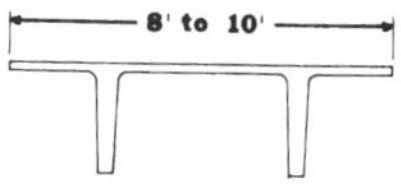	Floor	28″ to 34″	60 to 80	50 to 80
	Roof	12″ to 24″	30 to 50	40
	Wall	Width 8′	Up to 55′ high	Wind
Multiple Tee 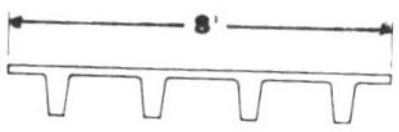	Roof	8″ to 12″	15 to 40	40
	Floor	8″ to 12″	15 to 30	100
Plank 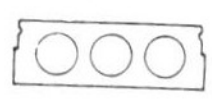 or 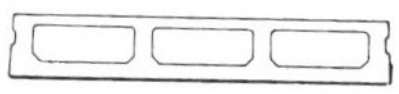	Roof or Floor	4″ 6″ 8″ 10″ 12″	Roof: 13, 22, 26, 33, 42 Floor: 12, 18, 25, 29, 32	40 for Roof 100 for Floor
Single Tee 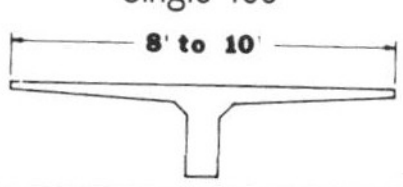	Roof	28″ 32″ 36″ 48″	40 80 100 120	40
AASHO Girder 	Bridges	Type 4 5 6	100 110 125	Highway
Box Beam 	Bridges	15″ 27″ 33″	40 to 100	Highway

The majority of precast projects today utilize double tees rather than single tees because of speed and ease of installation. As a result casting beds at manufacturing plants are normally formed for double tees. Single tee projects will therefore require an initial set up charge to be spread over the individual single tee costs.

For floors, a 2″ to 3″ topping is field cast over the shapes. For roofs, insulating concrete or rigid insulation is placed over the shapes.

Member lengths up to 40′ are standard haul, 40′ to 60′ require special permits and lengths over 60′ must be escorted. Over width and/or over length can add up to 100% on hauling costs.

Large heavy members may require two cranes for lifting which would increase erection costs by about 45%. An eight man crew can install 12 to 20 double tees, or 45 to 70 quad tees or planks per day.

Grouting of connections must also be included.

Several system buildings utilizing precast members are available. Heights can go up to 22 stories for apartment buildings. Optimum design ratio is 3 S.F. of surface to 1 S.F. of floor area.

R034136-90 Prestressed Concrete, Post-tensioned

In post-tensioned concrete the steel tendons are tensioned after the concrete has reached about 3/4 of its ultimate strength. The cableways are grouted after tensioning to provide bond between the steel and concrete. If bond is to be prevented, the tendons are coated with a corrosion-preventative grease and wrapped with waterproofed paper or plastic. Bonded tendons are usually used when ultimate strength (beams & girders) are controlling factors.

High strength concrete is used to fully utilize the steel, thereby reducing the size and weight of the member. A plasticizing agent may be added to reduce water content. Maximum size aggregate ranges from 1/2″ to 1-1/2″ depending on the spacing of the tendons.

The types of steel commonly used are bars and strands. Job conditions determine which is best suited. Bars are best for vertical prestresses since they are easy to support. The trend is for steel manufacturers to supply a finished package, cut to length, which reduces field preparation to a minimum.

Bars vary from 3/4″ to 1-3/8″ diameter. Table below gives time in labor-hours per tendon for placing, tensioning and grouting (if required) a 75′ beam. Tendons used in buildings are not usually grouted; tendons for bridges usually are grouted. For strands the table indicates the labor-hours per pound for typical prestressed units 100′ long. Simple span beams usually require one-end stressing regardless of lengths. Continuous beams are usually stressed from two ends. Long slabs are poured from the center outward and stressed in 75′ increments after the initial 150′ center pour.

Labor Hours per Tendon and per Pound of Prestressed Steel

Length	**100′ Beam**		**75′ Beam**		**100′ Slab**	
Type Steel	**Strand**		**Bars**		**Strand**	
Diameter	**0.5″**		**3/4″**	**1-3/8″**	**0.5″**	**0.6″**
Number	**4**	**12**	**1**	**1**	**1**	**1**
Force in Kips	**100**	**300**	**42**	**143**	**25**	**35**
Preparation & Placing Cables	3.6	7.4	0.9	2.9	0.9	1.1
Stressing Cables	2.0	2.4	0.8	1.6	0.5	0.5
Grouting, if required	2.5	3.0	0.6	1.3		
Total Labor Hours	8.1	12.8	2.3	5.8	1.4	1.6
Prestressing Steel Weights (Lbs.)	215	640	115	380	53	74
Labor-hours per Lb. Bonded	0.038	0.020	0.020	0.015		
Non-bonded					0.026	0.022

Flat Slab construction — 4000 psi concrete with span-to-depth ratio between 36 and 44. Two way post-tensioned steel averages 1.0 lb. per S.F. for 24′ to 28′ bays (usually strand) and additional reinforcing steel averages .5 lb. per S.F.

Pan and Joist construction — 4000 psi concrete with span-to-depth ratio 28 to 30. Post-tensioned steel averages .8 lb. per S.F. and reinforcing steel about 1.0 lb. per S.F. Placing and stressing averages 40 hours per ton of total material.

Beam construction — 4000 to 5000 psi concrete. Steel weights vary greatly.

Labor cost per pound goes down as the size and length of the tendon increase. The primary economic consideration is the cost per kip for the member.

Post-tensioning becomes feasible for beams and girders over 30′ long; for continuous two-way slabs over 20′ clear; also in transferring upper building loads over longer spans at lower levels. Post-tension suppliers will provide engineering services at no cost to the user. Substantial economies are possible by using post-tensioned Lift Slabs.

R034513-10 Precast Concrete Wall Panels

Panels are either solid or insulated with plain, colored or textured finishes. Transportation is an important cost factor. Prices shown in the unit cost section of the book are based on delivery within 50 miles of a plant including fabricators' overhead and profit. Engineering data is available from fabricators to assist with construction details. Usual minimum job size for economical use of panels is about 5000 S.F. Small jobs can double the prices shown. For large, highly repetitive jobs, deduct up to 15% from the prices shown.

2″ thick panels cost about the same as 3″ thick panels, and maximum panel size is less. For building panels faced with granite, marble or stone, add the material prices from those unit cost sections to the plain panel price shown. There is a growing trend toward aggregate facings and broken rib finish rather than plain gray concrete panels.

No allowance has been made in the unit cost section for supporting steel framework. On one story buildings, panels may rest on grade beams and require only wind bracing and fasteners. On multi-story buildings panels can span from column to column and floor to floor. Plastic-designed steel-framed structures may have large deflections which slow down erection and raise costs.

Large panels are more economical than small panels on a S.F. basis. When figuring areas include all protrusions, returns, etc. Overhangs can triple erection costs. Panels over 45′ have been produced. Larger flat units should be prestressed. Vacuum lifting of smooth finish panels eliminates inserts and can speed erection.

Concrete | R0347 | Site-Cast Concrete

R034713-20 Tilt Up Concrete Panels

The advantage of tilt up construction is in the low cost of forms and the placing of concrete and reinforcing. Panels up to 75′ high and 5-1/2″ thick have been tilted using strongbacks. Tilt up has been used for one to five story buildings and is well-suited for warehouses, stores, offices, schools and residences.

The panels are cast in forms on the floor slab. Most jobs use 5-1/2″ thick solid reinforced concrete panels. Sandwich panels with a layer of insulating materials are also used. Where dampness is a factor, lightweight aggregate is used. Optimum panel size is 300 to 500 S.F.

Slabs are usually poured with 3000 psi concrete which permits tilting seven days after pouring. Slabs may be stacked on top of each other and are separated from each other by either two coats of bond breaker or a film of polyethylene. Use of high early-strength cement allows tilting two days after a pour. Tilting up is done with a roller outrigger crane with a capacity of at least 1-1/2 times the weight of the panel at the required reach. Exterior precast columns can be set at the same time as the panels; interior precast columns can be set first and the panels clipped directly to them. The use of cast-in-place concrete columns is diminishing due to shrinkage problems. Structural steel columns are sometimes used if crane rails are planned. Panels can be clipped to the columns or lowered between the flanges. Steel channels with anchors may be used as edge forms for the slab. When the panels are lifted the channels form an integral steel column to take structural loads. Roof loads can be carried directly by the panels for wall heights to 14′.

Requirements of local building codes may be a limiting factor and should be checked. Building floor slabs should be poured first and should be a minimum of 5″ thick with 100% compaction of soil or 6″ thick with less than 100% compaction.

Setting times as fast as nine minutes per panel have been observed, but a safer expectation would be four panels per hour with a crane and a four-man setting crew. If crane erects from inside building, some provision must be made to get crane out after walls are erected. Good yarding procedure is important to minimize delays. Equalizing three-point lifting beams and self-releasing pick-up hooks speed erection. If panels must be carried to their final location, setting time per panel will be increased and erection costs may approach the erection cost range of architectural precast wall panels. Placing panels into slots formed in continuous footers will speed erection.

Reinforcing should be with #5 bars with vertical bars on the bottom. If surface is to be sandblasted, stainless steel chairs should be used to prevent rust staining.

Use of a broom finish is popular since the unavoidable surface blemishes are concealed.

Precast columns run from three to five times the C.Y. price of the panels only.

Concrete | R0352 | Lightweight Concrete Roof Insulation

R035216-10 Lightweight Concrete

Lightweight aggregate concrete is usually purchased ready mixed, but it can also be field mixed.

Vermiculite or Perlite comes in bags of 4 C.F. under various trade names. Weight is about 8 lbs. per C.F. For insulating roof fill use 1:6 mix. For structural deck use 1:4 mix over gypsum boards, steeltex, steel centering, etc., supported by closely spaced joists or bulb trees. For structural slabs use 1:3:2 vermiculite sand concrete over steeltex, metal lath, steel centering, etc., on joists spaced 2′-0″ O.C. for maximum L.L. of 80 P.S.F. Use same mix for slab base fill over steel flooring or regular reinforced concrete slab when tile, terrazzo or other finish is to be laid over.

For slabs on grade use 1:3:2 mix when tile, etc., finish is to be laid over. If radiant heating units are installed use a 1:6 mix for a base. After coils are in place, cover with a regular granolithic finish (mix 1:3:2) to a minimum depth of 1-1/2″ over top of units.

Reinforce all slabs with 6 x 6 or 10 x 10 welded wire mesh.

Masonry | R0401 | Maintenance of Masonry

R040130-10 Cleaning Face Brick

On smooth brick a person can clean 70 S.F. an hour; on rough brick 50 S.F. per hour. Use one gallon muriatic acid to 20 gallons of water for 1000 S.F. Do not use acid solution until wall is at least seven days old, but a mild soap solution may be used after two days.

Time has been allowed for clean-up in brick prices.

Masonry | R0405 | Common Work Results for Masonry

R040513-10 Cement Mortar (material only)

Type N - 1:1:6 mix by volume. Use everywhere above grade except as noted below. - 1:3 mix using conventional masonry cement which saves handling two separate bagged materials.

Type M - 1:1/4:3 mix by volume, or 1 part cement, 1/4 (10% by wt.) lime, 3 parts sand. Use for heavy loads and where earthquakes or hurricanes may occur. Also for reinforced brick, sewers, manholes and everywhere below grade.

Mix Proportions by Volume and Compressive Strength of Mortar

Where Used	Allowable Proportions by Volume					Compressive Strength @ 28 days
	Mortar Type	Portland Cement	Masonry Cement	Hydrated Lime	Masonry Sand	
Plain Masonry		1	1	—	6	
	M	1	—	1/4	3	2500 psi
		1/2	1	—	4	
	S	1	—	1/4 to 1/2	4	1800 psi
		—	1	—	3	
	N	1	—	1/2 to 1-1/4	6	750 psi
		—	1	—	3	
	O	1	—	1-1/4 to 2-1/2	9	350 psi
	K	1	—	2-1/2 to 4	12	75 psi
Reinforced Masonry	PM	1	1	—	6	2500 psi
	PL	1	—	1/4 to 1/2	4	2500 psi

Note: The total aggregate should be between 2.25 to 3 times the sum of the cement and lime used.

The labor cost to mix the mortar is included in the productivity and labor cost of unit price lines in unit cost sections for brickwork, blockwork and stonework.

The material cost of mixed mortar is included in the material cost of those same unit price lines and includes the cost of renting and operating a 10 C.F. mixer at the rate of 200 C.F. per day.

There are two types of mortar color used. One type is the inert additive type with about 100 lbs. per M brick as the typical quantity required. These colors are also available in smaller-batch-sized bags (1 lb. to 15 lb.) which can be placed directly into the mixer without measuring. The other type is premixed and replaces the masonry cement. Dark green color has the highest cost.

R040519-50 Masonry Reinforcing

Horizontal joint reinforcing helps prevent wall cracks where wall movement may occur and in many locations is required by code. Horizontal joint reinforcing is generally not considered to be structural reinforcing and an unreinforced wall may still contain joint reinforcing.

Reinforcing strips come in 10′ and 12′ lengths and in truss and ladder shapes, with and without drips. Field labor runs between 2.7 to 5.3 hours per 1000 L.F. for wall thicknesses up to 12″.

The wire meets ASTM A82 for cold drawn steel wire and the typical size is 9 ga. sides and ties with 3/16″ diameter also available. Typical finish is mill galvanized with zinc coating at .10 oz. per S.F. Class I (.40 oz. per S.F.) and Class III (.80 oz per S.F.) are also available, as is hot dipped galvanizing at 1.50 oz. per S.F.

R042110-10 Economy in Bricklaying

Have adequate supervision. Be sure bricklayers are always supplied with materials so there is no waiting. Place best bricklayers at corners and openings.

Use only screened sand for mortar. Otherwise, labor time will be wasted picking out pebbles. Use seamless metal tubs for mortar as they do not leak or catch the trowel. Locate stack and mortar for easy wheeling.

Have brick delivered for stacking. This makes for faster handling, reduces chipping and breakage, and requires less storage space. Many dealers will deliver select common in 2′ x 3′ x 4′ pallets or face brick packaged. This affords quick handling with a crane or forklift and easy tonging in units of ten, which reduces waste.

Use wider bricks for one wythe wall construction. Keep scaffolding away from wall to allow mortar to fall clear and not stain wall.

On large jobs develop specialized crews for each type of masonry unit.

Consider designing for prefabricated panel construction on high rise projects.

Avoid excessive corners or openings. Each opening adds about 50% to labor cost for area of opening.

Bolting stone panels and using window frames as stops reduces labor costs and speeds up erection.

R042110-20 Common and Face Brick

Common building brick manufactured according to ASTM C62 and facing brick manufactured according to ASTM C216 are the two standard bricks available for general building use.

Building brick is made in three grades; SW, where high resistance to damage caused by cyclic freezing is required; MW, where moderate resistance to cyclic freezing is needed; and NW, where little resistance to cyclic freezing is needed. Facing brick is made in only the two grades SW and MW. Additionally, facing brick is available in three types; FBS, for general use; FBX, for general use where a higher degree of precision and lower permissible variation in size than FBS is needed; and FBA, for general use to produce characteristic architectural effects resulting from non-uniformity in size and texture of the units.

In figuring the material cost of brickwork, an allowance of 25% mortar waste and 3% brick breakage was included. If bricks are delivered palletized with 280 to 300 per pallet, or packaged, allow only 1-1/2% for breakage. Packaged or palletized delivery is practical when a job is big enough to have a crane or other equipment available to handle a package of brick. This is so on all industrial work but not always true on small commercial buildings.

The use of buff and gray face is increasing, and there is a continuing trend to the Norman, Roman, Jumbo and SCR brick.

Common red clay brick for backup is not used that often. Concrete block is the most usual backup material with occasional use of sand lime or cement brick. Building brick is commonly used in solid walls for strength and as a fire stop.

Brick panels built on the ground and then crane erected to the upper floors have proven to be economical. This allows the work to be done under cover and without scaffolding.

R042110-50 Brick, Block & Mortar Quantities

Running Bond					
Number of Brick per S.F. of Wall - Single Wythe with 3/8″ Joints				C.F. of Mortar per M Bricks, Waste Included	
Type Brick	Nominal Size (incl. mortar) L H W	Modular Coursing	Number of Brick per S.F.	3/8″ Joint	1/2″ Joint
Standard	8 x 2-2/3 x 4	3C=8″	6.75	10.3	12.9
Economy	8 x 4 x 4	1C=4″	4.50	11.4	14.6
Engineer	8 x 3-1/5 x 4	5C=16″	5.63	10.6	13.6
Fire	9 x 2-1/2 x 4-1/2	2C=5″	6.40	550 # Fireclay	—
Jumbo	12 x 4 x 6 or 8	1C=4″	3.00	23.8	30.8
Norman	12 x 2-2/3 x 4	3C=8″	4.50	14.0	17.9
Norwegian	12 x 3-1/5 x 4	5C=16″	3.75	14.6	18.6
Roman	12 x 2 x 4	2C=4″	6.00	13.4	17.0
SCR	12 x 2-2/3 x 6	3C=8″	4.50	21.8	28.0
Utility	12 x 4 x 4	1C=4″	3.00	15.4	19.6

For Other Bonds Standard Size Add to S.F. Quantities in Table to Left		
Bond Type	Description	Factor
Common	full header every fifth course	+20%
	full header every sixth course	+16.7%
English	full header every second course	+50%
Flemish	alternate headers every course	+33.3%
	every sixth course	+5.6%
Header = W x H exposed		+100%
Rowlock = H x W exposed		+100%
Rowlock stretcher = L x W exposed		+33.3%
Soldier = H x L exposed		—
Sailor = W x L exposed		-33.3%

Concrete Blocks Nominal Size	Approximate Weight per S.F.		Blocks per 100 S.F.	Mortar per M block, waste included	
	Standard	Lightweight		Partitions	Back up
2″ x 8″ x 16″	20 PSF	15 PSF	113	27 C.F.	36 C.F.
4″	30	20		41	51
6″	42	30		56	66
8″	55	38		72	82
10″	70	47		87	97
12″	85	55		102	112

R042210-20 Concrete Block

The material cost of special block such as corner, jamb and head block can be figured at the same price as ordinary block of same size. Labor on specials is about the same as equal-sized regular block.

Bond beam and 16″ high lintel blocks are more expensive than regular units of equal size. Lintel blocks are 8″ long and either 8″ or 16″ high.

Use of motorized mortar spreader box will speed construction of continuous walls.

Hollow non-load-bearing units are made according to ASTM C129 and hollow load-bearing units according to ASTM C90.

R042210-30 Fully Grouted Reinforced Masonry Wall Capacities Per L.F. (Kips & In-Kips)

Earthquake Zones 1, 2 & 3					Allowable Vertical Wall Loads									Allowable Wall Moments Without Vertical Wall Loads	
Thk.	Length Or Height		Type Wall	Rebar	Eccentric Loads				Without Wind or Eccentric Loads		With Wind			Not Wind or Earthquake	
					7.0 in-K/Ft.		3.5 in-K/Ft.				Inspection				
T (Nom.) (in)	h′ (Ft.)	h′/T (in/in)	Brick Grout Conc. M.U.	Size and Spacing (in. O.C.)	Inspection No (K/Ft.)	Inspection Yes (K/Ft.)	Inspection No (K/Ft.)	Inspection Yes (K/Ft.)	Inspection No (K/Ft.)	Inspection Yes (K/Ft.)	No 15 psf (K/Ft.)	No 30 psf (K/Ft.)	Yes 15 & 30 (K/Ft.)	Inspection No (in.-K/Ft.)	Inspection Yes (in.-K/Ft.)
8″	8′	12	4″0″4″	#5@32	10.10	23.35	11.70	24.50	13.30	26.55	13.30	13.30	26.55	6.90	9.65
			Solid CMU		12.75	28.70	14.35	30.30	15.95	31.90	15.95	15.95	31.90	7.80	9.75
	12′	18	4″0″4″		9.30	21.60	10.80	23.05	12.25	24.55	12.25	12.25	24.55	6.90	9.65
			Solid CMU		16.65	26.50	18.15	28.00	14.70	29.45	14.70	14.70	29.45	7.80	9.75
	16′	24	4″0″4″		7.80	18.10	9.05	19.35	10.30	20.60	10.30	9.65	20.60	6.90	9.65
			Solid CMU		9.85	22.20	11.10	23.45	12.35	24.70	12.35	12.35	12.70	7.80	9.75
10″	8′	9.6	4″0″6″	#5@24	14.45	31.50	15.75	32.80	17.05	34.10	17.05	17.05	34.10	11.20	16.20
			Solid CMU		17.85	38.35	19.15	39.65	20.45	40.90	20.45	20.45	40.90	12.65	16.35
			4″2″4″		14.45	31.50	15.75	32.80	17.05	34.10	17.05	17.05	34.10	11.20	16.20
	12′	14.4	4″0″6″		13.90	30.35	15.15	31.60	16.40	32.80	16.40	16.40	32.80	11.20	16.20
			Solid CMU		17.20	36.90	18.45	38.15	19.70	39.40	19.70	19.70	39.40	12.65	16.35
			4″2″4″		13.90	30.35	15.15	31.60	16.40	32.80	16.40	16.40	32.80	11.20	16.20
	16′	19.2	4″0″6″		12.85	28.05	14.00	29.20	15.15	30.35	15.15	15.15	30.35	11.20	16.20
			Solid CMU		15.90	34.10	17.05	35.25	18.20	36.40	13.20	13.20	36.40	12.65	16.35
			4″2″4″		12.85	28.05	14.00	29.20	15.15	30.35	15.15	15.15	30.35	11.20	16.20
	20′	24	4″0″6″		11.15	24.25	12.10	25.25	13.10	26.20	13.10	12.40	26.20	11.20	16.20
			Solid CMU		13.75	29.50	14.75	30.50	15.75	31.50	15.75	15.75	31.50	12.65	16.35
			4″2″4″		11.15	24.25	12.10	25.25	13.10	26.20	13.10	12.40	26.20	11.20	16.20
12″	8′	8	4″0″8″	#8@48	18.60	39.35	19.65	40.40	20.75	41.50	20.75	20.75	41.50	16.65	24.90
			Solid CMU		22.75	47.65	23.80	48.70	24.90	49.80	24.90	24.90	49.80	18.80	25.10
			4″2″6″	#5@20	18.60	39.35	19.65	40.40	20.75	41.50	20.75	20.75	41.50	11.90	18.75
	12′	12	4″0″8″	#8@48	18.20	38.50	19.25	39.55	20.30	40.60	20.30	20.30	40.60	16.65	24.90
			Solid CMU		22.25	46.65	23.30	47.70	24.35	48.75	24.35	24.35	48.75	18.80	25.10
			4″2″6″	#5@20	18.20	38.50	19.25	39.55	20.30	40.60	20.30	20.30	40.60	11.90	18.75
	16′	16	4″0″8″	#8@48	17.45	36.90	18.45	37.90	19.45	38.90	19.45	19.45	38.90	16.65	24.90
			Solid CMU		21.30	44.70	22.35	45.70	23.35	46.70	23.25	23.35	46.70	18.80	25.10
			4″2″6″	#5@20	17.45	36.90	18.45	37.90	19.45	38.90	19.45	19.45	38.90	11.90	18.75
	24′	24	4″0″8″	#8@48	14.30	30.25	15.10	31.10	15.95	31.90	15.95	15.15	31.90	16.65	24.90
			Solid CMU		17.15	36.65	18.30	37.45	19.15	38.30	19.15	19.15	38.30	18.80	25.10
			4″2″6″	#5@20	14.30	30.25	15.10	31.10	15.95	31.90	15.95	15.15	31.90	11.90	18.75
16″	8′	6	4″0″12″	#8@32	26.40	54.40	27.20	55.20	28.00	56.05	28.00	28.00	56.05	31.20	49.10
			Solid CMU		32.00	65.65	32.80	66.45	33.60	67.25	33.60	33.60	67.25	35.25	50.40
			4″2″10″	#5@15	26.40	54.40	27.20	55.20	28.00	56.05	28.00	28.00	56.05	14.75	24.10
	12′	9	4″0″12″	#8@32	26.15	53.95	26.95	54.75	27.75	55.55	27.75	27.75	55.55	31.20	49.10
			Solid CMU		31.70	66.05	32.50	65.85	33.30	66.65	33.00	33.30	66.65	35.25	50.40
			4″2″10″	#5@15	26.15	53.95	26.95	54.75	27.75	55.55	27.75	27.75	55.55	14.75	24.10
	16′	12	4″0″12″	#8@32	25.70	53.05	26.50	53.80	28.30	54.60	27.30	27.30	54.60	31.20	49.10
			Solid CMU		31.20	63.95	31.95	64.75	32.75	65.50	32.75	32.75	65.50	35.25	50.40
			4″2″10″		25.70	53.05	26.50	53.80	27.30	54.60	27.30	27.30	59.60	14.75	24.10
	32′	24	4″0″12″	#8@32	20.35	41.95	20.95	42.55	21.60	43.20	18.75*	18.75*	43.20	31.20	49.10
			Solid CMU		24.65	50.60	25.30	51.20	25.90	51.85	24.55*	24.55*	51.85	35.25	50.40
			4″2″10″	#5@15	20.35	41.95	20.95	42.55	21.60	43.20	18.75*	18.75*	43.20	14.75	24.10

*Zone 3 Only

R042210-30 Unreinforced Masonry Wall Capacities Per L.F. (Kips & In-Kips)

Thk. T (Nom.) (in)	Earthquake Zones 0 & 1 Only			Allowable Vertical Wall Loads					Allowable Wall Moments (Without Vertical Wall Loads)			
	Length Or Height			Eccentric Loads		Without Wind or Eccentric Loads	With Wind		Not Wind or Earthquake		Wind or Earthquake	
									Inspection		Inspection	
	h′ (Ft.)	h′/t (in/in)	Type of Wall	7.0 in-K/Ft. (K/Ft.)	3.5 in-K/Ft. (K/Ft.)	(K/Ft.)	15 psf (K/Ft.)	30 psf (K/Ft.)	No (in-K/Ft.)	Yes (in-K/Ft.)	No (in-K/Ft.)	Yes (in-K/Ft.)
8″	8′	12	Solid Brick	4.85	7.50	10.15	10.15	10.15	1.15	2.30	1.55	3.10
			Solid CM Units	8.85	11.50	14.15	14.15	14.15	.70	1.40	.95	1.85
			Hollow CM Units	7.95	10.60	13.30	13.30	13.30	.70	1.40	.60	1.25
	10′	15	Solid Brick	4.70	7.30	9.85	9.85	9.85	1.15	2.30	1.55	3.10
			Solid CM Units	8.55	11.15	13.75	13.75	13.75	.70	1.40	.90	1.85
			Hollow CM Units	—	4.45	6.30	6.30	6.05	.45	.95	.60	1.25
	12′	18	Solid Brick	—	6.95	9.40	9.40	7.95	1.15	2.30	1.55	3.10
			Solid CM Units	8.15	10.60	13.10	13.10	12.90	.70	1.40	.90	1.85
			Hollow CM Units	—	4.25	6.00	6.00	4.75	.45	.95	.60	1.25
12″	8′	8	Solid Brick	12.30	14.10	15.90	15.90	15.90	2.70	5.40	3.60	7.20
			Solid CM Units	18.50	20.30	22.10	22.10	22.10	1.60	3.20	2.15	4.30
			Hollow CM Units	7.85	9.10	10.40	10.40	10.40	1.10	2.25	1.50	3.00
			Brick & Hollow CMU	5.50	7.80	10.05	10.05	10.05	1.35	2.70	1.80	3.60
	12′	12	Solid Brick	12.05	13.80	15.55	15.55	15.55	2.70	5.40	3.60	7.20
			Solid CM Units	13.45	14.75	16.05	16.05	16.05	1.60	3.20	2.15	4.30
			Hollow CM Units	7.65	8.90	10.15	10.15	10.15	1.10	2.25	1.50	3.00
			Brick & Hollow CMU	5.40	7.65	9.85	9.85	9.05	1.35	2.70	1.80	3.60
	16′	16	Solid Brick	11.55	13.20	14.90	14.90	14.35	2.70	5.40	3.60	7.20
			Solid CM Units	17.35	19.05	20.70	20.70	20.70	1.60	3.20	2.15	4.30
			Hollow CM Units	7.35	8.55	9.75	9.75	9.05	1.10	2.25	1.50	3.00
			Brick & Hollow CMU	5.20	7.30	9.45	9.10	—	1.35	2.70	1.80	3.60
16″	12′	9	Solid Brick	18.60	19.95	21.30	21.30	21.30	4.85	9.75	6.50	13.00
			Solid CM Units	26.95	28.30	29.60	29.60	29.60	2.90	5.85	3.90	7.80
			Hollow CM Units	10.85	12.10	13.30	13.30	13.30	1.50	3.05	2.00	4.05
			Brick & Hollow CMU	9.30	11.10	12.85	12.85	12.85	2.20	4.40	2.90	5.85
	16′	12	Solid Brick	18.30	19.60	20.90	20.90	20.90	4.85	9.75	6.50	13.00
			Solid CM Units	26.50	27.80	29.10	29.10	29.10	2.90	5.85	3.90	7.80
			Hollow CM Units	10.70	11.90	13.10	13.10	13.10	1.50	3.05	2.00	4.05
			Brick & Hollow CMU	9.15	10.90	12.65	12.65	11.10	2.20	4.40	2.90	5.85
	20′	15	Solid Brick	17.80	19.05	20.30	20.30	20.30	4.85	9.75	6.50	13.00
			Solid CM Units	25.75	27.00	28.25	28.25	28.25	2.90	5.85	3.90	7.80
			Hollow CM Units	10.40	11.55	12.70	12.70	10.95	1.50	3.05	2.00	4.05
			Brick & Hollow CMU	8.90	10.60	12.30	12.00	—	2.20	4.40	2.90	5.85
20″	12′	7.2	Solid Brick	24.80	25.85	26.90	26.90	26.90	7.70	15.40	10.25	20.50
			Solid CM Units	35.30	36.40	37.45	37.45	37.45	4.60	9.20	6.15	12.30
			Hollow CM Units	16.15	17.15	18.15	18.15	18.15	2.50	5.05	3.35	6.75
			Brick & Hollow CMU	16.10	17.25	18.45	18.45	18.45	4.05	8.10	5.40	10.80
	16′	9.6	Solid Brick	24.55	25.60	26.70	26.70	26.70	7.70	15.40	10.25	20.50
			Solid CM Units	35.00	36.10	37.15	37.15	37.15	4.60	9.20	6.15	12.30
			Hollow CM Units	16.00	17.00	18.00	18.00	18.00	2.50	5.05	3.35	6.75
			Brick & Hollow CMU	15.95	17.10	18.30	18.30	18.30	4.05	8.10	5.40	10.80
	24′	14.4	Solid Brick	23.70	24.70	25.75	25.75	25.75	7.70	15.40	10.25	20.50
			Solid CM Units	33.80	34.80	35.80	35.80	35.80	4.60	9.20	6.15	12.30
			Hollow CM Units	15.45	16.40	17.40	17.40	16.05	2.50	5.05	3.35	6.75
			Brick & Hollow CMU	15.40	16.50	17.65	17.65	15.25	4.05	8.10	5.40	10.80

Metals | R0505 | Common Work Results for Metals

R050516-30 Coating Structural Steel

On field-welded jobs, the shop-applied primer coat is necessarily omitted. All painting must be done in the field and usually consists of red oxide rust inhibitive paint or an aluminum paint. The table below shows paint coverage and daily production for field painting.

See Division 05950-650 for hot-dipped galvanizing and for field-applied cold galvanizing and other paints and protective coatings.

See Division 05910-500 for steel surface preparation treatments such as wire brushing, pressure washing and sand blasting.

Type Construction	Surface Area per Ton	Coat	One Gallon Covers		In 8 Hrs. Person Covers		Average per Ton Spray	
			Brush	Spray	Brush	Spray	Gallons	Labor-hours
Light Structural	300 S.F. to 500 S.F.	1st	500 S.F.	455 S.F.	640 S.F.	2000 S.F.	0.9 gals.	1.6 L.H.
		2nd	450	410	800	2400	1.0	1.3
		3rd	450	410	960	3200	1.0	1.0
Medium	150 S.F. to 300 S.F.	All	400	365	1600	3200	0.6	0.6
Heavy Structural	50 S.F. to 150 S.F.	1st	400	365	1920	4000	0.2	0.2
		2nd	400	365	2000	4000	0.2	0.2
		3rd	400	365	2000	4000	0.2	0.2
Weighted Average	225 S.F.	All	400	365	1350	3000	0.6	0.6

R050521-20 Welded Structural Steel

Usual weight reductions with welded design run 10% to 20% compared with bolted or riveted connections. This amounts to about the same total cost compared with bolted structures since field welding is more expensive than bolts. For normal spans of 18′ to 24′ figure 6 to 7 connections per ton.

Trusses — For welded trusses add 4% to weight of main members for connections. Up to 15% less steel can be expected in a welded truss compared to one that is shop bolted. Cost of erection is the same whether shop bolted or welded.

General — Typical electrodes for structural steel welding are E6010, E6011, E60T and E70T. Typical buildings vary between 2# to 8# of weld rod per ton of steel. Buildings utilizing continuous design require about three times as much welding as conventional welded structures. In estimating field erection by welding, it is best to use the average linear feet of weld per ton to arrive at the welding cost per ton. The type, size and position of the weld will have a direct bearing on the cost per linear foot. A typical field welder will deposit 1.8# to 2# of weld rod per hour manually. Using semiautomatic methods can increase production by as much as 50% to 75%.

R050523-10 High Strength Bolts

Common bolts (A307) are usually used in secondary connections (see Division 05090-150).

High strength bolts (A325 and A490) are usually specified for primary connections such as column splices, beam and girder connections to columns, column bracing, connections for supports of operating equipment or of other live loads which produce impact or reversal of stress, and in structures carrying cranes of over 5-ton capacity.

Allow 20 field bolts per ton of steel for a 6 story office building, apartment house or light industrial building. For 6 to 12 stories allow 25 bolts per ton, and above 12 stories, 30 bolts per ton. On power stations, 20 to 25 bolts per ton are needed.

Metals | R0512 | Structural Steel Framing

R051223-10 Structural Steel

The bare material prices for structural steel, shown in the unit cost sections of the book, are for 100 tons of shop-fabricated structural steel and include:

1. Mill base price of structural steel
2. Mill scrap/grade/size/length extras
3. Mill delivery to a metals service center (warehouse)
4. Service center storage and handling
5. Service center delivery to a fabrication shop
6. Shop storage and handling
7. Shop drafting/detailing
8. Shop fabrication
9. Shop coat of primer paint
10. Shop listing
11. Shop delivery to the job site

In unit cost sections of the book that contain items for field fabrication of steel components, the bare material cost of steel includes:

1. Mill base price of structural steel
2. Mill scrap/grade/size/length extras
3. Mill delivery to a metals service center (warehouse)
4. Service center storage and handling
5. Service center delivery to the job site

R051223-15 Structural Steel Estimating for Repair and Remodeling Projects

The correct approach to estimating structural steel is dependent upon the amount of steel required for the particular project. If the project is a sizable addition to a building, the data can be used directly from the cost data book. This is not the case however if the project requires a small amount of steel, for instance to reinforce existing roof or floor structural systems. To better understand this, please refer to the unit price line for a W16x31 beam with bolted connections. Assume your project requires the reinforcement of the structural members of the roof system of a 3 story building required for the installation a new piece of HVAC equipment. The project will need 4 pieces of W16x31 that are 30′ long. After pricing this 120 L.F. job using the unit prices for a W16x31, an analysis will reveal that the price is wholly inadequate.

The first problem is apparent if you examine the amount of steel that can be installed per day. The unit price line indicates 900 linear feet per day can be installed by a 5 man crew with an 90 ton crane. This productivity is correct for new construction but certainly is not for repair and remodeling work. Installation of new structural steel considers that each member is installed from the foundation to the roof of the structure in a planned and systematic manner with a crane having unrestricted access to all parts of the project. Additionally each connection is planned and detailed with full field access for fit-up and final bolting. The erection is planned and progresses such that interferences and conflicts with other structural members are minimized, if not completely eliminated. All of these assumptions are clearly not the case with a repair and remodeling job, and a significant decrease in the stated productivity will be observed.

A crane will certainly be needed to lift the members into the general area of the project but in most cases will not be able to place the beams into their final position. An opening in the existing roof may not be large enough to permit the beams to pass through and it may be necessary to bring them into the building through existing windows or doors. Moving the beams to the actual area where they will be installed may involve hand labor and the use of dollies. Finally, hoists and/or jacks may be needed for final positioning.

The connection of new members to existing can often be accomplished by field bolting with accurate field measurements and good planning but in many cases access to both sides of existing members is not possible and field welding becomes the only alternative. In addition to the cost of the actual welding, protection of existing finishes, systems and structure and fire protection must be considered.

New beams can never be installed tight to the existing decks or floors which they must support and the use of shims and tack welding becomes necessary. Additionally, further planning and cost is involved in assuring that existing loads are minimized during the installation and shimming process.

It is apparent that installation of structural steel as part of a repair and remodeling project involves more than simply installing the members and estimating the cost in the same manner as new construction. The best procedure for estimating the total cost is adequate planning and coordination of each process and activity that will be needed. Unit costs for the materials, labor and equipment can then be attached to each needed activity and a final, complete price can be determined.

R051223-20 Steel Estimating Quantities

One estimate on erection is that a crane can handle 35 to 60 pieces per day. Say the average is 45. With usual sizes of beams, girders, and columns, this would amount to about 20 tons per day. The type of connection greatly affects the speed of erection. Moment connections for continuous design slow down production and increase erection costs.

Short open web bar joists can be set at the rate of 75 to 80 per day, with 50 per day being the average for setting long span joists.

After main members are calculated, add the following for usual allowances: base plates 2% to 3%; column splices 4% to 5%; and miscellaneous details 4% to 5%, for a total of 10% to 13% in addition to main members.

The ratio of column to beam tonnage varies depending on type of steels used, typical spans, story heights and live loads.

It is more economical to keep the column size constant and to vary the strength of the column by using high strength steels. This also saves floor space. Buildings have recently gone as high as ten stories with 8″ high strength columns. For light columns under W8X31 lb. sections, concrete filled steel columns are economical.

High strength steels may be used in columns and beams to save floor space and to meet head room requirements. High strength steels in some sizes sometimes require long lead times.

Round, square and rectangular columns, both plain and concrete filled, are readily available and save floor area, but are higher in cost per pound than rolled columns. For high unbraced columns, tube columns may be less expensive.

Below are average minimum figures for the weights of the structural steel frame for different types of buildings using A36 steel, rolled shapes and simple joints. For economy in domes, rise to span ratio = .13. Open web joist framing systems will reduce weights by 10% to 40%. Composite design can reduce steel weight by up to 25% but additional concrete floor slab thickness may be required. Continuous design can reduce the weights up to 20%. There are many building codes with different live load requirements and different structural requirements, such as hurricane and earthquake loadings which can alter the figures.

Structural Steel Weights per S.F. of Floor Area

Type of Building	No. of Stories	Avg. Spans	L.L. #/S.F.	Lbs. Per S.F.	Type of Building	No. of Stories	Avg. Spans	L.L. #/S.F.	Lbs. Per S.F.
Steel Frame Mfg.	1	20′x20′	40	8	Apartments	2-8	20′x20′	40	8
		30′x30′		13		9-25			14
		40′x40′		18	Office	to 10	Various	80	10
Parking garage	4	Various	80	8.5		20			18
Domes (Schwedler)*	1	200′	30	10		30			26
		300′		15		over 50			35

R051223-25 Common Structural Steel Specifications

ASTM A992 (formerly A36, then A572 Grade 50) is the all-purpose carbon grade steel widely used in building and bridge construction.

The other high-strength steels listed below may each have certain advantages over ASTM A992 stuctural carbon steel, depending on the application. They have proven to be economical choices where, due to lighter members, the reduction of dead load and the associated savings in shipping cost can be significant.

ASTM A588 atmospheric weathering, high-strength low-alloy steels can be used in the bare (uncoated) condition, where exposure to normal atmosphere causes a tightly adherant oxide to form on the surface protecting the steel from further oxidation. ASTM A242 corrosion-resistant, high-strength low-alloy steels have enhanced atmospheric corrosion resistance of at least two times that of carbon structural steels with copper, or four times that of carbon structural steels without copper. The reduction or elimination of maintenance resulting from the use of these steels often offsets their higher initial cost.

Steel Type	ASTM Designation	Minimum Yield Stress in KSI	Shapes Available
Carbon	A36	36	All structural shape groups, and plates & bars up thru 8″ thick
	A529	42	Structural shape group 1, and plates & bars up thru 1/2″ thick
High-Strength Low-Alloy Manganese-Vanadium	A441	40	Plates & bars over 4″ up thru 8″ thick
		42	Structural shape groups 4 & 5, and plates & bars over 1-1/2″ up thru 4″ thick
		46	Structural shape group 3, and plates & bars over 3/4″ up thru 1-1/2″ thick
		50	Structural shape groups 1 & 2, and plates & bars up thru 3/4″ thick
High-Strength Low-Alloy Columbium-Vanadium	A572	42	All structural shape groups, and plates & bars up thru 6″ thick
		50	All structural shape groups, and plates & bars up thru 4″ thick
		60	Structural shape groups 1 & 2, and plates & bars up thru 1-1/4″ thick
		65	Structural shape group 1, and plates & bars up thru 1-1/4″ thick
High-Strength Low-Alloy Columbium-Vanadium	A992	50	All structural shape groups
Corrosion-Resistant High-Strength Low-Alloy	A242	42	Structural shape groups 4 & 5, and plates & bars over 1-1/2″ up thru 4″ thick
		46	Structural shape group 3, and plates & bars over 3/4″ up thru 1-1/2″ thick
		50	Structural shape groups 1 & 2, and plates & bars up thru 3/4″ thick
Weathering High-Strength Low-Alloy	A588	42	Plates & bars over 5″ up thru 8″ thick
		46	Plates & bars over 4″ up thru 5″ thick
		50	All structural shape groups, and plates & bars up thru 4″ thick
Quenched and Tempered Low-Alloy	A852	70	Plates & bars up thru 4″ thick
Quenched and Tempered Alloy	A514	90	Plates & bars over 2-1/2″ up thru 6″ thick
		100	Plates & bars up thru 2-1/2″ thick

R051223-30 High Strength Steels

The mill price of high strength steels may be higher than A992 carbon steel but their proper use can achieve overall savings thru total reduced weights. For columns with L/r over 100, A992 steel is best; under 100, high strength steels are economical. For heavy columns, high strength steels are economical when cover plates are eliminated. There is no economy using high strength steels for clip angles or supports or for beams where deflection governs. Thinner members are more economical than thick.

The per ton erection and fabricating costs of the high strength steels will be higher than for A992 since the same number of pieces, but less weight, will be installed.

R051223-35 Common Steel Sections

The upper portion of this table shows the name, shape, common designation and basic characteristics of commonly used steel sections. The lower portion explains how to read the designations used for the above illustrated common sections.

Shape & Designation	Name & Characteristics	Shape & Designation	Name & Characteristics
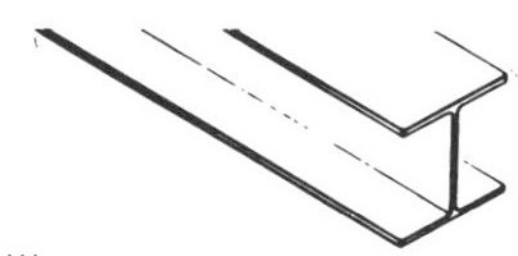 W	W Shape Parallel flange surfaces	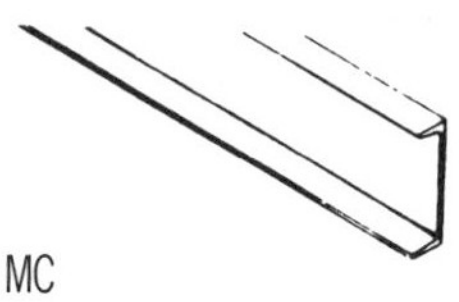 MC	Miscellaneous Channel Infrequently rolled by some producers
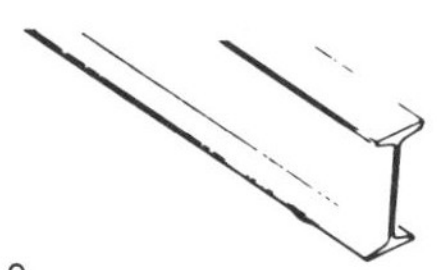 S	American Standard Beam (I Beam) Sloped inner flange	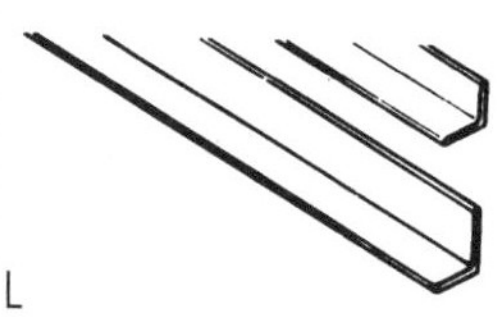 L	Angle Equal or unequal legs, constant thickness
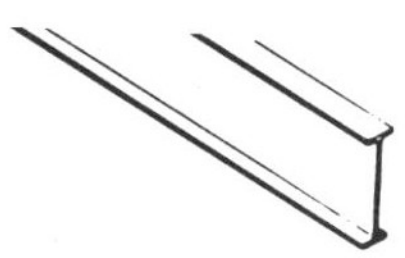 M	Miscellaneous Beams Cannot be classified as W, HP or S; infrequently rolled by some producers	 T	Structural Tee Cut from W, M or S on center of web
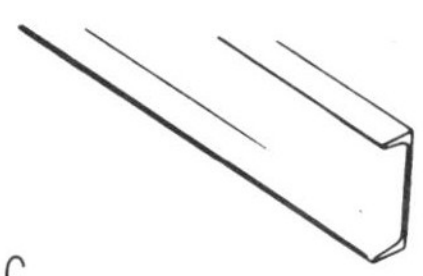 C	American Standard Channel Sloped inner flange	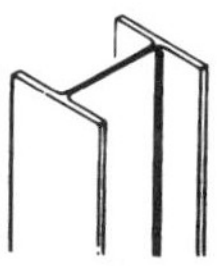 HP	Bearing Pile Parallel flanges and equal flange and web thickness

Common drawing designations follow:

W Shape → **W 18 x 35**
- 18 — Nominal Depth in Inches (Actual 17-3/4″)
- 35 — Weight in Pounds per Foot

American Standard Beam → **S 12 x 31.8**
- 12 — Depth in Inches
- 31.8 — Weight in Pounds per Foot

Miscellaneous Beam → **M 8 x 6.5**
- 8 — Depth in Inches
- 6.5 — Weight in Pounds per Foot

American Standard Channel → **C 8 x 11.5**
- 8 — Depth in Inches
- 11.5 — Weight in Pounds per Foot

Miscellaneous Channel → **MC 8 x 22.8**
- 8 — Depth in Inches
- 22.8 — Weight in Pounds per Foot

Angle → **L 6 x 3-1/2 x 3/8**
- 6 — Length of Long Leg in Inches
- 3-1/2 — Length of Other Leg in Inches
- 3/8 — Thickness of Each Leg in Inches

Tee Cut from W16 x 100 → **WT 8 x 50**
- 8 — Nominal Depth in Inches (Actual 8-1/2″)
- 50 — Weight in Pounds per Foot

Tee Cut from S12 x 35 → **ST 6 x 17.5**
- 6 — Depth in Inches (Actual 6-1/4″)
- 17.5 — Weight in Pounds per Foot

Tee Cut from M10 x 9 → **MT 5 x 4.5**
- 5 — Depth in Inches
- 4.5 — Weight in Pounds per Foot

Bearing Pile → **HP 12 x 84**
- 12 — Nominal Depth in Inches (Actual 12-1/4″)
- 84 — Weight in Pounds per Foot

R051223-45 Installation Time for Structural Steel Building Components

The following tables show the expected average installation times for various structural steel shapes. Table A presents installation times for columns, Table B for beams, Table C for light framing and bolts, and Table D for structural steel for various project types.

Table A

Description	Labor-Hours	Unit
Columns		
Steel, Concrete Filled		
3-1/2″ Diameter	.933	Ea.
6-5/8″ Diameter	1.120	Ea.
Steel Pipe		
3″ Diameter	.933	Ea.
8″ Diameter	1.120	Ea.
12″ Diameter	1.244	Ea.
Structural Tubing		
4″ x 4″	.966	Ea.
8″ x 8″	1.120	Ea.
12″ x 8″	1.167	Ea.
W Shape 2 Tier		
W8 x 31	.052	L.F.
W8 x 67	.057	L.F.
W10 x 45	.054	L.F.
W10 x 112	.058	L.F.
W12 x 50	.054	L.F.
W12 x 190	.061	L.F.
W14 x 74	.057	L.F.
W14 x 176	.061	L.F.

Table B

Description	Labor-Hours	Unit	Labor-Hours	Unit
Beams, W Shape				
W6 x 9	.949	Ea.	.093	L.F.
W10 x 22	1.037	Ea.	.085	L.F.
W12 x 26	1.037	Ea.	.064	L.F.
W14 x 34	1.333	Ea.	.069	L.F.
W16 x 31	1.333	Ea.	.062	L.F.
W18 x 50	2.162	Ea.	.088	L.F.
W21 x 62	2.222	Ea.	.077	L.F.
W24 x 76	2.353	Ea.	.072	L.F.
W27 x 94	2.581	Ea.	.067	L.F.
W30 x 108	2.857	Ea.	.067	L.F.
W33 x 130	3.200	Ea.	.071	L.F.
W36 x 300	3.810	Ea.	.077	L.F.

Table C

Description	Labor-Hours	Unit
Light Framing		
Angles 4″ and Larger	.055	lbs.
Less than 4″	.091	lbs.
Channels 8″ and Larger	.048	lbs.
Less than 8″	.072	lbs.
Cross Bracing Angles	.055	lbs.
Rods	.034	lbs.
Hanging Lintels	.069	lbs.
High Strength Bolts in Place		
3/4″ Bolts	.070	Ea.
7/8″ Bolts	.076	Ea.

Table D

Description	Labor-Hours	Unit	Labor-Hours	Unit
Apartments, Nursing Homes, etc.				
1-2 Stories	4.211	Piece	7.767	Ton
3-6 Stories	4.444	Piece	7.921	Ton
7-15 Stories	4.923	Piece	9.014	Ton
Over 15 Stories	5.333	Piece	9.209	Ton
Offices, Hospitals, etc.				
1-2 Stories	4.211	Piece	7.767	Ton
3-6 Stories	4.741	Piece	8.889	Ton
7-15 Stories	4.923	Piece	9.014	Ton
Over 15 Stories	5.120	Piece	9.209	Ton
Industrial Buildings				
1 Story	3.478	Piece	6.202	Ton

R051223-50 Subpurlins

Bulb tee subpurlins are structural members designed to support and reinforce a variety of roof deck systems such as precast cement fiber roof deck tiles, monolithic roof deck systems, and gypsum or lightweight concrete over formboard. Other uses include interstitial service ceiling systems, wall panel systems, and joist anchoring in bond beams. See Unit Price section for pricing on a square foot basis at 32-5/8″ O.C. Maximum span is based on a 3-span condition with a total allowable vertical load of 40 psf.

R051223-80 Dimensions and Weights of Sheet Steel

Gauge No.	Approximate Thickness				Weight		
	Inches (in fractions)	Inches (in decimal parts)		Millimeters			
	Wrought Iron	Wrought Iron	Steel	Steel	per S.F. in Ounces	per S.F. in Lbs.	per Square Meter in Kg.
0000000	1/2″	.5	.4782	12.146	320	20.000	97.650
000000	15/32″	.46875	.4484	11.389	300	18.750	91.550
00000	7/16″	.4375	.4185	10.630	280	17.500	85.440
0000	13/32″	.40625	.3886	9.870	260	16.250	79.330
000	3/8″	.375	.3587	9.111	240	15.000	73.240
00	11/32″	.34375	.3288	8.352	220	13.750	67.130
0	5/16″	.3125	.2989	7.592	200	12.500	61.030
1	9/32″	.28125	.2690	6.833	180	11.250	54.930
2	17/64″	.265625	.2541	6.454	170	10.625	51.880
3	1/4″	.25	.2391	6.073	160	10.000	48.820
4	15/64″	.234375	.2242	5.695	150	9.375	45.770
5	7/32″	.21875	.2092	5.314	140	8.750	42.720
6	13/64″	.203125	.1943	4.935	130	8.125	39.670
7	3/16″	.1875	.1793	4.554	120	7.500	36.320
8	11/64″	.171875	.1644	4.176	110	6.875	33.570
9	5/32″	.15625	.1495	3.797	100	6.250	30.520
10	9/64″	.140625	.1345	3.416	90	5.625	27.460
11	1/8″	.125	.1196	3.038	80	5.000	24.410
12	7/64″	.109375	.1046	2.657	70	4.375	21.360
13	3/32″	.09375	.0897	2.278	60	3.750	18.310
14	5/64″	.078125	.0747	1.897	50	3.125	15.260
15	9/128″	.0713125	.0673	1.709	45	2.813	13.730
16	1/16″	.0625	.0598	1.519	40	2.500	12.210
17	9/160″	.05625	.0538	1.367	36	2.250	10.990
18	1/20″	.05	.0478	1.214	32	2.000	9.765
19	7/160″	.04375	.0418	1.062	28	1.750	8.544
20	3/80″	.0375	.0359	.912	24	1.500	7.324
21	11/320″	.034375	.0329	.836	22	1.375	6.713
22	1/32″	.03125	.0299	.759	20	1.250	6.103
23	9/320″	.028125	.0269	.683	18	1.125	5.490
24	1/40″	.025	.0239	.607	16	1.000	4.882
25	7/320″	.021875	.0209	.531	14	.875	4.272
26	3/160″	.01875	.0179	.455	12	.750	3.662
27	11/640″	.0171875	.0164	.417	11	.688	3.357
28	1/64″	.015625	.0149	.378	10	.625	3.052

R053100-10 Decking Descriptions

General - All Deck Products

Steel deck is made by cold forming structural grade sheet steel into a repeating pattern of parallel ribs. The strength and stiffness of the panels are the result of the ribs and the material properties of the steel. Deck lengths can be varied to suit job conditions, but because of shipping considerations, are usually less than 40 feet. Standard deck width varies with the product used but full sheets are usually 12″, 18″, 24″, 30″, or 36″. Deck is typically furnished in a standard width with the ends cut square. Any cutting for width, such as at openings or for angular fit, is done at the job site.

Deck is typically attached to the building frame with arc puddle welds, self-drilling screws, or powder or pneumatically driven pins. Sheet to sheet fastening is done with screws, button punching (crimping), or welds.

Composite Floor Deck

After installation and adequate fastening, floor deck serves several purposes. It (a) acts as a working platform, (b) stabilizes the frame, (c) serves as a concrete form for the slab, and (d) reinforces the slab to carry the design loads applied during the life of the building. Composite decks are distinguished by the presence of shear connector devices as part of the deck. These devices are designed to mechanically lock the concrete and deck together so that the concrete and the deck work together to carry subsequent floor loads. These shear connector devices can be rolled-in embossments, lugs, holes, or wires welded to the panels. The deck profile can also be used to interlock concrete and steel.

Composite deck finishes are either galvanized (zinc coated) or phosphatized/painted. Galvanized deck has a zinc coating on both the top and bottom surfaces. The phosphatized/painted deck has a bare (phosphatized) top surface that will come into contact with the concrete. This bare top surface can be expected to develop rust before the concrete is placed. The bottom side of the deck has a primer coat of paint.

Composite floor deck is normally installed so the panel ends do not overlap on the supporting beams. Shear lugs or panel profile shape often prevent a tight metal to metal fit if the panel ends overlap; the air gap caused by overlapping will prevent proper fusion with the structural steel supports when the panel end laps are shear stud welded.

Adequate end bearing of the deck must be obtained as shown on the drawings. If bearing is actually less in the field than shown on the drawings, further investigation is required.

Roof Deck

Roof deck is not designed to act compositely with other materials. Roof deck acts alone in transferring horizontal and vertical loads into the building frame. Roof deck rib openings are usually narrower than floor deck rib openings. This provides adequate support of rigid thermal insulation board.

Roof deck is typically installed to endlap approximately 2″ over supports. However, it can be butted (or lapped more than 2″) to solve field fit problems. Since designers frequently use the installed deck system as part of the horizontal bracing system (the deck as a diaphragm), any fastening substitution or change should be approved by the designer. Continuous perimeter support of the deck is necessary to limit edge deflection in the finished roof and may be required for diaphragm shear transfer.

Standard roof deck finishes are galvanized or primer painted. The standard factory applied paint for roof deck is a primer paint and is not intended to weather for extended periods of time. Field painting or touching up of abrasions and deterioration of the primer coat or other protective finishes is the responsibility of the contractor.

Cellular Deck

Cellular deck is made by attaching a bottom steel sheet to a roof deck or composite floor deck panel. Cellular deck can be used in the same manner as floor deck. Electrical, telephone, and data wires are easily run through the chase created between the deck panel and the bottom sheet.

When used as part of the electrical distribution system, the cellular deck must be installed so that the ribs line up and create a smooth cell transition at abutting ends. The joint that occurs at butting cell ends must be taped or otherwise sealed to prevent wet concrete from seeping into the cell. Cell interiors must be free of welding burrs, or other sharp intrusions, to prevent damage to wires.

When used as a roof deck, the bottom flat plate is usually left exposed to view. Care must be maintained during erection to keep good alignment and prevent damage.

Cellular deck is sometimes used with the flat plate on the top side to provide a flat working surface. Installation of the deck for this purpose requires special methods for attachment to the frame because the flat plate, now on the top, can prevent direct access to the deck material that is bearing on the structural steel. It may be advisable to treat the flat top surface to prevent slipping.

Cellular deck is always furnished galvanized or painted over galvanized.

Form Deck

Form deck can be any floor or roof deck product used as a concrete form. Connections to the frame are by the same methods used to anchor floor and roof deck. Welding washers are recommended when welding deck that is less than 20 gauge thickness.

Form deck is furnished galvanized, prime painted, or uncoated. Galvanized deck must be used for those roof deck systems where form deck is used to carry a lightweight insulating concrete fill.

Wood, Plastics & Composites | R0611 | Wood Framing

R061110-30 Lumber Product Material Prices

The price of forest products fluctuates widely from location to location and from season to season depending upon economic conditions. The bare material prices in the unit cost sections of the book show the National Average material prices in effect Jan. 1 of this book year. It must be noted that lumber prices in general may change significantly during the year.

Availability of certain items depends upon geographic location and must be checked prior to firm-price bidding.

Wood, Plastics & Composites | R0616 | Sheathing

R061636-20 Plywood

There are two types of plywood used in construction: interior, which is moisture-resistant but not waterproofed, and exterior, which is waterproofed.

The grade of the exterior surface of the plywood sheets is designated by the first letter: A, for smooth surface with patches allowed; B, for solid surface with patches and plugs allowed; C, which may be surface plugged or may have knot holes up to 1″ wide; and D, which is used only for interior type plywood and may have knot holes up to 2-1/2″ wide. "Structural Grade" is specifically designed for engineered applications such as box beams. All CC & DD grades have roof and floor spans marked on them.

Underlayment-grade plywood runs from 1/4″ to 1-1/4″ thick. Thicknesses 5/8″ and over have optional tongue and groove joints which eliminate the need for blocking the edges. Underlayment 19/32″ and over may be referred to as Sturd-i-Floor.

The price of plywood can fluctuate widely due to geographic and economic conditions.

Typical uses for various plywood grades are as follows:

AA-AD Interior — cupboards, shelving, paneling, furniture

BB Plyform — concrete form plywood

CDX — wall and roof sheathing

Structural — box beams, girders, stressed skin panels

AA-AC Exterior — fences, signs, siding, soffits, etc.

Underlayment — base for resilient floor coverings

Overlaid HDO — high density for concrete forms & highway signs

Overlaid MDO — medium density for painting, siding, soffits & signs

303 Siding — exterior siding, textured, striated, embossed, etc.

Thermal & Moisture Protection | R0731 | Shingles & Shakes

R073126-20 Roof Slate

16″, 18″ and 20″ are standard lengths, and slate usually comes in random widths. For standard 3/16″ thickness use 1-1/2″ copper nails. Allow for 3% breakage.

Thermal & Moisture Protection | R0751 | Built-Up Bituminous Roofing

R075113-20 Built-Up Roofing

Asphalt is available in kegs of 100 lbs. each; coal tar pitch in 560 lb. kegs. Prepared roofing felts are available in a wide range of sizes, weights and characteristics. However, the most commonly used are #15 (432 S.F. per roll, 13 lbs. per square) and #30 (216 S.F. per roll, 27 lbs. per square).

Inter-ply bitumen varies from 24 lbs. per sq. (asphalt) to 30 lbs. per sq. (coal tar) per ply, MF4@ 25%. Flood coat bitumen also varies from 60 lbs. per sq. (asphalt) to 75 lbs. per sq. (coal tar), MF4@ 25%. Expendable equipment (mops, brooms, screeds, etc.) runs about 16% of the bitumen cost. For new, inexperienced crews this factor may be much higher.

Rigid insulation board is typically applied in two layers. The first is mechanically attached to nailable decks or spot or solid mopped to non-nailable decks; the second layer is then spot or solid mopped to the first layer. Membrane application follows the insulation, except in protected membrane roofs, where the membrane goes down first and the insulation on top, followed with ballast (stone or concrete pavers). Insulation and related labor costs are NOT included in prices for built-up roofing.

Thermal & Moisture Protection | R0752 | Modified Bituminous Membrane Roofing

R075213-30 Modified Bitumen Roofing

The cost of modified bitumen roofing is highly dependent on the type of installation that is planned. Installation is based on the type of modifier used in the bitumen. The two most popular modifiers are atactic polypropylene (APP) and styrene butadiene styrene (SBS). The modifiers are added to heated bitumen during the manufacturing process to change its characteristics. A polyethylene, polyester or fiberglass reinforcing sheet is then sandwiched between layers of this bitumen. When completed, the result is a pre-assembled, built-up roof that has increased elasticity and weatherablility. Some manufacturers include a surfacing material such as ceramic or mineral granules, metal particles or sand.

The preferred method of adhering SBS-modified bitumen roofing to the substrate is with hot-mopped asphalt (much the same as built-up roofing). This installation method requires a tar kettle/pot to heat the asphalt, as well as the labor, tools and equipment necessary to distribute and spread the hot asphalt.

The alternative method for applying APP and SBS modified bitumen is as follows. A skilled installer uses a torch to melt a small pool of bitumen off the membrane. This pool must form across the entire roll for proper adhesion. The installer must unroll the roofing at a pace slow enough to melt the bitumen, but fast enough to prevent damage to the rest of the membrane.

Modified bitumen roofing provides the advantages of both built-up and single-ply roofing. Labor costs are reduced over those of built-up roofing because only a single ply is necessary. The elasticity of single-ply roofing is attained with the reinforcing sheet and polymer modifiers. Modifieds have some self-healing characteristics and because of their multi-layer construction, they offer the reliability and safety of built-up roofing.

Thermal & Moisture Protection | R0784 | Firestopping

R078413-30 Firestopping

Firestopping is the sealing of structural, mechanical, electrical and other penetrations through fire-rated assemblies. The basic components of firestop systems are safing insulation and firestop sealant on both sides of wall penetrations and the top side of floor penetrations.

Pipe penetrations are assumed to be through concrete, grout, or joint compound and can be sleeved or unsleeved. Costs for the penetrations and sleeves are not included. An annular space of 1″ is assumed. Escutcheons are not included.

Metallic pipe is assumed to be copper, aluminum, cast iron or similar metallic material. Insulated metallic pipe is assumed to be covered with a thermal insulating jacket of varying thickness and materials.

Non-metallic pipe is assumed to be PVC, CPVC, FR Polypropylene or similar plastic piping material. Intumescent firestop sealant or wrap strips are included. Collars on both sides of wall penetrations and a sheet metal plate on the underside of floor penetrations are included.

Ductwork is assumed to be sheet metal, stainless steel or similar metallic material. Duct penetrations are assumed to be through concrete, grout or joint compound. Costs for penetrations and sleeves are not included. An annular space of 1/2″ is assumed.

Multi-trade openings include costs for sheet metal forms, firestop mortar, wrap strips, collars and sealants as necessary.

Structural penetrations joints are assumed to be 1/2″ or less. CMU walls are assumed to be within 1-1/2″ of metal deck. Drywall walls are assumed to be tight to the underside of metal decking.

Metal panel, glass or curtain wall systems include a spandrel area of 5′ filled with mineral wool foil-faced insulation. Fasteners and stiffeners are included.

Openings | R0813 | Metal Doors

R081313-20 Steel Door Selection Guide

Standard steel doors are classified into four levels, as recommended by the Steel Door Institute in the chart below. Each of the four levels offers a range of construction models and designs, to meet architectural requirements for preference and appearance, including full flush, seamless, and stile & rail. Recommended minimum gauge requirements are also included.

For complete standard steel door construction specifications and available sizes, refer to the Steel Door Institute Technical Data Series, ANSI A250.8-98 (SDI-100), and ANSI A250.4-94 Test Procedure and Acceptance Criteria for Physical Endurance of Steel Door and Hardware Reinforcements.

Level		Model	Construction	For Full Flush or Seamless		
				Min. Gauge	Thickness (in)	Thickness (mm)
I	Standard Duty	1 2	Full Flush Seamless	20	0.032	0.8
II	Heavy Duty	1 2	Full Flush Seamless	18	0.042	1.0
III	Extra Heavy Duty	1 2 3	Full Flush Seamless *Stile & Rail	16	0.053	1.3
IV	Maximum Duty	1 2	Full Flush Seamless	14	0.067	1.6

*Stiles & rails are 16 gauge; flush panels, when specified, are 18 gauge

Openings | R0851 | Metal Windows

R085123-10 Steel Sash

Ironworker crew will erect 25 S.F. or 1.3 sash unit per hour, whichever is less.

Mechanic will point 30 L.F. per hour.

Painter will paint 90 S.F. per coat per hour.

Glazier production depends on light size.

Allow 1 lb. special steel sash putty per 16″ x 20″ light.

Openings | R0852 | Wood Windows

R085216-10 Window Estimates

To ensure a complete window estimate, be sure to include the material and labor costs for each window, as well as the material and labor costs for an interior wood trim set.

Openings | R0853 | Plastic Windows

R085313-20 Replacement Windows

Replacement windows are typically measured per United Inch.

United Inches are calculated by rounding the width and height of the window opening up to the nearest inch, then adding the two figures.

The labor cost for replacement windows includes removal of sash, existing sash balance or weights, parting bead where necessary and installation of new window.

Debris hauling and dump fees are not included.

Openings | R0871 | Door Hardware

R087120-10 Hinges

All closer equipped doors should have ball bearing hinges. Lead lined or extremely heavy doors require special strength hinges. Usually 1-1/2 pair of hinges are used per door up to 7′-6″ high openings. Table below shows typical hinge requirements.

Use Frequency	Type Hinge Required	Type of Opening	Type of Structure
High	Heavy weight ball bearing	Entrances Toilet Rooms	Banks, Office buildings, Schools, Stores & Theaters Office buildings and Schools
Average	Standard weight ball bearing	Entrances Corridors Toilet Rooms	Dwellings Office buildings and Schools Stores
Low	Plain bearing	Interior	Dwellings

Door Thickness	Weight of Doors in Pounds per Square Foot				
	White Pine	Oak	Hollow Core	Solid Core	Hollow Metal
1-3/8″	3psf	6psf	1-1/2psf	3-1/2— 4psf	6-1/2psf
1-3/4″	3-1/2	7	2-1/2	4-1/2—5-1/4	6-1/2
2-1/4″	4-1/2	9	—	5-1/2—6-3/4	6-1/2

Openings | R0881 | Glass Glazing

R088110-10 Glazing Productivity

Some glass sizes are estimated by the "united inch" (height + width). The table below shows the number of lights glazed in an eight-hour period by the crew size indicated, for glass up to 1/4″ thick. Square or nearly square lights are more economical on a S.F. basis. Long slender lights will have a high S.F. installation cost. For insulated glass reduce production by 33%. For 1/2″ float glass reduce production by 50%. Production time for glazing with two glaziers per day averages: 1/4″ float glass 120 S.F.; 1/2″ float glass 55 S.F.; 1/2″ insulated glass 95 S.F.; 3/4″ insulated glass 75 S.F.

Glazing Method	United Inches per Light							
	40″	60″	80″	100″	135″	165″	200″	240″
Number of Men in Crew	1	1	1	1	2	3	3	4
Industrial sash, putty	60	45	24	15	18	—	—	—
With stops, putty bed	50	36	21	12	16	8	4	3
Wood stops, rubber	40	27	15	9	11	6	3	2
Metal stops, rubber	30	24	14	9	9	6	3	2
Structural glass	10	7	4	3	—	—	—	—
Corrugated glass	12	9	7	4	4	4	3	—
Storefronts	16	15	13	11	7	6	4	4
Skylights, putty glass	60	36	21	12	16	—	—	—
Thiokol set	15	15	11	9	9	6	3	2
Vinyl set, snap on	18	18	13	12	12	7	5	4
Maximum area per light	2.8 S.F.	6.3 S.F.	11.1 S.F.	17.4 S.F.	31.6 S.F.	47 S.F.	69 S.F.	100 S.F.

R092000-50 Lath, Plaster and Gypsum Board

Gypsum board lath is available in 3/8″ thick x 16″ wide x 4′ long sheets as a base material for multi-layer plaster applications. It is also available as a base for either multi-layer or veneer plaster applications in 1/2″ and 5/8″ thick–4′ wide x 8′, 10′ or 12′ long sheets. Fasteners are screws or blued ring shank nails for wood framing and screws for metal framing.

Metal lath is available in diamond mesh pattern with flat or self-furring profiles. Paper backing is available for applications where excessive plaster waste needs to be avoided. A slotted mesh ribbed lath should be used in areas where the span between structural supports is greater than normal. Most metal lath comes in 27″ x 96″ sheets. Diamond mesh weighs 1.75, 2.5 or 3.4 pounds per square yard, slotted mesh lath weighs 2.75 or 3.4 pounds per square yard. Metal lath can be nailed, screwed or tied in place.

Many **accessories** are available. Corner beads, flat reinforcing strips, casing beads, control and expansion joints, furring brackets and channels are some examples. Note that accessories are not included in plaster or stucco line items.

Plaster is defined as a material or combination of materials that when mixed with a suitable amount of water, forms a plastic mass or paste. When applied to a surface, the paste adheres to it and subsequently hardens, preserving in a rigid state the form or texture imposed during the period of elasticity.

Gypsum plaster is made from ground calcined gypsum. It is mixed with aggregates and water for use as a base coat plaster.

Vermiculite plaster is a fire-retardant plaster covering used on steel beams, concrete slabs and other heavy construction materials. Vermiculite is a group name for certain clay minerals, hydrous silicates or aluminum, magnesium and iron that have been expanded by heat.

Perlite plaster is a plaster using perlite as an aggregate instead of sand. Perlite is a volcanic glass that has been expanded by heat.

Gauging plaster is a mix of gypsum plaster and lime putty that when applied produces a quick drying finish coat.

Veneer plaster is a one or two component gypsum plaster used as a thin finish coat over special gypsum board.

Keenes cement is a white cementitious material manufactured from gypsum that has been burned at a high temperature and ground to a fine powder. Alum is added to accelerate the set. The resulting plaster is hard and strong and accepts and maintains a high polish, hence it is used as a finishing plaster.

Stucco is a Portland cement based plaster used primarily as an exterior finish.

Plaster is used on both interior and exterior surfaces. Generally it is applied in multiple-coat systems. A three-coat system uses the terms scratch, brown and finish to identify each coat. A two-coat system uses base and finish to describe each coat. Each type of plaster and application system has attributes that are chosen by the designer to best fit the intended use.

Gypsum Plaster Quantities for 100 S.Y.	2 Coat, 5/8″ Thick		3 Coat, 3/4″ Thick		
	Base	Finish	Scratch	Brown	Finish
	1:3 Mix	2:1 Mix	1:2 Mix	1:3 Mix	2:1 Mix
Gypsum plaster	1,300 lb.		1,350 lb.	650 lb.	
Sand	1.75 C.Y.		1.85 C.Y.	1.35 C.Y.	
Finish hydrated lime		340 lb.			340 lb.
Gauging plaster		170 lb.			170 lb.

Vermiculite or Perlite Plaster Quantities for 100 S.Y.	2 Coat, 5/8″ Thick		3 Coat, 3/4″ Thick		
	Base	Finish	Scratch	Brown	Finish
Gypsum plaster	1,250 lb.		1,450 lb.	800 lb.	
Vermiculite or perlite	7.8 bags		8.0 bags	3.3 bags	
Finish hydrated lime		340 lb.			340 lb.
Gauging plaster		170 lb.			170 lb.

Stucco-Three-Coat System Quantities for 100 S.Y.	On Wood Frame	On Masonry
Portland cement	29 bags	21 bags
Sand	2.6 C.Y.	2.0 C.Y.
Hydrated lime	180 lb.	120 lb.

R092910-10 Levels of Gypsum Drywall Finish

In the past, contract documents often used phrases such as "industry standard" and "workmanlike finish" to specify the expected quality of gypsum board wall and ceiling installations. The vagueness of these descriptions led to unacceptable work and disputes.

In order to resolve this problem, four major trade associations concerned with the manufacture, erection, finish and decoration of gypsum board wall and ceiling systems have developed an industry-wide *Recommended Levels of Gypsum Board Finish*.

The finish of gypsum board walls and ceilings for specific final decoration is dependent on a number of factors. A primary consideration is the location of the surface and the degree of decorative treatment desired. Painted and unpainted surfaces in warehouses and other areas where appearance is normally not critical may simply require the taping of wallboard joints and 'spotting' of fastener heads. Blemish-free, smooth, monolithic surfaces often intended for painted and decorated walls and ceilings in habitated structures, ranging from single-family dwellings through monumental buildings, require additional finishing prior to the application of the final decoration.

Other factors to be considered in determining the level of finish of the gypsum board surface are (1) the type of angle of surface illumination (both natural and artificial lighting), and (2) the paint and method of application or the type and finish of wallcovering specified as the final decoration. Critical lighting conditions, gloss paints, and thin wallcoverings require a higher level of gypsum board finish than do heavily textured surfaces which are subsequently painted or surfaces which are to be decorated with heavy grade wallcoverings.

The following descriptions were developed jointly by the Association of the Wall and Ceiling Industries-International (AWCI), Ceiling & Interior Systems Construction Association (CISCA), Gypsum Association (GA), and Painting and Decorating Contractors of America (PDCA) as a guide.

Level 0: No taping, finishing, or accessories required. This level of finish may be useful in temporary construction or whenever the final decoration has not been determined.

Level 1: All joints and interior angles shall have tape set in joint compound. Surface shall be free of excess joint compound. Tool marks and ridges are acceptable. Frequently specified in plenum areas above ceilings, in attics, in areas where the assembly would generally be concealed or in building service corridors, and other areas not normally open to public view.

Level 2: All joints and interior angles shall have tape embedded in joint compound and wiped with a joint knife leaving a thin coating of joint compound over all joints and interior angles. Fastener heads and accessories shall be covered with a coat of joint compound. Surface shall be free of excess joint compound. Tool marks and ridges are acceptable. Joint compound applied over the body of the tape at the time of tape embedment shall be considered a separate coat of joint compound and shall satisfy the conditions of this level. Specified where water-resistant gypsum backing board is used as a substrate for tile; may be specified in garages, warehouse storage, or other similar areas where surface appearance is not of primary concern.

Level 3: All joints and interior angles shall have tape embedded in joint compound and one additional coat of joint compound applied over all joints and interior angles. Fastener heads and accessories shall be covered with two separate coats of joint compound. All joint compound shall be smooth and free of tool marks and ridges. Typically specified in appearance areas which are to receive heavy- or medium-texture (spray or hand applied) finishes before final painting, or where heavy-grade wallcoverings are to be applied as the final decoration. This level of finish is not recommended where smooth painted surfaces or light to medium wallcoverings are specified.

Level 4: All joints and interior angles shall have tape embedded in joint compound and two separate coats of joint compound applied over all flat joints and one separate coat of joint compound applied over interior angles. Fastener heads and accessories shall be covered with three separate coats of joint compound. All joint compound shall be smooth and free of tool marks and ridges. This level should be specified where flat paints, light textures, or wallcoverings are to be applied. In critical lighting areas, flat paints applied over light textures tend to reduce joint photographing. Gloss, semi-gloss, and enamel paints are not recommended over this level of finish. The weight, texture, and sheen level of wallcoverings applied over this level of finish should be carefully evaluated. Joints and fasteners must be adequately concealed if the wallcovering material is lightweight, contains limited pattern, has a gloss finish, or any combination of these finishes is present. Unbacked vinyl wallcoverings are not recommended over this level of finish.

Level 5: All joints and interior angles shall have tape embedded in joint compound and two separate coats of joint compound applied over all flat joints and one separate coat of joint compound applied over interior angles. Fastener heads and accessories shall be covered with three separate coats of joint compound. A thin skim coat of joint compound or a material manufactured especially for this purpose, shall be applied to the entire surface. The surface shall be smooth and free of tool marks and ridges. This level of finish is highly recommended where gloss, semi-gloss, enamel, or nontextured flat paints are specified or where severe lighting conditions occur. This highest quality finish is the most effective method to provide a uniform surface and minimize the possibility of joint photographing and of fasteners showing through the final decoration.

R096613-10 Terrazzo Floor

The table below lists quantities required for 100 S.F. of 5/8″ terrazzo topping, either bonded or not bonded.

Description	Bonded to Concrete 1-1/8″ Bed, 1:4 Mix	Not Bonded 2-1/8″ Bed and 1/4″ Sand
Portland cement, 94 lb. Bag	6 bags	8 bags
Sand	10 C.F.	20 C.F.
Divider strips, 4′ squares	50 L.F.	50 L.F.
Terrazzo fill, 50 lb. Bag	12 bags	12 bags
15 Lb. tarred felt		1 C.S.F.
Mesh 2 x 2 #14 galvanized		1 C.S.F.
Crew J-3	0.77 days	0.87 days

2′ x 2′ panels require 1.00 L.F. divider strip per S.F.

3′ x 3′ panels require 0.67 L.F. divider strip per S.F.

4′ x 4′ panels require 0.50 L.F. divider strip per S.F.

5′ x 5′ panels require 0.40 L.F. divider strip per S.F.

6′ x 6′ panels require 0.33 L.F. divider strip per S.F.

R097223-10 Wall Covering

The table below lists the quantities required for 100 S.F. of wall covering.

Description	Medium-Priced Paper	Expensive Paper
Paper	1.6 dbl. rolls	1.6 dbl. rolls
Wall sizing	0.25 gallon	0.25 gallon
Vinyl wall paste	0.6 gallon	0.6 gallon
Apply sizing	0.3 hour	0.3 hour
Apply paper	1.2 hours	1.5 hours

Most wallpapers now come in double rolls only.
To remove old paper, allow 1.3 hours per 100 S.F.

R099100-20 Painting

Item	Coat	One Gallon Covers			In 8 Hours a Laborer Covers			Labor-Hours per 100 S.F.		
		Brush	Roller	Spray	Brush	Roller	Spray	Brush	Roller	Spray
Paint wood siding	prime	250 S.F.	225 S.F.	290 S.F.	1150 S.F.	1300 S.F.	2275 S.F.	.695	.615	.351
	others	270	250	290	1300	1625	2600	.615	.492	.307
Paint exterior trim	prime	400	—	—	650	—	—	1.230	—	—
	1st	475	—	—	800	—	—	1.000	—	—
	2nd	520	—	—	975	—	—	.820	—	—
Paint shingle siding	prime	270	255	300	650	975	1950	1.230	.820	.410
	others	360	340	380	800	1150	2275	1.000	.695	.351
Stain shingle siding	1st	180	170	200	750	1125	2250	1.068	.711	.355
	2nd	270	250	290	900	1325	2600	.888	.603	.307
Paint brick masonry	prime	180	135	160	750	800	1800	1.066	1.000	.444
	1st	270	225	290	815	975	2275	.981	.820	.351
	2nd	340	305	360	815	1150	2925	.981	.695	.273
Paint interior plaster or drywall	prime	400	380	495	1150	2000	3250	.695	.400	.246
	others	450	425	495	1300	2300	4000	.615	.347	.200
Paint interior doors and windows	prime	400	—	—	650	—	—	1.230	—	—
	1st	425	—	—	800	—	—	1.000	—	—
	2nd	450	—	—	975	—	—	.820	—	—

R131113-20 Swimming Pools

Pool prices given per square foot of surface area include pool structure, filter and chlorination equipment, pumps, related piping, ladders/steps, maintenance kit, skimmer and vacuum system. Decks and electrical service to equipment are not included.

Residential in-ground pool construction can be divided into two categories: vinyl lined and gunite. Vinyl lined pool walls are constructed of different materials including wood, concrete, plastic or metal. The bottom is often graded with sand over which the vinyl liner is installed. Vermiculite or soil cement bottoms may be substituted for an added cost.

Gunite pool construction is used both in residential and municipal installations. These structures are steel reinforced for strength and finished with a white cement limestone plaster.

Municipal pools will have a higher cost because plumbing codes require more expensive materials, chlorination equipment and higher filtration rates.

Municipal pools greater than 1,800 S.F. require gutter systems to control waves. This gutter may be formed into the concrete wall. Often a vinyl/stainless steel gutter or gutter/wall system is specified, which will raise the pool cost.

Competition pools usually require tile bottoms and sides with contrasting lane striping, which will also raise the pool cost.

R133113-10 Air Supported Structures

Air supported structures are made from fabrics that can be classified into two groups: temporary and permanent. Temporary fabrics include nylon, woven polyethylene, vinyl film, and vinyl coated dacron. These have lifespans that range from five to fifteen plus years. The cost per square foot includes a fabric shell, tension cables, primary and back-up inflation systems and doors. The lower cost structures are used for construction shelters, bulk storage and pond covers. The more expensive are used for recreational structures and warehouses.

Permanent fabrics are teflon coated fiberglass. The life of this structure is twenty plus years. The high cost limits its application to architectural designed structures which call for a clear span covered area, such as stadiums and convention centers. Both temporary and permanent structures are available in translucent fabrics which eliminates the need for daytime lighting.

Areas to be covered vary from 10,000 S.F. to any area up to 1000 foot wide by any length. Height restrictions range from a maximum of 1/2 of width to a minimum of 1/6 of the width. Erection of even the largest of the temporary structures requires no more than a week.

Centrifugal fans provide the inflation necessary to support the structure during application of live loads. Airlocks are usually used at large entrances to prevent loss of static pressure. Some manufacturers employ propeller fans which generate sufficient airflow (30,000 CFM) to eliminate the need for airlocks. These fans may also be automatically controlled to resist high wind conditions, regulate humidity (air changes), and provide cooling and heat.

Insulation can be provided with the addition of a second or even third interior liner, creating a dead air space with an "R" value of four to nine. Some structures allow for the liner to be collapsed into the outer shell to enable the internal heat to melt accumulated snow. For cooling or air conditioning, the exterior face of the liner can be aluminized to reflect the sun's heat.

R133113-90 Seismic Bracing

Sometimes referred to as anti-sway bracing, this support system is required in earthquake areas. The individual components must be assembled to make a required system.

Example

15070-400-0040	C-Clamp 3/8″ rod	2 ea.
15070-400-0330	Rod, continuous thread 3/8″	10 L.F.
15070-400-5030	Field, weld 1″	2 ea.

Additionally, height factors must be taken ainto account. Add the following percentages to labor for elevated installations:

15′ to 20′ high	10%
21′ to 25′ high	20%
26′ to 30′ high	30%
31′ to 35′ high	40%
36′ to 40′ high	50%
41′ to 50′ high	60%

Special Construction | R1334 | Fabricated Engineered Structures

R133419-10 Pre-engineered Steel Buildings

These buildings are manufactured by many companies and normally erected by franchised dealers throughout the U.S. The four basic types are: Rigid Frames, Truss type, Post and Beam and the Sloped Beam type. Most popular roof slope is low pitch of 1″ in 12″. The minimum economical area of these buildings is about 3000 S.F. of floor area. Bay sizes are usually 20′ to 24′ but can go as high as 30′ with heavier girts and purlins. Eave heights are usually 12′ to 24′ with 18′ to 20′ most typical.

Material prices shown in the Unit Price section are bare costs for the building shell only and do not include floors, foundations, anchor bolts, interior finishes or utilities. Costs assume at least three bays of 24′ each, a 1″ in 12″ roof slope, and they are based on 30 psf roof load and 20 psf wind load (wind load is a function of wind speed, building height, and terrain characteristics; this should be determined by a Registered Structural Engineer) and no unusual requirements. Costs include the structural frame, 26 ga. non-insulated colored corrugated or ribbed roofing and siding panels, fasteners, closures, trim and flashing but no allowance for insulation, doors, windows, skylights, gutters or downspouts. Very large projects would generally cost less for materials than the prices shown. For roof panel substitutions and wall panel substitutions, see appropriate Unit Price sections.

Conditions at the site, weather, shape and size of the building, and labor availability will affect the erection cost of the building.

R133423-30 Dome Structures

Steel — The four types are Lamella, Schwedler, Arch and Geodesic. For maximum economy, rise should be about 15 to 20% of diameter. Most common diameters are in the 200′ to 300′ range. Lamella domes weigh about 5 P.S.F. of floor area less than Schwedler domes. Schwedler dome weight in lbs. per S.F. approaches .046 times the diameter. Domes below 125′ diameter weigh .07 times diameter and the cost per ton of steel is higher. See R051223-20 for estimating weight.

Wood — Small domes are of sawn lumber, larger ones are laminated. In larger sizes, triaxial and triangular cost about the same; radial domes cost more. Radial domes are economical in the 60′ to 70′ diameter range. Most economical range of all types is 80′ to 200′ diameters. Diameters can run over 400′. All costs are quoted above the foundation. Prices include 2″ decking and a tension tie ring in place.

Plywood — Stock prefab geodesic domes are available with diameters from 24′ to 60′.

Fiberglass — Aluminum framed translucent sandwich panels with spans from 5′ to 45′ are commercially available.

Aluminum — Stressed skin aluminum panels form geodesic domes with spans ranging from 82′ to 232′. An aluminum space truss, triangulated or nontriangulated, with aluminum or clear acrylic closure panels can be used for clear spans of 40′ to 415′.

R142000-10 Freight Elevators

Capacities run from 2,000 lbs. to over 100,000 lbs. with 3,000 lbs. to 10,000 lbs. most common. Travel speeds are generally lower and control less intricate than on passenger elevators. Unit prices in division 14210-200 are for hydraulic and geared elevators.

R142000-20 Elevator Selective Costs

See R142000-40 for cost development.

	Passenger		Freight		Hospital	
A. Base Unit	Hydraulic	Electric	Hydraulic	Electric	Hydraulic	Electric
Capacity	1,500 lb.	2,000 lb.	2,000 lb.	4,000 lb.	4,000 lb.	4,000 lb.
Speed	100 F.P.M.	200 F.P.M.	100 F.P.M.	200 F.P.M.	100 F.P.M.	200 F.P.M.
#Stops/Travel Ft.	2/12	4/40	2/20	4/40	2/20	4/40
Push Button Oper.	Yes	Yes	Yes	Yes	Yes	Yes
Telephone Box & Wire	"	"	"	"	"	"
Emergency Lighting	"	"	No	No	"	"
Cab	Plastic Lam. Walls	Plastic Lam. Walls	Painted Steel	Painted Steel	Plastic Lam. Walls	Plastic Lam. Walls
Cove Lighting	Yes	Yes	No	No	Yes	Yes
Floor	V.C.T.	V.C.T.	Wood w/Safety Treads	Wood w/Safety Treads	V.C.T.	V.C.T.
Doors, & Speedside Slide	Yes	Yes	Yes	Yes	Yes	Yes
Gates, Manual	No	No	No	No	No	No
Signals, Lighted Buttons	Car and Hall	Car and Hall	Car and Hall	Car and Hall	Car and Hall	Car and Hall
O.H. Geared Machine	N.A.	Yes	N.A.	Yes	N.A.	Yes
Variable Voltage Contr.	"	"	N.A.	"	"	"
Emergency Alarm	Yes	"	Yes	"	Yes	"
Class "A" Loading	N.A.	N.A.	"	"	N.A.	N.A.

R142000-30 Passenger Elevators

Electric elevators are used generally but hydraulic elevators can be used for lifts up to 70′ and where large capacities are required. Hydraulic speeds are limited to 200 F.P.M. but cars are self leveling at the stops. On low rises, hydraulic installation runs about 15% less than standard electric types but on higher rises this installation cost advantage is reduced. Maintenance of hydraulic elevators is about the same as electric type but underground portion is not included in the maintenance contract.

In electric elevators there are several control systems available, the choice of which will be based upon elevator use, size, speed and cost criteria. The two types of drives are geared for low speeds and gearless for 450 F.P.M. and over.

The tables on the preceding pages illustrate typical installed costs of the various types of elevators available.

R142000-40 Elevator Cost Development

To price a new car or truck from the factory, you must start with the manufacturer's basic model, then add or exchange optional equipment and features. The same is true for pricing elevators.

Example:

Requirement: One-passenger elevator, five-story hydraulic, 2,500 lb. capacity, 12′ floor to floor, speed 150 F.P.M., emergency power switching and maintenance contract.

Description	Adjustment
A. Base Elevator: Hydraulic Passenger, 1500 lb. Capacity, 100 fpm, 2 Stops, Standard Finish	1 Ea.
B. Capacity Adjustment (2,500 lb.)	1 Ea.
C. Excess Travel Adjustment: 48′ Total Travel (4 x 12′) minus 12′ Base Unit Travel =	36 V.L.F.
D. Stops Adjustment: 5 Total Stops minus 2 Stops (Base Unit) =	3 Stops
E. Speed Adjustment (150 F.P.M.)	1 Ea.
F. Options:	
1. Intercom Service	1 Ea.
2. Emergency Power Switching, Automatic	1 Ea.
3. Stainless Steel Entrance Doors	5 Ea.
4. Maintenance Contract (12 Months)	1 Ea.
5. Position Indicator for main floor level (none indicated in Base Unit)	1 Ea.

Conveying Equipment | R1431 | Escalators

R143110-10 Escalators

Moving stairs can be used for buildings where 600 or more people are to be carried to the second floor or beyond. Freight cannot be carried on escalators and at least one elevator must be available for this function.

Carrying capacity is 5,000 to 8,000 people per hour. Power requirement is 2 to 3 KW per hour and incline angle is 30°.

Conveying Equipment | R1432 | Moving Walks

R143210-20 Moving Ramps and Walks

These are a specialized form of conveyor 3′ to 6′ wide with capacities of 3,600 to 18,000 persons per hour. Maximum speed is 140 F.P.M. and normal incline is 0° to 15°.

Local codes will determine the maximum angle. Outdoor units would require additional weather protection.

Fire Suppression | R2112 | Fire-Suppression Standpipes

R211226-10 Standpipe Systems

The basis for standpipe system design is National Fire Protection Association NFPA 14, however, the authority having jurisdiction should be consulted for special conditions, local requirements and approval.

Standpipe systems, properly designed and maintained, are an effective and valuable time saving aid for extinquishing fires, especially in the upper stories of tall buildings, the interior of large commercial or industrial malls, or other areas where construction features or access make the laying of temporary hose lines time consuming and/or hazardous. Standpipes are frequently installed with automatic sprinkler systems for maximum protection.

There are three general classes of service for standpipe systems:
Class I for use by fire departments and personnel with special training for heavy streams (2-1/2″ hose connections).
Class II for use by building occupants until the arrival of the fire department (1-1/2″ hose connector with hose).
Class III for use by either fire departments and trained personnel or by the building occupants (both 2-1/2″ and 1-1/2″ hose connections or one 2-1/2″ hose valve with an easily removable 2-1/2″ by 1-1/2″ adapter).

Standpipe systems are also classified by the way water is supplied to the system. The four basic types are:
Type 1: Wet standpipe system having supply valve open and water pressure maintained at all times.
Type 2: Standpipe system so arranged through the use of approved devices as to admit water to the system automatically by opening a hose valve.
Type 3: Standpipe system arranged to admit water to the system through manual operation of approved remote control devices located at each hose station.
Type 4: Dry standpipe having no permanent water supply.

R211226-20 NFPA 14 Basic Standpipe Design

Class	Design-Use	Pipe Size Minimums	Water Supply Minimums
Class I	2 1/2″ hose connection on each floor All areas within 150′ of an exit in every exit stairway Fire Department Trained Personnel	Height to 100′, 4″ dia. Heights above 100′, 6″ dia. (275′ max. except with pressure regulators 400′ max.)	For each standpipe riser 500 GPM flow For common supply pipe allow 500 GPM for first standpipe plus 250 GPM for each additional standpipe (2500 GPM max. total) 30 min. duration 65 PSI at 500 GPM
Class II	1 1/2″ hose connection with hose on each floor All areas within 130′ of hose connection measured along path of hose travel Occupant personnel	Height to 50′, 2″ dia. Height above 50′, 2 1/2″ dia.	For each standpipe riser 100 GPM flow For multiple riser common supply pipe 100 GPM 300 min. duration, 65 PSI at 100 GPM
Class III	Both of above. Class I valved connections will meet Class III with additional 2 1/2″ by 1 1/2″ adapter and 1 1/2″ hose.	Same as Class I	Same as Class I

*Note: Where 2 or more standpipes are installed in the same building or section of building they shall be interconnected at the bottom.

Combined Systems

Combined systems are systems where the risers supply both automatic sprinklers and 2-1/2″ hose connection outlets for fire department use. In such a system the sprinkler spacing pattern shall be in accordance with NFPA 13 while the risers and supply piping will be sized in accordance with NFPA 14. When the building is completely sprinklered the risers may be sized by hydraulic calculation.The minimum size riser for buildings not completely sprinklered is 6″.
The minimum water supply of a completely sprinklered, light hazard, high-rise occupancy building will be 500 GPM while the supply required for other types of completely sprinklered high-rise buildings is 1000 GPM.

General System Requirements

1. Approved valves will be provided at the riser for controlling branch lines to hose outlets.
2. A hose valve will be provided at each outlet for attachment of hose.
3. Where pressure at any standpipe outlet exceeds 100 PSI a pressure reducer must be installed to limit the pressure to 100 PSI. Note that the pressure head due to gravity in 100′ of riser is 43.4 PSI. This must be overcome by city pressure, fire pumps, or gravity tanks to provide adequate pressure at the top of the riser.
4. Each hose valve on a wet system having linen hose shall have an automatic drip connection to prevent valve leakage from entering the hose.
5. Each riser will have a valve to isolate it from the rest of the system.
6. One or more fire department connections as an auxiliary supply shall be provided for each Class I or Class III standpipe system. In buildings having two or more zones, a connection will be provided for each zone.
7. There will be no shutoff valve in the fire department connection, but a check valve will be located in the line before it joins the system.
8. All hose connections street side will be identified on a cast plate or fitting as to purpose.

R211313-10 Sprinkler Systems (Automatic)

Sprinkler systems may be classified by type as follows:

1. **Wet Pipe System.** A system employing automatic sprinklers attached to a piping system containing water and connected to a water supply so that water discharges immediately from sprinklers opened by a fire.
2. **Dry Pipe System.** A system employing automatic sprinklers attached to a piping system containing air under pressure, the release of which as from the opening of sprinklers permits the water pressure to open a valve known as a "dry pipe valve". The water then flows into the piping system and out the opened sprinklers.
3. **Pre-Action System.** A system employing automatic sprinklers attached to a piping system containing air that may or may not be under pressure, with a supplemental heat responsive system of generally more sensitive characteristics than the automatic sprinklers themselves, installed in the same areas as the sprinklers; actuation of the heat responsive system, as from a fire, opens a valve which permits water to flow into the sprinkler piping system and to be discharged from any sprinklers which may be open.
4. **Deluge System.** A system employing open sprinklers attached to a piping system connected to a water supply through a valve which is opened by the operation of a heat responsive system installed in the same areas as the sprinklers. When this valve opens, water flows into the piping system and discharges from all sprinklers attached thereto.
5. **Combined Dry Pipe and Pre-Action Sprinkler System.** A system employing automatic sprinklers attached to a piping system containing air under pressure with a supplemental heat responsive system of generally more sensitive characteristics than the automatic sprinklers themselves, installed in the same areas as the sprinklers; operation of the heat responsive system, as from a fire, actuates tripping devices which open dry pipe valves simultaneously and without loss of air pressure in the system. Operation of the heat responsive system also opens approved air exhaust valves at the end of the feed main which facilitates the filling of the system with water which usually precedes the opening of sprinklers. The heat responsive system also serves as an automatic fire alarm system.
6. **Limited Water Supply System.** A system employing automatic sprinklers and conforming to these standards but supplied by a pressure tank of limited capacity.
7. **Chemical Systems.** Systems using halon, carbon dioxide, dry chemical or high expansion foam as selected for special requirements. Agent may extinguish flames by chemically inhibiting flame propagation, suffocate flames by excluding oxygen, interrupting chemical action of oxygen uniting with fuel or sealing and cooling the combustion center.
8. **Firecycle System.** Firecycle is a fixed fire protection sprinkler system utilizing water as its extinguishing agent. It is a time delayed, recycling, preaction type which automatically shuts the water off when heat is reduced below the detector operating temperature and turns the water back on when that temperature is exceeded. The system senses a fire condition through a closed circuit electrical detector system which controls water flow to the fire automatically. Batteries supply up to 90 hour emergency power supply for system operation. The piping system is dry (until water is required) and is monitored with pressurized air. Should any leak in the system piping occur, an alarm will sound, but water will not enter the system until heat is sensed by a firecycle detector.

Area coverage sprinkler systems may be laid out and fed from the supply in any one of several patterns as shown below. It is desirable, if possible, to utilize a central feed and achieve a shorter flow path from the riser to the furthest sprinkler. This permits use of the smallest sizes of pipe possible with resulting savings.

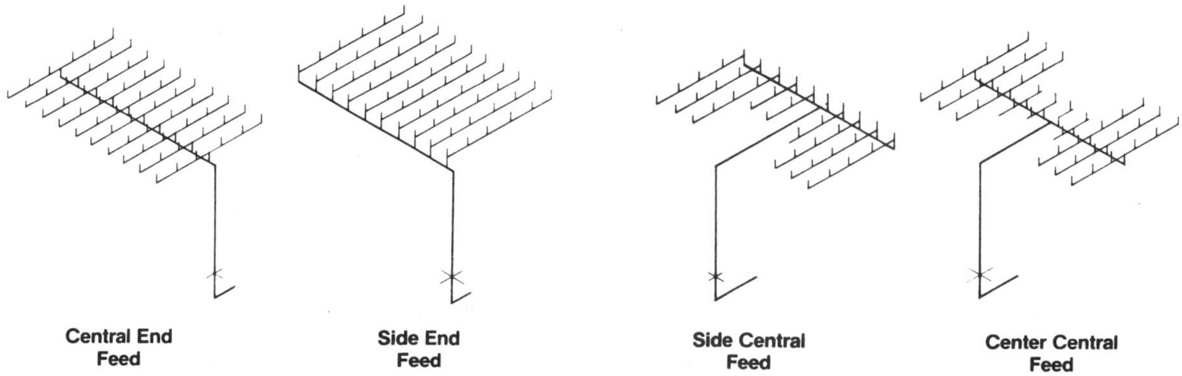

R211313-20 System Classification

System Classification

Rules for installation of sprinkler systems vary depending on the classification of occupancy falling into one of three categories as follows:

Light Hazard Occupancy

The protection area allotted per sprinkler should not exceed 225 S.F., with the maximum distance between lines and sprinklers on lines being 15′. The sprinklers do not need to be staggered. Branch lines should not exceed eight sprinklers on either side of a cross main. Each large area requiring more than 100 sprinklers and without a sub-dividing partition should be supplied by feed mains or risers sized for ordinary hazard occupancy.
Maximum system area = 52,000 S.F.

Included in this group are:

Churches
Clubs
Educational
Hospitals
Institutional
Libraries
(except large stack rooms)
Museums
Nursing Homes
Offices
Residential
Restaurants
Theaters and Auditoriums
(except stages and prosceniums)
Unused Attics

Ordinary Hazard Occupancy

The protection area allotted per sprinkler shall not exceed 130 S.F. of noncombustible ceiling and 130 S.F. of combustible ceiling. The maximum allowable distance between sprinkler lines and sprinklers on line is 15′. Sprinklers shall be staggered if the distance between heads exceeds 12′. Branch lines should not exceed eight sprinklers on either side of a cross main.
Maximum system area = 52,000 S.F.

Included in this group are:

Group 1
Automotive Parking and Showrooms
Bakeries
Beverage manufacturing
Canneries
Dairy Products Manufacturing/Processing
Electronic Plans
Glass and Glass Products Manufacturing
Laundries
Restaurant Service Areas

Group 2
Cereal Mills
Chemical Plants—Ordinary
Confectionery Products
Distilleries
Dry Cleaners
Feed Mills
Horse Stables
Leather Goods Manufacturing
Libraries—Large Stack Room Areas
Machine Shops
Metal Working
Mercantile
Paper and Pulp Mills
Paper Process Plants
Piers and Wharves
Post Offices
Printing and Publishing
Repair Garages
Stages
Textile Manufacturing
Tire Manufacturing
Tobacco Products Manufacturing
Wood Machining
Wood Product Assembly

Extra Hazard Occupancy

The protection area allotted per sprinkler shall not exceed 100 S.F. of noncombustible ceiling and 100 S.F. of combustible ceiling. The maximum allowable distance between lines and between sprinklers on lines is 12′. Sprinklers on alternate lines shall be staggered if the distance between sprinklers on lines exceeds 8′. Branch lines should not exceed six sprinklers on either side of a cross main.
Maximum system area:
Design by pipe schedule = 25,000 S.F.
Design by hydraulic calculation = 40,000 S.F.

Included in this group are:

Group 1
Aircraft hangars
Combustible Hydraulic Fluid Use Area
Die Casting
Metal Extruding
Plywood/Particle Board Manufacturing
Printing (inkls with flash points < 100 degrees F
Rubber Reclaiming, Compounding, Drying, Milling, Vulcanizing
Saw Mills
Textile Picking, Opening, Blending, Garnetting, Carding, Combing of Cotton, Synthetics, Wood Shoddy, or Burlap
Upholstering with Plastic Foams

Group 2
Asphalt Saturating
Flammable Liquids Spraying
Flow Coating
Manufactured/Modular Home Building Assemblies (where finished enclosure is present and has combustible interiors)
Open Oil Quenching
Plastics Processing
Solvent Cleaning
Varnish and Paint Dipping

R211313-30 Sprinkler Quantities for Various Sizes and Types of Pipe

Sprinkler Quantities: The table below lists the usual maximum number of sprinkler heads for each size of copper and steel pipe for both wet and dry systems. These quantities may be adjusted to meet individual structural needs or local code requirements. Maximum area on any one floor for one system is: light hazard and ordinary hazard 52,000 S.F., extra hazardous 25,000 S.F.

Pipe Size	Light Hazard Occupancy		Ordinary Hazard Occupancy		Extra Hazard Occupancy	
Diameter	Steel Pipe	Copper Pipe	Steel Pipe	Copper Pipe	Steel Pipe	Copper Pipe
1″	2 sprinklers	2 sprinklers	2 sprinklers	2 sprinklers	1 sprinklers	1 sprinklers
1-1/4″	3	3	3	3	2	2
1-1/2″	5	5	5	5	5	5
2″	10	12	10	12	8	8
2-1/2″	30	40	20	25	15	20
3″	60	65	40	45	27	30
3-1/2″	100	115	65	75	40	45
4″			100	115	55	65
5″			160	180	90	100
6″			275	300	150	170

Dry Pipe Systems: A dry pipe system should be installed where a wet pipe system is impractical, as in rooms or buildings which cannot be properly heated.

The use of an approved dry pipe system is more desirable than shutting off the water supply during cold weather.

Not more than 750 gallons of system capacity should be controlled by one dry pipe valve. Where two or more dry pipe valves are used, systems should preferably be divided horizontally.

R211313-40 Adjustment for Sprinkler/Standpipe Installations

Quality/Complexity Multiplier (For all installations)

Economy installation, add	0 to 5%
Good quality, medium complexity, add	5 to 15%
Above average quality and complexity, add	15 to 25%

R220105-10 Demolition (Selective vs. Removal for Replacement)

Demolition can be divided into two basic categories.

One type of demolition involves the removal of material with no concern for its replacement. The labor-hours to estimate this work are found under "Selective Demolition" in the Fire Protection, Plumbing and HVAC Divisions. It is selective in that individual items or all the material installed as a system or trade grouping such as plumbing or heating systems are removed. This may be accomplished by the easiest way possible, such as sawing, torch cutting, or sledge hammer as well as simple unbolting.

The second type of demolition is the removal of some item for repair or replacement. This removal may involve careful draining, opening of unions, disconnecting and tagging of electrical connections, capping of pipes/ducts to prevent entry of debris or leakage of the material contained as well as transport of the item away from its in-place location to a truck/dumpster. An approximation of the time required to accomplish this type of demolition is to use half of the time indicated as necessary to install a new unit. For example; installation of a new pump might be listed as requiring 6 labor-hours so if we had to estimate the removal of the old pump we would allow an additional 3 hours for a total of 9 hours. That is, the complete replacement of a defective pump with a new pump would be estimated to take 9 labor-hours.

R220523-80 Valve Materials

VALVE MATERIALS

Bronze:
Bronze is one of the oldest materials used to make valves. It is most commonly used in hot and cold water systems and other non-corrosive services. It is often used as a seating surface in larger iron body valves to ensure tight closure.

Carbon Steel:
Carbon steel is a high strength material. Therefore, valves made from this metal are used in higher pressure services, such as steam lines up to 600 psi at 850°F. Many steel valves are available with butt-weld ends for economy and are generally used in high pressure steam service as well as other higher pressure non-corrosive services.

Forged Steel:
Valves from tough carbon steel are used in service up to 2000 psi and temperatures up to 1000°F in Gate, Globe and Check valves.

Iron:
Valves are normally used in medium to large pipe lines to control non-corrosive fluid and gases, where pressures do not exceed 250 psi at 450° or 500 psi cold water, oil or gas.

Stainless Steel:
Developed steel alloys can be used in over 90% corrosive services.

Plastic PVC:
This is used in a great variety of valves generally in high corrosive service with lower temperatures and pressures.

VALVE SERVICE PRESSURES

Pressure ratings on valves provide an indication of the safe operating pressure for a valve at some elevated temperature. This temperature is dependent upon the materials used and the fabrication of the valve. When specific data is not available, a good "rule-of-thumb" to follow is the temperature of saturated steam on the primary rating indicated on the valve body. Example: The valve has the number 150S printed on the side indicating 150 psi and hence, a maximum operating temperature of 367°F (temperature of saturated steam and 150 psi).

DEFINITIONS

1. "WOG" – Water, oil, gas (cold working pressures).
2. "SWP" – Steam working pressure.
3. 100% area (full port) – means the area through the valve is equal to or greater than the area of standard pipe.
4. "Standard Opening" – means that the area through the valve is less than the area of standard pipe and therefore these valves should be used only where restriction of flow is unimportant.
5. "Round Port" – means the valve has a full round opening through the plug and body, of the same size and area as standard pipe.
6. "Rectangular Port" – valves have rectangular shaped ports through the plug body. The area of the port is either equal to 100% of the area of standard pipe, or restricted (standard opening). In either case it is clearly marked.
7. "ANSI" – American National Standards Institute.

R220523-90 Valve Selection Considerations

INTRODUCTION: In any piping application, valve performance is critical. Valves should be selected to give the best performance at the lowest cost.

The following is a list of performance characteristics generally expected of valves.

1. Stopping flow or starting it.
2. Throttling flow (Modulation).
3. Flow direction changing.
4. Checking backflow (Permitting flow in only one direction).
5. Relieving or regulating pressure.

In order to properly select the right valve, some facts must be determined.

A. What liquid or gas will flow through the valve?
B. Does the fluid contain suspended particles?
C. Does the fluid remain in liquid form at all times?
D. Which metals does fluid corrode?
E. What are the pressure and temperature limits? (As temperature and pressure rise, so will the price of the valve.)
F. Is there constant line pressure?
G. Is the valve merely an on-off valve?
H. Will checking of backflow be required?
I. Will the valve operate frequently or infrequently?

Valves are classified by design type into such classifications as Gate, Globe, Angle, Check, Ball, Butterfly and Plug. They are also classified by end connection, stem, pressure restrictions and material such as bronze, cast iron, etc. Each valve has a specific use. A quality valve used correctly will provide a lifetime of trouble- free service, but a high quality valve installed in the wrong service may require frequent attention.

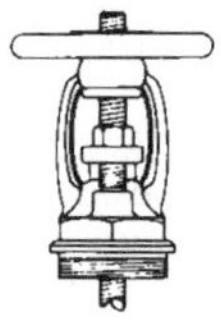

STEM TYPES
(OS & Y)—Rising Stem-Outside Screw and Yoke

Offers a visual indication of whether the valve is open or closed. Recommended where high temperatures, corrosives, and solids in the line might cause damage to inside-valve stem threads. The stem threads are engaged by the yoke bushing so the stem rises through the hand wheel as it is turned.

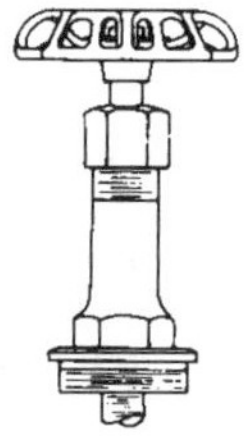

(R.S.)—Rising Stem-Inside Screw

Adequate clearance for operation must be provided because both the hand wheel and the stem rise.
The valve wedge position is indicated by the position of the stem and hand wheel.

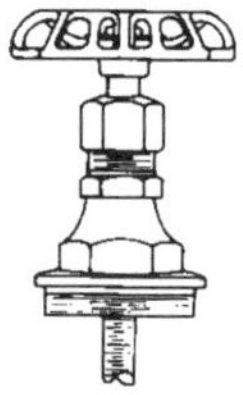

(N.R.S.)—Non-Rising Stem-Inside Screw

A minimum clearance is required for operating this type of valve. Excessive wear or damage to stem threads inside the valve may be caused by heat, corrosion, and solids. Because the hand wheel and stem do not rise, wedge position cannot be visually determined.

VALVE TYPES
Gate Valves

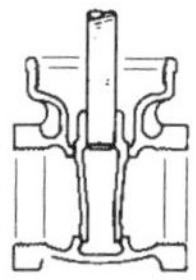

Provide full flow, minute pressure drop, minimum turbulence and minimum fluid trapped in the line.
They are normally used where operation is infrequent.

Globe Valves

Globe valves are designed for throttling and/or frequent operation with positive shut-off. Particular attention must be paid to the several types of seating materials available to avoid unnecessary wear. The seats must be compatible with the fluid in service and may be composition or metal. The configuration of the Globe valve opening causes turbulence which results in increased resistance. Most bronze Globe valves are rising stem-inside screw, but they are also available on O.S. & Y.

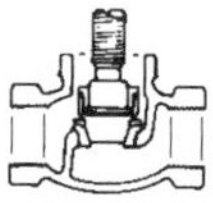

Angle Valves

The fundamental difference between the Angle valve and the Globe valve is the fluid flow through the Angle valve. It makes a 90° turn and offers less resistance to flow than the Globe valve while replacing an elbow. An Angle valve thus reduces the number of joints and installation time.

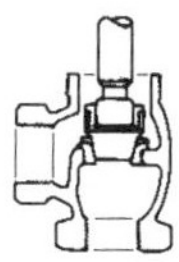

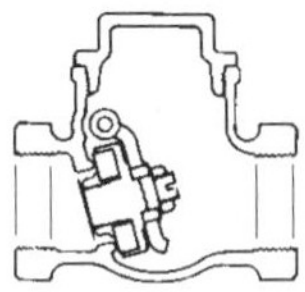

Check Valves

Check valves are designed to prevent backflow by automatically seating when the direction of fluid is reversed.
Swing Check valves are generally installed with Gate-valves, as they provide comparable full flow. Usually recommended for lines where flow velocities are low and should not be used on lines with pulsating flow. Recommended for horizontal installation, or in vertical lines only where flow is upward.

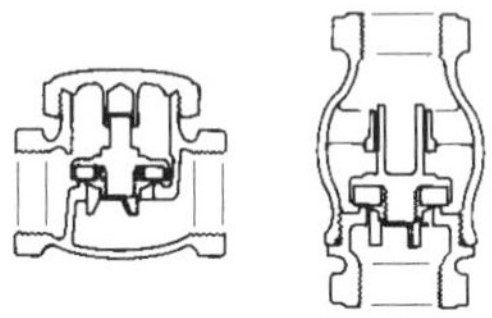

Lift Check Valves

These are commonly used with Globe and Angle valves since they have similar diaphragm seating arrangements and are recommended for preventing backflow of steam, air, gas and water, and on vapor lines with high flow velocities. For horizontal lines, horizontal lift checks should be used and vertical lift checks for vertical lines.

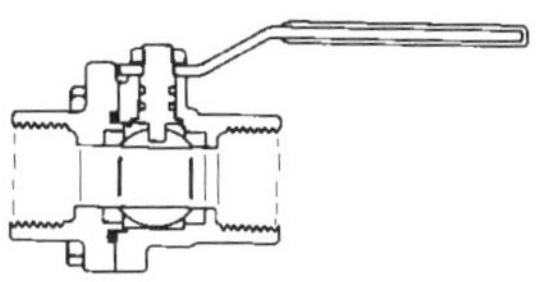

Ball Valves

Ball valves are light and easily installed, yet because of modern elastomeric seats, provide tight closure. Flow is controlled by rotating up to 90° a drilled ball which fits tightly against resilient seals. This ball seats with flow in either direction, and valve handle indicates the degree of opening. Recommended for frequent operation readily adaptable to automation, ideal for installation where space is limited.

Butterfly Valves

Butterfly valves provide bubble-tight closure with excellent throttling characteristics. They can be used for full-open, closed and for throttling applications.

The Butterfly valve consists of a disc within the valve body which is controlled by a shaft. In its closed position, the valve disc seals against a resilient seat. The disc position throughout the full 90° rotation is visually indicated by the position of the operator.

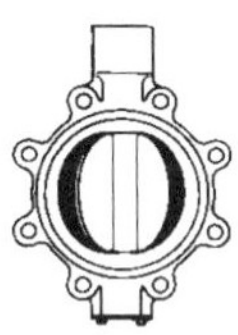

A Butterfly valve is only a fraction of the weight of a Gate valve and requires no gaskets between flanges in most cases. Recommended for frequent operation and adaptable to automation where space is limited.

Wafer and Lug type bodies when installed between two pipe flanges, can be easily removed from the line. The pressure of the bolted flanges holds the valve in place.
Locating lugs makes installation easier.

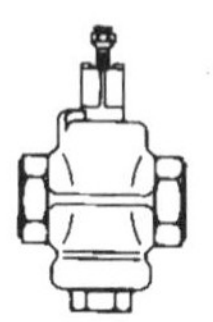

Plug Valves

Lubricated plug valves, because of the wide range of service to which they are adapted, may be classified as all purpose valves. They can be safely used at all pressure and vacuums, and at all temperatures up to the limits of available lubricants. They are the most satisfactory valves for the handling of gritty suspensions and many other destructive, erosive, corrosive and chemical solutions.

R221113-40 Plumbing Approximations for Quick Estimating

Water Control

Water Meter; Backflow Preventer, Shock Absorbers; Vacuum Breakers; Mixer. .... 10 to 15% of Fixtures

Pipe And Fittings .... 30 to 60% of Fixtures

Note: Lower percentage for compact buildings or larger buildings with plumbing in one area.
Larger percentage for large buildings with plumbing spread out.
In extreme cases pipe may be more than 100% of fixtures.
Percentages **do not** include special purpose or process piping.

Plumbing Labor

1 & 2 Story Residential .... Rough-in Labor = 80% of Materials

Apartment Buildings .... Rough-in Labor = 90 to 100% of Materials

Labor for handling and placing fixtures is approximately 25 to 30% of fixtures

Quality/Complexity Multiplier (for all installations)

Economy installation, add .... 0 to 5%

Good quality, medium complexity, add .... 5 to 15%

Above average quality and complexity, add .... 15 to 25%

R221113-50 Pipe Material Considerations

1. Malleable fittings should be used for gas service.
2. Malleable fittings are used where there are stresses/strains due to expansion and vibration.
3. Cast fittings may be broken as an aid to disassembling of heating lines frozen by long use, temperature and minerals.
4. Cast iron pipe is extensively used for underground and submerged service.
5. Type M (light wall) copper tubing is available in hard temper only and is used for nonpressure and less severe applications than K and L.
6. Type L (medium wall) copper tubing, available hard or soft for interior service.
7. Type K (heavy wall) copper tubing, available in hard or soft temper for use where conditions are severe. For underground and interior service.
8. Hard drawn tubing requires fewer hangers or supports but should not be bent. Silver brazed fittings are recommended, however soft solder is normally used.
9. Type DMV (very light wall) copper tubing designed for drainage, waste and vent plus other non-critical pressure services.

Domestic/Imported Pipe and Fittings Cost

The prices shown in this publication for steel/cast iron pipe and steel, cast iron, malleable iron fittings are based on domestic production sold at the normal trade discounts. The above listed items of foreign manufacture may be available at prices of 1/3 to 1/2 those shown. Some imported items after minor machining or finishing operations are being sold as domestic to further complicate the system.

Caution: Most pipe prices in this book also include a coupling and pipe hangers which for the larger sizes can add significantly to the per foot cost and should be taken into account when comparing "book cost" with quoted supplier's cost.

R221113-70 Piping to 10′ High

When taking off pipe, it is important to identify the different material types and joining procedures, as well as distances between supports and components required for proper support.

During the takeoff, measure through all fittings. Do not subtract the lengths of the fittings, valves, or strainers, etc. This added length plus the final rounding of the totals will compensate for nipples and waste.

When rounding off totals always increase the actual amount to correspond with manufacturer's shipping lengths.

A. Both red brass and yellow brass pipe are normally furnished in 12′ lengths, plain end. The Unit Price section includes in the linear foot costs two field threads and one coupling per 10′ length. A carbon steel clevis type hanger assembly every 10′ is also prorated into the linear foot costs, including both material and labor.

B. Cast iron soil pipe is furnished in either 5′ or 10′ lengths. For pricing purposes, the Unit Price section features 10′ lengths with a joint and a carbon steel clevis hanger assembly every 5′ prorated into the per foot costs of both material and labor.

Three methods of joining are considered: lead and oakum poured joints or push-on gasket type joints for the bell and spigot pipe and a joint clamp for the no-hub soil pipe. The labor and material costs for each of these individual joining procedures are also prorated into the linear costs per foot.

C. Copper tubing covers types K, L, M, and DWV which are furnished in 20′ lengths. Means pricing data is based on a tubing cut each length and a coupling and two soft soldered joints every 10′. A carbon steel, clevis type hanger assembly every 10′ is also prorated into the per foot costs. The prices for refrigeration tubing are for materials only. Labor for full lengths may be based on the type L labor but short cut measures in tight areas can increase the installation labor-hours from 20 to 40%.

D. Corrosion-resistant piping does not lend itself to one particular standard of hanging or support assembly due to its diversity of application and placement. The several varieties of corrosion-resistant piping do not include any material or labor costs for hanger assemblies (See the Unit Price section for appropriate selection).

E. Glass pipe is furnished in standard lengths either 5′ or 10′ long, beaded on one end. Special orders for diverse lengths beaded on both ends are also available. For pricing purposes, R.S. Means features 10′ lengths with a coupling and a carbon steel band hanger assembly every 10′ prorated into the per foot linear costs.

Glass pipe is also available with conical ends and standard lengths ranging from 6″ through 3′ in 6″ increments, then up to 10′ in 12″ increments. Special lengths can be customized for particular installation requirements.

For pricing purposes, Means has based the labor and material pricing on 10′ lengths. Included in these costs per linear foot are the prorated costs for a flanged assembly every 10′ consisting of two flanges, a gasket, two insertable seals, and the required number of bolts and nuts. A carbon steel band hanger assembly based on 10′ center lines has also been prorated into the costs per foot for labor and materials.

F. Plastic pipe of several compositions and joining methods are considered. Fiberglass reinforced pipe (FRP) is priced, based on 10′ lengths (20′ lengths are also available), with coupling and epoxy joints every 10′. FRP is furnished in both "General Service" and "High Strength." A carbon steel clevis hanger assembly, 3 for every 10′, is built into the prorated labor and material costs on a per foot basis.

The PVC and CPVC pipe schedules 40, 80 and 120, plus SDR ratings are all based on 20′ lengths with a coupling installed every 10′, as well as a carbon steel clevis hanger assembly every 3′. The PVC and ABS type DWV piping is based on 10′ lengths with solvent weld couplings every 10′, and with carbon steel clevis hanger assemblies, 3 for every 10′. The rest of the plastic piping in this section is based on flexible 100′ coils and does not include any coupling or supports.

This section ends with PVC drain and sewer piping based on 10′ lengths with bell and spigot ends and 0-ring type, push-on joints.

G. Stainless steel piping includes both weld end and threaded piping, both in the type 304 and 316 specification and in the following schedules, 5, 10, 40, 80, and 160. Although this piping is usually furnished in 20′ lengths, this cost grouping has a joint (either heli-arc butt-welded or threads and coupling) every 10′. A carbon steel clevis type hanger assembly is also included at 10′ intervals and prorated into the linear foot costs.

H. Carbon steel pipe includes both black and galvanized. This section encompasses schedules 40 (standard) and 80 (extra heavy).

Several common methods of joining steel pipe — such as thread and coupled, butt welded, and flanged (150 lb. weld neck flanges) are also included.

For estimating purposes, it is assumed that the piping is purchased in 20′ lengths and that a compatible joint is made up every 10′. These joints are prorated into the labor and material costs per linear foot. The following hanger and support assemblies every 10′ are also included: carbon steel clevis for the T & C pipe, and single rod roll type for both the welded and flanged piping. All of these hangers are oversized to accommodate pipe insulation 3/4″ thick through 5″ pipe size and 1-1/2″ thick from 6″ through 12″ pipe size.

I. Grooved joint steel pipe is priced both black and galvanized, in schedules 10, 40, and 80, furnished in 20′ lengths. This section describes two joining methods: cut groove and roll groove. The schedule 10 piping is roll-grooved, while the heavier schedules are cut-grooved. The labor and material costs are prorated into per linear foot prices, including a coupled joint every 10′, as well as a carbon steel clevis hanger assembly.

Notes:

The pipe hanger assemblies mentioned in the preceding paragraphs include the described hanger; appropriately sized steel, box-type insert and nut; plus a threaded hanger rod. On average, the distance from the pipe center line to the insert face is 2′.

C clamps are used when the pipe is to be supported from steel shapes rather than anchored in the slab. C clamps are slightly less costly than inserts. However, to save time in estimating, it is advisable to use the given line number cost, rather than substituting a C clamp for the insert.

Add to piping labor for elevated installation:

10′ to 15′ high	10%	30′ to 35′ high	40%
15′ to 20′ high	20%	35′ to 40′ high	50%
20′ to 25′ high	25%	Over 40′ high	55%
25′ to 30′ high	35%		

When using the percentage adds for elevated piping installations as shown above, bear in mind that the given heights are for the pipe supports, even though the insert, anchor, or clamp may be several feet higher than the pipe itself.

An allowance has been included in the piping installation time for testing and minor tightening of leaking joints, fittings, stuffing boxes, packing glands, etc. For extraordinary test requirements such as x-rays, prolonged pressure or demonstration tests, a percentage of the piping labor, based on the estimator's experience, must be added to the labor total. A testing service specializing in weld x-rays should be consulted for pricing if it is an estimate requirement. Equipment installation time includes start-up with associated adjustments.

R221316-10 Drainage Requirements

Drainage lines must have a slope to maintain flow for proper operation. This slope should not be less than 1/4″ per foot for 3″ diameter or smaller pipe and not less than 1/8″ per foot for 4″ diameter or larger pipe. The capacity of building drainage systems is calculated on a basis of "drainage fixture units" (d.f.u.) as per the following chart.

Type of Fixture	d.f.u. Value	Type of Fixture	d.f.u. Value
Automatic clothes washer (2″ standpipe)	3	Service sink (trap standard)	3
Bathroom group (water closet, lavatory and		Service sink (P trap)	2
bathtub or shower) tank type closet	6	Urinal, pedestal, syphon jet blowout	6
Bathtub (with or without overhead shower)	2	Urinal, wall hung	4
Clinic sink	6	Urinal, stall washout	4
Combination sink & tray with food disposal	4	Wash sink (cir. or mult.) per faucet set	2
Dental unit or cuspidor	1	Water closet, tank operated	4
Dental lavatory	1	Water closet, valve operated	6
Drinking fountain	1/2	Fixtures not listed above	
Dishwasher, domestic	2	Trap size 1-1/4″ or smaller	1
Floor drains with 2″ waste	3	Trap size 1-1/2″	2
Kitchen sink, domestic with one 1-1/2″ trap	2	Trap size 2″	3
Kitchen sink, domestic with food disposal	2	Trap size 2-1/2″	4
Lavatory with 1-1/4″ waste	1	Trap size 3″	5
Laundry tray (1 or 2 compartment)	2	Trap size 4″	6
Shower stall, domestic	2		

For continuous or nearly continuous flow into the sytem from a pump, air conditioning equipment or other item, allow 2 fixture units for each gallon per minute of flow.

When the "drainage fixture units" (d.f.u.) for each horizontal branch or vertical stack is computed from the table above, the appropriate pipe size for each branch or stack is determined from the table below.

R221316-20 Allowable Fixture Units (d.f.u.) for Branches and Stacks

Pipe Diam.	Horiz. Branch (not incl. drains)	Stack Size for 3 Stories or 3 Levels	Stack size for Over 3 levels	Maximum for 1 Story building Stack
1-1/2″	3	4	8	2
2″	6	10	24	6
2-1/2″	12	20	42	9
3″	20*	48*	72*	20*
4″	160	240	500	90
5″	360	540	1100	200
6″	620	960	1900	350
8″	1400	2200	3600	600
10″	2500	3800	5600	1000
12″	3900	6000	8400	1500
15″	7000			

*Not more than two water closets or bathroom groups within each branch interval nor more than six water closets or bathroom groups on the stack.

Stacks sized for the total may be reduced as load decreases at each story to a minimum diameter of 1/2 the maximum diameter.

R224000-10 Hot Water Consumption Rates

Type of Building	Size Factor	Maximum Hourly Demand	Average Day Demand
Apartment Dwellings	No. of Apartments:		
	Up to 20	12.0 Gal. per apt.	42.0 Gal. per apt.
	21 to 50	10.0 Gal. per apt.	40.0 Gal. per apt.
	51 to 75	8.5 Gal. per apt.	38.0 Gal. per apt.
	76 to 100	7.0 Gal. per apt.	37.0 Gal. per apt.
	101 to 200	6.0 Gal. per apt.	36.0 Gal. per apt.
	201 up	5.0 Gal. per apt.	35.0 Gal. per apt.
Dormitories	Men	3.8 Gal. per man	13.1 Gal. per man
	Women	5.0 Gal. per woman	12.3 Gal. per woman
Hospitals	Per bed	23.0 Gal. per patient	90.0 Gal. per patient
Hotels	Single room with bath	17.0 Gal. per unit	50.0 Gal. per unit
	Double room with bath	27.0 Gal. per unit	80.0 Gal. per unit
Motels	No. of units:		
	Up to 20	6.0 Gal. per unit	20.0 Gal. per unit
	21 to 100	5.0 Gal. per unit	14.0 Gal. per unit
	101 Up	4.0 Gal. per unit	10.0 Gal. per unit
Nursing Homes		4.5 Gal. per bed	18.4 Gal. per bed
Office buildings		0.4 Gal. per person	1.0 Gal. per person
Restaurants	Full meal type	1.5 Gal./max. meals/hr.	2.4 Gal. per meal
	Drive-in snack type	0.7 Gal./max. meals/hr.	0.7 Gal. per meal
Schools	Elementary	0.6 Gal. per student	0.6 Gal. per student
	Secondary & High	1.0 Gal. per student	1.8 Gal. per student

For evaluation purposes, recovery rate and storage capacity are inversely proportional. Water heaters should be sized so that the maximum hourly demand anticipated can be met in addition to allowance for the heat loss from the pipes and storage tank.

R224000-20 Fixture Demands in Gallons Per Fixture Per Hour

Table below is based on 140°F final temperature except for dishwashers in public places (*) where 180°F water is mandatory.

Fixture	Apartment House	Club	Gym	Hospital	Hotel	Indust. Plant	Office	Private Home	School
Bathtubs	20	20	30	20	20			20	
Dishwashers, automatic	15	50-150*		50-150*	50-200*	20-100*		15	20-100*
Kitchen sink	10	20		20	30	20	20	10	20
Laundry, stationary tubs	20	28		28	28			20	
Laundry, automatic wash	75	75		100	150			75	
Private lavatory	2	2	2	2	2	2	2	2	2
Public lavatory	4	6	8	6	8	12	6		15
Showers	30	150	225	75	75	225	30	30	225
Service sink	20	20		20	30	20	20	15	20
Demand factor	0.30	0.30	0.40	0.25	0.25	0.40	0.30	0.30	0.40
Storage capacity factor	1.25	0.90	1.00	0.60	0.80	1.00	2.00	0.70	1.00

To obtain the probable maximum demand multiply the total demands for the fixtures (gal./fixture/hour) by the demand factor. The heater should have a heating capacity in gallons per hour equal to this maximum. The storage tank should have a capacity in gallons equal to the probable maximum demand multiplied by the storage capacity factor.

R224000-30 Minimum Plumbing Fixture Requirements

Minimum Plumbing Fixture Requirements

Type of Building or Occupancy (2)	Water Closets (14) (Fixtures per Person)		Urinals (5,10) (Fixtures per Person)		Lavatories (Fixtures per Person)		Bathtubs or Showers (Fixtures per Person)	Drinking Fountains (Fixtures per Person) (3, 13)
	Male	Female	Male	Female	Male	Female		
Assembly Places-Theatres, Auditoriums, Convention Halls, etc.-for permanent employee use	1: 1 - 15 2: 16 - 35 3: 36 - 55	1: 1 - 15 2: 16 - 35 3: 36 - 55	0: 1 - 9 1: 10 - 50		1 per 40	1 per 40		
	Over 55, add 1 fixture for each additional 40 persons		Add one fixture for each additional 50 males					
Assembly Places-Theatres, Auditoriums, Convention Halls, etc. - for public use	1: 1 - 100 2: 101 - 200 3: 201 - 400	3: 1 - 50 4: 51 - 100 8: 101 - 200 11: 201 - 400	1: 1 - 100 2: 101 - 200 3: 201 - 400 4: 401 - 600		1: 1 - 200 2: 201 - 400 3: 401 - 750	1: 1 - 200 2: 201 - 400 3: 401 - 750		1: 1 - 150 2: 151 - 400 3: 401 - 750
	Over 400, add 1 fixture for each additional 500 males and 1 for each additional 125 females		Over 600, add 1 fixture for each additional 300 males		Over 750, add 1 fixture for each additional 500 persons			Over 750, add one fixture for each additional 500 persons
Dormitories (9) School or Labor	1 per 10	1 per 8	1 per 25		1 per 12	1 per 12	1 per 8	1 per 150 (12)
	Add 1 fixture for each additional 25 males (over 10) and 1 for each additional 20 females (over 8)		Over 150, add 1 fixture for each additional 50 males		Over 12 add 1 fixture for each additional 20 males and 1 for each 15 additional females		For females add 1 bathtub per 30. Over 150, add 1 per 20	
Dormitories-for Staff Use	1: 1 - 15 2: 16 - 35 3: 36 - 55	1: 1 - 15 3: 16 - 35 4: 36 - 55	1 per 50		1 per 40	1 per 40	1 per 8	
	Over 55, add 1 fixture for each additional 40 persons							
Dwellings: Single Dwelling Multiple Dwelling or Apartment House	1 per dwelling 1 per dwelling or apartment unit				1 per dwelling 1 per dwelling or apartment unit		1 per dwelling 1 per dwelling or apartment unit	
Hospital Waiting rooms	1 per room				1 per room			1 per 150 (12)
Hospitals-for employee use	1: 1 - 15 2: 16 - 35 3: 36 - 55	1: 1 - 15 3: 16 - 35 4: 36 - 55	0: 1 - 9 1: 10 - 50		1 per 40	1 per 40		
	Over 55, add 1 fixture for each additional 40 persons		Add 1 fixture for each additional 50 males					
Hospitals: Individual Room Ward Room	1 per room 1 per 8 patients				1 per room 1 per 10 patients		1 per room 1 per 20 patients	1 per 150 (12)
Industrial (6) Warehouses Workshops, Foundries and similar establishments-for employee use	1: 1 -10 2: 11 - 25 3: 26 - 50 4: 51 - 75 5: 76 - 100	1: 1 -10 2: 11 - 25 3: 26 - 50 4: 51 - 75 5: 76 - 100			Up to 100, per 10 persons Over 100, 1 per 15 persons (7, 8)		1 shower for each 15 persons exposed to excessive heat or to skin contamination with poisonous, infectious or irritating material	1 per 150 (12)
	Over 100, add 1 fixture for each additional 30 persons							
Institutional - Other than Hospitals or Penal Institutions (on each occupied floor)	1 per 25	1 per 20	0: 1 - 9 1: 10 - 50		1 per 10	1 per 10	1 per 8	1 per 150 (12)
			Add 1 fixture for each additional 50 males					
Institutional - Other than Hospitals or Penal Institutions (on each occupied floor)-for employee use	1: 1 - 15 2: 16 - 35 3: 36 - 55	1: 1 - 15 3: 16 - 35 4: 36 - 55	0: 1 - 9 1: 10 - 50		1 per 40	1 per 40	1 per 8	1 per 150 (12)
	Over 55, add 1 fixture for each additional 40 persons		Add 1 fixture for each additional 50 males					
Office or Public Buildings	1: 1 - 100 2: 101 - 200 3: 201 - 400	3: 1 - 50 4: 51 - 100 8: 101 - 200 11: 201 - 400	1: 1 - 100 2: 101 - 200 3: 201 - 400 4: 401 - 600		1: 1 - 200 2: 201 - 400 3: 401 - 750	1: 1 - 200 2: 201 - 400 3: 401 - 750		1 per 150 (12)
	Over 400, add 1 fixture for each additional 500 males and 1 for each additional 150 females		Over 600, add 1 fixture for each additional 300 males		Over 750, add 1 fixture for each additional 500 persons			
Office or Public Buildings - for employee use	1: 1 - 15 2: 16 - 35 3: 36 - 55	1: 1 - 15 3: 16 - 35 4: 36 - 55	0: 1 - 9 1: 10 - 50		1 per 40	1 per 40		
	Over 55, add 1 fixture for each additional 40 persons		Add 1 fixture for each additional 50 males					

R224000-30 Minimum Plumbing Fixture Requirements (cont.)

Minimum Plumbing Fixture Requirements

Type of Building or Occupancy	Water Closets (14) (Fixtures per Person)		Urinals (5, 10) (Fixtures per Person)		Lavatories (Fixtures per Person)		Bathtubs or Showers (Fixtures per Person)	Drinking Fountains (Fixtures per Person) (3, 13)
	Male	Female	Male	Female	Male	Female		
Penal Institutions - for employee use	1: 1 - 15 2: 16 - 35 3: 36 - 55	1: 1 - 15 3: 16 - 35 4: 36 - 55	0: 1 - 9 1: 10 - 50		1 per 40	1 per 40		1 per 150 (12)
	Over 55, add 1 fixture for each additional 40 persons		Add 1 fixture for each additional 50 males					
Penal Institutions - for prison use								
Cell	1 per cell				1 per cell			1 per cellblock floor
Exercise room	1 per exercise room		1 per exercise room		1 per exercise room			1 per exercise room
Restaurants, Pubs and Lounges (11)	1: 1 - 50 2: 51 - 150 3: 151 - 300	1: 1 - 50 2: 51 - 150 4: 151 - 300	1: 1 - 150		1: 1 - 150 2: 151 - 200 3: 201 - 400	1: 1 - 150 2: 151 - 200 3: 201 - 400		
	Over 300, add 1 fixture for each additional 200 persons		Over 150, add 1 fixture for each additional 150 males		Over 400, add 1 fixture for each additional 400 persons			
Schools - for staff use All Schools	1: 1 - 15 2: 16 - 35 3: 36 - 55	1: 1 - 15 3: 16 - 35 4: 36 - 55	1 per 50		1 per 40	1 per 40		
	Over 55, add 1 fixture for each additional 40 persons							
Schools - for student use: Nursery	1: 1 - 20 2: 21 - 50	1: 1 - 20 2: 21 - 50			1: 1 - 25 2: 26 - 50	1: 1 - 25 2: 26 - 50		1 per 150 (12)
	Over 50, add 1 fixture for each additional 50 persons				Over 50, add 1 fixture for each additional 50 persons			
Elementary	1 per 30	1 per 25	1 per 75		1 per 35	1 per 35		1 per 150 (12)
Secondary	1 per 40	1 per 30	1 per 35		1 per 40	1 per 40		1 per 150 (12)
Others (Colleges, Universities, Adult Centers, etc.	1 per 40	1 per 30	1 per 35		1 per 40	1 per 40		1 per 150 (12)
Worship Places Educational and Activities Unit	1 per 150	1 per 75	1 per 150		1 per 2 water closets			1 per 150 (12)
Worship Places Principal Assembly Place	1 per 150	1 per 75	1 per 150		1 per 2 water closets			1 per 150 (12)

Notes:

1. The figures shown are based upon one (1) fixture being the minimum required for the number of persons indicated or any fraction thereof.
2. Building categories not shown on this table shall be considered separately by the Administrative Authority.
3. Drinking fountains shall not be installed in toilet rooms.
4. Laundry trays. One (1) laundry tray or one (1) automatic washer standpipe for each dwelling unit or one (1) laundry trays or one (1) automatic washer standpipes, or combination thereof, for each twelve (12) apartments. Kitchen sinks, one (1) for each dwelling or apartment unit.
5. For each urinal added in excess of the minimum required, one water closet may be deducted. The number of water closets shall not be reduced to less than two-thirds (2/3) of the minimum requirement.
6. As required by ANSI Z4.1-1968, Sanitation in Places of Employment.
7. Where there is exposure to skin contamination with poisonous, infectious, or irritating materials, provide one (1) lavatory for each five (5) persons.
8. Twenty-four (24) lineal inches of wash sink or eighteen (18) inches of a circular basin, when provided with water outlets for such space shall be considered equivalent to one (1) lavatory.
9. Laundry trays, one (1) for each fifty (50) persons. Service sinks, one (1) for each hundred (100) persons.
10. General. In applying this schedule of facilities, consideration shall be given to the accessibility of the fixtures. Conformity purely on a numerical basis may not result in an installation suited to the need of the individual establishment. For example, schools should be provided with toilet facilities on each floor having classrooms.
    a. Surrounding materials, wall and floor space to a point two (2) feet in front of urinal lip and four (4) feet above the floor, and at least two (2) feet to each side of the urinal shall be lined with non-absorbent materials.
    b. Trough urinals shall be prohibited.
11. A restaurant is defined as a business which sells food to be consumed on the premises.
    a. The number of occupants for a drive-in restaurant shall be considered as equal to the number of parking stalls.
    b. Employee toilet facilities shall not to be included in the above restaurant requirements. Hand washing facilities shall be available in the kitchen for employees.
12. Where food is consumed indoors, water stations may be substituted for drinking fountains. Offices, or public buildings for use by more than six (6) persons shall have one (1) drinking fountain for the first one hundred fifty (150) persons and one additional fountain for each three hundred (300) persons thereafter.
13. There shall be a minimum of one (1) drinking fountain per occupied floor in schools, theaters, auditoriums, dormitories, offices of public building.
14. The total number of water closets for females shall be at least equal to the total number of water closets and urinals required for males.

R224000-40 Plumbing Fixture Installation Time

Item	Rough-In	Set	Total Hours	Item	Rough-In	Set	Total Hours
Bathtub	5	5	10	Shower head only	2	1	3
Bathtub and shower, cast iron	6	6	12	Shower drain	3	1	4
Fire hose reel and cabinet	4	2	6	Shower stall, slate		15	15
Floor drain to 4 inch diameter	3	1	4	Slop sink	5	3	8
Grease trap, single, cast iron	5	3	8	Test 6 fixtures			14
Kitchen gas range		4	4	Urinal, wall	6	2	8
Kitchen sink, single	4	4	8	Urinal, pedestal or floor	6	4	10
Kitchen sink, double	6	6	12	Water closet and tank	4	3	7
Laundry tubs	4	2	6	Water closet and tank, wall hung	5	3	8
Lavatory wall hung	5	3	8	Water heater, 45 gals. gas, automatic	5	2	7
Lavatory pedestal	5	3	8	Water heaters, 65 gals. gas, automatic	5	2	7
Shower and stall	6	4	10	Water heaters, electric, plumbing only	4	2	6

Fixture prices in front of book are based on the cost per fixture set in place. The rough-in cost, which must be added for each fixture, includes carrier, if required, some supply, waste and vent pipe connecting fittings and stops. The lengths of rough-in pipe are nominal runs which would connect to the larger runs and stacks. The supply runs and DWV runs and stacks must be accounted for in separate entries. In the eastern half of the United States it is common for the plumber to carry these to a point 5′ outside the building.

R224000-50 Water Cooler Application

Type of Service	Requirement
Office, School or Hospital	12 persons per gallon per hour
Office, Lobby or Department Store	4 or 5 gallons per hour per fountain
Light manufacturing	7 persons per gallon per hour
Heavy manufacturing	5 persons per gallon per hour
Hot heavy manufacturing	4 persons per gallon per hour
Hotel	.08 gallons per hour per room
Theatre	1 gallon per hour per 100 seats

Heating, Ventilating & Air Cond. | R2305 | Common Work Results for HVAC

R230500-10 Subcontractors

On the unit cost pages of the R.S. Means Cost Data books, the last column is entitled "Total Incl. O&P". This is normally the cost of the installing contractor. In the HVAC Division, this is the cost of the mechanical contractor. If the particular work being estimated is to be performed by a sub to the mechanical contractor, the mechanical's profit and handling charge (usually 10%) is added to the total of the last column.

R233100-10 Loudness Levels for Moving Air thru Fans, Diffusers, Register, Etc. (Measured in Sones)

Recommended Loudness Levels (Sones)

Area	Very Quiet	Quiet	Noisy	Area	Very Quiet	Quiet	Noisy	Area	Very Quiet	Quiet	Noisy
Auditoriums				**Hotels**				**Offices**			
Auditorium lobbies	3	4	6	Banquet Rooms	1.5	3	6	Conference rooms	1	1.7	3
Concert and opera halls	0.8	1	1.5	Individual rooms, suites	1	2	4	Drafting	2	4	8
Courtrooms	2	3	4	Kitchens and laundries	4	7	10	General open offices	2	4	8
Lecture halls	1.5	2	3	Lobbies	2	4	8	Halls and corridors	2.5	5	10
Movie theaters	1.5	2	3	**Indoor Sports**				Professional offices	1.5	3	6
Churches and Schools				Bowling alleys	3	4	6	Reception room	1.5	3	6
Kitchens	4	6	8	Gymnasiums	2	4	6	Tabulation & computation	3	6	12
Laboratories	2	4	6	Swimming pools	4	7	10	**Public Buildings**			
Libraries	1.5	2	3	**Manufacturing Areas**				Banks	2	4	6
Recreation halls	2	4	8	Assembly lines	5	12	30	Court houses	2.5	4	6
Sanctuaries	1	1.7	3	Foreman's office	3	5	8	Museums	2	3	4
Schools and classrooms	1.5	2.5	4	Foundries	10	20	40	Planetariums	1.5	2	3
Hospital and Clinics				General storage	5	10	20	Post offices	2.5	4	6
Laboratories	2	4	6	Heavy machinery	10	25	60	Public libraries	1.5	2	4
Lobbies, waiting rooms	2	4	6	Light machinery	5	12	30	Waiting rooms	3	5	8
Halls and corridors	2	4	6	Tool maintenance	4	7	10	**Restaurants**			
Operating rooms	1.5	2.5	4	**Stores**				Cafeterias	3	6	10
Private rooms	1	1.7	3	Department stores	3	6	8	Night clubs	2.5	4	6
Wards	1.5	2.5	4	Supermarkets	4	7	10	Restaurants	2	4	8

R233100-20 Ductwork

Duct Weight in Pounds per L.F., Straight Runs

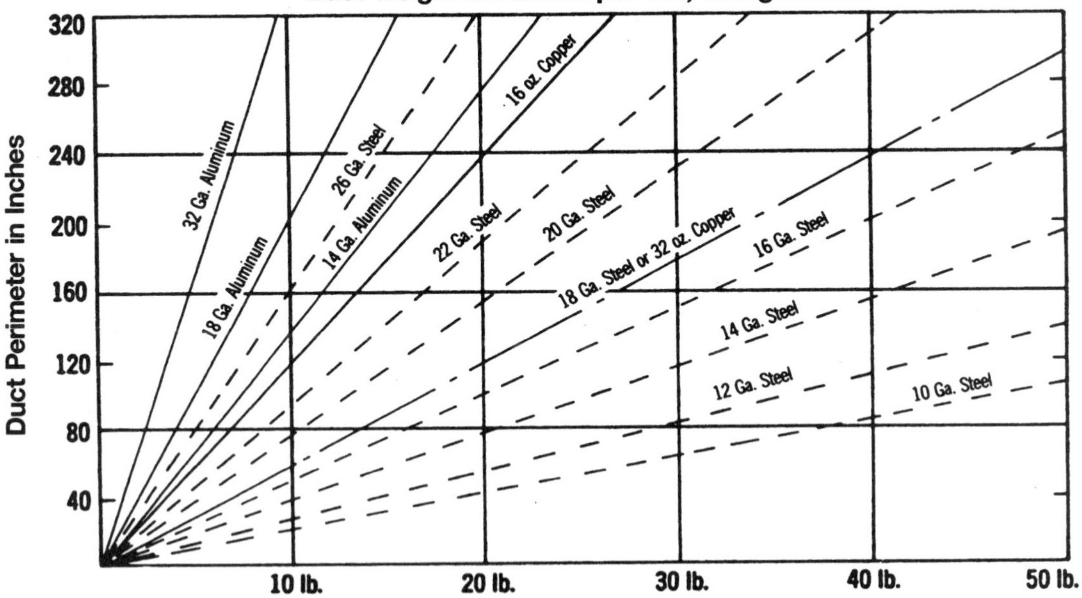

Add to the above for fittings; 90° elbow is 3 L.F.; 45° elbow is 2.5 L.F.; offset is 4 L.F.; transition offset is 6 L.F.; square-to-round transition is 4 L.F.; 90° reducing elbow is 5 L.F. For bracing and waste, add 20% to aluminum and copper, 15% to steel.

R233100-30 Duct Fabrication/Installation

The labor cost for fabricated sheet metal duct includes both the cost of fabrication and installation of the duct. The split is approximately 40% for fabrication, 60% for installation. It is for this reason that the percentage add for elevated installation of fabricated duct is less than the percentage add for prefabricated/preformed duct.

Example: assume a piece of duct cost $100 installed (labor only)

Sheet Metal Fabrication = 40% = $40

Installation = 60% = $60

The add for elevated installation is:

Based on total labor (fabrication & installation) = $100 x 6% = $6.00

Based on installation cost only (Material purchased prefabricated) = $60 x 10% = $6.00

The $6.00 markup (10′ to 15′ high) is the same.

R233100-40 Sheet Metal Calculator (Weight in Lb./Ft. of Length)

Gauge	26	24	22	20	18	16
Wt.-Lb./S.F.	.906	1.156	1.406	1.656	2.156	2.656
SMACNA Max. Dimension – Long Side		30″	54″	84″	85″ Up	
Sum-2 sides						
2	.3	.40	.50	.60	.80	.90
3	.5	.65	.80	.90	1.1	1.4
4	.7	.85	1.0	1.2	1.5	1.8
5	.8	1.1	1.3	1.5	1.9	2.3
6	1.0	1.3	1.5	1.7	2.3	2.7
7	1.2	1.5	1.8	2.0	2.7	3.2
8	1.3	1.7	2.0	2.3	3.0	3.6
9	1.5	1.9	2.3	2.6	3.4	4.1
10	1.7	2.2	2.5	2.9	3.8	4.5
11	1.8	2.4	2.8	3.2	4.2	5.0
12	2.0	2.6	3.0	3.5	4.6	5.4
13	2.2	2.8	3.3	3.8	4.9	5.9
14	2.3	3.0	3.5	4.1	5.3	6.3
15	2.5	3.2	3.8	4.4	5.7	6.8
16	2.7	3.4	4.0	4.6	6.1	7.2
17	2.8	3.7	4.3	4.9	6.5	7.7
18	3.0	3.9	4.5	5.2	6.8	8.1
19	3.2	4.1	4.8	5.5	7.2	8.6
20	3.3	4.3	5.0	5.8	7.6	9.0
21	3.5	4.5	5.3	6.1	8.0	9.5
22	3.7	4.7	5.5	6.4	8.4	9.9
23	3.8	5.0	5.8	6.7	8.7	10.4
24	4.0	5.2	6.0	7.0	9.1	10.8
25	4.2	5.4	6.3	7.3	9.5	11.3
26	4.3	5.6	6.5	7.5	9.9	11.7
27	4.5	5.8	6.8	7.8	10.3	12.2
28	4.7	6.0	7.0	8.1	10.6	12.6
29	4.8	6.2	7.3	8.4	11.0	13.1
30	5.0	6.5	7.5	8.7	11.4	13.5
31	5.2	6.7	7.8	9.0	11.8	14.0
32	5.3	6.9	8.0	9.3	12.2	14.4
33	5.5	7.1	8.3	9.6	12.5	14.9
34	5.7	7.3	8.5	9.9	12.9	15.3
35	5.8	7.5	8.8	10.2	13.3	15.8
36	6.0	7.8	9.0	10.4	13.7	16.2
37	6.2	8.0	9.3	10.7	14.1	16.7
38	6.3	8.2	9.5	11.0	14.4	17.1
39	6.5	8.4	9.8	11.3	14.8	17.6
40	6.7	8.6	10.0	11.6	15.2	18.0
41	6.8	8.8	10.3	11.9	15.6	18.5
42	7.0	9.0	10.5	12.2	16.0	18.9
43	7.2	9.2	10.8	12.5	16.3	19.4
44	7.3	9.5	11.0	12.8	16.7	19.8
45	7.5	9.7	11.3	13.1	17.1	20.3
46	7.7	9.9	11.5	13.3	17.5	20.7
47	7.8	10.1	11.8	13.6	17.9	21.2
48	8.0	10.3	12.0	13.9	18.2	21.6
49	8.2	10.5	12.3	14.2	18.6	22.1
50	8.3	10.7	12.5	14.5	19.0	22.5
51	8.5	11.0	12.8	14.8	19.4	23.0
52	8.7	11.2	13.0	15.1	19.8	23.4
53	8.8	11.4	13.3	15.4	20.1	23.9
54	9.0	11.6	13.5	15.7	20.5	24.3
55	9.2	11.8	13.8	16.0	20.9	24.8

Gauge	26	24	22	20	18	16
Wt.-Lb./S.F.	.906	1.156	1.406	1.656	2.156	2.656
SMACNA Max. Dimension – Long Side		30″	54″	84″	85″ Up	
Sum-2 Sides						
56	9.3	12.0	14.0	16.2	21.3	25.2
57	9.5	12.3	14.3	16.5	21.7	25.7
58	9.7	12.5	14.5	16.8	22.0	26.1
59	9.8	12.7	14.8	17.1	22.4	26.6
60	10.0	12.9	15.0	17.4	22.8	27.0
61	10.2	13.1	15.3	17.7	23.2	27.5
62	10.3	13.3	15.5	18.0	23.6	27.9
63	10.5	13.5	15.8	18.3	24.0	28.4
64	10.7	13.7	16.0	18.6	24.3	28.8
65	10.8	13.9	16.3	18.9	24.7	29.3
66	11.0	14.1	16.5	19.1	25.1	29.7
67	11.2	14.3	16.8	19.4	25.5	30.2
68	11.3	14.6	17.0	19.7	25.8	30.6
69	11.5	14.8	17.3	20.0	26.2	31.1
70	11.7	15.0	17.5	20.3	26.6	31.5
71	11.8	15.2	17.8	20.6	27.0	32.0
72	12.0	15.4	18.0	20.9	27.4	32.4
73	12.2	15.6	18.3	21.2	27.7	32.9
74	12.3	15.8	18.5	21.5	28.1	33.3
75	12.5	16.1	18.8	21.8	28.5	33.8
76	12.7	16.3	19.0	22.0	28.9	34.2
77	12.8	16.5	19.3	22.3	29.3	34.7
78	13.0	16.7	19.5	22.6	29.6	35.1
79	13.2	16.9	19.8	22.9	30.0	35.6
80	13.3	17.1	20.0	23.2	30.4	36.0
81	13.5	17.3	20.3	23.5	30.8	36.5
82	13.7	17.5	20.5	23.8	31.2	36.9
83	13.8	17.8	20.8	24.1	31.5	37.4
84	14.0	18.0	21.0	24.4	31.9	37.8
85	14.2	18.2	21.3	24.7	32.3	38.3
86	14.3	18.4	21.5	24.9	32.7	38.7
87	14.5	18.6	21.8	25.2	33.1	39.2
88	14.7	18.8	22.0	25.5	33.4	39.6
89	14.8	19.0	22.3	25.8	33.8	40.1
90	15.0	19.3	22.5	26.1	34.2	40.5
91	15.2	19.5	22.8	26.4	34.6	41.0
92	15.3	19.7	23.0	26.7	35.0	41.4
93	15.5	19.9	23.3	27.0	35.3	41.9
94	15.7	20.1	23.5	27.3	35.7	42.3
95	15.8	20.3	23.8	27.6	36.1	42.8
96	16.0	20.5	24.0	27.8	36.5	43.2
97	16.2	20.8	24.3	28.1	36.9	43.7
98	16.3	21.0	24.5	28.4	37.2	44.1
99	16.5	21.2	24.8	28.7	37.6	44.6
100	16.7	21.4	25.0	29.0	38.0	45.0
101	16.8	21.6	25.3	29.3	38.4	45.5
102	17.0	21.8	25.5	29.6	38.8	45.9
103	17.2	22.0	25.8	29.9	39.1	46.4
104	17.3	22.3	26.0	30.2	39.5	46.8
105	17.5	22.5	26.3	30.5	39.9	47.3
106	17.7	22.7	26.5	30.7	40.3	47.7
107	17.8	22.9	26.8	31.0	40.7	48.2
108	18.0	23.1	27.0	31.3	41.0	48.6
109	18.2	23.3	27.3	31.6	41.4	49.1
110	18.3	23.5	27.5	31.9	41.8	49.5

Example: If duct is 34″ x 20″ x 15′ long, 34″ is greater than 30″ maximum, for 24 ga. so must be 22 ga. 34″ + 20″ = 54″ going across from 54″ find 13.5 lb. per foot. 13.5 x 15′ = 202.5 lbs. For S.F. of surface area 202.5 ÷ 1.406 = 144 S.F.

Note: Figures include an allowance for scrap.

Heating, Ventilating & Air Cond. | R2331 | HVAC Ducts & Casings

R233100-50 Ductwork Packages (per Ton of Cooling)

System	Sheet Metal	Insulation	Diffusers	Return Register
Roof Top Unit Single Zone	120 Lbs.	52 S.F.	1	1
Roof Top Unit Multizone	240 Lbs.	104 S.F.	2	1
Self-contained Air or Water Cooled	108 Lbs.	—	2	—
Split System Air Cooled	102 Lbs.	—	2	—

Systems reflect most common usage.
Refer to system graphics for duct layout.

Heating, Ventilating & Air Cond. | R2334 | HVAC Fans

R233400-10 Recommended Ventilation Air Changes

Table below lists range of time in minutes per change for various types of facilities.

Facility	Minutes	Facility	Minutes	Facility	Minutes
Assembly Halls	2-10	Dance Halls	2-10	Laundries	1-3
Auditoriums	2-10	Dining Rooms	3-10	Markets	2-10
Bakeries	2-3	Dry Cleaners	1-5	Offices	2-10
Banks	3-10	Factories	2-5	Pool Rooms	2-5
Bars	2-5	Garages	2-10	Recreation Rooms	2-10
Beauty Parlors	2-5	Generator Rooms	2-5	Sales Rooms	2-10
Boiler Rooms	1-5	Gymnasiums	2-10	Theaters	2-8
Bowling Alleys	2-10	Kitchens-Hospitals	2-5	Toilets	2-5
Churches	5-10	Kitchens-Restaurant	1-3	Transformer Rooms	1-5

CFM air required for changes = Volume of room in cubic feet ÷ Minutes per change.

Heating, Ventilating & Air Cond. | R2337 | Air Outlets & Inlets

R233700-60 Diffuser Evaluation

CFM = V × An × K where V = Outlet velocity in feet per minute. An = Neck area in square feet and K = Diffuser delivery factor. An undersized diffuser for a desired CFM will produce a high velocity and noise level. When air moves past people at a velocity in excess of 25 FPM, an annoying draft is felt. An oversized diffuser will result in low velocity with poor mixing. Consideration must be given to avoid vertical stratification or horizontal areas of stagnation.

R235000-10 Heating Systems

Heating Systems

The basic function of a heating system is to bring an enclosed volume up to a desired temperature and then maintain that temperature within a reasonable range. To accomplish this, the selected system must have sufficient capacity to offset transmission losses resulting from the temperature difference on the interior and exterior of the enclosing walls in addition to losses due to cold air infiltration through cracks, crevices and around doors and windows. The amount of heat to be furnished is dependent upon the building size, construction, temperature difference, air leakage, use, shape, orientation and exposure. Air circulation is also an important consideration. Circulation will prevent stratification which could result in heat losses through uneven temperatures at various levels. For example, the most efficient use of unit heaters can usually be achieved by circulating the space volume through the total number of units once every 20 minutes or 3 times an hour. This general rule must, of course, be adapted for special cases such as large buildings with low ratios of heat transmitting surface to cubical volume. The type of occupancy of a building will have considerable bearing on the number of heat transmitting units and the location selected. It is axiomatic, however, that the basis of any successful heating system is to provide the maximum amount of heat at the points of maximum heat loss such as exposed walls, windows, and doors. Large roof areas, wind direction, and wide doorways create problems of excessive heat loss and require special consideration and treatment.

Heat Transmission

Heat transfer is an important parameter to be considered during selection of the exterior wall style, material and window area. A high rate of transfer will permit greater heat loss during the wintertime with the resultant increase in heating energy costs and a greater rate of heat gain in the summer with proportionally greater cooling cost. Several terms are used to describe various aspects of heat transfer. However, for general estimating purposes this book lists U values for systems of construction materials. U is the "overall heat transfer coefficient." It is defined as the heat flow per hour through one square foot when the temperature difference in the air on either side of the structure wall, roof, ceiling or floor is one degree Fahrenheit. The structural segment may be a single homogeneous material or a composite.

Total heat transfer is found using the following equation:

$Q = AU(T_2 - T_1)$ where

- **Q** = Heat flow, BTU per hour
- **A** = Area, square feet
- **U** = Overall heat transfer coefficient
- $(T_2 - T_1)$ = Difference in temperature of air on each side of the construction component. (Also abbreviated TD)

Note that heat can flow through all surfaces of any building and this flow is in addition to heat gain or loss due to ventilation, infiltration and generation (appliances, machinery, people).

R235000-20 Heating Approximations for Quick Estimating

Oil Piping & Boiler Room Piping

Small System	20 to 30% of Boiler
Complex System with Pumps, Headers, Etc.	80 to 110% of Boiler
Breeching With Insulation:	
Small	10 to 15% of Boiler
Large	15 to 25% of Boiler
Coils:	15 to 30% of Containing Unit
Balancing (Independent)	1/2% of H.V.A.C. Estimate

Quality/Complexity Adjustment: For all heating installations add these adjustments to the estimate to more closely allow for the equipment and conditions of the particular job under consideration.

Economy installation, add	0 to 5% of System
Good quality, medium complexity, add	5 to 15% of System
Above average quality and complexity, add	15 to 25% of System

R235000-30 The Basics of a Heating System

The function of a heating system is to achieve and maintain a desired temperature in a room or building by replacing the amount of heat being dissipated. There are four kinds of heating systems: hot-water, steam, warm-air and electric resistance. Each has certain essential and similar elements with the exception of electric resistance heating.

The basic elements of a heating system are:

A. A **combustion chamber** in which fuel is burned and heat transferred to a conveying medium.

B. The **"fluid"** used for conveying the heat (water, steam or air).

C. **Conductors** or pipes for transporting the fluid to specific desired locations.

D. A means of disseminating the heat, sometimes called **terminal units.**

A. The **combustion chamber** in a furnace heats air which is then distributed. This is called a warm-air system.

The combustion chamber in a boiler heats water which is either distributed as hot water or steam and this is termed a hydronic system.

The maximum allowable working pressures are limited by ASME "Code for Heating Boilers" to 15 PSI for steam and 160 PSI for hot water heating boilers, with a maximum temperature limitation of 250°F. Hot water boilers are generally rated for a working pressure of 30 PSI. High pressure boilers are governed by the ASME "Code for Power Boilers" which is used almost universally for boilers operating over 15 PSIG. High pressure boilers used for a combination of heating/process loads are usually designed for 150 PSIG.

Boiler ratings are usually indicated as either Gross or Net Output. The Gross Load is equal to the Net Load plus a piping and pickup allowance. When this allowance cannot be determined, divide the gross output rating by 1.25 for a value equal to or greater than the net heat loss requirement of the building.

B. Of the three **fluids** used, steam carries the greatest amount of heat per unit volume. This is due to the fact that it gives up its latent heat of vaporization at a temperature considerably above room temperature. Another advantage is that the pressure to produce a positive circulation is readily available. Piping conducts the steam to terminal units and returns condensate to the boiler.

The **steam system** is well adapted to large buildings because of its positive circulation, its comparatively economical installation and its ability to deliver large quantities of heat. Nearly all large office buildings, stores, hotels, and industrial buildings are so heated, in addition to many residences.

Hot water, when used as the heat carrying fluid, gives up a portion of its sensible heat and then returns to the boiler or heating apparatus for reheating. As the heat conveyed by each pound of water is about one-fiftieth of the heat conveyed by a pound of steam, it is necessary to circulate about fifty times as much water as steam by weight (although only one-thirtieth as much by volume). The hot water system is usually, although not necessarily, designed to operate at temperatures below that of the ordinary steam system and so the amount of heat transfer surface must be correspondingly greater. A temperature of 190°F to 200°F is normally the maximum. Circulation in small buildings may depend on the difference in density between hot water and the cool water returning to the boiler; circulating pumps are normally used to maintain a desired rate of flow. Pumps permit a greater degree of flexibility and better control.

In **warm-air** furnace systems, cool air is taken from one or more points in the building, passed over the combustion chamber and flue gas passages and then distributed through a duct system. A disadvantage of this system is that the ducts take up much more building volume than steam or hot water pipes. Advantages of this system are the relative ease with which humidification can be accomplished by the evaporation of water as the air circulates through the heater, and the lack of need for expensive disseminating units as the warm air simply becomes part of the interior atmosphere of the building.

C. Conductors (pipes and ducts) have been lightly treated in the discussion of conveying fluids. For more detailed information such as sizing and distribution methods, the reader is referred to technical publications such as the American Society of Heating, Refrigerating and Air-Conditioning Engineers "Handbook of Fundamentals."

D. Terminal units come in an almost infinite variety of sizes and styles, but the basic principles of operation are very limited. As previously mentioned, warm-air systems require only a simple register or diffuser to mix heated air with that present in the room. Special application items such as radiant coils and infrared heaters are available to meet particular conditions but are not usually considered for general heating needs. Most heating is accomplished by having air flow over coils or pipes containing the heat transporting medium (steam, hot-water, electricity). These units, while varied, may be separated into two general types, (1) radiator/convectors and (2) unit heaters.

Radiator/convectors may be cast, fin-tube or pipe assemblies. They may be direct, indirect, exposed, concealed or mounted within a cabinet enclosure, upright or baseboard style. These units are often collectively referred to as "radiatiors" or "radiation" although none gives off heat either entirely by radiation or by convection but rather a combination of both. The air flows over the units as a gravity "current." It is necessary to have one or more heat-emitting units in each room. The most efficient placement is low along an outside wall or under a window to counteract the cold coming into the room and achieve an even distribution.

In contrast to radiator/convectors which operate most effectively against the walls of smaller rooms, **unit heaters** utilize a fan to move air over heating coils and are very effective in locations of relatively large volume. Unit heaters, while usually suspended overhead, may be floor mounted. They also may take in fresh outside air for ventilation. The heat distributed by unit heaters may be from a remote source and conveyed by a fluid or it may be from the combustion of fuel in each individual heater. In the latter case the only piping required would be for fuel, however, a vent for the products of combustion would be necessary.

The following list gives may of the advantages of unit heaters for applications other than office or residential:

a. Large capacity so smaller number of units are required,
b. Piping system simplified, **c.** Space saved where they are located overhead out of the way, **d.** Rapid heating directed where needed with effective wide distribution, **e.** Difference between floor and ceiling temperature reduced, **f.** Circulation of air obtained, and ventilation with introduction of fresh air possible, **g.** Heat output flexible and easily controlled.

R235000-35 Heating (42° degrees latitude)

$$\text{Approximate S.F. radiation} = \frac{\text{S.F. sash}}{2} + \frac{\text{S.F. wall + roof — sash}}{20} + \frac{\text{C.F. building}}{200}$$

R235000-50 Factor for Determining Heat Loss for Various Types of Buildings

General: While the most accurate estimates of heating requirements would naturally be based on detailed information about the building being considered, it is possible to arrive at a reasonable approximation using the following procedure:

1. Calculate the cubic volume of the room or building.
2. Select the appropriate factor from Table 1 below. Note that the factors apply only to inside temperatures listed in the first column and to 0°F outside temperature.
3. If the building has bad north and west exposures, multiply the heat loss factor by 1.1.
4. If the outside design temperature is other than 0°F, multiply the factor from Table 1 by the factor from Table 2.
5. Multiply the cubic volume by the factor selected from Table 1. This will give the estimated BTUH heat loss which must be made up to maintain inside temperature.

Table 1 — Building Type	Conditions	Qualifications	Loss Factor*
Factories & Industrial Plants General Office Areas at 70°F	One Story	Skylight in Roof	6.2
		No Skylight in Roof	5.7
	Multiple Story	Two Story	4.6
		Three Story	4.3
		Four Story	4.1
		Five Story	3.9
		Six Story	3.6
	All Walls Exposed	Flat Roof	6.9
		Heated Space Above	5.2
	One Long Warm Common Wall	Flat Roof	6.3
		Heated Space Above	4.7
	Warm Common Walls on Both Long Sides	Flat Roof	5.8
		Heated Space Above	4.1
Warehouses at 60°F	All Walls Exposed	Skylights in Roof	5.5
		No Skylight in Roof	5.1
		Heated Space Above	4.0
	One Long Warm Common Wall	Skylight in Roof	5.0
		No Skylight in Roof	4.9
		Heated Space Above	3.4
	Warm Common Walls on Both Long Sides	Skylight in Roof	4.7
		No Skylight in Roof	4.4
		Heated Space Above	3.0

*Note: This table tends to be conservative particularly for new buildings designed for minimum energy consumption.

Table 2 — Outside Design Temperature Correction Factor (for Degrees Fahrenheit)

Outside Design Temperature	50	40	30	20	10	0	-10	-20	-30
Correction Factor	.29	.43	.57	.72	.86	1.00	1.14	1.28	1.43

R235000-70 Transmission of Heat

R235000-70 and R235000-80 provide a way to calculate heat transmission of various construction materials from their U values and the TD (Temperature Difference).

1. From the Exterior Enclosure Division or elsewhere, determine U values for the construction desired.
2. Determine the coldest design temperature. The difference between this temperature and the desired interior temperature is the TD (temperature difference).
3. Enter R235000-70 or R235000-80 at correct U Value. Cross horizontally to the intersection with appropriate TD. Read transmission per square foot from bottom of figure.
4. Multiply this value of BTU per hour transmission per square foot of area by the total surface area of that type of construction.

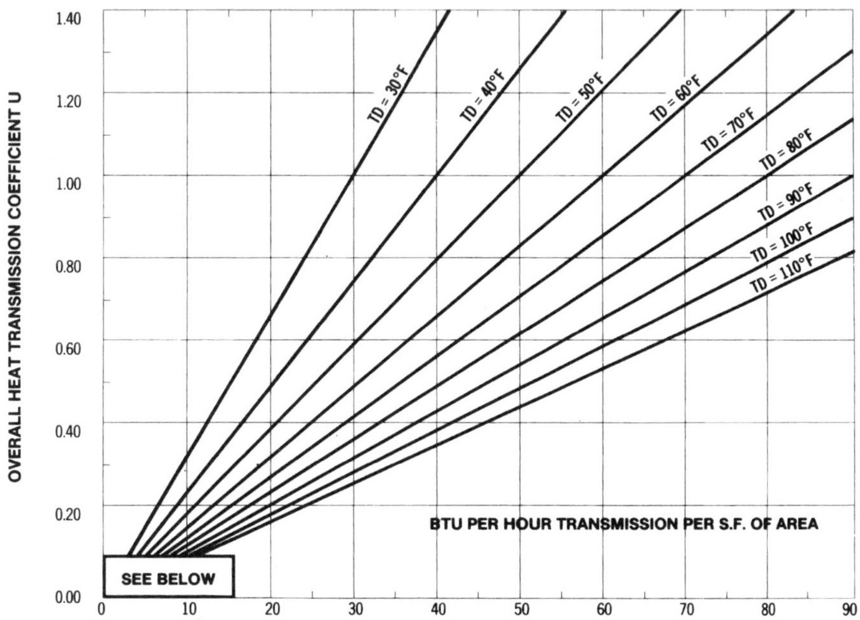

R235000-80 Transmission of Heat (Low Rate)

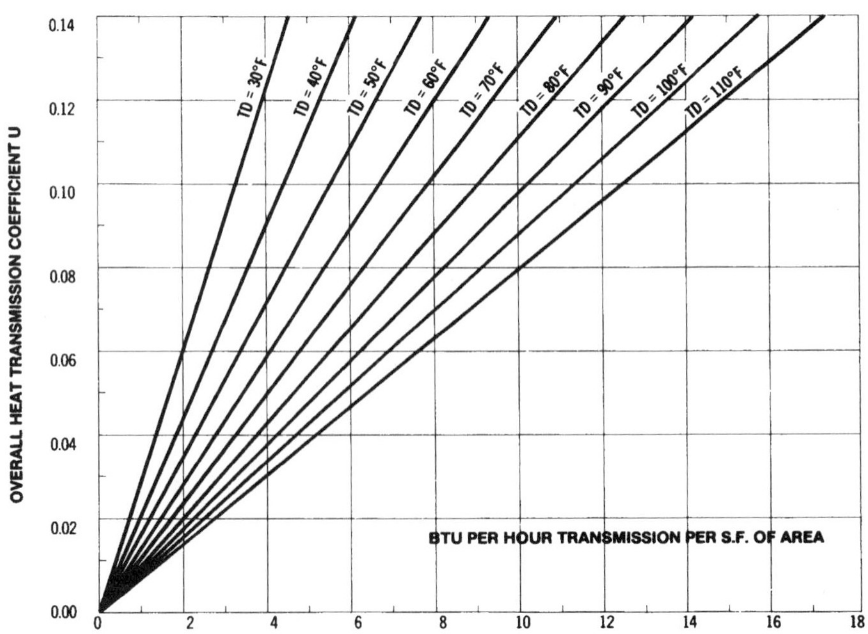

Heating, Ventilating & Air Cond. | R2356 | Solar Energy Heating Equipment

R235616-60 Solar Heating (Space and Hot Water)

Collectors should face as close to due South as possible, however, variations of up to 20 degrees on either side of true South are acceptable. Local climate and collector type may influence the choice between east or west deviations. Obviously they should be located so they are not shaded from the sun's rays. Incline collectors at a slope of latitude minus 5 degrees for domestic hot water and latitude plus 15 degrees for space heating.

Flat plate collectors consist of a number of components as follows: Insulation to reduce heat loss through the bottom and sides of the collector. The enclosure which contains all the components in this assembly is usually weatherproof and prevents dust, wind and water from coming in contact with the absorber plate. The cover plate usually consists of one or more layers of a variety of glass or plastic and reduces the reradiation by creating an air space which traps the heat between the cover and the absorber plates.

The absorber plate must have a good thermal bond with the fluid passages. The absorber plate is usually metallic and treated with a surface coating which improves absorptivity. Black or dark paints or selective coatings are used for this purpose, and the design of this passage and plate combination helps determine a solar system's effectiveness.

Heat transfer fluid passage tubes are attached above and below or integral with an absorber plate for the purpose of transferring thermal energy from the absorber plate to a heat transfer medium. The heat exchanger is a device for transferring thermal energy from one fluid to another.

Piping and storage tanks should be well insulated to minimize heat losses.

Size domestic water heating storage tanks to hold 20 gallons of water per user, minimum, plus 10 gallons per dishwasher or washing machine. For domestic water heating an optimum collector size is approximately 3/4 square foot of area per gallon of water storage. For space heating of residences and small commercial applications the collector is commonly sized between 30% and 50% of the internal floor area. For space heating of large commercial applications, collector areas less than 30% of the internal floor area can still provide significant heat reductions.

A supplementary heat source is recommended for Northern states for December through February.

The solar energy transmission per square foot of collector surface varies greatly with the material used. Initial cost, heat transmittance and useful life are obviously interrelated.

Heating, Ventilating & Air Cond. | R2360 | Central Cooling Equipment

R236000-10 Air Conditioning

General: The purpose of air conditioning is to control the environment of a space so that comfort is provided for the occupants and/or conditions are suitable for the processes or equipment contained therein. The several items which should be evaluated to define system objectives are:

- Temperature Control
- Humidity Control
- Cleanliness
- Odor, smoke and fumes
- Ventilation

Efforts to control the above parameters must also include consideration of the degree or tolerance of variation, the noise level introduced, the velocity of air motion and the energy requirements to accomplish the desired results.

The variation in **temperature** and **humidity** is a function of the sensor and the controller. The controller reacts to a signal from the sensor and produces the appropriate suitable response in either the terminal unit, the conductor of the transporting medium (air, steam, chilled water, etc.), or the source (boiler, evaporating coils, etc.).

The **noise level** is a by-product of the energy supplied to moving components of the system. Those items which usually contribute the most noise are pumps, blowers, fans, compressors and diffusers. The level of noise can be partially controlled through use of vibration pads, isolators, proper sizing, shields, baffles and sound absorbing liners.

Some **air motion** is necessary to prevent stagnation and stratification. The maximum acceptable velocity varies with the degree of heating or cooling which is taking place. Most people feel air moving past them at velocities in excess of 25 FPM as an annoying draft. However, velocities up to 45 FPM may be acceptable in certain cases. Ventilation, expressed as air changes per hour and percentage of fresh air, is usually an item regulated by local codes.

Selection of the system to be used for a particular application is usually a trade-off. In some cases the building size, style, or room available for mechanical use limits the range of possibilities. Prime factors influencing the decision are first cost and total life (operating, maintenance and replacement costs). The accuracy with which each parameter is determined will be an important measure of the reliability of the decision and subsequent satisfactory operation of the installed system.

Heat delivery may be desired from an air conditioning system. Heating capability usually is added as follows: A gas fired burner or hot water/steam/electric coils may be added to the air handling unit directly and heat all air equally. For limited or localized heat requirements the water/steam/electric coils may be inserted into the duct branch supplying the cold areas. Gas fired duct furnaces are also available.

Note: When water or steam coils are used the cost of the piping and boiler must also be added. For a rough estimate use the cost per square foot of the appropriate sized hydronic system with unit heaters. This will provide a cost for the boiler and piping, and the unit heaters of the system would equate to the approximate cost of the heating coils.

R236000-20 Air Conditioning Requirements

BTU's per hour per S.F. of floor area and S.F. per ton of air conditioning.

Type of Building	BTU per S.F.	S.F. per Ton	Type of Building	BTU per S.F.	S.F. per Ton	Type of Building	BTU per S.F.	S.F. per Ton
Apartments, Individual	26	450	Dormitory, Rooms	40	300	Libraries	50	240
Corridors	22	550	Corridors	30	400	Low Rise Office, Exterior	38	320
Auditoriums & Theaters	40	300/18*	Dress Shops	43	280	Interior	33	360
Banks	50	240	Drug Stores	80	150	Medical Centers	28	425
Barber Shops	48	250	Factories	40	300	Motels	28	425
Bars & Taverns	133	90	High Rise Office—Ext. Rms.	46	263	Office (small suite)	43	280
Beauty Parlors	66	180	Interior Rooms	37	325	Post Office, Individual Office	42	285
Bowling Alleys	68	175	Hospitals, Core	43	280	Central Area	46	260
Churches	36	330/20*	Perimeter	46	260	Residences	20	600
Cocktail Lounges	68	175	Hotel, Guest Rooms	44	275	Restaurants	60	200
Computer Rooms	141	85	Corridors	30	400	Schools & Colleges	46	260
Dental Offices	52	230	Public Spaces	55	220	Shoe Stores	55	220
Dept. Stores, Basement	34	350	Industrial Plants, Offices	38	320	Shop'g. Ctrs., Supermarkets	34	350
Main Floor	40	300	General Offices	34	350	Retail Stores	48	250
Upper Floor	30	400	Plant Areas	40	300	Specialty	60	200

*Persons per ton
12,000 BTU = 1 ton of air conditioning

R236000-30 Psychrometric Table

Dewpoint or Saturation Temperature (F)

Relative humidity (%)															
100	32	35	40	45	50	55	60	65	70	75	80	85	90	95	100
90	30	33	37	42	47	52	57	62	67	72	77	82	87	92	97
80	27	30	34	39	44	49	54	58	64	68	73	78	83	88	93
70	24	27	31	36	40	45	50	55	60	64	69	74	79	84	88
60	20	24	28	32	36	41	46	51	55	60	65	69	74	79	83
50	16	20	24	28	33	36	41	46	50	55	60	64	69	73	78
40	12	15	18	23	27	31	35	40	45	49	53	58	62	67	71
30	8	10	14	18	21	25	29	33	37	42	46	50	54	59	62
20	6	7	8	9	13	16	20	24	28	31	35	40	43	48	52
10	4	4	5	5	6	8	9	10	13	17	20	24	27	30	34
	32	**35**	**40**	**45**	**50**	**55**	**60**	**65**	**70**	**75**	**80**	**85**	**90**	**95**	**100**

Dry bulb temperature (F)

This table shows the relationship between RELATIVE HUMIDITY, DRY BULB TEMPERATURE AND DEWPOINT.

As an example, assume that the thermometer in a room reads 75°F, and we know that the relative humidity is 50%. The chart shows the dewpoint temperature to be 55°. That is, any surface colder than 55°F will "sweat" or collect condensing moisture. This surface could be the outside of an uninsulated chilled water pipe in the summertime, or the inside surface of a wall or deck in the wintertime. After determining the extreme ambient parameters, the table at the left is useful in determining which surfaces need insulation or vapor barrier protection.

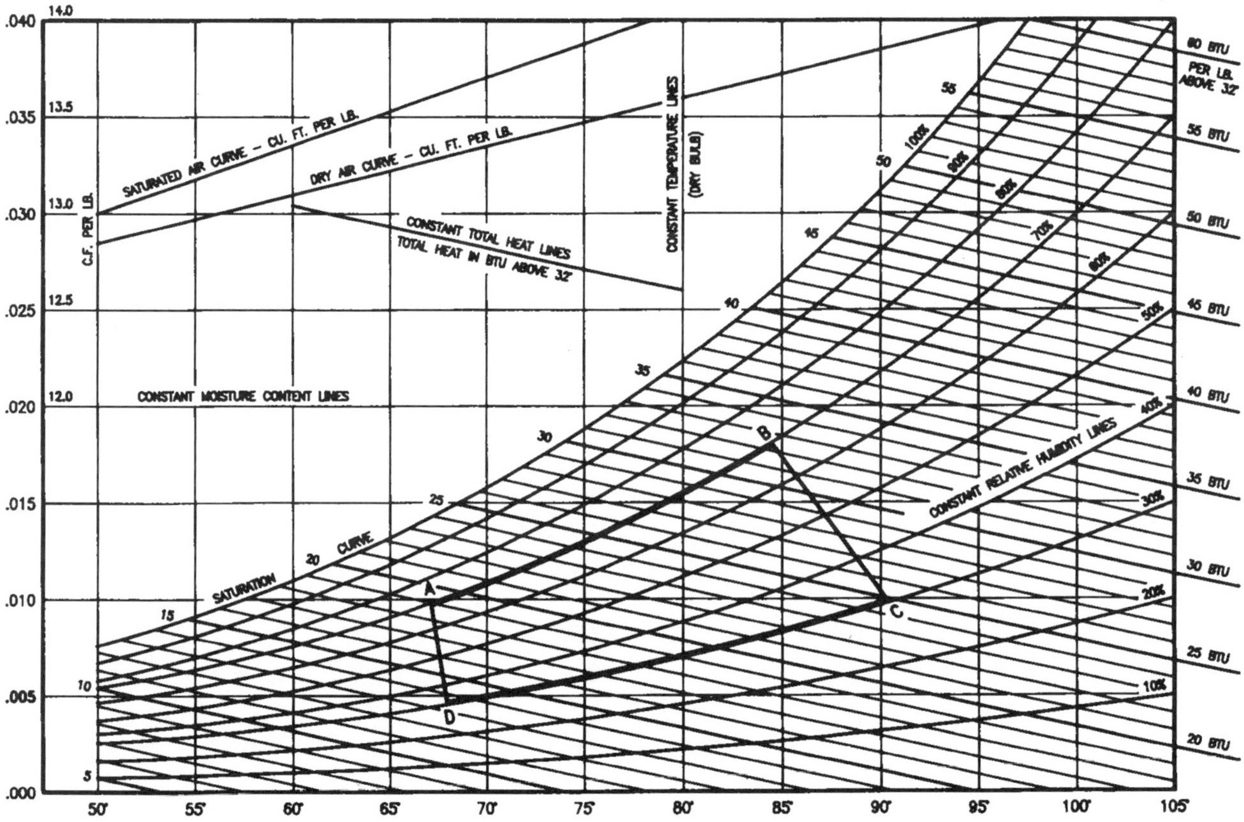

TEMPERATURE DEGREES FAHRENHEIT
TOTAL PRESSURE = 14.696 LB. PER SQ. IN. ABS.

Psychrometric chart showing different variables based on one pound of dry air. Space marked A B C D is temperature-humidity range which is most comfortable for majority of people.

R236000-90 Quality/Complexity Adjustment for Air Conditioning Systems

Economy installation, add	0 to 5%
Good quality, medium complexity, add	5 to 15%
Above average quality and complexity, add	15 to 25%

Add the above adjustments to the estimate to more closely allow for the equipment and conditions of the particular job under consideration.

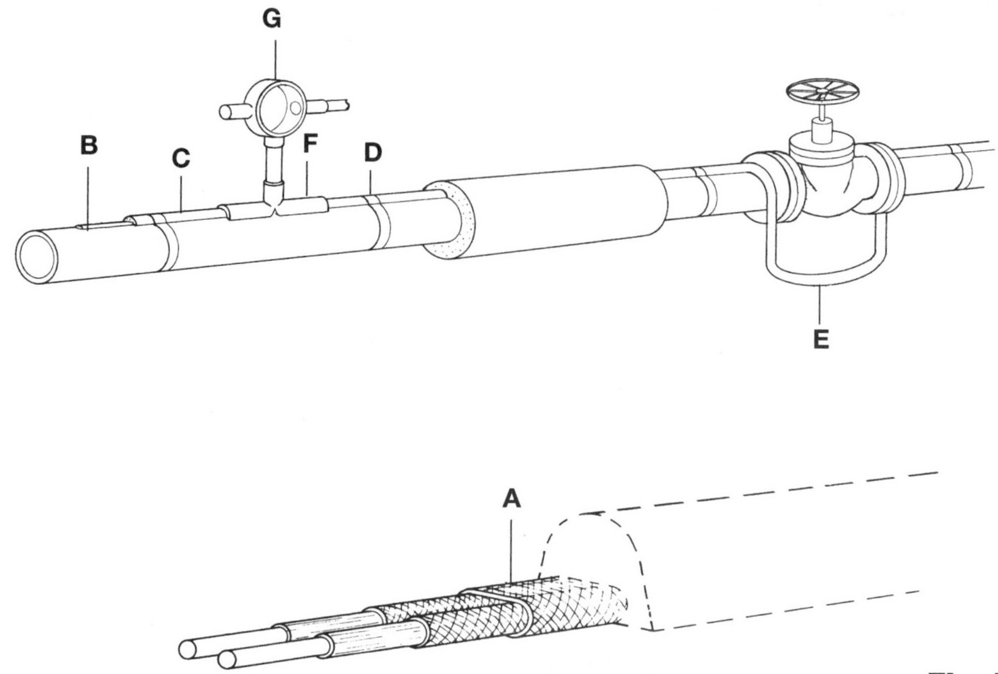

Fig. R238313-11

R238313-10 Heat Trace Systems

Before you can determine the cost of a HEAT TRACE installation the method of attachment must be established. There are (4) common methods:

1. Cable is simply attached to the pipe with polyester tape every 12′.
2. Cable is attached with a continuous cover of 2″ wide aluminum tape.
3. Cable is attached with factory extruded heat transfer cement and covered with metallic raceway with clips every 10′.
4. Cable is attached between layers of pipe insulation using either clips or polyester tape.

Example: Components for method 3 must include:

A. Heat trace cable by voltage and watts per linear foot.
B. Heat transfer cement, 1 gallon per 60 linear feet of cover.
C. Metallic raceway by size and type.
D. Raceway clips by size of pipe.

When taking off linear foot lengths of cable add the following for each valve in the system. (E)

In all of the above methods each component of the system must be priced individually.

SCREWED OR WELDED VALVE:

Size		Length
1/2″	=	6″
3/4″	=	9″
1″	=	1′ -0″
1-1/2″	=	1′ -6″
2″	=	2′
2-1/2″	=	2′ -6″
3″	=	2′ -6″
4″	=	4′ -0″
6″	=	7′ -0″
8″	=	9′ -6″
10″	=	12′ -6″
12″	=	15′ -0″
14″	=	18′ -0″
16″	=	21′ -6″
18″	=	25′ -6″
20″	=	28′ -6″
24″	=	34′ -0″
30″	=	40′ -0″

FLANGED VALVE:

Size		Length
1/2″	=	1′ -0″
3/4″	=	1′ -6″
1″	=	2′ -0″
1-1/2″	=	2′ -6″
2″	=	2′ -6″
2-1/2″	=	3′ -0″
3″	=	3′ -6″
4″	=	4′ -0″
6″	=	8′ -0″
8″	=	11′ -0″
10″	=	14′ -0″
12″	=	16′ -6″
14″	=	19′ -6″
16″	=	23′ -0″
18″	=	27′ -0″
20″	=	30′ -0″
24″	=	36′ -0″
30″	=	42′ -0″

BUTTERFLY VALVES:

Size		Length
1/2″	=	0′
3/4″	=	0′
1″	=	1′ -0″
1-1/2″	=	1′ -6″
2″	=	2′ -0″
2-1/2″	=	2′ -6″
3″	=	2′ -6″
4″	=	3′ -0″
6″	=	3′ -6″
8″	=	4′ -0″
10″	=	4′ -0″
12″	=	5′ -0″
14″	=	5′ -6″
16″	=	6′ -0″
18″	=	6′ -6″
20″	=	7′ -0″
24″	=	8′ -0″
30″	=	10′ -0″

R238313-10 Heat Trace Systems (cont.)

Add the following quantities of heat transfer cement to linear foot totals for each valve:

Nominal Valve Size	Gallons of Cement per Valve
1/2"	0.14
3/4"	0.21
1"	0.29
1-1/2"	0.36
2"	0.43
2-1/2"	0.70
3"	0.71
4"	1.00
6"	1.43
8"	1.48
10"	1.50
12"	1.60
14"	1.75
16"	2.00
18"	2.25
20"	2.50
24"	3.00
30"	3.75

The following must be added to the list of components to accurately price HEAT TRACE systems:

1. Expediter fitting and clamp fasteners (F)
2. Junction box and nipple connected to expediter fitting (G)
3. Field installed terminal blocks within junction box
4. Ground lugs
5. Piping from power source to expediter fitting
6. Controls
7. Thermostats
8. Branch wiring
9. Cable splices
10. End of cable terminations
11. Branch piping fittings and boxes

Deduct the following percentages from labor if cable lengths in the same area exceed:

150′ to 250′	10%	351′ to 500′	20%
251′ to 350′	15%	Over 500′	25%

Add the following percentages to labor for elevated installations:

15′ to 20′ high	10%	31′ to 35′ high	40%
21′ to 25′ high	20%	36′ to 40′ high	50%
26′ to 30′ high	30%	Over 40′ high	60%

R238313-20 Spiral-Wrapped Heat Trace Cable (Pitch Table)

In order to increase the amount of heat, occasionally heat trace cable is wrapped in a spiral fashion around a pipe; increasing the number of feet of heater cable per linear foot of pipe.

Engineers first determine the heat loss per foot of pipe (based on the insulating material, its thickness, and the temperature differential across it). A ratio is then calculated by the formula:

$$\text{Feet of Heat Trace per Foot of Pipe} = \frac{\text{Watts/Foot of Heat Loss}}{\text{Watts/Foot of the Cable}}$$

The linear distance between wraps (pitch) is then taken from a chart or table. Generally, the pitch is listed on a drawing leaving the estimator to calculate the total length of heat tape required. An approximation may be taken from this table.

Feet of Heat Trace Per Foot of Pipe																
	Nominal Pipe Size in Inches															
Pitch In Inches	1	1¼	1½	2	2½	3	4	6	8	10	12	14	16	18	20	24
3.5	1.80															
4	1.65															
5	1.46	1.60	1.80													
6	1.34	1.45	1.55	1.75												
7	1.25	1.35	1.43	1.57	1.75											
8	1.20	1.28	1.34	1.45	1.60	1.80										
9	1.16	1.23	1.28	1.37	1.51	1.68										
10	1.13	1.19	1.24	1.32	1.44	1.57	1.82									
15	1.06	1.08	1.10	1.15	1.21	1.29	1.42	1.78								
20	1.04	1.05	1.06	1.08	1.13	1.17	1.25	1.49	1.73							
25		1.04	1.04	1.06	1.08	1.11	1.17	1.33	1.51	1.72						
30				1.04	1.05	1.07	1.12	1.24	1.37	1.54	1.70	1.80				
35					1.06	1.06	1.09	1.17	1.28	1.42	1.54	1.64	1.78			
40						1.05	1.07	1.14	1.22	1.33	1.44	1.52	1.64	1.75		
50							1.05	1.09	1.15	1.22	1.29	1.35	1.44	1.53	1.64	1.83
60								1.06	1.11	1.16	1.21	1.25	1.31	1.39	1.46	1.62
70								1.05	1.08	1.12	1.17	1.19	1.24	1.30	1.35	1.47
80									1.06	1.09	1.13	1.15	1.19	1.24	1.28	1.38
90									1.04	1.06	1.10	1.13	1.16	1.19	1.23	1.32
100										1.05	1.08	1.10	1.13	1.15	1.19	1.23

Note: Common practice would normally limit the lower end of the table to 5% of additional heat and above 80% an engineer would likely opt for two (2) parallel cables.

R260105-30 Electrical Demolition (Removal for Replacement)

The purpose of this reference number is to provide a guide to users for electrical "removal for replacement" by applying the rule of thumb: 1/3 of new installation time (typical range from 20% to 50%) for removal. Remember to use reasonable judgment when applying the suggested percentage factor. For example:

Contractors have been requested to remove an existing fluorescent lighting fixture and replace with a new fixture utilizing energy saver lamps and electronic ballast:

In order to fully understand the extent of the project, contractors should visit the job site and estimate the time to perform the renovation work in accordance with applicable national, state and local regulation and codes.

The contractor may need to add extra labor hours to his estimate if he discovers unknown concealed conditions such as: contaminated asbestos ceiling, broken acoustical ceiling tile and need to repair, patch and touch-up paint all the damaged or disturbed areas, tasks normally assigned to general contractors. In addition, the owner could request that the contractors salvage the materials removed and turn over the materials to the owner or dispose of the materials to a reclamation station. The normal removal item is 0.5 labor hour for a lighting fixture and 1.5 labor-hours for new installation time. Revise the estimate times from 2 labor-hours work up to a minimum 4 labor-hours work for just fluorescent lighting fixture.

For removal of large concentrations of lighting fixtures in the same area, apply an "economy of scale" to reduce estimating labor hours.

R260519-20 Armored Cable

Armored Cable – Quantities are taken off in the same manner as wire.

Bx Type Cable – Productivities are based on an average run of 50′ before terminating at a box fixture, etc. Each 50′ section includes field preparation of (2) ends with hacksaw, identification and tagging of wire. Set up is open coil type without reels attaching cable to snake and pulling across a suspended ceiling or open face wood or steel studding, price does not include drilling of studs.

Cable in Tray – Productivities are based on an average run of 100 L.F. with set up of pulling equipment for (2) 90° bends, attaching cable to pull-in means, identification and tagging of wires, set up of reels. Wire termination to breakers equipment, etc., are not included.

Job Conditions – Productivities are based on new construction to a height of 15′ using rolling staging in an unobstructed area. Material staging is assumed to be within 100′ of work being performed.

R260519-80 Undercarpet Systems

Takeoff Procedure for Power Systems: List components for each fitting type, tap, splice, and bend on your quantity takeoff sheet. Each component must be priced separately. Start at the power supply transition fittings and survey each circuit for the components needed. List the quantities of each component under a specific circuit number. Use the floor plan layout scale to get cable footage.

Reading across the list, combine the totals of each component in each circuit and list the total quantity in the last column. Calculate approximately 5% for scrap for items such as cable, top shield, tape, and spray adhesive. Also provide for final variations that may occur on-site.

Suggested guidelines are:

1. Equal amounts of cable and top shield should be priced.
2. For each roll of cable, price a set of cable splices.
3. For every 1 ft. of cable, price 2-1/2 ft. of hold-down tape.
4. For every 3 rolls of hold-down tape, price 1 can of spray adhesive.

Adjust final figures wherever possible to accommodate standard packaging of the product. This information is available from the distributor.

Each transition fitting requires:

1. 1 base
2. 1 cover
3. 1 transition block

Each floor fitting requires:

1. 1 frame/base kit
2. 1 transition block
3. 2 covers (duplex/blank)

Each tap requires:

1. 1 tap connector for each conductor
2. 1 pair insulating patches
3. 2 top shield connectors

Each splice requires:

1. 1 splice connector for each conductor
2. 1 pair insulating patches
3. 3 top shield connectors

Each cable bend requires:

1. 2 top shield connectors

Each cable dead end (outside of transition block) requires:

1. 1 pair insulating patches

Labor does not include:

1. Patching or leveling uneven floors.
2. Filling in holes or removing projections from concrete slabs.
3. Sealing porous floors.
4. Sweeping and vacuuming floors.
5. Removal of existing carpeting.
6. Carpet square cut-outs.
7. Installation of carpet squares.

Takeoff Procedures for Telephone Systems: After reviewing floor plans identify each transition. Number or letter each cable run from that fitting. Start at the transition fitting and survey each circuit for the components needed. List the cable type, terminations, cable length, and floor fitting type under the specific circuit number. Use the floor plan layout scale to get the cable footage. Add some extra length (next higher increment of 5 feet) to preconnectorized cable.

Transition fittings require:

1. 1 base plate
2. 1 cover
3. 1 transition block

Floor fittings require:

1. 1 frame/base kit
2. 2 covers
3. Modular jacks

Reading across the list, combine the list of components in each circuit and list the total quantity in the last column. Calculate the necessary scrap factors for such items as tape, bottom shield and spray adhesive. Also provide for final variations that may occur on-site.

Adjust final figures whenever possible to accommodate standard packaging. Check that items such as transition fittings, floor boxes, and floor fittings that are to utilize both power and telephone have been priced as combination fittings, so as to avoid duplication.

Make sure to include marking of floors and drilling of fasteners if fittings specified are not the adhesive type.

Labor does not include:

1. Conduit or raceways before transition of floor boxes
2. Telephone cable before transition boxes
3. Terminations before transition boxes
4. Floor preparation as described in power section

Be sure to include all cable folds when pricing labor.

Takeoff Procedure for Data Systems: Start at the transition fittings and take off quantities in the same manner as the telephone system, keeping in mind that data cable does not require top or bottom shields.

The data cable is simply cross-taped on the cable run to the floor fitting.

Data cable can be purchased in either bulk form in which case coaxial connector material and labor must be priced, or in preconnectorized cut lengths.

Data cable cannot be folded and must be notched at 1 inch intervals. A count of all turns must be added to the labor portion of the estimate. (Note: Some manufacturers have prenotched cable.)

Notching required:

1. 90 degree turn requires 8 notches per side.
2. 180 degree turn requires 16 notches per side.

Floor boxes, transition boxes, and fittings are the same as described in the power and telephone procedures.

Since undercarpet systems require special hand tools, be sure to include this cost in proportion to number of crews involved in the installation.

Job Conditions: Productivity is based on new construction in an unobstructed area. Staging area is assumed to be within 200′ of work being performed.

R260519-90 Wire

Wire quantities are taken off by either measuring each cable run or by extending the conduit and raceway quantities times the number of conductors in the raceway. Ten percent should be added for waste and tie-ins. Keep in mind that the unit of measure of wire is C.L.F. not L.F. as in raceways so the formula would read:

$$\frac{\text{(L.F. Raceway x No. of Conductors) x 1.10}}{100} = \text{C.L.F.}$$

Price per C.L.F. of wire includes:

1. Set up wire coils or spools on racks
2. Attaching wire to pull in means
3. Measuring and cutting wire
4. Pulling wire into a raceway
5. Identifying and tagging

Price does not include:

1. Connections to breakers, panelboards or equipment
2. Splices

Job Conditions: Productivity is based on new construction to a height of 15′ using rolling staging in an unobstructed area. Material staging is assumed to be within 100′ of work being performed.

Economy of Scale: If more than three wires at a time are being pulled, deduct the following percentages from the labor of that grouping:

4-5 wires	25%
6-10 wires	30%
11-15 wires	35%
over 15	40%

If a wire pull is less than 100′ in length and is interrupted several times by boxes, lighting outlets, etc., it may be necessary to add the following lengths to each wire being pulled:

Junction box to junction box	2 L.F.
Lighting panel to junction box	6 L.F.
Distribution panel to sub panel	8 L.F.
Switchboard to distribution panel	12 L.F.
Switchboard to motor control center	20 L.F.
Switchboard to cable tray	40 L.F.

Measure of Drops and Riser: It is important when taking off wire quantities to include the wire for drops to electrical equipment. If heights of electrical equipment are not clearly stated, use the following guide:

	Bottom A.F.F.	Top A.F.F.	Inside Cabinet
Safety switch to 100A	5′	6′	2′
Safety switch 400 to 600A	4′	6′	3′
100A panel 12 to 30 circuit	4′	6′	3′
42 circuit panel	3′	6′	4′
Switch box	3′	3′6″	1′
Switchgear	0′	8′	8′
Motor control centers	0′	8′	8′
Transformers - wall mount	4′	8′	2′
Transformers - floor mount	0′	12′	4′

R260519-91 Maximum Circuit Length (approximate) for Various Power Requirements Assuming THW, Copper Wire @ 75° C, Based Upon a 4% Voltage Drop

Maximum Circuit Length: Table R260519-91 indicates typical maximum installed length a circuit can have and still maintain an adequate voltage level at the point of use. The circuit length is similar to the conduit length.

If the circuit length for an ampere load and a copper wire size exceeds the length obtained from Table R260519-91, use the next largest wire size to compensate voltage drop.

		Maximum Circuit Length in Feet				
	Wire	2 Wire, 1 Phase		3 Wire, 3 Phase		
Amperes	Size	120V	240V	240V	480V	600V
15	14*	50	105	120	240	300
	14	50	100	120	235	295
20	12*	60	125	145	290	360
	12	60	120	140	280	350
30	10*	65	130	155	305	380
	10	65	130	150	300	375
50	8	60	125	145	285	355
65	6	75	150	175	345	435
85	4	90	185	210	425	530
115	2	110	215	250	500	620
130	1	120	240	275	555	690
150	1/0	130	260	305	605	760
175	2/0	140	285	330	655	820
200	3/0	155	315	360	725	904
230	4/0	170	345	395	795	990
255	250	185	365	420	845	1055
285	300	195	395	455	910	1140
310	350	210	420	485	975	1220
380	500	245	490	565	1130	1415

*Solid Conductor

Note: The circuit length is the one-way distance between the origin and the load.

Example: A 130 ampere load at 480 volts, 3 phase, 3 wire with No. 1 wire can be run a maximum of 555 L.F. and provide satisfactory operation. If the same load is to be wired at the end of a 625 L.F. circuit, then a larger wire must be used.

R260519-92 Minimum Copper and Aluminum Wire Size Allowed for Various Types of Insulation

Minimum Wire Sizes										
	Copper		Aluminum			Copper		Aluminum	
Amperes	THW THWN or XHHW	THHN XHHW *	THW XHHW	THHN XHHW *	Amperes	THW THWN or XHHW	THHN XHHW *	THW XHHW	THHN XHHW *
15A	#14	#14	#12	#12	195	3/0	2/0	250kcmil	4/0
20	#12	#12	#10	#10	200	3/0	3/0	250kcmil	4/0
25	#10	#10	#10	#10	205	4/0	3/0	250kcmil	4/0
30	#10	#10	# 8	# 8	225	4/0	3/0	300kcmil	250kcmil
40	# 8	# 8	# 8	# 8	230	4/0	4/0	300kcmil	250kcmil
45	# 8	# 8	# 6	# 8	250	250kcmil	4/0	350kcmil	300kcmil
50	# 8	# 8	# 6	# 6	255	250kcmil	4/0	400kcmil	300kcmil
55	# 6	# 8	# 4	# 6	260	300kcmil	4/0	400kcmil	350kcmil
60	# 6	# 6	# 4	# 6	270	300kcmil	250kcmil	400kcmil	350kcmil
65	# 6	# 6	# 4	# 4	280	300kcmil	250kcmil	500kcmil	350kcmil
75	# 4	# 6	# 3	# 4	285	300kcmil	250kcmil	500kcmil	400kcmil
85	# 4	# 4	# 2	# 3	290	350kcmil	250kcmil	500kcmil	400kcmil
90	# 3	# 4	# 2	# 2	305	350kcmil	300kcmil	500kcmil	400kcmil
95	# 3	# 4	# 1	# 2	310	350kcmil	300kcmil	500kcmil	500kcmil
100	# 3	# 3	# 1	# 2	320	400kcmil	300kcmil	600kcmil	500kcmil
110	# 2	# 3	1/0	# 1	335	400kcmil	350kcmil	600kcmil	500kcmil
115	# 2	# 2	1/0	# 1	340	500kcmil	350kcmil	600kcmil	500kcmil
120	# 1	# 2	1/0	1/0	350	500kcmil	350kcmil	700kcmil	500kcmil
130	# 1	# 2	2/0	1/0	375	500kcmil	400kcmil	700kcmil	600kcmil
135	1/0	# 1	2/0	1/0	380	500kcmil	400kcmil	750kcmil	600kcmil
150	1/0	# 1	3/0	2/0	385	600kcmil	500kcmil	750kcmil	600kcmil
155	2/0	1/0	3/0	3/0	420	600kcmil	500kcmil		700kcmil
170	2/0	1/0	4/0	3/0	430		500kcmil		750kcmil
175	2/0	2/0	4/0	3/0	435		600kcmil		750kcmil
180	3/0	2/0	4/0	4/0	475		600kcmil		

*Dry Locations Only

Notes:

1. Size #14 to 4/0 is in AWG units (American Wire Gauge).
2. Size 250 to 750 is in kcmil units (Thousand Circular Mils).
3. Use next higher ampere value if exact value is not listed in table.
4. For loads that operate continuously increase ampere value by 25% to obtain proper wire size.
5. Refer to Table R260519-91 for the maximum circuit length for the various size wires.
6. Table R260519-92 has been written for estimating purpose only, based on ambient temperature of 30 °C (86°F); for ambient temperature other than 30 °C (86°F), ampacity correction factors will be applied.

R260519-93 Metric Equivalent, Wire

U.S. vs. European Wire – Approximate Equivalents			
United States		European	
Size AWG or kcmil	Area Cir. Mils.(cmil) mm²	Size mm²	Area Cir. Mils.
18	1620/.82	.75	1480
16	2580/1.30	1.0	1974
14	4110/2.08	1.5	2961
12	6530/3.30	2.5	4935
10	10,380/5.25	4	7896
8	16,510/8.36	6	11,844
6	26,240/13.29	10	19,740
4	41,740/21.14	16	31,584
3	52,620/26.65	25	49,350
2	66,360/33.61	–	–
1	83,690/42.39	35	69,090
1/0	105,600/53.49	50	98,700
2/0	133,100/67.42	–	–
3/0	167,800/85.00	70	138,180
4/0	211,600/107.19	95	187,530
250	250,000/126.64	120	236,880
300	300,000/151.97	150	296,100
350	350,000/177.30	–	–
400	400,000/202.63	185	365,190
500	500,000/253.29	240	473,760
600	600,000/303.95	300	592,200
700	700,000/354.60	–	–
750	750,000/379.93	–	–

R260519-94 Size Required and Weight (Lbs./1000 L.F.) of Aluminum and Copper THW Wire by Ampere Load

Amperes	Copper Size	Aluminum Size	Copper Weight	Aluminum Weight
15	14	12	24	11
20	12	10	33	17
30	10	8	48	39
45	8	6	77	52
65	6	4	112	72
85	4	2	167	101
100	3	1	205	136
115	2	1/0	252	162
130	1	2/0	324	194
150	1/0	3/0	397	233
175	2/0	4/0	491	282
200	3/0	250	608	347
230	4/0	300	753	403
255	250	400	899	512
285	300	500	1068	620
310	350	500	1233	620
335	400	600	1396	772
380	500	750	1732	951

R260526-80 Grounding

Grounding When taking off grounding systems, identify separately the type and size of wire.

Example:

Bare copper & size
Bare aluminum & size
Insulated copper & size
Insulated aluminum & size

Count the number of ground rods and their size.

Example:

1. 8′ grounding rod – 5/8″ dia. 20 Ea.
2. 10′ grounding rod – 5/8″ dia. 12 Ea.
3. 15′ grounding rod – 3/4″ dia. 4 Ea.

Count the number of connections; the size of the largest wire will determine the productivity.

Example:

Braze a #2 wire to a #4/0 cable
The 4/0 cable will determine the L.H. and cost to be used.

Include individual connections to:

1. Ground rods
2. Building steel
3. Equipment
4. Raceways

Price does not include:

1. Excavation
2. Backfill
3. Sleeves or raceways used to protect grounding wires
4. Wall penetrations
5. Floor cutting
6. Core drilling

Job Conditions: Productivity is based on a ground floor area, using cable reels in an unobstructed area. Material staging area assumed to be within 100′ of work being performed.

R260533-20 Conduit To 15′ High

List conduit by quantity, size, and type. Do not deduct for lengths occupied by fittings, since this will be allowance for scrap. Example:

A. Aluminum — size
B. Rigid galvanized — size
C. Steel intermediate (IMC) — size
D. Rigid steel, plastic-coated 20 Mil. — size
E. Rigid steel, plastic-coated 40 Mil. — size
F. Electric metallic tubing (EMT) — size
G. PVC Schedule 40 — size

Types (A) thru (E) listed above contain the following per 100 L.F.:

1. (11) Threaded couplings
2. (11) Beam-type hangers
3. (2) Factory sweeps
4. (2) Fiber bushings
5. (4) Locknuts
6. (2) Field threaded pipe terminations
7. (2) Removal of concentric knockouts

Type (F) contains per 100 L.F.:

1. (11) Set screw couplings
2. (11) Beam clamps
3. (2) Field bends on 1/2″ and 3/4″ diameter
4. (2) Factory sweeps for 1″ and above
5. (2) Set screw steel connectors
6. (2) Removal of concentric knockouts

Type (G) contains per 100 L.F.:

1. (11) Field cemented couplings
2. (34) Beam clamps
3. (2) Factory sweeps
4. (2) Adapters
5. (2) Locknuts
6. (2) Removal of concentric knockouts

Labor-hours for all conduit to 15′ include:

1. Unloading by hand
2. Hauling by hand to an area up to 200′ from loading dock
3. Setup of rolling staging
4. Installation of conduit and fittings as described in Conduit models (A) thru (G)

Not included in the material and labor are:

1. Staging rental or purchase
2. Structural modifications
3. Wire
4. Junction boxes
5. Fittings in excess of those described in conduit models (A) thru (G)
6. Painting of conduit

Fittings

Only those fittings listed above are included in the linear foot totals, although they should be listed separately from conduit lengths, without prices, to ensure proper quantities for material procurement.

If the fittings required exceed the quantities included in the model conduit runs, then material and labor costs must be added to the difference. If actual needs per 100 L.F. of conduit are: (2) sweeps, (4) LBs and (1) field bend, then, (4) LBs and (1) field bend must be priced additionally.

R260533-21 Hangers

It is sometimes desirable to substitute an alternate style of hanger if the support being used is not the type described in the conduit models.

One approach is the substitution method:

1. Find the cost of the type hanger described in the conduit model.
2. Calculate the cost of the desired type hanger (it may be necessary to calculate individual components such as drilling, expansion shields, etc.).
3. Calculate the cost difference (delta) between the two types of hangers.
4. Multiply the cost delta by the number of hangers in the model.
5. Divide the total delta cost for hangers in the model by the length of the model to find the delta cost per L.F. for that model.
6. Modify the given unit costs per L.F. for the model by the delta cost per L.F.

Another approach to hanger configurations would be to start with the conduit only and add all the supports and any other items as separate lines. This procedure is most useful if the project involves racking many runs of conduit on a single hanger, for instance, a trapeze type hanger.

Example: Five (5) 2″ RGS conduits, 50 L.F. each, are to be run on trapeze hangers from one pull box to another. The run includes one 90° bend.

1. List the hangers' components to create an assembly cost for each 2′-wide trapeze.
2. List the components for the 50′ conduit run, noting that 6 trapeze supports will be required.

Job Conditions: Productivities are based on new construction to 15′ high, using scaffolding in an unobstructed area. Material storage is assumed to be within 100′ of work being performed.

Add to labor for elevated installations:

15′ to 20′ High–10%	30′ to 35 High–30%
20′ to 25′ High–20%	35′ to 40′ High–35%
25′ to 30′ High–25%	Over 40′ High–40%

Add these percentages to the L.F. labor cost, but not to fittings. Add these percentages only to quantities exceeding the different height levels, rather than the total conduit quantities.

Linear foot price for labor does not include penetrations in walls or floors and must be added to the estimate

R260533-22 Conductors in Conduit

Table below lists maximum number of conductors for various sized conduit using THW, TW or THWN insulations.

Copper Wire Size	Conduit Size																										
	1/2″			3/4″			1″			1-1/4″			1-1/2″			2″			2-1/2″			3″		3-1/2″		4″	
	TW	THW	THWN	TW	THW	THWN	TW	THW	THWN	TW	THW	THWN	TW	THW	THWN	TW	THW	THWN	TW	THW	THWN	THW	THWN	THW	THWN	THW	THWN
#14	9	6	13	15	10	24	25	16	39	44	29	69	60	40	94	99	65	154	142	93		143		192			
#12	7	4	10	12	8	18	19	13	29	35	24	51	47	32	70	78	53	114	111	76	164	117		157			
#10	5	4	6	9	6	11	15	11	18	26	19	32	36	26	44	60	43	73	85	61	104	95	160	127		163	
#8	2	1	3	4	3	5	7	5	9	12	10	16	17	13	22	28	22	36	40	32	51	49	79	66	106	85	136
#6		1	1		2	4		4	6		7	11		10	15		16	26		23	37	36	57	48	76	62	98
#4		1	1		1	2		3	4		5	7		7	9		12	16		17	22	27	35	36	47	47	60
#3		1	1		1	1		2	3		4	6		6	8		10	13		15	19	23	29	31	39	40	51
#2		1	1		1	1		2	3		4	5		5	7		9	11		13	16	20	25	27	33	34	43
#1					1	1		1	1		3	3		4	5		6	8		9	12	14	18	19	25	25	32
1/0					1	1		1	1		2	3		3	4		5	7		8	10	12	15	16	21	21	27
2/0					1	1		1	1		1	2		3	3		5	6		7	8	10	13	14	17	18	22
3/0					1	1		1	1		1	1		2	3		4	5		6	7	9	11	12	14	15	18
4/0						1		1	1		1	1		1	2		3	4		5	6	7	9	10	12	13	15
250 kcmil								1	1		1	1		1	1		2	3		4	4	6	7	8	10	10	12
300								1	1		1	1		1	1		2	3		3	4	5	6	7	8	9	11
350									1		1	1		1	1		1	2		3	3	4	5	6	7	8	9
400											1	1		1	1		1	1		2	3	4	5	5	6	7	8
500											1	1		1	1		1	1		1	2	3	4	4	5	6	7
600												1		1	1		1	1		1	1	3	3	4	4	5	5
700														1	1		1	1		1	1	2	3	3	4	4	5
750 ▼														1	1		1	1		1	1	2	2	3	3	4	4

R260533-23 Metric Equivalent, Conduit

U.S. vs. European Conduit – Approximate Equivalents			
United States		European	
Trade Size	Inside Diameter Inch/mm	Trade Size	Inside Diameter mm
½	.622/15.8	11	16.4
¾	.824/20.9	16	19.9
1	1.049/26.6	21	25.5
1¼	1.380/35.0	29	34.2
1½	1.610/40.9	36	44.0
2	2.067/52.5	42	51.0
2½	2.469/62.7		
3	3.068/77.9		
3½	3.548/90.12		
4	4.026/102.3		
5	5.047/128.2		
6	6.065/154.1		

R260533-24 Conduit Weight Comparisons (Lbs. per 100 ft.) Empty

Type	1/2″	3/4″	1″	1-1/4″	1-1/2″	2″	2-1/2″	3″	3-1/2″	4″	5″	6″
Rigid Aluminum	28	37	55	72	89	119	188	246	296	350	479	630
Rigid Steel	79	105	153	201	249	332	527	683	831	972	1314	1745
Intermediate Steel (IMC)	60	82	116	150	182	242	401	493	573	638		
Electrical Metallic Tubing (EMT)	29	45	65	96	111	141	215	260	365	390		
Polyvinyl Chloride, Schedule 40	16	22	32	43	52	69	109	142	170	202	271	350
Polyvinyl Chloride Encased Burial						38		67	88	105	149	202
Fibre Duct Encased Burial						127		164	180	206	400	511
Fibre Duct Direct Burial						150		251	300	354		
Transite Encased Burial						160		240	290	330	450	550
Transite Direct Burial						220		310		400	540	640

R260533-25 Conduit Weight Comparisons (Lbs. per 100 ft.) with Maximum Cable Fill*

Type	1/2″	3/4″	1″	1-1/4″	1-1/2″	2″	2-1/2″	3″	3-1/2″	4″	5″	6″
Rigid Galvanized Steel (RGS)	104	140	235	358	455	721	1022	1451	1749	2148	3083	4343
Intermediate Steel (IMC)	84	113	186	293	379	611	883	1263	1501	1830		
Electrical Metallic Tubing (EMT)	54	116	183	296	368	445	641	930	1215	1540		

*Conduit & Heaviest Conductor Combination

R260533-60 Wireway

When "taking off" Wireway, list by size and type.

Example:
1. Screw cover, unflanged + size
2. Screw cover, flanged + size
3. Hinged cover, flanged + size
4. Hinged cover, unflanged + size

Each 10′ length on Wireway contains:
1. 10′ of cover either screw or hinged type
2. (1) Coupling or flange gasket
3. (1) Wall type mount

All fittings must be priced separately.

Substitution of hanger types is done the same as described in R260533-21, "HANGERS," keeping in mind that the wireway model is based on 10′ sections instead of a 100′ conduit run.

Labor-hours for wireway include:
1. Unloading by hand
2. Hauling by hand up to 100′ from loading dock
3. Measuring and marking
4. Mounting wall bracket using (2) anchor type lead fasteners
5. Installing wireway on brackets, to 15′ high (For higher elevations use factors in R260533-20)

Job Conditions: Productivity is based on new construction, to a height of 15′ using rolling staging in an unobstructed area.

Material staging area is assumed to be within 100′ of work being performed.

R260533-65 Outlet Boxes

Outlet boxes should be included on the same takeoff sheet as branch piping or devices to better explain what is included in each circuit.

Each unit price in this section is a stand alone item and contains no other component unless specified. For example, to estimate a duplex outlet, components that must be added are:
1. 4″ square box
2. 4″ plaster ring
3. Duplex receptacle
4. Device cover

The method of mounting outlet boxes is (2) plastic shield fasteners.

Outlet boxes plastic, labor-hours include:
1. Marking box location on wood studding
2. Mounting box

Economy of Scale – For large concentrations of plastic boxes in the same area deduct the following percentages from labor-hour totals:

1	to	10	0%
11	to	25	20%
26	to	50	25%
51	to	100	30%
	over	100	35%

Note: It is important to understand that these percentages are not used on the total job quantities, but only areas where concentrations exceed the levels specified.

R260533-70 Pull Boxes and Cabinets

List cabinets and pull boxes by NEMA type and size.

Example:	**TYPE**	**SIZE**
	NEMA 1	6″W x 6″H x 4″D
	NEMA 3R	6″W x 6″H x 4″D

Labor-hours for wall mount (indoor or outdoor) installations include:
1. Unloading and uncrating
2. Handling of enclosures up to 200′ from loading dock using a dolly or pipe rollers
3. Measuring and marking
4. Drilling (4) anchor type lead fasteners using a hammer drill
5. Mounting and leveling boxes

Note: A plywood backboard is not included.

Labor-hours for ceiling mounting include:
1. Unloading and uncrating
2. Handling boxes up to 100′ from loading dock
3. Measuring and marking
4. Drilling (4) anchor type lead fasteners using a hammer drill
5. Installing and leveling boxes to a height of 15′ using rolling staging

Labor-hours for free standing cabinets include:
1. Unloading and uncrating
2. Handling of cabinets up to 200′ from loading dock using a dolly or pipe rollers
3. Marking of floor
4. Drilling (4) anchor type lead fasteners using a hammer drill
5. Leveling and shimming

Labor-hours for telephone cabinets include:
1. Unloading and uncrating
2. Handling cabinets up to 200′ using a dolly or pipe rollers
3. Measuring and marking
4. Mounting and leveling, using (4) lead anchor type fasteners

R260533-75 Weight Comparisons of Common Size Cast Boxes in Lbs.

Size NEMA 4 or 9	Cast Iron	Cast Aluminum	Size NEMA 7	Cast Iron	Cast Aluminum
6″ x 6″ x 6″	17	7	6″ x 6″ x 6″	40	15
8″ x 6″ x 6″	21	8	8″ x 6″ x 6″	50	19
10″ x 6″ x 6″	23	9	10″ x 6″ x 6″	55	21
12″ x 12″ x 6″	52	20	12″ x 6″ x 6″	100	37
16″ x 16″ x 6″	97	36	16″ x 16″ x 6″	140	52
20″ x 20″ x 6″	133	50	20″ x 20″ x 6″	180	67
24″ x 18″ x 8″	149	56	24″ x 18″ x 8″	250	93
24″ x 24″ x 10″	238	88	24″ x 24″ x 10″	358	133
30″ x 24″ x 12″	324	120	30″ x 24″ x 10″	475	176
36″ x 36″ x 12″	500	185	30″ x 24″ x 12″	510	189

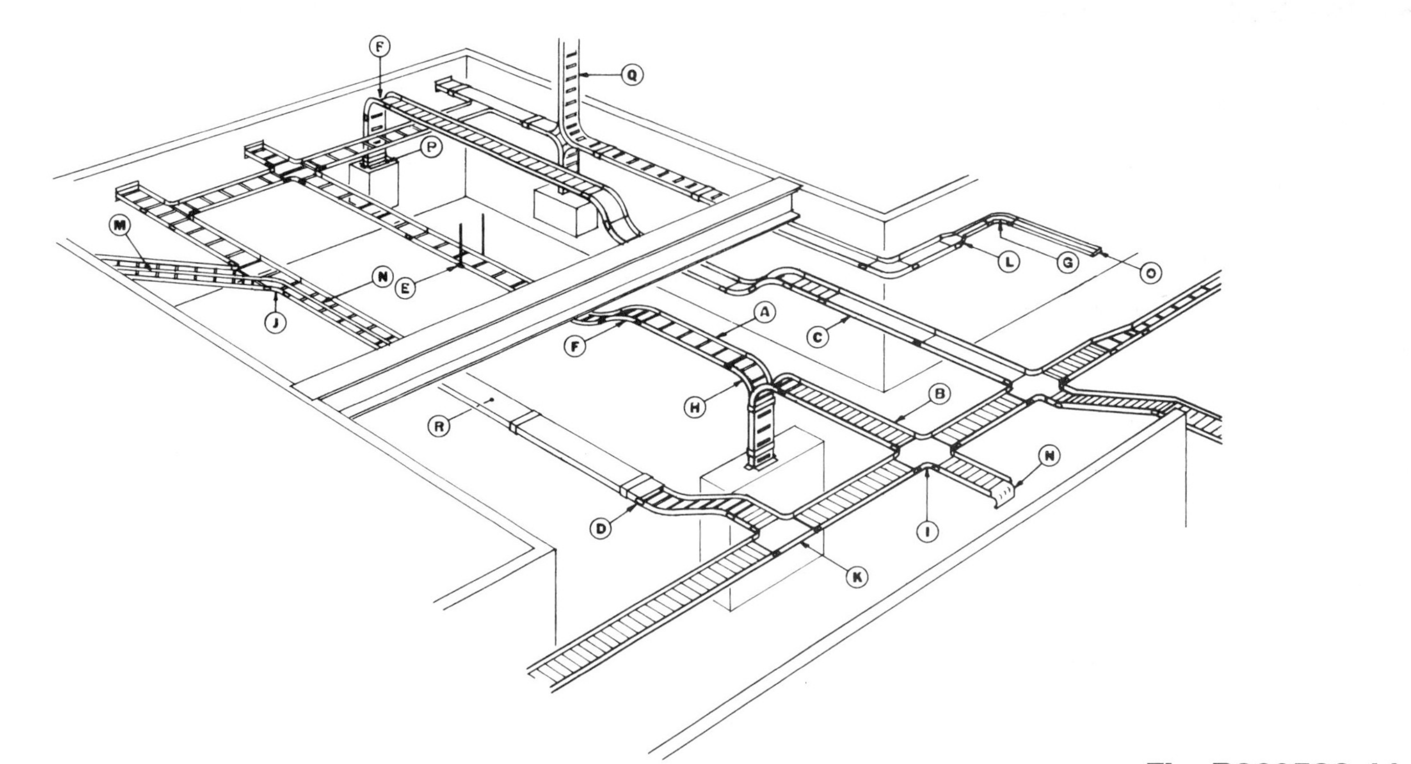

Fig. R260536-11

R260536-10 Cable Tray

Cable Tray - When taking off cable tray it is important to identify separately the different types and sizes involved in the system being estimated. (Fig. R260536-11)

A. – Ladder Type, galvanized or aluminum
B. – Trough Type, galvanized or aluminum
C. – Solid Bottom, galvanized or aluminum

The unit of measure is calculated in linear feet; do not deduct from this footage any length occupied by fittings, this will be the only allowance for scrap. Be sure to include all vertical drops to panels, switch gear, etc.

Hangers – Included in the linear footage of cable tray is

D. – 1 – Pair of connector plates per 12 L.F.
E. – 1 – Pair clamp type hangers and 4′ of 3/8″ threaded rod per 12 L.F.

Not included are structural supports, which must be priced in addition to the hangers.

Fittings – Identify separately the different types of fittings
1.) Ladder Type, galvanized or aluminum
2.) Trough Type, galvanized or aluminum
3.) Solid Bottom Type, galvanized or aluminum

The configuration, radius and rung spacing must also be listed. The unit of measure is "Ea."

F. – Elbow, vertical Ea.
G. – Elbow, horizontal Ea.
H. – Tee, vertical Ea.
I. – Cross, horizontal Ea.
J. – Wye, horizontal Ea.
K. – Tee, horizontal Ea.
L. – Reducing fitting Ea.

Depending on the use of the system other examples of units which must be included are:

M. – Divider strip L.F.
N. – Drop-outs Ea.
O. – End caps Ea.
P. – Panel connectors Ea.

Wire and cable are not included and should be taken off separately, see Division 161.

Job Conditions – Unit prices are based on a new installation to a work plane of 15′ using rolling staging.

Add to labor for elevated installations:

15′ to 20′ High	10%
20′ to 25′ High	20%
25′ to 30′ High	25%
30′ to 35′ High	30%
35′ to 40′ High	35%
Over 40′ High	40%

Add these percentages for L.F. totals but not to fittings. Add percentages to only those quantities that fall in the different elevations, in other words, if the total quantity of cable tray is 200′ but only 75′ is above 15′ then the 10% is added to the 75′ only.

Linear foot costs do not include penetrations through walls and floors which must be added to the estimate.

Cable Tray Covers

Covers – Cable tray covers are taken off in the same manner as the tray itself, making distinctions as to the type of cover. (Fig. R260536-11)

Q. – Vented, galvanized or aluminum
R. – Solid, galvanized or aluminum

Cover configurations are taken off separately noting type, specific radius and widths.

Note: Care should be taken to identify from plans and specifications exactly what is being covered. In many systems only vertical fittings are covered to retain wire and cable.

R260539-30 Conduit In Concrete Slab

List conduit by quantity, size and type.

Example:

A. Rigid galvanized steel + size
B. P.V.C. + size

Rigid galvanized steel (A) contains per 100 L.F.:

1. (20) Ties to slab reinforcing
2. (11) Threaded steel couplings
3. (2) Factory sweeps
4. (2) Field threaded conduit terminations
5. (2) Fiber bushings + locknuts
6. (2) Removal of concentric knockouts

P.V.C. (B) contains per 100 L.F.:

1. (20) Ties to slab reinforcing
2. (11) Field cemented couplings
3. (2) Factory sweeps
4. (2) Adapters
5. (2) Removal of concentric knockouts

R260539-40 Conduit In Trench

Conduit in trench is galvanized steel and contains per 100 L.F.:

1. (11) Threaded couplings
2. (2) Factory sweeps
3. (2) Fiber bushings + (4) locknuts
4. (2) Field threaded conduit terminations
5. (2) Removal of concentric knockouts

Note:

Conduit in Unit Price sections do not include:

1. Floor cutting
2. Excavation or backfill
3. Grouting or patching

Conduit fittings in excess of those listed in the above Conduit model must be added. (Refer to R260533-20 for Procedure example.)

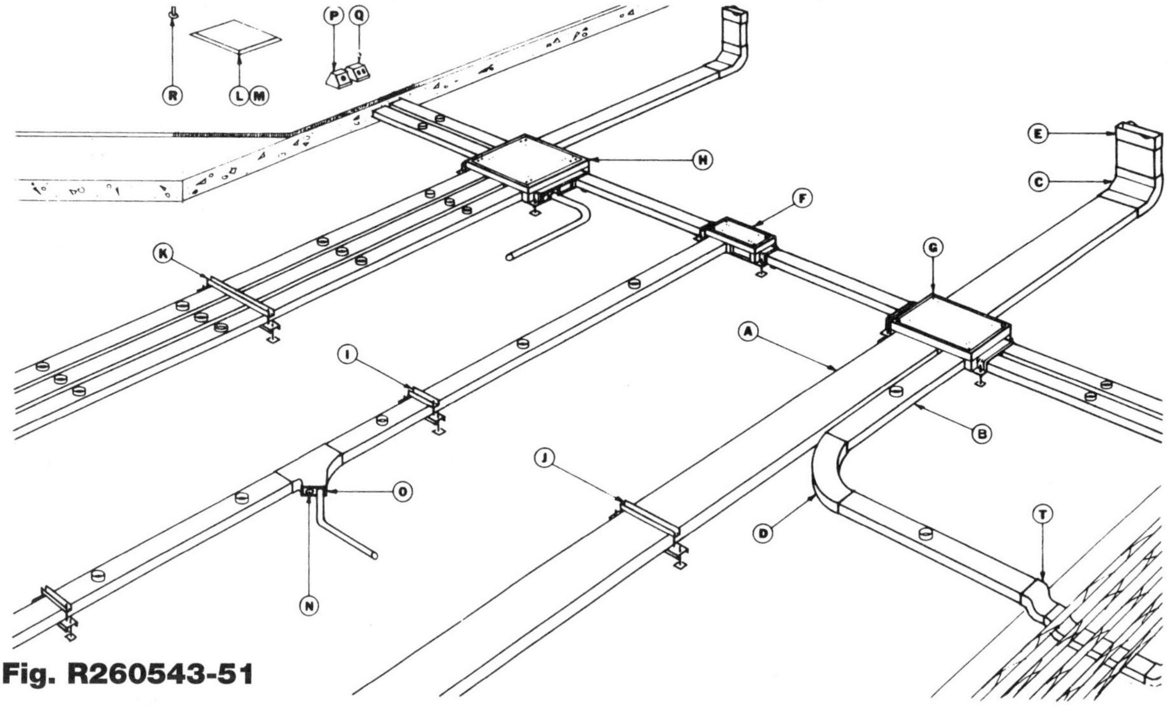

Fig. R260543-51

R260543-50 Underfloor Duct

When pricing Underfloor Duct it is important to identify and list each component, since costs vary significantly from one type of fitting to another. Do not deduct boxes or fittings from linear foot totals; this will be your allowance for scrap.

The first step is to identify the system as either:

FIG. R260543-51 Single Level

FIG. R260543-52 Dual Level

Single Level System

Include on your "takeoff sheet" the following unit price items, making sure to distinquish between Standard and Super duct:

A. Feeder duct (blank) in L.F.
B. Distribution duct (Inserts 2′ on center) in L.F.
C. Elbows (Vertical) Ea.
D. Elbows (Horizontal) Ea.
E. Cabinet connector Ea.
F. Single duct junction box Ea.
G. Double duct junction box Ea.
H. Triple duct junction box Ea.
I. Support, single cell Ea.
J. Support, double cell Ea.
K. Support, triple cell Ea.
L. Carpet pan Ea.
M. Terrazzo pan Ea.
N. Insert to conduit adapter Ea.
O. Conduit adapter Ea.
P. Low tension outlet Ea.
Q. High tension outlet Ea.
R. Galvanized nipple Ea.
S. Wire per C.L.F.
T. Offset (Duct type) Ea.

Dual Level System + Labor
see next page

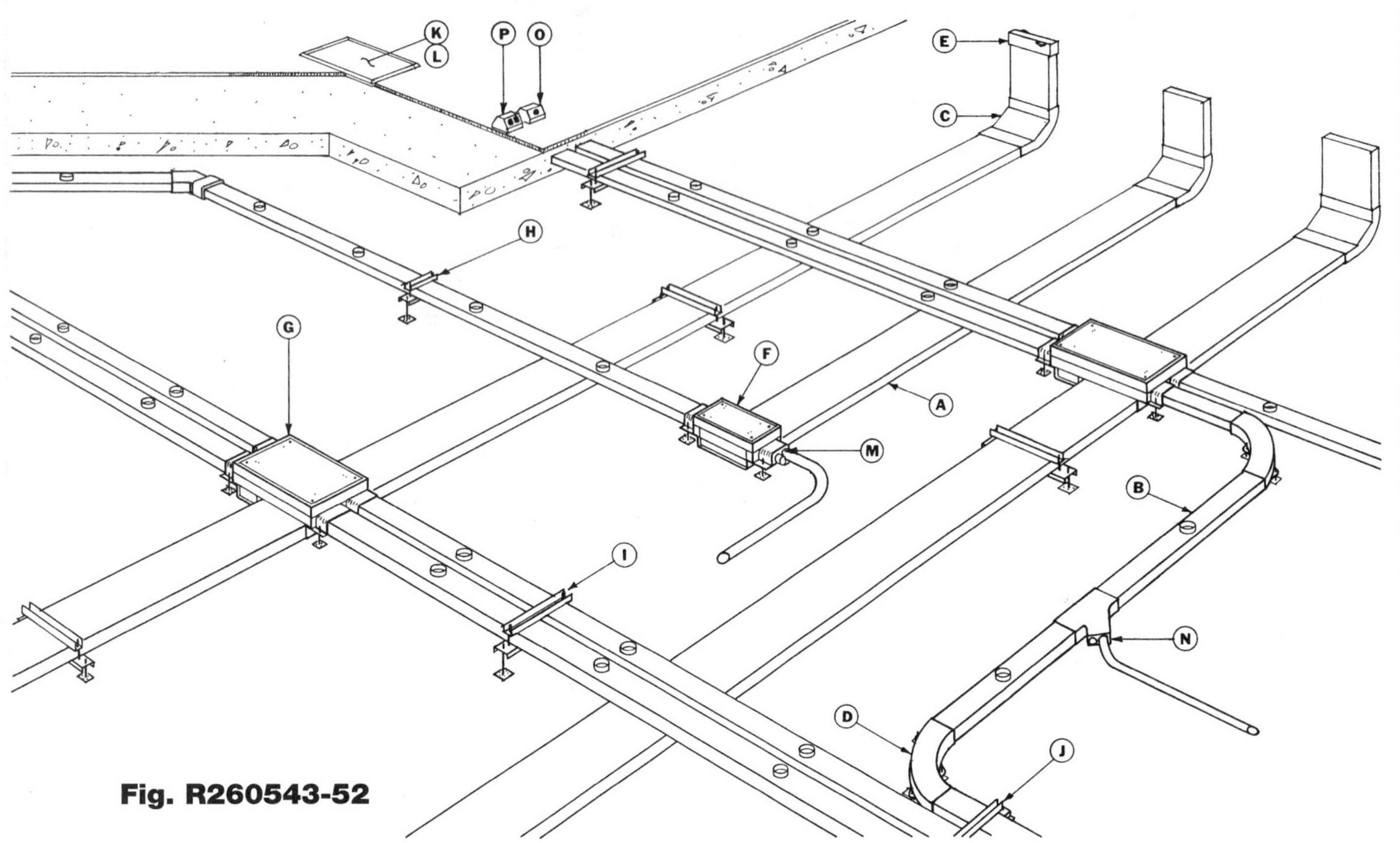

Fig. R260543-52

R260543-50 Underfloor Duct (cont.)

Dual Level

Include the following when "taking off" Dual Level systems:

Distinguish between Standard and Super duct.

A. Feeder duct (blank) in L.F.
B. Distribution duct (Inserts 2′ on center) in L.F.
C. Elbows (Vertical) Ea.
D. Elbows (Horizontal) Ea.
E. Cabinet connector Ea.
F. Single duct, 2 level, junction box Ea.
G. Double duct, 2 level, junction box Ea.
H. Support, single cell Ea.
I. Support, double cell Ea.
J. Support, triple cell Ea.
K. Carpet pan Ea.
L. Terrazzo pan Ea.
M. Insert to conduit adapter Ea.
N. Conduit adapter Ea.
O. Low tension outlet Ea.
P. High tension outlet Ea.
Q. Wire per C.L.F.

Note: Make sure to include risers in linear foot totals. High tension outlets include box, receptacle, covers and related mounting hardware.

Labor-hours for both Single and Dual Level systems include:

1. Unloading and uncrating
2. Hauling up to 200′ from loading dock
3. Measuring and marking
4. Setting raceway and fittings in slab or on grade
5. Leveling raceway and fittings

Labor-hours do not include:

1. Floor cutting
2. Excavation or backfill
3. Concrete pour
4. Grouting or patching
5. Wire or wire pulls
6. Additional outlets after concrete is poured
7. Piping to or from Underfloor Duct

Note: Installation is based on installing up to 150′ of duct. If quantities exceed this, deduct the following percentages:

1. 150′ to 250′ - 10%
2. 250′ to 350′ - 15%
3. 350′ to 500′ - 20%
4. over 500′ - 25%

Deduct these percentages from labor only.

Deduct these percentages from straight sections only.

Do not deduct from fittings or junction boxes.

Job Conditions: Productivity is based on new construction.

Underfloor duct to be installed on first three floors.

Material staging area within 100′ of work being performed.

Area unobstructed and duct not subject to physical damage.

R260580-75 Motor Connections

Motor connections should be listed by size and type of motor. Included in the material and labor cost is:

1. (2) Flex connectors
2. 18″ of flexible metallic wireway
3. Wire identification and termination
4. Test for rotation
5. (2) or (3) Conductors

Price does not include:

1. Mounting of motor
2. Disconnect Switch
3. Motor Starter
4. Controls
5. Conduit or wire ahead of flex

Note: When "Taking off" Motor connections, it is advisable to list connections on the same quantity sheet as Motors, Motor Starters and controls.

R260590-05 Typical Overhead Service Entrance

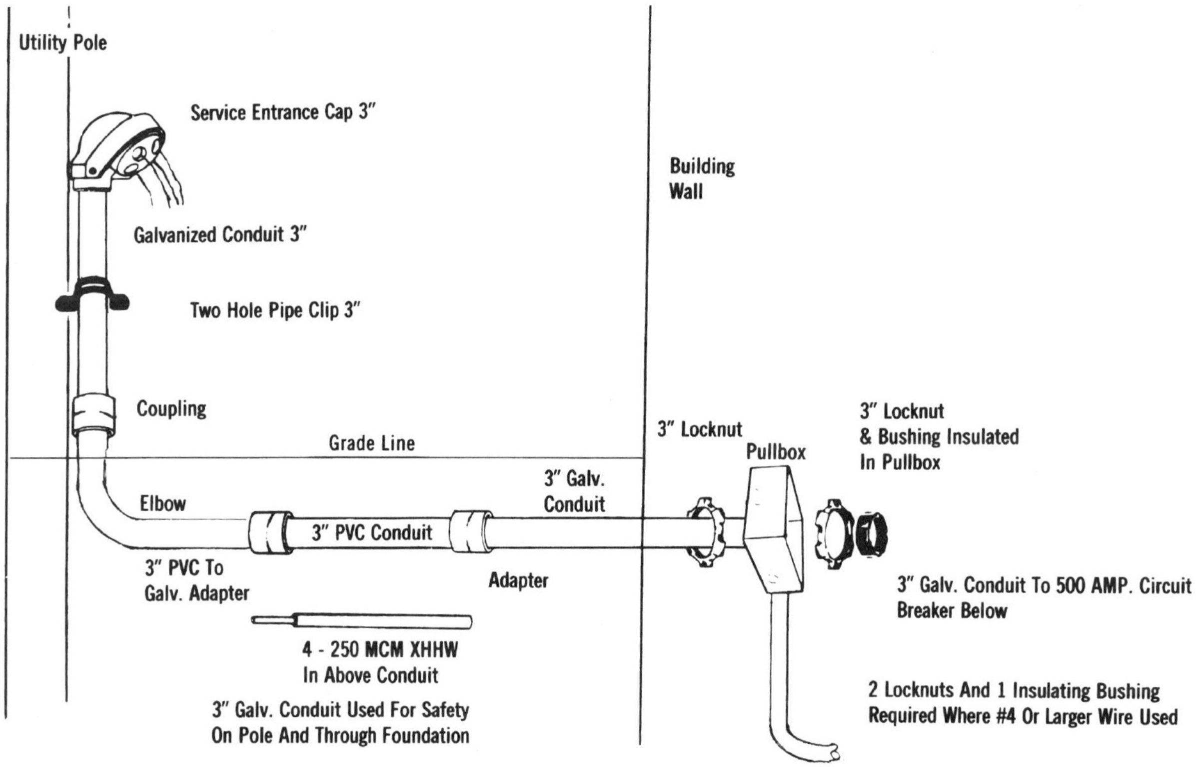

R260913-80 Switchboard Instruments

Switchboard instruments are added to the price of switchboards according to job specifications. This equipment is usually included when ordering "Gear" from the manufacturer and will arrive factory installed.

Included in the labor cost is:
1. Internal wiring connections
2. Wire identification
3. Wire tagging

Transition sections include:
1. Uncrating
2. Hauling sections up to 100′ from loading dock
3. Positioning sections
4. Leveling sections
5. Bolting enclosures
6. Bolting vertical bus bars

Price does not include:
1. Equipment pads
2. Steel channels embedded or grouted in concrete
3. Special knockouts
4. Rigging

Job Conditions: Productivity is based on new construction, equipment to be installed on the first floor within 200′ of the loading dock.

R262213-10 Electric Circuit Voltages

General: The following method provides the user with a simple non-technical means of obtaining comparative costs of wiring circuits. The circuits considered serve the electrical loads of motors, electric heating, lighting and transformers, for example, that require low voltage 60 Hertz alternating current.

The method used here is suitable only for obtaining estimated costs. It is **not** intended to be used as a substitute for electrical engineering design applications.

Conduit and wire circuits can represent from twenty to thirty percent of the total building electrical cost. By following the described steps and using the tables the user can translate the various types of electric circuits into estimated costs.

Wire Size: Wire size is a function of the electric load which is usually listed in one of the following units:

1. Amperes (A)
2. Watts (W)
3. Kilowatts (kW)
4. Volt amperes (VA)
5. Kilovolt amperes (kVA)
6. Horsepower (HP)

These units of electric load must be converted to amperes in order to obtain the size of wire necessary to carry the load. To convert electric load units to amperes one must have an understanding of the voltage classification of the power source and the voltage characteristics of the electrical equipment or load to be energized. The seven A.C. circuits commonly used are illustrated in Figures R262213-11 thru R262213-17 showing the tranformer load voltage and the point of use voltage at the point on the circuit where the load is connected. The difference between the source and point of use voltages is attributed to the circuit voltage drop and is considered to be approximately 4%.

Motor Voltages: Motor voltages are listed by their point of use voltage and not the power source voltage.

For example: 460 volts instead of 480 volts
200 instead of 208 volts
115 volts instead of 120 volts

Lighting and Heating Voltages: Lighting and heating equipment voltages are listed by the power source voltage and not the point of wire voltage.

For example: 480, 277, 120 volt lighting
480 volt heating or air conditioning unit
208 volt heating unit

Transformer Voltages: Transformer primary (input) and secondary (output) voltages are listed by the power source voltage.

For example: Single phase 10 kVA
Primary 240/480 volts
Secondary 120/240 volts

In this case, the primary voltage may be 240 volts with a 120 volts secondary or may be 480 volts with either a 120V or a 240V secondary.

For example: Three phase 10 kVA
Primary 480 volts
Secondary 208Y/120 volts

In this case the transformer is suitable for connection to a circuit with a 3 phase 3 wire or 3 phase 4 wire circuit with a 480 voltage. This application will provide a secondary circuit of 3 phase 4 wire with 208 volts between phase wires and 120 volts between any phase wire and the neutral (white) wire.

R262213-11

3 Wire, 1 Phase, 120/240 Volt System

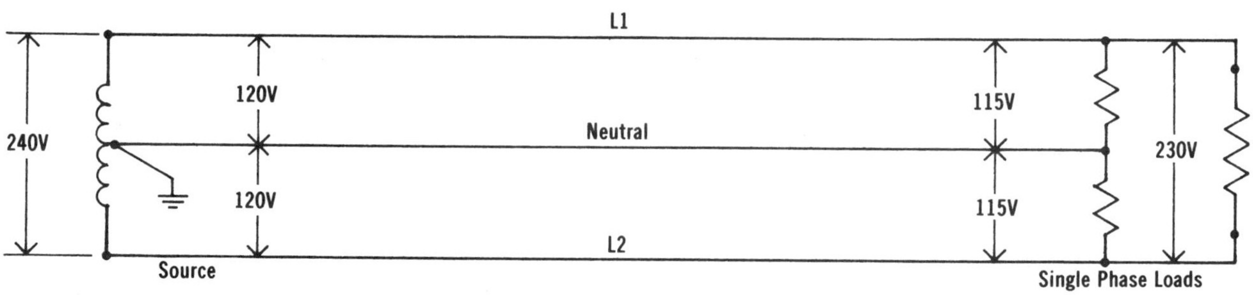

R262213-12

4 wire, 3 Phase, 208Y/120 Volt System

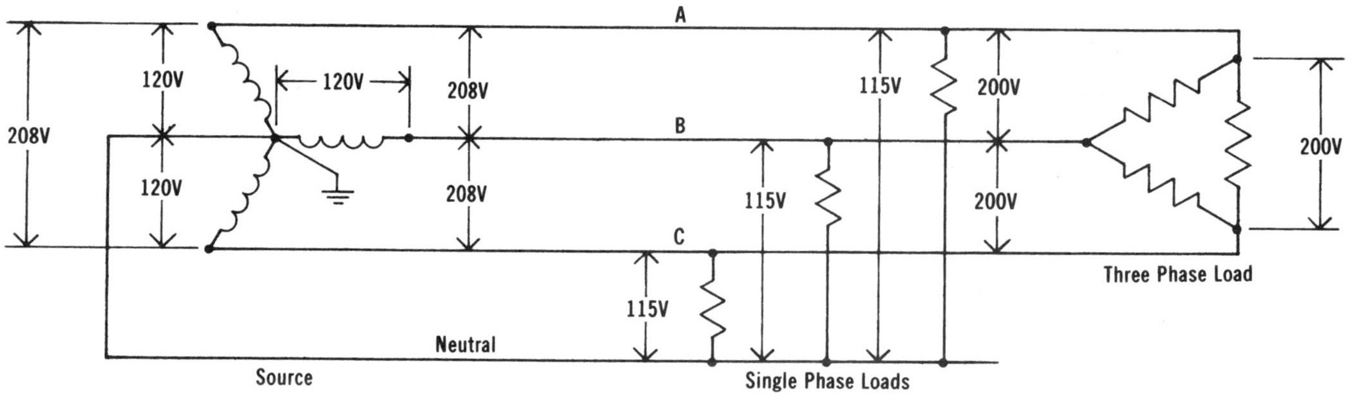

R262213-13 Electric Circuit Voltages (cont.)

3 Wire, 3 Phase 240 Volt System

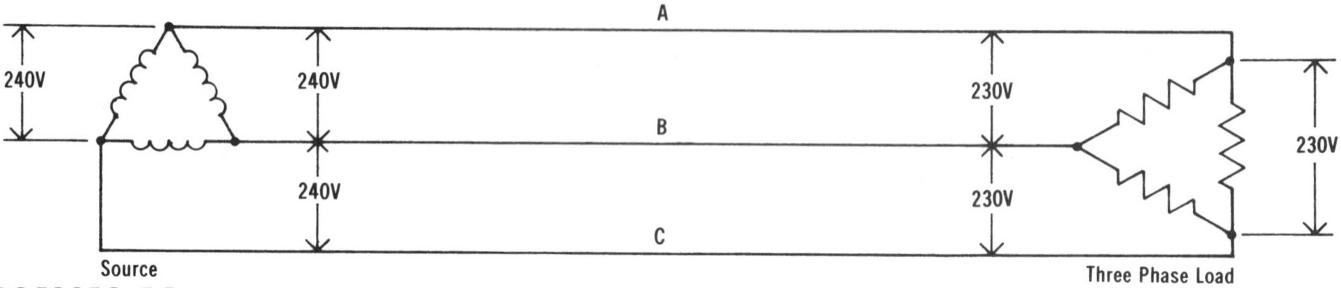

R262213-14

4 Wire, 3 Phase, 240/120 Volt System

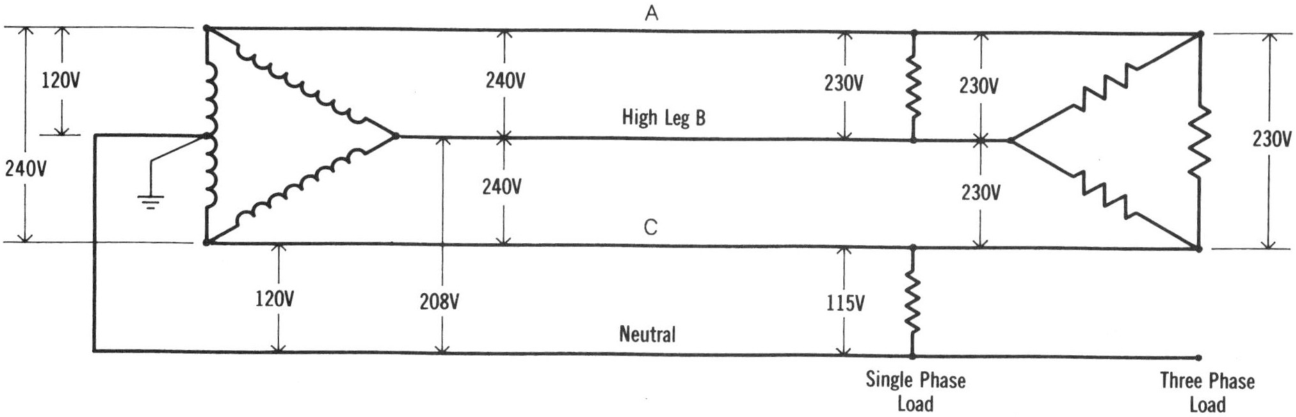

R262213-15

3 Wire, 3 Phase 480 Volt System

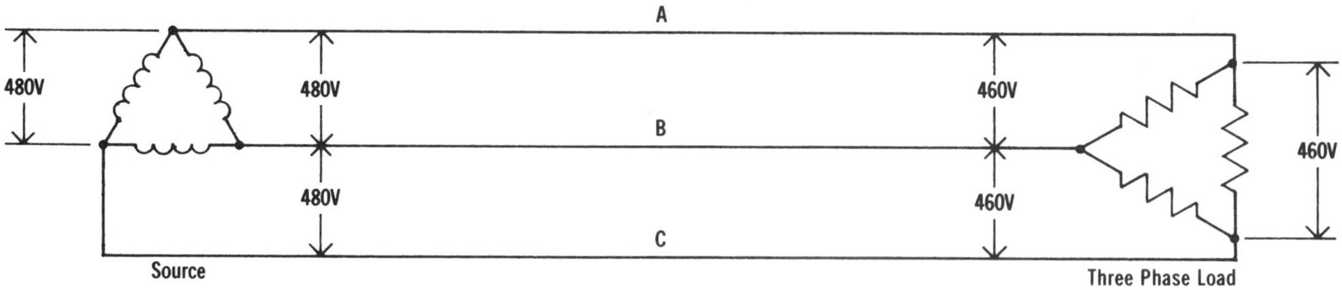

R262213-16

4 Wire, 3 Phase, 480Y/277 Volt System

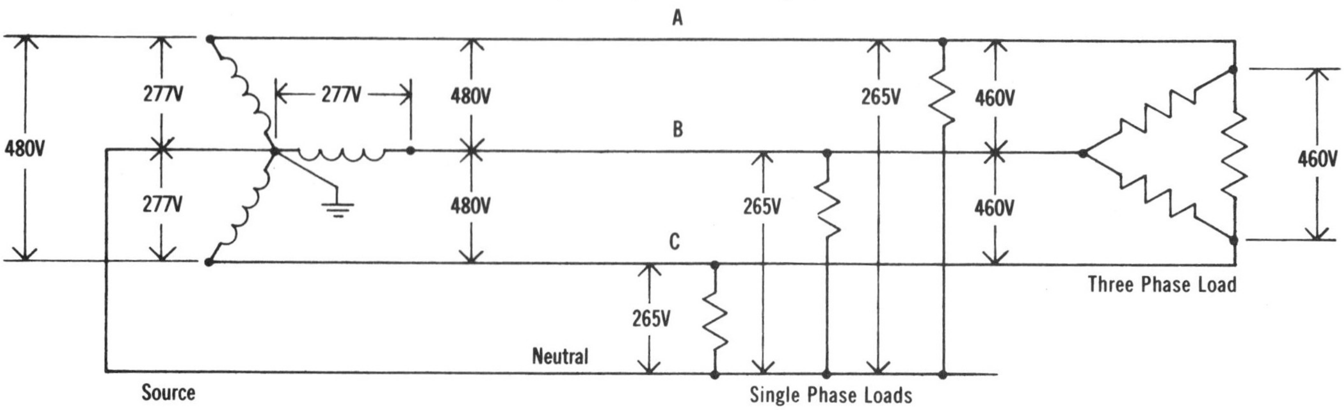

R262213-17 Electric Circuit Voltages (cont.)

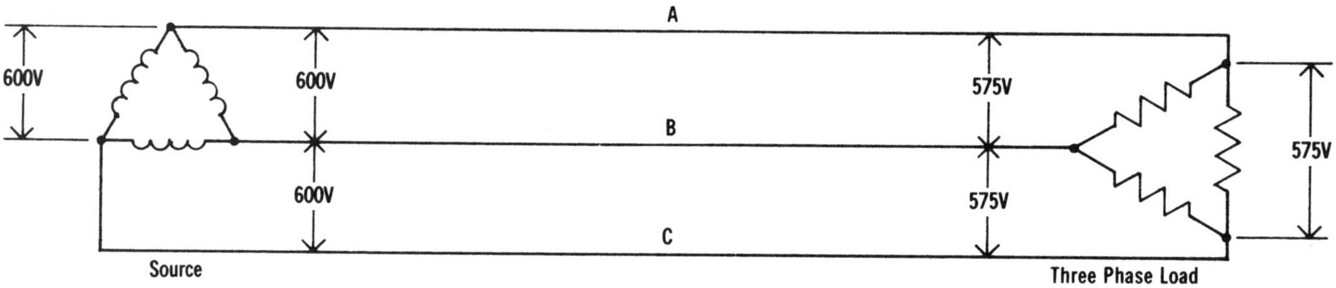

R262213-20 kW Value/Cost Determination

General: Lighting and electric heating loads are expressed in watts and kilowatts.

Cost Determination:

The proper ampere values can be obtained as follows:

1. Convert watts to kilowatts (watts 1000 ÷ kilowatts)
2. Determine voltage rating of equipment.
3. Determine whether equipment is single phase or three phase.
4. Refer to Table R262213-21 to find ampere value from kW, Ton and Btu/hr. values.
5. Determine type of wire insulation – TW, THW, THWN.
6. Determine if wire is copper or aluminum.
7. Refer to Table R260519-92 to obtain copper or aluminum wire size from ampere values.
8. Next refer to Table R260533-22 for the proper conduit size to accommodate the number and size of wires in each particular case.
9. Next refer to unit cost data for the per linear foot cost of the conduit.
10. Next refer to unit cost data for the per linear foot cost of the wire. Multiply cost of wire per L.F. x number of wires in the circuits to obtain total wire cost per L.F.
11. Add values obtained in Step 9 and 10 for total cost per linear foot for conduit and wire x length of circuit = Total Cost.

Notes:

1. 1 Phase refers to single phase, 2 wire circuits.
2. 3 Phase refers to three phase, 3 wire circuits.
3. For circuits which operate continuously for 3 hours or more, multiply the ampere values by 1.25 for a given kw requirement.
4. For kW ratings not listed, add ampere values.

   For example: Find the ampere value of 9 kW at 208 volt, single phase.

   4 kW = 19.2A
   5 kW = 24.0A
   9 kW = 43.2A
5. "Length of Circuit" refers to the one way distance of the run, not to the total sum of wire lengths.

R262213-21 Ampere Values as Determined by kW Requirements, BTU/HR or Ton, Voltage and Phase Values

			Ampere Values						
			120V	208V		240V		277V	480V
kW	Ton	BTU/HR	1 Phase	1 Phase	3 Phase	1 Phase	3 Phase	1 Phase	3 Phase
0.5	.1422	1,707	4.2A	2.4A	1.4A	2.1A	1.2A	1.8A	0.6A
0.75	.2133	2,560	6.2	3.6	2.1	3.1	1.9	2.7	.9
1.0	.2844	3,413	8.3	4.9	2.8	4.2	2.4	3.6	1.2
1.25	.3555	4,266	10.4	6.0	3.5	5.2	3.0	4.5	1.5
1.5	.4266	5,120	12.5	7.2	4.2	6.3	3.1	5.4	1.8
2.0	.5688	6,826	16.6	9.7	5.6	8.3	4.8	7.2	2.4
2.5	.7110	8,533	20.8	12.0	7.0	10.4	6.1	9.1	3.1
3.0	.8532	10,239	25.0	14.4	8.4	12.5	7.2	10.8	3.6
4.0	1.1376	13,652	33.4	19.2	11.1	16.7	9.6	14.4	4.8
5.0	1.4220	17,065	41.6	24.0	13.9	20.8	12.1	18.1	6.1
7.5	2.1331	25,598	62.4	36.0	20.8	31.2	18.8	27.0	9.0
10.0	2.8441	34,130	83.2	48.0	27.7	41.6	24.0	36.5	12.0
12.5	3.5552	42,663	104.2	60.1	35.0	52.1	30.0	45.1	15.0
15.0	4.2662	51,195	124.8	72.0	41.6	62.4	37.6	54.0	18.0
20.0	5.6883	68,260	166.4	96.0	55.4	83.2	48.0	73.0	24.0
25.0	7.1104	85,325	208.4	120.2	70.0	104.2	60.0	90.2	30.0
30.0	8.5325	102,390		144.0	83.2	124.8	75.2	108.0	36.0
35.0	9.9545	119,455		168.0	97.1	145.6	87.3	126.0	42.1
40.0	11.3766	136,520		192.0	110.8	166.4	96.0	146.0	48.0
45.0	12.7987	153,585			124.8	187.5	112.8	162.0	54.0
50.0	14.2208	170,650			140.0	208.4	120.0	180.4	60.0
60.0	17.0650	204,780			166.4		150.4	216.0	72.0
70.0	19.9091	238,910			194.2		174.6		84.2
80.0	22.7533	273,040			221.6		192.0		96.0
90.0	25.5975	307,170					225.6		108.0
100.0	28.4416	341,300							120.0

R262213-25 kVA Value/Cost Determination

General: Control transformers are listed in VA. Step-down and power transformers are listed in kVA.

Cost Determination:

1. Convert VA to kVA. Volt amperes (VA) ÷ 1000 = Kilovolt amperes (kVA).
2. Determine voltage rating of equipment.
3. Determine whether equipment is single phase or three phase.
4. Refer to Table R262213-26 to find ampere value from kVA value.
5. Determine type of wire insulation – TW, THW, THWN.
6. Determine if wire is copper or aluminum.
7. Refer to Table R260519-92 to obtain copper or aluminum wire size from ampere values.
8. Next refer to Table R260533-22 for the proper conduit size to accommodate the number and size of wires in each particular case.
9. Next refer to unit price data for the per linear foot cost of the conduit.
10. Next refer to unit price data for the per linear foot cost of the wire. Multiply cost of wire per L.F. x number of wires in the circuits to obtain total wire cost.
11. Add values obtained in Step 9 and 10 for total cost per linear foot for conduit and wire x length of circuit = Total Cost.

Example: A transformer rated 10 kVA 480 volts primary, 240 volts secondary, 3 phase has the capacity to furnish the following:

1. Primary amperes = 10 kVA x 1.20 = 12 amperes (from Table R262213-26)
2. Secondary amperes = 10 kVA x 2.40 = 24 amperes (from Table R262213-26)

Note: Transformers can deliver generally 125% of their rated kVA. For instance, a 10 kVA rated transformer can safely deliver 12.5 kVA.

R262213-26 Multiplier Values for kVA to Amperes Determined by Voltage and Phase Values

Volts	Multiplier for Circuits	
	2 Wire, 1 Phase	3 Wire, 3 Phase
115	8.70	
120	8.30	
230	4.30	2.51
240	4.16	2.40
200	5.00	2.89
208	4.80	2.77
265	3.77	2.18
277	3.60	2.08
460	2.17	1.26
480	2.08	1.20
575	1.74	1.00
600	1.66	0.96

R262213-27 Central Air Conditioning Watts per S.F., BTUs per Hour per S.F. of Floor Area and S.F. per Ton of Air Conditioning

Type Building	Watts per S.F.	BTUH per S.F.	S.F. per Ton	Type Building	Watts per S.F.	BTUH per S.F.	S.F. per Ton	Type Building	Watts per S.F.	BTUH per S.F.	S.F. per Ton
Apartments, Individual	3	26	450	Dormitory, Rooms	4.5	40	300	Libraries	5.7	50	240
Corridors	2.5	22	550	Corridors	3.4	30	400	Low Rise Office, Ext.	4.3	38	320
Auditoriums & Theaters	3.3	40	300/18*	Dress Shops	4.9	43	280	Interior	3.8	33	360
Banks	5.7	50	240	Drug Stores	9	80	150	Medical Centers	3.2	28	425
Barber Shops	5.5	48	250	Factories	4.5	40	300	Motels	3.2	28	425
Bars & Taverns	15	133	90	High Rise Off.-Ext. Rms.	5.2	46	263	Office (small suite)	4.9	43	280
Beauty Parlors	7.6	66	180	Interior Rooms	4.2	37	325	Post Office, Int. Office	4.9	42	285
Bowling Alleys	7.8	68	175	Hospitals, Core	4.9	43	280	Central Area	5.3	46	260
Churches	3.3	36	330/20*	Perimeter	5.3	46	260	Residences	2.3	20	600
Cocktail Lounges	7.8	68	175	Hotels, Guest Rooms	5	44	275	Restaurants	6.8	60	200
Computer Rooms	16	141	85	Public Spaces	6.2	55	220	Schools & Colleges	5.3	46	260
Dental Offices	6	52	230	Corridors	3.4	30	400	Shoe Stores	6.2	55	220
Dept. Stores, Basement	4	34	350	Industrial Plants, Offices	4.3	38	320	Shop'g. Ctrs., Sup. Mkts.	4	34	350
Main Floor	4.5	40	300	General Offices	4	34	350	Retail Stores	5.5	48	250
Upper Floor	3.4	30	400	Plant Areas	4.5	40	300	Specialty Shops	6.8	60	200

*Persons per ton

12,000 BTUH = 1 ton of air conditioning

R262213-60 Oil Filled Transformers

Transformers in this section include:

1. Rigging (as required)
2. Rental of crane and operator
3. Setting of oil filled transformer
4. (4) Anchor bolts, nuts and washers in concrete pad

Price does not include:

1. Primary and secondary terminations
2. Transformer pad
3. Equipment grounding
4. Cable
5. Conduit locknuts or bushings

Transformers in Unit Price sections for dry type, back-boost and isolating transformers include:

1. Unloading and uncrating
2. Hauling transformer to within 200′ of loading dock
3. Setting in place
4. Wall mounting hardware
5. Testing

Price does not include:

1. Structural supports
2. Suspension systems
3. Welding or fabrication
4. Primary & secondary terminations

Add the following percentages to the labor for ceiling mounted transformers:

10′ to 15′	= + 15%
15′ to 25′	= + 30%
Over 25′	= + 35%

Job Conditions: Productivities are based on new construction. Installation is assumed to be on the first floor, in an obstructed area to a height of 10′. Material staging area is within 100′ of final transformer location.

Electrical | R2622 | Low Voltage Transformers

R262213-65 Transformer Weight (Lbs.) by kVA

Oil Filled 3 Phase 5/15 KV To 480/277			
kVA	**Lbs.**	**kVA**	**Lbs.**
150	1800	1000	6200
300	2900	1500	8400
500	4700	2000	9700
750	5300	3000	15000
Dry 240/480 To 120/240 Volt			
1 Phase		**3 Phase**	
kVA	**Lbs.**	**kVA**	**Lbs.**
1	23	3	90
2	36	6	135
3	59	9	170
5	73	15	220
7.5	131	30	310
10	149	45	400
15	205	75	600
25	255	112.5	950
37.5	295	150	1140
50	340	225	1575
75	550	300	1870
100	670	500	2850
167	900	750	4300

Electrical | R2624 | Switchboards & Panelboards

R262416-50 Load Centers and Panelboards

When pricing Load Centers list panels by size and type. List Breakers in a separate column of the "Quantity Sheet," and define by phase and ampere rating.

Material and Labor prices include breakers; for example: a 100A, 3-Wire, 102/240V, 18 circuit panel w/ main breaker, as described in the unit cost section, contains 18 single pole 20A breakers.

If you do not choose to include a full panel of single pole breakers, use the following method to adjust material and labor costs.

Example: In an 18 circuit panel only 16 single pole breakers are desired, requiring that the cost of 2 breakers be subtracted from the panel cost.

1. Go to the appropriate unit cost section of the book to find the unit prices of the given circuit breaker type.
2. Modify those costs as follows: Bare material price x 0.50 Bare labor cost x 0.60.
3. Multiply those modified bare costs by 2 (breakers in this example).
4. Subtract the modified costs for the 2 breakers from the given cost of the panel.

Labor-hours for Load Center installation includes:

1. Unloading, uncrating, and handling enclosures 200′ from unloading area
2. Measuring and marking
3. Drilling (4) lead anchor type fasteners using a hammer drill
4. Mounting and leveling panel to a height of 6′
5. Preparation and termination of feeder cable to lugs or main breaker
6. Branch circuit identification
7. Lacing using tie wraps
8. Testing and load balancing
9. Marking panel directory

Not included in the material and labor are:

1. Modifications to enclosure
2. Structural supports
3. Additional lugs
4. Plywood backboards
5. Painting or lettering

Note: Knockouts are included in the price of terminating pipe runs and need not be added to the Load Center costs.

Job Conditions: Productivity is based on new construction to a height of 6′, in an unobstructed area. Material staging area is assumed to be within 100′ of work being performed.

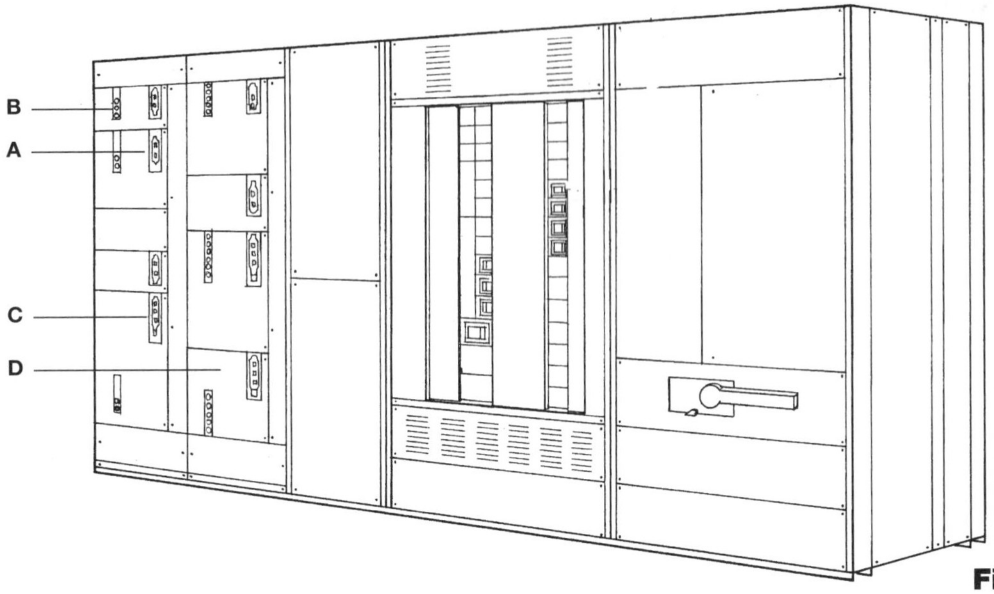

Fig. R262419-61

R262419-60 Motor Control Centers

When taking off Motor Control Centers, list the size, type and height of structures.

Example:

1. 600A, 22,000 RMS, 72″ high
2. 600A, back to back, 72″ high

Next take off individual starters; the number of structures can also be determined by adding the height in inches of starters divided by the height of the structure, and list on the same quantity sheet as the structures. Identify starters by Type, Horsepower rating, Size and Height in inches.

Example:

A. Class I, Type B, FVNR starter, 25 H.P., 18″ high.
   Add to the list with Starters, factory installed controls.

Example:

B. Pilot lights
C. Push buttons
D. Auxiliary contacts

Identify starters and structures as either copper or aluminum and by the NEMA type of enclosure.

When pricing starters and structures, be sure to add or deduct adjustments, using lines in the unit price section.

Included in the cost of Motor Control Structures are:

1. Uncrating
2. Hauling to location within 100′ of loading dock
3. Setting structures
4. Leveling
5. Aligning
6. Bolting together structure frames
7. Bolting horizontal bus bars

Labor-hours do not include:

1. Equipment pad
2. Steel channels embedded or grouted in concrete
3. Pull boxes
4. Special knockouts
5. Main switchboard section
6. Transition section
7. Instrumentation
8. External control wiring
9. Conduit or wire

Material for Starters includes:

1. Circuit breaker or fused disconnect
2. Magnetic motor starter
3. Control transformer
4. Control fuse and fuse block

Labor-hours for Starters include:

1. Handling
2. Installing starter within structure
3. Internal wiring connections
4. Lacing within enclosure
5. Testing
6. Phasing

Job Conditions: Productivity is based on new construction. Motor control location assumed to be on first floor in an unobstructed area. Material staging area within 100′ of final location.

Note: Additional labor-hours must be added if M.C.C. is to be installed on other than the first floor, or if rigging is required.

R262419-65 Motor Starters and Controls

Motor starters should be listed on the same "Quantity Sheet" as Motors and Motor Connections. Identify each starter by:

1. Size
2. Voltage
3. Type

Example:

A. FVNR, 480V, 2HP, Size 00
B. FVNR, 480V, 5HP, Size 0, Combination type
C. FVR, 480V, Size 2

The NEMA type of enclosure should also be identified.

Included in the labor-hours are:

1. Unloading, uncrating and handling of starters up to 200′ from loading area
2. Measuring and marking
3. Drilling (4) anchor type lead fasteners, using a hammer drill
4. Mounting and leveling starter
5. Connecting wire or cable to line and load sides of starter (when already lugged)
6. Installation of (3) thermal type heaters
7. Testing

The following is not included unless specified in the unit price description.

1. Control transformer
2. Controls, either factory or field installed
3. Conduit and wire to or from starter
4. Plywood backboard
5. Cable terminations

The following material and labor has been included for Combination type starters:

1. Unloading, uncrating and handling of starter up to 200′ of loading dock
2. Measuring and marking
3. Drilling (4) anchor type lead fasteners using a hammer drill
4. Mounting and leveling
5. Connecting prepared cable conductors
6. Installation of (3) dual element cartridge type fuses
7. Installation of (3) thermal type heaters
8. Test for rotation

MOTOR CONTROLS

When pricing motor controls make sure you consider the type of control system being utilized. If the controls are factory installed and located in the enclosure itself, then you would add to your starter price the items 5200 thru 5800 in the Unit Price section. If control voltage is different from line voltage, add the material and labor cost of a control transformer.

For external control of starters include the following items:

1. Raceways
2. Wire & terminations
3. Control enclosures
4. Fittings
5. Push button stations
6. Indicators

Job Conditions: Productivity is based on new construction to a height of 10′.

Material staging area is assumed to be within 100′ of work being performed.

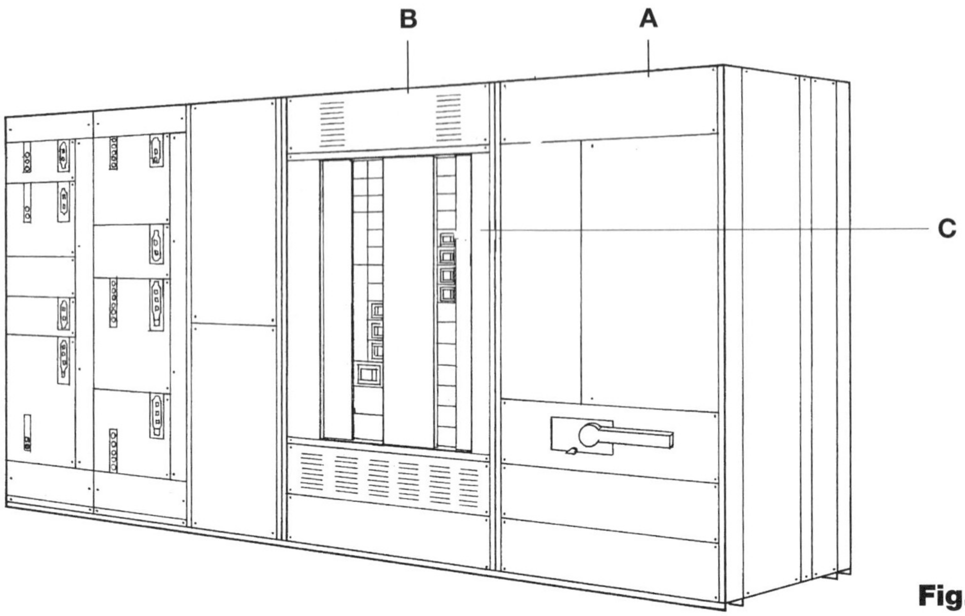

Fig. R262419-81

R262419-80 Distribution Section

After "Taking off" the Switchboard section, include on the same "Quantity sheet" the Distribution section; identify by:

1. Voltage
2. Ampere rating
3. Type

Example:

(Fig. R262419-81) (B) Distribution Section

Included in the labor costs of the Distribution section is:

1. Uncrating
2. Hauling to 200′ of loading dock
3. Setting of distribution panel
4. Leveling & shimming
5. Anchoring of equipment to pad or floor
6. Bolting of horizontal bus bars between
7. Testing of equipment

Not included in the Distribution section is:

1. Breakers (C)
2. Equipment pads
3. Steel channels embedded or grouted in concrete
4. Pull boxes
5. Special knockouts
6. Transition section
7. Conduit or wire

R262419-82 Feeder Section

List quantities on the same sheet as the Distribution section. Identify breakers by: (Fig. R262419-81)

1. Frame type
2. Number of poles
3. Ampere rating

Installation includes:

1. Handling
2. Placing breakers in distribution panel
3. Preparing wire or cable
4. Lacing wire
5. Marking each phase with colored tape
6. Marking panel legend
7. Testing
8. Balancing

R262419-84 Switchgear

It is recommended that "Switchgear" or those items contained in the following sections be quoted from equipment manufacturers as a package price.

Switchboard instruments

Switchboard distribution sections

Switchboard feeder sections

Switchboard Service Disc.

Switchboard In-plant Dist.

Included in these sections are the most common types and sizes of factory assembled equipment.

The recommended procedure for low voltage switchgear would be to price (Fig. R262419-81) (A) Main Switchboard

Identify by:
1. Voltage
2. Amperage
3. Type

Example:
1. 120/208V, 4-wire, 600A, nonfused
2. 277/480V, 4-wire, 600A, nonfused
3. 120/208V, 4-wire, 400A w/fused switch & CT compartment
4. 277/480V, 4-wire, 400A, w/fused switch & CT compartment
5. 120/208V, 4-wire, 800A, w/pressure switch & CT compartment
6. 277/480V, 4-wire, 800A, w/molded CB & CT compartment

Included in the labor costs for Switchboards are:
1. Uncrating
2. Hauling to 200′ of loading dock
3. Setting equipment
4. Leveling and shimming
5. Anchoring
6. Cable identification
7. Testing of equipment

Not included in the Switchboard price is:
1. Rigging
2. Equipment pads
3. Steel channels embedded or grouted in concrete
4. Special knockouts
5. Transition or Auxiliary sections
6. Instrumentation
7. External control
8. Conduit and wire
9. Conductor terminations

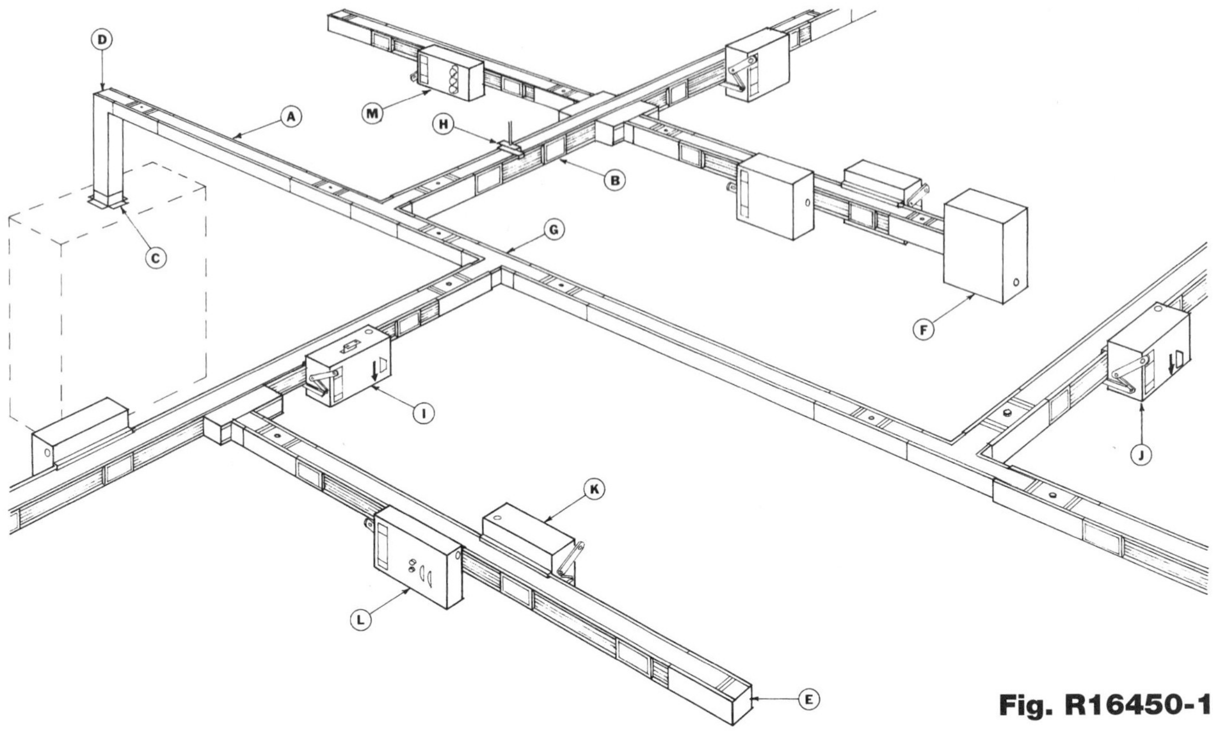

Fig. R16450-101

R262513-10 Aluminum Bus Duct

When taking off bus duct identify the system as either:

1. Aluminum
2. Copper

List straight lengths by type and size

A. Plug-in — 800 A
B. Feeder — 800 A

Do not measure thru fittings as you would on conduit, since there is no allowance for scrap in bus duct systems.

If upon taking off linear foot quantities of bus duct you find your quantities are not divisible evenly by 10 ft., then the remainder must be priced as a special item and quoted from the manufactuer. Do not use the bare material cost per L.F. for these special items. You can, however, safely use the bare labor cost per L.F. for the entire length.

Identify fittings by type and ampere rating.

Example:

C. Switchboard stub 800 A

D. Elbows 800 A

E. End box 800 A

F. Cable tap box 800 A

G. Tee Fittings 800 A

H. Hangers

Plug-in Units – List separately plug-in units and identify by type and ampere rating

I. Plug-in switches 600 Volt 3 phase 60 A

J. Plug-in molded case C.B. 60 A

K. Combination starter FVNR NEMA 1

L. Combination contactor & fused switch NEMA 1

M. Combination fusible switch & lighting control 60 A

Labor-hours for feeder and plug-in sections include:

1. Unloading and uncrating
2. Hauling up to 200 ft. from loading dock
3. Measuring and marking
4. Setup of rolling staging
5. Installing hangers
6. Hanging and bolting sections
7. Aligning and leveling
8. Testing

Labor-hours do not include:

1. Modifications to existing structure for hanger supports
2. Threaded rod in excess of 2 ft.
3. Welding
4. Penetrations thru walls
5. Staging rental

Deduct the following percentages from labor only:

150 ft. to 250 ft. — 10%
251 ft. to 350 ft. — 15%
351 ft. to 500 ft. — 20%
Over 500 ft. — 25%

Deduct percentage only if runs are contained in the same area.

Example: If the job entails running 100 ft. in 5 different locations do not deduct 20%, but if the duct is being run in 1 area and the quantity is 500 ft. then you would deduct 20%.

Deduct only from straight lengths, not fittings or plug-in units.

R262513-10 Aluminum Bus Duct (cont.)

Add to labor for elevated installations:

15 ft. to 20 ft. high	10%
21 ft. to 25 ft. high	20%
26 ft. to 30 ft. high	30%
31 ft. to 35 ft. high	40%
36 ft. to 40 ft. high	50%
Over 40 ft. high	60%

Bus Duct Fittings:

Labor-hours for fittings include:
1. Unloading and uncrating
2. Hauling up to 200 ft. from loading dock
3. Installing, fitting, and bolting all ends to in-place sections

Plug-in units include:
1. Unloading and uncrating
2. Hauling up to 200 ft. from loading dock
3. Installing plug-in into in-place duct
4. Setup of rolling staging
5. Connecting load wire to lugs
6. Marking wire
7. Checking phase rotation

Labor-hours for plug-ins do not include:
1. Conduit runs from plug-in
2. Wire from plug-in
3. Conduit termination

Economy of Scale – For large concentrations of plug-in units in the same area deduct the following percentages:

11	to	25	15%
26	to	50	20%
51	to	75	25%
76	to	100	30%
100	and	over	35%

Job Conditions: Productivities are based on new construction in an unobstructed first floor area to a height of 15 ft. using rolling staging.

Material staging area is within 100 ft. of work being performed.
Add to the duct fittings and hangers:
- Plug-in switches (fused)
- Plug-in breakers
- Combination starters
- Combination contactors
- Combination fusible switch and lighting control

Labor-hours for duct and fittings include:
1. Unloading and uncrating
2. Hauling up to 200 ft. from loading dock
3. Measuring and marking
4. Installing duct runs
5. Leveling
6. Sound testing

Labor-hours for plug-ins include:
1. Unloading and uncrating
2. Hauling up to 200 ft. from loading dock
3. Installing plug-ins
4. Preparing wire
5. Wire connections and marking

R262513-15 Weight (Lbs./L.F.) of 4 Pole Aluminum and Copper Bus Duct by Ampere Load

Amperes	Aluminum Feeder	Copper Feeder	Aluminum Plug-In	Copper Plug-In
225			7	7
400			8	13
600	10	10	11	14
800	10	19	13	18
1000	11	19	16	22
1350	14	24	20	30
1600	17	26	25	39
2000	19	30	29	46
2500	27	43	36	56
3000	30	48	42	73
4000	39	67		
5000		78		

R262716-40 Standard Electrical Enclosure Types

NEMA Enclosures

Electrical enclosures serve two basic purposes; they protect people from accidental contact with enclosed electrical devices and connections, and they protect the enclosed devices and connections from specified external conditions. The National Electrical Manufacturers Association (NEMA) has established the following standards. Because these descriptions are not intended to be complete representations of NEMA listings, consultation of NEMA literature is advised for detailed information.

The following definitions and descriptions pertain to NONHAZARDOUS locations.

NEMA Type 1: General purpose enclosures intended for use indoors, primarily to prevent accidental contact of personnel with the enclosed equipment in areas that do not involve unusual conditions.

NEMA Type 2: Dripproof indoor enclosures intended to protect the enclosed equipment against dripping noncorrosive liquids and falling dirt.

NEMA Type 3: Dustproof, raintight and sleet-resistant (ice-resistant) enclosures intended for use outdoors to protect the enclosed equipment against wind-blown dust, rain, sleet, and external ice formation.

NEMA Type 3R: Rainproof and sleet-resistant (ice-resistant) enclosures which are intended for use outdoors to protect the enclosed equipment against rain. These enclosures are constructed so that the accumulation and melting of sleet (ice) will not damage the enclosure and its internal mechanisms.

NEMA Type 3S: Enclosures intended for outdoor use to provide limited protection against wind-blown dust, rain, and sleet (ice) and to allow operation of external mechanisms when ice-laden.

NEMA Type 4: Watertight and dust-tight enclosures intended for use indoors and out – to protect the enclosed equipment against splashing water, see page of water, falling or hose-directed water, and severe external condensation.

NEMA Type 4X: Watertight, dust-tight, and corrosion-resistant indoor and outdoor enclosures featuring the same provisions as Type 4 enclosures, plus corrosion resistance.

NEMA Type 5: Indoor enclosures intended primarily to provide limited protection against dust and falling dirt.

NEMA Type 6: Enclosures intended for indoor and outdoor use – primarily to provide limited protection against the entry of water during occasional temporary submersion at a limited depth.

NEMA Type 6R: Enclosures intended for indoor and outdoor use – primarily to provide limited protection against the entry of water during prolonged submersion at a limited depth.

NEMA Type 11: Enclosures intended for indoor use – primarily to provide, by means of oil immersion, limited protection to enclosed equipment against the corrosive effects of liquids and gases.

NEMA Type 12: Dust-tight and driptight indoor enclosures intended for use indoors in industrial locations to protect the enclosed equipment against fibers, flyings, lint, dust, and dirt, as well as light splashing, see page, dripping, and external condensation of noncorrosive liquids.

NEMA Type 13: Oil-tight and dust-tight indoor enclosures intended primarily to house pilot devices, such as limit switches, foot switches, push buttons, selector switches, and pilot lights, and to protect these devices against lint and dust, see page, external condensation, and sprayed water, oil, and noncorrosive coolant.

The following definitions and descriptions pertain to HAZARDOUS, or CLASSIFIED, locations:

NEMA Type 7: Enclosures intended to use in indoor locations classified as Class 1, Groups A, B, C, or D, as defined in the National Electrical Code.

NEMA Type 9: Enclosures intended for use in indoor locations classified as Class 2, Groups E, F, or G, as defined in the National Electrical Code.

R262726-90 Wiring Devices

Wiring devices should be priced on a separate takeoff form which includes boxes, covers, conduit and wire.

Labor-hours for devices include:

1. Stripping of wire
2. Attaching wire to device using terminators on the device itself, lugs, set screws etc.
3. Mounting of device in box

Labor-hours do not include:

1. Conduit
2. Wire
3. Boxes
4. Plates

Economy of Scale – for large concentrations of devices in the same area deduct the following percentages from labor-hours:

1	to	10	0%
11	to	25	20%
26	to	50	25%
51	to	100	30%
	over	100	35%

R262726-90 Wiring Devices (cont.)

NEMA No.	15 R	20 R	30 R	50 R	60 R
1 125V 2 Pole, 2 Wire	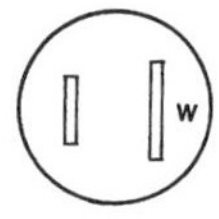				
2 250V 2 Pole, 2 Wire		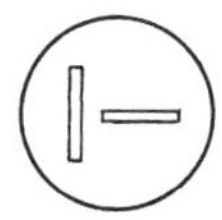			
5 125V 2 Pole, 3 Wire	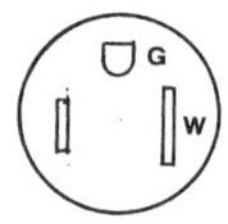	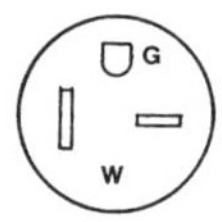	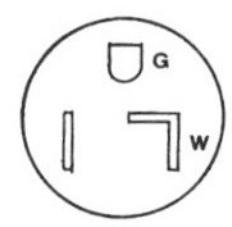	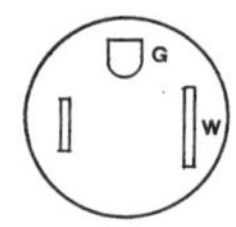	
6 250V 2 Pole, 3 Wire	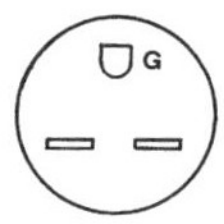	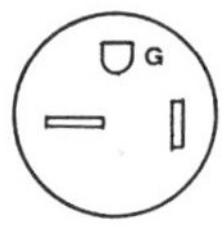	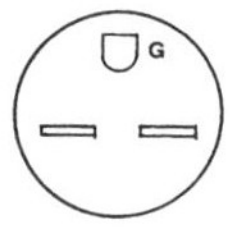	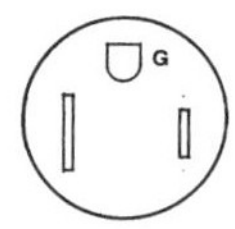	
7 277V, AC 2 Pole, 3 Wire	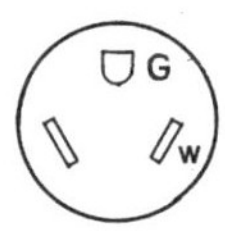			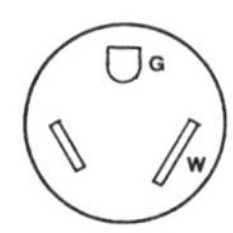	
10 125/250V 3 Pole, 3 Wire		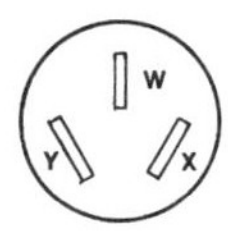	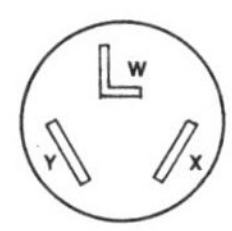	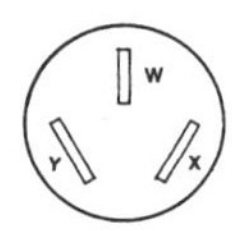	
11 3 Phase 250V 3 Pole, 3 Wire	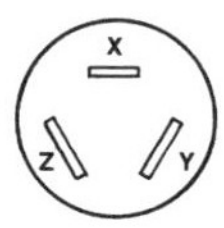	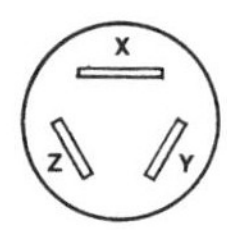	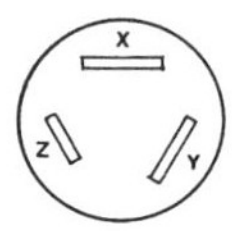	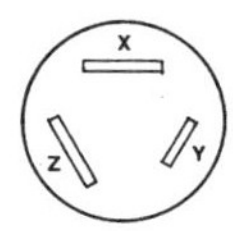	
14 125/250V 3 Pole, 4 Wire	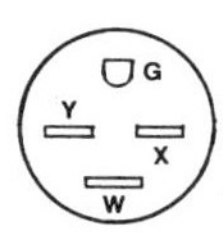	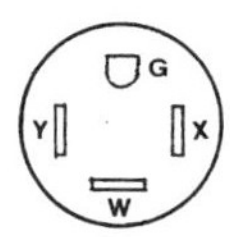	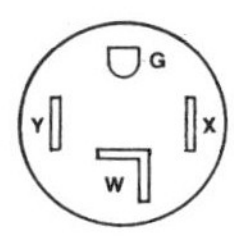	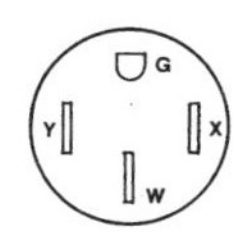	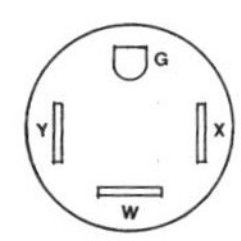
15 3 Phase 250V 3 Pole, 4 Wire	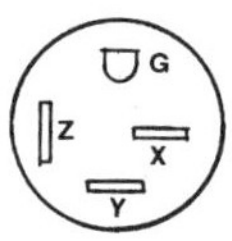	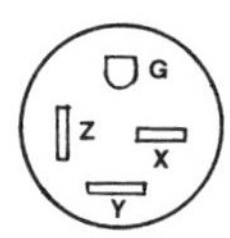	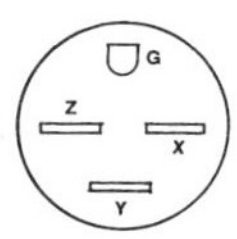	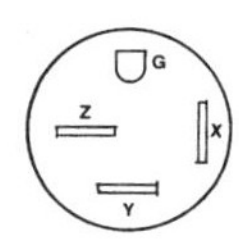	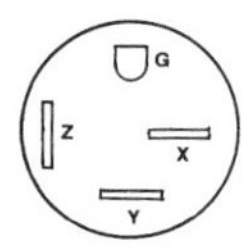
18 3 Phase 208Y/120V 4 Pole, 4 Wire	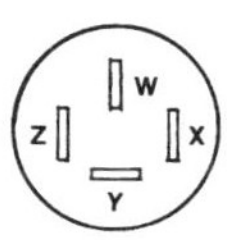	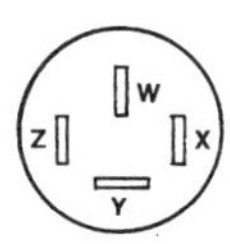	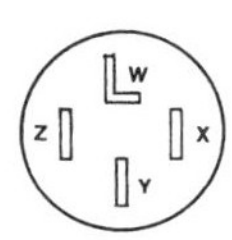	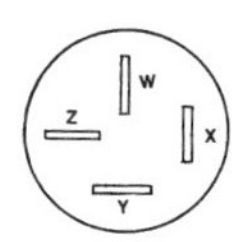	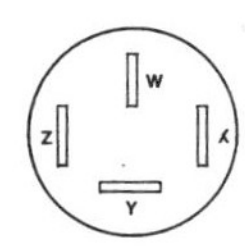

R262726-90 Wiring Devices (cont.)

NEMA No.	15 R	20 R	30 R
L 1 125V 2 Pole, 2 Wire			
L 2 250V 2 Pole, 2 Wire		15A	
L 5 125 V 2 Pole, 3 Wire			
L 6 250 V 2 Pole, 3 Wire			
L 7 227 V, AC 2 Pole, 3 Wire			
L 8 480 V 2 Pole, 3 Wire			
L 9 600 V 2 Pole, 3 Wire			
L 10 125 /250V 3 Pole, 3 Wire			
L 11 3 Phase 250 V 3 Pole, 3 Wire			
L 12 3 Phase 480 V 3 Pole, 3 Wire			
L 13 3 Phase 600V 3 Pole, 3 Wire			
L 14 125/250V 3 Pole, 4 Wire			
L 15 3 Phase 250 V 3 Pole, 4 Wire			
L 16 3 Phase 480V 3 Pole, 4 Wire			
L 17 3 Phase 600 V 3 Pole, 4 Wire			
L 18 3 Phase 208Y/120V 4 Pole, 4 Wire			
L 19 3 Phase 480Y/277V 4 Pole, 4 Wire			
L 20 3 Phase 600Y/347V 4 Pole, 4 Wire			
L 21 3 Phase 208Y/120V 4 Pole, 5 Wire			
L 22 3 Phase 480Y/277V 4 Pole, 5 Wire			
L 23 3 Phase 600Y/347V 4 Pole, 5 Wire			

R262816-80 Safety Switches

List each Safety Switch by type, ampere rating, voltage, single or three phase, fused or nonfused.

Example:

A. General duty, 240V, 3-pole, fused
B. Heavy Duty, 600V, 3-pole, nonfused
C. Heavy Duty, 240V, 2-pole, fused
D. Heavy Duty, 600V, 3-pole, fused

Also include NEMA enclosure type and identify as:

1. Indoor
2. Weatherproof
3. Explosionproof

Installation of Safety Switches includes:

1. Unloading and uncrating
2. Handling disconnects up to 200′ from loading dock
3. Measuring and marking location
4. Drilling (4) anchor type lead fasteners, using a hammer drill
5. Mounting and leveling Safety Switch
6. Installing (3) fuses
7. Phasing and tagging line and load wires

Price does not include:

1. Modifications to enclosure
2. Plywood backboard
3. Conduit or wire
4. Fuses
5. Termination of wires

Job Conditions: Productivities are based on new construction to an installed height of 6′ above finished floor.

Material staging area is assumed to be within 100′ of work in progress.

R263213-45 Generator Weight (Lbs.) by kW

3 Phase 4 Wire /480 Volt			
Gas		Diesel	
kW	Lbs.	kW	Lbs.
7.5	600	30	1800
10	630	50	2230
15	960	75	2250
30	1500	100	3840
65	2350	125	4030
85	2570	150	5500
115	4310	175	5650
170	6530	200	5930
		250	6320
		300	7840
		350	8220
		400	10750
		500	11900

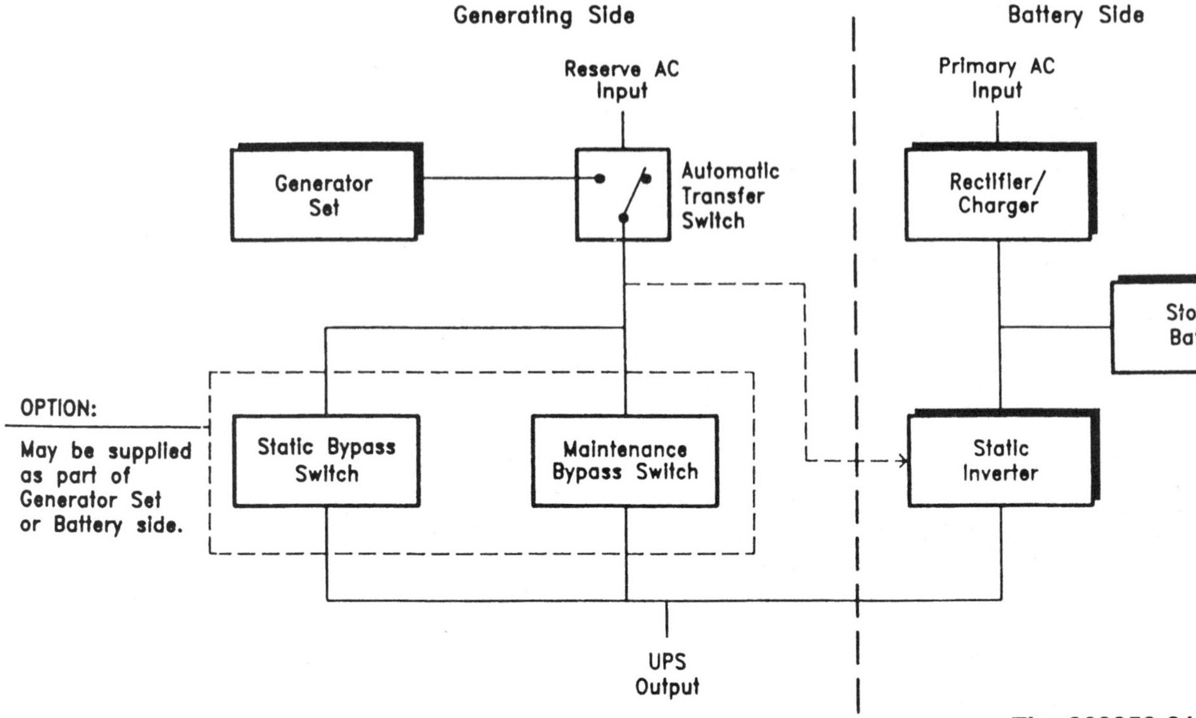

Fig. 263353-81

R263353-80 Uninterruptible Power Supply Systems

General: Uninterruptible Power Supply (UPS) Systems are used to provide power for legally required standby systems. They are also designed to protect computers and provide optional additional coverage for any or all loads in a facility. Figure R263353-81 shows a typical configuration for a UPS System with generator set.

Cost Determination:

It is recommended that UPS System material costs be obtained from equipment manufacturers as a package. Installation costs should be obtained from a vendor or contractor.

The recommended procedure for pricing UPS Systems would be to identify:

1. Frequency – by Hertz (Hz.)
   For example: 50, 60, and 400/415 Hz.
2. Apparent power and real power
   For example: 200 kVA/170 kW
3. Input/Output voltage For example: 120V, 208V or 240V
4. Phase – either single or three-phase
5. Options and accessories – For extended run time more batteries would be required. Other accessories include battery cabinets, battery racks, remote control panel, power distribution unit (PDU), and warranty enhancement plans.

Note:

1. Larger systems can be configured by paralleling two or more standard small size single modules.
   For example: two 15 kVA modules, combined together and configured to 30 kVA.
2. Maximum input current during battery recharge is typically 15% higher than normal current.
3. UPS Systems weights vary depending on options purchased and power rating in kVA.

R263413-30 HP Value/Cost Determination

General: Motors can be powered by any of the seven systems shown in Figure R262213-11 thru Figure R262213-17 provided the motor voltage characteristics are compatible with the power system characteristics.

Cost Determination:

Motor Amperes for the various size H.P. and voltage are listed in Table R263413-31. To find the amperes, locate the required H.P. rating and locate the amperes under the appropriate circuit characteristics.

For example:

A. 100 H.P., 3 phase, 460 volt motor = 124 amperes (Table R263413-31)

B. 10 H.P., 3 phase, 200 volt motor = 32.2 amperes (Table R263413-31)

Motor Wire Size: After the amperes are found in Table R263413-31 the amperes must be increased 25% to compensate for power losses. Next refer to Table R260519-92. Find the appropriate insulation column for copper or aluminum wire to determine the proper wire size.

For example:

A. 100 H.P., 3 phase, 460 volt motor has an ampere value of 124 amperes from Table R263413-31

B. 124A x 1.25 = 155 amperes

C. Refer to Table R260519-92 for THW or THWN wire insulations to find the proper wire size. For a 155 ampere load using copper wire a size 2/0 wire is needed.

D. For the 3 phase motor three wires of 2/0 size are required.

Conduit Size: To obtain the proper conduit size for the wires and type of insulation used, refer to Table R260533-22.

For example: For the 100 H.P., 460V, 3 phase motor, it was determined that three 2/0 wires are required. Assuming THWN insulated copper wire, use Table R260533-22 to determine that three 2/0 wires require 1-1/2″ conduit.

Material Cost of the conduit and wire system depends on:

1. Wire size required
2. Copper or aluminum wire
3. Wire insulation type selected
4. Steel or plastic conduit
5. Type of conduit raceway selected.

Labor Cost of the conduit and wire system depends on:

1. Type and size of conduit
2. Type and size of wires installed
3. Location and height of installation in building or depth of trench
4. Support system for conduit.

R263413-31 Ampere Values Determined by Horsepower, Voltage and Phase Values

H.P.	Amperes							
	Single Phase			Three Phase				
	115V	208V	230V	200V	208V	230V	460V	575V
1/6	4.4A	2.4A	2.2A					
1/4	5.8	3.2	2.9					
1/3	7.2	4.0	3.6					
1/2	9.8	5.4	4.9	2.5A	2.4A	2.2A	1.1A	0.9A
3/4	13.8	7.6	6.9	3.7	3.5	3.2	1.6	1.3
1	16	8.8	8	4.8	4.6	4.2	2.1	1.7
1-1/2	20	11	10	6.9	6.6	6.0	3.0	2.4
2	24	13.2	12	7.8	7.5	6.8	3.4	2.7
3	34	18.7	17	11.0	10.6	9.6	4.8	3.9
5	56	30.8	28	17.5	16.7	15.2	7.6	6.1
7-1/2	80	44	40	25.3	24.2	22	11	9
10	100	55	50	32.2	30.8	28	14	11
15				48.3	46.2	42	21	17
20				62.1	59.4	54	27	22
25				78.2	74.8	68	34	27
30				92	88	80	40	32
40				120	114	104	52	41
50				150	143	130	65	52
60				177	169	154	77	62
75				221	211	192	96	77
100				285	273	248	124	99
125				359	343	312	156	125
150				414	396	360	180	144
200				552	528	480	240	192
250							302	242
300							361	289
350							414	336
400							477	382

R263413-30 Cost Determination (cont.)

Magnetic starters, switches, and motor connection:

To complete the cost picture from H.P. to Costs additional items must be added to the cost of the conduit and wire system to arrive at a total cost.

1. Assembly Table D5020-160 Magnetic Starters Installed Cost lists the various size starters for single phase and three phase motors.
2. Assembly Table D5020-165 Heavy Duty Safety Switches Installed Cost lists safety switches required at the beginning of a motor circuit and also one required in the vicinity of the motor location.
3. Assembly Table D5020-170 Motor Connection lists the various costs for single and three phase motors.

Worksheet to obtain total motor wiring costs:

It is assumed that the motors or motor driven equipment are furnished and installed under other sections for this estimate and the following work is done under this section:

1. Conduit
2. Wire (add 10% for additional wire beyond conduit ends for connections to switches, boxes, starters, etc.)
3. Starters
4. Safety switches
5. Motor connections

Figure R263413-32

Worksheet for Motor Circuits

Item	Type	Size	Quantity	Cost Unit	Cost Total
Wire					
Conduit					
Switch					
Starter					
Switch					
Motor Connection					
Other					
Total Cost					

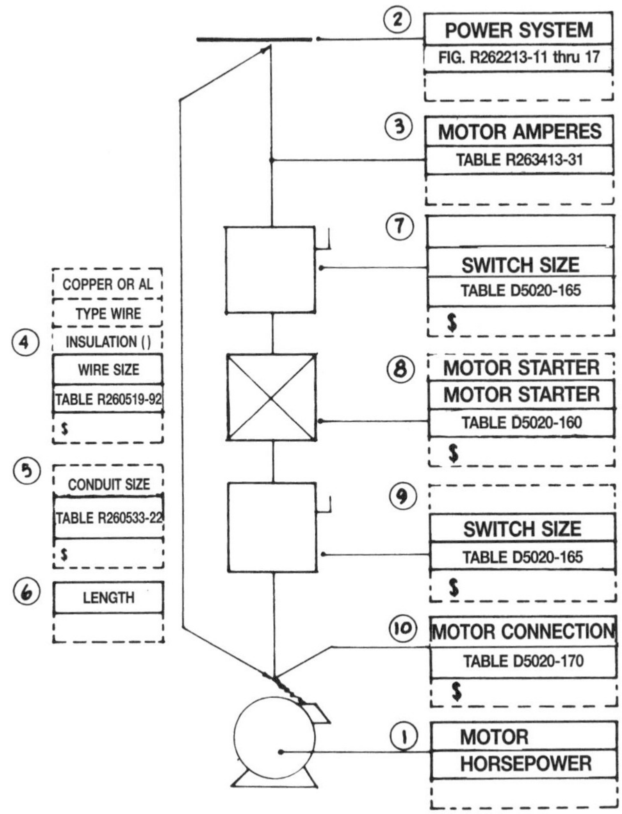

R263413-33 Maximum Horsepower for Starter Size by Voltage

Starter Size	Maximum HP (3φ) 208V	240V	480V	600V
00	1½	1½	2	2
0	3	3	5	5
1	7½	7½	10	10
2	10	15	25	25
3	25	30	50	50
4	40	50	100	100
5		100	200	200
6		200	300	300
7		300	600	600
8		450	900	900
8L		700	1500	1500

Electrical | R2636 | Transfer Switches

R263623-60 Automatic Transfer Switches

When taking off Automatic transfer switches identify by voltage, amperage and number of poles.

Example: Automatic transfer switch, 480V, 3 phase, 30A, NEMA 1 enclosure

Labor-hours for transfer switches include:

1. Unloading, uncrating and handling switches up to 200′ from loading dock
2. Measuring and marking
3. Drilling (4) lead type anchors, using a hammer drill
4. Mounting and leveling
5. Circuit identification
6. Testing and load balancing

Labor-hours do not include:

1. Modifications in enclosure
2. Structural supports
3. Additional lugs
4. Plywood backboard
5. Painting or lettering
6. Conduit runs to or from transfer switch
7. Wire
8. Termination of wires

Electrical | R2651 | Interior Lighting

R265113-40 Interior Lighting Fixtures

When taking off interior lighting fixtures, it is advisable to set up your quantity work sheet to conform to the lighting schedule as it appears on the print. Include the alpha-numeric code plus the symbol on your work sheet.

Take off a particular section or floor of the building and count each type of fixture before going on to another type. It would also be advantageous to include on the same work sheet the pipe, wire, fittings and circuit number associated with each type of lighting fixture. This will help you identify the costs associated with any particular lighting system and in turn make material purchases more specific as to when and how much to order under the classification of lighting.

By taking off lighting first you can get a complete "WALK THRU" of the job. This will become helpful when doing other phases of the project.

Materials for a recessed fixture include:

1. Fixture
2. Lamps
3. 6′ of jack chain
4. (2) S hooks
5. (2) Wire nuts

Labor for interior recessed fixtures include:

1. Unloading by hand
2. Hauling by hand to an area up to 200′ from loading dock
3. Uncrating
4. Layout
5. Installing fixture
6. Attaching jack chain & S hooks
7. Connecting circuit power
8. Reassembling fixture
9. Installing lamps
10. Testing

Material for surface mounted fixtures includes:

1. Fixture
2. Lamps
3. Either (4) lead type anchors, (4) toggle bolts, or (4) ceiling grid clips
4. (2) Wire nuts

Material for pendent mounted fixtures includes:

1. Fixture
2. Lamps
3. (2) Wire nuts
4. Rigid pendents as required by type of fixtures
5. Canopies as required by type of fixture

Labor hours include the following for both surface and pendent fixtures:

1. Unloading by hand
2. Hauling by hand to an area up to 200′ from loading dock
3. Uncrating
4. Layout and marking
5. Drilling (4) holes for either lead anchors or toggle bolts using a hammer drill
6. Installing fixture
7. Leveling fixture
8. Connecting circuit power
9. Installing lamps
10. Testing

Labor for surface or pendent fixtures does not include:

1. Conduit
2. Boxes or covers
3. Connectors
4. Fixture whips
5. Special support
6. Switching
7. Wire

Economy of Scale: For large concentrations of lighting fixtures in the same area deduct the following percentages from labor:

25	to	50	fixtures	15%
51	to	75	fixtures	20%
76	to	100	fixtures	25%
101	and over			30%

Job Conditions: Productivity is based on new construction in a unobstructed first floor location, using rolling staging to 15′ high.

Material staging is assumed to be within 100′ of work being performed.

Add the following percentages to labor for elevated installations:

15′	to	20′	high	10%
21′	to	25′	high	20%
26′	to	30′	high	30%
31′	to	35′	high	40%
36′	to	40′	high	50%
41′	and over			60%

R265723-05 Comparison - Operation of High Intensity Discharge Lamps

Lamp Type	Wattage	Life (Hours)	Circuit Wattage [1]	Average Initial Lumens	L.L.D. [2]	Mean Lumens [3]
M.V.	100 DX	24,000	125	4,000	61%	2,440
L.P.S.	SOX-35	18,000	65	4,800	100%	4,800
H.P.S.	LU-70	12,000	84	5,800	90%	5,220
M.H.	No Equivalent					
M.V.	175 DX	24,000	210	8,500	66%	5,676
M.V.	250 DX	24,000	295	13,000	66%	7,986
L.P.S.	SOX-55	18,000	82	8,000	100%	8,000
H.P.S.	LU-100	12,000	120	9,500	90%	8,550
M.H.	No Equivalent					
M.V.	400 DX	24,000	465	24,000	64%	14,400
L.P.S.	SOX-90	18,000	141	13,500	100%	13,500
H.P.S.	LU-150	16,000	188	16,000	90%	14,400
M.H.	MH-175	7,500	210	14,000	73%	10,200
M.V.	No Equivalent					
L.P.S.	SOX-135	18,000	147	22,500	100%	22,500
H.P.S.	LU-250	20,000	310	25,500	92%	23,205
M.H.	MH-250	7,500	295	20,500	78%	16,000
M.V.	1000 DX	24,000	1,085	63,000	61%	37,820
L.P.S.	SOX-180	18,000	248	33,000	100%	33,000
H.P.S.	LU-400	20,000	480	50,000	90%	45,000
M.H.	MH-400	15,000	465	34,000	72%	24,600
H.P.S.	LU-1000	15,000	1,100	140,000	91%	127,400

1. Includes ballast losses and average lamp watts
2. Lamp lumen depreciation (% of initial light output at 70% rated life)
3. Lamp lumen output at 70% rated life (L.L.D. x initial)
4. Cost of yearly operation = Cicuit Wattage/1000 x annual operating hours x average cost per KWH

M.V. = Mercury Vapor
L.P.S. = Low pressure sodium
H.P.S. = High pressure sodium
L.H. = Metal halide

R265723-10 For Other than Regular Cool White (CW) Lamps

Multiply Material Costs as Follows:					
Regular Lamps	Cool white deluxe (CWX)	x 1.35	Energy Saving Lamps	Cool white (CW/ES)	x 1.35
	Warm white deluxe (WWX)	x 1.35		Cool white deluxe (CWX/ES)	x 1.65
	Warm white (WW)	x 1.30		Warm white (WW/ES)	x 1.55
	Natural (N)	x 2.05		Warm white deluxe (WWX/ES)	x 1.65

R265723-20 Lamp Comparison Chart with Enclosed Floodlight, Ballast, & Lamp for Pole Mounting

Type	Watts	Initial Lumens	Lumens per Watt	Lumens @ 40% Life	Life (Hours)
Incandescent	150	2,880	19	85	750
	300	6,360	21	84	750
	500	10,850	22	80	1,000
	1,000	23,740	24	80	1,000
	1,500	34,400	23	80	1,000
Tungsten Halogen	500	10,950	22	97	2,000
	1,500	35,800	24	97	2,000
Fluorescent Cool White	40	3,150	79	88	20,000
	110	9,200	84	87	12,000
	215	16,000	74	81	12,000
Deluxe Mercury	250	12,100	48	86	24,000
	400	22,500	56	85	24,000
	1,000	63,000	63	75	24,000
Metal Halide	175	14,000	80	77	7,500
	400	34,000	85	75	15,000
	1,000	100,000	100	83	10,000
	1,500	155,000	103	92	1,500
High Pressure Sodium	70	5,800	83	90	20,000
	100	9,500	95	90	20,000
	150	16,000	107	90	24,000
	400	50,000	125	90	24,000
	1,000	140,000	140	90	24,000
Low Pressure Sodium	55	4,600	131	98	18,000
	90	12,750	142	98	18,000
	180	33,000	183	98	18,000

Color: High Pressure Sodium — Slightly Yellow
Low Pressure Sodium — Yellow
Mercury Vapor — Green-Blue
Metal Halide — Blue White

Note: Pole not included.

R265723-25 Energy Efficiency Rating for Luminaires

The energy efficiency program for luminaires recommends the use of a metric called the Luminaire Efficacy Rating (LER). The LER value expresses the total luminaires generated by a lamp, to the watts consumed by the lamp.

Lamp Type	Lumens per Watt
Incandescent	17
Tungsten Halogen	14 - 20
Fluorescent	50 - 104
Metal Halide	64 - 96
High Pressure Sodium	76 - 116

Communications | R2713 | Communications Backbone Cabling

R271323-40 Fiber Optics

Fiber optic systems use optical fiber such as plastic, glass, or fused silica, a transparent material, to transmit radiant power (i.e., light) for control, communication, and signaling applications. The types of fiber optic cables can be nonconductive, conductive, or composite. The composite cables contain fiber optics and current-carrying electrical conductors. The configuration for one of the fiber optic systems is as follows:

The transceiver module acts as transmitting and receiving equipment in a common house, which converts electrical energy to light energy or vice versa.

Pricing the fiber optic system is not an easy task. The performance of the whole system will affect the cost significantly. New specialized tools and techniques decrease the installing cost tremendously. In the fiber optic section of Means Electrical Cost Data, a benchmark for labor-hours and material costs is set up so that users can adjust their costs according to unique project conditions.

Units for Measure: Fiber optic cable is measured in hundred linear feet (C.L.F.) or industry units of measure - meter (m) or kilometer (km). The connectors are counted as units (EA.)

Material Units: Generally, the material costs include only the cable. All the accessories shall be priced separately.

Labor Units: The following procedures are generally included for the installation of fiber optic cables:

- Receiving
- Material handling
- Setting up pulling equipment
- Measuring and cutting cable
- Pulling cable

These additional items are listed and extended: Terminations

Takeoff Procedure: Cable should be taken off by type, size, number of fibers, and number of terminations required. List the lengths of each type of cable on the takeoff sheets. Total and add 10% for waste. Transfer the figures to a cost analysis sheet and extend.

Communications | R2715 | Communications Horizontal Cabling

R271513-75 High Performance Cable

There are several categories used to describe high performance cable. The following information includes a description of categories CAT 3, 5, 5e, 6, and 7, and details classifications of frequency and specific standards. The category standards have evolved under the sponsorship of organizations such as the Telecommunication Industry Association (TIA), the Electronic Industries Alliance (EIA), the American National Standards Institute (ANSI), the International Organization for Standardization (ISO), and the International Electrotechnical Commission (IEC), all of which have catered to the increasing complexities of modern network technology. For network cabling, users must comply with national or international standards. A breakdown of these categories is as follows:

Category 3: Designed to handle frequencies up to 16 MHz.

Category 5: (TIA/EIA 568A) Designed to handle frequencies up to 100 MHz.

Category 5e: Additional transmission performance to exceed Category 5.

Category 6 (draft): Development by TIA and other international groups to handle frequencies of 250 MHz.

Category 7 (draft): Under development to handle a frequency range from 1 to 600 MHz.

R312316-40 Excavating

The selection of equipment used for structural excavation and bulk excavation or for grading is determined by the following factors.

1. Quantity of material
2. Type of material
3. Depth or height of cut
4. Length of haul
5. Condition of haul road
6. Accessibility of site
7. Moisture content and dewatering requirements
8. Availability of excavating and hauling equipment

Some additional costs must be allowed for hand trimming the sides and bottom of concrete pours and other excavation below the general excavation.

When planning excavation and fill, the following should also be considered.

1. Swell factor
2. Compaction factor
3. Moisture content
4. Density requirements

A typical example for scheduling and estimating the cost of excavation of a 15′ deep basement on a dry site when the material must be hauled off the site is outlined below.

Assumptions:

1. Swell factor, 18%
2. No mobilization or demobilization
3. Allowance included for idle time and moving on job
4. No dewatering, sheeting, or bracing
5. No truck spotter or hand trimming

Number of B.C.Y. per truck = 1.5 C.Y. bucket x 8 passes = 12 loose C.Y.

$$= 12 \text{ x } \frac{100}{118} = 10.2 \text{ B.C.Y. per truck}$$

Truck Haul Cycle:

Load truck, 8 passes	=	4 minutes
Haul distance, 1 mile	=	9 minutes
Dump time	=	2 minutes
Return, 1 mile	=	7 minutes
Spot under machine	=	1 minute
		23 minute cycle

Fleet Haul Production per day in B.C.Y.

$$4 \text{ trucks x } \frac{50 \text{ min. hour}}{23 \text{ min. haul cycle}} \text{ x 8 hrs. x 10.2 B.C.Y.}$$

$$= 4 \text{ x } 2.2 \text{ x } 8 \text{ x } 10.2 = 718 \text{ B.C.Y./day}$$

Add the mobilization and demobilization costs to the total excavation costs. When equipment is rented for more than three days, there is often no mobilization charge by the equipment dealer. On larger jobs outside of urban areas, scrapers can move earth economically provided a dump site or fill area and adequate haul roads are available. Excavation within sheeting bracing or cofferdam bracing is usually done with a clamshell and production is low, since the clamshell may have to be guided by hand between the bracing. When excavating or filling an area enclosed with a wellpoint system, add 10% to 15% to the cost to allow for restricted access. When estimating earth excavation quantities for structures, allow work space outside the building footprint for construction of the foundation and a slope of 1:1 unless sheeting is used.

R312316-45 Excavating Equipment

The table below lists THEORETICAL hourly production in C.Y./hr. bank measure for some typical excavation equipment. Figures assume 50 minute hours, 83% job efficiency, 100% operator efficiency, 90° swing and properly sized hauling units, which must be modified for adverse digging and loading conditions. Actual production costs in the front of the book average about 50% of the theoretical values listed here.

Equipment	Soil Type	B.C.Y. Weight	% Swell	1 C.Y.	1-1/2 C.Y.	2 C.Y.	2-1/2 C.Y.	3 C.Y.	3-1/2 C.Y.	4 C.Y.
Hydraulic Excavator "Backhoe" 15′ Deep Cut	Moist loam, sandy clay	3400 lb.	40%	85	125	175	220	275	330	380
	Sand and gravel	3100	18	80	120	160	205	260	310	365
	Common earth	2800	30	70	105	150	190	240	280	330
	Clay, hard, dense	3000	33	65	100	130	170	210	255	300
Power Shovel Optimum Cut (Ft.)	Moist loam, sandy clay	3400	40	170	245	295	335	385	435	475
				(6.0)	(7.0)	(7.8)	(8.4)	(8.8)	(9.1)	(9.4)
	Sand and gravel	3100	18	165	225	275	325	375	420	460
				(6.0)	(7.0)	(7.8)	(8.4)	(8.8)	(9.1)	(9.4)
	Common earth	2800	30	145	200	250	295	335	375	425
				(7.8)	(9.2)	(10.2)	(11.2)	(12.1)	(13.0)	(13.8)
	Clay, hard, dense	3000	33	120	175	220	255	300	335	375
				(9.0)	(10.7)	(12.2)	(13.3)	(14.2)	(15.1)	(16.0)
Drag Line Optimum Cut (Ft.)	Moist loam, sandy clay	3400	40	130	180	220	250	290	325	385
				(6.6)	(7.4)	(8.0)	(8.5)	(9.0)	(9.5)	(10.0)
	Sand and gravel	3100	18	130	175	210	245	280	315	375
				(6.6)	(7.4)	(8.0)	(8.5)	(9.0)	(9.5)	(10.0)
	Common earth	2800	30	110	160	190	220	250	280	310
				(8.0)	(9.0)	(9.9)	(10.5)	(11.0)	(11.5)	(12.0)
	Clay, hard, dense	3000	33	90	130	160	190	225	250	280
				(9.3)	(10.7)	(11.8)	(12.3)	(12.8)	(13.3)	(12.0)
				Wheel Loaders				Track Loaders		
				3 C.Y.	4 C.Y.	6 C.Y.	8 C.Y.	2-1/4 C.Y.	3 C.Y.	4 C.Y.
Loading Tractors	Moist loam, sandy clay	3400	40	260	340	510	690	135	180	250
	Sand and gravel	3100	18	245	320	480	650	130	170	235
	Common earth	2800	30	230	300	460	620	120	155	220
	Clay, hard, dense	3000	33	200	270	415	560	110	145	200
	Rock, well-blasted	4000	50	180	245	380	520	100	130	180

R312319-90 Wellpoints

A single stage wellpoint system is usually limited to dewatering an average 15′ depth below normal ground water level. Multi-stage systems are employed for greater depth with the pumping equipment installed only at the lowest header level. Ejectors with unlimited lift capacity can be economical when two or more stages of wellpoints can be replaced or when horizontal clearance is restricted, such as in deep trenches or tunneling projects, and where low water flows are expected. Wellpoints are usually spaced on 2-1/2′ to 10′ centers along a header pipe. Wellpoint spacing, header size, and pump size are all determined by the expected flow as dictated by soil conditions.

In almost all soils encountered in wellpoint dewatering, the wellpoints may be jetted into place. Cemented soils and stiff clays may require sand wicks about 12″ in diameter around each wellpoint to increase efficiency and eliminate weeping into the excavation. These sand wicks require 1/2 to 3 C.Y. of washed filter sand and are installed by using a 12″ diameter steel casing and hole puncher jetted into the ground 2′ deeper than the wellpoint. Rock may require predrilled holes.

Labor required for the complete installation and removal of a single stage wellpoint system is in the range of 3/4 to 2 labor-hours per linear foot of header, depending upon jetting conditions, wellpoint spacing, etc.

Continuous pumping is necessary except in some free draining soil where temporary flooding is permissible (as in trenches which are backfilled after each day's work). Good practice requires provision of a stand-by pump during the continuous pumping operation.

Systems for continuous trenching below the water table should be installed three to four times the length of expected daily progress to ensure uninterrupted digging, and header pipe size should not be changed during the job.

For pervious free draining soils, deep wells in place of wellpoints may be economical because of lower installation and maintenance costs. Daily production ranges between two to three wells per day, for 25′ to 40′ depths, to one well per day for depths over 50′.

Detailed analysis and estimating for any dewatering problem is available at no cost from wellpoint manufacturers. Major firms will quote "sufficient equipment" quotes or their affiliates offer lump sum proposals to cover complete dewatering responsibility.

Description for 200′ System with 8″ Header		Quantities
Equipment & Material	Wellpoints 25′ long, 2″ diameter @ 5′ O.C.	40 Each
	Header pipe, 8″ diameter	200 L.F.
	Discharge pipe, 8″ diameter	100 L.F.
	8″ valves	3 Each
	Combination jetting & wellpoint pump (standby)	1 Each
	Wellpoint pump, 8″ diameter	1 Each
	Transportation to and from site	1 Day
	Fuel for 30 days x 60 gal./day	1800 Gallons
	Lubricants for 30 days x 16 lbs./day	480 Lbs.
	Sand for points	40 C.Y.
Labor	Technician to supervise installation	1 Week
	Labor for installation and removal of system	300 Labor-hours
	4 Operators straight time 40 hrs./wk. for 4.33 wks.	693 Hrs.
	4 Operators overtime 2 hrs./wk. for 4.33 wks.	35 Hrs.

R312323-30 Compacting Backfill

Compaction of fill in embankments, around structures, in trenches, and under slabs is important to control settlement. Factors affecting compaction are:

1. Soil gradation
2. Moisture content
3. Equipment used
4. Depth of fill per lift
5. Density required

Production Rate:

$$\frac{1.75' \text{ plate width x 50 F.P.M. x 50 min./hr. x .67' lift}}{\text{27 C.F. per C.Y.}} = 108.5 \text{ C.Y./hr.}$$

Production Rate for 4 Passes:

$$\frac{108.5 \text{ C.Y.}}{4 \text{ passes}} = 27.125 \text{ C.Y./hr. x 8 hrs.} = 217 \text{ C.Y./day}$$

Example:

Compact granular fill around a building foundation using a 21″ wide x 24″ vibratory plate in 8″ lifts. Operator moves at 50 F.P.M. working a 50 minute hour to develop 95% Modified Proctor Density with 4 passes.

Earthwork | R3141 | Shoring

R314116-40 Wood Sheet Piling

Wood sheet piling may be used for depths to 20′ where there is no ground water. If moderate ground water is encountered Tongue & Groove sheeting will help to keep it out. When considerable ground water is present, steel sheeting must be used.

For estimating purposes on trench excavation, sizes are as follows:

Depth	Sheeting	Wales	Braces	B.F. per S.F.
To 8′	3 x 12′s	6 x 8′s, 2 line	6 x 8′s, @ 10′	4.0 @ 8′
8′ x 12′	3 x 12′s	10 x 10′s, 2 line	10 x 10′s, @ 9′	5.0 average
12′ to 20′	3 x 12′s	12 x 12′s, 3 line	12 x 12′s, @ 8′	7.0 average

Sheeting to be toed in at least 2′ depending upon soil conditions. A five person crew with an air compressor and sheeting driver can drive and brace 440 SF/day at 8′ deep, 360 SF/day at 12′ deep, and 320 SF/day at 16′ deep. For normal soils, piling can be pulled in 1/3 the time to install. Pulling difficulty increases with the time in the ground. Production can be increased by high pressure jetting.

R314116-45 Steel Sheet Piling

Limiting weights are 22 to 38#/S.F. of wall surface with 27#/S.F. average for usual types and sizes. (Weights of piles themselves are from 30.7#/L.F. to 57#/L.F. but they are 15″ to 21″ wide.) Lightweight sections 12″ to 28″ wide from 3 ga. to 12 ga. thick are also available for shallow excavations. Piles may be driven two at a time with an impact or vibratory hammer (use vibratory to pull) hung from a crane without leads. A reasonable estimate of the life of steel sheet piling is 10 uses with up to 125 uses possible if a vibratory hammer is used. Used piling costs from 50% to 80% of new piling depending on location and market conditions. Sheet piling and H piles can be rented for about 30% of the delivered mill price for the first month and 5% per month thereafter. Allow 1 labor-hour per pile for cleaning and trimming after driving. These costs increase with depth and hydrostatic head. Vibratory drivers are faster in wet granular soils and are excellent for pile extraction. Pulling difficulty increases with the time in the ground and may cost more than driving. It is often economical to abandon the sheet piling, especially if it can be used as the outer wall form. Allow about 1/3 additional length or more for toeing into ground. Add bracing, waler and strut costs. Waler costs can equal the cost per ton of sheeting.

Earthwork | R3145 | Vibroflotation & Desification

R314513-90 Vibroflotation and Vibro Replacement Soil Compaction

Vibroflotation is a proprietary system of compacting sandy soils in place to increase relative density to about 70%. Typical bearing capacities attained will be 6000 psf for saturated sand and 12,000 psf for dry sand. Usual range is 4000 to 8000 psf capacity. Costs in the front of the book are for a vertical foot of compacted cylinder 6′ to 10′ in diameter.

Vibro replacement is a proprietary system of improving cohesive soils in place to increase bearing capacity. Most silts and clays above or below the water table can be strengthened by installation of stone columns.

The process consists of radial displacement of the soil by vibration. The created hole is then backfilled in stages with coarse granular fill which is thoroughly compacted and displaced into the surrounding soil in the form of a column.

The total project cost would depend on the number and depth of the compacted cylinders. The installing company guarantees relative soil density of the sand cylinders after compaction and the bearing capacity of the soil after the replacement process. Detailed estimating information is available from the installer at no cost.

R316000-20 Pile Caps, Piles and Caissons

General: The function of a reinforced concrete pile cap is to transfer superstructure load from isolated column or pier to each pile in its supporting cluster. To do this, the cap must be thick and rigid, with all piles securely embedded into and bonded to it.

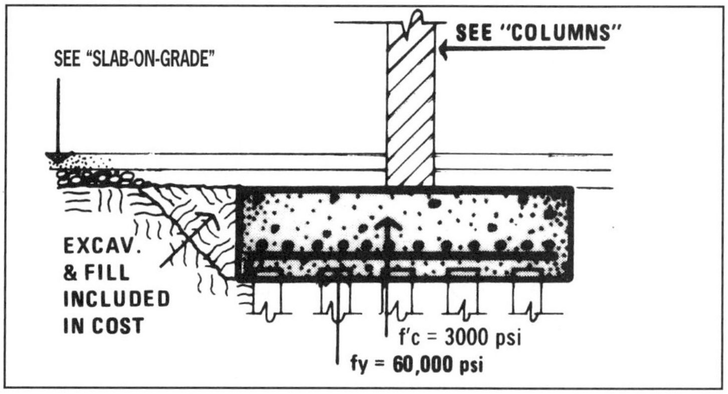

Figure 1.1-331 Section Through Pile Cap

Table 1.1-332 Concrete Quantities for Pile Caps

Load	Number of Piles @ 3′-0″ O.C. Per Footing Cluster									
Working (K)	2 (CY)	4 (CY)	6 (CY)	8 (CY)	10 (CY)	12 (CY)	14 (CY)	16 (CY)	18 (CY)	20 (CY)
50	(.9)	(1.9)	(3.3)	(4.9)	(5.7)	(7.8)	(9.9)	(11.1)	(14.4)	(16.5)
100	(1.0)	(2.2)	(3.3)	(4.9)	(5.7)	(7.8)	(9.9)	(11.1)	(14.4)	(16.5)
200	(1.0)	(2.2)	(4.0)	(4.9)	(5.7)	(7.8)	(9.9)	(11.1)	(14.4)	(16.5)
400	(1.1)	(2.6)	(5.2)	(6.3)	(7.4)	(8.2)	(13.7)	(11.1)	(14.4)	(16.5)
800		(2.9)	(5.8)	(7.5)	(9.2)	(13.6)	(17.6)	(15.9)	(19.7)	(22.1)
1200			(5.8)	(8.3)	(9.7)	(14.2)	(18.3)	(20.4)	(21.2)	(22.7)
1600				(9.8)	(11.4)	(14.5)	(19.5)	(20.4)	(24.6)	(27.2)
2000				(9.8)	(11.4)	(16.6)	(24.1)	(21.7)	(26.0)	(28.8)
3000						(17.5)		(26.5)	(30.3)	(32.9)
4000								(30.2)	(30.7)	(36.5)

Table 1.1-333 Concrete Quantities for Pile Caps

Load	Number of Piles @ 4′-6″ O.C. Per Footing					
Working (K)	2 (CY)	3 (CY)	4 (CY)	5 (CY)	6 (CY)	7 (CY)
50	(2.3)	(3.6)	(5.6)	(11.0)	(13.7)	(12.9)
100	(2.3)	(3.6)	(5.6)	(11.0)	(13.7)	(12.9)
200	(2.3)	(3.6)	(5.6)	(11.0)	(13.7)	(12.9)
400	(3.0)	(3.6)	(5.6)	(11.0)	(13.7)	(12.9)
800			(6.2)	(11.5)	(14.0)	(12.9)
1200				(13.0)	(13.7)	(13.4)
1600						(14.0)

R316000-20 Pile Caps, Piles and Caissons (cont.)

General: Piles are column-like shafts which receive superstructure loads, overturning forces, or uplift forces. They receive these loads from isolated column or pier foundations (pile caps), foundation walls, grade beams, or foundation mats. The piles then transfer these loads through shallower poor soil strata to deeper soil of adequate support strength and acceptable settlement with load.

Be sure that other foundation types aren't better suited to the job. Consider ground and settlement, as well as loading, when reviewing. Piles usually are associated with difficult foundation problems and substructure condition. Ground conditions determine type of pile (different pile types have been developed to suit ground conditions.) Decide each case by technical study, experience and sound engineering judgment—not rules of thumb. A full investigation of ground conditions, early, is essential to provide maximum information for professional foundation engineering and an acceptable structure.

Piles support loads by end bearing and friction. Both are generally present; however, piles are designated by their principal method of load transfer to soil.

Boring should be taken at expected pile locations. Ground strata (to bedrock or depth of 1-1/2 building width) must be located and identified with appropriate strengths and compressibilities. The sequence of strata determines if end bearing or friction piles are best suited.

End bearing piles have shafts which pass through soft strata or thin hard strata and tip bear on bedrock or penetrate some distance into a dense, adequate soil (sand or gravel.)

Friction piles have shafts which may be entirely embedded in cohesive soil (moist clay), and develop required support mainly by adhesion or "skin-friction" between soil and shaft area.

Piles pass through soil by either one of two ways:

1. Displacement piles force soil out of the way. This may cause compaction, ground heaving, remolding of sensitive soils, damage to adjacent structures, or hard driving.
2. Non-displacement piles have either a hole bored and the pile cast or placed in hole, or open ended pipe (casing) driven and the soil core removed. They tend to eliminate heaving or lateral pressure damage to adjacent structures of piles. Steel "HP" piles are considered of small displacement.

Placement of piles (attitude) is most often vertical; however, they are sometimes battered (placed at a small angle from vertical) to advantageously resist lateral loads. Seldom are piles installed singly but rather in clusters (or groups). Codes require a minimum of three piles per major column load or two per foundation wall or grade beam. Single pile capacity is limited by pile structural strength or support strength of soil. Support capacity of a pile cluster is almost always less than the sum of its individual pile capacities due to overlapping of bearing the friction stresses.

Large rigs for heavy, long piles create large soil surface loads and additional expense on weak ground.

Fewer piles create higher costs per pile.

Pile load tests are frequently required by code, ground situation, or pile type.

R316000-20 Pile Caps, Piles and Caissons (cont.)

General: Caissons, as covered in this section, are drilled cylindrical foundation shafts which function primarily as short column-like compression members. They transfer superstructure loads through inadequate soils to bedrock or hard stratum. They may be either reinforced or unreinforced and either straight or belled out at the bearing level.

Shaft diameters range in size from 20″ to 84″ with the most usual sizes beginning at 34″. If inspection of bottom is required, the minimum diameter practical is 30″. If handwork is required (in addition to mechanical belling, etc.) the minimum diameter is 32″. The most frequently used shaft diameter is probably 36″ with a 5′ or 6′ bell diameter. The maximum bell diameter practical is three times the shaft diameter.

Plain concrete is commonly used, poured directly against the excavated face of soil. Permanent casings add to cost and economically should be avoided. Wet or loose strata are undesirable. The associated installation sometimes involves a mudding operation with bentonite clay slurry to keep walls of excavation stable (costs not included here).

Reinforcement is sometimes used, especially for heavy loads. It is required if uplift, bending moment, or lateral loads exist. A small amount of reinforcement is desirable at the top portion of each caisson, even if the above conditions theoretically are not present. This will provide for construction eccentricities and other possibilities. Reinforcement, if present, should extend below the soft strata. Horizontal reinforcement is not required for belled bottoms.

There are three basic types of caisson bearing details:

1. Belled, which are generally recommended to provide reduced bearing pressure on soil. These are not for shallow depths or poor soils. Good soils for belling include most clays, hardpan, soft shale, and decomposed rock.

   Soils requiring handwork include hard shale, limestone, and sandstone.

   Soils not recommended include sand, gravel, silt, and igneous rock. Compact sand and gravel above water table may stand. Water in the bearing strata is undesirable.
2. Straight shafted, which have no bell but the entire length is enlarged to permit safe bearing pressures. They are most economical for light loads on high bearing capacity soil.
3. Socketed (or keyed), which are used for extremely heavy loads. They involve sinking the shaft into rock for combined friction and bearing support action. Reinforcement of shaft is usually necessary. Wide flange cores are frequently used here.

Advantages include:

A. Shafts can pass through soils that piles cannot
B. No soil heaving or displacement during installation
C. No vibration during installation
D. Less noise than pile driving
E. Bearing strata can be visually inspected & tested

Uses include:

A. Situations where unsuitable soil exists to moderate depth
B. Tall structures
C. Heavy structures
D. Underpinning (extensive use)

See R033053-10 for Soil Bearing Capacities.

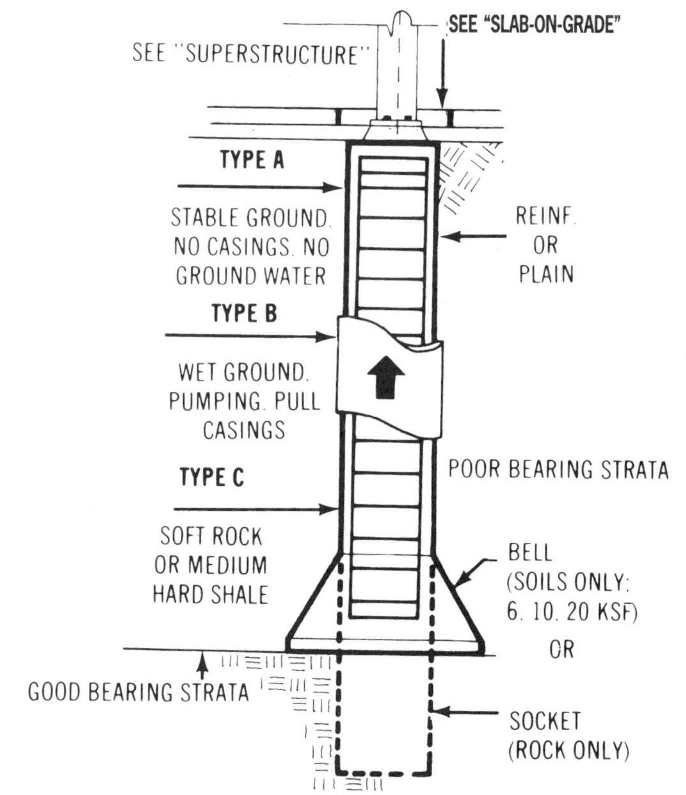

Figure 1.4-201 Design Assumptions

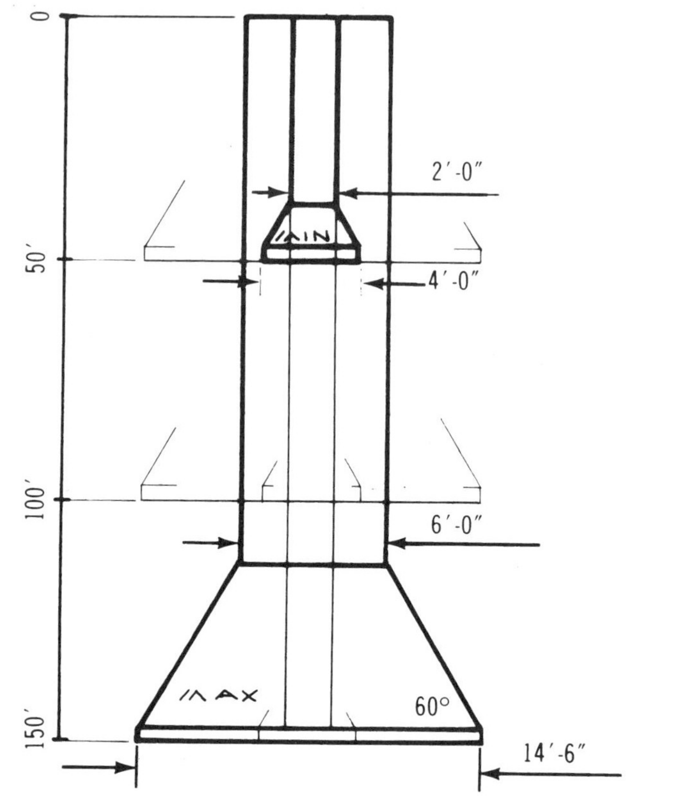

Figure 1.4-202 Size Range

Earthwork | R3163 | Drilled Caissons

R316326-60 Caissons

The three principal types of cassions are:

(1) Belled Caissons, which except for shallow depths and poor soil conditions, are generally recommended. They provide more bearing than shaft area. Because of its conical shape, no horizontal reinforcement of the bell is required.

(2) Straight Shaft Caissons are used where relatively light loads are to be supported by caissons that rest on high value bearing strata. While the shaft is larger in diameter than for belled types this is more than offset by the saving in time and labor.

(3) Keyed Caissons are used when extremely heavy loads are to be carried. A keyed or socketed caisson transfers its load into rock by a combination of end-bearing and shear reinforcing of the shaft. The most economical shaft often consists of a steel casing, a steel wide flange core and concrete. Allowable compressive stresses of .225 f'c for concrete, 16,000 psi for the wide flange core, and 9,000 psi for the steel casing are commonly used. The usual range of shaft diameter is 18″ to 84″. The number of sizes specified for any one project should be limited due to the problems of casing and auger storage. When hand work is to be performed, shaft diameters should not be less than 32″. When inspection of borings is required a minimum shaft diameter of 30″ is recommended. Concrete caissons are intended to be poured against earth excavation so permanent forms which add to cost should not be used if the excavation is clean and the earth sufficiently impervious to prevent excessive loss of concrete.

Soil Conditions for Belling		
Good	**Requires Handwork**	**Not Recommended**
Clay	Hard Shale	Silt
Sandy Clay	Limestone	Sand
Silty Clay	Sandstone	Gravel
Clayey Silt	Weathered Mica	Igneous Rock
Hard-pan		
Soft Shale		
Decomposed Rock		

Earthwork | R3171 | Tunnel Excavation

R317100-10 Tunnel Excavation

Bored tunnel excavation is common in rock for diameters from 4 feet for sewer and utilities, to 60 feet for vehicles. Production varies from a few linear feet per day to over 200 linear feet per day. In the smaller diameters, the productivity is limited by the restricted area for mucking or the removal of excavated material.

Most of the tunnels in rock today are excavated by boring machines called moles. Preparation for starting the excavation or setting up the mole is very costly. Shafts must be excavated to the invert of the proposed tunnel and the mole must be lowered into the shaft. If excavating a portal tunnel, that is starting at an open face, the cost is reduced considerably both for mobilization and mucking.

In soft ground and mixed material, special bucket excavators and rotary excavators are used inside a shield. Tunnel liners must follow directly behind the shield to support the earth and prevent cave-ins.

Traditional muck haulage operations are performed by rail with locomotives and muck cars. Sometimes conveyors are more economical and require less ventilation of the tunnel.

Ventilation and air compression are other important cost factors to consider in tunnel excavation. Continuous ventilation ducts are sometimes fabricated at the tunnel site.

Tunnel linings are steel, cast in place reinforced concrete, shotcrete, or a combination of these. When required, contact grouting is performed by pumping grout between the lining and the excavation. Intermittent holes are drilled into the lining and separate costs are determined for drilling per hole, grout pump connecting per hole, and grout per cubic foot.

Consolidation grouting and roof bolts may also be required where the excavation is unstable or faulting occurs.

Tunnel boring is usually done 24 hours per day. A typical crew for rock boring is:

Tunneling Crew based on three 8 hour shifts

1 Shifter
1 Walker
1 Machine Operator for mole
1 Oiler
1 Mechanic
3 Locomotives with operators
5 Miners for rails, vent ducts, and roof bolts
1 Electrician
2 Pumps
2 Laborers for hoisting
1 Hoist operator for muck removal
1 Oiler

Surface Crew Based on normal 8 hour shift

2 Shop Mechanics
1 Electrician
1 Shifter
2 Laborers
1 Operator with 18 ton cherry picker
1 Operator with front end loader

R320113-70 Pavement Maintenance

Routine pavement maintenance should be performed to keep a paved surface from deteriorating under the normal forces of nature and traffic.

The msot important maintenance function is the early detection and repair of minor pavement defects. Cracks and other surface breaks can develop into serious defects if not repaired in their earliest stages. For these reasons a pavement preventive maintenance program should include frequent close inspections of pavement surfaces. When suspicious areas are detected, a detailed investigation should be undertaken to determine the appropriate repair. Where subsurface or pavement deterioration is detected, the Benkelman Beam can be used to make deflection measurements under normal traffic stresses. This is done to determine the extent of the affected area.

Patching or resurfacing work should be done during warm (10°C and above) and dry weather. Adequate compaction is dificult to achieve when hot or warm mixtures are placed on cold pavements. Asphalt and asphalt mixtures usually do not bond well to damp surfaces.

Exterior Improvements | R3292 | Turf & Grasses

R329219-50 Seeding

The type of grass is determined by light, shade and moisture content of soil plus intended use. Fertilizer should be disked 4″ before seeding. For steep slopes disk five tons of mulch and lay two tons of hay or straw on surface per acre after seeding. Surface mulch can be staked, lightly disked or tar emulsion sprayed. Material for mulch can be wood chips, peat moss, partially rotted hay or straw, wood fibers and sprayed emulsions. Hemp seed blankets with fertilizer are also available. For spring seeding, watering is necessary. Late fall seeding may have to be reseeded in the spring. Hydraulic seeding, power mulching, and aerial seeding can be used on large areas.

R329343-10 Plant Spacing Chart

This chart may be used when plants are to be placed equidistant from each other, staggering their position in each row.

Plant Spacing (Inches)	Row Spacing (Inches)	Plants Per (CSF)	Plant Spacing (Feet)	Row Spacing (Feet)	Plants Per (MSF)
			4	3.46	72
6	5.20	462	5	4.33	46
8	6.93	260	6	5.20	32
10	8.66	166	8	6.93	18
12	10.39	115	10	8.66	12
15	12.99	74	12	10.39	8
18	15.59	51.32	15	12.99	5.13
21	18.19	37.70	20	17.32	2.89
24	20.78	28.87	25	21.65	1.85
30	25.98	18.48	30	25.98	1.28
36	31.18	12.83	40	34.64	0.72

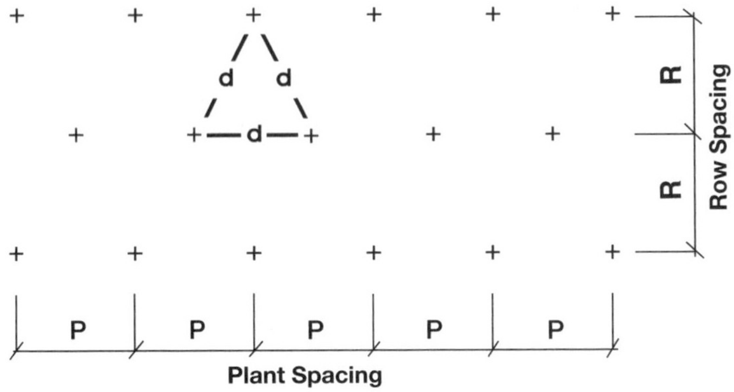

R329343-20 Trees and Plants by Environment and Purposes

Dry, Windy, Exposed Areas
Barberry
Junipers, all varieties
Locust
Maple
Oak
Pines, all varieties
Poplar, Hybrid
Privet
Spruce, all varieties
Sumac, Staghorn

Lightly Wooded Areas
Dogwood
Hemlock
Larch
Pine, White
Rhododendron
Spruce, Norway
Redbud

Total Shade Areas
Hemlock
Ivy, English
Myrtle
Pachysandra
Privet
Spice Bush
Yews, Japanese

Cold Temperatures of Northern U.S. and Canada
Arborvitae, American
Birch, White
Dogwood, Silky
Fir, Balsam
Fir, Douglas
Hemlock
Juniper, Andorra
Juniper, Blue Rug
Linden, Little Leaf
Maple, Sugar
Mountain Ash
Myrtle
Olive, Russian
Pine, Mugho
Pine, Ponderosa
Pine, Red
Pine, Scotch
Poplar, Hybrid
Privet
Rosa Rugosa
Spruce, Dwarf Alberta
Spruce, Black Hills
Spruce, Blue
Spruce, Norway
Spruce, White, Engelman
Yellow Wood

Wet, Swampy Areas
American Arborvitae
Birch, White
Black Gum
Hemlock
Maple, Red
Pine, White
Willow

Poor, Dry, Rocky Soil
Barberry
Crownvetch
Eastern Red Cedar
Juniper, Virginiana
Locust, Black
Locust, Bristly
Locust, Honey
Olive, Russian
Pines, all varieties
Privet
Rosa Rugosa
Sumac, Staghorn

Seashore Planting
Arborvitae, American
Juniper, Tamarix
Locust, Black
Oak, White
Olive, Russian
Pine, Austrian
Pine, Japanese Black
Pine, Mugho
Pine, Scotch
Privet, Amur River
Rosa Rugosa
Yew, Japanese

City Planting
Barberry
Fir, Concolor
Forsythia
Hemlock
Holly, Japanese
Ivy, English
Juniper, Andorra
Linden, Little Leaf
Locust, Honey
Maple, Norway, Silver
Oak, Pin, Red
Olive, Russian
Pachysandra
Pine, Austrian
Pine, White
Privet
Rosa Rugosa
Sumac, Staghorn
Yew, Japanese

Bonsai Planting
Azaleas
Birch, White
Ginkgo
Junipers
Pine, Bristlecone
Pine, Mugho
Spruce,k Engleman
Spruce, Dwarf Alberta

Street Planting
Linden, Little Leaf
Oak, Pin
Ginkgo

Fast Growth
Birch, White
Crownvetch
Dogwood, Silky
Fir, Douglas
Juniper, Blue Pfitzer
Juniper, Blue Rug
Maple, Silver
Olive, Autumn
Pines, Austrian, Ponderosa, Red Scotch and White
Poplar, Hybrid
Privet
Spruce, Norway
Spruce, Serbian
Texus, Cuspidata, Hicksi
Willow

Dense, Impenetrable Hedges
Field Plantings:
- Locust, Bristly,
- Olive, Autumn
- Sumac

Residential Area:
- Barberry, Red or Green
- Juniper, Blue Pfitzer
- Rosa Rugosa

Food for Birds
Ash, Mountain
Barberry
Bittersweet
Cherry, Manchu
Dogwood, Silky
Honesuckle, Rem Red
Hawthorn
Oaks
Olive, Autumn, Russian
Privet
Rosa Rugosa
Sumac

Erosion Control
Crownvetch
Locust, Bristly
Willow

R329343-30 Zones of Plant Hardiness

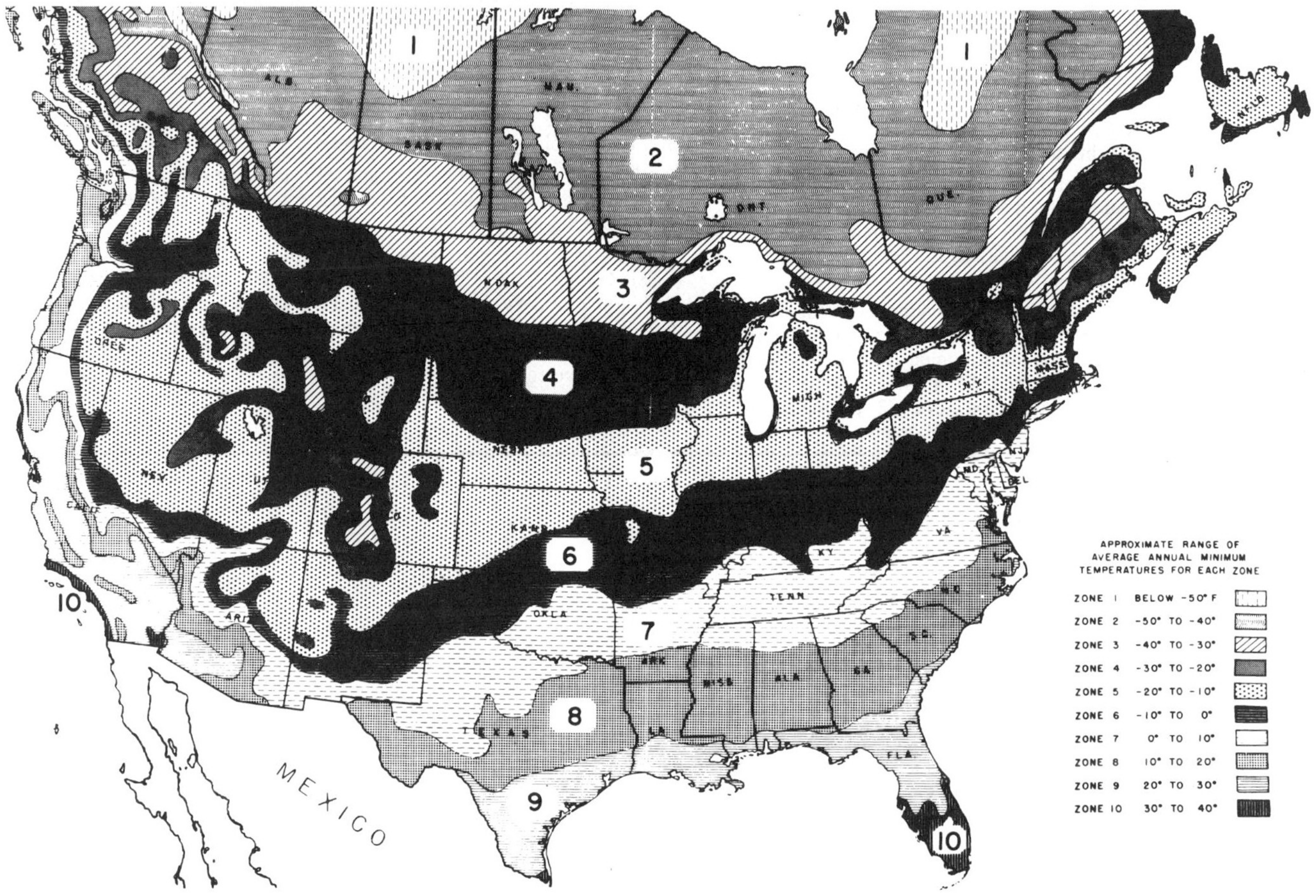

R331113-80 Piping Designations

There are several systems currently in use to describe pipe and fittings. The following paragraphs will help to identify and clarify classifications of piping systems used for water distribution.

Piping may be classified by schedule. Piping schedules include 5S, 10S, 10, 20, 30, Standard, 40, 60, Extra Strong, 80, 100, 120, 140, 160 and Double Extra Strong. These schedules are dependent upon the pipe wall thickness. The wall thickness of a particular schedule may vary with pipe size.

Ductile iron pipe for water distribution is classified by Pressure Classes such as Class 150, 200, 250, 300 and 350. These classes are actually the rated water working pressure of the pipe in pounds per square inch (psi). The pipe in these pressure classes is designed to withstand the rated water working pressure plus a surge allowance of 100 psi.

The American Water Works Association (AWWA) provides standards for various types of **plastic pipe.** C-900 is the specification for polyvinyl chloride (PVC) piping used for water distribution in sizes ranging from 4″ through 12″. C-901 is the specification for polyethylene (PE) pressure pipe, tubing and fittings used for water distribution in sizes ranging from 1/2″ through 3″. C-905 is the specification for PVC piping sizes 14″ and greater.

PVC pressure-rated pipe is identified using the standard dimensional ratio (SDR) method. This method is defined by the American Society for Testing and Materials (ASTM) Standard D 2241. This pipe is available in SDR numbers 64, 41, 32.5, 26, 21, 17, and 13.5. Pipe with an SDR of 64 will have the thinnest wall while pipe with an SDR of 13.5 will have the thickest wall. When the pressure rating (PR) of a pipe is given in psi, it is based on a line supplying water at 73 degrees F.

The National Sanitation Foundation (NSF) seal of approval is applied to products that can be used with potable water. These products have been tested to ANSI/NSF Standard 14.

Valves and strainers are classified by American National Standards Institute (ANSI) Classes. These Classes are 125, 150, 200, 250, 300, 400, 600, 900, 1500 and 2500. Within each class there is an operating pressure range dependent upon temperature. Design parameters should be compared to the appropriate material dependent, pressure-temperature rating chart for accurate valve selection.

R337116-60 Average Transmission Line Material Requirements (Per Mile)

Terrain:		Flat				Rolling				Mountain			
Item		69kV	161kV	161kV	500kV	69kV	161kV	161kV	500kV	69kV	161kV	161kV	500kV
Pole Type	Unit	Wood	Wood	Steel	Steel	Wood	Wood	Steel	Steel	Wood	Wood	Steel	Steel
Conductor: 397,500 – Cir. Mil., 26/7–ACSR													
Structures	Ea.	12[1]				9[2]				7[2]			
Poles	Ea.	12				18				14			
Crossarms	Ea.	24[3]				9[4]				7[4]			
Conductor	Ft.	15,990				15,990				15,990			
Insulators[5]	Ea.	180				135				105			
Ground Wire	Ft.	5,330				10,660				10,660			
Conductor: 636,000 – Cir. Mil., 26/7–ACSR													
Structures	Ea.	13[1]	11[6]			10[2]	9[2]	6[9]		8[2]	8[2]	6[9]	
Excavation	C.Y.	–	–			–	–	120		–	–	120	
Concrete	C.Y.	–	–			–	–	10		–	–	10	
Steel Towers	Tons	–	–			–	–	32		–	–	32	
Poles	Ea.	13	11			20	18	–		16	16	–	
Crossarms	Ea.	26[3]	33[7]			10[4]	9[8]	–		8[4]	8[8]	–	
Conductor	Ft.	15,990	15,990			15,990	15,990	15,990		15,990	15,990	15,990	
Insulators[5]	Ea.	195	165			150	297	297		120	264	297	
Ground Wire	Ft.	5330	5330			10,660	10,660	10,660		10,660	10,660	10,660	
Conductor: 954,000 – Cir. Mil., 45/7–ACSR													
Structures	Ea.	14[1]	12[6]		4[11]	10[2]	9[2]	6[9]	4[11]	8[2]	8[2]	6[9]	4[13]
Excavation	C.Y.	–	–		200	–	–	125	214	–	–	125	233
Concrete	C.Y.	–	–		20	–	–	10	21	–	–	10	21
Steel Towers	Tons	–	–		57	–	–	33	57	–	–	33	63
Poles	Ea.	14	12		–	20	18	–	–	16	16	–	–
Crossarms	Ea.	28[10]	36[7]		–	10[4]	9[8]	–	–	8[4]	8[8]	–	–
Conductor	Ft.	15,990	15,990		47,970[12]	15,990	15,990	15,990	47,970[12]	15,990	15,990	15,990	47,970[12]
Insulators[5]	Ea.	210	180		288	150	297	297	288	120	264	297	576
Ground Wire	Ft.	5330	5330		10,660	10,660	10,660	10,660	10,660	10,660	10,660	10,660	10,660
Conductor: 1,351,500 – Cir. Mil., 45/7–ACSR													
Structures	Ea.			8[14]				8[14]					
Excavation	C.Y.			220				220					
Concrete	C.Y.			28				28					
Steel Towers	Ton			46				46					
Conductor	Ft.			31,680[15]				31,680[15]					
Insulators	Ea.			528				528					
Ground Wire	Ft.			10,660				10,660					

1. Single pole two-arm suspension type construction
2. Two-pole wood H-frame construction
3. 4¾″ x 5¾″ x 8′ and 4¾″ x 5¾″ x 10′ wood crossarm
4. 6″ x 8″ x 26′-0″ wood crossarm
5. 5¾″ x 10″ disc insulator
6. Single pole construction with 3 fiberglass crossarms (5 fog- type insulators per phase)
7. 7′-0″ fiberglass crossarms
8. 6″ x 10″ x 35′-0″ wood crossarm
9. Laced steel tower, single circuit construction
10. 5″ x 7″ x 8′ and 5″ x 7″ x 10′ wood crossarm
11. Laced steel tower, single circuit 500-kV construction
12. Bundled conductor (3 sub-conductors per phase)
13. Laced steel tower, single circuit restrained phases (500-kV)
14. Laced steel tower, double circuit construction
15. Both sides of double circuit strung

Note: To allow for sagging, a mile (5280 Ft.) of transmission line uses 5330 Ft. of conductor per wire (called a wire mile).

Utilities | R3371 | Electrical Utility Transmission & Distribution

R337119-30 Concrete for Conduit Encasement

Table below lists C.Y. of concrete for 100 L.F. of trench. Conduits separation center to center should meet 7.5″ (N.E.C.).

Number of Conduits	1	2	3	4	6	8	9	Number of Conduits
Trench Dimension	11.5″ x 11.5″	11.5″ x 19″	11.5″ x 27″	19″ x 19″	19″ x 27″	19″ x 38″	27″ x 27″	Trench Dimension
Conduit Diameter 2.0″	3.29	5.39	7.64	8.83	12.51	17.66	17.72	Conduit Diameter 2.0″
2.5″	3.23	5.29	7.49	8.62	12.19	17.23	17.25	2.5″
3.0″	3.15	5.13	7.24	8.29	11.71	16.59	16.52	3.0″
3.5″	3.08	4.97	7.02	7.99	11.26	15.98	15.84	3.5″
4.0″	2.99	4.80	6.76	7.65	10.74	15.30	15.07	4.0″
5.0″	2.78	4.37	6.11	6.78	9.44	13.57	13.12	5.0″
6.0″	2.52	3.84	5.33	5.74	7.87	11.48	10.77	6.0″

Transportation | R3472 | Railway Construction

R347216-10 Single Track R.R. Siding

The costs for a single track RR siding in the Unit Price section include the components shown in the table below.

Description of Component	CSI Line Item	Qty. per L.F. of Track	Unit
Ballast, 1-1/2″ crushed stone	02060-150-0320	.667	C.Y.
6″ x 8″ x 8′-6″ Treated timber ties, 22″ O.C.	05655-700-1600	.545	Ea.
Tie plates, 2 per tie	05655-750-0300	1.091	Ea.
Track rail	05655-750-1000	2.000	L.F.
Spikes, 6″, 4 per tie	05655-750-0200	2.182	Ea.
Splice bars w/ bolts, lock washers & nuts, @ 33′ O.C.	05655-750-0100	.061	Pair
Crew B-14 @ 57 L.F./Day		.018	Day

R347216-20 Single Track, Steel Ties, Concrete Bed

The costs for a R.R. siding with steel ties and a concrete bed in the Unit Price section include the components shown in the table below.

Description of Component	CSI Line Item	Qty. per L.F. of Track	Unit
Concrete bed, 9′ wide, 10″ thick	03310-240-3950	.278	C.Y.
Ties, W6x16 x 6′-6″ long, @ 30″ O.C.	05120-640-0120	.400	Ea.
Tie plates, 4 per tie	05655-750-0300	1.600	Ea.
Track rail	05655-750-1000	2.000	L.F.
Tie plate bolts, 1″, 8 per tie	05655-750-0020	3.200	Ea.
Splice bars w/bolts, lock washers & nuts, @ 33′ O.C.	05655-750-0100	.061	Pair
Crew B-14 @ 22 L.F./Day		.045	Day

Change Orders

Change Order Considerations

A Change Order is a written document, usually prepared by the design professional, and signed by the owner, the architect/engineer and the contractor. A change order states the agreement of the parties to: an addition, deletion, or revision in the work; an adjustment in the contract sum, if any; or an adjustment in the contract time, if any. Change orders, or "extras" in the construction process occur after execution of the construction contract and impact architects/engineers, contractors and owners.

Change orders that are properly recognized and managed can ensure orderly, professional and profitable progress for all who are involved in the project. There are many causes for change orders and change order requests. In all cases, change orders or change order requests should be addressed promptly and in a precise and prescribed manner. The following paragraphs include information regarding change order pricing and procedures.

The Causes of Change Orders

Reasons for issuing change orders include:

- Unforeseen field conditions that require a change in the work
- Correction of design discrepancies, errors or omissions in the contract documents
- Owner-requested changes, either by design criteria, scope of work, or project objectives
- Completion date changes for reasons unrelated to the construction process
- Changes in building code interpretations, or other public authority requirements that require a change in the work
- Changes in availability of existing or new materials and products

Procedures

Properly written contract documents must include the correct change order procedures for all parties—owners, design professionals and contractors—to follow in order to avoid costly delays and litigation.

Being "in the right" is not always a sufficient or acceptable defense. The contract provisions requiring notification and documentation must be adhered to within a defined or reasonable time frame.

The appropriate method of handling change orders is by a written proposal and acceptance by all parties involved. Prior to starting work on a project, all parties should identify their authorized agents who may sign and accept change orders, as well as any limits placed on their authority.

Time may be a critical factor when the need for a change arises. For such cases, the contractor might be directed to proceed on a "time and materials" basis, rather than wait for all paperwork to be processed—a delay that could impede progress. In this situation, the contractor must still follow the prescribed change order procedures, including but not limited to, notification and documentation.

All forms used for change orders should be dated and signed by the proper authority. Lack of documentation can be very costly, especially if legal judgments are to be made and if certain field personnel are no longer available. For time and material change orders, the contractor should keep accurate daily records of all labor and material allocated to the change. Forms that can be used to document change order work are available in *Means Forms for Building Construction Professionals.*

Owners or awarding authorities who do considerable and continual building construction (such as the federal government) realize the inevitability of change orders for numerous reasons, both predictable and unpredictable. As a result, the federal government, the American Institute of Architects (AIA), the Engineers Joint Contract Documents Committee (EJCDC) and other contractor, legal and technical organizations have developed standards and procedures to be followed by all parties to achieve contract continuance and timely completion, while being financially fair to all concerned.

In addition to the change order standards put forth by industry associations, there are also many books available on the subject.

Pricing Change Orders

When pricing change orders, regardless of their cause, the most significant factor is *when* the change occurs. The need for a change may be perceived in the field or requested by the architect/engineer *before* any of the actual installation has begun, or may evolve or appear *during* construction when the item of work in question is partially installed. In the latter cases, the original sequence of construction is disrupted, along with all contiguous and supporting systems. Change orders cause the greatest impact when they occur *after* the installation has been completed and must be uncovered, or even replaced. Post-completion changes may be caused by necessary design changes, product failure, or changes in the owner's requirements that are not discovered until the building or the systems begin to function.

Specified procedures of notification and record keeping must be adhered to and enforced regardless of the stage of construction: *before, during,* or *after* installation. Some bidding documents anticipate change orders by requiring that unit prices including overhead and profit percentages—for additional as well as deductible changes—be listed. Generally these unit prices do not fully take into account the ripple effect, or impact on other trades, and should be used for general guidance only.

When pricing change orders, it is important to classify the time frame in which the change occurs. There are two basic time frames for change orders: *pre-installation change orders*, which occur before the start of construction, and *post-installation change orders*, which involve reworking after the original installation. Change orders that occur between these stages may be priced according to the extent of work completed using a combination of techniques developed for pricing *pre-* and *post-installation* changes.

The following factors are the basis for a check list to use when preparing a change order estimate.

Factors To Consider When Pricing Change Orders

As an estimator begins to prepare a change order, the following questions should be reviewed to determine their impact on the final price.

General

- Is the change order work *pre-installation* or *post-installation*?

  Change order work costs vary according to how much of the installation has been completed. Once workers have the project scoped in their mind, even though they have not started, it can be difficult to refocus. Consequently they may spend more than the normal amount of time understanding the change. Also, modifications to work in place such as trimming or refitting usually take more time than was initially estimated. The greater the amount of work in place, the more reluctant workers are to change it. Psychologically they may resent the change and as a result the rework takes longer than normal. Post-installation change order estimates must include demolition of existing work as required to accomplish the change. If the work is performed at a later time, additional obstacles such as building finishes may be present which must be protected. Regardless of whether the change occurs pre-installation or post-installation, attempt to isolate the identifiable factors and price them separately. For example, add shipping costs that may be required pre-installation or any demolition required post-installation. Then analyze the potential impact on productivity of psychological and/or learning curve factors and adjust the output rates accordingly. One approach is to break down the typical workday into segments and quantify the impact on each segment. The following chart may be useful as a guide:

Activities (Productivity) Expressed as Percentages of a Workday			
Task	Means Mechanical Cost Data (for New Construction)	Pre-Installation Change Orders	Post-Installation Change Orders
1. Study plans	3%	6%	6%
2. Material procurement	3%	3%	3%
3. Receiving and storing	3%	3%	3%
4. Mobilization	5%	5%	5%
5. Site movement	5%	5%	8%
6. Layout and marking	8%	10%	12%
7. Actual installation	64%	59%	54%
8. Clean-up	3%	3%	3%
9. Breaks—non-productive	6%	6%	6%
Total	100%	100%	100%

Change Order Installation Efficiency

The labor-hours expressed (for new construction) are based on average installation time, using an efficiency level of approximately 60-65%. For change order situations, adjustments to this efficiency level should reflect the daily labor-hour allocation for that particular occurrence.

If any of the specific percentages expressed in the above chart do not apply to a particular project situation, then those percentage points should be reallocated to the appropriate task(s). Example: Using data for new construction, assume there is no new material being utilized. The percentages for Tasks 2 and 3 would therefore be reallocated to other tasks. If the time required for Tasks 2 and 3 can now be applied to installation, we can add the time allocated for *Material Procurement* and *Receiving and Storing* to the *Actual Installation* time for new construction, thereby increasing the Actual Installation percentage.

This chart shows that, due to reduced productivity, labor costs will be higher than those for new construction by 5% to 15% for pre-installation change orders and by 15% to 25% for post-installation change orders. Each job and change order is unique and must be examined individually. Many factors, covered elsewhere in this section, can each have a significant impact on productivity and change order costs. All such factors should be considered in every case.

- Will the change substantially delay the original completion date?

  A significant change in the project may cause the original completion date to be extended. The extended schedule may subject the contractor to new wage rates dictated by relevant labor contracts. Project supervision and other project overhead must also be extended beyond the original completion date. The schedule extension may also put installation into a new weather season. For example, underground piping scheduled for October installation was delayed until January. As a result, frost penetrated the trench area, thereby changing the degree of difficulty of the task. Changes and delays may have a ripple effect throughout the project. This effect must be analyzed and negotiated with the owner.

- What is the net effect of a deduct change order?

  In most cases, change orders resulting in a deduction or credit reflect only bare costs. The contractor may retain the overhead and profit based on the original bid.

Materials

- Will you have to pay more or less for the new material, required by the change order, than you paid for the original purchase?

  The same material prices or discounts will usually apply to materials purchased for change orders as new construction. In some instances, however, the contractor may forfeit the advantages of competitive pricing for change orders. Consider the following example:

  A contractor purchased over $20,000 worth of fan coil units for an installation, and obtained the maximum discount. Some time later it was determined the project required an additional matching unit. The contractor has to purchase this unit from the original supplier to ensure a match. The supplier at this time may not discount the unit because of the small quantity, and the fact that he is no longer in a competitive situation. The impact of quantity on purchase can add between 0% and 25% to material prices and/or subcontractor quotes.

- If materials have been ordered or delivered to the job site, will they be subject to a cancellation charge or restocking fee?

  Check with the supplier to determine if ordered materials are subject to a cancellation charge. Delivered materials not used as result of a change order may be subject to a restocking fee if returned to the supplier. Common restocking charges run between 20% and 40%. Also, delivery charges to return the goods to the supplier must be added.

Labor

- How efficient is the existing crew at the actual installation?

  Is the same crew that performed the initial work going to do the change order? Possibly the change consists of the installation of a unit identical to one already installed; therefore the change should take less time. Be sure to consider this potential productivity increase and modify the productivity rates accordingly.

- If the crew size is increased, what impact will that have on supervision requirements?

  Under most bargaining agreements or management practices, there is a point at which a working foreman is replaced by a nonworking foreman. This replacement increases project overhead by adding a nonproductive worker. If additional workers are added to accelerate the project or to perform changes while maintaining the schedule, be sure to add additional supervision time if warranted. Calculate the hours involved and the additional cost directly if possible.

- What are the other impacts of increased crew size?

  The larger the crew, the greater the potential for productivity to decrease. Some of the factors that cause this productivity loss are: overcrowding (producing restrictive conditions in the working space), and possibly a shortage of any special tools and equipment required. Such factors affect not only the crew working on the elements directly involved in the change order, but other crews whose movement may also be hampered.

  As the crew increases, check its basic composition for changes by the addition or deletion of apprentices or nonworking foreman and quantify the potential effects of equipment shortages or other logistical factors.

- As new crews, unfamiliar with the project, are brought onto the site, how long will it take them to become oriented to the project requirements?

  The orientation time for a new crew to become 100% effective varies with the site and type of project. Orientation is easiest at a new construction site, and most difficult at existing, very restrictive renovation sites. The type of work also affects orientation time. When all elements of the work are exposed, such as concrete or masonry work, orientation is decreased. When the work is concealed or less visible, such as existing electrical systems, orientation takes longer. Usually orientation can be accomplished in one day or less. Costs for added orientation should be itemized and added to the total estimated cost.

- How much actual production can be gained by working overtime?

  Short term overtime can be used effectively to accomplish more work in a day. However, as overtime is scheduled to run beyond several weeks, studies have shown marked decreases in output. The following chart shows the effect of long term overtime on worker efficiency. If the anticipated change requires extended overtime to keep the job on schedule, these factors can be used as a guide to predict the impact on time and cost. Add project overhead, particularly supervision, that may also be incurred.

Days per Week	Hours per Day	Production Efficiency					Payroll Cost Factors	
		1 Week	2 Weeks	3 Weeks	4 Weeks	Average 4 Weeks	@ 1-1/2 Times	@ 2 Times
5	8	100%	100%	100%	100%	100%	100%	100%
	9	100	100	95	90	96.25	105.6	111.1
	10	100	95	90	85	91.25	110.0	120.0
	11	95	90	75	65	81.25	113.6	127.3
	12	90	85	70	60	76.25	116.7	133.3
6	8	100	100	95	90	96.25	108.3	116.7
	9	100	95	90	85	92.50	113.0	125.9
	10	95	90	85	80	87.50	116.7	133.3
	11	95	85	70	65	78.75	119.7	139.4
	12	90	80	65	60	73.75	122.2	144.4
7	8	100	95	85	75	88.75	114.3	128.6
	9	95	90	80	70	83.75	118.3	136.5
	10	90	85	75	65	78.75	121.4	142.9
	11	85	80	65	60	72.50	124.0	148.1
	12	85	75	60	55	68.75	126.2	152.4

Effects of Overtime

Caution: Under many labor agreements, Sundays and holidays are paid at a higher premium than the normal overtime rate.

The use of long-term overtime is counterproductive on almost any construction job; that is, the longer the period of overtime, the lower the actual production rate. Numerous studies have been conducted, and while they have resulted in slightly different numbers, all reach the same conclusion. The figure above tabulates the effects of overtime work on efficiency.

As illustrated, there can be a difference between the *actual* payroll cost per hour and the *effective* cost per hour for overtime work. This is due to the reduced production efficiency with the increase in weekly hours beyond 40. This difference between actual and effective cost results from overtime work over a prolonged period. Short-term overtime work does not result in as great a reduction in efficiency, and in such cases, effective cost may not vary significantly from the actual payroll cost. As the total hours per week are increased on a regular basis, more time is lost because of fatigue, lowered morale, and an increased accident rate.

As an example, assume a project where workers are working 6 days a week, 10 hours per day. From the figure above (based on productivity studies), the average effective productive hours over a four-week period are:

$$0.875 \times 60 = 52.5$$

Depending upon the locale and day of week, overtime hours may be paid at time and a half or double time. For time and a half, the overall (average) *actual* payroll cost (including regular and overtime hours) is determined as follows:

$$\frac{40 \text{ reg. hrs.} + (20 \text{ overtime hrs.} \times 1.5)}{60 \text{ hrs.}} = 1.167$$

Based on 60 hours, the payroll cost per hour will be 116.7% of the normal rate at 40 hours per week. However, because the effective production (efficiency) for 60 hours is reduced to the equivalent of 52.5 hours, the effective cost of overtime is calculated as follows:

For time and a half:

$$\frac{40 \text{ reg. hrs.} + (20 \text{ overtime hrs.} \times 1.5)}{52.5 \text{ hrs.}} = 1.33$$

Installed cost will be 133% of the normal rate (for labor).

Thus, when figuring overtime, the actual cost per unit of work will be higher than the apparent overtime payroll dollar increase, due to the reduced productivity of the longer workweek. These efficiency calculations are true only for those cost factors determined by hours worked. Costs that are applied weekly or monthly, such as equipment rentals, will not be similarly affected.

Equipment

- What equipment is required to complete the change order?

  Change orders may require extending the rental period of equipment already on the job site, or the addition of special equipment brought in to accomplish the change work. In either case, the additional rental charges and operator labor charges must be added.

Summary

The preceding considerations and others you deem appropriate should be analyzed and applied to a change order estimate. The impact of each should be quantified and listed on the estimate to form an audit trail.

Change orders that are properly identified, documented, and managed help to ensure the orderly, professional and profitable progress of the work. They also minimize potential claims or disputes at the end of the project.

Square Foot Costs

Estimating Tips

- The cost figures in this Square Foot Cost section were derived from approximately 11,200 projects contained in the RSMeans database of completed construction projects. They include the contractor's overhead and profit but do not generally include architectural fees or land costs. The figures have been adjusted to January of the current year. New projects are added to our files each year, and outdated projects are discarded. For this reason, certain costs may not show a uniform annual progression. In no case are all subdivisions of a project listed.
- These projects were located throughout the U.S. and reflect a tremendous variation in square foot (S.F.) and cubic foot (C.F.) costs. This is due to differences, not only in labor and material costs, but also in individual owners' requirements. For instance, a bank in a large city would have different features than one in a rural area. This is true of all the different types of buildings analyzed. Therefore, caution should be exercised when using these Square Foot costs. For example, for court houses, costs in the database are local court house costs and will not apply to the larger, more elaborate federal court houses. As a general rule, the projects in the 1/4 column do not include any site work or equipment, while the projects in the 3/4 column may include both equipment and site work. The median figures do not generally include site work.
- None of the figures "go with" any others. All individual cost items were computed and tabulated separately. Thus, the sum of the median figures for Plumbing, HVAC, and Electrical will not normally match the total Mechanical and Electrical costs arrived at by separate analysis and tabulation of the projects.
- Each building was analyzed as to total and component costs and percentages. The figures were arranged in ascending order with the results tabulated as shown. The 1/4 column shows that 25% of the projects had lower costs and 75% had higher. The 3/4 column shows that 75% of the projects had lower costs and 25% had higher. The median column shows that 50% of the projects had lower costs and 50% had higher.
- There are two times when square foot costs are useful. The first is in the conceptual stage when no details are available. Then square foot costs make a useful starting point. The second is after the bids are in and the costs can be worked back into their appropriate units for information purposes. As soon as details become available in the project design, the square foot approach should be discontinued and the project priced as to its particular components. When more precision is required or for estimating the replacement cost of specific buildings, the current edition of *RSMeans Square Foot Costs* should be used.
- When using the figures in this section, it is recommended that the median column be used for preliminary figures if no additional information is available. The median figures, when multiplied by the total city construction cost index figures (see City Cost Indexes) and then multiplied by the project size modifier at the end of this section, should present a fairly accurate base figure, which would then have to be adjusted in view of the estimator's experience, local economic conditions, code requirements, and the owner's particular requirements. There is no need to factor the percentage figures as these should remain constant from city to city. All tabulations mentioning air conditioning had at least partial air conditioning.
- The editors of this book would greatly appreciate receiving cost figures on one or more of your recent projects which would then be included in the averages for next year. All cost figures received will be kept confidential except that they will be averaged with similar projects to arrive at Square Foot cost figures for next year's book. See the last page of the book for details and the discount available for submitting one or more of your projects.

	50 17 00 | S.F. Costs		UNIT	UNIT COSTS 1/4	UNIT COSTS MEDIAN	UNIT COSTS 3/4	% OF TOTAL 1/4	% OF TOTAL MEDIAN	% OF TOTAL 3/4	
01	0010	**APARTMENTS** Low Rise (1 to 3 story)	S.F.	56.50	71	94				01
	0020	Total project cost	C.F.	5.05	6.70	8.25				
	0100	Site work	S.F.	4.81	6.55	11.55	6.05%	10.55%	14.05%	
	0500	Masonry		1.11	2.57	4.45	1.54%	3.67%	6.35%	
	1500	Finishes		5.95	8.20	10.15	9.05%	10.75%	12.85%	
	1800	Equipment		1.83	2.79	4.14	2.73%	4.03%	5.95%	
	2720	Plumbing		4.38	5.65	7.15	6.65%	8.95%	10.05%	
	2770	Heating, ventilating, air conditioning		2.79	3.44	5.05	4.20%	5.60%	7.60%	
	2900	Electrical		3.25	4.32	5.80	5.20%	6.65%	8.40%	
	3100	Total: Mechanical & Electrical		11.25	14.35	17.90	15.90%	18.05%	23%	
	9000	Per apartment unit, total cost	Apt.	52,500	80,000	118,000				
	9500	Total: Mechanical & Electrical	"	9,900	15,600	20,400				
02	0010	**APARTMENTS** Mid Rise (4 to 7 story)	S.F.	74.50	90	110				02
	0020	Total project costs	C.F.	5.80	8.05	11				
	0100	Site work	S.F.	2.98	5.90	11.65	5.25%	6.70%	9.20%	
	0500	Masonry		4.96	6.85	9.75	5.15%	7.40%	10.50%	
	1500	Finishes		9.40	13.10	15.20	10.55%	13.45%	17.70%	
	1800	Equipment		2.45	3.49	4.48	2.55%	3.48%	4.31%	
	2500	Conveying equipment		1.69	2.08	2.48	2.05%	2.29%	2.69%	
	2720	Plumbing		4.37	7	7.75	5.70%	7.20%	8.95%	
	2900	Electrical		4.92	7	8.20	6.65%	7.20%	8.95%	
	3100	Total: Mechanical & Electrical		15.75	19.70	24.50	18.85%	21%	25%	
	9000	Per apartment unit, total cost	Apt.	84,500	99,500	165,000				
	9500	Total: Mechanical & Electrical	"	15,900	18,400	23,300				
03	0010	**APARTMENTS** High Rise (8 to 24 story)	S.F.	84.50	102	117				03
	0020	Total project costs	C.F.	8.20	10.05	12.20				
	0100	Site work	S.F.	3.07	4.96	6.95	2.58%	4.84%	6.15%	
	0500	Masonry		4.89	8.90	11.05	4.74%	9.65%	11.05%	
	1500	Finishes		9.40	11.75	13.85	9.75%	11.80%	13.70%	
	1800	Equipment		2.72	3.34	4.43	2.78%	3.49%	4.35%	
	2500	Conveying equipment		1.92	2.92	4.16	2.23%	2.78%	3.37%	
	2720	Plumbing		6.25	7.35	10.30	6.80%	7.20%	10.45%	
	2900	Electrical		5.80	7.35	9.90	6.45%	7.65%	8.80%	
	3100	Total: Mechanical & Electrical		17.40	22	26.50	17.95%	22.50%	24.50%	
	9000	Per apartment unit, total cost	Apt.	88,000	97,000	134,500				
	9500	Total: Mechanical & Electrical	"	19,000	21,700	23,000				
04	0010	**AUDITORIUMS**	S.F.	87	119	170				04
	0020	Total project costs	C.F.	5.50	7.65	11.50				
	2720	Plumbing	S.F.	5.55	7.70	9.75	5.85%	7.20%	8.70%	
	2900	Electrical		7.10	9.85	13.20	6.80%	8.95%	11.30%	
	3100	Total: Mechanical & Electrical		13.60	19.30	39	24.50%	30.50%	31.50%	
05	0010	**AUTOMOTIVE SALES**	S.F.	63.50	88	108				05
	0020	Total project costs	C.F.	4.28	5.15	6.65				
	2720	Plumbing	S.F.	2.97	5.15	5.60	2.89%	6.05%	6.50%	
	2770	Heating, ventilating, air conditioning		4.58	7	7.55	4.61%	10%	10.35%	
	2900	Electrical		5.25	8.05	11.15	7.40%	9.95%	12.40%	
	3100	Total: Mechanical & Electrical		14.60	20.50	26	19.15%	20.50%	26%	
06	0010	**BANKS**	S.F.	127	158	200				06
	0020	Total project costs	C.F.	9.10	12.40	16.35				
	0100	Site work	S.F.	14.55	22	33	7.85%	12.95%	17%	
	0500	Masonry		6.65	12.45	23	3.36%	6.95%	10.35%	
	1500	Finishes		11.30	15.45	19.85	5.85%	8.45%	11.25%	
	1800	Equipment		5.10	10.50	22.50	1.34%	5.95%	10.65%	
	2720	Plumbing		4	5.70	8.35	2.82%	3.90%	4.93%	
	2770	Heating, ventilating, air conditioning		7.60	10.15	13.50	4.86%	7.15%	8.50%	
	2900	Electrical		12.05	16.10	21	8.20%	10.20%	12.20%	
	3100	Total: Mechanical & Electrical		29	38	46	16.55%	19.45%	23%	
	3500	See also division 11020 & 11030								

50 17 | Square Foot Costs

	50 17 00 | S.F. Costs		UNIT	UNIT COSTS 1/4	UNIT COSTS MEDIAN	UNIT COSTS 3/4	% OF TOTAL 1/4	% OF TOTAL MEDIAN	% OF TOTAL 3/4	
13	0010	**CHURCHES**	S.F.	86	109	141				13
	0020	Total project costs	C.F.	5.30	6.70	8.85				
	1800	Equipment	S.F.	1.09	2.46	5.20	1.03%	2.24%	4.50%	
	2720	Plumbing		3.34	4.67	6.90	3.51%	4.96%	6.25%	
	2770	Heating, ventilating, air conditioning		7.80	10.20	14.45	7.50%	10%	12%	
	2900	Electrical		7.20	9.90	13.25	7.30%	8.75%	10.90%	
	3100	Total: Mechanical & Electrical	↓	22	29	40	18.25%	22%	24%	
	3500	See also division 11040								
15	0010	**CLUBS, COUNTRY**	S.F.	92	111	140				15
	0020	Total project costs	C.F.	7.40	9.05	12.50				
	2720	Plumbing	S.F.	5.90	8.25	18.80	5.60%	7.90%	10%	
	2900	Electrical		7.25	10.35	13.65	7%	8.95%	11%	
	3100	Total: Mechanical & Electrical	↓	22	38.50	48	19%	26.50%	29.50%	
17	0010	**CLUBS, SOCIAL** Fraternal	S.F.	73.50	105	142				17
	0020	Total project costs	C.F.	4.59	6.95	8.30				
	2720	Plumbing	S.F.	4.62	5.75	8.70	5.60%	6.90%	8.55%	
	2770	Heating, ventilating, air conditioning		6.65	8.05	10.35	8.20%	9.25%	14.40%	
	2900	Electrical		5.85	9.10	11	6.50%	9.50%	10.55%	
	3100	Total: Mechanical & Electrical	↓	16.30	31	39.50	21%	23%	23.50%	
18	0010	**CLUBS, Y.M.C.A.**	S.F.	92.50	125	152				18
	0020	Total project costs	C.F.	4.25	7.10	10.60				
	2720	Plumbing	S.F.	5.85	11.60	13	5.65%	7.60%	10.85%	
	2900	Electrical		7	9.55	13.60	6.05%	7.80%	10.20%	
	3100	Total: Mechanical & Electrical	↓	26	31.50	35	18.40%	21.50%	28.50%	
19	0010	**COLLEGES** Classrooms & Administration	S.F.	103	136	184				19
	0020	Total project costs	C.F.	7.45	10.60	16.70				
	0500	Masonry	S.F.	6.85	12.90	15.85	5.65%	8.25%	10.50%	
	2720	Plumbing		5.05	10.10	18.15	5.10%	6.60%	8.95%	
	2900	Electrical		8.45	12.95	16.10	7.70%	9.85%	12%	
	3100	Total: Mechanical & Electrical	↓	31	43.50	51.50	24%	28%	31.50%	
21	0010	**COLLEGES** Science, Engineering, Laboratories	S.F.	173	203	247				21
	0020	Total project costs	C.F.	9.95	14.50	16.45				
	1800	Equipment	S.F.	9.65	22	24	2%	6.45%	12.65%	
	2900	Electrical		14.30	19.50	31	7.10%	9.40%	12.10%	
	3100	Total: Mechanical & Electrical	↓	53	63	97.50	28.50%	31.50%	41%	
	3500	See also division 11600								
23	0010	**COLLEGES** Student Unions	S.F.	110	155	182				23
	0020	Total project costs	C.F.	6.15	8.10	10.40				
	3100	Total: Mechanical & Electrical	S.F.	29	45	53	23.50%	26%	29%	
25	0010	**COMMUNITY CENTERS**	"	90	113	152				25
	0020	Total project costs	C.F.	5.95	8.50	11.05				
	1800	Equipment	S.F.	2.29	3.90	6.15	1.87%	3.12%	6%	
	2720	Plumbing		4.43	7.55	11.05	4.94%	7%	9.10%	
	2770	Heating, ventilating, air conditioning		7.30	10.50	14.45	6.95%	10.65%	13.05%	
	2900	Electrical		7.75	9.95	14.50	7.35%	9.10%	10.85%	
	3100	Total: Mechanical & Electrical	↓	27	32.50	46.50	22.50%	26.50%	32.50%	
28	0010	**COURT HOUSES**	S.F.	131	151	176				28
	0020	Total project costs	C.F.	10.10	12.20	15.20				
	2720	Plumbing	S.F.	6.25	8.75	12.55	5.95%	7.45%	8.20%	
	2900	Electrical		12.90	15.25	18.40	8.55%	9.95%	11.50%	
	3100	Total: Mechanical & Electrical	↓	35.50	41	51.50	22.50%	29.50%	30.50%	
30	0010	**DEPARTMENT STORES**	S.F.	48.50	66	83				30
	0020	Total project costs	C.F.	2.61	3.54	4.59				
	2720	Plumbing	S.F.	1.59	1.92	2.91	1.82%	4.21%	5.90%	
	2770	Heating, ventilating, air conditioning	↓	4.43	6.85	10.30	8.20%	9.10%	14.80%	

	50 17 00 | S.F. Costs		UNIT	UNIT COSTS			% OF TOTAL			
				1/4	MEDIAN	3/4	1/4	MEDIAN	3/4	
30	2900	Electrical	S.F.	5.60	7.65	9.05	9.05%	12.15%	14.95%	30
	3100	Total: Mechanical & Electrical		9.85	12.55	22	13.20%	21.50%	50%	
31	0010	**DORMITORIES** Low Rise (1 to 3 story)	S.F.	83	117	146				31
	0020	Total project costs	C.F.	5.50	8.45	12.65				
	2720	Plumbing	S.F.	5.55	7.45	9.40	8.05%	9%	9.65%	
	2770	Heating, ventilating, air conditioning		5.85	7.05	9.35	4.61%	8.05%	10%	
	2900	Electrical		5.95	8.95	12.45	6.55%	8.90%	9.55%	
	3100	Total: Mechanical & Electrical		31	32.50	34	22%	26%	29%	
	9000	Per bed, total cost	Bed	39,100	43,500	93,000				
32	0010	**DORMITORIES** Mid Rise (4 to 8 story)	S.F.	113	148	181				32
	0020	Total project costs	C.F.	12.50	13.70	16.45				
	2900	Electrical	S.F.	12.05	13.70	17.80	8.20%	10.20%	11.10%	
	3100	Total: Mechanical & Electrical	″	31	35	68	19.50%	30.50%	37.50%	
	9000	Per bed, total cost	Bed	16,100	36,800	76,000				
34	0010	**FACTORIES**	S.F.	43	64	98.50				34
	0020	Total project costs	C.F.	2.75	4.11	6.80				
	0100	Site work	S.F.	4.90	8.95	14.15	6.95%	11.45%	17.95%	
	2720	Plumbing		2.31	4.31	7.10	3.73%	6.05%	8.10%	
	2770	Heating, ventilating, air conditioning		4.50	6.45	8.70	5.25%	8.45%	11.35%	
	2900	Electrical		5.35	8.45	12.85	8.10%	10.50%	14.20%	
	3100	Total: Mechanical & Electrical		12.70	20.50	31	21%	28.50%	35.50%	
36	0010	**FIRE STATIONS**	S.F.	85	116	151				36
	0020	Total project costs	C.F.	4.98	6.85	9.10				
	0500	Masonry	S.F.	12.50	22.50	29.50	8.60%	11.65%	16.45%	
	1140	Roofing		2.78	7.50	8.55	1.90%	4.94%	5.05%	
	1580	Painting		2.28	3.22	3.29	1.37%	1.57%	2.07%	
	1800	Equipment		1.80	2.55	6.15	.90%	2.50%	4.07%	
	2720	Plumbing		5.40	7.95	11.70	5.85%	7.35%	9.50%	
	2770	Heating, ventilating, air conditioning		4.64	7.40	11.50	4.86%	7.25%	9.25%	
	2900	Electrical		6.05	10.40	13.70	6.80%	8.75%	10.65%	
	3100	Total: Mechanical & Electrical		30	35	41.50	19.60%	23%	27%	
37	0010	**FRATERNITY HOUSES** and Sorority Houses	S.F.	85	110	150				37
	0020	Total project costs	C.F.	8.50	8.85	11.35				
	2720	Plumbing	S.F.	6.45	7.35	13.50	6.80%	8%	10.85%	
	2900	Electrical		5.60	12.15	14.85	6.60%	9.90%	10.65%	
	3100	Total: Mechanical & Electrical		15	21.50	26		15.10%	15.90%	
38	0010	**FUNERAL HOMES**	S.F.	90	122	222				38
	0020	Total project costs	C.F.	9.15	10.20	19.60				
	2900	Electrical	S.F.	3.97	7.25	8.60	3.58%	4.44%	5.95%	
	3100	Total: Mechanical & Electrical	″	14.10	21	28.50	12.90%	12.90%	12.90%	
39	0010	**GARAGES, COMMERCIAL** (Service)	S.F.	53	80	109				39
	0020	Total project costs	C.F.	3.42	4.95	7.20				
	1800	Equipment	S.F.	2.87	6.45	10	2.69%	4.62%	6.80%	
	2720	Plumbing		3.52	5.45	9.90	5.45%	7.85%	10.65%	
	2730	Heating & ventilating		4.63	6.30	8.70	5.25%	6.85%	8.20%	
	2900	Electrical		4.90	7.50	10.70	7.15%	9.25%	10.85%	
	3100	Total: Mechanical & Electrical		10.95	20.50	30.50	13.60%	17.40%	27%	
40	0010	**GARAGES, MUNICIPAL** (Repair)	S.F.	74.50	99.50	140				40
	0020	Total project costs	C.F.	4.67	5.90	10.20				
	0500	Masonry	S.F.	7	13.70	21.50	5.60%	9.15%	12.50%	
	2720	Plumbing		3.35	6.45	12.15	3.59%	6.70%	7.95%	
	2730	Heating & ventilating		5.75	8.30	16	6.15%	7.45%	13.50%	
	2900	Electrical		5.55	8.70	12.50	6.65%	8.15%	11.15%	
	3100	Total: Mechanical & Electrical		17.75	35.50	51.50	21.50%	25.50%	28.50%	

50 17 00 | S.F. Costs

		50 17 00 | S.F. Costs	UNIT	UNIT COSTS 1/4	UNIT COSTS MEDIAN	UNIT COSTS 3/4	% OF TOTAL 1/4	% OF TOTAL MEDIAN	% OF TOTAL 3/4	
41	0010	**GARAGES, PARKING**	S.F.	29	42.50	73				41
	0020	Total project costs	C.F.	2.73	3.71	5.40				
	2720	Plumbing	S.F.	.82	1.28	1.97	1.72%	2.70%	3.85%	
	2900	Electrical		1.59	1.96	3.07	4.33%	5.20%	6.30%	
	3100	Total: Mechanical & Electrical		3.26	4.54	5.95	7%	8.90%	11.05%	
	3200									
	9000	Per car, total cost	Car	12,300	15,400	19,700				
43	0010	**GYMNASIUMS**	S.F.	80.50	107	138				43
	0020	Total project costs	C.F.	4.03	5.45	6.70				
	1800	Equipment	S.F.	1.92	3.60	6.90	2.07%	3.35%	6.70%	
	2720	Plumbing		5.10	6.30	7.85	4.95%	6.75%	7.75%	
	2770	Heating, ventilating, air conditioning		5.50	8.40	16.85	5.80%	9.80%	11.10%	
	2900	Electrical		6.15	8.20	10.40	6.60%	8.30%	10.30%	
	3100	Total: Mechanical & Electrical		22	31	36.50	20.50%	26%	29.50%	
	3500	See also division 11480								
46	0010	**HOSPITALS**	S.F.	155	190	288				46
	0020	Total project costs	C.F.	11.75	14.65	21				
	1800	Equipment	S.F.	3.94	7.55	13.05	1.10%	2.68%	5%	
	2720	Plumbing		13.35	18.70	24	7.60%	9.10%	10.85%	
	2770	Heating, ventilating, air conditioning		19.60	25	34	7.80%	12.95%	16.65%	
	2900	Electrical		16.95	22	34	9.85%	11.75%	14%	
	3100	Total: Mechanical & Electrical		47.50	63	103	26.50%	33%	36.50%	
	9000	Per bed or person, total cost	Bed	127,000	203,500	272,000				
	9900	See also division 11700								
48	0010	**HOUSING** For the Elderly	S.F.	76.50	96.50	119				48
	0020	Total project costs	C.F.	5.45	7.55	9.65				
	0100	Site work	S.F.	5.65	8.40	12.10	5.05%	7.90%	12.10%	
	0500	Masonry		2.33	8.70	12.70	1.30%	6.05%	11%	
	1800	Equipment		1.84	2.54	4.04	1.88%	3.23%	4.43%	
	2510	Conveying systems		1.86	2.50	3.38	1.78%	2.20%	2.81%	
	2720	Plumbing		5.65	7.25	9.45	8.15%	9.55%	10.50%	
	2730	Heating, ventilating, air conditioning		2.91	4.12	6.15	3.30%	5.60%	7.25%	
	2900	Electrical		5.70	7.70	9.90	7.30%	8.50%	10.25%	
	3100	Total: Mechanical & Electrical		19.60	23.50	31	18.10%	22.50%	29%	
	9000	Per rental unit, total cost	Unit	71,000	83,000	92,500				
	9500	Total: Mechanical & Electrical	"	15,800	18,200	21,200				
50	0010	**HOUSING** Public (Low Rise)	S.F.	64	89	116				50
	0020	Total project costs	C.F.	5.70	7.10	8.85				
	0100	Site work	S.F.	8.15	11.75	19.05	8.35%	11.75%	16.50%	
	1800	Equipment		1.75	2.85	4.34	2.26%	3.03%	4.24%	
	2720	Plumbing		4.63	6.10	7.75	7.15%	9.05%	11.60%	
	2730	Heating, ventilating, air conditioning		2.33	4.51	4.94	4.26%	6.05%	6.45%	
	2900	Electrical		3.88	5.80	8	5.10%	6.55%	8.25%	
	3100	Total: Mechanical & Electrical		18.40	23.50	26.50	14.50%	17.55%	26.50%	
	9000	Per apartment, total cost	Apt.	70,500	80,000	100,500				
	9500	Total: Mechanical & Electrical	"	15,000	18,500	20,500				
51	0010	**ICE SKATING RINKS**	S.F.	57	128	141				51
	0020	Total project costs	C.F.	4.03	4.12	4.75				
	2720	Plumbing	S.F.	2.05	3.84	3.93	3.12%	3.23%	5.65%	
	2900	Electrical		5.85	9	9.55	6.30%	10.15%	15.05%	
	3100	Total: Mechanical & Electrical		9.80	13.80	17.25	18.95%	18.95%	18.95%	
52	0010	**JAILS**	S.F.	167	216	278				52
	0020	Total project costs	C.F.	15.75	21	26				
	1800	Equipment	S.F.	6.50	19.25	32.50	2.80%	5.55%	11.90%	
	2720	Plumbing		17	21.50	28.50	7%	8.90%	13.35%	
	2770	Heating, ventilating, air conditioning		15.05	20	39	7.50%	9.45%	17.75%	
	2900	Electrical		17.80	23	28.50	8.20%	11.55%	14.70%	

50 17 00 | S.F. Costs

		50 17 00 | S.F. Costs	UNIT	UNIT COSTS 1/4	UNIT COSTS MEDIAN	UNIT COSTS 3/4	% OF TOTAL 1/4	% OF TOTAL MEDIAN	% OF TOTAL 3/4
52	3100	Total: Mechanical & Electrical	S.F.	44.50	83	98.50	27.50%	30%	34%
53	0010	**LIBRARIES**	S.F.	106	132	174			
	0020	Total project costs	C.F.	7.20	9.05	11.55			
	0500	Masonry	S.F.	8.25	14.60	24.50	5.80%	8%	12.35%
	1800	Equipment		1.44	3.88	6.05	.41%	1.50%	4.16%
	2720	Plumbing		3.90	5.65	7.70	3.43%	4.60%	5.70%
	2770	Heating, ventilating, air conditioning		8.65	14.60	19	7.80%	10.95%	12.80%
	2900	Electrical		10.75	13.85	17.50	8.35%	10.40%	11.95%
	3100	Total: Mechanical & Electrical		32	40	50	20.50%	23%	26.50%
54	0010	**LIVING, ASSISTED**	S.F.	97.50	115	135			
	0020	Total project costs	C.F.	8.20	9.60	10.90			
	0500	Masonry	S.F.	2.87	3.42	4.02	2.37%	3.16%	3.86%
	1800	Equipment		2.22	2.58	3.30	2.12%	2.45%	2.66%
	2720	Plumbing		8.20	10.95	11.35	6.05%	8.15%	10.60%
	2770	Heating, ventilating, air conditioning		9.70	10.15	11.10	7.95%	9.35%	9.70%
	2900	Electrical		9.55	10.55	12.20	9%	10%	10.70%
	3100	Total: Mechanical & Electrical		27.50	32	36	26%	29%	31.50%
55	0010	**MEDICAL CLINICS**	S.F.	99	122	155			
	0020	Total project costs	C.F.	7.30	9.85	12.50			
	1800	Equipment	S.F.	2.69	5.65	8.80	1.10%	3.13%	6.35%
	2720	Plumbing		6.60	9.30	12.45	6.15%	8.40%	10.10%
	2770	Heating, ventilating, air conditioning		7.85	10.30	15.15	6.65%	8.85%	11.35%
	2900	Electrical		8.35	12.10	15.80	8.10%	10%	12.20%
	3100	Total: Mechanical & Electrical		27	37	51	22.50%	27%	33.50%
	3500	See also division 11700							
57	0010	**MEDICAL OFFICES**	S.F.	93.50	115	141			
	0020	Total project costs	C.F.	6.95	9.45	12.80			
	1800	Equipment	S.F.	3.25	6.10	8.65	.98%	5.10%	7.05%
	2720	Plumbing		5.15	7.95	10.75	5.60%	6.80%	8.50%
	2770	Heating, ventilating, air conditioning		6.25	9.15	11.90	6.15%	8.05%	9.70%
	2900	Electrical		7.50	10.85	15.20	7.65%	9.80%	11.70%
	3100	Total: Mechanical & Electrical		20	29	43	19.35%	23%	30.50%
59	0010	**MOTELS**	S.F.	59	87	113			
	0020	Total project costs	C.F.	5.25	7.05	11.50			
	2720	Plumbing	S.F.	6	7.60	9.10	9.45%	10.60%	12.55%
	2770	Heating, ventilating, air conditioning		3.64	5.45	9.75	5.60%	5.60%	10%
	2900	Electrical		5.60	7.15	9.10	7.45%	9.20%	10.80%
	3100	Total: Mechanical & Electrical		18.95	23.50	40.50	18.50%	21%	25.50%
	5000								
	9000	Per rental unit, total cost	Unit	30,000	57,000	61,500			
	9500	Total: Mechanical & Electrical	"	5,850	8,850	10,300			
60	0010	**NURSING HOMES**	S.F.	93.50	119	146			
	0020	Total project costs	C.F.	7.25	9.10	12.40			
	1800	Equipment	S.F.	2.91	3.86	6.45	2.02%	3.62%	4.99%
	2720	Plumbing		7.95	12	14.45	8.75%	10.10%	12.70%
	2770	Heating, ventilating, air conditioning		8.35	12.65	16.80	9.70%	11.45%	11.80%
	2900	Electrical		9.30	11.50	15.60	9.50%	10.60%	12.50%
	3100	Total: Mechanical & Electrical		22	30.50	51	26%	29.50%	30.50%
	9000	Per bed or person, total cost	Bed	41,000	51,500	66,000			
61	0010	**OFFICES** Low Rise (1 to 4 story)	S.F.	77.50	99.50	130			
	0020	Total project costs	C.F.	5.55	7.65	10.10			
	0100	Site work	S.F.	6.20	10.50	15.35	6.20%	9.70%	13.60%
	0500	Masonry	"	2.59	5.85	10.60	2.55%	5.40%	8%
	1800	Equipment	S.F.	.82	1.62	4.42	.77%	1.50%	3.69%
	2720	Plumbing		2.78	4.30	6.25	3.66%	4.50%	6.10%
	2770	Heating, ventilating, air conditioning		6.15	8.70	12.55	7.20%	10.40%	11.70%
	2900	Electrical		6.35	9.05	12.75	7.45%	9.65%	11.40%

50 17 00 | S.F. Costs

		50 17 00 | S.F. Costs	UNIT	UNIT COSTS 1/4	UNIT COSTS MEDIAN	UNIT COSTS 3/4	% OF TOTAL 1/4	% OF TOTAL MEDIAN	% OF TOTAL 3/4	
61	3100	Total: Mechanical & Electrical	S.F.	16.50	23	34.50	18.25%	22.50%	27%	61
62	0010	**OFFICES** Mid Rise (5 to 10 story)	S.F.	82.50	100	132				62
	0020	Total project costs	C.F.	5.85	7.45	10.55				
	2720	Plumbing	S.F.	2.49	3.86	5.55	2.83%	3.74%	4.50%	
	2770	Heating, ventilating, air conditioning		6.25	8.95	14.30	7.65%	9.40%	11%	
	2900	Electrical		6.10	7.85	10.85	6.35%	7.80%	10%	
	3100	Total: Mechanical & Electrical	↓	15.65	19.95	39	18.95%	21%	27.50%	
63	0010	**OFFICES** High Rise (11 to 20 story)	S.F.	101	128	157				63
	0020	Total project costs	C.F.	7.10	8.85	12.70				
	2900	Electrical	S.F.	6.15	7.50	11.15	5.80%	7.85%	10.50%	
	3100	Total: Mechanical & Electrical	"	19.80	26.50	45	16.90%	23.50%	34%	
64	0010	**POLICE STATIONS**	S.F.	122	156	199				64
	0020	Total project costs	C.F.	9.65	11.85	16.20				
	0500	Masonry	S.F.	11.40	20	26.50	6.70%	10.55%	11.35%	
	1800	Equipment		1.93	8.50	13.45	1.43%	4.07%	6.70%	
	2720	Plumbing		7	13.55	16.85	5.65%	6.90%	10.75%	
	2770	Heating, ventilating, air conditioning		10.60	14.10	19.15	5.85%	10.55%	11.70%	
	2900	Electrical		13.25	19.80	25	9.80%	11.85%	14.80%	
	3100	Total: Mechanical & Electrical	↓	43.50	52.50	71	28.50%	32%	32.50%	
65	0010	**POST OFFICES**	S.F.	96	118	151				65
	0020	Total project costs	C.F.	5.75	7.30	8.65				
	2720	Plumbing	S.F.	4.32	5.35	6.75	4.24%	5.30%	5.60%	
	2770	Heating, ventilating, air conditioning		6.75	8.35	9.25	6.65%	7.15%	9.35%	
	2900	Electrical		7.90	11.15	13.20	7.25%	9%	11%	
	3100	Total: Mechanical & Electrical	↓	23	30	34	16.25%	18.80%	22%	
66	0010	**POWER PLANTS**	S.F.	665	880	1,625				66
	0020	Total project costs	C.F.	18.30	40	85.50				
	2900	Electrical	S.F.	47	99.50	149	9.30%	12.75%	21.50%	
	8100	Total: Mechanical & Electrical	"	117	380	855	32.50%	32.50%	52.50%	
67	0010	**RELIGIOUS EDUCATION**	S.F.	77.50	101	119				67
	0020	Total project costs	C.F.	4.29	6.15	7.70				
	2720	Plumbing	S.F.	3.23	4.58	6.45	4.40%	5.30%	7.10%	
	2770	Heating, ventilating, air conditioning		8.15	9.25	13.05	10.05%	11.45%	12.35%	
	2900	Electrical		6.15	8.65	11.35	7.60%	9.10%	10.35%	
	3100	Total: Mechanical & Electrical	↓	25	32.50	38.50	22%	23%	26%	
69	0010	**RESEARCH** Laboratories and facilities	S.F.	115	166	242				69
	0020	Total project costs	C.F.	9.05	17.40	20.50				
	1800	Equipment	S.F.	5.15	10.10	24.50	.90%	4.58%	8.80%	
	2720	Plumbing		11.60	14.85	24	6.15%	8.30%	10.80%	
	2770	Heating, ventilating, air conditioning		10.40	35	41.50	7.25%	16.50%	17.50%	
	2900	Electrical		13.30	22.50	38	9.45%	11.15%	15.40%	
	3100	Total: Mechanical & Electrical	↓	41	78	111	29.50%	37%	45.50%	
70	0010	**RESTAURANTS**	S.F.	111	143	188				70
	0020	Total project costs	C.F.	9.45	12.35	16.25				
	1800	Equipment	S.F.	7.25	17.80	27	6.10%	13%	15.65%	
	2720	Plumbing		8.90	10.75	14.15	6.10%	8.15%	9%	
	2770	Heating, ventilating, air conditioning		11.30	15.60	20.50	9.20%	12%	12.40%	
	2900	Electrical		11.85	14.60	19	8.35%	10.55%	11.55%	
	3100	Total: Mechanical & Electrical	↓	36.50	38.50	50	19.25%	24%	29.50%	
	9000	Per seat unit, total cost	Seat	4,125	5,475	6,500				
	9500	Total: Mechanical & Electrical	"	1,025	1,375	1,625				
72	0010	**RETAIL STORES**	S.F.	52	70.50	92.50				72
	0020	Total project costs	C.F.	3.55	5.05	7.05				
	2720	Plumbing	S.F.	1.90	3.17	5.40	3.26%	4.60%	6.80%	
	2770	Heating, ventilating, air conditioning		4.10	5.60	8.45	6.75%	8.75%	10.15%	
	2900	Electrical		4.70	6.45	9.30	7.25%	9.90%	11.65%	
	3100	Total: Mechanical & Electrical	↓	12.05	16.05	21.50	17.05%	21%	23.50%	

50 17 00 | S.F. Costs

		50 17 00 | S.F. Costs	UNIT	UNIT COSTS 1/4	UNIT COSTS MEDIAN	UNIT COSTS 3/4	% OF TOTAL 1/4	% OF TOTAL MEDIAN	% OF TOTAL 3/4	
74	0010	**SCHOOLS** Elementary	S.F.	84.50	104	128				74
	0020	Total project costs	C.F.	5.60	7.15	9.25				
	0500	Masonry	S.F.	7.55	12.50	18.65	5.45%	10.65%	14.70%	
	1800	Equipment		2.49	4.23	7.80	1.90%	3.33%	5%	
	2720	Plumbing		4.90	6.95	9.30	5.70%	7.15%	9.35%	
	2730	Heating, ventilating, air conditioning		7.35	11.75	16.40	8.15%	10.80%	15.20%	
	2900	Electrical		7.95	10.40	13.05	8.40%	10.05%	11.70%	
	3100	Total: Mechanical & Electrical	↓	28	35	42.50	24%	27.50%	30%	
	9000	Per pupil, total cost	Ea.	9,825	14,600	43,200				
	9500	Total: Mechanical & Electrical	"	2,775	3,500	12,500				
76	0010	**SCHOOLS** Junior High & Middle	S.F.	86	108	126				76
	0020	Total project costs	C.F.	5.60	7.25	8.10				
	0500	Masonry	S.F.	9.45	13.45	16.20	8%	11.60%	14.30%	
	1800	Equipment		2.81	4.53	7	1.81%	3.26%	5.15%	
	2720	Plumbing		5.60	6.35	8.20	5.50%	6.90%	8.15%	
	2770	Heating, ventilating, air conditioning		6.45	12.40	17.30	8.75%	12.75%	17.45%	
	2900	Electrical		8.50	10.35	13.30	7.90%	9.25%	10.60%	
	3100	Total: Mechanical & Electrical	↓	25.50	36	44	22.50%	25.50%	29.50%	
	9000	Per pupil, total cost	Ea.	11,200	15,100	19,700				
78	0010	**SCHOOLS** Senior High	S.F.	90	110	139				78
	0020	Total project costs	C.F.	5.80	8.15	13.50				
	1800	Equipment	S.F.	2.41	5.70	8.40	1.88%	3.29%	4.80%	
	2720	Plumbing		5.25	7.80	14.10	5.70%	7%	8.35%	
	2770	Heating, ventilating, air conditioning		10.50	12.05	23	8.95%	11.60%	15%	
	2900	Electrical		9.10	11.70	18.75	8.35%	10.10%	12.35%	
	3100	Total: Mechanical & Electrical	↓	31	35.50	60	23%	26.50%	28.50%	
	9000	Per pupil, total cost	Ea.	8,650	17,600	22,000				
80	0010	**SCHOOLS** Vocational	S.F.	74	105	132				80
	0020	Total project costs	C.F.	4.61	6.60	9.15				
	0500	Masonry	S.F.	4.36	10.75	16.45	3.53%	4.61%	10.95%	
	1800	Equipment	"	2.22	3.15	8	1.24%	3.13%	4.68%	
	2720	Plumbing	S.F.	4.75	7.25	10.85	5.40%	6.90%	8.55%	
	2770	Heating, ventilating, air conditioning		6.65	12.35	20.50	8.60%	11.90%	14.65%	
	2900	Electrical		7.75	10.55	14.90	8.45%	10.95%	13.20%	
	3100	Total: Mechanical & Electrical	↓	27	29.50	51	23.50%	29.50%	31%	
	9000	Per pupil, total cost	Ea.	10,300	27,700	41,300				
83	0010	**SPORTS ARENAS**	S.F.	65	86.50	134				83
	0020	Total project costs	C.F.	3.52	6.30	8.15				
	2720	Plumbing	S.F.	3.76	5.70	12.05	4.35%	6.35%	9.40%	
	2770	Heating, ventilating, air conditioning		8.10	9.60	13.30	8.80%	10.20%	13.55%	
	2900	Electrical		6.75	9.15	11.85	8.60%	9.90%	12.25%	
	3100	Total: Mechanical & Electrical	↓	16.80	30	39	21.50%	25%	27.50%	
85	0010	**SUPERMARKETS**	S.F.	60	69.50	84				85
	0020	Total project costs	C.F.	3.34	4.03	6.10				
	2720	Plumbing	S.F.	3.35	4.21	4.91	5.40%	6%	7.45%	
	2770	Heating, ventilating, air conditioning		4.92	6.55	7.95	8.60%	8.65%	9.60%	
	2900	Electrical		7.50	8.60	10.20	10.40%	12.45%	13.60%	
	3100	Total: Mechanical & Electrical	↓	19.25	21	29	20.50%	26.50%	31%	
86	0010	**SWIMMING POOLS**	S.F.	97	163	345				86
	0020	Total project costs	C.F.	7.80	9.70	10.55				
	2720	Plumbing	S.F.	8.95	10.25	14.20	4.80%	9.70%	20.50%	
	2900	Electrical		7.30	11.80	17.20	6.50%	7.25%	7.60%	
	3100	Total: Mechanical & Electrical	↓	17.80	45	61.50	11.15%	14.10%	23.50%	
87	0010	**TELEPHONE EXCHANGES**	S.F.	129	189	239				87
	0020	Total project costs	C.F.	8	12.85	17.70				
	2720	Plumbing	S.F.	5.45	8.60	12.30	4.52%	5.80%	6.90%	
	2770	Heating, ventilating, air conditioning	↓	12.60	25.50	31.50	11.80%	16.05%	18.40%	

		50 17 00 | S.F. Costs	UNIT	UNIT COSTS 1/4	UNIT COSTS MEDIAN	UNIT COSTS 3/4	% OF TOTAL 1/4	% OF TOTAL MEDIAN	% OF TOTAL 3/4	
87	2900	Electrical	S.F.	13.10	21	37	10.90%	14%	17.85%	87
	3100	Total: Mechanical & Electrical	↓	38.50	73.50	104	29.50%	33.50%	44.50%	
91	0010	**THEATERS**	S.F.	81	101	153				91
	0020	Total project costs	C.F.	3.74	5.55	8.15				
	2720	Plumbing	S.F.	2.70	2.93	11.95	2.92%	4.70%	6.80%	
	2770	Heating, ventilating, air conditioning		7.85	9.55	11.80	8%	12.25%	13.40%	
	2900	Electrical		7.10	9.55	19.45	8.05%	9.95%	12.25%	
	3100	Total: Mechanical & Electrical	↓	18.25	27	56	23%	26.50%	27.50%	
94	0010	**TOWN HALLS** City Halls & Municipal Buildings	S.F.	95	121	153				94
	0020	Total project costs	C.F.	8.25	10.15	13.90				
	2720	Plumbing	S.F.	3.77	7.05	13	4.31%	5.95%	7.95%	
	2770	Heating, ventilating, air conditioning		6.85	13.55	19.80	7.05%	9.05%	13.45%	
	2900	Electrical		8.60	12.55	16.75	8.05%	9.50%	12.05%	
	3100	Total: Mechanical & Electrical	↓	29	34	45	22%	26.50%	31%	
97	0010	**WAREHOUSES** And Storage Buildings	S.F.	35.50	51	72				97
	0020	Total project costs	C.F.	1.94	2.77	4.57				
	0100	Site work	S.F.	3.48	6.90	10.45	6.05%	12.95%	19.85%	
	0500	Masonry		2.10	4.79	10.35	3.73%	7.40%	12.30%	
	1800	Equipment		.54	1.17	6.55	.91%	1.82%	5.55%	
	2720	Plumbing		1.12	2.02	3.77	2.90%	4.80%	6.55%	
	2730	Heating, ventilating, air conditioning		1.35	3.62	4.86	2.41%	5%	8.90%	
	2900	Electrical		2.05	3.76	6.25	5.15%	7.20%	10.10%	
	3100	Total: Mechanical & Electrical	↓	5.55	8.55	18.70	12.75%	18.90%	26%	
99	0010	**WAREHOUSE & OFFICES** Combination	S.F.	41.50	55.50	74.50				99
	0020	Total project costs	C.F.	2.12	3.08	4.56				
	1800	Equipment	S.F.	.72	1.39	2.15	.52%	1.21%	2.40%	
	2720	Plumbing		1.60	2.84	4.16	3.74%	4.76%	6.30%	
	2770	Heating, ventilating, air conditioning		2.53	3.96	5.55	5%	5.65%	10.05%	
	2900	Electrical		2.78	4.13	6.50	5.85%	8%	10%	
	3100	Total: Mechanical & Electrical	↓	7.75	11.80	18.85	14.40%	19.95%	24.50%	

Square Foot Project Size Modifier

One factor that affects the S.F. cost of a particular building is the size. In general, for buildings built to the same specifications in the same locality, the larger building will have the lower S.F. cost. This is due mainly to the decreasing contribution of the exterior walls plus the economy of scale usually achievable in larger buildings. The Area Conversion Scale shown below will give a factor to convert costs for the typical size building to an adjusted cost for the particular project.

Example: Determine the cost per S.F. for a 100,000 S.F. Mid-rise apartment building.

$$\frac{\text{Proposed building area} = 100{,}000 \text{ S.F.}}{\text{Typical size from below} = 50{,}000 \text{ S.F.}} = 2.00$$

Enter Area Conversion scale at 2.0, intersect curve, read horizontally the appropriate cost multiplier of .94. Size adjusted cost becomes .94 x $90.00 = $84.60 based on national average costs.

Note: For Size Factors less than .50, the Cost Multiplier is 1.1
For Size Factors greater than 3.5, the Cost Multiplier is .90

The Square Foot Base Size lists the median costs, most typical project size in our accumulated data and the range in size of the projects.

The Size Factor for your project is determined by dividing your project area in S.F. by the typical project size for the particular Building Type. With this factor, enter the Area Conversion Scale at the appropriate Size Factor and determine the appropriate cost multiplier for your building size.

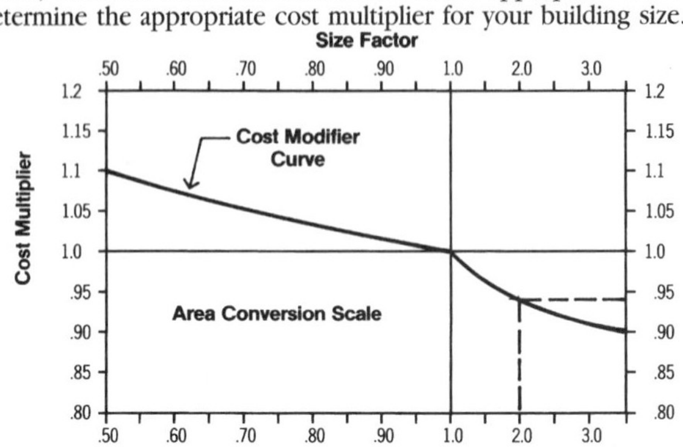

Square Foot Base Size

Building Type	Median Cost per S.F.	Typical Size Gross S.F.	Typical Range Gross S.F.	Building Type	Median Cost per S.F.	Typical Size Gross S.F.	Typical Range Gross S.F.
Apartments, Low Rise	$ 71.00	21,000	9,700 - 37,200	Jails	$216.00	40,000	5,500 - 145,000
Apartments, Mid Rise	90.00	50,000	32,000 - 100,000	Libraries	132.00	12,000	7,000 - 31,000
Apartments, High Rise	102.00	145,000	95,000 - 600,000	Living, Assisted	115.00	32,300	23,500 - 50,300
Auditoriums	119.00	25,000	7,600 - 39,000	Medical Clinics	122.00	7,200	4,200 - 15,700
Auto Sales	88.00	20,000	10,800 - 28,600	Medical Offices	115.00	6,000	4,000 - 15,000
Banks	158.00	4,200	2,500 - 7,500	Motels	87.00	40,000	15,800 - 120,000
Churches	109.00	17,000	2,000 - 42,000	Nursing Homes	119.00	23,000	15,000 - 37,000
Clubs, Country	111.00	6,500	4,500 - 15,000	Offices, Low Rise	99.50	20,000	5,000 - 80,000
Clubs, Social	105.00	10,000	6,000 - 13,500	Offices, Mid Rise	100.00	120,000	20,000 - 300,000
Clubs, YMCA	125.00	28,300	12,800 - 39,400	Offices, High Rise	128.00	260,000	120,000 - 800,000
Colleges (Class)	136.00	50,000	15,000 - 150,000	Police Stations	156.00	10,500	4,000 - 19,000
Colleges (Science Lab)	203.00	45,600	16,600 - 80,000	Post Offices	118.00	12,400	6,800 - 30,000
College (Student Union)	155.00	33,400	16,000 - 85,000	Power Plants	880.00	7,500	1,000 - 20,000
Community Center	113.00	9,400	5,300 - 16,700	Religious Education	101.00	9,000	6,000 - 12,000
Court Houses	151.00	32,400	17,800 - 106,000	Research	166.00	19,000	6,300 - 45,000
Dept. Stores	66.00	90,000	44,000 - 122,000	Restaurants	143.00	4,400	2,800 - 6,000
Dormitories, Low Rise	117.00	25,000	10,000 - 95,000	Retail Stores	70.50	7,200	4,000 - 17,600
Dormitories, Mid Rise	148.00	85,000	20,000 - 200,000	Schools, Elementary	104.00	41,000	24,500 - 55,000
Factories	64.00	26,400	12,900 - 50,000	Schools, Jr. High	108.00	92,000	52,000 - 119,000
Fire Stations	116.00	5,800	4,000 - 8,700	Schools, Sr. High	110.00	101,000	50,500 - 175,000
Fraternity Houses	110.00	12,500	8,200 - 14,800	Schools, Vocational	105.00	37,000	20,500 - 82,000
Funeral Homes	122.00	10,000	4,000 - 20,000	Sports Arenas	86.50	15,000	5,000 - 40,000
Garages, Commercial	80.00	9,300	5,000 - 13,600	Supermarkets	69.50	44,000	12,000 - 60,000
Garages, Municipal	99.50	8,300	4,500 - 12,600	Swimming Pools	163.00	20,000	10,000 - 32,000
Garages, Parking	42.50	163,000	76,400 - 225,300	Telephone Exchange	189.00	4,500	1,200 - 10,600
Gymnasiums	107.00	19,200	11,600 - 41,000	Theaters	101.00	10,500	8,800 - 17,500
Hospitals	190.00	55,000	27,200 - 125,000	Town Halls	121.00	10,800	4,800 - 23,400
House (Elderly)	96.50	37,000	21,000 - 66,000	Warehouses	51.00	25,000	8,000 - 72,000
Housing (Public)	89.00	36,000	14,400 - 74,400	Warehouse & Office	55.50	25,000	8,000 - 72,000
Ice Rinks	128.00	29,000	27,200 - 33,600				

Abbreviations

A	Area Square Feet; Ampere
ABS	Acrylonitrile Butadiene Stryrene; Asbestos Bonded Steel
A.C.	Alternating Current; Air-Conditioning; Asbestos Cement; Plywood Grade A & C
A.C.I.	American Concrete Institute
AD	Plywood, Grade A & D
Addit.	Additional
Adj.	Adjustable
af	Audio-frequency
A.G.A.	American Gas Association
Agg.	Aggregate
A.H.	Ampere Hours
A hr.	Ampere-hour
A.H.U.	Air Handling Unit
A.I.A.	American Institute of Architects
AIC	Ampere Interrupting Capacity
Allow.	Allowance
alt.	Altitude
Alum.	Aluminum
a.m.	Ante Meridiem
Amp.	Ampere
Anod.	Anodized
Approx.	Approximate
Apt.	Apartment
Asb.	Asbestos
A.S.B.C.	American Standard Building Code
Asbe.	Asbestos Worker
A.S.H.R.A.E.	American Society of Heating, Refrig. & AC Engineers
A.S.M.E.	American Society of Mechanical Engineers
A.S.T.M.	American Society for Testing and Materials
Attchmt.	Attachment
Avg.	Average
A.W.G.	American Wire Gauge
AWWA	American Water Works Assoc.
Bbl.	Barrel
B&B	Grade B and Better; Balled & Burlapped
B.&S.	Bell and Spigot
B.&W.	Black and White
b.c.c.	Body-centered Cubic
B.C.Y.	Bank Cubic Yards
BE	Bevel End
B.F.	Board Feet
Bg. cem.	Bag of Cement
BHP	Boiler Horsepower; Brake Horsepower
B.I.	Black Iron
Bit.; Bitum.	Bituminous
Bk.	Backed
Bkrs.	Breakers
Bldg.	Building
Blk.	Block
Bm.	Beam
Boil.	Boilermaker
B.P.M.	Blows per Minute
BR	Bedroom
Brg.	Bearing
Brhe.	Bricklayer Helper
Bric.	Bricklayer
Brk.	Brick
Brng.	Bearing
Brs.	Brass
Brz.	Bronze
Bsn.	Basin
Btr.	Better
BTU	British Thermal Unit
BTUH	BTU per Hour
B.U.R.	Built-up Roofing
BX	Interlocked Armored Cable
c	Conductivity, Copper Sweat
C	Hundred; Centigrade
C/C	Center to Center, Cedar on Cedar
Cab.	Cabinet
Cair.	Air Tool Laborer
Calc	Calculated
Cap.	Capacity
Carp.	Carpenter
C.B.	Circuit Breaker
C.C.A.	Chromate Copper Arsenate
C.C.F.	Hundred Cubic Feet
cd	Candela
cd/sf	Candela per Square Foot
CD	Grade of Plywood Face & Back
CDX	Plywood, Grade C & D, exterior glue
Cefi.	Cement Finisher
Cem.	Cement
CF	Hundred Feet
C.F.	Cubic Feet
CFM	Cubic Feet per Minute
c.g.	Center of Gravity
CHW	Chilled Water; Commercial Hot Water
C.I.	Cast Iron
C.I.P.	Cast in Place
Circ.	Circuit
C.L.	Carload Lot
Clab.	Common Laborer
Clam	Common maintenance laborer
C.L.F.	Hundred Linear Feet
CLF	Current Limiting Fuse
CLP	Cross Linked Polyethylene
cm	Centimeter
CMP	Corr. Metal Pipe
C.M.U.	Concrete Masonry Unit
CN	Change Notice
Col.	Column
CO_2	Carbon Dioxide
Comb.	Combination
Compr.	Compressor
Conc.	Concrete
Cont.	Continuous; Continued
Corr.	Corrugated
Cos	Cosine
Cot	Cotangent
Cov.	Cover
C/P	Cedar on Paneling
CPA	Control Point Adjustment
Cplg.	Coupling
C.P.M.	Critical Path Method
CPVC	Chlorinated Polyvinyl Chloride
C.Pr.	Hundred Pair
CRC	Cold Rolled Channel
Creos.	Creosote
Crpt.	Carpet & Linoleum Layer
CRT	Cathode-ray Tube
CS	Carbon Steel, Constant Shear Bar Joist
Csc	Cosecant
C.S.F.	Hundred Square Feet
CSI	Construction Specifications Institute
C.T.	Current Transformer
CTS	Copper Tube Size
Cu	Copper, Cubic
Cu. Ft.	Cubic Foot
cw	Continuous Wave
C.W.	Cool White; Cold Water
Cwt.	100 Pounds
C.W.X.	Cool White Deluxe
C.Y.	Cubic Yard (27 cubic feet)
C.Y./Hr.	Cubic Yard per Hour
Cyl.	Cylinder
d	Penny (nail size)
D	Deep; Depth; Discharge
Dis.;Disch.	Discharge
Db.	Decibel
Dbl.	Double
DC	Direct Current
DDC	Direct Digital Control
Demob.	Demobilization
d.f.u.	Drainage Fixture Units
D.H.	Double Hung
DHW	Domestic Hot Water
Diag.	Diagonal
Diam.	Diameter
Distrib.	Distribution
Dk.	Deck
D.L.	Dead Load; Diesel
DLH	Deep Long Span Bar Joist
Do.	Ditto
Dp.	Depth
D.P.S.T.	Double Pole, Single Throw
Dr.	Driver
Drink.	Drinking
D.S.	Double Strength
D.S.A.	Double Strength A Grade
D.S.B.	Double Strength B Grade
Dty.	Duty
DWV	Drain Waste Vent
DX	Deluxe White, Direct Expansion
dyn	Dyne
e	Eccentricity
E	Equipment Only; East
Ea.	Each
E.B.	Encased Burial
Econ.	Economy
E.C.Y	Embankment Cubic Yards
EDP	Electronic Data Processing
EIFS	Exterior Insulation Finish System
E.D.R.	Equiv. Direct Radiation
Eq.	Equation
Elec.	Electrician; Electrical
Elev.	Elevator; Elevating
EMT	Electrical Metallic Conduit; Thin Wall Conduit
Eng.	Engine, Engineered
EPDM	Ethylene Propylene Diene Monomer
EPS	Expanded Polystyrene
Eqhv.	Equip. Oper., Heavy
Eqlt.	Equip. Oper., Light
Eqmd.	Equip. Oper., Medium
Eqmm.	Equip. Oper., Master Mechanic
Eqol.	Equip. Oper., Oilers
Equip.	Equipment
ERW	Electric Resistance Welded
E.S.	Energy Saver
Est.	Estimated
esu	Electrostatic Units
E.W.	Each Way
EWT	Entering Water Temperature
Excav.	Excavation
Exp.	Expansion, Exposure
Ext.	Exterior
Extru.	Extrusion
f.	Fiber stress
F	Fahrenheit; Female; Fill
Fab.	Fabricated
FBGS	Fiberglass
F.C.	Footcandles
f.c.c.	Face-centered Cubic
f'c.	Compressive Stress in Concrete; Extreme Compressive Stress
F.E.	Front End
FEP	Fluorinated Ethylene Propylene (Teflon)
F.G.	Flat Grain
F.H.A.	Federal Housing Administration
Fig.	Figure
Fin.	Finished
Fixt.	Fixture
Fl. Oz.	Fluid Ounces
Flr.	Floor
F.M.	Frequency Modulation; Factory Mutual
Fmg.	Framing
Fndtn.	Foundation

Abbreviations

Fori.	Foreman, Inside
Foro.	Foreman, Outside
Fount.	Fountain
FPM	Feet per Minute
FPT	Female Pipe Thread
Fr.	Frame
F.R.	Fire Rating
FRK	Foil Reinforced Kraft
FRP	Fiberglass Reinforced Plastic
FS	Forged Steel
FSC	Cast Body; Cast Switch Box
Ft.	Foot; Feet
Ftng.	Fitting
Ftg.	Footing
Ft. Lb.	Foot Pound
Furn.	Furniture
FVNR	Full Voltage Non-Reversing
FXM	Female by Male
Fy.	Minimum Yield Stress of Steel
g	Gram
G	Gauss
Ga.	Gauge
Gal.	Gallon
Gal./Min.	Gallon per Minute
Galv.	Galvanized
Gen.	General
G.F.I.	Ground Fault Interrupter
Glaz.	Glazier
GPD	Gallons per Day
GPH	Gallons per Hour
GPM	Gallons per Minute
GR	Grade
Gran.	Granular
Grnd.	Ground
H	High; High Strength Bar Joist; Henry
H.C.	High Capacity
H.D.	Heavy Duty; High Density
H.D.O.	High Density Overlaid
Hdr.	Header
Hdwe.	Hardware
Help.	Helper Average
HEPA	High Efficiency Particulate Air Filter
Hg	Mercury
HIC	High Interrupting Capacity
HM	Hollow Metal
H.O.	High Output
Horiz.	Horizontal
H.P.	Horsepower; High Pressure
H.P.F.	High Power Factor
Hr.	Hour
Hrs./Day	Hours per Day
HSC	High Short Circuit
Ht.	Height
Htg.	Heating
Htrs.	Heaters
HVAC	Heating, Ventilation & Air-Conditioning
Hvy.	Heavy
HW	Hot Water
Hyd.;Hydr.	Hydraulic
Hz.	Hertz (cycles)
I.	Moment of Inertia
I.C.	Interrupting Capacity
ID	Inside Diameter
I.D.	Inside Dimension; Identification
I.F.	Inside Frosted
I.M.C.	Intermediate Metal Conduit
In.	Inch
Incan.	Incandescent
Incl.	Included; Including
Int.	Interior
Inst.	Installation
Insul.	Insulation/Insulated
I.P.	Iron Pipe
I.P.S.	Iron Pipe Size
I.P.T.	Iron Pipe Threaded
I.W.	Indirect Waste
J	Joule
J.I.C.	Joint Industrial Council
K	Thousand; Thousand Pounds; Heavy Wall Copper Tubing, Kelvin
K.A.H.	Thousand Amp. Hours
KCMIL	Thousand Circular Mils
KD	Knock Down
K.D.A.T.	Kiln Dried After Treatment
kg	Kilogram
kG	Kilogauss
kgf	Kilogram Force
kHz	Kilohertz
Kip.	1000 Pounds
KJ	Kiljoule
K.L.	Effective Length Factor
K.L.F.	Kips per Linear Foot
Km	Kilometer
K.S.F.	Kips per Square Foot
K.S.I.	Kips per Square Inch
kV	Kilovolt
kVA	Kilovolt Ampere
K.V.A.R.	Kilovar (Reactance)
KW	Kilowatt
KWh	Kilowatt-hour
L	Labor Only; Length; Long; Medium Wall Copper Tubing
Lab.	Labor
lat	Latitude
Lath.	Lather
Lav.	Lavatory
lb.; #	Pound
L.B.	Load Bearing; L Conduit Body
L. & E.	Labor & Equipment
lb./hr.	Pounds per Hour
lb./L.F.	Pounds per Linear Foot
lbf/sq.in.	Pound-force per Square Inch
L.C.L.	Less than Carload Lot
L.C.Y.	Loose Cubic Yard
Ld.	Load
LE	Lead Equivalent
LED	Light Emitting Diode
L.F.	Linear Foot
Lg.	Long; Length; Large
L & H	Light and Heat
LH	Long Span Bar Joist
L.H.	Labor Hours
L.L.	Live Load
L.L.D.	Lamp Lumen Depreciation
lm	Lumen
lm/sf	Lumen per Square Foot
lm/W	Lumen per Watt
L.O.A.	Length Over All
log	Logarithm
L-O-L	Lateralolet
L.P.	Liquefied Petroleum; Low Pressure
L.P.F.	Low Power Factor
LR	Long Radius
L.S.	Lump Sum
Lt.	Light
Lt. Ga.	Light Gauge
L.T.L.	Less than Truckload Lot
Lt. Wt.	Lightweight
L.V.	Low Voltage
M	Thousand; Material; Male; Light Wall Copper Tubing
M^2CA	Meters Squared Contact Area
m/hr; M.H.	Man-hour
mA	Milliampere
Mach.	Machine
Mag. Str.	Magnetic Starter
Maint.	Maintenance
Marb.	Marble Setter
Mat; Mat'l.	Material
Max.	Maximum
MBF	Thousand Board Feet
MBH	Thousand BTU's per hr.
MC	Metal Clad Cable
M.C.F.	Thousand Cubic Feet
M.C.F.M.	Thousand Cubic Feet per Minute
M.C.M.	Thousand Circular Mils
M.C.P.	Motor Circuit Protector
MD	Medium Duty
M.D.O.	Medium Density Overlaid
Med.	Medium
MF	Thousand Feet
M.F.B.M.	Thousand Feet Board Measure
Mfg.	Manufacturing
Mfrs.	Manufacturers
mg	Milligram
MGD	Million Gallons per Day
MGPH	Thousand Gallons per Hour
MH, M.H.	Manhole; Metal Halide; Man-Hour
MHz	Megahertz
Mi.	Mile
MI	Malleable Iron; Mineral Insulated
mm	Millimeter
Mill.	Millwright
Min., min.	Minimum, minute
Misc.	Miscellaneous
ml	Milliliter, Mainline
M.L.F.	Thousand Linear Feet
Mo.	Month
Mobil.	Mobilization
Mog.	Mogul Base
MPH	Miles per Hour
MPT	Male Pipe Thread
MRT	Mile Round Trip
ms	Millisecond
M.S.F.	Thousand Square Feet
Mstz.	Mosaic & Terrazzo Worker
M.S.Y.	Thousand Square Yards
Mtd.	Mounted
Mthe.	Mosaic & Terrazzo Helper
Mtng.	Mounting
Mult.	Multi; Multiply
M.V.A.	Million Volt Amperes
M.V.A.R.	Million Volt Amperes Reactance
MV	Megavolt
MW	Megawatt
MXM	Male by Male
MYD	Thousand Yards
N	Natural; North
nA	Nanoampere
NA	Not Available; Not Applicable
N.B.C.	National Building Code
NC	Normally Closed
N.E.M.A.	National Electrical Manufacturers Assoc.
NEHB	Bolted Circuit Breaker to 600V.
N.L.B.	Non-Load-Bearing
NM	Non-Metallic Cable
nm	Nanometer
No.	Number
NO	Normally Open
N.O.C.	Not Otherwise Classified
Nose.	Nosing
N.P.T.	National Pipe Thread
NQOD	Combination Plug-on/Bolt on Circuit Breaker to 240V.
N.R.C.	Noise Reduction Coefficient
N.R.S.	Non Rising Stem
ns	Nanosecond
nW	Nanowatt
OB	Opposing Blade
OC	On Center
OD	Outside Diameter
O.D.	Outside Dimension
ODS	Overhead Distribution System
O.G.	Ogee
O.H.	Overhead
O&P	Overhead and Profit
Oper.	Operator
Opng.	Opening
Orna.	Ornamental
OSB	Oriented Strand Board

Abbreviations

O.S.&Y.	Outside Screw and Yoke
Ovhd.	Overhead
OWG	Oil, Water or Gas
Oz.	Ounce
P.	Pole; Applied Load; Projection
p.	Page
Pape.	Paperhanger
P.A.P.R.	Powered Air Purifying Respirator
PAR	Parabolic Reflector
Pc., Pcs.	Piece, Pieces
P.C.	Portland Cement; Power Connector
P.C.F.	Pounds per Cubic Foot
P.C.M.	Phase Contrast Microscopy
P.E.	Professional Engineer; Porcelain Enamel; Polyethylene; Plain End
Perf.	Perforated
Ph.	Phase
P.I.	Pressure Injected
Pile.	Pile Driver
Pkg.	Package
Pl.	Plate
Plah.	Plasterer Helper
Plas.	Plasterer
Pluh.	Plumbers Helper
Plum.	Plumber
Ply.	Plywood
p.m.	Post Meridiem
Pntd.	Painted
Pord.	Painter, Ordinary
pp	Pages
PP; PPL	Polypropylene
P.P.M.	Parts per Million
Pr.	Pair
P.E.S.B.	Pre-engineered Steel Building
Prefab.	Prefabricated
Prefin.	Prefinished
Prop.	Propelled
PSF; psf	Pounds per Square Foot
PSI; psi	Pounds per Square Inch
PSIG	Pounds per Square Inch Gauge
PSP	Plastic Sewer Pipe
Pspr.	Painter, Spray
Psst.	Painter, Structural Steel
P.T.	Potential Transformer
P. & T.	Pressure & Temperature
Ptd.	Painted
Ptns.	Partitions
Pu	Ultimate Load
PVC	Polyvinyl Chloride
Pvmt.	Pavement
Pwr.	Power
Q	Quantity Heat Flow
Quan.;Qty.	Quantity
Q.C.	Quick Coupling
r	Radius of Gyration
R	Resistance
R.C.P.	Reinforced Concrete Pipe
Rect.	Rectangle
Reg.	Regular
Reinf.	Reinforced
Req'd.	Required
Res.	Resistant
Resi.	Residential
Rgh.	Rough
RGS	Rigid Galvanized Steel
R.H.W.	Rubber, Heat & Water Resistant; Residential Hot Water
rms	Root Mean Square
Rnd.	Round
Rodm.	Rodman
Rofc.	Roofer, Composition
Rofp.	Roofer, Precast
Rohe.	Roofer Helpers (Composition)
Rots.	Roofer, Tile & Slate
R.O.W.	Right of Way
RPM	Revolutions per Minute
R.S.	Rapid Start
Rsr	Riser
RT	Round Trip
S.	Suction; Single Entrance; South
SC	Screw Cover
SCFM	Standard Cubic Feet per Minute
Scaf.	Scaffold
Sch.; Sched.	Schedule
S.C.R.	Modular Brick
S.D.	Sound Deadening
S.D.R.	Standard Dimension Ratio
S.E.	Surfaced Edge
Sel.	Select
S.E.R.; S.E.U.	Service Entrance Cable
S.F.	Square Foot
S.F.C.A.	Square Foot Contact Area
S.F. Flr.	Square Foot of Floor
S.F.G.	Square Foot of Ground
S.F. Hor.	Square Foot Horizontal
S.F.R.	Square Feet of Radiation
S.F. Shlf.	Square Foot of Shelf
S4S	Surface 4 Sides
Shee.	Sheet Metal Worker
Sin.	Sine
Skwk.	Skilled Worker
SL	Saran Lined
S.L.	Slimline
Sldr.	Solder
SLH	Super Long Span Bar Joist
S.N.	Solid Neutral
S-O-L	Socketolet
sp	Standpipe
S.P.	Static Pressure; Single Pole; Self-Propelled
Spri.	Sprinkler Installer
spwg	Static Pressure Water Gauge
S.P.D.T.	Single Pole, Double Throw
SPF	Spruce Pine Fir
S.P.S.T.	Single Pole, Single Throw
SPT	Standard Pipe Thread
Sq.	Square; 100 Square Feet
Sq. Hd.	Square Head
Sq. In.	Square Inch
S.S.	Single Strength; Stainless Steel
S.S.B.	Single Strength B Grade
sst	Stainless Steel
Sswk.	Structural Steel Worker
Sswl.	Structural Steel Welder
St.;Stl.	Steel
S.T.C.	Sound Transmission Coefficient
Std.	Standard
STK	Select Tight Knot
STP	Standard Temperature & Pressure
Stpi.	Steamfitter, Pipefitter
Str.	Strength; Starter; Straight
Strd.	Stranded
Struct.	Structural
Sty.	Story
Subj.	Subject
Subs.	Subcontractors
Surf.	Surface
Sw.	Switch
Swbd.	Switchboard
S.Y.	Square Yard
Syn.	Synthetic
S.Y.P.	Southern Yellow Pine
Sys.	System
t.	Thickness
T	Temperature; Ton
Tan	Tangent
T.C.	Terra Cotta
T & C	Threaded and Coupled
T.D.	Temperature Difference
Tdd	Telecommunications Device for the Deaf
T.E.M.	Transmission Electron Microscopy
TFE	Tetrafluoroethylene (Teflon)
T. & G.	Tongue & Groove; Tar & Gravel
Th.;Thk.	Thick
Thn.	Thin
Thrded	Threaded
Tilf.	Tile Layer, Floor
Tilh.	Tile Layer, Helper
THHN	Nylon Jacketed Wire
THW.	Insulated Strand Wire
THWN;	Nylon Jacketed Wire
T.L.	Truckload
T.M.	Track Mounted
Tot.	Total
T-O-L	Threadolet
T.S.	Trigger Start
Tr.	Trade
Transf.	Transformer
Trhv.	Truck Driver, Heavy
Trlr	Trailer
Trlt.	Truck Driver, Light
TTY	Teletypewriter
TV	Television
T.W.	Thermoplastic Water Resistant Wire
UCI	Uniform Construction Index
UF	Underground Feeder
UGND	Underground Feeder
U.H.F.	Ultra High Frequency
U.L.	Underwriters Laboratory
Unfin.	Unfinished
URD	Underground Residential Distribution
US	United States
USP	United States Primed
UTP	Unshielded Twisted Pair
V	Volt
V.A.	Volt Amperes
V.C.T.	Vinyl Composition Tile
VAV	Variable Air Volume
VC	Veneer Core
Vent.	Ventilation
Vert.	Vertical
V.F.	Vinyl Faced
V.G.	Vertical Grain
V.H.F.	Very High Frequency
VHO	Very High Output
Vib.	Vibrating
V.L.F.	Vertical Linear Foot
Vol.	Volume
VRP	Vinyl Reinforced Polyester
W	Wire; Watt; Wide; West
w/	With
W.C.	Water Column; Water Closet
W.F.	Wide Flange
W.G.	Water Gauge
Wldg.	Welding
W. Mile	Wire Mile
W-O-L	Weldolet
W.R.	Water Resistant
Wrck.	Wrecker
W.S.P.	Water, Steam, Petroleum
WT., Wt.	Weight
WWF	Welded Wire Fabric
XFER	Transfer
XFMR	Transformer
XHD	Extra Heavy Duty
XHHW; XLPE	Cross-Linked Polyethylene Wire Insulation
XLP	Cross-linked Polyethylene
Y	Wye
yd	Yard
yr	Year
Δ	Delta
%	Percent
~	Approximately
Ø	Phase
@	At
#	Pound; Number
<	Less Than
>	Greater Than

A

B

C

D

E

F

G

H

I

J

K

L

M

INDEX

T

U

V

W

X

Y

Z

Notes

Notes

Notes

Notes

Notes

Division Notes

		CREW	DAILY OUTPUT	LABOR-HOURS	UNIT	2006 BARE COSTS MAT.	LABOR	EQUIP.	TOTAL	TOTAL INCL O&P

Division Notes

		CREW	DAILY OUTPUT	LABOR-HOURS	UNIT	2006 BARE COSTS MAT.	LABOR	EQUIP.	TOTAL	TOTAL INCL O&P

Division Notes

		CREW	DAILY OUTPUT	LABOR-HOURS	UNIT	2006 BARE COSTS MAT.	LABOR	EQUIP.	TOTAL	TOTAL INCL O&P

Reed Construction Data, Inc.

Reed Construction Data, Inc., a leading worldwide provider of total construction information solutions, is comprised of three main product groups designed specifically to help construction professionals advance their businesses with timely, accurate and actionable project, product and cost data. Reed Construction Data is a division of Reed Business Information, a member of the Reed Elsevier plc group of companies.

The Project, Product and Cost & Estimating divisions offer a variety of innovative products and services designed for the full spectrum of design, construction and manufacturing professionals. Through its International companies, Reed Construction Data's reputation for quality construction market data is growing worldwide.

Cost Information

RSMeans, the undisputed market leader and authority on construction costs, publishes current cost and estimating information in annual cost books and on the *Means CostWorks* CD-ROM. RSMeans furnishes the construction industry with a rich library of corresponding reference books and a series of professional seminars that are designed to sharpen professional skills and maximize the effective use of cost estimating and management tools. RSMeans also provides construction cost consulting for owners, manufacturers, designers, and contractors.

Project Data

Reed Construction Data provides complete, accurate and relevant project information through all stages of construction. Customers are supplied industry data through leads, project reports, contact lists, plans and specifications surveys, market penetration analyses and sales evaluation reports. Any of these products can pinpoint a county, look at a state, or cover the country. Data is delivered via paper, e-mail, CD-ROM or the Internet.

Building Product Information

The First Source suite of products is the only integrated building product information system offered to the commercial construction industry for comparing and specifying building products. These print and online resources include *First Source,* CSI's SPEC-DATA™, CSI's MANU-SPEC™, First Source CAD, and Manufacturer Catalogs. Written by industry professionals and organized using CSI's MasterFormat™, construction professionals use this information to make better design decisions. FirstSourceONL.com combines Reed Construction Data's project, product and cost data with news and information from Reed Business Information's *Building Design & Construction* and *Consulting-Specifying Engineer magazines.* This industry-focused site offers easy and unlimited access to vital information for all construction professionals.

International

Reed Construction Data Canada serves the Canadian construction market with reliable and comprehensive project and product information services that cover all facets of construction. Core services include: *BuildSource, BuildSpec, BuildSelect,* product selection and specification tools available in print and on the Internet; Building Reports, a national construction project lead service; CanaData, statistical and forecasting information; *Daily Commercial News,* a construction newspaper reporting on news and projects in Ontario; and *Journal of Commerce,* reporting news in British Columbia and Alberta.

BIMSA/Mexico provides construction project news, product information, cost data, seminars and consulting services to construction professionals in Mexico. Its subsidiary, PRISMA, provides job costing software.

Byggfakta Scandinavia AB, founded in 1936, is the parent company for the leaders of customized construction market data for Denmark, Estonia, Finland, Norway and Sweden. Each company fully covers the local construction market and provides information across several platforms including via subscription, on an ad-hoc basis, electronically and on paper.

Cordell Building Information Services, with its complete range of project and cost and estimating services, is Australia's specialist in the construction information industry. Cordell provides in-depth and historical information on all aspects of construction projects and estimation, including several customized reports, construction and sales leads and detailed cost information.

For more information, please visit our Web site at www.reedconstructiondata.com.

Reed Construction Data, Inc., Corporate Office
30 Technology Parkway South
Norcross, GA 30092-2912
(800) 322-6996
(800) 895-8661 (fax)
info@reedbusiness.com
www.reedconstructiondata.com

Reed Construction Data®

Means Project Cost Report

By filling out this report, your project data will contribute to the database that supports the Means Project Cost Square Foot Data. When you fill out this form, Means will provide a 30% discount off one of the Means products advertised in the following pages. Please complete the form including all items where you have cost data, and all the items marked (✔).

$30.00 Discount per product for each report you submit.

Project Description (No remodeling projects, please.)

✔ Building Use (Office, School...) ______

✔ Address (City, State) ______

______

✔ Frame (Wood, Steel...) ______

✔ Exterior Wall (Brick, Tilt-up...) ______

✔ Basement: (check one) ☐ Full ☐ Partial ☐ None

✔ Number Stories ______

✔ Floor-to-Floor Height ______

% Air Conditioned ______ Tons ______

Comments ______

Total Project Cost $ ______

Owner ______

Architect ______

General Contractor ______

✔ Bid Date ______

Typical Bay Size ______

✔ Labor Force: ____ % Union ____ % Non-Union

✔ Project Description (Circle one number in each line)

1. Economy 2. Average 3. Custom 4. Luxury
1. Square 2. Rectangular 3. Irregular 4. Very Irregular

A	✔	General Conditions	$
B	✔	Site Work	$
C	✔	Concrete	$
D	✔	Masonry	$
E	✔	Metals	$
F	✔	Wood & Plastics	$
G	✔	Thermal & Moisture Protection	
GR		Roofing & Flashing	$
H	✔	Doors and Windows	$
J	✔	Finishes	$
JP		Painting & Wall Covering	$

K	✔	Specialties	$
L	✔	Equipment	$
M	✔	Furnishings	$
N	✔	Special Construction	$
P	✔	Conveying Systems	$
Q	✔	Mechanical	$
QP		Plumbing	$
QB		HVAC	$
R	✔	Electrical	$
S	✔	Mech./Elec. Combined	$

Please specify the Means product you wish to receive.
Complete the address information.

Product Name ______

Product Number ______

Your Name ______

Title ______

Company ______

☐ Company

☐ Home Street Address ______

City, State, Zip ______

Return by mail or Fax 888-492-6770.

Method of Payment:

Credit Card # ______

Expiration Date ______

Check ______

Purchase Order ______

Reed Construction Data/RSMeans
Square Foot Costs Department
P.O. Box 800
Kingston, MA 02364-9988

For more information visit Means Web site at www.rsmeans.com

Reed Construction Data/RSMeans . . . a tradition of excellence in Construction Cost Information and Services since 1942.

Table of Contents

Book Selection Guide

The following table provides definitive information on the content of each cost data publication. The number of lines of data provided in each unit price or assemblies division, as well as the number of reference tables and crews is listed for each book. The presence of other elements such as an historical cost index, city cost indexes, square foot models or cross-referenced index is also indicated. You can use the table to help select the Means' book that has the quantity and type of information you most need in your work.

Unit Cost Divisions	Building Construction Costs	Mechanical	Electrical	Repair & Remodel.	Square Foot	Site Work Landsc.	Assemblies	Interior	Concrete Masonry	Open Shop	Heavy Construc.	Light Commercial	Facil. Construc.	Plumbing	Western Construction Costs	Residential
1	554	273	358	465		496		297	465	553	502	210	979	325	553	161
2	2560	1345	460	1457		8234		543	1383	2466	5287	695	4835	1593	2543	768
3	1393	112	99	721		1250		191	1774	1388	1395	229	1315	81	1388	262
4	854	18	0	666		660		576	1075	830	597	432	1076	0	840	352
5	1876	239	193	965		817		967	738	1844	1077	835	1856	316	1816	759
6	1487	82	78	1465		450		1385	317	1477	607	1603	1574	47	1824	1734
7	1252	159	71	1240		472		485	401	1251	355	963	1306	169	1252	760
8	1893	28	0	1965		313		1682	642	1870	51	1329	2095	0	1852	1263
9	1595	47	0	1425		232		1688	360	1547	157	1337	1811	47	1586	1217
10	871	47	25	508		193		710	170	873	0	404	917	233	871	217
11	1018	322	169	502		137		813	44	925	107	218	1173	291	925	104
12	313	0	0	46		213		1422	27	304	0	65	1442	0	304	63
13	1317	1016	385	659		451		884	75	1292	340	484	1931	929	1183	194
14	345	36	0	258		36		292	0	344	30	12	362	35	343	6
15	1976	13019	636	1752		1542		1192	58	1985	1729	1225	10787	9669	2005	829
16	1279	470	10114	1002		739		1108	55	1297	760	1083	9885	415	1228	554
Totals	**20583**	**17213**	**12588**	**15096**		**16235**		**14235**	**7584**	**20246**	**12994**	**11124**	**43344**	**14150**	**20513**	**9243**

Assembly Divisions	Building Construction Costs	Mechanical	Elec-trical	Repair & Remodel.	Square Foot	Site Work Landsc.	Assemblies	Interior	Concrete Masonry	Open Shop	Heavy Construc.	Light Commercial	Facil. Construc.	Plumbing	Western Construction Costs	Asm Div	Resi-dential
A	612	19	0	192	150	540	612	0	550		542	149	24	0		1	374
B	5584	0	0	809	2479	0	5584	333	1914		0	2023	144	0		2	217
C	1208	0	0	635	853	0	1208	1555	132		0	757	238	0		3	588
D	2392	1014	780	693	1776	0	2392	752	0		0	1262	1010	890		4	867
E	294	0	0	85	257	0	294	5	0		0	257	5	0		5	393
F	124	0	0	0	123	0	124	0	0		0	123	3	0		6	357
G	584	465	172	330	111	1844	584	0	471		423	110	113	559		7	299
																8	760
																9	80
																10	0
																11	0
																12	0
Totals	**10798**	**1498**	**952**	**2744**	**5749**	**2384**	**10798**	**2645**	**3067**	**0**	**965**	**4681**	**1537**	**1449**			**3935**

Reference Section	Building Construction Costs	Mechanical	Electrical	Repair & Remodel.	Square Foot	Site Work Landsc.	Assemblies	Interior	Concrete Masonry	Open Shop	Heavy Construc.	Light Commercial	Facil. Construc.	Plumbing	Western Construction Costs	Resi-dential
Tables	237	237	237	237	4	237	219	237	237	237	237	237	237	237	237	35
Models					105							46				32
Crews	452	452	452	434		452		452	452	431	452	431	434	452	452	431
City Cost Indexes	yes	yes	yes	yes	yes	yes	yes	yes	yes	yes	yes	yes	yes	yes	yes	yes
Historical Cost Indexes	yes	yes	yes	yes	yes	yes	yes	yes	yes	yes	yes	yes	yes	yes	yes	no

Means Building Construction Cost Data 2006

Available in Both Softbound and Looseleaf Editions

The "Bible" of the industry comes in the standard softcover edition or the looseleaf edition.

Many customers enjoy the convenience and flexibility of the looseleaf binder, which increases the usefulness of Means *Building Construction Cost Data 2006* by making it easy to add and remove pages. You can insert your own cost information pages, so everything is in one place. Copying pages for faxing is easier also. Whichever edition you prefer, softbound or the convenient looseleaf edition, you'll be eligible to receive *The Change Notice* FREE. Current subscribers receive *The Change Notice* via e-mail.

$157.95 per copy, Looseleaf
Catalog No. 61016

Means Building Construction Cost Data 2006

Offers you unchallenged unit price reliability in an easy-to-use arrangement. Whether used for complete, finished estimates or for periodic checks, it supplies more cost facts better and faster than any comparable source. Over 23,000 unit prices for 2006. The City Cost Indexes cover over 930 areas, for indexing to any project location in North America. Order and get *The Change Notice* FREE. You'll have year-long access to the Means Estimating **HOTLINE** FREE with your subscription. Expert assistance when using Means data is just a phone call away.

$126.95 per copy
Over 700 pages, illustrated, available Oct. 2005
Catalog No. 60016

Means Metric Construction Cost Data 2006

An 880+ page compendium of all the data from both the 2006 *Building Construction Cost Data* AND the *Heavy Construction Cost Data*, in **metric** format! Access all of this vital information from one complete source. It contains more than 600 pages of unit costs and 40 pages of assemblies costs. The Reference Section contains over 200 pages of tables, charts and other estimating aids. A great way to stay in step with today's construction trends and rapidly changing costs.

$149.95 per copy
Over 800 pages, illus., available Dec. 2005
Catalog No. 63016

For more information visit Means Web site at www.rsmeans.com

Annual Cost Guides

Means Mechanical Cost Data 2006

• HVAC Controls

Total unit and systems price guidance for mechanical construction... materials, parts, fittings, and complete labor cost information. Includes prices for piping, heating, air conditioning, ventilation, and all related construction.

Plus new 2006 unit costs for:

- Over 2500 installed HVAC/controls assemblies
- "On Site" Location Factors for over 930 cities and towns in the U.S. and Canada
- Crews, labor, and equipment

$126.95 per copy
Over 600 pages, illustrated, available Oct. 2005
Catalog No. 60026

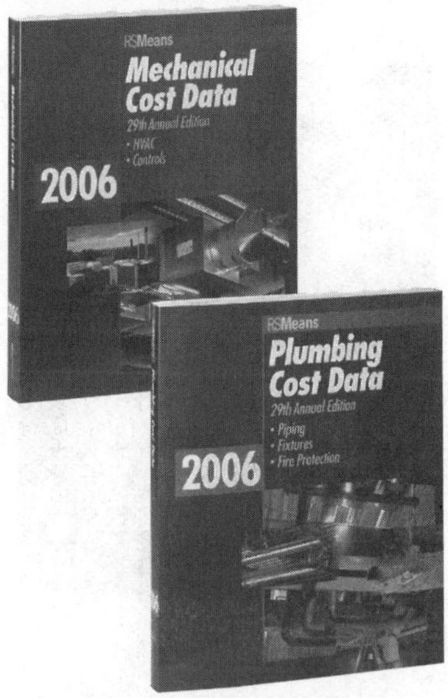

Means Plumbing Cost Data 2006

Comprehensive unit prices and assemblies for plumbing, irrigation systems, commercial and residential fire protection, point-of-use water heaters, and the latest approved materials. This publication and its companion, Means *Mechanical Cost Data*, provide full-range cost estimating coverage for all the mechanical trades.

$126.95 per copy
Over 550 pages, illustrated, available Oct. 2005
Catalog No. 60216

Means Electrical Cost Data 2006

Pricing information for every part of electrical cost planning. More than 15,000 unit and systems costs with design tables; clear specifications and drawings; engineering guides and illustrated estimating procedures; complete labor-hour and materials costs for better scheduling and procurement; and the latest electrical products and construction methods.

- A variety of special electrical systems including cathodic protection
- Costs for maintenance, demolition, HVAC/ mechanical, specialties, equipment, and more

$126.95 per copy
Over 450 pages, illustrated, available Oct. 2005
Catalog No. 60036

Means Electrical Change Order Cost Data 2006

Means *Electrical Change Order Cost Data* 2006 provides you with electrical unit prices exclusively for pricing change orders based on the recent, direct experience of contractors and suppliers. Analyze and check your own change order estimates against the experience others have had doing the same work. It also covers productivity analysis and change order cost justifications. With useful information for calculating the effects of change orders and dealing with their administration.

$126.95 per copy
Over 450 pages, available Oct. 2005
Catalog No. 60236

Means Square Foot Costs 2006

It's Accurate and Easy To Use!

- **Updated 2006 price information**, based on nationwide figures from suppliers, estimators, labor experts, and contractors
- "How-to-Use" sections, with **clear examples** of commercial, residential, industrial, and institutional structures
- Realistic graphics, offering true-to-life illustrations of building projects
- Extensive information on using square foot cost data, including sample estimates and alternate pricing methods

$137.95 per copy
Over 450 pages, illustrated, available Nov. 2005
Catalog No. 60056

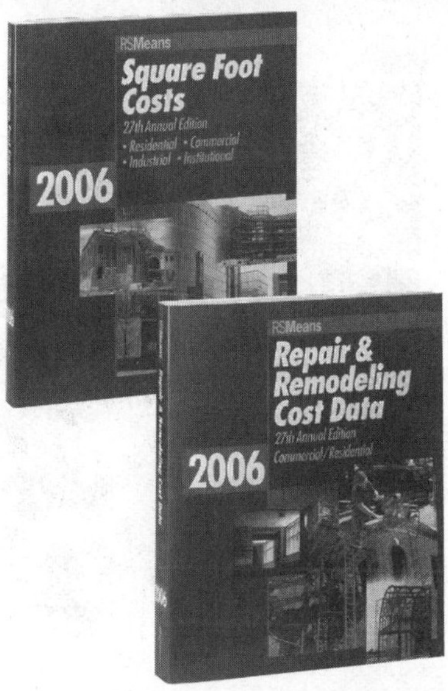

Means Repair & Remodeling Cost Data 2006

Commercial/Residential

Use this valuable tool to estimate commercial and residential renovation and remodeling.

Includes: New costs for hundreds of unique methods, materials, and conditions that only come up in repair and remodeling. PLUS:

- Unit costs for over 16,000 construction components
- Installed costs for over 90 assemblies
- Costs for 300+ construction crews
- Over 930 "On Site" localization factors for the U.S. and Canada.

$107.95 per copy
Over 650 pages, illustrated, available Nov. 2005
Catalog No. 60046

Annual Cost Guides

For more information visit Means Web site at www.rsmeans.com

Means Facilities Construction Cost Data 2006

For the maintenance and construction of commercial, industrial, municipal, and institutional properties. Costs are shown for new and remodeling construction and are broken down into materials, labor, equipment, overhead, and profit. Special emphasis is given to sections on mechanical, electrical, furnishings, site work, building maintenance, finish work, and demolition.

More than 45,000 unit costs, plus assemblies and reference sections are included.

$299.95 per copy
Over 1200 pages, illustrated, available Nov. 2005
Catalog No. 60206

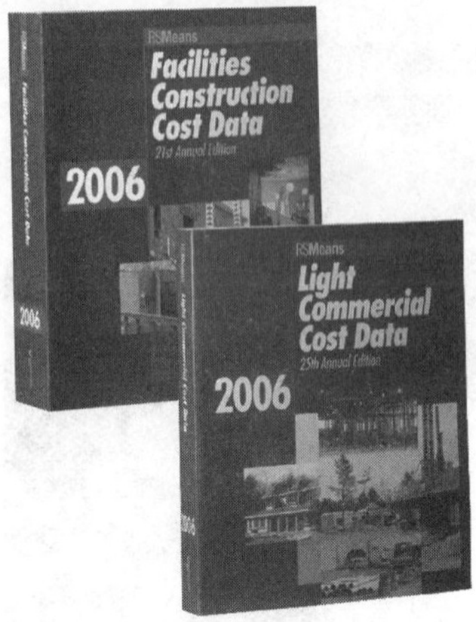

Means Light Commercial Cost Data 2006

Specifically addresses the light commercial market, which is an increasingly specialized niche in the industry. Aids you, the owner/designer/contractor, in preparing all types of estimates, from budgets to detailed bids. Includes new advances in methods and materials.

Assemblies section allows you to evaluate alternatives in the early stages of design/planning.

Over 13,000 unit costs for 2006 ensure you have the prices you need... when you need them.

$107.95 per copy
Over 650 pages, illustrated, available Nov. 2005
Catalog No. 60186

Means Residential Cost Data 2006

Contains square foot costs for 30 basic home models with the look of today, plus hundreds of custom additions and modifications you can quote right off the page. With costs for the 100 residential systems you're most likely to use in the year ahead. Complete with blank estimating forms, sample estimates and step-by-step instructions.

$107.95 per copy
Over 600 pages, illustrated, available Oct. 2005
Catalog No. 60176

Means Site Work & Landscape Cost Data 2006

Means *Site Work & Landscape Cost Data* 2006 is organized to assist you in all your estimating needs. Hundreds of fact-filled pages help you make accurate cost estimates efficiently.

Updated for 2006!

- Demolition features—including ceilings, doors, electrical, flooring, HVAC, millwork, plumbing, roofing, walls, and windows
- State-of-the-art segmental retaining walls
- Flywheel trenching costs and details
- Updated wells section
- Landscape materials, flowers, shrubs, and trees

$126.95 per copy
Over 600 pages, illustrated, available Nov. 2005
Catalog No. 60286

Means Assemblies Cost Data 2006

Means *Assemblies Cost Data* 2006 takes the guesswork out of preliminary or conceptual estimates. Now you don't have to try to calculate the assembled cost by working up individual components costs. We've done all the work for you.

Presents detailed illustrations, descriptions, specifications and costs for every conceivable building assembly—240 types in all—arranged in the easy-to-use UNIFORMAT II system. Each illustrated "assembled" cost includes a complete grouping of materials and associated installation costs, including the installing contractor's overhead and profit.

$206.95 per copy
Over 600 pages, illustrated, available Oct. 2005
Catalog No. 60066

Means Open Shop Building Construction Cost Data 2006

The latest costs for accurate budgeting and estimating of new commercial and residential construction... renovation work... change orders... cost engineering.

Means *Open Shop BCCD* will assist you to:

- Develop benchmark prices for change orders
- Plug gaps in preliminary estimates, and budgets
- Estimate complex projects
- Substantiate invoices on contracts
- Price ADA-related renovations

$126.95 per copy
Over 700 pages, illustrated, available Dec. 2005
Catalog No. 60156

For more information
visit Means Web site
at www.rsmeans.com

Annual Cost Guides

Means Building Construction Cost Data 2006

Western Edition

This regional edition provides more precise cost information for western North America. Labor rates are based on union rates from 13 western states and western Canada. Included are western practices and materials not found in our national edition: tilt-up concrete walls, glu-lam structural systems, specialized timber construction, seismic restraints, and landscape and irrigation systems.

$126.95 per copy
Over 650 pages, illustrated, available Dec. 2005
Catalog No. 60226

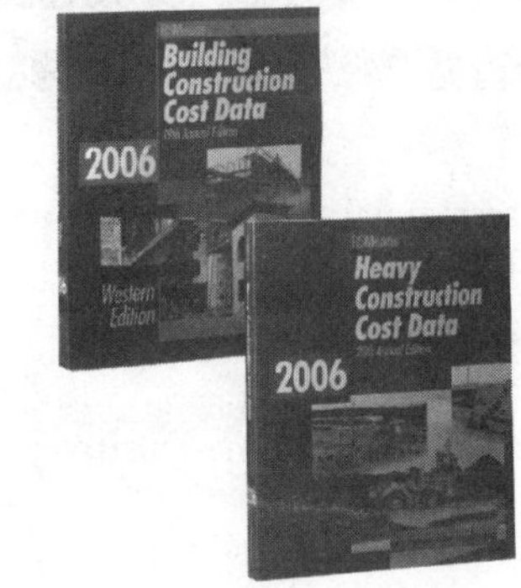

Means Heavy Construction Cost Data 2006

A comprehensive guide to heavy construction costs. Includes costs for highly specialized projects such as tunnels, dams, highways, airports, and waterways. Information on different labor rates, equipment, and material costs is included. Features unit price costs, systems costs, and numerous reference tables for costs and design. Valuable not only to contractors and civil engineers, but also to government agencies and city/town engineers.

$126.95 per copy, Over 450 pages, illustrated, available Nov. 2005
Catalog No. 60166

Means Construction Cost Indexes 2006

Who knows what 2006 holds? What materials and labor costs will change unexpectedly? By how much?

- Breakdowns for 316 major cities
- National averages for 30 key cities
- Expanded five major city indexes
- Historical construction cost indexes

$272.00 per year
$68.00 individual quarters
Catalog No. 60146 A,B,C,D

Means Interior Cost Data 2006

Provides you with prices and guidance needed to make accurate interior work estimates. Contains costs on materials, equipment, hardware, custom installations, furnishings, labor costs... every cost factor for new and remodel commercial and industrial interior construction, including updated information on office furnishings, plus more than 50 reference tables. For contractors, facility managers, and owners.

$126.95 per copy, Over 600 pages, illustrated, available Oct. 2005
Catalog No. 60096

Means Concrete & Masonry Cost Data 2006

Provides you with cost facts for virtually all concrete/masonry estimating needs, from complicated formwork to various sizes and face finishes of brick and block, all in great detail. The comprehensive unit cost section contains more than 8,500 selected entries. Also contains an assemblies cost section, and a detailed reference section that supplements the cost data.

$114.95 per copy
Over 450 pages, illustrated, available Dec. 2005
Catalog No. 60116

Means Labor Rates for the Construction Industry 2006

Complete information for estimating labor costs, making comparisons, and negotiating wage rates by trade for over 300 U.S. and Canadian cities. With 46 construction trades listed by local union number in each city, and historical wage rates included for comparison. No similar book is available through the trade.

Each city chart lists the county and is alphabetically arranged with handy visual flip tabs for quick reference.

$274.95 per copy
Over 300 pages, available Dec. 2005
Catalog No. 60126

Means Facilities Maintenance & Repair Cost Data 2006

Means *Facilities Maintenance & Repair Cost Data* gives you a complete system to manage and plan your facility repair and maintenance costs and budget efficiently. Guidelines for auditing a facility and developing an annual maintenance plan. Budgeting is included, along with reference tables on cost and management and information on frequency and productivity of maintenance operations.

The only nationally recognized source of maintenance and repair costs. Developed in cooperation with the Army Corps of Engineers.

$274.95 per copy
Over 600 pages, illustrated, available Dec. 2005
Catalog No. 60306

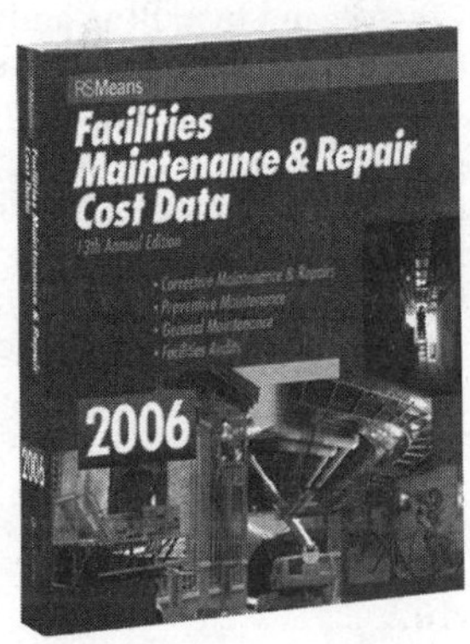

Builder's Essentials: Estimating Building Costs for the Residential & Light Commercial Contractor

By Wayne J. DelPico

Step-by-step estimating methods for residential and light commercial contractors. Includes a detailed look at every construction specialty—explaining all the components, takeoff units, and labor needed for well-organized, complete estimates. Covers:

- Correctly interpreting plans and specifications
- Developing accurate and complete labor and material costs
- Understanding direct and indirect overhead costs... and accounting for time-sensitive costs
- Using historical cost data to generate new project budgets
- Allocating the right amount for profit and contingencies

$29.95 per copy
Over 400 pages, illustrated, Softcover
Catalog No. 67343

Green Building: Project Planning & Cost Estimating

By RSMeans and Contributing Authors

Written by a team of leading experts in sustainable design, this book is a complete guide to planning and estimating green building projects, a growing trend in building design and construction. It explains:

- All the different criteria for "green-ness"
- What criteria your building needs to meet to get a LEED, ENERGYSTAR®, or other recognized rating for green buildings
- How the project team works differently on a green versus a traditional building project
- How to select and specify green products
- How to evaluate the cost and value of green products versus conventional ones—not only for their initial installation cost, but their cost over time in maintenance and operation

Features an extensive Green Building Cost Data section

$89.95 per copy
350 pages, illustrated, Hardcover
Catalog No. 67338

Means ADA Compliance Pricing Guide

New Second Edition

By Adaptive Environments and RSMeans

Completely updated and revised to the new 2004 *Americans with Disabilities Act Accessibility Guidelines*, this book features more than 70 of the most commonly needed modifications for ADA compliance—their design requirements, suggestions, and final cost. Projects range from installing ramps and walkways, widening doorways and entryways, and installing and refitting elevators, to relocating light switches and signage, and remodeling bathrooms and kitchens. Also provided are:

- Detailed cost estimates for budgeting modification projects, including estimates for each of 260 alternates.
- An assembly estimate for every project, with detailed cost breakdown including materials, labor-hours, and contractor's overhead.
- 3,000 additional ADA compliance-related unit cost line items.
- Costs that are easily adjusted to over 900 cities and towns.

$79.95 per copy
Over 350 pages, illustrated, Softcover
Catalog No. 67310A

For more information
visit Means Web site
at www.rsmeans.com

Reference Books

Value Engineering: Practical Applications . . . For Design, Construction, Maintenance & Operations
By Alphonse Dell'Isola, PE

A tool for immediate application—for engineers, architects, facility managers, owners, and contractors. Includes: making the case for VE—the management briefing, integrating VE into planning, budgeting, and design, conducting life cycle costing, using VE methodology in design review and consultant selection, case studies, VE workbook, and a life cycle costing program on disk.

$79.95 per copy
Over 450 pages, illustrated, Softcover
Catalog No. 67319

Facilities Operations & Engineering Reference
By the Association for Facilities Engineering and RSMeans

An all-in-one technical referance for planning and managing facility projects and solving day-to-day operations problems. Selected as the official Certified Plant Engineer reference, this handbook covers financial analysis, maintenance, HVAC and energy efficiency, and more.

$109.95 per copy
Over 700 pages, illustrated, Hardcover
Catalog No. 67318

The Building Professional's Guide to Contract Documents
3rd Edition

By Waller S. Poage, AIA, CSI, CVS

A comprehensive reference for owners, design professionals, contractors, and students.

- Structure your documents for maximum efficiency.
- Effectively communicate construction requirements.
- Understand the roles and responsibilities of construction professionals.
- Improve methods of project delivery.

$32.48 per copy, 400 pages
Diagrams and construction forms, Hardcover
Catalog No. 67261A

Building Security: Strategies & Costs
By David Owen

This comprehensive resource will help you evaluate your facility's security needs, and design and budget for the materials and devices needed to fulfill them.

Includes over 130 pages of Means cost data for installation of security systems and materials, plus a review of more than 50 security devices and construction solutions.

$44.98 per copy
350 pages, illustrated, Hardcover
Catalog No. 67339

Cost Planning & Estimating for Facilities Maintenance

In this unique book, a team of facilities management authorities shares their expertise at:

- Evaluating and budgeting maintenance operations
- Maintaining and repairing key building components
- Applying Means *Facilities Maintenance & Repair Cost Data* to your estimating

Covers special maintenance requirements of the ten major building types.

$89.95 per copy
Over 475 pages, Hardcover
Catalog No. 67314

Life Cycle Costing for Facilities
By Alphonse Dell'Isola and Dr. Steven Kirk

Guidance for achieving higher quality design and construction projects at lower costs! Cost-cutting efforts often sacrifice quality to yield the cheapest product. Life cycle costing enables building designers and owners to achieve both. The authors of this book show how LCC can work for a variety of projects – from roads to HVAC upgrades to different types of buildings.

$99.95 per copy
450 pages, Hardcover
Catalog No. 67341

Building & Renovating Schools

This all-inclusive guide covers every step of the school construction process—from initial planning, needs assessment, and design, right through moving into the new facility. A must-have resource for anyone concerned with new school construction or renovation, including architects and engineers, contractors and project managers, facility managers, school administrators and school board members, building committees, community leaders, and anyone else who wants to ensure that the project meets schools' needs in a cost-effective, timely manner. With square foot cost models for elementary, middle, and high school facilities and real-life case studies of recently completed school projects.

The contributors to this book—architects, construction project managers, contractors, and estimators who specialize in school construction—provide start-to-finish, expert guidance on the process.

$99.95 per copy
Over 425 pages, Hardcover
Catalog No. 67342

Reference Books

For more information visit Means Web site at www.rsmeans.com

Interior Home Improvement Costs, New 9th Edition

Estimates for the most popular remodeling and repair projects—from small, do-it-yourself jobs to major renovations and new construction. Includes: Kitchens & Baths; New Living Space from your Attic, Basement, or Garage; New Floors, Paint, and Wallpaper; Tearing Out or Building New Walls; Closets, Stairs, and Fireplaces; New Energy-Saving Improvements, Home Theaters, and More!

$24.95 per copy
250 pages, illustrated, Softcover
Catalog No. 67308E

Exterior Home Improvement Costs New 9th Edition

Estimates for the most popular remodeling and repair projects—from small, do-it-yourself jobs, to major renovations and new construction. Includes: Curb Appeal Projects—Landscaping, Patios, Porches, Driveways, and Walkways; New Windows and Doors; Decks, Greenhouses, and Sunrooms; Room Additions and Garages; Roofing, Siding, and Painting; "Green" Improvements to Save Energy & Water.

$24.95 per copy
Over 275 pages, illustrated, Softcover
Catalog No. 67309E

Builder's Essentials: Plan Reading & Material Takeoff

By Wayne J. DelPico

For Residential and Light Commercial Construction

A valuable tool for understanding plans and specs, and accurately calculating material quantities. Step-by-step instructions and takeoff procedures based on a full set of working drawings.

$35.95 per copy
Over 420 pages, Softcover
Catalog No. 67307

Builder's Essentials: Framing & Rough Carpentry, 2nd Edition

By Scot Simpson

Develop and improve your skills with easy-to-follow instructions and illustrations. Learn proven techniques for framing walls, floors, roofs, stairs, doors, and windows. Updated guidance on standards, building codes, safety requirements, and more. Also available in Spanish!

$24.95 per copy
Over 150 pages, Softcover
Catalog No. 67298A
Spanish Catalog No. 67298AS

Builder's Essentials: Best Business Practices for Builders & Remodelers

An Easy-to-Use Checklist System

By Thomas N. Frisby

A comprehensive guide covering all aspects of running a construction business, with more than 40 user-friendly checklists. Provides expert guidance on: increasing your revenue and keeping more of your profit, planning for long-term growth, keeping good employees, and managing subcontractors.

$29.95 per copy
Over 220 pages, Softcover
Catalog No. 67329

Unit Price Estimating Methods

New 3rd Edition

This new edition includes up-to-date cost data and estimating examples, updated to reflect changes to the CSI numbering system and new features of Means cost data. It describes the most productive, universally accepted ways to estimate, and uses checklists and forms to illustrate shortcuts and timesavers. A model estimate demonstrates procedures. A new chapter explores computer estimating alternatives.

$59.95 per copy
Over 350 pages, illustrated, Hardcover
Catalog No. 67303A

Means Landscape Estimating Methods

4th Edition

By Sylvia H. Fee

This revised edition offers expert guidance for preparing accurate estimates for new landscape construction and grounds maintenance. Includes a complete project estimate featuring the latest equipment and methods, and chapters on *Life Cycle Costing* and *Landscape Maintenance Estimating*.

$62.95 per copy
Over 300 pages, illustrated, Hardcover
Catalog No. 67295B

Means Environmental Remediation Estimating Methods, 2nd Edition

By Richard R. Rast

Guidelines for estimating 50 standard remediation technologies. Use it to prepare preliminary budgets, develop estimates, compare costs and solutions, estimate liability, review quotes, negotiate settlements.

A valuable support tool for Means Environmental Remediation Unit Price and Assemblies books.

$99.95 per copy
Over 750 pages, illustrated, Hardcover
Catalog No. 64777A

For more information
visit Means Web site
at www.rsmeans.com

Reference Books

Means Illustrated Construction Dictionary, Condensed, 2nd Edition

By RSMeans

Recognized in the industry as the best resource of its kind. This essential tool has been further enhanced with updates to existing terms and the addition of hundreds of new terms and illustrations—in keeping with recent developments. For contractors, architects, insurance and real estate personnel, homeowners, and anyone who needs quick, clear definitions for construction terms.

$59.95 per copy
Over 500 pages, Softcover
Catalog No. 67282A

Means Repair & Remodeling Estimating New 4th Edition

By Edward B. Wetherill & RSMeans

This important reference focuses on the unique problems of estimating renovations of existing structures and helps you determine the true costs of remodeling through careful evaluation of architectural details and a site visit.

New section on disaster restoration costs.

$69.95 per copy
Over 450 pages, illustrated, Hardcover
Catalog No. 67265B

Facilities Planning & Relocation

New, lower price and user-friendly format.

By David D. Owen

A complete system for planning space needs and managing relocations. Includes step-by-step manual, over 50 forms, and extensive reference section on materials and furnishings.

$89.95 per copy
Over 450 pages, Softcover
Catalog No. 67301

Means Square Foot & Assemblies Estimating Methods, 3rd Edition

Develop realistic square foot and assemblies costs for budgeting and construction funding. The new edition features updated guidance on square foot and assemblies estimating using UNIFORMAT II. An essential reference for anyone who performs conceptual estimates.

$69.95 per copy
Over 300 pages, illustrated, Hardcover
Catalog No. 67145B

Means Electrical Estimating Methods 3rd Edition

Expanded new edition includes sample estimates and cost information in keeping with the latest version of the CSI MasterFormat and UNIFORMAT II. Complete coverage of fiber optic and uninterruptible power supply electrical systems, broken down by components and explained in detail. Includes a new chapter on computerized estimating methods. A practical companion to Means *Electrical Cost Data.*

$64.95 per copy
Over 325 pages, Hardcover
Catalog No. 67230B

Means Mechanical Estimating Methods 3rd Edition

This guide assists you in making a review of plans, specs, and bid packages with suggestions for takeoff procedures, listings, substitutions and pre-bid scheduling. Includes suggestions for budgeting labor and equipment usage. Compares materials and construction methods to allow you to select the best option.

$64.95 per copy
Over 350 pages, illustrated, Hardcover
Catalog No. 67294A

Means Spanish/English Construction Dictionary

By RSMeans, The International Conference of Building Officials (ICBO), and Rolf Jensen & Associates (RJA)

Designed to facilitate communication among Spanish- and English-speaking construction personnel—improving performance and job-site safety. Features the most common words and phrases used in the construction industry, with easy-to-follow pronunciations. Includes extensive building systems and tools illustrations.

$22.95 per copy
250 pages, illustrated, Softcover
Catalog No. 67327

Project Scheduling & Management for Construction

New 3rd Edition

By David R. Pierce, Jr.

A comprehensive yet easy-to-follow guide to construction project scheduling and control—from vital project management principles through the latest scheduling, tracking, and controlling techniques. The author is a leading authority on scheduling with years of field and teaching experience at leading academic institutions. Spend a few hours with this book and come away with a solid understanding of this essential management topic.

$64.95 per copy
Over 300 pages, illustrated, Hardcover
Catalog No. 67247B

Reference Books

For more information visit Means Web site at www.rsmeans.com

Concrete Repair and Maintenance Illustrated

By Peter Emmons

Hundreds of illustrations show users how to analyze, repair, clean, and maintain concrete structures for optimal performance and cost effectiveness. From parking garages to roads and bridges to structural concrete, this comprehensive book describes the causes, effects, and remedies for concrete wear and failure. Invaluable for planning jobs, selecting materials, and training employees, this book is a must-have for concrete specialists, general contractors, facility managers, civil and structural engineers, and architects.

$34.98 per copy
300 pages, illustrated, Softcover
Catalog No. 67146

Plumbing Estimating Methods 3rd Edition

By Joseph Galeno and Sheldon Greene

Updated and revised! This practical guide walks you through a plumbing estimate, from basic materials and installation methods through change order analysis. *Plumbing Estimating Methods* covers residential, commercial, industrial, and medical systems, and features sample takeoff and estimate forms and detailed illustrations of systems and components.

$29.98 per copy
330+ pages, Softcover
Catalog No. 67283B

Building Spec Homes Profitably

By Kenneth V. Johnson

Professional guidance from a veteran spec home builder… This complete system covers all aspects of spec home building—from market research and site selection to financing and design to scheduling and supervision, and more!

You'll learn to reduce risk and ensure profits in spec home building, no matter what the economic climate.

$29.95 per copy
200 pages, Softcover
Catalog No. 67312

How to Estimate with Means Data & CostWorks

New 2nd Edition

By RSMeans and Saleh Mubarak

Learn estimating techniques using Means cost data. Includes an instructional version of Means *CostWorks* CD-ROM with sample building plans. The step-by-step guide takes you through all the major construction items. Over 300 sample estimating problems are included.

$59.95 per copy
Over 190 pages, Softcover
Catalog No. 67324A

Means Productivity Standards for Construction

3rd Edition

By RSMeans

Completely updated, with well over 3,000 new work items added to cover topics in current demand! This comprehensive reference for the productivity of construction crews, equipment, and labor includes precise descriptions of each construction task organized according to the Construction Specifications Institute's MasterFormat.

$49.98 per copy
800+ pages, Hardcover
Catalog No. 67236A

Total Productive Facilities Management

By Richard W. Sievert, Jr.

Today, facilities are viewed as strategic resources… elevating the facility manager to the role of asset manager supporting the organization's overall business goals. Now, Richard Sievert Jr., in this well-articulated guidebook, sets forth a new operational standard for the facility manager's emerging role… a comprehensive program for managing facilities as a true profit center.

$29.98 per copy
275 pages, Softcover
Catalog No. 67321

Preventive Maintenance Guidelines for School Facilities

By John C. Maciha

A complete PM program for K-12 schools that ensures sustained security, safety, property integrity, user satisfaction, and reasonable ongoing expenditures.

Includes schedules for weekly, monthly, semiannual, and annual maintenance with hard copy and electronic forms.

$149.95 per copy
Over 225 pages, Hardcover
Catalog No. 67326

Preventive Maintenance for Higher Education Facilities

By Applied Management Engineering, Inc.

An easy-to-use system to help facilities professionals establish the value of PM, and to develop and budget for an appropriate PM program for their college or university. Features interactive campus building models typical of those found in different-sized higher education facilities, and PM checklists linked to each piece of equipment or system in hard copy and electronic format.

$149.95 per copy
150 pages, Hardcover
Catalog No. 67337

Historic Preservation: Project Planning & Estimating

By Swanke Hayden Connell Architects

Expert guidance on managing historic restoration, rehabilitation, and preservation building projects and determining and controlling their costs. Includes:

- How to determine whether a structure qualifies as historic
- Where to obtain funding and other assistance
- How to evaluate and repair more than 75 historic building materials

$99.95 per copy
Over 675 pages, Hardcover
Catalog No. 67323

Means Illustrated Construction Dictionary

Unabridged 3rd Edition

Long regarded as the Industry's finest, the Means Illustrated Construction Dictionary is now even better. With the addition of over 1,000 new terms and hundreds of new illustrations, it is the clear choice for the most comprehensive and current information. The companion CD-ROM that comes with this new edition adds many extra features: larger graphics, expanded definitions, and links to both CSI MasterFormat numbers and product information.

$99.95 per copy
Over 790 pages, illustrated, Hardcover
Catalog No. 67292A

Designing & Building with the IBC, 2nd Edition

By Rolf Jensen & Associates, Inc.

This updated comprehensive guide helps building professionals make the transition to the 2003 International Building Code®. Includes a side-by-side code comparison of the IBC 2003 to the IBC 2000 and the three primary model codes, a quick-find index, and professional code commentary. With illustrations, abbreviations key, and an extensive Resource section.

$99.95 per copy
Over 875 pages, Softcover
Catalog No. 67328A

Residential & Light Commercial Construction Standards, 2nd Edition

By RSMeans and Contributing Authors

For contractors, subcontractors, owners, developers, architects, engineers, attorneys, and insurance personnel, this book provides authoritative requirements and recommendations compiled from the nation's leading professional associations, industry publications, and building code organizations.

$59.95 per copy
600 pages, illustrated, Softcover
Catalog No. 67322A

Builder's Essentials: Advanced Framing Methods

By Scot Simpson

A highly illustrated, "framer-friendly" approach to advanced framing elements. Provides expert, but easy to interpret, instruction for laying out and framing complex walls, roofs, and stairs, and special requirements for earthquake and hurricane protection. Also helps bring framers up to date on the latest building code changes, and provides tips on the lead framer's role and responsibilities, how to prepare for a job, and how to get the crew started.

$24.95 per copy
250 pages, illustrated, Softcover
Catalog No. 67330

Means Estimating Handbook

2nd Edition

By RSMeans

Updated new Second Edition answers virtually any estimating technical question—all organized by CSI MasterFormat. This comprehensive reference covers the full spectrum of technical data required to estimate construction costs. The book includes information on sizing, productivity, equipment requirements, code-mandated specifications, design standards, and engineering factors.

$99.95 per copy
Over 900 pages, Hardcover
Catalog No. 67276A

Seminars

Means CostWorks Training

This one-day seminar course has been designed with the intention of assisting both new and existing users to become more familiar with the *CostWorks* program. The class is broken into two unique sections: (1) A one-half day presentation on the function of each icon; and each student will be shown how to use the software to develop a cost estimate. (2) Hands-on estimating exercises that will ensure that each student thoroughly understands how to use *CostWorks*. You must bring your own laptop computer to this course.

CostWorks Benefits/Features:

- Estimate in your own spreadsheet format
- Power of Means National Database
- Database automatically regionalized
- Save time with keyword searches
- Save time by establishing common estimate items in "Bookmark" files
- Customize your spreadsheet template
- Hot key to Product Manufacturers' listings and specs
- Merge capability for networking environments
- View crews and assembly components
- AutoSave capability
- Enhanced sorting capability

Unit Price Estimating

This interactive two-day seminar teaches attendees how to interpret project information and process it into final, detailed estimates with the greatest accuracy level.

The single most important credential an estimator can take to the job is the ability to visualize construction in the mind's eye, and thereby estimate accurately.

Some Of What You'll Learn:

- Interpreting the design in terms of cost
- The most detailed, time-tested methodology for accurate "pricing"
- Key cost drivers—material, labor, equipment, staging, and subcontracts
- Understanding direct and indirect costs for accurate job cost accounting and change order management

Who Should Attend: Corporate and government estimators and purchasers, architects, engineers... and others needing to produce accurate project estimates.

Square Foot and Assemblies Cost Estimating

This two-day course teaches attendees how to quickly deliver accurate square foot estimates using limited budget and design information.

Some Of What You'll Learn:

- How square foot costing gets the estimate done faster
- Taking advantage of a "systems" or "assemblies" format
- The Means "building assemblies/square foot cost approach"
- How to create a very reliable preliminary and systems estimate using bare-bones design information

Who Should Attend: Facilities managers, facilities engineers, estimators, planners, developers, construction finance professionals... and others needing to make quick, accurate construction cost estimates at commercial, government, educational, and medical facilities.

Repair and Remodeling Estimating

This two-day seminar emphasizes all the underlying considerations unique to repair/remodeling estimating and presents the correct methods for generating accurate, reliable R&R project costs using the unit price and assemblies methods.

Some Of What You'll Learn:

- Estimating considerations—like labor-hours, building code compliance, working within existing structures, purchasing materials in smaller quantities, unforeseen deficiencies
- Identify problems and provide solutions to estimating building alterations
- Rules for factoring in minimum labor costs, accurate productivity estimates and allowances for project contingencies
- R&R estimating examples are calculated using unit prices and assemblies data

Who Should Attend: Facilities managers, plant engineers, architects, contractors, estimators, builders... and others who are concerned with the proper preparation and/or evaluation of repair and remodeling estimates.

Mechanical and Electrical Estimating

This two-day course teaches attendees how to prepare more accurate and complete mechanical/electrical estimates, avoiding the pitfalls of omission and double-counting, while understanding the composition and rationale within the Means Mechanical/Electrical database.

Some Of What You'll Learn:

- The unique way mechanical and electrical systems are interrelated
- M&E estimates, conceptual, planning, budgeting, and bidding stages
- Order of magnitude, square foot, assemblies, and unit price estimating
- Comparative cost analysis of equipment and design alternatives

Who Should Attend: Architects, engineers, facilities managers, mechanical and electrical contractors... and others needing a highly reliable method for developing, understanding, and evaluating mechanical and electrical contracts.

Plan Reading and Material Takeoff

This two-day program teaches attendees to read and understand construction documents and to use them in the preparation of material takeoffs.

Some of What You'll Learn:

- Skills necessary to read and understand typical contract documents—blueprints and specifications
- Details and symbols used by architects and engineers
- Construction specifications' importance in conjunction with blueprints
- Accurate takeoff of construction materials and industry-accepted takeoff methods

Who Should Attend: Facilities managers, construction supervisors, office managers... and others responsible for the execution and administration of a construction project including government, medical, commercial, educational, or retail facilities.

Facilities Maintenance and Repair Estimating

This two-day course teaches attendees how to plan, budget, and estimate the cost of ongoing and preventive maintenance and repair for existing buildings and grounds.

Some Of What You'll Learn:

- The most financially favorable maintenance, repair, and replacement scheduling and estimating
- Auditing and value engineering facilities
- Preventive planning and facilities upgrading
- Determining both in-house and contract-out service costs; annual, asset-protecting M&R plan

Who Should Attend: Facility managers, maintenance supervisors, buildings and grounds superintendents, plant managers, planners, estimators... and others involved in facilities planning and budgeting.

Scheduling and Project Management

This two-day course teaches attendees the most current and proven scheduling and management techniques needed to bring projects in on time and on budget.

Some Of What You'll Learn:

- Crucial phases of planning and scheduling
- How to establish project priorities and develop realistic schedules and management techniques
- Critical Path and Precedence Methods
- Special emphasis on cost control

Who Should Attend: Construction project managers, supervisors, engineers, estimators, contractors... and others who want to improve their project planning, scheduling, and management skills.

Advanced Project Management

This two-day seminar will teach you how to effectively manage and control the entire design-build process and allow you to take home tangible skills that will be immediately applicable on existing projects.

Some Of What You'll Learn:

- Value engineering, bonding, fast-tracking, and bid package creation
- How estimates and schedules can be integrated to provide advanced project management tools
- Cost engineering, quality control, productivity measurement, and improvement
- Front loading a project and predicting its cash flow

Who Should Attend: Owners, project managers, architectural and engineering managers, construction managers, contractors... and anyone else who is responsible for the timely design and completion of construction projects.

For more information visit Means Web site at www.rsmeans.com

Seminars

2006 Means Seminar Schedule

Note: Call for exact dates and details.

Location	Dates
Las Vegas, NV	March
Washington, DC	April
Phoenix, AZ	April
Denver, CO	May
San Francisco, CA	June
Philadelphia, PA	June
Washington, DC	September
Dallas, TX	September
Las Vegas, NV	October
Orlando, FL	November
Atlantic City, NJ	November
San Diego, CA	December

Registration Information

Register Early... Save up to $100! Register 30 days before the start date of a seminar and save $100 off your total fee. *Note: This discount can be applied only once per order. It cannot be applied to team discount registrations or any other special offer.*

How to Register Register by phone today! Means' toll-free number for making reservations is: **1-800-334-3509.**

Individual Seminar Registration Fee $895. Individual CostWorks Training Registration Fee $349. To register by mail, complete the registration form and return with your full fee to: Seminar Division, Reed Construction Data, RSMeans Seminars, 63 Smiths Lane, Kingston, MA 02364.

Federal Government Pricing All federal government employees save 25% off regular seminar price. Other promotional discounts cannot be combined with Federal Government discount.

Team Discount Program Two to four seminar registrations: Call for pricing.

Multiple Course Discounts When signing up for two or more courses, call for pricing.

Refund Policy Cancellations will be accepted up to ten days prior to the seminar start. There are no refunds for cancellations received later than ten working days prior to the first day of the seminar. A $150 processing fee will be applied for all cancellations. Written notice of cancellation is required. Substitutions can be made at any time before the session starts. **No-shows are subject to the full seminar fee.**

AACE Approved Courses The RSMeans Construction Estimating and Management Seminars described and offered to you here have each been approved for 14 hours (1.4 recertification credits) of credit by the AACE International Certification Board toward meeting the continuing education requirements for recertification as a Certified Cost Engineer/Certified Cost Consultant.

AIA Continuing Education We are registered with the AIA Continuing Education System (AIA/CES) and are committed to developing quality learning activities in accordance with the CES criteria. RSMeans seminars meet the AIA/CES criteria for Quality Level 2. AIA members will receive (14) learning units (LUs) for each two-day RSMeans Course.

NASBA CPE Sponsor Credits We are part of the National Registry of CPE Sponsors. Attendees may be eligible for (16) CPE credits.

Daily Course Schedule The first day of each seminar session begins at 8:30 A.M. and ends at 4:30 P.M. The second day is 8:00 A.M.–4:00 P.M. Participants are urged to bring a hand-held calculator since many actual problems will be worked out in each session.

Continental Breakfast Your registration includes the cost of a continental breakfast, a morning coffee break, and an afternoon break. These informal segments will allow you to discuss topics of mutual interest with other members of the seminar. (You are free to make your own lunch and dinner arrangements.)

Hotel/Transportation Arrangements RSMeans has arranged to hold a block of rooms at each hotel hosting a seminar. To take advantage of special group rates when making your reservation, be sure to mention that you are attending the Means Seminar. You are, of course, free to stay at the lodging place of your choice. **(Hotel reservations and transportation arrangements should be made directly by seminar attendees.)**

Important Class sizes are limited, so please register as soon as possible.

Note: *Pricing subject to change.*

Registration Form **Call 1-800-334-3509 to register or FAX 1-800-632-6732. Visit our Web site www.rsmeans.com**

Please register the following people for the Means Construction Seminars as shown here. Full payment or deposit is enclosed, and we understand that we must make our own hotel reservations if overnight stays are necessary.

☐ Full payment of $______________________ enclosed.

☐ Bill me

Name of Registrant(s)
(To appear on certificate of completion)
____
____
____

P.O. #: ______________________________
GOVERNMENT AGENCIES MUST SUPPLY PURCHASE ORDER NUMBER

Firm Name ______________________________

Address ______________________________

City/State/Zip ______________________________

Telephone No. Fax No. ______________________________

E-mail Address ______________________________

Charge our registration(s) to: ☐ MasterCard ☐ VISA ☐ American Express ☐ Discover

Account No. ____________________ Exp. Date ____________

Cardholder's Signature ______________________________

Seminar Name City Dates
____
____

Please mail check to: Seminar Division, Reed Construction Data, RSMeans Seminars, 63 Smiths Lane, P.O. Box 800, Kingston, MA 02364 USA

MeansData™

CONSTRUCTION COSTS FOR SOFTWARE APPLICATIONS

Your construction estimating software is only as good as your cost data.

A proven construction cost database is a mandatory part of any estimating package. The following list of software providers can offer you MeansData™ as an added feature for their estimating systems. See the table below for what types of products and services they offer (match their numbers). Visit online at **www.rsmeans.com/demosource/** for more information and free demos. Or call their numbers listed below.

1. **3D International**
   713-871-7000
   venegas@3di.com
2. **4Clicks-Solutions, LLC**
   719-574-7721
   mbrown@4clicks-solutions.com
3. **Aepco, Inc.**
   301-670-4642
   blueworks@aepco.com
4. **Applied Flow Technology**
   800-589-4943
   info@aft.com
5. **ArenaSoft Estimating**
   888-370-8806
   info@arenasoft.com
6. **BSD - Building Systems Design, Inc.**
   888-273-7638
   bsd@bsdsoftlink.com
7. **CMS - Computerized Micro Solutions**
   800-255-7407
   cms@proest.com
8. **Corecon Technologies, Inc.**
   714-895-7222
   sales@corecon.com
9. **CorVet Systems**
   301-622-9069
   sales@corvetsys.com
10. **Estimating Systems, Inc.**
    800-967-8572
    esipulsar@adelphia.net
11. **MC² - Management Computer**
    800-225-5622
    vkeys@mc2-ice.com
12. **Maximus Asset Solutions**
    800-659-9001
    assetsolutions@maximus.com
13. **Shaw Beneco Enterprises, Inc.**
    877-719-4748
    inquire@beneco.com
14. **Timberline Software Corp.**
    800-628-6583
    product.info@timberline.com
15. **US Cost, Inc.**
    800-372-4003
    sales@uscost.com
16. **Vanderweil Facility Advisors**
    617-451-5100
    info@VFA.com
17. **WinEstimator, Inc.**
    800-950-2374
    sales@winest.com

TYPE	1	2	3	4	5	6	7	8	9	10	11	12	13	14	15	16	17
BID					•	•	•		•		•			•			•
Estimating		•			•	•	•	•	•	•	•		•	•	•		•
DOC/JOC/SABER		•			•	•			•	•		•	•	•			•
ID/IQ		•							•	•		•	•				•
Asset Mgmt.												•				•	•
Facility Mgmt.	•		•									•				•	
Project Mgmt.	•	•						•	•				•	•			
TAKE-OFF					•		•	•	•		•			•	•		
EARTHWORK								•			•						
Pipe Flow				•													
HVAC/Plumbing					•		•		•								
Roofing					•		•		•								
Design	•				•								•	•			•
Other Offers Links:																	
Accounting/HR		•			•								•	•			
Scheduling					•									•	•		•
CAD														•			•
PDA														•	•		•
Lt. Versions		•					•							•	•		•
Consulting	•	•			•		•		•			•		•	•	•	•
Training		•			•	•	•	•	•	•		•	•	•	•	•	•

Qualified re-seller applications now being accepted. Call Carol Polio, Ext. 5107.

FOR MORE INFORMATION
CALL 1-800-448-8182, EXT. 5107 OR FAX 1-800-632-6732

ORDER TOLL FREE 1-800-334-3509
OR FAX 1-800-632-6732.

2006 Order Form

Qty.	Book No.	COST ESTIMATING BOOKS	Unit Price	Total
	60066	Assemblies Cost Data 2006	$206.95	
	60016	Building Construction Cost Data 2006	126.95	
	61016	Building Const. Cost Data–Looseleaf Ed. 2006	157.95	
	60226	Building Const. Cost Data–Western Ed. 2006	126.95	
	60116	Concrete & Masonry Cost Data 2006	114.95	
	60146	Construction Cost Indexes 2006	272.00	
	60146A	Construction Cost Index–January 2006	68.00	
	60146B	Construction Cost Index–April 2006	68.00	
	60146C	Construction Cost Index–July 2006	68.00	
	60146D	Construction Cost Index–October 2006	68.00	
	60346	Contr. Pricing Guide: Resid. R & R Costs 2006	39.95	
	60336	Contr. Pricing Guide: Resid. Detailed 2006	39.95	
	60326	Contr. Pricing Guide: Resid. Sq. Ft. 2006	39.95	
	60236	Electrical Change Order Cost Data 2006	126.95	
	60036	Electrical Cost Data 2006	126.95	
	60206	Facilities Construction Cost Data 2006	299.95	
	60306	Facilities Maintenance & Repair Cost Data 2006	274.95	
	60166	Heavy Construction Cost Data 2006	126.95	
	60096	Interior Cost Data 2006	126.95	
	60126	Labor Rates for the Const. Industry 2006	274.95	
	60186	Light Commercial Cost Data 2006	107.95	
	60026	Mechanical Cost Data 2006	126.95	
	63016	Metric Construction Cost Data 2006	149.95	
	60156	Open Shop Building Const. Cost Data 2006	126.95	
	60216	Plumbing Cost Data 2006	126.95	
	60046	Repair and Remodeling Cost Data 2006	107.95	
	60176	Residential Cost Data 2006	107.95	
	60286	Site Work & Landscape Cost Data 2006	126.95	
	60056	Square Foot Costs 2006	137.95	
		REFERENCE BOOKS		
	67310A	ADA Compliance Pricing Guide, 2nd Ed.	79.95	
	67273	Basics for Builders: How to Survive and Prosper	34.95	
	67330	Bldrs Essentials: Adv. Framing Methods	24.95	
	67329	Bldrs Essentials: Best Bus. Practices for Bldrs	29.95	
	67298A	Bldrs Essentials: Framing/Carpentry 2nd Ed.	24.95	
	67298AS	Bldrs Essentials: Framing/Carpentry Spanish	24.95	
	67307	Bldrs Essentials: Plan Reading & Takeoff	35.95	
	67261A	Bldg. Prof. Guide to Contract Documents 3rd Ed.	32.48	
	67342	Building & Renovating Schools	99.95	
	67339	Building Security: Strategies & Costs	44.98	
	67312	Building Spec Homes Profitably	29.95	
	67146	Concrete Repair & Maintenance Illustrated	34.98	
	67314	Cost Planning & Est. for Facil. Maint.	89.95	
	67317A	Cyberplaces: The Internet Guide 2nd Ed.	59.95	
	67328A	Designing & Building with the IBC, 2nd Ed.	99.95	
	67230B	Electrical Estimating Methods 3rd Ed.	64.95	
	64777A	Environmental Remediation Est. Methods 2nd Ed.	$ 99.95	
	67343	Estimating Bldg. Costs for Resi. & Lt. Comm.	29.95	
	67160	Estimating for Contractors	17.98	
	67276A	Estimating Handbook 2nd Ed.	99.95	
	67249	Facilities Maintenance Management	86.95	
	67246	Facilities Maintenance Standards	79.95	

Qty.	Book No.	REFERENCE BOOKS (Cont.)	Unit Price	Total
	67318	Facilities Operations & Engineering Reference	109.95	
	67301	Facilities Planning & Relocation	89.95	
	67231	Forms for Building Const. Professional	47.48	
	67260	Fundamentals of the Construction Process	25.98	
	67338	Green Building: Proj. Planning & Cost Est.	89.95	
	67323	Historic Preservation: Proj. Planning & Est.	99.95	
	67308E	Home Improvement Costs–Int. Projects 9th Ed.	24.95	
	67309E	Home Improvement Costs–Ext. Projects 9th Ed.	24.95	
	67324A	How to Est. w/Means Data & CostWorks 2nd Ed.	59.95	
	67304	How to Estimate with Metric Units	9.98	
	67306	HVAC: Design Criteria, Options, Select. 2nd Ed.	84.95	
	67281	HVAC Systems Evaluation	84.95	
	67282A	Illustrated Construction Dictionary, Condensed	59.95	
	67292A	Illustrated Construction Dictionary, w/CD-ROM	99.95	
	67348	Job Order Contracting	89.95	
	67347	Kitchen & Bath Project Costs	29.95	
	67295B	Landscape Estimating 4th Ed.	62.95	
	67341	Life Cycle Costing	99.95	
	67302	Managing Construction Purchasing	19.98	
	67294A	Mechanical Estimating 3rd Ed.	64.95	
	67283B	Plumbing Estimating Methods 3rd Ed.	29.98	
	67345	Pract. of Cost Segregation Analysis	99.95	
	67326	Preventive Maint. Guidelines for School Facil.	149.95	
	67337	Preventive Maint. for Higher Education Facilities	149.95	
	67346	Preventive Maint. for Multi-Family Housing	89.95	
	67236A	Productivity Standards for Constr.–3rd Ed.	49.98	
	67247B	Project Scheduling & Management for Constr. 3rd Ed.	64.95	
	67265B	Repair & Remodeling Estimating 4th Ed.	69.95	
	67322A	Resi. & Light Commercial Const. Stds. 2nd Ed.	59.95	
	67327	Spanish/English Construction Dictionary	22.95	
	67145B	Sq. Ft. & Assem. Estimating Methods 3rd Ed.	69.95	
	67287	Successful Estimating Methods	32.48	
	67313	Successful Interior Projects	24.98	
	67233	Superintending for Contractors	35.95	
	67321	Total Productive Facilities Management	29.98	
	67284	Understanding Building Automation Systems	29.98	
	67303A	Unit Price Estimating Methods 3rd Ed.	59.95	
	67319	Value Engineering: Practical Applications	79.95	
		MA residents add 5% state sales tax		
		Shipping & Handling**		
		Total (U.S. Funds)*		

Prices are subject to change and are for U.S. delivery only. *Canadian customers may call for current prices. **Shipping & handling charges: Add 7% of total order for check and credit card payments. Add 9% of total order for invoiced orders.

Send Order To: **ADDV-1001**

Name (Please Print) ______________________

Company ______________________

☐ **Company**

☐ **Home** Address ______________________

____

City/State/Zip ______________________

Phone # ____________ P.O. # ____________

(Must accompany all orders being billed)

Mail To: **RSMeans** P.O. Box 800, Kingston, MA 02364-0800